AUTOMOTIVE TECHNOLOGY

A SYSTEMS APPROACH

5th Edition

Jack Erjavec

DELMAR
CENGAGE Learning™

Australia • Brazil • Japan • Korea • Mexico • Singapore • Spain • United Kingdom • United States

DELMAR
CENGAGE Learning™

AUTOMOTIVE TECHNOLOGY:
A Systems Approach, 5e
Jack Erjavec

Vice President, Career and Professional Editorial: Dave Garza

Director of Learning Solutions: Sandy Clark

Executive Editor: David Boelio

Managing Editor: Larry Main

Senior Product Manager: Matthew Thouin

Editorial Assistant: Lauren Stone

Vice President, Career and Professional Marketing: Jennifer McAvey

Executive Marketing Manager: Deborah S. Yarnell

Senior Marketing Manager: Jimmy Stephens

Marketing Specialist: Mark Pierro

Production Director: Wendy Troeger

Production Manager: Mark Bernard

Content Project Manager: Cheri Plasse

Art Director: Benj Gleeksman

Library of Congress Control Number: 2008934340

ISBN-13: 978-1428311497

ISBN-10: 1428311491

Delmar
5 Maxwell Drive
Clifton Park, NY 12065-2919
USA

Cengage Learning products are represented in Canada by Nelson Education, Ltd.

For your lifelong learning solutions, visit **delmar.cengage.com**

Visit our corporate website at **www.cengage.com.**

Printed in the United States of America
4 5 XX 12 11

CONTENTS

SECTION 3 ELECTRICITY 433

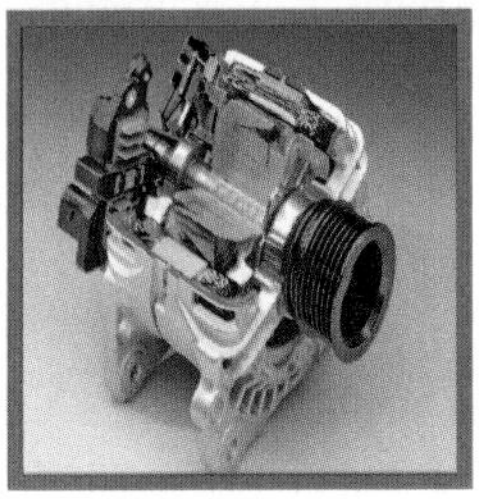

SECTION 8 BRAKES 1424

SECTION 9 PASSENGER COMFORT 1535

PHOTO SEQUENCES

PREFACE

ABOUT THE BOOK

All of the changes to the various systems of an automobile and the integration of those systems have made becoming a successful technician more challenging than ever before. This book, *Automotive Technology: A Systems Approach,* was designed and written to prepare students for those challenges. With students having so much to learn in a short time, why fill the pages of a textbook with information they do not need? The emphasis of this book is on those things that students need to know about the vehicles of yesterday, today, and tomorrow.

This does not mean that the pages are filled with fact after fact. Rather, each topic is explained in a logical way, slowly but surely. After many years of teaching, I have a good sense of how students read and study technical material. I also know what things draw their interest into a topic and keep it there. These things have been incorporated in the writing and features of the book.

This new edition of *Automotive Technology: A Systems Approach* represents the many changes that have taken place in the automotive industry over the past few years. With each new edition, a new challenge (for me) presents itself. What should I include and what should I delete? I hope that I made the right choices. Of course, if I did, I give much of the credit to the feedback I have received from users of the fourth edition and those individuals who reviewed this new edition while it was in the making. They all did a fantastic job and showed that they are truly dedicated to automotive education.

NEW TO THIS EDITION

This new edition is not the previous edition with a new cover and some new chapters. Although much of the information from the fourth edition was retained, each chapter has been updated in response to the changing industry. In addition, there are some new chapters that should be helpful to students and their instructors.

The first section of chapters, which give an overview of the automotive industry, careers, working as a technician, tools, diagnostic equipment, and basic automotive systems, has a totally new look. In fact, the content of these chapters has been rearranged to better match the responsibilities and career of automotive technicians. Chapter 1 explores the career opportunities in the automotive industry. This discussion has been expanded to include more about alternative careers, including the parts distribution world. Chapter 2 covers workplace skills and the ways to go about seeking and selecting a job in the automotive field. This chapter goes through the process of getting a job and keeping it. It also covers some of the duties common to all automotive technicians.

Chapter 3 covers the basic systems of the automobile in a very basic approach and has been updated to include hybrid vehicles. Chapters 4 through 6 cover very important issues regarding hand tools, shop equipment, and safety issues (including bloodborne pathogens). Throughout these chapters, there is a strong emphasis on safely working on today's vehicles and the correct tools required to do so. Chapter 5 gives a brief look at the special and diagnostic tools required for working in each of the eight primary ASE certification areas. The tools discussed include all of the required tools for each area as defined by NATEF.

Chapter 7 is new. It goes through the procedures involved in common safety inspections and preventive maintenance programs. These are the things that students ought to know before they enter into an entry-level position as a technician.

Chapter 8 has been refined and covers the science and math principles that are the basis for the operating principles of an automobile. Too often, we as instructors assume that our students know these basics. I have included this chapter to serve as a reference for those students who want to be good technicians, and to do that they need a better understanding of why things happen the way they do. The rest of Section 1 has been updated with more coverage of current trends, safety shop equipment, and preventive maintenance services.

Section 2, which includes the chapters on engines, has been changed to include more coverage on the latest engine designs and technologies. There is more coverage on the theory, diagnosis, and service to alloy engines and overhead camshaft engines. There are also discussions on the latest trends, including variable valve timing and lift and variable compression ratios. An emphasis on light-duty diesel engines and those engines used in hybrid vehicles is also part of the entire section.

It seems that everyone appreciated the coverage of basic electricity and electronics in the previous edition. As a result, nothing was deleted from those chapters. However, the organization of the chapters is different. I moved the coverage of electronics and electronic-related equipment later in the section. Hopefully this will allow students to gain a solid understanding of basic electricity before moving on to electronics. Each of the chapters in Section 3 covers the parts and pieces that are part of the latest hybrid systems. I feel that it is better to talk about them as the main topic appears rather than to treat them as something special. For example, although lead-acid batteries are the norm for conventional vehicles, the discussion on batteries includes the various types of batteries used in hybrids. This discussion includes the operation of batteries and fuel cells that may be used in the future. This approach was also used in the discussion of motors and charging systems.

Coverage of all the major electrical systems has been increased to include new technologies. This includes high-voltage systems, new exterior lighting systems, new restraint systems, adaptive systems (such as cruise control), and many new accessories. The rest of the section has been brought up to date with additional coverage on body computers and the use of lab scopes and graphing meters.

The entire Engine Performance section (Section 4) has been updated. New to this edition are introductory chapters that deal with overall engine performance theories and testing. This change represents the approach taken by most experienced technicians. It is hoped that students will be able to grasp a global look at these systems and can become better diagnosticians. The revision of the section covers the individual engine performance systems, their operation, and how to test them with current diagnostic equipment. Added emphasis on diagnostics was the main goal of the revision of the rest of this section.

Sections 5 and 6 cover transmissions and drivelines. All of the chapters in these sections have been updated to include more coverage on electronic controls. A new chapter (Chapter 41) has been added to the automatic transmission section. This chapter is all about electronically controlled transmissions, which are the standard transmissions of today's vehicles. There is also more coverage on six-, seven-, and eight-speed transmissions, automatic manual transmissions, new differential designs, and electronic automatic transmissions and transaxles. In addition, there is complete coverage on the transmissions used in today's hybrid vehicles.

The suspension and steering systems section has increased coverage on electronic controls and systems. This includes the new magneto-rheological shock absorbers and four-wheel steering systems. Chapter 47, Wheel Alignment, has been updated to include the latest techniques for performing a four-wheel alignment.

The Brakes section has also been updated to reflect current technology. This includes the latest antilock brake, stability control, and traction control systems.

Heating and air-conditioning systems are covered in Section 9. The content in Chapters 52 and 53 was totally revised and include hybrid systems as well as future systems using CO_2 as a refrigerant.

ORGANIZATION AND GOALS OF THIS EDITION

This edition is still a comprehensive guide to the service and repair of our contemporary automobiles. It is still divided into nine sections that relate to the specific automotive systems. The chapters within each section describe the various subsystems and individual components. Diagnostic and service procedures that are unique to different automobile manufacturers also are included in these chapters. Because many automotive systems are integrated, the chapters explain these important relationships in great detail.

Effective diagnostic skills begin with learning to isolate the problem. The exact cause is easier to pinpoint by identifying the system that contains the problem. Learning to think logically about troubleshooting problems is crucial to mastering this essential skill. Therefore, logical troubleshooting techniques are discussed throughout this text. Each chapter describes ways to isolate the problem system and then the individual components of that system.

This *systems approach* gives the student important preparation opportunities for the ASE certification exams. These exams are categorized by the automobile's major systems. The book's sections are outlined to match the ASE test specifications and competency task lists. The review questions at the end of every chapter give students practice in answering ASE-style review questions.

More importantly, a *systems approach* allows students to have a better understanding of the total vehicle. With this understanding, they have a good chance for a successful career as an automotive technician. That is the single most important goal of this text.

ACKNOWLEDGMENTS

I would like to acknowledge and thank the following dedicated and knowledgeable educators for their comments, criticisms, and suggestions during the review process:

Rob Barszcz
Porter and Chester Institute
Chicopee, MA

Bill Bentley
Coosa Valley Technical College
Rome, GA

David Borelli
Lincoln Technical Institute
Philadelphia, PA

Ron Chappell
Santa Fe Community College (retired)
Gainesville, FL

Kevin Cornell
Porter and Chester Institute
Rocky Hill, CT

Harold Davis
Porter and Chester Institute
Stratford, CT

Robert Day
Lincoln College of Technology
Grand Prairie, TX

Michael Dommer
Indian Hills Community College
Ottumwa, IA

Dan Encinas
Los Angeles Trade Tech College
Los Angeles, CA

Douglas Frye
Porter and Chester Institute
Rocky Hill, CT

Deno Garzolini
Area 30 Career Center
Greencastle, IN

Gary Grote
Porter and Chester Institute
Rocky Hill, CT

Carl Hader
Grafton High School
Grafton, WI

James Harper
Lincoln College of Technology
Grand Prairie, TX

Ron Harris
Lincoln College of Technology
Grand Prairie, TX

Glenn Hartland
Porter and Chester Institute
Westborough, MA

James Haun
Walla Walla Community College
Walla Walla, WA

David E. Hayman
Lincoln Technical Institute
Queens, NY

Joe Jackson
Ranken Technical College
St. Louis, MO

Ed Lingle
Lincoln College of Technology
Grand Prairie, TX

Sherri McDougal
Clintonville High School
Clintonville, WI

James McGuire
Lincoln College of Technology
Grand Prairie, TX

Rick Mercado
Intellitec College
Colorado Springs, CO

Robert Michelini
Muncie Area Career Center
Muncie, IN

David Mitchell
Prosser School of Technology
Sellersburg, IN

Calvin Motley
Central New Mexico Community College
Laguna, NM

Cliff Owen
Griffin Technical College (retired)
Griffin, GA

Donald Paulak
Hennepin Technical College
Eden Prairie, MN

John Reinwald
Baran Institute of Technology
East Windsor, CT

Michael Riel
Porter and Chester Institute
Chicopee, MA

Jeremiah Rossmanith
Iowa Central Community College
Fort Dodge, IA

Harold Rowzee
Jones County Junior College
Ellisville, MS

Scott Scheife
Milwaukee Area Technical College
Mequon, WI

Gregory Schmidt
Henry Ford Community College
Dearborn, MI

Philip Shastid
Lincoln College of Technology
Grand Prairie, TX

Wendell Soucy
Porter and Chester Institute
Enfield, CT

Rob Thompson
South-Western Career Academy
Grove City, OH

Mark Whitehead
Lincoln College of Technology
Grand Prairie, TX

John Wood
Ranken Technical College
St. Louis, MO

Charles Woodard
Broward Community College
Miramar, FL

ABOUT THE AUTHOR

Jack Erjavec has become a fixture in the automotive textbook publishing world. He has many years of experience as a technician, educator, author, and editor and has authored or coauthored more than thirty automotive textbooks and training manuals. Mr. Erjavec holds a Master of Arts degree in Vocational and Technical Education from Ohio State University. He spent 20 years at Columbus State Community College as an instructor and administrator and has also been a long-time affiliate of the North American Council of Automotive Teachers, including serving on the board of directors and as executive vice-president. Jack is also associated with ATMC, SAE, ASA, ATRA, AERA, and other automotive professional associations.

FEATURES OF THE TEXT

Learning how to maintain and repair today's automobiles can be a daunting endeavor. To guide the readers through this complex material, we have built in a series of features that will ease the teaching and learning processes.

OBJECTIVES

Each chapter begins with the purpose of the chapter, stated in a list of objectives. Both cognitive and performance objectives are included in the lists. The objectives state the expected outcome that will result from completing a thorough study of the contents in the chapters.

CAUTIONS AND WARNINGS

Instructors often tell us that shop safety is their most important concern. Cautions and warnings appear frequently in every chapter to alert students to important safety concerns.

SHOP TALK

These features are sprinkled throughout each chapter to give practical, common-sense advice on service and maintenance procedures.

CUSTOMER CARE

Creating a professional image is an important part of shaping a successful career in automotive technology. The customer care tips were written to encourage professional integrity. They give advice on educating customers and keeping them satisfied.

USING SERVICE INFORMATION

Learning to use available service information is critical to becoming a successful technician. The source of information varies from printed material to online materials. The gathering of information can be a time-consuming task but nonetheless is extremely important. We have included a feature that points the student in the right direction to find the right information.

PERFORMANCE TIPS

This feature introduces students to the ideas and theories behind many performance-enhancing techniques used by professionals.

"GO TO" FEATURE

This new feature is used throughout the chapters and tells the student where to go for prerequisite and additional information on the topic.

PHOTO SEQUENCES

Step-by-step photo sequences illustrate practical shop techniques. The photo sequence focus on techniques that are common, need-to-know service and maintenance procedures. These photo sequences give students a clean, detailed image of what to look for when they perform these procedures. This was a popular feature of the previous editions, so we now have a total of 53.

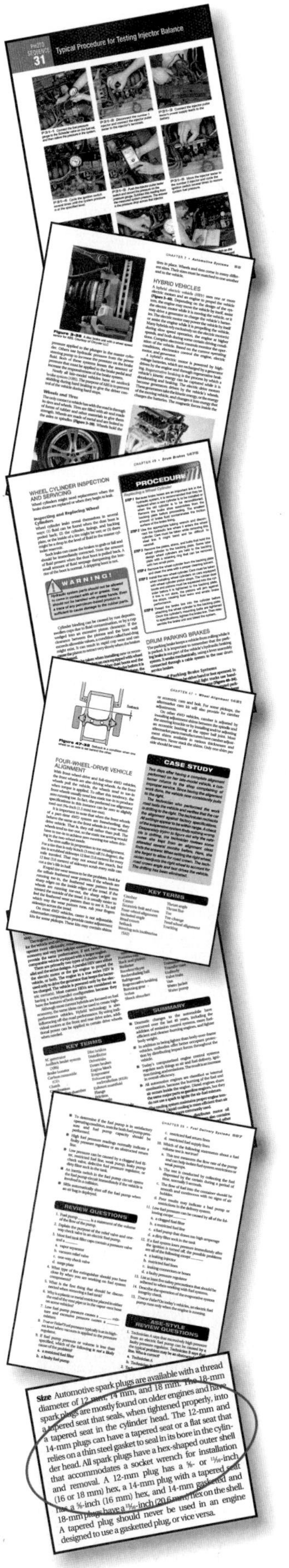

HYBRID VEHICLES

Abundant content on hybrid vehicles and related technologies is presented throughout the textbook within the specific system areas that are being discussed.

PROCEDURES

This feature gives detailed, step-by-step instructions for important service and maintenance procedures. These hands-on procedures appear frequently and are given in great detail because they help to develop good shop skills and help to meet competencies required for ASE certification.

CASE STUDIES

Case studies highlight our emphasis on logical troubleshooting. A service problem is outlined at the end of the chapters, and then a technician's solution is described. This gives the student a practical example of logical troubleshooting.

KEY TERMS

Each chapter ends with a list of the terms that were introduced in the chapter. These terms are highlighted in the text when they are first used, and many are defined in the glossary.

SUMMARY

Highlights and key bits of information from the chapter are listed at the end of each chapter. This listing is designed to serve as a refresher for the reader.

REVIEW QUESTIONS

A combination of short-answer essay, fill-in-the-blank, and multiple-choice, questions make up the end-of-chapter review questions. Different question types are used to challenge the reader's understanding of the chapter's contents. The chapter objectives are used as the basis for the review questions.

ASE-STYLE REVIEW QUESTIONS

In any chapter that relates to one of the ASE certification areas, there are ten ASE-style review questions that relate to that area. Some are quite challenging and others are a simple review of the contents of the chapter.

METRIC EQUIVALENTS

Throughout the text, all measurements are given in UCS and metric increments.

SUPPLEMENTS

The *Automotive Technology* package offers a full complement of supplements.

Student Online Companion

Each textbook provides access to the new *Student Online Companion,* which includes the following components:

- **PowerPoint**—Study outlines with images for each textbook chapter.
- **ASE-Style Practice Questions**—Over 350 questions to help students review chapter material and get familiar with the question types they will see on certification exams.
- **Web Links and Activities**—Links to industry Web resources/reference material with related research activities for many.
- **Challenging Concepts**—45 videos, narration, and questions to help students comprehend more challenging topics.
- **Interactive Online Game**—A self-review Q&A game to help students comprehend the chapter material.

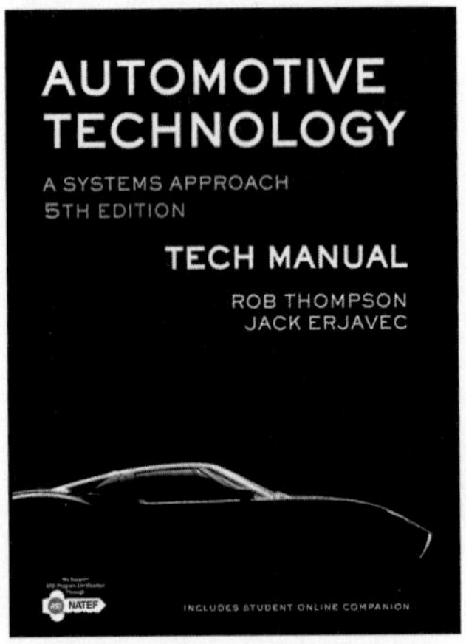

Tech Manual

The *Tech Manual* (ISBN 1428311505) offers students opportunities to strengthen their comprehension of key concepts and to develop their hands-on, practical shop experience. Each chapter includes Concept Activities and Job Sheets, many of which are directly correlated to specific NATEF tasks. Service manual report sheets and review questions also are included to offer a rounded approach to each lesson.

Instructor Resources

The *Instructor Resources* DVD (ISBN 1428311521) for the fifth edition includes the following components to help minimize instructor prep time and engage students:

- **PowerPoint**—Chapter outlines with images, animations, and video clips for each textbook chapter.
- **Computerized Test Bank in Exam View**—Hundreds of modifiable questions for exams, quizzes, in-class work, or homework assignments. All applicable questions are correlated to the 2008 NATEF Automobile Standards.
- **Image Library**—A searchable database of hundreds of images from the textbook that can be used to easily customize the PowerPoint outlines.
- **Photo Sequences**—Each of the 53 photo sequences from the textbook are provided within PowerPoint for easy classroom projection.
- **End-of-Chapter Review Questions**—Word files of all textbook review questions are provided on the DVD.
- **Instructor's Manual**—An electronic version of the Instructor's Manual is included in the Instructor Resources.
- **NATEF Correlations**—The 2008 NATEF Automobile Standards are correlated to the chapter and page numbers of the core text and all relevant Tech Manual job sheets.
- **Job Sheet Template**—For instructors who develop their own job sheets, a template is provided to help with their formatting.

Instructor's Manual

This comprehensive guide provides lecture outlines with teaching hints, answers to review questions from the textbook, and answers to *Tech Manual* questions, as well as guidelines for using the *Tech Manual.* A correlation chart to the 2008 NATEF Task List provides references to topic coverage in both the text and *Tech Manual.*

CAREERS IN THE AUTOMOTIVE INDUSTRY

CHAPTER 1

OBJECTIVES

■ Describe the reasons why today's automotive industry is considered a global industry. ■ Explain how computer technology has changed the way vehicles are built and serviced. ■ Explain why the need for qualified automotive technicians is increasing. ■ Describe the major types of businesses that employ automotive technicians. ■ List some of the many job opportunities available to people with a background in automotive technology. ■ Describe the different ways a student can gain work experience while attending classes. ■ Describe the requirements for ASE certification as an automotive technician and as a master auto technician.

THE AUTOMOTIVE INDUSTRY

Each year millions of new cars and light trucks are produced and sold in North America **(Figure 1–1)**. The automotive industry's part in the total economy of the United States is second only to the food industry. Manufacturing, selling, and servicing these vehicles are parts of an incredibly large, diverse, and expanding industry.

Figure 1–1 Ford's F-150 pickup has been the best selling vehicle in America for many years.

Thirty years ago, America's "big three" automakers—General Motors Corporation, Ford Motor Company, and Chrysler Corporation—dominated the auto industry. This is no longer true. The industry is now a global industry **(Table 1–1)**. Automakers from Japan, Korea, Germany, Sweden, and other European and Asian countries compete with companies in the United States for domestic and foreign sales.

Several foreign manufacturers, such as Honda, Toyota, and BMW, operate assembly plants in the United States and Canada. Automobile manufacturers have joined together, or merged, to reduce costs and increase market share. In addition, many smaller auto manufacturers have been bought by larger companies to form larger global automobile companies. Most often the ownership of a company is not readily identifiable by the brand name. An example of this is Ford Motor Company; Ford brands include Ford, Mercury, Lincoln, Volvo, and Mazda.

There are also a number of vehicles built jointly by the United States and foreign manufacturers. These vehicles are built and sold in North America or exported to other countries.

TABLE 1–1 WORLDWIDE UNIT SALES OF PASSENGER CARS, LIGHT-, MEDIUM-, AND HEAVY-DUTY TRUCKS

Manufacturer	Country of Origin	Approx. Units Sold Annually	Notes
Toyota Motor Corp.	Japan	8.8 million	Includes Lexus, Scion, Daihatsu, and Hino
General Motors	U.S.	8.7 million	Includes GM/Daewoo, Holden, Hummer, Opel, and Saab
Ford Motor Co.	U.S.	6.0 million	Includes Volvo Car Corp.
Volkswagen AG	Germany	5.7 million	Includes Audi, Bentley, Bugatti, Lamborghini, Skoda, and Seat
Hyundai-Kia Automotive	Korea	3.8 million	Includes Hyundai and Kia
Honda Motor Co.	Japan	3.6 million	Includes Acura
Nissan Motor Co.	Japan	3.5 million	Includes Infiniti
PSA/Peugeot-Citroen SA	France	3.4 million	Includes Citroen and Peugeot
Daimler Benz AG	Germany	2.7 million	Includes Mercedes-Benz, Smart, EvoBus, Freightliner, and Mitsubishi Fuso
Renault SA	France	2.4 million	Includes Dacia and Renault-Samsung Motors
Fiat S.p.A.	Italy	2.3 million	Includes Ferrari, Alfa Romeo, Iveco, Lancia, and Maserati
Suzuki Motor Corp.	Japan	2.2 million	
Chrysler LLC	U.S.	2.0 million	
BMW Group 11	Germany	1.4 million	Includes Rolls-Royce and Cooper
Mitsubishi Motors Corp.	Japan	1.3 million	
Mazda Motor Corp.	Japan	1.2 million	
AutoVaz	Russia	860 thousand	
China FAW Group Corp.	China	682 thousand	
Isuzu Motors Ltd.	Japan	651 thousand	
Fuji Heavy Industries Ltd.	Japan	602 thousand	Includes Subaru
Dongfeng Motor Corp.	China	460 thousand	
Chongqing Changan	China	456 thousand	
Tata Motors Ltd.	India	454 thousand	Includes Jaguar and Land Rover
Shanghai Automotive	China	420 thousand	Includes Wuling
Beijing Automobile Works	China	337 thousand	
Chery Automobile Co.	China	304 thousand	
Hafei Motor Co.	China	231 thousand	
Volvo Truck Group	Sweden	230 thousand	Includes Mack Trucks, RVI, Volvo Buses, and Nissan Diesel Motor Co.
AutoGaz	Russia	224 thousand	
Zhejiang Geely	China	204 thousand	
Anhui Jianghuai	China	172 thousand	
Paccar	U.S.	167 thousand	Includes DAF, Kenworth, Leyland, Peterbilt, and Foden
Navistar International	U.S.	158 thousand	

TABLE 1-1 *Continued*

Manufacturer	Country of Origin	Approx. Units Sold Annually	Notes
Mahindra & Mahindra	India	148 thousand	
Iran Khodro	Iran	132 thousand	
Proton	Malaysia	126 thousand	Includes Lotus
Shenyang Brilliance Jinbei	China	126 thousand	
SsangYong Motor Co.	Korea	110 thousand	
Porsche AG	Germany	96 thousand	
Jiangxi Changhe	China	89 thousand	
MAN Nutzfahrzeuge	Germany	87 thousand	
Scania	Sweden	65 thousand	
MG Rover Group	U.K.	53 thousand	
Others	China	122 thousand	Includes Great Wall Motor Co. and Southeast (Fujian) Motor Co.
Others	India	116 thousand	Includes Ashok Leyland, Eicher Motors, Force Motors/Bajaj Tempo, and Hindustan Motors Ltd

This cooperation between manufacturers has given customers an extremely wide selection of vehicles to choose from. This variety has also created new challenges for automotive technicians, based on one simple fact: Along with the different models come different systems.

The Importance of Auto Technicians

The automobile started out as a simple mechanical beast. It moved people and things with little regard to the environment, safety, and comfort. Through the years these concerns have provided the impetus for design changes. One area that has affected automobile design the most is the same area that has greatly influenced the rest of our lives, electronics. Today's automobiles are sophisticated electronically controlled machines. To provide comfort and safety while being friendly to the environment, today's automobiles use the latest developments of many different technologies—mechanical and chemical engineering, hydraulics, refrigeration, pneumatics, physics, and, of course, electronics.

An understanding of electronics is a must for all automotive technicians **(Figure 1–2)**. The needed level of understanding is not that of an engineer; rather, technicians need a practical understanding of electronics. In addition to having the mechanical skills needed to remove, repair, and replace faulty or damaged components, today's technician also must be able to diagnose and service complex electronic systems.

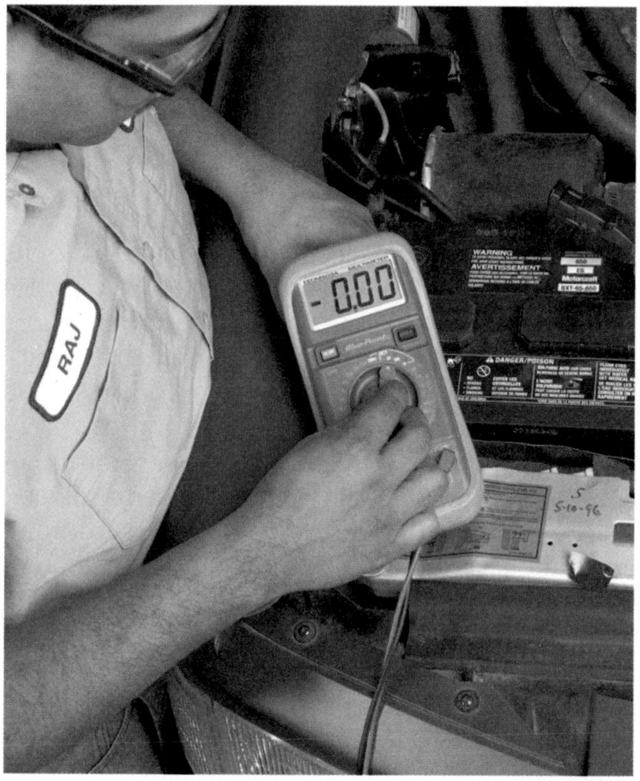

Figure 1–2 An understanding of electronics is a must for all automotive technicians.

Computers and electronic devices are used to control the operation of an engine. Because of these controls, today's automobiles use less fuel, perform

Figure 1-3 The Toyota Prius is the best selling hybrid vehicle. *Courtesy of Dewhurst Photography and Toyota Motor Sales, U.S.A., Inc.*

better, and run cleaner than those in the past. Electronic controls also are used in nearly all systems of an automobile. The number of electronically controlled systems on cars and trucks increases each year. There are many reasons for the heavy insurgence of electronics into automobiles. Electronics are based on electricity and electricity moves at the speed of light. This means the operation of the various systems can be monitored and changed very quickly. Electronic components have no moving parts, are durable, do not require periodic adjustments, and are very light. All of these allow today's automobiles to be more efficient, cleaner, safer, and better performing than vehicles of the past.

The application of electronics has also led to the success of hybrid vehicles **(Figure 1–3)**. A hybrid vehicle has two separate sources of power. Those power sources can work together to move the vehicle or can power the vehicle on their own. Today's hybrid vehicles are moved by electric motors and/or a gasoline engine. Hybrid vehicles are complex machines and all who work on them must be properly trained.

The design of today's automobiles is also influenced by legislation. Throughout history, automobile manufacturers have been required to respond to new laws designed to make automobiles safer and cleaner-running. In response to these laws, new systems and components are introduced. Anyone desiring to be a good technician must regularly update his or her skills to keep up with the technology.

Legislation has not only influenced the design of gasoline-powered vehicles, it has also led to a wider use of diesel engines in passenger vehicles. By mandating cleaner diesel fuels, the laws have opened the door for clean burning and highly efficient diesel engines. These new engines are also fit with electronic controls.

Figure 1-4 Good technicians are able to follow specific manufacturers' diagnostic charts and interpret the results of diagnostic tests.

Many states have laws that require owners to have their vehicles' exhaust tested on an annual basis. Some states require automobiles to pass an annual or biannual **Inspection/Maintenance (I/M)** test. Today's automotive technician must be able to find the cause of test failures and correct them.

The Need for Quality Service

The need for good technicians continues to grow. Currently there is a great shortage of qualified automotive technicians. This means there are, and will be, excellent career opportunities for good technicians. Good technicians are able to diagnose and repair problems in today's automobiles **(Figure 1–4)**.

Car owners demand that when things go wrong, they should be "fixed right the first time." The primary reason some technicians are unable to fix a particular problem is simply that they cannot find the cause of the problem. Today's vehicles are complex and a great amount of knowledge and understanding is required to diagnose them. Today's technicians must have good **diagnostic skills**. Technicians who can identify and solve problems the first time the vehicle is brought into the shop are wanted by the industry and have many excellent career opportunities.

The Need for Ongoing Service

Electronic controls have not eliminated the need for routine service and scheduled maintenance **(Figure 1–5)**. In fact, they have made it more important than ever. Although electronic systems can make adjustments to compensate for some problems, a computer cannot replace worn parts. A computer

Figure 1-5 Regular preventive maintenance (PM) is important for keeping electronic control systems operating correctly. A common part of PM is changing the engine's oil and filter.

cannot tighten loose belts or change dirty coolant or engine oil. Simple problems such as these can set off a chain of unwanted events in an engine control system. Electronic controls are designed to help a well-maintained vehicle operate efficiently. They are not designed to repair systems.

Electronic systems are based on the same principles as a computer. In fact, these systems rely on a computer to control the operation of a component or system. Instead of a keyboard, automotive electronic systems rely on sensors or inputs. These send information to the computer. The computer receives the inputs and through computer logic causes a component to change the way it is operating. These controlled outputs are similar to your computer screen or printer.

Each automobile manufacturer recommends that certain maintenance services be performed according to a specific schedule. These maintenance procedures are referred to as **preventive maintenance (PM)** because they are designed to prevent problems. Scheduled PM normally includes oil and filter changes; coolant and lubrication services; replacement of belts and hoses; and replacement of spark plugs, filters, and worn electrical parts (**Figure 1–6**).

If the owner fails to follow the recommended maintenance schedule, the vehicle's warranty might not cover problems that result. For example, if the engine fails during the period covered by the warranty, the warranty may not cover the engine if the owner does not have proof that the engine's oil was changed according to the recommended schedule.

Warranties A new car **warranty** is an agreement by the auto manufacturer to have its authorized dealers repair, replace, or adjust certain parts if they become defective. This agreement typically lasts until the vehicle has been driven 36,000 miles (58,000 km), and/or has been owned for 3 years.

The details of most warranties vary with the manufacturer, vehicle model, and year. Most manufacturers also provide a separate warranty for the powertrain (engine, transmission, and so on) that covers these parts for a longer period than the basic warranty. There are also additional warranties for other systems or components of the vehicle.

Often, according to the terms of the warranty the owner must pay a certain amount of money, called the **deductible**. The manufacturer pays for all repair costs over the deductible amount.

Battery and tire warranties are often prorated, which means that the amount of the repair bill covered by the warranty decreases over time. Some warranties are held by a third party, such as the manufacturer of the battery or tires. Although the manufacturer sold the vehicle with the battery or set of tires, their warranty is the manufacturer's responsibility.

There are also two government-mandated warranties: the Federal Emissions Defect Warranty and the Federal Emissions Performance Warranty. The Federal Emissions Defect Warranty ensures that the vehicle meets all required emissions regulations and that the vehicle's emission control system works as designed and will continue to do so for 2 years or 24,000 miles. The warranty does not cover problems caused by accidents, floods, misuse, modifications, poor maintenance, or the use of leaded fuels. The systems typically covered by this warranty are:

- Air induction
- Fuel metering
- Ignition
- Exhaust
- Positive crankcase ventilation
- Fuel evaporative control
- Emission control system sensors

The Federal Emissions Performance Warranty covers the catalytic converter(s) and engine control module for a period of 8 years or 80,000 miles. If the owner properly maintains the vehicle and it fails an emissions test approved by the Environmental Protection Agency (EPA), an authorized service facility will repair or replace the emission-related parts covered by the warranty at no cost to the owner. Some states, such as California, require the manufacturers to offer additional or extended warranties.

5,000 MILES OR 6 MONTHS

- Replace engine oil and oil filter
- Rotate tires
- Visually inspect brake linings

Additional maintenance items for special operating conditions

- Inspect ball joints and dust covers
- Inspect drive shaft boots
- Inspect air filter
- Inspect steering linkage and boots
- Retorque drive shaft bolt
- Tighten nuts and bolts on chassis

10,000 MILES OR 12 MONTHS

(Same as 5,000 miles and 6 months)

Additional maintenance items for special operating conditions

(Same as 5,000 miles and 6 months)

15,000 MILES OR 18 MONTHS

(Same as 5,000 miles and 6 months)

Plus:

- Clean cabin air filter
- Inspect the following:
  - Ball joints and dust covers
  - Drive shaft boots
  - Engine air filter
  - Steering linkage and boots
  - Retorque drive shaft bolt
  - Tighten nuts and bolts on chassis

20,000 MILES OR 24 MONTHS

(Same as 5,000 miles and 6 months)

Plus:

- Replace cabin filter

Additional maintenance items for special operating conditions

(Same as 5,000 miles and 6 months)

25,000 MILES OR 30 MONTHS

(Same as 5,000 miles and 6 months)

Additional maintenance items for special operating conditions

(Same as 5,000 miles and 6 months)

30,000 MILES OR 36 MONTHS

(Same as 5,000 miles and 6 months)

Plus:

- Replace cabin filter
- Rotate tire
- Replace engine air filter
- In addition, inspect the following:
  - Brake lines and hoses
  - Differential oil
  - Engine coolant
  - Exhaust pipes and mountings
  - Fuel lines and connections, fuel tank band, and fuel tank vapor system hoses
  - Fuel tank cap gasket
  - Radiator core and condenser
  - Steering gear box
  - Steering linkage and boots
  - Transmission fluid or oil

Additional maintenance items for special operating conditions

(Same as 5,000 miles and 6 months)

35,000 MILES OR 42 MONTHS

(Same as 5,000 miles and 6 months)

Additional maintenance items for special operating conditions

(Same as 5,000 miles and 6 months)

40,000 MILES OR 48 MONTHS

(Same as 20,000 miles and 24 months)

Additional maintenance items for special operating conditions

(Same as 20,000 miles and 24 months)

45,000 MILES OR 54 MONTHS

(Same as 15,000 miles and 18 months)

Additional maintenance items for special operating conditions

(Same as 15,000 miles and 18 months)

50,000 MILES OR 60 MONTHS

(Same as 5,000 miles and 6 months)

Additional maintenance items for special operating conditions

(Same as 5,000 miles and 6 months)

55,000 MILES OR 66 MONTHS

(Same as 20,000 miles and 24 months)

Additional maintenance items for special operating conditions

(Same as 20,000 miles and 24 months)

60,000 MILES OR 72 MONTHS

(Same as 15,000 miles and 18 months)

Plus:

- Inspect:
  - Drive belts
  - Engine valve clearance

Additional maintenance items for special operating conditions

(Same as 15,000 miles and 18 months)

Plus:

- Replace front differential oil
- Replace transmission oil or fluid

Figure 1-6 A typical preventive maintenance schedule.

Figure 1-7 Automotive service technicians can enjoy careers in many different automotive businesses.

All warranty information can be found in the vehicle's owner's manual. Whenever there are questions about the warranties, carefully read that section in the owner's manual. If you are working on a vehicle and know that the part or system is covered under a warranty, make sure to tell the customer before proceeding with your work. Doing this will save the customer money and you will earn his or her trust.

Career Opportunities

Automotive service technicians can enjoy careers in many different types of automotive businesses **(Figure 1–7)**. Because of the skills required to be a qualified technician, there are also career opportunities for those who do not want to repair automobiles the rest of their lives. There are also many opportunities for good technicians who want to change careers. The knowledge required to be a good service technician can open many doors of opportunity.

Dealerships New car dealerships **(Figure 1–8)** serve as the link between the vehicle manufacturer and the customer. They are privately owned businesses. Most dealerships are franchised operations, which means the owners have signed a contract with particular auto manufacturers and have agreed to sell and service their vehicles.

The manufacturer usually sets the sales and service policies of the dealership. Most warranty repair work is done at the dealership. The manufacturer then pays the dealership for making the repair. The manufacturer also provides the service department at the dealership with the training, special tools, equipment, and information needed to repair its vehicles. The manufacturers also help the dealerships get service business. Often, their commercials stress the importance of using their replacement parts and promote their technicians as the most qualified to work on their products.

Working for a new car dealership can have many advantages. Technical support, equipment, and the opportunity for ongoing training are usually excellent. At a dealership, you have a chance to become very skillful in working on the vehicles you service. However, working on one or two types of vehicles does not appeal to everyone. Some technicians want diversity.

Figure 1-8 Dealerships sell and service vehicles made by specific auto manufacturers.

Figure 1-9 Full-service gasoline stations are not as common as they used to be, but they are a good example of an independent service shop.

Independent Service Shops Independent shops **(Figure 1–9)** may service all types of vehicles or may specialize in particular types of cars and trucks or specific systems of a car. Independent shops outnumber dealerships by six to one. As the name states, an independent service shop is not associated with any particular automobile manufacturer. Many independent shops are started by technicians eager to be their own boss and run their own business.

An independent shop may range in size from a two-bay garage with two to four technicians to a multiple-bay service center with twenty to thirty technicians. A **bay** is simply a work area for a complete vehicle **(Figure 1–10)**. The amount of equipment in an independent shop varies; however, most are well equipped to do the work they do best. Working in an independent shop may help you develop into a well-rounded technician.

Figure 1-10 A bay in an independent service shop.

Specialty shops specialize in areas such as engine rebuilding, transmission/transaxle overhauling, and air conditioning, brake, exhaust, cooling, emissions, and electrical work. A popular type of specialty shop is the "quick lube" shop, which takes care of the PM of vehicles. It hires lubrication specialists who change fluids, belts, and hoses in addition to checking certain safety items on the vehicle.

The number of specialty shops that service and repair only one or two systems of the automobile has steadily increased over the past 10 to 20 years. Technicians employed by these shops have the opportunity to become very skillful in one particular area of service.

Franchise Repair Shop A great number of jobs are available at service shops that are run by large companies such as Firestone, Goodyear, and Midas. These shops do not normally service and repair all of the systems of the automobile. However, their customers do come in with a variety of service needs. Technicians employed by these shops have the opportunity to become very proficient in many areas of service and repair.

Some independent shops may look like they are part of a franchise but are actually independent. Good examples of this type of shop are the NAPA service centers **(Figure 1–11)**. These centers are not controlled by NAPA, nor are they franchises of NAPA. They are called NAPA service centers because the facility has met NAPA's standards of quality and the owner has agreed to use NAPA as the primary source of parts and equipment.

Figure 1-11 NAPA service centers are a good example of an independent repair shop that has affiliated with a large business. In these arrangements, the shops are still run independently.

Store-Associated Shops Other major employers of auto technicians are the service departments of department stores. Many large stores that sell automotive parts often offer certain types of automotive services, such as brake, exhaust system, and wheel and tire work.

Fleet Service and Maintenance Any company that relies on several vehicles to do its business faces an ongoing vehicle service and PM problem. Small fleets often send their vehicles to an independent shop for maintenance and repair. Large fleets, however, usually have their own PM and repair facilities and technicians **(Figure 1–12)**.

Utility companies (such as electric, telephone, or cable TV), car rental companies, overnight delivery services, and taxicab companies are good examples of businesses that usually have their own service departments. These companies normally purchase their vehicles from one manufacturer. Technicians who work on these fleets have the same opportunities and benefits as technicians in a dealership. In fact, the technicians of some large fleets are authorized to do warranty work for the manufacturer. Many good career opportunities are available in this segment of the auto service industry.

Figure 1-12 Large fleets usually have their own preventive maintenance and repair facilities and technicians.

JOB CLASSIFICATIONS

The automotive industry offers numerous types of employment for people with a good understanding of automotive systems.

Service Technician

A **service technician (Figure 1–13)** assesses vehicle problems, performs all necessary diagnostic tests, and competently repairs or replaces faulty components. The skills to do this job are based on a sound understanding of auto technology, on-the-job experience, and continuous training in new technology as it is introduced by auto manufacturers.

Individuals skilled in automotive service are called technicians, not mechanics. There is a good reason for this. *Mechanic* stresses the ability to repair and service mechanical systems. While this skill is still very much needed, it is only part of the technician's overall job. Today's vehicles require mechanical knowledge plus an understanding of other technologies, such as electronics, hydraulics, and pneumatics.

Figure 1-13 A service technician troubleshoots problems, performs all necessary diagnostic tests, and competently repairs or replaces faulty components.

Figure 1-14 Specialty technicians work on only one vehicle system, such as brakes.

A technician may work on all systems of the car or may become specialized. Specialty technicians concentrate on servicing one system of the automobile, such as electrical, brakes **(Figure 1–14)**, or transmission. These specialties require advanced and continuous training in that particular field.

In many automotive shops, the technician also has the responsibility for diagnosing the concerns of the customer and preparing a cost estimate for the required services.

Often individuals begin their career as a technician in a new car dealership by performing new car preparation, commonly referred to as "new car prep." The basic purpose of new car prep is to make a vehicle ready to be delivered to a customer. Each dealership has a list of items and services that are performed prior to delivery. Some of the services may include removing the protective plastic from the vehicle's exterior and interior and installing floor mats. At times, new car prep includes tightening certain bolts that may have been intentionally left loose for shipping. New car prep is a great way for a new technician to become familiar with the vehicles sold at the dealership.

Shop Foreman

The **shop foreman** is the one who helps technicians with more difficult tasks and serves as the quality control expert. In some shops, this is the role of the **lead tech**. For the most part, both jobs are the same. Some shops have technician teams. On these teams, there are several technicians, each with a different level of expertise. The lead tech is sort of the shop foreman of the team. Lead techs and shop foremen have a good deal of experience and excellent diagnostic skills.

Service Advisor

The person who greets customers at a service center is the **service advisor** **(Figure 1–15)**, sometimes called a service writer or consultant. Service advisors need to have an understanding of all major systems of an automobile and be able to identify all major

Figure 1-15 A service advisor's main job is to record the customer's concerns.

components and their locations. They also must be able to describe the function of each of those components and be able to identify related components. A good understanding of the recommended service and maintenance intervals and procedures is also required. With this knowledge they are able to explain the importance and complexity of each service and are able to recommend other services.

A thorough understanding of warranty policies and procedures is also a must. Service advisors must be able to explain and verify the applicability of warranties, service contracts, service bulletins, and campaign/recalls procedures.

Service advisors also serve as the liaison between the customer and the technician in most dealerships. They have responsibility for explaining the customer's concerns and/or requests to the technician plus keeping track of the progress made by the technician so the customer can be informed. This monitoring is also important because it impacts the completion of service on the vehicles of other customers.

Often automotive technicians or students of automotive service programs realize a need to change career choices but desire to stay in the service industry. Becoming a service writer, advisor, or consultant is a good alternative. This job is good for those who have the technical knowledge but lack the desire or physical abilities to physically work on automobiles.

Many of the requirements for being a successful technician apply to being a successful service consultant. However, being a service consultant requires greater skill levels in customer relations, internal communication and relations, and sales. Service consultants must communicate well with customers, over the telephone or in person, in order to satisfy their needs or concerns. Most often this satisfaction involves the completion of a repair order, which contains customer information, instructions to the technicians, and a cost estimate.

Accurate estimates are not only highly appreciated by the customer, but they are also required by law in most states. Writing an accurate estimate requires a solid understanding of the automobile, good communications with the customers and technicians, and good reading and math skills.

Most shops use computers to generate the repair orders and estimates and to schedule the shop's workload. Therefore, having solid computer skills is an asset for service advisors.

Service Manager

The service manager is responsible for the operation of the entire service department at a large dealership or independent shop. Normally, customer concerns and complaints are handled by the service manager. Therefore, a good service manager has good people skills in addition to organizational skills and a solid automotive background.

In a dealership, the service manager makes sure the manufacturers' policies on warranties, service procedures, and customer relations are carried out. The service manager also arranges for technician training and keeps all other shop personnel informed and working together.

Service Director

Large new car dealerships often have a **service director** who oversees the operation of the service and parts departments as well as the body shop. The service director has the main responsibility of keeping the three departments profitable. The service director coordinates the activities of these separate departments to ensure efficiency.

Many service directors began their career as technicians. As technicians they demonstrated a solid knowledge of the automotive field and had outstanding customer relations skills and good business sense. The transition from technician to director typically involves promotion to various other managerial positions first.

Parts Counterperson

A parts counterperson (**Figure 1–16**) can have different duties and is commonly called a parts person or specialist. Parts specialists are found in nearly all automotive dealerships and auto parts retail and wholesale stores. They sell auto parts directly to customers and issue materials and supplies to auto repair specialists working in automotive services facilities and body shops. A parts counterperson must be friendly, professional, and efficient when working with all customers, both on the phone and in person.

Figure 1–16 A parts counterperson has an important role in the operation of a store or dealership.

Depending on the parts store or department, duties may also include delivering parts, purchasing a variety of automotive parts, maintaining inventory levels, and issuing parts to customers and technicians. Responsibilities include preparing purchase orders, scheduling deliveries, assisting in the receipt and storage of parts and supplies, and maintaining contact with vendors. An understanding of automotive terminology and systems is a must for good parts counterpersons.

This career is an excellent alternative for those who know about cars but would rather not work on them. Much of the knowledge required to be a technician is also required for a parts person. However, a parts specialist requires a different set of skills. Most automotive parts specialists acquire the sales and customer service skills needed to be successful primarily through on-the-job experience and training. They may also gain the necessary technical knowledge on the job or through educational programs and/or experience. To better understand the world of the parts industry refer to **Figure 1–17**, which defines the common terms used by parts personnel.

Parts Manager

The parts manager is in charge of ordering all replacement parts for the repairs the shop performs. The ordering and timely delivery of parts is extremely important for the smooth operation of the shop. Delays in obtaining parts or omitting a small but crucial part from the initial parts order can cause frustrating holdups for both the service technicians and customers.

Most dealerships and large independent shops keep an inventory of commonly used parts, such as filters, belts, hoses, and gaskets. The parts manager is responsible for maintaining this inventory.

An understanding of automotive systems and their parts, thoroughness, attention to detail, and the ability to work with people face to face and over the phone are essential for a parts manager.

RELATED CAREER OPPORTUNITIES

In addition to careers in automotive service, there are many other job opportunities directly related to the automotive industry.

Parts Distribution

The **aftermarket** refers to the network of businesses (**Figure 1–18**) that supplies replacement parts to independent service shops, car and truck dealerships, fleet operations, and the general public.

Vehicle manufacturers and independent parts manufacturers sell and supply parts to approximately a thousand warehouse distributors throughout the United States. These **warehouse distributors (WDs)** carry substantial inventories of many part lines.

Warehouse distributors serve as large distribution centers. WDs sell and supply parts to parts wholesalers, commonly known as jobbers.

Jobbers sell parts and supplies to shops and do-it-yourselfers. Jobbers often have a delivery service that gets the desired parts to a shop shortly after it ordered them. Some parts stores focus on individual or walk-in customers. These businesses offer the do-it-yourselfers repair advice, and some even offer testing of old components. Selling good parts at a reasonable price and offering extra services to their customers are the characteristics of successful parts stores. Many jobbers operate machine shops that offer another source of employment for skilled technicians. Jobbers or parts stores can be independently owned and operated. They can also be part of a larger national chain (**Figure 1–19**). Auto manufacturers have also set up their own parts distribution systems to their dealerships and authorized service outlets. Parts manufactured by the original vehicle manufacturer are called **original equipment manufacturer (OEM)** parts.

Opportunities for employment exist at all levels in the parts distribution network, from warehouse distributors to the counterpeople at local jobber outlets.

Marketing and Sales

Companies that manufacture equipment and parts for the service industry are constantly searching for knowledgeable people to represent and sell their products. For example, a sales representative working for an aftermarket parts manufacturer should have a good knowledge of the company's products. The sales representative also works with WDs, jobbers, and service shops to make sure the parts are being sold and installed correctly. They also help coordinate training and supply information so that everyone using their products is properly trained and informed.

Other Opportunities

Other career possibilities for those trained in automotive service include automobile and truck recyclers, insurance company claims adjusters, auto body shop technicians, and trainers for the various manufacturers or instructors for an automotive program (**Figure 1–20**). The latter two careers require solid experience and a thorough understanding of the automobile. It is not easy being an instructor or

ACCOUNTS RECEIVABLE Money due from a customer.

ALPHANUMERIC A numbering system commonly used in parts catalogs and price listings. This system uses a combination of letters and numbers. They are placed in order starting from the left digit and working across to the right.

BACK ORDER Parts ordered from a supplier that have not been shipped to the store or shop because the supplier has none in its inventory.

BILL OF LADING A shipping document acknowledging receipt of goods and stating terms of delivery.

CATALOGING The process of looking up the needed parts in a parts catalog.

CORE CHARGE A charge that is added when a customer buys a remanufactured part. Core charges are refunded to the customer when he or she returns a rebuildable part.

CORRECTION BULLETIN A bulletin that corrects catalog errors due to printing errors or inaccurately assigned part numbers.

CUSTOMER RELATIONS A description of how a salesperson interacts with the customer.

DEALERS The jobber's wholesale customers, such as service stations, garages, and vehicle dealers, who install parts in their customers' vehicles.

DISCOUNT The amount of savings being offered to a customer, normally expressed as a percentage.

DISTRIBUTOR A large-volume parts-stocking business that sells to wholesalers.

FREIGHT CHARGE A charge added to special order parts to cover their transportation to the store.

GROSS PROFIT The selling price of a part minus its cost.

HIGH-VOLUME Describes a popular item, which is sold in large numbers.

INDIVIDUALLY PRICED The condition of having each part of a display priced for customer convenience.

INVENTORY The parts a store or shop has in its possession for resale.

INVENTORY CONTROL A method of determining amounts of merchandise to order based on supplies on hand and past sales of the item.

INVOICE The record of a sale to a customer.

JOBBER The owner or operator of an auto parts store usually wholesaling products to volume purchasers such as dealers, fleet owners, and businesses. They also may sell retail to do-it-yourselfers.

LIST PRICE The suggested selling price for an item.

MARGIN Same as gross profit.

MARKUP The amount a business charges for a part above the actual cost of the part.

NET PRICE A business's profit after deducting the cost of all of its merchandise and all expenses involved in operating the business.

NO-RETURN POLICY A store policy that certain parts cannot be returned after purchase. It is common to have a no-return policy on electrical and electronic parts.

ON HAND The quantity of an item that the store or shop has in its possession.

PERPETUAL INVENTORY A method of keeping a continuous record of stock on hand through sales receipts and/or invoices.

PHYSICAL INVENTORY The process whereby each part is manually counted and the number on hand is written on a form or entered into a computer.

PROFIT The amount received for goods or services above the shop's or store's cost for the part or service.

PURCHASE ORDER A form giving someone the authority to purchase goods or services for a company.

REMANUFACTURED PART A part that has been reconditioned to its original specifications and standards.

RESTOCKING FEE The fee charged by a store or supplier for having to handle a returned part.

RETAIL Selling merchandise to walk-in trade (do-it-yourselfers).

RETURN POLICY A policy regarding the return of unwanted and unneeded parts. Return policies may include restocking fees or prohibit the return of certain parts.

SELLING PRICE The price at which a part is sold. This price will vary according to the type of customer (retail or wholesale) who is purchasing the part.

SPECIAL ORDER An order placed whenever a customer purchases an item not normally kept in stock.

STOCK ORDER A process by which the store orders more stock from its suppliers in order to maintain its inventory.

STOCK ROTATION Selling the older stock on hand before selling the newer stock.

SUPERSESSION BULLETIN A bulletin sent by the parts supplier that lists part numbers that now replace (supersede) previous part numbers.

TURNOVER The number of times each year that a business buys, sells, and replaces a part.

VENDOR The supplier.

WARRANTY RETURN A defective part returned to the supplier due to failure during its warranty period.

WAREHOUSE DISTRIBUTOR The jobber's supplier who is the link between the manufacturer and the jobber.

WHOLESALE The business's price to large-volume customers.

Figure 1-17 Some of the common terms used by parts personnel.

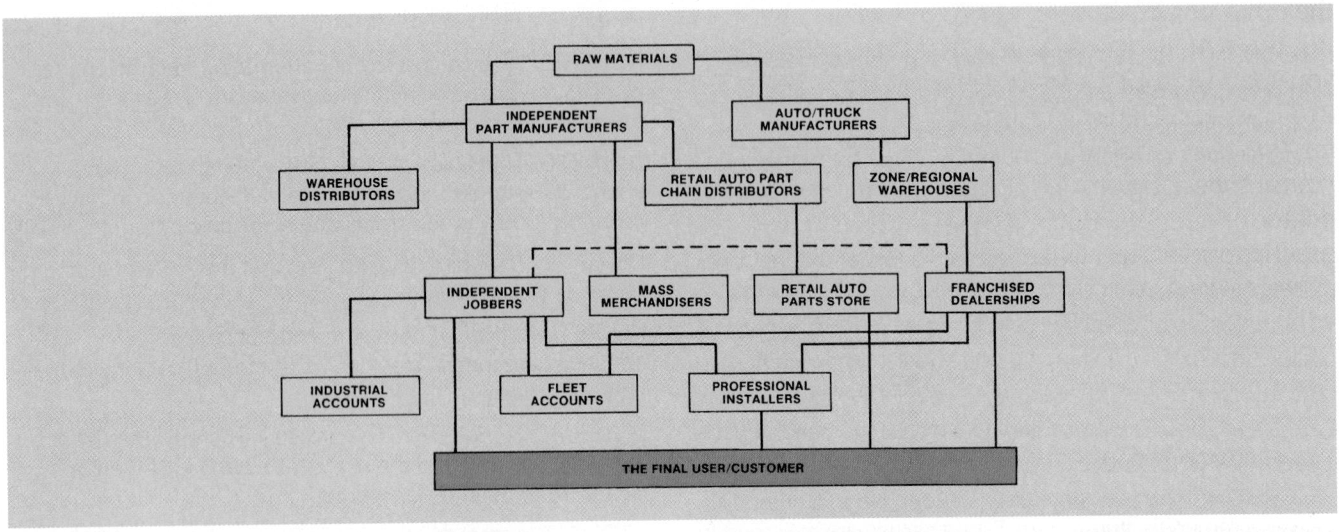

Figure 1-18 The auto parts supply network.

Figure 1-19 Many parts stores are part of a national corporation with stores located across the country.

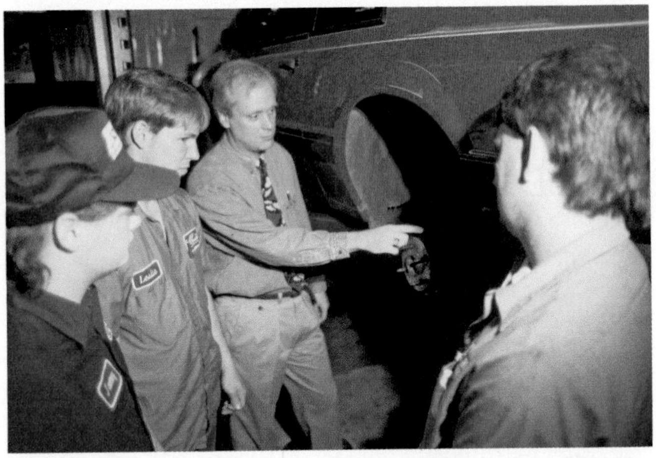

Figure 1-20 A career possibility for an experienced technician is that of a trainer for the various manufacturers or instructors for an automotive program.

trainer; however, passing on knowledge can be very rewarding. Undoubtedly, there is no other career that can have as much impact on the automotive service industry as that of a trainer or instructor.

TRAINING FOR A CAREER IN AUTOMOTIVE SERVICE

Those interested in a career in auto service can receive training in formal school settings—secondary, postsecondary, and vocational schools; and technical or community colleges, both private and public.

Student Work Experience

There are many ways to gain work experience while you are a student. You may already be involved in one of the following; if not, consider becoming involved in one of these possibilities.

Job Shadowing Program In this program you follow an experienced technician or service writer. The primary program objective is to expose you to the "real world," to see what it takes to be a successful technician or service writer. By job shadowing, you will also become familiar with the total operation of a service department.

Mentoring Program This program has the lowest participation rate of all these programs but can be one of the most valuable. In a mentoring program, you have someone who is successful to use as an expert. Your mentor has agreed to stay in contact with you, to answer questions, and to encourage you. When you have a good mentor, you have someone who may be able to explain things a little differently than the way things are explained in class. A mentor may also be able to give real life examples of why some of the things you need to learn are important.

Cooperative Education This type of program is typically 2 years in length. One year is spent in school and the other in a dealership or service facility. This does not mean that 1 solid year is spent in school; rather you spend 8 to 12 weeks at school, and then work for 8 to 12 weeks. The switching back and forth continues for 2 years. Not only do you earn an hourly wage while you are working, you also earn credit toward your degree or diploma. While at work, you get a chance to practice and perfect what you learned in school. Your experiences at work are carefully coordinated with your experiences at school; therefore, it is called a cooperative program—industry cooperates with education. Examples of this type of program are the Chrysler CAPS, Ford ASSET, GM ASEP, and Toyota T-Ten (in Canada these are called T-TEP) programs.

Apprenticeship Program Similar to a cooperative education program, an apprenticeship program combines work experiences with education. The primary difference between the two programs is that in an apprenticeship program students attend classes in the evening after completing a day's work. During this rigorous training program, you receive a decent hourly wage and plenty of good experience. You start the program as a helper to an experienced technician and can begin to do more on your own as you progress through the program. Most apprenticeship programs take 2 years to complete. Automobile manufacturers and dealers often sponsor these programs.

Part-Time Employment The success of this experience depends on you and your drive to learn. Working part-time will bring you good experience, some income, and a good start in getting a great full-time position after you have completed school. The best way to approach this is to find a position and service facility that will allow you to grow. You need to start at a right level and be able to take on more difficult tasks when you are ready. The most difficult challenge when working part-time is to keep up with your education while you are working. Many times work may get in the way, but if you truly want to learn, you will find a way to fit your educational needs around your work schedule.

Postgraduate Education A few manufacturer programs are designed for graduates of postsecondary schools. These programs train individuals to work on particular vehicles. For example, BMW's Service Technician Education Program (STEP) is a scholarship program for the top graduates of automotive postsecondary schools. Students in the program apply what they learned in their 2-year program and learn to diagnose and service BMW products. BMW says this program is the most respected and intense training program of its kind in the world. For more information go to http://www.bmwstep.com.

The Need for Continuous Learning

Training in automotive technology and service does not end with graduation. Nor does the *need to read* end. A professional technician constantly learns and keeps up to date. In order to maintain your image as a professional and to keep your knowledge and skills up to date, you need to do what you can to learn new things. You need to commit yourself to lifelong learning.

There are many ways in which you can keep up with the changing technology. Short courses on specific systems or changes are available from the manufacturers and a number of companies that offer formal training, such as such as Federal Mogul, NAPA, AC Delco, and local parts jobbers. There are also several on-line courses available. It is wise to attend update classes as soon as you can. If you wait too long, you may have a difficult time catching up with the ever-changing technologies.

In addition to taking classes, you can learn by reading automotive magazines or the newest editions of automotive textbooks. A good technician takes advantage of every opportunity to learn.

ASE CERTIFICATION

The National Institute for **Automotive Service Excellence (ASE)** has established a voluntary certification program for automotive, heavy-duty truck, auto body repair, and engine machine shop technicians. In addition to these programs, ASE also offers individual testing in the areas of automotive and heavy-duty truck parts, service consultant, alternate fuels, advanced engine performance, and a variety of other areas. This certification system combines voluntary testing with on-the-job experience to confirm that technicians have the skills needed to work on today's more complex vehicles. ASE recognizes two distinct levels of service capability—the automotive technician and the master automotive technician. The master automotive technician is certified by ASE in all major automotive systems. The automotive technician may have certification in only several areas.

To become ASE certified, a technician must pass one or more tests that stress system diagnosis and repair procedures. The eight basic certification areas in automotive repair follow:

1. Engine repair
2. Automatic transmission/transaxle
3. Manual transmissions and drive axles
4. Suspension and steering

Figure 1–21 ASE certification shoulder patches worn by (*left*) automotive technicians and (*right*) master automotive technicians.

5. Brakes
6. Electrical systems
7. Heating and air conditioning
8. Engine performance (driveability)

After passing at least one exam and providing proof of 2 years of hands-on work experience, the technician becomes ASE certified. Retesting is necessary every 5 years to remain certified. A technician who passes one examination receives an automotive technician shoulder patch. The master automotive technician patch is awarded to technicians who pass all eight of the basic automotive certification exams (**Figure 1–21**).

ASE also offers advanced-level certification in some areas. The most commonly sought advanced certification for automobile technicians is the L1 or Advanced Engine Performance. Individuals seeking this certification must be certified in Electricity and Engine Performance before taking this exam.

ASE also offers specialist certifications. For example, to become a certified Undercar Specialist, you must have certification in Suspension and Steering, Brake, and Exhaust Systems (a speciality test). Certification is also available for Parts Counterperson and Service Consultants.

As mentioned, ASE certification requires that you have 2 years of full-time, hands-on working experience as an automotive technician. You may receive credit toward this 2-year experience requirement by completing formal training in one or a combination of the following:

- High school training
- Post–high school training
- Short courses
- Apprenticeship programs

Each certification test consists of forty to eighty multiple-choice questions. The questions are written by a panel of technical service experts, including domestic and import vehicle manufacturers, repair and test equipment and parts manufacturers, working automotive technicians, and automotive instructors. All questions are pretested and quality checked on a national sample of technicians before they are included in the actual test. Many test questions force the student to choose between two distinct repair or diagnostic methods.

For further information on the ASE certification program, go to its Web site at http://www.ase.com.

KEY TERMS

Aftermarket
Automotive Service Excellence (ASE)
Bay
Deductible
Diagnostic skills
Independent shops
Inspection/ Maintenance (I/M)
Jobbers
Lead tech
Original equipment manufacturer (OEM)
Preventive maintenance (PM)
Service advisor
Service director
Service technician
Shop foreman
Warehouse distributors (WDs)
Warranty

SUMMARY

- The modern auto industry is a global industry involving vehicle and parts manufacturers from many countries.
- Electronic computer controls are found on many auto systems, such as engines, ignition systems, transmissions, steering systems, and suspensions. The use of electronics in automobiles is increasing rapidly.
- Preventive maintenance is extremely important in keeping today's vehicles in good working order.
- New car dealerships, independent service shops, specialty service shops, fleet operators, and many other businesses are in great need of qualified service technicians.
- A solid background in auto technology may be the basis for many other types of careers within the industry. Some examples are parts management, collision damage appraisal, sales, and marketing positions.
- Training in auto technology is available from many types of secondary, vocational, and technical

schools. Auto manufacturers also have cooperative programs with schools to ensure that graduates understand modern systems and the equipment to service them.

- The National Institute for Automotive Service Excellence (ASE) actively promotes professionalism within the industry. Its voluntary certification program for automotive technicians and master auto technicians helps guarantee a high level of quality service.
- The ASE certification process involves both written tests and credit for on-the-job experience. Testing is available in many areas of auto technology.

REVIEW QUESTIONS

1. Give a brief explanation of why electronics are so widely used on today's vehicles.
2. Explain the basic requirements for becoming a successful automotive technician.
3. List at least five different types of businesses that hire service technicians. Describe the types of work these businesses handle and the advantages and disadvantages of working for them.
4. Name four ways that you can gain work experience while you are a student.
5. *True or False?* In most cooperative education programs, students work at an automotive service facility during the day and attend classes in the evenings.
6. Individuals often begin a career as an automotive technician in a new car dealership by ______, which is a good way for a new technician to become familiar with the vehicles sold at the dealership.
   a. working at the parts counter
   b. performing new car prep
   c. being a service advisor
   d. serving as the lead tech
7. *True or False?* A hybrid vehicle uses two different power sources.
8. The government-mandated warranty that specifically covers the catalytic converter(s) and engine control module is the ______.
   a. Federal Emissions Defect Warranty
   b. Federal Powertrain Warranty
   c. Federal Emissions Performance Warranty
   d. Extended Federal Exhaust Warranty
9. Which of the following is typically included in a scheduled preventive maintenance program?
   a. oil and filter changes
   b. coolant and lubrication services
   c. replacement of filters
   d. all of the above
10. In a large new car dealership, the individual who oversees the operation of the service department, parts department, and body shop is the ______.
    a. service manager
    b. service director
    c. shop foreman
    d. parts manager
11. Repair work performed on vehicles still under the manufacturer's warranty is usually performed by ______.
    a. independent service shops
    b. dealerships
    c. specialty shops
    d. either a or b
12. Which of the following businesses perform work on only one or two automotive systems?
    a. dealerships
    b. independent service shops
    c. specialty shops
    d. fleet service departments
13. Normally, whose job is it to greet the customer and complete the repair or work order?
    a. service manager
    b. parts manager
    c. automotive technician
    d. service advisor
14. Technician A says that all an individual needs to do in order to become certified by ASE in a particular area is to pass the certification exam in that area. Technician B says that all of the questions on an ASE exam force the test taker to choose between two distinct repair methods. Who is correct?
    a. Technician A
    b. Technician B
    c. Both A and B
    d. Neither A nor B
15. To be successful, today's automotive technician must have ______.
    a. an understanding of electronics
    b. the ability to repair and service mechanical systems

c. the dedication to always be learning something new

d. all of the above

16. A technician must have a minimum of ______ year(s) of hands-on work experience to get ASE certification.

a. 1
b. 2
c. 3
d. 4

17. A technician who passes all eight basic ASE automotive certification tests is certified as a(n) ______.

a. automotive technician
b. master automotive technician
c. service manager
d. parts manager

18. Technician A says battery warranties are often prorated. Technician B says prorated warranties have a deductible. Who is correct?

a. Technician A
b. Technician B
c. Both A and B
d. Neither A nor B

19. Wholesale auto parts stores that sell aftermarket parts and supplies to service shops and the general public are called ______.

a. warehouse distributors
b. mass merchandisers
c. jobbers
d. freelancers

20. Ongoing technical training and support is available from ______.

a. aftermarket parts manufacturers
b. auto manufacturers
c. jobbers
d. all of the above

WORKPLACE SKILLS

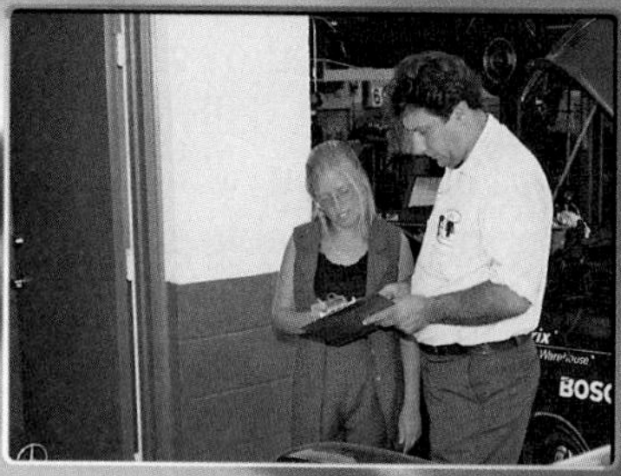

CHAPTER 2

OBJECTIVES

■ Develop a personal employment plan. ■ Seek and apply for employment. ■ Prepare a resume and cover letter. ■ Prepare for an employment interview. ■ Accept employment. ■ Understand how automotive technicians are compensated. ■ Understand the proper relationship between an employer and an employee. ■ Explain the key elements of on-the-job communications. ■ Be able to use critical thinking and problem-solving skills. ■ Explain how you should look and act to be regarded as a professional. ■ Explain how fellow workers and customers should be treated.

This chapter gives an overview of what you should do to get a job and how to keep it. The basis for this discussion is respect—respect for yourself, your employer, fellow employees, your customers, and everyone else. Also included in this discussion are the key personal characteristics required of all seeking to be successful automotive technicians and employees.

SEEKING AND APPLYING FOR EMPLOYMENT

Becoming employed, especially in the field in which you want a career, involves many steps. As with many things in life, you must be adequately prepared before taking the next step toward employment. This discussion suggests ways you can prepare and what to expect while taking these steps.

Employment Plan

An **employment plan** is nothing more than an honest appraisal of yourself and your career hopes. The plan should include your employment goals, a timetable for reaching those goals, and a prioritized list of potential employers or types of employers. You may need to share your employment plan with someone while you are seeking employment, so make sure it is complete. Even if no one else will see it, you should be as thorough as possible because it will help keep you focused during your quest for employment.

Think about the type of job you want and do some research to find out what is required to get that job. Evaluate yourself against those requirements. If you do not meet the requirements, set up a plan for obtaining the needed skills. Also, consider the working conditions of that type of job. Are you willing and able to be a productive worker in those conditions? If not, find a job that is similar to your desires and pursue that career.

Self-Appraisal To begin the self-appraisal part of your employment plan, ask yourself:

- Why am I looking for a job?
- What specifically do I hope to gain by having a job?
- What do I like to do?
- What am I good at?
- Which of my skills would I like to use in my job?

By honestly answering these questions, you should be able to identify the jobs that will help you meet your goals. If you are just seeking a job to pay bills or buy a car and have no intention of turning this job into a career, be honest with yourself and your potential employer. If you are hoping to begin a successful career, realize you will probably start at the bottom of the ladder to success. You must also realize that how quickly you climb the ladder is your responsibility. An employer's responsibility is merely to give you a fair chance to climb it.

Identifying Your Skills Honestly evaluate yourself and your life to determine what skills you have. Even if you have never had a job, you still have skills and talents that can make you a desirable employee. Make a list of all of the things you have learned from your school, friends, and family and through television, volunteering, books, hobbies, and so on. You may be surprised by the number of skills you have. Identify these skills as being either technical or personal skills.

Technical skills include things you can do well and enjoy, such as:

- Using a computer
- Working with tools, machines, or equipment
- Playing video games
- Doing math problems
- Maintaining or fixing things
- Figuring out how things work
- Making things with your hands
- Working with ideas and information
- Solving puzzles or problems
- Studying or reading
- Doing experiments or researching a topic
- Expressing yourself through writing

Personal skills are also called **soft skills** and are things that are part of your personality. These are things you are good at or enjoy doing, such as:

- Working with people
- Caring for or helping people
- Working as a member of a team and independently
- Leading or supervising others
- Following orders or instructions
- Persuading people
- Negotiating with others

By identifying these skills you will have created your personal skills inventory. From the inventory you match your skills and personal characteristics to the needs and desires of potential employers. The inventory will also come in handy when marketing yourself for a job, such as when preparing your resume and cover letter and during an interview.

Identifying Job Possibilities

One of the things you identified in your employment plan was your preferred place to work. This may have been a specific business or a type of business, such as a new car dealership or independent shop. Now your task is to identify the companies that are looking for

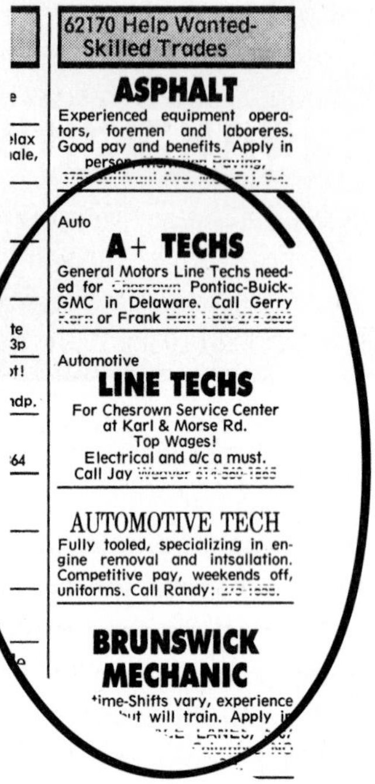

Figure 2-1 Check the help-wanted ads in your local newspaper for businesses that are looking for technicians.

someone. To do this, look through the help-wanted section in the newspaper **(Figure 2–1)**. You can also check your school's job posting board or ask people you know who already work in the business. If there is nothing available in the business you prefer, look for openings in the type of business that was second on your priority list.

Carefully look at the description of the job. Make sure you meet the qualifications for the job before you apply. For example, if you have a drug problem and the ad states that all applicants will be drug tested, you should not bother applying and should concentrate on breaking the habit. Even if the ad says nothing about testing for drug use, you should know that there is no place for drugs at work and continued drug use will only jeopardize your career.

Driving Record Your driving record is something you must also be aware of, and you probably are. If you have a poor record, you may not be considered for a job that requires operating a vehicle. In the same way that a driving record affects your personal car insurance, the employer's insurance costs can also increase because of your poor driving record. A bad driving record or the loss of a driver's license can jeopardize getting or keeping a job.

Preparing Your Resume

Your **resume** and cover letter are your own personal marketing tools and may be the first look at you an employer has. Although not all employers require a resume, you should prepare one for those that do. Preparing a resume also forces you to look at your qualifications for a job. That alone justifies having a resume.

Keep in mind that although you may spend hours writing and refining your resume, an employer may only take a minute or two from his or her busy schedule to look it over. With this in mind, put together a resume that tells the employer who you are in such a way that he or she wants to interview you.

A resume normally includes your contact information, career objective, skills and/or accomplishments, work experience, education, and a statement about references. There are different formats you can follow when designing your resume. If you have limited work experience, make sure the resume emphasizes your skills and accomplishments rather than work history. Even if you have no work experience, you can sell yourself by highlighting some of the skills and attributes you identified in your employment plan.

When listing or mentioning your attributes and skills, express them in a way that shows how they relate to the job you are seeking. For instance, if you practice every day at your favorite sport so you can make the team, you may want to describe yourself as being persistent, determined, motivated, and goal-oriented. Another example is if you have ever pulled an all-nighter to get an assignment done on time, it can mean that you work well under pressure and always get the job done. Another example would be if you keep your promises and do what you said you would do, you may want to describe yourself as reliable, a person who takes commitment seriously.

Identifying your skills may be a difficult task, so have your family and/or friends help you. Keep in mind that you have qualities and skills that employers want. You need to recognize them, put them in a resume, and tell them to your potential employer. Do not put the responsibility of figuring out who you are on the employers—tell them.

Figure 2–2 is an example of a basic resume for an individual seeking an entry-level position as a technician.

Putting Together an Effective Resume Follow these guidelines while preparing and writing your resume:

- Your resume must be typewritten. If you do not have access to a computer or a typewriter, your local library probably has them available for public use.
- Make sure it is neat, uncluttered, and easy to read.
- Use quality white paper.
- Keep it short—a maximum of one or two pages.
- Use dynamic words to describe your skills and experience, such as accomplished, achieved, communicated, completed, created, delivered, designed, developed, directed, established, founded, instructed, managed, operated, organized, participated, prepared, produced, provided, repaired, and supervised.
- Choose your words carefully; remember that the resume is a look at you.
- Make sure all information is accurate.
- Make sure the information you think is the most important stands out and is positioned near the top of the page.
- Design your resume with a clean letter type (font) and wide margins (1½ inches on both sides is good) so that it is easy on the eyes.
- Only list the "odd" jobs you had if they are related to the job you are applying for.
- Do not repeat information.
- Proofread the entire resume to catch spelling and grammatical errors. If you find them, fix them and print a new, clean copy.
- Do not make handwritten corrections or use correction fluid to cover mistakes.
- Make sure your resume is not dirty and wrinkled when you deliver it.

References

A **reference** is someone who will be glad to tell a potential employer about you. A reference can be anyone who knows you, other than a family member or close friend. Employers contact references to verify or complete their picture of you. Make a list of three to five people you can use as references, including their contact information. If you do not supply references, the potential employer may assume that you cannot find anyone who has anything nice to say about you. You probably will not be considered for the job.

Choose your references wisely. Teachers (past and present), coaches, and school administrators are good examples of who you can ask to be a reference. People you have worked for or have helped are also good references. Try also to get someone whose opinion is respected, such as a priest, minister, or elder in your church or someone you know well who holds a high position.

Always talk to your references first, and get permission to give their names and telephone numbers to an employer. If they do not seem comfortable with

Jack Erjavec
1234 My Street
Somewhere, OZ 99902
123-456-7890

Performance-oriented student, with an excellent reputation as a responsible and hard-working achiever, seeking a position as an entry-level automotive technician in a new car dealership.

Skills and Attributes

- People oriented
- Motivated
- Committed
- Strong communication and teamwork skills
- Honest
- Reliable
- Organized
- Methodical
- Creative problem-solver
- Good hand skills

Work Experience

2006–2008 Somewhere Soccer Association (Assistant coach)

- Instructed and supervised junior team
- Performed administrative tasks as the Coach required

2004–2006 Carried out various odd jobs within the community

- Washing and waxing cars, picking up children from school, raking leaves, cutting grass

Education

Somewhere Senior High School, graduated in 2008
Somewhere Community College, currently enrolled in the Automotive Technology Program

Extracurricular Activities

1999–2008 Active member of the video game club
1999–2008 Member of the varsity soccer team

Hobbies and Activities

Reading auto-related magazines, going to races, doing puzzles, working on cars with family and friends.

References

Available upon request.

Figure 2-2 A sample of a resume for someone who has little work experience.

giving you a reference, take the hint and move on to someone else. If someone is willing to provide you with a written reference, make several copies of the letter so you can attach them to your resume and/or job application. Give copies of your resume to those on your reference list. Make sure to bring your reference list when applying for a job.

Preparing Your Cover Letter

A cover letter **(Figure 2–3)** should be sent with every resume you mail, e-mail, fax, or personally deliver. A cover letter gives you a chance to point out exactly why you are perfect for the job. You should not send out the same cover letter to all potential employers. Adjust the letter to match the company and position

Jack Erjavec
1234 My Street
Somewhere, OZ 99902

March 24, 2008

Mrs. Need Someone
Service Manager, Exciting New Cars
56789 Big Dealer Avenue
Somewhere, OZ 99907

Re: application for an entry-level automotive technician position

Dear Mrs. Someone:

Your ad in the March 14 edition of the *Dogpatch* for an automotive technician greatly interested me, as this position is very much in line with my immediate career objective—a career position as an automotive technician in a new car dealership. Because of the people and cars featured at your dealership, I know working there would be exciting.

I have tinkered with cars for most of my life and am currently enrolled in the Automotive Technology program at Somewhere Community College. I chose this program because you are on the advisory council and I knew it must be a good program. I have good hand skills and work hard to be successful. My being on the varsity soccer team for four years should attest to that. I also enjoy working with people and have developed excellent communication skills. The position you have open is a perfect fit for me. A resume detailing my skills and work experience is attached for your review.

I would appreciate an opportunity to meet with you to further discuss my qualifications. In the meantime, many thanks for your consideration, and I look forward to hearing from you soon. I can be reached by phone at 123-456-7890, most weekdays after 2:30 PM. If I am unable to answer the phone when you call, please leave a message and I will return your call as soon as I can. Thanks again.

Sincerely,

Jack Erjavec

encl.

Figure 2-3 An example of a cover letter that can be sent with the resume in Figure 2–2.

you are applying for. Yes, this means a little more work, but it will be worth it. For example, if you state in one letter that you have always been a "Chevy nut" that may help at a Chevrolet dealership but will not at a Toyota or Ford dealership.

A good cover letter is normally made up of three paragraphs, each with its own purpose.

First Paragraph In the first paragraph, tell the employer that you are interested in working for the company, the position you are interested in, and why. Make sure you let the employer know that you know something about the company and what the job involves. Also include a statement of how you found out about the open position, which could be a help-wanted ad, a job posting at school, and/or a referral by someone who works for the company.

Second Paragraph In this paragraph, sell yourself by mentioning one or two of your job qualifications and describe them in more detail than you did in your resume. Make sure you expand on the material in your resume rather than simply repeat it. Point out any special training or experience you have that directly relates to the job. When doing this, give a summary without listing the places and dates. This information is listed in your resume so simply refer to the resume for details. This summary is another opportunity for you to let the employer know that you understand what they do, what the job involves, and how you can help them.

Third Paragraph Typically this paragraph is the end or closing of the letter. Make sure you thank the employer for taking the time to review your resume and ask him or her to contact you to make an appointment for an interview. Make sure you give a phone number where you can be reached. If you have particular times when it is best to contact you, put those times in this paragraph. Make sure you have a clear and understandable message on your telephone's answering machine, just in case you miss the employer's call. Also, have an organized work area around the phone so you can accurately schedule any interview appointments.

Guidelines for Writing an Effective Cover Letter Follow these guidelines while preparing and writing your cover letter:

- Address the letter to a person, not just a title. If you do not know the person's name, call the company and ask for the correct spelling of the person's name and his or her title.
- Make sure the words you use in the letter are upbeat.
- Use a natural writing style, keeping it professional but friendly.
- Try hard not to start every sentence with "I"; make some "you" statements.
- Check the letter for spelling and grammatical errors.
- Type the letter on quality paper and make sure it is neat and clean.
- Make sure you sign the letter before sending it.

Contacting Potential Employers

Unless the help-wanted ad or job posting tells you otherwise, it is best to drop your resume and cover letter off in person (preferably to the person who does the hiring). When you are doing this, make sure you tell the employer who you are and the job you want. Make sure you are prepared for what happens next. You may be given an interview right then. You may be asked to fill out an application. If so, fill it out.

Before you leave, thank the employer and ask if you can call back in a few days if you do not hear from him. If you do not hear back within a week, call to make sure the employer received your resume, reminding him of who you are and what job you applied for. If he tells you that the job is filled or that no jobs are available, politely thank him for considering you and tell him you will stay in touch in case there is a future job opening.

Employment Application

An application for employment is a legal document that summarizes who you are. It is also another marketing tool for you. Filling out the application is the first task the employer has asked you to do, so do it thoroughly and carefully. Make sure you are prepared to fill out an application before you go. Take your own pen and a paperclip so that you can attach your resume to the application. Make sure you have your reference list. When filling out the application, neatly print your answers.

Read over the entire application before filling it out. Make sure you follow the directions carefully. Too often applicants try to rush through the application and make mistakes or provide the wrong information. Also by reading through the application before you fill in the blanks, you have a better chance of filling it out neatly. A messy application or one with crossed out or poorly erased information tells employers you may not care about the quality of your work.

By following the directions and providing the employer with the information asked for, you are demonstrating that you have the ability to read, understand, and follow written instructions, rules, and procedures. When answering the questions, be honest.

Make sure you completely fill out the application. Doing this shows the employer that you can complete a task. Answer every question. Write N/A (nonapplicable) if a question does not apply to you. If lines are left blank, the employer may think you do not pay attention to the details of a job or are a bit lazy. When you have completed the application, sign it and attach your cover letter and resume to it.

The Interview

Typically if employers are interested in you, you will be contacted to come in for an interview. This is a good sign. If they were not impressed with what they know of you so far, they will not ask for an interview. Knowing this should give you some confidence as you prepare for the interview.

Although an interview does not last very long, it is a time when you can either get the job or lose it.

Get ready for the interview by taking time to learn as much as you can about the company. Think of some of the reasons the company should hire you. When doing this, think of how both of you would benefit. Think of questions you might ask the interviewer to show you are interested in the job and the business. Then make a list of questions that you think the employer might ask. Think about how you should answer each of them and practice the answers with your family and friends. Some of the more common interview questions include:

- What can you tell me about yourself?
- Why are you interested in this job?
- What are your strengths and weaknesses?
- If we employ you, what will you do for us?
- Do you have any questions about us or the job?

Tips for a Successful Interview

- Before the interview, think about what days and hours you can work and when you can start working.
- Make sure you take your social security card (or SIN card), extra copies of your resume, a list of your references and their contact information, as well as copies of any letters of recommendation you may have.
- Take paper and a pen to the interview so you can take notes. Often the interviewer will be doing the same.
- Try to relax right before the interview.
- Be on time (early is good) for the interview. If you are not exactly sure how to get to the business or what types of problems you may face getting there (such as traffic jams or construction), make a trip there 1 or 2 days before the interview. If you must be late, or if you cannot make it to the interview, call the employer as soon as possible and explain why. Ask if you can arrange for a new interview time.
- Show up looking neat and professional. Wear something more formal than what you would wear on the job.
- When you are greeted by the interviewer, introduce yourself and be ready to shake hands. Do it firmly!
- Listen closely to the interviewer and look at the interviewer while he or she talks.
- Answer all questions carefully and honestly. If you do not have an immediate answer, think about it before you open your mouth. If you do not understand the question, restate the question in the way you understand it. The interviewer will then know what question you are answering.
- Never answer questions with a simple "yes" or "no." Answer all questions with examples or explanations that show your qualities or skills.
- Market yourself but do not lie about or exaggerate your abilities.
- Show your desire and enthusiasm for the job, but try to be yourself; that is, not too shy or too aggressive.
- Never say anything negative about other people or past employers.
- Do not be overly familiar with the interviewer and do not use slang during the interview, even if the interviewer does.
- Restate your interest in the job and summarize your good points at the end of the interview.
- Ask the interviewer if you can call back in a few days.

After the Interview

After the interview, go to a quiet place and reflect on what just took place. Think about what you did well and what you could have done better. Write these down so you can refer to them when you are preparing for your next interview.

Within 3 days after the interview, write a letter to the interviewer, thanking him or her for his or her time. Make sure you remind the interviewer of your interest and qualifications. Take advantage of this additional chance to market yourself but do not be overly aggressive when doing this.

Remember, finding a job takes time and seldom do you land a job on your first attempt. If you do not get a job offer as a result of a first interview, do not give up. Do your best not to feel depressed or dejected. Simply realize that, although you are qualified, someone with more experience was chosen. Send a thank-you letter anyway; this may prompt the interviewer to think of you the next time a similar job becomes available.

Review your cover letter, resume, and interview experience. Identify anything that can improve your marketing tools. Do not feel shy about asking the employer who did not hire you what you could have done better. Discuss your job hunt with your family and friends who will provide support and encouragement. Keep in touch with people you know who are working and who may have job leads. Explore other options. Do not rule out volunteering or job shadowing as a means of connecting with the workplace.

If you do get a job offer, do not be afraid to discuss the terms and conditions before accepting. Find out, or confirm, things such as what you will be doing, the hours you will be working, how you will be paid, and what to do when you report to work the first day. If you have any concerns, do not hesitate to share them with someone whose opinion you respect before committing yourself to the job. Do not commit to the job and then change your mind a few days later. If you have any doubts about the job, think seriously about it before you accept or decline.

ACCEPTING EMPLOYMENT

When you accept the job, you are entering into an agreement with the employer. That agreement needs to be honored. Make sure you are ready to start working. You need to have transportation to and from work and the required tools and clothes for the job.

You will also need a social security (or, in Canada, a social insurance) number. If you do not already have one, you need one quickly. In the United States, you can apply for a social security number if you are a legal citizen or if you have a nonimmigrant visa status and have permission to work in the United States.

To apply for a social security number, you must appear in person at a Social Security office to complete the application form. You must take your birth certificate or valid passport with the necessary cards and authorizations to be employed. Once the forms are completed and submitted, it may take more than 2 weeks for you to receive a card with your number on it.

Typically before you begin to work, or at least before you get paid, you will fill out state and federal income tax forms. These forms give the company authorization to deduct income taxes from your wages. When you are an employee for a company, the company must deduct those taxes. One form you will fill out is the employees withholding allowance certificate form, called the W-4. This form tells the employer how much, according to a scale, should be deducted from your pay for taxes. Basically the form asks how many exemptions you would like to claim. What you should claim depends on many things, and it is best that you seek advice from someone before you fill this in. In fact, do this well before you arrive to fill out the form.

Compensation

Automotive technicians **(Figure 2–4)** can be paid in a number of ways. When deciding on whether or not to accept a job, make sure you understand how you will be paid. Keep in mind that the employer agrees to pay you in exchange for your work, the quality of which is unknown before you start to work. When you accept employment, you accept the terms of compensation offered to you. Do not show up on the first day of work demanding more. After you have started working, progressed on the job, and made the company money, you can ask for more.

Hourly Wages Most often, new or apprentice technicians are paid a fixed wage for every hour they work. The amount of pay per hour depends on the business, your skill levels, and the work you will be doing. While collecting an hourly wage, you have a chance to learn the trade and the business. Time is usually spent working with a master technician or doing low-skilled

Figure 2-4 Automotive technicians can be paid in a number of ways.

jobs. As you learn more and become more productive, you can earn more. Many shops pay a good hourly rate to their productive technicians. Some have bonus plans that allow technicians to make more when they are highly productive. Nearly all service facilities for fleets pay their technicians an hourly wage.

Commission When technicians are paid on a **commission** basis, they receive a minimum hourly wage plus a percentage of what the shop receives from the customers for performing various services. This pay system can work well for technicians who are employed in a shop whose business fluctuates through the year. This system, along with the "flat-rate" system, is often referred to as incentive pay systems.

Flat Rate **Flat rate** is a pay system in which a technician is paid for the amount of work he or she does. The flat-rate system favors technicians who work in a shop that has a large volume of work. Although this pay plan offers excellent wages, it is not recommended for new and inexperienced technicians.

Every conceivable service to every different model of vehicle has a flat-rate time. These times are assigned by the automobile manufacturers. The times are based on the average time it takes a team of technicians to perform the service on new vehicle models. Flat-rate times **(Figure 2–5)** are listed in a **labor guide**, which can be a manual or be available on a computer. When you are paid on a flat-rate basis, your pay is based on that time, regardless of how long it took to complete the job. Flat-rate times are also used to determine the amount of money the dealership will receive from the manufacturer for making warranty repairs.

To explain how this system works, suppose a technician is paid $15.00 per hour flat rate. If a job has a flat rate time of 3 hours, the technician will be paid $45.00 for the job, regardless of how long it took to complete it. Experienced technicians beat the flat-rate time nearly all of the time. Their weekly pay is based on the time "turned in," not on the time spent. If the technician turns in 60 hours of work in a 40-hour workweek, he or she actually earns $22.50 each hour worked. However, if the technician turns in only 30 hours in the 40-hour week, the hourly pay is $11.25.

The flat-rate times from the manufacturers are used primarily for warranty repairs. Once a vehicle gets a little older, it takes a little longer to service it. This is because dirt, rust, and other conditions make the services more difficult. Because of this, the flat-rate times for older vehicles are longer. Because nondealership service facilities normally work on "out-of-warranty" vehicles, therefore older vehicles, the flat-rate times are about 20% higher than those used in a dealership. These flat-rate times are given in flat-rate manuals published by Chilton, Motor, and Mitchell.

At times, a flat-rate technician will be paid for the amount of time spent on the job. This is commonly referred to as "straight" or "clock" time. Straight time is paid when a service procedure is not listed in the flat-rate manual and when the customer's concern requires more than normal diagnostic time.

Benefits Along with the pay, the employer may offer benefits, sometimes called "fringe benefits." The cost of the benefits may be paid for by the business, or you may need to pay a share or all of the costs. There is no common benefit package for automotive technicians. Common benefits include:

- Health insurance
- Retirement plans
- Paid vacations
- Paid sick days
- Uniforms and uniform cleaning services
- Update training

When accepting employment, make sure you understand the benefits and seek help in choosing which you should participate in.

Total Earnings Depending on the business, you may be paid weekly or twice a month. The total amount of what you earn is called your **gross pay**. This is not your "take home" or **net pay**. Your net pay is the result of subtracting all taxes and benefit costs from your gross pay. These deductions may include:

- Federal income taxes
- State income taxes
- City income taxes
- Federal Insurance Contribution Act (FICA) taxes—this is commonly known as social security taxes
- Your contribution toward health insurance
- Uniform costs

WORKING AS A TECHNICIAN

Once you have the job, you need to keep it. Your performance during the first few weeks will determine how long you will stay employed and how soon you will get a raise or a promotion. Make sure you arrive to work on time. If you are going to be late or absent, call

HON-6

ACCORD : CIVIC : CIVIC DEL SOL : CRX : INSIGHT : PRELUDE : S2000

	LABOR TIME	SEVERE SERVICE
Fuel Cut-off Solenoid, Replace (B)		
1981-91 Accord	.7	*.7*
1981-91 Civic	.5	*.5*
CRX	.5	*.5*
1981-91 Prelude	.7	*.7*
Fuel Tank, Replace (B)		
Includes: Drain and refill.		
Accord		
1981-95 (.9)	1.3	*1.6*
1996-97 (.8)	1.2	*1.5*
1998-02		
disc brakes	4.0	*4.3*
drum brakes	4.6	*4.9*
2003-05 (3.1)	4.5	*4.8*
Civic (1.0)	1.5	*1.8*
CRX, Civic del Sol	1.7	*2.0*
Insight (1.2)	1.8	*2.1*
Prelude		
1981-82	1.4	*1.7*
1983-87	2.3	*2.6*
1988-96	1.3	*1.6*
1997-01	2.1	*2.4*
S2000 (4.0)	5.8	*6.1*
w/NGV Civic GX add	.5	*.5*
Fuel Pump, Replace (B)		
Includes: Testing.		
1981-86		
electric	1.2	*1.5*
manual	.7	*.9*
Accord		
1987-89	.8	*1.1*
1990-97	1.6	*1.8*
1998-05 (.6)	1.0	*1.3*
Civic, CRX, Civic del Sol		
1988-91	1.6	*1.9*
1992-05 (.6)	1.0	*1.3*
Civic HB, IMA (.4)	.6	*.8*
Insight (1.4)	2.2	*2.5*
Prelude	1.0	*1.2*
S2000 (1.5)	2.4	*2.7*
Fuel Pump Relay, Replace (B)		
1981-01	.5	*.5*
Intake Manifold and/or Gasket, Replace (B)		
Includes: Adjustments.		
Accord		
1981-97	3.8	*4.3*
1998-05		
4 cyl.	1.6	*1.8*
V6	.8	*.9*
Civic		
1981-87	3.6	*3.9*
1988-00	3.8	*4.3*
2001-05		
1.3L, 1.7L (.9)	1.4	*1.5*
2.0L (1.4)	2.1	*2.4*
CRX		
1984-87	3.0	*3.3*
1988-91	4.1	*4.4*
Insight (1.2)	1.8	*2.1*
Civic del Sol, Prelude	3.8	*4.3*
S2000 (2.0)	3.1	*3.4*
Replace		
injector plate 2001-05 Civic add	.3	*.3*
manifold add	.5	*.7*
INJECTION		
Fast Idle Valve, Replace (B)		
1985-01 (.4)	.7	*.7*
Fuel Injectors, Replace (B)		
Accord		
1985-93	.7	*1.0*
1994-03		
4 cyl. (.4)	.7	*1.0*
V6	2.3	*2.6*
Civic		
1985-00	.7	*1.0*
2001-03		
1.3L, 2.0L (.4)	.7	*1.0*
1.7L (.7)	1.1	*1.2*
CRX	1.7	*2.0*
Civic del Sol	1.2	*1.5*
Insight, S2000 (.4)	.7	*1.0*
1985-01 Prelude	.7	*1.0*
Fuel Pressure Regulator, Replace (B)		
1985-05 (.3)	.5	*.7*
w/1998-01 Accord V6 add	1.1	*1.2*
w/2001-05 Civic 1.3L, 1.7L add	.2	*.2*
Idle Air Control (IAC) Valve, Replace (B)		
Accord		
1986-93	.5	*.5*
1994-02	1.2	*1.4*
2003-05 (.5)	.8	*.9*
Civic		
1987-95	.3	*.3*
1996-00		
exc. Si	.9	*1.0*
Si	.3	*.3*
2001-05 (.5)	.9	*1.0*
CRX, Civic del Sol	.3	*.3*
Insight (.7)	1.1	*1.2*
Prelude	.5	*.5*
S2000 (.3)	.5	*.5*
Throttle Body Assy., Replace (B)		
Accord		
1985-89	1.8	*1.9*
1990-97		
4 cyl.	.8	*.9*
V6	1.6	*1.7*
1998-05 (.5)	.8	*.9*
Civic, CRX		
1985-91	1.5	*1.6*
1992-05 (.5)	.8	*.9*
Civic del Sol	1.2	*1.3*
Insight (.6)	1.0	*1.1*
Prelude		
1985-91	1.8	*1.9*
1992-01	.8	*.9*
S2000 (.5)	.8	*.9*
EXHAUST		
Catalytic Converter, R&I or Replace (B)		
Accord		
1981-85	1.2	*1.3*
1986-05 (.5)	.8	*.9*
Civic, CRX		
1981-87	1.2	*1.3*
1988-05 (.5)	.8	*.9*
Civic del Sol	1.2	*1.3*
Insight (.5)	.8	*.9*
Prelude		
1981-87	1.2	*1.3*
1988-01 (.5)	.8	*.9*
S2000 (.5)	.8	*.9*
Center Muffler, Replace (B)		
1986-87	1.7	*1.8*
Exhaust Manifold or Gasket, Replace (B)		
Accord		
1981-83	3.7	*4.0*
1984-89	1.5	*1.8*
1990-97	.8	*1.1*
1998-02		
4 cyl.	.8	*1.1*
V6		
one side	1.8	*2.1*
both sides	2.3	*2.6*
2003-05 (.4)	.6	*.7*
Civic, CRX		
1981-87	3.7	*4.0*
1988-91	1.2	*1.5*
1992-05 (.5)	.8	*1.1*
Civic del Sol	1.2	*1.5*
Insight (.3)	.5	*.8*
Prelude		
1981-82	3.7	*4.0*
1983-87	1.7	*2.0*
1988-91	2.6	*2.9*
1992-01	.8	*1.1*
S2000 (.7)	1.1	*1.3*
Exhaust Pipe or Crossover Pipe Flange Gasket or Seal, Replace (B)		
Exc. Insight each (.6)	.9	*1.0*
Insight each (.3)	.5	*.5*
Front or Rear Exhaust Pipe, Replace (B)		
Each (.8)	1.2	*1.3*
Muffler, Replace (B)		
1981-05 (.5)	.8	*.9*
One-Piece Exhaust System, Replace (B)		
CRX (1.1)	1.7	*1.8*
Tail Pipe, Replace (B)		
1981-05 (.2)	.3	*.4*

Figure 2-5 When you are paid flat rate, you are paid for the times listed in a labor guide. *Courtesy of* Chilton, *an imprint of Cengage Learning Inc.*

the employer as soon as you can. Once you are at work:

- Be cheerful and cooperative with those around you.
- Do not spend a lot of time talking when you should be working.
- Find out what is expected of you and do your best to meet those expectations.
- Make sure you ask about anything you are not sure of, but try to think things out for yourself whenever you can.
- Show that you are willing to learn and to help out in emergencies.

A successful automotive technician has a good understanding of how the various automotive systems work, has good hand skills, has a desire to succeed, and has a commitment to be a good employee. The required training is not just in the automotive field. Because good technicians spend a great deal of time working with service manuals, good reading skills are a must. Technicians must also be able to accurately describe what is wrong to customers and the service advisor. Often these descriptions are done in writing; therefore, a technician also needs to be able to write well.

Technicians also should have a basic knowledge of computers. Computers not only control the major systems of today's vehicles, they also are used for diagnostics, tracking customers, and record keeping and as sources for information. If you have little or no experience with computers, take a computer course and spend time using a computer. If you do not have access to one, go to your local library.

Employer-Employee Relationships

Being a good employee requires more than job skills. When you become an employee, you sell your time, skills, and efforts. In return, your employer has certain responsibilities:

- **Instruction and Supervision.** You should be told what is expected of you. Your work should be observed and you should be told if your work is satisfactory and offered ways to improve your performance.
- **Good Working Conditions.** An employer should provide a clean and safe work area as well as a place for personal cleanup.
- **Wages.** You should know how much you are to be paid and what your pay will be based on. Will you be paid by the hour, by the amount of work completed, or by a combination of these two? Your employer should pay you on designated paydays.
- **Benefits.** When you were hired, you were told what fringe benefits you can expect. The employer should provide these when you are eligible to receive them.
- **Opportunity and Fair Treatment.** You should be given a chance to succeed and possibly advance within the company. You and all other employees should be treated equally, without prejudice or favoritism.

On the other side of this business relationship, you have responsibilities to the employer, including:

- **Regular Attendance.** A good employee is reliable. Businesses cannot operate successfully unless their workers are on the job. One of the first things a potential employer will ask an instructor is about the student's attendance.
- **Following Directions.** As an employee, you are part of a team. Doing things your way may not serve the best interests of the company.
- **Team Membership.** A good employee works well with others and strives to make the business successful.
- **Responsibility.** Be willing to answer for your behavior and work habits.
- **Productivity.** Remember that you are paid for your time as well as your skills and effort. You have a duty to be as effective as possible when you are at work.
- **Loyalty.** Loyalty is expected by any employer. This means you are expected to act in the best interests of your employer, both on and off the job.

COMMUNICATIONS

Employers value employees who can communicate. Effective communications include listening, reading, speaking, and writing. Communication is a two-way process. The basics of communication are simply sending a message and receiving a response.

To be successful, you should carefully follow all oral and written directions that pertain to your job. If you do not fully understand them, ask for clarification. You also need to be a good listener. Like other things in life, messages can appear to be good, bad, or have little worth to you. Regardless of how you rate the message, you should show respect to the person giving the message. Look at the person while he or she is speaking and listen to the message before you respond. In order to totally understand the message, you may need to ask questions and gather as many

GENERAL SPECIFICATIONS	
DISPLACEMENT	1.9L
NUMBER OF CYLINDERS	1–4
BORE AND STROKE	
1 9L	82 × 88 (3.23 × 3.46)
FIRING ORDER	1-3–4–2
OIL PRESSURE (HOT 2000 RPM)	240–450 kPa (35–65 psi)
DRIVE BELT TENSION	178–311 (40–70 Lb-Ft)

CYLINDER HEAD AND VALVE TRAIN 1 2	
COMBUSTION CHAMBER VOLUME (oc)	EFI-HO 55 ± 1.6
VALUE GUIDE BORE DIAMETER	EFI 39.9 ± 0.8
Intake	13.481-13 519 mm (0.531-0.5324 in.)
Exhaust	13.481-13.519 mm (0.531-0.532 in.)
VALVE GUIDE I.D.	
Intake and Exhaust	8.063–8.094 mm (.3174–.3187 in.)
Width — Intake & Exhaust	1.75–2.32 mm (0.069–0.091 in.)
Angle	45°
Runout (T.I.R.)	0.076 mm (0.003 in.) MAX.
Bore Diameter (Insert Counterbore Diameter)	
Intake	(EFI-HO) 43.763 mm (1.723 in.) MIN.
	43.788 mm (1.724 in.) MAX.
	(EFI) 39.940 mm (1.572 in.) MIN.
	39.965 mm (1.573 in.) MAX.
Exhaust	(EFI-HO) 38.263 mm (1.506 in.) MIN.
	38.288 mm (1.507 in.) MAX.
	(EFI) 34.940 mm (1.375 in.) MIN.
	39.965 mm (1.573 in) MAX.
GASKETS SURFACE FLATNESS	0.04 mm (0.0016 in.)/26 mm (1 in.)
	0.08 mm (0.003 in.)/156 mm (6 in.)
	0.15 mm (0.006 in.) Total
HEAD FACE SURFACE FINISH	0.7/2.5 0.8 (28/100 .030)
VALVE STEM TO GUIDE CLEARANCE	
Intake	0.020–0.069 mm (0.0008–0.0027 in.)
Exhaust	0.046–0.095 mm (0.0018–0.0037 in.)
VALVE HEAD DIAMETER	
Intake	42.1–41.9 mm (1.66–1.65 in.)
Exhaust	37.1–36.9 mm (1.50–1.42 in.)
VALVE FACE RUNOUT	
LIMIT	Intake & Exhaust 05 mm (0.002 in.)
VALVE FACE ANGLE	45.6°
VALVE STEM DIAMETER (Std.)	
Intake	8.043–8.025 mm (0.3167–0.3159 in.)

Figure 2-6 Being able to read and understand the information and specifications given in service information is a must for automotive technicians. *Courtesy of Ford Motor Company*

details as possible. *Hint:* Try to put yourself in the other person's shoes and listen without bias.

Obviously, when you read something, you are receiving a message without the advantage of seeing the message sender. Therefore, you must take what you read at face value. This is important because being able to read and understand the information and specifications given in service information is necessary for automotive technicians **(Figure 2–6)**.

Do your best to think through the words you use to convey a message to the customer or your supervisor. Pay attention to how they are listening and adjust your words and mannerisms accordingly. When writing a response, think about to whom the message is going and adjust your words to match their abilities and attitudes. Also, keep in mind that more than one person may read it, so think of others' needs as well.

Working in an automotive facility requires speaking to your supervisors, fellow employees, and customers. Always keep in mind that communication is a two-way street; do not try to totally control the conversation, and give listeners a chance to speak.

Proper telephone etiquette is also important. Most businesses will tell you how to answer the phone, typically involving the name of the company followed by your name. Make sure you listen carefully to the person calling. When you are the one making the call, make sure you introduce yourself and state the overall purpose of the phone call. Again, the key to proper phone etiquette is respect.

You will also be required to write things, such as warranty reports and work orders. You may also need to speak with or write to customers, parts suppliers, and supervisors to clarify an issue. Take your time and write clear, concise, complete, and grammatically correct sentences and paragraphs. Doing this will not only help you to get your message across but will also make you a more prized employee.

Nonverbal Communication

In all communications, some of the true meaning is lost in the transmission of a message from a sender to a receiver. In many cases, the heard message is often far different from the one intended. Because the words spoken are not always understood or are interpreted wrongly because of personal feelings, you can alter the meaning of words significantly by changing the tone of your voice. Think of how many ways you can say "no"; you could express mild doubt, terror, amazement, anger, and other emotions.

It is important that you realize that a major part of communication is nonverbal. **Nonverbal communication** is the things you do while communicating. Pay attention to your nonverbal communication as well as to that of others.

Nonverbal communication includes such things as body language and tone. Body language includes facial expression, eye movement, posture, and gestures. All of us read people's faces; we interpret what they say or feel. We also look at posture to give us a glimpse of how the other person feels about the message. Posture can indicate self-confidence, aggressiveness, fear, guilt, or anxiety. Similarly, we look at how they place their hands or give a handshake.

Posture and other aspects of body language have been identified as important keys to communication. Many scholars have studied them and defined what they indicate. Some divide postures into two basic groups:

- *Open/closed* is the most obvious. People with their arms folded, legs crossed, and bodies turned away are signaling that they are rejecting or are closed to messages, whereas people fully facing you with open hands and both feet planted on the ground are saying they are open to and accepting the message.

- *Forward/back* indicates whether people are actively or passively reacting to the message. When they are leaning forward and pointing toward you, they are actively accepting or rejecting the message. When they are leaning back, looking at the ceiling, doodling on a pad, or cleaning their glasses, they are either passively absorbing or ignoring the message.

SOLVING PROBLEMS AND CRITICAL THINKING

Anyone who can think critically and logically to evaluate situations is considered very desirable by the industry. **Critical thinking** is the art of being able to judge or evaluate something without bias or prejudice. When diagnosing an automotive problem, critical thinkers are able to locate the cause of the problem by responding to what is known, not what is supposed!

Good critical thinkers begin solving problems by carefully observing what is and what is not happening. Based on these observations, something is declared as a fact. For example, if the right headlamp of a vehicle does not light and the left headlamp does, a critical thinker will be quite sure that the source of the problem is related to the right headlamp and not the left one. Therefore, all testing will be centered on the right headlamp. The critical thinker then studies the circuit and determines the test points. Prior to conducting any test, the critical thinker knows what to test and what the possible test results would indicate.

Critical thinkers solve problems in an orderly way and do not depend on chance. They come to conclusions based on a sound reasoning. They also understand that if a specific problem exists only during certain conditions, there are a limited number of causes. They further understand the relationship between how often the problem occurs and the probability of accurately predicting the problem. Also, they understand that one problem may cause other problems and they know how to identify the connection between the problems.

Solving problems is something we do every day. Often the problems are trivial, such as deciding what to watch on television. Other times they are critical and demand much thought. At these times, thinking critically will really pay off. Although it is impossible to guarantee that critical thinking will lead to the correct decision, it will lead to good decisions and solutions.

Diagnosis

The word **diagnosis** is used to define one of the major duties of a technician. Diagnosis is a way of looking at systems that are not functioning properly and finding out why. It is not guessing, and it is more than following a series of interrelated steps in order to find the solution to a specific problem. Solid diagnosis is based on an understanding of the purpose and operation of the system that is not working properly.

In service manuals there are diagnostic aids given for many different problems. These are either symptom based or flow charts. **Flow charts** or decision trees **(Figure 2–7)** guide you through a step-by-step process. As you answer the questions given at each step, you are told what your next step should be. Symptom-based diagnostic charts **(Figure 2–8)** focus on a definition of the problem and offer a list of possible causes of the problem. Sometimes the diagnostic aids are a combination of the two—a flow chart based on clearly defined symptoms.

When these diagnostic aids are not available or prove to be ineffective, most good technicians conduct a good visual inspection and then take a logical approach to finding the cause of the problem. This relies on critical thinking skills as well as system knowledge. Logical diagnosis follows these steps:

1. **Gather information about the problem.** Find out when and where the problem happens and what exactly happens.
2. **Verify that the problem exists.** Take the vehicle for a road test and try to duplicate the problem, if possible.
3. **Thoroughly define what the problem is and when it occurs.** Pay strict attention to the conditions present when the problem happens. Also pay attention to the entire vehicle; another problem may be evident to you that was not evident to the customer.
4. **Research all available information** to determine the possible causes of the problem. Try to match the exact problem with a symptoms chart or think about what is happening and match a system or some components to the problem.
5. **Isolate the problem** by testing. Narrow down the probable causes of the problem by checking the obvious or easy-to-check items.

DTC B3138 DOOR LOCK CIRCUIT (HIGH)

STEP	ACTION	YES	NO
1	Was a BCM diagnostic check performed?	Go to step 2	See BCM diagnostics
2	* Check for current DTCs with a scan tool. * Disconnect power to the LH door lock switch. * Does the scan tool display B3138 as a current code?	Go to step 3	Go to step 6
3	* Check for current DTCs with a scan tool. * Disconnect power to the RH door lock switch. * Does the scan tool display B3138 as a current code?	Go to step 4	Go to step 7
4	* Disconnect the brown BCM (C1) connector. * Backprobe connectors with a digital multimeter. * Measure voltage between A4 (LT BLU) and ground. * Does the multimeter show battery voltage?	Go to step 5	Go to step 8
5	Locate and repair the short to battery voltage in CKT 195 (LT BLU) between the BCM and the LOCK relay, or the left or right front door switches.	Go to step 9	—
6	Replace the LH power door lock switch. Is the repair complete?	Go to step 9	—
7	Replace the RH power door lock switch. Is the repair complete?	Go to step 9	—
8	* Replace the BCM. * Program the BCM with proper calibrations. * Perform the learn procedure. Is the repair complete?	Go to step 9	—
9	* Reconnect all disconnected components. * Clear the DTCs. Is the action complete?	System OK	—

Figure 2-7 A typical decision tree for diagnostics.

6. Continue testing to **pinpoint the cause of the problem**. Once you know where the problem should be, test until you find it!
7. **Locate and repair the problem, then verify the repair.** Never assume that your work solved the original problem. Make sure the problem is history before returning it to the customer.

PROFESSIONALISM

The key to effective communications is respect. Like communication, respect is a two-way process. You should respect others and others should respect you. However, respect cannot be commanded; it must be earned. As a technician, you can earn respect in many ways. All of these result from the amount of professionalism you display. Professionalism is best shown by having a positive attitude, displaying good behavior, and accepting responsibility.

A good technician is a highly skilled and knowledgeable individual. A professional technician is a good technician who dresses and acts appropriately. A professional demonstrates the following:

- Self-esteem, pride, and confidence
- Honesty, integrity, and personal ethics

Symptom	Probable cause	Remedy
Engine will not start or hard to start	Vacuum hose disconnected or damaged	Repair or replace
	EGR valve is not closed	Repair or replace
	Malfunction of the EVAP Canister Purge	Repair or replace
	Solenoid Valve	Repair or replace
Rough idle or engine stalls	Vacuum hose disconnected or damaged	Repair or replace
	EGR valve is not closed	Repair or replace
	Malfunction of the PCV valve	Replace
	Malfunction of the EVAP Canister Purge System	Check the system; if there is a problem, check its component parts
Excessive oil consumption	Positive crankcase ventilation line clogged	Check positive crankcase ventilation system
Poor fuel mileage	Malfunction of the exhaust gas recirculation	Check the system; if there is a problem, check its component parts

Figure 2-8 A symptom-based diagnostic chart.

- A positive attitude toward learning, growth, and personal health
- Initiative, energy, and persistence to get the job done
- Respect for others
- A display of initiative and assertiveness
- The ability to set goals and priorities in work and personal life
- The ability to plan and manage time, money, and other resources to achieve goals
- The willingness to follow rules, regulations, and policies
- The willingness to fulfill the responsibilities of your job
- Assuming responsibility and accountability for your decisions and actions
- The ability to apply ethical reasoning

Coping with Change

Your professionalism is also evident by how you react to change. Unfortunately, work environments never stay the same. New rules and regulations, supervisors, fellow employees, business owners, and vehicle systems are all potential sources of stress. Rather than focusing on the negatives of these changes, you should identify the positives. This will help you minimize stress. If you feel stress, do what you can to relieve it. Activities such as walking, running, or playing sports help reduce stress. When you are stressed, whether it is caused by something at work or in your personal life, it is difficult to be a productive worker. Therefore, do your best to put things in perspective and do some critical thinking to identify what you can do to change the situation that is causing the stress.

When the source of stress is related to your job, spend time to decide whether the stressful situation can be changed or not. If it cannot and you feel you can no longer cope with it, it may be wise to find employment elsewhere. This stress can be the result of change or can be caused by a realization that you do not like what you are doing or do not have the ability to do what is required of you.

If you decide that leaving your job is the best solution, do it professionally. Do not simply stop showing up for work or walk up to the employer and say "I quit!" The best way to quit a job is to write a letter of resignation and personally present it to the employer. The letter should state why you are leaving the company. Be careful not to attack the business, the employer, or fellow workers. You can simply say you are looking at other opportunities or have found another job. Bad-mouthing the business is a sure way of losing a good work reference—one that you may need for your next job. The letter should also include the last day you intend to work. Your last day should be approximately 2 weeks after you notify the employer. At the end of your letter of resignation, thank the employer for the opportunity to work for him or her and for the personal growth experiences they provided for you.

INTERPERSONAL RELATIONSHIPS

As an employee, you have responsibilities to your fellow workers. You are a member of a team. Teamwork means cooperating with and caring about other workers. All members of the team should understand and contribute to the goals of the business. Keep in mind that if the business does not make money, you may not have a job in the future. Your responsibility is more than simply doing your job. You should also:

- Suggest improvements that may make the business more successful.
- Display a positive attitude.
- Work with team members to achieve common goals.
- Respect the thoughts and opinions of your fellow workers and your employer.
- Exercise "give and take" for the benefit of the business.
- Value individual diversity.
- Respond to praise or criticism in a professional way.
- Provide constructive praise or criticism.
- Channel and control emotional reactions to situations.
- Resolve conflicts in a professional way.
- Identify and react to any intimidation or harassment.

Customer Relations

Good customer relations are important for all members of the team. You should make sure you listen and communicate clearly (**Figure 2–9**). Be polite and organized, particularly when dealing with customers on the telephone. Always be as honest as you possibly can.

Figure 2-9 Good customer relations are important; make sure you always listen and communicate clearly.

Present yourself as a professional. Professionals are proud of what they do and they show it. Always dress and act appropriately and watch your language, even when you think no one is near.

Respect the vehicles you are working on. They are important to the lives of your customers. Always return the vehicles to their owners clean and in an undamaged condition. Remember, a vehicle is the second largest expense a customer has. Treat it that way. It does not matter if you like the vehicle. It belongs to the customer; treat it respectfully.

Explain the repair process to the customer in understandable terms. Whenever you are explaining something to a customer, make sure you do it in a simple way without making the customer feel stupid. Always show customers respect and be courteous to them. Not only is this the right thing to do but it also leads to loyal customers.

KEY TERMS

Commission
Critical thinking
Diagnosis
Employment plan
Flat rate
Flow charts
Gross pay
Labor guide
Net pay
Nonverbal communication
Reference
Resume
Soft skills

SUMMARY

- An employment plan is an honest appraisal of yourself and your career hopes.
- A reference is someone who will be glad to tell a potential employer about you and your work habits.
- A resume and cover letter are personal marketing tools and may be the first look at you an employer has.
- A resume normally includes your contact information, career objective, skills and/or accomplishments, work experience, education, and a statement about references.
- A cover letter gives you a chance to point out exactly why you are perfect for a particular job.
- An application form is a legal document that summarizes who you are.
- Good preparation for an employment interview will result in a good experience.

- Automotive technicians are typically paid an hourly wage or on the flat-rate system.
- A successful automotive technician has good training, a desire to succeed, and a commitment to be a good technician and a good employee.
- As part of an employment agreement your employer also has certain responsibilities to you, and you have responsibilities to the employer.
- Effective communications include listening, reading, speaking, and writing.
- Nonverbal communication is a key part of sending and receiving a message and includes such things as body language and tone.
- Employers value someone who can think critically and act logically to evaluate situations and who has the ability to solve problems and make decisions.
- Diagnosis means finding the cause or causes of a problem. It requires a thorough understanding of the purpose and operation of the various automotive systems.
- Diagnostic charts found in service manuals can aid in diagnostics.
- Professionalism is best displayed by having a positive attitude, displaying good behavior, and accepting responsibility.
- New rules and regulations, supervisors, fellow employees, vehicle systems, and vehicles are all potential sources of stress, and your professionalism will be measured by how well you cope.
- Teamwork means cooperating with and caring about other workers.
- Good customer relations is a quality of good technicians and is based on respect.

REVIEW QUESTIONS

1. Which of the following is *not* a recommended step for accurate diagnosis of a problem?
   a. Gather as much information as you can about the problem.
   b. Thoroughly define the problem.
   c. Replace system components and identify the cause of a problem through the process of elimination.
   d. Research all available information and knowledge to determine the possible causes of the problem.
2. What type of information should go into your employment plan?
3. What does it mean to be paid based on flat rate?
4. What should be included in the three main paragraphs of a cover letter?
5. *True or False?* When you feel stress from your job and also feel that the situation will not change, you should seek employment elsewhere.
6. Which of the following is *not* a desired characteristic of a good resume?
   a. It is neat, uncluttered, and easy to read.
   b. It has a list of all jobs you have done, whether for pay or just to help someone out.
   c. It is only one or two pages long.
   d. Important information appears near the top of the paper.
7. Which of the following behaviors does *not* show that you are a responsible person?
   a. having set goals and priorities in your work and personal life
   b. showing a willingness to follow rules, regulations, and policies
   c. showing a willingness to share the consequences of your mistakes with others
   d. using ethical reasoning when making decisions
8. *True or False?* If you decide that leaving your job is the best way to relieve stress at work, you should stop showing up for work and send your employer a letter stating why you left the company.
9. *True or False?* When you are filling out an application, do not attempt to answer the questions that do not pertain to you and your situation.
10. Which of the following is *not* the right thing to do when you are being interviewed for a job?
    a. Show up looking neat and in the clothing you would wear on the job.
    b. Never hesitate with an answer to a question, because hesitation may indicate you are a shy person.
    c. To avoid saying too much or offending the interviewer, answer as many questions as you can with a simple yes or no.
    d. Listen closely to the interviewer and look at the interviewer while he or she talks.
11. Which of the following is *not* a characteristic of a good employee?
    a. reliable
    b. responsible
    c. overly sociable
    d. loyal

12. Applicant A goes to a quiet place immediately after an interview and reflects on what just took place. Applicant B sends a letter of thanks to the interviewer if he has not heard back from the employer within 2 weeks. Who is doing the right thing?
    a. A only
    b. B only
    c. Both A and B
    d. Neither A nor B
13. Technician A always looks at people while they are speaking and listens to their message before responding. Technician B always looks at customers when they are speaking to show she is interested in what they are saying. Who is correct?
    a. Technician A only
    b. Technician B only
    c. Both A and B
    d. Neither A nor B
14. Technician A always speaks to customers with his arms folded across his chest because he does not know what else to do with them. Technician B always tries to fully comprehend the message by asking questions about it and gathering as many details as possible. Who is correct?
    a. Technician A only
    b. Technician B only
    c. Both A and B
    d. Neither A nor B
15. Which of the following would *not* be considered a soft skill?
    a. enjoying solving puzzles or problems
    b. caring for or helping people
    c. the ability to work independently
    d. taking care to follow orders or instructions
16. Describe the seven basic steps for logical diagnosis.
17. *True or False?* Critical thinking is the art of being able to judge or evaluate something with bias or prejudice.
18. *True or False?* The business you decide to work for will pay for all of the benefits you receive.
19. Define the term "diagnosis" as it applies to the duties of an automotive technician.
20. When identifying individuals to use as references while seeking employment, consider all of the following *except:*
    a. someone who knows and whose opinion is respected, such as a priest, a minister, or an elder in your church
    b. a family member or close friend
    c. past and present teachers, coaches, and school administrators
    d. people you have worked for or have helped

AUTOMOTIVE SYSTEMS

CHAPTER 3

OBJECTIVES

- Explain the major events that have influenced the development of the automobile during the last 40 years.
- Explain the difference between unitized vehicles and body-over-frame vehicles.
- Describe the manufacturing process used in a modern automated automobile assembly plant.
- List the basic systems that make up an automobile and name their major components and functions.

HISTORICAL BACKGROUND

The automobile has changed quite a bit since the first horseless carriage went down an American street. In 1896, both Henry Ford and Ransom Eli Olds test drove their first gasoline-powered vehicles. Prior to this time, other individuals were making their own automobiles (**Figure 3–1**). Most were powered by electricity or steam. The year 1896 marks the beginning of the automotive industry, not because of what Ford or Olds did, but because of the Duryea Brothers, who, by 1896, had made thirteen cars in the first factory that made cars for customers.

In the beginning, the automobile looked like the horse-drawn carriage it was designed to replace. In 1919, 90% of the cars had carriagelike open bodies. These early cars had rear-mounted engines and very tall tires. They were designed to move people down dirt roads.

Figure 3-1 The 1886 Benz Patent Motor Wagen, one of the first automobiles made. © *Courtesy of Daimler AG*

The automobile changed when the roads became paved, more people owned cars, manufacturers tried to sell more cars, concerns for safety and the environment grew, and new technology was developed. All of these changes resulted in automobiles that are more practical, more affordable, safer, more comfortable, more dependable, and faster. Although many improvements have been made to the original design, the basics of the automobile have changed very little:

- Nearly all of today's cars still use gasoline engines to drive two or more wheels.
- A steering system is used to control the direction of the car.
- A brake system is used to slow down and stop the car.
- A suspension system is used to absorb road shocks and help the driver maintain control on bumpy roads.
- These major systems are mounted on steel frames and the frame is covered with body panels.
- The body panels give the car its shape and protect those inside from the weather and dirt.
- The body panels also offer some protection for the passengers if the automobile is in an accident.

Although these basics have changed little in the past 100 years, the design of the systems has greatly changed. The entire automobile is technologically light-years ahead of Ford's and Olds's early models. New technologies have changed the slow, unreliable, user-hostile vehicles of the early 1900s into vehicles that can travel at very high speeds, operate trouble-free for thousands of miles, and provide comforts that even the very rich had not dreamed of in 1896.

Social and political pressures have had a great influence on automobile design for the past 40-plus years. In 1965, laws were passed that limited the amount of harmful gases emitted by an automobile. Although this had little immediate effect on the industry, the automobile manufacturers were forced to focus on the future. They needed to build cleaner-burning engines. In the following years, stricter emissions laws were passed and manufacturers were required to develop new emission control systems.

World events in the 1970s continued to shape the development of the automobile. An oil embargo by Arab nations in 1973 caused the price of gasoline to quickly increase to four times its normal price. This event caused most Americans to realize that the supply of gasoline and other nonrenewable resources was limited. Car buyers wanted cars that were not only kind to the environment but that also used less fuel.

The **Corporate Average Fuel Economy (CAFE)** standards were set in 1975. These required automakers to build more fuel-efficient vehicles. Under the CAFE standards, different models from each manufacturer are tested for the number of miles they can be driven on a gallon of gas. The fuel efficiencies of these vehicles are averaged together to arrive at a corporate average. The CAFE standards have increased many times since it was established. A manufacturer that does not meet CAFE standards for a given model year faces heavy fines.

While trying to produce more fuel-efficient vehicles, manufacturers replaced large eight-cylinder engines with four-cylinder and other small engines. Basic engine systems like carburetors and ignition breaker points were replaced by electronic fuel injection and electronic ignition systems.

By the mid-1980s, all automobiles were equipped with some type of electronic control system. These systems did, and still do, monitor the engine's operation and provide increased power outputs while minimizing fuel consumption and emissions. Electronic sensors are used to monitor the engine and many other systems. Computerized engine control systems control air and fuel delivery, ignition timing, emission systems operation, and a host of other related operations. The result is a clean-burning, fuel-efficient, and powerful engine **(Figure 3–2)**.

Figure 3-2 A cutaway of a late-model V-10 gasoline engine.

DESIGN EVOLUTION

Not too long ago, nearly every car and truck was built with body-over-frame construction, rear-wheel drive, and symmetrical designs. Today, nearly all cars do not have a separate frame; instead, the frame and body are built as a single unit, called a unibody. Most pickup trucks and large SUVs still are built on a frame.

Another major influence to design was the switch from rear-wheel drive to front-wheel drive. Making this switch accomplished many things, the most notable being improved traction at the drive wheels, increased interior space, shorter hood lines, and a very compact driveline. Because of the weight and loads that pickup trucks are designed to move, most remain rear-wheel drive.

Perhaps the most obvious design change through the years has been body styles. Body styles have changed to respond to the other design considerations and to trends of the day. For example, in the '50s America had a strange preoccupation with the unknown, outer space, which led to cars that had rocketlike fins. Since then fins have disappeared and body styles have become more rounded to reduce air drag.

Body-Over-Frame Construction

In body-over-frame construction, the frame is the vehicle's foundation. The body and all major parts of the vehicle are attached to the frame. The frame must be strong enough to keep the rest of the vehicle in alignment should a collision occur.

The frame is an independent, separate component that is not rigidly attached to the vehicle's body. The body is generally bolted to the frame **(Figure 3–3)**. Large, specially designed rubber mounts are placed between the frame and body to reduce noise and vibration from entering the passenger compartment. Quite often two layers of rubber are used in these pads

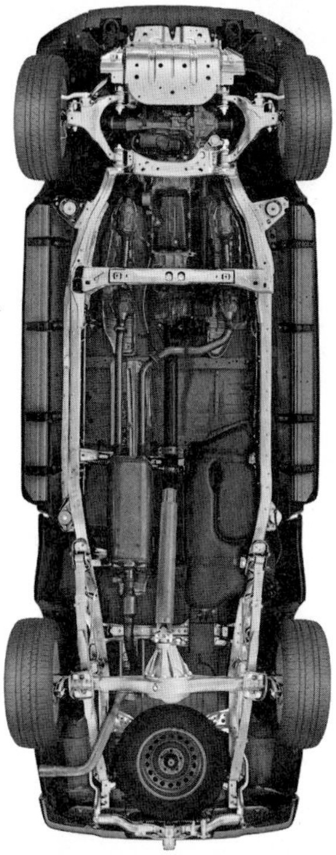

Figure 3-3 In body-over-frame construction, the frame is the vehicle's foundation. *Courtesy of Rob Lawrence and Toyota Motor Sales, U.S.A., Inc.*

to provide a smoother ride. The frame rails are made of stamped steel, which are welded together. Some frames are made by a **hydroforming** process, which uses high-pressure water, rather than heat, to shape the steel into the desired shape.

Unitized Construction

A **unibody** **(Figure 3–4)** is a stressed hull structure in which each of the body parts supplies structural support and strength to the entire vehicle. Unibody vehicles tend to be tightly constructed because the major parts are all welded together. This helps protect the occupants during a collision. However, it causes damage patterns that differ from those of body-over-frame vehicles. Rather than localized damage, the stiffer sections used in unibody design tend to transmit and distribute impact energy throughout more of the vehicle.

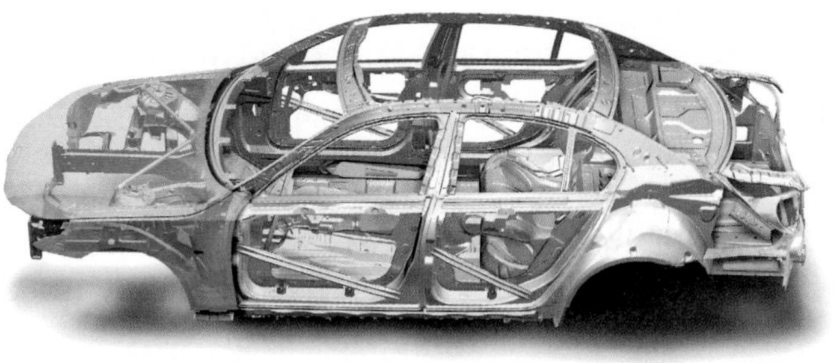

Figure 3-4 The structure of a unibody car. *Courtesy of BMW of North America, LLC*

Nearly all unibodies are constructed from steel. A few cars, such as the Audi A8, use aluminum instead. An aluminum car body and frame can weigh up to 40% less than an identical body made of steel. Most front-wheel-drive unibody vehicles have a **cradle** or partial frame that is used to support the powertrain and suspension for the front wheels.

BODY SHAPES

Various methods of classifying vehicles exist. Vehicles may be classified by engine type, body/frame construction, fuel consumption structure, type of drive, or the classifications most common to consumers, which are body shape, seat arrangement, and number of doors. Eight basic body shapes are used today:

- **Sedan.** A vehicle with front and back seats that accommodates four to six persons is classified as either a two- or four-door sedan **(Figure 3–5)**. Often, a two-door sedan is called a coupe **(Figure 3–6)**. If the vehicle's B pillars do not extend up through the side windows, the car is called a hardtop.

Figure 3-5 This Toyota Camry is an example of a typical late-model sedan. *Courtesy of Toyota Motor Sales, U.S.A., Inc.*

Figure 3-6 This Honda Civic SI is a coupe.

Figure 3-7 A BMW 3 series convertible. *Courtesy of BMW of North America, LLC*

Figure 3-8 This Aston Martin is an English sports car.

Figure 3-9 A Toyota Prius is an example of a hatchback. Its rear luggage compartment is an extension of the passenger compartment. *Courtesy of Toyota Motor Sales, U.S.A., Inc.*

- **Convertibles.** Convertibles have vinyl roofs that can be raised or lowered. A few late-model convertibles feature a folding metal roof that tucks away in the trunk when it is down **(Figure 3–7)**. Some convertibles have both front and rear seats. Those without rear seats are commonly referred to as sports cars **(Figure 3–8)**.
- **Liftback** or **hatchback.** The distinguishing feature of this vehicle is its rear luggage compartment, which is an extension of the passenger compartment. Access to the luggage compartment is gained through an upward opening hatch-type door **(Figure 3–9)**. This design car can be a three- or a five-door model. The third or fifth door is the rear hatch.
- **Station wagon.** A station wagon is characterized by its roof, which extends straight back, allowing a spacious interior luggage compartment in the rear. The rear door, which can be opened in various ways depending on the model, provides access to the luggage compartment. Station wagons typically have four doors and can have space for up to nine passengers.
- **Pickups.** Pickup truck body designs have an open cargo area behind the driver's compartment. There are many varieties available today: there are compact, medium-sized, full-sized, and heavy-duty pickups. They also can be had in two-, three-, or four-door models. Some have extended cab areas with seats in back of the front seat. They are available in two-wheel drive, four-wheel drive (4X4), or all-wheel drive.
- **Vans.** The van body design has a tall roof and a totally enclosed large cargo or passenger area. Vans can seat from two to twelve passengers, depending on size and design. Basically, there are

Figure 3-10 This Dodge Caravan is an example of a late-model mini-van. Full-size vans are also available. *Courtesy of Chrysler LLC*

two sizes of vans: mini- and full-size. The most common are mini-vans **(Figure 3–10)**.

- **Sport utility vehicles (SUVs)**. SUVs are best described as multipurpose vehicles that can carry a wide range of passengers, depending on their size and design. A good majority of SUVs have four-wheel drive, although some do not. Most small SUVs are based on automobile platform and take on many different looks and features **(Figure 3–11)**. Mid-size SUVs are larger and typically offer more features and comfort. There are many large SUVs available **(Figure 3–12)**. These vehicles can seat up to nine adults and tow up to 6 tons.
- **Crossover vehicles**. These automobiles look like an SUV but are built lighter and offer fuel efficiency. They are actually a combination of a station wagon and an SUV. They have SUV features but are not quite the same size. The basic construction of a crossover vehicle leads to a less trucklike ride than a normal SUV **(Figure 3–13)**. They also are not designed to tow heavy loads or for off-the-road use.

Figure 3-11 This Honda Element is considered a small SUV and has many unique features including anytime four-wheel drive.

Figure 3-12 This Lincoln Navigator is an excellent example of a large SUV, and it is based on a truck chassis.

Figure 3-13 This is an Audi crossover vehicle.

THE BASIC ENGINE

The engine provides the power to drive the wheels of the vehicle. All automobile engines, both gasoline and diesel, are classified as internal combustion engines because the combustion or burning that creates energy takes place inside the engine. **Combustion** is the burning of an air and fuel mixture. As a result of combustion, large amounts of pressure are generated in the engine. This pressure or energy is used to power the car. The engine must be built strong enough to hold the pressure and temperatures formed by combustion.

Diesel engines have been around a long time and are mostly found in big heavy-duty trucks. However, they are also used in some pickup trucks and will become more common in automobiles in the future **(Figure 3–14)**. Although the construction of gasoline and diesel engines is similar, their operation is quite different.

A gasoline engine relies on a mixture of fuel and air that is ignited by a spark to produce power. A diesel engine also uses fuel and air, but it does not need a spark to cause ignition. A diesel engine is often called a compression ignition engine. This is because its incoming air is tightly compressed, which greatly raises its temperature. The fuel is then injected into

Figure 3–14 A four-cylinder automotive diesel engine.

the compressed air. The heat of the compressed air ignites the fuel and combustion takes place. The following sections cover the basic parts and the major systems of a gasoline engine.

Cylinder Block

The biggest part of the engine is the **cylinder block**, which is also called an **engine block** **(Figure 3–15)**. The cylinder block is a large casting of metal (cast iron or aluminum) that is drilled with holes to allow for the passage of lubricants and coolant through the block and provide spaces for movement of mechanical parts. The block contains the cylinders, which are round passageways fitted with pistons. The block houses or holds the major mechanical parts of the engine.

Cylinder Head

The **cylinder head** fits on top of the cylinder block to close off and seal the top of the cylinders **(Figure 3–16)**. The **combustion chamber** is an area into which the

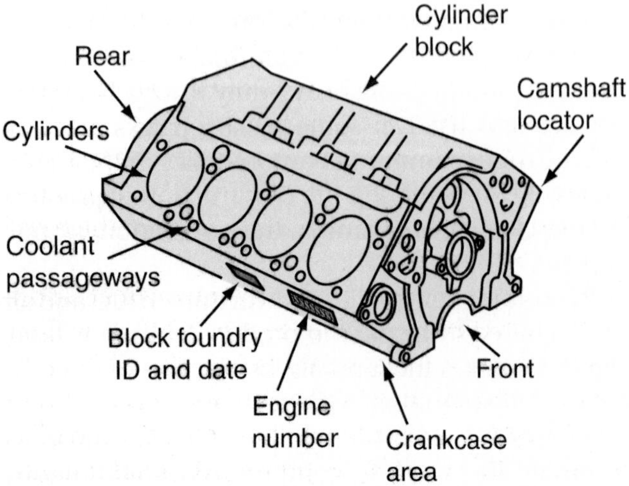

Figure 3–15 An engine block for a V8 engine.

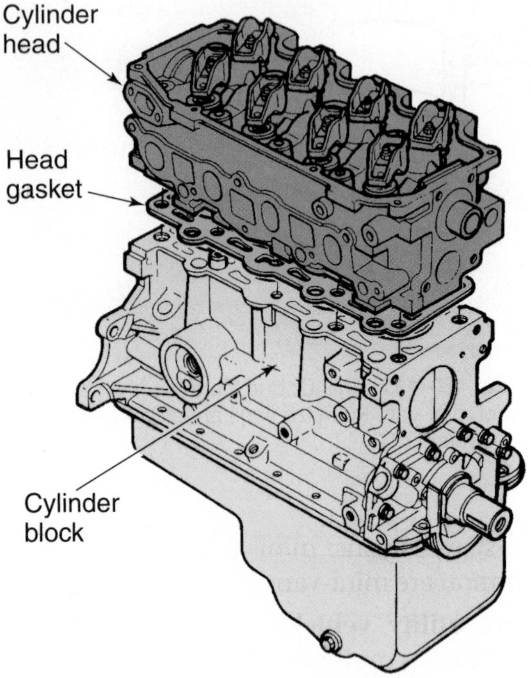

Figure 3–16 The two major units of an engine, the cylinder block and the cylinder head, are sealed together with a gasket and are bolted together. *Courtesy of Federal-Mogul Corporation*

air-fuel mixture is compressed and burned. The cylinder head contains all or most of the combustion chamber. The cylinder head also contains **ports**, which are passageways through which the air-fuel mixture enters and burned gases exit the cylinder. A cylinder head can be made of cast iron or aluminum.

Piston

The burning of air and fuel takes place between the cylinder head and the top of the piston. The **piston** is a can-shaped part closely fitted inside the cylinder **(Figure 3–17)**. In a four-stroke cycle engine, the piston moves through four different movements or strokes to complete one cycle. These four are the intake, compression, power, and exhaust strokes. On the intake stroke, the piston moves downward, and a charge of air-fuel mixture is introduced into the cylinder. As the piston travels upward, the air-fuel mixture is compressed in preparation for burning. Just before the piston reaches the top of the cylinder, ignition occurs and combustion starts. The pressure of expanding gases forces the piston downward on its power stroke. When it reciprocates, or moves upward again, the piston is on the exhaust stroke. During the exhaust stroke, the piston pushes the burned gases out of the cylinder.

Connecting Rods and Crankshaft

The reciprocating motion of the pistons must be converted to rotary motion before it can drive the wheels

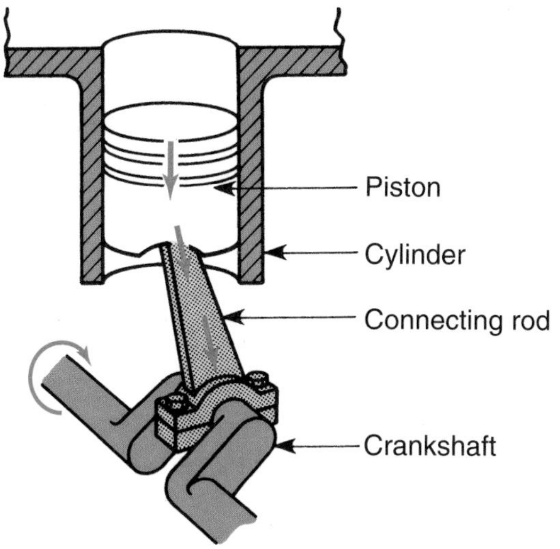

Figure 3–17 The engine's pistons fit tightly in the cylinders and are connected to the engine's crankshaft with connecting rods.

of a vehicle. This conversion is achieved by linking the piston to a **crankshaft** with a **connecting rod**. As the piston is pushed down on the power stroke, the connecting rod pushes on the crankshaft, causing it to rotate. The end of the crankshaft is connected to the transmission to continue the power flow through the drivetrain and to the wheels.

Valve Train

A **valve train** is a series of parts used to open and close the intake and exhaust ports. A valve is a movable part that opens and closes a passageway. A camshaft controls the movement of the valves **(Figure 3–18)**, causing them to open and close at the proper time. Springs are used to help close the valves.

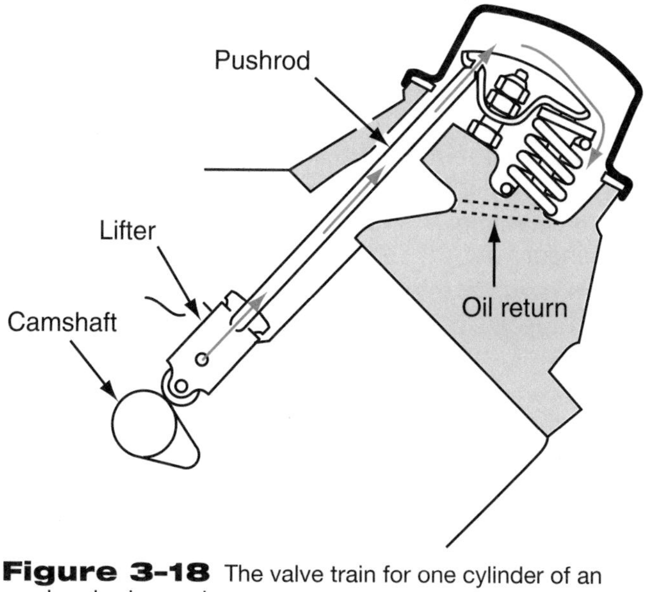

Figure 3–18 The valve train for one cylinder of an overhead valve engine.

Figure 3–19 The blue manifold is the intake manifold and the red manifold is for the exhaust.

Manifolds

A **manifold** is metal ductwork assembly used to direct the flow of gases to or from the combustion chambers. Two separate manifolds are attached to the cylinder head **(Figure 3–19)**. The **intake manifold** delivers a mixture of air and fuel to the intake ports. The **exhaust manifold** mounts over the exhaust ports and carries exhaust gases away from the cylinders.

ENGINE SYSTEMS

The following sections present a brief explanation of the systems that help an engine run and keep running.

Lubrication System

The moving parts of an engine need constant lubrication. Lubrication limits the amount of wear and reduces the amount of friction in the engine. **Friction** is heat generated when two objects rub against each other.

Motor or engine oil is the fluid used to lubricate the engine. Several quarts of oil are stored in an **oil pan** bolted to the bottom of the engine block. The oil pan is also called the crankcase or **oil sump**. When the engine is running, an oil pump draws oil from the pan and forces it through oil galleries. These galleries are small passageways that direct the oil to the moving parts of the engine.

Oil from the pan passes through an oil filter before moving through the engine **(Figure 3–20)**. The filter removes dirt and metal particles from the oil. Premature wear and damage to parts can result from dirt in the oil. Regular replacement of the oil filter and oil is an important step in a preventive maintenance program.

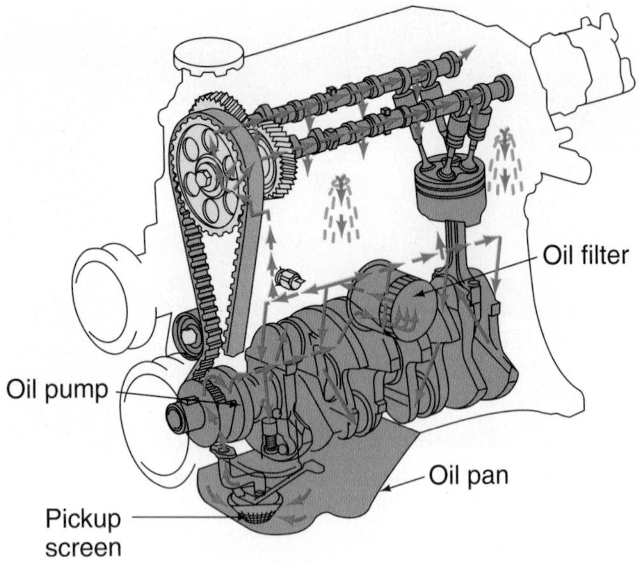

Figure 3-20 Oil flow in a typical engine's lubrication system.

Cooling System

The burning of the air-fuel mixture in the combustion chambers of the engine produces large amounts of heat. This heat must not be allowed to build up and must be reduced. This heat can easily damage and warp the metal parts of an engine. To prevent this, engines have a cooling system **(Figure 3–21)**.

The most common way to cool an engine is to circulate a liquid coolant through passages in the engine block and cylinder head. An engine can also be cooled by passing air over and around the engine. Few air-cooled engines are used in automobiles today because it is very difficult to maintain a constant temperature at the cylinders. If the engine is kept at a constant temperature, it will run more efficiently.

A typical cooling system relies on a **water pump**, which circulates the coolant through the system. The pump is typically driven by the engine. The coolant is a mixture of water and antifreeze. The coolant is pushed through passages, called **water jackets**, in the cylinder block and head to remove heat from the area around the cylinders' combustion chambers. The heat picked up by the coolant is sent to the **radiator**. The radiator transfers the coolant's heat to the outside air as the coolant flows through its tubes. To help remove the heat from the coolant, a cooling fan is used to pull cool outside air through the fins of the radiator.

To raise the boiling point of the coolant, the cooling system is pressurized. To maintain this pressure, a radiator or **pressure cap** is fitted to the radiator. A **thermostat** is used to block off circulation in the system until a preset temperature is reached. This allows the engine to warm up faster. The thermostat also

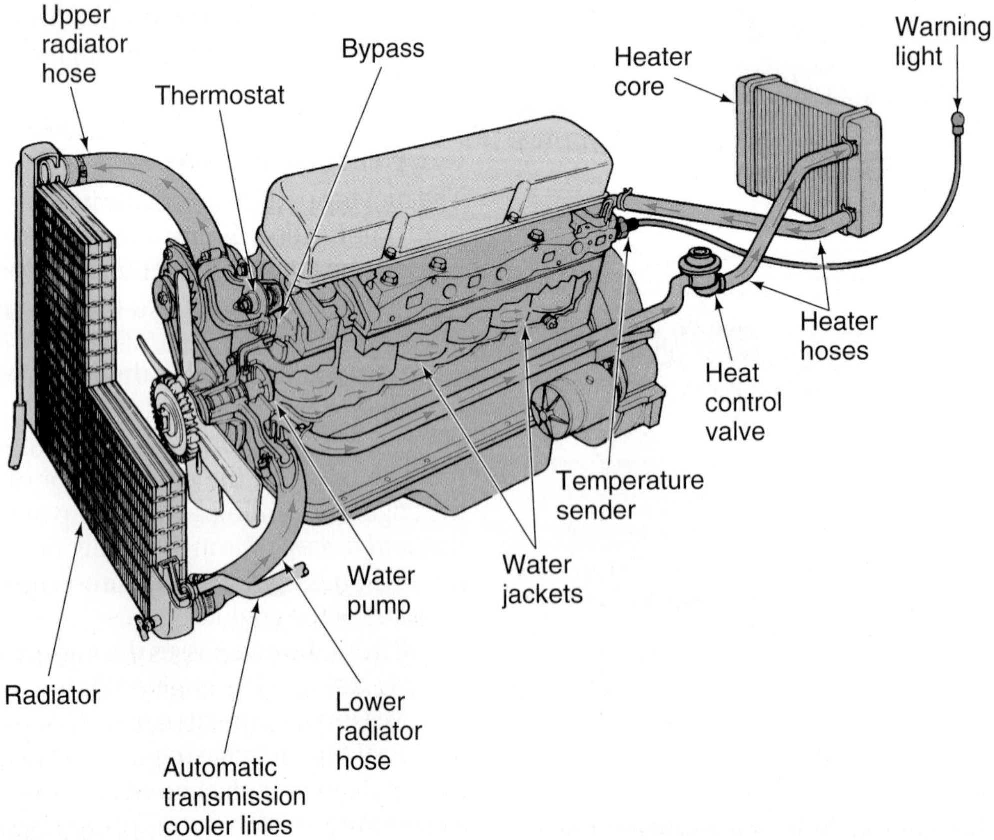

Figure 3-21 A typical engine cooling system.

keeps the engine temperature at a predetermined level. Because parts of the cooling system are located in various spots under the vehicle's hood, hoses are used to connect these parts and keep the system sealed.

Fuel and Air System

The fuel and air system is designed to supply the correct amount of fuel mixed with the correct amount of air to the cylinders of the engine. This system also:

- Stores the fuel for later use
- Delivers fuel to a device that will control the amount of fuel going to the engine
- Collects and cleans the outside air
- Delivers outside air to the individual cylinders
- Changes the fuel and air ratios to meet the needs of the engine during different operating conditions

The fuel system is made up of different parts. A fuel tank stores the liquid gasoline. Fuel lines carry the liquid from the tank to the other parts of the system. A pump moves the gasoline from the tank through the lines. A filter removes dirt or other particles from the fuel. A fuel pressure regulator keeps the pressure below a specified level. An air filter cleans the outside air before it is delivered to the cylinders. Fuel injectors mix fuel with the air for delivery to the cylinders or directly to the cylinders. An intake manifold directs the air to each of the cylinders **(Figure 3–22)**.

Emission Control System

In the past one of the chief contributors to air pollution was the automobile. For some time now, engines have been engineered to emit very low amounts of certain pollutants. The pollutants that have been drastically reduced are **hydrocarbons (HC)**, **carbon monoxide (CO)**, and **oxides of nitrogen (NO_X)**. The Environmental Protection Agency establishes emissions standards that limit the amount of these pollutants a vehicle can emit.

To meet these standards, many changes have been made to the engine itself. There also have been systems developed and added to the engines to reduce the pollutants they emit. A list of the most common pollution control devices follows:

- ***Positive crankcase ventilation (PCV) system.*** This system reduces HC emissions by drawing fuel and oil vapors from the crankcase and sends them into the intake manifold where they are delivered to and burned in the cylinders. This system prevents the pressurized vapors from escaping the engine and entering into the atmosphere.

Figure 3-22 The intake system for a V-10 in a Dodge Viper. *Courtesy of Chrysler LLC*

- *Evaporative emission control system.* This system reduces HC emissions by drawing fuel vapors from the fuel system and releases them into the intake air to be burned. This system stops these vapors from leaking into the atmosphere.
- ***Exhaust gas recirculation (EGR)*** *system.* This system introduces exhaust gases into the intake air to reduce the temperatures reached during combustion. This reduces the chances of forming NO_X during combustion.
- *Catalytic converter.* Located in the exhaust system, it allows for the burning or converting of HC, CO, and NO_X into harmless substances, such as water.
- *Air injection system.* This system reduces HC emissions by introducing fresh air into the exhaust stream to cause minor combustion of the HC in the engine's exhaust.

Exhaust System

During the exhaust stroke, the engine's pistons move up and push the burned air-fuel mixture, or exhaust, out of the combustion chamber and into the exhaust manifold. From the manifold, the gases travel through the other parts of the exhaust system until they are expelled into the atmosphere **(Figure 3–23)**. The exhaust system is designed to direct toxic exhaust fumes away from the passenger compartment, to quiet the sound of the exhaust pulses, and to burn or

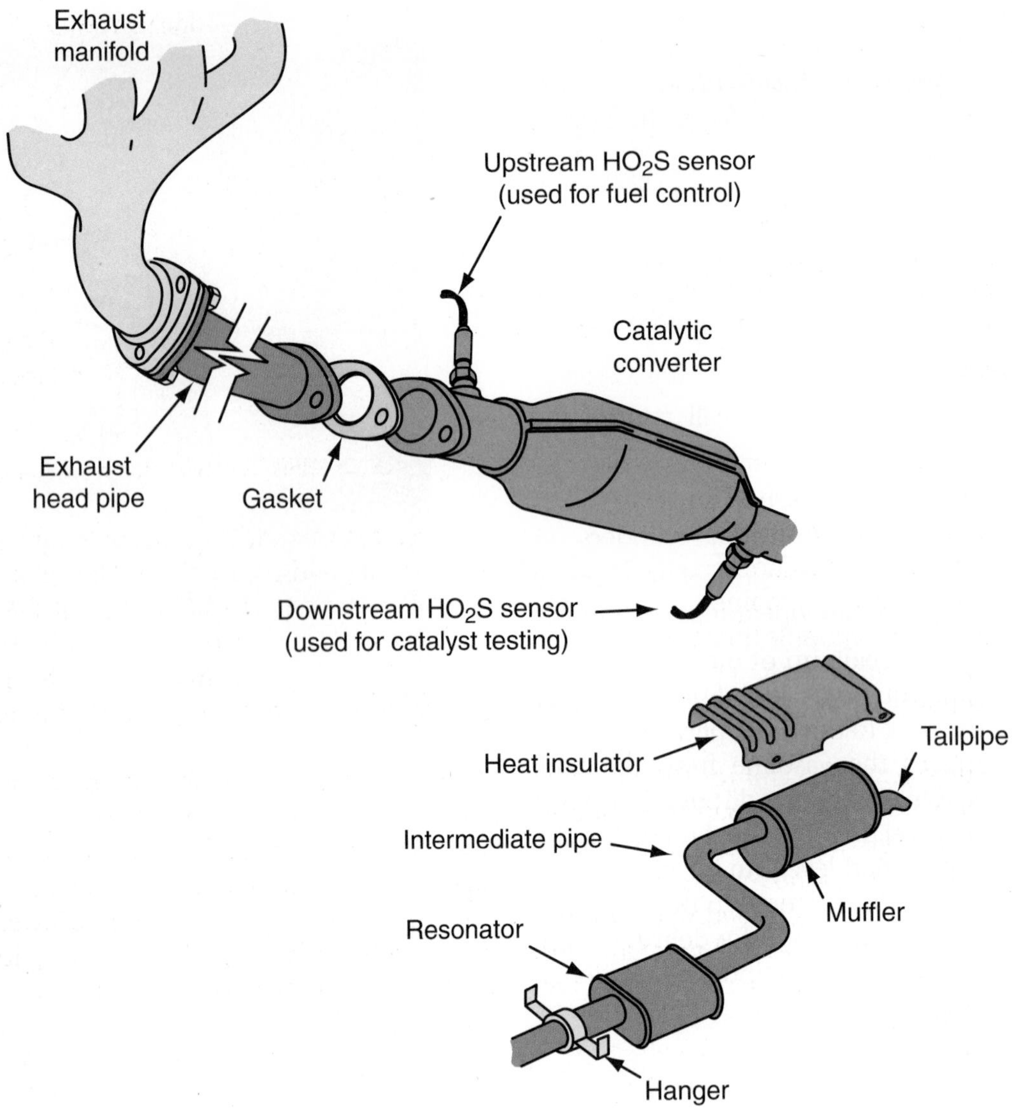

Figure 3-23 A typical exhaust system on a late-model car.

catalyze pollutants in the exhaust. A typical exhaust system contains the following components:

- Exhaust manifold and gasket
- Exhaust pipe, seal, and connector pipe
- Intermediate pipes
- Catalytic converter(s)
- Muffler and resonator
- Tailpipe
- Heat shields
- Clamps, gaskets, and hangers

ELECTRICAL AND ELECTRONIC SYSTEMS

Automobiles have many circuits that carry electrical current from the battery to individual components. The total electrical system includes such major subsystems as the ignition system, starting system, charging system, and the lighting and other electrical systems.

Ignition System

After the air-fuel mixture has been delivered to the cylinder and compressed by the piston, it must be ignited. A gasoline engine uses an electrical spark to ignite the mixture. Generating this spark is the role of the ignition system.

The **ignition coil** generates the electricity that creates this spark **(Figure 3–24)**. The coil transforms the low voltage of the battery into a burst of 30,000 to 100,000 volts. This burst is what ignites the mixture. The mixture must be ignited at the proper time in order for complete combustion to occur. Although the exact proper time varies with engine design, ignition must occur at a point before the piston has completed its compression stroke.

On most engines, the motion of the piston and the rotation of the crankshaft are monitored by a

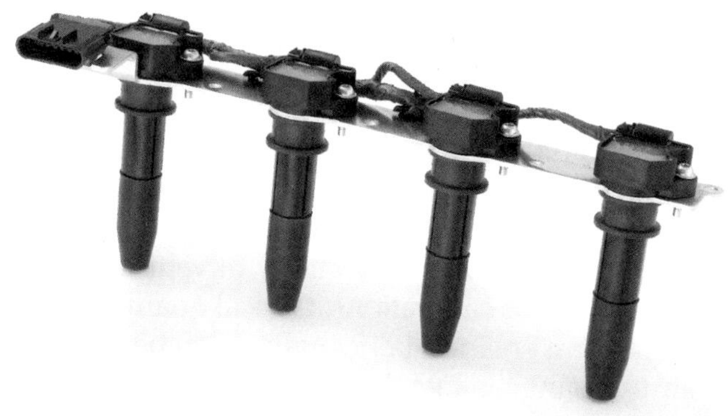

Figure 3-24 An ignition module and coil assembly for four cylinders.

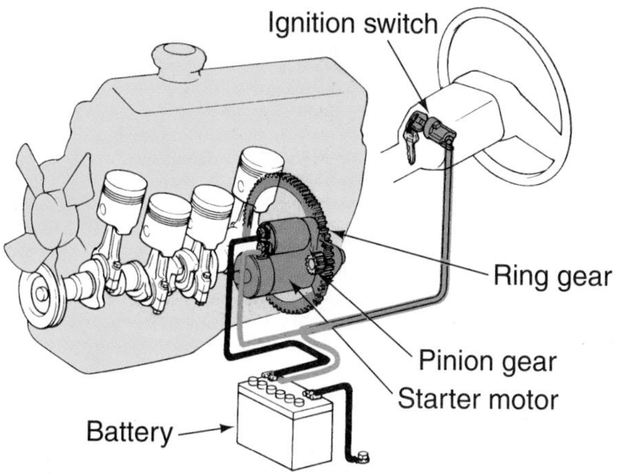

Figure 3-25 A typical starting system.

crankshaft position sensor. The sensor electronically tracks the position of the crankshaft and relays that information to an ignition control module. Based on input from the crankshaft position sensor, and, in some systems, the electronic engine control computer, the ignition control module then turns the battery current to the coil on and off at just the precise time so that the voltage surge arrives at the cylinder at the right time.

The voltage surge from the coil must be distributed to the correct cylinder because only one cylinder is fired at a time. In earlier systems, this was the job of the **distributor**. A distributor is driven by a gear on the camshaft at one-half the crankshaft speed. It transfers the high-voltage surges from the coil to spark plug wires in the correct firing order. The spark plug wires then deliver the high voltage to the spark plugs, which are screwed into the cylinder head. The voltage jumps across a space between two electrodes on the end of each **spark plug** and causes a spark. This spark ignites the air-fuel mixture.

Today's ignition systems do not use a distributor. Instead, these systems have several ignition coils—one for each spark plug or pair of spark plugs. When a coil is activated by the electronic control module, high voltage is sent through a spark plug circuit. The electronic control module has total control of the timing and distribution of the spark-producing voltage to the various cylinders.

Starting and Charging Systems

The starting system is responsible for getting the engine started **(Figure 3–25)**. When the ignition key is turned to the start position, a small amount of current flows from the battery to a **solenoid** or relay. This activates the solenoid or relay and closes another electrical circuit that allows full battery voltage to reach the starter motor. The starter motor then rotates the flywheel mounted on the rear of the crankshaft. As the crankshaft turns, the pistons move through their strokes. At the correct time for each cylinder, the ignition system provides the spark to ignite the air-fuel mixture. If good combustion takes place, the engine will now rotate on its own without the need of the starter motor. The ignition key is now allowed to return to the "on" position. From this point on, the engine will continue to run until the ignition key is turned off.

The electrical power for the engine and the rest of the car comes from the car's battery. The battery is especially important for the operation of the starting system. While the starter is rotating the crankshaft, it uses a lot of electricity. This tends to lower the amount of power in the battery. Therefore, a system is needed to recharge the battery so that engine starts can be made in the future.

The charging system is designed to recharge and maintain the battery's state of charge. It also provides electrical power for the ignition system, air conditioner, heater, lights, radio, and all electrical accessories when the engine is running.

The charging system includes an AC generator (alternator), voltage regulator, indicator light, and the necessary wiring. Rotated by the engine's crankshaft through a drive belt, the **AC generator** **(Figure 3–26)** converts mechanical energy into electrical energy. The AC is converted into direct current (DC) within the alternator. When the output or electrical current from the charging system flows back to the battery, the battery is being charged. When the current flows out of the battery, the battery is said to be discharging.

Electronic Engine Controls

Nearly all vehicles on the road have an electronic engine control system. This is a system comprised of many electronic and electromechanical parts. The system is designed to continuously monitor the operation of the engine and to make adjustments that will cause the engine to run more efficiently. Electronic

Figure 3-26 The major components of a late-model AC generator. *Courtesy of Robert Bosch GmbH, www.bosch-presse.de*

engine control systems have dramatically improved fuel mileage, engine performance, and driveability and have greatly reduced exhaust emissions.

Electronic control systems have three main types of components: input sensors, a computer, and output devices **(Figure 3–27)**. The computer analyzes data from the input sensors. Then, based on the inputs and the instructions held in its memory, the computer directs the output devices to make the necessary changes in the operation of some engine systems. Electronic control systems have fewer moving parts than old-style mechanical and vacuum controls. Therefore, the engine and other support systems can maintain their calibration almost indefinitely.

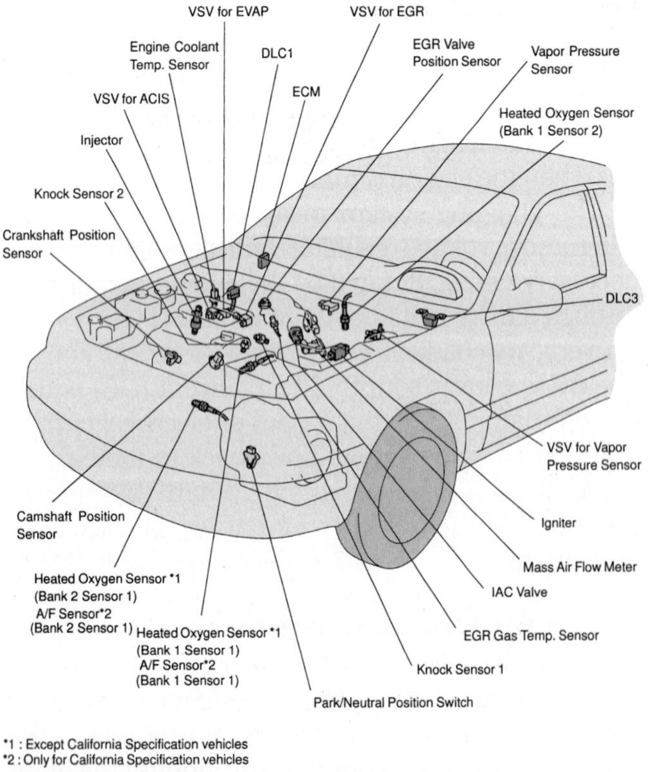

Figure 3-27 Late-model electronic engine control systems are made up of many different sensors and actuators and a central computer or control module.

As an added advantage, an electronic control system is very flexible. Because it uses computers, it can be programmed to meet a variety of vehicle engine combinations or calibrations. Critical quantities that determine an engine's performance can be changed easily by changing data that is stored in the computer's memory.

On-Board Diagnostics

Today's engine control systems are **on-board diagnostic (OBD II)** second-generation systems. These systems were developed to ensure proper emission control system operation for the vehicle's lifetime by monitoring emission-related components and systems for deterioration and malfunction. This monitoring includes also a check of the tank ventilation system for vapor leaks. The OBD system consists of the engine and transmission control modules and their sensors and actuators along with the diagnostic software.

The computer **(Figure 3–28)** can detect system problems even before the driver notices a driveability problem because many problems that affect emissions can be electrical or even chemical in nature.

When the OBD system determines that a problem exists, a corresponding "diagnostic trouble code" is stored in the computer's memory. The computer also illuminates a yellow dashboard light indicating "check engine" or "service engine soon" or displays an engine symbol. This light informs the driver of the need for service, not of the need to stop the vehicle.

A blinking or flashing dashboard lamp indicates a rather severe level of engine misfire. When this occurs, the driver should reduce engine speed and load and

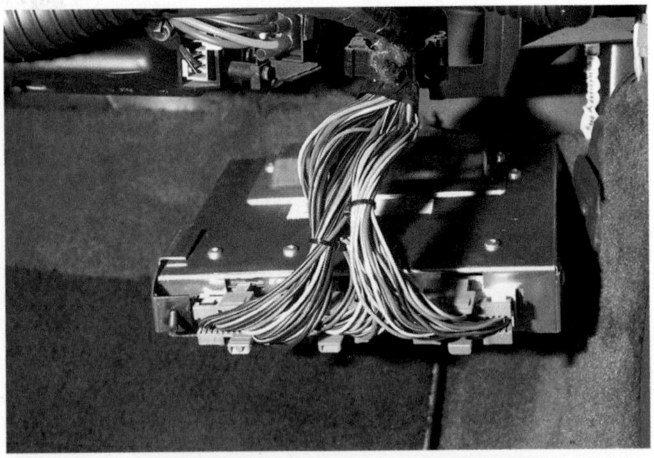

Figure 3-28 A typical automotive computer.

have the vehicle serviced as soon as possible. After the problem has been fixed, the dashboard lamp will be turned off.

HEATING AND AIR-CONDITIONING SYSTEMS

Heating and air-conditioning systems do little for the operation of a vehicle; they merely provide comfort for the passengers of the vehicle. Both systems are dependent on the proper operation of the engine. The heating system basically adds heat to the vehicle's interior, whereas air conditioning removes heat. To do this, the systems rely on many parts to put basic theories to work.

Heating Systems

To meet federal safety standards, all vehicles must be equipped with passenger compartment heating and windshield defrosting systems. The main components of an automotive heating system are the heater core, the heater control valve, the blower motor and the fan, and the heater and defroster ducts. The heating system works with the engine's cooling system and converts the heat from the coolant circulating inside the engine to hot air, which is blown into the passenger compartment. A heater hose transfers hot coolant from the engine to the heater control valve and then to the heater core inlet. As the coolant circulates through the core, heat is transferred from the coolant to the tubes and fins of the core. Air blown through the core by the blower motor and fan then picks up the heat from the surfaces of the core and transfers it into the passenger compartment. After giving up its heat, the coolant is then pumped out through the heater core outlet, where it is returned to the engine's cooling system to be heated again.

Transferring heated air from the heater core to the passenger compartment is the job of the heater and defroster ducts. The ducts are typically part of a large plastic shell that connects to the necessary inside and outside vents. Contained inside the duct are also the doors required to direct air to the floor, dash, and/or windshield.

Air-Conditioning Systems

An air-conditioning (A/C) system is designed to pump heat from one point to another. In an automotive A/C system, heat is removed from the passenger compartment and moved to the outside of the vehicle.

The substance used to remove heat from the inside of a vehicle is called the **refrigerant**. To understand how a refrigerant is used to cool the interior of a vehicle, the effects of pressure and temperature on it must be first understood. If the pressure of the refrigerant is high, so is its temperature. Likewise if the pressure is low, so is its temperature. Therefore, changing its pressure can change the refrigerant's temperature.

To absorb heat, the temperature and pressure of the refrigerant are kept low. To get rid of the heat, the temperature and pressure are high. As the refrigerant absorbs heat, it evaporates or changes from a liquid to a vapor. As it dissipates heat, it condenses and changes from a vapor to a liquid. These two changes of state occur continuously as the refrigerant circulates through the system.

An A/C system is a closed, pressurized system. It consists of a compressor, condenser, receiver/dryer or accumulator, expansion valve or orifice tube, and an evaporator. The best way to understand the purpose of the components is to divide the system into two sides: the high side and the low side. High side refers to the side of the system that is under high pressure and high temperature. Low side refers to the low-pressure, low-temperature side **(Figure 3–29)**.

Compressor The **compressor** separates the high and low sides of the system. Its primary purpose is to draw the low-pressure and low-temperature vapor from the evaporator and compress this vapor into high-temperature, high-pressure vapor. The secondary purpose of the compressor is to circulate or pump the refrigerant through the system. The compressor is located on the engine and is driven by the engine's crankshaft via a drive belt.

Compressors are equipped with an electromagnetic clutch as part of the compressor pulley

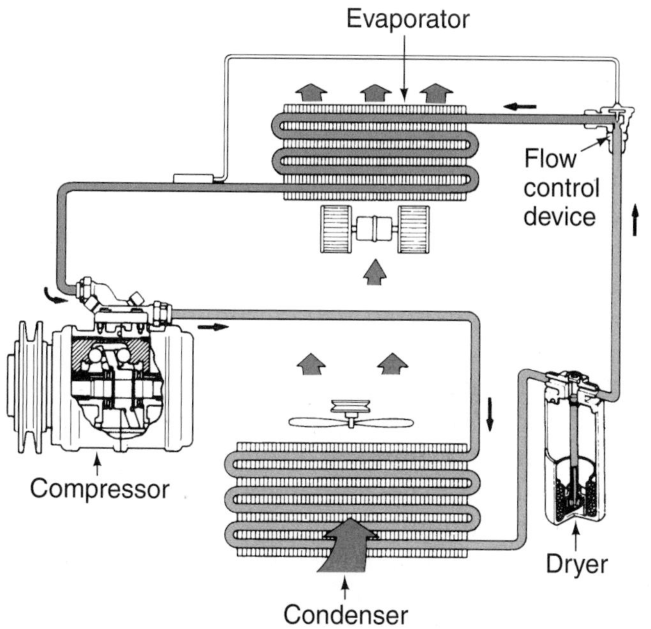

Figure 3-29 A simple look at an air conditioning system. The blue signifies low pressure and the red is high pressure.

assembly. The clutch is designed to engage the pulley to the compressor shaft when the clutch coil is energized. When the clutch is not engaged, the compressor shaft does not rotate, and the pulley freewheels. The clutch provides a way for turning the compressor on or off.

Condenser The **condenser** consists of coiled tubing mounted in a series of thin cooling fins to provide maximum heat transfer in a minimum amount of space. The condenser is normally mounted just in front of the vehicle's radiator. It receives the full flow of ram air from the movement of the vehicle or airflow from the radiator fan when the vehicle is standing still.

The condenser condenses or liquefies the high-pressure, high-temperature vapor coming from the compressor. To do so, it must give up its heat. Very hot, high-pressure refrigerant vapor enters the inlet at the top of the condenser, and as the hot vapor passes down through the condenser coils, heat moves from the refrigerant into the cooler air that flows across the coils and fins. This loss of heat causes the refrigerant to change from a high-pressure hot vapor to a high-pressure warm liquid. The high-pressure warm liquid flows from the bottom of the condenser to the receiver/dryer or to the refrigerant metering device if an accumulator is used.

Receiver/Dryer The **receiver/dryer** is a storage tank for the liquid refrigerant from the condenser. The refrigerant flows into the receiver tank, which contains a bag of desiccant (moisture-absorbing material). The desiccant absorbs unwanted water and moisture in the refrigerant.

Accumulator Most late-model systems have an accumulator rather than a receiver/dryer. The accumulator is connected into the low side at the outlet of the evaporator. The accumulator contains a desiccant and is designed to store excess refrigerant. If liquid refrigerant flows out of the evaporator, it will be collected by and stored in the accumulator. The main purpose of an accumulator is to prevent liquid from entering the compressor.

Thermostatic Expansion Valve/Orifice Tube The refrigerant flow to the evaporator must be controlled to obtain maximum cooling while ensuring complete evaporation of the liquid refrigerant within the evaporator. This is the job of a **thermostatic expansion valve (TEV or TXV)** or a fixed orifice tube. The TEV is mounted at the inlet to the evaporator and separates the high-pressure side of the system from the low-pressure side. The TEV regulates the refrigerant flow to the evaporator by balancing the inlet flow to the outlet temperature.

Like the TEV, the **orifice tube** is the dividing point between the high- and low-pressure sides of the system. However, its metering or flow rate control does not depend on comparing evaporator pressure and temperature. It is a fixed orifice. The flow rate is determined by pressure difference across the orifice and by the additional cooling of the refrigerant in the bottom of the condenser after it has changed from vapor to liquid.

Evaporator The **evaporator**, like the condenser, consists of a refrigerant coil mounted in a series of thin cooling fins. The evaporator is usually located beneath the dashboard or instrument panel.

The low-pressure, low-temperature liquid refrigerant from the TEV or orifice tube enters the evaporator as a spray. The heat at the evaporator causes the refrigerant to boil and change into a vapor. The transfer of heat from the evaporator to the refrigerant causes the evaporator to get cold. Because hot air always moves toward cold air, the hot air from inside the vehicle moves across the evaporator. As the process of heat loss from the air to the evaporator core surface is taking place, any moisture in the air condenses on the outside of the evaporator core and is drained off as water. This dehumidification of air adds to passenger comfort.

Refrigerant Lines There are three major refrigerant lines. Suction lines are located between the outlet side of the evaporator and the inlet side or suction side of the compressor. They carry the low-pressure, low-temperature vapor to the compressor. The discharge or high-pressure line connects the compressor to the condenser. The liquid lines connect the condenser to the receiver/dryer and the receiver/dryer to the inlet side of the expansion valve. Through these lines, the refrigerant travels in its path from a gas state (compressor outlet) to a liquid state (condenser outlet) and then to the inlet side of the expansion valve, where it vaporizes at the evaporator.

DRIVETRAIN

The **drivetrain** is made up of all components that transfer power from the engine to the driving wheels of the vehicle. The exact components used in a vehicle's drivetrain depend on whether the vehicle is equipped with rear-wheel drive, front-wheel drive, or four-wheel drive.

Today, most cars are front-wheel drive (FWD). Some larger luxury and performance cars are rear-wheel drive (RWD). Most pickup trucks, minivans, and SUVs are also RWD vehicles. Power flow in an RWD vehicle passes through the **clutch** or **torque converter**, manual or automatic transmission, and the driveline (drive shaft assembly). Then it goes through the rear differential, the rear-driving axles, and onto the rear wheels.

Power flow through the drivetrain of FWD vehicles passes through the clutch or torque converter, then moves through a front differential, the driving axles, and onto the front wheels.

Four-wheel-drive (4WD) or all-wheel-drive (AWD) vehicles combine features of both rear- and front-wheel-drive systems so that power can be delivered to all wheels either on a permanent or on-demand basis. Typically if a truck, pickup or SUV, has 4WD, the system is based on an RWD and a front drive axle is added. When a car has AWD or 4WD, the drivetrain is a modified FWD system. Modifications include a rear drive axle and an assembly that transfers some of the power to the rear axle.

Clutch

A clutch is used with manual transmissions/transaxles. It mechanically connects the engine's flywheel to the transmission/transaxle input shaft **(Figure 3–30)**. This is accomplished by a special friction plate that is splined to the input shaft of the transmission. When the clutch is engaged, the friction plate contacts the flywheel and transfers power to the input shaft.

When stopping, starting, and shifting from one gear to the next, the clutch is disengaged by pushing down on the clutch pedal. This moves the clutch plate away from the flywheel, stopping power flow to the transmission. The driver can then shift gears without damaging the transmission or transaxle. Releasing the clutch pedal reengages the clutch and allows power to flow from the engine to the transmission.

Manual Transmission

A manual or standard transmission is one in which the driver manually selects the gear of choice. Proper gear selection allows for good driveability and requires some driver education.

Whenever two or three gears have their teeth meshed together, a gearset is formed. The movement of one gear in the set will cause the others to move. If any of the gears in the set are a different size than the others, the gears will move at different speeds. The size ratio of a gearset is called the **gear ratio** of that gearset.

A manual transmission houses a number of individual gearsets, which produce different gear ratios **(Figure 3–31)**. The driver selects the desired operating gear or gear ratio. A typical manual transmission has four or five forward gear ratios, neutral, and reverse.

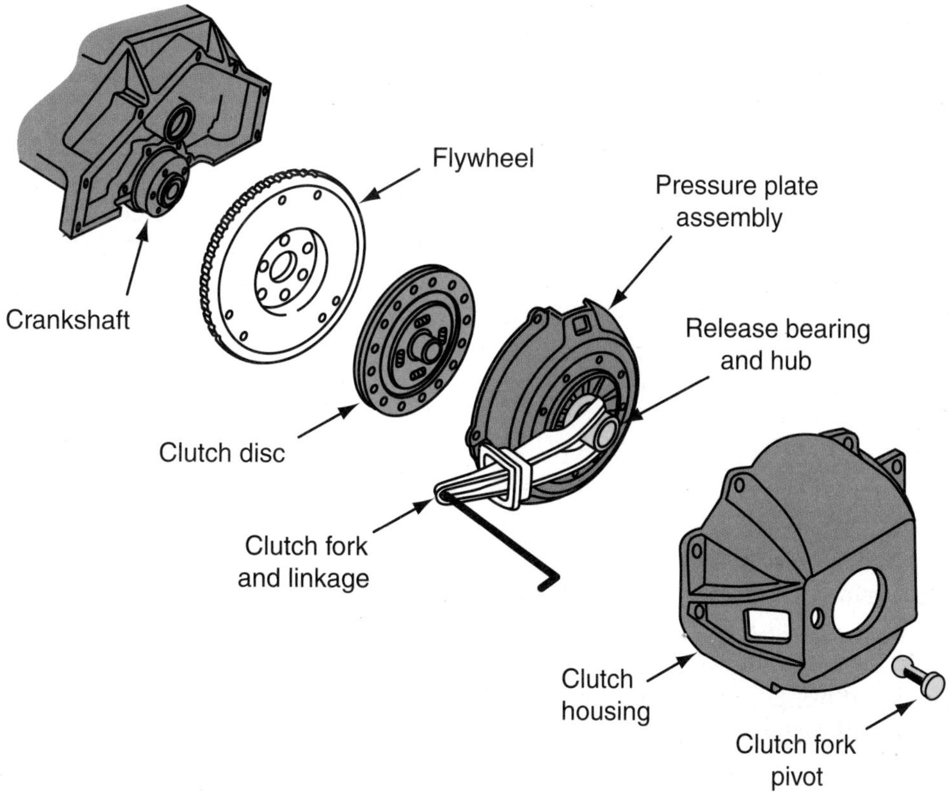

Figure 3–30 The major components of a clutch assembly for a manual transmission.

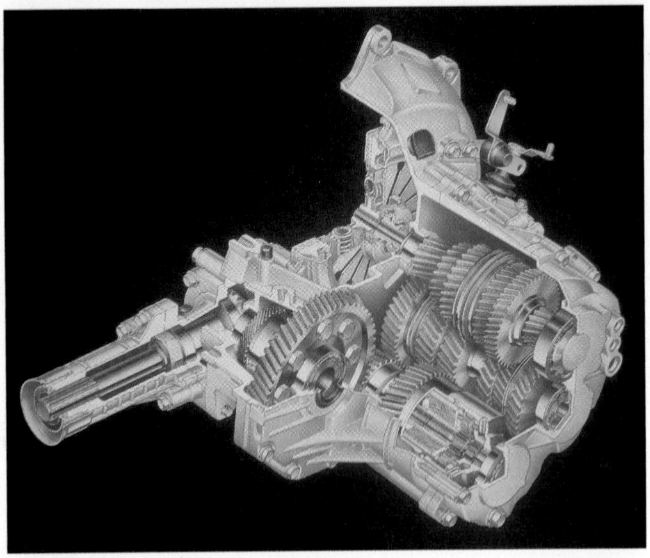

Figure 3-31 A typical manual transaxle. *Courtesy of Chrysler LLC*

Automatic Transmission

An automatic transmission does not need a clutch pedal and shifts through the forward gears without the control of the driver. Instead of a clutch, it uses a torque converter to transfer power from the engine's flywheel to the transmission input shaft. The torque converter allows for smooth transfer of power at all engine speeds **(Figure 3–32)**.

Shifting in an automatic transmission is controlled by a hydraulic and/or electronic control system. In a hydraulic system, an intricate network of valves and other components use hydraulic pressure to control the operation of planetary gearsets. These gearsets provide the three or four forward speeds, neutral, park, and reverse gears normally found in automatic transmissions. Newer electronic shifting systems use electric solenoids to control shifting mechanisms. Electronic shifting is precise and can be varied to suit certain operating conditions. All automatic transmission-equipped vehicles with OBD II have electronic shifting.

Figure 3-32 A cutaway of a six-speed automatic transmission shown with the torque converter in the housing. *Courtesy of BMW of North America, LLC*

Driveline

Drivelines are used on RWD vehicles and 4WD vehicles. They connect the output shaft of the transmission to the gearing in the rear axle housing. They are also used to connect the output shaft to the front and rear drive axles on a 4WD vehicle.

A driveline consists of a hollow drive or propeller shaft that is connected to the transmission and drive axle differential by universal joints (U-joints). These U-joints allow the drive shaft to move with movement of the rear suspension, preventing damage to the shaft.

Differential

On RWD vehicles, the drive shaft turns perpendicular to the forward motion of the vehicle. The differential gearing in the rear axle housing is designed to turn the direction of the power so that it can be used to drive the wheels of the vehicle. The power flows into the **differential**, where it changes direction, then flows to the rear axles and wheels **(Figure 3–33)**.

The gearing in the differential also multiplies the torque it receives from the drive shaft by providing a final gear reduction. Also, it divides the torque between the left and right driving axles and wheels so that a differential wheel speed is possible. This means one wheel can turn faster than the other when going around turns.

Driving Axles

Driving axles are solid steel shafts that transfer the torque from the differential to the driving wheels. A separate axle shaft is used for each driving wheel. In an RWD vehicle, the driving axles are part of the

Figure 3-33 The driveline connects the output from the transmission to the differential unit and drive axles. *Courtesy of BMW of North America, LLC*

differential and are enclosed in an axle housing that protects and supports these parts. Some rear drive axle units are mounted to an independent suspension and the drive axle assembly is similar to that of a FWD vehicle.

Each drive axle is connected to the side gears in the differential. The inner ends of the axles are splined to fit into the side gears. As the side gears are turned, the axles to which they are splined turn at the same speed.

The drive wheels are attached to the outer ends of the axles. The outer end of each axle has a flange mounted to it. A **flange** is a rim for attaching one part to another part. The flange, fitted with studs, at the end of an axle holds the wheel in place. **Studs** are threaded shafts, resembling bolts without heads. One end of the stud is screwed or pressed into the flange. The wheel fits over the studs, and a nut, called the **lug nut**, is tightened over the open end of the stud. This holds the wheel in place.

The differential carrier supports the inner end of each axle. A bearing inside the axle housing supports the outer end of the axle shaft. This bearing, called the axle bearing, allows the axle to rotate smoothly inside the axle housing.

Transaxle

A **transaxle** is used on FWD vehicles. It is made up of a transmission and differential housed in a single unit **(Figure 3–34)**. The gearsets in the transaxle provide the required gear ratios and direct the power flow into the differential. The differential gearing provides the final gear reduction and splits the power flow between the left and right drive axles.

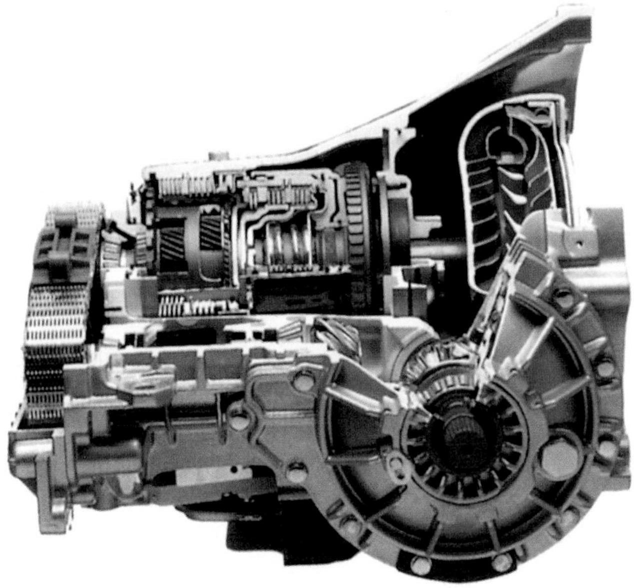

Figure 3–34 A cutaway of an automatic transaxle. *Courtesy of Chrysler LLC*

The drive axles extend from the sides of the transaxle. The outer ends of the axles are fitted to the hubs of the drive wheels. **Constant velocity (CV) joints** mounted on each end of the drive axles allow for changes in length and angle without affecting the power flow to the wheels.

Four-Wheel-Drive System

4WD or AWD vehicles combine the features of RWD transmissions and FWD transaxles. Additional **transfer case** gearing splits the power flow between a differential driving the front wheels and a rear differential that drives the rear wheels. This transfer case can be a housing bolted directly to the transmission/transaxle, or it can be a separate housing mounted somewhere in the driveline. Most RWD-based 4WD vehicles have a drive shaft connecting the output of the transmission to the rear axle and another connecting the output of the transfer case to the front drive axle. Typically, AWD cars have a center differential, which splits the torque between the front and rear drive axles.

RUNNING GEAR

The **running gear** of a vehicle includes those parts that are used to control the vehicle, which includes the wheels and tires and the suspension, steering, and brake systems.

Suspension System

The suspension system **(Figure 3–35)** includes such components as the springs, shock absorbers, MacPherson struts, torsion bars, axles, and connecting linkages. These components are designed to support the body and frame, the engine, and the drivelines. Without these systems, the comfort and ease of driving the vehicle would be reduced.

Springs or **torsion bars** are used to support the axles of the vehicle. The two types of springs commonly used are the coil spring and the leaf spring. Torsion bars are also used and are long spring steel rods. One end of the rod is connected to the frame, whereas the other end is connected to the movable parts of the axles. As the axles move up and down, the rod twists and acts as a spring.

Shock absorbers dampen the upward and downward movement of the springs. This is necessary to limit the car's reaction to a bump in the road.

Steering System

The steering system allows the driver to control the direction of the vehicle. It includes the steering wheel, steering gear, steering shaft, and steering linkage.

Two basic types of steering systems are used today: the **rack-and-pinion** and **recirculating ball** systems

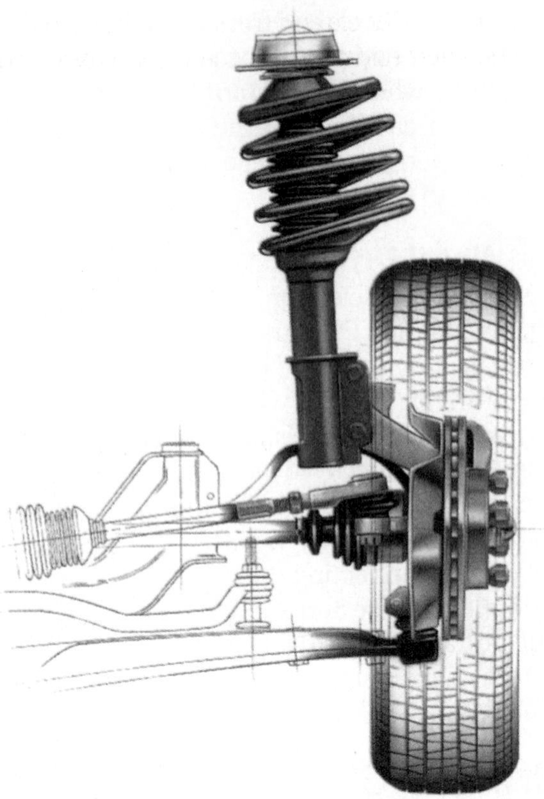

Figure 3-35 A strut assembly of a typical suspension system. *Courtesy of Ford Motor Company*

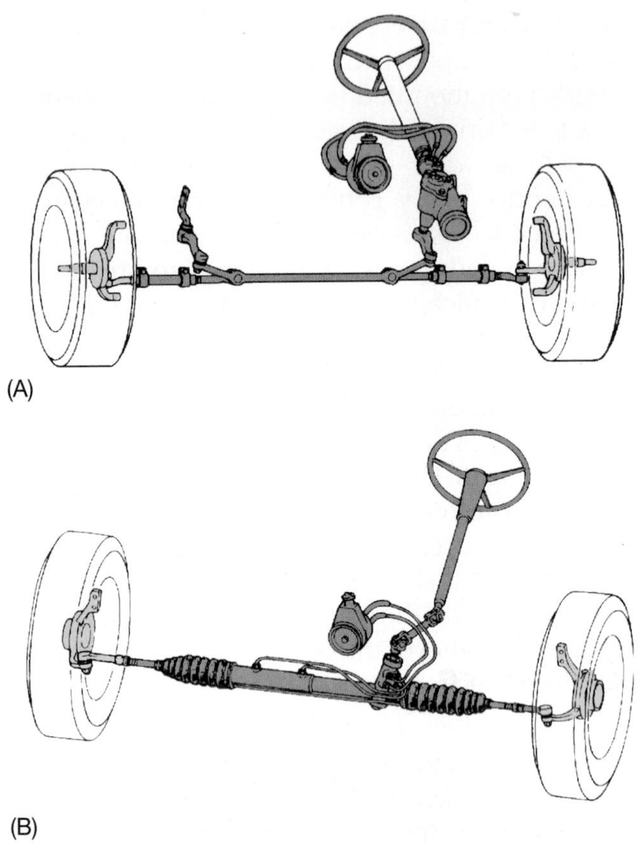

Figure 3-36 (A) A parallelogram-type steering system. (B) A rack and pinion steering system. *Courtesy of Federal-Mogul Corporation*

(Figure 3–36). The rack-and-pinion system is commonly used in passenger cars. The recirculating ball system is normally used only on pickup trucks, SUVs, and full-size luxury cars.

Steering gears provide a gear reduction to make changing the direction of the wheels easier. On most vehicles, the steering gear is also power assisted to ease the effort of turning the wheels. In a power-assisted system, a pump provides hydraulic fluid under pressure to the steering gear. Pressurized fluid is directed to one side or the other of the steering gear to make it easier to turn the wheels.

Some vehicles are equipped with speed-sensitive power steering systems. These change the amount of power assist according to vehicle speed. Power assist is the greatest when the vehicle is moving slowly and decreases as speed increases.

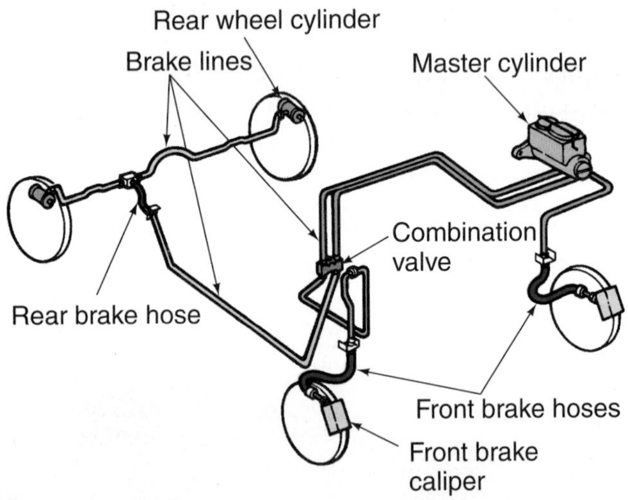

Figure 3-37 A typical hydraulic brake system with disc brakes at the front and rear wheels.

Brakes

Obviously, the brake system is used to slow down and stop a vehicle **(Figure 3–37)**. Brakes, located at each wheel, use friction to slow and stop a vehicle.

The brakes are activated when the driver presses down on the brake pedal. The brake pedal is connected to a plunger in a **master cylinder**, which is filled with hydraulic fluid. As pressure is put on the brake pedal, a force is applied to the hydraulic fluid in the master cylinder. This force is increased by the master cylinder and transferred through brake hoses and lines to the four brake assemblies.

Two types of brakes are used—**disc brakes** and **drum brakes**. Many vehicles use a combination of the two types: disc brakes at the front wheels **(Figure 3–38)** and drum brakes at the rear wheels; others have disc brakes at all wheels.

Most vehicles have power-assisted brakes. Many vehicles use a vacuum **brake booster** to increase the

Figure 3-38 A disc brake unit with a wheel speed sensor for ABS. *Courtesy of Chrysler LLC*

pressure applied to the plunger in the master cylinder. Others use hydraulic pressure from the power steering pump to increase the pressure on the brake fluid. Both of these systems lessen the amount of pressure that must be applied to the brake pedal and increase the responsiveness of the brake system.

Nearly all late-model vehicles have an **antilock brake system (ABS)**. The purpose of ABS is to prevent skidding during hard braking to give the driver control of the vehicle during hard stops.

Wheels and Tires

The only contact a vehicle has with the road is through its tires and wheels. Tires are filled with air and made of forms of rubber and other materials to give them strength. Wheels are made of metal and are bolted to the axles or spindles **(Figure 3–39)**. Wheels hold the tires in place. Wheels and tires come in many different sizes. Their sizes must be matched to one another and to the vehicle.

Figure 3-39 An alloy wheel with high-performance tires.

HYBRID VEHICLES

A **hybrid electric vehicle (HEV)** uses one or more electric motors and an engine to propel the vehicle **(Figure 3–40)**. Depending on the design of the system, the engine may move the vehicle by itself, assist the electric motor while it is moving the vehicle, or it may drive a generator to charge the vehicle's batteries. The electric motor may power the vehicle by itself or assist the engine while it is propelling the vehicle. Many hybrids rely exclusively on the electric motor(s) during slow speed operation, the engine at higher speeds, and both during some certain driving conditions. Complex electronic controls monitor the operation of the vehicle. Based on the current operating conditions, electronics control the engine, electric motor, and generator.

A hybrid's electric motor is powered by high-voltage batteries, which are recharged by a generator driven by the engine and through regenerative braking. **Regenerative braking** is the process by which a vehicle's kinetic energy can be captured while it is decelerating and braking. The electric drive motors become generators driven by the vehicle's wheels. These generators take the kinetic energy, or the energy of the moving vehicle, and changes it into energy that charges the batteries. The magnetic forces inside the

Figure 3-40 The electric motor in this hybrid arrangement fits between the engine and the transmission. *Courtesy of American Honda Motor Co., Inc.*

generator cause the drive wheels to slow down. A conventional brake system brings the vehicle to a safe stop.

The engines used in hybrids are specially designed for the vehicle and electric assist. Therefore, they can operate more efficiently, resulting in very good fuel economy and very low tailpipe emissions. HEVs can provide the same performance, if not better, as a comparable vehicle equipped with a larger engine.

There are primarily two types of hybrids: the parallel and the series designs. A parallel HEV uses either the electric motor or the gas engine to propel the vehicle, or both. The engine in a true series HEV is used only to drive the generator that keeps the batteries charged. The vehicle is powered only by the electric motor(s). Most current HEVs are considered as having a series/parallel configuration because they have the features of both designs.

Although most current hybrids are focused on fuel economy, the same ideas can be used to create high-performance vehicles. Hybrid technology is also influencing off-the-road performance. By using individual motors at the front and rear drive axles, additional power can be applied to certain drive wheels when needed.

KEY TERMS

AC generator
Antilock brake system (ABS)
Brake booster
Carbon monoxide (CO)
Clutch
Combustion
Combustion chamber
Compressor
Condenser
Connecting rod
Constant velocity (CV) joint
Convertible
Corporate Average Fuel Economy (CAFE)
Cradle
Crankshaft
Crankshaft position sensor
Crossover vehicles
Cylinder block
Cylinder head
Differential
Disc brakes
Distributor
Drivetrain
Drum brakes
Engine block
Evaporator
Exhaust gas recirculation (EGR)
Exhaust manifold
Flange
Friction
Gear ratio
Hatchback
Hybrid electric vehicle (HEV)
Hydrocarbons (HC)
Hydroforming
Ignition coil
Intake manifold
Liftback
Lug nut
Manifold
Master cylinder
Oil pan
Oil sump
On-board diagnostics (OBD II)
Orifice tube
Oxides of nitrogen (NO_X)
Pickup
Piston
Port
Positive crankcase ventilation (PCV)
Pressure cap
Rack and pinion
Radiator
Receiver/dryer
Recirculating ball
Refrigerant
Regenerative braking
Running gear
Sedan
Shock absorber
Solenoid
Spark plug
Sport utility vehicle (SUV)
Spring
Station wagon
Steering gear
Stud
Thermostat
Thermostatic expansion valve (TEV or TXV)
Torque converter
Torsion bar
Transaxle
Transfer case
Unibody
Valve train
Van
Water jacket
Water pump

SUMMARY

- Dramatic changes to the automobile have occurred over the last 40 years, including the addition of emission control systems, more fuel-efficient and cleaner-burning engines, and lighter body weight.
- In addition to being lighter than body-over-frame vehicles, unibodies offer better occupant protection by distributing impact forces throughout the vehicle.
- Today's computerized engine control systems regulate such things as air and fuel delivery, ignition timing, and emissions. The result is an increase in overall efficiency.
- All automotive engines are classified as internal combustion, because the burning of the fuel and air occurs inside the engine. Diesel engines share the same major parts as gasoline engines, but they do not use a spark to ignite the air-fuel mixture.
- The cooling system maintains proper engine temperatures. Liquid cooling is more efficient than air cooling and it is more commonly used.
- The lubrication system distributes motor oil throughout the engine. This system also contains the oil filter necessary to remove dirt and other foreign matter from the oil.
- The fuel system is responsible not only for fuel storage and delivery, but also for atomizing and mixing it with the air in the correct proportion.

- The exhaust system has three primary purposes: to channel toxic exhaust away from the passenger compartment, to quiet the exhaust pulses, and to burn the emissions in the exhaust.
- The electrical system of an automobile includes the ignition, starting, charging, and lighting systems. Electronic engine controls regulate these systems very accurately through the use of computers.
- Modern automatic transmissions use a computer to match the demand for acceleration with engine speed, wheel speed, and load conditions. It then chooses the proper gear ratio and, if necessary, initiates a gear change.
- The running gear is critical to controlling the vehicle. It consists of the suspension system, braking system, steering system, and wheels and tires.
- Hybrid electric vehicles have an internal combustion engine and an electrical motor that are used to propel the vehicle together or separately.

REVIEW QUESTIONS

1. Under the CAFE standards, for what are vehicles tested?
2. Define internal combustion.
3. In addition to the battery, what does the charging system include?
4. Which of the following is *not* a typical emission control system?
   a. EGR
   b. PCV
   c. EPA
   d. air injection
5. Automatic transmissions use a ______ instead of a clutch to transfer power from the flywheel to the transmission's input shaft.
   a. differential
   b. U-joint
   c. torque converter
   d. constant velocity joint
6. Which of the following is *not* one of the strokes of a four-cycle engine?
   a. compression
   b. exhaust
   c. intake
   d. combustion
7. Technician A says that the PCV system is designed to limit CO emissions. Technician B says that catalytic converters reduce HC emissions at the tailpipe. Who is correct?
   a. Technician A
   b. Technician B
   c. Both A and B
   d. Neither A nor B
8. Which type of engine is classified as internal combustion?
   a. gasoline
   b. diesel
   c. both a and b
   d. neither a nor b
9. What does the valve train do?
   a. It delivers fuel to a device that controls the amount of fuel going to the engine.
   b. It houses the major parts of the engine.
   c. It converts reciprocating motion to rotary motion.
   d. It opens and closes the intake and exhaust ports of each cylinder.
10. Technician A says that liquid cooling an engine maintains a constant operating temperature. Technician B says that oil is circulated through the cooling system to remove heat from the engine's parts. Who is correct?
    a. Technician A
    b. Technician B
    c. Both A and B
    d. Neither A nor B
11. An engine will not start and no spark is found at the spark plugs when the engine is turned over by the starter. Technician A says that the problem is probably the battery. Technician B says that the ignition system is most likely at fault. Who is correct?
    a. Technician A
    b. Technician B
    c. Both A and B
    d. Neither A nor B
12. Which emission control system introduces exhaust gases into the intake air to reduce the formation of NO_X in the combustion chamber?
    a. evaporative emission controls
    b. exhaust gas recirculation
    c. air injection
    d. early fuel evaporation
13. Technician A says that many vehicles use an AC generator as the charging unit. Technician B says that many vehicles use an alternator as the charging unit. Who is correct?
    a. Technician A
    b. Technician B
    c. Both A and B
    d. Neither A nor B
14. Technician A says that a transaxle delivers torque to the front and the rear drive axles. Technician B says that a transaxle is most commonly found in 4WD pickups and SUVs. Who is correct?
    a. Technician A
    b. Technician B
    c. Both A and B
    d. Neither A nor B

15. Which of the following is *not* part of the running gear?
    a. differential
    b. steering
    c. suspension
    d. brakes
16. Which of the following statements about unibody construction is *not* true?
    a. A unibody is a stressed hull structure in which each of the body parts supplies structural support and strength to the entire vehicle.
    b. Unibody vehicles tend to be tightly constructed because the major parts are all welded together.
    c. All unibodies are constructed from steel.
    d. Most front-wheel-drive unibody vehicles have a cradle or partial frame.
17. A four-cylinder engine can have ______ ignition coils.
    a. one
    b. two
    c. four
    d. one, two, or four
18. Which of the following components are used to dampen the upward and downward movement of a vehicle's springs?
    a. torsion bars
    b. shock absorbers
    c. constant velocity joints
    d. connecting linkages
19. *True or False?* A hybrid electric vehicle uses regenerative braking to safely stop.
20. What two major engine components work together to change the reciprocating motion of the pistons into a rotary motion?
    a. crankshaft and connecting rod
    b. camshaft and crankshaft
    c. camshaft and connecting rod
    d. crankshaft and valve train
21. Technician A says that an air-conditioning system removes heat from inside the vehicle. Technician B says that the heating system adds heat to the inside of the vehicle. Who is correct?
    a. Technician A
    b. Technician B
    c. Both A and B
    d. Neither A nor B
22. The boiling point of the coolant in an engine's cooling system is raised by the ______.
    a. thermostat
    b. pressure cap
    c. radiator
    d. water pump
23. While discussing the operation of air-conditioning systems: Technician A says that in order for the refrigerant to release heat, it must be at a low temperature and pressure. Technician B says that in order for the refrigerant to absorb heat, it must be at a high temperature and pressure. Who is correct?
    a. Technician A
    b. Technician B
    c. Both A and B
    d. Neither A nor B
24. While discussing HEVs: Technician A says that in a series hybrid, the engine is only used to power a generator to recharge the batteries. Technician B says that in a parallel hybrid, the engine always supplies some of the power to move the vehicle. Who is correct?
    a. Technician A
    b. Technician B
    c. Both A and B
    d. Neither A nor B
25. Which of the following is *not* accomplished by the fuel and air system of a gasoline engine?
    a. stores the fuel for later use
    b. collects and cleans the outside air
    c. delivers fuel to a device that will control the amount of fuel going to the engine
    d. keeps the fuel and air ratio constant regardless of the conditions under which the engine is operating

HAND TOOLS AND SHOP EQUIPMENT

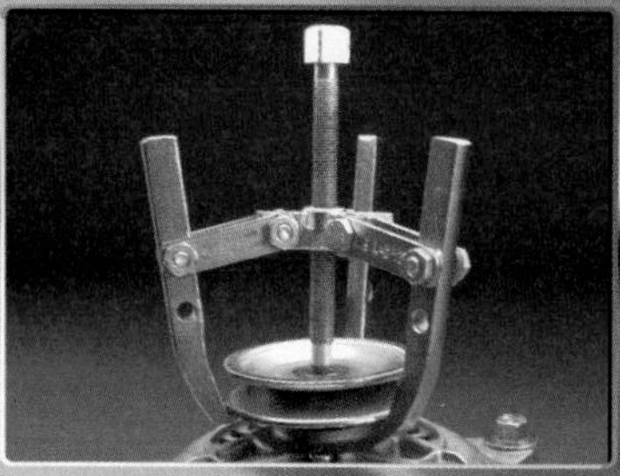

CHAPTER 4

OBJECTIVES

■ List the basic units of measure for length, volume, and mass in the two measuring systems. ■ Describe the different types of fasteners used in the automotive industry. ■ List the various mechanical measuring tools used in the automotive shop. ■ Describe the proper procedure for measuring with a micrometer. ■ List some of the hand tools used in auto repair. ■ List the common types of shop equipment and state their purpose. ■ Describe the use of common pneumatic, electrical, and hydraulic power tools found in an automotive service department. ■ Describe the different sources for service information that are available to technicians.

Repairing the modern automobile requires the use of various tools. Many of these tools are common hand and power tools used every day by a technician. Other tools are very specialized and are only for specific repairs on specific systems and/or vehicles. This chapter presents some of the more commonly used hand and power tools with which every technician must be familiar. Because units of measurement play such an important part in tool selection and in diagnosing automotive problems, this chapter begins with a presentation of measuring systems. Prior to the discussion on tools, there is a discussion on another topic that relates very much to measuring systems—fasteners.

MEASURING SYSTEMS

Two systems of weights and measures exist side by side in the United States—the Imperial or U.S. customary system and the international or metric system.

The basic unit of linear measurement in the Imperial system is the inch. The basic unit of linear measurement in the metric system is the meter. The meter is easily broken down into smaller units, such as the centimeter (1/100 meter) and millimeter (1/1,000 meter).

All units of measurement in the metric system are related to each other by a factor of 10. Every metric unit can be multiplied or divided by the factor of 10 to get larger units (multiples) or smaller units (submultiples). This makes the metric system much easier to use, with less chance of math errors than when using the Imperial system **(Figure 4–1)**.

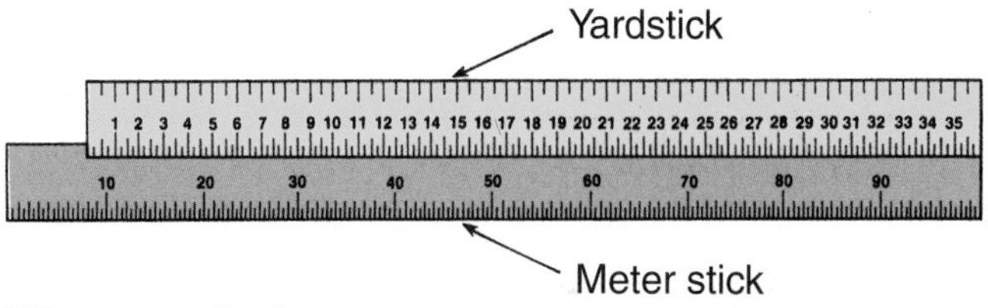

Figure 4–1 A meter stick has 1,000 increments known as millimeters and is slightly longer than a yardstick.

The United States passed the Metric Conversion Act in 1975 in an attempt to get American industry and the general public to use the metric system, as the rest of the world does. While the general public has been slow to drop the customary measuring system of inches, gallons, and pounds, many industries, led by the automotive industry, have now adopted the metric system for the most part.

Nearly all vehicles are now built to metric standards. Technicians must be able to measure and work with both systems of measurement. The following are some common equivalents in the two systems:

Linear Measurements

1 meter (m) = 39.37 inches (in.)

1 centimeter (cm) = 0.3937 inch

1 millimeter (mm) = 0.03937 inch

1 inch = 2.54 centimeters

1 inch = 25.4 millimeters

1 mile = 1.6093 kilometers

Square Measurements

1 square inch = 6.452 square centimeters

1 square centimeter = 0.155 square inch

Volume Measurements

1 cubic inch = 16.387 cubic centimeters

1,000 cubic centimeters = 1 liter (l)

1 liter (l) = 61.02 cubic inches

1 gallon = 3.7854 liters

Weight Measurements

1 ounce = 28.3495 grams

1 pound = 453.59 grams

1,000 grams = 1 kilogram

1 kilogram = 2.2046 pounds

Temperature Measurements

1°Fahrenheit (F) = $\frac{9}{5}$C + 32°

1°Celsius (C) = $\frac{5}{9}$(F – 32°)

Pressure Measurements

1 pound per square inch (psi) = 0.07031 kilogram (kg) per square centimeter

1 kilogram per square centimeter = 14.22334 pounds per square inch

1 bar = 14.504 pounds per square inch

1 pound per square inch = 0.06895 bar

1 atmosphere = 14.7 pounds per square inch

Torque Measurements

10 foot-pounds (ft.-lb) = 13.558 Newton meters (N-m)

1 N-m = 0.7375 ft.-lb

1 ft.-lb = 0.138 m kg

1 cm kg = 7.233 ft.-lb

10 cm kg = 0.98 N-m

FASTENERS

Fasteners are used to secure or hold different parts together or to mount a component. Many types and sizes of fasteners are used in automobiles. Each fastener is designed for a specific purpose and condition. The most commonly used is the threaded fastener.

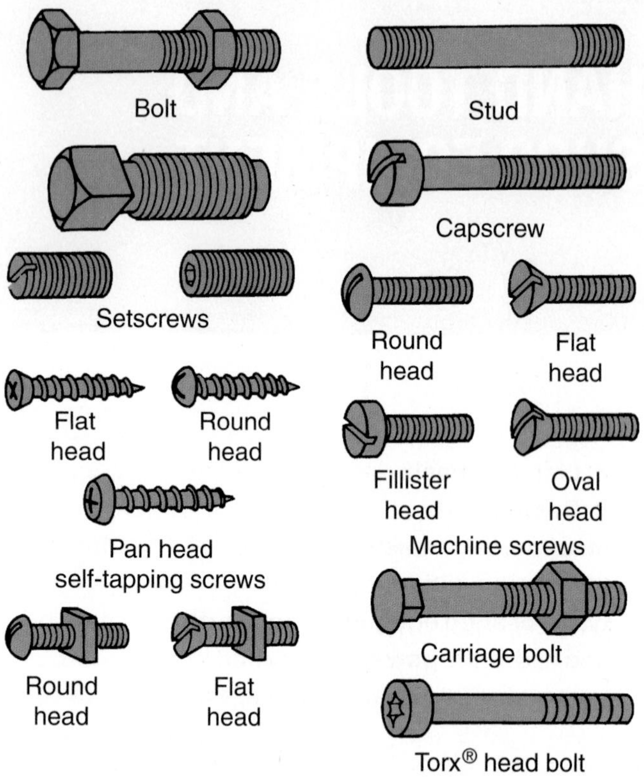

Figure 4–2 Common automotive threaded fasteners.

Threaded fasteners include bolts, nuts, screws, and similar items that allow for easy removal and installation of parts **(Figure 4–2)**.

The threads can be cut or rolled into the fastener. Rolled threads are 30% stronger than cut threads. They also offer better fatigue resistance because there are no sharp notches to create stress points. There are four classifications for the threads of Imperial fasteners: Unified National Coarse (UNC), Unified National Fine (UNF), Unified National Extrafine (UNEF), and Unified National Pipe Thread (UNPT or NPT). Metric fasteners are also available in fine and coarse threads.

NPT is the standard thread design for joining pipes and fittings. There are two basic designs: tapered and straight cut threads. Straight cut pipe thread is used to join pipes but it does not provide a good seal at the joining point. Tapered pipe threads provide a good seal because the internal and external threads compress against each other as the joint is tightened. Most often a sealant is used on pipe threads to provide a better seal. Pipe threads are commonly used at the ends of hoses and lines that carry a liquid or gas **(Figure 4–3)**.

Coarse (UNC) threads are used for general-purpose work, especially where rapid assembly and disassembly are required. Fine threads (UNF) are used where greater holding force is necessary. They are also used where greater vibration resistance is desired.

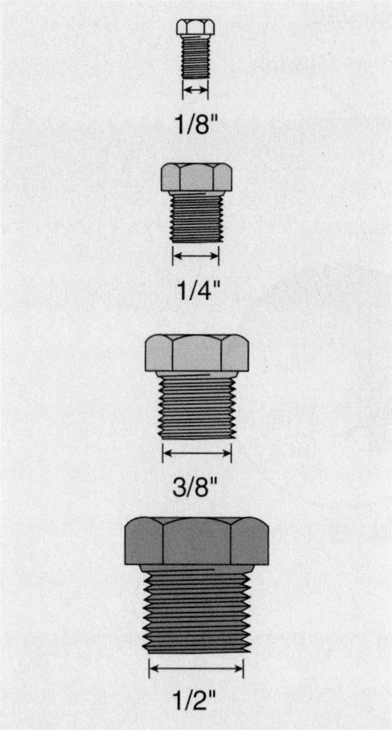

Figure 4–3 Various sizes of pipe fittings used with lines and hoses.

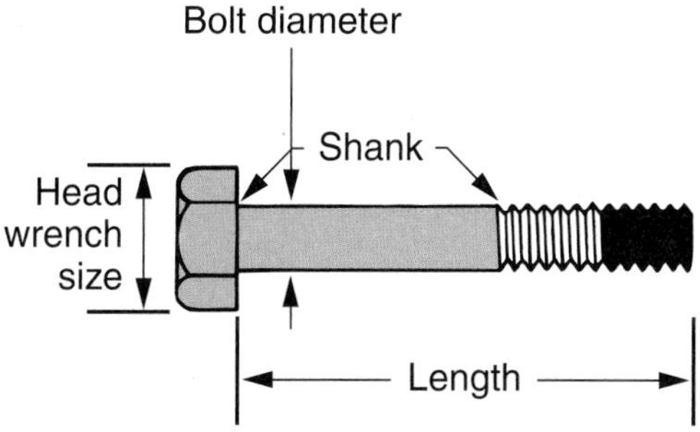

Figure 4–4 Basic terminology for bolt identification.

TABLE 4–1 STANDARD BOLT HEAD SIZES

Common English (U.S. Customary) Head Sizes	Common Metric Head Sizes
Wrench Size (inches)*	**Wrench Size (millimeters)***
3/8	9
7/16	10
1/2	11
9/16	12
5/8	13
11/16	14
3/4	15
13/16	16
7/8	17
15/16	18
1	19
1 1/16	20
1 1/8	21
1 3/16	22
1 1/4	23
1 5/16	24
1 3/8	26
1 7/16	27
1 1/2	29
	30
	32

**This does not suggest equivalency.*

Bolts

Bolts have a head on one end and threads on the other. They are identified by their head size, shank (shoulder) diameter, thread pitch, length **(Figure 4–4)**, and grade. The threads of a bolt travel from below the shank to the end of the bolt.

The **bolt head** is used to loosen and tighten the bolt; a socket or wrench fits over the head and is used to screw the bolt in or out. The size of the bolt head varies with the bolt's diameter and is available in Imperial and metric wrench sizes. Many confuse the size of the head with the size of the bolt. The size of a bolt is the diameter of its shank. **Table 4–1** lists the most common bolt head sizes.

Bolt diameter is the measurement across the diameter of the threaded area or **bolt shank**. The length of a bolt is measured from the bottom surface of the head to the end of the threads.

The **thread pitch** of a bolt in the Imperial system is the number of threads that are in 1 inch of the threaded length and is expressed in number of threads per inch. A UNF bolt with a 3/8-inch diameter is marked as a 3/8 × 24 bolt. It has 24 threads per inch. Likewise a 3/8-inch UNC bolt is called a 3/8 × 16 bolt.

The distance, in millimeters, between two adjacent threads determines the thread pitch in the metric system. This distance will vary between 1.0 and 2.0 and depends on the diameter of the bolt. The lower the number, the closer the threads are placed and the finer the threads are.

The bolt's tensile strength, or grade, is the amount of stress or stretch it is able to withstand before it breaks. The material that the bolt is made of and the diameter of the bolt determine its grade. In the Imperial system, the tensile strength of a bolt is identified by the number of radial lines (**grade marks**) on the

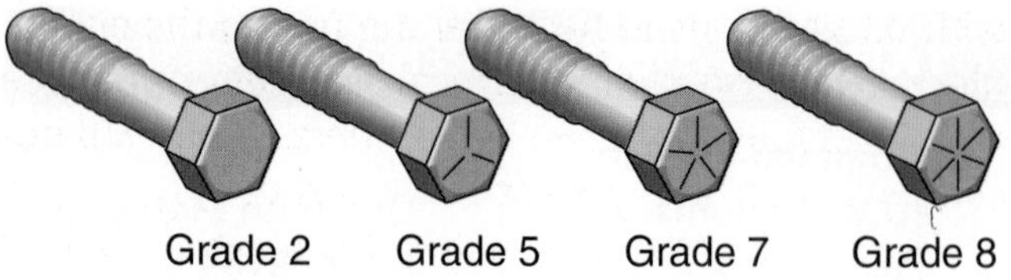

Customary (inch) bolts—identification marks correspond to bolt strength—increasing numbers represent increasing strength.

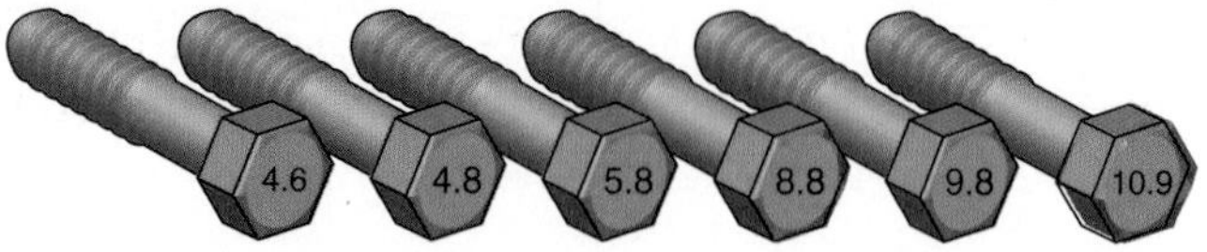

Metric bolts—identification class numbers correspond to bolt strength—increasing numbers represent increasing strength.

Figure 4–5 Bolt grade markings.

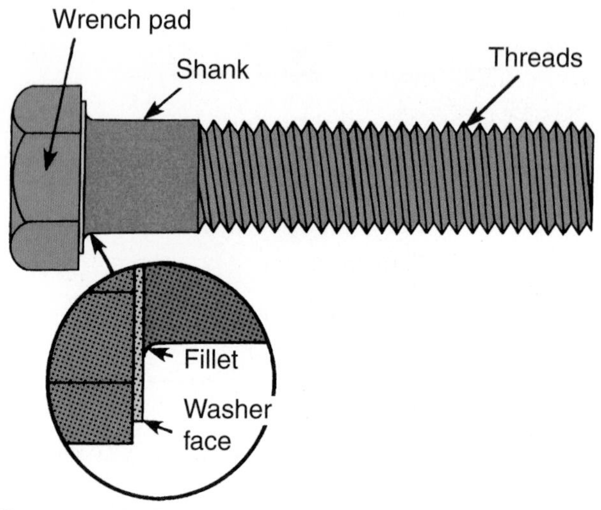

Figure 4–6 Bolt fillet detail.

bolt's head. More lines mean higher tensile strength **(Figure 4–5)**. Count the number of lines and add 2 to determine the grade of a bolt.

On metric bolts, a property class number on the bolt head identifies its grade. The property class is expressed with two numbers. The first represents the tensile strength of the bolt. The higher the number, the greater the tensile strength. The second is a percentage rating of the bolt's yield strength. This denotes how much stress the bolt can take before it is not able to return to its original shape. For example, a 10.9 bolt has a tensile strength of 1,000 MPa (145,000 psi) and a yield strength of 900 MPa (90% of 1,000). A 10.9 metric bolt is similar in strength to a grade 8 bolt.

Nuts are graded to match their respective bolts **(Table 4–2)**; a grade 8 nut must be used with a grade 8 bolt. If a grade 5 nut is used, a grade 5 connection would result.

SHOP TALK

Bolt heads can pop off because of fillet damage. The fillet is the slightly curved area where the shank flows into the bolt head (Figure 4–6). Scratches in this area introduce stress to the bolt head, causing failure. Replace all bolts that are damaged.

Tightening Bolts Any fastener is near worthless if it is not as tight as it should be. When a bolt is properly tightened, it will be "spring loaded" against the part it is holding. This spring effect is caused by the stretch of the bolt. Normally a properly tightened bolt is stretched to 70% of its elastic limit. The elastic limit of a bolt is that point of stretch from which the bolt will

TABLE 4–2 STANDARD NUT STRENGTH MARKINGS

Inch System		Metric System	
Grade	**Identification**	**Class**	**Identification**
Hex Nut Grade 5	 3 Dots	Hex Nut Property Class 9	 Arabic 9
Hex Nut Grade 8	 6 Dots	Hex Nut Property Class 10	 Arabic 10
Increasing dots represent increasing strength.		Can also have blue finish or paint dab on hex flat. Increasing numbers represent increasing strength.	

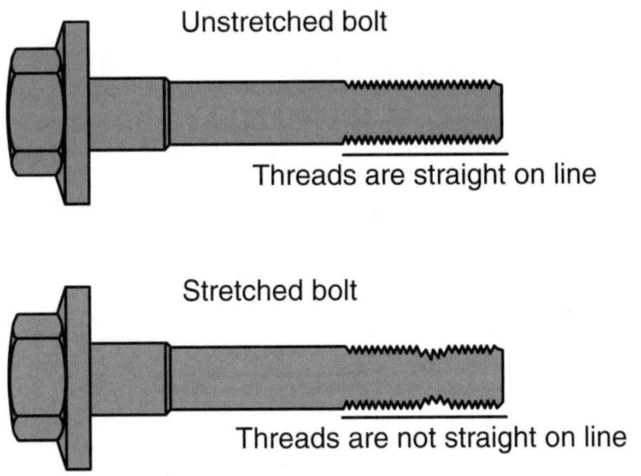

Figure 4–7 A comparison of a stretched and an unstretched bolt.

not return to its original shape when it is loosened. Not only will an overtightened or stretched bolt not have sufficient clamping force, it also will have distorted threads. The stretched threads will make it more difficult to screw and unscrew the bolt or a nut on the bolt **(Figure 4–7)**.

Fatigue breaks are the most common causes of bolt failure. A bolt becomes fatigued when it is able to move in its bore due to being undertightened.

Washers

Many different types of washers are used with fasteners. Flat washers are used to spread out the load of tightening a nut or bolt. This stops the bolt head or nut from digging into the surface as it is tightened. Also place flat washers with their rounded, punched side against the bolt head. Soft flat washers, sometimes called compression washers, are also used to spread the load of tightening and help seal one component to another. Copper washers are often used with oil pan bolts to help seal the pan to the engine block. Grade 8 and other critical applications require the use of fully hardened flat washers. These will not dish out when tightened like soft washers.

Lock washers are used to lock the head of a bolt or nut to the workpiece to keep them from coming loose and to prevent damage to softer metal parts.

Other Common Fasteners

The use of other fastener designs depends on the purpose of the fastener. Some of the more commonly used fasteners are described here.

Nuts Nuts are used with other threaded fasteners. Many different designs of nuts are found on today's cars **(Figure 4–8)**. The most common one is the hex nut, which is used with studs and bolts and tightened with a wrench.

Locknuts are often used in places where vibration may tend to loosen a nut. Locking nuts are standard nuts with nylon inserted into a section of the threads. The nylon cushions the vibrations.

Studs Studs are rods with threads on both ends. Most often, the threads on one end are coarse while the threads on the other end are fine. One end of the stud is screwed into a threaded bore. A bore in the part that will be mounted by the stud is fit over the stud and held in place with a nut that is screwed onto the stud. Studs are used when the clamping pressures of a fine thread are needed and a bolt will not work. If the stud is being screwed into soft (such as aluminum) or granular (such as cast iron) material, that end of the stud will have coarse threads. The opposite end will have fine threads. As a result, a coarse thread is used to hold the stud in a component and a fine-threaded

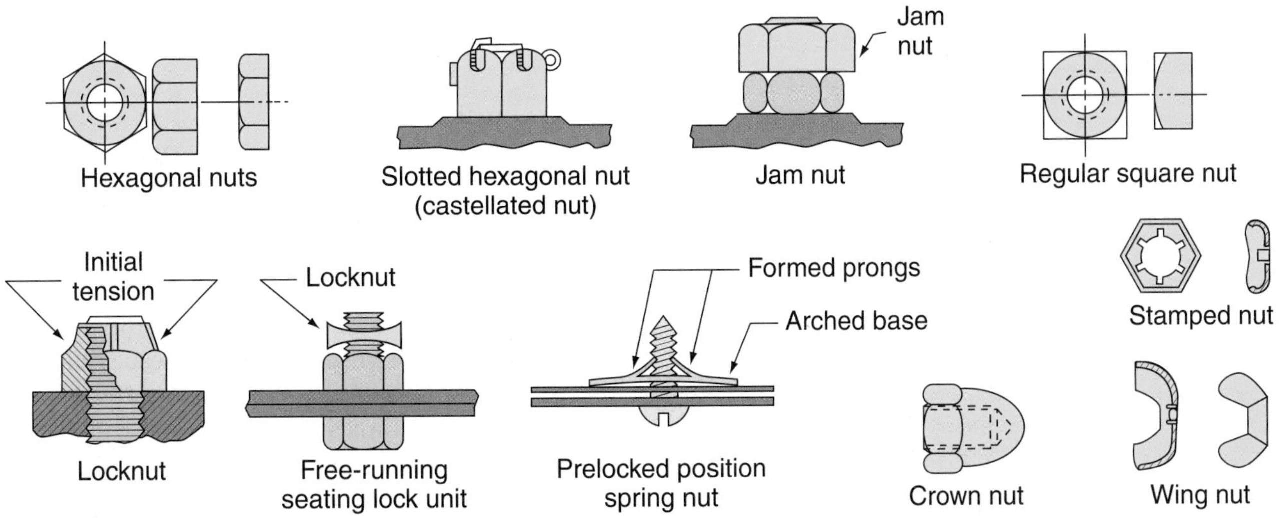

Figure 4–8 Many different types of nuts are used on automobiles. Each type has a specific purpose.

nut is used to clamp the other part to it. This provides for the clamping force of fine threads and the holding power of coarse threads.

Cap Screws Cap screws are similar to bolts, but they do not have a shoulder. The threads travel from the head to the end of the screw. Never use a cap screw in place of a bolt.

Setscrews Setscrews are used to prevent rotary motion between two parts, such as a pulley and shaft. Setscrews have a square head and are moved with a wrench, or they are headless and require an Allen wrench or screwdriver to move them.

Machine Screws The length of machine screws is entirely threaded. These screws have a head on one end and a flat bottom on the other. Machine screws are used to mount one component to another that has a threaded bore. They are also used with a nut to hold parts together. Machine screws can have a round, flat, Torx, oval, or fillister head.

Self-Tapping Screws Self-tapping screws are used to fasten sheet metal parts or to join together light metal with wood or plastic parts. These screws form their own threads in the material into which they are screwed.

Thread Lubricants and Sealants

It is often recommended that the threads of a bolt or stud be coated with a sealant or lubricant. The most commonly used lubricant is antiseize compound. Antiseize compound is used where a bolt might become difficult to remove after a period of time—for example, in an aluminum engine block. The amount of torque required to properly tighten a bolt treated with antiseize compound should be reduced. Thread lubricants may also cause hydrostatic lock; oil can be trapped in a blind hole. When the bolt contacts the oil, it cannot compress it; therefore, the bolt cannot be properly tightened or fully seated.

Thread sealants are used on bolts that are tightened into an oil cavity or coolant passage. The sealant prevents the liquid from seeping past the threads. Teflon tape is often used as a sealant. Another commonly used thread chemical, called threadlocker **(Figure 4–9)**, prevents a bolt from working loose as the engine or another part vibrates.

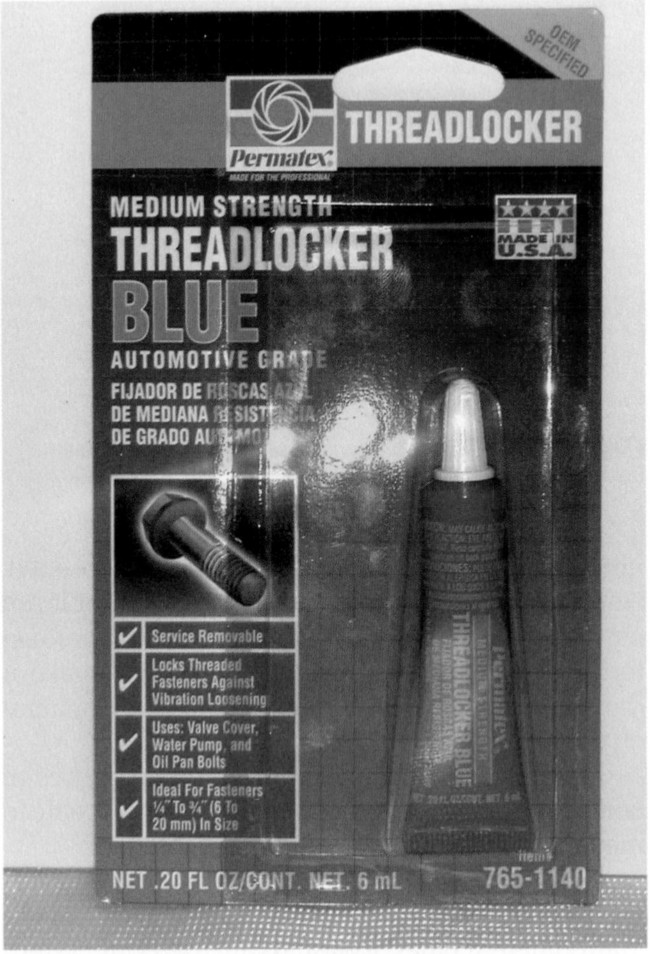

Figure 4–9 A container of threadlocker. *Courtesy of Permatex, Inc.*

Thread Pitch Gauge

The use of a **thread pitch gauge** provides a quick and accurate way to check the thread pitch of a fastener. The leaves of the tool are marked with the various pitches. To check the pitch of threads, simply match the teeth of the gauge with the threads of the fastener. Then read the pitch from the leaf. Thread pitch gauges are available for the various threads used by the automotive industry.

Taps and Dies

The hand **tap** is a small tool used for hand cutting internal threads **(Figure 4–10)**. An internal thread is cut on the inside of a part, such as a thread on the inside of a nut. Taps are also available that only clean and restore threads that were previously cut. Taps are selected by size and thread pitch. Photo Sequence 1 goes through the correct procedure for repairing damaged threads with a tap.

When tapping a bore, rotate the tap in a clockwise direction. Then, turn the tap counterclockwise about a quarter turn to break off any metal chips that may have accumulated in the threads. These small metal pieces can damage the threads as you continue to tap. These metal chips are gathered in the tap's flutes, which are recessed areas between the cutting teeth of the tap **(Figure 4–11)**. After backing off the tap, continue rotating the tap clockwise. Remember

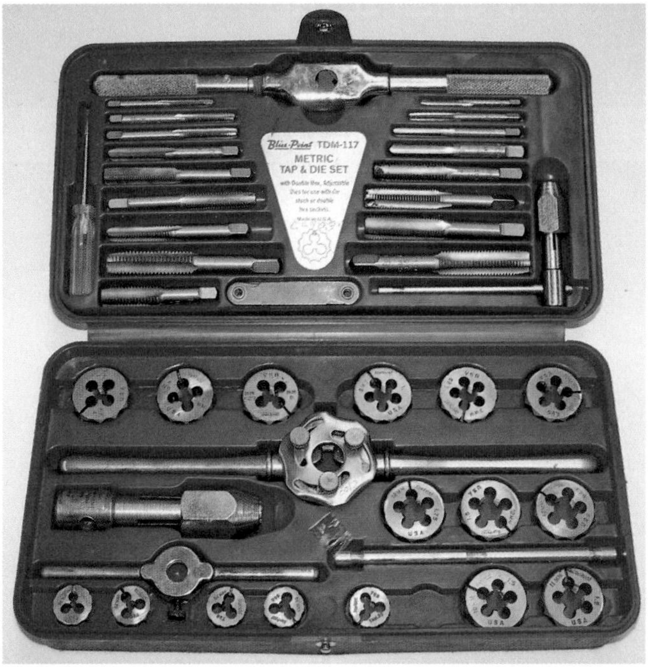

Figure 4-10 A tap and die set.

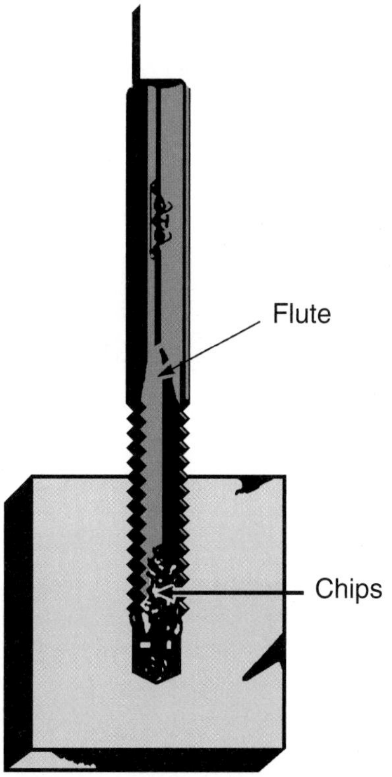

Figure 4-11 Metal chips are gathered into the flutes of a tap.

to back off the tap periodically and make sure all of the existing threads in the bore have been recut by the tap.

Hand-threading **dies** are the opposite of taps because they cut external (outside) threads on bolts, rods, and pipes rather than internal threads. Dies are made in various sizes and shapes, depending on the particular work for which they are intended. Dies may be solid (fixed size), split on one side to permit adjustment, or have two halves held together in a collet that provides for individual adjustments. Dies fit into holders called die stocks.

Threaded Inserts

When the threads in a bore are excessively damaged, it is better to replace them than try to tap them. A thread insert can be used to restore the original threads. Inserts require drilling the bore to a larger diameter and tapping that bore to allow the insert to be screwed into it. The inner threaded diameter of the insert will provide fresh threads for the bolt **(Figure 4–12)**.

Spark Plug Thread Repair Sometimes when spark plugs are removed from a cylinder head, the threads have traces of metal on them. This happens more often with aluminum heads. When this occurs, the spark plug bore must be corrected by installing thread inserts.

> **SHOP TALK**
>
> Never change spark plugs when the cylinder head is hot. The bores for the plugs can take on an oval shape as the cylinder head cools without spark plugs in the bores.

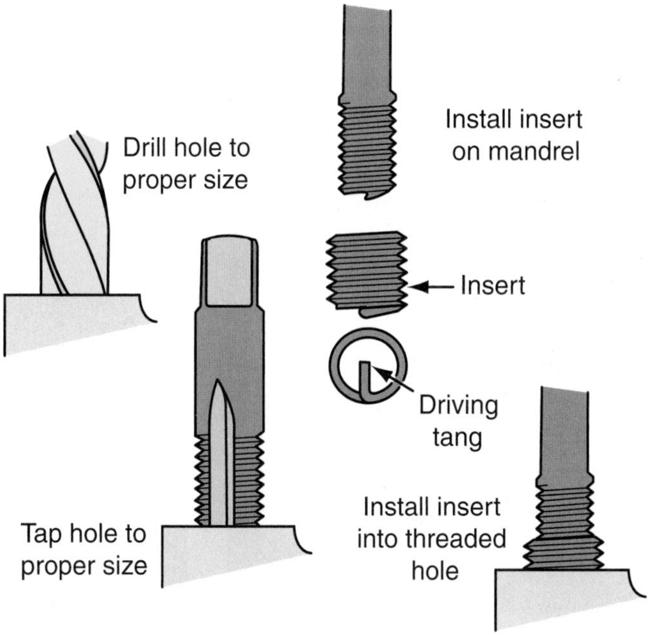

Figure 4-12 Using a threaded insert (heli-coil®) to repair damaged threads. *Courtesy of Emhart Fastening Teknologies*

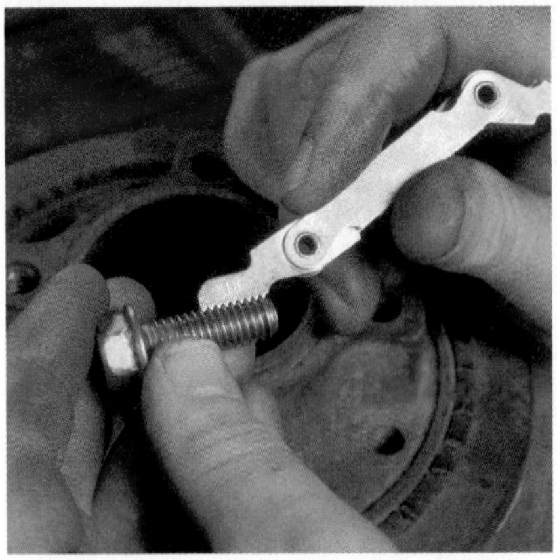

P1-1 *Using a thread pitch gauge, determine the thread size of the fastener that should fit into the damaged internal threads.*

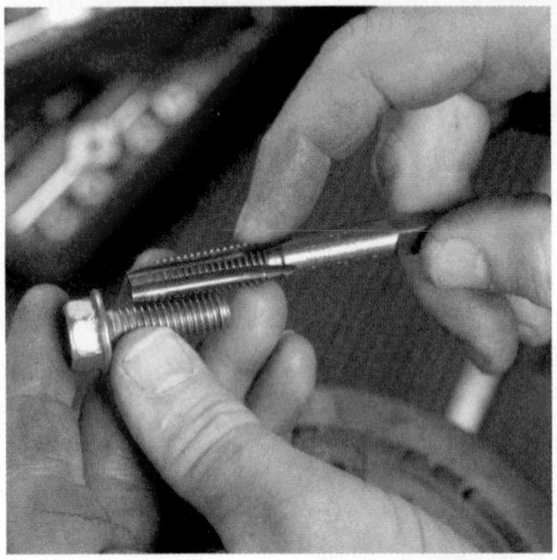

P1-2 *Select the correct size and type of tap for the threads and bore to be repaired.*

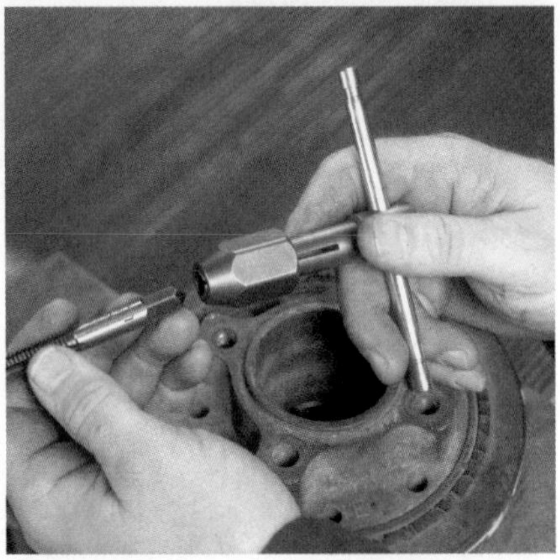

P1-3 *Install the tap into a tap wrench.*

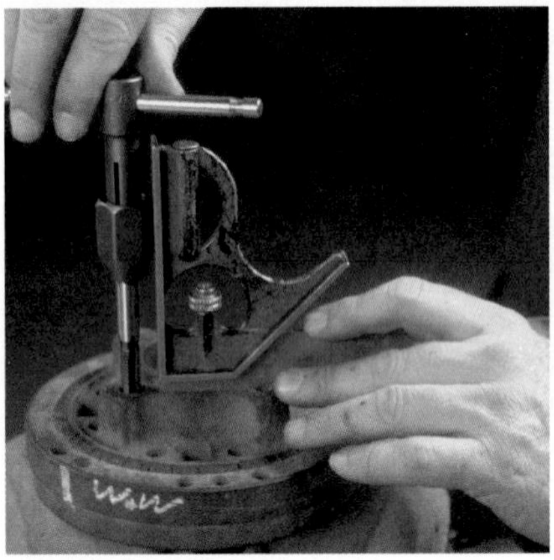

P1-4 *Start the tap squarely in the threaded hole using a machinist square as a guide.*

P1-5 *Rotate the tap clockwise into the bore until the tap has run through the entire length of the threads. While doing this, periodically turn the tap backward to clean the threads. This prevents breaking the tap.*

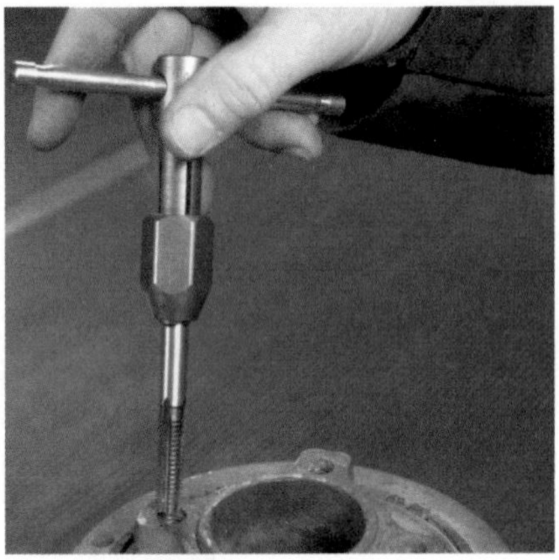

P1-6 *Drive the tap back out of the hole by turning it counterclockwise.*

P1-7 *Clean the metal chips left by the tap out of the hole.*

P1-8 *Inspect the threads left by the tap to be sure they are acceptable.*

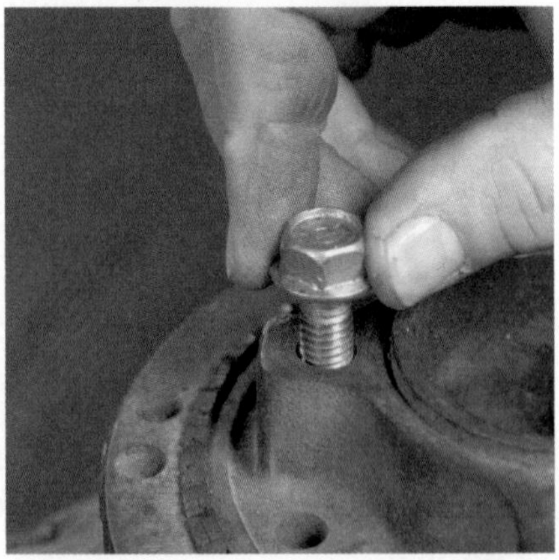

P1-9 *Test the threads by threading the correct fastener into the threaded hole.*

When installing spark plugs, if the plugs cannot be installed easily by hand, the threads in the cylinder head may need to be cleaned with a thread-chasing tap. There are special taps for spark plug bores, simply called **spark plug thread taps**. Be especially careful not to cross-thread the plugs when working with aluminum heads. Always tighten the plugs with a torque wrench and the correct spark plug socket, following the vehicle manufacturer's specifications. Also, when changing spark plugs in aluminum heads, the temperature of the heads should be ambient temperature before attempting to remove the plugs.

MEASURING TOOLS

Some service work, such as engine repair, requires very exact measurements, often in ten-thousandths (0.0001) of an inch or thousandths (0.001) of a millimeter. Accurate measurements with this kind of precision can only be made by using precise measuring devices.

Measuring tools are precise and delicate instruments. In fact, the more precise they are, the more delicate they are. They should be handled with great care. Never pry, strike, drop, or force these instruments. They may be permanently damaged.

Precision measuring instruments, especially micrometers, are extremely sensitive to rough handling. Clean them before and after every use. All measuring should be performed on parts that are at room temperature to eliminate the chance of measuring something that has contracted because it was cold or has expanded because it was hot.

> **SHOP TALK**
>
> Check measuring instruments regularly against known good equipment to ensure that they are operating properly and are capable of accurate measurement. Always refer to the appropriate material for the correct specifications before performing any service or diagnostic procedures. The close tolerances required for the proper operation of some automotive parts make using the correct specifications and taking accurate measurements very important. Even the slightest error in measurement can be critical to the durability and operation of an engine and other systems.

Machinist's Rule

The **machinist's rule** looks very much like an ordinary ruler. Each edge of this basic measuring tool is

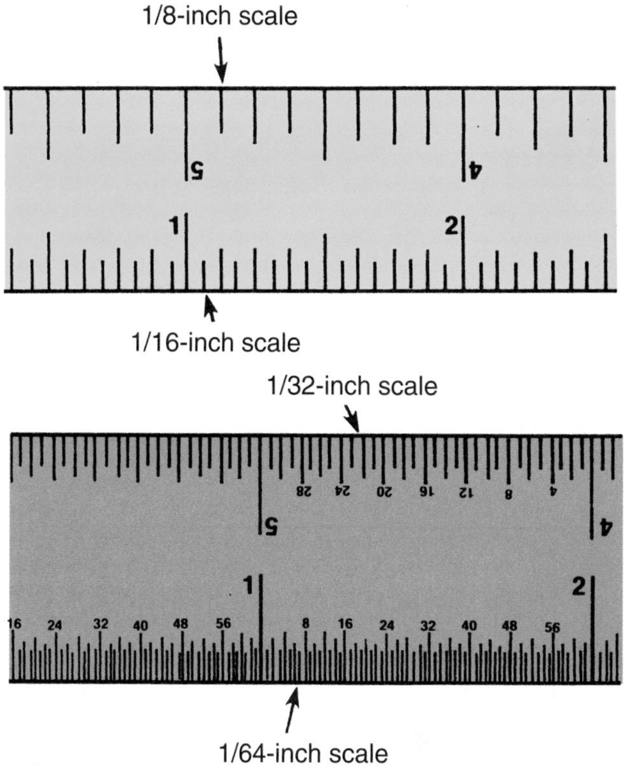

Figure 4–13 Graduations on a typical machinist's rule.

divided into increments based on a different scale. As shown in **Figure 4–13**, a typical machinist's rule based on the Imperial system of measurement may have scales based on 1/8-, 1/16-, 1/32-, and 1/64-inch intervals. Of course, metric machinist rules are also available. Metric rules are usually divided into 0.5 mm and 1 mm increments.

Some machinist's rules may be based on decimal intervals. These are typically divided into 1/10-, 1/50-, and 1/1,000-inch increments. Decimal machinist's rules are very helpful when measuring dimensions specified in decimals; they make such measurements much easier.

Vernier Caliper

A **vernier caliper** is a measuring tool that can make inside, outside, or depth measurements. It is marked in both British Imperial and metric divisions called a vernier scale. A vernier scale consists of a stationary scale and a movable scale, in this case the vernier bar to the vernier plate. The length is read from the vernier scale.

A vernier caliper has a movable scale that is parallel to a fixed scale (**Figure 4–14**). These precision measuring instruments are capable of measuring outside and inside diameters and most will even measure depth. Vernier calipers are available in both Imperial and metric scales. The main scale of the caliper is divided into inches; most measure up to 6 inches. Each inch is divided into 10 parts, each equal to

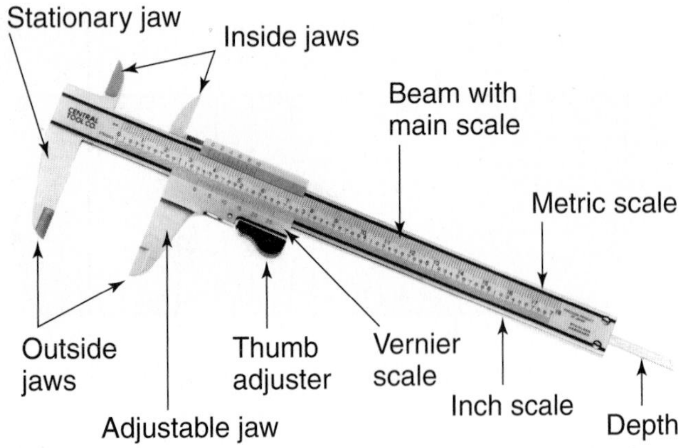

Figure 4–14 A vernier caliper. *Courtesy of Central Tools, Inc.*

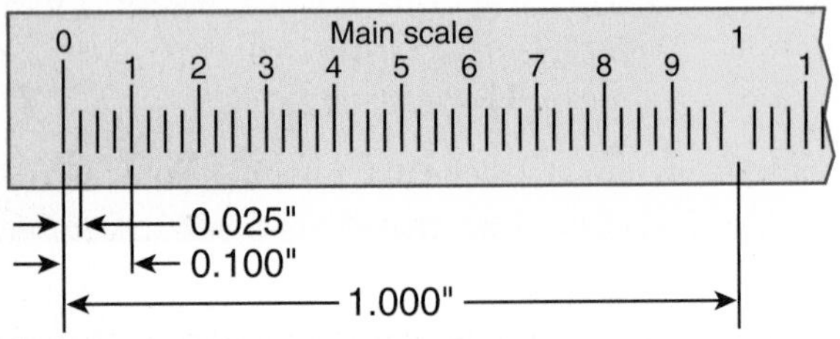

Figure 4–15 Each line of the main scale equals 0.025 inch.

0.100 inch. The area between the 0.100 marks is divided into four. Each of these divisions is equal to 0.025 inch **(Figure 4–15)**.

The vernier scale has 25 divisions, each one representing 0.001 inch. Measurement readings are taken by combining the main and vernier scales. At all times, only one division line on the main scale will line up with a line on the vernier scale **(Figure 4–16)**. This is the basis for accurate measurements.

To read the caliper, locate the line on the main scale that lines up with the zero (0) on the vernier scale. If the zero lined up with the 1 on the main scale, the reading would be 0.100 inch. If the zero on the vernier scale does not line up exactly with a line on the main scale, then look for a line on the vernier scale that does line up with a line on the main scale.

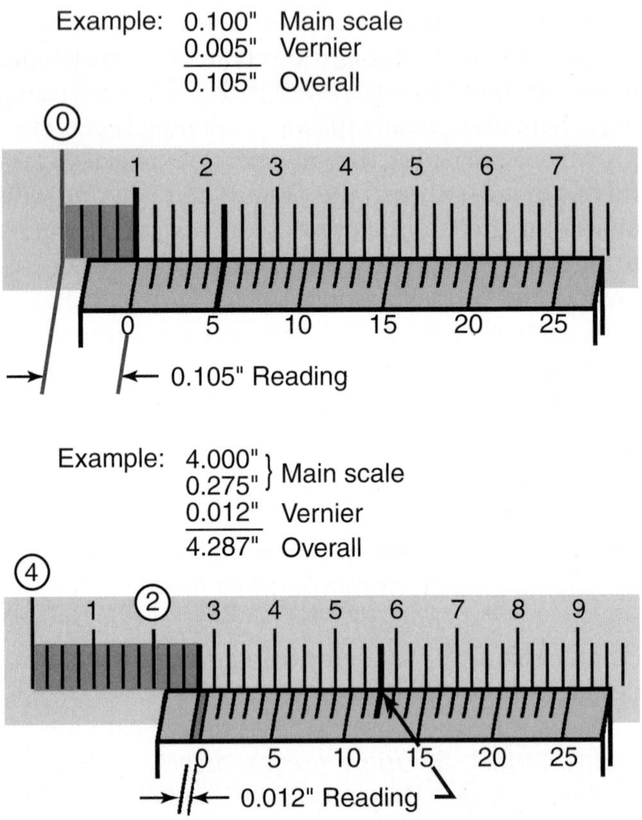

Figure 4–16 To get a final measurement, line up the vernier scale line that is exactly aligned with any line on the main scale.

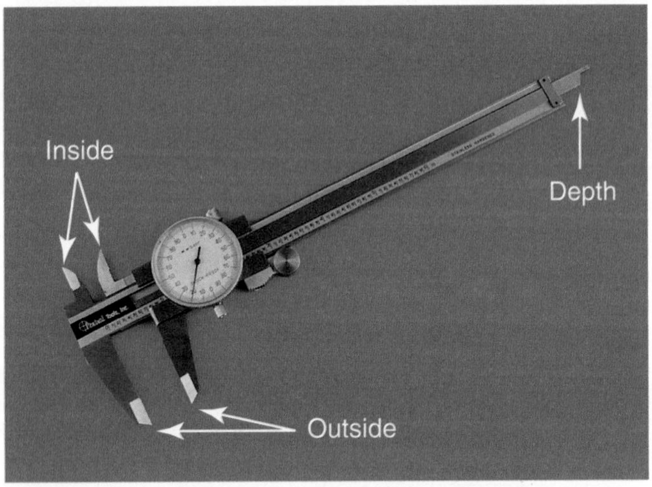

Figure 4–17 A dial vernier caliper. *Courtesy of Central Tools, Inc.*

Dial Caliper

The **dial caliper (Figure 4–17)** is an easier-to-use version of the vernier caliper. Imperial calipers commonly measure dimensions from 0 to 6 inches (0 to 150 mm). Metric dial calipers typically measure from 0 to 150 mm in increments of 0.02 mm. The dial caliper features a depth scale, bar scale, dial indicator, inside measurement jaws, and outside measurement jaws.

The main scale of a British Imperial dial caliper is divided into one-tenth (0.1) inch graduations. The dial indicator is divided into one-thousandth (0.001) inch graduations. Therefore, one revolution of the dial indicator needle equals one-tenth inch on the bar scale.

A metric dial caliper is similar in appearance; however, the bar scale is divided into 2 mm increments. Additionally, on a metric dial caliper, one revolution of the dial indicator needle equals 2 mm.

Both English and metric dial calipers use a thumb-operated roll knob for fine adjustment. When you use a dial caliper, always move the measuring jaws backward and forward to center the jaws on the object being measured. Make sure the caliper jaws lay flat on or around the object. If the jaws are tilted in any way, you will not obtain an accurate measurement.

Although dial calipers are precision measuring instruments, they are only accurate to plus or minus two-thousandths (±0.002) of an inch. Micrometers are preferred when extremely precise measurements are desired.

Micrometers

The **micrometer** is used to measure linear outside and inside dimensions. Both outside and inside micrometers are calibrated and read in the same manner. Measurements on both are taken with the measuring points in contact with the surfaces being measured.

The major components and markings of a micrometer include the frame, anvil, spindle, locknut, sleeve, sleeve numbers, sleeve long line, thimble marks, thimble, and ratchet **(Figure 4–18)**. Micrometers are calibrated in either inch or metric graduations and are available in a range of sizes. The proper procedure for measuring with an inch-graduated outside micrometer is outlined in Photo Sequence 2.

Most micrometers are designed to measure objects with accuracy to 0.001 (one-thousandth) inch. Micrometers are also available to measure in 0.0001 (ten-thousandths) of an inch. This type of micrometer should be used when the specifications call for this much accuracy. Digital micrometers are also available **(Figure 4–19)**. These eliminate the need to do math and still receive a precise measurement.

Reading a Metric Outside Micrometer The metric micrometer is read in the same manner as the inch-

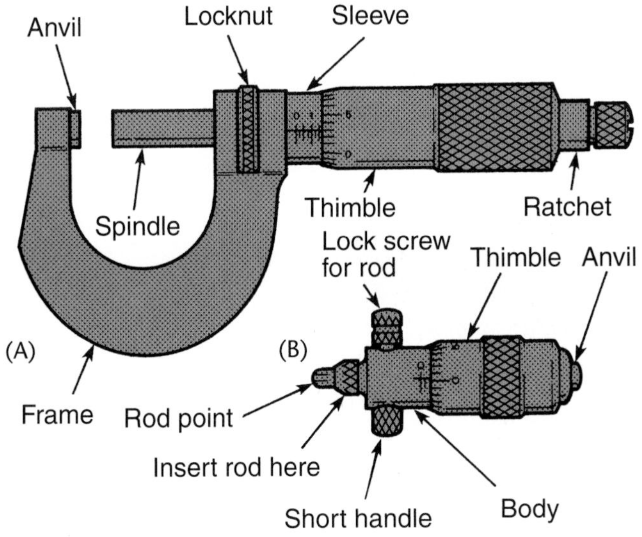

Figure 4-18 Major components of (A) an outside and (B) an inside micrometer.

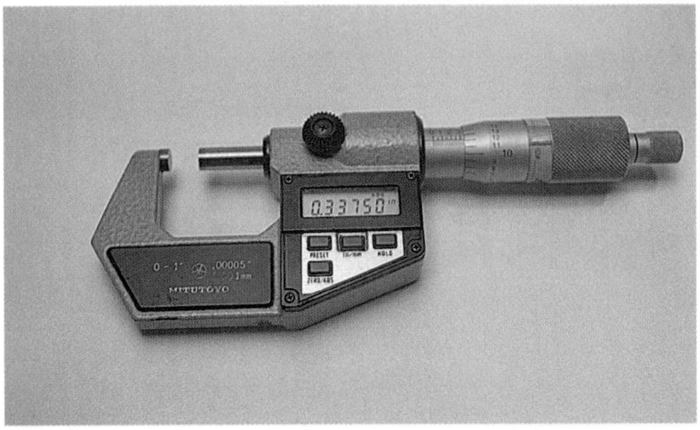

Figure 4-19 A digital micrometer eliminates the need to do math.

graduated micrometer, except the graduations are expressed in the metric system of measurement. Readings are obtained as follows:

- Each number on the sleeve of the micrometer represents 5 millimeters (mm) or 0.005 meter (m) **(Figure 4–20A)**.
- Each of the 10 equal spaces between each number, with index lines alternating above and below the horizontal line, represents 0.5 mm or five-tenths of a mm. One revolution of the thimble changes the reading one space on the sleeve scale or 0.5 mm **(Figure 4–20B)**.

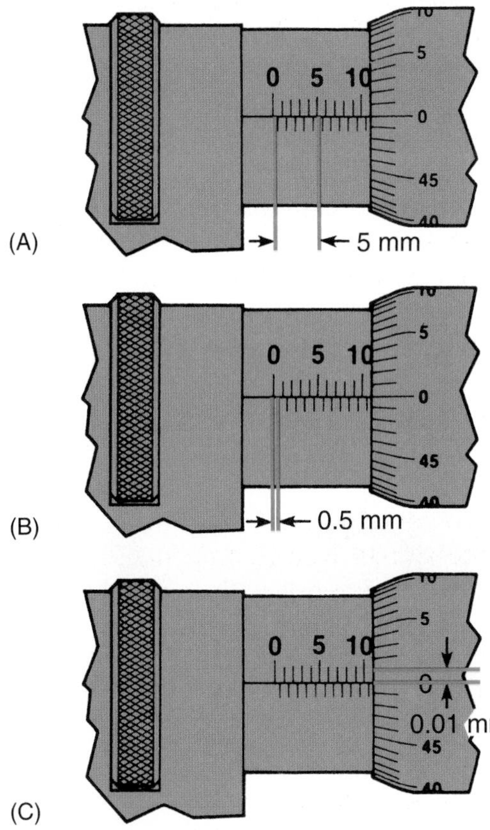

Figure 4-20 Reading a metric micrometer: (A) 10 mm plus (B) 0.5 mm plus (C) 0.01 mm equals 10.51 mm.

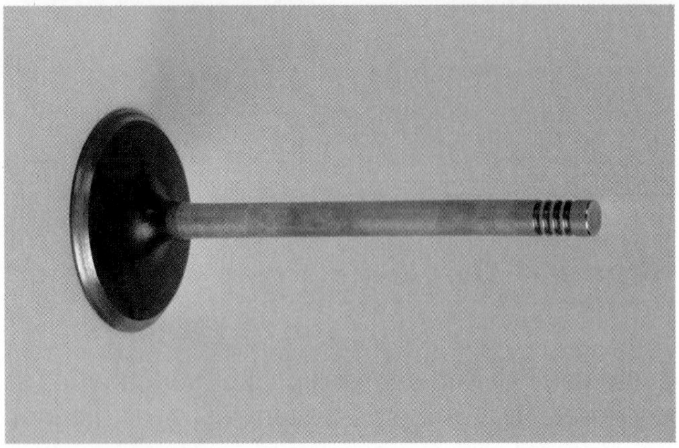

P2-1 *Micrometers can be used to measure the diameter of many different objects. By measuring the diameter of a valve stem in two places, the wear of the stem can be determined.*

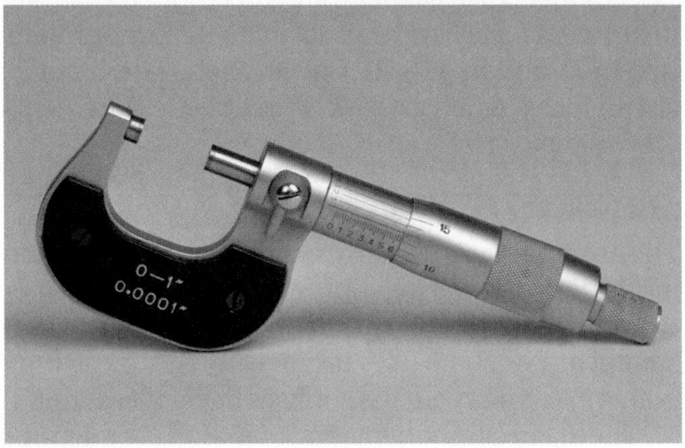

P2-2 *Because the diameter of a valve stem is less than 1 inch, a 0-to-1-inch outside micrometer is used.*

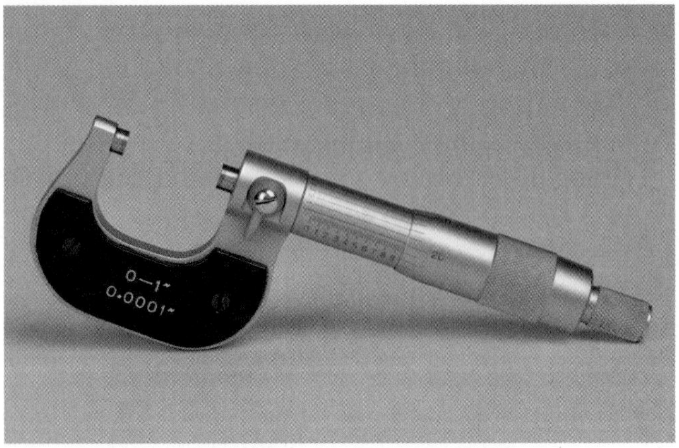

P2-3 *The graduations on the sleeve each represent 0.025 inch. To read a measurement on a micrometer, begin by counting the visible lines on the sleeve and multiplying them by 0.025.*

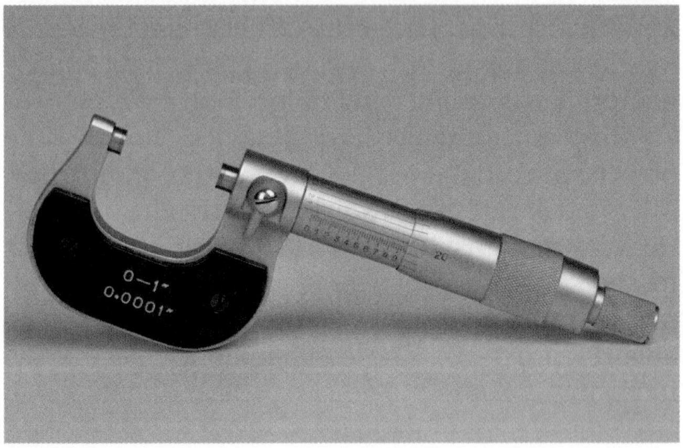

P2-4 *The graduations on the thimble assembly define the area between the lines on the sleeve. The number indicated on the thimble is added to the measurement shown on the sleeve.*

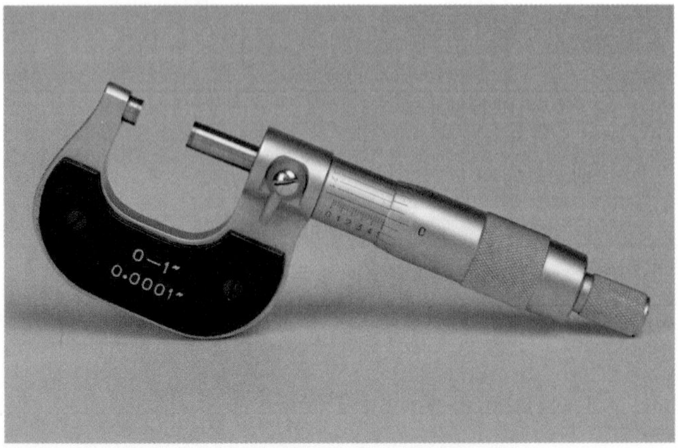

P2-5 *A micrometer reading of 0.500 inch.*

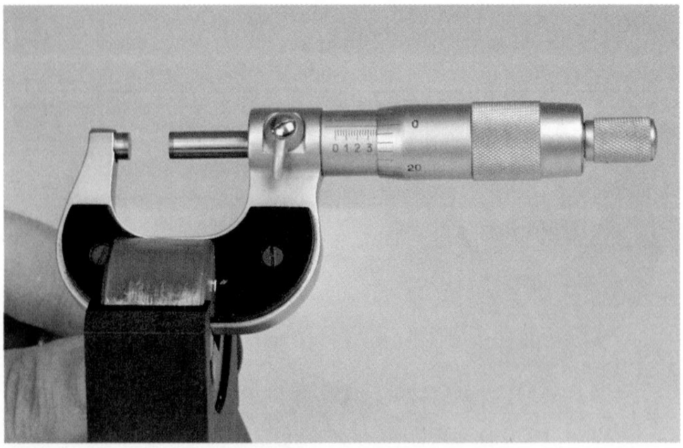

P2-6 *A micrometer reading of 0.375 inch.*

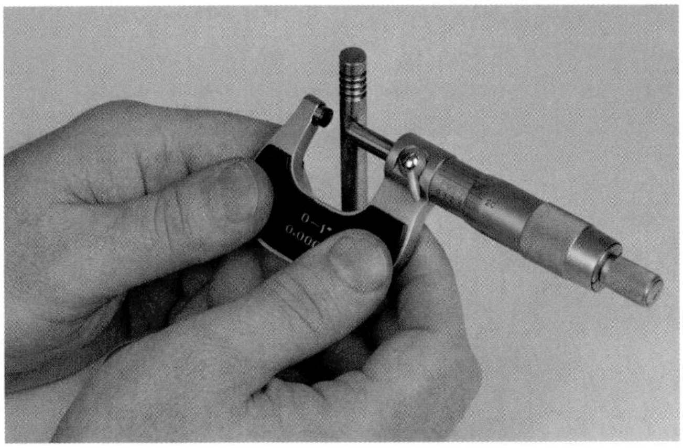

P2-7 *Normally, little stem wear is evident directly below the keeper grooves. To measure the diameter of the stem at that point, close the micrometer around the stem.*

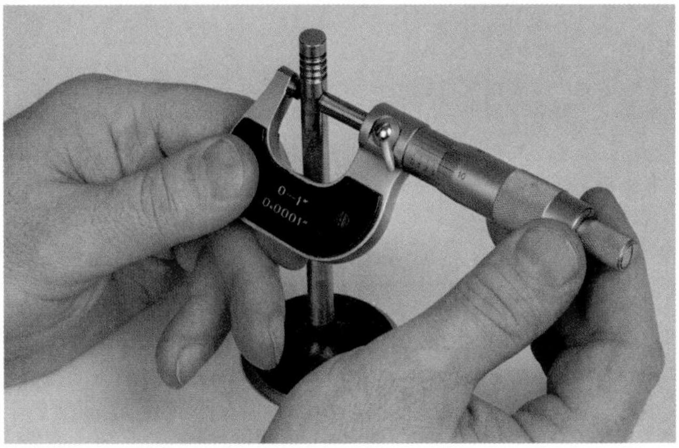

P2-8 *To get an accurate reading, slowly close the micrometer until a slight drag is felt while passing the valve in and out of the micrometer.*

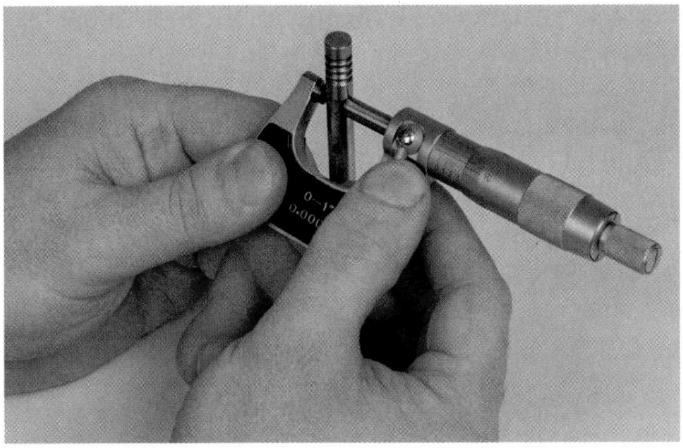

P2-9 *To prevent the reading from changing while you move the micrometer away from the stem, use your thumb to activate the lock lever.*

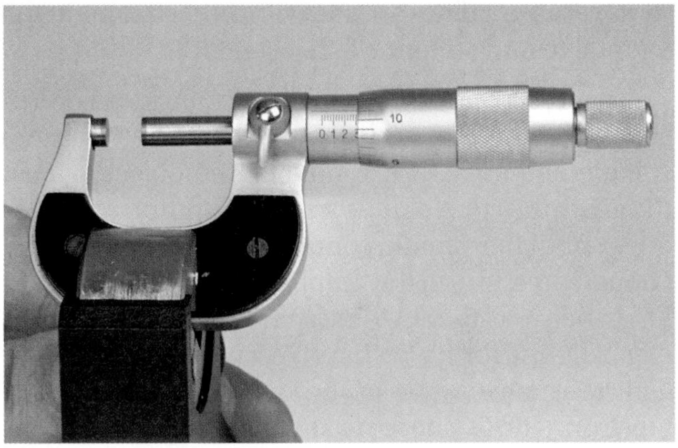

P2-10 *This reading (0.311 inch) represents the diameter of the valve stem at the top of the wear area.*

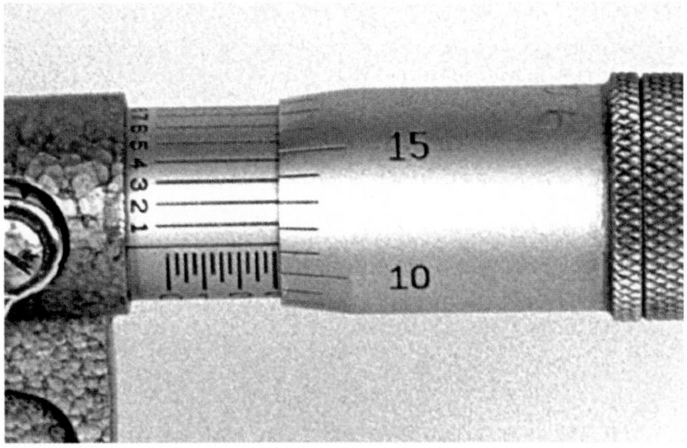

P2-11 *Some micrometers are able to measure in 0.0001 (ten-thousandths) of an inch. Use this type of micrometer if the specifications call for this much accuracy. Note that the exact diameter of the valve stem is 0.3112 inch.*

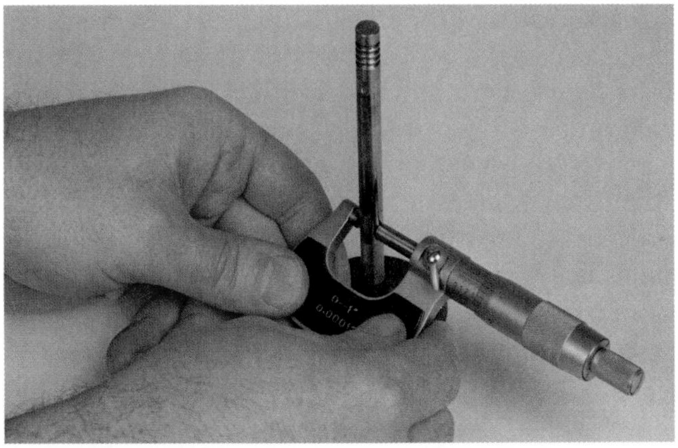

P2-12 *Most valve stem wear occurs above the valve head. The diameter here should also be measured. The difference between the diameter of the valve stem just below the keepers and just above the valve head represents the amount of valve stem wear.*

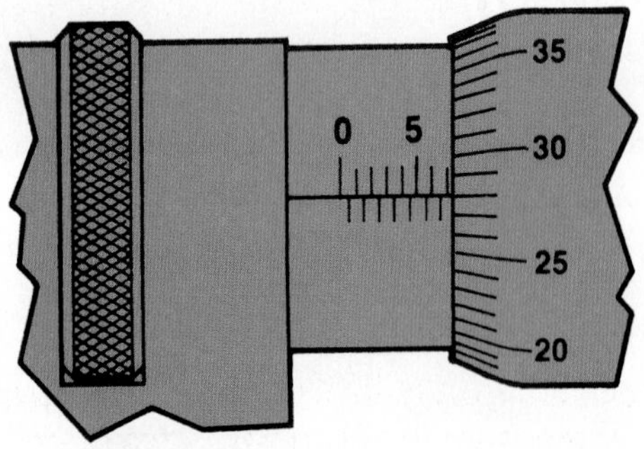

Figure 4-21 The total reading on this micrometer is 7.28 mm.

- The beveled edge of the thimble is divided into 50 equal divisions with every fifth line numbered: 0, 5, 10, . . . 45. Since one complete revolution of the thimble advances the spindle 0.5 mm, each graduation on the thimble is equal to one hundredth of a millimeter **(Figure 4–20C)**.
- As with the inch-graduated micrometer, the three separate readings are added together to obtain the total reading **(Figure 4–21)**.

To measure small objects with an outside micrometer, open the tool and slip the object between the spindle and anvil. While holding the object against the anvil, turn the thimble with your thumb and forefinger until the spindle contacts the object. Use only enough pressure on the thimble to allow the object to just fit between the anvil and spindle. Slip the micrometer back and forth over the object until you feel a very light resistance, while at the same time rocking the tool from side to side to make certain the spindle cannot be closed any further **(Figure 4–22)**. After your final adjustment, lock the micrometer and read the measurement.

Micrometers are available in different sizes. The size is dictated by the smallest to the largest measurement it can make. Examples of these sizes are the 0-to-1-inch, 1-to-2-inch, 2-to-3-inch, and 3-to-4-inch micrometers.

Reading an Inside Micrometer Inside micrometers are used to measure the inside diameter of a bore or hole. The tool is placed into the bore and extended until each end touches the bore's surface. If the bore is large, it might be necessary to use an extension rod to increase the micrometer's range. These extension rods come in various lengths.

To get a precise measurement, keep the anvil firmly against one side of the bore and rock the micrometer back and forth and side to side. This centers the micrometer in the bore. Make sure there is

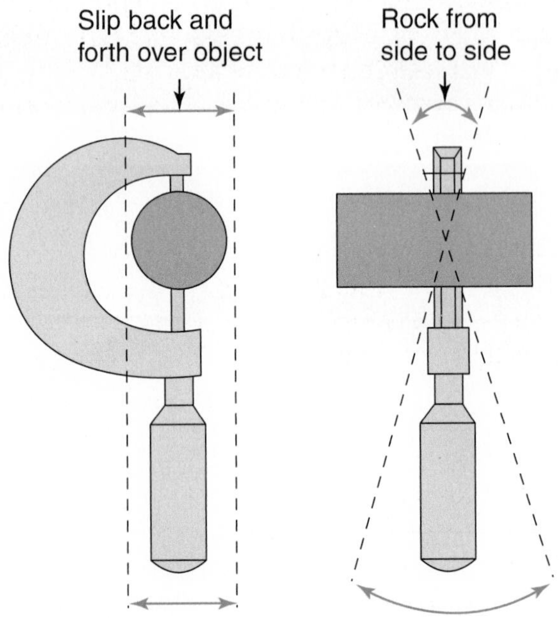

Figure 4-22 Slip the micrometer over the object and rock it from side to side.

correct resistance on both ends of the tool before taking a reading.

Reading a Depth Micrometer A depth micrometer **(Figure 4–23)** is used to measure the distance between two parallel surfaces. It operates and is read in the

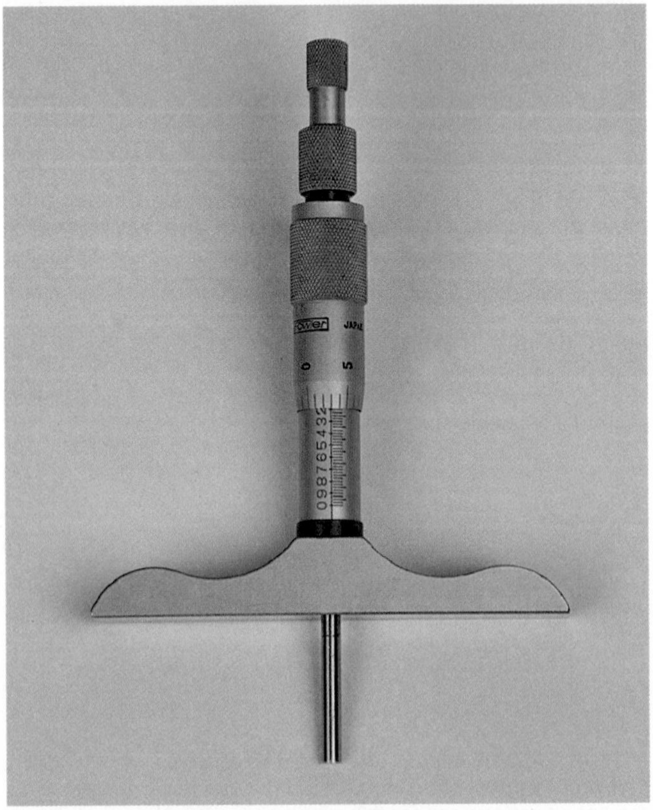

Figure 4-23 A depth micrometer. *Courtesy of Central Tools, Inc.*

same way as other micrometers. If a depth micrometer is used with a gauge bar, it is important to keep both the bar and the micrometer from rocking. Any movement of either part will result in an inaccurate measurement.

SHOP TALK

Measurements with any micrometer will be reliable only if the micrometer is calibrated correctly. To calibrate a micrometer, close the micrometer over a micrometer standard. If the reading differs from that of the standard, the micrometer should be adjusted according to the instructions provided by the tool manufacturer. Proper care of a micrometer is also important to ensure accurate measurements. This care includes:

- Always clean the micrometer before using it.
- Do not touch the measuring surfaces.
- Store the tool properly. The spindle face should not touch the anvil face; a change in temperature might spring the micrometer.
- Clean the micrometer after use. Wipe it clean of any oil, dirt, or dust using a lint-free cloth.
- Never use the tool as a clamp or tighten the jaws too tightly around an object.
- Do not drop the tool.
- Check the calibration weekly.

Telescoping Gauge

Telescoping gauges (**Figure 4–24**) are used for measuring bore diameters and other clearances. They may also be called **snap gauges**. They are available in sizes ranging from fractions of an inch through 6 inches (150 mm). Each gauge consists of two telescoping plungers, a handle, and a lock screw. Snap gauges are normally used with an outside micrometer.

To use the telescoping gauge, insert it into the bore and loosen the lock screw. This will allow the plungers to snap against the bore. Once the plungers have expanded, tighten the lock screw. Then, remove the gauge and measure the expanse with a micrometer.

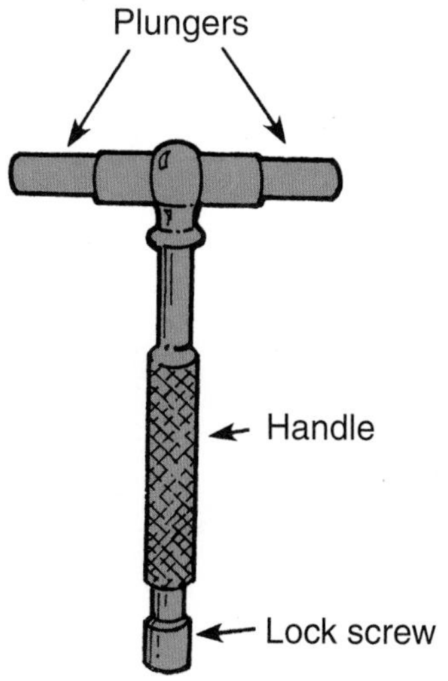

Figure 4-24 Parts of a telescoping gauge.

Small Hole Gauge

A small hole or **ball gauge** works just like a telescoping gauge. However, it is designed to be used on small bores. After it is placed into the bore and expanded, it is removed and measured with a micrometer (**Figure 4–25**). Like the telescoping gauge, the small hole gauge consists of a lock, a handle, and an expanding end. The end expands or retracts by turning the gauge handle.

Feeler Gauge

A **feeler gauge** is a thin strip of metal or plastic of known and closely controlled thickness. Several of these metal strips are often assembled together as a feeler gauge set that looks like a pocket knife (**Figure 4–26**). The desired thickness gauge can be pivoted away from others for convenient use. A steel feeler gauge pack usually contains strips or leaves of 0.002- to 0.010-inch thickness (in steps of 0.001 inch) and leaves of 0.012- to 0.024-inch thickness (in steps of 0.002 inch). Metric feeler gauges are also available.

A feeler gauge can be used by itself to measure piston ring side clearance, piston ring end gap, connecting rod side clearance, crankshaft end play, and other distances.

Round wire feeler gauges are often used to measure spark plug gap. The round gauges are designed to give a better feel for the fit of the gauge in the gap.

Straightedge

A **straightedge** is no more than a flat bar machined to be totally flat and straight, and to be effective it must be flat and straight. Any surface that should be flat can be checked with a straightedge and feeler gauge set. The straightedge is placed across and at angles on the surface. At any low points on the surface, a feeler gauge can be placed between the straightedge and the surface (**Figure 4–27**). The size gauge that fills in the gap indicates the amount of warpage or distortion.

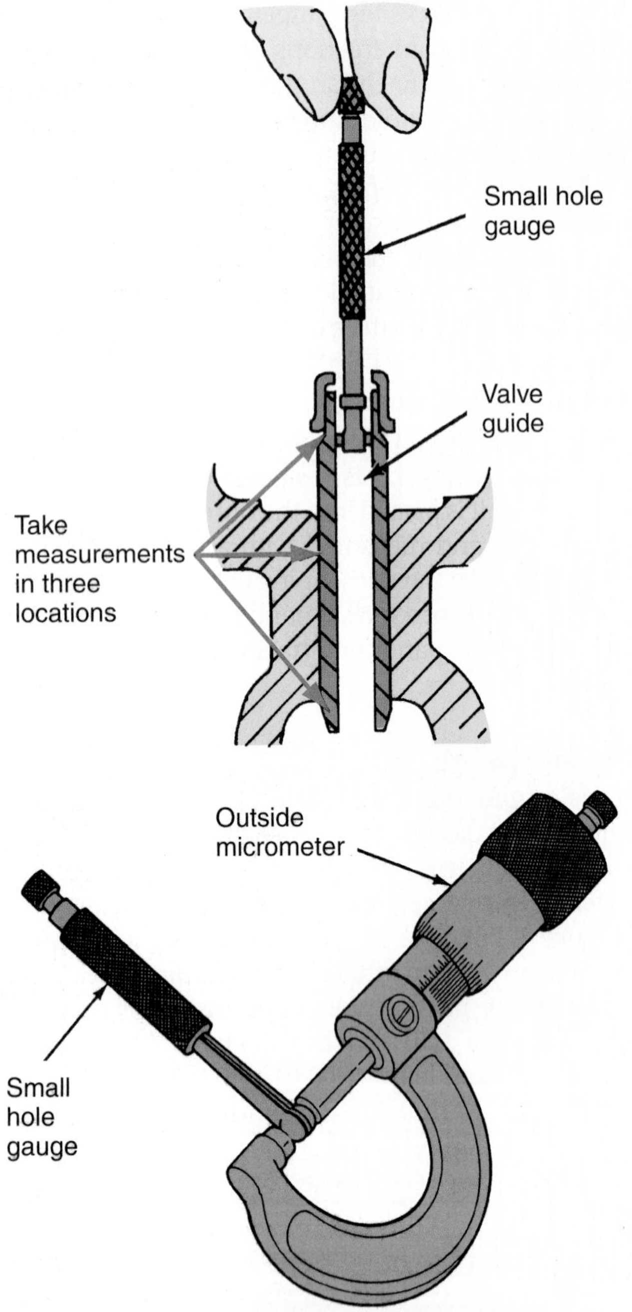

Hole gauge method
to measure valve guide wear

Figure 4–25 Insert the ball gauge into the bore to be measured. Then expand it, lock it, and remove it. Now measure it with an outside micrometer. *Courtesy of Ford Motor Company*

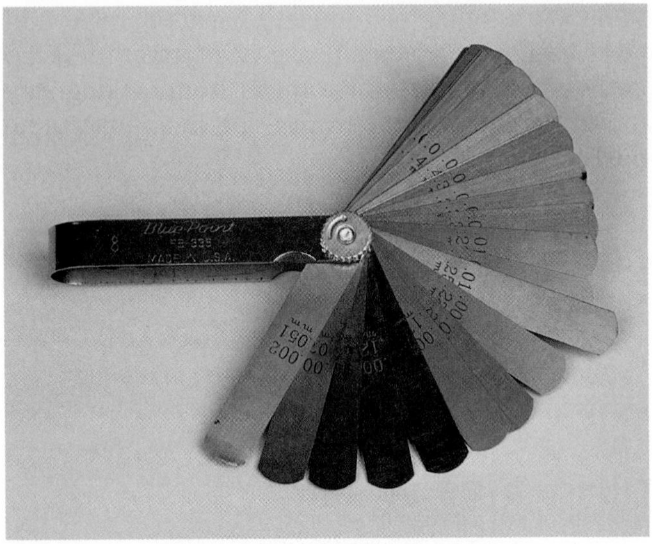

Figure 4–26 Typical feeler gauge set.

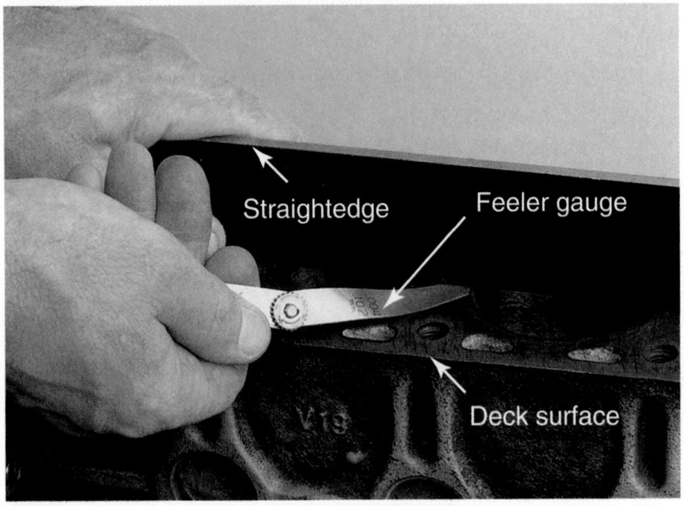

Figure 4–27 Using a feeler gauge and precision straightedge to check for warpage.

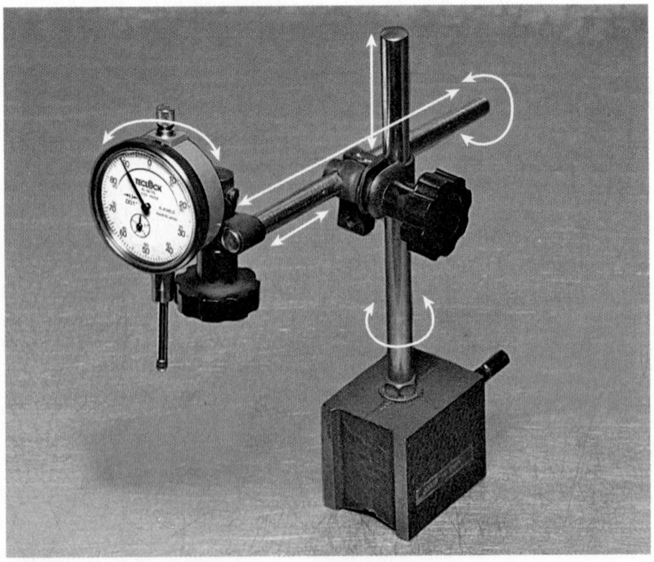

Figure 4–28 A dial indicator with a highly adaptive holding fixture.

Dial Indicator

The **dial indicator (Figure 4–28)** is calibrated in 0.001-inch (one-thousandth inch) increments. Metric dial indicators are also available. Both types are used to measure movement. Common uses of the dial indicator include measuring valve lift, journal concentricity, flywheel or brake rotor runout, gear backlash, and crankshaft end play. Dial indicators are available with various face markings and

Figure 4-29 This dial indicator setup will measure the amount this axle can move in and out.

measurement ranges to accommodate many measuring tasks.

To use a dial indicator, position the indicator rod against the object to be measured. Then, push the indicator toward the work until the indicator needle travels far enough around the gauge face to permit movement to be read in either direction **(Figure 4–29)**. Zero the indicator needle on the gauge. Always be sure the range of the dial indicator is sufficient to allow the amount of movement required by the measuring procedure. For example, never use a 1-inch indicator on a component that will move 2 inches.

HAND TOOLS

Most service procedures require the use of hand tools. Therefore, technicians need a wide assortment of these tools. Each has a specific job and should be used in a specific way. Most service departments and garages require their technicians to buy their own hand tools.

Wrenches

The word *wrench* means twist. A wrench is a tool for twisting and/or holding bolt heads or nuts. Nearly all bolt heads and nuts have six sides; the jaw of a wrench fits around these sides to turn the bolt or nut. All technicians should have a complete collection of **wrenches**. This includes both metric and SAE wrenches in a variety of sizes and styles **(Figure 4–30)**. The width of the jaw opening determines its size. For example, a ½-inch wrench has a jaw opening (from face to face) of ½ inch. The size is actually slightly larger than its nominal size so the wrench fits around a nut or bolt head of equal size.

SHOP TALK

Metric and SAE wrenches are not interchangeable. For example, a 9⁄16-inch wrench is 0.02 inch larger than a 14-millimeter nut. If the 9⁄16-inch wrench is used to turn or hold a 14-millimeter nut, the wrench will probably slip. This may cause rounding of the points of the nut and possibly skinned knuckles as well.

The following is a brief discussion of the types of wrenches used by automotive technicians.

Open-End Wrench The jaws of the open-end wrench **(Figure 4–31)** allow the wrench to slide around two sides of a bolt or nut head where there might be insufficient clearance above or on one side of the nut to accept a box wrench.

Box-End Wrench The end of the box-end wrench is boxed or closed rather than open. The jaws of the wrench fit completely around a bolt or nut, gripping each point on the fastener. The box-end wrench is not likely to slip off a nut or bolt. It is safer than an open-end wrench. Box-end wrenches are available as 6 point and 12 point **(Figure 4–32)**. The 6-point box end grips the screw more securely than a 12-point box-end wrench can and avoids damage to the bolt head.

Combination Wrench The combination wrench has an open-end jaw on one end and a box-end on the other. Both ends are the same size. Every auto technician should have two sets of wrenches: one for holding and one for turning. The combination wrench is probably the best choice for the second set. It can be used with either open-end or box-end wrench sets and can be used as an open-end or box-end wrench.

Flare Nut (Line) Wrenches Flare nut or line wrenches should be used to loosen or tighten brake line or tubing fittings. Using open-ended wrenches on these fittings tends to round the corners of the nut, which are typically made of soft metal and can distort easily. Flare nut wrenches surround the nut and provide a better grip on the fitting. They have a section cut out so that the wrench can be slipped around the brake or fuel line and dropped over the flare nut.

Allen Wrench Setscrews are used to fasten door handles, instrument panel knobs, engine parts, and even brake calipers. A set of fractional and metric hex head

Figure 4–30 A technician needs many different sets of wrenches. *Reproduced under license from Snap-on Incorporated. All of the marks are marks of their owners.*

wrenches, or Allen wrenches **(Figure 4–33)**, should be in every technician's toolbox. An Allen wrench can be L-shaped or can be mounted in a socket driver and used with a ratchet.

Adjustable-End Wrench An adjustable-end wrench (commonly called a crescent wrench) has one fixed jaw and one movable jaw. The wrench opening can be adjusted by rotating a helical adjusting screw that is mated to teeth in the lower jaw. Because this type of wrench does not firmly grip a bolt's head, it is likely to slip. Adjustable wrenches should be used carefully and *only* when it is absolutely necessary. Be sure to put all of the turning pressure on the fixed jaw.

Sockets and Ratchets

A set of Imperial and metric sockets combined with a ratchet handle and a few extensions should be

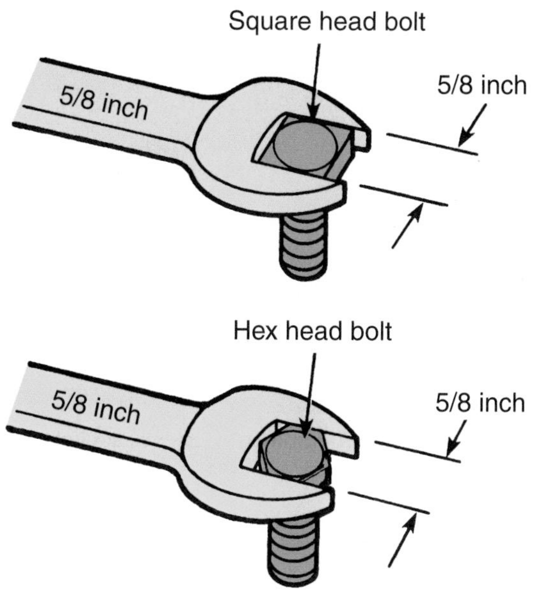

Figure 4-31 An open-end wrench grips only two sides of a fastener.

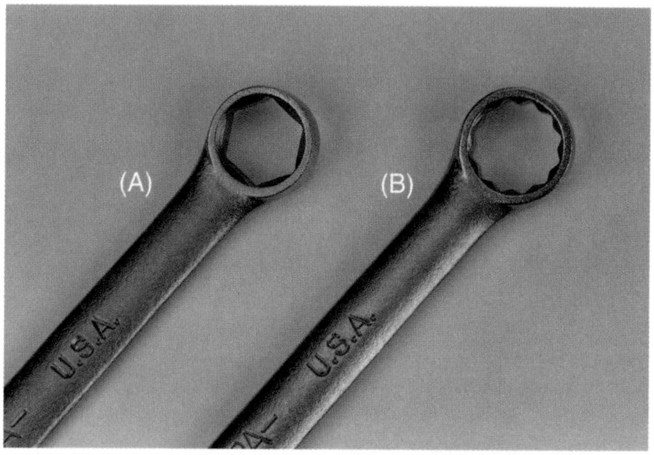

Figure 4-32 Six-point and twelve-point box-end wrenches are available.

included in your tool set. The ratchet allows you to turn the socket in one direction with force and in the other direction without force, which allows you to tighten or loosen a bolt without removing and resetting the wrench after you have turned it. In many situations, a socket wrench is much safer, faster, and easier to use than any other wrench. In fact, sometimes it is the only wrench that will work.

The basic socket wrench set consists of a ratchet handle and several barrel-shaped sockets. The socket fits over and around a bolt or nut (**Figure 4–34**). Inside, it is shaped like a box-end wrench. Sockets are available in 6, 8, or 12 points. A 6-point socket has stronger walls and improved grip on a bolt compared to a normal 12-point socket. However, 6-point sockets have half the positions of a 12-point socket. Six-point sockets are mostly used on fasteners that are rusted or rounded. Eight-point sockets are available to use on square nuts or square-headed bolts. Some axle and transmission assemblies use square-headed plugs in the fluid reservoir.

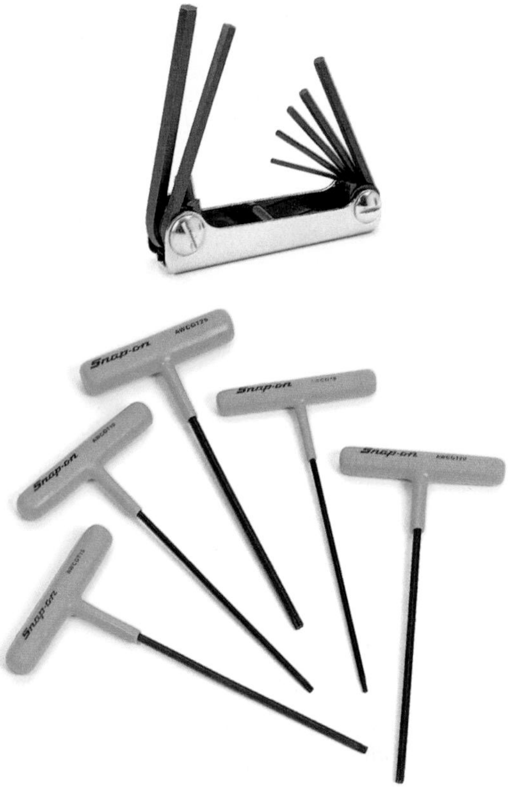

Figure 4-33 Top: A handy tool containing many different Allen wrenches. Bottom: Tee-handle Allen wrenches designed for better gripping and easier torque application. *Reproduced under license from Snap-on Incorporated. All of the marks are marks of their owners.*

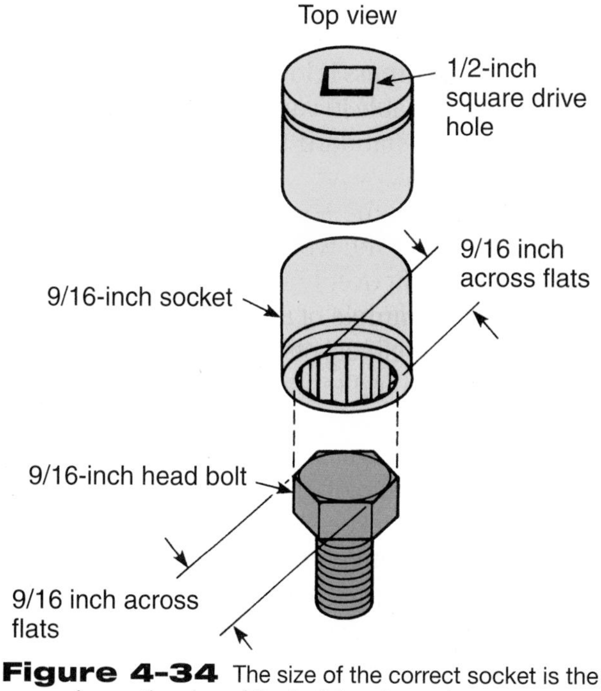

Figure 4-34 The size of the correct socket is the same size as the size of the bolt head or nut.

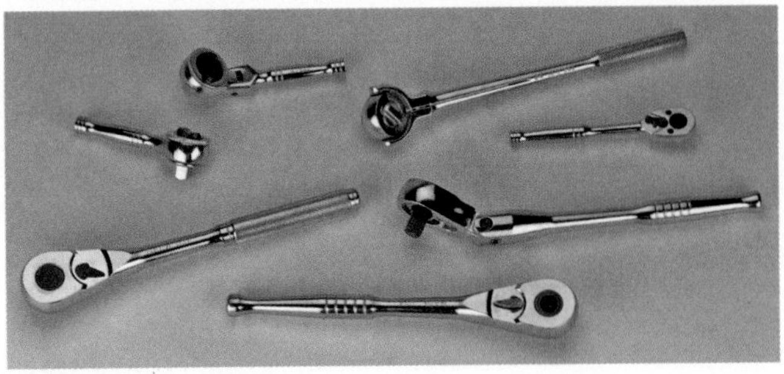

Figure 4–35 An assortment of ratchets.

The top side of a socket has a square hole that accepts a square lug on the socket handle. This square hole is the drive hole. The size of the hole and handle lug (¼ inch, ⅜ inch, ½ inch, and so on) indicates the drive size of the socket wrench. One handle fits all the sockets in a set. On better-quality handles, a spring-loaded ball in the square drive lug fits into a depression in the socket. This ball holds the socket to the handle. An assortment of socket (ratchet) handles is shown in **Figure 4–35**.

Not all socket handles are ratcheting. Some, called breaker bars, are simply long arms with a swivel drive used to provide extra torque onto a bolt to help loosen it. These are available in a variety of lengths and drive sizes. Sometimes nut drivers are used. These handles look like screwdrivers and have a drive shaft on the end of the shaft. Sockets and/or various attachments are inserted on the drive lug. These drivers are only used when bolt tightness is low.

Sockets are available in various sizes, lengths, and bore depths. Both standard SAE and metric socket wrench sets are necessary for automotive service. Normally, the larger the socket size, the longer the socket or the deeper the well. Deep-well sockets are made extra long to fit over bolt ends or studs. A spark plug socket is an example of a special purpose deep-well socket. Deep-well sockets are also good for reaching nuts or bolts in limited-access areas. Deep-well sockets should not be used when a regular-size socket will do the job. The longer socket develops more twist torque and tends to slip off the fastener.

Heavier-walled sockets are designed for use with an impact wrench and are called impact sockets. Most sockets are chrome-plated, except for impact sockets, which are not **(Figure 4–36)**.

WARNING!

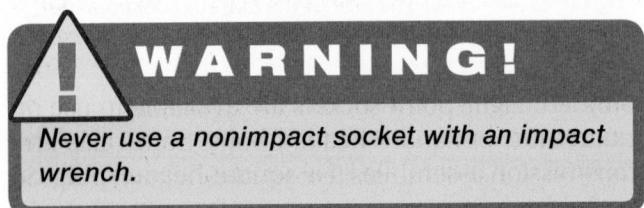

Never use a nonimpact socket with an impact wrench.

Figure 4–36 A chromed deep-well socket and an impact socket.

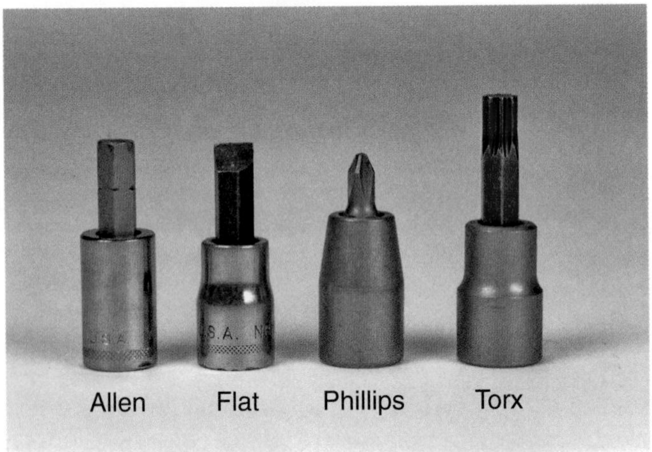

Figure 4–37 A typical set of socket drivers.

Special Sockets Screwdriver (including Torx® driver) and Allen wrench attachments are also available for use with a socket wrench. **Figure 4–37** shows a typical set of specialty socket drivers. These socket wrench attachments are very handy when a fastener cannot be loosened with a regular screwdriver. The leverage given by the ratchet handle is often just what it takes to break a stubborn screw loose.

Swivel sockets are also available. These sockets are fitted with a flexible joint that accommodates odd angles between the socket and the ratchet handle. These sockets are often used to work bolts that are difficult to reach.

Although crowfoot sockets are not really sockets, they are used with a ratchet or breaker bar. These sockets are actually the end of an open-end or line wrench made with a drive bore, which allows a ratchet to move the socket.

Extensions An extension is commonly used to separate the socket from the ratchet or handle. The extension moves the handle away from the bolt and makes the use of a ratchet more feasible. Extensions are

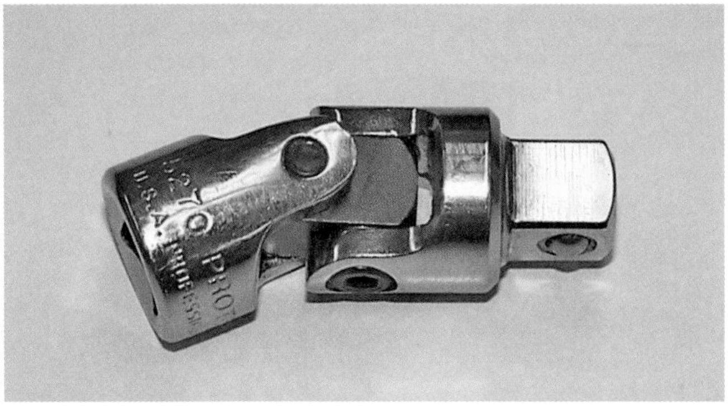

Figure 4-38 A flexible adapter is used when direct access to the bolt is not possible.

available in all common drive sizes and in a variety of lengths. The most common lengths are 1 inch, 3 inches, 6 inches, and 10 inches; however 2- and 3-foot extensions are also quite common. Flexible adapters are used with extensions to gain access to bolts that cannot be directly tightened or loosened **(Figure 4–38)**.

Wobble extensions allow a socket to pivot slightly at the drive connection. This type of extension provides for a more positive connection to the socket than swivel joints but only allows approximately 16 degrees of flexibility.

Socket Adapters When sockets of a different drive size must be used with a particular ratchet or handle, an adapter can be inserted between the socket and the drive on the handle. An example of a common adapter is one that allows for the use of a $\frac{1}{4}$-inch drive socket on a $\frac{3}{8}$-inch drive ratchet.

Torque Wrenches

Torque wrenches (Figure 4–39) measure how tight a nut or bolt is. Many of a car's nuts and bolts should be tightened to a certain amount and have a torque specification that is expressed in foot-pounds (U.S.)

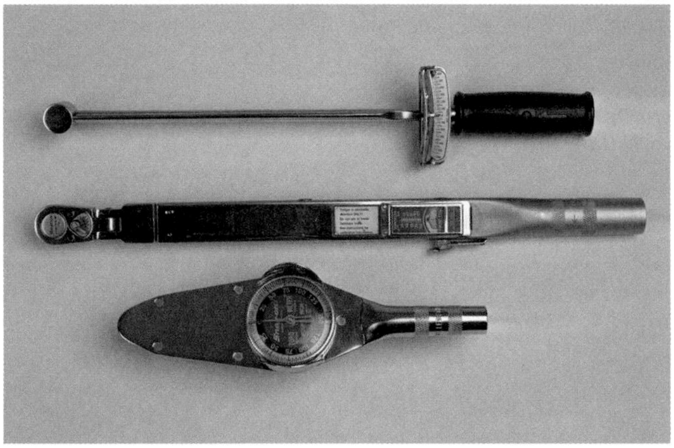

Figure 4-39 The common types of torque wrenches.

or Newton-meters (metric). A foot-pound is the work or pressure accomplished by a force of 1 pound through a distance of 1 foot. A Newton-meter is the work or pressure accomplished by a force of 1 kilogram through a distance of 1 meter.

A torque wrench is basically a ratchet or breaker bar with some means of displaying the amount of torque exerted on a bolt when pressure is applied to the handle. Torque wrenches are available with the various drive sizes. Sockets are inserted onto the drive and then placed over the bolt. As pressure is exerted on the bolt, the torque wrench indicates the amount of torque.

SHOP TALK

Following torque specifications is critical. However, there is a possibility that the torque spec is wrong as printed. (In other words, someone made a mistake.) If the torque spec seems way too tight or loose for the size of bolt, find the torque spec in a different source. If the two specs are the same, use it. If they are different, use the one that seems right.

The common types of torque wrenches are available with inch-pound and foot-pound increments.

- A beam torque wrench is not highly accurate. It relies on a beam metal that points to the torque reading.
- A "click"-type torque wrench clicks when the desired torque is reached. The handle is twisted to set the desired torque reading.
- A dial torque wrench has a dial that indicates the torque exerted on the wrench. The wrench may have a light or buzzer that turns on when the desired torque is reached.
- A digital readout type displays the torque and is commonly used to measure turning effort as well as for tightening bolts. Some designs of this type torque wrench have a light or buzzer that turns on when the desired torque is reached.

The correct torque provides the tightness and stress that the manufacturer has found to be the most desirable and reliable. For example, engine-bearing caps that are too tight distort the bearings, causing excessive wear and incorrect oil clearance. This often results in rapid wear of other engine parts due to decreased oil flow. Insufficient torque can result in out-of-round bores and subsequent failure of the parts.

PROCEDURE

When using a torque wrench, follow these steps to get an accurate reading:

1. Locate the torque specs and procedures in a service manual.
2. Mentally divide the torque specification by three.
3. Hold the wrench so that it is at a 90-degree angle from the fastener being tightened.
4. Tighten the bolt or nut to one-third of the specification.
5. Then tighten the bolt to two-thirds of the spec.
6. Now tighten the bolt to within 10 foot-pounds of the spec.
7. Tighten the bolt to the specified torque.
8. Recheck the torque.

Screwdrivers

A screwdriver drives a variety of threaded fasteners used in the automotive industry. Each fastener requires a specific kind of screwdriver, and a well-equipped technician has several sizes of each.

SHOP TALK

A screwdriver should not be used as a chisel, punch, or pry bar. Screwdrivers were not made to withstand blows or bending pressures. When misused in such a fashion, the tips will wear, become rounded, and tend to slip out of the fastener. At that point, a screwdriver becomes unusable. Remember a defective tool is a dangerous tool.

Screwdrivers are defined by their sizes, their tips **(Figure 4–40)**, and the types of fasteners they should be used with. Your tool set should include both blade and Phillips drivers in a variety of lengths from 2-inch "stubbies" to 12-inch screwdrivers. You also should have an assortment of special screwdrivers, such as those with a Torx® head design.

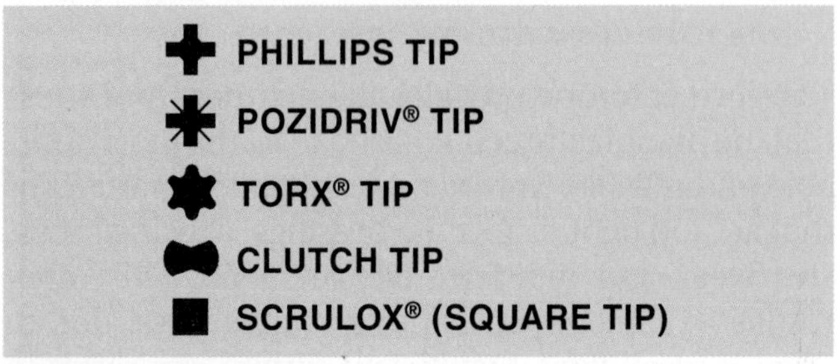

Figure 4–40 The various screwdriver tips that are available. *Reproduced under license from Snap-on Incorporated. All of the marks are marks of their owners.*

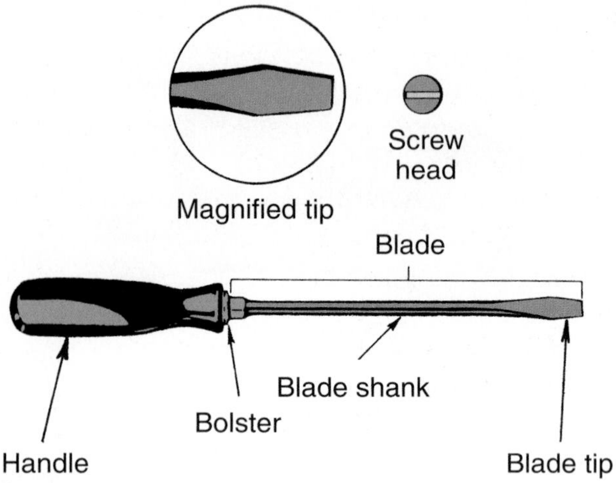

Figure 4–41 The blade tip screwdriver is used with slotted head fasteners.

- *Standard Tip Screwdriver:* A slotted screw accepts a screwdriver with a standard or blade-type tip. The standard tip screwdriver is probably the most common type **(Figure 4–41)**. It is useful for turning carriage bolts, machine screws, and sheet metal screws. The width and thickness of the blade determine the size of a standard screwdriver. Always use a blade that fills the slot in the fastener.
- *Phillips Screwdriver:* The tip of a **Phillips screwdriver** has four prongs that fit the four slots in a Phillips head screw **(Figure 4–42)**. The four surfaces enclose the screwdriver tip so it is less likely that the screwdriver will slip out of the fastener. Phillips screwdrivers come in sizes #0 (the smallest), #1, #2, #3, and #4 (the largest).
- *Reed and Prince Screwdriver:* The tip of a Reed and Prince screwdriver is like a Phillips except that the prongs come to a point rather than to a blunt end.
- *Pozidriv® Screwdriver:* The **Pozidriv screwdriver** is like a Phillips but its tip is flatter and blunter. The squared tip grips the screw's head and slips less than a Phillips screwdriver.

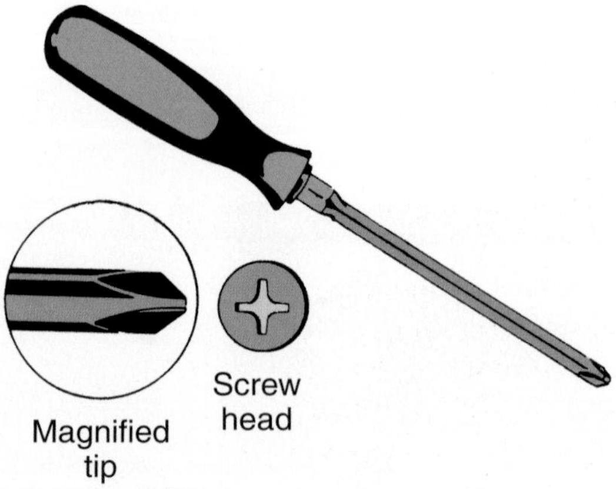

Figure 4–42 The tip of a Phillips screwdriver has four prongs that provide a good grip in the fastener.

- *Torx® Screwdriver:* The **Torx screwdriver** is used to secure headlight assemblies, mirrors, and luggage racks. Not only does the six-prong tip provide greater turning power and less slippage, but the Torx fastener also provides a measure of tamper resistance. Torx drivers come in sizes T15 (the smallest), T20, T25, and T27 (the largest).
- *Clutch Driver:* Fasteners that require a clutch driver are normally used in non-load-bearing places. Clutch head fasteners offer a degree of tamper resistance and offer less slippage than a standard slot screw. The clutch head design has been called a butterfly or figure-eight. Automotive technicians do not often use these drivers.
- *Scrulox® Screwdriver:* The Scrulox screwdriver has a square tip. The tip fits into a square recess in the top of a fastener. This type of fastener is commonly used on truck bodies, campers, and boats.

Impact Screwdriver

An impact screwdriver is used to loosen stubborn screws. Impact screwdrivers have interchangeable heads and bits that allow the handles of the tools to be used with various screw head designs.

To use an impact screwdriver **(Figure 4–43)**, select the correct bit and insert it into the driver's head. Then hold the bit against the screw slot while firmly twisting the handle in the desired direction. Strike the handle with a hammer. The force of the hammer will exert a downward force on the screw and, at the same time, exert a twisting force on the screw.

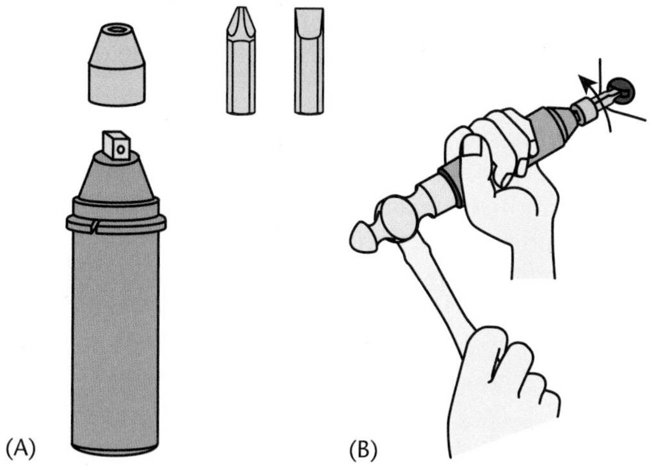

Figure 4-43 (A) An impact screwdriver set. (B) An impact screwdriver automatically tries to rotate the screw when it is struck with a hammer.

Pliers

Pliers (Figure 4–44) are gripping tools used for working with wires, clips, and pins. At a minimum, an auto technician should own several types: standard pliers for common parts and wires, needle nose for small

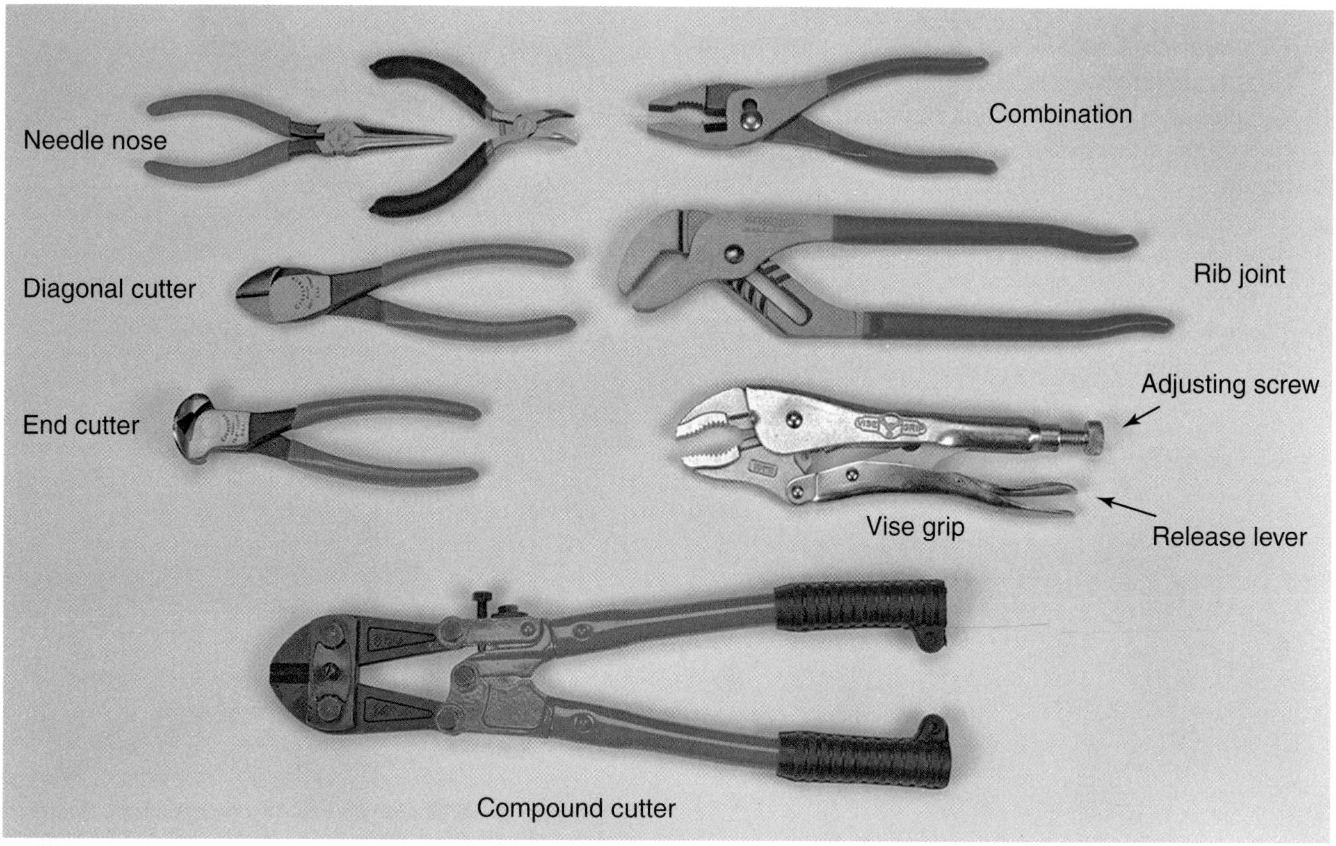

Figure 4-44 Various types of pliers.

parts, and large, adjustable pliers for large items and heavy-duty work. A brief discussion on the different types of pliers follows:

- *Combination pliers* are the most common type of pliers and are frequently used in many kinds of automotive repair. The jaws have both flat and curved surfaces for holding flat or round objects. Also called slip-joint pliers, the combination pliers have many jaw-opening sizes. One jaw can be moved up or down on a pin attached to the other jaw to change the size of the opening.
- *Adjustable pliers,* commonly called *channel locks,* have a multiposition slip joint that allows for many jaw-opening sizes.
- *Needle nose pliers* have long, tapered jaws. They are great for holding small parts or for reaching into tight spots. Many needle nose pliers also have wire-cutting edges and a wire stripper. Curved needle nose pliers allow you to work on a small object around a corner.
- *Locking pliers,* or *vise grips,* are similar to the standard pliers, except they can be tightly locked around an object. They are extremely useful for holding parts together. They are also useful for getting a firm grip on a badly rounded fastener that is impossible to turn with a wrench or socket. Locking pliers come in several sizes and jaw configurations for use in many auto repair jobs.
- *Diagonal-cutting pliers,* or cutters, are used to cut electrical connections, cotter pins, and wires on a vehicle. Jaws on these pliers have extra-hard cutting edges that are squeezed around the item to be cut.
- *Snap- or lock ring pliers* are made with a linkage that allows the movable jaw to stay parallel throughout the range of opening **(Figure 4–45)**. The jaw surface is usually notched or toothed to prevent slipping.

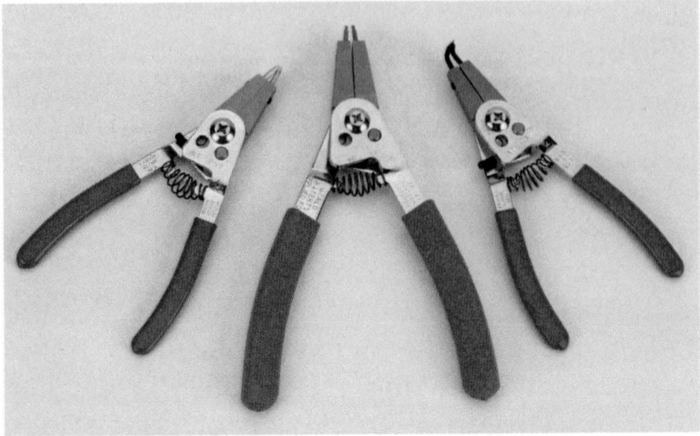

Figure 4–45 Snapring and retaining ring pliers.

- *Retaining ring pliers* are identified by their pointed tips that fit into holes in retaining rings. Retaining ring pliers come in fixed sizes but are also available in sets with interchangeable jaws.

Hammers

Hammers are identified by the material and weight of the head. There are two groups of hammer heads: steel and soft faced **(Figures 4–46** and **4–47)**. Your tool set should include at least three hammers: two

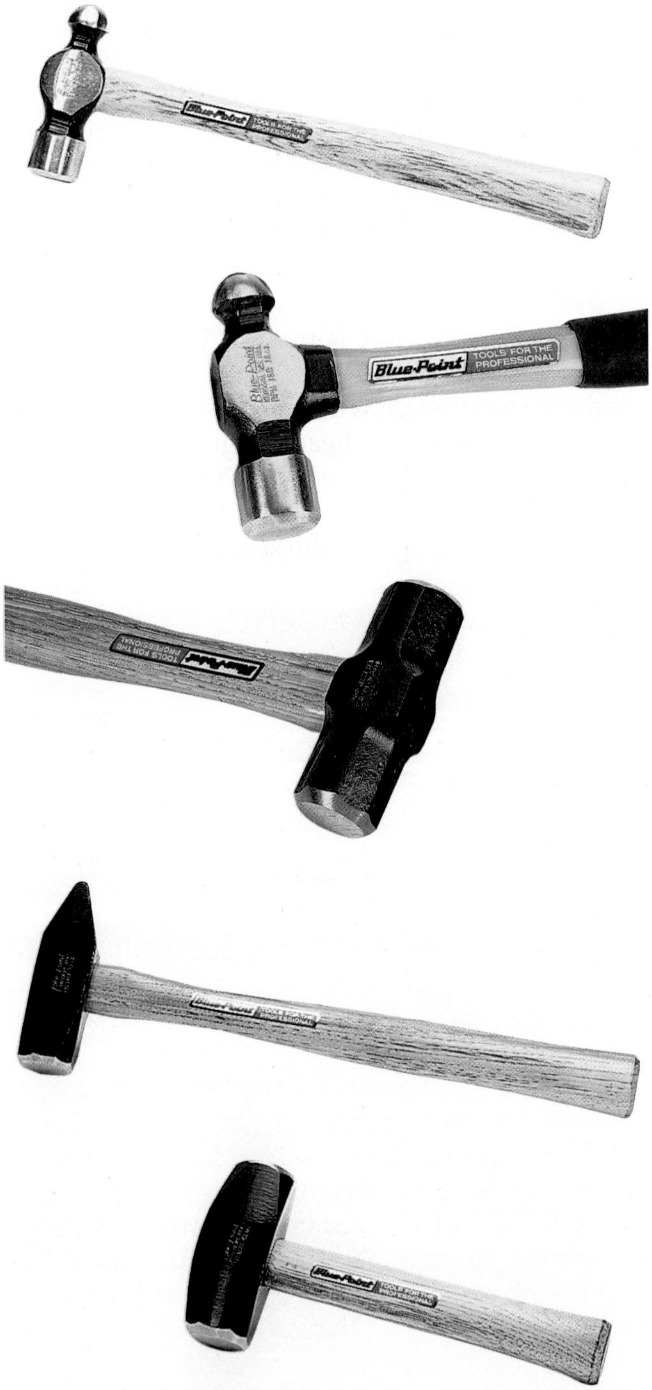

Figure 4–46 Various steel-faced hammers. *Reproduced under license from Snap-on Incorporated. All of the marks are marks of their owners.*

Figure 4-47 Soft-faced hammers. *Reproduced under license from Snap-on Incorporated. All of the marks are marks of their owners.*

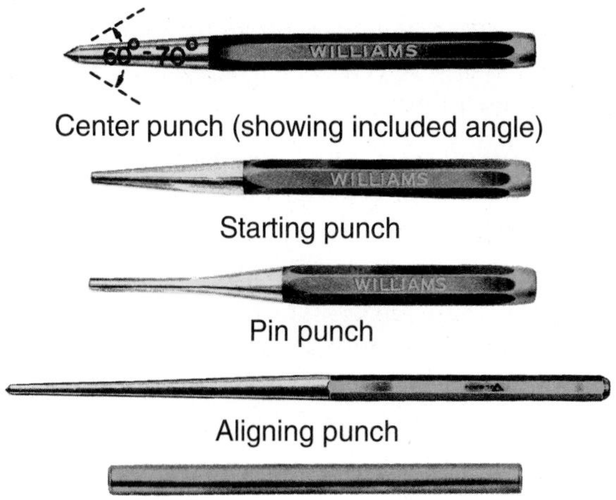

Figure 4-48 Punches are defined by their shape and the diameter of the point.

ball-peen hammers, one 8-ounce and one 12- to 16-ounce hammer, and a small sledgehammer. You should also have a plastic and lead or brass-faced mallet. The heads of steel-faced hammers are made from high-grade alloy steel. The steel is deep forged and heat treated to a suitable degree of hardness. Soft-faced hammers have a surface that yields when it strikes an object. Soft-faced hammers should be used on machined surfaces and when marring a finish is undesirable. For example, a brass hammer should be used to strike gears or shafts because it will not damage them.

Chisels and Punches

Chisels are used to cut metal by driving them with a hammer. Automotive technicians use a variety of chisels for cutting sheet metal, shearing off rivet and bolt heads, splitting rusted nuts, and chipping metal. A variety of chisels are available, each with a specific purpose, including flat, cape, round-nose cape, and diamond point chisels.

Punches **(Figure 4–48)** are used for driving out pins, rivets, or shafts; aligning holes in parts during assembly; and marking the starting point for drilling a hole. Punches are designated by their point diameter and punch shape. Drift punches are used to remove drift and roll pins. Some drifts are made of brass; these should be used whenever you are concerned about possible damage to the pin or to the surface surrounding the pin. Tapered punches are used to line up bolt holes. Starter or center punches are used to make an indent before drilling to prevent the drill bit from wandering.

Removers

Rust, corrosion, and prolonged heat can cause automotive fasteners, such as cap screws and studs, to become stuck. A box wrench or socket is used to loosen cap screws. A special gripping tool is designed to remove studs. However, if the fastener breaks off, special extracting tools and procedures must be employed.

One type of stud remover is shown in **Figure 4–49**. These tools are also used to install studs. Stud removers have hardened, knurled, or grooved eccentric rollers or jaws that grip the stud tightly when operated. Stud removers/installers are turned by a socket wrench drive handle, a socket, or wrench.

Extractors are used on screws and bolts that are broken off below the surface. Twist drills, fluted extractors, and hex nuts are included in a screw extractor set **(Figure 4–50)**. This type of extractor lessens the tendency to expand the screw or stud that has

Figure 4–49 Stud installation/removal tool.

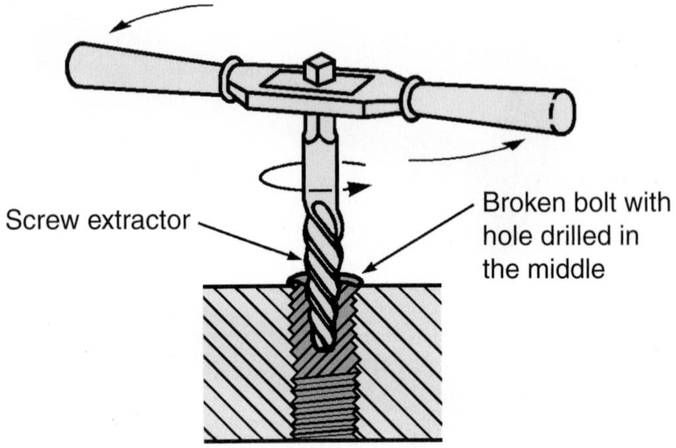

Figure 4–51 Using a screw extractor to remove a broken bolt.

Figure 4–50 Screw extractors.

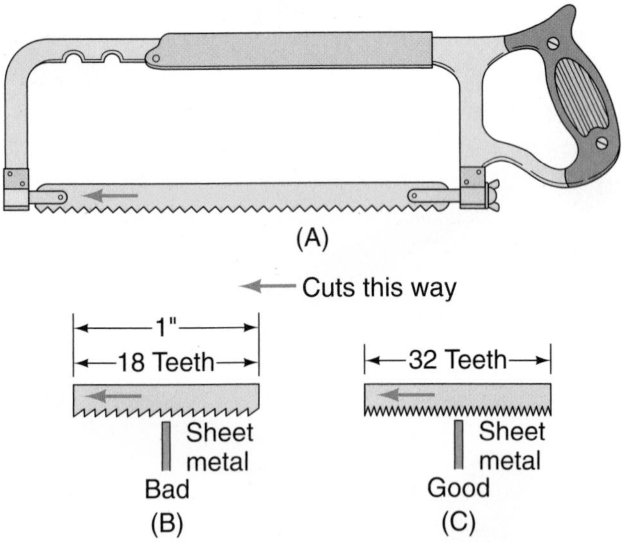

Figure 4–52 (A) The teeth on the blade in a hacksaw should face forward. (B) A coarse blade should not be used with sheet metal. (C) A fine blade will work well with sheet metal.

been drilled out by providing gripping power along the full length of the stud.

Screw extractors are often called easy outs. To use an extractor, the bolt must be drilled and the extractor forced into that bore. The teeth of the extractor grip the inside of the drilled bore and allow the bolt to be turned out **(Figure 4–51)**. Easy outs typically have the size of the required drill bit stamped on one side.

At times a broken bolt can be loosened and removed from its bore by driving it in a counterclockwise direction with a chisel and hammer. A bolt broken off above the surface may be able to be removed with locking pliers.

Hacksaws

A hacksaw is used to cut metal **(Figure 4–52)**. The blade only cuts on the forward stroke. The teeth of the blade should always face away from the saw's handle. The number of teeth on the blade determines the type of metal the saw can be used on. A fine-toothed blade is best for thin sheet metal, whereas a coarse blade is used on thicker metals.

When using a hacksaw, never bear down on the blade while pulling it toward you; this will dull the blade. Use the entire blade while cutting.

Files

Files are commonly used to shape or smooth metal edges. Files typically have square, triangular, rectangular (flat), round, or half-round shapes **(Figure 4–53)**. They also vary in size and coarseness. The most commonly used files are the half-round and flat with either single-cut or double-cut designs. A single-cut file has its cutting grooves lined up diagonally across the face of the file. The cutting grooves of a double-cut file run diagonally in both directions across the

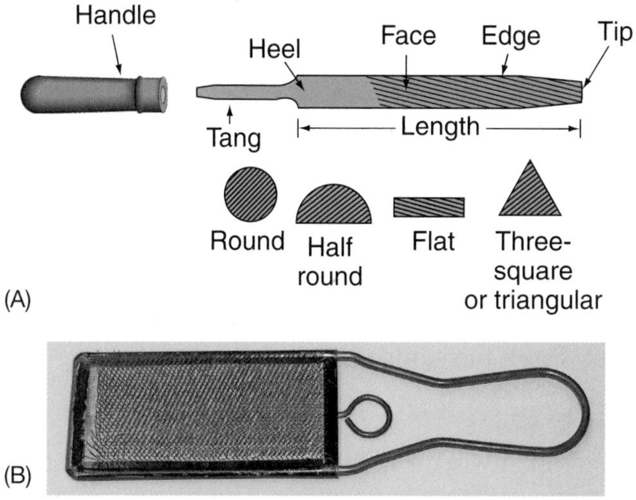

Figure 4-53 (A) Files come in a variety of shapes. *Courtesy of Ford Motor Company* (B) A file card.

face. Double-cut files are considered first cut or roughening files because they can remove large amounts of metal. Single-cut files are considered finishing files because they remove small amounts of metal.

To avoid personal injury, files should always be used with a plastic or wooden handle. Like hacksaws, files only cut on the forward stroke. Coarse files are used for soft metals, and smoother, or finer, files are used to work steel and other hard metals.

Keep files clean, dry, and free of oil and grease. To clean filings from the teeth of a file, use a special tool called a file card.

Gear and Bearing Pullers

Many precision gears and bearings have a slight interference fit (**press fit**) when installed on a shaft or housing. For example, the inside diameter of a bore is 0.001 inch smaller than the outside diameter of a shaft. When the shaft is fitted into the bore it must be pressed in to overcome the 0.001-inch interference fit. This press fit prevents the parts from moving on each other. The removal of gears and bearings must be done carefully. Prying or hammering can break or bind the parts. A puller with the proper jaws and adapters should be used when applying force to remove gears and bearings. Using proper tools, the force can be applied with a slight and steady motion.

Pullers are available in many different designs and therefore are designed for specific purposes. Most pullers come with various jaw lengths and shapes to allow them to work in a number of different situations.

Some pullers are fitted to the end of a slide hammer **(Figure 4–54)** and are used to remove slightly press-fit items. After the mounting plate of the puller is secure in or on the object to be removed, the weight on the tool's hammer is slid back with force against the handle of the tool, generating a pulling force and jerking the object out of its bore.

Figure 4-54 Using a slide hammer-type puller to remove a drive axle.

To pull something out of a bore, the puller must be designed to expand its jaws outward. The jaws also must be small enough to reach into the bore without damaging the bore while still firmly gripping the object that is being removed. This type puller is commonly used to remove seals, bushings, and bearing cups.

Jaw-type pullers are used to pull an object off a shaft. These pullers are available with two or three jaws **(Figure 4–55)**. Jaw-type pullers are commonly used to remove bearings, pulleys, and gears.

Some pullers are actually pushers. A push-puller is used to push a shaft out of its bore in a housing. It is often difficult to grip the end of the shaft with a puller, so a push-puller is used to move the shaft out of the bore.

Bearing, Bushing, and Seal Drivers

Another commonly used group of special tools includes the various designs of bearing, bushing, and

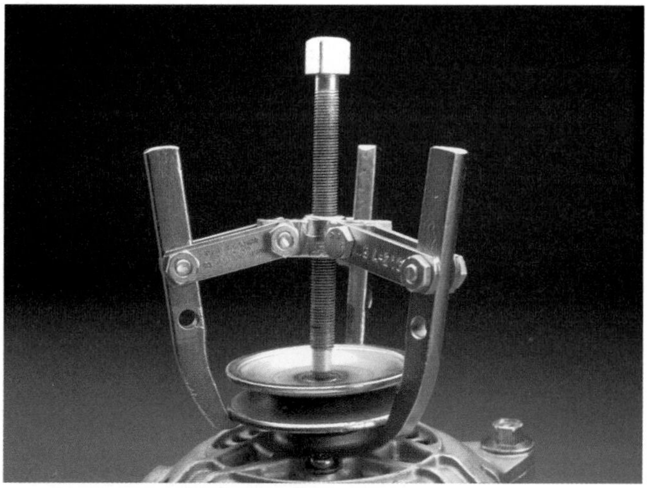

Figure 4-55 The jaws on this puller are reversible to allow for inside and outside pulls.

seal drivers. Auto manufacturers supply their dealerships with drivers for specific components. However, universal sets of drivers are also available. These sets include a variety of driver plates, each of a different diameter. The plates are often reversible. The flat side of the plate is used to install seals and the tapered side is used to install tapered bearing races. A driver handle is threaded into the appropriate plate. The bearing or seal is driven into place by tapping on the driver hammer.

Always make sure you use the correct tool for the job; bushings and seals are easily damaged if the wrong tool or procedure is used. Car manufacturers and specialty tool companies work closely together to design and manufacture special tools required to repair cars.

Trouble Light

Adequate light is necessary when working under and around automobiles. A **trouble light** can be battery powered (like a flashlight) or need to be plugged into a wall socket. Some shops have trouble lights that pull down from a reel suspended from the ceiling. Trouble lights should have LED or fluorescent bulbs. Incandescent bulbs should not be used because they can pop and burn. Take extra care when using a trouble light. Make sure its cord does not get caught in a rotating object. The bulb or tube should be surrounded by a cage or enclosed in clear plastic to prevent accidental breaking and burning.

Creeper

Rather than crawl on your back to work under a vehicle, use a creeper **(Figure 4–56)**. A creeper is a platform with small wheels. It allows you to slide under a vehicle and easily maneuver while working. To protect yourself and others around you, never have the creeper lying on the floor when you are not using it.

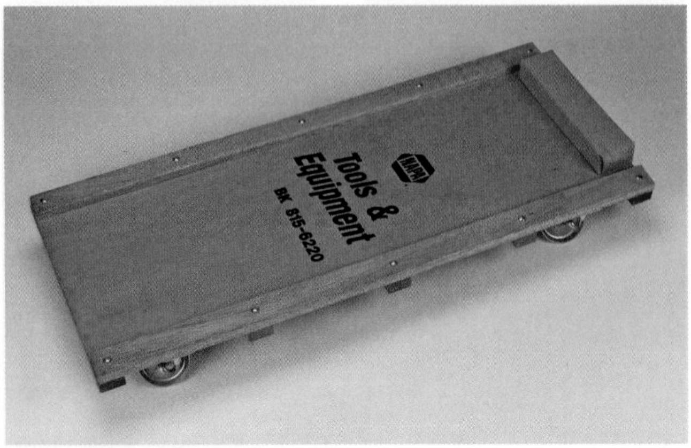

Figure 4–56 A creeper allows you to work comfortably and safely under a vehicle.

Accidentally stepping on it can result in a serious fall. Always keep it standing on its end when it is not being used.

SHOP EQUIPMENT

Some tools and equipment are supplied by the service facility and few technicians have these as part of their tool assortment. These tools are commonly used but there is no need for each technician to own them. Many shops have one or two of each.

Bench Vises

Often repair work is completed with a part or assembly removed from the vehicle. The repairs are typically safely and quickly made by securing the assembly. Small parts are usually secured with a bench vise. The vise is bolted to a workbench to give it security. The object to be held is placed into the tool's jaws and the jaws are tightened around the object. If the object could be damaged or marred by the jaws, brass jaw caps are installed over the jaws before the object is placed between them.

Bench Grinder

This electric power tool is generally bolted to a workbench. The grinder should have safety shields and guards. Always wear face protection when using a grinder. A bench grinder is classified by wheel size. Six- to ten-inch wheels are the most common in auto repair shops. Three types of wheels are available with this bench tool:

1. Grinding wheel, for a wide variety of grinding jobs from sharpening cutting tools to deburring
2. Wire wheel brush, for general cleaning and buffing, removing rust, scale, and paint, deburring, and so forth
3. Buffing wheel, for general purpose buffing, polishing, and light cutting

Presses

Many automotive jobs require the use of powerful force to assemble or disassemble parts that are press fit together. Removing and installing piston pins, servicing rear axle bearings, pressing brake drum and rotor studs, and performing transmission assembly work are just a few examples. Presses can be hydraulic, electric, air, or hand driven. Capacities range up to 150 tons of pressing force, depending on the size and design of the press. Smaller arbor and C-frame presses can be bench or pedestal mounted, while high-capacity units are freestanding or floor mounted **(Figure 4–57)**.

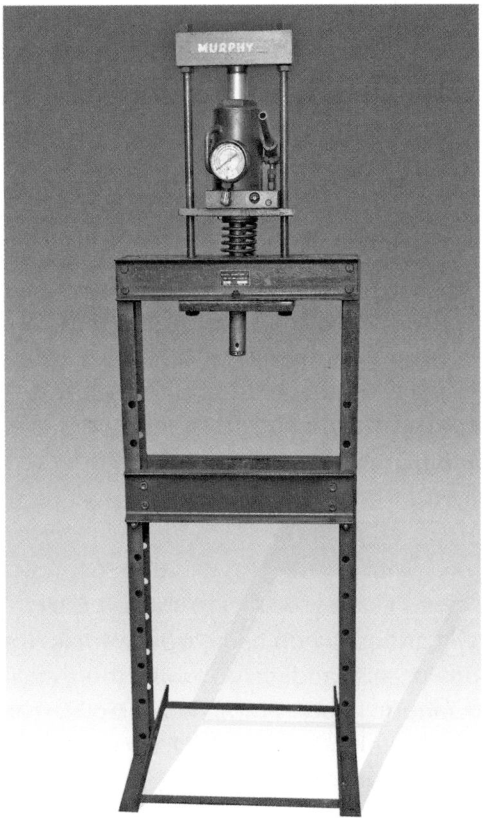

Figure 4-57 A floor-mounted hydraulic press.

WARNING!

Always wear safety glasses when using a press.

Grease Guns

Some shops are equipped with air-powered grease guns, while in others, technicians use a manually operated grease gun. Both types can force grease into a grease fitting. Hand-operated grease guns are often preferred because the pressure of the grease can be controlled by the technician. However, many shops use low air pressure to activate a pneumatic grease gun. The suspension and steering system may have several grease or zerk fittings.

Oxyacetylene Torches

Oxyacetylene torches **(Figure 4–58)** have many purposes. In the automotive service industry they are used to heat metal when two parts are difficult to separate, to cut metal (such as when replacing exhaust system parts), and to weld or connect two metal parts together.

Oxyacetylene welding and cutting equipment uses the combustion of acetylene in oxygen to produce a flame temperature of about 5,600°F (3,100°C). Acetylene is used as the fuel and oxygen is used to aid in the combustion of the fuel.

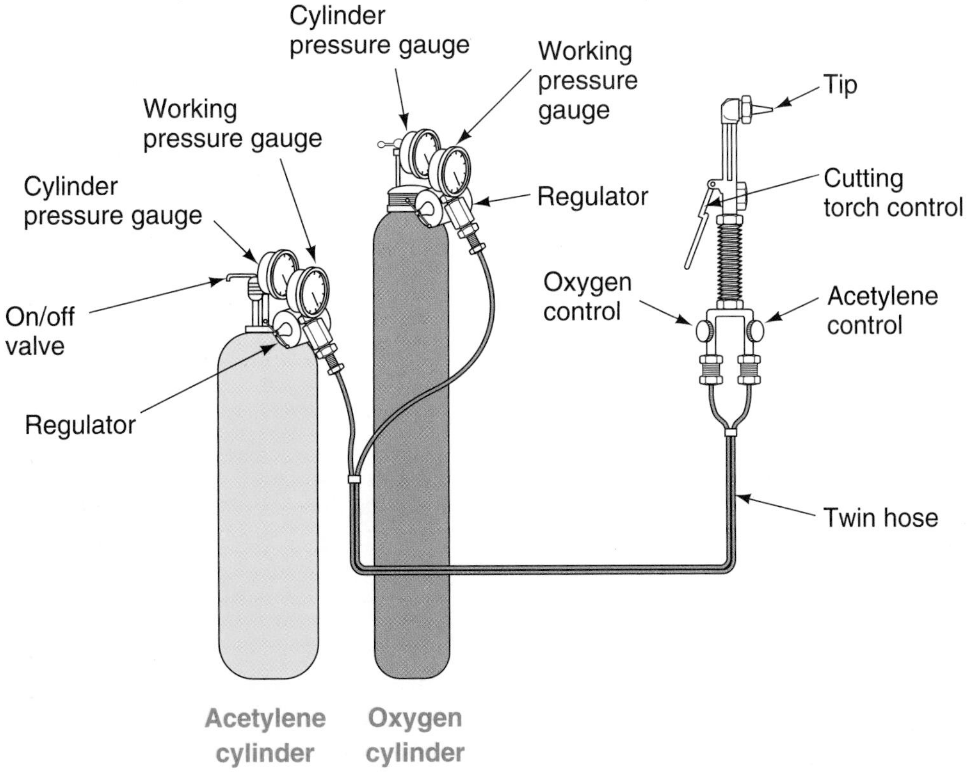

Figure 4-58 Oxyacetylene welding equipment shown with a cutting torch.

The equipment includes cylinders of oxygen and acetylene, two pressure regulators, two flexible hoses (one for each cylinder), and a torch. The torches are selected for the job being done—welding torch for welding, brazing, soldering, and heating and cutting torch for cutting metal.

There are three sets of valves for each gas: the tank valve, the regulator valve, and the torch valve. The oxygen hose is colored green, and the acetylene hose is red. The acetylene connections have left-hand threads and the oxygen connectors have right-hand threads.

Welding and Heating Torch The hoses connect the cylinders to the torch. On the torch there are separate valves for each gas. The torch is comprised of the valves, a handle, a mixing chamber (where the fuel gas and oxygen mix), and a tip (where the flame forms). Many different tips can be used with a welding torch. Always select the correct size for the job.

Cutting Torch A cutting torch is used to cut metal. It is similar to a welding torch. However, the cutting torch has a third tube from the valves to the mixing chamber. It carries high-pressure oxygen, which is controlled by a large lever on the torch. During cutting, the metal is heated until it glows orange, and then a lever on the torch is pressed to pass a stream of oxygen through the heated metal to burn it away where the cut is desired.

Precautions Never use oxyacetylene equipment unless you have been properly trained to do so. Also, adhere to all safety precautions, including:

- Always pay attention to what you are doing: While using a torch, severe and fatal burns and violent explosions can result from inattention and carelessness.
- Before using an oxyacetylene torch, make sure that all flammable materials such as grease, oil, paint, sawdust, and so on are cleared from the area.
- Keep oxygen away from all combustibles.
- Wear approved shaded goggles with enclosed sides, or a shield with a shaded lens to protect your eyes from glare and sparks.
- Wear leather gloves to protect your hands from burns.
- Wear clothes and shoes/boots appropriate for welding. They should be free of grease and oil.
- Make sure that the gas cylinders are securely fastened upright to a wall or a post or a portable cart.
- Never move an oxygen tank around without its valve cap screwed in place.
- Never lay an acetylene tank on its side while being used.
- Never oil an oxygen regulator.

POWER TOOLS

Power tools make a technician's job easier. They operate faster and with more torque than hand tools. However, power tools require greater safety measures. Power tools do not stop unless they are turned off. Power is furnished by air (pneumatic), electricity, or hydraulic fluid. *Power tools should only be used for loosening nuts and/or bolts.*

SHOP TALK

Safety is critical when using power tools. Carelessness or mishandling of power tools can cause serious injury. Do not use a power tool without obtaining permission from your instructor. Be sure you know how to operate the tool properly before using it. Prior to using a power tool, read the instructions carefully.

Impact Wrench

An **impact wrench** **(Figure 4–59)** is a portable hand-held reversible wrench. A heavy-duty model can deliver up to 450 foot-pounds (607.5 N-m) of torque. When triggered, the output shaft, onto which the impact socket is fastened, spins freely at 2,000 to 14,000 rpm, depending on the wrench's make and model. When the impact wrench meets resistance, a small spring-loaded hammer situated near the end of the tool strikes an anvil attached to the drive shaft onto which the socket is mounted. Each impact moves the socket around a little until torque equilibrium is reached, the fastener breaks, or the trigger is

Figure 4-59 A typical air impact wrench.

released. Torque equilibrium occurs when the torque of the bolt equals the output torque of the wrench. Impact wrenches can be powered either by air or by electricity.

SHOP TALK

When using an air impact wrench, it is important that only impact sockets and adapters be used. Other types of sockets and adapters, if used, might shatter and fly off, endangering the safety of the operator and others in the immediate area.

An impact wrench uses compressed air or electricity to hammer or impact a nut or bolt loose or tight. Light-duty impact wrenches are available in three drive sizes—¼, ⅜, and ½ inch—and two heavy-duty sizes—¾ and 1 inch.

WARNING!

Impact wrenches should not be used to tighten critical parts or parts that may be damaged by the hammering force of the wrench.

Air Ratchet

An air ratchet, like the hand ratchet, has a special ability to work in hard-to-reach places. Its angle drive reaches in and loosens or tightens where other hand or power wrenches just cannot work **(Figure 4–60)**. The air ratchet looks like an ordinary ratchet but has a fat handgrip that contains the air vane motor and drive mechanism. Air ratchets usually have a ⅜-inch drive. Air ratchets are not torque sensitive; therefore, a torque wrench should be used on all fasteners after snugging them up with an air ratchet.

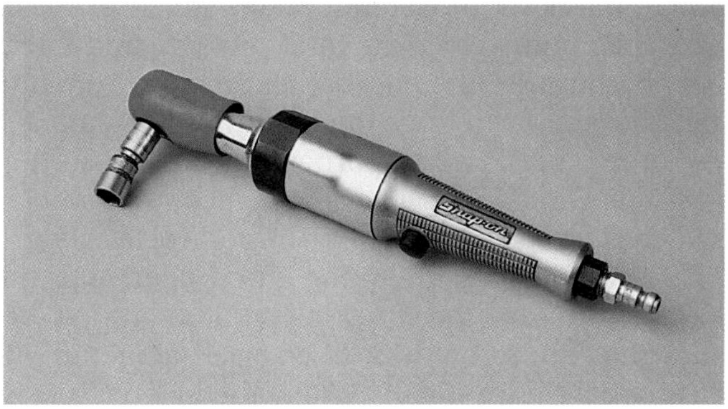

Figure 4-60 An air ratchet.

Air Drill

Air drills are usually available in ¼-, ⅜-, and ½-inch sizes. They operate in much the same manner as an electric drill, but are smaller and lighter. This compactness makes them a great deal easier to use for drilling operations in auto work.

Blowgun

Blowguns are used for blowing off parts during cleaning. Never point a blowgun at yourself or anyone else. A blowgun **(Figure 4–61)** snaps into one end of an air hose and directs airflow when a button is pressed. Always use an OSHA-approved air blowgun. Before using a blowgun, be sure it has not been modified to eliminate air-bleed holes on the side.

JACKS AND LIFTS

Jacks are used to raise a vehicle off the ground and are available in two basic designs and in a variety of sizes. The most common jack is a hydraulic floor jack, which is classified by the weights it can lift: 1½, 2, and 2½ tons, and so on. These jacks are controlled by moving the handle up and down. The other design of portable floor jack uses compressed air. Pneumatic jacks are operated by controlling air pressure at the jack.

CAUTION!

Before lifting a vehicle with air suspension, turn off the system. The switch is usually in the trunk.

The hydraulic floor lift is the safest lifting tool and is able to raise the vehicle high enough to allow you to walk and work under it. Various safety features prevent a hydraulic lift from dropping if a seal does leak or if air pressure is lost. Before lifting a vehicle, make sure the lift is correctly positioned.

Floor Jack

A floor jack is a portable unit mounted on wheels. The lifting pad on the jack is placed under the chassis of the vehicle, and the jack handle is operated with a pumping action. This forces fluid into a hydraulic cylinder in the jack, and the cylinder extends to force the jack lift pad upward and to lift the vehicle. Always be sure that the lift pad is positioned securely under one of the car manufacturer's recommended lifting points. To release the hydraulic pressure and lower the vehicle, the handle or release lever must be turned slowly.

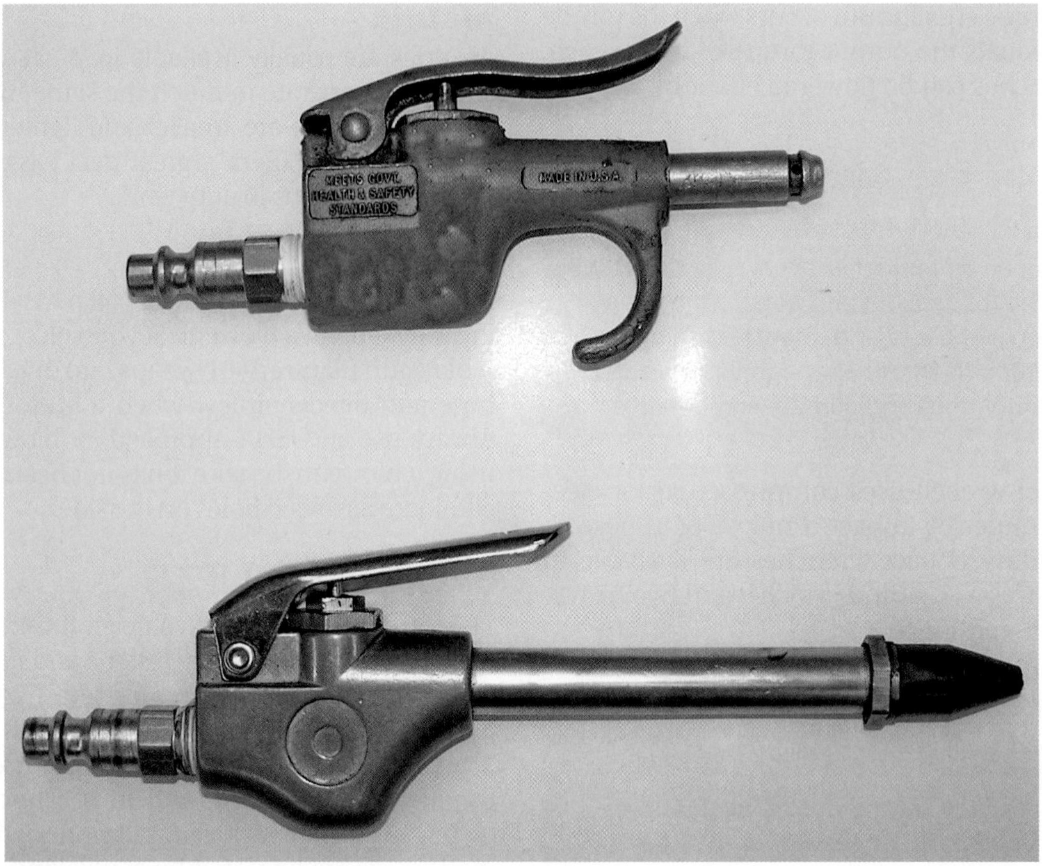

Figure 4-61 Two types of air nozzles (blowguns).

The maximum lifting capacity of the floor jack is usually written on the jack decal. Never lift a vehicle that exceeds the jack lifting capacity. This action may cause the jack to break or collapse, resulting in vehicle damage or personal injury.

When a vehicle is raised by a floor jack, it should be supported by safety stands **(Figure 4–62)**. Never work under a car with only a jack supporting it; always use safety stands. Hydraulic seals in the jack can let go and allow the vehicle to drop.

Figure 4-62 Whenever you have raised a vehicle with a floor jack, the vehicle should be supported with jack stands.

Lift

A lift is used to raise a vehicle so the technician can work under the vehicle. The lift arms must be placed under the car manufacturer's recommended lifting points prior to raising a vehicle. There are three basic types of lifts: frame contact **(Figure 4–63)**, wheel contact, and axle engaging. These categories define where the frame contact points align with the vehicle.

Twin posts are used on some lifts **(Figure 4–64)**, whereas other lifts have a single post **(Figure 4–65)**. Some lifts have an electric motor, which drives a hydraulic pump to create fluid pressure and force the lift upward. Other lifts use air pressure from the shop air supply to force the lift upward. If shop air pressure is used for this purpose, the air pressure is applied to fluid in the lift cylinder. A control lever or switch is placed near the lift. The control lever supplies shop air pressure to the lift cylinder, and the switch turns on the lift pump motor. Always be sure that the safety lock is engaged after the lift is raised **(Figure 4–66)**. When the safety lock is released, a release lever is operated slowly to lower the vehicle.

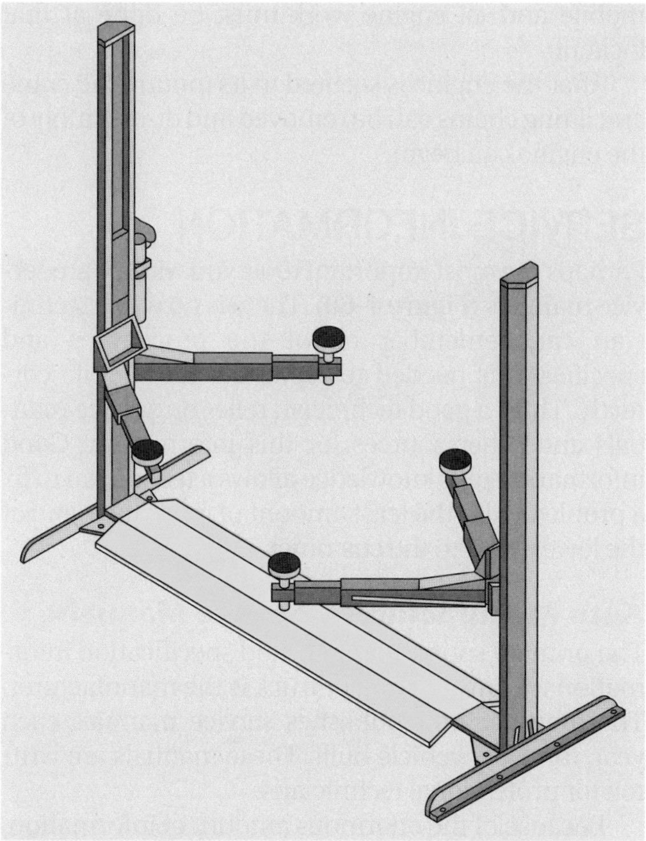

Figure 4-63 An aboveground or surface mount frame-contact lift. *Courtesy of Automotive Lift Institute*

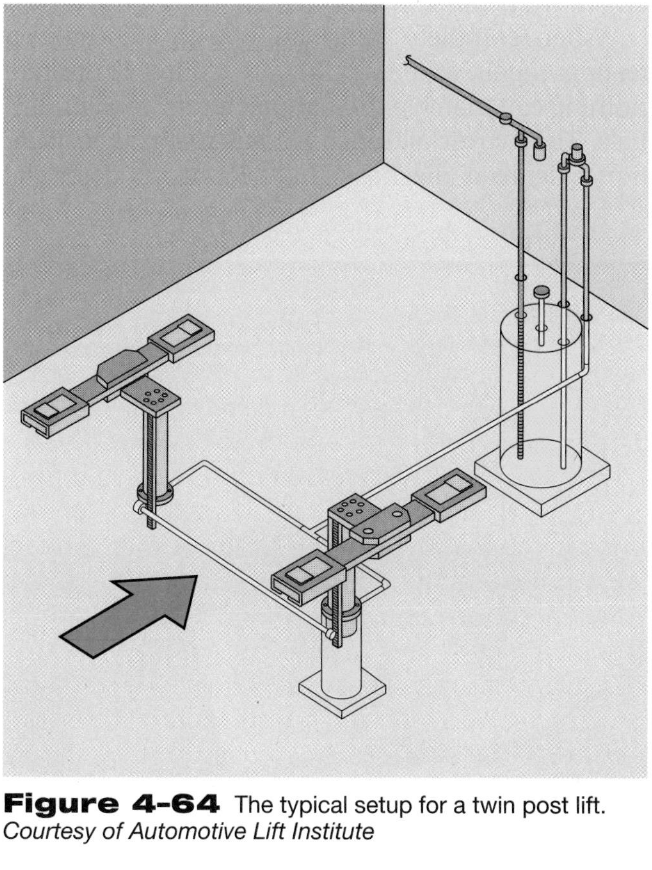

Figure 4-64 The typical setup for a twin post lift. *Courtesy of Automotive Lift Institute*

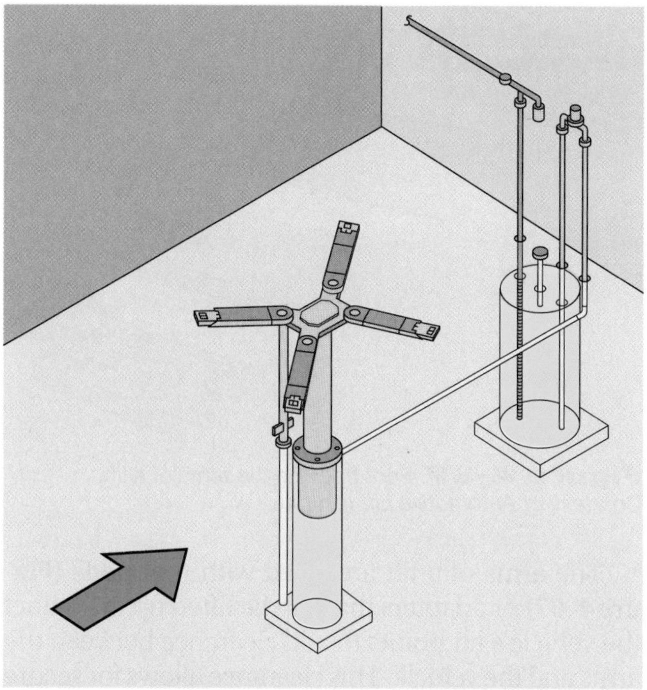

Figure 4-65 The typical setup for a single post lift. *Courtesy of Automotive Lift Institute*

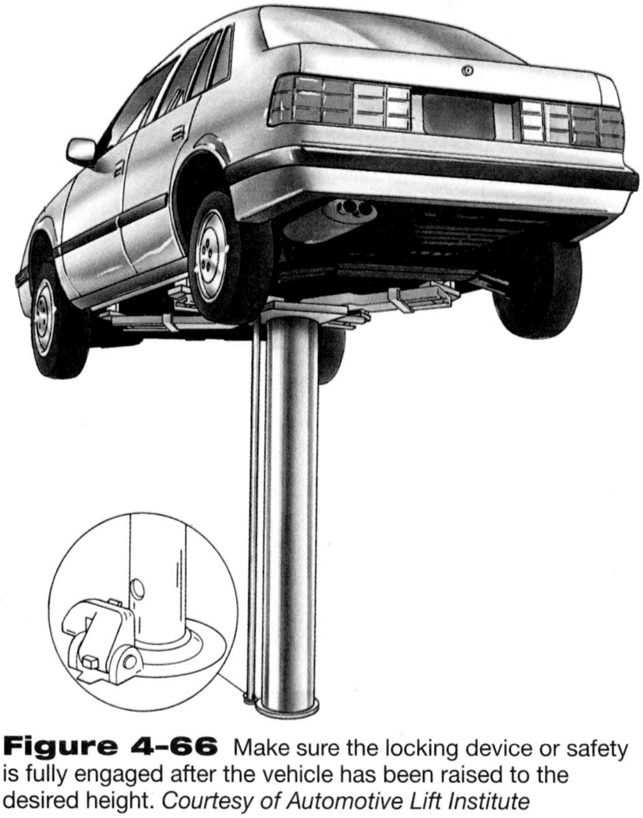

Figure 4-66 Make sure the locking device or safety is fully engaged after the vehicle has been raised to the desired height. *Courtesy of Automotive Lift Institute*

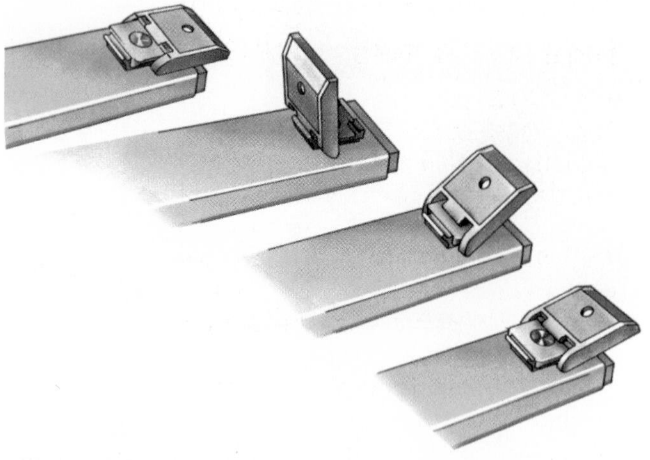

Figure 4-67 Foot pads on the arms of a lift.
Courtesy of Automotive Lift Institute

The arms of a lift are fitted with **foot pads (Figure 4–67)** or adapters that can be lifted up to contact the vehicle's lift points to add clearance between the arms and the vehicle. This clearance allows for secure lifting without damaging any part of the body or underbody of the vehicle.

Portable Crane

To remove and install an engine, a portable crane, frequently called a cherry picker, is used. To lift an engine, attach a pulling sling or chain to the engine. Some engines have eye plates for use in lifting. If they are not available, the sling must be bolted to the engine. The sling attaching bolts must be large enough to support the engine and must thread into the block a minimum of 1½ times the bolt diameter. Connect the crane to the chain. Raise the engine slightly and make sure the sling attachments are secure. Carefully lift the engine out of its compartment.

Lower the engine close to the floor so the transmission and torque converter or clutch can be removed from the engine, if necessary.

Engine Stands/Benches

After the engine has been removed, use the crane to raise the engine. Position the engine next to an engine stand. Most stands use a plate with several holes or adjustable arms. The engine must be supported by at least four bolts that fit solidly into the engine. The engine should be positioned so that its center is in the middle of the engine's stand adapter plate. The adapter plate can swivel in the stand. By centering the engine, the engine can be easily turned to the desired working positions.

Some shops have engine mounts bolted to the top of workbenches. The engine is suspended off the side of the workbench. These have the advantage of a good working space next to the engine, but they are not mobile and all engine work must be done at that location.

After the engine is secured to its mount, the crane and lifting chains can be removed and disassembly of the engine can begin.

SERVICE INFORMATION

Perhaps the most important tools you will use are service manuals **(Figure 4–68)**. There is no way a technician can remember all of the procedures and specifications needed to repair an automobile correctly. Thus, a good technician relies on service manuals and other sources for this information. Good information plus knowledge allows a technician to fix a problem with the least amount of frustration and at the lowest cost to the customer.

Auto Manufacturers' Service Manuals

The primary source of repair and specification information for any car, van, or truck is the manufacturer. The manufacturer publishes service manuals each year, for every vehicle built. These manuals are written for professional technicians.

Because of the enormous amount of information, some manufacturers publish more than one manual per year per car model. They may be separated into sections such as chassis, suspension, steering, emission control, fuel systems, brakes, basic maintenance, engine, transmission, body, and so on **(Figure 4–69)**.

When complete information with step-by-step testing, repair, and assembly procedures is desired, nothing can match auto manufacturers' repair manuals. They cover all repairs, adjustments, specifications, detailed diagnostic procedures, and special

Figure 4-68 One of your most important tools is a service manual.

TABLE OF CONTENTS	SECTION NUMBER
GENERAL INFO. AND LUBE	
General Information	0A
Maintenance and Lubrication	0B
HEATING AND AIR COND.	
Heating and Vent.(nonA/C)	1A
Air Conditioning System	1B
V-5 A/C Compressor Overhaul	1D3
BUMPERS AND FRONT BODY PANELS	
Bumpers (See 10-4)	
Fr. End Body Panels (See 10-5)	
STEERING, SUSPENSION, TIRES, AND WHEELS	
Diagnosis	3
Wheel Alignment	3A
Power Steering Gear & Pump	3B1
Front Suspension	3C
Rear Suspension	3D
Tires and Wheels	3E
Steering Col. On-Vehicle Service	3F
Steering Col. - Std. Unit Repair	3F1
Steering Col. - Tilt, Unit Repair	3F2
DRIVE AXLES	
Drive Axles	4D
BRAKES	
General Info. - Diagnosis and On-Car Service	5
Compact Master Cylinder	5A1
Disc Brake Caliper	5B2
Drum Brake - Anchor Plate	5C2
Power Brake Booster Assembly	5D2
ENGINES	
General Information	6
2.0 Liter I-4 Engine	6A1
3.1 Liter V6 Engine	6A3
Cooling System	6B
Fuel System	6C
Engine Electrical - General	6D
Battery	6D1
Cranking System	6D2
Charging System	6D3
Ignition System	6D4
Engine Wiring	6D5
Driveability & Emissions - Gen.	6E
Driveability & Emissions - TBI	6E2
Driveability & Emissions - PFI	6E3
Exhaust System	6F

TABLE OF CONTENTS	SECTION NUMBER
TRANSAXLE	
Auto. Transaxle On-Car Serv.	7A
Auto. Trans. - Hydraulic Diagnosis	3T40-HD
Auto. Trans. - Unit Repair	3T40
Man. Trans. On-Car Service	7B
5-Sp. 5TM40 Man. Trans. Unit Repair	7B1
5-Sp. Isuzu Man. Trans. Unit Repair	7B2
Clutch	7C
CHASSIS ELECTRICAL, INSTRUMENT PANEL & WASHER WIPER	
Electrical Diagnosis	8A
Lighting and Horns	8B
Instrument Panel and Console	8C
Windshield Wiper/Washer	8E5
ACCESSORIES	
Audio System	9A
Cruise Control	9B
Engine Block Heater	9C
BODY SERVICE	
General Body Service	10-1
Stationary Glass	10-2
Underbody	10-3
Bumpers	10-4
Body Front End	10-5
Doors	10-6
Rear Quarters	10-7
Body Rear End	10-8
Roof & Convertible Top	10-9
Seats	10-10
Safety Belts	10-11
Body Wiring	10-12
Unibody Collision Repair	11-1
Welded Panel Replacement	11-2
INDEX	
Alphabetical Index	

Figure 4-69 The main index of a factory service manual showing that the manual is divided by major vehicle systems.

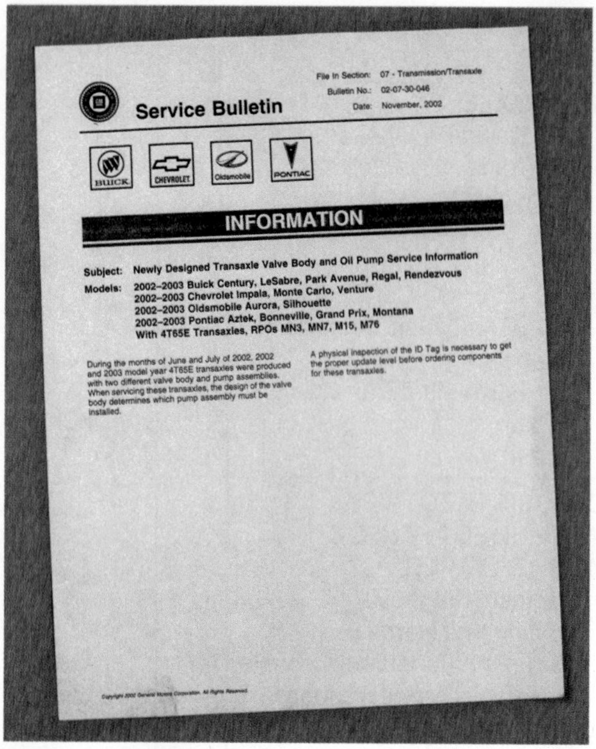
File In Section: 07 - Transmission/Transaxle
Bulletin No.: 02-07-30-046
Date: November, 2002

Service Bulletin

BUICK CHEVROLET Oldsmobile PONTIAC

INFORMATION

Subject: Newly Designed Transaxle Valve Body and Oil Pump Service Information

Models: 2002–2003 Buick Century, LeSabre, Park Avenue, Regal, Rendezvous
2002–2003 Chevrolet Impala, Monte Carlo, Venture
2002–2003 Oldsmobile Aurora, Silhouette
2002–2003 Pontiac Aztek, Bonneville, Grand Prix, Montana
With 4T65E Transaxles, RPOs MN3, MN7, M15, M76

During the months of June and July of 2002, 2002 and 2003 model year 4T65E transaxles were produced with two different valve body and pump assemblies. When servicing these transaxles, the design of the valve body determines which pump assembly must be installed.

A physical inspection of the ID Tag is necessary to get the proper update level before ordering components for these transaxles.

Figure 4–70 A technical service bulletin.

tools required. They can be purchased directly from the automobile manufacturer.

To help you learn how to effectively use service manuals, you will find service manual references and tips throughout this text.

Since many technical changes occur on specific vehicles each year, manufacturers' service manuals need to be constantly updated. Updates are published as service bulletins (often referred to as technical service bulletins or TSBs) that show the changes in specifications and repair procedures during the model year **(Figure 4–70)**. These changes do not appear in the service manual until the next year. The car manufacturer provides these bulletins to dealers and repair facilities on a regular basis.

Automotive manufacturers also publish a series of technician reference books. The publications provide general instructions about the service and repair of the manufacturers' vehicles and also indicate their recommended techniques.

General and Specialty Repair Manuals

Service manuals are also published by independent companies rather than the manufacturers. However, they pay for and get most of their information from the car makers. The manuals contain component information, diagnostic steps, repair procedures, and specifications for several makes of automobiles in one book. Information is usually condensed and is more general than the manufacturers' manuals. The condensed format allows for more coverage in less space and, therefore, is not always specific. They may also contain several years of models as well as several makes in one book.

Finding Information in Service Manuals

Although the manuals from different publishers vary in presentation and arrangement of topics, all service manuals are easy to use after you become familiar with their organization. Most shop manuals are divided into a number of sections, each covering different aspects of the vehicle. The beginning sections commonly provide vehicle identification and basic maintenance information. The remaining sections deal with each different vehicle system in detail, including diagnostic, service, and overhaul procedures. Each section has an index indicating more specific areas of information.

To obtain the correct system specifications and other information, you must first identify the exact system you are working on. The best source for vehicle identification is the VIN. The code can be interpreted through information given in the service manual. The manual may also help you identify the system through identification of key components or other identification numbers and/or markings.

To use a service manual:

1. Select the appropriate manual for the vehicle being serviced.
2. Use the table of contents to locate the section that applies to the work being done.
3. Use the index at the front of that section to locate the required information.
4. Carefully read the information and study the applicable illustrations and diagrams.
5. Follow all of the required steps and procedures given for that service operation.
6. Adhere to all of the given specifications and perform all measurement and adjustment procedures with accuracy and precision.

Aftermarket Suppliers' Guides and Catalogs

Many of the larger parts manufacturers have excellent guides on the various parts they manufacture or supply. They also provide updated service bulletins on their products. Other sources for up-to-date technical information are trade magazines and trade associations.

Lubrication Guides

These specially designed service manuals contain information on lubrication, maintenance, capacities, and underhood service. The lubrication guide includes lube and maintenance instructions, lubrication diagrams and specifications, vehicle lift points, and preventive maintenance mileage/time intervals. The capacities listed include cooling system, air conditioning, cooling system air bleed locations, wheel and tire specifications, and wheel lug torque specifications. The underhood information includes specifications for tune-up; mechanical, electrical, and fuel systems; diagrams; and belt tension.

Owner's Manuals

An owner's manual comes with the vehicle when it is new. It contains operating instructions for the vehicle and its accessories. It also contains valuable information about checking and adding fluids, safety precautions, a complete list of capacities, and the specifications for the various fluids and lubricants for the vehicle.

Flat-Rate Manuals

Flat-rate manuals contain standards for the length of time a specific repair is supposed to require. Normally, they also contain a parts list with approximate or exact prices of parts. They are excellent for making cost estimates and are published by the manufacturers and independents.

Computer-Based Information

Most technicians no longer rely on printed copies of service manuals. They access the same information, as well as service bulletins, electronically on **compact disk–read-only memory (CD-ROMs)** **(Figure 4–71)**, digital video disks (DVDs), and the Internet. Computer-based information eliminates the need for a huge library of printed manuals. Using electronics to find information is also easier and quicker. The disks are normally updated monthly and not only contain the most recent service bulletins but also engineering and field service fixes. DVDs hold more information than CDs; therefore, there are fewer disks with systems that use DVDs.

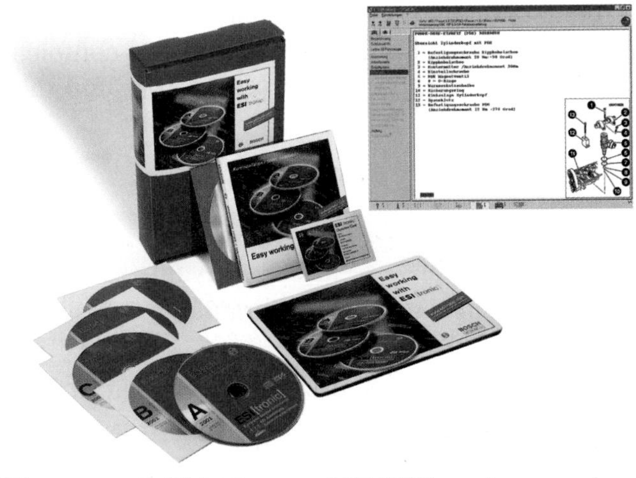

Figure 4-71 The use of CD-ROMs and a computer makes accessing information quick and easy. *Courtesy of Robert Bosch GmbH, www.bosch-presse.de*

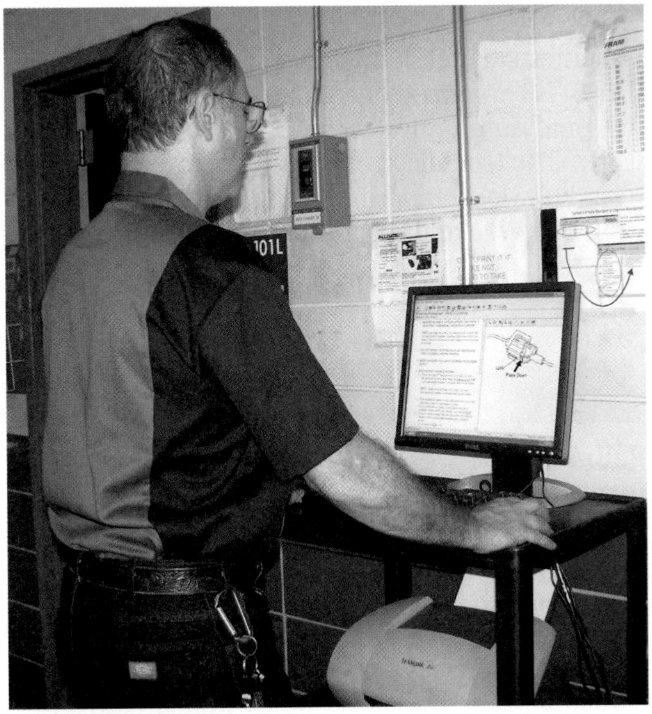

Figure 4-72 The computer screen can display everything that would be in a printed service manual, but computer systems are quicker and using them makes finding information easier.

A technician enters vehicle information and then selects the appropriate part or system **(Figure 4–72)**. The appropriate information then appears on the computer's screen. Online data can be updated instantly and requires no space for physical storage. These systems are easy to use and the information is quickly accessed and displayed. Once the information is retrieved, a tech can read it off the screen or print it out and take it to the service bay.

Hotline Services

Hotline services provide answers to service concerns by telephone. Manufacturers provide help by telephone for technicians in their dealerships. There are subscription services for independents to be able to get repair information by phone. Some manufacturers also have a phone modem system that can transmit computer information from the car to another location. The vehicle's diagnostic link is connected to the modem. The technician in the service bay runs a test sequence on the vehicle. The system downloads the latest updated repair information on that particular model of car. If that does not repair the problem, a technical specialist at the manufacturer's location will review the data and propose a repair.

iATN

The International Automotive Technician's Network (iATN) is comprised of a group of thousands of professional automotive technicians from around the world. The technicians in this group exchange technical knowledge and information with other members. The Web address for this group is http://www.iatn.net.

KEY TERMS

Ball gauge
Blowgun
Bolt diameter
Bolt head
Bolt shank
Chisel
Compact disk–read-only memory (CD-ROM)
Dial caliper
Dial indicator
Die
Extractor
Feeler gauge
Fillet
Foot pads
Grade marks
Impact wrench
Machinist's rule
Micrometer
Phillips screwdriver
Pliers
Pozidriv screwdriver
Press fit
Punch
Snap gauge
Spark plug thread tap
Straightedge
Tap
Telescoping gauge
Thread pitch
Thread pitch gauge
Torque wrench
Torx screwdriver
Trouble light
Vernier caliper
Wrench

SUMMARY

- Repairing the modern automobile requires the use of many different hand and power tools. Units of measurement play a major role in tool selection. Therefore, it is important to be knowledgeable about the Imperial and the metric systems of measurement.
- Measuring tools must be able to measure objects to a high degree of precision. They should be handled with care at all times and cleaned before and after every use.
- A micrometer can be used to measure the outside diameter of shafts and the inside diameter of holes. It is calibrated in either inch or metric graduations.
- Telescoping gauges are designed to measure bore diameters and other clearances. They usually are used with an outside micrometer. Small hole gauges are used in the same manner as the telescoping gauge, usually to determine valve guide diameter.
- The screw pitch gauge provides a fast and accurate method of measuring the threads per inch (pitch) of fasteners. This is done by matching the teeth of the gauge with the fastener threads and reading the pitch directly from the leaf of the gauge.
- It is crucial to use the proper amount of torque when tightening nuts or cap screws on any part of a vehicle, particularly the engine. A torque-indicating wrench makes it possible to duplicate the conditions of tightness and stress recommended by the manufacturer.
- Metric and SAE size wrenches are not interchangeable. An auto technician should have a variety of both types.
- A screwdriver, no matter what type, should never be used as a chisel, punch, or pry bar.
- The hand tap is used for hand cutting internal threads and for cleaning and restoring previously cut threads. Hand-threading dies cut external threads and fit into holders called die stocks.
- Carelessness or mishandling of power tools can cause serious injury. Safety measures are needed when working with such tools as impact and air ratchet wrenches, blowguns, bench grinders, lifts, hoists, and hydraulic presses.
- The primary source of repair and specification information for any vehicle is the manufacturer's service manual. Updates are published as service bulletins and include changes made during the model year, which will not appear in the manual until the following year.
- Flat-rate manuals are ideal for making cost estimates. Published by manufacturers and independent companies, they contain figures showing how long specific repairs should take to complete, as well as a list of the necessary parts and their prices.

REVIEW QUESTIONS

1. How often should the calibration of a micrometer be checked?
2. List some common uses of the dial indicator.
3. Wrenches are marked with their size. What does the size represent?
4. *True or False?* The same information available in service manuals and bulletins is also available

electronically: on compact disks (CD-ROMs), digital video disks (DVDs), and the Internet.

5. How do manufacturers inform technicians about changes in vehicle specifications and repair procedures during the model year?
6. Which of the following wrenches is the best choice for turning a bolt?
   a. open-end
   b. box-end
   c. combination
   d. none of the above
7. Technician A says that a vernier caliper can be used to measure the outside diameter of something. Technician B says that a vernier caliper can be used to measure the inside diameter of a bore. Who is correct?
   a. Technician A
   b. Technician B
   c. Both A and B
   d. Neither A nor B
8. Technician A says that a tap cuts external threads. Technician B says that a die cuts internal threads. Who is correct?
   a. Technician A
   b. Technician B
   c. Both A and B
   d. Neither A nor B
9. Which of the following screwdrivers is like a Phillips but has a flatter and blunter tip?
   a. standard
   b. Torx®
   c. Pozidriv®
   d. clutch head
10. Which of the following types of pliers is best for grasping small parts?
    a. adjustable
    b. needle nose
    c. retaining ring
    d. snapring
11. Technician A uses a punch to align holes in parts during assembly. Technician B uses a punch to drive out rivets. Who is correct?
    a. Technician A
    b. Technician B
    c. Both A and B
    d. Neither A nor B
12. An extractor is used for removing broken ______.
    a. seals
    b. bushings
    c. pistons
    d. bolts
13. Which of the following statements about items that are press fit is *not* true?
    a. Many precision gears and bearings have a slight interference fit when installed on a shaft or housing.
    b. The press fit allows slight motion between the parts and therefore prevents wear.
    c. The removal of gears and bearings that are press fit must be done carefully to avoid breaking or binding the parts.
    d. A puller with the proper jaws and adapters should be used when applying force to remove press-fit gears and bearings.
14. Technician A uses a blowgun to blow off parts during cleaning. Technician B uses a blowgun to clean off his uniform after working. Who is correct?
    a. Technician A
    b. Technician B
    c. Both A and B
    d. Neither A nor B
15. Technician A says that flare nut or line wrenches should be used to loosen or tighten brake line or tubing fittings. Technician B says that open-end wrenches will surround the fitting's nut and provide a positive grip on the fitting. Who is correct?
    a. Technician A
    b. Technician B
    c. Both A and B
    d. Neither A nor B
16. Technician A uses a dial caliper to take inside and outside measurements. Technician B uses a dial caliper to take depth measurements. Who is correct?
    a. Technician A
    b. Technician B
    c. Both A and B
    d. Neither A nor B
17. For a measurement that must be made within one ten-thousandth of an inch, Technician A uses a machinist's rule. For the same accuracy, Technician B uses a standard micrometer. Who is correct?
    a. Technician A
    b. Technician B
    c. Both A and B
    d. Neither A nor B
18. Technician A says that portable floor jacks are operated by hydraulics. Technician B says that portable floor jacks may be operated by compressed air. Who is correct?
    a. Technician A
    b. Technician B
    c. Both A and B
    d. Neither A nor B
19. When using an air impact wrench, Technician A uses impact sockets and adapters. Technician B uses chrome-plated sockets. Who is correct?
    a. Technician A
    b. Technician B
    c. Both A and B
    d. Neither A nor B
20. While discussing thread sealants and lubricants: Technician A says that antiseize compound is used where a bolt might become difficult to remove after a period of time. Technician B says that thread sealants are used on bolts that are

tightened into an oil cavity or coolant passage. Who is correct?

a. Technician A
b. Technician B
c. Both A and B
d. Neither A nor B

21. The metric equivalent to 10 miles is ______.

a. 0.06895 bar
b. 10.6895 bars
c. 1.6093 kilometers
d. 16.093 kilometers

22. *True or False?* A metric bolt with the marking of 8.9 has more yield strength than a bolt marked 10.5.

23. List the steps that should be followed to precisely tighten a bolt to specifications.

24. Which of the following thread designs typically has tapered threads to provide for sealing at a joint?

a. UNC
b. NPT
c. UNF
d. UNEF

25. Which of the following is *not* measured with a feeler gauge set?

a. piston ring side clearance
b. flywheel runout
c. crankshaft end play
d. spark plug gap

DIAGNOSTIC EQUIPMENT AND SPECIAL TOOLS

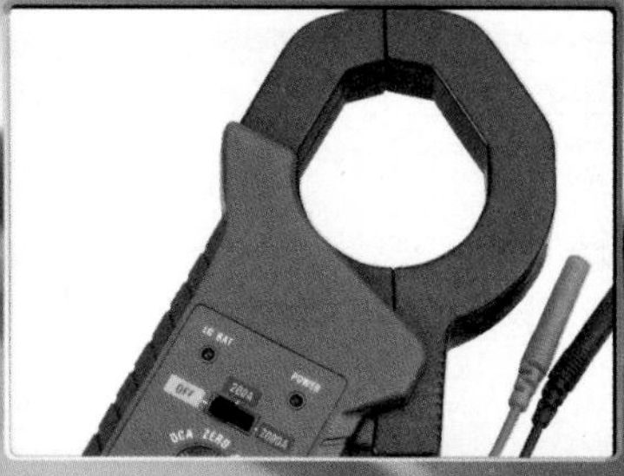

CHAPTER 5

OBJECTIVES

■ Describe the various diagnostic tools used to check an engine and its related systems. ■ Describe the common tools used to service an engine and its related systems. ■ Describe the various diagnostic tools used to check electrical and electronic systems. ■ Describe the common tools used to service electrical and electronic systems. ■ Describe the various diagnostic tools used to check engine performance systems. ■ Describe the common tools used to service engine performance systems. ■ Describe the various diagnostic tools used to check hybrid vehicles. ■ Describe the various diagnostic tools used to check a vehicle's drivetrain. ■ Describe the common tools used to service a vehicle's drivetrain. ■ Describe the various diagnostic tools used to check a vehicle's running gear for wear and damage. ■ Describe the common tools used to service a vehicle's running gear. ■ Describe the various diagnostic tools used to check a vehicle's heating and air-conditioning system. ■ Describe the common tools used to service a vehicle's heating and air-conditioning system.

Diagnosing and servicing the various systems of an automobile require many different tools. Tools that are used to check the performance of a system or component are commonly referred to as diagnostic tools. Tools designed for a particular purpose or system are referred to as special tools. This chapter looks at the common diagnostic and special tools required to service the different systems of a vehicle.

ENGINE REPAIR TOOLS

Engine repair **(Figure 5–1)** and diagnostic tools are discussed in the following paragraphs. This discussion does not cover all of the tools you may need: Only the most commonly used are discussed. Details of when and how to use these tools are presented in Section 2 of this book.

Compression Testers

The operation of an engine depends on the compression of the air-fuel mixture within its cylinders. If the combustion chamber leaks, some of the mixture will escape while it is being compressed, resulting in a loss of power and a waste of fuel.

A **compression gauge** is used to check cylinder compression. The dial face on the gauge indicates pressure in both pounds per square inch (psi) and metric kilopascals (kPa). The range is usually 0 to 300 psi and 0 to 2,100 kPa. There are two basic types of compression gauges: the push-in gauge **(Figure 5–2)** and the screw-in gauge.

The push-in type has a short stem that is either straight or bent at a 45-degree angle. The stem ends in a tapered rubber tip that fits any size spark plug hole. After the spark plugs have been removed, the rubber tip is placed in the spark plug hole and held there while the engine is cranked through several compression cycles. Although simple to use, the push-in gauge may give inaccurate readings if it is not held tightly in the hole.

The screw-in gauge has a long, flexible hose that ends in a threaded adapter **(Figure 5–3)**. This type of compression tester is often used because its flexible hose can reach into areas that are difficult to reach with a push-in-type tester. The threaded adapters are changeable and come in several thread sizes to fit 10 mm, 12 mm, 14 mm, and 18 mm diameter holes. The adapters screw into the spark plug holes in place of the spark plugs.

Most compression gauges have a vent valve that holds the highest pressure reading on its meter.

Figure 5-1 Proper engine rebuilding requires many different tools.

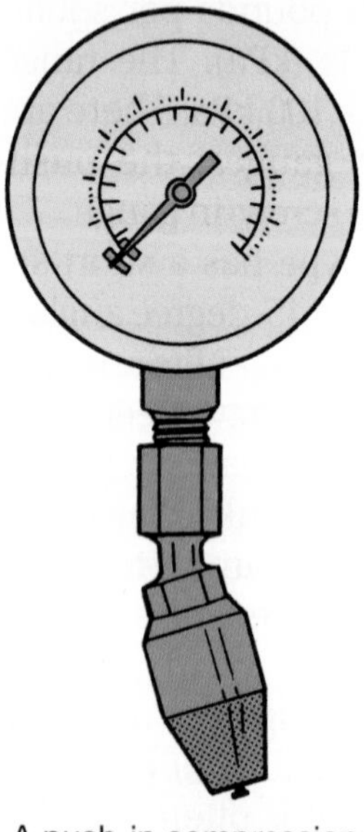

Figure 5-2 A push-in compression gauge.

Opening the valve releases the pressure when the test is complete.

Cylinder Leakage Tester

If a compression test shows that any of the cylinders are leaking, a cylinder leakage test can be performed to measure the percentage of compression lost and help locate the source of leakage.

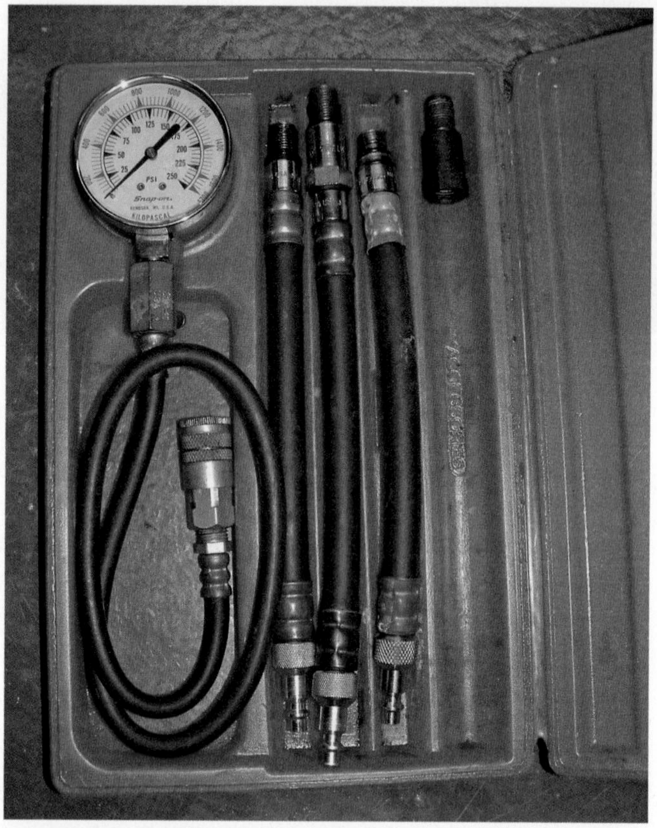

Figure 5-3 A screw-in compression gauge set.

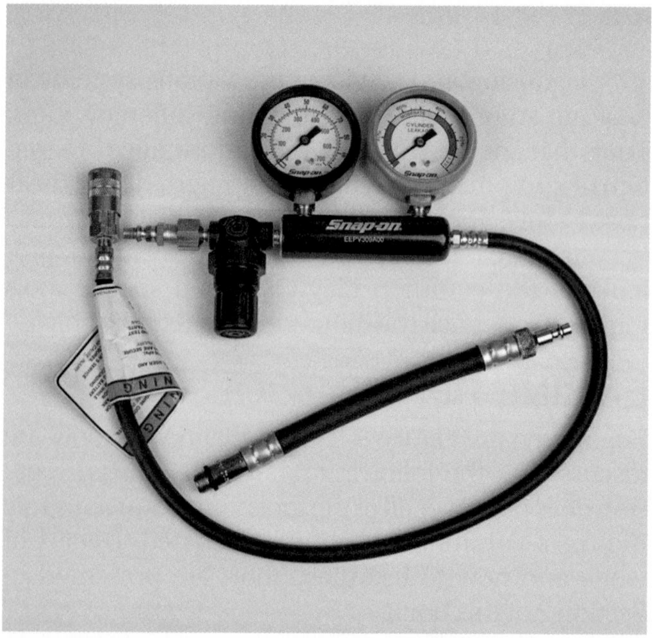

Figure 5-4 A cylinder leakage tester.

A **cylinder leakage tester (Figure 5–4)** applies compressed air to a cylinder, with the piston at the top of its bore, through the spark plug hole. A threaded adapter on the end of the air pressure hose screws into the spark plug hole. A pressure regulator in the tester controls the pressure applied to the cylinder. A gauge registers the percentage of air pressure lost from the cylinder when the compressed air is applied. The scale on the dial face reads 0% to 100%.

A zero reading means there is no leakage in the cylinder. A reading of 100% indicates that the cylinder will not hold any pressure. The location of the compression leak can be found by listening and feeling around various parts of the engine.

Oil Pressure Gauge

Checking the engine's oil pressure gives information about the condition of the oil pump, the pressure regulator, and the entire lubrication system. Lower-than-normal oil pressures can be caused by excessive engine bearing clearances. Oil pressure is checked at the sending unit passage with an externally mounted mechanical oil pressure gauge. Various fittings are usually supplied with the oil pressure gauge to fit different openings in the lubrication system.

Stethoscope

A **stethoscope** is used to locate the source of engine and other noises. The stethoscope pickup is placed on the suspected component, and the stethoscope receptacles are placed in the technician's ears **(Figure 5–5)**. Some sounds can be heard easily without using a listening device, but others are impossible to hear unless amplified, which is what a stethoscope does. It can also help you distinguish between normal and abnormal noise. The best results, however, are obtained with an electronic listening device. With this tool you can tune into the noise, which allows you to eliminate all other noises that might distract or mislead you.

Transaxle Removal and Installation Equipment

The engines of some FWD vehicles are removed by lifting them from the top. Others must be removed from the bottom and this requires special equipment. The required equipment varies with manufacturer and vehicle model; however, most accomplish the same thing.

To remove the engine from under the vehicle, the vehicle must be raised. A crane and/or support fixture is used to hold the engine and transaxle assembly in place while the engine is being readied for removal. Then the engine is lowered onto an engine cradle. The cradle is similar to a floor jack and lowers the engine further so it can be rolled out from under the vehicle.

Often a transverse-mounted engine is removed with the transaxle. The transaxle is separated from the engine after it has been removed.

Ridge Reamer

After many miles of use, a ridge is formed at the top of the engine's cylinders. Because the top piston ring stops traveling before it reaches the top of the cylinder, a ridge of unworn metal is left. This ridge must be removed to push the pistons out of the block without damaging them during engine rebuilding. This ridge is removed with a **ridge reamer (Figure 5–6)**. The tool is adjusted for the bore, inserted into it, and rotated with a wrench until the ridge is removed.

Ring Compressor

A **ring compressor** is used to install a piston into a cylinder bore. The compressor wraps around the rings to make their outside diameter smaller than the inside diameter of the bore. With the compressor tool adjusted properly, the piston assembly can be easily pushed into the bore without damaging the bore or piston.

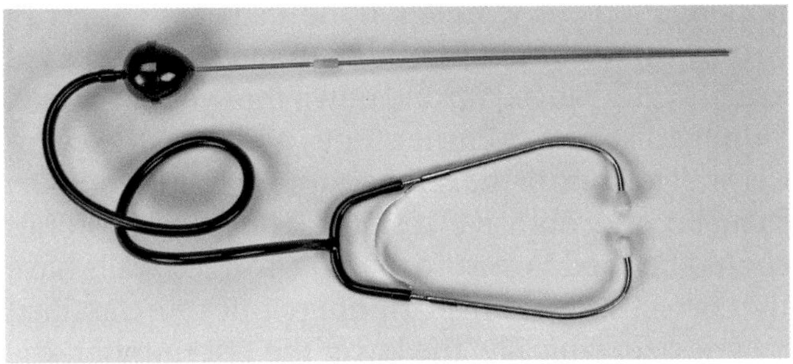

Figure 5-5 A stethoscope.

Figure 5-6 A ridge reamer.

Figure 5–7 A piston ring compressor. *Reproduced under license from Snap-on Incorporated. All of the marks are marks of their owners.*

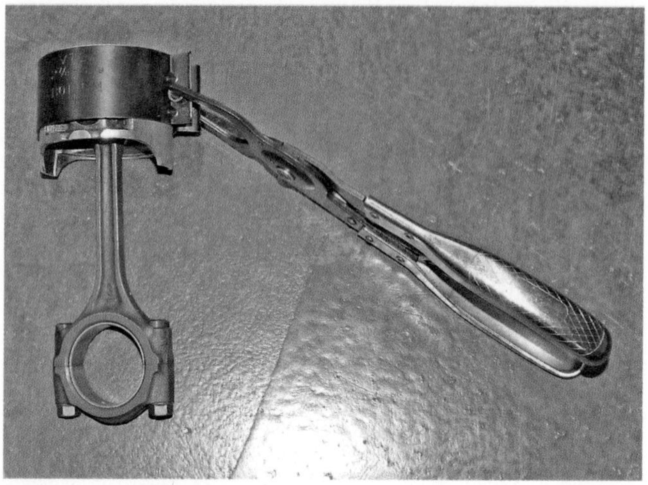

Figure 5–8 A piston ring compressor with a steel band and ratcheting pliers.

There are three basic types of ring compressors. One style has an adjustable band with a ratchet mechanism to tighten it around the piston **(Figure 5–7)**. Another style uses ratcheting pliers to tighten a steel band around the piston **(Figure 5–8)**. The bands are available in a variety of sizes. The third type has a single band that is wrinkled. The band is tightened by moving a lever. Once the rings are totally compressed into the piston, a thumbscrew is tightened to hold the band in position.

Ring Expander

To prevent damage to the piston rings during removal and installation, a **ring expander** should be used. To install a piston ring, the ring must be made large enough to fit over the piston. The rings fit into the jaws of the expander and the handle of the tool is squeezed to expand the ring **(Figure 5–9)**. Expand the rings only to the point where they can fit over the piston. Using an expander prevents the possibility of cracking or distorting the rings while they are being expanded. The tool also helps to prevent cut fingers caused by the edges of the rings.

Ring Groove Cleaner

Before installing piston rings onto a piston, the ring grooves should be cleaned. The carbon and other debris that may be present in the back of the groove will not allow the rings to compress evenly and com-

Figure 5–9 A piston ring expander.

pletely into the grooves. Piston ring grooves are best cleaned with a **ring groove cleaner**. This tool is adjustable to fit the width and depth of the groove. Make sure it is properly adjusted before using it and make sure you do not damage the piston while cleaning it.

Dial Bore Indicator

Cylinder bore taper and out-of-roundness can be measured with a micrometer and a telescoping gauge. However, most shops use a **dial bore gauge**. This gauge typically consists of a handle, guide blocks, a lock, an indicator contact, and an indicator. They also come with extensions that make them adaptable to various size bores. As the dial bore gauge is moved inside the bore, the indicator will show any change in the bore's diameter.

Cylinder Hone and Deglazer

The proper surface finish on a cylinder wall acts as a reservoir for oil to lubricate the piston rings and prevent piston and ring scuffing. On most late-model engines, sleeves or inserts into the engine block provide this finish, and when damaged or worn, the sleeves are replaced. On other engines, the cylinder walls can be refurbished. Always refer to the manufacturer's recommendation before servicing cylinder walls. When the walls have minor problems, the bore can be honed. Honing sands the walls to remove imperfections. A **cylinder hone** usually consists of two or three stones. The hone rotates at a selected speed and is moved up and down the cylinder's bore. Honing oil flows over the stones and onto the cylinder wall to control the temperature and flush out any metallic and abrasive residue. The correct stones should be used to ensure that the finished walls have the correct surface finish. Honing stones are classified by grit size; typically, the lower the grit number, the coarser the stone.

Cylinder honing machines are available in manual and automatic models. The major advantage of the automatic type is that it allows the technician to dial in the exact crosshatch angle needed.

If the cylinder walls have surface conditions, taper, and out-of-roundness that are within acceptable limits, the walls only need to be deglazed. Combustion heat, engine oil, and piston movement combine to form a thin residue on the cylinder walls that is commonly called glaze.

Most cylinder deglazers or **glaze breakers** use an abrasive with about 220 or 280 grit. The glaze breaker is installed in a slow-moving electric drill or in a honing machine. Many deglazers use round stones that extend on coiled wire from the center shaft. This type deglazer may also be used to lightly hone the bore. Various sizes of resilient-based hone-type brushes are available for honing and deglazing.

When cylinder surfaces are badly worn or excessively scored or tapered, a **boring bar** is used to cut the cylinders for oversize pistons or sleeves. A boring bar leaves a pattern similar to uneven screw threads. Therefore, the bore should be honed to the correct finish after it has been bored.

Cam Bearing Driver Set

The camshaft is supported by several friction-type bearings, or bushings. They are designed as one piece and are typically pressed into the camshaft bore in the cylinder head or block; however, some overhead camshaft (OHC) engines use split bearings to support the camshaft. Camshaft bearings are normally replaced during engine rebuilding. Cam bearings are normally press fit into the block or head using a bushing driver and hammer.

V-Blocks

The various shafts in an engine must be straight and not distorted. Visually it is impossible to see any distortions unless the shaft is severely damaged. Warped or distorted shafts will cause many problems, including premature wear of the bearings they ride on. The best way to check a shaft is to place the ends of the shaft onto V-blocks. These blocks will support the shaft and allow you to rotate the shaft. Place the plunger of a dial indicator on the journals of the shaft and rotate the shaft. Any movement of the indicator's needle suggests a problem.

Valve and Valve Seat Resurfacing Equipment

Whenever the valves have been removed from the cylinder head, the valve heads and valve seats should be resurfaced. The most critical sealing surface in the valvetrain is between the face of the valve and its seat in the cylinder head. Leakage between these surfaces reduces the engine's compression and power and can lead to valve burning. To ensure proper seating of the valve, the seat area on the valve face and seat must be the correct width, at the correct location, and concentric with the guide. These conditions are accomplished by renewing the surface of the valve face (**Figure 5–10**) and seat.

Figure 5–10 A valve grinding machine.

Valve and valve seat grinding or refacing is done by machining with a grinding stone or metal cutters to achieve a fresh, smooth surface on the valve faces and stem tips. Valve faces suffer from burning, pitting, and wear caused by opening and closing millions of times during the life of an engine. Valve stem tips wear because of friction from the rocker arms or actuators.

Valve Guide Repair Tools

The amount of valve guide wear can be measured with a ball gauge and micrometer. If wear or taper is excessive, the guide must be machined or replaced. If the original guide can be removed and a new one inserted, press out the old valve guide with a properly sized driver. Then install a new guide with a press and the same driver.

Some technicians knurl the old guide to restore the inside diameter dimension (ID) of a worn valve guide. Knurling raises the surface on the inside of the guide by plowing tiny furrows through the surface. This effectively decreases the ID of the guide. A burnisher is used to flatten the ridges and produce a proper-sized hole to restore the correct guide-to-stem clearance.

Reaming is often done to increase a guide's ID to accept an oversized valve stem or a guide insert. Some valve guide liners or inserts are not precut to length and the excess must be milled off before finishing.

Figure 5–11 A typical spring compressor for OHC valves.

Figure 5–12 A C-clamp valve spring compressor.

Valve Spring Compressor

To remove the valves from a cylinder head, the valve spring assemblies must be removed. To do this, the valve spring must be compressed enough to remove the valve keepers, then the retainer. There are many types of **valve spring compressors**. Some are designed to allow valve spring removal while the cylinder head is still on the engine block. Other designs are only used when the cylinder head is removed.

The prybar-type compressor is used while the cylinder head is still mounted to the block. With the cylinder's piston at TDC, shop air is fed into the cylinder to hold the valve up and prevent it falling into the cylinder.

Some OHC engines require the use of a special spring compressor **(Figure 5–11)**. Often these special tools can be used when the cylinder head is attached to the block and when it is on a bench. They bolt to the cylinder head and have a threaded plunger that fits onto the retainer. As the plunger is tightened down on the retainer, the spring compresses.

C-clamp-type spring compressors can only be used on cylinder heads after they have been removed **(Figure 5–12)**. These are normally equipped with interchangeable jaws and can be pneumatically or manually operated. One end of the clamp is positioned on the valve head and the other on the valve's retainer. After the compressor is adjusted, it is activated to squeeze down on the spring. Once the spring is compressed, the valve keepers can be removed. Then the tension of the compressor is slowly released and the valve retainer and spring can be removed.

Valve Spring Tester

Before valve springs are reused, they should be checked to make sure they are within specifications. This checking should include their freestanding height and squareness. If those two dimensions are good, the spring should be checked with a **valve spring tester**. A valve spring tester checks each valve's open and close pressure. Correct close pressure guarantees a tight seal. The open pressure overcomes valvetrain inertia and closes the valve when it should close. The tester's gauge reflects the pressure of the spring when it is compressed to the installed or valve-closed height. Read the pressure on the tester and compare this reading to specifications. Any pressure outside the pressure range given in the specifications indicates the spring should be replaced.

Torque Angle Gauge

Most manufacturers recommend the torque-angle method for tightening cylinder head bolts, which requires the use of a **torque angle gauge**. Typically two steps are involved: Tighten the bolt to the specified torque, and then tighten the bolt an additional amount. The latter is expressed in degrees. To accurately measure the number of degrees added to the bolt, a torque angle gauge **(Figure 5–13)** is attached to the wrench. The additional tightening will stretch the

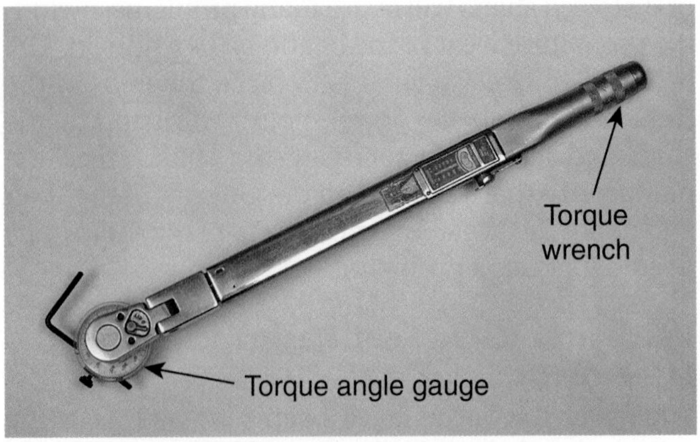

Figure 5–13 A torque angle gauge attached to the drive lug of a torque wrench.

bolt and produce a very reliable clamp load that is much higher than can be achieved just by torquing.

Oil Priming Tool

Prior to starting a freshly rebuilt engine, the oil pump must be primed. There are several ways to prelubricate, or prime, an engine. One method is to drive the oil pump with an electric drill. With some engines, it is possible to make a drive that can be chucked in an electric drill motor to engage the drive on the oil pump. Insert the fabricated oil pump drive extension into the oil pump through the distributor drive hole. To control oil splash, loosely set the valve cover(s) on the engine. After running the oil pump for several minutes, remove the valve cover and see whether there is any oil flow to the rocker arms. If oil reached the cylinder head, the engine's lubrication system is full of oil and is operating properly. If no oil reached the cylinder head, there is a problem either with the pump, with the alignment of an oil hole in a bearing, or perhaps with a plugged gallery.

Using a prelubricator **(Figure 5–14)**, which consists of an oil reservoir attached to a continuous air supply, is the best method of prelubricating an engine without running it. When the reservoir is attached to the engine and the air pressure is turned on, the prelubricator will supply the engine's lubrication system with oil under pressure.

Figure 5-14 An engine preluber kit. *Courtesy of SPX Service Solutions*

Figure 5-15 Cooling system pressure tester.

Cooling System Pressure Tester

A cooling system pressure tester **(Figure 5–15)** contains a hand pump and a pressure gauge. A hose is connected from the hand pump to a special adapter that fits on the radiator filler neck. This tester is used to pressurize the cooling system and check for coolant leaks. Additional adapters are available to connect the tester to the radiator cap. With the tester connected to the radiator cap, the pressure relief action of the cap may be checked.

Coolant Hydrometer

A coolant **hydrometer** is used to check the amount of antifreeze in the coolant. This tester contains a pickup hose, a coolant reservoir, and a squeeze bulb. The pickup hose is placed in the radiator coolant. When the squeeze bulb is squeezed and released, coolant is drawn into the reservoir. As coolant enters the reservoir, a float moves upward with the coolant level. A pointer on the float indicates the freezing point of the coolant on a scale located on the reservoir housing.

Refractory Testers For many shops, the preferred way to check coolant is with a refractometer **(Figure 5–16)**. This tester works on the principle that light bends as it passes through a liquid. A sample of the coolant is placed in the tester. As light passes through the sample of coolant, it bends and shines on a scale in the tester. A reading is taken at the point on the scale where there a separation of light and dark. Most refractory coolant testers can also check the electrolyte in a battery.

Measuring pH Acids produced by bacteria and other contaminants can reduce the effectiveness of coolant. Some shops measure the pH of coolant to determine deterioration of the coolant. The pH is measured

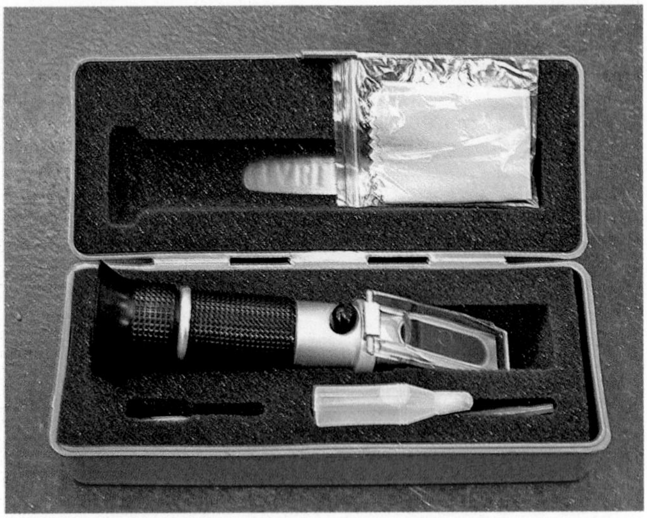

Figure 5-16 A refractometer checks the condition of the engine's coolant.

by placing test strips or a digital pH tester into the coolant.

Coolant Recovery and Recycle System

A coolant recovery and recycle machine **(Figure 5–17)** typically can drain, recycle, fill, flush, and pressure

Figure 5-17 A coolant recycling machine that drains, back flushes, and fills the cooling system.

test a cooling system. Usually additives are mixed into the used coolant during recycling. These additives either bind to contaminants in the coolant so they can be easily removed, or they restore some of the chemical properties in the coolant.

ELECTRICAL/ELECTRONIC SYSTEM TOOLS

Electrical system service and diagnostic tools are discussed in the following paragraphs. This discussion does not cover all of the tools you may need; rather, these tools are the most commonly used by the service industry. Many automotive systems are electrically controlled and operated; therefore, these tools are also used in those systems. Details of when and how to use these tools are presented in Section 3 of this book, as well as in the sections that discuss the various other automotive systems.

Computer Memory Saver

Memory savers are an external power source used to maintain the memory circuits in electronic accessories and the engine, transmission, and body computers when the vehicle's battery is disconnected. The saver is plugged into the vehicle's cigar lighter outlet. It can be powered by a 9- or 12-volt battery **(Figure 5–18)**.

Circuit Tester

Circuit testers **(Figure 5–19)** are used to check for voltage in an electrical circuit. A circuit tester, commonly called a **testlight**, looks like a stubby ice pick. Its handle is transparent and contains a light bulb. A probe extends from one end of the handle and a ground clip and wire from the other end. When the ground clip is attached to a good ground and the probe touched to a live connector, the bulb in the

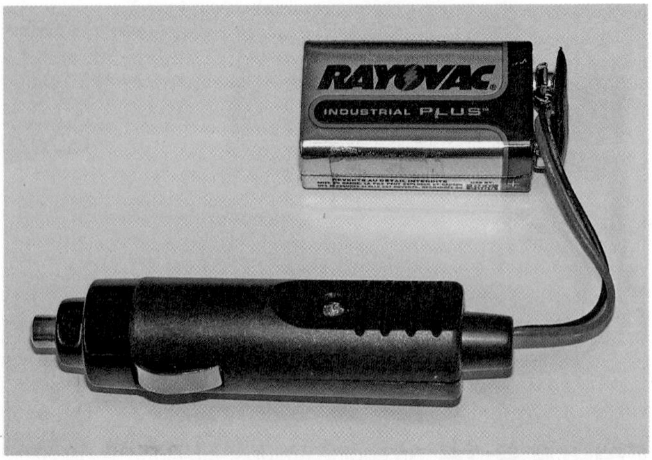

Figure 5-18 A simple 9-volt memory saver.

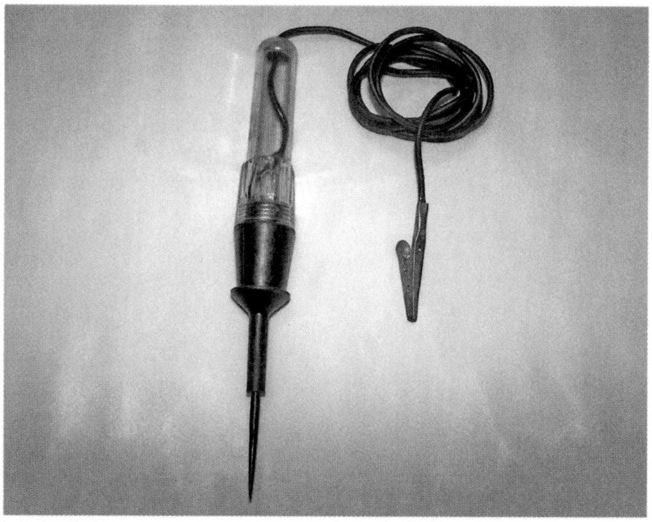

Figure 5-19 Typical circuit tester, commonly called a testlight.

handle will light up. If the bulb does not light, voltage is not available at the connector.

> **WARNING!**
>
> *Never use a 12 V testlight to diagnose components and wires in computer systems. The current draw of these testlights may damage the computer system components. High-impedance testlights are available for diagnosing computer systems.*

A self-powered testlight is called a **continuity tester**. It is used on open circuits. It looks like a regular testlight but has a small internal battery. When the ground clip is attached to one end of the wire or circuit and the probe touched to the other end, the lamp will light if there is continuity in the circuit. If an open circuit exists, the light will not illuminate.

> **WARNING!**
>
> *Do not use any type of testlight or circuit tester to diagnose automotive air bag systems. Use only the vehicle manufacturer's recommended equipment on these systems.*

Voltmeter

A **voltmeter** has two leads: a red positive lead and a black negative lead. The red lead should be connected to the positive side of the circuit or component. The black should be connected to ground or to the negative side of the component. Voltmeters should be connected across the circuit being tested.

A voltmeter measures the voltage available at any point in an electrical system. A voltmeter can also be used to test voltage drop across an electrical circuit, component, switch, or connector. A voltmeter can also be used to check for proper circuit grounding.

Ohmmeter

An **ohmmeter** measures resistance to current flow in a circuit. In contrast to the voltmeter, which operates by the voltage available in the circuit, an ohmmeter is battery powered. The circuit being tested must have no power applied. If the power is on in the circuit, the ohmmeter will be damaged.

The two leads of the ohmmeter are placed across or in parallel with the circuit or component being tested. The red lead is placed on the positive side of the circuit and the black lead is placed on the negative side of the circuit. The meter sends current through the component and determines the amount of resistance based on the voltage dropped across the load. The scale of an ohmmeter reads from 0 to infinity (∞). A 0 reading means there is no resistance in the circuit and may indicate a short in a component that should show a specific resistance. An infinity reading indicates a number higher than the meter can measure, which usually indicates an open circuit.

Ammeter

An **ammeter** measures current flow in a circuit. Current is measured in amperes. Unlike the voltmeter and ohmmeter, the ammeter must be placed into the circuit or in series with the circuit being tested. Normally, this requires disconnecting a wire or connector from a component and connecting the ammeter between the wire or connector and the component. The red lead of the ammeter should always be connected to the side of the connector closest to the positive side of the battery and the black lead should be connected to the other side.

It is much easier to test current using an ammeter with an inductive pickup (**Figure 5–20**). The pickup clamps around the wire or cable being tested. The ammeter determines amperage based on the magnetic field created by the current flowing through the wire. This type of pickup eliminates the need to separate the circuit to insert the meter.

Volt/Ampere Tester

A volt/ampere tester (VAT), shown in **Figure 5–21**, is used to test batteries, starting systems, and charging systems. The tester contains a voltmeter, ammeter, and carbon pile. The carbon pile is a variable resistor.

Figure 5-20 An ammeter with an inductive pickup is called a current probe. *Courtesy of SPX Service Solutions*

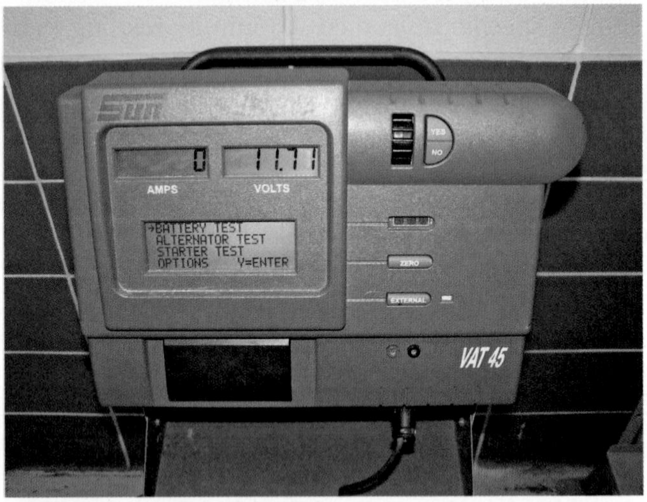

Figure 5-21 The volt/amp tester checks batteries and the starting and charging systems.

When the tester is attached to the battery and turned on, the carbon pile draws current out of the battery. The ammeter will read the amount of current draw. The maximum current draw from the battery, with acceptable voltage, is compared to the rating of the battery to see if the battery is okay. A VAT also measures the current draw of the starter and current output from the charging system.

Multimeters

A **multimeter** is a must for diagnosing the individual components of an electrical system. Multimeters

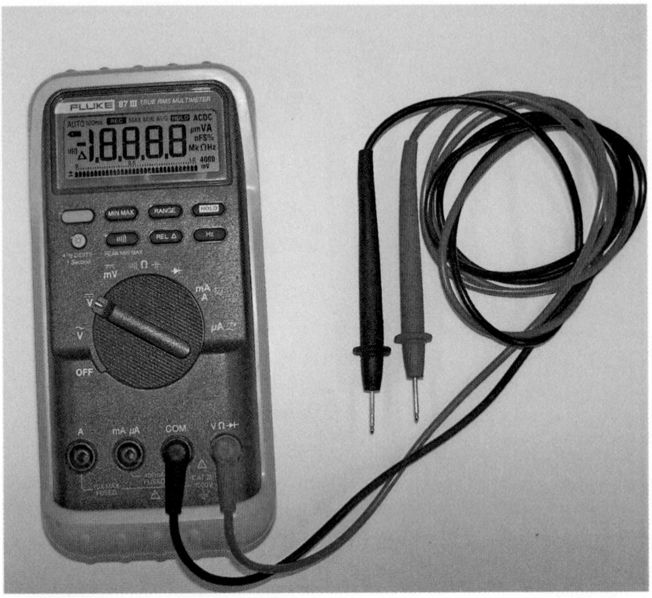

Figure 5-22 Typical multifunctional, low-impedance multimeter.

have different names, depending on what they measure and how they function. A volt-ohm-milliamp meter is referred to as a VOM or DVOM, if it is digital. A **digital multimeter (DMM)** can measure many more things than volts, ohms, and low current.

Most multimeters **(Figure 5–22)** measure direct current (dc) and alternating current (ac) amperes, volts, and ohms. More advanced multimeters may also measure diode continuity, frequency, temperature, engine speed, and dwell, and/or duty cycle.

Multimeters are available with either digital or analog displays. DMMs provide great accuracy by measuring volts, ohms, or amperes in tenths, hundredths, or thousandths of a unit. Several test ranges are usually provided for each of these functions. Some meters have multiple test ranges that must be manually selected; others are autoranging.

Analog meters use a sweeping needle against a scale to display readings and are not as precise as digital meters. Analog meters have low input impedance and should not be used on sensitive electronic circuits or components. Digital meters have high impedance and can be used on electronic circuits as well as electrical circuits.

Lab Scopes

An oscilloscope or **lab scope** is a visual voltmeter **(Figure 5–23)**. A lab scope converts electrical signals to a visual image representing voltage changes over a specific period of time. This information is displayed in the form of a continuous voltage line called a **waveform** or **trace**. With a scope, precise measurement is possible. A scope displays any change in voltage as it occurs.

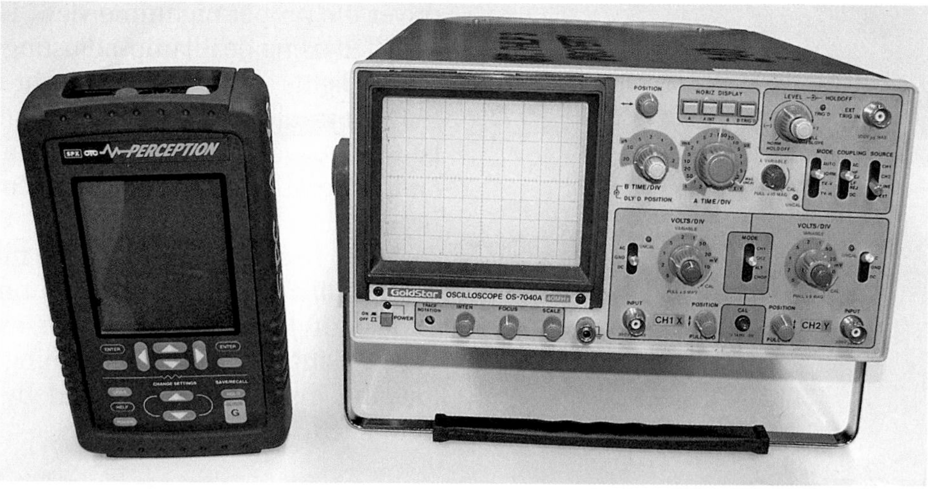

Figure 5-23 Two types of lab scopes.

An upward movement of the voltage trace on an oscilloscope indicates an increase in voltage, and a downward movement of this trace represents a decrease in voltage. As the voltage trace moves across an oscilloscope screen, it represents a specific length of time.

The size and clarity of the displayed waveform is dependent on the voltage scale and the time reference selected. Most scopes are equipped with controls that allow voltage and time interval selection. It is important when choosing the scales to remember that a scope displays voltage over time.

Dual-trace scopes can display two different waveform patterns at the same time. This type scope is especially important for diagnosing intermittent problems.

Most new lab scopes can display more than two waveforms and are hand-held units. Some scopes have an electronic library of known good signals, which allow technicians to compare what they see with what they should be seeing. Some also include wiring diagrams and additional diagnostic and testing information.

Graphing Multimeter

One of the latest trends in diagnostic tools is a graphing digital multimeter. These meters display readings over time, similar to a lab scope. The graph displays the minimum and maximum readings on a graph, as well as displaying the current reading **(Figure 5–24)**. By observing the graph, a technician can detect any undesirable changes during the transition from a low reading to a high reading, or vice versa. These glitches are some of the more difficult problems to identify without a graphing meter or a lab scope.

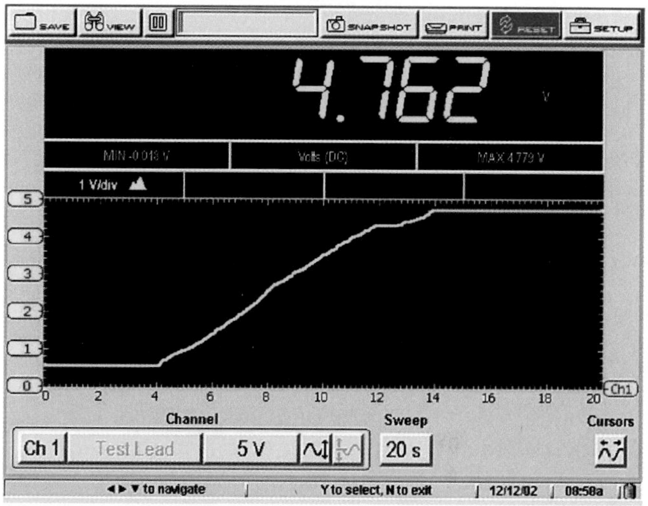

Figure 5-24 The screen of a graphing multimeter. *Reproduced under license from Snap-on Incorporated. All of the marks are marks of their owners.*

Battery Hydrometer

On unsealed batteries, the specific gravity of the electrolyte can be measured to give a fairly good indication of the battery's state of charge. A hydrometer **(Figure 5–25)** is used to perform this test. A battery hydrometer consists of a glass tube or barrel, rubber bulb, rubber tube, and a glass float or hydrometer with a scale built into its upper stem. The glass tube encases the float and forms a reservoir for the test electrolyte. Squeezing the bulb pulls electrolyte into the reservoir.

When filled with test electrolyte, the hydrometer float bobs in the electrolyte. The depth to which the glass float sinks in test electrolyte indicates its relative weight compared to water. The reading is taken off the scale by sighting along the level of the electrolyte.

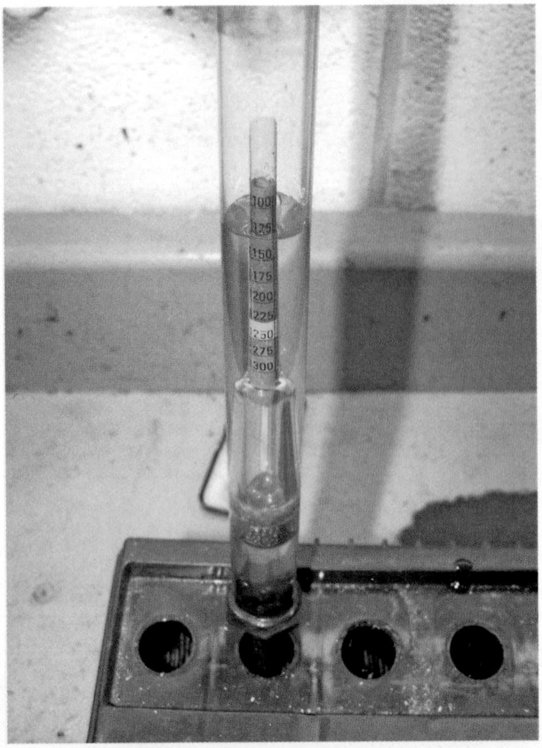

Figure 5-25 A hydrometer is used to measure the specific gravity of a battery's electrolyte.

> **WARNING!**
>
> *Be careful not to allow the battery's electrolyte to drip on you or the vehicle. The electrolyte is a very strong acid and can burn your skin and damage the vehicle's paint.*

Wire and Terminal Repair Tools

Many automotive electrical problems can be traced to faulty wiring. Loose or corroded terminals; frayed, broken, or oil-soaked wires; and faulty insulation are the most common causes.

Wires and connectors are often repaired or replaced. Sometimes an entire length of wire is replaced; other times only a section is. In either case, the wire must have the correct terminal or connector to work properly in the circuit. Wire cutters, stripping tools, terminal crimpers, and **connector picks** are the most commonly used tools for wire repair. Also, soldering equipment is used to provide the best electrical connection for a wire to another wire and for a wire to a connector.

Headlight Aimers

Headlights must be kept in adjustment to obtain maximum illumination. Sealed beams that are properly adjusted cover the correct range and afford the driver the proper nighttime view. Headlights can be adjusted using headlamp-adjusting tools or by shining the lights on a chart. Headlight-aiming tools give the best results with the least amount of work. Many late-model vehicles have levels built into the headlamp assemblies that are used to correctly adjust the headlights.

Most headlight aimers use mirrors with split images, like split-image finders on some cameras, and spirit levels to determine exact adjustment. When using any headlight-aiming equipment, follow the instructions provided by the equipment manufacturer.

ENGINE PERFORMANCE TOOLS

Diagnostic and special tools for the air, fuel, ignition, emission, and engine-control systems are discussed in the following paragraphs. This discussion does not cover all of the tools you may need; rather, these tools are the most commonly used by the service industry. Some are also used when diagnosing or servicing the controls of other automotive systems. Details of when and how to use these tools are covered in Section 4 of this book, as well as in other sections where necessary.

Scan Tools

A **scan tool (Figure 5–26)** is a microprocessor designed to communicate with the vehicle's computer. Connected to the computer through diagnostic connectors, a scan tool can access diagnostic trouble codes (DTCs), run tests to check system operations, and monitor the activity of the system. Trouble codes and test results are displayed on a screen or printed out on the scanner printer.

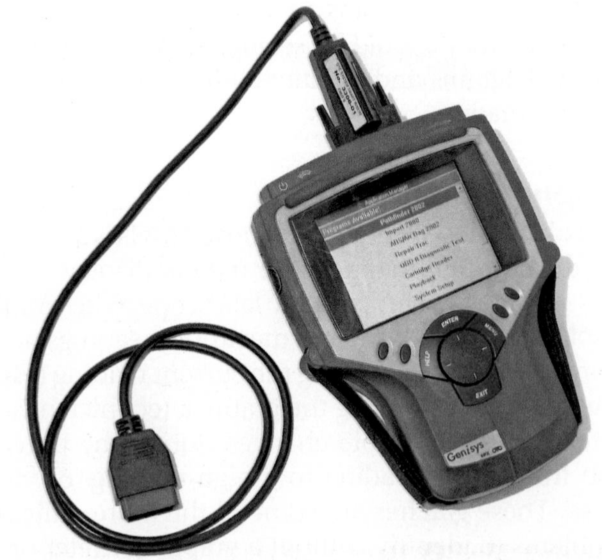

Figure 5-26 A Genysis scan tool.

Figure 5-27 Various scan tool connectors for OBD-II systems.

The scan tool is connected to specific diagnostic connectors on the vehicle. Some manufacturers have one diagnostic connector. This connects the data wire from each computer to a specific terminal in this connector. Other manufacturers have several diagnostic connectors on each vehicle, and each of these connectors may be connected to one or more computers **(Figure 5–27)**. The scan tool must be programmed for the model year, make of vehicle, and type of engine.

With OBD-II, the diagnostic connectors (commonly called data link connectors, or DLCs) are located in the same place on all vehicles. Also, any scan tool designed for OBD-II will work on all OBD-II systems; therefore, the need to have designated scan tools or cartridges is eliminated. Most OBD-II scan tools have the ability to store, or "freeze," data during a road test **(Figure 5–28)** and then play back this data when the vehicle is returned to the shop.

There are many different scan tools available. Some are a combination of other diagnostic tools,

Figure 5-28 Using a scan tool during a road test.
Courtesy of SPX Service Solutions

such as a lab scope and graphing multimeter. These may have the following capabilities:

- Retrieve DTCs.
- Monitor system operational data.
- Reprogram the vehicle's electronic control modules.
- Perform systems diagnostic tests.
- Display appropriate service information, including electrical diagrams.
- Display TSBs.
- Display troubleshooting instructions.
- Perform easy tool updating through a personal computer (PC).

Some scan tools work directly with a PC through uncabled communication links, such as Bluetooth. Others use a Personal Digital Assistant (PDA). These are small hand-held units that allow you to read DTCs, monitor the activity of sensors, and view inspection/maintenance system test results to quickly determine what service the vehicle requires. Most of these scan tools also have the ability to:

- Perform system and component tests.
- Report test results of monitored systems.
- Exchange files between a PC and a PDA.
- View and print files on a PC.
- Print DTC/freeze frame.
- Generate emissions reports.
- View IM/Mode 6 information.
- Display relative TSBs.
- Display full diagnostic code descriptions.
- Observe live sensor data.
- Update the scan tool as a manufacturer's interfaces change.

Engine Analyzers

When performing a complete engine performance analysis, an engine analyzer is used **(Figure 5–29)**. An engine analyzer houses all of the necessary test equipment. With an engine analyzer, you can perform tests on the battery, starting system, charging system, primary and secondary ignition circuits, electronic control systems, fuel system, emissions system, and engine assembly. The analyzer is connected to these systems by a variety of leads, inductive clamps, probes, and connectors. The data received from these connections is processed by several computers within the analyzer.

Most engine analyzers have both manual and automatic test modes. In the manual modes, any

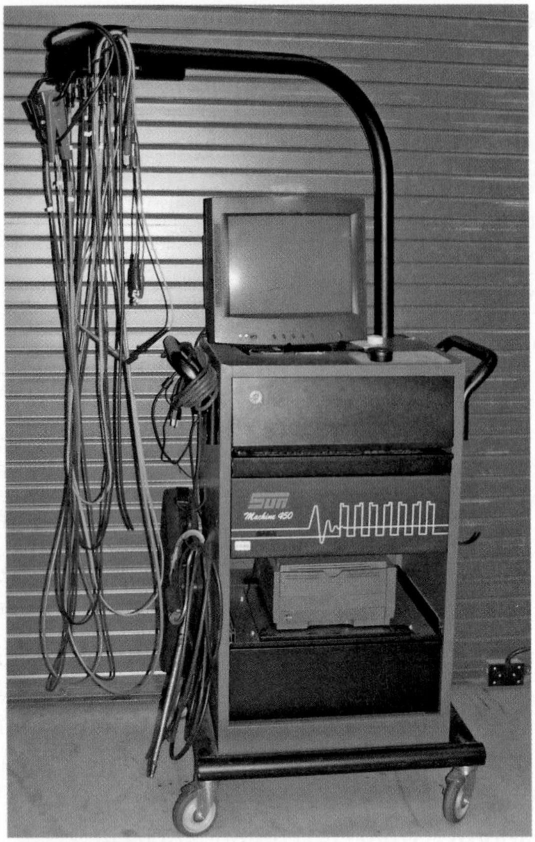

Figure 5-29 An engine analyzer.

single test, such as cylinder compression or generator output, can be performed. When the automatic test mode is selected, specific tests are automatically performed in a specific sequence.

The analyzer may compare the test results to the vehicle manufacturer's specifications. When the test series is completed, the analyzer prints a summary report. Many analyzers also provide diagnostic assistance for the problems indicated by their measurements.

Fuel Pressure Gauge

A fuel **pressure gauge** is essential for diagnosing fuel injection systems **(Figure 5–30)**. These systems rely on very high fuel pressures, from 35 to 70 psi. A drop in fuel pressure reduces the amount of fuel delivered to the injectors and results in a lean air-fuel mixture.

A fuel pressure gauge is used to check the discharge pressure of fuel pumps, the regulated pressure of fuel injection systems, and injector pressure drop. This test can identify faulty pumps, regulators, or injectors and can identify restrictions present in the fuel delivery system. Restrictions are typically caused by a dirty fuel filter, collapsed hoses, or damaged fuel lines.

Some fuel pressure gauges also have a valve and outlet hose for testing fuel pump discharge volume.

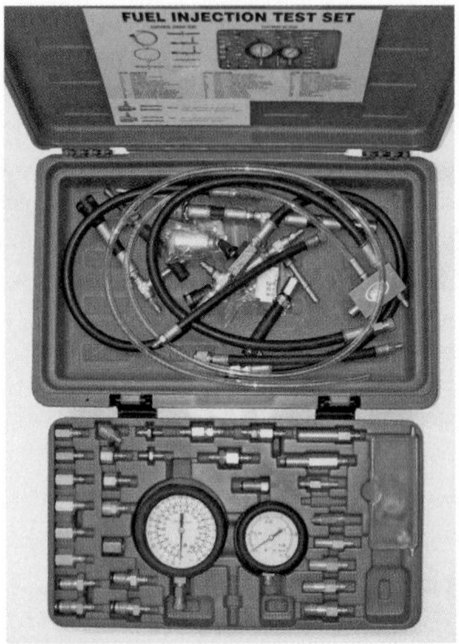

Figure 5-30 A fuel pressure gauge and adapters.

The manufacturer's specification for discharge volume is given as a number of pints or liters of fuel that should be delivered in a certain number of seconds.

CAUTION!

While testing fuel pressure, be careful not to spill gasoline. Gasoline spills may cause explosions and fires, resulting in serious personal injury and property damage.

Injector Balance Tester

The injector balance tester **(Figure 5–31)** is used to test the injectors in a port fuel injected engine for proper operation. A fuel pressure gauge is also used during the injector balance test. The injector balance tester contains a timing circuit, and some injector balance testers have an off-on switch. A pair of leads on the tester must be connected to the battery with the correct polarity. The injector terminals are

Figure 5-31 A fuel injection balance tester.

disconnected, and a second double lead on the tester is attached to the injector terminals.

The fuel pressure gauge is connected to the Schrader valve on the fuel rail, and the ignition switch should be cycled two or three times until the specified fuel pressure is indicated on the pressure gauge. When the tester push button is depressed, the tester energizes the injector winding for a specific length of time, and the technician records the pressure decrease on the fuel pressure gauge. This procedure is repeated on each injector.

If the pressure drops very little, or if there is no pressure drop, the injector's orifice is restricted or the injector is faulty. If there is an excessive amount of pressure drop, the injector plunger is sticking open. Sticking injector plungers may result in a rich air-fuel mixture.

WARNING!

Electronic fuel injection systems are pressurized, and these systems require depressurizing prior to fuel pressure testing and other service procedures.

Injector Circuit Testlight

A special testlight called a **noid light** can be used to determine if a fuel injector is receiving its proper voltage pulse from the computer **(Figure 5–32)**. The wiring harness connector is disconnected from the injector and the noid light is plugged into the connector. After disabling the ignition to prevent starting, the engine is turned over by the starter motor. The noid light will flash rapidly if the voltage signal is present.

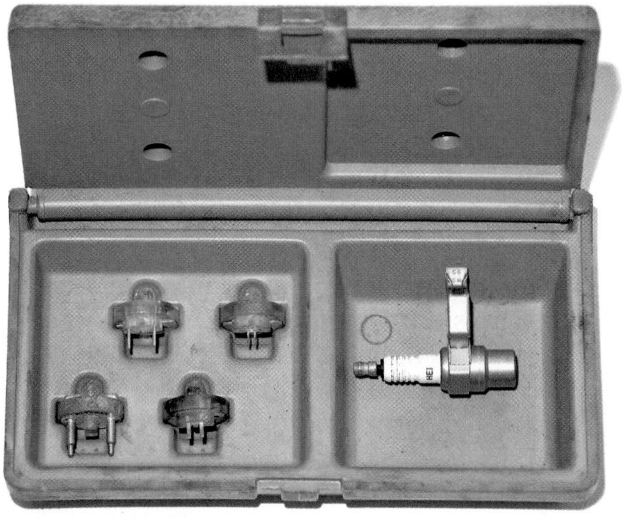

Figure 5-32 A set of noid lights and a test spark plug.

No flash usually indicates an open in the power feed or ground circuit to the injector.

Fuel Injector Cleaners

Fuel injectors spray a certain amount of fuel into the intake system. If the fuel pressure is low, not enough fuel will be sprayed. Low pressure may also occur if the fuel injector is dirty. Normally, clogged injectors are the result of inconsistencies in gasoline detergent levels and the high sulfur content of gasoline. When these sensitive fuel injectors become partially clogged, fuel flow is restricted. Spray patterns are altered, causing poor performance and reduced fuel economy.

The solution to a sulfated and/or plugged fuel injector is to clean it, not replace it. There are two kinds of fuel injector cleaners. One is a pressure tank. A mixture of solvent and unleaded gasoline is placed in the tank, following the manufacturer's instructions for mixing, quantity, and safe handling. The vehicle's fuel pump is disabled and, on some vehicles, the fuel line must be blocked between the pressure regulator and the return line. Then, the hose on the pressure tank is connected to the service port in the fuel system. The inline valve is then partially opened and the engine is started. It should run at approximately 2,000 rpm for about 10 minutes to clean the injectors thoroughly.

An alternative to the pressure tank is a pressurized canister in which the solvent solution is premixed. Use of the canister-type cleaner is similar to this procedure but does not require mixing or pumping. The canister is connected to the injection system's servicing fitting, and the valve on the canister is opened. The engine is started and allowed to run until it dies. Then, the canister is discarded.

Fuel Line Tools

Many vehicles are equipped with quick-connect line couplers. These work well to seal the connection but are nearly impossible to disconnect if the correct tools are not used. There is a variety of quick-connect fittings and tools **(Figure 5–33)**.

Pinch-Off Pliers

The need to pinch off a rubber hose is common during diagnostics and service. Special pliers are designed to do this without damaging the hose **(Figure 5–34)**. These pliers are much like vise-grip pliers in that they hold their position until they are released. The jaws of the pliers are flat and close in a parallel motion. Both of these features prevent damage to the hose.

Vacuum Gauge

Measuring intake manifold vacuum is another way to diagnose the condition of an engine. Manifold

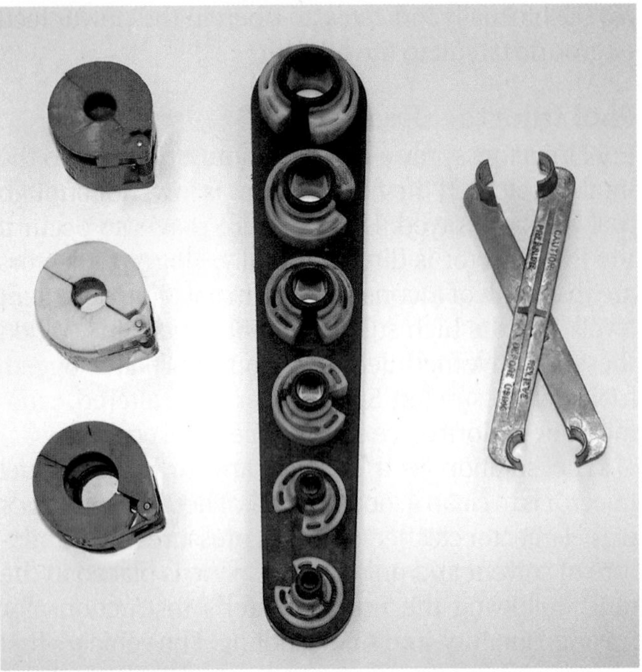

Figure 5-33 Various fuel line disconnect tools.

Figure 5-34 Pinch-off pliers closing off a vacuum hose.

vacuum is tested with a vacuum gauge. **Vacuum** is formed on a piston's intake stroke. As the piston moves down, it lowers the pressure of the air in the cylinder—if the cylinder is sealed. This lower cylinder pressure is called engine vacuum. If there is a leak, atmospheric pressure will force air into the cylinder and the resultant pressure will not be as low. The reason atmospheric pressure enters is simply that whenever there is a low and high pressure, the high pressure always moves toward the low pressure. Vacuum is measured in inches of mercury (in. Hg) and in kilopascals (kPa).

To measure vacuum, a flexible hose on the vacuum gauge **(Figure 5–35)** is connected to a source of manifold vacuum, either on the manifold or at a point below the throttle plates. The test is made with the engine cranking or running. A good vacuum reading is typically at least 16 in. Hg. However, a reading of 15 to 20 in. Hg (50 to 65 kPa) is normally acceptable. Since the intake stroke of each cylinder occurs at a different time, the production of vacuum occurs in pulses. If the amount of vacuum produced by each cylinder is the same, the vacuum gauge will show a steady reading. If one or more cylinders are producing different amounts of vacuum, the gauge will show a fluctuating reading.

Figure 5-35 A vacuum/pressure tester used to measure engine vacuum.

Vacuum Pump

There are many vacuum-operated devices and vacuum switches on cars. These devices use engine vacuum to cause a mechanical action or to switch something on or off. The tool used to test vacuum-actuated components is the vacuum pump. There are two types of vacuum pumps: an electrical-operated pump and a hand-held pump. The hand-held pump is most often used for diagnostics **(Figure 5–36)**. A hand-held vacuum pump consists of a hand pump, a vacuum gauge, and a length of rubber hose used to attach the pump to the component being tested. Tests with the vacuum pump can usually be performed without removing the component from the vehicle.

When the handles of the pump are squeezed together, a piston inside the pump body draws air out of the component being tested. The partial vacuum created by the pump is registered on the pump's vacuum gauge. While forming a vacuum in a component, watch the action of the component. The vacuum level needed to actuate a given component should be compared to the specifications given in the factory service manual.

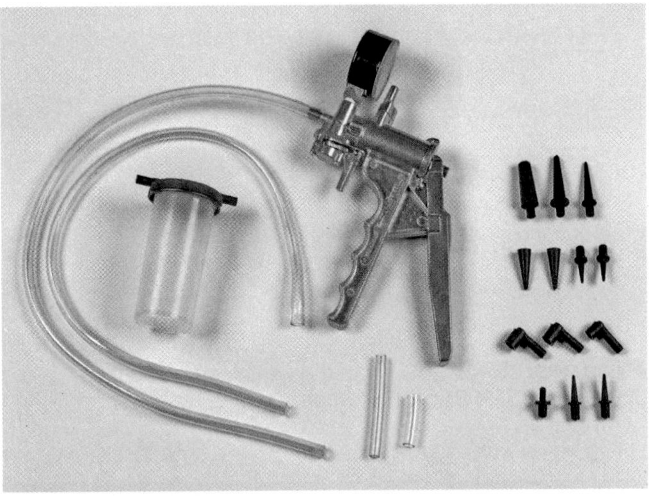

Figure 5-36 Typical hand-operated vacuum pump with accessories.

The vacuum pump is also used to locate vacuum leaks by connecting the vacuum pump to a suspect vacuum hose or component and applying vacuum. If the needle on the vacuum gauge begins to drop after the vacuum is applied, a leak exists somewhere in the system.

Vacuum Leak Detector

A vacuum or compression leak might be revealed by a compression check, a cylinder leakage test, or a manifold vacuum test. However, finding the location of the leak can often be very difficult.

A simple, but time-consuming, way to find leaks in a vacuum system is to check each component and vacuum hose with a vacuum pump. Simply apply vacuum to the suspected area and watch the gauge for any loss of vacuum. A good vacuum component holds the vacuum applied to it.

Another method of leak detection is done by using an ultrasonic leak detector. Air rushing through a vacuum leak creates a high-frequency sound higher than the range of human hearing. An ultrasonic leak detector is designed to hear the frequencies of the leak. When the tool is passed over a leak, the detector responds to the high-frequency sound by emitting a warning beep. Some detectors also have a series of light emitting diodes (LEDs) that light up as the frequencies are received. The closer the detector is moved to the leak, the more LEDs light up or the faster the beeping occurs, allowing the technician to zero in on the leak. An ultrasonic leak detector can sense leaks as small as 1/500 inch and accurately locate the leak to within 1/16 inch.

Tachometer

A **tachometer** is used to measure engine speed. Like other meters, tachometers are available in analog and digital types. Digital meters are the most common. Tachometers are connected to the ignition system to monitor ignition pulses, which are then converted to engine speed by the meter.

Several types of inductive pickup tachometers that simplify rpm testing are available. An inductive tachometer simply clamps over the number 1 spark plug wire. The digital display gives the engine rpm, based on the magnetic pulses created by the secondary voltage in the wire. This type of tachometer is suitable for distributorless ignition systems.

Timing Light

A **timing light** (**Figure 5–37**) is used to check ignition timing. The timing light is connected to the battery terminals and has an inductive clamp that fits over the number 1 spark plug wire. While the engine is running, the timing light emits a beam of light each time the spark plug fires. Many timing lights have a timing advance knob that may be used to check spark advance. Some timing lights electronically measure timing advance as the engine rpm is increased and displays it on an LED display.

Spark Tester

A **spark tester** (see Figure 5–32) is a fake spark plug. The tester is constructed like a spark plug but does not have a ground electrode. In place of the electrode

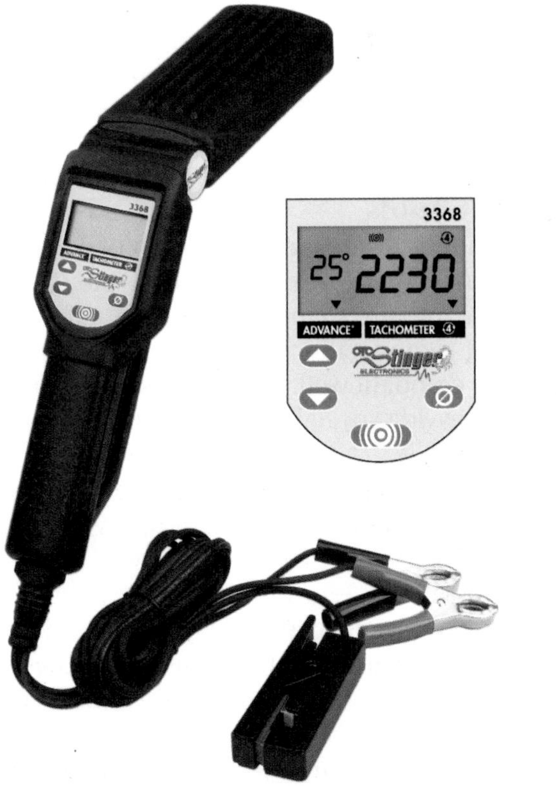

Figure 5-37 A digital timing light. *Courtesy of SPX Service Solutions*

there is a grounding clamp. Using test spark plugs is an easy way to determine if the ignition problem is caused by something in the primary or secondary circuit.

The spark tester is inserted in the spark plug end of an ignition cable. When the engine is cranked, a spark should be seen from the tester to a ground. Experience with these testers will also help you determine the intensity of the spark.

Logic Probes

In some circuits, pulsed or digital signals pass through the wires. These on-off digital signals either carry information or provide power to drive a component. Many sensors, used in a computer-controlled circuit, send digital information back to the computer. To check the continuity of the wires that carry digital signals, a logic probe can be used.

A **logic probe** is similar in appearance to a test-light. It contains three different-colored LEDs. A red LED lights when there is high voltage at the point being probed. A green LED lights to indicate low voltage. A yellow LED indicates the presence of a voltage pulse. The logic probe is powered by the circuit and reflects only the activity at the point being probed. When the probe's test leads are attached to a circuit, the LEDs display the activity.

If a digital signal is present, the yellow LED turns on. When there is no signal, the LED is off. If voltage is present, the red or green LEDs will light, depending on the amount of voltage. When there is a digital signal and the voltage cycles from low to high, the yellow LED will be lit and the red and green LEDs will cycle, indicating a change in the voltage.

Sensor Tools

Oxygen sensors are replaced as part of the preventive maintenance program and when they are faulty. Because they are shaped much like a spark plug with wires or a connector coming out of the top, ordinary sockets do not fit well. For this reason, tool manufacturers provide special sockets for these sensors (**Figure 5–38**).

Special sockets are also available for other sending units and sensors.

Static Strap

Because electronic components are sensitive to voltage, static electricity can destroy them. Static straps are available for technicians to wear while working on or around electronic components. These straps typically are worn around a wrist and connected to a known good ground on the vehicle. The straps send all static electricity to the ground of the vehicle, thereby eliminating the chance of this electricity going to the electronic components.

Figure 5–38 A heated oxygen sensor socket. *Courtesy of SPX Service Solutions*

Figure 5–39 A hand-held digital infrared temperature gauge.

Pyrometers

The converter should be checked for its ability to convert CO and HC into CO_2 and water by doing a delta temperature test. To conduct this test, use a hand-held digital **pyrometer** (**Figure 5–39**). By touching the pyrometer probe or placing it near to the exhaust pipe just ahead of and just behind the converter, there should be an increase of at least 100°F or 8% above the inlet temperature reading as the exhaust gases pass through the converter. If the outlet temperature is the same or lower, nothing is happening inside the converter. A pyrometer can also be used to measure the temperature of the coolant at various stages of its travel.

Spark Plug Sockets

Special sockets are available for the installation and removal of spark plugs (**Figure 5–40**). These sockets are deep sockets with a hex nut drive at the end to

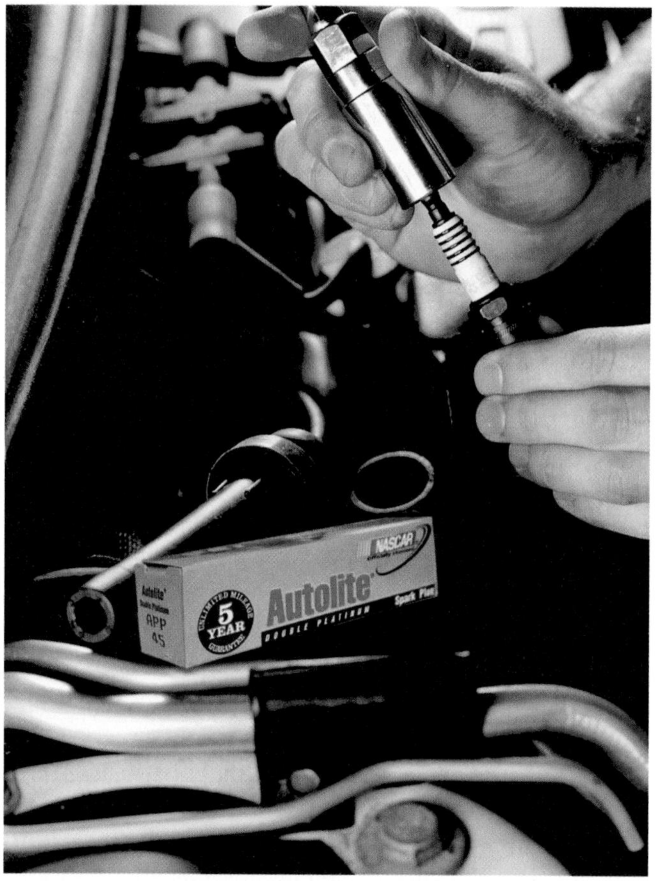

Figure 5-40 A spark plug socket. *Courtesy of Honeywell International Inc.*

Figure 5-41 A five-gas exhaust analyzer.

allow a technician to turn them with a ratchet or an open-end wrench. The sockets are available in the common sizes of spark plugs (5/8-inch, 9/16-inch, and 13/16-inch) and have a 3/8-inch drive. The socket is built with a rubber sleeve that surrounds the insulator part of the spark plug to prevent cracking or other damage to the plug while it is being removed or installed.

Exhaust Analyzers

Federal laws require that new cars and light trucks meet specific emissions levels. State governments have also passed laws requiring that owners maintain their vehicles so that the emissions remain below an acceptable level. Most states require an annual emissions inspection to meet that goal. Many shops have an exhaust analyzer for inspection purposes.

Exhaust analyzers **(Figure 5–41)** are also very valuable diagnostic tools. By looking at the quality of an engine's exhaust, a technician is able to look at the engine's combustion process and the efficiency of the vehicle's emission controls. Any defect will cause emission levels to increase. The amount and type of change is considered during diagnostics.

Exhaust analyzers measure the amount of HC and CO in the exhaust. HC is measured in parts per million (ppm) or grams per mile (g/mi) and CO is measured as a percent of the total exhaust. In addition to measuring HC and CO levels, an exhaust analyzer also monitors CO_2 and O_2 levels. Many exhaust analyzers also measure a fifth gas **(Figure 5–42)**, oxides of nitrogen (NO_x).

By measuring NO_x, CO_2, and O_2, in addition to HC and CO, a technician gets a better look at the engine's efficiency. There is a desired relationship among the

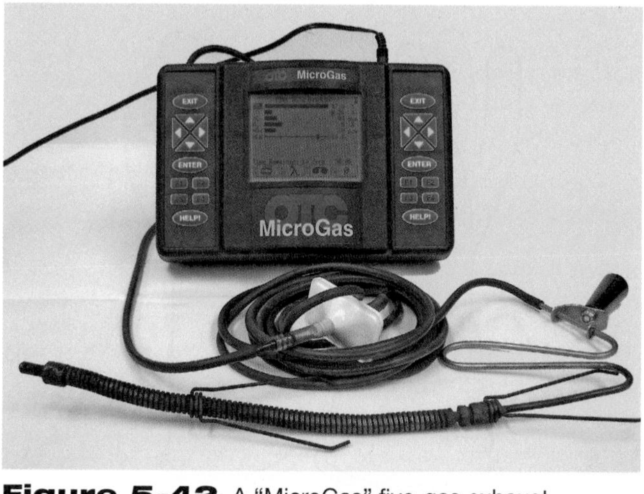

Figure 5-42 A "MicroGas" five-gas exhaust analyzer.

five gases. Any deviation from this relationship can be used to diagnose a driveability problem.

Chassis Dynamometer

A chassis **dynamometer**, commonly called a dyno, is used to simulate a road test. A vehicle can be driven through a wide assortment of operating conditions without leaving the shop. Because the vehicle is stationary, test equipment can be connected and monitored while the vehicle is driven under various loads. This is extremely valuable when diagnosing a problem. A chassis dyno can also be used for performance tuning.

The vehicle's drive wheels are positioned on large rollers. The electronically controlled rollers offer rotational resistance to simulate the various loads a vehicle may face.

Some performance shops have an engine dynamometer that directly measures the output from an engine. A chassis dynamometer measures the engine's output after it has passed through the driveline.

Hybrid Tools

A hybrid vehicle is an automobile and as such is subject to many of the same problems as a conventional vehicle. Most systems in a hybrid vehicle are diagnosed in the same way as well. However, a hybrid vehicle has unique systems that require special procedures and test equipment. It is imperative to have good information before attempting to diagnose these vehicles. Also, make sure you follow all test procedures precisely as they are given.

Gloves Always wear safety gloves when working on or around the high-voltage systems. These gloves must be class "0" rubber insulating gloves **(Figure 5–43)**, rated at 1,000 volts (these are commonly called "**lineman's gloves**"). Also, to protect the integrity of the insulating gloves, as well as you, wear leather gloves over the insulating gloves while doing a service.

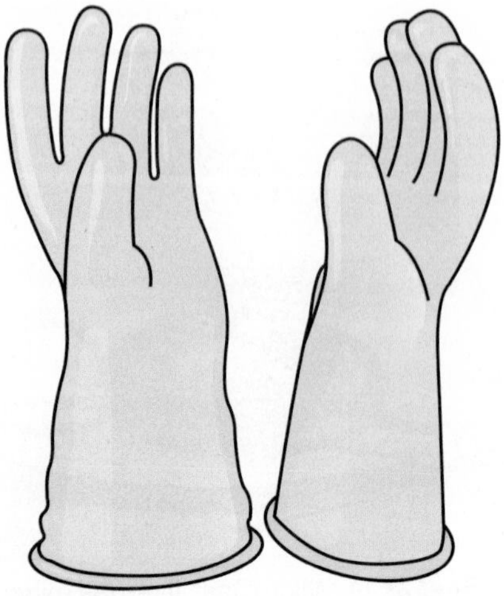

Figure 5-43 A pair of lineman's gloves.

CAUTION!

The condition of the gloves must be checked before each use. Make sure there are no tears or signs of wear. Electrons are very small and can enter through the smallest of holes in your gloves. To check the condition of the gloves, blow enough air into each one so they balloon out. Then fold the open end over to seal the air in. Continue to slowly fold that end of the glove toward the fingers. This will compress the air. If the glove continues to balloon as the air is compressed, it has no leaks. If any air leaks out, the glove should be discarded. All gloves, new and old, should be checked before they are used.

Test Equipment An important diagnostic tool is a DMM. However, this is not the same DMM used on a conventional vehicle. The meter used on hybrids (and EVs and FCEVs) should be classified as a category III meter. There are basically four categories for low-voltage electrical meters, each built for specific purposes and to meet certain standards. Low voltage, in this case, means voltages less than 1,000 volts. The categories define how safe a meter is when measuring certain circuits. The standards for the various categories are defined by the American National Standards Institute (ANSI), the International Electrotechnical Commission (IEC), and the Canadian Standards Association (CSA). A CAT III meter **(Figure 5–44)** is required for testing hybrid vehicles because of the high voltages, three-phase current, and the potential for high transient voltages. **Transient voltages** are voltage surges or spikes that occur in AC circuits. To be safe, you should have a CAT III 1000 V meter. A meter's voltage rating reflects its ability to withstand transient voltages. Therefore, a CAT III 1000 V meter offers much more protection than a CAT III meter rated at 600 volts.

Figure 5-44 Only meters with this symbol should be used on the high-voltage systems in a hybrid vehicle.

WARNING!

Always follow the test procedures defined by the manufacturers when using their equipment.

Another important tool is an **insulation resistance tester**. These can check for voltage leakage through the insulation of the high-voltage cables. Obviously no leakage is desired and any leakage can cause a safety hazard as well as damage to the vehicle. Minor leakage can also cause hybrid system-related driveability problems. This meter is not one commonly used by automotive technicians but should be for anyone who might service a damaged hybrid vehicle, such as doing body repair. This should also be a CAT III meter and may be capable of checking resistance and voltage of circuits like a DMM.

To measure insulation resistance, system voltage is selected at the meter and the probes placed at their test position. The meter will display the voltage it detects. Normally, resistance readings are taken with the circuit de-energized unless you are checking the effectiveness of the cable or wire insulation. In this case, the meter is measuring the insulation's effectiveness and not its resistance.

The probes for the meters should have safety ridges or finger positioners. These help prevent physical contact between your fingertips and the meter's test leads.

TRANSMISSION AND DRIVELINE TOOLS

The repair and diagnostic tools for manual and automatic transmissions, as well as those required for driveline service, are discussed in the following paragraphs. This discussion does not cover all of the tools you may need; rather, these tools are the most commonly used by the service industry. Details of when and how to use these tools are covered in Sections 5 and 6 of this book.

Transaxle Removal and Installation Equipment

The removal and replacement (R&R) of transaxles mounted to transversely mounted engines may require different tools than those needed to remove a transmission from a RWD vehicle. Make sure you follow the manufacturer's instructions before attempting to remove the engine and/or transaxle from a FWD vehicle.

To remove the engine and transmission from under the vehicle, the vehicle must be raised. A crane and/or support fixture is used to hold the engine and transaxle assembly in place while the assembly is being readied for removal. When everything is set for removal of the assembly, the crane is used to lower the assembly onto a cradle. The cradle is similar to a hydraulic floor jack and is used to lower the assembly further so it can be rolled out from under the vehicle. The transaxle can be separated from the engine once it has been removed from the vehicle.

When the transaxle is removed as a single unit, the engine must be supported while it is in the vehicle before, during, and after transaxle removal. Special fixtures **(Figure 5–45)** mount to the vehicle's upper frame or suspension parts. These supports have a bracket that is attached to the engine. With the bracket in place, the engine's weight is on the support fixture and the transmission can be removed.

Transmission/Transaxle Holding Fixtures

Special holding fixtures should be used to support the transmission or transaxle after it has been removed from the vehicle **(Figure 5–46)**. These holding fixtures may be stand-alone units or may be bench mounted.

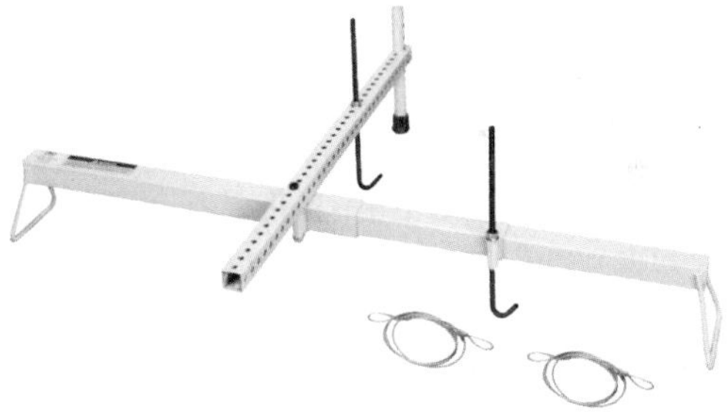

Figure 5-45 An engine support is used to hold the engine in place while the transaxle is removed on many FWD vehicles. *Courtesy of SPX Service Solutions*

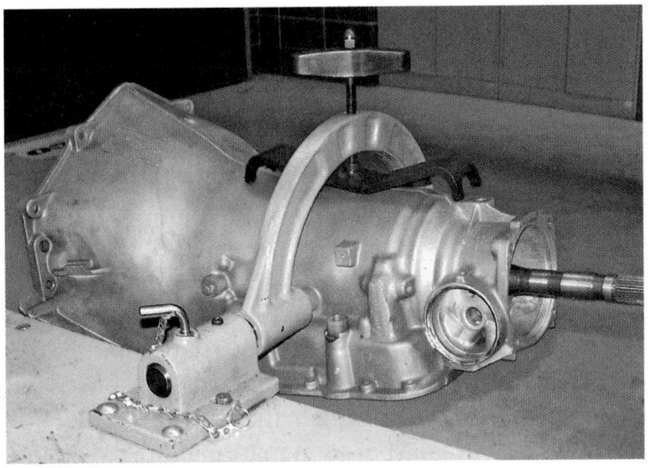

Figure 5-46 A transmission holding fixture.

Figure 5-47 A typical transmission jack.

They allow the transmission to be easily repositioned during repair work.

Transmission Jack

A transmission jack **(Figure 5–47)** is designed to help you while removing a transmission from under the vehicle. The weight of the transmission makes it difficult and unsafe to remove it without much assistance and/or a transmission jack. These jacks fit under the transmission and are usually equipped with hold-down chains, which are used to secure the transmission to the jack. The transmission's weight rests on the jack's saddle.

Transmission jacks are available in two basic styles. One is used when the vehicle is raised by a hydraulic jack and sitting on jack stands. The other style is used when the vehicle is raised on a lift.

Axle Pullers

Axle pullers are used to pull rear axles in RWD vehicles. Most rear axle pullers are slide hammer type.

Special Tool Sets

Vehicle manufacturers and specialty tool companies work closely together to design and manufacture special tools required to repair transmissions. Most of these special tools are listed in the appropriate service manuals and are part of each manufacturer's "essential tool kit."

Figure 5-48 A clutch alignment tool set with various sizes of pilots, adapters, and alignment cones. *Courtesy of SPX Service Solutions*

Clutch Alignment Tool

To keep the clutch disc centered on the flywheel while assembling the clutch, a clutch alignment tool is used. The tool is inserted through the input shaft opening of the pressure plate and is passed through the clutch disc. The tool then is inserted into the pilot bushing or bearing. The outside diameter (OD) of the alignment tool that goes into the pilot must be only slightly smaller than the ID of the pilot bushing. The OD of the tool that holds the disc in place must likewise be only slightly smaller than the ID of the disc's splined bore. The effectiveness of this tool depends on its diameter, so it is best to have various sizes of clutch alignment tools **(Figure 5–48)**.

Clutch Pilot Bearing/Bushing Puller/Installer

To remove and install a clutch pilot bearing or bushing, special tools are needed. These tools not only make the job easier but also prevent damage to the bore in the flywheel.

Universal Joint Tools

Although servicing universal joints can be done with hand tools and a vise, many technicians prefer the use of specifically designed tools. One such tool is a C-clamp modified to include a bore that allows the joint's caps to slide in while tightening the clamp over an assembled joint to remove it **(Figure 5–49)**. Other

Figure 5-49 A universal joint bearing press with adapters. *Reproduced under license from Snap-on Incorporated. All of the marks are marks of their owners.*

Figure 5-50 An angle gauge is used to check the angle of a drive shaft. *Reproduced under license from Snap-on Incorporated. All of the marks are marks of their owners.*

tools are the various drivers used with a press to press the joint in and out of its yoke.

Drive Shaft Angle Gauge

Critical to the durability of universal joints and vibration-free vehicle operation is the angle of the drive shaft. The angle of the drive shaft at the transmission should equal its angle at the drive axle. There are many ways to measure the angle; one way involves the use of an **inclinometer** or drive shaft angle gauge **(Figure 5–50)**.

Hydraulic Pressure Gauge Set

A common diagnostic tool for automatic transmissions is a hydraulic pressure gauge **(Figure 5–51)**. A pressure gauge measures pressure in pounds per square inch (psi) and/or kilopascals (kPa). The gauge is normally part of a kit that contains various fittings and adapters.

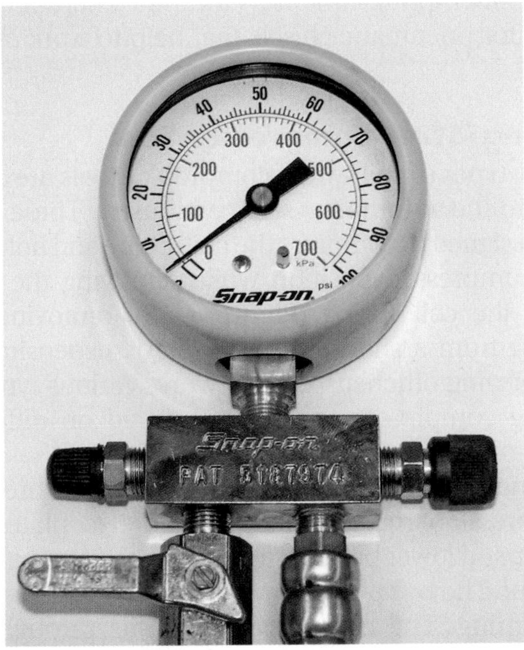

Figure 5-51 Pressure gauges are used to diagnose automatic transmissions and power steering systems.

SUSPENSION AND STEERING TOOLS

Suspension and steering repair and diagnostic tools as well as wheel alignment tools and equipment are discussed in the following paragraphs. This discussion does not cover all of the tools you may need; rather, these tools are the most commonly used by the service industry. Details of when and how to use these tools are covered in Section 7 of this book.

Tire Tread Depth Gauge

A tire tread depth gauge measures tire tread depth. This measurement should be taken at three or four locations around the tire's circumference to obtain an average tread depth. This gauge is used to determine the remaining life of a tire as well as for comparing wear of one tire to the other tires. It is also used when making tire warranty adjustments.

Power Steering Pressure Gauge

A power steering pressure gauge is used to test the power steering pump pressure. This test is also important when checking hydraulic boost brake systems. Because the power steering pump delivers extremely high pressure during this test, the recommended procedure in the vehicle manufacturer's service manual must be followed.

A pressure gauge with a shutoff valve is installed between the pump and the steering gear. Adapters are used to make good connections with the vehicle's power steering system.

Control Arm Bushing Tools

A variety of control arm bushing tools are available to remove and replace control arm bushings. Old bushings are pressed out of the control arm. A C-clamp tool can be used to remove the bushing. The C-clamp is installed over the bushing. An adapter is selected to fit on the bushing and push the bushing through the control arm. Turning the handle on the C-clamp pushes the bushing out of the control arm.

New bushings can be installed by driving or pressing them in place. Adapters are available for the C-clamp tool to install the new bushings. After the correct adapters are selected, position the bushing and tool on the control arm. Turning the C-clamp handle pushes the bushing into the control arm.

Tie-Rod End and Ball Joint Puller

Some car manufacturers recommend a tie-rod end and ball joint puller to remove tie-rod ends and pull ball joint studs from the steering knuckle.

Ball joint removal and pressing tools are designed to remove and replace pressed-in ball joints on front

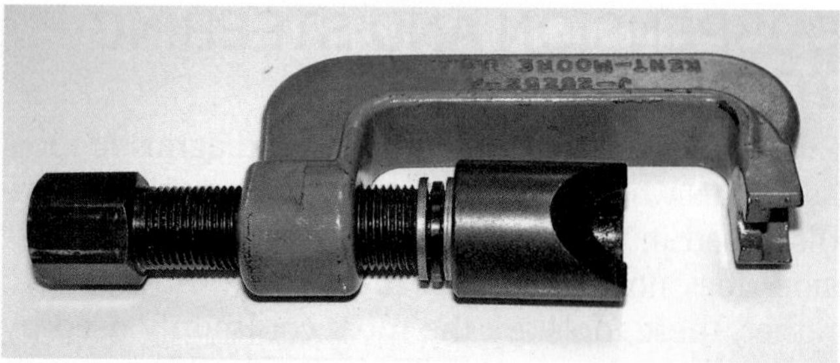

Figure 5-52 A ball joint removal tool.

suspension systems **(Figure 5–52)**. Often these tools are used in conjunction with a hydraulic press. The size of the removal and pressing tool must match the size of the ball joint.

Some ball joints are riveted to the control arm and the rivets are drilled out for removal.

Front Bearing Hub Tool

Front bearing hub tools are designed to remove and install front wheel bearings on FWD cars. These bearing hub tools are usually designed for a specific make of vehicle and the correct tools must be used for each application. Failure to do so may result in damage to the steering knuckle or hub. Also, the use of the wrong tool will waste quite a bit of your time.

Pitman Arm Puller

A pitman arm puller is a heavy-duty puller designed to remove the pitman arm from the pitman shaft **(Figure 5–53)**. These pullers can also be used to separate tie-rod ends and ball joints.

Tie-Rod Sleeve-Adjusting Tool

A tie-rod sleeve-adjusting tool **(Figure 5–54)** is required to rotate the tie-rod sleeves and perform some front wheel adjustments. Never use anything except a tie-rod adjusting tool to adjust the tie-rod sleeves. Tools such as pipe wrenches will damage the sleeves.

Figure 5-53 A pitman arm puller is designed to remove the pitman arm from the pitman shaft. *Courtesy of SPX Service Solutions*

Figure 5-54 A tie-rod sleeve-adjusting tool. *Courtesy of SPX Service Solutions*

Steering Column Special Tool Set

A wheel puller is used to remove the steering wheel from its shaft. Mount the puller over the wheel's hub after the horn button and air bag have been removed. Make sure you follow the recommendations exactly for air bag module removal. Screw the bolts into the threaded bores in the steering wheel. Then tighten the puller's center bolt against the steering wheel shaft until the steering wheel is free.

Special tools are also required to service the lock mechanism and ignition switch.

Shock Absorber Tools

Often shock absorbers can be removed with regular hand tools, but there are times when special tools may be necessary. The shocks are under the vehicle and are subject to dirt and moisture, which may make it difficult to loosen the mounting nut from the stud of the shock. Wrenches are available to hold the stud while attempting to loosen the nut. There are also tools for pneumatic chisels that help to work off the nut.

Spring/Strut Compressor Tool

Many types of coil spring compressor tools are available to the automotive service industry. These tools are designed to compress the coil spring and hold it in the compressed position while removing the strut from the coil spring **(Figure 5–55)**, removing the spring from a short-long arm (SLA) suspension, or performing other suspension work. Various types of spring compressor tools are required on different types of front suspension systems.

One type of spring compressor uses a threaded compression rod that fits through two plates, an upper and lower ball nut, a thrust washer, and a forcing nut. The two plates are positioned at either end of the spring. The compression rod fits through the plates with a ball nut at either end. The upper ball nut is pinned to the rod. The thrust washer and forcing

Figure 5-55 A spring compressor for a strut suspension.

nut are threaded onto the end of the rod. Turning the forcing nut draws the two plates together and compresses the spring.

There is a tremendous amount of energy in a compressed coil spring. Never disconnect any suspension component that will suddenly release this tension. This action could result in serious personal injury and vehicle or property damage.

Power Steering Pump Pulley Special Tool Set

When a power steering pump pulley must be replaced, it should never be hammered off or on. Doing so will cause internal damage to the pump. Normally the pulley can be removed with a gear puller, although special pullers are available. To install a pulley, a special tool is used to press the pulley on without a press or the need to drive the pulley in place.

Brake Pedal Depressor

A brake pedal depressor must be installed between the front seat and the brake pedal to apply the brakes

Figure 5-56 The technician is installing a brake pedal depressor; also note that a steering wheel lock is in place.

while checking some front wheel alignment angles to prevent the vehicle from moving **(Figure 5–56)**.

Wheel Alignment Equipment—Four Wheel

Many automotive shops are equipped with a computerized four-wheel alignment machine **(Figure 5–57)** that can check all front- and rear-wheel alignment angles quickly and accurately.

After vehicle information is keyed into the machine and the wheel units are installed, the machine must be compensated for wheel runout. When compensation is complete, alignment measurements are instantly displayed. Also displayed are the specifications for that vehicle. In addition to the normal alignment specifications, the screen may display asymmetric tolerances, different left- and right-side specifications, and cross specifications. (A difference is allowed between left and right sides.) Graphics and text on the screen show the technician where and

Figure 5-57 A computerized four-wheel alignment setup. *Courtesy of RTI Technologies, Inc.*

how to make adjustments. As the adjustments are made on the vehicle, the technician can observe the center block slide toward the target. When the block aligns with the target, the adjustment is within half the specified tolerance.

Tire Changer

Tire changers are used to demount and mount tires. A wide variety of tire changers are available, and each one has somewhat different operating procedures. Always follow the procedure in the equipment operator's manual and the directions provided by your instructor.

Wheel Balancer—Electronic Type

The most commonly used wheel balancer requires that the tire/wheel assembly be taken off and mounted on the balancer's spindle **(Figure 5–58)**. Weights are added to balance the tire/wheel assembly. The wheel assembly is rotated at high speed and the machine indicates the amount of weight to be added and the location where the weights should be placed.

Several electronic dynamic/static balancer units are available that permit balancing while the wheel and tire are on the vehicle. Often a strobe light flashes at the heavy point of the tire and wheel assembly.

Figure 5-58 An electronic wheel balancer.

Figure 5-59 Wheel weight pliers.

Wheel Weight Pliers

Wheel weight pliers are actually combination tools designed to install and remove clip-on lead wheel weights **(Figure 5–59)**. The jaws of the pliers are designed to hook into a hole in the weight's bracket. The pliers are then moved toward the outside of the wheel and the weight is pried off. On one side of the pliers is a plastic hammer head used to tap the weights onto the rim.

BRAKE SYSTEM TOOLS

The repair and diagnostic tools for brake service are discussed in the following paragraphs. This discussion does not cover all of the tools you may need; rather, these tools are the most commonly used by the service industry. Details of when and how to use these tools are presented in Section 8 of this book.

Cleaning Equipment and Containment Systems

Equipment should be used to safely contain asbestos while doing brake work. A negative-pressure enclosure and high-efficiency particulate air (HEPA) vacuum system allow you to clean and inspect brake assemblies while preventing the release of asbestos fibers into the air. A vacuum pump and a HEPA filter keep the enclosure under negative pressure as work is done.

Low-pressure wet cleaning systems wash dirt from the brake assembly and catch the contaminated cleaning agent in a basin. This system uses water mixed with an organic solvent or wetting agent. The brake assembly is gently flooded to prevent any asbestos-containing brake dust from becoming airborne.

Figure 5-60 Brake spring pliers.

Holddown Spring and Return Spring Tools

Brake shoe return springs used on drum brakes are very strong and require special tools for removal and installation. Most return spring tools have special sockets and hooks to release and install the spring ends. Some are built like pliers **(Figure 5–60)**.

Holddown springs for brake shoes are much lighter than return springs, and many such springs can be released and installed by hand. A holddown spring tool **(Figure 5–61)** looks like a cross between a screwdriver and a nut driver. A specially shaped end grips and rotates the spring retaining washer.

Figure 5-61 A holddown spring compressor tool.

Boot Drivers, Rings, and Pliers

Dust boots attach between the caliper bodies and pistons of disc brakes to keep dirt and moisture out of the caliper bores. A special driver is used to install a dust boot with a metal ring that fits tightly on the caliper body. The circular driver is centered on the boot placed against the caliper and then hit with a hammer to drive the boot into place. Other kinds of dust boots fit into a groove in the caliper bore before the piston is installed. Special rings or pliers are then needed to expand the opening in the dust boot and let the piston slide through it for installation.

Caliper Piston Removal Tools

A caliper piston can usually be slid or twisted out of its bore by hand. Rust and corrosion (especially where road salt is used in the winter) can make piston removal difficult. One simple tool that will help with the job is a set of special pliers that grips the inside of the piston and lets you move it by hand with more force. These pliers work well on pistons that are only mildly stuck.

For a severely stuck caliper piston, a hydraulic piston remover can be used. This tool requires that the caliper be removed from the car and installed in a holding fixture. A hydraulic line is connected to the caliper inlet and a hand-operated pump is used to apply up to 1,000 psi of pressure to loosen the piston. Because of the danger of spraying brake fluid, always wear eye protection when using this equipment.

Drum Brake Adjusting Tools

Although almost all drum brakes built during the past 30 years have some kind of self-adjuster, the brake shoes still require an initial adjustment after they are installed. The star wheel adjusters of many drum brakes can be adjusted with a flat-blade screwdriver. Brake adjusting spoons **(Figure 5–62)** and wire hooks designed for this specific purpose can make the job faster and easier, however.

Brake Cylinder Hones

Cylinder hones are used to clean light rust, corrosion, pits, and built-up residue from the bores of master cylinders, wheel cylinders, and calipers. A hone can be a very useful—sometimes necessary—tool when you have to overhaul a cylinder. A hone will not, however, save a cylinder with severe rust or corrosion.

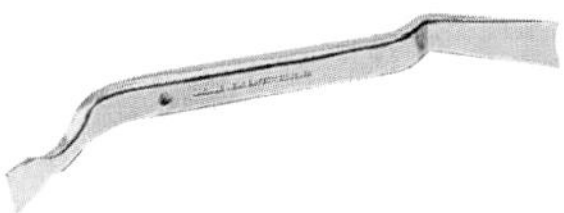

Figure 5-62 A drum brake adjustment tool. *Reproduced under license from Snap-on Incorporated. All of the marks are marks of their owners.*

The most common cylinder hones have two or three replaceable abrasive stones at the ends of spring-loaded arms. Spring tension usually is adjustable to maintain proper stone pressure against the cylinder walls. The other end of the hone is mounted in a drill motor for use, and the hone's flexible shaft lets the motor turn the hone properly without being precisely aligned with the cylinder bore.

Another kind of hone is the **brush** or **ball hone**. It has abrasive balls attached to flexible metal brushes that are, in turn, mounted on the hone's flexible shaft. In use, centrifugal force moves the abrasive balls outward against the cylinder walls; tension adjustment is not required. A brush hone provides a superior surface finish and is less likely to remove too much metal than a stone hone.

Tubing Tools

The rigid brake lines, or pipes, of the hydraulic system are made of steel tubing to withstand high pressure and to resist damage from vibration, corrosion, and work hardening. Brake lines often can be purchased in preformed lengths to fit specific locations on specific vehicles. Straight brake lines can also be purchased in many lengths and several diameters and bent to fit specific vehicle locations. Even with prefabricated lines available, you probably will have many occasions to cut and bend steel lines and form flared ends for installation. The common tools **(Figure 5–63)** you should have are:

Figure 5-63 A typical tubing tool set. *Reproduced under license from Snap-on Incorporated. All of the marks are marks of their owners.*

Figure 5-64 A digital caliper for measuring brake disc thickness. *Courtesy of Honeywell International Inc.*

- A tubing cutter and reamer
- Tube benders
- A double flaring tool for SAE flares
- An International Standards Organization (ISO) flaring tool for European-style ISO flares

Brake Disc Micrometer

A special micrometer should be used to check the thickness of a rotor accurately. A brake disc micrometer has pointed anvils that allow the tip to fit into grooves worn on the rotor. This type of micrometer is read in the same way as other micrometers but is made with a range from 0.300 to 1.300 inches. Digital calipers are also used to measure disc brake thickness **(Figure 5–64)**.

Drum Micrometer

A drum micrometer is a single-purpose instrument used to measure the inside diameter of a brake drum. A drum micrometer has two movable arms on a shaft **(Figure 5–65)**. One arm has a precision dial indicator; the other arm has an outside anvil that fits against the inside of the drum. In use, the arms are secured on the shaft by lock screws that fit into grooves every 1/8 inch (0.125) on the shaft. The dial indicator is graduated in 0.005-inch increments.

Metric drum micrometers work the same way except that the shaft is graduated in 1 cm major increments and the lock screws fit in notches every 2 mm.

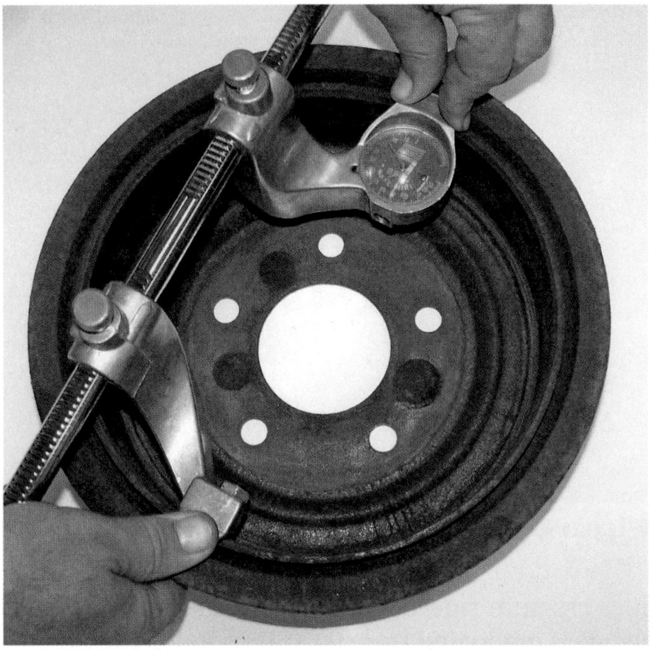

Figure 5-65 A drum micrometer.

Brake Shoe Adjusting Gauge (Calipers)

A brake shoe adjusting gauge is an inside-outside measuring device **(Figure 5–66)**. This gauge is often called a brake shoe caliper. During drum brake service, the inside part of the gauge is placed inside a newly surfaced drum and expanded to fit the drum diameter. The lock screw is then tightened and the gauge moved to the brake shoes installed on the backing plate. The brake shoes are then adjusted until the outside part of the gauge just slips over them. This action provides a rough adjustment of the brake shoes. Final adjustment must still be done after the drum is installed, but the brake shoe gauge makes the job faster.

Figure 5-66 A drum brake shoe adjusting gauge.

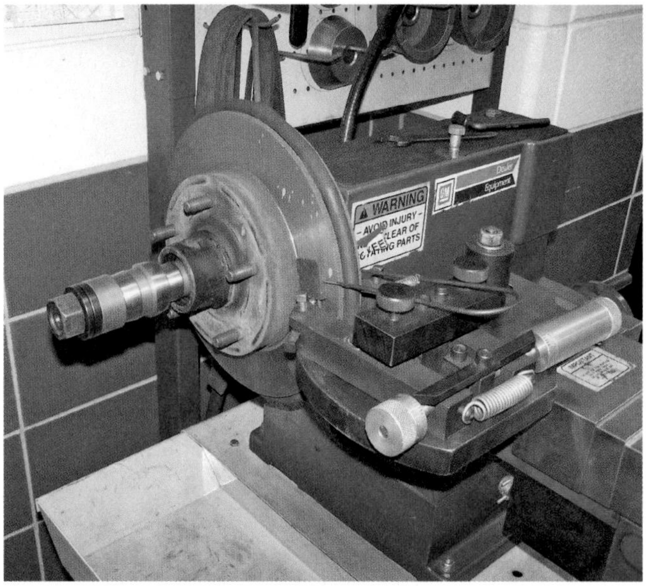

Figure 5-67 A bench brake lathe.

Brake Lathes

Brake lathes are special power tools used only for brake service. They are used to turn and resurface brake rotors and drums. Turning involves cutting away very small amounts of metal to restore the surface of the rotor or drum. The traditional brake lathe is an assembly mounted on a stand or workbench. The bench lathe requires that the drum or rotor be removed from the vehicle and mounted on the lathe for service **(Figure 5–67)**.

As the drum or rotor is turned on the lathe spindle, a carbide steel cutting bit is passed over the drum or rotor friction surface to remove a small amount of metal. The cutting bit is mounted rigidly on a lathe fixture for precise control as it passes across the friction surface.

An on-car lathe **(Figure 5–68)** is bolted to the vehicle suspension or mounted on a rigid stand to provide a stable mounting point for the cutting tool. The rotor may be turned by either the vehicle's engine or drive train (for a FWD vehicle) or by an electric motor and drive attachment on the lathe. As the rotor is turned, the lathe cutting tool is moved across both surfaces of the rotor to refinish it. An on-car lathe not only has the obvious advantage of speed, it also rotates the rotor on the vehicle wheel bearings and hub so that these sources of runout, or wobble, are compensated for during the refinishing operation.

Bleeder Screw Wrenches

Special bleeder screw wrenches often are used to open bleeder screws. Bleeder screw wrenches are small, 6-point box wrenches with strangely offset handles for access to bleeder screws in awkward

Figure 5-68 An on-vehicle disc brake lathe.
Courtesy of RTI Technologies, Inc.

locations. The 6-point box end grips the screw more securely than a 12-point box wrench can and avoids damage to the screw.

Bleeding Tools

Removing the air from the closed hydraulic brake system is very important. This is done by bleeding the system. Bleeding can be done manually, with a vacuum pump, or with a pressure bleeder. The latter two are preferred because they are quick and very efficient, and the technician can do without an assistant.

HEATING AND AIR-CONDITIONING TOOLS

The repair and diagnostic tools for the heating, ventilation, and air-conditioning (A/C) systems are discussed in the following paragraphs. This discussion does not cover all of the tools you may need; rather, these tools are the most commonly used by the service industry. Details of when and how to use these tools are covered in Section 9 of this book.

Manifold Gauge Set

A **manifold gauge set** **(Figure 5–69)** is used when discharging, charging, evacuating, and for diagnosing trouble in an A/C system. With the new legislation on handling refrigerants, all gauge sets are required to have a valve device to close off the end of the hose so that the fitting not in use is automatically shut.

The low-pressure gauge is graduated into pounds of pressure from 1 to 120 (with cushion to 250) in 1-pound graduations, and, in the opposite direction, in inches of vacuum from 0 to 30. This is the gauge that should always be used in checking pressure on the low-pressure side of the system. The high-pressure gauge is graduated from 0 to 500 pounds pressure in 10-pound graduations. This gauge is used for checking pressure on the high-pressure side of the system.

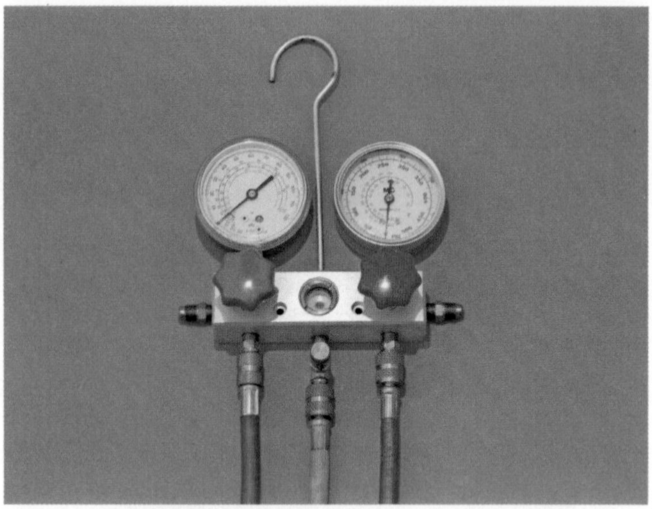

Figure 5-69 A manifold gauge set.

The gauge manifold is designed to control refrigerant flow. When the manifold test set is connected into the system, pressure is registered on both gauges at all times.

Because R-134a is not interchangeable with R-12, separate sets of hoses, gauges, and other equipment are required to service vehicles. All equipment used to service R-134a and R-12 systems must meet SAE standard J1991. The service hoses on the manifold gauge set must have manual or automatic backflow valves at the service port connector ends to prevent the refrigerant from being released into the atmosphere during connection and disconnection. Manifold gauge sets for R-134a can be identified by labels on the gauges and/or have a light blue color on the face of the gauges.

For identification purposes, R-134a service hoses must have a black stripe along their length and be clearly labeled. The low-pressure hose is blue with a black stripe. The high-pressure hose is red with black stripe and the center service hose is yellow with a black stripe. Service hoses for one type of refrigerant will not easily connect into the wrong system, as the fittings for an R-134a system are different than those used in an R-12 system.

Service Port Adapter Set

To connect a manifold gauge set to an A/C system, adapters are sometimes needed **(Figure 5–70)**. The high-side fitting on many vehicles with an R-12 system may require the use of a special adapter to connect the manifold gauge set to the service port. The service hoses of some manifold gauge sets are not

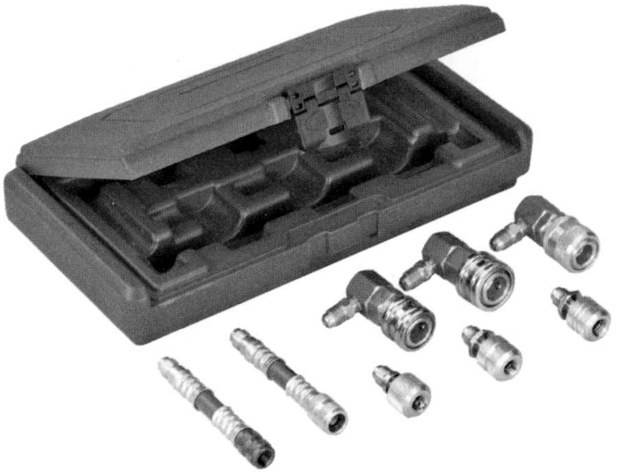

Figure 5-70 An A/C port adapter set. *Courtesy of SPX Service Solutions*

equipped with a Schrader valve-depressing pin. Therefore, when connecting this type hose to a Schrader valve, an adapter must be used. The manifold and gauge sets for R-12 and R-134a are not interchangeable; therefore, there are no suitable adapters for using R-12 gauges on an R-1134a system or vice versa.

Electronic Leak Detector

An electronic leak detector **(Figure 5–71)** is safe and effective and can be used with all types of refrigerants. A hand-held battery-operated electronic leak detector contains a test probe that is moved about 1 inch per second in areas of suspected leaks. Since refrigerant is heavier than air, the probe should be positioned below the test point. An alarm or a buzzer on the detector indicates the presence of a leak. On some models, a light flashes when refrigerant is detected.

Fluorescent Leak Tracer

To find a refrigerant leak using the fluorescent tracer system, first introduce a fluorescent dye into the air-conditioning system with a special infuser included with the detector equipment. Run the air conditioner for a few minutes, giving the tracer dye fluid time to circulate and penetrate. Wear the tracer protective goggles and scan the system with a black-light glow gun. Leaks in the system will shine under the black light as a luminous yellow-green **(Figure 5–72)**.

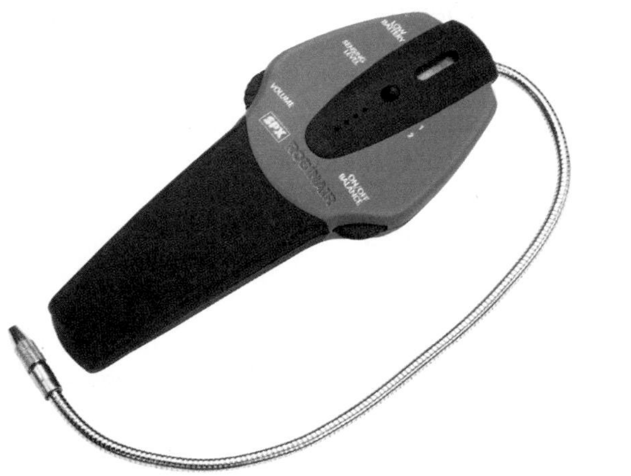

Figure 5-71 An electronic leak detector. *Courtesy of SPX Service Solutions*

Figure 5-72 With a fluorescent tracer system, refrigerant leaks will glow brightly when inspected with a UV/Blue light as a luminous yellow-green. *Courtesy of Tracer Products*

Refrigerant Identifier

A refrigerant identifier **(Figure 5–73)** is used to identify the type of refrigerant present in a system. This test should be done before any service work. The tester is used to identify the purity and quality of the refrigerant sample taken from the system.

Refrigerant Recycling/Charging Stations

A charging station **(Figure 5–74)** removes, evacuates, and recharges an A/C system. The amount of

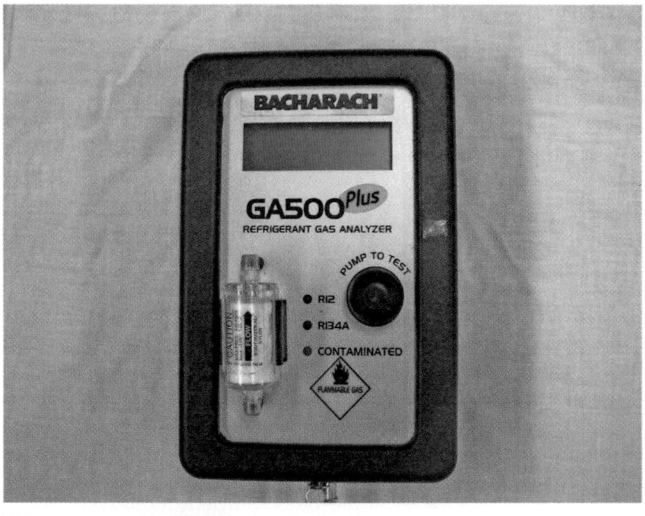

Figure 5-73 A refrigerant identifier (analyzer).

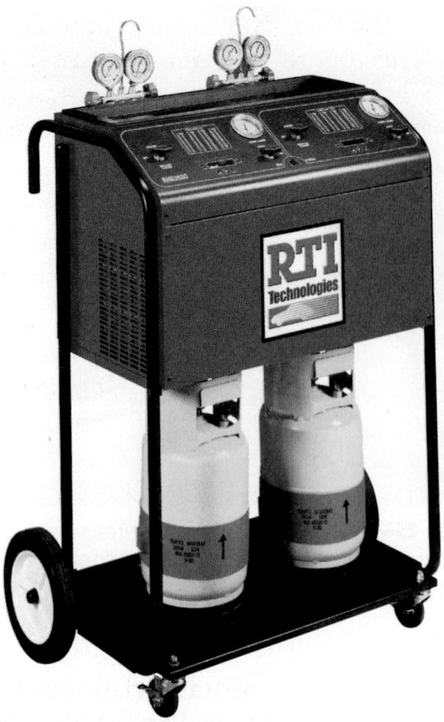

Figure 5–74 A dual (R-12 and R-134a) charging station. *Courtesy of RTI Technologies, Inc.*

refrigerant put into the system is adjusted through controls on the station.

Special equipment is used to remove and recycle the refrigerant in a system **(Figure 5–75)**. These machines draw the old refrigerant out of the system, filter it, separate the oil, remove moisture and air, and store the refrigerant for future use.

Figure 5–75 A single-pass refrigerant recovery and recycling machine. *Courtesy of RTI Technologies, Inc.*

Figure 5–76 A dial-type thermometer used to check A/C operation. *Courtesy of SPX Service Solutions*

All recycled refrigerant must be safely stored in DOT CFR Title 49 or UL-approved containers. Before recycled refrigerant can be used, it must be checked for noncondensable gases.

Thermometer

A digital readout or dial-type **thermometer (Figure 5–76)** is often used to measure the air temperature at the vent outlets, which indicates the overall performance of the system. The thermometer can also be used to check the temperature of refrigerant lines, hoses, and components while diagnosing a system. While doing the latter, an electronic pyrometer works best and is often used.

Compressor Tools

Although compressors are usually replaced when they are faulty, certain service procedures for them are standard practice. Most of these procedures focus on compressor clutch and shaft seal service and they require special tools. Clutch plate tools are required to gain access to the shaft seal. They are also needed to reinstall the clutch plate after service.

Many different tools are required to perform these services to a compressor. Typically to replace a shaft seal, you will need an adjustable or fixed spanner wrench, clutch plate installer/remover, ceramic seal installer/remover, seal assembly installer/remover, seal seat installer/remover, shaft seal protector, snap-ring pliers, O-ring remover, and O-ring installer. Some of these tools are for a specific model compressor; others are universal fit or have interchangeable parts to allow them to work on a variety of compressors.

Hose and Fitting Tools

An A/C system is a closed system, meaning outside air should never enter the system and the refrigerant in the system should never exit to the outside. To maintain this closed system, special fittings and hoses are used. Often, special tools, such as the spring-lock coupling tool set, are required when servicing the system's fittings and hoses. Without this tool, it is impossible to separate the connector and not damage it.

KEY TERMS

Ammeter
Ball hone
Boring bar
Brush hone
Compression gauge
Connector pick
Continuity tester
Cylinder hone
Cylinder leakage tester
Dial bore gauge
Digital multimeter (DMM)
Dual-trace
Dynamometer
Glaze breaker
Hydrometer
Inclinometer
Insulation resistance tester
Lab scope
Lineman's gloves
Logic probe
Manifold gauge set
Multimeter
Noid light
Ohmmeter
Pressure gauge
Pyrometer
Ridge reamer
Ring compressor
Ring expander
Ring groove cleaner
Scan tool
Spark tester
Stethoscope
Tachometer
Testlight
Thermometer
Timing light
Torque angle gauge
Trace
Transient voltage
Vacuum
Valve spring compressor
Valve spring tester
VAT
Voltmeter
Waveform

SUMMARY

- Common diagnostic tools used to check an engine and its related systems include a compression gauge, cylinder leakage tester, oil pressure gauge, stethoscope, dial bore indicator, valve spring tester, cooling system pressure tester, coolant hydrometer, engine analyzers, fuel pressure gauge, injector balance tester, injector circuit test light, vacuum gauge, vacuum pump, vacuum leak detector, spark tester, logic probes, pyrometers, and exhaust analyzers.
- Common tools used to service an engine and its related systems include transaxle removal and installation equipment, ridge reamer, ring compressor, ring expander, ring groove cleaner, cylinder deglazer, cylinder hone, boring bar, cam bearing driver set, V-blocks, valve and valve seat resurfacing equipment, valve guide repair tools, valve spring compressor, torque angle gauge, oil priming tool, a coolant recovery and recycle system, fuel injector cleaners, fuel line tools, pinch-off pliers, timing light, and spark plug sockets.
- Some of the common diagnostic tools for electronic and electrical systems include a testlight, continuity tester, voltmeter, ohmmeter, ammeter, volt/ampere tester, DMM, lab scope, scan tools, and battery hydrometer.
- Common electrical and electronic system service tools include a computer memory saver, wire and terminal repair tools, headlight aimers, static straps, and sensor tools.
- The tools required to work on hybrid vehicles are the same as those used on a conventional vehicle with the addition of tools designed for high voltages, such as linemen's gloves, CAT III test equipment, and insulation resistance testers.
- Diagnostic tools for a vehicle's drivetrain include a drive shaft angle gauge and hydraulic pressure gauge set.
- Tools required to service the drivetrain include transaxle removal and installation equipment, transmission/transaxle holding fixtures, transmission jack, axle pullers, special tool sets, clutch alignment tool, clutch pilot bearing/bushing puller/installer, and universal joint tools.
- The various diagnostic tools used on a vehicle's running gear include a tire tread depth gauge, power steering pressure gauge, wheel alignment equipment, brake disc micrometer, and drum micrometer.
- Some of the common tools used to service a vehicle's running gear include control arm bushing tools, tie-rod end and ball joint pullers, front bearing hub tool, pitman arm puller, tie-rod sleeve adjusting tool, steering column special tool set, shock absorber tools, spring/strut compressor tool, power steering pump pulley special tool set, brake pedal depressor, tire changer, wheel balancer, wheel weight pliers, brake cleaning equipment and containment systems, holddown spring and return spring tools, boot drivers and pliers, caliper piston removal tools, drum brake adjusting tools, brake cylinder hones, tubing tools, brake shoe adjusting gauge, brake lathes, bleeder screw wrenches, and pressure bleeders.
- Common tools used to check a vehicle's heating and air-conditioning system include a manifold gauge set, a service port adapter set, an electronic leak detector, a fluorescent leak tracer, and a thermometer.
- Tools used to service air-conditioning systems include a refrigerant identifier, refrigerant charging station, refrigerant recovery and recycling system, compressor tools, and hose and fitting tools.

REVIEW QUESTIONS

1. What are the two types of testlights and how do they differ?
2. *True or False?* Knurling is used to repair worn valve guides by increasing the inside diameter of the guide.
3. Name the two basic types of compression gauges.
4. What tool is used to test engine manifold vacuum?
5. Which of the following statements is *not* true?
   a. Exhaust analyzers allow a technician to look at the effectiveness of a vehicle's emission control systems.
   b. Most exhaust analyzers measure HC in parts per million or grams per mile.
   c. CO is measured as a percent of the total exhaust.
   d. Emission controls greatly alter O_2 and CO_2 emissions.
6. Technician A says that a brake disc micrometer has pointed anvils. Technician B says that a brake drum micrometer is a large inside micrometer that is read like any other micrometer. Who is correct?
   a. Technician A c. Both A and B
   b. Technician B d. Neither A nor B
7. Technician A says that a pyrometer measures temperature. Technician B says that a thermometer measures temperature. Who is correct?
   a. Technician A c. Both A and B
   b. Technician B d. Neither A nor B
8. *True or False?* A lab scope is a visual voltmeter that shows voltage over a period of time.
9. When using a voltmeter, Technician A connects it across the circuit being tested. Technician B connects the red lead of the voltmeter to the more positive side of the circuit. Who is correct?
   a. Technician A c. Both A and B
   b. Technician B d. Neither A nor B
10. Technician A uses a digital volt/ohmmeter to test voltage. Technician B uses the same tool to test resistance. Who is correct?
    a. Technician A c. Both A and B
    b. Technician B d. Neither A nor B
11. Technician A says that a charging station removes old refrigerant and recharges an A/C system. Technician B says that a charging cylinder meters out the desired amount of refrigerant by weight. Who is correct?
    a. Technician A c. Both A and B
    b. Technician B d. Neither A nor B
12. Which of the following statements about manifold gauge sets is *not* true?
    a. An adapter is required for using R-12 gauges on an R-1134a system.
    b. A manifold gauge set is used when discharging, charging, and evacuating and for diagnosing trouble in an A/C system.
    c. The gauge manifold is designed to control refrigerant flow. When the manifold test set is connected into the system, pressure is registered on both gauges at all times.
    d. R-134a service hoses have a black stripe along their length, the low-pressure hose is blue, and the high-pressure hose is red.
13. *True or False?* A brake shoe adjusting gauge is an inside-outside measuring device used to initially adjust the expanse of brake shoes before the brake drum is installed.
14. When conducting an oil pressure test: Technician A says that lower than normal pressure can be caused by a burned intake valve. Technician B says that lower than normal oil pressure can be caused by excessive engine bearing clearances. Who is correct?
    a. Technician A c. Both A and B
    b. Technician B d. Neither A nor B
15. Which of the following conditions can be revealed by fuel pressure readings?
    a. faulty fuel pump
    b. faulty fuel pressure regulator
    c. restricted fuel delivery system
    d. all of the above
16. Technician A uses a high-impedance testlight on the high-voltage systems in hybrid vehicles. Technician B uses only CAT-III test instruments to check high-voltage systems. Who is correct?
    a. Technician A c. Both A and B
    b. Technician B d. Neither A nor B
17. Technician A says that a sulfated and plugged fuel injector is caused by electrical problems. Technician B says that a sulfated and plugged fuel injector can be cleaned. Who is correct?
    a. Technician A c. Both A and B
    b. Technician B d. Neither A nor B

18. The tests conducted by a scan tool can also be done by some ______.
    a. fuel injector pulse testers
    b. exhaust analyzers
    c. engine analyzers
    d. digital volt/ohmmeters
19. When using a fuel injector pulse tester, Technician A says that little or no pressure drop indicates a plugged or defective injector. Technician B says that no pressure drop indicates an overly rich condition. Who is correct?
    a. Technician A
    b. Technician B
    c. Both A and B
    d. Neither A nor B
20. It is much easier to test current using an ammeter equipped with a(n) ______.
    a. continuity tester
    b. carbon pile
    c. inductive pickup
    d. tachometer
21. While discussing a clutch alignment tool: Technician A says that the part of the tool that fits into the clutch plate must have a slightly larger OD than the bore in the disc. Technician B says that the OD of the tool that goes into the pilot must be slightly smaller than the ID of the pilot bushing. Who is correct?
    a. Technician A
    b. Technician B
    c. Both A and B
    d. Neither A nor B
22. *True or False?* The angle of the drive shaft at the transmission should equal its angle at the drive axle.
23. Technician A says that ball joints may be pressed into the control arm. Technician B says that ball joints may be riveted to the control arm. Who is correct?
    a. Technician A
    b. Technician B
    c. Both A and B
    d. Neither A nor B
24. Which of the following is *not* a suitable way to bleed a hydraulic brake system?
    a. manual bleeding
    b. bench bleeding
    c. pressure bleeding
    d. vacuum bleeding
25. Hydraulic pressure gauges are used to diagnose all of these, except:
    a. power steering systems
    b. automatic transmissions
    c. hydraulic-boost brake systems
    d. engine compression

CHAPTER

6

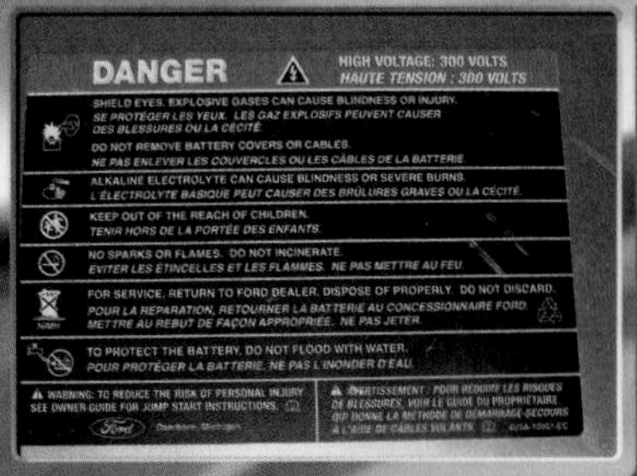

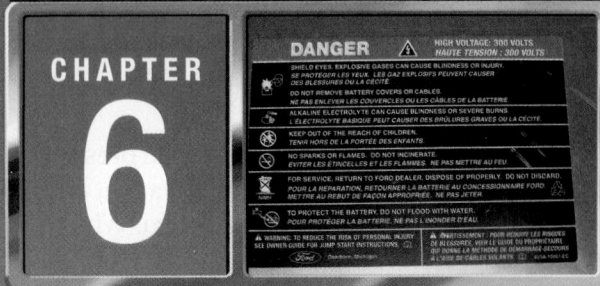

WORKING SAFELY IN THE SHOP

OBJECTIVES

■ Understand the importance of safety and accident prevention in an automotive shop. ■ Explain the basic principles of personal safety, including protective eye wear, clothing, gloves, shoes, and hearing protection. ■ Explain the procedures and precautions for safely using tools and equipment. ■ Explain the precautions that need to be followed to safely raise a vehicle on a lift. ■ Explain what should be done to maintain a safe working area in a shop. ■ Describe the unique safety considerations that must be adhered to when working on hybrid vehicles. ■ Describe the purpose of the laws concerning hazardous wastes and materials, including the Right-To-Know laws. ■ Describe your rights, as an employee and/or student, to have a safe place to work.

Working on automobiles can be dangerous. It can also be fun and very rewarding. To keep the fun and rewards rolling in, you need to try to prevent accidents by working safely. In an automotive repair shop, there is great potential for serious accidents, simply because of the nature of the business and the equipment used. When there is carelessness, the automotive repair industry can be one of the most dangerous occupations.

However, the chances of you being injured while working on a car are close to nil if you learn to work safely and use common sense. Shop safety is the responsibility of everyone in the shop—you, your fellow students or employees, and your employer or instructor. Everyone must work together to protect the health and welfare of all who work in the shop.

Unless you want to get hurt or want your fellow students or employees to get hurt, you should strive to work safely. Shop accidents can cause serious injury, temporary or permanent disability, and death.

This chapter covers many guidelines concerning personal, work area, tool and equipment, and hazardous material safety. In addition to this chapter, special warnings are given throughout this book to alert you to situations where carelessness could result in personal injury. When working on cars, always follow the safety guidelines given in service manuals and other technical literature. They are there for your protection.

PERSONAL SAFETY

To protect yourself from injuries, you must take precautions. This includes wearing protective gear, dressing appropriately, working professionally, and correctly handling tools and equipment.

Eye Protection

Your eyes can become infected or permanently damaged by many things in a shop. Consider the following:

- Dirt and sharp bits of rust can easily fall into your eyes while you are working under a vehicle.
- Some procedures, such as grinding, release tiny particles of metal and dust, which are thrown off at very high speeds. These particles can easily get into your eyes, scratching or cutting them.
- Pressurized gases and liquids escaping a ruptured hose or loose hose fitting can spray into your eyes and cause blindness.

To be safe, you should wear suitable eye protection whenever you are working in the shop. In most shops, this is not an option—you must wear eye protection. There are many types of eye protection available (**Figure 6–1**).

Safety glasses have lenses made of safety glass. They also offer some sort of side protection. To help

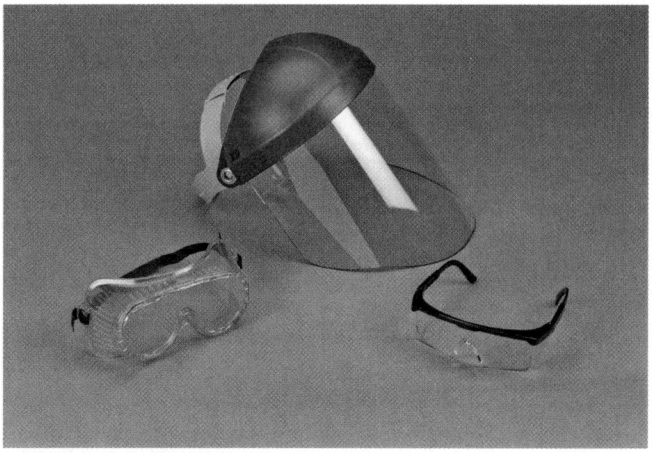

Figure 6–1 Various types of eye protection: safety (splash) goggles, face shield, and safety glasses.

develop the habit of wearing safety glasses, make sure the glasses fit well and feel comfortable.

Regular prescription glasses do not offer sufficient protection and, therefore, should not be worn as a substitute for safety glasses. Prescription glasses can be made with polycarbonate lenses and can be worn as safety glass if they are rated ANSI Z87 and have side shields fixed to the frame.

Some procedures may require that you wear additional eye protection. For example, when you are working around air conditioning systems, you should wear splash goggles and, when cleaning parts with a pressurized spray, you should wear a face shield. The face shield will also protect the rest of your face.

Eye First Aid If chemicals such as battery acid, fuel, or solvents get into your eyes, flush them continuously with clean water. Have someone call a doctor and get medical help immediately.

Many shops have eye wash stations or safety showers **(Figure 6–2)** that should be used whenever you or someone else has been sprayed or splashed with a chemical.

Clothing

Clothing that hangs out freely, such as shirttails, can create a safety hazard and cause serious injury. Nothing you wear should be allowed to dangle in the engine compartment or around equipment. Shirts should be tucked in and buttoned and long sleeves buttoned or carefully rolled up. Your clothing should be well fitted and comfortable but made with strong material. Some technicians prefer to wear coveralls or shop coats to protect their personal clothing. Your work clothes should offer you some protection but should not restrict your movement.

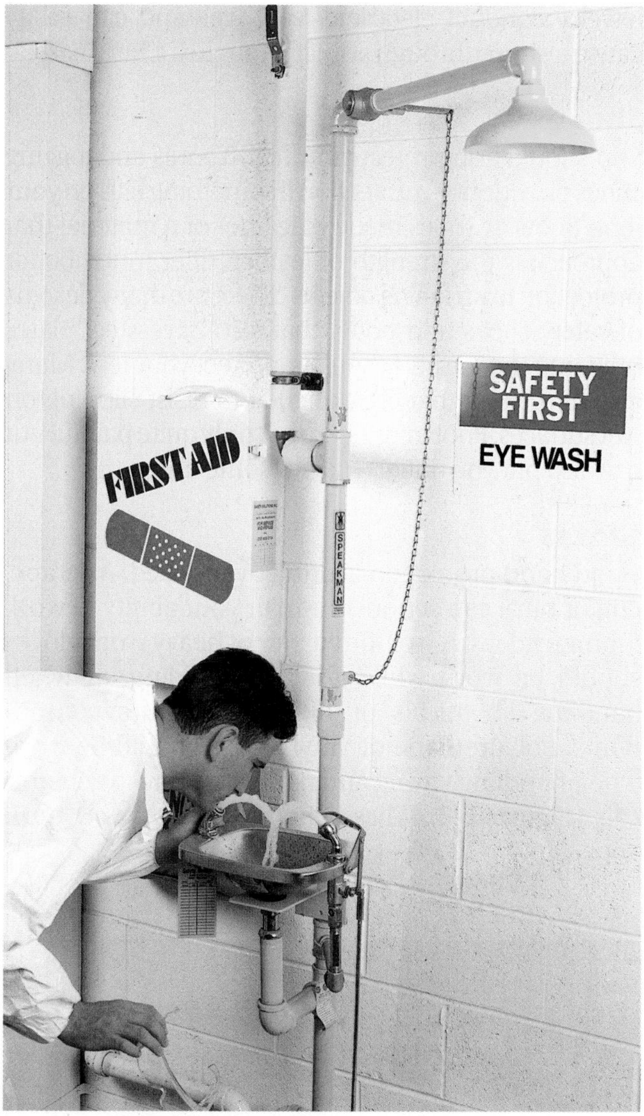

Figure 6–2 A combination eye wash and safety shower. *Courtesy of DuPont Performance Coatings*

Keep your clothing clean. If you spill gasoline or oil on yourself, change that item of clothing immediately. Oil against your skin for a prolonged period can produce rashes or other allergic reactions. Gasoline can irritate cuts and sores.

Hair and Jewelry

Long hair and loose, hanging jewelry can create the same type of hazard as loose-fitting clothing. They can get caught in moving engine parts and machinery. If you have long hair, tie it back or tuck it under a cap.

Rings, necklaces, bracelets, and watches should not be worn while working. A ring can rip your finger off, a watch or bracelet can cut your wrist, and a necklace can choke you. This is especially true when working with or around electrical wires. The metals in most jewelry conduct electricity very well and can easily cause a short, through you, if it touches a bare wire.

Shoes

You should also protect your feet. Tennis and jogging shoes provide little protection if something falls on your foot. Boots or shoes made of leather or a material that approaches the strength of leather, offer much better protection from falling objects. There are many designs of safety shoes and boots that also have steel plates built into the toe and shank to protect your feet. Many also have soles that are designed to resist slipping on wet surfaces. Foot injuries are not only quite painful but can also put you out of work for some time.

Gloves

Good hand protection is often overlooked. A scrape, cut, or burn can seriously impair your ability to work for many days. A well-fitted pair of heavy work gloves should be worn while grinding, welding, or when handling chemicals or high-temperature components. Polyurethane or vinyl gloves should be worn when handling strong and dangerous caustic chemicals **(Figure 6–3)**. These chemicals can easily burn your skin.

Figure 6–3 Polyurethane or vinyl gloves should be worn when handling strong and dangerous caustic chemicals.

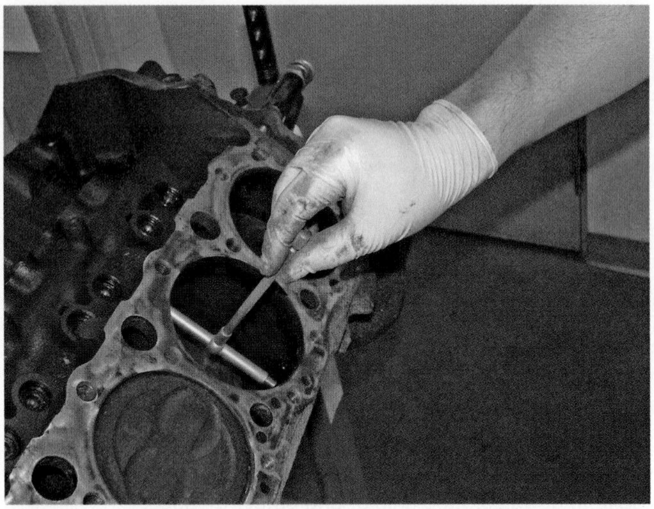

Figure 6–4 Many technicians wear thin, surgical-type latex or nitrile gloves whenever they are working on vehicles.

Many technicians wear thin, surgical-type latex **(Figure 6–4)** or nitrile gloves whenever they are working on vehicles. These offer little protection against cuts but do offer protection against disease and grease buildup under and around your fingernails. Latex gloves are more comfortable but weaken when they are exposed to gas, oil, and solvents. Nitrile gloves are not as comfortable but they are not affected by gas, oil, and solvents. Your choice of hand protection should be based on what you are doing.

Disease Prevention

When you are ill with something that may be contagious, see a doctor and do not go to work or school until the doctor says there is little chance of someone else contracting the illness from you. Doing this will protect others, and if others do this you will be protected.

You should also be concerned with and protect yourself and others from bloodborne pathogens. **Bloodborne pathogens** are pathogenic microorganisms that are present in human blood and can cause disease. These pathogens include, but are not limited to, staph infections caused by the bacteria *Staphylococcus aureus*, hepatitis B virus (HBV), and human immunodeficiency virus (HIV). For everyone's protection, any injury that causes bleeding should be dealt with as a threat to others. You should avoid contact with the blood of another. If you need to administer some form of first aid, make sure you wear hand protection before you do so. You should also wear gloves and other protection when handling the item that caused the cut. This item should be sterilized immediately. Most importantly, like all injuries, report the accident to your instructor or supervisor.

Figure 6–5 While working in a noisy environment, your ears can be protected with earmuffs or earplugs.

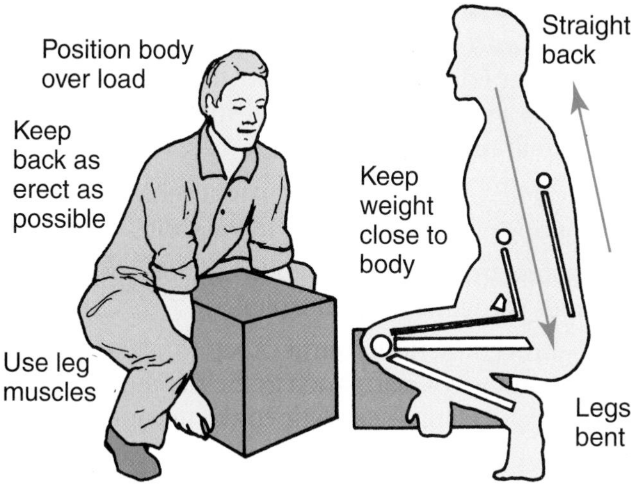

Figure 6–6 Use your leg muscles—*never* your back—to lift heavy objects.

Ear Protection

Exposure to very loud noise levels for extended periods can lead to hearing loss. Air wrenches, engines running under a load, and vehicles running in enclosed areas can all generate harmful levels of noise. Simple earplugs or earphone-type protectors **(Figure 6–5)** should be worn in environments that are constantly noisy.

Respiratory Protection

Technicians often work with chemicals that have toxic fumes. Air or respiratory masks should be worn whenever you will be exposed to toxic fumes. Cleaning parts with solvents and painting are the most common times when respiratory masks should be worn.

Masks should also be worn when handling parts that contain asbestos or when handling hazardous materials. The proper handling of these materials is covered in great detail later in this chapter.

Lifting and Carrying

At least once a week a technician will need to move something that is heavy. Knowing how to lift these heavy things can save your career. When lifting any object, follow these steps:

1. Place your feet close to the object. Position your feet so you will be able to maintain a good balance.
2. Keep your back and elbows as straight as possible. Bend your knees until your hands reach the best place to get a strong grip on the object **(Figure 6–6)**.
3. If the part is in a cardboard box, make sure the box is in good condition. Old, damp, or poorly sealed boxes will tear and the part will fall out.
4. Firmly grasp the object or container. Never try to change your grip as you move the load.
5. Keep the object close to your body, and lift it up by straightening your legs. Use your leg muscles, not your back muscles.
6. If you must change your direction of travel, never twist your body. Turn your whole body, including your feet.
7. When placing the object on a shelf or counter, do not bend forward. Place the edge of the load on the shelf and slide it forward. Be careful not to pinch your fingers.
8. When setting down a load, bend your knees and keep your back straight. Never bend forward. This strains the back muscles.
9. When lowering something heavy to the floor, set the object on blocks of wood to protect your fingers.

You should also use back-protection devices when you lift heavy objects. Always lift and work within your ability and ask others (or use a hoist) to help when you are not sure whether you can handle the size or weight of an object. Even small, compact parts can be surprisingly heavy or unbalanced. Think about how you are going to lift something before beginning.

CAUTION!

Trying to "muscle" something with your arms or back can result in severe damage to your back and may end your career and limit what you do the rest of your life!

Professional Behavior

Accidents can be prevented simply by the way you behave. The following list does not include everything you should or should not do; it merely gives some things to think about:

- Never smoke while working on a vehicle or while working with any machine in the shop.
- Playing around or "horseplay" is not fun when it sends someone to the hospital.
- To prevent serious burns, keep your skin away from hot metal parts such as the radiator, exhaust manifold, tailpipe, catalytic converter, and muffler.
- Always disconnect electric engine cooling fans when working around the radiator. Many of these will turn on without warning and can easily chop off a finger or hand. Make sure you reconnect the fan after you have completed your repairs.
- When working with a hydraulic press, make sure the pressure is applied in a safe manner. It is generally wise to stand to the side when operating the press.
- Properly store all parts and tools by putting them away in a place where people will not trip over them. This practice not only cuts down on injuries, it also reduces time wasted looking for a misplaced part or tool.
- Keep your work area clean and uncluttered. Make sure you clean up all spills before continuing to work.

TOOL AND EQUIPMENT SAFETY

An automotive technician must adhere to the following shop safety guidelines when using all tools and equipment.

Hand Tool Safety

Careless use of simple hand tools such as wrenches, screwdrivers, and hammers causes many shop accidents that could be prevented. Keep in mind the following tips when using hand tools:

- Keep all hand tools grease-free. Oily tools can slip out of your hand, causing broken fingers or at least cut or skinned knuckles.
- Inspect your tools for cracks, broken parts, or other dangerous conditions before you use them.
- Hand tools should only be used for the purpose they were designed for. Use the right tool for the job.
- Make sure the tool is of professional quality.
- Never use broken or damaged tools.
- When using a wrench, always pull it, not push it, toward you.
- Always use the correct size of wrench.
- Use a box-end or socket wrench whenever possible.
- Do not use deep-well sockets when a regular size socket will work. The longer socket develops more twist torque and tends to slip off the fastener.
- Use an adjustable wrench only when it is absolutely necessary; pull the wrench so that the force of the pull is on the nonadjustable jaw.
- When using an air impact wrench, always use impact sockets.
- Never use wrenches or sockets that have cracks or breaks.
- Never use a wrench or pliers as a hammer.
- Never use pliers to loosen or tighten a nut; use the correct wrench.
- Always be sure to strike an object with the full face of the hammerhead.
- Always wear safety glasses when using a hammer and/or chisel.
- Never strike two hammer heads together.
- Never use screwdrivers as chisels.
- Be careful when using sharp or pointed tools.
- Do not place sharp tools or other sharp objects into your pockets.
- If a tool is supposed to be sharp, make sure it is sharp. Dull tools can be more dangerous than sharp tools.
- Use knives, chisels, and scrapers in a motion that will keep the point or blade moving away from your body.
- Always hand a pointed or sharp tool to someone else with the handle toward the person to whom you are handing the tool.

Power Tool Safety

CAUTION!

Carelessness or mishandling of power tools can cause serious injury. Make sure you know how to operate a tool before using it.

Power tools are operated by an outside power source, such as electricity, compressed air, or hydraulic pressure. Always respect the tool and its power source. Carelessness can result in serious injury. Also, always wear safety glasses when using power tools. Never try to use a tool beyond its stated capacity.

Electrical Tools When using an electrically powered tool, make sure it is properly grounded. Check the wiring for insulation cracks, as well as bare wires, before using it. Also, when using electrical power tools, never stand on a wet or damp floor. Before plugging in any electric tool, make sure its switch is in the *off* position. When you are finished using the tool, turn it off and unplug it. Never leave a running power tool unattended.

When using power equipment on a small part, never hold the part in your hand. Always mount the part in a bench vise or use vise grip pliers.

When using a bench or floor grinding wheel, check the machine and the grinding wheels for signs of damage before using them. If the wheels are damaged, they should be replaced before using the machine. Make sure the speed rating of the replacement wheels match the speed of the machine. Be sure to place all safety guards in position **(Figure 6–7)**. A safety guard is a protective cover over a moving part. Although the safety guards are designed to prevent injury, you should still wear safety glasses and/or a face shield while using the machine. Make sure there are no people or parts around the machine before starting it. Keep your hands and clothing away from the moving parts. Maintain a balanced stance while using the machine.

Compressed Air Tools Tools that use compressed air are called **pneumatic tools**. Compressed air is used to inflate tires, apply paint, and drive tools, such as air ratchets and impact wrenches. Pneumatic tools must always be operated at the pressure recommended by the manufacturer. Before using a pneumatic tool, check all hose connections for leaks. Also check for air line damage.

When using an air nozzle, wear safety glasses and/or a face shield. Particles of dirt and pieces of metal, blown by the high-pressure air, can penetrate your skin or get into your eyes. Always hold an air nozzle or air control device securely when starting or shutting off the compressed air. A loose nozzle can whip suddenly and cause serious injury. Never point an air nozzle at anyone. Never use compressed air to blow dirt from your clothes or hair. Never use compressed air to clean the floor or workbench. Never spin bearings with compressed air. If the bearing is damaged, one of the steel balls or rollers might fly out and cause serious injury.

Lift Safety

Always be careful when raising a vehicle on a lift or a hoist. Adapters and hoist plates must be positioned correctly on twin post- and rail-type lifts to prevent damage to the underbody of the vehicle. There are specific lift points. These points allow the weight of the vehicle to be evenly supported by the adapters or hoist plates. The correct lift points can be found in the vehicle's service manual. **Figure 6–8** shows typical locations for unibody and frame cars. These diagrams are for illustration only. Always follow the manufacturer's instructions. Before operating any lift or hoist, you should already have been trained on the proper use of the lift. Always follow

Figure 6-7 A bench grinder with its safety shields in place.

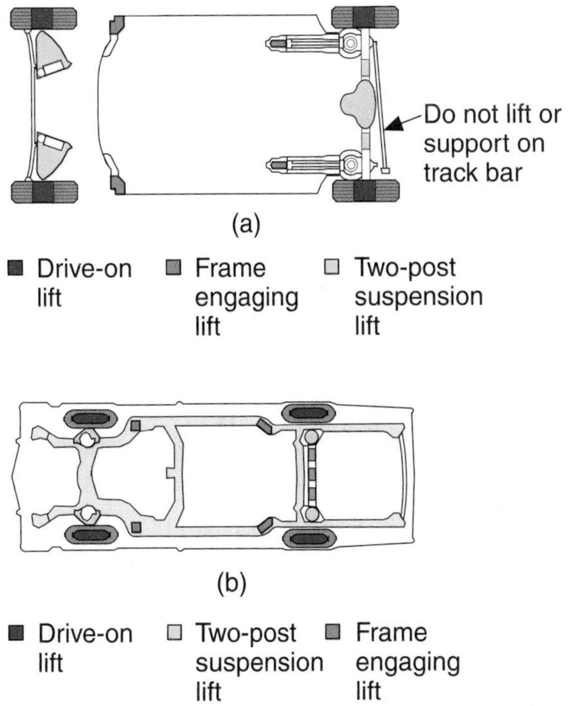

Figure 6-8 Typical lifting points: The correct ones for a vehicle are given in the service manual for that vehicle.

the recommended instructions for operating a particular lift.

Once you feel the lift supports are properly positioned under the vehicle, raise the lift until the supports contact the vehicle. Then, check the supports to make sure they are in full contact with the vehicle. Shake the vehicle to make sure it is securely balanced on the lift, then raise the lift to the desired working height.

WARNING!

Before working under a car, make sure the lift's locking device is engaged.

The Automotive Lift Institute (ALI) is an association concerned with the design, construction, installation, operation, maintenance, and repair of automotive lifts. Their primary concern is safety. Every lift approved by ALI has the label shown in **Figure 6–9**. It is a good idea to read through the safety tips included on that label before using a lift.

AUTOMOTIVE LIFT

SAFETY TIPS

Post these safety tips where they will be a constant reminder to your lift operator. For information specific to the lift, always refer to the lift manufacturer's manual.

1. Inspect your lift daily. Never operate if it malfunctions or if it has broken or damaged parts. Repairs should be made with original equipment parts.
2. Operating controls are designed to close when released. Do not block open or override them.
3. Never overload your lift. Manufacturer's rated capacity is shown on nameplate affixed to the lift.
4. Positioning of vehicle and operation of the lift should be done only by trained and authorized personnel.
5. Never raise vehicle with anyone inside it. Customers or by-standers should not be in the lift area during operation.
6. Always keep lift area free of obstructions, grease, oil, trash, and other debris.
7. Before driving vehicle over lift, position arms and supports to provide unobstructed clearance. Do not hit or run over lift arms, adapters, or axle supports. This could damage lift or vehicle.
8. Load vehicle on lift carefully. Position lift supports to contact at the vehicle manufacturer's recommended lifting points. Raise lift until supports contact vehicle. Check supports for secure contact with vehicle. Raise lift to desired working height. CAUTION: If you are working under vehicle, lift should be raised high enough for locking device to be engaged.
9. Note that with some vehicles, the removal (or installation) of components may cause a critical shift in the center of gravity and result in raised vehicle instability. Refer to the vehicle manufacturer's service manual for recommended procedures when vehicle components are removed.
10. Before lowering lift, be sure tool trays, stands, etc. are removed from under vehicle. Release locking devices before attempting to lower lift.
11. Before removing vehicle from lift area, position lift arms and supports to provide an unobstructed exit (See Item #7).

These "Safety Tips," along with "Lifting it Right," a general lift safety manual, are presented as an industry service by the Automotive Lift Institute. For more information on this material, write to: ALI, P.O. Box 1519, New York, NY 10101.

Look For This Label on all Automotive Service Lifts.

FOUNDED 1945

AUTOMOTIVE LIFT INSTITUTE MEMBER ® ALI

THIS AUTOMOTIVE LIFT WAS MANUFACTURED TO CONFORM TO THE REQUIREMENTS OF ANSI/ALI B153.1, A SAFETY STANDARD DEVELOPED COOPERATIVELY WITH THE INDUSTRY AND THOSE SUBSTANTIALLY CONCERNED WITH ITS SCOPE AND PROVISIONS. THE MANUFACTURER IS RESPONSIBLE FOR THE CONSTRUCTION OF THIS PRODUCT TO THIS STANDARD.

AUTOMOTIVE LIFT INSTITUTE, INC.

Figure 6–9 Automotive lift safety tips. *Courtesy of the Automotive Lift Institute*

Jack and Jack Stand Safety

A vehicle can be raised off the ground by a hydraulic jack **(Figure 6–10)**. A handle on the jack is moved up and down to raise part of a vehicle and a valve is turned to release the hydraulic pressure in the jack to lower the part. At the end of the jack is a lifting pad. The pad must be positioned under an area of the vehicle's frame or at one of the manufacturer's recommended lift points. Never place the pad under the floorpan or under steering and suspension components, because they can easily be damaged by the weight of the vehicle. Always position the jack so that the wheels of the vehicle can roll as the vehicle is being raised.

WARNING!

Never use a lift or jack to move something heavier than it is designed for. Always check the rating before using a lift or jack. If a jack is rated for 2 tons, do not attempt to use it for a job that requires a 5-ton jack. It is dangerous for you and the vehicle.

Safety stands, also called **jack stands (Figure 6–11)**, are supports of various heights that sit on the floor. They are placed under a sturdy chassis member, such as the frame or axle housing, to support the vehicle. Once the safety stands are in position, the hydraulic pressure in the jack should be slowly released until the weight of the vehicle is on the stands. Like jacks, jack stands also have a capacity rating. Always use a jack stand of the correct rating.

Never move under a vehicle when it is supported by only a hydraulic jack. Rest the vehicle on the safety stands before moving under the vehicle.

The jack should be removed after the jack stands are set in place. This eliminates a hazard, such as a jack handle sticking out into a walkway. A jack handle that is bumped or kicked can cause a tripping accident or cause the vehicle to fall.

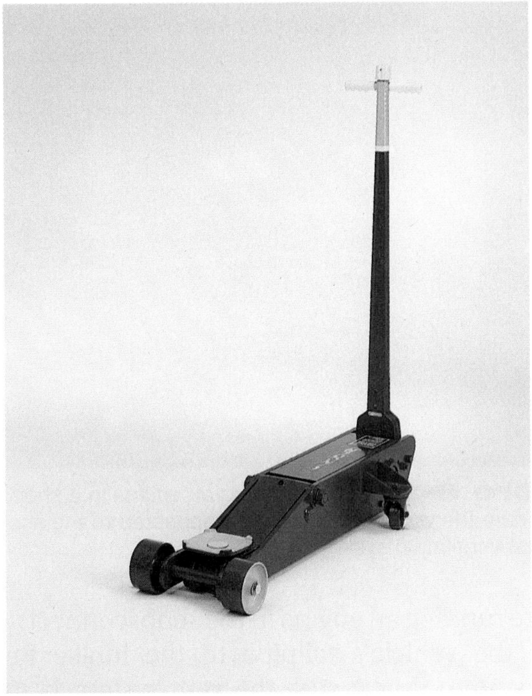

Figure 6–10 Typical hydraulic jack. *Courtesy of Lincoln Automotive Company*

Figure 6–11 Jack stands should be used to support the vehicle after it has been raised by a jack.

Chain Hoist and Crane Safety

Heavy parts of the automobile, such as engines, are removed by using chain hoists **(Figure 6–12)** or cranes. Another term for a chain hoist is chain fall. Cranes often are called cherry pickers.

To prevent serious injury, chain hoists and cranes must be properly attached to the parts being lifted. Always use bolts with enough strength to support the object being lifted. After you have attached the lifting chain or cable to the part that is being removed, have your instructor check it. Place the chain hoist or crane directly over the assembly. Then, attach the lifting chain or cable to the hoist.

Figure 6–12 A heavy-duty engine hoist. *Courtesy of Lincoln Automotive Company*

Figure 6–13 A solvent-based parts washer.

Cleaning Equipment Safety

Parts cleaning is a necessary step in most repair procedures. Cleaning automotive parts can be divided into three basic categories.

Chemical cleaning relies primarily on some type of chemical action to remove dirt, grease, scale, paint, or rust **(Figure 6–13)**. A combination of heat, agitation, mechanical scrubbing, or washing may be used to help remove dirt. Chemical cleaning equipment includes small parts washers, hot/cold tanks, pressure washers, spray washers, and salt baths.

Thermal cleaning relies on heat, which bakes off or oxidizes the dirt. Thermal cleaning leaves an ash residue on the surface that must be removed by an additional cleaning process, such as airless shot blasting or spray washing.

Abrasive cleaning relies on physical abrasion to clean the surface. This includes everything from a wire brush to glass bead blasting, airless steel shot blasting, abrasive tumbling, and vibratory cleaning. Chemical in-tank solution sonic cleaning might also be included here because it relies on the scrubbing action of ultrasonic sound waves to loosen surface contaminants.

Vehicle Operation

When a customer brings a vehicle in for service, certain driving rules should be followed to ensure your safety and the safety of those working around you. For example, before moving a car into the shop, buckle your seat belt. Make sure no one is near, the way is clear, and there are no tools or parts under the car before you start the engine.

Check the brakes before putting the vehicle in gear. Then, drive slowly and carefully in and around the shop.

When road testing the car, obey all traffic laws. Drive only as far as is necessary to check the automobile and verify the customer's complaint. Never make excessively quick starts, turn corners too quickly, or drive faster than conditions allow.

If the engine must be kept running while you are working on the car, block the wheels to prevent the vehicle from moving. Place the transmission in park for automatic transmissions or in neutral for manual transmissions. Set the parking (emergency) brake. Never stand directly in front of or behind a running vehicle.

When parking a vehicle in the shop, always roll the windows down. This allows for access if the doors accidentally lock.

Venting the Engine's Exhaust Whenever you need to have the engine running for diagnosis or service, the engine's exhaust must be vented to the outside. Carbon monoxide (CO) is present in the exhaust. CO is an odorless, tasteless, and colorless deadly gas. Inhaling CO can cause brain damage and, in severe cases, death. Early symptoms of CO poisoning include headaches, nausea, and fatigue.

Most shops have an exhaust ventilation system **(Figure 6–14)**; always use it. These systems collect the engine's exhaust and release it to the outside air.

Figure 6–14 When running an engine in a shop; make sure the vehicle's exhaust is connected to the shop's exhaust ventilation system.

Before running an engine in the shop, connect a hose from the vehicle's tailpipe to the intake for the vent system. Make sure the vent system is turned on before running the engine. If the work area does not have an exhaust venting system, use a hose to direct the exhaust out of the building.

Electrical Safety

Much of your work on an automobile will be around or with the vehicle's electrical system. To prevent personal injury or damage to the vehicle, you should always take the necessary precautions before working. When possible, you should disconnect the vehicle's battery before disconnecting any electrical wire or component. This prevents the possibility of a fire or electrical shock. It also eliminates the possibility of an accidental short, which can ruin the car's electrical system. Disconnect the negative or ground cable first **(Figure 6–15)**, then disconnect the positive cable.

Figure 6–15 Before doing any electrical work or working around the battery, disconnect the negative lead of the battery.

Because electrical circuits require a ground to be complete, by removing the ground cable you eliminate the possibility of a circuit accidentally becoming completed. When reconnecting the battery, connect the positive cable first, then the negative.

Also, remove wristwatches and rings before servicing any part of the electrical system. This helps prevent the possibility of electrical arcing and burns. When disconnecting electrical connectors, do not pull on the wires. When reconnecting the connectors, make sure they are securely connected.

Battery Precautions Because the vehicle's electrical power is stored in a battery or battery pack, special handling precautions must be followed when working with or near batteries. Hybrid and other electric vehicles have very high voltages; therefore, special precautions apply to these vehicles and are given following the general precautions for batteries.

- Make sure you are wearing safety glasses (preferably a face shield) and protective clothing when working around and with batteries.
- Keep all flames, sparks, and excessive heat away from the battery at all times, especially when it is being charged.
- Never smoke near the top of a battery and never use a lighter or match as a flashlight.
- Never lay metal tools or other objects on the battery because a short circuit across the terminals can result.
- All batteries have an electrolyte, which is very corrosive. It can cause severe injuries if it comes in contact with your skin or eye. If electrolyte gets on you, immediately wash with baking soda and water. If the acid gets in your eyes, immediately flush them with cool water for a minimum of 15 minutes and get immediate medical attention.
- Lead-acid batteries use sulfuric acid as the electrolyte. Sulfuric acid is poisonous, is highly corrosive, and produces gases that can explode in high heat.
- Acid from the battery damages a vehicle's paint and metal surfaces and harms shop equipment. Neutralize any electrolyte spills during servicing.
- The most dangerous battery is one that has been overcharged. It is hot and has been, or still may be, producing large amounts of hydrogen. Allow the battery to cool before working with or around it. Also never use or charge a battery that has frozen electrolyte.
- Always use a battery carrier or lifting strap to make moving and handling batteries easier and safer.
- Always charge a battery in well-ventilated areas.
- Never connect or disconnect charger leads when the charger is turned on. This generates a dangerous spark.
- Never recharge the battery when the system is on.
- Turn off all accessories before charging the battery and correct any parasitic drain problems.
- Make sure the charger's power switch is off when you are connecting or disconnecting the charger cables to the battery.
- Always double-check the polarity of the battery charger's connections before turning the charger on. Incorrect polarity can damage the battery or cause it to explode.
- Never attempt to use a charger as a boost to start the engine.

High-Voltage Systems Electric drive vehicles (battery-operated, hybrid, and fuel cell electric vehicles) have high-voltage electrical systems (from 42 volts to 650 volts). These high voltages can kill you! Fortunately, most high-voltage circuits are identifiable by size and color. The cables have thicker insulation and are typically colored orange **(Figure 6–16)**. The connectors are also colored orange. On some vehicles, the high-voltage cables are enclosed in an orange shielding or casing; again the orange indicates high voltage. In addition, the high-voltage battery pack and most high-voltage components have "High Voltage" caution labels **(Figure 6–17)**. Be careful not to touch these wires and parts. There are other safety precautions that should always be adhered to when working on an electric drive vehicle:

Figure 6–16 The high-voltage cables on this Civic hybrid are colored orange and are enclosed in orange casing.

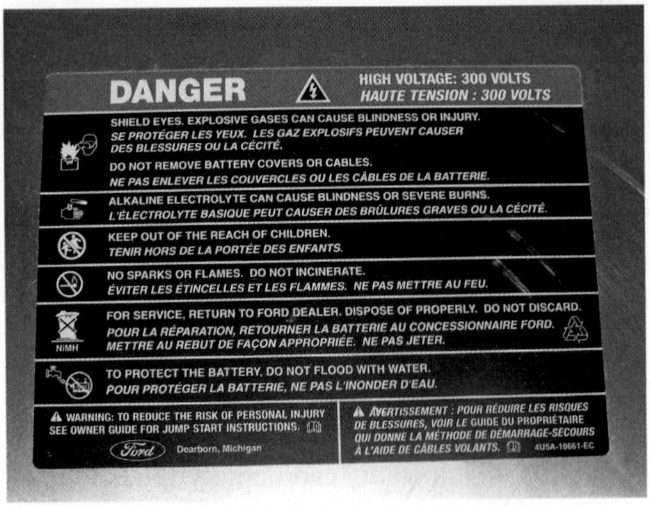

Figure 6-17 Most high-voltage components in a hybrid vehicle have "High Voltage" caution labels.

- Always adhere to the safety guidelines given by the vehicle's manufacturer.
- Obtain the necessary training before working on these vehicles.
- Be sure to perform each repair operation correctly.
- Disable or disconnect the high-voltage system before performing services to those systems. Do this according to the procedures given by the manufacturer.
- Any time the engine is running in a hybrid vehicle, the generator is producing high-voltage. Take care to prevent being shocked.
- Before doing any service to an electric drive vehicle, make sure the power to the electric motor is disconnected or disabled.
- Systems may have a high-voltage capacitor that must be discharged after the high-voltage system has been isolated. Make sure to wait the prescribed amount of time (normally about 10 minutes) before working on or around the high-voltage system.
- After removing a high-voltage cable, cover the terminal with vinyl electrical tape.
- Always use insulated tools.
- Use only the tools and test equipment specified by the manufacturer and follow the test procedures defined by the equipment manufacturer.
- Alert other technicians that you are working on the high-voltage systems with a warning sign such as "High-voltage work: Do not touch."
- Always follow the instructions given by the manufacturer for removing high-voltage battery packs.
- Wear insulating gloves, commonly called "lineman's gloves," when working on or around the high-voltage system. Make sure they have no tears, holes, or cracks and that they are dry. The integrity of the gloves should be checked before using them.
- Always install the correct type of circuit protection device into a high-voltage circuit.
- Many electric motors have a strong permanent magnet in them; do not handle these parts if you have a pacemaker.
- When an electric drive vehicle needs to be towed into the shop for repairs, make sure it is not towed on its drive wheels. Doing this will drive the generator(s), which can overcharge the batteries and cause them to explode. Always tow these vehicles with the drive wheels off the ground or move them on a flat bed.

Rotating Pulleys and Belts

Be very careful around belts, pulleys, wheels, chains, or any other rotating mechanism. When working around an engine's drive belts and pulleys, make sure your hands, shop towels, or loose clothing do not come in contact with the moving parts. Hands and fingers can be quickly pulled into a revolving belt or pulley even at engine idle speeds.

> **WARNING!**
>
> *Be careful when working around electric engine cooling fans. These fans are controlled by a thermostat and can come on without warning, even when the engine is not running. Whenever you must work around these fans, disconnect the electrical connector to the fan motor before reaching into the area around the fan.*

WORK AREA SAFETY

The floor of your work area and bench tops should be kept clean, dry, and orderly. Any oil, coolant, or grease on the floor can make it slippery. Slips can result in serious injuries. To clean up oil, use commercial oil absorbent. Also, keep all water off the floor. Water is slippery on smooth floors, and electricity flows well through water. Aisles and walkways should be kept clean and wide enough to easily move through. Make sure the work areas around machines are large enough to safely operate the machine.

Make sure all drain covers are snugly in place. Open drains or covers that are not flush to the floor can cause toe, ankle, and leg injuries.

Keep an up-to-date list of emergency telephone numbers clearly posted next to the telephone. These numbers should include a doctor, hospital, and fire and police departments. Also, the work area should have a first-aid kit for treating minor injuries and eye-flushing kits readily available. You should know where these items are kept.

Fire Hazards and Prevention

Gasoline is a highly flammable volatile liquid. Something that is flammable catches fire and burns easily. A volatile liquid is one that vaporizes very quickly. Flammable volatile liquids are potential fire bombs. Always keep gasoline, ethanol, or diesel fuel in an approved safety can **(Figure 6–18)**, and never use gasoline to clean your hands or tools.

The presence of gasoline is so common that its dangers are often forgotten. A slight spark or an increase in heat can cause a fire or explosion. Gasoline fumes are heavier than air. Therefore, when an open container of gasoline is sitting about, the fumes spill out over the sides of the container. These fumes are more flammable than liquid gasoline and can easily explode.

> **CAUTION!**
>
> *Never siphon gasoline or diesel fuel with your mouth. These liquids are poisonous and can make you sick or fatally ill.*

Never smoke around gasoline or in a shop filled with gasoline fumes. If the vehicle has a gasoline leak or you have caused a leak by disconnecting a fuel line, wipe it up immediately and stop the leak. Make sure that any grinding or welding that may be taking place in the area is stopped until the spill is totally cleaned up and the floor has been flushed with water. The rags used to wipe up the gasoline should be taken outside to dry, then stored in an approved dirty rag container. If vapors are present in the shop, have the doors open and turn on the ventilating system. It takes only a small amount of fuel mixed with air to cause combustion.

Ethanol Most commonly found as E85 (15% gasoline mixed with 85% ethanol), ethanol is a very volatile liquid. Ethanol is a non-petroleum-based fuel and is used as an alternative fuel to gasoline. Ethanol is also used as an additive to increase the octane rating of gasoline. Handle and store E85 in the same way as gasoline.

Diesel Fuel Diesel fuel is not as volatile as gasoline but should be stored and handled in the same way.

Figure 6-18 Flammable liquids should stored in safety-approved containers.

It is also not as refined as gasoline and tends to be a very dirty fuel. It normally contains many impurities, including active microscopic organisms that can be highly infectious. If diesel fuel happens to get on an open cut or sore, thoroughly wash it immediately.

Solvents Cleaning solvents are also not as volatile as gasoline, but they are still flammable. These should be stored and treated in the same way as gasoline. Handle all solvents (or any liquids) with care to avoid spillage. Keep all solvent containers closed, except when pouring. Proper ventilation is very important in areas where volatile solvents and chemicals are used. Solvent and other combustible materials must be stored in approved and designated storage cabinets or rooms **(Figure 6–19)**. Storage rooms should have adequate ventilation.

Discard or clean all empty solvent containers. Solvent fumes in the bottom of these containers are very flammable. Never light matches or smoke near flammable solvents and chemicals, including battery acids.

Rags Oily or greasy rags can also be a source for fires. These rags should be stored in an approved container **(Figure 6–20)** and never thrown out with normal trash. When these oily, greasy, or paint-soaked rags are left lying about or are not stored properly, they

Figure 6-19 Store combustible materials in approved safety cabinets.

Figure 6-20 Dirty rags and towels should be kept in an approved container. *Courtesy of the DuPont Company*

Figure 6-21 Know the location and types of fire extinguishers that are available in the shop.

can cause spontaneous combustion. Spontaneous combustion results in a fire that starts by itself, that is, without a match.

Fire Extinguishers

You should know the location of all fire extinguishers **(Figure 6–21)** and fire alarms in the shop and you should also know how to use them before you need one. You should also be aware of the different types of fires and fire extinguishers **(Table 6–1)**. All extinguishers are marked with a symbol or letter signifying the class of fire for which they are intended. Using the wrong type of extinguisher may cause the fire to grow instead of being put out.

If a fire extinguisher is not handy, a blanket or fender cover may be used to smother the flames. Be careful when doing this because the heat of the fire may burn you and the blanket. If the fire is too great to smother, move everyone away from the fire and call the local fire department. A simple under-the-hood fire can cause the total destruction of the car and the building and can take some lives. You must be able to respond quickly and precisely to avoid a disaster.

Using a Fire Extinguisher Remember, never open doors or windows during a fire unless it is absolutely necessary; the extra draft will only make the fire worse. Make sure the fire department is contacted before or during your attempt to extinguish a fire. To extinguish a fire, stand 6 to 10 feet from the fire. Before releasing the agent from the extinguisher, hold the extinguisher firmly in an upright position. Aim the nozzle at the base and use a side-to-side motion, sweeping the

TABLE 6-1 GUIDE TO EXTINGUISHER SELECTION

	Class of Fire	Typical Fuel Involved	Type of Extinguisher
Class A Fires (green)	**For Ordinary Combustibles** Put out a class A fire by lowering its temperature or by coating the burning combustibles.	Wood Paper Cloth Rubber Plastics Rubbish Upholstery	Water*[1] Foam* Multipurpose dry chemical[4]
Class B Fires (red)	**For Flammable Liquids** Put out a class B fire by smothering it. Use an extinguisher that gives a blanketing, flame-interrupting effect; cover whole flaming liquid surface.	Gasoline Oil Grease Paint Lighter fluid	Foam* Carbon dioxide[5] Halogenated agent[6] Standard dry chemical[2] Purple K dry chemical[3] Multipurpose dry chemical[4]
Class C Fires (blue)	**For Electrical Equipment** Put out a class C fire by shutting off power as quickly as possible and by always using a nonconducting extinguishing agent to prevent electric shock.	Motors Appliances Wiring Fuse boxes Switchboards	Carbon dioxide[5] Halogenated agent[6] Standard dry chemical[2] Purple K dry chemical[3] Multipurpose dry chemical[4]
Class D Fires (yellow)	**For Combustible Metals** Put out a class D fire of metal chips, turnings, or shavings by smothering or coating with a specially designed extinguishing agent.	Aluminum Magnesium Potassium Sodium Titanium Zirconium	Dry powder extinguishers and agents only

**Cartridge-operated water, foam, and soda-acid types of extinguishers are no longer manufactured. These extinguishers should be removed from service when they become due for their next hydrostatic pressure test.*

Notes:

(1) Freezes in low temperatures unless treated with antifreeze solution, usually weighs more than 20 pounds (9 kg), and is heavier than any other extinguisher mentioned.

(2) Also called ordinary or regular dry chemical (sodium bicarbonate).

(3) Has the greatest initial fire-stopping power of the extinguishers mentioned for class B fires. Be sure to clean residue immediately after using the extinguisher so sprayed surfaces will not be damaged (potassium bicarbonate).

(4) The only extinguishers that fight A, B, and C classes of fires. However, they should not be used on fires in liquefied fat or oil of appreciable depth. Be sure to clean residue immediately after using the extinguisher so sprayed surfaces will not be damaged (ammonium phosphates).

(5) Use with caution in unventilated, confined spaces.

(6) May cause injury to the operator if the extinguishing agent (a gas) or the gases produced when the agent is applied to a fire is inhaled.

entire width of the fire **(Figure 6–22)**. Stay low to avoid inhaling the smoke. If it gets too hot or too smoky, get out. Remember, never go back into a burning building for anything. To help remember how to use an extinguisher, remember the word “PASS.”

Pull the pin from the handle of the extinguisher.

Aim the extinguisher’s nozzle at the base of the fire.

Squeeze the handle.

Sweep the entire width of the fire with the contents of the extinguisher.

Figure 6-22 Aim the nozzle at the base of the fire and sweep the entire width of the fire.

MANUFACTURERS' WARNINGS AND GOVERNMENT REGULATIONS

A typical shop contains many potential health hazards for those working in it. These hazards can cause injury, sickness, health impairments, discomfort, and even death. These hazards can be classified as:

Chemical hazards—caused by high concentrations of vapors, gases, or dust

Hazardous wastes—those substances that are the result of a service

Physical hazards—include excessive noise, vibration, pressures, and temperatures

Ergonomic hazards—conditions that impede normal body position and motion

Many government agencies have the responsibility to ensure safe work environments for all workers. Federal agencies include the **Occupational Safety and Health Administration (OSHA)**, Mine Safety and Health Administration (MSHA), and National Institute for Occupational Safety and Health (NIOSH). These agencies, as well as state and local governments, have instituted regulations that must be understood and followed. Everyone in a shop has the responsibility to adhere to these regulations.

OSHA

In 1970, OSHA was formed to "assure safe and healthful working conditions for working men and women; by authorizing enforcement of the standards developed under the Act; by assisting and encouraging the States in their efforts to assure safe and healthful working conditions by providing research, information, education, and training in the field of occupational safety and health."

The established safety standards are consistent across the country. It is the employer's responsibility to provide a place of employment free from all recognized hazards. All automotive industry safety and health issues are controlled by OSHA.

RIGHT-TO-KNOW LAW

OSHA also regulates the use of many potentially hazardous materials. The **Environmental Protection Agency (EPA)** regulates their disposal.

Servicing and maintaining a vehicle involves the handling and managing of a wide variety of materials and wastes. Some of these wastes can be toxic to fish, wildlife, and humans when improperly managed. It is to the shop's legal and financial advantage to manage the wastes properly and, even more importantly, to prevent the pollution of our natural resources.

An important part of a safe work environment is the employee's knowledge of potential hazards. All employees in a shop are protected by **Right-To-Know Laws** concerning all potentially hazardous materials. OSHA's Hazard Communication Standard was originally intended for chemical companies and manufacturers that require employees to handle potentially hazardous materials. Since then federal courts decided that these regulations should apply to all companies, including auto repair shops.

The general intent of Right-To-Know Laws is for employers to provide their employees with a safe working place. All employees must be trained about their rights under the legislation, the nature of the hazardous chemicals in their workplace, and the contents of the labels on the chemicals. All of the information about each chemical must be posted on **material safety data sheets (MSDS)** and must be accessible. The manufacturer of the chemical must provide these sheets upon request **(Figure 6-23)**. They detail the chemical composition and precautionary information for all products that can present a health or safety hazard.

An MSDS lists the product's ingredients, potential health hazards, physical description, explosion and fire data, reactivity and stability data, and protection data including first aid and proper handling.

All hazardous materials must be properly labeled, indicating what health, fire, or reactivity hazard they pose and what protective equipment is necessary when handling each. The manufacturer of the hazardous materials must provide all warnings and

HEXANE

MSDS Safety Information

Ingredients

Name: HEXANE (N_HEXANE)
% Wt: >97
OSHA PEL: 500 PPM
ACGIH TLV: 50 PPM
EPA Rpt Qty: 1 LB
DOT Rpt Qty: 1 LB

Health Hazards Data

LD50 LC50 Mixture: LD50:(ORAL RAT) 28.7 KG/MG
Route Of Entry Inds _ Inhalation: YES
Skin: YES
Ingestion: YES
Carcinogenicity Inds _ NTP: NO
IARC: NO
OSHA: NO
Effects of Exposure: ACUTE:INHALATION AND INGESTION ARE HARMFUL AND MAY BE FATAL. INHALATION AND INGESTION MAY CAUSE HEADACHE, NAUSEA, VOMITING, DIZZINESS, IRRITATION OF RESPIRATORY TRACT, GASTROINTESTINAL IRRITATION AND UNCONSCIOUSNESS. CONTACT W/SKIN AND EYES MAY CAUSE IRRITATION. PROLONGED SKIN MAY RESULT IN DERMATITIS (EFTS OF OVEREXP)
Signs And Symptoms Of Overexposure: HLTH HAZ:CHRONIC:MAY INCLUDE CENTRAL NERVOUS SYSTEM DEPRESSION.
Medical Cond Aggravated By Exposure: NONE IDENTIFIED.
First Aid: CALL A PHYSICIAN. INGEST:DO NOT INDUCE VOMITING. INHAL:REMOVE TO FRESH AIR. IF NOT BREATHING, GIVE ARTIFICIAL RESPIRATION. IF BREATHING IS DIFFICULT, GIVE OXYGEN. EYES:IMMED FLUSH W/PLENTY OF WATER FOR AT LEAST 15 MINS. SKIN:IMMED FLUSH W/PLENTY OF WATER FOR AT LEAST 15 MINS WHILE REMOVING CONTAMD CLTHG & SHOES. WASH CLOTHING BEFORE REUSE.

Handling and Disposal

Spill Release Procedures: WEAR NIOSH/MSHA SCBA & FULL PROT CLTHG. SHUT OFF IGNIT SOURCES:NO FLAMES, SMKNG/FLAMES IN AREA. STOP LEAK IF YOU CAN DO SO W/OUT HARM. USE WATER SPRAY TO REDUCE VAPS. TAKE UP W/SAND OR OTHER NON-COMBUST MATL & PLACE INTO CNTNR FOR LATER (SU PDAT)
Neutralizing Agent: NONE SPECIFIED BY MANUFACTURER.
Waste Disposal Methods: DISPOSE IN ACCORDANCE WITH ALL APPLICABLE FEDERAL, STATE AND LOCAL ENVIRONMENTAL REGULATIONS. EPA HAZARDOUS WASTE NUMBER:D001 (IGNITABLE WASTE).
Handling And Storage Precautions: BOND AND GROUND CONTAINERS WHEN TRANSFERRING LIQUID. KEEP CONTAINER TIGHTLY CLOSED.
Other Precautions: USE GENERAL OR LOCAL EXHAUST VENTILATION TO MEET TLV REQUIREMENTS. STORAGE COLOR CODE RED (FLAMMABLE).

Fire and Explosion Hazard Information

Flash Point Method: CC
Flash Point Text: _9F,_23C
Lower Limits: 1.2%
Upper Limits: 77.7%
Extinguishing Media: USE ALCOHOL FOAM, DRY CHEMICAL OR CARBON DIOXIDE. (WATER MAY BE INEFFECTIVE).
Fire Fighting Procedures: USE NIOSH/MSHA APPROVED SCBA & FULL PROTECTIVE EQUIPMENT (FP N).
Unusual Fire/Explosion Hazard: VAP MAY FORM ALONG SURFS TO DIST IGNIT SOURCES & FLASH BACK. CONT W/STRONG OXIDIZERS MAY CAUSE FIRE. TOX GASES PRDCED MAY INCL:CARBON MONOXIDE, CARBON DIOXIDE.

Figure 6–23 Material safety data sheets are an important part of employee training and should be readily accessible.

precautionary information, which must be read and understood by the user before using the product. Also, a list of all hazardous materials used in the shop must be posted for the employees to see.

Shops must also keep records of all training programs, records of accidents or spill incidents, satisfaction of employee requests for specific chemical information via the MSDS, and a general right-to-know compliance procedure manual utilized within the shop.

Hazardous Wastes

WARNING!

When handling any hazardous waste material, be sure to wear the proper safety equipment recommended by the MSDS. Follow all required procedures. This includes the use of approved respirator equipment.

Many repair and service procedures generate **hazardous wastes**, such as dirty solvents. Something is classified as a hazardous waste by the EPA if it is on its list of known harmful materials. A complete EPA list of hazardous wastes can be found in the Code of Federal Regulations. It should be noted that a material is only considered a hazardous waste when the shop is ready to dispose of it.

Regulations on the generation and handling of hazardous waste have led to the development of equipment found in shops. Examples of these are thermal cleaning units, close-loop steam cleaners, waste oil furnaces, oil filter crushers, refrigerant recycling machines, engine coolant recycling machines, and highly absorbent cloths.

WARNING!

The shop is ultimately responsible for the safe disposal of hazardous waste, even after it leaves the shop. Only licensed waste removal companies should dispose of the waste. In addition to hauling the waste away, they will also take care of all the paperwork, deal with the various government agencies, and advise the shop on how to recover the disposal costs. If there is a hazardous waste spill, contact the National Response Center (1-800-424-8802) immediately. Failure to do so can result in a $10,000 fine or a year in jail, or both.

Always keep hazardous waste separate from other wastes. Make sure they are properly labeled and sealed in the recommended containers. The storage area should be covered and may need to be fenced and locked if vandalism could be a problem.

Guidelines for Handling Shop Wastes

Some of the common hazardous wastes, along with what you should do with them follows:

Oil Recycle oil. Set up equipment, such as a drip table or screen table with a used-oil collection bucket, to collect oil that drips off parts. Place drip pans underneath vehicles that are leaking fluids onto the storage area. Do not mix other wastes with used oil, except as allowed by your recycler. Used oil generated by a shop (and/or oil received from household do-it-yourself generators) may be burned on site in a commercial space heater. Also, used oil may be burned for energy recovery. Contact state and local authorities to determine requirements and to obtain necessary permits.

Oil Filters Drain for at least 24 hours, crush (**Figure 6–24**) and recycle used oil filters.

Figure 6-24 A hydraulic single oil filter crusher.
Courtesy of SPX Service Solutions

Batteries Recycle batteries by sending them to a reclaimer or back to the distributor. Keeping shipping receipts can demonstrate that you have recycled. Store batteries in a watertight, acid-resistant container. Inspect batteries for cracks and leaks when they come in. Treat a dropped battery as if it were cracked. Acid residue is hazardous because it is corrosive and may contain lead and other toxins. Neutralize spilled acid by covering it with baking soda or lime, and dispose of all hazardous material.

Metal Residue from Machining Collect metal filings when machining metal parts. Keep separate and recycle if possible. Prevent metal filings from falling into a storm sewer drain.

Refrigerants Recover and/or recycle refrigerants during the servicing and disposal of motor vehicle air conditioners and refrigeration equipment. It is not allowable to knowingly vent refrigerants into the atmosphere. Recovery and/or recycling during servicing must be performed by an EPA-certified technician using certified equipment and following specified procedures.

Solvents Replace hazardous chemicals with less toxic alternatives that have equal performance. For example, substitute water-based cleaning solvents for petroleum-based solvent degreasers. To reduce the amount of solvent used when cleaning parts, use a two-stage process: dirty solvent followed by fresh solvent. Hire a hazardous waste management service to clean and recycle solvents. (Some spent solvents must be disposed of as hazardous waste, unless recycled properly.) Store solvents in closed containers to prevent evaporation. Evaporation of solvents contributes to ozone depletion and smog formation. In addition, the residue from evaporation must be treated as a hazardous waste. Properly label spent solvents and store on drip pans or in diked areas and only with compatible materials.

Containers Cap, label, cover, and properly store aboveground outdoor liquid containers and small tanks within a diked area and on a paved impermeable surface to prevent spills from running into surface or ground water.

Other Solids Store materials such as scrap metal, old machine parts, and worn tires under a roof or tarpaulin to protect them from the elements and to prevent potential contaminated runoff. Consider recycling tires by retreading them.

Liquid Recycling Collect and recycle coolants from radiators. Store transmission fluids, brake fluids, and solvents containing chlorinated hydrocarbons separately, and recycle or dispose of them properly.

Shop Towels/Rags Keep waste towels in a closed container marked "contaminated shop towels only." To reduce costs and liabilities associated with disposal of used towels, which can be classified as hazardous wastes, investigate using a laundry service that is able to treat the wastewater generated from cleaning the towels.

Asbestos has been identified as a health hazard. *Asbestos* is a term used to describe a number of naturally occurring fibrous materials. It is a carcinogen that causes a number of diseases that result in cancer. Asbestos-caused cancer, or mesothelioma, is a form of lung cancer. When breathed in, the asbestos fibers cause scarring of the lungs and/or damage to the lung's air passages. The injuries and scars become an effective holding place for the asbestos. Obviously, you want to avoid breathing in asbestos dust and fibers. Be careful when working with asbestos materials, such as brake pads, clutch discs, and some engine gaskets. All asbestos waste must be disposed of in accordance with OSHA and EPA regulations.

For more on work environment safety, contact the U.S. EPA Office of Compliance at http://es.inel.gov.

KEY TERMS

Abrasive cleaning
Asbestos
Bloodborne pathogens
Chemical cleaning
Environmental Protection Agency (EPA)
Hazardous waste
Jack stands
Material safety data sheets (MSDS)
Occupational Safety and Health Administration (OSHA)
Pneumatic tools
Power tools
Right-To-Know Laws
Safety glasses
Safety stands
Thermal cleaning

SUMMARY

- Dressing safely for work is very important. Wear snug-fitting clothing, eye and ear protection, protective gloves, steel-toed shoes, and caps to cover long hair.
- When choosing eye protection, make sure it has safety glass and offers side protection.
- A respirator should be worn whenever you are working around toxic fumes or excessive dust.
- When shop noise exceeds safe levels, protect your ears by wearing earplugs or earmuffs.
- Safety while using any tool is essential, and even more so when using power tools. Before plugging in a power tool, make sure the power switch is off. Disconnect the power before servicing the tool.
- Always observe all relevant safety rules when operating a vehicle lift or hoist. Jacks, jack stands, chain hoists, and cranes can also cause injury if not operated safely.
- Use care whenever it is necessary to move a vehicle in the shop. Carelessness and playing around can lead to a damaged vehicle and serious injury.
- Carbon monoxide (CO) gas is a poisonous gas present in engine exhaust fumes. Exhaust must be properly vented from the shop using tailpipe hoses or other reliable methods.
- Adequate ventilation is also necessary when working with any volatile solvent or material.
- Gasoline and diesel fuel are highly flammable and should be kept in approved safety cans.
- Never light matches near any combustible materials.
- It is important to know when to use each of the various types of fire extinguishers. When fighting a fire, aim the nozzle at the base and use a side-to-side sweeping motion.
- Right-To-Know Laws came into effect in 1983 and are designed to protect employees who must handle hazardous materials and wastes on the job.
- Material safety data sheets (MSDS) contain important chemical information and must be furnished to all employees annually. New employees should be given the sheets as part of their job orientation.
- All hazardous and asbestos waste should be disposed of according to OSHA and EPA regulations.

REVIEW QUESTIONS

1. What is the correct way to dispose of used oil filters?
2. Where in the shop should a list of emergency telephone numbers be posted?
3. Which of the following offer(s) the least protection for your eyes?
   a. face shield
   b. safety glasses
   c. splash goggles
   d. prescription glasses
4. Which of the following statements about latex and nitrile gloves is *not* true?
   a. The gloves offer protection against cuts.
   b. The gloves offer protection against disease and grease buildup under and around your fingernails.
   c. Latex gloves are more comfortable but weaken when they are exposed to gas, oil, and solvents.
   d. Nitrile gloves are not as comfortable but they are not affected by gas, oil, and solvents.
5. Describe the correct process for lifting a heavy object.
6. What are bloodborne pathogens and why should technicians be concerned about them?
7. List at least five things you should remember when using hand tools.
8. List at least five precautions that must be adhered to while working with or around a vehicle's battery.
9. How should a class B fire be extinguished?
10. Where can complete EPA lists of hazardous wastes be found?

11. Which of the following statements about safety glasses is true?
    a. They should offer side protection.
    b. The lenses should be made of a shatterproof material.
    c. Some service operations require that additional eye protection be worn with safety glasses.
    d. All of the above statements are true.
12. Gasoline is ______.
    a. highly volatile
    b. highly flammable
    c. dangerous, especially in vapor form
    d. all of the above
13. Technician A says that it is recommended that you wear shoes with nonslip soles in the shop. Technician B says that steel-toed shoes offer the best foot protection. Who is correct?
    a. Technician A c. Both A and B
    b. Technician B d. Neither A nor B
14. Technician A says that used engine coolant should be collected and recycled. Technician B says that all oil-based waste materials can be collected in the same container if an approved waste disposal company is hired to rid the shop of the oil. Who is correct?
    a. Technician A c. Both A and B
    b. Technician B d. Neither A nor B
15. There are many ways to clean parts while they are being serviced. These methods can be grouped into three separate categories. What are they?
16. Federal Right-To-Know Laws concern ______.
    a. auto emission standards
    b. hazards associated with chemicals used in the workplace
    c. employee benefits
    d. hiring practices
17. Which of the following is/are important when working in an automotive shop?
    a. using the proper tool for the job
    b. avoiding loose-fitting clothes
    c. wearing steel-toed shoes
    d. all of the above
18. Technician A says that the volatility of a substance is a statement of how easily the substance vaporizes or explodes. Technician B says that the flammability of a substance is a statement of how well the substance supports combustion. Who is correct?
    a. Technician A c. Both A and B
    b. Technician B d. Neither A nor B
19. Which of the following is *not* recommended for use when trying to extinguish flammable liquid fires?
    a. foam c. water
    b. carbon dioxide d. dry chemical
20. List at least five precautions that must be adhered to while working on a vehicle with a high-voltage system.
21. What is the correct procedure for using a fire extinguisher to put out a fire?
22. Technician A ties his long hair behind his head while working in the shop. Technician B covers her long hair with a brimless cap. Who is correct?
    a. Technician A c. Both A and B
    b. Technician B d. Neither A nor B
23. Technician A uses compressed air to blow dirt from his clothes and hair. Technician B uses compressed air to clean off the top of a workbench. Who is correct?
    a. Technician A c. Both A and B
    b. Technician B d. Neither A nor B
24. Heavy protective gloves should be worn when ______.
    a. welding
    b. grinding metal
    c. working with caustic cleaning solutions
    d. all of the above
25. Proper disposal of oil filters includes ______.
    a. recycling used filters
    b. draining them for at least 24 hours
    c. crushing them
    d. all of the above

PREVENTIVE MAINTENANCE AND BASIC SERVICES

CHAPTER 7

OBJECTIVES

■ Describe the information that should be included on a repair order. ■ Explain how repair costs can be estimated. ■ Explain how the vehicle and its systems can be defined by deciphering its VIN. ■ Explain the importance of preventive maintenance, and list at least six examples of typical preventive maintenance. ■ Understand the differences between the fluids required for preventive maintenance and know how to select the correct one for a particular vehicle. ■ Explain how the design of a vehicle determines what preventive maintenance procedures must be followed.

Preventive services are those services performed not to correct problems but rather to prevent them. These and other basic services are covered in this chapter. All of these services may be performed by technicians in many different types of service facilities—dealerships, independents, and specialty shops. Regardless of what type of shop, the first thing a tech needs to worry about is the repair order.

REPAIR ORDERS

A **repair order (RO)** is written for every vehicle brought into the shop for service. ROs may also be called service or work orders. ROs contain information about the customer, the vehicle, the customer's concern or request, an estimate of the cost for the services, and the time the services should be completed **(Figure 7–1)**. ROs are legal documents that are used for many other purposes, such as payroll and general record keeping **(Figure 7–2)**. Legally, an RO protects the shop and the customer.

Although every shop may enter different information onto the original RO, most ROs contain the following information:

- Complete customer information
- Complete vehicle identification
- The service history of the vehicle
- The customer's complaint
- The preliminary diagnosis of the problem
- An estimate of the amount of time required for the service
- An estimate of the costs of the parts involved in the service
- The time the services should be completed
- The name or other identification of the technician assigned to perform the services
- The actual services performed with their cost
- The parts replaced during the services
- Recommendations for future services
- The total cost of the services

Figure 7–1 Service facilities run smoothly when there is good communications between the customer and the technician.

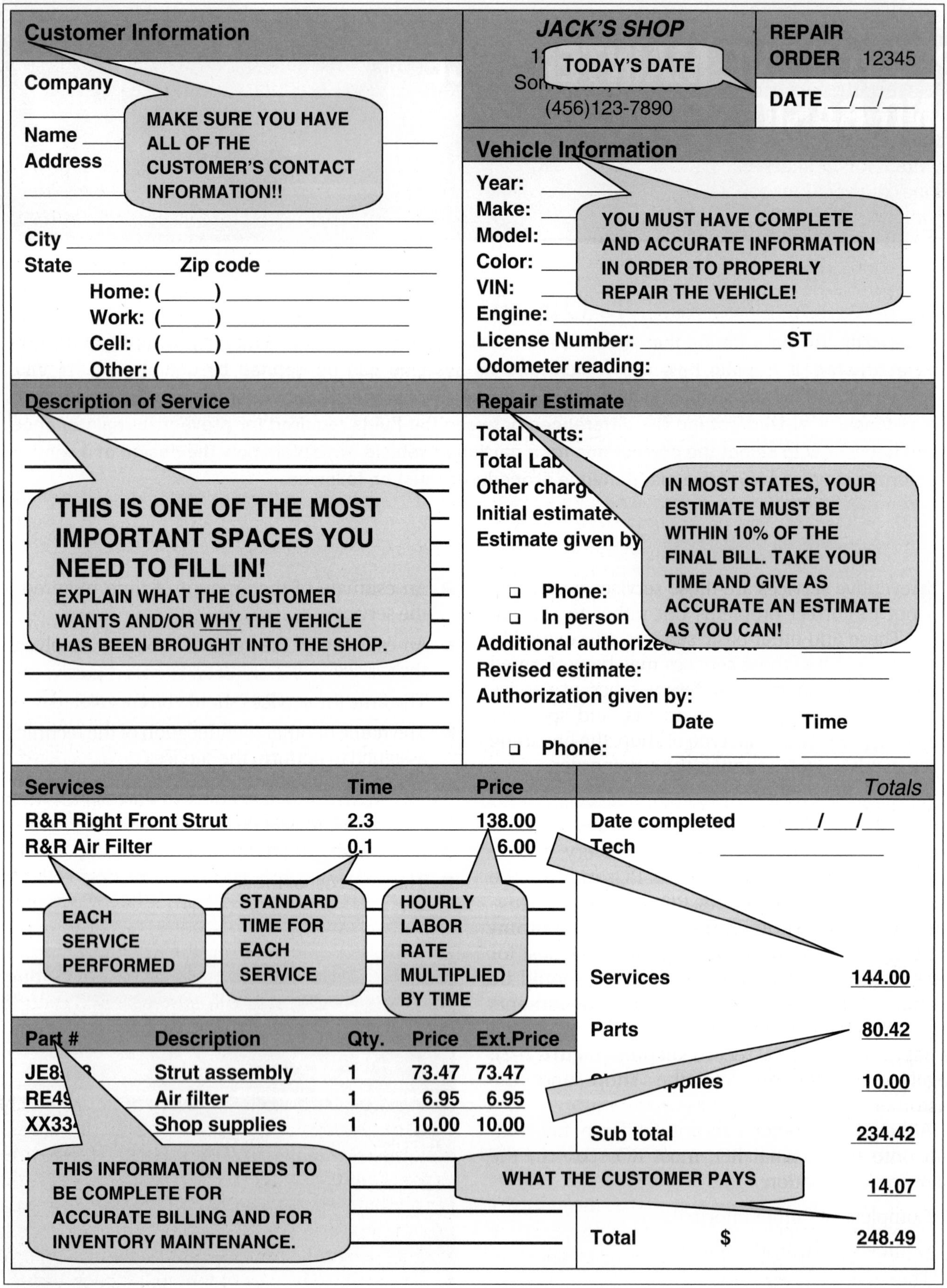

Customer Information

Company ______

Name ______

Address ______

City ______

State ______ Zip code ______

Home: (_____) ______

Work: (_____) ______

Cell: (_____) ______

Other: (_____)

JACK'S SHOP

(456)123-7890

REPAIR ORDER 12345

DATE __/__/__

Vehicle Information

Year: ______

Make: ______

Model: ______

Color: ______

VIN: ______

Engine: ______

License Number: ______ ST ______

Odometer reading:

Description of Service

Repair Estimate

Total Parts: ______

Total Lab...

Other charg...

Initial estimate:

Estimate given by

❑ Phone:

❑ In person

Additional authorized amount: ______

Revised estimate: ______

Authorization given by:

	Date	Time
❑ Phone:	______	______

Services	Time	Price
R&R Right Front Strut	2.3	138.00
R&R Air Filter	0.1	6.00

Part #	Description	Qty.	Price	Ext.Price
JE8...	Strut assembly	1	73.47	73.47
RE4...	Air filter	1	6.95	6.95
XX33...	Shop supplies	1	10.00	10.00

Totals

Date completed __/__/__

Tech ______

Services	144.00
Parts	80.42
Shop supplies	10.00
Sub total	234.42
	14.07
Total $	248.49

Figure 7-2 A completed repair order.

An RO is signed by the customer, who in doing so authorizes the service and accepts the terms noted on the RO. The customer also agrees to pay for the services when they are completed. Many states require a customer signature to begin repair work and for a change in the original estimate. If a signature is not required for changes in the original estimate, all phone conversations concerning the estimate should be noted on the RO.

In most cases when a customer signs the RO, he or she acknowledges the shop's right to impose a mechanic's lien. This lien basically says that the shop may gain possession of the vehicle if the customer does not pay for the agreed-upon services and the vehicle remains at the shop for a period of 90 or more days. This clause ensures that the shop will receive some compensation for the work performed, whether or not the customer pays the bill.

After the work has been completed and the RO filed, it becomes a legal service record. Service records are kept by the shop to maintain the vehicle's service history and for legal purposes. Evidence of repairs and recommended repairs is very important for settling potential legal disputes with the vehicle's owner in the future.

Computerized Shop Management Systems

Today, most service facilities use computerized shop management software. The information for the completion of an RO is input on the computer's keyboard. The software package also helps in the estimation of repair costs. The software also takes information from the RO and saves it in various files. These files are used for many purposes, such as schedule reminders, bookkeeping, vehicle/owner history, and tracking employee productivity. Notes can also be added to the RO (these do not appear on the RO). These personal notes can be used to remind the shop of commitments made to the customer, any special information about the customer and/or the vehicle, and any abnormal events that took place during the customer's last visit to the shop.

When the customer arrives at the shop, the computer can quickly recall all pertinent information about the vehicle. Typically, all the service writer needs to do is key in the vehicle's license number, the vehicle's identification number, or the owner's name. If the customer has been to the shop before, all information will be available to the service writer. Also, most shop management software relies on numerical codes to denote what services have been and will be performed. These codes serve as shortcuts so the service writer does not need to key in the description of each service. The codes are designated by the software company or the vehicle's manufacturer. At a dealership, these link directly to the warranty reimbursement file.

Parts Replacement

Very often when a service is performed, parts are replaced. This appears on the RO as "R&R," which stands for "remove and replace." In a dealership, nearly all of the replacement parts are original equipment manufacturer (OEM) parts obtained through the parts department. Some replacement parts installed by a dealership and nearly all parts installed by other service facilities are from the aftermarket. Other replacement parts may be rebuilt or remanufactured units. These are based on parts that have failed and have been rebuilt or restored to specifications. Normally, remanufactured parts are totally tested, disassembled, cleaned, and machined, and all of the weak or dysfunctional parts replaced. If this process is completed correctly, the remanufactured part will be as reliable as an original equipment (OE) part.

Parts that are destined to be rebuilt have a value to them. Therefore, when they are replaced, the original (replaced) part has a core value and the customer is charged a "**core charge**." The replaced part is called a **core**, because it is the basis for rebuilding. The core charge represents the value of the failed part. Core charges are built into the RO and can be negated when the shop or customer returns the core. Core charges are typical for replacement engines, transmissions, starters, generators, and brake shoes.

If the replaced part has no core value, the shop disposes of the part. However, many shops offer the part to the customer as proof that the part was removed and a new one installed. At times, the customer will insist that the part be given them. Always place the part in plastic or another container before putting it inside the vehicle. This will prevent any dirt on the part from getting on anything inside the vehicle.

Sublet Repairs

Service facilities typically do not perform all possible services. Often another business will be contracted to perform a service or part of the service. This is referred to as subletting. **Sublet repairs** are sent to shops that specialize in certain repairs, such as radiator repairs. Often a repair, or part of the repair, is performed by another person or company outside the dealership or service facility. The cost of the subletting is added to the costs of the services performed by the service facility. Often the customer is billed slightly more than the actual cost of the sublet repair.

When estimating the cost of a repair or service:

- Make sure you have the correct contact information for the customer.
- Make sure you have the correct information about the vehicle.
- Always use the correct labor and parts guide or database for that specific vehicle.
- Locate the exact service for that specific vehicle in the guide or database.
- Using the guidelines provided in the guide or database, choose the proper time allocation listed for the service.
- Multiply the allocated time by the shop's hourly labor rate.
- If any sublet repairs are anticipated, list this service as a sublet repair and add the cost to the labor costs.
- Using the information given in the guide or database, identify the parts that will be replaced for that service.
- Locate the cost of the parts in the guide or database or in the catalogs used by the shop.
  - Repeat the process for all other services required or requested by the customer.
  - Multiply the time allocations by the shop's hourly flat rate.
  - Add all of the labor costs together; this sum is the labor estimate for those services.
  - Add the cost of all the parts together; this sum is the estimate for the parts required for the services.
  - Add the total labor and parts costs together. If the shop charges a standard fee for shop supplies, add it to the labor and parts total. This sum is the cost estimate to present to the customer.

Figure 7-3 Guidelines for estimating the cost of repairs.

Estimating Repair Costs

For legal reasons and to establish good customer relations, estimated repair costs must be calculated with as much accuracy as possible. The customer is protected against being charged more than the estimate given on the RO, unless he or she later authorizes a higher amount. Some states allow shops to be within 10% of the estimate, whereas others hold the shop to the amount that was estimated. **Figure 7–3** lists some things to follow when estimating the cost of services and repairs.

VEHICLE IDENTIFICATION

Before any service is done to a vehicle, it is important for you to know exactly what type of vehicle you are working on. The best way to do this is to refer to the **vehicle's identification number (VIN)**. The VIN is given on a plate behind the lower corner of the driver's side of the windshield as well as other locations on the vehicle. The VIN is made up of seventeen characters and contains all pertinent information about the vehicle. The use of the seventeen number and letter code became mandatory beginning with 1981 vehicles and is used by all manufacturers of vehicles both domestic and foreign. Most new vehicles have a scan code below the VIN **(Figure 7–4)**.

Each character of a VIN has a particular purpose. The first character identifies the country where the vehicle was manufactured; for example:

- 1 or 4 – U.S.A.
- 2 – Canada

Figure 7-4 A vehicle identification plate.

- 3 – Mexico
- J – Japan
- K – Korea
- S – England
- W – Germany

The second character identifies the manufacturer; for example:

- A – Audi
- B – BMW
- C – Chrysler
- D – Mercedes Benz
- F – Ford
- G – General Motors
- H – Honda
- N – Nissan
- T – Toyota

The third character identifies the vehicle type or manufacturing division (passenger car, truck, bus, and so on). The fourth through eighth characters identify the features of the vehicle, such as the body style, vehicle model, and engine type.

The ninth character is used to identify the accuracy of the VIN and is a check digit. The tenth character identifies the model year; for example:

- S – 1995
- V – 1997
- W – 1998
- Y – 2000
- 1 – 2001
- 3 – 2003
- 5 – 2005
- 7 – 2007
- 9 – 2009

The eleventh character identifies the plant where the vehicle was assembled, and the twelfth to seventeenth characters identify the production sequence of the vehicle as it rolled off the manufacturer's assembly line.

PREVENTIVE MAINTENANCE

Preventive maintenance (PM) involves performing certain services to a vehicle on a regularly scheduled basis before there is any sign of trouble. Regular inspection and routine maintenance can prevent major breakdowns and expensive repairs. It also keeps cars and trucks running efficiently and safely.

A recent survey of 2,375 vehicles conducted during National Car Care Month found that more than 90% of the cars looked at needed some form of service. The cars were inspected for exhaust emissions, fluid levels, tire pressure, and other safety features. The results indicated that 34% of the cars had restricted air filters; 27% had worn belts; 25% had clogged PCV filters; 14% had worn hoses; and 20% had bad batteries, battery cables, or terminals.

During the fluid and cooling system inspection, 39% failed due to bad or contaminated transmission or power steering fluid, 36% had worn-out or dirty engine oil, 28% had inadequate cooling system protection, and 8% had a faulty radiator cap.

In the safety category, 50% failed due to worn or improperly inflated tires, 32% had inoperative headlights or brake lights, and 14% had worn wipers.

A typical PM schedule recommends particular service at mileage or time intervals. Driving habits and conditions should also be used to determine the frequency of PM service intervals. For example, vehicles that frequently are driven for short distances in city traffic may require more frequent oil changes due to the more rapid accumulation of condensation and unburned fuel in the oil. Most manufacturers also specify more frequent service intervals for vehicles that are used to tow a trailer or those that operate in extremely dusty or unusual conditions.

Safety Inspections

Several states and provinces require annual or biennial vehicle safety inspections. The intent of these inspections is to improve road safety. Research shows that states with annual safety inspection programs have 20% fewer accidents than states without safety inspections. These inspections consist of a series of safety-related checks of various systems and areas of a vehicle. For example, some common checks are shown in **Figure 7–5**. The exact systems and subsystems that are inspected vary. The inspections are part of the vehicle registration process. Often automobile dealers are required to complete a safety inspection on all used vehicles before they are sold and report the results to the customer.

INSPECT WINDSHIELD AND OTHER GLASS FOR:

Cloudiness, distortion, or other obstruction to vision.
Cracked, scratched, or broken glass.
Window tinting.
Operation of front door glass.

INSPECT WINDSHIELD WIPER/WASHER FOR:

Operating condition.
Condition of blade.

INSPECT WINDSHIELD DEFROSTER FOR:

Operating condition.

INSPECT MIRRORS FOR:

Rigidity of mounting.
Condition of reflecting surface.
View of road to rear.

INSPECT HORN FOR:

Electrical connections, mounting, and horn button.
Emits a sound audible for a minimum of 200 feet.

INSPECT DRIVER'S SEAT FOR:

Anchorage.
Location.
Condition.

INSPECT SEAT BELTS FOR:

Condition.

INSPECT HEADLIGHTS FOR:

Approved type, aim, and output.
Condition of wiring and switch.
Operation of beam indicator.

INSPECT OTHER LIGHTS FOR:

Operation of all lamps, lens color, and condition of lens.
Aim of fog and driving lamps.

INSPECT SIGNAL DEVICE FOR:

Correct operation of indicators (visual or audible).
Illumination of all lamps, lens color, and condition of lens.

INSPECT FRONT DOORS FOR:

Handle or opening device permits the opening of the door from the outside and inside of the vehicle.
Latching system that holds door in its proper closed position.

INSPECT HOOD FOR:

Operating condition of hood latch.

INSPECT FLUIDS FOR:

Levels that are below the proper level.

INSPECT BELTS AND HOSES FOR:

Belt tension, wear, or absence.
Hose damage.

INSPECT POLLUTION CONTROL SYSTEM FOR:

Presence of emissions system—evidence that no essential parts have been removed, rendered inoperative, or disconnected.

INSPECT BATTERY FOR:

Proper anchorage.
Loose or damaged connections.

INSPECT FUEL SYSTEM FOR:

Any part that is not securely fastened.
Liquid fuel leakage.
Fuel tank filler cap for presence.

INSPECT EXHAUST SYSTEM FOR:

Damaged exhaust—manifold, gaskets, pipes, mufflers, connections, etc.
Leakage of gases at any point from motor to point discharged from system.

INSPECT STEERING AND SUSPENSION FOR:

Play in steering wheel.
Wear in bushings, kingpins, ball joints, wheel bearings, tie-rod ends.
Looseness of gear box on frame, condition of drag link, and steering arm.
Wheel alignment and axle alignment.
Broken spring leaves and worn shackles.
Shock absorbers.
Broken frame.
Broken or missing engine mounts.
Lift blocks.

INSPECT FLOOR PAN FOR:

Holes that allow exhaust gases to enter occupant compartment.
Conditions that create a hazard to the occupants.

INSPECT BRAKES FOR:

Worn, damaged, or missing parts.
Worn, contaminated, or defective linings or drums.
Leaks in system and proper fluid level.
Worn, contaminated, or defective disc pads or discs.
Excessive pedal play.

INSPECT PARKING BRAKE FOR:

Proper adjustment.

INSPECT TIRES, WHEELS, AND RIMS FOR:

Proper inflation.
Loose or missing lug nuts.
Condition of tires, including tread depth.
Mixing radials and bias ply tires.
Wheels that are cracked or damaged so as to cause unsafe operation.

Figure 7-5 A safety inspection may include these items.

BASIC SERVICES

The services required for a PM program are generally noncorrective procedures. However, often while performing PM on a vehicle, a technician notices the need for a minor repair. Both PM and those basic minor services are covered in the rest of this chapter.

Customer Care

Whenever you do any service to a vehicle, use fender covers (Figure 7–6) and do not leave fingerprints on the exterior or interior of the car. If oil or grease gets on the car, clean it off.

Engine Oil

Engine oil is a clean or refined form of **crude oil**. Crude oil, when taken out of the ground, is dirty and does not work well as a lubricant for engines. Crude oil must be refined to meet industry standards. Engine oil (often called motor oil) is just one of the many products that come from crude oil. Engine oil is specially formulated so that it:

- Can flow easily through the engine
- Provides lubrication without foaming
- Reduces friction and wear
- Prevents the formation of rust and corrosion
- Cools the engine parts it flows over
- Keeps internal engine parts clean

Engine oil contains many additives, each intended to improve the effectiveness of the oil. The **American Petroleum Institute** (**API**) classifies engine oil as standard or S-class for passenger cars and light trucks and as commercial or C-class for heavy-duty commercial applications. The various types of oil within each class are further rated according to their ability to meet the engine manufacturers' warranty specifications (**Table 7–1**). Engine oils can be classified as **energy-conserving** (fuel-saving) **oils**. These are designed to reduce friction, which in turn reduces fuel consumption. Friction modifiers and other additives are used to achieve this.

Figure 7-6 Fender covers should be used when working under the hood.

TABLE 7-1 ENGINE OIL SERVICE RATINGS

RATING	COMMENTS
SA	Straight mineral oil (no additives), not suitable for use in any engine
SB	Non-detergent oil with additives to control wear and oil oxidation
SC	Obsolete since 1964
SD	Obsolete since 1968
SE	Obsolete since 1972
SF	Obsolete since 1980
SG	Obsolete since 1988
SH	Obsolete since 1993
SJ	Obsolete since 1997
SL	Started in 2001
SM	Started in 2005

In addition to the API rating, oil **viscosity** is important in selecting engine oil. The ability of oil to resist flowing is its viscosity. The thicker the oil, the higher its viscosity rating. Viscosity is affected by temperature; hot oil flows faster than cold oil. Oil flow is important to the life of an engine. Because an engine operates under a wide range of temperatures, selecting the correct viscosity is very important.

The **Society of Automotive Engineers** (**SAE**) has established an oil viscosity classification system that is accepted throughout the industry. This system is a numeric rating in which the higher viscosity, or heavier weight, oils receive the higher numbers. For example, oil classified as SAE 50 weight oil is heavier and flows slower than SAE 10 weight oil. Heavyweight oils are best suited for use in high-temperature regions. Low-weight oils work best in low-temperature operations.

Figure 7-7 The SAE classification and the API rating are displayed in this way on a container of oil.

Although single viscosity oils are available, most engine oils are **multiviscosity** oils. These oils carry a combined classification such as 10W-30. This rating says the oil has the viscosity of both a 10- and a 30-weight oil. The "W" after the 10 notes that the oil's viscosity was tested at 0°F (–18°C). This is commonly referred to as the "winter grade." Therefore, the 10W means the oil has a viscosity of 10 when cold. The 30 rating is the hot rating. This rating was the result of testing the oil's viscosity at 212°F (100°C). To formulate multiviscosity oils, polymers are blended into the oil. Polymers expand when heated. With the polymers, the oil maintains its viscosity to the point where it is equal to 30-weight oil. The SAE classification and the API rating are displayed on the container of oil **(Figure 7–7)**.

ISLAC Oil Ratings The International Lubrication Standardization and Approval Committee (ISLAC) has developed an oil rating that combines SAE viscosity ratings and the API service rating. If engine oil meets the standards, a "sunburst" symbol is displayed on the container **(Figure 7–8)**. This means the oil is suitable for use in nearly any gasoline engine.

SHOP TALK

Many engines have very specific requirements. Always install the type of oil specified by the manufacturer. Never assume that a particular type of oil can be used in an engine.

Figure 7-8 The ILSAC certification mark, commonly referred to as "the Starburst."

Synthetic Oils Synthetic oils are considered synthetic because the finished product does not occur naturally and it was made through a chemical, not natural, process. The introduction of synthetic oils dates back to World War II. Synthetic oils have many advantages over mineral oils, including better fuel economy and engine efficiency by reducing friction; they have low viscosity in low temperatures and a higher viscosity in warm temperatures, and they tend to have a longer useful life. Synthetic oils cost much more than mineral oils, which is the biggest drawback for using them. Engine oils that are blends of mineral oils and synthetics to keep the cost down are available but offer many of the advantages of synthetic oil.

Maintenance Perhaps the PM service that is best known to the public is changing the engine's oil and filter. Because oil is the lifeblood of an engine, it is critical that the oil and filter are changed on a regular basis. Photo Sequence 3 shows the steps involved in changing the engine oil and oil filter. Whenever doing this, make sure the oil is the correct rating for the vehicle.

In between oil and filter changes, the level of the oil should be periodically checked. When doing this, make sure the vehicle is parked on level ground. Locate and remove the oil **dipstick**. With a clean rag, wipe the oil from the dipstick and reinsert it all the way in its tube. Remove it again and check the level of the oil **(Figure 7–9)**. If the level is at the "full" mark, the level is okay. If the level is at the "add" mark, this means the level is about 1 quart low. Regardless of the level, examine the oil for evidence of dirt. If the oil is contaminated, it must be changed.

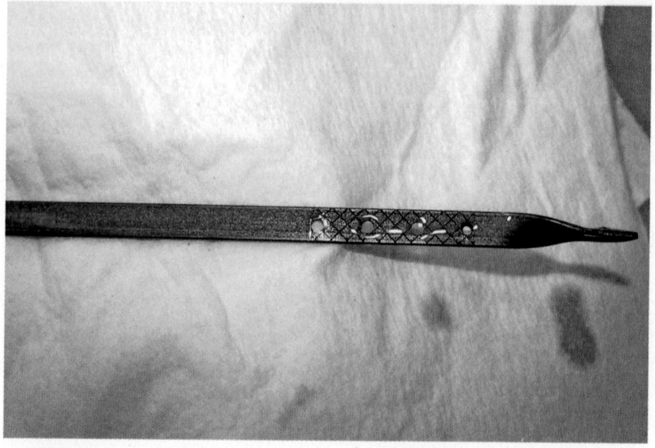

Figure 7-9 Check the engine's oil level with the dipstick.

P3-1 *Always make sure the vehicle is positioned safely on a lift or supported by jack stands before working under it. Before raising the vehicle, allow the engine to run awhile. After it is warm, turn off the engine.*

P3-2 *The tools and other items needed to change the engine's oil and oil filter are rags, a funnel, an oil filter wrench, safety glasses, and a wrench for the drain plug.*

P3-3 *Place the oil drain pan under the drain plug before beginning to drain the oil.*

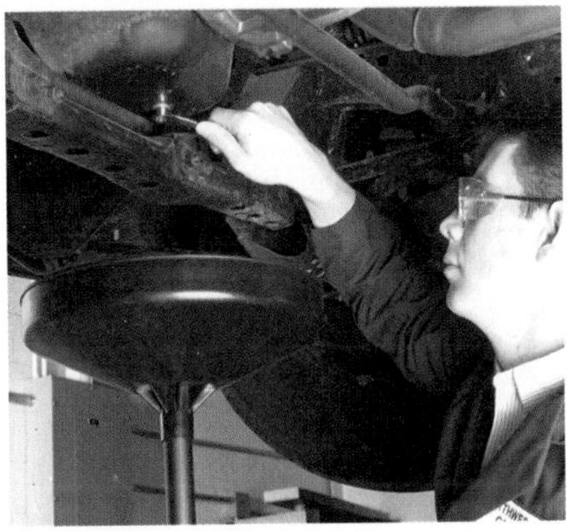

P3-4 *Loosen the drain plug with the appropriate wrench. After the drain plug is loosened, quickly remove it so the oil can freely drain from the oil pan.*

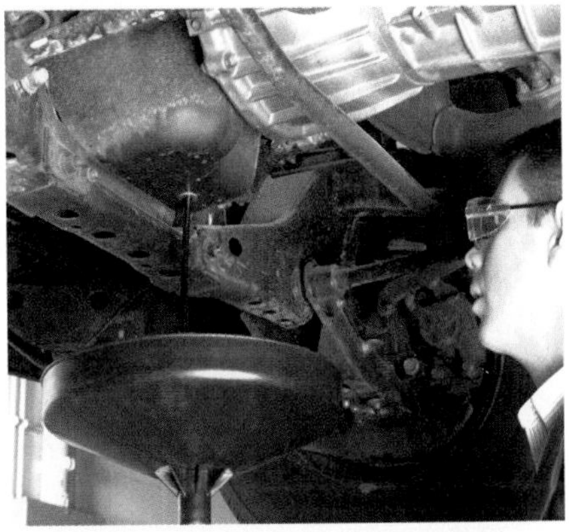

P3-5 *Make sure the drain pan is positioned so it can catch all of the oil.*

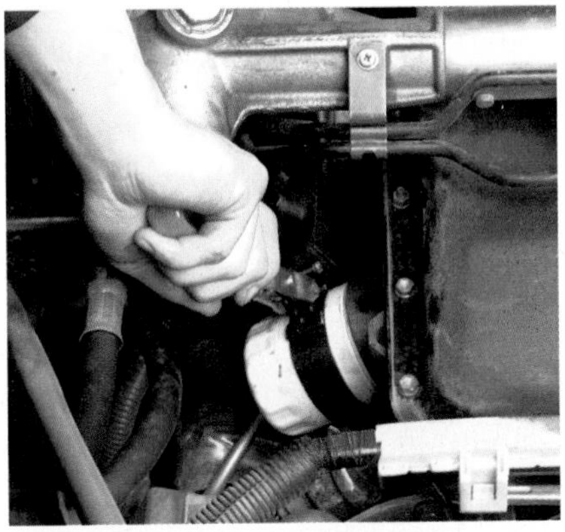

P3-6 *While the oil is draining, use an oil filter wrench to loosen and remove the oil filter.*

P3-7 *Make sure the oil filter seal came off with the filter. Then place the filter into the drain pan so it can drain. After it has completely drained, discard the filter according to local regulations.*

P3-8 *Wipe off the oil filter sealing area on the engine block. Then apply a coat of clean engine oil onto the new filter's seal.*

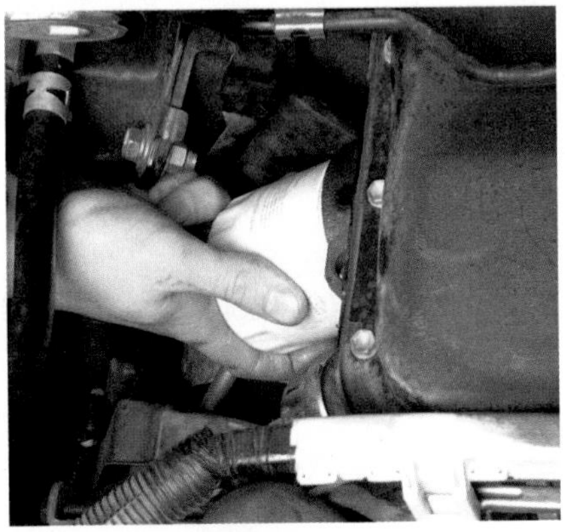

P3-9 *Install the new filter and hand-tighten it. Oil filters should be tightened according to the directions given on the filter.*

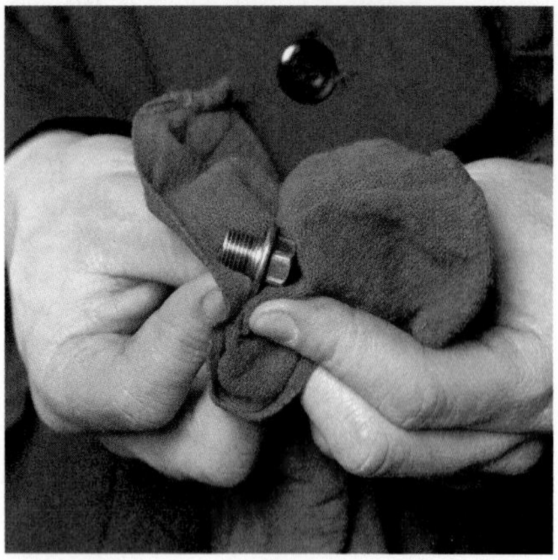

P3-10 *Prior to installing the drain plug, wipe off its threads and sealing surface with a clean rag.*

P3-11 *Tighten the drain plug according to the manufacturer's recommendations. Overtightening can cause thread damage, whereas undertightening can cause an oil leak.*

P3-12 *With the oil filter and drain plug installed, lower the vehicle and remove the oil filler cap.*

P3-13 *Carefully pour the oil into the engine. The use of a funnel usually keeps oil from spilling on the engine.*

P3-14 *After the recommended amount of oil has been put in the engine, check the oil level.*

P3-15 *Start the engine and allow it to reach normal operating temperature. While the engine is running, check the engine for oil leaks, especially around the oil filter and drain plug. If there is a leak, shut down the engine and correct the problem.*

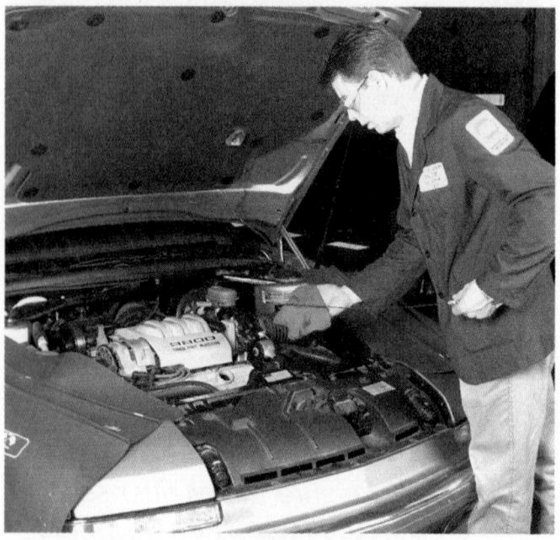

P3-16 *After the engine has been turned off, recheck the oil level and correct it as necessary.*

Cooling System

Whenever you change an engine's oil, you should also do a visual inspection of the different systems under the hood, including the cooling system. Inspect all cooling system hoses for signs of leakage and/or damage. Replace all hoses that are swollen, cracked, or show signs of leakage. The radiator should also be checked for signs of leaks; if any are evident the radiator should be repaired or replaced. Also, check the front of the radiator for any buildup of dirt and bugs **(Figure 7–10)**. This can restrict airflow through the radiator and should be removed by thorough cleaning.

The level and condition of the engine's coolant should also be checked. Check the coolant's level at the coolant recovery tank **(Figure 7–11)**. It should be between the "low" and "full" lines. If the level is too low, more coolant should be added through the cap of the tank, not the radiator. Bring the level up to the "full" line. Always use the correct type of coolant when topping off or replacing it. Look at the color of the coolant when checking the level. It should be green, or perhaps orange, but it should not look rusty or cloudy. If the coolant looks contaminated, the cooling system should be flushed and new coolant put into the system.

Figure 7-10 A buildup of dirt and bugs can restrict airflow through the radiator.

Figure 7-11 The level of coolant in the cooling system should be checked at the coolant recovery tank.

SHOP TALK

Recycle all used antifreeze/coolant or take it to an authorized collection point. Do not dump old coolant into a sewage drain, the ground, or any body of water.

CAUTION!

Never remove the radiator cap when the coolant is hot. Because the system is pressurized, the coolant can be hotter than boiling water and will cause severe burns. Wait until the top radiator hose is not too hot to touch. Then press down on the cap and slowly turn it until it hits the first stop. Now slowly let go of the cap. If there is any built-up pressure in the system, it will be released when the cap is let up. After all pressure has been exhausted, turn the radiator cap to remove it.

Coolant Engine **coolant** is a mixture of water and antifreeze/coolant. Water alone has a boiling point of 212°F (100°C) and a freezing point of 32°F (0°C) at sea level. A mixture of 67% antifreeze and 33% water will raise the boiling point of the mixture to 235°F (113°C) and lower the freezing point to −92°F (−69°C). As can be seen in **Figure 7–12**, antifreeze in excess of 67% will actually raise the freezing point of the mixture. Normally, the recommended mixture is a 50/50 solution of water and antifreeze/coolant. Some coolant suppliers offer a mixture of pure water and

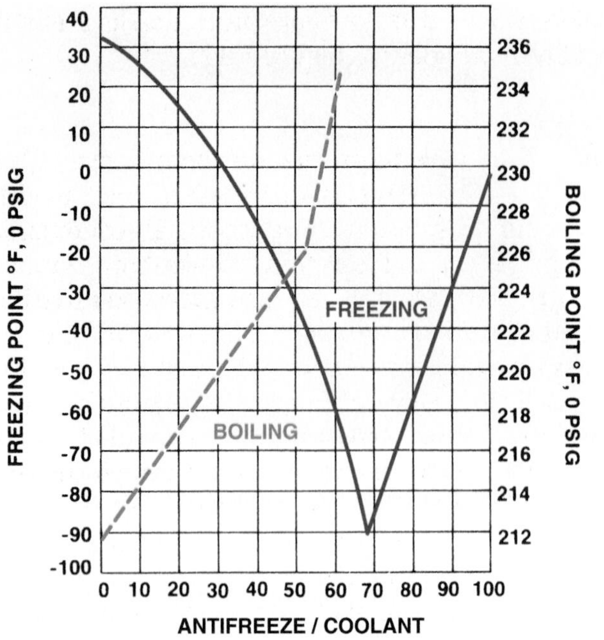

Figure 7–12 The relationship of the percentage of antifreeze in the coolant to the coolant's freezing and boiling points.

antifreeze that can be used to top off a cooling system when the level is low (**Figure 7–13**).

The antifreeze concentration must always be a minimum of 44% all year and in all climates. If the percentage is lower than 44%, engine parts may be eroded by cavitation, and cooling system components may be severely damaged by corrosion.

Five types of coolant are commonly available:

- **Ethylene glycol**—This is the most commonly used antifreeze/coolant. It is green in color and provides good protection regardless of climate, but it is poisonous.

CAUTION!

Never leave ethylene glycol or propylene glycol coolant out and lying around. Both children and animals will drink it because of its sweet taste. The coolant is poisonous and can cause death.

- **Propylene glycol**—This type has the same basic characteristics as ethylene glycol-based coolant but is not sweet tasting and is less harmful to animals and children. Propylene glycol-based coolants should not be mixed with ethylene glycol.
- **Phosphate-free**—This is ethylene glycol-based coolant that has no phosphates, which makes it more environmentally friendly. Phosphate-free coolant is recommended by some auto manufacturers.
- **Organic acid technology (OAT)**—This coolant is also environmentally friendly and contains zero phosphates or silicones. This orange coolant is often referred to by a brand name "DEX-COOL" and is used in all late-model GM vehicles (**Figure 7–14**).
- **Hybrid organic acid technology (HOAT)**—This is similar to OAT coolant but has been enhanced with additives that make the coolant less abrasive to water pumps.

Coolant Condition A coolant hydrometer is used to check the amount of antifreeze in the coolant. This tester contains a pickup hose, coolant reservoir, and squeeze bulb. The pickup hose is placed in the

Figure 7–13 Topping off the cooling system with a 50/50 mixture of antifreeze and water.

Figure 7–14 Ethylene glycol is the most commonly used antifreeze/coolant and is green in color. OAT coolant is orange and is often referred to by a brand name "DEX-COOL."

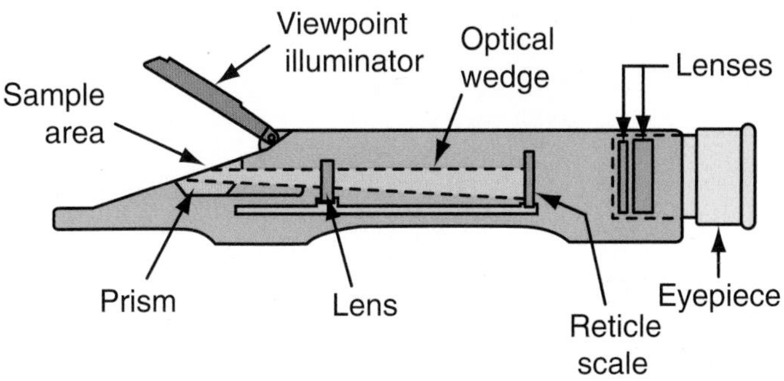

Figure 7-15 A refractometer that tests coolant condition and battery electrolyte.

radiator coolant. When the squeeze bulb is squeezed and released, coolant is drawn into the reservoir. As coolant enters the reservoir, a pivoted float moves upward with the coolant level. A pointer on the float indicates the freezing point of the coolant on a scale located on the reservoir housing.

A refractometer **(Figure 7–15)** offers a precise way to check coolant condition. Most refractometers can also measure the specific gravity of battery electrolyte and test the condition of brake fluid. A sample of the fluid is placed in the sample area of the meter and as light passes through the sample, a line is cast on the meter's scale. The line shows the concentration of the antifreeze in the coolant **(Figure 7–16)**.

Test strips are also used to gain a precise evaluation of coolant. The test strips are immersed into a sample of coolant. After about 5 minutes the strip will change color. The color of the strip is then compared to a scale on the container of strips. Matching the colors will indicate the freeze protection level and the acidity of the coolant **(Figure 7–17)**.

Drive Belts

Drive belts have been used for many years. **V-belts** and **V-ribbed** (serpentine) **belts** are used to drive water pumps, power steering pumps, air-conditioning compressors, generators, and emission control pumps **(Figure 7–18)**. Heat has adverse effects on drive belts and they tend to overcure due to excessive heat. This causes the rubber to harden and crack. Excessive heat normally comes from slippage. Slippage can be caused by improper belt tension or oily conditions. When there is slippage, heat also travels through the drive pulley and down the shaft to the support bearing of the component it is driving. These bearings may be damaged if the slippage is allowed to continue.

V-belts ride in a matching groove in the engine's pulleys. The angled sides of the belt contact the inside of the pulleys' grooves **(Figure 7–19)**. This point of contact is where motion is transferred. As a V-belt wears, it begins to ride deeper in the groove. This reduces its tension and promotes slippage. Because this is a normal occurrence, periodic adjustment of belt tension is necessary.

Drive belts can be used to drive a single part or a combination of parts. An engine can have three or more V-belts. In some cases, two matched belts are used on the same pulley set. This increases the strength of the belt and pulley connection and provides redundancy in case a belt breaks.

Most late-model vehicles use a **serpentine belt** to drive all or most accessories. Serpentine belts are long and follow a complex path that weaves around the

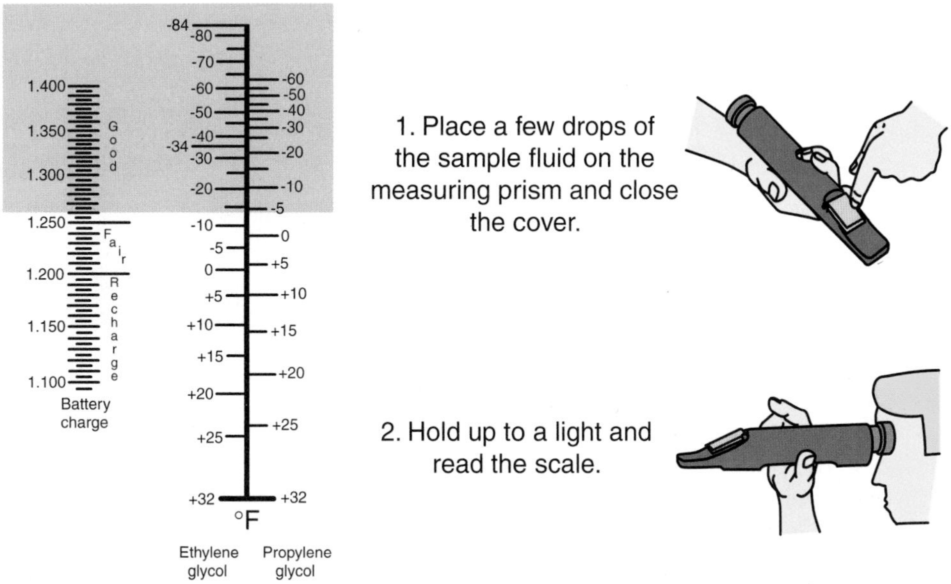

Figure 7-16 Measuring antifreeze and battery electrolyte levels with a refractometer.

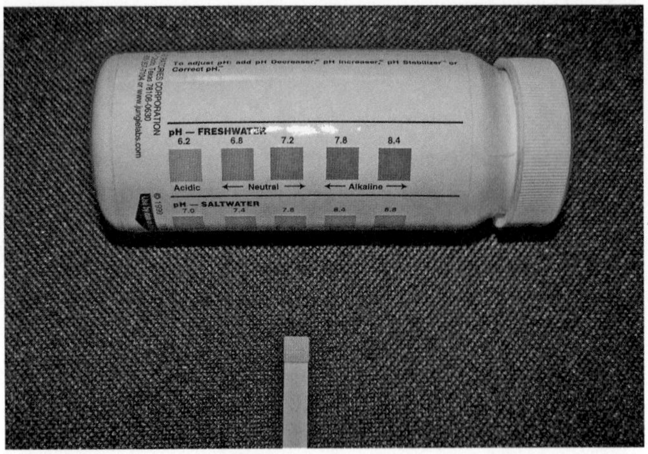

Figure 7-17 Matching the color of a test strip to the scale on its container will indicate the freeze protection level and the acidity of the coolant.

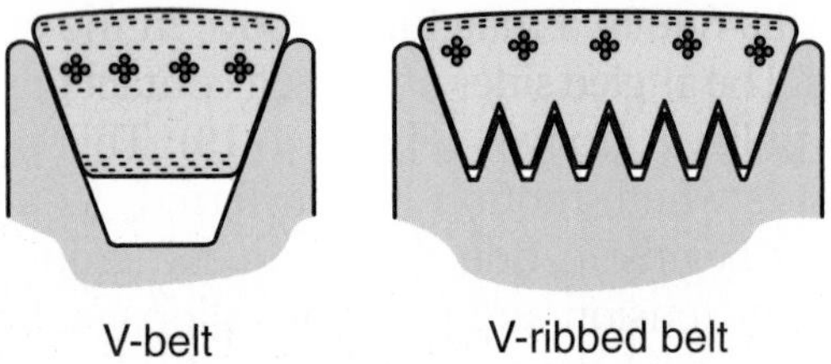

Figure 7-18 A V-belt rides in a single groove, whereas a V-ribbed belt rides in several grooves.

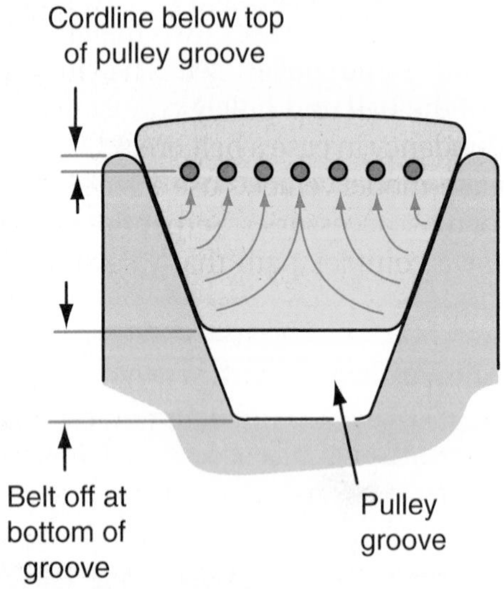

Figure 7-19 The sides of a V-belt contact the grooves of the pulleys.

various pulleys (**Figure 7–20**). Proper tension is critical on a serpentine belt due to the complex routing. Serpentine belts are flat on the outside and have a series of continuous ribs on the inside. These ribs fit into matching grooves in the pulleys. Both the ribbed side and the flat side of the belt can be used to transfer power. Over time, the belts will stretch and lose their tension. To compensate for this and to keep a proper amount of belt tension, most serpentine belt systems

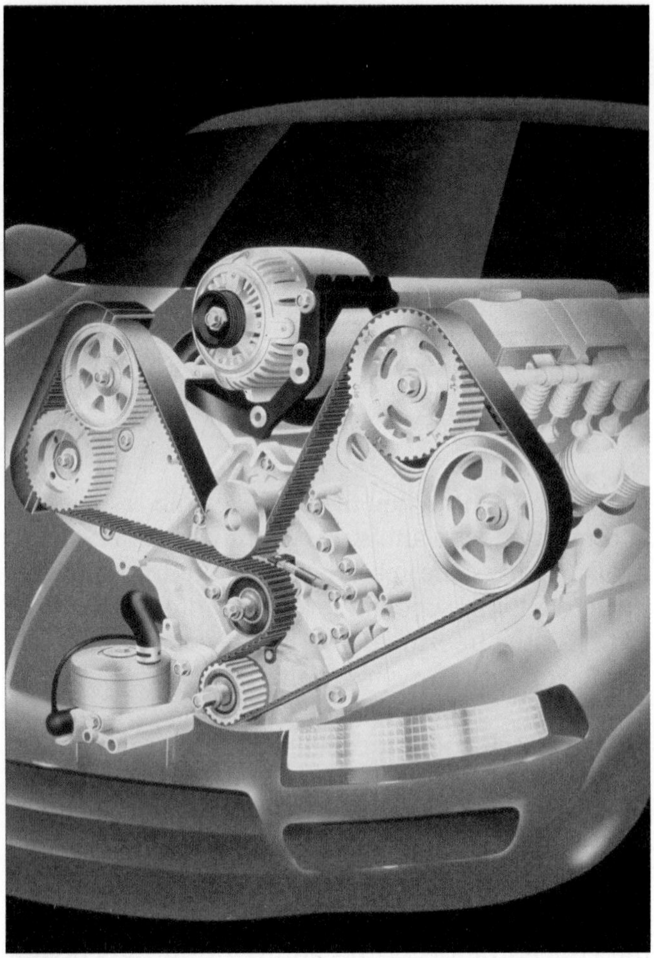

Figure 7-20 A serpentine drive belt. *Courtesy of Gates Corporation*

Figure 7-21 A belt tensioner for a serpentine belt. *Courtesy of Gates Corporation*

have a belt tensioner pulley. This pulley is a spring-loaded pulley (**Figure 7–21**) that exerts a predetermined amount of the pressure on the belt to keep it at the desired tension.

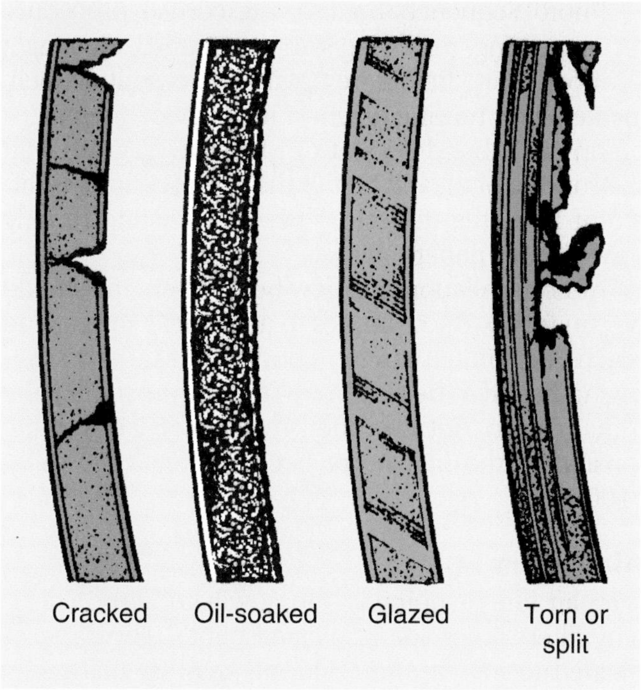

Figure 7-22 Drive belts should be inspected.
Courtesy of Chrysler LLC

If a belt does not have the proper tension, it may squeal or chirp, it may roll off a pulley, or it may slip. Excessive tension may put unwanted forces on the pulleys and the shafts they are attached to, leading to noise, belt breakage, glazing, and damage to the bearings and bushings in the driven components.

Inspection Even the best drive belts last only an average of 4 years. That time can be shortened by several things; most of these can be found by inspecting the belts. Check the condition of all of the drive belts on the engine. Carefully look to see if they have worn or glazed edges, tears, splits, and signs of oil soaking **(Figure 7–22)**. If these conditions exist, the belt should be replaced. Also inspect the grooves of the drive pulleys for rust, oil, wear, and other damage. If a pulley is damaged, it should be replaced. Rust, dirt, and oil should be cleaned off the pulley before installing a new belt.

Misalignment of the pulleys reduces the belt's service life and brings about rapid pulley wear, which causes thrown belts and noise. Undesirable side or end thrust loads can also be imposed on pulley or pump shaft bearings. Check alignment with a straightedge. Pulleys should be in alignment within 1⁄16 inch (1.59 mm) per foot of the distance across the face of the pulleys.

Belt Tension A quick check of a belt's tension can be made by locating the longest span of the belt between two pulleys. With the engine off, press on the belt midway through that distance. If the belt moves more than ½ inch per foot of free span, the belt should be adjusted. Keep in mind that different belts require different tensions. The belt's tension should be checked with a belt tension gauge **(Figure 7–23)**. The tension should meet the manufacturer's specifications. Many engines are now equipped with a ribbed V-belt, which has an automatic tensioning pulley; therefore, a tension adjustment is not required.

Figure 7-23 Check the tension of a drive belt with a belt tension gauge.

USING SERVICE INFORMATION

Proper belt tightening procedures and specifications are given in the specification section of most service manuals.

The exact procedure for adjusting belt tension depends on what the belt is driving. Normally, the mounting bracket for the component driven by the belt and/or its tension adjusting bolt is loosened. The mounting brackets on generators, power steering pumps, and air compressors are designed to be adjustable. Some brackets have a hole or slot to allow the use of a prybar. Other brackets have a ½-inch square opening in which a breaker bar can be installed to move the component and tighten the belt. Other engines have an adjusting bolt, sometimes called a jackscrew, that can be tightened to correct the belt tension. Loosen the mounting bolts and hold the component in the position that provides for the correct tension. Be careful not to damage the part you are prying against. Then tighten the mounting bolts or tension adjusting bolt to keep the tension on the belt. Once tightened, recheck the belt tension with the tension gauge.

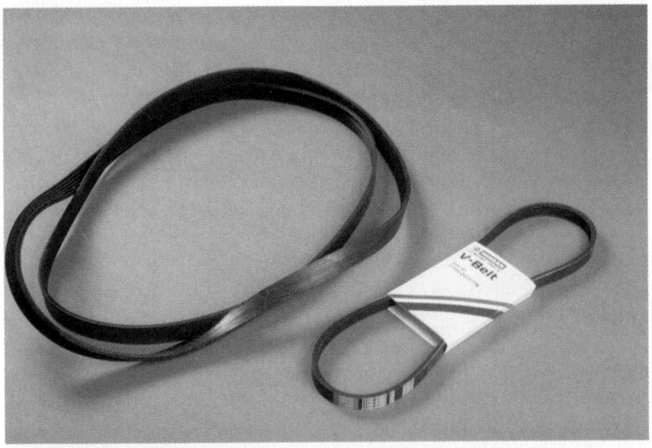

Figure 7-24 The size and part number of a new belt are given on the belt container. The size can be verified by physically comparing the old with the new belt.

Belt Replacement If a drive belt is damaged, it should be replaced. If there is more than one drive belt, all should be replaced even if only one is bad. Always use an exact replacement belt. The size of a new belt is typically given, along with the part number, on the belt container **(Figure 7–24)**. You can verify that the new belt is a replacement for the old one by physically comparing the two. This, however, does not account for any belt stretch that may have occurred. Therefore, only use this comparison as verification. The best way to select the correct replacement belt is with a parts catalog and/or by matching the numbers on the old belt to the numbers on the new belt.

To replace a V-belt on some engines, it may be necessary to remove the fan, fan pulley, and other accessory drive belts to gain access to belts needing replacement. Also, before removing the old drive belt, disconnect the electric cooling fan at the radiator, if the vehicle has one. Remove the old belt by loosening the components that have adjusting slots for belt tension. Then slip the old belt off. Check the condition and alignment of the pulleys. Correct any problems before installing the new belt. Place the new belt around the pulleys. Once in place, loosely tighten the bolts that were loosened during belt removal. Then adjust the tension of the belt and retighten all mounting hardware.

SHOP TALK

It is never advisable to pry a belt onto a pulley. Obtain enough slack so the belt can be slipped on without damaging either the V-belts or a pulley. Some power steering pumps have a ½-inch drive socket to aid in adjusting belts to the proper tension without prying against any accessory.

Photo Sequence 4 shows the correct procedure for inspecting, removing, replacing, and adjusting a V-ribbed belt. Before removing a serpentine belt, locate a belt routing diagram in a service manual or on an underhood decal. Compare the diagram with the routing of the old belt. If the actual routing is different from the diagram, draw the existing routing on a piece of paper.

After installation of a new belt, the engine should be run for 10 to 15 minutes to allow belts to seat and reach their initial stretch condition. Modern steel-strengthened V-belts do not stretch much after the initial run-in, but it is often recommended that the tension of the belt be rechecked after 5,000 miles (8,000 km).

Air Filters

If an air filter is doing its job, it will get dirty. This is why filters are made of pleated paper. The paper is pleated to increase the filtering area. By increasing the area, the amount of time it will take for dirt to plug the filter becomes longer. As a filter gets dirty, the amount of air that can flow through it is reduced. This is not a problem until less air than what the engine needs can get through the filter. Without the proper amount of air, the engine will not be able to produce the power it should; nor will it be as fuel efficient as it should be.

Included in the PM plan for all vehicles is the periodic replacement of the air filter. This mileage or time interval is based on normal vehicle operation. If the vehicle is used, or has been used, in heavy dust, the life of the filter is shorter. Always use a replacement filter that is the same size and shape as the original. An air filter should be periodically checked for excessive dirt or blockage **(Figure 7–25)**. The best way to do this is to remove it and hold it up against a light. If little

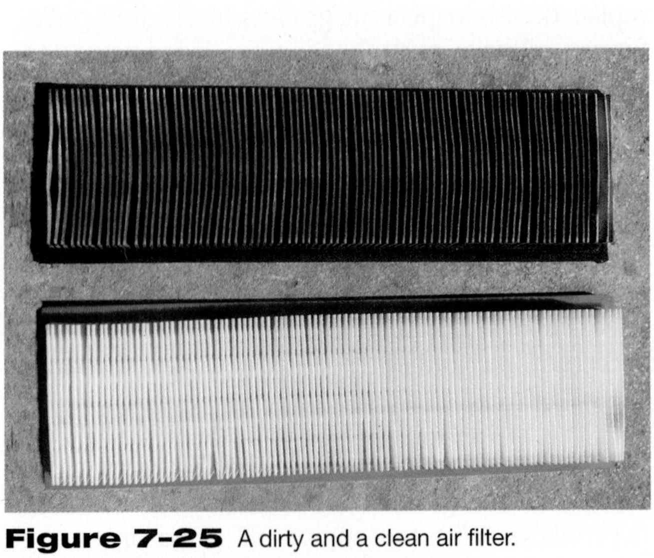

Figure 7-25 A dirty and a clean air filter.

PHOTO SEQUENCE 4

Typical Procedure for Inspecting, Removing, Replacing, and Adjusting a Drive Belt

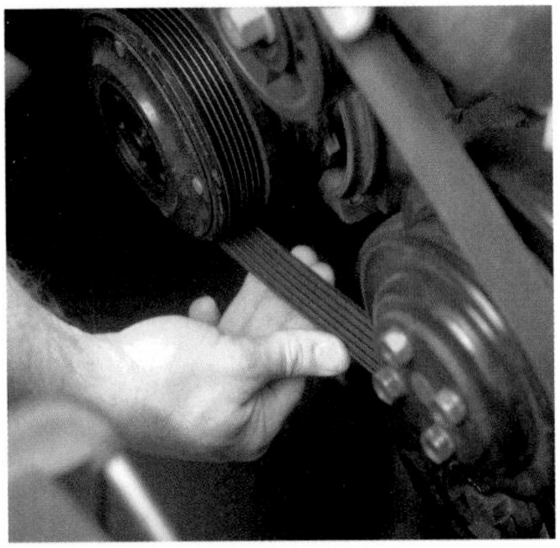

P4-1 *Inspect the belt by looking at both sides.*

P4-2 *Look for signs of glazing.*

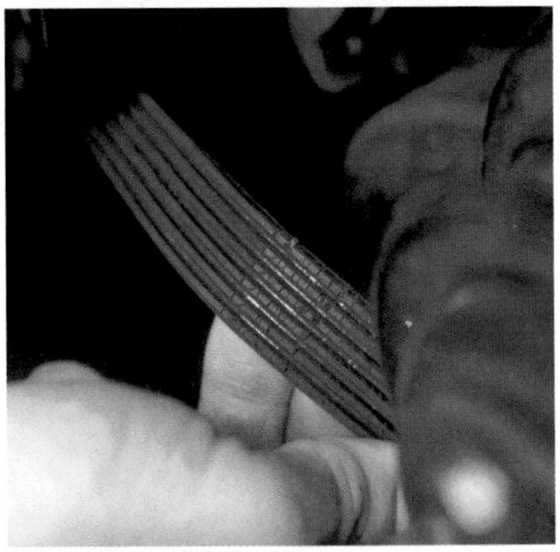

P4-3 *Look for signs of tearing or cracking.*

P4-4 *To replace a worn belt, locate the tensioner or generator pulley.*

P4-5 *Loosen the hold-down fastener for the tensioner or generator pulley.*

P4-6 *Pry the tensioner or generator pulley inward to release the belt tension and remove the belt.*

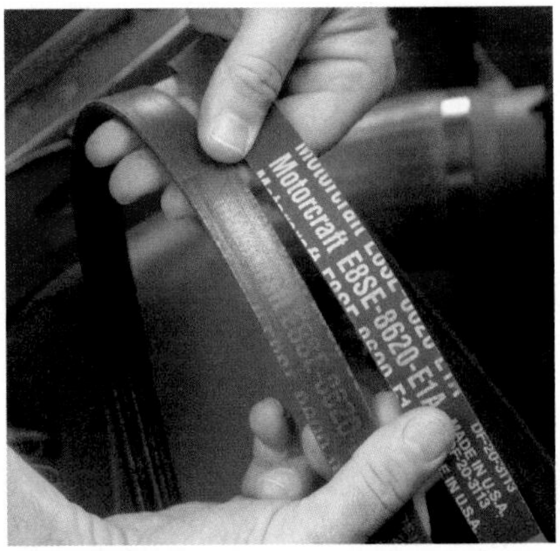

P4-7 *Match the old belt up for size with the new replacement belt.*

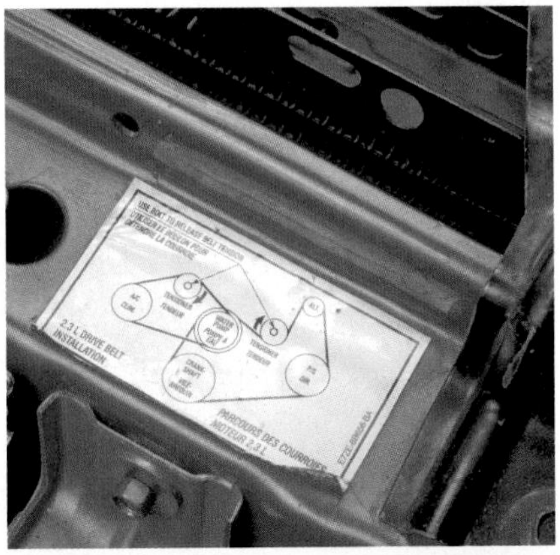

P4-8 *Observe the belt routing diagram in the engine compartment.*

PHOTO SEQUENCE 4

Typical Procedure for Inspecting, Removing, Replacing, and Adjusting a Drive Belt *(continued)*

P4-9 *Install the new belt over each of the drive pulleys. Often the manufacturer recommends a sequence for feeding the belt around the pulleys.*

P4-10 *Pry out the tensioner or generator pulley to put tension on the belt.*

P4-11 *Install the belt squarely in the grooves of each pulley.*

P4-12 *Measure the belt deflection in its longest span. If a belt tension gauge is available, use it and compare the tension to specifications.*

P4-13 *Pry the tensioner or generator pulley to adjust the belt to specifications.*

P4-14 *Tighten the tensioner or generator pulley fastener.*

P4-15 *Start the engine and check the belt for proper operation.*

or no light passes through the filter, it should be replaced. Air filters are typically replaced every 30,000 miles (50,000 km).

When replacing the filter element, carefully remove all dirt from the inside of the housing. Large pieces or dirt and stones accumulate here. It would be disastrous if that dirt got into the cylinders. Also make sure that the air cleaner housing is properly aligned and closed around the filter to ensure good airflow of clean air. If the filter does not seal well in the housing, dirt and dust can be pulled into the airstream to the cylinders. The shape and size of the air filter element depends on its housing; the filter must be the correct size for the housing or dirt will be drawn into the engine.

Battery

The battery is the main source of electrical energy for the vehicle. It is very important that it is inspected and checked on a regular basis.

SHOP TALK

It should be noted that disconnecting the battery on late-model cars removes some memory from the engine's computer and the car's accessories. Besides losing the correct time on its clock or the programmed stations on the radio, the car might run roughly. If this occurs, allow the engine to run for a while before shutting it off.

PROCEDURE

1. Visually inspect the battery cover and case for dirt and grease.
2. Check the electrolyte level (if possible).
3. Inspect the battery for cracks, loose terminal posts, and other signs of damage.
4. Check for missing cell plug covers and caps.
5. Inspect all cables for broken or corroded wires, frayed insulation, or loose or damaged connectors.
6. Check the battery terminals, cable connectors, metal parts, holddowns, and trays for corrosion damage or buildup—a bad connection can cause reduced current flow.
7. Check the heat shield for proper installation on vehicles so equipped.

If the battery or any of the associated parts are dirty **(Figure 7–26)** or corroded, they should be removed and cleaned. Photo Sequence 5 shows the

Figure 7-26 A really dirty battery.

correct procedure for cleaning a battery, a battery tray, and battery cables.

SHOP TALK

When removing or installing a battery, always use the built-in battery strap or a battery lifting tool to lift the battery in or out of its tray.

Transmission Fluid

The oil **(Figure 7–27)** used in automatic transmissions is called **automatic transmission fluid (ATF)**. This special fluid is dyed red so that it is not easily confused with engine oil. Before checking the fluid,

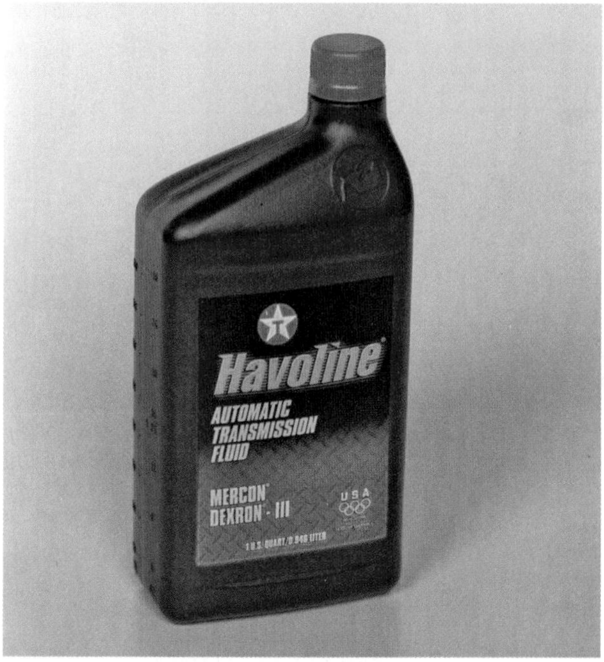

Figure 7-27 Automatic transmission fluid (ATF).

P5-1 *Loosen the battery negative terminal clamp.*

P5-2 *Use a terminal clamp puller to remove the negative cable.*

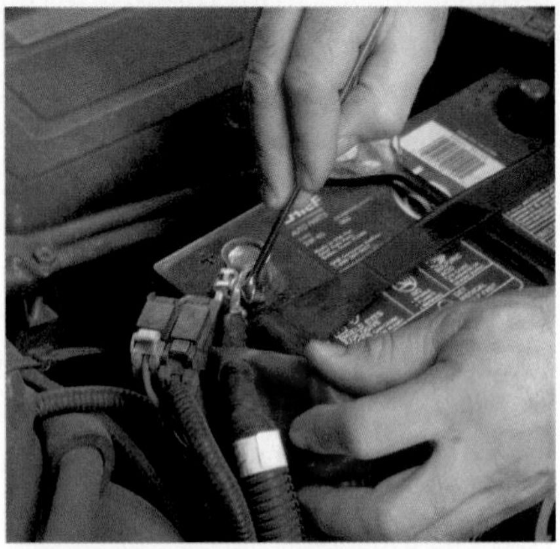
P5-3 *Loosen the battery positive terminal clamp.*

P5-4 *Use a terminal clamp puller to remove the positive clamp.*

P5-5 *Remove the battery hold-down hardware and any heat shields.*

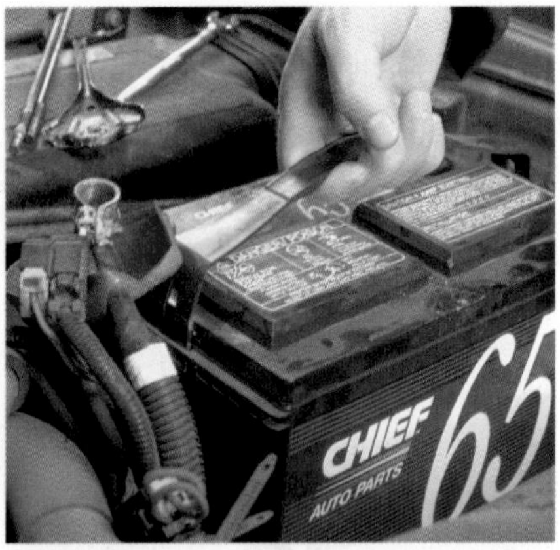

P5-6 *Remove the battery from the tray.*

P5-7 *Mix a solution of baking soda and water.*

P5-8 *Brush the baking soda solution over the battery case, but do not allow the solution to enter the cells of the battery.*

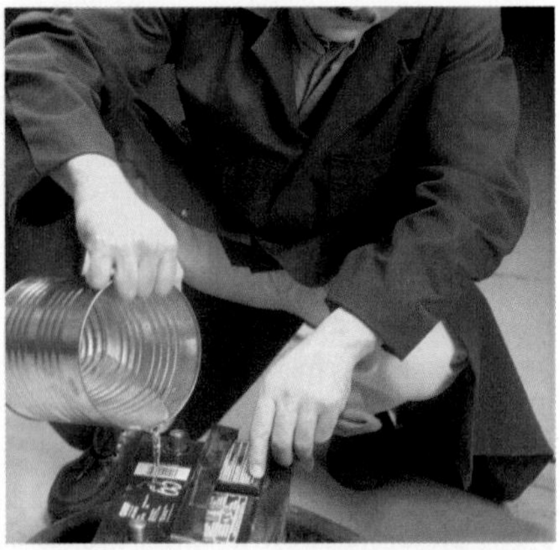
P5-9 *Flush the baking soda off with water.*

PHOTO SEQUENCE 5

Typical Procedure for Cleaning a Battery Case, Tray, and Cables *(continued)*

P5-10 *Use a scraper and wire brush to remove corrosion from the hold-down hardware.*

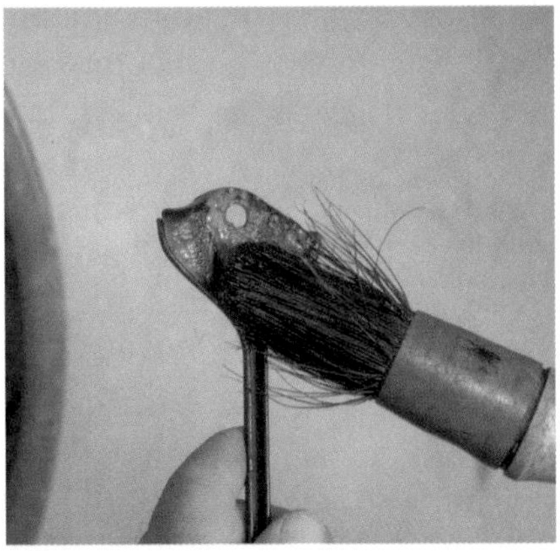

P5-11 *Brush the baking soda solution over the hold-down hardware and then flush with water.*

P5-12 *Allow the hardware to dry, then paint it with corrosion-proof paint.*

P5-13 *Use a terminal cleaner brush to clean the battery cables.*

P5-14 *Use a terminal cleaner brush to clean the battery posts.*

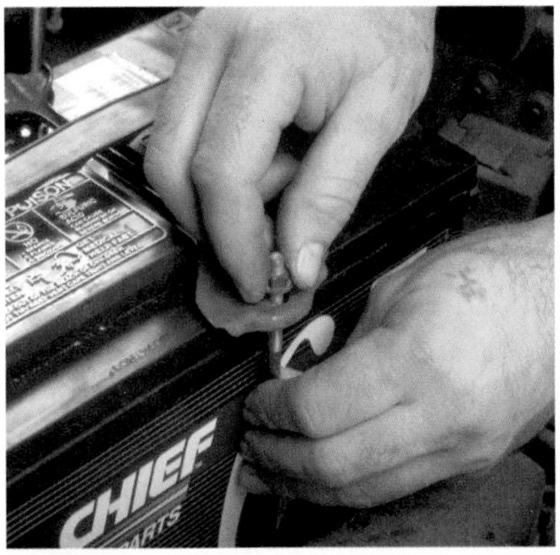

P5-15 *Install the battery back into the tray. Also install the hold-down hardware.*

P5-16 *Install the positive battery cable. Then install the negative cable.*

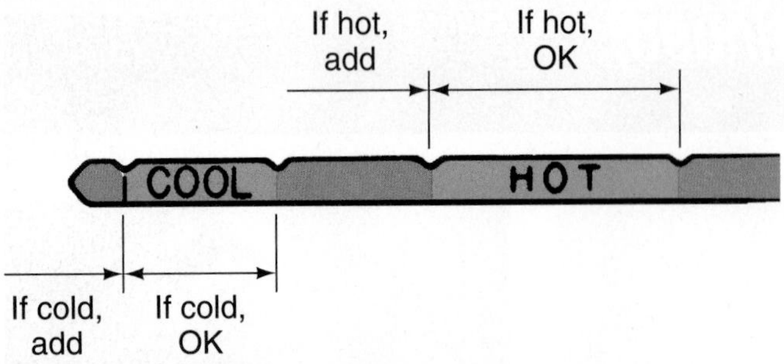

Figure 7-28 Automatic transmission fluid should be checked regularly. Normally the level is checked when the engine is warm. The normal cold level is well below the normal hot level.

make sure the engine is warm and the vehicle is level. Then set the parking brake and allow the engine to idle. Sometimes the manufacturer recommends that the ATF level be checked when the transmission is placed into Park; however, some may require some other gear. Make sure you follow those requirements. Locate the fluid dipstick (normally located to the rear of the engine) and pull it out of its tube. Check the level of the fluid on the dipstick **(Figure 7–28)**. If the level is low, add only enough to bring the level to full. Make sure you only use the fluid recommended by the manufacturer.

The condition of the fluid should be checked while checking the fluid level. The normal color of ATF is pink or red. If the fluid has a dark brownish or blackish color and/or a burned odor, the fluid has been overheated. A milky color indicates that engine coolant has been leaking into the transmission's cooler in the radiator.

After checking the ATF level and color, wipe the dipstick on absorbent white paper and look at the stain left by the fluid. Dark particles are normally band and/or clutch material, whereas silvery metal particles are normally caused by the wearing of the transmission's metal parts. If the dipstick cannot be wiped clean, it is probably covered with varnish, which results from fluid oxidation. Varnish will cause the transmission's valves to stick, causing improper shifting speeds. Varnish or other heavy deposits indicate the need to change the transmission's fluid and filter.

The exact fluid that should be used in an automatic transmission depends on the transmission design and the year the transmission was built. It is very important that the correct type of ATF be used. Always refer to the service or owner's manual for the correct type of fluid to use. Some transmission dipsticks are also marked with the type of ATF required.

Although there are many types of ATF available, the following are the most common:

- **ATF +3 (Chrysler Specification MS-7176E)**—This is a fluid formulated for Chrysler automatic transmissions where ATF+, ATF+2, or ATF +3 is recommended.
- **Type F**—This fluid is typically recommended for Ford and some imported vehicle automatic transmissions built prior to the 1977 model year as well as some 1977 through 1982 models. Do not assume that all Ford vehicles use type F; they do not and it has been a long time since they did!
- **Dexron® VI/Mercon®**—This fluid is sometimes referred to as multipurpose ATF because it is recommended for all GM and Ford automatic transmissions (since 1983) requiring Dexron or Mercon transmission fluids. It also is suitable for most Mercedes-Benz passenger car automatic transmissions.
- **Multivehicle ATF**—This ATF is specially formulated to meet the requirements of a broad range of automatic transmission specifications. It can be safely used in most U.S. vehicles but should not be used in a few pre-1986 vehicles where type F fluids are specified, in vehicles requiring Dexron VI, or in some recent vehicles equipped with continuous variable transmissions (CVTs).

Some transmissions require the use of fluids not mentioned here. CVTs require a fluid that is much different from that used in automatic transmissions. Always use the fluid recommended by the manufacturer. The use of the wrong fluid may cause the transmission to operate improperly and/or damage the transmission.

Manual Transmissions Manual transmissions, transaxles, and drive axle units require the use of specific lubricants or oils, and the levels need to be checked according to the manufacturer's recommended service intervals. Some manufacturers recommend that the fluids be changed periodically. Most repair shops have an air-operated dispenser for these fluids; others rely on a hand-operated oil pump **(Figure 7–29)**.

Power-Steering Fluid

Now locate the power-steering pump. The level of power-steering fluid is checked with the engine off. The filler cap on the power-steering pump normally has a dipstick. Unscrew the cap and check the level **(Figure 7–30)**. The level of the fluid is normally checked when the engine is warm. If the fluid is cold,

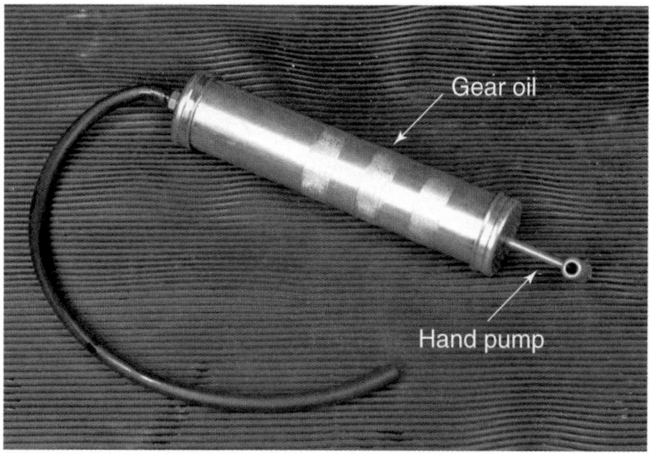

Figure 7-29 A hand-operated pump used to fill transmissions and drive axles with lubricant.

Figure 7-30 The filler cap on a power-steering pump normally has a dipstick to check the fluid level.

it will read lower than normal. Add fluid as necessary. Sometimes the fluid used in these systems is ATF; check the service manual for the proper fluid type before adding fluid.

Brake Fluid

Brake fluid levels are checked at the master cylinder. Older master cylinders are made of cast iron or aluminum and have a metal bail that snaps over the master cylinder cover to hold it in place. Normally the bail can be moved in only one direction. Once moved out of the way, the master cylinder cover can be removed. Once removed, the fluid levels can be checked.

Figure 7-31 Translucent brake fluid reservoirs allow the fluid level to be observed from the outside.

Newer master cylinders have a metal or plastic reservoir mounted above the cylinder. The reservoir will have one or two caps. To check the fluid level in the metal reservoir, the cap must be removed. Most often the caps are screwed on. The caps on some plastic reservoirs have snaps to hold them. Unsnap the cap to check the fluid. It is important to clean the area around the caps before removing them. This prevents dirt from falling into the reservoir. A rubber diaphragm attached to the inside of the caps is designed to stop dirt, moisture, and air from entering into the reservoir. Make sure the diaphragm is not damaged.

Most new plastic reservoirs are translucent and allow the fluid level to be observed from the outside **(Figure 7–31)**.

While checking the fluid level, look at the color of the fluid. Brake fluid tends to absorb moisture and its color gives clues as to the moisture content of the fluid. Dark- or brown-colored fluid indicates contamination; the system must be flushed and the fluid replaced.

When it is necessary to add brake fluid, make sure the fluid is the correct type and is fresh and clean. There are basically four types of brake fluids: DOT 3, DOT 4, DOT 5, and DOT 5.1 **(Figure 7–32)**. The specifications for all automotive brake fluids are defined by Society of Automotive Engineers (SAE) Standard J1703 and Federal Motor Vehicle Safety Standard (FMVSS) 116. Fluids classified according to FMVSS 116 are assigned Department of Transportation (DOT) numbers. Basically, the higher the DOT number, the more rigorous the specifications for the fluid. Domestic automakers specify DOT 3 fluid for their vehicles. However, Ford calls for a heavy-duty variation, which meets the basic specifications for

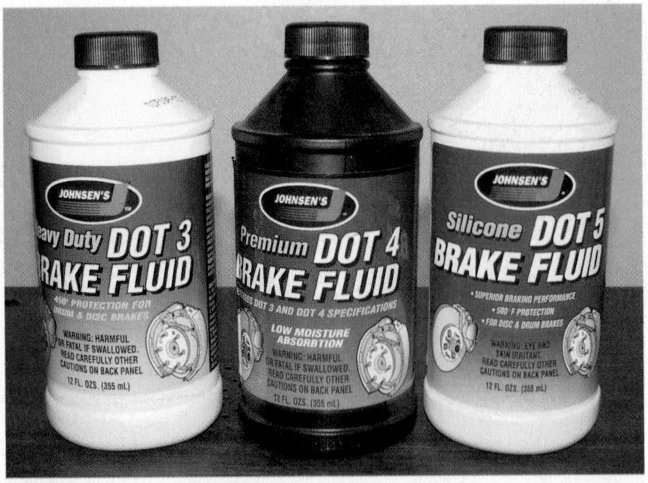

Figure 7–32 The three types of brake fluid: DOT 3 is the most commonly used.

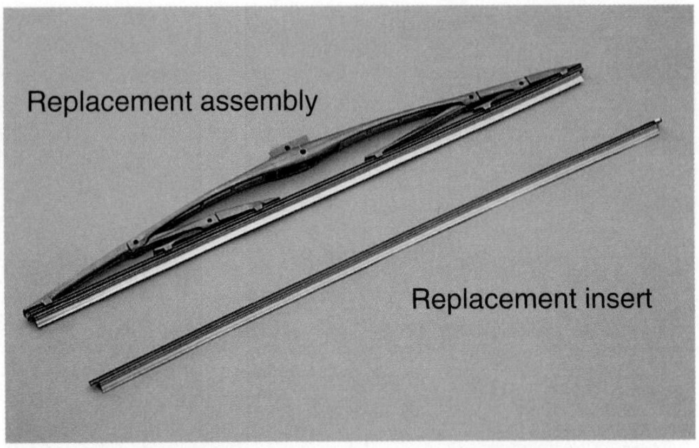

Figure 7–33 Windshield wiper blades are replaced as a complete assembly, or blade inserts are fitted into the blade.

DOT 3 but has the higher boiling point of DOT 4. Import manufacturers are about equally divided between DOT 3 and DOT 4.

DOT 3, DOT 4, and DOT 5.1 fluids are polyalkylene-glycol-ether mixtures, called "**polyglycol**" for short. The color of both DOT 3 and DOT 4 fluid ranges from clear to light amber. DOT 5 fluids are all silicone based because only silicone fluid—so far—can meet the DOT 5 specifications. No vehicle manufacturer, however, recommends DOT 5 fluid for use in its brake systems. Although all three fluid grades are compatible they do not combine well if mixed together in a system. Therefore, the best rules are to use the fluid type recommended by the manufacturer and to never mix fluid types in a system.

Clutch Fluid On some vehicles with a manual transmission, there is another but smaller master cylinder close to the brake master cylinder. This is the clutch master cylinder. Its fluid level needs to be checked, which is done in the same way as brake fluid. In most cases, the clutch master cylinder uses the same type of fluid as the brake master cylinder. However, check this out before adding any fluid.

Windshield Wipers

Check the condition of the windshield wipers. Wiper blades can become dull, torn, or brittle. If they are, they should be replaced. Also, check the condition of the wiper arms. Look for signs of distortion or damage. Also, check the spring on the arm. This spring is designed to keep the wiper blade fairly tight against the windshield. If the spring is weak or damaged, the blade will not do a respectable job cleaning the glass.

Most wiper blade assemblies have replaceable blades or inserts (**Figure 7–33**). To replace the blades, grab hold of the assembly and pivot it away from the windshield. Once the arm is moved to its maximum position, it should stay there until it is pivoted back to the windshield. Doing this will allow you to easily replace the wiper blades without damaging the vehicle's paint or glass.

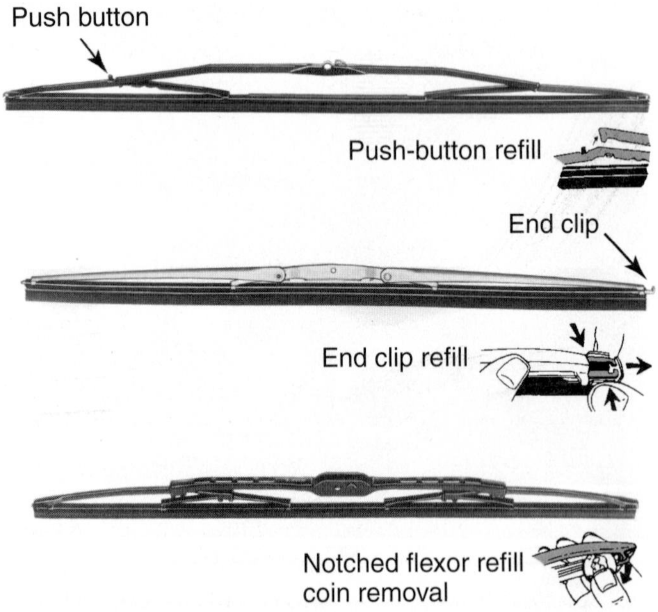

Figure 7–34 Examples of the different ways that wiper blade inserts are secured to the blade assembly. *Courtesy of Federal-Mogul Corporation*

There are three basic types of wiper blade inserts (**Figure 7–34**). Look carefully at the old blade to determine which one to install. Remove the old insert and install the new one. After installation, pull on the insert to make sure it is properly secured. If the insert comes loose while the wipers are moving across the windshield, the wiper arm could scratch the glass.

Most often wiper blades are replaced as an assembly. There are several methods used to secure the blades to the wiper arm (**Figure 7–35**). Most

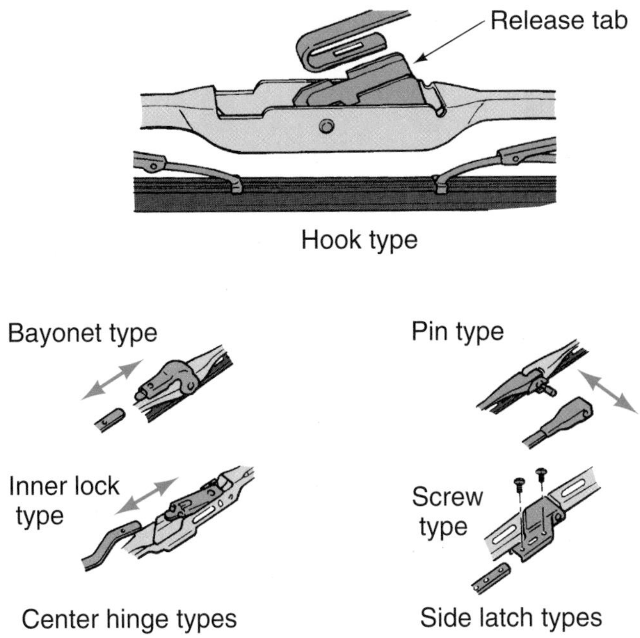

Figure 7-35 Examples of the different ways windshield wiper blades are secured to the wiper arm. *Courtesy of Chrysler LLC*

replacement blades come with the necessary adapters to secure the blade to the arm.

When it is necessary to replace the wiper arm or the entire assembly, they must be removed. Wiper arms are either mounted onto a threaded shaft and held in place by a nut, or they are pressed over a splined shaft. Some shaft-mounted arms are held in place by a clip that must be released before the arm can be pulled off. When installing wiper arms, make sure they are positioned so the blades do not hit the frame of the windshield while they are operating. When checking the placement and operation of the wipers, wet the windshield before turning on the wipers. The water will serve as a lubricant for the wipers.

Windshield Washer Fluid The last fluid level to check is the windshield washer fluid (**Figure 7–36**). Visually check the level and add as necessary. Always use windshield washer fluid and never add water to the washer fluid reservoir, especially in cold weather. The water can freeze and crack the tank or clog the washer hoses and nozzles.

Tires

The vehicle's tires should be checked for damage and wear. Tires should have at least 1⁄16″ of tread remaining. Any less and the tire should be replaced. Tires have "tread wear indicators" molded into them. When the wear bar shows across the width of the tread, the tire is worn beyond its limits. Most shops use a tire wear gauge, which gives an accurate measurement of the tread depth (**Figure 7–37**). Also, check the tires for bulges, nails, tears, and other damage. All of these indicate the tire should be replaced.

Figure 7-36 Check the level of the windshield washer fluid at the reservoir.

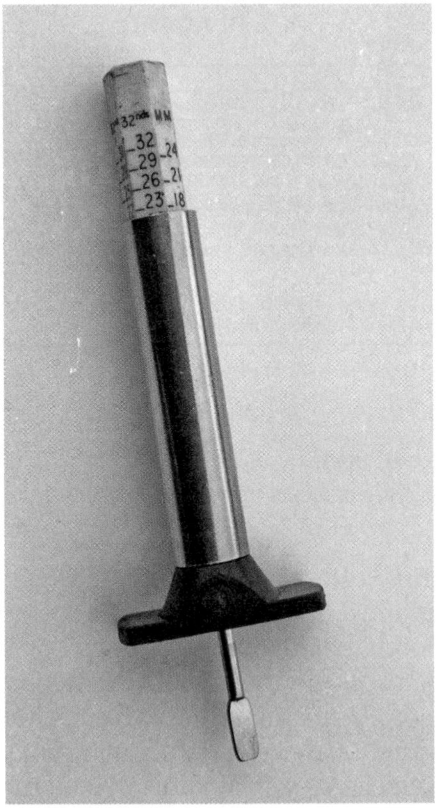

Figure 7-37 A tire tread depth gauge.

Inflation Check the inflation of the tires. To do this, use a tire pressure gauge (**Figure 7–38**). Press the gauge firmly onto the tire's valve stem. The air pressure in the tire will push the scale out of the tool. The highest number shown on the scale is the air pressure of the tire. Compare this reading with the specifications for the tire.

Figure 7-38 Check the tires and wheels for damage and proper inflation.

TIRE PLACARD

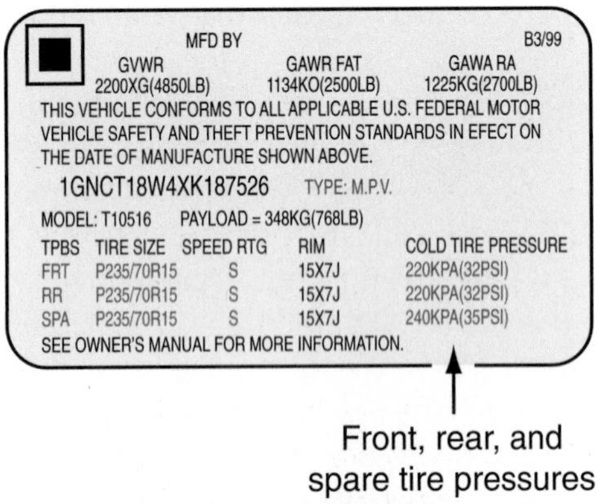
MFD BY
B3/99
GVWR
2200XG(4850LB)
GAWR FAT
1134KO(2500LB)
GAWA RA
1225KG(2700LB)
THIS VEHICLE CONFORMS TO ALL APPLICABLE U.S. FEDERAL MOTOR
VEHICLE SAFETY AND THEFT PREVENTION STANDARDS IN EFFECT ON
THE DATE OF MANUFACTURE SHOWN ABOVE.
1GNCT18W4XK187526 TYPE: M.P.V.
MODEL: T10516 PAYLOAD = 348KG(768LB)
TPBS TIRE SIZE SPEED RTG RIM COLD TIRE PRESSURE
FRT P235/70R15 S 15X7J 220KPA(32PSI)
RR P235/70R15 S 15X7J 220KPA(32PSI)
SPA P235/70R15 S 15X7J 240KPA(35PSI)
SEE OWNER'S MANUAL FOR MORE INFORMATION.
Front, rear, and
spare tire pressures

Figure 7-39 The tire placard gives the recommended cold tire pressure for that vehicle.

The correct tire pressure is listed in the vehicle's owner's manual or on a decal (placard) stuck on the driver's doorjamb **(Figure 7–39)**. The air pressure rating on the tire is not the amount of pressure the tire should have. Rather this rating is the maximum pressure the tire should ever have when it is cold.

New vehicles are fit with tire inflation monitoring systems. These systems have an air pressure sensor attached to the inside of each wheel. When the pressure is below or above a specified range, the vehicle's computer causes a warning light on the dash to illuminate. This alerts the driver of a problem. These monitors can be, and should be, checked, because tire pressure is important to the safety of the vehicle's occupants, and false monitor readings can cause many hardships.

Tire Rotation To equalize tire wear, most car and tire manufacturers recommend that the tires be rotated. Front and rear tires perform different jobs and can wear differently, depending on driving habits and the type of vehicle. In an RWD vehicle, for instance, the front tires usually wear along the outer edges, primarily because of the scuffing and slippage encountered in cornering. The rear tires wear in the center because of acceleration thrusts. To equalize wear, it is recommended that tires be rotated as illustrated in **Figure 7–40**. Bias ply and bias-belted tires should be rotated about every 6,000 miles. Radial tires should be initially rotated at 7,500 miles and then at least every 15,000 miles thereafter. It is important that directional tires are kept rotating in the direction they are designed for. This means the tires may need to be dismounted from the wheel, flipped, and reinstalled on the rim before being put on the other side of the car.

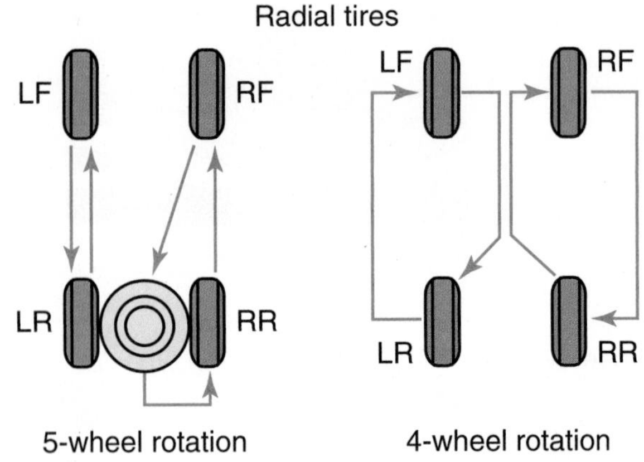

Figure 7-40 Rotation sequence for radial tires.

Lug Nut Torque Obviously, to rotate the tires you must remove the tire/wheel assemblies and then reinstall them. Before reinstalling a tire/wheel assembly on a vehicle, make sure the wheel studs are clean and not damaged, then clean the axle/rotor flange and wheel bore with a wire brush or steel wool. Coat the axle pilot flange with disc brake caliper slide grease or an equivalent. Place the wheel on the hub. Install the lug nuts, and tighten them alternately to draw the wheel evenly against the hub. They should be tightened to a specified torque **(Figure 7–41)** and sequence to avoid distortion. Many tire technicians snug up the lug nuts, then when the car is lowered to the floor, they use a torque wrench for the final tightening.

WARNING!

Overtorquing of the lug nuts is the most common cause of disc brake rotor distortion. Also, an overtorqued lug distorts the threads of the lug and could lead to premature failure.

Figure 7-41 Wheel lugs should be tightened to the specified torque.

Figure 7-42 Torque sticks are color coded to indicate their torque setting.

Some technicians use a torque absorbing adapter, also called a torque stick **(Figure 7–42)**, to tighten the lug nuts. Make sure you use the correct stick for the recommended torque. Then check the actual torque of the lug nuts with a torque wrench.

Chassis Lubrication

A PM procedure that is becoming less common because of changing technology is chassis lubrication. However, all technicians should know how to do this. During the lubrication procedure, grease is forced between two surfaces that move or rub against each other. The grease reduces the friction produced by the movement of the parts. During a chassis lube, grease is forced into a pivot point or joint through a grease fitting. Grease fittings are found on steering and suspension parts, which need lubrication to prevent wear and noise caused by their action during vehicle operation.

Grease fittings are called **zerk fittings** and are threaded into the part that should be lubricated. A fitting at the end of a manual or pneumatic grease gun fits over the zerk to inject the lubricant. Older vehicles have zerk fittings in many locations, whereas newer vehicles use permanently lubricated joints. Some of these joints have threaded plugs that can be removed to lubricate the joint. A special adapter is threaded onto the grease gun and into the plug's bore to lubricate the joint. After grease has been injected into the joint, the plug should be reinstalled or a zerk fitting installed. On some vehicles, rubber or plastic plugs are installed at the factory; they should never be reused.

Before lubricating the chassis, refer to the service manual and identify the lubrication points for the vehicle. Then raise the vehicle. Locate the lubrication points and wipe the fittings clean with a shop towel. Zerk fittings have a one-way spring-loaded check valve that allows grease into the joint but prevents it from leaking out. Dirt can plug the valve, allowing grease to leak out and water and dirt to leak in.

Carefully look at the joints to see if the joint boots are sealed or not. Some joints, such as tie-rod ends and ball joints, are sealed with rubber boots. If the boots are good, push the grease gun's nozzle straight onto a zerk fitting and pump grease slowly into the joint **(Figure 7–43)**. If the joint has a sealed boot, put just enough grease into the joint to cause the boot to slightly expand. If the boot is not sealed, put in enough grease to push the old grease out. Then wipe off the old grease and any excess grease. Repeat this at all lubrication points.

Greases Greases are made from oil blended with thickening agents. There are a few synthetic greases available that meet the same standards as petroleum greases. The thickening agent increases the viscosity

Figure 7-43 A grease gun forces lubrication into a joint through a zerk fitting.

NLGI Grade	Worked Penetration after 60 strokes at 77°F(25°C)	Appearance
000	44.5–47.5 mm	fluid
00	4.00–4.30 mm	fluid
0	3.55–3.85 mm	very soft
1	3.10–3.40 mm	soft
2	2.65–2.95 mm	moderately soft
3	2.20–2.50 mm	semifluid
4	1.75–2.05 mm	semihard
5	1.30–1.60 mm	hard
6	0.85–1.15 mm	very hard

Figure 7-44 The table shows the NLGI grades and the worked penetration ranges.

of the grease. Greases are categorized by a **National Lubricating Grease Institute (NLGI)** number and by the thickeners and additives that are in the grease, such as lithium, molybdenum disulfide, calcium, aluminum, barium, or sodium. Some greases are also labeled with an "EP," which means they have extreme pressure additives. The number assigned by the NLGI is based on test results and the specifications set by the American Society for Testing Materials (ASTM).

The ASTM specifies the consistency of grease using a penetration test. During this test, the grease is heated to 77°F (25°C) and placed below the tip of the test cone. The cone is dropped into the grease. The distance the cone is able to penetrate the grease is measured. The cone will penetrate deeper into soft grease. The NLGI number represents the amount of penetration **(Figure 7–44)**. The higher the NLGI number, the thicker the grease is. NLGI #2 is typically specified for wheel bearings and chassis lubrication.

The NLGI also specifies grease by its use and has established two categories for automotive use. Chassis lubricants are identified with the prefix "L," and wheel bearing lubricants have a prefix of "G." Greases are further defined within those groups by their overall performance. Chassis greases are classified as either LA or LB, and there are three classifications for wheel bearing greases (GA, GB, and GC). LB and GC have the highest performance ratings and are the greases specified for chassis and wheel bearing lubrication. Many types of greases are labeled as both GC and LB and are acceptable for both. These are often referred to as multipurpose greases **(Figure 7–45)**. The NLGI certification mark is included on the grease's container **(Figure 7–46)**.

Class	Purpose
GA	Mild duty—wheel bearings
GB	Mild to moderate duty—wheel bearings
GC	Mild to severe duty—wheel bearings
LA	Mild duty—chassis parts and universal joints
LB	Mild to severe duty—chassis parts and universal joints

Figure 7-45 ASTM grease designation guide.

HYBRID VEHICLES

Hybrid vehicles are maintained and serviced in the same way as conventional vehicles, except for the hybrid components. The latter includes the high-voltage battery pack and circuits, which must be respected when doing any service on the vehicles. Other services to hybrid vehicles are normal services that must be completed in a different way.

For the most part, service to the hybrid system is not something that is done by technicians, unless they are certified to do so by the automobile manufacturer. Keep in mind that a hybrid has nearly all of the basic systems as a conventional vehicle and these are diagnosed and serviced in the same way. Through

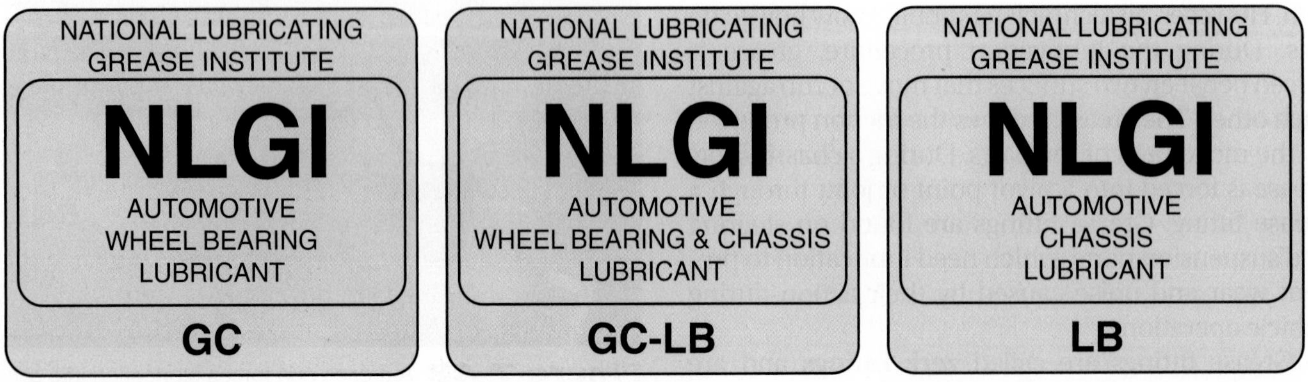

Figure 7-46 NLGI identification symbols.

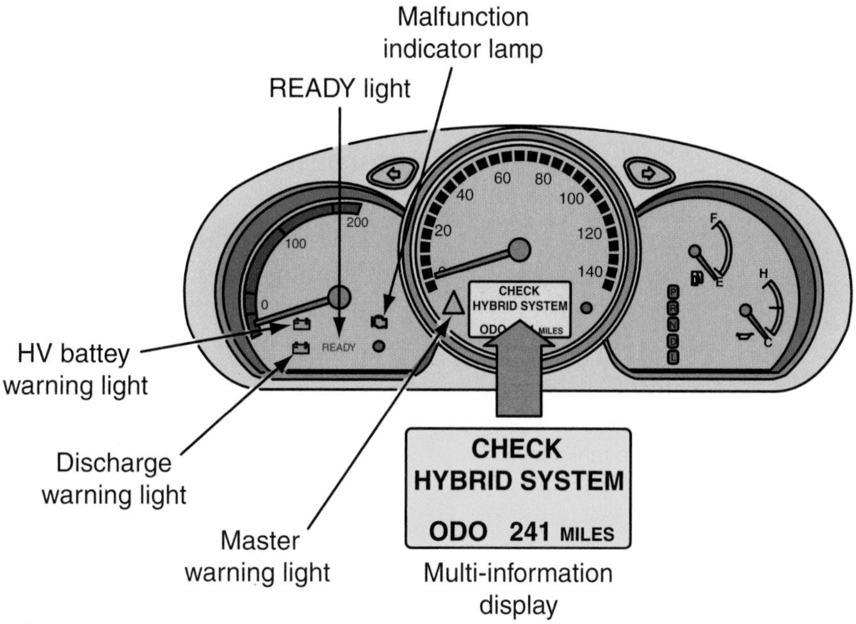

Figure 7-47 An example of some of the warning lights in a hybrid vehicle.

an understanding of how the hybrid vehicle operates, you can safely service them.

One of the things to pay attention to is the stop-start feature. You need to know when the engine will normally shut down and restart. Without this knowledge, or the knowledge of how to prevent this, the engine may start on its own when you are working under the hood. Needless to say, this can create a safety hazard. There is a possibility that your hands or something else can be trapped in the rotating belts or hit by a cooling fan. Unless the system is totally shut down, the engine may start at any time when its control system senses that the battery needs to be recharged.

In addition, there is a possibility that the system will decide to power the vehicle by electric only. When it does this, there is no noise, just a sudden movement of the vehicle. This can scare you and can be dangerous. To prevent both of these incidences, always remove the key from the ignition. Make sure the "READY" lamp in the instrument cluster is off; this lets you know the system is also off **(Figure 7–47)**.

Maintenance

Maintenance of a hybrid vehicle is much the same as a conventional one. Care needs to be taken to avoid anything orange while carrying out the maintenance procedures.

The computer-controlled systems are extremely complex, especially in assist and full hybrids, and are very sensitive to voltage changes. This is why the manufacturers recommend a thorough inspection of the auxiliary battery and connections every 6 months.

The engines used in hybrids are modified versions of engines found in other models offered by the manufacturer. Other than fluid checks and changes, there is little maintenance required on these engines. However, there is less freedom in deciding the types of fluids that can be used and the parts that can replace the original equipment. Hybrids are not very forgiving. Always use the exact replacement parts and the fluids specified by the manufacturer.

Typically, the weight of the engine oil used in a hybrid is very light. If heavier oil is used, the computer may see this as a problem and prevent the engine from starting. The heavier oil may cause an increase in the current required to crank the engine. If the computer senses very high current draw while attempting to crank the engine, it will open the circuit in response.

Special coolants are required in most hybrids because the coolant cools not only the engine, but also the inverter assembly. Cooling the inverter is important and checking its coolant condition and level is an additional check during PM. The cooling systems used in some hybrids feature electric pumps and storage tanks **(Figure 7–48)**. The tanks store heated coolant and can cause injury if you are not aware of how to carefully check them.

The battery cooling system may need to be serviced at regular intervals. There is a filter in the ductwork from the outside of the vehicle to the battery box. This filter needs to be periodically changed. If the filter becomes plugged, the temperature of the battery will rise to dangerous levels. In fact, if the computer senses high temperatures it may shut down the system.

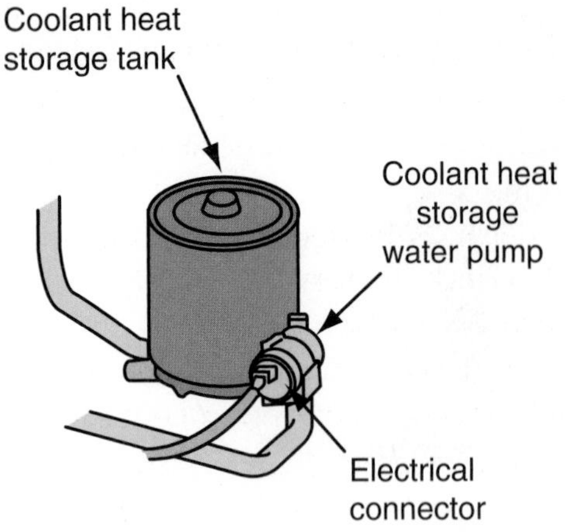

Figure 7-48 The hot coolant storage tank for Toyota hybrids.

A normal part of PM is checking power steering and brake fluids. The power steering systems used by the manufacturers vary; some have a belt-driven pump, some have an electrically driven pump, and others have a pure electric and mechanical steering gear. Each variety requires different care; therefore, always check the service manual for the specific model before doing anything to these systems. Also, keep in mind that some hybrids use the power steering pump as the power booster for the brake system.

Hybrids are all about fuel economy and reduced emissions. Everything that would affect these should be checked on a regular basis. Items such as tires, brakes, and wheel alignment can have a negative effect, and owners of hybrids will notice the difference. These owners are constantly aware of their fuel mileage due to the displays on the instrument panel.

ADDITIONAL PM CHECKS

The following PM checks are in addition to those items specified by the manufacturer. These should be performed at these suggested time intervals to help ensure safe and dependable vehicle operation.

Time: While operating the vehicle

- Pay attention to and note any changes in the sound of the exhaust or any smell of exhaust fumes in the vehicle.
- Check for vibrations in the steering wheel. Notice any increased steering effort or looseness in the steering wheel.
- Notice if the vehicle constantly turns slightly or pulls to one side of the road.
- When stopping, listen and check for strange sounds, pulling to one side, increased brake pedal travel, or hard-to-push brake pedal.
- If any slipping or changes in the operation of the transmission occur, check the transmission fluid level.
- Check for fluid leaks under the vehicle. (Water dripping from the air-conditioning system after use is normal.)
- Check the automatic transmission's park function.
- Check the parking brake.

Time: At least monthly

- Check the operation of all exterior lights, including the brake lights, turn signals, and hazard warning flashers.

Time: At least twice a year

- Check the pressure in the spare tire.
- Check headlight alignment.
- Check the muffler, exhaust pipes, and clamps.
- Inspect the lap/shoulder belts for wear.
- Check the radiator, heater, and air-conditioning hoses for leaks or damage.

Time: At least once a year

- Lubricate all hinges and all outside key locks.
- Lubricate the rubber weather strips for the doors.
- Clean the body's water drain holes.
- Lubricate the transmission controls and linkage.

KEY TERMS

American Petroleum Institute (API)
Automatic transmission fluid (ATF)
Coolant
Core
Core charge
Crude oil
Dipstick
Energy-conserving oil
Multiviscosity
National Lubricating Grease Institute (NLGI)
Polyglycol
Repair order (RO)
Serpentine belt
Society of Automotive Engineers (SAE)
Sublet repair
V-belts
Vehicle identification number (VIN)
Viscosity
V-ribbed belts
Zerk fitting

SUMMARY

- A repair order (RO) is a legal document used for many purposes.
- An RO includes a cost estimate for the repairs. By law, this estimate must be quite accurate.
- Preventive maintenance (PM) involves regularly scheduled service on a vehicle to keep it operating efficiently and safely. Professional technicians should stress the importance of PM to their customers.
- Engine oil is a clean or refined form of crude oil. It contains many additives, each intended to improve the effectiveness of the oil.
- The American Petroleum Institute (API) classifies engine oil according to its ability to meet the engine manufacturers' warranty specifications.
- The Society of Automotive Engineers (SAE) has established an oil viscosity classification system that has a numeric rating in which the higher viscosity, or heavier weight oils, receives the higher numbers.
- Changing the engine's oil and filter should be done on a regular basis.
- Whenever the engine's oil is changed, a thorough inspection of the cooling systems should be done.
- Normally, the recommended mixture for engine coolant is a 50/50 solution of water and antifreeze/coolant.
- V-belts and V-ribbed (serpentine) belts are used to drive water pumps, power steering pumps, air-conditioning compressors, generators, and emission control pumps.
- If a belt does not have the proper tension, it may produce squealing and chirping noises; allow the belt to roll off a pulley; or slip, which reduces the power that drives the component.
- Excessive belt tension may put unwanted forces on the pulleys and the shafts they are attached to, leading to noise, belt breakage, glazing, and damage to the bearings and bushings in water pumps, generators, and power steering pumps.
- The air filter should be periodically checked for excessive dirt or blockage and a replacement filter should be the same size and shape as the original.
- The battery is the main source of electrical energy for the vehicle. It is very important that it is checked on a regular basis.
- If the battery or any of the associated parts are dirty or corroded, remove the battery and clean them.
- The condition of the automatic transmission fluid (ATF) should be checked while checking the fluid level.
- Normally the fluid used in power-steering systems is ATF. Check the service manual for the proper fluid type before adding fluid.
- Check the level of the brake fluid and make sure the fluid is the correct type and is fresh and clean.
- There are basically four types of brake fluids: DOT 3, DOT 4, DOT 5, and DOT 5.1. Most automakers specify DOT 3 fluid for their vehicles.
- Check the windshield wipers for signs of dullness, tears, and hardness. Also check the spring on the wiper arm.
- The vehicle's tires should be checked for damage and wear as well as for proper inflation.
- To equalize tire wear, most car and tire manufacturers recommend that the tires be rotated after a specified mileage interval.
- Several parts of a vehicle may need periodic lubrication; always use the correct type of grease when doing this.
- Some PM procedures are unique to hybrid vehicles; always follow the recommendations of the manufacturer.
- When servicing a hybrid vehicle, always respect its high-voltage system.

REVIEW QUESTIONS

1. Describe the information found in a VIN.
2. Technician A stresses the need to follow the manufacturer's recommendations for preventive maintenance to his customers. Technician B says that the proper PM service intervals depends on the customer's driving habits and typical driving conditions. Who is correct?
   a. Technician A
   b. Technician B
   c. Both A and B
   d. Neither A nor B
3. Technician A says that tires should have a tread depth of at least 1/16 of an inch. Technician B says that tires have a tread wear indicator built into them. Who is correct?
   a. Technician A
   b. Technician B
   c. Both A and B
   d. Neither A nor B

4. When working on a hybrid electric vehicle, which of the following statements does *not* reflect things you should be aware of?
   a. The engine will normally shut down and restart when stopped.
   b. The engine may start at any time when its control system senses that the battery needs to be recharged.
   c. The system may decide to power the vehicle, while it is parked, by the engine anytime the load is great.
   d. Make sure the "READY" lamp in the instrument cluster is off; this lets you know the system is off.
5. *True or False?* Legally, an RO protects the shop and the customer.
6. While examining the color of an engine's coolant: Technician A says that because it is orange, the cooling system should be flushed and new coolant put into the system. Technician B says that the coolant looks rusty and the cooling system should be flushed and new coolant put into the system. Who is correct?
   a. Technician A c. Both A and B
   b. Technician B d. Neither A nor B
7. While checking the condition of a vehicle's ATF: Technician A says that if the fluid has a dark brownish or blackish color, the fluid has been overheated. Technician B says that a milky color indicates that engine coolant has been leaking into the transmission's cooler in the radiator. Who is correct?
   a. Technician A c. Both A and B
   b. Technician B d. Neither A nor B
8. Technician A says that if a drive belt does not have the proper tension, it may produce squealing and chirping noises. Technician B says that excessive tension may cause noise, belt breakage, and glazing. Who is correct?
   a. Technician A c. Both A and B
   b. Technician B d. Neither A nor B
9. List five things that engine oil is formulated to do.
10. Technician A says that the rubber diaphragm attached to the inside of the master cylinder caps is designed to stop dirt from entering the reservoir. Technician B says that the rubber diaphragm is designed to stop air from entering the reservoir. Who is correct?
    a. Technician A c. Both A and B
    b. Technician B d. Neither A nor B
11. Why should you wipe off the outside of a zerk fitting before injecting grease into it?
12. *True or False?* All engine coolants contain some phosphates, which make them unfriendly to the environment.
13. List at least five things that should be checked while inspecting a vehicle's battery.
14. *True or False?* If brake fluid is dark or brown colored, the system must be flushed and the fluid replaced.
15. Which of the following statements about drive belt slippage is *not* true?
    a. Excessive heat normally comes from slippage.
    b. As a V-belt slips, it begins to ride deeper in the pulley groove.
    c. Slippage can be caused by improper belt tension or oily conditions.
    d. When there is slippage, heat travels through the drive pulley and down the shaft to the support bearing of the component it is driving.
16. *True or False?* ATF labeled as "Multivehicle ATF" can safely be used in all automatic transmissions.
17. Which of the following statements about an oil's viscosity is *not* true?
    a. The ability of oil to flow is its viscosity.
    b. Viscosity is affected by temperature; hot oil flows faster than cold oil. Oil flow is important to the life of an engine.
    c. In the API system of oil viscosity classification, the lighter oils receive a higher number.
    d. Heavyweight oils are best suited for use in high-temperature regions. Low-weight oils work best in low temperature operations.
18. While discussing the readings on a built-in hydrometer in a battery: Technician A says that the green dot means the battery is charged enough for testing. Technician B says that the red dot means the battery is completely discharged and must be replaced. Who is correct?
    a. Technician A c. Both A and B
    b. Technician B d. Neither A nor B
19. Which of the following is *not* a true statement about a mechanic's lien?
    a. This lien states that the shop may gain possession of the vehicle if the customer does not pay for the agreed-upon services.

   b. The right to impose a mechanic's lien can be exercised by a shop within 30 days after the services have been completed.
   c. In most cases, the shop's right to impose a lien on the vehicle being serviced must be acknowledged by the customer prior to beginning any services to the vehicle.
   d. This clause ensures that the shop will receive some compensation for the work performed, whether or not the customer pays the bill.

20. *True or False?* An oil with the classification of 5W-30 has the viscosity of both a 10- and 30-weight oil. The 5W means the oil has a viscosity of 5 when warm and a 30 rating when it is cold.
21. While discussing the effects of overtorquing wheel lugs: Technician A says that this can cause the threads of the lugs and/or studs to distort. Technician B says that this is the most common cause of disc brake rotor distortion. Who is correct?
   a. Technician A
   b. Technician B
   c. Both A and B
   d. Neither A nor B
22. The most commonly used brake fluid in domestic vehicles is:
   a. DOT 2
   b. DOT 3
   c. DOT 4
   d. DOT 5
23. Which of the following greases are best suited for lubricating automotive wheel bearings?
   a. LA
   b. LB
   c. GA
   d. GC
24. While discussing the special preventive maintenance items for a hybrid vehicle: Technician A says that special coolants are required in most hybrids because the coolant not only cools the engine, but also the inverter assembly. Technician B says that the battery cooling system may have a filter in the ductwork from the outside of the vehicle to the battery box and that this filter needs to be periodically changed. Who is correct?
   a. Technician A
   b. Technician B
   c. Both A and B
   d. Neither A nor B
25. How does a technician determine the proper inflation of a vehicle's tires?

CHAPTER 8

BASIC THEORIES AND MATH

OBJECTIVES

■ Describe the states in which all matter exists. ■ Explain what energy is and how energy is converted. ■ Calculate the volume of a cylinder. ■ Explain the forces that influence the design and operation of an automobile. ■ Describe and apply Newton's laws of motion to an automobile. ■ Define friction and describe how it can be minimized. ■ Describe the various types of simple machines. ■ Explain the difference between torque and horsepower. ■ Differentiate between a vibration and a sound. ■ Explain Pascal's law and give examples of where it is applied to an automobile. ■ Explain the behavior of gases. ■ Explain how heat affects matter. ■ Describe what is meant by the chemical properties of a substance. ■ Explain the difference between oxidation and reduction. ■ Describe the origin and practical applications of electromagnetism.

This chapter contains many of the things you have learned or will learn in other courses. The material is not intended to take the place of those other courses but rather to emphasize the knowledge you need to gain employment and be successful in an automotive career. Many of the facts presented in this chapter will be addressed again in greater detail according to the topic. Make sure you understand the contents of this chapter.

MATTER

Matter is anything that occupies space. All matter exists as a gas, liquid, or solid. Gases and liquids are considered fluids because they move or flow easily and easily respond to pressure. A gas has neither a shape nor volume of its own and tends to expand without limits. A liquid takes a shape and has volume. A solid is matter that does not flow.

Atoms and Molecules

All matter is made up of countless tiny particles called **atoms**. A substance with only one type of atom is referred to as an **element**. Over 100 elements are known to exist; 92 occur naturally and the rest have been manufactured in laboratories **(Figure 8–1)**. The atom is the smallest particle of an element and has all of the chemical characteristics of the element.

Small, positively charged particles called protons are located in the center, or nucleus, of each atom. In most atoms, the nucleus also contains neutrons. Neutrons have no electrical charge, but they add weight to the atom. The positively charged protons tend to repel each other, and this repelling force could destroy the nucleus. The presence of the neutrons with the protons cancels the repelling action and keeps the nucleus together. Electrons are small, very light particles with a negative electrical charge. Electrons move in orbits around the atom's nucleus. Elements are listed on the atomic scale, or periodic chart, according to their number of protons and electrons. For example, hydrogen is number 1 on this scale, and copper is number 29.

A proton is about 1,840 times heavier than an electron. Therefore, electrons are easier to move than protons. While the electrons are orbiting, centrifugal force tends to move them away from the nucleus. However, the attraction between the positively charged protons and the negatively charged electrons holds the electrons in their orbits.

Atoms of different elements have different numbers of protons, electrons, and neutrons. Some of the lighter elements have the same number of protons and neutrons, but many of the heavier elements have more neutrons than protons.

Group:	1	2	3	4	5	6	7	8	9	10	11	12	13	14	15	16	17	18
Period	**1A**	**2A**	**3B**	**4B**	**5B**	**6B**	**7B**		**8B**		**1B**	**2B**	**3A**	**4A**	**5A**	**6A**	**7A**	**8A**
1	1 H																	2 He
2	3 Li	4 Be											5 B	6 C	7 N	8 O	9 F	10 Ne
3	11 Na	12 Mg											13 Al	14 Si	15 P	16 S	17 Cl	18 Ar
4	19 K	20 Ca	21 Sc	22 Ti	23 V	24 Cr	25 Mn	26 Fe	27 Co	28 Ni	29 Cu	30 Zn	31 Ga	32 Ge	33 As	34 Se	35 Br	36 Kr
5	37 Rb	38 Sr	39 Y	40 Zr	41 Nb	42 Mo	[43] Tc	44 Ru	45 Rh	46 Pd	47 Ag	48 Cd	49 In	50 Sn	51 Sb	52 Te	53 I	54 Xe
6	55 Cs	56 Ba	*	72 Hf	73 Ta	74 W	75 Re	76 Os	77 Ir	78 Pt	79 Au	80 Hg	81 Tl	82 Pb	83 Bi	84 Po	85 At	86 Rn
7	87 Fr	88 Ra	**	[104] Unq	[105] Unp	[106] Unh	[107] Uns	[108] Uno	[109] Une	[110] Uun	[111] Uuu	[112] Uub	[113] Uut	[114] Uuq	[115] Uup	[116] Uuh	[117] Uus	[118] Uuo

***Lanthanides:**	57 La	58 Ce	59 Pr	60 Nd	[61] Pm	62 Sm	63 Eu	64 Gd	65 Tb	66 Dy	67 Ho	68 Er	69 Tm	70 Yb	71 Lu
****Actinides:**	89 Ac	90 Th	91 Pa	92 U	[93] Np	[94] Pu	[95] Am	[96] Cm	[97] Bk	[98] Cf	[99] Es	[100] Fm	[101] Md	102 No	103 Lr

LEGEND:

Alkali Metals	Noble Gases
Alkaline Earth Metals	Halogens
Other Metals	Other Nonmetals
Semiconductors	No Data Available

Figure 8-1 The periodic table of the elements with each element's natural state shown.

A hydrogen (H) atom is the simplest atom. It has one proton and one electron **(Figure 8–2)**. A copper (CU) atom has 29 protons and 29 electrons. The electrons orbit in four different rings around the nucleus. Because two, eight, and eighteen electrons are the maximum number of electrons in the first three electron rings next to the nucleus, the fourth ring has one electron **(Figure 8–3)**. The outer ring of electrons is called the valence ring, and the number of electrons in the valence ring determines the electrical characteristics of the element.

A single atom of some elements does not exist. An example of this is oxygen, whose symbol is O. Pure oxygen exists only as a pair of oxygen atoms and has a symbol of O_2. This is a molecule of oxygen. A **molecule** is the smallest particle of an element or compound. A molecule can be made of one type or different types of atoms. A compound contains two or more different types of atoms. Oxygen atoms readily combine with atoms of other elements to form a compound. Many other atoms also have this same characteristic.

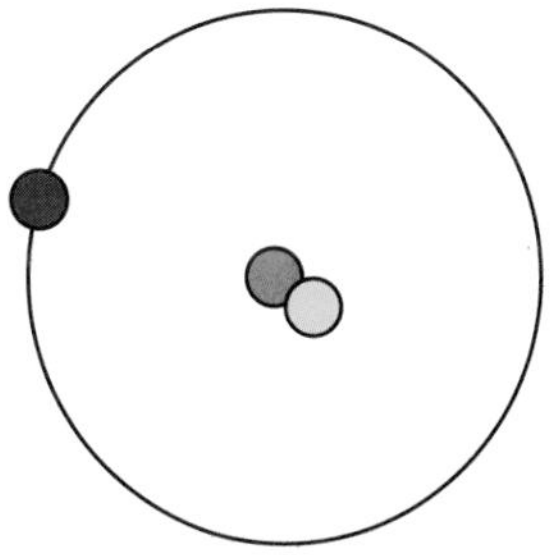

Figure 8-2 A hydrogen atom has one proton (red), one neutron (yellow), and one electron (green).

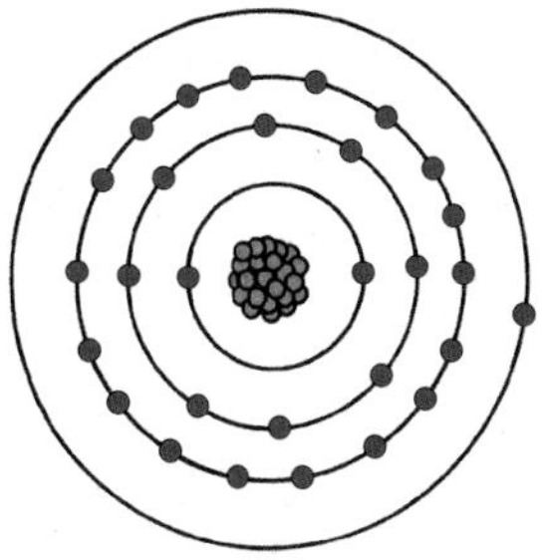

Figure 8-3 A copper atom.

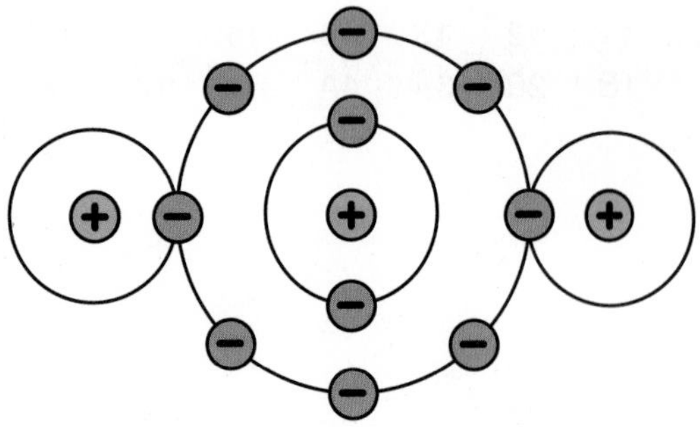

Figure 8-4 A molecule of water.

Water is a compound of oxygen and hydrogen atoms. The chemical symbol for water is H_2O. This symbol indicates that each molecule of water contains two atoms of hydrogen and one atom of oxygen **(Figure 8–4)**.

Ions

An **ion** is an atom or molecule that has lost or gained one or more electrons. As a result, it has a negative or positive electrical charge. A negatively charged ion has more electrons than it has protons. The opposite is true of positively charged ions, which have fewer electrons than protons. Ions are denoted in the same way as other atoms and molecules except for a superscript symbol or number that shows the electrical charge and the number of electrons gained or lost. For example, hydrogen with a positive charge (H^+) and oxygen with a negative charge (O^{2-}) is called an **oxide**.

Plasma Considered by scientists as the fourth state of matter, **plasma** refers to an ionized gas that has about an equal amount of positive ions and electrons. The electrons travel with the nucleus of the atoms but can move freely and are not bound to it. The gas at this point no longer behaves as a gas. It now has electrical properties and creates a magnetic field, which radiates light and other forms of electromagnetic energy. It typically takes the form of gaslike clouds and is the basis of most stars. In fact, our sun is really just a large piece of plasma. Plasmas are the most common form of matter in the universe. Plasma in the stars and in the space between them occupies nearly 99% of the visible universe.

Plasma does not exist as a solid, liquid, or gas; it is different and has a much different temperature range. Plasma is more dense than other states of matter.

Behavior of the States

The particles of a solid are held together in a rigid structure. When a solid dissolves into a liquid, its particles break away from this structure and mix evenly in the liquid, forming a **solution**. When they are heated, most liquids **evaporate**, which means atoms or molecules break free from the body of the liquid to become gas particles. When all of the liquid has evaporated, a solid is left behind. The particles of the solid are normally arranged in a structure called a crystal.

Absorption and Adsorption Not all solids dissolve in a liquid; rather, the liquid is either absorbed or adsorbed. The action of a sponge is the best example of absorption. When a dry sponge is put into water, the water is absorbed by the sponge. The sponge does not dissolve; the water merely penetrates into the sponge and the sponge becomes filled with water. There is no change to the atomic structure of the sponge, nor does the structure of the water change. If we take a glass and put it into water, the glass does not absorb the water. The glass, however, still gets wet as a thin layer of water adheres to the glass. This is adsorption. Materials that *absorb* fluids are **permeable** substances. **Impermeable** substances, such as glass, *adsorb* fluids. Some materials are impermeable to most fluids, whereas others are impermeable to just a few.

ENERGY

Energy may be defined as the ability to do work. Because all matter consists of atoms and molecules in constant motion, all matter has energy. Energy is not matter, but it affects the behavior of matter. Everything that happens requires energy, and energy comes in many forms.

Each form of energy can change into other forms. However, the total amount of energy never changes; it can only be transferred from one form to another, not created or destroyed. This is known as the "principle of the conservation of energy."

> **A Look at History** *Albert Einstein, in his theory of relativity, proposed an equation for energy that many have heard of but few understand. He stated that energy equals mass times the speed of light squared or $E = m \times c^2$.*

Engine Efficiency

Engine efficiency is a measurement of the amount of energy put into the engine and the amount of energy available from the engine. It is expressed in a percentage. The formula for determining efficiency is: (output energy ÷ input energy) × 100.

Other aspects of the engine are expressed in efficiency. These include mechanical efficiency, volumetric efficiency, and thermal efficiency. They

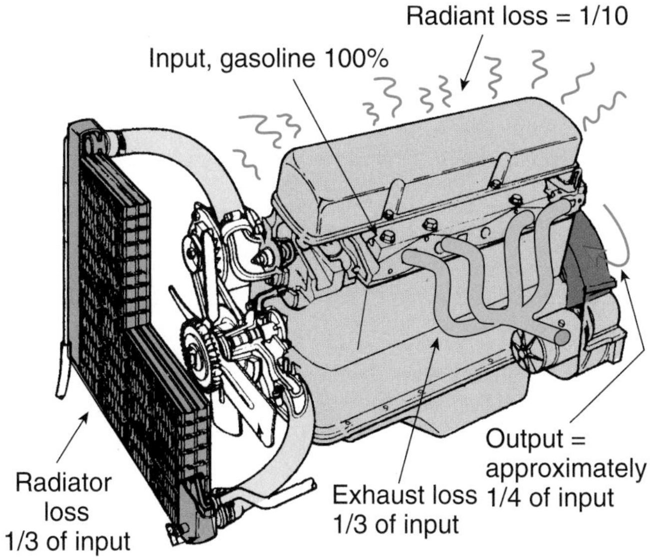

Figure 8–5 A gasoline engine wastes or loses most of the energy it receives.

are expressed as a ratio of input (actual) to output (maximum or theoretical). Efficiencies are always less than 100%. The difference between the efficiency and 100% is the percentage lost. For example, if 100 units of energy are put into an engine and 28 units were used to power the vehicle, the efficiency is 28%. This means 72% of the energy received was wasted or lost **(Figure 8–5)**.

Kinetic and Potential Energy

When energy is released to do work, it is called **kinetic energy**. Kinetic energy may also be referred to as energy in motion **(Figure 8–6)**. Stored energy is called **potential energy**.

There are many automotive systems that have potential energy and, at times, kinetic energy. The ignition system is a source of high electrical energy.

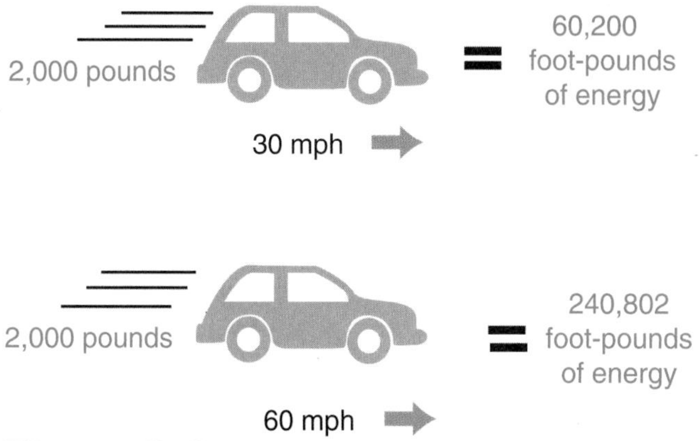

Figure 8–6 The kinetic energy of a moving vehicle increases exponentially with its speed.

The heart of the ignition system is the ignition coil, which has much potential energy. When it is time to fire a spark plug, that energy is released and becomes kinetic energy as it creates a spark across the gap of a spark plug.

Energy Conversion

Energy conversion occurs when one form of energy is changed to another. Because energy is not always in the desired form, it must be converted to a form that can be used. Some of the most common energy conversions are discussed here.

Chemical to Thermal Energy Chemical energy in gasoline or diesel fuel is converted to thermal energy when the fuel burns in the engine cylinders.

Chemical to Electrical Energy The chemical energy in a battery **(Figure 8–7)** is converted to electrical energy to power many of the accessories on an automobile.

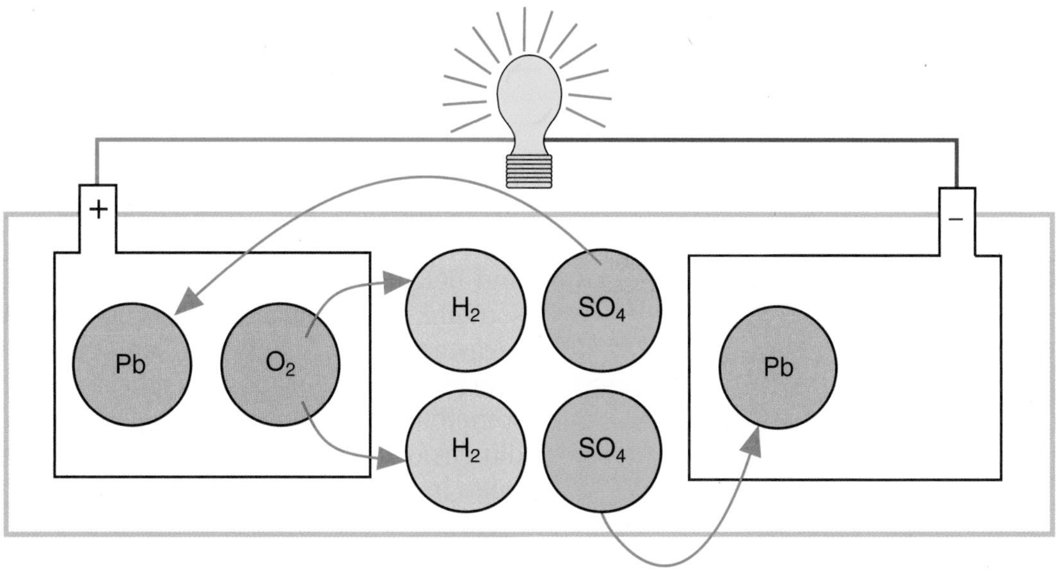

Figure 8–7 Chemical energy is converted to electrical energy in a battery.

Electrical to Mechanical Energy In an automobile, the battery supplies electrical energy to the starting motor, and this motor converts the electrical energy to mechanical energy to crank the engine.

Thermal to Mechanical Energy The thermal energy that results from the burning of the fuel is converted to mechanical energy, which is used to move the vehicle.

Mechanical to Electrical Energy The generator is driven by the mechanical energy of the engine. The generator converts this energy to electrical energy to power the vehicle's electrical accessories and recharges the battery.

Electrical to Radiant Energy Radiant energy is light energy. Electrical energy is converted to thermal energy to heat up a filament inside a light bulb to illuminate it and release radiant energy.

Kinetic to Mechanical to Electrical Energy Hybrid vehicles have a system, called regenerative braking, that uses the energy of the moving vehicle (kinetic) to rotate a generator. The mechanical energy used to operate the generator is used to provide electrical energy to charge the batteries **(Figure 8–8)** or power the electric drive motor.

Mass and Weight

Mass is the amount of matter in an object. **Weight** is a force and is measured in pounds or kilograms. Gravitational force gives the mass its weight. As an example, a spacecraft can weigh 500 tons (one million pounds) here on earth where it is affected by the earth's gravitational pull. In outer space, beyond the earth's gravity and atmosphere, the spacecraft is nearly weightless but its mass remains unchanged **(Figure 8–9)**.

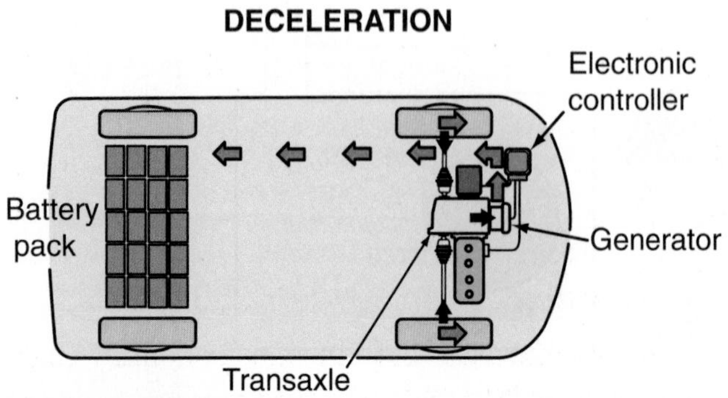

Figure 8-8 Regenerative braking captures some of the vehicle's kinetic energy to charge the batteries or power the electric drive motor.

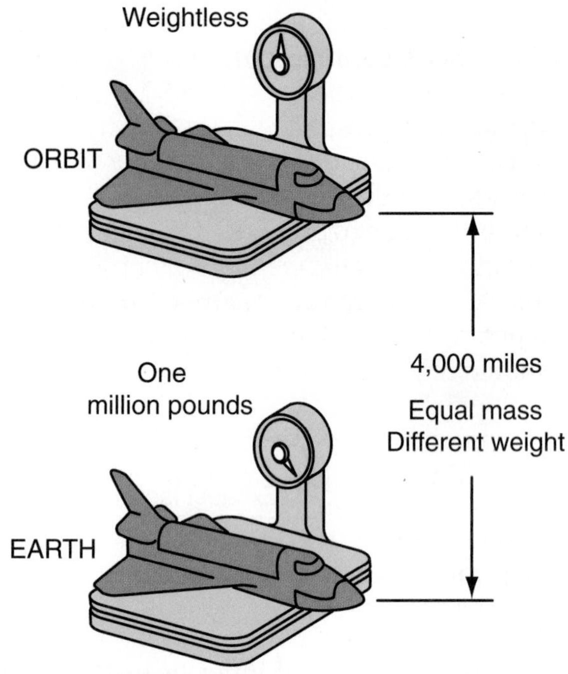

Figure 8-9 The difference in weight of a space shuttle on earth and in space.

Automobile specifications list the weight of a vehicle primarily in two ways. **Gross weight** is the total weight of the vehicle when it is fully loaded with passengers and cargo. **Curb weight** is the weight of the vehicle when it is not loaded with passengers or cargo.

Metric Conversion To convert kilograms into pounds, simply multiply the weight in kilograms by 2.2046. For example, if something weighs 5 kilograms, then $5 \times 2.2046 = 11.023$ pounds. To express the answer in pounds and ounces, convert the 0.023 pound into ounces. Because there are 16 ounces in a pound, multiply 16 by 0.023 ($16 \times 0.023 = 0.368$ ounce). Therefore, 5 kilograms is equal to 11 pounds and 0.368 ounce.

Size

The size of something is related to its mass. An object's size defines how much space it occupies. Size dimensions are typically stated in terms of its length, width, and height. Length is a measurement of how long something is from one end to another. Width is a measurement of how wide something is from one side to another. Obviously height is the distance from something's bottom to its top. All three of these dimensions are measured in inches, feet, yards, and miles in the English system and meters in the metric system.

Sometimes distance measurements are made with a rule that has fractional, rather than decimal, increments. Most automotive specifications are given decimally; therefore, fractions must be converted into decimals. It is also easier to add and subtract dimensions if they are expressed in decimal form rather than fractions. If you want to find the rolling circumference of a tire and the diameter of the tire is 20⅜ inches, convert the fraction to decimals before going further. The distance around the tire is the circumference and is equal to the diameter multiplied by a constant called pi (π). Pi is equal to approximately 3.14; therefore, the circumference of the tire is equal to the diameter multiplied by 3.14. Convert the 20⅜ inches into a whole number and a decimal. To convert the ⅜ to a decimal, divide 3 by 8 ($3 \div 8 = 0.375$). Therefore, the diameter of the tire is 20.375 inches. Now multiply the diameter by π ($20.375 \times 3.14 = 63.98$). The circumference of the tire is nearly 64 inches.

Metric Conversion To convert meters into feet, multiply the number of meters by 3.281. To convert feet into inches, multiply the number of feet by 12. For example, to convert 0.01 mm to inches, begin by converting 0.01 mm into meters. Because 1 mm is equal to 0.001 meter, multiply 0.01 by 0.001 ($0.001 \times 0.01 = 0.00001$). Then multiply 0.00001 meter by 3.281 ($0.00001 \times 3.281 = 0.00003281$ foot). Now convert feet into inches by multiplying by 12 ($0.00003281 \times 12 = 0.00039372$ inch).

To do this easier, recognize that 1 mm is equal to 0.03937 inch. Then multiply 0.01 mm by 0.03937 ($0.01 \times 0.03937 = 0.0003937$ inch).

VOLUME

Volume is also a measurement of size and is related to mass and weight. Volume is the amount of space occupied by an object in three dimensions: length, width, and height. For example, a pound of gold and a pound of feathers both have the same weight, but the pound of feathers occupies a much larger volume. In the English system, volume is measured in cubic inches, cubic feet, cubic yards, or gallons. The measurement for volume in the metric system is cubic centimeters or liters **(Figure 8–10)**.

The volume of a container is calculated by taking an object's measured length, width, and height and multiplying them. For example, if a box has a length of 2 inches, a width of 3 inches, and a height of 4 inches ($2 \times 3 \times 4 = 24$), its volume equals 24 cubic inches. Different shapes have different formulas for calculating volume but all consider the object's three dimensions.

The volume of an engine's cylinders is expressed as displacement. This size does *not* reflect the external (length, width, and height) of the engine. Cylinder **displacement** is the maximum volume of a cylinder. A piston's travel from its lowest point (BDC) to its highest point (TDC) within the cylinder is called the stroke of the piston **(Figure 8–11)**. A cylinder's bore is the diameter of the cylinder.

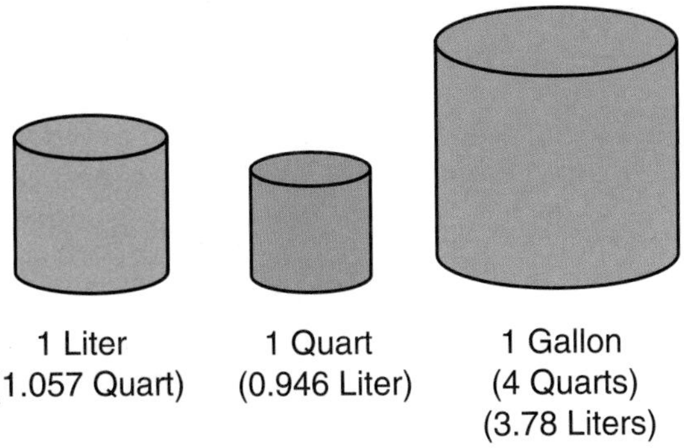

Figure 8-10 A comparison of metric and English units of volume.

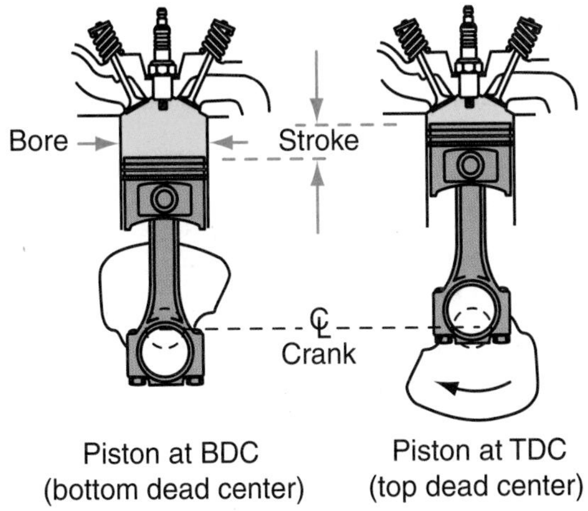

Figure 8-11 The bore and stroke of an engine.

Displacement is usually measured in cubic inches, cubic centimeters, or liters. The total displacement of an engine (including all cylinders) is a rough indicator of its power output. Total displacement is the sum of displacements for all cylinders in an engine. Engine cubic inch displacement (CID) may be calculated as follows:

$$CID = \pi \times R^2 \times L \times N$$

where $\pi = 3.1416$

R = the radius of the cylinder or (the diameter, the bore, ÷ 2)

L = length of stroke

N = number of cylinders in the engine

Example: Calculate the CID of a six-cylinder engine with a 3.7 in. bore and 3.4 in. stroke.

$$CID = 3.1416 \times 1.85^2 \times 3.4 \times 6$$

$$CID = 219.66$$

SHOP TALK

Engine displacement can also be calculated by using this formula: 0.7854 × Bore × Bore × Stroke × Number of cylinders = Displacement

Most of today's engines are listed by their metric displacement. Cubic centimeters and liters are determined by using metric measurements in the displacement formula.

Example: Calculate the metric displacement of a four-cylinder engine with a 78.9 mm stroke and a 100 mm bore. Before you use the formula to find the displacement in cubic centimeters, convert the millimeter measurements to centimeters: 78.9 mm = 7.89 cm and 100 mm = 10 cm.

$\text{Displacement} = 3.1416 \times 5^2 \times 7.89 \times 4$

Displacement = 2479 cubic centimeters (cc) or approximately 2.5 liters (L)

Ratios

Often automotive features are expressed as ratios. A ratio expresses the relationship between two things. If something is twice as large as another, there is a ratio of 2:1. Sometimes ratios are used to compare the movement of an object. For example, if a 1-inch movement by something causes something else to move 2 inches, there is a travel ratio of 1:2.

An engine's **compression ratio** expresses how much the air and fuel mixture is compressed as a cylinder's piston moves from the bottom (BDC) to the top (TDC) of the cylinder. The compression ratio is defined as the ratio of the volume in the cylinder above the piston when the piston is at BDC to the volume in the cylinder is at TDC **(Figure 8–12)**. The formula for calculating the compression ratio is as follows:

volume above the piston at BDC ÷
volume above the piston at TDC

or

total cylinder volume ÷
total combustion chamber volume

In many engines, the top of the piston is at the top of the cylinder block during TDC. The combustion chamber is the cavity in the cylinder head above the piston. This may be modified slightly by the shape of the top of the piston. The volume of the combustion chamber must be added to each volume in the formula in order to get an accurate calculation of compression ratio.

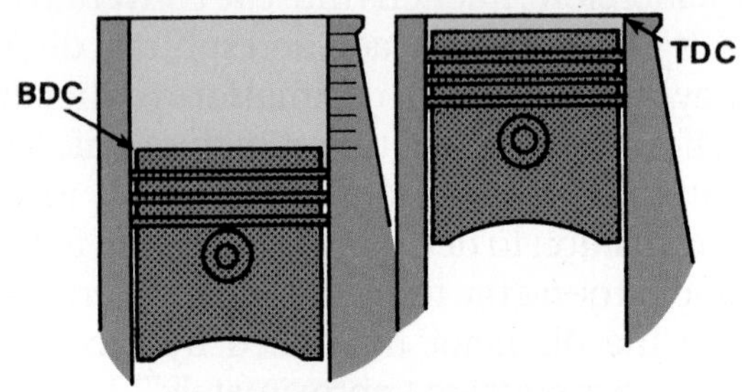

Figure 8-12 An engine's compression ratio indicates the amount the air and fuel mixture is compressed during the compression stroke of a piston.

Example: Calculate the compression ratio if the total piston displacement is 45 cubic inches and the combustion chamber volume is 5.5 cubic inches.

$$45 + 5.5 \div 5.5 = 9.1$$

Therefore, the compression ratio is 9.1 to 1 or 9.1:1.

Proportions

Ratios are also used to express the correct mixture for something. For example, engine coolant should typically be mixed with 50% coolant and 50% water when the cooling system is refilled **(Figure 8–13)**. This is a 1:1 ratio. This ratio allows for maximum hot and cold protection.

Consider a cooling system that has a capacity of 9.5 liters. Because most coolant is sold in gallon containers, to determine the amount of coolant that should be put in the system, first convert the liter

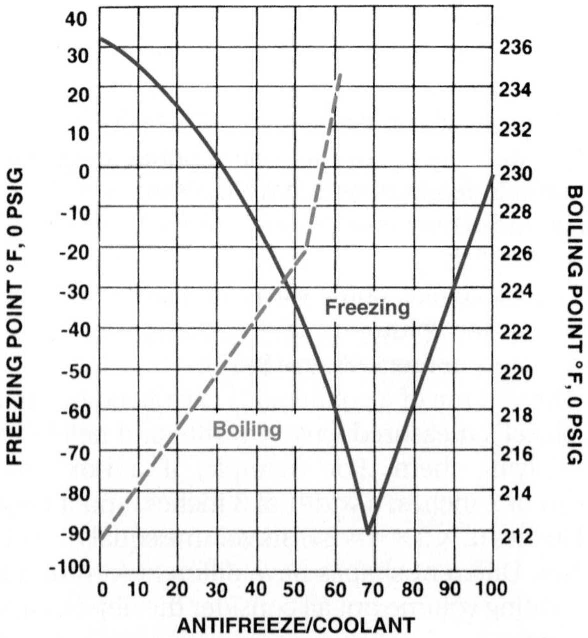

Figure 8-13 The relationship of the percentage of antifreeze to the freezing and boiling points of the engine's coolant.

capacity to gallons. Because 1 gallon equals 3.7854 liters, divide 9.5 liters by 3.7854 (9.5 ÷ 3.7854 = 2.5097). Therefore, the total capacity of the cooling system is a little more than 2.5 gallons. To determine how much coolant or antifreeze to put in the system, divide the total capacity by 2 (2.5 ÷ 2 = 1.25). Therefore, the correct mixture is 1¼ gallons of coolant mixed with 1¼ gallons of water.

FORCE

A **force** is a push or pull and can be large or small. Force can be applied by direct contact or from a distance. Gravity and electromagnetism are examples of forces that are applied from a distance. Forces can be applied from any direction and with any intensity. For example, if a pulling force is twice that of the pushing force, the object will be pulled at one-half of the pulling force. When two or more forces are applied to an object, the combined force is called the resultant. The resultant is the sum of the amount and direction of the forces. For example, when a mass is suspended by two lengths of wire, each wire will carry half the weight of the mass. If the attachment of the wires is moved so they are now at an angle to the mass, the wires will carry more force. They now carry the force of the mass plus the force that pulls against the other wire.

Automotive Forces

When a vehicle is sitting still, gravity exerts a downward force on the vehicle. The ground exerts an equal and opposite upward force and supports the vehicle. When the engine is running and its power is transferred to the drive wheels, the wheels exert a force against the ground in a horizontal direction. This causes the vehicle to move but it is opposed by the mass of the vehicle **(Figure 8–14)**. To move the vehicle faster, the force supplied by the wheels must increase beyond the opposing forces. As the vehicle does move faster, it pushes against the air as it travels. This becomes a growing opposing force, and the force at the drive wheels must overcome that force in order for the vehicle to increase speed. After the vehicle has achieved the desired speed, no additional force is required at the drive wheels.

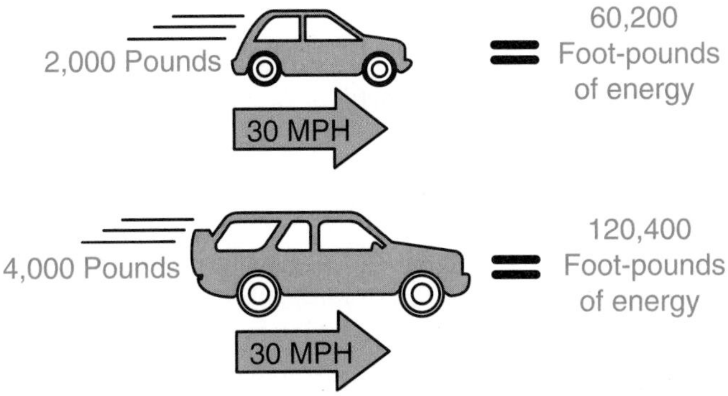

Figure 8-14 The amount of energy required to move a vehicle depends on its mass.

Force—Balanced and Unbalanced When the applied forces are balanced and there is no overall resultant force, the object is in **equilibrium**. An object sitting on a solid flat surface is in equilibrium. If the surface is set at a slight angle, the forces will cause the object to slowly slide down the surface. If the surface is at a severe angle, the downward force will cause the object to quickly slide down the slope. In both cases, the surface is still supplying the force needed to support the object but the pull of gravity is greater and the resultant force causes the object to slide down.

Turning Forces Forces can cause rotation as well as straight line motion. A force acting on an object that is free to rotate will have a turning effect, or turning force. This force is equal to the force multiplied by the distance of the force from the turning point around which it acts.

Forces on Tires and Wheels

If you roll a cone-shaped object on a smooth surface, the cone will not roll in a straight line. Rather it will move in the direction of the cone's tilt **(Figure 8–15)**. Riding a bicycle is an example of this. To turn left, it is easier to tilt the bicycle to the left. A tilted, rolling wheel tends to move in the direction of the tilt. Similarly, if a vehicle's tire and wheel are tilted, the tire and

Figure 8-15 A tire at an angle will roll in the same way as a cone would.

wheel will tend to move in the direction of the tilt. This is a principle considered during wheel alignment.

While riding a bicycle, your weight is projected through the bicycle's front fork to the road surface. The centerline of the front fork is tilted rearward in relation to the vertical centerline of the wheel. When the handle bars are turned, the tire pivots on the wheel's vertical centerline. Because the tire's pivot point is behind where your weight is projected against the surface of the road, the front wheel tends to return to the straight-ahead position after a turn. The wheel also tends to remain in the straight-ahead position as the bicycle is driven. This principle of resultant forces is the basis for precise wheel alignment.

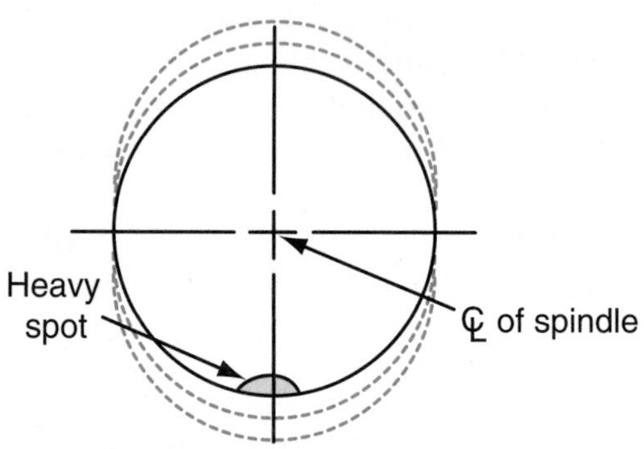

Figure 8–16 Wheel tramp is the result of a tire and wheel assembly being statically unbalanced.

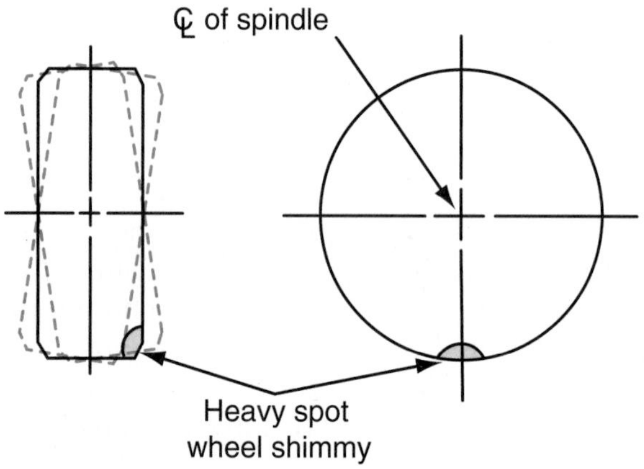

Figure 8–17 Dynamic imbalance causes wheel shimmy.

Centrifugal/Centripetal Forces

When an object moves in a circle, its direction is continuously changing. All directional changes require a force. The forces required to maintain a circular motion are called **centripetal** and **centrifugal forces.** The required forces depend on the size of the circle and the object's mass and speed.

Centripetal force tends to pull the object toward the center of the circle. Centrifugal force tends to push the object away from the center. The centripetal force that keeps an object whirling around on the end of a string is caused by **tension** in the string. If the string breaks, there is no longer string tension and the object will fly off in a straight line because of the centrifugal force on it. Gravity is the centripetal force that keeps the planets orbiting around the sun. Without this centripetal force, the earth would move in a straight line through space.

Wheel and Tire Balance

When the weight of a wheel and tire assembly is distributed equally around the center of wheel rotation, the wheel and tire has proper static balance. Being statically balanced, the wheel and tire assembly will not tend to rotate by itself, regardless of the wheel position. If the weight is not distributed equally, the wheel and tire assembly is statically unbalanced. As the wheel and tire rotate, centrifugal force acts on this static unbalance and causes the wheel to "tramp" or "hop" **(Figure 8–16)**.

Dynamic balance exists when the weight thrown to the sides of a rotating tire and wheel assembly are equal **(Figure 8–17)**. To illustrate this, assume we have a bar with a ball attached by string to both ends of the bar. If we rotate the bar, the balls will turn with the bar and the centripetal and centrifugal forces will keep the balls in an orbit around the rotating bar. If the two balls weigh the same and are at an equal distance from the bar, the bar will rotate smoothly. However, if one of the balls is heavier, the bar will wobble as it rotates. The greater the difference in weight, the greater the wobble. The wobble can eventually destroy the mechanism used to rotate the bar.

Now if we add some weight to the end of the bar that has the lighter ball, the weights and forces can be equalized and the wobble removed. This is basically how we dynamically balance a wheel and tire assembly **(Figure 8–18)**.

When we think of all the parts of an automobile that rotate, it is easy to see why proper balance is important. Improper balance can cause premature wear or destruction of parts.

Pressure

Pressure is a force applied against an object and is measured in units of force per unit of surface area (pounds per square inch or kilograms per square centimeter). Mathematically, pressure is equal to the applied force divided by the area over which the force acts. Consider two 10-pound weights sitting on a table; one occupies an area of 1 square inch and the

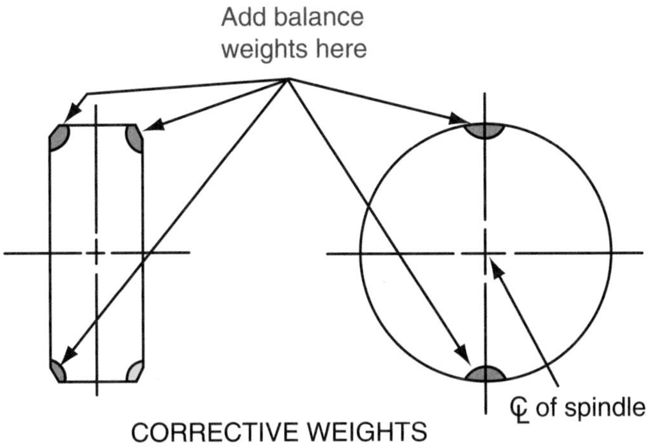

Figure 8-18 Adding a weight to counteract with the heavy spot of a tire and wheel assembly.

other an area of 4 square inches. The pressure exerted by the first weight would be 10 pounds per 1 square inch or 10 psi. The other weight, although it weighs the same, will exert only 2.5 psi (10 pounds per 4 square inches = 10 ÷ 4 = 2.5). This shows that a force acting over a large area will exert less **pressure** than the same force acting over a small area.

Because pressure is a force, all principles of force apply to pressure. If more than one pressure is applied to an object, the object will respond to the resultant force. Also, all matter (liquids, gases, and solids) tend to move from an area of high pressure to an area of low pressure.

TIME

The word **time** is used to mean many things. For this discussion, time is defined as a measurement of the duration of something that has happened, is happening, or will happen. Time is measured by the increments of a clock: seconds, minutes, and hours. Often an automotive technician is concerned with how long something occurs, such as the length of time a spark plug fires to cause combustion. This time, called spark duration, is typically about 3 milliseconds (0.003 second) and is measured with a lab scope because it would be very difficult to measure that short of a time with a clock.

Technicians also monitor how many times a cycle is repeated within a period of time, such as a minute. A tachometer, which measures engine revolutions per minute, is an often used diagnostic tool.

MOTION

When the forces on an object do not cancel each other out, they will change the object's speed or direction of motion, or both. The greater the object's mass, the greater the force must be to change its motion. This resistance to change is called **inertia**. Inertia is the tendency of an object at rest to remain at rest or the tendency of an object in motion to stay in motion in one direction. The inertia of an object at rest is called static inertia, whereas the inertia of an object in motion is called dynamic inertia. Inertia exists in liquids, solids, and gases. When you push and move a parked vehicle, you overcome the static inertia of the vehicle. If you catch a ball in motion, you overcome the dynamic inertia of the ball.

When a force overcomes static inertia and moves an object, the object gains momentum. **Momentum** is the product of an object's weight and speed. Momentum is a type of mechanical energy. An object loses momentum if another force overcomes the dynamic inertia of the moving object.

Rates

Speed is the distance an object travels in a set amount of time. It is calculated by dividing the distance traveled by the time it took to travel that distance. We refer to the speed of a vehicle in miles per hour (mph) or kilometers per hour (km/h). **Velocity** is the speed of an object in a particular direction. **Acceleration** is the rate of increase in speed. Acceleration is calculated by dividing the change in speed by the time it took for that change. **Deceleration** is the reverse of acceleration; it is the rate of a decrease in speed.

Newton's Laws of Motion

How forces change an object's motion was first explained by Sir Isaac Newton in what is known as Newton's laws. Newton's first law of motion is called the law of inertia. It states that an object at rest tends to remain at rest and an object in motion tends to remain in motion, unless some force acts on it. When a car is parked on a level street, it remains stationary unless it is driven or pushed.

Newton's second law states that when a force acts on an object, the motion of the object will change. This change is equal to the size of the force divided by the object's mass. Trucks have a greater mass than cars. Because a large mass requires a larger force to produce a given acceleration, a truck needs a larger engine than a car.

Newton's third law says that for every action there is an equal and opposite reaction. A practical application of this occurs when the wheel strikes a bump in the road. This action drives the wheel and suspension upward with a certain force, and a specific amount of energy is stored in the spring. After this, the spring forces the wheel and suspension downward with a force equal to the initial upward force caused by the bump.

Friction

Friction is a force that slows or prevents the motion of two objects or surfaces that touch. Friction may occur in solids, liquids, and gases. It is the joining or bonding of the atoms at each of the surfaces that causes the friction. When you attempt to pull an object across a surface, the object will not move until these bonds are overcome. Smooth surfaces produce little friction; therefore, only a small amount of force is needed to break the bonds of the atoms. Rougher surfaces produce a larger friction force because there are stronger bonds between the two surfaces **(Figure 8–19)**. To move an object over a rough surface, such as sandpaper, a great amount of force is required.

Friction is put to good use in disc brakes **(Figure 8–20)**. The friction between the brake disc and pad slows the rotation of the wheel, reducing the vehicle's speed. In doing so, it converts the kinetic energy of the vehicle into heat.

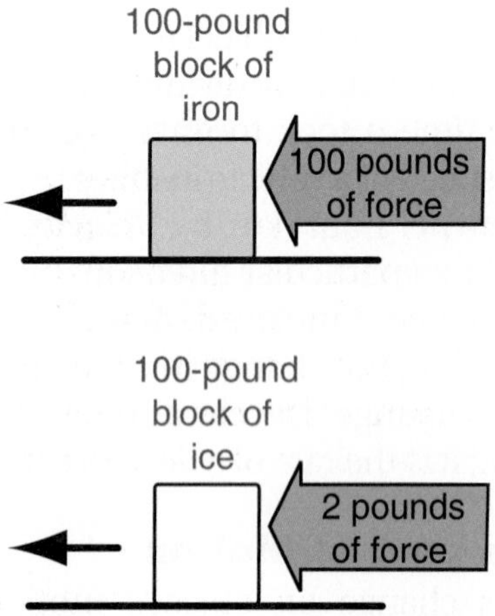

Figure 8-19 Sliding ice across a surface produces less friction than sliding a rougher material, such as iron, across a surface.

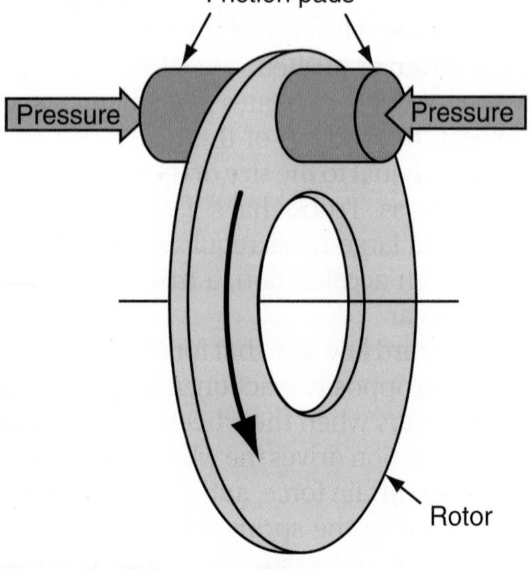

Figure 8-20 As pressure is applied to the friction pads, the pads attempt to stop the rotor to which the tire and wheel are attached.

Lubrication Friction can be reduced in two main ways: by lubrication or by the use of rollers. The presence of oil or another fluid between two surfaces keeps the surfaces apart. Because fluids (liquids and gases) flow, they allow movement between surfaces. The fluid keeps the surfaces apart, allowing them to move smoothly past one another **(Figure 8–21)**.

Rollers Rollers placed between two surfaces also keep the surfaces apart. An object placed on rollers will move smoothly if pushed or pulled. Instead of sliding against one another, the surfaces produce turning forces, which cause each roller to spin. This leaves very little friction to oppose motion. Bearings are a type of roller used to reduce the friction between moving parts such as a wheel and its axle **(Figure 8–22)**. As the wheel turns on the axle, the balls in the bearing roll around inside the bearing, drastically reducing the friction between the wheel and axle.

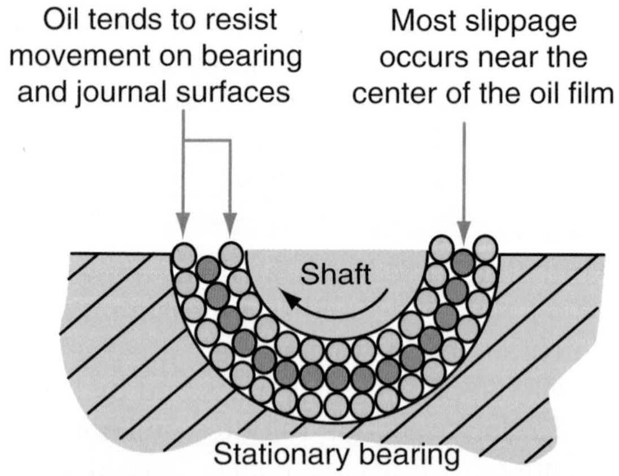

Figure 8-21 Oil separates the rotating shaft from the stationary bearing.

Figure 8-22 An assortment of tapered roller bearings and their races.

Air Resistance

When a vehicle is driven, resistance occurs between the air and the vehicle's body. This resistance or friction opposes the momentum, or mechanical energy, of the moving vehicle. The mechanical energy from the engine must overcome the vehicle's inertia and the friction of the air striking the vehicle. The faster an object moves, the greater the air resistance.

Body design, obviously, affects the amount of friction developed by the air striking the vehicle. The total resistance to motion caused by friction between a moving vehicle and the air is referred to as coefficient of drag (Cd). At 45 miles per hour (72 kilometers per hour), half of the engine's mechanical energy can be used to overcome air resistance. Therefore, reducing a vehicle's Cd is a very effective way to improve fuel economy. Cd may also be called aerodynamic drag.

Aerodynamics is the study of the effects of air on a moving object **(Figure 8–23)**. The basics of this science are fairly easy to understand. The larger the area facing the moving air is, the more the air will tend to hold back or resist forward motion.

Needless to say, the less air a vehicle pushes out of its way, the less power it needs to move at a given speed. If engineers want a vehicle that uses less fuel and emits fewer pollutants, they do whatever it takes to make the engine work less hard. Aerodynamics is one of those things.

Most aerodynamic design work is done initially on a computer; the design is then checked and modified by placing the vehicle in a wind tunnel **(Figure 8–24)**. A wind tunnel is a carefully constructed facility with a large fan at one end. Inside the tunnel, the movement of air over, under, and around the vehicle is studied.

Ideally, the air moved by the vehicle will follow the contours of the vehicle. This prevents the air from doing funny things as it is pushed away. If the air that moves under the vehicle has a place to push up, the vehicle will tend to lift. This creates poor handling, a situation that can be very unsafe. Air can also be trapped under the vehicle, which increases the vehicle's air drag. If air moving over the top pushes against the vehicle, there is an increase in air drag. To help direct the air and make it usable, air dams and spoilers or wings are used.

Figure 8-24 This wind tunnel can generate winds as high as 150 miles per hour. *Courtesy of Chrysler LLC*

WORK

When a force moves a mass a specific distance, **work** is done. When work is accomplished, a mass may be lifted or pushed against a resistance or opposing force **(Figure 8–25)**. Work is equal to the applied force multiplied by the distance the object moved

Figure 8-23 The movement of air as it goes over this car. *Courtesy of Chrysler LLC*

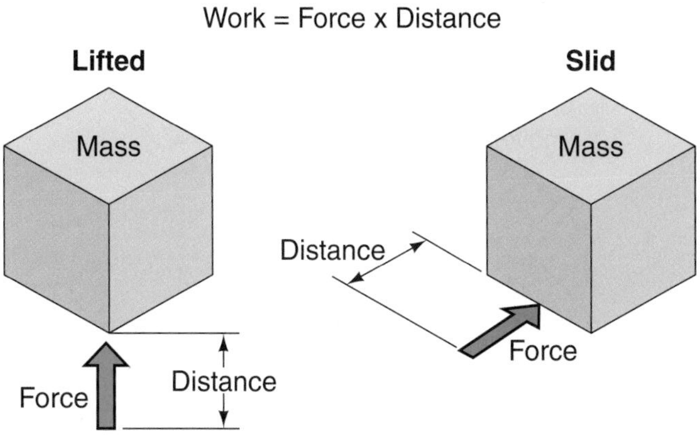

Figure 8-25 When work is performed, a mass is moved a certain distance.

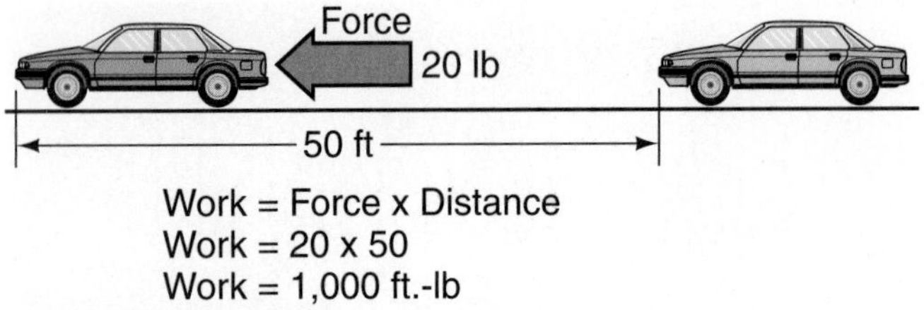

Figure 8-26 1,000 foot-pounds of work.

(force × distance = work) and is measured in foot-pounds **(Figure 8–26)**, watts, or Newton-meters. For example, if a force moves a 3,000-pound car 50 feet, 150,000 foot-pounds of work was done.

During work, a force acts on an object to start, stop, or change its direction. It is possible to apply a force and not move the object. For example, you may push with all your strength on a car stuck in a ditch and not move it. This means no work was done. Work is only accomplished when an object is started, stopped, or redirected by a force.

Simple Machines

A machine is any device used to transmit a force and, in doing so, changes the amount of force and/or its direction. A common example of a simple machine that does both is a valve rocker arm. One end of a rocker arm is pushed up by the action of the engine's camshaft. When this happens, the other end of the rocker arm pushes down on a valve to open it. A rocker arm is also designed to change the size of the force applied to it. Rocker arms provide more movement on the valve side or output than the input side. This is referred to as the rocker arm's ratio. If a rocker arm has a ratio of 1.5:1, one end of it will move 1.5 times more than the other **(Figure 8–27)**. For example, if the camshaft causes one end of the rocker arm to move ½ inch, the other end will move ¾ of an inch.

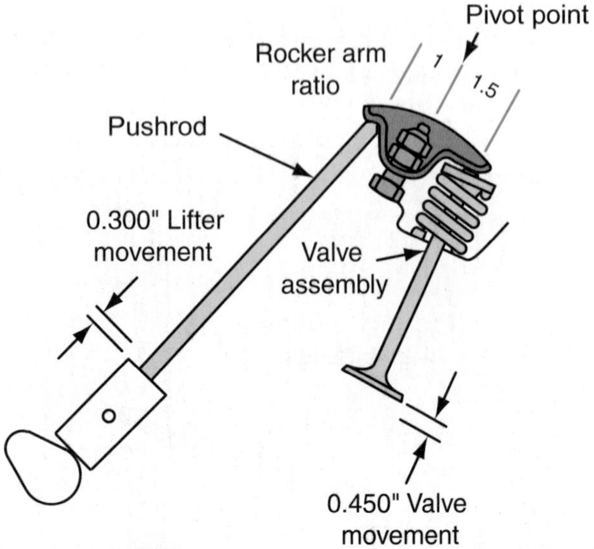

Figure 8-27 A rocker arm with a ratio of 1:1.5.

Figure 8-28 It takes less energy to pull a mass up an inclined plane than would be required to lift the mass vertically.

The force applied to a machine is called the effort, while the force it overcomes is called the **load**. The effort is often smaller than the load, because a small effort can overcome a heavy load if the effort is moved a larger distance. The machine is then said to give a mechanical advantage. Although the effort will be smaller when using a machine, the amount of work done, or energy used, will be equal to or greater than that without the machine.

Inclined Plane The force required to drag an object up a slope **(Figure 8–28)** is less than that required to lift it vertically. However, the overall distance moved by the object is greater when pulled up the slope than if it were lifted vertically. A screw is an inclined plane wrapped around a shaft. The force that turns the screw is converted to a larger one, which moves a shorter distance and drives the screw in.

Pulleys A **pulley** is a wheel with a grooved rim in which a rope, belt, or chain runs to move something by pulling on the other end of the rope, belt, or chain. A simple pulley changes the direction of a force but not its size. Also, the distance the force moves does not change. By using several pulleys connected together as in a block and tackle, the size of the force can be changed too, so that a heavy load can be lifted using a small force. With a double pulley, the required applied force to move an object can be reduced by one-half but the distance the force must be moved is doubled. A quadruple pulley can reduce the force by four times but the distance will be increased by four times. Pulleys of different sizes can change the required applied force as well as the speed or distance the pulley needs to travel to accomplish work **(Figure 8–29)**.

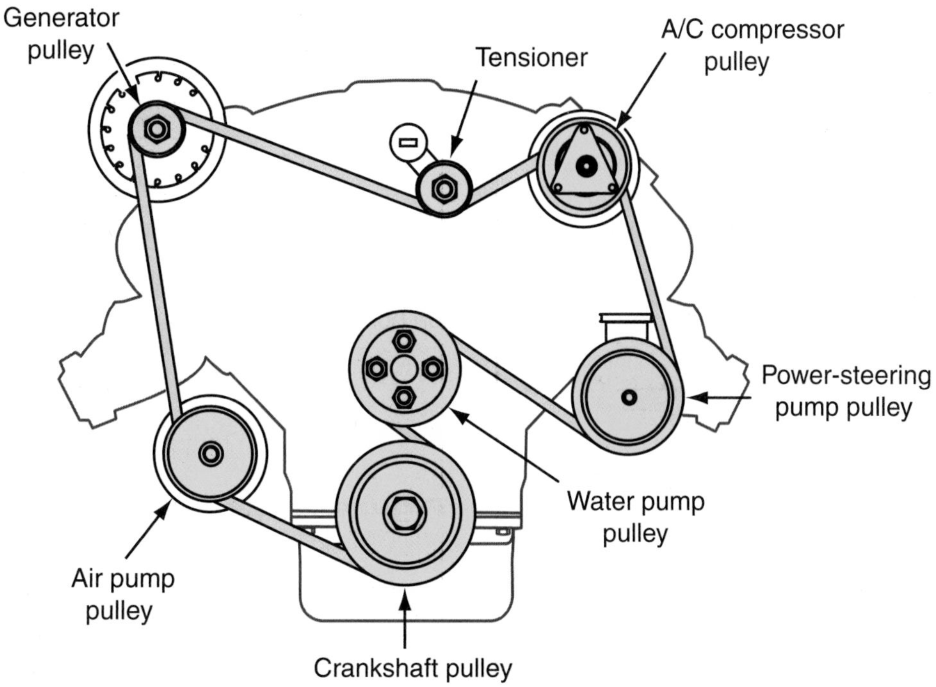

Figure 8-29 Accessories are driven by a common drive belt but they rotate at different speeds because of the differences in pulley size.

Levers A **lever** is a device made up of a bar moving on a fixed pivot point called the fulcrum. A lever uses a force applied at one point to move a mass on the other end of the bar. Levers are divided into classes. In a class one lever, the fulcrum is between the effort and the load **(Figure 8–30)**. The load is larger than the effort, but it moves a smaller distance. A pair of pliers is an example of a class one lever. In a class two lever, the load is between the fulcrum and the effort. Here again, the load is greater than the effort and moves through a smaller distance **(Figure 8–31)**. In a class three lever, the effort is between the fulcrum and the load. In this case, the load is less than the effort but it moves through a greater distance.

Gears A **gear** is a toothed wheel that becomes a machine when it is meshed with another gear. The action of one gear is the same as a rotating lever and moves the other gear with it. Based on the size of the gears, the amount of force applied from one gear to the other can be changed. Keep in mind that this does not change the amount of work performed by the gears because the change in force is accompanied by a change in the distance of travel **(Figure 8–32)**. The relationship of force and distance is inverse. Gear ratios express the mathematical relationship (diameter and number of teeth) of one gear to another.

Wheels and Axles The most obvious application of a wheel and axle is a vehicle's tires and wheels. These revolve around an axle and limit the amount of area that contacts the road. Wheels function as rollers to reduce the friction between a vehicle and the road. Basically, the larger the wheel, the less force is required to turn it. However, the wheel moves farther as it gets larger. An example of this is a steering wheel. A steering wheel that is twice the size of another will require one-half the force to turn it but will also require twice the distance to accomplish the same work.

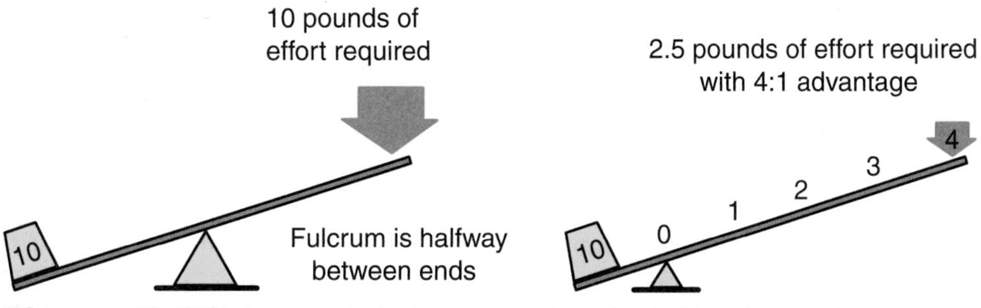

Figure 8-30 A mechanical advantage can be gained with a class one lever.

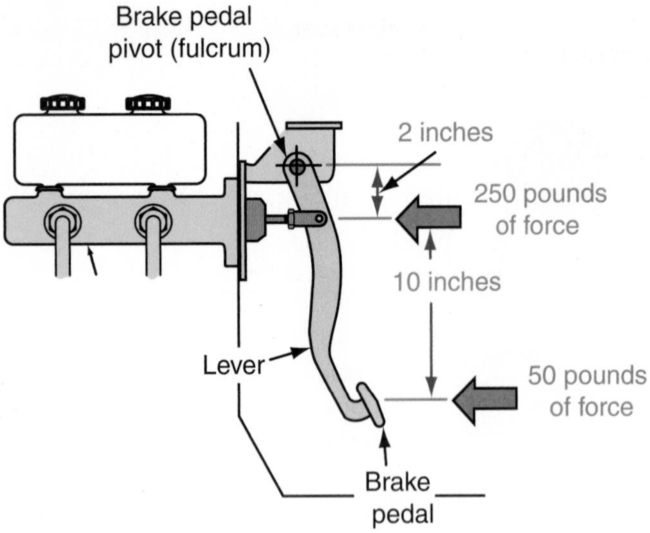

Figure 8-31 A brake pedal assembly is an example of a class two lever.

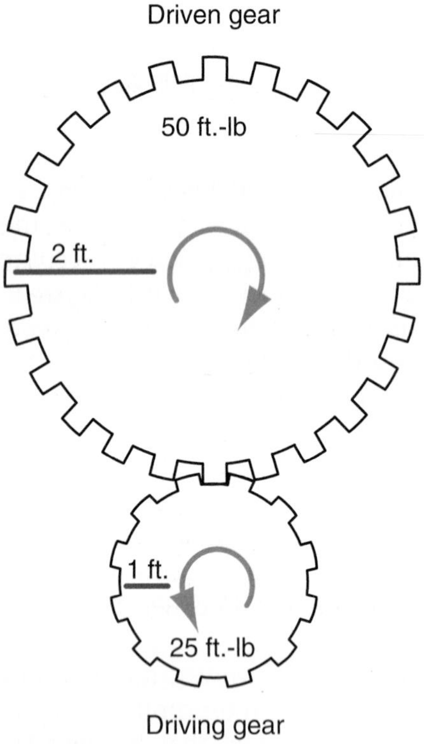

Figure 8-32 When a small gear drives a larger gear, the larger gear turns with more force but travels less; therefore, the amount of work stays the same.

Torque

Torque is a force that tends to rotate or turn things and is measured by the force applied and the distance traveled. The technically correct unit of measurement for torque is pounds per foot (lb-ft.). However, it is rather common to see torque stated as foot-pounds (ft.-lb). In the metric, or SI, system, torque is stated in Newton-meters (N-m) or kilogram-meters (kg-m).

An engine creates torque and uses it to rotate the crankshaft. The combustion of gasoline and air creates pressure against the top of a piston. That pressure creates a force on the piston and pushes it down.

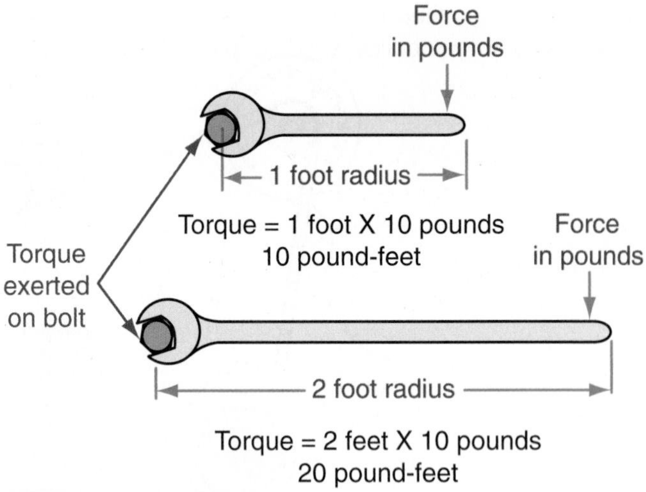

Figure 8-33 The amount of torque applied to a wrench is changed by the length of the wrench.

The force is transmitted from the piston to the connecting rod and from the connecting rod to the crankshaft. The engine's crankshaft rotates with a torque that is sent through the drivetrain to turn the drive wheels of the vehicle.

Torque is force times leverage, the distance from a pivot point to an applied force. Torque is generated any time a wrench is turned with force. If the wrench is a foot long, and you put 20 pounds of force on it, 20 pounds per foot are being generated. To generate the same amount of torque while exerting only 10 pounds of force, the wrench needs to be 2 feet long **(Figure 8–33)**. To have torque, it is not necessary to have movement. When you pull a wrench to tighten a bolt, you supply torque to the bolt. If you pull on a wrench to check the torque on a bolt and the bolt torque is sufficient, torque is applied to the bolt but no movement occurs. If the bolt turns during torque application, work is done. When a bolt does not rotate during torque application, no work is accomplished.

Torque Multiplication

When gears with different numbers of teeth mesh, each rotates at a different speed and force. Torque is calculated by multiplying the force by the distance from the center of the shaft to the point where the force is exerted.

The distance from the center of a circle to its outside edge is its radius. On a gear, the radius is the distance from the center of the gear to the point on its teeth where force is applied.

If a tooth on the driving gear is pushing against a tooth on the driven gear with a force of 25 pounds and the force is applied at a distance of 1 foot (the radius of the driving gear), a torque of 25 ft.-lb is applied to the driven gear. The 25 pounds of force from the teeth of the smaller (driving) gear is applied to the teeth of the larger (driven) gear. If that same force were applied at a distance of 2 feet from the center, the torque on the shaft of the driven gear would be 50 ft.-lb (see Figure 8-32). The same force is acting at twice the distance from the shaft center.

The amount of torque that can be applied from a power source is proportional to the distance from the center at which it is applied. If a fulcrum or pivot point is placed closer to the object being moved, more torque is available to move the object, but the lever must move farther than if the fulcrum were farther away from the object. The same principle is used for gears in mesh: A small gear will drive a large gear more slowly but with greater torque.

A drivetrain consisting of a driving gear with eleven teeth and a radius of 1 inch and a driven gear with forty-four teeth and a radius of 4 inches will have a torque multiplication factor of 4 and a speed reduction of ¼. Thus, the larger gear will turn with four times the torque but one-fourth the speed **(Figure 8–34)**. The radii between the teeth of a gear act as levers.

Gear ratios express the mathematical relationship of one gear to another. Gear ratios can vary by changing the diameter and number of teeth of the gears in mesh. A gear ratio also expresses the amount of torque multiplication between two gears. The ratio is obtained by dividing the diameter or number of teeth of the driven gear by the diameter or teeth of the drive gear. If the smaller driving gear had eleven teeth and the larger gear had forty-four teeth, the ratio is 4:1.

Power

Power is a measurement of the rate, or speed, at which work is done. The metric unit for power is the watt. A watt is equal to one Newton-meter per second. Power is a unit of speed combined with a unit of force. For example, if you were pushing something with a force of 1 N and it moved at 1 meter per second, the power output would be 1 watt.

In electrical terms, 1 watt is equal to the amount of electrical power produced by a current of 1 ampere across a potential difference of 1 volt. This is expressed as Power (P) = Voltage (E) × Current (I) or $P = E \times I$.

Horsepower

Horsepower is the rate at which torque is produced. James Watt is credited with being the first person to calculate horsepower and power. He measured the amount of work a horse could do within a specific time. A horse could move 330 pounds 100 feet in 1 minute **(Figure 8–35)**. Therefore, he determined that one horse could do 33,000 ft.-lb of work in 1 minute. Thus, 1 horsepower is equal to 33,000 ft.-lb per minute, or 550 ft.-lb per second. Two horsepower could do this same amount of work in ½ minute. If you push a 3,000-pound (1,360-kilogram) car for 11 feet (3.3 meters) in ¼ minute, you produce 4 horsepower.

An engine producing 300 ft.-lb of torque at 4,000 rpm produces 228 horsepower at 4,000 rpm. This is based on the fact that horsepower is equal to torque multiplied by engine speed, and that quantity is divided by 5252 ([torque × engine speed] ÷ 5252 = horsepower). The constant, 5252, is used to convert the rpm into revolutions per second.

SHOP TALK

Manufacturers are now rating their engines' outputs in watts. One horsepower is equal to approximately 746 watts. Therefore, a 228 hp engine is rated at 170,088 watts or about 170 kW.

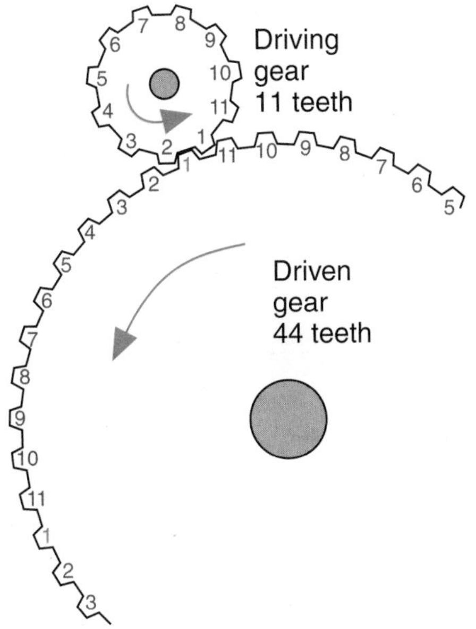

Figure 8–34 The driving gear must rotate four times to rotate the driven gear once.

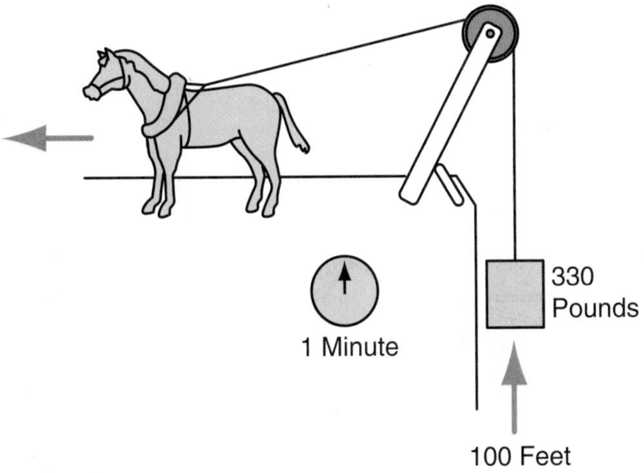

Figure 8–35 This is how James Watt defined 1 horsepower.

WAVES AND OSCILLATIONS

An **oscillation** is the back-and-forth movement of an object between two points. When that motion travels through matter or space, it becomes a **wave**. A mass suspended by a spring, for example, is acted on by two forces: gravity and the tension in the spring. At the point of equilibrium, the resultant of these forces is zero. When the mass is given a downward push, the tension of the spring exceeds the weight of the mass. The resultant upward force accelerates the mass back up toward its original position and its momentum carries it farther upward. When the weight exceeds the spring's tension, the mass moves down again and the oscillation repeats itself until the mass is at equilibrium. As the mass oscillates toward the equilibrium position, the size of the oscillation decreases. As the mass oscillates, the air around it moves and becomes an air wave.

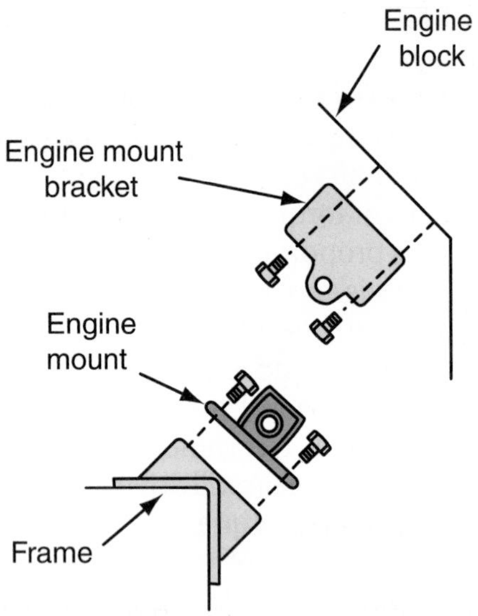

Figure 8-37 An engine mount holds the engine in place and isolates engine vibrations from the rest of the vehicle.

Vibrations

When an object oscillates, it vibrates **(Figure 8–36)**. To prevent the vibration of one object from causing a vibration in other objects, the oscillating mass must be isolated from other objects. This is often a difficult task. For example, all engines vibrate as they run. To reduce the transfer of engine vibrations to the rest of the vehicle, the engine is held by special mounts. The materials used in the mounts must keep the engine in place and must be elastic enough to absorb the engine's vibrations **(Figure 8–37)**. If the engine was mounted solidly, the vibrations would be felt throughout the vehicle.

Vibration control is also important for the reliability of components. If the vibrations are not controlled, the object could shake itself to destruction. Vibration control is the best justification for always mounting parts in the way they were designed to be mounted.

Unwanted and uncontrolled vibrations typically result from one component vibrating at a different frequency than another part. When two waves or vibrations meet, they add up or interfere. This is called the principle of superposition and is common to all waves. Making unwanted vibrations tolerable can be done by canceling them with equal and opposite vibrations. This approach to vibration reduction is best illustrated by the use of balance shafts in an engine. These shafts are designed to counter the vibrations caused by the rotation of the engine's crankshaft and pistons **(Figure 8–38)**. The balance shaft spins and creates an equal but opposite vibration to cancel the vibrations of the crankshaft.

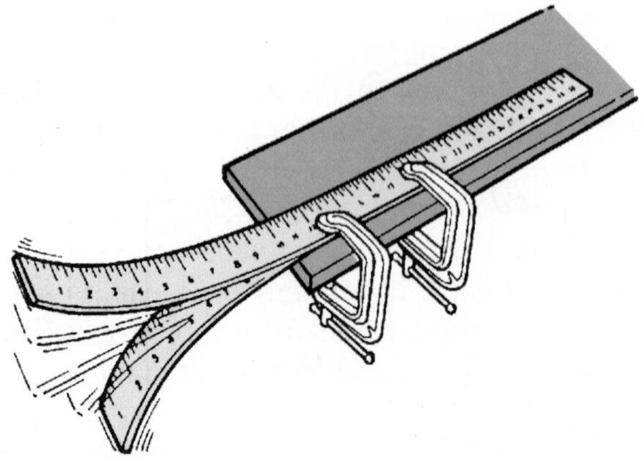

Figure 8-36 Vibrations happen in cycles; notice how the yardstick moves up and down in cycles.

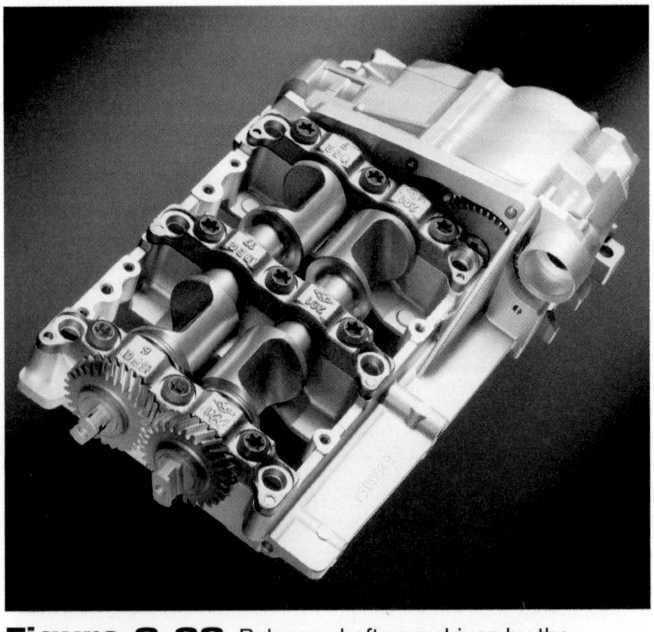

Figure 8-38 Balance shafts are driven by the crankshaft and work to counter crankshaft pulses and vibrations by acting with an equal force but in the opposite direction. *Courtesy of BMW of North America, LLC*

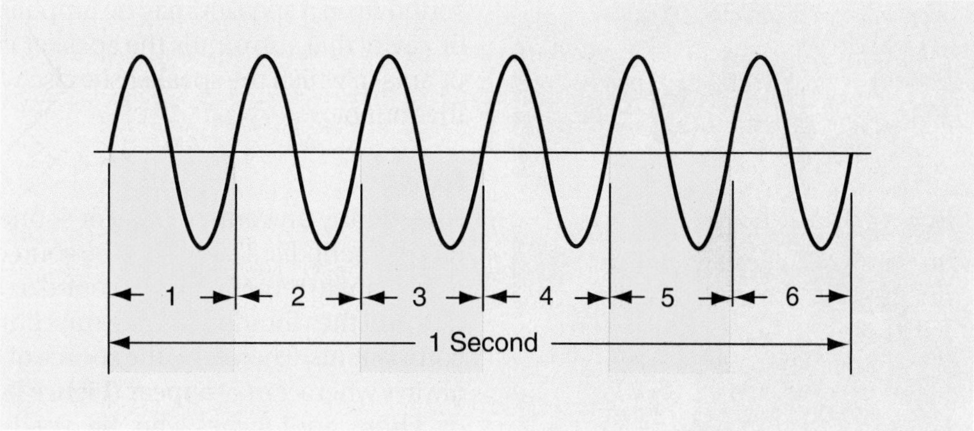

Figure 8-39 Frequency is a statement of how many cycles occur in a second.

How many times the vibration occurs in 1 second is called **frequency**. Frequency **(Figure 8–39)** is most often expressed in **hertz** (Hz). One hertz is equal to one cycle per second. The name is in honor of Heinrich Hertz, an early German investigator of radio wave transmission. The **amplitude** of a vibration is its intensity or strength **(Figure 8–40)**. The velocity of a vibration is the result of its amplitude and its frequency. All materials have a unique resonant or natural vibration frequency.

Sound

Vibration results in the phenomenon of sound. In air, the vibrations that cause sound are transmitted as a wave between air molecules; many other substances transmit sound in a similar way. A vibrating object causes pressure variations in the surrounding air. Areas of high and low pressure, known as compressions and rarefactions, move through the air as sound waves. Compression makes the sound waves denser, whereas rarefaction makes them less dense. The distance between each compression of a sound wave is called its **wavelength**. Sound waves with a short wavelength have a high frequency and a high-pitched sound.

When the rapid variations in pressure occur between about 20 Hz and 20 kHz, sound is audible. Audible sound is the sensation (as detected by the ear) of very small rapid changes in the air pressure above and below atmospheric pressure.

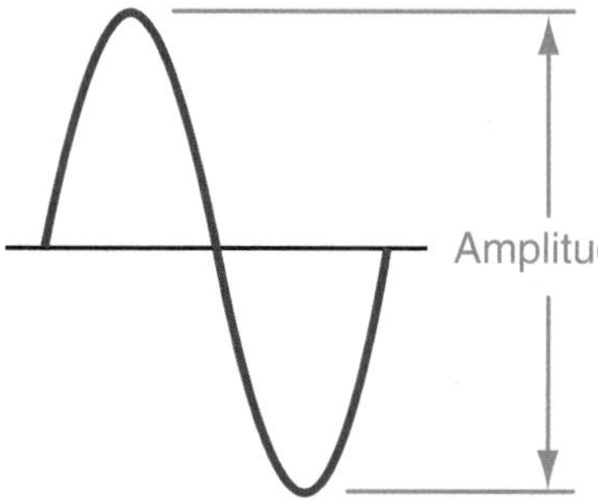

Figure 8-40 Amplitude is a measurement of a vibration's intensity.

Certain terms are used to describe sound:

- The pitch of a sound is based on its frequency. The greater the frequency, the higher the pitch.
- A decibel is a numerical expression of the loudness of a sound.
- Intensity is amount of energy in a sound wave.
- An overtone is an additional tone that is heard because of the air waves of the original tone.
- Harmonics result from the presence of two or more tones at the same time.
- Resonance is produced when the natural vibration of a mass is greatly increased by vibrations at the same or nearly the same frequency of another source or mass. A cavity has certain resonant frequencies. These frequencies depend on the shape and size of the cavity and the velocity of sound within the cavity.

During diagnostics, you often need to listen to the sound of something. You will be paying attention to the type of sound and its intensity and frequency. The tone of the sound usually indicates the type of material that is causing the noise. If there is high pitch, you know that the source of the sound is something that is vibrating quickly. This means the source is less rigid than something that vibrates with a low pitch. Although pitch is dependent on the sound's frequency, the frequency itself can identify the possible sources of the sound. For example, if a sound from an engine increases with an increase in engine speed, you know that the source of the sound must be something that is moving faster as a result of the increase in engine speed. If the frequency of the sound appears to be at one-half the speed of the engine, you know that the source of the sound is something that is rotating at that speed, such as the camshaft.

Figure 8-41 A variety of speakers. *Courtesy of Robert Bosch GmbH, www.bosch-presse.de*

Speakers A speaker **(Figure 8–41)** converts electrical energy into sound energy or waves. A constantly changing electrical signal is fed to the coil of a speaker, which lies within the magnetic field of a permanent magnet. The signal in the coil causes it to behave like an electromagnet, making it push against the field of the permanent magnet. The speaker cone is then pushed in and out by the coil in time with the electrical signal. As the cone moves forward, the air immediately in front of it is compressed, causing a slight increase in air pressure; it then moves back past its rest position and causes a reduction in the air pressure (rarefaction). This process continues so that a wave of alternating high and low pressure is radiated away from the speaker cone at the speed of sound. These changes in air pressure are actually sound. The sound from a speaker may be amplified by the space or cavity that surrounds the speaker cone. The room or area in which the speaker sits also works to amplify the sound.

Noise

Noise is any unwanted signal or sound. It can be random or periodic. To identify the source of a noise, it is important to remember that sound or noise is a vibration and the vibration may be traveling through other components. Therefore, the source of the noise is not always where it may appear **(Figure 8–42)**.

Three approaches can be used to prevent or reduce noise. The most effective way is to intervene at the design stage to make a noisy component produce less noise. A relatively new technique of noise reduction is antinoise or active noise control. This involves producing a sound that is similar to, but out of phase with, the noise. This effectively cancels the original noise. More obvious methods of noise reduction, or passive noise control, involve the use of filters, insulation, and noise barriers.

A filter is an electrical circuit that allows signals in certain frequency ranges to pass through and blocks all other frequencies. Sound insulation prevents sound from traveling from one place to another. Heavy materials like concrete are the most effective materials for sound insulation. Sound insulation or deadening materials are placed strategically throughout a modern automobile. Some sound deadening materials actually absorb sounds. These materials are able to vibrate without creating sound.

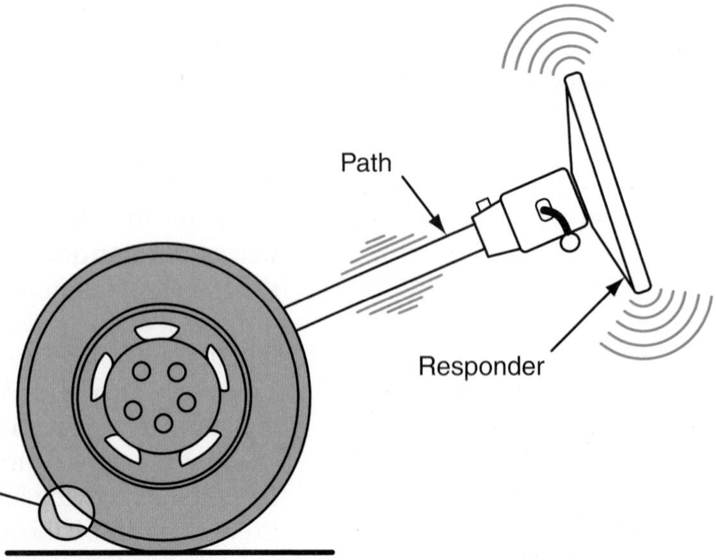

Figure 8-42 A vibration and/or noise will easily move through components so that it appears that the responder (in this case, the steering wheel) is the cause of the noise or vibration.

LIGHT

Light is a form of electromagnetic radiation. In free space, it travels in a straight line at 300 million meters per second. When a beam of light meets an object, a proportion of the rays may be reflected. Some light may also be absorbed, and some transmitted. Without reflection, we would only be able to see objects that give out their own light. Light always reflects from a surface at the same angle at which it strikes. Therefore, parallel rays of light reflecting off a very flat surface will remain parallel. A beam of light reflecting from an irregular surface will scatter in all directions. Light that passes through an object is bent or refracted. The angle of refraction depends on the angle at which the light meets the object and the material it passes through. Lenses and mirrors can cause light rays to diverge or converge. When light rays converge, they can reach a point of focus.

These principles are the basis for fiber-optic lighting. With fiber optics, the light from a single lamp moves through one or more fiber cables to illuminate a point away from the source lamp. The fiber cables are designed to allow the light to travel without losing intensity and the light can be delivered to many locations at the same time.

Photo Cells

Radiation is produced in the sun's core during its nuclear reactions and is the source of most of the earth's energy. A transfer of energy, from electromagnetic radiation to electrical energy, takes place in a photovoltaic (photo) cell, or solar cell. When no light falls on it, it can supply no electricity.

LIQUIDS

A fluid is something that does not have a definite shape; therefore, liquids and gases are fluids. A characteristic of all fluids is that they will conform to the shape of their container. A major difference between a gas and a liquid is that a gas will always fill a sealed container, whereas a liquid may not. A gas will also readily expand or compress according to the pressure exerted on it. Liquids are basically incompressible, which gives them the ability to transmit force **(Figure 8–43)**. Liquids also always seek a common level. A liquid may also change to a gas in response to temperature increases.

Liquids exert pressure on immersed objects, resulting in an upward resultant force called upthrust. The upthrust is equal to the weight of the liquid displaced by the immersed object. If the upthrust on an object is greater than the weight of the object, then the object will float. Large ships float because they displace huge amounts of water, producing a large upthrust.

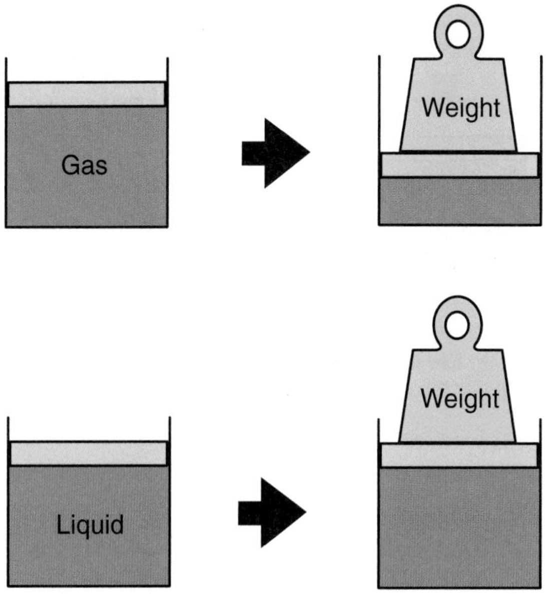

Figure 8-43 Gases compress, whereas liquids do not.

Laws of Hydraulics

Hydraulics is the study of liquids in motion. Liquids will predictably respond to pressures put on them. This allows hydraulics to do work. A simple hydraulic system has liquid, a pump, lines to carry the liquid, control valves, and an output device. The liquid must be available from a continuous source, such as an oil pan or sump. A pump is used to move the liquid through the system. The lines that carry the liquid may be pipes, hoses, or a network of internal bores or passages in a housing. Control valves regulate hydraulic pressure and direct the flow of the liquid. The output device is the unit that uses the pressurized liquid to do work.

Over 300 years ago a French scientist, Blaise Pascal, determined that if you had a liquid-filled container with only one opening and applied force to the liquid through that opening, the force would be evenly distributed throughout the liquid. This explains how pressurized liquid is used to operate and control systems, such as the brake system and automatic transmissions.

Pascal constructed the first known hydraulic device, which consisted of two sealed containers connected by a tube. The cylinders' pistons are sealed against the walls of each cylinder to prevent the liquid from leaking out and to prevent air from entering into the cylinder. When the piston in the first cylinder has a force applied to it, the pressure moves everywhere within the system. The force is transmitted through

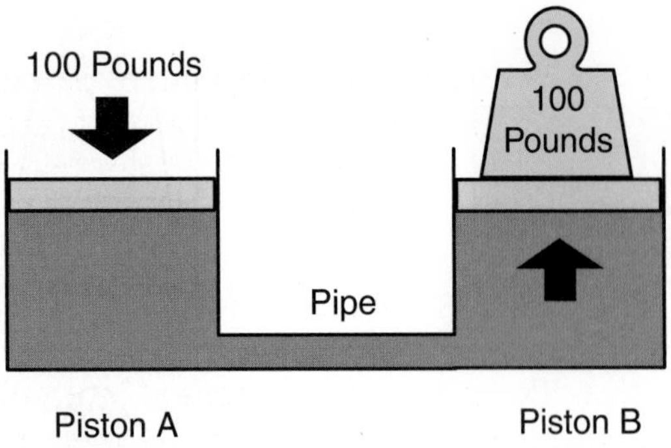

Figure 8-44 In a hydraulic circuit, pressure is transferred equally throughout the system.

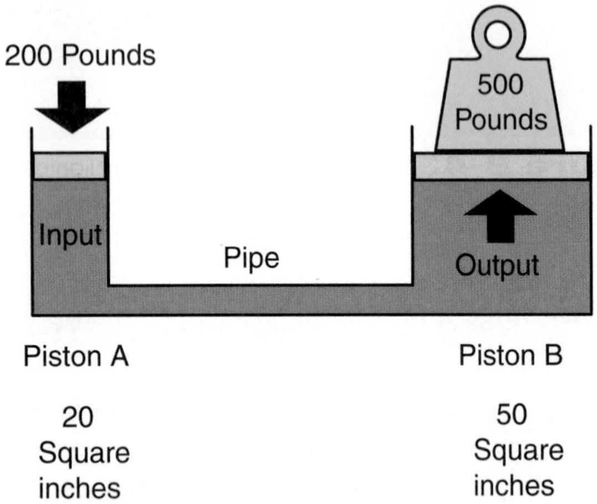

Figure 8-45 The force available to do work can be increased by increasing the size of the piston doing the work.

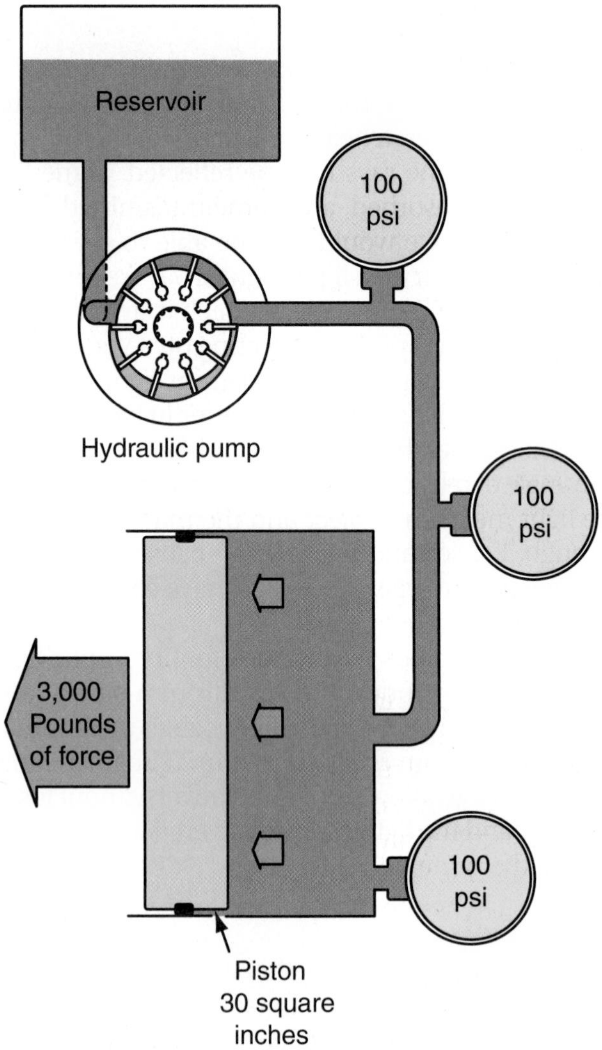

Figure 8-46 A pressure applied to a liquid is transmitted equally and acts with equal force at every point within the hydraulic circuit.

the connecting tube to the second cylinder. The pressurized fluid in the second cylinder exerts force on the bottom of the second piston, moving it upward and lifting the load on the top of it **(Figure 8–44)**. By using this device, Pascal found he could increase the force available to do work **(Figure 8–45)**, just as could be done with levers or gears.

Pascal determined that force applied to liquid creates pressure, or the transmission of the force through the liquid. These experiments revealed two important aspects of a liquid when it is confined and put under pressure. The pressure applied to it is transmitted equally in all directions and this pressure acts with equal force at every point in the container. If a liquid is confined and a force applied, pressure is produced. In order to pressurize a liquid, the liquid must be in a sealed container. Any leak in the container will decrease the pressure.

Mechanical Advantage with Hydraulics

Hydraulics are used to do work in the same way as a lever or gear. These systems transmit energy. Because energy cannot be created or destroyed, these systems only redirect energy to perform work. They do not create more energy. If a hydraulic pump provides 100 psi, there will be 100 pounds of pressure on every square inch of the system **(Figure 8–46)**. If the system included a piston with an area of 50 square inches, each square inch receives 100 pounds of pressure. This means there will be 5,000 pounds of force applied to that piston **(Figure 8–47)**. The use of the larger piston gives the system a **mechanical advantage** as it increases the force available to do work. The multiplication of force through a hydraulic system is directly proportional to the difference in the piston sizes throughout the system.

By changing the size of the pistons in a hydraulic system, force is multiplied, and as a result, low amounts of force can be used to move heavy objects. The mechanical advantage of a hydraulic system can be increased further by the use of levers to increase the force applied to a piston.

Although the force available to do work is increased by using a larger piston in one cylinder, the

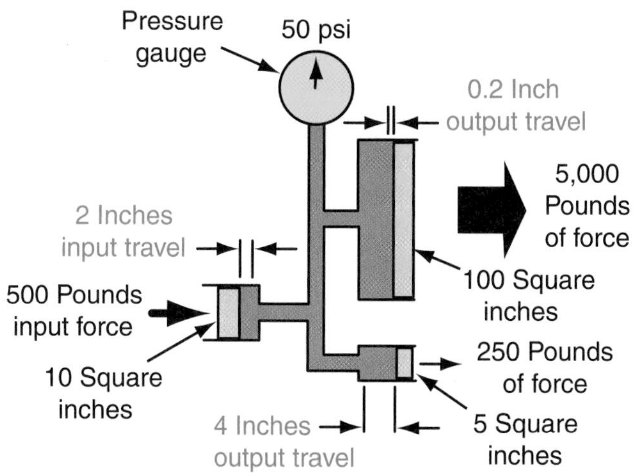

Figure 8-47 Hydraulic systems can provide an increase in force (mechanical advantage), but the output's travel will decrease proportionally.

total movement of the larger piston is less than that of the smaller one. A hydraulic system with two cylinders, one with a 1-inch piston and the other with a 2-inch, will double the force at the second piston. However, the total movement of the larger piston will be half the distance of the smaller one.

The use of hydraulics to gain a mechanical advantage is similar to the use of levers or gears. Hydraulics is preferred when the size and shape of the system is of concern. In hydraulics, the force applied to one piston will transmit through the fluid and the opposite piston will have the same force on it. The distance between the two pistons in a hydraulic system does not affect the force in a static system. Therefore, the force applied to one piston can be transmitted without change to another piston located somewhere else.

A hydraulic system responds to the pressure or force applied to it. The mere presence of different-sized pistons does not always result in fluid power. Either the pressure applied to the pistons or the size of the pistons must be different to cause fluid power. If an equal amount of pressure is exerted onto the pistons in a system and the pistons are the same size, neither piston will move, and the system is balanced or is at equilibrium. The pressure in a balanced hydraulic system is called **static pressure** because there is no fluid motion.

When an unequal amount of pressure is exerted on the pistons, the piston with the least amount of pressure on it will move in response to the difference between the two pressures. Likewise, if the size of the two pistons is different and an equal amount of pressure is exerted on the pistons, the fluid will move. The pressure of the fluid while it is in motion is called **dynamic pressure**.

GASES

A gas is a fluid made up of independent particles—atoms or molecules—that are in constant, random motion. This means that a gas will fill any container into which it is placed. The random movement of gas particles also ensures that any two gases sharing the same container will totally mix. This is **diffusion**.

The kinetic energy of atoms and molecules increases as the temperature increases. Molecules in solids move slowly compared to those in liquids or gases. Gas molecules move quickly compared to liquid molecules. At higher temperatures, gas molecules spread out more, whereas at lower temperatures, gas molecules move closer together. The bombardment of particles against the sides of the container produces pressure.

Behavior of Gases

Three simple laws describe the predictable behavior of gases: Boyle's law, Charles' law, and the pressure (ideal gas) law. Each of these laws describes a relationship between the pressure, volume, and temperature of a gas.

Boyle's law states that the volume and pressure of a mass of gas at a fixed temperature are inversely proportional. If the pressure on a gas increases, its volume will decrease; likewise if the volume is increased, the pressure will decrease.

Charles' law states that the volume of a mass of gas depends on its temperature. Therefore, at a constant pressure, the volume of a gas will increase or decrease in relationship to its temperature increase or decrease. Increasing the temperature of the gas will increase its pressure. If the volume cannot change, the pressure of the gas will. Therefore, the pressure and temperature of a gas are also directly related. If you increase one, you also increase the other. This explains why cold air is denser than warm air.

In the event that all three variables—pressure, volume, and temperature—are changed, the ideal gas law allows for changes of two variables to utilize either Boyle's or Charles' law.

Air Pressure

Because air is gaseous matter with mass and weight, it exerts pressure on the earth's surface. A 1-square-inch column of air extending from the earth's surface to the outer edge of the atmosphere weighs 14.7 psi at sea level. Therefore, atmospheric pressure is 14.7 psi at sea level (**Figure 8–48**). **Atmospheric pressure** may be defined as the total weight of the earth's atmosphere. Pressure greater than atmospheric pressure may be measured in psi gauge (psig). Using

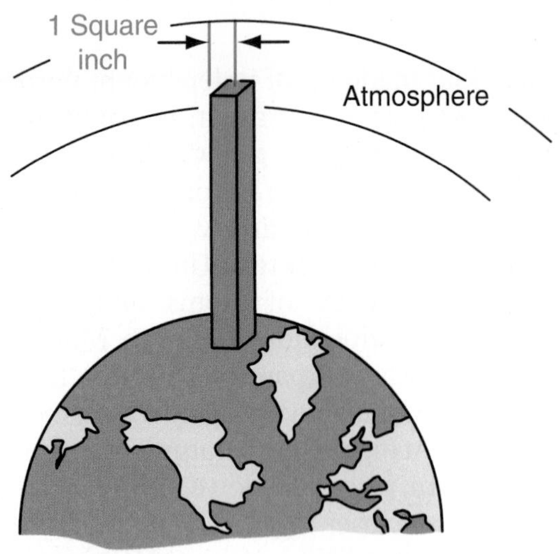

Figure 8-48 One square inch of air equals 14.7 pounds per square inch of pressure at sea level.

a standard pressure gauge, air pressure is compared to that of normal atmospheric pressure. When the actual pressure is 19.7 psi, the gauge will read 5 psi, showing the pressure differential **(Figure 8–49)**. The actual pressure is referred to as psi absolute (psia).

When air becomes hotter, it expands, and this hotter air is lighter compared to an equal volume of cooler air. This hotter, lighter air exerts less pressure on the earth's surface compared to cooler air. This means the weight of the atmosphere changes with weather. This change is rather slight. As the weight changes, so does the atmospheric pressure. The change in atmospheric pressure is measured with a barometer and is called **barometric pressure**. Barometric pressure at normal atmospheric pressure is 29.92 inches of mercury. The increments for measuring barometric pressure are based on the increments of a barometer. A barometer is a "J"-shaped tube with mercury in it. One end of the tube is exposed to normal atmospheric pressure and the other end to current atmospheric pressure. When the current atmospheric pressure equals normal atmospheric pressure, the level of the mercury will be 29.92 inches up the tall part of the "J." When the current atmospheric pressure is lower than normal, the normal atmospheric pressure pushes the mercury down. Likewise, when current atmospheric pressure is higher than normal, it will push the mercury up the tube. The amount of mercury movement reflects the difference in the two pressures. This corresponds with a universal law that states a high pressure always moves toward a lower pressure.

Although the pressure of the atmosphere only changes slightly, the impact of these changes can be critical to the overall operation of an engine. The combustion process depends on having the correct amount of air enter into the cylinders. If the calibrations for the air and the accompanying amount of fuel did not consider the changes in atmospheric pressure, the engine would most often not receive the correct mixture of air and fuel. Today's engines are equipped with a sensor to monitor barometric pressure.

To further consider the law that states a high pressure always moves to a lower pressure, look at what happens when a nail punctures an automotive tire. The high-pressure air in the tire leaks out until the pressure inside the tire is equal to the atmospheric pressure outside the tire. When the tire is repaired and inflated, air with a pressure higher than atmospheric is forced into the tire.

When you climb above sea level, atmospheric pressure decreases. The weight of a column of air is less at an elevation of 5,000 feet (1,524 meters) than it is at sea level. As altitude continues to increase, atmospheric pressure and weight continue to decrease. At an altitude of several hundred miles above sea level, the earth's atmosphere ends, and there is no pressure beyond that point **(Figure 8–50)**.

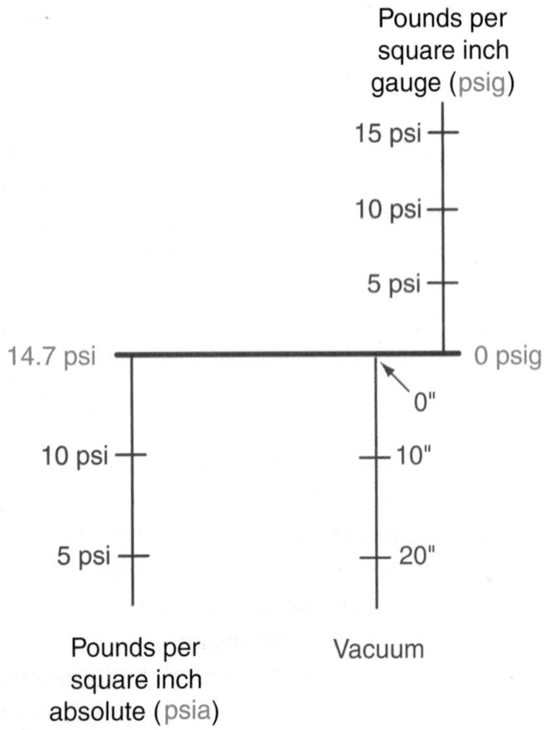

Figure 8-49 The relationship between psia and psig.

Vacuum Scientifically, **vacuum** is defined as the absence of atmospheric pressure. However, it is commonly used to refer to any pressure less than atmospheric pressure. Vacuum may also be referred to as low or negative pressure simply because it is a pressure lower than atmospheric pressure.

Altitude in feet	Altitude in meters	Atmospheric pressure
Sea level	Sea level	14.7 psi
18,000	5,486.3	7.35 psi
52,926	16,131.9	1.47 psi
101,381	30,900.9	0.147 psi
159,013	48,467.2	0.0147 psi
227,889	69,463.6	0.00147 psi
283,076	96,281.6	0.000147 psi

Figure 8-50 The higher the altitude, the lower the pressure of the atmosphere.

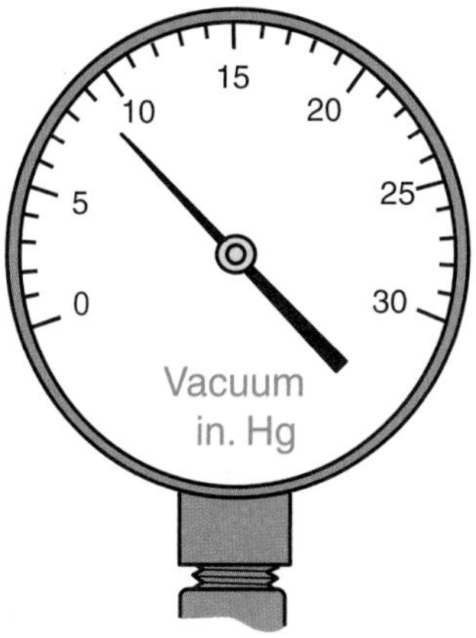

Figure 8-51 A vacuum gauge measures pressures below atmospheric pressure in units of inches of mercury.

Vacuum could be measured in psig or psia, but inches of mercury (in. Hg) is most commonly used for this measurement **(Figure 8–51)**. Let us assume that a plastic "U" tube is partially filled with mercury, and atmospheric pressure is allowed to enter one end of the tube. If vacuum is supplied to the other end of the "U" tube, the mercury is forced downward by the atmospheric pressure. When this movement occurs, the mercury also moves upward on the side where the vacuum is supplied. For example, if the mercury moves downward 10 inches, or 25.4 centimeters (cm) where the atmospheric pressure is supplied, and upward 10 in. (25.4 cm) where the vacuum is supplied, the mercury moved a total of 20 in. and the vacuum is rated as 20 in. Hg. The highest possible, or perfect, vacuum is approximately 29.9 in. Hg.

HEAT

Heat is a form of energy. The main sources of heat are the sun, the earth, chemical reactions, electricity, friction, and nuclear energy. Heat results from the kinetic energy that is present in all matter; therefore, everything has heat. Cold objects have low kinetic energy because their atoms and molecules are moving very slowly, whereas hot objects have more kinetic energy because their atoms and molecules are moving fast.

From	To °F	To °C	To °K
°F	°F	(°F − 32)/1.8	(°F − 32) × 5/9 + 273.15
°C	(°C × 1.8) + 32	°C	°C + 273.15
°K	(°K − 273.15) × 9/5 + 32	°K − 273.15	°K

Figure 8-52 Conversion guidelines for °F, °C, and °K.

Temperature is an indication of an object's kinetic energy. Temperature is measured with a thermometer, which has a Fahrenheit (F), Celsius (Centigrade) (C), or Kelvin (K) scale **(Figure 8–52)**. At absolute zero (−459.4°F, −273°C, or 0°Kelvin), particles of matter do not vibrate, but at all other temperatures, particles have motion. The temperature of an object is a statement of how cold or hot an object is. Heat and temperature are not the same thing: Heat is the movement of kinetic energy from one object to another, whereas temperature is an indication of the amount of kinetic energy something has. Energy from something hot will always move to an object that is colder until both are at the same temperature. The greater the difference in temperature between the two objects, the faster the heat will flow from one to the other.

Heat is measured in British thermal units (Btus) and calories. One Btu is the amount of heat required to heat 1 pound of water by 1 degree Fahrenheit. One calorie is equal to the amount of heat needed to raise the temperature of 1 gram of water 1 degree Celsius.

Heat Transfer

Heat transfers between two substances that have different temperatures through convection, conduction, or radiation. **Convection** is the transfer of heat by the movement of a heated object. Convection can be easily seen by watching a pot of water on a stove. The water on the bottom of the pot is the first to be heated by the stove. As the water at the bottom becomes hotter, it expands and becomes lighter than the water at the top of the pot. This causes the heavier water to sink toward the bottom and push the warmer water up. This continues until all of the water in the pot is at the same temperature.

Conduction is the movement of heat through a material. The immense heat that results from combustion is absorbed by the engine and is used to push the pistons down. The engine's cooling system uses conduction to move the heat from the parts to help cool the engine. Because heat energy moves from something hot toward something colder, the engine's heat moves to the engine's coolant circulating within the engine. Heat can be conducted to a liquid, gas, or solid.

Radiation does not rely on another material to transfer heat. The moving atoms and molecules within an object create waves of radiant energy. These waves are typically called infrared rays. Hot objects give off more infrared rays than colder objects. Therefore, a hot object will give off infrared rays to anything around it that is colder. No movement is necessary to transfer this heat. You can feel radiation in action by simply putting your hand near something that is hot. The hot object cools as it radiates its heat energy. In an engine's cooling system, the radiator uses radiation to transfer heat from the coolant to the surrounding air.

The Effects of Temperature Change

Anytime the temperature of an object changes, a transfer of heat has occurred. The change in temperature can also cause the object to change size or its state of matter. As heat moves in and out of a mass, the movement of atoms and molecules in that mass increases or slows down. With an increase in motion, the size of the mass tends to get bigger or expand. This is commonly called **thermal expansion**. **Thermal contraction** takes place when a mass has heat removed from it and the atoms and molecules slow down. All gases and most liquids and solids expand when heated, with gases expanding the most. Solids, because they are not fluid, expand and contract at a much lower rate. It is important to realize that all materials do not expand and contract at the same rate. For example, an aluminum component will expand at a faster rate than the same component made of iron. This explains why aluminum cylinder heads have unique service requirements and procedures when compared to iron cylinder heads.

Thermal expansion takes place every time fuel and air are burned in an engine. The sudden temperature increase inside the cylinder causes a rapid expansion of the gases, which pushes the piston downward.

Typically when heat is added to a mass, the temperature of the mass increases. This does not always happen, however. In some cases, the additional heat causes no increase in temperature but causes the mass to change its state (solid to liquid or liquid to gas). For example, if we take an ice cube and heat it to 32°F (0°C), it will begin to melt **(Figure 8–53)**. As heat is added to the ice cube, the temperature of the ice cube will not increase until it becomes a liquid. The heat added to the ice cube that did not raise its temperature but caused it to melt is called **latent heat** or the heat of fusion. Each gram of ice at 0°C requires 80 calories of heat to melt it to water at 0°C. As more heat is added to the 0°C water, the water's temperature will once again increase. This continues until the temperature of the water reaches 212°F (100°C). This is the boiling temperature of water. At this point, any additional heat applied to the water is latent heat, causing the water to change its state to that of a gas. This added heat is called the heat of evaporation.

To change the water gas back to liquid water, the same amount of heat required to change the liquid to a gas must be removed from the gas. At that point the gas condenses to a liquid. As additional heat is removed, the temperature will drop until enough

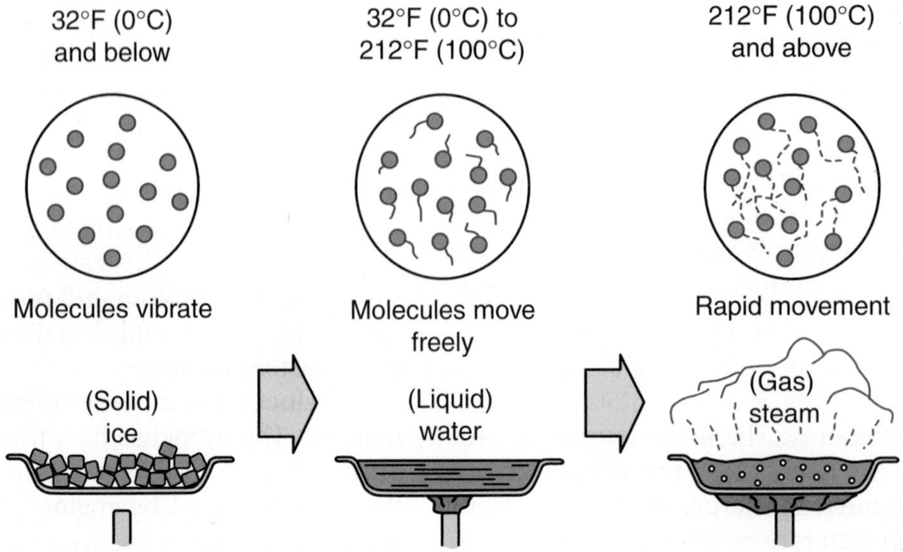

Figure 8-53 Water can exist in three different states of matter.

heat is removed to bring its temperature back down to freezing (melting in reverse) point. At that time latent heat must be removed from the liquid before the water turns to ice again.

Controlling Heat

There is a particular temperature range in which all parts of an automobile will operate best. Engineers strive to control those temperatures to ensure reliable and efficient operation. This is a major task because parts that do not perform well when hot are often mounted close to something that is very hot. High heat could transfer to the heat sensitive parts if insulation or passing outside air was not present. Although some parts tolerate extreme temperatures, they must still be protected from overheating. The combustion process can generate temperatures as great as 2,500°F (1,371°C). If that heat moved uncontrollably to other parts of the engine, those parts would expand to the point where they could no longer move or they may melt. This is why the engine's cooling system is so important. The cooling system is a controlled heat transfer system designed to protect the engine and allow it to run more efficiently.

CHEMICAL PROPERTIES

The properties of something describe or identify the characteristics of an object. Physical properties are characteristics that are readily observable, such as color, size, luster, and smell.

Chemical properties are only observable during a chemical reaction and describe how one type of matter reacts with another to form a new and different substance. Chemical properties are quite different from physical properties. A chemical property of some metals is the ability to combine with oxygen to form rust (iron and oxygen) or tarnish (silver and sulfur). Another example is hydrogen's ability to combine with oxygen to form water.

A solution is a mixture of two or more substances. Most solutions are liquids, but solutions of gases and solids are possible. An example of a gas solution is the air we breathe; it is composed of mostly oxygen and nitrogen. Brass is a good example of a solid solution because it is composed of copper and zinc. The liquid in a solution is called the **solvent**, and the substance added is the solute. If both are liquids, the one present in the smaller amount is usually considered the solute. Solutions can vary widely in terms of how much of the dissolved substance is actually present. A heavily diluted (much water) acid solution has very little acid and may not be noticeably acidic.

Specific Gravity

Specific gravity is the heaviness or relative **density** of a substance as compared to water. If something is 3.5 times as heavy as an equal volume of water, its specific gravity is 3.5. Its density is 3.5 grams per cubic centimeter, or 3.5 kilograms per liter. Specific gravity checks of a battery's electrolyte are an indication of the battery's state of charge **(Figure 8–54)**.

Density is a statement of how much mass there is in a particular volume. Water is denser than air; therefore, there will be less air in a given container than water in that same container **(Figure 8–55)**. The density of a material changes with temperature as well **(Figure 8–56)**. This is the reason an engine runs more efficiently with cool intake air.

Chemical Reactions

Chemical changes, or chemical reactions, result in the production of another substance, such as wood turning to carbon after it has been completely burned. A chemical reaction is always accompanied by a change in energy. This means energy is given off or taken in during the reaction. Some reactions that release energy need some energy to get the reaction

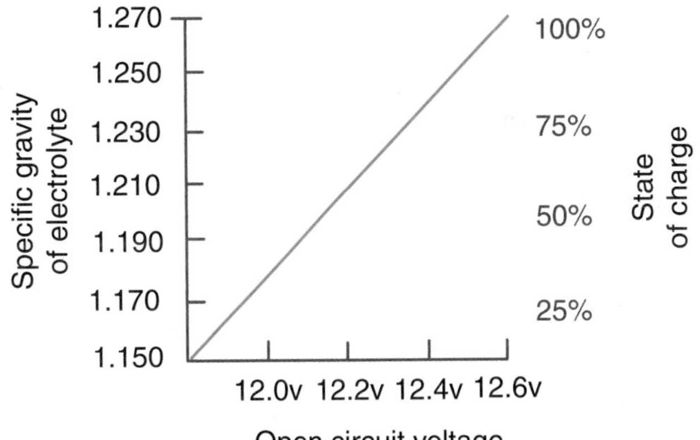

Figure 8–54 Specific gravity checks of a battery's electrolyte are an indication of the battery's state of charge.

Substance	Density in g/cm³
Air	0.0013
Ice	0.92
Water	1.00
Aluminum	2.70
Steel	7.80
Gold	19.30

Figure 8–55 A look at the density of different substances as compared to water.

Temp °F	Temp °C	Approx. Change In Density
200°	93°	−21%
180°	82°	−16.8%
160°	71°	−12.6%
140°	60°	−8.4%
120°	49°	−4.2%
100°	38°	—
80°	27°	+4.2%
60°	16°	+8.4%
40°	4°	+12.6%
20°	−7°	+16.8%
0°	−18°	+21%

Figure 8-56 The effect that temperature has on the density of air at atmospheric pressure.

started. A reaction takes place when two or more molecules interact and one of the following happens:

- A chemical change occurs.
- Single reactions occur as part of a large series of reactions.
- Ions, molecules, or pure atoms are formed.

Catalysts and Inhibitors

Reactions need a certain amount of energy to happen. A **catalyst** lowers the amount of energy needed to make a reaction happen. A catalyst is any substance that affects the speed of a chemical reaction without itself being consumed or changed. Catalysts tend to be highly specific, reacting with one substance or a small set of substances. In a car's catalytic converter, the platinum catalyst converts unburned hydrocarbons and nitrogen compounds into products that are harmless to the environment **(Figure 8–57)**. Water, especially salt water, catalyzes oxidation and corrosion. An inhibitor is the opposite of a catalyst and stops or slows the rate of a reaction.

Acids/Bases

An ion is an atom or group of atoms with one or more positive or negative electric charges. Ions are formed when electrons are added to or removed from neutral molecules or other ions. Ions are what make something an **acid** or a **base**.

Acids are compounds that break into hydrogen (H^+) ions and another compound when placed in an aqueous (water) solution. They have a sour taste, are corrosive, react with some metals to produce hydrogen, react with carbonates to produce carbon dioxide, change the color of litmus from blue to red, and become less acidic when combined with alkalis. Most acids are slow reacting, especially if they are weak acids. Acids also react with bases to form salts.

Alkalis (bases) are compounds that release hydroxide ions (OH^-) and react with H^+ ions to produce water, thus neutralizing each other. Most substances are neutral (not an acid or a base). Alkalis feel slippery, change the color of litmus from red to blue, and become less alkaline when they are combined with acids.

A hydroxide is any compound made up of one atom each of hydrogen and oxygen, bonded together and acting as the hydroxyl group or hydroxide anion (OH^-). An oxide is any chemical compound in which oxygen is combined with another element. Metal oxides typically react with water to form bases or with acids to form salts. Oxides of nonmetallic elements react with water to form acids or with bases to form salts.

A salt is a chemical compound formed when the hydrogen of an acid is replaced by a metal. Typically, an acid and a base react to form a salt and water.

pH The **pH scale** is used to measure how acidic or basic a solution is. Its name comes from the fact that pH is the absolute value of the power of the hydrogen ion concentration. The scale goes from 0 to 14. Distilled (pure) water is 7. Acids are found between 0 and 7 and bases are from 7 to 14. When the pH of a substance is low, the substance has many H^+ ions. When the pH is high, the substance has many OH^- ions. The pH value helps inform scientists, as well as technicians, of the nature, composition, or extent of reaction of substances.

The pH of something is typically checked with litmus paper. Litmus is a mixture of colored organic compounds obtained from several species of lichen. Lichen is a type of plant that is actually a combination of a fungus and algae. Litmus test strips can be used to check the condition of the engine's coolant **(Figure 8–58)**.

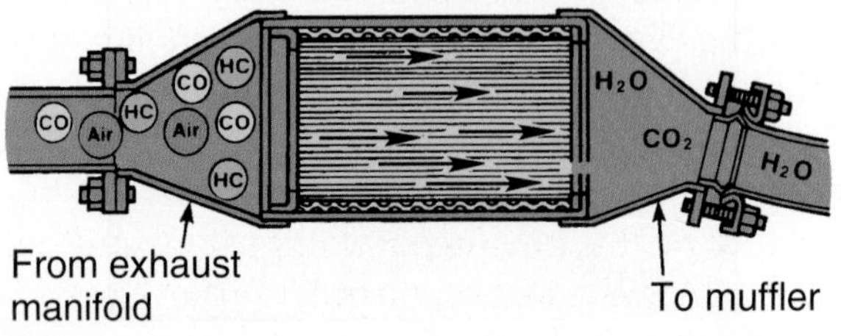

Figure 8-57 A basic catalytic converter that changes pollutants into chemicals that are good for the environment.

Reduction and Oxidation

Oxidation is a chemical reaction in which a substance combines with oxygen. Rapid oxidation produces heat fast enough to cause a flame. When fuel burns, it

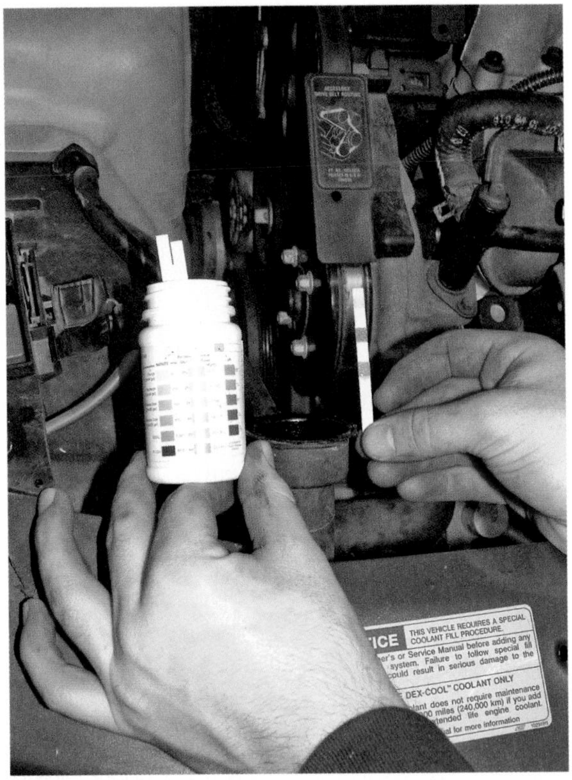

Figure 8-58 Litmus test strips can be used to check the condition of an engine's coolant.

combines with oxygen to form other compounds. This chemical reaction is combustion, which produces heat and fire.

The addition of hydrogen atoms or electrons is **reduction**. Oxidation and reduction always occur simultaneously: One substance is oxidized by the other, which is reduced. During oxidation, a molecule provides electrons. During reduction, a molecule accepts electrons. Oxidation and reduction reactions are usually called redox reactions. Redox is any chemical reaction in which electrons are transferred. Batteries, also known as voltaic cells, produce an electrical current at a constant voltage through redox reactions.

An oxidizing agent is something that accepts electrons and oxidizes something else while being reduced in the process. A reducing agent is something that provides electrons and reduces something else while being oxidized.

Every atom or ion has an oxidation number. This value compares the number of protons and electrons in that atom. In many cases, the oxidation number reflects the actual charge on the atom, but there are many cases in which it does not. The oxidation number is reduced during reduction by adding electrons. The oxidation number is increased during oxidation by removing electrons. All free, uncombined elements have an oxidation number of zero. Hydrogen, in all its compounds except hydrides, has an oxidation number of $+1$. Oxygen, in all its compounds except peroxides, has an oxidation number of -2.

Metallurgy

Metallurgy is the art and science of extracting metals from their ores and modifying them for a particular use. This includes the chemical, physical, and atomic properties and structures of metals and the way metals are combined to form alloys. An **alloy** is a mixture of two or more metals. Steel is an alloy of iron plus carbon and other elements.

Metals have one or more of the following properties:

- Good heat and electric conduction
- Malleability—can be hammered, pounded, or pressed into a shape without breaking
- Ductility—can be stretched, drawn, or hammered without breaking
- High light reflectivity—can make light bounce off its surface
- The capacity to form positive ions in a solution and hydroxides rather than acids when their oxides meet water

About three-quarters of the elements are metals. The most abundant metals are aluminum, iron, calcium, sodium, potassium, and magnesium.

Rust and Corrosion The rusting of iron is an example of oxidation. Unlike fire, rusting occurs so slowly that little heat is produced. Iron combines with oxygen to form rust. The rate at which this occurs depends on several factors: temperature, surface area (more iron exposed for oxygen to reach), and catalysts (speed up a reaction but do not react and change themselves).

Corrosion is the wearing away of a substance due to chemical reactions. It occurs whenever a gas or liquid chemically attacks an exposed surface. This action is accelerated by heat, acids, and salts. Some materials naturally resist corrosion; others can be protected by painting, coatings, galvanizing, or anodizing.

Galvanizing involves the coating of zinc onto iron or steel to protect it against exposure to the atmosphere. If galvanizing is properly applied, it can protect the metals for 15 to 30 years or more.

Metals can be anodized for corrosion resistance, electrical insulation, thermal control, abrasion resistance, sealing, improving paint adhesion, and decorative finishing. Anodizing is a process that electrically deposits an oxide film from an aqueous solution onto the surface of a metal, often aluminum. During the process, dyes can be added to the process to give the material a colored surface.

Hardness The hardness of something describes its resistance to scratching. **Hardening** is a process that increases the hardness of a metal, deliberately or accidentally, by hammering, rolling, carburizing, heat treating, tempering, or other processes. All of these deform the metal by compacting the atoms or molecules to make the material denser.

Carburizing hardens the surface of steel with heat. It increases the hardness of the outer surface while leaving the core relatively soft. The combination of a hard surface and soft interior withstands very high stress. It also has a low cost and offers flexibility for manufacturing. To carburize, the steel parts are placed in a carbonaceous environment (with charcoal, coke, and carbonates or carbon dioxide, carbon monoxide, methane, or propane) at a high temperature for several hours. The carbon diffuses into the surface of the steel, altering the crystal structure of the metal. Gears, ball and roller bearings, and piston pins are often carburized.

Heat treating changes the properties of a metal (including iron, steel, aluminum, copper, and titanium) by using heat. **Tempering** is the heat treating of metal alloys, particularly steel. For example, raising the temperature of hardened steel to 752°F (400°C) and holding it for a time before quenching it in oil decreases its hardness and brittleness and produces strong steel.

Solids under Tension

The atoms of a solid are closely packed, so solids have a greater density than most liquids and gases. The rigidity results from the strong attraction between its atoms. A force pulling on a solid moves these atoms farther apart, creating an opposing force called tension. If a force pushes on a solid, the atoms move closer together, creating compression. These are the principles of how springs function. Springs are used in many automotive systems, the most obvious of which are those used in suspension systems **(Figure 8–59)**.

Figure 8-59 A coil spring for a suspension system.

An elastic substance is a solid that gets larger under tension, gets smaller under compression, and returns to its original size when no force is acting on it. Most solids show some elastic behavior, but there is usually a limit to the force that the material can face. When excessive force is applied, the material will not return to its original size and it will be distorted or will break. The limit depends on the material's internal structure; for example, steel has a low elastic limit and can only be extended about 1% of its length, whereas rubber can be extended to about 1,000%. Another factor involved in elasticity is the cross-sectional area of the material.

Tensile strength is the ratio of the maximum load a material can support without breaking while being stretched. It is dependent on the cross-sectional area of the material. When stresses less than the tensile strength are removed, the material returns to its original size and shape. Greater stresses form a narrow, constricted area in the material, which is easily broken. Tensile strengths are measured in units of force per unit area.

Electrochemistry

Electrochemistry is concerned with the relationship between electricity and chemical change. Many spontaneous chemical reactions release electrical energy and some of these are used in batteries and fuel cells to produce electric power. The basis for electricity is the movement of electrons from one atom to another.

Electrolysis is an electrochemical process. During this process, electric current is passed through a substance, causing a chemical change. This change causes either a gain or loss of electrons. Electrolysis normally takes place in an electrolytic cell made of separated positive and negative electrodes immersed in an electrolyte.

An **electrolyte** is a substance that conducts current as a result of the breaking down of its molecules into positive and negative ions. The most familiar electrolytes are acids, bases, and salts that ionize when dissolved in solvents such as water and alcohol. Ions drift to the electrode of the opposite charge and are the conductors of current in electrolytic cells.

ELECTRICITY AND ELECTROMAGNETISM

All electrical effects are caused by electric charges. There are two types of electric charges: positive and negative. These charges exert electrostatic forces on each other due to the strong attraction of electrons to protons. An electric field is the area on which these forces have an effect. Protons carry a positive charge, while electrons carry a negative charge. Atoms are normally neutral, having an equal number of protons and electrons, but an atom can gain or lose electrons. Electricity has many similarities with magnetism. For example, the lines of the electric fields between charges take the same form as the lines of magnetic force, so magnetic fields can be said to be an equivalent to electric fields. Charges of the same type repel, while charges of a different type attract **(Figure 8–60)**.

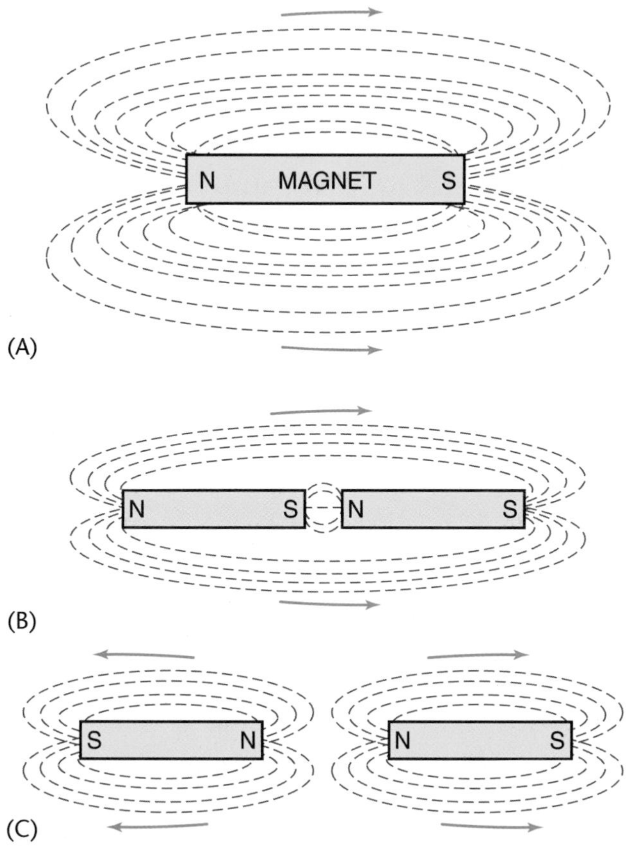

Figure 8-60 (A) In a magnet, lines of force emerge from the north pole and travel to the south pole before passing through the magnet back to the north pole. (B) Unlike poles attract, while (C) similar poles repel each other.

Electricity

An electric circuit is simply the path in which an electric current flows. Electrons can be moved around a circuit by electrostatic forces. A circuit usually consists of a conductive material, such as a metal, where the electrons are held very loosely to their atoms, thus making electron movement possible. The strength of the electrostatic force is the voltage. The movement of the electric charge is called an electric current. The higher the voltage, the greater the current will be. But the current also depends on the thickness, length, temperature, and nature of the materials used as a conductor. Electrical resistance opposes the flow of electric current. Good conductors have low resistance, which means that a small amount of voltage will produce high current. In batteries, the dissolving of an electrode causes the freeing of electrons, resulting in their movement to another electrode and the formation of a current.

Magnets

Some materials are natural magnets; however, most magnets are produced. The materials used to make a permanent magnet are commonly called ferromagnetic materials. These are made of mostly heated iron compounds. The heat causes the atoms to shift direction, and once all of them point in the same direction, the metal becomes a magnet. This sets up two distinct poles called the north and south poles. The poles are at the ends of the magnet and there is an attraction between the two separate poles.

The lines of a magnetic field form closed lines of force from the north to the south. If another iron or steel object enters into the magnetic field, it is pulled into the magnet. If another magnet is introduced into the magnetic field, it will either move into the field or push away from it. This is the result of the natural attraction of a magnet from north to south. If the north pole of one magnet is introduced to the north pole of another, the two poles will oppose each other and will push away. If the south pole is introduced to the north pole of another, the two magnets will join together because the opposite poles are attracted to each other.

The strength of the magnetic force is uniform around the outside of the magnet. The force is strongest at the surface of the magnet and weakens with distance. If you double the distance from a magnet, the force is reduced by ¼.

The strength of a magnetic field is typically measured with a magnetometer and in units of Gauss (G).

Electromagnetism

Electrical current will produce magnetism that affects other objects in the same way as permanent magnets. The arrangement of force lines around a current-carrying conductor, its magnetic field, is circular. The magnetic effect of electrical current is increased by making the current-carrying wire into a coil **(Figure 8–61)**.

When a coil of wire is wrapped around an iron bar, it is called an **electromagnet**. The magnetic field produced by the coil magnetizes the iron bar, strengthening the magnetic field. The strength of the magnetism produced depends on the number of windings in the coil and the amount of the current flowing in the coil.

Producing Electrical Energy

There are many ways to generate electricity. The most common is to use coils of wire and magnets in a generator. When a wire and magnet are moved relative to each other, a voltage is produced **(Figure 8–62)**. In a generator, the wire is wound into a coil and the coil rotates within the field of the magnet. The more turns in the coil and the faster the coil moves, the greater the voltage. The coils or magnets spin around at high speed, typically turned by steam pressure. The steam is usually generated by burning coal or oil, a process that releases unwanted pollutants. Renewable sources of electricity, such as hydroelectric power, wind power, solar energy, and geothermal power, produce only heat as an emission. Automotive generators are driven by the engine's crankshaft via a belt. In a generator, the kinetic energy of a spinning object is converted into electrical energy.

A solar cell converts the energy of light directly into electrical energy, using layers of semiconductors. Electricity is produced by causing electrons to leave the atoms in a semiconductor material. Each electron leaves behind a hole or gap. Other electrons move into the hole, leaving holes in their atoms. This process continues and forms a moving chain of electrons, which is electrical current.

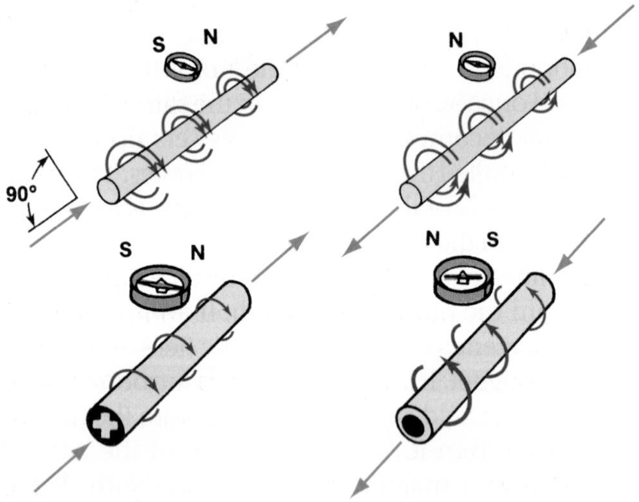

Figure 8-61 When current is passed through a conductor, such as a wire, magnetic lines of force are generated around the wire at right angles to the direction of the current flow.

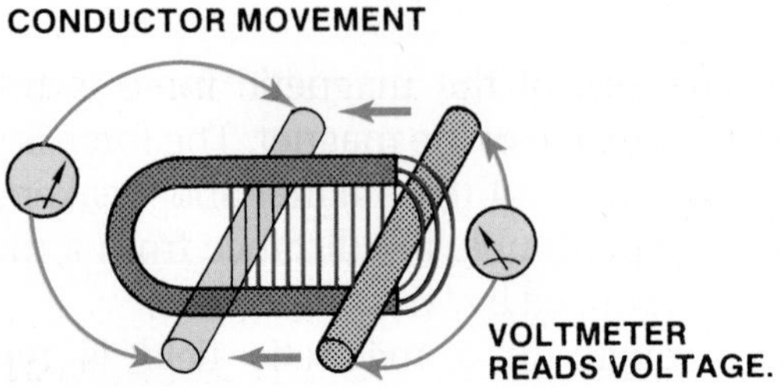

Figure 8-62 Moving a conductor across magnetic lines of force induces a voltage in the conductor.

Radio Waves

Electricity and magnetism are directly related. A changing electric field will produce a changing magnetic field, and vice versa. Whenever an electric charge accelerates, it gives out energy in the form of electromagnetic radiation. For example, electrons moving up and down a radio antenna produce a type of radiation known as radio waves. Electromagnetic radiation consists of oscillating electric and magnetic fields. There is a wide range of different types of electromagnetic radiation, called the electromagnetic spectrum, extending from low-energy radio waves to high-energy, short-wavelength gamma rays. This includes visible light and X-rays.

KEY TERMS

Acceleration
Acid
Aerodynamics
Alloy
Amplitude
Atmospheric pressure
Atoms
Barometric pressure
Base
Carburizing
Catalyst
Centrifugal force
Centripetal force
Compression ratio
Conduction
Convection
Curb weight
Deceleration
Density
Diffusion
Displacement
Dynamic pressure
Electrolysis
Electrolyte
Electromagnet
Element
Engine efficiency
Equilibrium
Evaporate
Force
Frequency
Gear
Gross weight
Hardening

Heat
Heat treating
Hertz (Hz)
Horsepower
Impermeable
Inertia
Ion
Kinetic energy
Latent heat
Lever
Load
Mass
Matter
Mechanical advantage
Molecule
Momentum
Oscillation
Oxidation
Oxide
Permeable
pH scale
Plasma
Potential energy
Power
Pressure
Pulley
Radiation
Reduction
Solution
Solvent
Specific gravity
Speed
Static pressure
Tempering
Tensile strength
Tension
Thermal contraction
Thermal expansion
Time
Torque
Vacuum
Velocity
Volume
Wave
Wavelength
Weight
Work

SUMMARY

- Matter is anything that occupies space, and it exists as a gas, liquid, or solid.
- All matter is made of many tiny particles called atoms and molecules.
- When a solid dissolves in a liquid, a solution is formed. Not all solids will dissolve; rather, the liquid is either absorbed or adsorbed.
- Materials that *absorb* fluids are permeable substances. Impermeable substances *adsorb* fluids.
- Energy is the ability to do work and all matter has energy.
- The total amount of energy never changes; it can only be transferred from one form to another, not created or destroyed.
- When energy is released to do work, it is called kinetic energy. Stored energy is called potential energy.
- Energy conversion occurs when one form of energy is changed to another form.
- Mass is the amount of matter in an object. Weight is a force and is measured in pounds or kilograms. Gravitational force gives the mass its weight.
- Volume is the amount of space occupied by an object.
- The volume of an engine's cylinders determines its size, expressed as displacement.
- The compression ratio of an engine is the ratio of the volume in the cylinder above the piston when the piston is at the bottom of its travel to the cylinder's volume above the piston when the piston is at its uppermost position.
- A force is a push or pull, which can be large or small, and can be applied to something by direct contact or from a distance.
- When an object moves in a circle, its direction is continuously changing and all changes in direction require a force. The forces required to maintain circular motion are called centripetal and centrifugal forces.
- Pressure is a force applied against an object and is measured in units of force per unit of surface area (pounds per square inch or kilograms per square centimeter).
- The greater the mass of an object, the greater the force needed to change its motion.
- When a force overcomes static inertia and moves an object, the object gains momentum. Momentum is the product of an object's weight times its speed.
- Speed is the distance an object travels in a set amount of time. Velocity is the speed of an object in a particular direction. Acceleration is the rate of speed increase. Deceleration is the reverse of acceleration, because it is the rate of the decrease in speed.
- Newton's laws of motion are: (1) an object at rest tends to remain at rest and an object in motion tends to remain in motion; (2) when a force acts on an object, the motion of the object will change; and (3) for every action there is an equal and opposite reaction.
- Friction is a force that slows or prevents motion of two moving objects that touch.
- Friction can be reduced in two main ways: by lubrication or by the use of rollers.
- Aerodynamics is the study of the effects of air on a moving object.
- When a force moves a certain mass a specific distance, work is done.
- A machine is any device used to transmit a force and, in doing so, changes the amount of force and/or its direction. Examples of simple machines are inclined planes, pulleys, levers, gears, and wheels and axles.

- Torque is a force that tends to rotate or turn things and is measured by the force applied and the distance traveled.
- Gear ratios express the mathematical relationship of one gear to another.
- Power is a measurement of the rate at which work is done and is measured in watts.
- Horsepower is the rate at which torque is produced.
- An oscillation is any single swing of an object back and forth between the extremes of its travel. When that motion travels through matter or space, it becomes a wave.
- How many times the vibration occurs in 1 second is called frequency and is commonly expressed in hertz (Hz), which is equal to one cycle per second. The amplitude of a vibration is its intensity or strength.
- Noise is any unwanted signal or sound and can be random or periodic.
- Light is a form of electromagnetic radiation. It travels in a straight line at 300 million meters per second.
- A gas will always fill a sealed container, whereas a liquid may not. A gas will also readily expand or compress according to the pressure exerted on it. Liquids are basically incompressible, which gives them the ability to transmit force.
- Hydraulics is the study of liquids in motion.
- Pascal constructed the first known hydraulic device and established what is known as Pascal's law of hydraulics.
- The pressure inside the hydraulic system is called static pressure because there is no fluid motion. The pressure of the fluid while it is in motion is called dynamic pressure.
- Boyle's law states that the volume and pressure of gas at a fixed temperature is inversely proportional.
- The pressure law states that the pressure exerted by a gas increases as the temperature of the gas is increased.
- Charles' law states that the volume of a mass of gas depends on its temperature.
- Atmospheric pressure is the total weight of the earth's atmosphere. Pressure greater than atmospheric pressure may be measured in psi gauge (psig); actual pressure is measured in psi absolute (psia).
- Scientifically, vacuum is defined as the absence of atmospheric pressure. However, it is commonly used to refer to any pressure less than atmospheric pressure.
- Heat is a form of energy caused by the movement of atoms and molecules and is measured in British thermal units (Btus) and calories.
- Temperature is an indication of an object's kinetic energy and is measured with a thermometer.
- Convection is the transfer of heat by the movement of a heated object.
- Conduction is the movement of heat through a material.
- Through radiation, heat is transferred by radiant energy.
- As heat moves in and out of a mass, the size of the mass tends to change.
- Sometimes additional heat causes no increase in temperature but causes the mass to change its state; this heat is called latent heat.
- The liquid in a solution is called the solvent, and the substance added is the solute.
- Specific gravity is the heaviness or density of a substance compared to that of water.
- A catalyst is a substance that affects a chemical reaction without being consumed or changed.
- An ion is an atom or group of atoms with one or more positive or negative electric charges. Ions are formed when electrons are added to or removed from neutral molecules or other ions. Ions are what make something an acid or a base.
- The pH scale is used to measure how acidic or basic a solution is.
- Oxidation is a chemical reaction in which a substance combines with oxygen. The addition of hydrogen atoms or electrons is called reduction.
- Hardening is a process that increases the hardness of a metal by hammering, rolling, carburizing, heat treating, tempering, or other processes.
- An elastic substance is a solid that gets larger under tension, smaller under compression, and returns to its original size when no force is acting on it.
- Tensile strength represents the maximum load a material can support without breaking when being stretched and is dependent on the cross-sectional area of the material.
- Electrolysis is an electrochemical process in which electric current is passed through a substance, causing a chemical change.
- An electrolyte is a substance or compound that conducts electric current as a result of the breaking

down of its molecules into positively and negatively charged ions.

- Any electrical current will produce magnetism. When a coil of wire is wrapped around an iron bar, it is called an electromagnet.
- The most common way to produce electricity is to use coils of wire and magnets in a generator.

REVIEW QUESTIONS

1. Describe Newton's first law of motion and give an application of this law in automotive theory.
2. In what four states does matter exist?
3. Explain Newton's second law of motion and give an example of how this law is used in automotive theory.
4. Describe six different forms of energy.
5. Describe four different types of energy conversion.
6. Explain why a rotating, tilted wheel moves in the direction of the tilt.
7. Why are gases and liquids considered fluids?
8. Describe the effects of static and dynamic balance.
9. Describe the effect of temperature on the volume of a gas.
10. The nucleus of an atom contains ______ and ______.
11. Which of the following is the correct formula used to calculate engine displacement?
    a. Displacement $= \pi \times R^2 \times L \times N$
    b. Displacement $= \pi^2 \times R \times L \times N$
    c. Displacement $= \pi \times D \times L \times N$
    d. Displacement $= \pi \times D \times L^2 \times N$
12. Work is calculated by multiplying ______ by ______.
13. Energy may be defined as the ability to do ______.
14. Name three types of simple machines.
15. When one object is moved over another object, the resistance to motion is called ______.
16. Weight is the measurement of the earth's ______ ______ on an object.
17. Torque is a force that does work with a ______ action.
18. How are engines mounted in a vehicle and why?
19. Vacuum is defined as the absence of ______ ______.
20. While discussing different types of energy: Technician A says that when energy is released to do work, it is called potential energy. Technician B says that stored energy is referred to as kinetic energy. Who is correct?
    a. Technician A only
    b. Technician B only
    c. Both A and B
    d. Neither A nor B
21. While discussing friction in matter: Technician A says that friction creates heat. Technician B says that friction occurs in liquids, solids, and gases. Who is correct?
    a. Technician A only
    b. Technician B only
    c. Both A and B
    d. Neither A nor B
22. While discussing mass and weight: Technician A says that mass is the measurement of an object's inertia. Technician B says that mass and weight may be measured in cubic inches. Who is correct?
    a. Technician A only
    b. Technician B only
    c. Both A and B
    d. Neither A nor B
23. When applying the principles of work and force, ______.
    a. work is accomplished when force is applied to an object that does not move
    b. in the metric system the measurement for work is cubic centimeters
    c. no work is accomplished when an object is stopped by mechanical force
    d. if a 50-pound object is moved 10 feet, 500 ft.-lb of work are produced
24. All these statements about energy and energy conversion are true, *except* ______.
    a. thermal energy may be defined as light energy
    b. chemical to thermal energy conversion occurs when gasoline burns
    c. mechanical energy is defined as the ability to do work
    d. mechanical to electrical energy conversion occurs when the engine drives the generator
25. Which of the following is *not* a true statement about heat?
    a. Whenever the temperature of something changes, a transfer of heat has occurred.
    b. All matter expands when heated.
    c. A change in temperature can cause the object to change size.
    d. A change in temperature can cause the object to change its state of matter.

CHAPTER 9

AUTOMOTIVE ENGINE DESIGNS AND DIAGNOSIS

OBJECTIVES

■ Describe the various ways in which engines can be classified. ■ Explain what takes place during each stroke of the four-stroke cycle. ■ Outline the advantages and disadvantages of the inline and V-type engine designs. ■ Define important engine measurements and performance characteristics, including bore and stroke, displacement, compression ratio, engine efficiency, torque, and horsepower. ■ Outline the basics of diesel, stratified, and Miller-cycle engine operation. ■ Explain how to evaluate the condition of an engine. ■ List and describe nine abnormal engine noises.

INTRODUCTION TO ENGINES

The engine (**Figure 9–1**) provides the power to drive the vehicle's wheels. All automobile engines, both gasoline and diesel, are classified as internal-combustion engines because the combustion or burning that creates energy takes place inside the engine.

The biggest part of the engine is the cylinder block (**Figure 9–2**). The cylinder block is a large casting of metal that is drilled with holes to allow for the passage of lubricants and coolant through the block and provide spaces for movement of mechanical parts. The block contains the cylinders, which are round passageways fitted with pistons. The block houses or holds the major mechanical parts of the engine.

The cylinder head fits on top of the cylinder block to close off and seal the top of the cylinder (**Figure 9–3**). The combustion chamber is an area into

Figure 9–1 Today's engines are complex, efficient machines.

Figure 9–2 A cylinder block for an eight-cylinder engine.

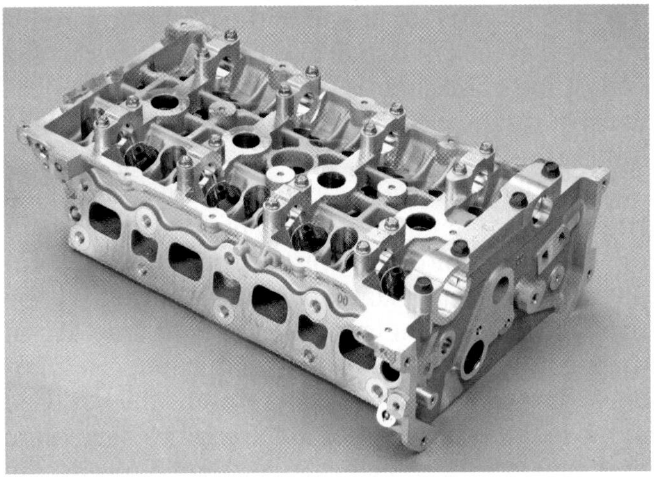

Figure 9-3 A cylinder head for a late-model inline four-cylinder engine. *Courtesy of Chrysler LLC*

which the air-fuel mixture is compressed and burned. The cylinder head contains all or most of the combustion chamber. The cylinder head also contains ports through which the air-fuel mixture enters and burned gases exit the cylinder and the bore for the spark plug.

The valve train is a series of parts used to open and close the intake and exhaust ports. A valve is a movable part that opens and closes the ports. A camshaft controls the movement of the valves. Springs are used to help close the valves.

The up-and-down motion of the pistons must be converted to rotary motion before it can drive the wheels of a vehicle. This conversion is achieved by linking the piston to a crankshaft with a connecting rod. The upper end of the connecting rod moves with the piston. The lower end of the connecting rod is attached to the crankshaft and moves in a circle. The end of the crankshaft is connected to the flywheel or flexplate.

Engine Construction

Modern engines are highly engineered power plants. These engines are designed to meet the performance and fuel efficiency demands of the public. Modern engines are made of lightweight engine castings and stampings; noniron materials (for example, aluminum, magnesium, fiber-reinforced plastics); and fewer and smaller fasteners to hold things together. These fasteners are made possible through computerized joint designs that optimize loading patterns. Each of these newer engine designs has its own distinct personality, based on construction materials, casting configurations, and design **(Figure 9–4)**.

These modern engine-building techniques have changed how engine repair technicians make a living.

Figure 9-4 A typical late-model engine. *Courtesy of American Honda Motor Co., Inc.*

Before these changes can be explained, it is important to explain the "basics" of engine design and operation.

ENGINE CLASSIFICATIONS

Today's automotive engines can be classified in several ways depending on the following design features:

- Operational cycles. Most technicians will generally come in contact with only four-stroke engines. However, a few older cars have used and some cars in the future will use a two-stroke engine.
- Number of cylinders. Current engine designs include 3-, 4-, 5-, 6-, 8-, 10-, and 12-cylinder engines.
- Cylinder arrangement. An engine can be flat (opposed), inline, or V-type. Other more complicated designs have also been used.
- Valve train type. Engine valve trains can be either the **overhead camshaft (OHC)** type or the camshaft in-block **overhead valve (OHV)** type. Some engines separate camshafts for the intake and exhaust valves. These are based on the OHC design and are called **double overhead camshaft (DOHC)** engines. V-type DOHC engines have four camshafts—two on each side.
- Ignition type. There are two types of ignition systems: spark and compression. Gasoline engines use a spark ignition system. In a spark ignition system, the air-fuel mixture is ignited by an electrical

spark. Diesel engines, or compression ignition engines, have no spark plugs. A diesel engine relies on the heat generated as air is compressed to ignite the air-fuel mixture for the power stroke.

- Cooling systems. There are both air-cooled and liquid-cooled engines in use. Nearly all of today's engines have liquid-cooling systems.
- Fuel type. Several types of fuel currently used in automobile engines include gasoline, natural gas, methanol, diesel, and propane. The most commonly used is gasoline although new fuels are being tested.

Four-Stroke Gasoline Engine

In a passenger car or truck, the engine provides the rotating power to drive the wheels through the transmission and driving axles. All automotive engines, both gasoline and diesel, are classified as internal combustion because the combustion or burning takes place inside the engine. These systems require an air-fuel mixture that arrives in the combustion chamber at the correct time and an engine constructed to withstand the temperatures and pressures created by the burning of thousands of fuel droplets.

The **combustion chamber** is the space between the top of the piston and the cylinder head. It is an enclosed area in which the fuel and air mixture is burned. The piston fits into a hollow metal tube, called a cylinder. The piston moves up and down in the cylinder.

This reciprocating motion must be converted to a rotary motion before it can drive the wheels of a vehicle. This change of motion is accomplished by connecting the piston to a crankshaft with a connecting rod **(Figure 9–5)**. The upper end of the connecting rod moves with the piston as it moves up and down in the cylinder. The lower end of the connecting rod is attached to the crankshaft and moves in a circle. The end of the crankshaft is connected to the flywheel, which transfers the engine's power through the drivetrain to the wheels.

In order to have complete combustion in an engine, the right amount of fuel must be mixed with the right amount of air. This mixture must be compressed in a sealed container, then shocked by the right amount of heat (spark) at the right time. When these conditions exist, all the fuel that enters a cylinder is burned and converted to power, which is used to move the vehicle. Automotive engines have more than one cylinder. Each cylinder should receive the same amount of air, fuel, and heat, if the engine is to run efficiently.

Although the combustion must occur in a sealed cylinder, the cylinder must also have some means of allowing heat, fuel, and air into it. There must also be a means to allow the burnt air-fuel mixture out so a fresh mixture can enter and the engine can continue to run. To accommodate these requirements, engines are fitted with valves.

There are at least two valves at the top of each cylinder. The air-fuel mixture enters the combustion chamber through an intake valve and leaves (after having been burned) through an exhaust valve **(Figure 9–6)**. The valves are accurately machined plugs that fit into machined openings. A valve is said to be seated or closed when it rests in its opening. When the valve is pushed off its seat, it opens.

A rotating camshaft, driven and timed to the crankshaft, opens and closes the intake and exhaust valves. **Cams** are raised sections of a shaft that have high spots called **lobes**. Cam lobes are oval shaped. The placement of the lobe on the shaft determines when the valve will open. The height and shape of the lobe determines how far the valve will open and how

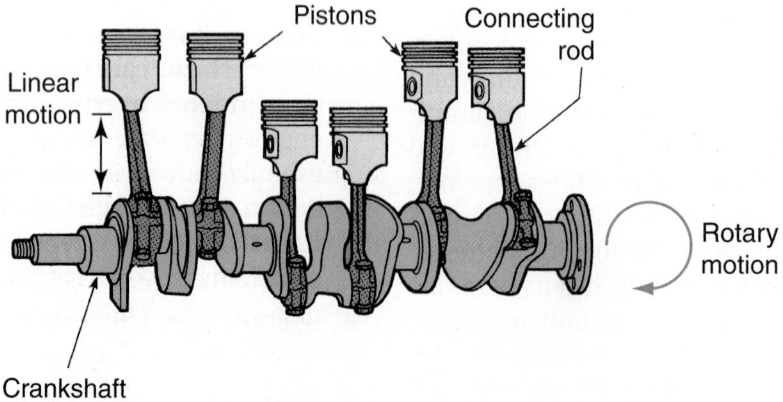

Figure 9-5 The linear (reciprocating) motion of the pistons is converted to rotary motion by the crankshaft.

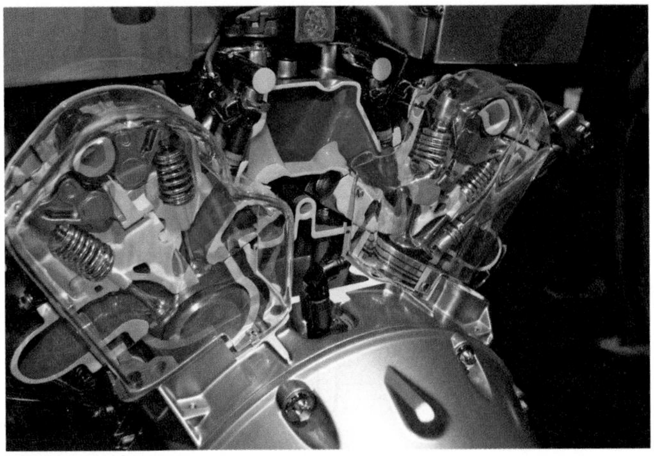

Figure 9-6 A cutaway of an engine showing the intake passages (blue) and valve and exhaust passage (red) and valve.

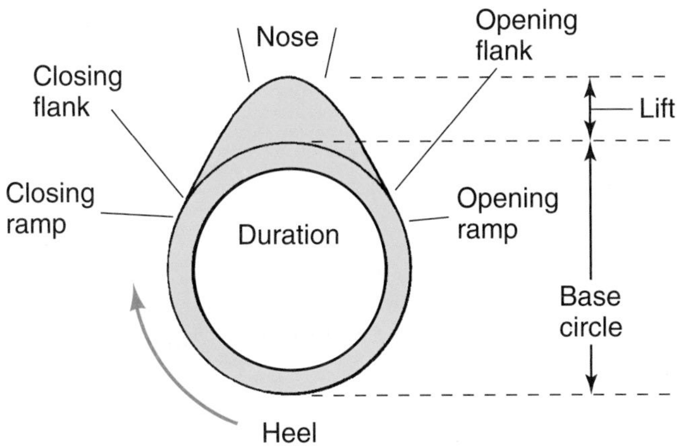

Figure 9-7 The height and width of a cam lobe determine when and for how long a valve will be open.

long it will remain open in relation to piston movement **(Figure 9–7)**.

As the camshaft rotates, the lobes rotate and push the valve open by pushing it away from its seat. Once the cam lobe rotates out of the way, the valve, forced by a spring, closes. The camshaft can be located either in the cylinder block or in the cylinder head.

When the action of the valves and the spark plug is properly timed to the movement of the piston, the combustion cycle takes place in four strokes of the piston: the intake stroke, the compression stroke, the power stroke, and the exhaust stroke. The camshaft is driven by the crankshaft through gears, or sprockets, and a cogged belt, or timing chain. The camshaft turns at half the crankshaft speed and rotates one complete turn during each complete four-stroke cycle.

Four-Stroke Cycle A **stroke** is the full travel of the piston either up or down in a cylinder's bore. The reciprocal movement of the piston during the four strokes is converted to a rotary motion by the crankshaft. It takes two full revolutions of the crankshaft to complete the four-stroke cycle. One full revolution of the crankshaft is equal to 360 degrees of rotation; therefore, it takes 720 degrees to complete the four-stroke cycle. During one piston stroke, the crankshaft rotates 180 degrees.

Flywheel The piston moves by the pressure produced during combustion, but this moves the piston only about half a stroke or one-quarter of a revolution of the crankshaft. This explains why a flywheel is needed. The flywheel stores some of the power produced by the engine. This power is used to keep the pistons in motion during the rest of the four-stroke cycle. A heavy flywheel is only found on engines equipped with a manual transmission. Engines with automatic transmissions have a flexplate and a torque converter. The weight and motion of the fluid inside the torque converter serve as a flywheel.

Intake Stroke The first stroke of the cycle is the intake stroke. As the piston moves away from **top dead center (TDC)**, the intake valve opens **(Figure 9–8A)**. The downward movement of the piston increases the volume of the cylinder above it, reducing the pressure in the cylinder. This reduced pressure, commonly referred to as engine vacuum, causes the atmospheric pressure to push a mixture of air and fuel through the open intake valve. (Some engines are equipped with a super- or turbocharger that pushes more air past the valve.) As the piston reaches the bottom of its stroke, the reduction in pressure stops, causing the intake of air-fuel mixture to slow down. It does not stop because of the weight and movement of the air-fuel mixture. It continues to enter the cylinder until the intake valve closes. The intake valve closes after the piston has reached **bottom dead center (BDC)**. This delayed closing of the valve increases the volumetric efficiency of the cylinder by packing as much air and fuel into it as possible.

Compression Stroke The compression stroke begins as the piston starts to move from BDC. The intake valve closes, trapping the air-fuel mixture in the cylinder **(Figure 9–8B)**. The upward movement of the piston compresses the air-fuel mixture, thus heating it up. At TDC, the piston and cylinder walls form a combustion chamber in which the fuel will be burned. The volume of the cylinder with the piston at BDC compared to the volume of the cylinder with the piston at TDC determines the compression ratio of the engine.

Power Stroke The power stroke begins as the compressed fuel mixture is ignited **(Figure 9–8C)**. With

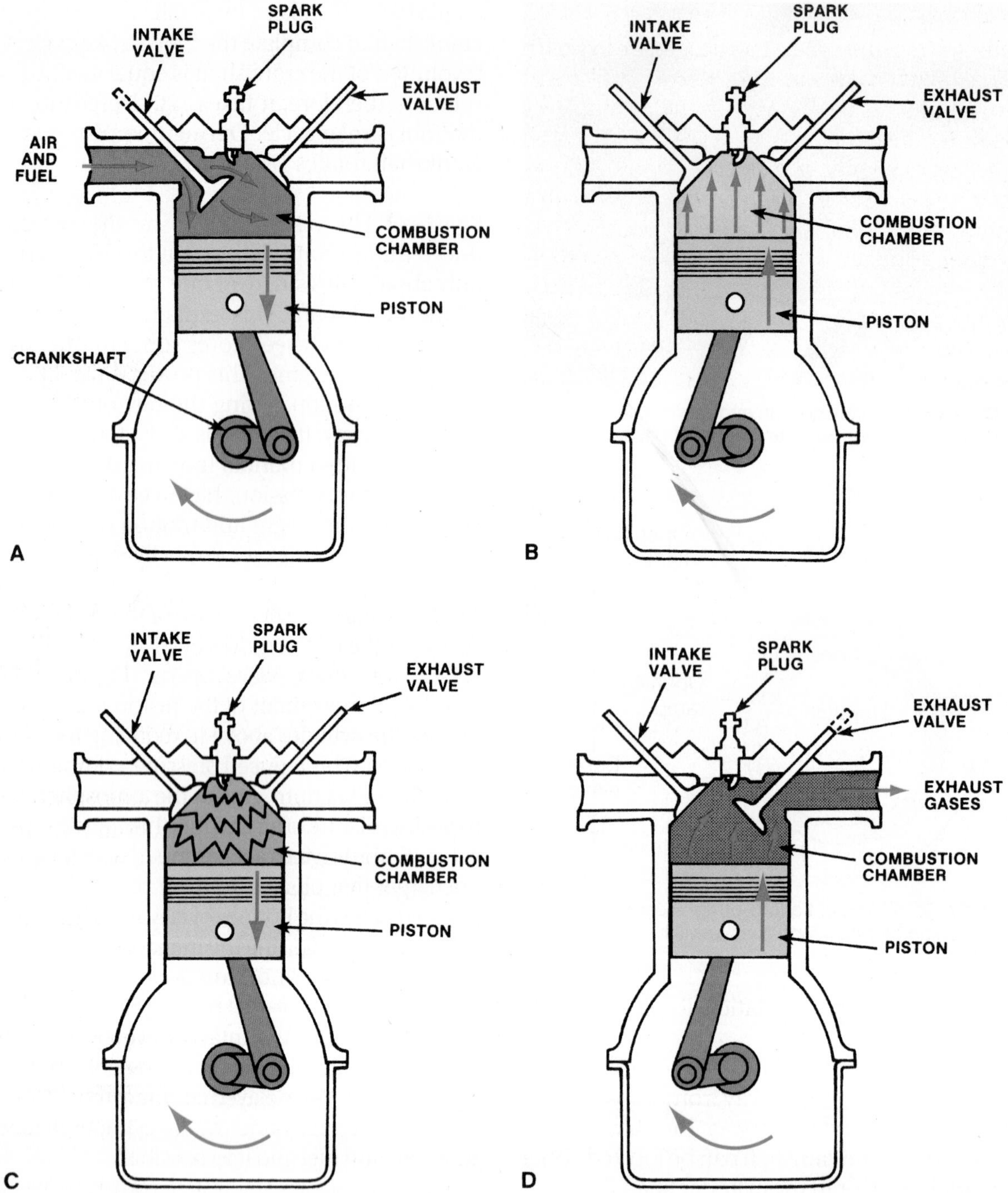

Figure 9-8 (A) Intake stroke, (B) compression stroke, (C) power stroke, and (D) exhaust stroke.

the valves still closed, an electrical spark across the electrodes of a spark plug ignites the air-fuel mixture. The burning fuel rapidly expands, creating a very high pressure against the top of the piston. This drives the piston down toward BDC. The downward movement of the piston is transmitted through the connecting rod to the crankshaft.

Exhaust Stroke The exhaust valve opens just before the piston reaches BDC on the power stroke (**Figure 9–8D**). Pressure within the cylinder causes the exhaust gas to rush past the open valve and into the exhaust system. Movement of the piston from BDC pushes most of the remaining exhaust gas from the cylinder. As the piston nears TDC, the exhaust valve begins to close as the intake valve starts to open. The exhaust stroke completes the four-stroke cycle. The opening of the intake valve begins the cycle again. This cycle occurs in each cylinder and is repeated over and over, as long as the engine is running.

Firing Order

An engine's **firing order** states the sequence in which an engine's pistons are on their power stroke and therefore the order in which the cylinders' spark plugs fire. The firing order also indicates the position of all of the pistons in an engine when a cylinder is firing. For example, consider a four-cylinder engine with a firing order of 1-3-4-2. The sequence begins with piston #1 on the compression stroke. During that time, piston #3 is moving down on its intake stroke, #4 is moving up on its exhaust stroke, and #2 is moving down on its power stroke. These events are identified by what needs to happen in order for #3 to be ready to fire next, and so on.

The firing order of an engine is determined by its design and manufacturer's preference. An engine's firing order can be found on the engine or on the engine's emissions label and in service manuals. **Figure 9–9** shows some of the common cylinder arrangements and their associated firing orders.

COMMON CYLINDER NUMBERING AND FIRING ORDER	
IN-LINE	
4-Cylinder	**6-Cylinder**
① ② ③ ④	① ② ③ ④ ⑤ ⑥
Firing Order 1–3–4–2 1–2–4–3	Firing Order 1–5–3–6–2–4
V CONFIGURATION	
V6	**V8**
⑤ ③ ① Right Bank ⑥ ④ ② Left Bank Firing Order 1–4–5–2–3–6	① ② ③ ④ Right Bank ⑤ ⑥ ⑦ ⑧ Left Bank Firing Order 1–5–4–8–6–3–7–2
② ④ ⑥ Right Bank ① ③ ⑤ Left Bank Firing Order 1–6–5–4–3–2	① ② ③ ④ Right Bank ⑤ ⑥ ⑦ ⑧ Left Bank Firing Order 1–5–4–2–6–3–7–8
① ② ③ Right Bank ④ ⑤ ⑥ Left Bank Firing Order 1–2–3–4–5–6	② ④ ⑥ ⑧ Right Bank ① ③ ⑤ ⑦ Left Bank Firing Order 1–8–4–3–6–5–7–2
① ② ③ Right Bank ④ ⑤ ⑥ Left Bank Firing Order 1–4–2–3–5–6	② ④ ⑥ ⑧ Right Bank ① ③ ⑤ ⑦ Left Bank Firing Order 1–8–7–2–6–5–4–3

Figure 9-9 Examples of cylinder numbering and firing orders.

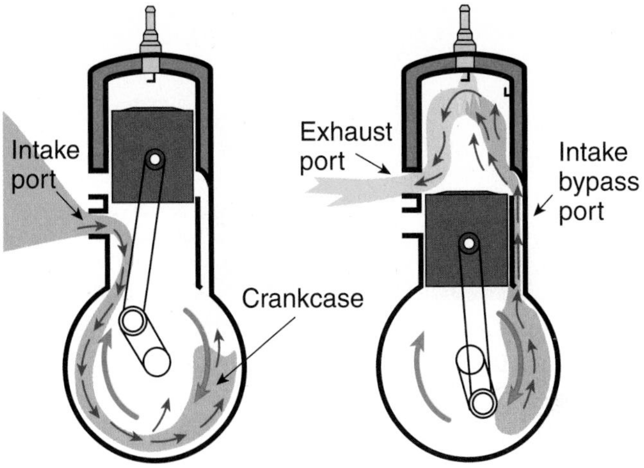

Figure 9-10 A two-stroke cycle.

Two-Stroke Gasoline Engine

In the past, several imported vehicles have used two-stroke engines. As the name implies, this engine requires only two strokes of the piston to complete all four operations: intake, compression, power, and exhaust **(Figure 9–10)**. This is accomplished as follows:

1. Movement of the piston from BDC to TDC completes both intake and compression.
2. When the piston nears TDC, the compressed air-fuel mixture is ignited, causing an expansion of the gases. During this time, the intake and exhaust ports are closed.
3. Expanding gases in the cylinder force the piston down, rotating the crankshaft.
4. With the piston at BDC, the intake and exhaust ports are both open, allowing exhaust gases to leave the cylinder and air-fuel mixture to enter.

Although the two-stroke-cycle engine is simple in design and lightweight because it lacks a valve train, it has not been widely used in automobiles. It tends to be less fuel efficient and releases more pollutants into the atmosphere than four-stroke engines. Oil is often in the exhaust stream because these engines require constant oil delivery to the cylinders to keep the piston lubricated. Some of these engines require a certain amount of oil to be mixed with the fuel.

Engine Rotation To meet the standards set by the SAE, nearly all engines rotate in a counterclockwise direction. This can be confusing because its apparent direction changes with what end of the engine you look at. If one looks at the front of the engine, it rotates in a clockwise direction. The standards are based on the rotation of the flywheel, which is at the rear of

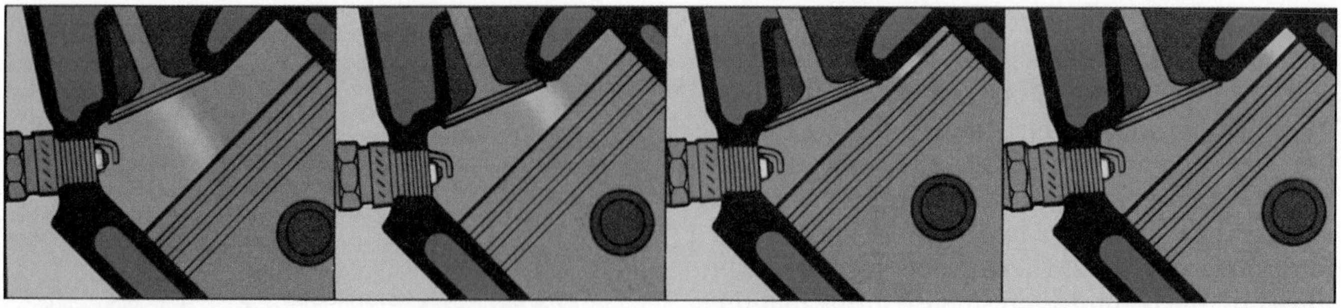

Figure 9-11 Normal combustion. *Courtesy of Federal-Mogul Corporation*

the engine, and there the engine rotates counterclockwise.

Combustion

Although many different things and events can affect combustion in the engine's cylinders, the ignition system has the responsibility for beginning and maintaining the combustion process. Obviously when combustion does not occur in all of the cylinders, the engine will not run. If combustion occurs in all but one or two cylinders, the engine may start and run but will run poorly. The lack of combustion is not always caused by the ignition system. Poor combustion can also be caused by problems in the engine, air-fuel system, or the exhaust system.

When normal combustion occurs, the burning process moves from the gap of the spark plug across the compressed air-fuel mixture. The movement of this flame front should be rapid and steady and should end when all of the air-fuel mixture has been burned **(Figure 9–11)**. During normal combustion, the rapidly expanding gases push down on the piston with a powerful but constant force.

When all of the air and fuel in the cylinder are involved in the combustion process, complete combustion has occurred. When something prevents this, the engine will misfire or experience incomplete combustion. Misfires cause a variety of driveability problems, such as a lack of power, poor gas mileage, excessive exhaust emissions, and a rough running engine.

Engine Configurations

Depending on the vehicle, either an inline, V-type, slant, or opposed cylinder design can be used. The most popular designs are inline and V-type engines.

Inline Engine In the inline engine design **(Figure 9–12)**, the cylinders are all placed in a single row. There is one crankshaft and one cylinder head for all of the cylinders. The block is cast so that all cylinders are located in an upright position.

Figure 9-12 The cylinder block for an inline engine. *Courtesy of Chrysler LLC*

Inline engine designs have certain advantages and disadvantages. They are easy to manufacture and service. However, because the cylinders are positioned vertically, the front of the vehicle must be higher. This affects the aerodynamic design of the car. Aerodynamic design refers to the ease with which the car can move through the air. When equipped with an inline engine, the front of a vehicle cannot be made as low as it can with other engine designs.

V-Type Engine The V-type engine design has two rows of cylinders **(Figure 9–13)** located 60 to 90 degrees away from each other. A V-type engine uses one crankshaft, which is connected to the pistons on both sides of the V. This type of engine has two cylinder heads, one over each row of cylinders.

One advantage of using a V-configuration is that the engine is not as high or long as one with an inline

Figure 9–13 A V-type engine. *Courtesy of Chrysler LLC*

Figure 9–14 A horizontally opposed cylinder engine, commonly called a boxer engine.

configuration. The front of a vehicle can now be made lower. This design improves the outside aerodynamics of the vehicle. If eight cylinders are needed for power, a V-configuration makes the engine much shorter, lighter, and more compact. Many years ago, some vehicles had an inline eight-cylinder engine. The engine was very long and its long crankshaft also caused increased torsional vibrations in the engine.

A variation of the V-type engine is the W-type engine. These engines are basically two V-type engines joined together at the crankshaft. This design makes the engine more compact. They are commonly found in late-model Volkswagens.

Slant Cylinder Engine Another way of arranging the cylinders is in a slant configuration. This arrangement is much like an inline engine, except the entire block has been placed at a slant. The slant engine was designed to reduce the distance from the top to the bottom of the engine. Vehicles using the slant engine can be designed more aerodynamically.

Opposed Cylinder Engine In this design, two rows of cylinders are located opposite the crankshaft **(Figure 9–14)**. These engines have a common crankshaft and a cylinder head on each bank of cylinders. Porsches and Subarus use this style of engine, commonly called a boxer engine. Boxer engines have a low center of gravity and tend to run smoothly during all operating conditions.

Camshaft and Valve Location

The valves in all modern engines are placed in the cylinder head above the top of the piston. The valves in many older engine designs were placed to the side of the piston. Camshafts are located inside the engine

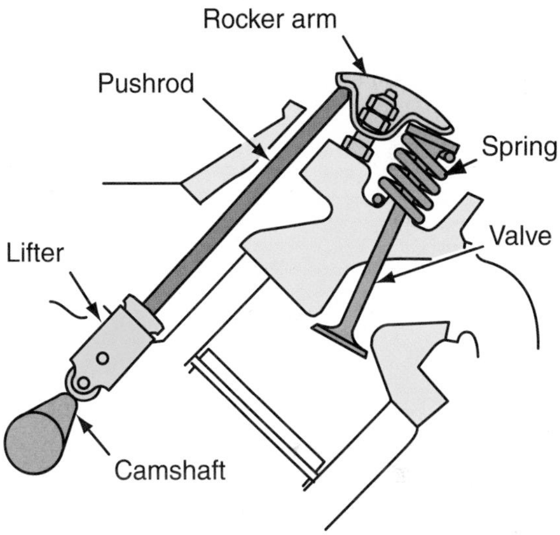

Figure 9–15 The basic valve train for an overhead valve engine.

block or above the cylinder head. The placement of the camshaft further describes an engine.

Overhead Valve (OHV) As the same implies, the intake and exhaust valves in an OHV engine are mounted in the cylinder head and are operated by a camshaft located in the cylinder block. This arrangement requires the use of valve lifters, pushrods, and rocker arms to transfer camshaft rotation to valve movement **(Figure 9–15)**.

Overhead Cam (OHC) An OHC engine also has the intake and exhaust valves located in the cylinder head. But as the name implies, the cam is located in the cylinder head. In an OHC engine, the valves are operated directly by the camshaft or through cam followers or tappets **(Figure 9–16)**. Engines with one camshaft above a cylinder are often referred to as single overhead camshaft (SOHC) engines.

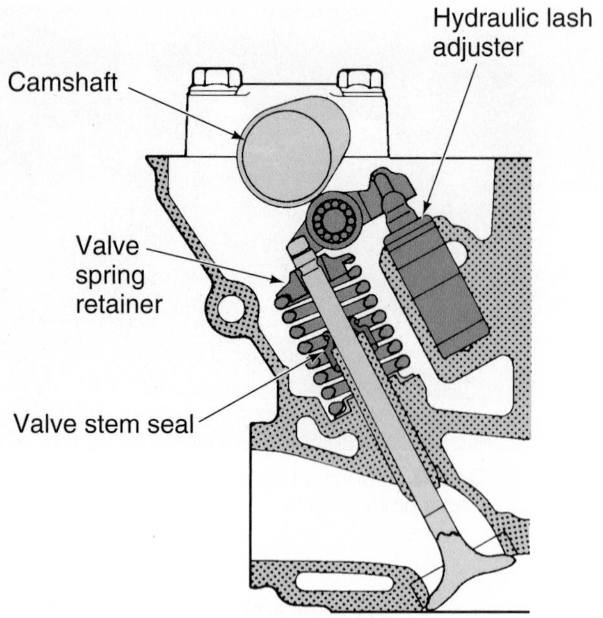

Figure 9-16 Basic valve and camshaft placement in an overhead camshaft engine. *Courtesy of Hyundai Motor America*

Engine Location

The engine is usually placed in one of three locations. In most vehicles, it is located at the front of the vehicle, in front of the passenger compartment. Front-mounted engines can be positioned either longitudinally or transversely with respect to the vehicle.

The second engine location is a mid-mount position between the passenger compartment and rear suspension. Mid-mount engines are normally transversely mounted. The third, and least common, engine location is the rear of the vehicle. The engines are typically opposed-type engines.

Each of these engine locations offers advantages and disadvantages.

Front Engine Longitudinal In this type of vehicle, the engine, transmission, front suspension, and steering equipment are installed in the front of the body, and the differential and rear suspension are installed in the rear of the body. Most front engine longitudinal vehicles are rear-wheel drive. Some front-wheel-drive cars with a transaxle have this configuration, and most four-wheel-drive vehicles are equipped with a transfer case and have the engine mounted longitudinally in the front of the vehicle.

Total vehicle weight can be evenly distributed between the front and rear wheels with this configuration. This lightens the steering force and equalizes the braking load. With this design, it is possible to independently remove and install the engine, propeller shaft, differential, and suspension. Longitudinally mounted engines require large engine compartments. The need for a rear-drive propeller shaft and differential also cuts down on passenger compartment space.

Front Engine Transverse Front engines that are mounted transversely sit sideways in the engine compartment. They are used with transaxles that combine transmission and differential gearing into a single compact housing, fastened directly to the engine. Transversely mounted engines reduce the size of the engine compartment and overall vehicle weight.

Transversely mounted front engines allow for down-sized, lighter vehicles with increased interior space. However, most of the vehicle weight is toward the front of the vehicle. This provides for increased traction by the drive wheels. The weight also places a greater load on the front suspension and brakes.

Mid-Engine Transverse In this design, the engine and drivetrain are positioned between the passenger compartment and rear axle. Mid-engine location is used in smaller, rear-wheel-drive, high-performance sports cars for several reasons. The central location of heavy components results in a center of gravity very near the center of the vehicle, which vastly improves steering and handling. Since the engine is not under the hood, the hood can be sloped downward, improving aerodynamics and increasing the driver's field of vision. However, engine access and cooling efficiency are reduced. A barrier is also needed to reduce the transfer of noise, heat, and vibration to the passenger compartment.

ENGINE MEASUREMENT AND PERFORMANCE

Many of the engine measurements and performance characteristics a technician should be familiar with were discussed in Chapter 8. What follows are some of the important facts of each.

Bore and Stroke

The **bore** of a cylinder is simply its diameter measured in inches (in.) or millimeters (mm). The stroke is the length of the piston travel between TDC and BDC. Between them, bore and stroke determine the displacement of the cylinders. When the bore and stroke are of equal size, the engine is called a *square engine.* Engines that have a larger bore than stroke are called oversquare and engines with a larger stroke than bore are referred to as being undersquare. **Oversquare** engines offer the opportunity to fit larger valves in the

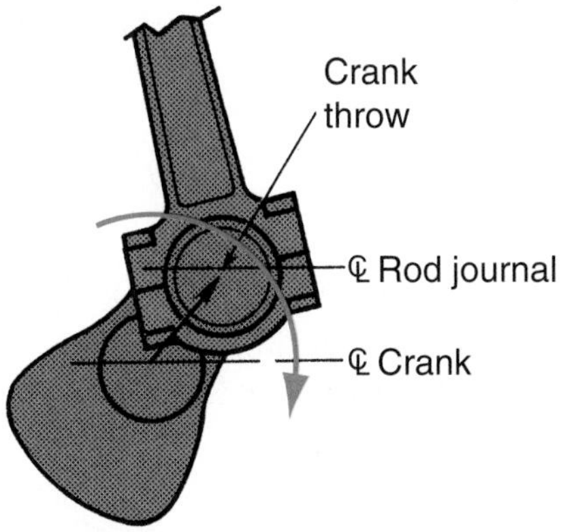

Figure 9-17 The stroke of an engine is equal to twice the crank throw.

combustion chamber and use longer connecting rods, which means oversquare engines are capable of running at higher engine speeds. But because of the size of the bore, the engines tend to be physically larger than undersquare engines. **Undersquare** engines have short connecting rods that aid in the production of more power at lower engine speeds. A square engine is a compromise between the two designs.

The **crank throw** is the distance from the crankshaft's main bearing centerline to the connecting rod journal centerline. The stroke of any engine is twice the crank throw (**Figure 9–17**).

Displacement

A cylinder's displacement is the volume of the cylinder when the piston is at BDC. An engine's displacement is the sum of the displacements of each of the engine's cylinders (**Figure 9–18**). Typically, an engine with a larger displacement produces more torque than a smaller displacement engine; however, many other factors influence an engine's power output. Engine displacement can be changed by changing the size of the bore and/or stroke of an engine.

Calculation of an engine's displacement is given in Chapter 8.

Performance TIP

The throw of a crankshaft determines the stroke. The length of the connecting rod only determines where the piston will be as it travels through the stroke. Therefore, it is possible that the piston may reach out above its bore if a crankshaft with a longer stroke is installed with standard connecting rods. The correct combination of pistons with a higher piston pin hole must be used to prevent damage to the engine.

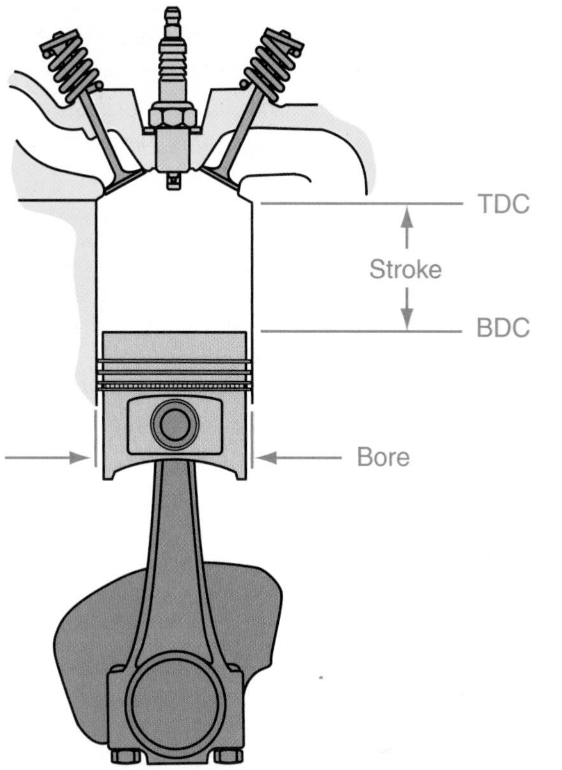

Figure 9-18 Displacement is the volume the cylinder holds between TDC and BDC.

Compression Ratio

An engine's stated compression ratio is a comparison of a cylinder's volume when the piston is at BDC to the cylinder's volume when the piston is at TDC. The compression ratio is a statement of how the air-fuel mixture is compressed during the compression stroke. It is important to keep in mind that this ratio can change through wear and carbon and dirt buildup in the cylinders. For example, if a great amount of carbon collects on the top of the piston and around the combustion chamber, the volume of the cylinder changes. This buildup of carbon will cause the compression ratio to increase because the volume at TDC will be smaller.

The higher the compression ratio, the more power an engine theoretically can produce. Also, as the compression ratio increases, the heat produced by the compression stroke also increases. Gasoline with a low-octane rating burns fast and may explode rather than burn when introduced to a high-compression ratio, which can cause preignition. The higher a gasoline's octane rating, the less likely it is to explode.

As the compression ratio increases, the octane rating of the gasoline should also be increased to prevent abnormal combustion.

Performance TIP *Often the bore of an engine is cut larger to incorporate larger pistons and to increase the engine's displacement. Doing this increases the power output of the engine. However, this will also increase the engine's compression ratio. The compression ratio may also be increased by removing metal from the mating surface of the cylinder head and/or the engine block or by installing a thinner head gasket. Care must be taken not to raise the compression too high. High-compression ratios require high-octane fuels and if the required fuel is not available, any performance gains can be lost. Use this formula to determine the exact compression ratio of an engine after modifications have been made:*

CR = total cylinder volume with the piston at BDC ÷ the total cylinder volume with the piston at TDC

The volume at BDC is equal to the cylinder's volume when the piston is at BDC plus the volume of the combustion chamber plus the volume of the head gasket. The volume of the head gasket is calculated by multiplying its thickness by the square of the bore and 0.7854.

The volume at TDC is equal to the volume in the cylinder when the piston is at TDC plus the volume of the combustion chamber plus the volume of the head gasket.

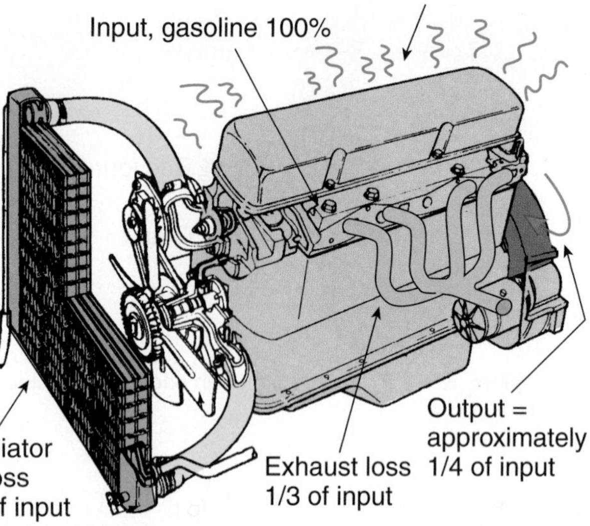

Figure 9–19 A gasoline engine is only about 25% thermal efficient.

Engine Efficiency

One of the dominating trends in automotive design is increasing an engine's efficiency. **Efficiency** is simply a measure of the relationship between the amount of energy put into an engine and the amount of energy available from the engine. Other factors, or efficiencies, affect the overall efficiency of an engine.

Volumetric Efficiency Volumetric efficiency describes the engine's ability to have its cylinders filled with air-fuel mixture. If the engine's cylinders are able to be filled with air-fuel mixture during its intake stroke, the engine has a volumetric efficiency of 100%. Typically, engines have a volumetric efficiency of 80% to 100% if they are not equipped with a turbo- or supercharger. Basically, an engine becomes more efficient as its volumetric efficiency is increased.

Thermal Efficiency Thermal efficiency is a measure of how much of the heat formed during the combustion process is available as power from the engine. Typically only one-fourth of the heat is used to power the vehicle. The rest is lost to the surrounding air and engine parts and to the engine's coolant (**Figure 9–19**). Obviously, when less combustion heat is lost, the engine is more efficient.

Mechanical Efficiency Mechanical efficiency is a measure of how much power is available once it leaves the engine compared to the amount of power that was exerted on the pistons during the power stroke. Power losses occur because of the friction generated by the moving parts. Minimizing friction increases mechanical efficiency.

Torque versus Horsepower

Torque is a twisting or turning force. Horsepower is the rate at which torque is produced. An engine produces different amounts of torque based on the rotational speed of the crankshaft and other factors. A mathematical representation, or graph, of the relationship between the horsepower and torque of an engine is shown in **Figure 9–20**.

This graph shows that torque begins to decrease when the engine's speed reaches about 1,700 rpm. Brake horsepower increases steadily until about 3,500 rpm. Then it drops. The third line on the graph indicates the horsepower needed to overcome the friction or resistance created by the internal parts of the engine rubbing against each other.

Brake horsepower is a term used to express the amount of horsepower measured on a dynamometer. This measurement represents the amount of horsepower an engine provides when it is held at a specific speed at full throttle. Horsepower is also

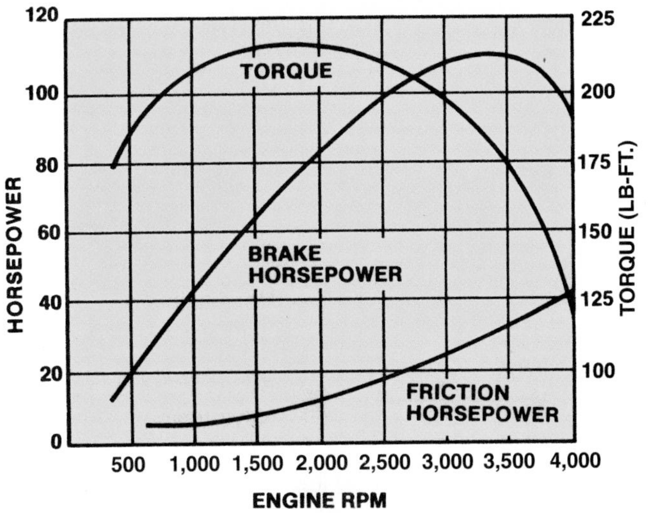

Figure 9-20 The relationship between horsepower and torque.

expressed as SAE gross horsepower, which is the maximum amount of power an engine produces at a specified speed with some of its accessories disconnected or removed. SAE net horsepower represents the power produced by an engine at a specified speed when all of its accessories are operating.

Atkinson Cycle Engines

An **Atkinson cycle** engine is a four-stroke cycle engine in which the intake valve is held open longer than normal during the compression stroke **(Figure 9–21)**. As the piston is moving up, the mixture is being compressed and some of it pushed back into the intake manifold. As a result, the amount of mixture in the cylinder and the engine's effective displacement and compression ratio are reduced. Typically there is a "surge tank" in the intake manifold to hold the mixture that is pushed out of the cylinder during the Atkinson compression stroke. Often the Atkinson cycle is referred to as a five-stroke cycle because there are two distinct cycles during the compression stroke. The first is while the intake valve is open and the second is when the intake valve is closed. This two-stage compression stroke creates the "fifth" cycle.

In a conventional engine, much engine power is lost due to the energy required to compress the mixture during the compression stroke. The Atkinson cycle reduces this power loss and this leads to greater engine efficency. The Atkinson cycle also effectively changes the length of time the mixture is being compressed. Most Atkinson cycle engines have a long piston stroke. Keeping the intake valve open during compression effectively shortens the stroke. However, because the valves are closed during the power stroke, that stroke is long. The longer power stroke allows the combustion gases to expand more and reduces the amount of heat that is lost during the exhaust stroke. As a result, the engine runs more efficently than a conventional engine.

Although these engines provide improved fuel economy and lower emissions, they also produce less power. The lower power results from the lower operating displacement and compression ratio. Power also is lower because these engines take in less air than a conventional engine.

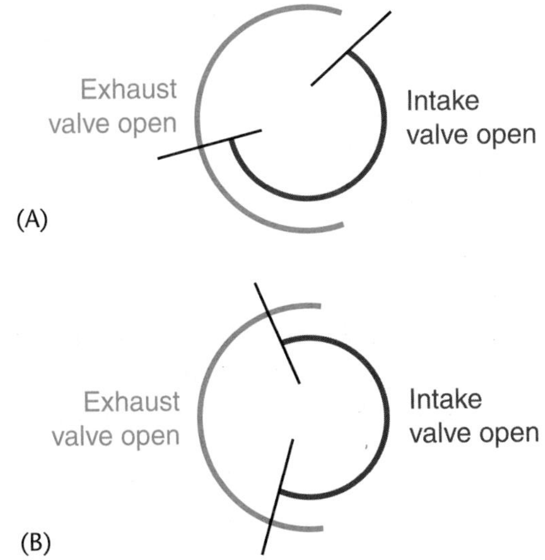

Figure 9-21 (A) Typical valve timing for an Atkinson cycle engine. (B) Typical valve timing for a conventional four-stroke cycle engine. Notice that the intake valve in the Atkinson cycle engine opens and closes later.

Hybrid Engines Many hybrid vehicles have Atkinson cycle engines. The low-power output from the engine is supplemented with the power from the electric motors. This combination offers good fuel economy, low emissions, and normal acceleration.

Some Toyota Atkinson cycle engines use variable valve timing to allow the engine to run with low displacement (Atkinson cycle) or normal displacement. The opening and closing of the intake valves is controlled by the engine control system **(Figure 9–22)**. While the valve is open during the compression stroke, the effective displacement of the engine is reduced. When the displacement is low, fuel consumption is minimized, as are exhaust emissions. The engine runs with normal displacement when the intake valves close earlier. This action provides for more power output. The control unit adjusts valve timing according to engine speed, intake air volume, throttle position, and water temperature. Because this system responds to operating conditions, the displacement of the engine changes accordingly.

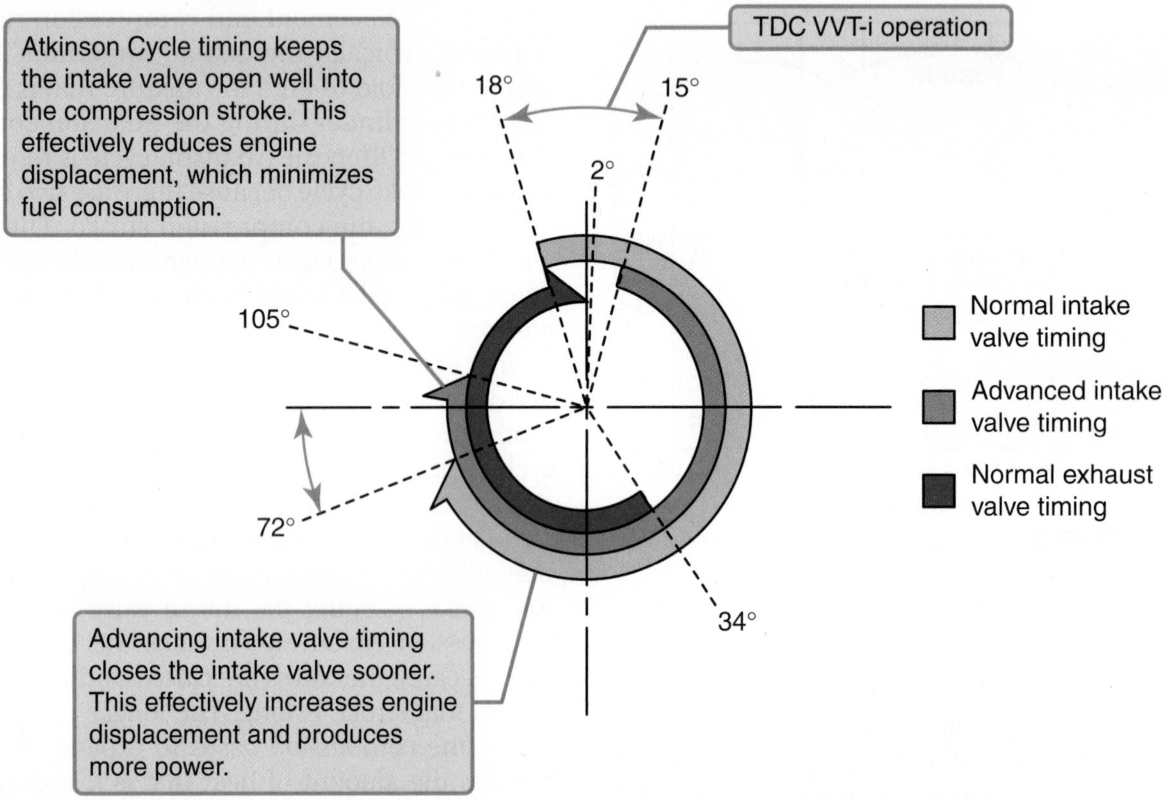
Atkinson Cycle timing keeps the intake valve open well into the compression stroke. This effectively reduces engine displacement, which minimizes fuel consumption.
TDC VVT-i operation
18°
15°
2°
105°
72°
34°
Normal intake valve timing
Advanced intake valve timing
Normal exhaust valve timing
Advancing intake valve timing closes the intake valve sooner. This effectively increases engine displacement and produces more power.

Figure 9-22 Toyota's VVT-i (variable valve timing with intelligence) changes the engine from a conventional four-stroke cycle to an Atkinson cycle according to the vehicle's operating conditions.

In response to these inputs, the control unit sends commands to the camshaft timing oil control valve. A controller at the end of the camshaft is driven by the crankshaft. The control unit regulates the oil pressure sent to the controller. A change in oil pressure changes the position of the camshaft and the timing of the valves. The camshaft timing oil control valve is duty cycled by the control unit to advance or retard intake valve timing. The controller rotates the intake camshaft in response to the oil pressure. An advance in timing results when oil pressure is applied to the timing advance chamber. When the oil control valve is moved and the oil pressure is applied to the timing retard side vane chamber (**Figure 9–23**), the timing is retarded.

Miller Cycle Engines An Atkinson cycle engine with forced induction (supercharging) is called a Miller cycle engine. The decrease of intake air and resulting low power is compensated by the supercharger. The supercharger forces air into the cylinder during the compression stroke. Keep in mind that the actual compression stroke in an Atkinson cycle engine does not begin until the intake valve closes. The supercharger in a Miller cycle engine forces more air past the valve and, therefore, there is more air in the cylinder when the intake closes.

The Miller cycle is efficient only if the supercharger uses less energy to compress the mixture than the piston would normally need to compress it during a normal compression stroke. This is an obstacle for engineers because to drive a supercharger requires approximately 10% to 20% of the engine's output. The latest Miller cycle engines control the action of the supercharger so that it is only used when it is better for compression and is shut down when piston compression is best.

DIESEL ENGINES

Diesel engines represent tested, proven technology with a long history of success. Invented by Dr. Rudolph Diesel, a German engineer, and first marketed in 1897, the diesel engine is now the dominant power plant in heavy-duty trucks, construction equipment, farm equipment, buses, and marine applications. Diesel engines in cars and light trucks will become more common soon. There are many reasons for this, one of which is that low-sulfur diesel fuel will be available in the United States. Diesel vehicles are very common in Europe and other places where cleaner fuels are available (**Figure 9–24**).

The operation of a **diesel engine** is comparable to a gasoline engine. They also have a number of

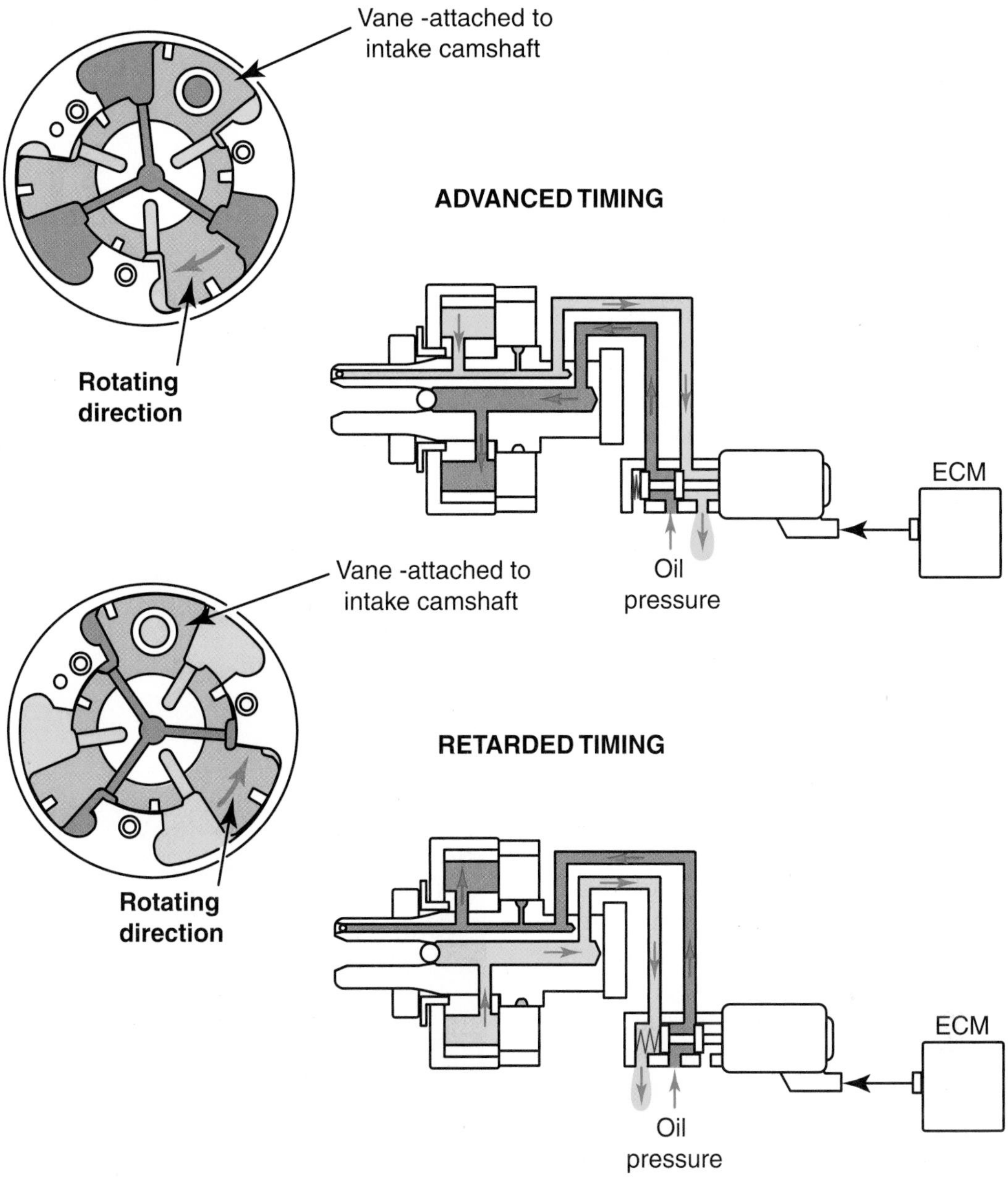

Figure 9–23 Oil flow for the VVT-i as it advances and retards the valve timing.

Figure 9–24 A European four-cylinder passenger car diesel engine.

components in common, such as the crankshaft, pistons, valves, camshaft, and water and oil pumps. They both are available as two- or four-stroke combustion cycle engines. However, diesel engines have compression ignition systems **(Figure 9–25)**. Rather than relying on a spark for ignition, a diesel engine uses the heat produced by compressing air in the combustion chamber to ignite the fuel. The compression ratio of diesel engines is typically three times (as high as 25:1) that of a gasoline engine. As intake air is compressed, its temperature rises to 1,300°F to 1,650°F (700°C to 900°C). Just before the air is fully compressed, a fuel injector sprays a small amount of diesel fuel into the cylinder. The high temperature of the compressed air instantly ignites the fuel. The combustion causes increased heat in

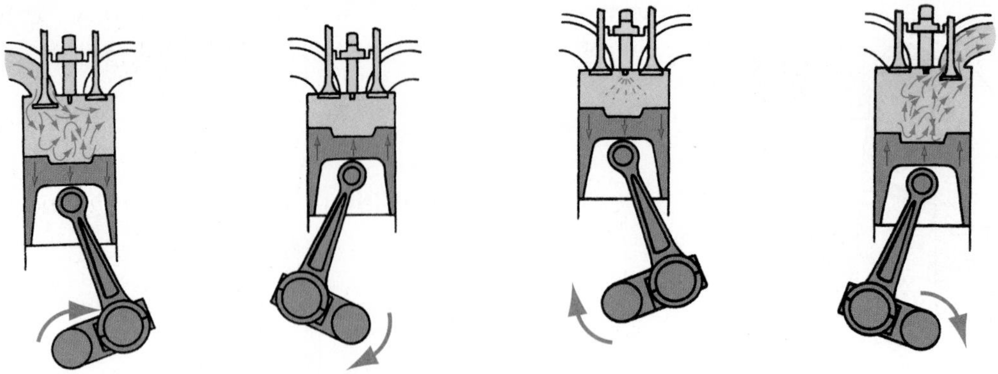

Figure 9-25 A four-stroke diesel engine cycle.

the cylinder and the resulting high pressure moves the piston down on its power stroke.

Construction

Diesel engines are heavier than gasoline engines of the same power. A diesel engine must be made stronger to contain the extremely high compression and combustion pressures. A diesel engine also produces less horsepower than a same-sized gasoline engine. Therefore, to provide the required power, the displacement of the engine is increased. This results in a physically larger engine. Diesels have high torque outputs at very low engine speeds but do not run well at high engine speeds. On many diesel engines, turbochargers and intercoolers are used to increase their power output **(Figure 9–26)**.

Figure 9-26 The high-output Cummins turbo diesel I-6 engine used in Dodge Ram heavy-duty trucks. *Courtesy of Chrysler LLC*

Diesel combustion chambers are different from gasoline combustion chambers because diesel fuel burns differently. Three types of combustion chambers are used in diesel engines: open combustion chamber, precombustion chamber, and turbulence combustion chamber. The open combustion chamber is located directly inside the piston. Diesel fuel is injected directly into the center of the chamber. The shape of the chamber and the quench area produce turbulence. The precombustion chamber is a smaller, second chamber connection to the main combustion chamber. On the power stroke, fuel is injected into the small chamber. Combustion is started there and then spreads to the main chamber. This design allows for lower fuel injection pressures and simpler injection systems. The turbulence combustion chamber creates an increase in air velocity or turbulence in the combustion chamber. The fuel is injected into the turbulent air and burns more completely.

Fuel injection is used on all diesel engines. Older diesel engines had a distributor-type injection pump driven and regulated by the engine. The pump supplied fuel to injectors that sprayed the fuel into the engine's combustion chamber. Newer diesel engines are equipped with common rail systems **(Figure 9–27)**. Common rail systems are **direct injection (DI)** systems. The injectors' nozzles are placed inside the combustion chamber. The piston top has a depression where initial combustion takes place. The injector must be able to withstand the temperature and pressure inside the cylinder and must be able to deliver a fine spray of fuel into those conditions. These systems have a high-pressure (14,500+ psi or 1,000+ bar) fuel rail connected to individual solenoid-type injectors.

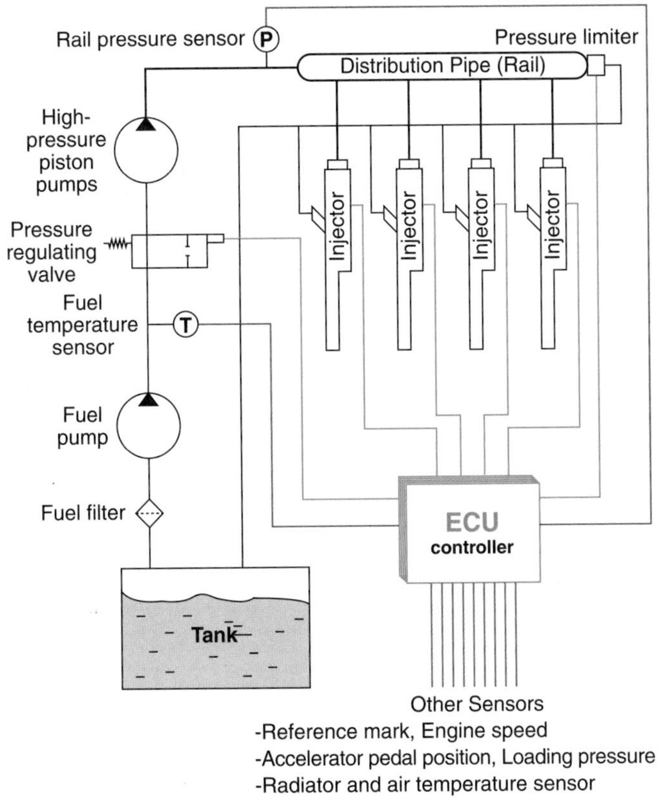

Figure 9-27 A common rail fuel injection system.

The injectors are controlled by a computer that attempts to match injector operation to the operating conditions of the engine. Newer diesel fuel injectors rely on stacked piezoelectric crystals rather than solenoids. Piezo crystals quickly expand when electrical current is applied to them. The crystals allow the injectors to respond very quickly to the needs of the engine. With this new-style injector, diesel engines are quieter, more fuel efficient, cleaner, and have more power.

Diesel engines are also available in two-stroke-cycle models. Most diesels generally use the four-stroke cycle, while some larger diesels operate with the two-stroke cycle. Two-stroke diesels must use forced induction from either a turbocharger or a supercharger. These engines are ideal for some applications because they provide high torque for their displacement.

Advantages

When compared to gasoline engines, diesel engines offer many advantages. They are more efficient and use less fuel than a gasoline engine of the same size. Diesel engines are very durable. This is due to stronger construction and the fact that diesel fuel is a better lubricant than gasoline. This means that the fuel is less likely to remove the desired film of oil on the cylinder walls and piston rings of the engine. Diesel engines are also better suited for moving heavy loads at low speeds.

Disadvantages

The primary disadvantages of using diesel engines in passenger cars and light trucks include:

- Low power output
- Difficult cold weather starting
- Noise
- Exhaust emissions

Many diesel engines are fit with a turbocharger to increase their power. Combining turbochargers with common rail injection systems have resulted in more horsepower.

In cold weather, diesel engines can be difficult to start because the cold air cannot become hot enough to cause combustion, in spite of the high compression ratios. This problem is compounded by the fact that the cold metal of the cylinder block and head absorbs the heat generated during the compression stroke. Some diesel engines use **glow plugs** to help ignite fuel during cold starting. These small electrical heaters are placed inside the cylinder and are used only to warm the combustion chamber when the engine is cold. Other diesels have a resistive grid heater in the intake manifold to warm the air until the engine reaches operating temperature.

A characteristic of a diesel engine is its sound. This noise, knock or clatter, is caused by the sudden ignition of the fuel as it is injected into the combustion chamber. Through the use of electronically controlled common rail injector systems, manufacturers have been able to minimize the noise.

Emissions have always been an obstacle for diesel cars and new stricter emissions standards will go into effect shortly. Cleaner, low-sulfur, diesel fuel has been available in the United States since 2007. With new technologies and the cleaner fuel, the emissions levels from a diesel engine should be able to run as clean as most gasoline engines. Many diesel vehicles have an assortment of traps and filters to clean the exhaust before it enters the atmosphere. Some diesel engines have diesel particulate filters and catalytic converters (**Figure 9–28**). Particulate filters catch the black soot (unburned carbon compounds) that is typically expelled from a diesel vehicle's exhaust. Most diesel cars will have **selective catalytic reduction** (SCR) systems to reduce NO_x emissions. SCR is a process wherein a substance is injected into the exhaust stream and then absorbed onto a catalyst. This action breaks down the exhaust's NO_x to form H_2O and N_2. Others will use NO_x traps. Diesel engines produce

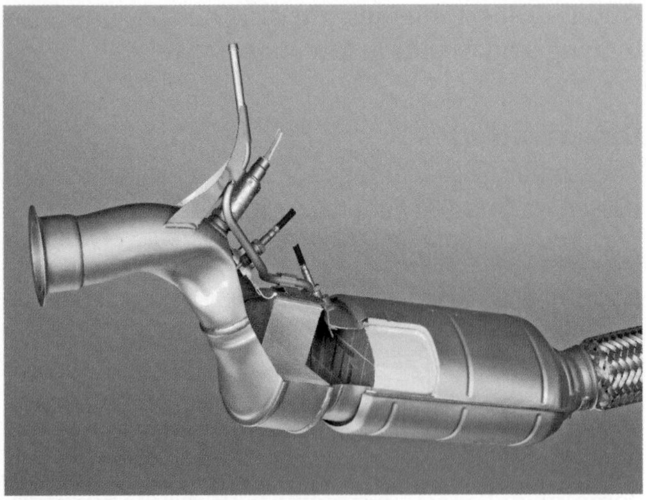

Figure 9-28 A catalytic converter and particulate trap for a diesel engine. *Courtesy of BMW of North America, LLC*

very little carbon monoxide because they run with an abundance of air.

OTHER AUTOMOTIVE POWER PLANTS

In an attempt to reduce fuel consumption and harmful exhaust emissions, many manufacturers are supplementing or modifying the basic internal combustion engine. Many of these power plants were developed during the early days of automobiles. Due to the advancements made in electronic controls, they are becoming a viable alternative to the conventional gasoline engine.

Hybrids

A hybrid vehicle has at least two different types of power or propulsion systems. Today's hybrid vehicles have an internal combustion engine and an electric motor (some vehicles have more than one electric motor). A hybrid's electric motor is powered by batteries and/or ultracapacitors, which are recharged by a generator that is driven by the engine **(Figure 9–29)**. They are also recharged through regenerative braking. The engine may use gasoline, diesel, or an alternative fuel. Complex electronic controls monitor the operation of the vehicle. Based on the current operating conditions, electronics control the engine, electric motor, and generator.

Depending on the design of the hybrid vehicle, the engine may power the vehicle, assist the electric motor while it is propelling the vehicle, or drive a generator to charge the vehicle's batteries. The electric motor may propel the vehicle by itself, assist the engine while it is propelling the vehicle, or act as a

Figure 9-29 The Honda Civic Hybrid has a 1.3-liter gasoline engine and a 20-horsepower electric motor. *Courtesy of American Honda Motor Co., Inc.*

generator to charge the batteries. Many hybrids rely exclusively on the electric motor(s) during slow-speed operation, on the engine at higher speeds, and on both during some certain driving conditions.

Often hybrids are categorized as series or parallel designs. In a series hybrid, the engine never directly powers the vehicle. Rather it drives a generator, and the generator either charges the batteries or directly powers the electric motor that drives the wheels **(Figure 9–30)**. Currently there are no true series hybrids manufactured. A parallel hybrid vehicle uses either the electric motor or the gas engine to propel the vehicle, or both **(Figure 9–31)**. Most current hybrids can be considered as having a series/parallel configuration because they have the features of both designs.

Although most current hybrids are focused on fuel economy, the same construction is used to create high-performance vehicles. The added power of the electric motor boosts the performance levels provided by the engine. Hybrid technology also enhances off-the-road performance. By using individual motors at the front and rear drive axles, additional power can be applied to certain drive wheels when needed.

The engines used in hybrids are specially designed for fuel economy and low emissions. The engines tend to be small displacement engines that use variable valve timing and the Atkinson cycle to

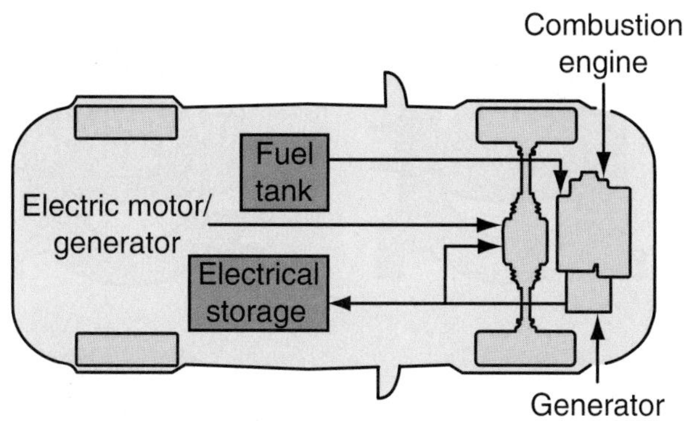

Figure 9-30 The configuration of a series hybrid vehicle.

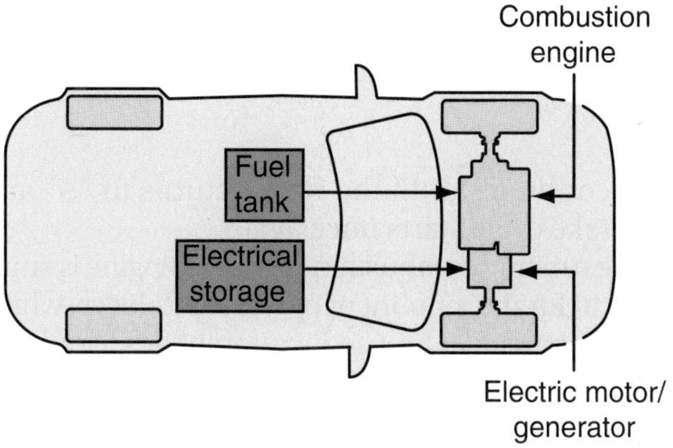

Figure 9-31 The configuration of a parallel hybrid vehicle.

provide low fuel consumption. These advanced engines, however, cannot produce the power needed for reasonable acceleration by themselves. The electric motor provides additional power for acceleration and for overcoming loads.

Battery-Operated Electric Vehicles A battery-operated electric vehicle, sometimes referred to as an EV, uses one or more electric motors to turn its drive wheels. The electricity for the motors is stored in batteries that must be recharged by an external electrical power source. Normally they are recharged by plugging them into an outlet at home or other locations. The recharging time varies with the type of charger, the size and type of battery, and other factors. Normal recharge time is 4 to 8 hours.

An electric motor is quiet and has few moving parts. It starts well in the cold, is simple to maintain, and does not burn petroleum products to run. The disadvantages of an EV are limited speed, power, and range as well as the need for heavy, costly batteries. However, an EV is much more efficient than a conventional gasoline-fueled vehicle. EVs are considered zero emissions vehicles because they do not directly pollute the air. The only pollution associated with them is the result of creating the electricity to charge their batteries.

In the early days of the automobile, electric cars outnumbered gasoline cars. Today, there are few EVs on the road but they are commonly used in manufacturing, shipping, and other industrial plants, where the exhaust of an internal combustion engine could cause illness or discomfort to the workers in the area. They are also used on golf courses, where the quiet operation adds to the relaxing atmosphere. Some auto manufacturers are still studying their use. Whether battery-operated EVs return to the market really depends on the development of new batteries and motors. To be practical, EVs need to have much longer driving ranges between recharges and must be able to sustain highway speeds for great distances.

Fuel Cell Electric Vehicles Although just experimental at this time, there is much promise for fuel cell EVs. These vehicles are powered solely by electric motors, but the energy for the motors is produced by fuel cells. Fuel cells rely on hydrogen to produce the electricity. A fuel cell generates electrical power through a chemical reaction. A fuel cell EV uses the electricity produced by the fuel cell to power motors that drive the vehicle's wheels **(Figure 9–32)**. The batteries in these vehicles do not need to be charged by an external source.

Fuel cells convert chemical energy to electrical energy by combining hydrogen with oxygen. The hydrogen can be supplied directly as pure hydrogen gas or through a "fuel reformer" that pulls hydrogen from hydrocarbon fuels such as methanol, natural gas, or gasoline. Simply put, a fuel cell is comprised of two electrodes (the anode and the cathode) located on either side of an electrolyte. As the hydrogen enters the fuel cell, the hydrogen atoms give up electrons at the anode and become hydrogen ions in the electrolyte. The electrons that were released at the anode move through an external circuit to the

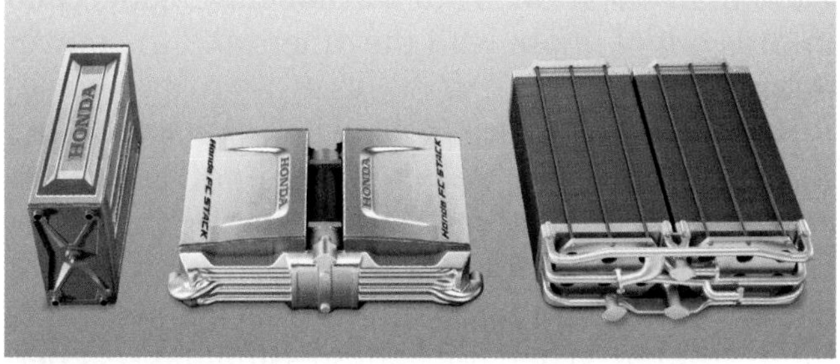

Figure 9-32 The sources of power for a fuel cell electric vehicle: fuel cell stack (left), power control unit (center), and lithium ion battery pack (right). *Courtesy of American Honda Motor Co., Inc.*

cathode. As the electrons move toward the cathode, they can be diverted and used to power the vehicle. When the electrons and hydrogen ions combine with oxygen molecules at the cathode, water and heat are formed. There are no smog-producing or greenhouse gases produced. Although vehicles equipped with reformers emit some pollutants, those that run on pure hydrogen are true zero-emission vehicles.

Rotary Engines

The **rotary** engine, or **Wankel** engine, is somewhat similar to the standard piston engine in that it is a spark ignition, internal combustion engine. Its design, however, is quite different. For one thing, the rotary engine uses a rotating motion rather than a reciprocating motion. In addition, it uses ports rather than valves for controlling the intake of the air-fuel mixture and the exhaust of the combusted charge.

The main part of a rotary engine is a roughly triangular rotor that rotates within an oval-shaped housing. The rotor has three convex faces and each face has a recess in it. These recesses increase the overall displacement of the engine. The tips of the rotor are always in contact with the walls of the housing as the rotor moves to seal the sides (chambers) to the walls. As the rotor rotates, it creates three separate chambers of gas. Also, as it rotates, the volume between the sides of the rotor and the housing continuously changes. During rotor rotation, the volume of the gas in each chamber alternately expands and contracts. It is how a rotary engine rotates through the basic four-stroke cycle.

The rotor "walks" around a rigidly mounted gear in the housing. The rotor is connected to the crankshaft through additional gears that allow every rotation of the rotor to rotate the crankshaft three times. This means that the output shaft only rotates three times for every revolution of the rotor, which allows only one power stroke for each revolution of the output shaft. This is why a rotary engine produces less power than a conventional four-stroke engine. When more than one rotor is fitted inside the engine, each rotor is out of phase with the others and the power output is increased.

Referring to **Figure 9–33**, when the side of the rotor is in position "A," the intake port is uncovered and the air-fuel mixture is entering the upper chamber. As the rotor moves to "B," the intake port closes and the upper chamber reaches its maximum volume. When full compression has reached "C," the two spark plugs fire, one after the other, to start the power stroke. At "D," the side of the rotor uncovers the exhaust port and exhaust begins. This cycle continues until the rotor returns to "A" and the intake cycle starts once again.

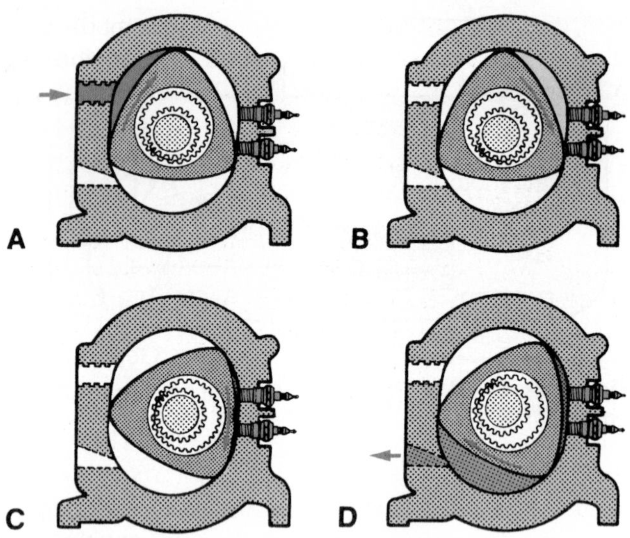

Figure 9-33 A rotary engine cycle.

The rotating combustion chamber engine is small and light for the amount of power it produces, which makes it attractive for use in automobiles. However, the rotary engine at present cannot compete with a piston gasoline engine in terms of durability, exhaust emissions, and economy. After a few years of not offering a rotary engine, Mazda has released a version of the engine, called the Renesis, that produces lower emissions and has two rotors.

Stratified Charge Engines

The stratified charge engine **(Figure 9–34)** combines the features of gasoline and diesel engines. It differs

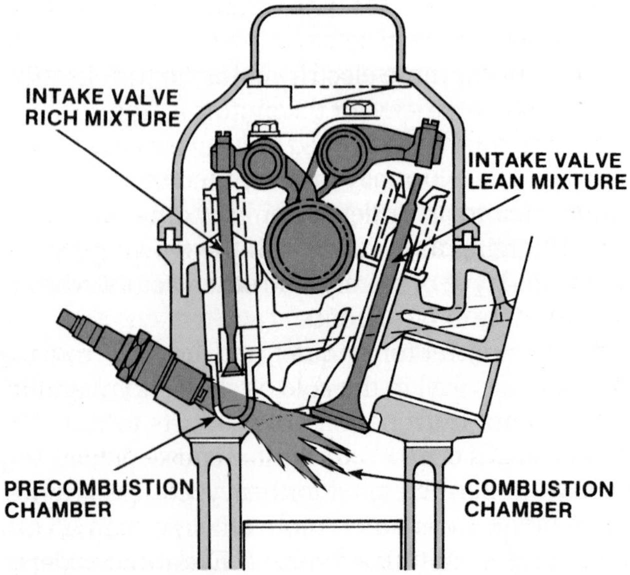

Figure 9-34 A typical stratified charge engine.

from the conventional gasoline engine in that the air-fuel mixture is deliberately stratified to produce a small rich mixture at the spark plug while providing a leaner, more efficient and cleaner burning main mixture. In addition, the air-fuel mixture is swirled to provide for more complete combustion.

A large amount of very lean mixture is drawn through the main intake valve on the intake stroke to the main combustion chamber. At the same time, a small amount of rich mixture is drawn through the auxiliary intake valve into the precombustion chamber. At the end of the compression stroke, the spark plug fires the rich mixture in the precombustion chamber. As the rich mixture ignites, it in turn ignites the lean mixture in the main chamber. The lean mixture minimizes the formation of carbon monoxide during the power stroke. In addition, the peak temperature stays low enough to minimize the formation of NO_x, and the mean temperature is held high enough and long enough to reduce hydrocarbon emissions.

The Honda CVCC engine uses a stratified charge design. This engine uses a third valve to release the initial charge. The stratified charge combustion chamber has three important advantages: It produces good part-load fuel economy, it can run efficiently on low-octane fuel, and it has low exhaust emissions.

Homogeneous Charge Compression Ignition Engines

Within the next few years, some automobiles will be equipped with **homogeneous charge compression ignition (HCCI)** engines. HCCI engines offer the high efficiency and torque of a diesel engine while providing the low emissions and power of a gasoline engine. Basically these engines have a combustion process that allows a gasoline or diesel engine to operate with either compression ignition or spark ignition. With spark ignition the air and fuel are mixed (homogenized) before ignition and ignition is caused by a spark. In a diesel engine the air and fuel are never mixed. The air is compressed and ignition occurs when fuel is sprayed into the high-temperature air. In an HCCI engine, the air and fuel are mixed and ignition occurs as the mixture is compressed. During compression, the mixture gets hot enough to "autoignite." HCCI is also referred to as controlled auto-ignition (CAI).

In an HCCI engine, combustion immediately and simultaneously begins at several points within the mixture. This means the combustion process occurs rapidly and is controlled by the quality and temperature of the compressed mixture. This spontaneous combustion produces a flameless release of energy to drive the piston down.

The HCCI engine runs on a lean, diluted mixture of fuel, air, and exhaust gases. Only the heat inside the cylinder determines when ignition will occur. This fact makes it hard to control ignition timing. The temperature of the mixture at the beginning of the compression stroke must be increased to autoignition temperatures at the end of the compression stroke. Autoignition usually occurs when the temperature reaches 1,430°F to 1,520°F (777°C to 827°C) for gasoline. The engine's control unit must supply the correct amount of fuel mixed with the correct amount of air in order for combustion to occur at the right time. In addition, the control unit must provide a mixture that is hot enough to be able to autoignite at the end of the compression stroke. Therefore, it must be able to vary the compression ratio, the temperature of the intake air, the pressure of the intake air, or the amount of retained or reinducted exhaust gas. The role of the control unit is extremely important for proper operation.

Dual Mode A practical application of an HCCI engine would be one with dual mode capabilities. The spark ignition mode could be used when high power is required, and the compression ignition mode would be used during steady loads and speeds. To do this, the engine must be able to smoothly switch from the HCCI mode to the spark ignition mode from one cylinder firing to the next. This would require precise control of valve timing, air and fuel metering, and spark plug timing.

Benefits A gasoline HCCI engine could deliver almost the same fuel economy as a diesel engine and at a much lower cost. GM estimates that HCCI could improve gasoline engine fuel efficiency by 20%, while emitting near-zero amounts of NO_x and particulate matter. In fact, HCCI engines emit extremely low levels of NO_x without a catalytic converter.

However, a gasoline engine running in the HCCI mode produces more noise and vibrations than a conventional engine. Also, they tend to experience incomplete combustion, which leads to hydrocarbon and carbon monoxide emissions. To rectify this, HCCI engines are fitted with typical emission control systems, including an oxidizing catalytic converter.

Variable Compression Ratio Engines

Variable compression engines are being explored, not only for use with HCCI, but for use in conventional engines. Changing the compression ratio is

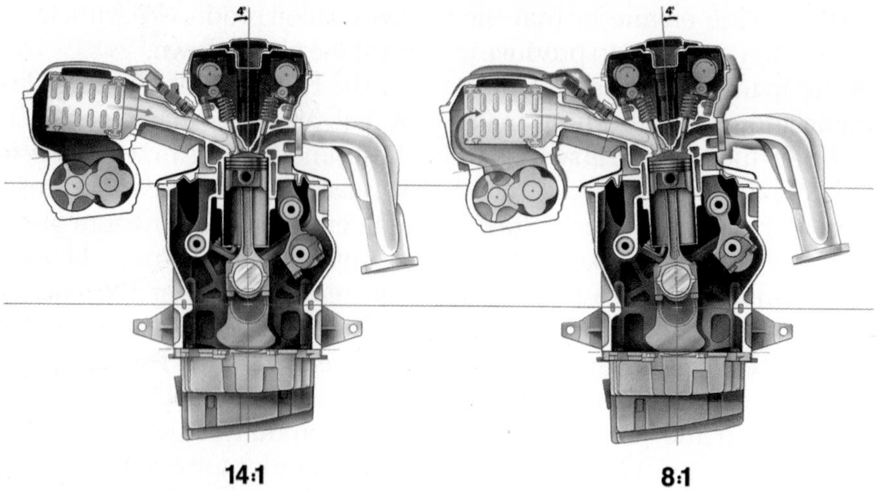

Figure 9-35 The SVC can vary the engine's compression ratio from 8:1 to 14:1.
Courtesy of Saab Automobile AB

one way to provide power when needed and minimizing fuel consumption. One way to do this is through changes in valve timing. The process is similar to the modifications made for the Atkinson cycle. Another way is to change the volume of the combustion chamber in response to the engine's operating conditions.

Saab has developed such an engine, called Saab variable compression (SVC), that has a cylinder head constructed with integrated cylinders. The compression ratio is altered by changing the slope of the cylinder head in relation to the engine block. This changes the volume of the combustion chamber **(Figure 9–35)**. The cylinder head is pivoted at the crankshaft by a hydraulic actuator and can be as much as 4 degrees. The engine management system adjusts the angle in response to engine speed, engine load, and fuel quality. The cylinder head is sealed to the engine block by a rubber bellows.

ENGINE IDENTIFICATION

USING SERVICE INFORMATION

Normally, information used to identify the size of an engine is given in service manuals at the beginning of the section covering that particular manufacturer.

By referring to the VIN, much information about the vehicle can be determined. Identification numbers are also found on the engine. Some manufacturers use tags or stickers attached at various places, such as the valve cover or oil pan. Blocks often have a serial number stamped into them **(Figure 9–36)**. Service manuals typically give the location of the code for a particular engine. The engine code is generally found beside the serial number. A typical

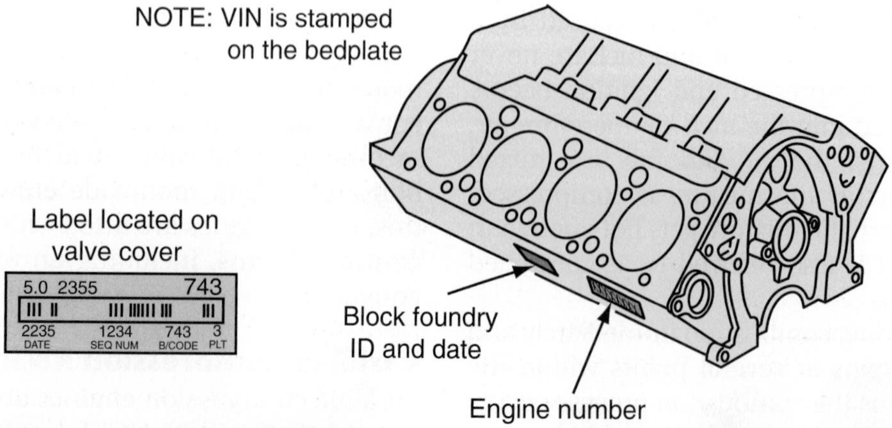

Figure 9-36 Examples of the various identification numbers found on an engine.

engine code might be DZ or MO. These letters indicate the horsepower rating of the engine, whether it was built for an automatic or manual transmission, and other important details. The engine code will help you determine the correct specifications for that particular engine.

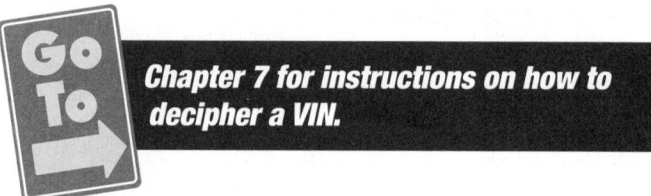

Go To ***Chapter 7 for instructions on how to decipher a VIN.***

Engine ID Tags Many engines have ID tags or stickers attached to various places on the engine, such as the valve cover or oil pan. The tags include the displacement, assembly plant, model year, change level, engine code, and date of production. Service manuals normally note the location of these stickers or tags on a particular engine.

Casting Numbers Whenever an engine part such as an engine block or head is cast, a number is put into the mold to identify the casting and the date when the part was made. This date does not indicate when the engine was assembled or placed into the vehicle at the factory. A part made during one year may be installed in the vehicle in the following year; therefore, the casting date may not match the model year of the vehicle. Casting numbers should not be used for identifying the displacement of an engine. They only indicate the basic design of an engine. The same block or head can be used with a variety of different displacement engines.

Underhood Label Vehicles produced since 1972 have an underhood emission control label that contains such useful information as ignition timing specifications, emission control devices, engine size, vacuum hose routing, and valve adjustment specifications.

ENGINE DIAGNOSTICS

As the trend toward the integration of ignition, fuel, and emission systems progresses, diagnostic test equipment must also keep up with these changes. New tools and techniques are constantly being developed to diagnose electronic engine control systems. However, not all engine performance problems are related to electronic control systems; therefore, technicians still need to understand basic engine tests. These tests are an important part of modern engine diagnosis.

Compression Test

Internal combustion engines depend on compression of the air-fuel mixture to maximize the power produced by the engine. The upward movement of the piston on the compression stroke compresses the air-fuel mixture within the combustion chamber. The air-fuel mixture gets hotter as it is compressed. The hot mixture is easier to ignite, and when ignited it generates much more power than the same mixture at a lower temperature.

If the combustion chamber leaks, some of the air-fuel mixture will escape when it is compressed, resulting in a loss of power and a waste of fuel. The leaks can be caused by burned valves, a blown head gasket, worn rings, slipped timing belt or chain, worn valve seats, a cracked head, and more.

An engine with poor compression (lower compression pressure due to leaks in the cylinder) will not run correctly. If a symptom suggests that the cause of a problem may be poor compression, a compression test is performed.

A compression gauge is used to check cylinder compression. The dial face on the typical compression gauge indicates pressure in both pounds per square inch (psi) and metric kilopascals (kPa). Most compression gauges have a vent valve that holds the highest pressure reading on its meter. Opening the valve releases the pressure when the test is complete. The steps for conducting a cylinder compression test are shown in Photo Sequence 6.

Ford, Toyota, and other hybrids use Atkinson cycle engines. These engines delay the closing of the intake valve, which means that the overall compression ratio and displacement of the engine are reduced. Therefore, when conducting a compression test on these engines, expect a slightly lower reading than what you would expect from a conventional engine.

To conduct a compression test on a Ford Escape, you must use a scan tool and the one from Ford is preferred. The scan tool allows you to enter into the engine cranking diagnostic mode. This mode allows the engine to crank with the fuel injection system disabled. It also makes sure that the starter motor/generator is not activated (except for activating the starter motor to crank the engine), which not only is good for safety purposes, it is also good because the load of the generator cannot affect the test results because it is not energized. Always follow the sequence as stated in the service manual. Failure to do so will result in bad readings.

P6-1 *Before conducting a compression test, disable the ignition and the fuel injection system. Most manufacturers recommend that the engine be warm when testing.*

P6-2 *Prop the throttle plate into a wide-open position to allow an unrestricted amount of air to enter the cylinders during the test.*

P6-3 *Remove all of the engine's spark plugs.*

P6-4 *Connect a remote starter button to the starter system.*

P6-5 *Many types of compression gauges are available. The screw-in type tends to be the most accurate and easiest to use.*

P6-6 *Carefully install the gauge into the spark plug hole of the first cylinder.*

P6-7 *Connect a battery charger to the car to allow the engine to crank at consistent and normal speeds needed for accurate test results.*

P6-8 *Depress the remote starter button and observe the gauge's reading after the first engine revolution.*

P6-9 *Allow the engine to turn through four revolutions, and observe the reading after the fourth. The reading should increase with each revolution.*

Conducting a Cylinder Compression Test *(continued)*

P6-10 *Readings observed should be recorded. After all cylinders have been tested, a comparison of cylinders can be made.*

P6-11 *Before removing the gauge from the cylinder, release the pressure from it using the release valve on the gauge.*

P6-12 *Each cylinder should be tested in the same way.*

P6-13 *After completing the test on all cylinders, compare them. If one or more cylinders is much lower than the others, continue testing those cylinders with the wet test.*

P6-14 *Squirt a small amount of oil into the weak cylinder(s).*

P6-15 *Reinstall the compression gauge into that cylinder and conduct the test.*

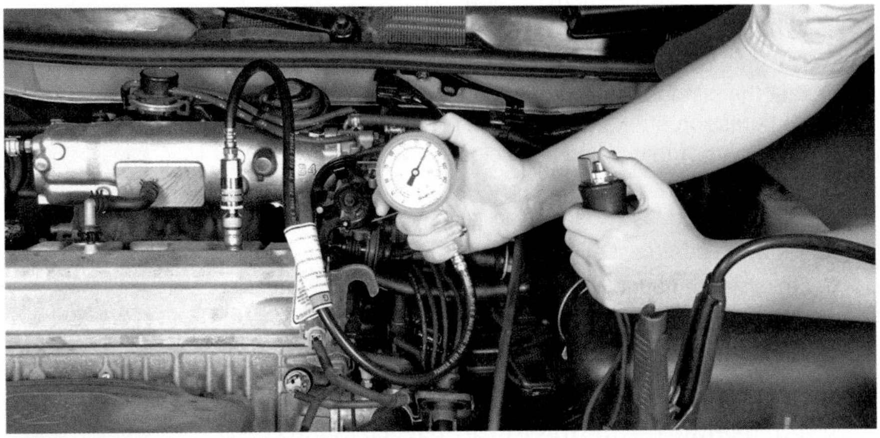

P6-16 *If the reading increases with the presence of oil in the cylinder, the most likely cause of the original low readings was poor piston ring sealing. Using oil during a compression test is normally referred to as a wet test.*

Wet Compression Test Because many things can cause low compression, it is advisable to conduct a wet compression test on the low cylinders. This test allows you to identify if it is caused by worn or damaged piston rings. To conduct this test, add two squirts of oil into the low cylinders. Then measure the compression of that cylinder. If the readings are higher, it is very likely that the piston rings are the cause of the problem. The oil temporarily seals the piston to the cylinder walls, which is why the readings increased. If the readings do not increase, or increase only slightly, the cause of the low readings is probably the valves.

Figure 9-37 Rotate the engine so that the piston of the cylinder that will be tested is at TDC before checking leakage.

Cylinder Leakage Test

If a compression test shows that any of the cylinders are leaking, a cylinder leakage test can be performed to measure the percentage of compression lost and to help locate the source of leakage. A cylinder leakage tester applies compressed air to a cylinder through the spark plug hole. The source of the compressed air is normally the shop's compressed air system. The tester's pressure regulator controls the pressure applied to the cylinder. A gauge registers the percentage of air pressure lost when the compressed air is applied to the cylinder. The scale on the gauge typically reads 0% to 100%. The amount and location of the air that escapes give a good idea of the engine's condition and can pinpoint where compression is lost.

CAUTION!

Always follow the precautions given by the manufacturer when conducting a compression test or other engine-related tests, especially when doing this on a hybrid vehicle. In most hybrids, the engine is cranked by a high-voltage motor. Because this motor is required to run the test, the high-voltage system cannot be isolated. Therefore, extreme care must be taken and all appropriate safety precautions must be followed.

PROCEDURE

1. Make sure the engine is at operating condition.
2. Remove the radiator cap, oil filler cap, dipstick tube, air filter cover, and all spark plugs.
3. Rotate the crankshaft with a remote starter button so that the piston of the tested cylinder is at TDC on its compression stroke **(Figure 9–37)**. This ensures that the valves of that cylinder are closed.
4. Insert the threaded adapter on the end of the tester's air pressure hose into the spark plug hole.
5. Allow the compressed air to enter the cylinder.
6. Observe the gauge reading **(Figure 9–38)**.
7. Listen and feel to identify the source of any escaping air.

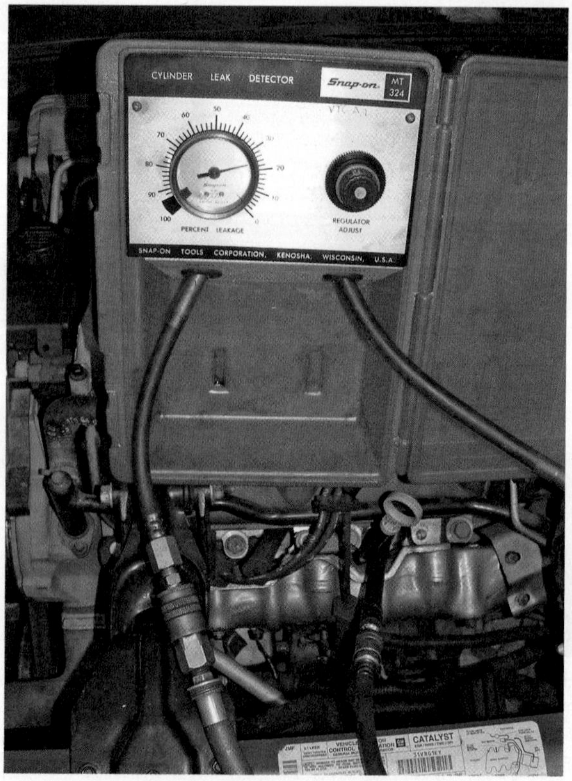

Figure 9-38 The reading on the tester is the percentage of air that leaked out during the test.

A zero reading means there is no leakage in the cylinder. Readings of 100% indicate that the cylinder will not hold any pressure. Any reading that is more than 0% indicates there is some leakage **(Figure 9–39)**. Most engines, even new ones, experience some leakage

Measured Leakage	Conclusion
Less than 10%	Good
Between 10 and 20%	Acceptable
Between 20 and 30%	**Worn engine**
Above 30%	Definite problem
100%	Serious problem

Figure 9-39 Cylinder leakage test results.

around the rings. Up to 20% is considered acceptable. When the engine is running, the rings will seal much better and the actual leakage will be lower.

SHOP TALK

Some leakage testers read in the opposite way; a reading of 100% may indicate a totally sealed cylinder, whereas 0% indicates a very serious leak. Always refer to the manufacturer's literature before using test equipment.

The location of the compression leak can be found by listening and feeling around various parts of the engine **(Figure 9–40)**.

Cylinder Power Balance Test

The cylinder power balance test is used to check if all of the engine's cylinders are producing the same amount of power. Ideally, all cylinders will produce the same amount. To check an engine's power balance, each cylinder is disabled, one at a time, and the change in engine speed is recorded. If all of the cylinders are producing the same amount of power, engine speed will drop the same amount as each cylinder is disabled. Unequal cylinder power balance can be caused by the following problems:

- Defective ignition coil
- Defective spark plug wire
- Defective or worn spark plug
- Damaged head gasket
- Worn piston rings
- Damaged piston
- Damaged or burned valves
- Broken valve spring
- Worn camshaft
- Defective lifters, pushrods, and/or rocker arms
- Leaking intake manifold
- Faulty fuel injector

Source of Leakage	Probable Cause
Radiator	Faulty head gasket Cracked cylinder head Cracked engine block
Throttle body	Damaged intake valve
Tailpipe	Damaged exhaust valve
Oil filler or dipstick tube	Worn piston rings
Adjacent spark plug hole	Faulty head gasket Cracked cylinder head

Figure 9-40 Sources of cylinder leakage and the probable causes.

A power balance test is performed quickly and easily using an engine analyzer, because the firing of the spark plugs can be automatically controlled or manually controlled by pushing a button. Some vehicles have a power balance test built into the engine control computer. This test is either part of a routine self-diagnostic mode or must be activated by the technician.

WARNING!

On some computer-controlled engines, certain components must be disconnected before attempting the power balance test. Because of the wide variations from manufacturer to manufacturer, always check the appropriate service manual. On all vehicles with an electric cooling fan, override the controls by using jumper wires to make the fan run constantly. If the fan control cannot be bypassed, disconnect the fan. Be careful not to run the engine with a disabled cylinder for more than 15 seconds. The unburned fuel in the exhaust can build up in the catalytic converter and create an unsafe situation. Also run the engine for at least 10 seconds between testing individual cylinders.

Connect the engine analyzer's leads according to the manufacturer's instructions. Turn the engine on and allow it to reach normal operating temperature. Set the engine speed at 1,000 rpm and connect a vacuum gauge to the intake manifold. As each cylinder is shorted, note and record the rpm drop and the change in vacuum.

As each cylinder is shorted, a noticeable drop in engine speed should be noted. Little or no decrease in speed indicates a weak cylinder. If all of the readings are fairly close to each other, the engine is in good condition. If the readings from one or more cylinders differ from the rest, there is a problem. Further testing

Figure 9-41 The vacuum gauge is connected to the intake manifold where it reads engine vacuum.

may be required to identify the exact cause of the problem.

Vacuum Tests

Measuring intake manifold vacuum is another way to diagnose the condition of an engine. Vacuum is formed by the downward movement of the pistons during their intake stroke. If the cylinder is sealed, a maximum amount will be formed.

Manifold vacuum is tested with a vacuum gauge. The gauge's hose is connected to a vacuum fitting on the intake manifold (**Figure 9–41**). Normally a "tee" fitting and short piece of vacuum hose are used to connect the gauge.

Vacuum gauge readings (**Figure 9–42**) can be interpreted to identify many engine conditions, including the ability of the cylinder to seal, the timing of the opening and closing of the engine's valves, and ignition timing.

Ideally each cylinder of an engine will produce the same amount of vacuum; therefore, the vacuum gauge reading should be steady and give a reading of at least 17 inches of mercury (in. Hg). If one or more cylinders produce more or less vacuum than the others, the needle of the gauge will fluctuate. The intensity of the fluctuation indicates the severity of the problem. For example, if the reading on the vacuum gauge fluctuates between 10 and 17 in. Hg we should look at the rhythm of the needle. If the needle seems to stay at 17 most of the time but drops to 10 and quickly rises, we know that the reading is probably caused by a problem in one cylinder. Fluctuating or low readings can indicate many different problems. For example, a low, steady reading might be caused by retarded ignition timing or incorrect valve timing. A sharp vacuum drop at regular intervals might be caused by a burned intake valve. Other conditions that can be revealed by vacuum readings follow:

- Stuck or burned valves
- Improper valve or ignition timing
- Weak valve springs
- Faulty PCV, EGR, or other emission-related system
- Uneven compression
- Worn rings or cylinder walls
- Leaking head gaskets
- Vacuum leaks
- Restricted exhaust system
- Ignition defects

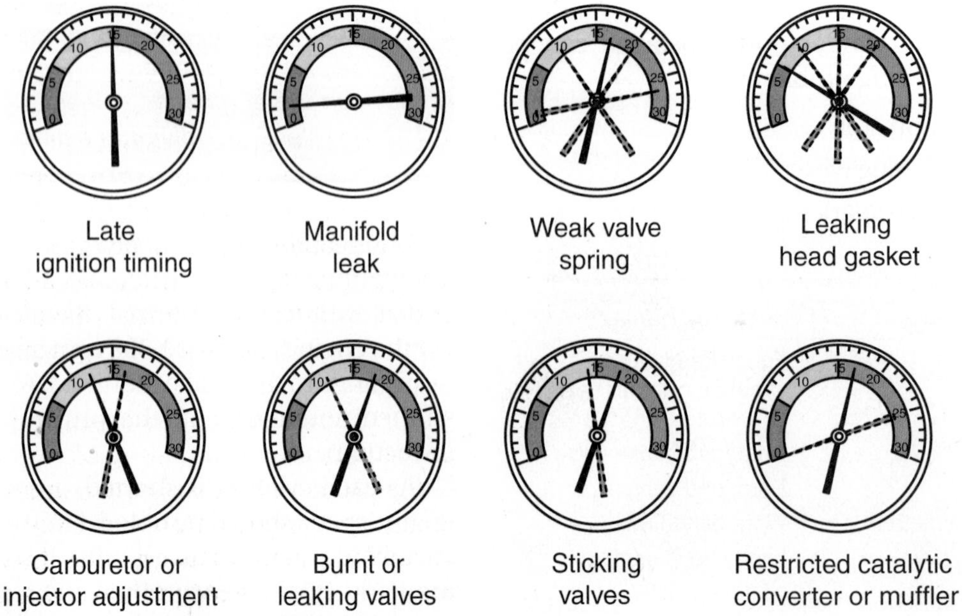

Figure 9-42 Vacuum gauge readings and the engine condition indicated by each.

Oil Pressure Testing

An oil pressure test is used to determine the wear of an engine's parts. The oil pressure test is performed with an oil pressure gauge, which measures the pressure of the oil as it circulates through the engine. Basically, the pressure of the oil depends on the efficiency of the oil pump and the clearances through which the oil flows. Excessive clearances, most often caused by wear between a shaft and its bearings, will cause a decrease in oil pressure.

Loss of performance, excessive engine noise, and poor starting can be caused by abnormal oil pressure. When the engine's oil pressure is too low, premature wear of its parts will result.

An oil pressure tester is a gauge with a high-pressure hose attached to it. The scale of the gauge typically reads from 0 to 100 psi (0 to 690 kPa). Using the correct fittings and adapters, the hose is connected to an oil passage in the engine block. The test normally includes the following steps:

1. Remove the oil pressure sensor **(Figure 9–43)** and tighten the threaded end of the gauge's hose into that bore.
2. Run the engine until it reaches normal operating temperature.
3. Observe the gauge reading while the engine is running at about 1,000 rpm and at 2,500 rpm (or the specified engine speed).
4. Compare the readings to the manufacturer's specifications.

Excessive bearing clearances are not the only possible causes for low oil pressure readings; others are oil pump-related problems, a plugged oil pickup screen, weak or broken oil pressure relief valve, low oil level, contaminated oil, or low oil viscosity.

Figure 9-43 The oil pressure gauge is installed into the oil pressure sending unit's bore in the engine block.

Higher than normal readings can be caused by too much oil, cold oil, high oil viscosity, restricted oil passages, and a faulty pressure regulator.

Oil Pressure Warning Lamp The instrument panel of most vehicles has an oil pressure warning lamp that lights when the oil pressure drops below a particular amount. This lamp should turn on when the ignition key is initially turned to the *on* position and the engine is not running. Once the engine starts, the lamp should go out. If the lamp fails to turn off, there may be an oil pressure problem or a fault in the warning lamp electrical circuit. To determine if the problem is the engine, conduct an oil pressure test. If there is normal oil pressure, the cause of the lamp staying on is an electrical problem.

EVALUATING THE ENGINE'S CONDITION

Once the compression, cylinder leakage, vacuum, and power balance tests are performed, a technician is ready to evaluate the engine's condition. For example, an engine with good relative compression but high cylinder leakage past the rings is typical of a high-mileage worn engine. This engine would have these symptoms: excessive blowby, lack of power, poor performance, and reduced fuel economy.

If these same compression and leakage test results are found on an engine with comparatively low mileage, the problem is probably stuck piston rings that are not expanding properly. If this is the case, try treating the engine with a combustion chamber cleaner, oil treatment, or engine flush. If this fails to correct the problem, an engine overhaul is required.

A cylinder that has poor compression but minimal leakage indicates a valve train problem. Under these circumstances, a valve might not be opening at the right time, might not be opening enough, or might not be opening at all. This condition can be confirmed on engines with a pushrod-type valve train by pulling the rocker covers and watching the valves operate while the engine is cycled. If one or more valves fail to move, either the lifters are collapsed or the cam lobes are worn. If all of the cylinders have low compression with minimal leakage, the most likely cause is incorrect valve timing.

If compression and leakage are both good, but the power balance test reveals weak cylinders, the cause of the problem is outside the combustion chamber. Assuming there are no ignition or fuel problems, check for broken, bent, or worn valve train components, collapsed lifters, leaking intake manifold, or excessively leaking valve guides. If the latter is suspected,

Figure 9-44 Oil leaking from around the oil pan gasket.

squirt some oil on the guides. If they are leaking, blue smoke will be seen in the exhaust.

Fluid Leaks

When inspecting the engine, check it for leaks (**Figure 9–44**). There are many different fluids under the hood of an automobile so care must be taken to identify the type of fluid that is leaking (**Figure 9–45**). Carefully look at the top and sides of the engine, and note any wet residue that may be present. Sometimes road dirt will mix with the leaking fluid and create a heavy coating. Also look under the vehicle for signs of leaks or drips; make sure you have good lighting. Note the areas around the leaks and identify the possible causes. Methods for positively identifying the source of leaks from various components are covered later in this section. All leaks should be corrected because they can result in more serious problems.

Description	Probable Source
Honey or dark greasy fluid	Engine oil
Honey or dark thick fluid with a chestnut smell	Gear oil
Green, sticky fluid	Engine coolant
Slippery clear or yellowish fluid	Brake fluid
Slippery red fluid	Transmission or power-steering fluid
Bluish watery fluid	Washer fluid

Figure 9-45 Identification of fluid leaks.

Sometimes smell will identify the fluid. Gasoline evaporates when it leaks out and may not leave any residue, but it is easy to identify by its smell.

Exhaust Smoke Diagnosis

Examining and interpreting the vehicle's exhaust can give clues of potential engine problems. Basically there should be no visible smoke coming out of the tailpipe. There is an exception to this rule, however, on a cold day after the vehicle has been idling for awhile, it is normal for white smoke to come out of the tailpipe. This is nothing else but the water that has condensed in the exhaust system becoming steam. However, the steam should stop once the engine reaches normal operating temperature. If it does not, a problem is indicated. The color of the exhaust is used to diagnose engine concerns (**Figure 9–46**).

Engine Type	Visible Sign	Diagnosis	Probable Causes
Gasoline	Gray or black smoke	Incomplete combustion or excessively rich A/F mixture	■ Clogged air filter ■ Faulty fuel injection system ■ Faulty emission control system ■ Ignition problem ■ Restricted intake manifold
Diesel	Gray or black smoke	Incomplete combustion	■ Clogged air filter ■ Faulty fuel injection system ■ Faulty emission control system ■ Wrong grade of fuel ■ Engine overheating
Gasoline and Diesel	Blue smoke	Burning engine oil	■ Oil leaking into combustion chamber ■ Worn piston rings, cylinder walls, valve guides, or valve stem seals ■ Oil level too high
Gasoline	White smoke	Coolant/water is burning in the combustion chamber	■ Leaking head gasket ■ Cracked cylinder head or block
Diesel	White smoke	Fuel is not burning	■ Faulty injection system ■ Engine overheating

Figure 9-46 Exhaust analysis.

NOISE DIAGNOSIS

More often than not, malfunction in the engine will reveal itself first as an unusual noise. This can happen before the problem affects the driveability of the vehicle. Problems such as loose pistons, badly worn rings or ring lands, loose piston pins, worn main bearings and connecting rod bearings, loose vibration damper or flywheel, and worn or loose valve train components all produce telltale sounds. Unless the technician has experience in listening to and interpreting engine noises, it can be very hard to distinguish one from the other.

Customer Care

When attempting to diagnose the cause of abnormal engine noise, it may be necessary to temper the enthusiasm of a customer who thinks they have pinpointed the exact cause of the noise using nothing more than their own two ears. While the owner's description may be helpful (and should always be asked for), it must be stressed that one person's "rattle" can be another person's "thump." You are the professional. The final diagnosis is up to you. If customers have been proved correct in their diagnosis, make it a point to tell them so. Everyone feels better about dealing with an automotive technician who listens to them.

When correctly interpreted, engine noise can be a very valuable diagnostic aid. For one thing, a costly and time-consuming engine teardown might be avoided. Always make a noise analysis before doing any repair work. This way, there is a much greater likelihood that only the necessary repair procedures will be done.

Using a Stethoscope

Some engine sounds can be easily heard without using a listening device, but others are impossible to hear unless amplified. A stethoscope **(Figure 9–47)** is very helpful in locating engine noise by amplifying the sound waves. It can also distinguish between normal and abnormal noise. The procedure for using a stethoscope is simple. Use the metal prod to trace the sound until it reaches its maximum intensity. Once the precise location has been discovered, the sound can be better evaluated. A sounding stick, which is nothing more than a long, hollow tube, works on the same principle, though a stethoscope gives much clearer results.

Figure 9-47 Using a stethoscope helps to identify the source of an abnormal noise.

The best results, however, are obtained with an electronic listening device. With this tool you can tune into the noise. Doing this allows you to eliminate all other noises that might distract or mislead you.

WARNING!

Be very careful when listening for noises around moving belts and pulleys at the front of the engine. Keep the end of the hose or stethoscope probe away from the moving parts. Physical injury can result if the hose or stethoscope is pulled inward or flung outward by moving parts.

Common Noises

Figure 9–48 gives examples of abnormal engine noises, including a description of the sound, and their likely causes. An important point to keep in mind is that insufficient lubrication is the most common cause of engine noise. For this reason, always check the fluid levels first before moving on to other areas of the vehicle. Some noises are more pronounced on a cold engine because clearances are greater when parts are not expanded by heat. Remember that aluminum and iron expand at different rates as temperatures rise. For example, a knock that disappears as the engine warms up probably is piston slap or knock. An aluminum piston expands more than the iron block, allowing the piston to fit more closely as engine temperature rises.

Also keep in mind that loose accessories, cracked flexplates, loose bolts, bad belts, broken mechanical fuel pump springs, and other noninternal engine problems can be mistaken for more serious internal

Type	Sound	Mostly Heard During	Possible Causes
Ring noise	High-pitched rattle or clicking	Acceleration	■ Worn piston rings ■ Worn cylinder walls ■ Broken piston ring lands ■ Insufficient ring tension
Piston slap	Hollow, bell-like	Cold engine operation and is louder during acceleration	■ Worn piston rings ■ Worn cylinder walls ■ Collapsed piston skirts ■ Misaligned connecting rods ■ Worn bearings ■ Excessive piston to wall clearance ■ Poor lubrication
Piston pin knock	Sharp, metallic rap	Hot engine operation at idle	■ Worn piston pin ■ Worn piston pin boss ■ Worn piston pin bushing ■ Lack of lubrication
Main bearing noise	Dull, steady knock	Louder during acceleration	■ Worn bearings ■ Worn crankshaft
Rod bearing noise	Light tap to heavy knocking or pounding	Idle speeds and low load higher speeds	■ Worn bearings ■ Worn crankshaft ■ Misaligned connecting rod ■ Lack of lubrication
Thrust bearing noise	Heavy thumping	Irregular sound, may be heard only during acceleration	■ Worn thrust bearing ■ Worn crankshaft ■ Worn engine saddles
Tappet noise	Light regular clicking	Mostly heard during idle	■ Improper valve adjustment ■ Worn or damaged valve train ■ Dirty hydraulic lifters ■ Lack of lubrication
Timing chain noise	Severe knocking	Increases with increase in engine speed	■ Loose timing chain

Figure 9-48 Common engine noises.

engine problems. Always attempt to identify the exact source before completing your diagnosis. In most cases, the source of internal engine noises is best identified by tearing down the engine and inspecting all parts.

Abnormal Combustion Noises

Detonation and preignition noises are caused by abnormal engine combustion. **Detonation** knock or **ping** is a noise most noticeable during acceleration with the engine under load and running at normal temperature. Detonation occurs when part of the air-fuel mixture begins to ignite on its own. This results in the collision of two flame fronts (**Figure 9–49**). One flame front is the normal front moving from the spark plug tip. The other front begins at another point in the combustion chamber. The air-fuel mixture at that point is ignited by heat, not by the spark. The colliding flame fronts cause high-frequency shock waves (heard as a knocking or pinging sound) that could cause physical damage to the pistons, valves, bearings, and spark plugs. Excessive detonation can be very harmful to the engine.

Detonation is usually caused by excessively advanced ignition timing, engine overheating, excessively lean mixtures, or the use of gasoline with too low of an octane rating. A malfunctioning EGR valve can also cause detonation and even rod knock.

Another condition that also causes pinging or spark knocking is called **preignition**, which occurs when combustion begins before the spark plug fires

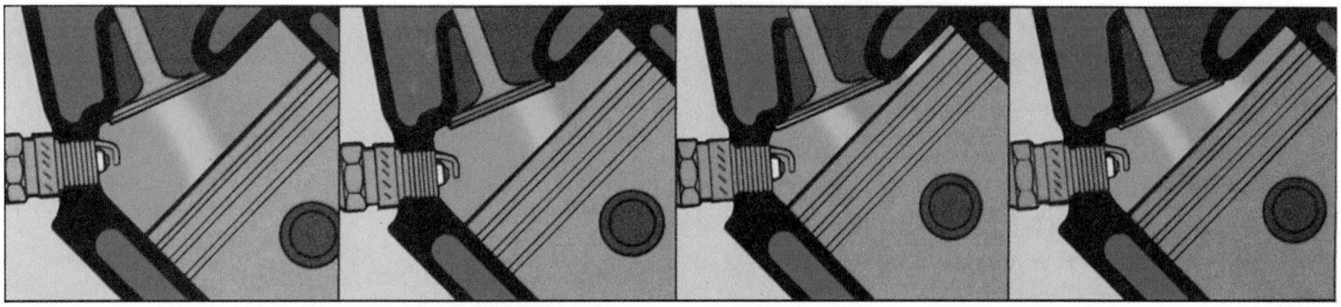

Figure 9–49 Detonation. *Courtesy of Federal-Mogul Corporation*

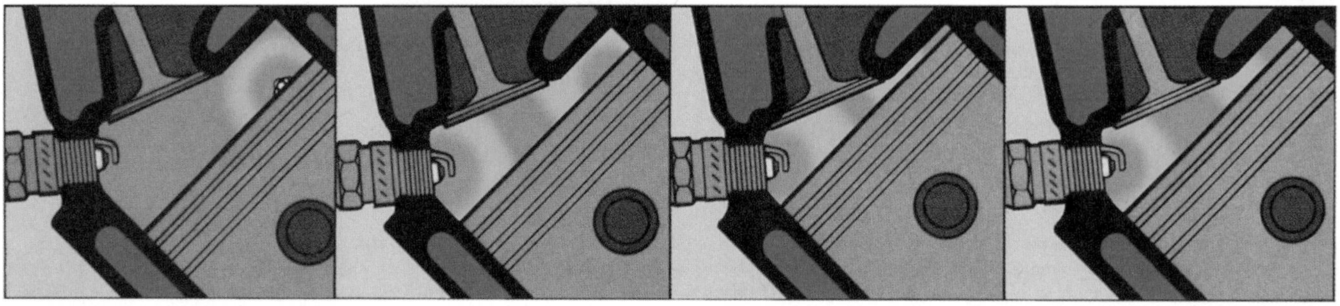

Figure 9–50 Preignition. *Courtesy of Federal-Mogul Corporation*

(**Figure 9–50**). Any hot spot within the combustion chamber can cause preignition. Common causes of preignition are incandescent carbon deposits in the combustion chamber, a faulty cooling system, too hot of a spark plug, poor engine lubrication, and cross firing. Preignition can lead to detonation; however, preignition and detonation are two separate events. Preignition normally does not cause engine damage; detonation does.

Sometimes abnormal combustion causes engine parts to make an abnormal noise. For example, *rumble* is a term that is used to describe the knock or noise resulting from abnormal ignition. Rumble is a vibration of the crankshaft and connecting rods that is caused by multisurface ignition. This is a form of preignition in which several flame fronts occur simultaneously from overheated deposit particles.

Multisurface ignition causes a tremendous sudden pressure rise near TDC. It has been reported that the rate of pressure rise during rumble is five times the rate of normal combustion.

Cleaning Carbon Deposits A buildup of carbon on the top of the piston, intake valve, or in the combustion chamber can cause a number of driveability concerns, including preignition. There are a number of techniques used to remove or reduce the amount of carbon inside the engine. One way, of course, is to disassemble the engine and remove the carbon with a scraper or wire wheel. Two other methods are more commonly used. One is simply adding chemicals to the fuel. These chemicals work slowly so do not expect quick results.

The other method requires more labor but is more immediately effective. This uses a carbon blaster, which is a machine that uses compressed air to force crushed walnut shells into the cylinders. The shells beat on the piston top and combustion chamber walls to loosen and remove the carbon. Basically to use a carbon blaster, the intake manifold and spark plugs are removed. The output hose of the blaster is attached to a cylinder's intake port or inserted into the bore for the fuel injector. A hose is inserted into the spark plug bore; this is where the shells and carbon exit the cylinder. Once connected to the cylinder, the blaster forces a small amount of shells in and out of the cylinder. Hopefully, the carbon deposits leave with the shells. To help remove any remaining bits of shells, compressed air is applied to the cylinder. This operation is done at each cylinder. It is important to note that any remaining shell bits will be burned once the engine is run again.

CASE STUDY

A customer complains that the engine has a rough idle and does not have as much power as it used to. The customer also says that the engine is difficult to start when it is cold and runs better while cruising at highway speeds. On examining the car, it is found that it has a four-cylinder engine with nearly 65,000 miles on it. Driving the car verifies the customer's complaint.

Diagnosis begins with a visual inspection that reveals only that the car has been well maintained. Because there is an endless list of possible causes for these problems, the first test is for engine vacuum. With the vacuum gauge connected to the manifold and the engine at idle speed, the gauge's needle constantly drops from a normal reading to 10 in. Hg. The rhythm of the drop matches the rhythm of the idle. The behavior of the gauge indicates a probable problem in one cylinder.

To verify this diagnosis, a cylinder power balance test is conducted. The results show an engine speed drop of 100 to 125 rpm when cylinders 1, 2, and 4 are shorted, and a drop of only 10 rpm when cylinder 3 is shorted. Based on this test, cylinder 3 is identified as having the problem.

To identify the exact fault, further testing is required. The spark plugs are removed and inspected. All look normal, including the one from cylinder 3. Next, a compression test is taken. All cylinders have normal readings. A cylinder leakage test is then conducted and it too shows normal conditions. The results of the power balance, compression, and cylinder leakage tests lead to the conclusion that the cause has to be in the valve train. Something is preventing a valve from opening.

Removing the cam cover for a visual inspection leads to the discovery of the fault: The intake lobe for cylinder 3 on the camshaft is severely worn. A replacement of the camshaft and matching lifter will correct the problem. The worn lobe only affects the opening of the valve and does not prevent it from sealing, which is why the compression and cylinder leakage test results were normal. Cylinder power and vacuum are affected by the valve not opening fully.

KEY TERMS

Atkinson cycle
Bore
Bottom dead center (BDC)
Cam
Combustion chamber
Crank throw
Detonation
Diesel engine
Direct injection (DI)
Double overhead camshaft (DOHC)
Efficiency
Firing order
Glow plug
Homogeneous charge compression ignition (HCCI)
Lobe
Overhead camshaft (OHC)
Overhead valve (OHV)
Oversquare
Ping
Preignition
Rotary
Selective catalytic reduction (SCR)
Stroke
Top dead center (TDC)
Undersquare
Wankel

SUMMARY

- Automotive engines are classified by several different design features such as operational cycles, number of cylinders, cylinder arrangement, valve train type, valve arrangement, ignition type, cooling system, and fuel system.
- The basis of automotive gasoline engine operation is the four-stroke cycle. This includes the intake stroke, compression stroke, power stroke, and exhaust stroke. The four strokes require two full crankshaft revolutions.
- The most popular engine designs are the inline (in which all the cylinders are placed in a single row) and V-type (which features two rows of cylinders). The slant design is much like the inline, but the entire block is placed at a slant. Opposed cylinder engines use two rows of cylinders located opposite the crankshaft.
- The two basic valve and camshaft placement configurations currently in use on four-stroke engines are the overhead valve and overhead cam.
- Bore is the diameter of a cylinder, and stroke is the length of piston travel between top dead center (TDC) and bottom dead center (BDC). Together these two measurements determine the displacement of the cylinder.
- Compression ratio is a measure of how much the air and fuel are compressed during the compression stroke.

- In an Atkinson cycle engine the intake valve is held open longer than normal during the compression stroke. As a result, the amount of mixture in the cylinder and the engine's effective displacement and compression ratio are reduced.
- Diesel engines have compression ignition systems. Rather than relying on a spark for ignition, a diesel engine uses the heat produced by compressing the intake air to ignite the fuel.
- A hybrid electric vehicle has two different types of power or propulsion systems: an internal combustion engine and an electric motor.
- A compression test is conducted to check a cylinder's ability to seal and therefore its ability to compress the air-fuel mixture inside the cylinder.
- A cylinder leakage test is performed to measure the percentage of compression lost and to help locate the source of leakage.
- A cylinder power balance test reveals whether all of an engine's cylinders are producing the same amount of power.
- Vacuum gauge readings can be interpreted to identify many engine conditions, including the ability of the engine's cylinders to seal, the timing of the opening and closing of the engine's valves, and ignition timing.
- An oil pressure test measures the pressure of the engine's oil as it circulates throughout the engine. This test is very important because abnormal oil pressures can cause a host of problems, including poor performance and premature wear.
- Carefully observing the exhaust can aid engine diagnosis.
- An engine malfunction often reveals itself as an unusual noise. When correctly interpreted, engine noise can be a very helpful diagnostic aid.

REVIEW QUESTIONS

1. What occurs in the combustion chamber of a four-stroke engine?
2. Name the four strokes of a four-stroke cycle engine.
3. As an engine's compression ratio increases, what should happen to the octane rating of the gasoline?
4. What test can be performed to check the efficiency of individual cylinders?
5. Describe tappet noise.
6. Which of the following statements about engines is *not* true?
   a. The engine provides the rotating power to drive the wheels through the transmission and driving axle.
   b. Only gasoline engines are classified as internal combustion.
   c. The combustion chamber is the space between the top of the piston and the cylinder head.
   d. For the combustion in the cylinder to take place completely and efficiently, air and fuel must be combined in the right proportions.
7. Which stroke in the four-stroke cycle begins as the compressed fuel mixture is ignited in the combustion chamber?
   a. power stroke
   b. exhaust stroke
   c. intake stroke
   d. compression stroke
8. *True or False?* In an HCCI engine, combustion immediately and simultaneously produces a steady flame across the mixture.
9. Which of the following is *not* a true statement about diesel engines?
   a. The operation and main components of a diesel engine are comparable to those of a gasoline engine.
   b. Diesel and gasoline engines are available as four-stroke combustion cycle engines.
   c. Diesel engines rely on glow plugs instead of spark plugs to initiate ignition.
   d. The compression ratio of diesel engines is typically three times that of a gasoline engine.
10. *True or False?* When looking at the front of a running engine, it will be rotating in a counterclockwise direction.
11. Which engine system removes burned gases and limits noise produced by the engine?
    a. exhaust system
    b. emission control system
    c. ignition system
    d. air-fuel system
12. In a six-cylinder engine with a firing order of 1-4-5-2-3-6, what stroke is piston #5 on when #1 is on its compression stroke?
13. The stroke of an engine is ____________ the crank throw.
    a. half
    b. twice
    c. four times
    d. equal to

14. *True or False?* The camshaft for a two-stroke cycle engine is always located in the engine block.
15. Which of the following is an expression of how much of the heat formed during the combustion process is available as power from the engine?
    a. mechanical efficiency
    b. engine efficiency
    c. volumetric efficiency
    d. thermal efficiency

ASE-STYLE REVIEW QUESTIONS

1. While diagnosing the cause for black smoke from the tailpipe of a gasoline engine: Technician A says that a faulty fuel injection system is a likely cause. Technician B says that it is most likely caused by oil leaking into the combustion chamber. Who is correct?
    a. Technician A   c. Both A and B
    b. Technician B   d. Neither A nor B
2. While discussing piston slap: Technician A says that it is a high-pitched clicking that becomes louder during deceleration caused by detonation. Technician B says that it is the noise made by the piston when it contacts the cylinder wall due to excessive clearances. Who is correct?
    a. Technician A   c. Both A and B
    b. Technician B   d. Neither A nor B
3. While determining the cause for air leaking out of the oil dipstick tube during a cylinder leakage test: Technician A says that a burnt exhaust or intake valve is indicated. Technician B says that a warped cylinder head or bad head gasket is indicated. Who is correct?
    a. Technician A   c. Both A and B
    b. Technician B   d. Neither A nor B
4. While diagnosing the cause for an engine having good results from a compression test and cylinder leakage test but poor results from a cylinder power balance test: Technician A says that incorrect valve timing is the most likely cause. Technician B says that a collapsed lifter is a likely cause. Who is correct?
    a. Technician A   c. Both A and B
    b. Technician B   d. Neither A nor B
5. While looking at the results of an oil pressure test: Technician A says that higher than normal readings can be caused by a defective pressure regulator. Technician B says that higher than normal readings can be expected on a cold engine. Who is correct?
    a. Technician A   c. Both A and B
    b. Technician B   d. Neither A nor B
6. A vehicle is producing a sharp, metallic rapping sound originating in the upper portion of the engine. It is most noticeable during idle. Technician A diagnoses the problem as piston pin knock. Technician B says that the problem is most likely a loose crankshaft thrust bearing. Who is correct?
    a. Technician A   c. Both A and B
    b. Technician B   d. Neither A nor B
7. While conducting an engine vacuum test: Technician A says that a steady low vacuum reading can be caused by a burned intake valve. Technician B says that an overall low vacuum reading is caused by something that affects all of the engine's cylinders. Who is correct?
    a. Technician A   c. Both A and B
    b. Technician B   d. Neither A nor B
8. While determining the most likely problem of an engine with poor compression test results but acceptable cylinder leakage readings: Technician A says that the problem may be incorrect valve timing. Technician B says that the problem is a leaking valve guide. Who is correct?
    a. Technician A   c. Both A and B
    b. Technician B   d. Neither A nor B
9. When a customer refers to the engine component that opens and closes the intake and exhaust valves: Technician A believes that the customer is referring to the camshaft. Technician B thinks that the component in question is the intake manifold. Who is correct?
    a. Technician A   c. Both A and B
    b. Technician B   d. Neither A nor B
10. Technician A says that if an engine had good results from a compression test, it will have good results from a cylinder leakage test. Technician B says that if an engine had good results from a cylinder leakage test, it will have good results from a compression test. Who is correct?
    a. Technician A   c. Both A and B
    b. Technician B   d. Neither A nor B

ENGINE DISASSEMBLY AND CLEANING

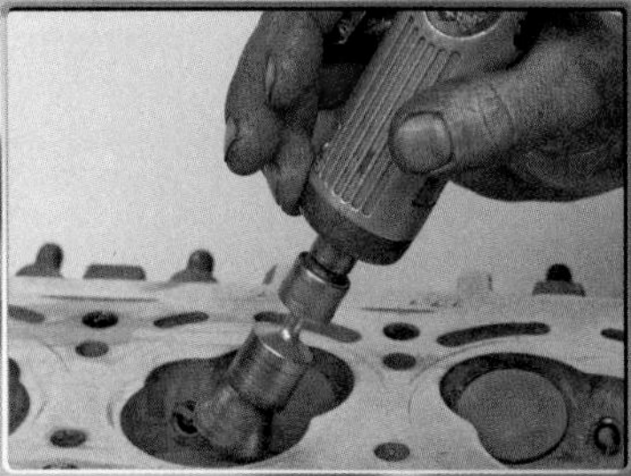

CHAPTER 10

OBJECTIVES

■ Prepare an engine for removal. ■ Remove an engine from a FWD and a RWD vehicle. ■ Separate the engine into its basic components. ■ Name the three basic cleaning processes.

This chapter covers the removal and installation of an engine. It also covers basic disassembly and general cleaning of the components. The material is presented so that it applies not only to engine rebuilding, but also to the replacement of individual parts when a total rebuild is not necessary. Complete disassembly and assembly of the engine block and cylinder head are covered in Chapters 11 and 12.

Before removing the engine, clean it and the area around it. Also, check the service manual for the correct procedure for removing the engine from a particular vehicle. Make sure you adhere to all precautions given by the manufacturer.

REMOVING AN ENGINE

Make sure you have the tools and equipment required for the job before you begin. In addition to hand tools and some special tools, you will need an engine hoist or crane **(Figure 10–1)** and a jack.

Figure 10–1 To pull an engine out of a vehicle, the chain on the lifting crane is attached to another chain secured to the engine.

The basic procedures for engine removal vary, depending on whether the engine is removed from the bottom of the vehicle or through the hood opening. Many FWD vehicles require removal of the engine from the bottom, whereas most RWD vehicles require the engine to come out from the hood opening. The engine exit point is something to keep in mind while you are disconnecting and removing items in preparation for engine removal.

General Procedures

When removing an engine, setting the vehicle on a frame contact hoist is recommended. When the vehicle is sitting on the floor, block the wheels so it does not move while you are working. Open the hood and put fender covers on both front fenders **(Figure 10–2)**. Once the vehicle is in position, relieve the pressure in the fuel system using the procedures given by the manufacturer.

> **Customer Care**
>
> Make sure your hands, shoes, and clothing are clean before getting into a customer's car. Disposable seat and floor coverings should be used to help protect the interior.

Figure 10–2 Before doing anything, put covers on the fenders.

Battery Install a memory saver before you disconnect the battery to prevent the vehicle's computers and other devices from losing what they have stored in their memory. Disconnect all negative battery cables **(Figure 10–3)**, tape their connectors, and place them away from the battery. If the battery will be in the way of engine removal, remove the positive cables and the battery.

Hood The vehicle's hood will get in the way during engine removal. If the hinges allow the hood to be set straight up above the engine compartment, prop it in that position with wood or a broom. Make sure the hood is secure before proceeding. In many cases, the hood should be removed and set aside in a safe place on fender covers or cardboard. Make sure not to damage the vehicle's paint while doing this. Before removing the hood, mark the location of the hinges on the hood. Then unbolt and remove the hood with the help of someone else.

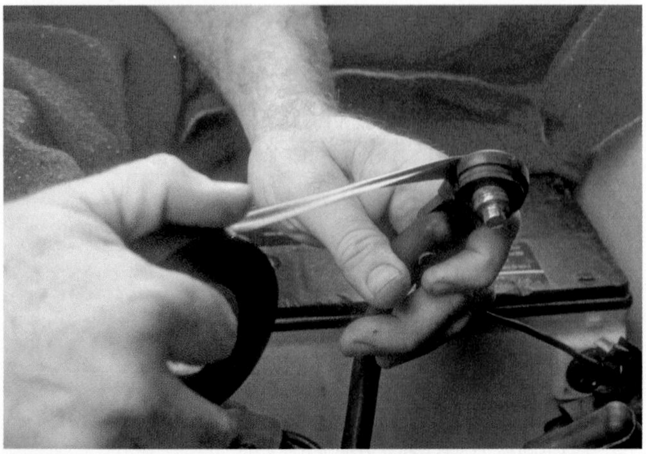

Figure 10–3 After the negative battery cable is disconnected, tape the terminal end to prevent it from accidentally touching the battery.

Figure 10–4 Drain the engine's coolant and recycle it.

SHOP TALK

Often the safest place to store the hood is on the vehicle's roof.

Fluids Drain the engine's oil and remove the oil filter. Then drain the coolant from the radiator **(Figure 10–4)** and engine block, if possible. To increase the flow of the coolant out of the cooling system, remove the radiator cap. Make sure the engine is cool before opening the coolant drain and before removing the radiator cap. After collecting the old coolant, recycle it. If the transmission will be removed with the engine, drain its fluid.

Underbody Connections While you are under the vehicle to drain the fluids, disconnect the shift linkage, transmission cooling lines, all electrical connections, vacuum hoses, and clutch linkages from the transmission **(Figure 10–5)**. If the clutch is

Figure 10–5 If the transmission will be removed with the engine, disconnect all linkages, lines, and electrical connectors.

hydraulically operated, unbolt the slave cylinder and set it aside, if possible. If this is not possible, disconnect and plug the line to the cylinder.

Air-Fuel System Remove the air intake ducts and air cleaner assembly. Disconnect and plug the fuel line at the fuel rail. If the engine is equipped with a return fuel line from the fuel pressure regulator, disconnect that as well **(Figure 10–6)**. Make sure all fuel lines are closed off with pinch pliers or the appropriate plug or cap. Most late-model fuel lines have quick-connect fittings that are separated by squeezing the retainer tabs together and pulling the fitting off the fuel line nipple.

Disconnect all vacuum lines at the engine. Make sure these are labeled before disconnecting them. Most automobiles have a vacuum wiring diagram decal under the hood **(Figure 10–7)**. The diagram and the labels (these could be masking tape with the connecting point written on it) will make it easier to reconnect the hoses when the engine is reinstalled.

Now disconnect the throttle linkage at the throttle body and the electrical connector to the throttle position (TP) sensor.

SHOP TALK

Some technicians are using instant cameras or video recording cameras to help recall the locations of underhood items by taking pictures before work is started. This technique can be quite valuable considering how complex the underhood systems of current cars have become.

Accessories Remove all drive belts **(Figure 10–8)**. Unbolt and move the power steering pump and

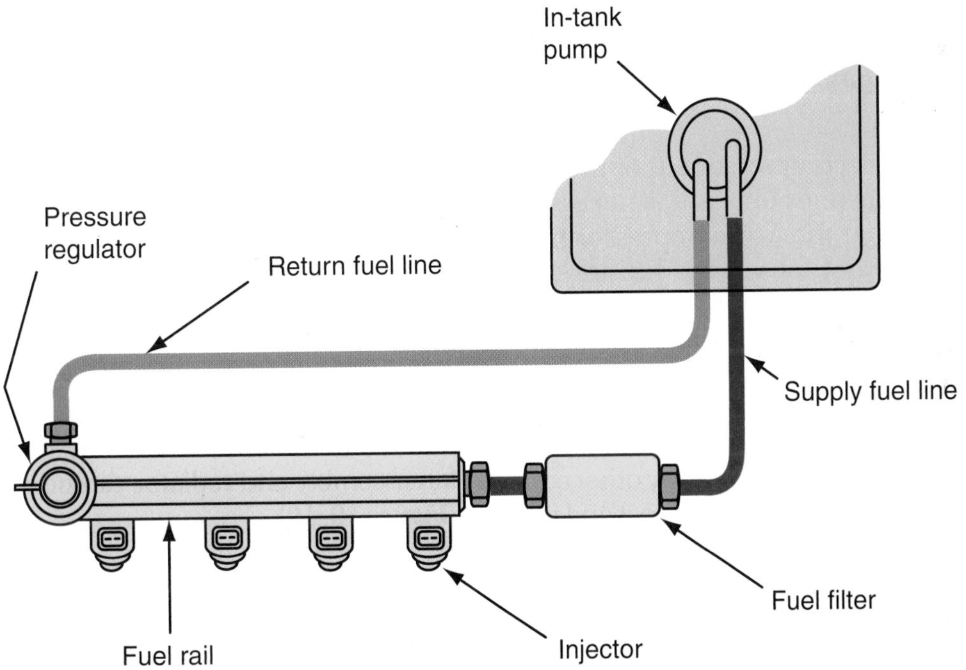

Figure 10–6 Disconnect and plug the fuel lines at the fuel rail and the pressure regulator.

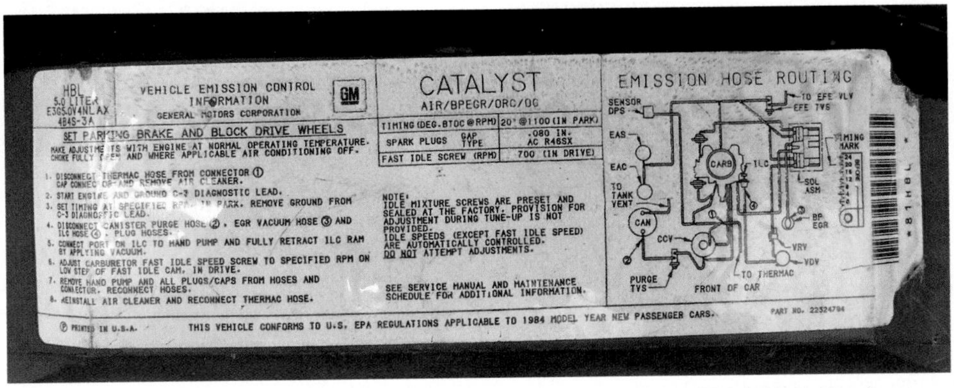

Figure 10–7 An underhood decal showing the routing of vacuum lines for the vehicle.

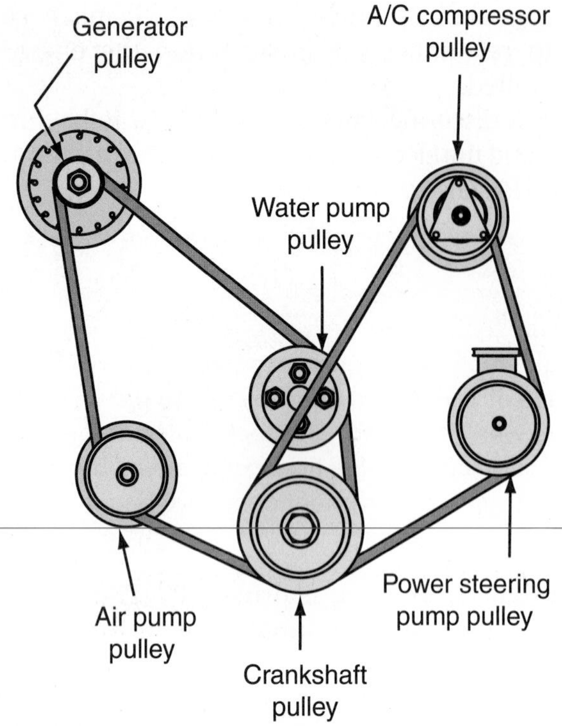

Figure 10–8 Before removing the drive belts, pay attention to the routing of each belt.

Figure 10–9 Before unplugging the electrical wires between the engine and the vehicle, use masking tape as a label to identify all disconnected wires.

air-conditioning (A/C) compressor out of the way. Do not disconnect the lines unless it is necessary. If the pressure hoses at the A/C compressor must be disconnected, do not loosen the fittings until the refrigerant has been captured by a refrigerant reclaimer/recycling machine. Once disconnected, plug the lines and connections at the compressor to prevent dirt and moisture from entering.

Remove or move the A/C compressor bracket, power steering pump, air pump, and any other components attached to the engine. Disconnect and plug all transmission and oil cooler lines.

SHOP TALK

When removing the fasteners, pay close attention to their size and type. The brackets used to secure accessories use different size fasteners. Mark and organize the fasteners so their proper location can be easily identified later. It is a good idea to store the fasteners in different containers, one for each system or section of the engine.

Electrical Connections Unplug all electrical wires between the engine and the vehicle. Use masking tape as a label to identify all wires that are disconnected **(Figure 10–9)**. Some engines have a crankshaft position sensor attached above the flywheel or flexplate. This sensor must be removed before separating the engine from the bell housing. Make sure the engine ground strap is disconnected, preferably at the engine.

Cooling System Disconnect the heater inlet and outlet hoses. Then disconnect the upper and lower radiator hoses. If the radiator is fitted with a fan shroud, carefully remove it along with the cooling fan. If the vehicle is equipped with an electric cooling fan, disconnect the wiring to the cooling fan. Then unbolt and remove the radiator mounting brackets and remove the radiator. Normally the electric cooling fan assembly and radiator can be removed as a unit **(Figure 10–10)**.

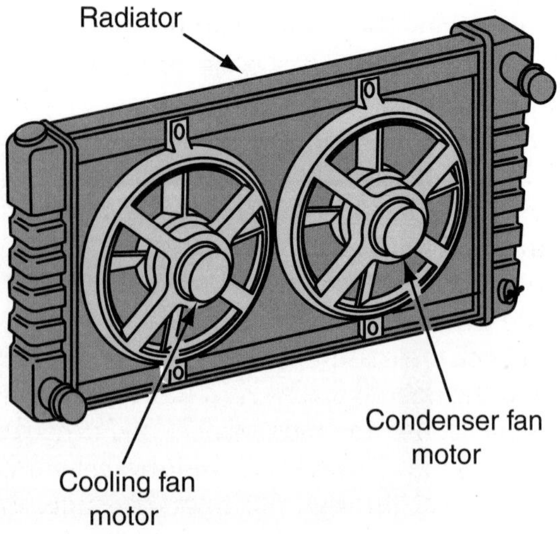

Figure 10–10 Typically the electric cooling fans can be removed as a unit with the radiator.

Figure 10–11 The exhaust system must be disconnected to remove the engine. Often this is the condition of the exhaust, so take care not to damage anything.

Miscellaneous Stuff Disconnect the exhaust system; attempt to do this at the exhaust manifold **(Figure 10–11)**. When disconnecting the exhaust system, make sure the wires connected to the exhaust sensors are disconnected before the system is moved. Remove any heat shields that may be in the way of moving or removing the exhaust system. Now carefully check under the hood to find and remove anything that may interfere with engine removal.

Removing the engine from a RWD vehicle is more straightforward than removing one from a FWD model, because there is typically easy access to the cables, wiring, and bell housing bolts. Engines in FWD cars can be more difficult to remove because large assemblies such as engine cradles, suspension components, brake components, splash shields, or other pieces may need to be disassembled or removed.

FWD Vehicles

Before removing the engine, identify any special tool needs and precautions that are recommended by the manufacturer. Most often the engine in a FWD vehicle is removed through the bottom of the vehicle. Special tools may be required to hold the transaxle and/or engine in place as it is being disconnected from the vehicle **(Figure 10–12)**. Always refer to the service manual before proceeding to remove the transaxle. You will waste much time and energy if you do not check the manual first.

When the engine is removed through the bottom of the vehicle, use an engine cradle and dolly to support the engine. If the manufacturer recommends engine removal through the hood opening, use an engine hoist. Regardless of the method of removal, the engine and transaxle are usually removed as a unit. The transaxle can be separated from the engine once it has been lifted out of the vehicle.

Figure 10–12 A transverse engine support bar provides the necessary support when removing an engine from a FWD vehicle. *Courtesy of SPX Service Solutions*

Figure 10–13 Use a large breaker bar to loosen the axle shaft hub nuts.

Drive Axles Using a large breaker bar, loosen and remove the axle shaft hub nuts **(Figure 10–13)**. It is recommended that these nuts be loosened with the vehicle on the floor and the brakes applied.

Raise the vehicle so you can comfortably work under it. Then remove the wheel and tire assemblies from the front wheels. Tap the splined CV joint shaft with a soft-faced hammer to see if it is loose. Most will come loose with a few taps. Many Ford FWD cars use an interference fit spline at the hub. You will need a special puller for this type of CV joint; the tool pushes the shaft out, and on installation pulls the shaft back into the hub.

Disconnect all suspension and steering parts that need to be removed according to the service manual.

Figure 10-14 Pull the steering knuckle outward to allow the CV joint shaft to slide out of the hub.

Index the parts so wheel alignment will be close after reassembly. Normally the lower ball joint must be separated from the steering knuckle. The ball joint will either be bolted to the lower control arm or the ball joint will be held into the knuckle with a pinch bolt. Once the ball joint is loose, the control arm can be pulled down and the knuckle can be pushed outward to allow the CV joint shaft to slide out of the hub **(Figure 10–14)**.

The inboard joint can then either be pried out or will slide out. Some transaxles have retaining clips that must be removed before the inner joint can be removed. Others have a flange-type mounting. These must be unbolted to remove the shafts. In some cases, flange-mounted drive shafts may be left attached to the wheel and hub assembly and only unbolted at the transmission flange. The free end of the shafts should be supported and placed out of the way.

Pull the drive axles out of the transaxle. While removing the axles, make sure the brake lines and hoses are not stressed. Suspend them with wire to relieve the weight on the hoses and to keep them out of the way.

Transaxle Connections Disconnect all electrical connectors and the speedometer cable at the transaxle. Then disconnect the shift linkage or cables and the clutch cable.

Starter Now, remove the starter. The starter wiring may be left connected, or you can also completely remove the starter from the vehicle to get it totally out of the way. The starter should never be left to hang by the wires attached to it. The weight of the starter can damage the wires or, worse, break the wires and allow the starter to fall, possibly on you or someone else. Always securely support the starter and position it out of the way after you have unbolted it from the engine.

Figure 10-15 Some engines are equipped with eye plates to which the hoist can be safely attached.

Removing the Engine through the Hood Opening Connect the engine sling or lifting chains to the engine. Use the lifting hooks on the engine **(Figure 10–15)** or fasten the sling to the points given in the service manual. Connect the sling to the crane and raise the crane just enough to support the engine.

From under the vehicle, remove the cross member. Then remove the mounting bolts for the engine at the engine and transmission mounts. With the transmission jack supporting the transmission, remove the transaxle mounts.

From under the hood, remove all remaining mounts. Raise the engine slightly to free it from the mounts. Then slowly raise the engine from the engine compartment. Guide the engine around all wires and hoses to make sure nothing gets damaged. Once the engine is cleared from the vehicle, prepare to separate it from the transaxle.

Removing the Engine from under the Vehicle Position the engine cradle and dolly under the engine. Adjust the pegs of the cradle so they fit into the recesses on the bottom of the engine, and secure the engine.

Remove all engine and transmission mount bolts. If required, remove the frame member from the vehicle. It may also be necessary to disconnect the steering gear from the frame. Double-check to ensure that all wires and hoses are disconnected from the engine. With the transmission jack supporting the transmission, remove the transaxle mounts.

Slowly raise the vehicle, lifting it slightly away from the engine. As the vehicle is lifted, the engine remains on the cradle. During this process, continually check for interference with the engine and the body of the vehicle. Also watch for any wires and hoses that may still be attached to the engine. Once the vehicle is clear of the engine, prepare to separate the engine from the transaxle.

RWD Vehicles

The engine is removed through the hood opening with an engine hoist on most RWD vehicles. Refer to the service manual to determine the proper engine lift points. Attach a pulling sling or chain to the engine. Some engines have eye plates for lifting. If they are not available, the sling must be bolted to the engine. The sling attaching bolts must be large enough to support the engine and must thread into the block a minimum of 1½ times the bolt diameter.

If the transmission is being removed with the engine, position the hook of the engine hoist to the lifting chain so that the engine tips a little toward the transmission. Lift the engine slightly and check for any additional things behind and under the engine that should be disconnected.

Transmission If the engine and transmission must be separated before engine removal, remove all clutch (bell) housing bolts. If the vehicle has an automatic transmission, remove the torque converter mounting bolts.

If the transmission is being removed with the engine, place a drain pan under the transmission and drain the fluid from the transmission. Once the fluid is out, move the drain pan under the rear of the transmission. Use chalk to index the alignment of the rear U-joint and the pinion flange **(Figure 10–16)**. Then remove the drive shaft.

Disconnect all electrical connections and the speedometer cable at the transmission. Make sure you place these away from the transmission so they are not damaged during transmission removal or installation.

Disconnect and remove the transmission and clutch linkage. Disconnect the parts of the exhaust system that may get in the way. It is best to do this by disconnecting as little as possible.

Use a transmission jack to securely support the transmission **(Figure 10–17)** and unbolt the motor mounts. If the engine is removed with its transmission, the front of the engine must come straight up as the transmission moves away from the bottom of the vehicle. Remove the transmission mount and cross member **(Figure 10–18)**.

Removing the Engine Center the boom of the crane (hoist) directly over the engine and raise the engine slightly. Make sure the engine is securely fastened to the chain and that nothing else is still attached to the engine. Continue raising the engine while pulling it forward. Make sure that the engine does not bind or damage anything in the engine compartment while doing this. When the engine is high enough to clear the front of the vehicle, roll the crane and engine away from the vehicle.

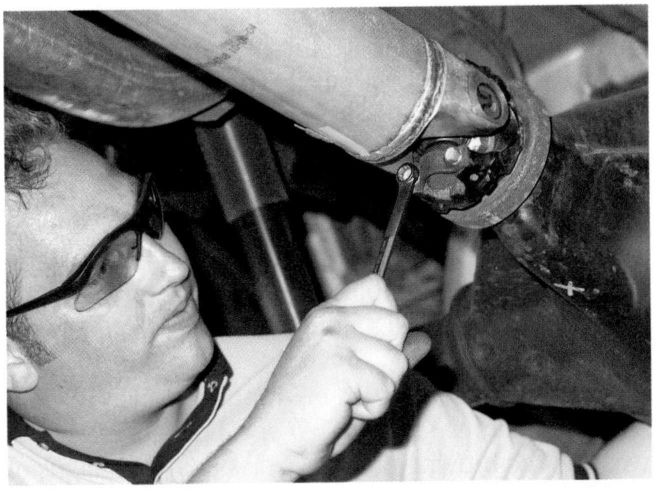

Figure 10-16 Mark the alignment of the rear U-joint and the pinion flange before removing the drive shaft.

Figure 10-17 Use a transmission jack to securely support the transmission before removing the motor mounts.

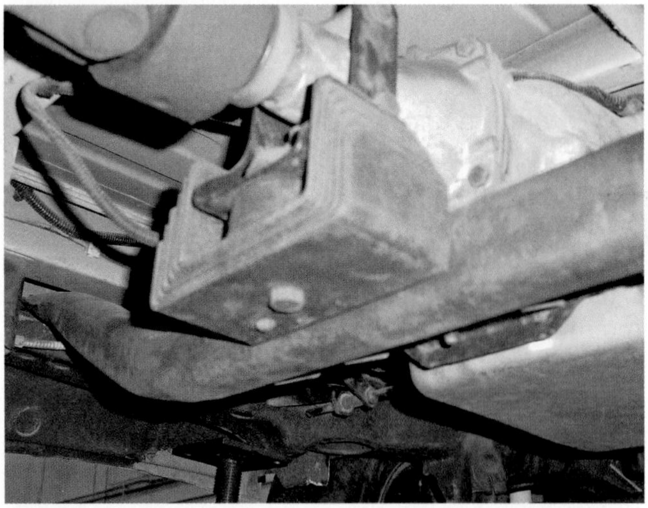

Figure 10-18 If the engine is removed with its transmission, remove the transmission mount and cross member.

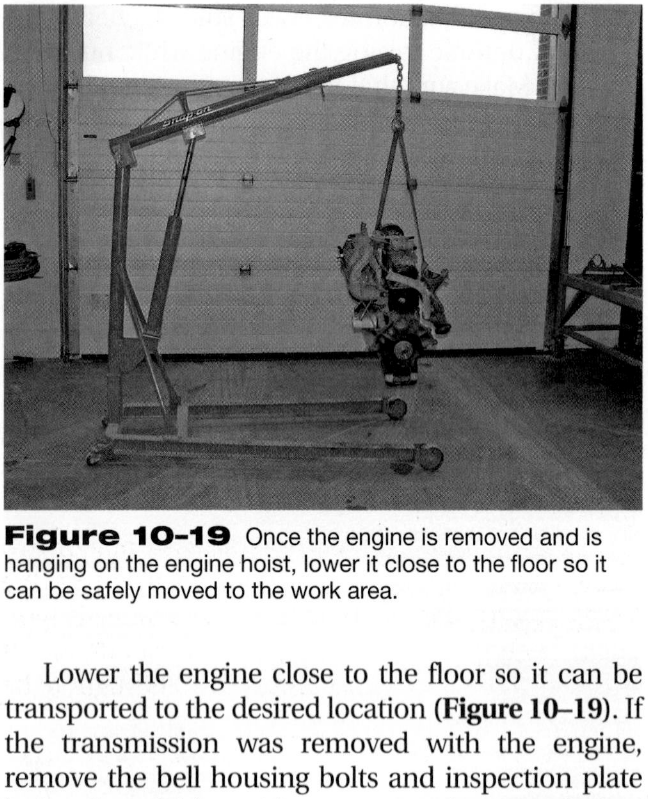

Figure 10-19 Once the engine is removed and is hanging on the engine hoist, lower it close to the floor so it can be safely moved to the work area.

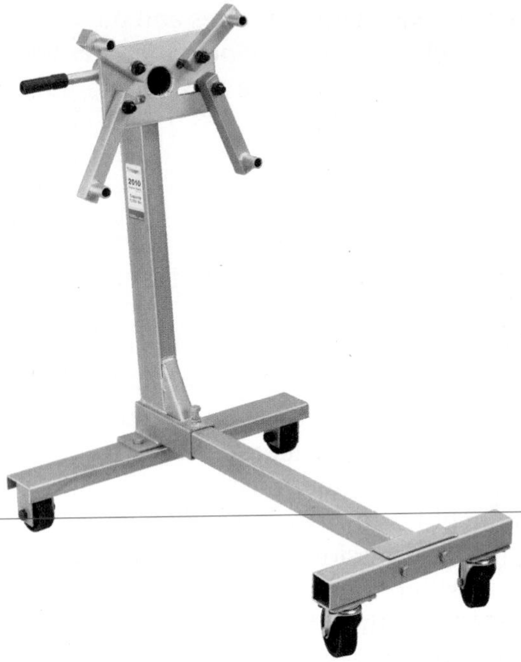

Figure 10-20 A typical engine stand. *Courtesy of SPX Service Solutions*

Lower the engine close to the floor so it can be transported to the desired location **(Figure 10–19)**. If the transmission was removed with the engine, remove the bell housing bolts and inspection plate bolts. On vehicles with an automatic transmission, also remove the torque converter-to-flexplate bolts. Use a C-clamp or other brace to prevent the torque converter from falling. Also mark the location of the torque converter in relation to the flexplate.

ENGINE DISASSEMBLY AND INSPECTION

Raise the engine and position it next to an engine stand **(Figure 10–20)**. Mount the engine to the engine stand with bolts. Most stands use a plate with several holes or adjustable arms. The engine must be supported by at least four bolts that fit solidly into it. The engine should be positioned so that its center is in the middle of the engine stand's adapter plate. This will ensure that the engine is not too heavy when rotated on the engine stand.

In most cases, the flywheel or flexplate must be removed to mount the engine on its stand. Mark the position of the flywheel on the crankshaft. This aids the reassembly of the engine. To do this, loosen—but do not remove—the attaching bolts in a "star" pattern to reduce the chance of distorting the flywheel. At times, the flywheel will turn with the wrench as the bolts are being loosened. When this happens, a flywheel lock should be used to stop the flywheel from turning. Once all of the bolts are loosened, take hold of the flywheel while removing the bolts. The flywheel can be quite heavy and if it falls, you can be injured or the flywheel can be damaged. The flywheel for manual transmissions should be inspected for possible damage and for signs of clutch problems. Place the flywheel on a flat surface.

Once the engine is securely mounted to the engine stand, remove the sling or lifting chain. The engine can now be disassembled and cleaned. Always refer to the service manual before you start to disassemble an engine.

Slowly disassemble the engine and visually inspect each part for any signs of damage. Look for excessive wear on the moving parts. Check all parts for signs of overheating, unusual wear, and chips. Look for signs of gasket and seal leakage.

USING SERVICE INFORMATION

Look up the specific model car and engine prior to disassembling the engine.

Cylinder Head Removal

The first step in disassembly of an engine is usually the removal of the intake and exhaust manifolds. On some inline engines, the intake and exhaust manifolds are often removed as an assembly.

SHOP TALK

It is important to let an aluminum cylinder head cool before removing it.

To start cylinder head removal, remove the valve cover or covers and disassemble the rocker arm components **(Figure 10–21)** according to the guidelines given by the manufacturer. Check the rocker area for sludge. Excessive buildup can indicate a poor maintenance schedule and is a signal to look for wear on other components.

Figure 10-21 Remove the valve cover and disassemble the rocker arm components. Check the rocker area for sludge. Keep the rocker arms or rocker arm assemblies in the order they were installed.

On OHC engines, the timing belt cover must be removed. Under the cover is the timing belt or chain and sprockets. In the service information, there will be a description of the type and location of the timing marks on the crankshaft and camshaft sprockets. If possible, rotate the crankshaft to check the alignment of the sprockets. If the shafts are not aligned, make note of this for later reference. The valves will hit the pistons on some engines when the timing belt or chain slips, skips, or breaks. These engines are commonly called **interference engines**. When the valves hit the piston, they will bend. The valves in **freewheeling engines** will not hit the piston when valve timing is off. However, the keys and keyways in the camshaft sprocket may be damaged.

Interference engines typically have a decal on the cam cover that states the belt must be changed at a particular mileage interval. Potential valve and/or piston damage is the reason why timing belt replacement is recommended.

The belt or chain must be removed before removing the cylinder head. Locate and move the belt's tensioner pulley to remove its tension on the belt. Slip the belt off the camshaft and crankshaft sprockets, if possible.

When removing the cylinder head, keep the pushrods and rocker arms or rocker arm assemblies in exact order. Use an organizing tray or label the parts with a felt-tipped marker to keep them together and labeled accurately. This type of organization greatly aids in diagnosing valve-related problems. Remove the lifters from the block and place them in the order they were installed.

The cylinder head bolts are loosened one or two turns each, following the pattern specified by the manufacturer **(Figure 10–22)**. The sequence is typically the opposite of the tightening sequence. If there is no specified procedure, the bolts ought to be

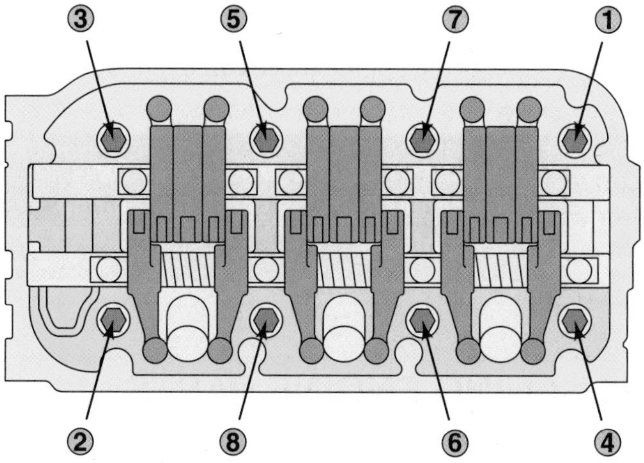

Figure 10-22 When loosening cylinder head bolts; follow the sequence given by the manufacturer.

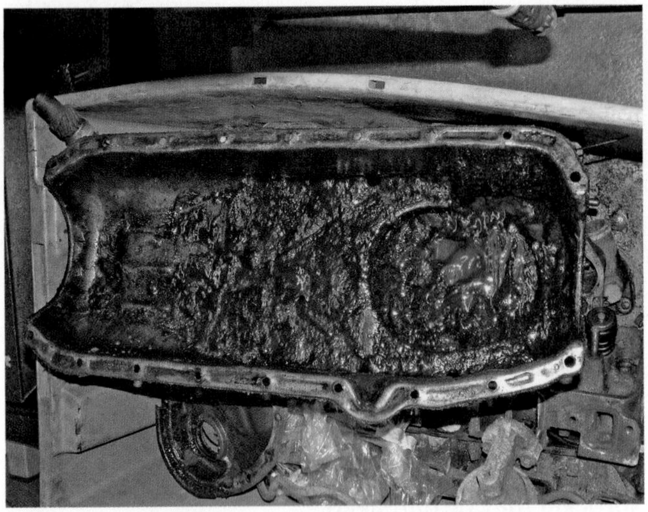

Figure 10–23 A major buildup of sludge on the bottom of this oil pan.

Figure 10–24 From (A) grime to (B) shine.

loosened one or two turns, beginning in the ends and working toward the center. This prevents the distortion that can occur if bolts are all loosened at once. The bolts are then removed and the cylinder head can be lifted off. The cylinder head gasket should be saved to compare with the new head gasket during reassembly. Set the cylinder head(s) on cardboard or another soft surface to prevent damage to the sealing surfaces.

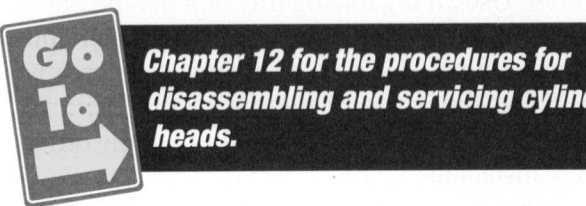

Go To *Chapter 12 for the procedures for disassembling and servicing cylinder heads.*

The water pump is normally mounted to the front of the engine. Unbolt and remove it. Rotate the engine block so the oil pan is up. Remove the pan's attaching bolts. Then lift off the oil pan. Once the pan is removed, look inside for metal shavings and sludge **(Figure 10–23)**. Both of these are indications of problems. Disassembly of the engine block can begin now.

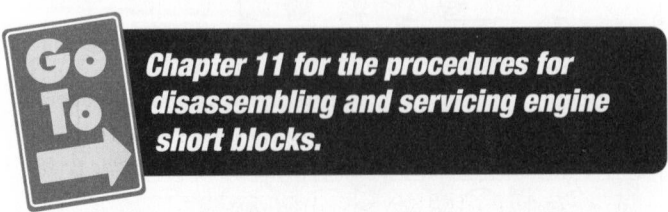

Go To *Chapter 11 for the procedures for disassembling and servicing engine short blocks.*

CLEANING ENGINE PARTS

After the component that needs service has been disassembled, its parts should be thoroughly cleaned **(Figure 10–24)**. The cleaning method depends on the component and the type of equipment available. An incorrect cleaning method or agent can often be more harmful than no cleaning at all. For example, using caustic soda to clean aluminum parts will dissolve the part. **Caustic soda** is a strong detergent that is commonly found in solvents that are effective in removing carbon.

WARNING!

Always wear the appropriate eye protection and gloves when working with cleaning solvents.

Only after all components have been thoroughly and properly cleaned can an effective inspection be made or proper machining be done.

Types of Contaminants

Being able to recognize the type of dirt you are to clean will save you time and effort. Basically there are four types of dirt.

Water-Soluble Soils The easiest dirt to clean is **water-soluble** soils, which includes dirt, dust, and mud.

Organic Soils **Organic soils** contain carbon and cannot be effectively removed with plain water. There are three distinct groupings of organic soils:

- Petroleum by-products derived from crude oil, including tar, road oil, engine oil, gasoline, diesel fuel, grease, and engine oil additives

- By-products of combustion, including carbon, varnish, gum, and sludge
- Coatings, including such items as rust-proofing materials, gasket sealers and cements, paints, waxes, and sound-deadener coatings

Rust Rust is the result of a chemical reaction that takes place when iron and steel are exposed to oxygen and moisture. Corrosion, like rust, results from a similar chemical reaction between oxygen and metal containing aluminum. If left unchecked, both rust and corrosion can physically destroy metal parts quite rapidly. In addition to metal destruction, rust also acts to insulate and prevent proper heat transfer inside the cooling system.

Scale When water containing mineral and deposits is heated, suspended minerals and impurities tend to dissolve, settle out, and attach to the surrounding hot metal surfaces. This buildup of minerals and deposits inside the cooling system is known as scale. Over a period of time, scale can accumulate to the extent that passages become blocked, cooling efficiency is compromised, and metal parts start to deteriorate.

Cleaning with Chemicals

There are three basic processes for cleaning automotive engine parts. The first process that is discussed is chemical cleaning.

This method of cleaning uses chemical action to remove dirt, grease, scale, paint, and/or rust.

CAUTION!

When working with any type of cleaning solvent or chemical, be sure to wear protective gloves and goggles and work in a well-ventilated area. Prolonged immersion of the hands in a solvent can cause a burning sensation. In some cases a skin rash might develop. There is one caution to mention about all manufactured cleaning materials that cannot be overemphasized: Read the labels carefully before mixing or using.

Unfortunately, the most traditional line of defense against soils involves the use of cleaning chemicals. Chlorinated hydrocarbons and mineral spirits may have some health risks associated with their use through skin exposure and inhalation of vapors. Hydrocarbon cleaning solvents are also flammable. The use of water-based nontoxic chemicals can eliminate such risks.

WARNING!

Prior to using any chemical, read through all of the information given on the material safety data sheet (MSDS) or the Canadian workplace hazardous materials information systems sheets (WHMIS) for that chemical. Become aware of the health hazards presented by the various chemicals.

Hydrocarbon solvents are labeled hazardous or toxic and require special handling and disposal procedures. Many water-based cleaning solutions are biodegradable. Once the cleaning solution has become contaminated with grease and grime, it too becomes a hazardous or toxic waste that can be subject to the same disposal rules as a hydrocarbon solvent.

Some manufacturers offer waste-handling and solvent recycling services. The old solvent is recycled by a distillation process to separate the sludge and contaminants. The solvent is then returned to service and the contaminants disposed of. Independent services for maintaining hot tanks and spray washers are also available.

CAUTION!

Care needs to be taken with alloy blocks with sleeves or liners. The different metals react differently to chemicals. Make sure to check with the service manual before using a cleaning solution on these parts. The wrong chemical can cause damage to the block and/or sleeves.

Chemical Cleaning Machines

Parts Washers Parts washers (often called solvent tanks) are one of the most widely used and inexpensive methods of removing grease, oil, and dirt from the metal surfaces of a seemingly infinite variety of automotive components and engine parts. A typical washer setup **(Figure 10–25)** might consist of a tank to hold a given volume of solvent cleaner and some method of applying the solvent. These methods include soaking, soaking and agitation, solvent streams, and spray gun applicators.

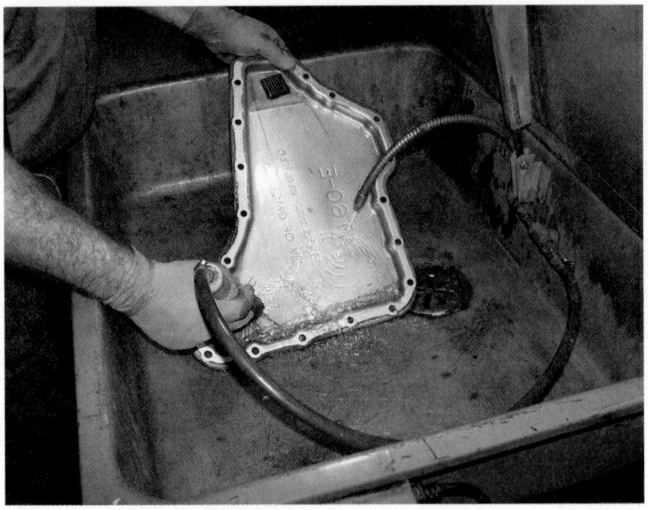

Figure 10-25 A typical parts washer.

Soak Tanks There are two types of soak tanks: cold and hot. Cold soak tanks are commonly used to clean carburetors, throttle bodies, and aluminum parts. A typical cold soak unit consists of a tank to hold the cleaner and a basket to hold the parts to be cleaned. After soaking with or without gentle agitation is complete, the parts are removed, flushed with water, and blown dry with compressed air.

Cleaning time is short, about 20 to 30 minutes, when the chemical cleaner is new. The time becomes progressively longer as the chemical ages. Agitation by raising and lowering the basket (usually done mechanically) will reduce the soak period to about 10 minutes. Some more elaborate tanks are agitated automatically.

Hot soak tanks are actually heated cold tanks. The source of heat is electricity, natural gas, or propane. The solution inside the hot tanks usually ranges from 160°F to 200°F (71°C to 93°C). Most tanks are generally large enough to hold an entire engine block and its related parts.

Hot tanks use a simple immersion process that relies on a heated chemical to lift the grease and grime off the surface. Liquid or parts agitation may also be used to speed up the job. Agitation helps shake the grime loose and also helps the liquid penetrate blind passageways and crevices in the part **(Figure 10–26)**. Generally speaking, it takes one to several hours to soak most parts clean.

Figure 10-26 A hot soak (d.p) tank.

Hot Spray Tanks The hot spray tank **(Figure 10–27)** works like a large automatic dishwasher and removes organic and rust soils from a variety of automotive parts. As with the hot soak method, spray washers soak the parts, but they also have the benefit of moderate pressure cleaning.

Figure 10-27 A hot spray cleaning machine.
Courtesy of Better Engineering Mfg., Inc.

Using a hot jet spray washer can cut cleaning time to less than 10 minutes. Normally, a strong soap solution is used as the cleaning agent. The speed of this system, along with lower operating costs, makes it popular with many machine shop owners.

SHOP TALK

Caustic soda, also known as sodium hydroxide, can be a very dangerous irritant to the eyes, skin, and mucous membranes. These chemicals should be used and handled with care. Because of the accumulation of heavy metals, it is considered a hazardous waste material and must be disposed of in accordance with EPA guidelines.

Spray washers are often used to preclean engine parts prior to disassembly. A pass-through spray washer is fully automatic once the parts have been loaded, and the cabinet prevents the runoff from going down the drain or onto the ground (which is not permitted in many areas because of local waste disposal regulations). Spray washers are also useful for post-machining cleanup to remove machine oils and metal chips.

Thermal Cleaning

The second basic process for cleaning engine parts is thermal cleaning. This process relies on heat to bake off or oxidize dirt and other contaminants.

Thermal cleaning ovens **(Figure 10–28)**, especially the pyrolytic type, have become increasingly popular. The main advantage of thermal cleaning is a total reduction of all oils and grease on and in blocks, heads, and other parts. The high temperature inside the oven (generally 650°F to 800°F [343°C to 426°C]) oxidizes all the grease and oil, leaving behind a dry, powdery ash on the parts. The ash must then be removed by shot blasting or washing. The parts come out dry, which makes subsequent cleanup with shot blast or glass beads easier because the shot will not stick.

Figure 10-28 A cleaning furnace. *Courtesy of Pollution Control Products Co.*

SHOP TALK

A slow cooling rate is recommended to prevent distortion that could be caused by unequal cooling rates within complex castings.

One of the major attractions of cleaning ovens is that they offer a more environmentally acceptable process than chemical cleaning. But although there is no solvent or sludge to worry about with an oven, the ash residue that comes off the cleaned parts must still be handled according to local disposal regulations.

Abrasive Cleaners

The third process used to clean engine parts involves the use of abrasives. Most abrasive cleaning machines are used in conjunction with other cleaning processes rather than as a primary cleaning process itself.

Cleaning by Hand Some manual cleaning is inevitable. Heavy buildups of grease and/or carbon should initially be removed by scraping or wire brushing. Cleaning aluminum and other soft metals with either technique should be done with extreme care, especially while using a steel scraper or brush. Steel or plastic scrapers are used to remove old gasket material from a surface and heavy sludge. Power tools with a small sanding disc (normally emery cloth) are available **(Figure 10–29)**. These are designed to remove all soft materials without damaging the hard metal surface. After the item has been scraped, an additional cleaning method is used to finalize cleaning.

Carbon can be removed with a handheld wire brush or a wire wheel driven by an electric or air drill motor **(Figure 10–30)**. Moving the wire wheel in a

Figure 10-29 Using a power scraper pad will prevent any metal from being removed.

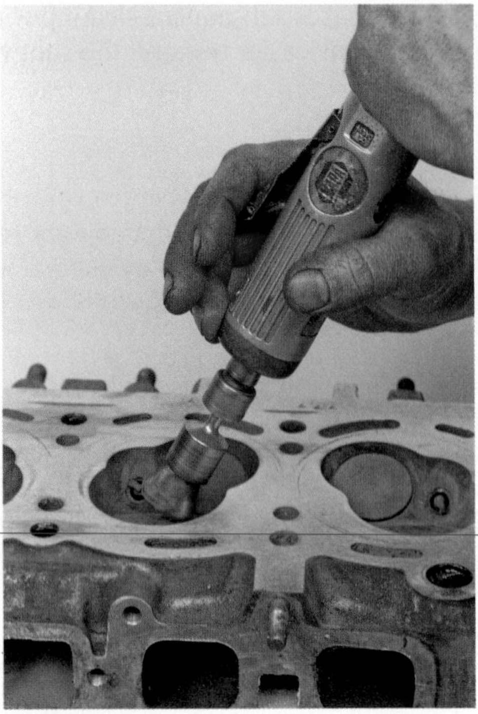

Figure 10-30 Carbon can be removed with a wire wheel driven by an electric or air drill motor.

light circular motion against the carbon helps to crack and dislodge the carbon. Some shops use a wire brush in addition to another cleaning method.

Wire brushes are also used to clean the inside of oil and coolant galleries. The brushes are soaked in a cleaning solvent and then passed through the passages in the block. To do this, the gallery plugs must be removed (**Figure 10–31**).

Abrasive Blaster Compressed air shot and grit blasters are best used on parts that will be machined after they have been cleaned. Two basic types of media are available: shot and grit. Shot is round; grit is angular in shape. Parts must be dry and grease-free when they go into an abrasive blast machine. Otherwise, the shot or beads will stick. Steel shot and glass beads are used for cleaning and/or peening the part's surfaces. **Peening** is a process of hammering on the surface. This packs the molecules tighter to increase the part's resistance to fatigue and stress. Steel shot is normally used with airless wheel blast equipment, which hurls the shot at the part by the centrifugal force of the spinning wheel. Glass beads are blown through a nozzle by compressed air in an enclosure (**Figure 10–32**).

Figure 10-31 It is often necessary to remove the gallery plugs and hand clean the oil galleries.

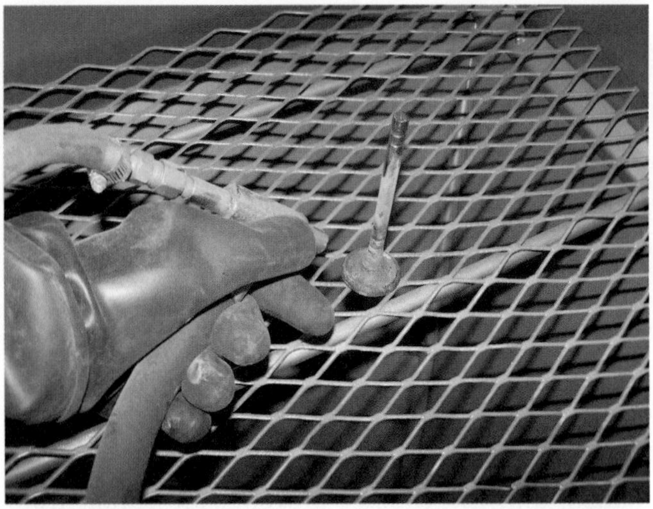

Figure 10-32 Using a blast nozzle to clean the backside of a valve.

Grit is primarily used for aggressive cleaning or on surfaces that need to be etched to improve paint adhesion. However, it removes metal, which can lead to some changes in tolerances. Grit blasting also chews out pits in the surface into which pollutants and blast residue can settle. This leads to stress corrosion unless the surface is painted or treated. These tiny crevices can also form surface stresses in the metal, which can lead to cracking in highly loaded parts. Grit should never be used for peening. Steel and aluminum oxide are the two most common types of grit.

Parts Tumbler A cleaning alternative that can save considerable labor when cleaning small parts such as engine valves is a tumbler. Various cleaning media can be used in a tumbler to scrub the parts clean. This saves considerable hand labor and eliminates dust. In some tumblers, all parts are rotated and tilted at the same time.

Vibratory Cleaning Shakers, as they are frequently called, use a vibrating tub filled with ceramic, steel, porcelain, or aluminum abrasive to scrub parts clean (**Figure 10–33**). Most shakers flush the tub with solvent to help loosen and flush away the dirt and grime.

Figure 10–33 A vibratory parts cleaner. *Courtesy of C & M TOPLINE*

The solvent drains out the bottom and is filtered to remove the sludge.

Alternative Cleaning Methods

Three of the most popular alternatives to traditional chemical cleaning systems are ultrasonic cleaning, citrus chemicals, and salt baths.

Ultrasonic Cleaning This cleaning process has been used for a number of years to clean small parts like jewelry, dentures, and medical instruments. Recently, however, the use of larger ultrasonic units has expanded into small engine parts cleaning. **Ultrasonic cleaning** **(Figure 10–34)** utilizes high-frequency sound waves to create microscopic bubbles that burst into energy to loosen soil from parts. Because the tiny bubbles do all the work, the chemical content of the cleaning solution is minimized, making waste disposal less of a problem. At the present time, however, the initial cost and handling capacity of ultrasonic equipment is its major disadvantage.

Citrus Chemicals Some chemical producers are starting to develop citrus-based cleaning chemicals as a replacement for the more hazardous solvent and alkaline-based chemicals currently used. Because of their citrus origin, these chemicals are safer to handle, easier to dispose of, and even smell good.

Salt Bath The **salt bath** is a unique process that uses high-temperature molten salt to dissolve organic materials, including carbon, grease, oil, dirt, paint, and some gaskets. For cast iron and steel, the salt bath operates at about 700°F to 850°F (371°C to 454°C). For aluminum or combinations of aluminum and iron, a different salt solution is used at a lower temperature (about 600°F [315°C]). The contaminants precipitate out of the solution and sink to the bottom of the tank, where they must be removed periodically. The salt bath itself lasts indefinitely as long as the salt is maintained properly. Cycling times with a salt bath are fairly quick, averaging 20 to 30 minutes. Like a hot tank, the temperature of the salt bath is maintained continuously.

CRACK DETECTION

Once engine parts have been cleaned, everything should be carefully inspected. This inspection should include a check for cracks, especially in the engine block and cylinder head. If cracks in the metal casting are discovered during the inspection, they should be repaired or the part replaced.

Cracks in metal castings are the result of stress or strain in a section of the casting. This stress or strain finds a weak point in that section of the casting and causes it to distort or separate at that point **(Figure 10–35)**. Such stresses or strains in castings can develop from the following:

- Pressure or temperature changes during the casting procedure may cause internal material structure defects, inclusion, or voids.
- Fatigue may result from fluctuating or repeated stress cycles. It might begin as small cracks and progress to larger ones under the action of the stress.
- Flexing of the metal may result due to its lack of rigidity.
- Impact damage may occur by a solid, hard object hitting a component.

Figure 10–34 An ultrasonic parts cleaner.

Figure 10-35 Examples of stress cracks.

- Constant impacting of a valve against a hardened seat may produce vibrations that could possibly lead to fracturing a thin-walled casting.
- Chilling of a hot engine by a sudden rush of cold water or air over the surface may happen.
- Excessive overheating is possible due to improper operation of an engine system.

Methods

Cracks can be found by visual inspection; however, many are not easily seen **(Figure 10–36)**. Therefore, engine rebuilders use special equipment to detect cracks, especially if there is reason to suspect a crack.

Pressure Checks Pressure checking a cylinder block or head is done in the same way a tire is checked for leaks. All of the coolant passages are plugged with rubber stoppers or gaskets. Compressed air is injected into a water jacket and the point of air entry is sealed. The block or head is then submerged into water. Bubbles will form in the water if there is a leak. The spot where the bubbles are forming is the location of the leak.

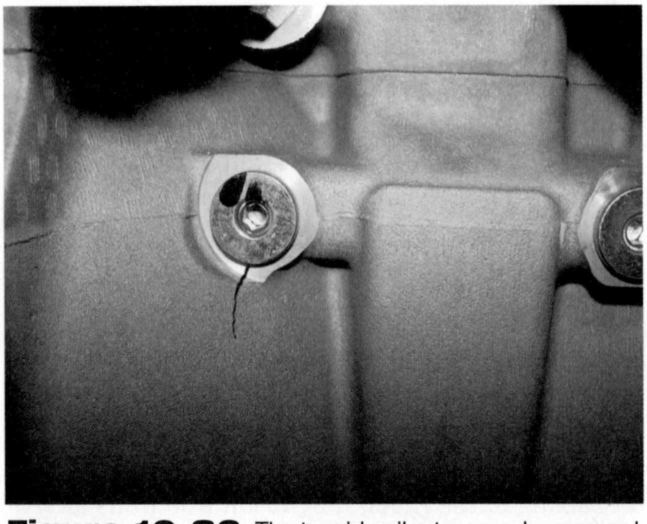

Figure 10-36 The topside oil artery crack appeared when an oxyacetylene flame was passed over the casting. Carbon in the flame was trapped in the crack, highlighting it.

Figure 10-37 MPI testing passes a magnetic field through the iron item being checked.

Magnetic Checks Magnetic particle inspection (MPI) uses a permanent or electromagnet to create a magnetic field in a cast iron unit **(Figure 10–37)**. When the legs of the detector tool are placed on the metal, the magnetic field travels through the metal. Iron filings are sprinkled in the surface to detect a secondary magnetic field resulting from a crack **(Figure 10–38)**. Because the secondary magnetic field will not form if the crack is in the same direction as the magnet, the magnet must be rotated and the metal checked in both directions.

Dye Penetrant Another common way to detect cracks is by using three separate chemicals:

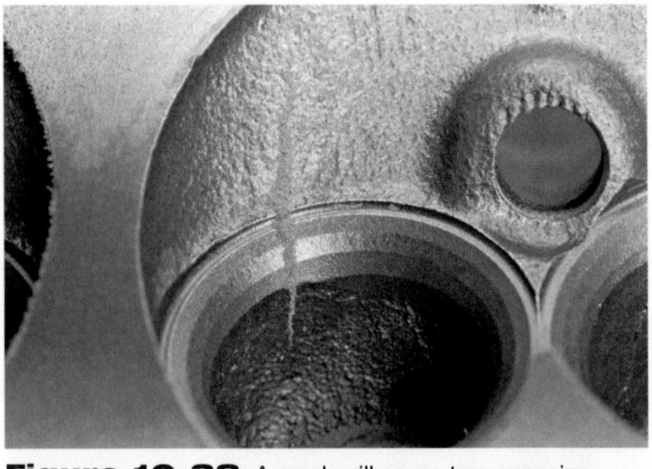

Figure 10-38 A crack will cause two opposing magnetic poles to form on each side; the iron filings used with the magnet will show these fields.

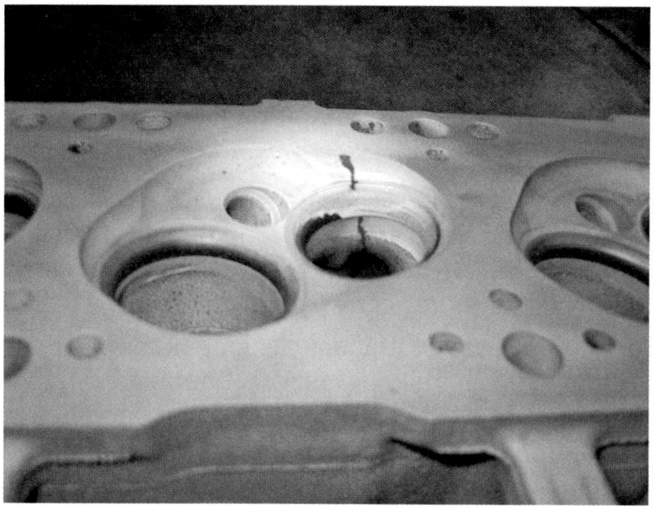

Figure 10-39 Cracks appear as red lines when a dye penetrant is used. *Courtesy of LOCK-N-STITCH Inc.*

penetrant, cleaner, and developer. The part to be checked must be clean and dry. This check must be done according to the following sequence:

1. Spray or brush the penetrant onto the surface.
2. Wait 5 minutes.
3. Spray the cleaner onto a clean cloth.
4. Wipe off all visible penetrant.
5. Spray the developer on the tested area.
6. Wait until the developer is totally dry.
7. Inspect the area. Cracks will appear as a red line **(Figure 10–39)**.

Crack Repair

If a crack is found, a decision must be made as to whether the part should be replaced or repaired. This decision should be based on the cost of repair as well as any other repair that the part may need. Further inspection of the cylinder block and head and the service and repair procedures for each are covered in the Chapters 11 and 12.

CASE STUDY

A four-cylinder engine is brought into the shop. The customer complains of excessive oil consumption and oil leaks. Compression and cylinder leakage tests indicate that the cylinders are sealing well, and a power balance test indicates that all of the cylinders are producing about the same amount of power. Based on these results, the technician assumes that the problem is leaking valve seals. The initial plan is to replace the seals and regasket the engine.

It is odd that the engine has both of these problems. It has less than 50,000 miles on it. Not really sure if the problems are related, the technician proceeds to disassemble the engine. Upon removing the valve cover, large amounts of sludge are evident throughout the valve train. This is normally a sign that the engine has been neglected. However, a review of the files indicates that the oil has recently been changed. In fact, the car has been well maintained. Is the sludging related to the oil consumption and leaks?

The oil pan is removed and additional sludge is found. The cylinder head is then removed from the block. The piston tops and the combustion chamber are covered with a thick black carbon coating. Is this buildup related to other problems?

The cylinder head is disassembled and each of the valve seals is found to be deteriorated. What could cause the deterioration of rubber parts, leaking gaskets, sludging, and carbon buildup in the cylinders? After careful thought, the technician pays attention to the parts taken off the engine during initial disassembly. A thorough inspection is made of the PCV system and it is discovered that the hose that connects the valve to the manifold is plugged solid. The valve is also found to be plugged.

The PCV system is designed to remove crankcase fumes and pressure from the crankcase. These fumes can cause rapid sludging of the oil and deterioration of rubber parts. Excessive crankcase pressure can cause leaks, as the pressure seeks to relieve itself. A faulty PCV valve can cause all of the problems exhibited by this engine. In fact, it is the cause of the problems.

The engine is resealed and new valve stem seals are installed. The engine is then installed with a new PCV valve and hose. Not only is the customer's complaint taken care of, but so is the cause of the problem.

KEY TERMS

Caustic soda
Freewheeling engines
Interference engines
Organic soils
Peening
Rust
Salt bath
Scale
Solvent tank
Ultrasonic cleaning
Water soluble

SUMMARY

- When preparing an engine for removal and disassembly, it is important to always follow the specific service manual procedures for the particular vehicle being worked on.
- A hoist and chain are needed to lift an engine out of its compartment. Mount the engine to an engine stand with a minimum of four bolts, or set it securely on blocks.
- While an engine teardown of both the cylinder head and block is a relatively standard procedure, exact details vary among engine types and styles. The vehicle's service manual should be considered as the final word.
- An understanding of specific soil types can save time and effort during the engine cleaning process. The main categories of contaminants include water-soluble and organic soils, rust, and scale.
- Protective gloves and goggles should be worn when working with any type of cleaning solvent or chemical. Read the label carefully before using as well as all of the information provided on material safety data sheets.
- Parts washers, or solvent tanks, are a popular and inexpensive means of cleaning the metal surfaces of many automotive components and engine parts. Regardless of the type of solvent used, it usually requires some brushing, scraping, or agitation to increase the cleaning effectiveness.
- Cold soak tanks are used to clean carburetors, throttle bodies, and aluminum parts. Hot soak tanks, which can accommodate an entire engine block, use a heated cleaning solution to boil out dirt. Hot heat spray washers have the added benefit of moderate pressure cleaning.
- Alternatives to caustic chemical cleaning have emerged in recent years, including ultrasonic cleaning, salt baths, and citrus chemical cleaning. These methods are all growing in popularity.
- The main advantage of thermal cleaning is its total reduction of all oils and grease. The high temperatures inside the oven leave a dry, powdery ash on the parts. This is then removed by shot blasting or washing.
- Steel shot and glass beads are used for cleaning operations where etching or material removal is not desired. Grit, the other type of abrasive blaster, is used for more aggressive cleaning jobs.
- Some degree of manual cleaning is necessary in any engine rebuilding job. Very fine abrasive paper should be used to remove surface irregularities. A handheld or power wire brush is also helpful, though it can be time-consuming to work with.
- There are three common methods for detecting cracks in the metal casting of engine parts: using a magnet and magnetic powder (iron filings), using penetrant dye (especially for aluminum heads and blocks), and pressuring with air.

REVIEW QUESTIONS

1. What should be worn when working with any type of cleaning solvent or chemical?
2. *True or False?* Most engines in a RWD vehicle must be removed with the transmission still attached.
3. What is the best way to lift a vehicle when preparing to remove an engine?
   a. frame contact hoist
   b. drive on lift
   c. hydraulic jack and safety stands
   d. engine hoist
4. *True or False?* The first step in disassembling an engine is usually the removal of the intake and exhaust manifolds.
5. Which of the following statements is *not* true?
   a. When the engine is removed through the bottom of the vehicle, use an engine cradle and dolly to support the engine.
   b. If the manufacturer recommends engine removal through the hood opening, use an engine hoist.
   c. Regardless of the method of removal, the engine and transaxle in a FWD vehicle are usually removed as a unit.
   d. The transaxle can be separated from the engine after the engine is off its mounts.

6. The buildup of minerals and deposits inside the cooling system is called ______.
   a. organic soil
   b. scale
   c. rust
   d. grime
7. Hydrocarbon solvents are ______.
   a. flammable
   b. toxic
   c. both a and b
   d. neither a nor b
8. *True or False?* On many FWD vehicles, the suspension system must be partially disassembled to remove the radiator.
9. Which cleaning method uses high-frequency sound waves to create microscopic bubbles that loosen dirt from parts?
   a. ultrasonic
   b. salt bath
   c. thermal
   d. caustic
10. Parts must be ______ when they go into an abrasive blast machine.
    a. wet
    b. dry
    c. grease-free
    d. both b and c
11. An engine block should be mounted to an engine stand using a minimum of ______ bolts.
    a. four
    b. six
    c. three
    d. five
12. Which of the following is *not* considered part of the organic soil grouping?
    a. petroleum by-products derived from crude oil, including tar, road oil, engine oil, gasoline, diesel fuel, grease, and engine oil additives
    b. rust that is a product of coolant and aluminum
    c. by-products of combustion, including carbon, varnish, gum, and sludge
    d. coatings, including such items as rust-proofing materials, gasket sealers and cements, paints, waxes, and sound-deadener coatings
13. Why should a memory saver be installed before disconnecting a vehicle's battery?
14. Which of the following statements is *not* true about thermal cleaning?
    a. The main advantage of thermal cleaning is a total reduction of all oils and grease on and in blocks, heads, and other parts.
    b. After a part has been thermally cleaned, it should be submerged in water to cool it quickly.
    c. Thermal cleaning leaves behind a dry, powdery ash on the parts.
    d. After a part has been thermally cleaned, it should be washed or blasted with shot.
15. Which of the following is *not* a common way to identify the location of cracks in the engine block or cylinder head?
    a. pressure checks
    b. vacuum test
    c. magnetic particle inspection
    d. penetrant dye

ASE-STYLE REVIEW QUESTIONS

1. While working on an engine with an excessive amount of sludge buildup: Technician A says that the presence of sludge is a signal to look for wear on other components. Technician B says that excessive buildup can indicate a poor maintenance schedule. Who is correct?
   a. Technician A
   b. Technician B
   c. Both A and B
   d. Neither A nor B
2. While discussing the common causes for cracks developing in a cylinder block or head: Technician A says that the chilling of a hot engine by a sudden rush of cold water or air over the surface may cause cracking. Technician B says that excessive overheating is a common cause. Who is correct?
   a. Technician A
   b. Technician B
   c. Both A and B
   d. Neither A nor B
3. Technician A uses a crane to remove an engine from its compartment. Technician B uses an engine cradle to remove an engine from its compartment. Who is correct?
   a. Technician A
   b. Technician B
   c. Both A and B
   d. Neither A nor B
4. While removing a cylinder head: Technician A keeps all rocker arms and pushrods in order. Technician B loosens each head bolt, starting with the center bolts and moving toward the ends. Who is correct?
   a. Technician A
   b. Technician B
   c. Both A and B
   d. Neither A nor B

5. While discussing abrasive cleaners: Technician A says that shot is angular in shape and is used for aggressive cleaning. Technician B says that grit is an angular-shaped media and is used to peen metal surfaces. Who is correct?
   a. Technician A c. Both A and B
   b. Technician B d. Neither A nor B
6. Technician A labels or marks all electrical wires before disconnecting them. Technician B labels or marks all vacuum hoses and verifies the connections to the underhood decal before disconnecting them. Who is correct?
   a. Technician A c. Both A and B
   b. Technician B d. Neither A nor B
7. While discussing cleaning engine parts: Technician A says that the cleaning method used depends on the component to be cleaned and the type of cleaning equipment available. Technician B says that sometimes it is best to clean parts by hand with soap and warm water. Who is correct?
   a. Technician A c. Both A and B
   b. Technician B d. Neither A nor B
8. While loosening the axle shaft hub nuts on a FWD vehicle: Technician A says that a large breaker bar should be used to prevent damage to the bearings. Technician B says that these nuts should be loosened with the vehicle on the floor and the brakes applied. Who is correct?
   a. Technician A c. Both A and B
   b. Technician B d. Neither A nor B
9. While preparing to remove an engine: Technician A disconnects the refrigerant lines at the air-conditioning compressor and allows the refrigerant to totally leak out before removing the compressor. Technician B installs plugs in the ends of the refrigerant hoses after they have been disconnected. Who is correct?
   a. Technician A c. Both A and B
   b. Technician B d. Neither A nor B
10. After removing the vehicle's hood in preparation for removing the engine: Technician A places the hood on the roof of the vehicle. Technician B sets the hood aside in a safe place on fender covers or cardboard. Who is correct?
    a. Technician A c. Both A and B
    b. Technician B d. Neither A nor B

LOWER END THEORY AND SERVICE

CHAPTER 11

OBJECTIVES

■ Describe how to disassemble and inspect an engine. ■ List the parts that make up a short block and briefly describe their operation. ■ Describe the major service and rebuilding procedures performed on cylinder blocks. ■ Describe the purpose, operation, and location of the camshaft. ■ Describe the four types of camshaft drives. ■ Inspect the camshaft and timing components. ■ Describe how to install a camshaft and its bearings. ■ Explain crankshaft construction, inspection, and rebuilding procedures. ■ Explain the function of engine bearings, flywheels, and harmonic balancers. ■ Explain the common service and assembly techniques used in connecting rod and piston servicing. ■ Explain the purpose and design of the different types of piston rings. ■ Describe the procedure for installing pistons in their cylinder bores. ■ Inspect, service, and install an oil pump.

The lower end of an engine is the cylinder block assembly. This includes the block, camshaft, crankshaft, bearings, pistons, piston rings, and oil pump. Many of these parts are made by casting or forging. To **cast** is to form molten metal into a particular shape by pouring it into a mold. To **forge** is to form metal into a shape by heating it and pressing into a mold. Some forging is done with cold metals. These manufactured parts then undergo a number of machining operations. Following are a few examples:

- The top of the block must be perfectly smooth so that the cylinder head can seal it.
- The bottom of the block is also machined to allow for proper sealing of the oil pan.
- The cylinder bores must be smooth and have the correct diameter to accept the pistons.
- The main bearing area of the block must be align bored (cut a series of holes in a straight line) to a diameter that will accept the crankshaft. Camshaft bearing bores must also be aligned.

When there is a major engine failure, shops either rebuild or replace the engine **(Figure 11–1)**. Most often the **short block** is repaired or replaced as an assembly. A basic short block consists of a cylinder block, crankshaft, bearings, connecting rods, pistons and rings, and oil gallery and core plugs. Parts related to the short block but not necessarily included with it are the flywheel and harmonic balancer. A short block may also include the engine's camshaft and timing gear. A **long block** is basically a short block with cylinder heads. These terms are commonly used when purchasing replacement engines.

Figure 11–1 A cutaway showing the fit of the piston assemblies and crankshaft in an engine block.

SHORT BLOCK DISASSEMBLY

This chapter begins with the assumption that the engine has been removed and the cylinder head(s) have been separated from the cylinder (engine) block. However, few parts have been removed from the short block. If the oil pan and water pump are still attached to the block, remove them before proceeding.

Remove the **harmonic balancer**, also called a vibration damper. The harmonic balancer is an assembly of an inner hub bonded with rubber to an outer ring. Its purpose is to absorb the torsional vibrations of the crankshaft. Removal of the balancer requires the correct type puller **(Figure 11–2)**. If a jawed puller is used, it is very likely that the rubber bonding will be damaged. This would make the balancer useless and can cause engine vibrations and crankshaft damage. Once the balancer is removed, carefully check the rubber for tears or other damage. If there are any faults, replace the balancer.

On overhead valve engines, remove the timing cover. Under the cover are the timing gears. The timing sprocket on the crankshaft snout has a slight interference fit and can normally be pulled off by hand. However, the camshaft sprocket and chain must be removed with the crankshaft sprocket. The camshaft sprocket is either press-fit or bolted to the camshaft.

Before removing the gears and chain, check the deflection of the timing chain. Depress the chain at its midway point between the gears and measure the amount that the chain can be deflected. If the deflection measurement exceeds specifications, the timing chain and gears should be replaced.

Loosen the camshaft sprocket and pull the timing gears and chain from the engine. Be careful not to lose the keys in each shaft or any shims that may be behind the sprockets.

Often the timing chain assembly has tensioners and guides. The timing gear assembly is normally replaced during an engine overhaul. The tensioners and guides wear and should be replaced as well.

The timing belt or chain on OHC engines has already been loosened before the cylinder head was removed. However, it may still be dangling around the crankshaft sprocket. Slip it off, if possible. Unbolt the timing chain or belt cover and gently pry it away from the engine block and cylinder head. Remove the crankshaft position sensor, timing chain guide, chain tensioner slipper, and chain. Pull the timing gear or sprocket off the crankshaft. On some engines, the oil pump is driven by the crankshaft at the front of the engine. The pump should be unbolted and removed. Rotate the crankshaft counterclockwise to align the timing marks on the oil pump sprocket with the mark on the oil pump. Remove the attachment bolt for the sprocket and the chain tensioner plate and spring. Then remove the oil pump sprockets and chain. Remove the oil screen then the oil pump **(Figure 11–3)**. On some engines, the alignment of the inner and outer gears of the pump should be marked before removing the pump.

If the lifters have not been removed, do so now. Place them on a bench in the order they were removed. Carefully pull the camshaft out of the block. Support the camshaft during removal to avoid dragging its lobes over the surfaces of the camshaft bearings. This can damage the bearings and lobes. Some engines require the removal of a thrust plate before removing the camshaft.

Figure 11–2 The harmonic balancer should be removed with the correct puller.

Figure 11–3 On some engines, the oil pump is driven by the crankshaft at the front of the engine.

Cylinder Block Disassembly

Rotate the engine on its stand so that the bottom is facing up. Remove the oil pan if it was not removed previously. Then remove the oil pump as directed in the service manual. Be careful not to lose the drive shaft while pulling the pump off the engine.

If the engine has balance shafts, check the thrust clearance of the shafts before removing the assembly. Set a dial indicator so it can read the back-and-forth movement (end play) of the shaft. Measure the total distance that the shaft is able to move in the housing. Compare that reading to specifications. If the reading is more than the specified maximum, the balance shaft housing and bearings should be replaced. Unbolt the housing following the sequence given in the shop manual **(Figure 11–4)**.

Remove the housing cover. Then lift the balance shaft(s) out of the housing. Inspect the bearings for unusual wear or damage. Keep the bearings in their original location or mark them for identification. Check the journals of the balance shafts for scratches, pitting, and other damage. If a bearing or journal is damaged, replace the bearings and/or balance shaft.

Rotate the engine so the bottom is at the bottom again. Rotate the crankshaft so that the piston of one cylinder is at BDC. Carefully remove the cylinder ridge with a ridge reamer tool. Rotate the tool clockwise with a wrench to remove the ridge **(Figure 11–5)**. Do not cut too deeply, because an indentation may be left in the bore. Remove just enough metal to allow the piston assembly to slip out of the bore without causing damage to the bore.

The ridge is formed at the top of the cylinder. Because the top ring stops traveling before it reaches

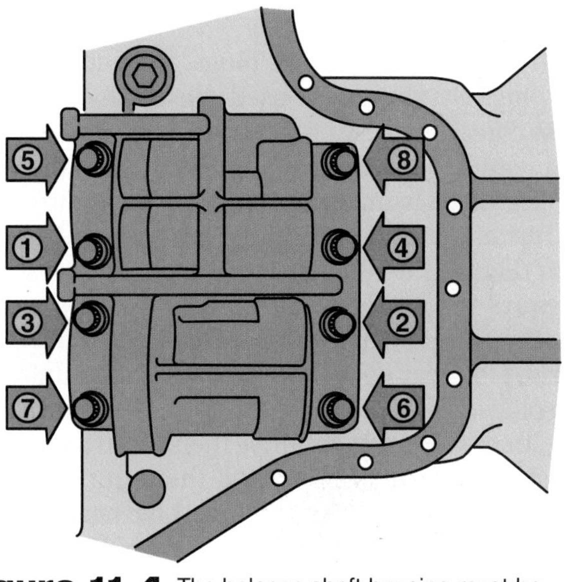

Figure 11–4 The balance shaft housing must be unbolted in the prescribed order to prevent shaft and housing warpage.

Figure 11–5 A ridge reamer should be used on all cylinders before removing the pistons.

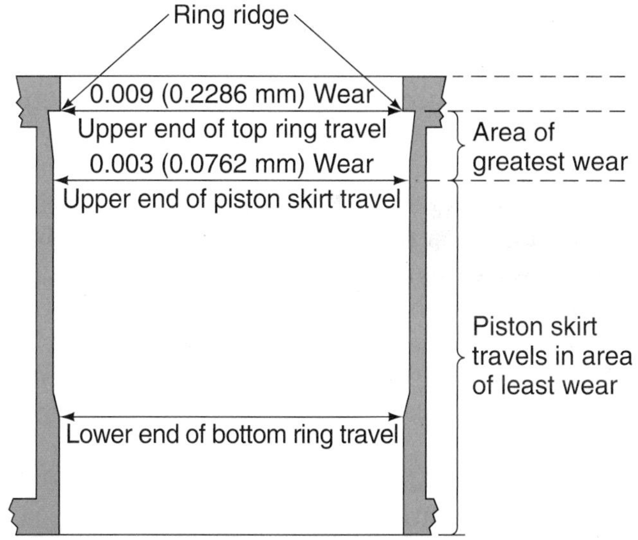

Figure 11–6 Normal cylinder wear. *Courtesy of Dana Corporation*

the top of the cylinder, a ridge of unworn metal is left **(Figure 11–6)**. Carbon also builds up above this ridge, adding to the problem. If the ridge is not removed, the piston's ring lands may be damaged as the piston is driven out of its bore.

Repeat the process on all cylinders. After removing the ridges, use an oily rag to wipe the metal cuttings out of the cylinder. The cuttings will stick to it.

Rotate the engine to put the bottom side up. Check all connecting rods and main bearing caps for correct position and numbering. If the numbers

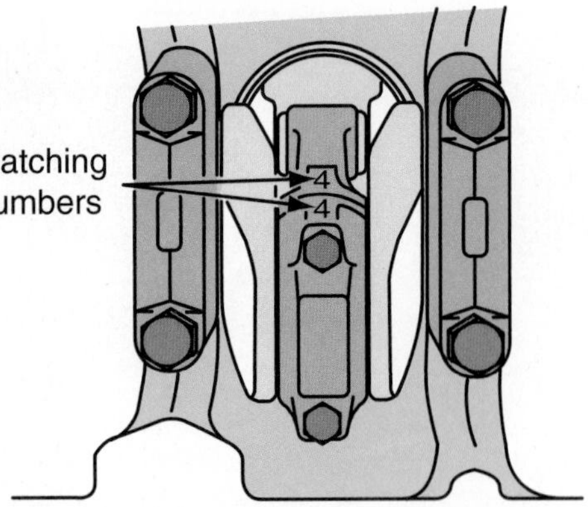

Figure 11–7 Check all connecting rods and main bearing caps for correct position and numbering. If the numbers are not visible, use a center punch or number stamp to number them.

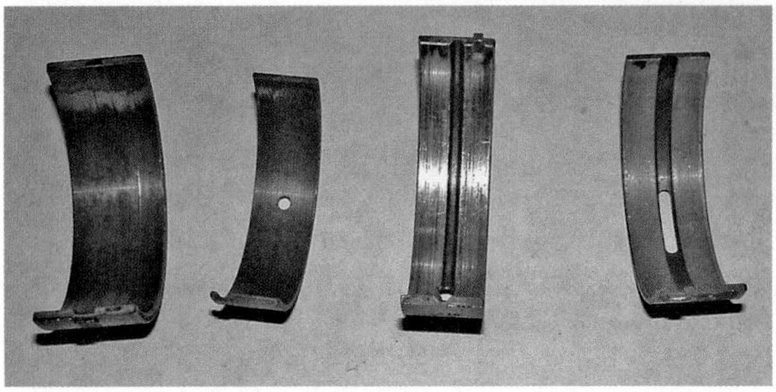

Figure 11–8 The bearings on the left have no babitt left, and the ones on the right are slightly worn and scored.

Figure 11–9 Inspect each crankshaft journal for damage and wear. *Courtesy of Dana Corporation*

are not visible, use a center punch or number stamp to number them (**Figure 11–7**). Caps and rods should be stamped on the external flat surface. If the rods are already numbered or marked, make sure the marks designate the cylinder where the rods should be installed. They may have been installed incorrectly or they may have been taken from another engine. Re-mark them to show their current location.

To remove the piston and rod assemblies, position the crankshaft throw at the bottom of its stroke. Remove the connecting rod nuts and cap. Remember that the caps and rods must remain as a set. Tap the cap lightly with a soft hammer or wood block to aid in cap removal. Cover the rod bolts with protectors to avoid damage to the crankshaft journals. Carefully push the piston and rod assembly out with the wooden hammer handle or wooden drift and support the piston by hand as it comes out of the cylinder. Be sure that the connecting rod does not damage the cylinder wall during removal. With the bearing inserts in the rod and cap, replace the cap (numbers on the same side) and install the nuts. Repeat the procedure for all piston and rod assemblies.

In the specified order, loosen and remove the main bearing cap bolts and main bearing cap. Keeping the main bearing caps in order is very important. The location and position of each main bearing cap should be marked. Most high-performance engines have a main bearing girdle or bearing support. These use at least four bolts at each main bearing. It is important that the recommended bolt loosening sequence be followed.

After removing the main bearing caps, carefully take out the crankshaft by lifting both ends equally to avoid bending and damage. Store the crankshaft in a vertical position to avoid damage.

Remove the rear main oil seal and main bearings from the block and caps. Examine the bearing inserts for signs of abnormal engine conditions such as embedded metal particles, lack of lubrication, antifreeze contamination, oil dilution, and uneven wear (**Figure 11–8**). Also inspect them for any unusual signs of wear, and check them for indications that they are over- or undersized. Then carefully inspect the main journals on the crankshaft for damage (**Figure 11–9**).

Engine blocks have **core plugs**, also called expansion plugs. Sand cores are used when a block is made. These cores are partly broken and dissolved when the hot metal is poured into the mold. To get the sand out, the block is made with holes. These core holes are machined and core plugs are installed to seal them (**Figure 11–10**). Blocks are also made with passageways for oil. These are machined in after the block is made and are sealed with plugs.

The block cannot be thoroughly cleaned unless all core plugs and oil plugs are removed. To remove cup-type "freeze"/core plugs, drive them in on a slant and use channel lock pliers to pull them out. Flat-type plugs can be removed by drilling a hole near the center and inserting a slide hammer to pull it out. On some engines, the cup-type plug can be easily removed by driving the plug out from the backside with a long rod.

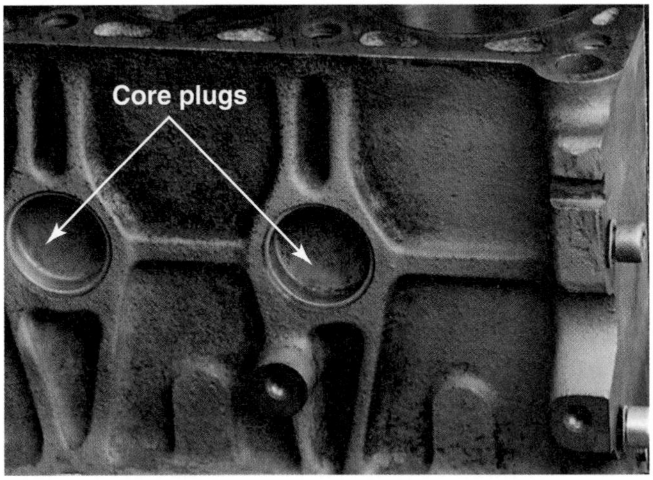

Figure 11–10 Core plugs in an engine block.

Sometimes removing threaded front and rear oil gallery plugs can be difficult. Using a drill and screw extractor can help.

> **SHOP TALK**
>
> Using heat to melt paraffin into the threads of oil plugs will make removal much easier. As the part is heated, it will expand and the paraffin will leak down between the threads. Because the paraffin serves as a lubricant, you will be able to loosen the plug. Hot paraffin burns, so wear gloves when handling it.

After cleaning, the block and its parts must be visually checked for cracks or other damage.

CYLINDER BLOCK

The cylinder block houses the areas where combustion takes place. It is normally a one-piece casting that is machined so that all the parts contained in it fit properly. Blocks may be cast from different materials such as iron, aluminum **(Figure 11–11)**, magnesium, or possibly, in the future, plastic. A few engine blocks have an exterior made of magnesium and the interior (cylinders, coolant passages, etc.) of a cast-iron insert. Some late-model blocks are made of two pieces: an upper unit that contains the cylinders and a lower one that surrounds the crankshaft **(Figure 11–12)**. Other blocks have main cap girdles (also referred to as the bedplate) integrated in the housing to provide added strength.

Cast-iron blocks are very strong but heavy. Many engines have an aluminum block to reduce the vehicle's overall weight. Certain materials are added to aluminum to make it stronger and less likely to warp from the heat of combustion. The cylinder walls of

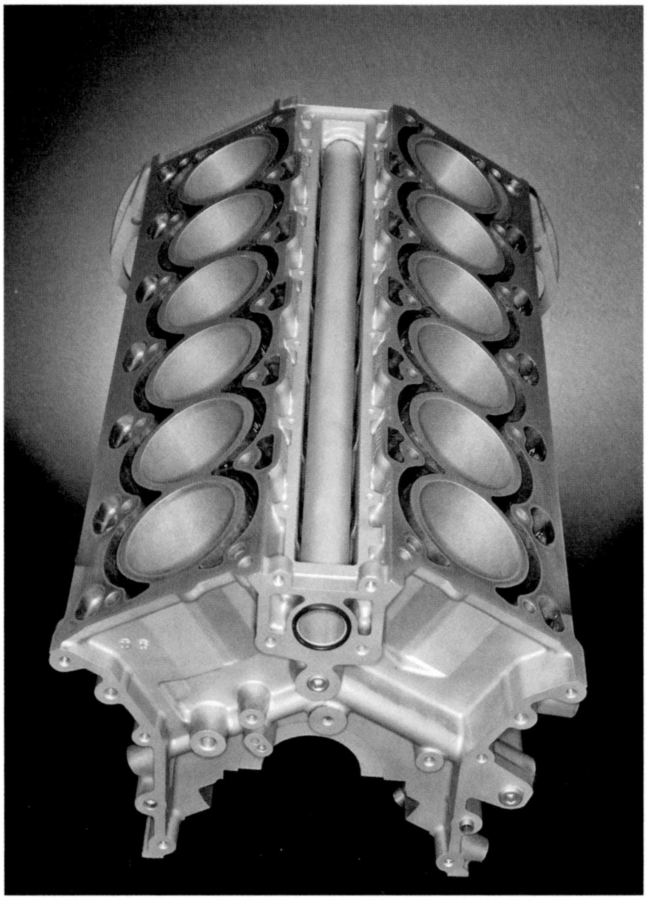

Figure 11–11 An engine block for a 12-cylinder engine. *Courtesy of BMW of North America, LLC*

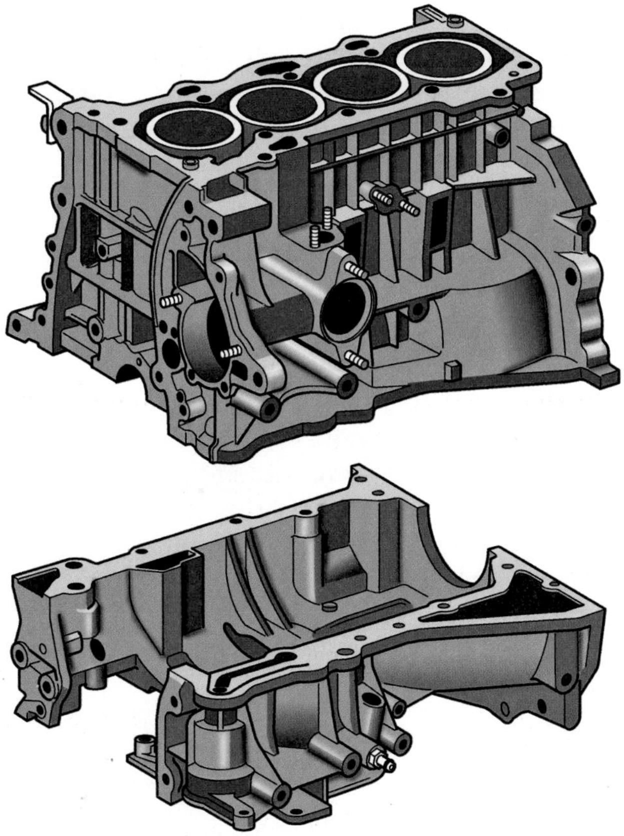

Figure 11–12 Aluminum engine blocks are often two-piece units. *Courtesy of Toyota Motor Sales, U.S.A., Inc.*

aluminum blocks may be treated with a special coating or may have a sleeve or liner to serve as cylinder walls.

Cylinder Liners

Most aluminum blocks have cylinder liners (**Figure 11–13**). The liners are normally made of a cast-iron alloy. On some engines, the liners can be replaced and/or machined if they are damaged. Most are very thin and cannot be serviced and the block must be replaced if the walls are damaged. Liners are pressed into the block or are placed in the mold before the block is cast. The liners have ribs on their outside diameter. These ribs hold the liner in place and increase its ability to dissipate heat to the block. There are two types of sleeves: wet and dry. The dry sleeve is supported from top to bottom by the block. The wet sleeve is supported only at the top and bottom. Coolant touches the center part of a wet sleeve.

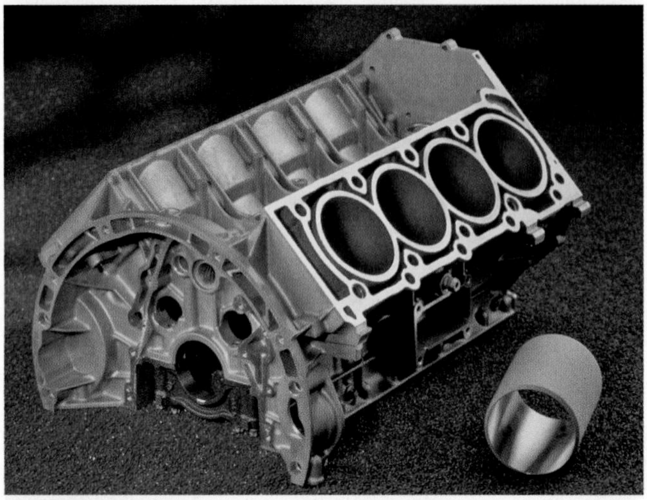

Figure 11-13 A cylinder block and cylinder liner for a late-model aluminum V8 engine. *Courtesy of Chrysler LLC*

Lubrication and Cooling

A cylinder block contains a series of oil passages that allow engine oil to be pumped through the block and crankshaft and on to the cylinder head. Water jackets are also cast in the block around the cylinder bores. Coolant circulates through these jackets to transfer heat away.

Some engine blocks are cast with a plastic spacer for the water jackets **(Figure 11–14)**. The spacers provide a uniform distribution of heat throughout the cylinders by directing the flow of coolant toward the normally hotter areas. For example, a spacer is used to direct coolant away from the center of the cylinder

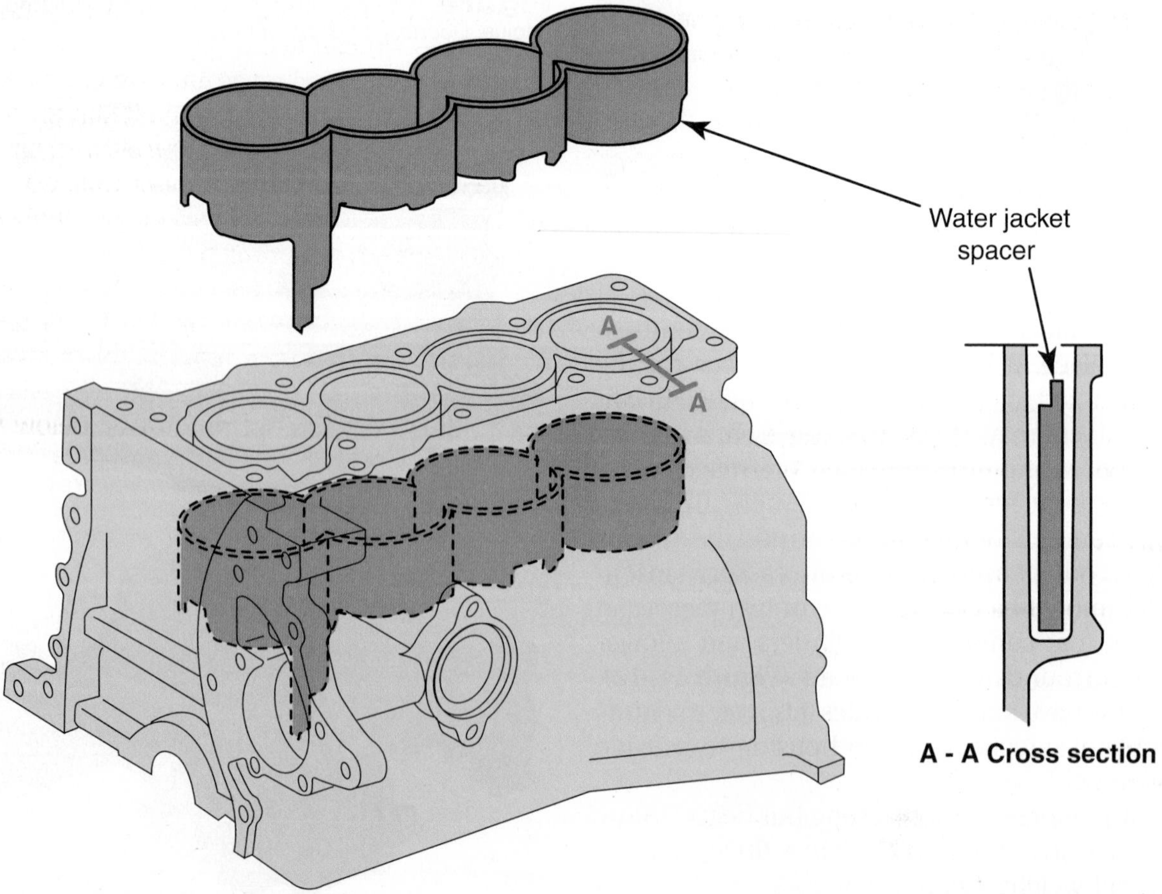

Figure 11-14 If the engine has water jacket spacers, they must be removed before cleaning the block and to prevent damage to them.

bore to the top and bottom of the bore. The top and bottom of the bore are normally the hotter areas, so suppressing the coolant flow to the center may result in the center having the same temperature as the top and bottom.

Other Features

An engine block has other features. Many of these are machined areas on which to mount other parts. These areas typically include threaded bores. Brackets and housings may also be cast onto the basic block. Some blocks have air passage holes near the bottom to allow for any pressure buildup caused by the pumping action of the pistons. The bottoms of the pistons are moving with pumping action at the same time the tops are moving through their strokes.

CYLINDER BLOCK RECONDITIONING

Before doing any service to the block, clean all threaded bores with the correct-size thread chaser to remove any burrs or dirt. Use a bottoming tap in any blind holes. These bores should be slightly **chamfered** to eliminate thread pulls and jagged edges. If there is damage to the threads, they should be repaired. To restore damaged threads in an aluminum part, a threaded insert should be installed in the bore.

Check the block for cracks and other damage. Cast-iron blocks can be checked by magnafluxing. Aluminum blocks are checked for cracks with penetrant dye and a black light. Some cracks can be repaired; however, if they are in critical areas, the block should be replaced.

Deck Flatness

The top of the block where the cylinder head mounts is called the **deck**. To check deck warpage, use a precision straightedge and feeler gauge. With the straightedge positioned diagonally across the deck, the amount of warpage is determined by the size of feeler gauge that fits into the gap between the deck and the straightedge (**Figure 11–15**).

Some engines have special deck flatness requirements. Always refer to the manufacturer's specifications. If the deck is warped beyond limits, the block should be decked or replaced. Decking requires a special grinder that will shave off small amounts of metal, leaving a flat surface. Some manufacturers do not recommend decking, especially if the block is aluminum. If the block has more than one deck surface (such as a V-type engine), each deck should be machined to the same height. If the deck is warped and not corrected, the valve seats can distort when the head is tightened to the block. Coolant and combustion leakage can also occur.

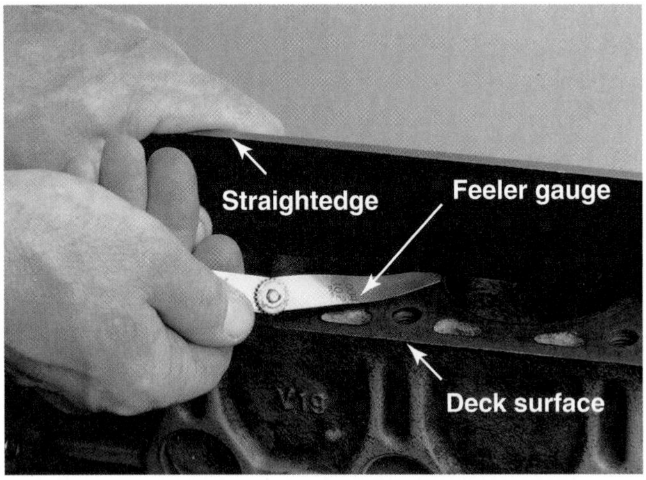

Figure 11–15 Checking for deck warpage with a straightedge and feeler gauge.

Cylinder Bore Inspection

Inspect the cylinder walls for scoring, roughness, or other signs of wear. Dirt can accelerate ring and cylinder wall wear. It also can get caught in the piston rings and can grind away at the metal surfaces.

Scuffed or scored pistons, rings, and cylinder walls (**Figure 11–16**) can act as passages for oil to bypass the rings and enter the combustion chamber. Scuffing and scoring occur when the oil film on the cylinder wall is ruptured, allowing metal-to-metal contact of the piston rings on the cylinder wall. Cooling system hot spots, oil contamination, and fuel wash are typical causes of this problem.

Most cylinder wear occurs at the top of the ring travel. Pressure on the top ring is at a peak and lubrication at a minimum when the piston is at the top of its stroke. A ridge of unworn material will remain above the upper limit of ring travel. Below the ring

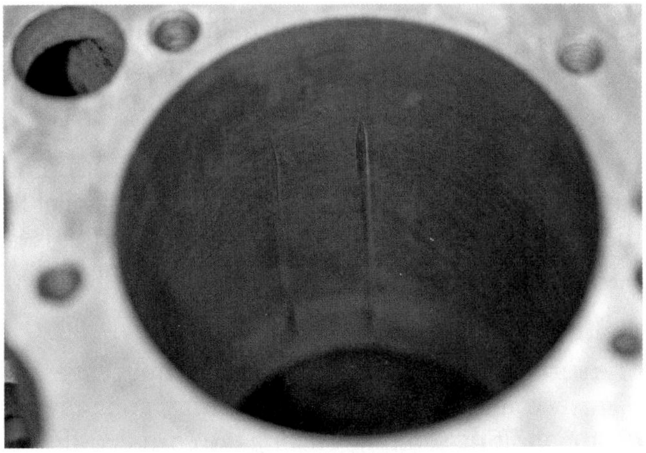
Figure 11–16 A scored cylinder wall.

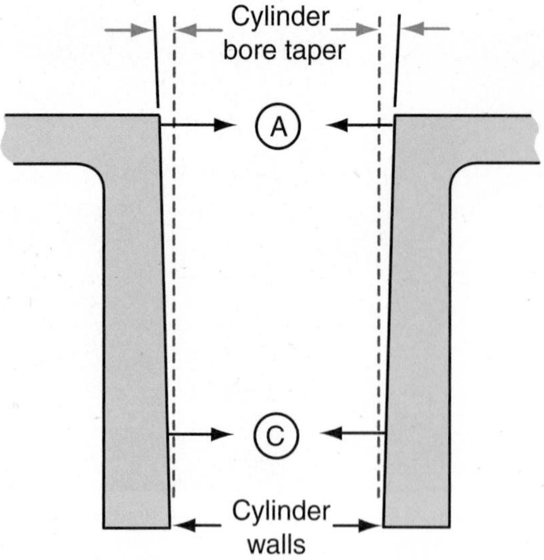

Figure 11–17 To check for taper, measure the diameter of the cylinder at A and C. The difference between the two readings is the amount of taper.

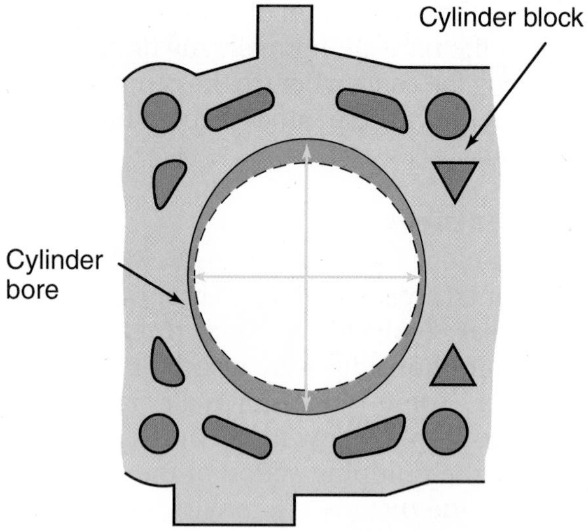

Figure 11–18 To check cylinder out-of-roundness, measure the bore in different locations. *Courtesy of Ford Motor Company*

travel area, wear is negligible because only the piston skirt contacts the cylinder wall.

A properly reconditioned cylinder must have the correct diameter, have no taper or out-of-roundness, and the surface finish must be such that the piston rings will seat to form a seal that will control oil and minimize blowby.

Taper is the difference in diameter between the bottom of the cylinder bore and the top of the bore just below the ridge (**Figure 11–17**). Subtracting the smaller diameter from the larger one gives the cylinder taper. Some taper is permissible, but normally not more than 0.006 inch (0.1524 mm). If the taper is less than that, reboring the cylinder is not necessary.

Cylinder out-of-roundness is the difference of the cylinder's diameter when measured parallel with the crank and then perpendicular to the crank (**Figure 11–18**). Out-of-roundness is measured at the top of the cylinder just below the ridge. Typically the maximum allowable out-of-roundness is 0.0015 inch (0.0381 mm). Normally a cylinder bore is checked for out-of-roundness with a dial bore gauge (**Figure 11–19**). However, a telescoping gauge can also be used.

When using a dial bore gauge or a telescoping gauge, make sure the measuring arms are parallel to the plane of the crankshaft. The best way to do this is to rock the gauge until the smallest reading is obtained.

Figure 11–19 Cylinder bore is checked for out-of-roundness with a dial bore gauge.

Cylinder Bore Surface Finish

The surface finish on a cylinder wall should act as an oil reservoir to lubricate the piston rings and prevent piston and ring scuffing. Having the correct cylinder wall finish is important. Piston ring faces can be damaged and wear prematurely if the wall is too rough. A surface that is too smooth will not hold enough oil and scuffing may occur.

The desired finish has many small crisscross grooves (**Figure 11–20**). Ideally, these grooves cross at 50- to 60-degree angles, although anything from 20 to 60 degrees is acceptable. This finish leaves millions of tiny diamond-shaped areas to serve as oil reservoirs (**Figure 11–21**). It also provides flat areas or plateaus on which an oil film can form to separate the rings from the wall. If the angle of the crosshatch is too steep, the oil film will be too thin, causing ring and cylinder scuffing. If the angle is too flat, the pistons may hydroplane and excessive oil consumption will result.

Cylinder Deglazing If the inspection and measurements of the cylinder wall show that surface conditions, taper, and out-of-roundness are within

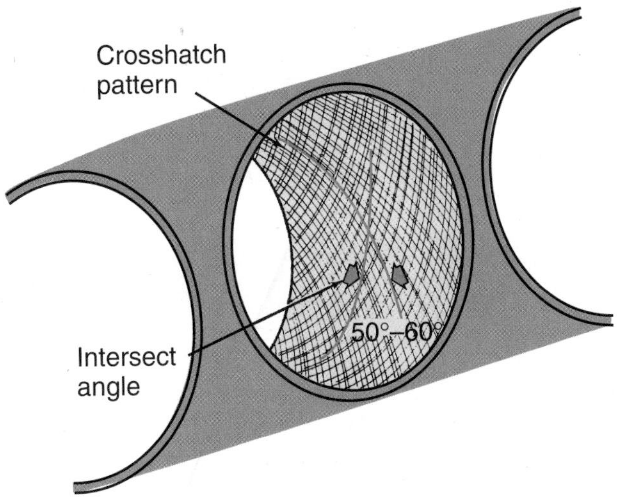

Figure 11-20 Ideal crosshatch pattern for cylinder walls. *Courtesy of Chrysler LLC*

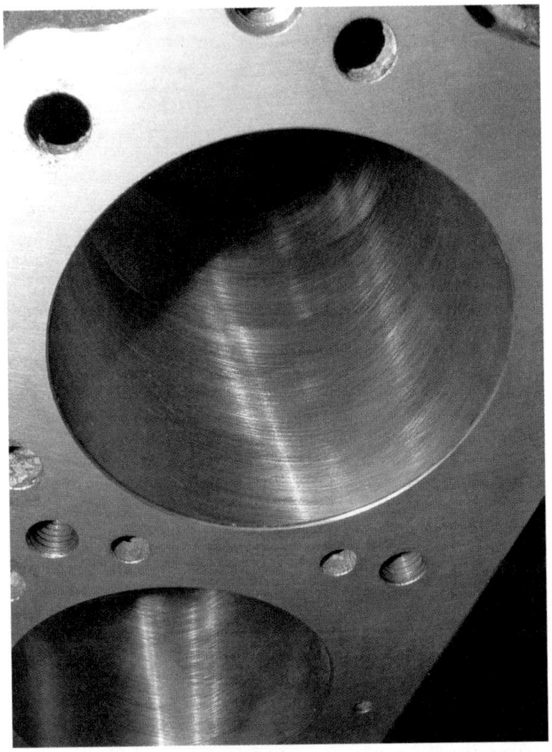

Figure 11-21 The desired cylinder wall finish for most types of piston rings.

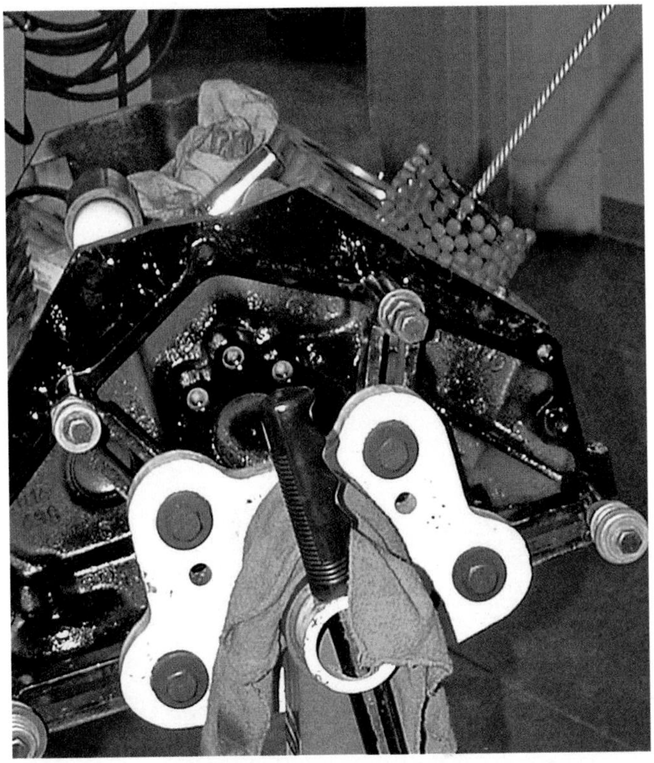

Figure 11-22 Using a resilient-based, hone-type brush, commonly called a ball hone.

acceptable limits, the cylinder walls may only need to be deglazed. Combustion heat, engine oil, and piston movement combine to form a thin residue, **glaze**, in the cylinders. Glazed walls allow the piston rings to slide over them and prevent a positive seal between the walls and the rings.

It is easy to confuse glaze with the polished surface that appears in the cylinders after the engine has some miles on it. Often, glazing can be removed by wiping the cylinders with denatured alcohol or lacquer thinner. Fine honing stones can also remove glaze and leave the walls with a desired finish. Often, technicians use a ball hone to deglaze **(Figure 11–22)** and create the desired pattern on the walls of the cylinder. On many newer engines, deglazing with a ball hone or stones is not recommended. Always check the manufacturer's recommendations before deglazing or honing the cylinder walls.

Cylinder Honing A cylinder **hone** usually has three or four stones. The hone is spun by an electric motor and is moved up and down the cylinder's bore. Springs push the stones against the walls. As they rotate, a small amount of metal is removed from the walls. Honing oil flows over the stones and onto the cylinder wall to control the temperature and to flush out any metallic and abrasive residue. To achieve the correct wall finish, the correct type of stones should be used. Stones are classified by grit size: The lower the grit number, the coarser the stone. The type of piston ring that will be installed normally dictates the desirable grit.

Cylinder honing machines are available in manual and automatic models **(Figure 11–23)**. Automatic models allow the technician to dial in the desired crosshatch angle. When honing a cylinder by hand, use a slow-speed (200–450 rpm) electric drill. Mount the honing tool into the drill and insert it into the bore. Adjust the stones so they fit snugly to the narrowest section of the cylinder. Move the drill and the hone up and down in the bore with short strokes.

Figure 11-23 Automatic cylinder hone machine.

Figure 11-24 Torque plates are fastened to the block during cylinder boring and honing to prevent block distortion during the machining process. *Courtesy of Jasper Engines & Transmissions*

Never remain in one spot too long. Squirt some honing oil on the walls and occasionally stop honing and clean the stones. Continue until the desired results are achieved.

Cylinder Boring When cylinder surfaces are badly worn or excessively scored or tapered, a **boring bar** or boring machine is used to cut the cylinders for oversize pistons or sleeves. A boring bar leaves a pattern on the cylinder wall similar to uneven screw threads. These can cause poor oil control and excessive blowby. Therefore, always hone a cylinder after it has been bored.

CAUTION!

Always wear eye protection when operating deglazing, honing, or boring equipment.

When engines are bored, oversized pistons and rings are used. These are available as 0.020″ (0.50 mm), 0.030″ (0.77 mm), 0.040″ (1.0 mm), and 0.060″ (1.5 mm) oversized.

Torque plates simulate the weight and structure of a cylinder head. They are used by engine rebuilding shops and are fastened to the cylinder block to equalize or prevent twist and distortion when honing or boring a cylinder (**Figure 11–24**).

After servicing the walls, use plenty of hot, soapy water; a stiff-bristle brush; and a soft, lint-free cloth to clean the walls. Then rinse the block with water and allow it to thoroughly dry. Put a light coat of clean engine oil on the walls to prevent rust.

Lifter Bores

Carefully check each valve lifter bore for cracks and evidence of excessive wear. Oblong or egg-shaped bores indicate wear. If the bores are rusted, glazed, or have burrs and high spots, they can be honed with a brake wheel cylinder hone. Be careful not to remove more than 0.0005 inch of metal while honing. If the bores exceed allowable wear limits or are damaged, the engine block should be replaced.

Checking Crankshaft Saddle Alignment

If the block is warped and its main bearing bores are out of alignment (**Figure 11–25**), the crankshaft will bend as it rotates. This causes bearing failure and possibly a broken crankshaft. Blocks that are not severely

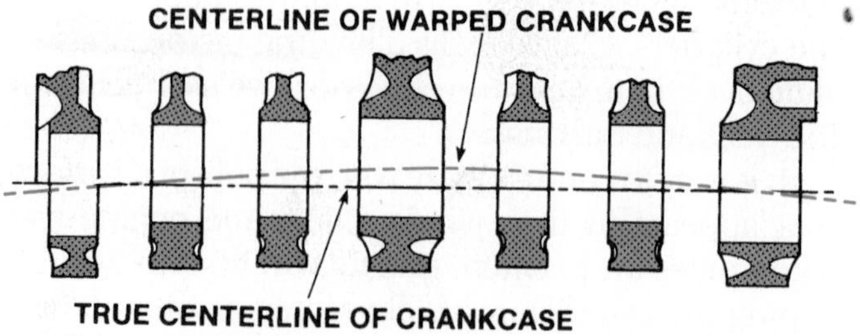

Figure 11-25 An exaggerated view of crankcase housing misalignment.

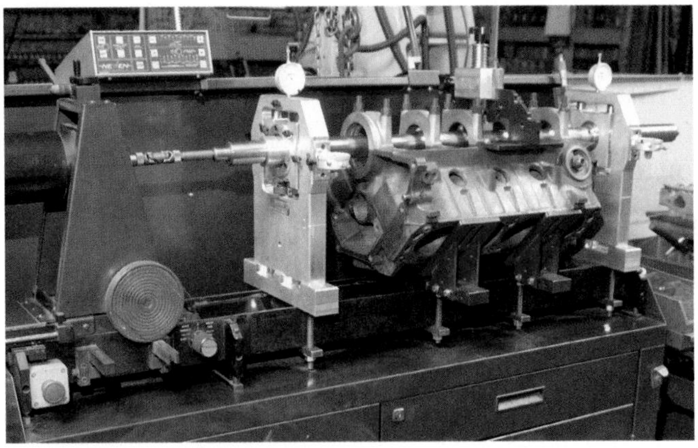

Figure 11-26 A line boring machine for correct crankshaft saddle alignment.

warped can be repaired by **line boring** them. This is a special machining operation in which the main bearing bores are cut to an oversize in order to keep perfect alignment **(Figure 11–26)**. Badly warped blocks are replaced.

The alignment of the crankshaft saddle bore can be checked with a precisely ground arbor placed into the bearing bores. The arbor is rotated in the bores; the effort required to rotate it indicates the alignment of the bores.

Saddle alignment can also be checked with a metal straightedge **(Figure 11–27)**. Place the straightedge in the saddles. Attempt to slide a feeler gauge that is half the maximum specified oil clearance under the straightedge. If this can be done at any saddle, the saddles are out of alignment and the block must be line-bored.

The roundness of the bearing saddles should be checked with a dial bore gauge or telescoping gauge and micrometer. To do this, install and tighten the main bearing caps, then measure the inside diameter in many places in each bore.

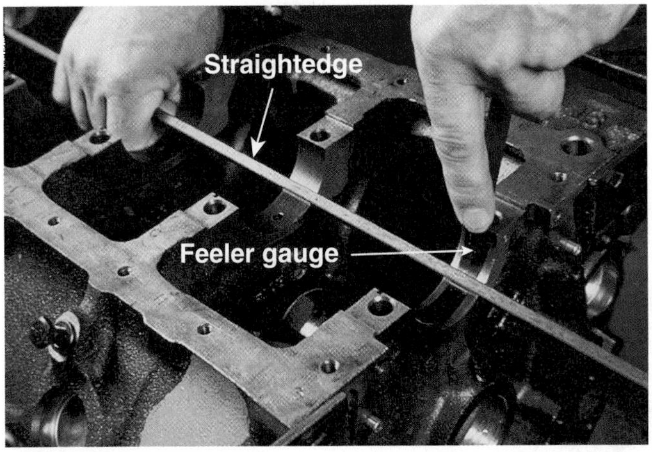

Figure 11-27 Checking bore alignment with a straightedge and feeler gauge. *Courtesy of Federal-Mogul Corp.*

Installing Core Plugs

After the block has been serviced and cleaned, new oil and core plugs should be installed. The plugs' bores should be inspected for any damage that would interfere with the proper sealing. Make sure the plugs are the correct size and type. Coat the plug or bore lightly with a nonhardening oil-resistant (oil gallery) or water-resistant (cooling jacket) sealer. The oil plugs are threaded into their bores. If the threads are damaged, run a tap through the bore.

If the bore for a core plug is damaged, it should be bored out for an oversized plug. Oversize (OS) plugs are identified by the OS stamped on the plug. The correct way to install a core plug depends on the type of plug.

Disc- or Dished-Type These fit in a bore with the dished side facing out **(Figure 11–28A)**. With a hammer, hit the center of the disc's crown and drive the plug in until just the crown becomes flat. This allows the plug to expand properly and have a tight fit.

Cup-Type These are installed with the flanged edge outward **(Figure 11–28B)**. The bore for these plugs has its largest diameter at the outer (sealing) edge. The outside of the cup must be positioned behind the chamfered edge of the bore to effectively seal the bore.

Expansion-Type These are installed with the flanged edge inward **(Figure 11–28C)**. The bore for these plugs has its largest diameter at the back of the bore. The base of the plug must be positioned at the rear of the bore to seal it.

CAMSHAFTS

A camshaft **(Figure 11–29)** has a cam for each exhaust and intake valve. Each cam has a lobe that controls the opening of the valves. The height of the lobe is proportional to the amount the valve will open. The camshaft in older engines had an eccentric to operate the fuel pump and a gear to drive the distributor and oil pump. Some diesel engines have cam lobes for fuel injectors, fuel injection pumps, and/or air starting valves. The camshaft fits in a bore above the crankshaft. The bore is either in the center of the block or slightly off to one side, unless the engine has overhead camshafts. Many camshafts are cast iron or steel; others are forged steel units. To reduce weight and add strength, the camshaft in many late-model engines is a steel tube with lobes welded to it.

On OHV engines, the camshaft lobes work with lifters, pushrods, and rocker arms to open the valves.

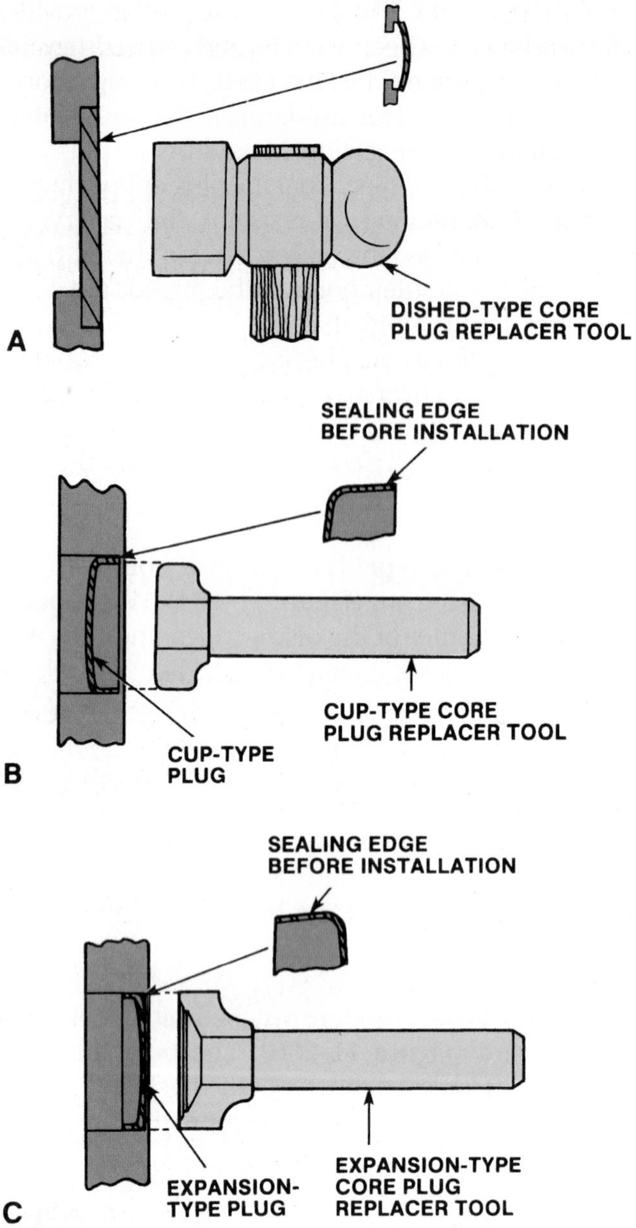

Figure 11-28 Core plug installation methods: (A) dished, (B) cup, and (C) expansion.

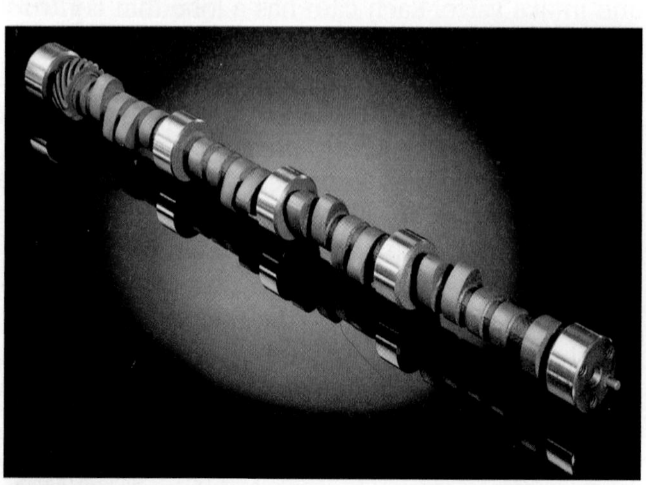

Figure 11-29 A camshaft for a V8 engine. *Courtesy of Melling Automotive Products*

As the cam lobe rotates, it pushes up on the lifter, which moves the pushrod and one end of the rocker arm up. Since a rocker arm is a lever, as that end goes up, the other moves down. This action pushes down on the valve to open it. As the cam continues to rotate, the valve spring closes the valve while maintaining contact between the valve and the rocker arm, thereby keeping the pushrod and the lifter in contact with the rotating cam.

OHC engines may have separate camshafts for the intake and exhaust valves. As these camshafts rotate, the lobes directly open the valves or open them indirectly with cam followers, rocker arms, or bucket-type tappets. The closing of the valves is still the responsibility of the valve springs. Service to the camshaft(s) in an OHC engine is usually part of the procedures for reconditioning the engine's cylinder head.

A camshaft is driven by the crankshaft at half its speed. This is accomplished through the use of a camshaft drive gear or drive sprocket that is twice as large as the crankshaft sprocket. During one full rotation of the camshaft, the intake and exhaust valves open and close once. To synchronize the opening and closing of the valves with the position and movement of the pistons, the camshaft is timed to the crankshaft.

In the valve timing diagram **(Figure 11–30)** valve action is shown in relation to crankshaft rotation. The

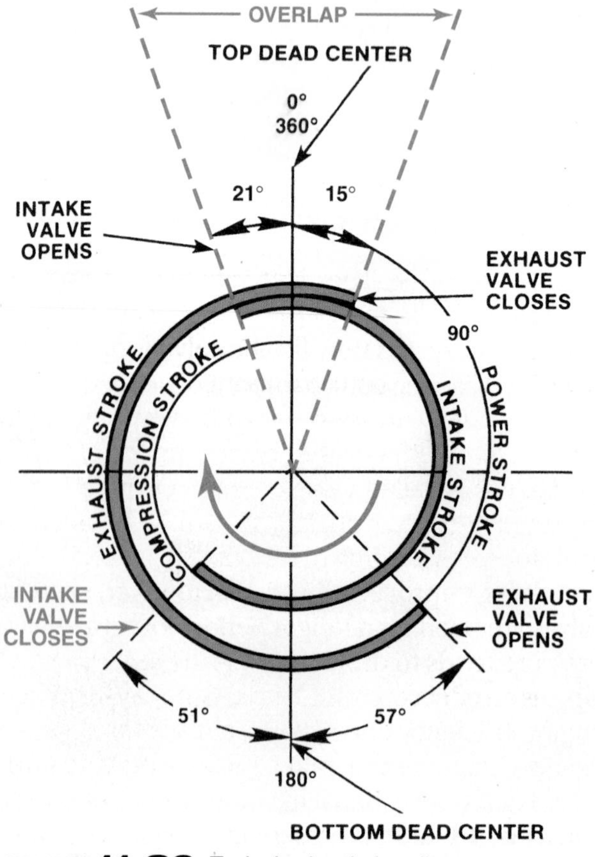

Figure 11-30 Typical valve timing diagram.

intake valve starts to open at 21 degrees BTDC and remains open until it has traveled 51 degrees past BDC. The number of degrees between the valve's opening and closing is called intake valve duration time (252 degrees). The exhaust stroke begins at 57 degrees before BDC and continues until 15 degrees ATDC. The total exhaust valve duration is 252 degrees. The specifications for the camshaft used for the figure show the duration of the intake valve to be the same as the exhaust. This is typical; however, some camshafts are designed with different durations for the intake and exhaust valves. These camshafts are called dual-pattern cams. Different engine designs require different valve opening and closing times. Therefore, each engine design has a unique camshaft.

The actual design of the cams and lobes varies with the type of lifter or follower used in the engine. There four distinct types of lifters: solid nonroller, hydraulic nonroller, solid roller, and hydraulic roller. Camshafts designed for solid and hydraulic nonroller cams are often called "flat-tappet" cams. A camshaft must be matched with the type of lifter for which it was designed.

Camshaft Terminology

Many different terms are used to define the specifications of a camshaft. The actual shape of a cam lobe is called the cam profile. The profile determines the duration and lift provided by the camshaft. Valve overlap is controlled by the placement of the lobes on the camshaft.

Camshaft **duration** is how long the cam holds the valves open. It is expressed in crankshaft degrees. The width of the lobe determines the cam's duration. Lift is the distance the lobe moves the lifter or follower to open the valve. Maximum lift is determined by the height of the lobe. The valve is fully open only when the lifter is at the top of the lobe. Camshaft lift does not always express how far the valve is open. Rocker arms increase the actual amount of valve opening.

Both valves are open slightly at the end of the exhaust stroke and the beginning of the intake stroke; this is called valve **overlap**. Overlap is critical to exhaust gas scavenging. A camshaft with a long overlap helps empty the cylinders at high engine speeds for improved efficiency. However, since both valves are open for a longer period, low-rpm cylinder pressure tends to drop. Because the amount of overlap has an effect on cylinder pressure, it affects overall engine efficiency and exhaust emissions. Valve overlap also helps get the intake mixture moving into the cylinder. As the exhaust gases move out of the cylinder, a low pressure is present in the cylinder. This low pressure causes atmospheric pressure to push the intake charge into the cylinder. Less overlap provides for more pressure in the cylinder at low speeds, resulting in more torque at lower speeds.

Performance TIP

Installing a camshaft with a different profile is a popular way to increase the performance of an engine. Choosing the right cam for an engine can be tricky. There are many things that must be considered, such as performance goals, type of transmissions, the weight of the vehicle, and other modifications made to the engine. Most camshaft suppliers offer guidelines for choosing the right cam. Basically, opening the valves more and keeping them open longer allows more air and fuel into the engine but allows more mixture to escape through the exhaust. Longer duration camshafts improve high engine speed performance but weaken low-speed power. Lower duration improves low-speed torque, but it limits high-speed power. Typically if the vehicle is driven mostly on the street, a camshaft with a modest increase in duration and higher lift is best.

Lobe Terminology Other terms are used to describe a camshaft's profile:

- *Base Circle*—The base circle is the cam without its lobe. It is also the part of the cam where valve adjustments are made.
- *Nose*—The nose is the highest portion of the cam lobe measured from the base circle. This point provides for the maximum amount of lift.
- *Ramp (Flank)*—The ramps are the sides of a cam lobe that lie between the nose and base circle. The ramp on one side is for valve opening and the other for valve closing. How quickly a valve will open and close depends on the steepness of the ramp.
- *Clearance Ramps*—Clearance ramps are at the very beginning or end of the opening and closing ramps. The opening clearance ramp removes any slack in the valve train before applying pressure against the valve spring to open the valve. The closing clearance ramp restores that clearance. These clearance ramps are important to the durability of the valve train.
- *Lobe Separation*—Lobe separation is the angle formed by the centerlines of the intake and exhaust lobes. This angle is measured at the camshaft and the degree rating reflects the angle at the camshaft and has nothing to do with crankshaft rotation. If the maximum lift points of the

intake and exhaust lobes are 105 degrees apart, the camshaft has a 105-degree lobe separation angle. Lobe separation angle has a direct relationship to amount of overlap. The larger or wider the separation, the less overlap there is.

Timing Mechanisms

The camshaft and crankshaft must always remain in the same relative position to each other. They must also be aligned to each other. The alignment is initially set by matching marks on the gears for both shafts; these are called timing marks. The following are the basic configurations for driving the camshaft.

Belt Drive Sprockets on the crankshaft and the camshaft are linked by a continuous neoprene belt. The belt has square-shaped internal teeth that mesh with teeth on the sprockets. The timing belt is reinforced with nylon or fiberglass to give it strength and prevent stretching. This drive configuration is limited to OHC engines. Most technicians replace the belt any time it has been removed.

Chain Drive Sprockets on the camshaft and the crankshaft are linked by a continuous chain. The sprocket on the crankshaft is steel. The sprocket on the camshaft is steel on heavy-duty applications. When quiet operation is a goal, an aluminum sprocket with nylon teeth is used. Nearly all OHV engines use a chain drive system. Chain drives are also used on many OHC engines, especially DOHCs. Often multiple chains are used and arranged in an elaborate fashion. Most chain drives have a chain tensioner to maintain proper tightness and different silencing pads to reduce chain noise.

Gear Drive A gear on the crankshaft meshes directly with a gear on the camshaft. The crankshaft gear is usually iron or steel. The camshaft gear is steel on heavy-duty applications, or aluminum or pressed fiber when quiet operation is a major consideration. The gears are helical. Helical gears are strong and tend to push the camshaft backward to help prevent it from walking out of the block.

Tensioners There is a tensioner on long chain and belt drives. The tensioner may be spring loaded and/or hydraulically operated. Its purpose is to keep the belt or chain under the correct tension as it wears and stretches. All belts and chains have a drive or tension side and a slack side. The tension side is the side that is always being pulled. The tensioner is positioned on the slack side and presses in on the belt or chain.

Variable Valve Timing Previously, variable intake and exhaust timing was only possible with overhead cams. DaimlerChrysler became the first manufacturer to produce a cam-in-block engine with independent control of exhaust camshaft timing. This system was introduced in the 2008 Dodge Viper SRT10's 8.4-liter engine. This is the first production pushrod equipped engine with true **variable valve timing (VVT)**. The VVT system electronically adjusts valve overlap by changing exhaust valve opening times in response to engine speed and load. The system provides an increase in horsepower and torque. It also reduces fuel consumption and exhaust emissions.

This VVT system uses a special camshaft and phaser. The phaser is attached to the end of the camshaft. Inside the phaser are vanes that move within a fixed cavity inside a sealed hub. The movement of the vanes is controlled by oil pressure. The applied oil pressure is controlled by the powertrain control module (PCM). The PCM transmits a signal to a solenoid to move a valve spool that regulates the flow of oil to the phaser cavity. As the applied oil pressure increases, the vanes move against spring pressure. Each vane can rotate a total of 22.5 degrees inside its chamber.

The camshaft is actually two camshafts: an inner shaft and an outer hollow tube-type shaft. It is a camshaft within a camshaft. The exhaust lobes are attached to the outer shaft and the intakes are pinned to the inner camshaft through slots in the outer tube (**Figure 11–31**). Locking pins pass through the slots and are driven through the intake cam lobe assemblies. The exhaust lobes are pressed into position on the hollow outer shaft.

The phaser hub is fit with an external gear that is driven, via a chain, by the crankshaft. The vanes are connected to the outer tube and the hub. As the vanes

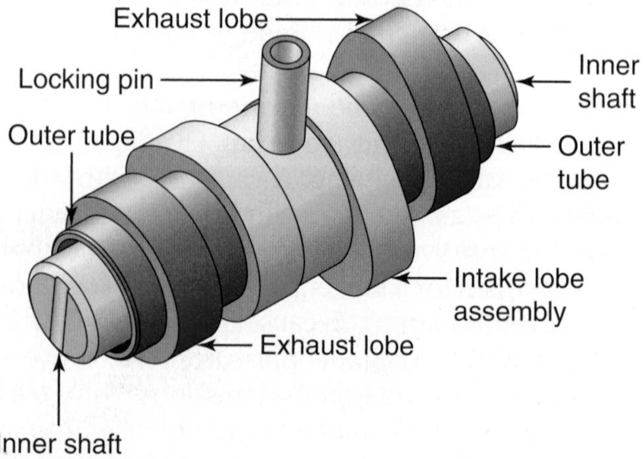

Figure 11–31 The basic construction of the camshaft within a camshaft used for variable valve timing on an OHV engine.

rotate, the position of the exhaust lobes, in relationship to the intake lobes, changes. The amount the exhaust lobes can move is limited by the size of the oil cavity and the slots in the outer tube.

Although it is possible to control both the intake and exhaust valves with this VVT system, Chrysler chose to alter only the exhaust valve timing. Its goal was to increase horsepower without causing rough idle and high emissions. By controlling exhaust valve timing, Chrysler was able to use high-performance camshaft profiles without affecting low-speed operation.

Camless Engines In the very near future, four-stroke cycle engines will not have a camshaft. In a camless engine, the valves are opened and closed electronically or with electrohydraulic devices. The electrohydraulic system uses an electromagnetic field to control hydraulic solenoids. Camless technology will allow total control of the valves' lift, duration, and overlap. This would make engines much more efficient, and there would be less engine power loss due to friction. A conventional engine loses much power through the camshaft and its drive gears and chains or belts. It is said that this technology will improve fuel economy about 20% and provide nearly 20% more low-speed torque.

Valve Lifters

Valve lifters, sometimes called **cam followers** or **tappets**, follow the contour or shape of the cam lobe. Lifters are either mechanical (solid) or hydraulic. Solid valve lifters provide a rigid connection between the camshaft and the valves. Hydraulic valve lifters provide the same connection but use oil to absorb the shock that results from the movement of the valve train.

Hydraulic lifters are designed to automatically compensate for the effects of engine temperature. Changes in temperature cause valve train components to expand and contract. Hydraulic lifters automatically maintain a direct connection between valve train parts.

Solid lifters do not have this built-in feature and require a clearance between the parts of the valve train. This clearance allows for component expansion as the engine gets hot. Periodic adjustment of this clearance must be made. Excessive clearance might cause a clicking noise. This clicking noise is also an indication of the hammering of valve train parts against one another, which results in reduced camshaft and lifter life.

Nonroller lifters rotate in their bore. The contact area between the cam lobe and lifter is one of the highest stressed areas in an engine. Lifter rotation places this stress at different areas at the bottom of the lifter and prevents excessive wear. Cam lobes are ground with a slight taper **(Figure 11–32)**, approximately 0.001 inch (0.254 mm). The diameter of the front of the lobe's base circle is different from the diameter at the rear of the lobe. The speed at which the lifter rotates in its bore depends on the amount of taper. The interface of the lifter and lobe taper also prevents the camshaft from moving to the front or rear while the engine is running.

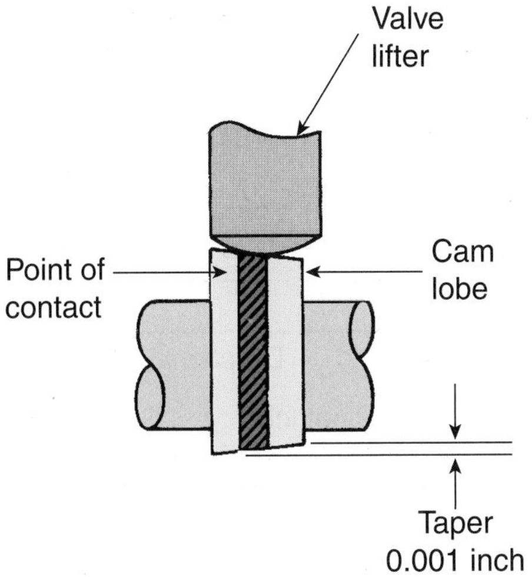

Figure 11-32 The camshaft's lobe is tapered to cause the lifters to rotate.

Roller Lifters In an effort to reduce the friction of the lifter rubbing against the cam lobes—and the resulting power loss—manufacturers often use roller-type lifters. Roller lifters **(Figure 11–33)** have a roller on the camshaft end of the lifter. The roller acts as a wheel and allows the lifter to follow the contour of the cam lobe with little friction. The lifter rolls along the surface of the lobe rather than rub against it. Roller lifters may be solid or hydraulic. The hydraulic part of a

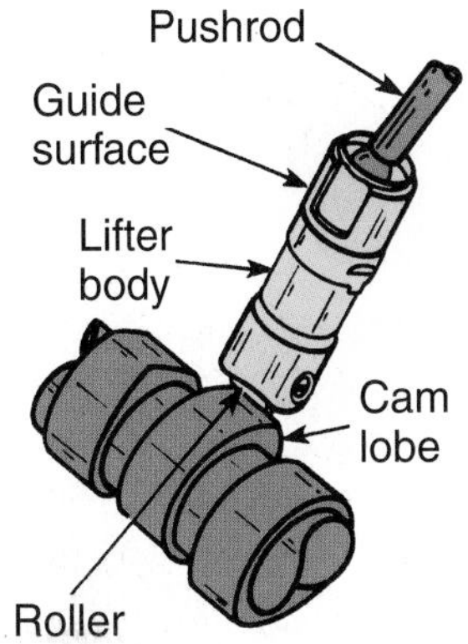

Figure 11-33 A typical roller lifter.

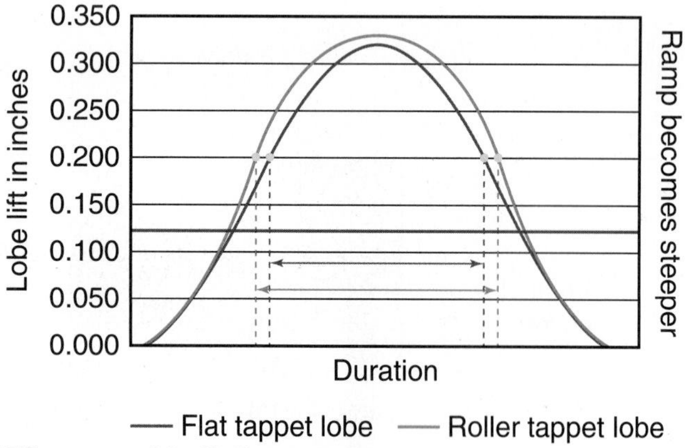

Figure 11-34 Roller lifters are able to follow very aggressive cam lobe designs that provide more lift at a given duration than a flat tappet.

roller lifter works in the same way as a flat hydraulic tappet.

The rollers also allow the lifter to follow aggressive cam lobe designs that provide more lift at a given duration than a flat tappet **(Figure 11–34)**. The lobes on a roller camshaft have steeper ramps and a blunt nose. If the same lobe was used with flat tappets, the edge of the lifter would contact the lobe. This would cause serious damage to the lifter and cam lobe **(Figure 11–35)**. Another advantage of roller lifters is that as long as they are not damaged, they can be reused on a different roller camshaft.

Roller lifters do not and should not rotate in their bores. Therefore, the lobes on roller camshafts are not tapered. To prevent the lifters from rotating, a pair of lifters may be connected by a bar. The bar prevents the lifters from rotating but allows each lifter to move up and down independently. Some manufacturers do not use a tie-bar; rather, they use special fixtures mounted on the block to prevent lifter rotation.

Since the lobes of the camshaft are not tapered, a roller cam will tend to walk toward the front or rear of the engine block. Cam walk can cause a number of problems to the lifters and camshaft. To prevent this problem, thrust washers or cam buttons are fitted at the ends of the camshaft. These keep the camshaft in the correct position under the lifters.

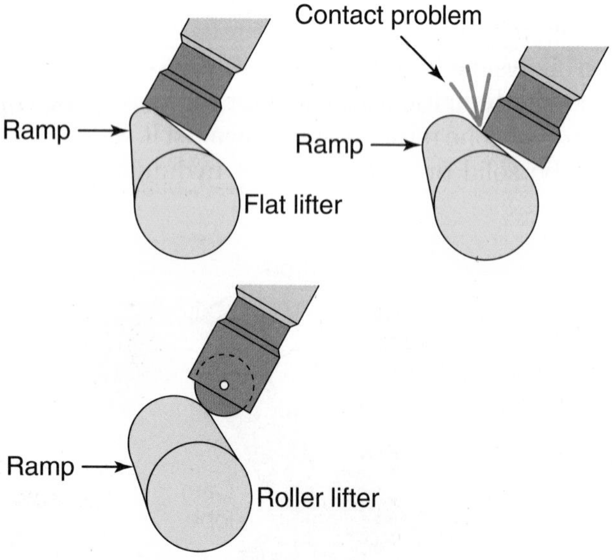

Figure 11-35 The lobes on a roller camshaft can hold the valves open longer at higher lifts. If the same lobe was used with flat tappets, the edge of the lifter would contact the lobe.

Operation of Hydraulic Valve Lifters A hydraulic lifter has a plunger, oil metering valve, pushrod seat, check valve spring, and a plunger return spring housed in a hardened iron body.

When the lifter is resting on the cam's basic circle, the valve is closed and the lifter maintains a zero clearance in the valve train. Oil is fed to the lifter through feed holes in its bore. Oil pressure forces the lifter's check valve closed to keep the oil inside the lifter. This forms a rigid connection between the lifter and pushrod. When there is some clearance in the valve train, a spring between the plunger and lifter body pushes the plunger up to eliminate the clearance. As the cam lobe turns and opens a valve, the lifter's oil feed hole moves away from the oil feed in the lifter bore. New oil cannot enter the lifter. The effort to open the valve pushes the lifter's plunger down slightly. This allows a small amount of oil to leak out; this is called **leakdown**. When the lifter returns to the base of the cam, oil can again fill the lifter **(Figure 11–36)**.

If a hydraulic lifter is not able to leak down or does not fill with oil, a noise will be heard from the engine.

Camshaft Bearings

The camshaft is supported by several bearings, or bushings. OHV camshaft bearings are one-piece plain bearings pressed into the camshaft bore **(Figure 11–37)**. The bearings are either aluminum or steel with a lining of babbitt. **Babbitt** is a soft slippery material made of mostly lead and tin. Alloys of aluminum are commonly found in late-model engines. Aluminum bearings have a longer service life because they are harder than babbitt, but they are more susceptible to damage from dirt and poor lubrication.

OHC camshafts may be supported by split plain bearings. These split bearings are similar to main and connecting rod bearings. Camshaft bearings are normally replaced during engine rebuilding. The old bearings should be inspected for signs of unusual wear that may indicate an oiling or bore alignment problem.

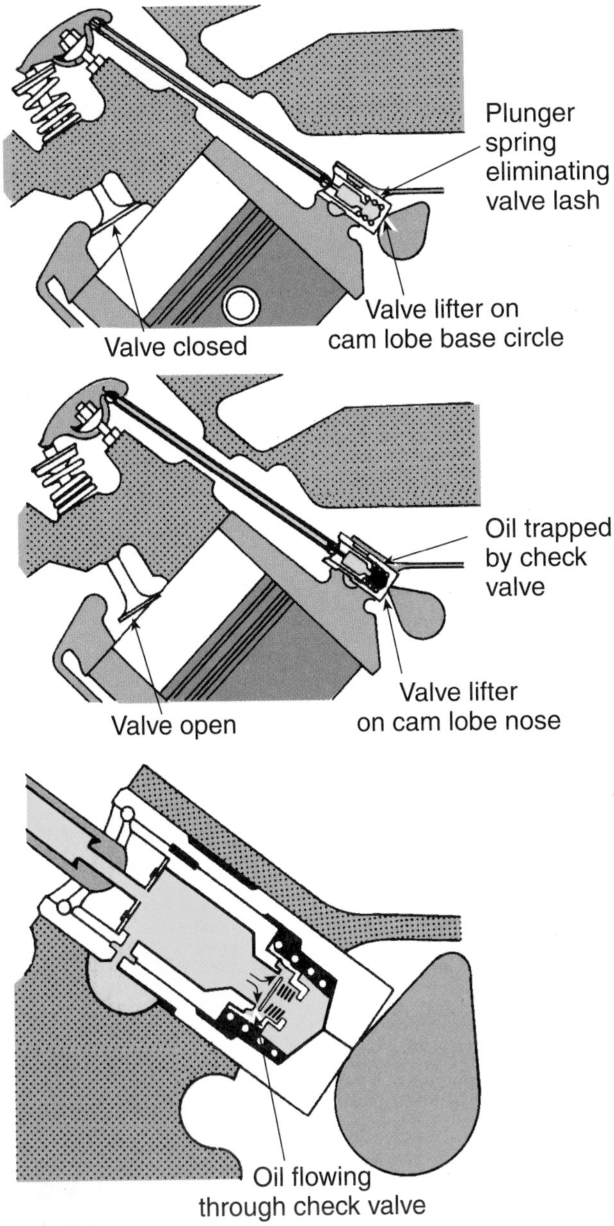

Figure 11-36 Hydraulic lifter operation.

> **SHOP TALK**
>
> Many OHC engines do not use camshaft bearings. Rather the cylinder head is machined to provide support for the camshaft and to serve as bearing surfaces. When these surfaces are damaged, the entire cylinder head is typically replaced.

Balance Shafts

Many late-model engines have one or more balance (silence) shafts to smoothen engine operation. An engine's crankshaft is one of the main sources of engine vibration because it is inherently out of balance. Balance shafts are designed to cancel out these vibrations.

Figure 11-37 The typical camshaft bearing is a full round design.

A balance shaft has counterweights designed to mirror the throws of the crankshaft. These weights are positioned to the opposite side of the weight of the crankshaft and are rotated in the opposite direction as the crankshaft. As the engine turns, the opposing weights mutually cancel out any vibrations. Balance shafts rotate at twice the speed of the crankshaft and are synchronized or timed to the rotation of the crankshaft. If the balance shaft(s) are not timed to the crankshaft, the engine may vibrate more than it would without a balance shaft.

Balance shafts are located in the engine block on one side of the crankshaft, in the camshaft bore, or in a separate assembly bolted to the engine block **(Figure 11–38)**.

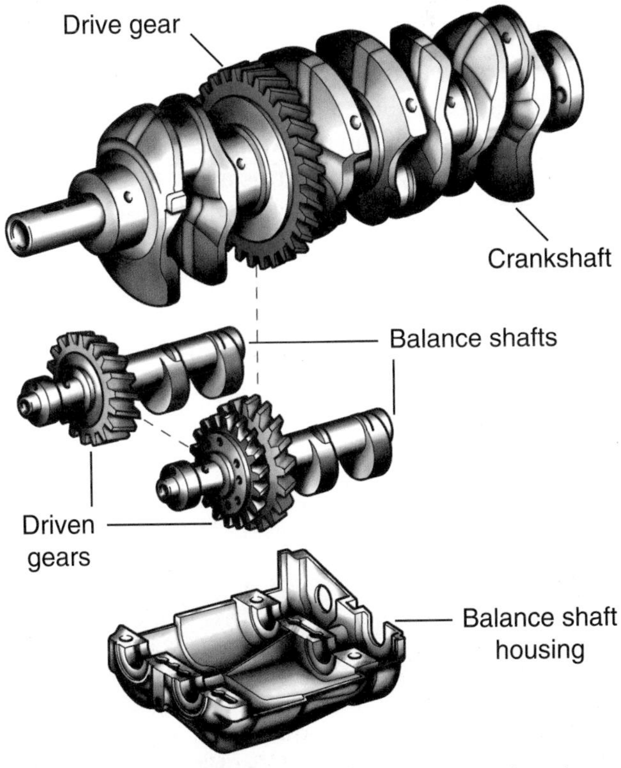

Figure 11-38 Balance shaft assemblies for a four-cylinder engine. *Courtesy of Toyota Motor Sales, U.S.A., Inc.*

Balance shafts are inspected and serviced as part of reconditioning or building a short block. The shafts' journals and bearings need to be checked for wear, damage, and proper oil clearances. Check the bolts for the housing. If they are damaged, replace them. Use a vernier caliper to measure the length of the bolts from their seat to the end. Compare the length to specifications. If the bolts are too long, replace them. The drive chain for the shaft(s) should also be checked for stretching. This is done by pulling on two ends of the chain and measuring that length. If the length is greater than specifications, the chain should be replaced.

CRANKSHAFT

Crankshafts **(Figure 11–39)** are made of cast iron, forged cast steel, or nodular iron, and then machined. At the centerline of the crankshaft are the main bearing journals. These journals are machined to a very close tolerance because the weight and movement of the crankshaft are supported at these points. The number of main bearings varies with engine design. V-type engines have fewer main bearings than an inline engine with the same number of cylinders. A V-type engine uses a shorter crankshaft.

Offset from the crankshaft's centerline are connecting rod journals. The degree of offset and number of journals depends on engine design **(Figure 11–40)**. An inline six-cylinder engine has six connecting rod journals. A V6 engine may have only three. Each journal has two rods attached to it, one from each side of the V. A connecting rod journal is also called a crank pin.

The position of the rod journals places the weight and pressure from the pistons away from the center of crankshaft. This creates an imbalanced condition. To overcome this imbalance, counterweights are added to the crankshaft. These are positioned opposite the connecting rod journals.

The main and rod bearing journals must have a very smooth surface. A clearance between the journal and bearing is needed to allow a film of oil to form. The crankshaft rotates on this film. If the crankshaft journals become out of round, tapered, or scored, the oil film will not form properly and the journal will contact the bearing surface. This causes early bearing or crankshaft failure. The main and rod bearings are generally made of lead-coated copper or tin, or aluminum. These materials are softer than the crankshaft. This means that wear will appear first on the bearings.

Figure 11–39 A crankshaft.

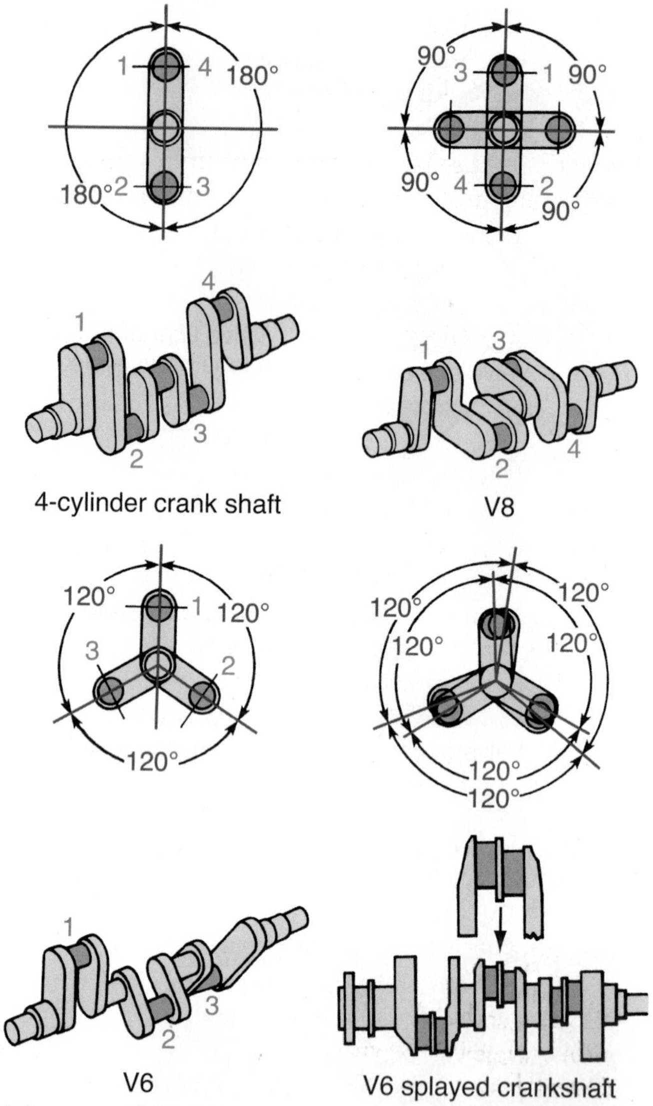

Figure 11–40 Various crankshaft configurations.

It is important that the journals receive an ample supply of clean oil. Each main bearing journal has a hole drilled into it with a connecting bore or bores leading to one or more rod bearing journals. Pressurized oil moves in, over, and out of the journals.

A crankshaft has two distinct ends. One is called the flywheel end and, as its name implies, this is where the flywheel is attached. The front end or belt drive end of the crankshaft has a snout for mounting the crankshaft timing gear and damper.

Crankshaft Torsional Dampers

Combustion causes an extreme amount of pressure in a cylinder (more than 2 tons each time a cylinder fires). This pressure is applied to the pistons and moves through the connecting rods to the crankshaft. This downward force causes the crankshaft to rotate. In an engine with more than one cylinder, this pressure is exerted at different places on the crankshaft and at different times. As a result, the crankshaft tends to twist and deflect, causing torsional harmonic vibrations. These vibrations constantly change but there are specific engine speeds where these harmonics are amplified. This increase in torsional vibration can cause damage to the crankshaft, the engine, and/or any accessories that are driven by the crankshaft.

These harmful vibrations are often limited by a torsional damper located at the front of the crankshaft. There are two common types of torsional dampers: harmonic balancers and fluid dampers. Both use friction to reduce crankshaft vibrations.

Harmonic Balancer A harmonic balancer (**Figure 11–41**), also called a vibration damper, is the most common. A harmonic balancer has a cast-iron hub and an outer cast-iron inertia ring that is connected to the hub with an elastomer (rubber) sleeve. The hub of the harmonic balancer is pressed onto the snout of the crankshaft. The inertia ring is heavy and is machined to serve as a counterweight for the crankshaft. As the crankshaft twists, the hub applies a force to the rubber. The rubber then applies this force to the inertia ring. The counterweight is snapped in the direction of crankshaft rotation to counterbalance the torsional vibrations from the pulsating crankshaft. The connecting rod journals also snap as they receive the high pressure from combustion and the snaps cause the counterweight to snap.

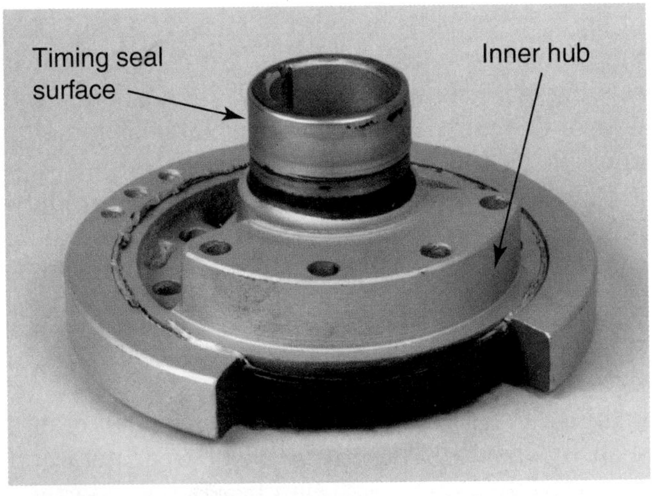

Figure 11-41 A vibration damper or harmonic balancer.

To allow the outer ring to move independently of the hub, the rubber sleeve deflects slightly. The condition of this sleeve is critical to the effectiveness of the balancer. Check the condition of the rubber; look for any broken areas or tears. If it looks good, press on the rubber; it should spring back. If the balancer fails the checks, it should be replaced.

Fluid Damper This type of torsional damper is seldom used by the OEM. However, it is commonly installed by the aftermarket. They are effective in a wide range of engine speeds, especially high speeds. Fluid-filled dampers have a hub surrounded by an inertia ring. Rather than connecting the two with rubber, the outer ring encases the hub. A high-viscosity silicone fluid surrounds the hub. As the outer ring snaps in response to the snapping of the connecting rods and crankshaft, the outer ring rubs against the hub. This rubbing creates friction. The friction is absorbed by the fluid and turned into heat. Therefore, the vibrations are changed to heat and the heat is dissipated from the damper.

Flywheel

The **flywheel** also helps the engine run smoother by applying a constant moving force to carry the crankshaft from one firing stroke to the next. Once the flywheel starts to rotate, its weight tends to keep it rotating. The flywheel's inertia keeps the crankshaft rotating smoothly in spite of the pulses of power from the pistons.

Because of its large diameter, the flywheel also makes a convenient point to connect the starter to the engine. The large diameter supplies good gear reduction for the starter, making it easy for it to turn the engine against its compression. The surface of a flywheel may be used as part of the clutch. On a vehicle with an automatic transmission, a lighter **flexplate** is used. The torque converter provides the weight required to attain flywheel functions.

Flywheel Inspection Check the runout of the flywheel and carefully inspect its surface. Replacement or resurfacing may be required. Excessive runout can cause vibrations, poor clutch action, and clutch slippage. With both manual shift and automatic transmissions, inspect the flywheel for a damaged or worn ring gear. Many ring gears can be removed and flipped over if they are damaged on one side.

CRANKSHAFT INSPECTION AND REBUILDING

Check the crankshaft for the following:

- Are the vibration damper and flywheel mounting surfaces eroded or fretted?
- Are there indications of damage from previous engine failures?
- Do any of the journal diameters show signs of heat checking or discoloration from high-operating temperatures?
- Are any of the sealing surfaces deeply worn, sharply ridged, or scored?
- Are there any signs of surface cracks or hardness distress?

If any or all of these conditions are present, the parts need to be repaired or replaced.

To measure the diameter of the journals, use an outside micrometer **(Figure 11–42)**. Measure them for size, out-of-roundness, and taper **(Figure 11–43)**. Taper is measured from one side of the journals to the other. The maximum taper is 0.001 inch.

Compare these measurements to specifications to determine if the crankshaft needs to be reground or replaced. If the journals are within specifications, the journal area needs only to be cleaned.

USING SERVICE INFORMATION

Crankshaft specifications can be found in the engine specification section of a service manual.

Crankshaft Reconditioning

If the crankshaft is severely damaged, it should be replaced. A crankshaft with journal taper or grooves, burnt marks, or small nicks in the journal surfaces may be reusable after the journals are refinished. This process grinds away some of the metal on the journals to provide an even and mar-free surface.

Minor journal damage may be corrected by polishing the journals with a very fine sandpaper. A polishing tool rotates a long loop of sandpaper against the journals as the crankshaft is rotated by a stand. The constant movement of the sandpaper and the rotation of the crankshaft prevent the creation of flat spots on the journals.

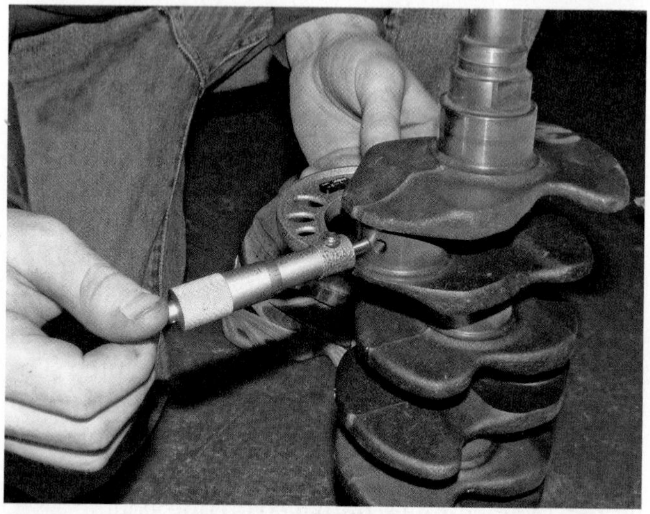

Figure 11-42 Measure the diameters of the crank journals with an outside micrometer.

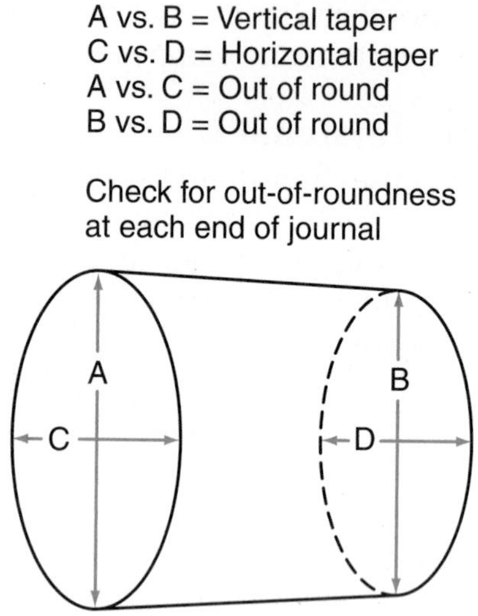

Figure 11-43 Checking crankshaft journals for out-of-roundness and taper.

Checking Crankshaft Straightness

To check the straightness of the crankshaft, ensure that it is supported on V-blocks positioned on the end main bearing journals. Position a dial indicator at the 3 o'clock position on the center main bearing journal.

Set the indicator at 0 (zero) and turn the crankshaft through one complete rotation. The total deflection of the indicator, the amount greater than zero plus the amount less than zero, is the **total indicator reading (TIR)**. Bow is 50% of the TIR. Compare the bow of the crankshaft to the acceptable alignment/bow specifications.

A special machine is used to straighten crankshafts but will only be found in serious engine rebuilding shops. In most cases if the crankshaft is warped, it is replaced.

Crankshaft Bearings

Bearings are used to carry the loads created by crankshaft movement. They are a major wear item and require close inspection. Main bearings support the crankshaft journals. Connecting rod bearings connect the crankshaft to the connecting rods.

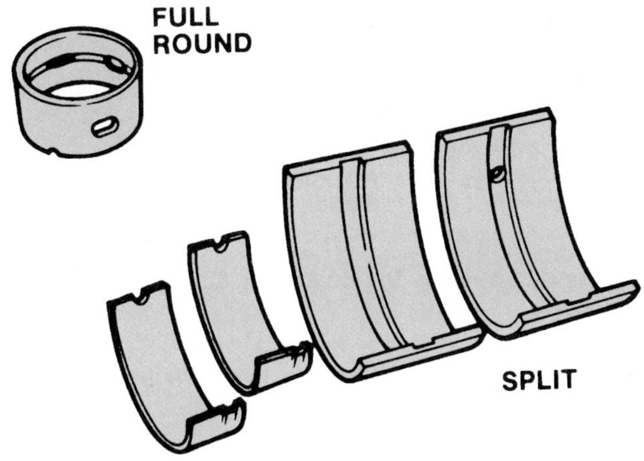

Figure 11-44 Full-round and split insert bearings.

Modern crankshaft bearings are known as insert bearings. There are two basic designs **(Figure 11–44)**. A **full-round** (one-piece) **bearing** is used in bores that allow the shaft's journals to be inserted into the bearing, such as a camshaft. A **split** (two halves) **bearing** is used where the bearing must be assembled around the journal. Crankshaft bearings are the split type.

Many crankshafts have a main bearing with flanged sides. This is called a thrust bearing and is used to control the horizontal movement or end play of the shaft. Most thrust main bearings are double flanged **(Figure 11–45)**. Some crankshafts use flat insert thrust bearings.

Some late-model engines do not have separate main bearing caps; rather, they are fitted with a lower engine block assembly. This assembly works like a bridge and contains the lower half of the main bearing bores. The assembly is torqued to the block.

On some engines, the main bearing caps and lower block assemblies are given additional strength

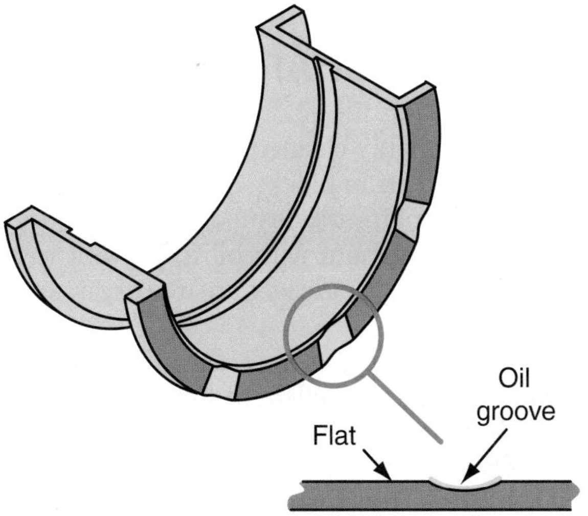

Figure 11-45 A thrust bearing with grooves cut into its flange to provide for better lubrication.

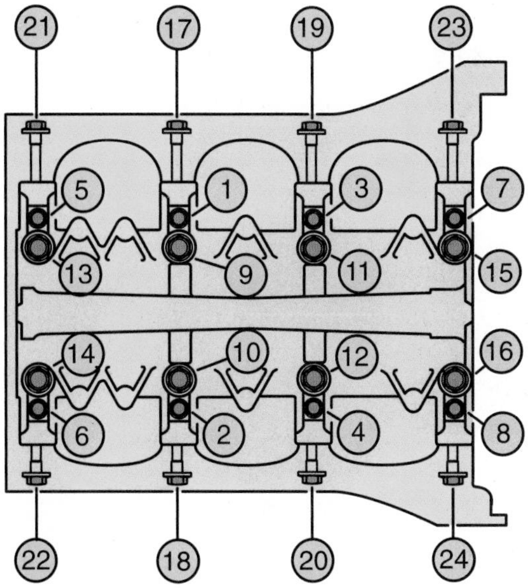

Figure 11-46 Six bolts secure each main cap in this engine. Each bolt must be tightened in correct sequence and to the correct torque.

through the use of additional bolts. The main caps are secured by four bolts, two on each side of the cap. Other designs may have additional side bolts to fasten the side of the cap to the engine block. Regardless of the number and position of the bolts, proper tightening sequences **(Figure 11–46)** must be followed.

Bearing Materials

Bearings can be made of aluminum, aluminum alloys, copper and lead alloys, and steel backings coated with babbitt. Each has advantages in terms of resistance to corrosion, rate of wear, and fatigue strength. Aluminum alloy bearings are the most commonly used. These bearings contain silicon, which helps to reduce wear. Some bearings use a combination of metals, such as a layer of copper-lead alloy on a steel backing, followed by a thin coating of babbitt **(Figure 11–47)**. This design takes advantage of the excellent properties of each metal.

Bearing Spread

Most main and connecting rod bearings have "spread." This means the distance across the outside parting edges of the bearing insert is slightly greater than the diameter of the housing bore. To position a bearing half with spread, it must be snapped into place **(Figure 11–48)**. This provides a good fit inside the bore and helps keep the bearings in place during assembly.

Bearing Crush

Each half of a split bearing is made slightly larger than an exact half. This can be seen quite easily when a half is snapped into place. The parting faces extend a little beyond the seat **(Figure 11–49)**. This extension is called crush. When the two halves are assembled and

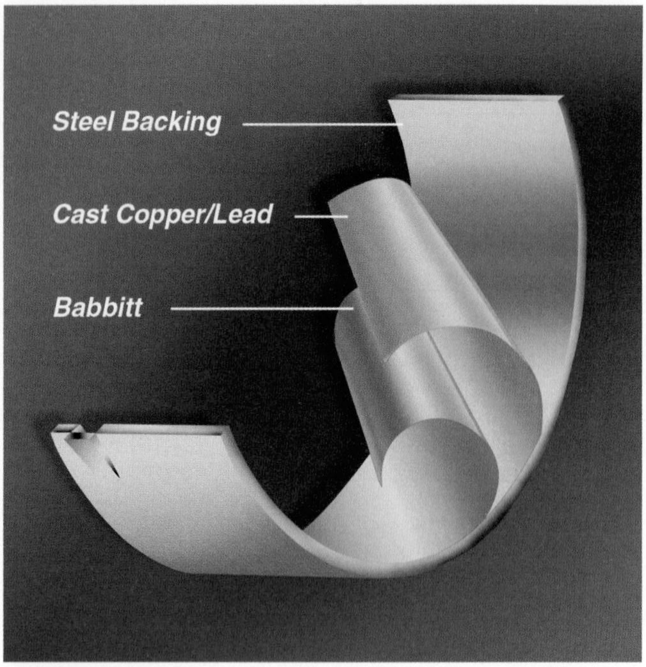

Figure 11-47 The basic construction of a bearing composed of three metals. *Courtesy of Dana Corporation*

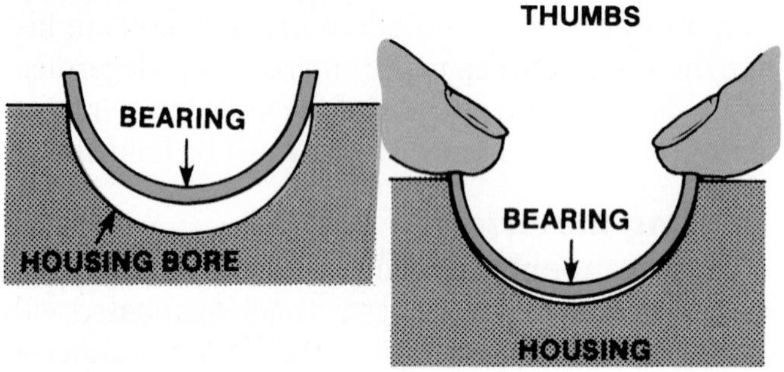

Figure 11-48 Spread requires a bearing to be lightly snapped into place.

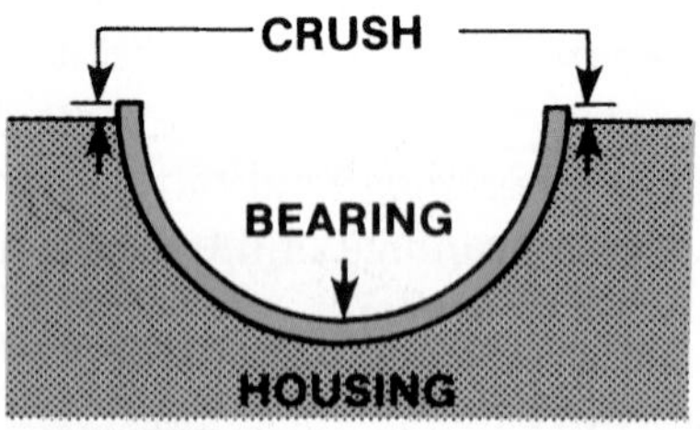

Figure 11-49 Crush ensures good contact between the bearing and the housing.

the cap tightened, the crush forces the bearing halves into the bore. Bearing crush increases the contact area between the bearing and bore, allowing for better heat transfer, and compensates for slight bore distortions.

Bearing Locating Devices

Engine bearings must not be able to rotate or shift sideways in their bores. Many different methods are used to keep the bearings in place. The most common way is the use of a locating lug. As shown in **Figure 11–50**, this is a protrusion at the parting

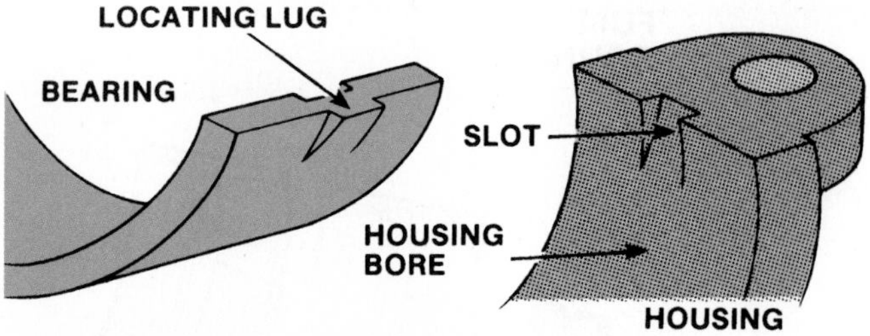

Figure 11-50 The locating lug fits into the slot in the housing.

face of the bearing. The lug fits into a slot in the bearing's bore.

Oil Grooves

To ensure an adequate oil supply to the bearing's surface, an oil groove is added to the bearing. Most OEM bearings have a full groove around the entire circumference of the bearing, and others have a groove only in the upper bearing half.

Oil Holes

Oil holes in the bearings allow oil to flow through the block and into the bearing's oil clearance. These holes control the amount of oil sent to the connecting rod bearings and other parts of the engine. For example, oil squirt holes in connecting rods are used to spray oil onto the cylinder walls. The oil hole normally lines up with the groove in a bearing. When installing bearings, make sure the oil holes in the block line up with holes in the bearings.

Bearing Failure and Inspection

As shown in **Figure 11–51**, bearings can fail for many reasons. Dirt and oil starvation are the major reasons for bearing failure. Other engine problems, such as bent or twisted crankshafts or connecting rods or out-of-shape journals, can also cause bearings to wear irregularly.

INSTALLING MAIN BEARINGS AND CRANKSHAFT

Before assembling the short block, make sure the engine is thoroughly cleaned. Hot water and detergent are used to clean blocks, crankshafts, and camshafts. After the parts are cleaned, blow them dry and immediately coat them with oil to prevent rusting. Also make sure all oil passages are free of dirt and foreign particles.

A gap or clearance between the outside diameter of the crankshaft journals and the inside diameter of its bearings is necessary. This clearance allows for the building and maintenance of an oil film. Proper lubrication and cooling of the bearing depend on correct crankshaft oil clearances. Scored

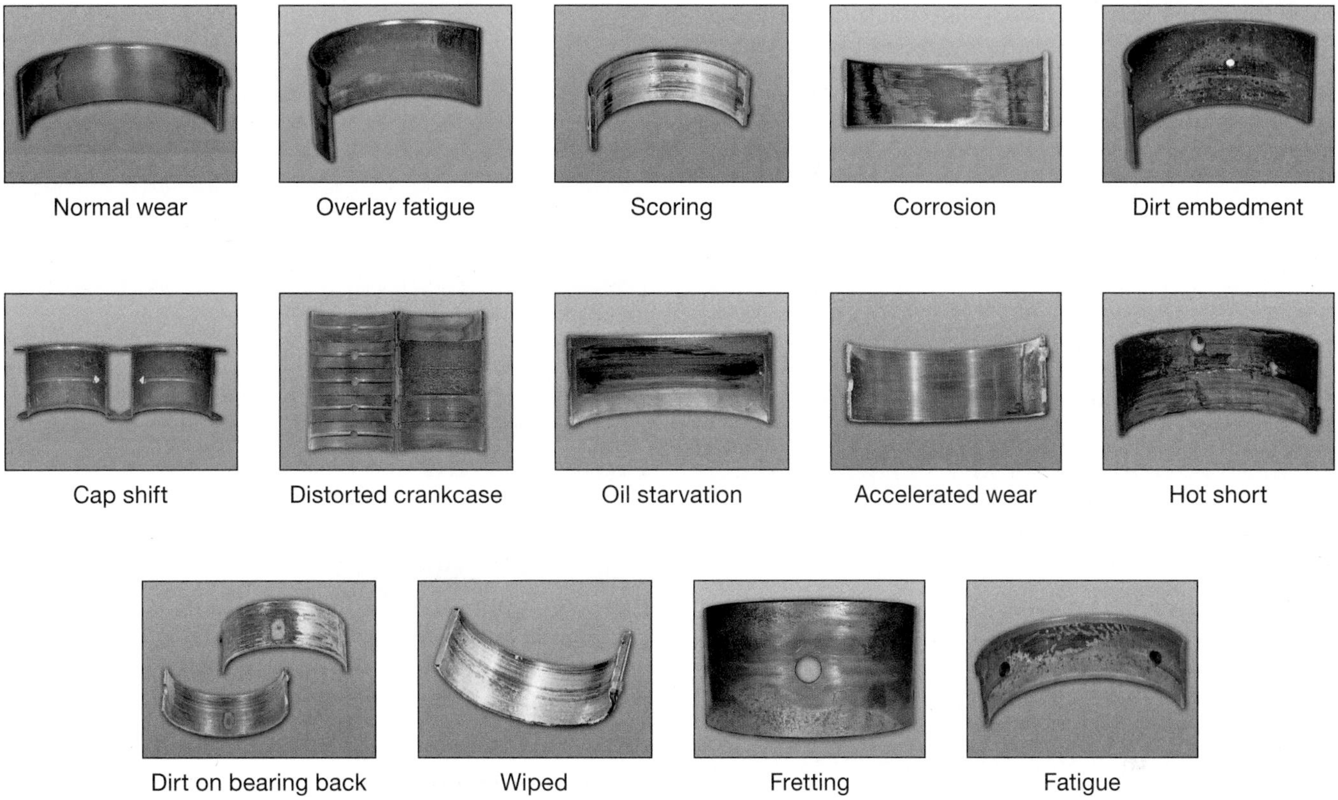

Figure 11–51 Common forms of bearing distress. *Courtesy of Dana Corporation*

bearings, worn crankshaft, excessive cylinder wear, stuck piston rings, and worn pistons can result from too small an oil clearance. If the oil clearance is too great, the crankshaft might pound up and down, overheat, and weld itself to the insert bearings.

During an engine rebuild, if there is little or no wear on the journals, the proper oil clearance can be restored with the installation of standard-size bearings. However, if the crankshaft is worn to the point where standard-size bearings will leave an excessive oil clearance, a bearing with a thicker wall must be used. Although these bearings are thicker, they are known as undersize because the journals and crankpins are smaller in diameter. In other words, they are under the standard size.

Undersize bearings are available in 0.001 inch (0.0254 mm) or 0.002 inch (0.0508 mm) sizes for shafts that are uniformly worn by that amount. Undersize bearings are also available in thicker sizes, such as 0.010 inch (0.2540 mm), 0.020 inch (0.5080 mm), and 0.030 inch (0.7620 mm), for crankshafts that have been refinished (or reground) to a standard undersize. The difference in bearing thickness is normally stamped onto the backside of the bearing. Bearings may also be color-coded to indicate their size. Often engines are manufactured with other than standard journal sizes. The crankshaft is marked to show the size of bearing used **(Figure 11–52)**.

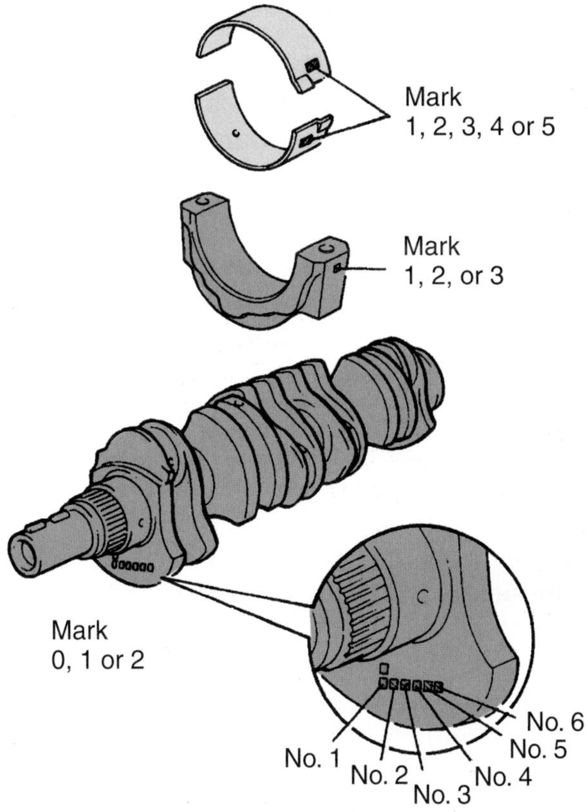

Figure 11–52 Size markings on a crankshaft, connecting rod, and rod bearing. *Courtesy of Toyota Motor Sales, U.S.A., Inc.*

If the housing bores have been machined larger by align boring or align honing, oversized bearings are used to take up this space.

Make sure the new bearings match the crankshaft journal diameters and main bearing bores. Before the bearings are installed, make sure the bore is clean and dry. Use a clean, lint-free cloth to wipe the bearing back and bore surface.

SHOP TALK

The critical bolts in most engines must be tightened with a torque-angle gauge after a specific torque has been reached. Make sure to do this. Also check the length and condition of all bolts before reusing them.

Put the new main bearings into each main bearing cap and the bearing bores in the cylinder block. Make sure all holes are aligned. The backs of the bearing inserts should never be oiled or greased. Place the crankshaft onto the bearings. The oil clearance is now measured with Plastigage. Make sure the journals are clean and free of oil. The presence of oil will give inaccurate clearance measurements.

Plastigage is fine, plastic string that flattens out as the caps are tightened on the crankshaft. The procedure for using Plastigage is shown in Photo Sequence 7.

One side of the Plastigage's package is used for inch measurements, the other for metric measurements. The string can be purchased to measure different clearance ranges. Usually, the smallest clearance range is required for engine work.

If the oil clearance is not within specifications, the crankshaft needs to be machined or replaced, or undersized bearings should be installed.

SHOP TALK

If the journals measure within specifications but they are pitted or gouged, polish the worst journal to determine whether or not grinding is necessary. If polishing achieves smoothness, then grinding is probably not necessary.

Crankshaft End Play

Crankshaft end play can be measured with a feeler gauge by prying the crankshaft rearward and measuring the clearance between the thrust bearing flange and a machined surface on the crankshaft. Insert the feeler gauge at several locations around the rear thrust bearing face (**Figure 11–53**). You may also position a dial indicator so that the fore and aft movement of the crankshaft can be measured.

Figure 11-53 Crankshaft end play can be checked with a feeler gauge. *Courtesy of Federal-Mogul Corp.*

If the end play is less than or greater than the specified limits, the main bearing with the thrust surface must be exchanged for one with a thicker or thinner thrust surface. If the engine has thrust washers or shims, thicker or thinner washers or shims must be used.

Most engines require the installation of main bearing seals during the final installation of the crankshaft.

Connecting Rod

The connecting rod is used to transmit the pressure applied on the piston to the crankshaft (**Figure 11–54**). Connecting rods are able to swivel at the piston and the crankshaft. This allows them to freely move the pistons up and down while they rotate around the crankshaft. A connecting rod faces great stress. The force applied during the power stroke is applied to the connecting rod as it moves through a variety of angles. The rod has great force applied to it from the top and has great resistance to movement at the bottom. The center section of a rod is basically an "I-beam." This provides maximum strength with minimum weight.

Connecting rods are kept as light as possible. They are generally forged from high-strength steel or made of nodular steel or cast iron. Cast iron is rarely used in automotive engines. Aluminum and titanium connecting rods are also used. Aluminium rods are light and have the ability to absorb high-pressure shocks, but they are not as durable as steel rods. Titanium rods are very strong and light but are rather expensive. Some late-model engines, such as the Ford 4.6-liter and the Chrysler 2.0-liter engines, have powdered (sintered) metal connecting rods.

SEQUENCE 7

Checking Main Bearing Clearance with Plastigage

P7-1 *Checking main bearing clearance begins with mounting the engine block upside down on an engine stand.*

P7-2 *Install main bearings into bores, being careful to properly seat them. Wipe the bearings with a clean lint-free rag.*

P7-3 *Wipe the crankshaft journals with a clean rag.*

P7-4 *Carefully install the crankshaft into the bearings. Try to keep the crankshaft from moving on the bearing surfaces.*

P7-5 *Place a piece of Plastigage on the journal. The piece should fit between the radius of the journal.*

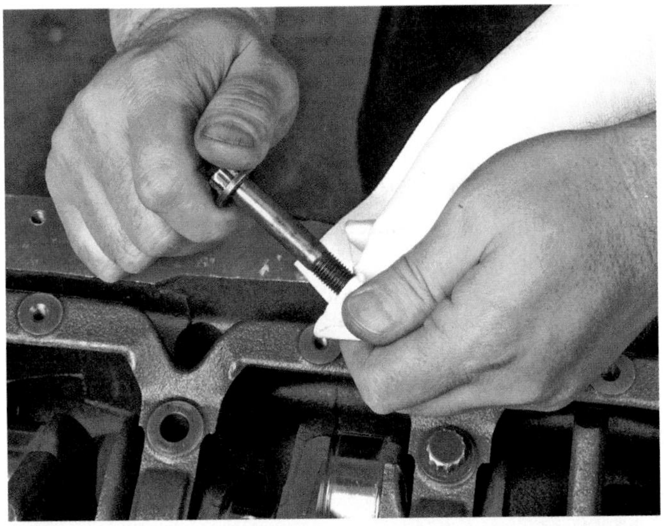

P7-6 *Install the main caps in their proper locations and directions. Wipe the threads of the cap bolts with a clean rag.*

P7-7 *Install the cap bolts and tighten them according to the manufacturer's recommendations.*

P7-8 *Remove the main caps and observe the spread of the Plastigage. If the gage did not spread, try again with a larger gage.*

P7-9 *Compare the spread of the gage with the scale given on the Plastigage container. Compare the clearance with the specifications.*

P7-10 *Carefully scrape the Plastigage off the journal surface.*

P7-11 *Wipe the journal clean with a rag.*

P7-12 *If the clearance was within the specifications, remove the crankshaft and apply a good coat of fresh engine oil to the bearings.*

P7-13 *Reinstall the crankshaft and apply a coat of oil to the journal surfaces.*

P7-14 *Reinstall the main caps and tighten according to specifications.*

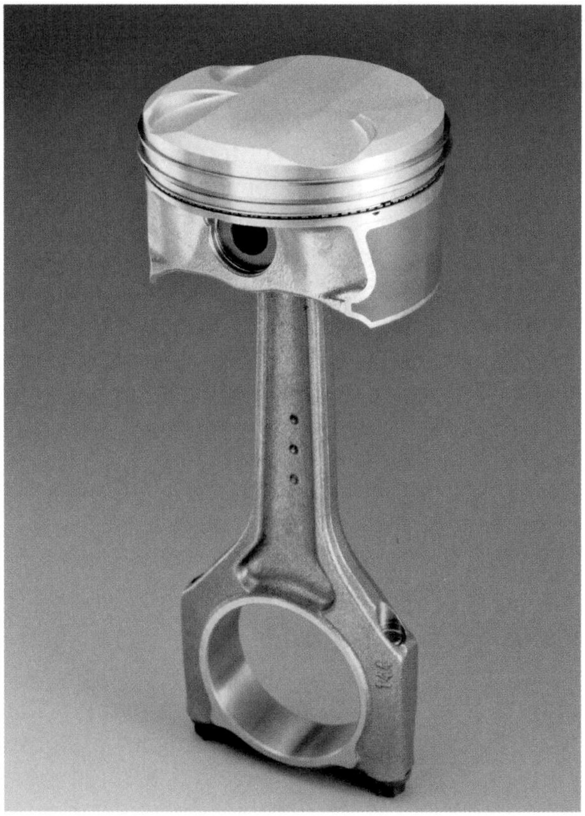

Figure 11-54 A piston and connecting rod assembly. *Courtesy of BMW of North America, LLC*

These rods are light and strong and are easily identified by their smoothness.

The small end or piston pin end is made to accept the **piston pin**, which connects the piston to the rod. The piston pin can be pressed-fit in the piston and free fit in the rod. When this is the case, the small end of the rod will be fitted with a bushing. The pin can also be a free fit in the piston and pressed-fit in the rod. In this case no bushings are used. The pin simply moves in the piston using the piston hole as a bearing surface.

The "big" end of the rod is used to attach it to the crankshaft. This end is made in two pieces. The upper half is part of the rod. The lower half is called the rod cap and is bolted to the rod. The connecting rod and its cap are manufactured as a unit and must always be kept together. During production, the rod caps are either machined off the rod or are scribed and broken off. Since the cap of a powdered metal rod is broken away from the rod during manufacturing, there is an uneven mating surface due to the break and the grain of the powdered metal. When the cap is assembled on the rod, these imperfections ensure that the cap is positioned exactly where it was before it was broken off. With other rods, there is the possibility that the rod and cap will be slightly misaligned when they are assembled.

The big end is fitted with bearing inserts made of the same material as the main bearings. Many connecting rods have a hole drilled through the big end to the bearing area. The bearing insert may have a hole that is aligned with the hole in the rod's bore. This hole supplies oil for lubricating. Some rods have an oil squirt hole. This hole sprays oil onto the cylinder wall to lubricate and cool the piston skirt. When the rod is properly installed, this squirt hole is pointed to the major thrust area of the cylinder wall.

Inspection Closely examine all piston skirts and bearings for unusual wear patterns that may indicate a twisted rod **(Figure 11–55)**. Rods suspected of being bent or distorted can be checked with a rod alignment checker. Normally a damaged rod is replaced, although equipment is available to straighten them and to rebore the small and big ends. Many manufacturers recommend a check of the rod bolts before reusing them. The typical procedure involves measuring the diameter of the bolt at its tension portion.

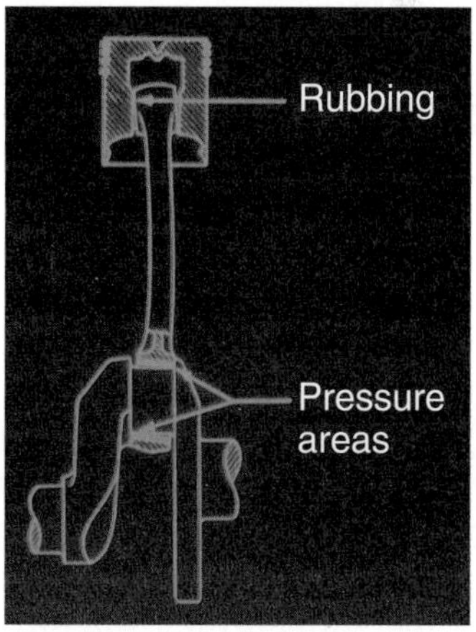

Major misalignment

A sign of a bent and twisted rod

Figure 11-55 The effects of rod misalignment on its bearings. *Courtesy of AE Clevite Engine Parts*

If the diameter is less than the minimum, it should be replaced.

PISTON AND PISTON RINGS

The piston forms the lower portion of the combustion chamber. The pressures from combustion are exerted against the top of the piston, called the **head** or **dome**. A piston must be strong enough to face this pressure; however, it should also be as light as possible. This is why most pistons are made of aluminum or aluminum alloys.

Aluminum pistons alloyed with copper, magnesium, nickel, and silicon are common. Silicon is the most common element mixed with aluminum to make pistons. Silicon makes the piston more resistive to corrosion and improves its strength, hardness, and wear resistance. It also helps to reduce the piston's weight.

There are three basic types of aluminum silicon alloys used in pistons: hypoeutectic, eutectic, and hypereutectic. Hypoeutectic pistons, common in earlier engines, have about 9% silicon. Most eutectic pistons have 11% to 12% silicon. Eutectic alloys provide good strength and are economical to make. **Hypereutectic** pistons have a silicon content above 12%. They offer low thermal expansion rates, improved groove wear, good resistance to high temperatures, and greater strength and scuff and seizure resistance.

The head of the piston can be flat, concave, convex, crowned, raised and relieved for valves, or notched for valves. Newer pistons are typically flat, flat with valve notches, or have a slightly dished crown. The dished crown concentrates the pressure of combustion at the thickest part of the piston head, right above the top of the piston pin boss. The piston pin boss is a built-up area around the bore for the piston pin, sometimes called the **wrist pin** (**Figure 11–56**). The pin bore is not always centered in the piston. It can be offset toward the major thrust side of the piston, which is the side that will contact the cylinder wall during the power stroke.

Piston heads are often coated with hard anodizing, ceramics, or electroplating. These coatings increase the hardness and resistance to corrosion, cracking, wear, and scratching. New ceramic coatings offer nearly three times the surface hardness of traditional hard-anodized coatings. Ceramic coatings also help protect against spontaneous detonation.

Just below the dome, around the sides of the piston, is a series of grooves. The grooves are used to hold the piston rings. The high parts between the grooves are called **ring lands**. Some pistons have a ceramic coating in the top ring groove to prevent the ring from being "welded" inside the groove. Normally there are three grooves: two compression and one oil control. The compression grooves are located toward the top of the piston. The depth of grooves varies with the size of the piston and the type of rings used. The oil control groove is the lowest groove on the piston. They are normally wider than the compression ring grooves and have holes or slots to allow oil to drain. The positions of the ring grooves vary with engine design. Many newer engines have the top compression ring as close as possible to the piston head. This reduces the amount of fuel that can drop down the sides of the piston before combustion. This hidden fuel is not involved in the combustion process but leaves as unburned hydrocarbons during the exhaust

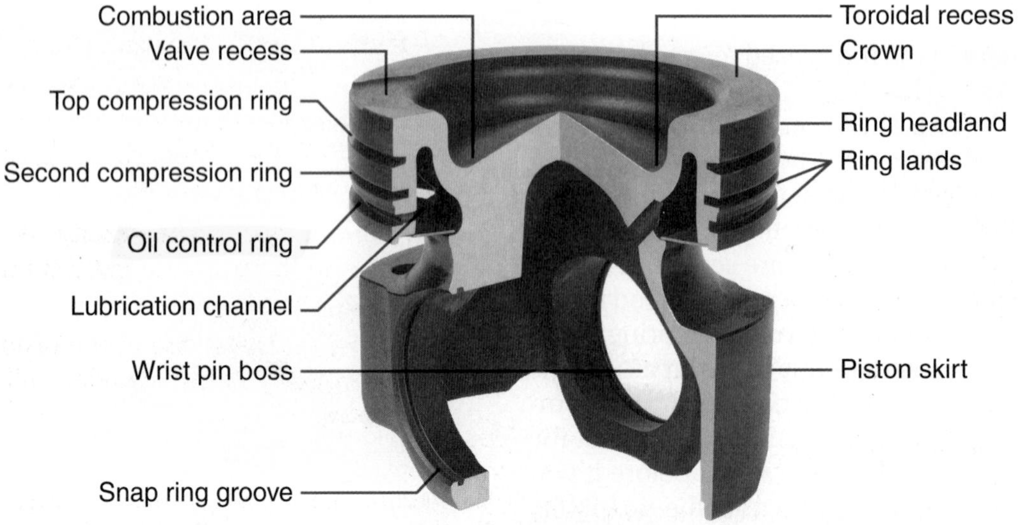

Figure 11-56 The features and terminology used to describe a piston. *Courtesy of MAHLE International GmbH*

stroke. In this design, all rings are placed close together. On a few pistons, the piston pin bore is very close to the piston head, behind the groove for the lower oil control ring.

The area below the piston pin is called the **piston skirt**. The area from just below the bottom ring groove to the tip of the skirt is the piston thrust surface. There are two basic types of piston skirts: the slipper and full skirt. The full skirt is used primarily in truck and commercial engines. The slipper type is used in automobile engines and allows the piston enough thrust surface for normal operation. A slipper skirt also allows the piston to be lighter and reduces piston expansion because there is less material to hold heat.

Late-model engines that are capable of running to fairly high rpm use lighter pistons. These pistons have skirts only on the thrust sides. Often the skirts are coated with molybdenum to prevent cylinder wall scuffing.

To ensure that the piston is installed correctly and has the correct offset, the top of the piston will have a mark. The most common mark is a notch that is machined into the top edge of the piston. Always check with the service manual for the correct direction and position of the mark. The front of the piston must match the front of the connecting rod **(Figure 11–57)**.

When an engine is designed, piston expansion determines how much piston clearance is required in the cylinder bore. Too little clearance will cause the piston to bind at operating temperatures. Too much will cause piston slap. The normal piston clearance for an engine is about 0.001 to 0.002 inch (0.0254 to 0.0508 mm). This clearance is measured between the piston skirt and the cylinder wall. Advances in piston technology have allowed manufacturers to build engines with about half that clearance. This leads to increased efficiency and lower emissions.

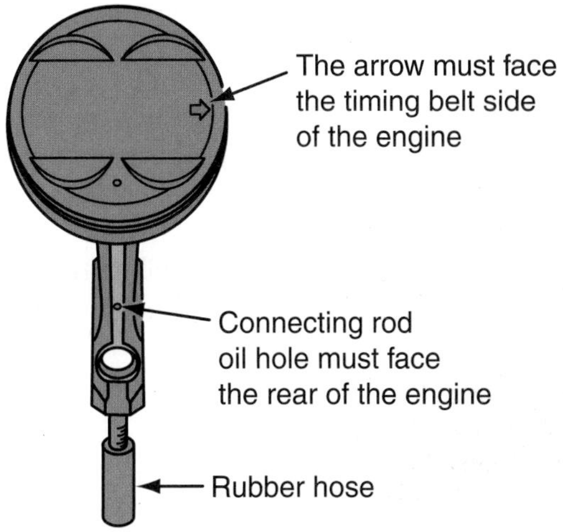
The arrow must face the timing belt side of the engine
Connecting rod oil hole must face the rear of the engine
Rubber hose

Figure 11-57 Always make sure that the markings on the piston and connecting rod are in the correct relationship to each other and they face the correct direction.

Piston Terminology

Many different terms are used to describe the design of a piston; these include:

- *Compression distance (or height)*—The distance from the center of the piston pin bore to the top of the piston.
- *Ring belt*—The area between the top of the piston and the pin bore where the piston rings are installed.
- *Heat dam*—A narrow groove cut in some pistons to reduce heat flow to the top ring groove. During engine operation, the groove fills with carbon and absorbs the heat of combustion.
- *Land diameter*—The diameter of the ring land. On some pistons, the land diameter will be same for each ring; on others it will increase from top to bottom.
- *Land clearance*—The difference between the diameters of the land and the cylinder.
- *Groove root diameter*—The diameter of a piston measured at the bottom of a ring groove. The root diameter of each groove may vary with the type of ring used.
- *Groove protector*—A steel or cast-iron insert placed into the top groove of an aluminum piston to extend the life of the top compression ring.
- *Top groove spacer*—A steel spacer installed above the ring in a reconditioned groove to bring the ring's side clearance within specifications.
- *Piston pin bushing*—Found primarily in cast-iron pistons, this bushing serves as a bearing for the piston pin. It is inserted into the piston pin bore.
- *Major thrust face*—The part of the piston skirt that has the greatest thrust load. This is typically the right side when looking at the engine from the flywheel end.
- *Minor thrust face*—The part of the piston skirt that is opposite the major thrust face.
- *Skirt clearance*—The difference between the diameter of the piston skirt diameter and the diameter of the cylinder.
- *Piston skirt taper*—The difference between the diameter of the piston at the top and bottom of the skirt.
- *Piston cam*—The shape of the piston skirt area, which provides correct cylinder wall contact and clearances.

Inspection Each piston should be carefully checked for damage and cracks. Pay attention to the ring lands and the pin boss area. Look for scuffing on the sides

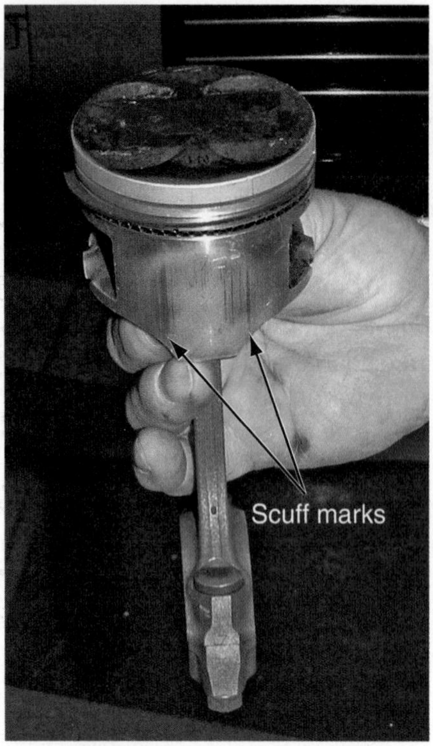

Figure 11-58 Each piston should be carefully checked for scuffing on the sides of the piston.

of the piston **(Figure 11–58)**. Light up and down scuffing is normal. Excessive, irregular, or diagonal scuff marks indicate lubrication, cooling system, or combustion problems. Scuffing may also be caused by a bent connecting rod, seized piston pin, or inadequate piston-to-wall clearance. If any damage is evident, the piston should be replaced.

Remove the piston rings. A ring expander should be used to remove the compression rings. Normally the oil control ring can be rolled off by hand. Remove the carbon from the top of the piston with a gasket scraper. Carbon and oil build up in the back part of the groove. This buildup must be removed. The dirt will prevent the rings from seating properly. Clean the grooves of the piston with a groove cleaning tool or a broken piston ring. When doing this, make sure no metal is scraped off. The oil control ring groove has slots or holes. These should also be cleaned. Use a drill bit or small brush. Once the grooves are clean, use a brush and solvent to thoroughly clean the piston. Do not use a wire brush.

Ring side clearance should be measured. Side clearance is the difference between the thickness of the ring and the width of its groove. To measure this, place a new ring in its groove and, with a feeler gauge, measure the clearance between the ring and the top of the groove **(Figure 11–59)**. If the clearance is not within the specified range, the piston should be replaced.

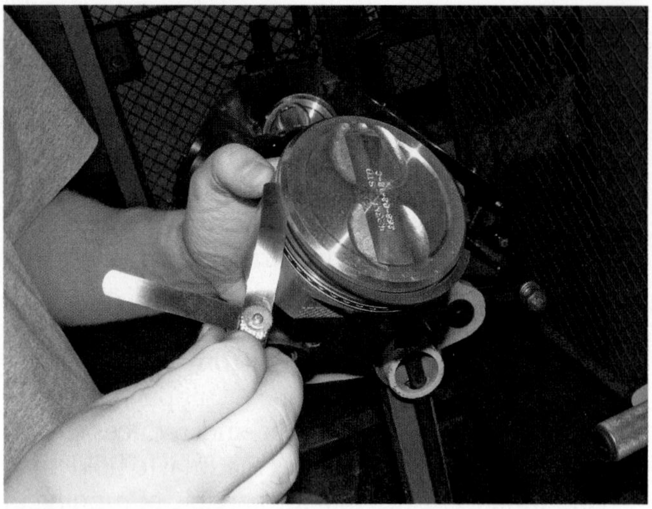

Figure 11-59 Ring side clearance should be checked on each piston.

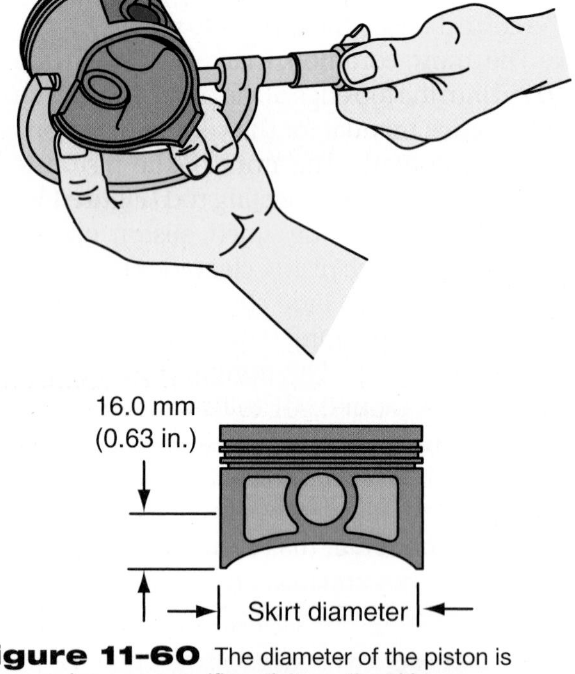

Figure 11-60 The diameter of the piston is measured across specific points on the skirt.

The diameter of the piston should be measured. This measurement is normally taken across specific points on the skirt **(Figure 11–60)**. If the diameter is not within specifications, the piston should be replaced. Some engine rebuilders will knurl the skirts if the diameter is slightly less than specifications.

Piston Pins

Piston pins are basically thick-walled hollow tubes. Like the rest of the piston and connecting rod assembly, it is built to be strong and light. Most are made from alloy steel and are plated with chrome, carburized, and/or heat-treated to provide good wear resistance. Piston pins are lubricated by oil fed through passages in the connecting rods, oil splashing in the crankcase, or spray nozzles in the rods or pistons.

A piston pin fits through the small end of the connecting rod and the piston's pin bore. The way the pin is retained is used to describe it.

- A stationary pin is pressed into the piston. The connecting rod swivels on the pin.
- A semifloating pin is pressed into the connecting rod. The piston swivels on the pin.
- A full-floating pin is able to move or rotate in the piston and connecting rod. The pins are retained by caps, plugs, snaprings, or spring clips inserted in the piston at the ends of the pin. Full-floating pins are the most commonly used.

Inspect the pin boss area on the piston for signs of pin wobble. Then remove the pin to inspect it. With full-floating pins, the retaining clips are removed and the pin pushed out.

A pin press is used to remove and install pressed-fit pins. When installing a piston pin, make sure the piston is facing the correct direction in regard to the connecting rod.

Inspect the pin closely for signs of wear. Full-floating pins should have an even wear pattern. Carefully inspect the pin bore in the piston. Since a piston is made from softer material than the pin, the piston will wear before the pin. If there are signs of uneven wear, suspect a lubrication or connecting rod problem.

Check the fit of the pin. It should move freely through the bores. Also attempt to move the pin up and down in its bores. Any movement means the piston bore or pin is worn. To determine if the bore or pin is worn, measure the diameter of the pin bore. If the bore is not within specifications, replace the piston. Then measure the diameter of the pin. If the pin is not within specifications, replace it. If the piston bore and pin meet specifications, measure the small end bore of the connecting rod (**Figure 11–61**). If the diameter is not within specifications, replace the connecting rod.

Some manufacturers recommend a check of the pin's oil clearance. To do this, subtract the diameter of the pin from the diameter of the piston pin bore. If the oil clearance exceeds specifications, replace the piston and pin. Now subtract the diameter of the pin from the diameter of the connecting rod's small end. If the oil clearance exceeds specifications, replace the connecting rod and/or the pin.

Connecting rods may have a piston pin bushing. Measure the inside diameter of the bushings and compare the reading to specifications. If the bushing is worn or damaged, it should be replaced. The bushing is pressed out of the rod with a pin press. Installing new bushings is also done with a press; some technicians heat the rods and freeze the pins before pressing them in. This makes installation easier. Before applying pressure on the pin, make sure it is set squarely above the bore.

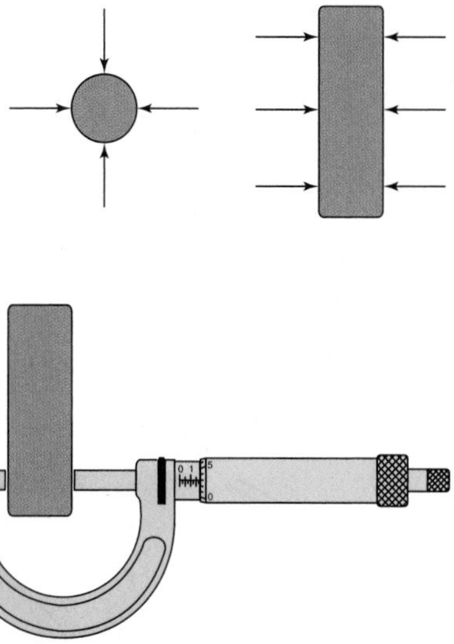

Figure 11-61 The piston pin is measured at a variety of spots and its diameter compared to the ID of the piston's pin bore and the small end of the connecting rod.

Piston Rings

Piston rings are used to fill the gap between the piston and cylinder wall. The piston rings seal the combustion chamber at the piston. Piston rings must also remove oil from the cylinder walls to prevent oil from entering into the combustion chamber. They also carry heat from the piston to the cylinder walls to help cool the piston.

In most engines, pistons are fitted with two compression rings and one oil control ring. The compression rings are found in the two upper grooves closest to the piston head. The oil ring is fitted to the groove just above the wrist pin. There are many different designs of rings; each has a specific application.

Compression Rings Compression rings are designed to use combustion pressure to force them against the cylinder wall. During the power stroke, the pressure caused by the expanding air-fuel mixture is applied between the inside of the ring and the piston ring groove. This forces the ring into full contact with the cylinder walls. The same force is applied to the top of the ring, forcing it against the bottom of the ring groove. These two actions help to form a tight ring seal.

Common compression rings are made of cast iron, cast iron coated with molybdenum (moly), and

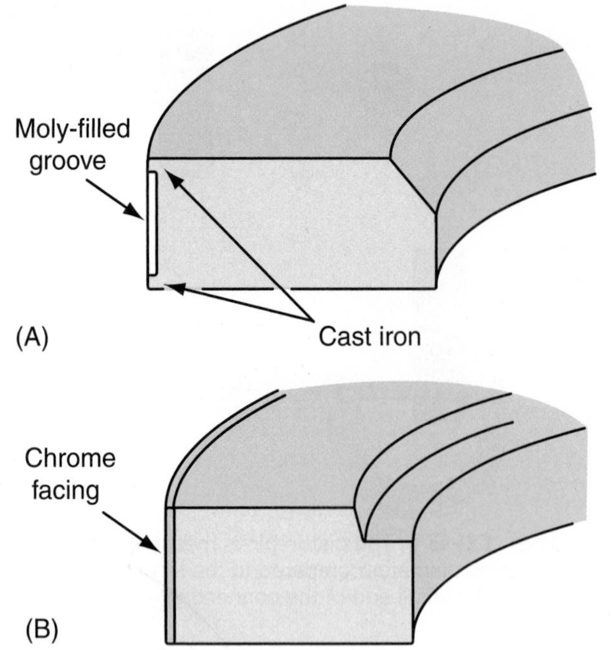

Figure 11-62 (A) A moly-coated compression ring. (B) A chrome-faced compression ring.

cast iron coated with chrome **(Figure 11–62)**. Cast iron offers a durable wear surface and costs less than a moly- or chrome-faced ring. These rings are ideal for normal driving. Moly coatings are quite porous and can hold oil. As a result, moly rings have a very high resistance to scuffing. These rings are used in engines that are run at continuous high speeds or severe load conditions. Chrome also has good resistance to scuffing but does not have the oil retention capabilities of moly. Chrome rings are recommended when driving conditions include frequent travel on dusty or unpaved roads. Chrome is very dense and hard and will push away any dirt that enters the cylinder on the intake stroke. Moly coatings, due to their porosity, will allow dirt to become embedded in a ring's face. Normally, a moly ring is used in the top ring groove with a cast iron or chrome ring in the second groove.

Other face coatings include ceramics, graphite, phosphate, and iron oxide. All coatings are designed to help in the wear-in process. Wear-in is the time required for the rings to conform to the shape and surface of the cylinder wall.

Oil Control Rings Oil is constantly being applied to the cylinder walls. The oil lubricates and cleans the cylinder wall and helps cool the piston. Controlling this oil is the primary purpose of the oil ring. The two common types of oil rings are the segmented oil ring and the cast-iron oil ring. Both are slotted so that excess oil from the cylinder wall can pass through the ring. The piston's oil ring groove is also slotted. Oil passes through the ring and slots in the piston and returns to the oil sump.

Segmented oil rings have an upper and a lower scraper rail and an expander. The scraper rings are often chrome rings. The expander pushes the two scrapers out against the cylinder wall. During installation, the end gaps of the three pieces must be staggered to prevent oil from escaping into the cylinder.

Installing the Rings

Some engines use low-tension piston rings; make sure the new rings are the correct ones for the engine. Before installing the rings onto a piston, check the rings' end gap. Place a compression ring into a cylinder. Use an upside-down piston to square the ring in the bore. Measure the gap between the ends of the ring with a feeler gauge. Compare the reading to specifications. If the gap exceeds limits, oversized rings should be used. If the gap is less than specifications, the ends of the ring can be filed with a special tool.

The preceding procedure for checking ring gap assumes that all taper and imperfections in the bore have been corrected. If the bore has some taper, the end gap should be checked in the cylinder at the lowest point of piston travel.

Piston ring gap is critical. Excessive gap will allow combustion gases to leak into the crankcase. This is commonly called blowby. Too little clearance can score the cylinder walls as the ends of the rings come in contact with each other as the engine heats up. The gap of the top compression ring allows some combustion pressure to leak onto the second compression ring. This helps the second ring seal.

Apply a light coat of oil on the rings. The oil control ring is installed first. Insert the expander; position the ends above a pin boss but do not allow them to overlap. Then install the rails. Stagger the ends of the three parts. The oil control ring assembly can be installed by hand. If the piston pin is set up into the oil ring groove, an oil ring support must be inserted in the ring groove. The support gives the oil rings a place to sit at the points of the groove where there is no piston material beneath the rings. The support has a dimple to prevent it from rotating around the piston. This keeps the ring gap at the desired spot at all times.

Use a piston expander to install the top and second compression rings **(Figure 11–63)**. Install the second ring first. Make sure the rings are installed in the right position. This includes making sure that the correct side of the ring is facing up. Rings have some sort of mark to show which side

Figure 11-63 Use a ring expander to install the compression rings.

should be up. Check the instructions of the ring manufacturer.

INSTALLING PISTONS AND CONNECTING RODS

The piston and connecting rod assemblies are installed next. Check the marks on the connecting rod caps and the connecting rods to make sure they are a match. The bearings for the connecting rods must be the correct size. If the crankshaft has been machined undersize, matching rod bearing inserts must be installed. The size of the bearing inserts is printed on their box and is stamped on the backs of the bearings or color coded.

Snap the bearing inserts into the connecting rods and rod caps. Make sure the tang on the bearing fits snugly into the matching notch. Check the clearance of the connecting rod bearings with Plastigage or a micrometer. Follow the same procedure used to check main bearing clearance. If the clearance is not within specifications, the connecting rod may need to be machined or replaced, or undersized bearings should be installed.

Wipe a coat of clean oil on the cylinder walls. Then install the piston and rod according to the procedure shown in Photo Sequence 8. Remember that connecting rods are numbered for proper assembly. Also make sure that the end gaps of the piston rings are staggered according to the manufacturer's recommendations **(Figure 11–64)** prior to installing the piston assembly.

Coat the crankshaft with clean lubricant or engine oil. After each piston is installed, rotate the crankshaft and check its freedom of movement. If the crankshaft is hard to rotate after a piston has been installed, remove it and look for signs of binding.

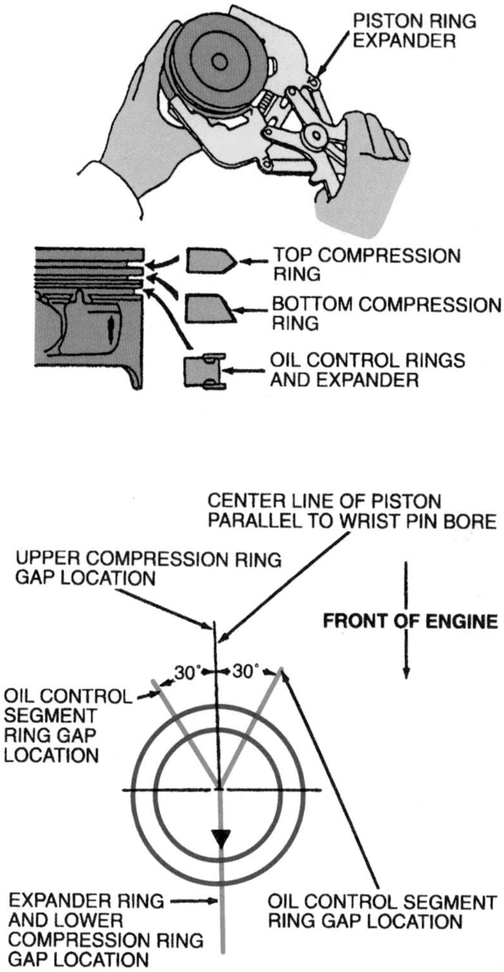

Figure 11-64 Install the piston rings onto the piston with a ring expander. Also make sure that the ring end gaps are arranged according to specifications. *Courtesy of Ford Motor Company*

> **SHOP TALK**
>
> Many technicians find it easier to use a wrinkle band ring compressor **(Figure 11–65)** to install pistons. These push the rings into their grooves but do not contact the entire outside of the piston. This makes it easier to push the piston out of the tool.

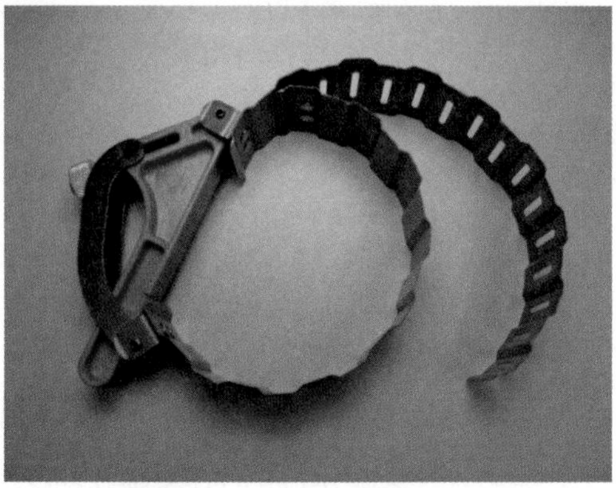

Figure 11-65 A wrinkle band ring compressor. *Reproduced under license from Snap-on Incorporated. All of the marks are marks of their owners.*

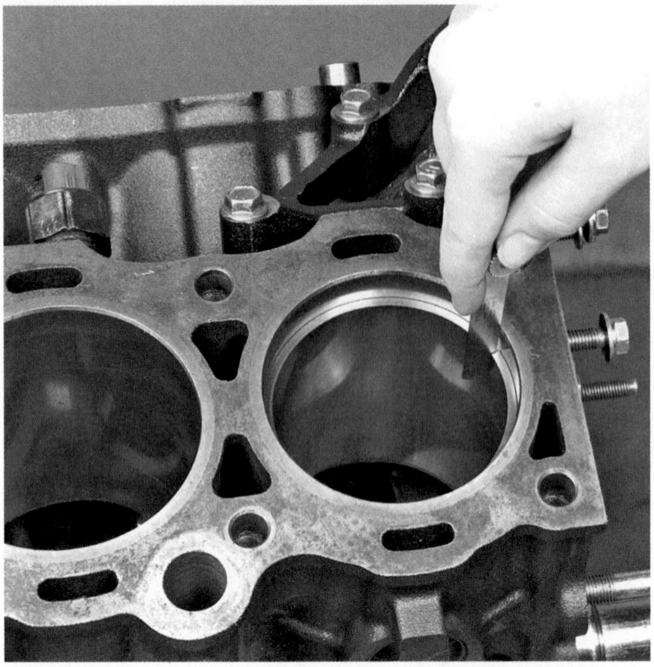

P8-1 *Insert a new piston ring into the cylinder. Use the head of the piston to position the ring so that it is square with the cylinder wall. Use a feeler gauge to check the end gap. Compare end-gap specifications with the measured gap. Correct as needed. Normally, the gaps of the piston rings are staggered to prevent them from being in line with each other. Piston rings are installed easily with a ring expander.*

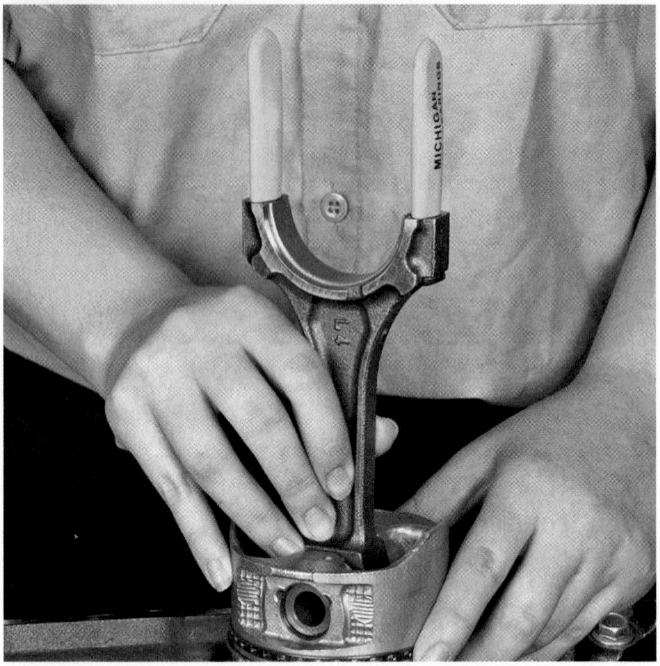

P8-2 *Before attempting to install the piston and rod assembly into the cylinder bore, place rubber or aluminum protectors or boots over the threaded section of the rod bolts. This will help prevent bore and crankpin damage.*

P8-3 *Lightly coat the piston, rings, rod bearings, cylinder wall, and crankpin with an approved assembly lubricant or a light engine oil. Some technicians submerge the piston in a large can of clean engine oil before it is installed.*

P8-4 *Stagger the ring end gaps and compress the rings with the ring compressor. This tool is expanded to fit around the piston rings. It is tightened to compress the piston rings. When the rings are fully compressed, the tool will not compress any further. The piston will fit snugly but not tightly.*

PHOTO SEQUENCE 8 Installing a Piston and Rod Assembly *(continued)*

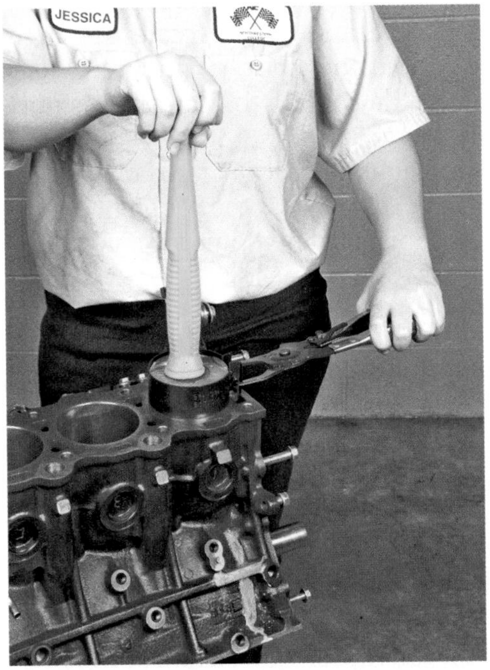

P8-5 *Rotate the crankshaft until the crankpin is at its lowest level (BDC). Then place the piston/rod assembly into the cylinder bore until the ring compressor contacts the cylinder block deck. Make sure that the piston reference mark is in correct relation to the front of the engine. Also, when installing the assembly, make certain that the rod threads do not touch or damage the crankpin.*

P8-6 *Lightly tap on the head of the piston with a mallet handle or block of wood until the piston enters the cylinder bore. Push the piston down the bore while making sure that the connecting rod fits into place on the crankpin. Remove the protective covering from the rod bolts.*

P8-7 *Position the matching connecting rod cap and finger tighten the rod nuts. Make sure the connecting rod blade and cap markings are on the same side. Gently tap each cap with a plastic mallet as it is being installed to properly position and seat it. Torque the rod nuts to the specifications given in the service manual. Repeat the piston/rod assembly procedure for each assembly.*

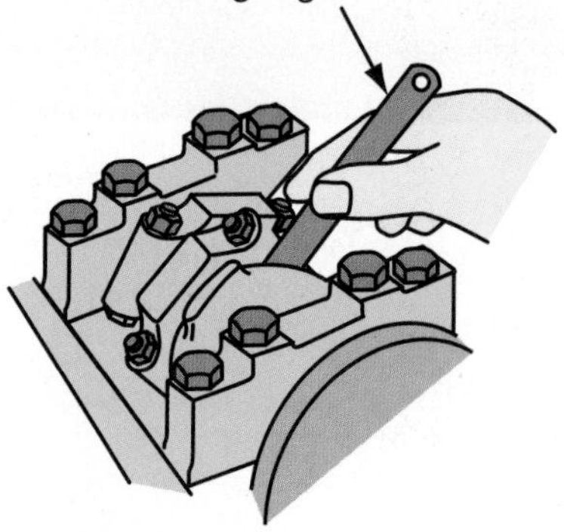

Figure 11-66 Measuring connecting rod side clearance.

After the pistons and rods have been installed, check the connecting rod side clearances **(Figure 11–66)**. Side clearance is the distance between the crankshaft and the side of the connecting rod. It is measured with a feeler gauge. If the clearance is not correct, the rods may need to be machined or replaced.

INSPECTION OF CAMSHAFT AND RELATED PARTS

After the camshaft has been cleaned, check each lobe **(Figure 11–67)** for scoring, scuffing, fractured surface, pitting, and signs of abnormal wear; also check for plugged oil passages.

Premature lobe and lifter wear is generally caused by metal-to-metal contact between the cam lobe and the lifter bottom due to inadequate lubrication. Wear normally begins within the first few minutes of operation. It is the result of insufficient lubrication.

To check cam lobe wear, measure the lobe height with a micrometer. Measure from the heel to the nose and then again 90 degrees from the original measurement. Record the measurement for each intake and exhaust lobe. Any variation in height indicates wear. Also check the measurements against the manufacturer's specifications.

Measure each camshaft journal in several places to determine if it is worn. If a journal is 0.001 inch (0.0254 mm) or more below the manufacturer's specifications, it should be replaced.

Figure 11-67 A worn lifter and camshaft. *Courtesy of Federal-Mogul Corporation*

The camshaft should also be checked for straightness with a dial indicator. Place the camshaft on V-blocks. Position the dial indicator on the center bearing journal and slowly rotate the camshaft. If the dial indicator shows runout (a 0.002-inch [0.0508 mm] deviation), the camshaft is not straight and must be replaced.

If the engine has a worn or damaged camshaft, identify the cause and fix the problem before installing a new camshaft, lifters, and/or followers.

Timing Components

The timing belt or chain and crankshaft/camshaft gears (sprockets) should be inspected and replaced if damaged or worn. A timing gear with cracks, spalling, or excessive tooth wear is an indication of improper **backlash** (either insufficient or excessive). Excess backlash causes the noise as the gears' teeth make violent contact. This overloading causes accelerated tooth wear and breakage. Insufficient backlash can cause the gears to bind. It can also cause high contact forces that can rupture the oil film between the teeth, causing spalling and wear.

Gear backlash is checked with a dial indicator. Check the movement between the camshaft and crankshaft gears at six equally spaced teeth. Hold the gear firmly against the block while making the check. Refer to specifications for backlash limits.

Lifters

When inspecting lifters, carefully check their bottoms and pushrod sockets. Wear should be off the center of the lobe and there should be no signs of edge contact. If they show abnormal wear, scoring, or pitting, they should be replaced. Any lifter that has its contact face worn flat or concave must also be replaced. Lifter bottoms are normally spherical.

Technically, normal lifter wear is referred to as adhesive or galling wear. It results from two solid surfaces (camshaft lobe and lifter face) rubbing on each other. The two surfaces tend to weld together. Fortunately, proper lubrication retards this process. However, excessive loading will accelerate the wear process **(Figure 11–68)**. Excessive loading is caused by incorrectly matched valve springs (too much spring pressure), old lifters on a new camshaft, or new lifters on an old camshaft.

Hydraulic lifters should be removed, disassembled, cleaned, and checked. Check the lifter's leakdown with a leakdown tester. Lifter leakdown rate is important. If the lifters leak down too quickly, noisy operation will result. If the lifters do not match

Figure 11-68 The possible wear patterns of a lifter that does not spin in its bore. *Courtesy of TRW, Inc.*

Figure 11-69 Cam bearings are normally press-fit into the block or head using a bushing driver and hammer.

specifications, they should be replaced, along with the camshaft. Never install old flat lifters with a new camshaft or new lifters on a used camshaft.

INSTALLATION OF CAMSHAFT AND RELATED PARTS

Before installing the camshaft and balance shafts with their bearings, coat all parts with assembly lubricant. This lubricant should have an extreme pressure (EP) lubricant rating. This helps prevent scuffing and galling during initial startup. It should also have excellent adhesion to prevent the lubricant from draining off the parts during engine reassembly.

Camshaft Bearings

Although the installation of camshaft bearings can be done after the short block is assembled, it may be easier to align the bearings' oil holes when the crankshaft is not installed. Keep in mind that any engine block that needed to have its main bearing bore alignment corrected due to distortion is likely to have camshaft bearing bore misalignment problems.

OHV cam bearings are normally press-fit into the block or head using a bushing driver and hammer (**Figure 11–69**). The bearing at the rear of the block should be installed first. The camshaft journals may have different diameters, with the smallest being on the rear of the block and each journal being progressively larger.

The new bearing is fit over the tool's expanding mandrel and the tool is set into the block. A guide cone is used to keep the tool centered in the bore. Once the bearing is at the outside of its bore, rotate the bearing to align the oil hole in the bearing with the oil hole in the block.

On blocks with grooves behind the bearings, the bearing should be installed with the oil hole at the 2 o'clock position as viewed from the front. While holding the centering cone against the outside bore, drive the bearing into its bore. If the cone and tool are allowed to move while inserting the bearings, the bearing can be damaged. While driving the bearings into their bore, be careful not to shave metal off the backs of the bearings. This galling may cause a buildup of metal between the outside of the bearing and the housing bore and result in less clearance. To prevent galling, housing bores should be chamfered.

Camshaft

Wipe off each cam bearing with a lint-free cloth, then thoroughly coat the camshaft lobes, bearing journals, and distributor drive gear (if there is one) with assembly lube. Also lubricate the lifters. Most premature cam wear develops within the first few minutes of operation. Prelubrication helps to prevent this. Use the lubricant recommended by the manufacturer.

The camshaft should be carefully installed to avoid damaging the bearings with the edge of a cam lobe or journal. Keep it straight while installing it. A threaded bolt in the front of the camshaft helps guide the cam in place. An alternative is to install the camshaft while the block rests on its end.

When the camshaft is in place, install the thrust plate and the timing gear. Be sure to align the thrust plate with the woodruff key. Never hammer a gear or sprocket onto the shaft. Heating steel gears helps with installation. Do not heat fiber gears. When installing the gears, be sure to keep the gear square and aligned with the keyway at all times.

Once the shaft is completely in the block, the shaft should be able to be turned by hand. If the cam does not turn, check for a damaged bearing, a nick on the cam's journal, or a slight misalignment of the block journals. If the bearing clearance is too small, some technicians ream away a slight amount of the bearing;

others hone the bearing. Both of these need to be done carefully. The best way to increase the clearance is to cut the camshaft journals on a lathe. Reaming or honing the inside diameter of cam bearings is not recommended because grit may become embedded in bearing surfaces, which will cause shaft wear.

CRANKSHAFT AND CAMSHAFT TIMING

USING SERVICE INFORMATION

Normally, camshaft timing marks are shown in the engine section of a service manual under the heading Timing Belt or Chain R&R.

During most engine rebuilds, a completely new timing assembly is installed. The camshaft drive must be installed so that the camshaft and crankshaft are in time with each other. Both sprockets are held in position by a key or possibly a pin. There are factory timing marks on the crankshaft and camshaft gears or sprockets **(Figure 11–70)**. The timing must be positioned according to the manufacturer's instructions.

SHOP TALK

Make sure the timing marks are precisely matched. If the gears are misaligned by one gear tooth, the timing will be off by about 17 degrees.

The chain is installed on the crankshaft gear first, then around the camshaft sprocket. Never wind a chain onto the gears. Also, a screwdriver, prybar, or hammer should never be used to force a chain into position. Prying or pounding on the chain will damage the links, causing the chain to stretch and fail. Carefully place the entire assembly as a unit onto the shafts by pressing both gears evenly, keeping the keyways aligned. While keeping the gears parallel and aligned, gently tap them in place.

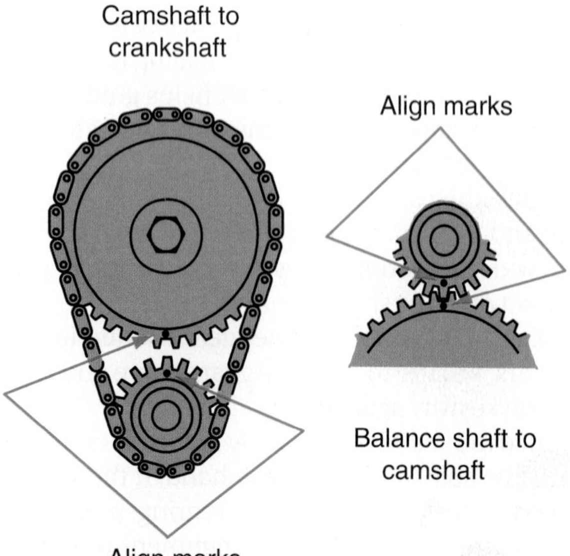

Figure 11–70 Camshaft-to-crankshaft and camshaft-to-balance shaft timing marks.

SHOP TALK

There are some OHV engines that are interference engines. If the crankshaft-to-camshaft timing is not correct, the pushrods normally bend when the engine is started. Double-check the alignment.

Some manufacturers recommend additional checks of the camshaft after it has been inserted in the block. These include a check of the clearance between the camshaft gear and the backing plate. This check is made with a feeler gauge. This clearance can often be corrected by repositioning the camshaft gear. Camshaft end play is measured with a dial indicator. As the camshaft is moved back and forth in the block, the indicator will read the amount of movement. Compare this to specifications. End play can be corrected by changing the size of the shim behind the thrust plate or by replacing the thrust plate. Some engines limit the end play with a cam button. A nylon button or Torrington bearing sets on a spring and is installed between the gear and timing chain cover.

Performance TIP

A camshaft is designed to open and close the valves at the correct time. The camshaft, crankshaft, timing gears or sprockets, and timing chains or belts are made to specifications. Those specifications will include a range of acceptable tolerances. This means there is no guarantee that the timing assembly will provide exactly what the camshaft was intended to do. Overall engine performance can be affected by a few degrees of misalignment. Many performance engine builders degree the camshaft to precisely match the position of the camshaft with the crankshaft. Degreeing a cam is highly recommended when installing a performance camshaft. An engine with a properly degreed camshaft will be more efficient than one without a degreed cam.

The exact procedure for degreeing camshafts varies with each engine design. The procedures must be followed exactly as stated by the cam manufacturer. Degreeing the cam requires a degree wheel, a stable pointer that can be attached to the block, a dial indicator with a mount, a positive TDC stop tool, the specification sheet for the camshaft, and the parts necessary to adjust the mounting of the timing gears. The latter are necessary to position the gears at their correct positions.

Balance Shafts

Balance shafts are serviced in the same way as crankshafts. Each bearing and journal should be cleaned and carefully inspected. All worn or damaged parts should be replaced. The oil clearance of the shafts and bearings should be checked with Plastigage. Place the balance shaft half bearings into the crankcase and then position the balance shaft(s) onto the bearings. Then lay a strip of Plastigage across each journal. Insert the matching bearing halves into the balance shaft housing and tighten the housing bolts to specifications and in the proper sequence. Now remove the housing bolts and housing. Measure the Plastigage at its widest point. Compare the readings to specifications. If the oil clearance is excessive, measure the shaft bore and the shaft(s) journals. Determine the cause of excessive clearances and replace the bearing or balance shaft. Remove the Plastigage completely after measuring.

To install the balance shaft assembly, install the bearings in the crankcase and housing. Apply a light coat of clean engine oil to the bearings. The shafts need to be properly timed to the crankshaft. On some engines there are two balance shafts; these must be timed to each other, in addition to the crankshaft **(Figure 11–71)**. Align the timing marks of the balance shaft(s). Set the shaft(s) into position in the block. Set the housing with bearings over the shafts. Install and tighten the housing bolts. In most cases these bolts should be tightened in two steps and in a prescribed sequence.

Lifters

Before installing the lifters into the lifter bores, generously coat the lifter bores with assembly lube. Coat the lifters and/or followers one at a time as they are being installed. After they have been installed, rotate the camshaft to check for binding or misalignment.

SHOP TALK

Some technicians submerge the hydraulic lifters in clean oil to allow them to fill with oil before installing them.

OIL PUMPS

The oil pump may be located in the oil pan or mounted to the front of the engine **(Figure 11–72)**. Its purpose is to supply oil to cool, clean, and lubricate the various moving parts in the engine. The

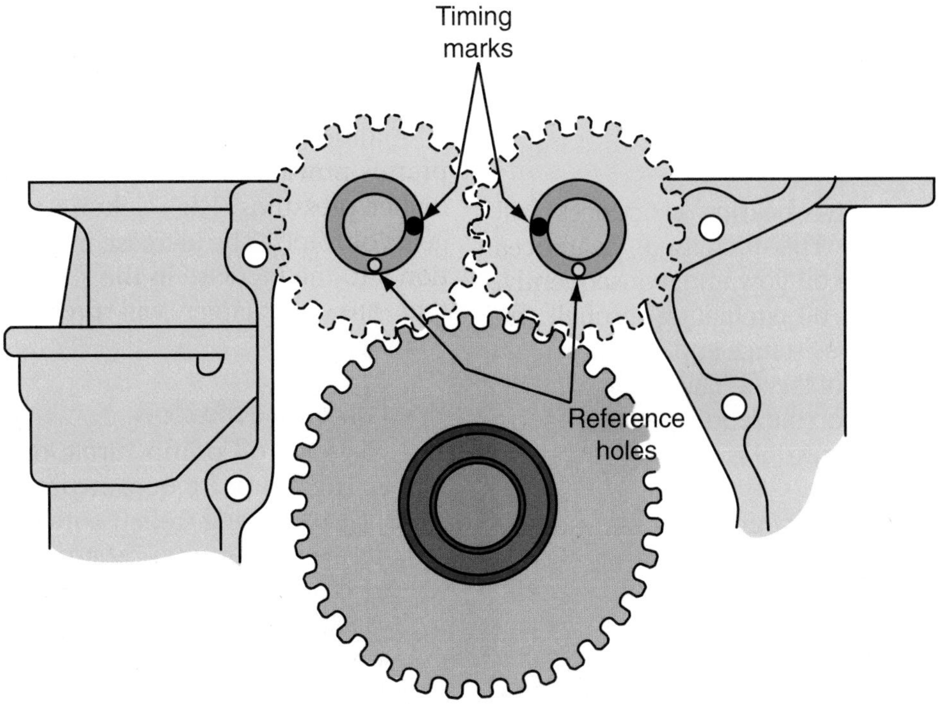

Figure 11-71 Balance shafts must be timed to each other and the crankshaft.

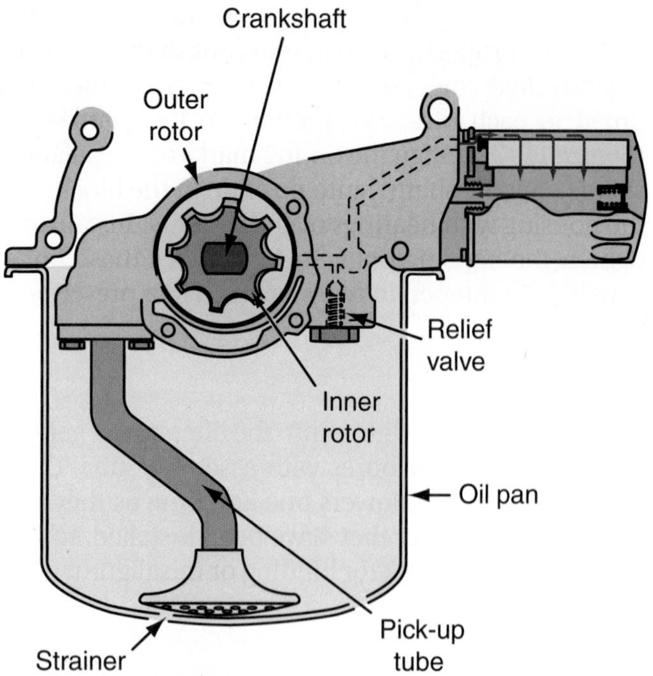

Figure 11-72 The different ways an oil pump is driven by the crankshaft.

pump is normally driven by the crankshaft and creates suction to draw oil from the oil pan through a strainer. The pump's suction creates pressure that forces the oil through the oil filter to various passages. The oil then returns to the oil pan.

An oil pump does not create oil pressure; it merely moves oil from one place to another. Oil pumps are positive displacement pumps; that is, the amount of oil that leaves the pump is the same amount that enters it. Output volume is proportional to pump speed. As engine rpm increases, pump output also increases. As the oil leaves the pump, it passes through many passages. These passages restrict oil flow. These restrictions are what cause oil pressure. Small passages cause the pressure to increase; larger ones decrease the pressure.

This is why excessive bearing clearances will decrease oil pressure. The increased clearances reduce the resistance to oil flow and, consequently, increase the volume of oil circulating through the engine. This decreased resistance and increased volume lower the pressure of the oil. The ability of an oil pump to deliver more than the required volume of oil is a safety measure to ensure lubrication of vital parts as the engine wears.

Oil pressure is also determined by the viscosity and temperature of the oil. High-viscosity oil has more flow resistance than low-viscosity oil.

Types of Oil Pumps

Oil pumps are driven by the camshaft or crankshaft. In newer engines, the pump is driven by the crankshaft. Integral pumps are driven directly by the crankshaft, and others are driven by a chain or gear. How the pump is driven is dictated by the location of the pump. Some oil pumps have an intermediate or drive shaft that is driven by a gear on the camshaft. Other pumps are driven by an auxiliary shaft meshed with the camshaft.

The most commonly used oil pumps are the rotor and gear types. A typical **rotor-type pump (Figure 11–73)** has an inner rotor and an outer rotor, which is driven by the inner rotor. The outer rotor always has one more lobe than the inner rotor. When the rotors turn and the rotors' lobes unmesh, oil is drawn into that space. As the rotors continue to turn, the oil becomes trapped between the lobes, cover plate, and top of the pump cavity. It is then forced out of the pump body by the meshing of the lobes. This squeezes the oil out and directs it through the engine. The amount of oil forced out of the pump depends on the diameter and thickness of the pump's rotors.

Gear-type pumps use a drive gear, which is connected to an input shaft, and a driven gear. Both gears trap oil between their teeth and the pump cavity wall. As the gears rotate, oil is forced out as the gear teeth unmesh. The output volume per revolution depends on the length and depth of the gear teeth. Another style of gear-type oil pump uses an idler gear with internal teeth that spins around the drive gear. In this style of pump, often called a crescent or trochoidal type, the gears are eccentric **(Figure 11–74)**. That is, as the larger gear turns, it walks around the smaller one, moving the oil in the space between.

The rotor type moves a greater volume of oil than the gear type because the space in the open lobe of the outer rotor is greater than the space between the teeth of the gears of a gear-type pump.

At times, high-volume pumps are installed. These pumps provide more oil flow. They do not provide higher pressures. High-volume pumps have larger gears or rotors. The increase in oil volume is proportional to the increase in the size of the gears. Gears that are 20% larger will provide 20% more oil volume.

Pressure Regulation

The faster an oil pump turns, the more oil it will move. Therefore, its output pressure increases. A pressure-regulating (relief) valve is used to control the pump's maximum pressure. Excessive oil pressure can lead to poor lubrication due to the oil blowing past parts rather than flowing over them. A pressure regulator valve is loaded with a closely calibrated spring that allows oil to bleed off at a given pressure. When the pressure from the pump reaches

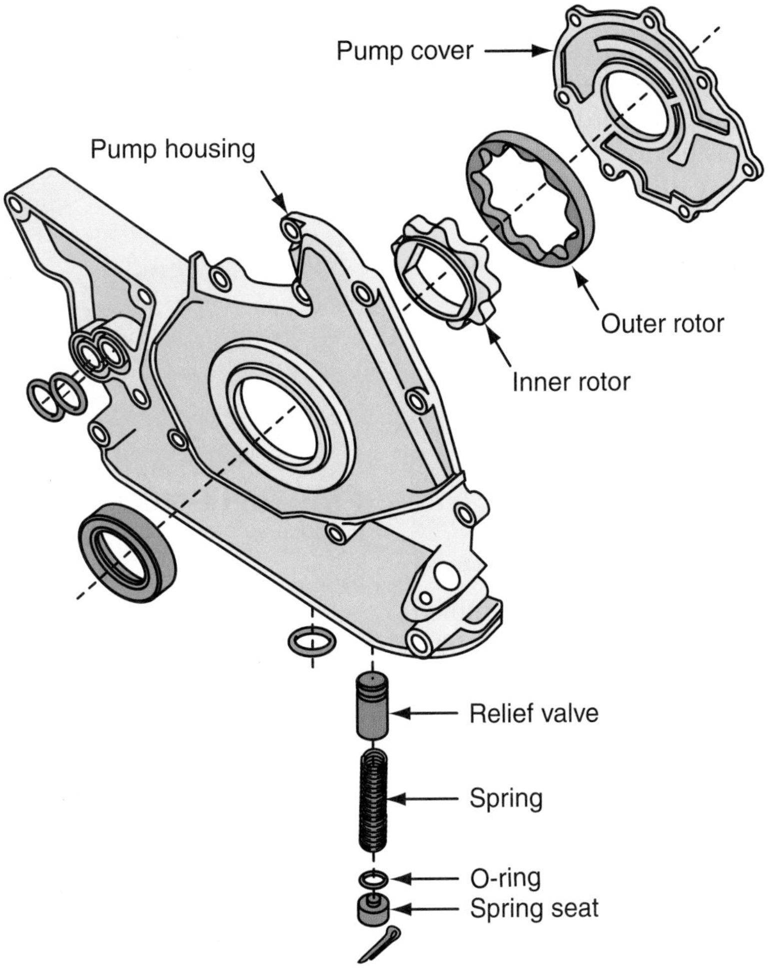

Figure 11-73 A rotor-type oil pump.

Figure 11-74 A gear-type oil pump.

a preset level, a check valve, ball, or plunger unseats and allows the oil to return to either the inlet side of the pump or to the crankcase.

High-pressure oil pumps are basically stock pumps with a stronger relief valve spring. The spring delays the opening of the valve until a higher pressure is reached.

OIL PUMP SERVICE

Many technicians install a new or rebuilt oil pump on each engine they rebuild. Although the oil pump is probably the best lubricated part of the engine, it is lubricated before the oil passes through the filter. Therefore, it can experience premature failure due to dirt or other materials entering the pump. Foreign particles can cause the following:

1. Fine abrasive particles gradually wear the surfaces, causing a reduction in efficiency.
2. Hard particles larger than the clearances can cause scoring as they pass through, finally resulting in seizure.
3. Large particles will physically lock up the pump. Of course, when the pump seizes or locks up, the gears or the drive shaft will be twisted or sheared off.

Inspection

The oil pump should be carefully inspected. Carefully remove the pressure relief valve and note the direction it is pointing to so it can be reinstalled in its proper position. If the relief valve is installed backward, the pump will not be able to build up pressure. Inspect the relief valve spring for signs of collapsing or wear. Check the tension of the spring according to specifications. Also check the piston for scores and free operation in its bore. Remove the housing bolts, then remove the cover from the housing.

Inspect the gears or rotors and the pump housing for scoring or other damage. Replace the oil pump if any are found.

CAUTION!

Always follow the manufacturer's procedure for servicing or replacing an oil pump.

Mark the gear teeth so they can be reassembled with the same tooth indexing (**Figure 11–75**). The gears and rotors of some pumps are marked by the factory. Once all parts have been removed, clean them and dry them off with compressed air.

SHOP TALK

Use a paint stick or other nonviolent means to mark the gears. If a center punch must be used, file down the raised material around the indent before reassembling the pump. The raised edges may cause interference and wear if they are not removed.

Inspect the pump gears or rotors for chipping, galling, pitting, or signs of abnormal wear. Examine the housing bores for similar signs of wear. If any part of the housing is scored or noticeably worn, replace the pump assembly.

Check the mating surface of the pump cover for wear. If the cover is excessively worn, scored, or scratched, replace the pump. Use a feeler gauge and straightedge to check its flatness. There are specifications for the amount of acceptable warpage. If the cover is excessively warped, replace the pump.

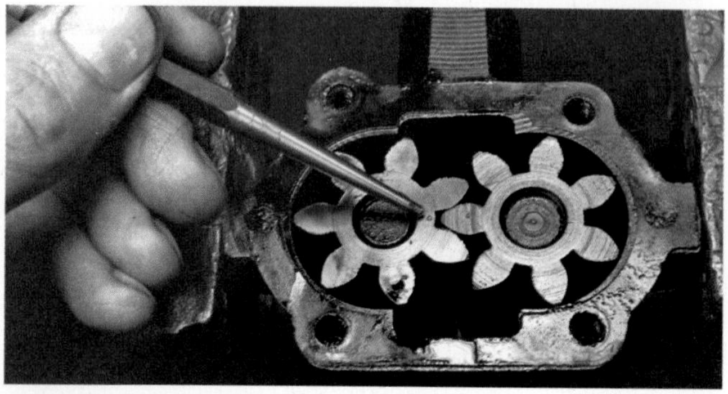

Figure 11–75 Mark the gear teeth so they can be reassembled with the same indexing.

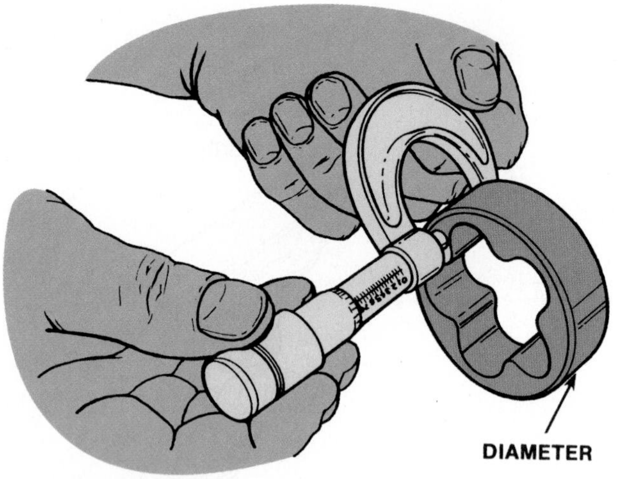

Figure 11–76 Measuring the outer rotor with an outside micrometer.

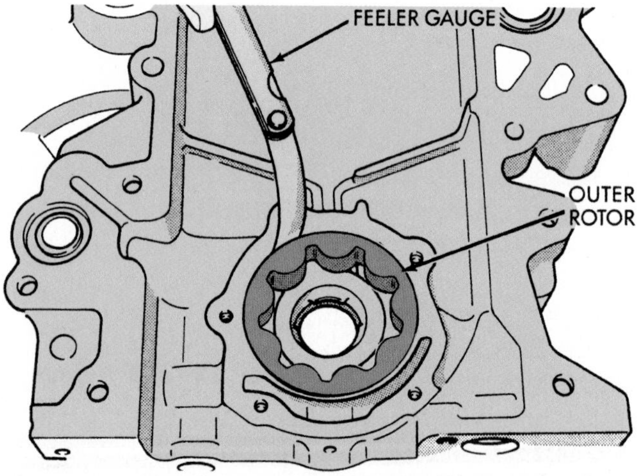

Figure 11–77 Checking clearance between the outer rotor and pump body. *Courtesy of Chrysler LLC*

Use a micrometer to measure the diameter and thickness of the rotors (**Figure 11–76**). If these dimensions are less than the specified amount, the rotors or the pump must be replaced.

With rotor pumps, reinstall the rotors into the pump body. Use a feeler gauge to check the clearance between the outer rotor and pump body (**Figure 11–77**). If the housing-to-rotor clearance exceeds specifications, replace the oil pump. If specifications are not available, replace the pump if the measured clearance is greater than 0.012 inch (0.305 mm).

Position the inner and outer rotor lobes so they face each other. Measure the clearance between them with a feeler gauge (**Figure 11–78**). A clearance of more than 0.010 inch (0.2540 mm) is unacceptable and the pump should be replaced. The timing case

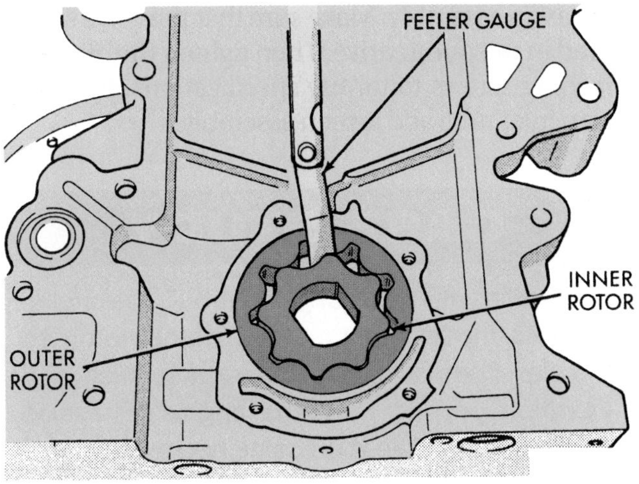

Figure 11-78 Measuring clearance between the inner and outer rotor lobes. *Courtesy of Chrysler LLC*

and gear thrust plate might also be worn. Excessive clearance can limit pump efficiency. Replace them as necessary.

On a gear-type pump, measure the clearance between the gear teeth and pump housing. Take several measurements at various locations around the housing and compare the readings. If the clearance at any point exceeds 0.005 inch (0.0762 mm), replace the pump.

On both gear-type or rotor-type oil pumps, make sure the housing surface is clean and free of gasket material before measuring the end clearance of the gears or rotors. Place a straightedge across the pump housing and measure the clearance between the straightedge and gears **(Figure 11–79)**. The desired end play clearance should not exceed 0.003 inch (0.1270 mm).

Some manufacturers recommend that liquid thread lock be applied to the pump housing bolts. Always check the service manual before doing this. Install the cover and tighten the bolts to specifications. The gasket that is used to seal the end housing is also designed to provide the proper clearance between the gears and the end plate. Do not substitute another gasket or make a gasket to replace the original one. If a precut gasket was not originally used, seal the end housing with a thin bead of anaerobic sealing material. Turn the input shaft or gear by hand. It should rotate easily. If it does not and the pump passed all tests, replace the pump.

Remove the old oil seal from the oil pump. Install a new seal into the pump housing. Make sure all mating surfaces of the pump are clean, undamaged, and dry.

If the pump uses a hexagonal drive shaft, inspect the pump drive and shaft to make sure that the corners are not rounded. Check the drive shaft-to-housing clearance by measuring the OD of the shaft and the ID of the drive.

The pickup screen **(Figure 11–80)** is normally replaced when an engine is rebuilt. It is important that the pickup is positioned properly. This will avoid oil pan interference and ensure that the pickup is always submerged in oil. When installing the pickup tube, be sure to use new gaskets and seals. Air leaks on the suction side of the oil pump can cause the pressure relief valve to hammer back and forth. Over time, this will cause the valve to fail. Air leaks can also cause oil aeration, foaming, marginal lubrication, and premature engine wear. Care should be taken to make sure that all parts on the suction side of the pump fit tightly and that there is no place for air leakage. Air leakage often comes from cracked seams in the pickup tube.

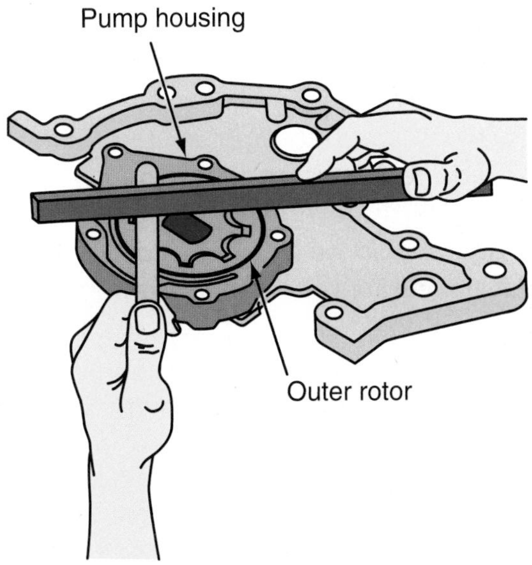

Figure 11-79 Check the housing-to-rotor clearance at the (A) rotors and (B) the housing.

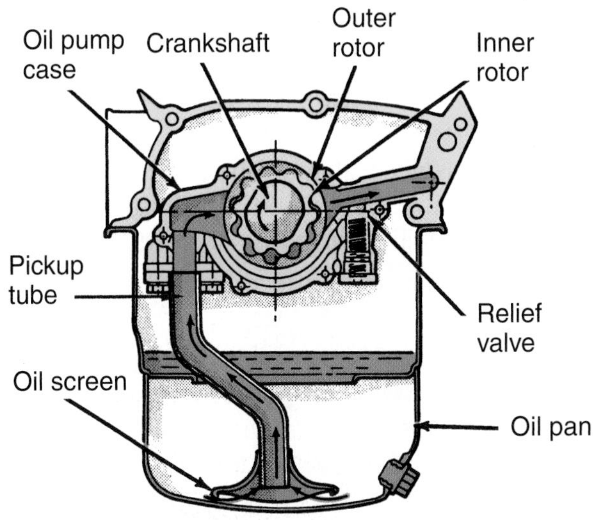

Figure 11-80 Oil pump pickup. *Courtesy of Chrysler LLC*

INSTALLING THE OIL PUMP

The pump should be primed before assembly. The preferred way to do this is to submerge the pump in clean engine oil. Make sure the inlet port is fully in the oil. Then turn the pump by hand until you see oil flow from the outlet of the pump. The use of heavy lubricants inside the oil pump is not recommended because they do not circulate easily and can cause oil starvation to critical parts when the engine is first started.

Crankshaft-Driven Pump

The installation of a typical crankshaft-driven oil pump is as follows:

1. Install a new oil seal into the pump.
2. Apply liquid gasket evenly to the pump's mating surfaces on the block.
3. Do not allow the gasket material to dry.
4. Coat the lip of the oil seal and the O-rings with oil.
5. Align the inner rotor with the crankshaft.
6. Install and tighten the oil pump.
7. Clean all excess grease on the snout of the crankshaft.
8. Install the oil pickup.

Cam-Driven Pumps

The installation of a typical camshaft-driven oil pump is as follows:

1. Apply a suitable sealant to the pump and block.
2. Make sure that the drive gears are properly meshed and that the drive shaft is seated in the pump.
3. Install the pump to its full depth and rotate it back and forth slightly to ensure proper positioning and alignment through the full surface of the pump and the block machined interface surfaces.
4. Once installed, tighten the bolts or screws. The pump must be held in a fully seated position while installing the bolts or screws.
5. Install the oil pump inlet tube and screen assembly.

Distributor-Driven Pump When installing an older-style, distributor-driven oil pump, position the drive shaft into the distributor socket. The stop on the shaft should touch the roof of the crankcase when the shaft is fully seated. With the stop in position, insert the drive shaft into the oil pump. Install the pump and shaft as an assembly. Make sure that the drive shaft is seated in the pump drive. Then tighten the oil pump, attaching screws to torque specifications. Install the pump inlet tube and screen assembly.

CASE STUDY

3,000 miles after the shop had rebuilt the engine in a late-model Toyota, the customer returned the car complaining of abnormal noise and vibration coming from the engine. She also said it seemed like the engine had a "miss."

The technician who did the engine work was experienced and was quite sure he did everything right. The car was taken on a road test and the problem was verified. The noise and vibration varied with engine speed and the engine just did not seem to run right. The vibration seemed to be caused by a loose, damaged, or broken engine mount. So a careful inspection was done. No problem was found with the mounts.

Next, the technician connected a scan tool to see if there were any hints as to why the engine vibrated or had a miss. No codes were retrieved. A thorough inspection was done on everything under the hood. Nothing was obviously broken or disconnected.

He started the engine and looked around it. This is when the potential problem was noticed. The harmonic balancer was wobbling on the shaft. Turning off the engine, he reached down to find that the attaching bolt was loose. He knew he had tightened it during the engine rebuild so the fact that it was loose was a mystery.

He proceeded to tighten the bolt, but checked the torque specifications first. This is when he discovered what he did wrong. The bolt should be tightened to a little over 200 ft.-lb. He knew he did not tighten the bolt that much. To do this, the balancer and crankshaft need to be held stationary while turning the bolt. The balancer, crankshaft sprocket, and woodruff key were inspected. These were not damaged yet. He borrowed the special tool to hold the crank and tightened the bolt. The engine was started and the problem disappeared.

The technician was actually lucky; the wobbling balancer could have caused serious damage to the crankshaft. On this engine, the crank sprocket and harmonic balancer share the same woodruff key. This is the reason for the required bolt torque. The engine miss was caused by the wobbling balancer. The rotation of the crankshaft sprocket is monitored by the crankshaft position sensor, which is located next to the sprocket. The sensor, reacting to the wobbling of the sprocket, sent erratic readings to the engine control module. These unstable signals were enough to cause the miss but were within the normal range so the system did not set a trouble code.

The technician learned once more that following the recommendations of the manufacturer can save time, frustration, and reputation.

KEY TERMS

Babbitt
Backlash
Boring bar
Cam follower
Cast
Chamfering
Core plug
Deck
Dome
Duration
Flexplate
Flywheel
Forge
Full-round bearing
Gear-type pump
Glaze
Harmonic balancer
Head
Hone
Hypereutectic
Leakdown
Line boring
Long block
Overlap
Piston pin
Piston rings
Piston skirt
Plastigage
Ring lands
Rotor-type pump
Short block
Split bearing
Taper
Tappet
Total indicator reading (TIR)
Valve lifters
Variable valve timing (VVT)
Wrist pin

SUMMARY

- The basic short block assembly consists of the cylinder block, crankshaft, crankshaft bearings, connecting rods, pistons and rings, and oil gallery and core plugs. On OHV engines the camshaft and its bearings are also included.
- The cylinder block houses the areas where combustion occurs.
- A properly reconditioned cylinder must be of the correct diameter, have no taper or runout, and have a surface finish such that the piston rings can seal.
- Glaze is the thin residue that forms on cylinder walls due to a combination of heat, engine oil, and piston movement.
- Core plugs and oil gallery plugs are normally removed and replaced as part of cylinder block reconditioning. The three basic core plugs are the disc or dished type, cup type, and expansion type.
- A cam changes rotary motion into reciprocating motion. The part of the cam that controls the opening of the valves is the cam lobe. The closing of the valves is the responsibility of the valve springs.
- Solid valve lifters provide for a rigid connection between the camshaft and the valves. Hydraulic valve lifters do the same but use oil to absorb the shock resulting from movement of the valve train. Roller lifters are used to reduce friction and power loss.
- The camshaft is driven by the crankshaft at half its speed.
- The most common ways to measure cam lobe wear are with a dial indicator or an outside micrometer. The dial indicator test should be conducted with the camshaft in the engine. When using an outside micrometer, the camshaft must be out of the engine.
- Most premature cam wear develops within the first few minutes of operation.
- The crankshaft turns on a film of oil trapped between the bearing surface and the journal surface. The journals must be smooth and highly polished. The flywheel adds to an engine's smooth running by applying a constant moving force to carry the crankshaft from one firing stroke to the next. The flywheel surface may be used as part of the clutch.
- Crankshaft checks include saddle alignment, straightness, clearance, and end play.
- Bearings carry the critical loads created by crankshaft movement. Most bearings used today are insert bearings.
- Maintaining a specific oil clearance is critical to proper bearing operation. Bearings are available in a variety of undersizes.

- Aluminum pistons are light, yet strong enough to withstand combustion pressure.
- Piston rings are used to fill the gap between the piston and cylinder wall. Most of today's vehicle engines are fitted with two compression rings and one oil control ring.
- When installing a piston and connecting rod assembly, various markings can be used to make sure the installation is correct. Always check the service manual for exact locations.
- Connecting rod side clearance determines the amount of oil throw-off from the bearings and measured with a feeler gauge.
- The camshaft is supported in the cylinder block by friction-type bearings, or bushings, which are typically pressed into the camshaft bore in the block or head. Camshaft bearings are normally replaced during engine rebuilding.
- During most engine rebuilds, a new timing assembly is installed. When installing the timing gears, make sure they are aligned to specifications.
- After an engine is rebuilt, a new or rebuilt oil pump is often installed. If the old pump is to be reused, it should be carefully inspected for wear and thoroughly cleaned.

REVIEW QUESTIONS

1. Name the two most common ways to measure cam lobe wear.
2. What is the deck?
3. What is the name of the component in the lubrication system that prevents excessively high system pressures from occurring as engine speed increases?
4. Where does maximum cylinder bore wear occur?
5. What type of valve lifter automatically compensates for the effects of engine temperature?
   a. hydraulic
   b. solid
   c. roller
   d. all of the above
6. What is the purpose of compression rings?
7. Most pistons used today are made of ______.
   a. cast iron
   b. aluminum
   c. ceramic
   d. none of the above
8. Core plugs ______.
   a. are also called expansion plugs
   b. are used in all cast-iron cylinder blocks
   c. are a possible source for coolant leaks
   d. all of the above
9. Which of the following is not of concern when checking a rotor-type oil pump?
   a. cover flatness
   b. rotor thickness
   c. inner rotor-to-outer rotor clearance
   d. inner rotor-to-pump housing clearance
10. *True or False?* Oil pressure is determined by whether the pump is a gear type or rotor type and by oil clearances, the pump's pressure regulator valve, and oil viscosity.
11. How is the oil pump driven?
12. Each half of a split bearing is made slightly larger than an exact half. What is this called?
    a. spread
    b. crush
    c. both a and b
    d. neither a nor b
13. The connecting rod journal is also called the ______.
    a. balancer shaft
    b. vibration damper
    c. Plastigage
    d. crankpin
14. What device in the valve train changes rotary motion into reciprocating motion?
    a. eccentric
    b. cam
    c. bushing
    d. mandrel
15. Which type of oil ring is slotted so that excess oil can pass through it?
    a. cast iron
    b. segmented
    c. both a and b
    d. neither a nor b

ASE-STYLE REVIEW QUESTIONS

1. Technician A uses a feeler gauge and straightedge to determine the pump cover flatness. Technician B uses an outside micrometer to measure the diameter and thickness of the outer rotor. Who is correct?

   a. Technician A c. Both A and B
   b. Technician B d. Neither A nor B

2. After installing cam bearings: Technician A checks that the oil holes in the bearings are properly aligned with those in the block by squirting oil into the holes. Technician B checks for proper alignment by inserting a wire through the holes. Who is correct?

   a. Technician A c. Both A and B
   b. Technician B d. Neither A nor B

3. Technician A says that a cylinder wall with too smooth a surface will prevent the piston rings from seating properly. Technician B says that a cylinder wall should have a crosshatch honing pattern. Who is correct?

   a. Technician A c. Both A and B
   b. Technician B d. Neither A nor B

4. Technician A uses a prybar to stretch the timing chain onto its sprockets. Technician B cools the timing sprockets so that the timing chain will slip over the teeth. Who is correct?

   a. Technician A c. Both A and B
   b. Technician B d. Neither A nor B

5. Technician A installs a cup-type core plug with its flanged edge outward. Technician B installs a dish-type core plug with the dished side facing inward. Who is correct?

   a. Technician A c. Both A and B
   b. Technician B d. Neither A nor B

6. Technician A uses a micrometer to measure the connecting rod journal for taper. Technician B uses a micrometer to measure the connecting rod journal for out-of-roundness. Who is correct?

   a. Technician A c. Both A and B
   b. Technician B d. Neither A nor B

7. Technician A says that piston ring end gaps should be the same for each ring on a piston. Technician B says that piston ring end gaps should be staggered before installing the piston into its bore. Who is correct?

   a. Technician A c. Both A and B
   b. Technician B d. Neither A nor B

8. Technician A checks crankshaft end play or end clearance with a feeler gauge. Technician B uses a dial indicator. Who is correct?

   a. Technician A c. Both A and B
   b. Technician B d. Neither A nor B

9. When installing an oil pump in an engine: Technician A packs the pump with petroleum jelly. Technician B submerges the pump in clean oil and rotates the pump shaft to fill the pump body. Who is correct?

   a. Technician A c. Both A and B
   b. Technician B d. Neither A nor B

10. Technician A says that counterweights are added to crankshafts to offset the weight of the connecting rods and pistons. Technician B says that the connecting rod bearings are fed a fresh supply of oil through holes drilled in the crankshaft. Who is correct?

    a. Technician A c. Both A and B
    b. Technician B d. Neither A nor B

CHAPTER 12

UPPER END THEORY AND SERVICE

OBJECTIVES

■ Describe the purpose of an engine's cylinder head, valves, and related valve parts. ■ Explain why there are special service procedures for aluminum and OHC heads. ■ Describe the types of combustion chamber shapes found on modern engines. ■ Describe the different ways that manufacturers vary valve timing. ■ Describe the procedures for inspecting valve train parts. ■ Explain the procedures involved in reconditioning cylinder heads. ■ Explain the procedures involved in reconditioning valve guides, valve seats, and valve faces. ■ Explain the steps in cylinder head and valve reassembly.

CYLINDER HEAD

The cylinder head **(Figure 12–1)** is made of cast iron or aluminum. On overhead valve engines, the cylinder head contains the valves, valve seats, valve guides, valve springs, rocker arm supports, and a recessed area that makes up the top portion of the combustion chamber. On overhead cam engines, the cylinder head contains these items, plus the supports for the camshaft and camshaft bearings **(Figure 12–2)**.

All cylinder heads contain passages that match passages in the cylinder block. These passages allow coolant to circulate in the head and allow oil to drain back into the oil pan. Oil also moves through some of the passages to lube the camshaft and valve train. The cylinder head also contains tapped holes in the combustion chamber to accept the spark plugs.

The sealing surface of the head must be flat and smooth. To aid in the sealing, a gasket is placed between the head and block. This gasket, called the head gasket, is made of special material that can withstand high temperatures, high pressures, and the expansion of the metals around it. The head also serves as the mounting point for the intake and exhaust manifolds and contains the intake and exhaust ports.

Cylinder head design is one of the most influential factors that affects the overall performance of an engine. The size and shape of the intake and exhaust ports affect the velocity and volume of the mixture entering and leaving the cylinders.

Figure 12–1 A typical late-model cylinder head.

Figure 12–2 An OHC aluminum cylinder head. *Courtesy of Chrysler LLC*

Openings in the cylinder head allow coolant to pass through the head. Coolant must circulate throughout the cylinder head to remove excess heat. The coolant flows from passages in the cylinder block through the head gasket and into the cylinder head. The coolant then passes back to other parts of the cooling system.

Ports

Intake and exhaust ports are cast into the cylinder head. One port is normally used for each valve. However, on engines with more than two valves per cylinder, the ports for the intake or exhaust valves may be combined. These ports are called **Siamese ports** (**Figure 12–3**). With Siamese ports, individual ports around each valve mesh together to form a larger single port that is connected to a manifold. **Cross-flow ports** are used on some engines and have intake and exhaust ports on opposite sides of the combustion chamber. Heads of this design are called cross-flow heads.

Aluminum Heads

Attempts to lower overall engine weight have resulted in many lightweight cylinder head designs. An aluminum head weighs roughly half as much as a cast-iron head. Eliminating anywhere from 20 to 40 pounds of weight is a plus for fuel economy, but these heads bring new concerns for technicians.

Aluminum expands and contracts almost twice as much as cast iron in response to temperature changes. When an aluminum head is mated to an iron block, the difference in thermal expansion between head and block creates a scrubbing stress on the head gasket. Unless the gasket is engineered to take this, leakage and premature gasket failure can result.

The difference in thermal expansion also creates a lot of stress throughout the head. The head wants to expand in all directions as it heats up, but the head bolts keep it from going sideways or lengthwise. The only way it can expand is up, so the head tends to bow up in the middle.

Engines with overhead cams and aluminum heads often have the cam journals machined in the head itself. Aluminum makes a fairly good bearing material. It is soft and provides good embedability to foreign particles. But it lacks the support and rigidity of a conventional steel-backed bearing in an iron saddle. Because of this, overhead cam bores experience more flex than their cast-iron counterparts. The result is usually accelerated wear and egg-shaped bores. If the head overheats and warps, alignment through the cam bores is destroyed. In some instances, the misalignment can be so bad that the cam will not turn once the head is unbolted from the engine.

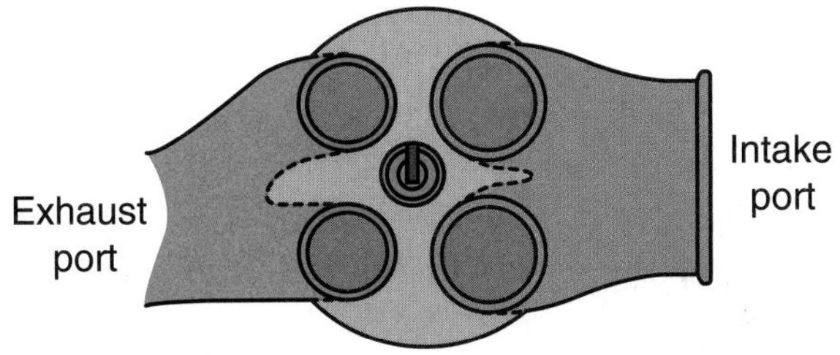

Figure 12–3 Siamese ports.

COMBUSTION CHAMBER

The performance of an engine, its fuel efficiency, and its exhaust emissions all depend to a large extent on the shape of the combustion chamber. An efficient combustion chamber must be compact to minimize the surface area through which heat is lost to the engine's cooling system. The point of ignition (the nose of the spark plug) should be at the center of the combustion chamber to minimize the flame path, or the distance from the spark to the furthermost point in the chamber. The shorter the flame path, the more evenly the air-fuel mixture will burn.

Manufacturers have designed several shapes of combustion chambers. Before looking at the popular combustion chamber designs, two terms should be defined.

1. **Turbulence** is a very rapid movement of gases. Turbulence causes better combustion because the air and fuel are mixed better.
2. **Quenching** is the mixing of gases by pressing them into a thin area. This area is called the quench area.

Wedge Chamber

In the **wedge-type combustion chamber**, the spark plug is located at the wide part of the wedge (**Figure 12–4**). The spark travels from the large area in

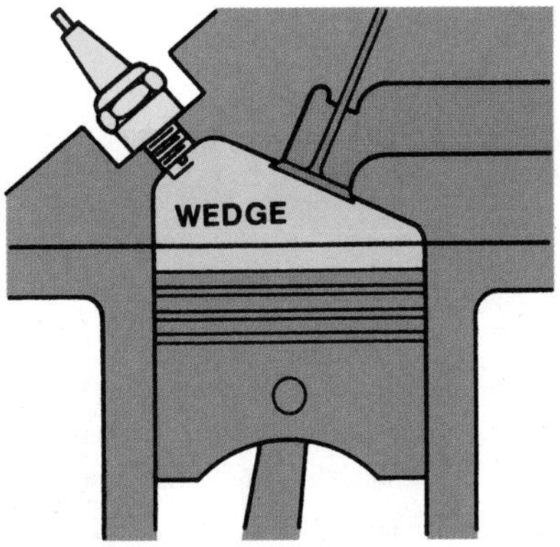

Figure 12–4 A typical wedge combustion chamber.

the chamber to a smaller one. This allows for rapid and even combustion. At the end of the chamber's wedge is a very narrow area. The **quench area**, also called the squish area, causes the air and fuel to be thoroughly mixed before combustion. The mixture in the quench area is squeezed out at high speed as the piston moves up. This causes turbulence in the chamber. The turbulence breaks down the fuel into small particles and helps mix the air and fuel. Both of these contribute to efficient combustion and reduce the chances of detonation. The turbulence also helps the flame front move evenly throughout the chamber. The wedge-shaped combustion chamber is also called a turbulence-type combustion chamber.

Hemispherical Chamber

The **hemispherical combustion chamber** gets its name from its basic shape. Hemi is defined as half, and spherical means circle. The combustion chamber is shaped like a half circle. This type of cylinder head is also called the **hemi-head**. The piston top forms the base of the hemisphere, and the valves are inclined at an angle of 60 to 90 degrees to each other, with the spark plug positioned between them **(Figure 12–5)**.

This design has several advantages. The flame path from the spark plug to the piston head is short, which gives efficient burning. The cross-flow arrangement of the inlet and exhaust valves allows for a relatively free flow of gases in and out of the chamber. The result is that the engine can breathe deeply, meaning that it can draw in a large volume of mixture for the space available and give a high power output.

The hemispherical combustion chamber is considered a nonturbulence-type combustion chamber. The mixture is compressed evenly on the compression stroke. Combustion radiates evenly from the spark plug, completely burning the air-fuel mixture.

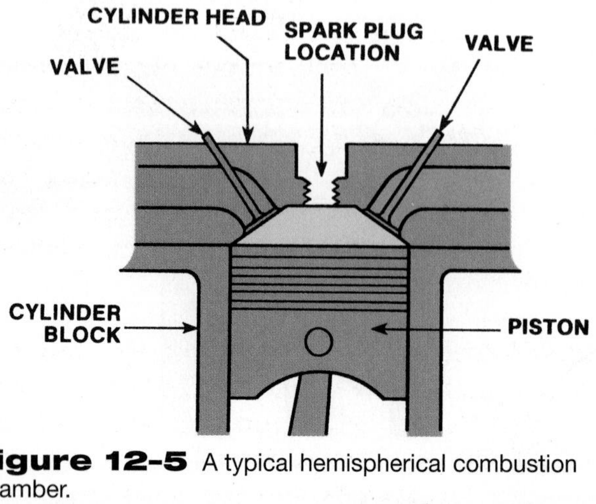

Figure 12-5 A typical hemispherical combustion chamber.

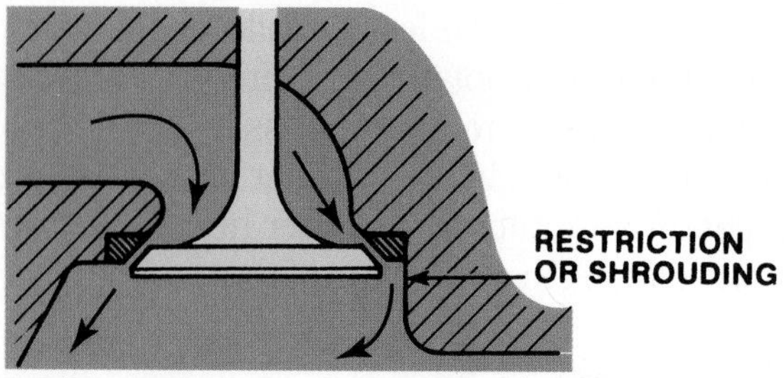

Figure 12-6 Shrouding is a restriction in the flow of gases caused by the location of the valves in the combustion chamber.

Figure 12-7 A cylinder head with pentroof combustion chambers.

One of the more important advantages of the hemispherical combustion chamber is that air and fuel can enter the chamber very easily. The wedge combustion chamber restricts the flow of air and fuel to a certain extent. This is called **shrouding**. **Figure 12–6** shows the valve very close to the side of the combustion chamber, which causes the flow of air and fuel to be restricted. Volumetric efficiency is reduced. Hemispherical combustion chambers do not have this restriction.

Pentroof Chamber

Many of today's engines have a **pentroof** combustion chamber. This design is a modified hemispherical chamber. It is mostly found in engines with four valves per cylinder. The spark plug is located in the center of the chamber and the intake and exhaust valves are on opposite sides of the chamber **(Figure 12–7)**. Pentroof chambers have a squish area around the entire cylinder.

INTAKE AND EXHAUST VALVES

The intake and exhaust valves are commonly called **poppet** valves **(Figure 12–8)**. They tend to pop open and close. When they open, they allow intake air to flow into the combustion chamber or allow the exhaust to leave it. When closed, they must (along with the cylinder head gasket, piston rings, and spark plug) seal the chamber. The heads of the intake and exhaust

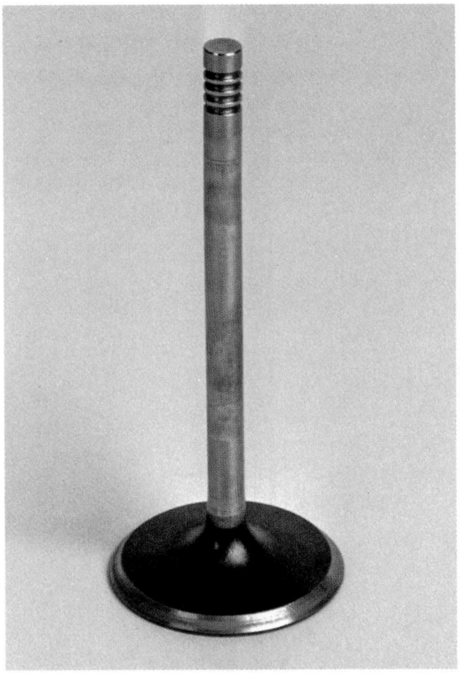

Figure 12–8 A poppet valve.

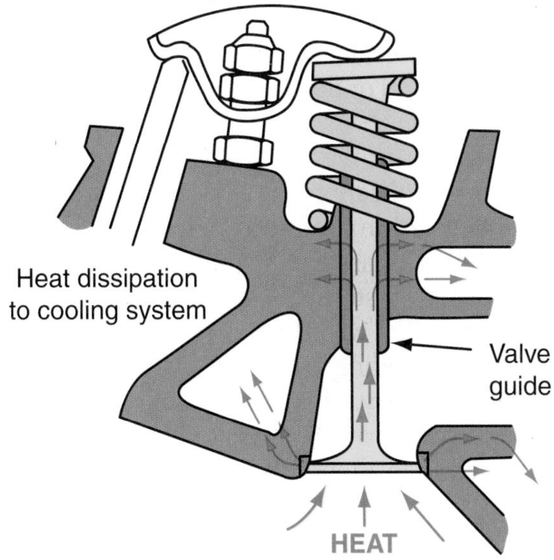

Figure 12–9 Valves cool by transferring heat to the liquid passages in the cylinder head.

valves have different diameters. The intake valve is the larger of the two. An exhaust valve can be smaller because exhaust gases move easier than intake air.

Valve Construction

Today, most valves are made from special hardened steel, steel alloys, or stainless steel. Other metals are often used in high-performance valves. Heat is an important factor in the design and construction of a valve. The material used to make a valve must be able to withstand high temperatures and be able to dissipate the heat quickly. Most of the heat is dissipated through the contact of the valve face and seat. The heat then moves through the cylinder head to its coolant passages. Heat is also transferred through the valve stem to the valve guide and again to the cylinder head **(Figure 12–9)**.

Intake and exhaust valves are typically made with different materials. Intake valves are typically low-alloy steels or heat- and corrosion-resistant high-alloy steels. The alloy used in a typical exhaust valve is chromium for oxidation resistance with small amounts of nickel, manganese, silicon, and/or nitrogen. Heat resistance is critical for exhaust valves because they face temperatures of 1,500°F to 4,000°F (816°C to 2,204°C). Intake valves need less heat resistance because the intake air and fuel tend to cool them. Intake valves also need less corrosion protection because they are not exposed to the corrosive action of the hot exhaust gases.

A valve can be made as a single piece or two pieces. Two-piece valves allow the use of different metals for the valve head and stem. The pieces are spin welded together. These valves typically have a stainless steel head and a high-carbon steel stem. The stems are often chrome plated so the weld is not visible. One-piece valves run cooler because the weld of a two-piece valve inhibits the flow of heat up the stem.

Today's engines require higher quality valves that contain more nickel to withstand the heat. Most exhaust valves and some intake valves have 4% nickel content. The intake valves with high nickel are used in turbocharged engines. Older valves are alloyed with 2% nickel. The alloys used to make valves depend on the intended use and the design of the engine. Newer engines also tend to have lighter valves than what was used in the past. The lighter weight decreases the amount of power lost moving the valves and allows for higher engine speeds.

Stainless Steel Valves Stainless steel is commonly used to make valves. **Stainless steel** is an iron-carbon alloy with a minimum of 10.5% chromium content. Stainless steel does not stain, corrode, or rust as easily as ordinary steel. There are different types of stainless steels used to make valves. Austenitic stainless steels contain a maximum of 0.15% carbon, a minimum of 16% chromium, and nickel and/or manganese to give it strength and improve its heat resistance. Stainless steel is nonmagnetic.

Inconel Valves An alloy that is being used by many manufacturers is Inconel. Inconel has a nickel base with 15% to 16% chromium and 2.4% to 3.0% titanium. This alloy is normally used in high-temperature applications and has good oxidation and corrosion resistance. Inconel is difficult to machine; therefore, Inconel valves are replaced when they are deformed or damaged.

Stellite Valves Another alloy that is used in high-temperature applications is stellite. **Stellite** is an alloy of nickel, chromium, and tungsten and is nonmagnetic. Stellite is a hard-facing material that is welded to valve faces and stems. It may also be used on the stem tip for added wear resistance. It comes in various grades depending on the mix of ingredients that are used in the alloy. This alloy has high resistance to wear, corrosion, erosion, abrasion, and galling. Stellite is available in many different grades, which are determined by the materials used in the alloy.

Sodium Filled Some exhaust valves have a hollow stem. The hollow section of the stem is partially filled with sodium (**Figure 12–10**). Sodium is a silver-white alkaline metallic substance that transfers heat much better than steel. At operating temperatures, the sodium becomes a liquid. When the valve opens, the sodium splashes down toward the head and absorbs heat. Then as the valve moves up, the sodium moves away from the head and up the stem. The heat absorbed by the sodium is then transferred to the guide where it moves to the coolant passages in the head. Sodium-filled valves should not be machined.

WARNING!

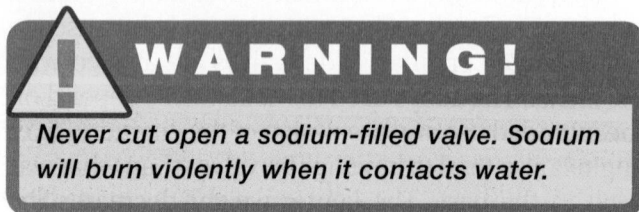

Never cut open a sodium-filled valve. Sodium will burn violently when it contacts water.

Titanium Valves Titanium alloys are added to valves to lighten them. Some high-performance engines have titanium valves. These valves dissipate heat well, are durable, and are very light. A titanium valve weighs less than half of a comparable steel valve.

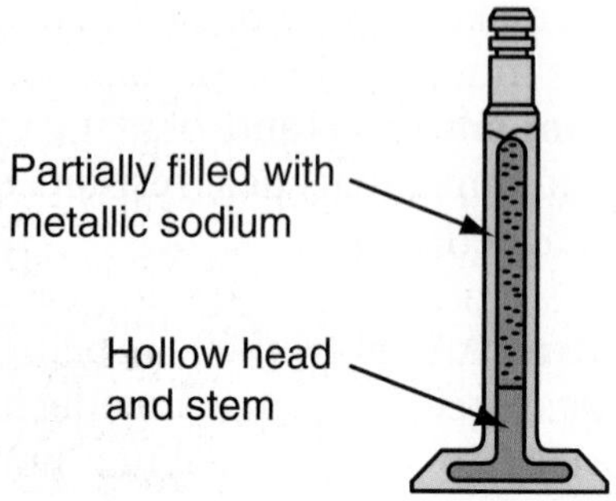

Figure 12–10 Some exhaust valves are partially filled with sodium to help cooling.

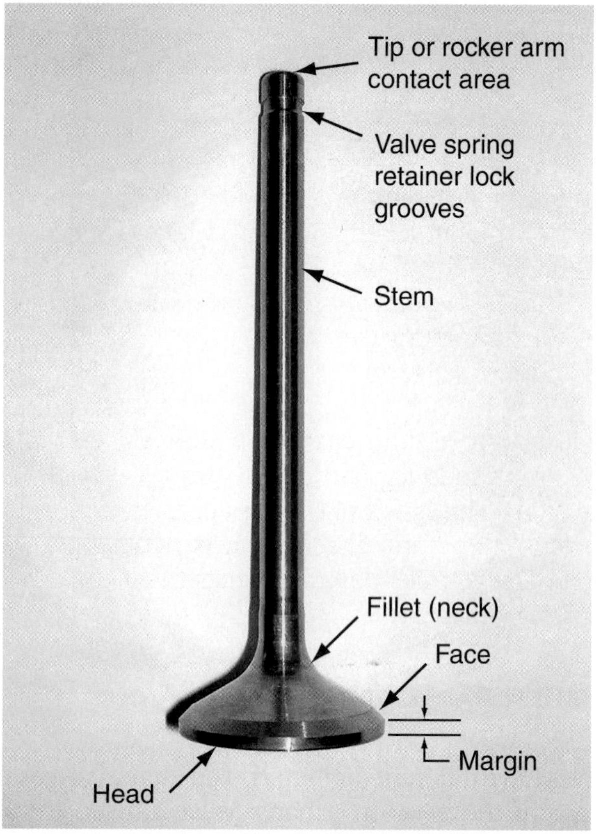

Figure 12–11 The parts of a typical valve.

Ceramic Valves Ceramic valves are being tested for future use. Ceramic materials weigh less than half of what a comparable size steel valve weighs and can withstand extreme temperatures without weakening or becoming deformed.

Valve Terminology Valves (**Figure 12–11**) have a round head with a tapered face used to seal the intake or exhaust port. This seal is made by the **valve face** contacting the valve seat. The angle of the taper depends on the design and manufacturer of the engine. The distance between the valve face and the head of the valve is called the **margin**. The **valve stem** guides the valve during its up-and-down movement and serves to connect the valve to its spring through its valve spring retainers and keepers. The keepers are fit into a machined slot at the top of the stem, called the valve keeper groove. The stem moves within a **valve guide** that is either machined into (integral type) or pressed into the head (insert type).

Valve Stems

Little oil passes through the clearance between the stem and valve guide. Therefore, the surfaces of the guide and the stem are designed to minimize friction. Valve stems have two common types of coating to prevent wear and reduce friction: chrome plating and black nitriding. In addition to these coatings, the tips of the stem are hardened or stellited to resist damage from the constant hammering they face as the stems are pushed open.

Chrome-plated stems help prevent valve stem scuffing and galling and provide protection against wear during initial engine starts when no oil is present on the valve stem. Chrome coating is also widely used on high-performance valves. The thickness of the chrome plating can vary from 0.0002 to 0.001 inch (0.0051 to 0.0254 mm).

Many foreign manufacturers use a black nitride coating rather than chrome plating on the valves. Black nitride is applied to the entire valve, not just the stem. The finish of the surface is smoother than chrome; therefore, less friction is produced by the stem. The nitride coating protects the stems against scuffing and wear.

Valve Seats

The valve seat **(Figure 12–12)** is the area of the cylinder head contacted by the face of the valve. The seat may be part of the casting and machined in the head (integral type) or it may be pressed into the head (insert type). Insert seats are always used in aluminum cylinder heads. They are also used to replace damaged integral seats.

Valve seats provide a sealing area for the valves. They also absorb the valve's heat and transfer it to the cylinder head. Seats must be hard enough to withstand the constant closing of the valve. Due to corrosive products found in exhaust gas, seats must be highly resistant to corrosion. When the head is made of cast iron, it has integral seats because cast iron meets those requirements. Cast iron is also used to make seat inserts. Most are induction hardened. These are hardened through electromagnetism, which heats the surface of the seat.

Many late-model engines with aluminum heads have sintered powder metal (tungsten carbide) seats. Powder metal seats are harder and more durable than cast-iron seats.

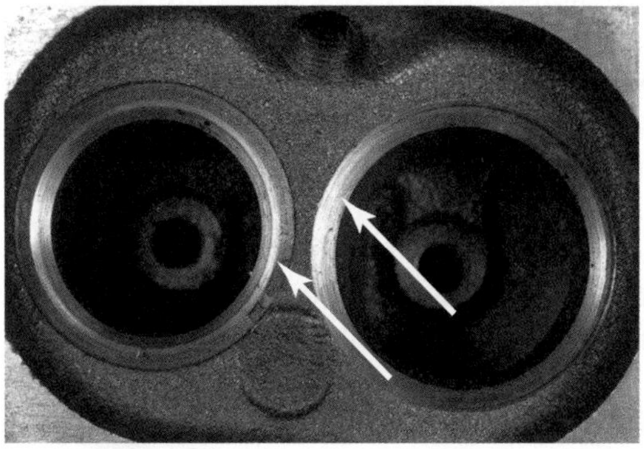

Figure 12-12 Valve seats.

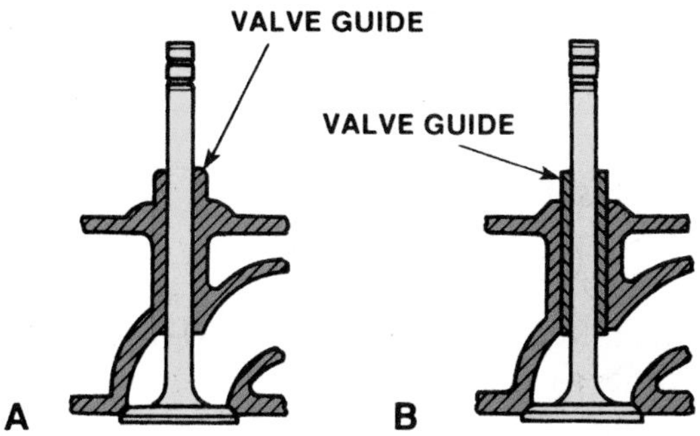

Figure 12-13 (A) Integral and (B) removable valve guides.

Valve Guides

Valve guides support the valves in the head and prevent the valves from moving in any direction other than up and down. The inside diameter of a guide is machined to provide a very small clearance with the valve stem. This close clearance is important for the following reasons:

- It keeps oil from being drawn into the combustion chamber past the intake valve stem during the intake stroke, and it keeps oil from leaking out to the exhaust port when the pressure in the exhaust port is lower than the pressure in the crankcase.
- It keeps exhaust gases from leaking into the crankcase area past the exhaust valve stems during the exhaust stroke.
- It keeps the valve face in perfect alignment with the valve seat.

Valve guides can be cast integrally with the head, or they can be removable **(Figure 12–13)**. Removable or insert guides are press-fit into the head. Aluminum heads are fitted with insert guides. Guides are made from materials that provide low friction and can transfer heat well. Cast-iron guides are mixed or coated with phosphorus and/or chrome. Bronze alloys are also used. These may contain some aluminum, silicon, nickel, and/or zinc.

Valve Spring Retainers and Oil Seals

The valve assembly is completed by the spring, retainer, and seal. An oil seal is placed over the top of the valve stem. The seal acts like an umbrella to keep oil from running down the valve stem and into the combustion chamber. The valve spring, which keeps the valve in a normally closed position, is held in place by the retainer. The retainer locks onto the valve stem with two wedge-shaped parts that are called **valve keepers**. **Figure 12–14** shows the components that make up a valve spring assembly.

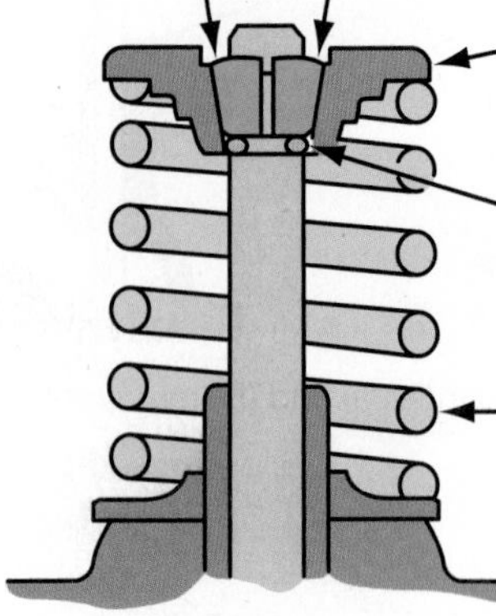

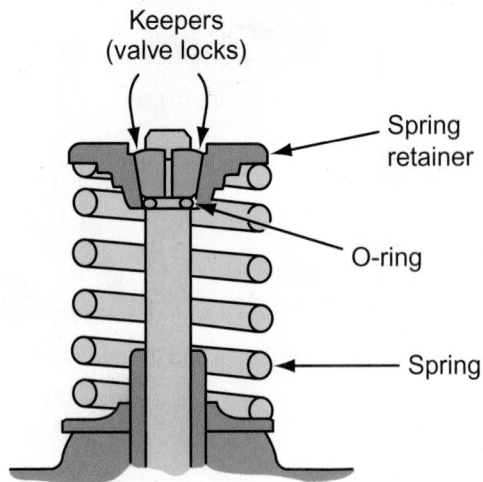

Figure 12–14 Valve assembly with spring, retainer, seals, and keepers.

Valve Rotators Some engines are equipped with retainers that cause the exhaust valves to rotate. These rotators prevent carbon from building up between the valve face and seat. Carbon buildup can hold the valve partially open, causing it to burn.

Valve Springs

A valve spring closes the valve and maintains valve train contact during valve opening and closing. Some engines have one spring per valve. Others use two or three springs. Often the second or third spring is a flat spring called a **damper spring**, which is designed to control vibrations. To dampen spring vibrations and increase total spring pressure, some engine manufacturers use a reverse wound secondary spring inside the main spring.

Low spring pressure may allow the valve to float during high-speed operation. Too much pressure will cause premature valve train or camshaft lobe wear and can also lead to valve breakage.

Other Valve-Related Parts

Other parts are associated with the valves.

Rocker Arms Rocker arms change the direction of the cam's lifting force. As the lifter and pushrod move upward, the rocker arm pivots at its center point. This causes a change in direction on the valve side and pushes the valve down. Rocker arms also allow the valve to open farther than the actual lift of the cam lobe. This is done by having different distances from the pivot point to the ends of the rocker arm. The difference in length from the valve end of the rocker arm and the center of the pivot point (shaft or stud) compared to the pushrod or cam end of the rocker arm and the pivot point (shaft or stud) is expressed as a ratio. Usually, rocker arm ratios range from 1:1 to 1:1.75. A ratio larger than 1:1 results in the valve opening farther than the actual lift of the cam lobe.

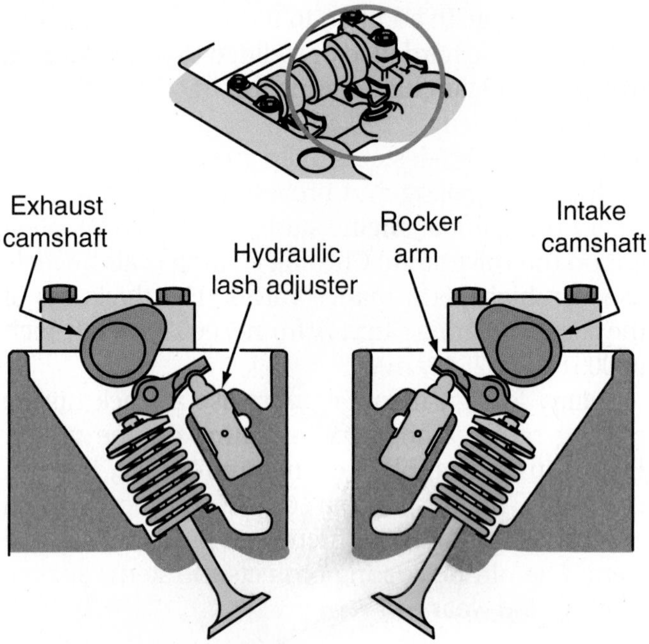

Figure 12–15 The camshaft in this setup rides on a rocker arm that has a hydraulic lash adjuster.

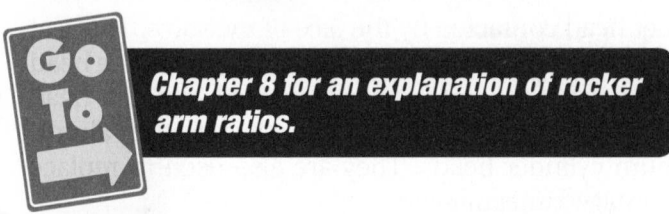

Go To
Chapter 8 for an explanation of rocker arm ratios.

The camshaft in some OHC engines rides directly on the rocker arm. One end of the rocker arm fits over a cam follower or lifter and the other end is over the valve stem **(Figure 12–15)**. Often OHC cylinder heads have a complex arrangement of rocker arms **(Figure 12–16)**. Other OHC engines have no rocker arms and the camshaft rides directly on top of the valves.

Rocker arms are made of stamped steel, cast aluminum, or cast iron. Cast adjustable rocker arms are attached to a rocker arm shaft that is mounted to the head by rocker arm brackets **(Figure 12–17)**. Cast-iron rockers are used in large, low-speed engines. They almost always pivot on a common shaft. Aluminum rockers, on the other hand, are generally used on high-performance applications and pivot on needle bearings to reduce friction.

Some domestic engines are equipped with a stamped steel rocker arm for each valve. The rocker arm is mounted to a stud that is either pressed or threaded into the cylinder head and must be replaced if worn, bent, broken, or loose. On some engines, the studs are drilled for an oil passage to the rocker arms.

Pushrods Pushrods are the connecting link between the rocker arm and the valve lifter. Pushrods transmit cam action to the valves. Often the pushrods have a

Figure 12–16 This arrangement has three rocker arms per valve. *Courtesy of American Honda Motor Co., Inc.*

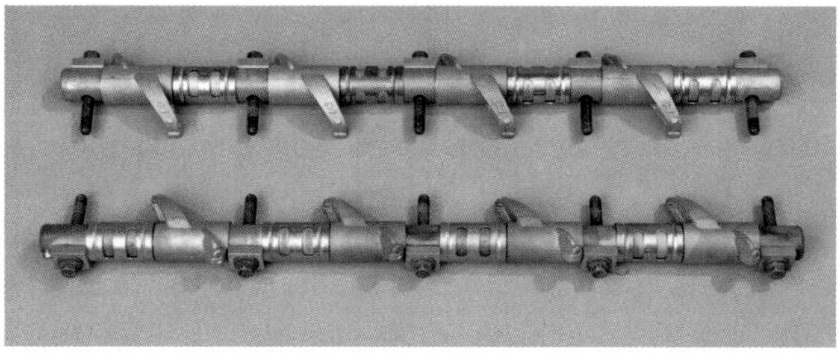

Figure 12–17 Rocker arm shaft assemblies. *Courtesy of Chrysler LLC*

hole in the center to allow oil to pass from the hydraulic lifter to the rocker arm assembly **(Figure 12–18)**. Some engines use pushrod **guide plates** to limit the side movement of the pushrods.

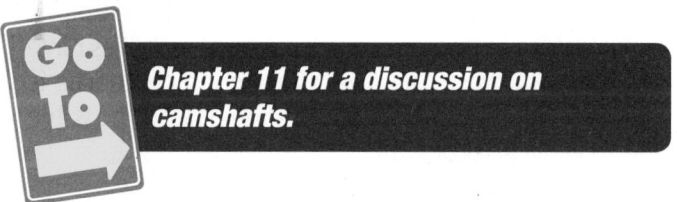

Go To **Chapter 11 for a discussion on camshafts.**

Cam Followers Found on some OHC engines, cam followers are cups that sit on top of the valve assembly. They provide a larger surface for the cam lobes to move the valves. Some followers have a hydraulic unit that fits under the cup to maintain proper valve clearance. Others require periodic adjustment. Most use metal shims between the cup and cam lobe. To adjust valve clearance, a shim with a different thickness is inserted.

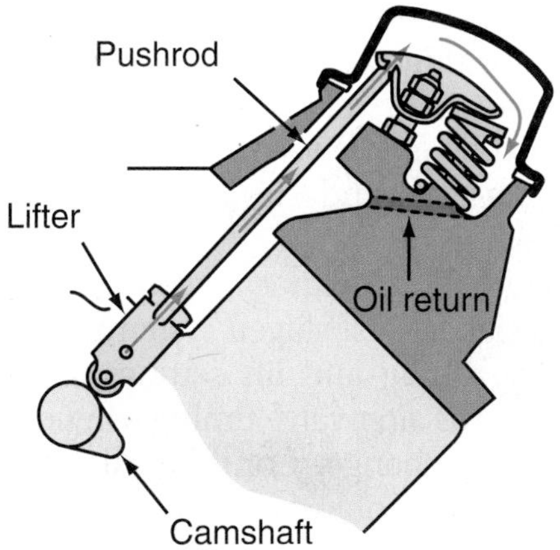

Figure 12–18 Most pushrods have a hole through the center to allow oil flow from the hydraulic lifter to the rocker arm assembly.

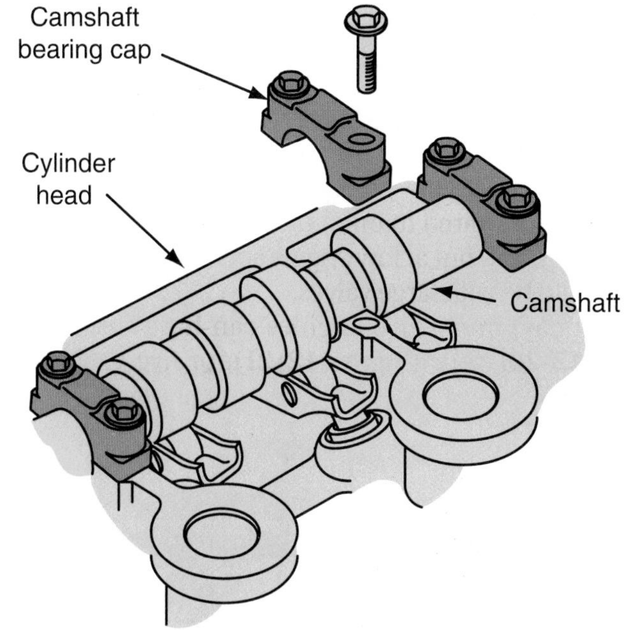

Figure 12–19 Many OHC cylinder heads are machined to accept one or two camshafts above the valves and have bearing caps to secure the camshaft.

Camshaft Bearings The camshaft is part of the cylinder head assembly in all OHC-type engines. The unit that holds the camshaft(s) may be a separate unit bolted to the cylinder head or the camshaft's bore is machined into the upper part of the head. In the most common design, the cylinder heads are machined to accept one or two camshafts and have caps that secure the camshaft **(Figure 12–19)**.

Multivalve Engines

Many newer engines use multivalve arrangements. One of the first cars to use four valves per cylinder as a way to enhance gas flow and increase horsepower was the 1918 dual-valve Pierce Arrow.

The basic idea behind using more than one intake and/or exhaust valve is simple—better efficiency. To improve efficiency, engineers need to improve the flow in and out of the combustion chamber. In the past, this was attempted by making the valves larger and by changing valve timing. Larger valves allowed more air in and more exhaust out, but the bigger valves weighed more and therefore required stronger springs to close them. The stronger springs held the valves closed tighter but required more engine power to open them. Also, when an engine is running at low speeds, the air moving past a large valve has a lower velocity than it would have if it flowed past a small valve. This reduces engine torque at low engine speeds.

Although two small valves weigh as much or more than one valve, each valve weighs less and, therefore, the spring tension on each is less. This means less power is required to open them. Also, the velocity of the air in and out at low engine speeds is quicker than it would be with large valves.

Today, multivalve engines can have three **(Figure 12–20)**, four **(Figure 12–21)**, or five valves per cylinder **(Figure 12–22)**. The most common arrangement is four valves per cylinder with two intake and two exhaust valves. All multivalve engines have cross-flow heads.

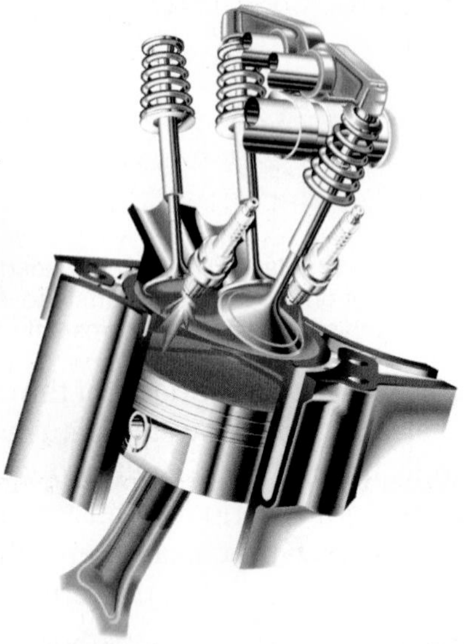

Figure 12–20 An engine with two spark plugs, two intake valves, and one exhaust valve for each cylinder. *© Courtesy of Daimler AG*

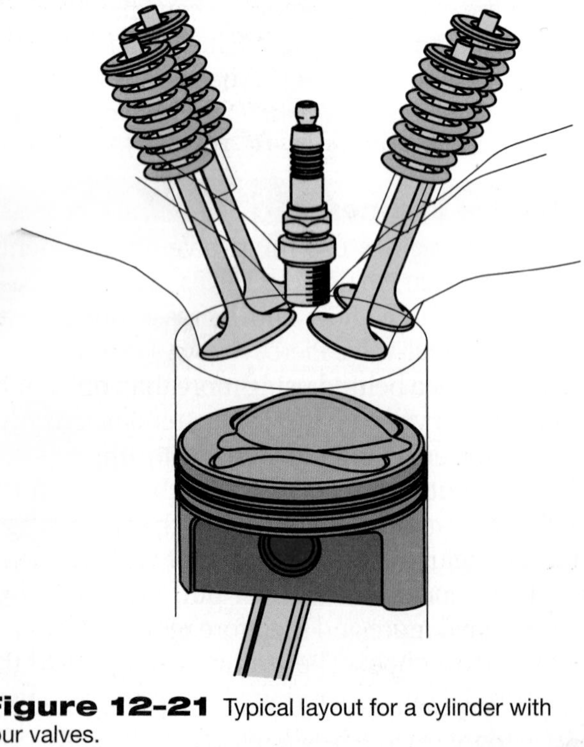

Figure 12–21 Typical layout for a cylinder with four valves.

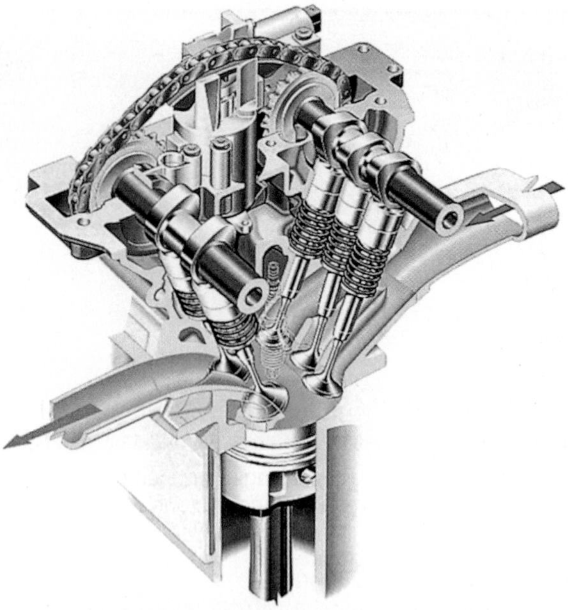

Figure 12–22 This five-valve per cylinder arrangement has three intake valves and two exhaust valves. *Courtesy of Chrysler LLC*

Using two intake and one or two exhaust valves increases the volume of the intake and exhaust ports. Therefore, more air can move in or out of the cylinder. This results in a more complete combustion, which reduces the chances of misfire and detonation. It also results in better fuel efficiency, cleaner exhaust, and increased power output. Two smaller valves have less mass than one big one, so mechanical inertia is reduced, making a higher engine speed possible before valve float occurs.

VARIABLE VALVE TIMING

Changing valve timing in response to driving conditions improves driveability and lowers fuel consumption and emission levels. There are many different variable valve timing (VVT) systems used on today's engines. Many systems only vary the timing of the intake camshaft, some vary the timing of the intake and exhaust valves, and a few vary the timing and lift of all the engine's valves.

VVT systems are either staged or continuously variable designs. Most staged systems allow two different valve timing and lift settings. Continuously variable systems alter valve timing whenever operating conditions change. Continuously variable systems change the phasing or timing of a valve's duration **(Figure 12–23)**. These systems provide a wider torque curve, reduction in fuel consumption,

Advancing	Retarding
Begins Intake Event Sooner	Delays Intake Event
Lengthens Valve Overlap	Shortens Valve Overlap
Builds More Low-End Torque	Builds More High-RPM Power
Decreases Piston to Intake Valve Clearance	Increases Piston-to-Intake Valve Clearance

Figure 12-23 The effects of changing intake valve timing.

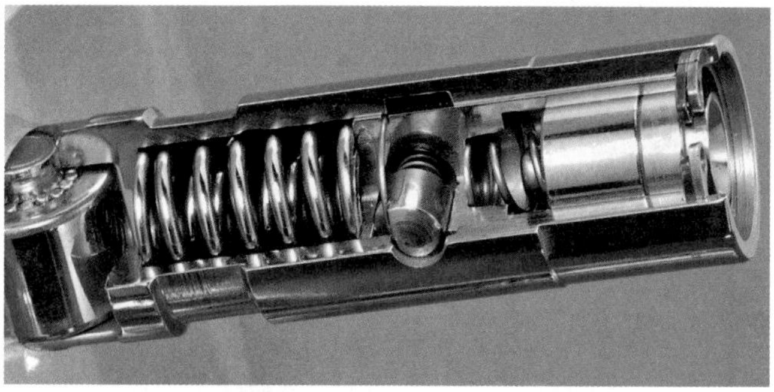

Figure 12-24 The lifter used in Chrysler's multidisplacement system. When the pin in the center is moved, the piston inside the lifter is disconnected from the lifter body. *Courtesy of Chrysler LLC*

improved power at high speeds, and a reduction in hydrocarbon and NO_x emissions. On some engines, VVT has eliminated the need for an exhaust gas recirculation (EGR) valve.

Cylinder Deactivation

Cylinder deactivation works by keeping the intake and exhaust valves constantly closed in a group of the engine's cylinders. This decreases the working displacement of the engine and provides an increase in fuel economy and reduced emissions. The systems are designed to make the deactivation and activation of the cylinders unnoticeable to the driver. This is accomplished by controlling the fuel injectors, ignition timing, throttle opening, and valve timing. The exact system used for cylinder deactivation varies with the engine design and the manufacturer.

OHC engines typically have a pair of rocker arms at each valve. One of the rocker arms rides on the camshaft lobe and the other works the valve. When the two rocker arms are locked together, the valve moves according to the rotation of the camshaft. To disable a cylinder, the rocker arms are unlocked. The rocker arm on the cam lobe continues to work but does not transfer its movement to the other. The locking device is simply a pin that moves in response to oil pressure. A solenoid, controlled by the PCM, directs oil pressure to the pin.

Honda's variable cylinder management (VCM) system is an example of this. This system is based on the i-VTEC variable valve control system, which is a staged valve timing system.

OHV engines also use oil pressure to deactivate the cylinders. High pressure is sent to the hydraulic lifters to collapse them. The lifters then follow the cam lobes but do not move the pushrods and rocker arms. There are two ways this is done. Chrysler's multidisplacement system (MDS), found in some V8 engines, has an oil circuit controlled by four solenoids and eight unique hydraulic roller lifters. When conditions dictate that the vehicle does not need all eight cylinders, the PCM energizes the solenoids. Oil pressure is sent to the lifters. The pressure pushes on a small pin in the lifters **(Figure 12–24)**. As the pin moves, the piston inside the lifter is disconnected from the lifter body. The lifter body continues to move with the cam lobe but no motion is passed on to the rocker arms.

General Motors' displacement on demand (DoD) system, now called active fuel management (AFM), uses two-stage switching lifters. The lifters have an inner and outer body connected by a spring-loaded pin. High oil pressure, sent by solenoids, collapses the spring and the two lifter bodies disconnect.

Staged Valve Timing

Most staged valve timing systems switch between two or more different camshaft profiles based on operating conditions. An example of this is Honda's VTEC system used on multivalve engines. The camshaft has three lobes for each pair of intake valves. The third lobe is shaped for more valve lift and different open and close times **(Figure 12–25)**. There is a rocker arm

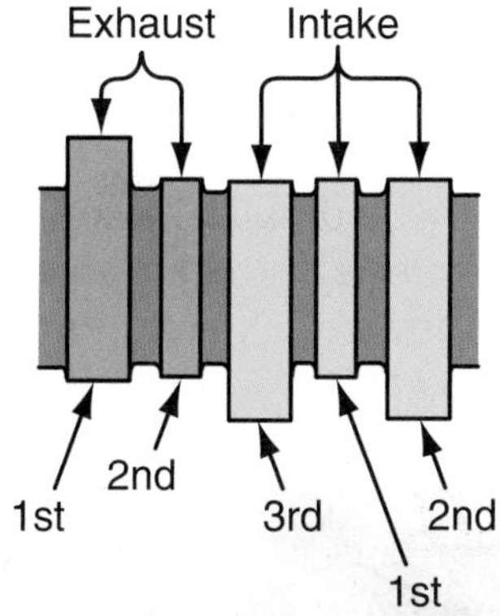

LOBE	INTAKE	EXHAUST
1st	1.1692 in. (29.700 mm)	1.1771 in. (29.900 mm)
2nd	1.4003 in. (35.568 mm)	1.4054 in. (35.699 mm)
3rd	1.4196 in. (36.060 mm)	

Figure 12-25 The size of the cam lobes for one cylinder for Honda's three-stage VTEC system.

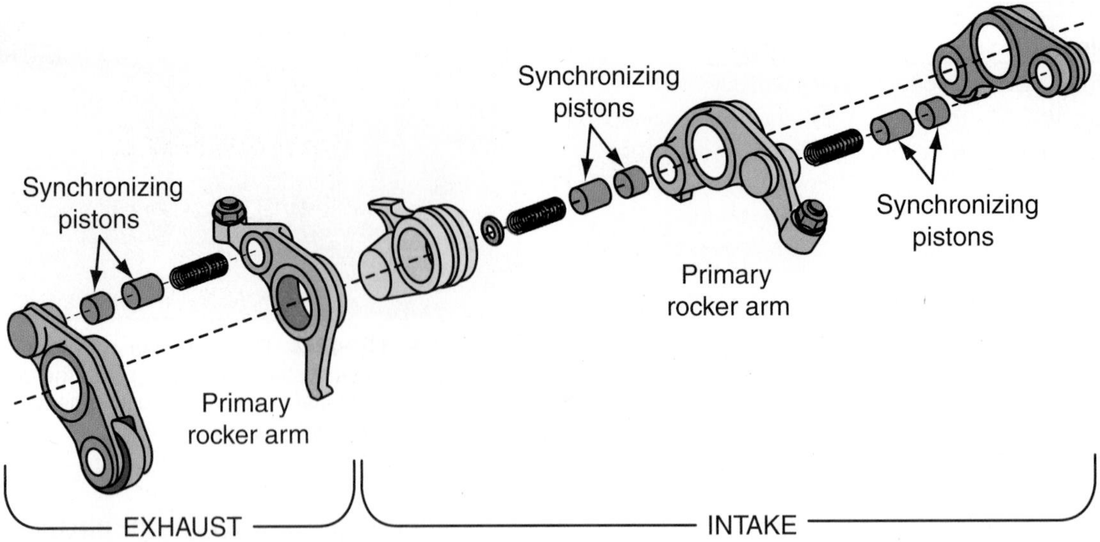

Figure 12–26 The synchronizing pin is controlled by oil pressure and locks the rocker arms together.

over each of the three lobes. At low engine speeds, only the second lobe's rocker arms move the valves. At high speed, a solenoid valve sends pressurized oil through the rocker shaft to a piston in the outer rocker arms **(Figure 12–26)**. This pushes the piston partly into the center rocker arm, locking the three rocker arms together. The valves now open according to the shape of the third lobe. When the solenoid valve closes, a spring pushes the pistons back into the outer rockers, and the engine runs with normal valve timing.

This system has been modified to include VCM, also called the cylinder idling system, on some engines. This system is used on Honda's hybrid to increase the regenerative braking capabilities of the vehicle and to minimize fuel consumption. The system allows normal and high-output valve timing, plus cylinder idling at all or some cylinders. The system is based on three rocker arms per valve. One rocker arm is used to activate the valve. The other two ride on camshaft lobes. A hydraulically controlled pin connects and disconnects the rocker arms. When there is no connection between the two cam-riding rockers and the valve rocker, the cylinder is idling or deactivated. There are three separate oil passages leading to the pin. As the pressure moves through a passage, it moves the pin. The PCM controls a spool valve that directs the pressurized oil to the appropriate passage. It also controls solenoids that control the amount of pressure.

Operation In a typical system, when the brake pedal is released and the accelerator depressed, the vehicle moves by both electric and engine power. At this time the engine is running in the economy mode with the valves opening by the low lift camshaft profile. When the driver is maintaining a very low cruising speed, the engine shuts off and the electric motor powers the car by itself. During this time, the engine's rocker arms are not opening the valves. During acceleration from a low speed, the engine runs in the economy mode. During heavy acceleration, the engine runs in its high-output mode and the electric motor assists the engine. During deceleration, the motor begins to work as a generator and the engine's valves close and remain closed. This allows for maximum regenerative braking and reduces fuel consumption.

Continuously Variable Timing

To provide continuously variable value timing, camshafts are fitted with a **phaser**. The phaser is mounted where a timing pulley, sprocket, or gear would be **(Figure 12–27)**. The phaser allows camshaft-to-crankshaft

Figure 12–27 The phaser assembly on the intake camshaft of a Lexus' VVT-i engine. *Courtesy of Toyota Motor Sales, U.S.A., Inc.*

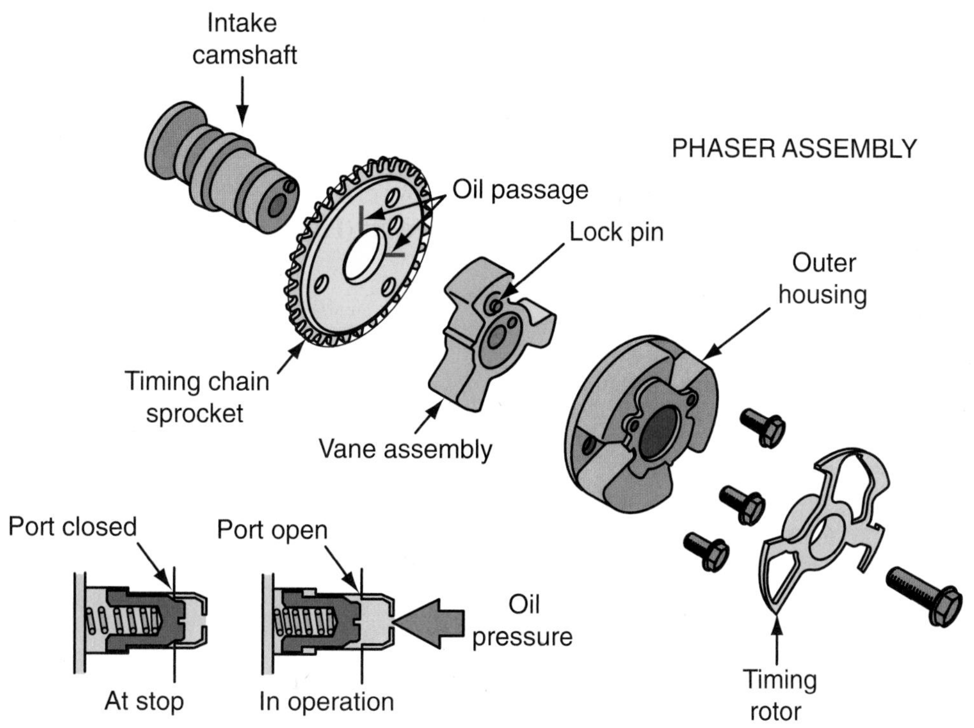

Figure 12-28 An exploded view of a phaser.

timing to change while the engine is running. Phasers can be electronically or hydraulically controlled. In a hydraulically controlled system, oil flow is controlled by the PCM. Electronic systems rely on stepper motors. A few engines have electronic/electric intake cam phasers with hydraulic exhaust cam phasers.

A phaser assembly is a sealed unit comprised of a hub and an internal vane assembly **(Figure 12–28)**. Around the hub is a timing gear that is connected by chain or belt to the crankshaft. The vane assembly is attached to the camshaft. As the vane assembly moves, the phasing of the camshaft changes. At the base of the hub are oil ports. Oil from control solenoids enters and exits through these ports. When the PCM determines a need to change valve timing, oil is sent to the correct port. The pressurized oil then pushes on the vanes and causes a change in valve timing. Changes in valve timing are made by changes in oil pressure on either side of the vane.

On a few engines, phasers are connected to the intake and exhaust camshafts, therefore altering the timing of both. It is important to realize that by altering the timing of both the intake and exhaust valves, valve overlap is also changed. Ford's 5.4-liter Triton V8 engine has a single camshaft in each cylinder bank. The engine also has VVT. This means the timing of the intake and exhaust valves are shifted in equal amounts. When more low-speed torque is required, the PCM orders earlier valve opening and closing. When more high-speed power is needed, the cam timing is retarded.

Prius VVT-i System The engine in a Prius, like other hybrid vehicles, operates on the Atkinson cycle. However, it also runs with a conventional four-stroke cycle. The switching between the two cycles is done by valve control. Toyota's VVT-i system is reprogrammed to allow the intake valve to close later for the Atkinson cycle. The Atkinson cycle effectively reduces the displacement of the engine and is in operation when there is low engine load.

The VVT-i system is controlled by the PCM. The PCM adjusts valve timing according to engine speed, intake air volume, throttle position, and water temperature. In response to these inputs, the ECM sends commands to the camshaft timing oil control valve **(Figure 12–29)**. A change in oil pressure changes the position of the camshaft and the timing of the valves. The camshaft timing oil control valve is duty cycled by the ECM to advance or retard intake valve timing.

Valvetronic System

Many BMW engines have infinitely variable intake and exhaust valve lift and timing control. Valvetronic is used with variable intake and exhaust valve timing to regulate the flow of air into the cylinders by controlling valve lift. By doing this, the engine has no need for a throttle plate. In fact, this is one of the biggest advantages of the

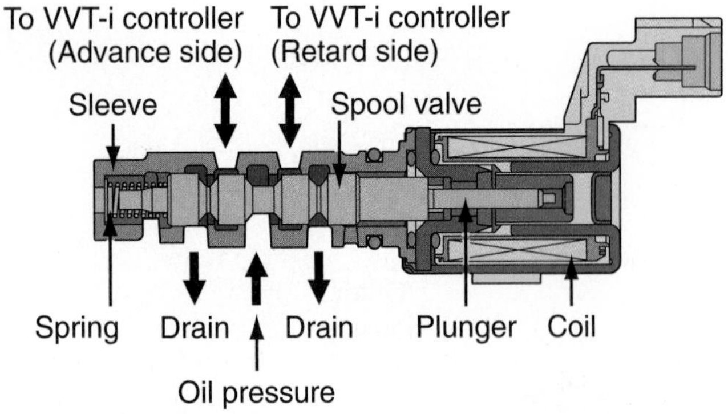

Figure 12-29 The spool valve in the camshaft timing oil control valve is duty cycled by the PCM. This allows oil pressure to be applied to the advance or retard side of the phaser. *Courtesy of Toyota Motor Sales, U.S.A., Inc.*

system. A throttle plate has a tendency to rob the engine of power, especially at low engine speeds.

In conventional engines the throttle plate regulates the flow of incoming air, while the lift and duration of the intake valve remain constant. During low speeds, the throttle plate is almost totally closed and blocks most of the air available for the cylinders. This results in a pumping loss. **Pumping loss** is a term used to describe the difficulty a piston has in moving air into the cylinder and moving it out on the exhaust stroke. Pumping losses are a major reason why engines consume a disproportionately large amount of fuel in city driving.

Valvetronic solves this problem by controlling the flow of incoming air at the intake valves. This system uses a conventional camshaft with a secondary eccentric shaft and a series of levers and roller followers that are activated by a stepper motor **(Figure 12–30)**. A computer changes the phase of the eccentric cam to change the action of the valves.

At high engine speeds the system dials in maximum lift, opening the ports for maximum flow to guarantee rapid filling of the cylinder **(Figure 12–31)**. At low engine speeds, the system reverts to minimal valve lift **(Figure 12–32)**. This reduces the amount of

Figure 12-30 The secondary eccentric shaft and stepper motor assembly for the Valvetronic system. *Courtesy of BMW of North America, LLC*

Figure 12-31 The position of the eccentric shaft to provide maximum lift in a Valvetronic system. *Courtesy of BMW of North America, LLC*

Figure 12-32 The position of the eccentric shaft to provide minimum lift in a Valvetronic system. *Courtesy of BMW of North America, LLC*

air entering the cylinder. The action of the valves becomes the engine's throttle plates.

Other VVT Systems

A unique setup for controlling intake and exhaust valve timing and lift relies on a conventional camshaft, which is ground for high performance, with hydraulic lifters fitted with ultra-high-speed valves to bleed off the fluid. The PCM changes the pressure in the lifter to delay valve openings, change valve duration, or prevent valves from opening. A solenoid is used to control the flow of oil into a piston in each lifter that effectively determines the tappet's height.

Another design uses pressurized oil to allow a four-valve-per-cylinder engine to operate as a three-valve engine at low speeds. At high speeds the engine uses the four valves. Below 2,500 rpm, each intake valve follows a separate camshaft lobe. The primary valve opens and closes normally, while the secondary intake opens just enough to keep the engine running smoothly. As the engine reaches 2,500 rpm, the PCM allows pressurized oil to move small pins that lock each pair of rocker arms together, causing both intake valves to follow the normal cam lobe. When the engine slows down, the pressurized oil is bled off and the pin releases, separating the two rocker arms.

The Volvo 3.2-liter engine uses a mix of staged and continuously variable cam timing. It is also unique in the fact that the camshaft is driven at the flywheel end of the engine. The camshaft's phasers are chain-driven. The lift of the intake valves varies with the cam lobe that the rocker arms are riding on. This change is done with a locking pin controlled by oil pressure.

CYLINDER HEAD DISASSEMBLY

On some OHC engines, the rocker arms must be removed before the head is disassembled. If the camshaft rides directly above the rocker arms, use the appropriate spring compressor and depress the valve enough to pull the rocker arm out **(Figure 12–33)**. Some have the rocker arms mounted on a separate shaft. The ends of the rocker arms do not directly contact the valves; rather a bridge rocker arm is used. The bridge rocker arm assembly is also mounted on a shaft. To remove these rocker arms, both shafts are unbolted **(Figure 12–34)**.

On all OHC engines, the camshaft must be removed before the cylinder head can be disassembled. Follow the specified order for loosening the camshaft bearing caps. Keep the caps in the order they were on the head. Also, draw a diagram of the arrangement of the cam follower assemblies and mark each part. This will ensure that each part is returned to the same position.

Figure 12–33 On some OHC heads, the valve springs must be slightly compressed to remove the rocker arms.

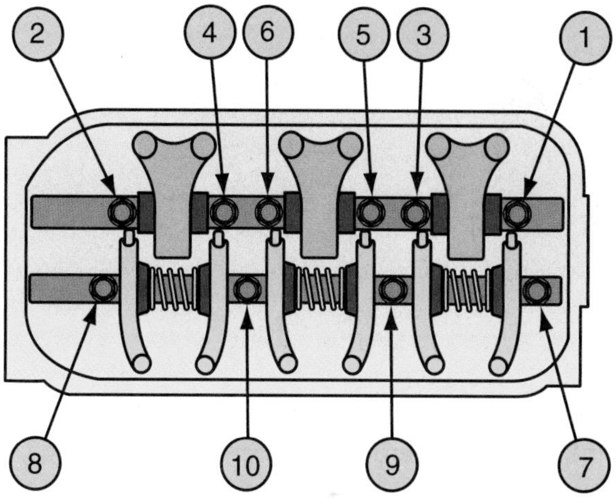

Figure 12–34 Rocker arm assemblies should always be removed by following the prescribed order for loosening the mounting bolts.

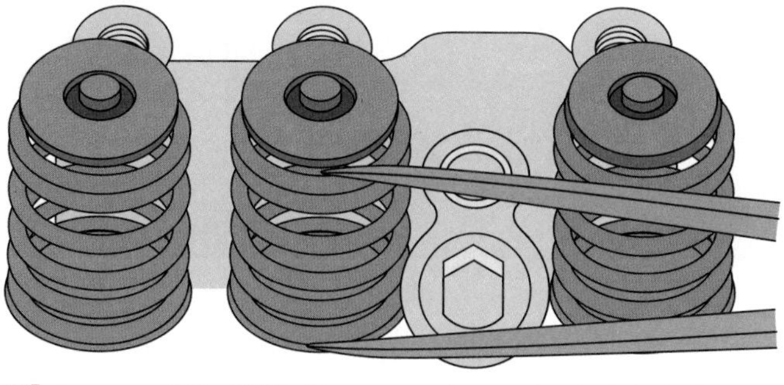

Figure 12–35 Measure valve spring height before disassembling the cylinder head. *Courtesy of Ford Motor Company*

Measure the installed spring height for each valve and record it **(Figure 12–35)**. This measurement will be needed during reassembly. To remove the valves from the cylinder head, use a valve spring compressor. First, select a socket that fits over the valve tip and onto the retainer. Tap the socket with a plastic mallet to loosen the valve keepers. Adjust the jaws of the compressor so they fit securely over the spring retainers. Compress the valve springs just enough to remove the keepers **(Figure 12–36)**.

Next, remove the valve oil seals and the valves. Keep all assemblies together according to the cylinder they were in. If a valve does not pass through its guide, the tip might be mushroomed or peened over and have a ridge around it. Do not drive the valve through the guide. It could score or crack the valve guide or head. Raise the stem and file off the ridge until the stem slides through the guide easily **(Figure 12–37)**.

Figure 12–36 With a spring compressor, compress the springs just enough to remove the keepers.

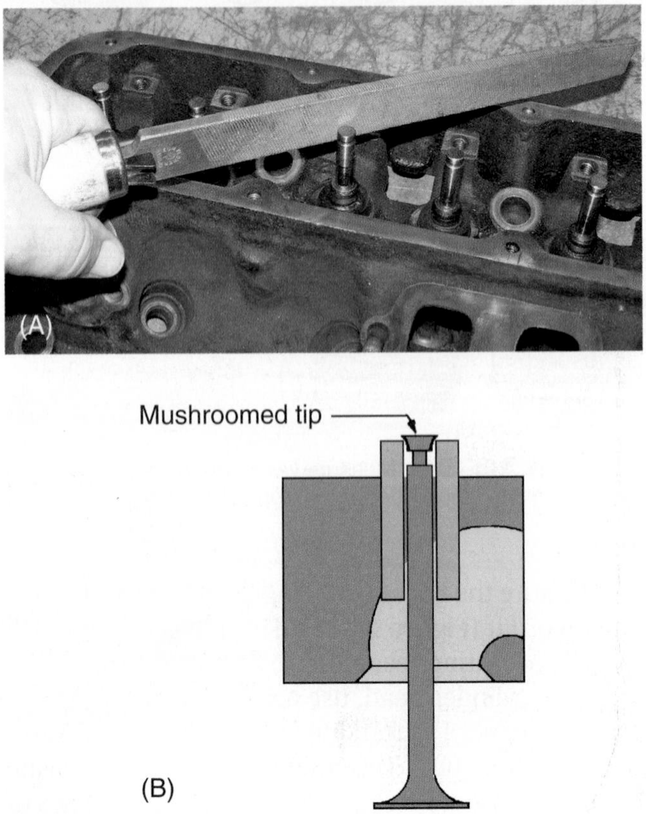

Figure 12–37 A mushroomed valve tip should be (A) filed before removing the valve from (B) its guide.

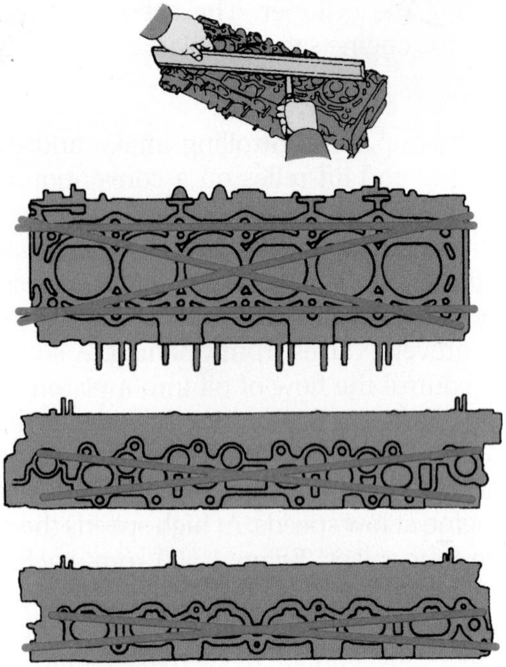

Figure 12–38 Checking a cylinder head for warpage.

Inspection of Head

Cylinder heads should be carefully inspected after they are cleaned. Any severe damage to the sealing and valve areas indicates that the head should be repaired or replaced. Also use the appropriate method for detecting any and all cracks. Check the heads for dents, scratches, and corrosion around water passages.

As engines undergo heating and cooling cycles over their life span, certain parts tend to warp. This is especially true of cylinder heads. By using a precision straightedge and feeler gauge, the amount of warpage can be measured. The surface should be checked both across the head as well as lengthwise **(Figure 12–38)**. In general, maximum allowable deformation is 0.004 inch (0.1016 mm). Check the manufacturer's recommendations for the maximum allowable warpage for the engine you are working on. Many manufacturers recommend head replacement if it is warped beyond allowable limits. Also check the flatness of the intake and exhaust manifold mounting surfaces.

Aluminum Cylinder Heads

WARNING!

Aluminum cylinder head bolts should never be loosened or tightened when the metal is hot. Doing so may cause the cylinder head to warp due to torque changes.

Figure 12–39 Lightweight aluminum heads are prone to cracking. This can lead to a recessed exhaust valve seat.

Aluminum heads experience fair amounts of thermal expansion, which can lead to cracking. The most crack-prone areas are usually the areas around the valve seats **(Figure 12–39)**. High-combustion temperatures and the constant pounding of the valve against its seat often cause cracking between the intake and exhaust seats or just under the exhaust seat.

Aluminum has another drawback—**porosity**. The casting process sometimes leaves microscopic pores in the metal, which can weep oil or coolant. In most instances, the problem is only cosmetic and does no real harm. But the customer may not agree. To the customer, a wet spot on the outside of a cylinder head looks like a leak.

Aluminum heads should be carefully checked for dents, scratches, and corrosion around water passages. Also, they should be checked for warpage. Warpage in an aluminum cylinder head is usually the result of overheating (low coolant, uneven coolant circulation within the head, a too lean fuel mixture, and incorrect ignition timing). For aluminum heads, the maximum allowable warpage is less than it is for cast-iron heads.

Always check the cylinder head thickness and specifications to be sure that material can be safely removed from the surface. Some manufacturers do not recommend any machining; rather they require head replacement if cylinder head flatness is not within specifications. Aluminum heads can be straightened through the use of heat and special clamping fixtures. However, some manufacturers do not recommend this.

Alignment of the cam bores in an OHC head should be checked with a straightedge and feeler gauge. If the bores are more than 0.002 to 0.003 inch (0.0508 to 0.0762 mm) out of alignment, corrective action is required.

Valves

Each valve face should be carefully checked for evidence of burning **(Figure 12–40)**. Also check the entire valve for signs of wear or distortion **(Figure 12–41)**. Replace any valves that are badly burned, worn, or bent. Also discard any valve that is badly burned, cracked, pitted, or shows signs of excessive wear. Carefully examine the facings on the stem. If the plating is flaking or chipped, the valve should be replaced.

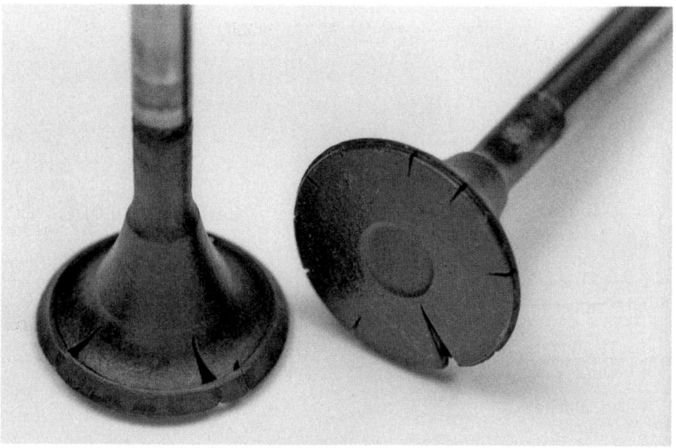

Figure 12–40 Severely burnt valves.

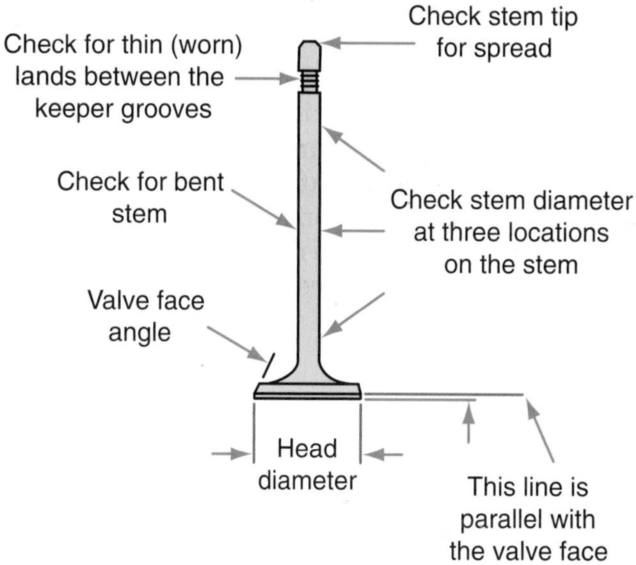

Figure 12–41 Parts of a valve that should be checked during your inspection.

Examine the backside of the intake valves. A black oily buildup in the neck and stem area indicates that oil is entering the cylinder through the intake valve **(Figure 12–42)** or valve guide seats. Measure the margin left on the valve. Valves that cannot be refaced without leaving at least a 1⁄32-inch (0.79 mm) valve margin should not be reused. Reusable valves are cleaned by soaking them in solvent, which will soften the carbon deposits. The deposits are then removed with a wire buffing wheel. Once the deposits are removed, the valve can be resurfaced.

When replacing a valve, make sure the new one is an exact replacement for the original. This includes the stem diameter, stem height, head diameter, and

Figure 12-42 An oily soot or heavy carbon buildup on the back of the valves indicates bad valve seals.

the material used to make the valve. The first three items can be measured and compared. However, it is not easy to identify the metals used for the valve. Replacement valves should be made from the same or a better alloy than the original valve. A good starting point to identifying the metal is to see if it is magnetic or not. Stainless steel is nonmagnetic, while carbon steel is magnetic.

Valve Seats

Valve seats should be checked for damage, cracks, burning, and deterioration. Also, check insert seats for looseness in their bores. This is done by prying on the inside of the seat. With moderate pressure there should be no movement. If any damage is found on the seat, a new insert seat can be installed. If an integral seat is damaged, the area around it will need to be cut out and an insert seat installed. If the seat appears to have sunk too deeply in its bore, it needs to be replaced.

A seat should also be replaced if its valve was broken or bent. This may have been caused by the seat not being concentric with the guide. This causes the valve stem to flex every time the valve closes.

Retainers and Keepers

Valve spring retainers and valve keepers hold the valve spring and valve in place. The retainer holds the spring in line with the valve stem. A worn retainer will allow the spring to move away from the centerline of the valve. This will affect valve operation because spring tension on the valve will not be evenly distributed. Each retainer should be carefully inspected for cracks, since a cracked retainer may result in serious damage to the engine. The inside shape of most retainers is a cone that matches the outside shape of the keepers.

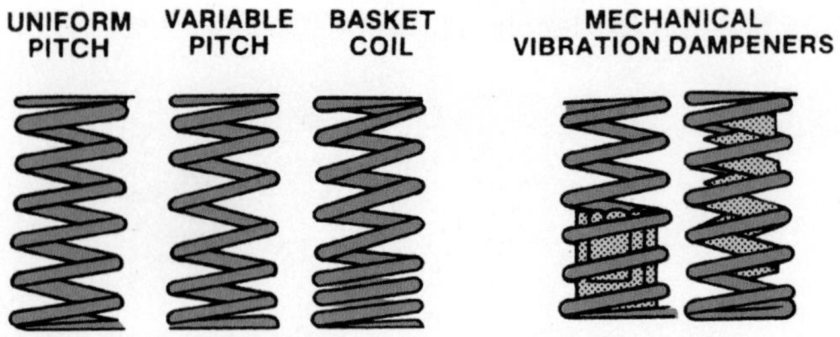

Figure 12-43 Common valve spring designs.

This must be a good fit in order to secure the keepers in their grooves on the valve stem. Both the retainers and keepers should be inspected for wear and damage. They should be replaced if a defect is found.

The valve stem grooves should match the inside shape of the keepers. Some valves have multiple keeper grooves. Others have only one. All of the valve stem grooves should be inspected for damage and fit by inserting a keeper in them.

Valve Rotators

When valves are refaced or replaced, the rotators should be replaced because they cannot be accurately inspected. Whether or not they rotate by hand is no indication of how they actually function. Uneven wear patterns on the valve stem tip are an indication that the rotators are not working properly.

Valve Springs

The valve spring assemblies **(Figure 12–43)**, including the damper springs, should be checked for signs of cracks, breaks, and damage. The high stresses and temperatures imposed on valve springs during operation cause them to weaken and sometimes break. Rust pits will also cause valve spring breakage. To determine if the spring can be reused, the following tests should be performed.

Freestanding Height Test Line up all the springs on a flat surface and place a straightedge across the tops. Measure from the table top to the straightedge. Throw away any spring that is not within specifications.

Spring Squareness Test A spring that is not square will cause side pressure on the valve stem and abnormal wear. To check squareness, set a spring upright against a square **(Figure 12–44)**. Turn the spring until a gap appears between the spring and the square. Measure the gap with a feeler gauge. If the gap is more than 0.060 inch (1.524 mm), the spring should be replaced.

Open/Close Spring Pressure Test The procedure for checking valve spring open and close spring pressure

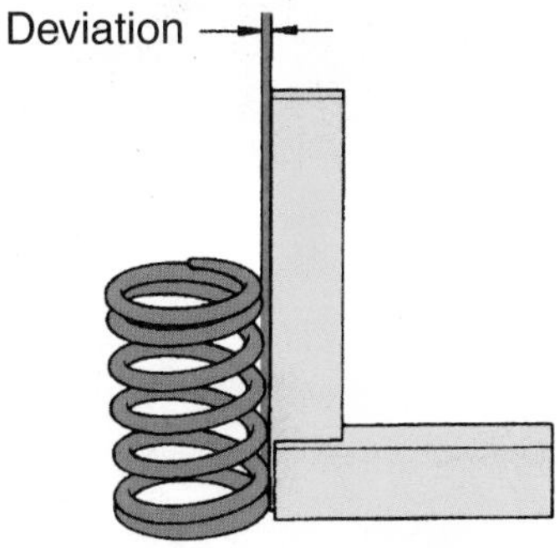

Figure 12-44 Spring squareness test.

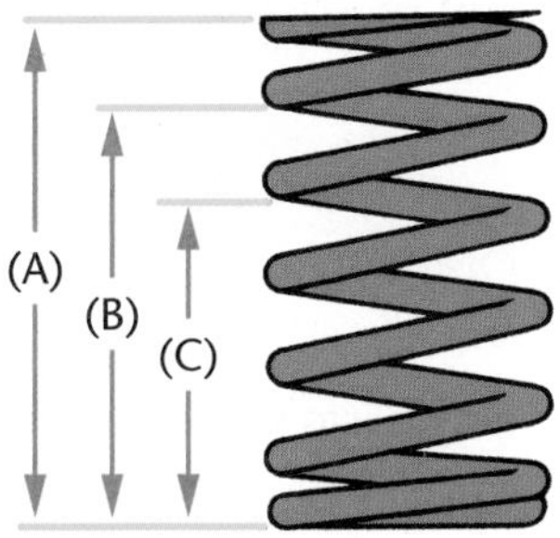

Figure 12-45 Valve spring height terminology: (A) free height, (B) valve closed spring height, and (C) valve open spring height.

is given in Photo Sequence 9. Close pressure guarantees a tight seal. The open pressure overcomes valve train inertia and closes the valve when it should close. Spring pressure specifications are listed according to spring height **(Figure 12–45)**. Any spring that does not meet specifications should be replaced.

Performance TIP *When making any change in hopes of increasing horsepower, especially a cam change, do not forget to install stiffer valve springs. Higher-tension springs will help keep the lifters in contact with the cam lobes and overcome the increased momentum of the valves and the valve train during high engine speeds. Match the springs' tension with the engine. Excessive spring tension is not good. It can put too much stress on the cam lobes, lifters, rocker arms, and so on, and cause them to wear prematurely. Always follow the recommendations of the camshaft manufacturer when replacing valve springs.*

INSPECTION OF THE VALVE TRAIN

When inspecting the valve train, each part should be carefully checked. Use the following guidelines when inspecting the components.

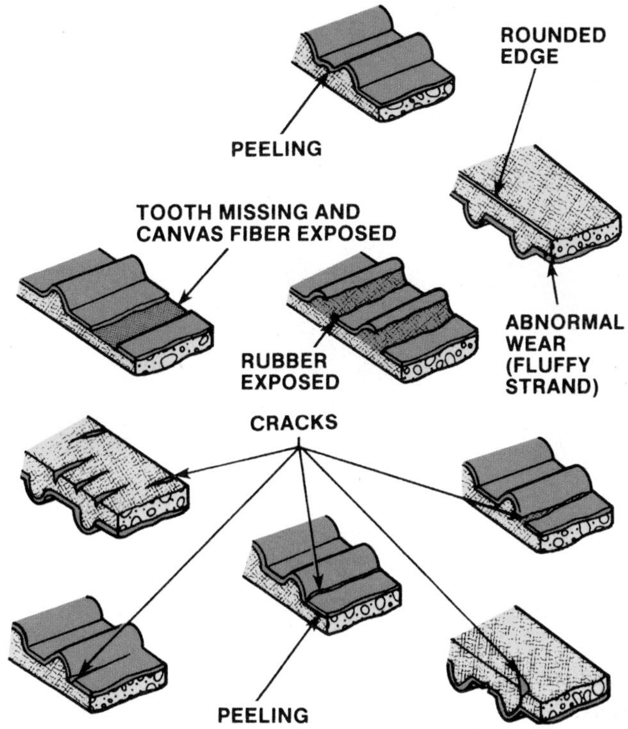

Figure 12-46 Various forms of timing belt wear.

Timing Belts

Most often, the belt is replaced when the engine or head is rebuilt. Stripped/broken belt failure is commonly due to insufficient tensioning, extended service life, abusive operation, or worn tensioners. Most manufacturers recommend timing belt replacement every 60,000 miles. Loose timing belts will jump across the teeth of the timing sprockets, causing shearing of the belt teeth. Check for cord separation and cracks on all surfaces **(Figure 12–46)** of the belt. Also check for evidence of exposure to oil or water. Both can cause deterioration of the belt. If the belts are contaminated or damaged, they should be replaced.

On many engines, severe engine damage will result from a broken timing belt. When a timing belt breaks, the camshaft no longer turns, but the crankshaft continues to rotate through inertia. When the camshaft stops, the valves stay where they were when the belt broke. This means some of the valves are open. As the pistons continue to move, they can strike the open valves. This results in bent or broken valves and/or damaged piston domes.

SHOP TALK

The condition of a rubber timing belt can frequently be checked by the fingernail test. Press your fingernail into the hardened backside of the belt. If no impression is left, the belt is too hard; this is caused by overheating and aging.

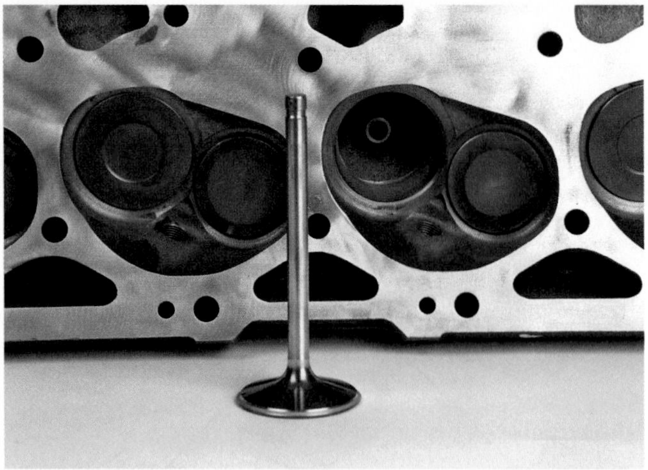

P9-1 *Prior to installing the valve and fitting valve springs, all other head work should be completed.*

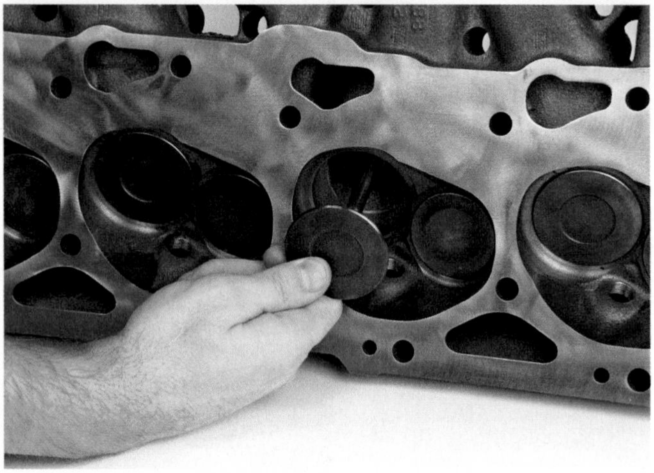

P9-2 *Install the valve into its proper valve guide.*

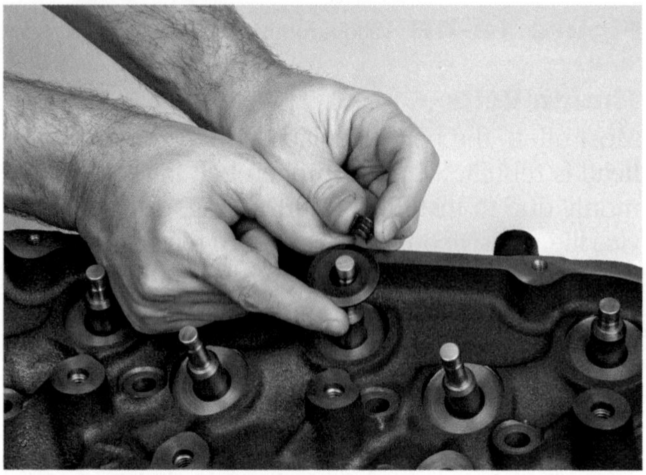

P9-3 *Install the valve retainer and keepers. Without the spring, these must be held in place.*

P9-4 *While pulling up on the retainer, measure the distance between the bottom of the retainer and the spring pad on the cylinder head with a divider.*

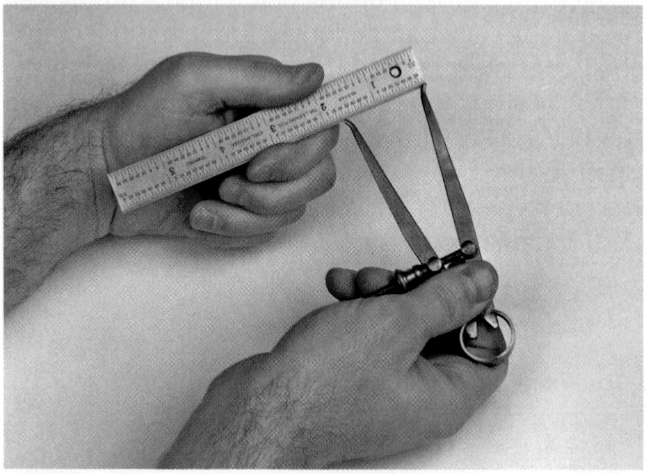

P9-5 *Use a scale to determine the measurement expressed by the divider.*

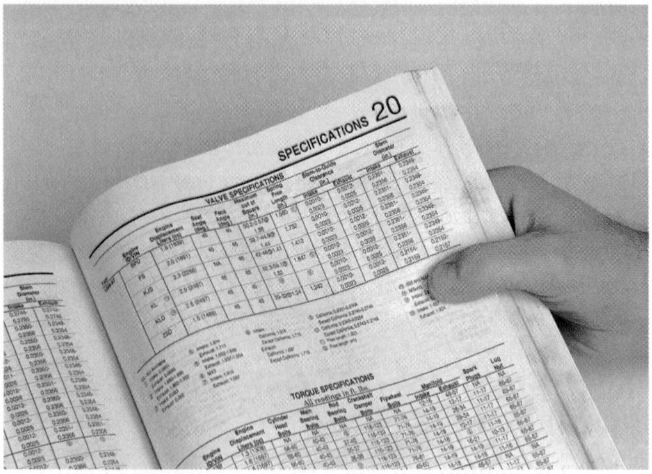

P9-6 *Compare this measurement with the specifications given in the service manual for installed spring height.*

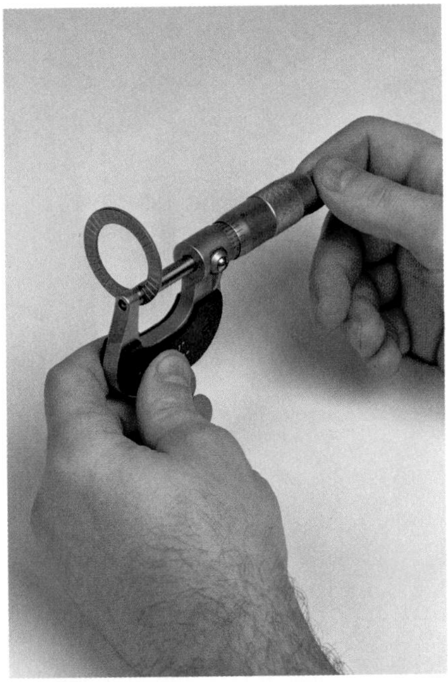

P9-7 *If measured installed height is greater than the specifications, a valve shim must be placed under the spring to correct the difference.*

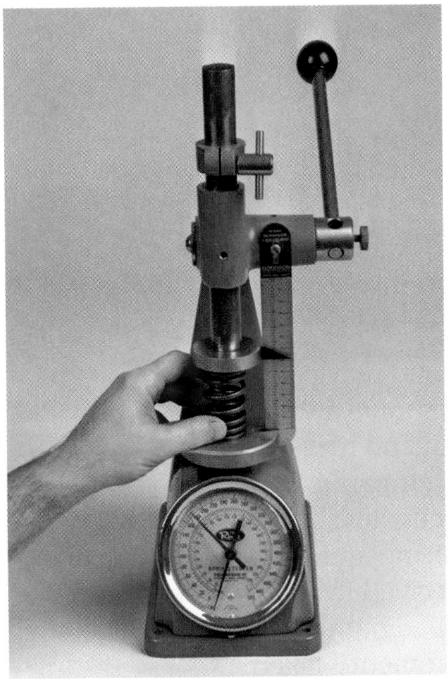

P9-8 *Spring tension must be checked at the installed spring height; therefore, if a shim is to be used, insert it under the spring on the valve spring tension gauge.*

P9-9 *Compress the spring into the installed height by pressing down on the tester's lever.*

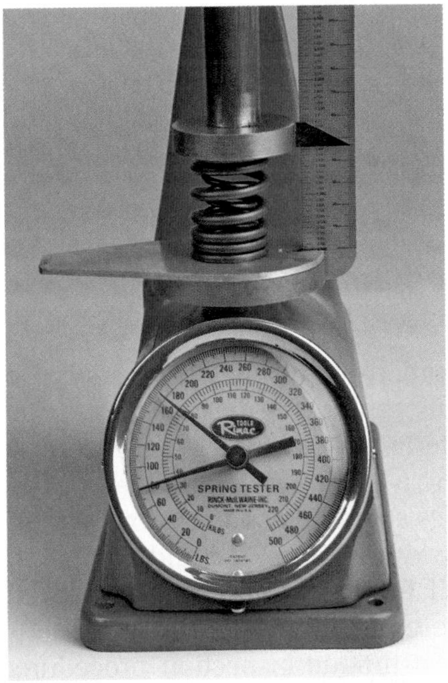

P9-10 *The tension gauge will reflect the pressure of the spring when compressed to the installed or valve closed height. Compare this reading to the specifications.*

P9-11 *Now compress the spring to the open height specification. Use the rule on the gauge or a scale to measure the compressed height.*

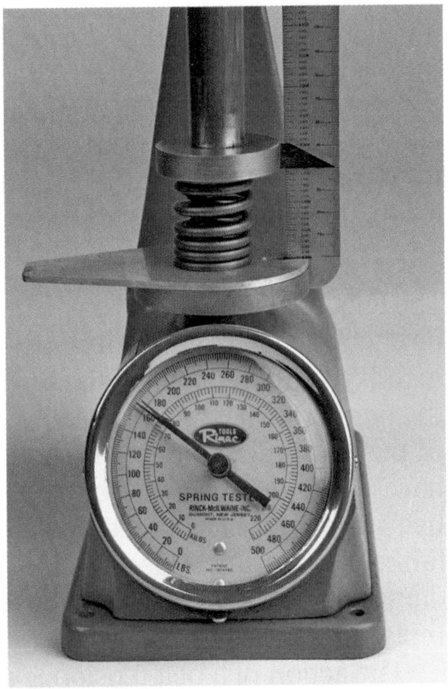

P9-12 *Compare this reading to specifications. Any pressure outside the pressure range given in the specifications indicates that the spring should be replaced. After the tension and height have been checked, the spring can be installed on the valve stem.*

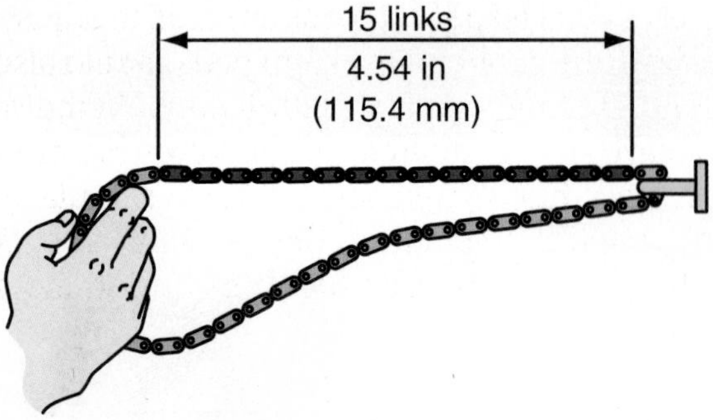

Figure 12-47 The timing chain on some engines should be measured in sections while it is being stretched.

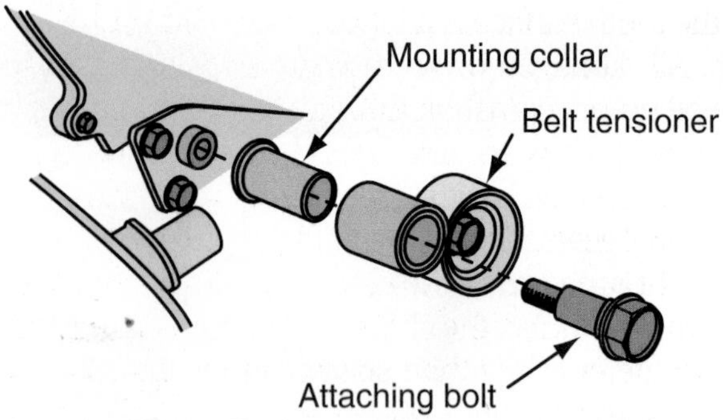

Figure 12-48 A timing belt tensioner assembly.

Timing Chains

Each drive chain should be inspected and replaced if it is damaged. The length of the chain should also be checked. Some manufacturers recommend measuring the entire length of the chain and comparing that to specifications. The chains on other engines should be measured in sections while they are being stretched. To do this, pull the chain with the specified tension. Then measure the length of the specified number of chain links **(Figure 12–47)**. This measurement is taken at three random sections of the chain. The average length is then compared to specifications. The chain should be replaced if it is not within specifications.

Belt Idler Pulley

All idler pulleys should be rotated by hand. They are okay if they move smoothly. The pulley should also be checked for signs of lubricant leakage. Check around the seal. If leakage is evident, the idler should be replaced.

Tensioners

The tensioners of belt and chain drive systems should be checked. There are many types of tensioners used in today's engines; refer to the service information for the correct inspection procedure. Check the surface of the tensioner's pulley. It should be smooth and have no buildup of grease or oil **(Figure 12–48)**. Most belt tensioners should also be checked for signs of lubricant leakage. Check around the seal. If any damage or leakage is evident, the tensioner should be replaced.

The action of a belt tensioner should be checked. Make sure the spring is free to move the tensioner pulley. If the tensioner spring is defective, replace the tensioner. On plunger-type tensioners, hold the tensioner with both hands and push the pushrod strongly against a flat surface. The pushrod should not move. If it does, replace the tensioner. Measure the distance the pushrod extends from the housing. Compare that distance to specifications. If this measurement is not within specifications, replace the tensioner.

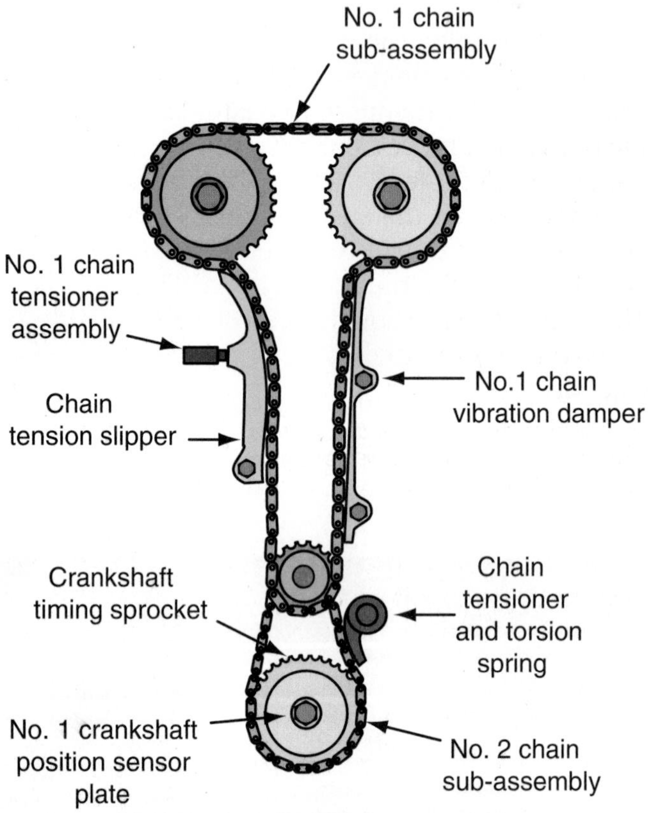

Figure 12-49 Timing chain components.

Chain drive systems have a variety of dampers and guides in addition to a tensioner **(Figure 12–49)**. The dampers and guides should be checked for wear. In most cases, their width is measured and compared to specifications. If they are worn, they should be replaced. Again, there are different types of chain tensioners, each with a unique inspection procedure. The plunger in ratchet-type tensioners should be able to be smoothly moved out by hand but should not be able to be pushed in by hand.

Gears and Sprockets

All timing gears and sprockets should be carefully inspected. A gear with cracks, spalling, or excessive wear on the tooth surface is an indication of improper backlash. All damaged or worn gears should be replaced. The oil pump, camshaft timing, crankshaft timing, and balance shaft gears and sprockets on some engines are measured with the drive chain wrapped around the individual gears. The diameter of the gear with the chain around it is measured with a vernier caliper. The caliper's jaws must connect the chain's rollers while doing this. If the diameter is less than specifications, the chain and gear or sprocket should be replaced.

Cam Phasers

Camshaft phasers are used in most VVT systems. The action of a hydraulic phaser can be checked while it is attached to the camshaft. Clamp the camshaft in a soft-jawed vise. Attempt to rotate the timing gear on the phaser. If it moves, the phaser must be replaced. Next, cover all of the oil ports on the phaser with electrical tape except the advance port **(Figure 12–50)**. Using an air nozzle with a rubber tip, apply the specified air pressure to the exposed port **(Figure 12–51)**. The timing gear should move counterclockwise. When the air is released, the timing gear should move clockwise. This check should be conducted several times and the timing gear should move smoothly. This process is then repeated at the retard port. In this case all ports, except the retard port, are sealed.

Cam Followers and Lash Adjusters

Overhead cam follower arm and lash adjuster assemblies should be carefully checked for broken or severely damaged parts. If pads are used to adjust the valve lash, the cups and the shim pads should also be carefully checked. A soft shim will not hold the valve at its correct lash; therefore, the hardness of each shim should be checked. You can do this by placing a shim on the base circle of the camshaft and, with your hand, press down on the shim. You should feel no give.

Rocker Arms

Inspect the rocker arms for wear, especially at points that contact the valve stem and pushrod. Make sure the oil feed bore in each rocker arm is clear and not plugged with dirt. The fit between a rocker arm and its shaft is checked by measuring the outside diameter of the shaft and comparing it to the inside diameter of the rocker arm **(Figure 12–52)**. Excessive clearance requires replacement of the rocker arm or the rocker shaft, or both. Another wear point that should be checked is the pivot area of the rocker arm. Also check for loose mounting studs and nuts or bolts. Replacement press-in studs are available in standard sizes and oversizes. The standard size is used to replace damaged or worn studs and the oversizes are used for loose studs.

Excessive wear on the valve pad occurs when the rocker arm repeatedly strikes the valve tip in a hammer-like fashion. This is caused by excessive valve lash due to improperly adjusted valves or bad hydraulic lifters. Worn rocker arm valve pads can also be caused by poor lubrication. Although a cast rocker arm can be resurfaced, a stamped nonadjustable rocker arm that is worn must be replaced.

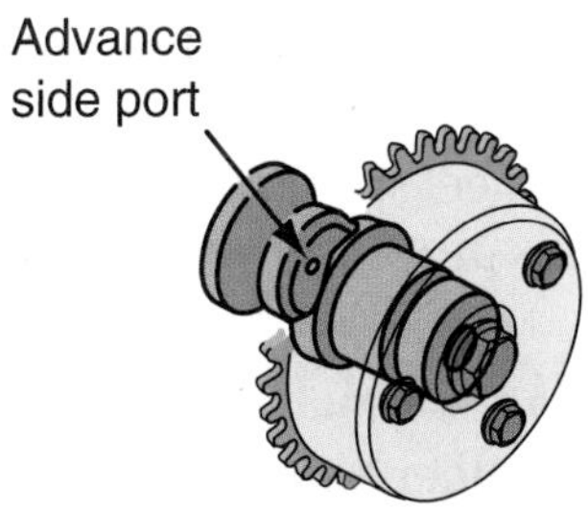

Figure 12-50 Location of the advance port on a camshaft phaser.

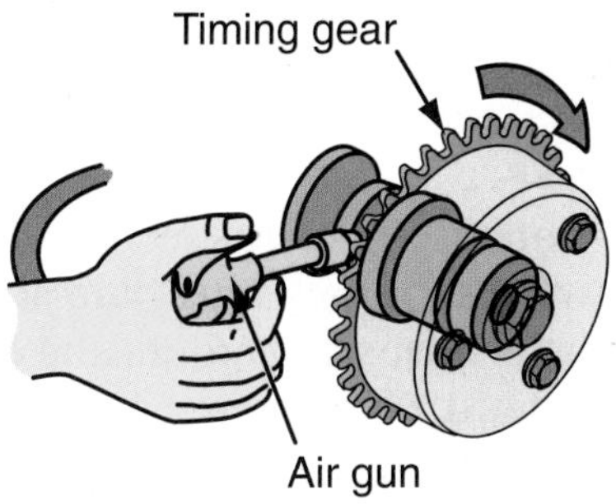

Figure 12-51 The action of the phaser is checked by applying air to the port and observing the timing gear.

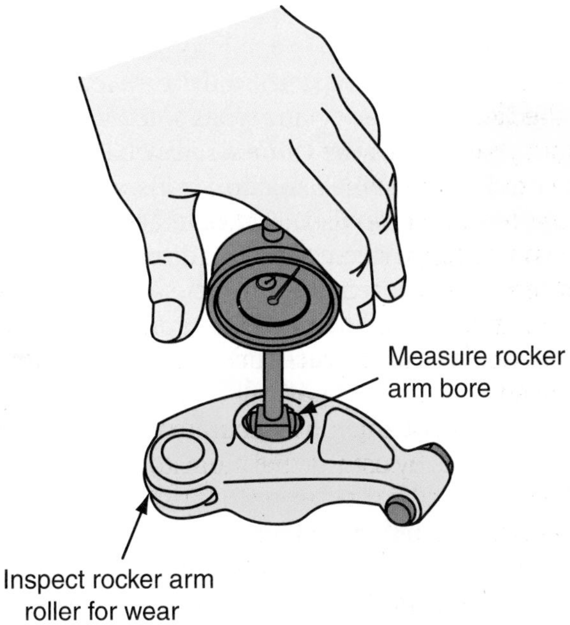

Figure 12-52 Measure the inside diameter of the rocker arm and check for an out-of-round condition.

Honda's Variable Cylinder Management Rocker Arms Honda uses oil pressure and two sets of rocker arms to alter cam timing and to deactivate cylinders. This system has unique inspection and service procedures. When inspecting them, keep all parts in order so they can be installed in their original location. Measure the outside diameter of each rocker shaft at the location of the first rocker arm. Then measure the inside diameter of the rocker arm. The difference between the two is the clearance. This clearance should be compared to specifications. Repeat this procedure for each rocker arm and shaft. If the clearance is beyond specifications, replace the shaft and all over-tolerance rocker arms. If one rocker arm needs replacement, all of the rocker arms on that shaft should be replaced.

Next, inspect the synchronizing pistons in the rocker arms. Slide them into the rocker arms; they should move smoothly. If they do not or are damaged, the rocker arm assembly should be replaced. The rocker arm oil control solenoid has a filter. This should be checked and replaced if it is clogged.

Pushrods

During inspection, some pushrods may be found to have a groove worn in the area in which they pass through the cylinder head, and some may have tip wear. Also, the ends of the pushrods should be checked for nicks, grooves, roughness, or signs of excessive wear. All damaged pushrods should be replaced. Hollow pushrods should be thoroughly cleaned so there are no blockages in the bore.

Check the straightness of each of the pushrods. Bent pushrods can be caused by incorrect valve timing, valve sticking, or improper valve adjustment. Bent or broken pushrods can also be caused by the use of incorrect valve springs or if the valve's installed height is less than specified. Also, insufficient valve-to-piston clearance can cause a collision between the valve and piston at high engine speeds.

Pushrods can be visually checked for straightness while they are installed in the engine by rotating them with the valve closed. With the pushrods out of the engine, they can be checked for straightness by rolling them over a flat surface such as a surface plate. If a pushrod is not straight, it will appear to hop as it is rolled. However, the most accurate way to check for straightness is by using a dial indicator.

Camshaft and Bearings

The camshafts in most OHC engines are secured to the cylinder head by bearing caps. Some ride on split bearings, whereas others ride on a machined surface in the cylinder head. The bearings or bearing surface should be carefully inspected for signs of unusual wear that may indicate an oiling or bore alignment problem. If the engine is equipped with camshaft bearings, they are normally replaced during engine rebuilding. If the camshaft bore is damaged, the cylinder head is normally replaced.

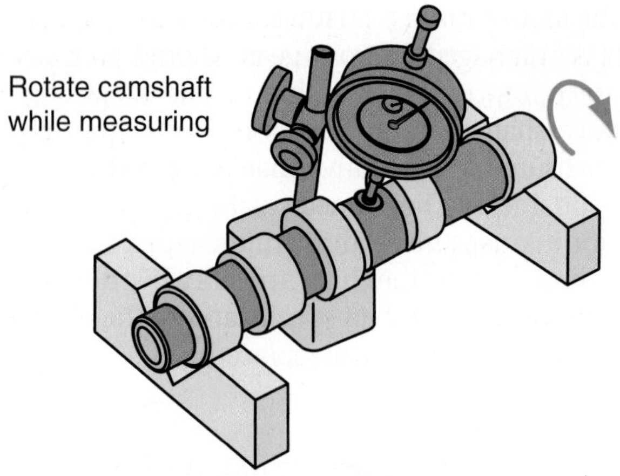

Figure 12-53 Camshafts should be checked for straightness.

Each lobe of the camshaft should be checked for wear, scoring, scuffing, fractured surface, pitting, and signs of abnormal wear. Also check for plugged oil passages.

The camshaft should also be checked for straightness. Place the camshaft on V-blocks. Place a dial indicator on one the middle journals. Rotate the camshaft and watch the dial indicator (**Figure 12–53**). Compare the highest reading to specifications. If the measurement exceeds specifications, replace the camshaft. With a micrometer, measure the height of each cam lobe and the diameter of each journal. If the readings do not meet specifications, replace the camshaft. Repeat this procedure on each camshaft.

SERVICING CYLINDER HEADS

Service to the cylinder head can involve many different procedures. These procedures vary with the metal used to make the head and the design of the engine. Always refer to the appropriate service information from the manufacturer before starting any work on the head.

Crack Repair

Common locations of cracks in a cylinder head include: between the spark plug bore and the valve seat, between the valve seats, around the valve guides, and in the exhaust ports. In most cases, a cracked head should be replaced. However, some cracks can be effectively repaired. Crack repair is normally done by specialty shops.

It is important to keep in mind that most cracks are caused by something other than a defect in the head. The cause of the cracking needs to be identified and corrected. No matter what caused the crack, the crack needs to be repaired if the head is reused. Crack repair is done by the cold process of pinning or the hot process of welding.

Furnace Welding Crack Repairs Furnace welding is considered the best way to repair cracks in a cast-iron head. To furnace weld, the head is first preheated in an oven. This minimizes thermal shock when the flame of a welding torch contacts the head. After the crack has been filled with metal, the head is allowed to slowly cool before it is used.

Flame Spray Welding Flame spray (powder) welding is also used to repair cast-iron heads. This process uses nickel-based powders and a special torch to fill the crack.

Repairing Aluminum Heads Defects in aluminum heads are commonly found as:

- Cracks between the valve seat rings
- Cracks in coolant passages
- Cracks across the main oil artery
- Detonation damage inside the combustion chamber
- Melted or deformed metal in the chamber
- Coolant-related metal erosion

In many cases, these problems result in head replacement. However, some heads are repairable. **Tungsten inert gas (TIG)** welding is the preferred way to repair aluminum heads **(Figure 12–54)**. Welding aluminum is often considered difficult because it welds differently than iron or steel. When exposed to air, aluminum forms an oxide coating on the surface that helps protect the metal against corrosion. The oxide layer makes welding difficult because it interferes with fusing and weakens the weld. A TIG welder prevents the formation of the oxide by bathing the weld with inert gas (normally argon).

Figure 12–54 TIG welding an aluminum head.

Pinning Cracks Pinning is commonly used to repair small accessible cracks in cast-iron and aluminum heads. Pins are used only when the metal is thick enough to secure the pins. Pinning is done with a drill, tap, and tapered or straight pins. Holes are drilled into the ends of the crack to keep it from spreading, then holes are drilled at various overlapping intervals along the length of the crack. After they are installed, the pins are peened to seal the surface. A sealant is not required when tapered pins are used; however, it must be applied over the repair area when straight pins are used.

Resurfacing Cylinder Heads

There are three reasons for resurfacing the deck surface of a cylinder head:

1. To make the surface flat so that the gasket seals properly
2. To raise the compression ratio
3. To square the deck to the main bores

Heads that are deformed beyond specifications must be replaced or resurfaced.

The refinished surface should not be too smooth. It must be rough enough to provide "bite" but not enough to cause a poor seal and leakage. For proper head gasket seating, the finish should have shallow scratches and small projections that allow for gasket support and sealing of voids.

WARNING!

Before operating any surfacing machine, be familiar with and follow all the cautions and warnings given in the machine's operation manual. Also, when operating these machines, wear safety glasses, goggles, or a face shield.

Belt Surfacers Belt surfacers resemble belt sanders. These machines are easy to set up and operate. An operator merely places the part to be surfaced on the belt. A restraint rail helps keep the part positioned **(Figure 12–55)**. Some machines have air-operated holddown fixtures.

Figure 12-55 (A) A typical belt surfacer and (B) how it is used.

Figure 12-56 Milling an aluminum cylinder head.

Resurfacing quality depends on operator skill and factors such as belt condition, machine horsepower, and the holddown pressure applied.

Milling Machines **Milling** machines cut away thin layers of metal to create a level, properly finished surface (**Figure 12–56**). Cutters remove up to 0.050 inch (1.27 mm) per pass. Both rough and finish cuts are usually made to create the desired finish.

Broaching Machines **Broaching** machines use an underside rotary cutter or broach. A block, cylinder head, or intake manifold is held in an inverted position as the broach passes underneath.

Surface Grinders Surface grinders use a grinding wheel to remove metal stock (**Figure 12–57**). They set up and operate similarly to milling machines.

Figure 12-57 A surface grinder.

Stock Removal Guidelines

The amount of stock removed from the head gasket surface must be limited. Excessive surfacing can lead to problems in the following areas.

Compression Ratio After resurfacing a cylinder head, the combustion chamber will be smaller. This will raise the compression ratio. How much depends on how much metal was removed from the surface and the type of chamber. A good rule of thumb is that when 0.010 inch (0.2540 mm) has been removed from the head, the compression ratio will increase by about 0.141:1 to 0.20:1.

To determine if the compression ratio increased too much, it may be necessary to measure the volume of the combustion chamber. Measuring the volume is called **cc-ing** the head. This is done with the valves and spark plugs installed. The cylinder head is mounted upside down, and a glass or plastic plate is installed over the combustion chamber. A graduated container called a burette is used to fill the combustion chamber with thin oil. The oil is poured through a hole in the plate, as shown in **Figure 12–58**. The

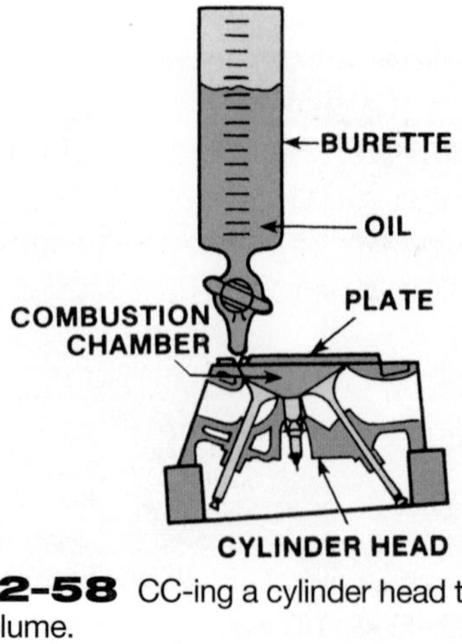

Figure 12-58 CC-ing a cylinder head to find the combustion volume.

amount of oil required to fill the chamber is equal to the volume (in cubic centimeters) of the chamber. The measured volume is compared to specifications.

If the chamber volumes are unequal, individual chambers can be matched by grinding the valve seats to sink the valves and by grinding and polishing metal from the combustion chamber surface. Either method can be used to equalize the volumes and adjust them to specifications.

On some engines, a thicker head gasket can be used to decrease the compression after the head has been resurfaced. There can be as much as 0.040-inch difference between various types and brands of gaskets. For instance, changing from a soft-faced to a steel or copper shim gasket can increase the compression ratio by as much as 0.50:1.

Valve Timing On many OHC engines, it is necessary to restore the distance between the camshaft and crankshaft gears after the head has been resurfaced. Special shims are used to raise the cylinder head, which raises the camshaft. If 0.030 inch (0.7620 mm) was removed from the head surface, the camshaft must be moved up 0.030 inch (0.7620 mm). If this is not done, valve timing will be altered.

Piston/Valve Interference When the block or head is surfaced, the piston-to-valve clearance becomes less. To prevent the valves from making contact with the piston, a minimum of 0.070 inch piston-to-valve clearance is recommended.

Misalignment Removing metal from the head or block also causes valve tips, rocker arms, and pushrods to be positioned closer to the camshaft. This causes a change in rocker arm geometry and can cause hydraulic lifters to bottom out.

Also, on V-type engines, the mountings for the manifolds will be lower. This can present a sealing problem with the intake manifold. Also, the ports might be mismatched and manifold bolts might not line up. In order to return the intake manifold to its original alignment, metal on the sealing surfaces of the manifold must be removed.

SHOP TALK

If one cylinder head on a V-type engine is resurfaced, the other one must be cut so equal amounts of metal are removed from each head.

RECONDITIONING VALVES

Whenever the valves have been removed from the cylinder head, the valve heads and valve seats should be resurfaced. The most critical sealing surface in the valve train is between the face of the valve and its seat when the valve is closed. Leakage between these surfaces reduces the engine's compression and power and can lead to valve burning. To ensure proper seating of the valve, the seat area on the valve face and seat must be the correct width, at the correct location, and concentric with the guide. These are accomplished by renewing the surface of the valve face and seat.

Valve grinding or refacing is done by machining a fresh, smooth surface on the valve faces and stem tips. Valve faces suffer from burning, pitting, and wear caused by opening and closing millions of times during the life of an engine. Valve stem tips wear because of friction from the rocker arms or actuators. Valve tips are machined after the valve face is refinished.

Valve Face

Valves can be refaced on either grinding or cutting **(Figure 12–59)** machines. Although grinding machines are common, they should not be used on some late-model valves. The manufacturer may recommend cutting the valve face. The process for grinding or cutting a valve face is much the same. However, grinding uses a stone and cutting uses a hardened blade.

USING SERVICE INFORMATION

Specifications for valve angles are listed in the engine specifications section of a service manual.

Figure 12–59 Cutter-type valve resurfacer. *Courtesy of Neway Manufacturing, Inc.*

Before using a stone to grind a valve face, it should be dressed **(Figure 12–60)**. A special diamond cutting tip is passed by the stone to clean it and provide an even surface. Place the valve as deeply as possible into the machine's chuck. This eliminates stem flexing while pressure is exerted on the valve. Set the angle of the stone or cutting bit to the desired angle.

SHOP TALK

Many manufacturers recommend that the angle of the valve seat and face have an interference angle of 1 degree. This provides a good seal through a wedging effect. The 1-degree difference can be made at the valve face or seat, or split between the two.

Take light cuts **(Figure 12–61)** using the full width of the grinding wheel. Make sure coolant is flowing

Figure 12-60 Before grinding a valve, the stone should be dressed.

Figure 12-61 Take light cuts on the valve using the full width of the grinding wheel.

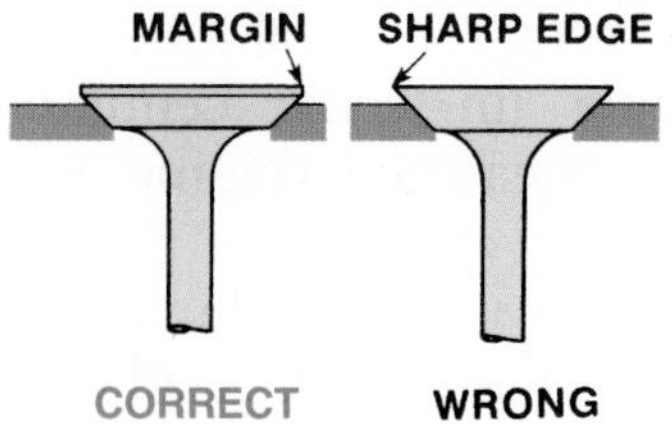

Figure 12-62 A sharp edge on the valve face is not recommended.

over the contact point between the valve face and the grinding wheel. Remove only enough metal to clean up the valve face. A knifelike edge will heat up and burn easily or might cause preignition **(Figure 12–62)**. The width of the edge of a valve head between the top of the valve and the edge of the face is the valve margin. As a general rule, it is not advisable to grind a valve face to a point where the margin is reduced by more than 25% or to where it is less than 0.045 inch (1.143 mm) on the exhaust and 0.030 inch (0.7620 mm) on the intake valves.

After grinding, check valve head runout. Use a dial indicator on the valve margin and rotate the valve while it is still in the chuck. Valve runout should not exceed 0.002 inch (0.0508 mm) TIR. The face should not show any chatter marks or unground areas. After grinding, examine the valve face for cracks. Sometimes fine cracks are visible only after grinding. Sometimes they occur during grinding due to inadequate coolant flow or excessive wheel pressure.

CAUTION!

Always wear eye protection when operating any type of grinding equipment.

Valve Stem

Removing metal from the valve face and/or seat will set the valve deeper into the port. As a result, more of the valve stem extends from the other side of the head. If the stem height is greater than that specified by the manufacturer, the valve stem tip must be ground down to bring the overall height of the stem back into specs. In some cases, if the stem height is excessive, the valve and/or valve seat must be replaced. Valve stem installed height is measured from the valve spring pad on the cylinder head to the top of the valve **(Figure 12–63)**. When the installed height is excessive, valve train geometry can be thrown off or there can be valve lash problems.

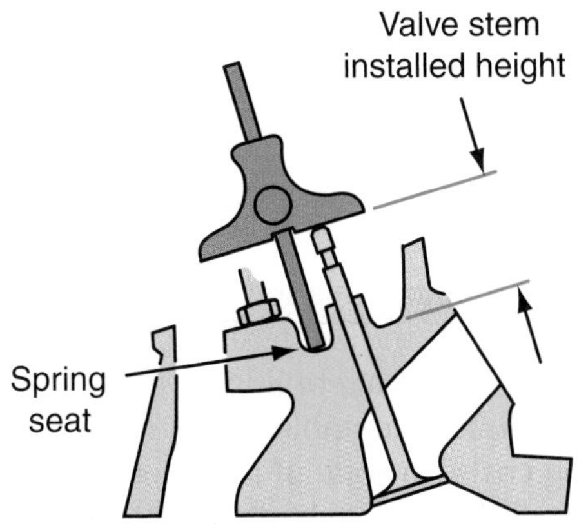

Figure 12-63 Measuring valve stem installed height.

The valve tip is ground so that it is exactly square with the stem. Because valve tips have hardened surfaces up to 0.030 inch (0.7620 mm) in depth, only 0.010 inch (0.2540 mm) can be removed during grinding. If more than 0.010 inch (0.2540 mm) is removed from the tip, the valve must be replaced. Follow the manufacturer's specifications for the allowable limits. After the tip has been ground, the edges should be chamfered. This is done by setting the valve at a specified angle to the grinding or cutting tool. The valve is then rotated until the entire edge is smooth.

VALVE GUIDE RECONDITIONING

Valve guide problems can be lumped into one of three basic categories: inadequate lubrication, valve geometry problems, and wrong valve stem-to-guide clearance.

Inadequate lubrication can be caused by oil starvation in the upper valve train due to low oil pressure, obstructed oil passages, improper operation of pushrods, and using the wrong type of valve seal. Insufficient lubrication results in stem scuffing, rapid stem and guide wear, possible valve sticking, and ultimately valve failure due to poor seating and overheating.

Geometry problems include an incorrectly installed valve height, off-square springs, and rocker arm tappet screws or rocker arms that push the valve sideways every time it opens. This causes uneven guide wear, leaving an egg-shaped hole. The wear leads to increased stem-to-guide clearance, poor valve seating, and premature valve failure.

A certain minimum amount of clearance is needed for lubrication and thermal expansion of the valve stem. Exhaust valves require more clearance than intakes because they run hotter. Clearance should also be close enough to prevent a buildup of varnish and carbon deposits on the stems, which could cause sticking. Insufficient clearance, however, can lead to rapid stem and guide wear, scuffing, and sticking, which prevents the valve from seating fully. This, in turn, causes the valve to run hot and burn.

Guide Wear

The amount of valve guide wear can be measured with a ball (small-bore) gauge and micrometer. Insert and expand the ball gauge at the top of the guide. Lock it to that diameter, remove it from the guide, and measure the ball gauge with an outside micrometer. Repeat this process with the ball gauge in the middle and the bottom of the guide **(Figure 12–64)**. Compare your measurements to the specifications for valve guide inside diameter. Then compare your measurements against each other. Any difference in reading shows a taper or wear inside the guide.

Another way to check for excessive guide wear is with a dial indicator. The accuracy of this check is directly dependent on the amount the valve is open during the check. Some manufacturers specify this amount or provide special spacers that are installed over the valve stem to ensure the proper height. Attach the dial indicator to the cylinder head and position it so the plunger is at a right angle to the valve stem being measured **(Figure 12–65)**. With the plunger in

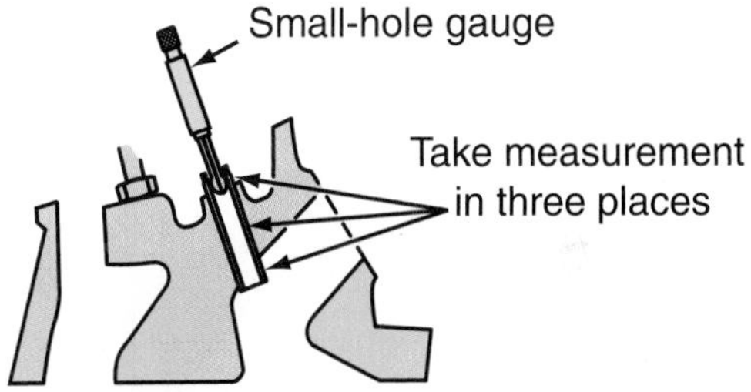

Figure 12-64 Valve guide wear can be measured with a small-hole gauge and a micrometer.

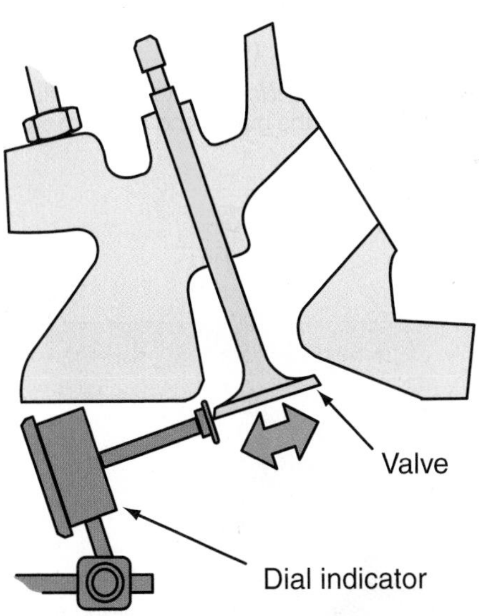

Figure 12-65 Checking for valve guide wear with a dial indicator.

contact with the valve head, move the valve toward the indicator and set the dial indicator to zero. Now move the valve from the indicator. Observe the reading on the dial while doing this. The reading on the indicator is the total movement of the valve and is indicative of the guide's wear. Compare the reading to specifications.

If the clearance is too great, oil can be drawn past both the intake and exhaust guides. Though oil consumption is more of a problem with sloppy or worn intake guides because the guides are constantly exposed to vacuum, oil can also be pulled down the exhaust guides by suction created in the exhaust port. The outflow of hot exhaust creates a venturi effect as it exits the exhaust port, creating enough vacuum to draw oil down a worn guide **(Figure 12–66)**.

Because it retains oil well, the antiseize and antiwear characteristics of bronze allow a bronze guide to last two to five times longer than a cast-iron guide.

Knurling

Knurling is one of the fastest ways to restore the inside diameter (ID) of a worn valve guide. The process raises the surface of the guide ID by plowing tiny furrows through the surface of the metal **(Figure 12–67)**. As the knurling tool cuts into the guide, metal is raised

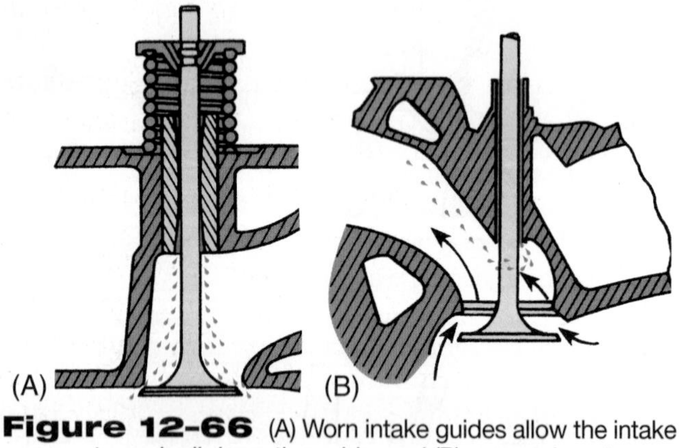

Figure 12-66 (A) Worn intake guides allow the intake vacuum to suck oil down the guide, and (B) worn exhaust guides can do the same.

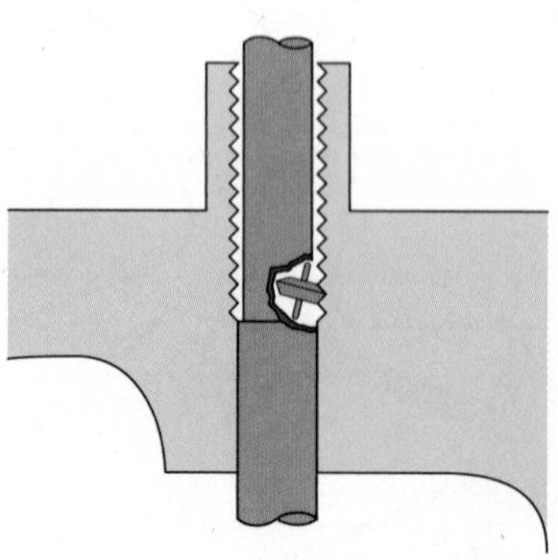

Figure 12-67 Knurling restores the inside diameter (ID) dimensions of a worn valve guide by raising the inside surface of the guide by plowing tiny furrows through the surface of the metal.

or pushed up on either side of the indentation. This effectively decreases the ID of the guide hole. A burnisher is used to press the ridges flat and is then used to shave off the peaks of these ridges to produce the proper-sized hole and restore the correct guide-to-stem clearance.

One of the main advantages of knurling is that it does not change the centerline of the valve stem. Knurling also allows a rebuilder to reuse the old valve if wear is within acceptable limits, helping to reduce rebuilding costs. In spite of its speed and simplicity, knurling is not a cure for badly worn guides.

Reaming and Oversized Valves

Reaming increases the guide hole size to take an oversize valve stem or restores the guide to its original diameter after installing inserts or knurling.

When reaming, limit the amount of metal removed per pass. Always reface the valve seat after the valve guide has been reamed and use a suitable scraper to break the sharp corner (ID) at the top and bottom of the valve guide.

The advantage of reaming for an oversized valve is that the finished product is totally new. The guide is straight, the valve is new, and the clearance is accurate. The use of oversized valve stems is generally considered to be superior to knurling. Yet, like knurling, it is relatively quick and easy. The only tool required is a reamer. Its use is limited to heads in which the guides are not worn beyond the limits of the available oversize valves.

Thin-Wall Guide Liners

Thin-wall guide liners **(Figure 12–68)** are often inserted into guides to restore them. They provide the benefits of a bronze guide surface. They can be used in integral and replaceable guides. It is faster, easier, and cheaper than installing new guides in heads with replaceable or integral guides, and it maintains guide centering with respect to the seats.

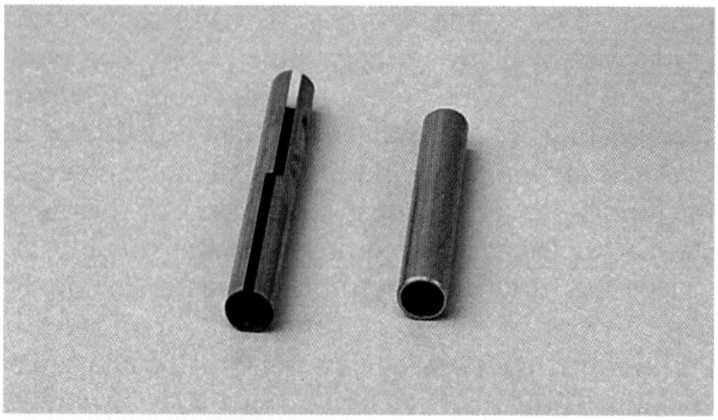

Figure 12-68 A thin-wall valve guide liner.

Phosphor-bronze or silicon-aluminum-bronze are used to make the liners. They are designed for a 0.002- to 0.0025-inch (0.0508 to 0.0635 mm) press-fit. A tight fit is necessary for proper heat transfer and to prevent the liner from working loose.

The original guides are bored to the required diameter for the liner. The liner is then pressed into the bore. It is important to note that some liners are not precut to the length of the original guide. When this is the case, the liner should be cleanly cut or milled to the correct length before installing it. After installation, guide-to-valve stem clearance should be checked and corrected by reaming or knurling.

Valve Guide Replacement

Replacing the valve guide is a repair option. However, pressing out the old guides and installing new ones can be difficult on aluminum heads. Cracking the head or galling the guide hole is always a risk.

Integral Guides To replace integral guides, bore the old guide out. The bore should allow for an interference fit of about 0.002 inch (0.0508 mm). Most technicians freeze the guides before attempting to install them. Lubricate the guide and press or drive it into the bore in the head. It is important to keep the guide straight while it is being installed.

Occasionally, the new guide will not be concentric with the valve seat. Install a new seat to correct the problem, and check the concentricity of the valve seat with a concentricity gauge **(Figure 12–69)**.

Insert Guides To press out an old valve guide, place a proper-sized driver so that its end fits snugly into the guide. The shoulder on the driver must also be slightly smaller than the OD of the guide. Use a heavy ball peen hammer to drive the guide out of the cylinder head.

This procedure should be done with care. There is always the danger of breaking the guide or damaging the guide bore in the head. Cast-iron guides in particular have a tendency to gall aluminum heads. If the bore becomes damaged, it must be rebored and an oversized guide installed. To help prevent these problems, freeze and lubricate the guides before installation. Also, use a press and the correct driver to install them. Keep the guide aligned in its bore and press straight down, not at an angle. If one end of the guide is cut at an angle, make sure the angle is properly placed within the valve port.

Valve Guide Height Guide height is important to avoid interference with the valve spring retainer. The manufacturer's specifications give the correct valve guide installed height, but it is wise to measure the height of the old guides and use this as a reference. As each guide is installed, the installed height should be measured. As each guide is installed, insert a valve. Check for any stem interference.

RECONDITIONING VALVE SEATS

To ensure proper seating of the valve, the valve seat must be the correct width **(Figure 12–70)**, in the correct location on the valve face, and concentric with the guide (less than 0.002-inch [0.0508 mm] runout).

The ideal seat width is 1⁄16 inch for intake valves and 3⁄32 inch (2.38 mm) for exhaust valves. This width ensures proper sealing and heat transfer. However, when an existing seat is refinished to make it smooth and concentric, it also becomes wider.

Wide seats cause problems. Seating pressure drops as seat width increases. Less force is available to crush carbon particles that stick to the seats, and seats run cooler, allowing deposits to build up on them.

The seat should contact the valve face 1⁄32 inch (0.79 mm) from the margin of the valve. When the engine reaches operating temperature, the valve expands slightly more than the seat. This moves the contact area down the valve face. Seats that make too low a contact with the valve face might lose partial contact at normal operating temperatures.

Valve seats can be reconditioned or repaired by one of two methods, depending on the seat type—machining a counterbore to install an insert seat, or grinding, cutting, or machining an integral seat.

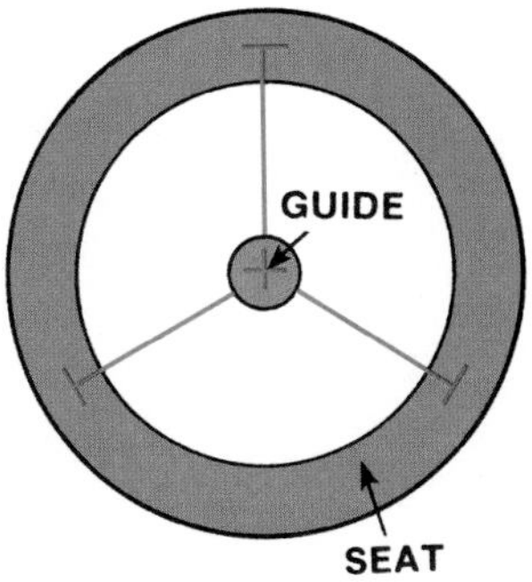

Figure 12–69 The centerline of the guide should be concentric with the seat.

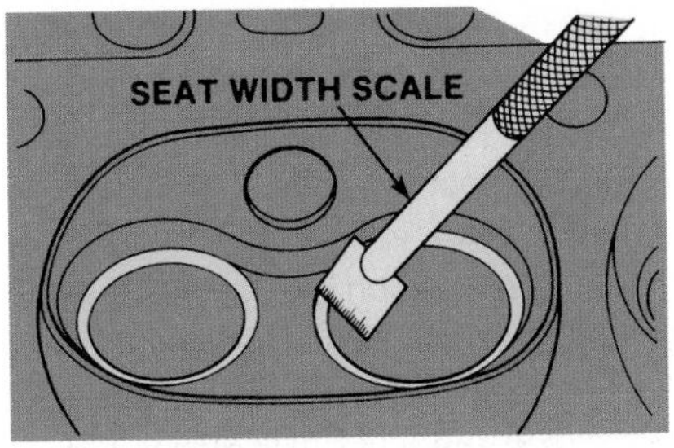

Figure 12–70 Checking valve seat width.

Cracked, heavily burnt, and loose seats should be replaced. If the seat is slightly blemished, it may only need to be refaced. Some rebuilders always replace the valve seats in an aluminum head when it is being reconditioned. This helps restore correct valve height and valve train geometry. Also, many engines have powdered metal seats. These become harder through use and can be very difficult to machine. Therefore, they are replaced with new powder metal or alloy seats.

Valve seats can be reconditioned or repaired by machining a counterbore to install an insert seat, or grinding, cutting, or machining an integral seat.

Seats should not be serviced until all other machining or repairs have been done to the head and the head has been thoroughly cleaned. This includes the reconditioning of the valve guides.

Installing Valve Seat Inserts

The following steps outline a typical procedure for valve seat insert removal and replacement.

PROCEDURE

Insert Valve Seat Removal and Replacement

STEP 1 To remove the damaged insert, use a puller or a prybar **(Figure 12–71)**.

STEP 2 After removal, clean up the counterbore or recut it to accommodate oversized inserts.

STEP 3 Insert the counterboring pilot into the valve guide. Then mount the base and ball shaft assembly to the gasket face angle of the cylinder head.

STEP 4 Use an outside micrometer to accurately expand the cutterhead to a predetermined size of the counterbore **(Figure 12–72)**. Remember that the counterbore should have a slightly smaller ID than the OD of the insert to provide for an interface fit.

STEP 5 Place the valve insert counterboring tool over the pilot and ball-shaft assembly. Preset the depth of the valve seat insert at the feed screw.

STEP 6 Cut the insert by turning the stop-collar until it reaches the present depth. Use a lubricant on the cutters for a smoother finish **(Figure 12–73)**.

STEP 7 To install the insert, heat the head in a parts cleaning oven to approximately 350°F to 400°F (176°C to 204°C). Chill the insert in a freezer or with dry ice before installation.

STEP 8 Press the seat with the proper interference fit using a driver.

STEP 9 When the installation is complete, the edge around the outside of the insert is staked as shown in **Figure 12–74**. By doing so, the insert will be secured more effectively in the counterbore.

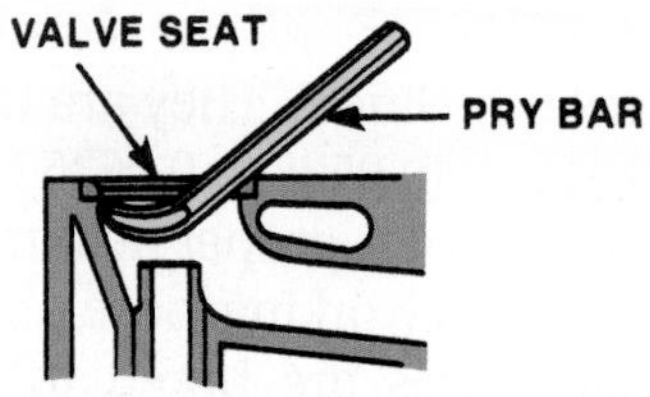

Figure 12–71 Using a prybar to remove a damaged insert seal.

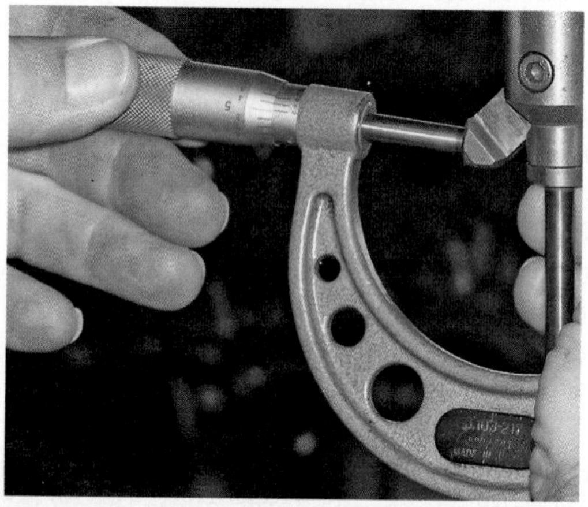

Figure 12–72 Using an outside micrometer to expand the cutter head to allow for an interference fit between the bore and the new valve seat.

Figure 12–73 The valve to openings of this head has been bored and is ready for new seat inserts.

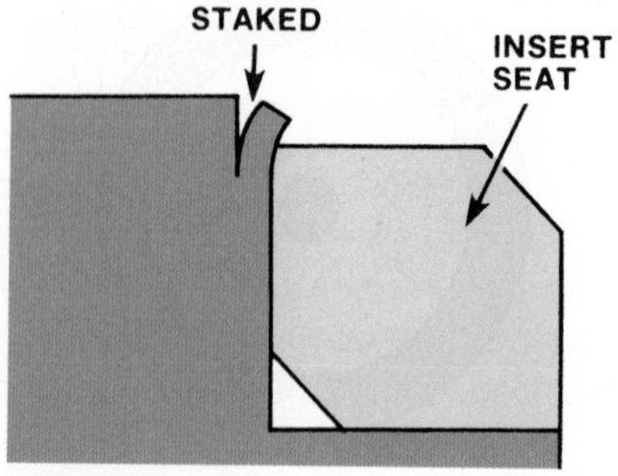

Figure 12–74 Staking the valve seat to the head can be done with a sharp chisel.

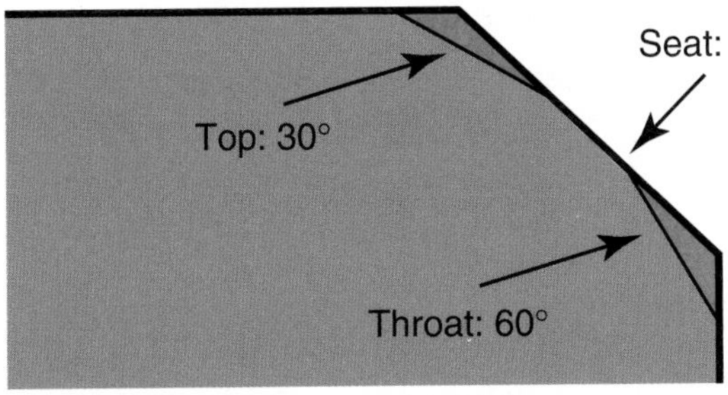

Figure 12-75 The three angles of a properly finished seat.

CAUTION!

Wear the proper gloves when handling dry ice.

Reconditioning Integral Seats

The average valve seat width is 0.060 inch (1.524 mm) and begins 0.030 inch (0.7620 mm) from the valve margin. A properly reconditioned seat has three angles: top, 30 or 15 degrees; seat, 45 or 30 degrees; and throat, 60 degrees. Using three angles provides the correct seat width and sealing position on the valve face **(Figure 12–75)**.

Integral valve seats can be reconditioned by grinding, cutting, or machining.

Grinding Valve Seats

When grinding a valve seat, it is very important to use the correct size pilot and grind stone. Hard seats use a soft stone and soft seats (cast iron) use a harder stone. The stone must be properly dressed and cutting oil used to aid in grinding.

SHOP TALK

Before grinding, many technicians clean the seats by placing a piece of fine emery cloth between the stone and the seat and giving the surface a hard rub. This will help prevent contamination of the seat grinding stone by any oil or carbon residue that might be present on the valve seat. Such contamination could cause glazing.

The grinding wheel is positioned and centered by inserting a properly sized pilot shaft into the valve guide **(Figure 12–76)**. All valve guide service must be completed before installing the pilot.

The seat is ground by continually and quickly raising and lowering the grinder unit on and off the seat. Grinding should only continue until the seat is clean and free of defects.

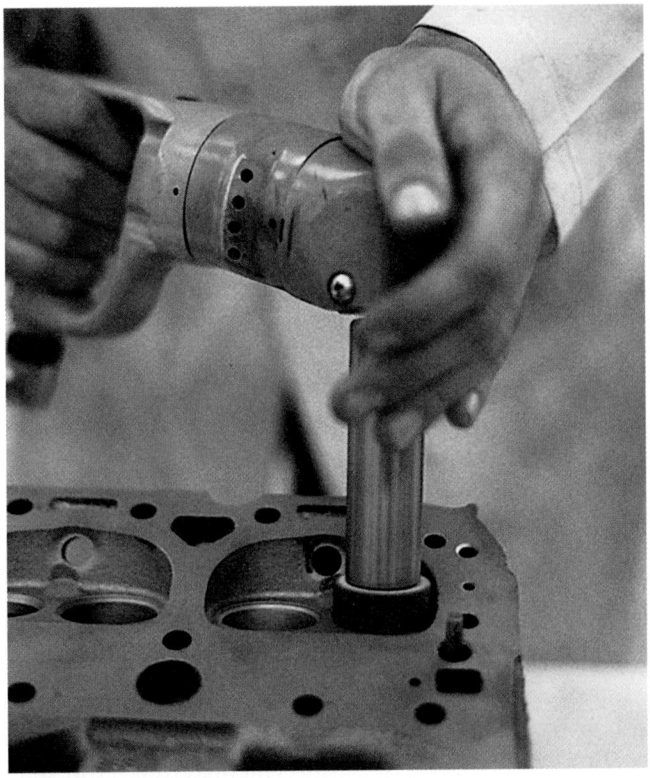

Figure 12-76 Grinding the seat.

Figure 12-77 A valve seat being cut at three angles at the same time. *Courtesy of Jasper Engines & Transmissions*

Cutting Valve Seats

Cutting valve seats differs from grinding only in the equipment used **(Figure 12–77)**. Hardened valve seat cutters replace grinding wheels. The basic seat cutting procedures are the same as those for grinding. Always check the fit of the valve face on the seat and adjust the seat as needed after initial cuts.

Checking Valve Fit

While cutting a 45-degree or other specified seat, make the seat wider than specifications. Bevel or chamfer the top of the seat with a **topping** stone (normally 30 degrees). Then bevel the bottom with a

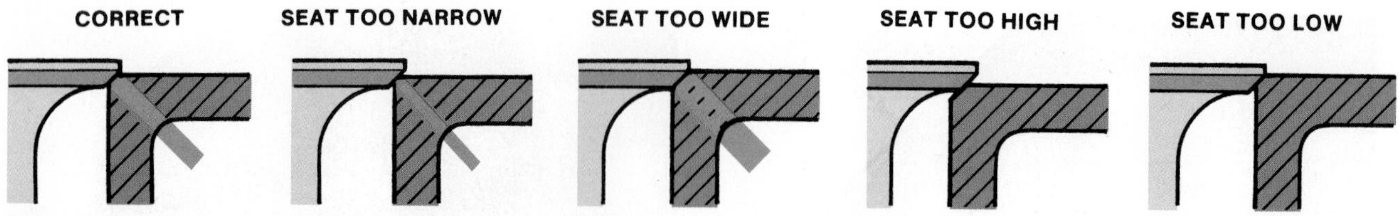

Figure 12–78 The fit of the valve face in the seat should be checked carefully.

throating stone (normally 60 degrees). Coat the face of the valve with machinist dye, which is a Prussian blue compound. Install the valve into its seat. Then open the valve and snap it closed against the seat several times. Remove the valve and look at the pattern on the valve face and seat.

The dye should show the face making full and even contact with the seat. The contact area should also be centered on the valve seat. If the pattern on the valve face is not even, the valve should be replaced. If the pattern on the seat is not even, check the concentricity of the guide and replace the guide and/or seat.

If the contact area is above the center of the seat, it needs to be lowered **(Figure 12–78)**. Cut the seat with the topping stone. This stone will also make the seat narrower. If the contact area is too low, cut the seat with the throating stone. When using the topping or throating stone or cutting bits to move the contact area, remove a little metal and recheck the contact area. Keep machining the seat until the contact area is the correct width and has the desired placement. After the contact area is acceptable, use the 45-degree stone to clean off any possible burrs caused by the cutting and grinding. Make sure all metal particles from machining the seat are totally removed before assembling the head.

Lapping Some recommend lapping the valves after the seat area has been adjusted. This procedure uses an abrasive compound to remove all small imperfections in the contact area between the valve face and the seat. The abrasive lapping compound is spread thinly over the face of the valve. With the face against the seat, the valve is moved back and forth with a lapping tool. A lapping tool is basically a stick with a suction cup on the end. The suction cup is placed on the outside of the valve head and the stick is rotated back and forth by hand. After several rotations, the valve is removed and its face and seat are wiped clean.

VALVE STEM SEALS

Valve stem seals are used to control the amount of oil allowed between the valve stem and the guide. The stems and guides will scuff and wear excessively if they do not have enough lubrication. Too much oil produces heavy deposits that build up on the intake valve and hard deposits on the head end of the exhaust valve stem. Worn valve stem seals can increase the oil consumption by as much as 70%.

There are basically three types of seals. **Positive seals** fit tightly around the top of the guide and scrape oil off the valve stem as it moves up and down. Deflector, splash, or **umbrella-type seals** ride up and down on the valve stems to deflect oil away from the guides. O-ring seals are used to prevent oil from moving into the guide when the valve is open.

Installing Positive Valve Seals

To install a positive valve seal **(Figure 12–79)**, place the plastic sleeve in the kit over the end of the valve

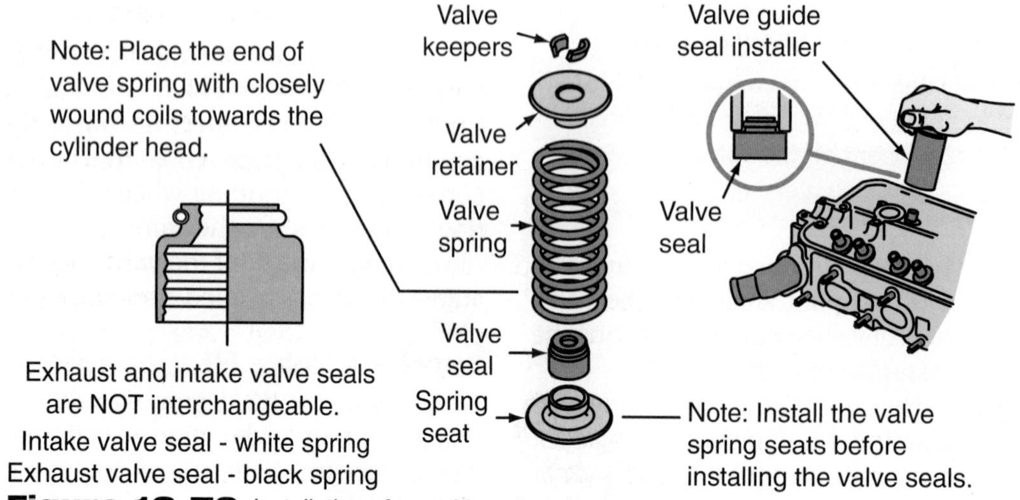

Figure 12–79 Installation of a positive oil seal onto a valve guide.

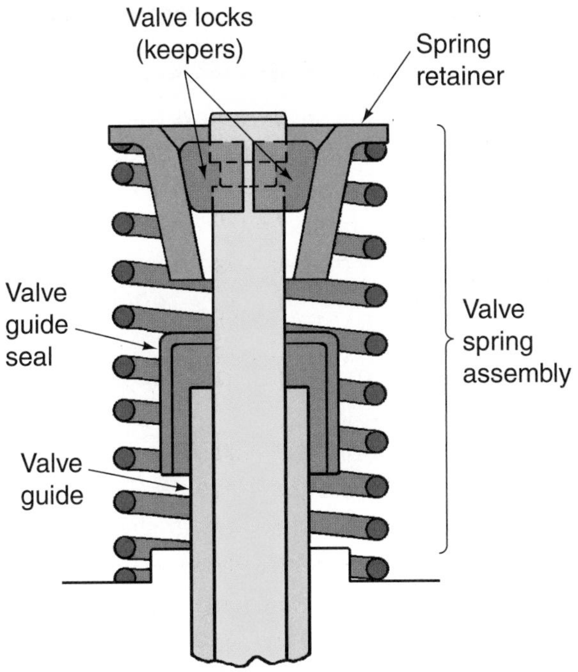

Figure 12-80 Valve assembly with an umbrella-type oil seal. *Courtesy of Federal-Mogul Corporation*

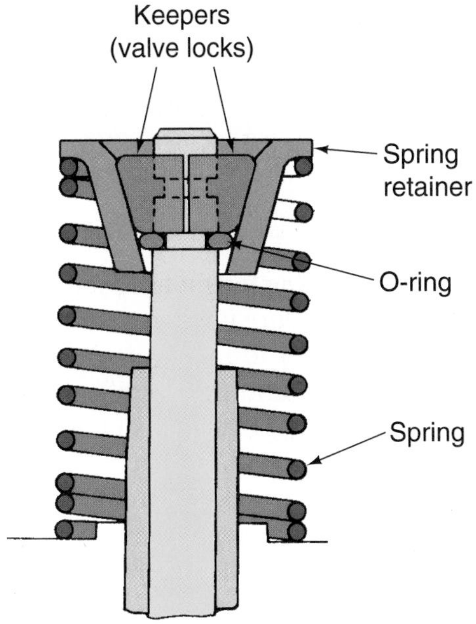

Figure 12-81 Valve assembly with an O-ring valve seal. *Courtesy of Federal-Mogul Corporation*

stem to protect the seal as it slides over the keeper grooves. Lightly lubricate the sleeve. Carefully place the seal on the cap over the valve stem and push the seal down until it touches the top of the valve guide. At this point, the installation cap can be removed and placed on the next valve. A special installation tool can be used to finish pushing the seal over the guide until the seal is flush with the top of the guide.

Installing Umbrella-Type Valve Seals

An umbrella-type seal is installed on the valve stem before the spring is installed. It is pushed down on the valve stem until it touches the valve guide boss **(Figure 12–80)**. It will be positioned correctly when the valve first opens.

Installing O-Rings

When installing O-rings, use engine oil to lightly lubricate the O-ring. Then install it in the lower groove of the lock section of the valve stem **(Figure 12–81)**. Make sure the O-ring is not twisted.

ASSEMBLING THE CYLINDER HEAD

Before a cylinder head is reassembled and installed, two critical measurements must be carefully checked: the installed stem height and the installed spring height.

Installed stem height is the distance between the spring seat and stem tip. A number of tools can be used to obtain an accurate stem height reading, including a depth micrometer, vernier caliper, and telescoping gauge.

USING SERVICE INFORMATION

Stem height specifications are often unavailable in service manuals. As a guide for assembly, record the stem heights for all valves during disassembly.

Another specification related to installed stem height is installed spring height. Installed spring height is measured from the spring seat to the underside of the retainer when it is assembled with keepers and held in place. This measurement can be made by using a set of dividers or scales, telescoping gauge, or spring height gauge.

If the spec for installed spring height for an exhaust valve is 1.600 inches (40.64 mm), and the measurement is 1.677 inches (42.54 mm), the increase in height is 0.077 inch (1.95 mm). This means the installed stem height has also been increased by 0.077 inch (1.95 mm).

Adjustments to valve spring height can be made with valve spring inserts, otherwise known as spring shims. Even though valve shims come in only three standard thicknesses—0.060, 0.030, and 0.015 inch (1.52, .7620, and .3810 mm)—using combinations of different shims gives the correct amount of compensation (within 0.005 or 0.010 inch [0.1270 or 0.2540 mm]).

SHOP TALK

Valve keepers should be replaced in pairs. If a new keeper is mated with a used one, the spring retainer may cock and break off the valve tip or allow the assembly to come apart.

By comparing spring height to specifications, the desired amount of spring tension correction can be easily determined. For example, if spring height is 0.180 inch (4.59 mm) and the specifications call for 0.149 inch (3.78 mm), a 0.030-inch shim (0.149 inch + 0.030 inch = 0.179 inch [3.78 mm + 0.7620 mm = 4.54 mm]) would be needed. If more than one shim is required, place the thickest one next to the spring. If one side of the shim is serrated or dimpled, place that side onto the spring seat.

The end of the spring that is wound the tightest should always be placed toward the head of the valve. This is also true of mechanical surge and vibration dampers.

With the valve inserted into its guide, position the valve spring inserts, valve spring, and retainer over the valve stem. Using a valve spring compressor, compress the spring just enough to install the valve keepers into their grooves. Excess pressure may cause the retainer to damage the oil seal. Release the spring compressor and tap the valve stem with a rubber mallet to seat the keepers. When doing this, the valve will open slightly. To prevent damage to the valves, never tap on the stems with the cylinder head lying flat on the bench. Turn the head on its side or raise it off the bench.

CAUTION!

If the keepers are not fully seated, the spring assembly could fly apart and cause personal injury (or serious damage to the engine if it occurs while the engine is running). For these reasons, it is good practice to assemble the valves with the retainers facing a wall and to wear eye protection.

OHC Engines

After the valves are installed and the cylinder head is assembled, the camshaft can be installed in the cylinder head. Some engines have a separate camshaft housing that bolts to the cylinder head. This should be installed with the proper seals and gaskets. Make sure the seals are properly seated in their grooves before tightening the housing to the head.

CASE STUDY

While traveling, the engine in a family's late-model car overheats. Water is added to the radiator. Then the car is taken to a service station to be diagnosed and repaired. Simple tests reveal that a heater hose has split and is leaking coolant. The hose is replaced and new coolant added to the system. With the car repaired, the family continues on their trip.

After driving only a short time, smoke is noticed coming out of the tailpipe. The engine shakes violently whenever it is placed in a load condition. The driver turns the car around and heads back to the shop that fixed the leak. Upon arrival, the driver says the technician must have knocked something loose or broken something while replacing the hose. The technician doubts this but agrees to take another look.

A cylinder power balance test would verify this, but it cannot be conducted because of the erratic idle. Instead, the spark plugs are removed and inspected. All look normal except for plug #2, which looks newer than the rest. A compression test is taken and cylinder #2 has very low dry and wet readings. Then a cylinder leakage test is conducted. Cylinder #2 has excessive leakage and air can be felt leaving the exhaust pipe.

These test results lead to the conclusion that the engine is running on one less cylinder because an exhaust valve is not sealing. The customer still feels the technician did something wrong but agrees to allow the technician to remove the cylinder head for further diagnosis. As soon as the head is off the block, the problem is identified. The exhaust valve seat for cylinder #2 came loose from the head and is preventing the valve from closing. The technician then explains that this undoubtedly happened when the engine overheated, due to the expansion of the metal. The customer is satisfied with the explanation and allows the technician to correct the problem. The family is then able to continue their trip.

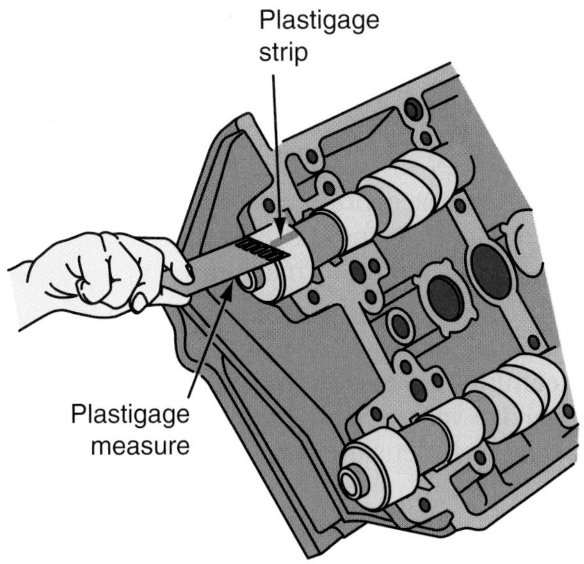

Figure 12–82 OHC camshaft bearing clearances can be checked with Plastigage.

Service to the camshaft bearings can now be done. Full-round insert bearings are pressed into the bores in the cylinder head. Special tools are designed to make this job easier. Never use a standard camshaft bearing driver and hammer to install these bearings. The hammering can easily break or damage the bearing supports. After each bearing is fully seated in its bore, double-check the alignment of the bearing's oil hole with the oil hole in the head.

Some OHCs do not have bearing inserts. The bores in the aluminum casting serve as a bearing surface. These surfaces can be cleaned up and/or align bored if needed.

Most late-model OHC engines use split bearings and bearing caps or have a separate housing for the camshaft. Working with split camshaft bearings is like working with crankshaft main bearings. This includes checking bearing clearances with Plastigage **(Figure 12–82)**.

Before installing the camshaft(s), install the timing gear. Make sure the alignment pin or key is aligned. Tighten the gear-to-shaft bolt to specifications. If the head has two camshafts, the manufacturer will give specific instructions for the alignment of the timing gears and chain. Normally one camshaft is installed with the chain wrapped around its gear. Once the camshaft shaft and gear are aligned with the timing reference marks, the bearing caps are installed and tightened in a specified order and torque. Then the other camshaft is installed by slipping the gear into the chain and then setting it into the bearing journals. After the timing marks are aligned on both shafts, the bearing caps for the second camshaft are tightened. When both camshafts are in place, the chain tensioner is installed.

Figure 12–83 Coating the camshaft with lubricant. *Courtesy of Dana Corporation*

Prior to installation, wipe off each cam bearing with a lint-free cloth, then thoroughly coat the camshaft lobes and bearing journals with the lube recommended by the manufacturer **(Figure 12–83)**.

Install the camshaft. Install new O-rings and gaskets, as required, while installing the camshaft. Some camshafts have an end thrust plate or shims to limit shaft end play. Make sure the originals are installed. After the camshaft is in place, check the end play and change the thickness of the shims as needed.

If the assembly has bearing caps, place them in their correct positions and tighten them according to specifications. Once tightened, the shaft should be able to rotate smoothly. On some engines, turning the camshaft causes the valves to open and close. If this is the case, stand the cylinder head on its end to prevent the valves from hitting the bench while you turn the camshaft. If the cam does not turn, binding might be the cause. Binding is the result of a damaged bearing, a nick on the cam's journal, or a slight misalignment of the block journals. The problem should be identified and corrected.

Gather the rocker arms, lash adjusters, pushrods, lifters, and other parts that transfer the motion of the camshaft to the valve stem. Coat all of these parts with clean engine oil. These parts should be installed according to the order given in the service manual. Some are installed before the camshaft is installed; others are installed after. Make sure all parts fit securely.

Normally the rocker arm assemblies are installed by turning the cam until the lobe for that valve faces away from a valve stem. The rocker arms can be slipped into position by depressing the valve spring slightly. Make sure the rocker arm mounting bolts are tightened to specifications. Follow the same procedure for all of the valves. The manufacturer may recommend lubricating some of the bolts, so check the manual.

Mount all manifolds and other parts that were on the head when it was removed from the block. Use new gaskets or the correct sealant where necessary. Adjust the valve lash (clearance) and mount the head onto the block.

KEY TERMS

Broaching
CC-ing
Cross-flow ports
Damper spring
Guide plate
Hemi-head
Hemispherical combustion chamber
Knurling
Margin
Milling
Pentroof
Phaser
Poppet
Porosity
Positive seal
Pumping loss
Quench area
Quenching
Reaming
Shrouding
Siamese ports
Stainless steel
Stellite
Throating
Topping
Tungsten inert gas (TIG)
Turbulence
Umbrella-type seal
Valve face
Valve guide
Valve keeper
Valve stem
Wedge-type combustion chamber

SUMMARY

- Pushrods are the connecting link between the rocker arm and the valve lifter.
- The rocker arm converts the upward movement of the valve lifter into a downward motion to open the valve.
- Aluminum cylinder heads are used on late-model engines because of their light weight. The thermal expansion characteristics of aluminum can lead to problems such as leaking and cracking.
- Every cylinder of a four-stroke engine contains at least one intake valve and one exhaust valve.
- Multivalve engines feature either three, four, or five valves per cylinder, which means better combustion and reduced misfire and detonation. These benefits are offset to some extent by a more complicated camshaft arrangement.
- The means of resurfacing the deck of a cylinder head include grinding, milling, belt surfacing, and broaching.
- The amount of stock removed from the cylinder head gasket surface must be limited. Excessive surfacing can create problems with the engine's compression ratio, not to mention piston/valve interference and misalignment.
- The two surfaces of a valve reconditioned by grinding are the face and the tip. Valves can be refaced on grinding or cutting machines.
- One of the fastest ways for restoring the inside diameter of a worn valve guide is knurling. Reaming increases the guide bore to take an oversize valve stem or restores the guide to its original dimension after knurling or installing inserts.
- Pressing out an old valve guide to install a new one can be difficult on some aluminum heads where the interference fit is considerable.
- To ensure proper seating of a valve, the seat must be the correct width, in the correct location on the valve face, and concentric with the guide.
- When grinding a valve seat, choosing the correct size pilot and stone is important.
- Valve stem seals are used to control the amount of oil between the valve stem and guide. Too much oil produces deposits, while insufficient lubrication leads to excessive wear.
- The valve spring closes the valve and maintains valve train contact during the opening and closing of the valve. To determine if a spring needs to be replaced, three tests are valuable: freestanding height, spring squareness, and open/close spring pressure.
- Two critical measurements that must be made before a cylinder head is reassembled are installed stem height and installed spring height. The first of these is determined by measuring the distance between the spring seat and the stem tip. The latter is measured by the spring seat to the underside of the retainer when it is assembled with keepers and held in place.

REVIEW QUESTIONS

1. What happens when spring tension is too low?
2. Define valve margin.
3. What usually causes warpage in an aluminum cylinder head?
4. What are the two ways pushrods can be checked for straightness?

5. What welding method is preferred for repairing aluminum heads?
6. Why do some technicians not consider knurling a long-term repair?
7. Which of the following is not true of knurling?
   a. It is one of the fastest techniques for restoring the ID dimensions of a worn valve guide.
   b. It reduces the amount of work necessary to reseat the valve.
   c. It is useful for restoring badly worn guides to their original condition.
   d. None of the above.
8. How does resurfacing the cylinder head of an OHC engine affect valve timing and what should be done when the head needs resurfacing?
9. To ensure proper seating of the valve, the valve seat must be ______.
   a. the correct width
   b. in the correct location on the valve face
   c. concentric with the guide
   d. all of the above
10. When grinding valve seats, ______.
    a. a pilot shaft is inserted into the valve guide
    b. a hard stone should be used on a hard seat
    c. a soft stone should be used on a soft seat
    d. all of the above
11. If the valve face and valve seat do not contact each other evenly after reconditioning, ______.
    a. regrind with the same stone
    b. regrind with stones of different angles
    c. discard the cylinder head
    d. none of the above
12. Many engines with VVT have a phaser mounted to the end of one or more camshafts. Which of the following statements is not true?
    a. A phaser is used to alter valve timing.
    b. Most phasers are controlled by oil pressure.
    c. Phasers are used to change valve lift.
    d. Phasers can be used to change valve overlap.
13. Multiple valve engines tend to be more efficient than two-valve-per-cylinder engines because they ______.
    a. allow for increased port areas
    b. have smaller valves
    c. provide less restriction to the airflow
    d. all of the above
14. Which type of surfacing machine uses underside rotary cutters?
    a. milling c. belt
    b. broaching d. grinding
15. Which of the following statements is not true?
    a. Normally the desired valve face-to-seat contact area for intake valves is 1/16 inch.
    b. Normally the desired valve face-to-seat contact area for exhaust valves is 1/16 inch.
    c. The average valve seat width is 0.060 inch and the average seat begins 0.030 inch from the valve margin.
    d. A properly reconditioned seat has the correct seat width and sealing position on the valve face.

ASE-STYLE REVIEW QUESTIONS

1. While discussing the reasons for resurfacing a cylinder head: Technician A says that it should be resurfaced to make it flat and very smooth so the head gasket can seal properly during engine assembly. Technician B says that it can be resurfaced to raise the compression. Who is correct?
   a. Technician A c. Both A and B
   b. Technician B d. Neither A nor B
2. Technician A says that bronze valve guides retain oil better than cast-iron ones. Technician B says that cast-iron valve guides are easier to machine than bronze guides. Who is correct?
   a. Technician A c. Both A and B
   b. Technician B d. Neither A nor B
3. Technician A says that positive valve stem seals fit tightly around the top of the stem. Technician B says that positive stem seals scrape oil off the valve as it moves up and down. Who is correct?
   a. Technician A c. Both A and B
   b. Technician B d. Neither A nor B
4. When fitting a freshly ground valve into a freshly ground seat: Technician A says that the seat should be ground more with the same stone if the margin is too small. Technician B says that the seat should be ground more with the same stone if the valve seat is too high. Who is correct?
   a. Technician A c. Both A and B
   b. Technician B d. Neither A nor B

5. Technician A says that low spring pressure may allow a valve to float. Technician B says that excessive spring tension may cause premature valve train wear. Who is correct?
   a. Technician A
   b. Technician B
   c. Both A and B
   d. Neither A nor B
6. Technician A says that it is not necessary to measure valve stem height unless the valves have been replaced. Technician B says that the valve stem height can be adjusted with shims. Who is correct?
   a. Technician A
   b. Technician B
   c. Both A and B
   d. Neither A nor B
7. Technician A uses a 30-degree stone for topping a 45-degree valve seat. Technician B uses a 60-degree stone to lower a 45-degree valve seat. Who is correct?
   a. Technician A
   b. Technician B
   c. Both A and B
   d. Neither A nor B
8. While discussing the Valvetronic system: Technician A says that the system has no need for a throttle plate. Technician B says that the system alters the duration and the lift of the intake valves. Who is correct?
   a. Technician A
   b. Technician B
   c. Both A and B
   d. Neither A nor B
9. While checking a hydraulic phaser after the camshaft has been removed from the engine: Technician A attempts to rotate the phaser's timing gear by hand and replaces it if it does move. Technician B applies air pressure to the advance and retard ports and replaces it if the gear does not move. Who is correct?
   a. Technician A
   b. Technician B
   c. Both A and B
   d. Neither A nor B
10. While servicing a warped aluminum cylinder head: Technician A checks the thickness of the head and the specifications to be sure that material can be safely removed from the surface. Technician B replaces the head if the warpage exceeds the specified allowable amount. Who is correct?
    a. Technician A
    b. Technician B
    c. Both A and B
    d. Neither A nor B

ENGINE SEALING AND REASSEMBLY

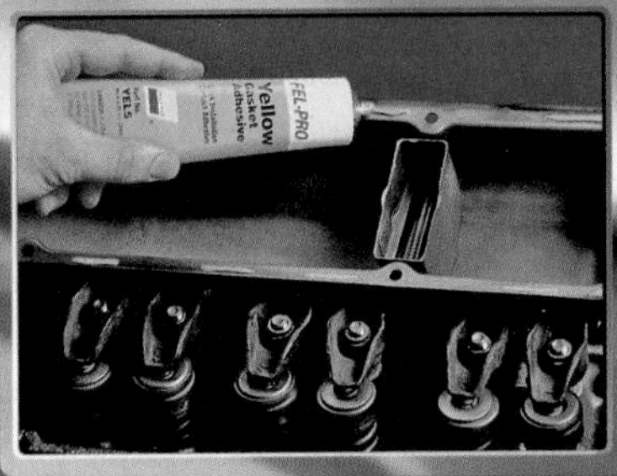

CHAPTER 13

OBJECTIVES

■ Explain the purpose of the various gaskets used to seal an engine. ■ Identify the major gasket types and their uses. ■ Explain general gasket installation procedures. ■ Describe the methods used to seal the timing cover and rear main bearing. ■ Reassemble an engine including core plugs, bearings, crankshaft, camshaft, pistons, connecting rods, timing components, cylinder head, valve train components, oil pump, oil pan, and timing covers. ■ Explain the ways to prelubricate a rebuilt engine. ■ Reinstall an engine and observe the correct starting and break-in procedures.

Proper sealing of an engine keeps the low-pressure liquids in the cooling system away from the cylinders and lubricating oil. It also keeps the high pressure of combustion in the cylinders. It prevents both internal and external oil leaks and suppresses and muffles noise.

TORQUE PRINCIPLES

All metals are elastic. **Elasticity** means a bolt can be stretched and compressed up to a certain point. This elastic, springlike property is what provides the clamping force when a bolt is threaded into a tapped hole or when a nut is tightened. As the bolt is stretched a few thousandths of an inch, clamping force or holding power is created due to the bolt's natural tendency to return to its original length **(Figure 13–1)**.

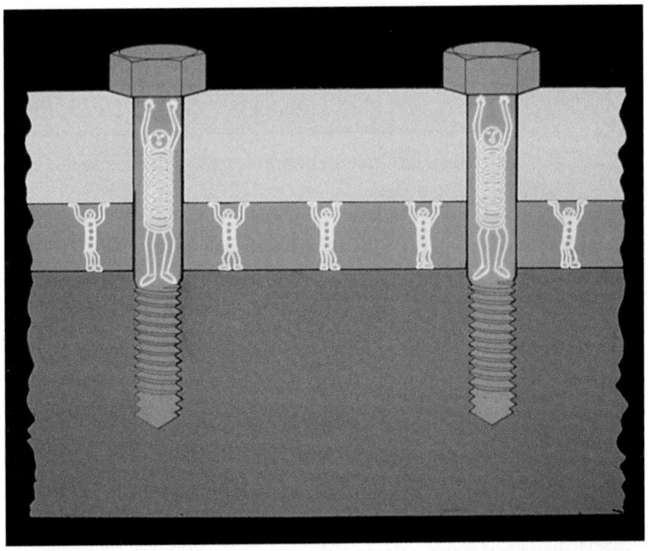

Figure 13–1 Clamping force results from bolt stretch. *Courtesy of Federal-Mogul Corporation*

Like a spring, the more a bolt is stretched, the tighter it becomes. However, a bolt can be stretched too far. This is obvious when the grip on the wrench feels "mushy." At this point, the bolt can no longer safely carry the load it was designed to support, called the modulus of elasticity.

If a bolt is stretched into **yield**, it takes a permanent set and never returns to normal **(Figure 13–2)**. The bolt will continue to stretch more each time it is used, just like a piece of taffy that is stretched until it breaks. Proper use of torque will avoid this yield condition.

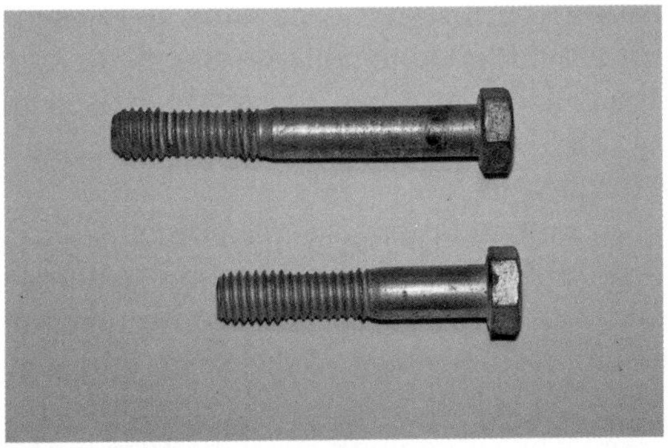

Figure 13–2 These bolts have been torqued beyond their yield. Note the soda bottle effect.

TABLE 13–1 THE EFFECT OF LUBRICATION ON BOLT TORQUE SPECIFICATIONS

Bolt Size	Thread	Lube	Torque	Change
5/16 –18	COARSE	None	29 ft.-lb	—
	COARSE	Plated and cleaned	19 ft.-lb	–34%
	COARSE	SAE 20 oil	18 ft.-lb	–38%
	COARSE	SAE 40 oil	17 ft.-lb	–41%
	COARSE	SAE 30 oil + plating	16 ft.-lb	–45%
	COARSE	Lithium grease	16 ft.-lb	–45%
	COARSE	Dry moly film	14 ft.-lb	–52%
	COARSE	Graphite and oil	13 ft.-lb	–55%
½ –13	COARSE	None	121 ft.-lb	—
	COARSE	Plated and cleaned	90 ft.-lb	–26%
	COARSE	SAE 20 oil	87 ft.-lb	–28%
	COARSE	SAE 40 oil	83 ft.-lb	–31%
	COARSE	SAE 30 oil + plating	79 ft.-lb	–35%
	COARSE	Lithium grease	79 ft.-lb	–35%
	COARSE	Dry moly film	66 ft.-lb	–45%
	COARSE	Graphite and oil	62 ft.-lb	–49%

*Use these lubrication percentages to decrease the torque specifications for other bolt sizes.

Appendix B gives standard bolt and nut torque specifications. If the manufacturer's torque specifications are available, follow them precisely.

Nonplated bolts have a rougher surface than plated finishes. Therefore, it takes more torque to produce the same clamping force as on a plated bolt.

Most printed torque values are for dry, plated bolts. Lubricants are beneficial when working with engines. They provide smoother surfaces and more consistent and evenly loaded connections. They also help reduce thread galling.

Reusing a dry nut will produce a connection with decreasing clamp force each time it is used. Nut threads are designed to collapse slightly to carry the bolt load. If dry nuts are reused, increased thread galling will result each time the nuts are reused at the same torque.

Lubrication of fasteners is recommended for consistency (**Table 13–1**). However, be sure to lubricate all the bolts with the same lubricant. Some lubricants are more slippery than others, which affects torque values. Lubricate the bolt, never the hole. Otherwise, the bolt may merely be tightening against the oil in the hole.

If a bolt with a reduced shank diameter (for example, a connecting rod bolt) is specified by the OEM, never replace it with a standard, straight shank bolt. A reduced shank diameter bolt looks "dog-boned." Its function is to reduce the stress on the threads by transferring it to the shank. A standard bolt under similar conditions would break very quickly at the threads.

Keep the following points in mind:

1. Visually inspect the bolts.
   - Threads must be clean and undamaged. Discard all bolts that are not acceptable.
   - Use liquid sealant or engine oil on the threads and seating face of cylinder head bolts to prevent seizure from rust and corrosion. This is particularly important for an aluminum block or head. Use sealant on those bolts that hold in coolant or oil.
   - Install bolts in their proper holes.
   - Run a nut over the bolt's threads by hand. Discard it if any binding occurs.
   - Clean bolt and cylinder block threads with a thread chaser or tap (**Figure 13–3**).
2. Apply a light coat of 10W engine oil to the threads and bottom face of the bolt head. A sealer is required for a bolt that enters a water jacket.

Figure 13-3 Cleaning bolt holes with tap. *Used with permission of Detroit Gaskets*

This will stop seepage around the bolt threads. Seeping coolant could get in the oil or cause corrosion that might damage parts, resulting in engine failure.

3. Tighten bolts in the recommended sequence. This is important to prevent warpage of the parts.
   - Use an accurate torque wrench.
   - Tighten bolts according to the recommended steps.
4. If bolt heads are not tight against the surface, the bolts should be removed and washers installed.
5. Make sure the bolt is the proper length (not too long).

Bolt hole threads often pull up, leaving a raised edge around the hole **(Figure 13–4)**. If a part has been resurfaced, the threads might run up to the surface. In either case, the bolt holes should be tapered at the surface by chamfering and the threads cleaned with an appropriate size bottoming tap. Always repair damaged threads.

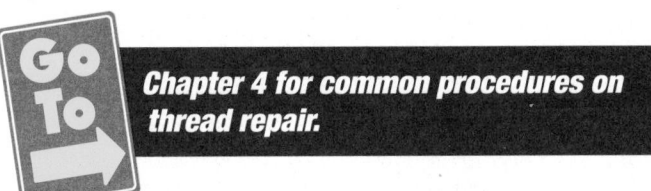

Go To ***Chapter 4 for common procedures on thread repair.***

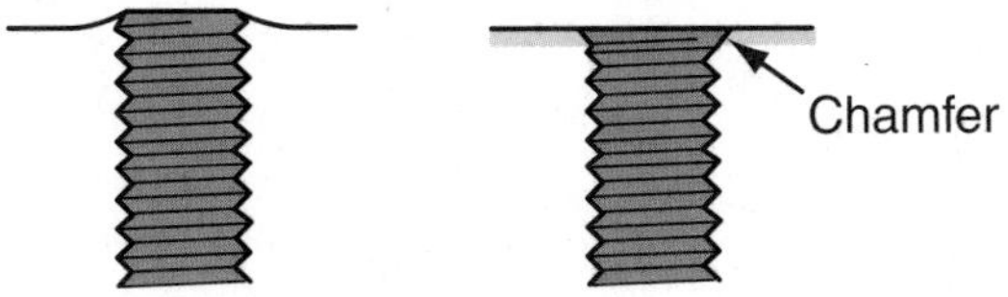

Figure 13-4 Bolt hole threads can pull up, leaving a raised edge. Also, if the block has been resurfaced, the threads may run up to the surface. In either case, the hole must be chamfered.

SHOP TALK

Impact wrenches should only be used to loosen nuts and bolts. Use other power or hand tools to tighten them. Final tightening should always be done with a torque wrench.

Thread Repair

A common fastening problem is threads stripping inside an engine block, cylinder head, or other structure. This problem is usually caused by overtorquing or by incorrectly threading the bolt into the hole. Rather than replacing the block or cylinder head, the threads can be replaced by the use of threaded inserts. The helically coiled insert is the most commonly used.

Go To ***Chapter 4 for the procedures for reconditioning fastener threads.***

Torque-to-Yield (TTY) Bolts

Some fasteners are intentionally torqued into a yield condition. These fasteners, known as **torque-to-yield (TTY)** bolts, are designed to stretch when properly tightened. When a bolt is stretched to its yield point, it exerts its maximum clamping force. TTY bolts are not ordinary bolts. The bolt shank is designed to stretch **(Figure 13–5)** and spring back up to its yield point when tightened. Once at the yield point, the bolt becomes permanently stretched and will not return to its original length. Therefore, TTY bolts should not be reused.

TTY bolts are commonly used as cylinder head bolts but may be used in other applications. Service

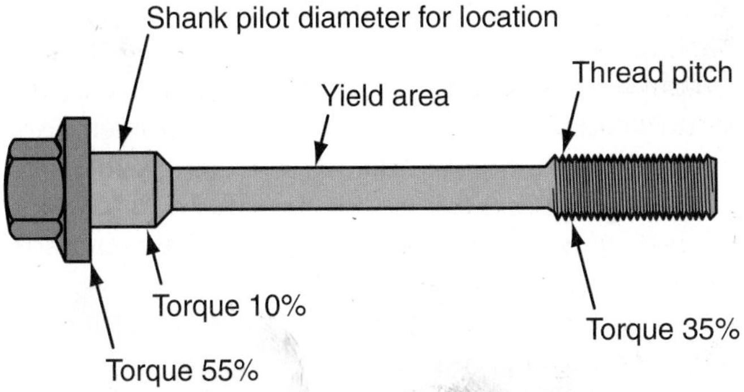

Figure 13-5 TTY bolts are designed with a reduced shank diameter; this is where the intended stretch occurs.

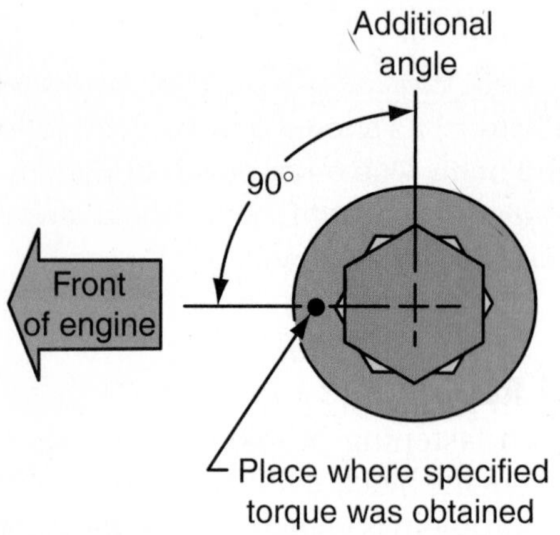

Figure 13-6 TTY bolts are tightened to a specified torque and then turned to an additional number of degrees.

manuals note the location of TTY bolts and give the proper tightening sequence for them.

Tightening a TTY bolt involves two distinct steps: Tighten the bolt to the specified torque, and then turn the bolt to a specified angle (measured in degrees) to load the bolt beyond its yield point (**Figure 13–6**). A torque angle gauge is required to do this. This gauge fits between the drive of the torque wrench and the socket. Once the specified torque is reached, the angle gauge is set to zero. Then the bolt is turned until the specified angle is read on the gauge.

GASKETS

When parts are bolted together, it is nearly impossible to obtain a positive seal between the parts. **Gaskets** are used to provide that seal. Some gaskets seal low-pressure fluids, such as oil and coolant, while others also seal high-combustion pressures, such as head gaskets. Gaskets also serve as spacers, wear insulators, and vibration dampers. Gaskets are only used between two stationary parts. Seals are used if one of the parts moves.

Gaskets can be made of paper, fiber, steel, cork, synthetic rubber, and combinations of these materials. Most late-model engines have molded silicone rubber gaskets and cut plastic gaskets with silicone sealing beads. The gaskets may incorporate flexible graphite cores, specialized surface coatings, asbestos-free materials, elastomeric beading, reinforced cork products, wire-ring combustion seals, flat-plate hoop-strength constructions, and many other design variations (**Figure 13–7**). The goal of all designs is provide long-term leak-free joints.

Cut Gaskets

Gaskets made of paper, fiber, and cork are normally called soft, cut gaskets (**Figure 13–8**). Each gasket is cut to the desired size and shape from a sheet of material.

Paper/Fiber Gaskets These are made of a fiber-reinforced paperlike material. This treated paper seals low-pressure, low-temperature areas well. For some applications, paper gaskets may be relatively thick. Paper gaskets are commonly used between the different housings and covers in a transmission or for some water pumps. These gaskets seldom need an additional sealant; however, a thin coating of adhesive may be used to hold the gasket in place while it is installed.

Cork Gaskets Cork gaskets are used to seal low-pressure areas like valve covers and oil pans. In most cases, cork gaskets are not used on today's engines. In fact, most gasket manufacturers substitute a rubber gasket for the original cork. Cork gaskets are very soft, easily distorted, and absorbent. In fact, they will weep some of the fluid they are sealing. They also tend to become brittle and crack over time. Composite gaskets, typically rubberized cork, are often used (**Figure 13–9**). These have a longer service life. A sealant or adhesive is often used to ease installation and to improve their effectiveness.

Molded Rubber Gaskets

Molded rubber gaskets provide excellent sealing and are commonly used on today's engines. These gaskets are made by injecting synthetic rubber (neoprene, nitrile, silicone, or other similar material) into a mold to form one-piece gaskets. Molded gaskets retain their flexibility and are durable. Molded gaskets are often used to seal intake manifolds, some thermostat or water pump housings, valve covers, and oil pans.

Some molded gaskets have a steel insert that adds stiffness and strength to the gasket (**Figure 13–10**). Some gaskets may also have reinforcements around the bolt holes (**Figure 13–11**) to limit the amount of crush when the parts are tightened together.

WARNING!

Do not use sealant or adhesive on rubber gaskets; they can prevent the gasket from sealing.

Manufacturers may not use premade gaskets when they are making their engines. Rather they use chemical gasketing. Robotic equipment applies a

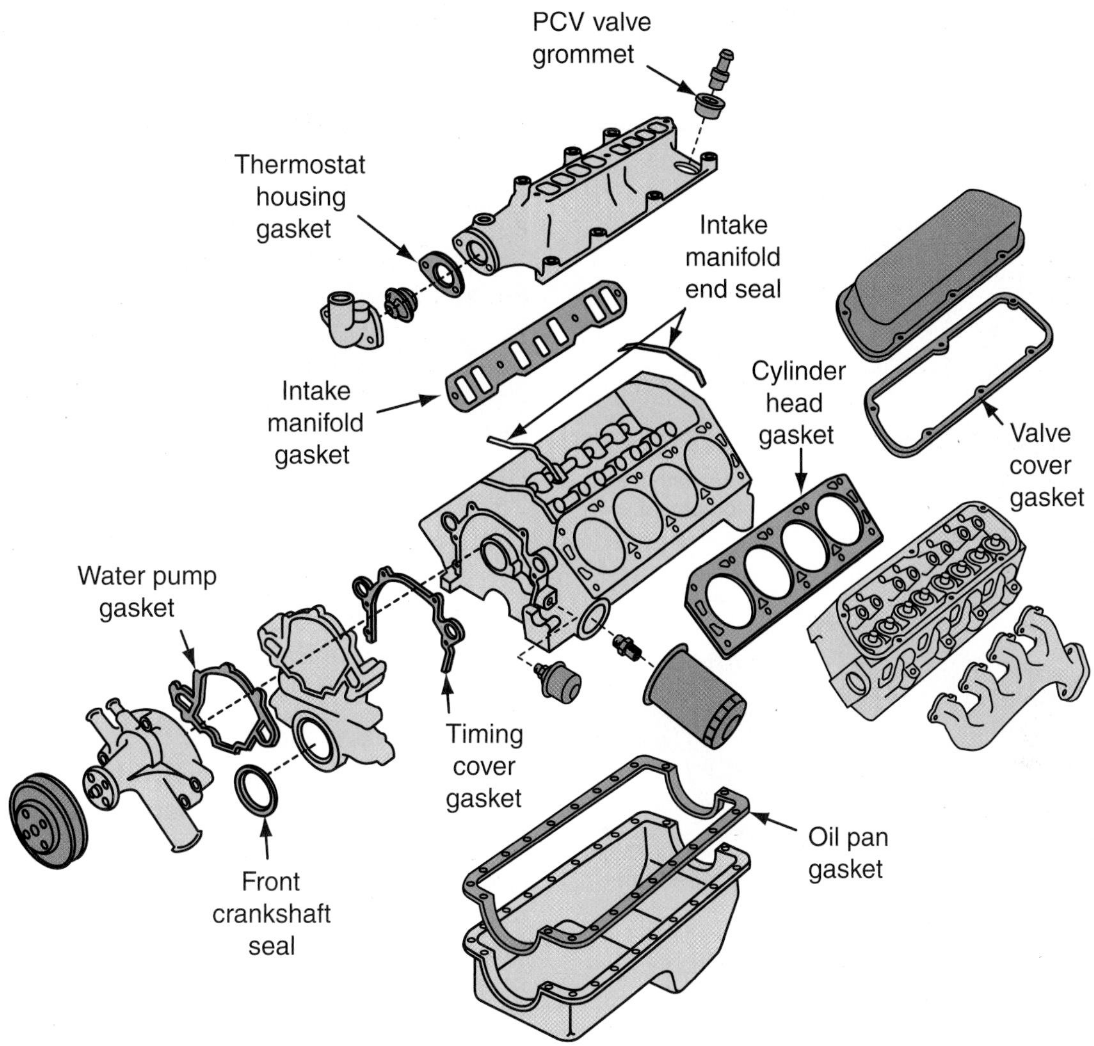

Figure 13-7 Typical engine gasket and seal locations.

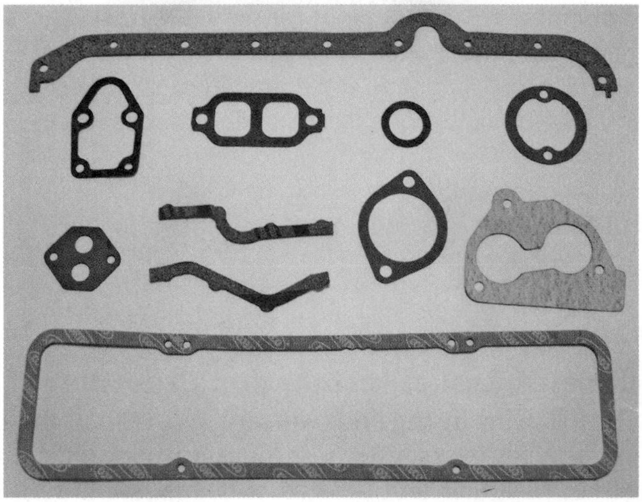

Figure 13-8 An assortment of cut gaskets.

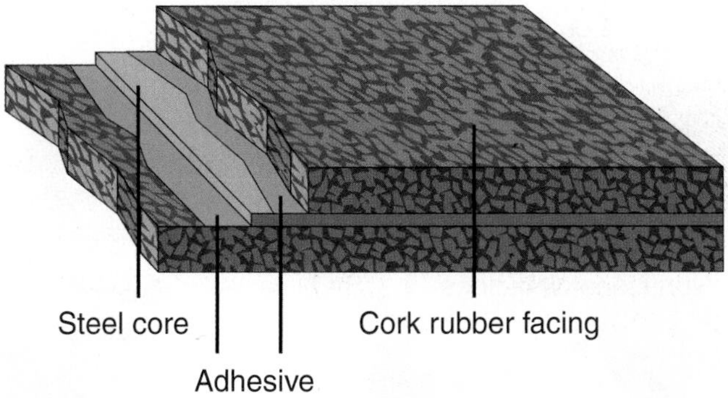

Figure 13-9 Composition of a cork/rubber gasket used mainly as valve cover, oil pan, and timing cover gaskets. *Courtesy of Dana Corporation*

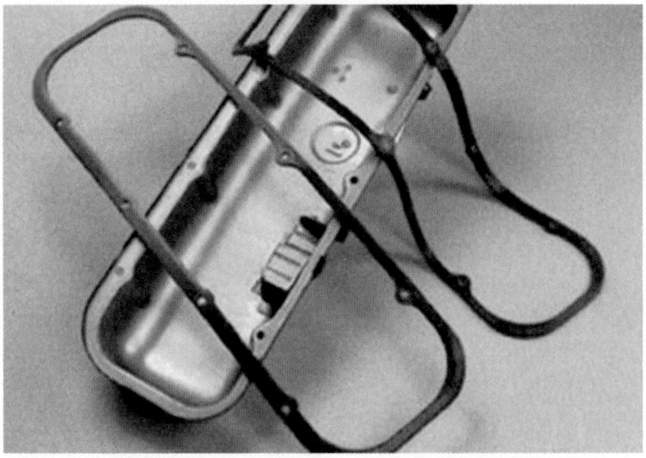

Figure 13–10 (Left) A valve cover gasket with a rigid core. (Right) A valve cover gasket without reinforcement. *Courtesy of Federal-Mogul Corporation*

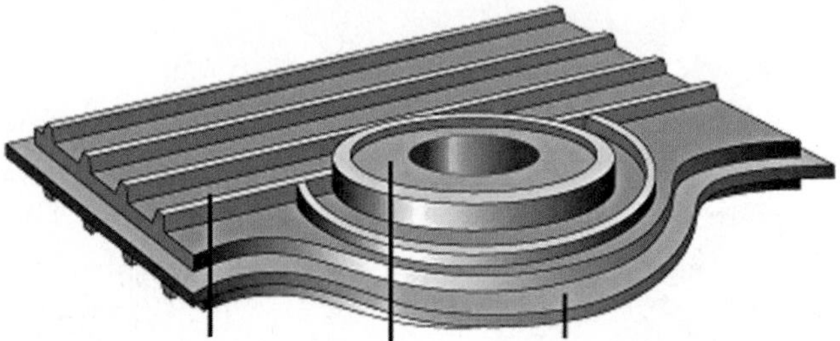

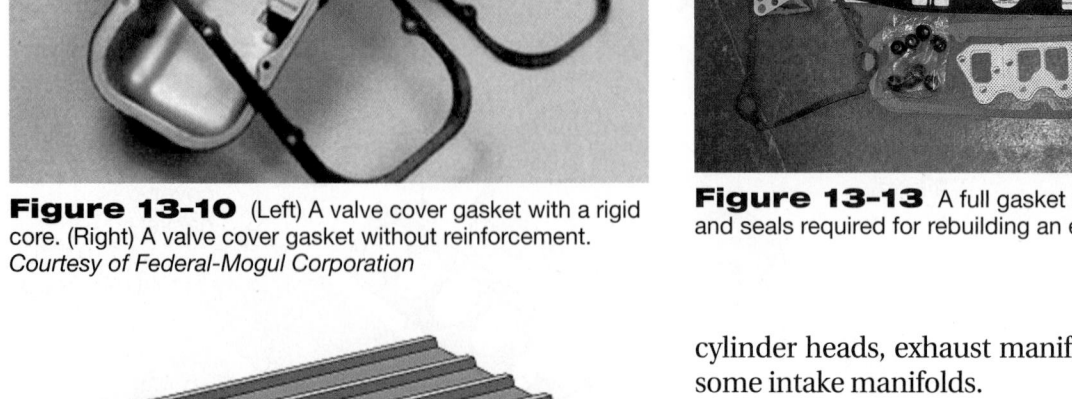

Figure 13–11 Composition of a molded rubber gasket with steel grommets used mainly as valve cover and oil pan gaskets. *Courtesy of Dana Corporation*

bead of sealant around the sealing area. The result is called a "formed-in-place" gasket.

Hard Gaskets

Hard gaskets are made from steel, stainless steel, copper, or a combination of metals and other materials. Often the metal is enclosed by a compressible and heat-resistant clay/fiber or Teflon® compound **(Figure 13–12)**. Hard gaskets are typically for

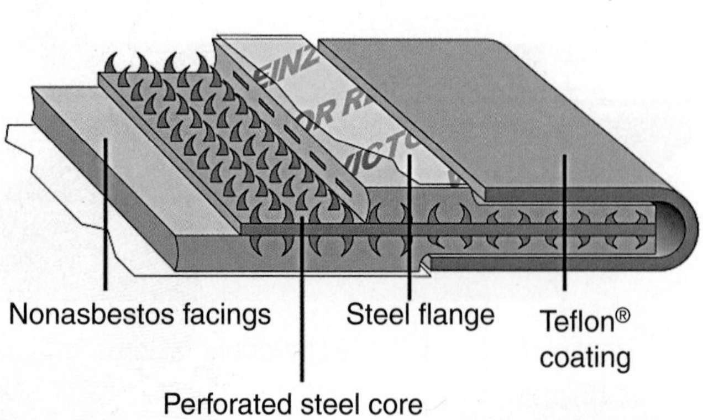

Figure 13–12 Composition of a Teflon®-coated perforated steel core gasket used mainly as head and intake manifold gaskets. *Courtesy of Dana Corporation*

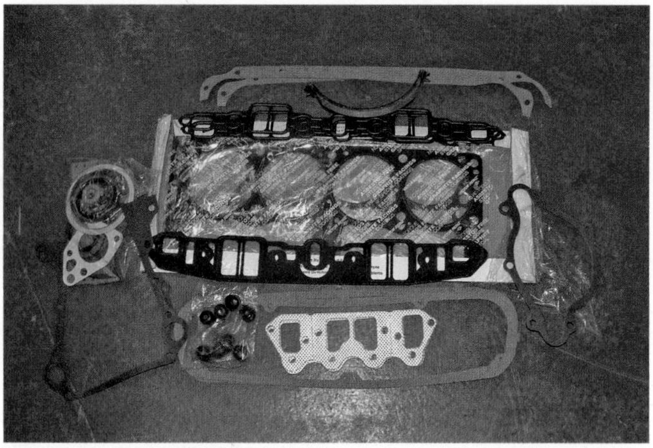

Figure 13–13 A full gasket set contains the gaskets and seals required for rebuilding an engine.

cylinder heads, exhaust manifolds, EGR valves, and some intake manifolds.

Replacement Gaskets

Gaskets can be purchased individually or in sets. Sets are often available for the service being performed on the engine. The most common are timing cover, head, manifold, oil pan, and full sets. A full set contains all gaskets and seals required for rebuilding an engine **(Figure 13–13)**. Normally there are more gaskets in a set than are needed. A particular engine may have been available with different equipment and the extra gaskets are for those variations. All gaskets need to be properly installed to provide a good, durable seal.

SHOP TALK

Before replacing a faulty gasket, check for any TSBs on the engine. There may be a recommended replacement for the OEM gasket. For example, some GM engines built between 1996 and 2002 have experienced premature intake manifold gasket failures. These failures are normally caused by the corrosive effects of organic acid technology (OAT) coolant. The cure is a replacement gasket that is less susceptible to OAT.

General Gasket Installation Procedures

The following instructions will serve as a helpful guide for installing gaskets. Because there are many different gasket materials and designs, it is impossible to list directions for every type of installation. Always follow any special directions provided in the instructions packed with most gasket sets.

1. Never reuse old gaskets. Even if the old gasket appears to be in good condition, it will never seal as well as a new one.
2. Handle new gaskets carefully. Be careful not to damage the new gaskets before placing them on the engine.

USING SERVICE INFORMATION

Always refer to the specific engine and engine part section of the service manual for the recommended procedures for using sealants.

SHOP TALK

Some technicians tend to use too much sealant on gaskets. Do not make this mistake. Because sealants have less strength than gasket materials, they create weaker joints. They can also prevent some gasket material from doing what it is supposed to do, which is to soak up oil and swell to make a tight seal.

3. Cleanliness is essential. New gaskets seal best when used on clean surfaces. Thoroughly clean all mating surfaces of dirt, oil deposits, rust, old sealer, and gasket material.
4. Use the right gasket in the right position. Always compare the new gasket to the mating surfaces to make sure it is the right gasket. Check that all bolt holes, dowel holes, coolant, and lubrication passages line up perfectly with the gasket. Some gaskets have directions such as "top," "front," or "this side up" stamped on one surface **(Figure 13–14)**. An upside-down or reversed gasket can easily cause a loss of oil pressure, overheating, and engine failure.
5. Use sealants and adhesives only when the engine or gasket manufacturer recommends their use. Some chemicals will react negatively with the gasket's coating. Normally if a gasket requires a sealant, it will be included with the gasket.
6. Make sure the threads of all fasteners are clean and undamaged.

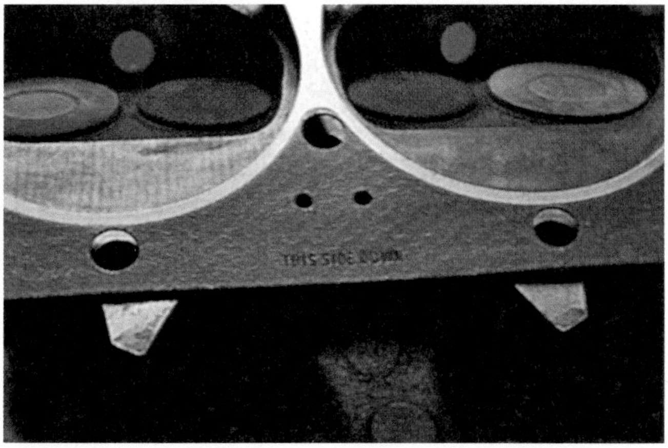

Figure 13–14 Some gaskets have installation directions stamped on them. *Courtesy of Detroit Gaskets*

SPECIFIC ENGINE GASKETS

There is a wide variety of gaskets used on engines, each with its own purpose, and each is designed for that application. The following is a discussion of the most common ones.

Cylinder Head Gaskets

Cylinder head gaskets seal the gap between the cylinder head and the block **(Figure 13–15)**. The head gasket has a very demanding job. It must seal the combustion chambers and the coolant and oil passages between the head and the block. The oil and coolant are low-temperature and low-pressure fluids. The gasket must prevent leakage inside and outside the engine. It must be able to do this while sustaining wide temperature ranges and pressures inside each cylinder. When a cold engine is first started, parts near the combustion chamber are very cold. Then, after only a few minutes of running, these same parts might reach 400°F (204°C). The inner edges of the head gasket are exposed to combustion temperatures from 2,000°F to 4,000°F (1,093°C to 2,204°C). Pressures in the combustion chamber also vary tremendously. On

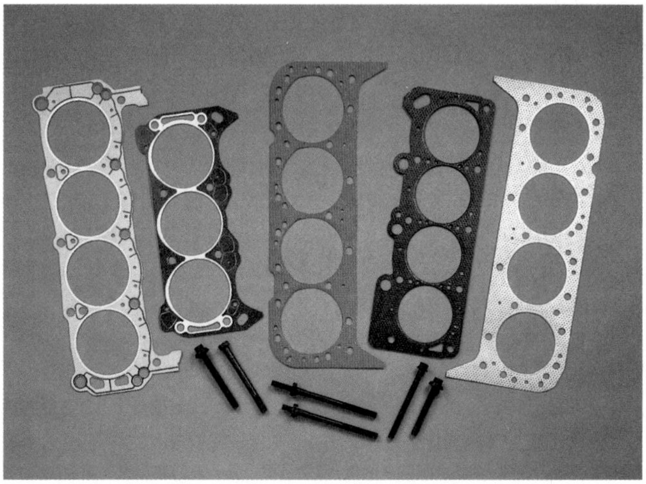

Figure 13–15 Different designs of cylinder head gaskets and replacement head bolts. *Courtesy of Detroit Gaskets*

the intake stroke a vacuum or negative pressure is present. After combustion, pressure peaks of approximately 1,000 psi (6,895 kPa) occur. This extreme change from low to high pressure happens in a fraction of a second.

Cylinder head gaskets must simultaneously do the following:

- Seal intake stroke vacuum, combustion pressure, and the heat of combustion.
- Prevent coolant leakage and resist rust, corrosion, and, in many cases, meter coolant flow.
- Seal oil passages through the block and head while resisting chemical action.
- Allow for lateral and vertical head movement as the engine heats and cools.
- Be flexible enough to seal minor surface warpage while being stiff enough to maintain adequate gasket compression.
- Fill small machining marks that could lead to serious gasket leakage and failure.
- Withstand forces produced by engine vibration.

Head gaskets for many late-model cast-iron engines have raised silicone, **Viton**, or fluoroelastomer sealing beads on their face to increase the clamping pressure around troublesome areas. Most head gaskets have a steel fire ring that surrounds the top of the cylinder. These protect the material used elsewhere in the gasket. The durability of a head gasket can also be improved by using strong, high-temperature fibers such as aramid and kevlar and by adding reinforcements around oil passages.

Bimetal Engine Requirements Most late-model engines have aluminum cylinder heads and cast-iron blocks. When heated, aluminum expands two to three times more than steel. This creates a back-and-forth scrubbing action on a head gasket as the engine temperature changes (**Figure 13–16**). This movement can tear a gasket apart if it is not designed to handle it. To reduce the chances of a gasket tearing, graphite or specially coated gaskets are used. **Graphite** is a relatively soft material that can withstand high temperatures. It is a natural lubricant, so graphite-faced gaskets are not coated. Teflon, molybdenum, and other similar slippery nonstick coatings are used on other gasket designs to prevent the gasket from sticking to either surface. This allows the head to expand and contract without destroying the gasket.

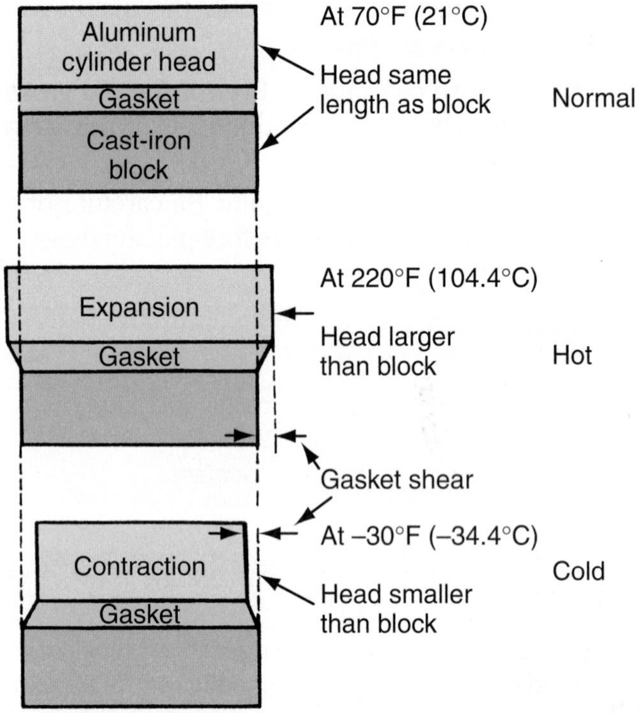

Figure 13–16 Thermal growth characteristics of bimetal engines.

Multilayer Steel (MLS) Many engines have **multilayer steel (MLS)** head gaskets. These gaskets are comprised of three to seven layers of steel. The outer layers are normally embossed stainless spring steel coated with a thin layer of an antifriction coating of Teflon, nitrile rubber, or Viton (**Figure 13–17**). The inner layers provide the necessary thickness. The use of an MLS gasket reduces the load on the head bolts and allows them to retain their shape after they are tightened in place. They are also very durable. MLS gaskets require an extremely smooth finish on the engine block and cylinder head. They are also used with TTY bolts.

SHOP TALK

One of the problems with resurfacing the head on an OHC engine is that cam timing is affected. This is caused by the fact that the camshaft is now closer to the crankshaft. Some aftermarket gasket companies offer an MLS gasket set made up of top and bottom layers and several steel spacer layers. The spacers fit in the middle of the outer layers to achieve the required thickness. These stackable head gaskets can also be used to lower compression when a turbocharger or supercharger is installed.

Head Gasket Failures When a head gasket has failed, it is important to correct the problem that caused it.

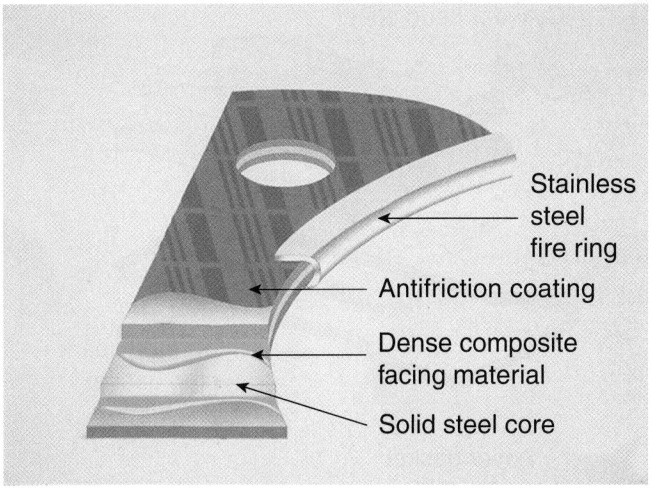

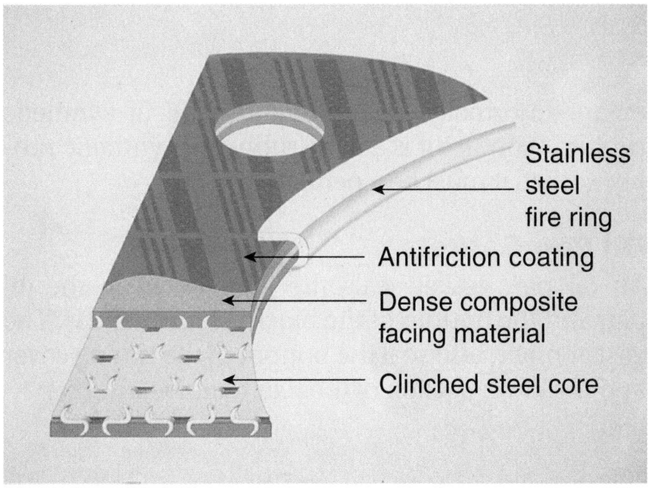

Figure 13–17 Two examples of the construction of an MLS head gasket. *Courtesy of Federal-Mogul Corporation*

Figure 13–18 shows the common causes of failure and the systems that should be checked.

Some engines will have hot spots that cannot be corrected. These engines typically have exhaust ports that are located next to each other. Heat builds up in these areas and causes the head to swell and crush the head gasket. Gasket manufacturers often incorporate reinforcements in those areas to resist the crushing.

SHOP TALK

It is important to find out why a head gasket has failed. When inspecting a gasket, measure its thickness at the damaged and undamaged areas. If the damaged area is thinner, the gasket failed because of overheating or a hot spot. If the fire ring around the cylinder bores is cracked or burnt, preignition or detonation caused the gasket to fail.

Problem	Probable Cause
Preignition/Detonation	Incorrect ignition timing
	Incorrect air-fuel mixture
	Vacuum leak
	Faulty cooling system
Overheating	Restricted radiator
	Cooling system leak
	Faulty thermostat
	Faulty water pump
	Inoperative cooling fan
	Faulty EGR system
Improper Installation	Wrong surface finish
	Incorrect bolt-tightening sequence
	Use of stretched or damaged bolts
	Improper use of sealant
	Use of incorrect gasket
	Dirty mating surfaces
Hot Spots	Use of incorrect gasket

Figure 13–18 Leading causes of cylinder head gasket failure.

Head Bolts Installation failures are commonly caused by head bolt problems. When installing a cylinder head and bolts, follow these guidelines:

- Make sure all head bolts are clean and have undamaged threads. Replace any bolts that are nicked, deformed, or worn.
- Make sure the correct length bolt is installed in the bore. Some engines use longer bolts in some locations.
- Check the length of each bolt and compare the measurements. Longer than normal bolts are stretched and should not be reused.
- Inspect the shank or top of the threaded area of the bolt for evidence of stretching.
- Never reuse TTY bolts.
- When installing an aluminum head, use hardened steel washers under the bolts. Place the washers so their rounded edge faces up.
- Clean the thread bores in the engine block with a bottoming tap.
- Make sure the top of each thread bore is chamfered.
- If the head has been resurfaced, make sure the bolts do not bottom out in their bores. If they do, install hardened steel washers under the bolt heads to raise them, or use a head gasket shim to bring the head to its proper height.

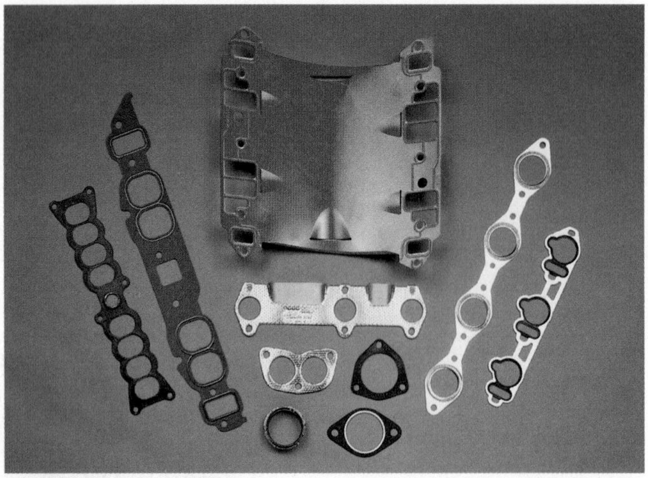
Figure 13-19 Assorted intake and exhaust manifold gaskets for diverse applications. *Courtesy of Detroit Gaskets*

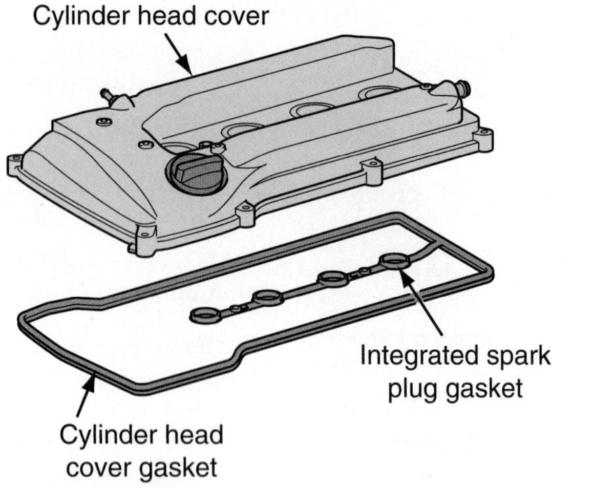

Figure 13-20 A cylinder cover gasket with integrated spark plug gaskets. *Courtesy of Toyota Motor Sales, U.S.A., Inc.*

- Lubricate the threads and the bottom of the bolt head with engine oil.
- Apply the correct type of thread sealant to all bolts that go into a coolant passage.

Manifold Gaskets

There are three basic types of manifold gaskets—the intake manifold, exhaust manifold, and a combination intake and exhaust. Combination gaskets are often used on inline engines without cross-flow heads. Manifold gaskets are made of a variety of materials, depending on the application. Each type of manifold gasket has its own sealing characteristics and problems **(Figure 13–19)**. Therefore, be sure to follow the manufacturer's instructions when installing them.

Before installing a manifold and its gasket, make sure that the mating surfaces on the head and manifold are flat and free of damage. Also, always follow the recommended bolt-tightening sequence and use the exact torque specs.

Valve Cover Gaskets

Valve cover gaskets must seal between a steel, aluminum, magnesium, or molded plastic cover and a cylinder head surface that might not be machined. On OHC engines, the valve cover may be called the cam or cylinder head cover. Cam covers are normally made of die-cast aluminum. Some cylinder cover gaskets have spark plug gaskets integrated into the gasket **(Figure 13–20)**. When installing these, make sure the gasket is perfectly aligned. Valve cover bolts are usually widely spaced so the gasket material must be able to seal without being tightly clamped. Valve cover gaskets must be able to withstand high temperatures and the caustic action of acids in the oil.

On older engines, the most commonly used material is a blend of cork and rubber particles. Late-model engines normally have gaskets made of synthetic rubber. Whether it is a cork/rubber or synthetic rubber gasket, it must be a perfect fit.

Oil Pan Gaskets

An oil pan gasket seals the joint between the oil pan and the bottom of the block **(Figure 13–21)**. The gasket might also seal the bottom of the timing cover and the lower section of the rear main bearing cap.

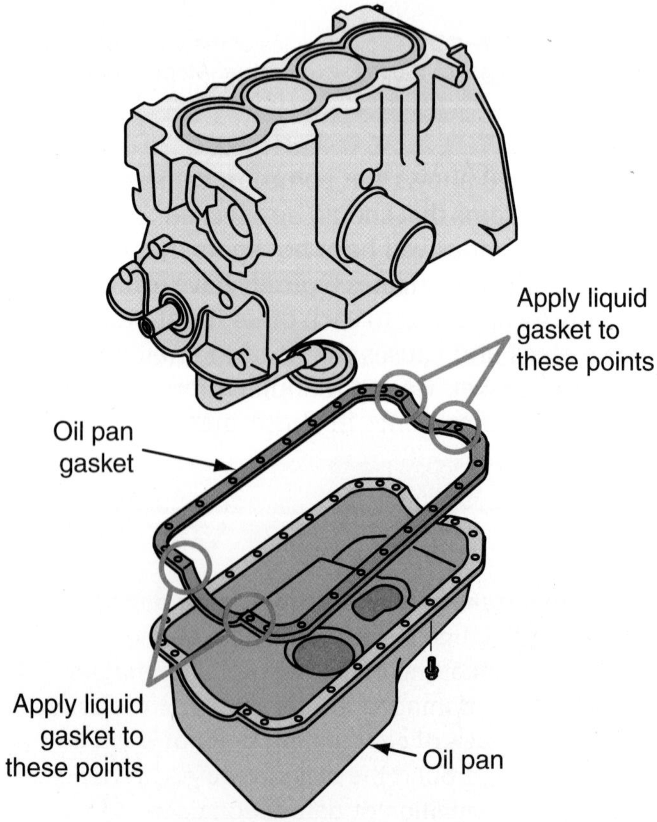

Figure 13-21 Oil pan gasket installation; note the points where a liquid gasket is required.

Like valve cover gaskets, the oil pan gasket must resist hot, thin engine oil. Oil pans are usually made of stamped steel, cast iron, or cast aluminum. Because of the added weight and splash of crankcase oil, the pan has many assembly bolts closely spaced. As a result, the clamping force on the oil pan gasket is great.

Oil pan gaskets are made of several types of material. A commonly used material is synthetic rubber, known for its long-term sealing ability. It is tough, durable, and resists hot engine oil. Synthetic rubber gaskets are easy to remove, so the sealing surfaces need less cleanup.

Many late-model engines use a hard gasket with a bead of sealant around the inside dimensions of the gasket **(Figure 13–22)**. The bead increases clamping pressures and provides a positive seal.

The oil pan on some engines is comprised of two parts: the upper unit, which is made of aluminum or magnesium, and the lower unit, which is made of stamped steel **(Figure 13–23)**. A gasket is used to seal the two units. Many engines also have a baffle assembly that must be bolted to the crankcase before the oil pan is installed.

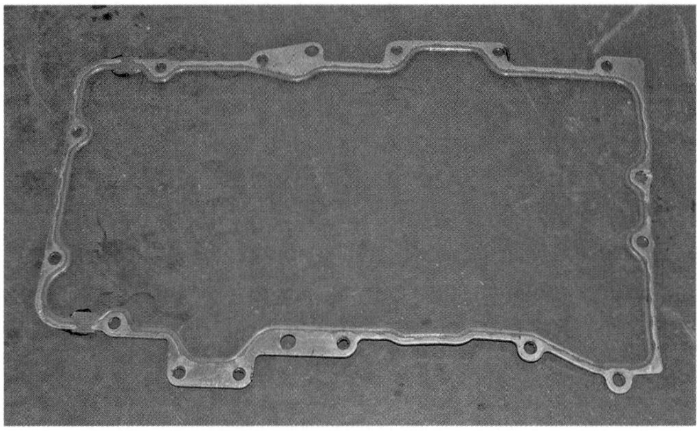

Figure 13-22 A hard oil pan gasket with a sealant bead.

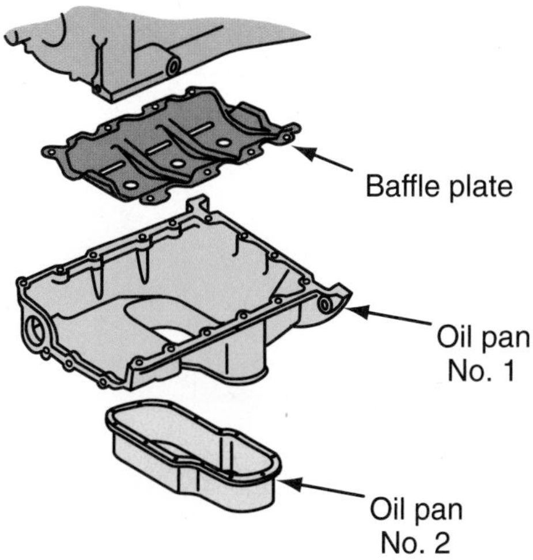

Figure 13-23 A two-piece oil pan assembly with a baffle plate.

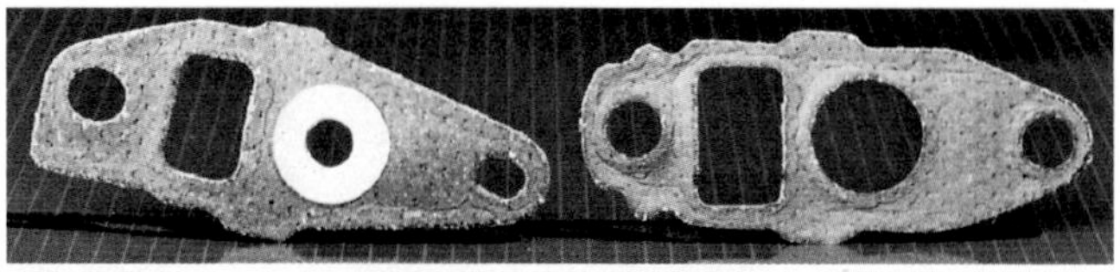

Figure 13-24 Typical EGR valve gaskets. *Courtesy of Federal-Mogul Corporation*

Carefully follow the recommendations of the OEM and gasket manufacturer. Before installing a stamped steel oil pan, make sure its flange is flat. Tighten the oil pan bolts according to specifications and in the sequence given in the service manual.

EGR Valve

The exhaust gas recirculation (EGR) valve takes a sample of the exhaust gases and introduces it back into the cylinders. This reduces combustion temperatures and prevents NO_x from forming. The sealing surface of the valve should be carefully inspected. Use a file to remove any minor imperfections that may prevent the valve from sealing. Also make sure that the new gasket is the correct size; some applications have specifically sized holes in the gasket that are used to regulate exhaust flow **(Figure 13–24)**. Using the wrong gasket could change how the engine performs.

ADHESIVES, SEALANTS, AND OTHER CHEMICAL SEALING MATERIALS

A number of chemicals can be used to reduce labor and ensure a good seal. Many gasket sets include a label with the recommended chemical to use with that gasket set. Some even include sealers when the OEM used a sealer to replace a gasket and a gasket cannot be manufactured for that application. They also include sealers when gasket unions need a sealant to ensure a good seal.

SHOP TALK

Chemical adhesives and sealants give added holding power and sealing ability where two parts are joined. Sealants usually are added to threads where fluid contact is frequent. Chemical thread retainers are either **aerobic** (cures in the presence of air) or **anaerobic** (cures in the absence of air). These chemical products are used in place of lock washers.

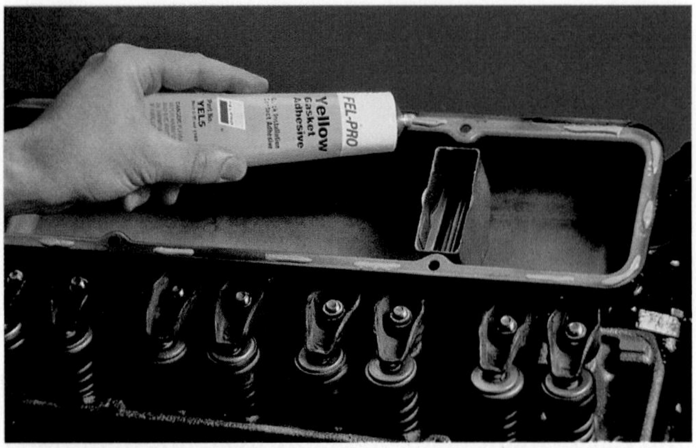

Figure 13–25 An adhesive is often used to hold a gasket in place during assembly. *Courtesy of Federal-Mogul Corporation*

Figure 13–26 Applying gasket sealer with a brush. *Courtesy of Federal-Mogul Corporation*

Adhesives

Gasket adhesives form a tough bond when used on clean, dry surfaces. Adhesives do not aid the sealing ability of the gasket. They are meant only to hold gaskets in place during assembly. Use small dabs. Do not assemble components until the adhesive is completely dry **(Figure 13–25)**.

Sealants

Manufacturers sometimes specify the use of sealants to assist a gasket or seal or to form a new gasket. These sealants should only be used when specified by the manufacturer. Also make sure to use the specific sealant recommended by the manufacturer.

WARNING!

Never use a sealant or liquid gasket material on exhaust manifolds.

General-Purpose Sealants General-purpose sealers are liquid and available in a brush type (known as brush tack) or an aerosol type (known as spray tack). General-purpose sealers **(Figure 13–26)** form a tacky, flexible seal when applied in a thin, even coat that aids in gasket sealing by helping to position the gasket during assembly.

WARNING!

Make sure every sealant you use on today's engines is oxygen sensor safe.

WARNING!

Never use a hard-drying sealant (such as shellac) on gaskets. It will make future disassembly extremely difficult and might damage the gasket.

Thread Sealants Bolts that pass through a liquid passage should be coated with Teflon thread **(Figure 13–27)** or a brush-on thread sealant. Some head bolts or water pump bolts tighten into a coolant passage and must be sealed or they will leak. These flexible sealants are nonhardening sealers that fill voids, preventing the fluid from running up the threads. They resist the chemical attack of lubricants, synthetic oils, detergents, antifreeze, gasoline, and diesel fuel.

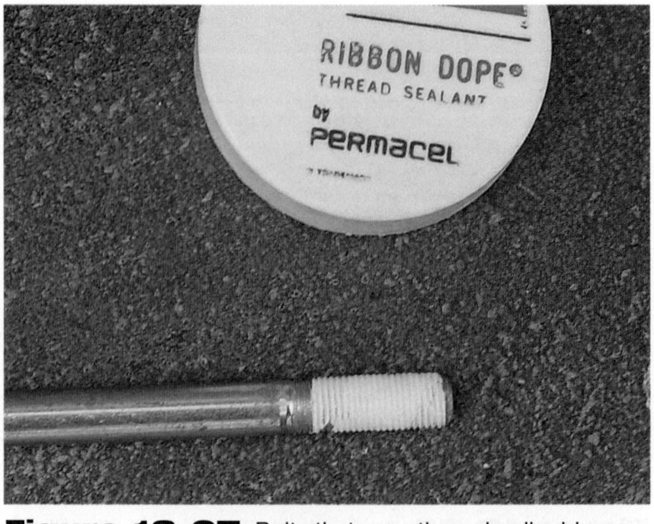

Figure 13–27 Bolts that pass through a liquid passage should be coated with Teflon or similar thread sealant.

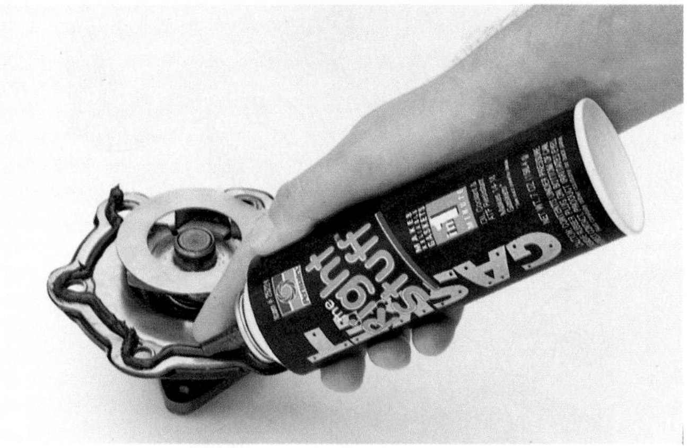

Figure 13-28 Applying a bead of RTV gasket material on a water pump. *Courtesy of Loctite Corporation*

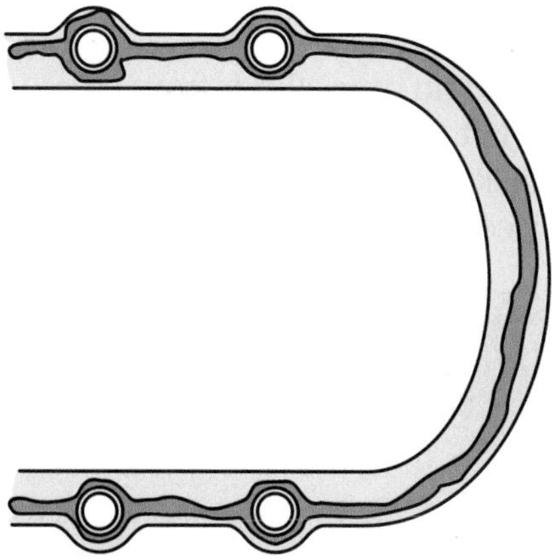

Figure 13-29 When applying RTV, make sure the bolt holes are encircled.

Silicone Sealants **Silicone** (or formed-in-place) **gaskets** are a liquid sealant applied directly to mating surfaces and allowed to cure in place. Many technicians use silicone gasketing to aid in sealing the corners, notches, or dovetails of gaskets. **Room temperature vulcanizing (RTV)** silicone sealing products are the most commonly used formed-in-place (FIP) gasket products. RTV is an aerobic sealant, which means it cures or hardens in the presence of air. RTV can be used to seal two stationary parts such as water pumps **(Figure 13–28)** and oil pans. It cannot be used as a head or exhaust gasket or in fuel systems. It should also not be used to coat gaskets. RTV comes in a variety of colors that denote the proper application. Black is for general purpose; blue is for special applications; and red is for high-temperature requirements. Always use the correct type for the application. Also make sure that the RTV is oxygen sensor safe. RTV silicone sealants are impervious to most fluids, are extremely resistant to oil, have great flexibility, and adhere very well to most materials.

WARNING!

Be careful not to use excessive amounts of RTV. If too much is applied, it can loosen up and get into the oil system where it can block an oil passage and cause severe engine damage.

To use RTV, make sure the mating surfaces are free of dirt, grease, and oil. Apply a continuous ⅛-inch bead on one surface only (preferably the cover side). Encircle all bolt holes **(Figure 13–29)**. Adjust the shape before a skin forms (in about 1 minute), then remove the excess RTV with a dry towel or paper towel. Press the parts together. Do not slide the parts together—this will disturb the bead. Tighten all retaining bolts to the specified torque.

CAUTION!

The uncured rubber contained in RTV silicone gasketing irritates the eyes. If any gets in your eyes, immediately flush with clean water or eyewash. If the irritation continues, see a doctor.

Anaerobic Sealants These materials are used for thread locking as well as gasketing. They are mostly used to hold sleeves, bearings, and locking screw nuts in places subject to much vibration. Anaerobic sealers are intended to be used between the machined surfaces of rigid castings, not on flexible stampings. Use anaerobic sealants only when specified.

SHOP TALK

Once hardened, an anaerobic bond is unbelievably tenacious and can withstand high temperatures. Different anaerobic sealants are for specific purposes and not readily interchangeable. For example, thread-locking products range from medium-strength antivibration agents to high-strength, weldlike retaining compounds. The inadvertent use of the wrong product could make future disassembly impossible.

Antiseize Compounds

Antiseize compounds prevent dissimilar metals from reacting with one another and seizing. This material is used on many fasteners, especially those used with aluminum parts. Always follow the manufacturer's recommendations when using this compound.

OIL SEALS

Seals keep oil and other fluids from escaping around a rotating shaft. There are three basic oil seal designs: the fiber packing, the two-piece split lip design, and the one-piece radial design **(Figure 13–30)**.

> **SHOP TALK**
>
> Whenever installing an oil seal, make sure its lip seal is lubricated with a light coating of grease. Also, make sure the lip portion of the seal is facing the direction that oil is coming against.

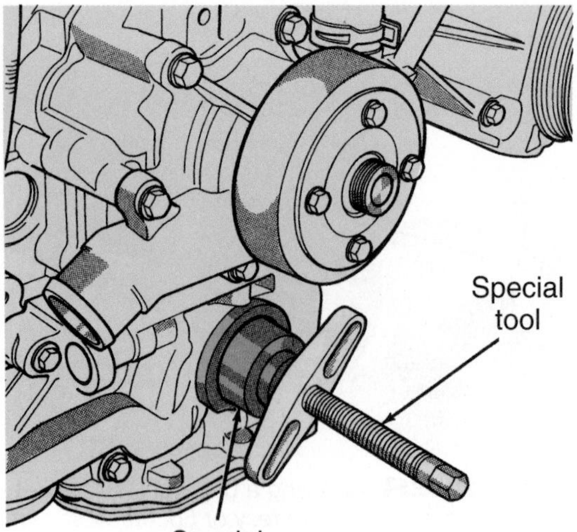

Figure 13-31 The installation of the timing cover oil seal requires the use of a special tool or driver. *Courtesy of Chrysler LLC*

Timing Cover Oil Seals

An oil seal in the timing cover prevents oil from leaking around the crankshaft. Its installation often requires the use of a special tool **(Figure 13–31)** or a driver. It is important that the seal be positioned squarely in the bore of the cover and the crankshaft be positioned in the center of the seal.

Rear Main Bearing Seals

Rear main bearing seals keep oil from leaking around the rear main bearing. There are two basic types: wick- or rope-type packing and molded synthetic rubber.

Wick- or rope-type packings are common on many older engines. Molded synthetic rubber lip-type seals are used on many newer engines. They do a good job of sealing as long as the surface of the shaft is very smooth. Synthetic rubber seals may be retrofitted to some older engines that have wick seals.

Three types of synthetic rubber are used for rear main bearing seals. **Polyacrylate** is commonly used because it is tough and abrasion resistant, with moderate temperature resistance to 350°F (177°C). Silicone synthetic rubber has a greater temperature range, but it has less resistance to abrasion and is more fragile than polyacrylate. Silicone seals must be handled carefully during installation to avoid damage. Viton has the abrasion resistance of polyacrylate and the temperature range of silicone, but it is the most expensive of the synthetic types. Synthetic rubber seals may be one piece **(Figure 13–32)** or two pieces **(Figure 13–33)**.

Always check the shaft for smoothness. Shafts should be free of nicks and burrs. Carefully remove any roughness with a very fine emery cloth and clean the shaft thoroughly. The shaft should have a highly polished appearance and a smooth feel. Also, be sure to check and clean the oil slinger and oil return channel in the bearing cap.

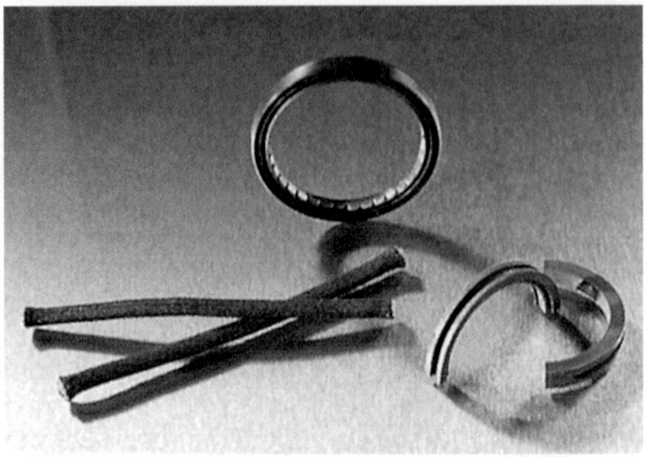

Figure 13-30 The three basic oil seal designs. *Courtesy of Detroit Gaskets*

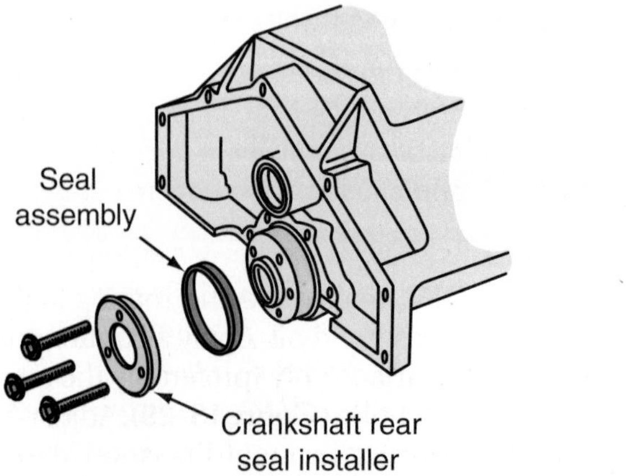

Figure 13-32 Installing a one-piece rubber rear crankshaft seal.

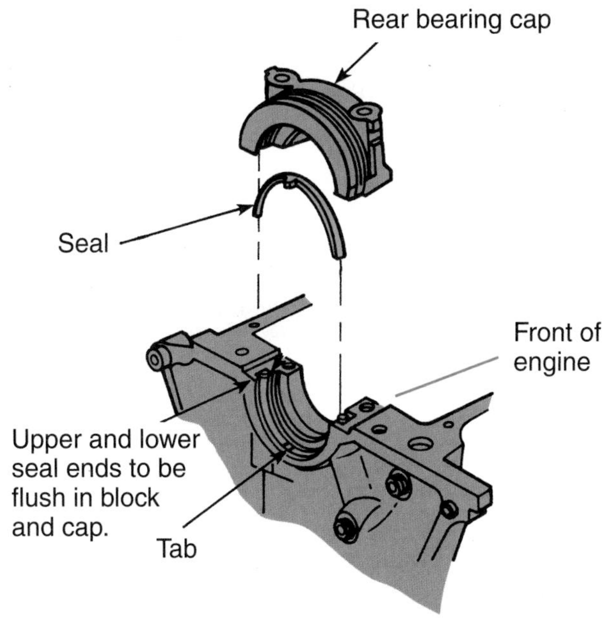

Figure 13-33 Typical two-piece rubber rear crankshaft seal.

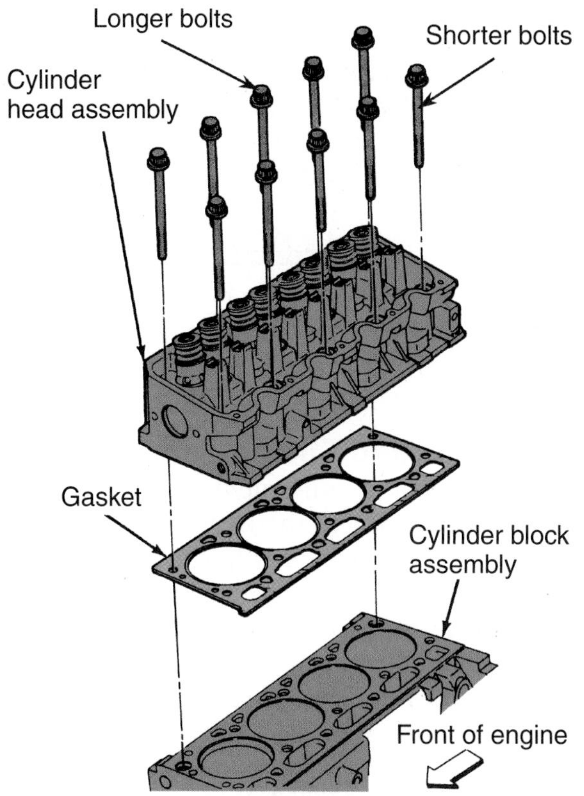

Figure 13-34 Head bolts are different lengths in some engines. *Courtesy of Ford Motor Company*

ENGINE REASSEMBLY

When reassembling an engine, the sequence is essentially the reverse of the teardown sequence given in a previous chapter. Always refer to the service manual before assembling an engine.

Installing the Cylinder Head and Valve train

Use a wire brush to clean the threads of the head bolts. Then check their condition and length. Many engines use head bolts of different lengths **(Figure 13–34)** and their location is given in the service manual. Lightly lubricate the threads with clean engine oil.

Position the head gasket on the block and make sure it matches the bores in the block. Place the cylinder head onto the block. Make sure that the dowel pins **(Figure 13–35)** are in place and that the head and block are properly aligned.

Tighten the head bolts according to the recommended sequence **(Figure 13–36)** and to the specified torque. Most heads are tightened in a sequence that starts in the middle and moves out to the ends. The bolts are generally tightened in two or three stages.

Before inserting the pushrods, apply some assembly lube to both ends. Liberally coat the rocker arms with assembly lube or clean engine oil. Then install the pushrods and rocker arms. Many OHV engines have positive stop rocker arm adjustments. This means that when torquing the rocker arms to spec, the plunger of the hydraulic lifter is properly positioned, giving the correct lifter adjustment.

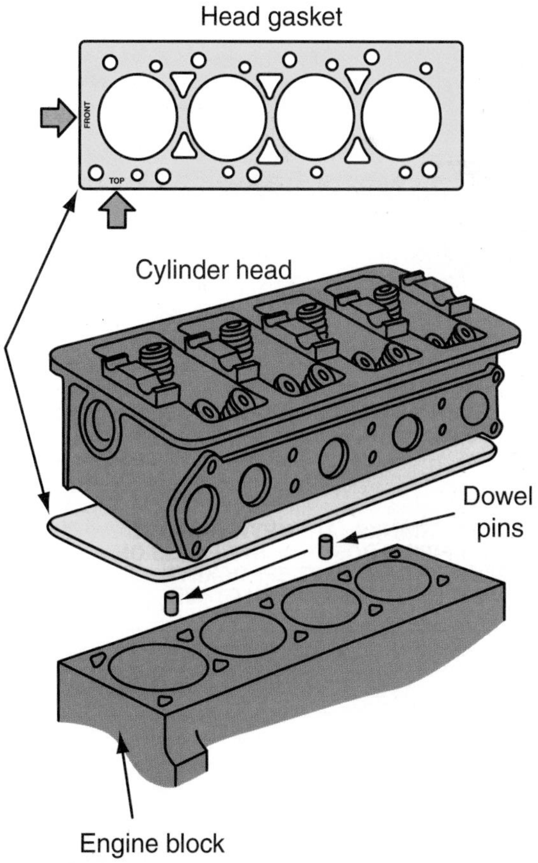

Figure 13-35 When installing a cylinder head onto an engine block, make sure that the dowel pins are securely installed in the block and that the head gasket is in the correct position.

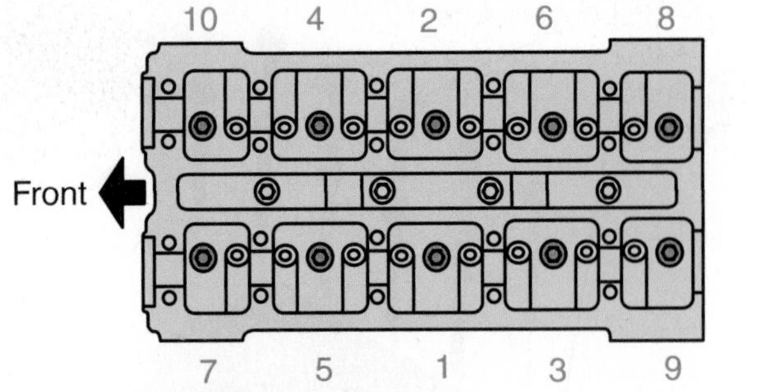

Figure 13-36 Always follow the specified tightening sequence when torquing cylinder head bolts.

Timing Belts and Chains

The alignment of the camshaft(s) with the crankshaft is critical. The alignment marks and the correct procedure for doing this vary with the engine design. Many engines have additional chains or belts for the balance shafts and oil pump. DOHC engines may have a chain that connects the two camshafts. All of these must be properly timed. Check the service manual for the correct alignment and tension adjustment procedures **(Figure 13–37)**. Make sure that the belt or chain tensioners are set properly.

> **USING SERVICE INFORMATION**
>
> Normally, camshaft timing marks are shown in the engine section of a service manual under the heading of Timing Belt or Chain R&R.

Timing belts can be replaced with the engine in the vehicle or when the engine is on a stand. Photo Sequence 10 shows a typical procedure for changing a timing belt with the engine in a vehicle.

After the belt is installed, adjust its tension according to the manufacturer's recommendations. Then, by turning the crankshaft, rotate the engine through two complete turns and recheck the tension. Now rotate the engine through at least more revolutions. Recheck the timing marks on the crankshaft and

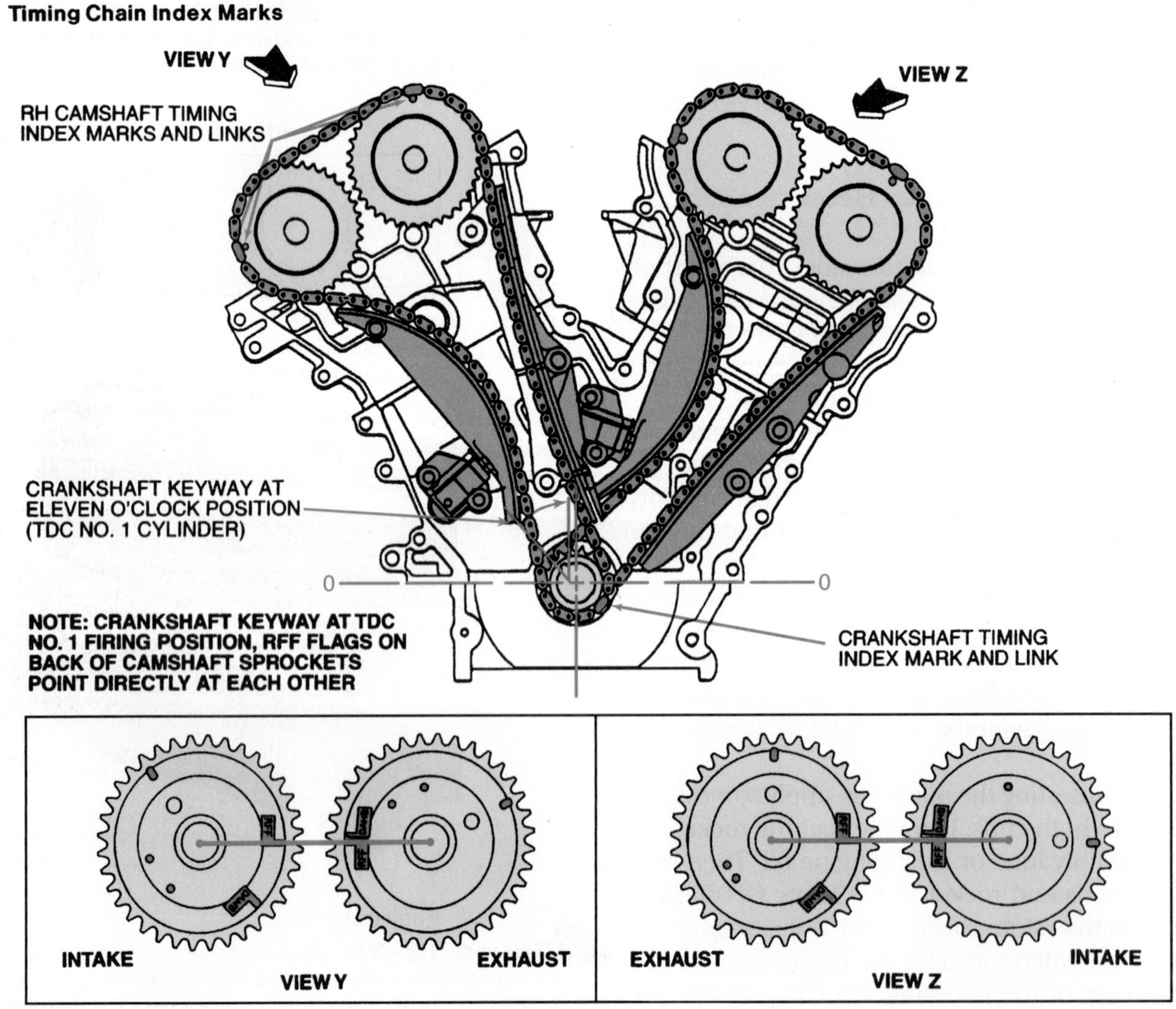

Figure 13-37 Note the timing marks and position of the chain tensioners on this DOHC engine.

P10-1 *Disconnect the negative cable from the battery prior to removing and replacing the timing belt.*

P10-2 *Carefully remove the timing cover. Be careful not to distort or damage it while pulling it up. With the cover removed, check the immediate area around the belt for wires and other obstacles. If some are found, move them out of the way.*

P10-3 *Align the timing marks on the camshaft's sprocket with the mark on the cylinder head. If the marks are not obvious, use a paint stick or chalk to clearly mark them.*

P10-4 *Carefully remove the crankshaft timing sensor and probe holder.*

P10-5 *Loosen the adjustment bolt on the belt tensioner pulley. It is normally not necessary to remove the tensioner assembly.*

P10-6 *Slide the belt off the crankshaft sprocket. Do not allow the crankshaft pulley to rotate while doing this.*

Replacing a Timing Belt on an OHC Engine *(continued)*

P10-7 *To remove the belt from the engine, the crankshaft pulley must be removed. Then the belt can be slipped off the crankshaft sprocket.*

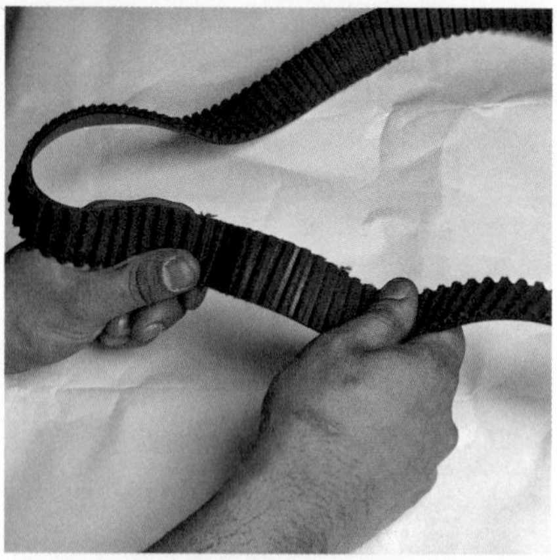

P10-8 *After the belt has been removed, inspect it for cracks and other damage. Cracks will become more obvious if the belt is twisted slightly. Any defects in the belt indicate it must be replaced.*

P10-9 *To begin reassembly, place the belt around the crankshaft sprocket. Then reinstall the crankshaft pulley.*

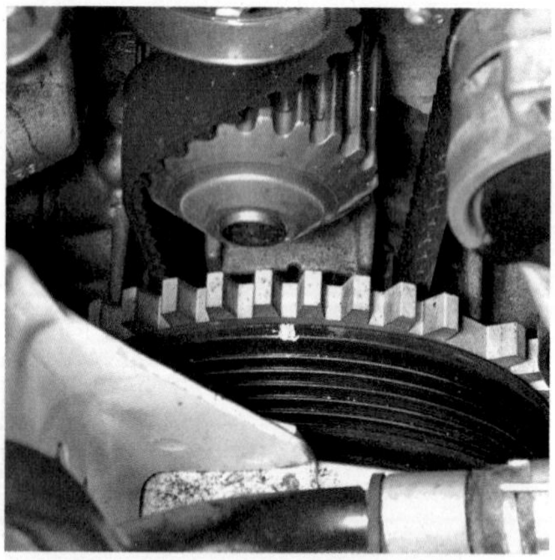

P10-10 *Make sure the timing marks on the crankshaft pulley are lined up with the marks on the engine block. If they are not, carefully rock the crankshaft until the marks are lined up.*

P10-11 *With the timing belt fitted onto the crankshaft sprocket and the crankshaft pulley tightened in place, the crankshaft timing sensor and probe can be reinstalled.*

P10-12 *Align the camshaft sprocket with the timing marks on the cylinder head. Then wrap the timing belt around the camshaft sprocket and allow the belt tensioner to put a slight amount of pressure on the belt.*

camshaft. If necessary, readjust the timing and tension.

Adjusting Valves

Nearly all engines must have their valve clearance set before starting after a rebuild. Engines with solid lifters also require periodic valve adjustments. Valve clearance is called **valve lash** and is checked by inserting a feeler gauge between the valve tip and the rocker arm or cam lobe. The proper amount of valve lash allows for the expansion of parts as the engine's temperature rises. This prevents excessive wear and/or damage to the valve train. It also minimizes valve train noise. Always refer to the service manual before adjusting valve lash.

Valve lash is adjusted in many different ways, depending on engine design. It is extremely important to make sure that the valves are totally closed before making an adjustment. Most OHV engines with hydraulic lifters are adjusted, one cylinder at a time, by bringing a cylinder to exact TDC. At this point the intake and exhaust valves are fully closed and the lifter is on the heel of the cam lobe. Rotate a valve's pushrod with your fingers while tightening the rocker arm's pivot nut. When the pushrod has some resistance, turn the nut an additional ⅛ to ½ turn. Once the valves of that cylinder are adjusted, the next cylinder in the firing order is brought to TDC and its valves adjusted. The process is repeated until all valves have been adjusted. Valve lash can be adjusted with the engine in or out of the vehicle.

SHOP TALK

Pay close attention to the valve lash specs. Typically the exhaust valves have a different lash spec than the intakes.

The procedure for adjusting an OHV with solid lifters is similar, except there must be a clearance between the top of the valve stem and the rocker arm tip. The lash is measured with a feeler gauge and an adjustment screw set once the desired lash is present. In most cases, a locking nut surrounds the screw. This nut is tightened while the screw is held in position so the adjustment does not change.

Similar lash adjustment procedures are used on some OHC engines. An adjustment screw at one end of the rocker arm is tightened or loosened to obtain the desired lash. To adjust the valves, make sure the valve is fully closed and loosen the adjuster locknut. Then turn the screw to achieve the proper clearance. While holding the screw and preventing it from turning, tighten the locknut. Follow the same procedure for each valve that needs to be adjusted.

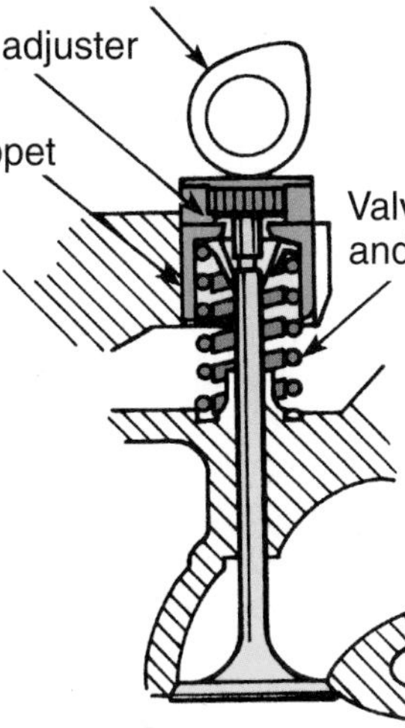

Figure 13-38 Some overhead cam engines feature a cam follower with an adjustment screw. *Courtesy of Ford Motor Company*

Some OHC engines have adjustable cam followers (**Figure 13–38**). Lash is changed by turning the adjuster.

Valve lash is controlled by select shims on many OHC engines. The shim sits between the cam lobe and the cam follower placed over the valve assembly. Valve lash is adjusted by inserting a shim of a different thickness into the follower.

PROCEDURE

Adjusting the valves on an engine that uses shims for valve clearance.

1. Rotate the crankshaft until piston #1 is at TDC (**Figure 13–39A**).
2. Check the camshaft alignment marks; if they are not aligned, rotate the crankshaft one full turn (**Figure 13–39B**) until they are aligned.
3. Refer to the service manual to identify the valves that are closed at this point. With a feeler gauge, measure and record the valve lash of those valves (**Figure 13–39C**).
4. Rotate the crankshaft one full turn with piston #1 again at TDC.
5. Measure and record the valve lash on the valves not measured in the previous step. Compare all measured clearances to specifications.
6. For any valves that do not have the proper lash, follow the rest of this procedure.
7. Rotate the camshaft so that the cam lobe for the valve needing adjustment is facing up.
8. Using a screwdriver, rotate the notch of the valve follower and shim assembly so that it is to the side of the camshaft.

9. While holding the camshaft in place, depress the valve lifter assembly.
10. Using a small screwdriver and a magnetic finger, remove the adjusting shim **(Figure 13–39D)**.
11. Measure the thickness of the shim with a micrometer **(Figure 13–39E)**.
12. Calculate the size of the required shim by adding the measured clearance to the size of the old shim. Then subtract the desired clearance from that total. That total is the required shim size. To correct excessive clearance, a thicker shim is installed. If reduced clearance is needed, a thinner shim must be installed.
13. Install the new shim and recheck the valve lash. Then move to the next valve that needs adjustment and repeat the process.
14. When all valves have been adjusted, reinstall the camshaft covers, timing belt cover, and anything else that has been removed.

Some engines rely on cam followers of different thicknesses to provide proper valve lash. To adjust lash on these engines, the valve clearance of all valves should be measured while the valves are closed. Once the lash adjustments have been determined, the camshaft needs to be removed. Then the thickness of the cam follower on the valves that were out of spec should be measured. The desired cam follower is determined in the same way as deciding the correct shim thickness. The thickness of the new follower should equal the thickness of the old follower plus the measured lash. The desired lash is then subtracted from that sum. The answer is the required thickness for the replacement follower.

New lifter thickness = (Old lifter thickness + measured valve lash) − desired valve lash

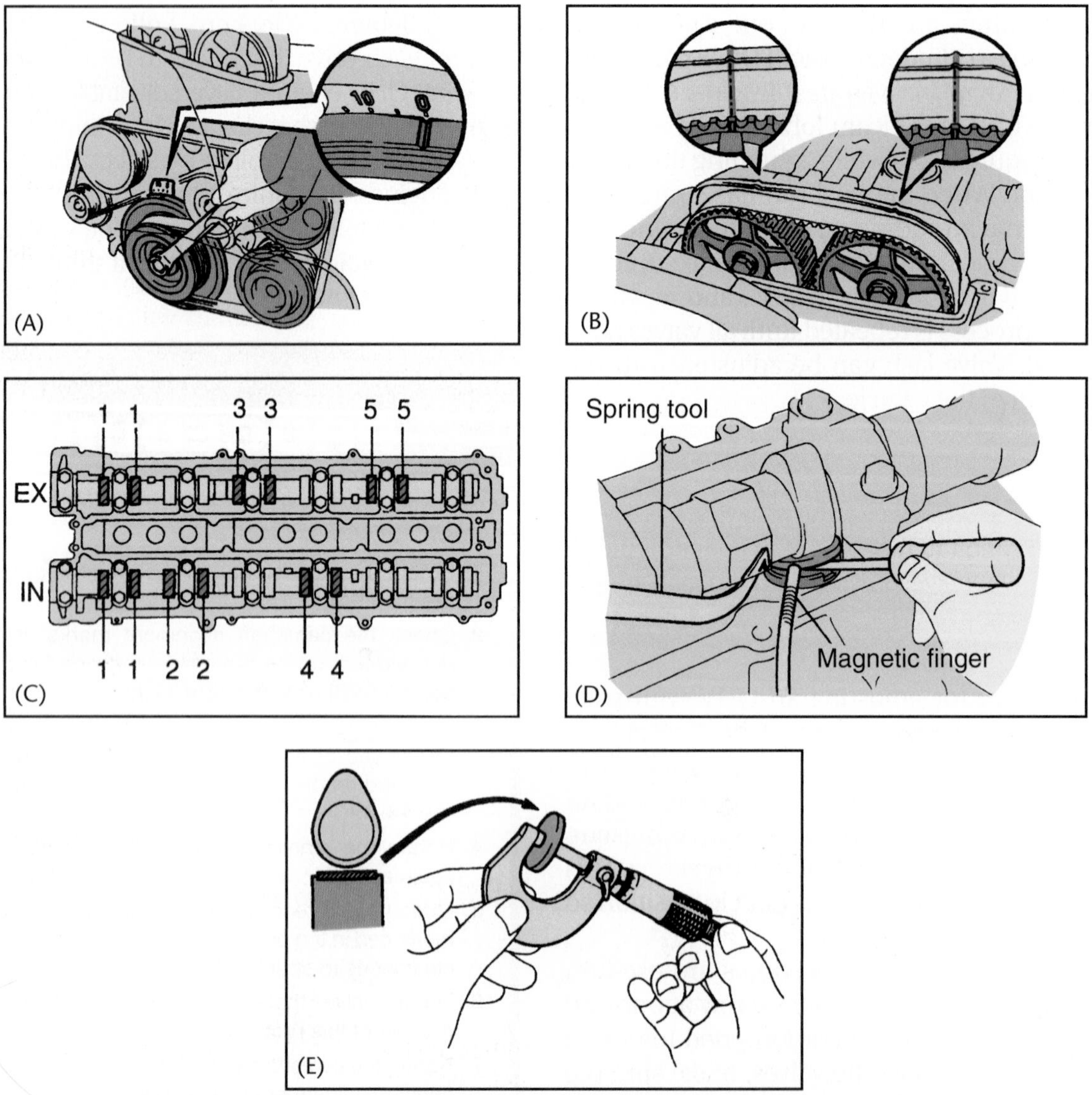

Figure 13-39 To adjust valve lash: (A) Rotate the crankshaft so that it is at TDC. (B) Check the camshaft alignment marks. (C) Measure and record the valve lash of the valves that are totally closed. (D) Use a small screwdriver and a magnetic finger to remove the adjusting shim. (E) Measure the thickness of the shim with a micrometer to determine the correct shim to install. *Courtesy of Toyota Motor Sales, U.S.A., Inc.*

Final Reassembly Steps

The final steps in engine assembly involve installing various covers, pulleys, sensors, and other related items that mount directly to the engine.

Coolant Drains and Plugs Make sure all coolant drains and plugs are installed in the block. Drains are normally threaded into the block with a thread sealant. This is also true for threaded plugs.

Timing Sensors Proper installation of the crankshaft and camshaft timing sensors is critical. Make sure to coat new O-rings with clean oil before they are installed **(Figure 13–40)**. Some sensors have a specified gap that must be set during installation. Also make sure that the trigger wheel for each sensor is properly aligned. Refer to the service manual.

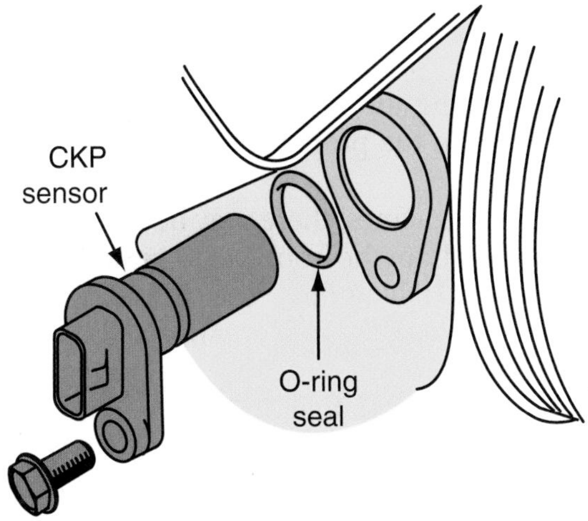

Figure 13-40 A crankshaft position sensor.

Timing Cover When replacing the timing cover, remove the old gaskets and seals from the timing cover and block. Install the new seals using a press, seal driver, or hammer and a clean block of wood. When installing seals, be sure to support the underside of the cover to prevent damage. The gasket for the cover can be made of a variety of materials. Some require that a light coat of adhesive be applied to hold the gasket in place during assembly. Do not do this unless the gasket manufacturer recommends doing so. Position the gasket on the cover. Mount the cover onto the block and torque the bolts to specifications **(Figure 13–41)**.

Vibration Damper Install the vibration damper onto the snout of the crankshaft using the proper tool. In most cases, the damper should bottom out against the oil slinger and timing sprocket. Make sure the woodruff key is in place. Some dampers are not pressed-fit onto the crankshaft. Be sure to install the large washer behind the damper-retaining bolt on these engines.

Mount the crankshaft pulley to the outside of the damper and tighten its mounting bolts to specifications. Often, the crankshaft will need to be held to prevent it from turning while the pulley's bolts are tightened. A special tool is normally required to do this.

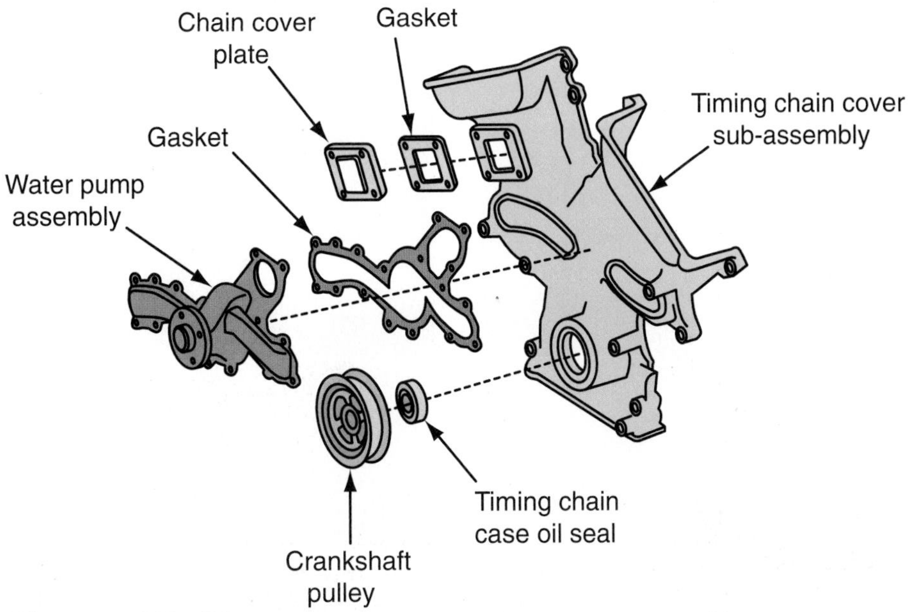

Figure 13-41 A timing chain cover and water pump assembly.

Valve (Cam) Cover Before installing a stamped valve cover, make sure the cover's sealing flange is flat, then apply contact adhesive to the valve cover's sealing surfaces in small dabs. Mount the valve cover gasket on the valve cover and align it in position. If the gasket has mounting tabs, use them in tandem with the contact adhesive. Allow the adhesive to dry completely before mounting the valve cover on the cylinder head. Torque the mounting bolts to specifications.

Cast plastic and aluminum cam and valve covers normally have a rubber gasket. Do not apply an adhesive or sealant to the gasket unless instructed to do so. Start all of the mounting bolts before beginning the tightening sequence. Torque the bolts to specifications.

Water Pump In many cases, a coating of good waterproof sealer should be applied to both sides of the new gasket. Position the pump against the block. Install the mounting bolts and tighten them, in a staggered sequence, to specifications. Careless tightening could cause the pump housing to crack. After tightening, check the pump to make sure it rotates freely. Then, install the drive pulley onto the pump's shaft.

Oil Pan Before installing a stamped oil pan, check the flanges for warpage. Use a straightedge **(Figure 13–42)** or lay the pan, flange side down, on a flat surface with a flashlight underneath it to spot uneven edges. Carefully check the flange around the bolt holes. Minor distortions can be corrected with a hammer and block of wood. If the flanges are too bent to be repaired in this manner, the pan should be replaced. Once it has been determined that the flanges are flat, install the oil pan with a new gasket.

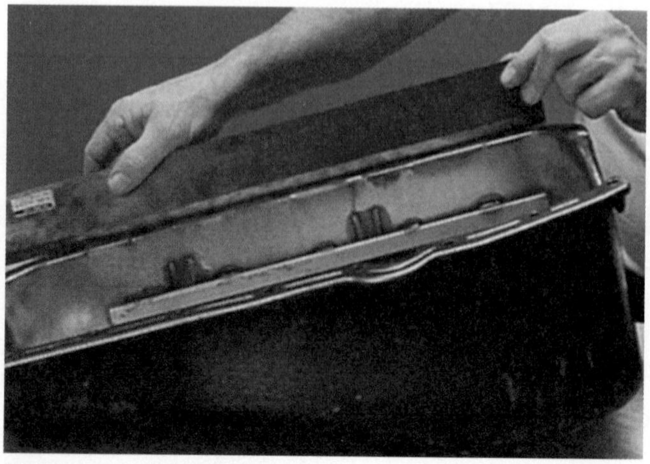

Figure 13–42 Checking the flatness of an oil pan flange. *Courtesy of Detroit Gaskets*

Cast pans are replaced if they are warped or damaged. Also, cast pans normally use a rubber gasket. The mounting bolts should be tightened in the proper sequence and to the specified torque.

Intake Manifold Before installing the manifold, thoroughly clean all of the sealing surfaces, bolt holes, and bolts. Inspect the surfaces for damage and repair or replace as necessary. Check the gaskets for any markings or installation instructions that may be stamped on them. Check the manufacturer's instructions for recommendations on the use of a supplementary sealant. Some intake manifold gaskets should be coated with a nonhardening sealer.

When installing an intake manifold on a V engine, it is wise to use guide bolts. These guides will make sure that the gaskets and the manifold are perfectly aligned before tightening them. They also prevent the manifold from shifting and rupturing the sealant. When tightening the bolts or nuts, make sure you tighten them to the proper torque and in the order specified.

On some V engines, there may be rubber or cork-rubber end seals for the front and the rear of the manifold. Before installing these seals, thoroughly clean all oil from the mating surfaces. Apply adhesive to the surface to hold the seals in place during installation. Once the intake manifold gaskets and seals are in place, apply a small bead (approximately ⅛ inch) of silicone RTV to the point where the seals meet the gasket **(Figure 13–43)**. Other engines may have a large one-piece intake manifold gasket with a splash guard. These are installed with the same care as other intake manifold gaskets.

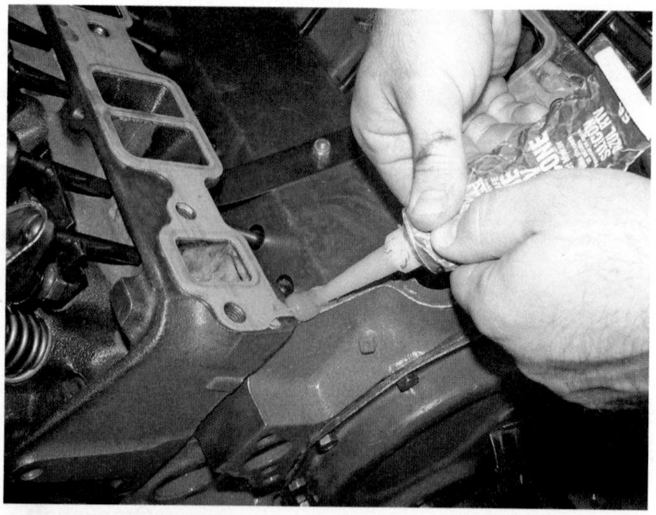

Figure 13–43 On some engines, after the intake manifold gaskets and seals are in place, a small bead of RTV should be applied at the point where the seals meet the gasket.

Make sure the correct gasket is installed. Some intake gaskets purposely block off coolant passages to allow coolant to flow in a predetermined path through the engine. If these ports are not blocked off, the engine can overheat.

Thermostat and Water Outlet Housing Install the thermostat with the temperature sensor facing into the block. If the thermostat is installed upside down, the engine will overheat. Also install any coolant pipes and hoses that route the coolant in and out of the engine.

Exhaust Manifold When installing the exhaust manifold(s), tighten the bolts in the center of the manifold first to prevent cracking it. If there are dowel holes in the exhaust manifold that align with dowels in the cylinder head, make sure these holes are larger than the dowels. If the dowels do not have enough clearance because of the buildup of foreign material, the manifold will not be able to expand properly and may crack. Be sure to install all heat shields **(Figure 13–44)**.

Flywheel or Flexplate Reinstall the engine sling. Raise the engine into the air on a suitable lift, and remove the engine stand mounting head. Set the assembled engine on the floor and support it on blocks of wood while attaching the flywheel or flexplate. Be sure to use the right flywheel bolts and lock washers. The bolts and washers are very thin. If normal bolts or washers are used, they may cause interference with the clutch disc or the torque converter. Manufacturers often recommend coating the bolts

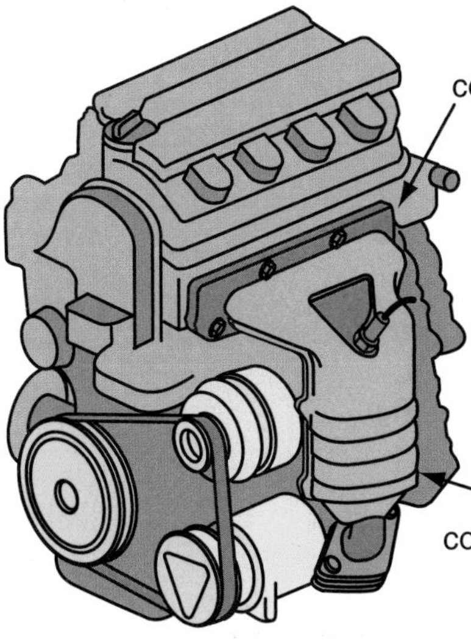

Figure 13–44 Exhaust manifold and heat shield for a four-cylinder engine.

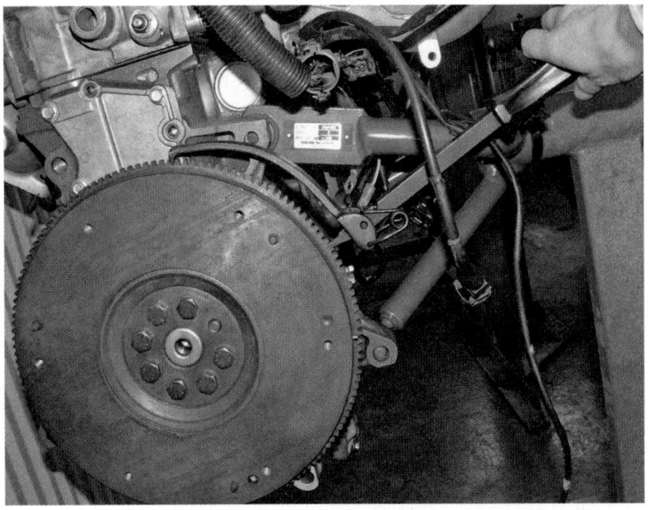
Figure 13–45 A flywheel holding and rotating tool.

with an adhesive. A flywheel holding tool is often required to tighten the bolts to specifications **(Figure 13–45)**. Always follow the specified sequence while tightening. These may be TTY bolts.

CAUTION!

On Honda and some other hybrids, extreme care must be taken when installing the rotor to the rear of the engine. The rotor assembly has very strong magnets. Anyone who has a pacemaker or other magnetically sensitive medical device should not handle the rotor assembly. The rotor should only be installed with the proper tools ***(Figure 13–46)*** *and not with your hands. During installation, the rotor may be suddenly drawn toward the stator with great force, which can cause serious hand or finger injury.*

Clutch Parts If the vehicle has a manual transmission, install the clutch assembly **(Figure 13–47)**. Make sure that the transmission's pilot bushing or bearing is in place in the rear of the crankshaft and that it is in good condition.

Using a clutch-aligning tool, or an old transmission input shaft, align the clutch disc. Then, tighten the disc and clutch pressure plate to the flywheel. Make sure the disc is installed in the right direction. There should be a marking on it that says "flywheel side." The flywheel may need to be held to tighten the bolts. Then install the bell housing if it was removed from the transmission. If the engine was removed with the transmission, reattach it now.

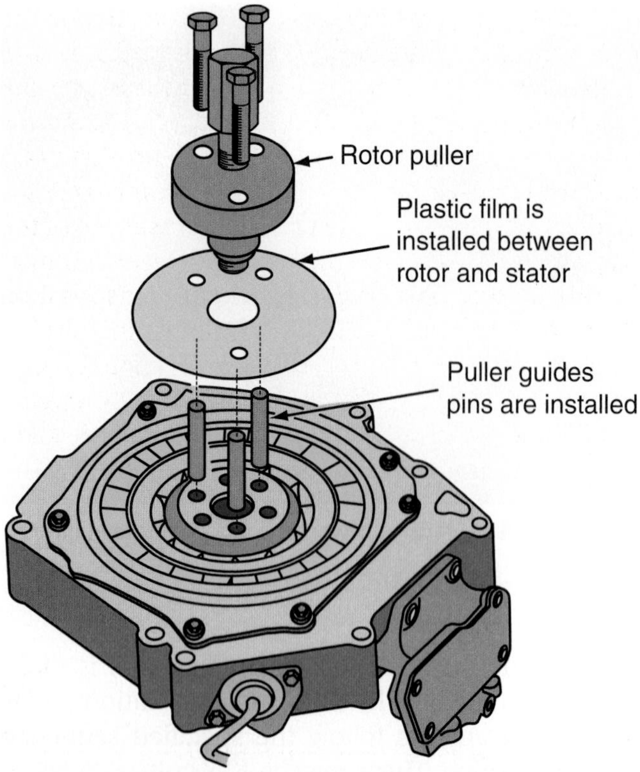

Figure 13-46 The special tools required to remove and install the rotor in a Honda hybrid.

Torque Converter On cars equipped with automatic transmissions, it is a good practice to replace the transmission's front pump seal before attaching it to the engine.

Install the torque converter, making sure that it is correctly engaged with the transmission's front pump. The drive lugs on the converter should be felt engaging the transmission front pump gear. Failure to correctly install the converter can result in damage to the transmission's front pump. If the engine was removed with the transmission, reattach it now.

Motor Mounts Check the condition of all motor mounts and replace any that are damaged. Loosely attach the mounts to their location on the engine. This allows for easy alignment of the mounts while installing the engine into the vehicle. After the engine is aligned in the vehicle, the bolts should be tightened to specifications.

Other Parts There are many other parts that can be reinstalled before the engine is put back into the vehicle. These vary with the model of car. Always check with the service information before installing anything onto the engine. The following are a few of things that may be installed with the engine out of the vehicle:

- Starter
- Oil dipstick guide
- Engine coolant temperature sensor
- Engine oil pressure switch
- Engine oil level sensor
- Knock sensor
- Fuel injector
- Fuel rail
- Fuel pump
- PCV valve
- Engine wiring harness
- Camshaft timing control valve
- Drive belt tensioner

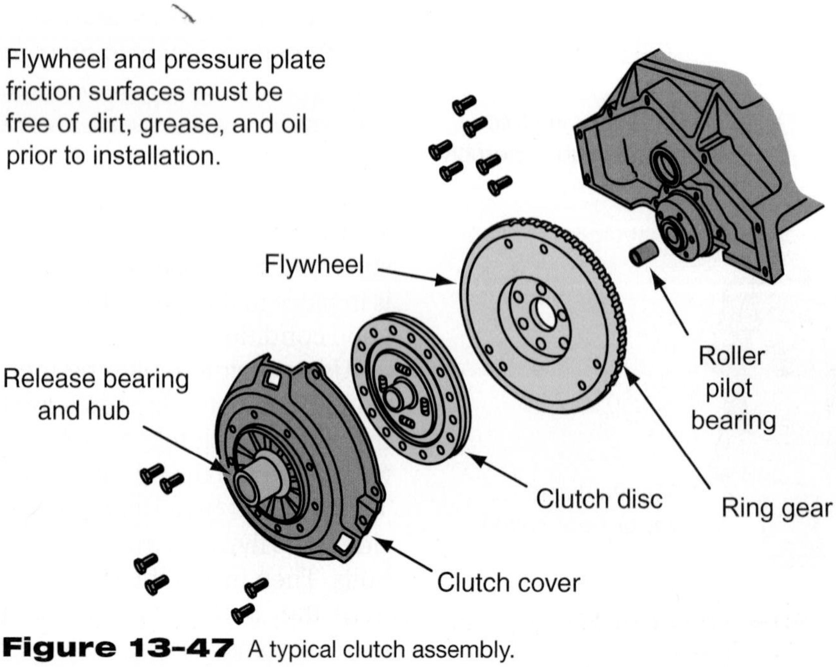

Figure 13-47 A typical clutch assembly.

- A/C compressor
- Generator assembly
- Drive belt
- Ignition coils
- Spark plugs
- Air cleaner assembly

INSTALLING THE ENGINE

Installing a computer-controlled engine can be a complex task requiring special procedures. Referring to the vehicle's service manual is absolutely essential for this procedure. Typically the procedure is the reverse of the removal procedure.

Installing an Engine into a FWD Vehicle

If the engine will be installed through the top, connect the engine to a sling and then connect the sling to the crane **(Figure 13–48)**. Slowly lower the engine into the engine compartment. Guide the engine around all wires and hoses to make sure nothing gets damaged. As the engine approaches its position in the engine compartment, align the engine and transmission mounts. Then, lower the engine so you can connect the engine and transmission mounts. Now raise the vehicle to a good working height.

If the engine must be installed from under the car, mount the engine onto the engine cradle and dolly. Lift the vehicle on a hoist and position the engine under the engine compartment. Carefully lower the vehicle until the engine and transmission are properly positioned in the engine compartment. While doing this, guide all wires and hoses out of the way and make sure that the vehicle does not contact or rest on any part of the engine or transmission. With an engine hoist, raise the engine/transmission into place. Align the engine and transmission mounts and tighten the bolts. Remove the engine cradle and dolly. Then raise the vehicle to a good working height.

SHOP TALK

The mounting bolts for some engines should be installed and tightened according to a specified sequence. Excessive noise and vibration may result from not following this sequence. Always refer to the appropriate service information when installing an engine.

While working under the vehicle, align and tighten all remaining engine/transmission mounts. Install the drive axle shafts and hubs. Connect the exhaust

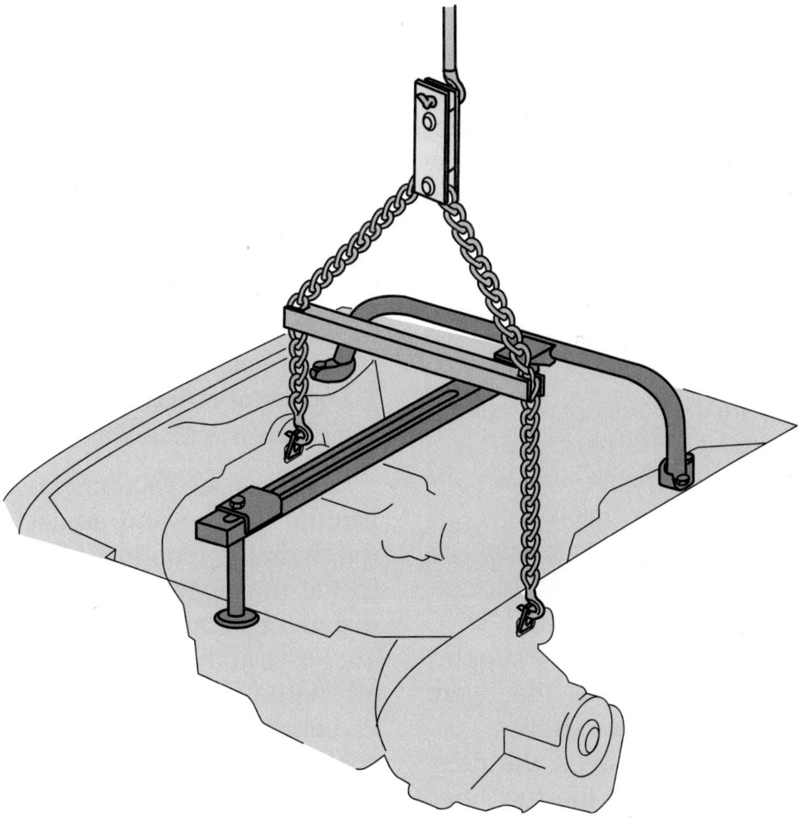

Figure 13-48 Equipment needed to install an engine from the top of a FWD vehicle.

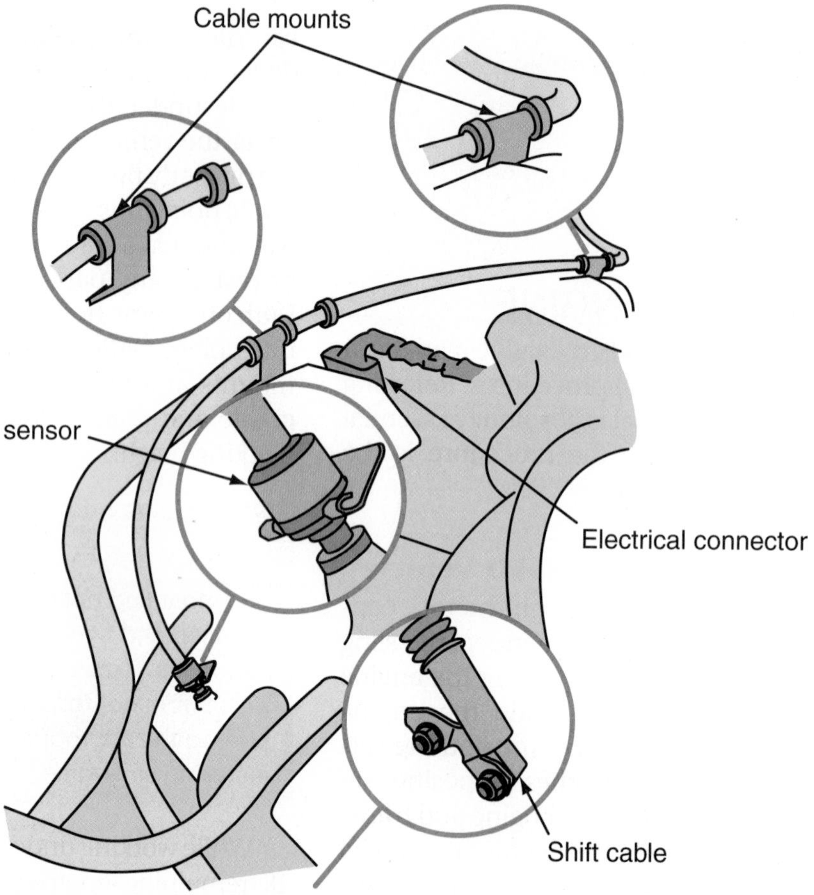

Figure 13-49 Various connections for a transaxle.

manifold to the exhaust system. Install any heat shields that were removed when the engine was removed. Connect all linkages, lines, hoses, and electrical wiring to the transmission **(Figure 13–49)**. Now reconnect all suspension and steering parts that were disconnected or removed. Check and replenish the fluid in the transaxle.

SHOP TALK

Many FWD vehicles require the powertrain cradle to be properly aligned after it is reconnected. This ensures correct steering and suspension alignment. After engine installation is completed, the vehicle's wheel alignment may need to checked and adjusted.

Lower the vehicle and remove the engine support bar or sling. Install all splash shields that were removed, then install the tire/wheel assemblies.

Lower the vehicle so its weight is on the tires. Tighten the axle hub nuts. Connect the fuel lines with their clamps. Install the canister purge valves and related hoses. Make sure all connections are secure. Now connect the engine wiring harness, ground straps, and all other electrical connectors and wires.

CAUTION!

*If working on a hybrid vehicle, make sure that the high-voltage system is isolated before making any connection. The high-voltage cables and connectors are orange in color **(Figure 13–50)**.*

Connect all vacuum hoses. Then connect the throttle linkage and adjust it if necessary. Install the radiator, cooling fan(s), and overflow tank. Install the upper and lower radiator hoses. Then install the heater hoses. Hybrid vehicles have coolant hoses at the inverter; make sure these are properly tightened.

Install the automatic transmission fluid (ATF) cooler hoses to the transmission. Now install the A/C compressor with the drive belt, condenser fan shroud, and the electrical connectors for the fan motor and compressor clutch. Also connect the wiring to the

Figure 13-50 The high-voltage connectors and cables in a hybrid are colored orange for identification.

steering linkage. Then, reinstall the engine compartment support strut.

CAUTION!

On some vehicles the A/C compressor is powered by high voltage; make sure the high-voltage system is off. Also make sure that the power cables are secure in their mounting clamps.

Now install the air induction system and connect any remaining items, including the battery and cables. Check the engine for fuel leaks. Do this by turning the ignition switch to the ON position and allowing the fuel pump to run for a few seconds. Turn the power off and check for signs of leaks. If there are any leaks, repair them before proceeding.

Make sure everything that was removed when the engine was pulled is reinstalled and secure. If the vehicle's hood was removed, carefully reattach it. Refill the radiator with coolant, and bleed air from the system with the heater valve open. Visually check for leaks. Prelubricate the engine and make sure the oil level is correct. On hybrid vehicles, the high-voltage system should be activated after everything is connected.

CAUTION!

Once the high-voltage system is activated, lineman gloves should be worn while doing any under-the-hood work.

Customer Care

After the engine has been started and everything checked out, the wheels must be aligned. Failure to do this can make a customer very unhappy.

Installing an Engine in a RWD Vehicle

Connect the engine to a sling and then connect the sling to the crane. Place a transmission jack under the transmission to hold it in position. Now slowly lower the engine into the engine compartment. Guide the engine around all wires and hoses to make sure nothing gets damaged. As the engine approaches its position in the engine compartment, align the engine to the input shaft of the transmission or the torque converter hub into the front pump. Carefully wiggle the engine until the input shaft slides through clutch disc splines or the torque converter seats fully into the transmission. Install and tighten the transmission to engine bolts. Start the engine mount bolts into their bores; you may need to wiggle the engine some to do this. Once the mount bolts are in place, tighten them and remove the transmission jack and engine sling.

CAUTION!

Do not force the engine and transmission together with the bolts.

Raise the vehicle to a good working height and install all remaining engine and transmission mounts. Connect the exhaust manifold to the exhaust system. Install any heat shields that were removed when the engine was removed. Reconnect the fuel line from the fuel tank to the engine. Install the drive shaft if it was removed. Make sure to align it with the index marks made during removal. Connect all electrical connectors, hoses, and linkages that are accessible from under the vehicle. Reinstall any suspension and steering part that was removed.

Lower the vehicle so you can work under the hood. Mount the A/C compressor and connect all lines and electrical connectors. Install the radiator and the cooling fan(s) and connect the rest of the hoses for the cooling system. Connect all vacuum and other hoses. Now connect the throttle linkage and adjust it if necessary. Connect all remaining electrical connectors.

Connect any remaining disconnected fuel lines. Make sure all connections are secure. Connect the radiator and heater hoses; use new clamps. Connect the engine electrical harness to the PCM or firewall

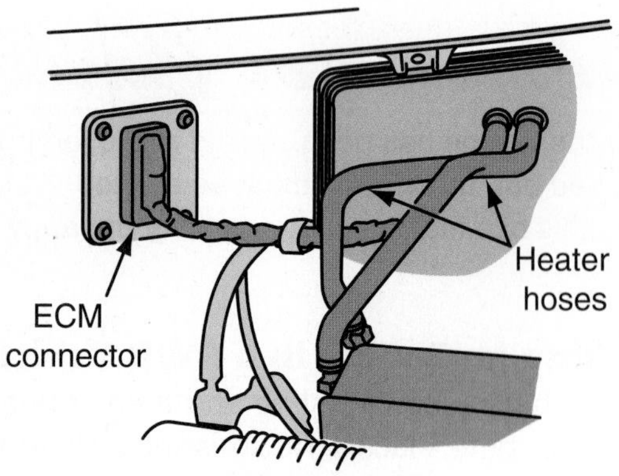

Figure 13-51 Heater hose connecting points and the ECM wiring harness on the firewall.

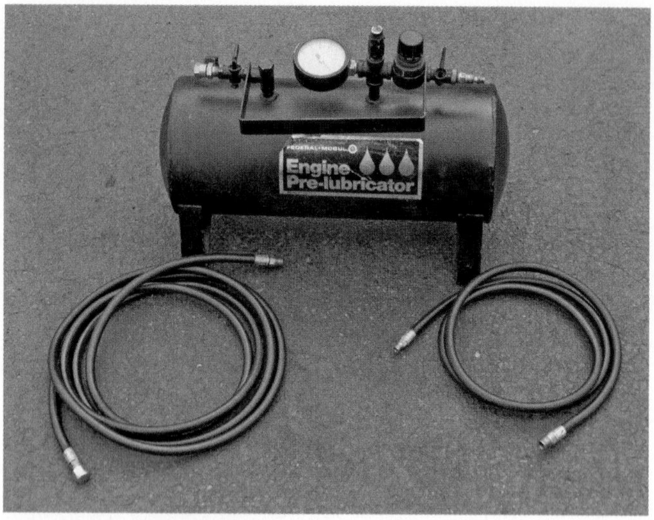

Figure 13-52 A prelubricator.

connector (**Figure 13–51**). Connect the engine ground straps. Install the battery and connect the battery cables. Turn the ignition switch to the ON position and allow the fuel pump to run for a few seconds. Turn the power off and check for signs of leaks. If there are any leaks, repair them before proceeding.

Now install the air induction system and connect any remaining items, including the throttle cables and sensor. Fill the radiator with coolant and visually check for leaks. If the engine does not have oil in it already, add the specified amount of the proper type oil. Prime the oil pump of the engine and prepare the engine for startup. Install and align the hood.

Prelubrication

Not prelubing a new or rebuilt engine before starting it can cause premature bearing failure due to poor lubrication. Other parts such as pistons, rings, and cylinder walls need immediate lubrication to prevent scuffing, scoring, and damage. It can take as long as 5 minutes after the engine has started before oil is distributed through all of the vital parts of an engine. It is claimed that more than 80% of all engine wear occurs when an engine is first started.

These problems can be prevented by lubricating the parts as they are assembled and by forcing oil into the oil galleries. This is the purpose of prelubing. A **prelubricator (Figure 13–52)** forces oil throughout the engine before it is started. There are several ways to prelubricate, or prime, an engine. One of the most common ways is to use an air-operated prelubricator. The procedure for using this type of prelubricator follows:

1. Fill the oil filter with clean engine oil and install it.
2. If the oil pressure sensor is installed, remove it.
3. Install an appropriately sized fitting for the preluber in the sensor's bore.
4. Connect the preluber hose to the fitting.
5. Open the valve on the preluber.
6. Fill the container for the preluber with at least 2 quarts of clean oil.
7. Pump the handle on the preluber until all of the oil in it has moved into the engine.
8. Close the valve and disconnect the hose from the fitting.
9. Remove the fitting and install the oil pressure sensor. Be sure to tighten it to specifications.
10. Check the oil level in the engine and add oil as needed.

SHOP TALK

While prelubing an engine, make sure there is a continuous flow of the correct type of oil. If the preluber runs out of oil during priming, an air pocket can form within the engine's lubrication system.

Distributor-Driven Oil Pumps On engines with a distributor-driven oil pump, the engine can be primed by driving the oil pump with an electric drill. A fabricated oil pump drive shaft is chucked in an electric drill motor and inserted through the distributor bore into the drive on the oil pump. Take the valve covers off but loosely set them over the valves to control oil splash. Drive the oil pump with the electric drill. After running the oil pump for several minutes, remove the valve cover and check for oil flow to the rocker arms. If oil reached the cylinder head, the engine's lubrication system is full of oil and is operating properly. If no oil reached the cylinder head, there is a problem with the pump, with an alignment of an oil hole in a bearing, or a plugged gallery.

Starting Procedure

On engines with an ignition distributor, set the ignition timing as accurately as possible before starting the engine. The timing can be properly set after it has been started by using an engine analyzer or timing light. Fill the gasoline tank with several gallons of fresh gasoline. Start the engine and run it at around 1,500 rpm. When the engine coolant reaches normal operating temperature, turn off the engine. Recheck all adjustments, ignition timing, and valve clearance. Look for signs of coolant or oil leaks.

Break-In Procedure

To prevent engine damage after it has been rebuilt and to ensure good initial oil control and long engine life, the proper **break-in** procedure must be followed. Make a test run at 30 mph and accelerate at full throttle to 50 mph. Repeat the acceleration cycle from 30 to 50 mph at least ten times. No further break-in is necessary. If traffic conditions will not permit this procedure, accelerate the engine rapidly several times through the intermediate gears during the road test. The objective is to apply a load to the engine for short periods and in rapid succession soon after the engine warms up. This action thrusts the piston rings against the cylinder wall with increased pressure and results in accelerated ring seating.

Performance TIP

If a performance flat tappet camshaft and dual valve springs were installed in the engine, special break-in procedures should be followed. The inner valve springs must be removed before the engine is started and run. Doing this makes it easier for the lifters to rotate in their bores and establish an acceptable wear pattern. Immediately after the engine starts, bring the engine speed up to 2,000 to 2,500 rpm and hold it there for about 30 minutes. After that time, shut down the engine and change the oil and filter to remove all contaminants from the engine. Then reinstall the inner valve springs.

Relearn Procedures

The computer in most late-model vehicles must undergo a relearn or initialization procedure after the battery has been reconnected. This procedure allows the computers to learn the condition of the engine and make adjustments according to the engine's restored condition. The last time the engine was run, the computer made adjustments based on the engine faults present. This procedure allows the computer to see that those faults were corrected. Initialization also resets the reference for the crankshaft position sensor and PCM. Always follow the manufacturer's procedures as outlined in the service manual.

Initialization is also necessary to reset the operating perimeters of many accessories, such as the power windows and the antitheft system. Devices like the clock and radio will need to be manually reset.

SHOP TALK

Hybrid vehicles have special initialization procedures. These systems automatically shut off the engine when the vehicle is stopped and restart it when certain conditions are present. In many cases, the PCM must know the exact position of the electric assist motor's rotor. If the PCM cannot recognize the rotor's position, the motor may not be able to run. In general, if initialization is not completed, the hybrid system will not work normally. The service information for the vehicle will give the procedures for each system that needs to be initialized.

Customer Care

After the engine has been totally checked over, return it to the owner with the following instructions:

1. Drive the vehicle normally but avoid sustained high speed during the first 500 miles (break-in period).
2. Do not allow the engine to sit idling for a long time.
3. Check the oil level frequently during the break-in period. It is not unusual to use 1 or 2 quarts of oil during this time.
4. The oil and oil filter should be changed at the end of the break-in time.
5. The cylinder head and intake manifold bolts may need to be retorqued.
6. Some adjustments, such as valve adjustments and ignition timing, may also need to be checked.

SHOP TALK

Some manufacturers recommend retorquing the cylinder head after a rebuilt engine has been run for the first time. With the engine at normal operating temperature, retighten each head bolt using the specified tightening sequence. If the engine has an aluminum head or block, do not retorque the bolts until the engine has cooled. On some engines it may be necessary to retorque the head, again, after a specified time or mileage interval. Always follow the manufacturer's recommendations.

CASE STUDY

A four-cylinder DOHC engine has just been rebuilt. During the initial running of the engine, it runs quite rough. This is normal until everything gets seated. However, the condition becomes worse the more the engine runs. Slight adjustments to the ignition timing do not help the condition. A visual inspection reveals nothing loose or disconnected. Each hose and wire is traced to make sure it is connected to the proper fitting and terminal.

Because it seems that the engine is running on only three cylinders, a cylinder leakage test is performed. The results indicate excessive leakage past the #4 intake valve. In an attempt to visually locate the problem, the intake cam cover is removed. A look at the camshaft reveals that the #4 intake lobe is worn. This was a new camshaft. Was it defective? Was it not hardened properly? Did something else cause this?

Further visual inspection reveals that the shim used to set valve lash was not fully seated in its cup. The edge of the shim was working like a knife, cutting off the metal from the lobe. The initial rough running of the engine was caused by the shim preventing the valve from fully closing. The condition worsened as the lobe was cut away.

The oil was changed and a new camshaft and shim were installed to correct the problem.

KEY TERMS

Aerobic
Anaerobic
Break-in
Elasticity
Gasket
Graphite
Multilayer steel (MLS)
Polyacrylate
Prelubricator
Room temperature vulcanizing (RTV)
Silicone gaskets
Torque-to-yield (TTY)
Valve lash
Viton
Yield

SUMMARY

- Elasticity means that a bolt can be stretched a certain amount and it returns to its original size when the load is reduced.
- Bolt yield means that a stretched bolt takes a permanent set and never returns to normal.
- Some fasteners are intentionally torqued into a yield condition. These torque-to-yield (TTY) bolts are designed to stretch when properly tightened. When a bolt is stretched to its yield point, it exerts its maximum clamping force. TTY bolts should not be reused.
- Gaskets serve as sealers, spacers, wear insulators, and vibration dampeners.
- Gaskets can be made of paper, fiber, steel, cork, synthetic rubber, and combinations of these materials.
- General recommendations for installing gaskets include: Never reuse old ones; handle new ones carefully; use sealants only when instructed; thoroughly clean all mating surfaces; and use the right gasket in the right position.
- Cylinder head gaskets must seal all of the combustion chambers and the coolant and oil passages between the head and the block.
- Common causes of cylinder head gasket failure include improper installation, overheating, hot spots, and detonation or preignition.
- Adhesives are used to hold a gasket in place during component assembly.
- Bolts that pass through a liquid passage should be coated with Teflon thread or a brush-on thread sealant.
- Silicone gasketing, of which RTV is the best known, is used on oil pans, valve covers, thermostat housing, timing covers, and water pumps.
- Anaerobic formed-in-place sealants are used for both thread locking and gasketing.

- Oil seals keep oil and other fluids from escaping around a rotating shaft.
- All engines with mechanical lifters have some method of adjustment to bring valve lash (clearance) back into specification.
- The steps in final engine assembly involve installing various engine covers, prelubing the engine, and installing manifolds and related items that mount directly to the engine assembly.
- The best method of prelubricating an engine under pressure without running it is to use a prelubricator, which consists of an oil reservoir attached to a continuous air supply.
- A proper break-in procedure is necessary to ensure good initial oil contact and long engine life.

REVIEW QUESTIONS

1. *True or False?* Make sure you locate and adhere to the "wet" torque specs when tightening a bolt that has been lubricated with oil or any other liquid.
2. What does it mean when a bolt is stretched into yield?
3. Name some applications of hard gaskets.
4. Where are flexible sealers most often used?
5. What are the major differences between aerobic and anaerobic sealants?
6. Which of the following statements is incorrect?
   a. Cylinder head bolts must be tightened to the proper amount of torque.
   b. Most cylinder head bolts are tightened in a sequence that starts on one end.
   c. On some engines, the head bolts are retorqued after the engine has been run and is hot.
   d. Many engines use head bolts of different lengths.
7. *True or False?* Valve lash is adjusted in many different ways; however, on most OHV engines with hydraulic lifters the clearance is set when the lifter is sitting on the nose of the cam lobe.
8. Which of the following statements about the purpose of a cylinder head gasket is *not* true?
   a. It seals intake stroke vacuum, combustion pressure, and the heat of combustion.
   b. It prevents coolant leakage and resists rust, corrosion and, in many cases, meter coolant flow.
   c. It allows for lateral and vertical head movement as the engine heats and cools.
   d. It meters lubricating oil onto the engine's cylinder walls.
9. *True or False?* The computer in most late-model vehicles must undergo a relearn or initialization procedure after the vehicle's battery has been disconnected and reconnected.
10. Graphite is ______.
    a. an anaerobic substance
    b. an RTV
    c. an aerobic substance
    d. none of these
11. Which of the following are considered soft, cut gaskets?
    a. paper gaskets
    b. cork gaskets
    c. cork/rubber gaskets
    d. all of the above
12. What material is typically not used to form a rear main bearing oil seal?
    a. polyacrylate
    b. RTV
    c. silicone synthetic rubber
    d. Viton
13. Which of the following statements about preparing to assemble an engine is *not* true?
    a. Discard and replace all bolts that are damaged.
    b. Identify what bolts go into the specific bores.
    c. Use a thread locker on all bolts.
    d. Clean bolt and cylinder block threads with a thread chaser or tap.
14. On nearly all hybrid vehicles, how can you identify the high-voltage cables and connectors?
15. *True or False?* Some crankshaft and camshaft timing sensors have a specified gap that must be set during installation.

ASE-STYLE REVIEW QUESTIONS

1. Technician A says that TTY bolts can be used after they were removed if there are no signs of distortion or stretching. Technician B says that head bolts that pass through a coolant passage

should be coated with a nonhardening sealer prior to installing them. Who is correct?

a. Technician A
b. Technician B
c. Both A and B
d. Neither A nor B

2. Technician A uses adhesives to hold gaskets in place while installing them. Technician B uses RTV sealant on all surfaces that are prone to leaks. Who is correct?

a. Technician A
b. Technician B
c. Both A and B
d. Neither A nor B

3. Technician A uses soft gaskets on valve covers. Technician B uses soft gasket on water pumps. Who is correct?

a. Technician A
b. Technician B
c. Both A and B
d. Neither A nor B

4. Technician A inspects all sealing surfaces for damage and uses a supplementary sealant on the gaskets if the surface is questionable. Technician B coats the intake gasket on some engines with a nonhardening sealer. Who is correct?

a. Technician A
b. Technician B
c. Both A and B
d. Neither A nor B

5. Technician A says that damaged bolt threads can be repaired with a tap and die set. Technician B says that if the bolt heads are not tight against the surface after they have been properly torqued, a washer should be installed under the bolt head. Who is correct?

a. Technician A
b. Technician B
c. Both A and B
d. Neither A nor B

6. Technician A says that aluminum expands at a different rate than cast iron. Technician B says that cast iron has a coefficient of thermal expansion two or three times greater than aluminum. Who is correct?

a. Technician A
b. Technician B
c. Both A and B
d. Neither A nor B

7. While discussing proper engine break-in: Technician A says that the engine should run at idle speed for at least 2 hours. Technician B says that the engine should be accelerated at full throttle to 50 mph and that this cycle should be repeated at least ten times. Who is correct?

a. Technician A
b. Technician B
c. Both A and B
d. Neither A nor B

8. While tightening a TTY bolt: Technician A initially tightens the bolt to the specified torque. Technician B turns the bolt an additional amount after the bolt has been tightened to the specified torque. Who is correct?

a. Technician A
b. Technician B
c. Both A and B
d. Neither A nor B

9. Before reusing head or other critical bolts: Technician A measures their length and compares the measurement to specifications. Technician B lubricates the threads and the bottom of the bolt head with engine oil before installing them. Who is correct?

a. Technician A
b. Technician B
c. Both A and B
d. Neither A nor B

10. While selecting the proper RTV for a particular application: Technician A uses RTV in a black tube for all high-temperature applications. Technician B uses RTV in a red tube for most general applications. Who is correct?

a. Technician A
b. Technician B
c. Both A and B
d. Neither A nor B

LUBRICATING AND COOLING SYSTEMS

CHAPTER 14

OBJECTIVES

■ Name and describe the components of a typical lubricating system. ■ Describe the purpose of a crankcase ventilation system. ■ Describe the operation of the cooling system. ■ List and describe the major components of the cooling system. ■ Describe the function of the water pump, radiator, radiator cap, and thermostat in the cooling system. ■ Diagnose the cause of engine overheating. ■ Explain why today's engines are concerned with engine operating temperatures. ■ Test and service the cooling system.

The life of an engine largely depends on its lubricating and cooling systems. If an engine does not have a supply of oil or cannot rid itself of high temperatures, it will be quickly destroyed.

LUBRICATION SYSTEM

An engine's lubricating system does several important things. It holds an adequate supply of oil to cool, clean, lubricate, and seal the engine. It also removes contaminants from the oil and delivers oil to all necessary areas of the engine. The main components of a typical lubricating system **(Figure 14–1)** are described here.

Engine Oil

Engine oil is specially formulated to lubricate and cool engine parts. The moving parts of an engine are fed a constant supply of oil. Engine oil is stored in the oil pan or sump. The oil pump draws the oil from the sump and passes it through a filter where dirt is removed. The oil is then moved throughout the engine via oil passages or galleries **(Figure 14–2)**. After circulating through the engine, the oil returns to the sump.

Go To *Chapter 7 for a detailed discussion of engine oil.*

Oil Pump

The oil pump is the heart of the lubricating system. Just as the heart in a human body circulates blood through veins, an engine's oil pump circulates oil through passages in the engine. The oil pump pickup is a line from the oil pump to the oil stored in the oil pan. It usually contains a filter screen, which is submerged in the oil at all times **(Figure 14–3)**. The screen serves to keep large particles from reaching the oil pump. This screen should be cleaned any time the oil pan is removed. The pickup may also contain a by-pass valve that allows oil to enter the pump if the screen becomes totally plugged.

Because the oil pump is a positive displacement pump, an oil **pressure relief valve** is used to prevent excessively high system pressures from occurring as engine speed is increased. Once oil pressure exceeds a preset limit, the spring-loaded pressure relief valve opens and allows the excess oil to bypass the rest of the system and return directly to the sump.

Go To *Chapter 11 for a detailed discussion of oil pumps.*

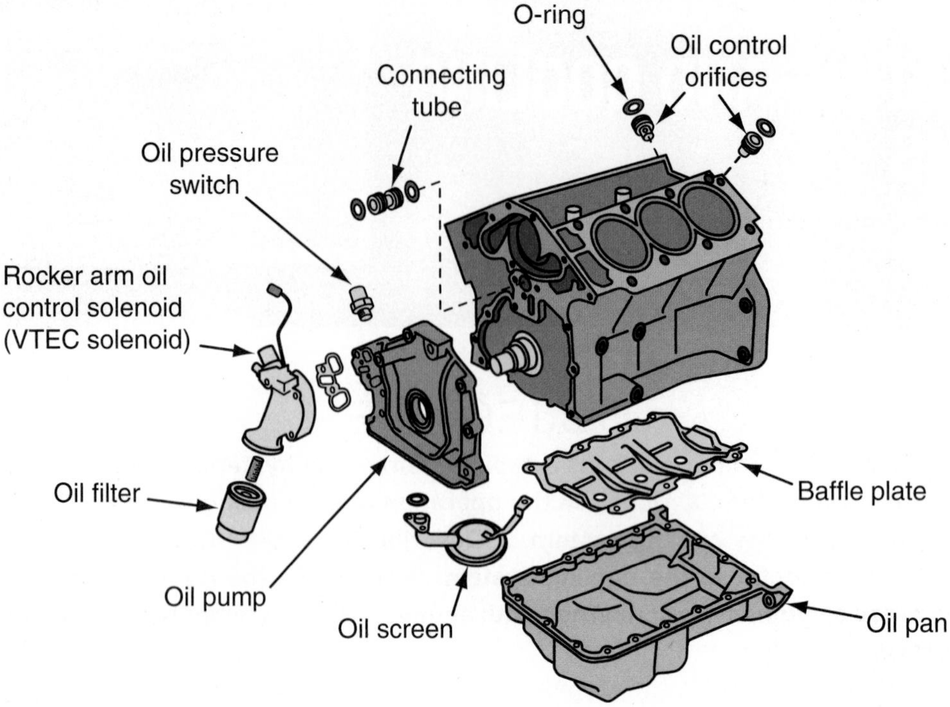

Figure 14-1 The major components of an engine's lubrication system.

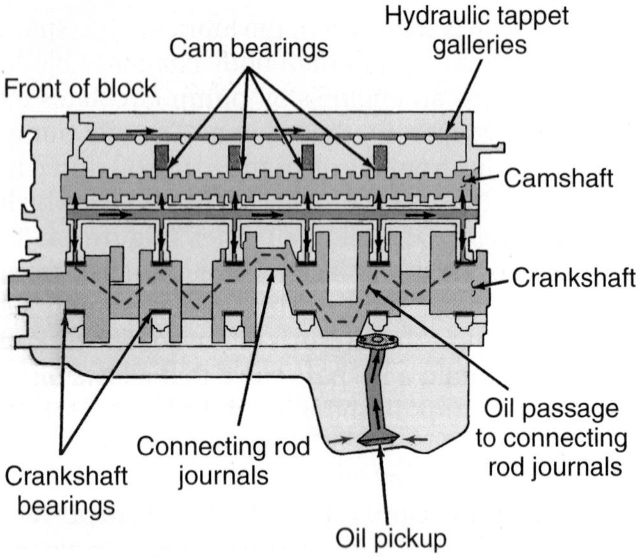

Figure 14-2 Direction of oil flow through this V-10 engine. *Courtesy of Chrysler LLC*

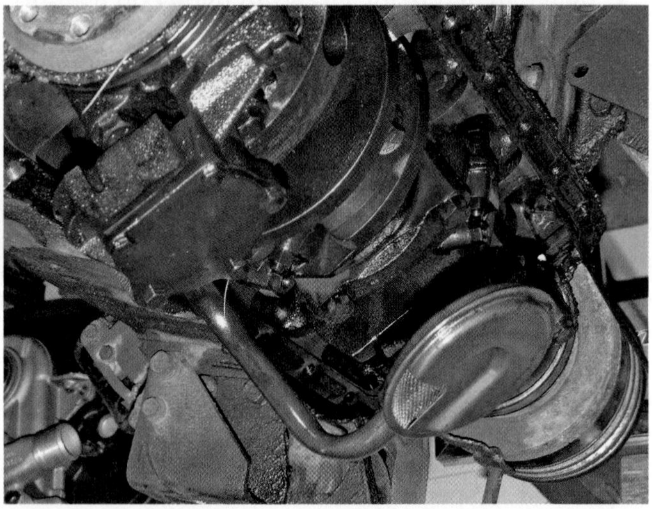

Figure 14-3 An oil pickup and screen.

Oil Pan or Sump

The oil pan is attached to the bottom of the engine block. It serves as the reservoir for the engine's oil. It is designed to hold the amount of oil that is needed to lubricate the engine when it is running, plus a reserve. The oil pan also helps to cool the oil through its contact with the outside air.

Oil is drawn out by the oil pump. The pump pressurizes the oil and sends it to the main oil gallery where it is channeled off to the bearing surfaces and the valve train. The oil drains back into the pan. This process maintains a flow of oil throughout the engine. This is often referred to as a **wet sump** oil system because the sump always has oil in it.

Pan Baffles In a wet sump system, the oil can slosh around during hard cornering or braking. During these times it is possible for the oil to move away from the oil pump's pickup. This will cause a temporary halt in oil flow through the engine, which can destroy it. Sloshing also can affect the rotation of the crankshaft. As the crankshaft rotates through a thick puddle of oil, it meets resistance and slows down. To help prevent sloshing, many engines have baffles in the oil pan to limit the movement of the oil.

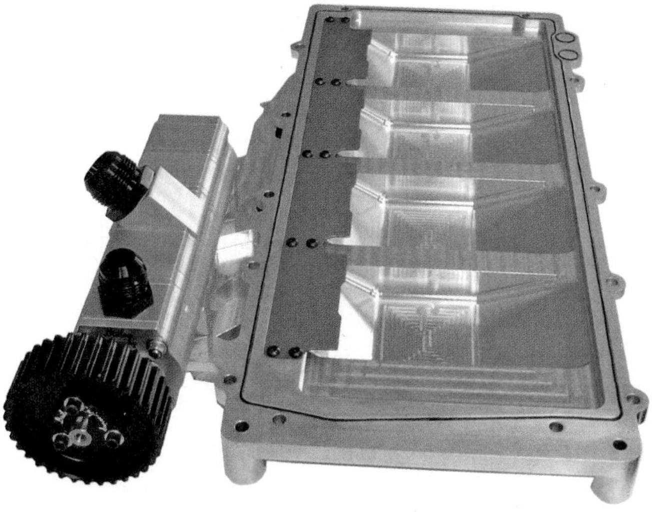

Figure 14–4 An oil pan and external pump for a dry sump system. *Courtesy of Dailey Engineering*

Dry Sump To eliminate the possibility of oil sloshing, some OEM engines are fitted with a **dry sump** oil system, as are most race engines. In a dry sump system, the oil pan does not store oil. The dry oil pan merely seals the bottom of the crankcase. The oil reservoir is a remote container set apart from the engine. Rather than having a single path for oil travel, dry sump pumps can feed oil directly to the crankshaft, valve train, and turbocharger. Normally one external oil pump **(Figure 14–4)** is used; however, some systems have two pumps. The second pump pulls all the oil out of the sump and returns it to the reservoir. This pump also lowers the pressure inside the crankcase.

Dry sump systems provide immediate oil delivery to critical areas of the engine. They also prevent oil starvation caused by acceleration, braking, and cornering forces. Because the dry sump is smaller than a wet sump, the engine can be placed lower in the frame to improve overall handling.

Oil Filter

All of the oil that leaves the oil pump is directed to an oil filter **(Figure 14–5)**. This ensures that small particles of dirt and metal suspended in the oil will not reach the close-fitting engine parts. This stops those impurities, which can cause premature wear, from circulating through the engine. Filtering also increases the usable life of the oil. The filter assembly threads directly onto the main oil gallery tube. The oil from the pump enters the filter and passes through the element of the filter. From the element, the oil flows back into the engine's main oil gallery.

An oil filter assembly is typically a disposable metal container filled with a special type of treated paper or other filter substance (cotton, felt, and the like). Some engines have a separate cartridge that fits into a metal casing **(Figure 14–6)**. The filter is usually mounted on

Figure 14–5 An oil filter being installed on an engine block. *Courtesy of Honeywell International Inc.*

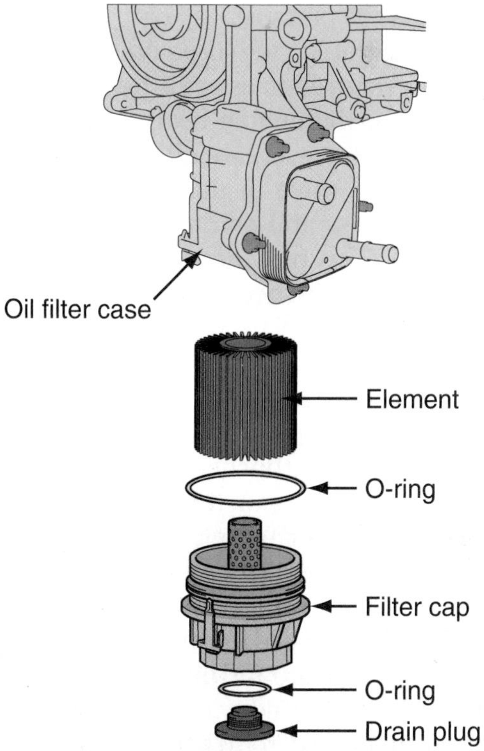

Figure 14–6 An oil filter assembly with a replaceable element. The oil in the oil filter can be drained by removing the drain plug before replacing the element. *Courtesy of Toyota Motor Sales, U.S.A., Inc.*

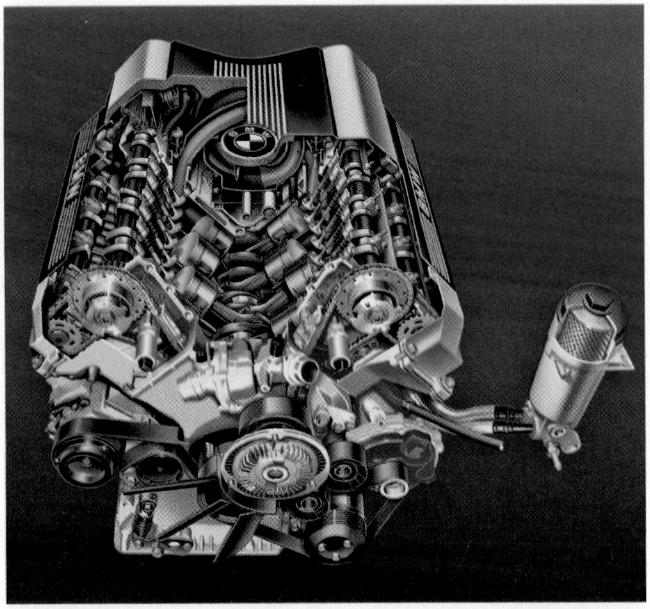

Figure 14–7 Oil lines carry the engine oil in and out of this remote filter before it moves through the engine to lubricate parts. *Courtesy of BMW of North America, LLC*

and sealed to an adapter bolted to the block. However, it may be attached to the timing cover or remotely mounted with oil lines connecting the filter to the oil galleries in the block **(Figure 14–7)**.

Some oil filters have an antidrainback valve. The purpose of this valve is to prevent oil drainage from the filter when the engine is not running. This allows for a supply of filtered oil as soon as the engine is started and has oil pressure.

The oil filtration system used on today's vehicles is commonly referred to as a full-flow system. All of the oil going to the engine bearings goes through the filter first. However, should the filter become plugged, a relief valve in the filter will open and allow oil to bypass and go directly to the bearings **(Figure 14–8)**. This provides the bearings and the rest of the engine with necessary, though unfiltered, lubrication.

Oil Coolers

To control oil temperature, many diesel, high-performance, and turbocharged engines have an external engine oil cooler **(Figure 14–9)**. Oil flows from the pump through the cooler and then to the engine. An oil cooler looks like a small radiator mounted near the front of the engine or within the radiator. Heat is removed from the oil as engine coolant flows through or around the cooler and air passes through it. Hot oil mixed with oxygen breaks down (oxidizes) and forms carbon and varnish. The higher the temperature, the faster these deposits build. An oil cooler helps keep the oil at its normal operating temperature.

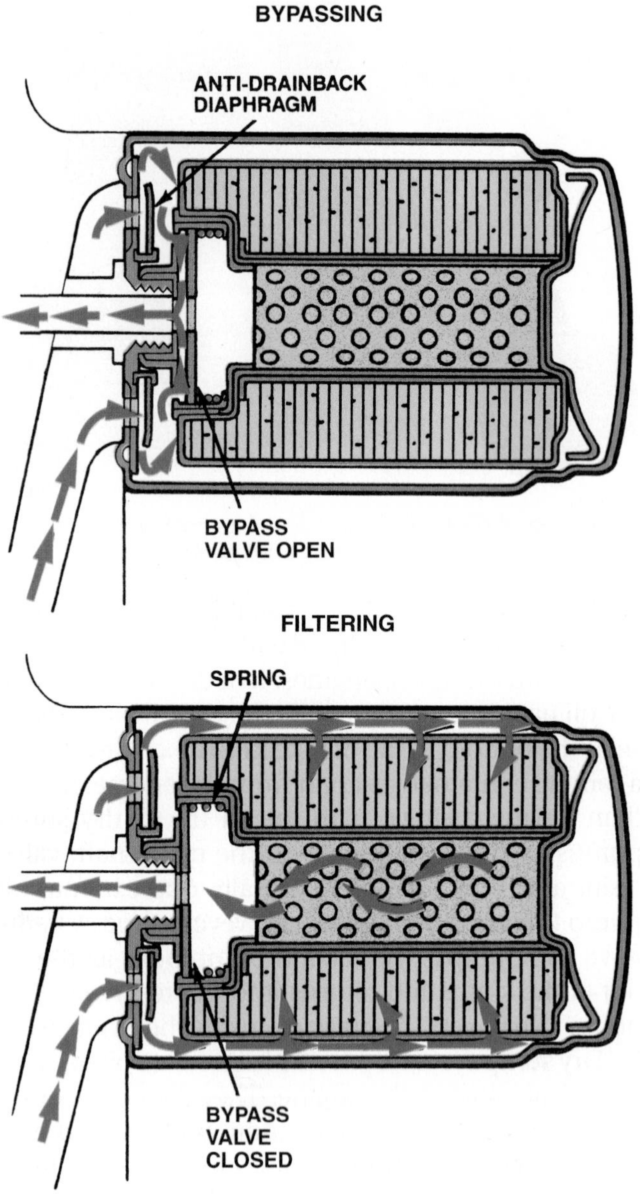

Figure 14–8 Oil flow through the filter. *Courtesy of Ford Motor Company*

Engine Oil Passages or Galleries

From the filter the oil flows into the engine oil galleries. These galleries consist of interconnecting passages that have been drilled completely through the engine block during manufacturing. The outside ends of the passages are blocked off so the oil can be routed through these galleries to various parts of the engine. The crankshaft also contains oil passages (oilways) to route the oil from the main bearings to the connecting rod bearing surfaces. Engines with a remote oil filter, an oil cooler, or a dry sump system have external oil lines that move the oil to designated areas.

Oil Seals and Gaskets

Seals and gaskets are used throughout the engine to prevent both external and internal oil leaks. The most

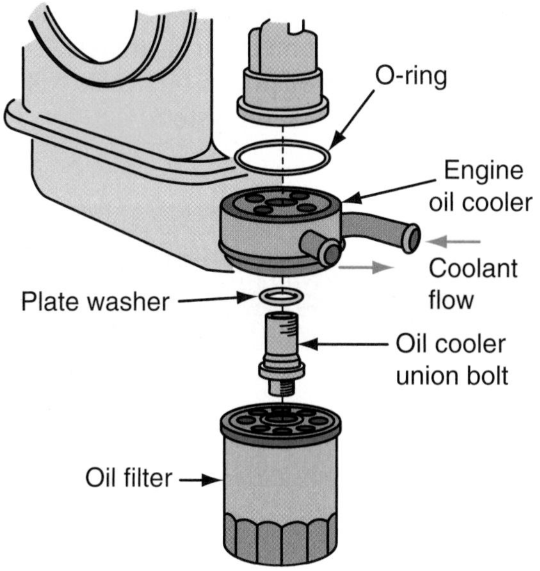

Figure 14–9 To control oil temperatures, some engines are fitted with a water-cooled oil cooler.

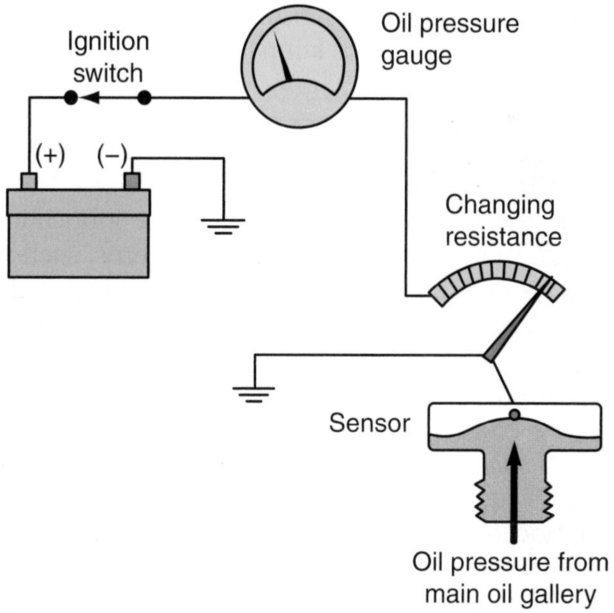

Figure 14–10 As oil pressure changes, the resistance in the oil pressure gauge circuit and the reading on the gauge change accordingly.

common materials used for sealing are synthetic rubber, soft plastics, fiber, and cork. In critical areas these materials might be bonded to a metal.

Dipstick

The dipstick is used to measure the level of oil in the oil pan. The end of the stick is marked to indicate when the engine oil level is correct. It also has a mark to indicate the need to add oil to the system.

Oil Pressure Indicator

All vehicles have an oil pressure gauge and/or a low-pressure indicator light. Oil gauges are either mechanically or electrically operated and display the actual oil pressure of the engine. The indicator light only warns the driver of low oil pressure.

In a mechanical gauge, oil travels up to the back of the gauge where a springy, flexible, hollow tube, called a Bourdon tube, uncoils as the pressure increases. A needle attached to the Bourdon tube moves over a scale to indicate the oil pressure.

Most pressure gauges are electrically controlled. An oil pressure sensor or sending unit is screwed into an oil gallery. As oil passes through an oil pressure sender **(Figure 14–10)**, it moves a diaphragm, which is connected to a variable resistor. This resistor changes the amount of current passing through the circuit. A gauge on the dashboard reacts to the current and moves a needle over a scale to indicate the oil pressure, or the current is translated into a digital reading on the gauge.

Warning light systems are basically simple electrical circuits. The indicator light comes on when the circuit is completed by a sensor. This sensor has a

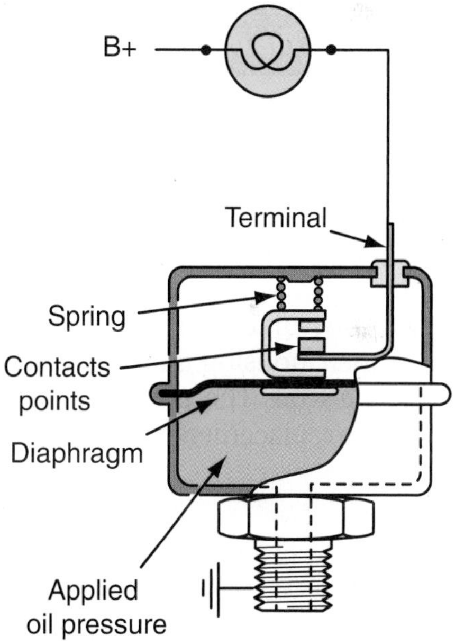

Figure 14–11 An oil pressure sensor for a warning gauge.

diaphragm connected to a switch inside the sensor. Under normal conditions, the sender switch is open. When oil pressure falls below a certain level, the reduction of pressure causes the diaphragm to move and close the sender switch **(Figure 14–11)**, which completes the electrical circuit. When this happens, electricity flows and turns on the warning light on the dashboard.

Oil Level Indicators Some late-model vehicles are equipped with an electronic oil level indicator. When the oil level in the oil pan drops below a predetermined level, a sensor completes the warning lamp circuit and the lamp lights.

ENGINE LUBRICATION DIAGNOSIS AND SERVICE

Other than engine destruction, engine lubrication problems can cause other engine concerns, such as noise, exhaust smoke, and the need to add oil to the crankcase.

Engine Oil

The level of the oil in an engine should be checked periodically. The condition of the oil should be noted at that time as well. Engine oil is specially formulated to lubricate and cool engine parts. Changing the engine's oil is a common preventive maintenance procedure.

Go To ***Chapter 7 for the correct procedure for changing the oil and oil filter.***

Oil Filter The oil filter is normally changed when the oil is changed. There are several designs of oil filters; always use the type of filter that is recommended for the vehicle you are working on. Some oil filters have a drain plug in their housing that should be used to drain the engine's oil. The drain pipe is normally included with the replacement filter cartridge.

> **SHOP TALK**
>
> On some engines, the oil filter should be removed before opening the drain plug at the sump. Doing this allows oil to flow freely from the crankcase.

> **WARNING!**
>
> *Used oil and oil filters must be disposed of properly and in accordance to local, state, and federal laws.*

Oil Pump

The oil pump seldom is the cause for lubrication problems. However, like any other engine part, it can go bad. The best way to check the performance of an oil pump is to conduct an oil pressure test. This test will not only check the oil pump, but it also will evaluate the rest of the lubrication system.

Go To ***Chapter 9 for the correct procedure for conducting and interpreting an oil pressure test.***

If the oil pump or its pressure regulator valve proves to be faulty, it should be replaced.

Go To ***Chapter 11 for the correct procedure for inspecting and installing an oil pump.***

Oil Passages, Galleries, and Lines

All oil passages, galleries, and lines should be thoroughly cleaned and flushed during and after an engine has been overhauled.

Oil Consumption

Excessive oil consumption can result from external and internal leaks or worn piston rings, valve seals, or valve guides. Internal leaks allow oil to enter the combustion chamber where it is burned. Blue exhaust smoke is an indication that an engine is burning oil.

If the valve guides are worn or the valve seals are worn, cracked, or improperly installed, oil will be drawn into the cylinder during the intake stroke. If there are worn or broken piston rings or worn cylinder walls, the affected cylinders will have low compression. The oil in the cylinder also tends to foul the spark plugs, which will cause misfires, high emissions, and possible damage to the catalytic converter.

External leaks are another common cause of excessive oil consumption. These leaks can occur at the valve or cam cover gasket, oil filter, front and rear seals, oil pan gasket, and timing gear cover. Fresh oil on the clutch housing, oil pan (**Figure 14–12**), edges of valve covers, external oil lines, crankcase filler tube, or at the bottom of the timing gear or chain cover usually indicates that the leak is close to or above that point.

When crankcase pressure is abnormally high, oil is forced out through joints that normally would not leak. Pressure develops when blowby becomes excessive or when a positive crankcase ventilation (PCV)

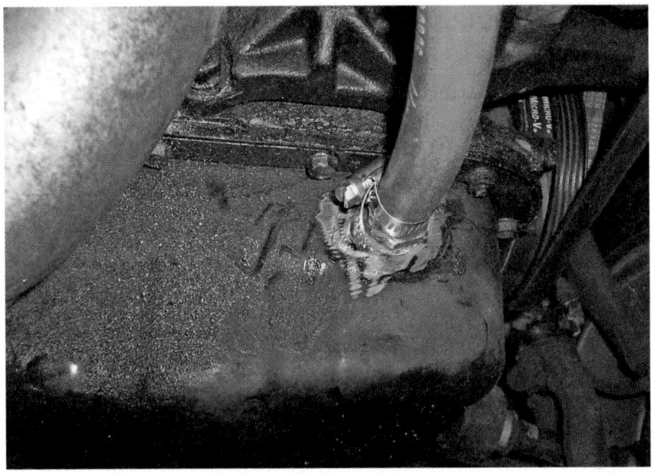

Figure 14-12 The presence of oil and dirt buildup around the oil pan indicates a leaky pan gasket.

system is not working properly. **Blowby** is a term used for the gases that escape the combustion chamber and enter the crankcase. These gases leak between the piston rings and cylinder walls. Blowby gases are normally pressurized air-fuel mixture and/or pressurized exhaust gases. The PCV system provides a continuous flow of fresh air through the crankcase to relieve the pressure and to prevent the formation of corrosive contaminants **(Figure 14–13)**. PCV valves are designed for a particular engine's operating characteristics. Using the wrong valve can cause oil consumption as well as other problems. If the PCV valve or connecting hoses become clogged, excessive pressure will develop in the crankcase. The pressure can cause oil to leak past gaskets and seals. It might also

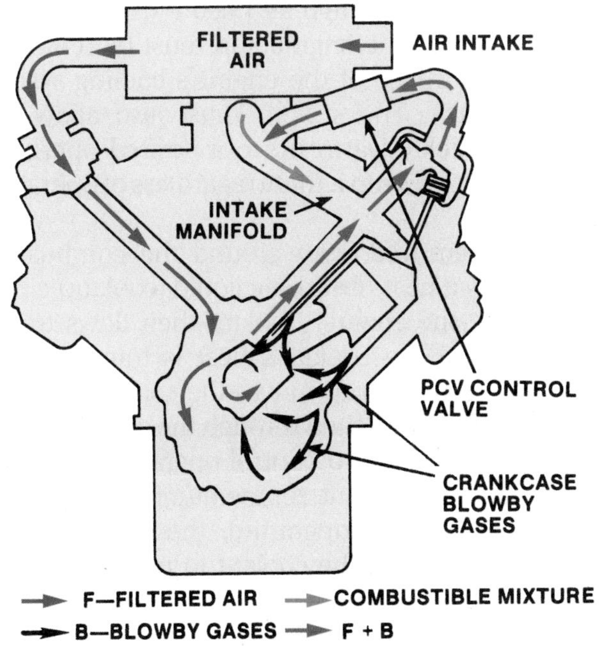

Figure 14-13 The operation of a PCV system.

force oil into the air cleaner or cause it to be drawn into the intake manifold.

Oil Usage Even the smallest oil leak can cause excessive oil consumption. Losing three drops of oil every 100 feet (30.48 m) equals 3 quarts (2.8 liters) of oil lost every 1,000 miles. Typically, new engines use less than a quart (0.946 L) of oil every 6,000 miles (9,656 km). As the engine wears, its oil usage increases. It is not unusual for a high-mileage engine to use a quart (0.946 L) of oil every 1,000 miles (1,609 km).

Sludge

The typical sign of poor maintenance is the buildup of yellow sludge inside the engine. **Sludge (Figure 14–14)** results from the oxidation of oil. When oil oxidizes, chemical compounds in the oil begin to chemically break down and solidify, forming a gel substance. The gelled oil cannot circulate through the engine and collects on engine parts. This buildup of sludge can also block normal lubrication paths. A sludged engine is a sick engine and parts will fail due to poor lubrication **(Figure 14–15)**.

Initial signs of sludge buildup include lower than normal oil pressure, increased fuel consumption, increased emissions, and poor driveability. These are caused by poor lubrication and the heat stored by the gel.

Sludge begins to form at the top of the engine and in the oil pan. As the engine is run with the broken-down oil, more sludge is formed and collects on the various parts of the engine.

A slight buildup of sludge on the inside of the oil filler cap is normal. This is caused by condensation. However, if there is quite a bit of sludge, there is probably sludge throughout the engine. Excessive sludge

Figure 14-14 Sludge buildup on the lower parts of an engine.

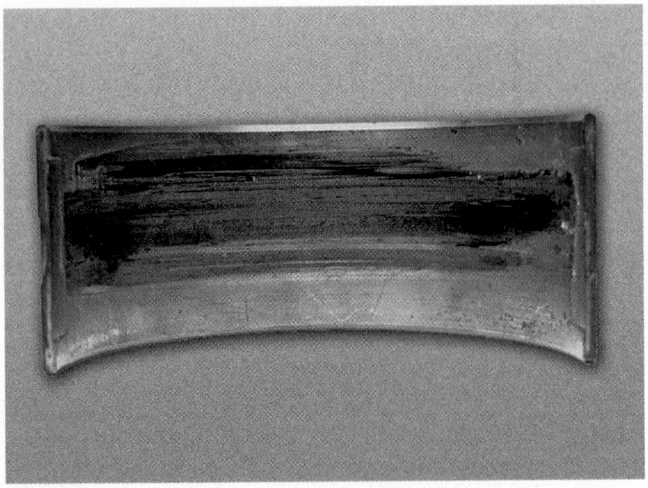

Figure 14-15 Damage to this main bearing was caused by a lack of oil. *Courtesy of Dana Corporation*

can also be caused by a plugged PCV hose or valve. Because the PCV purges the crankcase of vapor and moisture, a plugged system will allow condensation to build up and contaminate the oil. If the sludge on the filler cap is white, suspect a blown head gasket. The whitish gel is caused by coolant mixing with the oil. To verify this condition, use a combustion leak detector or run a cylinder leakage test.

Often, sludge can be removed by flushing the system. However, if the buildup is great, the engine should be torn down and cleaned.

SHOP TALK

Many late-model engines have a sludging problem. There are several reasons for this: These engines have long oil change intervals, use low-viscosity oils, have high operating temperatures, and run with a very lean mixture. Manufacturers of the affected engines have policies and procedures for dealing with these problem engines. Always check with the manufacturer when faced with a sludged engine.

Flushing the System

Flushing the lubrication system on a periodic basis or removing sludge is recommended by some manufacturers. However, there are others that recommend not to flush the engine. Flushing involves running a solvent through the engine and then draining the system. The ways to do this vary, as do the solvents used. Theoretically, flushing should help. The concern of those that do not recommend flushing is simply that the solvents may loosen up some dirt or sludge that will not drain out with the oil. These contaminants can block passages and restrict oil flow. Obviously, the manufacturers of the solvents and flushing equipment disagree.

Oil flushing solvents are available in containers that are added to the engine's oil before an oil change. The engine is run for about one-half hour and then the oil and filter are changed. Flushing machines connect to the filter housing and the drain plug port after the oil has been drained. A heated solution is pumped through the oil reservoir, passages, oil pump, and up into the valve train. The solvent back flushes the oil pump and pickup screen and breaks up and dissolves the sludge. The remains are drawn out by a vacuum. After flushing, a new oil filter is installed and clean oil is added to the engine.

Oil Cooler

If an engine has an external oil cooler, it and its lines should be inspected for leaks. If leaks are evident, the lines and/or cooler should be replaced. The cooler assembly should be flushed or replaced whenever there is a sludge buildup in the engine. If the engine was rebuilt, the cooler should be replaced and the lines cleaned. This is recommended because metal debris that may be trapped in the cooler can become dislodged after the engine has been run and cause oil starvation.

COOLING SYSTEMS

Today's engines create a tremendous amount of heat. Most of this heat is generated during combustion. Metal temperatures around the combustion chamber can run as high as 1,000°F (537.7°C). This heat can destroy the engine and must be removed. This is the purpose of the engine's cooling system **(Figure 14–16)**. The system must also allow the engine to quickly warm up to a desired operating temperature and keep it there regardless of operating conditions.

Heat is removed from around the combustion chambers by a heat-absorbing liquid (coolant) circulating inside the engine. Coolant then flows to the radiator where the coolant's heat is transferred to the outside air. A pump moves the coolant through the engine block and then through the cylinder head. A thermostat is used to control operating temperature. When the coolant temperature is below the desired operating temperature, the thermostat is closed, which allows the coolant to recycle through the engine. When normal operating temperatures are reached, the thermostat opens to allow the coolant to flow to the radiator. The coolant flows to the top of the radiator and loses heat as it flows down through the

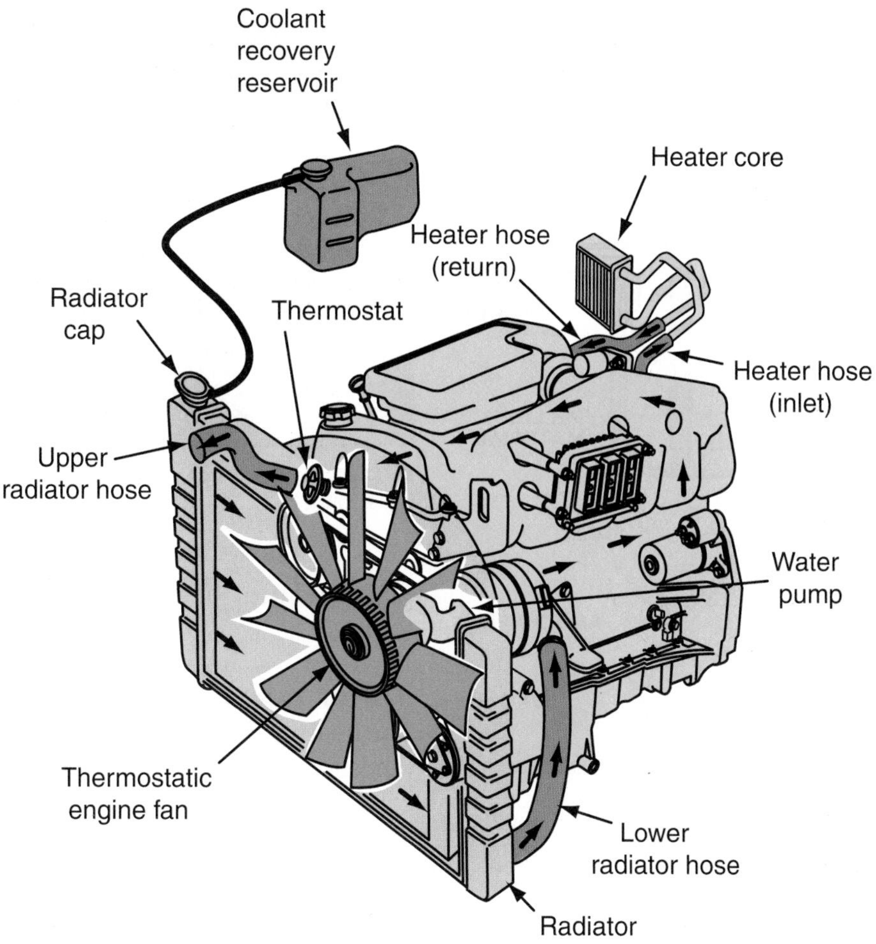

Figure 14-16 Major components of a liquid-cooling system. Arrows indicate the coolant flow.

radiator. Ram air and the airflow from the cooling fan move through the radiator and cool the coolant. The cooled coolant leaves the radiator and enters the water pump and then is sent back through the engine.

Coolant

Go To *Chapter 7 for a detailed discussion of engine coolant.*

Engine coolant is a mixture of water and antifreeze/coolant. Engine coolant has a higher boiling temperature and a lower freezing point than water. The exact boiling or freezing temperatures depend on the mixture. The typical recommended mixture is a 50/50 solution of water and antifreeze/coolant.

Thermostat

The **thermostat** controls the minimum operating temperature of the engine. The maximum operating temperature is controlled by the amount of heat produced by the engine and the cooling system's ability to dissipate the heat.

A thermostat is a temperature-responsive coolant flow control valve. It controls the temperature and amount of coolant entering the radiator. While the engine is cold, the thermostat remains closed **(Figure 14–17A)**, allowing coolant to only circulate in the engine. This allows the engine to uniformly warm up. When the coolant reaches a specified temperature, the thermostat begins to open and allows coolant to flow to the radiator. The hotter the coolant gets, the more the thermostat opens **(Figure 14–17B)**, sending more coolant to the radiator. Once the coolant moves through the radiator, it reenters the water pump. From there it is pushed through the passages in the engine and the cycle starts again.

The thermostat permits fast engine warmup. Slow warmup causes condensation in the crankcase, which can cause the formation of sludge. The thermostat also keeps the coolant above a specific minimum temperature to ensure efficient engine performance. Thermostats must start to open at a specified

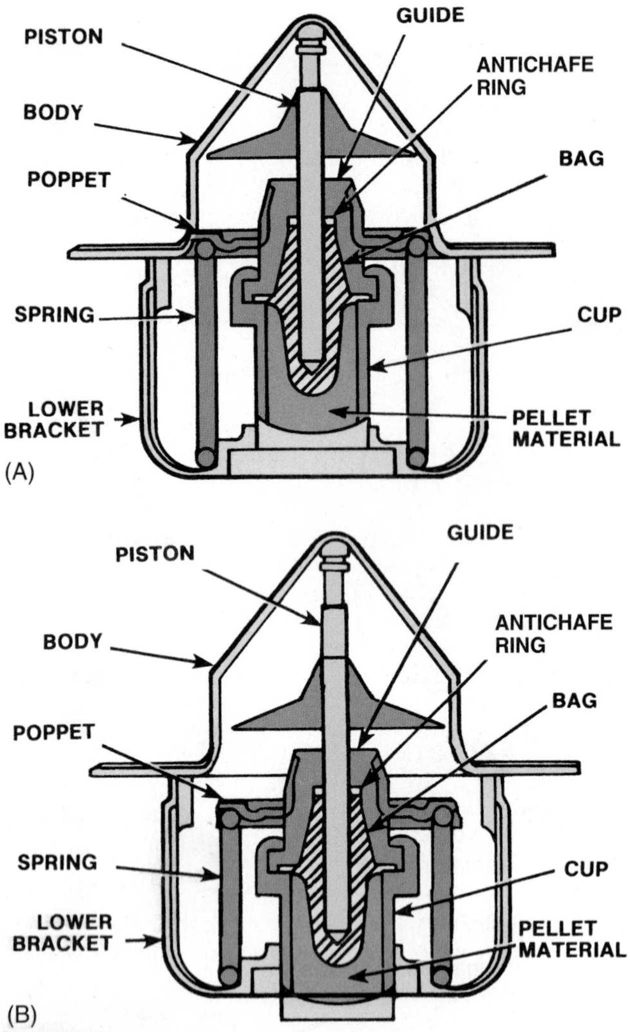

Figure 14-17 (A) Thermostat closed; (B) thermostat open.

temperature—normally 3°F (1.6°C) above or below its temperature rating. It must be fully opened at about 20°F (–7°C) above the "start to open" temperature.

Today's thermostats have a specially formulated wax and powdered metal pellet contained in a heat-conducting copper cup that has a piston inside a rubber boot. Heat causes the wax to expand, forcing the piston outward, which opens the thermostat's valve. The pellet responds to temperature changes and opens and closes the valve to control coolant temperature and flow. Today's thermostats are also designed to slow the flow of coolant when they are open. This prevents the coolant from moving too quickly through the engine. Fast-moving coolant may not have enough time to absorb heat and, therefore, overheating can result.

Most thermostats are located on the top and front of an engine **(Figure 14–18)**. The heat element fits into a recess in the block where it is exposed to hot coolant. The top of the thermostat is then covered by the water outlet housing, which holds it in place and provides a connection to the upper radiator hose.

Figure 14-18 A typical thermostat located in the water outlet.

Water Pump

The heart of the cooling system is the water pump. Its job is to move the coolant through the system. Typically the water pump is driven by the crankshaft through pulleys and a drive belt **(Figure 14–19** and **Figure 14–20)**. On some engines the pump is driven by the camshaft, timing belt or chain, or an electric motor. No matter how these units are driven, they all basically work the same way. The pumps are centrifugal-type pumps **(Figure 14–21)** with a rotating impeller to move the coolant. The shaft is mounted in the water pump housing and rotates on bearings. The pump has a seal to keep the coolant from passing through it. The inlet of the pump connects to the lower radiator hose, and its outlet connects to the engine block.

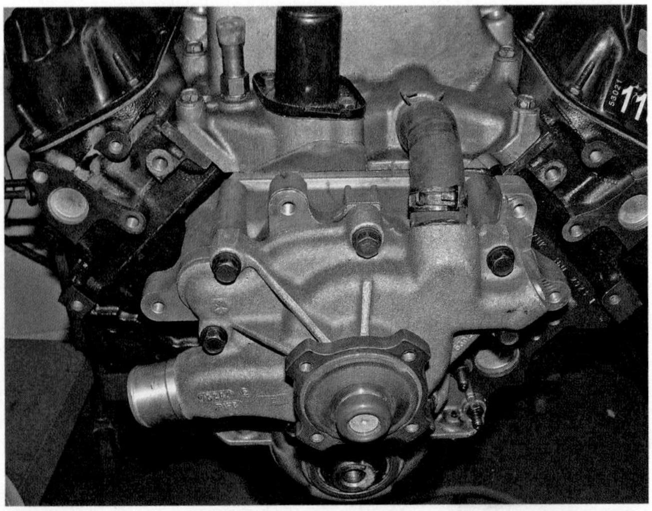

Figure 14-19 A water pump bolted to the front of an engine.

Figure 14-20 This water pump is driven by the timing belt.

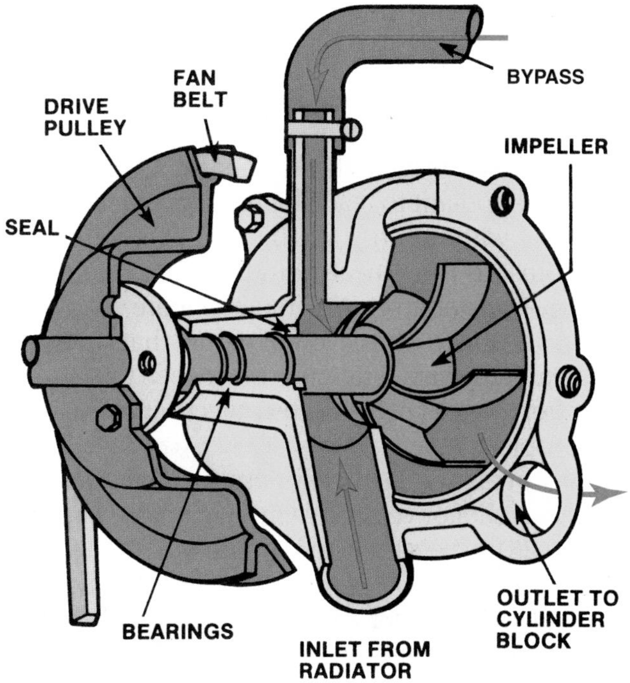

Figure 14-21 An impeller-type water pump.

Radiator

The radiator is basically a heat exchanger, transferring heat from the engine to the air passing through it. The radiator itself is a series of tubes and fins (collectively called the core) that expose the coolant's heat to as much surface area as possible (**Figure 14–22**). Attached to the sides or top and bottom of the core are plastic or aluminum tanks (**Figure 14–23**). One tank holds hot coolant and the other holds the cooled coolant. Cores are normally comprised of flattened aluminum tubes surrounded by thin aluminum fins. The fins conduct the heat from the tubes to the air flowing through the radiator. Most radiators have drain petcocks or plugs near the bottom. Coolant is added at the radiator cap or the recovery tank.

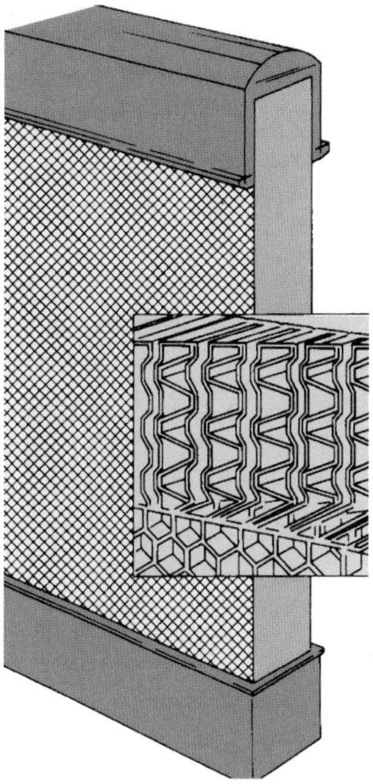

Figure 14-22 A radiator core is made of a series of tubes and fins that expose the coolant's heat to as much surface area as possible.

The efficiency of a radiator depends on its basic design, the area and thickness of the core, the amount of coolant going through the radiator, and the temperature of the cooling air. Today's radiators are designed to limit the amount of heat dissipated; this keeps the coolant somewhat hot at all times. Keeping engines operating at a high temperature is necessary to maintain low emission levels.

Radiators are normally based on one of two designs: cross flow or down flow. In a cross-flow radiator, coolant enters on one side, travels through tubes, and collects on the opposite side. In a down-flow radiator, coolant enters the top of the radiator and is drawn downward by gravity. Cross-flow radiators are seen most often on late-model cars because all the coolant flows through the fan's airstream, and the design allows for body designs with lower hood profiles.

Transmission Cooler Radiators used in vehicles with automatic transmissions have a sealed heat exchanger, or a form of radiator, located in the coolant outlet tank

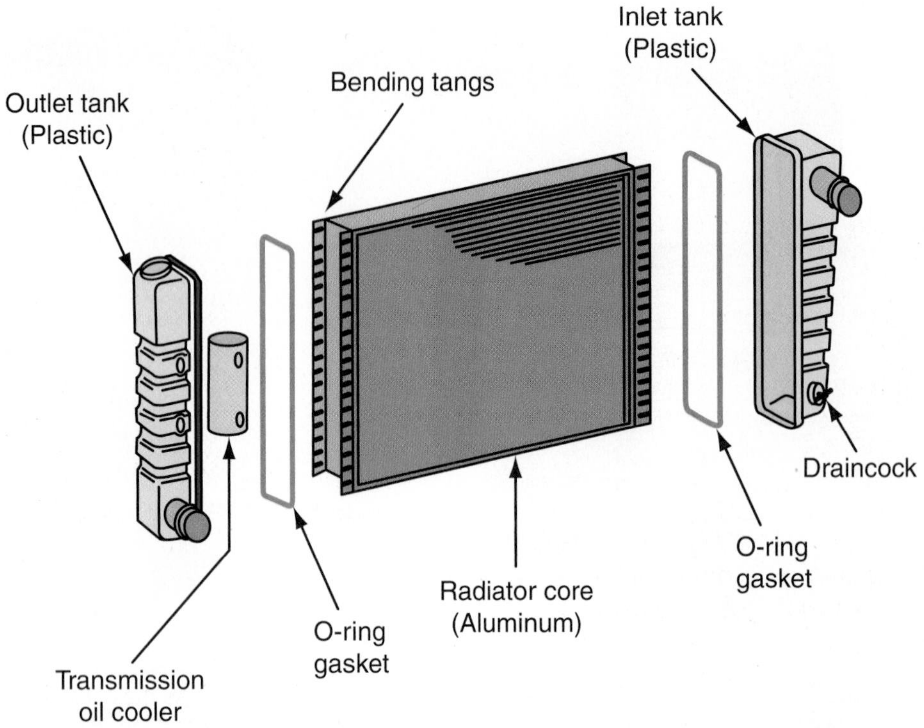

Figure 14–23 The core of a radiator is placed between plastic or aluminum tanks. One tank may contain the oil cooler for an automatic transmission and/or the engine.

of the engine's radiator. Metal or rubber hoses carry hot automatic transmission fluid to the oil cooler. The coolant passing over the sealed oil cooler cools the fluid, which is then returned to the transmission. Cooling the transmission fluid is essential to the efficiency and durability of an automatic transmission.

Radiator Pressure Cap

Radiator caps **(Figure 14–24)** keep the coolant from splashing out of the radiator. They also serve a very important role in keeping the coolant's temperature within a desired range. It does this by keeping the coolant pressurized to a specified level.

The pressure raises the boiling point of the coolant. For every pound of pressure put on the coolant, the boiling point is raised about 3¼ degrees Fahrenheit (1.8°C). Today's caps normally are designed to hold between 14 and 18 psi (93 and 123 kPa). This allows the coolant to reach higher-than-normal temperatures without boiling. This also allows the coolant to absorb more heat from the engine and more heat to transfer from the radiator core to outside air. This is due to a basic law of nature that states that the greater the heat difference is between two objects, the faster the heat of the hotter object will move to the cooler object.

The pressure in the system is regulated by a pressure relief or vent valve in the radiator cap **(Figure 14–25)**. When the cap is tightened on the radiator's filler neck, it seals the upper and lower sealing surfaces of the neck. The pressure relief valve is compressed against the lower seal. Coolant pressure builds up as the temperature of the coolant rises. When the pressure reaches the pressure rating of the cap, it pushes up on the spring in the pressure relief valve. This opens the valve and allows excess pressure to exit the radiator through a bore between the upper and lower seals. The bore is connected, by a tube, to the expansion or recovery tank. When enough pressure has been released

Figure 14–24 A radiator cap on a late-model engine.

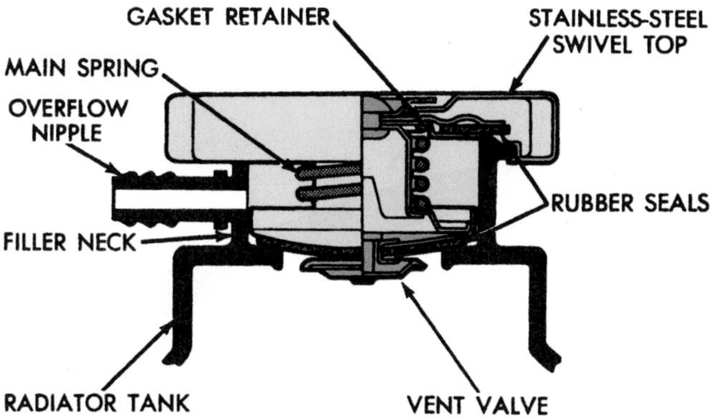

Figure 14–25 Parts of a radiator pressure cap assembly. *Courtesy of Chrysler LLC*

to drop system pressure below the cap's rating, the spring will close the pressure relief valve. When the coolant cools, its pressure drops. The low pressure opens the vacuum relief valve. The low pressure then draws coolant from the expansion tank to refill the radiator.

All radiator caps are designed to meet SAE standards for safety. This standard specifies that there shall be a detent or safety stop position allowing pressure to escape from the system without allowing the hot coolant to blow out of the radiator's neck. Only after all pressure has been relieved should the cap be removed from the filler neck.

WARNING!

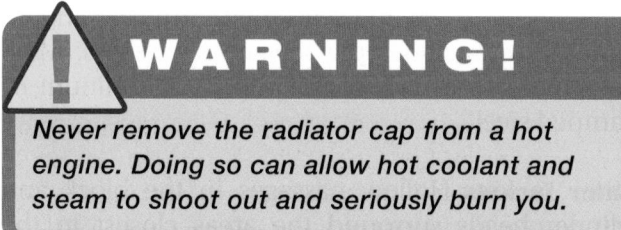

Never remove the radiator cap from a hot engine. Doing so can allow hot coolant and steam to shoot out and seriously burn you.

Radiator pressure cap specifications require that the cap must not leak below the low limit of the pressure range and must open above the high limit. They are labeled by the amount of pressure they should hold. For domestic vehicles, the pressure is stated in psi or kPa. Normally kPa ratings are expressed as a number times 100 kPa, for example: 1.3 × 100 kPa. This is a 130 kPa cap.

Radiator caps for some imported vehicles may have different markings. Some may be marked "0.9 Bar." This indicates that the pressure rating of the cap is 0.9 times normal atmospheric pressure. Because atmospheric pressure is 14.7 psi, a 0.9 cap has a pressure rating of about 13.2 psi (14.7 × 0.9). Another common rating is "100." The "100" indicates that the pressure rating of the cap is 100% of atmospheric pressure, or 14.7 psi.

Figure 14–26 A coolant expansion tank.

Expansion Tank All late-model cooling systems have an **expansion** or **recovery tank** (**Figure 14–26**). Cooling systems with expansion tanks are called closed-cooling systems. They are designed to catch and hold any coolant that passes through the pressure cap. As the engine warms up, the coolant expands. This eventually causes the pressure cap to release. The coolant passes to an expansion tank. When the engine is shut down, the coolant begins to shrink. Eventually, the vacuum spring inside the pressure cap opens and the coolant in the expansion tank is drawn back into the cooling system.

Hoses

Coolant flows from the engine to the radiator and from the radiator to the engine through radiator hoses. The hoses are usually made of butyl or neoprene rubber hoses to cushion engine vibrations and prevent damage to the radiator.

A hose is typically made up of three parts: an inner rubber tube, some reinforcement material, and an outer rubber cover. Different covers and reinforcements are used depending on the application of the hose. All three parts are bonded together. Basically the difference in hose construction lies in what the hose will carry, where it is located, and the temperature and pressure it will face. Cooling system hoses must be able to endure heavy vibrations and be resistant to oil, heat, abrasion, weathering, and pressure.

Most vehicles have at least four hoses in the cooling system; some have five or more (**Figure 14–27**). Two small diameter hoses send hot coolant from the water pump to the heater core and back. Two larger diameter hoses move the coolant from the water pump to the radiator and back into the engine block. The fifth hose is a small diameter bypass hose that

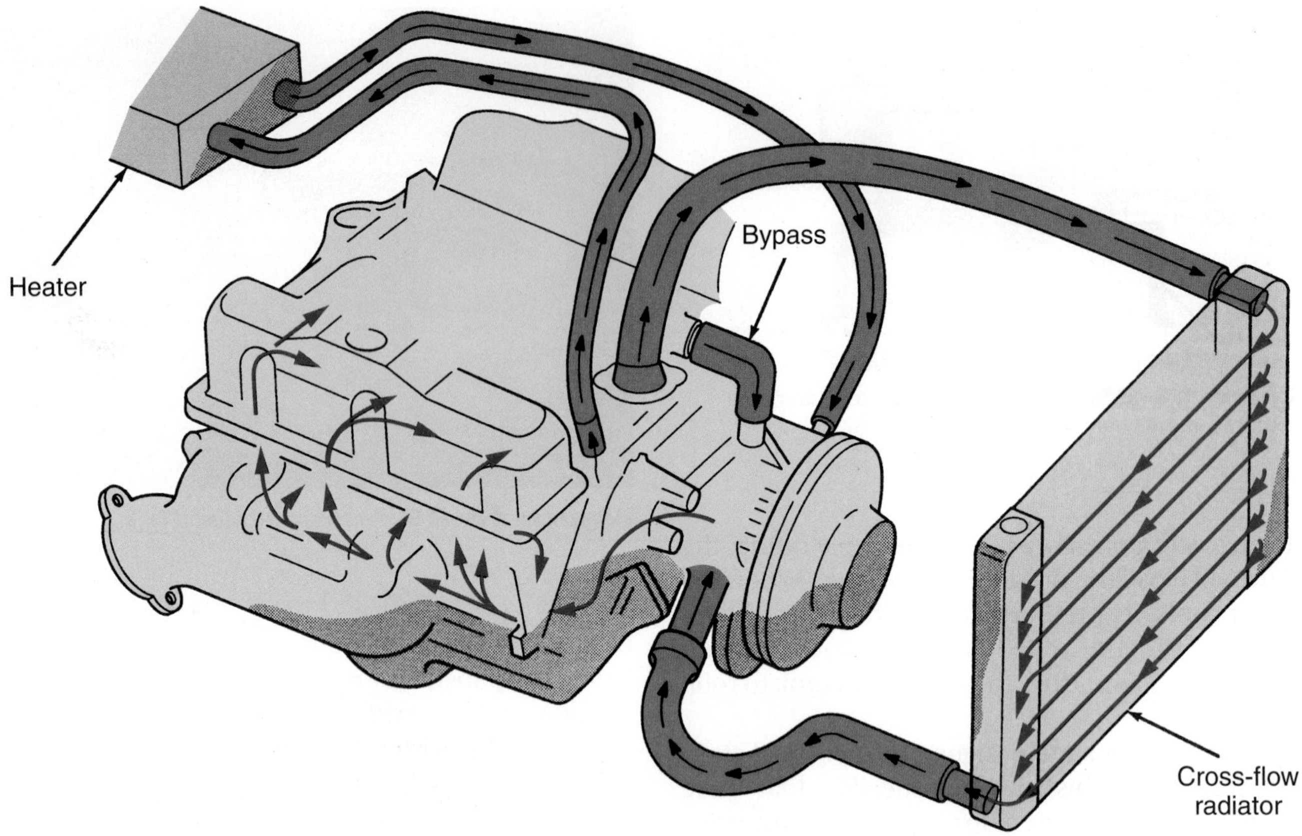

Figure 14–27 Routing of coolant through the cooling system hoses. *Courtesy of Chrysler LLC*

allows coolant to circulate within the engine when the thermostat is closed. This hose is not required on all engines because the bypass feature is built into the engine block or cylinder head.

Hoses are sized according to their inside diameter. For example, common heater hoses are ⅝ or ¾ inch. Radiator hoses are larger and have reinforcements that allow them to withstand about six times the normal operating pressure of the cooling system.

Radiator hoses are seldom straight tubes. They typically must bend or curve around parts to make a good connection without kinking. Straight hoses are not used because bending will cause them to collapse at the bend, causing a restriction. Most original equipment radiator hoses are molded to a specific shape to fit specific applications. Lower radiator hoses are normally reinforced with wire to prevent them from collapsing due to the suction of the water pump.

Heater hoses are made with reinforcements to help keep their shape. Some applications require a molded shape due to complex routing or curves. Some applications can use straight pieces of hose; however, most have bends.

Water Outlet The water outlet is the connection between the engine and the upper radiator hose through which hot coolant from the engine is pumped into the radiator. It has been called a gooseneck, an elbow, an inlet, an outlet, or a thermostat housing. Generally, it covers and seals the thermostat and, in some cases, includes the thermostat bypass. Most water outlets are made of cast iron, cast aluminum, or stamped steel.

Water Jackets Hollow passages in the block and cylinder heads surround the areas closest to the cylinders and combustion chambers **(Figure 14–28)**. Included in the water jackets are soft (core) plugs and a block drain plug. Some engines are equipped with plastic liners that direct coolant flow around critical areas. The soft plugs and drain plugs are usually removed during engine teardown. New ones are installed during reassembly. Core plugs are prone to rust and corrosion and, therefore, will weep coolant or rust through completely. When this happens, the core plugs should be replaced.

Hose Clamps

Hoses are attached to the engine and radiator with clamps **(Figure 14–29)**. Hose clamps are designed to apply clamping pressure around the outside of the hose at the point where it connects to the inlet and outlet connections at the radiator, engine block, water pump, or heater core. The pressure exerted on this

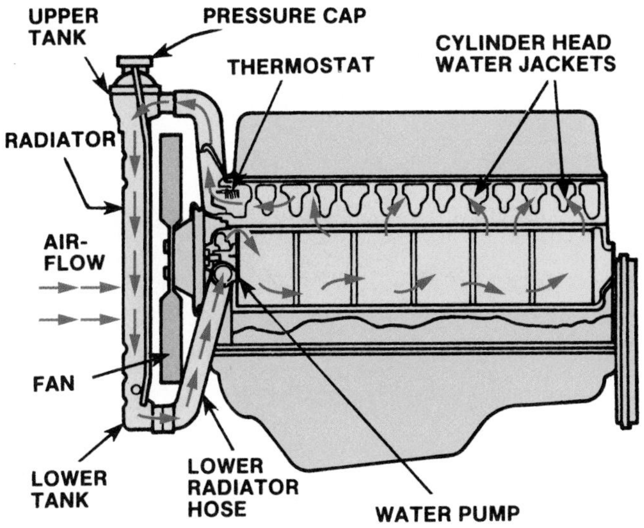

Figure 14-28 The cooling system circulates coolant through the engine's water jackets.

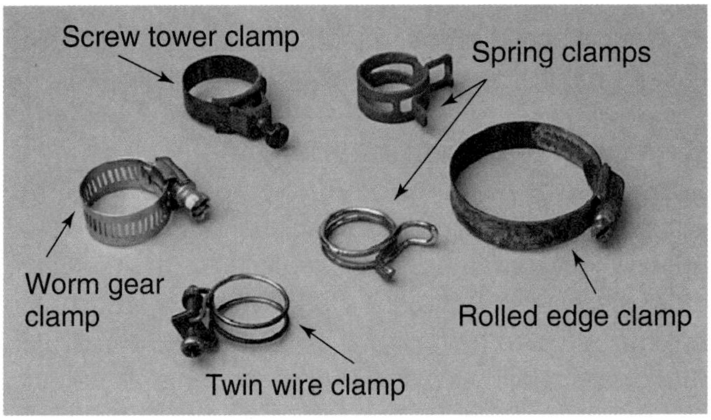

Figure 14-29 Common types of hose clamps.

connection is important to making and maintaining a seal at that point.

Belt Drives

Belt drives are used to power the water pump and/or cooling fan on many engines. The belts must be in good condition and properly tensioned in order to drive this at the correct speed.

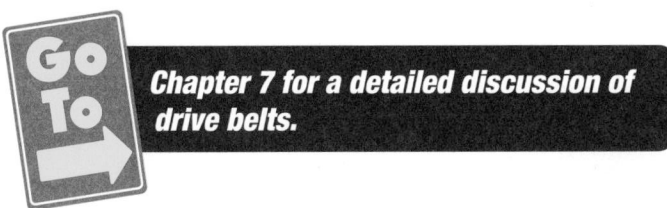

> **Go To** *Chapter 7 for a detailed discussion of drive belts.*

Heater System

A hot liquid passenger compartment heater is part of the engine's cooling system. Heated coolant flows from the engine through heater hoses and a heater control valve to a heater core located on either side of the fire wall **(Figure 14–30)**. Air is directed or blown

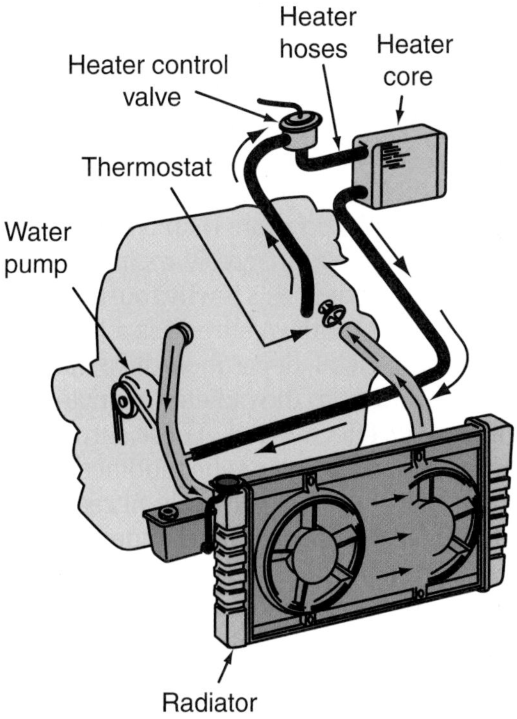

Figure 14-30 The coolant for the heater core is sent from the upper part of the engine through the heater core and back to the inlet side of the water pump.

over the heater core, and the heated air flows into the passenger compartment. Movable doors can be controlled to blend cool air with heated air for more or less heat.

Cooling Fans

The efficiency of the cooling system depends on the amount of heat that can be removed from the system and transferred to the air. At highway speeds, the ram air through the radiator should be sufficient to maintain proper cooling. At low speeds and idle, the system needs additional air. This air is delivered by a fan **(Figure 14–31)**. The fan may be driven by the engine,

Figure 14-31 A belt-driven cooling fan.

via a belt, or driven by an electric motor. A belt-driven fan is bolted to a pulley on the water pump and turns constantly with the engine. Thus, belt-driven fans always draw air through the radiator from the rear. The fan has several blades made of steel, nylon, or fiberglass, which are attached to a metal hub.

A fan that is placed more than 3 inches from the radiator is ineffective. It merely recirculates hot air around the fan blades. This is why most radiators are equipped with shrouds. A shroud is a large, circular piece of plastic, metal, or cardboardlike material that extends outward from the radiator to enclose the fan and increase its effectiveness. These shrouds should always be kept intact and not be modified.

Because fan air is usually only necessary at idle and low-speed operation, various design concepts are used to limit the fan's operation at higher speeds. Horsepower is required to turn the fan. Therefore, the operation of a cooling fan reduces the available horsepower to the drive wheels as well as the fuel economy of the vehicle. Fans are also very noisy at high speeds, adding to driver fatigue and total vehicle noise.

To eliminate this power drain during times when fan operation is not needed, many of today's belt-driven fans operate only when the engine and radiator heat up. This is accomplished by a **fan clutch** **(Figure 14–32)** located between the water pump pulley and the fan. When the engine and fan clutch are cold, the fan moves independently from the clutch and moves little air. The clutch locks the fan to its hub when the temperature of the air around the fan reaches a particular point. In most cases, the clutch slips at high speeds; therefore, it is not turning at full engine speed. The clutch assemblies rely on a thermostatic spring or silicone fluid.

Some engines have flexible blades or **flex-blades** that bend or change pitch based on engine speed. That is, at slower speeds the blade pitch is at the maximum. As engine speed increases, the blade pitch decreases as does the horsepower losses and noise levels.

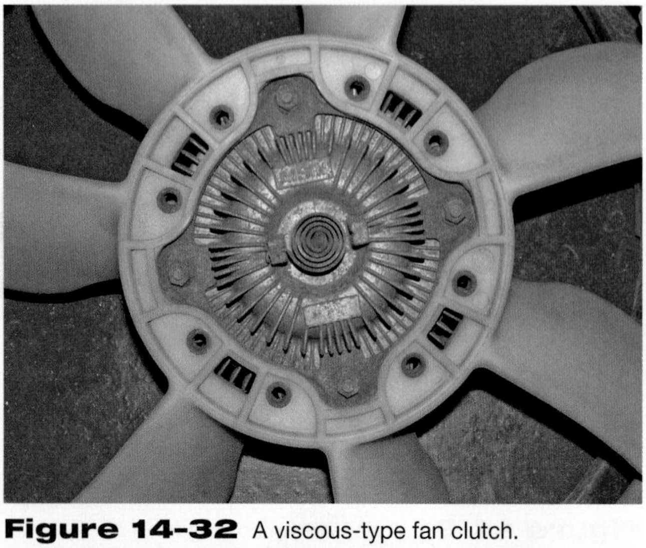

Figure 14–32 A viscous-type fan clutch.

Figure 14–33 This car has two separate electric cooling fans.

Electric Cooling Fans In most late-model applications, to save power and reduce the noise level, the conventional belt-driven, water pump–mounted engine cooling fan has been replaced with an electrically driven fan **(Figure 14–33)**. This fan and motor are mounted to a shroud. The 12-volt, motor-driven fan is electrically controlled by an engine coolant temperature switch and/or the air conditioner switch.

As the schematic in **Figure 14–34** shows, the cooling fan motor is connected to the battery through a normally open (NO) set of contacts in the cooling fan relay.

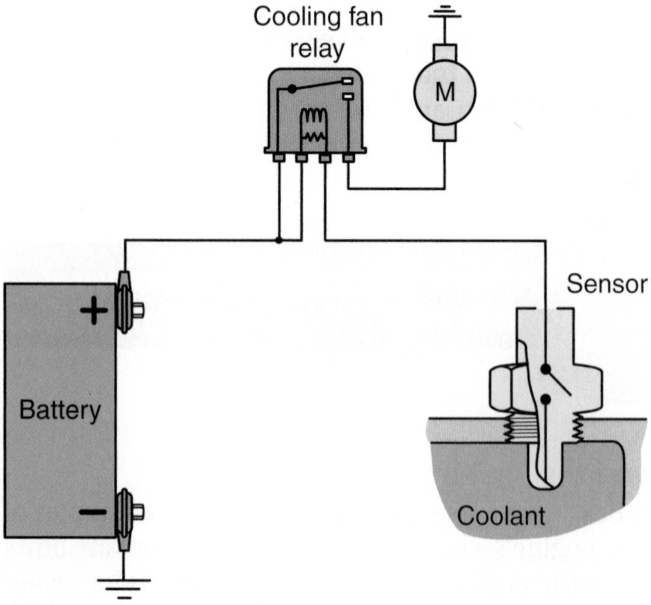

Figure 14–34 A simple schematic for an electric cooling fan.

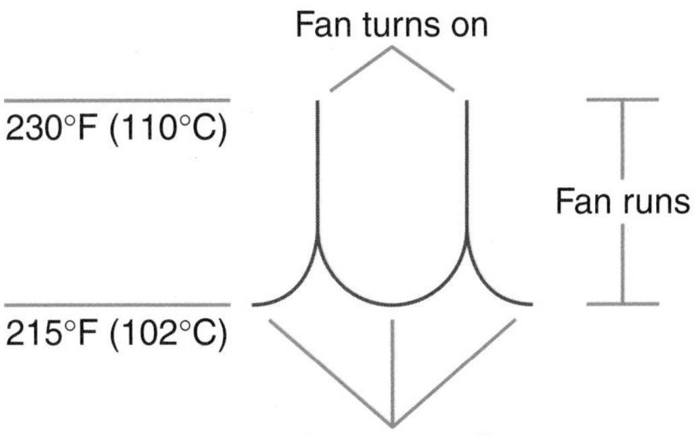

Figure 14–35 A simple graph showing when an electric cooling fan turns on and off.

During normal operation, with the air conditioner off and the engine coolant below a predetermined temperature of approximately 215°F (101.6°C), the relay contacts are open and the fan motor does not operate.

When coolant temperatures exceed approximately 230°F (110°C), the coolant temperature switch closes **(Figure 14–35)**. This energizes the fan relay coil, which in turn closes the relay contacts. The contacts provide 12 volts to the fan motor. When the air conditioner select switch is set to any cool position, regardless of engine temperature, a circuit is completed through the relay coil to ground. This closes the relay contacts to provide 12 volts to the fan motor. The fan then operates as long as the air conditioner and ignition switches are on.

There are many types of electric cooling fans. Some provide a cool-down period, which means the fan continues to run after the engine has been stopped and the ignition switch is turned off. These systems have a second temperature sensor. The fan stops only when the engine coolant falls to a predetermined temperature, usually about 210°F (98.8°C). In some systems, the fan does not start when the air conditioner is turned on unless the high side of the A/C system is above a predetermined temperature and/or pressure.

Some late-model cars control the cooling fan by completing the ground through the engine control computer. Check the service manual to see how an electric cooling fan is controlled before working with it.

CAUTION!

The electric cooling fan can come on at any time without warning even if the engine is not running. Always remove the negative terminal at the battery or the connector at the cooling fan motor while working around an electric fan. Make sure you reconnect the connector before giving the car back to the customer.

Figure 14–36 A coolant temperature sensor.

Temperature Sensors Proper electric cooling fan operation depends on the operation of a temperature sensor. A temperature sensor responds to changes in temperature. Some vehicles use more than one sensor to control the fans and to send engine temperature readings to the PCM. Based on this information, the PCM will adjust the fuel injection and ignition systems to provide efficient engine operation.

Temperature Indicators

Coolant temperature indicators alert the driver of an overheating condition. These indicators are a temperature gauge and/or a warning light. A temperature sensor is threaded into a bore in a water jacket **(Figure 14–36)**. Besides displaying coolant temperatures, temperature sensors supply some important information to the computers in engine control systems.

COOLING SYSTEM DIAGNOSIS

The cooling system must be inspected and serviced as a system. Replacing one damaged part while leaving others dirty or clogged will not increase system efficiency. Diagnosis of the system involves both a visual inspection of the parts, simple checks and tests, and leak testing. One of the first checks of the cooling system is the checking of coolant level and condition. These checks should be done during normal preventive maintenance and when there is a problem. All diagnosis of the cooling system should begin with a check of the coolant.

Go To *Chapter 7 for a general discussion of checking the coolant level and condition.*

There are marks on most expansion tanks that show where coolant levels should be when the car is hot and when it is cold. To check coolant levels on a car without a recovery tank, remove the radiator cap (when the engine is cold) and see if the coolant is up to where it should be. If there are no markings, make sure the coolant is covering the radiator core. If the coolant level is low after repeated filling, there is probably a leak in the cooling system. The condition or effectiveness of the coolant should be checked with a hydrometer, a refractometer, or alkaline test strips.

CAUTION!

When working on the cooling system, remember that at operating temperature the coolant is extremely hot. Touching the coolant or spilling it can cause serious body burns. Never remove the radiator cap when the engine is hot.

The proper mixture of water and antifreeze also reduces the amount of rust and lime deposits that can form in the system. These deposits tend to insulate the walls of the water jackets. As a result, the coolant is less able to absorb the engine's heat at the points where there is scale. This causes engine hot spots that result in increased component wear and make overheating more likely **(Figure 14–37)**.

Regardless of the mixture of the coolant or the type of antifreeze used, some lime, rust, and scale will always build in a cooling system. Any deposit on the walls of the water jackets will affect engine cooling. Changes in engine temperature cause the engine parts to expand and contract. Some of these deposits then break off and become suspended in the coolant. The coolant then becomes contaminated and the deposits may collect at a narrow passage, making the passage narrower. This restriction would further lessen the effectiveness of the cooling system. For these reasons and others, the engine's coolant should be replaced and the cooling system flushed every 1 or 2 years.

Testing for Electrolysis in Cooling Systems

Electrolysis is a process in which an electrical current is passed through water, causing the separation of hydrogen and oxygen molecules. In a cooling system, electrolysis removes the protective layer on the inside of the radiator tubes. It also can cause serious engine failures. Electrolysis occurs when there is improper grounding of electrical accessories and equipment or static electricity buildup somewhere in the vehicle. Checking for these should be a part of all checks of the coolant.

To check for the conditions prone to electrolysis, use a voltmeter that is capable of measuring AC and DC voltage. Set the meter so it can read in tenths of a volt DC. Attach the negative meter lead to a good ground. Place the positive lead into the coolant **(Figure 14–38)**. Look at the meter while the

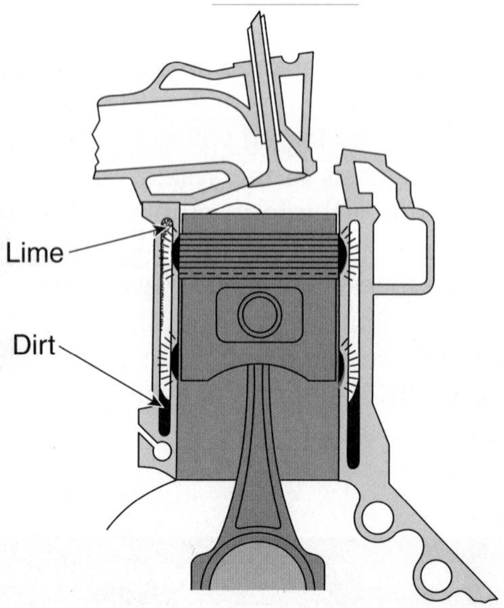

Figure 14-37 Lime and dirt buildup in the coolant passages tend to insulate the walls of the water jackets and can cause hot spots that result in increased wear and make overheating more likely.

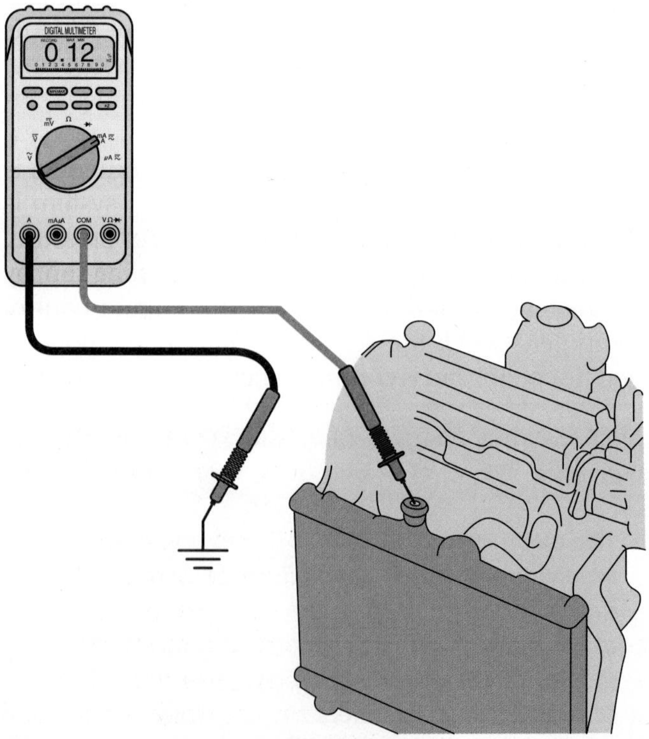

Figure 14-38 The setup for checking for conditions that can cause electrolysis.

engine is cranked with the starter and record the readings. Take another reading with the engine running and all accessories turned on. Record that reading. Voltage readings of 0 to 0.3 volt are normal for a cast-iron engine; normal readings for a bimetal or aluminum engine are half that amount. Repeat the test with the DMM in the AC voltage mode. There will be AC voltage if the problem is static electricity. Any readings above the normal indicate a problem.

To isolate the problem, look at where the high voltage was measured and think about the systems that were energized during that time. If the voltage was high when all of the accessories were turned on, turn them off one at a time until the voltage drops to normal. The circuit that was turned off prior to the drop in voltage has ground problems. After the electrical problems have been corrected, flush and replace the coolant.

INSPECTION OF COOLING SYSTEM

The most common cooling system problem is overheating. There are many reasons for this. Diagnosis of this condition involves many steps, simply because many things can cause this problem **(Figure 14–39)**. Basically, overheating can be caused by anything that decreases the cooling system's ability to absorb, transport, and dissipate heat: The first step is to determine whether the engine is indeed overheating.

SHOP TALK

A few late-model engines, such as GM's Northstar engine, will shut off cylinders to "air-cool" the engine and keep it running at reduced power when they overheat. This prevents the serious damage that may occur in other engines.

An overheating concern normally begins with high readings on the vehicle's temperature gauge or the illumination of the temperature warning lamp. These can be caused by a cooling system problem or a faulty temperature sensor, although when the engine is greatly overheating, it is obvious by the steam emitted by the system or by smell. The best way to check the accuracy of the temperature indicators is to measure the temperature of the coolant. If the indicators seem to be wrong, troubleshoot and repair the electrical circuit. Then recheck the system's temperature.

Condition	Cause
Overheats in heavy traffic or after idling for a long time	■ Low coolant level ■ Faulty radiator cap ■ Faulty thermostat ■ Cooling fan is not turning on ■ Restricted airflow through the radiator ■ Leaking head gasket ■ Restricted exhaust ■ Water pump impeller is corroded
Overheats when driving at speed, or after repeated heavy acceleration	■ Radiator and/or block are internally clogged with rust, scale, silt, or gel ■ Restricted airflow through the radiator ■ Faulty radiator cap ■ Faulty thermostat ■ Radiator fins are corroded and falling off ■ Water pump impeller is corroded ■ Collapsed lower radiator hose ■ Dragging brakes
Overheats any time or erratically	■ Low coolant level ■ Faulty radiator cap ■ Faulty thermostat ■ Temperature sender or related electrical problem ■ Cooling fan is not turning on
Overheats shortly after the engine is started	■ Temperature sender or related electrical problem
Seems slightly too hot all of the time; gauge nears the red zone at times	■ Radiator and/or block are internally clogged with rust, scale, silt, or gel ■ Restricted airflow through the radiator ■ Faulty radiator cap ■ Faulty thermostat ■ Radiator fins are corroded and falling off ■ Collapsed lower radiator hose ■ Cooling fan is not turning on
Bubbles in the coolant expansion tank	■ Faulty radiator cap ■ Failed head gasket
Air in the radiator but the expansion tank is full	■ Coolant leak ■ Faulty radiator cap ■ Air in the system ■ Faulty seal between the radiator cap and expansion tank ■ Failed head gasket

Figure 14-39 Common causes of engine overheating.

SHOP TALK

It is difficult to measure the actual temperature of the coolant on most late-model engines because the only access to the coolant is through the expansion tank. The coolant in the tank does not represent the coolant in the engine and it is not heated. Use a temperature probe or infrared sensor to measure the temperature of the inlet radiator tank. This will give an accurate measurement of system temperature.

Normal operating temperature for most engines is 195 to 220°F (91 to 104°C). To maintain this temperature, the coolant must circulate through the engine and radiator. Anything that will interfere with the movement of coolant, such as a faulty water pump or thermostat, or a loss of coolant, will cause overheating. Likewise anything that interferes with the passing of air through the radiator will also cause the engine to overheat.

Effects of Overheating

Engine overheating can cause the following problems:

- Detonation
- Preignition
- Blown head gasket
- Warped cylinder head
- OHC cam seizure and breakage
- Blown hoses
- Radiator leaks
- Cylinder damage due to swelling pistons
- Sticky exhaust valve stems
- Engine bearing damage

Temperature Test

A temperature test can be performed with an infrared temperature sensor, thermometer, or temperature probe. The latter may be a feature of a digital multimeter (DMM). A temperature test allows for monitoring temperature change through the cooling system. When a cold engine is started, the opening temperature of the thermostat can be observed as the engine warms. This can be compared to specifications to determine if the thermostat is bad. Once the engine has warmed up, the probe can be used to scan for cool spots in the radiator. These indicate an area where coolant is not flowing freely through the radiator. The cooling fan temperature switch can also be checked.

Radiator Checks

Cold spots on the radiator indicate internal restrictions. In most cases, this requires removal of the radiator so it can be deeply flushed or replaced. Normal cooling system flushing will normally not remove the restrictions. The restrictions are typically caused by internal corrosion or a buildup of scale and lime.

The radiator should also be inspected for external restrictions and for evidence of leaks. Dirt, bugs, and other debris on the surface of the radiator will block airflow. These should be removed by careful cleaning. Also, check for loose cooling fins. Salt and other road debris can corrode the solder used to attach the fins around the radiator's tubes. When the fins are not attached to the tubes, heat is not as easily transferred to the outside air.

Checking Hoses

Carefully check all cooling hoses for leakage, swelling, and chafing. Also replace any hose that feels mushy or extremely brittle when squeezed firmly **(Figure 14–40)**. When a hose becomes soft, it is deteriorating and should be replaced before more serious problems result. Hard hoses will resist flexing and may crack rather than bend and should be replaced.

Normally, hoses begin to deteriorate from the inside. Pieces of deteriorated hose will circulate through the system until they find a place to rest. This place is usually the radiator core, causing clogging. Deterioration can also cause leaks. Any external bulging or cracking of hoses is a definite sign of failure. When one hose fails, all of the others should be carefully inspected.

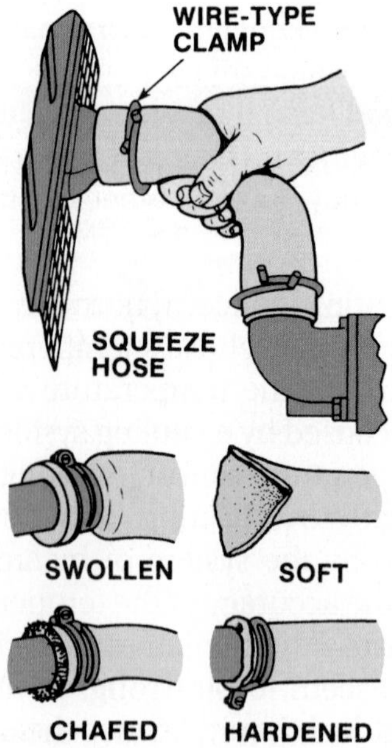

Figure 14–40 Defects in cooling hoses.

The upper radiator hose has the roughest service life of any hose in the cooling system. It must absorb more engine motion than any of the other hoses. It is exposed to the coolant at its hottest stage, and it is insulated by the hood during hot soak periods. These conditions make the upper hose the most likely to fail.

Check the firmness of the lower radiator hose to make sure that the internal reinforcement spring is in place and not damaged. Without this reinforcement, the hose can collapse under vacuum at high engine speeds and restrict coolant flow from the radiator to the engine.

Squeeze each hose and look for splits. These splits can burst wide open under pressure. Also look for rust stains around the clamps. Rust stains indicate a leak, possibly because the clamp has eaten into the hose. Loosen the clamp, slide it back, and check the hose for cuts.

A major cause of hose failure is an electrochemical attack on the rubber hose. This is known as **electrochemical degradation (ECD)**. It occurs because the hose, engine coolant, and the engine/radiator fittings form a galvanic (battery) cell. This chemical reaction causes very small cracks in the hose, allowing the coolant to attack and weaken the reinforcement in the hose **(Figure 14–41)**. ECD can cause pinhole leaks or hose rupture under normal operating pressures. The effects of ECD are accelerated by high temperatures and vibrations.

The best way to check hoses for ECD is to squeeze the hose near the clamps or connectors. ECD occurs within 2 inches of the ends of the hose—not in the middle. Compare the feel of the hose between the middle and the ends. Gaps can be felt along the length of the hose where it has been weakened by ECD. If the ends are soft and feel mushy, chances are the hose is under attack by ECD and the hose should be replaced.

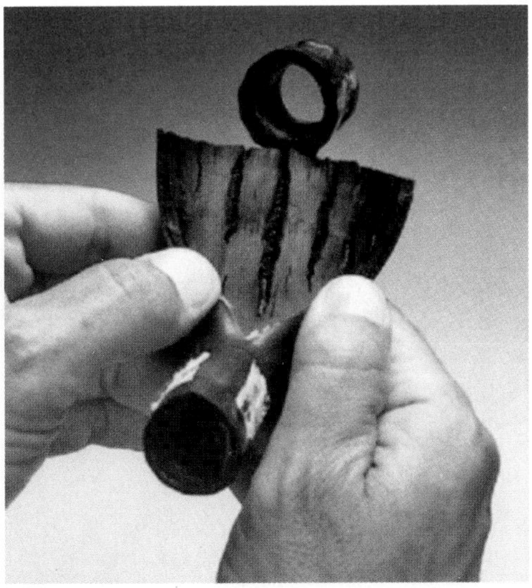

Figure 14–41 An ECD-damaged hose. *Courtesy of Gates Corporation*

Oil is another enemy to rubber hoses. A hose damaged by oil is swollen, soft, and sticky. If the oil leak is external, eliminate the oil leak or try to reroute the hose away from the oil leak to prevent oil damage to a new hose. At times, the oil damage occurs inside the hose. This can be caused by transmission fluid leaking into the coolant or an internal engine oil leak.

Belt Drives

Belts tend to overcure due to excessive heat. This causes the rubber to harden and crack. Excessive heat usually comes from slippage. Slippage can normally be attributed to lack of proper belt tension or oily conditions. When slippage occurs heat not only overcures the belt, it also travels through the drive pulley and down the shaft to the support bearings. These bearings might be damaged if the slippage is allowed to continue. As a V-belt wears, it begins to ride deeper in the pulley groove. This reduces its tension and promotes slippage. Carefully inspect all drive belts and replace them as necessary.

Go To
Chapter 7 for the correct procedure for changing a belt and adjusting its tension.

Checking Fans and Fan Clutches

The cooling fan is a very important part of the cooling system. Most engine-driven fans have a clutch, and most overheating problems are caused by a defective clutch. However, the fan itself as well as the fan shroud must be thoroughly checked. The fan shroud should be securely fastened to the radiator bracket. The shroud should also be checked for cracks and other damage. A damaged shroud should be replaced.

Any damage or distortion to the fan will cause it to be out of balance. This can cause major problems, including rapid and excessive water pump bearing and seal wear or damage to the radiator if the fan blades hit the radiator. A noticeable wobble as the fan spins means that the fan should be replaced.

Fan clutches are filled with silicone oil. The oil responds to speed and temperature. As the engine increases speed, the oil allows the fan to slip on its hub. This reduces the drag on the engine. If the clutch allows the fan to slip during low engine speeds, overheating can occur. A loss or deterioration of the oil will cause fan slippage. Over time, the oil begins to break down and offers less of a coupling between the fan and hub. It is said that the life of a fan clutch is about the same as that of a water pump and when one needs replacing, the other should also be replaced.

The clutch assembly should be carefully checked for leakage. Oily streaks radiating outward from the hub shaft mean that fluid has leaked out past the bearing seal. A leaking clutch should be replaced. The clutch should also be replaced if the fan can be spun with little or no resistance with the engine off, if the clutch wobbles when the fan is moved in or out, or if there are damaged or missing fins on the assembly.

Electric Cooling Fans The action of the electric cooling fan should be observed. However, before doing so, check the mounting of the fan assembly and the condition of the fan blades. The fan should be energized when the A/C is turned on and when the coolant reaches a specified temperature. If the fan does not turn on when it should, check the motor. Do this by connecting the motor directly to the battery. If the motor runs, the problem is in the motor's control circuit. Diagnosis of this problem should follow the prescribed sequence given by the manufacturer. If the jumped motor does not run, the problem is the fan motor **(Figure 14–42)**.

Testing the Thermostat

Thermostats are often the cause of overheating or poor heater performance. They can also cause an increase in fuel consumption and poor engine performance. Electronic engine control systems are programmed to deliver the ideal air-fuel mixture and ignition timing according to the engine's operating conditions. One of the conditions monitored by the PCM is engine temperature. The PCM also controls and regulates engine coolant temperatures to optimize efficiency.

A properly working thermostat is closed when the engine is cold and the coolant's temperature is lower than the thermostat's rating. Once the coolant reaches the opening temperature of the thermostat, the thermostat opens and allows coolant to flow to the radiator. The hotter the coolant gets, the more the thermostat opens, allowing more coolant to flow through the radiator. If the thermostat is stuck open, the coolant may not reach the desired temperature because it is cooled before it gets hot. If a thermostat is stuck closed because it failed or because there is a steam pocket below the thermostat, coolant will not flow between the engine and the radiator; the engine will, therefore, quickly overheat. The presence of steam results from low coolant levels or air in the system.

The easiest way to check thermostat operation is to measure engine temperature with an infrared thermometer **(Figure 14–43)**. However, you can use your hand to feel the temperature. Touch the upper and lower radiator hoses when the engine is first started. If the hoses do not become hot within several minutes, the thermostat is not opening.

Another way to test a thermostat is to remove it and totally submerge it in a small container of water so that it does not touch the bottom **(Figure 14–44)**. Place a thermometer in the water so that it does not touch the container and only measures water temperature. Heat the water. When the thermostat begins to open, read the thermometer. This is the

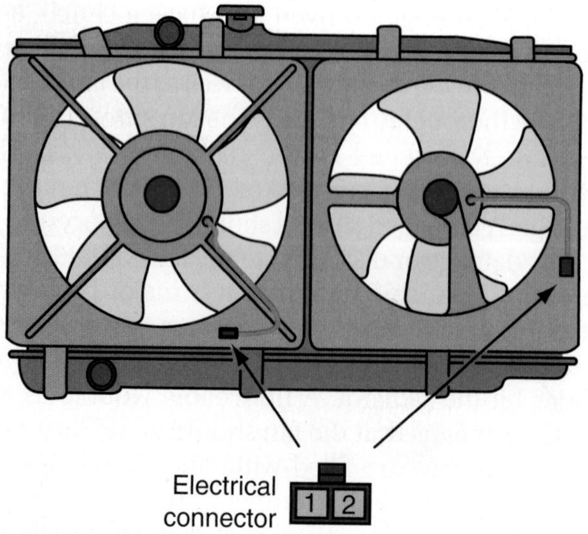

Figure 14–42 Test each fan motor by connecting battery power to terminal 1 and a ground to terminal 2. If either motor fails to run or does not run smoothly, replace it.

Figure 14–43 The operation of a thermostat can be observed with an infrared thermometer.

Figure 14–44 A thermostat can be tested by submerging it in water, heating the water, and observing the temperature at which the thermostat begins to open.

opening temperature of the thermostat. Then remove the thermostat from the water and allow it to cool. If the valve stays open after the thermostat is removed from the water, the thermostat is defective and must be replaced.

Water Pump Checks

Water pumps are driven by a belt, the crankshaft or camshaft, or by an electrical motor. A water pump pulls coolant from the radiator and pushes it through the engine. If the water pump is not working properly, the engine will quickly overheat. Seldom does a pump merely stop working. Problems with a pump are normally noise or leakage. An exception to this is electrically operated pumps. Electric problems can cause the pump to totally stop. The cause of these problems is identified by troubleshooting the circuit.

The majority of water pump failures are attributed to leaks. When the pump bearings and seals begin to fail, coolant will seep out of the weep hole in the casting or through the outer seal **(Figure 14–45)**. The seals may be worn out due to age or abrasives in the system, or they may have cracked because of thermal shock, such as adding cold water to an overheated engine.

CAUTION!

As soon as a water pump begins to leak, it should be replaced. Not only will the leak get worse and cause serious overheating, there is also the chance of the pump's shaft breaking and the fan moving into and destroying the radiator.

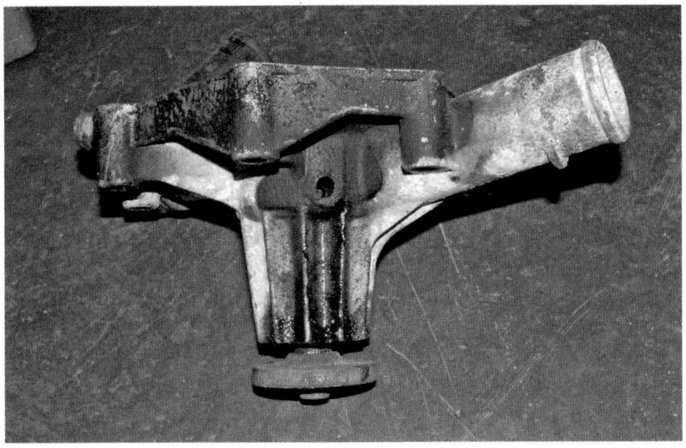

Figure 14–45 This pump shows signs of bad leaks.

Other failures can be attributed to bearing and shaft problems and an occasional cracked casting. Water pump bearing or seal failure can be caused by surprisingly small out-of-balance conditions that are difficult to spot. Look for the following: a bent fan, a piece of the fan missing, a cracked fan blade, a dirty fan-mounting surface, and a worn fan clutch. Any wobble of the pump shaft or fan means the pump and/or fan should be replaced.

Through time, the impeller blades may corrode or come loose from the shaft. Both of these situations will cause the pump not to work and it must be replaced. In less extreme cases, corrosion may cause the impeller to be loose on the shaft and not rotate all of the time. This can cause an intermittent cooling problem. Unfortunately, the only sure way of knowing the condition of the impeller is to remove the pump and inspect it.

To check the operation of the water pump, start the engine and allow it to warm up. Squeeze the upper radiator hose **(Figure 14–46)** and accelerate the

Figure 14–46 The operation of the water pump can be checked by squeezing the upper radiator hose while accelerating the engine. If a surge is felt in the hose, the pump is working.

engine. If a surge is felt in the hose, the pump is working.

Air in the cooling system can prevent the pump from working properly. To check for the presence of air, attach one end of a small hose to the radiator overflow outlet and put the other end into a jar of water. Then make sure all hose connections are tight, and the coolant level is correct. Run the engine and bring it to its normal operating temperature. Then run the engine at a fast idle. If a steady stream of bubbles appears in the jar, air is getting into the system.

The air may be entering the system because of a bad head gasket. Conduct a combustion leak or compression test. If the results of the compression test show that two adjacent cylinders tested have low compression, the gasket is bad or there is an air leak somewhere else in the cooling system.

On belt-driven pumps, turn the engine off and remove the drive belt and shroud. Grasp the fan and attempt to move it in and out and up and down. More than 1/16 inch (1.58 mm) of movement indicates worn bearings that require water pump replacement.

If the concern is excessive noise, start the engine and listen for a bad pump bearing, using a mechanic's stethoscope. Place the stethoscope on the bearing or pump shaft. If a louder-than-normal noise is heard, the bearing is defective.

CAUTION!

Whenever working near a running engine, keep your hands and clothing away from the moving fan, pulleys, and belts. Do not allow the stethoscope or rubber tubing to be caught by the moving parts.

TESTING FOR LEAKS

The most common cause of overheating is low coolant levels due to a leak. Leaks can occur anywhere in the system. The most common leak points include the hoses, radiator, heater core, water pump, thermostat housing, engine freeze plugs, transmission oil cooler, cylinder head(s), head gasket, and engine block. Often only a visual inspection of the cooling system and engine is necessary to find the source of a leak. The point of the leak may be wet or have a light gray color, the result of the coolant evaporating at that point.

Pressure Testing

A radiator pressure tester is the most common tool used to test a cooling system. It is extremely handy for finding the source of any leak in the cooling system. The tester applies pressure to the entire cooling system. A good cooling system should be able to hold about 14 psi (93 kPa) for 15 minutes or more without losing pressure. The procedure for using a pressure tester is given in Photo Sequence 11.

To use the tester, connect it to the radiator filler neck. Run the engine until it is warm, then pump the handle of the tester until the gauge reads the same pressure noted on the radiator cap. Watch the gauge. If the pressure drops, carefully check the hoses, radiator, heater core, and water pump for leaks. Often the leak will initially be obvious because coolant will spray out of the leak. Also look for bulges in the hoses. These indicate that the hose is soft and weak and should be replaced. If the pressure drops but there are no external leaks, suspect an internal leak. The source of internal leaks is found through additional testing.

An internal leak caused by a bad head gasket can be verified with the pressure tester. Relieve all pressure in the cooling system. Install the pressure tester onto the radiator filler neck and start the engine. Allow the engine to idle and watch the gauge on the tester. If the pressure begins to build, it is more than likely that the head gasket is blown and combustion gases are entering the coolant.

Leak Detection with Dye Another common way to identify the source of an external leak is to use dye penetrant and a black light. The dye is poured into the cooling system and the engine run until it reaches operating temperature. With the engine off, the engine and cooling system are inspected with the black light. Where the dyed coolant leaks, a bright or fluorescent green color will be seen **(Figure 14–47)**.

Figure 14-47 A common way to identify the source of an external leak is to use dye penetrant and a black light. Where the dyed coolant leaks, a bright or fluorescent green color will be seen. *Courtesy of Tracer Products*

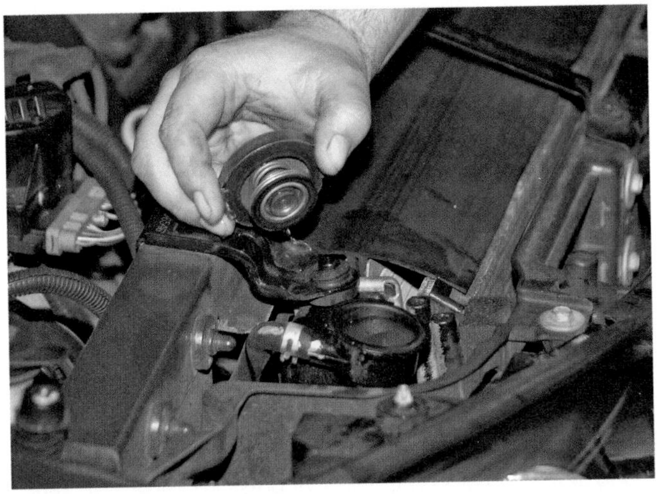

P11-1 *Remove the radiator cap.*

P11-2 *Top off the coolant level.*

P11-3 *Connect the tester to the radiator's filler tube and apply pressure to the system by pumping the handle of the tester.*

P11-4 *Once the pressure is equal to the rating of the pressure cap, watch for a pressure decrease and look for leaks.*

Using a Cooling System Pressure Tester *(continued)*

P11-5 *Using the appropriate adapter, mount the pressure cap to the tester.*

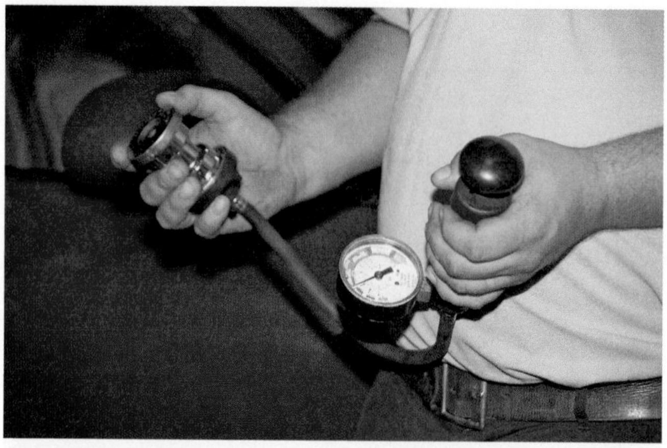

P11-6 *Apply pressure to the cap that is equal to the cap's rating. If the cap does not hold that pressure, replace it. Then apply more pressure. If the cap is good, the excess pressure will vent through the cap.*

P11-7 *Correct any leaks and top off the fluid level.*

Figure 14–48 A combustion leak tester.

Combustion Leak Check Internal leaks are suspected when there are no visible external leaks but the engine is losing coolant or could not hold pressure during the pressure test. These leaks are typically caused by a cracked cylinder head or block or a bad head gasket that is allowing coolant to leak into the cylinders or combustion gases leak into the cooling system. Sometimes steam, white smoke, or water in the exhaust, or coolant in the oil or oil in the coolant, will give a hint that there is a bad head gasket.

A combustion leak tester **(Figure 14–48)** is used to determine whether combustion gases are entering the cooling system. This tester is basically a glass tube with a rubber bulb. The tube is fit with a one-way valve at the bottom. To check for combustion gas leaks, follow this procedure:

PROCEDURE

1. Start the engine and allow it to reach normal operating temperature.
2. Fill the tester's glass tube with the test fluid (normally a blue liquid).
3. Carefully remove the radiator cap.
4. Make sure the coolant level in the radiator is below the lower sealing area of the filler neck.
5. Place the tester's tube into the filler neck.
6. Rapidly squeeze and release the rubber bulb. This will force air from the radiator up through the test fluid.
7. Observe the liquid. Combustion gases will change the color of the liquid to yellow. If the liquid remains blue, combustion gases were not present.
8. Dispose of the test fluid; never return used test fluid to its original container.

A bad head gasket normally results from another problem. Make sure that problem is corrected before replacing the gasket. Bad head gaskets should be replaced as soon as possible. The excessive heat and pressures resulting from this problem are far greater than what a cooling system can handle and can lead to serious engine problems.

SHOP TALK

The O_2 sensor should be replaced after a head gasket is replaced or if there was a crack in the head or block. Coolant contains silicone and silicates that may contaminate the sensor.

Testing the Radiator Pressure Cap

The opening pressure of the radiator cap should be measured and the cap checked for gasket cracking, brittleness, or deterioration **(Figure 14–49)**. Over time the spring in the cap can weaken and will lower the coolant's boiling point. This will cause the engine to lose coolant through the overflow tube whenever it gets hot. Also check the sealing surfaces in the radiator filler neck.

Install the correct cap adapter and radiator pressure cap to the tester head. Pump the tester until the gauge reads the pressure rating of the cap. The cap should hold that pressure for at least 1 minute. If it does not, replace it. Apply a pressure greater than the cap's rating. A good cap will vent the excess pressure. Remove the cap from the tester and visually inspect the condition of the cap's pressure valve and upper and lower sealing gaskets. If the gaskets are hard, brittle, or deteriorated, the cap may leak when exposed to hot, pressurized coolant. It should be replaced with a new cap in the same pressure range.

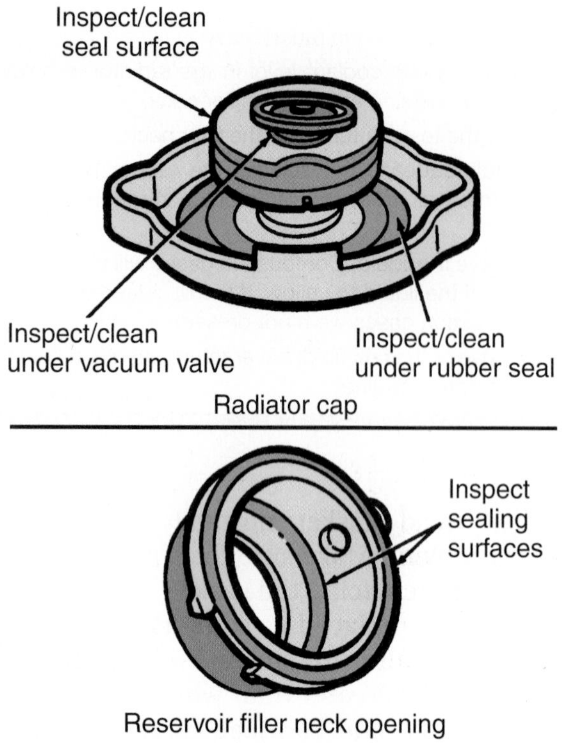

Figure 14-49 Radiator cap inspection. *Courtesy of Ford Motor Company*

> **WARNING!**
>
> *The radiator cap should never be removed when the radiator is hot to the touch. When the pressure in the radiator is suddenly released, the coolant's boiling temperature is reduced. This causes the coolant to immediately boil. Because coolant is a thick liquid, it will stick to your skin and can cause severe burns. The radiator should be allowed to cool, or you can force cool it by lightly spraying water on the radiator. When the cap is cool and the engine is shut off, use a cloth over the cap and turn it counterclockwise one-quarter turn to its safety stop and release the pressure. Let the cap remain in this position until all pressure subsides. When all discharge ends, use a cloth over the cap, press it down, and turn it counterclockwise to remove the cap.*

Water Outlet

Internal corrosion, contributes the failure of water outlets. Cast-iron water outlets are more resistant to corrosion than stamped steel or cast aluminum outlets. A more common cause of failure for a water outlet is the uneven torquing down of the mounting bolts, which can cause a mounting ear to break off. When this happens, the outlet will not seal and must be replaced.

COOLING SYSTEM SERVICE

Service to the cooling system entails replacing leaking or broken parts as well as emptying, flushing, and refilling the system. One of the common services is replacing the drive belt. Belts are a common wear item and should be installed at the correct tension. A loose belt will slip and prevent the water pump from moving coolant fast enough through the engine and/or turn the fan fast enough to cool the coolant.

The water pump on many late-model engines is driven by the engine's timing belt or chain. When replacing the water pump on these engines, always replace the timing belt. Make sure all pulleys and gears are aligned according to specifications when installing the belt.

Go To ***Chapter 7 for a detailed discussion of drive belts and the correct procedure for changing a belt and adjusting its tension.***

Hoses

> ***Customer Care***
>
> Technicians should do their customers a favor and remind them that all cooling hoses should be replaced every 2 to 4 years to prevent breakdowns.

Replacement hoses must be made of the right material and have the correct diameter, length, and shape. The hose's part number is often printed on the outside of the hose and replacement hoses should match or correspond to those numbers (**Figure 14–50**).

Nearly all OEM hoses are of the molded, curved design. Lower radiator hoses are normally wire reinforced to prevent collapse due to the suction of the water pump. Molded hoses designed to fit many applications are often sold. The hose is cut to fit a specific application. Some have printed cutoff marks on them to show where they should be cut. Others should be compared to the old hose for a cut reference.

Figure 14–50 Nearly all radiator hoses have a part number painted on the outside of the hose. These should be used when replacing a hose.

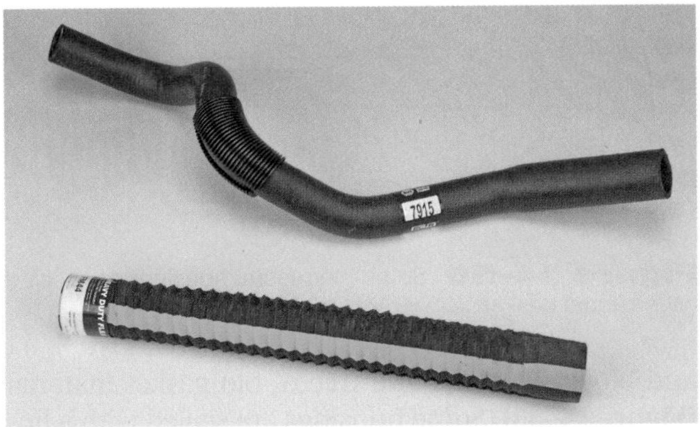

Figure 14–51 Aftermarket replacement hoses are either molded (top) or flex types (bottom).

Aftermarket hoses are either molded or flex types **(Figure 14–51)**. A flex-type hose allows greater vehicle coverage per part number. Flex hoses are available in different lengths and diameters. They can flex or bend into most required shapes without causing a restriction. Flex hoses may not be desirable for some systems that require radical bends and shapes.

Rather than replacing heater hoses with the molded type, formable heater hoses are available. This hose design has a wire spine. The wire spine allows the hose to bend into a curve without collapsing at the bend. Once the desired shape is obtained, the hose is cut to length and then installed.

All cooling system hoses are basically installed the same way. The hose is clamped onto inlet-outlet nipples on the radiator, water pump, and heater.

When replacing a hose, drain the coolant system below the level that is being worked on. Loosen or carefully cut the old clamp. With a knife, carefully cut the end of the old hose **(Figure 14–52)** so it can slide

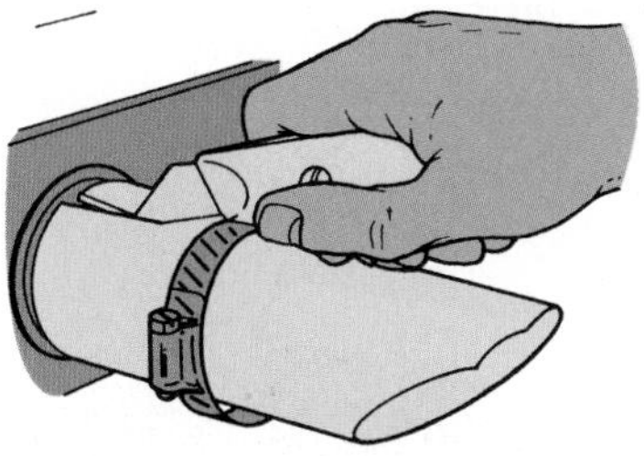

Figure 14–52 Cutting off an old hose.

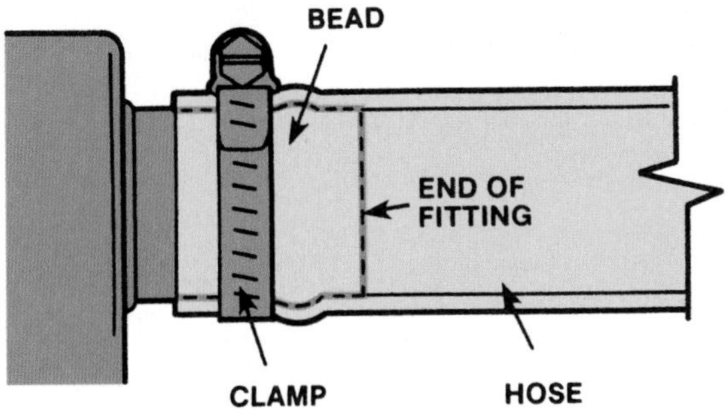

Figure 14–53 New clamps should be placed immediately after the bead of the fitting.

off its fitting. If the hose is stuck, do not pry it off. You could possibly damage the inlet/outlet nipple or the attachment between the end of the hose and the bead. Simply make more cuts into the hose.

Always clean the neck of the hose fitting or nipple with a wire brush or emery cloth after the old hose has been removed. Burrs or sharp edges can cut the hose and lead to premature failure. Dirty connections will prevent a good seal.

Dip the ends of the hose in coolant and slip the clamps over each end. Do not reuse spring-type clamps, even if they look good. Slip the hose over its fittings, engine end first. In cold weather, the hose may be stiff; it can be soaked in warm water to make it more flexible. If the hose does not fit properly, remove it and reverse the ends. Then slide the clamps to about ¼ inch from the end of the hose after it is properly positioned on the fitting **(Figure 14–53)**. Tighten the clamp securely but do not overtighten.

Hose Clamps Original equipment clamps are usually spring steel wires that are removed and replaced with special pliers. A worm drive hose clamp is often used as a replacement for many reasons. It provides even pressure around the outside of the hose. It is also easy to install and requires no special tools.

Rather than using steel clamps, some technicians prefer to use thermoplastic clamps **(Figure 14–54)**. These heat-sensitive clamps are installed on the hose ends and a heat gun is used to shrink the clamp. The shrinking of the clamp tightens the connection. As the engine runs, the heat of the coolant further tightens the connection.

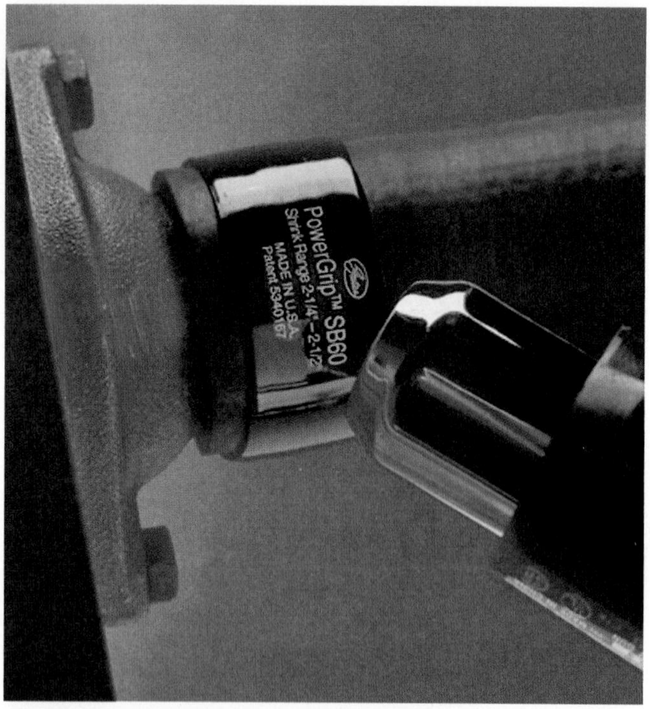

Figure 14-54 Thermoplastic clamps are tightened with a heat gun. *Courtesy of Gates Corporation*

Figure 14-55 Make sure all old gasket material is removed before installing a new thermostat gasket and housing.

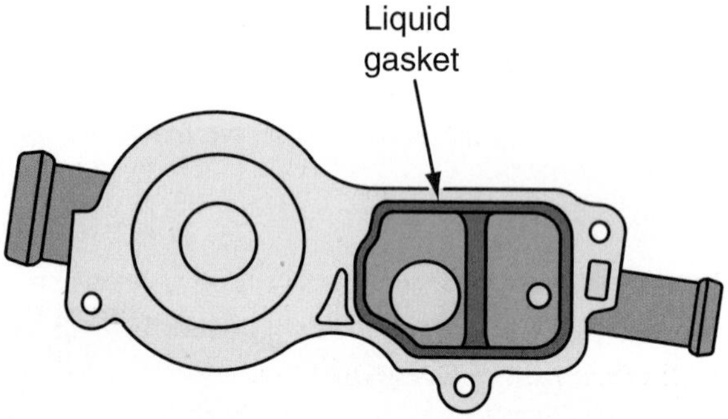

Figure 14-56 Some thermostat housings are sealed with a liquid gasket.

It is a good idea to readjust the clamp of a newly installed hose after a brief run-in period. The hose does not contract and expand at the same rate as the metal of the inlet/outlet nipple. Rubber hoses, warmed by the hot coolant and hot engine, will expand. When the engine cools, the fitting contracts more than the rubber, and the hose will not be as secure. This can result in cold coolant leaks at the inlet/outlet nipple. Retightening the clamp eliminates the problem.

Thermostat

When replacing the thermostat, the replacement should always have the same temperature rating as the original. Since 1971, nearly all engines require a thermostat with 192° or 195° ratings. Using a thermostat with a different opening temperature than was originally installed on a computerized engine will affect the operation of the fuel, ignition, and emissions control systems. This is due to the fact that the wrong thermostat can prevent the system from going into closed loop.

Replacement Markings on the thermostat normally indicate which end should face toward the radiator. Regardless of the markings, the sensored end must always be installed toward the engine.

When replacing the thermostat, also replace the gasket that seals the thermostat in place and is positioned between the thermostat housing and the block. Make sure the mating surfaces on the housing and block are clean and free of old gasket material (**Figure 14–55**). Some housings are sealed with a liquid gasket (**Figure 14–56**). A thin line of the liquid gasket should be applied around the water passages, and the housing installed before the liquid dries.

Normally there are locating pins on the thermostat that are used to position the thermostat in the block and housing (**Figure 14–57**). Most thermostats and housings are sealed with gaskets or rubber seals. The gaskets are normally composition fiber material cut to match the thermostat opening and mounting bolt configuration of the housing. Thermostat gaskets generally come with an adhesive backing. The backing holds the thermostat securely centered in the mounting flange. This makes it easier to properly align the housing to the block.

Repairing Radiators

SHOP TALK

Always refer to application charts or a service manual when replacing a pressure cap to make sure that the new cap has the same pressure range as the original cap.

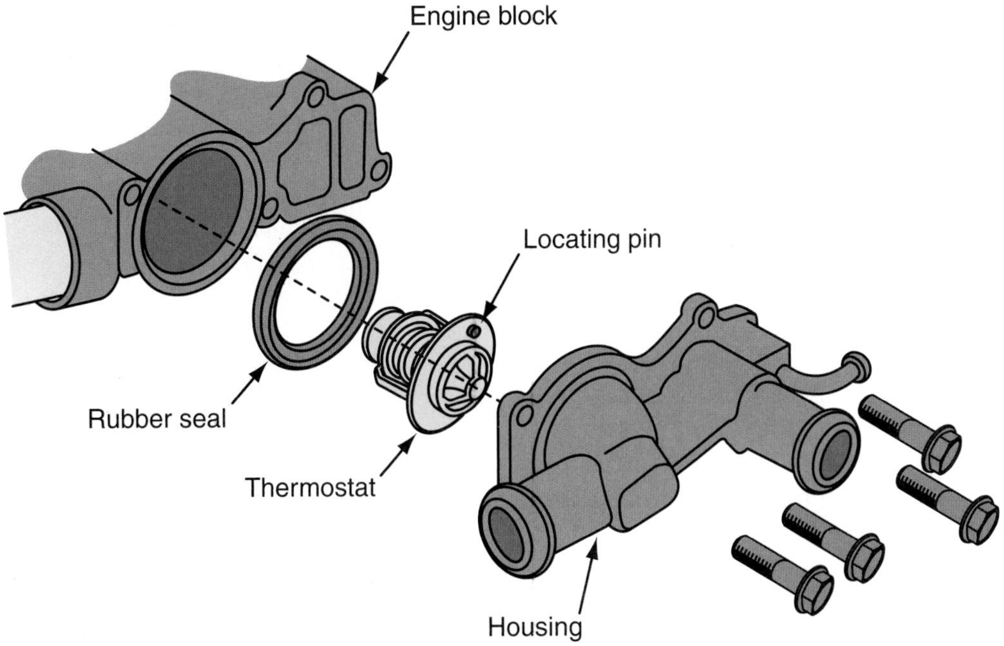

Figure 14-57 Normally there are locating pins on the thermostat that are used to position the thermostat in the block and housing.

Most radiator leak repairs require the removal of the radiator. Coolant must be drained and all hoses and oil cooler lines disconnected. Bolts holding the cooling fans and radiator are then loosened and removed.

The actual radiator repair procedures depend on its construction and the type of damage. Most repairs are made by radiator specialty shops whose technicians have knowledge of such work. Many of today's radiators have plastic tanks, which are not repaired. If these tanks leak, they are replaced. If the radiator is badly damaged, it should be replaced.

CAUTION!

When working on the coolant system, for example, replacing the radiator, thermostat, or water pump, a certain amount of coolant will spill on the floor. The antifreeze in the coolant causes it to be very slippery. Always immediately wipe up any coolant that spills to reduce or eliminate the chance of injury.

Replacing the Water Pump

When replacing a water pump, make sure that the replacement is the correct one. Using the wrong pump can cause engine overheating.

The water pump on some engines with a serpentine belt turns in the opposite direction to that used on the same engine with a V-belt.

Figure 14-58 Installing a water pump.

Before replacing a water pump, the cooling system should be drained. Remove all parts that will interfere with the removal of the pump. This includes the drive belts, fan, fan clutch, fan shroud, and pump pulley. Most pumps are bolted to the block as shown in **Figure 14–58**.

Loosen and remove the bolts in a crisscross pattern from the center outward. Insert a rag into the block opening and scrape off any remains of the old gasket. When replacing a water pump, always follow the procedures given by the manufacturer. Most often a coating of adhesive should be applied to a new gasket before it is placed into position on the water pump. Some pumps are sealed to the block with an O-ring. Make sure that the O-ring groove in the block is clean. When installing the O-ring, make sure it is totally inserted in the groove. Lubricate the O-ring to ensure it will not tear during installation. Position the pump against the block until it is properly seated. Install the mounting bolts and tighten them evenly in a staggered sequence to the torque specifications. Careless tightening can cause the pump housing to crack. Check the pump to make sure it rotates freely.

The water pumps on many late-model OHC engines are driven by the engine's timing belt. When replacing the water pump on these engines, always replace the timing belt. Make sure all pulleys and gears are aligned according to specifications when installing the belt.

Figure 14-59 Most engines have drain plugs that allow for draining coolant out of the block.

Draining the Coolant

An important part of a preventive maintenance program is changing the engine's coolant. This is done to prevent the coolant from breaking down chemically. When this happens, the coolant becomes too acidic. This should be done every 2 to 4 years. Obviously, draining the coolant is the first step in doing this. The coolant should also be drained when replacing cooling system parts.

Before draining the coolant, find the capacity of the system in the vehicle's specifications. This allows you to know what percentage of the coolant has been drained. Normally, 30% to 50% of the coolant will remain in the system.

Most radiators have a drain plug located in the lower part of a tank. Some have a petcock valve. Make sure the engine is cool before draining the coolant. Set the heater controls to the HOT position. Remove the radiator pressure cap. Remove the overflow reservoir and empty it into a catch can. Now, place the catch can under the drain plug and remove the plug. If the radiator has a petcock, open it fully. If the radiator does not have a drain plug, remove the lower radiator hose. Be careful not to force the hose away from the radiator. The tube on the radiator can be easily broken by excessive force. Twist the hose back and forth while pulling it off. If the hose is stuck, slip a thin screwdriver between the hose and the radiator tube to loosen it. After the coolant stops flowing out, install the drain plug or close the petcock.

Additional coolant can be drained through drain plugs in the engine block (**Figure 14–59**). Place the catch can below the drain plug and remove the plug. Once the coolant has drained, replace the plug. Make sure to apply sealer on the threads of the plug. Clean up any spilled coolant.

CAUTION!

Never pour engine coolant into a sewer or onto the ground. Used coolant is a hazardous waste and its disposal should be in accordance to local laws and regulations.

CAUTION!

Coolant can be very dangerous to children and animals. It has a sweet taste and can be deadly if ingested. Never keep coolant in an open container.

Coolant Recovery and Recycle System Whenever coolant is drained to service the cooling system, the used coolant should be drained and recycled by a coolant recovery and recycle machine. Typically additives are mixed into the used coolant during recycling. These additives either bind to contaminants in the coolant so they can be easily removed, or they restore some of the chemical properties in the coolant.

Flushing Cooling Systems

Whenever coolant is changed—and especially before a water pump is replaced—a thorough flushing should be performed. The bottom of the radiator will trap rust, dirt, and metal shavings. Draining the system will only remove the contaminants that are suspended in the drained fluid. Rust and scale will inevitably form in any cooling system. Buildups of these will affect the efficiency of the cooling system and can cause blockages inside the radiator. Flushing may not remove all debris. In fact if the radiator

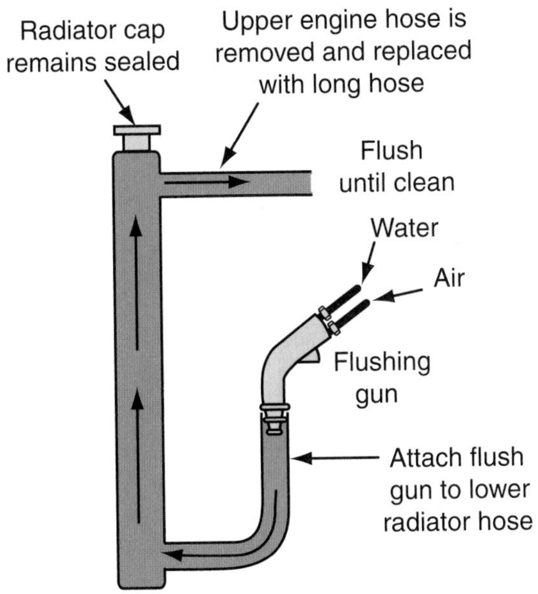

Figure 14-60 The typical setup for back flushing a cooling system.

Figure 14-61 Stop leak being added to the cooling system. *Courtesy of Honeywell International Inc.*

is plugged, the radiator should be removed and rebuilt or it should be replaced.

The system can be flushed in many ways. Power flushing equipment is often used to force the old coolant and contaminants out of the system. This function is often part of the operating cycle of a coolant exchanger. Back flushing forces clean water backward through the cooling system. The discharge of fluid carries away rust, scale, corrosion, and other contaminants. Flushing guns use compressed air to back flush the system and to break loose layers of dirt and other debris (**Figure 14–60**). This method of flushing the system is not recommended on systems that use plastic and aluminum radiators. Check the service manual for the proper way of cleaning out the cooling system in the vehicle being serviced.

A simple way to flush the system is to drain it and fill it with water. The engine is then run and brought to its operating temperature. At this point the engine is shut down and allowed to cool. The coolant is then drained. These steps are repeated until the discharged fluid is clear. The clear fluid indicates that all of the old coolant has been removed. Once this occurs, the system is drained again and refilled with coolant.

Flushing Chemicals Many different flush chemicals are available. Before using any chemical, make sure it is safe for the type of radiator. Coolant exchangers often use chemicals as part of their flushing function. The typical procedure for using flushing chemicals is to drain the system. Then pour the chemical into the radiator and top off the radiator with water. Install and tighten the radiator cap. Now start the car and set the heater control to its hottest position. Allow the engine to run until it reaches normal operating temperature. Then turn the engine off and allow it to cool. Once the engine has cooled, completely drain the radiator. If the flushing chemical requires a neutralizer, add it to the water remaining in the system and then refill the system with fresh coolant.

Customer Care

Many additives, inhibitors, and quick-fix remedies are available for cooling systems. These include, but are not limited to, stop leak, water pump lubricant, engine flush, and acid neutralizers. Stop leak or sealers (Figure 14–61) are often used to plug small leaks in the cooling system. These chemicals work to seal leaks in the radiator and engine (metal components). They do not seal leaking hoses and hose connections. Explain to your customers that caution should be exercised when using any additive. They must make sure that the chemicals are compatible with their cooling system. For example, caustic solutions must never be used in aluminum radiators. Alcohol-based remedies should never be used in any cooling system.

Refilling and Bleeding

After the cooling system has been drained and all services performed, the system needs to be refilled with the proper type and mixture of coolant. It is important that the correct type is put into the system. The color of the coolant normally indicates its purpose: Most coolants are green, extended-service coolants are orange, and special Toyota coolants are red. The additives in orange coolant are not chemically compatible with those in green or red coolant; therefore, the system must be completely drained and flushed before switching from orange to red or green, or vice versa.

When refilling a system, determine the total capacity of the system. Half of the capacity is the amount of undiluted coolant that should be put into the system. Then top off the system with water. Loosely install the radiator cap. Run the engine until it reaches operating temperature. Then turn it off and correct the level of the coolant. Tighten the radiator cap and run the engine again and check for leaks.

When refilling the cooling system, be sure you get it completely full. Some systems are difficult to fill without trapping air. Air pockets in the head(s), heater core, and below the thermostat can prevent proper coolant flow and cooling. If air is trapped in the engine block or cylinder head(s), it can also cause "hot spots." This can ruin the head gaskets, the cylinder walls, and the entire cooling system.

This is more of a problem on newer vehicles than older ones. In older cars, the top of the radiator sat higher than the rest of the cooling system. This positioning allowed air in the system to escape through the radiator cap. The radiator cap on many newer cars is much lower than the rest of the cooling system. This results from the lower hood lines. Because all liquids seek a natural level, that level may be above the cap and air gets trapped easily in other high places such as the block or head(s).

These new vehicles must be purged of the air after refilling. This can be done in many ways. Sometimes jacking up the front of the car will raise the radiator cap higher than the rest of the cooling system. With the radiator cap in its first lock position, start the engine. Allow it to run until the thermostat opens and the coolant circulates. Trapped air bubbles will naturally blow out the cap. When no more air escapes, shut off the engine and correct the coolant level. Then reinstall the radiator cap and tighten it to its fully locked position.

When refilling a system on a late-model vehicle, always check the service information for the correct coolant fill procedure. Each vehicle may have its own specific bleeding procedure. Some engines have air bleed valves located at the high point in the system **(Figure 14–62)**. Check the service manual for their location. These valves allow air to escape while the system is being filled.

To use the air bleeds, make sure the engine is warm and the heater is fully on. Connect a hose to the end of the valves and place the open end in a catch can. Open all bleed valves. Slowly put the required amount of coolant into the radiator until the coolant begins to leak out of the valves. Then close the valves and top off the system.

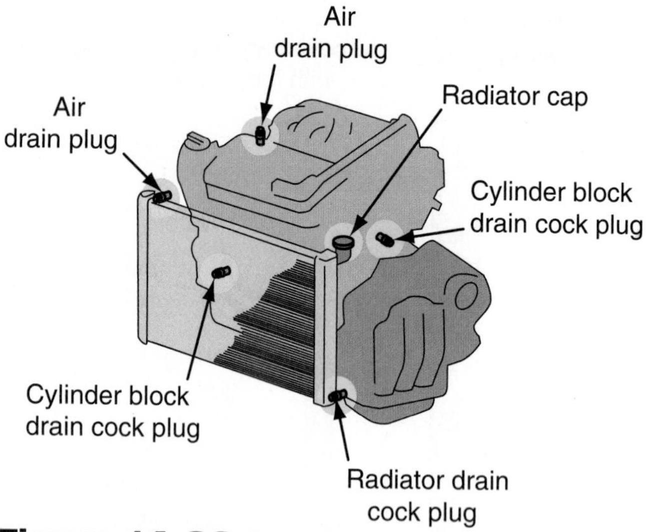

Figure 14-62 Some cooling systems are equipped with air bleed screws.

If the system does not have a bleeder valve, disconnect a heater hose at the highest point in the system. Once fluid flows steadily out of the hose, reconnect the hose. If the system was previously fit with a flushing Tee, remove its cap to purge the system. Always recheck the fluid level and make sure that all air is purged before putting the car back into service.

Special Precautions for Hybrid Vehicles

Special coolants are required in most hybrids because the coolant cools the engine and the converter/inverter assembly. Cooling the converter/inverter is important, and checking its coolant condition and level is an additional check during preventive maintenance. The cooling systems used in some hybrids feature electric pumps and storage tanks. The tanks store heated coolant and can cause injury if the technician is not aware of how to carefully check them.

Toyota hybrids have a system that heats a cold engine with retained hot coolant to provide reduced emissions levels. Hot coolant is stored in a container **(Figure 14–63)**. The coolant will circulate through the engine immediately after startup. The fluid also may circulate through the engine many hours after it is shut off. This fluid is under pressure and can cause serious burns to anyone who opens the system for inspection and/or repairs. To safely service this cooling system, the pump for the storage tank must be disconnected. The cooling system also is tied into the converter/inverter assembly. This also presents a potential problem, because it is easy to trap air in the cooling system due to the path of coolant flow **(Figure 14–64)**. To purge the system of air, there is a bleeder screw, and a scan tool is used to run the electrical water pump.

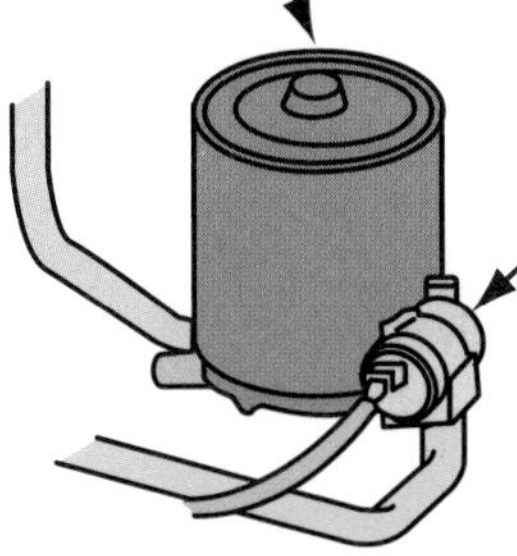

Figure 14-63 Toyota hybrids have a hot coolant storage container.

CAUTION!

The coolant in the coolant heat storage tank may be hot *when the engine and radiator are cold. Also,* never *remove the radiator cap when the engine or radiator is hot.*

PROCEDURE

The recommended procedure for draining and refilling the cooling system on a Toyota hybrid includes the following steps.

1. Remove the radiator's top cover and cap.
2. Disconnect the electrical connector on the water pump to prevent circulation of the coolant.
3. Connect a drain hose to the drain port on the bottom of the coolant heat storage tank, and then loosen the yellow drain plug on the tank.
4. Connect a drain hose to the drain port on the lower left corner of the radiator, and then loosen the yellow drain plug on the radiator.
5. Connect a drain hose to the drain port on the rear of the engine and loosen the drain plug.
6. After the coolant is drained, tighten the three drain plugs.
7. Reconnect the connector to the coolant heat storage tank's water pump.
8. Connect a hose to the radiator's bleeder valve port and place the other end of the hose into the coolant reservoir tank.
9. Loosen the radiator's bleeder plug.
10. Fill the radiator with the correct coolant.
11. Tighten the radiator's bleeder plug and install the radiator cap.
12. Connect the scan tool to DLC3.
13. Using the scan tool, run the water pump for the storage tank for 30 seconds.
14. Then loosen the radiator's bleeder plug.
15. Remove the radiator cap and top off the coolant in the radiator.
16. Repeat the refilling and bleeding sequence as often as necessary. Normally when no additional coolant is needed after the sequence, the system is bled.
17. Start the engine and allow it run for 1 to 2 minutes.
18. Turn off the engine and top off the fluid, if necessary.

Ford hybrids have two separate cooling systems: One is for engine cooling and the other is for hybrid components, called the Motor Electronics (M/E) cooling system. The engine cooling system is

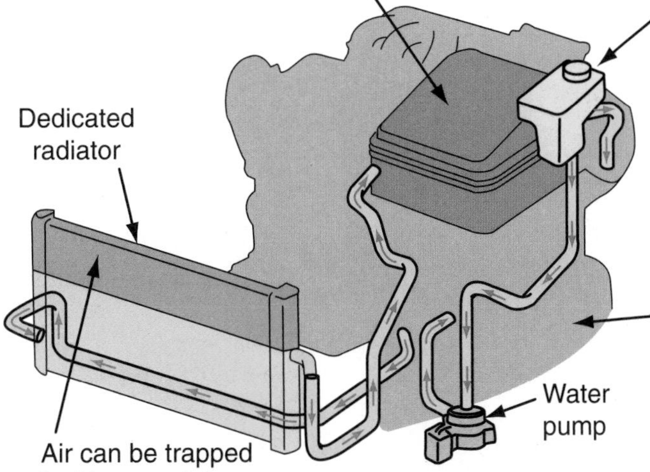

Figure 14-64 Since the reservoir tank and inverter is higher than the radiator cap, air is easily trapped during coolant refilling.

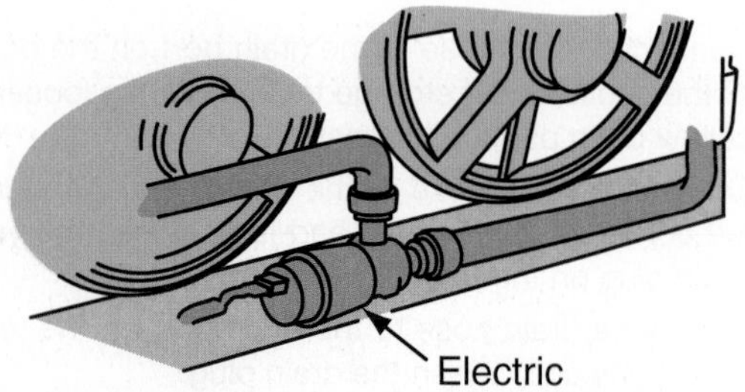

Figure 14–65 The electric water pump for the M/E cooling system on a Ford Escape Hybrid.

conventional. The M/E cooling system uses an electric water pump **(Figure 14–65)** to move coolant through the inverter, transmission, and a separate radiator mounted next to the conventional radiator **(Figure 14–66)**. The M/E coolant reservoir is located behind the engine coolant reservoir. Although the two systems operate similarly, the M/E cooling system typically operates at lower pressures and temperatures. The fluid levels in both cooling systems must be maintained.

It is easy to trap air in the M/E cooling system when filling the system. The system is fitted with a bleeder screw at the top of the inverter **(Figure 14–67)**. When servicing the system, make sure the high-voltage system is isolated by having the service connector in the SERVICING/SHIPPING position. Also, wear lineman's gloves because the bleeder screw is very close to the high-voltage cables.

Coolant Exchangers

A coolant exchanger **(Figure 14–68)** is used to remove old coolant from the system and to put in new coolant at the correct mixture. Some coolant exchangers will also leak test the system and flush it. Coolant exchangers are normally powered by shop air, although a few are powered by the vehicle's battery. Most exchangers move through their cycles with the engine off.

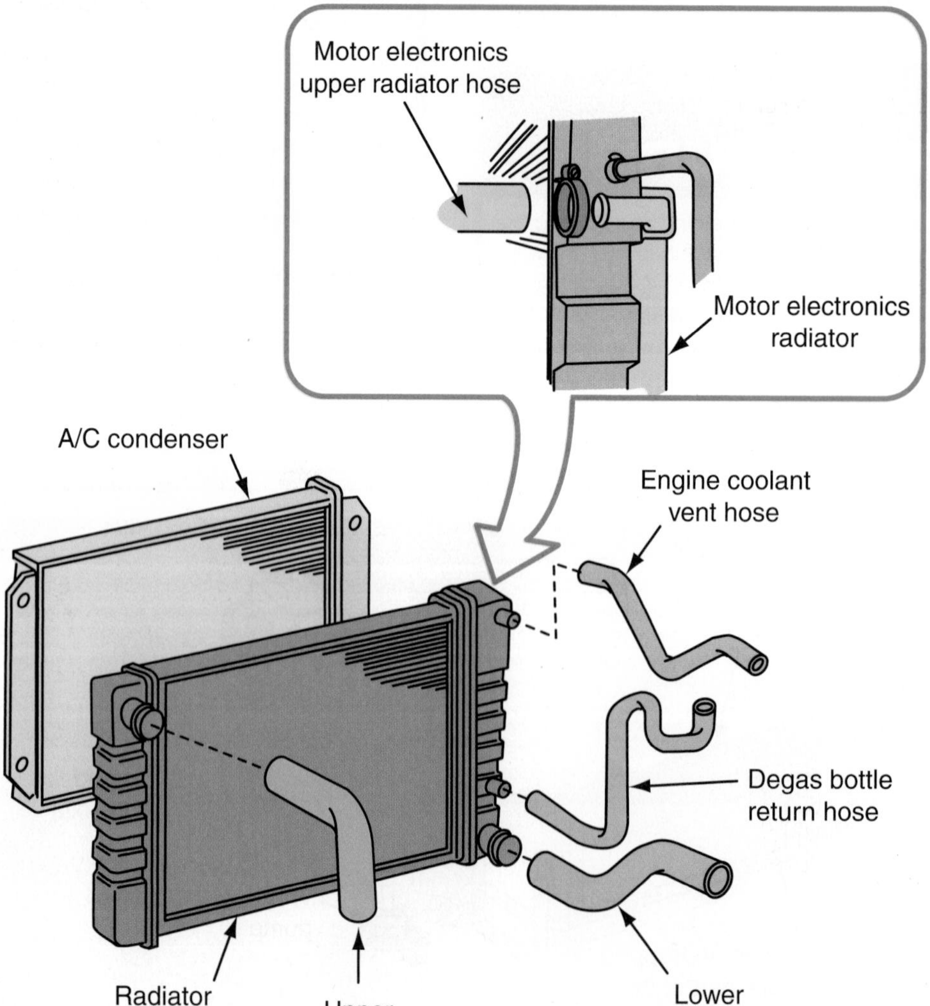

Figure 14–66 The radiator assembly for a Ford Escape Hybrid.

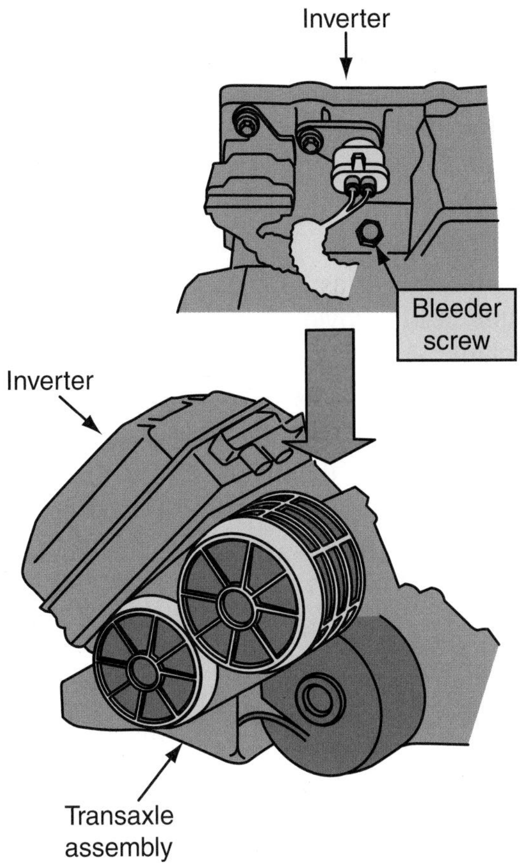

Figure 14-67 Location of the cooling system bleeder screw on a Ford Escape Hybrid.

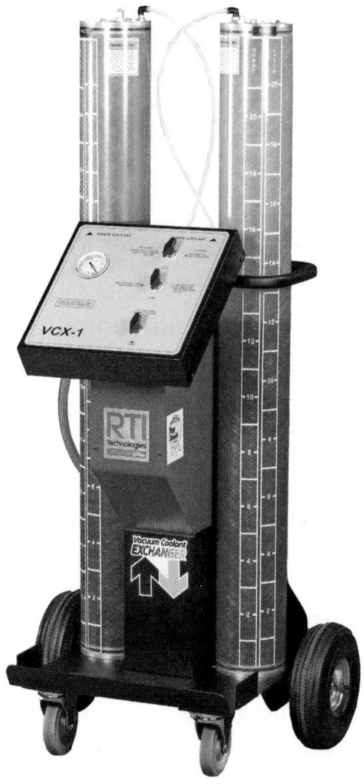

Figure 14-68 A coolant exchanger. *Courtesy of RTI Technologies, Inc.*

The machine is connected to the radiator filler neck, upper radiator hose, or a heater hose. The old coolant is siphoned out of the system and stored in a container. When all coolant is out, the low pressure in the cooling system siphons the fresh coolant mixture out of another container. The entire process, including flushing the system, takes only a few minutes. When complete, the system is free of air and full of coolant.

CASE STUDY

A late-model minivan's engine overheats while going up a mountain road. It is a very hot day so water is added to the radiator and the trip continues. After driving 30 miles, the engine overheats again. Again water is added. The engine is allowed to cool for an hour while the owner eats. After about 30 miles of driving, the engine overheats again. This time the van is taken to a service station to be diagnosed and repaired. It is late in the day and the owner agrees to stay overnight at a local motel. A fresh supply of water is added to the radiator and the owner drives about a mile to the motel.

In the morning the owner adds water to the system. The van is brought back to the service station for diagnosis. It is obvious that the engine is overheating and there is a loss of fluid. There is no need to verify the customer's complaint. A visual inspection shows little water in the radiator but no signs of external leakage is found. The oil level is found to be a little high and the oil appears to be thin.

A radiator tester is attached to the system and there is no sign of pressure or coolant leakage. The operation of the water pump is to be checked next. However, this test cannot be done because the engine will not start. The starter cannot turn the engine over. A quick check of the battery reveals that it is fully charged. Careful thought suggests that the problem is hydrostatic lockup. This is a condition in which a fluid fills the cylinder. Because a liquid cannot be compressed, the liquid prevents the piston from moving up.

To verify this, the spark plugs are removed. All plugs looked normal except cylinder #5.

It is wet with water. With the plugs out, the engine is cranked by the starter motor and water comes gushing out of the cylinder. The exact cause of the internal leak will be best identified by tearing down the engine. During disassembly, a crack in the engine block is found. This crack allowed the coolant to pass from a coolant passage in the block into the cylinder.

This leak is not detected by the radiator pressure test because when the engine is cold, little, if any, coolant leaks through the crack. The crack grows larger as the engine becomes hotter, allowing the cylinder to rapidly fill when it is hot. A new short block is put in the van and the problem is corrected.

KEY TERMS

Blowby
Dry sump
Electrochemical degradation (ECD)
Electrolysis
Expansion tank
Fan clutch
Flex-blades
Pressure relief valve
Recovery tank
Sludge
Thermostat
Wet sump

SUMMARY

- An engine's lubricating system must provide an adequate supply of oil to cool, clean, lubricate, and seal the engine. It also must remove contaminants from the oil and deliver oil to all necessary areas of the engine.
- Engine oil is stored in the oil pan or sump. The oil pump draws the oil from the sump and passes it through a filter where dirt is removed. The oil is then moved throughout the engine via oil passages or galleries. After circulating through the engine, the oil returns to the sump.
- The oil pump circulates oil through passages in the engine.
- An oil pump is a positive displacement pump and needs a pressure relief valve to prevent high pressures during high engine speeds.
- All of the oil that leaves the oil pump is directed to an oil filter.
- Excessive oil consumption can result from external and internal leaks or worn piston rings, valve seals, or valve guides. Internal leaks allow oil to enter the combustion chamber where it is burned.
- *Blowby* is a term used for the gases that escape the combustion chamber and enter the crankcase.
- When oil oxidizes, chemical compounds in the oil begin to chemically break down and solidify and form sludge.
- The fluid used as coolant today is a mixture of water and ethylene glycol-based antifreeze/coolant.
- Closed-cooling systems are cooling systems with an expansion or recovery tank.
- The purpose of the water pump is to move the coolant efficiently through the system.
- The radiator transfers heat from the engine to the air passing through it.
- The thermostat attempts to control the engine's operating temperature by routing the coolant either to the radiator or through the bypass or sometimes a combination of both.
- Radiator pressure caps keep the coolant from splashing out of the radiator and keep the coolant under pressure.
- For every pound of pressure put on the coolant, the boiling point is raised about 3°F (1.6°C).
- Coolant flows from the engine to the radiator and from the radiator to the engine through radiator hoses.
- A temperature indicator is mounted in the dashboard to alert the driver of an overheating condition.
- A hot liquid passenger heater is part of the engine's cooling system.
- Radiators for vehicles with automatic transmissions have a sealed oil cooler located in the coolant outlet tank.
- The cooling fan delivers additional air to the radiator to maintain proper cooling at low speeds and idle.
- To save power and reduce the noise levels, most late-model vehicles have an electrically driven cooling fan.
- The basic procedure for testing a vehicle's cooling system includes inspecting the radiator filler neck, inspecting the overflow tube for dents and other obstructions, testing for external leaks, and testing for internal leaks.
- Most radiator leak repairs require removing the radiator from the vehicle.
- The pressure cap should hold pressure in its range as indicated on the tester gauge dial for 1 minute.

- A thermostat can be tested in the engine or after it is removed.
- Hoses should be checked for leakage, swelling, and chafing. Any hose that feels mushy or extremely brittle or shows signs of splitting when it is squeezed should be replaced.
- The majority of water pump failures are attributed to leaks of some sort. Other failures can be attributed to bearing and shaft problems and an occasional cracked casting.
- Electrolysis removes the protective layer on the inside of the radiator tubes, which can lead to serious engine failures. It occurs when there is improper grounding of electrical accessories and equipment.
- Engine overheating can cause detonation, preignition, blown head gasket, OHC cam seizure and breakage, blown hoses, radiator leaks, cylinder damage due to swelling pistons, sticky exhaust valve stems, and engine bearing damage.
- A radiator pressure can help identify the source of any leak in the cooling system as well as check the operation of a pressure cap.
- Internal leaks can be verified with a combustion leak tester.
- When replacing a thermostat, the replacement should always have the same temperature rating as the original.
- The cooling system should be drained, flushed, and refilled with fresh coolant every 2 to 4 years.
- Hybrid vehicles typically require a special coolant because the cooling system also cools the inverter assembly. Hybrids also require special service procedures.

REVIEW QUESTIONS

1. What is the most efficient way to increase a radiator's efficiency?
2. What type of radiator pressure cap is found on today's vehicles?
3. Describe the simple test used to determine whether the water pump is causing good circulation.
4. Which of the following is a function of the engine's lubrication system?
   a. hold an adequate supply of oil
   b. remove contaminants from the oil
   c. deliver oil to all necessary areas of the engine
   d. all of the above
5. What is ECD and what can result from it?
6. *True or False?* To eliminate the drain of engine power during times when cooling fan operation is not needed, many of today's belt-driven fans have a fan clutch that prevents the operation of the fan when the engine and radiator are heated up.
7. For every pound of pressure put on engine coolant, the boiling point of the coolant is raised about ______.
   a. 2°F (1.1°C)
   b. 3°F (1.6°C)
   c. 4°F (2.2°C)
   d. 5°F (2.7°C)
8. In most automotive engines, the water pump is driven by the ______.
   a. geartrain
   b. oil pump
   c. crankshaft
   d. impeller
9. When must a thermostat be fully opened?
   a. at 3°F (1.6°C) above its temperature rating
   b. at 3°F (–16°C) below its temperature rating
   c. at 20°F (–6.6°C) above its opening temperature
   d. at 20°F (–6.6°C) below its opening temperature
10. Explain why a pressure relief valve is needed in an automotive oil pump.
11. List five problems that can be caused by engine overheating.
12. Cooling system hoses should be replaced when they are ______.
    a. hard (brittle)
    b. soft (spongy)
    c. swollen
    d. all of the above
13. *True or False?* The presence of steam in the cooling system results from low coolant levels or air in the system.
14. Which of the following is not a true statement about thermostats?
    a. Faulty thermostats are often the cause of air pockets in the coolant passages.
    b. They can also cause an increase in fuel consumption and poor engine performance.
    c. If the thermostat is stuck open, the coolant may not reach the desired temperature because it is cooled before it gets hot.
    d. If a thermostat is stuck closed, coolant will not flow between the engine and radiator and the engine will quickly overheat.
15. Describe the procedure for using air bleed valves to purge air out of a cooling system.

ASE-STYLE REVIEW QUESTIONS

1. While refilling an 8-quart (7.57 L) system with coolant: Technician A initially puts in 4 quarts (3.78 L) of straight coolant and tops it off with water. Technician B fills the system with the correct mixture, tightens the radiator cap, and then checks for leaks. Who is correct?
   a. Technician A c. Both A and B
   b. Technician B d. Neither A nor B
2. Technician A says that excessive engine oil consumption can be caused by low oil levels. Technician B says that excessive oil consumption can be caused by worn valve guides. Who is correct?
   a. Technician A c. Both A and B
   b. Technician B d. Neither A nor B
3. While checking for conditions that might cause coolant electrolysis: Technician A places the positive lead of a voltmeter into the coolant and connects the negative lead to a ground. Technician B checks for the presence of voltage in the coolant with the ignition off. Who is correct?
   a. Technician A c. Both A and B
   b. Technician B d. Neither A nor B
4. While checking the operation of a water pump: Technician A uses a pressure tester to apply normal operating pressure to the system. If the pressure increases, the water pump is working. Technician B starts the engine and allows it to warm up. The upper radiator hose is squeezed and the engine accelerated. If a surge is felt in the hose, the pump is working. Who is correct?
   a. Technician A c. Both A and B
   b. Technician B d. Neither A nor B
5. Technician A says that if the PCV valve or connecting hoses become clogged, excessive pressure will develop in the crankcase and can cause oil leaks. Technician B says that if the PCV valve or connecting hoses become clogged, oil may be pushed into the air cleaner or cause it to be drawn into the intake manifold. Who is correct?
   a. Technician A c. Both A and B
   b. Technician B d. Neither A nor B
6. While bleeding air from a cooling system: Technician A loosens a heater hose at its highest point. Technician B uses air bleed valves. Who is correct?
   a. Technician A c. Both A and B
   b. Technician B d. Neither A nor B
7. While diagnosing an engine that overheats after frequent hard accelerations: Technician A checks for a collapsed lower radiator hose. Technician B checks the cooling fan electrical circuit. Who is correct?
   a. Technician A c. Both A and B
   b. Technician B d. Neither A nor B
8. While performing a cooling system service on a hybrid vehicle: Technician A disconnects the electric water pump. Technician B disables the high-voltage system before serving the engine. Who is correct?
   a. Technician A c. Both A and B
   b. Technician B d. Neither A nor B
9. While discussing the causes of extreme engine bearing damage: Technician A says that contaminated and old engine oil could be the cause. Technician B says that a leaking radiator could be the cause. Who is correct?
   a. Technician A c. Both A and B
   b. Technician B d. Neither A nor B
10. While diagnosing the cause of air bubbles in the coolant recovery tank: Technician A checks for a bad head gasket. Technician B checks the radiator cap. Who is correct?
   a. Technician A c. Both A and B
   b. Technician B d. Neither A nor B

CHAPTER 15

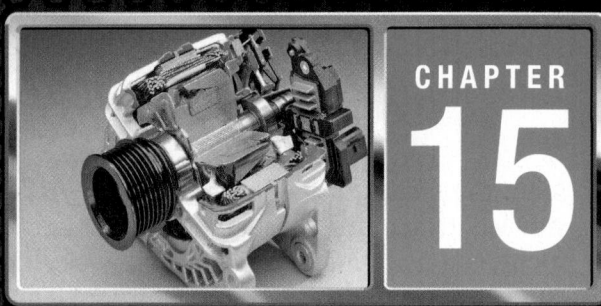

BASICS OF ELECTRICAL SYSTEMS

OBJECTIVES

- Explain the basic principles of electricity.
- Define the terms normally used to describe electricity.
- Use Ohm's law to determine voltage, current, and resistance.
- List the basic types of electrical circuits.
- Describe the differences between a series circuit and a parallel circuit.
- Name the various electrical components and their uses in electrical circuits.
- Describe the different kinds of automotive wiring.

There is often confusion concerning the terms *electrical* and *electronic.* In this book, electrical and electrical systems refer to wiring and electrical parts, such as generators, lights, and voltage regulators. **Electronics** means computers and other black box-type items used to control engine and vehicle systems.

A good understanding of electrical principles is important to proper diagnosis of any system that is monitored, controlled, or operated by electricity **(Figure 15–1)**.

BASICS OF ELECTRICITY

Perhaps the one reason why some people find it difficult to understand electricity is that they cannot see it. By actually knowing what it is and what it is *not,* you can easily understand it. Electricity is not magic! It is something that takes place or can take place in everything you know. It not only provides power for lights, TVs, stereos, and refrigerators, it is also the basis for the communications between our brain and the rest of our bodies. Although electricity cannot be

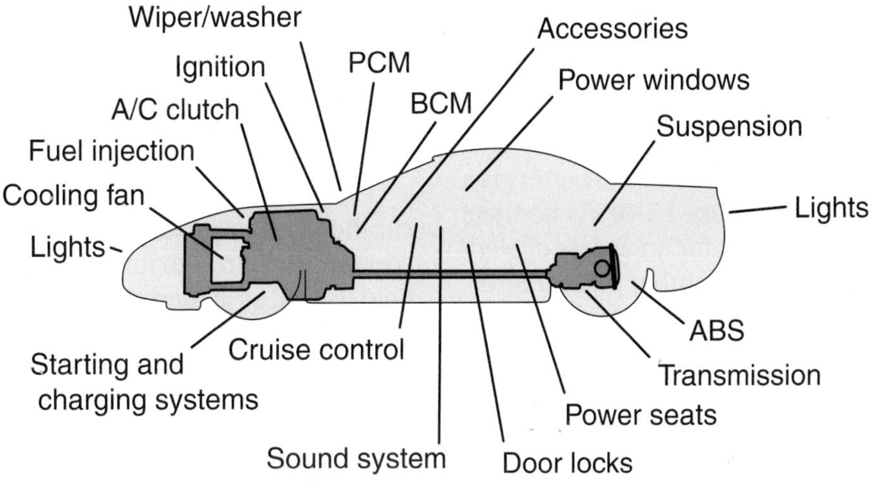

Figure 15-1 Basic overview of an electrical system for a late-model car.

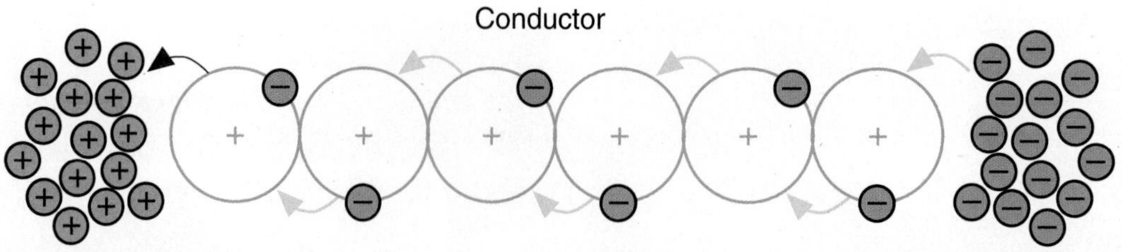

Figure 15–2 Electricity is the flow of electrons from one atom to another.

seen, the effects of it can be seen, felt, heard, and smelled. One of the most common displays of electricity is a lightning bolt. Lightning is electricity—a large amount of electricity. The power of lightning is incredible. Using the power from much smaller amounts of electricity to perform some work is the basis for an automobile's electrical system. Electricity cannot be seen because it is the movement of extremely small objects that move at the speed of light (186,282.397 miles [299,792,458 meters] per second).

Flow of Electricity

Electricity is the flow of electrons from one atom to another **(Figure 15–2)**. The release of energy as one electron leaves the orbit of one atom and jumps into the orbit of another is electrical energy. The key to creating electricity is to provide a reason for the electrons to move to another atom.

There is a natural attraction of electrons to protons. Electrons have a negative charge and are attracted to something with a positive charge. When an electron leaves the orbit of an atom, the atom then has a positive charge. An electron moves from one atom to another because the atom next to it appears to be more positive than the one it is orbiting around.

An electrical power source provides for a more positive charge and, to allow for a continuous flow of electricity, it supplies free electrons. To have a continuous flow of electricity, three things must be present: an excess of electrons in one place, a lack of electrons in another place, and a path between the two places.

Two power or energy sources are used in an automobile's electrical system. These are based on a chemical reaction and **magnetism**. A car's battery is a source of chemical energy **(Figure 15–3)**. A chemical reaction in the battery provides for an excess of electrons and a lack of electrons in another place. Batteries have two terminals, a positive and a negative. Basically, the negative terminal is the outlet for the electrons and the positive terminal is the inlet for the electrons to get to the protons. The chemical reaction in a battery causes a lack of electrons at the positive (+) terminal and an excess at the negative (−) terminal. This creates an electrical imbalance, causing the electrons to flow through the path provided by a wire.

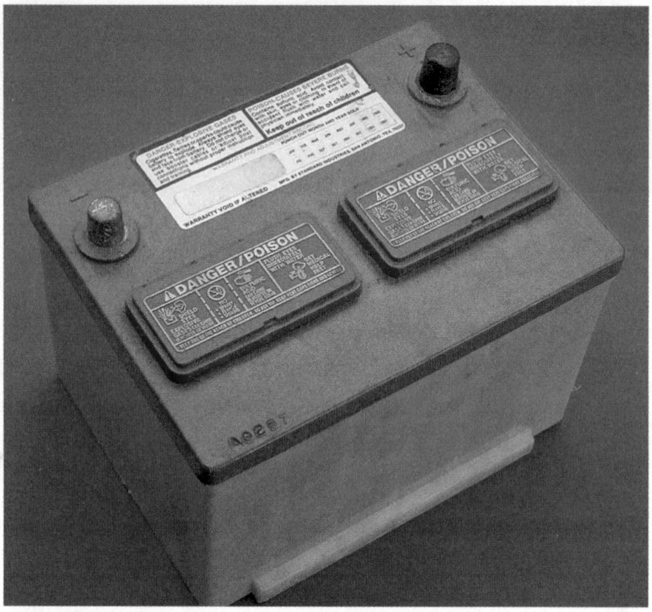

Figure 15–3 An automotive battery.

The chemical process in the battery continues to provide electrons until the chemicals become weak. At that time, either the battery has run out of electrons or all the protons are matched with an electron. When this happens, there is no longer a reason for the electrons to want to move to the positive side of the battery. To the electrons, it no longer looks more positive. Fortunately, the vehicle's charging system continuously restores the battery's supply of electrons. This allows the chemical reaction in the battery to continue indefinitely. In an electrical diagram, a battery is drawn as shown in **Figure 15–4**.

Electricity and magnetism are interrelated. One can be used to produce the other. Moving a wire (a conductor) through an already existing magnetic field (such as a permanent magnet) can produce electricity. This process of producing electricity

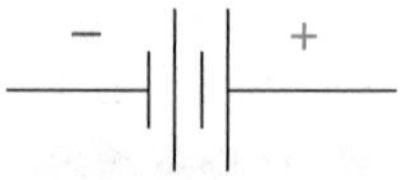

Figure 15–4 Symbol for a battery.

Figure 15-5 A late-model generator. *Courtesy of Robert Bosch GmbH, www.bosch-presse.de*

through magnetism is called **induction**. The heart of a vehicle's charging system is the AC generator **(Figure 15–5)**. A magnetic field, driven by the crankshaft via a drive belt, rotates through a coil of wire, producing electricity. The amount of electricity produced depends on a number of things: the strength of the magnetic field, the number of wires that are passed by the magnetic field, and the speed at which the magnetic field moves past the wires.

Electricity is also produced by chemical, photoelectrical, thermoelectrical, and piezoelectrical reactions. These sources of electricity are used throughout a modern automobile. Most are the basis of operation for electronic sensors and are discussed as those sensors are introduced in this book. The two most common ways of producing electricity are through electromagnetic induction and chemical reaction. These are main topics of the chapters in this section of the book.

ELECTRICAL TERMS

Electrical **current** describes the movement or flow of electricity. The greater the number of electrons flowing past a given point in a given amount of time, the more current the circuit has. The unit for measuring electrical current is the **ampere**, usually called an amp. The term *ampere* was assigned to units of current in honor of André Ampère, who studied the relationship between electricity and magnetism. The instrument used to measure electrical current flow in a circuit is called an ammeter.

When electricity flows, millions of electrons are moving past any given point at the speed of light. The electrical charge of any one electron is extremely small. It takes millions of electrons to make a charge that can be measured. In fact, 1 ampere of current means that 6.24 billion billion electrons are flowing past a given point in 1 second.

SHOP TALK

Engineers often express current flow in coulombs. One coulomb represents the amount of electric charge moved by a current of 1 ampere in 1 second. A coulomb represents 6.241 509 629 152 65 $\times$ 10^{18} electrons.

There are two types of current: **direct current (DC)** and **alternating current (AC)**. In direct current, the electrons flow in one direction only. In AC, the electrons change direction at a fixed rate. Typically, an automobile uses DC, while the current in homes and buildings is AC. Some components of the automobile generate or use AC. These are discussed in later chapters.

There are many theories about the direction of current flow. The conventional theory states that current flows from positive to negative. The electron theory says current moves from negative to positive. And the hole-flow theory basically says something is moving in both directions. Remember that a *theory* is *not* a *fact*. It is a concept that is yet to be proved wrong or right. Therefore, only one theory about current flow is correct and the rest are wrong. For the purposes of this book and for your own understanding of electricity, current flow **(Figure 15–6)** is described as moving from a point of higher potential (voltage) to a point of lower potential (voltage). This statement may not be absolutely correct, but it is sound and is based on what can be observed.

Voltage is electrical pressure **(Figure 15–7)**. It is the force developed by the attraction of the electrons to protons. The more positive one side of the circuit is, the more voltage is present in the circuit. Voltage does not flow; it is the pressure that causes current flow. To have current flow, some force is needed to move the electrons between atoms. This force is the

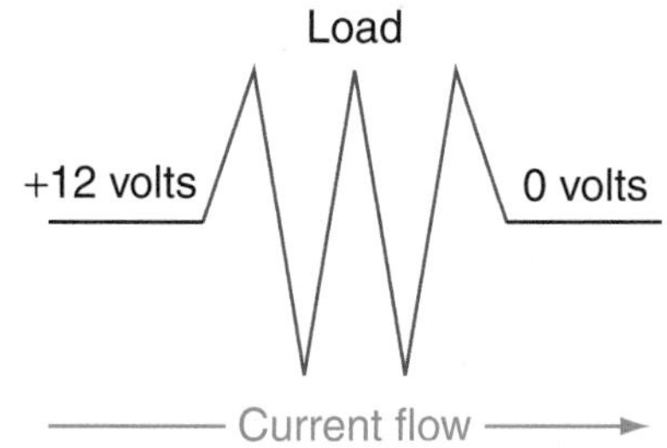

Figure 15-6 Current moves from a point of higher potential to a point of lower potential.

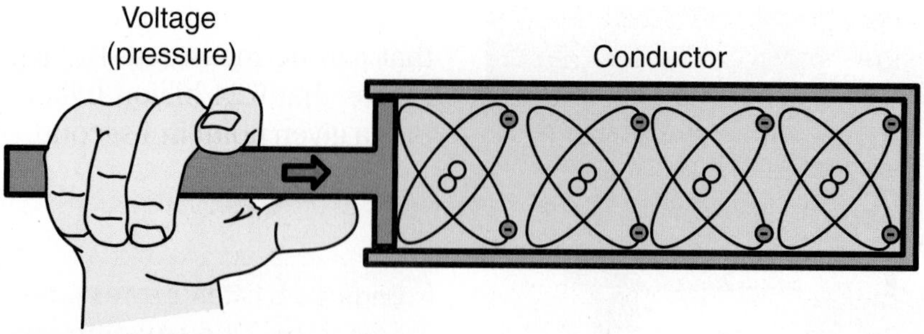

Figure 15-7 Voltage is electrical pressure.

Figure 15-8 Common symbol for an ohm.

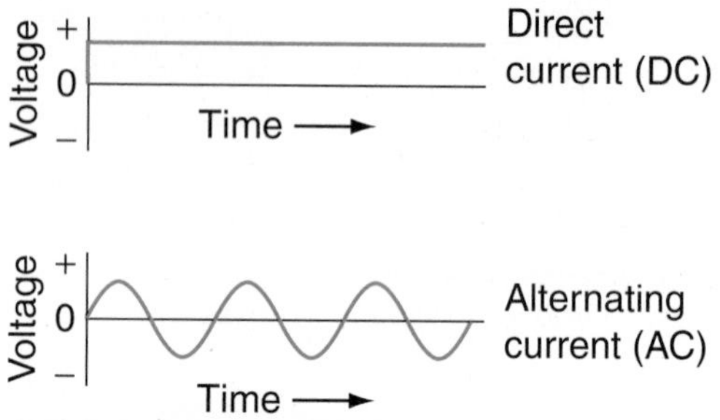

Figure 15-9 The difference between DC and AC.

pressure that exists between a positive and negative point within an electrical circuit. This force, also called **electromotive force (EMF)**, is measured in units called volts. One volt is the amount of pressure required to move 1 ampere of current through a resistance of 1 ohm. Voltage is measured by an instrument called a voltmeter. The unit of measurement for electrical pressure was so named to honor Alessandro Volta, who, in 1800, made the first electrical battery.

When any substance flows, it meets **resistance**. The resistance to electrical flow can be measured. The resistance to current flow produces heat. This heat can be measured to determine the amount of resistance. A unit of measured resistance is called an **ohm**. The common symbol for an ohm is shown in **Figure 15–8**. Resistance can be measured by an instrument called an ohmmeter.

Alternating Current

AC constantly changes in voltage and direction. If a graph is used to represent the amount of DC voltage available from a battery during a fixed period, the line on the graph will be flat, which represents a constant voltage. If AC voltage is shown on a graph, it will appear as a **sine wave (Figure 15–9)**. The sine wave shows AC changing in amplitude (strength) and direction. The highest positive voltage equals the highest negative voltage. The movement of the AC from its peak at the positive side of the graph to the negative side and then back to the positive peak is commonly referred to as "peak-to-peak" value. This value represents the amount of voltage available at a point. During each complete cycle of AC, there are always two maximum or peak values, one for the positive half-cycle and the other for the negative half-cycle. The difference between the peak positive value and the peak negative value is used to measure AC voltages **(Figure 15–10)**.

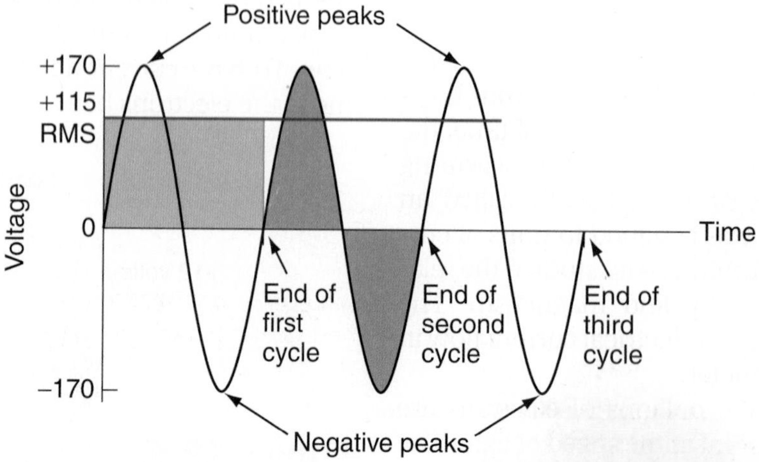

Figure 15-10 The action and measurement of alternating current.

AC does not have a constant value; therefore, as it passes through a resistance, nearly 29% less heat is produced when compared to DC. This is one reason that AC is preferred over DC when powering motors and other electrical devices.

The lack of heat also causes us to look at AC values differently than the same values in a DC circuit. AC has an effective value of 1 ampere when it produces heat in a given resistance at the same rate as does 1 ampere of DC. The effective value of an AC is equal to 0.707 times its maximum or peak current value. Because AC is caused by an alternating voltage, the ratio of the effective voltage value to the maximum voltage value is the same as the ratio of the effective current to the maximum current, or 0.707 times the maximum value. AC voltage measurements are often expressed in terms of root mean square (RMS) values.

AC voltage and current change constantly and their values must be viewed as average or effective. When AC is applied to a resistance, as the actual voltage changes in value and direction, so does the current. In fact, the change of current is in phase with the change in voltage. An "in-phase" condition exists when the sine waves of voltage and current are precisely in step with one another. The two sine waves go through their maximum and minimum points at the same time and in the same direction. In some circuits, several sine waves can be in phase with each other.

If a circuit has two or more voltage pulses but each has its own sine wave that begins and ends its cycle at a different time, the waves are "out of phase." If two sine waves are 180 degrees out of phase, they will cancel each other out if they have the same voltage and current. If two or more sine waves are out of phase but do not cancel each other, the effective voltage and current are determined by the position and direction of the sine wave at a given point within the circuit (**Figure 15–11**).

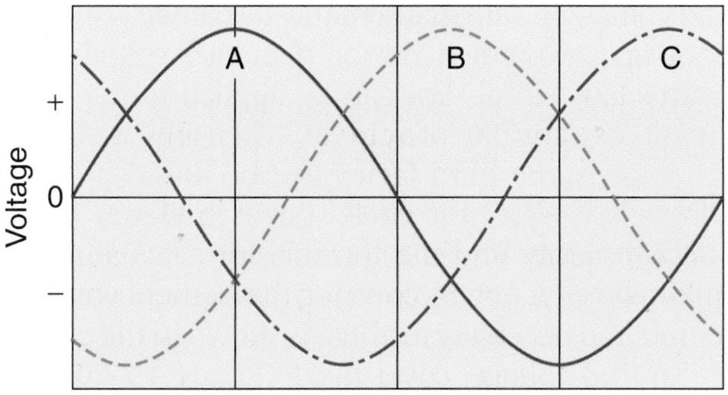

Figure 15-11 The sine waves of three-phase AC.

Circuit Terminology

An electrical circuit is considered complete when there is a path that connects the positive and negative terminals of the electrical power source. A completed circuit is called a **closed circuit**, whereas an incomplete circuit is called an **open circuit**. When a circuit is complete, there is **continuity**. Conductors are drawn in electrical diagrams as a line connecting two points, as shown in **Figure 15–12**.

In a complete circuit, the flow of electricity can be controlled and applied to do useful work, such as light a headlamp or turn over a starter motor. Components that use electrical power put a load on the circuit and consume electrical energy. These components are often referred to as electrical loads. Loads are drawn in electrical diagrams as a symbol representing the part or as a resistor. The typical drawing of a resistor is shown in **Figure 15–13**.

The amount of current that flows in a circuit is determined by the resistance in that circuit. As resistance goes up, the current goes down. The total resistance in a circuit determines how much current will flow through the circuit. The energy used by a load is measured in volts. Amperage stays constant in a circuit but the voltage drops as it powers a load. Measuring voltage drop tells how much energy is being consumed by the load.

The amount of electricity consumed by a load is normally called electrical power usage or watts. One watt is equal to 1 volt multiplied by 1 ampere. The formula for determining the amount of power consumed by a load is the amount of current through the load multiplied by the voltage drop across the load.

Grounding the Load Most automotive electrical circuits use the chassis as a conductor for the negative side of the battery through the battery's ground wire, as shown in **Figure 15–14**. Current passes from the battery, through the load, through the metal frame, and back to the battery. Using the frame as a return path or ground eliminates the need for a separate ground wire at each component. Without **grounding**

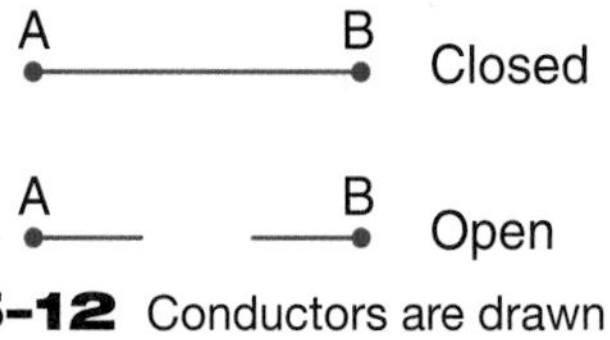

Figure 15-12 Conductors are drawn as lines from one point to another.

Figure 15-13 Symbol for a resistor.

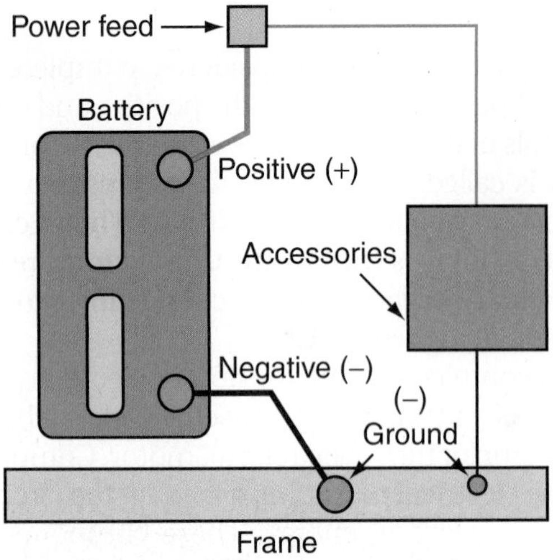

Figure 15-14 Most automotive electrical circuits use the chassis as the conductor for the negative side of the battery.

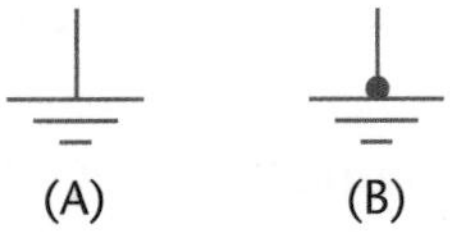

Figure 15-16 Symbols for grounds: (A) made through the component's mounting (B) made by a remote wire.

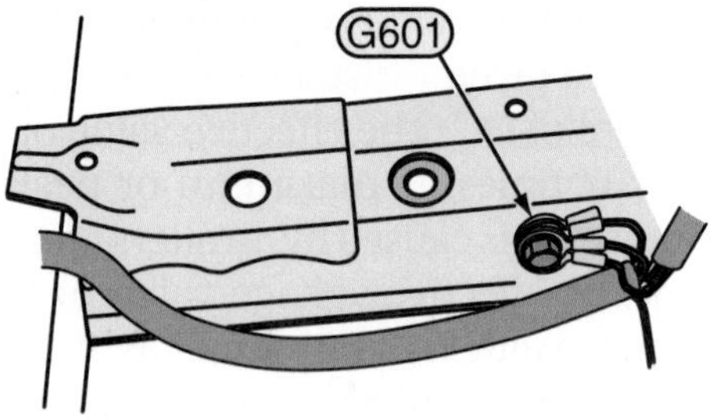

Figure 15-17 To ensure good grounding, manufacturers use a network of common grounding terminals and wires.

parts to the frame, hundreds of additional wires would be needed to complete the individual circuits.

Major components, such as the engine block and transmission case, also have a grounding wire connected to the frame. This provides a ground circuit for parts that are mounted directly to the block **(Figure 15–15)** or transmission. Other parts have a separate ground wire that connects the body of the part to the frame, engine, or transmission. These connections are called **chassis ground** connections. The wire that serves as the contact to the chassis is commonly called the **ground wire** or lead.

In wiring diagrams, chassis ground connections are drawn to show the normal type of ground connection for that part **(Figure 15–16)**. When the ground is made through the mounting of the component, the connection is represented in the drawing A. When the ground is made by a wire that connects to the chassis, the connection is shown as B.

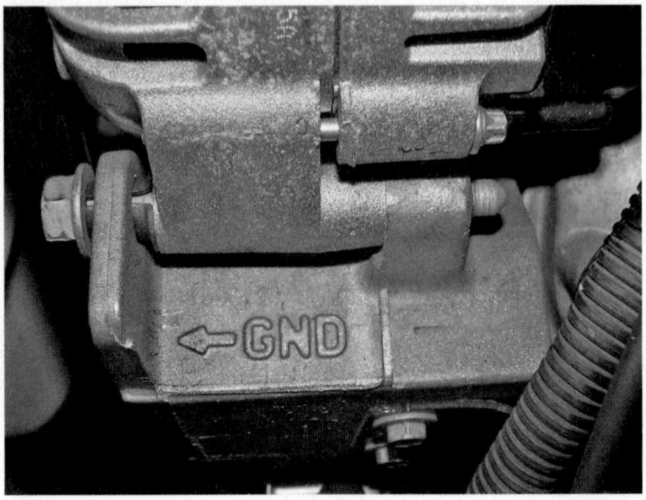

Figure 15-15 Many electrical parts are grounded through their mounting to the engine, transmission, or frame.

The increased use of plastics and other nonmetallic materials in body panels and engine parts has made electrical grounding more difficult. To ensure good grounding, some manufacturers use a network of common grounding terminals and wires **(Figure 15–17)**.

OHM'S LAW

In 1827, a German mathematics professor, Georg Ohm, published a book that included his explanation of the behavior of electricity. His thoughts have become the basis for a true understanding of electricity. He found it takes 1 volt of electrical pressure to push 1 ampere of electrical current through 1 ohm of resistance. This statement is the basic law of electricity. It is known as **Ohm's law**.

A simple electrical circuit is a load connected to a voltage source by conductors. The resistor could be a fog light, the voltage source could be a battery, and the conductor could be a copper wire **(Figure 15–18)**.

In any electrical circuit, current (I), resistance (R), and voltage (E) are mathematically related. This relationship is expressed in a mathematical statement of Ohm's law. Ohm's law can be applied to the entire circuit or to any part of a circuit. When any two factors are known, the third factor can be found by using Ohm's law. Using the circle shown in **Figure 15–19**, you can easily find the formula for calculating the unknown element. By covering the element you need to find, the necessary formula is shown in the circle.

To find voltage, cover the E **(Figure 15–20)**. The voltage (E) in a circuit is equal to the current (I) in amperes multiplied by the resistance (R) in ohms.

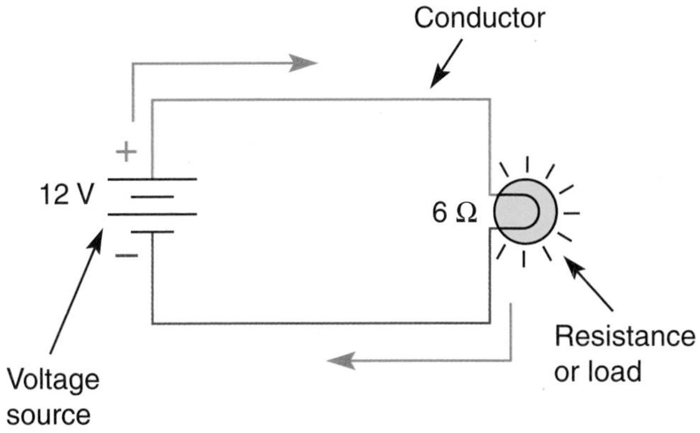

Figure 15-18 A simple circuit consists of a voltage source, conductors, and a resistance or load.

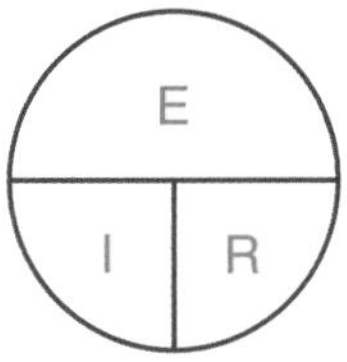

Figure 15-19 Ohm's law.

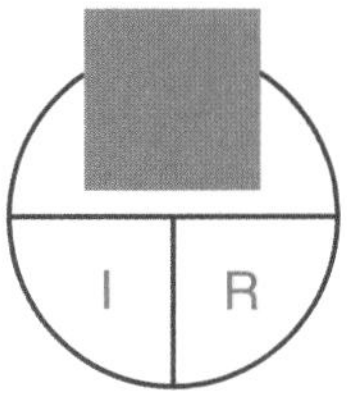

Figure 15-20 To find voltage, cover the *E* and use the exposed formula.

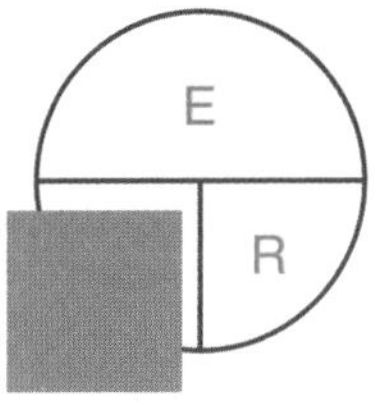

Figure 15-21 To find current, cover the *I* and use the exposed formula.

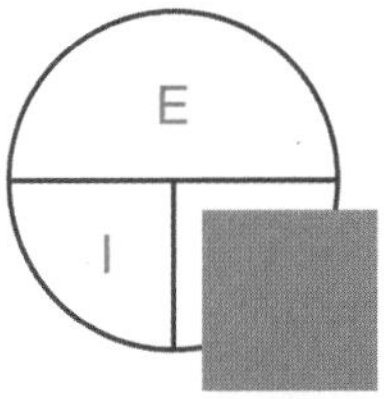

Figure 15-22 To find resistance, cover the *R* and use the exposed formula.

To find current, cover the *I* **(Figure 15–21)**. The current (amperage) in a circuit equals the voltage divided by the resistance (in ohms).

To find resistance, cover the *R* **(Figure 15–22)**. The resistance of a circuit (in ohms) equals the voltage divided by the current (in amperes).

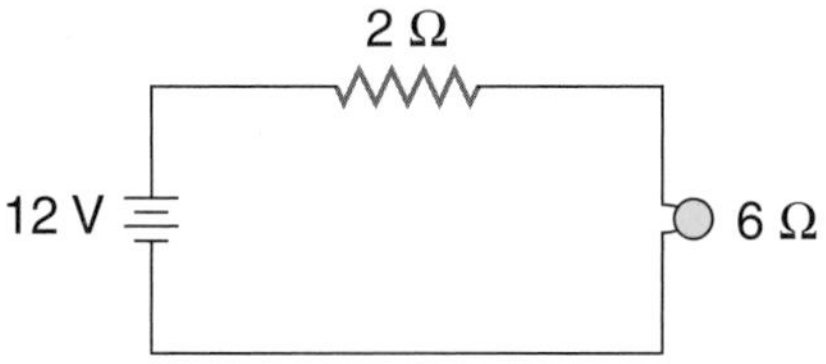

Figure 15-23 Same circuit as shown in Figure 15–18 but this one has a corroded wire, represented by the additional resistor.

It is very important for technicians to understand Ohm's law. It explains how an increase or decrease in voltage, resistance, or current affects a circuit.

For example, if the fog light in Figure 15–18 has a 6-ohm resistance, how many amperes does it use to operate? Since cars and light trucks have a 12-volt battery and you know two of the factors in the fog light circuit, it is simple to solve for the third.

$$I\text{(unknown)} = \frac{E\text{(12 volts)}}{R\text{(6 ohms)}}$$

$$I = \frac{12}{6}$$

$$I = 2 \text{ amperes}$$

In a clean, well-wired circuit, the fog lights will draw 2 amperes of current. What would happen if resistance in the circuit increases due to corroded or damaged wire or connections? If the corroded connections add 2 ohms of resistance to the fog light circuit, the total resistance is 8 ohms **(Figure 15–23)**. The amount of current flowing through the circuit for the lights decreases.

$$I = \frac{12}{6+2} = \frac{12}{8}$$

$$I = 1.5 \text{ amperes}$$

If the lights are designed to operate at 2 amperes, this decrease to 1.5 amperes causes them to burn dimly. Cleaning the corrosion away or installing new wires and connectors eliminates the unwanted resistance; the correct amount of current will flow through the circuit, allowing the lamp to burn as brightly as it should.

Voltage Drops

Voltage drop is the amount of voltage required to cause current flow through a load. Electrical energy is changed to another form of energy as it flows through a load and the voltage leaving the load is lower than it was when it entered. The amount of voltage dropped by a load depends on the circuit's current and the resistance of the load. Voltage drop can be calculated by using Ohm's law. If the resistance of the load and the circuit's current are known, the voltage drop is the product of the two. If there is only one load in the circuit, the voltage drop will equal the source voltage.

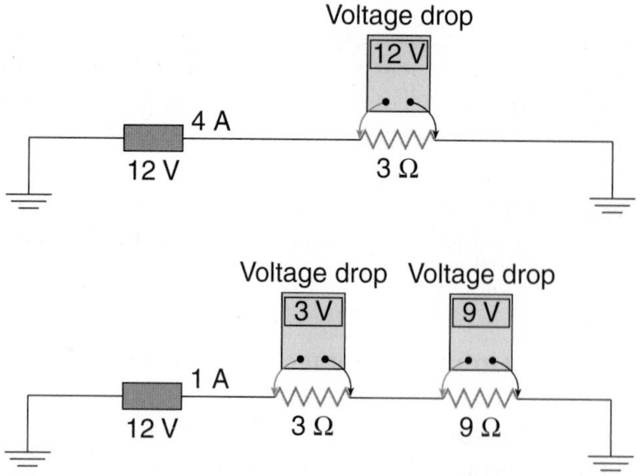

Figure 15-24 All of the available voltage is dropped by the loads in the circuit. The amount of drop depends on the resistance of the load.

When there is more than one load, the voltage drop will vary with the values of the loads. For example, if a 12-volt circuit has a 3-ohm load and a 9-ohm load, the total current will be 1 ampere and the total resistance will be 12 ohms. The voltage drop across the 3-ohm load will be 1×3, which equals 3 volts. The drop across the other load will be 1×9, or 9 volts **(Figure 15–24)**.

All of the available voltage in a circuit is dropped by the loads; therefore, 0 volts is present at the negative connection of the power source.

AC Circuits In AC circuits, the actual resistance of a load is called its **impedance**. Electrical impedance, like resistance, is measured in ohms and is simply defined as the operating resistance of a load. In addition to the static resistance of a load, additional opposition to AC current flow results from its interaction with magnetic fields within the conductor. Voltage drop in an AC circuit is the product of the current and the impedance (Z) of the circuit or load. It is expressed by the formula $E = I \times Z$ **(Figure 15–25)**.

Power and Watt's Law

Watt's law is the name given to the formula that calculates the electrical power (P) used by a load. The law is expressed by the mathematical formula of $P = I \times E$. In other words, power equals current multiplied by voltage. Electric power is expressed in **watts** and represents the rate at which electrical energy is converted to another form of energy. Electrical energy can be converted into sound, heat, light, or motion.

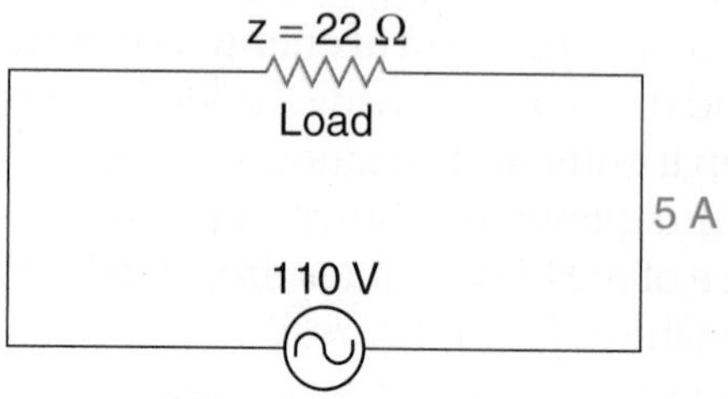

Figure 15-25 In alternating current circuits, the actual resistance of a load is called its impedance (Z).

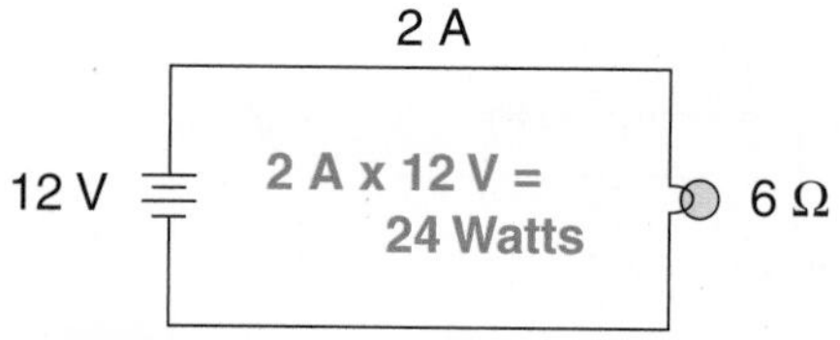

Figure 15-26 Electrical power is calculated by multiplying voltage by current.

If two loads have the same available voltage, the load with the lowest resistance will use the most power. For example, a 40-W household light bulb has a higher resistance than a 100-W bulb.

Although power measurements are rarely needed in automotive service, knowing the power requirements of light bulbs, electric motors, and other components is sometimes useful when troubleshooting electrical systems. Looking back at the example of the fog light circuit, the amount of power used by the fog light can be calculated **(Figure 15–26)**.

$$P = 12 \text{ volts} \times 2 \text{ amperes}$$
$$P = 24 \text{ watts}$$

The normal fog light uses 24 watts of power, whereas the corroded fog light circuit results in the following:

$$P = 12 \text{ volts} \times 1.5 \text{ amperes}$$
$$P = 18 \text{ watts}$$

This reduction in power explains the decrease in bulb brightness.

CIRCUITS

A complete electrical circuit exists when electrons flow along a path between two points. In a complete circuit, resistance must be low enough to allow the available voltage to push electrons between the two points. Most automotive circuits contain five basic parts.

1. Power sources, such as a battery or alternator, that provide the energy needed to cause electron flow
2. Conductors, such as copper wires, that provide a path for current flow
3. Loads, which are devices that use electricity to perform work, such as light bulbs, electric motors, or resistors

4. Controllers, such as switches or relays, that control or direct the flow of electrons
5. Circuit protection devices, such as fuses, circuit breakers, and fusible links

There are also three basic types of circuits used in automotive electrical systems: series circuits, parallel circuits, and series-parallel circuits. Each circuit type has its own characteristics regarding amperage, voltage, and resistance.

Series Circuits

A **series circuit** consists of one or more resistors connected to a voltage source with only one path for electron flow. For example, a simple series circuit consists of a resistor (2 ohms in this example) connected to a 12-volt battery **(Figure 15–27A)**. The current can be determined by applying Ohm's law.

$$I = \frac{E}{R} = \frac{12}{2} = 6 \text{ amperes}$$

Another series circuit may contain a 2-ohm resistor and a 4-ohm resistor connected to a 12-volt battery **(Figure 15–27B)**. The word *series* is given to a circuit in which the same amount of current is present throughout the circuit. The current that flows through one resistor also flows through other resistors in the circuit. As that amount of current leaves the battery, it flows through the conductor to the first resistor. At the resistor, some electrical energy or voltage is consumed as the current flows through it. The decreased amount of voltage is then applied to the next resistor as current flows to it. By the time the current is flowing in the conductor leading back to the battery, all voltage has been dropped. All of the source voltage available to the circuit is dropped by the resistors in the circuit.

In a series circuit, the total amount of resistance in the circuit is equal to the sum of all the individual resistors. In the circuit in Figure 15–27B, the total circuit resistance is 4 + 2 = 6 ohms. Based on Ohm's law, current is $I = E/R = 12/6 = 2$ amperes. In a series circuit, current is constant throughout the circuit. Therefore, 2 amps of current flows through the conductors and both resistors.

Ohm's law can be used to determine the voltage drop across parts of the circuit. For the 2-ohm resistor, $E = IR = 2 \times 2 = 4$ volts. For the 4-ohm resistor, $E = 2 \times 4 = 8$ volts. The sum of all voltage drops in a series circuit must equal the source voltage, or 4 + 8 = 12 volts.

An ammeter connected anywhere in this circuit will read 2 amperes, and a voltmeter connected across each of the resistors will read 4 volts and 8 volts, as shown in **Figure 15–28.**

All calculations for a series circuit work in the same way no matter how many resistors there are in series. Consider the circuit in **Figure 15–29**. This circuit has four resistors in series with each other. The total resistance is 12 ohms (5 ohms + 2 ohms + 4 ohms + 1 ohm). Using Ohm's law, we can see that the circuit current is 1 amp ($I = E/R = 12/12 = 1$ amp). We can also use Ohm's law to determine the voltage drop across each resistor in the circuit. For example, since the circuit current is 1 amp, 4 volts are dropped by the 4-ohm resistor ($E = I \times R = 1$ amp $\times$ 4 ohms = 4 volts).

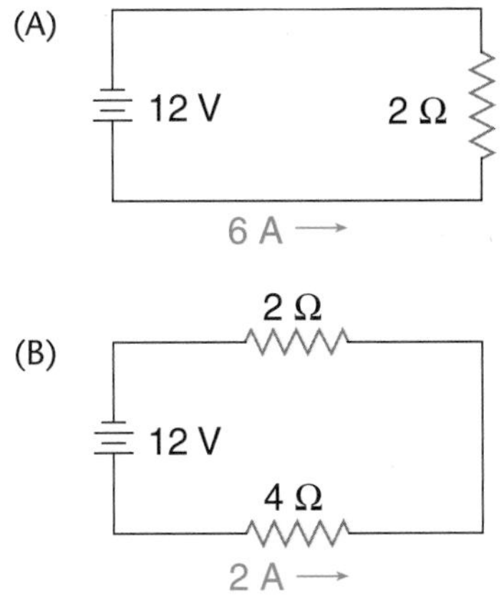

Figure 15–27 In a series circuit, the same amount of current flows through the entire circuit.

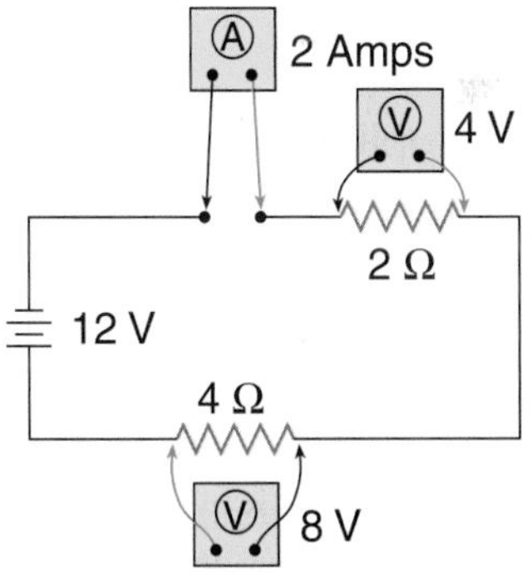

Figure 15–28 Measuring the current and voltage drops in the circuit.

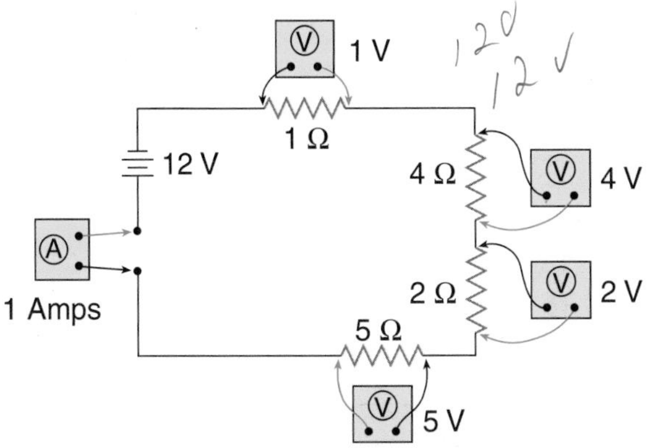

Figure 15–29 Values in the series circuit.

A series circuit is characterized by the following four facts:

1. The circuit's current is determined by the total amount of resistance in the circuit; it is constant throughout the circuit.
2. The voltage drops across each resistor are different if the resistance values are different.
3. The sum of the voltage drops equals the source voltage.
4. The total resistance is equal to the sum of all resistances in the circuit.

Parallel Circuits

A **parallel circuit** provides two or more different paths for the current to flow through. Each path has separate resistors (loads) and can operate independently of the other paths. The different paths for current flow are commonly called the legs of a parallel circuit.

A parallel circuit is characterized by the following facts:

1. Total circuit resistance is always lower than the resistance of the leg with the lowest total resistance.
2. The current through each leg will be different if the resistance values are different.
3. The sum of the current on each leg equals the total circuit current.
4. The voltage applied to each leg of the circuit will be dropped across the legs if there are no loads in series with the parallel circuit.

Consider the circuit shown in **Figure 15–30**. Two 3-ohm resistors are connected to a 12-volt battery. The resistors are in parallel with each other, since the battery voltage (12 volts) is applied to each resistor and they have a common negative lead. The current through each resistor or leg can be determined by applying Ohm's law. For the 3-ohm resistors, $I = E/R = 12/3 = 4$ amperes. Therefore, the total circuit current supplied by the battery is $4 + 4 = 8$ amperes. Using

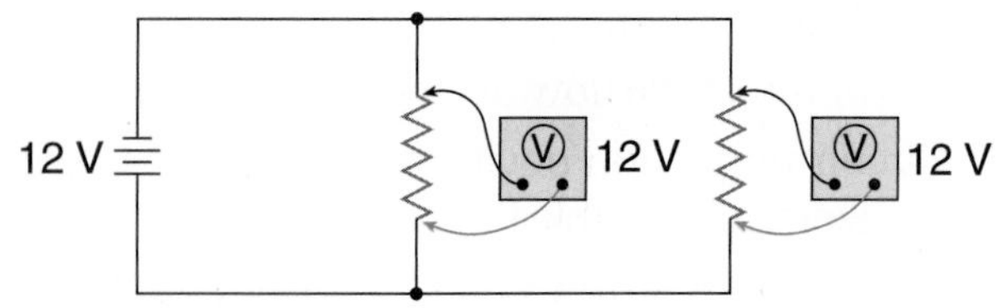

Figure 15-31 A parallel circuit with voltage drops shown.

Ohm's law, we find that 12 volts are dropped by both resistors (**Figure 15–31**).

Resistances are not added up to calculate the total resistance in a parallel circuit. Rather, they can be determined by dividing the product of their ohm values by the sum of their ohm values. This formula works when the circuit has two parallel legs.

$$\frac{3 \text{ ohms} \times 3 \text{ ohms}}{3 \text{ ohms} + 3 \text{ ohms}} = \frac{9}{6} = 1.5 \text{ ohms}$$

Total resistance can also be calculated by using Ohm's law if you know the total circuit current and the voltage ($R = E/I = 12/8 = 1.5$ ohms).

Consider another parallel circuit, **Figure 15–32**. In this circuit there are two legs and four resistors. Each leg has two resistors in series. One leg has a 4-ohm and a 2-ohm resistor. The total resistance on that leg is 6 ohms. The other leg has a 1-ohm and a 2-ohm resistor. The total resistance of that leg is 3 ohms. Therefore, we have 6 ohms in parallel with 3 ohms.

Current flow through the circuit can be calculated by different methods. Using Ohm's law, we know that $I = E/R$. If we take the total resistance of each leg and divide it into the voltage, we then know the current through that leg. Since total circuit current is equal to the sum of the current flows through each leg, we simply add the current across each leg together. This will give us total circuit current.

Leg 1: $I = E/R = 12/6 = 2$ amps

Leg 2: $I = E/R = 12/3 = 4$ amps

2 amps + 4 amps + 6 amps = total circuit current

Circuit current can also be determined by finding the total resistance of the circuit. To do this, the

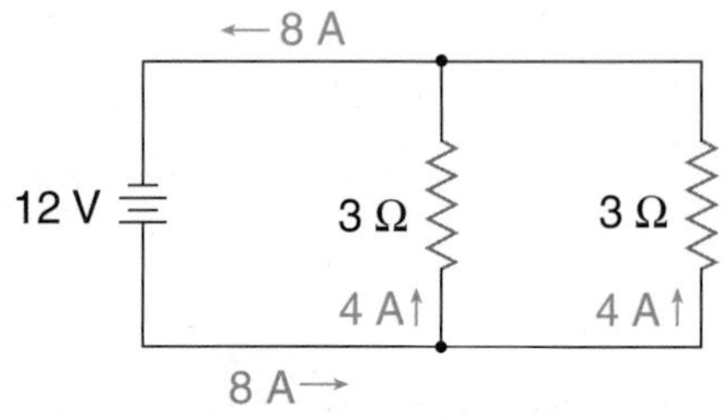

Figure 15-30 A simple parallel circuit.

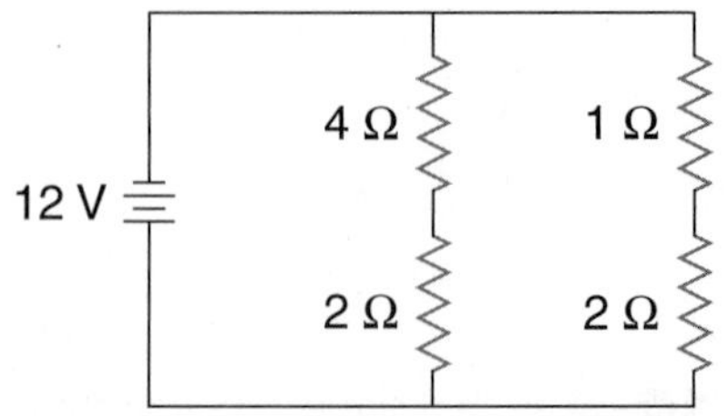

Figure 15-32 Series circuits within a parallel circuit.

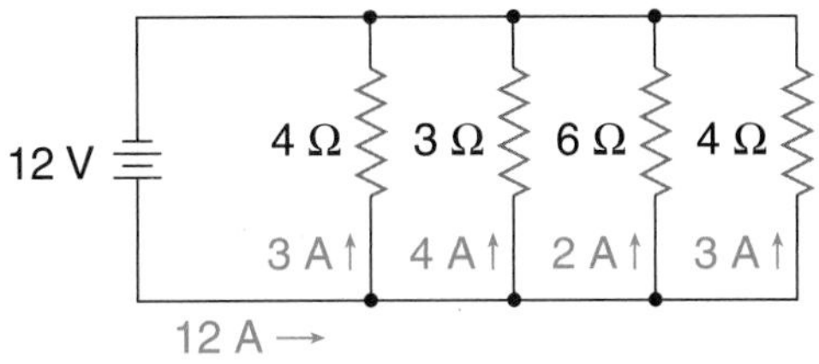

Figure 15-33 A parallel circuit with four legs.

product-over-sum formula is used. By dividing this total into the voltage, total circuit current is known.

$$\frac{\text{Leg 1} \times \text{Leg 2}}{\text{Leg 1} + \text{Leg 2}} = \frac{6 \times 3}{6 + 3} = \frac{18}{9} = 2 \text{ ohms}$$

since $I = E/R$, $I = {}^{12}\!/\!_{2}$, $I = 6$ amps (total circuit current)

When a circuit has more than two legs, the reciprocal formula should be used to determine total circuit resistance. The formula follows:

$$\frac{1}{\frac{1}{R_1} + \frac{1}{R_2} + \frac{1}{R_3} + \cdots \frac{1}{R_n}}$$

To demonstrate how to use this formula, consider the circuit in **Figure 15–33**. Here is a parallel circuit with four legs. The resistances across each leg are 4 ohms, 3 ohms, 6 ohms, and 4 ohms. Using the reciprocal formula, we will find that the total resistance of the circuit is 1 ohm. (Note that the total resistance is lower than the leg with the lowest resistance.)

$$\frac{1}{\frac{1}{4} + \frac{1}{3} + \frac{1}{6} + \frac{1}{4}} =$$

$$\frac{1}{\frac{3}{12} + \frac{4}{12} + \frac{2}{12} + \frac{3}{12}} = \frac{1}{\frac{12}{12}} = \frac{1}{1} = 1$$

The total of this circuit could also have been found by calculating the current across each leg then adding them together to get the total circuit current. Using Ohm's law, if you divide the voltage by the total circuit current, you will get total resistance.

Leg 1: $I = E/R = {}^{12}\!/\!_{4} = 3$ amps
Leg 2: $I = E/R = {}^{12}\!/\!_{3} = 4$ amps
Leg 3: $I = E/R = {}^{12}\!/\!_{6} = 2$ amps
Leg 4: $I = E/R = {}^{12}\!/\!_{4} = 3$ amps

Total circuit current $= 3 + 4 + 2 + 3 = 12$ amps; then,

$$R = E/I = {}^{12}\!/\!_{12} = 1 \text{ ohm}$$

Series-Parallel Circuits

In a **series-parallel circuit**, both series and parallel combinations exist in the same circuit. If you are faced with the task of calculating the values in a series-parallel circuit, determine all values of the parallel circuit(s) first. By looking carefully at a series-parallel circuit, you will find that it is nothing more than one or more parallel circuits in series with each other or in series with some other resistance.

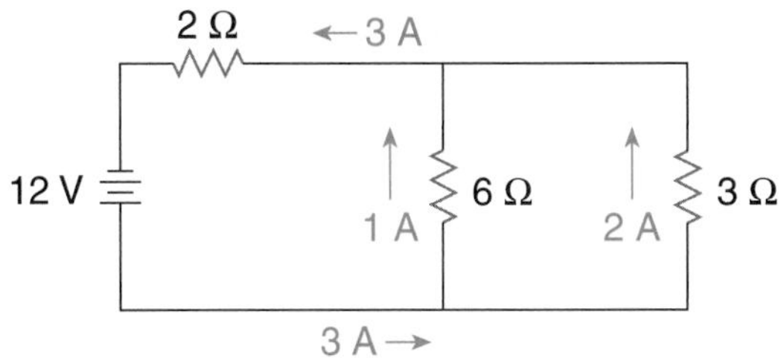

Figure 15-34 In a series-parallel circuit, the sum of the currents through the legs must equal the current through the series part of the circuit.

A series-parallel circuit is illustrated in **Figure 15–34**. The 6- and 3-ohm resistors are in parallel with each other and together are in series with the 2-ohm resistor.

The total current in this circuit is equal to the voltage divided by the total resistance. The total resistance can be determined as follows. The 6- and 3-ohm parallel resistors in **Figure 15–35** are equivalent to a 2-ohm resistor, since $(6 \times 3)/(6 + 3) = 2$. This equivalent 2-ohm resistor is in series with the other 2-ohm resistor. To find the total resistance, add the two resistance values together. This gives a total circuit resistance of 4 ohms ($2 + 2 = 4$ ohms). The total current, therefore, is $I = {}^{12}\!/\!_{4} = 3$ amperes. This means that 3 amps of current is flowing through the 2-ohm resistor in series and 3 amps are divided between the resistors in parallel. In series-parallel circuits, the sum of the currents, flowing in the parallel legs, must equal that of the series resistors' current.

To find the current through each of the resistors in parallel, find the voltage drop across those resistors first. With 3 amperes flowing through the 2-ohm resistor, the voltage drop across this resistor is $E = IR = 3 \times 2 = 6$ volts, leaving 6 volts across the 6- and 3-ohm resistors. The current through the 6-ohm resistor is $I = E/R = {}^{6}\!/\!_{6} = 1$ ampere, and through the 3-ohm resistor is $I = {}^{6}\!/\!_{3} = 2$ amperes. The sum of these two current values must equal the total circuit current, and it does,

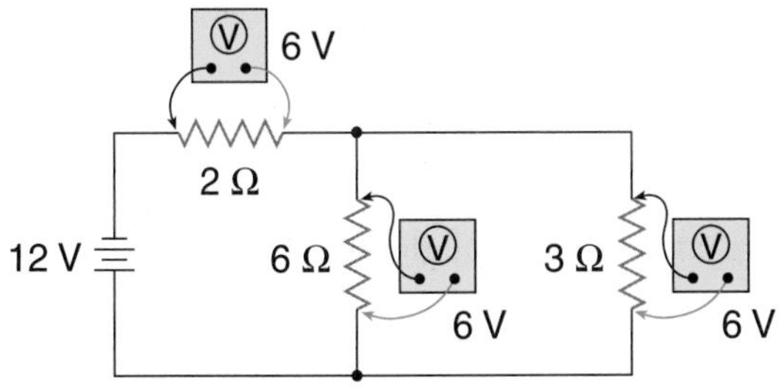

Figure 15-35 The circuit in Figure 15–34 with voltage drops shown.

1 + 2 = 3 amperes (Figure 15–35). The sum of the voltage drops across the parallel part of the circuit and the series part must also equal source voltage.

Kirchhoff's Law

In the 1840s, German physicist Gustav Kirchhoff described two laws on the behavior of electricity. Both of these laws relate to Ohm's law and help define the characteristics of series and parallel circuits.

Kirchhoff's "voltage law" basically states that the sum of the voltage drops in a closed circuit equals the voltage applied to that circuit. This directly describes the behavior of a series circuit. This law also describes the characteristics of more complex circuits in which there are positive and negative voltages. These are not discussed here.

Kirchhoff's "current law" explains the behavior of current in a parallel circuit. This law can be safely stated as: "At any junction point in a circuit, the current arriving to that point is equal to the current leaving." If 10 amperes of current arrives at the point in a circuit where a parallel leg is added to the circuit, the 10 amps will be divided according to the resistance of each leg, but a total of 10 amps will be present at the junction that ties the legs together again. In other words, the sum of the branch currents is equal to the total current entering the branches as well as the total current leaving the branches.

CIRCUIT COMPONENTS

Automotive electrical circuits contain a number of different electrical devices. The more common components are outlined in the following sections.

Power Sources

Most of today's vehicles operate with basically 12-volt electrical systems. The word *basically* is used because the battery is rated at 12 volts but stores 12.6 volts, and the charging system puts out about 14 volts while the engine is running. Because the charging system is the primary source of electrical power, it is fair to say that an automobile's electrical system is a 14-volt system.

In 1954, General Motors equipped its Cadillac with a 12-volt system. Prior to 1954, vehicles had 6-volt systems. The electrical demands of accessories, such as power windows and seats, put a severe strain on the 6-volt battery and charging system. With a 12-volt system, the charging system had to work less hard and there was plenty of electrical power for the electric accessories.

The increase in voltage also allowed wire sizes to decrease because the amperage required to power things was reduced. To explain this, consider an accessory that required 20 amps to operate in a 6-volt system (120 watts). When the voltage was increased to 12 volts, the system only drew 10 amperes.

Today we are faced with the same situation. The use of computers and the need to keep their memories fresh has put a drain on the battery even when the engine is not running. The number of electrical accessories also has and will continue to grow.

Today's vehicles are very sensitive to voltage change. In fact, their overall efficiency depends on a constant voltage. The demands of new technology make it difficult to maintain a constant voltage, and engineers have determined that system voltage must be increased to meet those demands. As vehicles evolve, emissions, fuel economy, comfort, convenience, and safety features will put more of a drain on the electrical system. This increased demand results from converting purely mechanical systems into electromechanical systems, such as steering, suspension, and braking systems, as well as new safety and communication systems. It has been estimated that in a few years the demand for continuous electrical power will be 3,000 to 7,000 W. Current 14-volt systems are rated at 800 to 1,500 W.

To meet these demands there are two possible solutions: increase the amperage capacity of the battery and charging system or increase system voltage.

Higher-amperage batteries and generators are only a Band-Aid solution. Because the generator is driven by engine power, more power from the engine will be required to keep the higher-capacity battery charged. Therefore, overall efficiency will decrease.

By moving to a higher system voltage, the battery will need to be larger and heavier. However, because system amperage will be lower, wire size will be smaller and perhaps the weight gain at the battery will be offset by the decreased weight of the wiring.

All of the advantages of moving from 6- to 12-volt systems apply to the move from 12 volts to 42 volts. But why 42 volts? Forty-two volts represent three 12-volt batteries. Engineers decided to take advantage of the fact that a 12-volt battery receives a 14-volt charge (3×14 volts $=$ 42 volts). Forty-two-volt systems are also desirable for safety reasons. Sixty volts can stop a person's heart; therefore, 42-volt systems allow for a margin of safety.

Forty-two-volt systems are based on a single 36-volt battery but have dual voltage systems (**Figure 15–36**). Part of the vehicle is powered by 12 to 14 volts and the rest by 36 to 42 volts. The battery has two positive connectors, one for each voltage, or the voltage is divided by a converter. The split voltage system provides 42 volts for high-voltage applications such as the starter/generator, power steering,

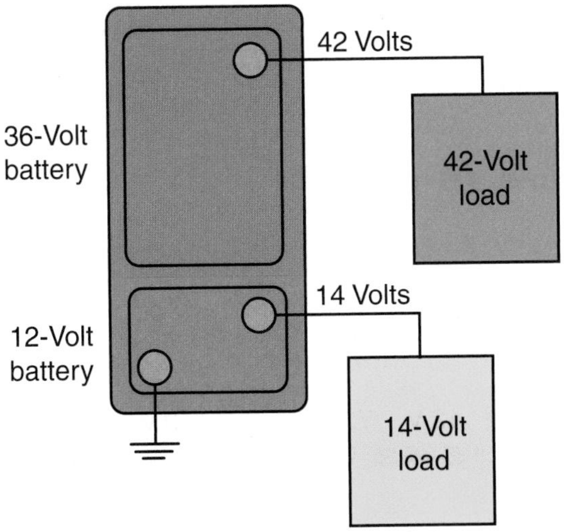

Figure 15-36 A split voltage arrangement providing 14 and 42 volts from a 36-volt battery.

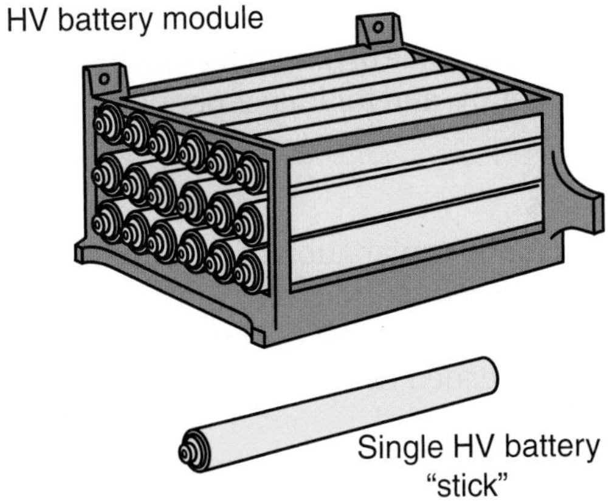

Figure 15-38 A high-voltage battery pack is made up of many individual battery cells.

air conditioning, traction control, brake, and engine cooling systems. The 14-volt system powers low-load systems, such as lights, power door locks, radios, and navigation systems.

High-Voltage Systems The battery in a hybrid or fuel cell vehicle serves a different purpose than a battery in a conventional vehicle. In a conventional vehicle, the battery's primary purpose is to provide a short powerful burst of power to start the engine. This type of battery is typically called a **starting battery**. In hybrid vehicles, the batteries provide continuous current to power electric motors and must be able to accept quick and frequent recharges.

A starting battery is also found in some hybrids in addition to a **high-voltage (HV) battery (Figure 15-37)**. The starting battery is used to start the engine, whereas the HV battery is used to power the electric motors. The HV system may also be used to power some accessories. In these vehicles, the starting battery is also used as the power source for normal accessories, such as lights.

Because there in no engine in a fuel cell electric vehicle (FCEV), there is no need for a starting battery. These vehicles are typically equipped with HV batteries.

Hybrid vehicles typically run on 42 to 600 volts. Hundreds of individual battery cells, each about the size of a flashlight battery, are connected together to store and provide the needed power **(Figure 15–38)**. The HV batteries used in hybrids are called battery packs. Many different types of batteries are available and under development to exceed the needs of these vehicles. In addition to batteries, some hybrid vehicles are equipped with ultracapacitors, which store and provide electrical energy.

HV is needed in hybrids to eliminate the need for extremely large cables and wiring. Also, keeping the required amperage low is easier on the batteries. A vehicle equipped with a starter/generator can be considered a mild hybrid and it relies on a 42-volt system. Functions such as stop-start, regenerative braking, and electrical assist are common to full and mild hybrid vehicles.

Only full hybrids have the ability to move in an electric-only mode. A full hybrid vehicle has a much

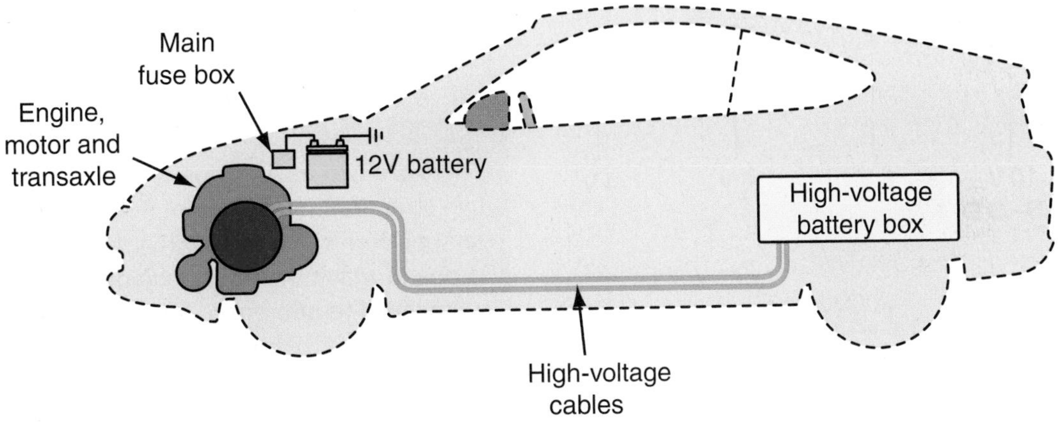

Figure 15-37 A hybrid vehicle with two battery packs.

higher voltage system than a mild hybrid; therefore, the battery is capable of providing much more power, more frequently, and for longer periods.

Resistors

As shown in the explanation of simple circuit design, resistors are used to limit current flow (and thereby voltage) in circuits where full current flow and voltage are not needed or desired. Resistors are devices specially constructed to put a specific amount of resistance into a circuit. In addition, some other components use resistance to produce heat and even light. An electric window defroster is a specialized type of resistor that produces heat. Electric lights are resistors that get so hot they produce light.

Automotive circuits typically contain these types of resistors: fixed value, stepped or tapped, and variable.

Fixed value resistors are designed to have only one rating, which should not change. These resistors are used to decrease the amount of voltage applied to a component, such as an ignition coil. Often manufacturers use a special wire, called **resistor wire**, to limit current flow and voltage in a circuit. This wire looks much like normal wire but is not a good conductor and is marked as a resistor.

Tapped or **stepped resistors** are designed to have two or more fixed values available by connecting wires to the several taps of the resistor. Heater motor resistor packs, which provide for different fan speeds, are an example of this type of resistor **(Figure 15–39)**.

Variable resistors are designed to have a range of resistances available through two or more taps and a control. Two examples of this type of resistor are **rheostats** and **potentiometers**. Rheostats **(Figure 15–40)** have two connections, one to the fixed end of a resistor and one to a sliding contact with the resistor. Moving the control moves the sliding contact away from or toward the fixed end tap, increasing or decreasing the resistance. Potentiometers **(Figure 15–41)** have three connections, one at each end of the resistance and one connected to a sliding contact with the resistor. Moving the control moves the sliding contact away from one end of the resistance but toward the other end. These are called potentiometers because different amounts of potential or voltage can be sent to another circuit. As the sliding contact moves, it picks up a voltage equal to the source voltage minus the amount dropped by the resistor, so far **(Figure 15–42)**.

Another type of variable resistor is the **thermistor**. This type of resistor is designed to change its resistance value as its temperature changes. Although most resistors are carefully constructed to maintain their rating within a few ohms through a range of temperatures, the thermistor is designed to change its rating. Thermistors are used to provide compensating voltage in components or to determine temperature. As a temperature sensor, the thermistor is connected to a voltmeter calibrated in degrees. As the temperature rises or falls, the resistance also changes, and so does the voltage in the circuit. These changes are read on the temperature gauge. Thermistors are also commonly used to sense temperature and send a signal back to a control unit. The control unit interprets the signal as a temperature value **(Figure 15–43)**.

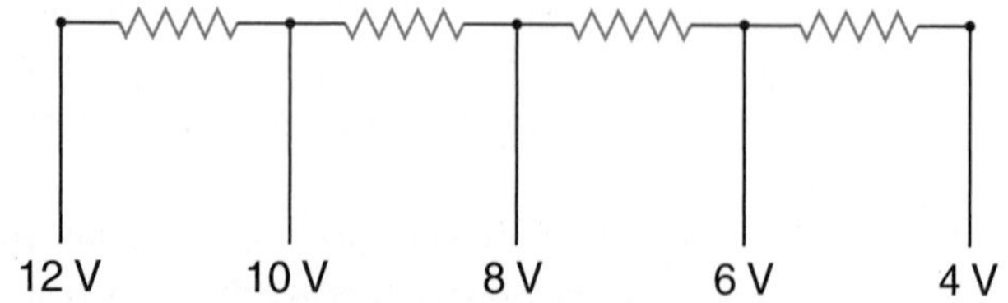

Figure 15-39 A stepped resistor.

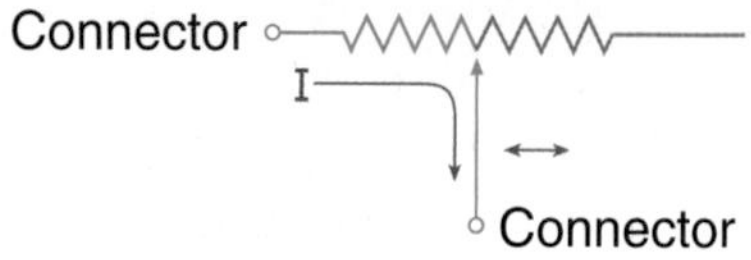

Figure 15-40 A rheostat.

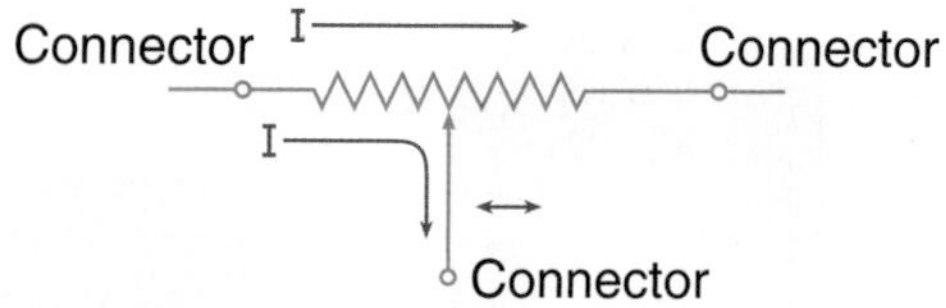

Figure 15-41 A potentiometer.

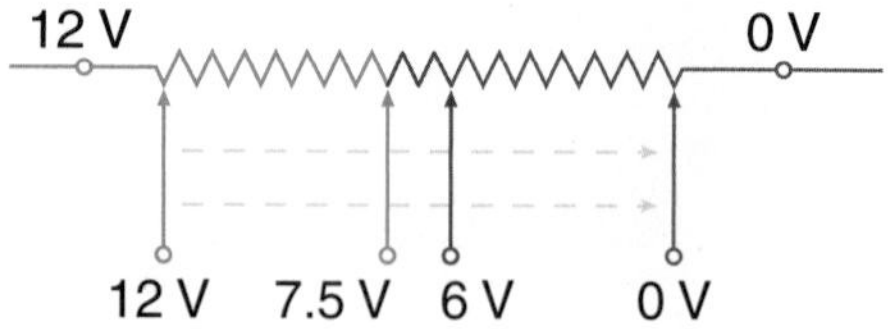

Figure 15-42 Voltage across a potentiometer.

SHOP TALK

There are two basic types of thermistors: negative temperature coefficient (NTC) and positive temperature coefficient (PTC). An NTC thermistor is one in which the resistance decreases with an increase in temperature. A PTC thermistor is one in which the resistance increases with an increase in temperature. To understand how a circuit works, begin by identifying the type used in the circuit.

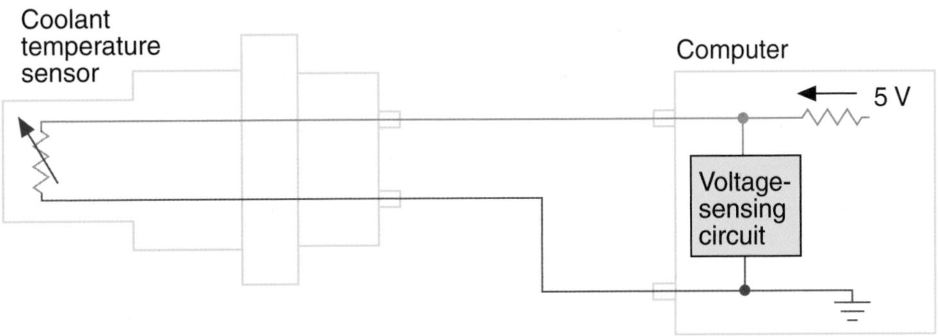

Figure 15-43 A thermistor is used to measure temperature. The sensing unit measures the change in resistance and translates this into a temperature value.

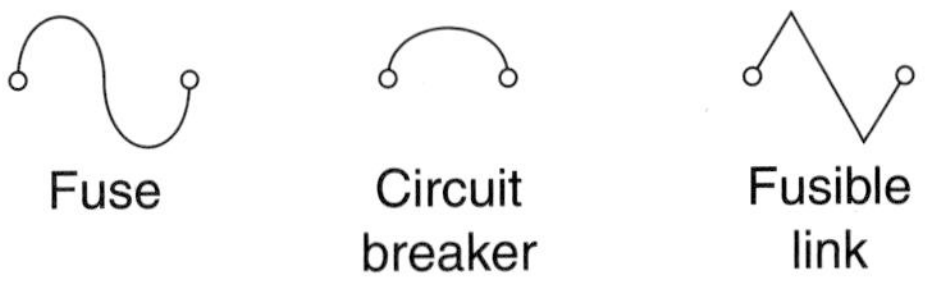

Figure 15-44 Electrical symbols for common circuit protection devices.

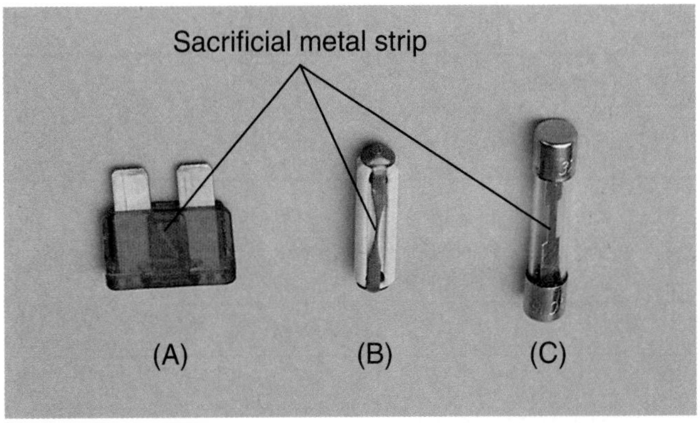

Figure 15-45 (A) Blade fuse, (B) ceramic fuse, and (C) cartridge fuse.

Circuit Protective Devices

When overloads or shorts in a circuit cause too much current to flow, the wiring in the circuit heats up, the insulation melts, and a fire can result, unless the circuit has some kind of protective device. Fuses, fuse links, maxi-fuses, and circuit breakers are designed to provide protection from high current. They may be used singly or in combination. Typical symbols for protection devices are shown in **Figure 15–44**.

WARNING!

Fuses and other protection devices normally do not wear out. They go bad because something went wrong. Never replace a fuse or fusible link, or reset a circuit breaker, without finding out why it went bad.

Fuses There are three basic types of fuses in automotive use: cartridge, blade, and ceramic **(Figure 15–45)**. The **cartridge fuse** is found on most older domestic cars and a few imports. It is composed of a strip of low temperature melting metal enclosed in a transparent glass or plastic tube. Late-model domestic vehicles and many imports use **blade** or **spade fuses**. The **ceramic fuse** was used on many European imports. The core is a ceramic insulator with a conductive metal strip along one side.

Fuses are rated by the current at which they are designed to blow. A three-letter code is used to indicate the types and sizes of fuses. Blade fuses have codes ATC or ATO. All glass SFE fuses have the same diameter, but the length varies with the current rating. Ceramic fuses are available in two sizes, code GBF (small) and the more common code GBC (large). The amperage rating is also embossed on the insulator. Codes, such as AGA, AGW, and AGC, indicate the length and diameter of the fuse. Fuse lengths in each of these series are the same, but the current rating can vary. The code and the current rating are usually stamped on the end cap.

The current rating for blade fuses is indicated by the color of the plastic case **(Table 15–1)**. In addition, it is usually marked on the top. The insulator on ceramic fuses is color coded to indicate different current ratings.

Three basic types of blade fuses are found on today's vehicles: the standard blade, the minifuse, and the maxi-fuse. The minifuse is the commonly used circuit protection device. Minifuses are available in ratings from 5 to 30 amps. The maxi-fuse is a serviceable replacement for a fusible link cable. It is used in circuits that have high operating current. Maxi-fuses are available in 2–100 amp ratings; the most common is the 30 amp.

TABLE 15-1 TYPICAL COLOR CODING OF PROTECTIVE DEVICES

Blade Fuse Color Coding	
Ampere Rating	**Housing Color**
4	pink
5	tan
10	red
15	light blue
20	yellow
25	natural
30	light green
Fuse Link Color Coding	
Wire Link Size	**Insulation Color**
20 GA	blue
18 GA	brown or red
16 GA	black or orange
14 GA	green
12 GA	gray
Maxi-Fuse Color Coding	
Ampere Rating	**Housing Color**
20	yellow
30	light green
40	amber
50	red
60	blue

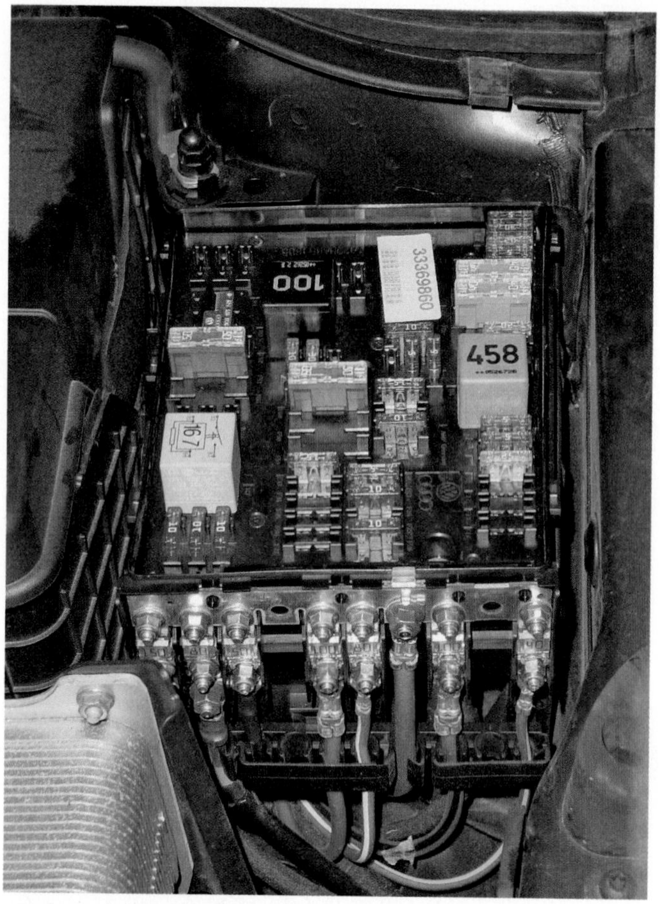

Figure 15-46 Typical fuse box or panel.

Fuses are located in a box or panel **(Figure 15–46)**, usually under the dashboard, behind a panel in the foot well, or in the engine compartment. Fuses are generally numbered, and the main components abbreviated. On late-model cars, there may be icons or symbols indicating which circuits they serve. This identification system is covered in more detail in the owner's and service manuals.

Fuse Links Fuse or **fusible links** are used in circuits when limiting the maximum current is not extremely critical. They are often installed in the positive battery lead to the ignition switch and other circuits that have power with the key off. Fusible links are normally found in the engine compartment near the battery. Fusible links are also used when it would be awkward to run wiring from the battery to the fuse panel and back to the load.

A fuse link **(Figure 15–47)** is a short length of small-gauge wire installed in a conductor. Because the fuse link is a lighter gauge of wire than the main conductor, it melts and opens the circuit before damage can occur in the rest of the circuit. Fuse link wire is covered with a special insulation that bubbles when it overheats, indicating that the link has melted.

CAUTION!

Always disconnect the battery ground cable prior to servicing any fuse link.

Maxi-Fuses Many late-model vehicles use **maxi-fuses** instead of fusible links. Maxi-fuses look and operate like two-prong, blade, or spade fuses, except that they are much larger **(Figure 15–48)** and can handle more current. Maxi-fuses are usually located in their own underhood fuse block.

Figure 15-47 A typical fuse link.

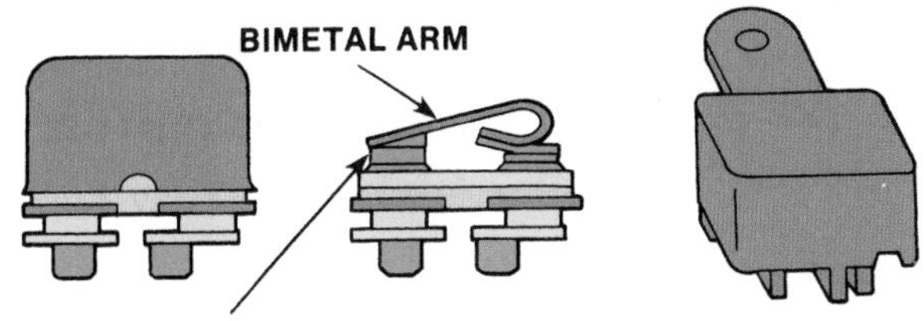

Figure 15-49 Cycling circuit breaker.

Maxi-fuses allow the vehicle's electrical system to be broken down into smaller circuits that are easy to diagnose and repair. For example, in some vehicles a single fusible link controls one-half or more of all circuitry. If it burns out, many electrical systems are lost. By replacing this single fusible link with several maxi-fuses, the number of systems lost due to a problem in one circuit is drastically reduced. This makes it easy to pinpoint the source of trouble.

Circuit Breakers Some circuits are protected by **circuit breakers** (abbreviated c.b. in the fuse chart of a service manual). They can be fuse panel mounted or inline. Like fuses, they are rated in amperes.

Each circuit breaker conducts current through an arm made of two types of metal bonded together (bimetal arm). If the arm starts to carry too much current, it heats up. As one metal expands faster than the other, the arm bends, opening the contacts. Current flow is broken. A circuit breaker can be cycling **(Figure 15–49)** or must be manually reset.

In the cycling type, the bimetal arm begins to cool once the current to it is stopped. Once it returns to its original shape, the contacts are closed and power is restored. If the current is still too high, the cycle of breaking the circuit is repeated.

Two types of noncycling or resettable circuit breakers are used. One is reset by removing the power from the circuit. There is a coil wrapped around a bimetal arm **(Figure 15–50A)**. When there is excessive current and the contacts open, a small current passes through the coil. This current through the coil is not enough to operate a load, but it does heat up both the coil and the bimetal arm. This keeps the arm in the open position until power is removed. The other type is reset by depressing a reset button. A spring pushes the bimetal arm down and holds the contacts together **(Figure 15–50B)**. When an overcurrent condition exists and the bimetal arm heats up, the bimetal arm bends enough to overcome the spring and the contacts snap open. The contacts stay open until the reset button is pushed, which snaps the contacts together again.

High-Voltage Systems HV systems require special circuit protection devices. Most circuit protection devices used in 12-volt systems are actually rated at 32 volts. If these protection devices were used in a

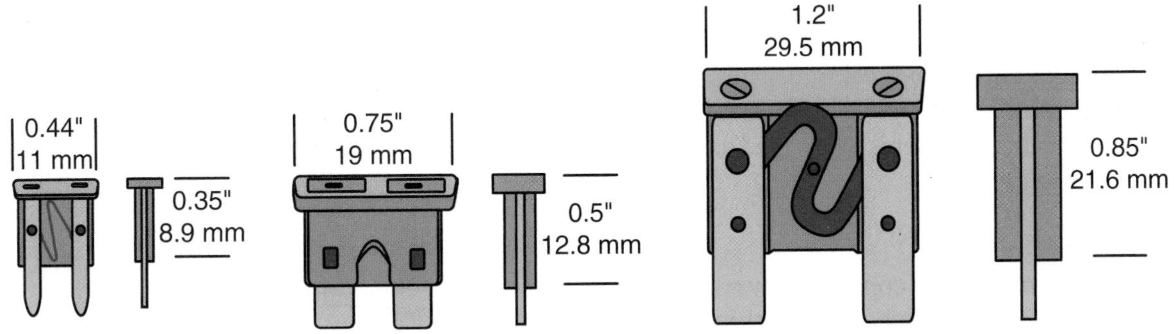

Figure 15-48 Size comparisons among a standard, mini, and maxi blade fuse.

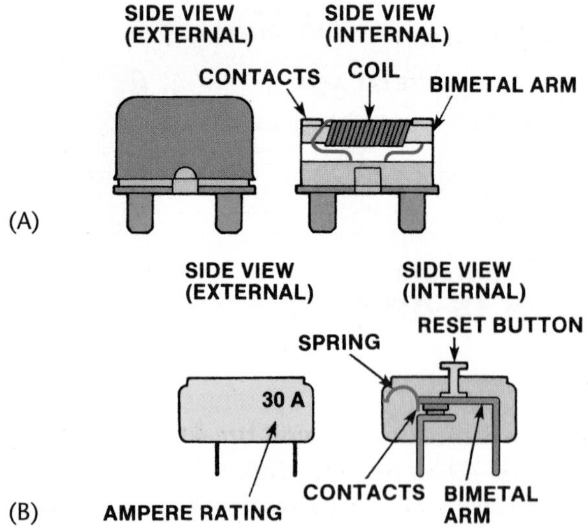

Figure 15-50 Resetting noncycling circuit breakers by (A) removing power from the circuit and (B) depressing a reset button.

42- or higher volt system, problems such as the burning of the vehicle's wiring and electrical components could result and also cause a fire.

Protection devices for the 42-volt systems are rated at 55 volts. This rating allows protection during times of HV spikes. The protection devices for 42-volt systems, and their receptacles, have a unique design to prevent the installation of the wrong type of fuse.

Higher voltage systems use a variety of electrical and electronic devices to protect the vehicle and circuits. Most have temperature sensors and constantly monitor the current in the system. If the temperature rises above the limit or if the current is higher than normal, the HV system automatically shuts down.

Voltage Limiter Some instrument panel gauges are protected against voltage fluctuations that could damage the gauges or give erroneous readings. A **voltage limiter** restricts voltage to the gauges to a particular amount. The limiter contains a heating coil, a bimetal arm, and a set of contacts. When the ignition is in the on or accessory position, the heating coil heats the bimetal arm, causing it to bend and open the contacts. When the arm cools down to the point that the contacts close, the cycle is repeated. The rapid opening and closing of the contacts produces a pulsating voltage at the output terminal averaging about 5 volts. A voltage limiter is also called an instrument voltage regulator (IVR).

Switches

Electrical circuits are usually controlled by some type of switch **(Figure 15–51)**. Switches do two things. They turn the circuit on or off, or they direct the flow of current in a circuit. Switches can be under the control of the driver or can be self-operating through a condition of the circuit, the vehicle, or the environment.

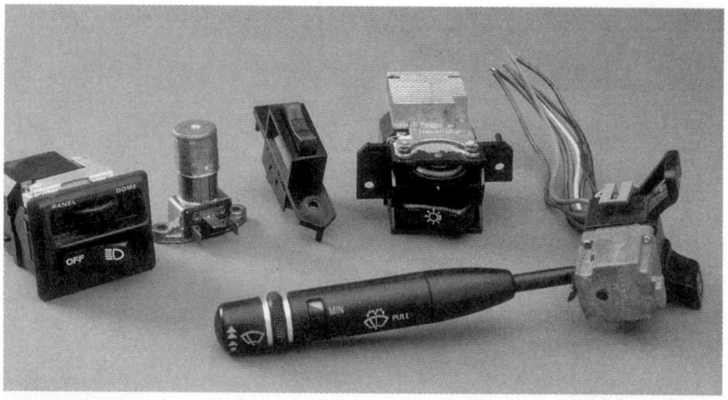

Figure 15-51 Examples of the various switches used in automobiles.

Contacts in a switch can be of several types, each named for the job they do or the sequence in which they work. A hinged-pawl switch **(Figure 15–52)** is the simplest type of switch. It either makes (allows for) or breaks (opens) current flow in a single conductor or circuit. This type of switch is a **single-pole, single-throw (SPST)** switch. The **throw** refers to the number of output circuits, and the **pole** refers to the number of input circuits made by the switch.

Another type of SPST switch is a momentary contact switch **(Figure 15–53)**. The spring-loaded contact on this switch keeps it from closing the circuit except when pressure is applied to the button. A horn switch is this type of switch. Because the spring holds the contacts open, the switch has a further designation: **normally open**. In the case where the contacts are held closed except when the button is pressed, the

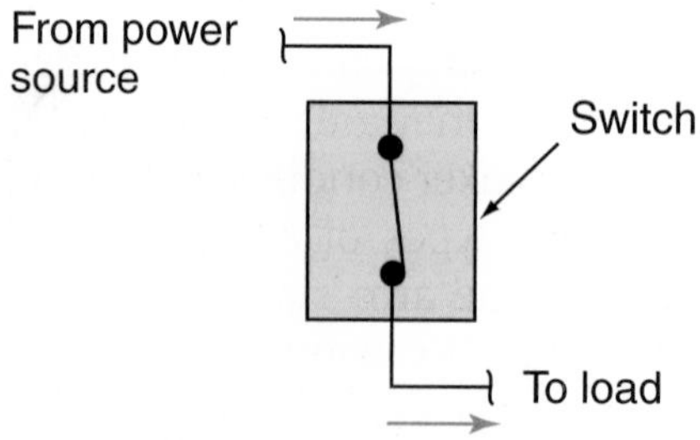

Figure 15-52 SPST hinged-pawl switch diagram.

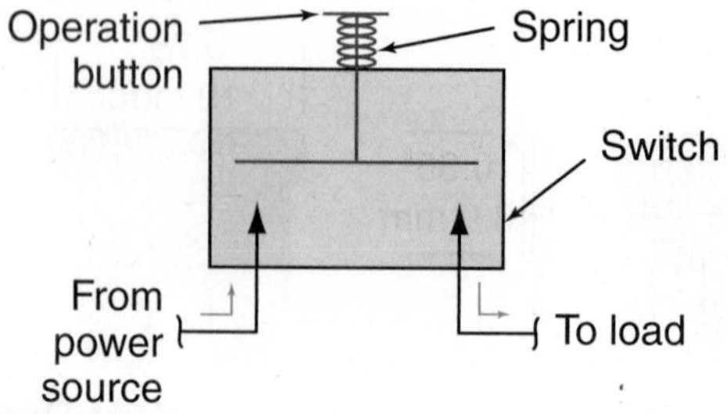

Figure 15-53 SPST momentary contact switch diagram.

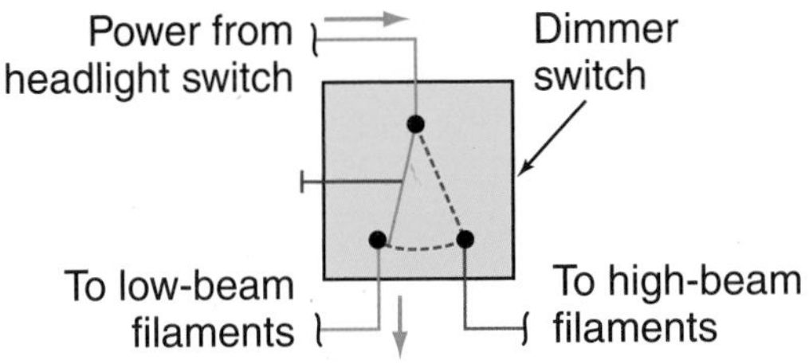

Figure 15–54 SPDT headlight dimmer switch.

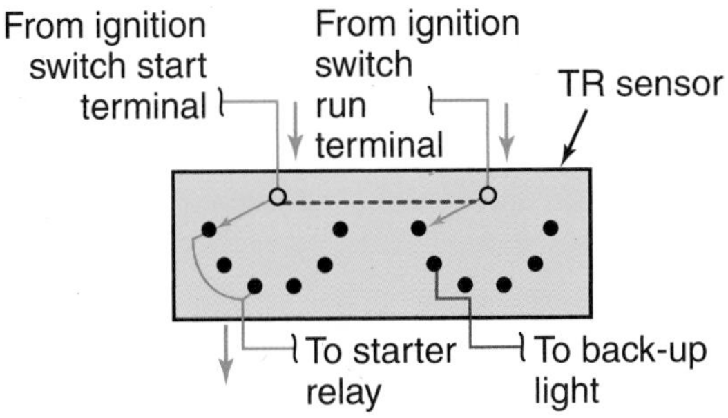

Figure 15–55 MPMT neutral start safety switch.

switch is designated **normally closed**. A normally closed momentary contact switch is the type of switch used to turn on the courtesy lights when one of the vehicle's doors is opened.

Single-pole, double-throw (SPDT) switches have one wire in and two wires out. **Figure 15–54** shows an SPDT hinged-pawl switch that feeds either the high-beam or low-beam headlight circuit. The dotted lines in the symbol show movement of the switch pawl from one contact to the other.

Switches can be designed with a great number of poles and throws. The transmission neutral start switch shown in **Figure 15–55**, for instance, has two poles and six throws and is referred to as a **multiple-pole, multiple-throw (MPMT)** switch. It contains two movable wipers that move in unison across two sets of terminals. The dotted line shows that the wipers are mechanically linked, or **ganged**. It is important to realize that current does not flow through the mechanical connection of the wipers. The switch closes a circuit to the starter in either P (park) or N (neutral) and to the back-up lights in R (reverse).

Most switches are combinations of hinged-pawl and push-pull switches, with different numbers of poles and throws. Some special switches are required, however, to satisfy the circuits of modern automobiles. A mercury switch is sometimes used to detect motion in a component, such as the one used in the engine compartment to turn on the compartment light.

Mercury is a very good conductor of electricity. In the mercury switch, a capsule is partially filled with mercury **(Figure 15–56)**. In one end of the capsule are two electrical contacts. The switch is attached to the

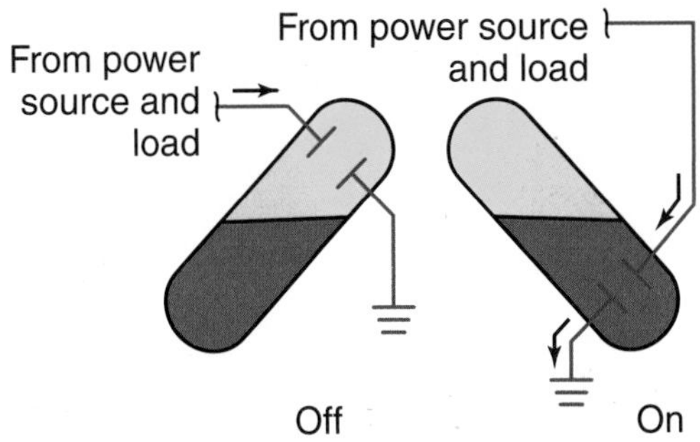

Figure 15–56 A typical mercury switch.

hood or luggage compartment lid. Normally, the mercury is in the end opposite to the contacts. When the lid is opened, the mercury flows to the contact end and provides a circuit between the electrical contacts.

A temperature-sensitive switch usually contains a bimetallic element heated either electrically or by some component in which the switch is used as a **sensor**. When engine coolant is below or at normal operating temperature, the engine coolant temperature sensor is in its normally open condition. If the coolant exceeds the temperature limit, the bimetallic element bends the two contacts together and the switch is closed to the indicator or the instrument panel. Other applications for heat-sensitive switches are time-delay switches and flashers.

Relays

A **relay** is an electric switch that allows a small amount of current to control a high-current circuit **(Figure 15–57)**. When the control circuit switch is open, no current flows to the coil of the relay, so the windings are de-energized. When the switch is closed, the coil is energized, turning the soft iron core into an electromagnet and drawing the armature down. This closes the power circuit contacts,

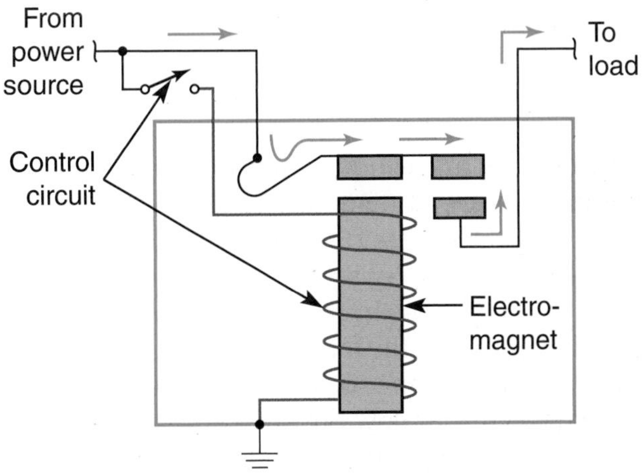

Figure 15–57 The basic way a relay works.

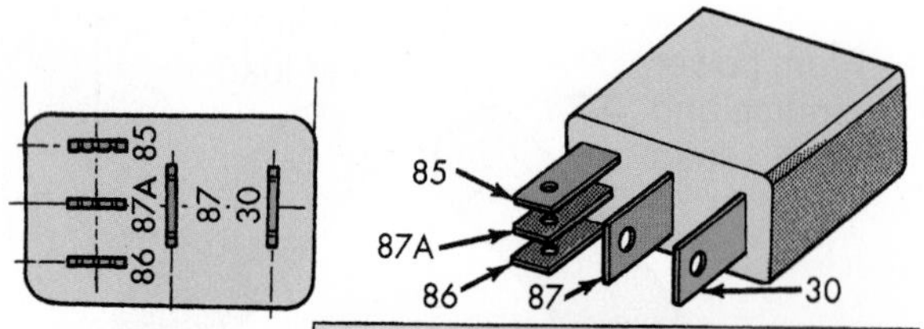

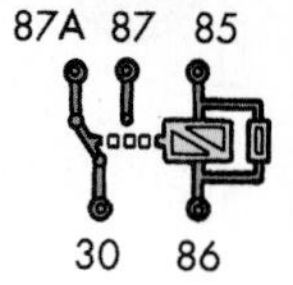

TERMINAL LEGEND	
NUMBER	IDENTIFICATION
30	COMMON FEED
85	COIL GROUND
86	COIL BATTERY
87	NORMALLY OPEN
87A	NORMALLY CLOSED

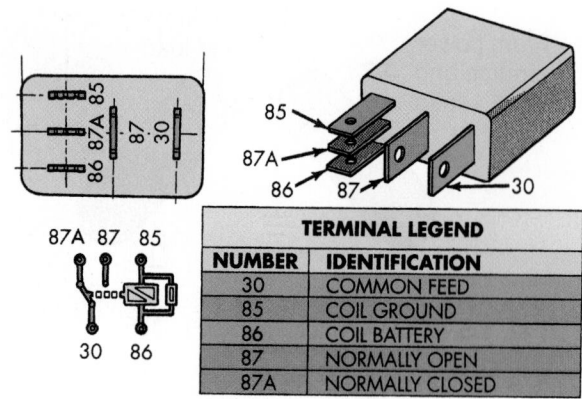

Figure 15-58 A typical electrical relay inner workings and connections. *Courtesy of Chrysler LLC*

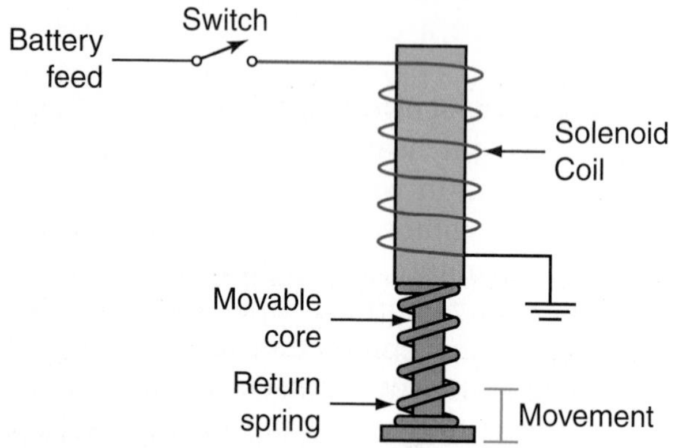

Figure 15-59 A solenoid is a device that has a movable electromagnetic core.

connecting power to the load circuit **(Figure 15–58)**. When the control switch is opened, current stops flowing in the coil and the electromagnet disappears. This releases the armature, which breaks the power circuit contacts.

Solenoids

Solenoids are also electromagnets with movable cores used to change electrical current flow into mechanical movement **(Figure 15–59)**. They are used in a wide variety of systems and can also close contacts, acting as a relay at the same time they mechanically cause something to happen.

Conductors and Insulators

Controlling and routing the flow of electricity requires the use of materials known as conductors and insulators **(Figure 15–60)**. **Conductors** are materials that have low resistance to current flow. If the number of electrons in the outer shell or ring of an atom is less than 4, the force holding them in place is weak. The voltage needed to move these electrons and create current flow is relatively small. These are good conductors. Most metals, such as copper, silver, and aluminum, are excellent conductors.

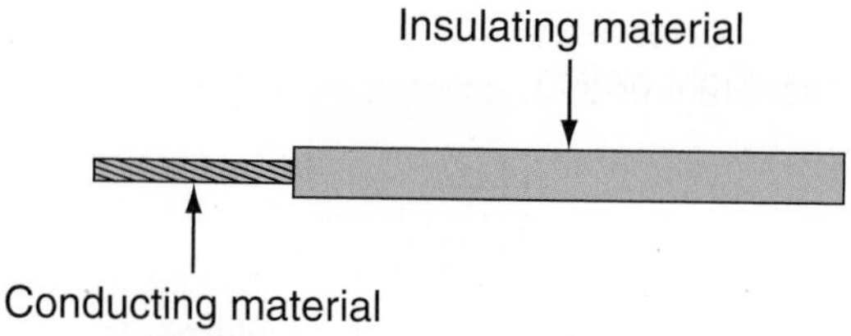

Figure 15-60 Basic construction of a wire.

When the number of electrons in the outer ring is greater than 4, the force holding them in orbit is very strong, and very high voltages are needed to move them. These materials are known as **insulators**. They resist the flow of current. Thermal plastics are the most common electrical insulators used today. They can resist heat, moisture, and corrosion without breaking down. Wire wound inside electrical units, such as ignition coils and generators, usually has a very thin baked-on insulating coating.

> **WARNING!**
>
> *Your body is a good conductor of electricity. Remember this when working on a vehicle's electrical system. Always observe all electrical safety rules.*

Copper wire is by far the most popular conductor. The resistance of a uniform, circular copper wire depends on the length, the diameter, and the temperature of the wire. If the length is doubled, the resistance between the wire ends is doubled. The longer the wire, the greater the resistance. If the diameter of a wire is doubled, the resistance for any given length is cut in half. The larger the wire's diameter, the lower the resistance.

The other important factor affecting the resistance of a copper wire is temperature. As the temperature increases, the resistance increases. Heat is developed in any wire-carrying current because of the resistance in the wire. If the heat becomes excessive, the insulation will be damaged. Resistance occurs when electrons collide as current flows through the conductor. These collisions cause friction, which in turn generates heat.

Wires Two basic types of wires are used: solid and stranded. **Solid wires** are single-strand conductors. **Stranded wires** are the most common and are made up of a number of small solid wires twisted together to form a single conductor.

Wire size standards were established by the Society of Automotive Engineers (SAE). These standards are called the **American wire gauge (AWG)** system. Sizes are identified by a numbering system ranging from number 0 to 40, with number 0 being the largest and number 40 the smallest in a cross-sectional area. Most automotive wiring range from number 10 to 24, with battery cables normally 4 gauge or lower. Battery cables must be large gauge wires in order to carry the heavy current draw of the starter motor.

Some manufacturers list wire size in metrics. Metric wiring sizes are based on the cross-sectional area of the wire. Most often wire size is listed in square millimeters (mm^2) rather than a metric wire gauge size. The stated size of a metric wire goes up as the wire's diameter increases. This is the opposite of the AWG system. **Table 15–2** gives the closest equivalent size cross references between metric and AWG wire sizes.

Automotive wiring can also be classified as primary or secondary. Primary wiring carries low voltage to all the electrical systems of the vehicle except to the secondary circuit of the ignition system. Secondary wire, also called **high-tension cable**, has extra thick insulation to carry high voltage from the ignition coil to the spark plugs. The conductor itself is designed for low currents.

Wires are commonly grouped together in harnesses. A single-plug harness connector may be the connection for four, six, or more circuits. Harnesses and harness connectors help organize the vehicle's electrical system and provide a convenient starting point for tracking and testing many circuits. Most major wiring harness connectors are located in a vehicle's dash or fire wall area **(Figure 15–61)**.

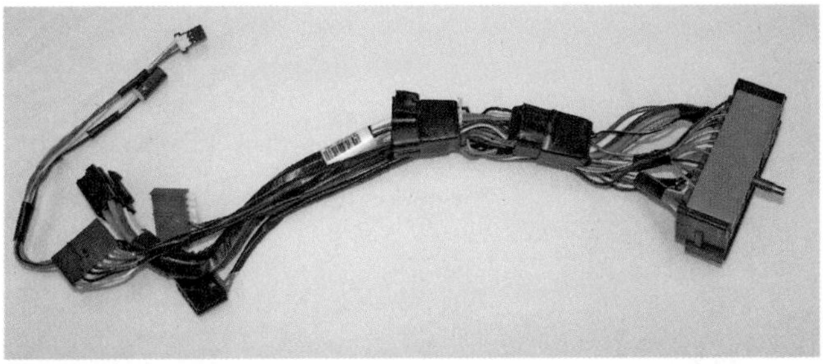

Figure 15-61 A typical partial wiring harness that connects to a connector on the fire wall.

TABLE 15-2 AWG WIRE GAUGE SIZE W/ METRIC EQUIVALENT

AWG Gauge #	Diameter in inches	Diameter in mm	Approx. Metric Equivalent Size
4	0.2043	5.189	19.0
6	0.1620	4.115	13.0
8	0.1285	3.264	8.0
10	0.1019	2.588	5.0
12	0.08081	2.053	3.0
14	0.06408	1.628	2.0
16	0.05082	1.291	1.0
18	0.04030	1.024	0.8
20	0.03196	0.8118	0.5
22	0.02535	0.6439	0.36
24	0.02010	0.5105	0.22

Flat Wiring As the number of electrical and electronic devices installed in vehicles increases, so does the need for more wiring. More wiring leads to more weight and more potential problems. The size and number of wiring harnesses also increases and spots to carefully tuck them away are limited. For example, a large wiring harness has a difficult time fitting between the roof and the head liner of the vehicle. To run wiring from the front of the vehicle to the rear via the roof, the harnesses are made of small groups of wires, which means there are more harnesses traveling along the roof. Another solution is to make the wiring harnesses flat.

Flat wiring reduces the bulge or thickness of a harness. The copper conductors inside these wires are flattened and no longer round in appearance. In a wiring harness, several flat wires are laid out next to each other and are covered with a plastic insulating material. The plastic offers protection and isolation to the conductors and keeps the harness flat and flexible. In the future, flat wiring may also have electronic components embedded in it. With this design, the wiring harness is not only easier to hide in body panels but also serves as a flexible printed circuit able to be located nearly anywhere in the vehicle.

Printed Circuits Many late-model vehicles use flexible printed circuits **(Figure 15–62)** and printed circuit boards. Both types of printed circuits allow for complete circuits to many components without having to run dozens of wires. Printed circuit boards are typically contained in a housing, such as the engine control module. These boards are not serviceable and in some cases not visible. When these boards fail, the entire unit is replaced.

A flexible printed circuit saves weight and space. It is made of thin sheets of nonconductive plastic onto which conductive metal, such as copper, has been

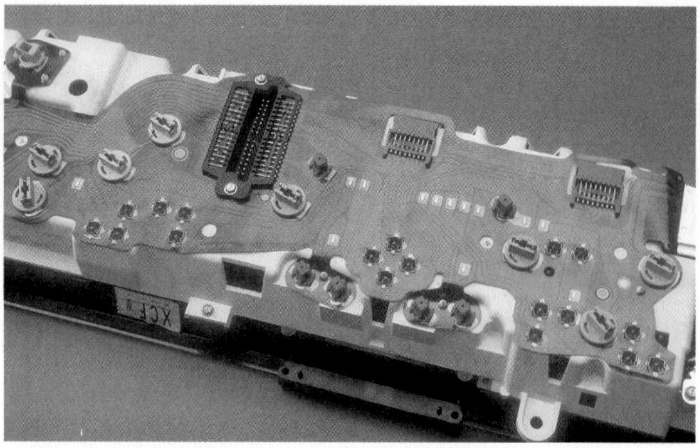

Figure 15-62 A typical printed circuit board.

deposited. Parts of the metal are then etched or eaten away by acid. The remaining metal lines form the conductors for the various circuits on the board. The printed circuit is normally connected to the power supply or ground wiring through the use of plug-in connectors mounted on the circuit sheet.

KEY TERMS

Alternating current (AC)
American wire gauge (AWG)
Ampere
Blade fuse
Cartridge fuse
Ceramic fuse
Chassis ground
Circuit breaker
Closed circuit
Conductor
Continuity
Current
Direct current (DC)
Electromotive force (EMF)
Electronics
Fixed value resistors
Fusible link
Ganged
Ground wire
Grounding
High-tension cable
High-voltage battery
Impedance
Induction
Insulator
Magnetism
Maxi-fuse
Multiple-pole, multiple throw (MPMT)
Normally closed
Normally open
Ohm
Ohm's law
Open circuit
Parallel circuit
Pole
Potentiometer
Relay
Resistance
Resistor wire
Rheostat
Sensor
Series circuit
Series-parallel circuit
Sine wave
Single-pole, double throw (SPDT)
Single-pole, single throw (SPST)
Solid wire
Spade fuse
Starting battery
Stepped resistor
Stranded wire
Tapped resistor
Thermistor
Throw
Variable resistor
Voltage
Voltage drop
Voltage limiter
Watt

SUMMARY

- Electricity results from the flow of electrons from one atom to another.
- The greater the number of electrons flowing past a given point in a given amount of time, the more current the circuit has. The unit for measuring electrical current is the ampere.
- There are two types of current: direct current (DC) and alternating current (AC). In direct current, the electrons flow in one direction only. In alternating current, the electrons change direction at a fixed rate.
- Voltage is electrical pressure or the force developed by the attraction of the electrons to protons.
- Resistance is measured in ohms, which is a measurement of the amount of the resistance something has to current flow.
- A completed electrical circuit is called a closed circuit, whereas an incomplete circuit is called an open circuit. When a circuit is complete, it is said to have continuity.
- The amount of electricity consumed by a load is normally called electrical power usage or watts.
- The mathematical relationship between current, resistance, and voltage is expressed in Ohm's law, $E = IR$.
- Voltage drop is the amount of voltage required to cause current flow through a load.
- Most automotive circuits contain five basic parts: power sources, conductors, loads, controllers, and circuit protection devices.
- There are three basic types of automotive electrical circuits: series circuits, parallel circuits, and series-parallel circuits.
- A series circuit consists of one or more resistors connected to a voltage source with only one path for electron flow.
- A parallel circuit provides two or more different paths for the current to flow through.
- In a series-parallel circuit, both series and parallel combinations exist in the same circuit.
- Resistors limit current flow.
- Fuses, fuse links, maxi-fuses, and circuit breakers protect circuits against overloads.

- Switches control on/off and direct current flow in a circuit.
- A relay is an electric switch.
- A solenoid is an electromagnet that translates current flow into mechanical movement.
- Controlling and routing the flow of electricity requires the use of materials known as conductors and insulators.
- Two basic types of wires are used: solid and stranded. Solid wires are single-strand conductors. Stranded wires are the most common and are made up of a number of small solid wires twisted together to form a single conductor.

REVIEW QUESTIONS

1. What is the name for the formula $E = IR$?
2. *True or False?* In a series circuit, circuit current is the same throughout the circuit.
3. What are the two types of wires and which is the most commonly used in automobiles?
4. Define voltage drop.
5. What is a thermistor?
6. What is a normally closed switch?
7. What happens in an electrical circuit when the resistance increases?
8. What is the difference between voltage and current?
9. What is an SPST switch?
10. Which type of resistor is commonly used in automotive circuits?
    a. fixed value
    b. stepped
    c. variable
    d. all of the above
11. Which of the following is a characteristic of all parallel circuits?
    a. Total circuit resistance is always higher than the resistance of the leg with the lowest total resistance.
    b. The current through each leg will be different if the resistance values are different.
    c. The sum of the resistance on each leg equals the total circuit resistance.
    d. The voltage applied to each leg of the circuit will be dropped across the legs if there are loads in series with the parallel circuit.
12. What is the current in a 12-volt circuit with two 6-ohm resistors connected in parallel?
    a. 2 amps
    b. 4 amps
    c. 6 amps
    d. 12 amps
13. Which of the following does not affect the resistance of a uniform, circular copper wire?
    a. the length of the wire
    b. the diameter of the wire
    c. the location of the wire
    d. the temperature of the wire
14. Which of the following is *not* a true statement about maxi-fuses?
    a. They are used in many late-model vehicles rather than fusible links.
    b. They automatically reset or can be manually reset and can handle more current than a blade fuse.
    c. They are normally in their own underhood fuse block.
    d. They allow the vehicle's electrical system to be broken down into small circuits.
15. Which of the following technologies is not used to produce electricity in an automobile?
    a. chemical reactions
    b. photoelectrical reactions
    c. thermonuclear reactions
    d. magnetic induction

ASE-STYLE REVIEW QUESTIONS

1. While discussing resistance: Technician A says that current will increase with a decrease in resistance. Technician B says that current will decrease with an increase in resistance. Who is correct?
    a. Technician A
    b. Technician B
    c. Both A and B
    d. Neither A nor B
2. Technician A says that a 12-volt light bulb that draws 12 amps has a power output of 1 watt. Technician B says that a motor that has 1 ohm of resistance has a power rating of 144 watts if it is connected to a 12-volt battery. Who is correct?
    a. Technician A
    b. Technician B
    c. Both A and B
    d. Neither A nor B

3. While discussing a 12-volt circuit with three resistors (a 3-ohm, a 6-ohm, and a 2-ohm) in parallel: Technician A says that the total resistance of the circuit is 1 ohm. Technician B says that the current flow through the 6-ohm resistor is 12 amps. Who is correct?
   a. Technician A
   b. Technician B
   c. Both A and B
   d. Neither A nor B
4. While discussing electrical grounds: Technician A says that a component may be grounded through its mounting to a major metal part. Technician B says that some components must be grounded by connecting its positive terminal to a metal section of the frame. Who is correct?
   a. Technician A
   b. Technician B
   c. Both A and B
   d. Neither A nor B
5. While discussing the batteries in hybrid vehicles: Technician A says that the battery packs must be able to supply bursts of electrical energy. Technician B says that the batteries must be able to accept quick and frequent recharges. Who is correct?
   a. Technician A
   b. Technician B
   c. Both A and B
   d. Neither A nor B
6. While discussing electrical relays: Technician A says that the relay's coil is energized when the circuit's control switch is open. This action connects power to the load circuit. Technician B says that a relay is an electric switch that allows a large amount of current to control a low-current circuit. Who is correct?
   a. Technician A
   b. Technician B
   c. Both A and B
   d. Neither A nor B
7. While discussing copper wire: Technician A says that if the length of wire is doubled, the resistance between the wire ends is doubled. Technician B says that if the diameter of a wire is doubled, the resistance between the wire ends is doubled. Who is correct?
   a. Technician A
   b. Technician B
   c. Both A and B
   d. Neither A nor B
8. While discussing current flow: Technician A says that current results from electrons being attracted to protons. Technician B says that for practical purposes, it is best to remember that current flows from a point of higher potential to a point of lower potential. Who is correct?
   a. Technician A
   b. Technician B
   c. Both A and B
   d. Neither A nor B
9. While discussing voltage: Technician A says that voltage is the force developed by the attraction of the electrons to protons. Technician B says that the force developed by the attraction of electrons to protons is called electromotive force and it is measured in volts. Who is correct?
   a. Technician A
   b. Technician B
   c. Both A and B
   d. Neither A nor B
10. While discussing the AWG system: Technician A says that most automotive wiring range from 0 to 4 gauge. Technician B says that the higher the gauge number, the larger the diameter of the wire. Who is correct?
   a. Technician A
   b. Technician B
   c. Both A and B
   d. Neither A nor B

GENERAL ELECTRICAL SYSTEM DIAGNOSTICS AND SERVICE

CHAPTER 16

OBJECTIVES

■ Describe the different possible types of electrical problems. ■ Read electrical automotive diagrams. ■ Perform troubleshooting procedures using meters, testlights, and jumper wires. ■ Describe how each of the major types of electrical test equipment are connected and interpreted. ■ Explain how to use a DMM for diagnosing electrical and electronic systems. ■ Explain how to use an oscilloscope for diagnosing electrical and electronic systems. ■ Test common electrical components. ■ Use wiring diagrams to identify circuits and circuit problems. ■ Diagnose common electrical problems. ■ Properly repair wiring and connectors.

Diagnosing nearly every system of a vehicle involves electrical and electronic systems. An understanding of how electrical/electronic systems work (**Figure 16–1**) and the knowledge of how to use the various types of test equipment are the keys to efficient diagnosis.

ELECTRICAL PROBLEMS

All electrical problems can be classified into one of three categories: opens, shorts, or high-resistance problems. Identifying the type of problem allows technicians to identify the correct tests to perform when diagnosing an electrical problem. An explanation of the different types of electrical problems follows.

Open Circuits

An **open** occurs when a circuit is incomplete. Without a completed path, current cannot flow (**Figure 16–2**) and the load or component will not work. An open circuit can be caused by a disconnected wire or connector, a broken wire, or a switch in the off position. When a circuit is off, it is open. When the circuit is on, it is closed. Switches open and close circuits, but at times a fault will cause an open. Opening a circuit stops current flow through the circuit. Voltage is still applied up to the open point, but there is no current flow. Without current flow, there are no voltage drops across the various loads.

Shorted Circuits

When a circuit has an unwanted path for current to follow, it has a **short**. When an energized wire accidentally contacts the frame or body of the car or another wire, circuit current can travel in unintended directions through the wires. This can cause uncontrollable circuits and high current through the circuits. Shorts are caused by damaged wire insulation, loose wires or connections, improper wiring, or careless installation of accessories.

A short creates an unwanted parallel leg or path in a circuit. As a result, circuit resistance decreases and current increases. The amount that the current will increase depends on the resistance of the short. The increased current flow can burn wires or components. Preventing this is the purpose of circuit protection devices. When a circuit protection device opens due to higher than normal current, a short is the likely cause. Also, if a connector or group of wires show signs of burning or insulation melting, high current is the cause, which is most likely caused by a short.

A short can be caused by a number of things and can be evident in a number of ways. It can be an unwanted path to ground. The short is often in

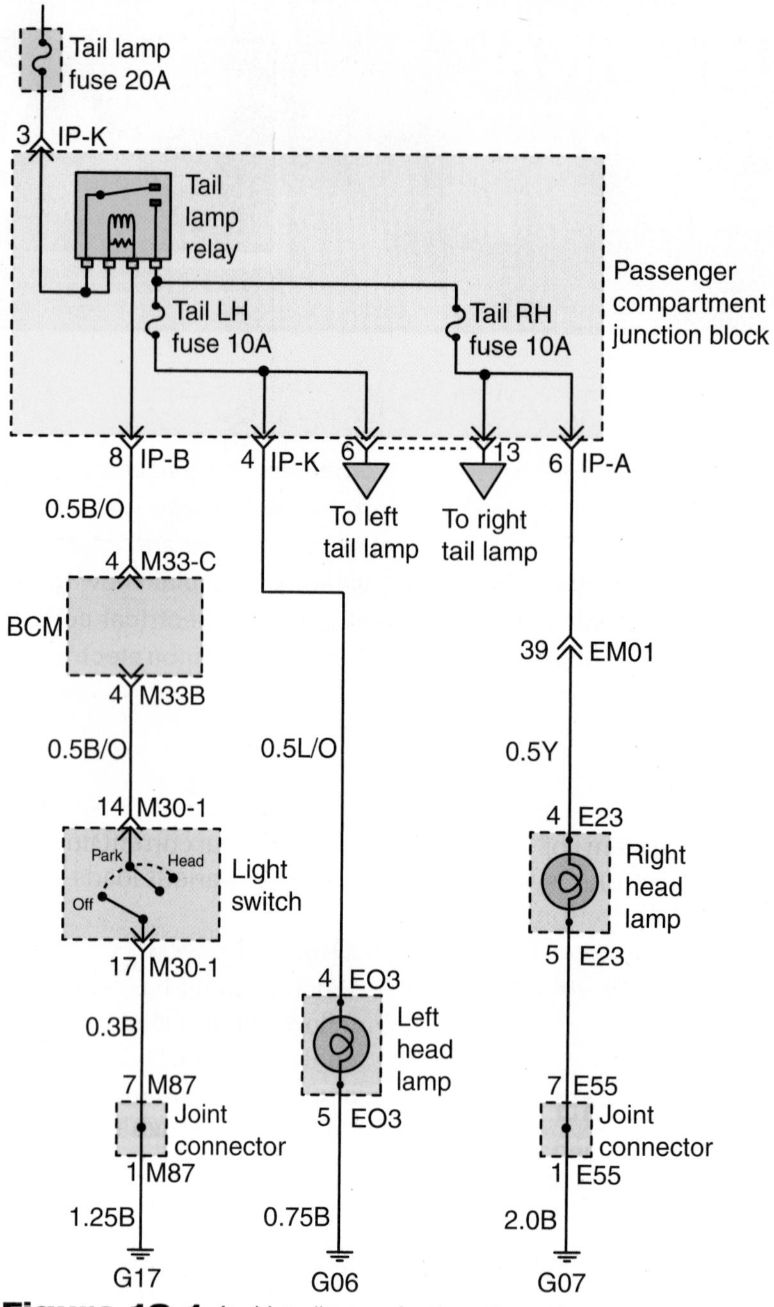

Figure 16–1 A wiring diagram for the tail, parking, and license lamps on a late-model vehicle.

parallel to a load and provides a low-resistance path to ground. Look at **Figure 16–3**; the short is probably caused by bad insulation that is allowing the power feed for the lamp to touch the ground for the same lamp. This problem creates a parallel circuit. **Figure 16–4** represents Figure 16–3 but is drawn to show the short as a parallel leg with very low resistance. The resistance assigned to the short may be more or less than an actual short, but the value 0.001 ohm is given to illustrate what happens. With the short, the circuit has three loads in parallel: 0.001, 3, and 6 ohms. The total resistance of this parallel circuit is less than the lowest resistance or 0.001 ohm. Using this value as the total resistance, circuit current is calculated to be more than 12,000 amps, which is much more than the fuse can handle. The high current will burn the fuse and the circuit will not work. Some call this problem a "grounded circuit" or a "copper-to-iron" short.

Sometimes two separate circuits become shorted together. When this happens, each circuit is controlled by the other. This may result in strange happenings, such as the horn sounding every time the brake pedal is depressed (**Figure 16–5**), or vice versa. In this case, the brake light circuit is shorted to the horn circuit. This is often called a "copper-to-copper" short.

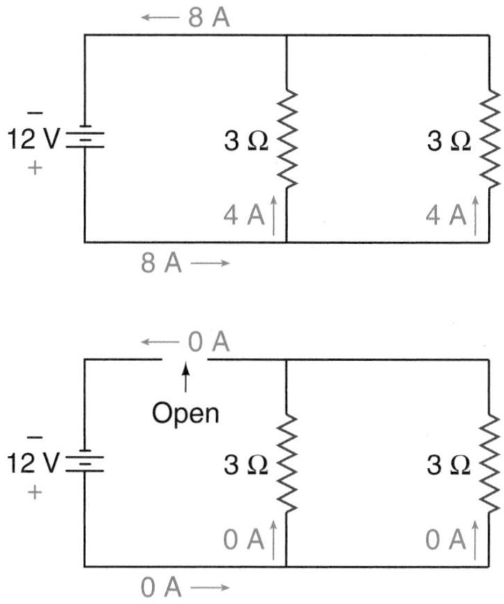

Figure 16–2 In an open circuit, there is no current flow. (A) is a normal circuit, and (B) is the same circuit with an open.

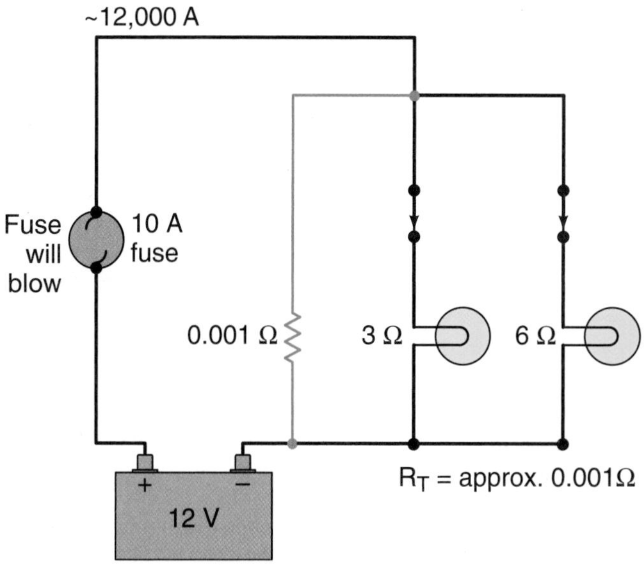

Figure 16–4 Ohm's law applied to Figure 16–3. Notice the rise in circuit amperage.

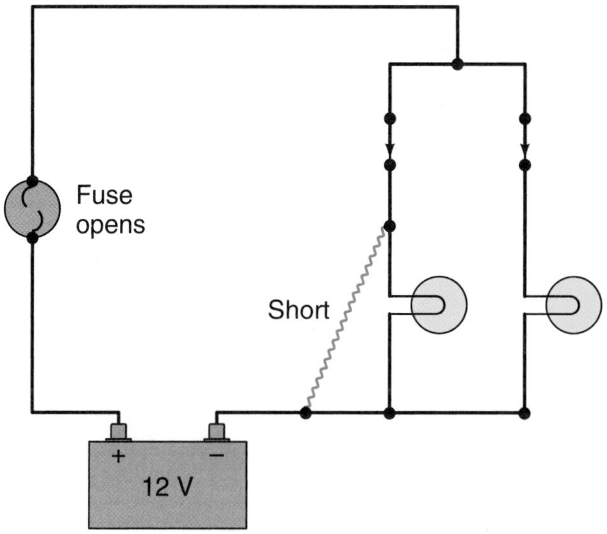

Figure 16–3 A short to ground.

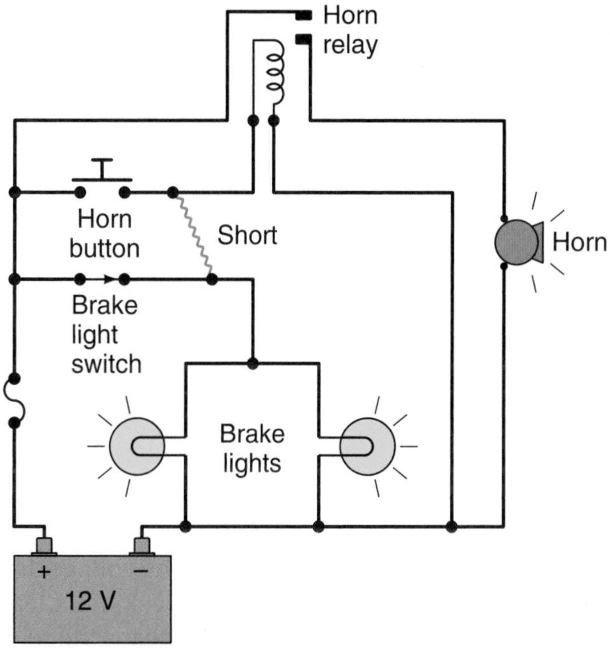

Figure 16–5 A wire-to-wire short.

High-Resistance Circuits

High-resistance problems occur when there is unwanted resistance in the circuit. The higher than normal resistance causes the current flow to be lower than normal and the components in the circuit are unable to operate properly. A common cause of this type of problem is corrosion at a connector. The corrosion becomes an additional load in the circuit **(Figure 16–6)**. This load uses some of the circuit's voltage, which prevents full voltage to the normal loads in the circuit.

Many sensors on today's vehicles are fed a 5-volt reference signal. The signal or voltage from the sensor is less than 5 volts depending on the condition it is measuring. A poor ground in the reference voltage circuit can cause higher than normal readings back to the computer. This seems to be contradictory to other high-resistance problems. However, if you look at a typical voltage divider circuit used to supply the reference voltage you will understand what is happening. Look at **Figure 16–7**. There are two resistors in series with the voltage reference tap between them. Because the total resistance in the circuit is 12 ohms, the circuit current is 1 amp. Therefore, the voltage drop across the 7-ohm resistor is 7 volts, leaving 5 volts at the tap.

Figure 16–8 is the same circuit, but a bad ground of 1 ohm was added. This low of resistance could be caused by corrosion at the connection. With the bad ground, the total resistance is now 13 ohms. This decreases our circuit current to approximately

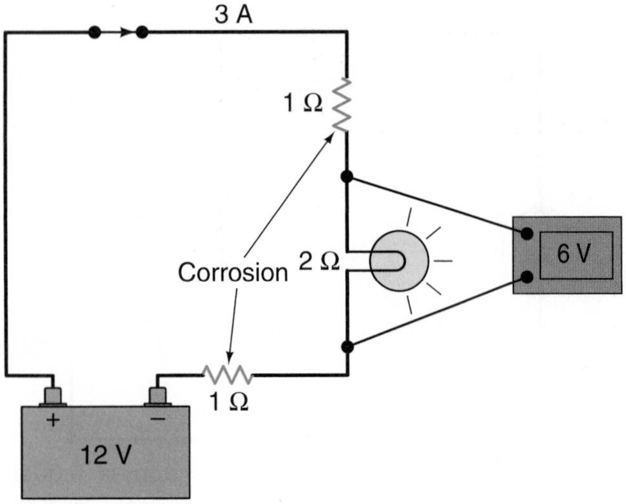

Figure 16-6 A simple light circuit with unwanted resistance. Notice the reduced voltage drop across the lamp and the reduced circuit current.

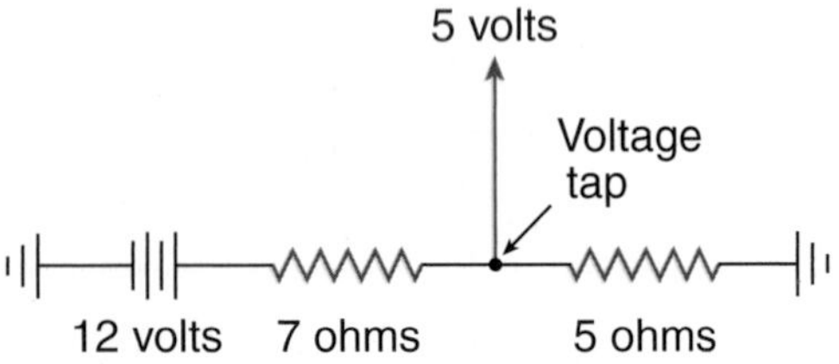

Figure 16-7 A voltage divider circuit.

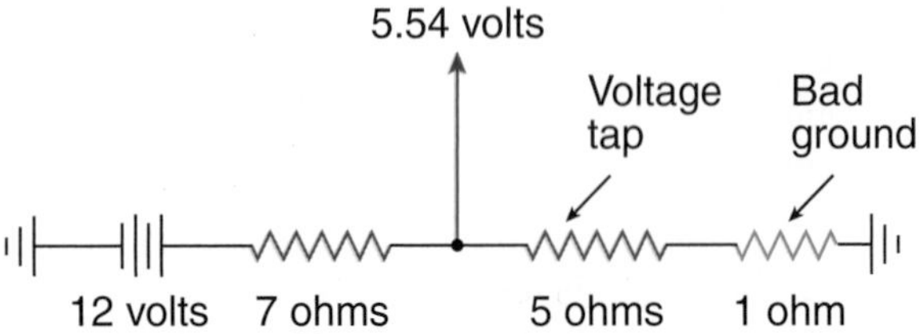

Figure 16-8 A voltage divider circuit with a bad ground.

0.92 amp. With this lower amperage, the voltage drop across the 7-ohm resistor is now about 6.46 volts, leaving 5.54 volts at the voltage tap. This means the reference voltage would be more than one-half volt higher than it should be. As a result, the computer will be receiving a return signal of at least one-half volt higher than it should. Depending on the sensor and the operating conditions of the vehicle, this could be critical.

ELECTRICAL WIRING DIAGRAMS

Wiring diagrams, sometimes called **schematics**, show how circuits are constructed. A typical service manual contains dozens of wiring diagrams that can be used to diagnose and repair a vehicle's electrical system.

A wiring diagram does not show the actual location of the parts or their appearance, nor does it indicate the length of the wire that runs between

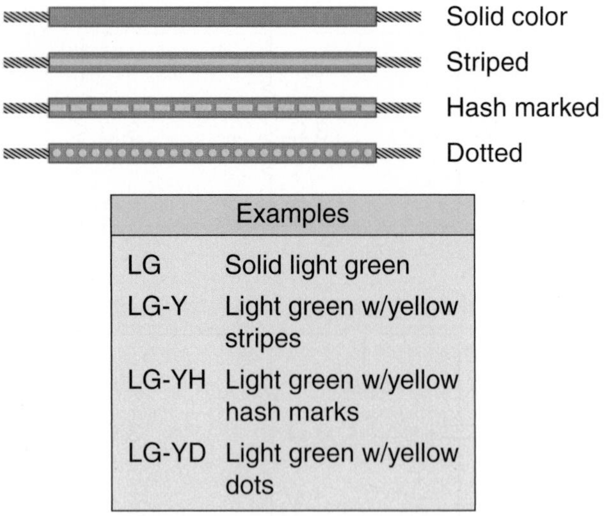

Figure 16-9 The different multicolor schemes of wires.

components. It usually indicates the color of the wire's insulation **(Figure 16–9)** and sometimes the wire gauge size. Typically the wire insulation is color coded as shown in **Table 16–1**. The first letter

TABLE 16-1 COMMON WIRE COLOR CODES

Color	Abbreviations			
Aluminum	AL			
Black	BLK	BK	B	SW
Blue	BL	BLU	L	
Dark Blue	BLU DK	DB	DK BLU	
Light Blue	BLU LT	LB	LT BLU	
Brown	BRN	BR	BN	
Glazed	GLZ	GL		
Gray	GRA	GR	G	GRY
Green	GN	GRN	G	
Dark Green	GRN DK	DG	DK GRN	
Light Green	GRN LT	LG	LT GRN	
Lilac	LI			
Maroon	MAR	M		
Orange	ORN	O	ORG	
Pink	PNK	PK	P	
Purple	PPL	PR	PUR	
Red	RED	R	RD	RO
Tan	TAN	T	TN	
Violet	VLT	V		
White	WHT	W	WH	WS
Yellow	YEL	Y	YL	GE

SYMBOLS USED IN WIRING DIAGRAMS			
+	Positive		Temperature switch
−	Negative		Diode
	Ground		Zener diode
	Fuse		Motor
	Circuit breaker	→≻ C101	Connector 101
	Condenser	→	Male connector
Ω	Ohms	≻	Female connector
	Fixed value resistor		Splice
	Variable resistor	S101	Splice number
	Series resistors		Thermal element
	Coil		Multiple connectors
	Open contacts	88:88	Digital readout
	Closed contacts		Single filament bulb
	Closed switch		Dual filament bulb
	Open switch		Light-emitting diode
	Ganged switch (N.O.)		Thermistor
	Single pole double throw switch		PNP bipolar transistor
	Momentary contact switch		NPN bipolar transistor
	Pressure switch		Gauge

Figure 16–10 Common electrical symbols used on wiring diagrams.

or set of letters usually indicates the base color of the insulation. The second set refers to the color of the stripe, hash marks, or dots on the wire, if there are any. Circuits are traced through a vehicle by identifying the beginning and end of a particular colored wire.

Many different symbols are used to represent components such as resistors, batteries, switches, transistors, and many other items. Some of these symbols have already been shown in earlier discussions. Other common symbols are shown in **Figure 16–10**. Connectors, splices, grounds, and other details are also given in a wiring diagram; a sample of these is shown in **Figure 16–11**.

Wiring diagrams can become quite complex. To avoid this, most diagrams usually illustrate one distinct system, such as the back-up light circuit, oil pressure indicator light circuit, or wiper motor circuit. In more complex ignition, electronic fuel injection, and computer-controlled systems, a diagram may be used to illustrate only part of the entire circuit.

USING SERVICE INFORMATION

Electrical symbols are not standardized throughout the automotive industry. Different manufacturers may have different methods of representing certain components, particularly the less common ones. Always refer to the symbol reference charts, wire color code charts, and abbreviation tables listed in the service information to avoid confusion when reading wiring diagrams.

ELECTRICAL TESTING TOOLS

With a basic understanding of electricity and simple circuits it is easier to understand the operation and purpose of the various types of electrical test equipment, which are described in the following sections. Several meters are used to test and diagnose electrical systems. These are the voltmeter, ohmmeter, ammeter, and volt/amp meter. These should be used along with jumper wires, testlights, and variable resistors (piles).

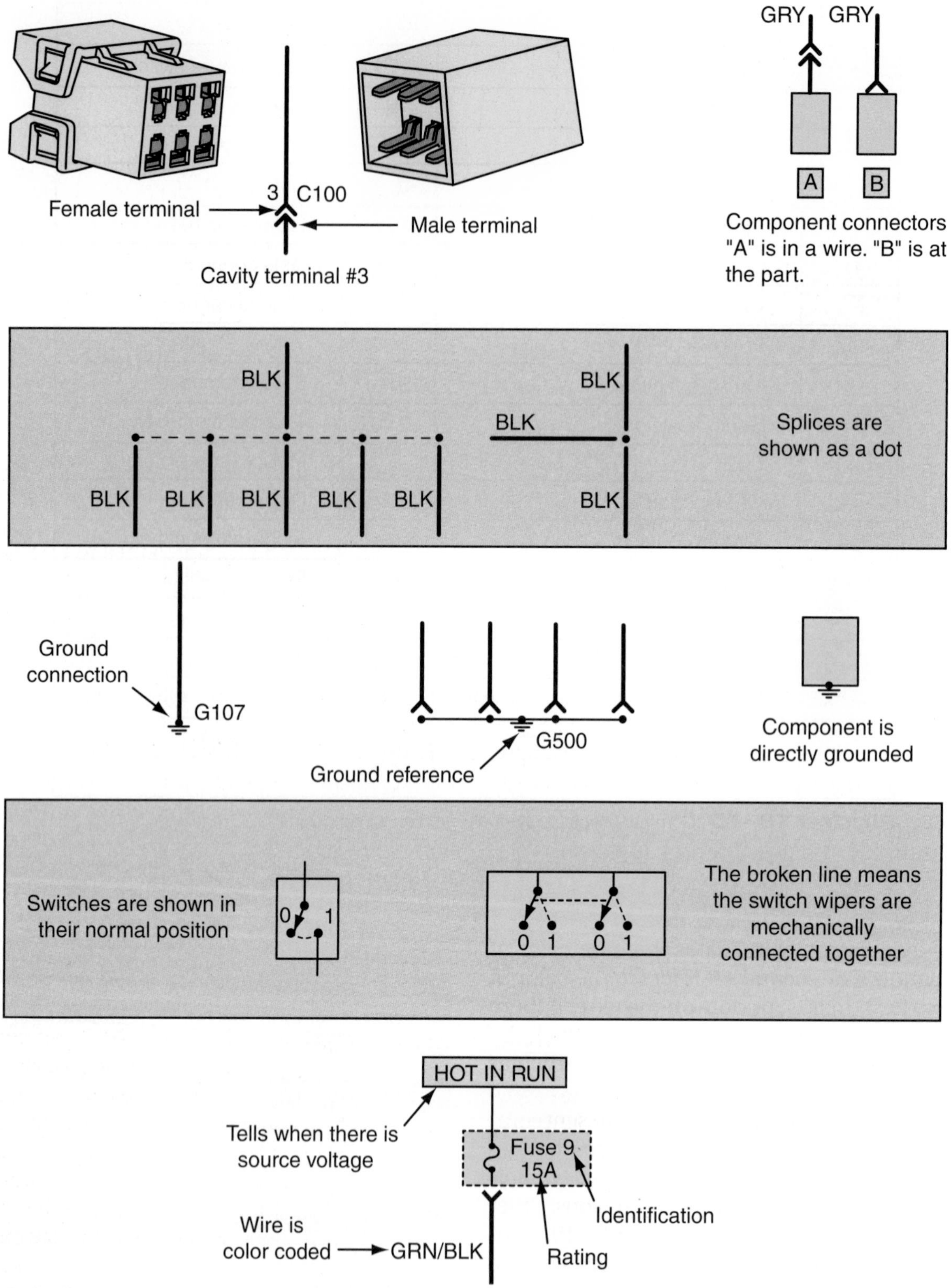

Figure 16-11 Examples of the information contained in a typical wiring diagram (schematic).

Go To *Refer to Chapter 5 for a description and basic uses for the various electrical test equipment.*

Circuit Testers

Circuit testers (testlights) are used to identify shorted and open circuits in electrical circuits. Low-voltage testers are used to troubleshoot 6- to 12-volt circuits. High-voltage circuit testers diagnose higher voltage systems, such as the secondary ignition circuit.

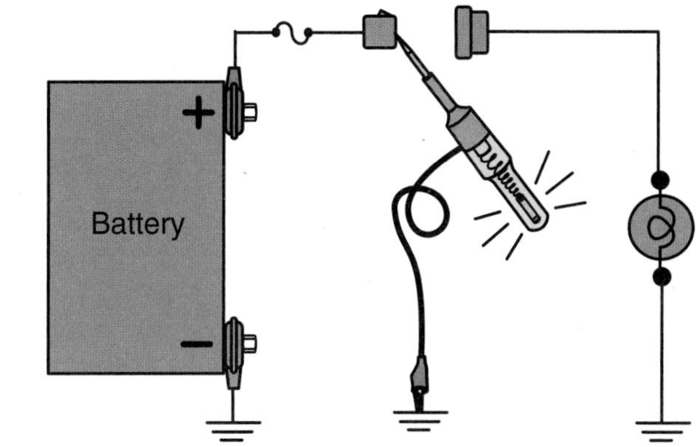

Figure 16-12 A testlight will illuminate when the probe is touched to part of the circuit that is powered.

High-impedance testlights are available for diagnosing electronic systems.

There are two basic types of testlights: nonpowered and powered. Nonpowered testlights are used to check for available voltage. With the wire lead connected to a good ground and the tester's probe at a point of voltage, the light turns on with the presence of voltage **(Figure 16–12)**. The amount of voltage determines the brightness of the light.

A self-powered testlight is used to check for continuity. Hooked across a circuit or component, the light turns on if the circuit is complete. A powered testlight should only be used if the power for the circuit or component has been disconnected.

Multimeter

A multimeter is one of the most important tools for troubleshooting electrical and electronic systems. A basic multimeter measures DC and AC voltage, current, and resistance. A multimeter eliminates the need for separate electrical meters. Most current multimeters also check engine speed, signal frequency, duty cycle, pulse width, diodes, temperature, and pressures. The desired test is selected by turning a control knob or depressing keys on the meter. These meters can be used to test simple electrical circuits, ignition systems, input sensors, fuel injectors, batteries, and starting and charging systems.

Lab Scopes

Lab scopes are fast reacting meters that measure and display voltages within a specific time frame. The voltage readings appear as a waveform or trace on the scope's screen. An upward movement of the trace indicates an increase in voltage, and a decrease in voltage results in a downward movement. These are especially valuable when watching the action of a circuit. The waveform displays what is happening and any problems can be observed when they happen.

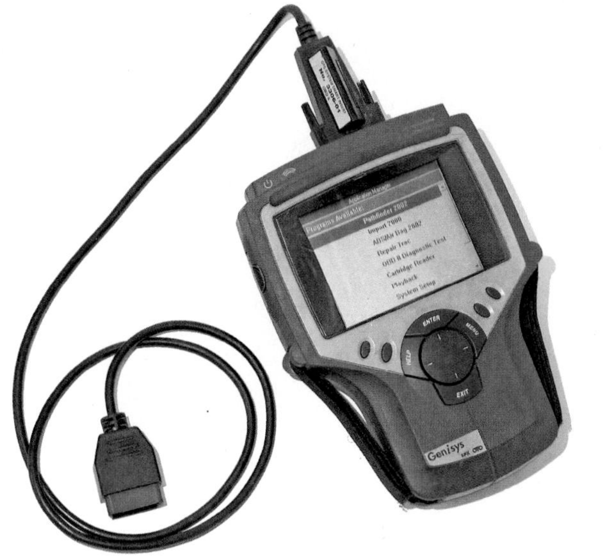

Figure 16-13 A typical scan tool.

This gives the technician a chance to observe what caused the change.

Scan Tools

When plugged into the vehicle's diagnostic connector, a scan tool can retrieve fault codes from a computer's memory and digitally display these codes **(Figure 16–13)**. A scan tool may also perform other diagnostic functions depending on the year and make of the vehicle. Many scan tools have removable modules that are updated each year. These modules are designed to test the computer systems on various makes of vehicles.

The scan tool must be programmed for the model year, make of vehicle, and type of engine. With some scan tools, this selection is made by pressing the appropriate buttons on the tester as directed by the digital tester display. On other scan tools, the appropriate memory card must be installed in the tester for the vehicle being tested. Some scan tools have a built-in printer to print test results, whereas others may be connected to an external printer.

Some scan tools display diagnostic information based on the trouble code. Service bulletins may also be indexed on the tool after vehicle information is entered into the tester. Other scan tools can also function as a multimeter. After retrieving related trouble codes, the tool allows for detailed testing. Most of these tools will display sensor specifications and instructions for further testing **(Figure 16–14)**.

Trouble codes are only set by the vehicle's computer when a voltage signal is entirely out of its normal range. The codes help technicians identify the cause of the problem. If a signal is within its normal range but it is still not correct, the vehicle's computer

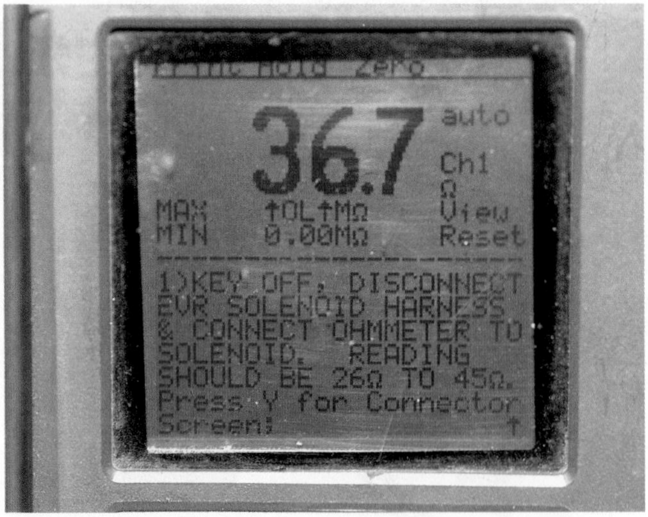

Figure 16–14 Some scan tools can also function as a multimeter and may display sensor specifications and instructions for further testing.

will not display a trouble code. However, a problem will still exist. As an aid to identify this type of problem, most manufacturers recommend that the signals to and from the computer be carefully looked at. This is done by observing the serial data stream or through the use of a breakout box.

Other Test Equipment

Other electrical test equipment may be needed to accurately diagnose an electrical circuit.

Jumper Wires Jumper wires are used to bypass individual wires, connectors, or components. Bypassing a component or wire helps to determine if that part is faulty **(Figure 16–15)**. If the symptom is no longer evident after the jumper wire is installed, the part bypassed is faulty. Technicians typically have jumper wires of various lengths; usually they have a fuse or circuit breaker in them to protect the circuits being tested.

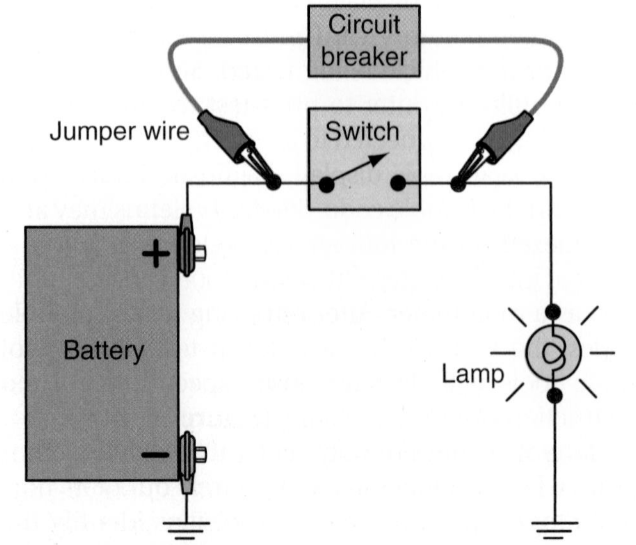

Figure 16–15 A jumper wire can be used to bypass a switch.

Computer Memory Saver Whenever a vehicle's battery needs to be disconnected, connect a memory saver to the vehicle. This saver will preserve the memory in the radio, electronic accessories, and the engine, transmission, and body computers.

Two types of memory savers are available or can be made for late-model vehicles. For a power source, use a 12-volt automotive battery or a 12-volt dry-cell lantern battery. If the vehicle's cigar lighter is continuously powered, a cigar lighter adapter plug with suitable wire leads and large alligator clips can be attached to the auxiliary battery. If the cigar lighter is controlled by the ignition switch, a set of jumper wires with alligator clips can be connected to the vehicle's electrical system.

To make either type of saver, use number 14 or 16 wire. Install a 5-ampere inline fuse in the positive lead, along with a diode **(Figure 16–16)**. The fuse will protect the memory saver and the electrical system from an accidental short. The diode will prevent current feedback from the vehicle's electrical system to the auxiliary battery.

If you connect the memory saver under the hood, do not connect it to the battery cable clamps. Removing and reinstalling the battery will likely dislodge the memory saver's alligator clips and make it useless. Instead, connect the saver's negative (−) lead to a good engine ground and the positive (+) lead to a point that is hot at all times, such as the battery connection at the generator or starter relay. Check a wiring diagram if you are unsure about the connection points.

If the memory saver must stay connected while the vehicle is raised and lowered on a hoist, securely place the battery out of the way of all moving parts. If

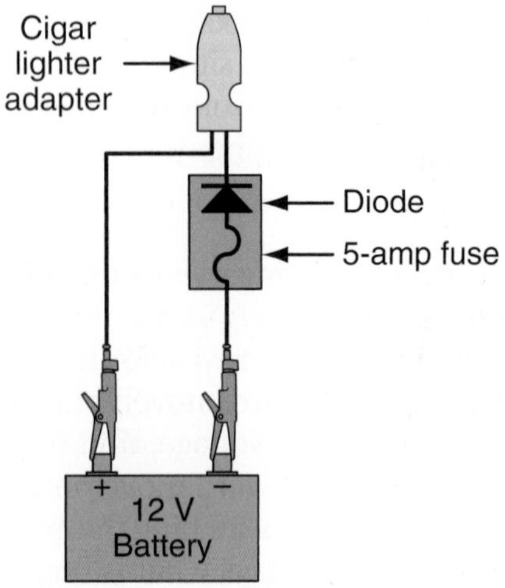

Figure 16–16 A homemade memory saver.

you place the battery inside the car, set it in a large plastic tray or tub to protect the vehicle from electrolyte or battery corrosion.

USING MULTIMETERS

Multimeters are available with either analog or digital displays. Analog meters **(Figure 16–17)** use a needle to point to a value on a scale. A digital meter shows the measured value in numbers or digits. The most commonly used meter is the digital multimeter (DMM), sometimes called a digital volt/ohmmeter (DVOM). Analog meters should not be used to test electronic components. They have low internal resistance (input impedance). The low input impedance allows too much current to flow through circuits and can damage delicate electronic devices. Digital meters, on the other hand, have high input impedance, 1 megohm (million ohms) to 10 megohms. In addition, metered voltage for resistance tests is well below 5 volts, reducing the risk of damage to sensitive components and delicate computer circuits.

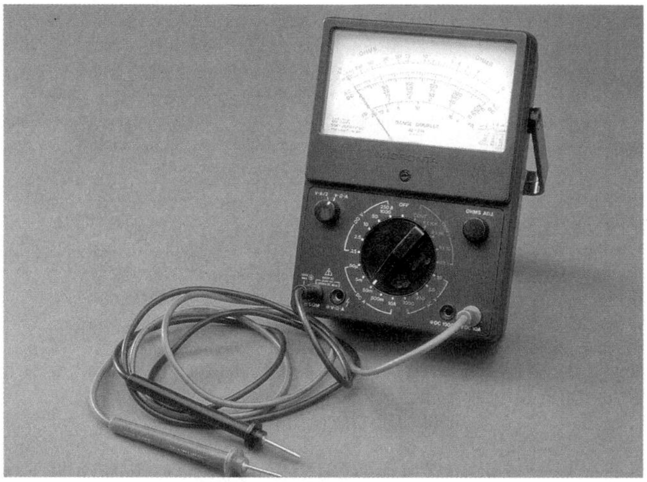

Figure 16–17 An analog multimeter.

Figure 16–18 Only meters with this symbol should be used on the high-voltage systems in a hybrid vehicle; it is preferable that the equipment is rated at 1,000 V.

> **SHOP TALK**
>
> The DMM used on high-voltage systems in hybrids should be classified as a category III meter. There are basically four categories for electrical meters, each built for specific purposes and to meet certain standards. The categories define how safe a meter is when measuring certain circuits. The standards for the various categories are defined by the American National Standards Institute (ANSI), the International Electrotechnical Commission (IEC), and the Canadian Standards Association (CSA). A CAT III meter is required for testing hybrid vehicles because of the high voltages, three-phase current, and the potential for high transient voltages. Transient voltages are voltage surges or spikes that occur in AC circuits. To be safe, you should have a CAT III 1,000 V meter. Within a particular category, meters have different voltage ratings. These reflect a meter's ability to withstand higher transient voltages. Therefore, a CAT III 1,000 V meter offers much more protection than a CAT III meter rated at 600 volts **(Figure 16–18)**.

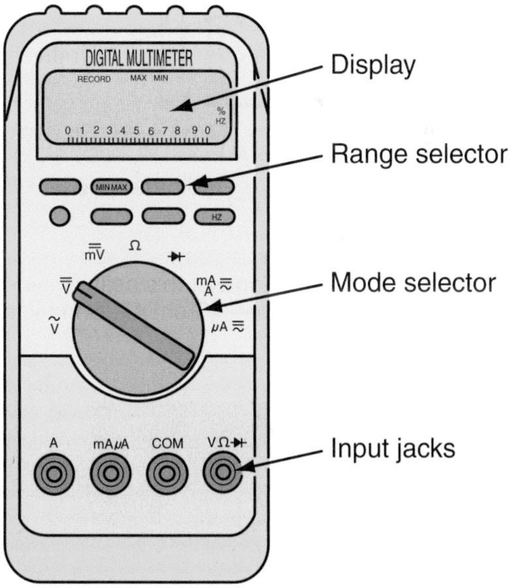

Figure 16–19 The front of a DMM normally has four distinct sections: the display area, range selectors, mode selector, and jacks for the test leads.

The front of a DMM is normally comprised of four distinct sections: the display area, range selectors, mode selector, and jacks for the test leads **(Figure 16–19)**. In the center of the display are large digits that represent the measured value. Normally there are four to five digits with a decimal point. To the right of the number, the measured units are displayed (V, A, or Ω). These units may be further defined by a symbol to denote a value of more or less than one. Examples of these are shown in **Figure 16–20**. In the upper-right-hand corner of the display, the type of voltage—AC or DC—is displayed. In the lower-right-hand corner of the display, the

Symbol	Name	Value
mV	millivolts	volts x 0.001
kV	kilovolts	volts x 1,000
mA	milliamps	amps x 0.001
μA*	microamps	amps x 0.000001
kΩ	kilo-ohms	ohms x 1,000
MΩ	megohms	ohms x 1,000,000

*Automotive technicians seldom use readings at the microamp level.

Figure 16-20 Symbols used to define the value of a measurement on a DMM.

DMM displays the measurement range; the range is automatically adjusted by the DMM or manually set by the technician **(Figure 16–21)**.

Setting the range on a DMM is important. If the measurement is beyond the set range, the meter will display a reading of "OL" or over limit **(Figure 16–22)**. The range on some DMMs is manually set, whereas others have an "auto range" feature, in which the appropriate scale is automatically selected by the meter. Meters with auto range allow you to disable it when you want to manually select ranges. Auto ranging is helpful when you do not know what value to expect. When using a meter with auto range, make sure you note the range being used by the meter. There is a big difference between 10 ohms and 10,000,000 (10 M) ohms.

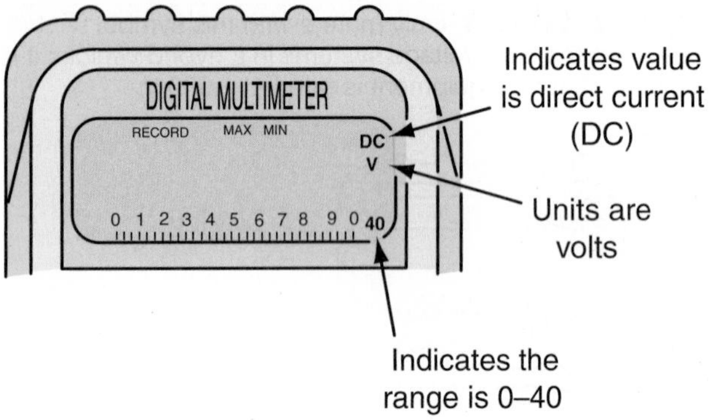

Figure 16-21 The measurement display also shows other important information. Some of this defines the measured value, the type of voltage—AC or DC—and the range selected on the meter.

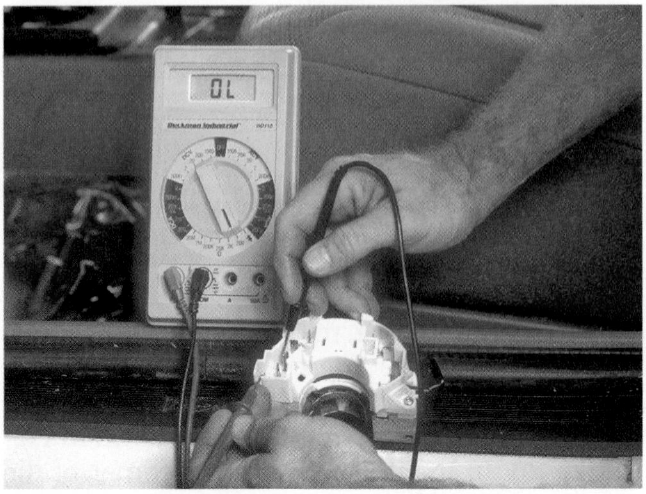

Figure 16-22 On some DMMs, an infinite reading is displayed as "OL."

WARNING!

Many DMMs with auto range display the measurement with a decimal point. Make sure you observe the decimal and the range being used by the meter (Figure 16–23). A reading of .972 K ohm equals 972 ohms. If you ignore the decimal point you will interpret the reading as 972,000 ohms.

The mode selector defines what the meter will be measuring **(Figure 16–24)**. The number of available modes varies with meter design, but nearly all have the following:

- Volts AC
- Volts DC
- Millivolts (mV) DC
- Resistance/Continuity (ohms)
- Diode Check
- Amps or Milliamps AC/DC

Most DMMs have two test leads and four input jacks. The black test lead always plugs into the COM

0.345 KΩ = 345 Ω

1025 mAmps = 1.025 Amps

Figure 16-23 Placement of decimal point and the scale should be noticed when measuring with a meter with auto range.

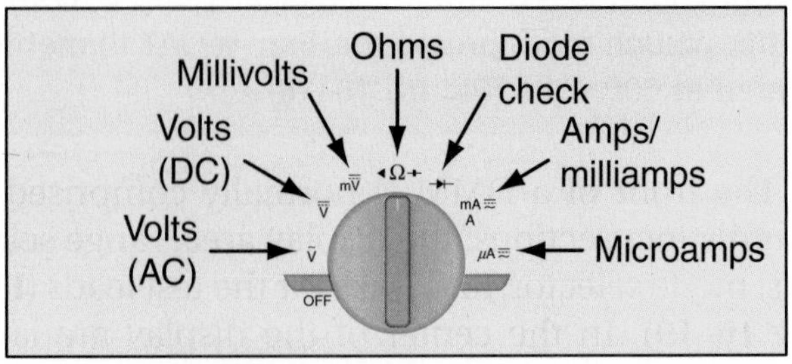

Figure 16-24 The mode selector defines what the meter will be measuring.

input jack and the red lead plugs into one of the other input jacks, depending on what is being measured. Often technicians have multiple sets of test leads, each for specific purposes. For example, a lead set with small tips is ideal for probing in hard-to-reach or tight spaces. Other test leads may be fitted with clips to hold the lead at a point during testing.

Typically, the three input jacks are:

- "A" for measuring up to 10 amps of current
- "A/mA" for measuring up to 400 mA of current
- "V/Ω/diode" for measuring voltage, resistance, conductance, capacitance, and checking diodes

Measuring Voltage

A DMM can measure source voltage, available voltage **(Figure 16–25)**, and voltage drops. Voltage is measured by placing the meter in parallel to the component or circuit being tested **(Figure 16–26)**. There are normally several voltage ranges available on a DMM. Most automotive circuits range from 50 mV to 15 volts.

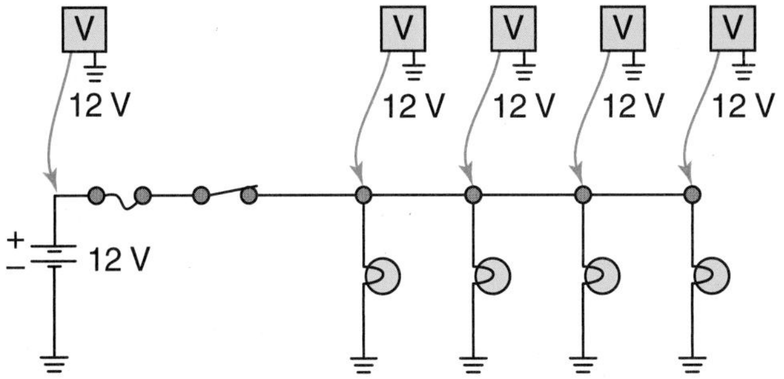

Figure 16-25 Checking available voltage at points within a circuit with a voltmeter.

> **WARNING!**
>
> *Hybrid vehicles have much higher voltage; always follow all safety precautions and service procedures when working with high-voltage circuits.*

A DMM can be used to check for proper circuit grounding. For example, if the voltmeter shows battery voltage at a lamp but no lighting is seen, the bulb

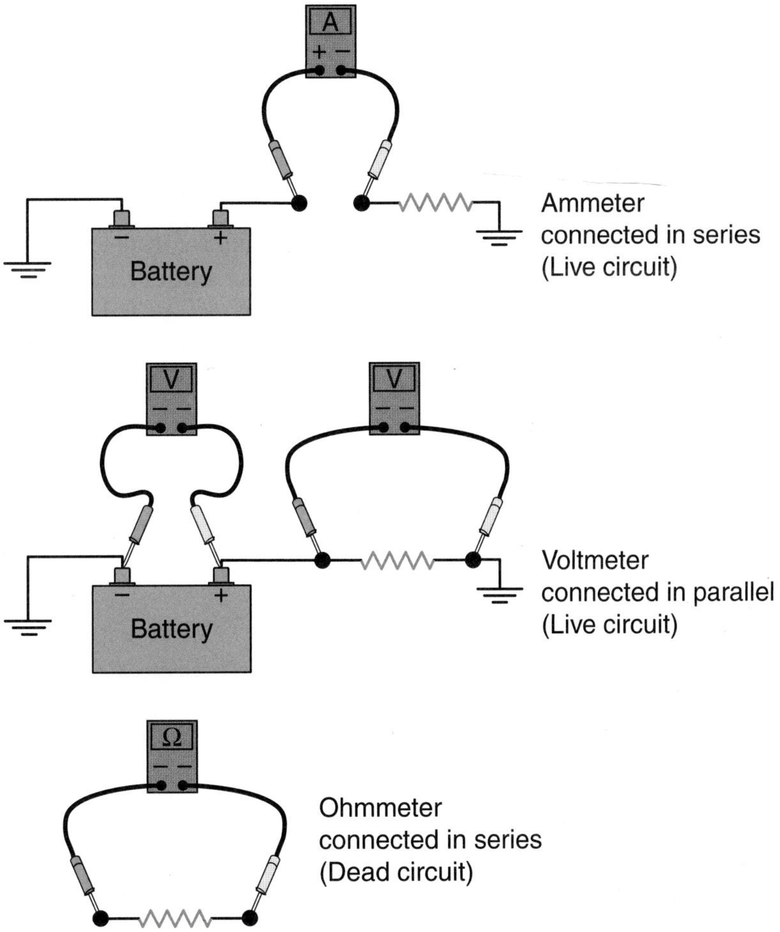

Figure 16-26 A voltmeter is connected in parallel or across a component or part of a circuit to measure available voltage and voltage drop.

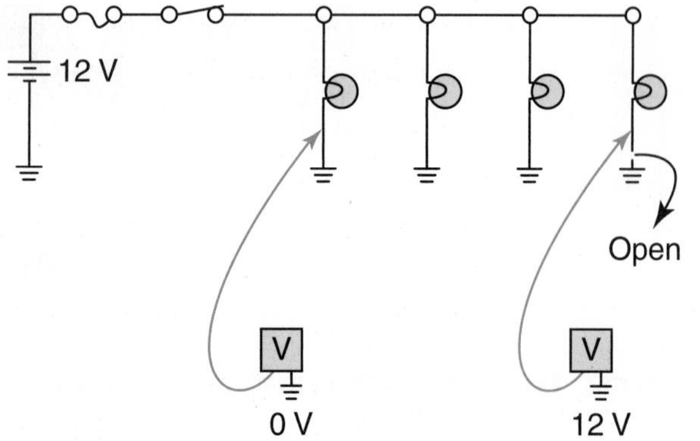

Figure 16-27 Using a voltmeter to check for open grounds.

or socket could be bad or the ground connection is faulty. An easy way to check for a defective bulb is to replace it with one known to be good. You can also use an ohmmeter to check for electrical continuity through the bulb.

If the bulbs are not defective, the problem is the light sockets or ground wires. Connect the voltmeter to the ground wire and a good ground as shown in **Figure 16–27**. If 0 volts is measured, move the positive meter lead to the power feed side of the bulb. If 0 volts is measured there, the light socket or feed wire is defective. In a normal light circuit, there should be 0 volts at the negative side of the bulb. If the socket is not defective and some voltage is measured at the ground, the ground circuit is faulty. The higher the voltage, the greater the problem.

PROCEDURE

To measure DC voltage:

1. Set the mode selector switch to Volts DC.
2. Select the auto range function, or manually set the range to match the anticipated value; normally the range is set to the closest to and higher than 12 volts.
3. Connect the test leads in parallel to the circuit or component being tested. The red lead should be connected first to the most positive side (side closest to the battery). The black lead is connected to a good ground.
4. Read the measurement on the display. If the reading is negative, it is likely that the leads are reversed.

Voltage Drop Test Measuring voltage drop is a very important test. It can identify circuits with unwanted resistances. This test is extremely valuable when working with electronic circuits; even the smallest loss of voltage will affect the performance of parts and

Figure 16-28 Measuring the voltage drop across the battery post and cable. It is easy to see why the voltage drop is high!

the vehicle as a whole. The test can be performed between any two points in a circuit and across any component, such as wires, switches, relay contacts and coils, and connectors **(Figure 16–28)**. A voltage drop test can find excessive resistance in a circuit that may not be detected using an ohmmeter. An ohmmeter works by passing a small amount of current through the component being tested while it is isolated from the circuit. A voltage drop test is conducted with the circuit operating with normal amounts of current.

Consider a simple circuit. If there are 12 volts available at the battery and the switch is closed, there should also be 12 volts available at each light. If, for example, less than 12 volts are measured, that means some additional resistance is somewhere else in the circuit. The lights may light but not as brightly as they should.

Figure 16–29 illustrates two headlights (2 ohms each) connected to a 12-volt battery using two wires. Each wire has an unwanted resistance of 0.05 ohm.

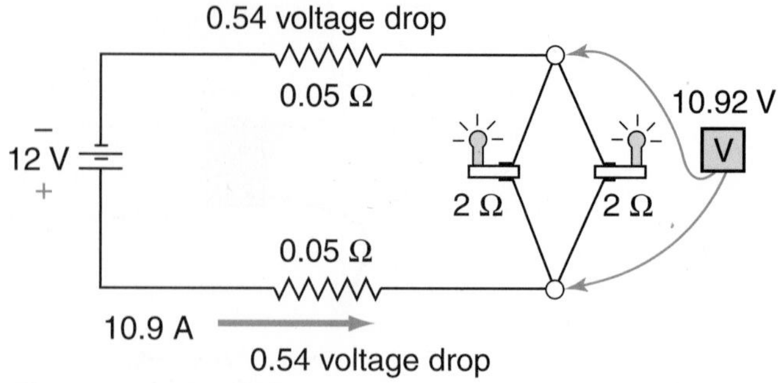

Figure 16-29 Wire resistance results in a slight voltage drop in the circuit.

The two headlights are wired parallel, and total resistance of the headlights is:

$$\frac{2 \text{ ohms} + 2 \text{ ohms}}{2 \text{ ohms} + 2 \text{ ohms}} = 1 \text{ ohm}$$

The total circuit resistance is:

$$1 \text{ ohm} + 0.05 \text{ ohm} + 0.05 \text{ ohm} = 1.1 \text{ ohms}$$

Therefore, the current in the circuit is

$$I = \frac{E}{R} = \frac{12}{1.1} = 10.9 \text{ amperes}$$

The voltage drop across each wire is:

$$E = I \div R$$
$$E = 10.9 \div 0.05 = 0.54 \text{ V}$$

This means there is a total of 1.08 volts dropped across the wires. When the voltage drop of the wires is subtracted from the 12-volt source voltage, 10.92 volts remain for the headlights.

Without the resistance in the wires, the headlights receive 12 amperes. With the resistance, the current flow was reduced to 10.9 amperes. The decreased current and voltage drop means the lights will not be as bright as normal.

PROCEDURE

The procedure for measuring voltage drop is shown in Photo Sequence 12. To measure voltage drop:

1. Set the DMM to the mV or V DC mode.
2. Set the DMM to auto range or a low-voltage range.
3. Connect the positive test lead to the most positive side of the part or circuit being tested.
4. Connect the negative test lead to the least positive side of the part or circuit being tested.
5. Power the circuit. Current must be flowing in order to have a voltage drop.
6. Read the meter. Excessive voltage drops indicate excessive resistance (an unwanted load) in that portion of the circuit.

Voltage drops should not exceed the following:

- 200 mV across a wire or connector
- 300 mV across a switch or relay contacts
- 100 mV at a ground connection
- 0 mV to < 50 mV across all sensor connections

Measuring AC Voltage There are two ways that multimeters display AC voltage: **root mean square (RMS)** and **average responding**. When an AC voltage signal is a true sine wave, both methods will display the same reading. Because most automotive sensors do not produce pure sine wave signals, it is important to know how the meter will display the AC voltage reading when comparing measured voltage to specifications. RMS meters convert the AC signal to a comparable DC voltage signal. Average responding meters display the average voltage peak. Always check the voltage specification to see if the specification is for RMS voltage; if it is, use an RMS meter (**Figure 16–30**).

Measuring Current

Testing the current through a circuit gives a true picture of what is happening in the circuit. This is because the circuit is being tested under load. Low current indicates that the circuit has higher than normal resistance, and high current means the circuit has lower than normal resistance. Many technicians do not check current because few specifications list what should be normal. This should not be a problem because current can be calculated if you know the resistance and voltage in the circuit. Remember, to have current the circuit must be complete and there must be voltage applied to it.

The circuit in **Figure 16–31** normally draws 6 amps and is protected by a 10 amp fuse. If the circuit constantly blows the fuse, a short exists somewhere in the circuit. Mathematically, each light should draw 1.5 amperes (6 ÷ 4 = 1.5). To find the short, disconnect all lights by removing them from their sockets. Then, close the switch and read the ammeter. With

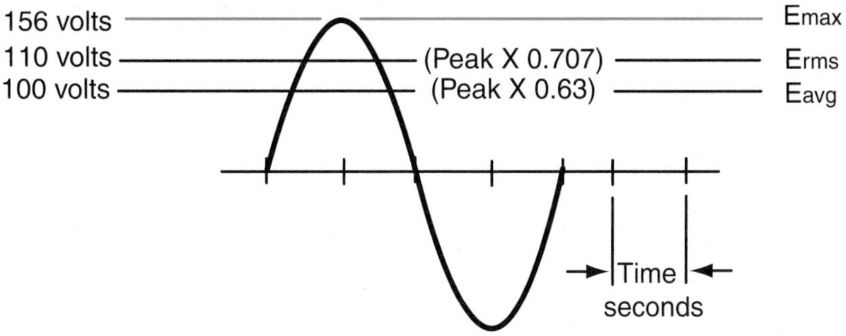

Figure 16–30 AC voltage: RMS.

P12-1 *Tools required to perform this task: voltmeter and fender covers.*

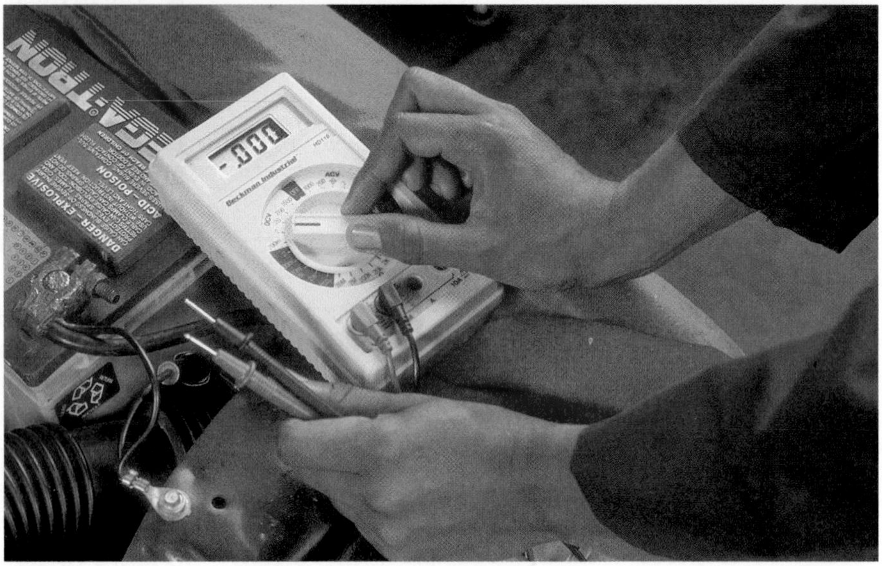

P12-2 *Set the voltmeter on its lowest DC volt scale.*

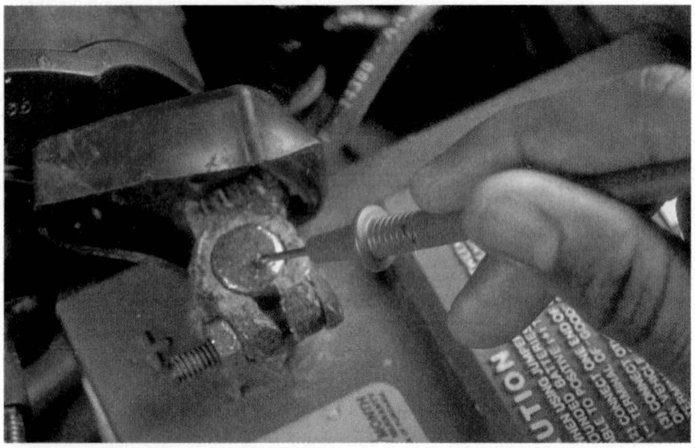

P12-3 *To test the voltage drop of the entire headlamp system, connect the positive (red) lead to the battery positive terminal.*

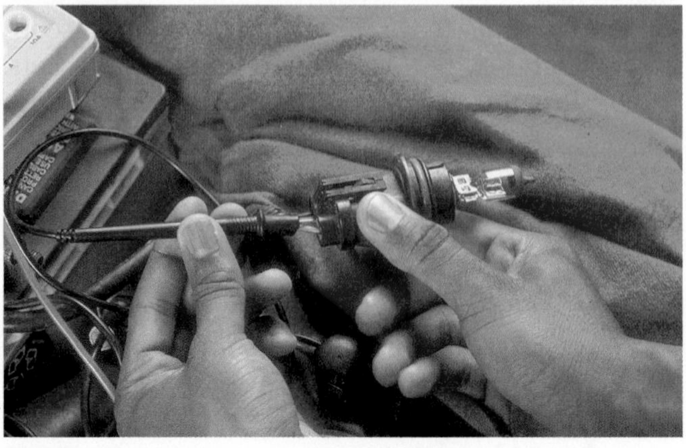

P12-4 *Connect the negative (black) lead to the low beam terminal of the headlight socket. Make sure you are connected to the battery "+" or power feed wire of the headlight.*

P12-5 *Turn on the headlights (low beam) and look at the voltmeter reading. The voltmeter will show the amount of voltage dropped between the battery and the headlight. This reading should be very low.*

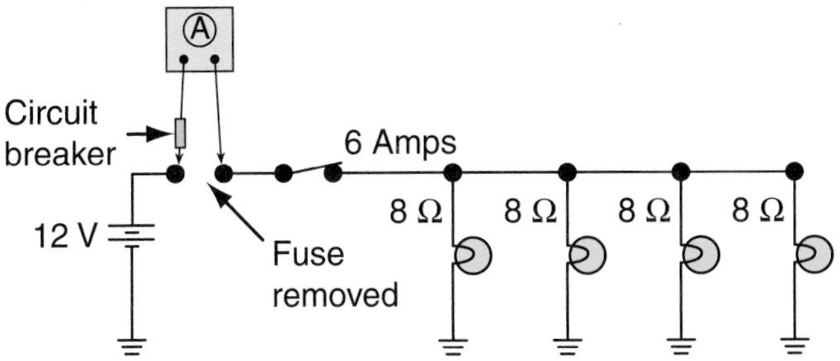

Figure 16-31 Checking a circuit using an ammeter.

the loads disconnected, the meter should read 0 ampere. If there is any reading, the wire between the fuse and the sockets is shorted to ground.

If 0 amp was measured, reconnect each light in sequence; the reading should increase 1.5 amperes with each bulb. If, when making any connection, the reading is higher than expected, the problem is in that part of the light circuit.

WARNING!

When testing for a short, always use a fuse. Never bypass the fuse with a wire. The fuse should be rated at no more than 50% higher capacity than specifications. This offers circuit protection and provides enough amperage for testing. After the problem is found and corrected, be sure to install a fuse with the specified rating.

Before checking current, make sure the meter is capable of measuring the suspected amount. To check the rating of the meter, look at the rating printed next to the DMM input jacks or check the rating of the meter's fuse (maximum current capacity is typically the same as the fuse rating). If you suspect that a measurement will have a current higher than the meter's maximum rating, use an inductive current probe.

An ammeter must be connected in series with the circuit; this allows circuit current to flow through the meter.

PROCEDURE

To measure current:

1. Turn off the circuit that will be tested.
2. Connect the test leads to the correct input jacks on the DMM.
3. Set the mode selector to the correct current setting (normally amps or milliamps).
4. Select the auto range function, or manually select the range for the expected current.
5. Open the circuit at a point where the meter can be inserted.
6. Connect a fused jumper wire to one of the test leads.
7. Connect the red lead to the most positive side of the circuit and the black lead to the other side.
8. Turn on the circuit.
9. Read the display on the DMM.
10. Compare the reading to specifications or your calculations.

WARNING!

Never place the leads of an ammeter across the battery or a load. This puts the meter in parallel with the circuit and will blow the fuse in the ammeter or possibly destroy the meter.

Inductive Current Probes Many DMMs have current probes (current clamps) that eliminate the need to insert the ammeter into the circuit. These probes read current by sensing the magnetic field formed in a wire by current flow **(Figure 16–32)**. Normally, to use a current probe, the DMM's mode selector is set to read millivolts (mV). The probe is then connected to the meter and turned on. Some probes must be zeroed prior to taking a measurement. This is done before the probe is clamped around a wire. The DMM may have a zero adjust control, which is turned until zero reads on the meter's display. The clamp is placed around a wire in the circuit being tested **(Figure 16–33)**. Make sure the arrow on the clamp is pointing in the direction of current flow. After the clamp is in place, the circuit is turned on and the

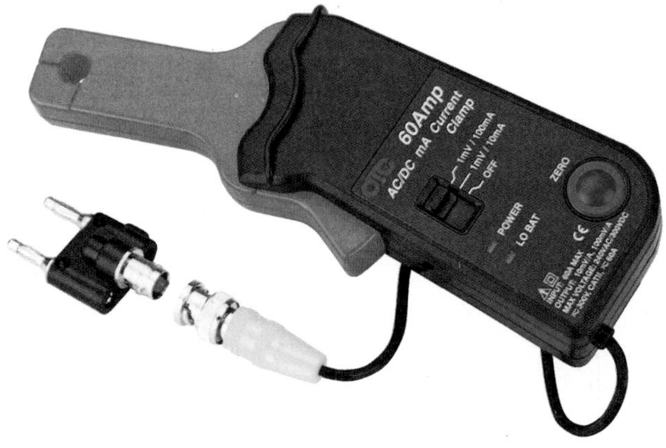

Figure 16-32 A low-amp current probe. *Courtesy of SPX Service Solutions*

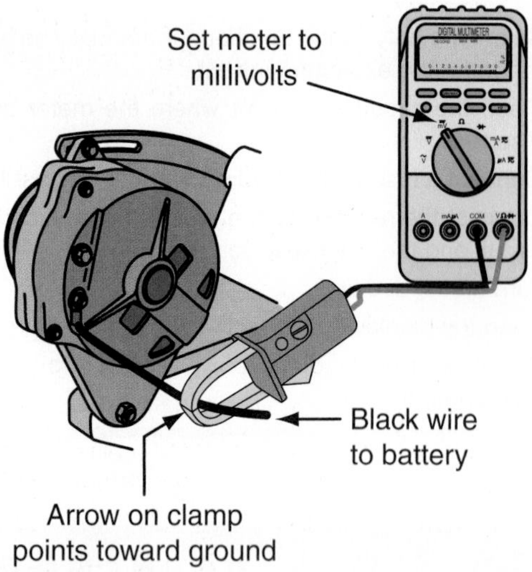

Figure 16-33 A current probe attached to a DMM allows for current measurements without breaking the circuit.

voltage read on the display. The voltage reading is then converted to an amperage reading—1 mV = 1 ampere.

Measuring Resistance

DMMs are used to test circuit continuity and resistance with no power applied. In other words, the circuit or component must first be disconnected from the power source **(Figure 16–34)**. Connecting an ohmmeter into a live circuit usually results in damage to the meter.

Checking the resistance can be used to check the condition of a component **(Figure 16–35)** or circuit. Often specifications list a normal range of resistance values for specific parts. If the resistance is too high, check for an open circuit or a faulty component. Excessive resistance can prevent a circuit from operating normally. Loose, damaged, or dirty connections are common causes of excessive resistance. If the resistance is too low, check for a shorted circuit or faulty component.

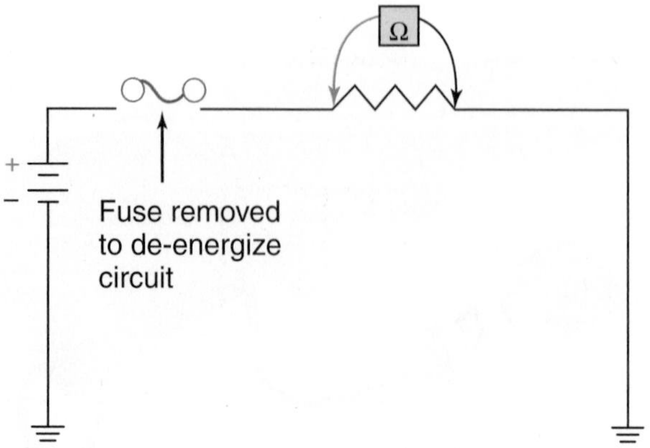

Figure 16-34 Measuring resistance using an ohmmeter. Note that the circuit fuse is removed to de-energize the circuit.

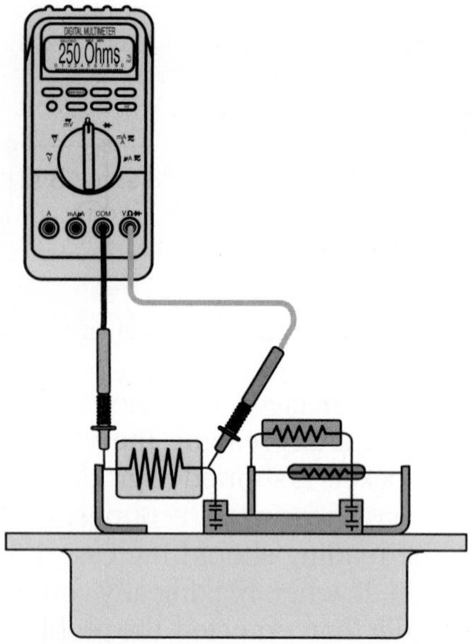

Figure 16-35 An ohmmeter can test the resistance of a component after it has been removed from the circuit.

CAUTION!

To avoid possible damage to the meter or to the equipment under test, disconnect the circuit power and discharge all high-voltage capacitors before measuring resistance. Always follow the manufacturer's test procedures when testing air bags.

Ohmmeters also are used to trace and check wires or cables. Connect one probe of the ohmmeter to the known wire at one end of the cable and touch the other probe to each wire at the other end of the cable. Any evidence of resistance indicates the correct wire. Using this same method, you can check for a defective wire. If low resistance is shown on the meter, the wire is sound. If no resistance is measured, the wire is open. Any indication of resistance means the wire may be shorted to another wire and the harness is defective.

PROCEDURE

To measure resistance:

1. Make sure the circuit or component to be tested is not connected to any power source.
2. Set the DMM mode selector to measure resistance.
3. Select the auto range feature or manually select the appropriate range.

4. Calibrate the meter by holding the two test leads together and adjusting the meter to zero. On some DMMs, the calibration should be checked whenever the range is changed.
5. Connect the meter leads in parallel to the component or part of the circuit that will be checked.
6. Read the measured value on the display. The DMM will show a zero or close to zero when there is good continuity. If there is no continuity, the meter will display an infinite or over-limit reading.

Continuity Tests Many DMMs have a continuity test mode that makes a beeping noise when there is continuity through the item being tested. This audible sound will continue as long as there is continuity. This feature can be handy for finding the cause of an intermittent problem. By connecting the DMM across a circuit and wiggling sections of the wiring harness, a problem can be noted when the beeping stops after a particular section or wire has been moved.

MIN/MAX Readings

Some DMMs also feature a MIN/MAX function. This displays the maximum, minimum, and average voltage recorded during the time of the test. This is valuable when checking sensors or when looking for electrical noise. Noise is primarily caused by radio frequency interference (RFI), which may come from the ignition system. RFI is an unwanted voltage signal that rides on a signal. This noise can cause intermittent problems with unpredictable results. The noise causes slight increases and decreases in the voltage. When a computer receives a voltage signal with noise, it will try to react to the minute changes. As a result, the computer responds to the noise rather than the voltage signal.

Other Measurements

Multimeters may also have the ability to measure duty cycle, pulse width, and frequency. Duty cycle **(Figure 16–36)** is measured as a percentage. A 60% duty cycle means that a device is on 60% of the time and off 40% of one cycle. When measuring duty cycle, you are looking at the amount of time something is on during one cycle.

Pulse width is normally measured in milliseconds. When measuring pulse width, you are looking at the amount of time something is on.

To accurately measure duty cycle, pulse width, and frequency, the meter's trigger level must be set. The trigger level tells the meter when to start counting. Trigger levels can be set at certain voltage levels or at a rise or fall in the voltage. Normally, meters have a built-in trigger level that corresponds with the voltage range setting. If the voltage does not reach the trigger level, the meter will not begin to recognize a cycle. On some meters you can select between a rise or fall in voltage to trigger the cycle count. Most technicians refer to this as a positive or negative slope trigger. A rise in voltage is a positive increase in voltage. This setting is used to monitor the activity of devices whose power feed is controlled by a computer. A fall in voltage is negative voltage. This setting is used to monitor ground-controlled devices.

Some DMMs can measure temperature. These meters are equipped with a thermocouple. Temperature readings can be made in Fahrenheit or Celsius. The thermocouple is connected to the DMM and placed on or near the object to be checked.

More elaborate DMMs have the ability to store and download data to a PC.

Safety Guidelines

To avoid possible electric shock or personal injury while using a DMM, follow these guidelines:

- Use a meter only as it was designed to be used; misuse may defeat the protection devices built into the meter.
- Never use a damaged meter.
- Make sure the battery is secure and totally enclosed with the meter.

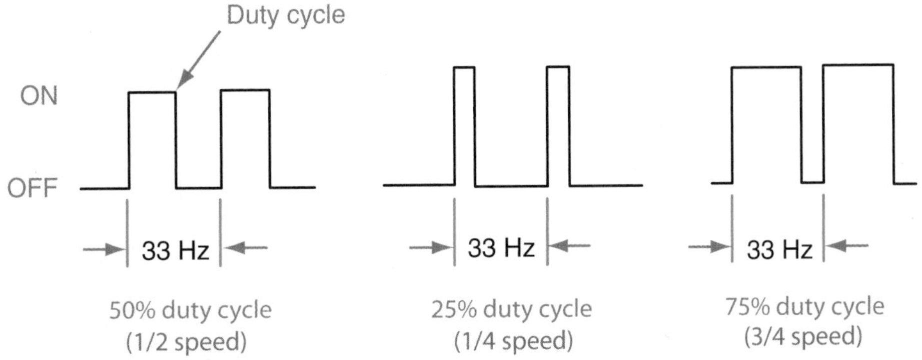

Figure 16–36 Duty cycle is expressed in a percentage.

- Inspect the test leads for damaged insulation or exposed metal. Replace damaged test leads before you use a meter.
- Never apply more than the rated voltage of the meter between the terminals or between any terminal and ground.
- Be extra careful when measuring high voltages.
- Use the proper connections, terminals, mode, and range for the measurements being taken.
- When measuring current, turn off the power to the circuit before connecting the meter in series with the circuit.
- When making electrical connections, connect the negative test lead before connecting the positive test lead; when disconnecting, disconnect the positive test lead before disconnecting the negative test lead.
- When using probes, keep your fingers behind the finger guards on the probes.
- Disconnect the power to the circuit and discharge all high-voltage capacitors before testing resistance, continuity, diodes, or capacitance.
- When testing HV components or circuits, wear high-voltage lineman's gloves.

USING LAB SCOPES

The lab scope has become the diagnostic tool of choice for many good technicians. When measuring voltage with an analog voltmeter, the meter only displays the average values at the point being probed. Digital voltmeters simply sample the voltage several times each second and update the meter's reading at a particular rate. If the voltage is constant, good measurements can be made with both types of voltmeters. A scope, however, will display any change in voltage as it occurs. This is important for diagnosing intermittent problems.

The screen is divided into small divisions of time and voltage **(Figure 16–37)**. These divisions set up a grid pattern on the screen. Time is represented by the horizontal movement of the trace. Voltage is measured with the vertical position of the trace. Because the scope displays voltage over time, the trace moves from the left (the beginning of measured time) to the right (the end of measured time). The value of the divisions can be adjusted to improve the view of the voltage trace. For example, the vertical scale can be adjusted so that each division represents 0.5 volt, and the horizontal scale can be adjusted so that each division equals 0.005 second (5 milliseconds). This allows for viewing small changes in voltage that occur during a very short period. The grid serves as a reference for measurements.

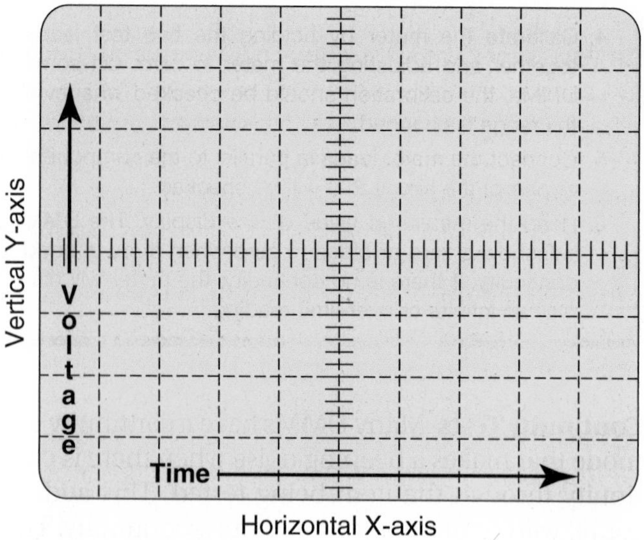

Figure 16–37 Grids on a scope screen that serve as a time and voltage reference.

Because a scope displays actual voltage, it will display any electrical noise or disturbances that accompany the voltage signal **(Figure 16–38)**. Electrical disturbances or **glitches** are momentary changes in the signal. These can be caused by intermittent shorts to ground, shorts to power, or opens in the circuit. These problems can occur for only a moment or may last for some time. A lab scope is handy for finding these and other causes of intermittent problems. By observing a voltage signal and wiggling or pulling a wiring harness, any looseness can be detected by a change in the voltage signal.

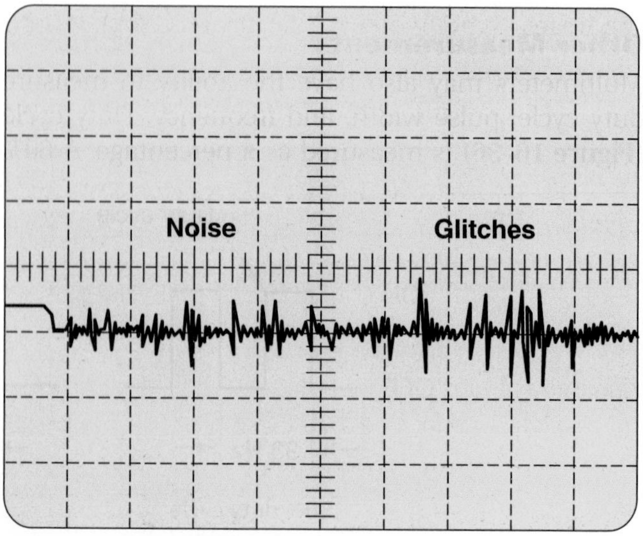

Figure 16–38 RFI noise and glitches may appear on a voltage signal.

Analog versus Digital Scopes

Analog scopes show the actual activity of a circuit and are called real-time scopes. This simply means that what is taking place at that time is what you see on the screen. Analog scopes have a fast update rate that allows for the display of activity without delay.

A digital scope, commonly called a **digital storage oscilloscope** or **DSO**, converts the voltage signal into digital information and stores it into its memory. Some DSOs send the signal directly to a computer or a printer or save it to a disk. To help in diagnostics, a technician can "freeze" the captured signal for close analysis. DSOs also have the ability to capture low-frequency signals. Low-frequency signals tend to flicker when displayed on an analog screen. To have a clean waveform on an analog scope, the signal must be repetitive and occurring in real time. The signal on a DSO is not quite real time. Rather it displays the signal as it occurred a short time before.

This delay is actually very slight. Most DSOs have a sampling rate of one million samples per second. This is quick enough to serve as an excellent diagnostic tool. This fast sampling rate allows slight changes in voltage to be observed. Slight and quick voltage changes cannot be observed on an analog scope.

Because digital signals are based on binary numbers, the trace appears to be slightly choppy when compared to an analog trace. However, the voltage signal is sampled more often, which results in a more accurate waveform. The waveform is constantly being refreshed as the signal is pulled from the scope's memory.

Both an analog and a digital scope can be dual trace **(Figure 16–39)** or multiple trace **(Figure 16–40)** scopes. This means they have the capability of displaying more than one trace at one time. By watching the traces simultaneously, the cause and effect of a sensor is observed and a good or normal waveform can be compared to the one being displayed.

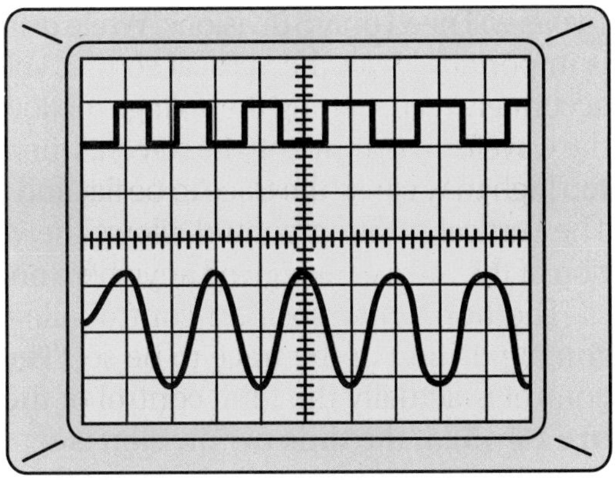

Figure 16-39 Two or more different signals can be observed on a multiple-channel scope. This is invaluable for diagnosis.

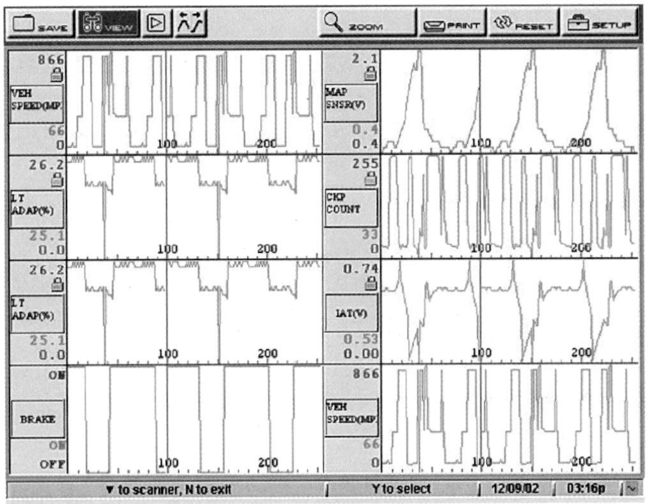

Figure 16-40 Some scopes and graphing scan tools can display many channels. *Reproduced under license from Snap-on Incorporated. All of the marks are marks of their owners.*

Waveforms

A waveform represents voltage over time. Any change in the amplitude of the trace indicates a change in the voltage. When the trace is a straight horizontal line, the voltage is constant. A diagonal line up or down represents a gradual increase or decrease in voltage. A sudden rise or fall in the trace indicates a sudden change in voltage.

Scopes can display AC and DC voltage, either one at a time or both at the same time, as in the case of noise caused by RFI. Noise results from AC voltage riding on a DC voltage signal. The consistent change of polarity and amplitude of the AC signal causes slight changes in the DC voltage signal. A normal AC signal changes its polarity and amplitude over time. The waveform created by AC voltage is typically called a sine wave **(Figure 16–41)**. One complete sine wave shows the voltage moving from zero to its positive peak then moving down through zero to its negative peak and returning to zero.

One complete sine wave is a cycle. The number of cycles that occur per second is the frequency of the signal. Frequency is measured in cycles per second (hertz or Hz). Checking frequency is one way of checking the operation of some electrical components. Input sensors are the most common components that produce AC voltage. Permanent magnet voltage generators produce an AC voltage that can be checked on a scope **(Figure 16–42)**. AC voltage waveforms should also be checked for noise and glitches. These may send false information to the computer.

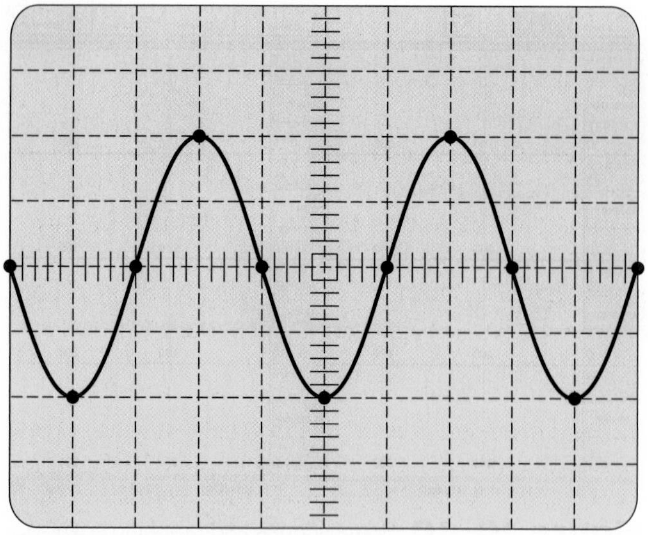

Figure 16-41 An AC voltage sine wave.

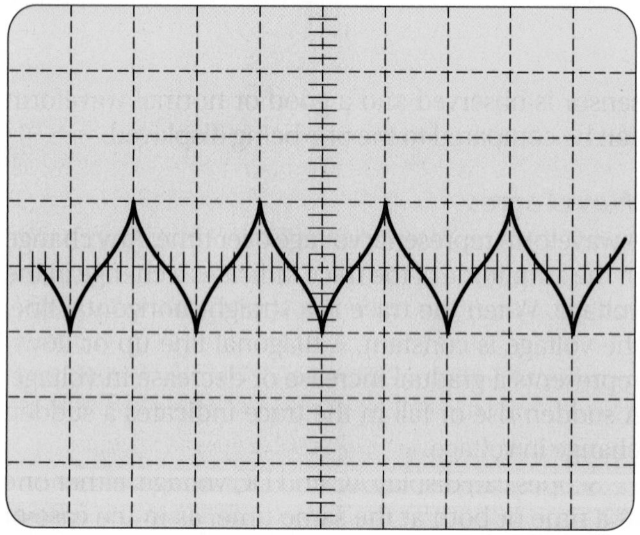

Figure 16-42 An AC voltage trace from a typical permanent magnet generator-type pickup or sensor.

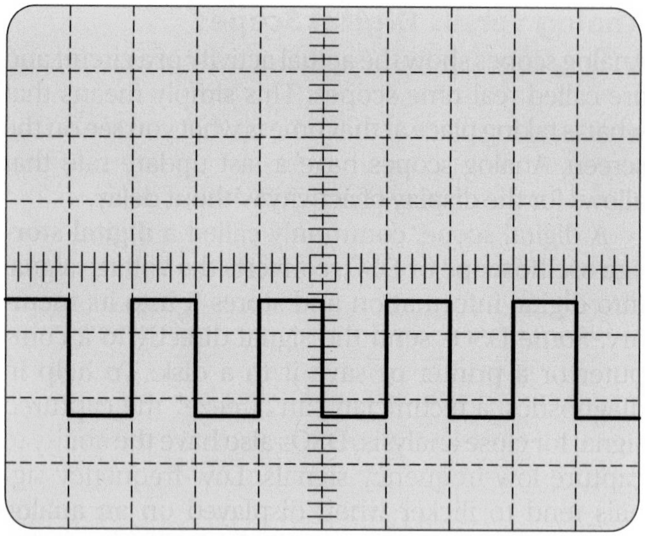

Figure 16-43 A typical square (on-off or high-low) wave.

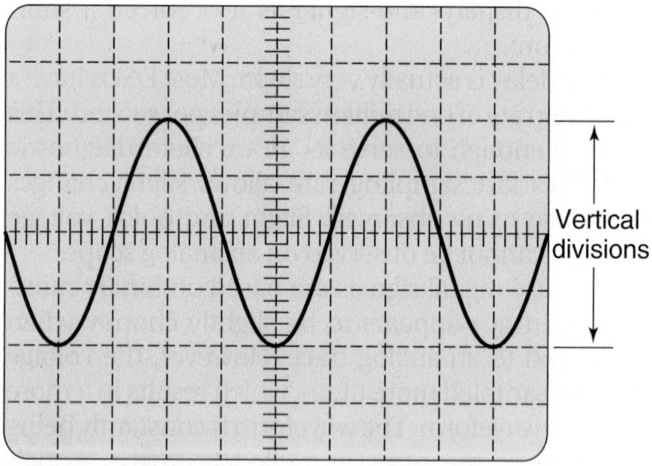

Figure 16-44 Vertical divisions represent voltage.

DC voltage waveforms may appear as a straight line or line showing a change in voltage. Sometimes a DC voltage waveform will appear as a square wave that shows voltage making an immediate change **(Figure 16–43)**. Square waves are identified by having straight vertical sides and a flat type. This type of wave represents voltage being applied (circuit being turned on), voltage being maintained (circuit remaining on), and no voltage applied (circuit is turned off). Of course, a DC voltage waveform may also show gradual voltage changes.

Scope Controls

Depending on manufacturer and model of the scope, the type and number of its controls will vary. However, nearly all scopes have intensity, vertical (Y-axis) adjustments, horizontal (X-axis) adjustments, and trigger adjustments. The intensity control is used to adjust the brightness of the trace. This allows for clear viewing regardless of the light around the scope screen.

The vertical adjustment actually controls the voltage that will be shown per division **(Figure 16–44)**. If the scope is set at 0.5 (500 milli) volt, a 5-volt signal will need 10 divisions. Likewise, if the scope is set to 1 volt, 5 volts will need only 5 divisions. While using a scope, it is important to set the vertical so that voltage can be accurately read. Setting the voltage too low may cause the waveform to move off the screen, whereas setting it too high may cause the trace to be flat and unreadable. The vertical position control allows the vertical position of the trace to be moved anywhere on the screen.

The horizontal position control allows the horizontal position of the trace to be set. The horizontal control is actually the time control of the trace **(Figure 16–45)**. If the time per division is set too low, the complete trace may not show across the screen. Also, if the time per division is set too high, the trace may be too crowded for detailed observation. The time

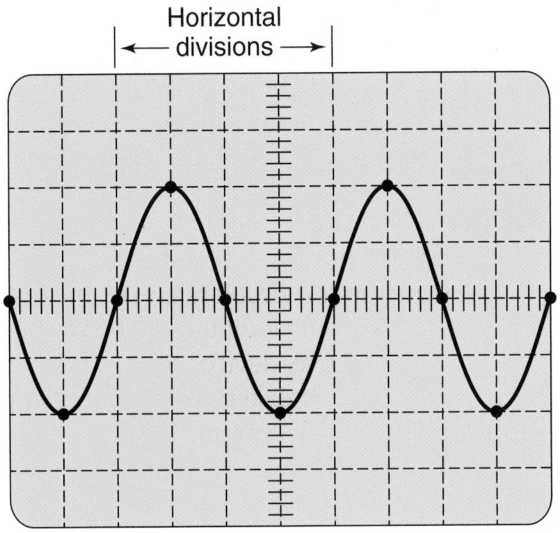

Figure 16-45 Horizontal divisions represent time.

per division (TIME/DIV) can be set from very short periods (millionths of a second) to full seconds.

Trigger controls tell the scope when to begin a trace across the screen. Setting the trigger is important when trying to observe the timing of something. Proper triggering will allow the trace to repeatedly begin and end at the same points on the screen. There are typically numerous trigger controls on a scope. The trigger mode selector has a NORM and an AUTO position. In the NORM setting, no trace will appear on the screen until a voltage signal occurs within the set time base. The AUTO setting will display a trace regardless of the time base.

Slope and level controls are used to define the actual trigger voltage. The slope switch determines whether the trace will begin on a rising or falling of the signal **(Figure 16–46)**. The level control sets when the time base will be triggered according to a certain point on the slope.

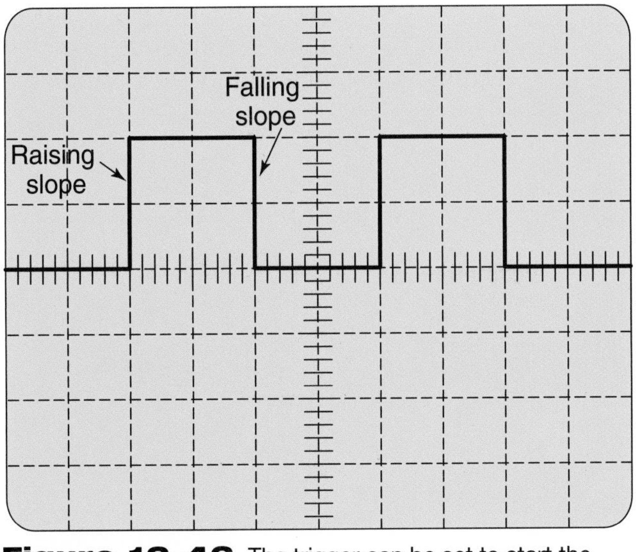

Figure 16-46 The trigger can be set to start the trace with a rise or fall of voltage.

A trigger source switch tells the scope which input signal to trigger on. This can be Channel 1, Channel 2, line voltage, or an external signal. External signal triggering is very useful when observing the trace of a component that may be affected by the operation of another component. An example of this would be observing fuel injector activity when changes in throttle position are made. The external trigger would be voltage changes at the Throttle Position Sensor. The displayed trace would be the cycling of a fuel injector.

Graphing Multimeters

One of the latest trends in diagnostic tools is a **graphing multimeter (GMM)**. A GMM is a DMM that displays voltage, resistance, current, and frequency as a waveform. The display shows the minimum and maximum readings as a graph, as well as all current measurements **(Figure 16–47)**. By observing the graph, a technician can see any undesirable changes during the transition from a low reading to a high reading, or vice versa. These glitches are some of the more difficult problems to identify without a graphing meter or a lab scope. The waveform on a DSO may miss a change in voltage, resistance, or current that occurs too quickly for the scope to detect and display it. DSOs do not display real-time data. There is always a slight delay between what it measures and what it displays, and depending on the refresh rate of the display it may miss some changes completely. A GMM is perhaps the best tool to use when trying to find the cause of intermittent problems in low-voltage DC circuits.

The capabilities of a GMM depend on the manufacturer and the model. Some of the features found on GMMs include a signal and data recorder, individual component tests, the ability to display measurements along with a graph, glitch capture, and an audible alarm. Some also have an electronic library of

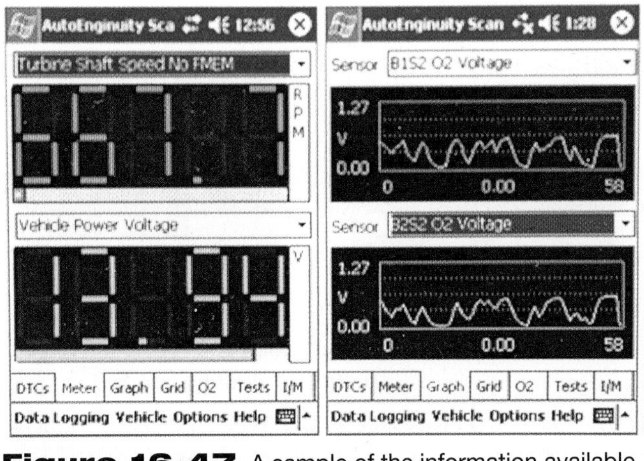

Figure 16-47 A sample of the information available from a GMM.

Figure 16-48 Many scan tools, DSOs, and GMMs allow for the transfer of captured information to a PC through a cable or wireless interface.

known good signals for comparison. These allow for a comparison of live patterns with expected values or known good waveforms. Some even have wiring diagrams and a vehicle-specific database of diagnostic and test information.

Transferring Data to a PC Many DSOs and GMMs allow for the transfer of captured information to a PC through a cable **(Figure 16–48)** or wireless interface. This feature allows for better viewing of the waveforms and other data and also allows for the creation of a personal library. The latter can be helpful in the future. There are several sources of waveforms readily available for download to a PC; these can be used as references and study guides.

TESTING BASIC ELECTRICAL COMPONENTS

All electrical components can fail. Testing them is the best way of determining if they are good or bad. For the most part, the proper way for checking electrical components is determined by what the component is supposed to do. If we think about what something is supposed to do, we can figure out how to test it. Often, removing the component and testing it on a bench is the best way to check it.

Protection Devices

When overloads or shorts in a circuit cause too much current, the wiring in the circuit heats up, the insulation melts, and a fire can result unless the circuit has some kind of protective device. Fuses, fuse links, maxi-fuses, and circuit breakers are designed to provide protection from high current. They may be used singularly or in combination.

WARNING!

Fuses and other protection devices normally do not wear out. They go bad because something went wrong. Never replace a fuse or fusible link, or reset a circuit breaker, without finding out why they went bad.

Circuit protection devices can be checked with an ohmmeter or testlight. If the fuse is good, there will be continuity through it. To test a circuit protection device with a voltmeter, check for available voltage at both terminals of the unit **(Figure 16–49)**. If the device is good, voltage will be present on both sides. A testlight can be used in place of a voltmeter.

Measuring voltage drop across a fuse or other circuit protection device tells more about its condition than its continuity. If a fuse, a fuse link, or a circuit breaker is in good condition, a voltage drop of zero will be measured. If 12 volts is read, the fuse is open. Any reading between 0 and 12 volts indicates that there is unwanted resistance and the fuse should be

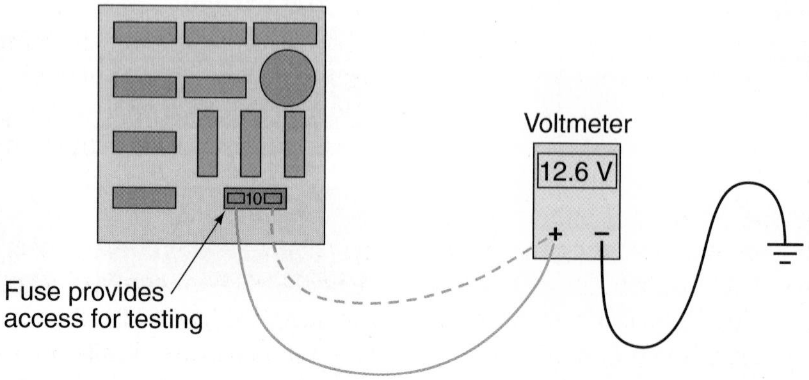

Figure 16-49 Circuit protection devices can be tested with a voltmeter. Make sure there is voltage present on both sides of the device.

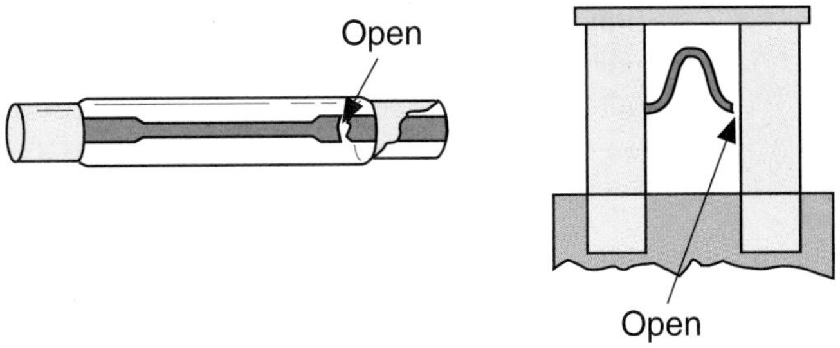

Figure 16–50 The condition of a fuse can often be checked visually.

replaced. Make sure you check the fuse holder for resistance as well.

Fuses A cartridge fuse can be visually checked by looking at the internal metal strip. Discoloration of the glass cover or glue bubbling around the metal end caps is an indication of overheating. Pull the fuse from its holder and look at the element through the transparent plastic housing. Look for internal breaks and discoloration **(Figure 16–50)**. A ceramic fuse is checked by looking for a break in the contact strip on the outside of the fuse.

Sometimes it is necessary to protect a device in a portion of a circuit even though the entire circuit is already protected by a fuse in the panel. This is done by installing an **inline fuse** in the power wire for the device. Inline fuses are primarily used on accessories that are very sensitive to power surges, such as radios and compact disk players. They are also used with driving lights and power antennas.

SHOP TALK

To calculate the correct fuse rating, use Watt's law: watts divided by volts equals amperes. For example, if you are installing a 55-watt pair of fog lights, divide 55 by the battery voltage to find how much current the circuit will draw. In this case, the current is approximately 5 amperes. To allow for current surges, the correct inline fuse should be rated slightly higher than the normal current flow. In this case, an 8- or 10-ampere fuse would do the job.

Fuse Links Fuse link wire is covered with a special insulation that bubbles when it overheats, indicating that the link has melted. If the insulation appears good, pull lightly on the wire. If the link stretches, the wire has melted. Of course, when it is hard to determine if the fuse link is burned out, check for continuity through the link with a testlight or ohmmeter.

WARNING!

Do not mistake a resistor wire for a fuse link. A resistor wire is generally longer and is clearly marked "Resistor do not cut or splice."

To replace a fuse link, cut the protected wire where it is connected to the fuse link. Then tightly crimp or solder a new fusible link of the same rating as the original link.

WARNING!

Always disconnect the battery ground cable prior to servicing any fuse link.

Maxi-Fuses Maxi-fuses are easier to inspect and replace than fuse links. To check a maxi-fuse, look at the fuse element through the transparent plastic housing. If there is a break in the element, it has blown. To replace it, pull it from its fuse box or panel. Always replace a blown maxi-fuse with one that has the same ampere rating.

Circuit Breakers Two types of circuit breakers are used. One is reset by removing the power from the circuit. The other type is reset by depressing a reset button. If a circuit breaker cannot be reset and remains open, replace it after making sure that there is not excessive current in the circuit.

Thermistors Some systems use a positive temperature coefficient (PTC) thermistor as a protection device. When there is high current, the resistance of the thermistor increases and causes a decrease in current flow. These can be checked with an ohmmeter **(Figure 16–51)**. If an infinite reading is displayed, the thermistor is open. Another way of checking a thermistor is to change its temperature and see if its resistance changes. A negative temperature coefficient thermistor works in the opposite way; that is, its resistance decreases as the temperature increases.

Switches

To check a switch, disconnect the connector at the switch. Check for continuity between the terminals of the switch **(Figure 16–52)** with the switch in the on and off positions. While in the off position, there should be no continuity between the terminals. With the switch on, there should be good continuity between

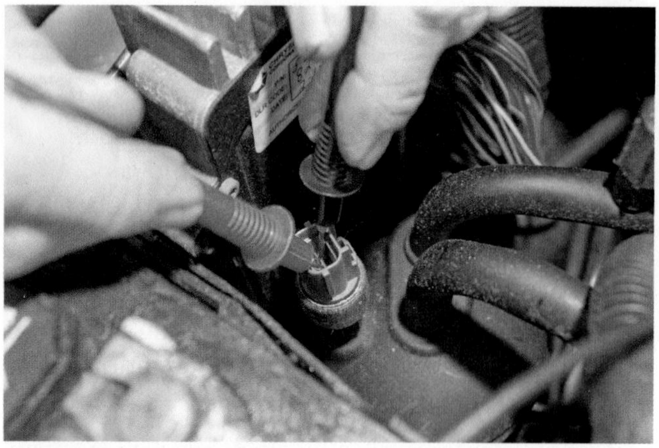

Figure 16–51 A temperature sensor (thermistor) can be checked with an ohmmeter.

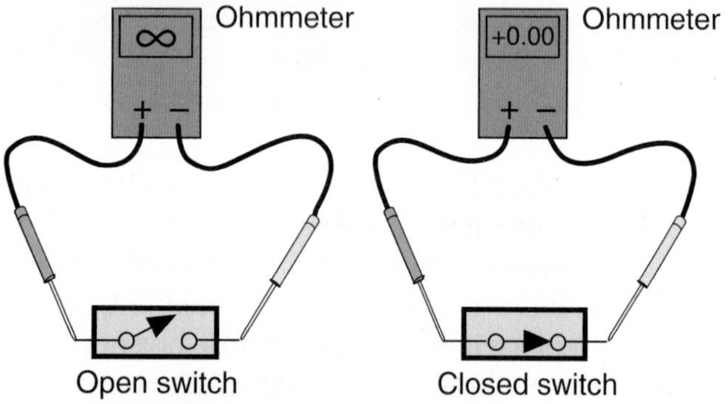

Figure 16–52 Checking a switch with an ohmmeter.

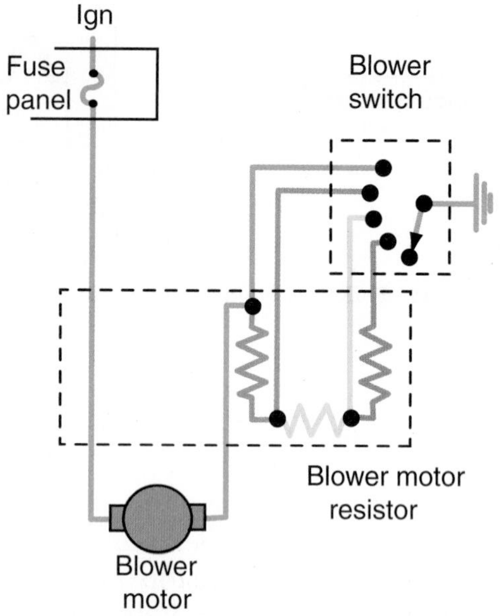

Figure 16–53 An MPMT switch should be checked in all of its possible positions. Use a wiring diagram to guide your tests.

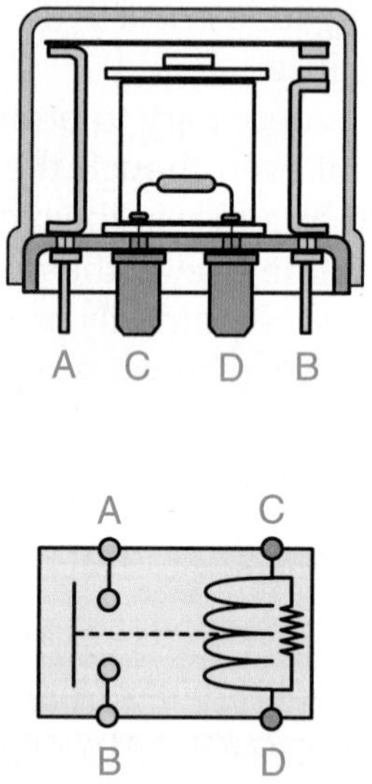

Figure 16–54 Use a wiring diagram to identify the terminals of a relay so it can be tested properly.

the terminals. If the switch is activated by something mechanical and does not complete the circuit when it should, check the adjustment of the switch (some switches are not adjustable). If the adjustment is correct, replace the switch. Another way to check a switch is to bypass it with a jumper wire. If the component works when the switch is jumped, the switch is bad. MPMT switches should be checked in each of its possible positions **(Figure 16–53)**. Use a wiring diagram to identify which terminals of the switch should have continuity during each switch position.

Voltage drop across switches should also be checked. Ideally when the switch is closed there should be no voltage drop. Excessive voltage drop indicates resistance, and the switch should be replaced.

Relays

A relay can be checked with a jumper wire, a voltmeter, an ohmmeter, or a testlight. If the terminals are accessible and the relay is *not* controlled by a computer, a jumper wire and testlight will be the quickest method to use. The schematic for a relay is typically shown on the outside of the relay. If not, check a wiring diagram to identify the terminals of the relay **(Figure 16–54)**. Also check the wiring diagram to determine if the relay is controlled by a power or ground switch.

PROCEDURE

If the relay is controlled on the ground side, follow this procedure to test the relay.

1. Use a testlight to check for voltage at the power feed to the relay. If voltage is not present, correct the fault in the battery feed to the relay. If there is voltage, continue testing.
2. Probe for voltage at the coil terminal. If voltage is not present, correct the fault in the battery feed to the relay. If voltage is present, continue testing.

3. Use a fused jumper wire to connect the control terminal to a good ground. If the relay works, the fault is in the control circuit. If the relay does not work, continue testing.
4. Connect the jumper wire from the battery to the output terminal of the relay. If the device operated by the relay works, the relay is bad. If the device does not work, the circuit between the relay and the device or the device's ground is faulty.

PROCEDURE

If the relay is controlled by a computer, do not use a testlight. Rather use a high-impedance voltmeter set to the 20 V DC scale, then:

1. Connect the negative lead of the voltmeter to a good ground.
2. Connect the positive lead to the output wire. If no voltage is present, continue testing. If there is voltage, disconnect the relay's ground circuit. The voltmeter should now read 0 volts. If it does, the relay is good. If voltage is still present, the relay is faulty.
3. Connect the positive voltmeter lead to the power input terminal. If near battery voltage is not measured there, the relay is faulty. If it is, continue testing.
4. Connect the positive lead to the control terminal. If near battery voltage is not measured there, check the circuit from the battery to the relay. If it is, continue testing.
5. Connect the positive lead to the relay ground terminal. If more than 1 volt is present, the circuit has a poor ground.

If the relay terminals are not accessible, remove the relay from its mounting. Use an ohmmeter to test for continuity between the relay coil terminals. If the meter indicates an infinite reading, replace the relay. If there is continuity, use a pair of jumper wires to energize the coil. Check for continuity through the relay contacts. If there is an infinite reading, the relay is faulty. If there is continuity, the relay is good and the circuits need to be tested.

Check the service manual for resistance specifications for the relay's coil and compare your readings to them. Low resistance indicates that the coil is shorted. If the coil is shorted, the transistors and/or driver circuits in the computer could be damaged due to excessive current flow.

Stepped Resistors

The best way to test a stepped resistor is with an ohmmeter. Remove the resistor and connect the ohmmeter leads to the two ends of the resistor. Compare the readings against specifications. Make sure the ohmmeter is set to the correct scale for the anticipated amount of resistance.

A stepped resistor can also be checked with a voltmeter. Measure the voltage after each part of the resistor block and compare the readings to specifications.

Variable Resistors

A common way to test a variable resistor is with an ohmmeter; however, it can also be checked by observing the output voltage. To test a rheostat, identify the input and output terminals and connect an ohmmeter across them. Rotate the control while observing the meter. If the resistance values do not match specifications, or if there is a sudden change in resistance as the control is moved, the unit is faulty. When using a voltmeter, the readings should be smooth and consistent as the control is moved.

To test a potentiometer, connect an ohmmeter across the resistor. The readings should be within the range listed in the specifications. Then move the leads to the input and output of the resistor. Manually change the resistance. The readings should sweep evenly and consistently within the specified resistance values. The condition of a potentiometer can also be checked with a voltmeter or a lab scope. Compare the voltage changes to the specifications.

Wiring

Wire insulation should be in good condition. Broken, frayed, or damaged insulation that exposes live wires can cause shorts (**Figure 16–55**). These conditions

Figure 16-55 Broken, frayed, or damaged insulation that exposes live wires can cause shorts.

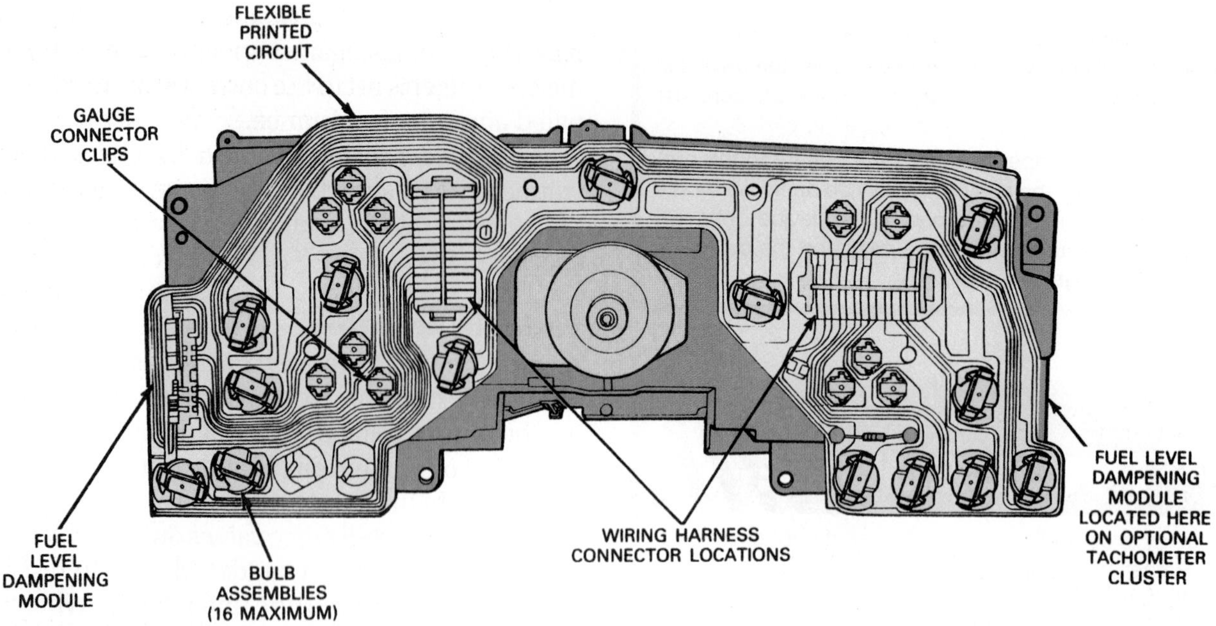

Figure 16-56 A typical printed circuit board.

can also create a safety hazard. Replace all wires that have damaged insulation.

When checking a circuit, make sure to check the ground connections, including the ground strap from the engine or other component to the chassis. An engine ground is typically a braided, flat cable. A bad ground cable can cause problems in many different circuits.

The best way to check a wire is to check the voltage drop across it. If the wire is in good shape, there should be very little or no voltage drop.

Wires are commonly grouped into a harness. A single-plug harness connector may form the connecting point for many circuits. Harnesses and harness connectors help organize the vehicle's electrical system and provide a convenient starting point for testing many circuits. Most major wiring harness connectors are located in the vehicle's dash or fire wall area.

Printed Circuits

Late-model vehicles use flexible printed circuits boards (**Figure 16–56**). Printed circuit boards are not serviceable and, in some cases, not visible. When these boards fail, the entire unit is replaced.

The following precautions should be observed when working with a printed circuit:

- Never touch the surface of the board. Dirt, salts, and acids on your fingers can etch the surface and set up a resistive condition.
- The copper conductors can be cleaned with a commercial cleaner or by lightly rubbing a pencil eraser across the surface.
- A printed circuit board is easily damaged because it is very thin. Be careful not to tear the surface, especially when plugging in connectors or bulbs.

TROUBLESHOOTING CIRCUITS

When troubleshooting an electrical problem, in any system, it is very important that a logical approach is taken. Making assumptions or jumping to conclusions can be very expensive and a total waste of time. The basic steps for diagnosis given in Chapter 2 should be followed. Here they are again, only modified to fit electrical problems:

PROCEDURE

To troubleshoot an electrical problem, follow these steps:

1. *Gather information about the problem.* From the customer find out when and where the problem happens and what exactly happens.
2. *Verify that the problem exists.* Take the vehicle for a road test or try the components of the customer's concerns and try to duplicate the problem, if possible.
3. *Thoroughly define what the problem is and when it occurs.* Pay attention to the conditions present when the problem happens. Also pay attention to the entire vehicle; another problem may be evident to you that was not evident to the customer. The most important thing is to fully understand what the problem is.
4. *Research all available information to determine the possible causes of the problem.* Look at all service bulletins and other information related to the problem

to see if this is a common concern. Study the wiring diagram of the system and match a system or some components to the problem.

5. *Isolate the problem.* Based on an understanding of the problem and circuit, make a list of probable causes. Narrow down this list of possible causes by checking the obvious or easy to check items. This includes a thorough visual inspection.
6. *Continue testing to pinpoint the cause of the problem.* Once you know where the problem should be, test until you find it! Begin testing to determine whether or not the most probable cause is the problem. If this is not the cause, move to the next most probable cause. Continue this until the problem is solved.
7. *Locate and repair the problem, then verify the repair.* Once you have determined the cause, make the necessary repairs. Never assume that your work solved the original problem. Operate all features of the circuit to be sure that the original problem has been corrected and that there are no other faults in the circuit.

Troubleshooting Logic

Remember there are three basic types of electrical problems. Knowing the type of problem that is causing the customer's concerns will dictate what tests should be conducted. If something does not work, the problem is most likely caused by a short or an open. If the fuse for that circuit is blown, the problem is a short. If the fuse is good, the problem is an open. If a part does not work correctly, such as a dim light bulb, the problem is high resistance.

Quick voltage checks will also help define the problem. Check for voltage at the part that is not working correctly. If source voltage is present at the part, the part is bad or the ground circuit is faulty. If less than source voltage is measured at the part, there is a fault in the power feed to the part. Also, measure the voltage drop across the part; this can indicate a problem with the part. If a faulty part is suspected, it should be checked or replaced.

When making any checks with a meter, follow all safety precautions. Try to take all measurements at a connector. Because the terminals at the connector can be damaged by inserting a meter's test leads into the connector, always use the correct adapter on the ends of the test leads. Adapters are available to match the size of the terminals. Using too large of an adapter can deform the terminals. When measurements are taken at the mating side (front) of a disconnected connector, this is called **front probing**. When measurements are taken at the back or wire side of a connected connector, this is called **back probing**. Front probing is the preferred way to take measurements. At times it may be necessary to make direct contact with a wire by piercing through the insulation. Make sure not to damage the wire and cover the pierced area with electrical tape or clear fingernail polish. This will prevent the copper wire from corroding.

The key to identifying the exact cause of the problem is limiting all testing to the components and circuits that could be causing the problem. An understanding of the problem, coupled with an understanding of the circuit, will lead to the fault. A wiring diagram will serve as the map to the problem. Your understanding and knowledge will tell you where you want to go and the wiring diagram will tell you how to get there.

Using Wiring Diagrams

During diagnosis, one of the most important sources of information is a wiring diagram. A wiring diagram shows the relationships of one circuit to the others. Based on an understanding of the diagram, electricity, and how a particular system is designed to work, testing points can be identified. Wiring diagrams are included in the electrical section of most service manuals.

Wiring diagrams contain much information about an electrical circuit **(Figure 16–57)**. They contain a comprehensive look at each circuit with all the connectors, wiring, signal connections (buses) between the devices, and electrical or electronic components of the circuit. The diagrams are drawn with lines that represent wires between the appropriate connectors. The distance between the connectors is not given. Most diagrams are drawn so the front of the car is on the left of the diagram and show the power source on the top of the wiring diagram and the ground source at the bottom.

Wiring diagrams typically show the following:

- *Wires by wire numbers or color coding.* All wires are identified by circuit number, color, and/or, in some cases, size. The wiring diagram also shows the color changes at a connector or splice. Often wire color changes at a connector, these changes are noted in the diagram.
- *Wire cross-section size.* Some manufacturers indicate the wire size along with the color code.
- *Ground connections.* Most diagrams show the point at which circuits are grounded. These are typically identified with the letter "G." Often there are common grounding points.
- *Wire connection points.* Individual connectors "C" are shown and listed by number. Note: The symbol for a connector varies with the manufacturer.
- *Reference of wire continuation.* At times, the part of a wiring diagram where a particular wire

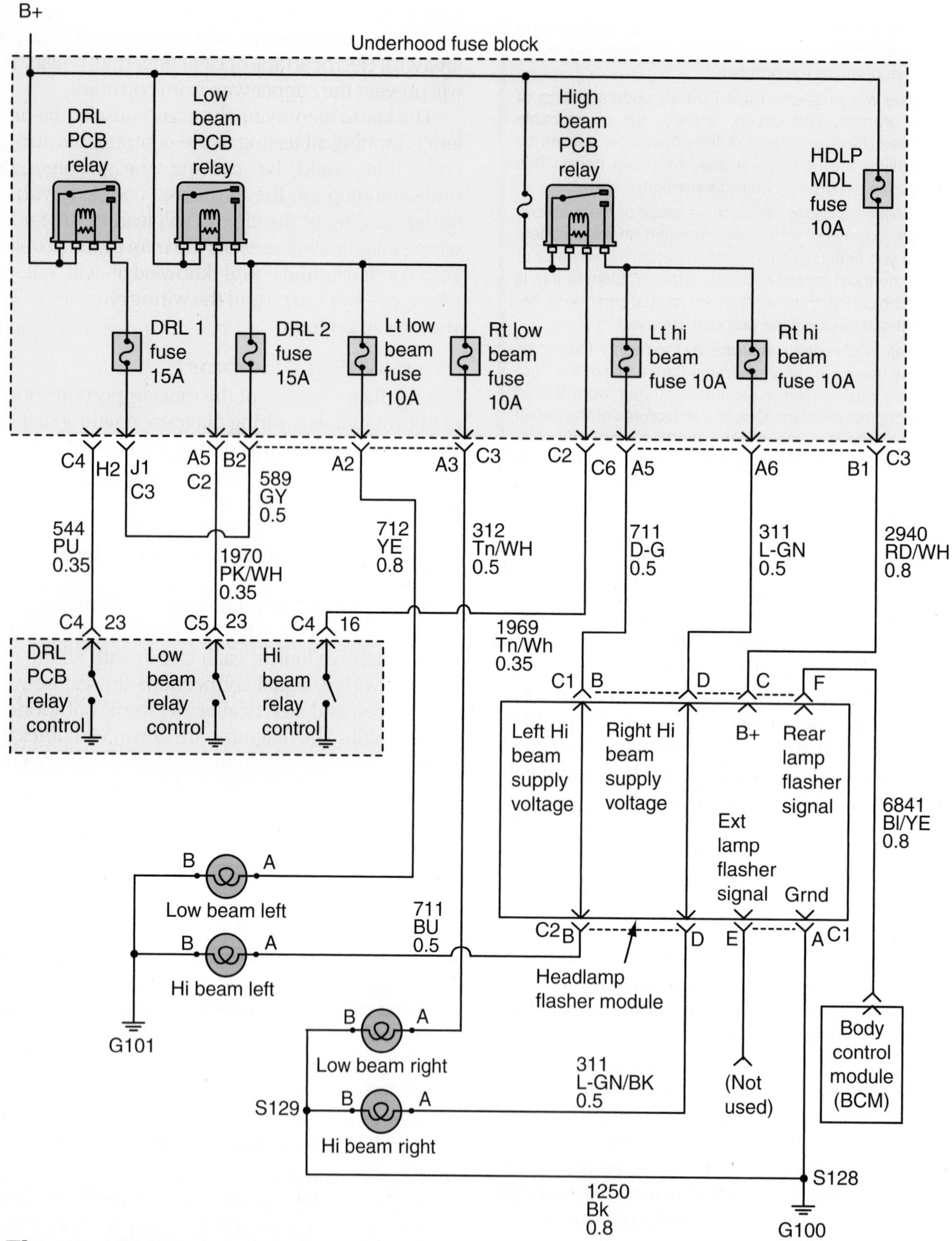

Figure 16-57 A wiring diagram of a headlight system.

continues is referenced to another diagram or a section area.

- *Location of splices.* Where a group of wires are electrically joined together is typically designated as "S."
- *Terminal designation.* A number or other label that shows where a particular wire is located in a multipin connector.
- *Component symbols.* A wiring diagram uses a set of symbols to represent electrical components or devices. These are quite standard but may vary with the manufacturer. In most cases, the components are drawn to show the way they are connected electrically.
- *Switches.* The placement of switches in the circuit are shown in their normal (NO or NC) position.
- *Fuse designation.* Fuses and other circuit protection devices are shown by their location and rating.
- *Relay information.* The location of all relays is shown as well as the connections to its terminals.
- *Continuation of circuit.* This may be noted by an arrow that notes where a wire or circuit is continued. The reference is normally to a seemingly unrelated circuit and is noted by an arrow.

SHOP TALK

To make it easier to identify where grounds, splices, and connectors are located on the vehicle, manufacturers use a numbering system that identifies the basic location of these. For example, many manufacturers, such as GM, have a three-digit number after the G, S, and C that represents their location. Typically, 100 through 199 means that the ground, splice, or connector is under the hood, 200–299 means they are under the dash, 300–399 means they are in the passenger compartment, and 400–499 means they are in the trunk.

Getting the Right Diagram Wiring diagrams are only valuable if they are the correct ones for the vehicle and show the circuit and/or components you wish to test. The wiring diagram should be for the exact year, make, and model of the vehicle. Most electronic service information systems will match the wiring diagram to the VIN. To retrieve a wiring diagram that will help in diagnosis of a problem, match the component to the index for the wiring diagram. The index will list a letter and number for each major component and many different connection points. Refer to those references. Most electronic information systems will automatically display the appropriate diagram once the component is selected.

For some vehicles, only total vehicle wiring diagrams are available. These can make it more difficult to locate the circuit you need. These diagrams also have an index that will identify the grid where a component can be found. The diagram is marked into equal sections by grid letters and numbers on the outside borders like a street map. If the wiring diagram is not indexed, you can locate the component by relating its general location in the vehicle to a general location on the wiring diagram.

Tracing a Circuit After you have the correct wiring diagram, identify all of the components, connectors, and wires that are directly related to that component. This is done by tracing through the circuit, starting at the component. Tracing through the circuit should identify the source of power for the component and its ground circuit. It also will identify all related controls. Tracing the circuit also allows you to understand the operation of the circuit. This is important because it will help determine where to test the circuit and what to expect at those test points. Remember, all electrical circuits have a power source, a load, and a path to ground. Make sure these are identified.

Tracing the circuit will also simplify complex wiring diagrams. Complex wiring diagrams are made up of many individual circuits. Some are directly interrelated and others are not. When the circuit of concern is pulled out of the wiring diagram, it is much easier to identify its wires and components. It also reduces the chances of being distracted by wires that probably are not the cause of the problem.

After you have traced the circuit, study it and make sure you know how the circuit is supposed to work. Then describe the problem and ask yourself what could cause this. Limit your answers to what is included in your traced wiring diagram. Also limit your answers to the description of the problem. It is wise to make a list of all probable causes of the problem then number them according to probability. For example, if all of the lights in the instrument panel do not work, it is most probable that the cause is not all bad bulbs. Rather it is more likely that the fuse or power feed is bad. After you have listed the probable causes, in order of probability, then look at the wiring diagram to identify how you can quickly test to find out if each is the cause.

Tracing the circuit can be done is one of two ways, neither of which involves marking a wiring diagram in a manual. Tracing can be done by drawing out the circuit on a piece of paper. It does not need to be pretty; it just needs to be accurate. Draw the component (can be a simple box) and draw all wires and controls that supply the power and ground; include all controls. Mark each wire with its color and make sure to note any change in color. Also label all controls and components that are included in the circuit.

Tracing can also be done with highlighters or markers. There are many ways to do this; the one here is just a suggestion. Developing your own method will work as long as you are consistent. On the wiring diagram, find the component of concern and outline it in yellow. Follow the wires toward the power source. Trace the power wire at the component to where it connects to a control or load. Color that wire red, it should have source voltage. Color the wire leading from the control to the component orange. If the control will pass a variable amount of voltage to the component, color the output wire green. Now look at the ground side of the component. If the path to ground is direct with no control, color the wire black. If the ground path has a control, color the wire to the control blue and the wire from the control to ground black. For an earlier and different approach to color-coding, see *"Mastering Complex Diagrams" (Motor Magazine, July, 1997)* and *"Wiring Diagram Color-Coding: More Than Meets the Eye" (Motor Magazine, December, 2008)*, both by Jorge Menchu.

Coloring the circuit as described allows for a simple reference as to what to expect at each wire in the circuit. The red wire(s) should always have source voltage. The orange wire will only have source voltage when the control is closed. The voltage on the green wire will vary with changes at the control. The black wire should have 0 volts at all times. However, the blue wire should have source voltage when the control is open and 0 volts when it is closed. Your testing should be based on this logic. Use the wiring diagram to identify the test points. **Figure 16–58** shows the circuit of the right low beam headlight. This is the same diagram as used in Figure 16–57, but only the parts of the circuit that could cause a problem with one headlamp are noted. Note that all wires and components that would affect more than that headlamp are not traced. Doing this certainly simplifies diagnosing the circuit.

If the power source for the component feeds more than one component or the ground is shared by others, check the operation of those components. If they operate normally, you know that the common power and ground circuits are good. Therefore, the problem must be between the common points. Likewise if the other components do not work correctly, you know that the problem is within the common part of the circuit.

TESTING FOR COMMON PROBLEMS

It would take thousands of pages to describe all of the possible combinations of electrical problems. But you are in luck. *All* problems can be boiled down to one of three problems. Identifying which one you are looking for will define what tests you need to conduct.

Testing for Opens

An open is evident by an inoperative component or circuit. Study the wiring diagram for the component. Begin your testing at the most accessible place in the circuit and work from there. Check for voltage at the positive side of the load. If there are 0 volts, move to the output of the control **(Figure 16–59)**. If there are at least 10.5 volts, the open is between the control and the load. If the reading is 10.5 volts or higher, check the ground side of the load. If the voltage there is 1 volt or lower and the load does not work, the load is bad. If the voltage at ground is greater than 1 volt, there is excessive resistance or an open in the ground circuit. If the voltage at the positive side of the load is less than 10.5 volts but above 0 volts, move the positive lead of the voltmeter toward the battery, testing all connections along the way. If 10.5 volts or more are present at any connector, there is an open or high resistance between that point and the point previously checked. If battery voltage is present at the ground of the load, there is an open in the ground circuit. Use a jumper wire to verify the location of the open.

Testing for Shorts

Use an ohmmeter to check for an internal short in a component. If the component is good, the meter will read the specified resistance or at least some resistance. If it is shorted, it will read lower than normal or zero resistance. Also, if the component has more than two terminals or pins, check for continuity across all combinations of these. Refer to the wiring diagram to see where there should be continuity. Any abnormal readings indicate an internal short.

When a fuse is blown, this probably is due to a wire-to-wire short or a short to ground. To test these circuits, a special jumper wire with a circuit breaker should be used as a substitute for the blown fuse. The jumper wire is fit with a 10- or 20-amp self-resetting circuit breaker. This will allow for testing the circuit

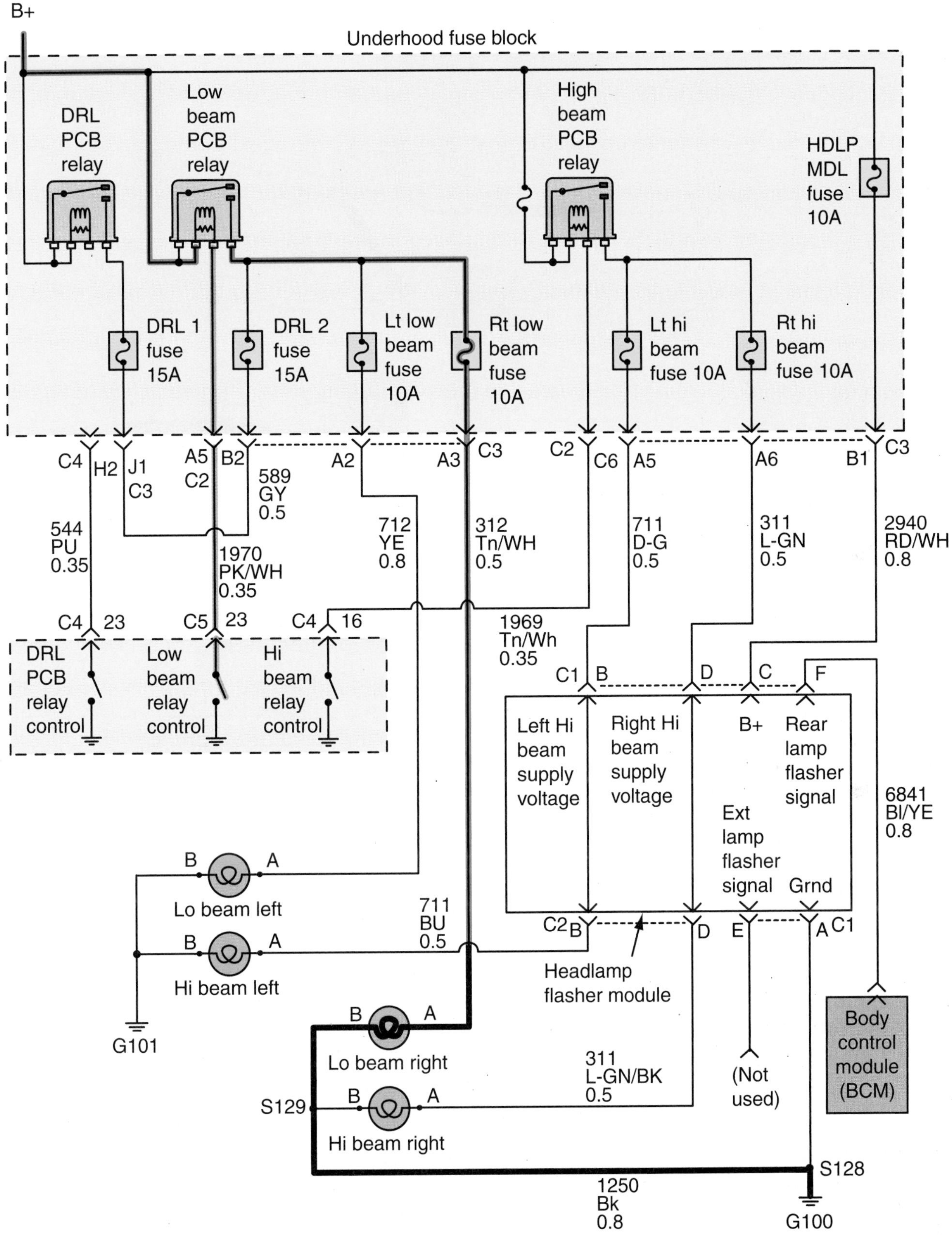

Figure 16-58 Figure 16–57 with the right low beam circuit traced.

without causing damage to the wires and components in the circuit. Some tool companies include a buzzer inline with the breaker.

When a wire-to-wire short is suspected, check the wiring diagram for all of the affected components. Identify all points where the affected circuits share a connector. Check the circuit protection devices for the circuits. High current due to the short will cause this. Check the wiring for signs of burned insulation and melted conductors. If a visual inspection does

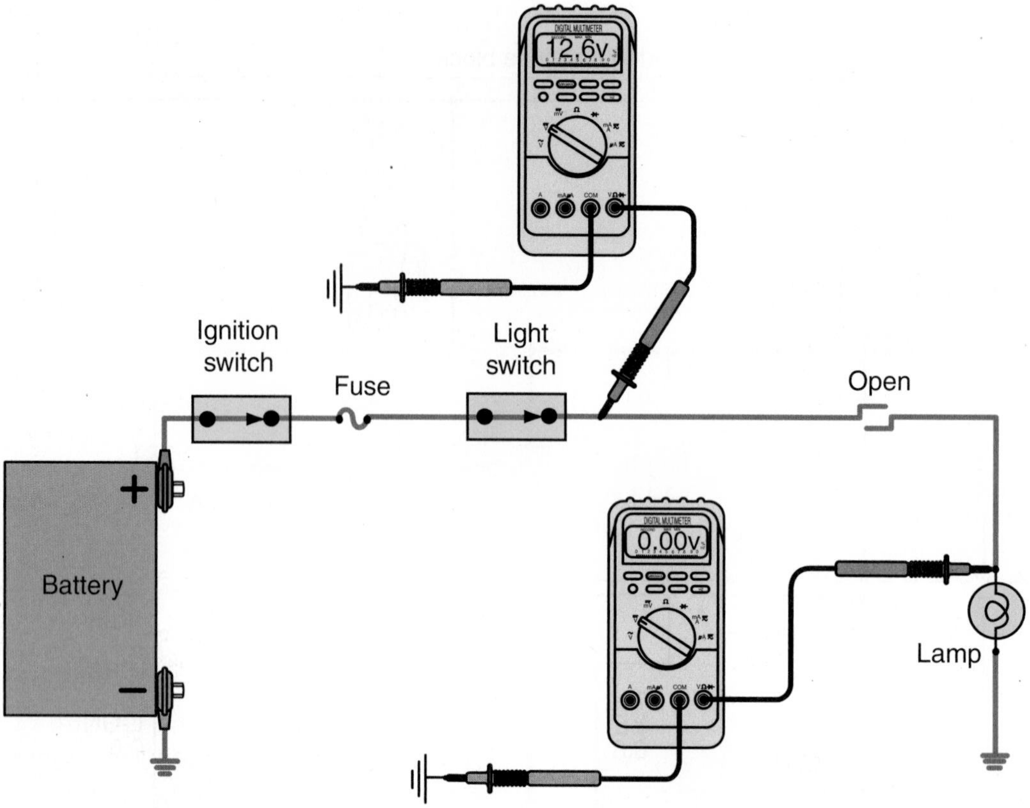

Figure 16-59 If you have 0 volts at the load, test the output of the switch. If there is power there, the open is between the switch and the load.

not identify the cause of the short, remove one of the fuses for the affected circuits. Install the special jumper wire across the fuse holder terminals. Activate that circuit and disconnect the loads that should be activated by the switch. This will create open circuits and, normally, current will not flow. If sound is coming from the buzzer in the jumper wire, current is still flowing somewhere in the circuit. Disconnect all connectors in the circuit one at a time. If the buzzer stops when a connector is disconnected, the short is in that circuit.

If the problem is a short to ground, the circuit's fuse or other protection device will be open. If the circuit is not protected, the wire, connector, or component will be burned or melted. To keep current flowing in the circuit so you can test it, connect the special jumper wire across the fuse holder. The circuit breaker will cycle open and closed, allowing you to test for voltage in the circuit. Connect a testlight in series with the cycling circuit breaker. Using the wiring diagram, identify the location of the connectors in the circuit. Starting at the ground end of the circuit, disconnect one connector at a time. Check the testlight after each connector. The short is in the circuit that was disconnected when the light went out.

An alternative to this method uses a DMM (**Figure 16–60**). Remove the bad fuse and disconnect the load. Connect the DMM across the fuse terminals. Refer to the wiring diagram to see if the ignition switch needs to be turned on to energize the circuit. If the meter reads a voltage, there must be a short before the load. Starting with the circuit's

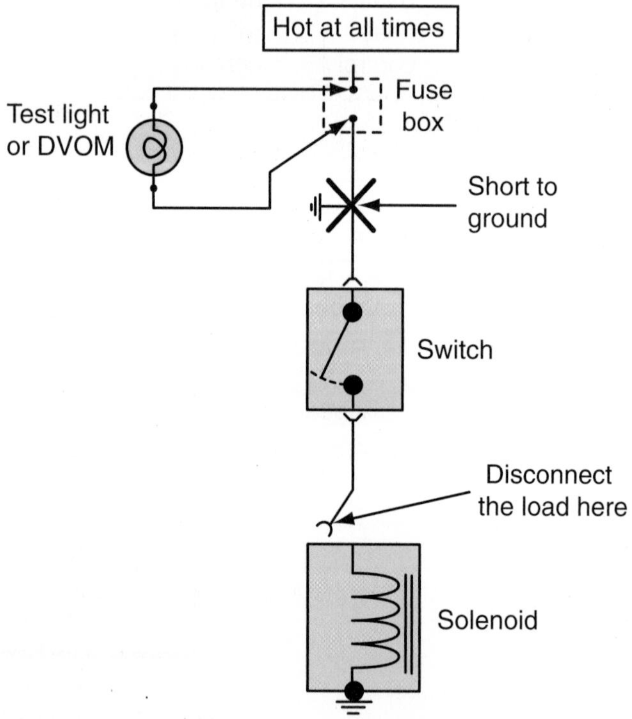

Figure 16-60 Through the process of elimination, the source of a short can be found.

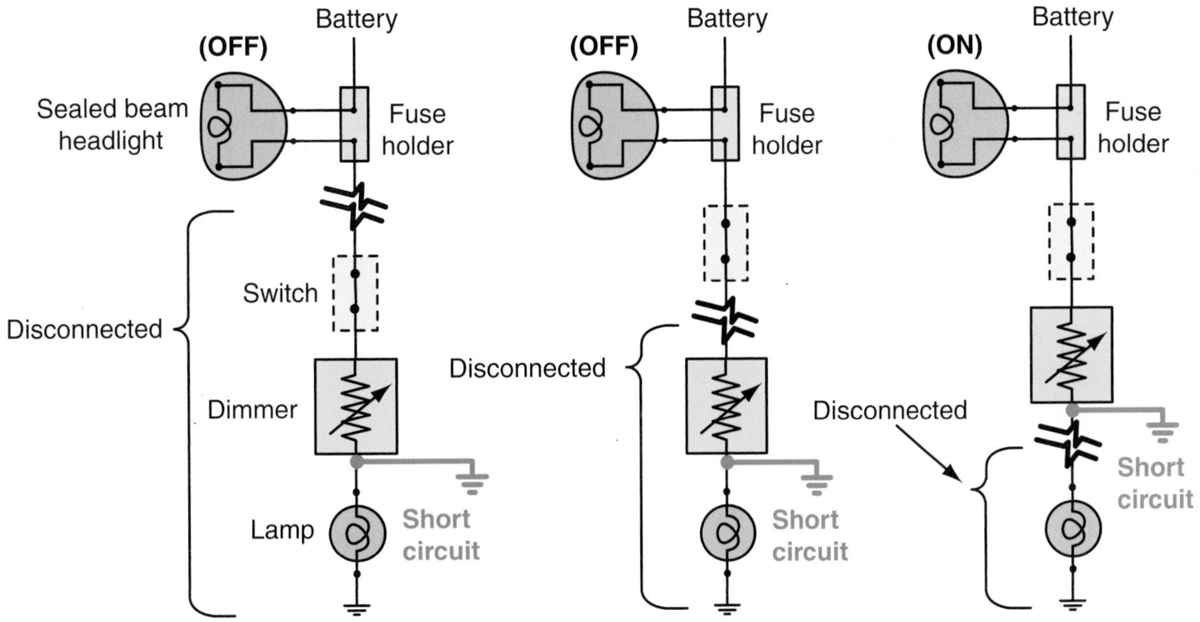

Figure 16-61 Many manufacturers do not recommend using a circuit breaker as a substitute for the fuse during testing; rather a sealed beam headlight is connected with jumper wires across the fuse holder.

wiring closest to the fuse, wiggle the wiring harness. Do this in short steps all along the wire. The point at which the voltage drops to zero is close to where the short is.

SHOP TALK

Many manufacturers do not recommend using a circuit breaker as a substitute for the fuse during testing; rather a sealed beam headlight (Figure 16–61) is connected with jumper wires across the fuse holder. The headlight serves a load and limits the current in the circuit. The headlight will light as long as current is flowing through the circuit.

Short Detector Some technicians use a compass or Gauss gauge to find the location of a short (**Figure 16–62**). A magnetic field is formed around a current-carrying conductor and a compass reacts to magnetic fields. The shorted circuit will have high current; therefore, a large magnetic field will be formed around the shorted circuit. With the wiring diagram and other service information, locate the routing of the wires in the affected circuit. Connect the jumper wire with a circuit breaker across the fuse holder for the blown fuse. Position the compass over or close to the wiring harness. The magnetic field in the wire will cause the compass' needle to move away from its north position. As the circuit breaker cycles, the needle will fluctuate. As the compass is slowly moved across the wire, it will continue to fluctuate until it passes the point where the short is. To find the exact location of the short, inspect the wire in that area. Look for signs of overheating and broken, cracked, exposed, or punctured wires.

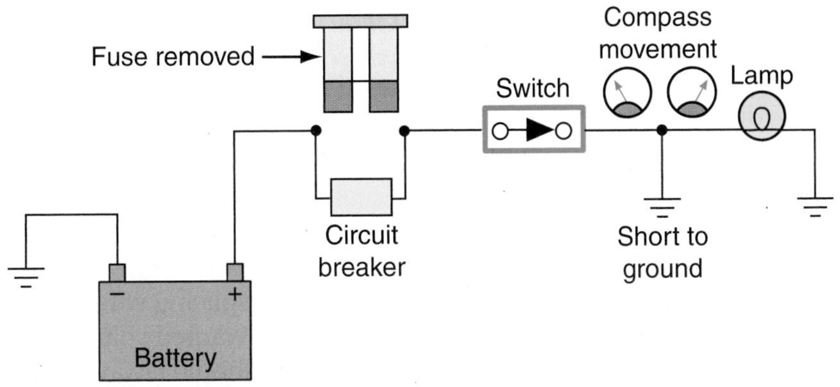

Figure 16-62 Use a compass to locate the cause of a short.

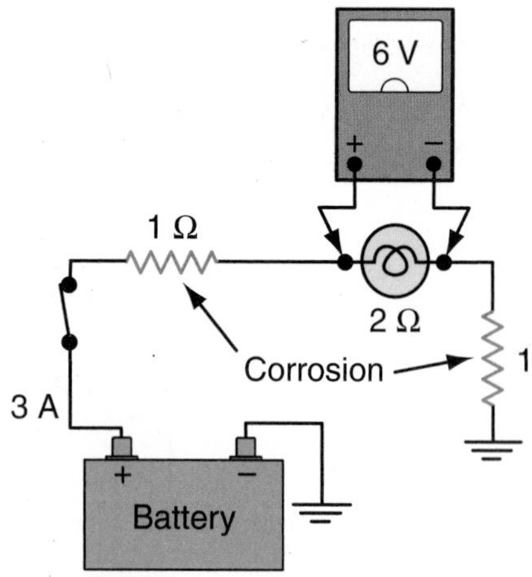

Figure 16-63 A simple light circuit with unwanted resistance. Notice the reduced voltage drop across the lamp and the reduced circuit current.

Testing for Unwanted Resistance

High-resistance problems are typically caused by corrosion on terminal ends, loose or poor connections, or frayed and damaged wires. Carefully inspect the affected circuit.

Whenever excessive resistance is suspected, both sides of the circuit should be checked. Begin by checking the voltage drop across the load **(Figure 16–63)**. This should be close to battery voltage unless the circuit contains a resistor located before the load. If the voltage is less than desired, check the voltage drop across the circuit from the switch to the load. If the voltage drop is excessive, that part of the circuit contains the unwanted resistance. If the voltage drop was normal, the high resistance is in the switch or in the circuit feeding the switch.

Check the voltage drop across the switch. If the voltage drop is excessive, the problem is the switch. If the voltage drop is normal, the high resistance is in the circuit feeding the switch. If battery voltage is present at the load, the ground circuit for the load should be checked. Connect the red voltmeter lead to the ground side of the load and the black lead to the grounding point for the circuit. If the voltage drop is normal, the problem is the grounding point. If the voltage drop is excessive, move the black meter lead toward the red. Check voltage drop at each step. Eventually you will read a high-voltage drop at one connector and then a low-voltage drop at the next. The point of high resistance is between those two test points. If the voltage drop is normal, the high resistance is in the switch or in the circuit feeding the switch.

CONNECTOR AND WIRE REPAIRS

Many electrical problems can be traced to faulty wiring or connections. Loose or corroded terminals; frayed, broken, or oil-soaked wires; and faulty insulation are the most common causes. Wires, fuses, and connections should be checked carefully. Keep in mind that a wire's insulation does not always appear to be damaged when the wire inside is broken. Also, a terminal may be tight but still may be corroded.

Check all connectors for corrosion, dirt, and looseness. Nearly all connectors have pushdown release-type locks **(Figure 16–64)**. Make sure these are not damaged. Many connectors have covers over them to protect them from dirt and moisture. Make sure these are properly installed to provide for that protection.

Never reroute wires when making repairs. Rerouting wires can result in induced voltages in nearby components. **Induced voltages** produce unwanted signals through magnetism rather than from the components within the circuit. These stray voltages can interfere with the function of electronic circuits.

WARNING!

When working with connectors, never pull on the wires to separate the connectors. This can create an intermittent contact and an intermittent problem that can be very difficult to find later. Always use the special tools designed for separating connectors to prevent this problem.

SHOP TALK

Apply dielectric grease to all connections before you assemble them. This will prevent future corrosion problems. Some manufacturers suggest using petroleum jelly at the connectors.

Replacement Wire Selection

Often electrical problems require the replacement of a wire or two. It is important that this is done in a way that corrects the original problem but does not create a new problem. All replacement wires should be of the same size or larger than the original. If adding an accessory, the new wire should be large enough to ensure safe and reliable performance. However, overly large wires add weight and expense, and add to the difficulty of splicing wires together. If the wire is too small, an unwanted voltage drop can occur. The two factors that should always be considered when determining the correct size of a wire are the total circuit

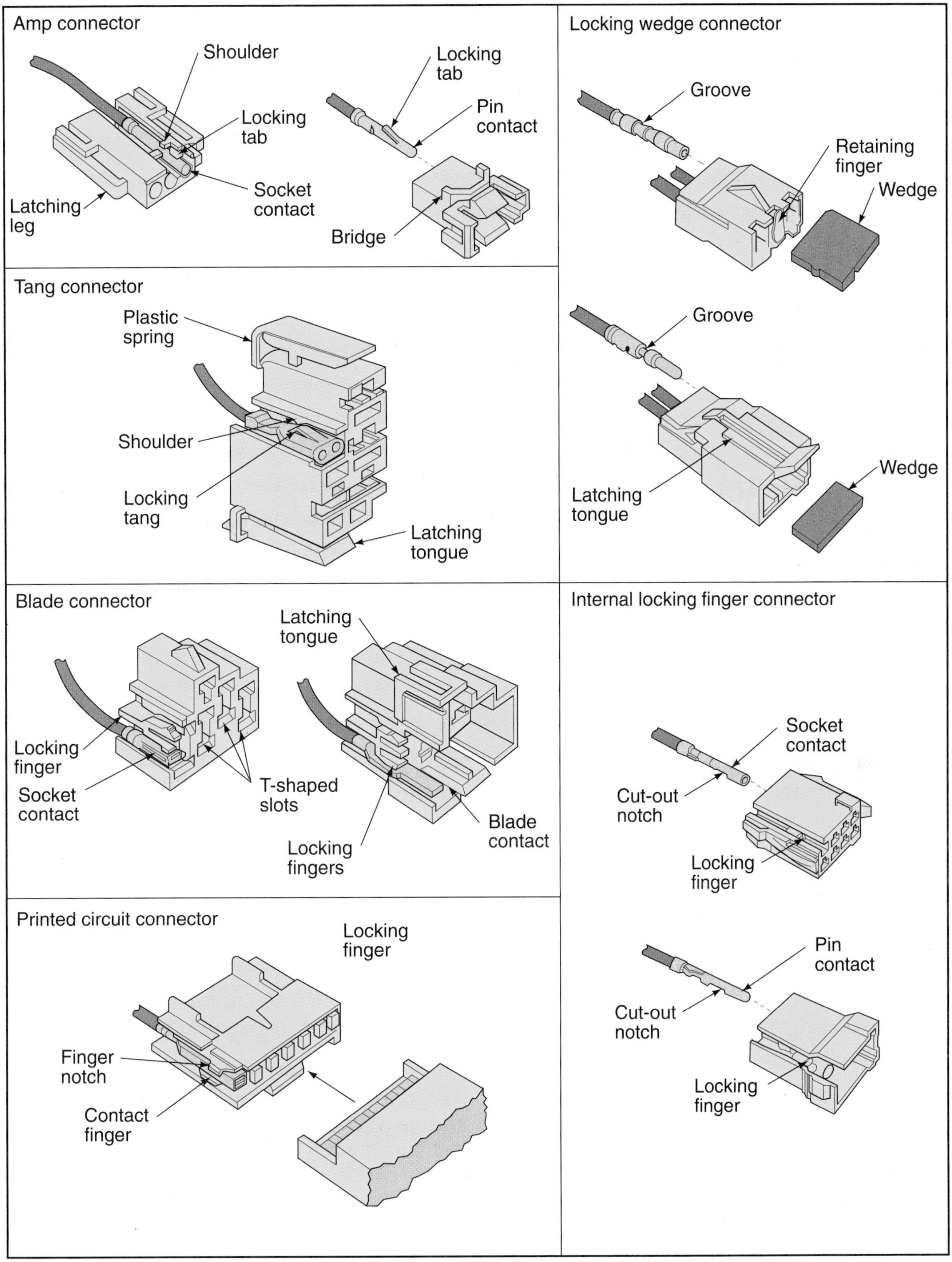

Figure 16-64 Different multiple-wire shell connectors and their locking mechanisms.

TABLE 16-2 AMPERAGE CAPACITY ACCORDING TO WIRE SIZE AND LENGTH

Approx. circuit current in amps at 12 V:	Required wire gauge per length in feet								
	3	5	7	10	15	20	25	30	40
1	18	18	18	18	18	18	18	18	18
2	18	18	18	18	18	18	18	18	18
4	18	18	18	18	18	18	18	16	16
6	18	18	18	18	18	18	16	16	16
8	18	18	18	18	16	16	16	16	16
10	18	18	18	18	16	16	16	14	12
15	18	18	18	18	14	14	12	12	12
20	18	18	16	16	14	12	10	10	10
30	18	16	16	14	10	10	10	10	10
40	18	16	14	12	10	10	8	8	6
50	12	12	10	10	6	6	4	4	4
100	10	10	8	8	4	4	2	2	2
200	10	8	8	6	4	4	2	2	1

amperage and the total length of wire (resistance increases with length) used in each circuit, including the ground. Allowance for the circuits, including grounds, has been computed in **Table 16–2**.

WARNING!

Supplemental restraint system (SRS) air bag harness insulation and the related connectors are usually color coded yellow or orange. Do not connect any accessories or test equipment to SRS-related wiring.

Connecting Wires

When a section of a wire needs to be replaced, it needs to be connected to an existing wire. Cut the damaged end of the wire from the main wire. Match the dimensions of the new wire to the old one. Measure the required length of the replacement wire; make sure it is slightly longer than the section that was removed. Then connect the two wires together. After the connection is made, the joint should be wrapped with tape or covered with heat-shrink tubing.

There are several ways that the original wire can be connected to the replacement wire. Some technicians use butt connectors; these can provide a good joint between the wires. However, the preferred way to connect wires or to install a connector is by soldering. Soldering joins two pieces of metal together by melting a lead and tin alloy and allowing it to flow into the joint. A soldering iron or gun is used to heat the solder. There are different types of solder, but only rosin-type or resin-type flux core solder should be used for electrical work.

Soldering is also used to mount components in circuit boards. Although this is not a typical procedure for an automotive technician, it should be noted that doing this requires great heat control. The heat from the soldering iron or gun can destroy electronic components. Also, when making wire repairs on or near an electronic component, use a heat sink to prevent the heat from traveling into the part.

Before using a soldering iron, make sure the tip is clean and tinned. The tip is made of copper, which corrodes through use. A corroded tip cannot transfer heat as it should. Use a file to remove all residue from the tip. When finished, the tip should be smooth and flat. Turn the iron on and allow it to heat. Then dip the hot tip into some soldering rosin flux. Remove the tip from the flux and immediately apply rosin core wire solder to all surfaces. The solder should flow over the tip. The tip is now tinned.

Photo Sequence 13 shows the procedure for soldering two copper wires together. Some manufacturers use aluminum in their wiring. Aluminum cannot be soldered. Follow the manufacturer's guidelines

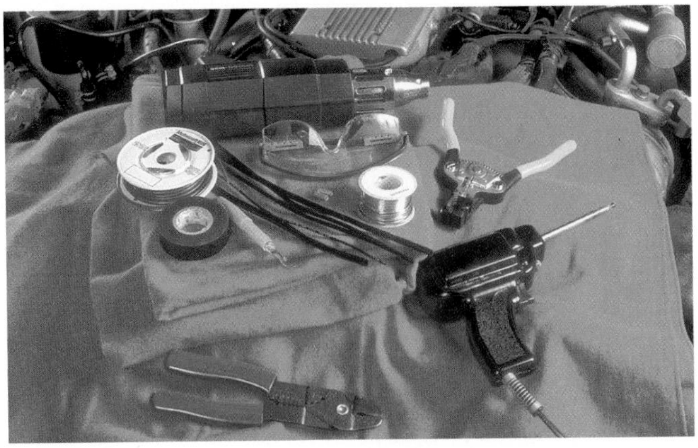

P13-1 *Tools required to solder copper wire: 100-watt soldering iron, 60/40 rosin core solder, crimping tool, splice clip, heat shrink tube, heating gun, and safety glasses.*

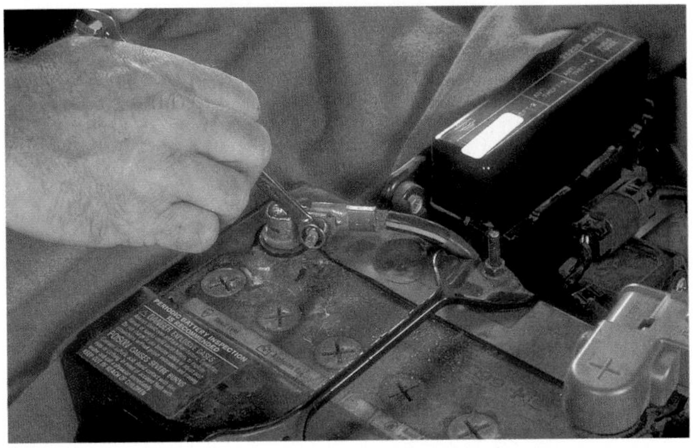

P13-2 *Disconnect the fuse that powers the circuit being repaired. Note: If the circuit is not protected by a fuse, disconnect the ground lead of the battery.*

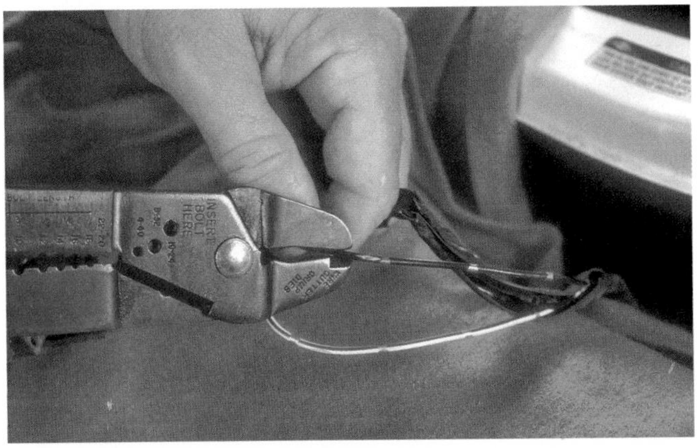

P13-3 *Cut out the damaged wire.*

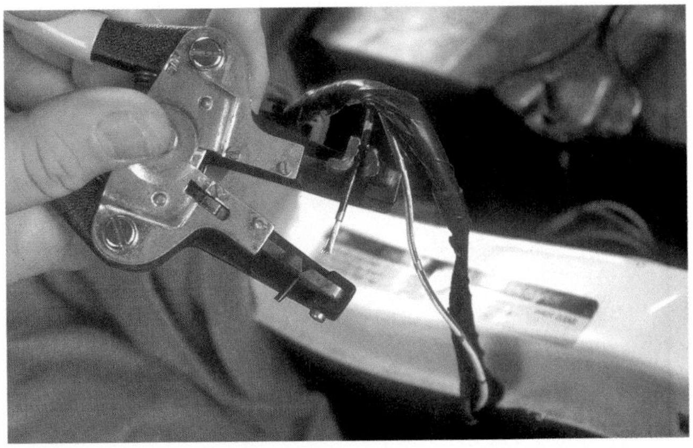

P13-4 *Using the correct size stripper, remove about ½ inch of the insulation from both wires.*

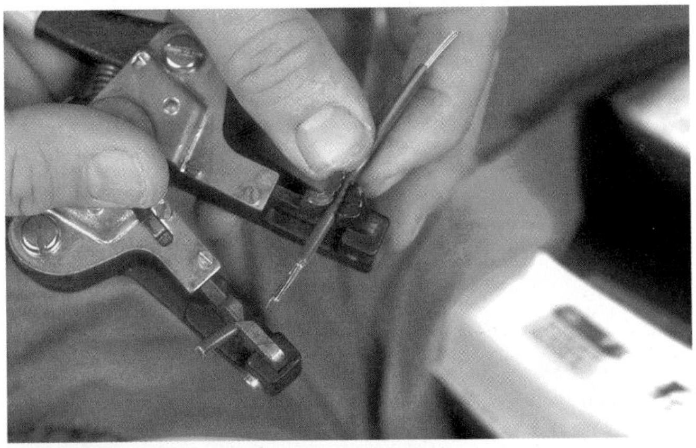

P13-5 *Now remove about ½ inch of the insulation from both ends of the replacement wire. The length of the replacement wire should be slightly longer than the length of the wire removed.*

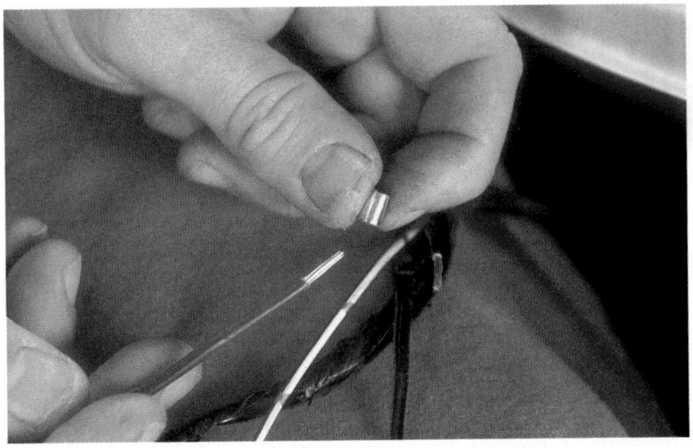

P13-6 *Select the proper size splice clip to hold the splice.*

Soldering Two Copper Wires Together *(continued)*

P13-7 *Place the correct size and length of heat shrink tube over the two ends of the wire.*

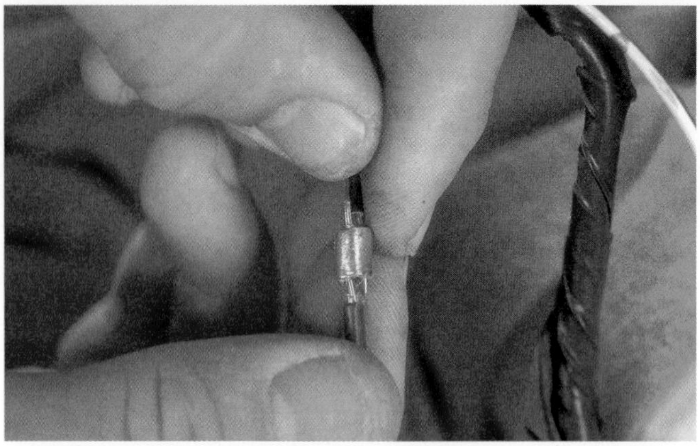

P13-8 *Overlap the two splice ends and center the splice clip around the wires, making sure that the wires extend beyond the splice clip in both directions.*

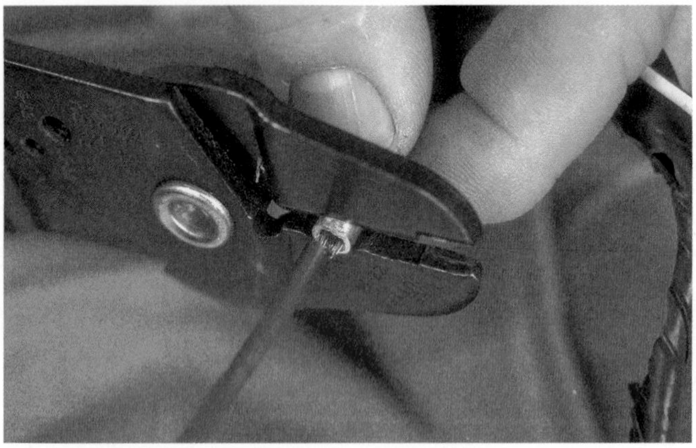

P13-9 *Crimp the splice clip firmly in place.*

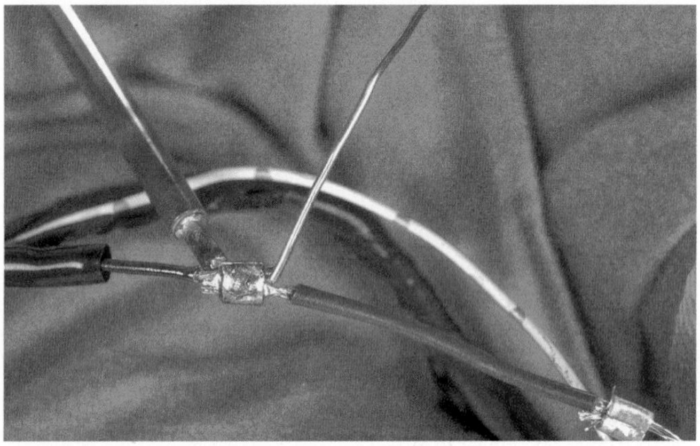

P13-10 *Apply the tip flat of the soldering iron against the splice to heat it. At the same time, apply solder to the opening of the clip. Do not apply solder to the iron. The iron should be 180 degrees away from the opening of the clip. As the splice and wires heat, the solder will flow through the splice.*

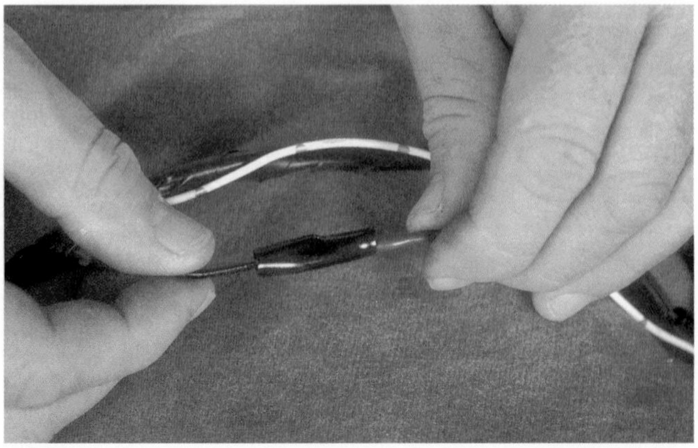

P13-11 *Place the hot soldering iron in its stand and unplug it. After the solder cools, slide the heat shrink tube over the splice.*

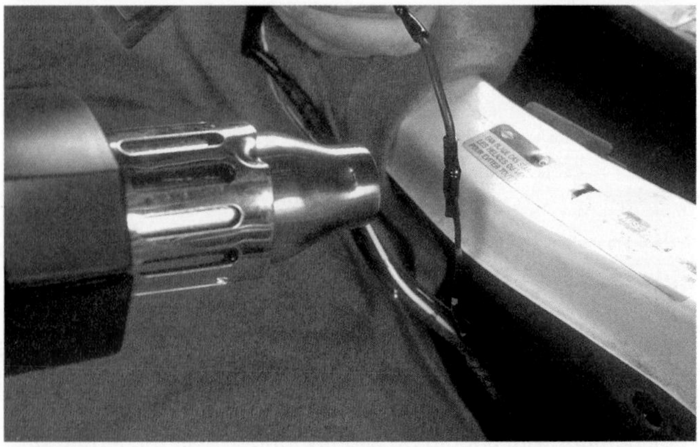

P13-12 *Heat the tube with the hot air gun until it shrinks around the splice. Do not overheat the heat shrink tube.*

and use the proper repair kits when repairing aluminum wiring.

WARNING!

Never use acid core solder. It creates corrosion and can damage electronic components.

Before splicing wires together, make sure the ends of the wires are clean. The correct size splice clip must be used. The size is based on the outside diameter of the wire, not the insulation. When inserting the wires into the joint, make sure that only the conductor enters the splice. Also make sure that the ends of the wires overlap each other inside the splice. Slip the jaws of the crimping tool over the center of the splice. Squeeze the crimping tool until the contact points of the crimper's jaws make contact with the splice. Then check the placement of the wires inside the splice and apply pressure to the crimping tool to form a tight crimp.

SHOP TALK

Rather than use a splice, some technicians twist the wire ends tightly together before soldering the joint. When doing this, it is important to realize that the solder does not provide for a mechanical joint. Therefore, it is important that the twisting provides a secure joint before soldering.

After a joint has been made, it must be insulated. This can be done with heat shrink tubing or tape. When using tape, place one end on the wire about 1 inch from the joint. Tightly wrap the tape around the wire. As the tape is being wrapped, about one-half of the previous wrap should be covered by the tape as it completes one turn around the wire. Once the wrapping has reached 1 inch beyond the joint, cut the tape. Firmly press on the tape at that end to form a good seal.

When using heat shrink tubing, make sure the tubing is slightly larger than the diameter of the splice. Cut a length of the tubing so that it is longer than the splice. Before joining the wires together, slip the tubing over one of the wires. Proceed to make the joint. After the wires are connected, move the shrink tubing over the splice. Use a heat gun and heat the tubing until it shrinks tightly around the splice. The tubing will only shrink a certain amount; therefore, do not continue to heat it after it is in place. Doing this can melt the tubing and/or the insulation of the wire.

Wire Terminals and Connectors

Many different types of connectors, terminals, and junction blocks are used on today's vehicles. In most cases, the type used in a particular application is shown in a wiring diagram. Wire end terminals are used as connecting points for wires. They are generally made of tin-plated copper and come in many shapes and sizes. They may be either soldered or crimped in place. When installing a terminal, select the appropriate size and type of terminal. Be sure it fits the unit's connecting post or prongs and it has enough current-carrying capacity for the circuit.

WARNING!

Always follow the manufacturer's wiring and terminal repair procedures. On some components and circuits, manufacturers recommend complete wiring harness replacement rather than making repairs to the wiring. For most vehicles, SRS air bag harness components, including wiring, insulation, and connectors, should not be repaired ***(Figure 16–65)****. Any SRS harness damage requires replacement of the related harness.*

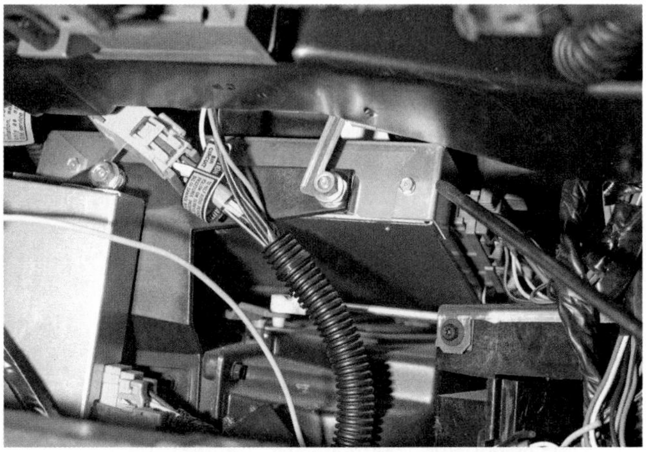

Figure 16-65 The wiring for SRS systems has yellow connectors, and all precautions and service procedures for dealing with these should be followed.

When a connector needs to be replaced because the original has melted or is otherwise damaged, attempt to replace it with the same type and size. Often this is difficult because so many different types are used and parts departments do not have all the various designs available. Normally the available connectors are based on common shapes with a common number of terminal cavities. Therefore, it is best to use a connector that meets the need; this may mean that some of the terminal cavities are left empty. For example, if the original connector has six terminals but the available replacement has eight, arrange the wire connections to the connector as if the connector had six. This will keep the wires in order for future diagnostics and leave the end pair blank. Of course, to do this, the male and female ends of the connector must be replaced. Sometimes, the replacement connector will require different terminal ends than the original. This requires the replacement of terminal ends on the wires for the male and female connectors.

Replacing a Terminal

Terminal ends are replaced when they are damaged or to accommodate the use of a connector. The replacement process must be done to provide for good continuity and to prevent electrical problems in the future.

PROCEDURE

To replace a terminal end on a wire, follow this procedure:

1. Use the service information to identify the type of terminal that should be used, the position of the locking clips, and the terminal unlocking procedures for that connector.
2. Use a small screwdriver or terminal pick **(Figure 16–66)** to unlock the secondary locking device.
3. Gently push the terminal into the connector and hold it there.
4. Insert the terminal pick into the connector and move the locking clip to the unlock position and hold it there.
5. Carefully pull the terminal from the connector by pulling the wire toward the rear of the connector. Do not use too much force.
6. Measure the diameter of the wire's insulation with a micrometer or vernier caliper.
7. Identify the type of terminal and use the measurement to select the correct size for the replacement terminal.
8. Select the correct size for the replacement wire.
9. Cut the old terminal from the wire in the harness.
10. Use the old wire as a guide and cut the replacement wire slightly longer. Be careful when doing this because if the wire is too short, there will be tension on the terminal, splice, or connector, which can lead to an open circuit. If the wire is too long, it may get pinched and cause a short.
11. Strip the insulation from the wire **(Figure 16–67)** in the harness and both ends of the replacement wire. Normally, ⅜ inch of insulation should be removed.
12. Make sure the strands of wire are not damaged while removing the insulation.
13. Place the ends of the wires into the terminal and connectors and crimp the terminal **(Figure 16–68)**. To get a proper crimp, place the open area of the connector facing toward the anvil of the tool. Make sure the wire is compressed under the crimp.

    WARNING! Do not crimp a terminal with the cutting edge of a pair of pliers. Although this method may crimp the terminal, it also weakens it.

14. If heat shrink will be used to seal the connections, slip the appropriate length of tubing over the end of the wire that will be spliced.
15. Install the terminal into the connector. Make sure the locking clip is in the proper position. If it is not, use the terminal pick to gently bend it back to its original shape.
16. Push the terminal into the connector until a click is heard.
17. Gently pull on the wire. If the terminal is locked in the connector, it will not move.
18. Connect both sides of the connector and secure the secondary locking device.
19. Tape the new wire to the wiring harness **(Figure 16–69)**. If the harness is contained in conduit, make sure it is fully enclosed and tape the outside of the conduit.

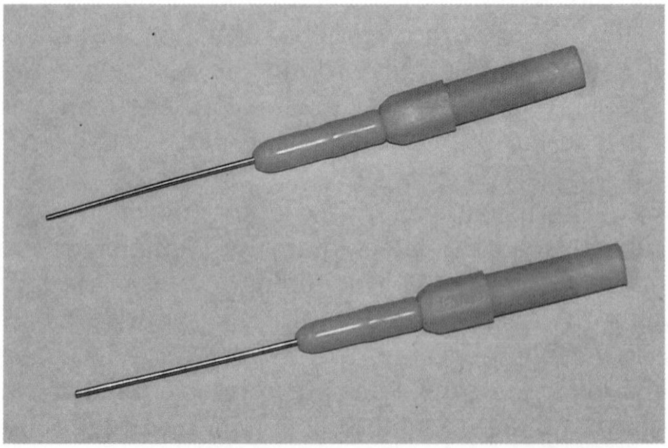

Figure 16–66 Terminal picks are used to unlock terminal and connector locks.

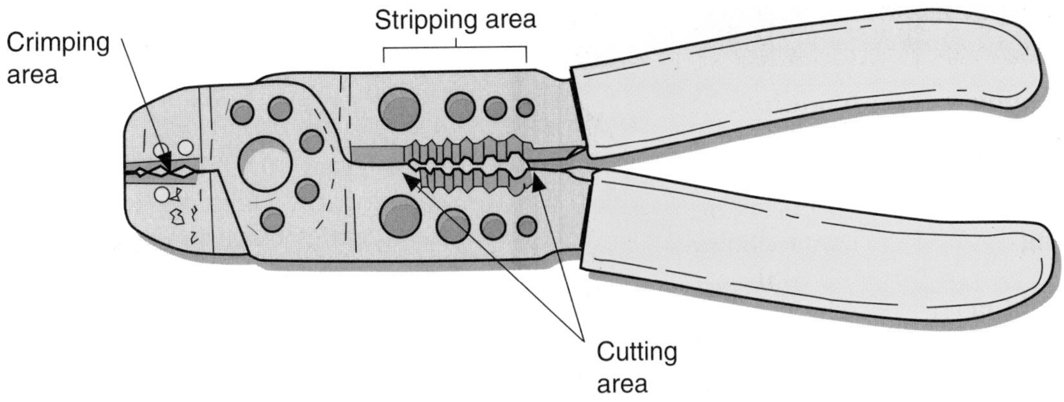

Figure 16-67 A typical crimping tool used for making electrical repairs.

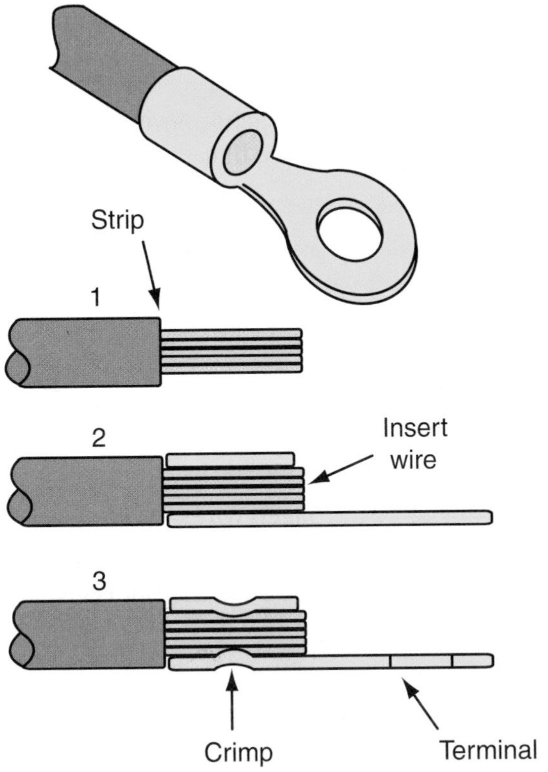

Figure 16-68 Placing a wire into and crimping it in a connector.

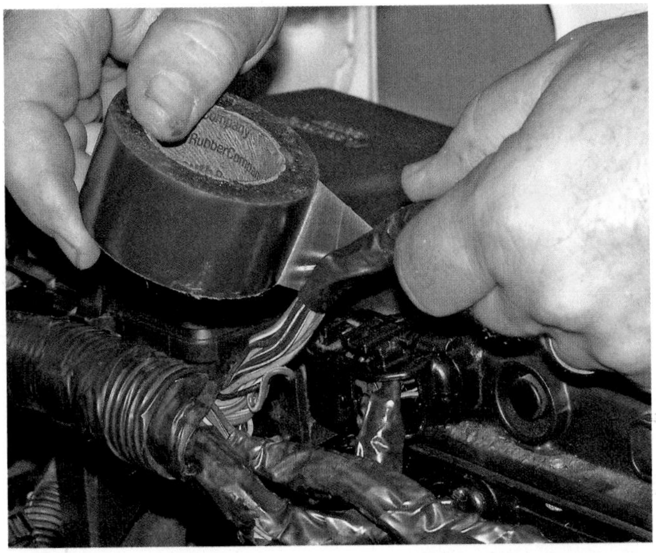

Figure 16-69 Tape all wire repairs to prevent corrosion and damage.

CASE STUDY

A customer tells the technician that the fuse for the windshield wiper blows as soon as he replaces it and turns on the wipers.

The technician removes the fuse and observes that it is black inside. The technician knows that these symptoms indicate a short to ground and so substitutes a 12-volt testlight for the blown fuse. With the testlight in place, the technician disconnects the wiper motor (a mechanically grounded load component), turns the ignition switch on, and notes the status of the testlight. Because the motor is disconnected, the circuit should be open and the testlight is lit. The technician knows that the problem is not in the motor. The problem is a short to ground in the wiring that leads to the motor. The technician continues the search for the short by separating circuit connectors one at a time. He starts at the connector that is farthest from the testlight and works toward the fuse panel. The light remains on when the technician unplugs the first connector so he knows that the short exists in some part of the circuit that is still intact. The light goes off when the technician unplugs the next connector one step closer to the fuse panel. The technician then realizes that the short is somewhere between the last two connectors unplugged because the testlight indicates that there is no longer a path to ground. The technician visually inspects the wiring between the two connectors and discovers and repairs a bare spot in the wire that was contacting ground.

KEY TERMS

Average responding
Back probing
Digital storage oscilloscope (DSO)
Front probing
Glitches
Graphing multimeter (GMM)
Induced voltage
Inline fuse
Open
Radio frequency interference (RFI)
Root mean square (RMS)
Schematics
Short

SUMMARY

- All electrical problems can be classified as an open, short, or high-resistance problem. Identifying the type of problem will allow for the identification of the correct tests to conduct when diagnosing an electrical circuit.
- Wiring diagrams show where wires are connected, the circuit's components, the color of the wires' insulation, and sometimes the wire gauge size.
- Voltmeters, ohmmeters, ammeters, and volt/amp meters are used to test and diagnose electrical systems. These are used with jumper wires, testlights, and variable resistors.
- Multimeters are multifunctional and can test DC and AC volts, ohms, and amperes. Some multimeters can also be used to measure engine rpm, duty cycle, pulse width, frequency, and temperature.
- There are two ways that DMMs display AC voltage: RMS and average responding.
- Some DMMs also feature a MIN/MAX function, which displays the maximum, minimum, and average voltage the meter recorded during the time of the test.
- On a lab scope, an upward movement of the trace indicates an increase in voltage, and a downward movement of this trace represents a decrease in voltage. As the trace moves across the screen, it represents a specific length of time.
- To troubleshoot a problem, begin by verifying the customer's complaint. Then operate the system and others to get a complete understanding of the problem. Use the correct wiring diagram and identify testing points and probable problem areas. Test and use logic to identify the cause of the problem.
- Wiring diagrams are invaluable for diagnostics. Tracing the diagram allows you to think about how the circuit should work and where it should be tested.
- Many automotive electrical problems can be traced to faulty wiring, such as loose or corroded terminals; frayed, broken, or oil-soaked wires; and faulty insulation.
- The preferred way to connect wires or to install a connector is by soldering. Never use acid core solder. It creates corrosion and can damage electronic components.

REVIEW QUESTIONS

1. How will an electrical circuit behave if there in an open in the circuit?
2. What happens to an electrical circuit when there is unwanted resistance in it?
3. An ammeter is always connected ______ with the circuit, whereas a voltmeter is connected in ______ with the circuit.
4. Which of the following is *not* a typical cause of unwanted or high resistance in a circuit?
   a. corrosion on terminal ends
   b. a power wire contacting the chassis
   c. loose or poor connections
   d. frayed and damaged wires
5. Refer to **Figure 16–70**. According to the color scheme used to trace a circuit that is given in the text, what color should wire A–B be?
   a. red
   b. blue
   c. green
   d. yellow
6. Refer to **Figure 16–71**. According to the color scheme used to trace a circuit that is given in the text, what wire is colored wrong?
   a. A–B
   b. B–C
   c. B–F
   d. D–E

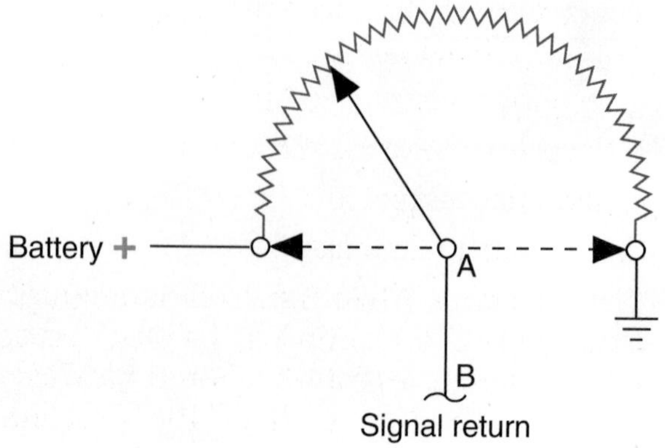

Figure 16–70

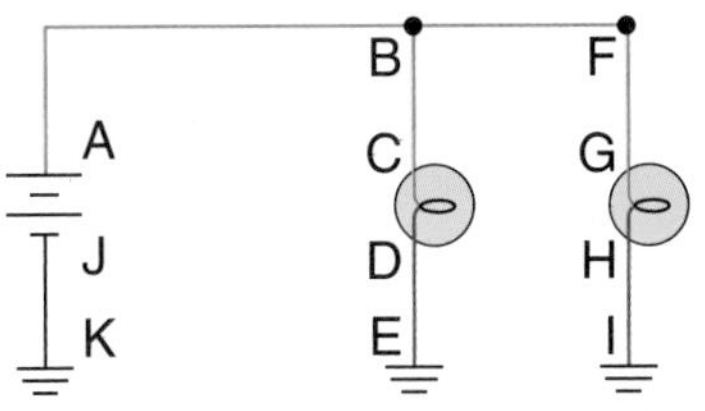

Figure 16-71

7. *True or False?* A zero reading on an ohmmeter means the circuit or component is open.
8. *True or False?* The maximum allowable voltage loss due to voltage drops across wires, connectors, and other conductors in a 12-volt circuit is 1.2 volts.
9. What type of solder should be used to repair electrical wiring?
10. Which of the following lab scope controls must be set when trying to observe the timing of something?
    a. intensity
    b. vertical
    c. horizontal
    d. trigger
11. Which of the following statements is true?
    a. A short to ground causes decreased current flow.
    b. An open causes unwanted voltage drops.
    c. High-resistance problems cause increased current flow.
    d. Both open and high-resistance problems may cause a load not to work.
12. *True or False?* When tracing a circuit in a wiring diagram, remember that all circuits have a power source, a load, and a path to ground and they need to be identified.
13. What would the results of a voltage drop test be if the circuit is open?
14. Which of the following information is not given in a wiring diagram?
    a. wire-by-wire color coding
    b. location of wire harness travel
    c. terminal designation
    d. component designations
15. *True or False?* While troubleshooting a problem, the key to identifying the exact cause of the problem is testing all of the vehicle's components and circuits until the problem is found.

ASE-STYLE REVIEW QUESTIONS

1. While discussing electricity: Technician A says that an open causes unwanted voltage drops. Technician B says that high-resistance problems cause increased current flow. Who is correct?
   a. Technician A c. Both A and B
   b. Technician B d. Neither A nor B
2. While discussing NTC thermistors: Technician A says that some systems use this type of thermistor as a protection device. Technician B says that when there is high current in a circuit, the resistance of the thermistor increases and causes a decrease in current flow. Who is correct?
   a. Technician A c. Both A and B
   b. Technician B d. Neither A nor B
3. While measuring the resistance of a wire with an ohmmeter: Technician A says that if low resistance is shown on the meter, the wire is basically good. Technician B says that if no resistance is measured, the wire is shorted. Who is correct?
   a. Technician A c. Both A and B
   b. Technician B d. Neither A nor B
4. While testing variable resistors: Technician A says that while checking a rheostat with a voltmeter, the voltage should change smoothly with a change in the control. Technician B says that a potentiometer should be checked with a testlight. Who is correct?
   a. Technician A c. Both A and B
   b. Technician B d. Neither A nor B
5. While using an ohmmeter to measure the resistance values of a component: Technician A says that if the component has less than the specified resistance, the part is open. Technician B says that if there is more resistance than called for, the part is shorted. Who is correct?
   a. Technician A c. Both A and B
   b. Technician B d. Neither A nor B
6. Technician A uses an ohmmeter to test circuit protection devices. Technician B uses a voltmeter to test circuit protection devices. Who is correct?
   a. Technician A c. Both A and B
   b. Technician B d. Neither A nor B
7. Technician A uses a testlight to detect resistance. Technician B uses a jumper wire to test circuit breakers, relays, and lights. Who is correct?
   a. Technician A c. Both A and B
   b. Technician B d. Neither A nor B

8. While diagnosing the location of a wire-to-wire short: Technician A checks the wiring of the affected circuits for signs of burned insulation and melted conductors. Technician B checks common connectors shared by the two affected circuits. Who is correct?
   a. Technician A
   b. Technician B
   c. Both A and B
   d. Neither A nor B
9. While measuring resistance: Technician A uses an ohmmeter to measure resistance of a component before disconnecting it from the circuit. Technician B uses a voltmeter to measure voltage drop. A circuit with very low resistance will drop zero or very little voltage. Who is correct?
   a. Technician A
   b. Technician B
   c. Both A and B
   d. Neither A nor B
10. While discussing how to test a switch: Technician A says that the action of the switch can be monitored by a voltmeter. Technician B says that continuity across the switch can be checked by measuring the resistance across the switch's terminals when the switch is in its different positions. Who is correct?
   a. Technician A
   b. Technician B
   c. Both A and B
   d. Neither A nor B

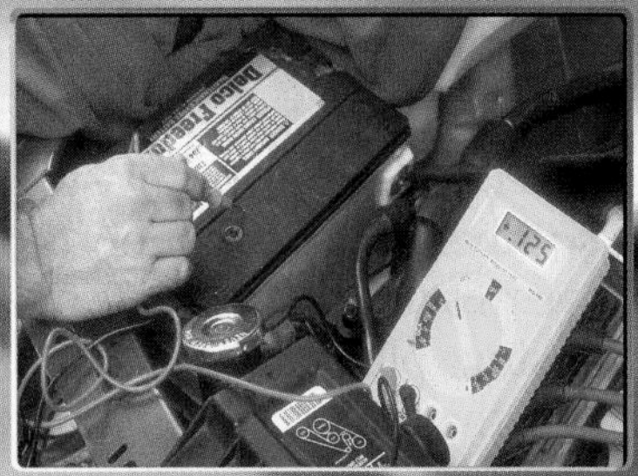

CHAPTER 17

BATTERIES: THEORY, DIAGNOSIS, AND SERVICE

OBJECTIVES

■ Describe how a battery works. ■ List the precautions that must be adhered to when working with or around batteries. ■ Describe the basic construction of an electrochemical cell. ■ Explain how electrochemical cells can be connected to increase voltage and current. ■ Explain the different methods used to recharge a battery. ■ List and describe the various ways a battery may be rated. ■ List and describe the various types of batteries according to their chemistries that may be used in automobiles. ■ List the precautions that must be adhered to when working with or around high-voltage systems. ■ Describe the construction and operation of a lead-acid battery. ■ Describe the various types of lead-acid batteries that are available today. ■ Describe the basic services and testing procedures for a lead-acid battery. ■ Describe the construction and operation of a nickel-metal hydride battery. ■ Describe the construction and operation of a lithium-based battery.

INTRODUCTION

The primary source for electrical power in all automobiles is the battery. The battery has undergone many changes through the years. The introduction of hybrid vehicles and the promise of fuel cell vehicles have drastically changed the basic design of an automotive battery. Many different types of batteries are available or under development to exceed the needs of hybrid or fuel cell vehicles. Lead-acid batteries have been, and continue to be, the power source for conventional vehicles. Each of these energy-storing devices is discussed in this chapter.

BASIC BATTERY THEORY

Electrical current is caused by the movement of electrons from something negative to something positive. The strength of the attraction of the electrons (negative) to the protons (positive) determines the amount of voltage present. When a path is not present for the electrons to travel through, voltage is still present but there is no current flow. When there is a path, the electrons move and there is current. This is the basic operation of batteries.

Basic Construction

Batteries are devices that convert chemical energy into electrical energy. Chemical reactions that produce electrons are called **electrochemical reactions**. A battery stores DC voltage and releases it when it is connected to a circuit. Inside the battery are two **electrodes** or **plates** surrounded by an electrolyte. These three elements make up an electrochemical cell (**Figure 17–1**). Batteries are normally made up of electrochemical cells connected together.

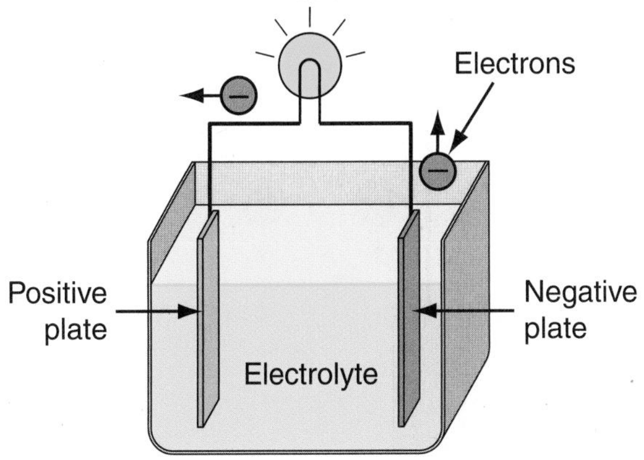

Figure 17–1 A simple electrochemical cell.

One of the plates has an abundance of electrons (negative plate) and the other has a lack of electrons (positive plate). The electrons want to move to the positive plate and do so when a circuit connects the two plates. Batteries have two terminals, a positive that is connected to the positive plate and a negative that is connected to the negative plate.

Electrolytes are chemical solutions that react with the metals used to construct the plates. These chemical reactions cause a lack of electrons on the positive electrode and an excess on the negative electrode. When connected into a circuit, the electrons move. The reactions continue to provide electrons for current flow until the circuit is opened or the chemicals inside the battery become weak. At that time, the battery has run out of electrons (the battery is worn out), the number of electrons on the positive and negative sides are equal, or all of the protons are matched with an electron. Recharging simply moves the electrons that moved to the positive electrode back to the negative electrode **(Figure 17–2)**.

Charging

Charging a battery restores the chemical nature of the cells. To do this, a chemical reaction takes place, causing current flow within the cells. Discharging allows for current flow outside the cell. To understand the charging process, remember that current flows from a higher potential (voltage) to a lower potential. If the voltage applied by an outside source to the battery is higher than the voltage of the battery, current will flow into the battery. This means the charging voltage must be higher than the battery's voltage in order to charge it.

Each battery design has its own charging requirements. It is very important to follow the correct procedure for the battery being charged. It is also important to prevent the battery from overheating during charging and to use the correct type of charger; these too vary with battery designs. Using the wrong charger can destroy the batteries or charger.

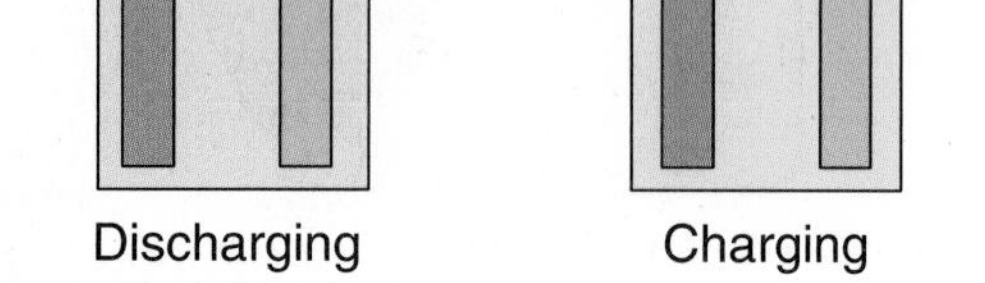

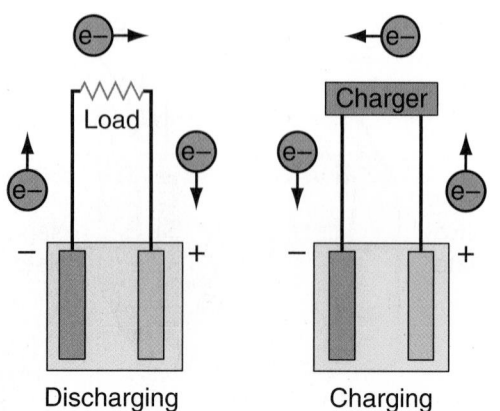

Figure 17–2 The flow of electrons in a battery while discharging and charging.

Cell Arrangements

The voltage produced by an individual battery cell varies with the chemicals and materials used to construct the cell. Most cells produce between 1.2 and 4 volts. To provide higher voltages, cells are connected together. In addition, there is a limited amount of current available from an individual cell, so to increase available current, cells are connected together. Cells can be connected in series or in parallel, or both.

Series Connections Cells are connected in series to provide higher voltages. In this arrangement, the total voltage is the sum of the voltages in each cell. For example, a lead-acid cell, commonly used in starting batteries, produces about 2.1 volts. By connecting six together in series, the battery has a voltage of 12.6 volts. Series connections have the positive terminal of one cell connected to the negative terminal of another, the positive terminal of that cell connected to the negative of another, and so on **(Figure 17–3)**. Individual batteries can also be connected in series. Forty-two-volt systems use a 36-volt battery pack, which can be made from three 12-volt batteries **(Figure 17–4)** or eighteen 2-volt cells connected in series. Forty-two volts are provided by the charging system.

Parallel Connections Cells are connected in parallel to increase the amperage of the pack of cells. The positive terminals are connected together, and all the negative terminals are connected together. The total amperage is the sum of amperages from each cell. The voltage is equal to the voltage of an individual cell.

Series-Parallel Connections In this arrangement, groups of cells are wired in parallel and then those groups are connected in series. This arrangement

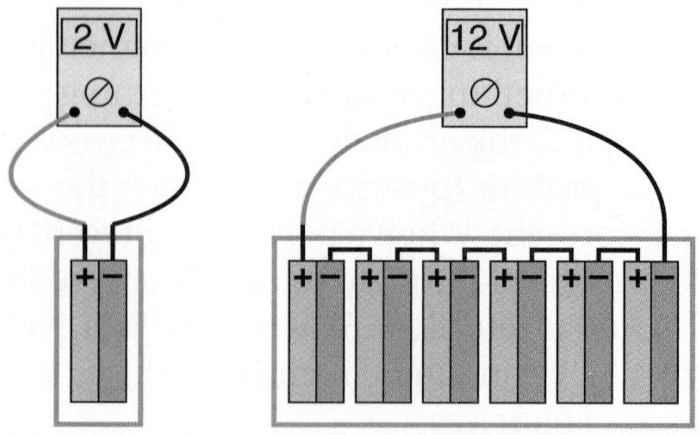

Figure 17–3 When individual cells are connected in series, the total voltage is equal to the sum of the cells.

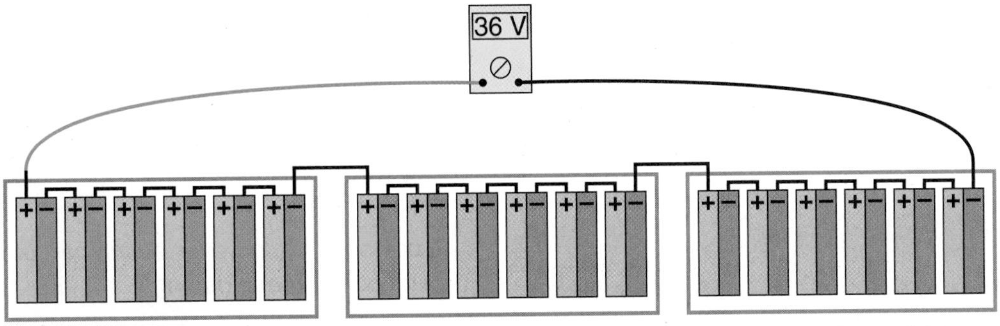

Figure 17-4 A 36-volt battery pack can be made from three 12-volt batteries or eighteen 2-volt cells connected in series.

provides for increases in voltage and amperage. Any number of cells can be connected in parallel as long as each group of parallel cells that are wired in series has the same power output.

Battery Hardware

In order to connect the battery to the vehicle's electrical system, battery cables are used. They must safely handle the voltage and current demands of the vehicle. Battery holddowns are used to prevent damage to the battery, and heat shields are sometimes used to keep battery temperatures down. Most high-voltage battery packs are enclosed in a box that serves to secure the pack and to keep it within a particular temperature range.

Battery Cables Battery cables must be able to carry the current required to meet all demands. Normal 12-volt cable size is 4 or 6 gauge. Various forms of clamps and terminals are used to ensure a good electrical connection at each end of the cable **(Figure 17–5)**. Connections must be clean and tight to prevent arcing and corrosion. The positive cable is normally red and the negative cable is black.

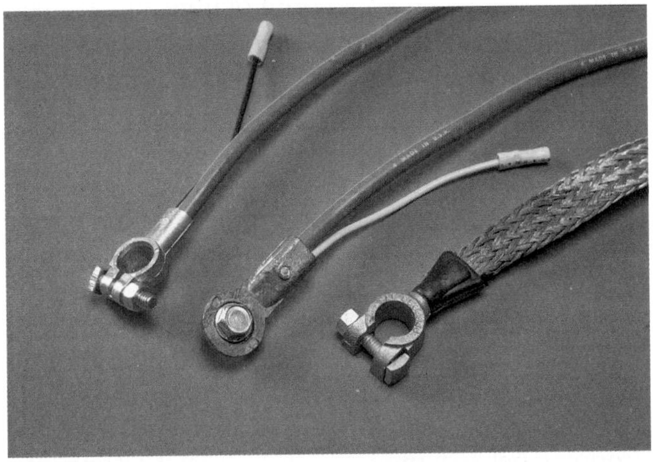

Figure 17-5 The battery cable is designed to carry the high current required to start the engine and supply the vehicle's electrical systems.

The high-voltage cables in nearly all hybrid vehicles are colored orange and have markings on them. Sometimes the cables are enclosed in an orange casing. It is important to remember that some hybrids power other accessories with high voltage; these cables are orange just like the battery cables **(Figure 17–6)**.

Battery Holddowns All batteries must be held securely in the vehicle to prevent damage to the battery and to prevent the terminals from shorting to the vehicle. Battery holddowns are made of metal or plastic **(Figure 17–7)**.

Cooling System The performance and durability of batteries, especially high-voltage battery packs, are

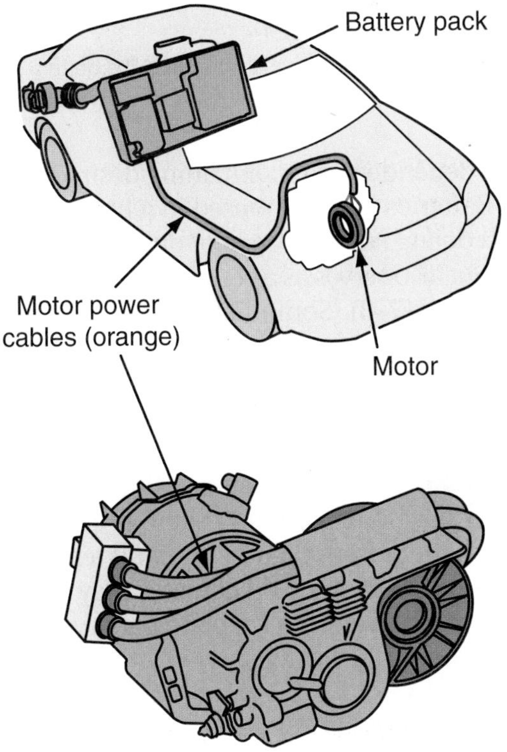

Figure 17-6 High-voltage cables are colored orange or are encased in an orange covering.

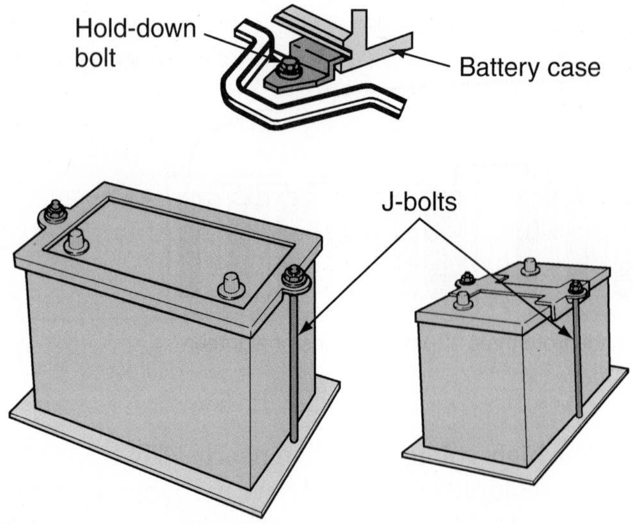

Figure 17-7 Examples of the different types of holddowns used with batteries. *Courtesy of Chrysler LLC*

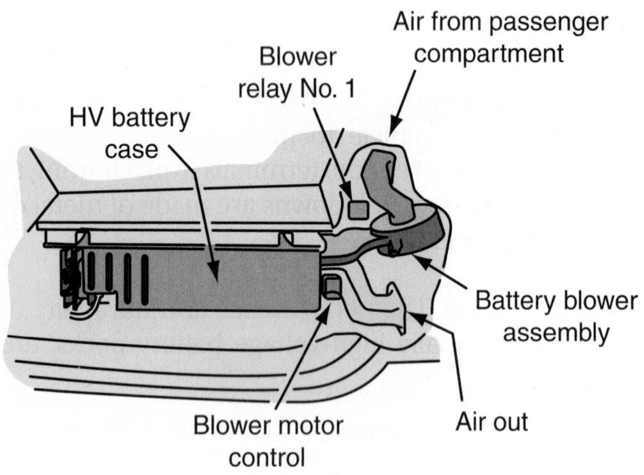

Figure 17-8 This assembly holds the battery pack and provides airflow to keep the batteries cool.

heavily dependent on maintaining desired temperatures. Batteries may be housed in a box or container with a cooling fan. The box not only secures the batteries, but also serves as a conduit for the air from the fan **(Figure 17–8)**. Some battery designs work best when they are warm. For these designs, the battery box also has a heater. Remember, each battery design has its own optimal temperature range.

Heat Shields Some batteries may have a heat shield made of plastic or another material to protect the starting battery from high underhood temperatures. Vehicles equipped for cold climates may have a battery blanket or heater to keep the battery warm during extremely cold weather.

Recycling Batteries

The materials used to make a battery can be used in the future through recycling. Batteries should not be discarded with regular trash because they contain metals and chemicals that are hazardous to the environment.

CAUTION!

A battery should never be incinerated; doing this can cause an explosion.

In 1994, the Rechargeable Battery Recycling Corporation (RBRC) was established to promote recycling of rechargeable batteries in North America. RBRC is a nonprofit organization that collects batteries from consumers and businesses and sends them to recycling companies. Collected batteries are sorted by their chemical makeup. Then they are broken apart and their elements separated. The chemicals or materials are further separated and then collected.

Ninety-eight percent of all lead-acid batteries are recycled. During the recycling process, the lead, plastic, and acids are separated. The electrolyte (sulfuric acid) can be reused or is discarded after it has been neutralized. The plastic casing is cut into small pieces, scrubbed, and melted to make new battery cases and other parts. The lead is also melted and poured into ingots to be used in new batteries.

BATTERY RATINGS

The voltage rating of a battery may be expressed as open circuit or operating voltage. **Open circuit voltage** is the voltage measured across the battery when there is no load on the battery. Operating voltage is the voltage measured across the battery when it is under a load.

The available current from a battery is expressed as the battery's capacity to provide a certain amount of current for a certain amount of time and at a certain temperature. Basically, a capacity rating expresses how much electrical energy a battery can store.

Ampere-Hour

A commonly used capacity rating is the ampere-hour rating. In the past, this was the common rating method for lead-acid batteries. However, these batteries are now rated otherwise. Other battery designs are still rated in ampere-hours or milliamp-hours.

The **ampere-hour (AH) rating** is the amount of steady current that a fully charged battery can supply for 20 hours at 80°F (26.7°) without the cell's voltage dropping below a predetermined level. For example, if a 12-volt battery can be discharged for 20 hours at a rate of 4.0 amperes before its voltage drops to

10.5 volts, it would be rated at 80 AH (20 hours × 4 amps = 80 AH). A 100 AH battery will provide 1 amp for 100 hours, or 100 amps for 1 hour.

Watt-Hour Rating

Some battery manufacturers rate their batteries in watt-hours. The **watt-hour rating** is determined at 0°F (–17.7°C) because the battery's capacity changes with temperature. The rating is calculated by multiplying a battery's AH rating by the battery's voltage. The watt-hour rating of a battery may be listed in units of kilowatts. If a battery can deliver 5 AH at 200 volts, it would be rated at 1 kilowatt-hour (5 AH × 200 volts = 1,000 watt-hour or 1 kilowatt-hour).

Cold Cranking Amps

The **cold cranking amps (CCA)** rating is the common method of rating most automotive starting batteries. It is determined by the load, in amperes, that a battery is able to deliver for 30 seconds at 0°F (–17.7°C) without its voltage dropping below a predetermined level. That voltage level for a 12-volt battery is 7.2 volts. The normal range for passenger car and light truck batteries is between 300 and 600 CCA; some batteries have a rating as high as 1,100 CCA.

Cranking Amps

The **cranking amps (CA) rating** is similar to CCA and is a measure of the current a battery can deliver at 32°F (0°C) for 30 seconds and maintain voltage at a predetermined level. Again, this level is 1.2 volts per cell (7.2 volts) for a 12-volt battery. This rating is more commonly used in climates that are not subject to extremely cold weather. Typically, the CCA rating of a battery is about 20% less than its CA rating.

Reserve Capacity

The **reserve capacity (RC) rating** is determined by the length of time, in minutes, that a fully charged starting battery at 80°F (26.7°C) can be discharged at 25 amperes before battery voltage drops below 10.5 volts. This rating gives an indication of how long the vehicle can be driven with the headlights on if the charging system fails. A battery with a reserve capacity of 120 would be able to deliver 25 amps for 120 minutes before its voltage drops below 10.5 volts.

COMMON TYPES OF BATTERIES

In addition to their use in automobiles, batteries are used in many other applications. As a result, there are many different types and designs of batteries available. Batteries differ in size, from small single cells to large battery packs, comprised of many cells. They also have different ratings (not always dependent on size) and service lives. The primary difference between batteries is the chemicals used in the cells.

Battery Chemistry

The following battery types can be or are being used in automobiles. Some of the batteries in the list that follows are discussed in more detail later in this chapter.

Lead-Acid Lead-acid batteries are the most commonly used starting battery. This type of battery is rechargeable. Several lead-acid batteries are connected in series to provide high voltage in some electric vehicles. There are many variations to the basic design, but all work and are constructed in the same way. The lead-acid cell has electrodes made of lead and lead-oxide with an electrolyte that is a strong acid. The lead-acid battery is one of the oldest battery designs.

Nickel-Cadmium (NiCad) NiCad batteries are mostly used in portable radios, emergency medical equipment, professional video cameras, and power tools. They provide great power and are normally the battery of choice for power tools. The electrodes in a NiCad cell are nickel hydroxide and cadmium. The electrolyte is potassium hydroxide **(Figure 17–9)**. NiCad batteries are economical and have a long service life. However, cadmium is an environmentally unfriendly metal, which is why NiCad batteries are being replaced by other designs.

Nickel-Metal Hydride (NiMH) Nickel-metal hydride batteries are rapidly replacing NiCad batteries

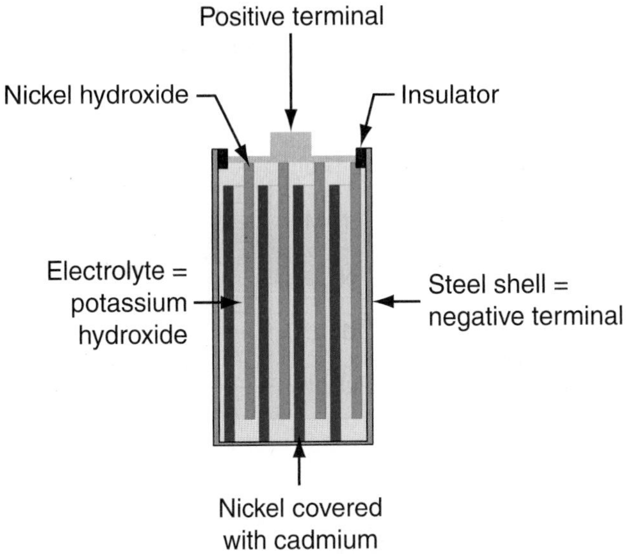

Figure 17-9 The basic construction of a NiCad battery.

because they are more environmentally friendly and are more capable of receiving a full recharge. NiMH batteries also have more capacity than a NiCad but have a reduced service life and a lower current capacity under load. These batteries are commonly used in today's hybrid vehicles. The cells have electrodes made of a metal hydride and nickel hydroxide. The electrolyte is potassium hydroxide.

Sodium-Sulfur (NaS) The electrodes in a sodium-sulfur battery cell are made of molten sodium (negative electrode) and liquid sulfur (positive electrode). The plates are separated by a solid ceramic electrolyte made of aluminum. The battery must be kept at about 570°F (300°C) to discharge and recharge. This design of battery is very efficient and is currently being researched for possible use in vehicles.

Sodium-Nickel-Chloride The electrodes in a sodium-nickel-chloride cell are made with nickel and iron powders and sodium chloride (table salt). The electrodes are separated by a ceramic electrolyte. Sodium-nickel-chloride batteries are also known as "ZEBRA" batteries **(Figure 17–10)**. These batteries have nearly five times the energy density as a lead-acid battery and are totally recyclable. However, they must operate at high temperatures, and the required thermal management system greatly raises their production cost. These batteries were designed to be used in automobiles and trains.

Lithium-Ion (Li-Ion) The electrodes in lithium-ion cells are made of a carbon compound (graphite) and a metal oxide. The electrodes are submersed in lithium salt. Overheating these cells may produce pure lithium in the cells. This metal is very reactive and can explode when hot. To prevent overheating, Li-Ion

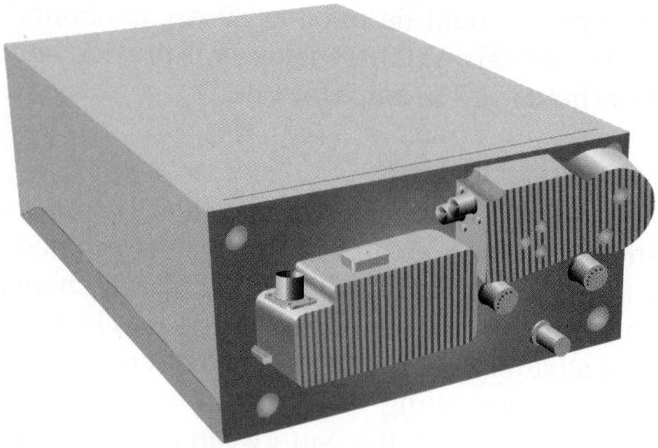

Figure 17–10 The sodium-nickel-chloride battery, made of nickel and iron powders, sodium chloride, and a ceramic electrolyte, is also known as the "ZEBRA" battery.

cells have built-in protective electronics and/or fuses to prevent reverse polarity and overcharging. Li-Ion batteries have very good power-to-weight ratios and are making their way into hybrid vehicles.

Lithium-Polymer (Li-Poly) The lithium-polymer battery is nearly identical to a Li-Ion battery. Like the Li-Ion, the electrodes are made of a carbon compound (graphite) and a metal oxide. However, the lithium salt electrolyte is held in a thin, solid, plasticlike polymer rather than as a liquid. The solid polymer electrolyte is not flammable; therefore, these batteries are less hazardous if they are mistreated. These batteries can store much more energy than a lead-acid battery. They also offer many other advantages and may be used in hybrid and fuel cell vehicles in the future **(Figure 17–11)**.

Nickel-Zinc Nickel-zinc battery cells are also being researched and tested for possible use in vehicles. They have high specific energy and power capability, have good deep cycle capability, can operate within a

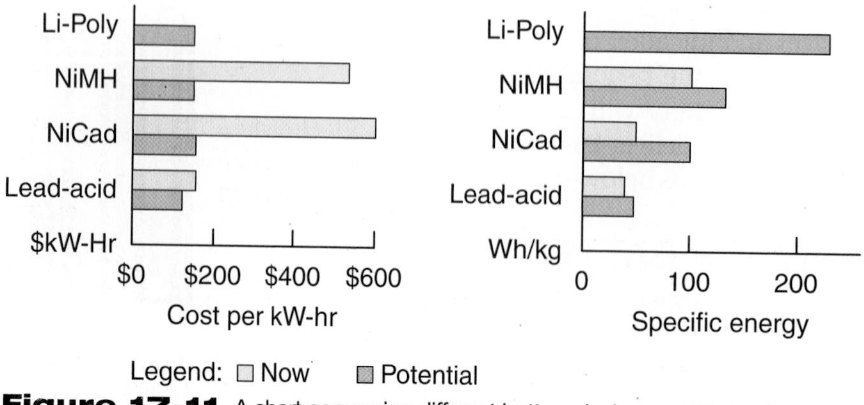

Figure 17–11 A chart comparing different battery designs and the main reasons Li-Poly batteries are being heavily researched for use in electric drive vehicles.

wide range of temperatures, are made of abundant low-cost materials, and are environmentally friendly. The nickel-zinc battery is an alkaline rechargeable system. These cells use a nickel/nickel-oxide electrode as the cathode and the zinc/zinc-oxide electrode as the anode. The electrolyte is normally potassium hydroxide.

HIGH-VOLTAGE BATTERIES

To power drive or traction motors, high voltages are necessary. High-voltage battery packs are assemblies of many cells. The cells are one of two configurations: cylindrical or prismatic **(Figure 17–12)**. In a **cylindrical cell**, the electrodes are rolled together and fit into a metal cylinder. A **separator** soaked in the electrolyte is placed between the plates. This design requires much storage space. When several cylinders are assembled together, there is much wasted space between the cylinders. **Prismatic cells** do not have this problem. They have flat electrodes placed into a box with separators placed between them. Prismatic cells tend to be more expensive to produce than cylindrical cells. Lead-acid batteries have prismatic cells, and some NiMH batteries found in hybrids are also constructed this way.

Applications

A vehicle equipped with a starter/generator can be considered a mild hybrid and relies on a 42-volt system. Functions such as stop-start, regenerative braking, and electrical assist are common to full and mild hybrid vehicles. However, only full hybrids have the ability to move in an electric-only mode. A full hybrid vehicle may have a battery pack that can supply more than 300 volts. Hybrid vehicles may also have a separate starting battery for the engine **(Figure 17–13)**. The starting battery also is the power source for the lights and other accessories.

Safety Issues

High-voltage circuits in hybrids are identifiable by size and color. The cables have thicker insulation and are colored orange. The connectors are also colored orange. On some vehicles, the high-voltage cables are enclosed in an orange shielding or casing; again the orange indicates high voltage. In addition, the high-voltage battery pack and other high-voltage components have "High Voltage" caution labels. It is important to remember that high voltage is also used to power some vehicle accessories. Always follow the safety precautions given by the manufacturer when working around or with high-voltage systems.

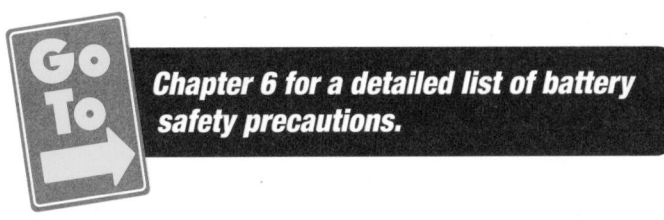

Go To ***Chapter 6 for a detailed list of battery safety precautions.***

LEAD-ACID BATTERIES

The most common automotive batteries are lead-acid designs. The wet cell, gel cell, absorbed glass mat (AGM), and valve regulated are versions of the lead-acid battery.

Basic Construction

A lead-acid battery consists of grids, positive plates, negative plates, separators, elements, electrolyte, a container, cell covers, vent plugs, and cell containers

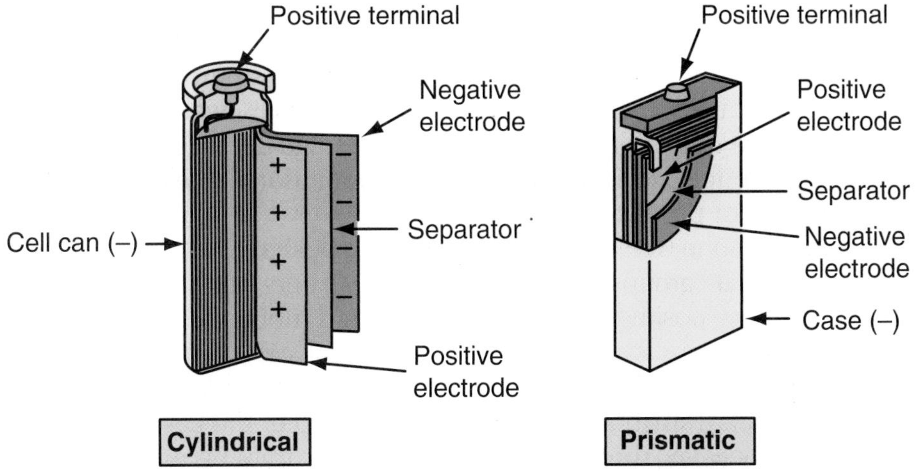

Figure 17–12 The basic construction of cylindrical and prismatic cells.

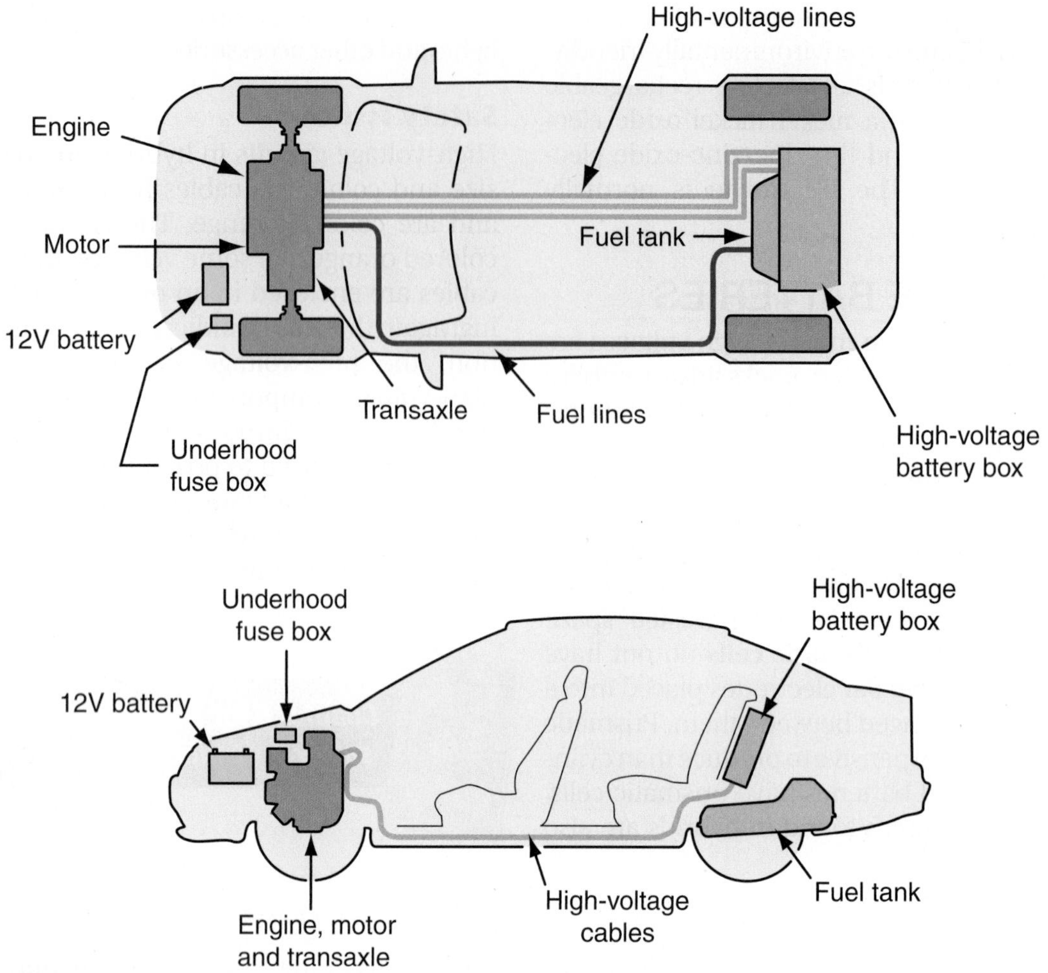

Figure 17-13 Location of the batteries in a Honda Civic hybrid.

(**Figure 17–14**). A **grid** is a lead alloy frame that supports the active material of each plate. Plates are typically flat, rectangular components that are either positive or negative, depending on the active material they hold.

The positive plate has a grid filled with its active material, lead peroxide. **Lead peroxide (PbO_2)** is a dark brown, crystalline material. The material pasted onto the grids of the negative plates is **sponge lead (Pb)**. Both plates are very porous and allow the liquid electrolyte to penetrate freely.

Each battery contains a number of elements. An **element** is a group of positive and negative plates (**Figure 17–15**). The plates are formed into a plate group, which holds a number of plates of the same polarity. The like-charged plates are welded to a lead alloy post or **plate strap**. The plate groups are placed alternately within the battery—positive, negative, positive, negative, and so on. There is usually one extra set of negative plates to balance the charge. To prevent the different plate groups from touching each other, separators are inserted between them. Separators are porous plastic sheets that allow for a transfer of ions between plates. When the element is placed into the battery case and immersed in electrolyte, it becomes a cell.

The electrolyte is a solution of sulfuric acid and water. The sulfuric acid supplies sulfate, which chemically reacts with both the lead and PbO_2 to release electrical energy. In addition, the sulfuric acid is the carrier for the electrons as they move inside the battery. To cause the required chemical reaction, the electrolyte must be the correct mixture of water and sulfuric acid. At 12.6 volts, the desired solution is 65% water and 35% sulfuric acid. Available voltage decreases when the percentage of acid in the solution decreases.

Casing Design The container or shell of the battery is usually a one-piece, molded assembly of polypropylene, hard rubber, or plastic. The case has a number of individual cell compartments. Cell connectors are used to join all cells of a battery in series.

The top of the battery is encased by a cell cover. The cover may be a one-piece design, or the cells might have their own individual covers. The cover

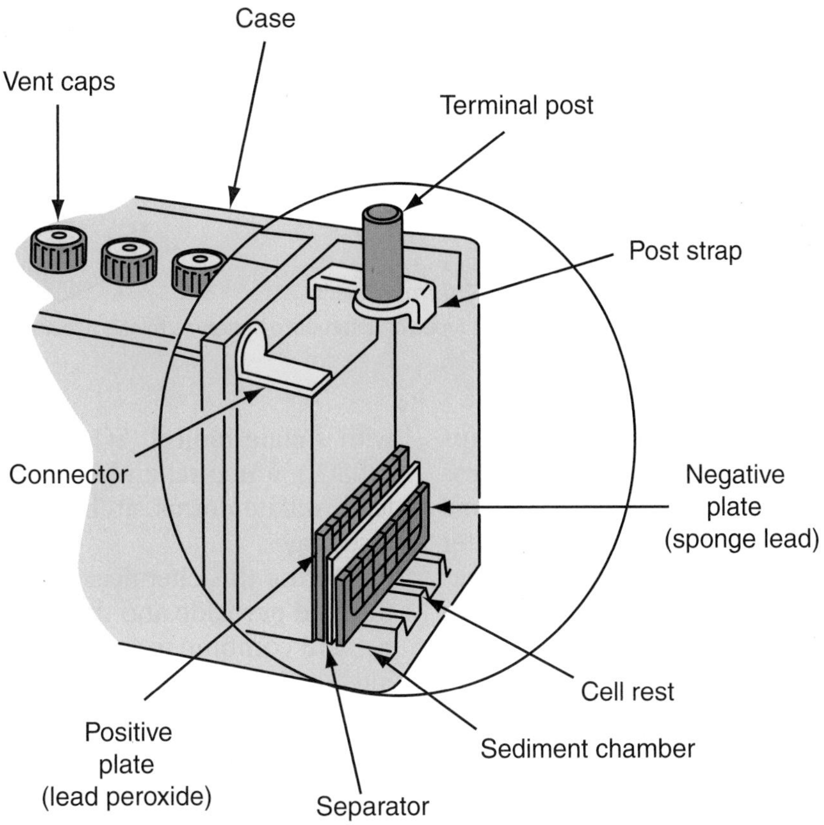

Figure 17–14 Components of a typical lead-acid storage battery.

must have vent holes to allow hydrogen and oxygen gases to escape. These gases are formed during charging and discharging. Battery vents can be permanently fixed to the cover or be removable, depending on the design of the battery. Vent plugs or caps are used on some batteries to close the openings in the cell cover and to allow for topping off the cells with electrolyte or water.

At the bottom of some battery casings is a sediment chamber. The chamber collects the materials that fall from the plates. If the sediments do not fall below the plates, they could cause a short between the plates. Some batteries do not have a sediment chamber; rather the separators are used to contain all sediments and keep them from contacting the plates.

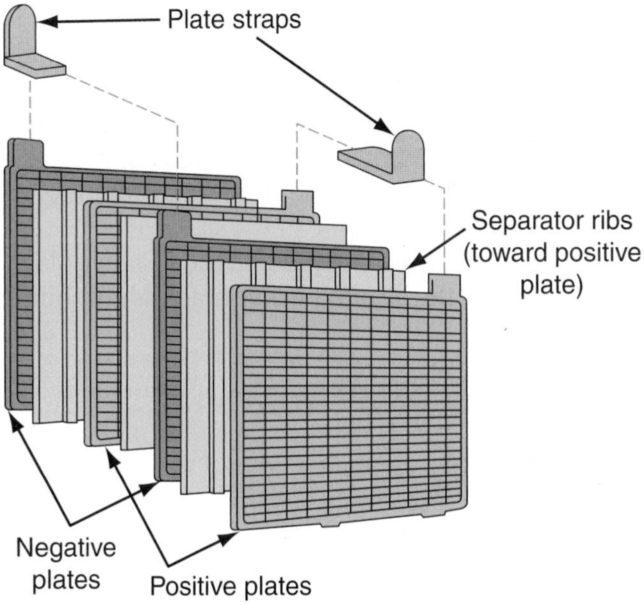

Figure 17–15 The parts of a typical battery element.

CAUTION!

When lifting a battery, excessive pressure on the end walls could cause acid to spew through the vent caps, resulting in personal injury. Lift with a battery carrier or with your hands on opposite corners.

Terminals The battery has two external terminals: a positive (+) and a negative (−). These terminals are two tapered posts, "L" terminals, threaded studs on top of the case, or two internally threaded connectors on the side **(Figure 17–16)**. The terminals have either a positive (+) or a negative (−) marking, depending on which end of the series they represent.

The size of the tapered terminals is specified by standards set by the **Battery Council International (BCI)** and Society of Automotive Engineers (SAE).

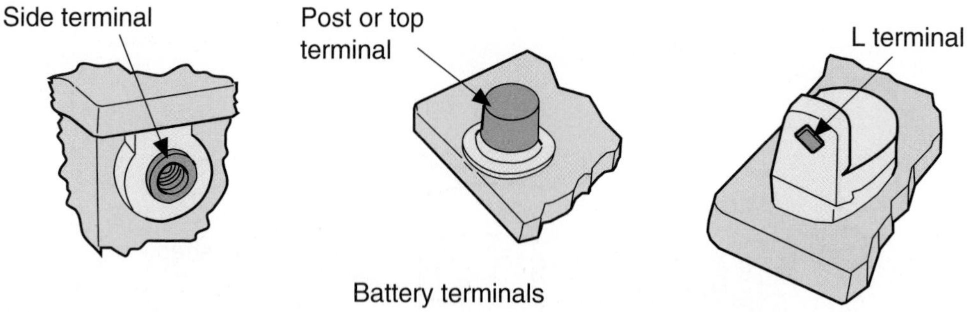

Figure 17-16 The most common types of automotive battery terminals. *Courtesy of Ford Motor Company*

This means that all positive and negative cable clamps will fit any corresponding battery terminal, regardless of the battery's manufacturer. The positive terminal is slightly larger, usually around 11/16 inch in diameter at the top, whereas the negative terminal usually has a 5/8-inch diameter. This minimizes, but does not prevent, the danger of installing the battery cables in reverse polarity.

Side terminals are positioned near the top of the battery case. These terminals are threaded and require a special bolt to connect the cables. Some batteries are fitted with both top and side terminals to allow them to be used in many different vehicles.

Discharging and Charging

The chemical reaction between the active materials on the positive and negative plates and the electrolyte releases electrical energy. When a battery discharges **(Figure 17–17)**, lead in the lead peroxide of the positive plate combines with the sulfate radical (SO_4) to form lead sulfate ($PbSO_4$).

A similar reaction takes place at the negative plate. The lead (Pb) of the negative active material combines with sulfate radical (SO_4) to also form lead sulfate ($PbSO_4$), a neutral and inactive material. Therefore, lead sulfate forms at both plates as the battery discharges.

During this chemical reaction, the oxygen from the lead peroxide and the hydrogen from the sulfuric acid combine to form water (H_2O). As discharging takes place, the electrolyte becomes weaker and the positive and negative plates become like one another.

The recharging process **(Figure 17–18)** is the reverse of discharging. Electricity from an outside source, such as the vehicle's generator or a battery recharger, is forced into the battery. The lead sulfate ($PbSO_4$) on both plates separates into lead (Pb) and sulfate (SO_4). As the sulfate (SO_4) leaves both plates, it combines with hydrogen in the electrolyte to form sulfuric acid (H_2SO_4). At the same time, the oxygen (O_2) in the electrolyte combines with the lead (Pb) at the positive plate to form lead peroxide (PbO_2). As a result, the negative plate returns to its original form of lead (Pb), and the positive plate reverts to lead peroxide (PbO_2).

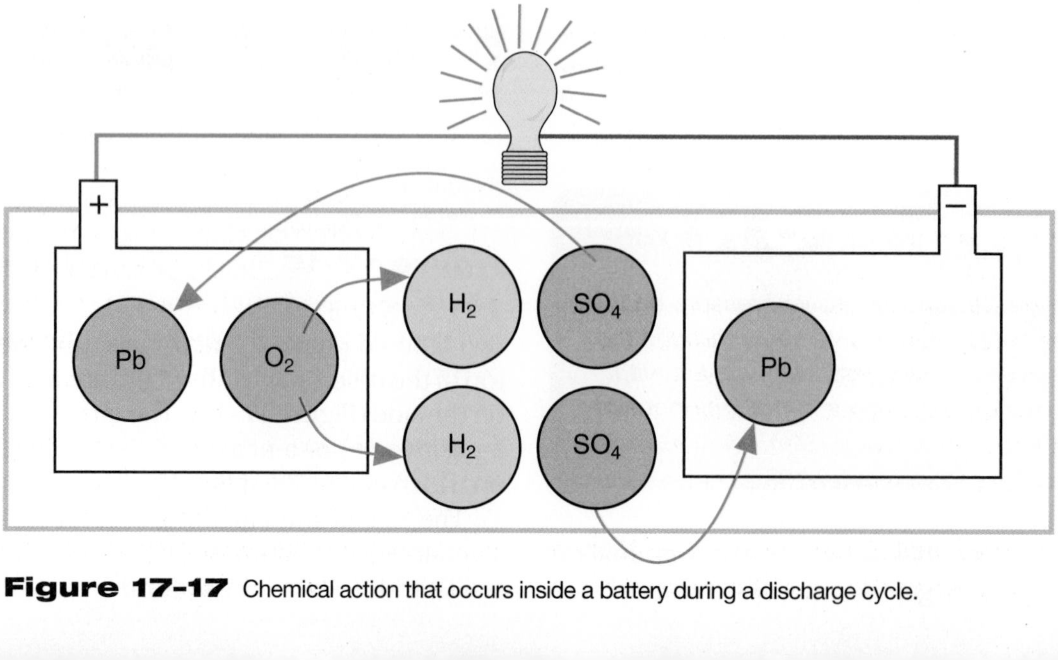

Figure 17-17 Chemical action that occurs inside a battery during a discharge cycle.

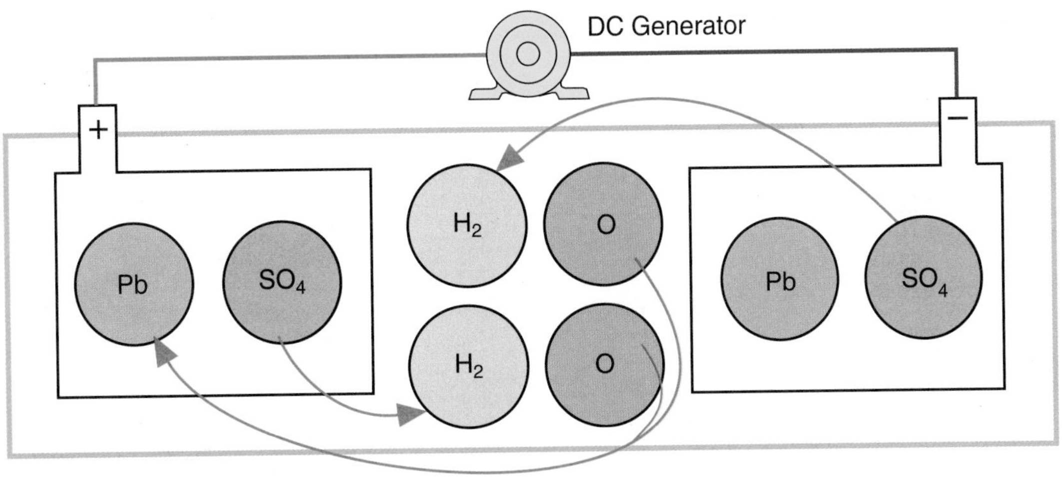

Figure 17-18 Chemical action inside a battery during the charge cycle.

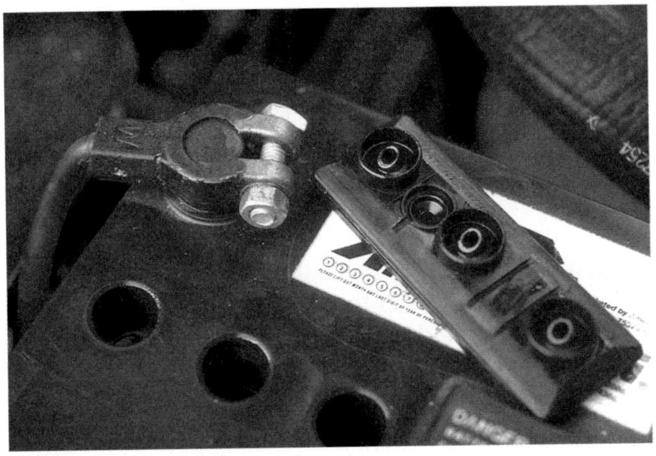

Figure 17-19 A battery with removable cell caps.

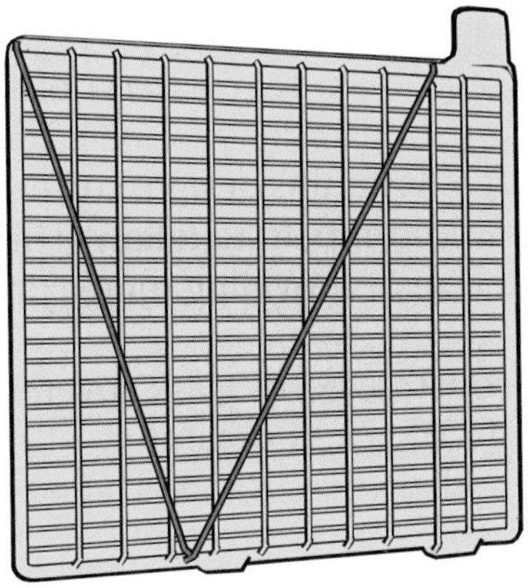

Figure 17-20 Maintenance-free battery grids with support bars give increased strength and faster electrical delivery. *Courtesy of Ford Motor Company*

An unsealed battery gradually loses water due to its conversion into hydrogen and oxygen; these gases escape the battery through the vent caps **(Figure 17–19)**. If the lost water is not replaced, the level of the electrolyte falls below the tops of the plates. This results in a high concentration of sulfuric acid in the electrolyte and permits the uncovered material of the plates to dry and harden. This will reduce the service life of a battery. This is why the electrolyte level in the battery must be frequently checked.

Designs

Lead-acid batteries can be designed as a starting battery or a deep cycle battery. Deep cycle batteries are designed to go through many charge and discharge cycles. They have thicker and fewer plates than a starting battery. The exact chemical composition of lead-acid batteries also depends on the designed purpose of the battery. However, all lead-acid batteries are based on the reaction of lead and acid.

Maintenance-Free and Low-Maintenance Batteries The majority of batteries installed in today's vehicles are low-maintenance or maintenance-free designs. A low-maintenance battery is a heavy-duty version of a normal lead-acid battery. Many of the parts are thicker and made with different, more durable materials. Low-maintenance batteries have vent holes and caps, which allow water to be added to the cells. However, a low-maintenance battery requires additional water substantially less often than a conventional battery.

Similar in construction but made with different plate materials, a **maintenance-free battery** experiences little gassing during discharge and charge cycles **(Figure 17–20)**. Therefore, maintenance-free batteries do not have external holes or caps **(Figure 17–21)**. They are equipped with small gas vents that prevent gas pressure buildup in the case. Water is never added to maintenance-free batteries.

Hybrid Batteries A **hybrid battery** can withstand six deep cycles and still retain 100% of its original reserve capacity. A battery experiences a deep cycle when it is near totally discharged and then recharged. The grid construction differs from other batteries in that the

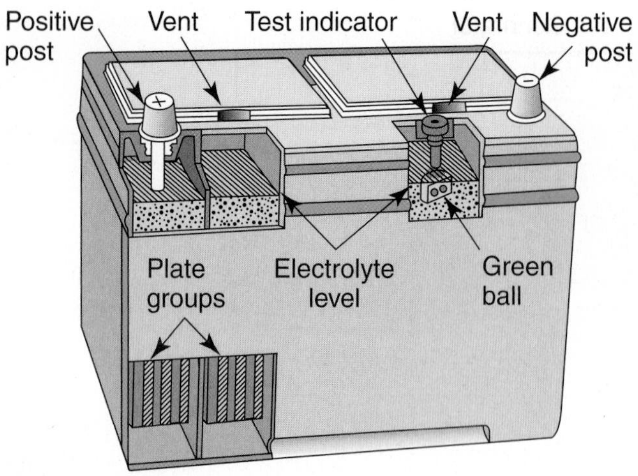

Figure 17-21 Construction of a maintenance-free battery showing the location of the gas vents.

plates have a lug located near the center of the grid. In addition, the vertical and horizontal grid bars are arranged in a radial design. With this design, current has less resistance and a shorter path to follow. This means the battery is capable of providing more current at a faster rate.

The separators are constructed of glass covered with a resin or fiberglass. The glass separators offer low electrical resistance with high resistance to chemical contamination.

Recombination Batteries A recombination or **recombinant battery** is a completely sealed maintenance-free battery that uses a gel-type electrolyte. In a gel cell battery, gassing is minimized and vents are not needed. During charging, the negative plates never reach a fully charged condition and therefore cause little or no release of hydrogen. Oxygen is released at the positive plates, but it passes through the separators and recombines with the negative plates.

Absorbed Glass Mat Batteries The electrolyte in **absorbed glass mat (AGM) batteries** is held in moistened fiberglass matting. The matting is sandwiched between the battery's plates, where it doubles as a vibration dampener. AGM batteries are recombinant batteries.

Rolls of high-purity lead plates are tightly compressed into six cells (**Figure 17–22**). The plates are separated by acid-permeated vitreous separators. Vitreous separators absorb acid in the same way a paper towel absorbs water. Each of the cells is enclosed in its own cylinder within the battery case, forming a sealed, closed system that resembles a six-pack of soda. During normal use, hydrogen and oxygen within the battery are captured and recombined to

Figure 17-22 The construction of an AGM battery. *Courtesy of Exide Technologies*

form water within the electrolyte. This eliminates the need to ever add water to the battery.

The spiral rolled plates and fiberglass mats are virtually impervious to vibration and impact. AGM batteries will never leak, have short recharging times, and have low internal resistance, which provides increased output.

Valve-Regulated Batteries **Valve-regulated lead-acid (VRLA) batteries** are similar to AGM batteries and are recombinant batteries. The oxygen produced on the positive plates is absorbed by the negative plate. That, in turn, decreases the amount of hydrogen produced at the negative plate. The combination of hydrogen and oxygen produces water, which is returned to the electrolyte. Therefore, this battery never needs to have water added to its electrolyte mixture.

One plate in a VRLA is made of a lead-tin-calcium alloy with porous lead dioxide; the other is also made of a lead-tin-calcium alloy but has spongy lead as the active material. The electrolyte is sulfuric acid that is absorbed into plate separators made of a glass-fiber fabric. The battery is equipped with a valve that opens to relieve any excessive pressure that builds up in the battery. At all other times the valve is closed and the battery is totally sealed.

Factors Affecting Battery Life

All storage batteries have a limited service life, but many conditions can decrease it.

Improper Electrolyte Levels With nonsealed batteries, water should be the only portion of the electrolyte lost due to evaporation during hot weather and gassing during charging. Maintaining an adequate electrolyte level is the basic step in extending battery life for these designs.

Temperature Batteries do not work well when they are cold. At 0°F (–17.8°C) a battery is only capable of working at 40% of its capacity. There is also the possibility of the battery freezing when it is very cold and its charge is low. When the battery is allowed to get too hot, the water in the electrolyte can evaporate. Batteries used in hot climates should have their electrolyte level checked very frequently and only distilled water should be added if necessary.

Corrosion Battery corrosion is commonly caused by spilled electrolyte or electrolyte condensation from gassing. In either case, the sulfuric acid corrodes, attacks, and can destroy not only connectors and terminals but holddown straps and the battery tray as well.

Corroded connections increase resistance at the battery terminals, which reduces the applied voltage to the vehicle's electrical system. Corrosion on the battery cover can also create a path for current, which can allow the battery to slowly discharge. Finally, corrosion can destroy the holddown straps and battery tray, which can result in physical damage to the battery and/or vehicle.

Overcharging Batteries can be overcharged by either the vehicle's charging system or a battery charger. In either case, the result is a violent chemical reaction within the battery that causes a loss of water in the cells. This can permanently reduce the capacity of the battery. Overcharging can also cause excessive heat, which can oxidize the positive plate grid material and even buckle the plates, resulting in a loss of cell capacity and early battery failure.

Undercharge/Sulfation The vehicle's charging system might not fully recharge the battery due to a fault in the charging system. This causes the battery to operate in a partially discharged condition. A battery in this condition will become sulfated when the sulfate normally formed in the plates becomes dense, hard, and chemically irreversible. This happens because the sulfate has been allowed to remain in the plates for a long period.

Sulfation of the plates causes two problems. First, it lowers the specific gravity levels of the electrolyte and increases the danger of freezing. Second, in cold weather, a sulfated battery often fails to crank the engine because of its lack of reserve power.

Poor Mounting Loose or missing holddown straps allow the battery to vibrate or bounce during vehicle operation. This can shake the active materials off the grid plates, severely shortening battery life. It can also loosen the plate connections to the plate strap, loosen cable connections, or even crack the battery case.

Cycling Heavy and repeated cycling can cause the positive plate material to break away from its grids and fall into the sediment chambers at the base of the case. This problem reduces battery capacity and can lead to premature short circuiting between the plates. Fortunately, the new envelope design found in many batteries reduces this problem.

SERVICING AND TESTING BATTERIES

CAUTION!

Always wear safety glasses or goggles when working with batteries. When a battery is charging or discharging, it gives off highly explosive hydrogen gas. Some hydrogen gas is present in the battery at all times. Any flame or spark can ignite this gas, causing the battery to violently explode, propelling the vent caps at a high velocity and spraying acid in a wide area. Sulfuric acid can cause severe skin burns. If electrolyte contacts your skin or eyes, flush the area with water for several minutes. When eye contact occurs, force your eyelid open and flush your eyes with eyewash solution. Do not rub your eyes or skin and call a doctor immediately.

Testing batteries is an important part of electrical system service. Prior to conducting any tests, make sure the battery is fully charged. Also remove the surface charge of the battery by turning on the headlights with the engine off. Keep the lights on for at least 3 minutes. Poor and inaccurate tests can lead to serious problems and expensive and unneeded repairs. Depending on the design of the battery, state of charge and capacity can be determined in several ways: specific gravity tests, visual inspection of batteries with a built-in hydrometer, open circuit voltage tests, and the capacity test.

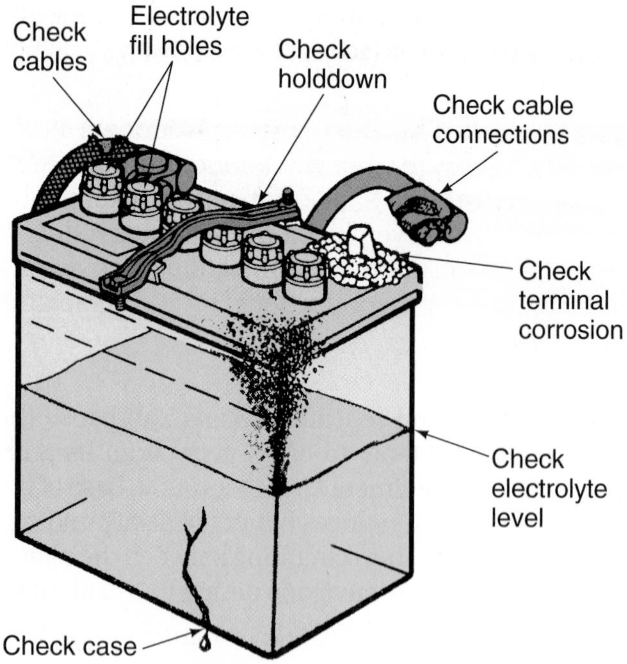

Figure 17-23 Batteries should be carefully checked for damage, dirt, and corrosion.

Inspection

Testing a lead-acid battery should begin with a thorough inspection of the battery and its terminals **(Figure 17–23)**. The following items should be checked:

1. Check the age of the battery by looking at the date code on the battery.
2. Check the condition of the case. A damaged battery should be replaced.
3. If the battery is not sealed, check the electrolyte levels in all cells and correct them as necessary. If water is added, charge the battery before conducting any test on the battery.
4. Check the condition of the battery terminals and cables. Clean any corrosion from the cable ends and terminals. Make sure the cable ends are tightly fastened to the terminals.
5. Make sure the battery holddowns are holding the battery securely in place.

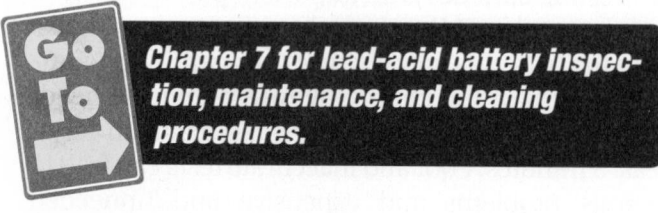

Go To ***Chapter 7 for lead-acid battery inspection, maintenance, and cleaning procedures.***

Battery Leakage Test

To perform a battery leakage test, set a voltmeter to a low DC volt range. Connect the negative lead to the battery's negative post. Then move the meter's positive lead across the top and sides of the battery case **(Figure 17–24)**. If some voltage is read on the meter, current is leaking out of the battery. The battery should be cleaned and then rechecked. If voltage is again measured, the battery should be replaced; the case is porous or cracked.

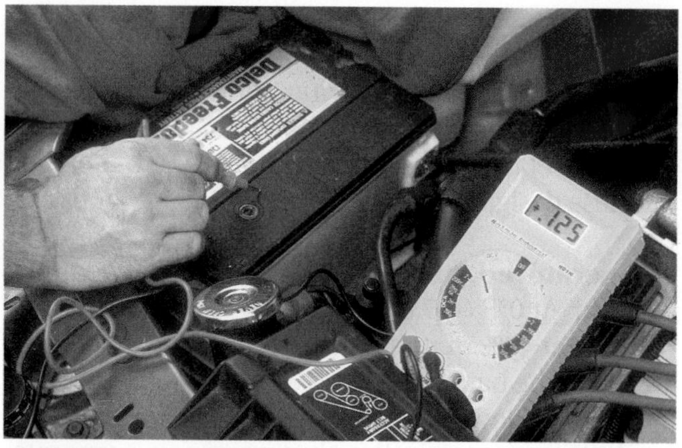

Figure 17-24 Performing a battery leakage test.

Cleaning the Battery and Terminals

Before removing the battery connectors or the battery, always neutralize any accumulated corrosion on terminals, connectors, and other metal parts. Apply a solution of baking soda and water or ammonia and water. Do not splash the corrosion onto the vehicle's paint, metal or rubber parts, or onto your hands and face. Be sure the solution cannot enter the battery cells. A stiff-bristle brush is ideal for removing heavy buildup. Dirt and accumulated grease can be removed with a detergent solution or solvent. After cleaning, rinse the battery and cable connections with clean water. Dry the components with a clean rag or low-pressure compressed air.

SHOP TALK

Remember to connect a memory saver and obtain security codes for the vehicle before disconnecting the battery.

To clean the inside surfaces of the connectors and the battery terminals, remove the cables. Always begin with the ground cable. Spring-type cable connectors are removed by squeezing the ends of their prongs together with wide-jaw, vise-gripping, channel lock, or battery pliers. This pressure expands the connector so it can be lifted off the terminal post.

For connectors tightened with nuts and bolts, loosen the nut using a box-end wrench and/or cable-clamp pliers **(Figure 17–25)**. Using ordinary pliers or an open-end wrench can cause problems. These tools

Figure 17–25 It is best to loosen a battery clamp with a box-end wrench and/or cable-clamp pliers.

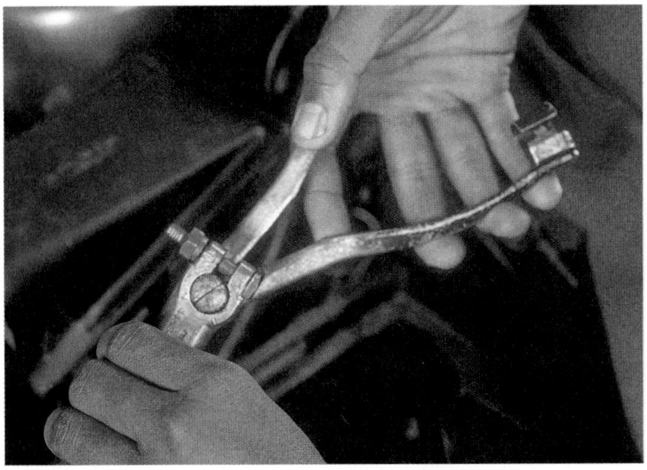

Figure 17–27 Before reinstalling the cable end onto the battery, use terminal end expanders to make sure the end does not need to be forced onto the battery post.

might slip off under pressure with enough force to break the cell cover or damage the casing.

Always grip the cable while loosening the nut. This reduces the pressure on the terminal post that could break it or loosen its mounting in the battery. If the connector does not lift easily off the terminal when loosened, use a clamp puller. Prying with a screwdriver or bar strains the terminal post and the plates attached to it. This can break the cell cover or pop the plates loose from the terminal post.

Once the connectors have been removed, open the connector using a connector-spreading tool. Neutralize any remaining corrosion by dipping it in a baking soda or ammonia solution. Next, clean the inside of the connectors and the posts using a wire brush with external and internal bristles **(Figure 17–26)**.

Begin reinstallation by expanding the opening of the clamp **(Figure 17–27)** so that force is not needed to place it over the post. Then position and tighten the terminal on its post. Do not overtighten any nuts or bolts; this could damage the post or terminal.

Battery Hydrometer

The electrolyte of a fully charged battery is usually about 64% water and 36% sulfuric acid. This corresponds to a specific gravity of 1.270. **Specific gravity** is the weight of a given volume of any liquid divided by the weight of an equal volume of water. Pure water has a specific gravity of 1.000, whereas battery electrolyte should have a specific gravity of 1.260 to 1.280 at 80°F (26.7°C). In other words, the electrolyte should be 1.260 to 1.280 times heavier than water.

The specific gravity of the electrolyte decreases as the battery discharges. This is why measuring the specific gravity of the electrolyte with a hydrometer can be a good indicator of how much charge a battery has lost **(Figure 17–28)**. Hydrometer readings should not vary more than 0.05 difference between cells. More variance is an indication that the battery is bad.

Figure 17–26 Combination external/internal wire brushes clean both terminals and inside cable connector surfaces.

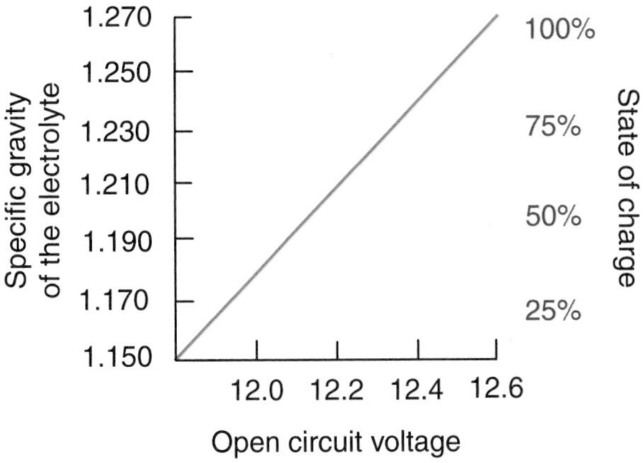

Figure 17–28 A graph showing the relationship between specific gravity, open circuit voltage, and the state of charge of a battery cell.

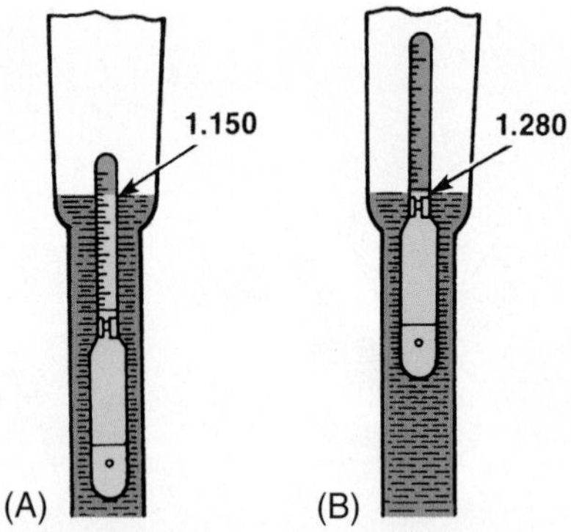

Figure 17-29 (A) When the scale sinks in the electrolyte, the specific gravity is low; (B) when it floats high, the specific gravity is high.

The electrolyte's specific gravity can be measured on unsealed batteries to give a fairly good indication of the battery's state-of-charge (SOC). A basic battery hydrometer uses a glass float or hydrometer in a glass tube to measure the electrolyte's specific gravity. Squeezing the hydrometer's bulb pulls electrolyte into the tube. When filled with test electrolyte, the hydrometer float bobs in the electrolyte. The depth at which the float sinks in the electrolyte indicates its relative weight compared to water. The reading is taken off the scale by sighting along the level of the electrolyte. If the hydrometer floats deeply in the electrolyte, the specific gravity is low **(Figure 17–29A)**. If the hydrometer floats high in the electrolyte, the specific gravity is high **(Figure 17–29B)**.

Because temperature affects the specific gravity of a substance, the specific gravity reading should be corrected by adding or subtracting 4 points (0.004) for each 10°F (–10°C), above or below the standard of 80°F (26.7°C). Most hydrometers have a built-in thermometer to measure the temperature of the electrolyte **(Figure 17–30)**. The hydrometer reading can be misleading if it is not adjusted. For example, a reading of 1.260 taken at 20°F (–6.6°C) would be 1.260 – (6 × 0.004 or 0.024) = 1.236. This lower reading means the cell has less charge than indicated. The specific gravity of the cells of a fully charged battery should be near 1.265 when corrected for electrolyte temperature. Recharge any battery if the specific gravity drops below an average of 1.230.

Refractometer In many cases, the same refractometer used to test engine coolant can be used to check the specific gravity of electrolyte. These tools also feature automatic temperature compensation. A refractometer uses a prism to analyze fluids.

Built-In Hydrometers

On some maintenance-free batteries, a special temperature-compensating hydrometer is built into the battery case **(Figure 17–31)**. A quick visual check indicates the battery's SOC **(Figure 17–32)**. It is important when observing the hydrometer that the battery has a clean top to see the correct indication. A flashlight may be required in poorly lit areas. Always look straight down when viewing the hydrometer.

A few battery designs have a charge indicator on the top of the case **(Figure 17–33)**. These batteries use a color display to note the SOC. Typically, green means "OK," gray means "Check for recharge," and white means "Change or replace."

Many batteries do not have a built-in hydrometer. A voltage check is the only way to check the battery's SOC. The specific gravity of these batteries cannot be checked because the batteries are sealed. Never pry the cell caps off to check the electrolyte; leave the battery sealed.

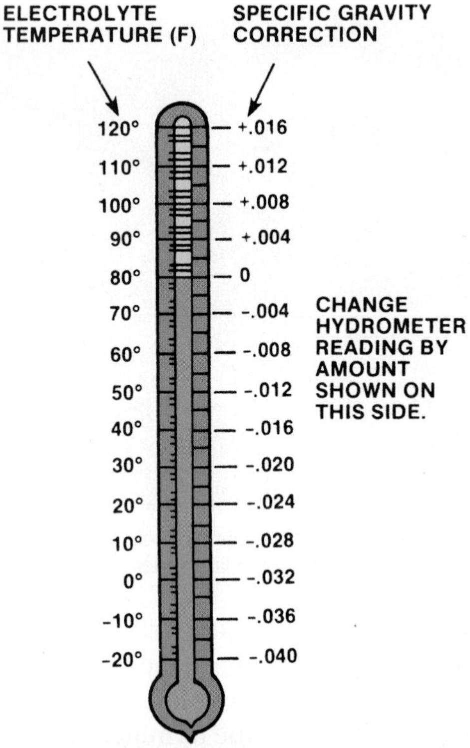

Figure 17-30 Hydrometers with thermometer correction scales make adjusting for electrolyte temperature easy.

Figure 17-31 Sight glass in a maintenance-free battery.

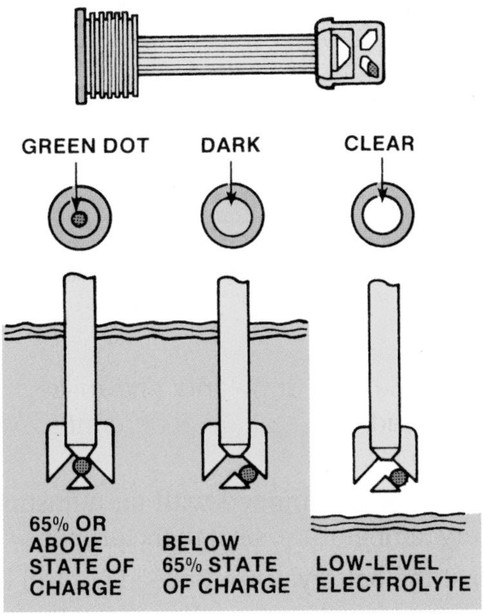

Figure 17-32 Design and operation of built-in hydrometers on maintenance-free sealed batteries.

Figure 17-33 A battery with a built-in charge indicator. *Courtesy of Robert Bosch GmbH, www.bosch-presse.de*

Open Circuit Voltage Test

An open circuit voltage check can be used as a substitute for the hydrometer specific gravity test on maintenance-free sealed batteries with no built-in hydrometer. As the battery is charged or discharged, slight changes occur in the battery's voltage. Therefore, battery voltage with no load applied can give some indication of the SOC.

The battery's temperature should be between 60° and 100°F (15.5° and 37.7°C). The voltage must be allowed to stabilize for at least 10 minutes with no load applied. On vehicles with high drains (computer controls, clocks, and accessories that always draw a small amount of current), it may be necessary to disconnect the battery ground cable. On batteries that have just been recharged, apply a heavy

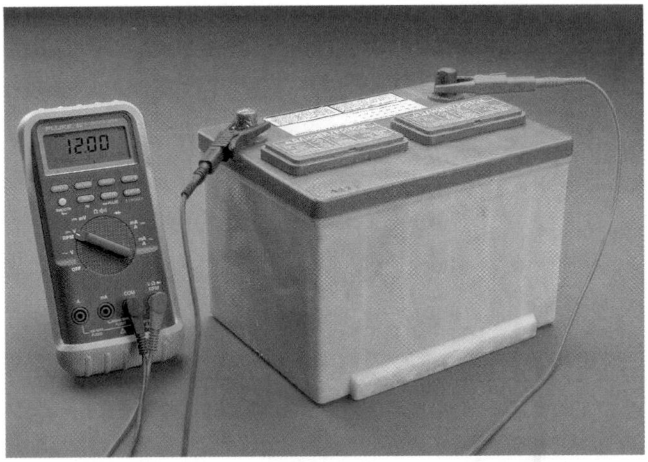

Figure 17-34 Measuring open circuit voltage across battery terminals using a voltmeter.

load for 15 seconds to remove the surface charge, then allow the battery to stabilize. Once voltage has stabilized, use a digital voltmeter to measure the battery voltage to the nearest one-tenth of a volt **(Figure 17–34)**. A fully charged 12-volt battery should have a terminal voltage of 12.6 volts. However, sealed AGM and gel cell batteries may have a slightly higher voltage (12.8–12.9 volts). Use Figure 17–28 to interpret the results. As you can see, minor changes in battery open circuit voltage can indicate major changes in SOC. If the test indicates an SOC below 75%, recharge the battery and perform the capacity test to determine battery condition.

Battery Load Test

The **load** or **capacity test** determines how well a battery performs under a load. It determines the battery's ability to furnish starting current and still maintain sufficient voltage to operate other systems. A battery load tester **(Figure 17–35)** or a volt/ampere tester (VAT) can be used for this test. A VAT can be

Figure 17-35 A battery load tester.

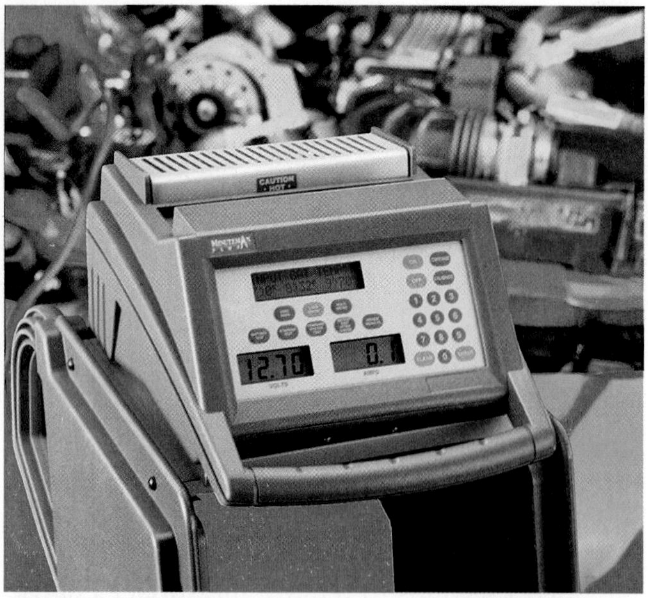

Figure 17–36 A typical automatic VAT. *Courtesy of SPX Service Solutions*

used to test batteries, starting systems, and charging systems. Both testers have a voltmeter, an ammeter, and a carbon pile. The carbon pile is a variable resistor. When the tester is attached to the battery and operated, the carbon pile will draw current from the battery. The ammeter will read the amount of current draw. The maximum current draw from the battery, with acceptable voltage, is compared to the rating of the battery to see if the battery is okay.

Some battery load testers and VATs automatically adjust the load or carbon pile. Battery information is inputted into these machines and it does the rest **(Figure 17–36)**.

The load or capacity test can be performed with the battery either in or out of the vehicle. The battery must be at or very near a full SOC. Use the specific gravity test or open circuit voltage test to determine the SOC, and recharge the battery if needed. For best results, the electrolyte should be as close to 80°F (26.7°C) as possible. Cold batteries show considerably lower capacity. Never load test a sealed battery if its temperature is below 60°F (15.5°C). Photo Sequence 14 shows the correct way to use a VAT-40 to conduct a load test.

During a load test, a load that simulates the current draw of a starting motor is put on the battery. The amount of current draw is determined by the rating of the battery and the battery's voltage is observed for 15 seconds.

When performing a battery load test, follow these guidelines:

- Make sure that the inductive pickup surrounds all of the wires from the battery's negative terminal.

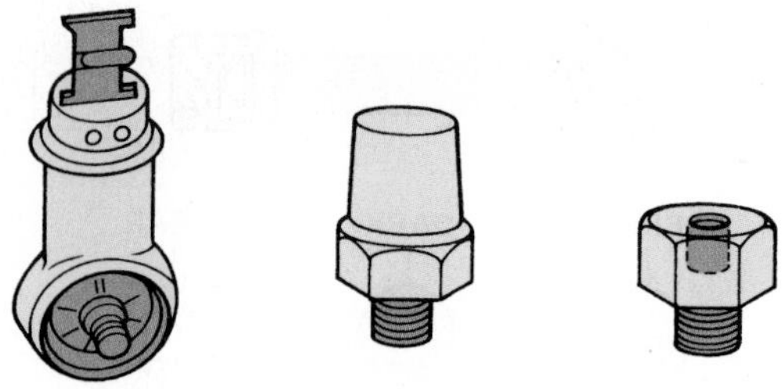

Figure 17–37 Adapters may be needed to test and change batteries with side-mount terminals.

- Observe the correct polarity and make sure that the test leads are making good contact with the battery posts.
- If the tester is equipped with an adjustment for battery temperature, set it to the proper setting.
- To test the battery, turn the load control knob to draw current at the rate of three times the battery's ampere-hour rating or one-half of its CCA rating.
- Discontinue the load after 15 seconds.

SHOP TALK

On some batteries with side terminals, obtaining a sound connection for the tester can be a problem. The best solution is to use the appropriate adapter **(Figure 17–37)**. If an adapter is not available, use a ⅜-inch coarse bolt with a nut on it. Bottom out the bolt. Back it off a turn. Then tighten the nut against the contact. Now attach the lead to the nut.

Interpreting Results If the voltage reading exceeds 10.5 volts, the battery is supplying sufficient current with a good margin to safety. If the reading is right on the spec, the battery might not have the reserve necessary to handle cranking during low temperatures.

After the load is turned off, if the voltage reads below the minimum, watch the voltmeter. If it rises above 12.4, the battery is bad—it can hold a charge but cannot supply the required current. The battery can be recharged and retested, but the results are likely to be the same.

If the voltage tests below the minimum and the voltmeter does not rise above 12.4 when the load is removed, the problem may only be a low SOC. Recharge the battery and load test again.

Battery Capacitance Test

Many manufacturers recommend that a **capacitance** or **conductance test** be performed on batteries. Conductance describes a battery's ability to conduct current. It is a measurement of the plate surface

P14-1 *To conduct this test, you need a VAT-40 or similar equipment. Before conducting the test, make sure the battery is fully charged.*

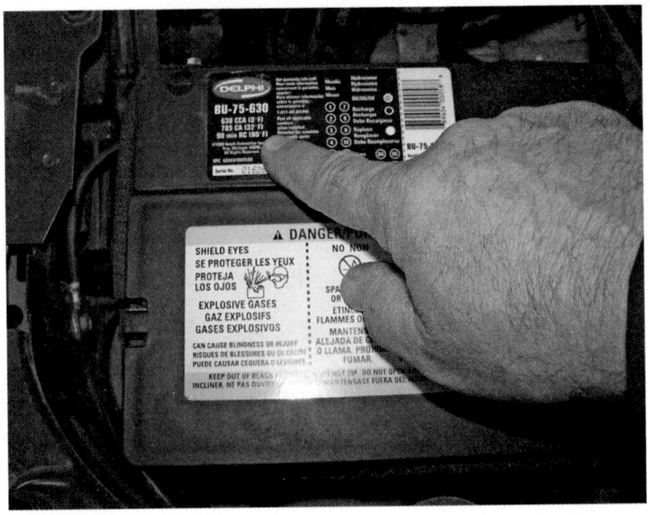

P14-2 *Check the markings on the battery to identify its CCA rating. This battery has a rating of 630 CCA.*

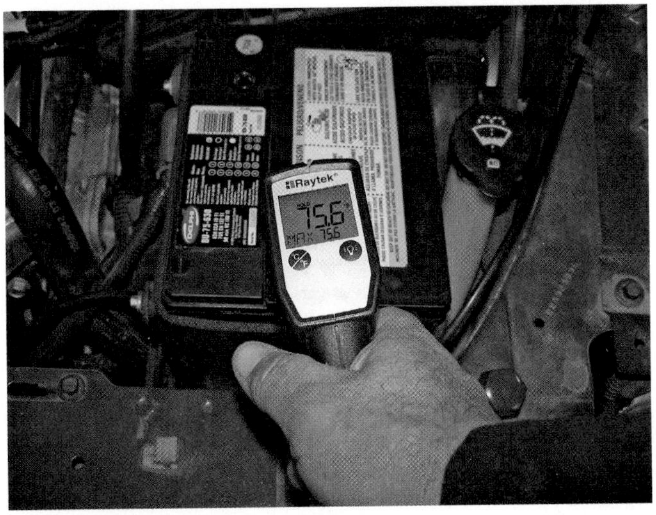

P14-3 *Check the temperature of the battery; it should be around 70°F (21°C).*

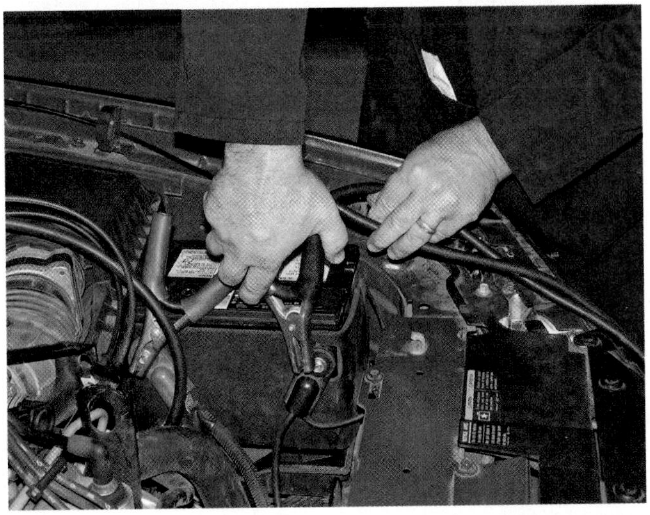

P14-4 *Attach the tester's battery leads to the correct terminals on the battery.*

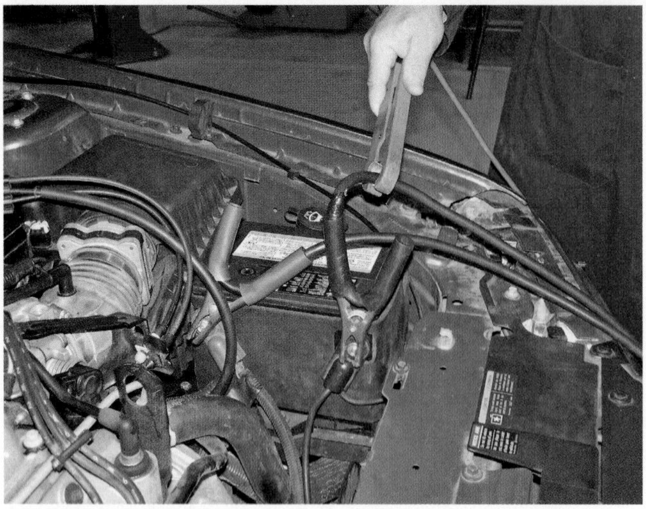

P14-5 *Clamp the tester's inductive lead around the tester's negative cable. Normally the arrow on the clamp should face away from the negative post of the battery.*

P14-6 *Turn the "Zero Adjust" knob on the tester until the ammeter reads zero.*

P14-7 *Rotate the load control until the ammeter reads one-half of the battery's CCA rating, which in this case is 315 amps.*

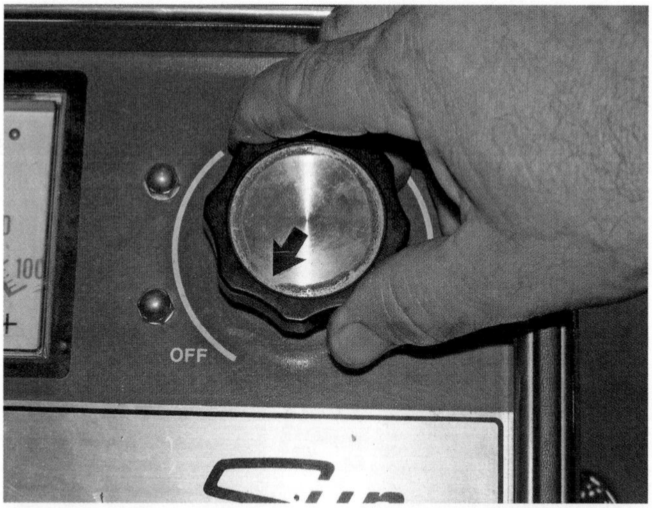

P14-8 *After 15 seconds, read the voltage of the battery and turn the load control* off. *Then disconnect the tester's leads.*

available in a battery for chemical reaction. Measuring conductance provides a reliable indication of a battery's condition and is correlated to battery capacity. Conductance can be used to detect cell defects, shorts, normal aging, and open circuits, which can cause the battery to fail.

A fully charged new battery will have a high conductance reading anywhere from 110% to 140% of its CCA rating. As a battery ages, the plate surface can sulfate or shed active material, which will lower its capacity and conductance. A conductance test is the only test that will yield accurate measurements of a battery with low SOC.

When a battery has lost a significant percentage of its cranking ability, the conductance reading will fall well below its rating and the test decision will be to replace the battery. Because conductance measurements can track the life of the battery, they are also effective for predicting end of life before the battery fails.

To measure conductance, the tester **(Figure 17–38)** creates a small signal that is sent through the battery and then measures a portion of the AC current response. The tester displays the service condition of the battery. The tester will indicate that the battery is good, needs to be recharged and tested again, has failed, or will fail shortly.

Many conductance testers will display a code at the end of the test. Be sure to look up this code in the operating manual before taking any action in response to the test results.

Figure 17-38 A conductance (capacitance) battery tester.

SHOP TALK

When testing an AGM battery with a typical capacitance tester, add 100 to the CCA rating of the battery when asked to enter the rating by the tester. If this is not done, all test results will be wrong.

Go To *Chapter 6 for a detailed list of battery safety precautions.*

Battery Drains

Many of today's vehicles have **parasitic loads**. Parasitic loads are current drains that exist when the key is off. This drain is caused by systems that operate when the engine is not running. A parasitic load is normal unless it exceeds specifications. These drains can cause a battery to lose its charge overnight or after a few days **(Table 17–1)**. The drains can also deplete the battery and cause various driveability problems. The

TABLE 17-1 A LOOK AT HOW LONG IT TAKES DIFFERENT CURRENT DRAINS TO DROP A TYPICAL BATTERY'S STATE-OF-CHARGE (SOC) TO 50%.

Constant Current Drain	50% SOC IN:
25 mA	30½ Days
50 mA	16½ Days
100 mA	8¼ Days
250 mA	3¼ Days
500 mA	1½ Days
750 mA	1 Day
1 A	19 Hours
2 A	12 Hours

computer may go into its back-up mode, set false codes, or raise idle speeds to compensate for the low battery voltage. A constantly low battery will also have a shortened service life.

Most manufacturers have a specification for the maximum allowable amount of parasitic drain. All vehicles will have some because small amounts of current are needed to maintain the memory in various systems. A drain of 30 mA is normal for most. This will not cause a battery to lose its charge quickly. Excessive current drains are caused by problems and they will run down a battery. The most common cause is a light that is not turning off—such as the glove box, trunk, or engine compartment light. The problem can also be in an electronic system. Many of these systems are designed to periodically monitor conditions; these episodes are called wake-up times. If an electronic unit wakes up but does not shut down soon afterward, it is malfunctioning and will drain the battery.

When a battery quickly loses its charge, a battery drain test should be conducted. The current drain can be measured with a DMM connected in series with the negative battery cable or by placing a low current probe around the cable. Current drain can also be measured with a high-current tester, such as a VAT. This requires the installation of a multiplying coil between the negative battery cable and the battery terminal post **(Figure 17–39)**. The tester's inductive probe is placed around the multiplying coil.

SHOP TALK

Remember to connect a memory saver before disconnecting the battery.

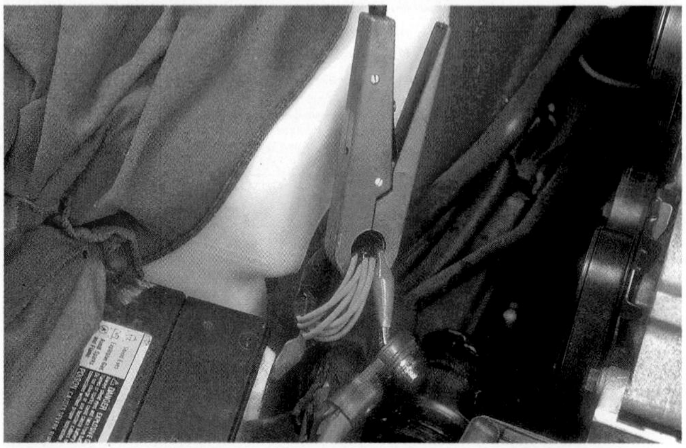

Figure 17-39 Using a multiplying coil to obtain accurate readings when measuring parasitic drains with a VAT-40 or similar tester.

PROCEDURE

To measure the parasitic drains on a battery:

1. Make sure the ignition is off and the key is out of the switch.
2. Turn off all accessories.
3. Place the ignition switch and all vehicle accessories in the off position. Some manufacturers recommend to wait 20 minutes after the ignition is off before testing.
4. Disconnect the underhood lamp, if equipped.
5. Set the DMM to read DC amps.
6. Zero the ammeter.
7. Place the probe around the negative battery cable or insert the meter in series with the cable.
8. Read the ammeter.
9. If the parasitic load is under 2 amps, change the DMM to read mAmps.
10. Read the ammeter.
11. If the parasitic drain is more than the specified maximum, check the trunk, glove box, and interior lights to see if they are on.
12. If a light was on, remove the bulb and watch the battery drain. If the drain is now okay, check that circuit.
13. If the cause of the drain is not the lights, find the problem circuit by removing one fuse at a time while watching the ammeter. The reading will drop when the fuse on the bad circuit is removed.
14. Test the components of that circuit to identify the cause of the drain.

Sometimes excessive battery drain will not be measured at the time of the test. If the battery runs down quickly and has passed all other tests, it may be necessary to monitor the drain overnight. Doing this will verify any problems related to the wake-up modes in the various systems. To do this, use the MIN/MAX feature of the DMM and monitor the parasitic drain overnight.

Battery Cables

Sometimes a battery's performance is affected by its cables. Poor connections and corrosion at the terminals and cable ends will cause voltage drops, which will reduce the available voltage at all of the vehicle's systems. A visual inspection may locate these problems but a voltage drop test across the connections is the best way to identify this problem. Heavily corroded terminals **(Figure 17–40)** are obvious, but sometimes corrosion forms a nearly invisible barrier between the battery terminals and cables. This can be unwanted resistance and will prevent the proper amount of current and voltage from going where they should.

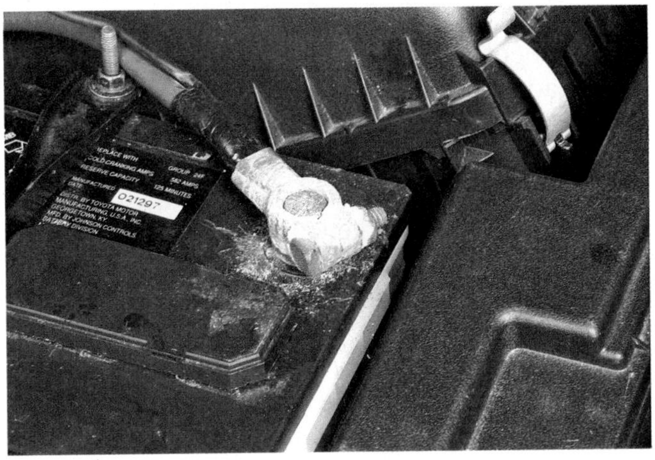

Figure 17-40 Corroded battery terminals reduce the efficiency of the battery and cause voltage drops.

The terminals may also be out of shape, loose, or damaged; these will cause excessive resistance. All battery cables should be inspected and tested for excessive voltage drops. All voltage drop tests should be done under a load. Operate a high-current system while measuring the voltage drop. In most cases, running the starter motor is the best load. Allow a drop of 0.1 volt for each cable connection; a total voltage drop across all battery connections should not exceed 0.2 volt.

Battery Chargers

Battery chargers are designed to supply a constant voltage or a constant current, or a mixture of the two. Constant voltage chargers provide a specific amount of voltage to the battery. The current varies with voltage of the battery. When the potential difference between the charger's voltage and the battery's voltage is great, the current is high. As the battery charges, its voltage increases and the charging current drops off. A constant current charger varies the voltage applied to the battery in order to maintain a constant current.

Both of these techniques work fine as long as the temperature of the battery is maintained. Some chargers have a thermometer to monitor battery temperature. These chargers reduce the charging voltage and/or current in response to rising temperatures.

There are many battery chargers that are "smart" or "intelligent." These chargers are designed to charge a battery in three basic steps: bulk, absorption, and float. During bulk charging, current is sent to the battery at a maximum safe rate until the voltage reaches approximately 80% of its capacity. Once the battery reaches that voltage level, the charger begins the absorption step. During this time, charging voltage is held constant while the current changes according to the battery's voltage. Once the battery is fully charged, the charger switches to the float step. During this step, the charger supplies a constant voltage equal to slightly more than the voltage of the battery. The current flow is very low. This step is a maintenance charge and is intended to keep a battery charged while it is not being used.

Chargers can also be designed to supply voltage and current to a battery according to the needs of the vehicle and, often, the needs of the customer. **Fast charging** quickly charges a battery. These supply large amounts of voltage and current. Although this charges the battery quickly, it also can overheat it if it is not closely monitored. This technique is best used when a battery is low on charge and will be installed into the vehicle in a short time. Batteries must be in good condition to accept a fast charge. Sulfation on the plates of the battery can lead to excessive gassing, boiling, and heat buildup during fast charging. Never fast-charge a battery that shows evidence of sulfation buildup or separator damage.

Slow or **trickle charging** applies low current to the battery and takes quite some time to fully charge it. However, it is unlikely that the battery will overheat and the battery has a good chance to be completely charged. Slow charging is the only safe way of charging sulfated batteries. The chemicals used in the construction of the battery should always be considered before fast charging or slow charging a battery.

Always charge a battery in a well-ventilated area away from sparks and open flames. The charger should be off before connecting or disconnecting the leads to the battery (**Figure 17–41**). Remember to wear eye protection, and never attempt to charge a frozen battery.

All battery chargers have manufacturer-specific characteristics and operating instructions that must be followed. When charging a battery in the vehicle,

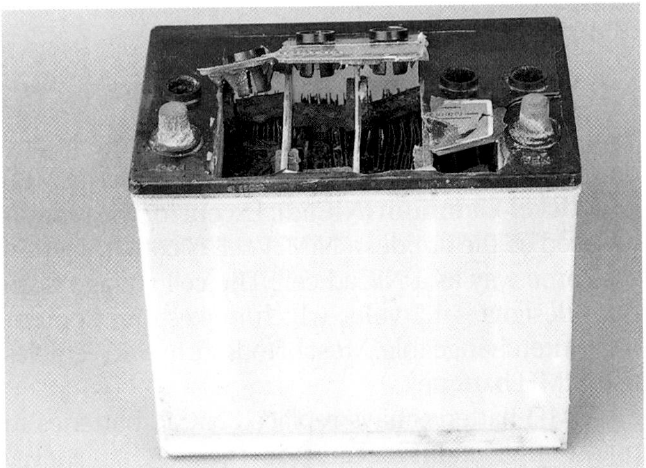

Figure 17-41 Careless use of a battery charger caused this battery to explode.

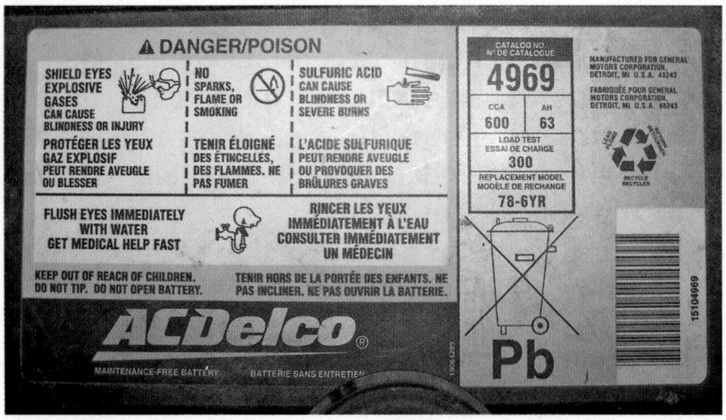

Figure 17-42 Battery sticker with identification and warnings.

always disconnect the battery cables to avoid damaging the generator or other electrical parts.

> **CAUTION!**
>
> *Do not exceed the manufacturer's battery-charging limits. Also never charge the battery if the built-in hydrometer is clear or light yellow. Replace the battery.*

Replacing a Battery

When a battery needs to be replaced, several things need to be considered when selecting the new battery: its capacity, ratings, and size. Make sure the new battery meets or exceeds the power requirement of the vehicle. The battery needs to fit the battery holding fixtures in the vehicle. The height of the battery is also important. The top of the battery and its terminals must fit safely under the hood without the possibility of shorting across the terminals. BCI group numbers are normally listed on the battery **(Figure 17–42)** and are used to state its physical size and rating.

NICKEL-BASED BATTERIES

Two types of nickel-based rechargeable batteries are commonly used: the **nickel-metal hydride (NiMH)** and **nickel-cadmium (NiCad)**. Except for the materials used as the anode, a NiMH cell is constructed in the same way as a NiCad cell. The cell voltage from both designs is 1.2 volts, which makes them potentially interchangeable. Most of today's hybrid vehicles use NiMH batteries.

NiMH batteries have replaced NiCad batteries in many applications because of their higher energy densities (more energy is available for a given amount of space), and they use environmentally friendly metals. Cadmium is considered to be harmful to the environment.

NiCad batteries also suffer from something called the "memory effect." This problem occurs when a battery is not fully discharged and then recharged. If the battery is consistently being recharged after it is only partially discharged, 50%, for example, the battery will eventually only accept and hold a 50% charge. NiMH batteries tend not to be affected by the memory effect.

Nickel-Metal Hydride Cells

NiMH cells are available in the cylindrical and prismatic designs. Both are used in hybrid vehicles. The prismatic design **(Figure 17–43)** requires less storage space but has less energy density than the cylindrical design **(Figure 17–44)**.

Figure 17-43 A high-voltage battery pack made with prismatic cells. *Courtesy of Toyota Motor Sales, U.S.A., Inc.*

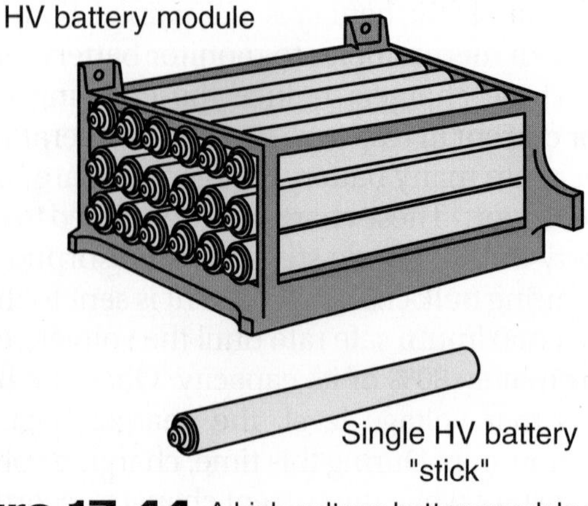

Figure 17-44 A high-voltage battery module made of several cylindrical battery cells.

These cells are still being studied and developed to overcome some of their weaknesses or limitations. They have a relatively short service life; however, most batteries used in hybrid vehicles have an 8-year warranty. Service life is reduced by subjecting the battery to many deep cycles of charging and discharging. NiMH cells generate high heat while being charged.

Chemical Reactions NiMH batteries have a positive plate that contains nickel hydroxide. The negative plate is made of hydrogen-absorbing metal alloys. The plates are separated by a sheet of fine fibers saturated with an aqueous and alkaline electrolyte—potassium hydroxide. The components of the cell are typically placed in a metal housing and the unit sealed. There is a safety vent that allows high pressures to escape, if needed.

The most commonly used hydrogen-absorbing alloys are compounds of titanium, vanadium, zirconium, nickel, cobalt, manganese, and aluminum. An alloy formed by the combination of two or three of these metals has the ability to absorb and store hydrogen. The amount of hydrogen that can be stored is many times greater than the actual volume of the alloy.

When an NiMH cell discharges, hydrogen moves from the negative to the positive plate (**Figure 17–45**). The electrolyte has no active role in the chemical reaction; therefore, the electrolyte level is not changed by the reaction. When the cell is recharged, hydrogen moves from the positive to negative electrode.

Characteristics There are certain characteristics of NiMH cells that must be understood and remembered, especially for diagnostic purposes and safety.

- NIMH batteries do not store well for long lengths of time, so if the vehicle has not been driven for a while, the battery may lose quite a bit of its normal capacity.
- These batteries have a limited number of charging cycles.
- All hybrids charge the battery while the vehicle is driven.
- The capacity of NiMH batteries decreases with an increase in heat.
- If the battery is exposed to intense heat, it is possible that hydrogen will be released from the battery, and hydrogen can be explosive when introduced to flame or sparks.
- The cells contain an electrolyte comprised mostly of potassium hydroxide. Exposure to the electrolyte can cause skin/eye irritation and or burns.
- The battery packs are very heavy.
- Used batteries should be disposed of according to local and federal laws.

Whenever a battery pack, cooling system components (**Figure 17–46**), or other hybrid parts must be removed, serviced, and/or replaced, always follow the specific procedures given by the manufacturer. If these are not followed exactly, injury to you, the vehicle, and others can result.

Recharging Recharging the high-voltage battery pack is best done by the vehicle itself; however, there are times when it may be necessary to recharge the battery in the shop. Doing so is not a typical procedure. Chances are your shop will not have the correct charger. For example, Toyota hybrid batteries require a special charger that is not sold to its dealerships (**Figure 17–47**). If there is a need for one, the dealership must contact the regional office and have one delivered, and only someone from that office is allowed to operate it. This charger has the normal connections plus a cable to power the battery's cooling system. The charger is designed to bring the battery pack to a 40% to 50% SOC within 3 hours. This is enough to start the vehicle and allow the engine to bring the battery back to full charge.

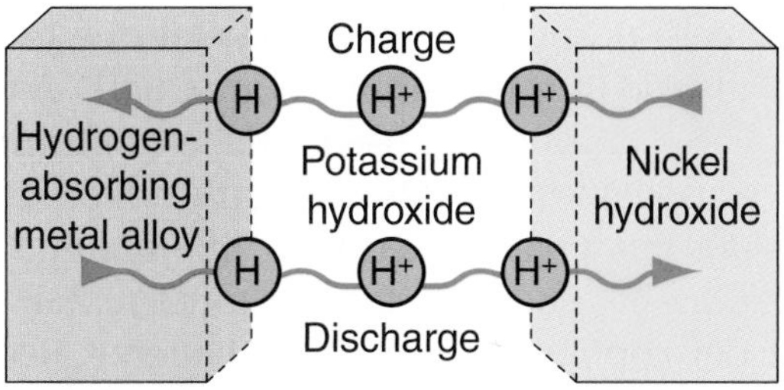

Figure 17-45 The chemical action inside an NiMH cell when it is discharging and charging.

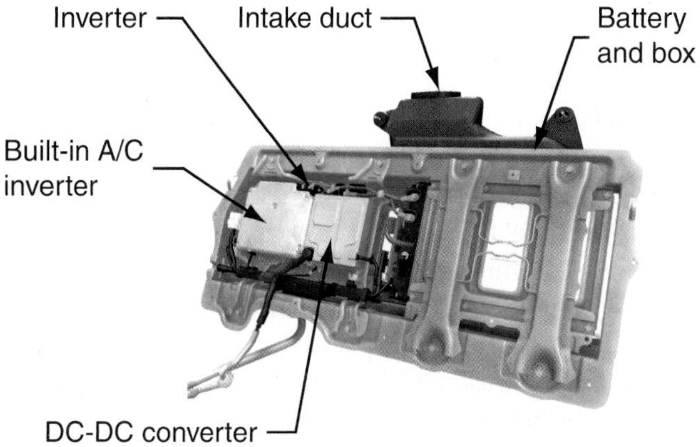

Figure 17-46 An assembly comprised of various hybrid parts plus the battery.

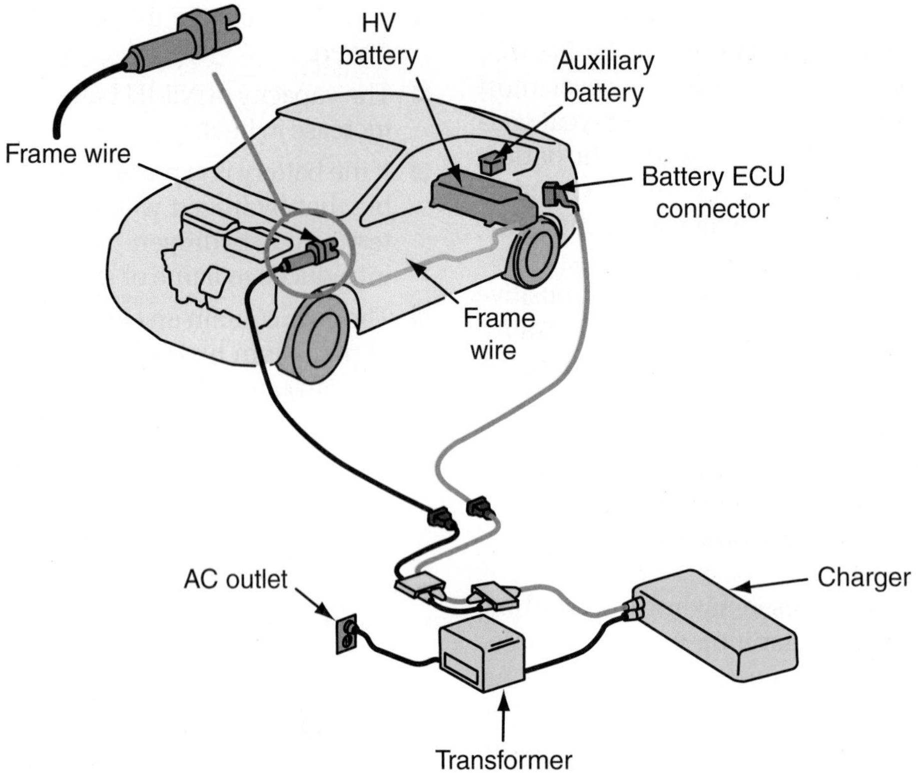

Figure 17-47 The high-voltage battery charger for Toyota hybrids. There are two output cables: one for the battery pack and the other operates the 12 V system for the battery cooling fans and computer.

Nickel-Cadmium Cells

NiCad (also referred to as NiCd) cells are commonly used in many appliances and may have a future in electric drive vehicles. They have a number of advantages over NiMH cells. One of these advantages is that they can withstand many deep cycles, about three times as many as NiMH cells. NiCad batteries can be produced at relatively low costs and have a long shelf and service life. NiCad batteries are especially good performers when high-energy boosts are required. Most NiCad batteries are of the cylindrical design.

On the downside, NiCad batteries have toxic metals, suffer from the memory effect, have low energy densities, and require charging after they have been unused for a while.

Chemical Reactions NiCad batteries use a nickel hydroxide positive electrode. This plate is made of a fiber mesh. The anode or negative plate is also made of fiber. The plate is covered with cadmium. An alkaline electrolyte, aqueous potassium hydroxide (KOH), serves as a conductor of ions and has only a slight role in the chemical reaction. During discharge, ions leave the anode and move to the cathode. During charging, the opposite occurs.

LITHIUM-BASED BATTERIES

Rechargeable lithium-based batteries are very similar in construction to nickel-based batteries. They also have high energy density, suffer less from the memory effect, and are environmentally friendlier. There are two major types of lithium-based cells: lithium-ion and lithium polymer.

Lithium is the lightest metal and provides the highest energy density of all known metals. Lithium oxidizes very rapidly and is highly flammable and slightly explosive when exposed to air and water. Lithium metal is also corrosive.

Lithium-Ion Battery

Battery cells do not use lithium metal due to safety issues; instead they use lithium compounds. A variety of lithium compounds is used. The term **lithium ion (Li-Ion)** applies to all batteries that use lithium regardless of the materials mixed with the lithium. These compounds have been the focus of much research during the development of lithium-ion batteries. Recently, a manganese lithium-ion battery has been developed that may last twice as long as an NiMH battery.

Li-Ion cells have a high voltage output, 3.6 volts. The compounds used in the cell determine the energy density of the cell. However, the higher density designs are more dangerous. Safety issues also surface when connecting Li-Ion cells, in series or parallel, to form a battery pack. Not all Li-Ion cells are designed to be used in a battery pack; only cells that meet tight voltage and capacity tolerances should be used. If the connected cells do not have the same output and capacity, the battery pack can be overcharged and cause a fire.

Li-Ion batteries are expensive to produce. This is mostly due to the required protection circuit. Because lithium metal is very reactive and can explode, the cycling of the battery must be monitored. The protection circuit limits the peak voltage of each cell during charging and prevents the voltage from dropping too low during discharge. The temperature of the battery pack is also monitored and charge and discharge activities are controlled to prevent high temperatures. The circuit also contains electronics and/or fuses to prevent polarity reversal.

Chemical Reactions As with most other rechargeable cells, ions move from the anode to the cathode when the cell is providing electrical energy, and during recharging the ions are moved back from the cathode to the anode **(Figure 17–48)**. The anode is made of graphite, a carbon compound. The cathode is mostly comprised of graphite and a lithium alloy oxide. The construction of the cathode is one of the areas that researchers are working on to produce a safer and stronger Li-Ion battery.

The electrolyte is also the target of much research. The basic electrolyte is a lithium salt mixed in a liquid. Polyethylene membranes are used to separate the plates inside the cells and, in effect, separate the ions from the electrons. The membranes have extremely small pores that allow the ions to move within the cell.

Charging Li-Ion batteries should be charged with a constant voltage. Charging current should change in response to the voltage of the cell. In other words, as the voltage of the cell increases, the charging current should decrease. Li-Ion batteries should not be fast-charged.

Due to the explosive nature of lithium, certain precautions must be followed when discharging and charging Li-Ion batteries:

- Never connect cells in parallel and/or series that are not designed to be connected together or do not have identical output voltages.
- Never charge or discharge the battery if it is not connected to its protection circuit.

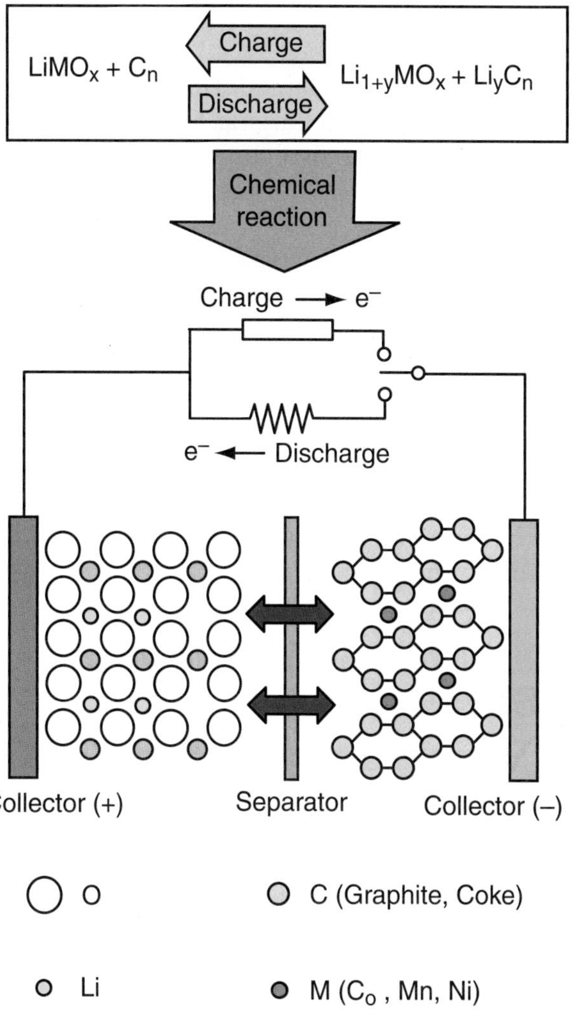

Figure 17–48 The chemical action inside a Li-Ion cell when it is supplying electrical energy and when it is being charged.

- If the protection circuit does not have a temperature sensor, carefully monitor the battery's temperature while charging and discharging.
- Never charge a Li-Ion battery that is physically damaged.

Lithium-Polymer Batteries

The **lithium-polymer (Li-Poly or LiPo) battery** was developed through the continuous research on the Li-Ion battery. The electrolyte used in Li-Poly cells is not a liquid; rather, the lithium salt is held in a solid polymer composite (such as polyacrylonitrile). The polymer electrolyte is not flammable and these batteries are less hazardous than Li-Ion batteries.

The dry polymer electrolyte does not conduct electricity; rather it allows ions to move between the anode and cathode. The polymer electrolyte also serves as the separator between the plates. The dry electrode has very high resistance and therefore cannot provide bursts of current for heavy loads.

Heating the cell to 140°F (60°C) or higher increases its efficiency. The voltage of a Li-Poly cell is about 4.23 volts when fully charged. The cells must be protected to prevent overcharging. However, these cells are more resistant to overcharging than Li-Ion cells and there is much less of a chance for electrolyte leakage.

Li-Poly cells are expensive to manufacturer and have a much higher cost-to-energy ratio than Li-Ion cells. However, because Li-Poly cells do not use a metal case, they are lighter and can be packaged in many ways. As a result of the flexible packaging, Li-Poly cells have a much higher energy density than Li-Ion, NiMH, and NiCad batteries.

In some Li-Poly cells, a gelled electrolyte has been added to enhance ion conductivity. These are called lithium-ion-polymer cells.

ISOLATING HIGH-VOLTAGE SYSTEMS

WARNING!

Careless handling of some hybrid components can lead to serious injury, including death. Always follow and adhere to the precautions given by the manufacturer. These precautions are clearly labeled in their service manuals. Here are just a few of those safety precautions:

- *Make sure the high-voltage system is shut down and isolated from the vehicle before working near or with any high-voltage component.*
- *When working on or near the high-voltage system, even when it is depowered, always use insulated tools.*
- *Never leave tools or loose parts under the hood or close to the battery pack. These can easily cause a short.*
- *Never wear anything metallic, such as rings, necklaces, watches, and earrings, when working on a hybrid vehicle.*
- *Warning and/or caution labels are attached to all high-voltage parts. Be careful not to touch these cables and parts without the correct protective gear, such as lineman's gloves.*
- *Always wear lineman's gloves during the process of depowering and powering the system back up again.*

The procedure for properly depowering and isolating the high-voltage system from the rest of the vehicle is very important and not very difficult. However, each manufacturer has its own procedure that must be followed in the order presented. With the correct information and by following the procedures you can safely work on a hybrid vehicle.

The following discussion covers the depowering procedures for the currently available assist and full hybrids. Other hybrids also have high-voltage systems. These systems should also be depowered before performing any service.

Honda Hybrids

Honda's procedure for depowering the high-voltage systems includes the following steps, which must be followed in the order they appear to avoid serious personal injury and damage to the vehicle's electrical system. Always wear undamaged electrically insulated gloves when inspecting and handling any high-voltage cable or component.

1. Turn the ignition switch off.
2. On many models, the rear seat back must be removed to gain access to the battery module.
3. Remove the switch lid from the IPU cover on the battery module.
4. Then remove the locking cover.
5. Turn the battery module switch off (**Figure 17–49**), and then reinstall the locking cover to secure the switch in the off position.
6. Wait for 5 or more minutes after turning the battery module switch off to allow the ultracapacitors to discharge.
7. Then disconnect the negative cable from the 12-volt battery.
8. Remove the IPU cover.

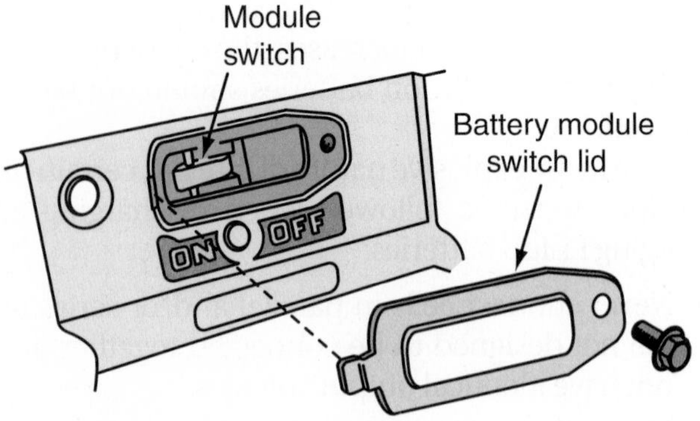

Figure 17–49 The battery module switch in a Honda hybrid.

9. Locate the terminals for the junction board.
10. Measure the voltage at those terminals. If the voltage is less than 30 volts, the system can be serviced. If the voltage is greater than 30 volts, there is a problem. This must be diagnosed before continuing with any service.
11. When the system is ready to be powered again, make sure all circuits have been securely reconnected, then reverse the depowering sequence.

Toyota Hybrids

All Toyota and Lexus hybrids have the same depowering sequence, although some use a much higher voltage than others. These higher voltage systems require only one additional precaution, that being the wait time for the ultracapacitors to discharge. Toyota calls for a wait of at least 5 minutes. On the vehicles with very high voltages, longer waiting times are safer. The basic procedure for depowering the high-voltage system includes:

1. Remove the key from the ignition switch. If the vehicle has a "smart" key, turn the smart key system off. This may be done by applying pressure to the brake pedal while depressing the start button for at least 2 seconds. If the READY lamp goes off, continue. If it does not, diagnose the problem before continuing.
2. Disconnect the negative (−) terminal cable from the auxiliary 12-volt battery. This should turn the high-voltage system off but does not complete the depowering process.
3. Make sure you are wearing insulated gloves and remove the service plug from the battery module **(Figure 17–50)**. To gain access to the plug, you may need to move the carpeting in the rear of the vehicle.
4. Put the service plug in your pocket to prevent others from reinstalling it before the system is ready or while you are working on the vehicle.
5. Put electrical insulating tape over the service plug connector.
6. Wait at least 5 minutes before proceeding or doing any work on or around the high-voltage system.
7. Prior to handling any high-voltage cable or part, check the voltage at the terminals. There should be less than 12 volts.
8. If a high-voltage cable must be disconnected for service, wrap its terminal with insulating tape to prevent a possible short.

Ford Motor Company Hybrids

The hybrids from Ford have a procedure similar to Toyota. However, the service disconnect plug can be

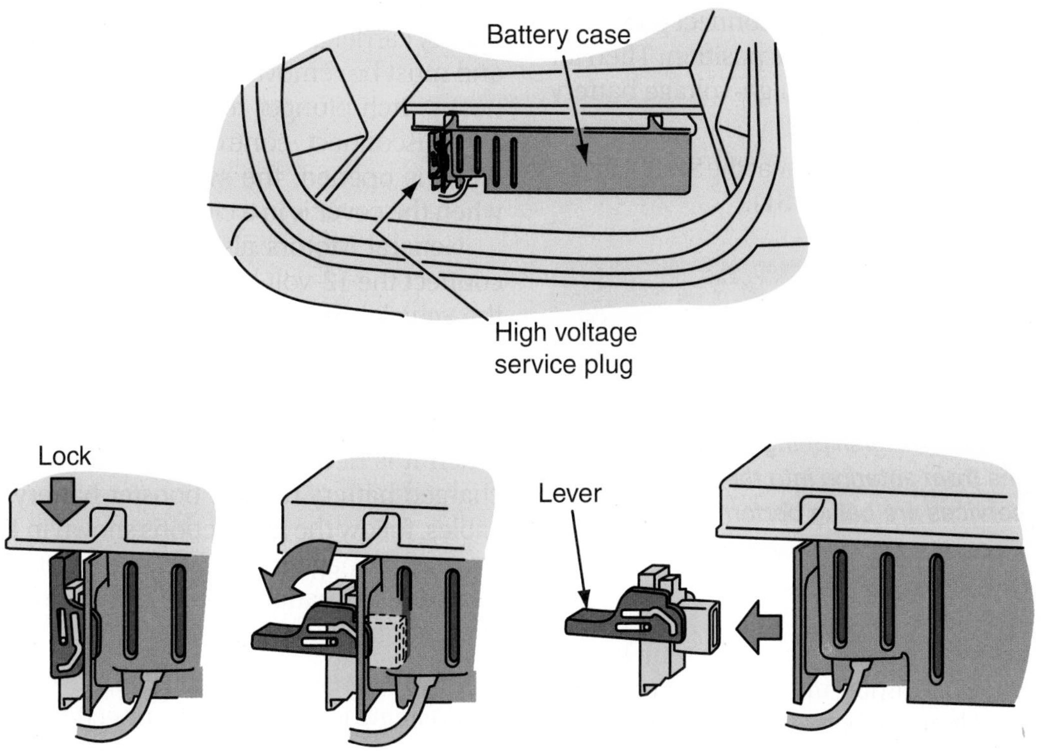

Figure 17–50 Removal of the service plug on a Toyota hybrid to disconnect the high-voltage circuits.

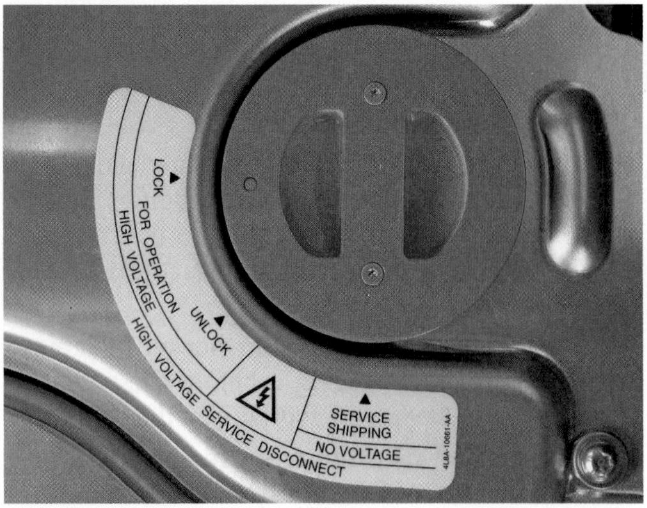

Figure 17-51 To isolate the high-voltage system, turn the service disconnect plug from the LOCK position (1) to the UNLOCK position (2). Then lift it out and reinstall it in the SERVICING/SHIPPING position (3).

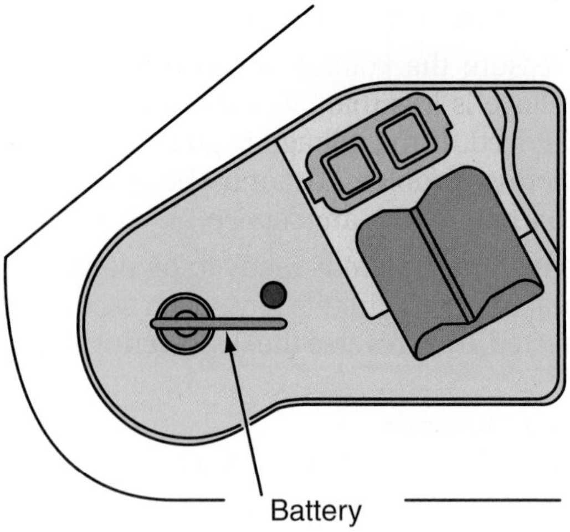

Figure 17-52 When the service disconnect switch for a GM hybrid switch is in the horizontal position, the switch is *off*.

installed in different positions, rather than simply be in or out. To isolate the high voltage:

1. Place the gear selector into the park position and remove the ignition key. This should isolate the high-voltage system, but for safety reasons continue through the procedure.
2. Disconnect the negative cable at the 12-volt battery; this should also isolate the high voltage, but continue through this procedure.
3. Lift up the carpeting behind the rear seat.
4. Locate and turn the service disconnect plug from the lock position to the unlock position. Then lift it out. This disconnects the high-voltage battery pack from the vehicle.
5. Reinstall the plug with its arrow at the SERVICING/SHIPPING mark **(Figure 17–51)**.

CAUTION!

Although the removal of the plug disconnects the battery pack, it should be returned to its bore but in the servicing/shipping position to prevent debris from entering into the battery pack while services are being performed.

GM's Hybrids

General Motors is currently using different hybrid systems. The most common of these have a 42-volt starter/generator located between the engine and transmission or at the front of the engine connected to the crankshaft via a belt.

To disconnect the 42-volt system from the vehicle there is a switch on the side of the battery pack called the energy storage box (ESB). The battery pack is located under the rear passenger seat. The service disconnect switch is located behind a removable cover on the battery pack. Once the cover is removed, the switch is turned to its horizontal or off position **(Figure 17–52)**. This disconnects the 42-volt system from the rest of the vehicle.

On some vehicles, the generator battery disconnect control module cover should be opened to disconnect the high-voltage system. This cover is a hinged panel attached to the rear of the generator battery carrier assembly. A bolt holds the cover closed and must be removed to open it. The service disconnect switch plunger rests against the generator battery disconnect control module cover. When the cover is opened, the switch will open; it will close when the cover is in a closed position.

General Motors also advises technicians to disconnect the 12-volt battery whenever working under the vehicle's hood and when working on or around any part related to the vehicle's hybrid system.

JUMP-STARTING

When it is necessary to jump-start a car with a discharged battery using a booster battery and jumper cables, follow the instructions shown in **Figure 17–53** to avoid damaging the charging system or creating a hazardous situation. Pay attention to the following precautions:

- Always wear eye protection when making or breaking jumper cable connections.
- Be sure the connections are done correctly and tightly.

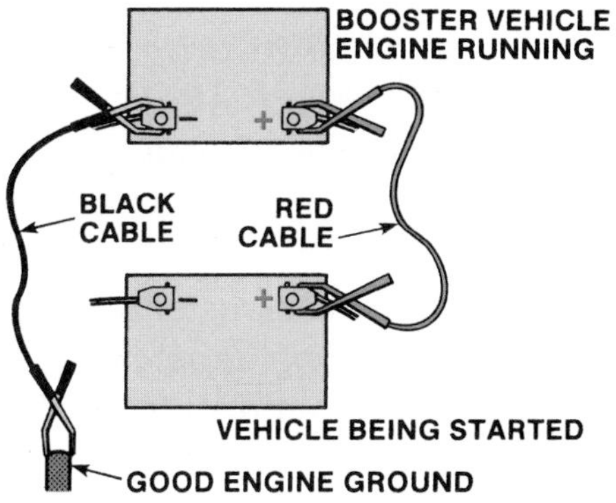

Figure 17-53 Proper setup and connections for jump-starting a vehicle with a low battery.

- Make sure all electrical accessories of both vehicles are off.
- When making the connections, do not lean over the battery or accidentally let the jumper cables or clamps touch anything except the correct battery terminals.
- Use only a 12-volt supply as the booster battery.
- The gases around the battery can explode if exposed to flames, sparks, or lit cigarettes.

PROCEDURE

To jump-start a typical hybrid vehicle with the battery of another vehicle, follow this procedure:

1. Park the booster vehicle close to the hood of the disabled vehicle, making sure that the two vehicles do not touch.
2. Set the parking brake on both vehicles.
3. Remove the ignition key and turn off accessories on both vehicles.
4. Open the hoods or trunks to gain access to the batteries. On some vehicles, there is a special jump-starting terminal under the hood.
5. Connect the clamp of the positive (red) jumper cable to the positive terminal on the weak battery.
6. Connect the clamp at the other end of the positive (red) jumper cable to the positive (+) terminal on the booster battery.
7. Connect the clamp of the negative (black) jumper cable to the negative (−) terminal on the booster battery.
8. Connect the clamp at the other end of the negative (black) jumper cable to an exposed metal part of the dead vehicle's engine, away from the battery and the fuel injection system. Never use a fuel line, engine camshaft or rocker arm covers, or the intake manifold for making this connection.
9. Make sure the jumper cables are clear of fan blades, belts, moving parts of both engines, or any fuel delivery system parts.
10. Run the engine of the booster vehicle at a medium speed for about 5 minutes.
11. Now attempt to start the disabled vehicle.
12. If the vehicle does not start, check the connections of the jumper cables and if a problem is found, correct it and try again. If the engine still does not start, the battery should be replaced.
13. Once the disabled vehicle has started, allow both engines to run for about 5 minutes.
14. Disconnect the negative cable from the disabled vehicle and then from the booster battery.
15. Next, disconnect the positive cable from the disabled vehicle and then from the booster battery.

High-Voltage Batteries

If the starting or high-voltage battery in a hybrid vehicle is discharged, the engine will not start nor will the vehicle be able to operate on electric power only. If the engine cranks but does not start, the high-voltage battery may need to be jumped. Manufacturers have built-in ways to jump-start these vehicles, if and when the batteries go dead. Some hybrid vehicles have separate procedures for jump-starting—one for the low-voltage battery and the other for the high-voltage battery. The basic connection from a booster battery to the dead battery is made in the same way as a conventional vehicle. However, the connecting points may be different and there are certain precautions to consider when jump-starting. Also refer to the service information prior to jump-starting a hybrid.

To jump-start, or get the engine running, on a Ford Escape or Mariner hybrid, make sure the ignition is *off* before proceeding. Then open the access panel on the driver's side foot well and press the jump-start button **(Figure 17–54)**. Wait at least 8 minutes before

Figure 17-54 To jump-start a Ford Escape or Mariner hybrid high-voltage battery, turn the ignition *off*, open the access panel on the driver's side foot well, and press the jump-start button.

continuing. This is important! If you continue sooner, the energy from the auxiliary battery will not be able to supply enough power to start the engine with the battery pack. When you depress the button, the system sends energy from the auxiliary battery to the battery pack. If the auxiliary battery has ample energy, it will be enough to start the engine. After the wait time, the warning lamp may blink for up to 2 minutes. After it stops, attempt to start the engine. If the engine still does not start, try again in a couple of minutes. If the engine still does not start, the low-voltage battery must be recharged or the vehicle jump-started through the 12-volt battery.

CASE STUDY I

A customer complains that the vehicle will not start without having to jump the battery. The technician learns that this happens every time the customer attempts to start the vehicle. The customer also says the voltmeter in the dash has been reading higher than normal.

The technician verifies the complaint. The engine turns over very slowly for a few seconds, then does not turn. After jumping the battery to get the vehicle into the shop, the technician makes a visual inspection of the battery and cables. The open circuit voltage test shows a voltage of 12.5 volts across the terminals. When a load test is conducted, the voltage drops to 7.8 volts at 80°F (26.7°C). After 10 minutes, the open circuit voltage is back up to 12.5 volts. The technician determines that the battery is sulfated and calls the customer with an estimate of repair costs. The customer agrees to have the battery replaced, which cures the problem.

CASE STUDY II

A very low mileage late-model Buick had been sitting in a garage for about 2 months. The owner no longer drove but went out every day or so and ran the engine for a while. He eventually sold the car and the day the new owner came to drive it away, the battery was totally dead. After a few wrong assumptions made as to why the battery was dead, a service truck was called in to jump the battery.

While connecting the booster battery to the vehicle, heavy sparking took place as the negative cable was connected. This was ignored and the owner started the car. It ran well but the new owner was concerned there was a problem and decided not to attempt to drive it home until he had someone follow him. That could happen in a few days. During this period, the owner went out daily to start and run the engine. For 3 days it started, but on the fourth day the battery was dead again. The owner contacted the new owner, who suggested that the vehicle be jumped and driven to a shop. Once at the shop, the technician tried to drive the car inside. The battery was dead. Again it was jumped and driven in for testing.

Battery tests were conducted and no problems identified. The technician called the new owner and reported the results with a suggestion that the battery probably just needed a slow charge. This was done and the owner had the car delivered back to him. The next day, he went to start the car and again the battery was dead. The shop was called and had the vehicle towed in. And again the battery needed to be jumped. This time, however, someone paid attention to the sparks when the jumper cables were connected. Once the car was in the shop, the battery was retested and was found to have low open circuit voltage. The technician then wiggled the battery and retested it. This time it had normal voltage. This confirmed the cause of the problem. It also explained why there was heavy sparking when it was connected to the booster battery. The battery had a loose plate and it was shorting out a battery cell sometimes. A new battery was installed and the new owner is still driving the hardly ever used Buick with a new battery.

KEY TERMS

Absorbed glass mat (AGM) battery
Ampere-hour (AH) rating
Battery Council International (BCI)
Capacitance test
Capacity test
Cold cranking amps (CCA)
Conductance test
Cranking amps (CA) rating
Cylindrical cell
Electrochemical reaction
Electrodes
Element
Fast charging
Grid
Hybrid battery
Lead peroxide (PbO_2)
Lithium-ion (Li-Ion)
Lithium-polymer (Li-Poly or LiPo) battery
Load test
Maintenance-free battery
Nickel-cadmium (NiCad)
Nickel-metal hydride (NiMH)
Open circuit voltage
Parasitic load
Plates
Plate strap
Prismatic cells
Recombinant battery
Reserve capacity (RC) rating
Separator
Slow charge
Specific gravity
Sponge lead (Pb)
Sulfation
Trickle charge
Valve-regulated lead-acid (VRLA) batteries
Watt-hour rating

SUMMARY

- The primary source for electrical power in all automobiles is the battery.
- Batteries are devices that convert chemical energy into electrical energy through electrochemical reactions.
- Batteries are normally made up of electrochemical cells connected together. Each cell has three major parts: an anode (negative plate or electrode), a cathode (positive plate or electrode), and electrolyte.
- Open circuit voltage is the voltage measured across the battery when there is no load on the battery.
- Operating voltage is the voltage measured across the battery when it is under a load.
- The ampere-hour (AH) rating is the amount of steady current that a fully charged battery can supply for 20 hours at 80°F (26.7°) without the cell's voltage dropping below a predetermined level.
- A battery's watt-hour rating is determined at 0°F (–17.7°C) and is calculated by multiplying a battery's amp-hour rating by the battery's voltage.
- The cold cranking amps (CCA) rating is determined by the load, in amperes, that a battery is able to deliver for 30 seconds at 0°F (–17.7°C) without its voltage dropping below a predetermined level.
- The cranking amps (CA) rating is a measure of the current a battery can deliver at 32°F (0°C) for 30 seconds and maintain voltage at a predetermined level.
- The reserve capacity (RC) rating is determined by the length of time, in minutes, that a fully charged starting battery at 80°F (26.7°) can be discharged at 25 amperes before battery voltage drops below 10.5 volts.
- There are many different types and designs of batteries available; the primary difference is the chemicals used in the cells, such as lead-acid, nickel-cadmium (NiCad), nickel-metal hydride (NiMH), sodium-sulfur (NaS), sodium-nickel-chloride, lithium-ion (Li-Ion), lithium-polymer (Li-Poly), and nickel-zinc.
- High-voltage battery packs are assemblies of cylindrical or prismatic cells.
- The most common automotive batteries are lead-acid designs. The wet cell, gel cell, absorbed glass mat (AGM), and valve regulated are versions of the lead-acid battery.
- A lead-acid battery consists of grids, positive plates, negative plates, separators, elements, electrolyte, a container, cell covers, vent plugs, and cell containers. Maintenance-free batteries have no holes or caps, but they do have gas vents. Sealed maintenance-free batteries do not require the gas vents used on other maintenance-free designs.
- Improper electrolyte levels, temperature, corrosion, overcharging, undercharging/sulfation, poor mounting, and cycling affect the service life of a battery.
- Depending on the design of the battery, its condition, SOC, and capacity can be measured by different tests: battery leakage test, specific gravity tests, built-in hydrometers, open circuit voltage test, capacity test, and capacitance or conductance test.
- Parasitic loads are current drains that exist when the key is off and may be normal or caused by a problem.
- To charge a battery, a given charging current is passed through the battery for a period of time. Fast chargers are more popular, but slow charging is the only safe way to charge a sulfated battery.
- Nickel-metal hydride (NiMH) cells are available in the cylindrical and prismatic designs and both are used in hybrid vehicles.

- Lithium-based batteries are similar in construction to nickel-based batteries but have higher energy density, suffer less from the memory effect, and are environmentally friendlier.
- There are two major types of lithium-based cells: lithium-ion and lithium-polymer.
- Always follow the correct procedure when jump-starting a vehicle.
- Some hybrid vehicles have separate procedures for jump-starting: one for the low-voltage battery and the other for the high-voltage battery.

REVIEW QUESTIONS

1. What is a parasitic load?
2. What design characteristics determine the current capacity of a battery?
3. How is a battery leakage test conducted?
4. Describe the difference between cylindrical and prismatic battery cells.
5. How can you identify the high-voltage system in most hybrid vehicles?
6. How many volts are present in a fully charged 12-volt battery?
7. What is meant by the term *specific gravity*?
8. What causes gassing in a battery?
9. List five things that shorten the life of a battery.
10. *True or False?* All electrochemical batteries have three major parts: an anode, a cathode, and an electrolyte.
11. *True or False?* A fully charged new battery will have a low conductance reading.
12. *True or False?* Batteries are recycled more than any other item.
13. *True or False?* Heating a lithium-polymer cell increases its efficiency.
14. Which of the following statements about NiMH cells is true?
    a. The cylindrical design requires less storage space but has less energy density than the prismatic design.
    b. They have a relatively long service life.
    c. Service life can be extended by frequent deep cycles.
    d. They do not respond well to overcharging; they should be trickle charged.
15. Which of the following statements about battery ratings is true?
    a. The ampere-hour rating is defined as the amount of steady current that a fully charged battery can supply for 1 hour at 80°F (26.7°C) without the cell voltage falling below a predetermined voltage.
    b. The cold cranking amps rating represents the number of amps that a fully charged battery can deliver at 0°F (–17.7°C) for 30 seconds while maintaining a voltage above 9.6 volts for a 12 V battery.
    c. The cranking amp rating expresses the number of amperes a battery can deliver at 32°F (0°C) for 30 seconds and maintain at least 1.2 volts per cell.
    d. The reserve capacity rating expresses the number of amperes a fully charged battery at 80°F can supply before the battery's voltage falls below 10.5 volts.

ASE-STYLE REVIEW QUESTIONS

1. Technician A uses a voltmeter to test the open circuit voltage of a battery. Technician B uses the results of an open circuit voltage test to determine the battery's state of charge. Who is correct?
   a. Technician A
   b. Technician B
   c. Both A and B
   d. Neither A nor B
2. Technician A says that the reserve rating of a battery is the amount of steady current that a fully charged battery can supply for 20 hours without the voltage falling below 10.5 volts. Technician B says that ampere-hour ratings state how many hours the battery is capable of supplying 25 amperes. Who is correct?
   a. Technician A
   b. Technician B
   c. Both A and B
   d. Neither A nor B
3. Technician A says that battery straps should be used whenever a battery is lifted. Technician B says that when a battery is removed, the positive cable should be disconnected first. Who is correct?
   a. Technician A
   b. Technician B
   c. Both A and B
   d. Neither A nor B

4. Technician A says that larger batteries have higher current capacities. Technician B says that BCI group numbers identify the features of a battery. Who is correct?

   a. Technician A c. Both A and B
   b. Technician B d. Neither A nor B

5. Technician A says that battery corrosion is commonly caused by spilled electrolyte. Technician B says that battery corrosion is commonly caused by electrolyte condensation from gassing. Who is correct?

   a. Technician A c. Both A and B
   b. Technician B d. Neither A nor B

6. Technician A says that before a battery is removed, a memory saver ought to be installed. Technician B says that before a battery is removed, the positive cable should be disconnected and the area around the battery should be cleaned. Who is correct?

   a. Technician A c. Both A and B
   b. Technician B d. Neither A nor B

7. Technician A says that nickel-metal hydride batteries have a negative plate that contains nickel hydroxide. Technician B says that the alkaline electrolyte has no active role in the chemical reaction. Who is correct?

   a. Technician A c. Both A and B
   b. Technician B d. Neither A nor B

8. While working with an NiMH battery: Technician A says that if the battery is exposed to extreme cold, it is possible that hydrogen will be released from the battery and hydrogen can be explosive when introduced to flame or sparks. Technician B says that the cells contain an electrolyte comprised mostly of sulfuric acid. Exposure to the electrolyte can cause skin/eye irritation and/or burns. Who is correct?

   a. Technician A c. Both A and B
   b. Technician B d. Neither A nor B

9. Technician A always wears safety glasses or goggles when working with batteries. Technician B says that there is no need to seek medical help if electrolyte gets in your eyes as long as you flush your eyes immediately after the electrolyte hits them. Who is correct?

   a. Technician A c. Both A and B
   b. Technician B d. Neither A nor B

10. While discussing the results of a load test: Technician A says that if a battery fails a load test but the voltage rises above 12.4 volts when the load is removed, the battery should be replaced. Technician B says that if a battery fails a load test but the voltage rises above 12.4 volts when the load is removed, the battery is able to hold a charge. Who is correct?

    a. Technician A c. Both A and B
    b. Technician B d. Neither A nor B

CHAPTER 18

STARTING AND TRACTION MOTOR SYSTEMS

OBJECTIVES

■ **Describe the basic operation of all electric motors.** ■ **Identify the major parts of a DC motor.** ■ **Explain the purpose of the starting system.** ■ **Describe the purpose and major components of the starting and starter control circuits.** ■ **Describe the different types of magnetic switches and starter drives.** ■ **Describe the different types of starting motors used in today's vehicles.** ■ **Inspect and test starter relays and solenoids and the switches, connectors, and wires of starter control circuits.** ■ **Perform starter circuit voltage drop tests and the starter current draw test.** ■ **Disassemble, clean, inspect, test, repair, and reassemble a starting motor.** ■ **Explain the basic operation of an AC motor.** ■ **Understand the characteristics of three-phase AC voltage and describe the operation of a three-phase AC motor.** ■ **Explain the purposes of a controller in a motor circuit.**

Electric motors are used to start the engine and, in hybrid and fuel cell vehicles, to move the vehicle. They also are used to operate many different accessories. For the most part, the basic difference between the motors used in these applications is the size and power output. Most of these motors are powered by DC voltage and are part of relatively simple circuits. Traction or drive motors use AC voltage and are controlled by rather complex circuits. Regardless of the voltage used to power the motor, the operation of all electric motors is based on the basic principles of magnetism.

BASICS OF ELECTROMAGNETISM

Electricity and magnetism are related. One can be used to create the other. Current flowing through a wire creates a magnetic field around the wire. Moving a wire through a magnetic field creates current flow in the wire.

Magnetism

A substance is said to be a magnet if it has the property of magnetism—the ability to attract such substances as iron, steel, nickel, or cobalt. A magnet has two points of maximum attraction, one at each end of the magnet. These points are designated the north pole and the south pole **(Figure 18–1A)**. When two magnets are brought together, opposite poles attract **(Figure 18–1B)**, while similar poles repel each other **(Figure 18–1C)**.

A magnetic field, called a **flux field**, exists around every magnet. The field consists of imaginary lines along which the magnetic force acts. These lines emerge from the north pole and enter the south pole, returning to the north pole through the magnet itself. All lines of force leave the magnet at right angles to the magnet. None of the lines cross each other and all are complete.

Electromagnets

Magnets can occur naturally in the form of a mineral called magnetite. Artificial magnets can be made by inserting a bar of magnetic material inside a coil of insulated wire and passing current through the coil. Another way of creating a magnet is by stroking the magnetic material with a bar magnet. Both methods force the randomly arranged molecules of the magnetic material to align themselves along north and south poles.

Artificial magnets can be either temporary or permanent. Temporary magnets are usually made of soft iron. They are easy to magnetize but quickly lose their

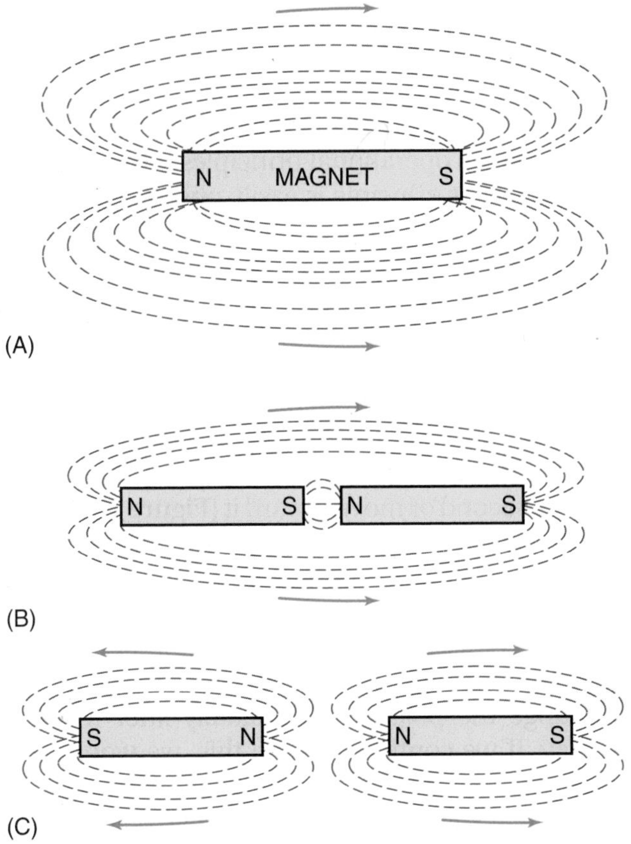

Figure 18–1 (A) In a magnet, lines of force emerge from the north pole and travel to the south pole before passing through the magnet back to the north pole. (B) Unlike poles attract, while (C) similar poles repel each other.

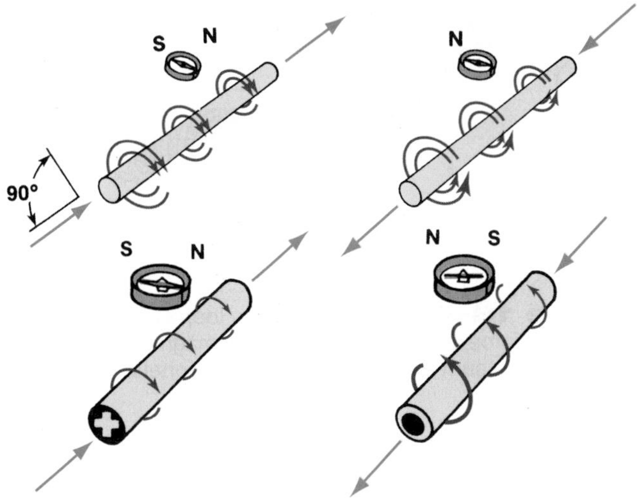

Figure 18–2 When current is passed through a conductor such as a wire, magnetic lines of force are generated around the wire at right angles to the direction of the current flow.

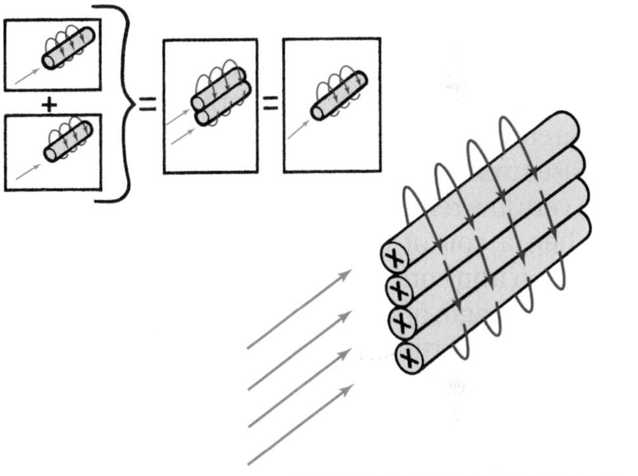

Figure 18–3 Increasing the number of conductors carrying current in the same direction increases the strength of the magnetic field around them.

magnetism when the magnetizing force is removed. Permanent magnets are difficult to magnetize. However, once magnetized they retain this property for very long periods.

The earth is a very large magnet, having a North Pole and a South Pole, with lines of magnetic force running between them. This is why a compass always aligns itself to straight north and south.

In 1820, a simple experiment discovered the existence of a magnetic field around a current-carrying wire. When a compass was held over the wire, its needle aligned itself at right angles to the wire **(Figure 18–2)**. The lines of magnetic force are concentric circles around the wire. The density of these circular lines of force is very heavy near the wire and decreases farther away from the wire. As is also shown in the same figure, the polarity of a current-carrying wire's magnetic field changes depending on the direction the current is flowing through the wire. These magnetic lines of force or flux lines do not move or flow around the wire. They simply have a direction as shown by their effect on a compass needle.

Flux Density The more flux lines, the stronger the magnetic field at that point. Increasing current will increase **flux density**. Also, two conducting wires lying side by side carrying equal currents in the same direction create a magnetic field equal in strength to one conductor carrying twice the current. Therefore, adding more wires increases the magnetic field **(Figure 18–3)**.

Coils Looping a wire into a coil concentrates the lines of flux inside the coil. The resulting magnetic field is the sum of the single-loop magnetic fields **(Figure 18–4)**. The overall effect is the same as placing many wires side by side, each carrying current in the same direction.

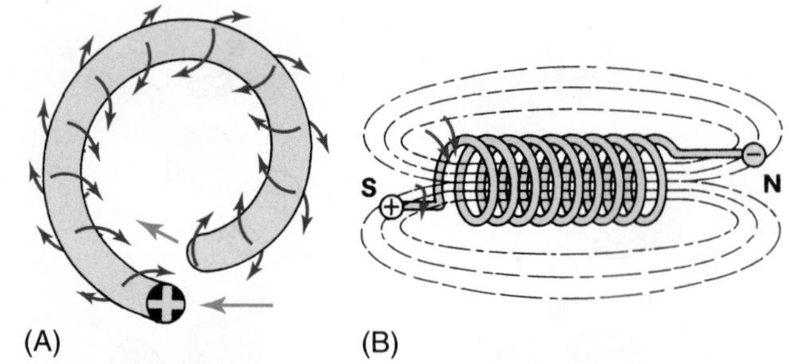

Figure 18–4 (A) Forming a wire loop concentrates the lines of force inside the loop. (B) The magnetic field of a wire coil is the sum of all the single-loop magnetic fields.

Magnetic Circuits and Reluctance

Just as current can only flow through a complete circuit, the lines of flux created by a magnet can only occupy a closed magnetic circuit. The resistance that a material offers to the passage of magnetic flux lines is called **reluctance**. Magnetic reluctance can be compared to electrical resistance.

Reconsider the coil of wire shown in Figure 18–4. The air inside the coil has very high reluctance and limits the magnetic strength produced. However, if an iron core is placed inside the coil, magnetic strength increases tremendously. This is because the iron core has a very low reluctance (**Figure 18–5**).

When a coil of current-carrying wire is wound around an iron core, it becomes a usable electromagnet. The strength of an electromagnet's magnetic field is directly proportional to the number of turns of wire and the current flowing through them.

The equation for an electromagnetic circuit is similar to Ohm's law for electrical circuits. It states that the number of magnetic lines is proportional to the ampere-turns divided by the reluctance. To summarize:

- The magnetic polarity of the coil depends on the direction of current flow through the loop.
- Field strength increases if current through the coil increases.
- Field strength increases if the number of coil turns increases.
- If reluctance increases, field strength decreases.

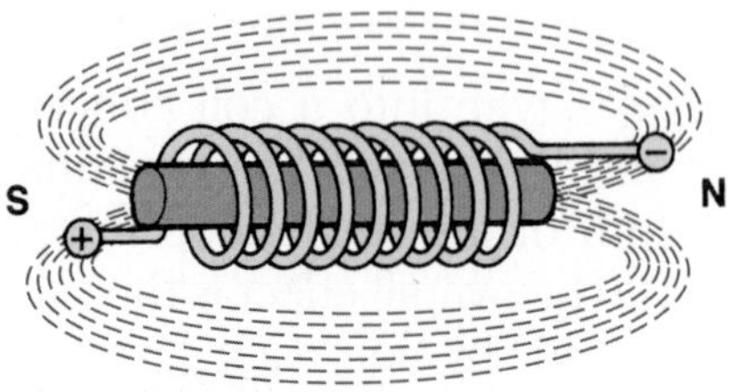

Figure 18–5 Placing a soft iron core inside a coil greatly reduces the reluctance of the coil and creates a usable electromagnet.

Motors

An electric motor converts electric energy into mechanical energy. Through the years, electric motors have changed substantially in design; however, the basic operational principles have remained the same. That principle is easily observed by taking two bar magnets and placing them end to end with each other. If the ends have the same polarity, they will push away from each other. If the ends have the opposite polarity, they will move toward each other and form one magnet.

If we put a pivot through the center of one of the magnets to allow it to spin and moved the other magnet toward it, the first magnet will either rotate away from the second or move toward it (**Figure 18–6**). This is basically how a motor works. Although we do not observe a complete rotation, we do see part of one, perhaps a half turn. If we could change the polarity of the second magnet, we would get another half turn. So in order to keep the first magnet spinning, we need to change the polarity immediately after it moves halfway. If we continued to do this, we would have a motor.

In a real motor, an electromagnet is fitted on a shaft. The shaft is supported by bearings or bushings to allow it to spin and to keep it in the center of the motor. Surrounding, but not touching, this inner magnet is a stationary permanent magnet or an electromagnet. Actually, there is more than one magnet or magnetic field in both components. The polarity of these magnetic fields is quickly switched and we have a constant opposition and attraction of magnetic fields. Therefore, we have a constantly rotating inner magnetic field, the shaft of which can do work due to the forces causing it to rotate. The torque of a motor varies with rotational speed, motor design, and the amount of current draw the motor has. The rotational

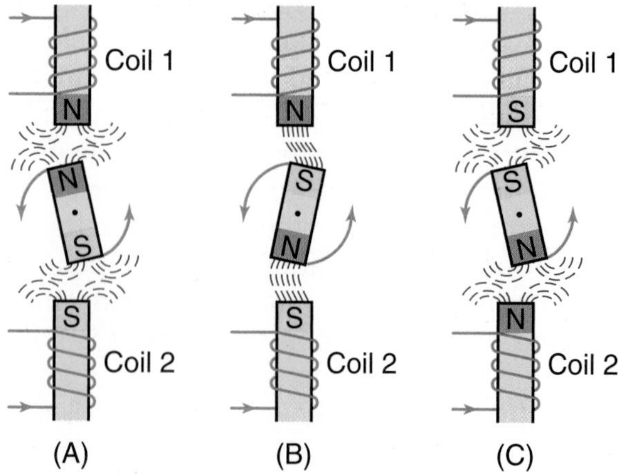

Figure 18–6 (A) Like poles repel. (B) Unlike poles are attracted to each other. Then if we change the polarity of the coils, (C) the like poles again repel.

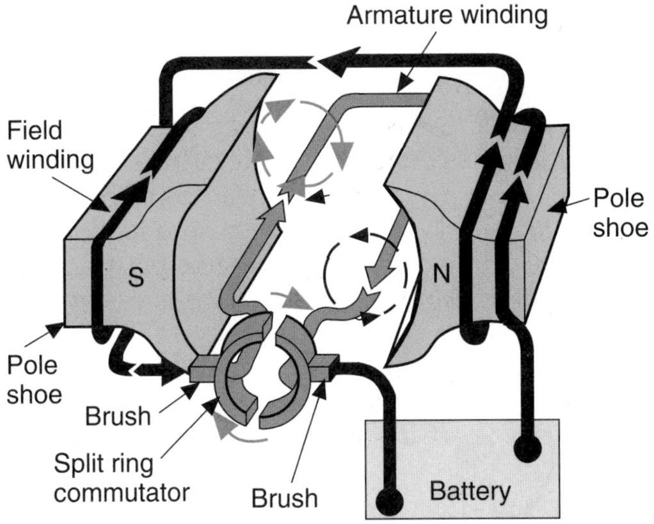

Figure 18-7 A simple electric motor.

Figure 18-8 A cutaway of a starter motor. *Courtesy of Robert Bosch GmbH, www.bosch-presse.de*

speed depends on the motor's current draw, the design of the motor, and the load on the motor's rotating shaft.

The basic components of a motor are the stator or field windings, which are the stationary parts of the motor, and the rotor or armature, which is the rotating part (**Figure 18–7**). The field windings are comprised of slotted cores made of thin sections of soft iron wound with insulated copper wire to form one or more pairs of magnetic poles. Some motors have the field windings wound around iron anchors, called pole shoes. The armature is comprised of loops of current-carrying wire. The loops are formed around a metal with low reluctance to increase the magnetic field. The magnetic fields around the armature are pushed away by the magnetic field of the field windings, causing the armature to rotate away from the windings' fields.

The field windings or the armature may be made with permanent magnets rather than electromagnets. Both cannot be permanent magnets. An electromagnet allows for a change in the polarity of the magnetic fields, which keeps the armature spinning. Changing the direction of current flow changes the magnetic polarities.

STARTING MOTORS

The starting motor (**Figure 18–8**) is a special type of electric motor designed to operate under great electrical loads and to produce great amounts of torque for short periods.

All starting motors are generally the same in design and operation. Basically the starter motor consists of a housing, field coils (windings), an armature, a commutator with brushes, and end frames (**Figure 18–9**).

The **starter housing** or **starter frame** encloses the internal parts and protects them from damage, moisture, and foreign materials. The housing also supports the field coils.

The **field coils** and their **pole shoes** (**Figure 18–10**) are securely attached to the inside of the housing. The field coils are insulated from the housing but are connected to a terminal that protrudes through the outer surface of the housing.

The field coils and pole shoes are designed to produce strong stationary electromagnetic fields within the starter when current flows to it. The magnetic fields are concentrated at the pole shoe. The coils are wound around respective pole shoes in opposite directions to create opposing magnetic fields.

The field coils connect in series with the armature winding through the starter **brushes**. This permits all current passing through the field coil circuit to also pass through the armature windings.

The **armature** is the rotating part of a starter. It is located between the starter drive and commutator end frames and the field windings. Current passing through the armature produces a magnetic field in each of its conductors. The reaction between the armature's magnetic field and the magnetic fields produced by the field coils causes the armature to rotate (**Figure 18–11**). This is the mechanical energy that is then used to crank the engine.

The armature has two main components: the armature windings and the **commutator**. Both mount to the armature shaft (**Figure 18–12**). The armature windings are made of heavy flat copper strips or wires that form a single loop and can handle the heavy current flow. The sides of these loops fit into slots in the armature core or shaft, but they are insulated from it.

The coils connect to each other and to the commutator so that current from the field coils flows

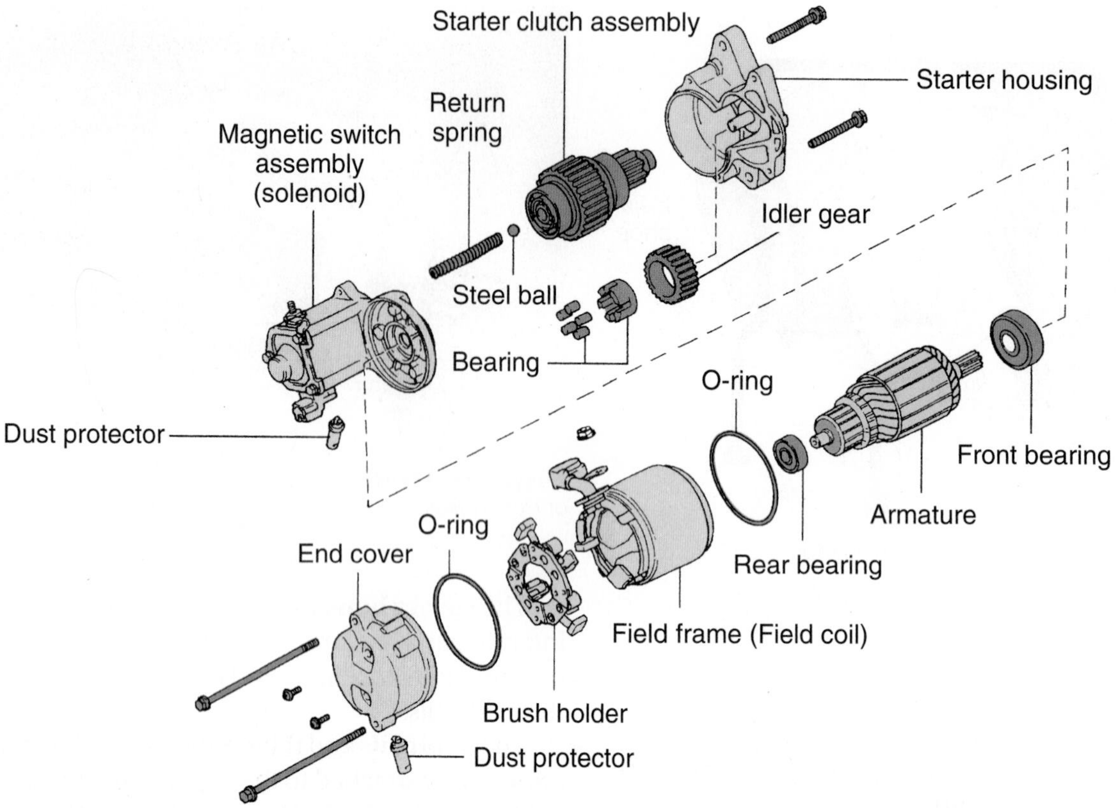

Figure 18-9 A typical starter motor assembly. *Courtesy of Toyota Motor Sales, U.S.A., Inc.*

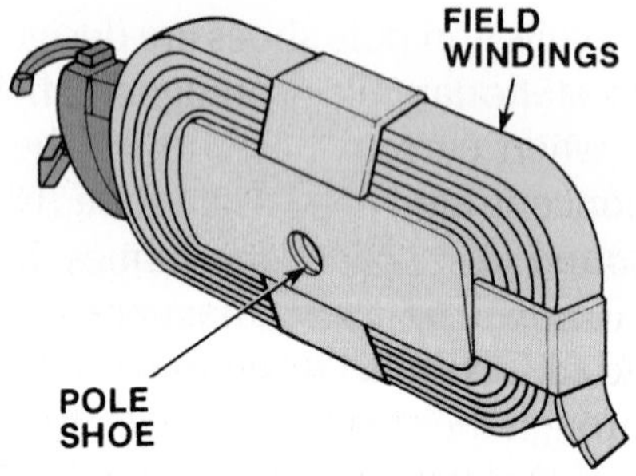

Figure 18-10 Example of a field coil and pole shoe.

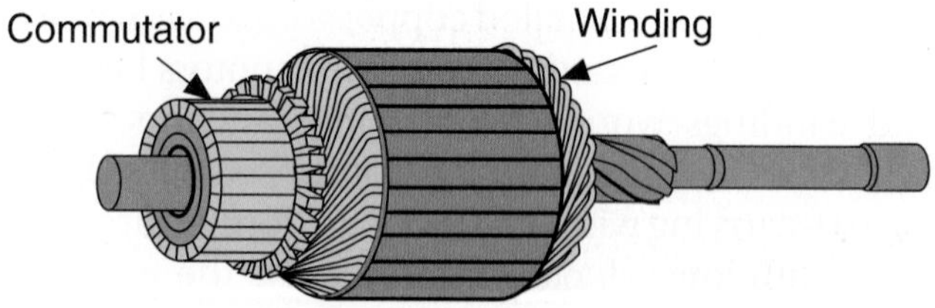

Figure 18-12 The armature of a starter motor.

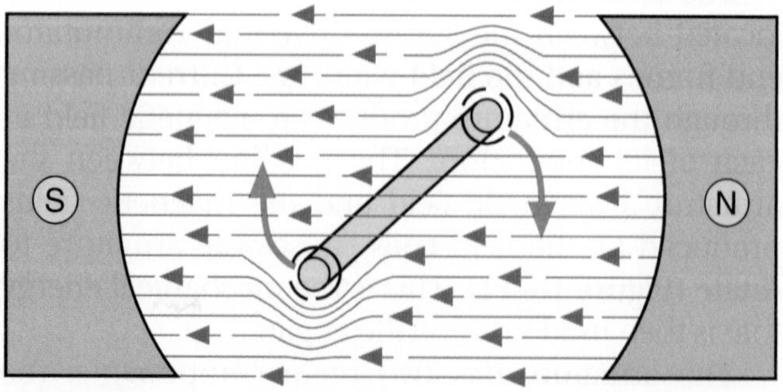

Figure 18-11 Rotation of the conductor is in the direction of the weaker magnetic field.

through all of the windings at the same time. This action generates a magnetic field around each armature winding, resulting in a repulsion force all around the conductor. This repulsion force causes the armature to turn.

The commutator assembly is made up of heavy copper segments separated from each other and the armature shaft by insulation. The commutator segments connect to the ends of the armature windings.

Starter motors have two to six brushes that ride on the commutator segments and carry the heavy current from the stationary field coils to the rotating armature windings. Each end of the armature windings is connected to one segment of the commutator. Carbon brushes are connected to one terminal of the power supply. The brushes contact the commutator segments conducting current to and from the armature coils.

The brushes mount on and operate in some type of holder, which may be a pivoting arm design inside the starter housing or frame (**Figure 18-13**). However, in many starters the brush holders are secured to the starter's end frame. Springs hold the brushes

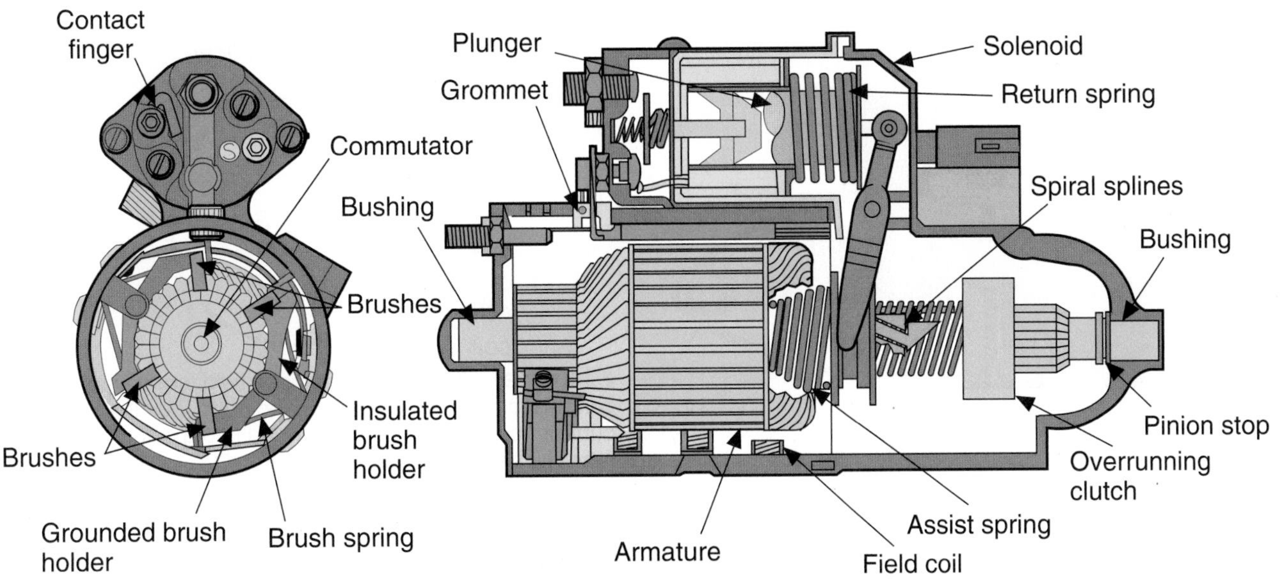

Figure 18–13 The location of the starter motor brushes and commutator.

against the commutator with the correct pressure. Finally, alternate brush holders are insulated from the housing or end frame. Those between the insulated holders are grounded.

The **end frame** is a metal plate that bolts to the commutator end of the starter housing. It supports the commutator end of the armature with a bushing or bearing.

Operating Principles

The starter motor converts current into torque through the interaction of magnetic fields. The magnetic field developed at the field windings and the armature have opposite polarities. When the armature windings are placed inside the field windings, part of the armature coil is pushed in one direction as the field opposes the field in the windings. This causes the armature to begin to rotate. As the armature moves, the contact between a brush and commutator segment is broken and the brush contacts a new segment. This causes a reverse in the polarity of the magnetic field around the armature. The new opposition of magnetic fields causes the armature to rotate more. This process continues and the armature continues to rotate until current stops flowing to the armature.

Many armature segments are used. This provides for a uniform turning motion because as one segment rotates past a brush, another immediately takes its place. This also provides for constant torque.

The number of coils and brushes may differ among starter motor models. The armature may be wired in series with the field coils (**series motor**); the field coils may be wired parallel or shunted across the armature (**shunt motors**); or a combination of series and shunt wiring (**compound motors**) may be used **(Figure 18–14)**.

A series motor develops its maximum torque at startup and develops less torque as speed increases. It is ideal for applications involving heavy starting loads.

Shunt or parallel-wound motors develop considerably less startup torque but maintain a constant speed at all operating loads. Compound motors combine the characteristics of good starting torque with constant speed. This design is particularly useful for applications in which heavy loads are suddenly applied. In a starter motor, a shunt coil is frequently used to limit the maximum free speed at which the starter can operate.

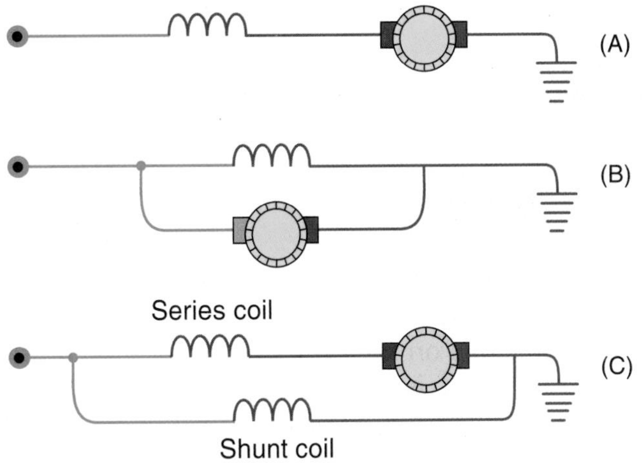

Figure 18–14 Starter motors are grouped according to how they are wired: (A) in series, (B) in parallel (shunt), or (C) a compound motor using both series and shunt coils.

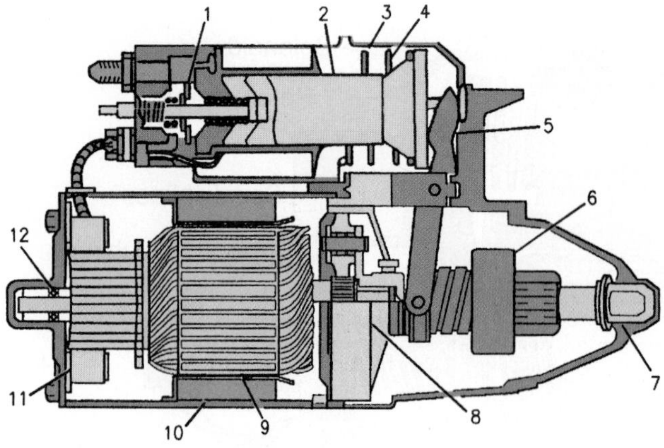

1 CONTACT DISC
2 PLUNGER
3 SOLENOID
4 RETURN SPRING
5 SHIFT LEVER
6 DRIVE ASSEMBLY
7 ROLLER BEARING
8 PLANETARY GEAR REDUCTION ASSEMBLY
9 ARMATURE
10 PERMANENT MAGNETS
11 BRUSH
12 BALL BEARINGS

Figure 18–15 Permanent magnet-type starter assembly.

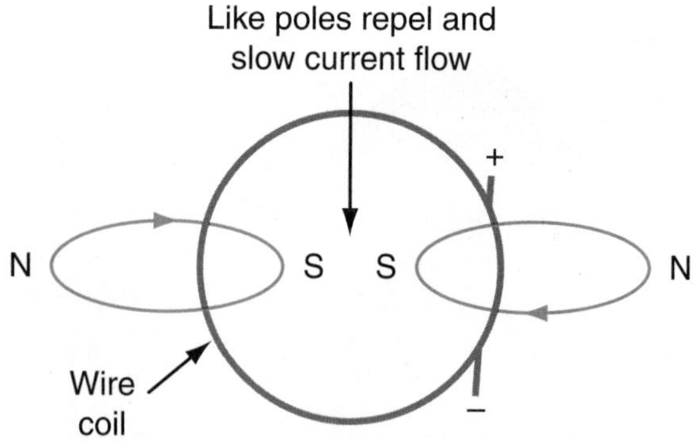

Figure 18–16 A magnetic field forms around a current-carrying conductor according to the direction of current flow. When a wire is formed into a coil, there are repelling poles that result in CEMF.

Permanent Magnet Motors A recent change in starter and accessory motors is the use of permanent magnets instead of electromagnets as field coils. Electrically, this type of motor is simpler. It does not require current for field coils. Current is delivered directly to the armature through the commutator and brushes **(Figure 18–15)**. With the exception of no electromagnets in the fields, this functions exactly as the other motors. Maintenance and testing procedures are the same as for other designs. Notice the use of a planetary gear reduction assembly on the front of the armature. This allows the armature to spin with increased torque, resulting in improved starter cold-cranking performance.

WARNING!

Permanent magnet motors require special handling because the permanent magnet material is quite brittle and can be destroyed with a sharp blow or if the motor is dropped.

Counter EMF The amount of torque from a starting motor depends on a number of things. One of the most important is current draw. The slower the motor turns, the more current it will draw. This is why a starter motor will draw large amounts of current when the engine is difficult to crank. A motor needs more torque to crank a difficult-to-start engine. The relationship between current draw and motor speed is explained by the principles of **counter EMF (CEMF)**. Electromotive force (EMF) is another name for voltage.

When the armature rotates within the field windings, conditions exist to induce a voltage in the armature. Voltage is induced anytime a wire is passed through a magnetic field. When the armature, which is a structure with many loops of wire, rotates past the field windings, a small amount of voltage is induced **(Figure 18–16)**. This voltage opposes the voltage supplied by the battery to energize the armature. As a result, less current is able to flow through the armature.

The faster the armature spins, the more induced voltage is present in the armature. The more voltage induced in the armature, the more opposition there is to normal current flow to the armature. The induced voltage in the armature opposes or is counter to the battery's voltage. This is why the induced voltage is called CEMF.

The effects of CEMF are quite predictable. When the armature of the motor turns slowly, low amounts of voltage are induced and, therefore, low amounts of CEMF are present. The low CEMF permits higher current flow. In fact, the only time a starter motor draws its maximum amount of current is when the armature is not rotating.

STARTING SYSTEM

The starting system is designed to turn or crank the engine until it can operate under its own power. To do this, the starter motor is engaged to the engine's flywheel. As it spins, it turns the engine's crankshaft. The sole purpose of the starting system is to crank the engine fast enough to run. The engine's ignition and fuel system provide the spark and fuel for engine operation, but they are not considered part of the starting system. They do affect how well an engine starts.

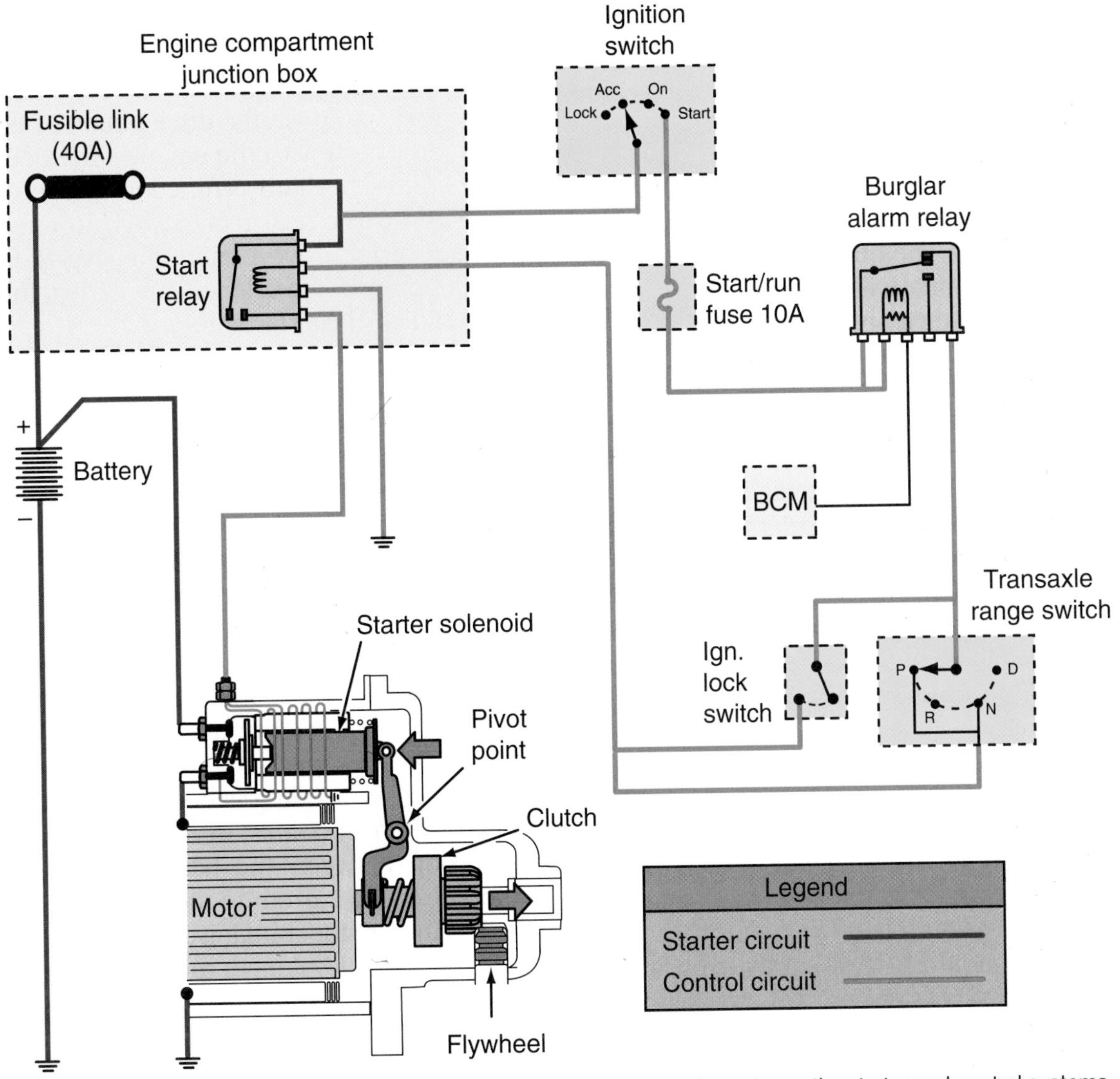

Figure 18–17 The starting system is made up of two separate systems: the starter and control systems.

> **SHOP TALK**
>
> A diesel engine does not have an ignition system. These engines run on the heat of compression. When they are cold and are being started, glow plugs are energized to heat the air-fuel mixture in the cylinders. Glow plugs are simple heaters and do not function as a spark plug. Once the engine is running and warmed up, the glow plug circuit is shut down. If the glow plug circuit is not operating properly, the engine may have a difficult time starting or it may not start at all.

A typical starting system has six basic components and two distinct electrical circuits. The components are the battery, ignition switch, battery cables, magnetic switch (either an electrical relay or a solenoid), starter motor, and the starter safety switch.

The starter motor draws a great deal of current from the battery. A large starter motor might require 250 or more amperes of current. This current flows through the large cables that connect the battery to the starter and ground.

The driver controls the flow of this current using the ignition switch normally mounted on the steering column. The battery cables are not connected to the switch. Rather, the system has two separate circuits: the starter circuit and the control circuit **(Figure 18–17)**. The starter circuit carries the heavy current from the battery to the starter motor through a magnetic switch in a relay or solenoid. The control circuit connects battery power at the ignition switch to the magnetic switch, which controls the high current to the starter motor.

STARTER CIRCUIT

The starter circuit carries the high current flow within the system and supplies power for the actual engine cranking. Components of the starter circuit are the battery, battery cables, magnetic switch or solenoid, and the starter motor.

Battery and Cables

Many of the problems associated with the starting system can be solved by troubleshooting the battery and its related components.

The starting circuit requires two or more heavy-gauge cables. One of these cables connects between the battery's negative terminal and the engine block or transmission case. The other cable connects the battery's positive terminal with the solenoid. On vehicles equipped with a **starter relay**, two positive cables are needed. One runs from the positive battery terminal to the relay and the second from the relay to the starter motor terminal. In any case, these cables carry the required heavy current from the battery to the starter and from the starter back to the battery.

Cables must be heavy enough to comfortably carry the required current load. Cranking problems can be created when undersized cables are installed. With undersized cables, the starter motor does not develop its greatest turning effort and even a fully charged battery might be unable to start the engine.

Magnetic Switches

Every starting system contains some type of magnetic switch that enables the control circuit to open and close the starter circuit. This magnetic switch can be one of several designs.

Solenoid The solenoid-actuated starter is by far the most common starter system used. A solenoid is an electromechanical device that uses the movement of a plunger to exert a pulling or holding force. As shown in **Figure 18–18**, the solenoid mounts directly on top of the starter motor.

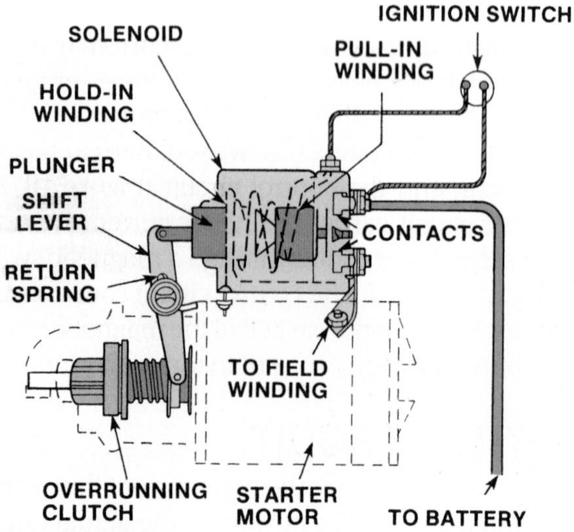

Figure 18–18 Example of a solenoid-actuated starter where the solenoid mounts directly to the starter motor.

In this type of starting system, the solenoid uses the electromagnetic field generated by its coil to perform two distinct jobs.

The first is to push the drive pinion of the starter motor into mesh with the engine's flywheel. This is the solenoid's mechanical function. The second job is to act as an electrical relay switch to energize the motor once the drive pinion is engaged. Once the contact points of the solenoid are closed, full battery current flows to the starter motor.

The solenoid assembly has two separate windings: a **pull-in winding** and a **hold-in winding**. The two windings have approximately the same number of turns but are wound from different size wire. Together these windings produce the electromagnetic force needed to pull the plunger into the solenoid coil. The heavier pull-in windings draw the plunger into the solenoid, while the lighter-gauge windings produce enough magnetic force to hold the plunger in this position.

Both windings are energized when the ignition switch is turned to the start position. When the plunger disc makes contact with this solenoid terminal, the pull-in winding is deactivated. At the same time, the plunger contact disc makes the motor feed connection between the battery and the starting motor, directing current to the field coils and starter motor armature for cranking power.

As the solenoid plunger moves, the shift fork also pivots on the pivot pin and pushes the starter drive pinion into mesh with the flywheel ring gear. When the starter motor receives current, its armature starts to turn. This motion is transferred through an overrunning clutch and pinion gear to the engine flywheel and the engine is cranked.

With this type of solenoid-actuated direct drive starting system, teeth on the **pinion gear** may not immediately mesh with the flywheel ring gear. If this occurs, a spring located behind the pinion compresses so the solenoid plunger can complete its stroke. When the starter motor armature begins to turn, the pinion teeth quickly line up with the flywheel teeth and the spring pressure forces them to mesh.

Starter Relay Starter relays (**Figure 18–19**) are similar to starter solenoids. However, they are not used to move the drive pinion into mesh. They are used as an electrical relay or switch. When current from the ignition switch arrives at the relay, a strong magnetic field is generated in the relay's coil. This magnetic force pulls the plunger contact disc up against the battery terminal and the starter terminal of the relay, allowing full current flow to the starter motor.

Figure 18-19 A starter relay/solenoid mounted on a vehicle.

A secondary function of the starter relay is to provide an alternate electrical path to the ignition coil during cranking. This current flow bypasses the resistance wire (or ballast resistor) in the ignition primary circuit. This is done when the plunger disc contacts the ignition by-pass terminal on the relay. Not all systems have an ignition by-pass setup.

All positive engagement starters use a relay in series with the battery cables to deliver current through the shortest possible battery cables. Some vehicles use both a starter relay and a starter motor-mounted solenoid. The relay controls current flow to the solenoid, which in turn controls current flow to the starter motor. This reduces the amount of current flowing through the ignition switch. In other words, it takes less current to activate the relay than to activate the solenoid.

Positive Engagement Movable Pole Shoe Drive Positive engagement movable pole shoe drive starters **(Figure 18–20)** are found mostly on older Ford products. In this design, the drive mechanism is an integral part of the motor, and the drive pinion is engaged with the flywheel before the motor is energized.

When the ignition switch is moved to the start position, the starter relay closes, and full battery current is delivered to the starter. This current runs through the winding of the movable pole shoe and through a set of contacts to ground. This generates a magnetic force that pulls down the movable pole shoe. It also forces the drive pinion to engage the flywheel ring gear using a lever action and opens the contacts. A small holding coil keeps the movable shoe

Figure 18-20 A positive engagement starter.

and lever assembly engaged during cranking. When the engine starts, an overrunning clutch prevents the flywheel from spinning the armature.

When the ignition switch is released from the start position, both the pole shoe and lever return to their original positions.

Starter Drives

The starter drive is the device that couples the armature with the flywheel. A pinion gear at one end of the armature meshes with the teeth on the outside of the flywheel **(Figure 18–21)**. The spinning armature then turns the flywheel. To prevent damage to the pinion gear or the ring gear on the flywheel, the pinion must mesh with the ring gear before the armature begins to spin. To help ensure smooth engagement, the end of the pinion gear is tapered **(Figure 18–22)**. To disengage the pinion from the flywheel, the pinion is mounted to the armature via an overrunning clutch.

Figure 18-21 Starter drive pinion gear is used to turn the engine's flywheel.

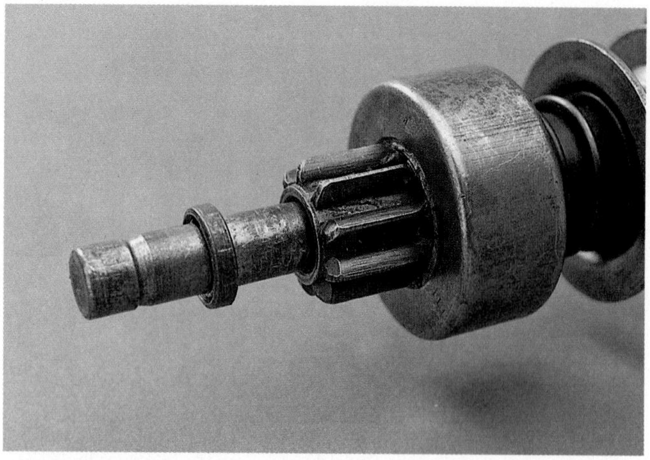

Figure 18–22 The pinion gear teeth are tapered to allow for smooth engagement.

Overrunning Clutch Once the engine starts, its speed increases. If the starter motor remains connected to the engine through the flywheel, it will spin at very high speeds, destroying the armature and other parts.

To prevent this, the starter drive must be disengaged as soon as the engine turns faster than the starter. In most cases, the pinion remains engaged until current stops flowing to the starter. To prevent the armature from spinning at engine speed, an overrunning clutch is used.

The clutch housing is internally splined to the armature shaft. The drive pinion turns freely on the armature shaft within the clutch housing. When the clutch housing is driven by the armature, the spring-loaded rollers are forced into the small ends of their tapered slots and wedged tightly against the pinion barrel. This locks the pinion and clutch housing solidly together, permitting the pinion to turn the flywheel and, thus, crank the engine.

When the engine starts **(Figure 18–23)**, the flywheel spins the pinion faster than the armature. This releases the rollers, unlocking the pinion gear from the armature shaft. The pinion then freely spins on the armature shaft. Once current flow is stopped, the pinion is then pulled away from the flywheel. The overrunning clutch is moved in and out of mesh by the starter drive linkage.

Gear Reduction Drive The armature of some starter motors does not directly drive the starter drive gear. Rather it drives a small gear that is permanently meshed with a larger gear **(Figure 18–24)**. This provides for a gear reduction and allows a small, high-speed motor to provide high torque at a satisfactory cranking speed. This starter design also tends to require lower current during engine startup. Some starters use a planetary gearset for gear reduction.

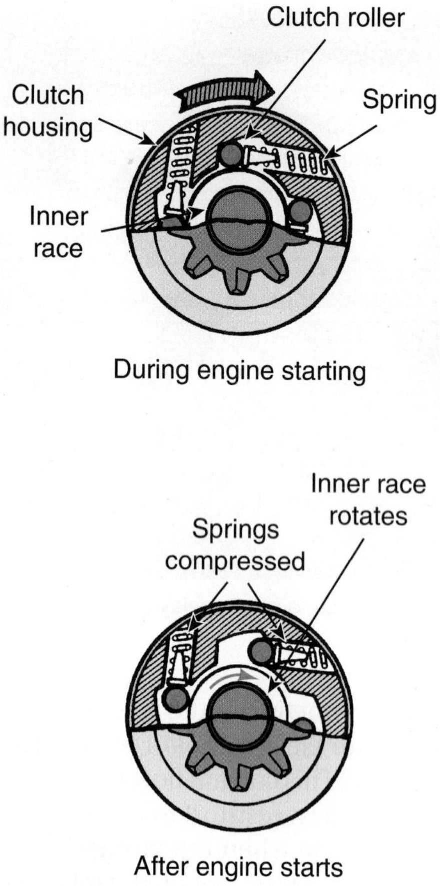

Figure 18–23 When the engine starts, the flywheel spins the pinion gear faster, which releases the rollers from the wedge.

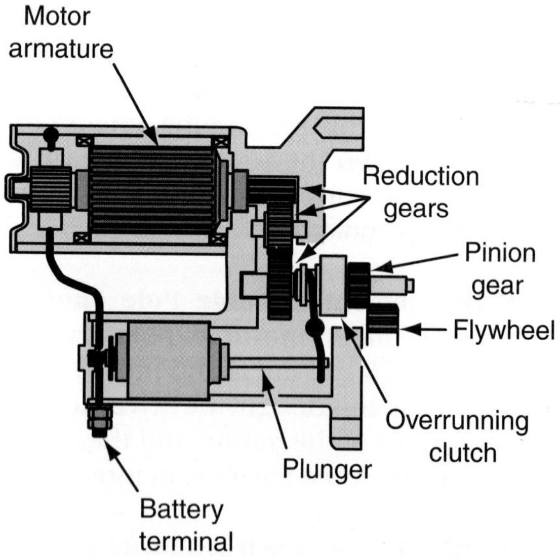

Figure 18–24 A gear reduction-drive starter.

CONTROL CIRCUIT

The control circuit allows the driver to use a small amount of battery current to control the flow of a large amount of current in the starting circuit.

The entire circuit usually consists of an ignition switch connected through normal-gauge wire to the battery and the magnetic switch (solenoid or relay). When the ignition switch is turned to the start position, a small amount of current flows through the coil of the magnetic switch, closing it and allowing full current to flow directly to the starter motor. The ignition switch performs other jobs besides controlling the starting circuit. It normally has at least four separate positions: accessory, off, on (run), and start.

Starting Safety Switch

The **starting safety switch**, often called the **neutral safety switch**, is a normally open switch that prevents the starting system from operating when the transmission is in gear. Starting safety switches can be located between the ignition switch and the relay or solenoid or the relay and ground.

The safety switch used with an automatic transmission is normally called a park/neutral position switch **(Figure 18–25)**. The switch contacts are wired in series with the control circuit so that no current can flow through the relay or solenoid unless the shift lever is in neutral or park. The switch is normally mounted on the transmission housing.

Mechanical safety switches for automatic transmissions physically block the movement of the ignition key when the transmission is in a gear. The ignition key can only be turned when the shift selector is in park or neutral. These are called interlock systems.

The safety switches used with manual transmissions are usually controlled by the clutch pedal. The clutch start switch serves the same purpose as a park/neutral position switch. The clutch start switch keeps the starter control circuit open until the clutch pedal is depressed **(Figure 18–26)**.

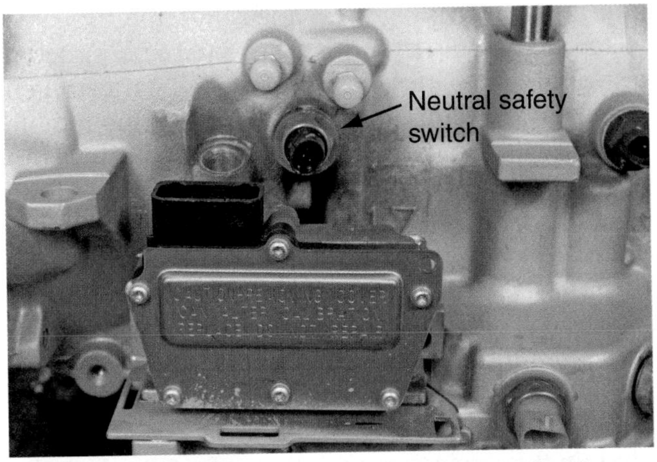

Figure 18-25 Neutral safety switch attached to an automatic transmission.

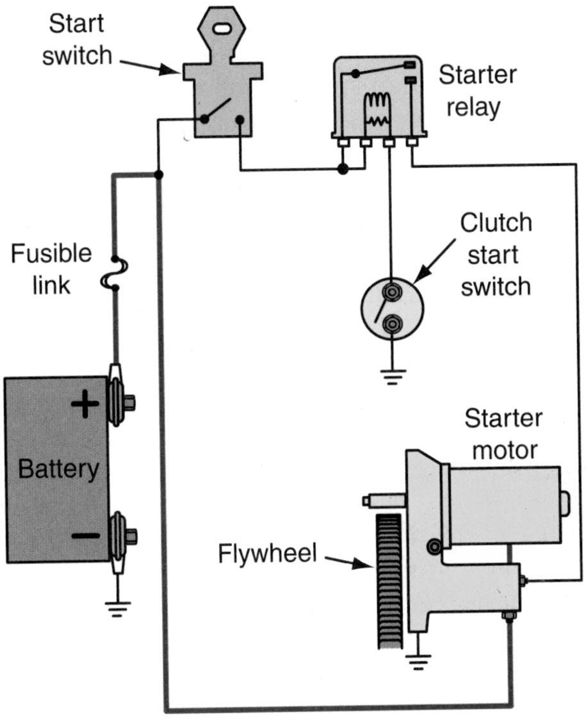

Figure 18-26 The clutch pedal must be fully depressed to close the clutch switch and complete the control circuit.

STARTING SYSTEM TESTING

As mentioned earlier, the starter motor is a special type of electrical motor designed for intermittent use only. During testing, it should never be operated for more than 15 seconds without resting for 2 minutes in between operation cycles to allow it to cool.

Preliminary Checks

The cranking output obtained from the motor is affected by the condition and charge of the battery, the circuit's wiring, and the engine's cranking requirement.

The battery should be checked and charged as needed before testing the starting system.

Check the wiring and cables for clean, tight connections. Loose or dirty connections will cause excessive voltage drops. Cables can be corroded by battery acid, and contact with engine parts and other metal surfaces can fray the cable insulation. Frayed insulation can cause a dead short that can seriously damage some of the electrical units of the vehicle.

Cables should also be checked to make sure they are not undersized or too long. When checking cables and wiring, always check the maxi-fuses and/or fusible links for the system. When one has failed, always troubleshoot the system and locate the cause before replacing the fuse or link.

Make certain that the engine is filled with proper weight oil as recommended by the vehicle manufacturer. Heavier-than-specified oil when coupled with low operating temperatures can drastically lower cranking speed to the point where the engine does not start and excessively high current is drawn by the starter.

Check the ignition switch for loose mounting, damaged wiring, sticking contacts, and loose connections. Check the wiring and mounting of the safety switch, if so equipped, and make certain the switch is properly adjusted. Check the mounting, wiring, and connections of the magnetic switch and starter motor. Also, be sure the starter pinion is properly adjusted.

Safety Precautions

Almost all starting system tests must be performed while the starter motor is cranking the engine. However, the engine must not start and run during the test or the readings will be inaccurate.

To prevent the engine from starting, the ignition switch can be bypassed with a remote starter switch that allows current to flow to the starting system but not to the ignition system.

During testing, be sure the transmission is out of gear during cranking and the parking brake is set. When servicing the battery, always follow safety precautions. Always disconnect the battery ground cable before making or breaking connections at the system's relay, solenoid, or starter motor.

SHOP TALK

Always approach starting system diagnosis in a logical way. This is the only way to identify the exact cause of a problem. Nearly 80% of starters returned as defective on warranty claims work perfectly when tested. This is often caused by poor or incomplete diagnosis of the starting and related charging systems.

Starter Solenoid Problems

A typical symptom of solenoid problems is a clicking noise when the ignition switch is turned to the start position. The noise is caused by the solenoid's plunger moving back and forth. Normally the plunger moves to the battery contacts and is held there by a magnetic field until the ignition switch is moved from the start position.

In order for the solenoid's plunger to move enough to complete the starter motor circuit and remain in that position, a strong magnetic field must be present around the solenoid's windings. The strength of the magnetic field depends on the current flowing through the windings. Therefore, anything that would reduce current flow would affect the operation of the solenoid. Common causes of the clicking are low battery voltage, low voltage available to the solenoid, or an open in the hold-in winding.

Checking voltage at the battery and to the solenoid will help you identify the cause of the problem. If the solenoid is bad, it can be replaced as a unit or with the starter motor.

Starting Safety Switches

Safety switches can be checked with a voltmeter or an ohmmeter. When the transmission is placed in park or neutral or when the clutch pedal is depressed, the switch should be closed. In other gear positions and when the clutch pedal is released, the switch should be open. Often these switches just need to be properly adjusted to correct their action. This is not possible on all vehicles. If adjustment does not correct the problem, the switch should be replaced.

Battery Load Test

A slow cranking engine is often caused by insufficient current from the battery. The battery must be able to crank the engine under all load conditions while maintaining enough voltage to supply ignition current for starting. Perform a battery load test before checking the starting systems.

Go To Chapter 17 for the correct procedures for conducting a battery load and other battery tests.

Cranking Voltage Test

The **cranking voltage test** measures the available voltage to the starter during cranking. To perform the test, disable the ignition or use a remote starter switch to bypass the ignition switch. Normally, the remote starter switch leads are connected to the positive terminal of the battery and the starter terminal of the solenoid or relay (**Figure 18–27**). Refer to the service manual for specific instructions on the model car being tested. Connect the voltmeter's negative lead to a good chassis ground. Connect the positive lead to the starter motor feed at the starter relay or solenoid. Activate the starter motor and observe the voltage reading. Compare the reading to the specifications given in the service manual. Normally, 9.6 volts is the minimum required.

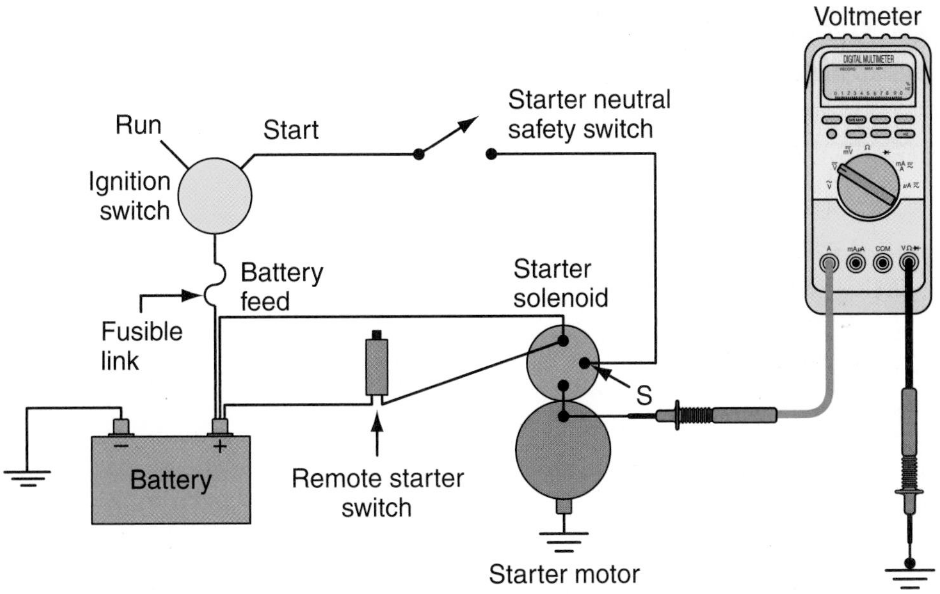

Figure 18-27 Using a remote starter switch to by pass the control circuit and ignition system.

> **SHOP TALK**
>
> Although most literature states that cranking voltage should be at least 9.6 volts, some engines will have a difficult time starting with voltages lower than 10.2 volts. Always check the manufacturer's specifications before coming to conclusions from this test.

Test Conclusions If the reading is above specifications but the starter motor still cranks poorly, the starter motor is faulty. If the voltage reading is lower than specifications, a cranking current test and circuit resistance test should be performed to determine if the problem is caused by high resistance in the starter circuit or an engine problem.

Cranking Current Test

The **cranking current test** measures the amount of current the starter circuit draws to crank the engine.

Nearly all starter current testers use an inductive pickup **(Figure 18–28)** to measure the current draw.

To conduct the cranking current test, connect a remote starter switch or disable the ignition prior to testing. Follow the instructions given with the tester when connecting the test leads. Crank the engine for no more than 15 seconds. Observe the voltmeter. If the voltage drops below 9.6 volts, a problem is indicated. Also, watch the ammeter and compare the reading to specifications.

Table 18–1 summarizes the most probable causes of too low or high starter motor current draw. If the problem appears to be caused by excessive resistance in the system, conduct an insulated circuit resistance test.

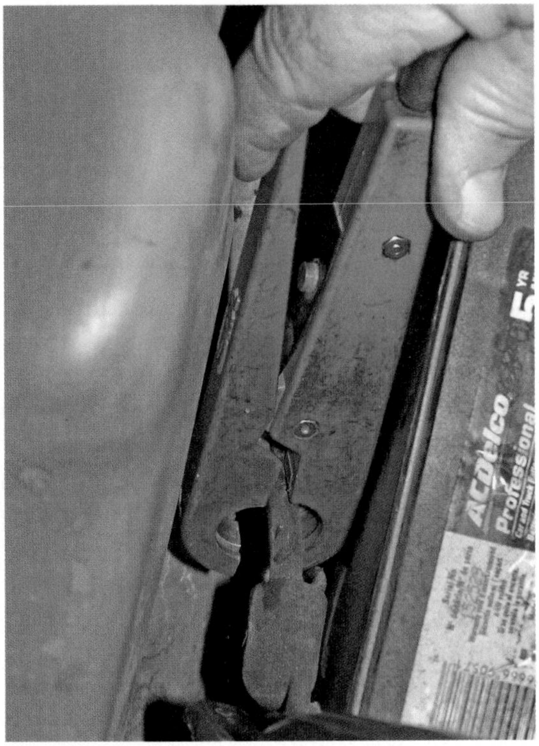

Figure 18-28 It is important to remember to move the inductive lead to the vehicle's system after checking the battery.

Insulated Circuit Resistance Test

The starter circuit is made up of the insulated circuit and the ground circuit. The insulated circuit includes all of the high current cables and connections from the battery to the starter motor.

TABLE 18–1 INTERPRETING THE RESULTS OF A CRANKING CURRENT TEST

Concern	Probable Cause
Low current draw	Undercharged or bad battery Excessive resistance in the starter circuit Excessive resistance in the starter or solenoid Excessive resistance at the various connections to the starter and/or solenoid
High current draw	The starter motor is shorted A short-to-ground in the starter circuit High mechanical resistance due to engine problems Misalignment of starter drive

To test the insulated circuit for high resistance, disable the ignition or bypass the ignition switch with a remote starter switch. Connect the positive (+) lead of the voltmeter to the battery's positive (+) terminal post or nut. Connect the negative (−) lead of the voltmeter to the starter terminal at the solenoid or relay. Crank the engine and record the voltmeter reading. If the reading is within specifications (usually 0.2 to 0.6 voltage drop), the insulated circuit does not have excessive resistance. Proceed to the ground circuit resistance test outlined in the next section. If the reading indicates a voltage loss above specifications, move the negative lead of the tester progressively closer to the battery, cranking the engine at each test point. Normally, a voltage drop of 0.1 volt is the maximum allowed across a length of cable.

Photo Sequence 15 goes through the correct procedure for conducting a voltage drop test on a typical starter circuit.

Test Conclusions When excessive voltage drop is observed, the trouble is located between that point and the preceding point tested. It is either a damaged cable or poor connection, an undersized wire, or possibly a bad contact assembly within the solenoid. Repair or replace any damaged wiring or faulty connections.

Starter Relay By-Pass Test

The starter relay by-pass test is a simple way to determine if the relay is operational. First, disable the ignition. Connect a heavy jumper cable between the battery's positive (+) terminal and the starter relay's starter terminal. This bypasses the relay. When the connection is made, the engine should crank.

CAUTION!

Make sure the vehicle's transmission is in park or neutral before doing this test. The starter motor can move the vehicle, which could injure you and others around you.

Test Conclusions If the engine cranks with the jumper installed and did not before the relay was bypassed, the starter relay is defective and should be replaced.

Ground Circuit Resistance Test

The ground circuit provides the return path to the battery for the current supplied to the starter by the insulated circuit. This circuit includes the starter-to-engine, engine-to-chassis, and chassis-to-battery ground terminal connections.

To test the ground circuit for high resistance, disable the ignition, or bypass the ignition switch with a remote starter switch. Refer to **Figure 18–29** for the proper test connection. Crank the engine and record the voltmeter reading.

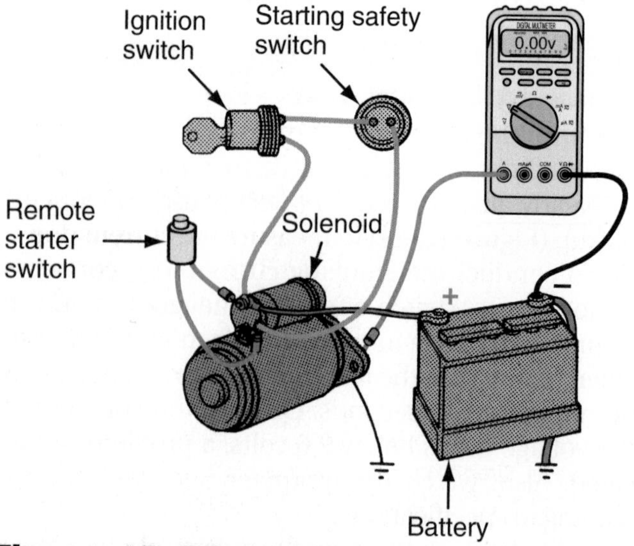

Figure 18–29 Setup for checking voltage drop across the ground circuit.

P15-1 *The tools required to measure voltage drop at various points within the starter circuit are fender covers, a DMM, and a remote starter switch. Make sure to apply the parking brake and set the transmission into neutral or park.*

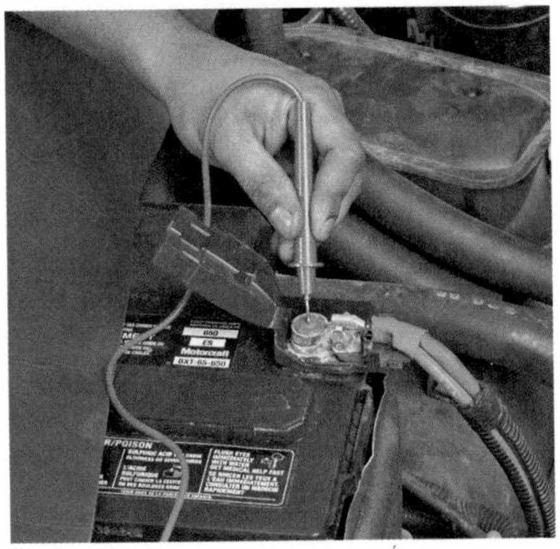

P15-2 *Connect the meter's positive lead to the positive battery post. If at all possible, do not connect it to the battery clamp.*

P15-3 *Connect the negative lead to the battery connection at the starter.*

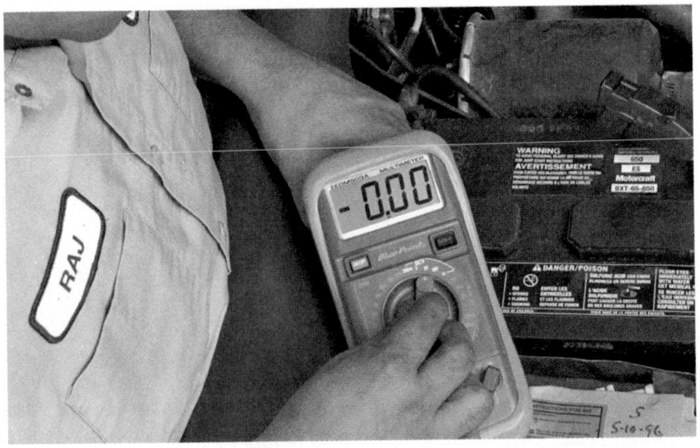

P15-4 *Set the voltmeter to the scale that is close to, but greater than, battery voltage.*

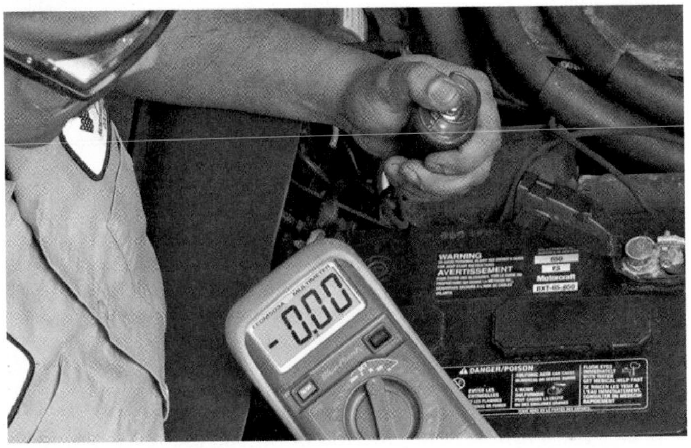

P15-5 *Disable the ignition and fuel injection and/or connect a remote starter switch.*

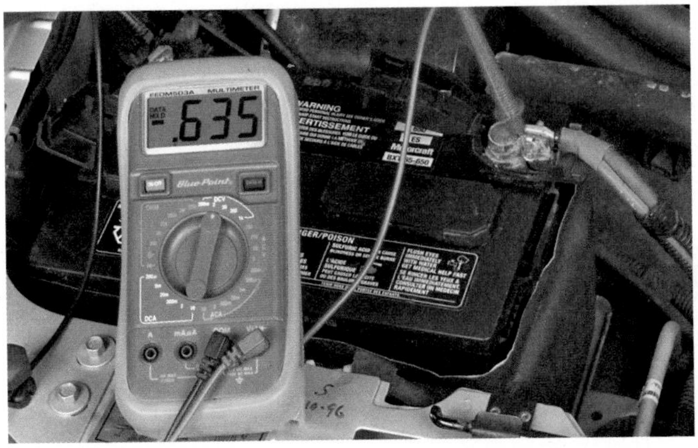

P15-6 *Crank the engine and read the voltmeter. This reading shows the voltage drop on the positive side of the starter circuit.*

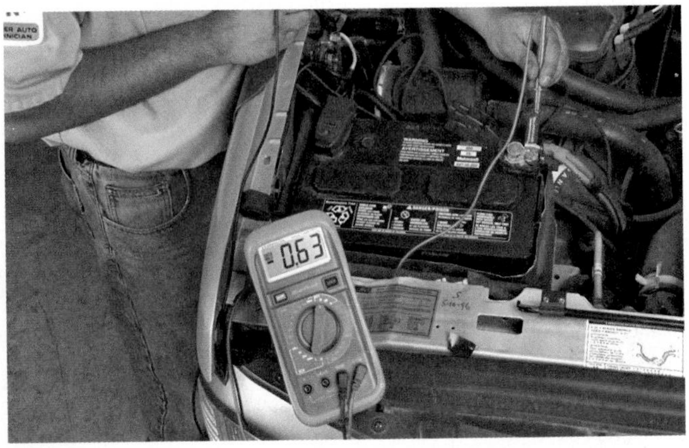

P15-7 *This reading shows excessive resistance. To locate the resistance, move the meter's negative lead to the next location toward the battery. In this case it is the starter side of the starter relay.*

Voltage Drop Testing of a Starter Circuit *(continued)*

P15-8 *Crank the engine and observe the reading on the meter. This is the voltage drop across the positive circuit from the battery to the output of the relay.*

P15-9 *There is still too much voltage drop; continue the test by moving the negative lead to the battery side of the relay.*

P15-10 *Crank the engine and observe the reading on the meter. This is the voltage drop across the cable from the battery to the relay. Notice that hardly any voltage was dropped. This cable is okay.*

P15-11 *Now connect the meter across the relay: the red lead on the battery side and the black lead on the starter side.*

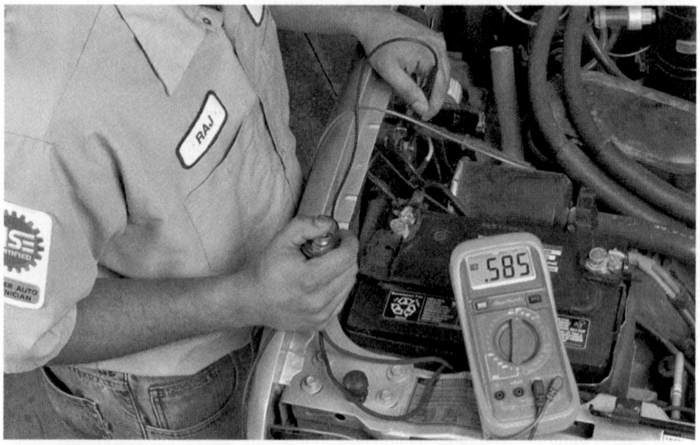

P15-12 *Crank the engine and observe the reading on the meter. This is the voltage drop across the contacts inside the relay.*

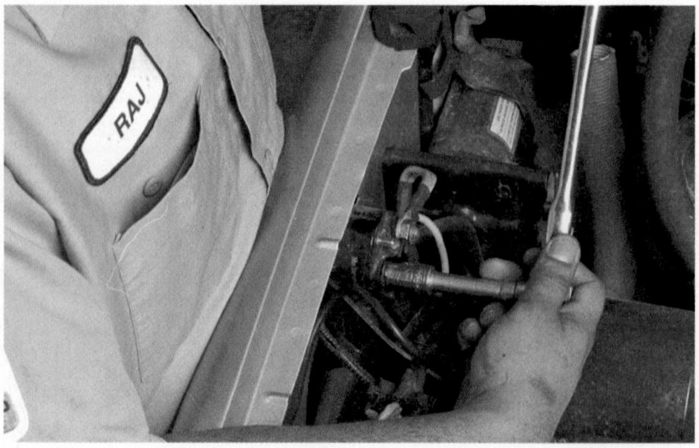

P15-13 *The reading is higher than normal; therefore, the starter relay has high resistance and needs to be replaced.*

Test Conclusions Good results would be less than a 0.2 volt drop for a 12-volt system. A voltage drop in excess of this indicates the presence of a poor ground circuit connection, resulting from a loose starter motor bolt, a poor battery ground terminal post connector, or a damaged or undersized ground system wire from the battery to the engine block. Isolate the cause of excessive voltage drop in the same manner as recommended in the insulated circuit resistance test by moving the positive (+) voltmeter lead progressively closer to the battery. If the ground circuit tests out satisfactorily and a starter problem exists, move on to the control circuit test.

Voltage Drop Test of the Control Circuit

The control circuit test examines all the wiring and components used to control the magnetic switch, whether it is a relay, a solenoid acting as a relay, or a starter motor-mounted solenoid.

High resistance in the solenoid switch circuit reduces current flow through the solenoid windings, which can cause improper functioning of the solenoid. In some cases of high resistance, it may not function at all. Improper functioning of the solenoid switch generally results in the burning of the solenoid switch contacts, causing high resistance in the starter motor circuit.

Check the vehicle wiring diagram, if possible, to identify all control circuit components. These normally include the ignition switch, safety switch, the starter solenoid winding, or a separate relay.

To perform the test, disable the ignition system. Connect the positive meter lead to the battery's positive terminal and the negative meter lead to the starter switch terminal on the solenoid or relay. Crank the engine and record the voltmeter reading.

Test Conclusions Generally, good results would be less than 0.5 volt, indicating that the circuit condition is good. If the voltage reading exceeds 0.5 volt, it is usually an indication of excessive resistance. However, on certain vehicles, a slightly higher voltage loss may be normal.

Identify the point of high resistance by moving the negative test lead back toward the battery's positive terminal, eliminating one wire or component at a time.

A reading of more than 0.1 volt across any one wire or switch is usually an indication of trouble. If a high reading is obtained across the safety switch used on an automatic transmission, check the adjustment of the switch according to the manufacturer's service manual. Clutch-operated safety switches cannot be adjusted. They must be replaced.

Test Starter Drive Components

This test detects a slipping starter drive without removing the starter from the vehicle. First, disable the ignition system or bypass the ignition switch with a remote starter switch. Turn the ignition switch to start and hold it in this position for several seconds. Repeat the procedure at least three times to detect an intermittent condition.

Test Conclusions If the starter cranks the engine smoothly, that is an indication that the starter drive is functioning properly. If the engine stops cranking and the starter spins noisily at high speed, the drive is slipping and should be replaced.

If the drive is not slipping, but the engine is not being cranked, inspect the flywheel for missing or damaged teeth. Remove the starter from the vehicle and check its drive components. Inspect the pinion gear teeth for wear and damage. Test the overrunning clutch mechanism. If good, the overrunning clutch should turn freely in one direction but not in the other. A bad clutch will turn freely in the overrun direction or not at all. If a drive locks up, it can destroy the starter by allowing the starter to spin at more than 15 times engine speed.

Removing the Starter Motor

If your testing indicates that the starter must be removed, the first step is to disconnect the negative cable at the battery and wrap the clamp with electrical tape. It may be necessary to place the vehicle on a lift to gain access to the starter. Before lifting the vehicle, disconnect all wires, fasteners, and so on that can be reached from under the hood.

Disconnect the wires leading to the solenoid terminals. To avoid confusion when reinstalling the starter, it is wise to mark the wires so they can be reinstalled on their correct terminals.

On some vehicles you may need to disconnect the exhaust system to be able to remove the starter. Loosen the starter mounting bolts and remove all but one. Support the starter while removing the remaining bolt. Then pull the starter out and away from the flywheel. Once the starter is free, remove the last bolt and the starter.

Once the starter is out, inspect the starter drive pinion gear and the flywheel ring gear **(Figure 18–30)**. When the teeth of the starter drive are abnormally worn, make sure you inspect the entire circumference of the flywheel. If the starter drive or the flywheel ring gear show signs of wear or damage, they must be replaced.

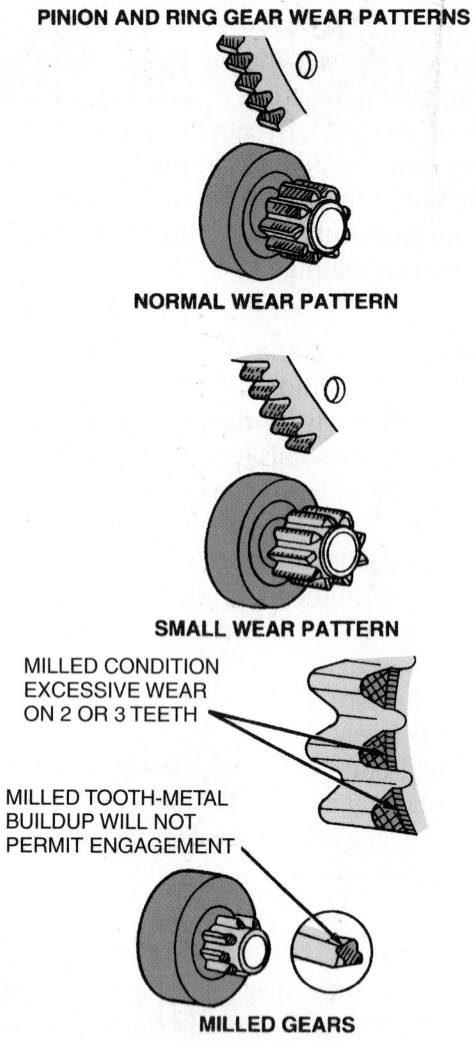

Figure 18-30 Starter drive and flywheel ring gear wear patterns. *Courtesy of Ford Motor Company*

Free Speed (No-Load) Test

Every starter should be bench tested after it is removed and before it is installed. To conduct a free speed or no-load test on a starter, follow these steps:

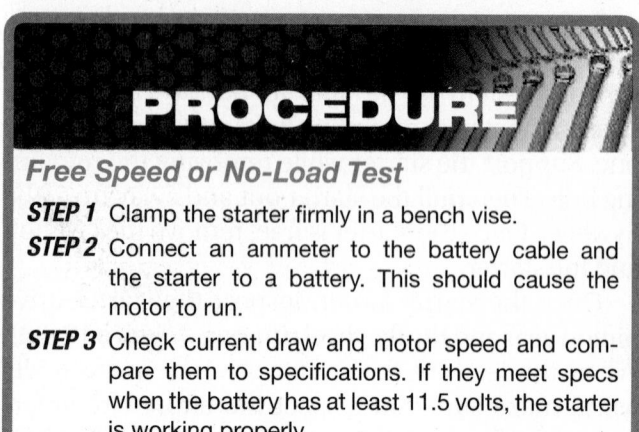

PROCEDURE

Free Speed or No-Load Test

STEP 1 Clamp the starter firmly in a bench vise.

STEP 2 Connect an ammeter to the battery cable and the starter to a battery. This should cause the motor to run.

STEP 3 Check current draw and motor speed and compare them to specifications. If they meet specs when the battery has at least 11.5 volts, the starter is working properly.

If the current draw was excessive or the motor speed too low, there is excessive physical resistance, which can be caused by worn bushings or bearings, a shorted armature, shorted field windings, or a bent armature.

If there was no current draw and the starter did not rotate, the problem could be caused by open field windings, open armature coils, broken brushes, or broken brush springs.

Low armature speed with low current draw indicates excessive resistance. There may be a poor connection between the commutator and the brushes, or the connections to the starter are bad. If the speed and current draw are both high, check for a shorted field winding.

STARTER MOTOR SERVICE

If the starter is bad, it should be replaced or rebuilt. Often technicians opt for replacing it rather than spending the time to repair or rebuild it. This decision, however, depends on a number of things, including the customer's desire, shop's policies, availability of repair parts, cost, and time.

Photo Sequence 16 covers the typical procedure for disassembling a Delco-Remy starter. Always refer to the manufacturer's procedures when repairing a starter.

WARNING!

Do not clean the starter motor in solvent. The residue left on the parts can ignite and cause a fire and/or destroy the starter. Use denatured alcohol, compressed air regulated to 25 psi (172.3 kPa), and/or clean rags to clean the unit and its parts.

The starter should be cleaned and inspected as it is disassembled. Inspect the end frame and drive housing for cracks or broken ends. Check the frame assembly for loose pole shoes and broken or frayed wires. Inspect the drive gear for worn teeth and proper overrunning clutch operation. The commutator should be free of flat spots and should not be excessively burned. Check the brushes for wear. Replace them if worn past specifications.

Starter Motor Component Tests

With the starter motor disassembled, tests can be conducted to determine the reason for failure. The armature and field coils should be checked for shorts and opens first. Normally, if the armature or coils are bad, the entire starter is replaced.

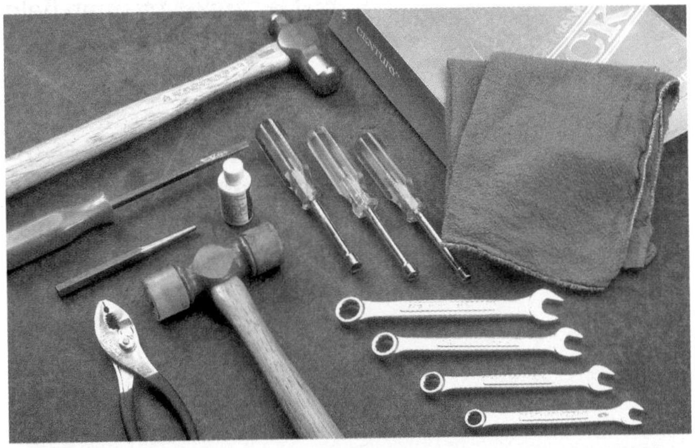

P16-1 *Always have a clean and organized work area. The tools required for disassembling this starter are: rags, assorted wrenches, snapring pliers, flat blade screwdriver, ball-peen hammer, plastic head hammer, punch, scribe, safety glasses, and small press.*

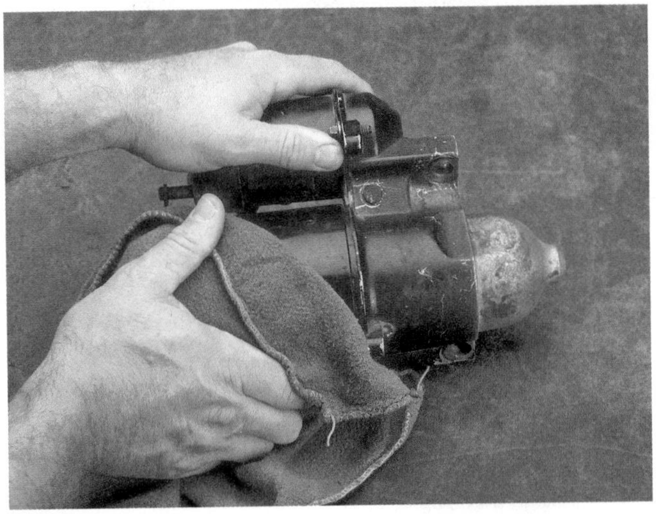

P16-2 *Clean the case.*

P16-3 *Scribe reference marks at each end of the starter end housings and the frame.*

P16-4 *Disconnect the field coil connection at the solenoid's M terminal.*

P16-5 *Remove the two screws that attach the solenoid to the starter drive housing.*

P16-6 *Rotate the solenoid until the locking flange of the solenoid is free. Then remove the solenoid.*

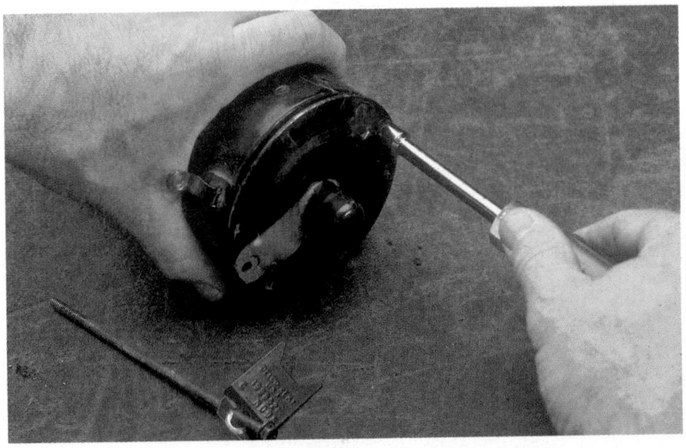

P16-7 *Remove the through-bolts from the end frame.*

Typical Procedure for Disassembling a Delco-Remy Starter *(continued)*

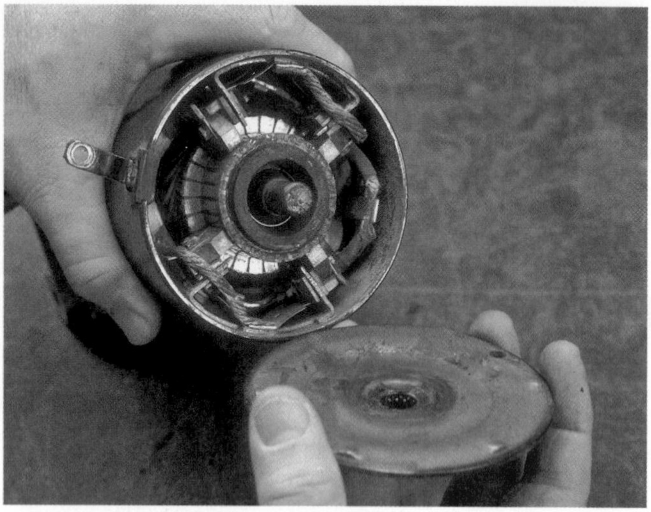

P16-8 *Remove the end frame.*

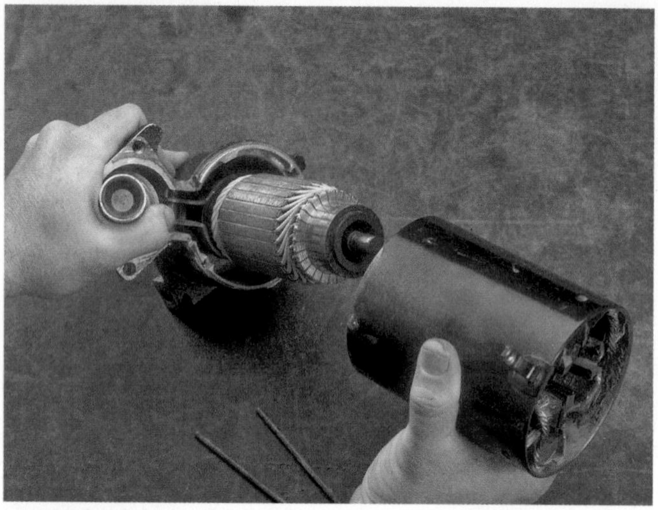

P16-9 *Remove the frame.*

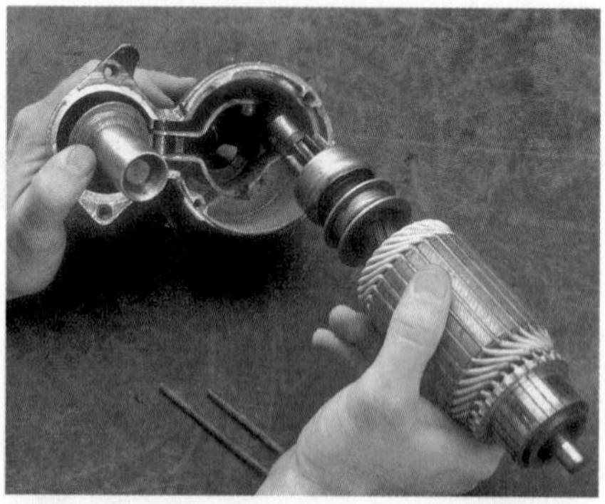

P16-10 *Remove the armature. Note: On some units it may be necessary to remove the shift lever from the drive housing before removing the armature.*

P16-11 *Place a ⅝-inch deep socket over the armature shaft until it contacts the retaining ring of the starter drive.*

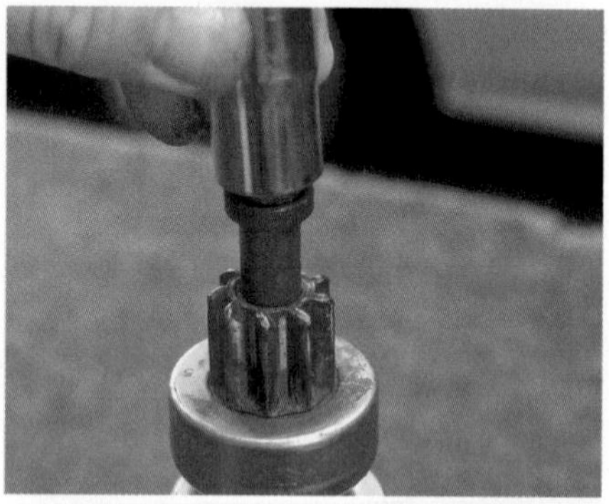

P16-12 *Tap the end of the socket with a plastic-faced hammer to drive the retainer toward the armature. Move it only far enough to access the snapring.*

P16-13 *Remove the snapring.*

P16-14 *Remove the retainer from the shaft and remove the clutch and spring from the shaft. Press out the drive housing bushing and the end-frame bushing.*

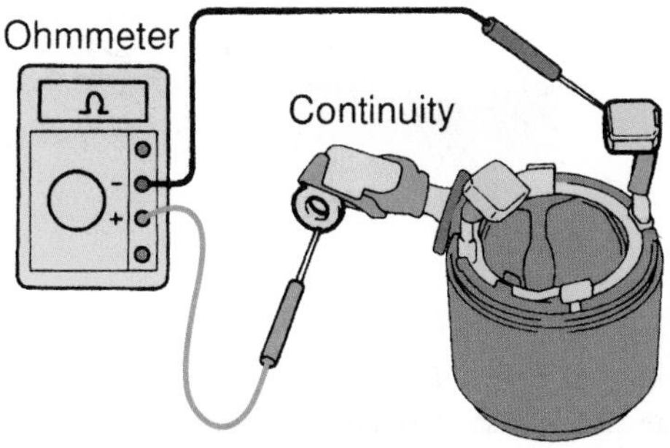

Figure 18–31 Checking a field coil for an open.
Courtesy of Toyota Motor Sales, U.S.A., Inc.

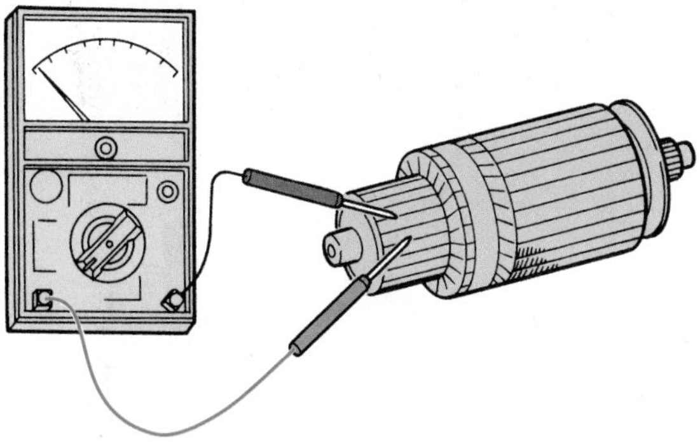

Figure 18–32 Checking the armature for an open.
Courtesy of American Isuzu Motors Inc.

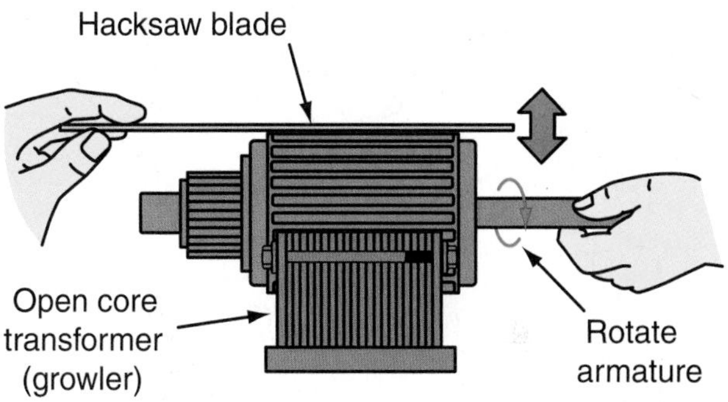

Figure 18–33 Testing an armature on a growler.

Field Coil Tests The field coil and frame assembly can be wired in a number of different ways. Accurate testing of the coils can only be done if you follow the specific guidelines of the manufacturer or if you know how the coils are wired. To do this, look at the wiring diagram and figure out where the coils get their power and where they ground. When you have this information, you will know if the coils are wired in series or parallel.

The usual way to check the field coils for opens is to connect an ohmmeter between the coils' power feed wire and the field coil brush lead **(Figure 18–31)**. If there is no continuity, the field is open. To check the field coil for a short to ground, connect the ohmmeter from the field coil brush lead and the starter (field frame) housing. If there is continuity, the field coil is shorted to the housing.

Armature Tests The armature should be inspected for wear or damage caused by contact with the permanent magnets or field windings. If there is wear or damage, check the pole shoes for looseness and repair as necessary. A damaged armature must be replaced.

Next, check the commutator of the armature. If the surface is dirty or burnt, clean it with emery cloth or cut it down with an armature lathe. Measure the diameter of the commutator with an outside micrometer or vernier caliper. If the diameter is less than specifications require, replace the armature.

Measure commutator runout by mounting the armature in V-blocks. Position a dial indicator so that it rides on the center of the commutator. If the runout is within specs, check the commutator for carbon dust or brass chips between the segments. If the commutator runout is beyond specs, replace the armature.

Check the depth of the insulating material (mica) between the commutator segments. Check each one and compare the depth with specifications. If necessary, undercut the mica with the proper tool or a hacksaw to achieve the proper depth. If the proper depth cannot be achieved, replace the armature.

Check for continuity between the segments of the commutator **(Figure 18–32)**. If an open circuit exists between any segments, replace the armature.

Place the armature in an armature tester, commonly called a **growler**. Hold a hacksaw blade on the armature core **(Figure 18–33)**. If the blade is attracted to the armature's core or vibrates while the core is turned, the armature is shorted and must be replaced.

With an ohmmeter, check the armature windings for a short to ground. Hold one meter lead to a commutator segment and the other on the armature core. Also check between the armature shaft and the commutator. If there is continuity at either of these two test points, the armature needs to be replaced.

Brush Inspection Brush inspection begins with an ohmmeter check of the brush holder. Connect one meter lead to a positive brush and the other lead to a negative brush. There should be no continuity between them. If there is, replace the brush holder. Install the brushes into the brush holder and slip the unit over the commutator. Using a spring scale, measure the spring tension of the holders at the moment the spring lifts off the brush. Compare the tension

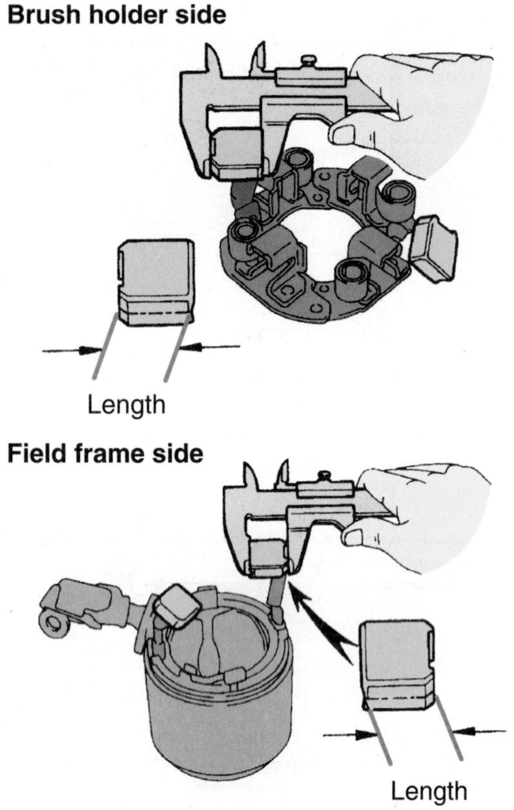

Figure 18–34 Measure the length of the brushes. *Courtesy of Toyota Motor Sales, U.S.A., Inc.*

with specs. If the tension is incorrect, replace the spring or the brush holder assembly.

Measure the length of the brushes **(Figure 18–34)**. If the brushes are not within specs, replace the brush or the brush holder assembly. To seat new brushes after installing them in the brush holder, slip a piece of fine sandpaper between the brush and the commutator. Then rotate the armature. This will put the contour of the commutator on the face of the brushes.

Bearings and Bushings Check each bearing and bushing by placing the armature into the bushing and paying attention to the fit and feel as the armature is rotated in the bushing. If the bushing or bearing feels too loose, tight, or rough, it should be replaced. Bushings can often be visually inspected for uneven and excessive wear. If the bushing is bad, replace it. Many bearings are held in the case by a retainer, while bushings are typically pressed out and into their bore.

Starter Drives and Clutches Carefully inspect the teeth on the starter drive. If the teeth are chipped, excessively worn, or damaged in any way, replace the drive assembly. Also check the teeth on the starter ring gear on the engine's flywheel. Often the same thing that caused damage to the starter drive will damage the teeth on the flywheel. If either or both are damaged, they should be replaced.

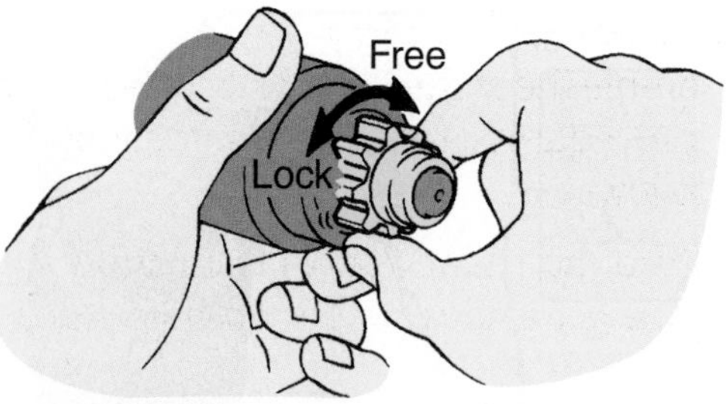

Figure 18–35 Check the overrunning clutch by attempting to rotate it in both directions.

To check the operation of the overrunning clutch, slide the drive and clutch assembly onto the armature shaft. Rotate the clutch in both directions. Check its movement. It should rotate smoothly in one direction and lock in the other **(Figure 18–35)**. If it does not lock in either direction or if it locks or barely moves in both directions, the assembly must be replaced.

CASE STUDY

A vehicle equipped with a solenoid-actuated direct-drive starting system is towed into the shop. The owner complains that the starter does not crank the engine when the ignition switch is turned to the start position.

The technician performs a battery load test to confirm that the battery is in good working order. It is. The technician then tests for voltage to the M or motor terminal on the solenoid using a voltmeter. Voltage reading at the M terminal is 12.5 volts.

Because there is voltage at the M terminal of the solenoid, inspection of individual connections and components, such as the starter safety switch, are not needed at this time. They are obviously allowing current to pass through the insulated circuit.

The technician then performs a ground circuit check to verify that the ground return path is okay. It is. Since the insulated and ground circuits have checked out okay, the only other source of an open circuit is the starter motor. The technician can now confidently pull the starter motor from the vehicle for rebuilding or replacement.

Starter Motor Reassembly

To reassemble the starter, basically reverse the disassembly procedures. Additional guidelines for reassembly follow:

- Lubricate the splines on the armature shaft that the drive gear rides on with a high-temperature grease.
- Lubricate the bearings and/or bushings with a high-temperature grease.
- Apply sealing compound to the solenoid flange before installing the solenoid to the starter motor.
- Check the pinion depth clearance.
- Perform a no-load test on the starter before installing it.

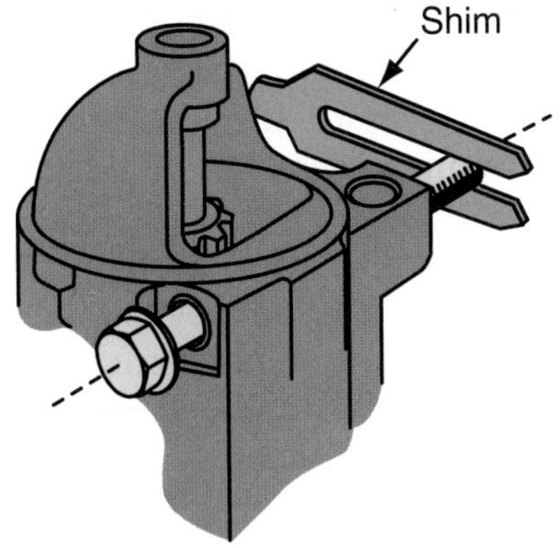

Figure 18–36 Shimming the starter to obtain proper pinion-to-ring gear clearance.

Installation

When installing a new or remanufactured starter, sand away the paint at the mounting point before installing it. Also, make sure you have a good hold on the starter while installing it. Tighten the retaining bolts to the specified torque and make sure the starter is fully seated before final tightening. Also, make sure that all electrical connections are tight. If the starter was installed with heat shields, make sure they are in place before tightening the bolts.

Some starters use shims between the starter and the mounting pad **(Figure 18–36)**. To check this clearance, install the starter and insert a flat blade screwdriver into the access slot on the side of the drive housing. Pry the drive pinion gear into the engaged position. Use a wire feeler gauge or a piece of 0.020-inch diameter wire to check the clearance between the gears **(Figure 18–37)**.

If the clearance between the two gears is incorrect, shims will need to be added or subtracted to bring the clearance within specs. If the clearance is excessive, the starter will produce a high-pitched whine while it is cranking the engine. If the clearance is too small, the starter will make a high-pitched whine after the engine starts and the ignition switch is returned to the RUN position.

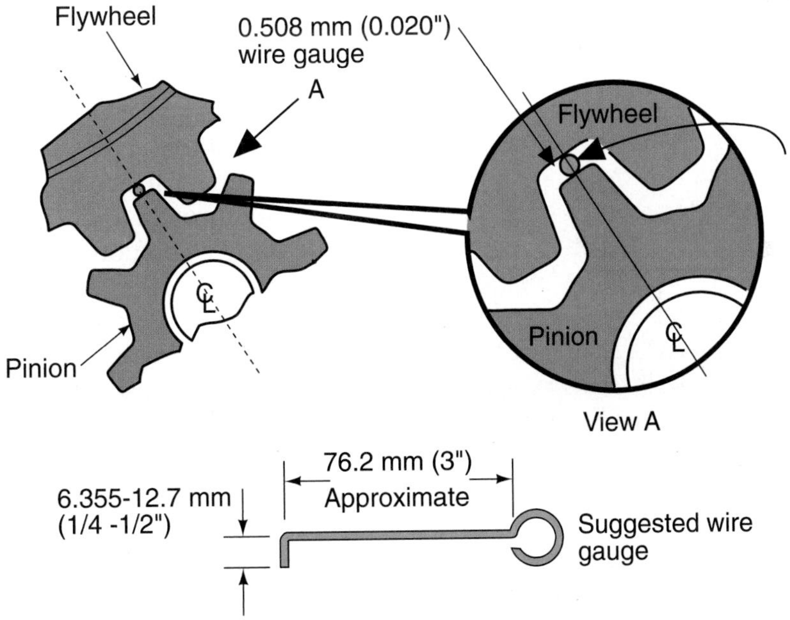

Figure 18–37 Checking the clearance between the pinion gear and the ring gear.

HIGH-VOLTAGE MOTORS

Hybrid and fuel cell vehicles use strong electric motors to move the vehicle and, in some cases, to start the engine as well. Often hybrid vehicles also have a normal starting motor, which may only serve as a back-up unit **(Figure 18–38)**. These starting motors are conventional and are not part of this discussion.

High-voltage motors are used for many purposes and in many different systems. They may be used to drive the air-conditioning compressor, work as a starting motor, or power the vehicle. The latter motors are referred to as traction motors.

Traction Motors

The placement and purpose of a **traction motor** depends on the design of the hybrid. All are powered by more than 12 volts. On some models, the high-voltage motor is not capable of moving the vehicle by itself. Instead it merely works as a starter motor and generator or adds power to the output of the engine. How these motors function as a generator is discussed in Chapter 19. Different motor designs are used as traction motors. A few are powered by DC voltage but most use AC.

The motors can be liquid or air cooled and are normally lubricated for life. Most production vehicles use AC motors, and many conversion electric vehicles use a DC motor. The latter is a consequence of cost. DC motors can be powered directly by the batteries, whereas AC motors require converters and inverters to change the DC voltage stored in the batteries into the AC required by the motors.

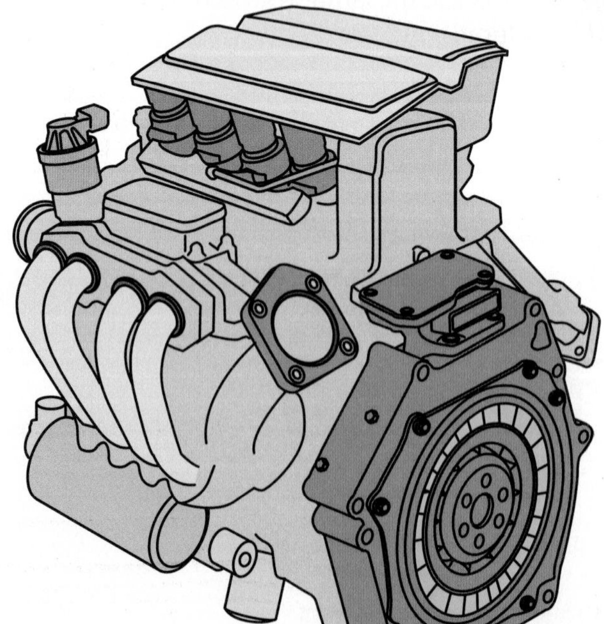

Figure 18-38 This electric motor, attached directly to the engine's crankshaft, starts the engine under normal conditions. When outside temperature is extremely low, when the battery state of charge is low, or if there is a problem with the hybrid system, the conventional starter starts the engine.

DC electric motors are quite reliable, but the brushes and commutator present some durability concerns. The carbon brushes spark and wear out, and the spring tension on the brushes must be kept within specifications. Excessive spring tension causes excessive friction and wear of the brushes and commutator. When the spring tension is too low, sparking occurs between the brush and the commutator, causing damage to both, and the brushes can bounce, which makes and breaks the circuit. Both of these problems can result in overheating the motor and decreasing reliability. Although brushless DC motors do not have this problem, most cannot provide enough power to move a vehicle.

DC motors are typically more expensive than comparable AC motors. In addition, the available torque from a DC motor is at its peak when the rotor or armature is not turning. The available torque decreases from that peak as armature speed increases. DC motors also tend to run hotter; therefore, they need proper cooling. DC motors also do not provide for regenerative braking unless they are separately excited (Sep-Ex) DC motors fitted with a special controller that allows for efficient switching of the motor to a generator and back to a motor.

AC motors are lighter than comparable DC motors. They are also very reliable. Because they have only one moving part, the shaft, they should last the life of the vehicle with little or no maintenance. In an AC motor, there is no commutator to distort or burn and no brushes to wear or spark.

AC systems typically operate at higher voltage and lower current than DC systems with the same power output. In an AC induction motor, the torque output is constant through a wide range of speeds. This provides even acceleration and often allows driving without the need of a transmission and different speed gears. There is no need for additional controllers or electronics to have regenerative braking.

The primary disadvantage of an AC motor is, again, the cost of the electronic systems required to convert (invert) the battery's DC to AC for the motor. However, most electronic equipment is becoming less expensive (as well as better), and this disadvantage has less merit than it did a few years ago and will be less of a consideration in the future. Manufacturers used AC motors simply because of their efficiency and lighter weight.

The addition of an electric motor or motors to the driveline is what makes a hybrid vehicle fuel efficient. The motor provides full torque at low speeds, which helps with good acceleration, and uses zero fuel when it is operating on its own. The intervention of the electric motor allows for the use of a smaller, more efficient engine. Although the engine is designed for fuel economy, it alone cannot provide adequate power for safe acceleration and load overtaking. The motors also have low production costs, low noise, and high efficiency. The electric motor can be configured, through electronics, to assist the engine during acceleration, passing, overtaking heavy loads, or hill climbing. Advanced electronics allow the motor to act as a generator when its power is not needed. The hybrid controller or computer-controlled system coordinates or synchronizes the action of both.

Brushless DC Motors

A brushless DC motor is much like a brushed DC motor, but the purpose of the rotor (armature) and stator (field windings) are reversed. The rotor has a set of permanent magnets and the stator has controllable electromagnets **(Figure 18–39)**. Obviously, a brushless motor has no brushes and no commutator. The normal electrical arcing that takes place between the brushes and commutator is eliminated with the brushless design. This arcing not only decreases the usable life of a motor, it also creates electromagnetic interference that is detrimental to advanced electronic systems.

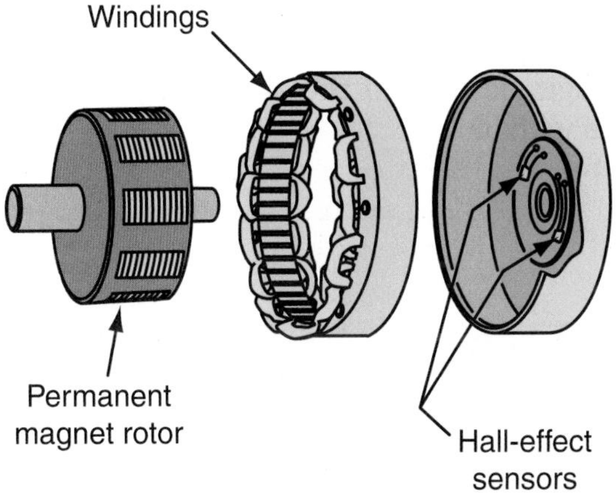

Figure 18-39 A brushless DC motor.

Rather than using brushes, an electronic circuit switches current flow to the different stator's windings to keep the rotor turning. The reversing of current flow through the windings is done by power transistors that switch according to the position of the rotor. Many brushless DC motors use Hall-effect sensors to monitor the position of the rotor. Other brushless motors monitor the CEMF in unexcited field windings to determine the position of the rotor. The current to the various stator windings is typically controlled by a **pulse width modulation (PWM)** frequency inverter. The voltage to the windings is changed by altering the duty cycle.

The **duty cycle** of something is the length of time the device is turned on compared to the time it is off **(Figure 18–40)**. Duty cycle can be expressed as a ratio or as a percentage. By quickly opening and closing the power circuit to the motor, the speed of the motor is controlled. This is called pulse width modulation. These power pulses vary in duration to change the speed of the motor. The longer the pulses, the faster the motor turns.

Brushless DC motors, when compared to brushed DC motors, are more reliable, more powerful, and more expensive. The expense is largely due to the cost of the electronic controls. High-output brushless DC motors are used in some electric drive vehicles.

AC Motors

With AC voltage, the direction of current flow changes but not immediately. Rather, as the current is getting ready to change directions, it decreases until it

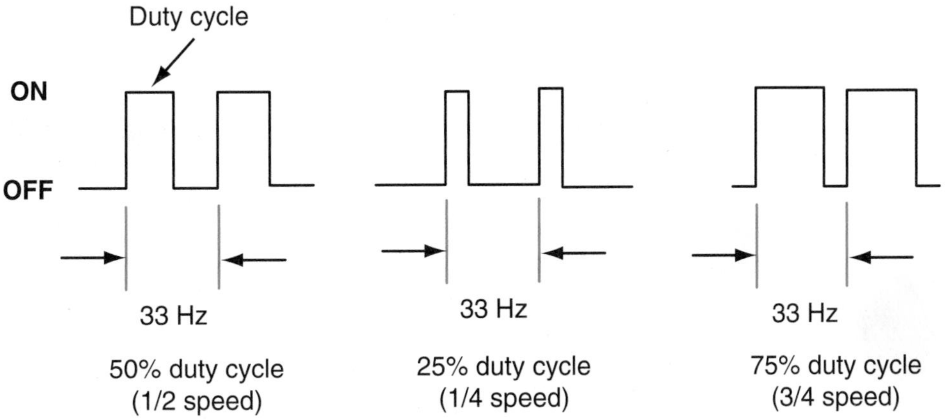

Figure 18-40 The action of various duty cycles.

reaches zero and then gradually builds up in the other direction. Therefore, the amount of current in an AC circuit always varies. When AC is given as a value, it is an average rating, not peak current. This average rating is referred to as a "root mean square" value.

Basic Construction An AC motor has two basic electrical parts: a stator and a rotor. The stator is comprised of individual electromagnets electrically connected to each other or connected in groups. The rotor is the rotating magnetic field and can be an electromagnet or a permanent magnet. The rotor is located within the stator fields. Like a DC motor, the rotor rotates as a result of the repulsion and attraction of the magnetic poles. The way this works is quite different.

Basic Operation Current is passed through the stator and rotor, causing the rotor to spin. Because the current is alternating, the polarity in the windings constantly changes. A synchronous AC motor will run at the frequency of the AC voltage. Many AC motors are induction types. In these motors, electrical current is induced in the rotor as it rotates, rather than have current delivered to it from an external source.

The rotor spins because it is pulled along by a rotating magnetic field in the stator. The stator does not physically move. The magnetic field does. If the windings of the stator are wired in series, current passes through them one at a time, and (because it is AC) the polarity and strength of the field around them are constantly changing **(Figure 18–41)**. The magnetic field of the rotor reacts and moves along with the "rotating magnetic field" of the stator.

To better understand this concept, let us look at a three-phase motor. Three-phase AC voltage is commonly used in motors because it provides a smoother and more constant supply of power. Three-phase AC voltage is much like having three independent AC power sources, each with the same amplitude and frequency but are 120 degrees out of phase with each other.

To produce a rotating magnetic field in the stator, each phase of the three-phase power source is connected to separate stator windings. Because each phase reaches its peak voltage at successively later times, the magnetic field is at its strongest point in each winding in succession as well. This creates the effect of the magnetic field continually moving around the stator. This rotating magnetic field will move around the stator once for every cycle of the voltage in each phase **(Figure 18–42)**. This means the field is rotating at the frequency of the source voltage. Remember that as the magnetic field moves, new magnetic polarities are present. As each polarity change is made, the poles of the rotor are attracted by the opposite poles on the stator. Therefore, as the magnetic field of the stator rotates, the rotor rotates with it.

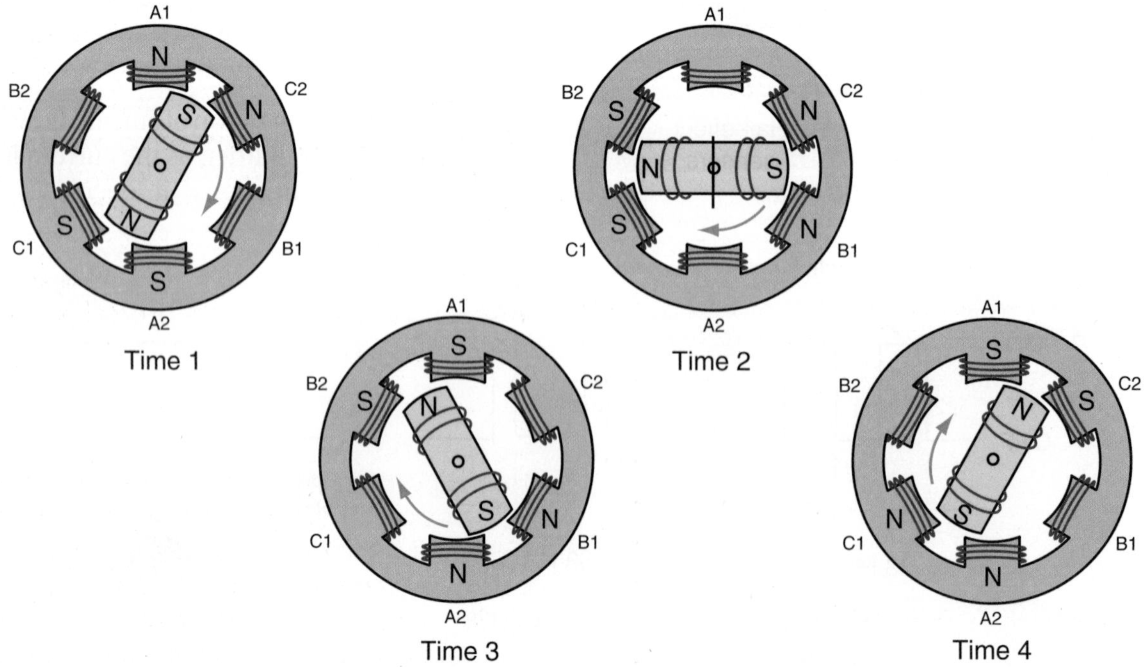

Figure 18-41 Notice how the polarity of the stator and rotor changes over time.

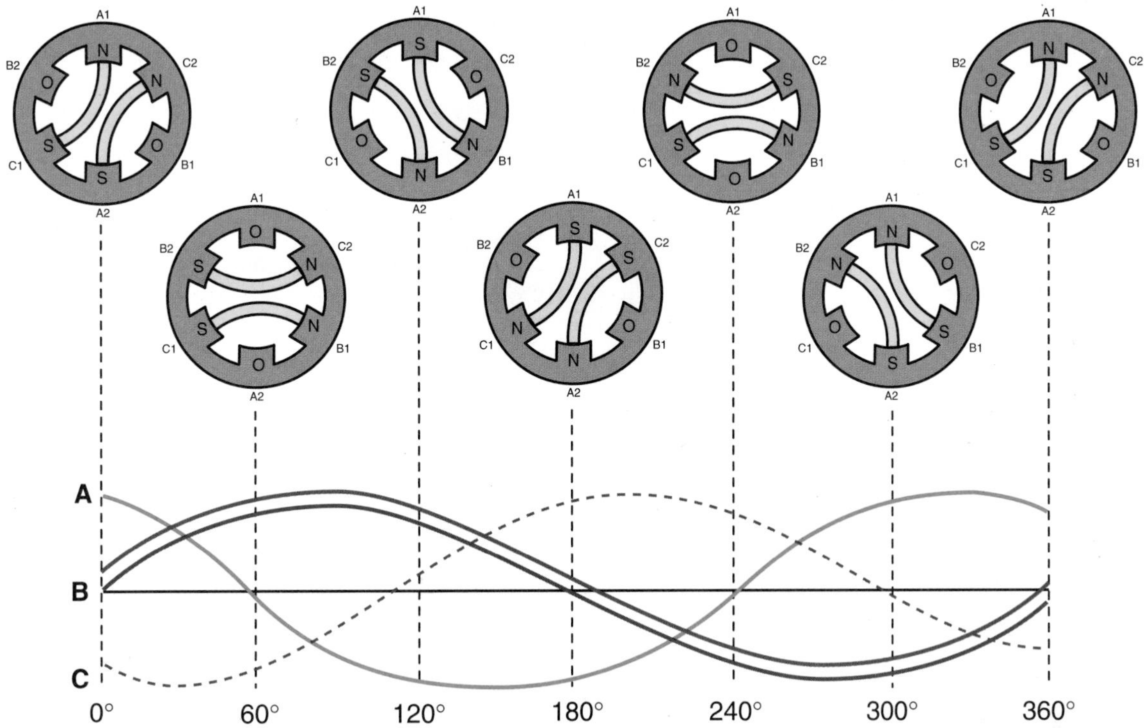

Figure 18-42 The rotor turns in response to the changing polarities caused by the three-phase AC.

Synchronous Motor

A synchronous motor **(Figure 18–43)** operates at a constant speed regardless of the load. Rotor speed is equal to the speed of the stator's rotating magnetic field. A synchronous motor is used when the exact speed of a motor must be maintained. Often, synchronous motors have magnets built into the rotor assembly. These magnets allow the rotor to easily align itself with the rotating magnetic field of the stator.

When three-phase AC is fed to the three sets of windings in the stator coil, a rotating magnetic field is present around the stator. The rotor simply rotates with that rotating magnetic field. The torque output of the rotor, therefore, is dependent on the strength of the magnetic field around the stator. The speed of the rotor is determined by the frequency of the AC input to the stator. Synchronous motors are available with outputs up to thousands of horsepower.

Figure 18-43 A synchronous motor assembly designed to fit between the engine's crankshaft and the transmission. *Courtesy of Robert Bosch GmbH, www.bosch-presse.de*

One of the disadvantages of synchronous motors is that they cannot be started by merely applying three-phase AC power to the stator. When AC is applied to the stator, a high-speed rotating magnetic field is immediately present. This field rushes past the rotor so quickly that the rotor cannot get started. The rotor is first repelled in one direction and then, very quickly, in another. There are many ways of addressing this, but for hybrid and electric vehicles the problem is solved by complex electronics that begin the rotating magnetic field in such a way and speed that the rotor simply follows the field **(Figure 18–44)**. Once the rotor is spinning, normal synchronous operation begins.

Asynchronous Motor

The most commonly used motor is the three-phase AC induction motor, which is most often found in hybrids. This motor has a low cost and a simple design. The stator is connected to the power source and the rotor. The three-phase AC sets up a rotating magnetic field around the stator. The rotor is not connected to an external source of voltage.

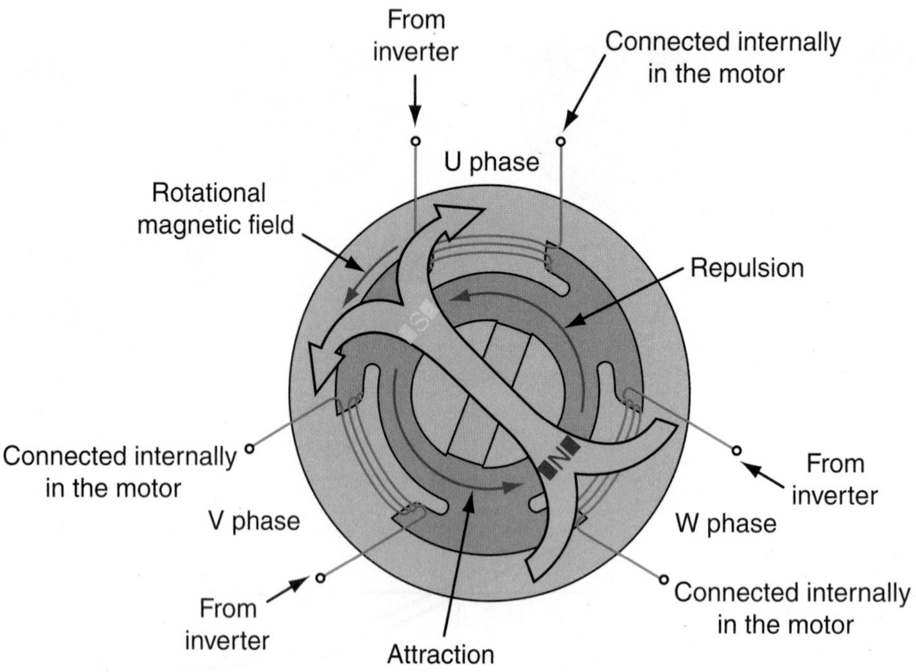

Figure 18-44 The synchronous motor used in Toyota's hybrid vehicles is controlled by the vehicle's inverter, which is part of the hybrid control system.

An induction motor generates its own rotor current. The current is induced in the rotor windings as the rotor cuts through the magnetic flux lines of the rotating stator field. The induced current causes each rotor conductor to act like the permanent magnet. As the magnetic field of the stator rotates, the magnetic field of the rotor follows the rotating magnetic field of the stator. It should be obvious that this type motor needs some rotation of the rotor before it can rotate on its own. Various methods are used to start these motors, including capacitors and separate starting windings.

It is impossible for the rotor of an induction motor to rotate at synchronous speed. If the rotor were to turn at the same speed as the rotating field, there would be no relative motion between the stator and rotor fields. As a result, no lines of force would be cut by the rotor's conductors, and there would be no induced voltage in the rotor. In an induction motor, the rotor must rotate at a speed slower than that of the rotating magnetic field. This is why these motors are called "**asynchronous motors**."

The difference between the synchronous speed and actual rotor speed is called **slip**. Slip is directly proportional to the load on the motor. When loads are on the rotor's shaft, the rotor tends to slow and slip increases. The slip then induces more current in the rotor and the rotor turns with more torque but at a slower speed; therefore, it produces less CEMF.

Asynchronous motors are used in most hybrid vehicles. Their placement can be between the engine and transmission, inside the transmission (**Figure 18–45**), or at the individual wheels. This motor offers many advantages for use in hybrid and fuel cell vehicles, one being the ease of controlling its speed. Basically its speed is determined by the frequency and amount of the voltage applied to it.

Figure 18-45 The transaxle bolted to the engine contains two separate electric motors. The object in the background is the battery pack. *Courtesy of Toyota Motor Sales, U.S.A., Inc.*

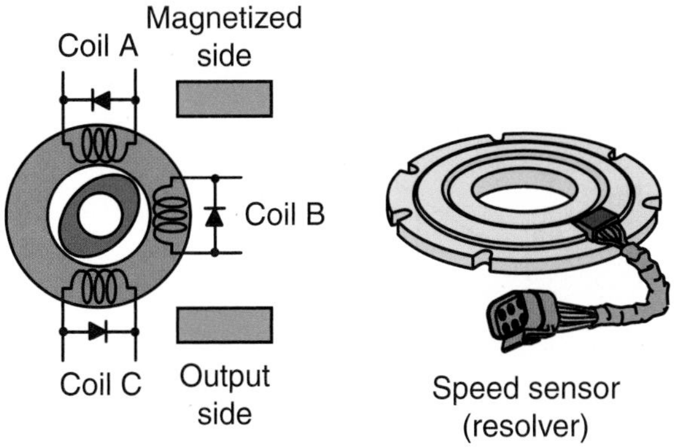

Figure 18–46 A resolver located in each motor precisely detects the magnetic pole position of the rotor and allows the control unit to monitor and control the activity of MG1 and MG2.

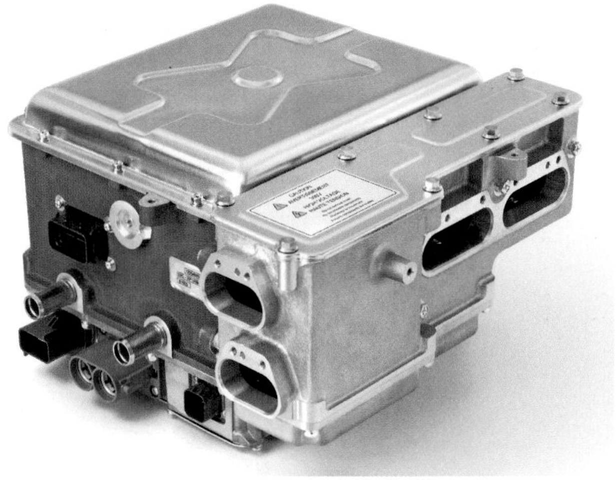

Figure 18–47 The power inverter for a two-mode hybrid system. *Courtesy of Chrysler LLC*

Rotor Sensors AC motors require a sensor to monitor the position of the rotor within the stator. It is necessary to time, or phase, the three-phase AC so it attracts the rotor's magnets and keeps it rotating and producing torque. AC creates a rotating magnetic field in the stator and the rotor chases that field. The control system monitors the position and speed of the rotor and controls the frequency of the stator's voltage, which in turn controls the torque and speed of the motor.

Toyota uses a sensor **(Figure 18–46)**, called a resolver, to monitor the position of the rotor in its traction motors. The sensor has three individual coil windings: one excitation coil and two detection coils. The detection coils are electrically staggered 90 degrees. The flow of AC into an excitation coil results in a constant frequency through the coil.

This frequency is sent to the detection coils and is altered by the presence of the rotating magnetic field. Because the rotor has a slight oval shape, the gap between the stator and rotor varies as the rotor rotates. A strengthening and weakening of the magnetic field occurs with the changing gap and alters the frequency at the detection coils. The control unit determines the exact position of the rotor according to the difference between the frequency values of the detection coils. The control unit also calculates the rotational speed of the rotor based on the speed at which the position changes.

Honda has a similar device but relies on three Hall-effect sensors rather than frequency detection coils.

Inverters

All AC motors require AC voltage. In a vehicle, this can be provided directly by an AC generator or through an inverter connected to the battery, which stores DC. Fuel cells produce DC voltage that can be used to power motors and lights and charge batteries without additional electronics.

A power **inverter** is used to change the battery's high-voltage DC into high-voltage, three-phase AC to power the electric motors **(Figure 18–47)**. It also changes the AC generated by the motors/generators to DC so it can recharge the battery pack. The inverter is basically a three-phase parallel circuit with each leg containing two power transistors connected in series. The electronic control unit controls the power transistors to ensure proper phasing of the AC to the motors. This circuit is connected between the motor/generator and the battery pack. The inverter assembly contains one of these circuits for each of the motor/generators in the vehicle. For additional power output from the motors, some vehicles have a boost converter that provides increased voltage to the motors. In many cases, the boost voltage is twice that of the battery's nominal voltage.

Controllers

A controller is used to manage the flow of electricity from the batteries to the motors, thereby controlling their speed. A sensor located by or connected to the throttle pedal sends the driver's input to the controller. The controller then sends the appropriate amount of voltage to the motor.

A simple controller is a variable resistor or potentiometer connected to the accelerator. When there is no pressure on the accelerator, resistance of the potentiometer is too high to allow voltage to the motor. When the accelerator is fully opened, the resistance is very low and full battery voltage is delivered to the motor. The various positions of the accelerator

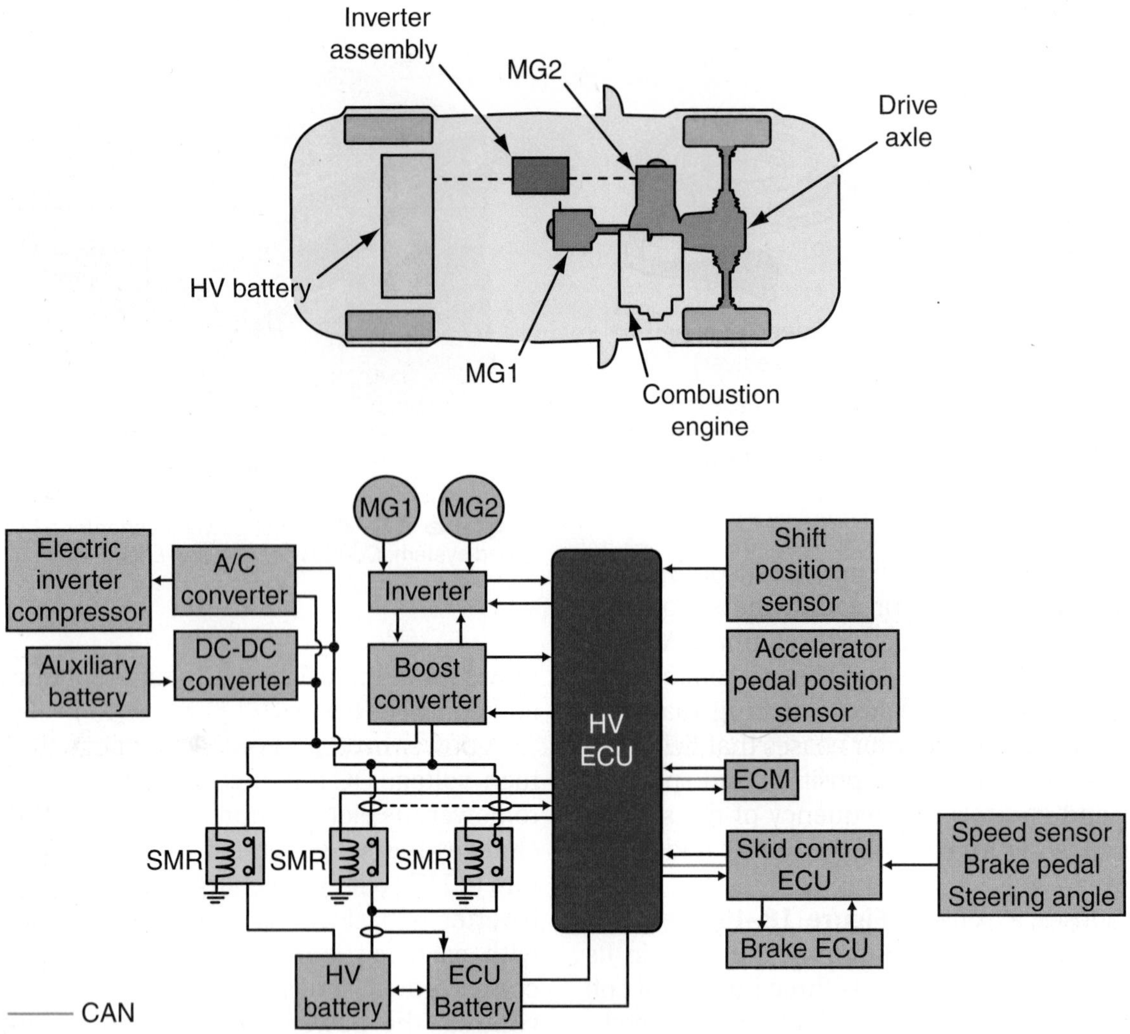

Figure 18–48 A diagram of Toyota's hybrid control system.

between closed and wide open allow corresponding amounts of voltage to the motor.

Using electronics, the same principle can be used but with more positive results. A sensor monitors the accelerator and sends information to a control unit. The control unit monitors that signal plus other inputs regarding the operating conditions of the vehicle **(Figure 18–48)**. The control unit then duty cycles the voltage from the inverter to the motor. In this way, the voltage is pulsed and more precise motor speed control is possible. Most controllers pulse the voltage more than 15,000 times per second.

Stop-Start Feature

The stop-start (Honda refers to this feature as idle stop) operational mode is initiated when the car is at a stop, the shift lever is placed in neutral, the clutch pedal is released, and/or when the brake pedal is held down **(Figure 18–49)**. Stop-start systems automatically shut down the engine to prevent the waste of fuel while the engine is idling. This feature can increase fuel economy by more than 5%. Stop-start does not occur if the voltage of the battery module is low. The benefit of these systems is most evident during stop-and-go driving. The engine is restarted automatically when the driver releases the brake pedal or when the control system senses the need. Typically the traction motor is used to quickly rotate the engine to restart it.

When the engine is restarted is a defining element of mild and full hybrids. Full hybrids can be powered by only the electric motor or the gasoline engine alone, or both. They have the stop-start feature but the engine does not automatically restart when the brakes are released. Because a full hybrid can accelerate by only the electric motor, the engine is not restarted until the engine's power is needed. This is determined by the control system and is based on the vehicle's load and the batteries' state of charge.

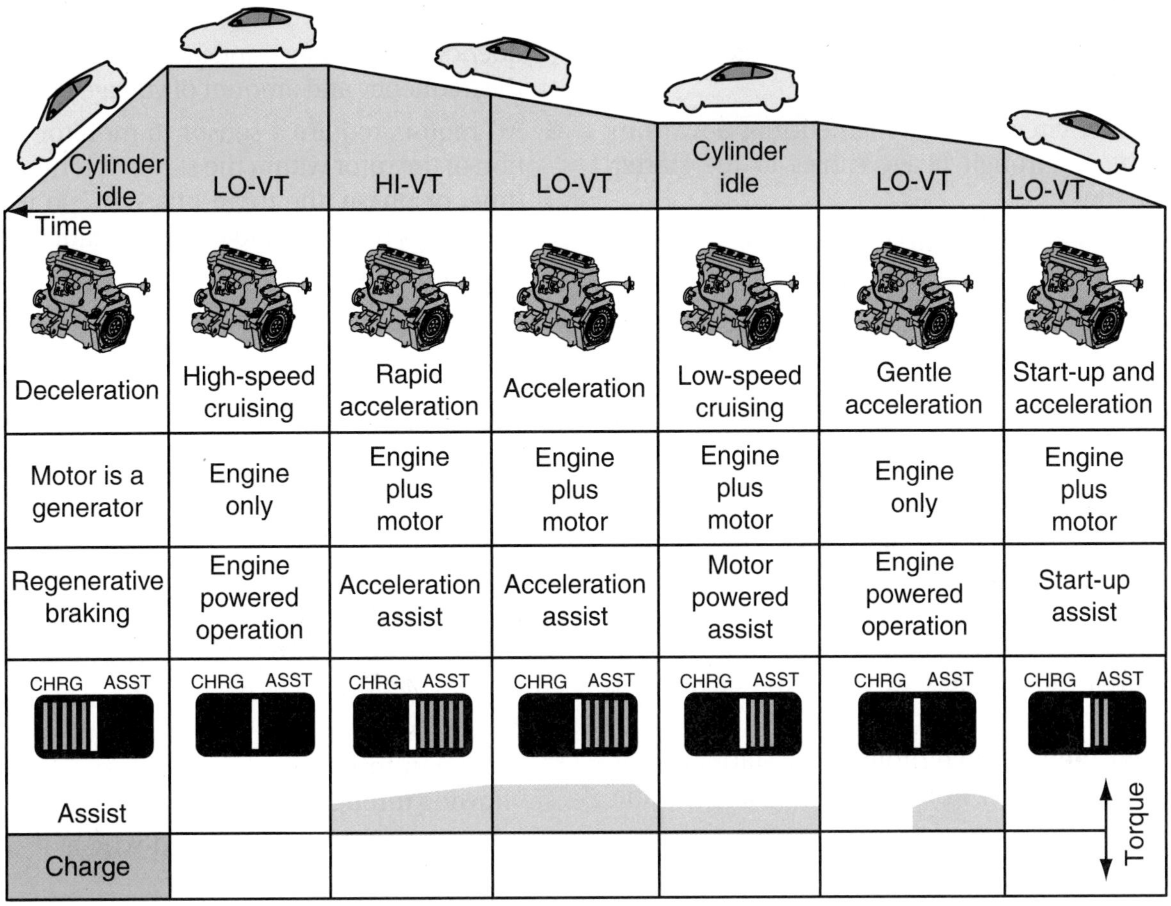

Cylinder idle
LO-VT
HI-VT
LO-VT
Cylinder idle
LO-VT
LO-VT
Time
Deceleration
High-speed cruising
Rapid acceleration
Acceleration
Low-speed cruising
Gentle acceleration
Start-up and acceleration
Motor is a generator
Engine only
Engine plus motor
Engine plus motor
Engine plus motor
Engine only
Engine plus motor
Regenerative braking
Engine powered operation
Acceleration assist
Acceleration assist
Motor powered assist
Engine powered operation
Start-up assist
CHRG ASST
Assist
Charge
Torque

Figure 18-49 The operational modes for a Honda Civic hybrid.

KEY TERMS

Armature
Asynchronous motor
Brushes
Commutator
Compound motor
Counter EMF (CEMF)
Cranking voltage test
Duty cycle
End frame
Field coil
Flux density
Flux field
Growler
Hold-in winding
Inverter
Neutral safety switch
Pinion gear
Pole shoes
Pull-in winding
Pulse width modulation (PWM)
Reluctance
Series motor
Shunt motor
Slip
Starter frame
Starter housing
Starter relay
Starting safety switch
Traction motor

SUMMARY

- Current flowing through a wire creates a magnetic field around the wire. Moving a wire through a magnetic field creates current flow in the wire.
- A magnetic field, called a flux field, exists around every magnet.
- The magnetic polarity of an electromagnet depends on the direction of current flow through the loop.
- The strength of the field around an electromagnet increases if current through the coil increases, the number of coil turns increases, and the reluctance decreases.
- The basic parts of a motor are the stator or field windings, which are the stationary parts of the motor, and the rotor or armature, which is the rotating part.
- A starting motor is a special type of electric motor designed to operate under great electrical loads and to produce great amounts of torque for short periods.
- All starting motors have a housing, field coils (windings), an armature, a commutator with brushes, and end frames.
- The amount of torque from a starting motor depends on its current draw, which is controlled by CEMF.

- The starting system has two distinct electrical circuits: the starter circuit and the control circuit.
- The starter circuit carries high current flow from the battery through heavy cables to the starter motor.
- The control circuit uses a small amount of current to operate a magnetic switch that opens and closes the starter circuit.
- The ignition switch is used to control current flow in the control circuit.
- Solenoids and relays are the two types of magnetic switches used in starting systems. Solenoids use electromagnetic force to pull a plunger into a coil to close the contact points. Relays use a hinged armature to open and close the circuit.
- The drive mechanism of the starter motor engages and turns the flywheel to crank the engine for starting.
- An overrunning clutch protects the starter motor from spinning too fast once the vehicle engine starts.
- Starting safety switches prevent the starting system from operating when the transmission is engaged.
- Battery load, cranking voltage, cranking current, insulated circuit resistance, starter relay bypass, ground circuit resistance, control circuit, and drive component tests are all used to troubleshoot the starting system.
- With the starter removed, inspect the starter drive pinion gear and the flywheel ring gear.
- The starter should be cleaned and inspected as it is disassembled.
- The housing, end frames, field coils, armature, brushes, and drive should be carefully inspected.
- Hybrid and fuel cell vehicles use strong electric motors to move the vehicle, and in some cases, to start the engine as well.
- The placement and purpose of a traction motor depends on the design of the hybrid.
- A brushless DC motor is much like a brushed DC motor, but the purpose of the rotor (armature) and stator (field windings) are reversed.
- In an AC motor, the polarity in the windings constantly changes. Most AC motors use three-phase AC.
- A synchronous AC motor will run at the frequency of the AC voltage.
- An asynchronous motor cannot run at the frequency of the voltage, and its speed is controlled by the frequency and amount of voltage applied to it.
- AC motors require a sensor to monitor the position of the rotor within the stator. It is necessary to time, or phase, the three-phase AC so it attracts the rotor's magnets and keeps it rotating and producing torque.
- An inverter is used to change the battery's high-voltage DC into high-voltage, three-phase AC to power the electric motors.
- The stop-start operational mode turns off the engine when the car is at a stop and quickly restarts it when it is needed.

REVIEW QUESTIONS

1. *True or False?* The strength of the magnetic poles in an electromagnet decreases with an increase in the number of turns of wire and the current flowing through them.
2. What is three-phase AC and why is it used for many motors?
3. Briefly explain the differences between a brushed and a brushless DC motor.
4. *True or False?* Many armature bearings are held in the case by a retainer, whereas bushings are typically pressed out and into their bore.
5. Which of the following is *not* part of the starter circuit?
   a. battery
   b. starting safety switch
   c. starter motor
   d. relay solenoid
6. Which of the following tests would *not* be performed to check for high resistance in the battery cables?
   a. cranking voltage test
   b. insulated circuit resistance test
   c. starter relay by-pass test
   d. ground circuit resistance test
7. When the starter spins but does not engage the flywheel, which of the following may be true?
   a. defective starter drive
   b. excessive resistance in the control circuit
   c. a faulty starter relay
   d. all of the above

8. If the solenoid clicks while trying to crank the engine with the starter, which of the following is *not* a probable cause?
   a. a faulty neutral safety switch
   b. low battery voltage
   c. low voltage available to the solenoid
   d. an open in the hold-in winding
9. The normal minimal cranking voltage specification is approximately ______.
   a. 9.6 volts
   b. 10.5 volts
   c. 11.0 volts
   d. 12.65 volts
10. If a ground circuit test reveals a voltage drop of more than 0.2 volt, the problem may be ______.
    a. a loose starter motor mounting bolt
    b. a poor battery ground terminal post connector
    c. a damaged battery ground cable
    d. all of the above
11. Which of the following would *not* cause excessive current draw by the starter motor?
    a. a short in the motor
    b. high resistance in the armature
    c. using oil that is too heavy in the engine
    d. very cold weather
12. The device that prevents the engine from turning the armature of the starter motor is the ______.
    a. overrunning clutch
    b. pinion gear
    c. flywheel
    d. pole shoe
13. The part of the armature that the brushes ride on is called the ______.
14. What is a synchronous motor?
15. A control circuit test may uncover which of the following conditions?
    a. high resistance in the solenoid switch circuit
    b. worn brushes
    c. loose battery cable connection
    d. short in the starter armature windings

ASE-STYLE REVIEW QUESTIONS

1. An engine cranks slowly: Technician A says that a possible cause of the problem is poor starter circuit connections. Technician B says that an engine problem could be the cause. Who is correct?
   a. Technician A
   b. Technician B
   c. Both A and B
   d. Neither A nor B
2. While discussing armature testing: Technician A says to test for shorts, place the armature in a growler, and hold a thin metal blade parallel to the armature. If the blade vibrates while the armature is turned, there is a short. Technician B says that an ohmmeter can be used to test for shorts. Who is correct?
   a. Technician A
   b. Technician B
   c. Both A and B
   d. Neither A nor B
3. Pinion gear-to-flywheel ring gear clearance is being discussed: Technician A says that if there is too much clearance there will be a high pitch while the engine cranks. Technician B says that if there is too little clearance, there will be a high pitch while the starter is cranking the engine. Who is correct?
   a. Technician A
   b. Technician B
   c. Both A and B
   d. Neither A nor B
4. Technician A says that a starter no-load test is conducted with the starter in a bench vise. Technician B says that the no-load test is similar to an open-circuit voltage test. Who is correct?
   a. Technician A
   b. Technician B
   c. Both A and B
   d. Neither A nor B
5. Technician A checks a starter's field coils for opens by connecting an ohmmeter between the coils' power feed wire and the field coil brush lead. Technician B checks a starter's field coil for a short to ground by connecting an ohmmeter across the field coil brush lead and the starter housing. Who is correct?
   a. Technician A
   b. Technician B
   c. Both A and B
   d. Neither A nor B
6. Technician A says that the purpose of the starter relay is to complete the circuit from the battery to the starter motor. Technician B says that the purpose of the solenoid is to complete the circuit from the battery to the starter motor. Who is correct?
   a. Technician A
   b. Technician B
   c. Both A and B
   d. Neither A nor B
7. While discussing the stop-start feature of hybrid vehicles: Technician A says that it is initiated when the car is at a stop, the shift lever is placed in neutral, the clutch pedal is released, and/or when the brake pedal is held down. Technician B says that stop-start does not occur if the voltage of the high-voltage battery's state-of-charge is low. Who is correct?
   a. Technician A
   b. Technician B
   c. Both A and B
   d. Neither A nor B

8. While discussing asynchronous motors: Technician A says that the speed of the motor stays constant regardless of the load put on it. Technician B says that their speed is determined by the frequency and amount of the voltage applied to them. Who is correct?

   a. Technician A
   b. Technician B
   c. Both A and B
   d. Neither A nor B

9. While discussing DC starting motors: Technician A says that the armature may be wired in series with the field coils. Technician B says that the field coils may be wired parallel with the armature. Who is correct?

   a. Technician A
   b. Technician B
   c. Both A and B
   d. Neither A nor B

10. While discussing electric motors: Technician A says that when there is high CEMF, there will be high current flow. Technician B says that if the reluctance of an electromagnet increases, its field strength increases. Who is correct?

    a. Technician A
    b. Technician B
    c. Both A and B
    d. Neither A nor B

CHARGING SYSTEMS

CHAPTER 19

OBJECTIVES

■ Explain the purpose of the charging system. ■ Identify the major components of the charging system. ■ Explain the purposes of the major parts of an AC generator. ■ Explain half- and full-wave rectification and how they relate to AC generator operation. ■ Identify the different types of AC voltage regulators. ■ Describe the two types of stator windings. ■ Explain the features enabled by the use of a starter/generator unit. ■ Perform charging system inspection and testing procedures using electrical test equipment.

The primary purpose of a charging system is to recharge the battery. After the battery has supplied the high current needed to start the engine, the battery, even a good battery, has a low charge. The charging system recharges the battery by supplying a constant and relatively low charge to the battery. Charging systems work on the principles of magnetism to change mechanical energy into electrical energy. This is done by inducing voltage.

Voltage is induced in a wire when it moves through a magnetic field. The wire or conductor becomes a source of electricity and has a polarity or distinct positive and negative ends. However, this polarity can be switched depending on the relative direction of movement between the wire and magnetic field **(Figure 19–1)**. This is why an AC generator produces alternating current **(Figure 19–2)**.

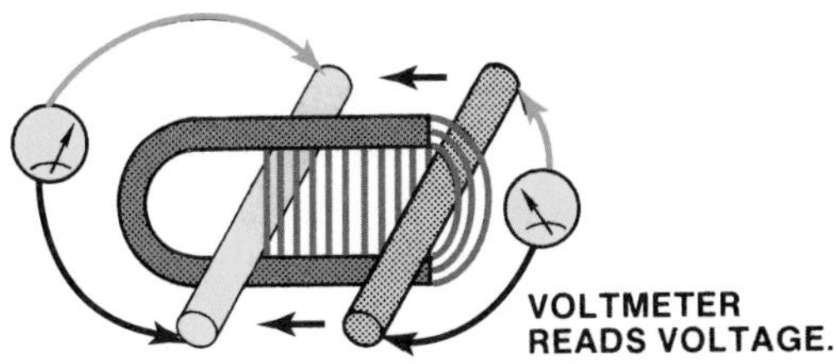

Figure 19–1 Moving a conductor so it cuts across the magnetic lines of force induces a voltage in the conductor.

Figure 19–2 An AC generator. *Courtesy of Robert Bosch GmbH, www.bosch-presse.de*

ALTERNATING CURRENT CHARGING SYSTEMS

During cranking, the battery supplies all of the vehicle's electrical power. However, once the engine is running, the charging system is responsible for producing enough energy to meet the demands of all of the loads in the electrical system while also recharging the battery. With all of the electrical and electronic devices found on today's vehicles, the charging system has a difficult job.

Several decades ago the charging system depended on a **DC generator**. The DC generator provided direct current (DC) and was similar to an electric motor in construction. The biggest difference between a generator and a motor is the wiring to the armature. In a motor, the armature receives current from the battery. This creates the magnetic field that opposes the magnetic fields in the motor's coils, which causes the armature to rotate. The armature in a DC generator is driven by the engine. It is not magnetized and the windings simply rotate through the stationary magnetic field of the field windings, inducing a voltage in the conductors inside the armature. A motor can become a generator by allowing current to flow from the armature instead of to it. In a DC generator, the placement of the brushes on the commutator changes the induced AC voltage to a DC voltage output.

DC generators had a very limited current output, especially at low speeds. They could not keep up with demands of the modern automobile and were replaced by AC generators. AC generators are capable of providing high current output even at low engine speeds.

SHOP TALK

With the implementation of OBD II, new terminology was given to many parts of an automobile. Prior to that law, an AC generator was called an **alternator**. In fact, in many cases and by many manufacturers an AC generator is still referred to as an alternator. To avoid confusion, just remember that an alternator is an AC generator and vice versa.

AC generators use a design that is basically the reverse of a DC generator. In an AC generator (**Figure 19–3**), a spinning magnetic field (called the rotor)

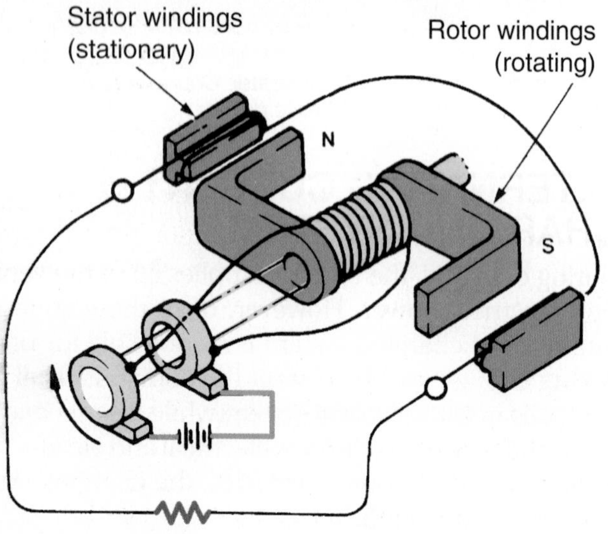

Figure 19-3 A simplified AC generator.

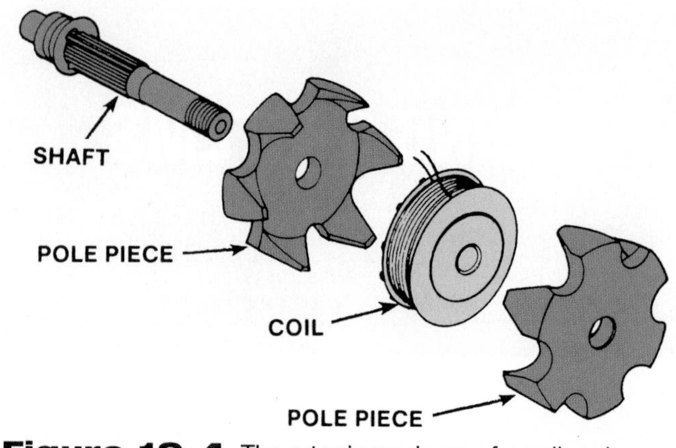

Figure 19-4 The rotor is made up of a coil, pole pieces, and a shaft.

rotates inside an assembly of stationary conductors (called the stator). As the spinning north and south poles of the magnetic field pass the conductors, they induce a voltage that first flows in one direction and then in the opposite direction (AC voltage). Because automobiles use DC voltage, the AC must be changed or rectified into DC. This is done through an arrangement of diodes that are placed between the output of the windings and the output of the AC generator.

AC Generator Construction

Rotor The rotor assembly consists of a drive shaft, coil, and two pole pieces (**Figure 19–4**). A pulley mounted on one end of the shaft allows the rotor to be spun by a belt driven by the crankshaft pulley.

The **rotor** is a rotating magnetic field inside the alternator. The field coil is simply a long length of insulated wire wrapped around an iron core. The core is located between the two sets of **pole pieces**. A magnetic field is formed by a small amount (4.0 to 6.5 amperes) of current passing through the coil winding. As current flows through the coil, the core is magnetized and the pole pieces assume the magnetic polarity of the end of the core that they touch. Thus, one pole piece has a north polarity and the other has a south polarity. The extensions of the pole pieces, known as **fingers**, form the actual magnetic poles. A typical rotor has fourteen poles, seven north and seven south, with the magnetic field between the pole pieces moving from the N poles to the adjacent S poles (**Figure 19–5**).

Slip Rings and Brushes Current to create the magnetic field is supplied to the coil from one of two sources: the battery or the AC generator itself. In either case, the current is passed through the AC generator's voltage regulator before it is applied to the coil. The voltage regulator varies the amount of current supplied. Increasing field current through the coil increases the strength of the magnetic field. This, in

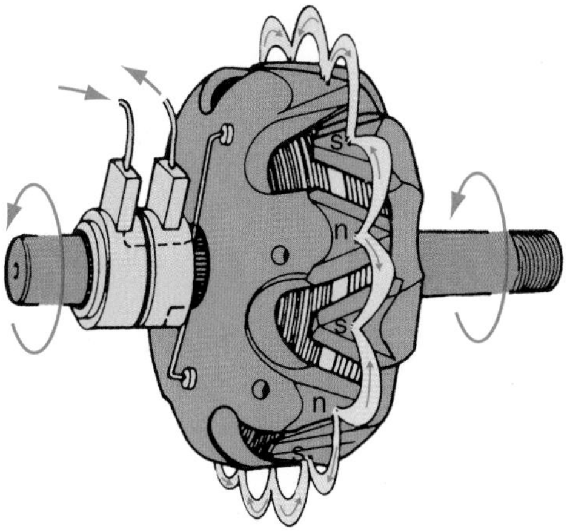

Figure 19-5 The magnetic field moves from the N poles, or fingers, to the S poles.

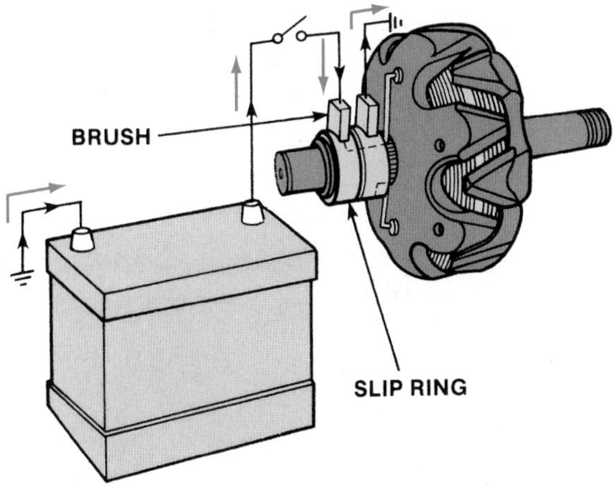

Figure 19-7 Current is carried by the brushes to the rotor windings via the slip rings.

turn, increases AC generator voltage output. Decreasing the field current to the coil has the opposite effect. Output voltage decreases.

Slip rings and brushes **(Figure 19–6)** conduct current to the spinning rotor. Most AC generators have two slip rings mounted directly on the rotor shaft. They are insulated from the shaft and each other. Each end of the **field coil** connects to one of the slip rings. A carbon brush located on each slip ring carries the current to the field coil. Current is transmitted from the field terminal of the voltage regulator through the first brush and slip ring to the field coil. Current passes through the field coil and the second slip ring and brush before returning to ground **(Figure 19–7)**.

Stator The **stator** is the stationary member of the generator. It is made up of a number of conductors,

Figure 19-6 The brushes ride on the slip rings.

or wires, into which the voltage is induced. Most AC generators use three windings to generate the required amperage output. They can be arranged in either a **delta** configuration or a **wye** configuration **(Figure 19–8)**. The delta winding **(Figure 19–9)** received its name because its shape resembles the Greek letter delta, Δ. The wye winding resembles the letter *Y*. Alternators use one or the other. Usually, a wye winding is used in applications in which high charging voltage at low engine speeds is required. AC generators with delta windings are capable of putting out higher amperages at high speeds but low engine speed output is poor.

The rotor rotates inside the stator. A small air gap between the two allows the rotor to turn without making contact with the stator.

Alternating current produces positive and negative pulses. The resultant waveform is a sine wave, which can be observed on a scope. The waveform starts at zero, goes positive, and then drops back to zero before turning negative. When the north pole magnetic field cuts across the stator, a positive voltage is generated. When the south polarity magnetic field cuts across the stator, a negative voltage is induced. A single loop of wire energized by a single north then a south result in a single-phase voltage. Remember that there are three overlapping stator windings. This produces overlapping sine waves **(Figure 19–10)** or **three-phase voltage**.

End Frame Assembly The end frame assembly, or housing, contains the bearings for the rotor shaft. Each end frame also has built-in ducts so the air from the rotor shaft fan can pass through the AC generator. Normally, a heat sink containing three positive rectifier diodes is attached to the rear end frame. Heat

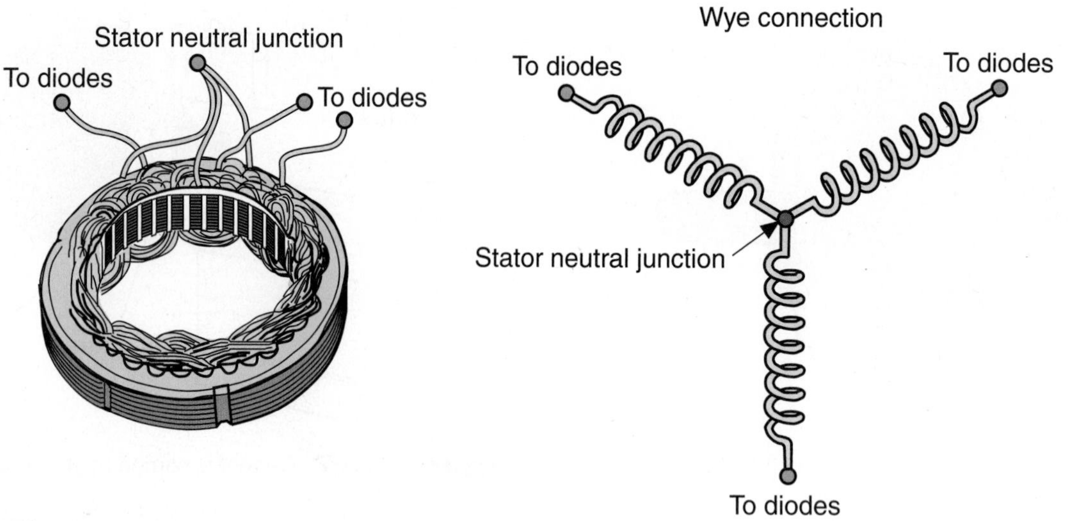

Figure 19-8 A wye-connected stator winding.

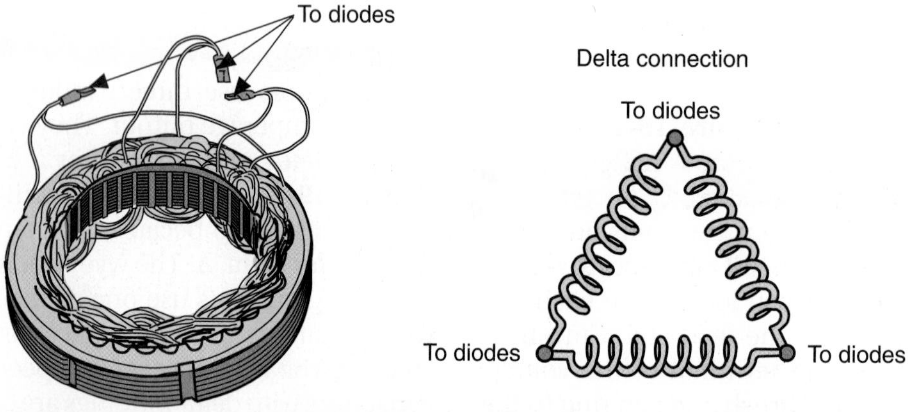

Figure 19-9 A delta-connected stator winding.

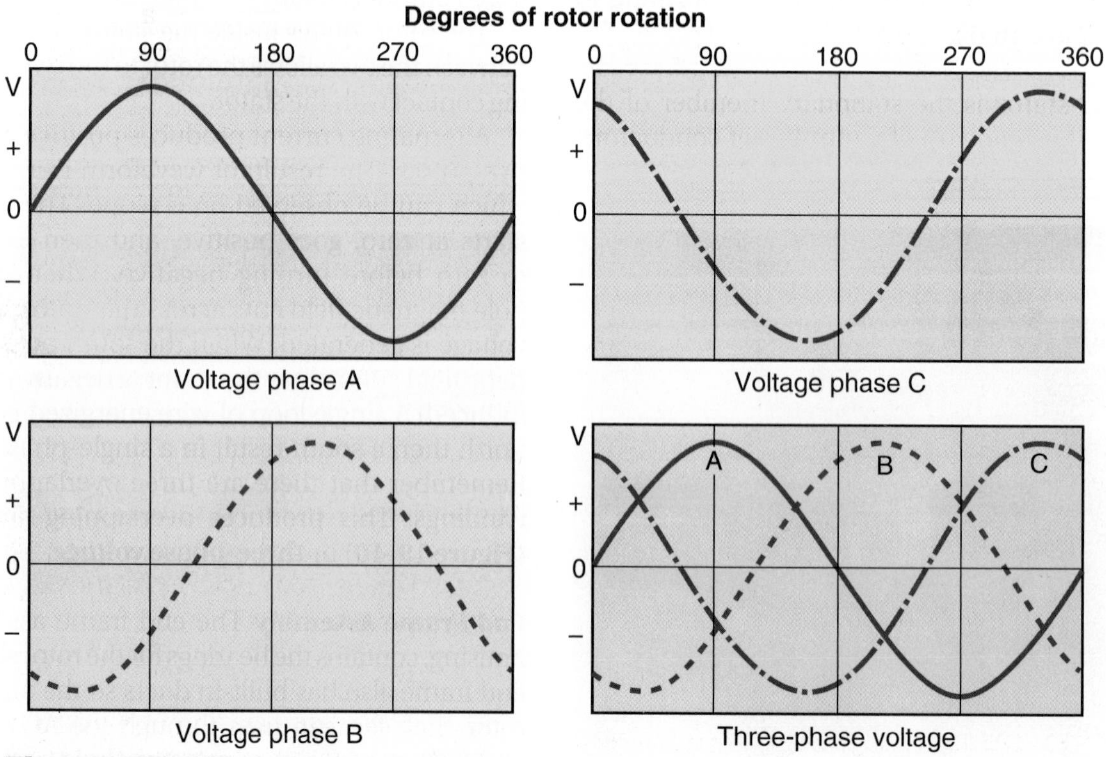

Figure 19-10 The voltage produced in each stator winding is added together to create a three-phase voltage.

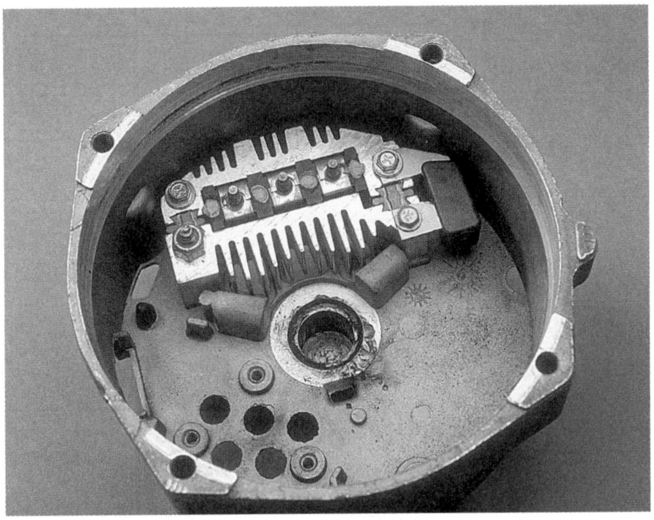

Figure 19-11 A bridge rectifier mounted in the end frame.

Figure 19-13 A water-cooled AC generator. *Courtesy of BMW of North America, LLC*

can pass easily from these diodes to the moving air (**Figure 19–11**). Three negative rectifier diodes are contained in the end frame itself. Because the end frames are bolted together and then bolted directly to the engine, the end frame assembly is part of the electrical ground path. This means that anything connected to the housing that is not insulated from the housing is grounded.

Cooling Fans Behind the drive pulley on most AC generators is a cooling fan that rotates with the rotor. This cooling fan draws air into the housing through the openings at the rear of the housing. The air leaves through openings behind the cooling fan (**Figure 19–12**). The moving air pulls heat from the diodes and their heat decreases.

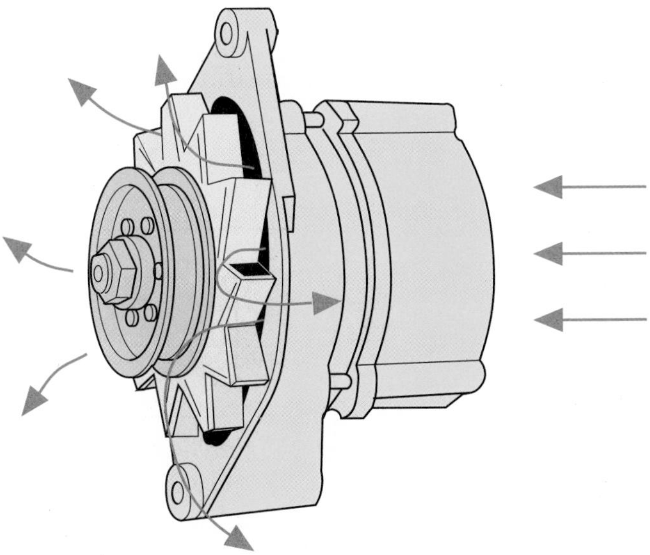

Figure 19-12 The cooling fan draws air in from the rear of the generator to keep the diodes cool. *Courtesy of Robert Bosch GmbH, www.bosch-presse.de*

Cooling the diodes is important for the efficiency and durability of an AC generator. Several generator designs have been introduced recently that increase the cooling efficiency of a generator. One of these is the AD-series generator from General Motors. The "A" stands for air cooled and the "D" means dual fans. This series is lighter than most other generators but capable of very high outputs. This type of generator does not have an external fan; instead, it has two internal fans.

Liquid-Cooled Generators Another recent design uses liquid cooling (**Figure 19–13**). Using water or coolant to cool a generator is a very efficient way to keep diode temperatures down. But the real reason for eliminating the fan and using liquid to cool the generator is to reduce noise. The rotating fan is a source of underhood noise that some automobile manufacturers want to eliminate. These new generators have water jackets cast into the housing. Hoses connect the housing to the engine's cooling system. Not only do these generators make less noise, they have higher power output and should last longer in the high-temperature environment of the engine compartment.

AC GENERATOR OPERATION

As mentioned earlier, AC generators produce alternating current that must be converted, or rectified, to DC. This is accomplished by passing the AC through diodes.

Diodes

A **diode** allows current to flow in one direction but not in the opposite direction. Therefore, it can function as a switch, acting as either conductor or insulator,

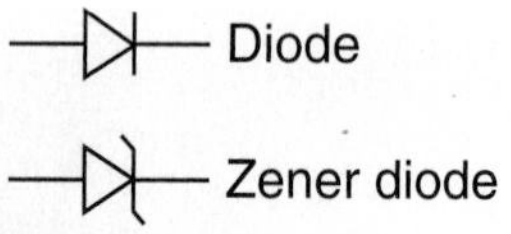

Figure 19–14 The schematic symbols for two common components of a generator and voltage regulator: a diode and a zener diode.

depending on the direction of current flow. In an AC generator, current is rectified (changed from AC to DC) through the use of diodes. The diodes are arranged so that current can leave the generator in one direction only (as DC).

A variation of the diode is the zener diode (**Figure 19–14**). This device functions like a standard diode until a certain voltage is reached. When the voltage level reaches this point, the zener diode allows current to flow in the reverse direction. Zener diodes are often used in electronic voltage regulators.

DC Rectification

Figure 19–15 shows that when AC passes through a diode, the negative pulses are blocked off to produce the scope pattern shown. If the diode is reversed, it blocks off current during the positive pulse and allows the negative pulse to flow (**Figure 19–16**). Because only half of the AC current pulses (either the positive or the negative) is able to pass, this is called **half-wave rectification.**

By adding more diodes to the circuit, more of the AC is rectified. When all of the AC is rectified, **full-wave rectification** occurs.

Full-wave rectification requires another circuit with similar characteristics. **Figure 19–17** shows a wye stator with two diodes attached to each winding. One diode is insulated, or positive, and the other is grounded, or negative. The center of the Y contains a common point for all windings. It can have a connection attached to it. It is called the stator neutral junction. At any time during the rotor movement, two windings are in series and the third coil is neutral and inactive. As the rotor revolves, it energizes the different sets of windings in different directions.

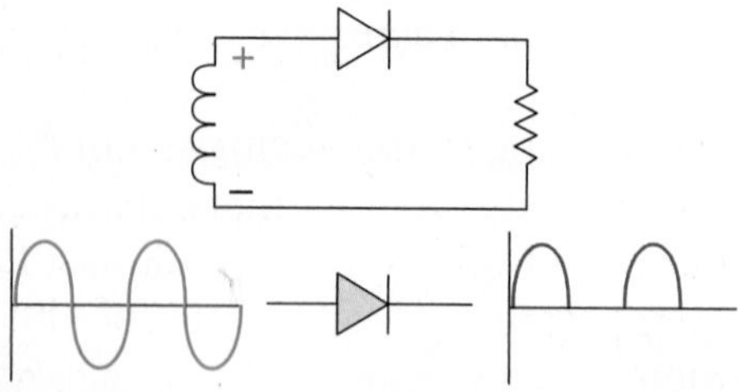

Figure 19–15 Half-wave rectification, diode positively biased.

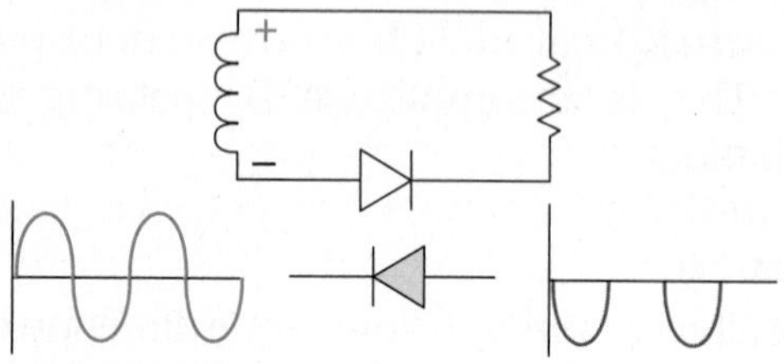

Figure 19–16 Half-wave rectification, diode negatively biased.

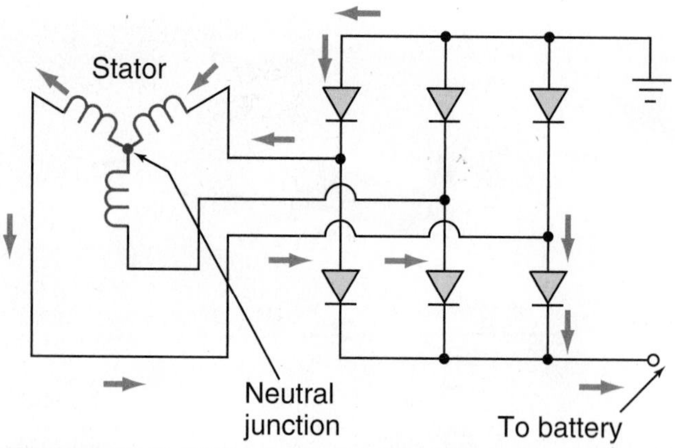

Figure 19–17 A wye stator wired to six diodes.

The diode action does not change when the stator and diodes are wired in a delta pattern. **Figure 19–18** shows the major difference. Instead of having two windings in series, the windings are in parallel. Thus, more current is available because the parallel paths allow more current to flow through the diodes. Nevertheless, the action of the diodes remains the same.

Many AC generators have an additional set of three diodes called the **diode trio**. The diode trio is used to rectify current from the stator so that it can be used to create the magnetic field in the rotor. Using the diode trio eliminates extra wiring. To control generator output, a voltage regulator regulates the current from the diode trio and to the rotor (**Figure 19–19**).

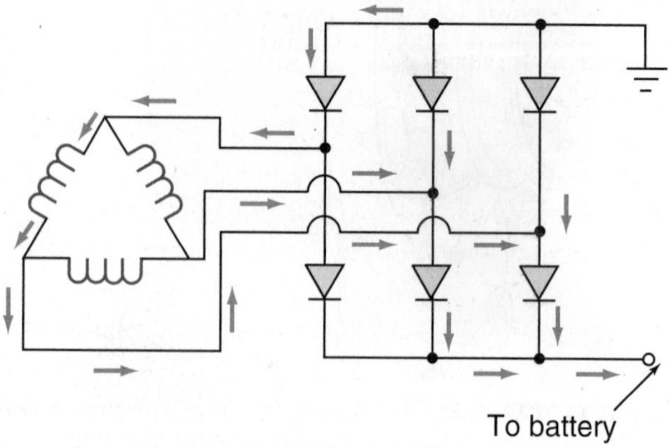

Figure 19–18 A delta stator wired to six diodes.

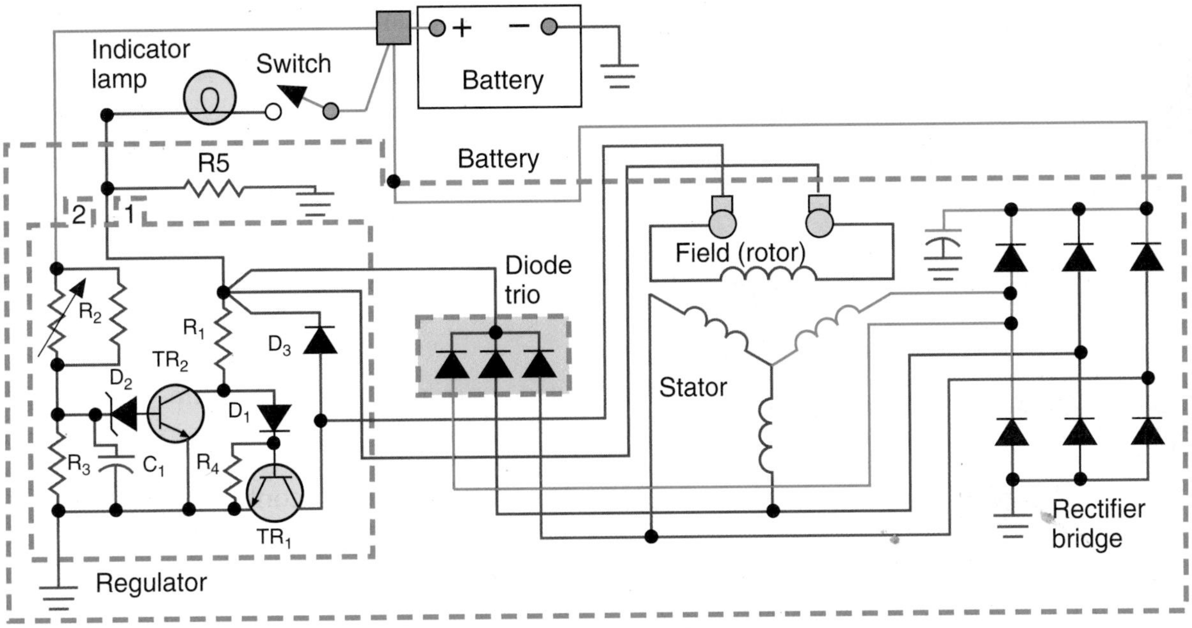

Figure 19–19 Wiring diagram of a charging circuit with a diode trio.

Factors Controlling Generator Output

Several factors determine the total output available from a generator other than the type of stator winding. These include:

- The rotational speed of the rotor. Higher speeds can lead to higher output.
- The number of windings in the rotor. Increased windings will increase output.
- The current flow through the rotor windings. Increased current through the rotor will increase output.
- The number of windings in the stator. An increase in the number of windings will increase output.

Performance TIP

Often owners will change the size of the generator's pulley to decrease the drag on the engine and, therefore, increase the engine's power output. Although the generator will spin at a lower speed, it also will have a lower output. Because the system will respond to the lower output by sending more current to the rotor, the magnetic field will be stronger and that alone will create more drag on the engine. Making the decision to change a generator pulley should be done only after much thought. If the battery is fully charged, there will be a net increase in engine output. If the battery is low on charge, there may be no gain at all.

VOLTAGE REGULATION

The output from an AC generator can reach as high as 250 volts if it is not controlled. The battery and the electrical system must be protected from this excessive voltage. Therefore, charging systems use a **voltage regulator** to control the generator's output. Voltage output is controlled by the voltage regulator as it varies the strength of the magnetic field in the rotor. Current output does not need to be controlled because an AC generator naturally limits the current output. To ensure that the battery stays fully charged, most regulators are set for a system voltage between 14.5 and 15.5 volts.

Voltage output is controlled by varying the field current through the rotor. The higher the field current, the higher the voltage output. By controlling the amount of resistance in series with the field coil, control of the field current and voltage output is obtained.

An input signal, called the **sensing voltage**, allows the regulator to monitor system voltage (**Figure 19–20**). If the sensing voltage is below the regulator setting, an increase in field current is allowed, which causes an increase in voltage output. Higher sensing voltage will result in a decrease in field current and voltage output. The regulator will reduce the charging voltage until it is at a level to run the ignition system while putting a low charge on the battery. If a heavy load is turned on, such as the headlights, the additional draw will cause a decrease in battery voltage. The regulator

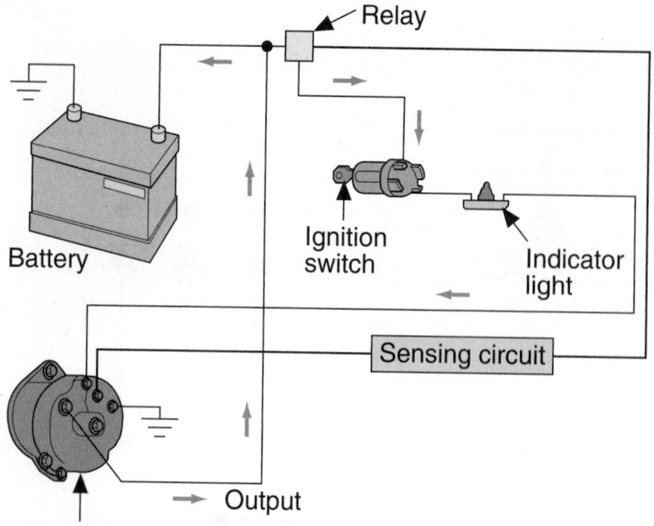

Figure 19-20 The voltage regulator adjusts the generator's output according to the voltage on the sensing circuit.

will sense the low voltage and will increase current to the rotor. When the load is turned off, the regulator senses the rise in system voltage and reduces the field current.

Another input that affects voltage regulation is temperature. Because ambient temperature influences the rate of charge that a battery can accept, regulators are temperature compensated. Temperature compensation is required because the battery is more reluctant to accept a charge at lower ambient temperatures. The regulator will increase the voltage output to force a charge on the battery.

Field Circuits

A regulator controls field current and is connected in one of three ways to the rotor. These are called field circuits. The A-circuit has the regulator on the ground side of the field coil. The battery feed (B+) for the field coil is picked up inside the AC generator (**Figure 19–21**). By placing the regulator on the ground side of the field coil, the regulator controls field current by varying the current flow to ground.

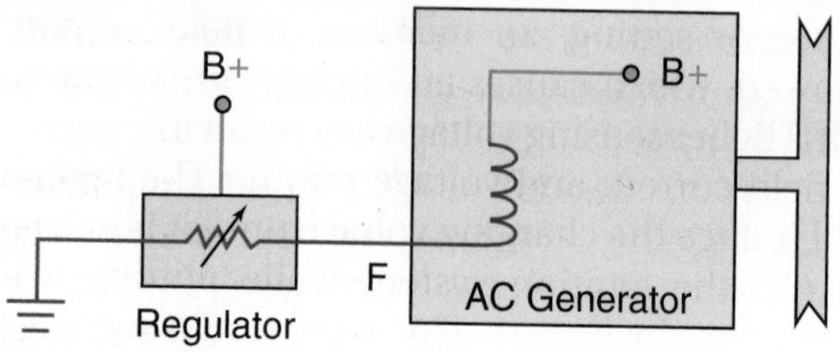

Figure 19-21 An A-circuit.

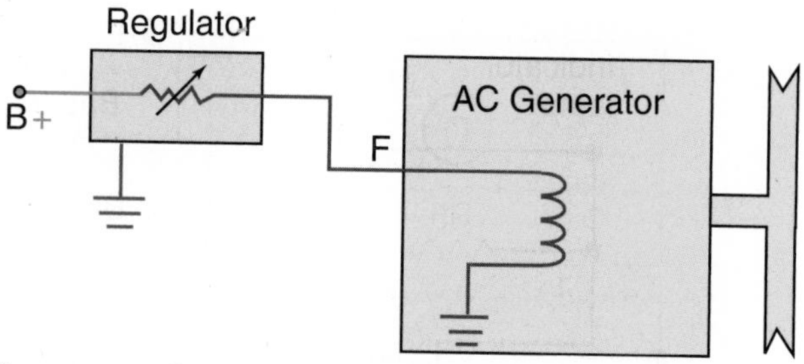

Figure 19-22 A B-circuit.

The second type of field circuit is the B-circuit. The voltage regulator controls the power side of the field circuit. The field coil is grounded inside the AC generator (**Figure 19–22**). Normally the B-circuit regulator is mounted outside the generator.

The third type of field circuit is called the isolated field. The AC generator has two field wires attached to the outside of the case. One wire is the ground; the other is the B+. The voltage regulator can be placed on either the ground or the B+ side of the circuit.

There are two basic types of regulators: electronic and electromechanical. Older vehicles were equipped with electromechanical regulators, whereas newer vehicles have an electronic regulator. Most newer vehicles do not have a separate voltage regulator; rather, they control charging system output through the PCM.

Electronic Regulators

Electronic regulators can be mounted outside the generator or be an integral part of the generator (**Figure 19–23**). Voltage output is controlled through the ground side of the field circuit (A-circuit control). Most electronic regulators have a zener diode that

Figure 19-23 An integrated voltage regulator.

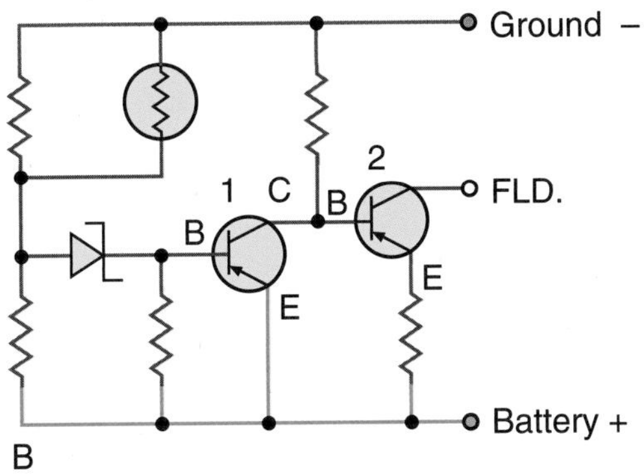

Figure 19-24 A simplified circuit of an electronic regulator with a zener diode.

Figure 19-25 Component locations of an AC generator with an internally mounted voltage regulator. *Courtesy of Robert Bosch GmbH, www.bosch-presse.de*

blocks current flow until a specific voltage is obtained, at which point it allows the current to flow. The schematic for an electronic voltage regulator with a zener diode is shown in **Figure 19–24**.

The generator's output is controlled by pulse width modulation. This varies the amount of time the field coil is energized in response to the vehicle's needs. For example, assume that a vehicle is equipped with a 100-ampere generator. If the electrical demand placed on the charging system requires 50 amps, the regulator would energize the field coil for 50% of the time. If the electrical system's demand were increased to 75 amps, the regulator would energize the field coil 75% of the cycle time.

Integrated circuit voltage regulators are used in many vehicles. These compact units are mounted either inside or on the back of the AC generator. Integrated circuit regulators are nonserviceable and must be replaced if defective.

The generator shown in **Figure 19–25** has an integrated regulator. The unit is mounted inside the generator's slip ring end frame along with the brush holder assembly. The regulator controls field current through a diode trio.

Fail-Safe Circuits To prevent simple electrical problems from causing high-voltage outputs that can damage delicate electronic components, many voltage regulators have **fail-safe circuits**.

If connections to the AC generator become corroded or accidentally disconnected, the regulator's fail-safe circuits may limit voltage output to prevent high-charging voltages. Under certain conditions, the fail-safe circuit may prevent the AC generator from charging at all. A fusible link or maxi-fuse in the fail-safe circuitry prevents high current from leaving the charging system and damaging the delicate electronic components in other vehicle systems.

Computer Regulation

Late-model vehicles do not have a separate voltage regulator. Instead voltage regulation is one of the many functions of the vehicle's PCM. The computer is used to control current to the field windings in the rotor.

This type of system switches or pulses field current on and off at a fixed frequency of about 400 cycles per second. A significant feature of this system is its ability to vary the amount of voltage according to vehicle requirements and ambient temperatures. This precise control allows the use of smaller, lighter storage batteries. It also reduces the magnetic drag of the AC generator, increasing engine output by several horsepower. Precise management of the charging rate also results in increased gas mileage.

The part of the PCM that controls charging output is typically called the power management system **(Figure 19–26)**. This system not only controls the field current, it also causes the engine's idle speed to increase when the battery is low and sends diagnostic messages to alert the driver of possible battery and generator problems. The condition of the battery is monitored with the ignition off and when it is on.

When the ignition is off, the battery's state-of-charge (SOC) is monitored by open-circuit voltage. Engine-off SOC is used as an indicator of the battery's condition. When the engine is running, the system continuously estimates SOC based on temperature,

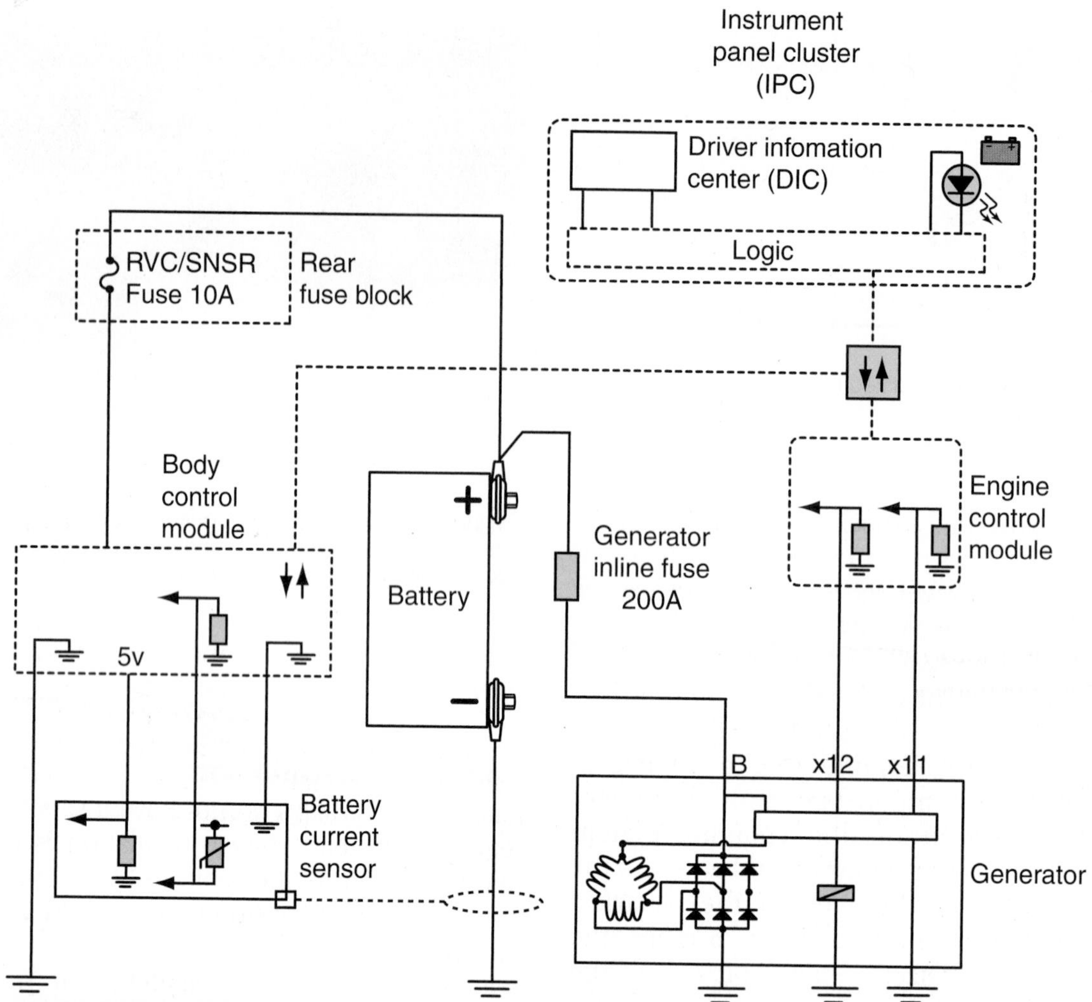

Figure 19-26 The main circuit for a PCM-controlled charging system.

battery capacity, the initial SOC, and the charging system's output.

NEW DEVELOPMENTS

In the quest to improve fuel economy, decrease emission levels, and make vehicles more reliable, engineers have applied advanced electronics to starters and generators.

Motor/Generators

Keep in mind that the main difference between a generator and a motor is that a motor has two magnetic fields that oppose each other, whereas a generator has one magnetic field and wires are moved through that field. Using electronics to control the current to and from the battery, a generator can also work as a motor. These units are called starter/generators or motor/generators. The construction of a motor/generator may be based on two sets of windings and brushes, a brushless design with a permanent magnet, or switched reluctance (**Figure 19–27**).

SHOP TALK

Although not commonly used today in automobiles, switched reluctance motors are common in other industries. Without going into great detail into their operation, the basics should be understood for the future. This design has a toothed stator and rotor and the rotor has no windings or magnets. The stator has slots containing a series of coil windings. A controller establishes a rotating magnetic field around the stator as it activates one coil set in the stator at one time. The timing of this activation is based on the position of the rotor. When one coil is energized, a magnetic field is formed around it. The rotor tooth that is closest to the magnetic field moves toward that field. When the tooth is close, current is switched to another stator winding and the tooth moves to it. As the current is sent to the consecutively placed windings, the rotor rotates.

Figure 19-27 A switched reluctance motor/generator. Note the design of the rotor. *Courtesy of Dana Corporation*

A motor/generator can be mounted externally to the engine and connected to the crankshaft by a drive belt. Belt-driven motor/generators have a belt tensioner that is mechanically or electrically controlled to allow it to drive or be driven by the engine's crankshaft. One system uses an electromagnetic clutch fitted to the crankshaft pulley. The clutch is engaged to allow the motor/generator to work as a generator or as a starter motor. When the engine is stopped, the crank pulley clutch disengages. Motor/generators can also be directly mounted to the crankshaft between the engine and transmission or integrated into the flywheel **(Figure 19–28)**. The most common hybrid vehicles have two motor/generators: one that serves as the engine starting motor and a generator and the other as a traction motor and generator.

Figure 19-28 An integrated motor/generator assembly built into the flywheel. *Courtesy of BMW of North America, LLC*

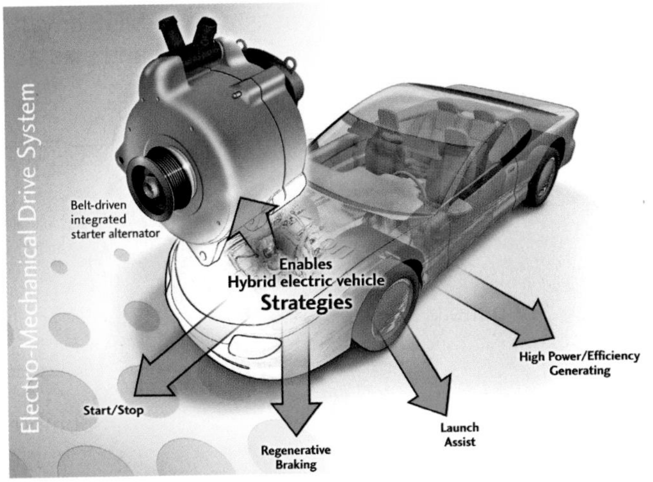

Figure 19-29 Motor/generators are capable of high charging outputs and can crank the engine at high speeds. They also allow for stop-start, regenerative braking, and electrical assist. *Courtesy of Visteon Corporation™*

Motor/generators are capable of high charging outputs and can crank the engine at high speeds. They also allow for other features that make the vehicle more efficient **(Figure 19–29)**:

- Stop-start
- Regenerative braking
- Electrical assist

CAUTION!

Always adhere to the precautions given by the manufacturer when working on or near high-voltage motor/generators and their related systems. Being careless can cause serious injuries, including death.

Regenerative Braking

When the brakes are applied in a conventional vehicle, friction at the wheel brakes converts the vehicle's kinetic energy into heat. With regenerative braking, that energy is used to recharge the batteries. The vehicle's kinetic energy is changed to electrical energy until the vehicle is stopped **(Figure 19–30)**. At that point, there is no kinetic energy. Through electronic controls, the motor/generator functions as a generator. The rotor is driven by the engine, which is driven by the vehicle's wheels, as the vehicle is slowing down. The amount of energy captured by a regenerative braking depends on many things, such as the SOC of the battery and the speed at which the generator's rotor is spinning.

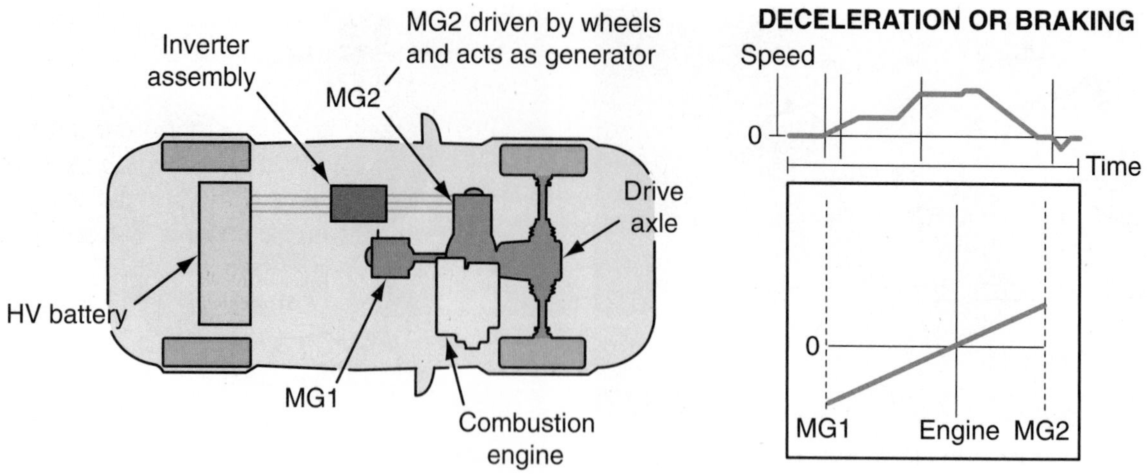

Figure 19-30 The power flow for a hybrid with two motor/generators during brake regeneration.

Regenerative braking not only recharges the battery pack, but the load of turning the generator's rotor helps slow and stop the vehicle. Regenerative braking is not used to completely stop the vehicle. A combination of conventional hydraulic brakes and regenerative braking is used. Hydraulic, friction-based brakes must be used when sudden and hard braking is needed. Regenerative braking systems are not a physical part of the brake system.

The system's controller regulates how fast the kinetic energy is converted, and thus regulates how fast braking will occur. The controller receives a signal from the brake pedal and other sensors. Based on programs, the controller determines the amount of regenerative and hydraulic braking needed. Regenerative braking works better when the generator's rotor spins fast. It works poorly at low speed; therefore, the controller must base the amount of regenerative braking on the vehicle's speed and the pressure applied to the brake pedal. The controller must also apply the regenerative braking smoothly so the vehicle does not jerk when it is engaged.

Regenerative braking is found on all hybrid vehicles. It can capture approximately 30% of the energy normally lost during braking in conventional vehicles. It is claimed that the electric energy resulting from regenerative braking supplies 20% of the energy used by a Toyota Prius. Regenerative braking also decreases brake wear and reduces maintenance costs.

42-Volt Generators

Nonhybrid vehicles equipped with a 42-volt electrical system will have an air- **(Figure 19–31)** or liquid-cooled generator capable of producing 42 volts and 5 to 10 kilowatts. Currently, a conventional 12-volt generator puts out about 1.5 kilowatts. Depending on the system, the vehicle may be fitted with a DC-to-DC converter that lowers some of the generator's high-voltage output to charge a 12-volt battery or power the vehicle's 12-volt loads **(Figure 19–32)**. Some new generators use switched reluctance techniques. These generators are very efficient at low speeds.

Figure 19-31 A belt-drive 42-volt motor/generator. *Courtesy of Toyota Motor Sales, U.S.A., Inc.*

PRELIMINARY CHECKS

The key to solving charging system problems is getting to the root of the trouble the first time. Once a customer drives away with the assurance that the problem is solved, another case of a dead battery is very costly—both in terms of a free service call and a damaged reputation. Add to this the many possible hours of labor trying to figure out why the initial repair failed, and the importance of a correct initial diagnosis becomes all too clear.

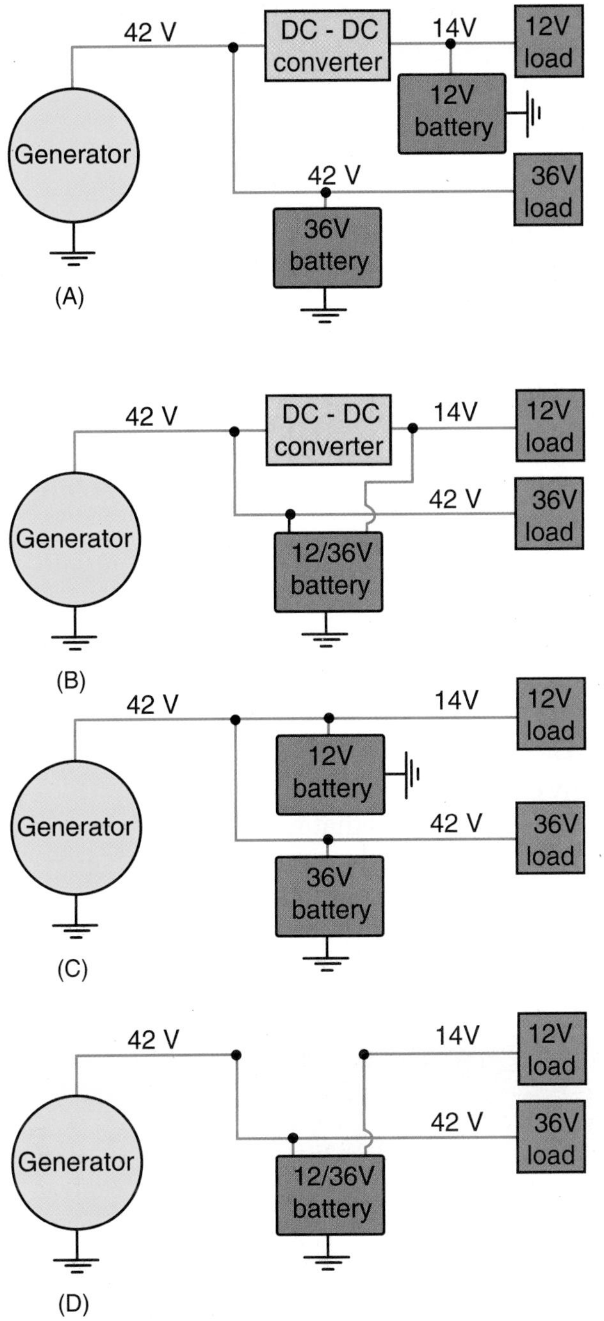

Figure 19-32 The different system layouts for 42-volt electrical systems: (A) a system with two batteries and a converter, (B) a system with one battery and a converter, (C) a two-battery system, and (D) a one-battery system.

Safety Precautions

- Disconnect the battery ground cable before removing any leads from the system. Do not reconnect the battery ground cable until all wiring connections have been made.
- Avoid contact with the AC generator output terminal. This terminal is hot (has voltage present) at all times when the battery cables are connected.
- The AC generator is not made to withstand a lot of force. Only the front housing is relatively strong. When adjusting belt tension, apply pressure only to the front housing to avoid damaging the stator and rectifier.
- When installing a battery, be careful to observe the correct polarity. Reversing the cables destroys the diodes. Proper polarity must also be observed when connecting a booster battery, positive to positive and negative to ground.
- Keep the tester's carbon pile off at all times, except during actual test procedures.
- Make sure all hair, clothing, and jewelry are kept away from moving parts.

Indicators

Part of all charging circuits is some sort of charging system warning device or indicator. The sole purpose of these is to alert the driver when there is a problem. They are also one of the first things to check when a charging system problem is suspected. Vehicles have an ammeter, a voltmeter, an indicator light, and/or a message center in the instrument panel that allows the driver to monitor the charging system.

Indicator Light This is the simplest and most common method of monitoring charging system performance. When the system fails to supply sufficient current, the light turns on. However, when the ignition switch is first activated, the light also comes on. This is due to the fact that the AC generator is not providing power to the battery and other electrical circuits. Thus, the only current path is through the ignition switch, indicator light, voltage regulator, part of the AC generator, and ground, and then back through the battery (**Figure 19–33**). Once enough power is provided, the lamp should turn off.

If, when the engine is running, the electrical load is greater than what can be handled by the charging system, the indicator light should turn on. This occasionally happens when the engine is idling. If there are no problems, the light should go out as the engine speed is increased. If it does not, either the AC generator or regulator is not working properly. If the lamp does not light when the ignition is placed in the on position, check the bulb and its circuitry.

SHOP TALK

On some late-model cars, the indicator light may be combined with the oil pressure warning light and is usually labeled "engine." If this light turns on while the engine is running, either the AC generator is not charging or the oil pressure is low—or both.

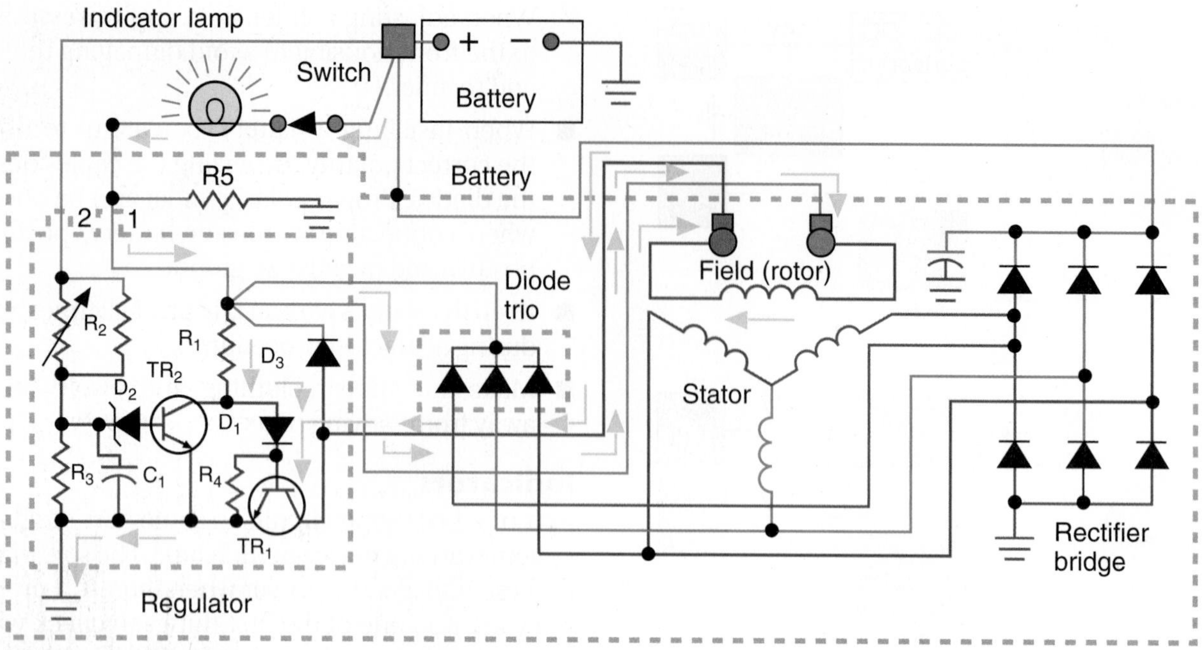

Figure 19-33 An electronic regulator with an indicator light "ON" due to no AC generator output.

Meters Some vehicles have an ammeter or voltmeter in their instrument cluster. The voltmeter displays the voltage at the battery. If the charging system is working fine, the voltmeter will read more than 12 volts. The ammeter monitors current flow in and out of the battery. When the generator is delivering current to the battery, the ammeter shows a positive (+) indication. When not enough current (or none at all) is being supplied, the result is a negative (−) indication. The latter situation indicates a problem with the charging system.

Warning Messages Warning messages are typically found on some vehicles with a PCM-controlled charging system. These messages are normally accompanied by the illumination of the charging system warning lamp. Like other warning lamp circuits, the lamp, along with a message, should be displayed when the ignition is first turned on. It should also turn off or clear once the engine is running, normally about 3 seconds. If the lamp and/or message remains on, the charging system needs to be diagnosed. If the lamp does not initially illuminate, that circuit needs to be diagnosed.

Normally the message center will display one of two messages when there is a faulty charging system. One message will say the battery is not being charged. An example for this would be: "Battery Not Charging Service Charging System." If the system senses a too low or too high charging voltage, the typical message will read: "Service Battery Charging System." Both of these messages will set a diagnostic trouble code that can be retrieved to help in diagnosis.

Inspection

Many charging system concerns are caused by easily repairable problems that reveal themselves during a visual inspection of the system. Remember to always look for the simple solution before performing more involved diagnostic procedures. Use the following inspection procedure when a problem is suspected.

PROCEDURE

Follow these steps while inspecting a charging system:

1. Inspect the battery. It might be necessary to charge the battery to restore it to a fully charged state. If the battery cannot be charged, it must be replaced. Also, make sure that the posts and cable clamps are clean and tight, because a bad connection can cause reduced current flow.
2. Inspect all system wiring and connections **(Figure 19–34)**. Many systems contain fusible links or maxi-fuses to protect against overloads; check them.
3. Inspect the generator and regulator mountings for loose or missing bolts. Replace or tighten as needed. Remember, most generators and regulators complete their ground through their mounting. If the mountings are not clean and tight, a high resistant ground will result.
4. Inspect the AC generator drive belt. Loose drive belts are a major source of charging problems. The correct procedure for inspecting, removing, replacing, and adjusting a drive belt is shown in Photo Sequence 4.
5. Before adjusting belt tension, check for proper pulley alignment.

Figure 19-34 Start your diagnosis with an inspection of the generator and its drive belt and wires.

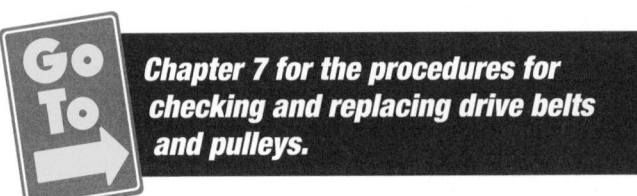

Go To *Chapter 7 for the procedures for checking and replacing drive belts and pulleys.*

PCM-Controlled Systems

With PCM-controlled systems, diagnosis should continue with a check for diagnostic trouble codes (DTCs). Normally there are two possible types of charging system DTCs: generator related and battery current sensor related. Always refer to the manufacturer's DTC charts and follow its prescribed diagnostic steps for those codes. On most systems, the voltage output can be monitored with a scan tool. If the voltage is not within the range specified by the manufacturer, check all connections before continuing with other tests.

In a typical PCM-controlled charging system, the PCM controls the generator by controlling the duty cycle of the field current. The PCM will also change charging rates according to existing conditions **(Figure 19–35)**. The PCM monitors a battery current sensor, the battery voltage, and the estimated battery temperature to determine the battery's SOC. The PCM can also increase the idle speed of the engine to raise charging rates in order to meet the current needs. The PCM sets a target charging output and adjusts the duty cycle of the field current to meet that target **(Table 19–1)** without overstressing the battery.

The battery current sensor is connected to the negative battery cable at the battery. The current sensor is a three-wire, Hall-effect sensor that directly feeds the PCM with information. The actual voltage regulator may be connected internally to the generator but it is controlled externally by the PCM. If the voltage regulator detects a charging system problem, it grounds the sensor's circuit, which signals the PCM that a problem exists. The PCM then controls the duty cycle based on information in its memory.

Several conditions will cause the PCM to increase or decrease the target charging output. These include the current operation of some heavy current draw accessories, low current demand at constant vehicle speeds, low speed after initial engine startup, and low or high temperatures. When monitoring the performance of these systems, it is important to note the operational conditions because these do affect the output of the generator.

At times it may be necessary to reprogram the computer with newer software. When doing this, always follow the instructions given by the manufacturer.

Go To *Chapter 22 for a discussion and typical procedures for reprogramming a vehicle's computer.*

Noise Diagnosis

Often a charging system concern is excessive noise. The cause of this noise can be something mechanical or electrical. Most often, electrical noise is normal and will vary with the load put on the charging system. However, a bad diode can cause an abnormal whirring noise and should be considered.

Most noises are caused by belts, bad bearings, or something rubbing against a pulley or belt. As with all problems, verify the noise and attempt to identify the area it is coming from. If a drive belt does not have the proper tension, it might produce a loud squealing sound. Check the condition and tension of the belt.

Check the mounting of the generator, its wiring harness, and all heater hoses, A/C lines, and other items that may be misrouted and making contact with belts or pulleys. If the cause of the noise is not identified, remove the drive belt. Spin the generator's pulley, idler pulley, and tensioner. If any of these do not spin freely, replace them. If no problem is found, check the tightness of the generator's mounting bolts and make sure the generator is

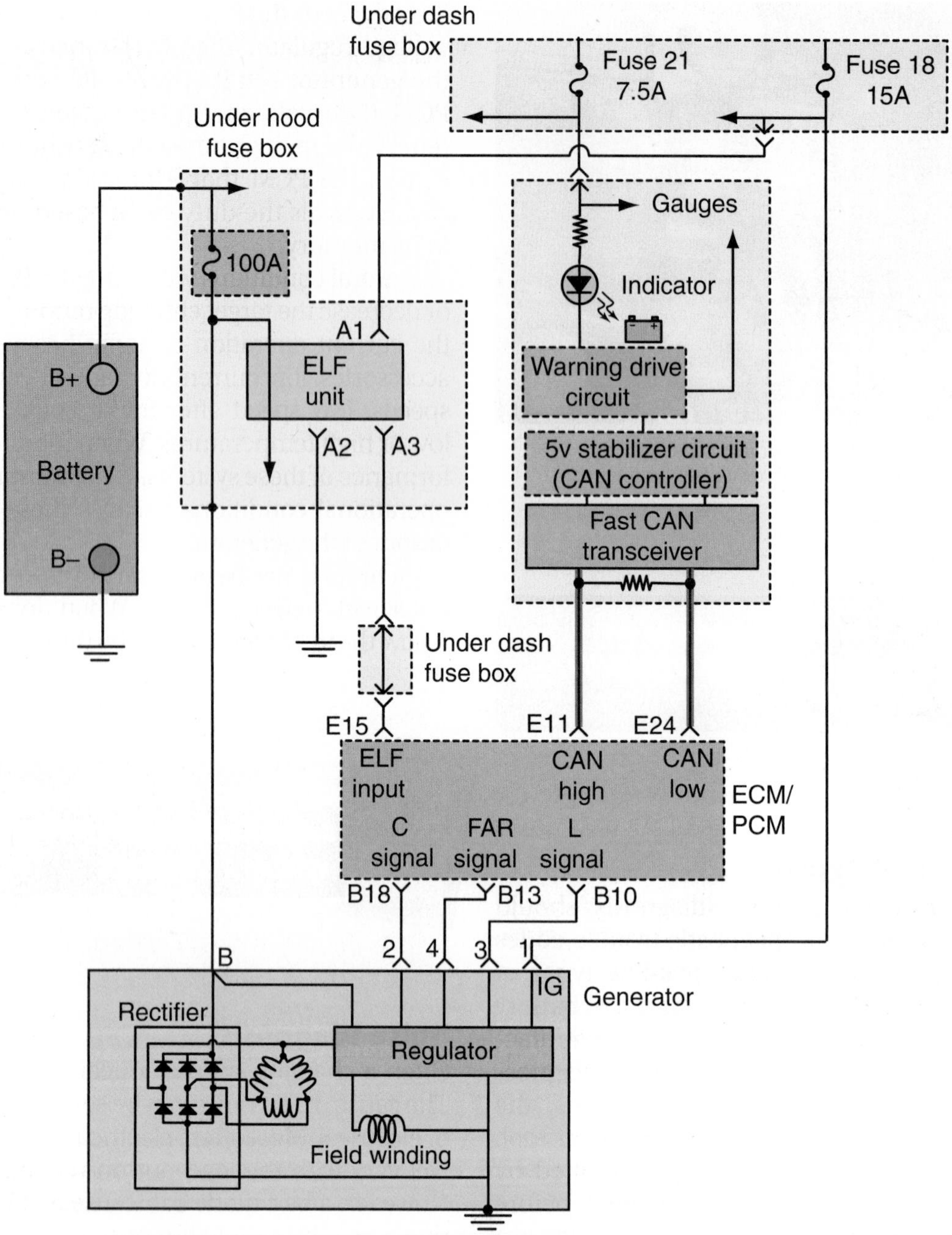

Figure 19-35 A schematic for a PCM-controlled charging system.

mounted correctly. If no problem is found, replace the generator.

GENERAL TESTING PROCEDURES

Diagnosing a charging system is a straightforward task. Tests can be conducted with a VAT, current probe, DMM, or a lab scope. Charging system tests for all cars are basically the same; however, it is very important to refer to the manufacturer's specifications. Even the most accurate test results are no good if they are not matched against the correct specs.

Regulator Tests

If the concern is excessive charge resulting in battery or component failure, the most likely cause is the voltage regulator. Seldom does a bad regulator cause low output, but it can cause no output. To determine if the regulator was the cause on older generators, the regulator was bypassed or full-fielded. This is not done with today's generators.

This requires knowledge of the type of field circuit the generator has. Then the field windings are given full current and the generator's output observed. Some generators had a window with a grounding tab

TABLE 19-1 THE RELATIONSHIP BETWEEN THE CONTROLLED OR COMMANDED FIELD CIRCUIT CURRENT DUTY CYCLE AND THE VOLTAGE OUTPUT OF THE GENERATOR

Commanded Duty Cycle	Output Voltage
10%	11 V
20%	11.6 V
30%	12.1 V
40%	12.7 V
50%	13.3 V
60%	13.8 V
70%	14.4 V
80%	15 V
90%	15.5 V

inside. By inserting a small screwdriver into the hole and grounding the tab, the regulator was bypassed. Also follow the manufacturer's procedures for full-fielding a generator.

Voltage Output Test

To check the charging system's voltage output, begin by measuring the battery's open circuit voltage. If the voltage is low, proceed to check the battery's condition. Then, start the engine and run it at the suggested rpm for this test (usually 1,500 rpm). With no electrical load, the voltage reading should be about 2 volts higher than the open circuit voltage.

A reading of less than 13 volts immediately after starting the engine indicates a charging problem. A reading of 16 or more volts indicates overcharging. A faulty voltage regulator or control voltage circuit is the most likely cause of overcharging.

If the unloaded charging system voltage is within specifications, test the output under a load. Increase the engine speed to about 2,000 rpm and turn on the headlights and other high current accessories. The voltage output while under heavy load should be about 0.5 volt above the battery's open circuit voltage.

Current Output Test

The output test looks at the current output of the system. A VAT can quickly check the amperage output. With the tester connected to the system **(Figure 19–36)**, the engine is run at a moderate speed (2,500 rpm) and the carbon pile adjusted to obtain maximum current output. This reading is compared

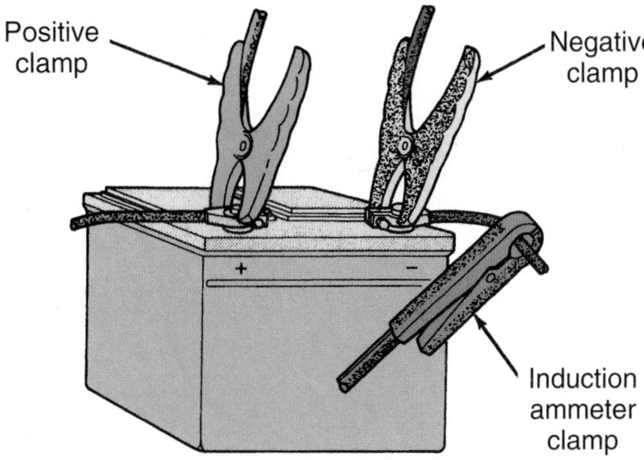

Figure 19-36 The lead connections for conducting a current output test with a VAT. Notice that the ammeter clamp is around the battery cable. *Courtesy of Chrysler LLC*

against the rated output. Normally, readings that are more than 10 amperes out of specifications indicate a problem and the generator should be repaired or replaced.

SHOP TALK

It is important to know the amperage rating of the generator before coming to any conclusions from the output test. The generator may have its rating stamped into its case, there may be a label on the case, and some have color markings that denote their rating. However, the appropriate service manual is the best way to identify the rating.

AC Leakage Test

It is important to remember that a battery cannot be charged with AC voltage. Therefore, it is important to make sure that no AC is leaking into the charging system from the AC generator. Leakage typically occurs at a faulty diode. To check for diode leakage, connect a DMM in series to the generator's output terminal while the engine is not running **(Figure 19–37)**. Set the meter to read milliamps. There should be no more than 0.5 milliamp. If there is more, check the diodes and/or replace the generator.

Diode Checks

The output of a generator is highly dependent on the condition of the diodes. Not only do the diodes rectify AC voltage to DC, they also prevent AC voltage from being present in the output. Bad diodes are indicated by the presence of more than 0.5 AC volt in the output wire. To check this, set the DMM to measure AC volts. Then connect the black meter lead to a

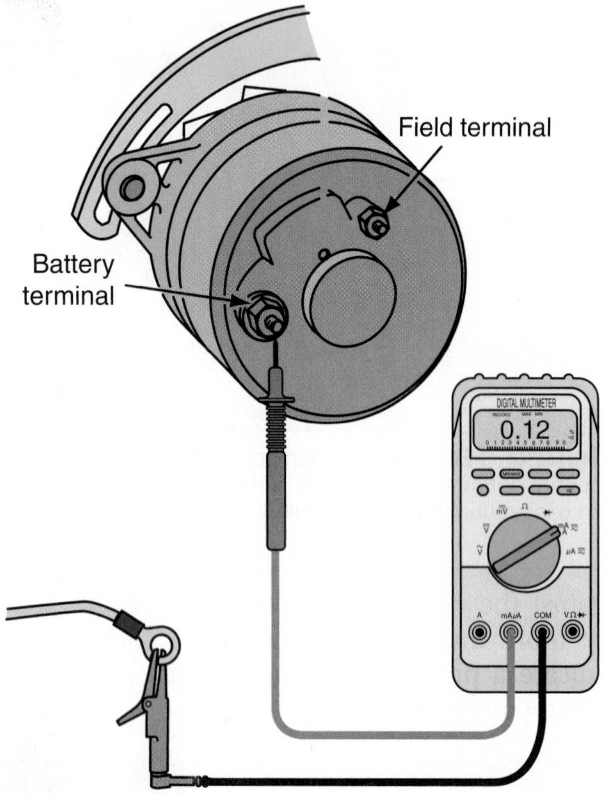

Figure 19-37 Setup for checking AC leakage.

good ground and the red lead to the generator's battery terminal.

Another check of the diodes while they are still in the generator is done with the engine off and with a low-amperage current probe. Measure the current on the generator's output wire. Any measurement greater than 0.5 milliamp indicates that one or more diodes are leaking and the generator or diodes need to be replaced.

Oscilloscope Checks

AC generator output can also be checked using an oscilloscope. **Figure 19–38** illustrates common AC generator voltage patterns for good and faulty generators. The correct pattern looks like the rounded top of a picket fence. A regular dip in the pattern indicates that one or more of the coil windings is grounded or open, or that a diode in the rectifier circuit of a diode trio circuit has failed. One or more bad or leaking diodes will decrease the output of a generator.

Generator Output and Generator Voltage Drop Tests

These tests measure voltage drop within the system wiring. They help pinpoint corroded connections or loose or damaged wiring.

Circuit resistance is checked by connecting a voltmeter to the positive battery terminal and the output, or battery terminal of the AC generator. The positive lead of the meter should be connected to the AC generator output terminal and the negative lead to the positive battery terminal **(Figure 19–39)**. To check the voltage drops across the ground circuit, connect the positive lead to the generator housing and the negative meter lead to the battery negative terminal. When measuring the voltage drop in these circuits, a sufficient amount of current must be flowing through the circuit. Therefore, turn on the headlights and other accessories to ensure that the AC generator is putting out at least 20 amps. If a voltage drop of more than 0.5 volt is measured in either circuit, there is a high-resistance problem in that circuit.

AC GENERATOR SERVICE

When the cause of charging system failure is the AC generator, it should be removed and replaced or rebuilt. Whether it is rebuilt or replaced depends on the type of generator it is, the time and cost required to rebuild it, your shop's policy, and your customer's desires. Most late-model AC generators are not rebuilt. They are traded in as a core toward the purchase of a new or remanufactured unit. Just in case you do need to rebuild one, Photo Sequence 17 is given as an example of what it takes to do so. This procedure is for a specific type and model of generator. Make sure you follow the procedures given by the manufacturer for the generator you are working on.

To test the components of an AC generator it must be removed and disassembled.

A faulty AC generator can be the result of many different types of internal problems. Diodes **(Figure 19–40)**, stator windings **(Figure 19–41)**, and field circuits **(Figure 19–42)** may be open, shorted **(Figure 19–43** and **Figure 19–44)**, or improperly grounded. The brushes or slip rings can become worn. The rotor shaft can become bent and the pulley can work loose or bend out of proper alignment.

Follow service manual procedures when removing and installing an AC generator. Remember, improper connections to an AC generator can destroy it.

Follow service manual procedures for disassembling, inspecting, testing, and rebuilding AC generators.

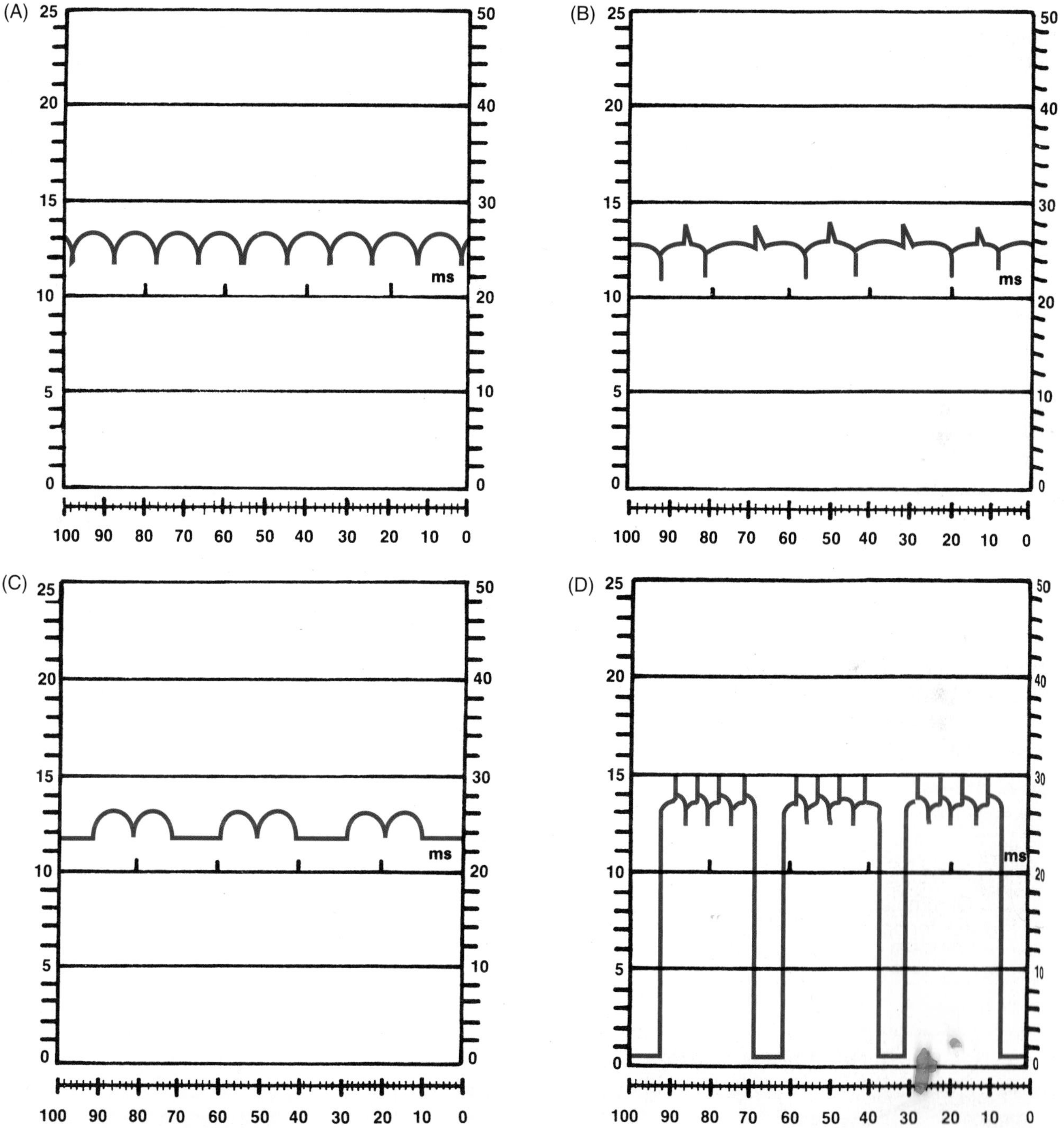

Figure 19-38 AC generator oscilloscope patterns: (A) good AC generator under full load, (B) good AC generator under no load, (C) shorted diode and/or stator winding under full load, and (D) open diode in diode trio.

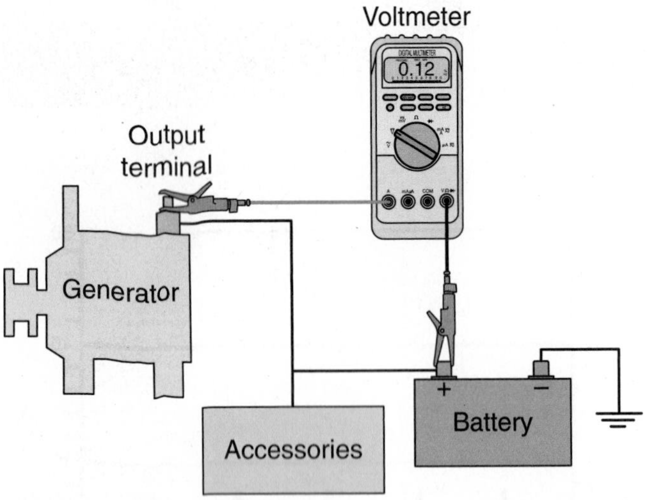

Figure 19-39 Check the resistance of the charging circuit by measuring the voltage drop from the positive side of the battery to the output terminal of the generator. Remember, the system must be working and current flowing to measure voltage drop.

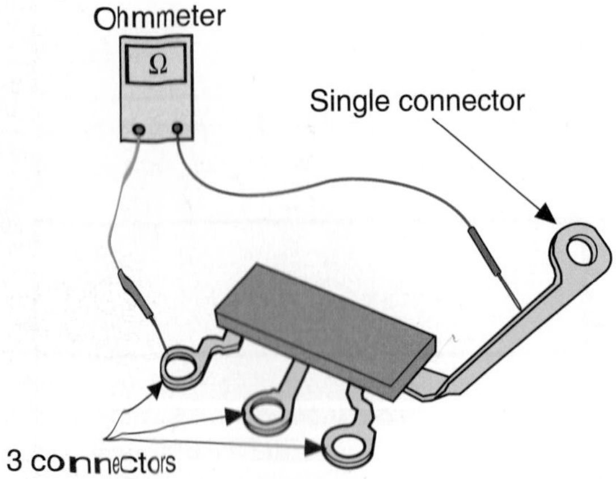

Figure 19-40 Using an ohmmeter to test a rectifier bridge or assembly.

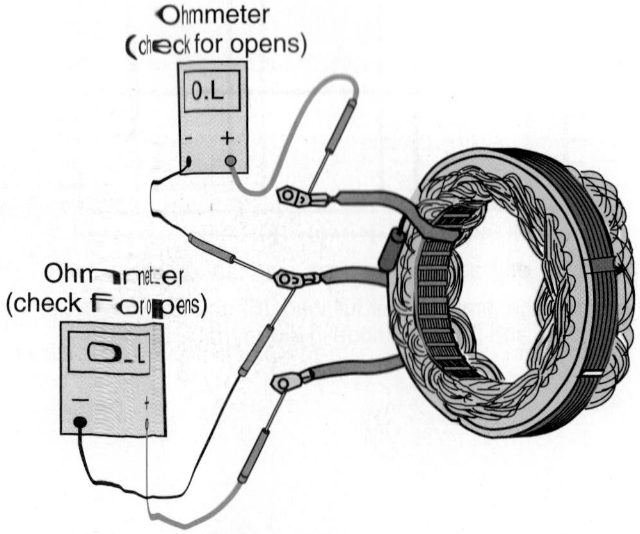

Figure 19-41 Testing a stator for opens.

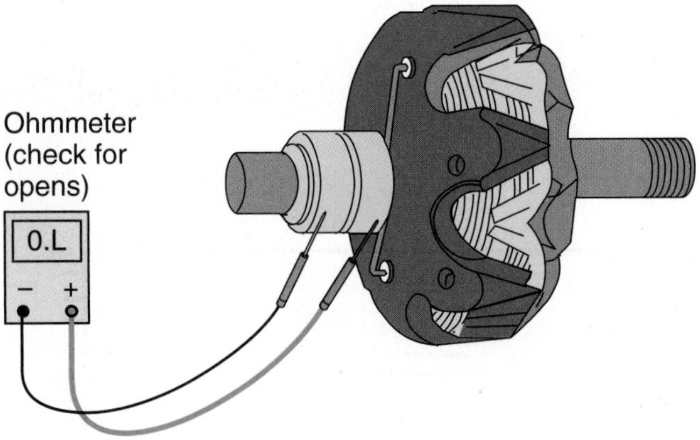

Figure 19-42 Testing a rotor for opens.

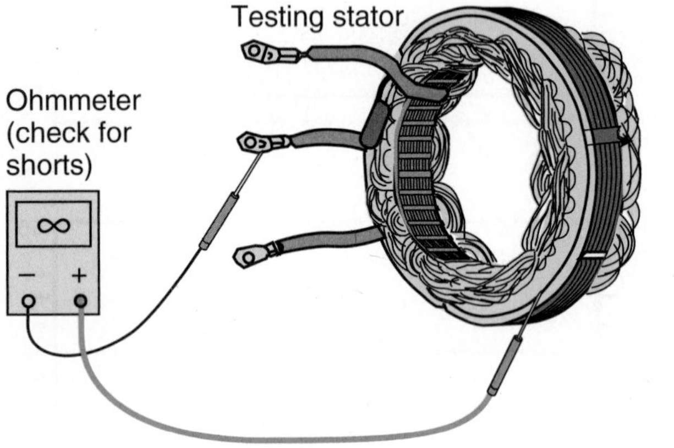

Figure 19-43 Testing a stator for a short to ground.

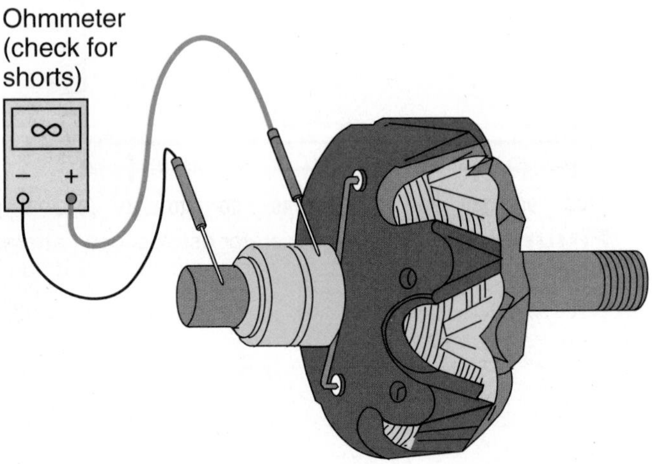

Figure 19-44 Testing a rotor for a short to ground.

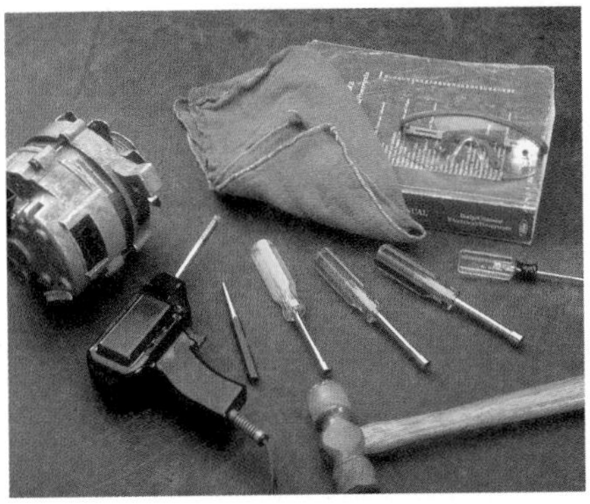

P17-1 *Always have a clean and organized work area. The tools required to disassemble a Ford IAR AC generator are rags, T20 torx wrench, plastic hammer, arbor press, 100-watt soldering iron, soft-jawed vise, safety glasses, and an assortment of sockets or nut drivers.*

P17-2 *Using the torx wrench, remove the four attaching screws that hold the regulator to the AC generator housing.*

P17-3 *Remove the regulator and brush assembly as a unit.*

P17-4 *Using the torx wrench, remove the two screws that attach the regulator to the brush holder. Then separate the regulator from the brush holder.*

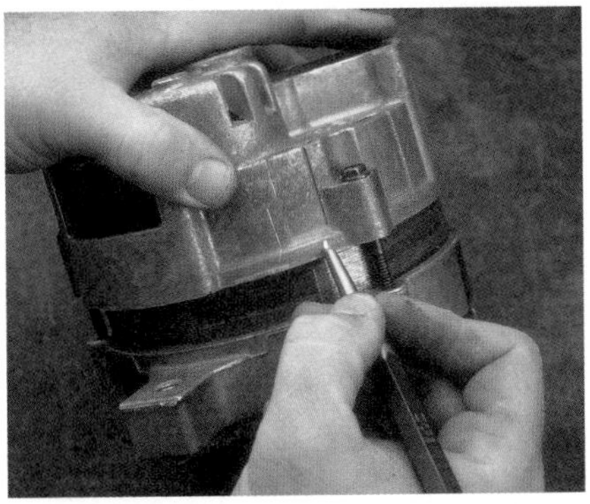

P17-5 *Scribe or mark the two end housings and the stator core for reference during reassembly.*

P17-6 *Remove the three through-bolts that secure the two housings.*

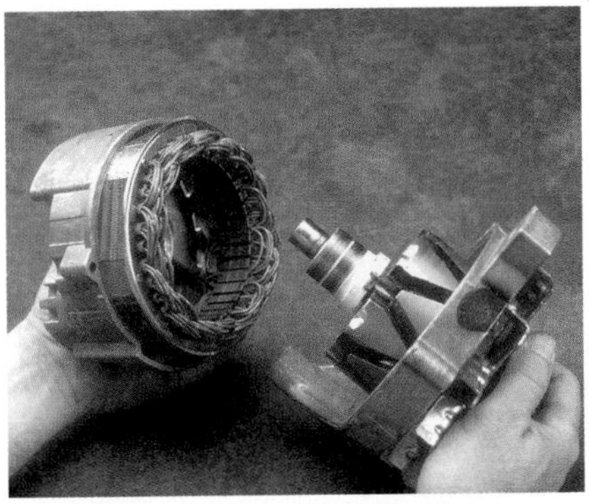

P17-7 *Separate the front housing from the rear housing. The rotor will come out with the front housing, and the stator will stay in the rear housing. It may be necessary to tap the front housing with the plastic hammer to get the two halves to separate.*

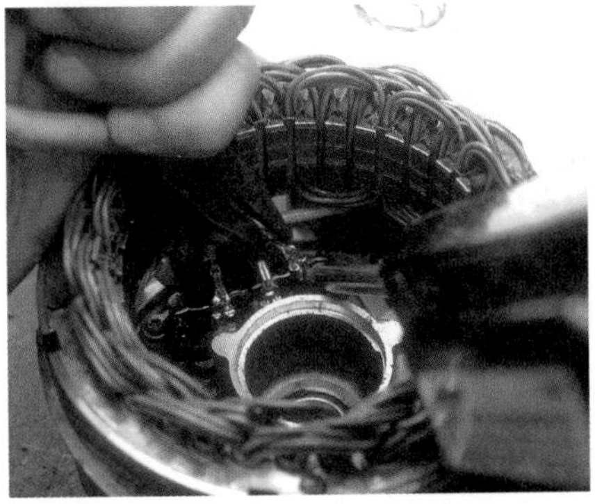

P17-8 *Separate the three stator lead terminals from the rectifier bridge.*

P17-9 *Remove the stator coil from the housing.*

P17-10 *Using the torx wrench, remove the four attaching bolts that hold the rectifier bridge.*

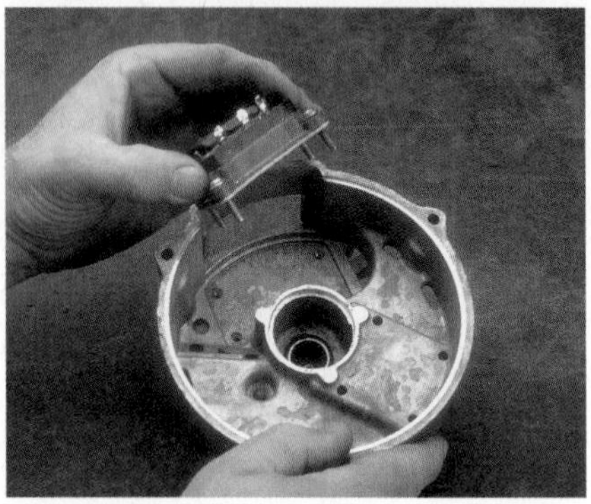

P17-11 *Remove the rectifier bridge from the housing.*

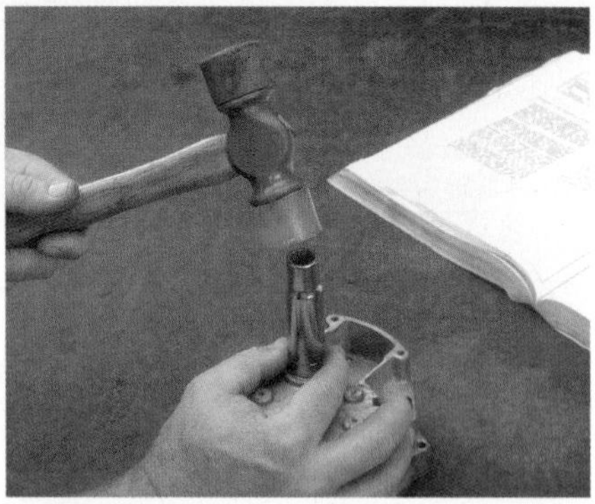

P17-12 *Use a socket or a bearing driver to tap out the bearing from the housing.*

P17-13 *Clamp the rotor in the vise.*

P17-14 *Remove the pulley-attaching nut, flat washer, drive pulley, fan, and fan spacer from the rotor shaft.*

P17-15 *Separate the front housing from the rotor. If the stop ring is damaged, remove it from the rotor. If not, leave it on the shaft.*

P17-16 *Remove the three screws that hold the bearing retainer to the front housing.*

P17-17 *Remove the bearing retainer.*

P17-18 *Remove the front bearing from the housing. Test and inspect all parts. Replace any defective ones. Reassembly is the reverse of this procedure.*

CASE STUDY

A customer brought his pickup truck into the dealership with a complaint that the truck must be jumped to get it started and once it is started, it runs for about 10 minutes then dies.

The technician verified the complaint, then did a visual inspection of the charging system. Based on the complaint, he knew that the battery was not being kept in charge by the AC generator or that the battery was unable to hold a charge. He found the drive belt extremely frayed and glazed. Knowing this could cause the problem, he replaced the belt. However, before releasing the truck back to the customer he tested the battery to make sure it was able to hold a charge if the AC generator was capable of charging it. The battery checked out fine. He then checked the AC generator output and found it to be within specifications. This was a capable technician; he knew that he should not assume that the obvious problem was the only problem. As a result, the customer was happy and should be for quite some time.

KEY TERMS

Alternator
DC generator
Delta
Diode
Diode trio
Fail-safe circuit
Field coils
Fingers
Full-wave rectification
Half-wave rectification
Integrated circuit voltage regulator
Pole piece
Rotor
Sensing voltage
Slip rings
Stator
Three-phase voltage
Voltage regulator
Wye

SUMMARY

- Inducing a voltage requires a magnetic field producing lines of force, conductors that can be moved, and movement between the conductors and the magnetic field so the lines of force are cut.
- Modern vehicles use an AC generator to produce electrical current in the charging system. Diodes in the generator change or rectify the alternating current to direct current.
- A voltage regulator keeps charging system voltage above battery voltage. Keeping the AC generator charging voltage above the 12.6 volts of the battery ensures that current flows into, not out of, the battery.
- Modern voltage regulators are completely solid-state devices that can be an integral part of the AC generator or mounted to the back of the generator housing. In some vehicles, voltage regulation is the job of the computer control module.
- Voltage regulators work by controlling current flow to the AC generator field circuit. This varies the strength of the magnetic field, which in turn varies current output.
- The driver can monitor charging system operation with indicator lights, a voltmeter, or an ammeter.
- Using electronics to control the current to and from battery, a generator can work as a motor or vice versa.
- Motor/generators are capable of high charging outputs. They can crank the engine at high speeds and allow for the stop-start feature and regenerative braking and may provide electrical assist to the engine.
- Regenerative braking captures some of the vehicle's kinetic energy to recharge the battery.
- Problems in the charging system can be as simple as worn or loose belts, faulty connections, or battery problems.
- Circuit resistance, current output, voltage output, field-current draw, and voltage regulator tests are all used to troubleshoot AC charging systems.

REVIEW QUESTIONS

1. *True or False?* A faulty voltage regulator can cause a no-charge or overcharge condition.
2. To protect electronic circuits, some voltage regulators have a ________ circuit built into them.
3. What is the purpose of a rectifier assembly?
4. How does a voltage regulator regulate the voltage output of an AC generator?
5. What are the four factors that determine how much voltage is induced by a magnet?
6. What would happen to the output of an AC generator if one of the stator windings has an open?
7. Describe the basic difference between a DC motor and a DC generator.

8. A rotating magnetic field inside a set of conducting wires is a simple description of a ______.
   a. DC generator
   b. AC generator
   c. voltage regulator
   d. none of the above
9. *True or False?* If the generator's voltage is within specifications, the generator is operating properly.
10. What part of the AC generator is the rotating magnetic field?
    a. stator
    b. rotor
    c. brushes
    d. poles
11. Which type of stator winding produces higher AC generator output at low speeds?
    a. wye
    b. delta
    c. trio
    d. series
12. Slip rings and brushes ______.
    a. mount on the rotor shaft
    b. conduct current to the rotor field coils
    c. are insulated from each other and the rotor shaft
    d. all of the above
13. The alternating current produced by the AC generator is rectified into DC, or direct current, through the use of ______.
    a. transistors
    b. electromagnetic relays
    c. diodes
    d. capacitors
14. A voltage drop over ______ indicates high resistance in either the positive or ground circuit of the charging system.
    a. 0.1
    b. 0.2
    c. 0.5
    d. 1.0
15. Voltage regulation by a PCM is done by ______.
    a. using a variable resistor to vary current flow to the rotor field windings
    b. pulsing current flow to the rotor field windings on and off to create a correct average field current supply
    c. changing the duty cycle of a current sensor to control the field's path to ground
    d. none of the above

ASE-STYLE REVIEW QUESTIONS

1. Technician A says that the waveform produced by an AC generator before rectification is called a sine wave. Technician B says that the waveform produced by the AC generator after the output moves through the diodes is a straight line because it is a constant DC voltage. Who is correct?
   a. Technician A
   b. Technician B
   c. Both A and B
   d. Neither A nor B
2. While discussing what is indicated by a vehicle's charge indicator: Technician A says that any positive reading on an ammeter shows that the battery is in good condition. Technician B says that if the voltmeter reading does not increase immediately after starting the engine, the battery is bad. Who is correct?
   a. Technician A
   b. Technician B
   c. Both A and B
   d. Neither A nor B
3. When checking the charging system on a late-model vehicle: Technician A connects a scan tool to monitor the system's voltage output. Technician B connects a scan tool to retrieve any DTCs that may be in the system. Who is correct?
   a. Technician A
   b. Technician B
   c. Both A and B
   d. Neither A nor B
4. While diagnosing the cause of a noise coming from a generator: Technician A says that the noise can be caused by something mechanical or electrical. Technician B says that the whirring noise is probably caused by a diode and it should be considered normal. Who is correct?
   a. Technician A
   b. Technician B
   c. Both A and B
   d. Neither A nor B
5. Technician A uses a current output test to check AC generator output. Technician B uses a voltage output test to check output. Who is correct?
   a. Technician A
   b. Technician B
   c. Both A and B
   d. Neither A nor B
6. Technician A says that many newer charging systems do not have a separate voltage regulator. Technician B says that most late-model charging systems are wired as A-type circuits and that the voltage is regulated through the ground of the field. Who is correct?
   a. Technician A
   b. Technician B
   c. Both A and B
   d. Neither A nor B

7. While diagnosing the braking system on a hybrid vehicle: Technician A says that the difficulty of the vehicle to come to a quick stop can be caused by the regenerative braking system. Technician B says that the difficulty of the vehicle to come to a quick stop can be caused by the hydraulic braking system. Who is correct?
   a. Technician A
   b. Technician B
   c. Both A and B
   d. Neither A nor B
8. Technician A says that if there was no output from the charging system until the AC generator was full-fielded, the voltage regulator is faulty. Technician B says that this indicates one or more leaking diodes in the generator. Who is correct?
   a. Technician A
   b. Technician B
   c. Both A and B
   d. Neither A nor B
9. When checking AC generator output on a late-model vehicle: Technician A tests the battery's open circuit voltage before testing the generator's output because the battery's SOC will affect the maximum output of the generator. Technician B checks the ambient temperature before performing any tests because the generator's output will be affected by temperature. Who is correct?
   a. Technician A
   b. Technician B
   c. Both A and B
   d. Neither A nor B
10. While testing a charging system: Technician A says that an overcharge condition is evident any time charging voltage is greater than 13 volts. Technician B says that the system should be able to supply its maximum current output when the engine is at idle and there is no load on the system. Who is correct?
    a. Technician A
    b. Technician B
    c. Both A and B
    d. Neither A nor B

CHAPTER

20

LIGHTING SYSTEMS

OBJECTIVES

■ Explain the operating principles of the various lighting systems. ■ Describe the different types of headlights and how they are controlled. ■ Understand the functions of turn, stop, and hazard warning lights. ■ Know how back-up lights operate. ■ Replace headlights and other burned-out bulbs. ■ Explain how to aim headlights. ■ Explain the purpose of auxiliary automotive lighting. ■ Describe the operation and construction of the various automotive lamps. ■ Diagnose lighting problems.

The lighting system provides power to both exterior and interior lights. It consists of the headlights, parking lights, marker lights, taillights, courtesy lights, dome/map lights, instrument illumination or dash lights, coach lights (if so equipped), headlight switch, and various other control switches **(Figure 20–1)**. Other lights, such as vanity mirror lights, the underhood light, the glove box light, and the trunk compartment light, are used on some vehicles and are also part of the lighting system.

Other lights that are not usually in the main lighting system are turn signal, hazard warning, back-up, and stoplights. These lights, as well as the pop-up headlights or retractable headlight covers found on some vehicles, are operated by separate control circuits and are covered later in this chapter.

Figure 20-1 Automotive lighting systems.

SHOP TALK

Lighting systems are largely regulated by federal laws, so the systems are similar among the various manufacturers. However, there are many variations. Before attempting to do any repairs on an unfamiliar system, you should always refer to the manufacturer's service manual.

LAMPS

A **lamp** generates light as current flows through the filament. This causes it to get very hot. The changing of electrical energy to heat energy in the resistive wire filament is so intense that the filament starts to glow and emits light. The glass envelope that encloses the filament is evacuated so that the filament "burns" in a vacuum. If air enters the envelope, the oxygen would cause the filament to oxidize and burn up.

It is important that any burned-out lamp be replaced with the correct lamp. You can determine what lamp to use by checking the lamp's standard trade number, usually present on the lamp's housing. Lamps are normally one of two types: a single filament **(Figure 20–2)** or a double filament **(Figure 20–3)**. Double-filament bulbs are designed to

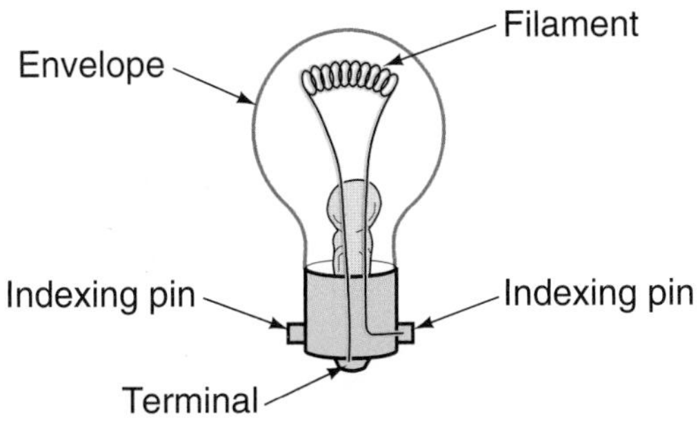

Figure 20-2 A single-filament bulb.

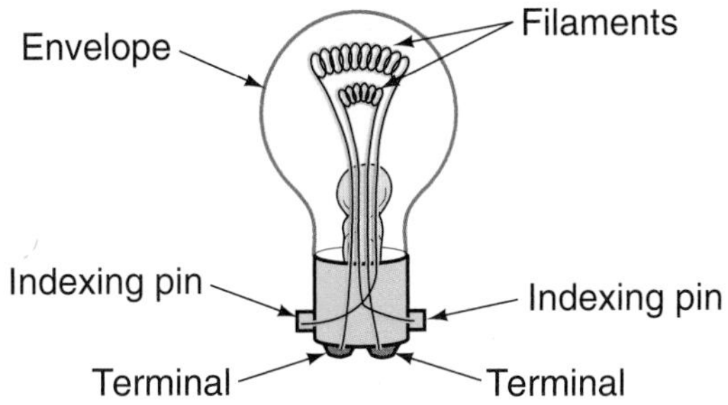

Figure 20-3 A double-filament bulb.

serve more than one function. They can be used as the lone bulb in the stoplight circuit, taillight circuit, and the turn signal circuit.

HEADLIGHTS

Headlights are mounted on the front of a vehicle to light the road ahead during darkness or other times when normal visibility is poor. Headlight designs and construction have been influenced by the changes in technology. In the past, all cars had two or four round or rectangular headlights. Now headlights are an integral part of a vehicle's overall design (**Figure 20–4**).

Sealed-Beam Headlights

Until recently, all vehicles had sealed-beam headlights. A **sealed-beam** headlamp is an air-tight

Figure 20-4 Today's headlights are an integral part of the appearance of vehicles.

assembly with a filament, reflector, and lens. The curved reflector is sprayed with vaporized aluminum and the inside of the lamp is normally filled with argon gas. The reflector intensifies the light produced by the filament, and the lens directs the light to form a broad flat beam (**Figure 20–5**). To direct the light, the surface of the glass lens has concave prisms.

Low- and high-beam filaments are placed at slightly different locations within a sealed-beam bulb (**Figure 20–6**). The filaments' location determines how light passes through the bulb's lens. This, in turn, determines the direction of the light beam. In a dual-filament lamp, the lower filament is used for the high beam and the upper filament is used for the low beam.

There are various ways to identify sealed-beam headlights, such as #1, #2, and the "halogen" or "H" marking molded on the front of the headlight lens. A type #1 is high beam only and has two electrical terminals. The type #2 has both low beam and high beam and three terminals. When a type #2 is switched to low beam, only one of its filaments is lit. When the

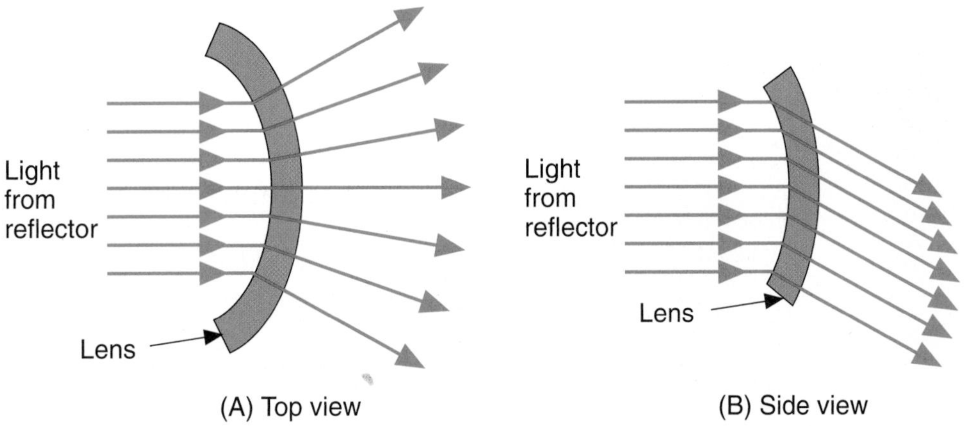

Figure 20-5 The reflector intensifies the light produced by the filament, and the lens directs the light to form a broad flat beam.

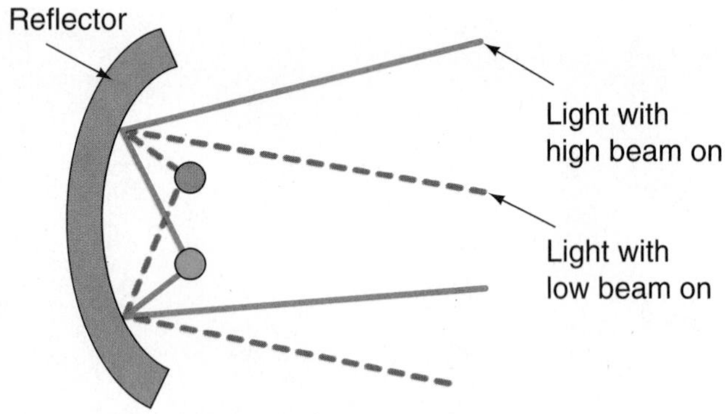

Figure 20-6 Filament placement controls the projection of the light beam.

Figure 20-8 A composite headlamp with a replaceable halogen bulb.

high beam is selected, the second filament lights in addition to the low beam.

When there are four headlights (two on each side), the single filament of the #1 lamp lights when the high beam is selected, along with both filaments of the #2 bulb. H1 and H2 quartz halogen bulbs are the most commonly used sealed-beam headlights.

Halogen Lamps The most commonly used sealed-beam headlight is the halogen type **(Figure 20–7)**. **Halogen** is a term used to identify a group of chemically related nonmetallic elements. These elements include chlorine, fluorine, and iodine. Most often a halogen headlamp is filled with iodine vapor. This headlamp is actually a bulb made of high-temperature resistant glass surrounding a tungsten filament. The bulb is inserted into a sealed glass or plastic housing. The tungsten filament is capable of withstanding higher temperatures than that of standard sealed-beam lamps because of the presence of the halogen. Therefore, because it can withstand higher temperatures, it can burn brighter.

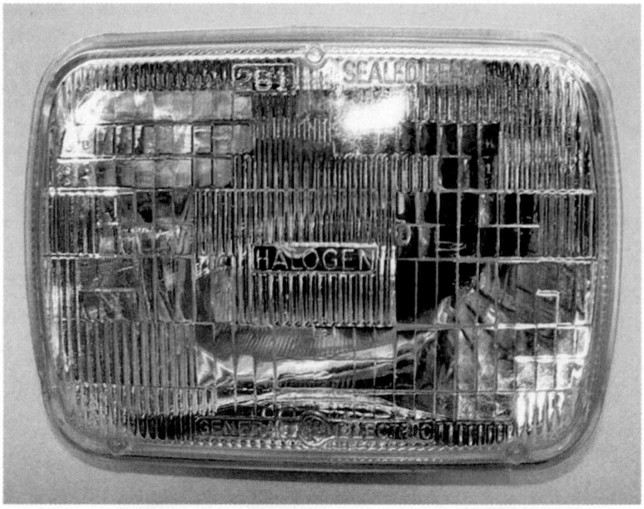

Figure 20-7 A sealed-beam halogen headlamp.

SHOP TALK

Because the filament is inside the inner bulb, cracking or breaking of the housing or lens does not prevent a halogen bulb from working. As long as the filament envelope has not been broken, the filament will continue to operate. However, a broken lens will result in poor light quality, so the lamp assembly should be replaced.

Composite Headlights

Many of today's vehicles have halogen headlight systems that use a replaceable bulb **(Figure 20–8)**. These are called **composite headlights**. By using these, manufacturers are able to produce any style of headlight lens they desire. This improves the aerodynamics, fuel economy, and styling of the vehicle.

Composite headlight housings are often vented to release some of the heat developed by the bulbs. The vents allow condensation to collect on the inside of the lens assembly. The condensation is not a problem and does not affect headlight operation. When the headlights are turned on, the heat generated by the halogen bulb quickly dissipates the condensation. Ford uses integrated nonvented composite headlights. On these vehicles, condensation is not considered normal. The assembly should be replaced.

WARNING!

Whenever you replace a composite lamp, be careful not to touch the lamp's envelope with your fingers. Staining the bulb with skin oil can substantially shorten the life of the bulb. Handle the bulb only by its base.

Figure 20-9 A headlamp assembly with cylindrical bulb housings.

Cylindrical Housings To provide more precise light beam patterns, many manufacturers are using headlamp assemblies that have cylindrical bulb housings **(Figure 20–9)**. The light beam is shot out of the cylinder. The aim of the cylinder projects the beam without much scattering.

HID Headlamps

High-intensity discharge (HID) or **xenon headlamps** use gas-discharge lamps and are electronically controlled. These lights are recognizable by the blue-white color of their light **(Figure 20–10)**. They have this color because the light's spectrum is much closer to daylight than that of a halogen bulb.

Instead of using a filament, an electrical arc is created between two electrodes that excite a gas (usually xenon) inside the headlamp **(Figure 20–11)**, which in turn vaporizes metallic salts that sustain the arc and emit light. The presence of an inert gas amplifies the light given off by this arcing.

Figure 20-10 HID (xenon) headlights are readily identifiable by their bluish light. *Courtesy of Chrysler LLC*

Figure 20-11 A xenon light bulb. *Courtesy of Chrysler LLC*

More than 15,000 volts are used to jump the gap between the electrodes. To provide this voltage, a voltage booster and controller is required. Once the high voltage bridges the gap, only about 80 volts are needed to keep current flowing across the gap. When the headlights are switched on, it takes approximately 15 seconds for the lamps to reach maximum intensity. However, even during ignition these lamps provide more than adequate light for safe driving.

Xenon headlights illuminate the area to the front and sides of the vehicle with a beam that is both brighter and much more consistent than the light generated by halogen headlamps. The great light output of these lamps allows the headlamp assembly to be smaller and lighter. Xenon lights also produce significantly less heat.

Xenon headlights produce about twice as much light as comparable halogen headlights **(Figure 20–12)** and make night driving safer and less tiring for the driver's eyes. Xenon headlamps also use about two-thirds less power to operate and will last two or three times as long.

Bi-Xenon Lights Some vehicles have bi-xenon headlamps that provide xenon lights for low and high beams. These may also be fitted with halogen lights that are used for the flash-to-pass feature. Bi-xenon lights rely on a mechanical shield plate, or shutter, that physically obstructs a portion of the overall light beam emitted by the arc. When the driver selects high beams, the shutter reacts and allows the headlights to project the complete, unobstructed light beam.

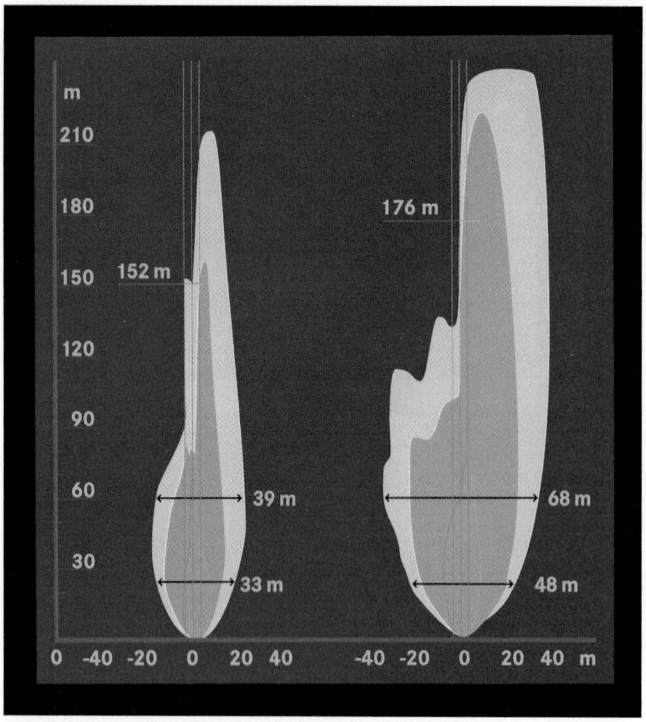

Figure 20–12 A comparison of the light pattern and intensity between a halogen (left) and a xenon (right) headlamp. *Courtesy of Chrysler LLC*

SHOP TALK

HID lamps offer great visibility but often drivers approaching a vehicle with HIDs become annoyed by the glare or brightness of the lamps. To eliminate these concerns, HID systems often have self-leveling and washer systems. Both of these are required in Europe and most European vehicles with HIDs have these features.

LED Headlights

Currently under development and perhaps available on some vehicles at this time are LED headlamps. **Light-emitting diodes (LEDs)** are semiconductors that change electrical energy into light. LEDs are used in many other systems and those are discussed later in this chapter. An individual LED does not produce enough light to serve as a headlamp. Ten to twenty LEDs are needed for ample forward lighting **(Figure 20–13)**. There are many reasons for using LEDs in headlights:

- LEDs do not require a vacuum bulb or high voltage to work.
- LED-based lighting sources require up to 40% less power than traditional lighting sources. This improves a vehicle's fuel economy.
- LEDs provide a whiter light than xenon.

Figure 20–13 This headlight assembly combines a xenon bulb with LEDs. The number of LEDs that are energized depends on the driving conditions.

- Prototype LED headlights have achieved a 1,000-lumen output in the low-beam mode; this is the same as a xenon headlamp.
- LEDs are mercury-free, making them environmentally friendly unlike some HID/xenon systems.
- The average operating life of an LED is twice that of the vehicle itself. This means the headlamp may never need to be replaced.
- LEDs are resistant to shock and vibration.
- LED headlamps reduce oncoming driver perception of glare.
- LED-based headlamps are up to 55% thinner than other designs, which gives designers more flexibility and freedom.

Incorporating LED technology into headlamps presents a problem common to all semiconductors, that being heat. The heat from the engine compartment and the lamp can cause failure. Therefore, LEDs need temperature controls and must be mounted to heat-retarding materials. LEDs also require precise current control, which requires complicated electronic circuitry. In spite of these obstacles, LED headlamps can open the door to other headlamp-related safety features. It is possible to have light beams that meet the current conditions automatically. This can be done by varying the number of LEDs powered or by the placement of special lenses in front of a particular group of LEDs. It is predicted that as LEDs become more powerful and less expensive, they may replace xenon headlamp systems.

Headlight Switches

Headlight switches are either mounted to the instrument panel **(Figure 20–14)** or are part of the multifunction switch on the steering column. The headlight

Figure 20-14 A headlight switch that also has positions for rear and front fog lights.

switch controls most of the exterior lighting for the vehicle. Most switches have three positions: off, park, and headlight. Most vehicles have a warning system that alerts the driver that the park or headlamps are on and that the driver's door is open.

When the headlight switch is in the OFF position, the open contacts prevent battery voltage from continuing on to the lamps **(Figure 20–15)**. When the switch is in the PARK position, battery voltage is applied to the parking lights, side markers, taillights, license plate lights, and instrument panel lamps. This circuit is usually protected by a 15- or 20-amp fuse that is separate from the headlight circuit.

When the switch is in the HEADLIGHT position, battery voltage is applied to the headlights. The lamps lit by the PARK position remain on. Normally, a self-resetting circuit breaker is installed between the battery feed and the headlights. The circuit breaker is designed to reset itself. If a problem causes the breaker to open, the lights will go off until the breaker resets. Then the lights will come back on. If there is a serious problem in the circuit, the headlights might flash as the breaker cycles. Some vehicles have a separate fuse for the headlight on each side of the vehicle. This allows one headlight to operate if there is a problem in the circuit for one side of the vehicle.

The instrument panel lights come on whenever the headlight switch is in the PARK or HEADLIGHT position. The brightness of these lamps is adjustable. A rheostat is used to allow the driver to control the brightness of the bulbs. This control may be part of the headlight switch, in which case the driver simply rotates the headlight switch knob to adjust the panel lights. Most vehicles have a separate control for instrument panel brightness **(Figure 20–16)**. Some of these display a brightness scale on the instrument panel **(Figure 20–17)** to show the driver where the lights are adjusted within their possible range.

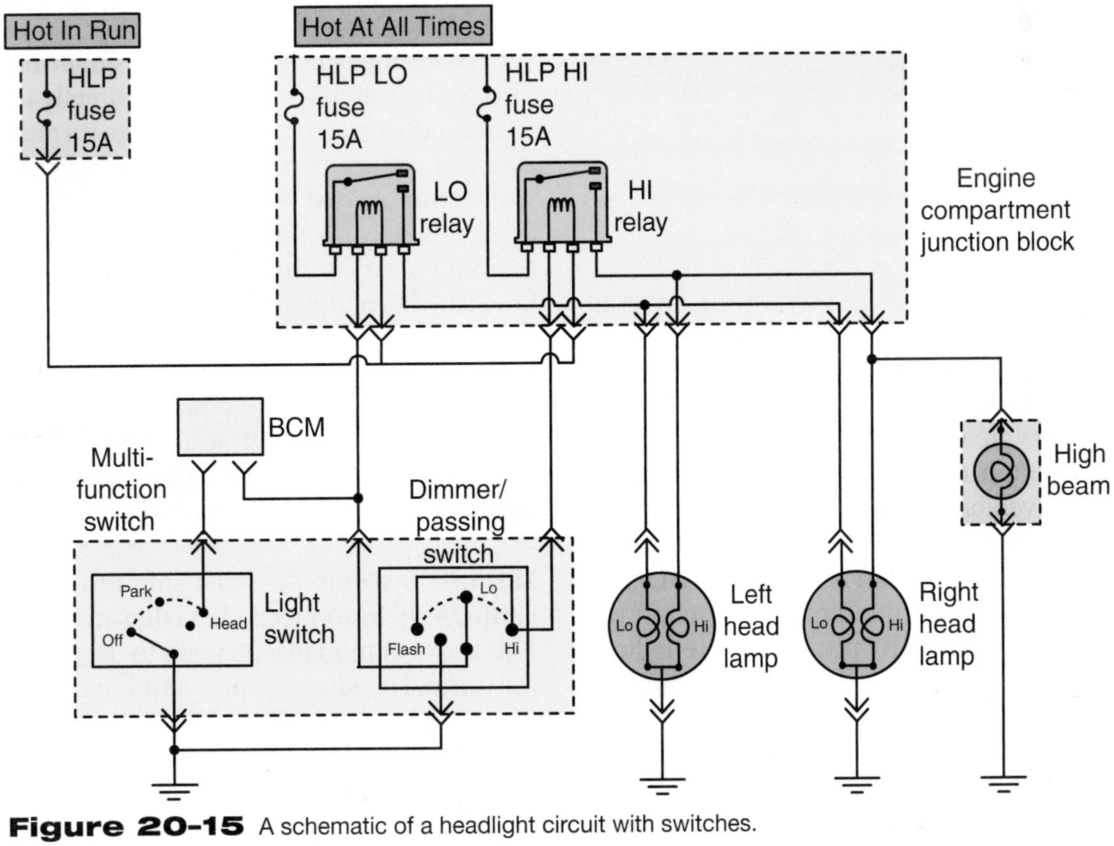

Figure 20-15 A schematic of a headlight circuit with switches.

Figure 20-16 An instrument panel light control switch. This switch also resets the trip odometer.

Figure 20-17 Some vehicles display a brightness scale on the instrument panel as the brightness control is moved.

Dimmer Switches

The **dimmer switch** provides a way for the driver to switch between high and low beams. A dimmer switch is connected in series with the headlight circuit and controls the current path to the headlights. The low-beam headlights are wired separately from the high-beam lamps. Often there is a relay for each set of beams. Newer vehicles have the dimmer switch on the steering column **(Figure 20–18)**.

When the headlight switch is in the HEADLIGHT position, current flows to the dimmer switch. If the dimmer switch is in the LOW position, current flows through the low-beam circuit. When the dimmer switch is in the HIGH position, current flows to the high-beam circuit (see Figure 20–15).

The circuits just discussed have switches that control battery voltage and the bulbs have a fixed ground. Many manufacturers use a system that relies on a groundside switch to control the headlights. In these systems, voltage is always available at the headlights. The headlight switch completes the circuit to ground and the headlights turn on. In these systems, the dimmer switch is also a ground control switch.

Figure 20-18 Most newer vehicles have the dimmer switch mounted on the steering column.

Daytime Running Lights

Canadian law requires that all new vehicles be equipped with **daytime running lights (DRL)** for added safety. This feature is also standard equipment on all new GM vehicles sold in North America. The system normally uses the vehicle's high-beam lights. The control circuit is connected directly to the vehicle's ignition switch so the lights are turned on whenever the vehicle is running. The circuit is equipped with a module that reduces 12-volt battery voltage to approximately 6 volts. This voltage reduction allows the high beams to burn with less intensity and prolongs the life of the bulbs. When the headlight switch is moved to the HEADLIGHT position, the module is deactivated and the lights work with their normal intensity and brightness. Applying the parking brake also deactivates the DRL system so the lights are not on when the vehicle is parked and the engine is running.

LED DRL Systems As a first step toward LED headlamps, Audi introduced LED DRLs in a separate unit. This move eliminates the need to drop the voltage to the normal headlamps for operation of the DRLs and therefore improves the vehicle's fuel economy. It is estimated that these lights consume more than 50% less power than the typical DRL. In addition, they use less space.

Concealed Headlights

Although not as common as they were a few years ago, concealed headlights are still found on some cars. Manufacturers use a concealed headlight system to improve the vehicle's aerodynamics and for style. Today, low profile headlight assemblies are being used instead of "pop-up" headlights.

When the headlight switch is moved to the HEADLIGHT position, the entire headlamp assembly pivots upward. This action is controlled by electric or vacuum motors.

The headlight switch for vacuum systems has vacuum motors attached to the headlight assemblies. When the switch is in the OFF position, engine vacuum is supplied to the motors to keep the headlight doors closed. When the headlight switch is moved to the HEADLIGHT position, a vacuum distribution valve vents the vacuum in the vacuum motors. This allows springs to open the doors. These systems are also equipped with a by-pass valve that allows the doors to be manually opened in case the system fails.

Electrically controlled systems use a torsion bar and a single motor to open both doors or have a separate motor for each headlight door. Current flow to the motors controls the opening and closing of the doors. Limit switches stop current to the motors when they are completely open or closed (**Figure 20–19**). The doors in electrically operated systems can also be manually opened.

Flash to Pass

Most steering column-mounted dimmer switches have an additional feature called flash to pass. This circuit illuminates the high-beam headlights even with the headlight switch in the OFF or PARK position. When the driver activates the flash to pass feature, the contacts of the dimmer switch complete the circuit to the high-beam filaments.

Automatic Light Systems

These systems provide light-sensitive, automatic on-off control of the light normally controlled by the regular headlight switch. It consists of a light-sensitive **photocell sensor/amplifier** assembly and a headlight control relay. Turning the regular headlight switch on overrides the automatic system. In other words, automatic operation is not possible until the regular headlight switch is turned off.

In normal operation, the photocell sensor/amplifier, which is usually mounted under a group of perforated holes in the upper instrument panel pad or slotted holes in the defroster grille panel, is exposed to ambient light. As the light level decreases, the light sensor's resistance increases. When resistance increases to a preset amount, the amplifier applies power to the headlight relay coil. The headlights, exterior lights, and instrument illumination lights turn on. The lights remain on until the system is turned off or the ambient light level increases.

Some systems have two sensors to monitor the ambient light. The light sensors monitor the intensity of the ambient light at an extended angle above the vehicle and in a narrow angle to the front of the vehicle.

Most automatic light systems have a headlamp delay system as well. This system allows the headlamps to stay on for a period of time after the vehicle is stopped and the ignition switch is turned off. A variable switch (**Figure 20–20**) allows the driver to set the amount of time the headlights should remain on after the ignition is turned off. The system can typically be adjusted to keep the headlights on for up to 3 minutes after the ignition is turned off. Of course, the driver can turn off the delay system and the headlamps will shut off as soon as the ignition is turned off.

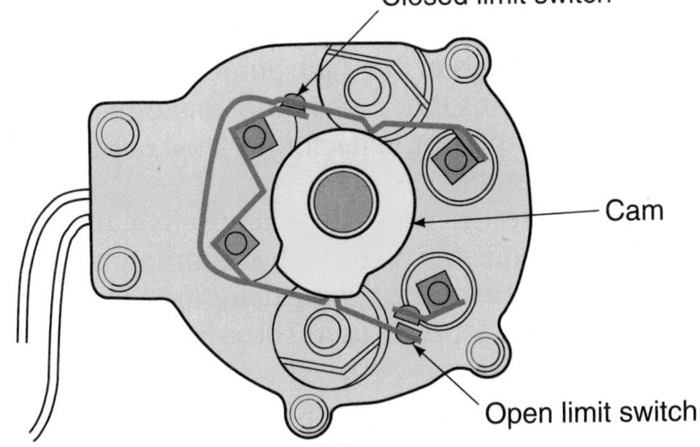

Figure 20–19 Most limit control switches for electric headlamp doors operate by a cam rotated by the motor. *Courtesy of Chrysler LLC*

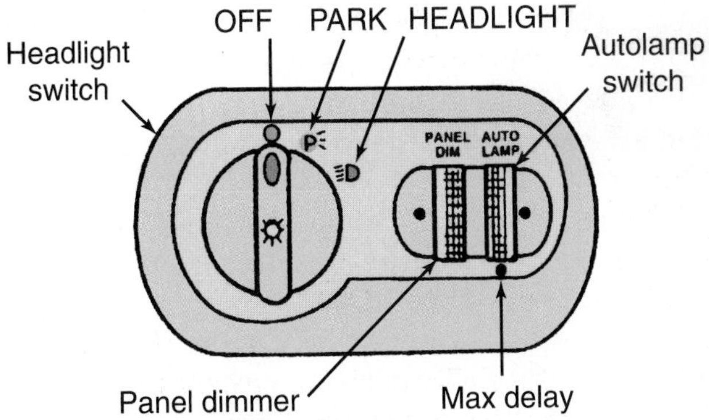

Figure 20–20 Headlight switch, instrument panel light control, and the auto lamp control. *Courtesy of Ford Motor Company*

Customer Care

If the customer's car is equipped with an automatic light control system, point out the location of the perforated holes or slots. Warn the customer not to place any items that may block light from the sensor/amplifier assembly. Blockage causes erratic operation of the system. The photocell must always be exposed to an outside light to function properly.

Figure 20-22 A comparison of how the road is lit up with a conventional (top) headlamp system and an adaptive (bottom) system. *Courtesy of Chrysler LLC*

High-Beam Detection This system automatically turns the high-beam headlights on or off according to conditions. A light sensor (commonly referred to as a camera) on the rearview mirror monitors the light in front of the vehicle **(Figure 20–21)**. When it is dark enough, the system will switch the high beams on. They will stay on until the sensor detects the headlights or taillights of another car. At that time the high beams are switched off until the lights of the other vehicle are no longer detected. The absence of those lights causes the system to switch back to high beams. The system also fades the change from low to high to prevent abrupt changes to the light on the road ahead. The system is programmed to ignore bright lights near the road and reflections from roadside signs. GM calls this system the "IntelliBeam Intelligent High-Beam Headlamp Control System."

Adaptive Headlights

Adaptive headlight systems are electronically controlled through inputs from the steering system and vehicle speed. They aim the lights in the direction the vehicle is turning **(Figure 20–22)**. Currently there are two basic system designs for adaptive lighting. Both are controlled by the vehicle's body control module (BCM). The systems respond in real time by responding to the vehicle's current steering angle, yaw rate, and road speed.

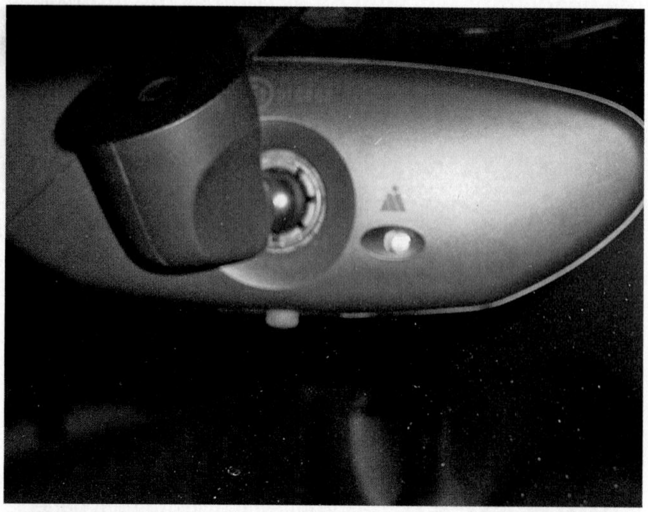

Figure 20-21 A photocell for an automatic lighting system.

One design swivels the entire headlamp assembly. Bidirectional electric motors attached at the base of the assembly **(Figure 20–23)** can move the left headlamp up to 15 degrees to the left and 5 degrees to the right, and the right headlamp up to 5 degrees to the left and 15 degrees to the right. Some manufacturers use a variation of this design and rotate the headlamp's light projector, reflector, or lens, rather than the entire assembly.

Another design uses extra lamps that turn on according to steering angles. These cornering lights may be housed in the headlamp or fog light assembly, or they can be a separate unit. These lamps normally only turn on during low-speed turning.

Adaptive headlights can also be controlled by GPS navigation and digital road maps. With information about the road ahead, the system can anticipate curves in the road and move the headlamps to

Figure 20-23 A headlamp assembly for an adaptive headlight system. *Courtesy of Chrysler LLC*

illuminate curves before the driver starts to turn the steering wheel.

Further advancements in adaptive headlights include variable beam patterns. These systems will automatically change the headlight beams for current conditions, such as rain, snow, and fog.

Auxiliary Lights

Although the car's headlights are adequate in normal driving circumstances, some customers desire auxiliary lights for special conditions, such as driving in heavy fog. In addition to fog lights, driving lights, and passing lights, there are off-road lights, worklights, rooflights, decklights, deckbars, and handheld spotlights.

Auxiliary lights are easy to install when the manufacturer's instructions are carefully followed. The mounting of auxiliary lights, of course, varies according to the type of lights being installed. Before beginning, become familiar with the local regulations for auxiliary lights and adhere to them. These regulations can be obtained from your state's department of transportation. Also, always follow all federal, state, and local laws when aiming headlights, fog lights, and driving lights.

Driving Lights Driving lights typically use an H3 or H4 quartz halogen bulb and a high-quality reflector and lens to project an intense, pencil-thin beam of light far down the road.

Proper aiming of these lights is extremely important. They are used to supplement the high beams and should be used in conjunction with them. Driving lights should be wired so that they are off when the high beams are off. One way to do this is to supply the controlling switch with current from the high-beam circuit, rather than from a circuit that is live all the time.

SHOP TALK

When adding auxiliary lights, make sure that the generator and wiring are heavy enough to handle the increased wattage requirement. Installing a higher output generator may be recommended, especially if other electrical accessories are also being installed. The choice of wire size should be based on the current of the load. Most light kits include a relay as a safeguard. A relay is used because the required current for these lights, especially halogen ones, can be quite high. It is not unusual that they require as much as 25 amperes.

Fog Lights Light does not penetrate fog well. Focus a powerful beam of light at the fog and all the driver sees is a powerful glare. To provide some light through the fog, fog lights are designed to send a flat, wide beam of light under the blanket of fog. This is why they are mounted low and are aimed low and parallel to the road. Because fog tends to reflect light back at the driver, fog lights are often fitted with yellow or amber lenses to reduce the discomfort caused by the glare. Some vehicles have OEM fog lights and are part of the normal lighting circuit.

HEADLIGHT SERVICE

When there is a headlight failure, it is typically caused by a burned-out bulb or lamp, especially if only one lamp fails. However, it possible that the circuit for that one lamp has an open or high resistance. Check for voltage at the bulb before replacing a bulb. If there is no voltage present, the circuit needs work and the original bulb may still be good. If more than one lamp (including the rear lights) is not working, carefully check the circuit. A problem there is much more likely than having a number of burned-out bulbs. Of course, if the charging system is not being regulated properly, the high voltage will cause lamps to burn out prematurely.

Headlight Replacement

There can be slight variations in procedure from one model to another when replacing headlights. For instance, on some models the turn signal light assembly must be removed before the headlight can be replaced. Overall, however, the procedure does not differ much from the following typical instructions.

Make sure the replacement bulb is the same type and part number as the one being replaced.

SHOP TALK

Because of the extremely high voltages involved, any work on xenon lighting should be done carefully and according to the manufacturer's recommendations.

PROCEDURE

Replacing Headlights

STEP 1 Remove the headlight bezel-retaining screws. Remove the bezel. If necessary, disconnect the turn signal lamp wires.

STEP 2 Remove the retaining ring screws from one or both lights.

STEP 3 Remove the retaining rings.

STEP 4 Remove the light from the housing. Disconnect the wiring connector from the back of the light **(Figure 20–24)**.

STEP 5 Push the wiring connector onto the prongs at the rear of the new light.

STEP 6 Place the new light in the headlight housing. Position it so the embossed number in the light lens is on the top.

STEP 7 Place the retaining ring over the light and install the retaining ring screws. Tighten them slightly.

STEP 8 Check the aim of the headlight and adjust it, if necessary.

STEP 9 Install the headlight bezel. Secure it with the retaining screws. Connect the turn signal lamp wiring (if it was disconnected).

SHOP TALK

Some manufacturers recommend coating the prongs and base of a new sealed beam with dielectric grease for corrosion protection. Use an electrical lubricant approved by the manufacturer.

HID Diagnosis and Service

Proper diagnosis of these systems depends on understanding how the circuit works. If only the low beams have a HID system, the headlights contain an arc tube and a transformer. The transformer increases the voltage so an arc can be established across the

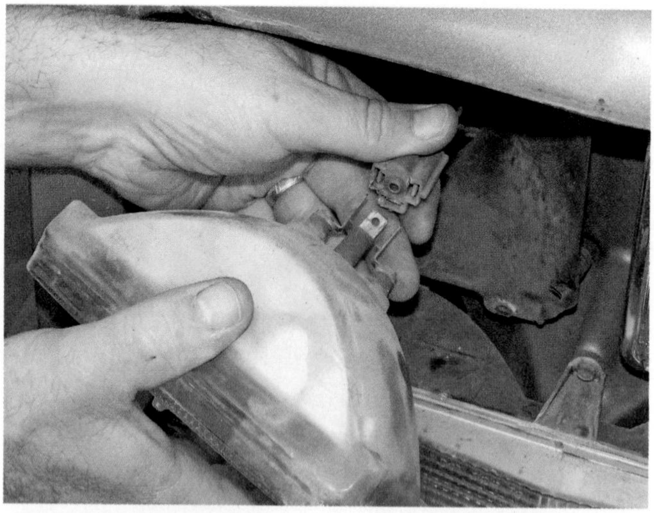

Figure 20–24 After unplugging the old bulb, check and clean the contacts in the connector before installing the new bulb.

electrodes. The arc tube is where the arcing takes place. When the headlight switch is moved to the HEADLIGHT position, the ground for the low-beam relay is completed by the control computer (BCM). This energizes the relay, and battery voltage is applied to the transformer in each headlamp assembly. The transformer increases the voltage and starts the high-voltage arcing to light the bulb.

In bi-xenon systems, there is no separate high-beam bulb. Rather a solenoid with a shutter is used to redirect the light beams from the bulb. When the headlamp dimmer switch is moved to the high-beam position, ground is applied through the dimmer switch to the BCM. In response to this signal, the BCM completes the ground for the high-beam relay, energizing the relay. Battery voltage is then applied to the left and right high-beam solenoids that move the shutter.

Normal Delay Often the wait period for xenon lights to become fully illuminated is considered a problem. This is not a problem; it is normal. The transformers need high current to start and keep the lamps illuminating. This takes a little time; it takes 2 seconds to establish the arc and then another 30 seconds to have full illumination. Obviously, if it takes longer than that, there is a problem.

Bulb Color Another normal concern is the color of the light. HIDs produce a bluish-white light. However, some produce a light beam that appears pure white. This is not a problem, even if one side appears to be a different color than the other side. The color of the light depends on many normal factors. A worn bulb may have a dim pinkish glow. Often replacement bulbs will have a yellowish-white look for the first 5 minutes when turned on. The lamps still provide the same amount of light as the originals and will provide a bluish light after about 100 hours of operation.

Bulb Replacement HID bulbs normally do not suddenly stop working. As the bulb wears, it will shut off and then turn back on. When the bulb is in the early stages of wear, this will happen very infrequently. As time goes on, the bulb will shut down and come on again rather frequently. This problem may progress to flickering until the system totally shuts down. Each manufacturer has a procedure for monitoring and diagnosing the system; always follow those.

Over time the resistance between the bulb's electrodes increases. This makes it harder for the transformer to maintain the arcing needed for light. When arcing is lost, the system will trigger the transformer to establish the arc again. This is what causes the light to flicker. It is best to replace the bulb before the lights constantly go on and off.

Every time the bulb shuts down, the transformer uses high current to reestablish the arc. The repetitive high current surges to the transformer damages the transformer. Eventually, the system will stop sending current to the transformer and the bulb will not work. When the lights have been flickering for a while the transformer should be replaced with a new bulb.

Headlight Adjustments

Headlights must be kept in adjustment to obtain the safest and best light beams on the road ahead. Headlights that are properly adjusted cover the correct range and afford the driver the proper nighttime view. Headlights that are out of adjustment can cause other drivers discomfort and sometimes create hazardous conditions.

Before adjusting or aiming a vehicle's headlights, make the following inspections to ensure that the vehicle is level. Any one of these conditions can result in an incorrect setting.

- The vehicle must be on a level floor.
- If the vehicle is coated with snow, ice, or mud, clean it, especially the underside. The additional weight can alter the riding height.
- Try to make the adjustment with the fuel tank half full; this should be the only load present on the vehicle.
- Worn or broken suspension components affect the setting, so check the springs or shock absorbers.

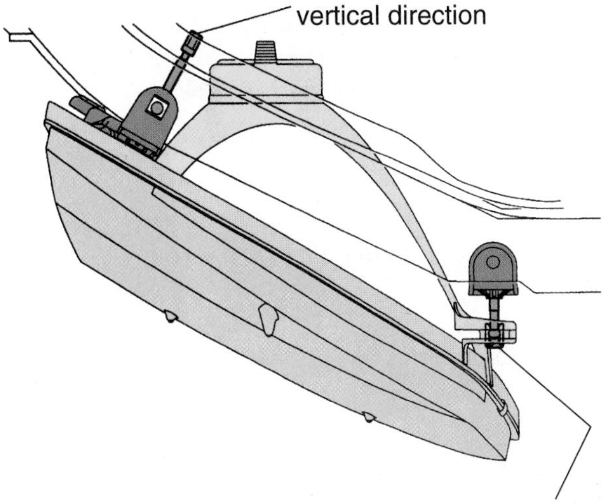

Figure 20-25 An example of headlight adjustment screws. *Courtesy of Toyota Motor Sales, U.S.A., Inc.*

- Inflate all tires to the recommended air pressure levels.
- Make sure the wheel alignment and rear axle tracking path are correct before adjusting the headlights.
- After placing the vehicle in position for the headlight test, push down on the front fender to settle the suspension.

Headlight assemblies have adjusting screws to move the headlight within its assembly to obtain correct headlight aim. Lateral or side-to-side adjustment is made by turning the adjusting screw at the side of the headlight (**Figure 20–25**). Vertical or up-and-down adjustment is made by turning the screw at the top of the headlight.

SHOP TALK

While adjusting headlight beams, make sure they meet the standards established by your local community, state, or province.

To properly adjust the headlights, their aim must be checked first. Many types of headlight aiming equipment are available (**Figure 20–26**). These aimers use mirrors with split images, like split-image finders on some cameras, and spirit levels to make exact adjustments. When using any mechanical aiming tool, follow the manufacturer's instructions. When headlight aiming equipment is not available, alignment can be checked by projecting the upper beam of each light on a screen or chart at a distance of about

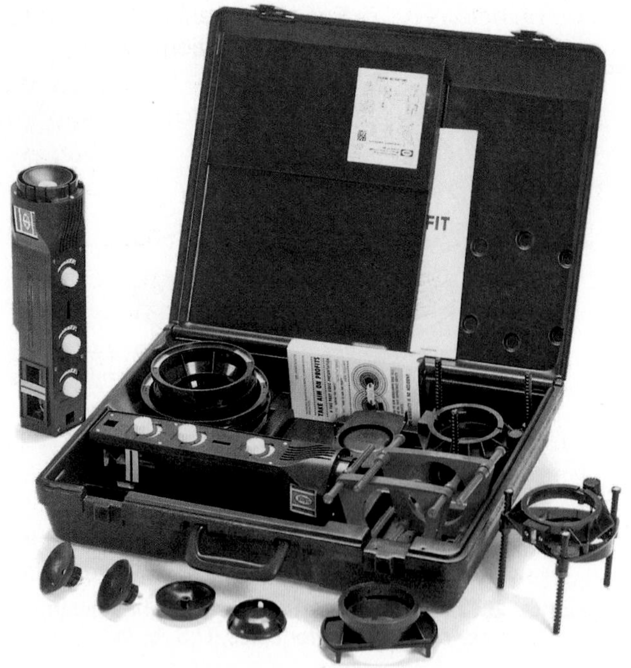

Figure 20-26 A typical headlight aiming kit. *Reproduced under license from Snap-on Incorporated. All of the marks are marks of their owners.*

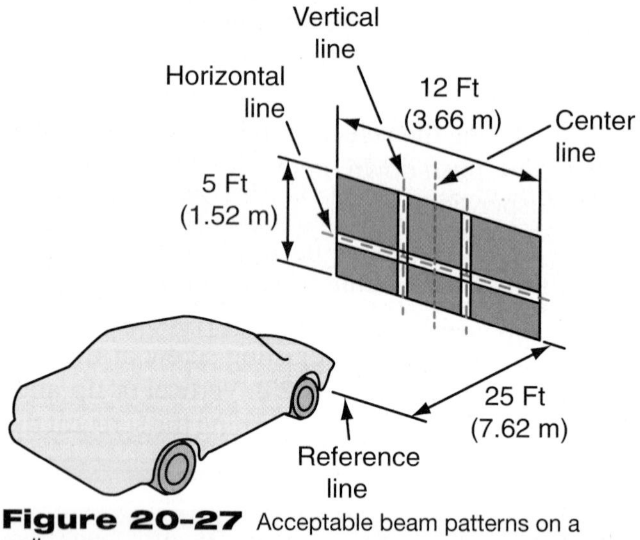

Figure 20-27 Acceptable beam patterns on a wall screen.

25 feet ahead of the headlight **(Figure 20–27)**. The vehicle must be exactly perpendicular to the chart.

To mark the chart for reference, measure the distance between the centers of the headlights. Use this measurement to draw two vertical lines that correspond to the center of the headlights. Then draw a vertical centerline halfway between the two vertical lines. Measure the distance from the floor to the centers of the headlights. Subtract 2 inches from this height and draw a horizontal line on the screen at this new height.

With the headlights on high beam, the hot spot of each projected beam should be centered on the point of intersection of the vertical and horizontal lines on the chart. If necessary, adjust the headlights vertically and laterally to obtain proper aim.

Some vehicles are equipped with indicators to help with adjustment. There is a horizontal indicator gear **(Figure 20–28)** at each headlamp assembly. Prior to making any adjustments, it is recommended that this indicator be at "0." A screwdriver is used to bring the gear back to "0." After this, the headlamps can be adjusted to specifications.

Many headlamps have a bubble level to aid vertical alignment. A horizontal gauge and magnifying window is located next to the bubble level as an aid for alignment **(Figure 20–29)**. The vertical bubble level is used to compensate for vehicle ride height changes due to heavy loads. The bubble level is calibrated to the earth's surface; therefore, the vehicle must be on

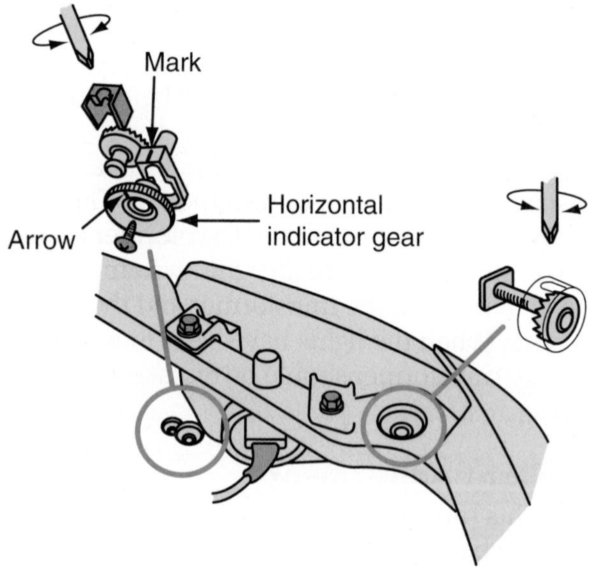

Figure 20-28 Honda's headlight adjustment screws with an indicator.

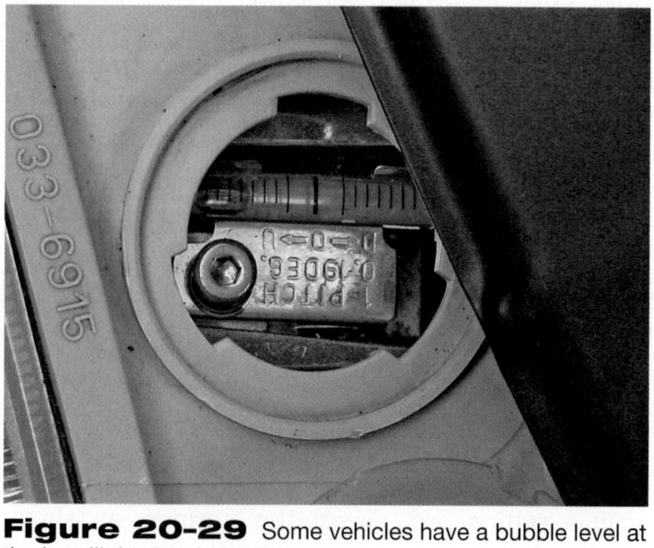

Figure 20-29 Some vehicles have a bubble level at the headlights to aid in alignment.

level ground when the headlights are aimed. If the headlight beam projection appears high to oncoming traffic, check the alignment on an alignment screen. If the beam pattern is above or to the left of the specified location on the screen, adjust the headlights, then recalibrate the bubble level and magnifying window. Ideally, if the headlights are aligned, the bubble level and magnifying window will be centered. Never change the calibration of the magnifying window or bubble level if the headlights are out of alignment.

Normally, a previously properly aimed headlight does not need to be reaimed after replacement of a bulb.

Auto-Leveling Headlamps

Some vehicles, primarily those equipped with xenon headlamps, have an automatic headlight leveling system. With this system there is no need to adjust the headlamps. However, if this system fails, the headlamps will not be properly aligned and the system must be diagnosed. Before diagnostics can take place, an understanding of how the system works is important.

The system keeps the headlights' aim within a prescribed range regardless of the vehicle's load and driving conditions. Each headlamp assembly has a headlamp leveling motor controlled by a headlamp control module. The control module is connected to motors at each headlamp and responds to demands by the BCM. The control module may be part of the lamp's transformer assembly. Based on inputs and its programming, the module determines the best headlight angle for the current conditions and orders that position to the electric motors. The motors, in turn, move the headlamp assembly.

The control module monitors inputs from position sensors at the control arms in the front and rear suspensions. As the suspension compresses and rebounds, the sensors' arms move and new signals are sent to the control module. The control module compares these inputs to information stored in its memory and sends out commands to adjust the aim of the headlamps, if needed.

The control module also receives inputs from a vehicle speed sensor. This allows the control module to know when the vehicle is accelerating or braking. During both of these events, the pitch or angle of the vehicle changes.

Diagnosis Diagnosis is done with a scan tool. The scan tool can be used to order the motors to move up and down. With the ignition on, do this and observe the headlamps. If they do not move, test the motors. If the motors move, check the voltage of the front and rear leveling position sensors. Normally the voltage should be between 0.5 and 4.9 volts. If the reading is not within the specified range, check the sensors.

The scan tool may also retrieve fault codes for the system. Possible fault codes can be set for a short to ground, open/high resistance, short to voltage, and abnormal signal performance for the following areas:

- Front and rear headlamp leveling sensor
- Front and rear headlamp leveling sensor 5-volt reference
- The headlamp leveling motor control circuits on both sides

The motor assembly is checked with an ohmmeter. With the ignition off, disconnect the electrical connector at each headlamp leveling motor. Take resistance readings across the various terminals of the connector and compare your reading to the specifications. The leveling sensors are checked with a voltmeter. First the reference voltage is checked at the specified terminals. If the voltage is higher than normal, there is a short to ground, an open, or high resistance in the circuit. If all circuits test normal, the appropriate headlamp leveling sensor may be bad.

Automatic Headlight System Diagnosis

Automatic headlight systems respond to changes in outside light levels. A photocell, normally in the vehicle's dash, sends a signal to the BCM. The BCM energizes the low-beam circuit when it is appropriate. If the headlamps turn on when the headlight switch is placed on the position but not when it is in "auto," the problem is in the auto circuit. Check the area around the photocell for anything that may be blocking light from the sensor. Check all connections in the system. The circuit normally contains a relay and amplifier along with the photocell. Each should be tested according to the procedures given by the manufacturer.

The photocell, or camera, must be properly aligned in order for the system to respond to the lights of oncoming vehicles and not lights coming from the far sides of the road. This typically is a very involved process and the procedures given by the manufacturer must be followed. This involves many measurements and conducting tests with a scan tool (**Figure 20–30**).

Most systems also keep the lights on for a short time after the vehicle has been parked. The controlling device for this feature is a potentiometer incorporated into the headlamp switch. When the delay feature is not working properly, begin testing at the timer control.

Adaptive Headlight Diagnosis

The operation of the adaptive headlight system is controlled by the headlamp control module.

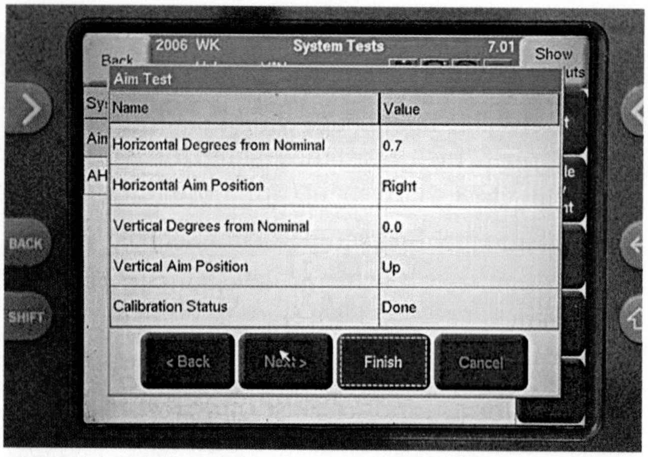

Figure 20-30 Calibration of the photocell includes a check with a scan tool.

Therefore, diagnosis begins with a scan tool. Because the system responds to many inputs, each one should be monitored.

The headlamp control module receives serial data from the engine control module, transmission control module, electronic brake control module, and body control module. The control module calculates the desired headlamp angle and orders the actuators or motors, at each headlight to move the headlamps. The control module also monitors the condition of the motor control circuits. If a problem is detected, a DTC will be set and a message displayed in the instrument panel to alert the driver of a problem.

OTHER LIGHT BULBS

There are several types of light bulbs used in today's vehicles (**Figure 20-31**). Each design has a unique purpose and a unique way it fits into its socket. This prevents the use of the wrong design. However, within a particular design, there may be different bulbs, each with its own power rating. Whenever replacing a bulb, make sure it is the exact replacement.

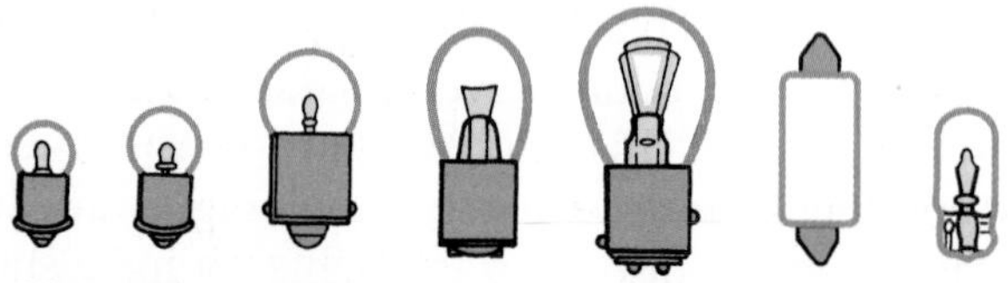

A,B MINIATURE BAYONET FOR INDICATOR AND INSTRUMENT LIGHTS
C—SINGLE CONTACT BAYONET FOR LICENSE AND COURTESY LIGHTS
D—DOUBLE CONTACT BAYONET FOR TRUNK AND UNDERHOOD LIGHTS
E—DOUBLE CONTACT BAYONET WITH STAGGERED INDEXING LUGS FOR STOP, TURN SIGNALS, AND BRAKE LIGHTS
F—CARTRIDGE TYPE FOR DOME LIGHTS
G—WEDGE BASE FOR INSTRUMENT LIGHTS

Figure 20-31 Common types of automotive bulbs.

SHOP TALK

The current trend is to use LEDs rather than standard bulbs in many different lighting systems. Taillights, turn signals, hazard lights, side lights, parking lights, and fog lamps may use an LED or a cluster of LEDs. When a light is not working, make sure you know if an LED is being used before getting a new bulb or beginning diagnostics.

Most bulbs fit into sockets and are held in place by spring tension or mechanical force. Bulbs are coded with numbers for replacement purposes. Bulbs with different code numbers might appear physically similar but have different wattage ratings.

Light Circuits

Light systems normally use one wire to the light, making use of the car body or frame to provide the ground. Because many of the manufacturers are now using plastic sockets and mounting plates (as well as plastic body parts) to reduce weight, many lights must now use two wires, the second one for the ground. Many bulbs have a single filament.

Double-filament bulbs are actually two separate bulbs in one casing; therefore, they have two hot wires. Some are grounded through their case and others require a third wire for ground.

Many light circuits are switched or controlled on the power side of their circuit; however, there are ground-controlled circuits. Knowing how the light is controlled is important for diagnostics.

INTERIOR LIGHT ASSEMBLIES

The types and numbers of interior light assemblies used vary significantly from one vehicle to another (**Figure 20-32**). Following are the more common ones.

Engine Compartment Light Operating the hood causes the engine compartment light mercury switch to close and light the underhood area. Some pickup trucks and SUVs are equipped with an underhood retractable magnetic base lamp mounted on a reel. The lamp can be used anywhere around the vehicle.

Glove Box Light Opening the glove box door closes the glove box light switch contacts and the light comes on.

Luggage Compartment Light The light is mounted in the underside of the trunk deck lid in the luggage compartment.

Figure 20-32 Full interior illumination is available with this light setup. *Courtesy of Chrysler LLC*

Trunk Lid Light Lifting the trunk lid causes the light mercury switch to close and the light to come on.

Vanity Light Pivoting the sun visor downward and opening the vanity mirror cover causes the vanity light switch contacts to close and the light to come on.

Courtesy Lights Vehicles may have courtesy lights in the door trim panels, under each side of the instrument panel, above the foot wells, or in the center of the headlining. Their sole purpose is to allow light inside the vehicle. They are illuminated when one of the doors is opened, by rotating the headlight switch to the full counterclockwise position, or by depressing the designated switch. **Figure 20–33** is a wiring diagram of a typical courtesy light circuit. The courtesy lights are also turned on by the illuminated entry or keyless entry systems, if the vehicle is equipped with one or both of these. LEDs are often used for interior lighting (**Figure 20–34**).

Typically, power is supplied from the fuse block to the courtesy lights. The ground is controlled by the position of the door switch. The door switches are held in an open position and do not provide for a ground circuit. When the door is opened, a spring

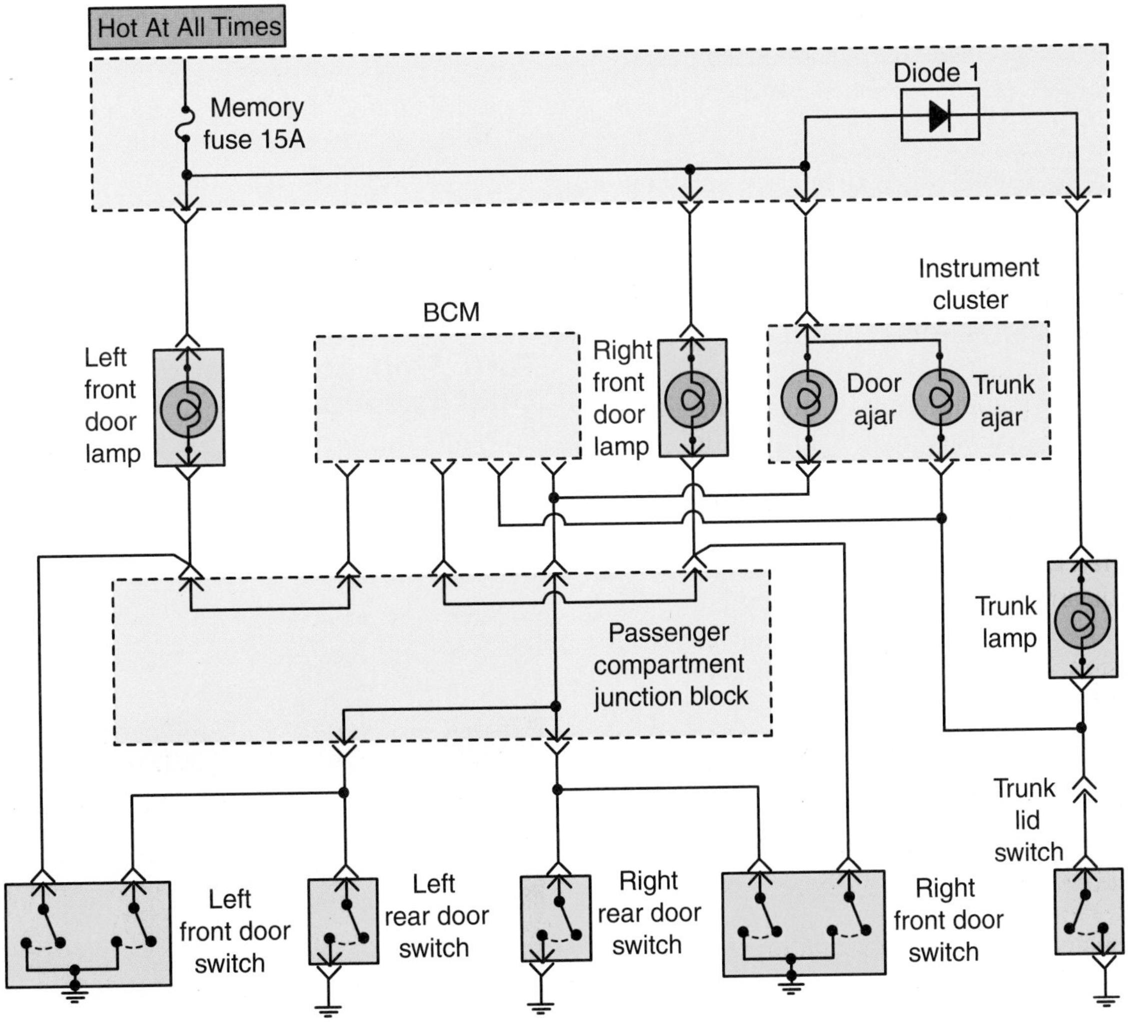

Figure 20-33 A typical courtesy light circuit.

Figure 20–34 An example of using LEDs for interior lights. *Courtesy of Chrysler LLC*

pushes the switch closed to ground the circuit, and the courtesy lights come on. Interior and courtesy lights rarely give any trouble. However, if they do not operate, check the fuse, bulb, switch, and wiring.

Dome/Map Light Map lights are typically located on each side of the dome/map light housing. They are operated independently of the dome light by switches located at each map light. The dome light **(Figure 20–35)** is actuated by a switch or along with the courtesy lights when the door is opened.

Illuminated Entry System This system assists vehicle entry during the hours of darkness by illuminating the door lock cylinder so it may be easily located for key insertion. The vehicle interior is also illuminated by the courtesy lights.

Most of these systems have an electronic actuator module, an illuminated door lock cylinder, a door latch switch, and wiring harness. The system is activated by raising the outside door handle, by depressing a key on the key fob, or by entering a code on the keyless entry system. This action momentarily closes a switch mounted on the door latch mechanism, which completes the ground circuit of the electronic actuator module and switches the system on. The vehicle interior lights turn on, and both front

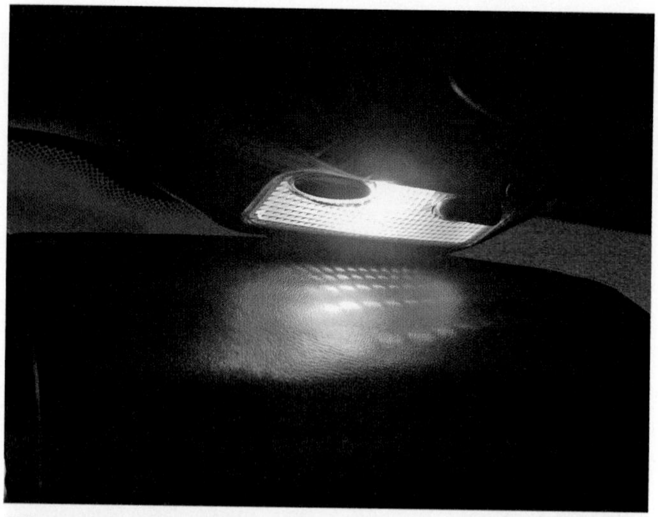

Figure 20–35 A dome light.

door lock cylinders are illuminated by a ring of light around the area where the key enters. This illumination remains on for approximately 25 seconds and then automatically turns off. During this 25-second period, the system can be manually deactivated by turning on the ignition switch.

REAR EXTERIOR LIGHTS

The rear of a vehicle has many different lights. These include the taillights, turn signal and hazard lamps, brake lights, a center high-mounted stoplight, side marker lamps, reverse or back-up lights, and license plate lamps. Many of them are incorporated in the taillight assembly. Rear lighting is definitely part of the overall style of the vehicle and most make a styling statement **(Figure 20–36)**.

Turn, Stop, and Hazard Warning Light Systems

Power for the turn (directional signal), stop, and hazard warning light systems is provided by the fuse panel **(Figure 20–37)**. Each system has a switch that

Figure 20–36 A stylish taillight assembly using LED technology.

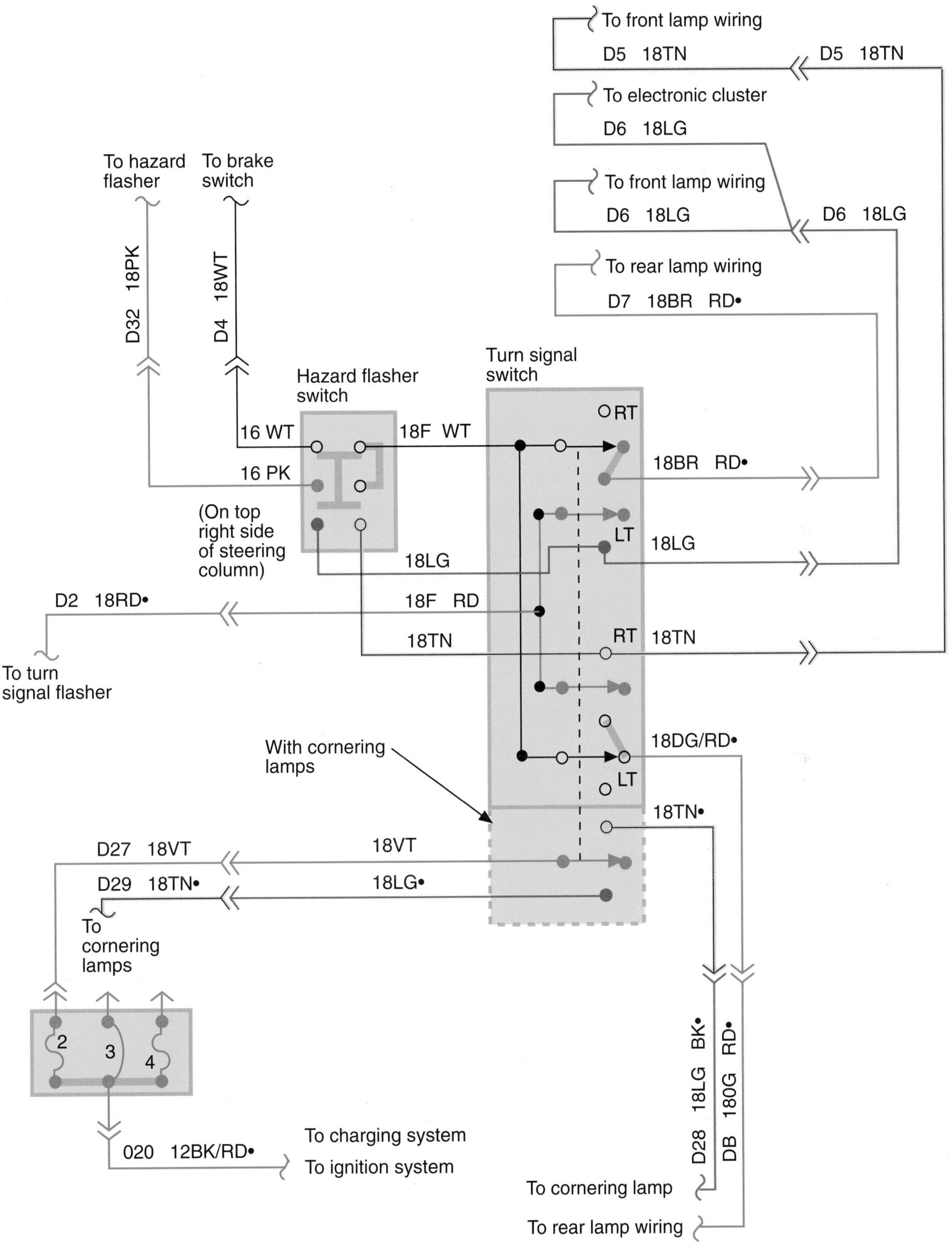

Figure 20-37 Turn signal circuit for a two-bulb system. *Courtesy of Chrysler LLC*

must close to turn on the lights in the circuit. Hazard lights are commonly referred to as 4-way flashers because the lights at all four corners of the vehicle will flash when the circuit is turned on.

The turn signal and hazard light switches on many current vehicles are part of a **multifunction switch**. When the turn or directional signal switch is activated, only one set of the switch contacts is closed—left or right. However, when the hazard switch is activated, all contacts are closed and all turn signal lights and indicators flash together and at the same time.

The power for the turn signals is provided through the fuse panel, but only when the ignition switch is on. The hazard lights are also powered through the fuse panel; however, they have power at all times regardless of ignition switch position.

Side markers are connected in parallel with the feed circuit (from the headlight switch) for the filaments of the front parking lights and rear taillights.

What a multifunction switch controls depends on the make, model, and year of the vehicle. Some control the directional signals and serve as the dimmer switch. Others control the turn and hazard signals and serve as the headlight, dimmer, windshield wiper, and windshield washer switch.

This switch is not repairable and must be replaced if defective. Photo Sequence 18 outlines the typical procedure for removing a multifunction switch. Some of the steps shown in this procedure may not apply to all types of vehicles; always refer to the service manual before removing this switch. Also carefully study the procedures beforehand and identify any special warnings that should be adhered to, especially those concerning the air bag.

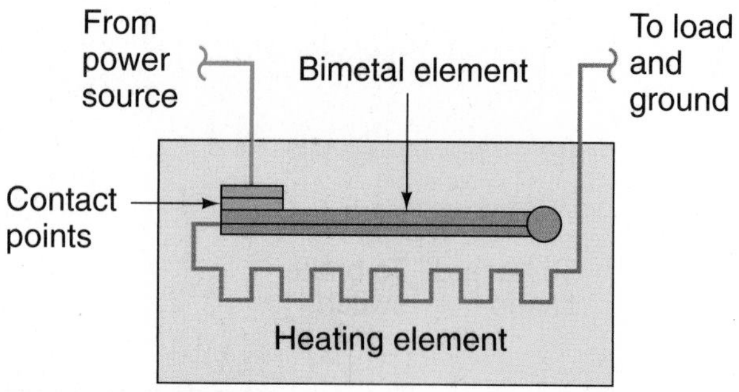

Figure 20-38 A typical turn signal flasher. *Courtesy of Ford Motor Company*

Flashers **Flashers** are components of both turn and hazard systems. They contain a temperature-sensitive bimetallic strip and a heating element **(Figure 20–38)**. The bimetallic strip is connected to one side of a set of contacts. Voltage from the fuse panel is connected to the other side. When the left turn signal switch is activated, current flows through the flasher unit to the turn signal bulbs. This current causes the heating element to emit heat, which in turn causes the bimetallic strip to bend and open the circuit. The absence of current flow allows the strip to cool and again close the circuit. This intermittent on/off interruption of current flow makes all left turn signal lights flash. Operation of the right turn is the same as the operation of the left turn signals.

Turn signal flashers are installed on the fuse panel on current car models and most current truck models **(Figure 20–39)**. However, on earlier models this is not

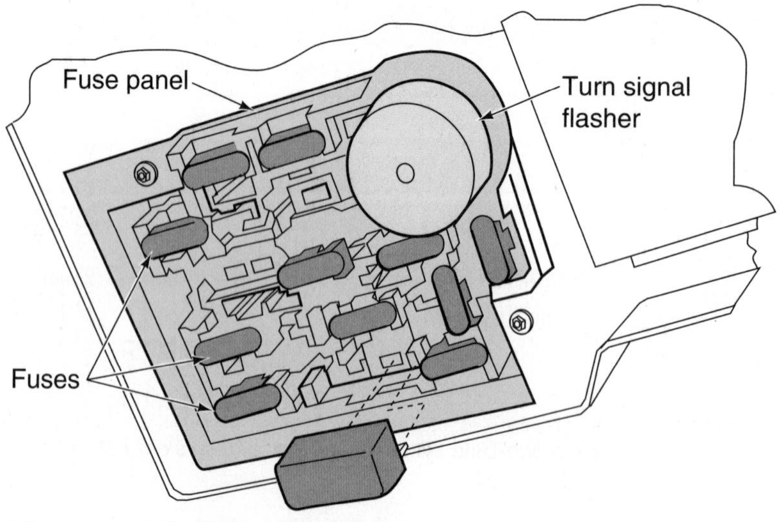

Figure 20-39 A common location for the flashers in the fuse panel. *Courtesy of Ford Motor Company*

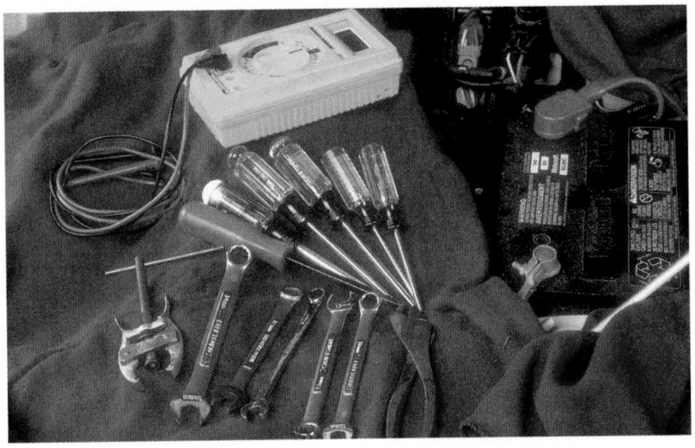

P18-1 *The tools required to test and remove a multifunction switch are fender covers, battery terminal pliers and pullers, assorted wrenches, Torx driver set, and an ohmmeter.*

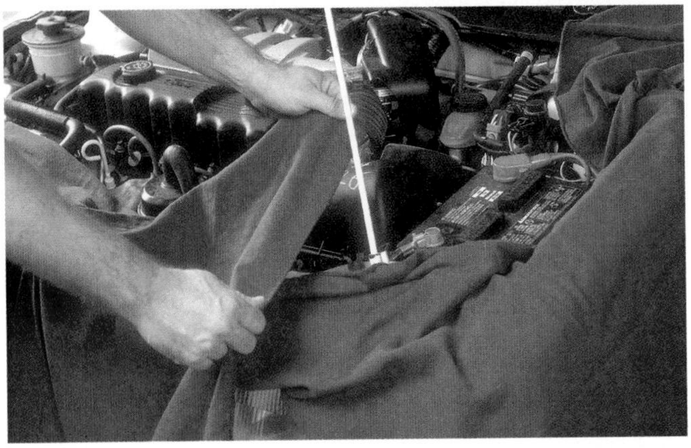

P18-2 *Place the fender covers over the fenders of the vehicle.*

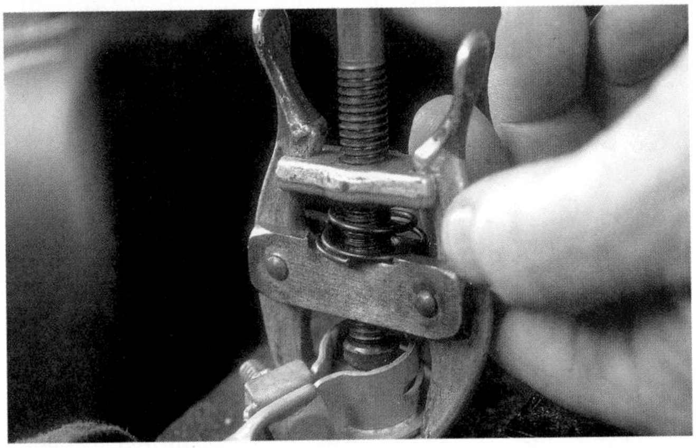

P18-3 *Loosen the negative battery clamp bolt and remove the battery clamp. Place the cable where it cannot contact the battery.*

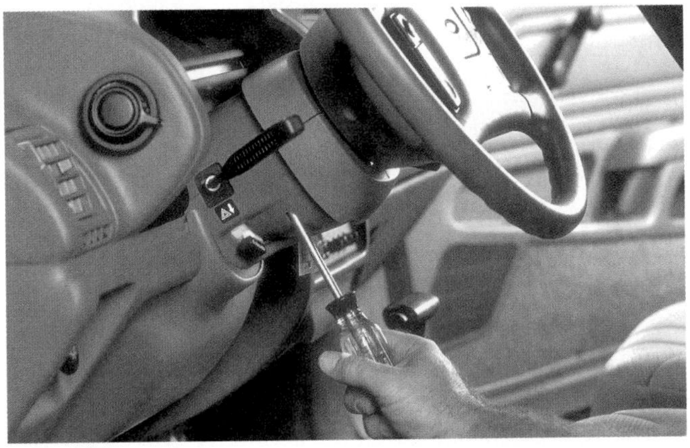

P18-4 *Remove the shroud retaining screws and remove the lower shroud from the steering column.*

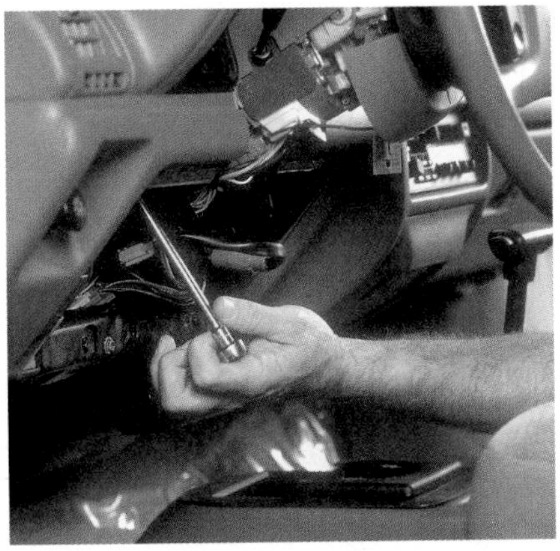

P18-5 *Loosen the steering column attaching nuts. Do not remove the nuts.*

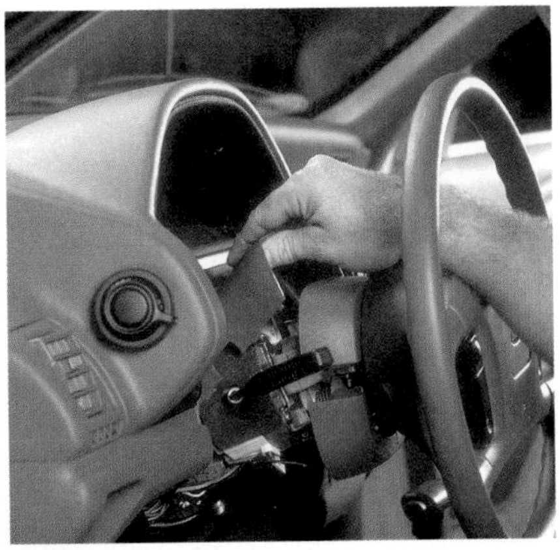

P18-6 *Lower the steering column just enough to remove the upper shroud.*

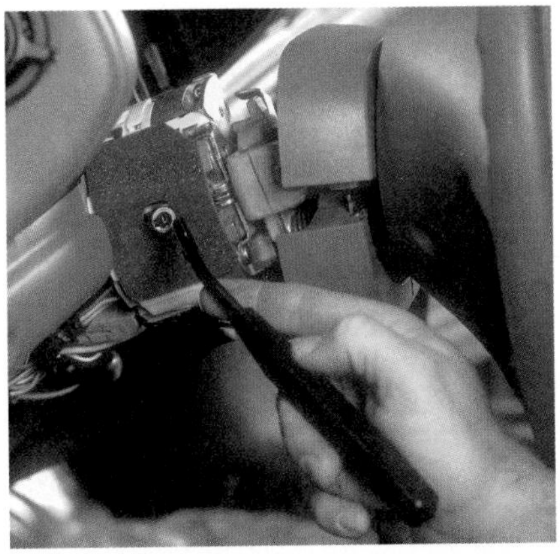

P18-7 *Remove the turn signal lever by simply rotating the outer end of the lever. Then pull it straight out.*

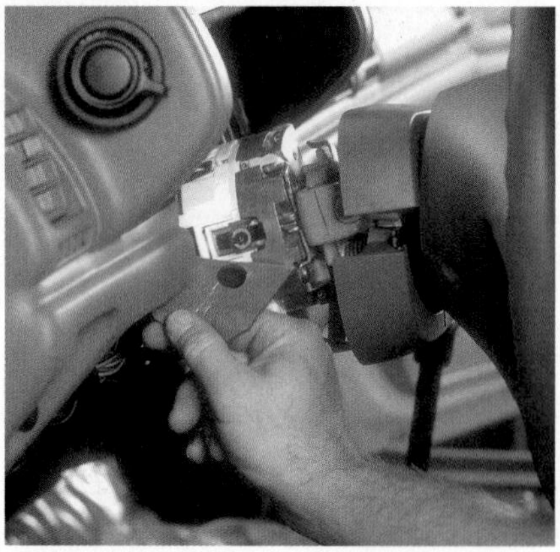

P18-8 *Peel back the foam shield from the turn signal switch.*

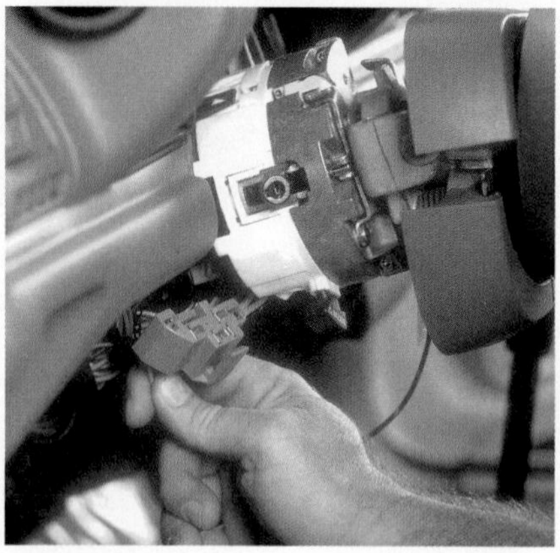

P18-9 *Disconnect the turn signal switch electrical connectors.*

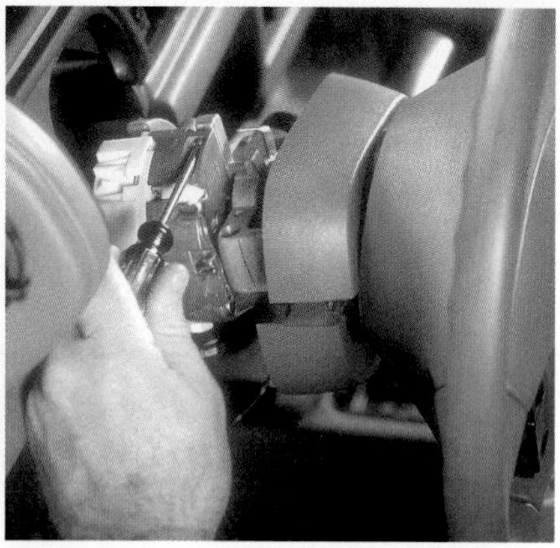

P18-10 *Remove the screws that attach the switch to the lock cylinder assembly.*

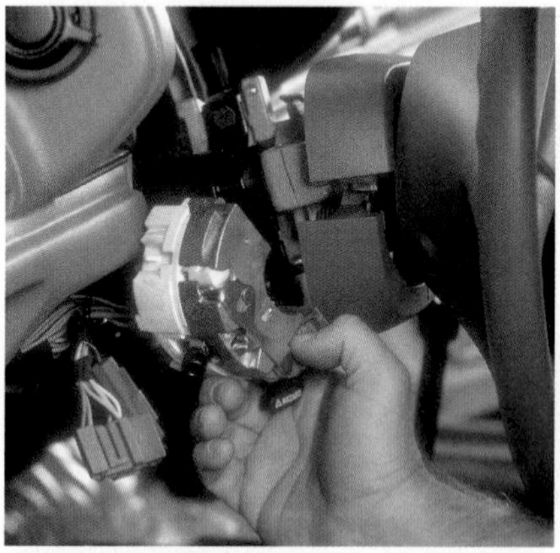

P18-11 *Disengage the switch from the lock assembly.*

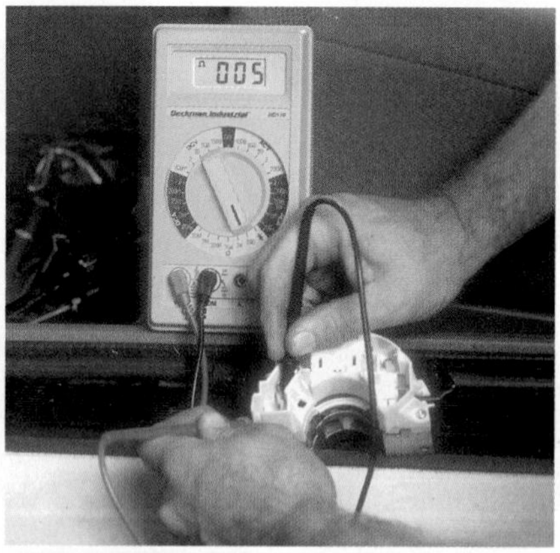

P18-12 *Use an ohmmeter to test the switch. Check for continuity when the dimmer switch is in the low-beam position.*

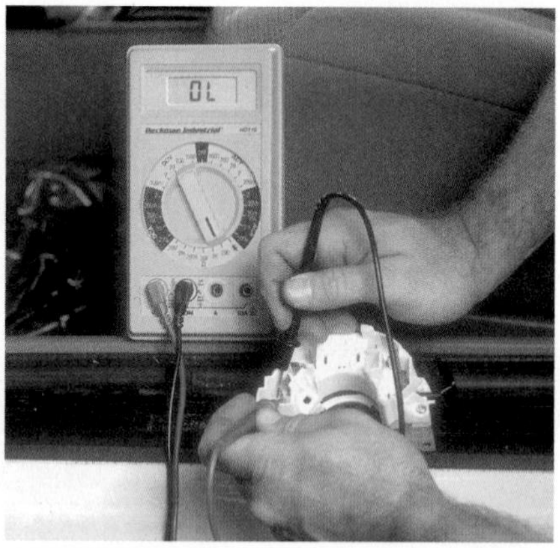

P18-13 *When the switch is in the low-beam position, the circuit should be open between the high-beam terminals.*

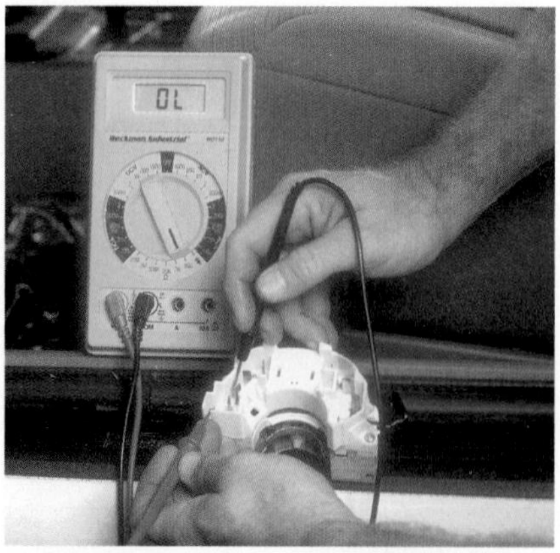

P18-14 *Also check the other terminals and circuits that should be open when the dimmer switch is in the low-beam position.*

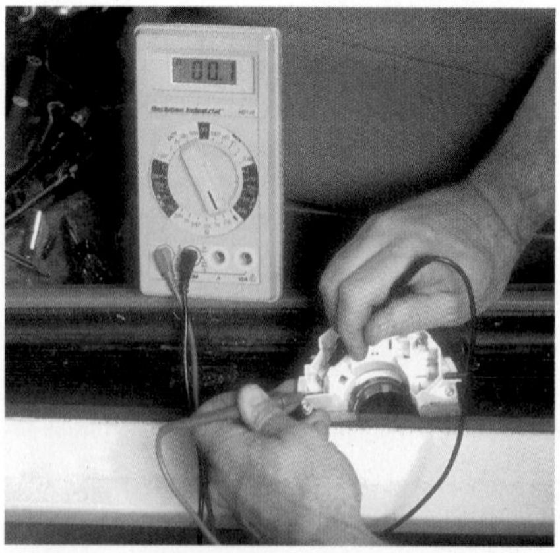

P18-15 *With the switch in the high-beam position, there should be continuity across the high-beam circuit. Also check for continuity across the other circuits that should be open when the switch is in the high-beam position.*

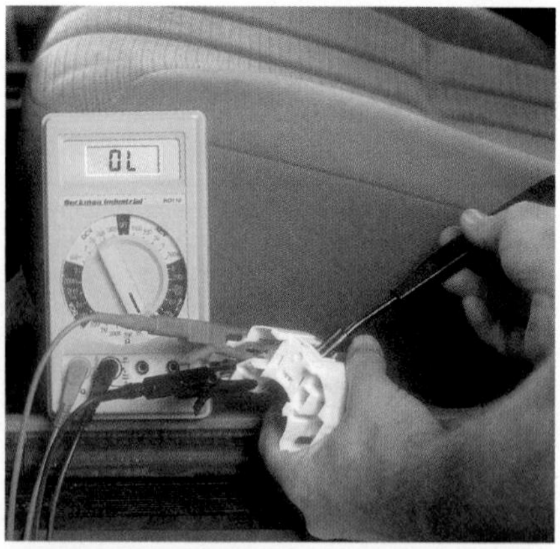

P18-16 *When the dimmer switch is placed in the flash-to-pass position, there should be continuity across those designated terminals and an open across the others.*

true. Hazard flashers are also mounted in various locations. Refer to the service manual for locations on the model being serviced.

A testlight can be used to determine which flasher is used for the turn signals and which is used for the hazard warning light. An easier way is to turn on both the directionals and the hazards. This activates both flasher units. By removing one of the flashers, the affected circuit no longer flashes. Therefore, that flash unit controls that particular circuit. If the turn signals fail to operate and the fuse is good, the flasher has probably failed.

Occasionally, the flasher does not flash as fast as it once did, or it flashes faster. This is also cause for replacement. If it flashes too slowly or not at all, check for a burned-out bulb first.

A flasher features two or three prongs that plug into a socket. Just pull the flasher out of the socket and replace it with a new one.

Flashers are designed to operate a specific number of bulbs to give a specific **candlepower** (brightness). If the candlepower on the turn signal bulbs is changed, or if additional bulbs are used (if a vehicle is hooked up to a trailer, for instance), a heavy-duty flasher must be used. This usually fits the socket without modifications. Although heavy-duty flashers will operate additional bulbs, they have one big disadvantage and should not be used unless it is necessary. These flashers will not cause the turn signals to flash slower if a bulb burns out. When a turn signal bulb fails, the driver has no idea that it did.

WARNING!

The flasher unit for turn signals should not be switched with a flasher unit for the hazard lights.

Some newer vehicles have a combination flasher unit that controls the flash rate of both the turn signals and the hazard lights. These combination flashers are electronic units **(Figure 20–40)**. The actual turning off and on of the lights is caused by the cycling of a transistor. This type of flasher also senses when a bulb is burned out and causes the remaining bulbs on that side to flash faster. Because this flasher is an electronic device, it cannot be tested with normal test equipment. The only test of the flasher is to substitute it with a known good one. If the lights flash normally, the original flasher unit was bad and needs to be replaced.

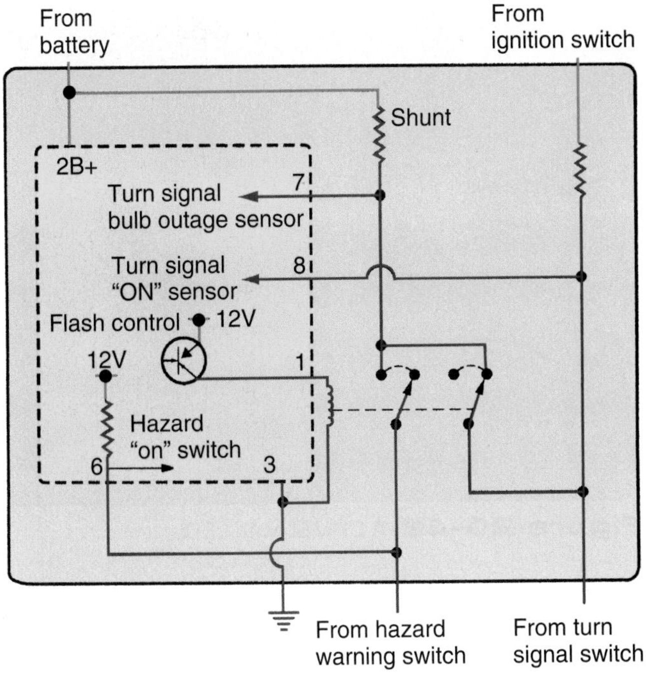

Figure 20–40 A typical electronic combination flasher unit. *Courtesy of Chrysler LLC*

Brake Lights

The brake (stop) lights are usually controlled by a stoplight switch that is normally mounted on the brake pedal arm **(Figure 20–41)**. Some cars are equipped with a brake or stoplight switch mounted on the master cylinder, which closes when hydraulic pressure increases as the brake pedal is depressed. In either case, voltage is present at the stoplight switch at all times. Depressing the brake pedal causes

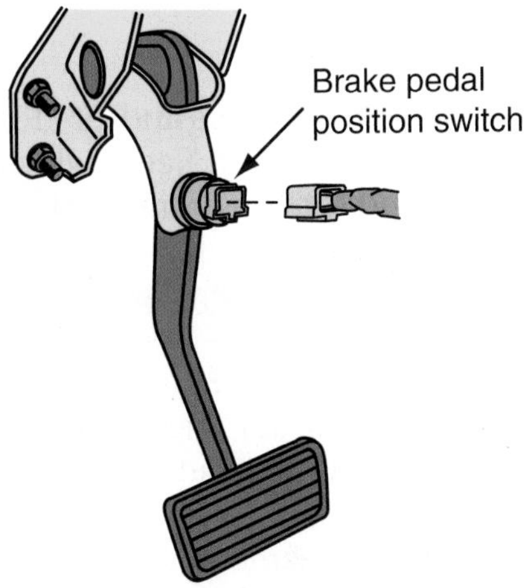

Figure 20–41 A stoplight switch located at the brake pedal assembly.

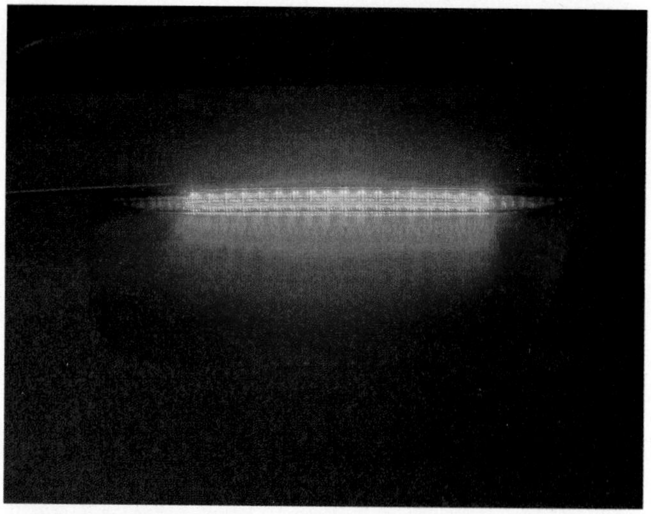

Figure 20-42 A CHMSL with LEDs.

the stoplight switch contacts to close. Current can then flow to the stoplight filament of the rear light assembly. These stay illuminated until the brake pedal is released.

In addition to the stoplights at the rear of the vehicle, all late-model vehicles have a center high-mounted stoplight (CHMSL) that provides an additional clear warning signal that the vehicle is braking (**Figure 20–42**). Federal studies have shown the additional stoplights to be effective in reducing the number and severity of rear collisions. The high-mounted stoplight is activated when current is applied to it from the stoplight switch. It stays illuminated until the brake pedal is released. When its contacts are closed, the stoplight switch can also provide current to the speed control amplifier, antilock brake control module, and the electric brake controller connector.

LED Lights

Some vehicles use neon lamps and/or LEDs for tail, brake, and turn signal lights. Neon lights are more energy efficient and turn on more quickly than regular lights. Because neon bulbs have no filament, the neon bulb will last longer than a conventional light bulb.

Whereas conventional bulbs take around 200 milliseconds to reach their full brightness, neon bulbs turn on within 3 milliseconds. The importance of this time difference is that it gives the driver behind the vehicle an earlier warning to stop. This early warning can give the approaching driver 19 more feet for stopping when driving at 60 miles per hour.

LEDs offer the same advantages as neon bulbs and turn on even quicker because they do not need to heat up to illuminate. LEDs achieve their full output in less than 1 millisecond. Several LEDs are placed behind the lens and are activated at the same time to give a bright illumination of the light assembly. LEDs also require a much smaller space so they are much less intrusive in the trunk. LEDs have a long operating life and provide a more precise contrast and signal pattern, thus attracting attention much more effectively.

Using the same basic technology as LEDs, laser-lit taillights consume seven times less power than incandescent sources. These savings are extremely important for electric vehicles. The light waves of a laser light beam move in the same direction and the light is all the same color. When used with rear exterior lights, fiber optics carry red light from a diode laser to a series of mirrors, which send the beam across a thin sheet of acrylic material.

Adaptive Brake Lights This system can select one of two available brake light areas for illumination: Moderate braking activates the standard brake lights incorporated within the taillight assemblies as well as the center high-mount brake light. Under intense braking and during all braking maneuvers with active ABS intervention, additional lamps are lit, thereby changing the size of the brake lights and their intensity (**Figure 20–43**). By increasing the brake lights' illuminated surface area, the system alerts drivers of following vehicles that the vehicle in front has started braking and decelerating at a rapid rate. This warning allows the driver of the following vehicle to react more quickly and reduces the danger of a rear impact.

An electronic control unit processes signals supplied by the speed sensor and the antilock brake system. It then uses these data to calculate the intensity of the braking as reflected by the vehicle's rate of deceleration.

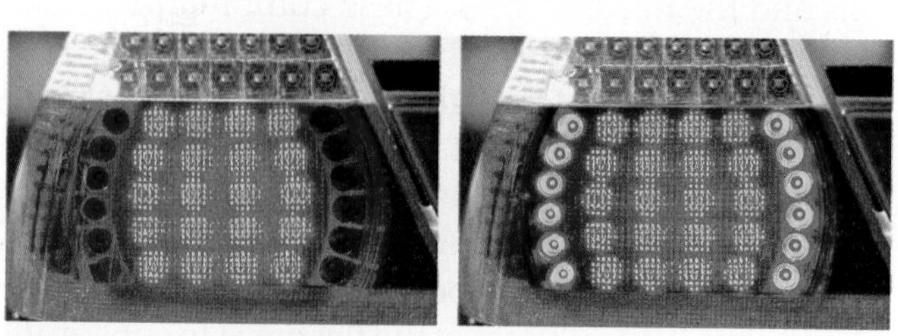

Figure 20-43 (Left) The normal illumination of the brake lights. (Right) The illumination during hard stops. *Courtesy of BMW of North America, LLC*

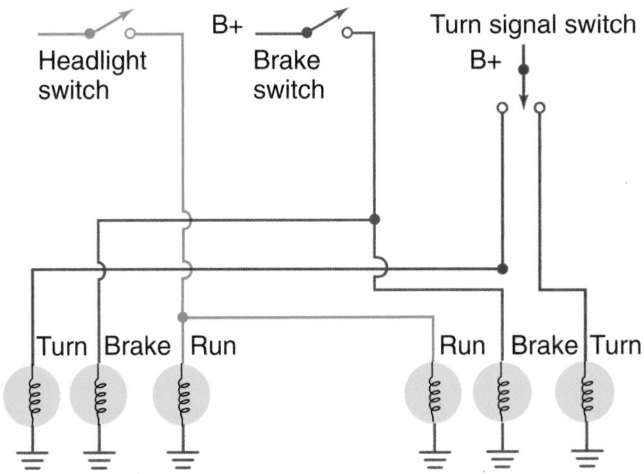

Figure 20-44 A typical three-bulb taillight circuit.

USING SERVICE INFORMATION

In addition to the taillight system, the rear of vehicles have many other lighting circuits. Most cars have brake lights, run lights, turn signals, and back-up lights. Let us look at these circuits **(Figure 20–44)** and see that their diagnosis is very simple once you are aware of how they appear in the service manual. Start with the brake lights. The easiest circuit to look at first is the three-bulb circuit found on many vehicles. The drawing shows a typical taillight circuit, which contains three separate filaments for each side of the rear of the vehicle. There is a separate filament for each function: brake, turn, and run. A constant source of fused B+ is made available to the brake switch. The brake switch is usually located on the brake pedal and is closed by pushing down on the brake pedal. B+ is now available to the bulbs, wired in parallel, at the rear of the vehicle. Releasing the brake pedal allows the spring-loaded normally open (NO) switch to open and turn the brake lights off. This is a simple circuit that requires only a 12-volt testlight or a voltmeter for diagnosis. The most common cause of failure is bulbs that burn out. Testing for B+ and ground at the bulb socket should verify the circuit. If B+ is not available at the socket, test for power at each connector, moving back toward the switch until it is found. Repair the open. Do not forget that the circuit is only hot if the brake pedal is depressed.

Back-Up Lights When the transmission is placed in reverse gear, back-up lights are turned on to illuminate the area behind the vehicle and to let other drivers know that the vehicle is in reverse. The major components in the system are the back-up light switch and the lights.

Power for the back-up light system is provided by the fuse panel. When the transmission is shifted to reverse, the back-up light switch closes and power flows to the back-up lights. That is, any time the transmission is in reverse, current flows from the fuse panel through the back-up light switch to the back-up lights. On many vehicles, the fuse that protects the back-up light system also protects the turn signal system.

In general, vehicles with a manual transmission have a separate switch. Those with an automatic transmission use a combination neutral start/back-up light switch. The combination neutral start/back-up light switch used with automatic transmissions is actually two switches combined in one housing. In park or neutral, current from the ignition switch is applied through the neutral start switch to the starting system. In reverse, current from the fuse panel is applied through the back-up light switch to the back-up lights.

The back-up light system is relatively easy to troubleshoot. On vehicles that use one fuse to protect both the turn signals and the back-up lights, the fuse can be checked. If the back-up lights are not working, check turn signal operation. If they work, the fuse is good. Check for power at the back-up light switch input and outlet with the transmission in reverse. (Make sure the parking brake is set.)

If the switch is okay, or there is no power to the switch, check the wiring—especially the connectors. If the back-up lights stay on when the transmission is not in reverse, suspect a short in the back-up light switch.

BASIC LIGHTING SYSTEM DIAGNOSIS

Light problems can be solved by simply replacing bulbs. A bulb can be quickly checked with a powered testlight or an ohmmeter. Connect the testlight across two of the bulb's terminals **(Figure 20–45)**. If there is continuity, the bulb is good.

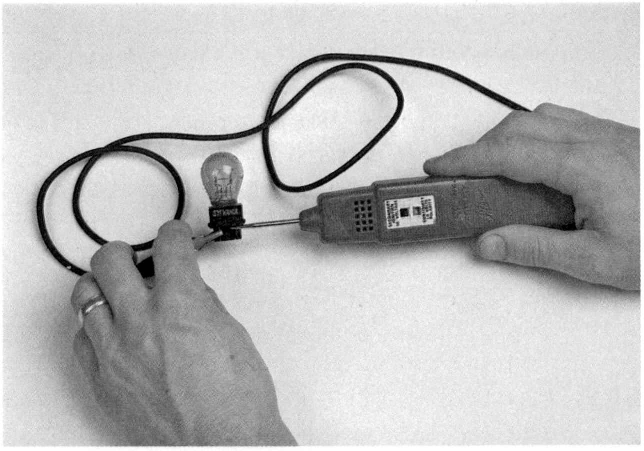

Figure 20-45 A bulb can be quickly checked with a continuity tester.

When a burnt-out bulb is not the problem, corroded or loose wiring is the next most common problem. All wiring connections should be clean and tight; light assemblies need to be securely mounted to provide a good ground. **Table 20–1** is a basic troubleshooting guide for lighting problems.

TABLE 20-1 GENERAL LIGHTING TROUBLESHOOTING GUIDE

Concern	Most Probable Cause
One light does not work	Bad bulb Open in that circuit
One light is dim	High resistance in the power or ground circuit
All lights in a circuit do not work	Open in the power or ground circuit Blown fuse Bad (control) switch
All lights in a circuit are dim	High resistance in the power or ground circuit
Flickering lights	Loose electrical connection A circuit breaker that is kicking out because of a short
One low-beam headlamp does not work	Bad bulb High resistance in the power or ground circuit
Both low beams do not work	Open in the power or ground circuit Bad (control) switch Blown fuse Bad bulbs due to overcharging
One high-beam headlamp does not work	Bad bulb High resistance in the power or ground circuit
Both high beams do not work	Open in the power or ground circuit Bad (control) switch Blown fuse Bad bulbs due to overcharging
Slow turn signal operation	Bad flasher unit High resistance in the power or ground circuit Incorrect bulb
Turn signal on one side does not work	Bad bulb High resistance in the power or ground circuit

Bulb Replacement

When replacing a bulb, make sure it is the correct replacement bulb. Using the wrong bulb can cause problems, such as an incorrect blink rate. Inspect the bulb socket. Often moisture gets into a bulb socket and causes corrosion at the electrical contacts in the socket. At times, the corrosion can be removed with sandpaper. Other times, the socket or light assembly should be replaced.

Light bulbs are contained in an assembly and access to the bulb is gained by reaching around the rear of the assembly and removing the socket. Other assemblies require the removal of the light's outside lens. With the lens removed, the bulb is removed from the front. While doing this, inspect the lens and gasket for damage and replace any damaged parts. If the lens is not sealed, dirt and moisture can enter the assembly and cause problems in the future.

On some light fixtures, the light assembly must be removed and the bulb and socket are removed from the rear of the assembly. Do not remove the lens from the assembly. It is a sealed unit and dust and other contaminants can cause serious damage to the reflector. Also, never attempt to clean the reflector. Wiping its surface can seriously reduce the light's brightness.

Bulbs are held in their sockets in a number of ways. Some bulbs are simply pushed into and pulled out of their sockets **(Figure 20–46)**, and some are screwed in and out. Some bulbs must be depressed and turned counterclockwise to remove them.

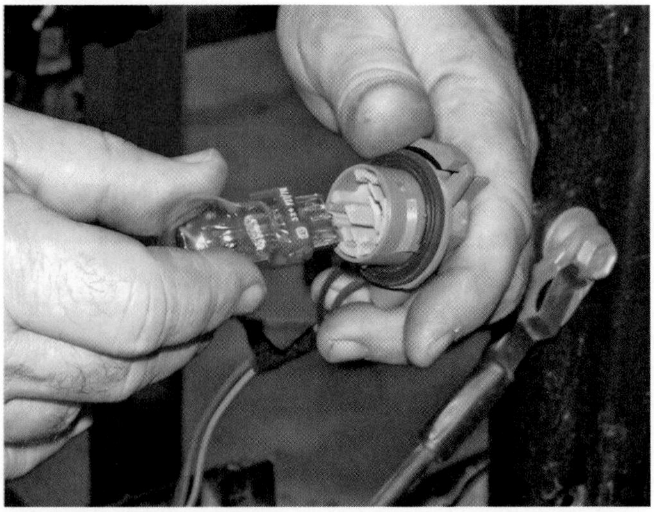

Figure 20-46 Many light bulbs are pulled out or pushed into their sockets.

CASE STUDY

A customer complains that the turn signals on his vehicle do not operate normally on the right side but work correctly on the left side.

The technician notices that the right front bulb lights but does not blink and the right rear does not work at all when the right turn signals are activated. He tests for power to the right rear with a testlight and the testlight blinks. He performs a voltage drop test and finds that the right rear bulb has an open circuit in its ground circuit. He repairs the ground and this system now works fine.

Turn signal circuits are frequent sources of difficulties. Their diagnosis, however, is not difficult and can usually be done with a testlight or DMM. Look at the common circuits, starting first with the flasher unit. The flasher is actually a type of circuit breaker, which is an overload protection device designed to open the circuit when excessive heat is developed from high current flow. Flashers are normally mounted in the fuse box and are made up of a fixed contact and a movable bimetallic contact. In the preceding case, not enough current was flowing through the right-side bulbs to operate the flasher when the right turn signals were activated. This was caused by the open ground circuit. The decreased current never heated the flasher unit enough to cause the front turn signal to blink. Of course, the rear bulb would not light because there was no ground.

KEY TERMS

Candlepower
Composite headlight
Daytime running lights (DRL)
Dimmer switch
Flasher
Halogen
High-intensity discharge (HID) headlamp
Lamp
Light-emitting diode (LED)
Multifunction switch
Photocell sensor/amplifier
Sealed-beam
Xenon headlamps

SUMMARY

- Headlight systems consist of two or four sealed-beam, halogen, or xenon bulbs.
- The headlight switch controls the headlights and all other exterior lights, with the exception of the turn signals, hazard warning, and stoplights.
- Dimmer switches allow the driver to select high or low headlight beams.
- Many vehicles have daytime running lights, which are typically normal headlamps powered by low voltage.
- Automatic headlight systems switch the headlights from high beam to low beam in response to current lighting conditions.
- Headlights must be kept in adjustment to obtain maximum illumination for safe driving.
- There are a number of interior lights found on today's vehicles. These vary with the make or model of the vehicle.
- The rear light assembly includes the taillights, turn signal/stop/hazard lights, high-mounted stoplight, rear side marker lights, back-up lights, and license plate lights.
- Flashers are used in turn signal, hazard warning, and side marker light circuits.
- The back-up light system illuminates the area behind the vehicle when it is put in reverse gear.
- Replacement light bulbs should be an exact match of the originals.

REVIEW QUESTIONS

1. What lighting systems are controlled by the headlight switch?
2. What is a CHMSL?
3. *True or False?* HID lamps produce more heat and white light than a halogen bulb.
4. *True or False?* The headlights on many adaptive headlight systems can swivel up to 5 degrees in the direction the vehicle is turning.
5. *True or False?* Many turn signal flasher units contain a temperature-sensitive bimetallic strip and a heating element.
6. List five things that should be done prior to aligning the headlights.
7. What are the most probable causes for one low-beam headlamp not working?

8. How many wires are normally connected to a double-filament light bulb?
9. Circuits that can energize both the high and low beams even if the headlight switch is off are known as ______ circuits.
   a. flash-to-pass
   b. mercury
   c. dimmer
   d. retractable
10. Why do some manufacturers protect the headlamp circuit with a circuit breaker instead of a fuse?
11. What kind of headlight does not have a filament?
12. What is the primary tool used to diagnose an auto-leveling headlamp system?
13. Which of the following would probably *not* cause the headlights on both sides of the vehicle not to work?
    a. an open in the power or ground circuit
    b. a bad (control) switch
    c. a blown fuse
    d. low generator output
14. The stoplight switch is normally mounted on the ______.
    a. instrument panel
    b. transmission
    c. brake pedal arm
    d. none of the above
15. Which of the following is *not* a true statement about LED-based lights?
    a. LEDs achieve their full output in about 200 milliseconds.
    b. LEDs require a much smaller space so they are much less intrusive in the trunk.
    c. LEDs have a long operating life.
    d. LEDs provide a more precise contrast and signal pattern, thus attracting attention much more effectively.

ASE-STYLE REVIEW QUESTIONS

1. While troubleshooting a headlight problem: Technician A says that when one headlamp does not work, the problem is probably a burned-out bulb or lamp. Technician B says that you should check for voltage at the bulb before replacing a bulb. Who is correct?
   a. Technician A
   b. Technician B
   c. Both A and B
   d. Neither A nor B
2. Technician A says that the condition of the vehicle's springs and shocks should be checked before aligning a headlight. Technician B says that the vehicle should have a full tank of fuel when its headlights are being adjusted. Who is correct?
   a. Technician A
   b. Technician B
   c. Both A and B
   d. Neither A nor B
3. Composite headlights are being discussed: Technician A says that they have replaceable bulbs. Technician B says that a cracked or broken lens will prevent the operation of a composite headlamp. Who is correct?
   a. Technician A
   b. Technician B
   c. Both A and B
   d. Neither A nor B
4. While troubleshooting a brake light problem: Technician A says that when the pedal is depressed the brake lights should come on. Technician B says that on some systems only part of the brake light should illuminate when there is slight pressure on the brake pedal. Who is correct?
   a. Technician A
   b. Technician B
   c. Both A and B
   d. Neither A nor B
5. Technician A says that a sealed-beam headlight that is cracked should be replaced. Technician B says that cracking or breaking of the housing or lens does not prevent a halogen bulb from working. Who is correct?
   a. Technician A
   b. Technician B
   c. Both A and B
   d. Neither A nor B
6. Technician A says that if the turn signals blink faster than normal, the problem may be a burned-out bulb on that side of the vehicle. Technician B says that if the turn signals blink slower, the problem may be a burned-out light bulb on that side of the vehicle. Who is correct?
   a. Technician A
   b. Technician B
   c. Both A and B
   d. Neither A nor B
7. Technician A says that flickering vehicle lights can be caused by a circuit breaker that is kicking out due to a short. Technician B says that flickering vehicle lights can be caused by a loose electrical connection. Who is correct?
   a. Technician A
   b. Technician B
   c. Both A and B
   d. Neither A nor B
8. The turn signals operate in the left direction only: Technician A says that the flasher unit is bad.

Technician B says that the fuse is probably blown. Who is correct?

a. Technician A
b. Technician B
c. Both A and B
d. Neither A nor B

9. The right back-up lamp is dim: Technician A says that the back-up lamp switch may have high resistance. Technician B says that the back-up lamp fuse may have corroded terminals. Who is correct?

a. Technician A
b. Technician B
c. Both A and B
d. Neither A nor B

10. When the headlights are switched from low beam to high beam, all headlights go out: Technician A says that the problem may be a bad dimmer switch. Technician B says that the ignition switch may be bad. Who is correct?

a. Technician A
b. Technician B
c. Both A and B
d. Neither A nor B

CHAPTER 21

ELECTRICAL INSTRUMENTATION

OBJECTIVES

■ **Describe the two types of instrument panel displays.** ■ **Know the purpose of the various gauges used in today's vehicles and how they function.** ■ **Describe the operation of the common types of gauges found in an instrument cluster.** ■ **List and explain the function of the various indicators found on today's vehicles.** ■ **List and explain the function of the various warning devices found on today's vehicles.** ■ **Explain the basics for diagnosing a gauge or warning circuit.**

Every vehicle has a number of electrical instruments. The number and type, and their related parts, vary from model to model and year to year. The appearance of the instrument panel also varies from the quite simple **(Figure 21–1)** to the elaborate **(Figure 21–2)**. No matter what they look like, the instrumentation must be easy to read and give accurate information. Instrument gauges, warning lights, and indicators provide valuable information to the driver concerning the vehicle's various systems.

Figure 21–1 A simple but functional approach to the layout of the essential gauges in an instrument panel. *Courtesy of Toyota Motor Sales, U.S.A., Inc.*

Figure 21–2 An elaborate arrangement of the essential gauges in an instrument panel. *Courtesy of Toyota Motor Sales, U.S.A., Inc.*

INSTRUMENT PANELS

Today's dashboard is more properly called the **instrument panel**. It contains an array of electrical gauges, switches, and controls connected to mazes of wiring, printed circuitry, and vacuum hoses beneath stylishly finished sheets of plastic or metal.

Displays

There are many different instrument panel designs and layouts. The two basic types of instrument panel displays are analog and digital. In a traditional analog display **(Figure 21–3)**, an indicator moves in front of a fixed scale to indicate a condition. The indicator is often a needle but it can also be a liquid crystal or graphic display. A digital display uses numbers instead of a needle or graphic symbol. Analog displays show relative change

Figure 21-3 An analog instrument panel.

Figure 21-4 A digital instrument panel. *Courtesy of Siemens VDO Automotive AG*

better than digital displays (**Figure 21–4**). They are useful when the driver must see something quickly and the exact reading is not important. For example, an analog tachometer shows the rise and fall of the engine speed better than a digital display. The driver does not need to know the exact engine speed. The most important thing is how fast the engine is reaching the red line on the gauge. A digital display is better for showing exact data such as miles. Many speedometer-odometer combinations are examples of both analog (speed) and digital (distance).

Three types of digital electronic displays are used today.

Light-Emitting Diode (LED) These displays are either used as single indicator lights or they can be grouped to show a set of letters or numbers. LED displays are commonly red, yellow, or green and can be hard to see in bright light.

Liquid Crystal Diode (LCD) These displays are made of sandwiches of special glass and liquid. A separate light source is required to make the display work. When there is no voltage, light cannot pass through the fluid. When voltage is applied, the light passes that point of the display. The action of LCDs slows down in cold weather. These displays are also very delicate and must be handled with care. Any rough handling of the display can damage it.

Vacuum Fluorescent These displays use glass tubes filled with argon or neon gas. The segments of the display are little fluorescent lights. When current is passed through the tubes, they glow very brightly. These displays are both durable and bright.

SHOP TALK

Most warning or instrument controls use **International Standards Organization (ISO)** symbols, which were developed by this organization to provide symbols easily recognizable throughout the world.

GAUGES

Gauges provide the driver with a scaled indication of the condition of a system. Two components are needed for the operation of an electrical gauge. These are instrument voltage regulators and sending units. **Instrument voltage regulators (IVRs)** are used to stabilize and limit voltage for accurate instrument operation (**Figure 21–5**). Vehicles equipped with an IVR also have a radio choke. A radio choke is a small coil of wire installed in the power lead to the IVR. The choke prevents radio interference caused by the pulsations of the IVR.

All gauges require an input from either a sending unit or a sensor. With sending or sensor units, a change or movement made by an external component causes a change in electrical resistance. Movement may be caused by pressure against a diaphragm, by heat, or by the motion of a float. In today's vehicles, the control computers and the gauges need the same information. The information passes through the computer and then to the gauge.

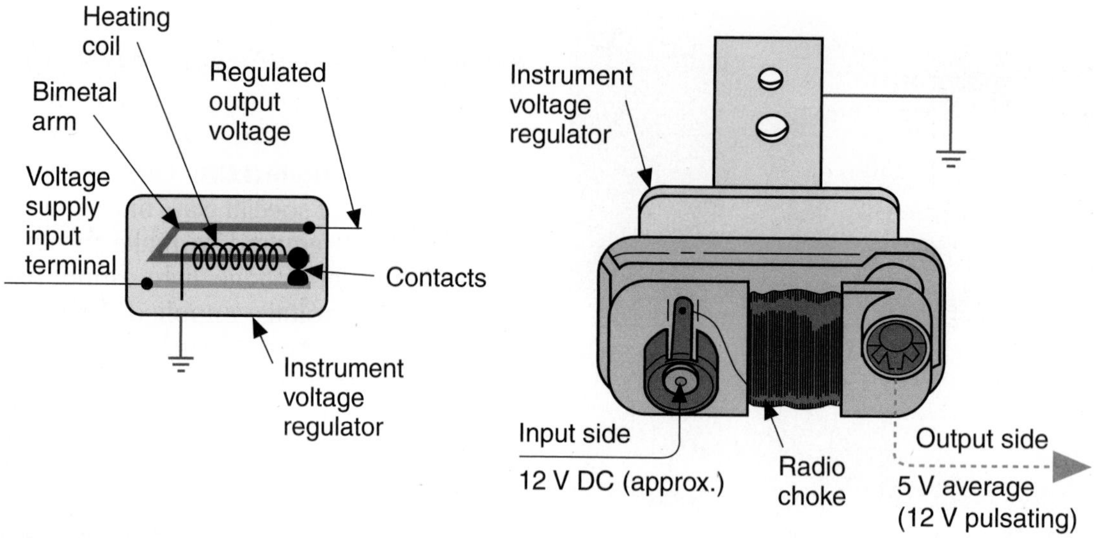

Figure 21-5 An IVR.

Magnetic Gauges

There are several types of magnetic gauges. The simplest form is the ammeter type **(Figure 21–6)**, in which a permanent magnet attracts a ferrous indicator needle connected to a pivot point and holds it centered on the gauge. An armature, or coil of wire, is wrapped around the base of the needle near the pivot point. When current flows through the armature, a magnetic field is formed. This magnetic field opposes that of the permanent magnet. Attractive or repulsive magnetic forces cause the needle to swing left or right. The direction the needle swings depends on the direction of current flow in the armature. The needle of the gauge pivots on the armature, often called a bobbin, which is why this type of gauge is often referred to as a **bobbin gauge**.

When sending unit resistance changes, current flow through the bobbin changes, causing the strength of the magnetic field created around the bobbin to change. As the resistance in the sending unit increases, the circuit's current decreases and a lower indicator position on the gauge results.

A **balancing coil** gauge also operates on principles of magnetic attraction and repulsion. However, a permanent magnet is not used. The base of the indicating arm is pivoted and includes an armature. Two coils are used to create magnetic fields **(Figure 21–7)**.

The two coils are connected so that electricity can flow through either one. When the resistance of the sending unit is low, the right-hand coil receives more current than the left-hand coil, attracting the armature. Thus, the gauge needle moves to the right.

When the resistance of the sending unit is high, the left-hand coil receives more current. More magnetic force is created in the left-hand coil and the needle swings to the left.

Thermal or Bimetallic Gauge

A bimetallic or **thermal gauge** operates through heat created by current flow **(Figure 21–8)**. A variable-resistance sending unit causes different

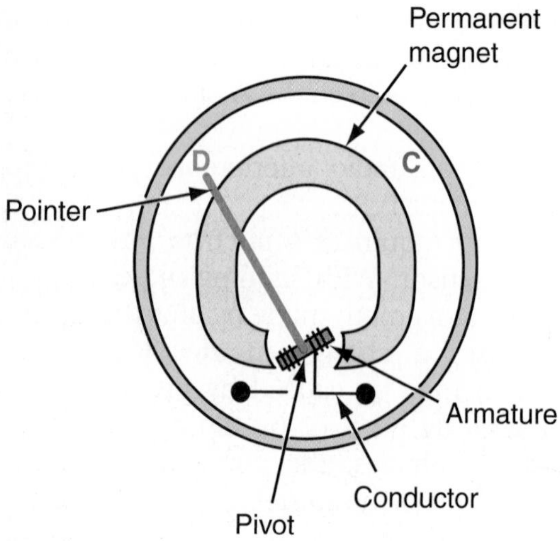

Figure 21-6 A simple ammeter that relies on magnetic principles.

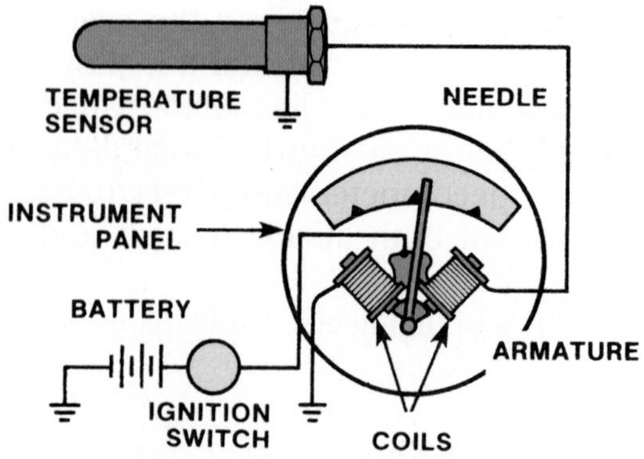

Figure 21-7 A balancing coil gauge. *Courtesy of Ford Motor Company*

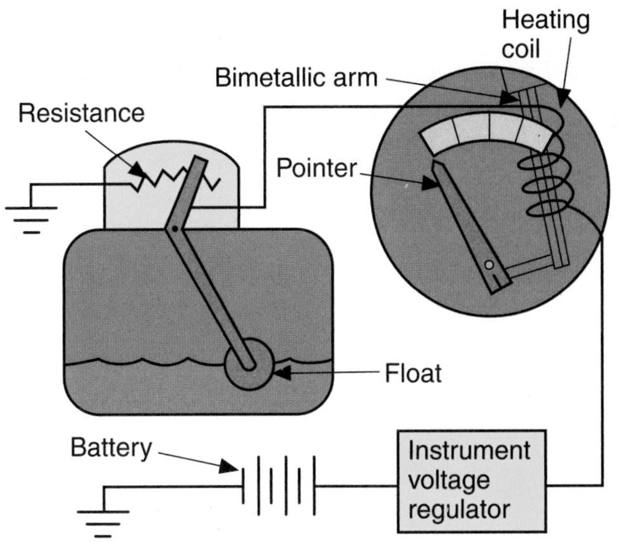

Figure 21-8 Typical bimetallic fuel gauge circuit.
Courtesy of Chrysler LLC

amounts of current to flow through a heating coil within a gauge. The heat acts on a bimetallic spring attached to a gauge needle. When more heat is created, the needle swings farther up the scale. When less heat is created, the needle moves down the scale.

Diagnosis

If all gauges fail to operate properly, begin by checking the circuit's fuse. Next, test for voltage at the last point common to all the malfunctioning gauges. If voltage is not present, work toward the battery to find the fault.

If the system uses an IVR, use a voltmeter to test for regulated voltage at a point common to all the gauges. If the voltage is out of specifications, check the ground circuit of the IVR. If that is good, replace the IVR. If there is no voltage present at the common point, check for voltage on the battery side of the IVR. If voltage is present at this point, replace the IVR. If regulated voltage is within specifications, test the circuit from the IVR to the gauges.

WARNING!

Many instruments and warning devices are linked to the vehicle's body control module and multiplexing network. Before troubleshooting a gauge or warning system, check the service manual to identify any special procedures or precautions.

BASIC INFORMATION GAUGES

The following gauges are found on nearly all instrument panels. The detailed operation of some gauges is described in other chapters of this book.

Speedometer

In the past, the speedometer was considered a nonelectrical or mechanical gauge. It had a drive cable attached to a gear in the transmission that turned a magnet inside a cup-shaped metal piece. The cup was attached to a needle, which was held at zero by a hairspring (a fine wire spring). As the cable rotated faster with an increase in speed, magnetic forces acted on the cup and forced it to move. As a result, the needle moved up the speed scale.

Electric speedometers are used in all late-model vehicles. While there are several systems in use, one of the most common types receives its speed information from the transmission-mounted vehicle speed sensor (**Figure 21–9**). This speed signal is also used by other modules in the vehicle, including the speed control module, ride control module, the engine control module, and others.

For each 40,000 pulses from the vehicle speed sensor (VSS), the trip and total odometers will increase by 1 mile. Speed is determined by dividing the input pulse frequency (in hertz) by 2.2 hertz/mph. The circuit electronics are calibrated to drive the pointer to a location in proportion to the speed input frequency. That is, as the pulse rate increases, the speedometer records it on an analog display. This display may be limited to a maximum of 85 mph or 199 km/h.

Most digital speedometers have a speed limit. If vehicle speed exceeds these values, the speedometer continues to display the top of its range. Vehicle speed is displayed whether the vehicle is moving forward or backward.

Odometer

A mechanical odometer is driven by a spiral gear cut on the speedometer's magnetic shaft. The odometer's numbered drums are geared so that when any one drum finishes a complete revolution, the drum to the left is turned one-tenth of a revolution. On some systems, a stepper motor is used to drive the odometer. The motor receives signals from the PCM and causes the odometer to move in steps.

An electronic odometer receives its information from the VSS. Odometers display seven digits, with the last digit in tenths of a unit. The accumulated mileage value of the digital display odometer is stored in a nonvolatile read-only memory (ROM) that retains the mileage even if the battery is disconnected.

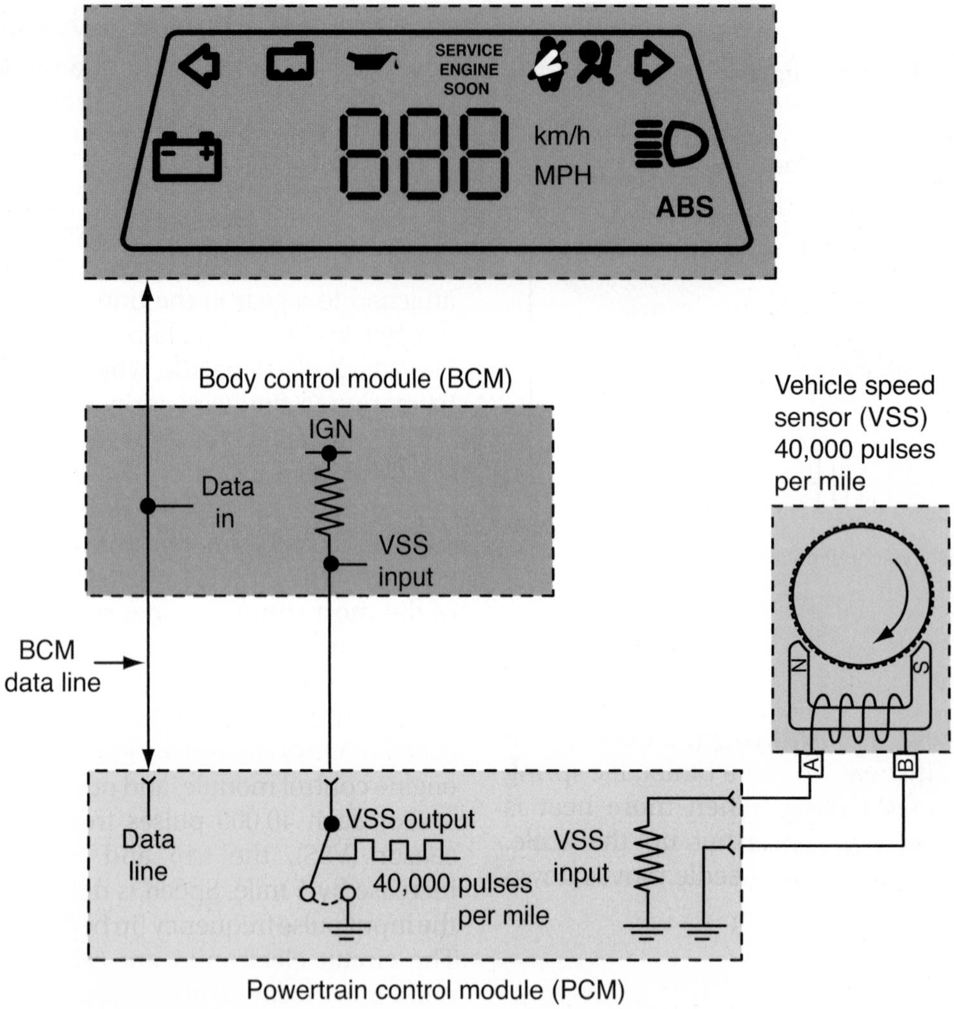

Figure 21-9 The signals from the VSS are shared by the instrument panel and the PCM.

Because the odometer records the number of miles or kilometers a vehicle has traveled, federal laws require that if a speedometer is replaced, the new odometer must be set to the same distance as was registered on the old one.

Trip Odometer This is part of the normal odometer. It records the distance traveled in intervals. The driver activates this gauge and it can be reset to zero whenever the driver desires.

Diagnosis If the speedometer or odometer is not working, the operation of the VSS should be checked before anything else. If the sensor is working fine, check the wiring from the sensor to the gauge. If all checks out, replace the speedometer/odometer assembly.

Oil Pressure Gauge

The oil pressure gauge indicates engine oil pressure. The oil pressure typically should be between 45 and 70 psi when the engine is running at a specified engine speed and at operating temperature. A lower pressure is normal at low idle speed.

With low oil pressure (or with the engine shut off), the oil pressure switch is open and no current flows through the gauge winding. The needle points to L. With oil pressure above a specific limit, the switch closes and current flows through the gauge winding to ground. A resistor limits current flow through the winding and ensures that the needle points to about mid-scale with normal oil pressure.

A piezoresistive sensor **(Figure 21–10)** is threaded into the oil delivery passage of the engine. The pressure of the oil causes a flexible diaphragm to move. This movement is transferred to a contact arm that slides down the resistor. The position of the sliding contact arm determines the resistance value and the amount of current flow through the gauge.

Diagnosis To test a piezoresistive-type sending unit, connect an ohmmeter to the sending unit's terminal and to ground. Check the resistance with the engine off and compare it to specifications. Start the engine

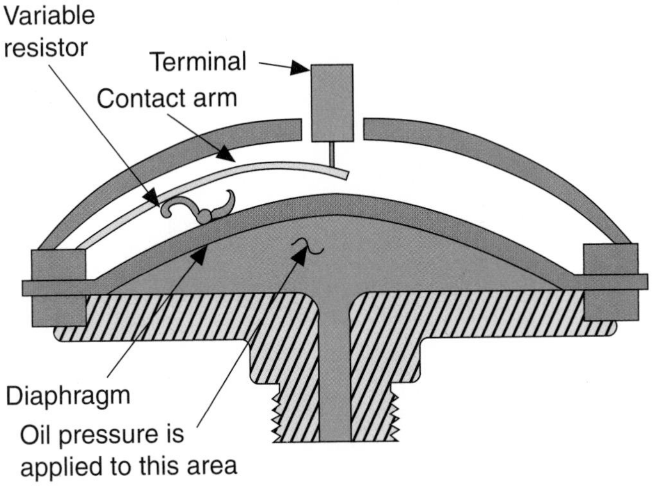

Figure 21-10 A piezoresistive sensor used for measuring engine oil pressure.

and allow it to idle. Check the resistance value and compare it to specifications. Before replacing the sending unit, connect a shop oil pressure gauge to confirm that the engine is producing adequate oil pressure.

This gauge indicates engine coolant temperature (ECT). It should normally indicate between C (cold) and H (hot). The sending unit is typically a negative temperature coefficient (NTC) thermistor. With low coolant temperature, sender resistance is high and the current flow to the temperature gauge winding is low **(Figure 21–11)**. The needle points to C. As coolant temperature increases, sender resistance decreases and current flow increases. The needle moves toward H.

The temperature gauge on a digital panel is of the bar type with a set number of segments. The number of illuminated bars varies according to the current flow from the sending unit. With low coolant temperature, few segments are lit. As coolant temperature increases, the number of illuminated segments increases.

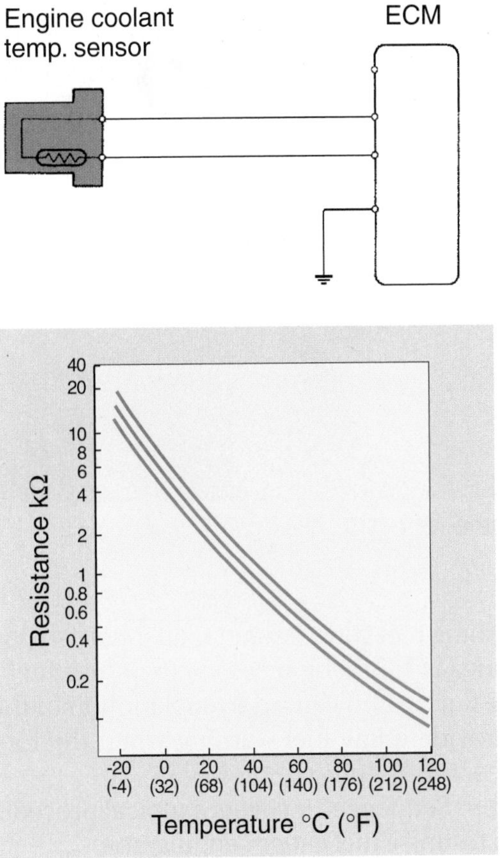

Figure 21-11 Most engine coolant temperature sensors use an NTC thermistor. The resistance of an NTC decreases with an increase in temperature.

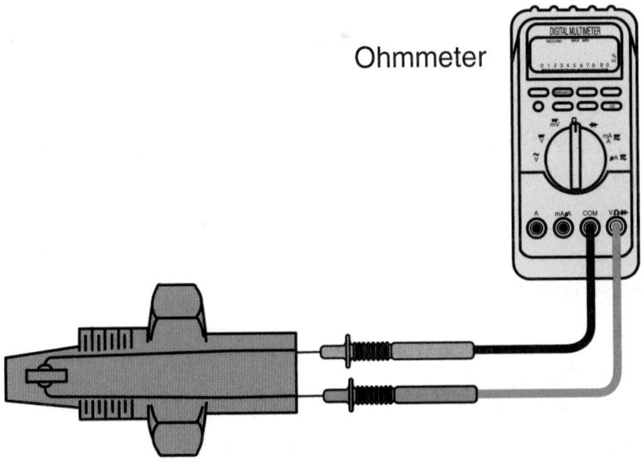

Figure 21-12 Testing a temperature sensor with an ohmmeter.

Diagnosis To test a coolant temperature sending unit, use an ohmmeter to measure resistance across the terminals **(Figure 21–12)**. The resistance value of the NTC thermistor decreases with an increase in temperature. Compare your measurements with the manufacturer's specifications. On most new vehicles, the action of an ECT can be monitored on a scan tool.

Fuel Level Gauge

The fuel level gauge indicates the fuel level in the fuel tank. It is a magnetic indicating system that can be found on either an analog (meter) or digital (bars) instrument panel.

The fuel sending unit is combined with the fuel pump assembly and consists of a variable resistor controlled by the level of an attached float in the fuel tank **(Figure 21–13)**. When the fuel level is low, resistance in the sender is low and movement of the gauge needle or number of lit bars is minimal (from empty position). When the fuel level is high, the resistance in the sender is high and movement of the gauge indicator (from the empty position) or number of lit bars on a digital display is greater.

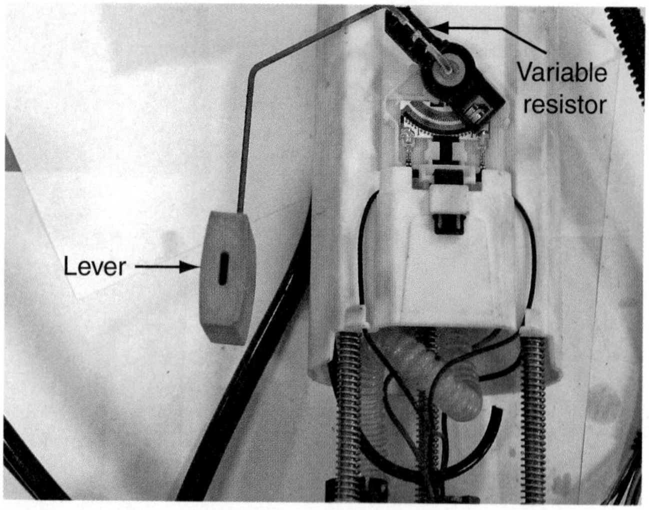

Figure 21-13 A fuel gauge sending unit.

In some fuel gauge systems, an antislosh/**low fuel warning (LFW)** module is used to reduce fuel gauge needle fluctuation caused by fuel motion in the tank and provide a low fuel warning when the fuel tank reaches ⅛ to 1/16 full.

Photo Sequence 19 covers a typical procedure for bench testing a fuel gauge sending unit.

SHOP TALK

Some technicians use a universal gauge tester (Figure 21–14) to diagnose fuel and other gauge problems. With this tester, it is possible to manually control the inputs to a gauge and observe its reaction. The tester can also check an IVR.

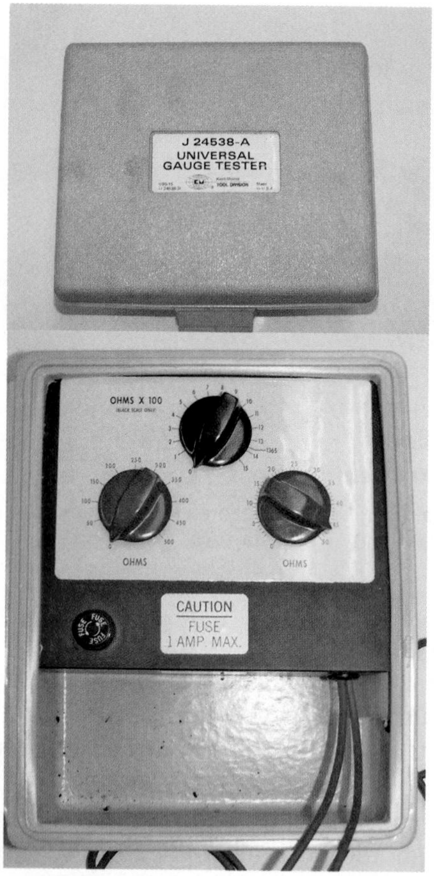

Figure 21-14 A universal gauge tester.

Go To *Chapter 19 for more information on charging indicators.*

Tachometer

The **tachometer** indicates engine rpm (engine speed). The electrical pulses to the tachometer typically come from the ignition module or PCM. The tachometer, using a balanced coil gauge, converts these impulses to rpm that can be read. The faster the engine rotates, the greater the number of impulses from the coil. Consequently, higher engine speeds are indicated.

In vehicles with digital instrumentation, the bar system is used with numbered segments. The numbers represent the engine's rpm times 1,000.

Charging Gauges

These allow the driver to monitor the charging system. Whereas a few older cars use an ammeter, most charging systems use a voltmeter or an indicator light. The indicator lamp will light when the voltage from the charging system is less than voltage at the battery.

INDICATOR AND WARNING DEVICES

Vehicles are equipped with several warning devices to alert the driver to problems. Most of these are safety or emissions related.

Indicator and Warning Lights

Indicator lights inform the driver that something that is normally off is now on, or vice versa. Consider the rear window defogger. When it is turned on, the indicator is lit. Another example is traction control. This system is normally on and when it is switched off, its indicator is lit. Warning lights notify the driver that something in the system is not functioning properly or that a situation exists that must be corrected. Some warning lights come on briefly when the engine is first

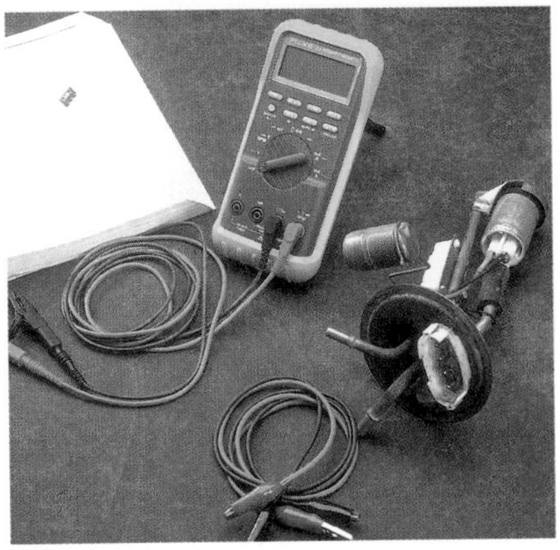

P19-1 *The tools required to perform this task are a DMM, jumper wires, and a service manual.*

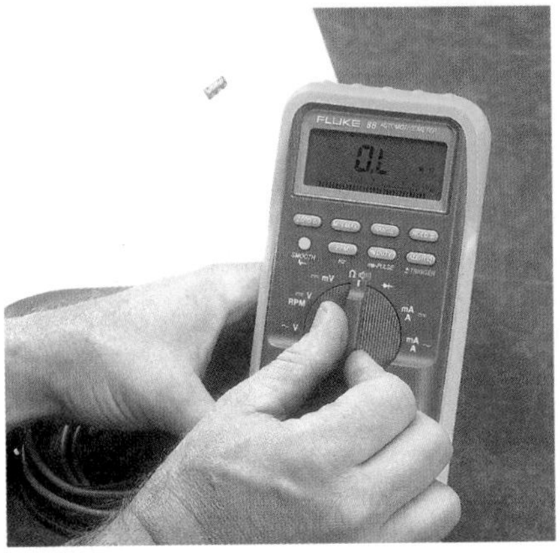

P19-2 *Select the ohmmeter function of the DMM.*

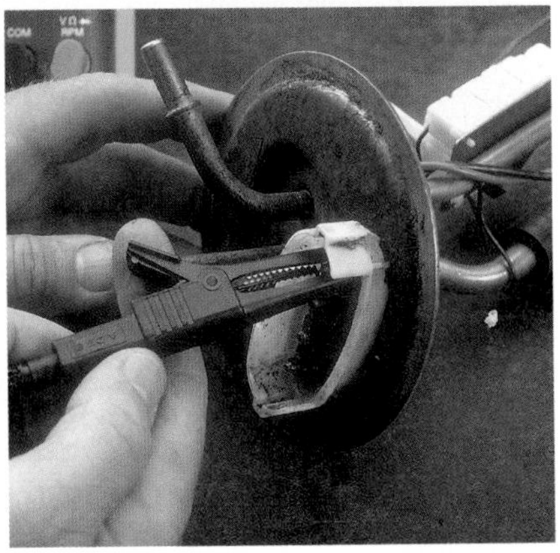

P19-3 *Connect the DMM's negative test lead to the ground terminal of the sending unit.*

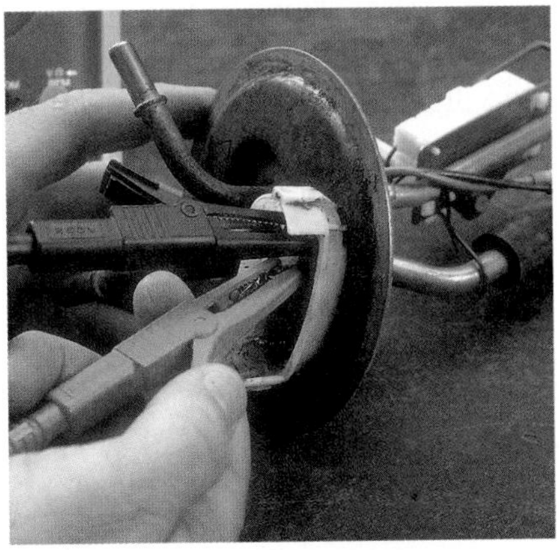

P19-4 *Connect the meter's positive lead to the variable resistor terminal.*

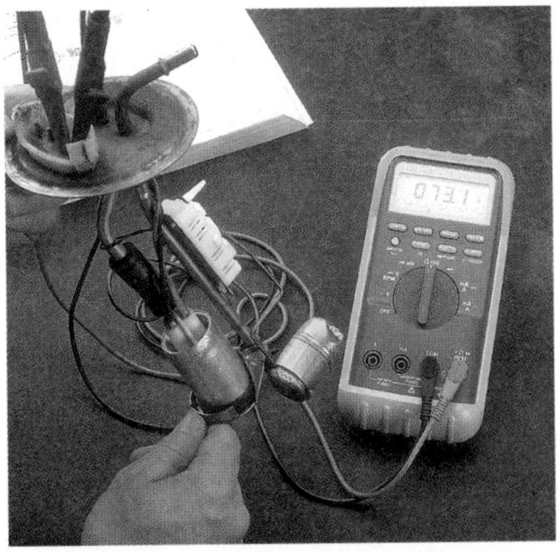

P19-5 *Holding the sender unit in its normal position, place the float rod against the empty stop.*

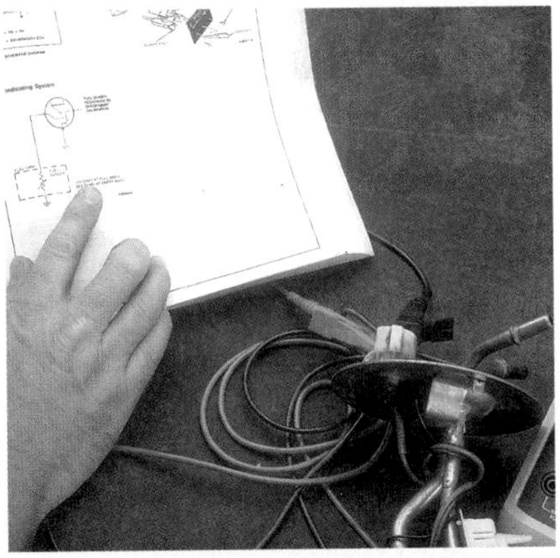

P19-6 *Read the ohmmeter and check the results with the specifications.*

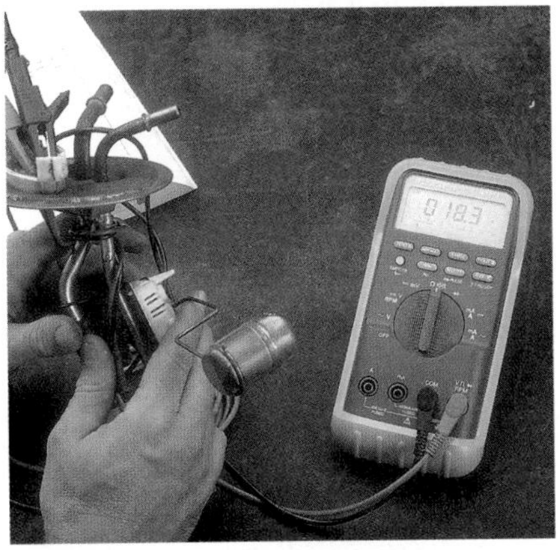

P19-7 *Slowly move the float toward the full stop while observing the ohmmeter. The resistance change should be smooth and consistent.*

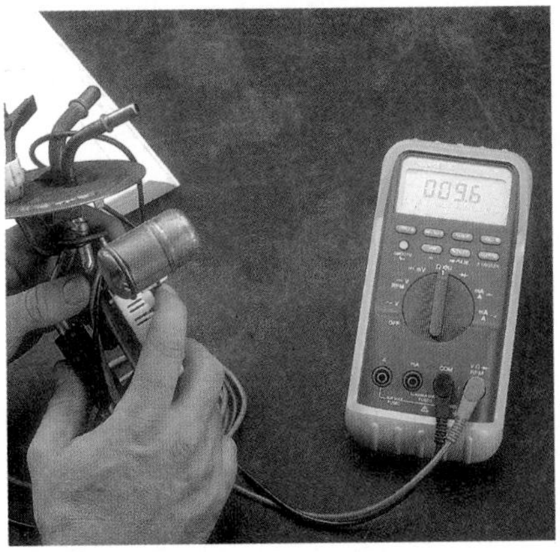

P19-8 *Check the resistance value while holding the float against the full stop. Check the results with the specifications.*

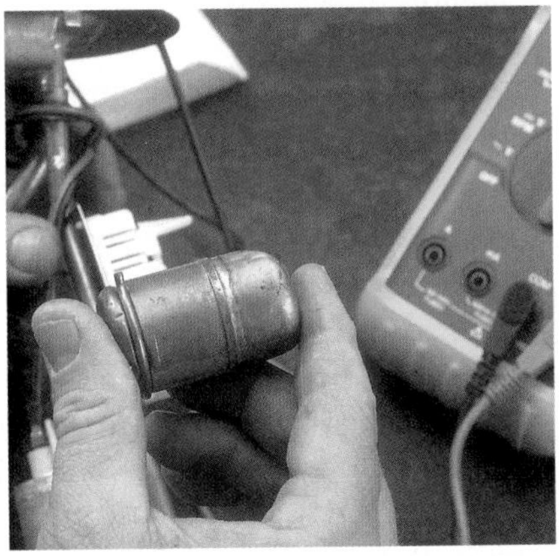

P19-9 *Check the float to be sure it is not filled with fuel, distorted, or loose.*

Figure 21–15 Most warning lights and indicators go through a bulb check when the ignition is first switched to the *on* position.

Figure 21–16 A TPM warning light signaling low tire pressure.

started; this is a bulb check **(Figure 21–15)**. Gauges and warning lights often work together to alert the driver to a problem.

Some of the more common indicator and warning lights are described here.

> **WARNING!**
>
> *Be aware that many of the warning and indicator lamps found on today's vehicles are triggered by a PCM or BCM. Often they are part of the multiplexed system. With this in mind, always refer to the testing methods recommended by the manufacturer before testing these systems. Using conventional testing methods on a computerized system may destroy part or all of the system.*

Air Bag Readiness Light The air bag readiness light lets the driver know the air bag system is working and ready to do its job. It lights briefly when the ignition is turned on. A malfunction in the air bag system will cause the light to stay on continuously or to flash.

Fasten Belts Indicator This lamp, typically accompanied by a chime, will illuminate when the driver has not fastened the seat belt. The chime and lamp will stay on until the belt is buckled. When the ignition switch is in the *on* position, the system is activated for about 5 seconds, whether or not the driver's belt is buckled.

Tire Pressure Monitor (TPM) When a low-inflated or flat tire is found, this warning light will be turned on. Some systems will illuminate this warning lamp in red or yellow. Red means there is an excessively low or flat tire and yellow means a tire has low pressure **(Figure 21–16)**. Some systems also emit a sound to alert the driver.

> **PROCEDURE**
>
> When the TPM light is on, follow these steps to identify why:
>
> 1. Check the tires for the proper inflation pressures.
> 2. Check the vehicle for nonstock accessories, such as tires, wheels, or devices that transmit radio waves, which could affect the operation of the TPM system.
> 3. Inspect the tires for physical damage or missing parts, such as the valve stem caps.
> 4. Check the tires for the presence of tire sealant. Tire sealants can block the sensor's pressure-sensing port.
> 5. With a scan tool, check the pressures that are being inputted to the receiver and compare those with the actual pressures.
> 6. Check the communication between the sensors and the receiver with a TPM diagnostic tool.

Brake Warning Light When this light is lit, it is an indication that the parking brake is engaged or there is a problem with the brake system. Vehicles built for Canada may have a separate warning lamp that indicates when the parking brake is applied.

Brake Fluid Level Warning Light The lamp is connected to a brake fluid level sensor in the brake fluid reservoir. If the brake fluid drops below a specified level, the sensor is activated and the light turns on while the engine is running.

Antilock Light If an antilock brake system fault is present, the antilock brake module grounds the indicator circuit, and the antilock light goes on. Vehicles built for Canada may have a different symbol on their warning lamp.

Brake Pad Indicator This lamp illuminates when the sensors at the wheel brake units sense thin brake pads. The sensors are typically embedded in the pads and when they contact the metal rotor, the warning light circuit is complete and the lamp is lit.

Stoplight Warning Light The light is controlled by a stoplight checker. The checker has a reed switch and magnetic coils. Magnetic fields form around the coils by the current flowing through each light when the stoplight switch is on. The magnetic fields cancel each other because the coils are wound in opposite directions. As a result, the reed switch and the warning light are off. If a stoplight fails, current only flows through one coil, and the resultant magnetic field causes the reed switch to close and the warning light to illuminate when the brake pedal is depressed.

Traction/Stability Control Lamp This lamp is illuminated in red when there is a problem with the traction and/or stability control systems. The red lamps will also be lit when the system is turned off. Yellow lamps are lit when the system is actively regulating drive torque and braking force **(Figure 21–17)**.

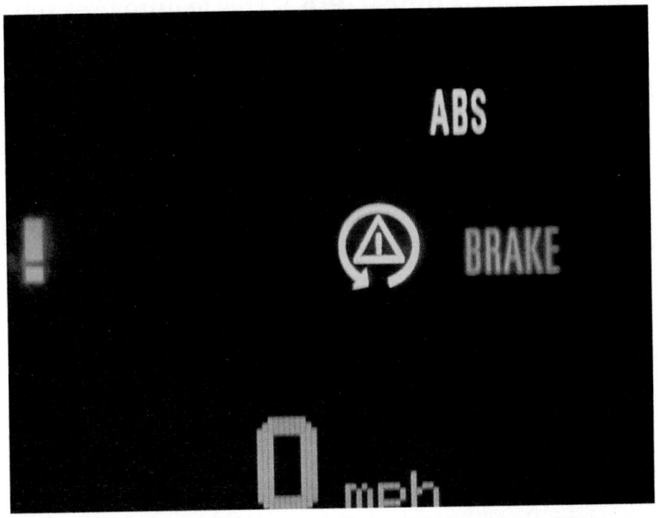

Figure 21-17 The traction control, brake system, and ABS indicators and warning lamps.

Check Engine Light This may be also labeled as "Service Engine Soon." This warning is primarily an emissions-related light. If something happens that will cause the vehicle to have higher emissions, the PCM will display this warning. This warning may also be triggered by oil pressure or coolant temperature. This warning may illuminate when the computer has stored a fault or diagnostic code in its memory.

Oil Pressure Light This lamp is operated by an oil pressure switch located in the engine's lubricating system. Some vehicles will illuminate this lamp in yellow or red to indicate the action the driver should take, red meaning the engine has an oil pressure problem and the engine should be shut down and yellow indicating the oil level is low and should be topped off as soon as possible.

Charge Indicator Light If there is something wrong with the charging system, this light comes on while the engine is running.

Check Filler Cap This lamp will be illuminated when the gas filler cap is not tight or off and when the engine control system senses a problem with the fuel system.

Low Fuel Warning This lamp will illuminate when the fuel level is below a quarter full. An electronic switch closes and power is applied to illuminate the lamp.

Add Coolant Lamp The purpose of this lamp is to inform the driver of low coolant levels in the cooling system.

Maintenance Reminder This lamp is triggered by mileage. It turns on to alert the driver that maintenance needs to be performed on the vehicle. This is not a warning; it is a reminder. Often the lamp will light for 2 seconds when the car is started, if the vehicle has been serviced within a specified interval. Normally, the required maintenance is an engine oil change. The name of this indicator includes "oil" on many vehicles. When the lamp is not reset during the specified distance, it will flash after the engine has started. If there was no service for about twice the recommended distance, the lamp will stay on continuously until it is reset.

The procedure for resetting this reminder varies by manufacturer. The following procedure should serve only as an example. To reset the reminder on a vehicle, refer to the service information for the vehicle being serviced.

PROCEDURE

To reset the maintenance reminder on a late-model Honda:

1. Turn the ignition switch on.
2. Press the ODO/TRIP switch until the trip meter A appears.
3. Turn the ignition switch off.
4. Then turn the ignition switch back on while pressing the trip odometer reset knob.
5. Hold the reset knob down for about 30 seconds. The reminder lamp should go out during that time.

Transmission Indicator This is part of an automatic transmission's control system. If the system detects a fault, it may operate the transmission in the fail-safe mode and will illuminate the warning light to inform the driver of the problem and to alert him or her that the transmission may not be working normally.

Drive Indicator Some four-wheel-drive vehicles have a lamp which, when lit, indicates that the vehicle is in four-wheel-drive mode of operation.

O/D Off Indicator This lamp is illuminated when the overdrive function of an automatic transmission has been switched off by the driver.

Rear Defrost Indicator When this light is lit, the defroster or deicer is operating.

High-Beam Indicator With the headlights turned on and the main light switch dimmer switch in the high-beam position, the indicator illuminates.

Left and Right Turn Indicators With the multifunction switch in the left or right turn position, voltage is applied to the circuit to illuminate the left or right turn indicator. The turn indicator flashes in unison with the exterior turn signal bulbs.

Fog Light Indicator This lamp is illuminated when the fog lights are turned on.

Lamp-Out Warning The lamp-out warning module is an electronic unit that is designed to measure small changes in voltage levels. An electronic switch in the module closes to complete a ground path for the indicator lights in the event of a bulb going out. The key to this system being able to detect one bulb out on a multibulb system is the use of the special resistance wires. With bulbs operating, the resistance wires provide 0.5 volt input to the light-out warning module. If one bulb in a particular system goes out, the input off the resistance wire drops to approximately 0.25 volt. The light-out warning module detects this difference and completes a ground path to the indicator light for the affected circuit.

Cruise Control Light This lamp is lit whenever the cruise control is turned on.

Air Suspension Light Voltage is present at the air suspension indicator at all times. If an air suspension fault exists, the ground of the light circuit is closed and the indicator illuminates.

Door Ajar Warning When the ignition is turned on and if the doors are left open or are ajar, this light comes on.

Add Washer Fluid Lamp Obviously, the purpose of this lamp is to inform the driver of a low level in the windshield washer fluid reservoir.

Sound Warning Devices

Various types of tone generators, including buzzers, chimes, and voice synthesizers, are used to remind drivers of a number of vehicle conditions. These warnings can include fasten seat belts, air bag operational, key left in ignition, door ajar, and light left on. **Figure 21–18** is a tone generator system schematic.

Park Distance Control (PDC) This feature uses sensors to measure the distance the front and/or rear of the vehicle is from an object **(Figure 21–19)**. An audible signal changes in frequency as the vehicle gets closer to an object. As the distance between the vehicle and object decreases, the intervals between the tones become shorter. When the object is very close, the tone is emitted continuously. The system uses four ultrasonic sensors at the rear and the front of the vehicle. Some systems include a visual indication of the distances to the obstacles, in addition to the audible warning.

Some systems allow the front sensors to be manually turned off in special situations such as stop-and-go traffic. The rear sensors automatically turn on when the transmission is placed into reverse.

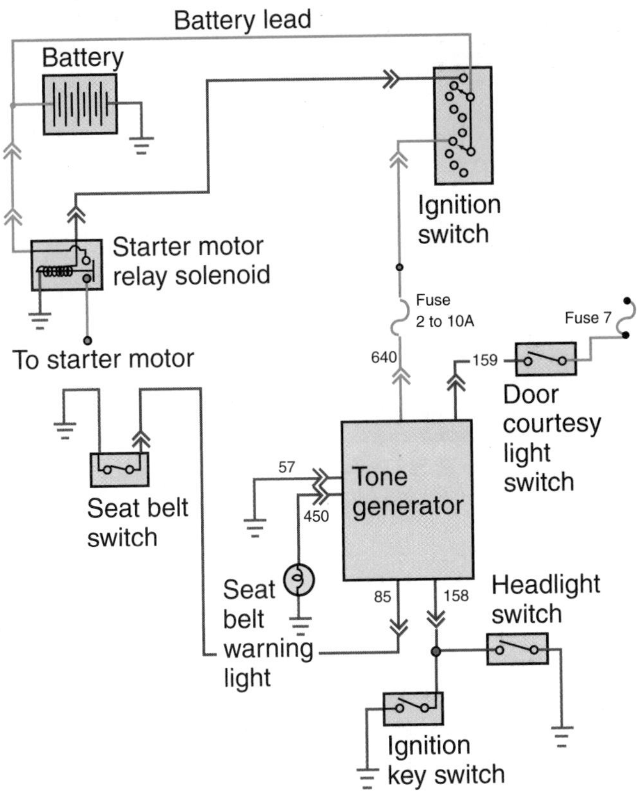

Figure 21-18 A tone generator warning system. *Courtesy of Ford Motor Company*

Figure 21-19 The sensors measuring the distance from the rear of the car to obstacles trigger the warning system (insert) inside the vehicle. *Courtesy of Robert Bosch GmbH, www.bosch-presse.de*

DRIVER INFORMATION CENTERS

The various gauges, warning devices, and comfort controls may be grouped together into a driver information center or instrument cluster. This may be simple **(Figure 21–20)** or an all-encompassing cluster of information. The message center keeps the driver alert of the information available. The types and extent of this information varies from one system to another.

Figure 21-20 Information for the driver appears across the bottom of the instrument cluster. *Courtesy of Toyota Motor Sales, U.S.A., Inc.*

In addition to standard warning signals, the information center may provide such vital data as fuel range, average or instantaneous fuel economy, fuel used since reset, time, date, estimated time of arrival (ETA), distance to destination, elapsed time since rest, average car speed, percent of oil life remaining, and various engine-operating parameters.

Other electronic displays and controls can be found on today's vehicles.

Heads-Up Display

A **heads-up display (HUD)** projects visual images onto the windshield using a vacuum fluorescent light source to complement in-dash instrumentation **(Figure 21–21)**. Because these images are projected in the driver's field of vision, the driver does not need to remove his or her eyes from the road to see certain

Figure 21-21 The heads-up display (HUD) can project the vehicle's speed onto the windshield, freeing the driver from looking down at the speedometer. *Courtesy of Siemens VDO Automotive AG*

pertinent information. Among the images HUD may display are vehicle speed, turn-signal indicators, low fuel warning, and a high-beam indicator. HUD systems use a prismatic mirror to project their images, a clean windshield, and dim ambient lighting to allow the driver to see the display.

Graphic Displays

Graphic displays are translucent drawings of the vehicle. They have small lamps located at various spots in the graphic. When a lamp is lit, the area indicated by the lamp has a problem. These indicators can note such things as the trunk is open or a light bulb is not working.

Hybrid Vehicles

Hybrid vehicles have some unique and interesting warnings, indicators, and displays. They all also have normal displays for the engine and other systems. The following is a brief look at the unique stuff they have.

Honda The instrument panel on most Honda hybrids displays the typical conditions for a gasoline engine, plus they display the operation of the hybrid (IMA) system and the car's fuel efficiency. Also, on cars with a manual transmission, the panel has upshift and downshift lights that are triggered by the PCM to inform the driver when it is most economical to shift gears.

The instrument panel has a meter that displays the status of the battery and IMA system **(Figure 21–22)**. A charge/assist indicator shows when the system's electric motor is assisting the engine. The amount of assist is indicated by bars. The number of bars illuminated indicates how much assist is being provided. This same display shows the amount of charge going to the batteries. When more bars are illuminated, the batteries are being recharged at a higher rate. Also on this side of the cluster is a state-of-charge indicator for the battery module. The entire cluster is designed to help the driver achieve maximum fuel economy.

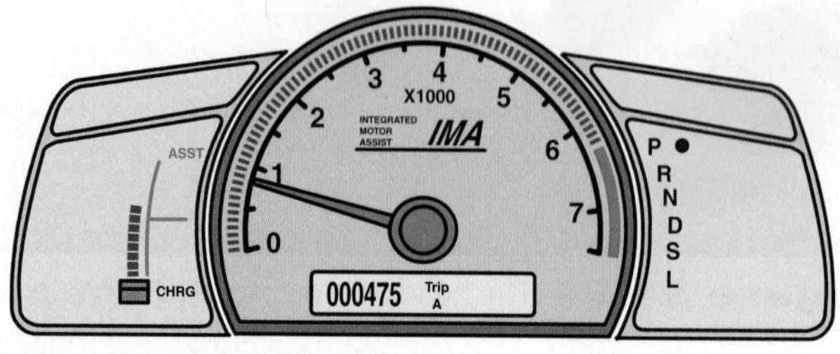

Figure 21–22 The instrument cluster in Honda hybrid vehicles has a meter that displays the status of the battery and IMA system.

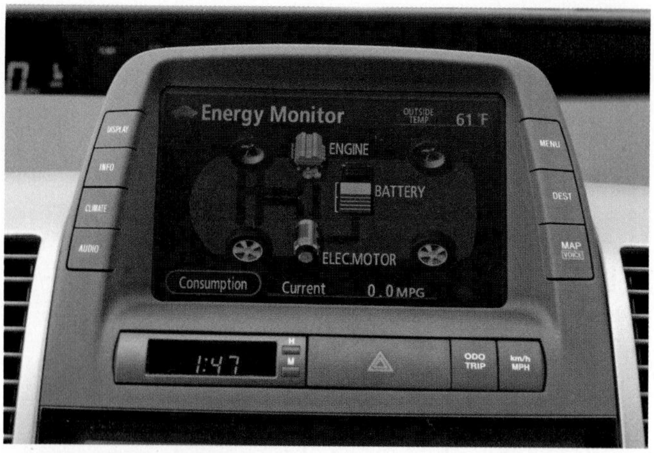

Figure 21–23 The energy monitor for a Toyota Prius. *Courtesy of Toyota Motor Sales, U.S.A., Inc.*

Toyota Most Toyota hybrids have a multi-information display in the center cluster panel. The display, a 7-inch LCD with a pressure sensitive touch panel, serves many functions. Many of these are typical, but some are unique to hybrid technologies. One is the fuel consumption screen. This display shows average fuel consumption, current fuel consumption, and the current amount of recovered energy.

Another unique display is the energy monitor screen **(Figure 21–23)**. This shows the direction and path of energy flow through the system in real time. By observing this display, drivers can alter their driving to achieve the most efficient operation for the current conditions.

Like other vehicles, these vehicles are equipped with a variety of warning lamps and indicators. Here are some of the indicators that are unique:

- "Ready" light—This lamp turns on when the ignition switch is turned to START to indicate that the car is ready to drive.
- Output control warning light—This lamp turns on when the temperature of the HV battery is too high or low. When this lamp is lit, the system's power output is limited.
- HV battery warning light—This lamp lights when the charge of the HV battery is too low.
- Hybrid system warning light—When the HV control unit detects a problem with the motor/generators, the inverter assembly, the battery pack, or the ECU itself, this lamp will be lit.
- Malfunction indicator light—This lamp is tied into the engine control system and will be lit when the PCM detects a fault within that control system.

- Discharge warning light—This lamp is tied to the 12-volt system and DC-DC converter. It will illuminate when there is problem in that circuit.

Hybrid four-wheel-drive SUVs show more information and include lamps for the four-wheel-drive option. The latter is monitored by a four-wheel-drive warning lamp that notifies the driver of any detected fault within the MGR (Motor/Generator—Rear) and rear transaxle. When this lamp is lit, a warning buzzer is also activated.

SHOP TALK

Starting or driving a hybrid is different. When the car is ready to be driven, the engine may not be running, but the "ready" lamp will be lit, which means the motor is ready to move the vehicle. If the "ready" lamp does not come on, the vehicle cannot be driven.

GENERAL DIAGNOSIS AND TESTING

Diagnosis should begin with a good visual inspection of the circuit. Check all sensors and actuators for physical damage. Check all connections to sensors, actuators, control modules, and ground points. Check the wiring for signs of burned or chafed spots, pinched wires, or contact with sharp edges or hot exhaust parts. After completing the visual checks, use a scan tool to retrieve any DTCs and monitor the data stream as necessary. Always refer to the manufacturer's recommended procedure before beginning to diagnose a circuit.

Items to be checked when a gauge or indicator system does not work right include:

- Fuses
- Indicator bulbs
- Detector switches (indicator systems)
- Sending units (gauge systems)
- IVR (gauge systems)
- Gauges (gauge systems)

Intermittent problems are usually caused by the following:

- Dirty or corroded terminals
- Wire chafing
- Poor wire-to-terminal connections
- Poor mating of connector halves or backed-out connector terminals
- Damaged connectors

CASE STUDY

A customer brings his pickup into the shop because the fuel gauge always shows empty, regardless of how much fuel is in the tank.

The technician verifies the problem and checks to make sure there is gasoline in the tank. Then she checks the fuse for the circuit. Finding all fuses in good shape, she refers to the service manual to determine if the gauge circuit has an IVR. This particular vehicle does not so she disconnects the wire to the fuel tank sending unit and checks it for voltage. Battery voltage is present when the ignition switch is turned on; therefore, she knows the power side of the circuit is good. She then connects a 10-ohm resistor to the wire and grounds the circuit. The fuel gauge now reads FULL. According to the service manual, this is what should happen. She now knows that the problem is either a faulty sending unit or a poor circuit ground. She reconnects the wire to the sending unit and then connects a jumper wire from the ground of the sending unit to a known good ground. The fuel gauge now shows a reading that seems to be accurate, so she cleans and corrects the ground circuit for the gauge. Then she verifies the repair by adding a few gallons of fuel and watching the response of the gauge.

KEY TERMS

Balancing coil
Bobbin gauge
Heads-up display (HUD)
Instrument panel
Instrument voltage regulator (IVR)
International Standards Organization (ISO)
Low fuel warning (LFW)
Tachometer
Thermal gauge

SUMMARY

- The two basic types of instrument panel displays are analog and digital. In an analog display, an indicator moves in front of a fixed scale to indicate a condition. A digital display uses numbers instead of a needle or graphic symbol.

- Three types of digital electronic displays are used today: light-emitting diode, liquid crystal diode, and vacuum fluorescent.
- A gauge circuit is often made of the gauge, a sending unit, and an instrument voltage regulator (IVR).
- Two types of electrical analog gauges—magnetic and thermal—are commonly used with sensors or sending units.
- Indicator lights and warning devices are generally activated by the closing of a switch.
- Various types of tone generators, including buzzers, chimes, and voice synthesizers, are used to inform drivers of vehicle conditions.
- Park distance control uses sensors to measure the distance the front and/or rear of the vehicle is from an object and emits an audible warning as the vehicle gets closer to an object.
- The various gauges, warning devices, and comfort controls may be grouped together into a driver information center or instrument cluster.
- A heads-up display projects visual images on the windshield by a vacuum fluorescent light source to complement existing traditional in-dash instrumentation.
- Diagnosis of gauges, indicators, and warning lights should begin with a good visual inspection of the circuit. Check all sensors and actuators; connections and wires to sensors, actuators, control modules, and ground points; and all vacuum hoses.
- Before troubleshooting a gauge or warning system, check the service manual to identify any special procedures or precautions.

REVIEW QUESTIONS

1. *True or False?* A heads-up display projects indicator and warning lights on the windshield rather than illuminate them in the instrument panel.
2. What is the purpose of an IVR?
3. Explain how a simple magnetic-type ammeter works.
4. What gauge indicates engine speed?
5. Describe the two types of instrument panel displays.
6. What is the device found in some fuel tanks to prevent fuel gauge fluctuations due to rough road surfaces?
7. What is the correct way to check a coolant temperature sensor?
8. What type of sending unit is typically used to monitor oil pressure?
9. What is the major difference between an indicator lamp and a warning light?
10. *True or False?* The traction/stability control lamp turns on whenever the system is actively regulating drive torque and braking force.
11. The indicator needle on a speedometer is held to the zero position by ______.
    a. magnetic force
    b. the weight of the needle
    c. the speedometer cable
    d. a hairspring
12. Which of the following is *not* a likely cause for an intermittent gauge problem?
    a. poor mating of connector halves or backed-out connector terminals
    b. an open ground wire
    c. connector body damage
    d. poor wire-to-terminal connections
13. What type of memory is used to store the accumulated mileage in an electronic odometer?
    a. RAM
    b. ROM
    c. PROM
    d. EPROM
14. Which of the following is *not* a true statement about lamp-out warning lights?
    a. The lamp-out warning module measures small changes in voltage levels.
    b. An electronic switch in the module completes a ground path for the indicator lights in the event of a bulb going out.
    c. A special resistance wire is used in multibulb systems.
    d. When a bulb burns out, the module senses the increased voltage drop and turns on the indicator lamp.
15. Which of the following statements about gauge sending units is *not* true?
    a. A switch can be used.
    b. A thermistor can be used.
    c. A piezoresistive sensor can be used.
    d. A variable resistor can be used.

ASE-STYLE REVIEW QUESTIONS

1. While discussing what the maintenance reminders base the time for the next service on: Technician A says that future service is based on the type of driving the vehicle has seen since the last service. Technician B says that the next interval is based on the number of miles or kilometers since the last service. Who is correct?
   a. Technician A c. Both A and B
   b. Technician B d. Neither A nor B
2. None of the engine's gauges works: Technician A checks the power to the IVR and the IVR's ground. Technician B begins by checking the fuse and then checks for voltage at the last point common to all the malfunctioning gauges. Who is correct?
   a. Technician A c. Both A and B
   b. Technician B d. Neither A nor B
3. The oil pressure light stays on whenever the engine is running, and the oil pressure has been checked and it meets specifications: Technician A says that a ground in the circuit between the indicator light and the pressure switch could be the cause. Technician B says that an open in the pressure switch could be the cause. Who is correct?
   a. Technician A c. Both A and B
   b. Technician B d. Neither A nor B
4. A digital speedometer constantly reads 0 mph: Technician A says that the problem may be the vehicle speed sensor. Technician B says that the problem may be the throttle position sensor. Who is correct?
   a. Technician A c. Both A and B
   b. Technician B d. Neither A nor B
5. All gauges read low: Technician A says that the connection to the instrument cluster may be corroded. Technician B says that the cluster's IVR may be open. Who is correct?
   a. Technician A c. Both A and B
   b. Technician B d. Neither A nor B
6. When testing an engine temperature sensor: Technician A uses a voltmeter to test the thermistor. Technician B says that the thermistor reacts to pressure and applies air pressure to the backside of the sensor. Who is correct?
   a. Technician A c. Both A and B
   b. Technician B d. Neither A nor B
7. The coolant temperature warning light stays on after the engine has been started: Technician A says that the wire to the sending unit may be shorted to ground. Technician B says that the sending unit may be electrically open. Who is correct?
   a. Technician A c. Both A and B
   b. Technician B d. Neither A nor B
8. While discussing the instrumentation on a Toyota hybrid: Technician A says that the "ready" light must be on before the vehicle is able to move. Technician B says that the engine must be running before the "ready" light is turned on. Who is correct?
   a. Technician A c. Both A and B
   b. Technician B d. Neither A nor B
9. The TPM warning lamp is lit: Technician A says that the cause could be low inflation. Technician B says that the customer may have used a tire sealant to seal a puncture. Who is correct?
   a. Technician A c. Both A and B
   b. Technician B d. Neither A nor B
10. The yellow traction control warning lamp comes on occasionally while the vehicle is driven: Technician A says that this means there is a problem with the system. Technician B says that the system is actively regulating drive torque and braking force when the lamp is lit. Who is correct?
    a. Technician A c. Both A and B
    b. Technician B d. Neither A nor B

CHAPTER 22

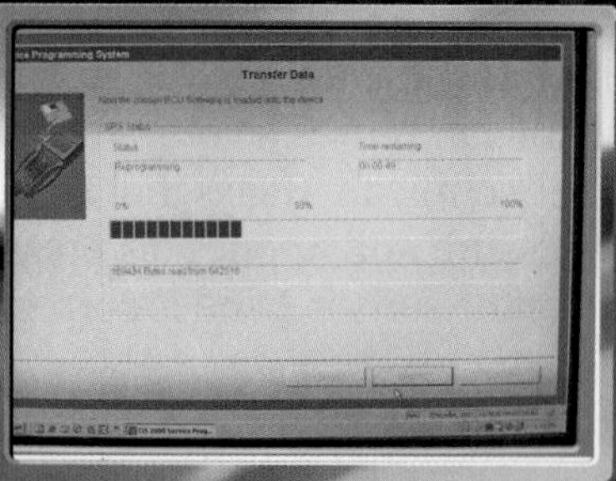

BASICS OF ELECTRONICS AND COMPUTER SYSTEMS

OBJECTIVES

■ Understand the purpose and operation of a capacitor. ■ Describe how semiconductors, diodes, and transistors work. ■ Explain the advantages of using electronic control systems. ■ Explain the basic function of the central processing unit (CPU). ■ List and describe the functions of the various sensors used by computers. ■ Explain the principle of computer communications. ■ Summarize the function of a binary code. ■ Name the various memory systems used in automotive computers. ■ List and describe the operation of output actuators. ■ Explain the principle of multiplexing. ■ Describe the precautions that must be taken when diagnosing electronic systems. ■ Perform a communications check on a multiplexed system. ■ Reprogram a control module in a vehicle.

Computerized engine controls and other features of today's vehicles would not be possible if it were not for electronics. For clarity, electronics is defined as the technology of controlling electricity. Capacitors, transistors, diodes, semiconductors, integrated circuits, and solid-state devices are all considered to be part of electronics rather than just electrical devices. But keep in mind that all the basic laws of electricity apply to electronic controls.

CAPACITORS

A **capacitor** is used to store and release electrical energy. Capacitors can be used to smooth out current fluctuations, store and release a high voltage, or block DC voltage. Capacitors are sometimes called condensers. Although a battery and a capacitor store electrical energy, the battery stores the energy chemically. A capacitor stores energy in an electrostatic field created between a pair of electrodes.

A capacitor can release all of its charged energy in an instant, whereas a battery slowly releases its charge. A capacitor is quick to discharge and quick to charge. A battery needs some time to discharge and charge but can provide continuous power. A capacitor only provides power in bursts.

Operation

A capacitor has a positive and a negative terminal **(Figure 22–1)**. Each terminal is connected to a thin electrode or plate (usually made of metal). The plates are parallel to each other and are separated by an insulating material called a dielectric. The dielectric can be paper, plastic, glass, or anything that does not conduct electricity. Placing a dielectric between the plates allows the plates to be placed close to each other without allowing them to touch.

When voltage is applied to a capacitor, the two electrodes receive equal but opposite charges **(Figure 22–2)**. The negative plate accepts electrons and stores them on its surface. The other plate loses electrons to the power source. This action charges the capacitor. Once the capacitor is charged, it has

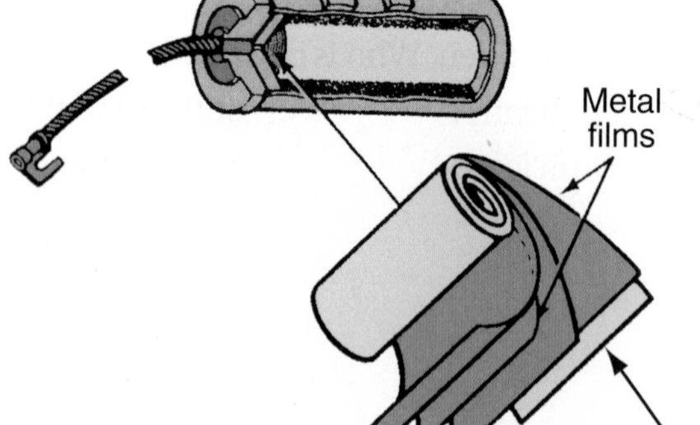

Figure 22-1 A construction of a condenser, which is a common name for a capacitor, that absorbs voltage spikes.

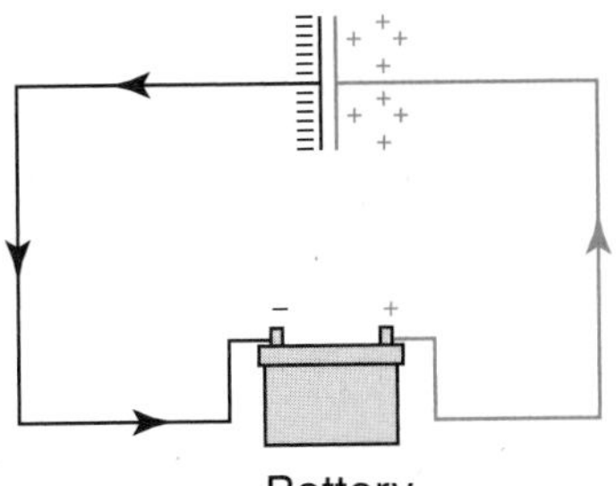

Figure 22-2 When voltage is applied to a capacitor, the two electrodes receive equal but opposite charges.

the same voltage as the power source. This energy is stored statically until the two terminals are connected together.

The ability of a capacitor to store an electric charge is called **capacitance**. The standard measure of capacitance is the **farad (F)**. A 1-farad capacitor can store one coulomb of charge at 1 volt. A coulomb is 6.25 *billion billion* electrons. One ampere equals the flow of 1 coulomb of electrons per second, so a 1-farad capacitor can hold 1 ampere-second of electrons at 1 volt. A capacitor's capacitance is directly proportional to the surface areas of the plates and the nonconductiveness of the dielectric and is inversely proportional to the distance between the plates. Most capacitors have a capacitance rating of much less than a farad and their values are expressed in one of these terms:

- Microfarads: μF ($1\ \mu F = 10^{-6}$ F)
- Nanofarads: nF ($1\ nF = 10^{-9}$ F)
- Picofarads: pF ($1\ pF = 10^{-12}$ F)

Capacitors oppose a change of voltage. If a battery is connected to a capacitor as shown in **Figure 22–2,** the capacitor will be charged when current flows from the battery to the plates. This current flow will continue until the plates have the same voltage as the battery. At this time, the current flow stops and the capacitor is charged.

The capacitor remains charged until a circuit is completed between the plates. If the charge is routed through a voltmeter, the capacitor will discharge with the same voltage as the battery that charged it. This statement explains why capacitors are commonly used to filter or clean up voltage signals, such as sound from a stereo. Current can only flow during the period that a capacitor is either charging or discharging.

Automotive capacitors are normally encased in metal. The grounded case provides a connection to one set of conductor plates and an insulated lead connects to the other set.

Variable capacitors are called **trimmers** or tuners and are rated very low in capacity because of the reduced size of their conducting plates. For this reason, they are only used in very sensitive circuits such as radios and other electronic applications.

Ultracapacitors

Ultra- or supercapacitors are used in hybrid vehicles and in some experimental fuel cell electric vehicles **(Figure 22–3)**. Ultracapacitors are capacitors with a large electrode surface area and a very small distance between the electrodes. These features give them very high capacitance; some are rated at 5,000 farads.

Ultracapacitors use an electrolyte rather than a dielectric. Although an ultracapacitor is an electrochemical device, no chemical reactions are involved in the storing of electrical energy.

Ultracapacitors are maintenance-free devices. They can withstand an infinite number of charge/discharge cycles without degrading and have a long service life. They also are very good at capturing the large amounts of energy from regenerative braking and can deliver power for traction motors quickly; this is needed for acceleration and overcoming heavy loads. Also, because they charge very quickly, they have energy available shortly after they have been discharged.

Ultracapacitors, however, cannot store as much total energy as batteries. To provide the required high voltages for electric vehicles, several capacitors must be connected in series. Each cell of an ultracapacitor can only store between 2 and 5 volts. Up to 500 cells are required to meet the needs of a typical electric drive vehicle.

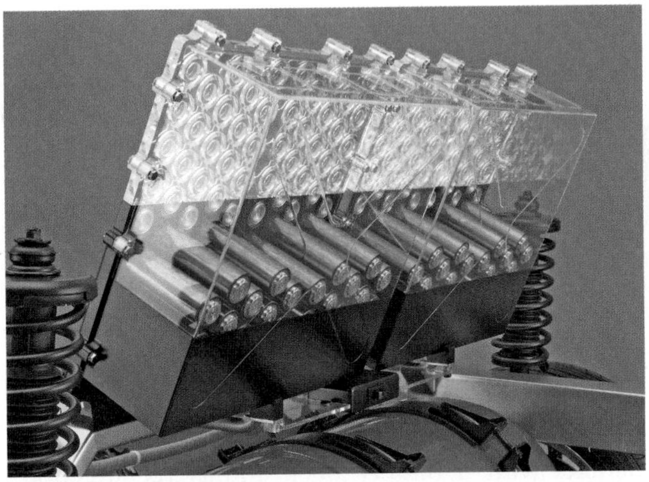

Figure 22-3 A super- or ultracapacitor assembly. *Courtesy of American Honda Motor Company, Inc.*

WARNING!

Part of the procedure for shutting down the high-voltage system in hybrid vehicles is to wait 5 or more minutes before touching anything. This is because the system needs to discharge the capacitors. Failure to do this can cause serious injury, including death.

Construction The electrodes of an ultracapacitor are typically made of carbon but can be made of a metal oxide or conducting polymers. The carbon surface of the electrodes is very coarse, with thousands of microscopic peaks and valleys. These irregularities increase the electrodes' surface area. In fact, an ounce (28.35 grams) of carbon provides nearly 13,500 square feet (1,250 sq. m) of surface area.

The plates are immersed in an electrolyte. The electrolyte is normally boric acid or sodium borate mixed in water and ethylene glycol or sugars to reduce the chances of evaporation. When voltage is applied to the capacitor, the electrolyte becomes polarized. The charge of the positive electrode attracts the negative ions in the electrolyte, and the charge of the negative electrode attracts the positive ions. The two layers of separated charges form a strong static charge.

A porous, dielectric separator is placed between the two electrodes to prevent the charges from moving between them. This separator is ultrathin and is the reason cell voltage must be kept low. High voltages would easily cause arcing across the plates.

Applications Ultracapacitors are used to store energy captured during deceleration and braking and to release that energy to assist the engine during acceleration (**Figure 22–4**). The energy in the capacitors may also be used to restart the engine during the stop/start sequence. While driving in city traffic, the stop/start feature can cycle many times. In very heavy traffic, this cycle can occur several times within a short period. The cycling is very hard on batteries. An ultracapacitor can be used to provide the power needed to start the engine. Recharged by regenerative braking and/or the generator, the ultracapacitor can be quickly charged and capable of providing enough energy for two engine restarts after it is charged.

Charging An ultracapacitor is normally placed in parallel with a battery pack and is recharged by regenerative braking. It can also be charged with a battery charger. A typical ultracapacitor needs about

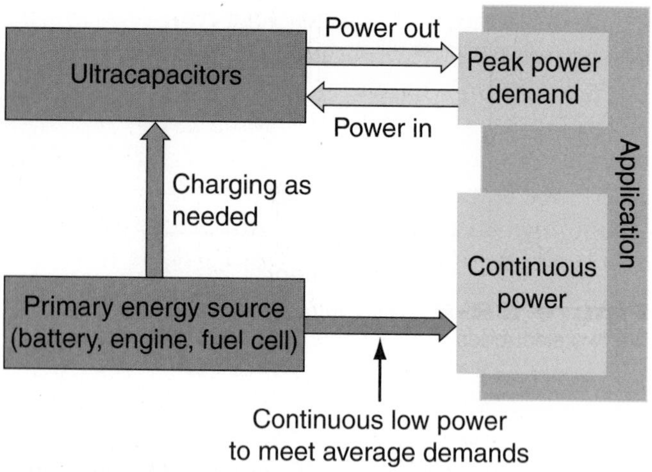

Figure 22–4 A look at the power flow for an ultracapacitor in a hybrid or fuel cell vehicle.

10 seconds to be fully recharged. An ultracapacitor is fully charged when it reaches the voltage of whatever is charging it. Ultracapacitors can be recharged and discharged an unlimited number of times.

SEMICONDUCTORS

A semiconductor is a material or device that can serve as a conductor or an insulator. Semiconductors have no moving parts; therefore, they seldom wear out or need adjustment. Semiconductors, or solid-state devices, are also small, require little power to operate, are reliable, and generate very little heat. For all these reasons, semiconductors are being used in many applications. However, current to them must be limited, as should heat.

Because a semiconductor can function as both a conductor and an insulator, it is often used as a switching device. How it behaves depends on what it is made of and which way current flows (or tries to flow) through it. Two common semiconductor devices are diodes and transistors. Diodes are used for isolation of components or circuits, clamping, or rectification of AC to DC. Transistors are used for amplification or switching.

Semiconductor materials have less resistance than an insulator but more resistance than a conductor. They also have a crystal structure. This means their atoms do not lose and gain electrons as conductors do. Instead, the atoms in semiconductors share outer electrons with each other. In this type of atomic structure, the electrons are tightly held and the element is stable. Common semiconductor materials are silicon (Si) and germanium (Ge).

Because the electrons are not free, the crystals cannot conduct current and are called **electrically inert materials**. In order to function as semiconductors, a small amount of trace element, called

impurities, must be added. The type of impurity determines the type of semiconductor.

N-type semiconductors have loose, or excess, electrons. They have a negative charge and can carry current. N-type semiconductors have an impurity with five electrons in the outer ring (called pentavalent atoms). Four of these electrons fit into the crystal structure, but the fifth is free. This excess of electrons produces the negative charge. **Figure 22–5** shows an example.

P-type semiconductors are positively charged materials. They are made by adding an impurity with three electrons in the outer ring (trivalent atoms). When this element is added to silicon or germanium, the three outer electrons fit into the pattern of the crystal, leaving a hole where a fourth electron would fit. This hole is actually a positively charged empty space. This hole carries the current in the P-type semiconductor. **Figure 22–6** shows an example of a P-type semiconductor.

Hole Flow

Understanding how semiconductors carry current without losing electrons requires an understanding of the concept of hole flow. The holes in a P-type semiconductor, being positively charged, attract electrons. Although the electrons cannot be freed from their atom, they can rearrange and fill a hole in a nearby atom. Whenever this happens, the place where the electron was is now a hole. This hole is then filled by another electron, and the process continues. The electrons move to the positive side of the structure, and the holes move to the negative side.

Diodes

Diodes are simple semiconductors. The most commonly used are regular diodes, LEDs, zener diodes, clamping diodes, and photo diodes. A diode allows current to flow in one direction; therefore, it can serve as a conductor or insulator, depending on the direction of current flow **(Figure 22–7)**. In a generator, voltage is rectified by diodes. Diodes are arranged so that current can leave the AC generator in one direction only (as direct current).

Inside a diode are positive and negative areas that are separated by a boundary area. The boundary area is called the PN junction. When the positive side of a diode is connected to the positive side of the circuit, it is said to have **forward bias (Figure 22–8)**.

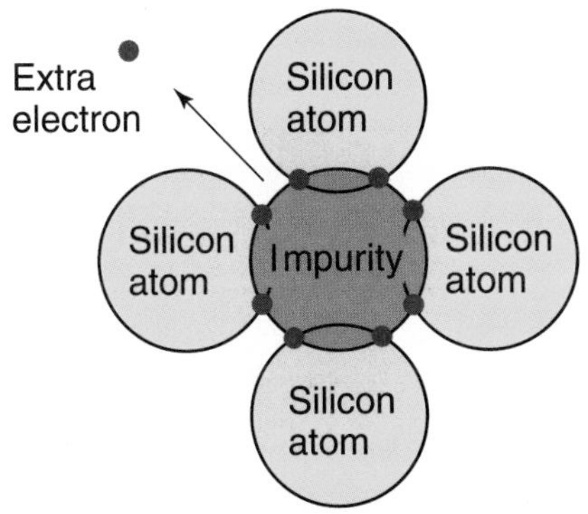

Figure 22-5 Atomic structure of an N-type silicon semiconductor.

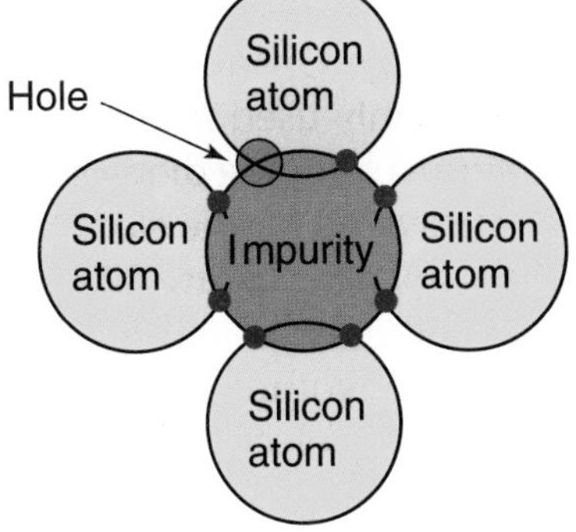

Figure 22-6 Atomic structure of a P-type silicon semiconductor.

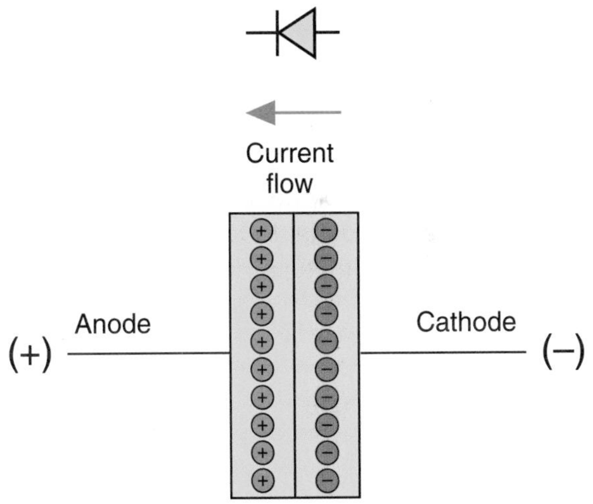

Figure 22-7 A diode and its schematical symbol.

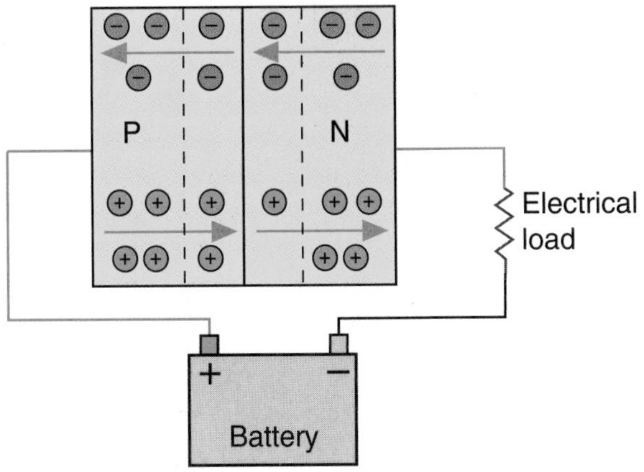

Figure 22-8 Forward biased voltage allows current flow through the diode.

Unlike electrical charges are attracted to each other and like charges repel each other. Therefore, the positive charge from the circuit is attracted to the negative side. The circuit's voltage is much stronger than the charges inside the diode, which causes the diode's charges to move. The diode's P material is repelled by the positive charge of the circuit and is pushed toward the N material and the N material toward the P. This causes the PN junction to become a conductor, allowing current to flow.

When **reverse bias** is applied to the diode, the P and N areas are connected to opposite charges. Because opposites attract, the P material moves toward the negative part of the circuit and the N material moves toward the positive part of the circuit. This empties the PN junction and current flow stops.

A **zener diode** works like a standard diode until a certain voltage is reached. When the voltage reaches this point, the diode allows current to flow in the reverse direction. Zener diodes are often used in electronic voltage regulators **(Figure 22–9)**.

LEDs emit light as current passes through them **(Figure 22–10)**. The color of the emitted light depends on the material used to make the LED. Typically LEDs are made of a variety of inorganic semiconductor materials that produce different colors **(Table 22–1)**.

Whenever the current flow through a coil of wire (such as used in a solenoid or relay) stops, a voltage surge or spike is produced. This surge results from the collapsing of the magnetic field around the coil. The movement of the field across the winding induces a very high-voltage spike, which can damage electronic components. In the past, a capacitor was used as a "shock absorber" to prevent component damage from this surge. On today's vehicles, a **clamping diode** is commonly used to prevent this voltage spike. By installing a clamping diode in parallel to the coil, a bypass is provided for the electrons during the time the circuit is opened **(Figure 22–11)**.

An example of the use of clamping diodes is on some air-conditioning compressor clutches. Because the clutch operates by electromagnetism, opening the clutch coil circuit produces a voltage spike. If the spike is left unchecked, it could damage the clutch coil relay contacts or the vehicle's computer.

TABLE 22-1 MATERIAL USED FOR DIFFERENT COLORED LEDS

Color	Semiconductor Material
Blue	Indium Gallium Nitride (Ingan) Sapphire (Al_2O_3) Silicon (Si) Silicon Carbide (Sic)
Bluish-Green	Indium Gallium Nitride (Ingan)
Green	Aluminium Gallium Indium Phosphide (Algainp) Aluminium Gallium Phosphide (Algap) Gallium Nitride (Gan) Gallium Phosphide (Gap)
Orange	Aluminium Gallium Indium Phosphide (Algainp) Gallium Arsenide Phosphide (Gaasp)
Orange-Red	Aluminium Gallium Indium Phosphide (Algainp) Gallium Arsenide Phosphide (Gaasp)
Red	Aluminium Gallium Arsenide (Algaas) Gallium Arsenide Phosphide (Gaasp) Gallium Phosphide (Gap)
White	Barrier of Aluminium Gallium Indium Nitride (Algainn)
Yellow	Aluminium Gallium Indium Phosphide (Algainp) Gallium Arsenide Phosphide (Gaasp) Gallium Phosphide (Gap)

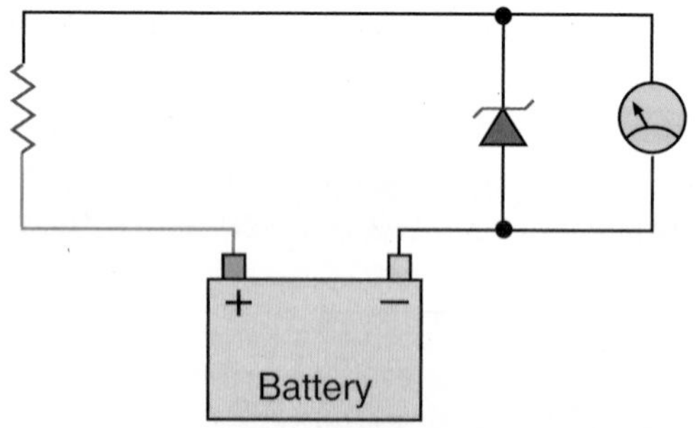

Figure 22-9 A simplified gauge circuit with a zener diode used to maintain a constant supply voltage to the gauge.

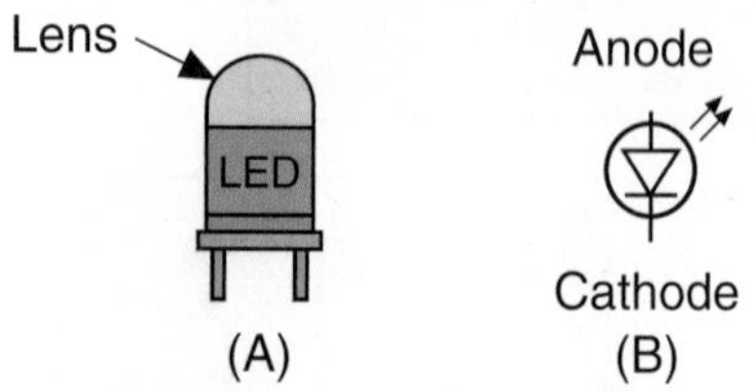

Figure 22-10 (A) An LED uses a lens to emit the light generated by current flow. (B) The schematical symbol for an LED.

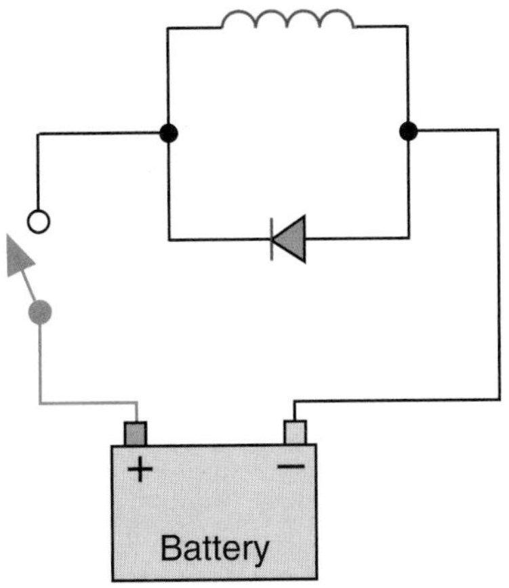

Figure 22-11 A clamping diode in parallel to a coil prevents voltage spikes when the switch is opened.

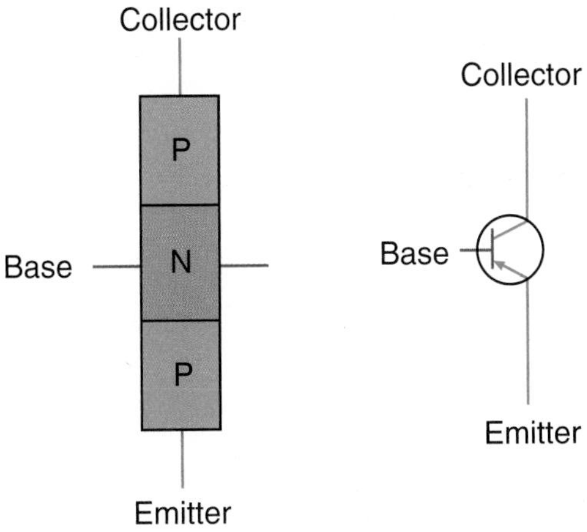

Figure 22-13 A PNP transistor and its schematical symbol.

The clamping diode is connected to the circuit in reverse bias.

Transistors

A transistor is produced by joining three sections of semiconductor materials. Like the diode, it is used as a switching device, functioning as either a conductor or an insulator. **Figure 22–12** shows some examples of transistors; there are many different sizes and types, depending on the application.

A transistor resembles a diode with an extra side. It can consist of two P-type materials and one N-type material or two N-type materials and one P-type material. These are called PNP and NPN types. In both types, junctions occur where the materials are joined. **Figure 22–13** shows a PNP transistor. Notice that each of the three sections has a lead connected to it. This allows any of the three sections to be connected to the circuit. The names for the legs are the **emitter**, **base**, and **collector**.

The center section is called the base and is the controlling part of the circuit or where the larger controlled part of the circuit is switched. The path to ground is through the emitter. A resistor is normally in the base circuit to keep current flow low. This prevents damage to the transistor. The emitter and collector make up the control circuit. When a transistor is drawn in an electrical schematic, the arrow on the emitter points to the direction of current flow.

The base of a PNP transistor is controlled by its ground. Current flows from the emitter through the base, then to ground. A negative voltage or ground must be applied to the base to turn on a PNP transistor. When the transistor is on, the circuit from the emitter to the collector is complete.

An NPN transistor is the opposite of a PNP. When positive voltage is applied to the base of an NPN transistor, the collector-to-emitter circuit is turned on **(Figure 22–14)**.

Transistors can also function as a variable switch. By varying the voltage applied to the base, the completeness of the emitter and collector circuit will also vary. This is done simply by the presence of a variable resistor in the base circuit. This principle is used in light-dimming circuits.

Figure 22-12 Typical transistors.

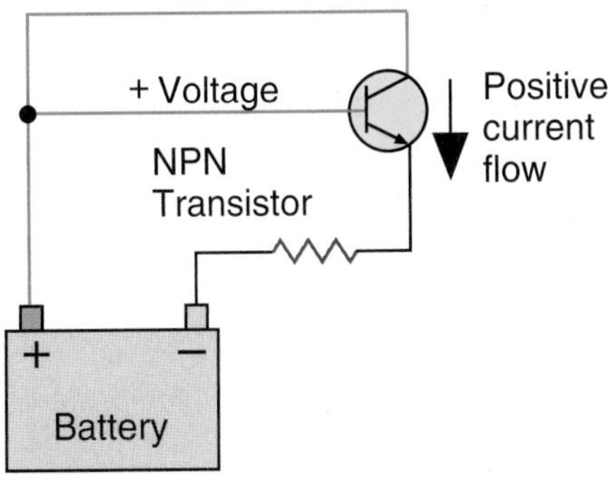

Figure 22-14 When positive voltage is applied to the base of an NPN transistor, current flows through the collector and the emitter.

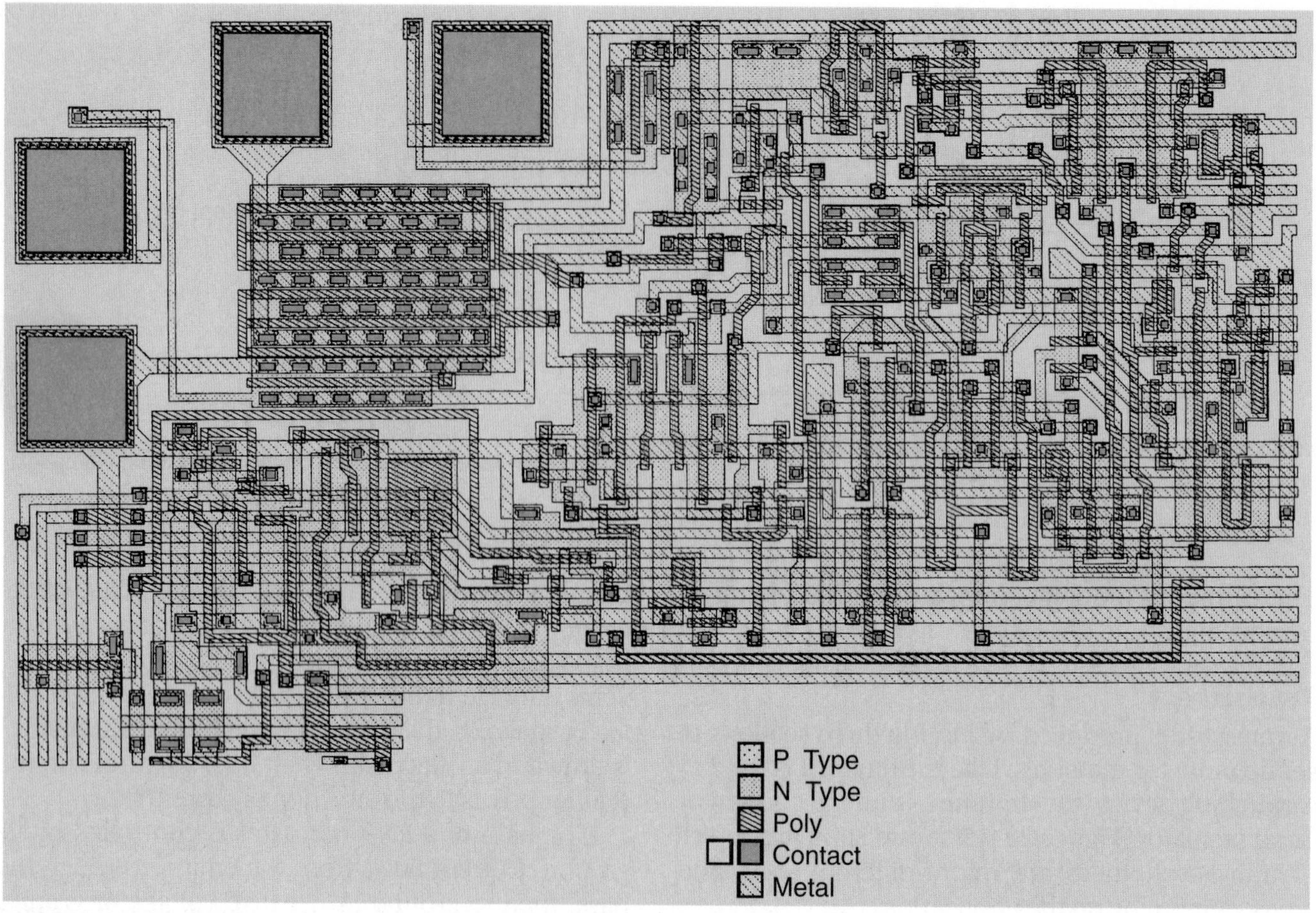

Figure 22-15 An enlarged illustration of an IC (chip) with thousands of transistors, diodes, resistors, and capacitors. The actual size of this chip can be less than 1/4 of an inch square.

Integrated Circuits

The ability of one transistor or diode is limited in its ability to do complex tasks. However, when many semiconductors are combined into a circuit, they can perform complex functions.

An integrated circuit is simply a large number of diodes, transistors, and other electronic components mounted on a single piece of semiconductor material **(Figure 22–15)**. This creates a very small package capable of performing many functions. Because of the size of an integrated circuit, many transistors, diodes, and other solid-state components can be installed in a car to make logic decisions and issue commands to other areas of the engine. This is the foundation of computerized engine control systems.

COMPUTER BASICS

Computers control nearly all of the systems in an automobile. Systems that once were controlled by vacuum, mechanical, and electromechanical devices are now controlled and operated by electronics. These are electronic control systems. They are made up of sensors, actuators, and a central processing unit, sometimes called a **microprocessor**. Electronic controls are designed to allow a system to operate in the most efficient way it can. They also provide for many driver conveniences. Although today's vehicles have several computers, they have two main computers—the powertrain control module (PCM) and body control module (BCM).

In addition to controlling various systems, the PCM and BCM continuously monitor operating conditions for possible system malfunctions. The computers compare system conditions against programmed parameters. If the conditions fall outside of these limits, the computers will detect the malfunction. A trouble code will be stored in the computers' memory to indicate the portion of the system at fault. A technician can access this code to aid in diagnosis.

The **central processing unit (CPU)** is basically thousands of transistors placed on a small chip. The CPU moves information in and out of the computer's memory **(Figure 22–16)**. Input information is processed in the CPU and checked against the programs stored in its memory. The CPU also checks for all other pertinent information held in memory. The CPU takes all of this information and uses computer

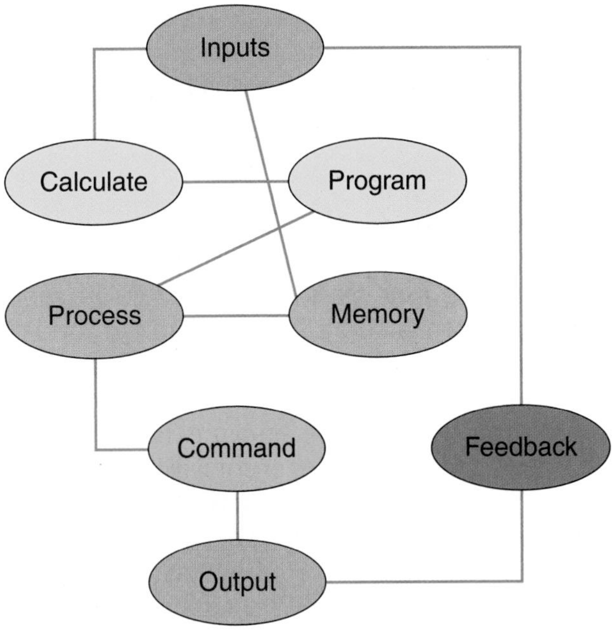

Figure 22-16 The basic information cycle for a computer.

logic to determine what should or should not happen. Once these decisions are made, the CPU sends out commands to make the required corrections or adjustments to the system.

A computer is an electronic device that stores and processes data. It relies on semiconductors and is really a group of integrated circuits. The four basic functions of a computer are:

1. Input: A signal sent from an input device. The device can be a sensor or a switch activated by the driver, technician, or a mechanical part.
2. Processing: The computer uses the input information and compares it to programmed instruction. This information is processed by logic circuits in the computer.
3. Storage: The program instructions are stored in the computer's memory. Some of the input signals are also stored for processing later.
4. Output: After the computer has processed the inputs and checked its programmed instructions, it will issue commands to various output devices. These output devices may be instrument panel displays or output actuators. The output of one computer may also be an input to other computers.

Inputs

The CPU receives inputs that it checks against programmed values. Depending on the input, the computer controls the actuator(s) until the programmed results are obtained. The inputs can come from other computers, the driver, the technician, or through a variety of sensors.

Driver input signals are usually provided by momentarily applying a ground through a switch. The computer receives this signal and performs the desired functions. For example, if the driver wishes to reset the trip odometer on a digital instrument panel, a reset button is depressed. This switch provides a momentary ground that the computer receives as an input and sets the trip odometer to zero.

Switches can be used as inputs for any operation that only requires a yes-no, or on-off, condition. Other inputs include those supplied by means of a sensor and those signals returned to the computer in the form of **feedback**. Feedback means that data concerning the effects of the computer's commands are fed back to the computer as an input signal.

If the computer sends a command signal to actuate an output device, a feedback signal may be sent back from the actuator to inform the computer that the task was performed. The feedback signal confirms both the position of the output device and the operation of the actuator. Another form of feedback is for the computer to monitor voltage when a switch, relay, or other actuator is activated. Changing positions of an actuator should result in predictable changes in the computer's voltage sensing circuit. The computer may set a diagnostic code if it does not receive the correct feedback signal.

All inputs have the same basic function. They detect a mechanical condition (movement or position), chemical state, or temperature condition and change it into an electrical signal that is used by the computer to make decisions. Each sensor has a specific job to do (for example, monitor throttle position, vehicle speed, and manifold pressure). Although there are many different sensor designs, they all are reference voltage sensors or voltage-generating sensors.

Reference Voltage Sensors To get an exact look at what is happening in a system, the computer sends a constant, predetermined voltage signal to a sensor. The sensor reacts to operating conditions and sends a voltage signal back to the computer. This type of sensor is called a **reference voltage (Vref) sensor**. The voltage sent out by the computer is called the reference voltage and normally has a value of 5 to 9 volts. The reference voltage is sent out to a sensor through a reference voltage regulator in the computer. The regulator keeps the reference voltage at a predetermined value. Because the computer knows that a certain voltage value has been sent out, it can indirectly interpret things like motion, temperature, and component position, based on what comes back.

Most reference voltage sensors are variable resistors or potentiometers. They modify the reference voltage and the return voltage represents a condition. The computer will use the return voltage to calculate the condition and order changes to system operation, if necessary.

For example, consider the operation of a throttle position sensor (TP sensor). During acceleration (from idle to wide-open throttle), the computer monitors throttle plate movement based on the changing reference voltage signal returned by the TP sensor **(Figure 22–17)**. The TP sensor is a rotary potentiometer that changes circuit resistance based on throttle shaft rotation. As the sensor's resistance varies, the computer responds in a specific manner to each corresponding voltage change.

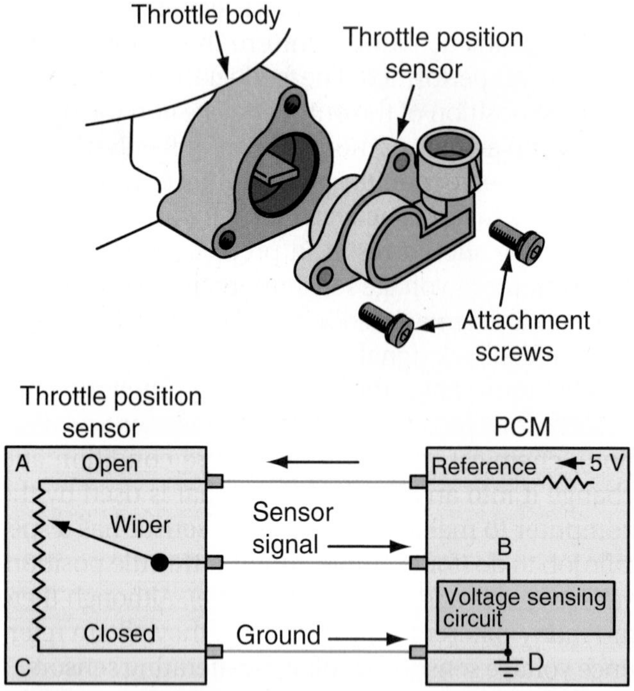

Figure 22-17 A TP switch is a voltage reference sensor.

Thermistors are also variable resistors and serve as voltage reference sensors. Inputs from a thermistor allow the computer to observe small changes in temperature. Thermistors are used to monitor the temperature of engine coolant, inside and outside ambient air, intake air, and many components.

Wheatstone bridges (Figure 22–18) are also used as variable resistance sensors. These are typically constructed of four resistors, connected in series-parallel between an input terminal and a ground terminal. Three of the resistors are kept at the same value. The fourth resistor is a sensing resistor. When all four of the resistors have the same value, the bridge is balanced and the voltage sensor will have a value of 0 volts. If the sensing resistor changes value, a change occurs in the circuit's balance. The sensing circuit will receive a voltage reading proportional to the amount of resistance change. If the Wheatstone bridge is used to measure temperature, temperature changes are indicated as a change in voltage by the sensing circuit. Wheatstone bridges are also used to measure pressure (**piezoresistive**) and mechanical strain.

In addition to variable resistors, another commonly used reference voltage sensor is a switch. By opening and closing a circuit, switches provide the necessary voltage information to the computer so vehicles can maintain the proper performance and driveability.

Voltage-Generating Sensors Although many sensors are variable resistors or switches, **voltage-generating sensors** are commonly used. These sensors include speed sensors, Hall-effect switches, oxygen sensors, and knock sensors. These are capable of producing an input voltage signal for the control system. This varying voltage signal allows the computer to monitor and immediately adjust the operation of a system to meet the current needs.

Magnetic pulse generators use the principle of magnetic induction to produce a voltage signal. They

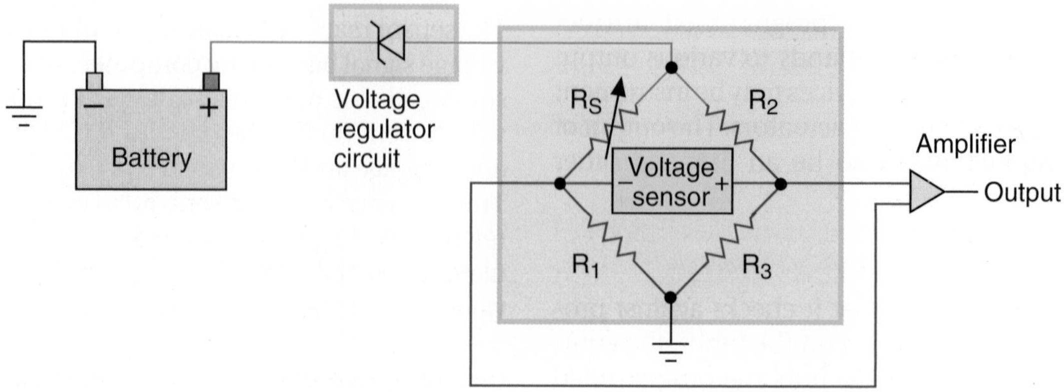

Figure 22-18 Wheatstone bridge.

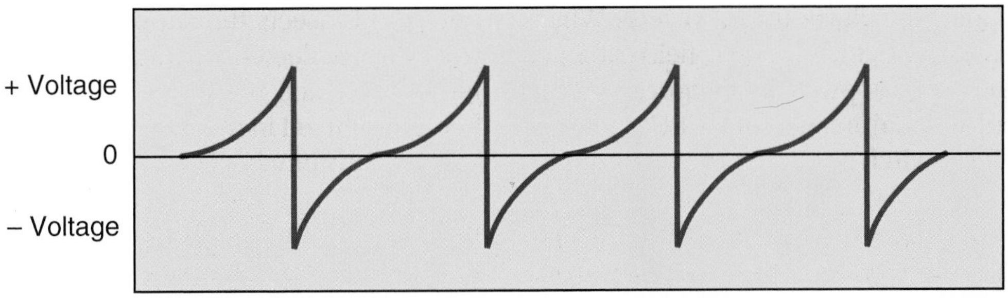

Figure 22-19 A pulsed voltage signal.

are also called permanent magnet (PM) generators. These sensors are often used to send data to the computer about the speed of the monitored component. These data provide information about vehicle speed, shaft speed, and wheel speed. The signals from speed sensors are used for instrumentation, cruise control systems, antilock brake systems, ignition systems, speed-sensitive steering systems, and automatic ride control systems. A magnetic pulse generator is also used to inform the computer about the position of a monitored device. This is common in engine controls where the CPU needs to know the position of the crankshaft in relation to rotational degrees.

The major components of a pulse generator are a timing disc and a pickup coil. The **timing disc** is attached to a rotating shaft or cable. The number of teeth on the timing disc is determined by the manufacturer and the application. If only the number of revolutions is required, the timing disc may have one tooth, whereas if it is important to track quarter revolutions, the timing disc needs at least four teeth. The teeth will generate a voltage that is constant per revolution of the shaft. For example, a vehicle speed sensor may be designed to deliver 4,000 pulses per mile. The number of pulses per mile remains constant regardless of speed. The computer calculates how fast the vehicle is going based on the frequency of the signal. The timing disc is also known as an armature, reluctor, trigger wheel, pulse wheel, or timing core.

The **pickup coil** is also known as a stator, sensor, or pole piece. It remains stationary while the timing disc rotates in front of it. The changes of magnetic lines of force generate a small voltage signal in the coil. A pickup coil consists of a permanent magnet with fine wire wound around it.

An air gap is maintained between the timing disc and the pickup coil. As the timing disc rotates in front of the pickup coil, the generator sends a pulse signal **(Figure 22–19)**. As a tooth on the timing disc aligns with the core of the pickup coil, it repels the magnetic field. The magnetic field is forced to flow through the coil and pickup core **(Figure 22–20)**. When the tooth passes the core, the magnetic field is able to expand **(Figure 22–21)**. This action is repeated every time a tooth passes the core. The moving lines of magnetic force cut across the coil windings and induce a voltage signal.

When a tooth approaches the core, a positive current is produced as the magnetic field begins to concentrate around the coil. When the tooth and core align, there is no more expansion or contraction of the

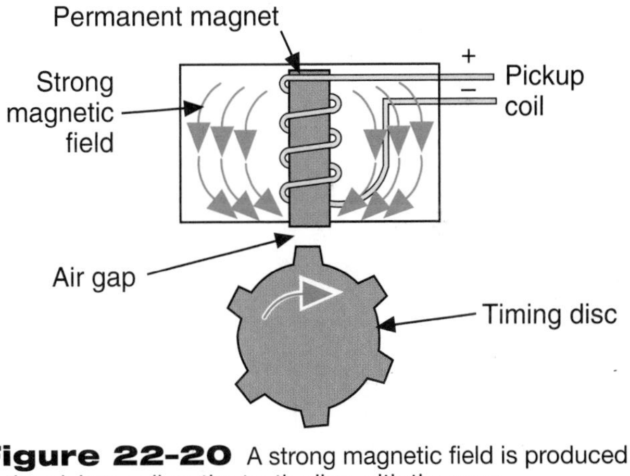

Figure 22-20 A strong magnetic field is produced in the pickup coil as the teeth align with the core.

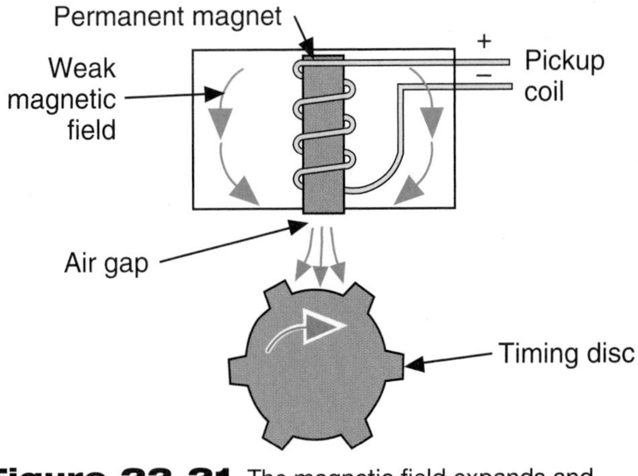

Figure 22-21 The magnetic field expands and weakens as the teeth pass the core.

magnetic field and the voltage drops to zero. When the tooth passes the core, the magnetic field expands and a negative current is produced **(Figure 22–22)**. The resulting pulse signal is sent to the CPU.

The **Hall-effect switch** performs the same function as a magnetic pulse generator. It operates on the principle that if a current is allowed to flow through thin conducting material exposed to a magnetic field, another voltage is produced **(Figure 22–23)**.

A Hall-effect switch contains a permanent magnet and a thin semiconductor layer made of gallium arsenate crystal (Hall layer) and a shutter wheel **(Figure 22–24)**. The Hall layer has a negative and a positive terminal connected to it. Two additional terminals located on either side of the Hall layer are used for the output circuit.

The permanent magnet is located directly across from the Hall layer so its lines of flux bisect the layer at right angles to the current flow. The permanent magnet is stationary, and a small air gap is between it and the Hall layer.

A steady current is applied to the crystal of the Hall layer. This produces a signal voltage perpendicular to the direction of current flow and magnetic flux. The signal voltage produced is a result of the effect the magnetic field has on the electrons. When the magnetic field bisects the supply current flow, the electrons are deflected toward the Hall layer negative terminal, which results in a weak voltage potential being produced in the Hall switch.

The shutter wheel consists of a series of alternating windows and vanes. It creates a magnetic shunt that changes the strength of the magnetic field from the permanent magnet. The shutter wheel is attached to a rotating component. As the wheel rotates, the vanes pass through the air gap. When a shutter vane enters the gap, it intercepts the magnetic field and shields the Hall layer from its lines of force. The electrons in the supply current are no longer disrupted and return to a normal state. This results in low voltage potential in the signal circuit of the Hall switch.

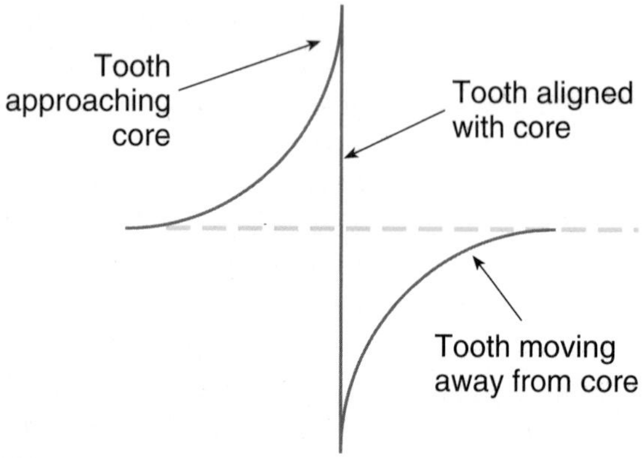

Figure 22-22 The waveform produced by a magnetic pulse generator.

Communication Signals

Voltage does not flow through a conductor; current flows while voltage is the pressure that pushes the current. However, voltage can be used as a signal; for example, difference in voltage levels, frequency of change, or switching from positive to negative values can be used as a signal.

A computer is capable of reading voltage signals. The programs used by the CPU are "burned" into IC chips using a series of numbers. These numbers represent various combinations of voltages that the computer can understand. The voltage signals to the computer can be either analog or digital. **Analog** means a voltage signal is infinitely variable, or can be changed, within a given range. **Digital** means a voltage signal that is in one of three states—either on-off, yes-no, or high-low. Digital signals produce a **square wave**. The wave represents the immediate change in the voltage signal. It is called a square wave because the digital signal creates a series of horizontal and vertical lines that connect to form a square-shaped pattern on a scope **(Figure 22–25)**. In a digital signal, voltage is represented by a series of digits, which create a **binary code**.

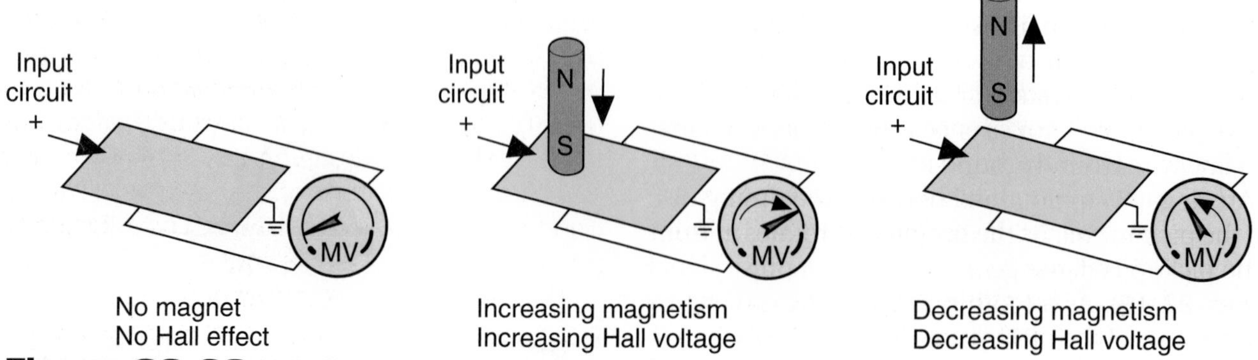

Figure 22-23 Hall-effect principles of voltage induction.

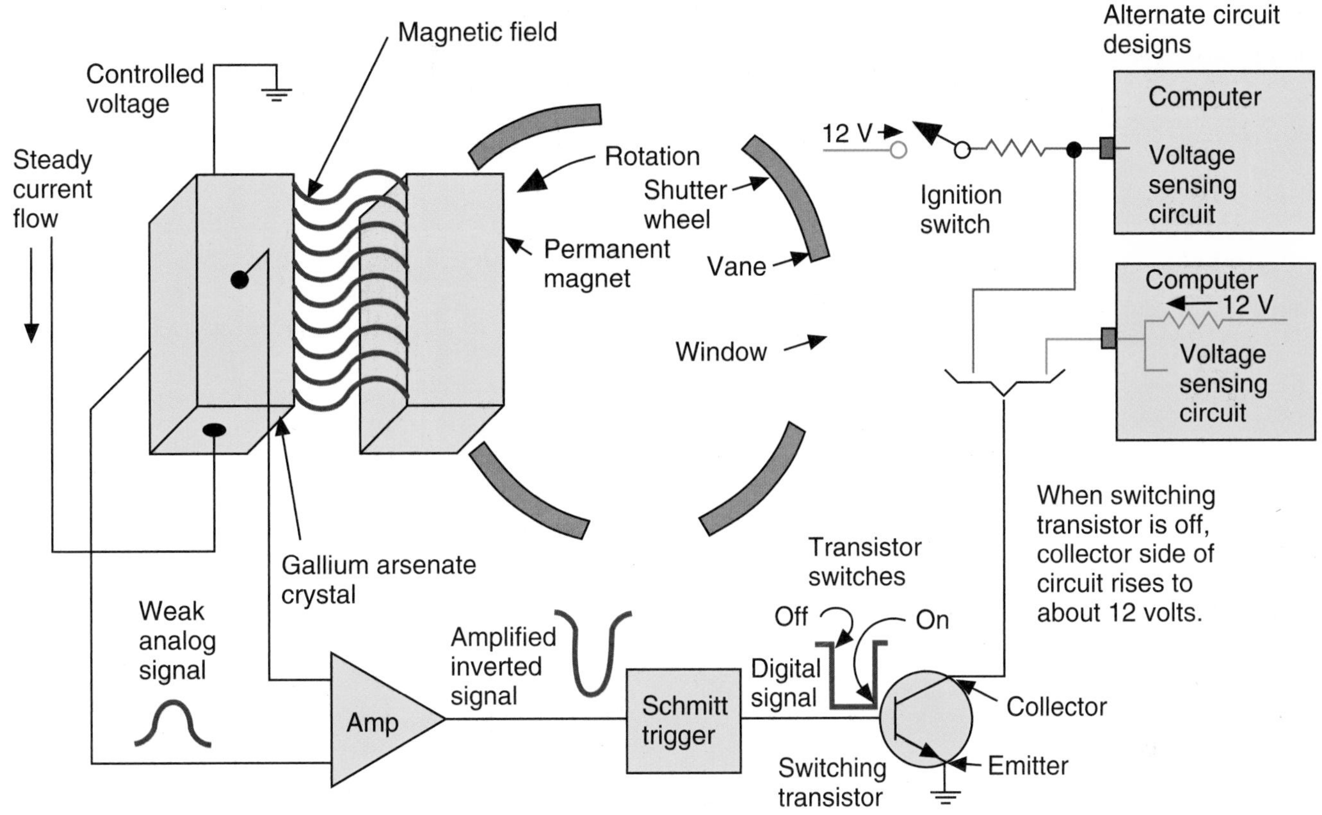

Figure 22-24 Typical circuit of a Hall-effect switch.

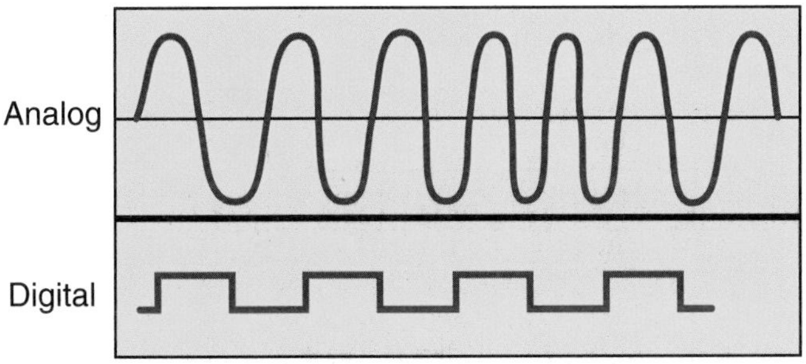

Figure 22-25 Analog signals can be constantly variable. Digital signals are either on/off or low/high.

Most input sensors produce a constantly variable voltage signal. For example, the voltage signal from an ambient temperature sensor never changes abruptly. The signal corresponds with a gradual increase or decrease in temperature and is therefore an analog signal.

A computer can only read a digital binary signal. To overcome this communication problem, all analog voltage signals are converted to a digital format by a device known as an analog-to-digital converter (**A/D converter**). The A/D converter **(Figure 22–26)** is located in the processor.

The A/D converter changes a series of signals to a binary number made up of 1s and 0s. Voltage above a given value converts to 1, and zero voltage converts to 0 **(Figure 22–27)**. Each 1 or 0 represents a **bit** of information. Eight bits equal a **byte** (sometimes referred to as a *word*). All communication between the CPU, the memories, and the interfaces is in binary code, with each information exchange in the form of a byte.

To get an idea of how binary coding works, let us see how signals from the coolant temperature sensor (CTS) are processed by the CPU. The CTS is an NTC thermistor that controls a reference signal based on temperature changes. Upon receiving the CTS's analog signals, the input conditioner immediately groups each signal value into a predetermined voltage range and assigns a numeric value to each range. In our example, use the following ranges and numeric values: 0 to 2 volts = 1, 2 to 4 volts = 2, and 4 to 5 volts = 3 (assuming a Vref of 5 volts).

When the CTS is hot, its resistance is low and the modified voltage signal it sends back falls into the high range (4 to 5 volts). Upon entering the A/D converter, the voltage value is assigned a numeric value of 3 (based on the ranges previously cited) and is ready for further translation into a binary code format. Binary numbers are represented by the numbers 0 and 1. Any number and word can be translated into a combination of binary 1s and 0s.

Without going into the finer points of binary numbering, the number 3 in binary is expressed as 11. To the thousands of tiny transistors and diodes that act

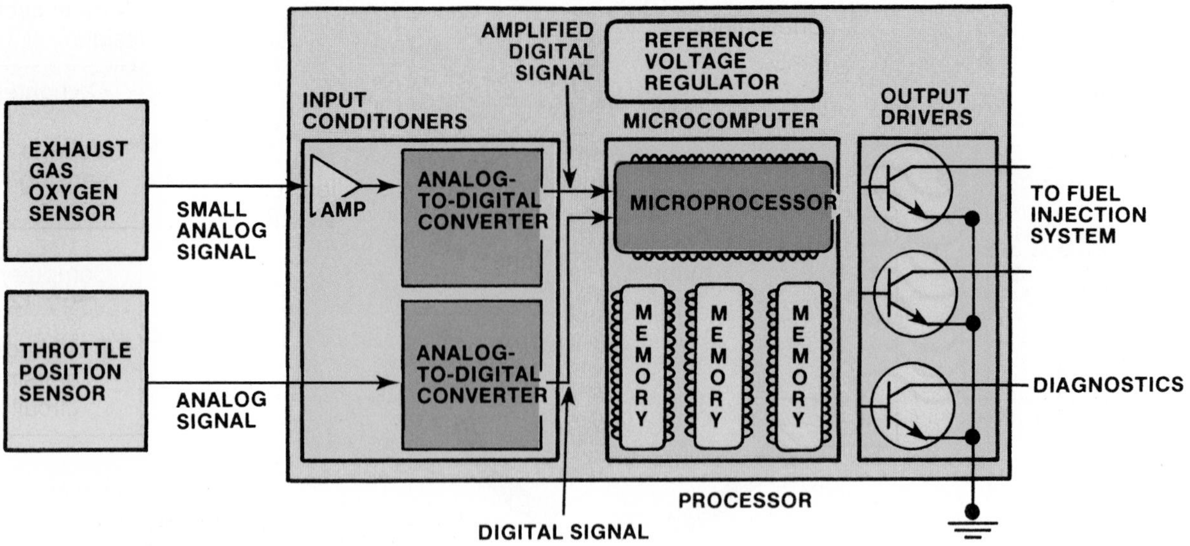

Figure 22-26 The A/D converter prepares input signals for the CPU.

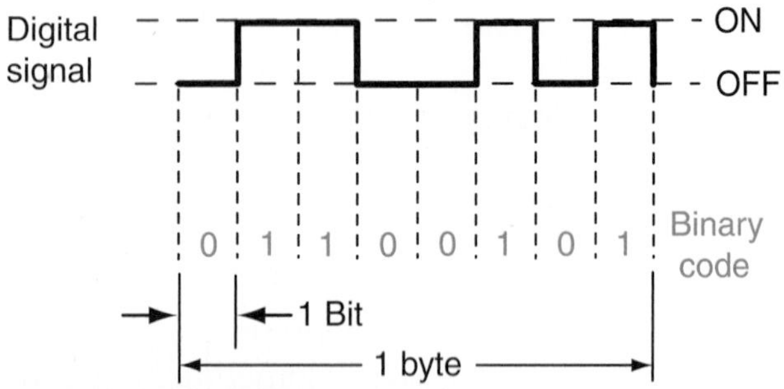

Figure 22-27 Each zero (0) and one (1) represents a bit of information. When eight bits are combined in specific sequence, they form a byte or word that makes up the basis of a computer's language.

as the on-off switches inside microprocessors, 11 instructs the computer to turn on or apply voltage to a specific circuit for a predetermined length of time (based on its program). **Table 22–2** illustrates how binary numbers can be converted into decimal or base ten numbers.

TABLE 22-2 BINARY NUMBER CODE

Binary Number	Base Ten Number
0000	0
0001	1
0010	2
0011	3
0100	4
0101	5
0110	6
0111	7
1000	8
1001	9
1010	10

Schmitt Trigger In addition to A/D conversion, some voltage signals require amplification before they can be processed by the computer. To do this, an input conditioner known as an amplifier is used to strengthen weak voltage signals. This is especially important for signals from Hall-effect switches. When the signal voltage leaves the sensor, it is a weak analog signal. The signal is amplified and inverted. It is then sent to a **Schmitt trigger**, which is a type of A/D converter, where it is digitized and conditioned into a clean square wave. The signal is then sent to a switching transistor that turns on and off in response to the signal and sets the frequency of the signal.

Clock Rate After an input signal has been generated, conditioned, and passed along to the computer, it is ready to be processed for performing some work and displaying information. The computer has a crystal oscillator or clock that delivers a constant time pulse. The crystal vibrates at a fixed rate when current, at a certain voltage level, is applied to it. The vibrations produce a very regular series of voltage pulses. The clock maintains an orderly flow of information throughout the computer's circuitry by transmitting one bit of binary code for each pulse. The clock enables the computer to know when one signal ends and when another begins.

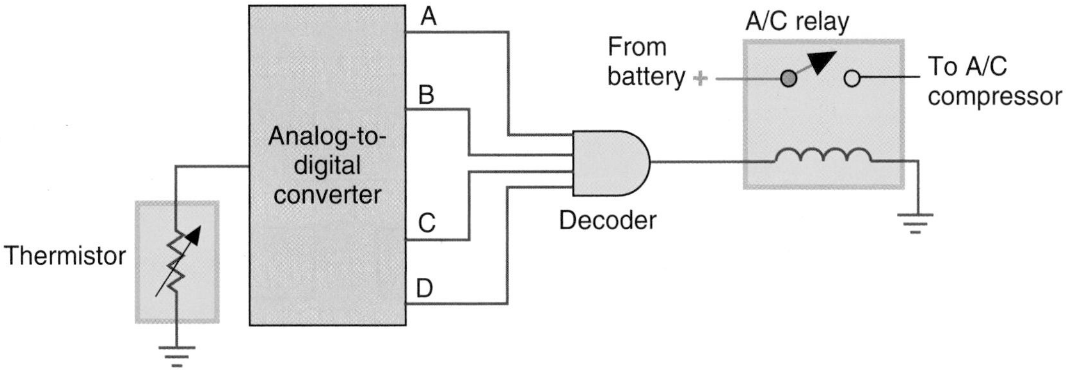

Figure 22-28 A simplified temperature sensing circuit that will turn on the A/C compressor when the inside temperature reaches a predetermined level.

Communication Rates The amount of information processed by a computer is dependent on its speed or **baud rate**. Baud rate is the speed of communication and is roughly equal to the number of bits per second a computer can process.

Logic Gates

Logic gates are the thousands of **field-effect transistors (FETs)** incorporated into the computer's circuitry. The FETs use the incoming voltage patterns to determine the pattern of pulses leaving the gate. These circuits are called logic gates because they act as gates to output voltage signals depending on different combinations of input signals. The following are the most common logic gates and their operation. The symbols in the figures represent functions and not electronic construction.

1. **NOT gate:** A NOT gate simply reverses binary 1s and 0s and vice versa. A high input results in a low output, and a low input results in a high output.
2. **AND gate:** The AND gate has at least two inputs and one output. The operation of the AND gate is similar to two switches in series with a load. The only way the load will turn on is if both switches are closed. Before current can be present at the output of the gate, current must be present at the base of both transistors.
3. **OR gate:** The OR gate operates similarly to two switches that are wired in parallel to a light. If one switch is closed, the light will turn on. A high signal to either input will result in a high output.
4. **NAND** and **NOR gates:** A NOT gate placed behind an OR or AND gate inverts the output signal.
5. **Exclusive-OR (XOR) gate:** A combination of gates that will produce a high output signal only if the inputs are different.

These different gates are combined to perform the processing function. The following are some of the most common combinations.

1. **Decoder circuits:** A combination of AND gates used to provide a certain output based on a given combination of inputs **(Figure 22–28)**. When the correct bit pattern is received by the decoder, it will produce the high-voltage signal to activate the relay coil.
2. **Multiplexer (MUX):** The basic computer is not capable of looking at all of the inputs at the same time. A multiplexer is used to examine one of many inputs depending on a programmed priority rating **(Figure 22–29)**.
3. **Demultiplexer (DEMUX):** Operates similarly to the MUX except that it controls the order of the outputs. The process that the MUX and DEMUX operate on is called sequential sampling. This means the computer deals with all the sensors and actuators one at a time.
4. **RS and clocked RS flip-flop circuits:** Flip-flop circuits remember previous inputs and do not change their outputs until they receive new input signals. The clocked flip-flop circuit has an inverted clock signal as an input so circuit operations occur in the proper order. Flip-flop circuits

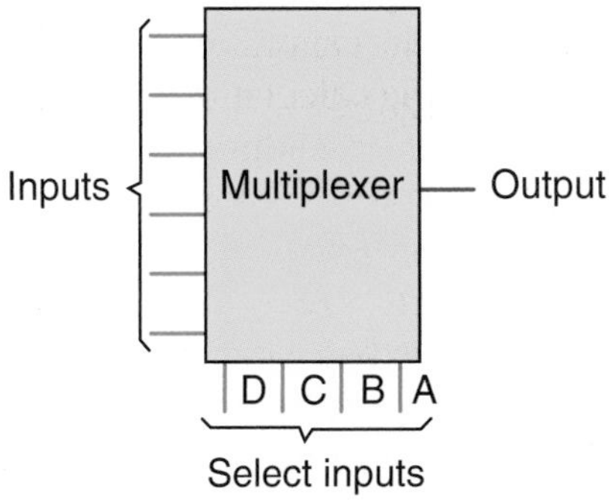

Figure 22-29 The selection of inputs at DCBA will determine which input data will be processed.

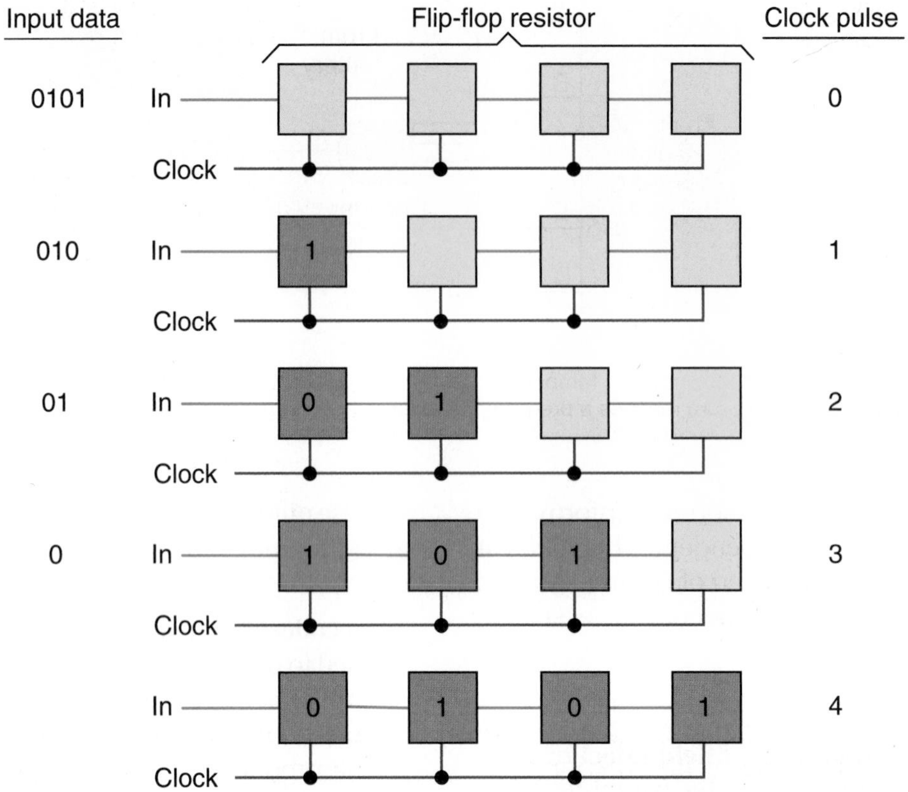

Figure 22-30 It takes four clock pulses to load four bits into the register.

are called sequential logic circuits because the output is determined by the sequence of inputs.

5. **Registers:** Used in the computer to temporarily store information. A register is a combination of flip-flops that transfers bits from one to another every time a clock pulse occurs **(Figure 22–30)**.
6. **Accumulators:** Registers designed to store the results of logic operations that can become inputs to other computers or modules.

Logic gates process input information to command output devices. The order of this logic or the instructions to the computer are held in the computer's memory.

Memories

A computer's memory holds the programs and other data, such as vehicle calibrations, which the CPU refers to while making calculations. The program is a set of instructions or procedures that it must follow. Included in the program are **look-up tables** that tell the computer when to retrieve an input (based on temperature, time, etc.), how to process it, and what to do with it after it has been processed. Look-up tables are sets of instructions and there is one for every possible condition the computer may detect.

The microprocessor works with memory in two ways: It can read information from memory or change information in memory by writing in or storing new information. To write information in memory, each memory location is assigned a number (written in binary code also) called an address. These addresses are sequentially numbered, starting with zero, and are used by the microprocessor to retrieve data and write new information into memory. During processing, the CPU often receives more data than it can immediately handle. In these instances, some information has to be temporarily stored or written into memory until the microprocessor needs it.

When ready, the microprocessor accesses the appropriate memory location (address) and is sent a copy of what is stored. By sending a copy, the memory retains the original information for future use.

Basically, three types of memory are used in automotive CPUs today **(Figure 22–31)**: read-only memory, programmable read-only memory, and random-accessmemory.

Read-Only Memory (ROM) Permanent information is stored in **read-only memory (ROM)**. Information in ROM cannot be erased, even if the system is turned off or the CPU is disconnected from the battery. As the name implies, information can only be read from ROM.

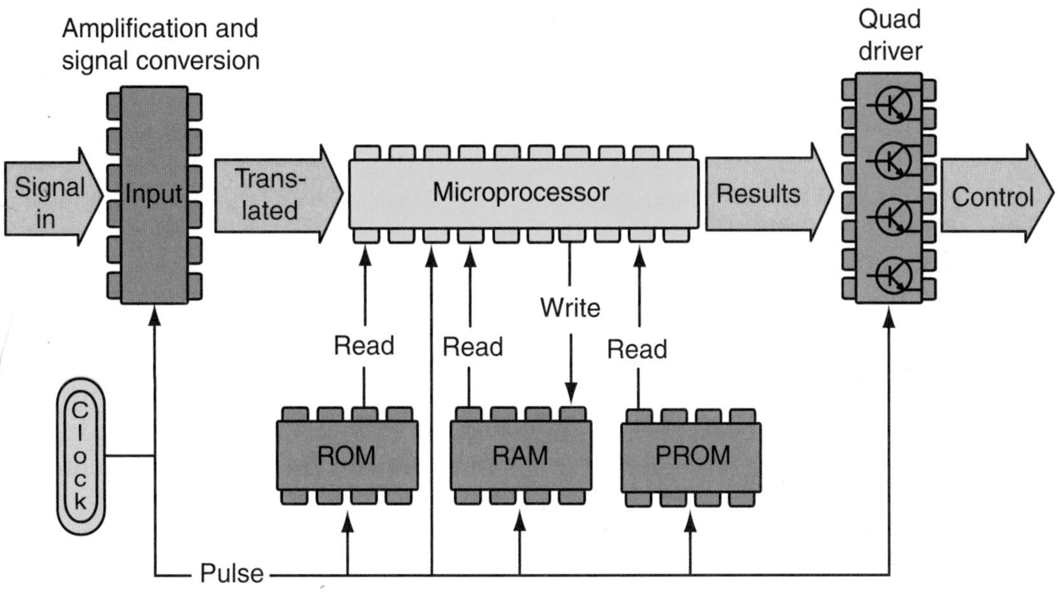

Figure 22-31 The three memories within a computer.

When making decisions, the microprocessor is constantly referring to the stored information and the input from sensors. By comparing information from these sources, the CPU makes informed decisions.

Programmable Read-Only Memory (PROM) The **programmable read-only memory (PROM)** differs from the ROM in that it plugs into the computer and may be reprogrammed or replaced with one containing a revised program. It contains program information specific to different vehicle model calibrations. The PROM in some computers is replaceable and can serve as a way to upgrade the system.

Erasable PROM (EPROM) is similar to PROM except that its contents can be erased to allow new data to be installed. A piece of Mylar tape covers a window. If the tape is removed, the memory circuit is exposed to ultraviolet light that erases its memory.

Electrically erasable PROM (EEPROM) allows changing the information electrically one bit at a time. Some manufacturers use this type of memory to store information concerning mileage, vehicle identification number, and options.

Random-Access Memory (RAM) The **random-access memory (RAM)** is used during computer operation to store temporary information. The CPU can write, read, and erase information from RAM in any order, which is why it is called random. One characteristic of RAM is that when the ignition key is turned off and the engine is stopped, information in RAM is erased. RAM is used to store information from the sensors, the results of calculations, and other data that are subject to constant change.

There are currently two other versions of RAM in use: volatile and nonvolatile. A volatile RAM, usually called **keep-alive memory (KAM)** has most of the features of RAM. Information can be written into KAM and can be read and erased from KAM. Unlike RAM, information in KAM is not erased when the ignition key is turned off and the engine is stopped. However, if battery power to the processor is disconnected, information in KAM is erased.

A **nonvolatile RAM (NVRAM)** does not lose its stored information if its power source is disconnected. Vehicles with digital display odometers usually store mileage information in nonvolatile RAM.

Actuators

Once the computer's programming determines that a correction or adjustment must be made in the controlled system, an output signal is sent to control devices called **actuators**. These actuators, which are solenoids, switches, relays, or motors, physically act on or carry out the command sent by the computer.

Actually, actuators are electromechanical devices that convert an electrical current into mechanical action. This mechanical action can then be used to open and close valves, control vacuum to other components, or open and close switches. When the CPU receives an input signal indicating a change in one or more of the operating conditions, the CPU determines the best strategy for handling the conditions. The CPU then controls a set of actuators to achieve a desired effect or strategy goal. For the computer to control an actuator, it must rely on a component called an **output driver**.

The output driver usually applies the ground circuit of the actuator. The ground can be applied steadily if the actuator must be activated for a selected amount of time, or the ground can be pulsed to activate the actuator in pulses.

Output drivers operate by the digital commands issued by the CPU. Basically, the output driver is nothing more than an electronic on-off switch used to control a specific actuator.

To illustrate this relationship, let us suppose the computer wants to turn on the engine's cooling fan. Once it makes a decision, it sends a signal to the output driver that controls the cooling fan relay (actuator). In supplying the relay's ground, the output driver completes the power circuit between the battery and cooling fan motor and the fan operates. When the fan has run long enough, the computer signals the output driver to open the relay's control circuit (by removing its ground), thus opening the power circuit to the fan.

SHOP TALK

Normally a computer will control an actuator with a low-side driver. These drivers complete the ground for the output device. Many newer systems use high-side drivers that control the outputs by varying the power to them. Most high-side drivers are metal oxide field effect transistors (MOSFET) controlled by another transistor. High-side drivers are used in circuits where a quick response to opens, shorts, and temperature changes is desired. Because a circuit's behavior depends on the driver being used, it is important to check the service information before diagnosing the system.

For actuators that cannot be controlled by a digital signal, the CPU must turn its digitally coded instructions back into an analog signal. This conversion is completed by the A/D converter.

Displays can be controlled directly by the CPU. They do not require digital-to-analog conversion or output drivers because they contain circuitry that decodes the microprocessor's digital signal. The decoded information is then used to indicate such things as vehicle speed, engine rpm, fuel level, or scan tool values.

Duty Cycle versus Pulse Width Often the computer controls the results of the output by controlling the duty cycle or pulse width of the actuator. Duty cycle is a measurement of the amount of time something is on compared to the time of one cycle and is measured in a percentage. When measuring duty cycle, you are looking at the amount of time something is on during one cycle. Pulse width is similar to duty cycle except that it is the exact time something is turned on and is measured in milliseconds.

Power Supply

The CPU also contains a power supply that provides the various voltages required by the microprocessor and internal clock that provides the clock pulse, which in turn controls the rate at which sensor readings and output changes are made. Also contained are protection circuits that safeguard the microprocessor from interference caused by other systems in the vehicle and diagnostic circuits that monitor all inputs and outputs and signal a warning light if any values go outside the specified parameters. This warning light is called the malfunction indicator lamp (MIL).

Awake/Sleep Modes

The control modules are able to control or perform all of their functions in the awake mode. They enter a sleep mode when normal control or monitoring of the system functions has stopped and a time limit has passed. There is still some activity during the sleep mode; what occurs depends on the system. Basically, during the sleep mode only enough power is used to maintain memory and for periodic monitoring of some systems. Once normal computer activity is called for, the computer wakes up and resumes its normal functions.

ON-BOARD DIAGNOSTICS

Although often thought of as something that only affects engine performance, on-board diagnostic (OBD) systems have shaped the operation of all electronic control modules. This is especially true of OBD II.

On-board diagnostic capabilities are incorporated into a vehicle's computer to monitor virtually every component that can affect emission performance. Each component is checked by a diagnostic routine to verify that it is functioning properly.

OBD I was the first generation of on-board diagnostic systems and was designed to monitor some of the vehicle's emission control components. Required on all 1991 and newer vehicles, OBD I systems monitored only a few of the emission-related components and were not calibrated to a specific level of emission performance. OBD II was developed to address these issues and to allow more accurate diagnosis by technicians.

The main goal of OBD II systems is to detect when engine or system wear or when component failure causes exhaust emissions to increase by 50% or more. According to the guidelines of OBD II, all vehicles have the following:

- A universal diagnostic test connector, known as the **data link connector (DLC)**, with dedicated pin assignments.
- A standard location for the DLC. It must be under the dash on the driver's side of the vehicle and must be visible.
- A standard list of **diagnostic trouble codes (DTCs)**.
- A standard communication protocol.
- The use of common scan tools on all vehicle makes and modes.
- Common diagnostic test modes.
- Vehicle identification must be automatically transmitted to the scan tool.
- Stored trouble codes must be able to be cleared from the computer's memory with the scan tool.
- The ability to record, and store in memory, a snapshot of the operating conditions that existed when a fault occurred.
- The ability to store a code whenever something goes wrong and affects exhaust quality.
- A standard glossary of terms, acronyms, and definitions must be used for all components in the electronic control systems.

To meet these standards, many new technologies were incorporated into the various control systems. Because nearly anything in an automobile can affect its emissions, the technologies carried over to all electronic control systems. The result is more efficient vehicles with more efficient engines and accessories.

By-Wire Technology

One of the things that has become a reality because of the high-powered computers used in today's vehicles is by-wire technology. Currently this technology has eliminated the mechanical connection from the throttle pedal and the fuel injection system **(Figure 22–32)**. Shift-by-wire technology is also used. In addition, it is being used by a few manufacturers in parking brake systems and adaptive cruise control. Soon it will be used in many other systems, such the brake and steering systems.

Drive-by-wire systems use sensors to translate the movement of pedals, the steering wheel, and other parts into electronic signals. The vehicle's computer receives these signals and commands electric motors to perform the function ordered by the driver. These systems respond much quicker than mechanical

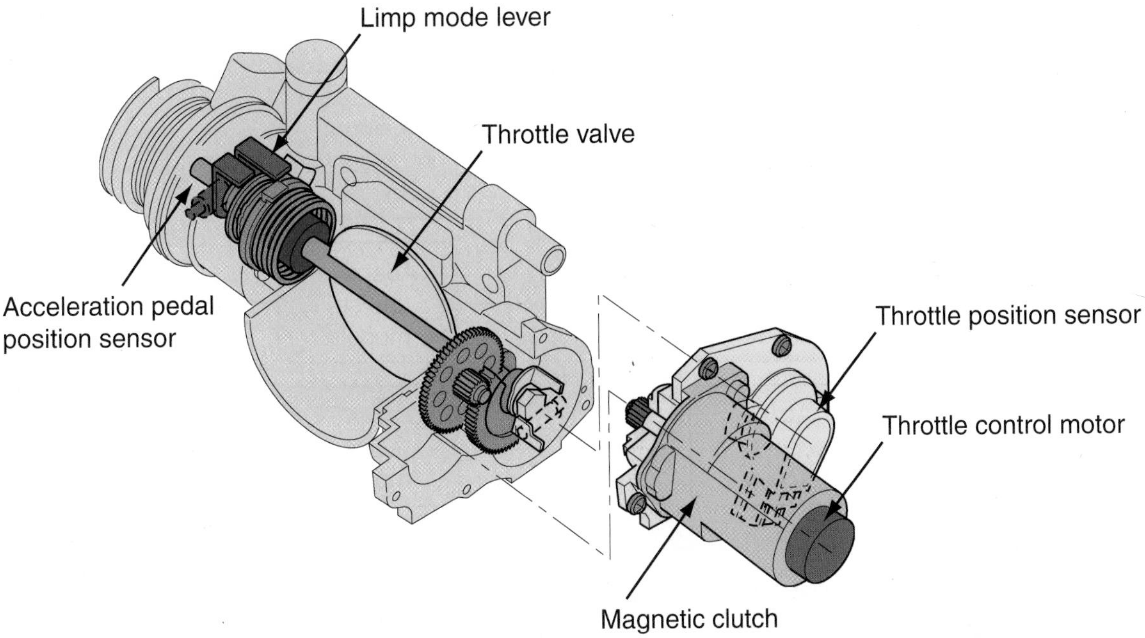

Figure 22-32 This electronic throttle control unit eliminates the need for a mechanical linkage between the throttle pedal and the fuel injection system. *Courtesy of Toyota Motor Sales, U.S.A., Inc.*

linkages and can send feedback to the computer as they operate.

MULTIPLEXING

Today's vehicles have hundreds of circuits, sensors, and other electrical parts. In order for the control systems to operate correctly, there must be some communication between them. Communication can take place through wires connecting each sensor and circuit to the appropriate control module. If more than one control module is involved, additional pairs of wires must connect the sensor or circuit to the other modules. The result of this communication network is miles of wires and hundreds of connectors. To eliminate the need for all of these wires, manufacturers are using multiplexing **(Figure 22–33)**.

Multiplexing, also called in-vehicle networking, provides efficient communications between vehicle systems. Multiplexing relies on one wire that allows many systems to communicate instead of many wires. A multiplex wiring system uses a **serial data bus** that connects different computers or control modules together. Each module can transmit and receive digital codes over the serial data bus, allowing one module to share information with other modules. For example, the signal relating to engine speed may be required by the engine control, transmission control, electronic brake control, and suspension control modules. Rather than have separate engine speed inputs for each module, the serial data bus carries the information, as well as other information, to all of the control modules.

Each sensor is wired directly to the control module that relies heavily on the sensor's signal. That control module sends the information, in binary code, to the serial data bus. Each control module has a code-reading device or chip that reads and sends messages on the serial data bus. Some chips can only send or only receive, depending on their purpose. All information on the serial data bus is available for all control modules. However, the chip of each device compares the coded message to its memory list to see whether the information is relevant to its own operation.

The chip is also used to prevent the signals from overlapping by allowing only one signal to be transmitted at a time. Each digital signal is preceded by an identification code that establishes its priority. If two modules attempt to send a message at the same time, the signal with the higher priority code is transmitted first. Because a control module processes only one input at a time, it orders the signals as it needs them.

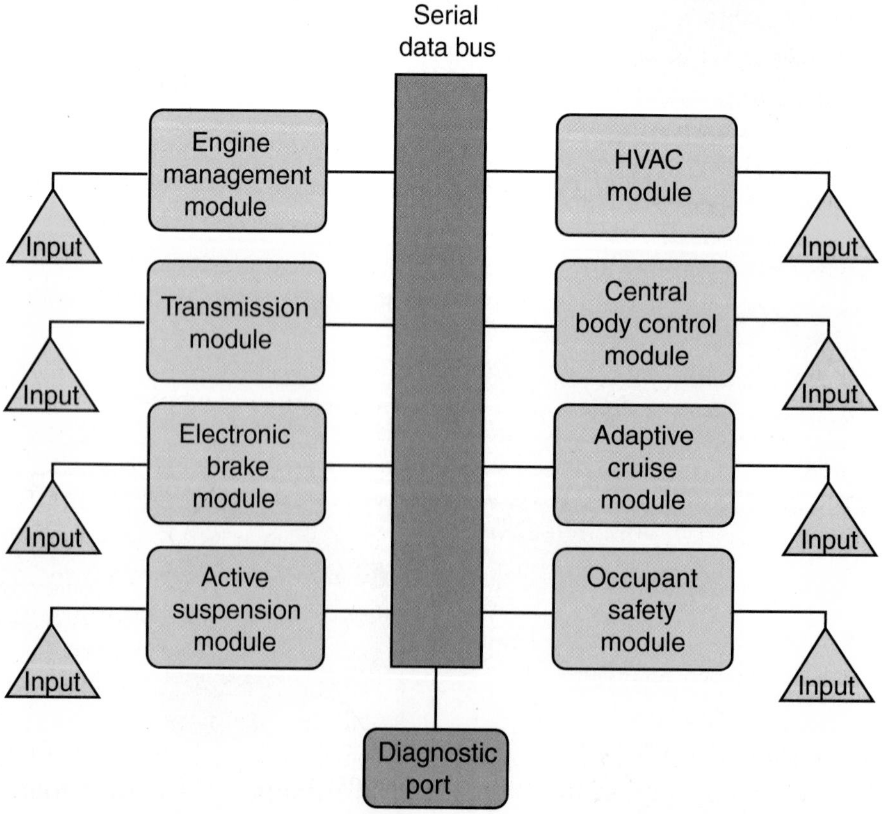

Figure 22-33 A multiplexed system uses a serial data bus to allow communications between the various control modules.

When one input is being received, the others are disregarded. Although it may appear to cause a time lapse, realize that the communication rate for most computers is between 10,000 and 1,000,000 bits per second.

Keep in mind that data are conveyed in binary numbers and therefore must be interpreted by some type of data processing before they become information. The stream of data across the bus is called serial data. It is essentially data that are transferred to and from a computer, one bit at a time. On many vehicles, serial data can be monitored with a scan tool connected to the vehicle's DLC. Monitoring serial data allows technicians to diagnose the various control modules and to check for DTCs.

The serial data bus is typically made of two wires: a ground wire and a serial bus transmission wire. These wires are twisted together to reduce magnetic interference, which can cause false information.

Advantages

Multiplexing offers many advantages over traditional wiring:

- The need for redundant sensors is eliminated because sensor data, such as vehicle speed and engine temperature, are available on the serial data bus where they can be used by several control modules.
- Accessories and vehicle features can be easily added to the vehicle through software changes.
- Fewer wires are required for the operation of each system, which means smaller wiring harnesses, lower cost and weight, and improved serviceability, reliability, and installation. Without multiplexing, it is necessary to add a ground, a power source, and control wires whenever an electronic component is added to the vehicle.
- Improved communications between control modules allows for more accurate recording and reporting of faults, which helps in locating and solving problems.

As the electrical content of today's vehicles continues to increase, the need for networking is even more evident.

Types of Multiplexing

There are four basic techniques used for multiplexing:

- Frequency division multiplexing (FDM). This is an analog technique in which each communications channel is assigned a frequency. The frequencies are stacked on top of each other and many frequencies can be sent at once. To prevent interference from other channels, each is separated by a small frequency.
- Time division multiplexing (TDM). This is a digital technique in which a sample of each channel is inserted into the data stream. The sample period is fast enough to sample all channels within a fixed period. It works like a very fast mechanical switch.
- Statistical time division multiplexing (STDM). This uses chips to allocate time only to channels when it is needed. This means more channels can be connected to the bus because the chips statistically compensate for times when a particular system is not being used.
- Wavelength division multiplexing (WDM). This is used in fiber-optic networks where multiple signals are transmitted as light is split into different wavelengths.

Communication Protocols

A protocol is the name for the language that computers speak when they are talking to each other. The differences in protocol are based on the speed and the technique used. The SAE has classified the different protocols by their speed and operation.

- Class A (low-speed communication): This is used for convenience systems, such as entertainment systems, audio, trip computer, seat controls, windows, and lighting. Most Class A functions require inexpensive, low-speed communication and use a generic universal asynchronous receiver/transmitter (UART). These functions are proprietary and have not been standardized by the industry.
- Class B (medium-speed communication): Class B multiplexing is used primarily with the instrument cluster, vehicle speed, and emissions data recording. Contained within this classification are different standards, designated by a number. The most commonly used is the SAE J1850 standard. Further, these standards are divided by their operation. One is a variable pulse width (VPW) type that uses a single bus wire. Another is a pulse width modulation (PWM) type that uses a two-wire differential bus.
- Class C (high-speed communication): This protocol is for real-time control of the powertrain, vehicle dynamics, and brake-by-wire. This protocol can use a twisted pair, but shielded coaxial cable or fiber optics may be used for less noise interference. The predominant class C protocol is CAN 2.0 (controller area network version 2.0). The CAN serial data line is a high-speed serial data bus

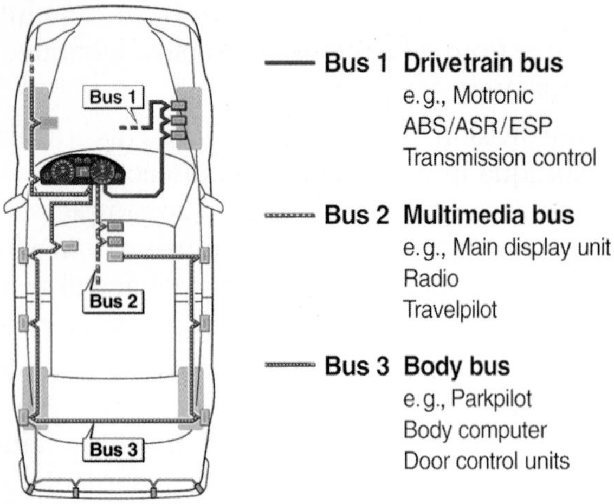

Figure 22-34 It is common to find a variety of protocols in a single vehicle. *Courtesy of Robert Bosch GmbH, www.bosch-presse.de*

that ensures that the required real-time response is maintained. CAN assigns a unique identifier to every message. The identifier classifies the content of the message and the priority of the message being sent. Each module processes only those messages whose identifiers are stored in the module's acceptance list.

It is common to find a variety of the different classes of multiplexing in a single vehicle **(Figure 22–34)**. Some systems, such as powertrain control and vehicle dynamics, require high-speed communications, whereas other systems do not.

Early multiplexed systems were often based on proprietary serial buses using generic UART or custom devices. This called for dedicated and specific scan tools, each with the ability to work only on specific systems. OBD II called for standardized diagnostic tools, which meant standard protocols had to be implemented. Currently, Class B communications is the standard protocol; however, many systems use Class C. Starting in 2008, Class C communications is the mandated protocol for diagnostics. CAN buses are found in nearly all late-model vehicles.

CAN Buses

The total network in most vehicles is comprised of two or three CAN buses. Each of these networks operates at different speeds. The different CAN buses are identified by a prefix or suffix. For example, a medium-speed bus may be called CAN B or MS-CAN. Likewise, a high-speed bus can be called CAN C or HS-CAN. Manufacturers are not consistent with these labels, so there are a variety of them.

Low- or medium-speed CANs are typically used for body functions, such as:

- Interior and exterior lights
- Horn
- Locks
- Windshield wipers
- Seats
- Window
- Sound systems

A high-speed bus is used for real-time functions such as:

- Engine management
- Antilock brake systems
- Transmission control
- Tire pressure monitoring systems
- Vehicle stability systems

These networks are integrated through the use of a gateway. A **gateway** module allows for data exchange between the different buses. It translates a message on one bus and transfers that message to another bus without changing the message **(Figure 22–35)**. The gateway interacts with each bus according to that bus's protocol. This is an important function; some information must be shared. In many vehicles; the BCM serves as the gateway for the different buses.

SHOP TALK

Although the BCM is designated as a body system, a vehicle may not start if the BCM is not operating correctly. This is due to its role as the gateway. If there is no communication between the security or antitheft system and the engine control system, the engine will not start. The PCM needs to know that it is okay to run the engine.

Each twisted wire in the CAN bus carries a different voltage **(Figure 22–36)**. Even the slightest change in voltage can affect the operation of one or more systems; therefore, all potential for voltage spikes, electrical noise, or induction must be eliminated. Twisting the wires eliminates the possibility of voltage being induced in one wire as current flows through the other. To eliminate other potential spikes and noise, two 120 Ω resistors are connected in parallel across the ends of the main CAN bus wires. These are called terminating resistors **(Figure 22–37)**. The location of the resistors varies. One may be located in the fuse block, and others can be internal to the ECM or PCM and the BCM.

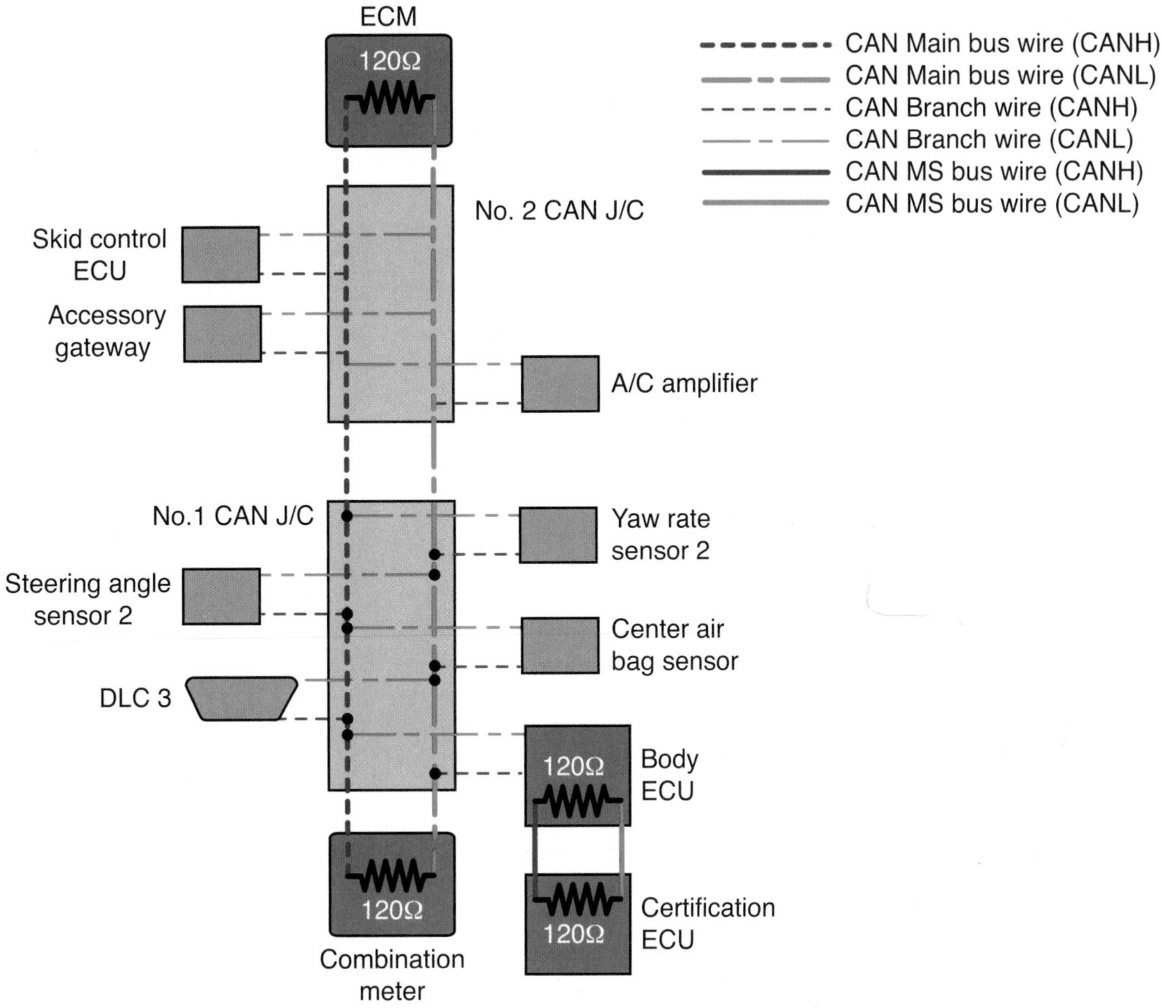

Figure 22-35 A basic look at a CAN communication system.

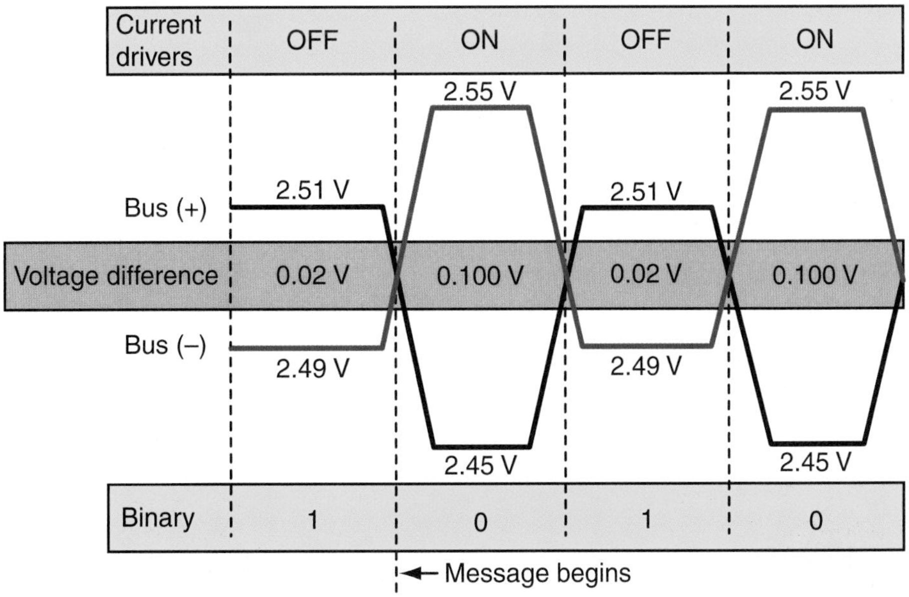

Figure 22-36 In order for a message to be transmitted, the drivers are energized to pull up the bias on the + bus and pull down on the – bus. There must be a voltage differential between the wires.

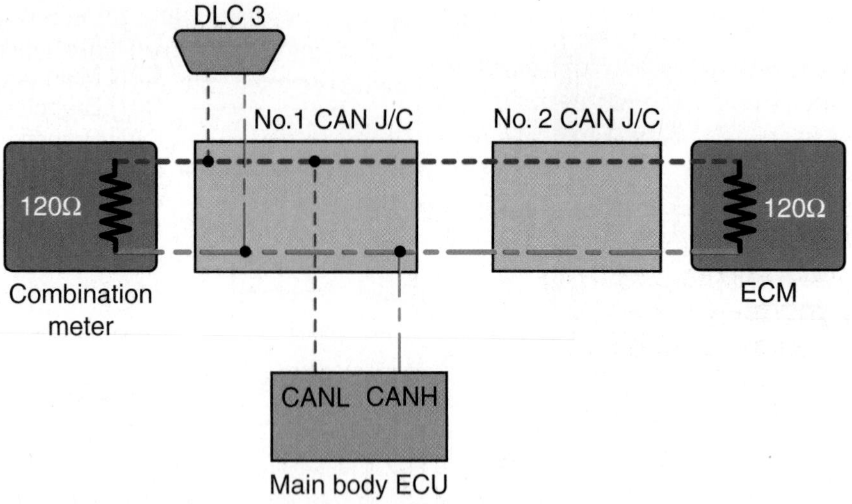

Figure 22-37 To eliminate potential voltage spikes and noise, two 120-Ω terminating resistors are connected in parallel across the ends of the main CAN bus wires.

PROTECTING ELECTRONIC SYSTEMS

The last thing a technician wants to do when a vehicle comes into the shop is create problems. This is especially true when it comes to electronic components. You should be aware of the ways to protect electrical systems and electronic components during storage and repair. Keep the following in mind at all times:

- Vehicle computer-controlled systems should avoid giving and receiving jump-starts due to the possibility of damage caused by voltage spikes.
- Do not connect or disconnect electronic components with the key on.
- Never touch the electrical contacts on any electrical or electronic part. Skin oils can cause corrosion and poor contacts.
- Be aware of any part that the manufacturer has marked with a code or symbol to warn technicians that it is sensitive to electrostatic discharge **(Figure 22–38)**.
- Before touching a computer, always touch a good ground first. This safely discharges any static electricity. Static electricity can generate up to 25,000 volts and can easily damage a computer.
- Tool companies offer static-proof work mats that allow work inside the vehicle without the fear of creating static electricity.
- Tool companies also have grounding wrist straps. A wire connects the wrist strap to a good ground.
- Never allow grease, lubricants, or cleaning solvents to touch the end of the sensor or its electrical connector.
- Be careful not to damage connectors and terminals when removing components. This may require special tools.
- When procedures call for connecting test leads, or wires, to electrical connections, use care and follow the manufacturer's instructions. Identify the correct test terminals before connecting the test leads.
- Do not connect jumper wires across a sensor unless indicated in the service manual to do so.
- Never apply 12 volts directly to an electronic component unless instructed to do so.
- Never use a testlight to test electronic ignition or any other computer-controlled system unless instructed to do so.
- Accidentally touching two terminals at the same time with a test probe can cause a short circuit.
- The sensor wires should never be rerouted. When replacing wiring, always check the service manual and follow the routing instructions.
- Disconnect any module that could be affected by welding, hammering, grinding, sanding, or metal straightening.

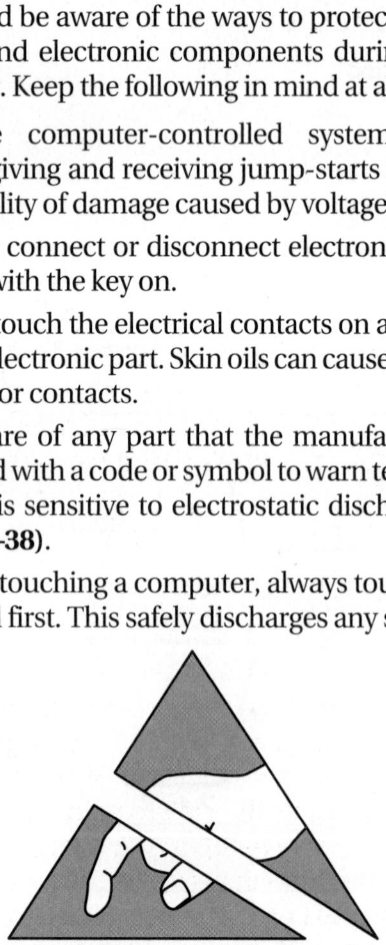

Figure 22-38 GM's electrostatic discharge (ESD) symbol warns technicians that a part or circuit is sensitive to static electricity.

DIAGNOSING BCMs

Before troubleshooting a system operated by a control module, check the service manual to identify any special procedures or precautions. When testing a system, you are basically trying to isolate a problem in one of the basic functions of the computer and its system.

The BCM continuously monitors operating conditions for possible system malfunctions. It compares system conditions against programmed parameters. If the conditions fall outside of these limits, it detects a malfunction and sets a trouble code that indicates the portion of the system that has the fault.

If the malfunction causes improper system operation, the computer may minimize the effects of the malfunction by using fail-safe action. During this the computer will control a system based on programmed values instead of the input signals. This allows the system to operate on a limited basis instead of shutting down completely.

Trouble Codes

There are as many ways to perform BCM diagnostics as there are automobile manufacturers. Nearly all vehicles require a scan tool to retrieve DTCs. The scan tool is plugged into the diagnostic connector for the system being tested. The technician chooses the system to be tested through the scan tool. Once the DTCs are retrieved, follow the appropriate diagnostic chart to isolate the fault. It is important to check the codes in the order recommended by the manufacturer.

On systems that do not retain codes after the ignition is switched off, operate the vehicle until the problem occurs again. Then retrieve the fault code before switching the ignition off. Remember, the trouble code does not necessarily indicate the faulty component; it only indicates the circuit of the system that is not operating properly. To locate the problem, follow the diagnostic procedure in the service manual for the code received.

Diagnosis should continue with a good visual inspection of the circuit involved with the code. Check all sensors and actuators for physical damage. Check all connections to sensors, actuators, control modules, and ground points. Check wiring for signs of burned or chafed spots, pinched wires, or contact with sharp edges or hot exhaust parts. Also check all vacuum hoses for pinches, cuts, or disconnects.

Communication Checks

Performing diagnostic checks on vehicles with a multiplex system should begin with a communications check. If the different control modules are not communicating with each other, there is no way to properly diagnose the systems. When a scan tool is installed, it will try to communicate with every module that could be in the vehicle. If an option is not there, the scan tool will display "No Comm" for that control module. That same message will appear if the module is present but not communicating. Therefore, always refer to the service information to identify what modules should be present before coming to any conclusions.

The system periodically checks itself for communication errors. The different buses send messages to each other immediately after it sends a message. The message between messages checks the integrity of the communication network. All of the modules in the network also receive a message within a specific time. If the message is not received, the control module will set a DTC stating that it did not receive the message.

There are three types of DTCs used by CAN buses:

- *Loss of communication.* Loss of communication (and Bus-off) DTCs are set when there is a problem with the communication between modules. This could be caused by bad connections, wiring, or the module. Note: In most cases, a lost communication DTC is set in modules other than the module with the communication problem.
- *Signal error.* The control modules can run diagnostics on some input circuits to determine if they are operating normally. If a circuit fails the test, a DTC will set.
- *Internal error.* The modules also run internal checks. If there is a problem, it will set an internal error DTC.

Bus Wire Service

If the bus wire needs repair due to an open, short, or high resistance, it must not be relocated or untwisted. The twisting serves an extremely important purpose **(Figure 22–39)**. After a bus wire has been repaired by soldering, wrap that part of the wire with vinyl tape. Never run the repair wire in such a way that it bypasses the twisted sections. CAN bus wires are likely to be influenced by noise if you bypass the twisted wires.

Reprogramming Control Modules

Before going deeply into diagnosing an electronic control circuit, it is wise to check all TSBs that may relate to the problem. Often there will be one that recommends reprogramming of the computer. This is typically called "**flashing**" the computer. When a computer is flashed, the old program is erased and a new one written in. Reprogramming is often necessary when the manufacturer discovers a common concern that can be solved through changing the system's software. New programs are downloaded into the scan tool and then downloaded into the computer through a dedicated circuit.

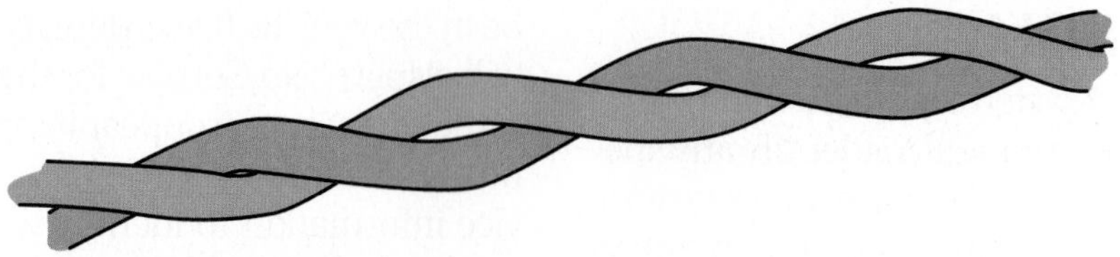

Figure 22-39 Electronic units use specially shielded wires for protection from unwanted induced voltages that can interfere or change voltage signals.

Each type of scan tool has a different procedure for flashing. Always follow the manufacturer's instructions. Some scan tools are connected to a PC and the software is transferred from a CD or a Web site. Photo Sequence 20 shows a typical procedure for flashing a BCM.

TESTING ELECTRONIC CIRCUITS AND COMPONENTS

Most electronic circuits can be checked in the same way as other electrical circuits. However, only high-impedance meters should be used. A lab scope is a valuable tool for diagnosing electronic circuits. The scope is primarily used to measure voltages, pulse, and duty cycle.

The scope displays voltage over time and the increments for voltage and time can be changed to provide a good look at the activity of a circuit. The time periods show the frequency of a voltage signal. *Frequency* is a term that describes how often a signal performs a complete cycle. Frequency is measured in hertz. To determine the frequency of something, divide the length of time it takes to complete one cycle into 1.

Whenever using a scope on a circuit, always follow the meter's instruction manual for hookup and proper settings. Most service manuals have illustrations of the patterns that are expected from the different electronic components. If the patterns do not match those in the manual, a problem is indicated. The problem may be in the component or in the circuit. Further testing is required to locate the exact cause of the problem.

Measuring Changing Voltages

A DMM may have AC selection modes for voltage and amperage. These modes are used to measure voltages and amperages that change polarity or levels very quickly. Most meters display the average voltage or current in an AC circuit. Some meters display root-mean-square (RMS) readings, which are very close to being average readings; however, there may be slight differences as this scale compensates for extreme fluctuations in voltage and current flow.

RMS refers to the effective or useful value of an AC signal. To determine the RMS value, peak voltage readings are multiplied by the RMS constant (0.707). The effective RMS value is calculated by squaring the instantaneous values of all the points on the sine wave, taking the average of these values, and extracting the square root. Therefore, the effective value is the root of the mean (average) square of the values along the sine wave. The resulting answer from this mathematical formula is the RMS.

Checking Diodes

Multimeters can be used to check diodes, including zener diodes and LEDs. Regardless of the bias of the diode, it should allow current flow in one direction only. Connect the meter's leads across the diode. Observe the reading on the meter. Then reverse the meter's leads and again observe the reading. The resistance in one direction should be very high or infinite and close to zero in the other direction **(Figure 22–40)**. If any other readings are observed, the diode is bad. A diode that has low resistance in both directions is shorted. A diode that has high resistance or an infinite reading in both directions is open.

You may run into problems when checking a diode with a high-impedance DMM. Because many diodes will not allow current flow through them unless the voltage is at least 0.6 volt, a digital meter may not be able to forward bias the diode. This will result in readings that indicate the diode is open, when in fact it may not be. Because of this problem, many multimeters are equipped with a diode testing feature. This allows for increased voltage at the test leads. Again, continuity should be present in one direction and not the other. Some meters will make a

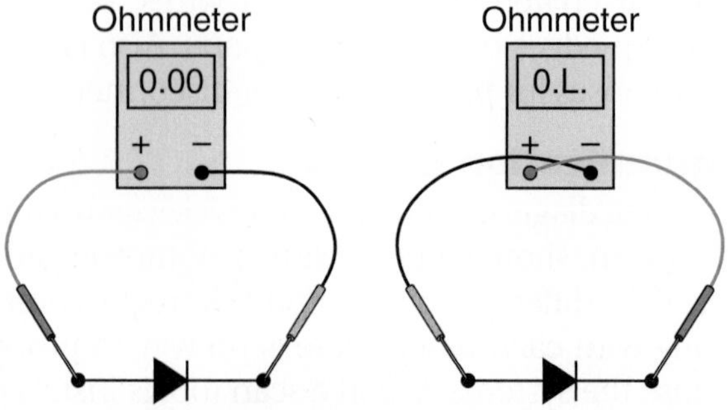

Figure 22-40 Testing a diode with an ohmmeter.

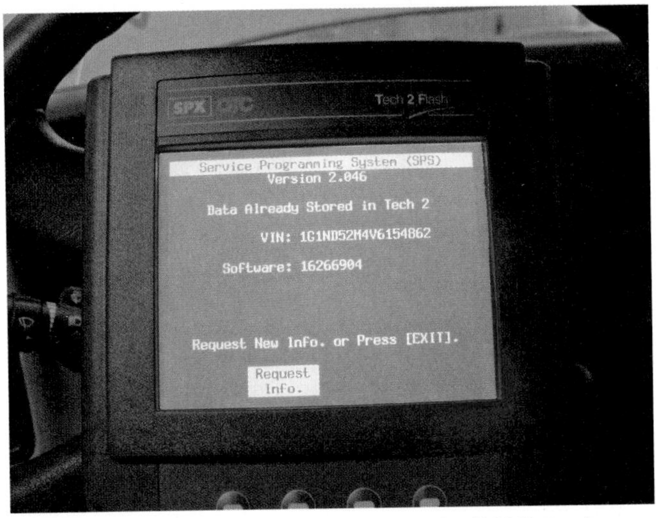

P20-1 *Use the scan tool to retrieve the BCM's part number or the vehicle's VIN and record it.*

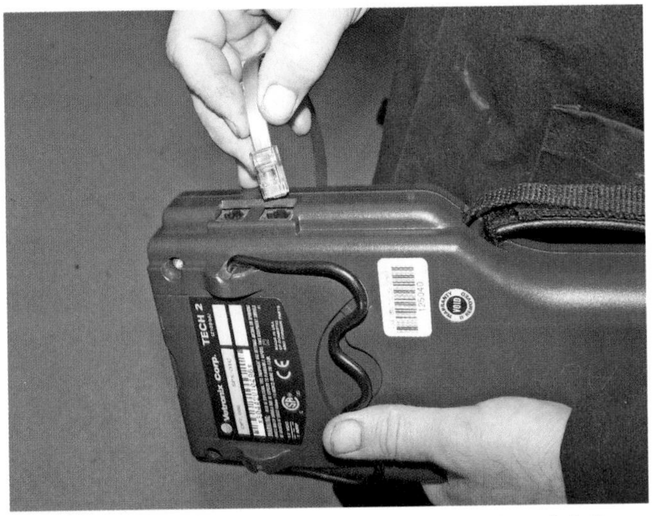

P20-2 *Connect the scan tool to a PC that can link the scan tool to the flash software. Some scan tools will connect directly to an Internet site.*

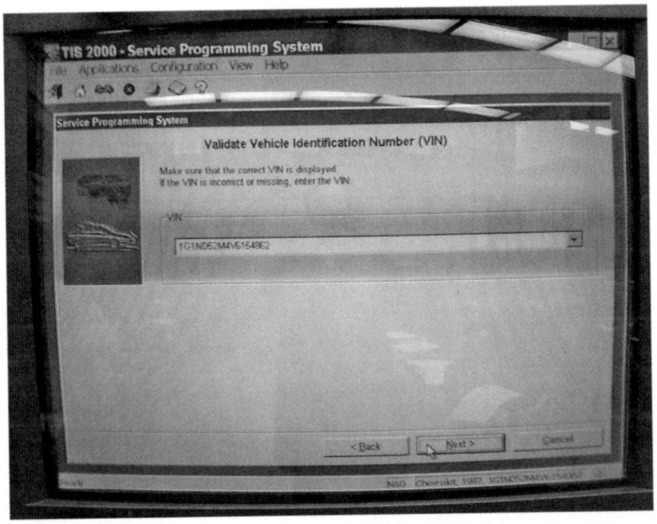

P20-3 *Enter the BCM part number in the appropriate field and select "Show Updates" on the menu.*

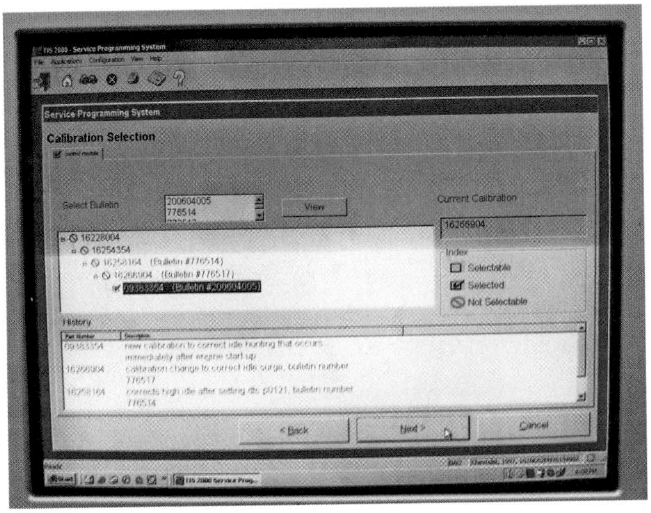

P20-4 *Select the desired flash line.*

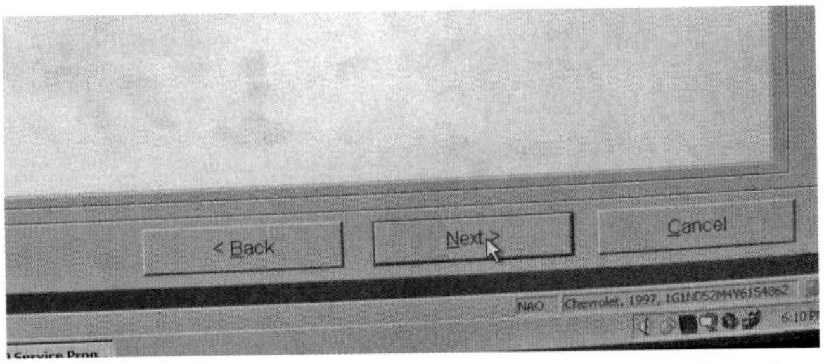

P20-5 *Hit the "Next" button to begin downloading of the software.*

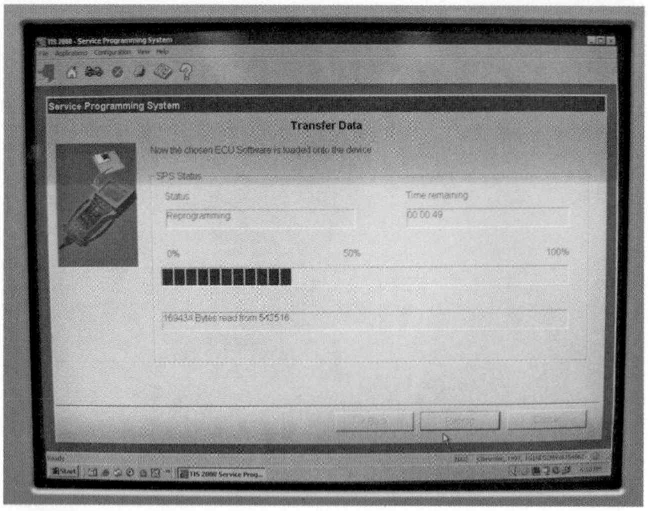

P20-6 *Monitor the progress of the downloading to the scan tool.*

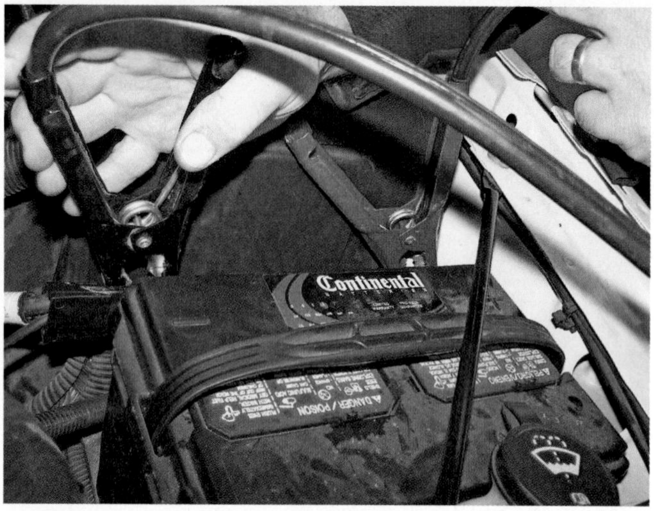

P20-7 *Once downloading is complete, connect a battery charger to the vehicle's battery. Turn it on and maintain about 13.5 volts at the battery.*

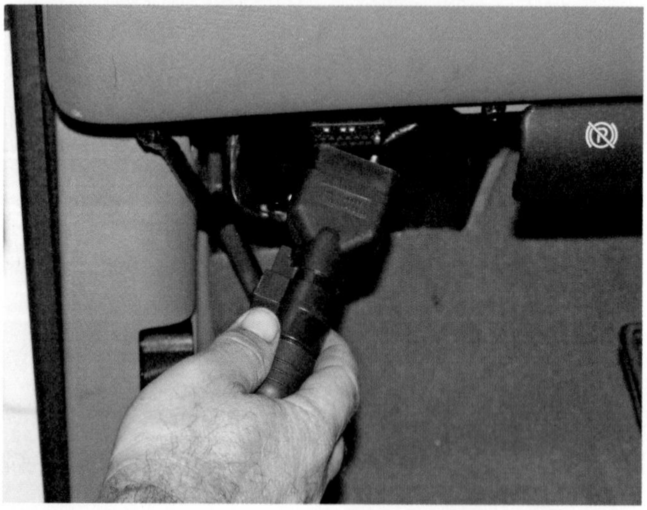

P20-8 *Connect the scan tool to the DLC and turn the scan tool on.*

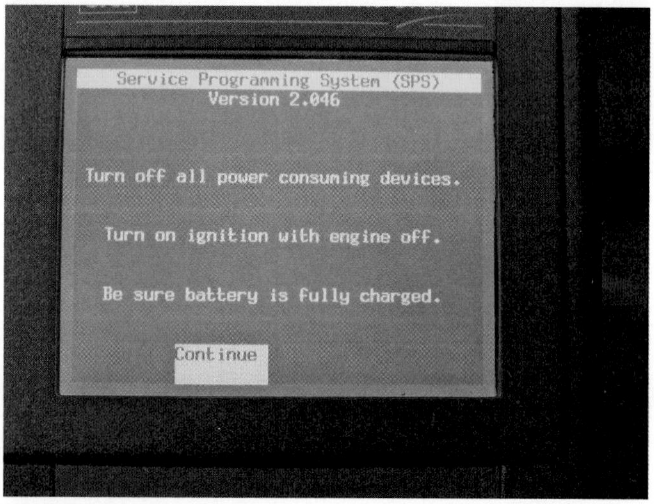

P20-9 *Move through the menus on the scan tool until the desired flash screen is shown. Then follow all instructions given on the tool.*

beeping noise when there is continuity during a diode check.

Diodes can also be tested with a voltmeter. Using the same logic as when testing with an ohmmeter, test the voltage drop across the diode. The meter should read low voltage in one direction and higher voltage in the other direction. Most automotive diodes will drop 500 to 650 mV.

CAUTION!

To avoid possible damage to the meter or to the part being tested, disconnect the circuit's power and discharge all high-voltage capacitors before testing.

Measuring Capacitance

Before checking a capacitor, make sure it is discharged. Use the DC voltage mode on a DMM to check it. Most DMMs have many different capacitance ranges **(Figure 22–41)**. Set the meter to read in the range that is the closest match to the expected reading. On many meters, if the range is too low for the capacitor, the display will show "diSC." Set the meter to read the capacitance and connect the meter leads across the capacitor. If there is no reading, reverse the leads before concluding that the capacitor is bad. Once your have a reading, compare that to specifications.

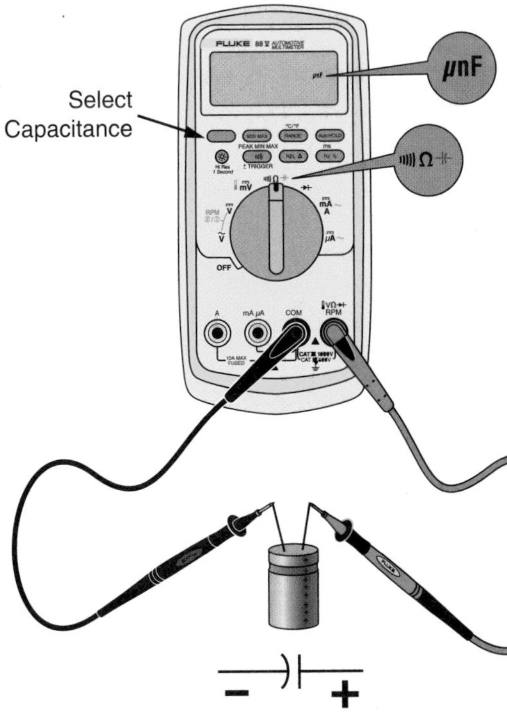

Figure 22–41 To measure capacitance, set up the DMM as shown. *Reproduced with the permission of Fluke*

KEY TERMS

A/D converter
Actuator
Base
Baud rate
Binary code
Bit
Byte
Capacitance
Capacitor
Central processing unit (CPU)
Clamping diode
Collector
Data link connector (DLC)
Diagnostic trouble code (DTC)
Electrically erasable PROM (EEPROM)
Electrically inert materials
Emitter
Erasable PROM (EPROM)
Farad (F)
Feedback
Field-effect transistors (FETs)
Flashing
Forward bias
Gateway
Hall-effect switch
Impurities
Keep-alive memory (KAM)
Logic gate
Look-up tables
Magnetic pulse generator
Microprocessor
Multiplexing
Nonvolatile RAM (VRAM)
Output driver
Pickup coil
Piezoresistive
Programmable read-only memory (PROM)
Random-access memory (RAM)
Read-only memory (ROM)
Reference voltage (Vref) sensor
Reverse bias
Schmitt trigger
Serial data bus
Square wave
Timing disc
Trimmers
Ultracapacitors
Voltage-generating sensors
Wheatstone bridge
Zener diode

SUMMARY

- All basic laws of electricity apply to electronic controls.
- A capacitor is used to store and release electrical energy.
- Ultra- or supercapacitors charge and discharge very quickly and have energy available shortly after they have been discharged.
- A diode allows current to flow in one direction but not in the opposite direction. It is formed by joining P-type semiconductor material with N-type semiconductor material.
- A transistor resembles a diode with an extra side. There are PNP and NPN transistors. They are used as switching devices. A very small current applied

to the base of the transistor controls a much larger current flowing through the entire transistor.

- Computers are electronic decision-making centers. Input devices called sensors feed information to the computer. The computer processes this information and sends signals to controlling devices.
- Most input sensors are reference voltage sensors or voltage-generating sensors.
- Computers work digitally; therefore, they must receive digital signals or convert analog signals to digital signals before processing them.
- A typical electronic control system is made up of sensors, actuators, a microcomputer, and related wiring.
- The microcomputer and its processors are the heart of the computerized engine controls.
- There are three types of computer memory used: ROM, PROM, and RAM.
- Output sensors or actuators are electromechanical devices that convert current into mechanical action.
- On-board diagnostic capabilities are incorporated into a vehicle's computer to monitor virtually every component that can affect emission performance. Each component is checked by a diagnostic routine to verify that it is functioning properly.
- Multiplexing provides communications between vehicle systems. It uses a serial data bus that connects different computers or control modules together.
- Controller area network (CAN) is the most commonly used network protocol in today's vehicles.
- Static electricity can generate up to 25,000 volts and do damage to components. Precautions for static discharge must be taken when handling electronic components.
- Diagnosis of a computer-controlled system includes an early check for diagnostic trouble codes.
- Communications between the various control modules in a vehicle is critical to the vehicle's overall operation.
- There are three types of DTCs used by CAN buses: loss of communication, signal error, and internal error.
- At times the manufacturer will recommend that a control module be reprogrammed; this is often called flashing the computer.
- Most electronic circuits can be checked in the same way as other electrical circuits. However, only high-impedance meters should be used.

REVIEW QUESTIONS

1. The three main types of memory in a computer are called ______, ______, and ______.
2. ______ signals show any change in voltage.
3. Digital signals are typically called ______ patterns.
4. ______ means that data concerning the effects of the computer's commands are fed back to the computer as an input signal.
5. The type of memory that contains specific information about the vehicle and can be replaced or reprogrammed is called ______.
6. A ______ is the simplest type of semiconductor.
7. What is meant by the term *pulse width?*
8. Variable capacitors are called ______.
9. How should you test a diode with a multimeter?
10. What is the major difference between ROM and RAM memory in a microprocessor?
11. Which of the following is not a type of information stored in ROM?
    a. strategy
    b. look-up tables
    c. sensor input
    d. none of the above
12. *True or False?* In a multiplex wiring system, a serial data bus is used to allow communication between the various control modules in a vehicle.
13. Which of the following is not made of semiconductor material?
    a. zener diode
    b. thermistor
    c. PM generator
    d. Hall-effect switch
14. Which of the following is a basic function of a computer?
    a. to store information
    b. to process information
    c. to send out commands
    d. all of the above
15. What is the name of the module that allows the different serial buses in a multiplexed system to communicate with each other?

ASE-STYLE REVIEW QUESTIONS

1. Technician A says that when positive voltage is present at the base of an NPN transistor, the transistor is turned on. Technician B says that when an NPN transistor is turned on, current flows through the collector and emitter of the transistor. Who is correct?
   a. Technician A
   b. Technician B
   c. Both A and B
   d. Neither A nor B
2. While checking an A/C compressor clutch coil spike-suppressing diode with an ohmmeter, the meter shows infinite resistance when the diode is measured in both directions: Technician A says that the diode is open. Technician B says that all diodes should behave this way until a specified voltage is applied to them. Who is correct?
   a. Technician A
   b. Technician B
   c. Both A and B
   d. Neither A nor B
3. Technician A says that some types of voltage sensors provide input to the computer by modifying or controlling a constant, predetermined voltage signal. Technician B says that another type of voltage sensor is a voltage-generating sensor. Who is correct?
   a. Technician A
   b. Technician B
   c. Both A and B
   d. Neither A nor B
4. Technician A says that multiplexing is a way to add wires without increasing the weight of the vehicle. Technician B says that multiplexing uses bus data links. Who is correct?
   a. Technician A
   b. Technician B
   c. Both A and B
   d. Neither A nor B
5. Technician A says that when a diode is placed in a circuit with the positive side of the circuit connected to the positive side of the diode and the negative side of the circuit connected to the negative side if the diode, the diode is said to have reverse bias. Technician B says that the diode's P material is repelled by the positive charge of the circuit and is pushed toward the N material and that the N material is pushed toward the P. Who is correct?
   a. Technician A
   b. Technician B
   c. Both A and B
   d. Neither A nor B
6. Technician A says that a capacitor always discharges with the same voltage it was charged with. Technician B says that current only flows through a capacitor when it is charging or discharging. Who is correct?
   a. Technician A
   b. Technician B
   c. Both A and B
   d. Neither A nor B
7. While discussing multiplex system protocols: Technician A says that CAN communications is based on variable pulse widths carried on a single bus wire. Technician B says that Class B communications offers real-time communications over a twisted pair of wires. Who is correct?
   a. Technician A
   b. Technician B
   c. Both A and B
   d. Neither A nor B
8. Technician A says that the signal from a Hall-effect switch needs to pass through a Schmitt trigger before the computer can process it. Technician B says that a Hall-effect switch produces a digital voltage signal. Who is correct?
   a. Technician A
   b. Technician B
   c. Both A and B
   d. Neither A nor B
9. While diagnosing an electronic control system: Technician A turns the ignition off before disconnecting a connector or component. Technician B uses a jumper wire to bypass a potentially faulty sensor and then observes the scan tool. Who is correct?
   a. Technician A
   b. Technician B
   c. Both A and B
   d. Neither A nor B
10. Technician A says that based on current conditions, a reference voltage sensor modifies a fixed voltage from the computer and sends that voltage back to the computer. Technician B says that most reference voltage sensors are variable resistors or potentiometers. Who is correct?
    a. Technician A
    b. Technician B
    c. Both A and B
    d. Neither A nor B

CHAPTER

23

ELECTRICAL ACCESSORIES

OBJECTIVES

■ Know the basic operation of electric windshield wiper and washer systems. ■ Explain the operation of power door locks, power windows, and power seats. ■ Determine how well the defroster system performs. ■ Identify the components of typical radio and audio systems. ■ Understand how cruise or speed control operates and the differences of various systems. ■ Describe the operation of keyless entry systems. ■ Identify the various security disabling devices. ■ Understand the operation of the various security alarms.

Electrical accessories make driving safer, easier, and more pleasant for the driver and passengers. This chapter covers many of the common accessories. Other automotive electric and electronic equipment, such as passive seat belts and air bags, are described elsewhere in this book. Most accessories are controlled by the BCM **(Figure 23–1)** and after verification of the problem, diagnostics begin with retrieving DTCs and observing data on a scan tool **(Figure 23–2)**.

Go To ***Chapter 22 for a discussion and the procedures for connecting and using a scan tool on a BCM.***

WINDSHIELD WIPER/WASHER SYSTEMS

There are several types of windshield wiper systems. Both rear and front systems can be found on a vehicle. Headlight wipers and washers are also available and work in unison with the windshield wipers. Windshield wiper motors can have electromagnetic fields or permanent magnetic fields.

Permanent Magnet Motor Circuits

In motors with permanent magnet fields, motor speed is controlled by the placement of the brushes on the commutator. Three brushes are used: common, high speed, and low speed. The common brush carries current whenever the motor is operating. The low-speed brush and high-speed brush are placed in different locations. Often, the high-speed and common brushes oppose each other, whereas the low-speed brush is offset **(Figure 23–3)**.

The placement of the brushes determines the number of armature windings connected in the circuit. When battery voltage is applied to fewer windings, there is less magnetism and less counter-EMF. With less CEMF, armature current is higher. This high current results in higher motor speeds. When more windings are energized, the magnetic field around the armature is greater and there is more CEMF. This results in lower current flow and slower motor speeds.

Park Switch A park switch is incorporated into the motor. It operates off a cam or latch arm on the motor's gear **(Figure 23–4)**. The switch supplies voltage to the motor after the wiper switch has been turned off. This allows the motor to continue running until it has reached the park position. The park switch changes position with each motor revolution. The switch remains in the run position for approximately 90% of the revolution. It is in the park position for another 10% of a revolution. This does not affect the operation of the motor until the wiper switch is placed in the park position.

When the wiper control is in the high-speed position, voltage is applied through the switch to the

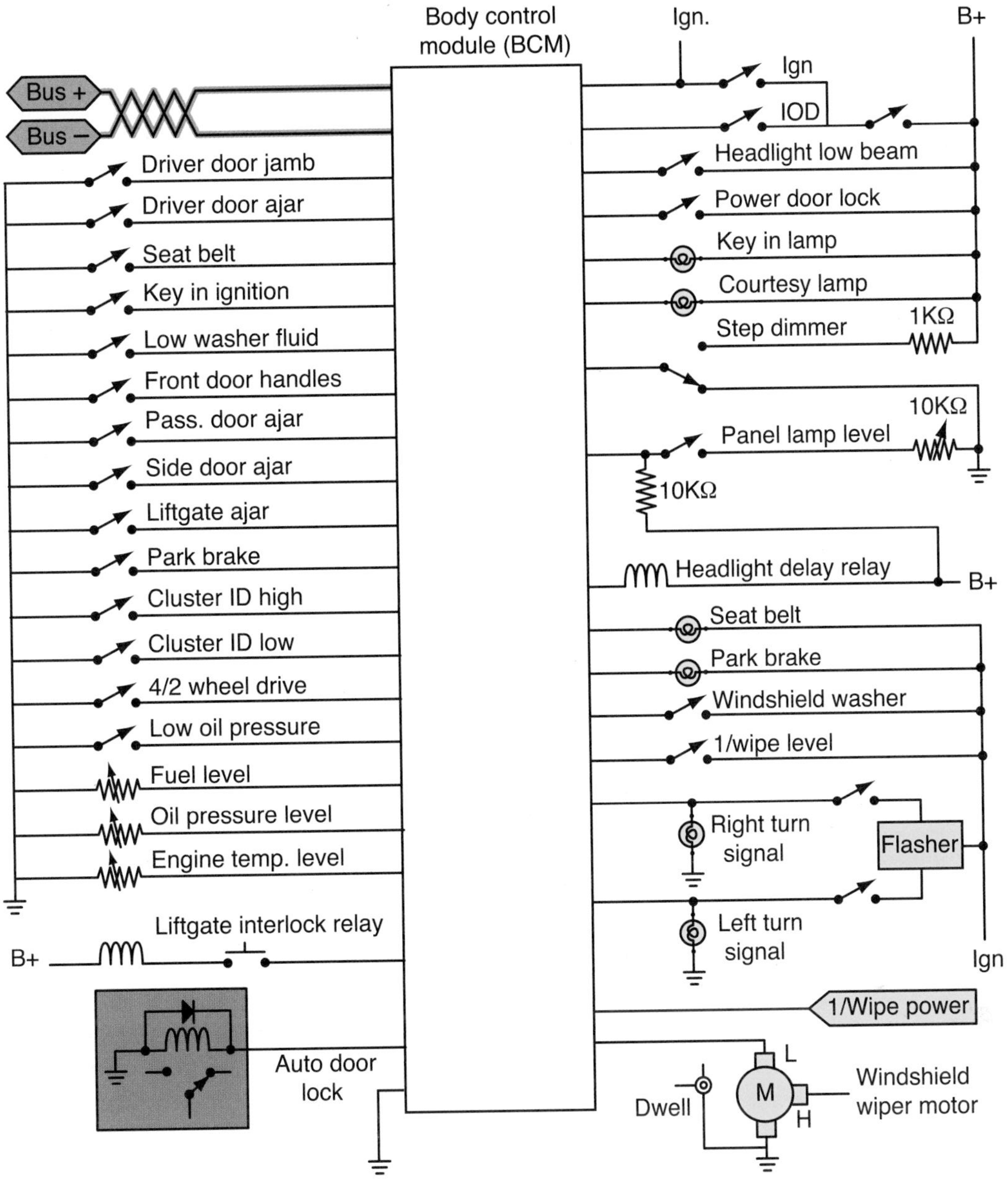

Figure 23-1 The BCM controls many of the vehicle's electrical systems. *Courtesy of Chrysler LLC*

Figure 23-2 The DLC is used to access diagnostic trouble codes and serial data.

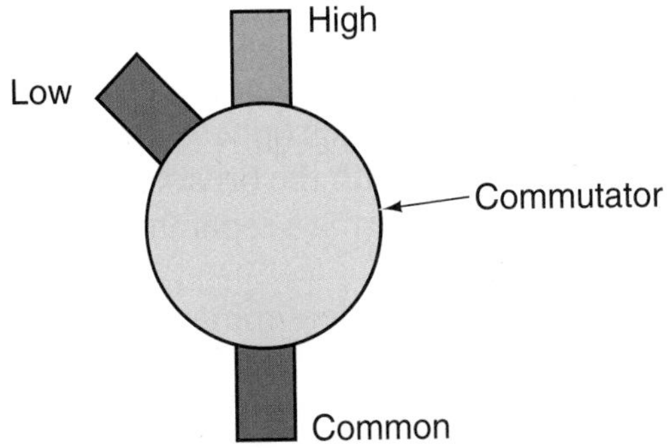

Figure 23-3 One style of brush arrangement has the high-speed brush opposite the common brush.

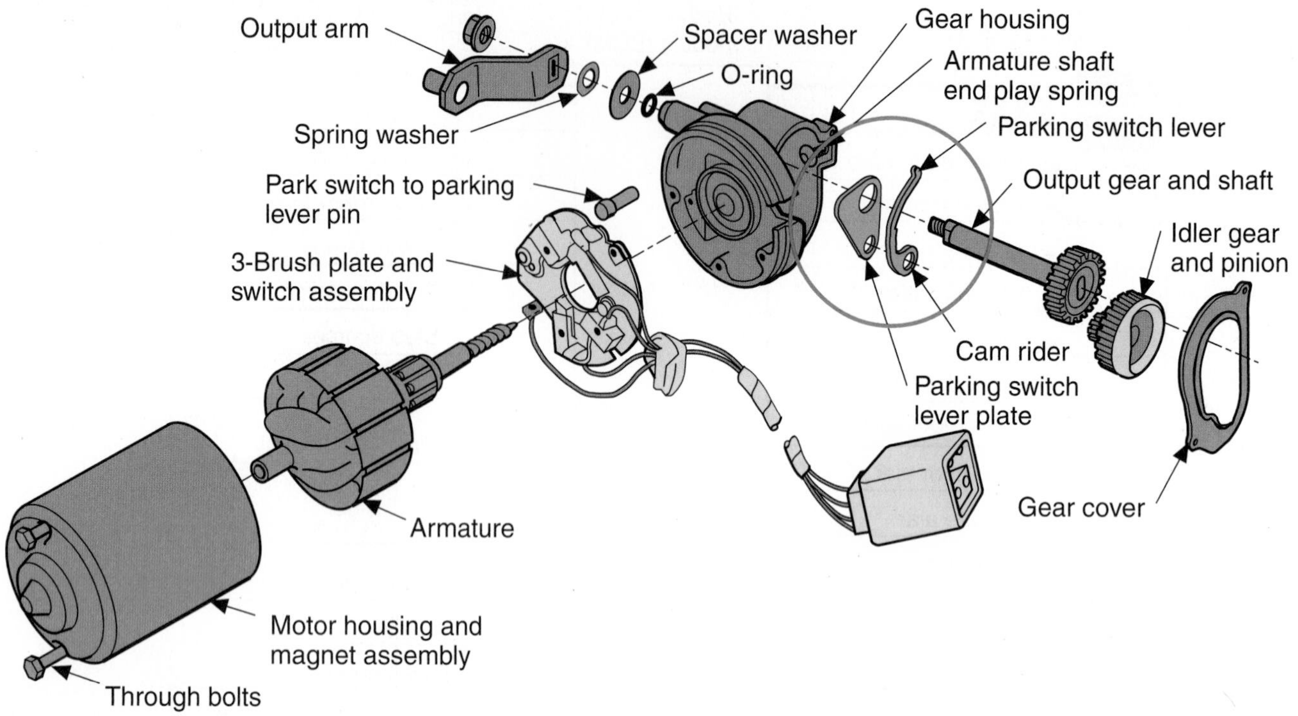

Figure 23–4 An exploded view of a wiper motor with a park mode.

high-speed brush **(Figure 23–5)**. Wiper 2 moves with wiper 1 but does not complete any circuits. When the switch is moved to the low-speed position, voltage is applied through wiper 1 to the low-speed brush. Wiper 2 also moves but does not complete any circuits.

When the switch is moved to the off position, wiper 1 opens. Voltage is applied to the park switch and wiper 2 allows current to flow to the low-speed brush. When the wiper blades reach their lowest position, the park switch is in the park position. This opens the circuit to the brush and the motor shuts off.

Electromagnetic Field Motor Circuits

Motors with electromagnetic fields have two brushes riding on the armature: a positive and a negative. The speed of the motor depends on the strength of the magnetic fields. Some two-speed and all three-speed wiper motors use two electromagnetic field windings. The two field coils are wound in opposite directions so that their magnetic fields oppose each other. The field is wired in series with the brushes and commutator. The shunt field forms a separate circuit off the series circuit to ground.

A ground side switch determines the path of current and the speed of the motor. One current path is directly to ground after the field coil and the other is to ground through a resistor. When the switch is placed in the low-speed position, the relay's contacts close and voltage is applied to the motor. The second wiper of the switch provides the path to ground for the shunt field. With no resistance in the shunt field coil, the shunt field is very strong and bucks the magnetic field of the series field. This results in slow motor operation.

When the switch is in the high-speed position, the shunt field finds its ground through the resistor. This results in low current and a weak magnetic field in the shunt coil; therefore, the armature turns at a higher speed.

Three-Speed Motors The control switch for a three-speed motor determines what resistors, if any, will be connected to the circuit of one of the fields **(Figure 23–6)**. When the switch is in the low-speed position, both field coils have the same amount of current flow. Therefore, the total magnetic field is weak and the motor runs slowly.

When the switch is in the medium-speed position, current flows through a resistor before going to the shunt field. This connection weakens the shunt coil and the motor's speed increases.

With the switch in the high-speed position, a resistor of greater value is connected to the shunt field. This connection weakens the magnetic strength of the coil and the motor runs faster.

Windshield Wiper Linkage and Blades Several arms and pivot shafts make up the linkage used to transmit the rotation of the motor to oscillate the windshield

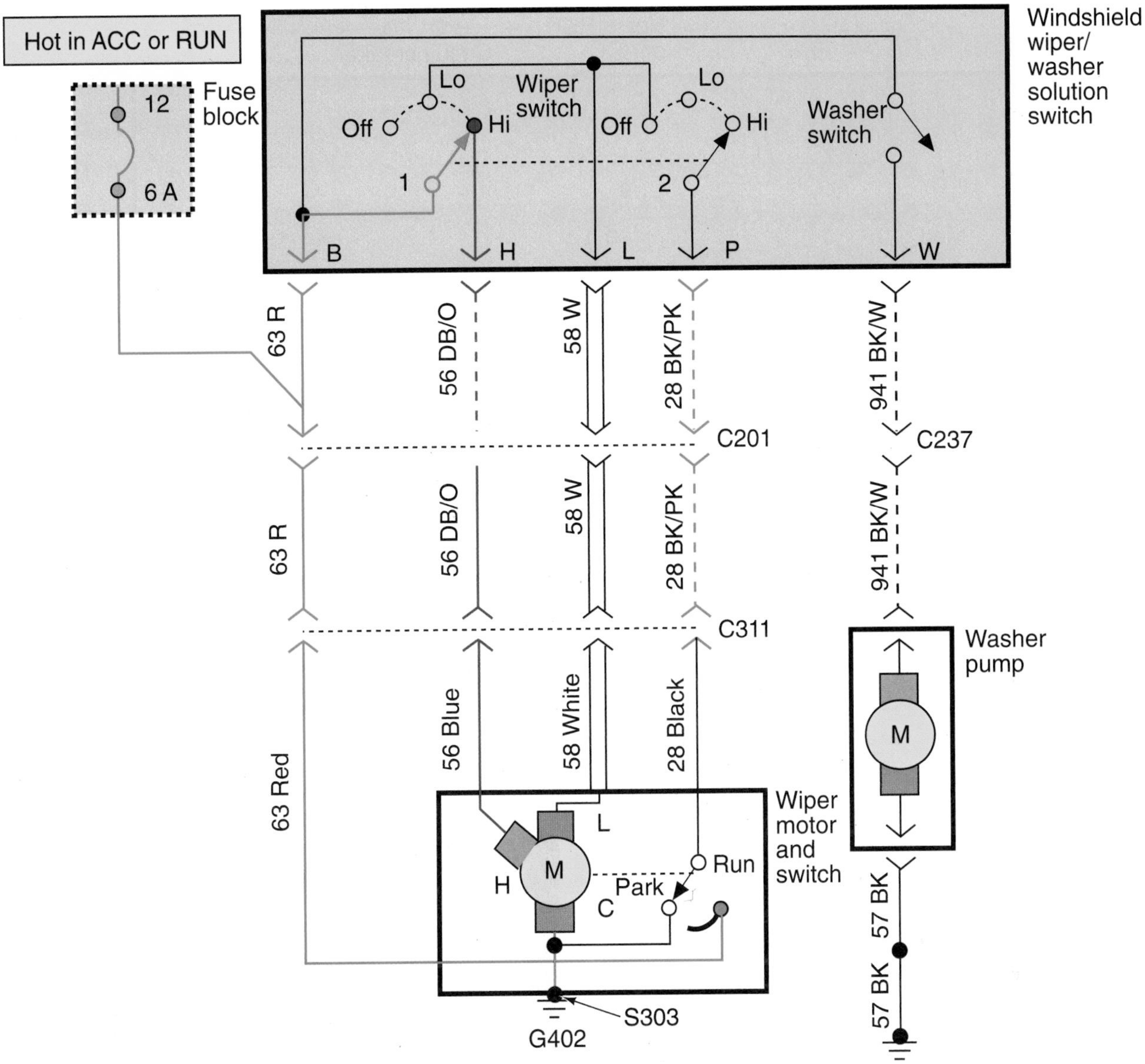

Figure 23-5 Current flow in the high-speed mode of a permanent magnet motor. *Courtesy of Ford Motor Company*

wipers. As the wiper motor runs, the linkage rotates the arms from left to right. The arrangement of the linkage causes the wipers' pivot points to oscillate. The wiper arms and blades are attached directly to the two pivot points.

A few wiper systems have two wiper motors that operate in opposite directions, thus creating the oscillation motion of the wipers **(Figure 23–7)**. These systems also occupy less space in the engine's cowl area.

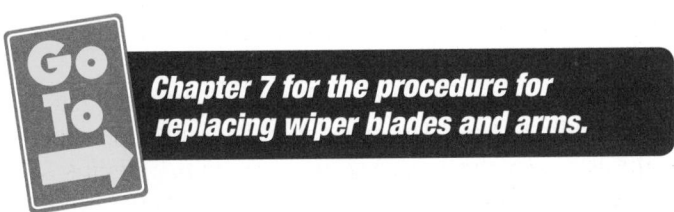

Go To *Chapter 7 for the procedure for replacing wiper blades and arms.*

Rear Window Wiper/Washer System This system is typically found on hatchbacks, vans, and SUVs **(Figure 23–8)** and has a separate switch to control the wiper motor. The parking function is completed within the rear window wiper motor and switch.

Intermittent Wiper Systems

Many wiper systems offer an intermittent mode that provides a variable interval between wiper sweeps. Many of these systems use a module, or governor, mounted near the steering column, or the system has a module connected to the BCM **(Figure 23–9)**.

The delay between wiper sweeps is controlled by a potentiometer. By rotating the intermittent control, the resistance value changes. The module contains a capacitor that is charged through the potentiometer. Once the capacitor is saturated, the electronic switch

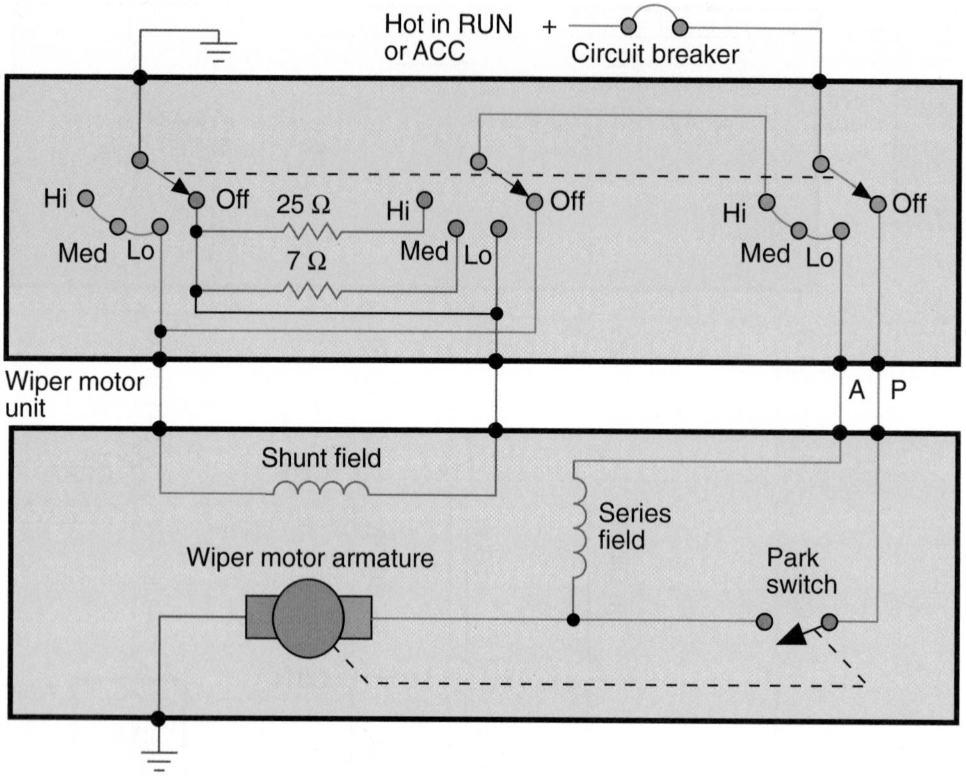

Figure 23-6 A three-speed wiper motor schematic.

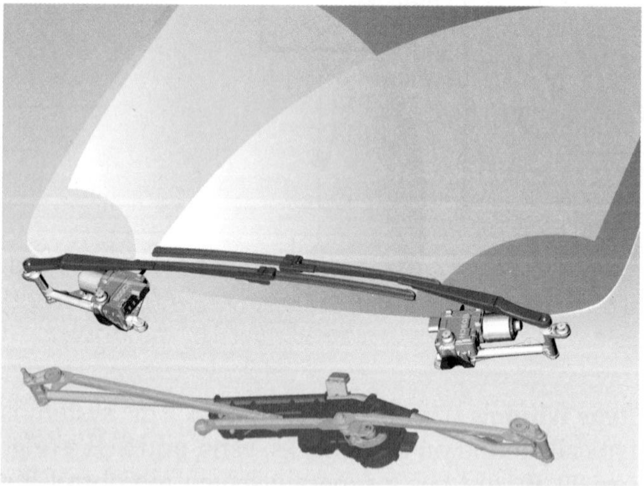

Figure 23-7 (Top) This system uses two motors that operate in the opposite direction. (Bottom) A typical single motor linkage. Note the differences in required space and complexity. *Courtesy of Robert Bosch GmbH, www.bosch-presse.de*

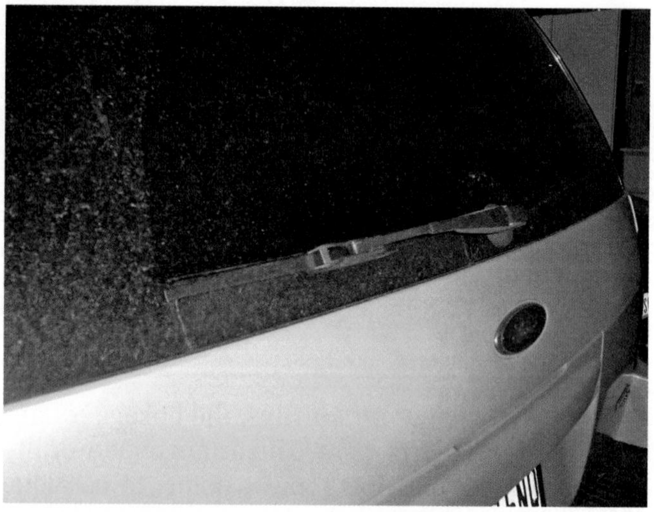

Figure 23-8 A rear window wiper.

is triggered and current flows to the wiper motor. The capacitor discharge is long enough to start the wiper operation and the park switch is returned to the run position. The wiper will continue to run until one sweep is completed and the park switch opens. The amount of time between sweeps is based on the length of time required to saturate the capacitor. As more resistance is added to the potentiometer, it takes longer to saturate the capacitor.

Rain-Sensing Wipers Some vehicles have a setting for windshield wiper operation that responds to water on the windshield. The sensor for these wipers is usually located in the center and at the top of the windshield behind the rear view mirror. The sensor transmits an infrared light onto the windshield's surface through a special optical element **(Figure 23–10)**. When the windshield is dry, all of the light is reflected back to the sensor. The windshield's ability to reflect light starts to change as soon as moisture begins to accumulate on the glass. This allows the infrared

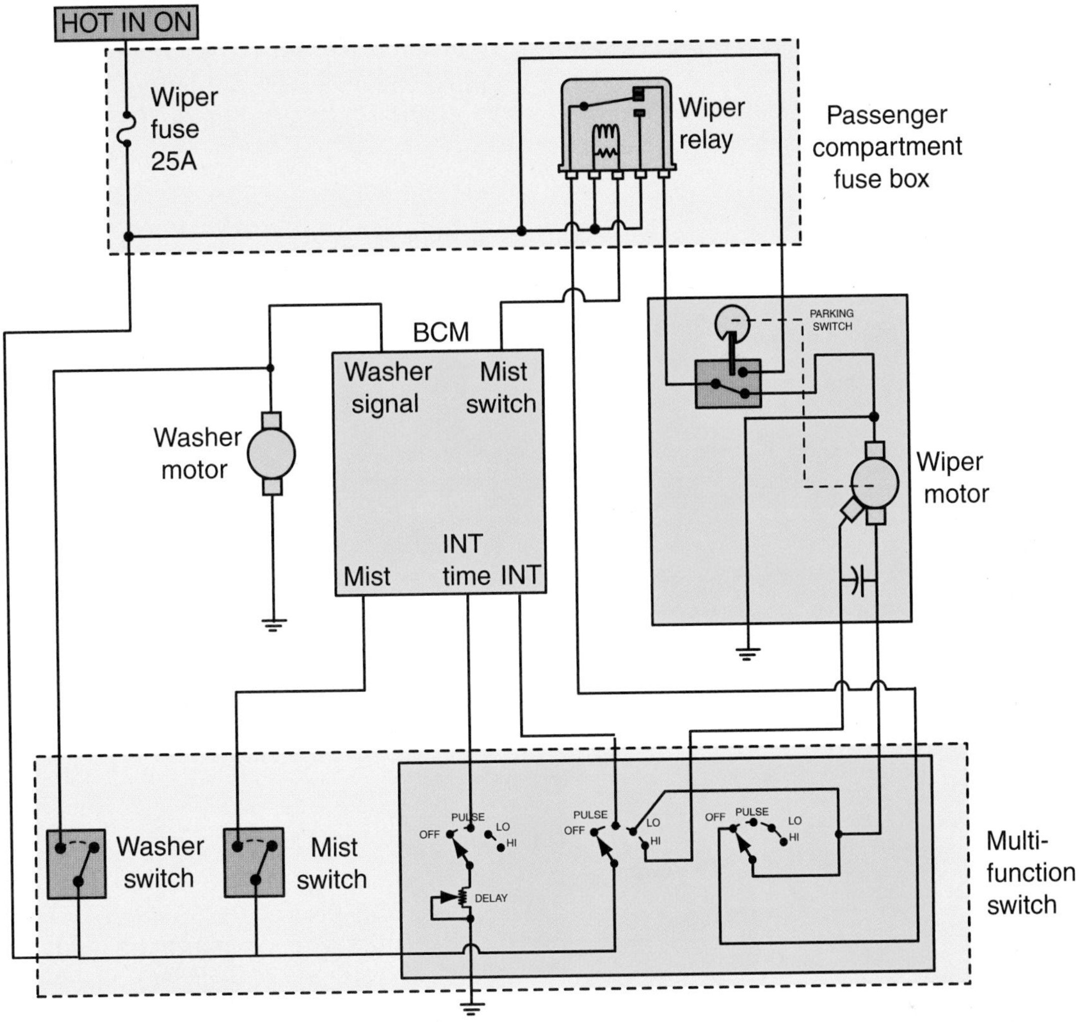

Figure 23-9 An intermittent wiper system.

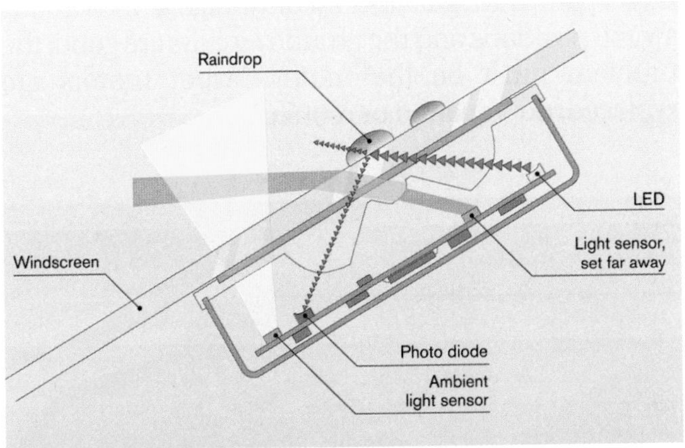

Figure 23-10 The basic operation of a rain sensor. *Courtesy of Robert Bosch GmbH, www.bosch-presse.de*

beam to penetrate through the windshield, thus reducing the amount of reflected light. This lower level of reflected light serves as an index indicating higher levels of moisture on the windshield's surface. The rain sensor uses all changes in reflected light as the basis for determining the intensity of the rain. In response, the number of sweeps made by the windshield wipers increases or decreases. The sensitivity level of this system can be adjusted by the driver.

Speed-Sensitive Wipers Some vehicles have speed-sensitive wipers that vary the speed or intermittent intervals according to vehicle speed. These systems are typically controlled by the BCM in response to inputs from the VSS.

Windshield Washers

Windshield washers spray a fluid onto the windshield and work in conjunction with the wiper blades to clean the windshield. Most systems have the washer pump installed in the fluid reservoir (**Figure 23–11**). A few systems, such as some from GM, use a pulse-type pump that operates off the wiper motor.

Washer systems are activated by holding the washer switch. If the wiper/washer system also has an intermittent control module, a signal is sent to the module when the washer switch is activated. An override circuit in the module operates the wipers at low speed for

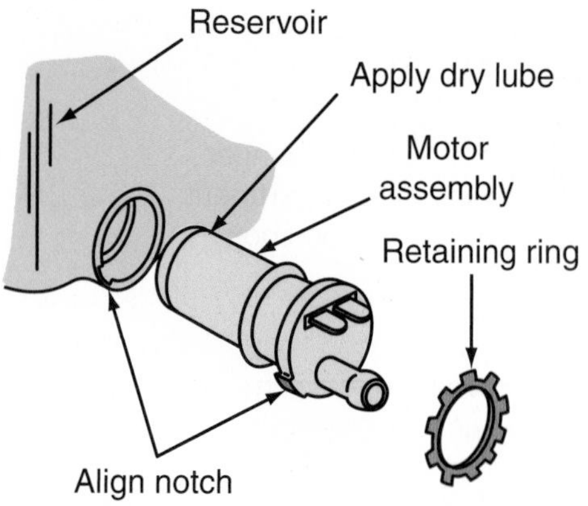

Figure 23-11 Installation of a washer pump and motor in a fluid reservoir.

Figure 23-12 Headlight washers. *Courtesy of Chrysler LLC*

a programmed length of time. The wipers either return to the park position or operate in intermittent mode, depending on the design of the system.

Some vehicles have washers that clean the headlights **(Figure 23–12)** and fog lights for maximum visibility. Headlight washer systems may operate from their own switch and pump or work along with the windshield washer system.

Some vehicles are equipped with a low fluid indicator. The washer fluid level switch closes when the fluid level in the reservoir drops below one-quarter full. Closing the switch allows power from the fuse panel to be applied to the indicator.

Wiper System Service

Customer complaints about windshield wiper operation can include poor wiping, no operation, intermittent operation, continuous operation, and wipers that do not park. Other complaints will be related to wiper arm adjustments, such as slapping the molding or one blade parks lower than the other.

When the wipers move as they should but do not wipe the glass surface the way they should or make noise while moving across the glass, the blades and/or arms should be replaced.

CAUTION!

Disconnect the battery's negative cable before removing or installing wiper linkages.

If the wipers work slower than expected, disconnect the wiper linkage at the motor **(Figure 23–13)**. Turn on the wiper system. If the motor runs properly, the problem is the linkage and is not electrical. If the motor runs slower than normal, check for resistance in the circuit.

If the motor does not run at a particular speed or at all, the problem is electrical. Carefully inspect the motor, wires, connectors, and switch. Your diagnosis should continue according to the guidelines given in the service manual. Pay attention to the circuits that could cause the problem, not the entire wiper circuit. Test for voltage at the motor in the various switch positions. Also check the ground circuit. If the motor is receiving the right amount of voltage at the various switch positions and the ground circuits are good, the problem must be the motor. Wiper motors are replaced, not repaired or rebuilt.

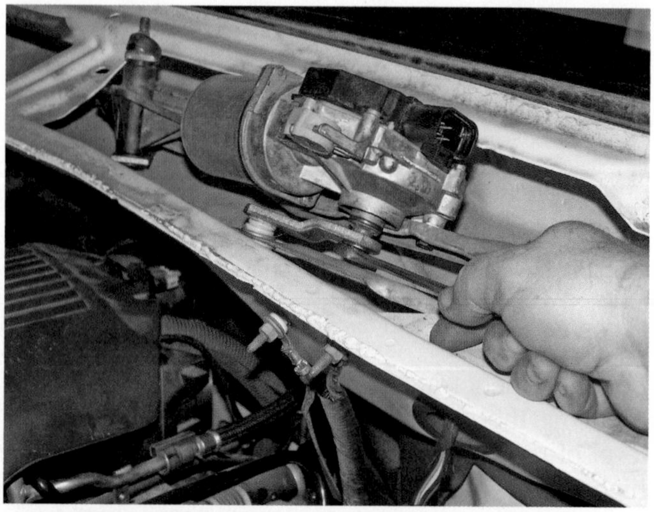

Figure 23-13 Disconnecting the wiper linkage arms at the wiper motor.

> **WARNING!**
>
> *Most wiper motors are the permanent magnet type, which can be quite delicate. Do not throw the motor around or hammer on the case. Both of these actions can destroy the magnetic fields.*

Washer System Service

Many washer problems are caused by restrictions in the fluid lines or nozzles. To check for restrictions, remove the hose from the pump and operate the system. If the pump ejects a stream of fluid, then the fault is in the delivery system. The exact location of the restriction can be found by reconnecting the fluid line to the pump and disconnecting the line at another location. If the fluid still streams out, the problem is after that new disconnect. If the fluid does not flow out, the problem is before where the hose was disconnected. Repeat this process until the problem is found.

If the pump does not spray out a steady stream of fluid, the problem is in the pump circuit. It should be tested in the same way as any other electrical circuit. Make sure it gets power from the switch when it should, then check the ground. If the power to the pump is good and there is a good ground, the problem is the pump. These are not rebuilt or repaired; they must be replaced.

HORNS/CLOCKS/CIGARETTE LIGHTER SYSTEMS

The purpose and operation of these systems are obvious and their circuits may vary from one model and year to another **(Figure 23–14)**. However, the overall operation remains the same.

Horns

Most horn systems are controlled by relays. When the horn button, ring, or padded unit is depressed, electricity flows from the battery through a horn lead, into an electromagnetic coil in the horn relay then to the ground. Low current through the coil creates a magnetic field that pulls on the movable arm. This action brings the relay's contacts together, causing the horn to sound.

Most vehicles have the horn switch mounted in the steering wheel assembly. Some vehicles are equipped with two horns, each designed to emit a different tone. The two horns provide a fuller sound than one horn can.

Horn Diagnosis If the horn does not work, check the fuse. If the fuse is bad, also check the relay. The relay is checked with a DMM. Refer to the wiring diagram

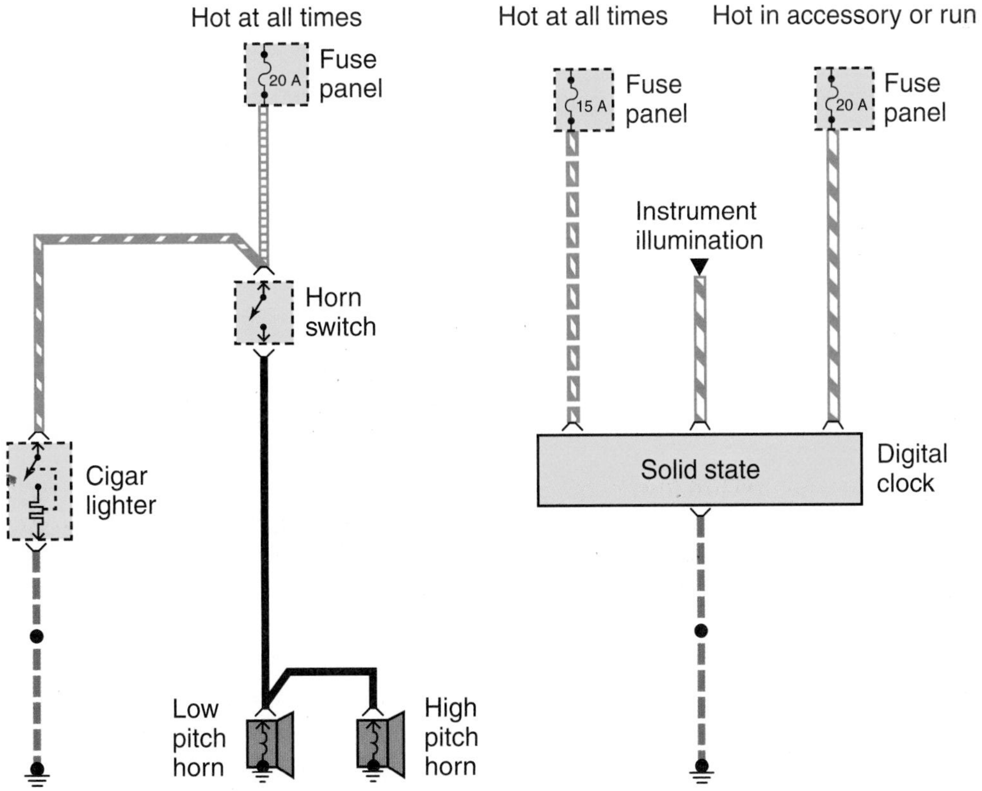

Figure 23–14 A typical horn/clock/cigar lighter circuit. *Courtesy of Ford Motor Company*

to identify the terminals that should have power when the horn switch is activated and deactivated. If power is not present, the problem may be that circuit or the relay. The relay can be checked for continuity with an ohmmeter.

If the relay is good, the switch should be checked if it or its circuit is suspect. Following the precautions for working around an air bag, disconnect the horn switch's connector. Connect the terminals of the switch together with a jumper wire. The horns should sound. If the horns work, replace the switch. If the horns do not work, check the cable reel behind the steering wheel. Jump across the cable's terminals. Again, the horns should sound. If they do not, check the wiring harness in the steering column for an open.

If all power circuits and controls check out, disconnect each horn. Test each horn by connecting battery power to the terminal and grounding the bracket. The horns should sound. If they do not, replace them.

Clock

The clock receives power directly from the fuse panel. Some clocks have additional functions. These are explained in the owner's guide for the particular vehicle.

Cigarette Lighter

The cigarette (cigar) lighter is a heating element that automatically releases itself from the pushed-in position when the appropriate heat level is reached.

For an inoperative system, first check the fuse(s). If the fuse is good, make certain that power is present at the lighter receptacle. If power is present, the lighter unit is probably bad. Refer to the service manual for additional troubleshooting information.

Many vehicles also have an additional power outlet that looks like an additional cigarette lighter receptacle. These can be used to power or recharge 12-volt appliances, such as cell phones.

CRUISE (SPEED) CONTROL SYSTEMS

Cruise control systems are designed to allow the driver to maintain a constant speed (usually above 30 mph or 48 km/h) without having to apply continual pressure on the accelerator pedal. Selected cruise speeds are easily maintained and can be easily changed. When engaged, the system sets the throttle position to the desired speed. The speed is maintained unless heavy loads and steep hills interfere. The cruise control switch is located on the end of the turn signal or near the center or sides of the steering wheel (**Figure 23–15**). There are usually several functions on the switch, including off-on, resume, and engage buttons. Cruise control is disengaged whenever the brake pedal or clutch pedal is depressed.

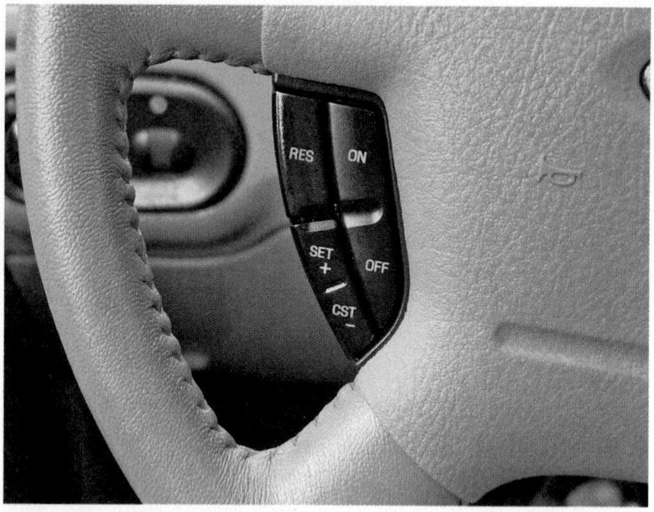

Figure 23-15 The cruise control switch is used to set or increase speed, resume speed, or turn the system off and on.

Vacuum Systems

Until recently, most cruise control systems relied on engine vacuum and mechanical linkages. The following are the common components in these systems (**Figure 23–16**):

- When the system is turned on, the **transducer** senses vehicle speed and controls the amount of vacuum applied to a vacuum servo. The amount of vacuum sets the servo position. Vehicle speed is sensed from the lower speedometer cable.
- The **servo unit** is connected to the throttle by a rod or linkage, chain, or cable. The servo unit maintains the desired speed by receiving a controlled amount of vacuum from the transducer. Variations in vacuum change the position of the throttle. When a vacuum is applied, the servo spring is compressed and the throttle is positioned correctly. When the vacuum is released, the servo spring is relaxed and the system is not operating.
- There are two brake-activated switches operated by the position of the brake pedal. When the pedal is depressed, the brake release switch disengages the system. A vacuum release valve is also used to disengage the system when the brake pedal is depressed.

Electronic Cruise Control Systems

Most late-model vehicles have electronic cruise control systems. These systems can use an electronic control module to move a servo unit or control an electric stepper motor. The motor moves a strap attached to the

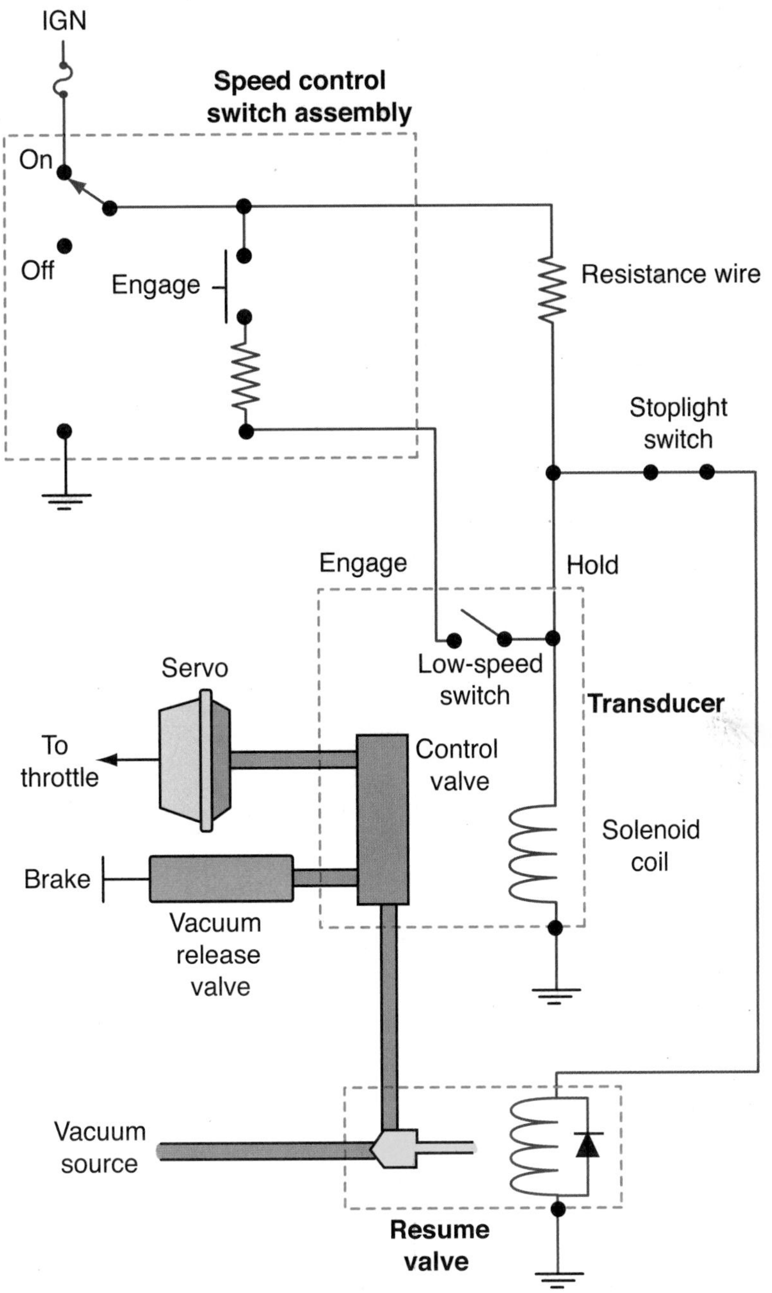

Figure 23–16 A cruise control circuit with vacuum and electrical circuits.

cruise control cable to move the throttle linkage. Engines with electronic throttle control (throttle-by-wire) do not need a separate cruise control module, stepper motor, or cable to control engine speed. The PCM has full control of the throttle and, therefore, the circuitry of the PCM operates the cruise control system.

A vehicle speed sensor (VSS) is used to monitor vehicle speed. The computer has several other inputs to help determine the operation of the system **(Figure 23–17)**. Most cruise control systems are connected to the vehicle's multiplexing network and should be serviced and tested according to the manufacturer's recommendations.

Adaptive Cruise Control

Like other cruise control systems, adaptive cruise control automatically maintains the desired speed of the vehicle, but it also maintains a safe distance between vehicles. The desired distance between vehicles is set by the driver. The system also adjusts the speed of the vehicle to mirror that of a slower vehicle in front of it, then maintains that speed. A laser or radar sensor **(Figure 23–18)** mounted near the front bumper serves as the eyes for the system. Other vehicles traveling within the sensor's range reflect the radar waves **(Figure 23–19)**, and the sensor picks up the returning signals. The control unit uses this

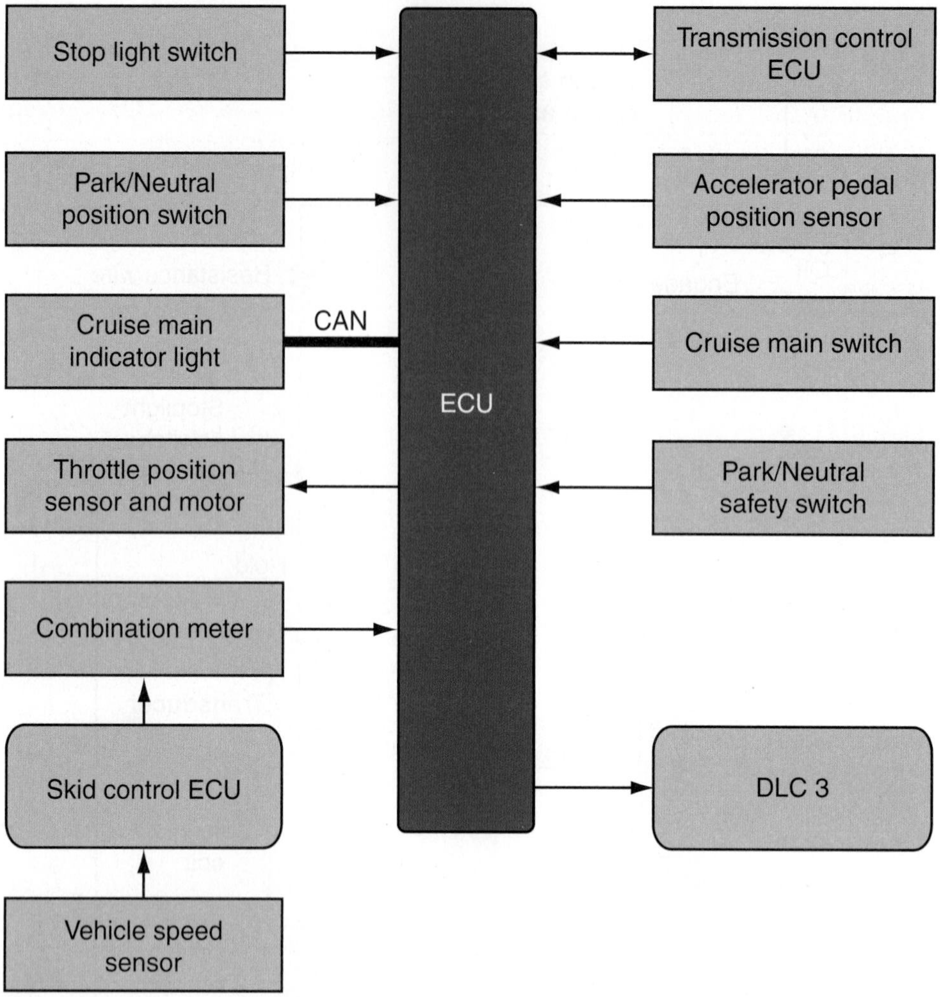

Figure 23-17 The main components of an electronic cruise control.

Figure 23-18 The distance sensor for an intelligent or adaptive cruise control system. *Courtesy of Chrysler LLC*

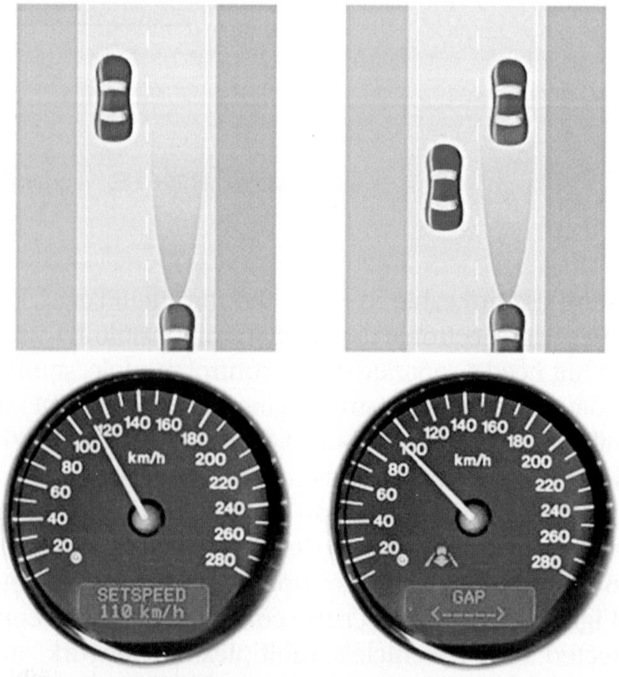

Figure 23-19 The response of an adaptive cruise control system. *Courtesy of Delphi Corporation*

information to determine the position and speed of the preceding vehicle. When the system detects a slower moving vehicle in the same lane, it reduces the throttle or gently applies the brakes to reduce speed. The vehicle then follows behind the preceding vehicle at the speed required to maintain a predefined distance. As soon as the vehicle in front has moved or increased speed, the system will accelerate the vehicle back up to the set and desired speed.

SHOP TALK

Most adaptive cruise control systems rely on radar sensors. *Radar* means "radio detection and ranging." Radio waves are transmitted from an antenna. When those waves contact another object, they bounce back to the antenna. The time it takes for the waves to return is used to determine how far away the object is. *Lidar* is "laser radar." This works the same way except light waves are used.

Cruise Control System Service

Most cruise control systems are controlled by the BCM and work with the electronic throttle control system. Diagnosis of these cruise systems is the same as for any other electronic system. Diagnostic work is done with a scan tool and service manual. Typically, on most late-model vehicles, cruise control problems are caused by faulty circuits, sensors, and/or switches **(Table 23–1)**.

Before connecting the scan tool, check the operation of the cruise indicator lamp. If it does not operate properly, test and repair that circuit before proceeding. If the light is blinking, the control module detects an electrical fault in the system. With the scan tool, verify that communication links on the CAN bus are good. Retrieve the DTCs and proceed with the recommended procedures for the codes.

TABLE 23–1 COMMON CRUISE CONTROL PROBLEMS AND THEIR PROBABLE CAUSES

Symptom	Likely Problem Area
Pushing the ON-OFF button does not turn the system on.	Stoplight switch circuit Clutch switch circuit (M/T) Vehicle speed sensor circuit Cruise control switch circuit Transmission range sensor circuit Control module
Vehicle speed cannot be set. (The CRUISE indicator is on)	Cruise control switch circuit Vehicle speed sensor circuit Stoplight switch Transmission range sensor circuit Clutch switch circuit Control module
Cruise control stops during operation.	Cruise control switch circuit Vehicle speed sensor circuit Stoplight switch Transmission range sensor circuit Clutch switch circuit Control module
Cruise control cannot be manually canceled.	Cruise control switch circuit Control module
Cruise control is not canceled when vehicle speed drops below the low-speed limit.	Vehicle speed sensor circuit Control module
Depressing the brake pedal does not cancel cruise control.	Stoplight switch circuit Control module
Depressing the clutch pedal does not cancel cruise control.	Clutch switch circuit Control module
Moving the shift lever does not cancel cruise control.	Transmission range sensor circuit Control module
Speed is not constant.	Vehicle speed sensor circuit Control module
CRUISE indicator is blinking.	Cruise control circuit Control module

Vacuum Systems Troubleshooting should begin with a thorough visual inspection. Check the system's fuses. Check all vacuum hoses for disconnects, pinches, loose connections, and so on. Inspect all wiring for tight and clean connections. Check and adjust the linkage according to specifications.

If the cruise control does not work, apply the brake pedal and verify brake light operation. If the lights do not work, repair them and see if this was the cause of the problem. If the vehicle has a manual transmission, check the clutch switch.

If the system does not keep a constant speed, disconnect the vacuum hose between the check valve and the servo. Apply 18 inches of vacuum to the check valve. The valve should hold the vacuum. If it does not, replace it. Check the vacuum dump valve, servo, and speed sensor according to the procedures

Figure 23-20 Check the linkage if the vehicle surges while cruise is activated.

outlined in the service manual. If everything checks out fine, replace the controller (amplifier).

When the cruise does not disengage, check the operation of the brake switch. Also, check the wires leading from the switch to the cruise control unit.

If the vehicle surges while cruise is activated, check the actuator linkage **(Figure 23–20)**. It should move smoothly. Then, check the speedometer cable for proper routing. Check the servo and vacuum dump valve with a handheld vacuum pump. If everything is okay, replace the controller. If the servo is found to be faulty, it must be replaced.

SOUND SYSTEMS

Sound systems are available in a wide variety of models. The complexity of the system varies significantly from the basic AM radio to more complex stereo systems **(Figure 23–21)** that include an AM/FM radio receiver, a stereo amplifier, compact disc (CD) player, cassette player, equalizer, several speakers, and a power antenna system.

A radio receives signals (radio waves) that are broadcast from radio station towers or antennas. Amplitude modulation (AM) waves travel far but cannot be used to broadcast in stereo. Also, AM does not have as good sound quality as frequency modulation (FM). Nearly all FM broadcasts are in stereo but the distance range for good reception is limited.

Sound quality depends on the basic system but especially on the quality of the speakers and their placement. Many sound systems are equipped with several speakers, each designed to produce a different range of sound. Matching speakers to the system is done by selecting and wiring speakers so they have the same impedance as the rest of the sound system. Impedance requirements are typically noted on the rear of the sound unit and/or in the installation instructions. The placement of the speakers is critical to good clean sound. Sound waves from the speaker will bounce off anything they hit, including other sound waves. This bouncing of sound can cause noise or distortion. To achieve a high-quality sound system, the speakers must be placed so that all bouncing is anticipated.

Many optional sound systems have very a high wattage output and use several amplifiers. Some even have automatic sound level systems that discreetly adjust the volume to compensate for changes in ambient noise and vehicle speed. An amplifier increases the volume of a sound without distorting it. Amplifiers are typically rated by the maximum power (watts) they can put out. In order to take advantage of the power output, speakers must be chosen that match the output.

CD players vary from being able to insert one or more CDs into the main unit to having an auxiliary unit where many CDs can be installed **(Figure 23–22)**.

Figure 23-21 An AM/FM radio with CD player.

Figure 23-22 An auxiliary CD player.

The control unit allows the operator to select the CD of choice. A few vehicles are fitted with MP3 systems.

Antenna

An antenna is designed to receive radio sound waves for both AM and FM stations. The design of the antenna must satisfy two basic requirements: for good AM reception the antenna should be as high as possible and for FM stations it should be 31 inches high. Therefore, most antennas are 31 inches high when they reach full height. Shorter antennas will have poor reception in both AM and FM. Many late-model vehicles are equipped with shorter antennas designed to enhance reception and therefore have no need to be long to provide good reception. The placement of these antennas is also selected to achieve optimal reception.

Power Antennas Many vehicles are equipped with electrically operated antennas that extend when the radio is turned on and lower when it is turned off. These antennas are powered by a small reversible electric motor. The motors are turned off by limit switches that open when the mast has extended or lowered to its desired height. Power antennas (even black-colored antennas) need to be cleaned with chrome polish on a regular basis to keep them working properly. Often when there is a problem with a power antenna, it is caused by dirt or a lack of lubricant on the telescoping mast. When there is a problem with the power unit, it is normally replaced as a unit.

Satellite Radio

To provide high-quality radio that is not interrupted by distance, some vehicles can be purchased with satellite radio. Satellite radios are also available as add-on items. These radios pick up sound waves from satellites many miles above the earth. Since the radio waves are transmitted by more than one satellite at all times, and each in their own orbit or place within the orbit, the same radio station can be heard from coast to coast **(Figure 23–23)**. Although distance does not hamper the reception, the radio waves cannot penetrate buildings, tunnels, or large groupings of trees. Therefore, to enjoy satellite radio, stay on the open road.

Figure 23-23 The reception of satellite radio systems is not affected by distance because the radio waves are transmitted by more than one satellite. *Courtesy of Delphi Corporation*

Phone and Multimedia Systems

There are many different systems that provide cell phone hookup and other entertainment features, including rear-seat DVD players. But the latest trend is to join all of these into a single system. These systems may also have voice control. The communication and entertainment systems are connected by Bluetooth technology or a standard USB port. Voice commands dial or answer the phone and choose the entertainment media. An example of this system is found in Ford products. The system is called "sync" and was co-developed with Microsoft.

Diagnosis

Internal inspection and service to the radio should be left to an authorized radio service center. However, technicians should be able to determine the cause of poor sound quality and radio reception. Most sound system problems are caused by the unit itself, the wiring in the circuit, the antenna, or the speakers.

If the radio system is not working, check the fuse. If the fuses are okay, refer to the service manual. If you determine the radio itself is the problem, remove it and send it to a qualified shop.

> **Customer Care**
>
> **Many customers do not understand the limitations of FM reception. Refer your customer to the owner's guide for information about the limitations of FM radio performance.**

If reception is bad and antenna height is correct, use an ohmmeter and check the ground of the antenna. Also connect one lead of the ohmmeter to the antenna's mast and its case; there should be no continuity between the two.

Poor speaker or sound quality is usually caused by one of the following:

- Damaged speaker cones, internal mountings, or wiring.
- Interference from the ignition system, neon signs, or electrical power lines.

- Distortion caused by the speaker, radio chassis, or wiring. If the concern is in the radio chassis, both speakers on the same side of the vehicle will have poor quality. Distortion caused by damaged wiring is most often accompanied by lower-than-normal sound output.
- Bent package tray sheet metal around the speaker opening, lack of mounting brackets, or missing or loose attaching hardware or speaker covers. Be careful not to overtighten hardware as this may bend or deform the speaker baskets, causing buzzes or distorted sound.

SHOP TALK

Antitheft audio systems have built-in devices that make the audio system soundless if stolen. If the power source for the audio system is cut, even if it is later reconnected, the audio system will not work unless its ID number is put in. When performing repairs on vehicles equipped with this system, before disconnecting the battery terminals or removing the audio system, ask the customer for the ID number so you can input the ID number afterward, or request that the customer input the ID number after the repairs are completed. With antitheft radio installation, there is a procedure in the service manual to obtain the factory backup code using a touch tone phone if the owner's code is not available. A memory saver or backup battery can be used to maintain the radio's code and settings during service.

POWER LOCK SYSTEMS

Although systems for automatically locking doors vary from one vehicle to another, the overall purpose is the same—to lock all outside doors. As a safety precaution against being locked in a car due to an electrical failure, power locks can be manually operated. Many late-model systems include automatic locking when the transmission selector is moved out of park or when the vehicle reaches a particular speed.

When either the driver's or passenger's control switch is activated (either locked or unlocked), power from the fuse panel is applied through the switch to a reversible motor. A rod that is part of the lock assembly moves up or down to lock or unlock the door. On some models the signal from the switch is applied to a relay that, when energized, applies an activating voltage to the door lock actuator. The door lock actuator consists of a motor and a built-in circuit breaker. Since the motors are reversible, each does not have its own ground. The ground for the lock circuits is at the master or door circuits (**Figure 23–24**).

Most models use a control switch mounted in the door arm rest or in the door trim panel. However, some models use switches controlled by the front door push button locks.

Power Trunk Release

The power trunk release system is a relatively simple electrical circuit that consists of a switch and a solenoid. When the trunk release switch is pressed, voltage is applied through the switch to the solenoid. With battery voltage on one side and ground on the other, the trunk release solenoid energizes and the trunk latch releases to open the trunk lid.

Diagnosis

Power door lock or trunk release systems rarely give trouble. Basic diagnosis includes:

- If none of the locks work, check the circuit protection devices. If they are okay, check the wiring.
- If the doors lock or unlock but not both, check the system relays. On vehicles without relays, check the wiring to and from the control switch on the nonworking side.
- If only one door lock does not operate, check the actuator for that door.
- If only one door lock switch does not work, check that switch.
- If the trunk release does not work, check the fuse first. Then check the switch. If the switch is okay, check for continuity through the trunk release solenoid to ground. If the solenoid is okay, check the wiring.

POWER WINDOWS

Obviously, the purpose of any power window system (**Figure 23–25**) is to raise and lower windows. The systems do not vary significantly from one model to another. The major components of a typical system are the master control switch, individual window control switches, and the window drive motors (**Figure 23–26**).

The master control switch provides overall system control. Power for the system comes directly from the fuse panel on two-door models and from an inline circuit breaker on four-door models. The window safety relay used on four-door models prevents operation of the system if the ignition switch is not in the run or accessory position. Power for the individual window control switches comes through the master control switch.

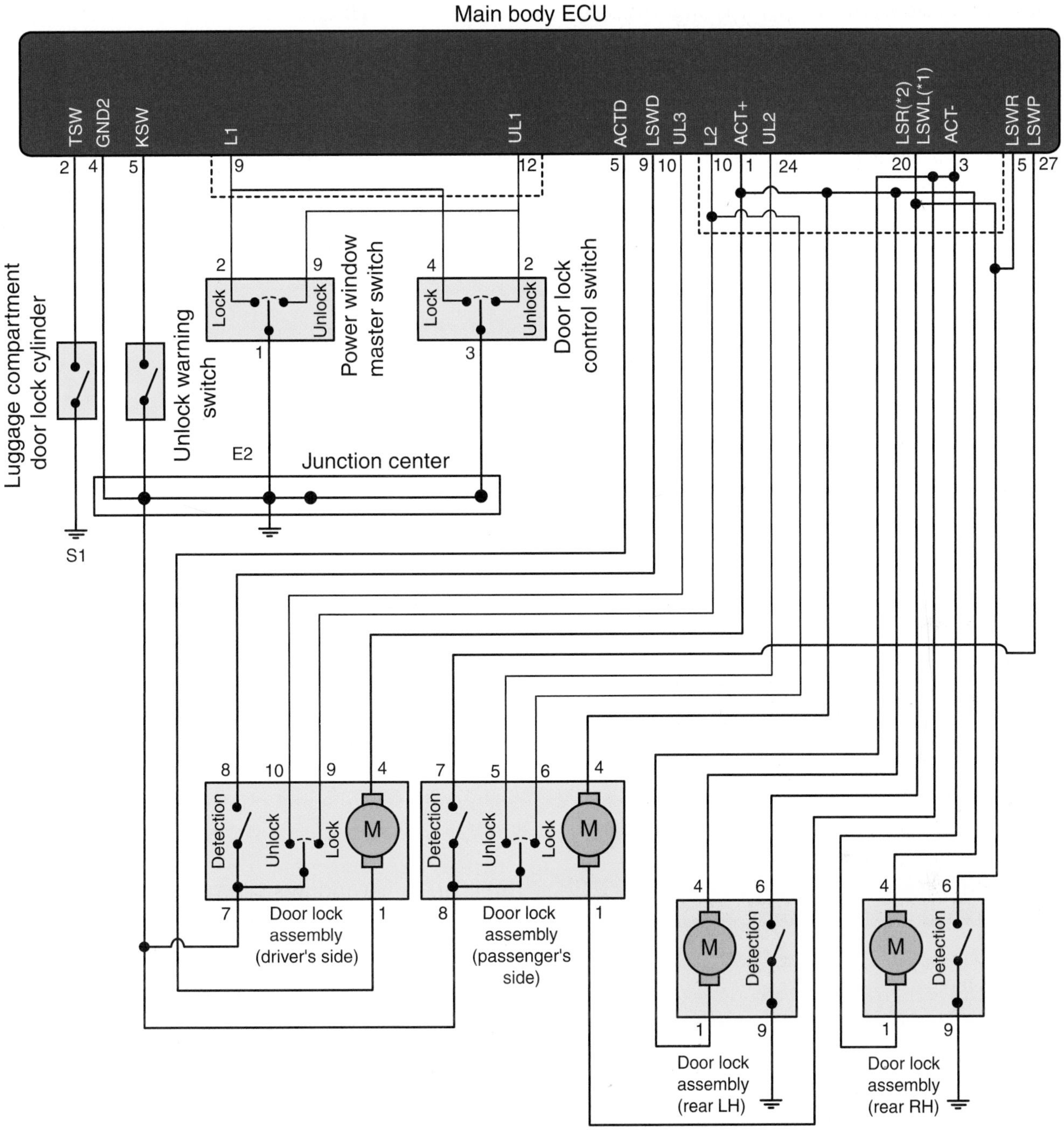

Figure 23-24 A wiring diagram for a power door lock system.

Four-door model master control switches usually have four segments **(Figure 23–27)** while two-door models have two. Each segment operates as a separate independent switch and controls power to a separate window motor. A lock switch included on four-door model master control switches is a safety device to prevent children from opening windows without the driver's knowledge.

Circuit Operation

Typically a permanent magnet motor operates each power window. Each motor raises or lowers the glass when voltage is applied to it. The direction that the motor moves the glass is determined by the polarity of the supply voltage.

Voltage is applied to a window motor when any UP switch in the master switch assembly is activated. The motor is grounded through the DOWN contact. Battery voltage is applied to the motor in the opposite direction when any DOWN switch in the master switch assembly is activated. The motor is then grounded through the master switch's UP contact.

The operation of the individual window switches is much the same. When the UP switch is activated,

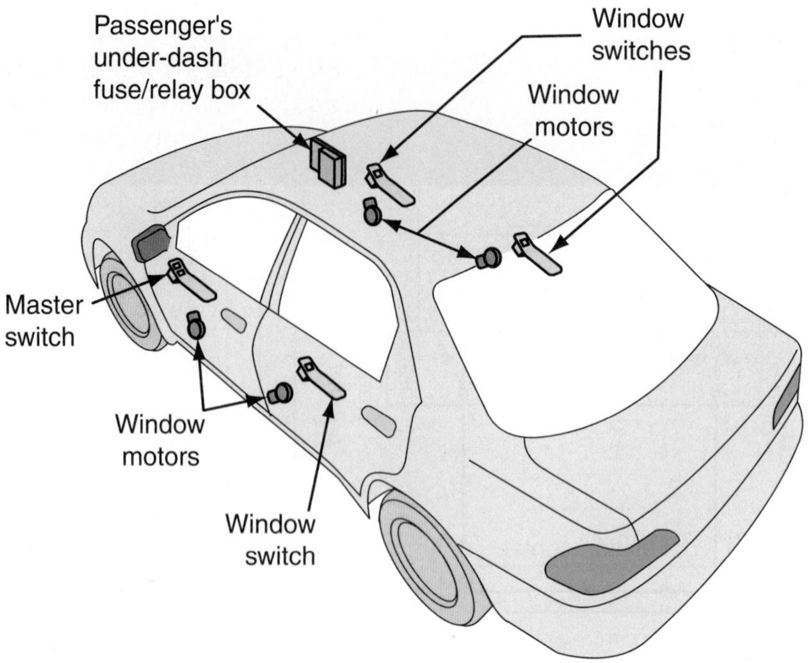

Figure 23-25 A power window system.

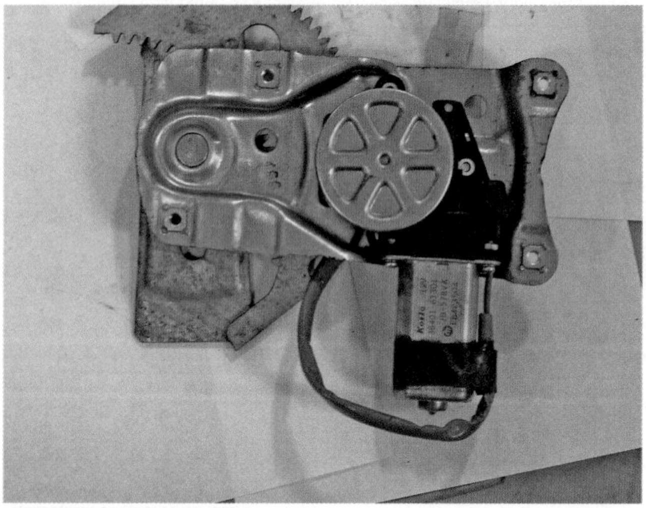

Figure 23-26 Electric window motor and regulator.

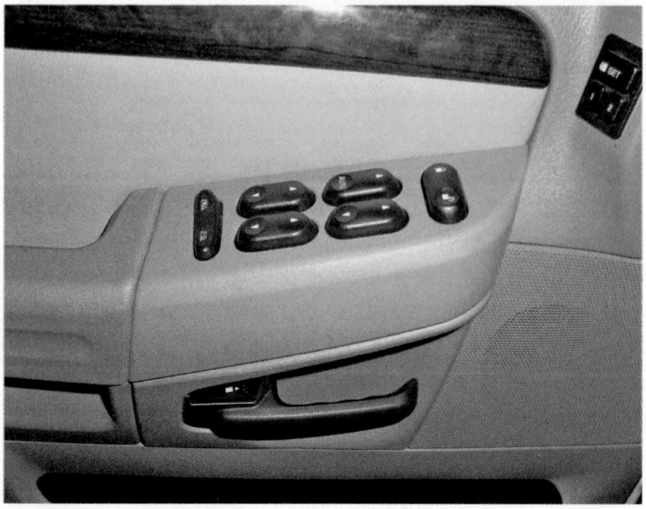

Figure 23-27 A master power window switch assembly.

voltage is applied to the window's motor. The motor is grounded through the DOWN contact at the switch and the DOWN contact at the master switch. When the DOWN switch in the window switch is activated, voltage is applied to the motor in the opposite direction. The motor is grounded through the UP contact at the window switch and the UP contact in the master switch. This runs the motor in the opposite direction.

Each motor is protected by an internal circuit breaker. If the window switch is held too long with the window obstructed or after the window is fully up or down, the circuit breaker opens the circuit.

Express Windows Some vehicles are equipped with an "express down" window feature. This feature fully opens the window by holding the window switch down for more than 0.3 second then releasing it. The window may be stopped at any time by depressing the switch. The express window option relies on an electronic module and a relay. When signaled, a control module energizes a relay, which completes the circuit to the motor. When the window is fully down, the module opens the relay, which stops the motor. The motor will also stop 10 to 30 seconds after the DOWN switch is depressed.

The express window circuit is part of the BCM and can be temporarily disabled if the battery is disconnected, if a fuse blows, or if parts are replaced. If this is the case, the system needs to be reinitialized. Follow the manufacturer's procedure for doing this.

Obstacle-Sensing Windows This feature prevents the window from closing if something is in the way, such as fingers. On some systems, if the window is

Figure 23-28 The infrared sensor for a smart window system. *Courtesy of Delphi Corporation*

closing and hits something, the window will reverse and open. Other systems do not require direct contact with an obstacle. Rather, they rely on infrared sensors **(Figure 23–28)**. When the light beams are broken by an obstacle, the window will reverse and go down.

Diagnosis

The first step in diagnosing a power window system is to determine whether the whole system or just one or two windows are not working correctly. If it is the whole system, the problem can be isolated to fuses, circuit breakers, or the master control switch. If only a portion of the system does not work, check the components used in the portion that is not working. Removing the door trim allows access to the motor and linkage. A special tool is often needed to release the retaining clips for the trim.

Guidelines Basic logic will help identify probable causes for window problems.

- When all windows do not operate, check the fuse and the wiring to the master switch, including the ground.
- When one window does not operate, check the wiring to the individual switch, the switch, and the motor. Also check the window for binding in its tracks.
- When both rear windows cannot be operated by their individual switches, check the lockout and master switch.
- When one window moves in one direction only, check the wiring between the master switch and the individual switch.

POWER SEATS

Power seats allow the driver or passenger to adjust the seat to the most comfortable position. The major components of the system are the seat control and the motors.

Power seats generally come in two configurations: four-way and six-way. However, some vehicles allow the seats to be adjusted in up to twelve directions. In a four-way system, the whole seat moves up or down, or forward and rearward. A six-way system **(Figure 23–29)** has the same adjustments, plus the capability to adjust the height of the front and rear of the seat. Generally, a four-way system is used on bench seats and a six-way system is used on split-bench and bucket seats. Some units also control the tilt, rear/forward movement, height, and angle of the seat back. The adjuster for the seat back may also control the height of the head rest or restraint.

Two motors are typically used on four-way systems, while three are used on a six-way system. The names of the motors identify their function. To raise or lower the entire seat on a six-way system, both the front height and the rear height motors are operated together. The motors are generally two-directional motor assemblies that include a circuit breaker to protect against circuit overload if the control switch is held in the actuate position for long periods **(Figure 23–30)**.

Diagnosis

Before testing a power seat system, conduct a visual inspection of the wires and connectors in the system. Two types of problems affect power seats: One is a tripped or constantly tripping circuit breaker and the other is the inability of the seat to move in a direction.

The resettable circuit breaker should protect the system from a short circuit or from high current due to an obstructed or stuck seat adjuster. The circuit breaker must be replaced if it is faulty. Before condemning or testing the circuit breaker, make sure the seat tracks are not damaged and that nothing is physically preventing the seat from moving.

When a seat does not move in a particular direction, turn on the dome light, then move the power seat switch in the problem direction. If the dome light dims, the seat may be binding or have some physical resistance on it. Check under the seat for binding or obstructions. If the dome light does not dim, test the system.

Disconnect the negative terminal at the battery. Remove the power seat switch from the seat or door armrest. Check for battery voltage to the switch. If

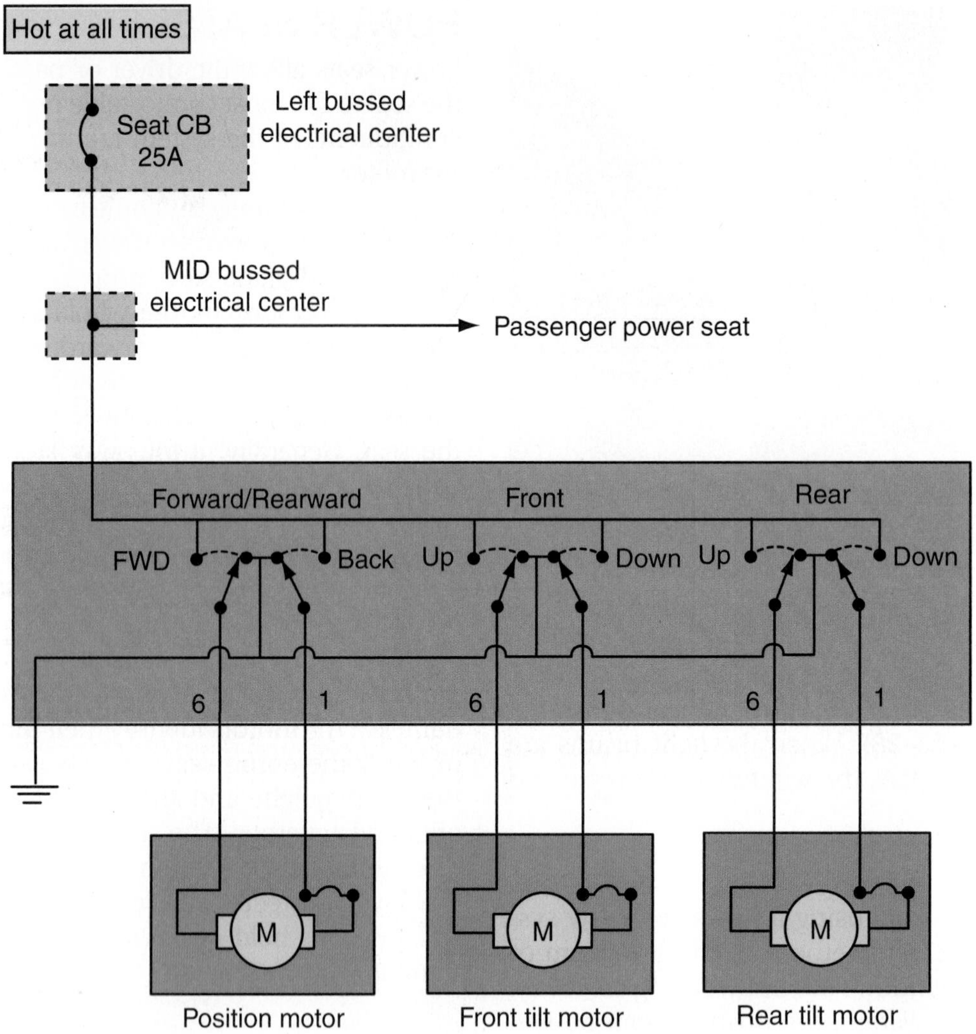

Figure 23-29 A power seat circuit.

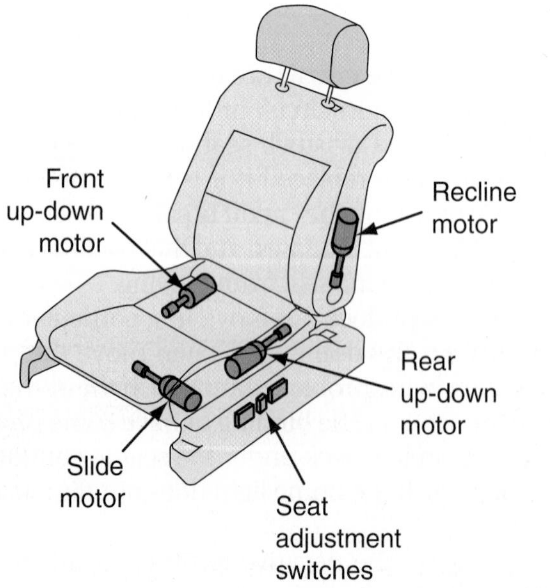

Figure 23-30 Location of the motors in a power seat.

there is no voltage and the circuit breaker is okay, check for an open in the power feed circuit. If voltage is present, check for continuity between the ground connection at the switch and a good ground. If there is no continuity, repair the ground circuit. If there is continuity, the switch must be tested. Use an ohmmeter to test the continuity of the switch in each position. Check the service manual for the expected continuity between the various terminals of the switch **(Figure 23–31)**. If the switch checks out, test the motor. If the switch is bad, replace it.

Other Seat Options

Many different options are available for the seats of a vehicle. Some of the following features are available on vehicles with manual seats; others are only available with the power seat option.

Temperature-Controlled Seats Some vehicles have an option that warms up the seats, an especially nice feature in cold climates. The system relies on heating coils in the seat cushion and back controlled by relays

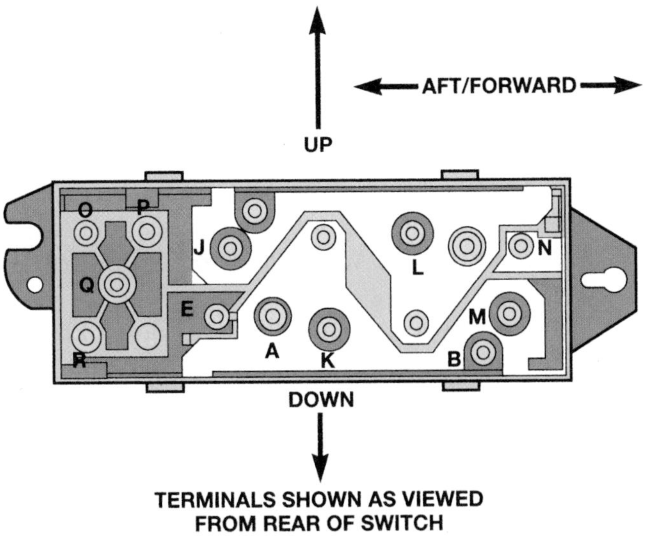

POWER SEAT SWITCH	
SWITCH POSITION	CONTINUITY BETWEEN
OFF	B-N, B-J, B-M, B-E, B-L, B-K
VERTICAL UP	A-E, A-M, B-N, B-J
VERTICAL DOWN	A-J, A-N, B-M, B-E
HORIZONTAL FORWARD	A-L, B-K
HORIZONTAL AFT	A-K, B-L
FRONT TILT UP	A-M, B-N
FRONT TILT DOWN	A-N, B-M
REAR TILT UP	A-E, B-J
REAR TILT DOWN	A-J, B-E
LUMBAR OFF	O-P, P-R
LUMBAR UP (INFLATE)	O-P, Q-R
LUMBAR DOWN (DEFLATE)	O-R, P-Q

Figure 23-31 The terminals of a power seat switch identified for continuity checks. *Courtesy of Chrysler LLC*

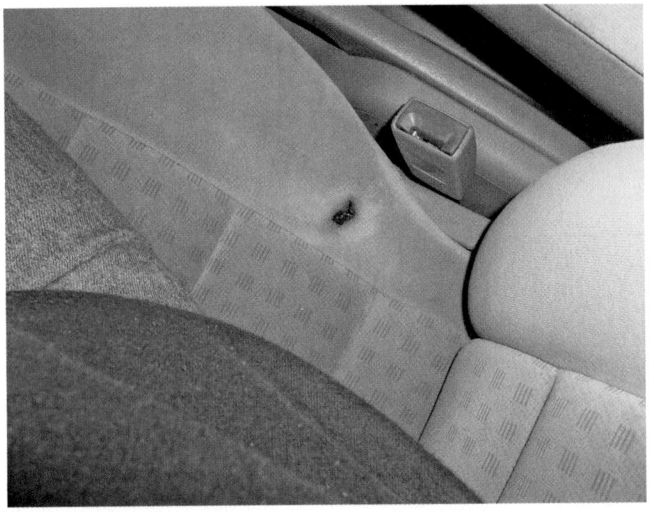

Figure 23-32 This heated seat became very hot because of a shorted heating coil.

and switches (**Figure 23–32**). In addition to warming the seats in cold weather, some vehicles allow the seats to cool by passing air through them during warm weather (**Figure 23–33**). The heating grids and switches are normally tested with an ohmmeter. Refer to the appropriate service manual for specifications and testing instructions.

Power Lumbar Supports A power lumbar support allows the driver to inflate or deflate a bladder located in the lower seat back. Adjusting this support improves the driver's comfort and gives support at the lower lumbar region of the spinal column.

Memory Seats The memory seat option allows for automatic positioning of the driver's seat to different, often up to three, programmable positions (**Figure 23–34**). This feature allows different drivers to have their desired seating position automatically adjusted by the system. It also allows an individual

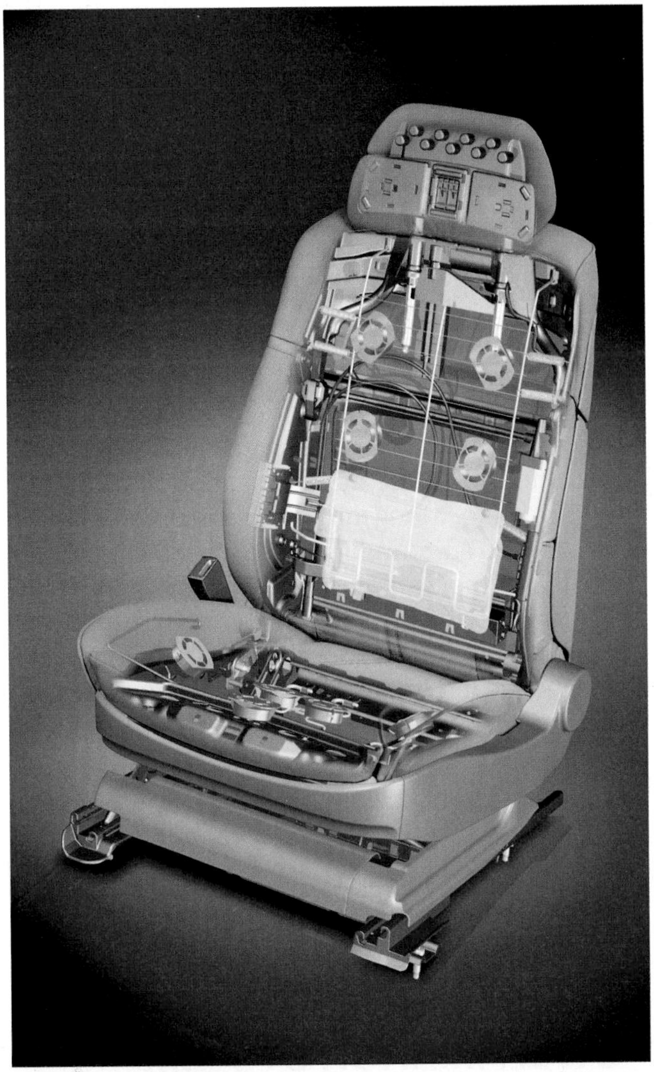

Figure 23-33 This seat features ventilation, heating, and multiple position adjustments, including the headrest and lumbar support; the small round things are fans. *Courtesy of the BMW of North America, Inc.*

Figure 23-34 Seat adjustment controls and seat memory buttons. *Courtesy of Chrysler LLC*

driver to set different positions for different driving situations.

Some systems with a remote key fob can be programmed to move the seat to its memory position whenever the unlock button of the key fob is depressed. Each driver can have his or her own key fob and their desired seat position will be selected when the door is unlocked. Also, other systems automatically adjust the power mirrors to a setting for each driver.

Adaptive and Active Seating Adaptive seating, a feature that moves the seat slightly as the driver shifts positions, is offered in some luxury vehicles. Moving the seat improves the driver's comfort and support while driving for a long time. Active seating stimulates the spine and surrounding muscles with continuous yet virtually imperceptible motion. This type of seating is designed to prevent the driver from getting saddle sore while sitting without moving for a long time. The right and left halves of the seat cushion move up and down at cyclical intervals. To do this, two pillows are integrated within the seat's upholstery. A hydraulic pump alternately inflates the two cavities with a mixture of water and glysantine.

Massaging Seats To help reduce driver fatigue, rows of rollers in the seat back move up and down when the driver depresses the control button. This massaging motion continues for about 10 seconds at a time.

POWER MIRROR SYSTEM

The power mirror system consists of a joystick-type control switch and a dual motor drive assembly located in each mirror assembly.

Rotating the power mirror switch to the left or right position selects one of the mirrors for adjustment. Moving the joystick control up and down or right and left moves the mirror to the desired position. The dual motor drive assembly is located behind the mirror glass. The position of the mirrors may be tied to the memory power seats and will automatically adjust when a seating position is selected.

A typical power mirror circuit **(Figure 23–35)** is an independent circuit unless it is tied into the convenience memory system. In that case, the BCM controls the mirrors along with the seats. The BCM in a few vehicles automatically tilts the passenger-side outside mirror downward whenever the transmission is placed into reverse. This allows the driver to see the area directly next to the vehicle.

Electrochromic Mirrors Many new vehicles are being equipped with electrochromic outside mirrors. These mirrors automatically adjust the amount of reflection based on the intensity of the light striking the mirror's surface. These mirrors use photo sensors to sense the light. When glare is heavy, the mirror darkens fully (down to 6% reflectivity). When glare is mild, the mirror provides 20% to 30% reflectivity. When glare subsides, the mirror changes to its clear daytime state.

Inside Mirrors

Some automatic day/night inside rearview mirrors use a thin layer of electrochromic material between two pieces of conductive glass. A switch located on the mirror allows the driver to turn the feature on or off. When it is turned on, the mirror switch is lighted by an LED. The self-dimming feature is disabled whenever the transmission is placed into reverse.

When the mirror is turned on, two photocell sensors monitor external light levels and adjust the reflection of the mirror **(Figure 23–36)**. The ambient photocell sensor detects the light levels outside and in front of the vehicle. The headlamp photocell faces rearward to detect the level of light coming in from the rear of the vehicle. When there is a difference in light levels between the two photocells, the mirror begins to darken.

The automatic day/night mirror cannot be repaired. If it is faulty, it must be replaced.

Some rearview mirrors are fitted with a directional compass display. The readings on the compass appear on the mirror's reflective surface, normally in the lower left corner.

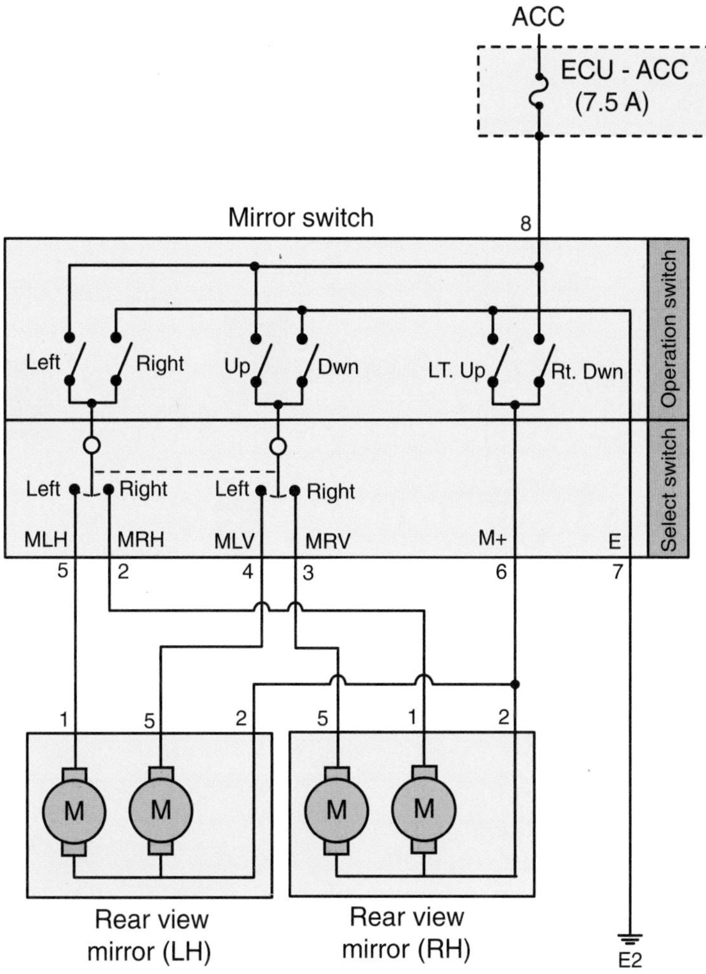

Figure 23-35 A wiring diagram for a power mirror circuit.

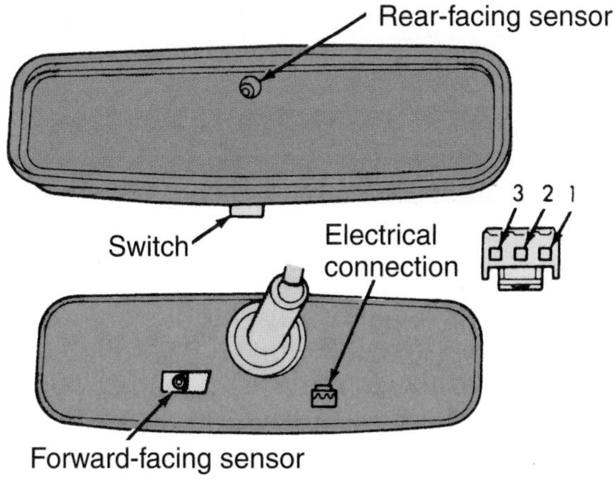

Figure 23-36 An automatic day/night mirror.
Courtesy of Chrysler LLC

REAR-WINDOW DEFROSTERS AND HEATED MIRROR SYSTEMS

The rear-window defroster (also called a **defogger** or **deicer**) heats the rear-window surface to remove moisture and ice from the window. On some vehicles, the same control heats the outside mirrors. The major components of a rear-window defroster include a switch, relay assembly, and the heating elements on the glass surface **(Figure 23–37)**.

Pressing the rear-window defroster switch momentarily energizes the relay. Battery voltage is then applied through closed contacts of the relay to the rear-window defroster grid. On models with a heated mirror, current also flows through a separate fuse to the mirror's heated grid.

After about 10 minutes, a time-delay circuit opens the ground path to the relay's coil and the coil de-energizes, shutting off power to the grids. The time-delay circuit prevents the system from remaining on during periods of extended driving. The system can also be manually turned off.

Diagnosis

One of the most common problems with rear-window defrosters is damage to the grids on the window. Damage can be caused by hard objects rubbing across the inside surface of the glass or by using harsh chemicals to clean the window. When a segment of the grid breaks, it opens the circuit. Often the customer's

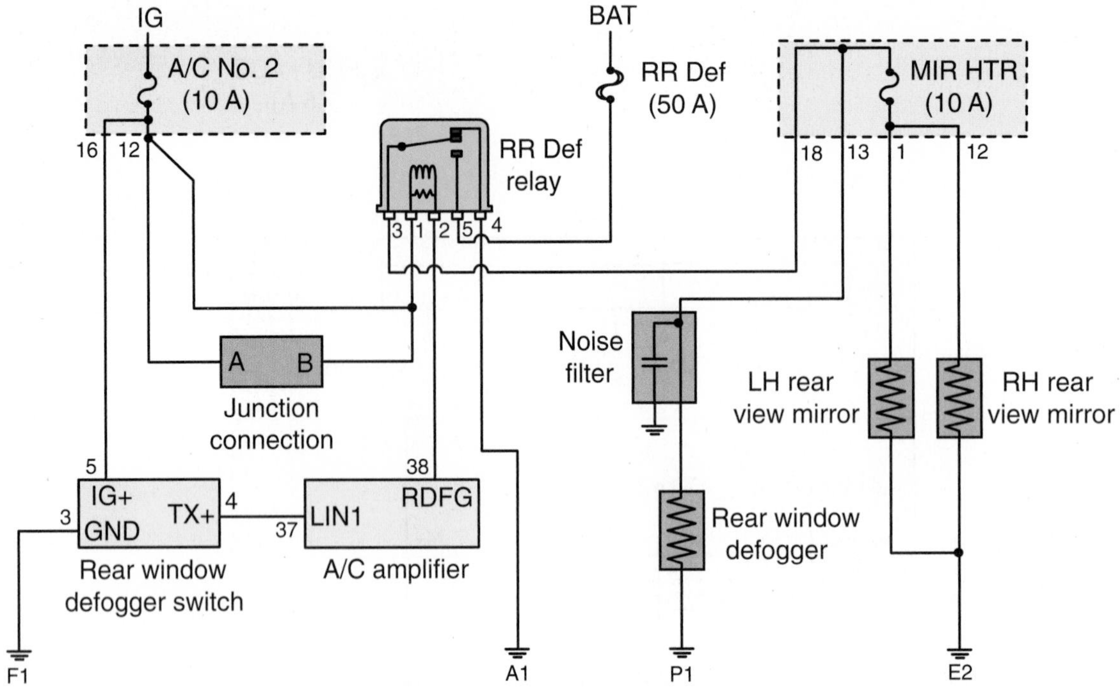

Figure 23-37 A wiring diagram for a rear-window defroster and heated mirror circuit.

complaint is that the unit does not defrost the entire window. Normally one or two lines of the grid are open. The open can be found by using a testlight. With the defroster turned on, voltage should be present at all points of the grid. If part of the grid does not have voltage, move the probe toward the positive side of the grid. Once voltage is present, you know the open is between those two points. Opens can be repaired by painting a special compound over the open. The correct procedure for doing this is shown in Photo Sequence 21.

If none of the grids heat up, check for voltage at the connection to the grids, and then check the ground circuit. If no voltage is present at the grids, check the circuit for the switch.

OTHER ELECTRONIC EQUIPMENT

Vehicles are being equipped with many electrical and electronic features. Examples of some of the more common newer accessories are discussed here.

Adjustable Pedals

Shorter drivers normally must move their seat very close, sometimes uncomfortably and unsafely close, to the steering wheel. By moving the pedals toward the driver, drivers may be able to adjust their seat position away from the steering wheel and still comfortably reach and use the pedals. An electric motor at the brake pedal (**Figure 23–38**), with a cable connection to the accelerator pedal, moves both pedals back and forth (up to 3 inches). A switch on the dash controls the motor. This feature may also be part of the seat memory system so the driver can quickly bring both the seat and pedals to the most comfortable position.

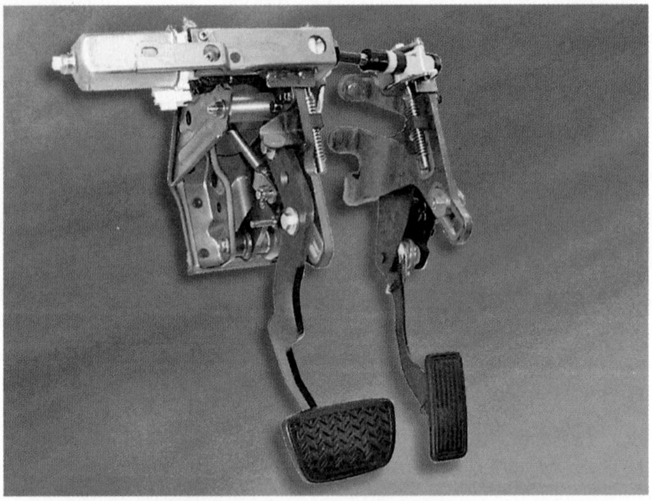

Figure 23-38 An adjustable pedal assembly. *Courtesy of Toyota Motor Sales, U.S.A., Inc.*

Heated Windshields

Heated, or self-defrosting, front windshield systems work like rear-window defrosters, heating the glass directly. However, instead of using a wire grid that could hinder the driver's vision, a microthin metallic coating inside the windshield is used. This coating or laminate is sandwiched between layers of glass. Bus bars are connected to the inner laminate or coating to provide power and ground circuits.

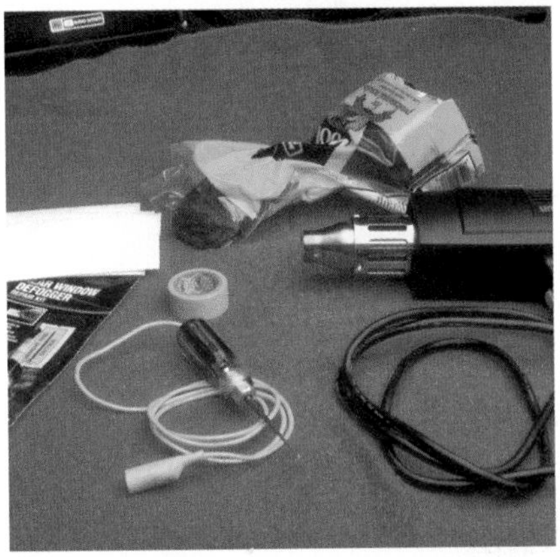

P21-1 *The tools required to perform this task include masking tape, repair kit, 500°F heat gun, testlight, steel wool, alcohol, and a clean cloth.*

P21-2 *Clean the grid line area to be repaired. Buff with fine steel wool. Wipe clean with a cloth dampened with alcohol. Clean an area about 1/4 inch (6 mm) on each side of the break.*

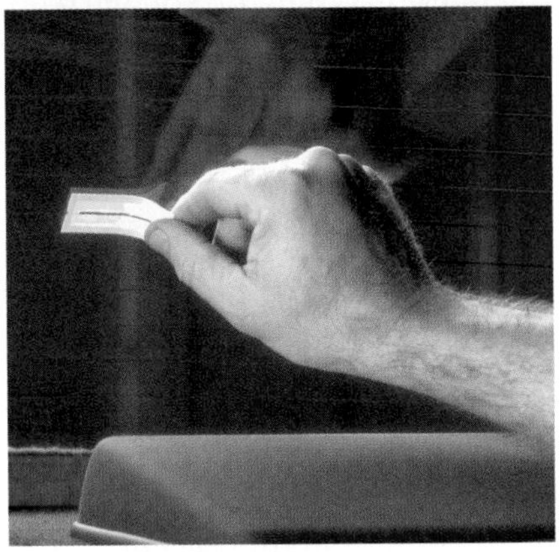

P21-3 *Position a piece of tape above and below the grid. The tape is used to control the width of the repair, so try to match the width with the original grid.*

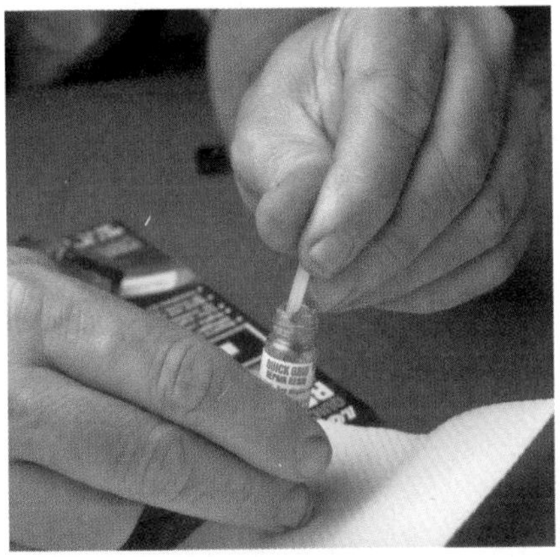

P21-4 *Mix the hardener and silver plastic thoroughly. If the hardener has crystallized, immerse the packet in hot water.*

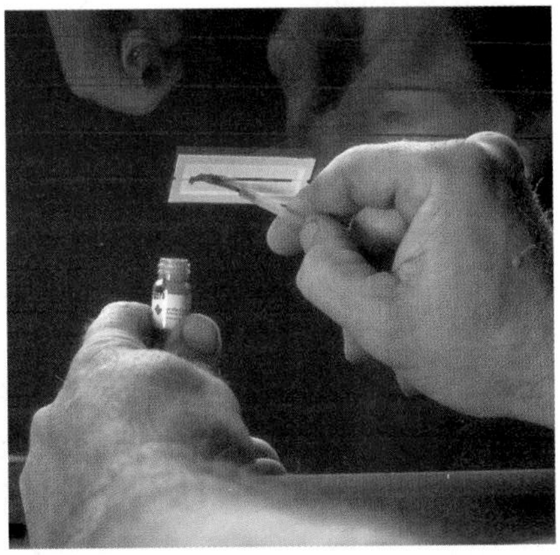

P21-5 *Apply the grid repair material to the repair area using a small stick.*

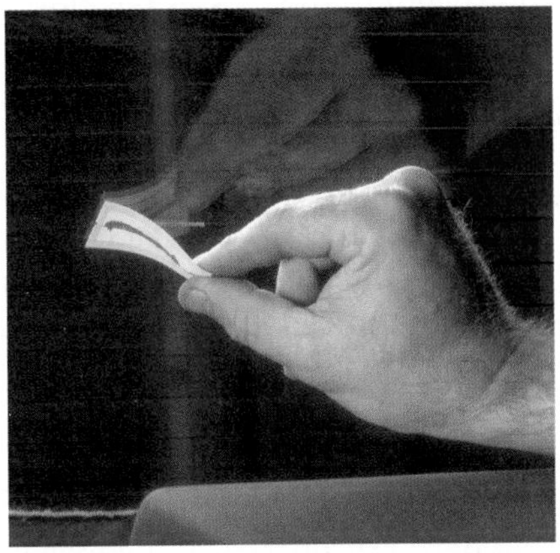

P21-6 *Carefully remove the tape.*

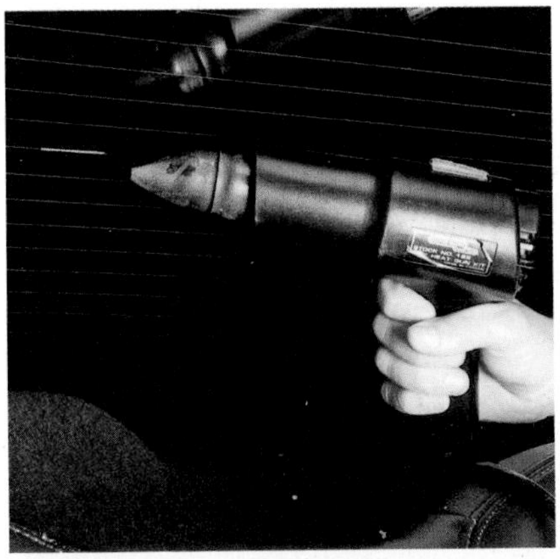

P21-7 *Apply heat to the repair area for 2 minutes. Hold the heat gun 1 inch (25 mm) away from the repair.*

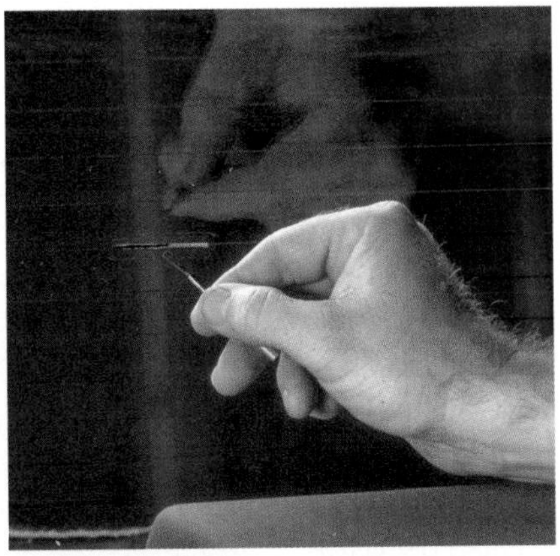

P21-8 *Inspect the repair. If it is discolored, apply a coat of tincture of iodine to the repair. Allow to dry for 30 seconds, then wipe off the excess with a cloth.*

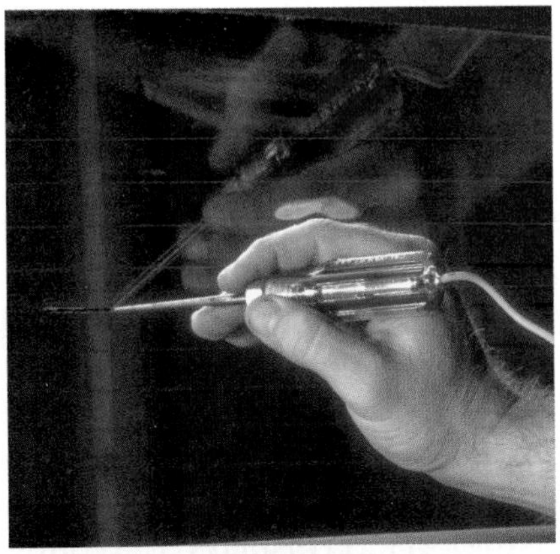

P21-9 *Test the repair with a testlight. Note: It takes 24 hours for the repair to fully cure.*

When the windshield is turned on, the generator's output is increased and redirected from the normal electrical system to the windshield circuit. This leaves all other electrical circuits to operate on power from the battery. When the voltage regulator senses a drop in battery voltage, the system is turned off and the generator's output is directed to the battery. Check the service manual for specific voltage details

Moon (Sun) Roof System

A power **moonroof** or **sunroof** can slide the roof panel open or closed. It can also tilt the panel up in the back to allow fresh air and natural light into the passenger compartment. The major components of any sunroof panel system are a relay, control switch, sliding roof panel, and motor. This circuit is normally protected by the inline circuit breaker.

When the two-position switch is moved to the open position, the roof panel moves into a storage area between the headliner and the roof. The panel stops moving any time the switch is released. Moving the switch to the closed position reverses the power flow through the motor.

If the system is not operating, check the fuse or circuit breaker. If these are okay, check for power at the switch with the ignition switch. If the voltage is present, the relay is okay. Check for power to the motor with the switch held in the open position. Refer to the service manual for additional diagnosis and testing information.

Figure 23-39 A retractable hardtop. *Courtesy of BMW of North America, Inc.*

Retractable Hardtops

A new trend in convertibles is retractable hardtops. The hard roof lowers and conceals itself in the vehicle's trunk. This system turns a regular coupe into a convertible. Most systems use electrohydraulic cylinders to move the hardtop **(Figure 23–39)**. In most cases, it takes about 30 seconds to stow or close the roof. Some of these hardtops have a power sliding glass sunroof. Diagnosis of these systems is conducted through the BCM.

iDrive

The iDrive feature offered on some BMWs gives a hint of where the industry may be headed with controls and information displays. The iDrive controller is a modified joystick installed on the center console **(Figure 23–40)**. It controls driving functions, including all active vehicle-control processes, such as those related to information displays. The controller is also used for the comfort and convenience vehicle's features. The controller can be pushed, turned, and slid to govern as many as 700 different vehicle functions. The selected functions are then displayed on the control display.

Figure 23-40 The iDrive controller. *Courtesy of BMW of North America, Inc.*

The controller is moved to access the main menus. To move to the submenus, it is rotated through electronic detents. The controller is pushed down to select a particular function or setting.

The control display of the iDrive system can also be used to go online to receive and send e-mails, make travel arrangements, get the latest news, and conduct a number of other online tasks. Online access is made through a special Internet portal via the wireless application protocol (WAP).

Blind Spot Detection

Blind-spot detection systems use bumper-mounted radar sensors that continuously monitor the rear blind spots on both sides of the vehicle. The sensors monitor the adjacent lanes behind the vehicle. When the vehicle enters the blind spot and there is a vehicle in one or both of them, a yellow light in the corresponding outside mirror lights up. If the driver puts on the turn signal for a lane change, the yellow light starts flashing as a more urgent warning that a vehicle is in that lane.

Rearview Cameras Some vehicles have a chip-based video camera mounted inside a small housing. The cameras consist of a silicon microchip assembly behind an optical lens. The view of the cameras is displayed on a screen to allow the driver to see what is behind the vehicle while traveling in reverse. This technology gives the driver a good view of what is behind the vehicle.

Parking Assist Similar technology is used to make parking easier. While parking the vehicle, or backing up, ultrasonic or radar-based sensors set off an audible warning when the driver gets too close to an object. The sensors are active only when the transmission is in reverse. They alert the driver as to the closeness of another vehicle or obstruction by a tone that becomes more rapid as the distance to the object decreases.

Self-Parking

A few vehicles offer an option for self-parking. This feature parallel parks a vehicle, without driver control, after the space is picked out by the driver. A variety of systems are used, but most calculate the size of the parking space and lets the driver know if the vehicle will fit into the space. This technology is only possible with vehicles that have electric steering.

The system uses ultrasonic sensors on either side of the front bumper, linked to a display in the instrument panel (**Figure 23–41**). It also relies on wheel-speed and steering-angle sensors to monitor the vehicle's movement and alert the driver of the remaining available space. Some systems rely on global positioning systems, a steering wheel sensor, and a rear-vision camera. In all systems, the driver lets go of the steering wheel and the system steers and moves the vehicle.

Figure 23-41 During self-parking, the driver must define the parking space on a screen in the instrument panel or the navigation display. *Courtesy of BMW of North America, LLC*

Night Vision

A few vehicles have a feature that uses military-style thermal imaging to allow drivers to see things they normally may not see until it is too late (**Figure 23–42**). The thermal-imaging system uses a camera with a fixed lens mounted behind the front grille and projects the image onto the bottom of the driver's side of the windshield by a HUD. The lens, which is an infrared sensor, is designed to operate at room temperature; therefore, the lens has its own heating and cooling system to keep it at the desired temperature.

The images seen on the HUD display are from the area in front of the light beams from the car's normal high-beam headlamps. The system allows the driver to see up to five times more of the road than with just headlights.

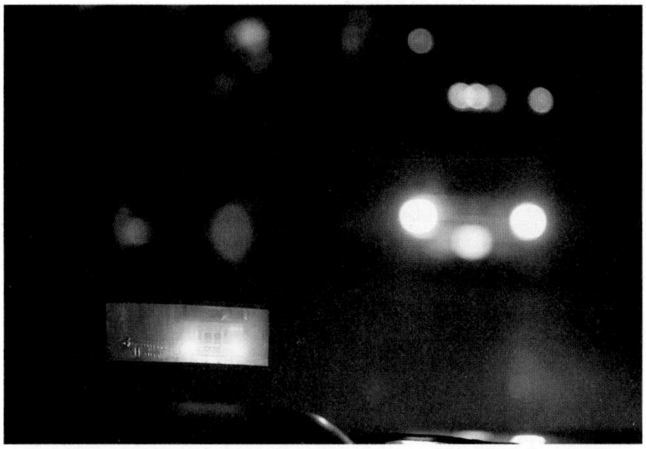

Figure 23-42 With night vision, driving is much safer. *Courtesy of Toyota Motor Sales, U.S.A., Inc.*

The system works by registering small differences in temperature and displaying them in sixteen different shades of gray on the HUD screen. It has the ability to display animals, or people, behind bushes or trees and can see through rain, fog, and smoke. Cold objects appear as dark images, whereas warm objects are white or a light color.

Navigation Systems

Navigation systems use global positioning satellites to help drivers make travel decisions. These global positioning systems (GPSs) set up a mathematical grid between the satellites and radio stations on the ground. The exact position of a vehicle can be plotted on the grid; therefore, the system knows exactly where the vehicle is **(Figure 23–43)**. GPS can display traffic and travel information on a display screen **(Figure 23–44)**. It can display a road map marking the exact location of the vehicle. It can plot out the best way of getting to a destination and can also tell the driver how many miles have been traveled and how many remain before reaching a destination. It can also display traffic information regarding traffic backups due to congestion, roadwork, and/or accidents, then display alternative routes so travel is not delayed.

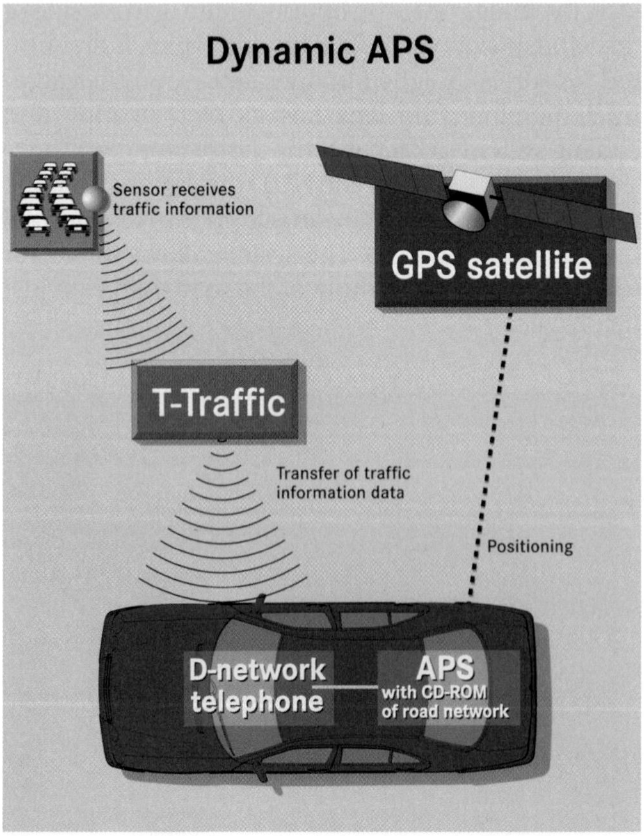

Figure 23-43 The communication network for GPS traffic and travel information. *Courtesy of Chrysler LLC*

Figure 23-44 The screen of a typical navigational center. *Courtesy of Toyota Motor Sales, U.S.A., Inc.*

A computer inside the vehicle compares the data it receives from the global positioning satellites with the information it has in its memory or what it reads from a CD or digital video disk (DVD). Periodically new disks are required to maintain the latest navigational information.

Many navigational systems give turn-by-turn guidance either by voice, on-screen displays, or both. Most of these systems feature touch-screen technology and can display and control other systems, such as air-conditioning, heating, and sound systems. Other systems allow you watch a movie on DVD while the vehicle is parked.

Vehicle Tracking Systems

Vehicle tracking systems can monitor the location of a vehicle if it has been stolen or lost. The system is based on the vehicle's navigational system or the cellular phone inside the vehicle. When the cell phone is the identifier, the tracking system is triggered when the thief attempts to place a phone call. If the correct code is not entered into the phone, the satellite begins to track the vehicle. This tracking signal is then monitored by an operator who can call the police in the area where the vehicle is being tracked. When the system relies on the GPS, a security or police officer can watch the movement of the vehicle on a remote computer screen.

Some systems automatically send a signal to the vehicle tracking operator if an air bag was deployed. This signal, in addition to the global positioning satellite network, lets the authorities immediately know when a serious accident has occurred and the emergency squad can respond without waiting to be called on the phone.

Systems, such as OnStar, can also provide the driver with emergency assistance and advice, roadside assistance, remote diagnostics (while the vehicle is driven), remote activation of the vehicle's horns and lights, remote door unlocking, news, and e-mail.

Voice Activation System

Voice activation or control systems allow for the control and operation of some accessories with voice commands. The voice commands are in addition to normal manual controls. Voice activation is commonly used for cell phone operation but can be used on other controls. Voice activation systems recognize the driver's voice and can respond with answers in response to the driver's questions. Once the system has understood the driver's request, it responds by carrying out the desired function, such as changing the stations on the radio. The system works through the microphones located near the driver. The voice activation system can recognize up to 2,000 commands and numeric sequences.

SECURITY AND ANTITHEFT DEVICES

Three basic types of antitheft devices are available: locking devices, disabling devices, and alarm systems. Many of the devices are available as optional equipment from the manufacturers; others are aftermarket installed.

Locks and Keys

Locks are designed to deny entry to the car as well as to prevent a thief from driving it away. Most locks move a mechanical block between the vehicle's body and the door. Keys simply move those blocks.

Manufacturers use specially cut keys that cannot be easily duplicated and lock mechanisms that are difficult to pick. The master key is often built into a remote control key handle. The master key can lock and unlock the doors, trunk, fuel filler door, and glove compartment all at the same time. The battery for the remote is charged each time the key is inserted into the ignition.

A special key, often called a valet key, only works in the doors and ignition, thereby preventing the valet from entering the trunk and glove compartment.

Many cars are equipped with special fuel filler doors that help to prevent the theft of gas from the fuel tank. Voltage is present at the fuel filler door release switch at all times. When the switch is closed, the door release solenoid is energized and the fuel door opens.

Passkey Systems

The passkey is a specially designed key, or transponder, that is programmed just for one vehicle. Although another key may fit into the ignition switch or door lock, the system will not allow the engine to start without the correct electrical signal from the key.

A **resistance key** **(Figure 23–45)** is a normal key with a small resistor bonded to it. When the key is inserted into the ignition switch, the circuit must recognize that resistance as being the correct amount for the vehicle before the engine will start.

Transponder key systems are based on a communication scheme between the vehicle's PCM and the transponder in the key **(Figure 23–46)**. Each time the key is inserted into the ignition switch, the PCM sends out a different radio signal. If the key's transponder is not capable of returning the same signal, the engine will not start.

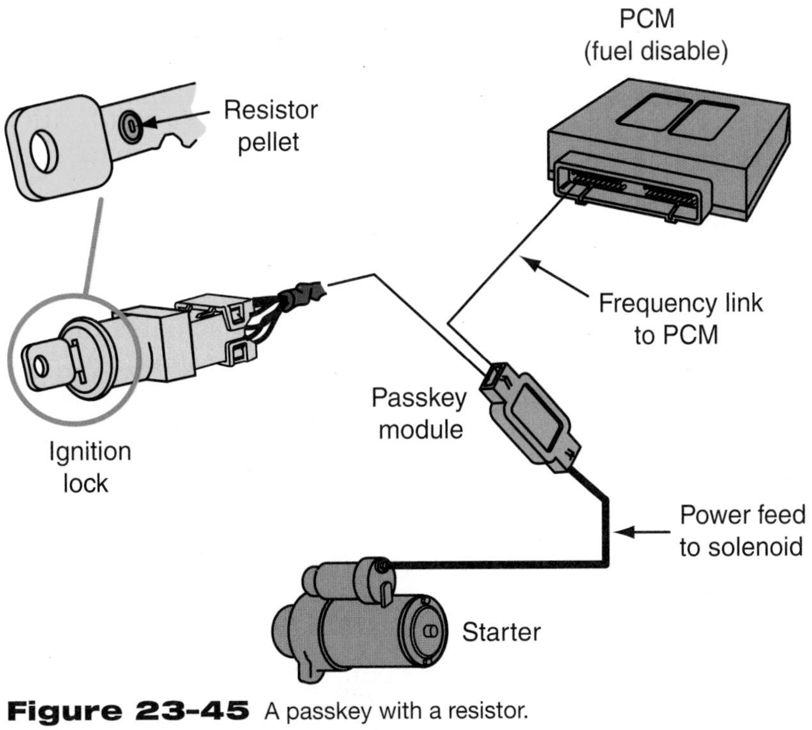

Figure 23-45 A passkey with a resistor.

Figure 23-46 An assortment of electronic keys and key fobs.

Some late-model vehicles use an ignition-kill system that prevents the vehicle from being hot-wired. In fact, when the ignition kill is activated, not even the ignition key starts the car. An ignition-kill system has a definite advantage as a theft deterrent. There is nothing to transport, no codes to remember, and nothing visible to mar the exterior of the car or to alert a thief.

Passwords On some passkey systems, a new password must be learned by the PCM or BCM when either has been replaced. The calibration is flashed into the new control module. After flashing, attempt to start the vehicle. It will not, but it will leave the ignition on until the theft system warning lamp turns off. Then turn the ignition off and attempt to start the engine. The engine should start this time.

Keyless Entry Systems

A keyless entry system allows the driver to unlock the doors or trunk lid from outside of the vehicle without using a key. It has two main components: an electronic control module and a coded-button keypad on the driver's door or a key fob. Some keyless systems also have an illuminated entry system.

The electronic control module typically can unlock all doors, unlock the trunk, lock all doors, lock the trunk, turn on courtesy lamps, and illuminate the keypad or keyhole after any button on the keypad is pushed or either front door handle is pulled.

Remote keyless entry systems rely on a handheld transmitter, frequently part of the key fob. With a press of the unlock button on the transmitter from 25 to 50 feet away (depending on the type of transmitter) in any direction range, the interior lights turn on, the driver's door unlocks, and the theft security system is disarmed. The trunk can also be unlocked. Pressing the lock button locks all doors and arms the security system. For maximum security, some remote units and their receiver change access codes each time the remote is used.

Some remote units can also open and close all of the vehicle's windows, including the sunroof. They may also be capable of setting off the alarm system in the case of panic.

Smart Keys

Many new vehicles do not come with a key; rather a transponder is used to perform all of the functions of a key. On some systems the transponder **(Figure 23–47A)** is inserted into a slot in the instrument panel **(Figure 23–47B)**. To start the engine, the transponder is pressed **(Figure 23–47C)**. On others, after the transponder is inserted into its slot, a start button is pressed. The system is based on a communication link between the transponder and the vehicle. If the codes of the transponder and the vehicle do not match, the engine will not start.

A few vehicles have smarter smart keys **(Figure 23–48)**. The transponder does not need to be inserted. It merely needs to be close to the vehicle. This system is normally called a smart access system. The system can perform many functions without inserting a key or pressing a button. It can lock and unlock the doors, allow the engine to start by pressing the engine switch while depressing the brake pedal, and open the trunk.

When the electronic key enters into zones around the vehicle, a certification control module certifies the ID code from the key. Once the signal is certified, the control module transmits an engine immobilizer deactivation signal to the ID code box and a steering unlock signal to the steering lock ECU. The BCM also receives a certification signal and actuates the door lock motor to unlock or lock the door.

The actuation zones are set up by several oscillators. Each oscillator transmits a signal every quarter of a second when the engine is off and the doors are locked. When a signal detects the electronic key, certification begins. The actuation zone is normally about 3 feet (1 meter).

The system can also trigger the BCM to restore the position of the driver's seat (driving position memory system), the shoulder belt anchor, the steering wheel, and the outside rearview mirror.

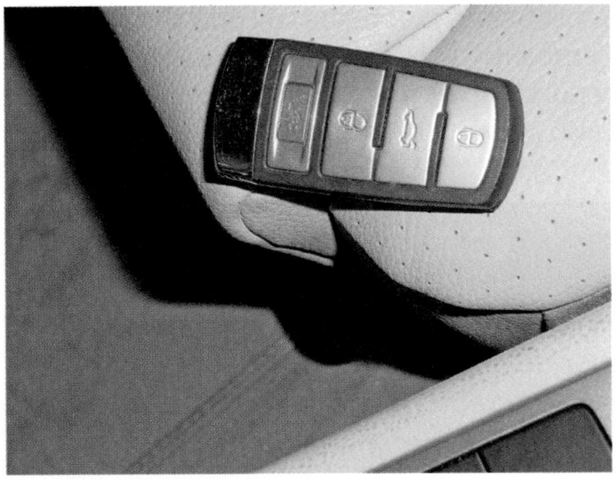

(A)

(B)

(C)

Figure 23-47 Some smart keys (A) are inserted into a slot (B), and then depressed (C) to start the engine.

Figure 23-48 A real smart key. *Courtesy of Toyota Motor Sales, U.S.A., Inc.*

SHOP TALK

A common service for antitheft systems is the replacement or addition of keys. Each manufacturer has a specific procedure for setting up communication with the vehicle and the key, transponder, and key fob. Without this communication link, the new key will not work.

Alarm Systems

The two methods for activating alarm systems are passive and active. Passive systems switch on automatically when the ignition key is removed or the doors are locked. Active systems are activated with a key fob transmitter, keypad, key, or toggle switch.

Switches similar to those used to turn on the courtesy lights as the doors are opened are often used. When a door, hood, or trunk is opened, the switch closes and the alarm sounds. It turns itself off automatically (provided the intruder has stopped trying to enter the car) to prevent the battery from being drained. It then automatically rearms itself.

Ultrasonic sensors are used to detect motion and will trigger the alarm if there is movement inside the vehicle. Current-sensitive sensors activate the alarm if there is a change in current within the electrical system, such as when a courtesy light goes on or the ignition starts. Motion detectors monitor changes in the vehicle's tilt, such as when someone is attempting to steal the tires.

Many alarm systems are designed to sound an alarm, turn on the hazard lights, and cause the high beams to flash along with the hazard lamps. Indicator lamps on the inside of the vehicle alert others that the alarm is set and also remind the driver to turn the alarm off before entering. To avoid false alarms, some systems allow for the disabling of particular sensors, such as the motion detector inside the vehicle that could be set off by a pet inside the vehicle.

Diagnosis There are many different types of antitheft systems used by the manufacturers; there also are many different aftermarket systems available. When diagnosing a problem, it is very important to have as much information as possible about the system. Most often a manufacturer specific scan tool is needed to diagnose an OEM antitheft system. In some cases, only dealership personnel have access to the necessary data for diagnosis. Before proceeding with a detailed diagnosis, inspect all fuses and relays in the system. Also check for loose wires, connectors, and components. This is especially important if the system works intermittently.

Antitheft system problems can prevent the engine from starting. If the system does not recognize the key or transponder, the engine will not start. The cause of the problem can be the key or transponder or the transponder control module. These same problems will also prevent the antitheft system from working.

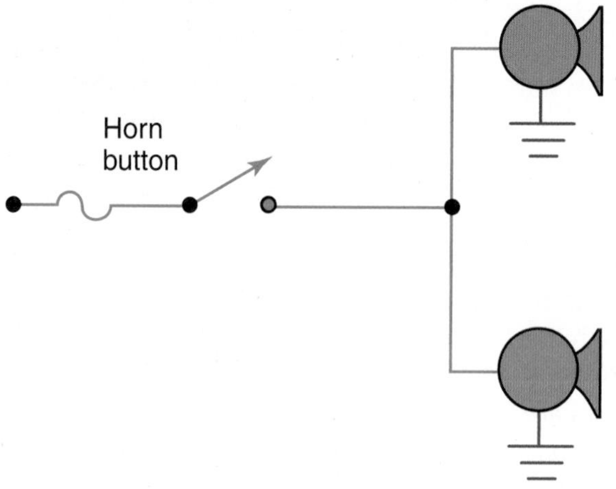

Figure 23-49 A two-horn circuit without a relay.

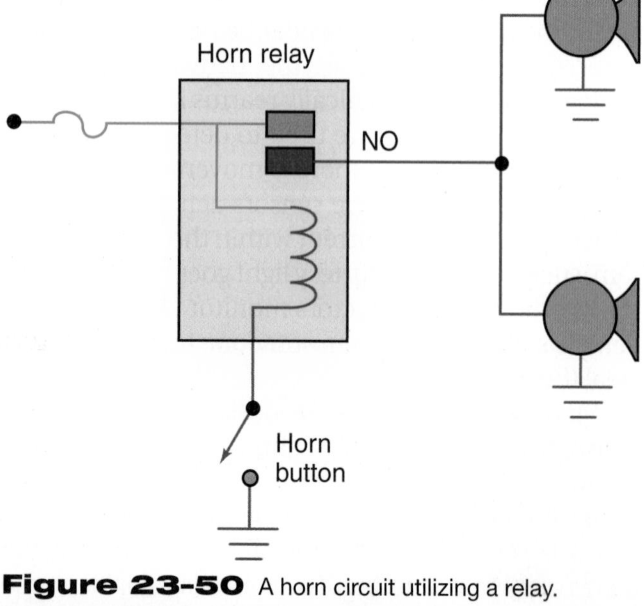

Figure 23-50 A horn circuit utilizing a relay.

CASE STUDY

A customer complains that the horn does not operate.

To correct this situation, the technician must determine whether the horn system has a relay. ***Figure 23–49*** *shows a two-horn circuit without a relay. Fused power goes up the steering column to the horn button, which is normally open (NO). Closing the switch closes the circuit, bringing power down the column to a common point where the two horns are connected in parallel. On the other side of the horn is a ground to complete the circuit. The circuit is easily diagnosed with a 12-volt test-light, following the usual procedure of looking for power with the horn button pushed in. The ground circuit should also be checked. Power in at the switch and no power out with the button depressed would be the only reason to replace the switch unit.*

Figure 23–50 *shows an adaptation of the simple horn circuit. Notice that fused power is applied to the horn relay rather than to the switch. An NO relay is used. Notice that the relay coil receives power from the same fused source as the contacts do. The other end of the relay coil runs up into the steering column where an NO switch is located. Closing the switch grounds the relay coil, which energizes the coil and causes the contacts to close. The closed contacts allow power to the common point for the two horns, which are grounded. Remember, any relay circuit must be diagnosed as two separate circuits. The relay coil is a separate circuit that needs a path for it to develop the magnetic field necessary to close the contacts. Closing the contacts allows current flow through the other circuit to the horns. Diagnose these circuits separately, as you would any relay circuit, and you are less likely to make mistakes. Use a DMM with a wiring diagram to trace the power through the circuit.*

KEY TERMS

Defogger
Deicer
Moonroof
Resistance key
Servo unit
Sunroof
Transducer
Transponder key

SUMMARY

- Many accessories may be controlled and operated by a body computer and, therefore, diagnosis may involve retrieving trouble codes from the computer.
- The basic designs of windshield wiper systems used on today's vehicles include a standard two- or three-speed system with or without an intermittent or rain-sensing feature.
- The motors can have electromagnetic fields or permanent magnetic fields. With permanent magnetic fields, motor speed is regulated by the placement of the brushes on the commutator. The speed of electromagnetic motors is controlled by the strength of the magnetic fields.
- Diagnosis of a wiper/washer system should begin by determining if the problem is mechanical or electrical.
- Most current vehicles have electronically regulated cruise control systems that are controlled with a separate control module or by the vehicle's PCM.
- Diagnosis of PCM-controlled cruise control systems is aided by a scan tool.
- Complex sound systems may include an AM/FM radio receiver, stereo amplifier, CD player, cassette player, MP3 player, equalizer, several speakers, and a power antenna system.
- Poor speaker or sound quality is usually caused by a bad antenna, damaged speakers, loose speaker mountings or surrounding areas, poor wiring, or damaged speaker housings.
- Diagnosis of power door locks, windows, and seats is best done by dividing the circuit into individual circuits and the total circuit and basing your testing on the symptoms.
- One of the most common problems with rear-window defrosters is damage to the grids on the window, which can usually be repaired.
- Navigation systems use global positioning satellites to help drivers make travel decisions while they are on the road.
- Three basic types of antitheft devices are available: locking devices, disabling devices, and alarm systems.
- Most ignition keys for late-model vehicles are either resistance or transponder passkeys that only work on the vehicle for which they were intended.
- Passive alarm systems switch on automatically when the ignition key is removed or the doors are locked. Active systems are activated manually.

REVIEW QUESTIONS

1. What is the primary purpose of adaptive cruise control?
2. Why are amplifiers added to sound systems?
3. What should be checked if none of the grids in a rear-window defogger works?
4. *True or False?* Rain-sensitive wiper systems respond to the weight of the water on the windshield to regulate the speed of the wipers.
5. What component in a wiper system makes it possible to adjust the wiper interval period? How does it work?
6. Name the two most common problems that occur with power seats.
7. *True or False?* On some passkey systems, when the PCM or BCM is replaced, the key must be reprogrammed with a new password.
8. *True or False?* Vehicles with electronic throttle control do not need a separate cruise control module, stepper motor, or cable to control engine speed. The PCM has full control of the throttle and, therefore, the circuitry of the PCM operates the cruise control system.
9. What could be the problem when the vehicle's dome light dims when a power seat is moved?
10. Rear defrosters generally have a relay with a timer. This allows ______.
    a. the defogger to shut down after a predetermined length of time
    b. the defogger to function just until the rear window is clear
    c. the defogger to be independent of the ignition switch
    d. none of the above
11. Which of the following can be sources of radio interference?
    a. ignition system wiring
    b. electrical power lines

c. neon signs
d. all of the above

12. Depressed part windshield wiper systems park the blades by generally ______.
    a. turning off power to the motor
    b. applying power to the motor
    c. ungrounding the switch
    d. reversing the motor direction
13. Which of the following parts or circuits could cause the cruise control not to cancel when the brake pedal is depressed?
    a. cruise control switch circuit
    b. vehicle speed sensor circuit
    c. stoplight switch
    d. transmission range sensor circuit
14. When diagnosing a vacuum/mechanical cruise control system that cannot maintain a constant speed, all of the following should be checked, except ______.
    a. the brake switch
    b. the vacuum lines
    c. the vacuum servo
    d. the vehicle speed sensor
15. Which of the following is not a true statement?
    a. When all power windows do not operate, check the fuse and the wiring to the master switch, including the ground.
    b. When one power window does not operate, check the wiring to the individual switch, the switch, and the motor. Also check the window for binging in its tracks.
    c. When both rear power windows cannot be operated by their individual switches, check the lockout and master switch.
    d. When one power window moves in one direction only, check the fuse and the wiring to the master switch, including the ground.

ASE-STYLE REVIEW QUESTIONS

1. While discussing power door locks: Technician A says that if none of the locks work, the actuators at each door should be checked first. Technician B says that if the power locks can lock or unlock but not both, the relays and wiring to the master switch should be checked. Who is correct?
    a. Technician A
    b. Technician B
    c. Both A and B
    d. Neither A nor B
2. Technician A says that horn buttons in circuits with relays are power switches. Technician B says that horn buttons in circuits without relays are grounding switches. Who is correct?
    a. Technician A
    b. Technician B
    c. Both A and B
    d. Neither A nor B
3. The right front power window does not work: Technician A says that the motor might have a bad ground. Technician B says that either the master or right door switch could be the problem. Who is correct?
    a. Technician A
    b. Technician B
    c. Both A and B
    d. Neither A nor B
4. The reasons for slower than normal wiper operation are being discussed: Technician A says that the problem may be in the mechanical linkage. Technician B says that there may be excessive resistance in the electrical circuit. Who is correct?
    a. Technician A
    b. Technician B
    c. Both A and B
    d. Neither A nor B
5. A six-way power seat does not work in any switch position: Technician A says to check the circuit breaker. Technician B says to use a continuity chart to test the switch. Who is correct?
    a. Technician A
    b. Technician B
    c. Both A and B
    d. Neither A nor B
6. Technician A says that some antitheft systems sound an alarm if someone enters the vehicle without a key. Technician B says that some antitheft devices prevent the engine from starting. Who is correct?
    a. Technician A
    b. Technician B
    c. Both A and B
    d. Neither A nor B
7. The circuit breaker at a power seat motor continuously trips: Technician A says that this could be caused by a seat track problem that is causing mechanical resistance. Technician B says that this could be caused by corrosion at the motor, which is causing electrical resistance. Who is correct?
    a. Technician A
    b. Technician B
    c. Both A and B
    d. Neither A nor B
8. Technician A says that a resistance key has a small thermistor bonded to it. Technician B says that a transponder key sends a radio signal to the PCM. Who is correct?
    a. Technician A
    b. Technician B
    c. Both A and B
    d. Neither A nor B

9. Technician A says that navigational systems rely on global positioning satellites. Technician B says that navigational systems rely on information programmed in its memory and information stored on a CD or DVD. Who is correct?
   a. Technician A
   b. Technician B
   c. Both A and B
   d. Neither A nor B

10. While diagnosing a cruise control system that does not turn on: Technician A checks the vehicle speed sensor and its input to the control module. Technician B checks the operation of the brake lights. Who is correct?
    a. Technician A
    b. Technician B
    c. Both A and B
    d. Neither A nor B

CHAPTER 24

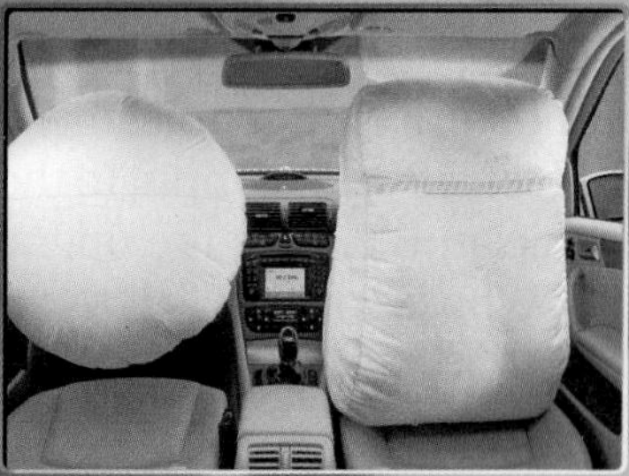

RESTRAINT SYSTEMS: THEORY, DIAGNOSIS, AND SERVICE

OBJECTIVES

■ **Identify and describe devices that contribute to automotive safety.** ■ **Explain the difference between active and passive restraint systems.** ■ **Know how to service and repair passive belt systems.** ■ **Describe the function and operation of air bags.** ■ **Identify the major parts of a typical air bag system.** ■ **Safely disarm and inspect an air bag assembly.** ■ **Know how to diagnose and service an air bag system.**

Safety is foremost in the minds of automobile manufacturers. According to a survey by the Insurance Institute for Highway Safety, occupant protection has emerged as a leading factor in determining which car people will buy. According to the institute, 68% of the households surveyed ranked the "degree to which the car protects people" as a very important purchase-decision factor.

Many safety features are now available as standard equipment or as options. Some of these include side impact barriers, crumple zone in the body, seat belts, antilock brakes, traction control, stability control, and air bags (**Figure 24–1**). There are many safety items that have been around for many years, such as laminated and tempered glass.

Common restraint systems—seat belts and air bags—are covered in this chapter. It is important for a technician to understand how these systems work and how to diagnose and service them.

An **active restraint system** is one that a vehicle's occupant must make a manual effort to use (**Figure 24–2**). For example, in most vehicles the passenger must fasten the seat belts for crash protection. A **passive restraint system** operates automatically. No action is required of the occupant to make it functional.

Passive safety equipment includes the safety belt system, the air bags, the rigid occupant cell, and the crumple zones at the front of the vehicle. The rear and sides of the body are among the most important safety features of today's cars and are designed to dissipate most of the impact energy for the protection of vehicle occupants.

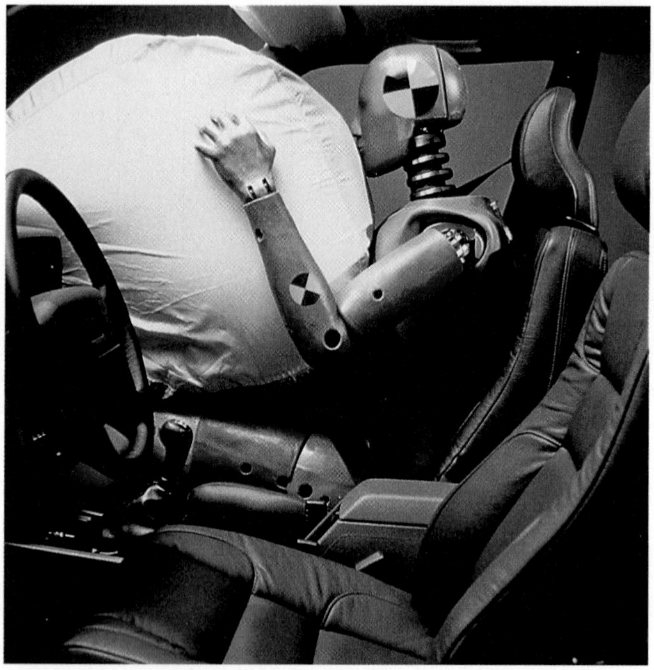

Figure 24-1 A passenger-side air bag. *Courtesy of Volvo Cars of North America*

SEAT BELTS

A passive seat belt system uses electric motors to automatically move shoulder belts across the driver and front seat passenger. The upper end of the belt is attached to a carrier that moves in a track at the top of the doorframe. The other end is secured to an inertia lock retractor mounted to the center console. When the door is opened, the outer end of the shoulder belt moves forward to allow for easy entry or exit. When the doors are closed and the

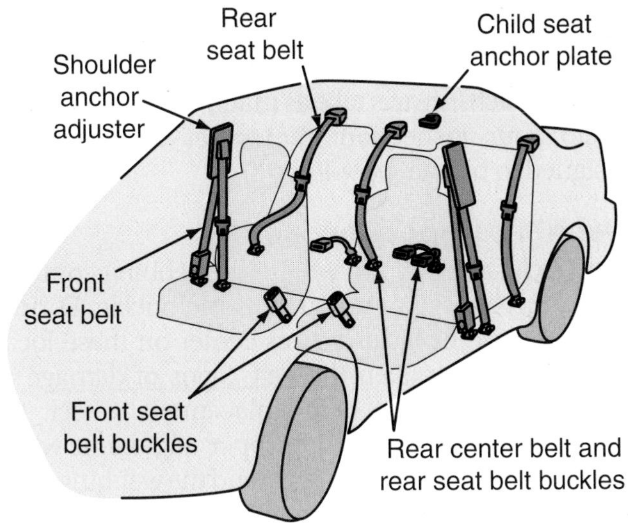

Figure 24-2 Active restraint systems.

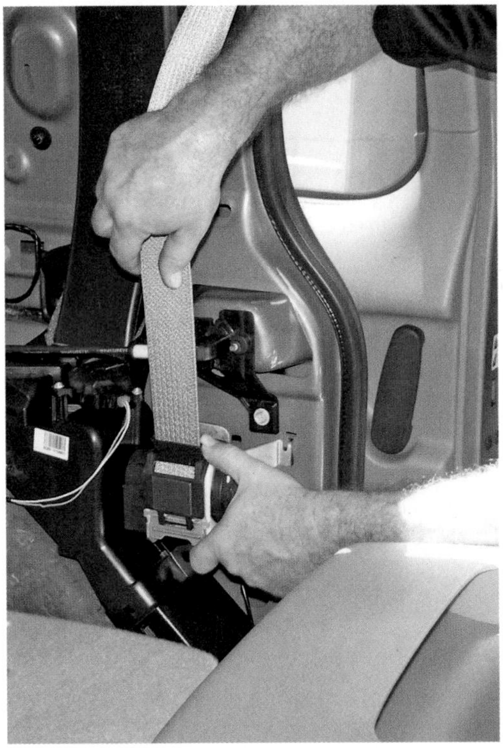

Figure 24-3 A seat belt retractor.

ignition is turned on, the belts move rearward and secure the occupants. The active lap belt is manually fastened and should be worn with the passive belt.

Most vehicles have two active belts. One is a lap belt that goes across the occupant's lap; the other is a shoulder belt that goes across the shoulder and chest. The two belts join together at a single point where they are inserted into a buckle anchored to the vehicle's floor.

If a seat belt will not buckle, use a flashlight and look inside the buckle. Often something is inside the buckle that can be safely removed. It does not take much to hamper the operation of the buckle. In most cases, the buckle should be replaced if something is lodged inside.

Seat Belt Retractors

When unbuckled, seat belts are stowed away by the seat belt retractors (**Figure 24–3**). The retractors may also work as pretensioners to take up the belt's slack during an accident to limit the forward movement of the occupant's body. **Inertia lock retractors** (**Figure 24–4**) prevent the belt from coming out of the retractor when there is a sudden pull on the belt. Some vehicles have electric or pyrotechnic-type **pretensioners**. Both of these are designed to quickly tighten the belt at the start of a crash.

Pyrotechnic pretensioners are the most common (**Figure 24–5**). When the pretensioner receives a signal from the control module, the pretensioner is ignited. There is a small explosion in the pretensioner that reverses its action. This puts a firm hold on the passenger. Many of these systems also have a mechanism that releases the pressure on the seat belt after it has been tightened by the pretensioner. When the pressure between the passenger's chest and the seat belt exceeds a particular point, the pressure on the seat belt is relaxed to prevent injury.

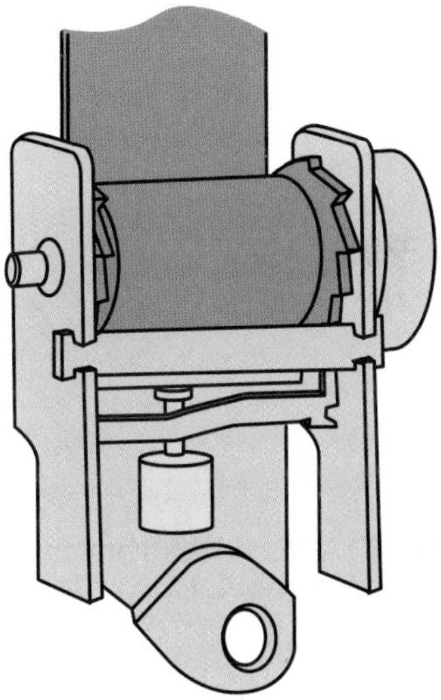

Figure 24-4 An inertia lock seat belt retractor.

On some vehicles, the action of the pretensioners varies with the weight of the person in the seat and the amount of force on the seat belt as that person is moving forward during an impact (**Figure 24–6**). Some vehicles are equipped with two-stage belt force limiters.

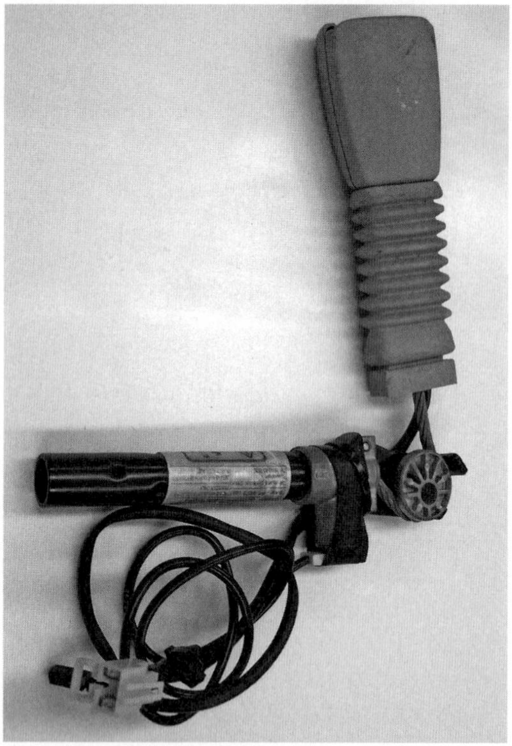

Figure 24–5 A buckle-mounted pretensioner.

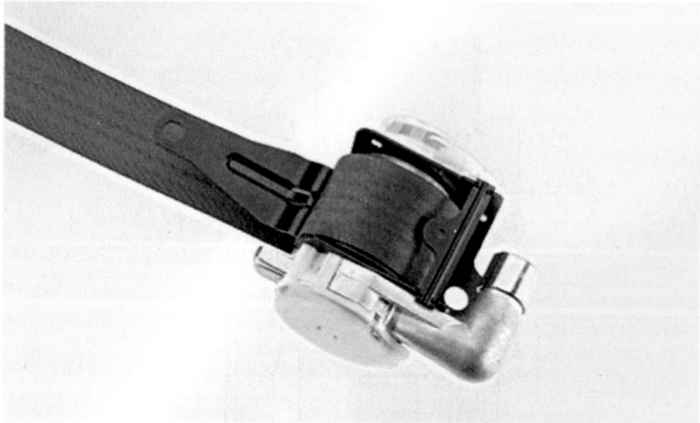

Figure 24–6 A mechanized pretensioning retractor/variable load-limiting seat belt retractor. *Courtesy of Delphi Corporation*

Warning Lights

All modern seat belt systems have a warning lamp and a buzzer or chime that is turned on when the vehicle is started to remind the occupants to buckle up. When the ignition is turned on, a signal is sent to the warning lamp **(Figure 24–7)**. If the seat belt is fastened, a signal is sent from the buckle switch to the indicator controller and the lamp turns off. If the belt is not buckled, the indicator and buzzer will alert the driver in intervals. There is a sensor in the front passenger seat that detects when someone is in the seat. This information is sent to the control module and the indicator lamp will blink until the seat belt is fastened.

SEAT BELT SERVICE

Inspecting seat belt systems should follow a systematic approach. Always take as much time as necessary to do your inspection; remember that they are designed to protect people.

Webbing Inspection

Pay special attention to where the webbing contacts maximum stress points, such as the buckle, D-ring, and retractor. Collision forces center on these locations and can weaken the belt. Signs of damage at these points require belt replacement. Check for twisted webbing due to improper alignment when connecting the buckle. Fully extend the webbing from the retractor. Inspect the webbing and replace it with a new assembly if the following conditions are noted **(Figure 24–8)**: cut or damaged webbing, broken or pulled threads, cut loops at the belt edge, color fading as a result of exposure to sun or chemical agents, or bowed webbing.

If the webbing cannot be pulled out of the retractor or will not retract to the stowed position, check for the following conditions and clean or correct as necessary: webbing soiled with gum, syrup, grease, or other material; twisted webbing; or the retractor or loop on the B-pillar out of position.

SHOP TALK

Never bleach or dye the belt webbing. Clean it with a mild soap solution and water.

Buckle Inspection

To determine if the buckle works or if the buckle housing has been damaged, insert the seat belt into the buckle until a click is heard. Pull back on the webbing quickly to ensure that the buckle is latched properly. Replace the seat belt assembly if the buckle does not latch. Depress the button on the buckle to release the belt. The belt should release with a pressure of approximately 2 pounds. Replace the seat belt assembly if the buckle cover is cracked, the push button is loose, or the pressure required to release the button is too high.

Retractor Inspection

Retractors for lap belts should lock automatically once the belt is fully out. Either webbing-sensitive or vehicle-sensitive seat belt retractors are used with passive seat belt systems. Webbing-sensitive retractors can be tested by grasping the seat belt and jerking it. The retractor should lock up; if it does not, replace the seat belt retractor.

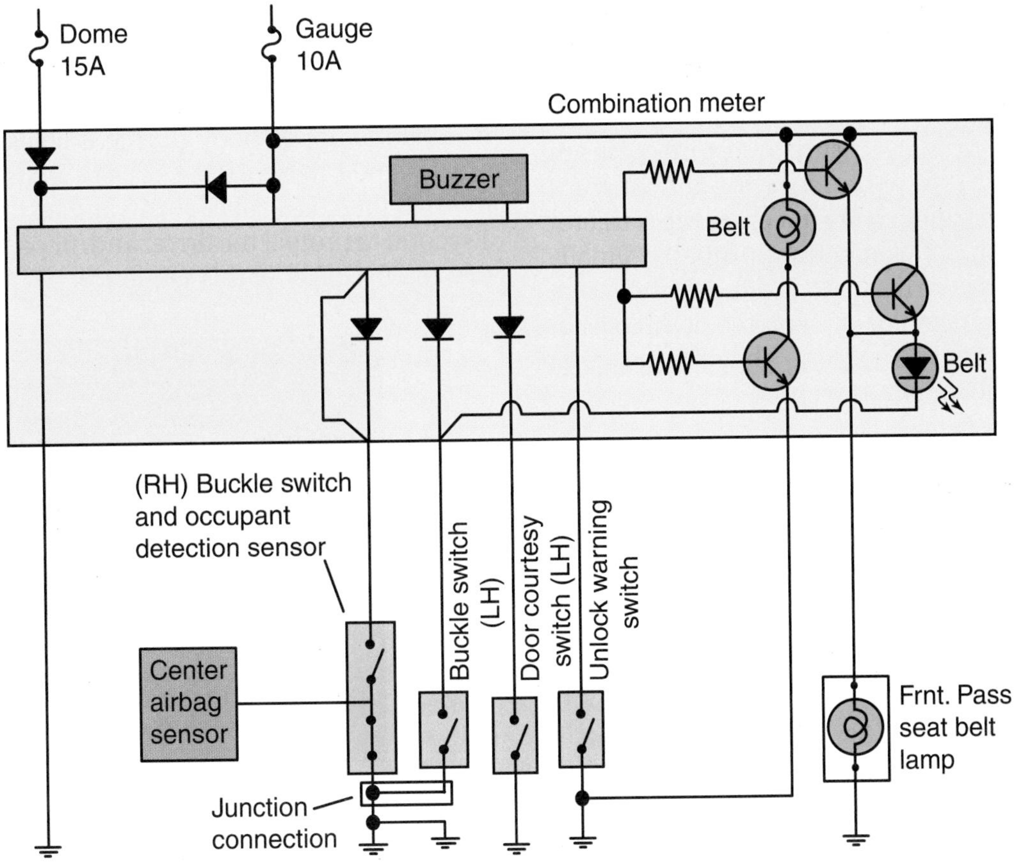

Figure 24-7 A wiring diagram for the seat belt warning and key reminder indicators.

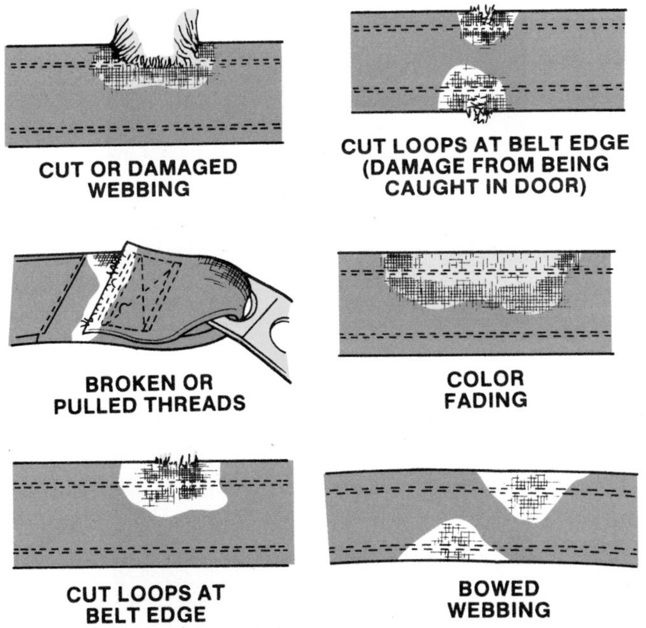

Figure 24-8 Examples of webbing defects.

Vehicle-sensitive belt retractors will not lock up using the same procedure. To test these belts, a braking test is required. Perform this test in a safe place. A helper is required to check the retractors on the passenger side and in the back if the vehicle is equipped with rear lap/shoulder belts.

Test each belt by driving the car at 5 to 8 mph and quickly applying the brakes. If a belt does not lock up, replace the seat belt assembly. During this test, it is important for the driver and helper to brace themselves in the event the retractor does not lock up.

Most retractors are not interchangeable. That is, an R marked on the retractor tab indicates that it is for the right side only, and an L should be used on the left side only.

CAUTION!

Never measure the resistance of a seat belt pretensioner. The voltage from the meter may accidentally ignite the pretensioner. This can result in serious personal injury.

Anchor Inspection

Carefully inspect the anchor areas and attaching bolts for the retractors. A buildup of dirt in the anchors can cause the seat belts to retract slowly. Wipe the inside of the loops with a clean cloth dampened in isopropyl alcohol. Loose bolts should be replaced and the new bolts tightened to specifications. Look for cracks and distortion in the metal at the anchor points. If there is damage to the metal in the mounting area, proper repairs, such as welding in reinforcement metal, must be completed before reattaching the anchor. Be sure to restore corrosion protection to the area. When spraying anticorrosion materials, make sure they do not enter the retractor. This can keep it from operating properly. Finally, look for dirt and corrosion around the anchor area.

Pretensioners should be replaced if a collision caused the air bags to deploy. They are explosive devices and good for one-time use.

Drive Track Assembly

Passive systems have a drive motor usually located at the base of the track assembly behind the rear seat side trim panel. The motor pulls the tape that positions the belt. If the motor is faulty, replace it. To service a motorized seat belt system, follow the instructions given in the service manual.

Rear Seat Restraint System

Rear seat belts are inspected in the same way as the front. However, some vehicles have a center seat belt. These belts do not have a retractor. Check the webbing, anchors, and the adjustable locking slide for the belt. Fasten the tongue to the buckle and adjust by pulling the webbing end at a right angle to the connector and buckle. Release the webbing and pull upward on the connector and buckle. If the slide lock does not hold, remove and replace that seat belt assembly.

Warning Light and Sound Systems

When the ignition is turned to the on or run position, the Fasten Seat Belt light should come on. There should also be a buzzer or chime. If these warning light and sound systems do not come on, check for a blown fuse or circuit breaker. If that checks out fine, and there is sound but no light, check for a damaged or burned-out bulb. If the bulb lights but there is no sound, check for damaged or loose wiring, switches, or buzzer (voice module).

Service Guidelines

Some guidelines for servicing lap and shoulder belts follow:

- Replace the seat belt with a new assembly if there is any abnormality.
- Never disassemble any part of the seat belt system.
- Never attempt repairs on lap or shoulder belt retractors or retractor covers. Replace them if necessary.
- Tighten all anchor bolts to specifications.

AIR BAGS

An air bag is much like a nylon balloon that quickly inflates to stop the forward movement of the occupant's upper body during a collision. Air bags are designed to be used *with* seat belts, not replace them. If there is a collision, an air bag takes less than 1 second to protect the driver and/or passengers **(Figure 24–9)**. Consider this sequence:

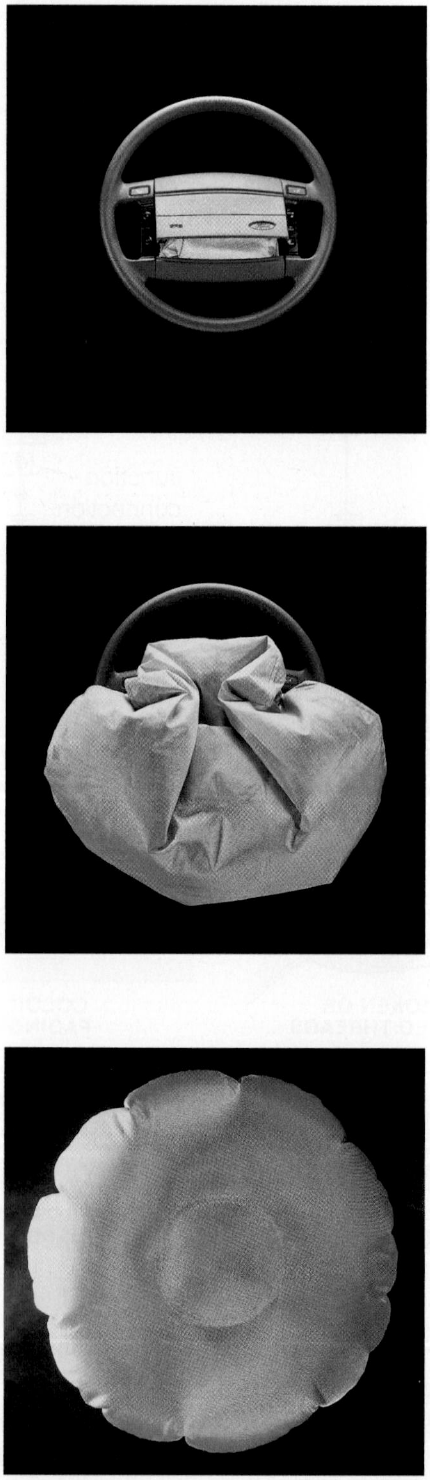

Figure 24-9 Various stages of air bag inflation.
Courtesy of TRW, Incorporated

- Time zero—Impact begins and the air bag system is doing nothing.
- Twenty milliseconds later—The sensors are sending an impact signal to the air bag module and the air bag begins to inflate.
- Three milliseconds later (total time from impact is now 23 milliseconds)—The air bag is inflated and is up against the occupant's chest. The occupant's body has not yet begun to move as a result of the impact.
- Seventeen milliseconds later (total time from impact is now 40 milliseconds)—The air bag is almost fully deployed and the occupant's body begins to move forward because of the impact.
- Thirty milliseconds later (total time from impact is just 70 milliseconds)—The air bag begins to absorb the forward movement of the occupant and the air bag begins to deflate through its vents. Once the air bag deflates, its job is over.

The systems and parts used to deploy an air bag vary with the year and manufacturer of the vehicle, as well as the location of the air bag. An air bag is inflated or deployed by rapid expansion (explosion) of a gas. The gas is fired by an igniter commonly called a **squib**.

Different manufacturers also call their air bag systems by different names, such as **supplemental inflatable restraint (SIR)** and **supplemental restraint system (SRS)**. All late-model vehicles have a driver-side and a passenger-side air bag **(Figure 24–10)**.

Figure 24-10 Driver-side and passenger-side air bags. *Courtesy of Chrysler LLC*

Passenger-side air bag modules are located in the vehicle's dash. These air bags are very similar in design and operation to those on the driver's side. However, many manufacturers use a different set of sensors. The actual capacity of gas required to inflate the passenger-side air bag is much greater because the bag must span the extra distance between the occupant and the dashboard. The steering wheel and column make up this difference on the driver's side.

A driver-side and passenger-side air bag system may include a knee diverter, also called a knee bolster. This unit is designed to help restrain the lower parts of the bodies of the driver and front passenger and prevent the driver from sliding under the air bag during a collision. It is located underneath the steering column and behind the steering column trim. Some vehicles have front and rear seat cushion air bags that inflate the front of the seat cushion to restrain the occupant's lower hip. This helps dampen the impact energy that acts on the occupant's upper body, including the head and chest.

Because of the concern for babies and small children, pickups and other two-seat vehicles either do not have a passenger-side SIR or have a switch that prevents it from deploying. The switch is typically operated with a key to activate or deactivate the SIR. An indicator light in the instrument panel shows the current status of the passenger-side SIR.

Side Air Bags

On some vehicles the occupants may be further protected by side air bags and/or side curtain air bags **(Figure 24–11)**. The rear passengers may be protected by air bags in the rear of the front seat backs, side air bags, and/or the side curtain air bags.

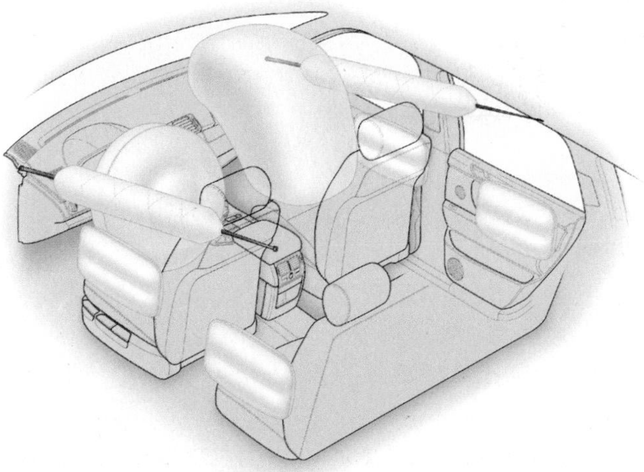

Figure 24-11 This vehicle has a total of eight air bags. *Courtesy of BMW of North America, Inc.*

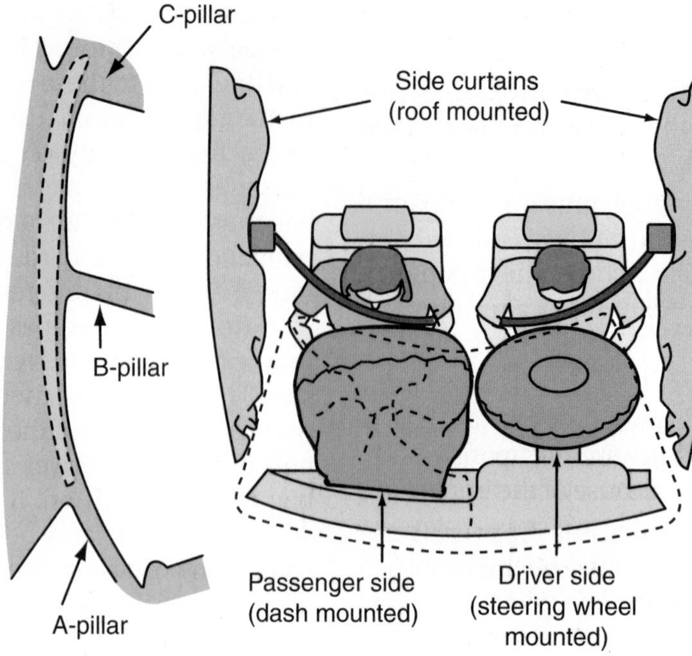

Figure 24–12 A side curtain.

Side air bags can take on many different shapes and are deployed from various locations. Side curtains **(Figure 24–12)** blanket the entire side of the car. Side air bags **(Figure 24–13)** are available for the front and rear doors on some cars. These air bags are deployed from the interior trim on the door or from the outside of the seat. Curtain air bags are located inside the headliner and extend from the driver's and front passenger's front pillars to the rear pillars behind the rear seat. Each air bag is a one-piece unit.

Side head protection systems inflate a long, narrow air bag that extends from the windshield area to the back of the front seat **(Figure 24–14)**.

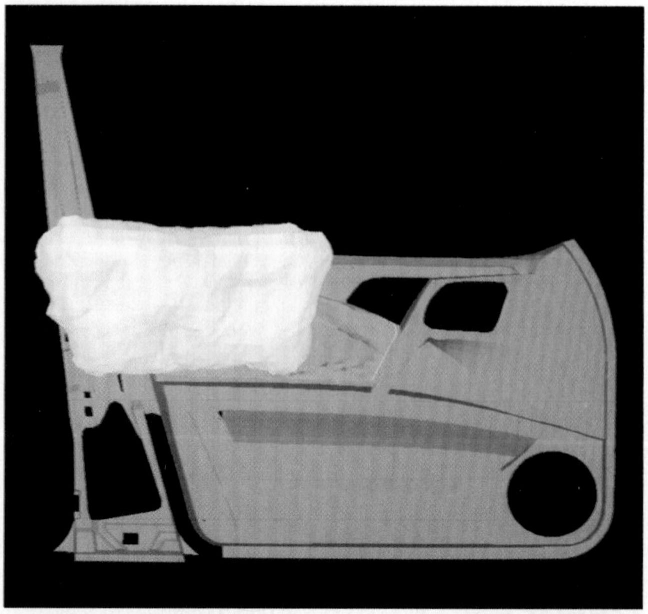

Figure 24–13 A side air bag. *Courtesy of Chrysler LLC*

Figure 24–14 A side impact head air bag. *Courtesy of BMW of North America, Inc.*

When a side impact air bag is deployed: the front side, rear side, and curtain shield air bags are deployed at the same time. Door-mounted side air bags must begin deploying in 5 to 6 milliseconds. This requirement is based on the fact that only a few inches separate the occupant from the other vehicle during a side impact. Seat-back-mounted side air bags do not need to operate at these great speeds. The head air bag is designed to stay inflated for about 5 seconds to offer protection against a second or third impact.

Current air bag systems work in conjunction with the seat belt pretensioners and retractors. When the air bag circuit is turned on, so is the pretensioner circuit. These actions limit the movement of the occupants.

Second-Generation Air Bags

Some vehicles are equipped with **second-generation air bags** that inflate with less force than earlier air bags. The air bags are depowered by reducing the peak inflation pressure and/or the force and speed at which an air bag inflates. These systems reduce the number of injuries caused by the air bag itself. Depending on the specific model vehicle, air bag size, and seat belt system, a second-generation air bag inflates with an average of 20% to 35% less energy.

Adaptive SRS Systems

All 2006 and newer vehicles must have a system that allows for air bag suppression when infants, children, or small adults are in the front passenger seat. This system uses a load sensor, seat belt tension sensor, and an electronic control unit **(Figure 24–15)**. The load sensor measures the weight on the seat and classifies the occupant as an adult or child and provides the classification to the air bag controller, which enables or suppresses passenger air bag deployment. A belt tension sensor identifies cinched child seats.

Some vehicles have "**smart**" or **adaptive air bags**. Many of these systems have two possible stages of air bag deployment. The force of the expanding air bag is controlled to match the severity of the impact and/or by the size and weight of the seat's occupant **(Figure 24–16)**. Other things considered are seat-track position and seat belt use. All of these factors require different deployment rates.

Two-stage air bags have twin chambers, comprised of two air bags, two containers of gas, and two squibs. When low-pressure deployment is desired, only one squib is fired. The air bag sensor assembly calculates the extent of impact, seat position, and status of the seat belts and controls the inflation times for the two chambers.

During a severe collision, the occupants need maximum protection and both squibs fire. The speed of deployment can also be controlled by the firing of the squibs. For rapid deployment, both squibs fire at the same time. To phase in full deployment, one squib is fired, then a few milliseconds later the other is. When the twin-chamber air bag is deployed, it forms the shape of two bags with a depression in the middle. The shape supports the occupant **(Figure 24–17)**.

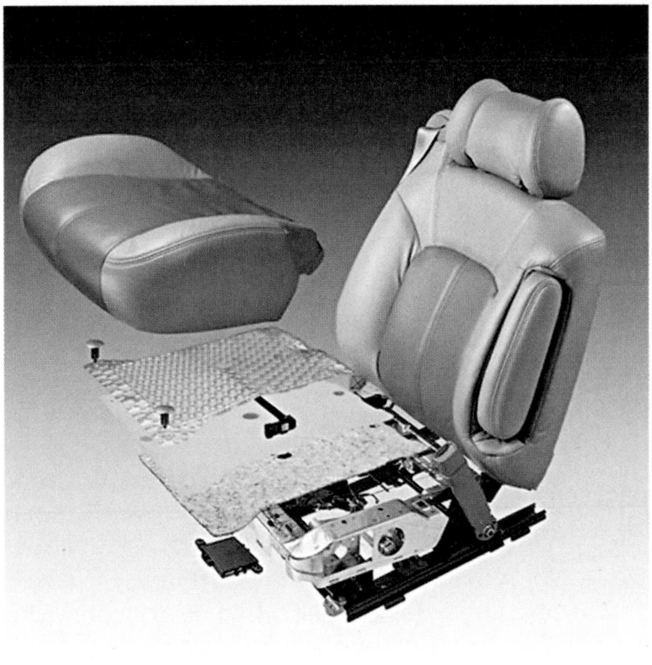

Figure 24–15 An occupant detection system with a seat belt tension sensor. *Courtesy of Delphi Corporation*

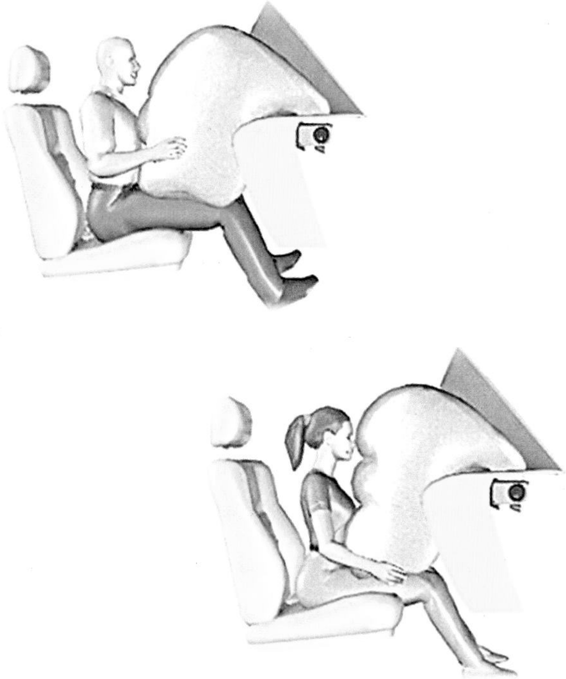

Figure 24–16 (Top) Full strength air bag deployment because of the weight of the occupant. (Bottom) Reduced air bag deployment because of the weight of the occupant. *Courtesy of Delphi Corporation*

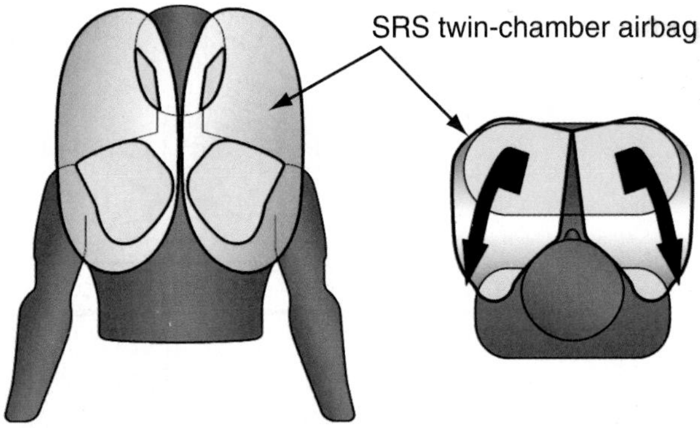

Figure 24–17 Full deployment of a twin-chamber air bag takes on the shape that surrounds the occupant. *Courtesy of Toyota Motor Sales, U.S.A., Inc.*

Mechanical Air Bag Systems

Mechanical air bag systems are found in some vehicles (early Toyotas, Volvos, and Jaguars). These systems are totally independent systems and do not rely on electricity or electronics to deploy. They have a mechanical trigger that ignites the propellant to deploy the air bag. The trigger is an impact sensor with a firing pin. During an impact, the firing pin moves and ignites a primer that ignites sodium azide pellets inside a gas generator. The gases released by the burning pellets inflate the air bag. This type of system is often used to equip a vehicle with air bags when it was not originally fitted with them.

Mechanical air bag systems can be used at any location in the vehicle. The number of gas generators used in each type of air bag depends on the size of the air bag. For example, a side impact air bag may have two generators and a side curtain can have as many as eight generators.

ELECTRICAL SYSTEM COMPONENTS

The electrical circuit of an air bag system includes impact sensors and an electronic control module. The electrical system conducts a system self-check to let the driver know that it is functioning properly, records DTCs for technician use, detects an impact, and sends a signal to deploy the air bags.

A vehicle can contain many different air bag modules **(Figure 24–18)**. The driver-side, passenger-side, side, and curtain air bag modules each has an inflator (igniter), air bag, and an ignition unit (cracker, igniter charger, gas). When the sensors send a signal to the module, current flows into the inflator and activates the ignition material to deploy the air bag.

To prevent accidental deployment, most systems require that at least two sensor switches be closed before an air bag is deployed **(Figure 24–19)**. The number of sensors used in a system depends on the design of the system. Normally, sensors are located in the engine and passenger compartments.

Sensors

Typically, ignition of the air bag only occurs when an outside (impact or crash) sensor and an inside (safing

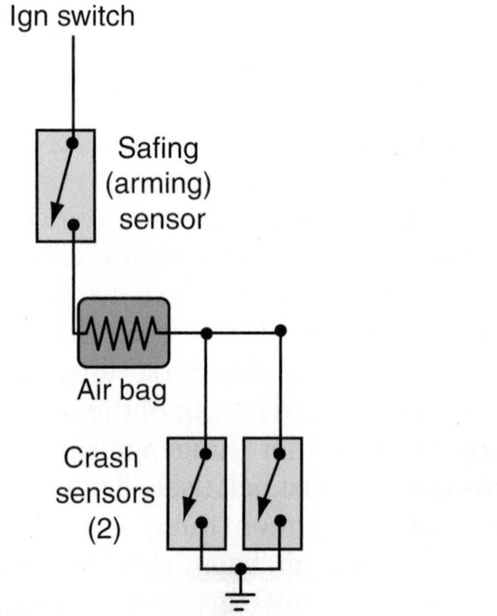

Figure 24-19 A simple air bag circuit with sensors.

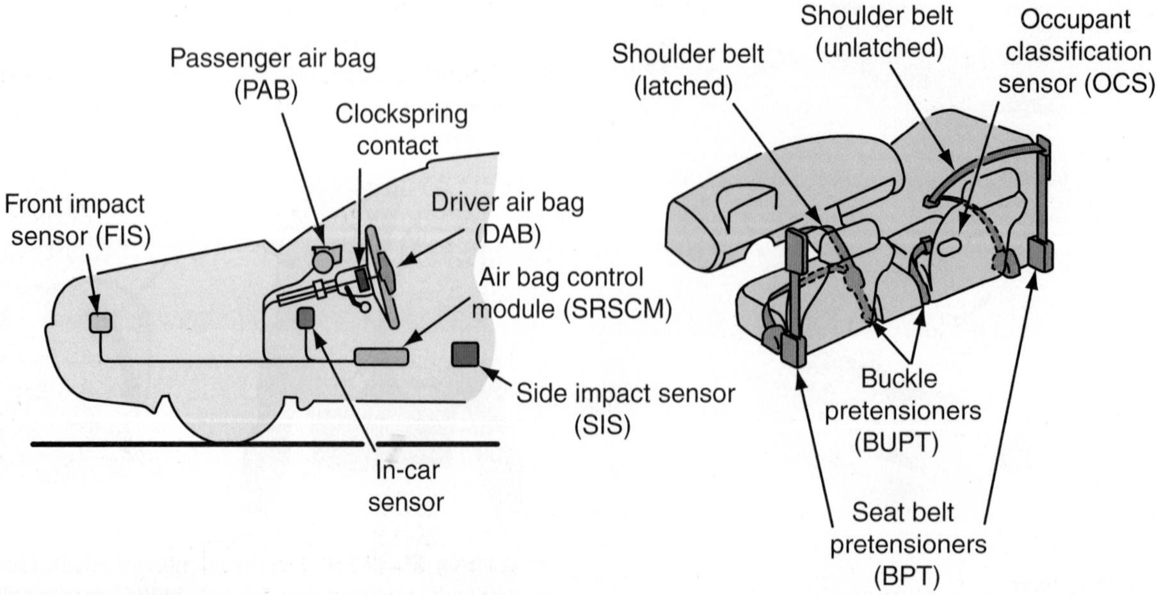

Figure 24-18 Location of common SRS components.

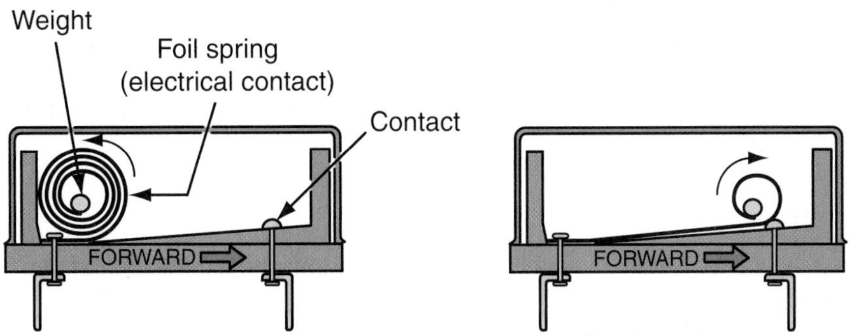

Figure 24–20 A roller-type air bag sensor.

or arming) sensor are closed. Once the two sensors are closed, the electrical circuit to the igniter is complete.

Impact—Safing Sensors There are three basic types of safing sensors:

Roller-Type—These have a roller located on a ramp **(Figure 24–20)**. One terminal of the sensor is connected to the ramp. The other terminal is connected to a spring contact extending through an opening in the ramp but not touching the ramp. Small springs hold the roller against a stop. During a heavy impact, the roller moves up the ramp and strikes the spring contact. This completes the circuit between the ramp and the spring, and the air bag deploys.

Mass-Type—This sensor has a normally open set of gold-plated switch contacts and ball. The ball is the sensing mass and is held in place by a magnet **(Figure 24–21)**. When there is sufficient force, the ball breaks loose and makes contact with the electrical contacts to complete the circuit.

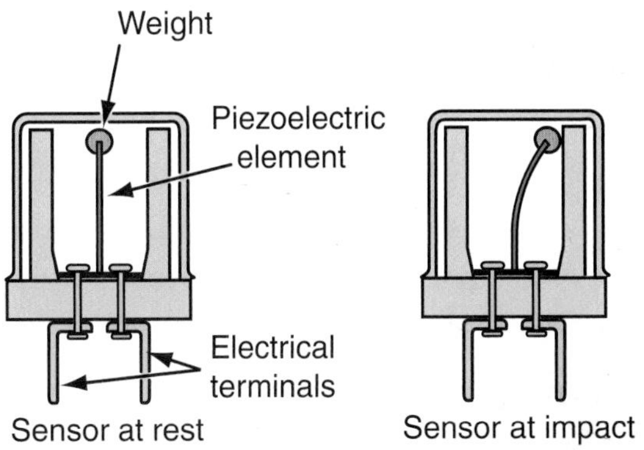

Figure 24–22 An accelerometer air bag sensor. *Courtesy of Chrysler LLC*

Accelerometer—This is a piezoelectric element that is distorted during a collision **(Figure 24–22)**. It generates an analog voltage that reflects the strength of the deceleration forces.

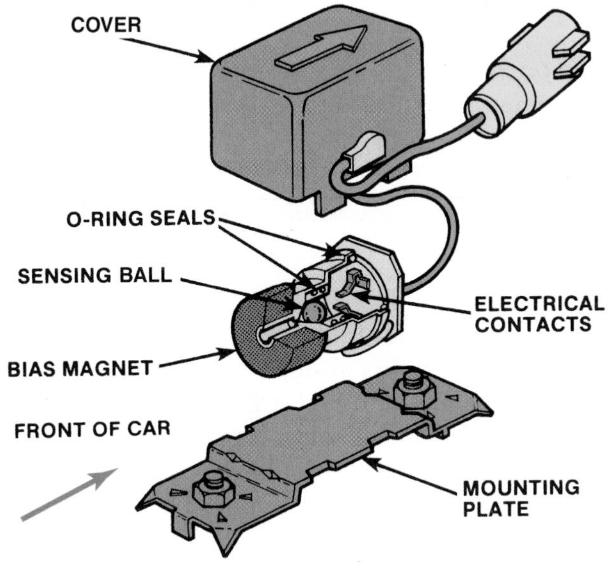

Figure 24–21 A typical ball and magnet sensor for an air bag system. *Courtesy of Chrysler LLC*

Seat Position Sensor The seat position sensor is mounted on the inner rail of the front seats **(Figure 24–23)**. The sensor is basically a magnet and a Hall-effect switch. The sensor is used to detect when the seat is in the forward position. When the seat is in a rearward position, the rail is close to the seat position sensor. When it is in the forward position, the distance between the rail and the sensor becomes larger. The position of the rail in relation to the Hall-effect switch determines the signal the sensor sends to the air bag sensor assembly. When the seat is close to the steering wheel, the air bag deploys with less pressure.

Seat Belt Buckle Switch The seat belt buckle switch detects whether or not the seat belt is fastened. It also

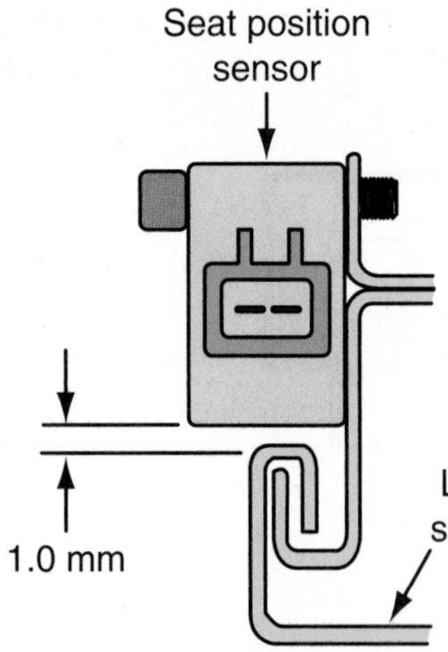

Figure 24-23 The seat position sensor has a specified gap between the sensor and the lower seat rail.

is a Hall-effect switch and magnet. The magnetic field changes as the seat belt is fastened and unfastened. The SRS system adjusts the force of the air bags according to this input.

Occupant Classification Sensor

To determine the force of the front and side passenger air bags, the weight of the occupant is measured. If there is no occupant, the air bags will not deploy during a collision. Most passenger occupant classification sensors use a bladder placed beneath the seat cushion. The bladder is connected to a pressure sensor that sends a signal to the control unit for the air bags. The bladder is a silicone-filled gel mat. When there is weight on the seat cushion, pressure is applied to the silicone in the mat. The pressure sensor measures that pressure and converts it to voltage signals **(Figure 24–24)**.

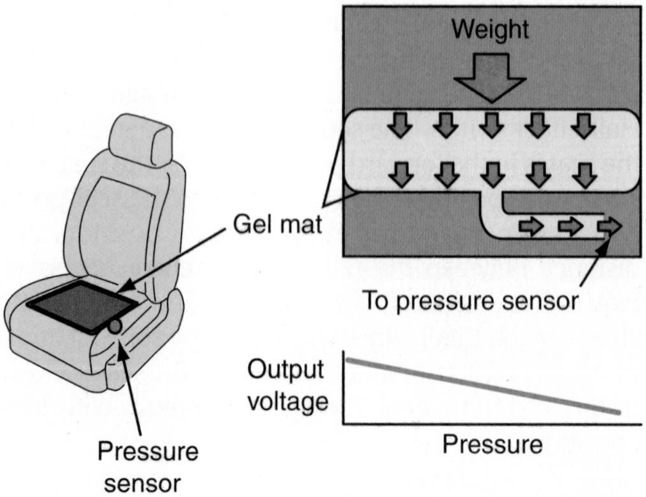

Figure 24-24 A gel mat-type occupant classification sensor.

Other occupant classification sensors are made up of two sheets of electrodes separated by a spacer. Weight on the seat causes the electrodes to contact each other through a hole in the spacer. The weight on the seat determines the contact area of the sheets. The amount of conductance between the sheets is used to determine weight.

The occupant classification sensor, seat belt tension sensor, and seat belt buckle switch send signals to the control unit. The control unit then determines if the seat is occupied, and if the seat is occupied by an adult, a child, or a child in a booster seat. The latter is detected by the seat belt tension sensor. When a child seat is sitting on the seat cushion, seat belt tension pulls the child seat down, adding pressure to the seat cushion. This pressure, plus the weight of the child and the child seat, are sensed by the weight detection sensor.

Diagnostic Monitor Assembly The **air bag sensing diagnostic monitor (ASDM)** constantly monitors the readiness of the SIR electrical system. If the module determines there is a fault, it will illuminate the warning lamp. Depending on the fault, the SIR system may be disarmed until the fault is corrected.

The diagnostic module also supplies back-up power to the air bag module in the event that the battery or cables are damaged during the accident. The stored charge can last up to 30 minutes after the battery has been disconnected.

> **WARNING!**
>
> *The back-up power supply must be depleted before any air bag service is performed. To deplete this back-up power, disconnect the positive battery cable and wait at least 30 minutes. Refer to the information in the service manual to determine exactly how long you should wait.*

Wiring Harness

For identification and safety purposes, the electrical harnesses of the SIR system typically have yellow connectors. Each connector has a special function and is designed specifically for the SRS. To increase their reliability, all SRS connectors have durable gold-plated terminals and are placed in specific locations. The connectors also have a shorting bar, called an activation prevention mechanism, that prevents accidental deployment while being serviced.

Single-stage air bags have one inflator and one pair of wires that connect to the air bag module. Two-stage air bags have two inflators and two pairs of wires connected to the air bag module.

Clockspring The **clockspring** allows for electrical contact to the air bag module at all times. Since the air bag module sits in the center of the steering wheel, the clockspring is designed to provide voltage to the module regardless of steering wheel position. The clockspring is located between the steering wheel and the steering column.

The clockspring's electrical connector contains a long conductive ribbon. The wires from the air bag's electrical system are connected from the underside of the clockspring to the conductive ribbon. The other end of the ribbon is connected to the air bag module. When the steering wheel is turned, the ribbon coils and uncoils without breaking the electrical connection.

SIR or Air Bag Readiness Light This light lets the driver know the air bag system is ready to do its job. The warning lamp is operated by the diagnostic module. The readiness lamp lights briefly when the driver turns the ignition key from off to run. The lamp should go out once the engine is running. A malfunction in the air bag system causes the light to stay on continuously or to flash. Some systems have a tone generator that sounds if there is a problem in the system or if the readiness light is not functioning.

Air Bag Module

The air bag module is the air bag and inflator assembly packaged into a single unit or module. The module is located in the steering wheel for the driver and in the dash panel for the front-seat passenger **(Figure 24–25)**. The various types of side protection air bags have the module located at the point where the bag is deployed.

The inflation of the air bag is typically accomplished through an explosive release of nitrogen gas. The igniter **(Figure 24–26)** is an integral part of the inflator assembly. It starts a chemical reaction to inflate the air bag. At the center of the igniter assembly is the squib, which contains **zeronic potassium perchlorate (ZPP)**. When voltage is supplied through

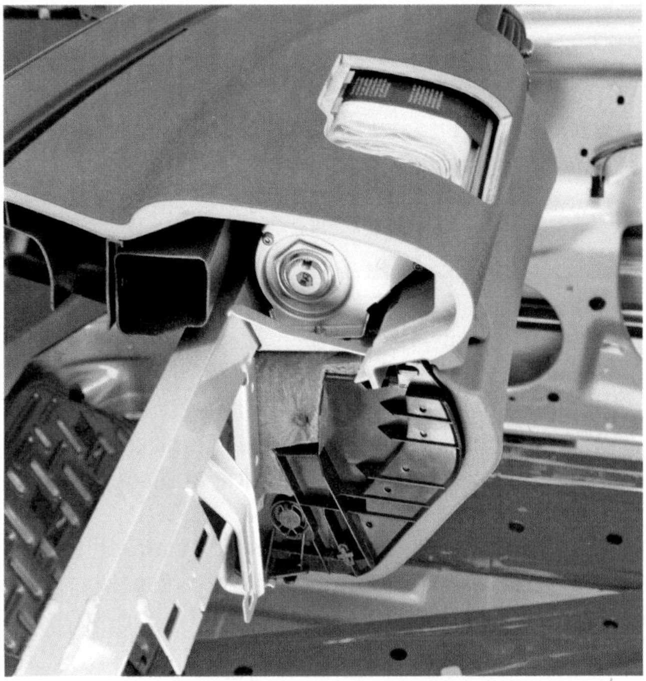

Figure 24-25 A cutaway view showing the complete passenger-side air bag module in the dash. *Courtesy of Chrysler LLC*

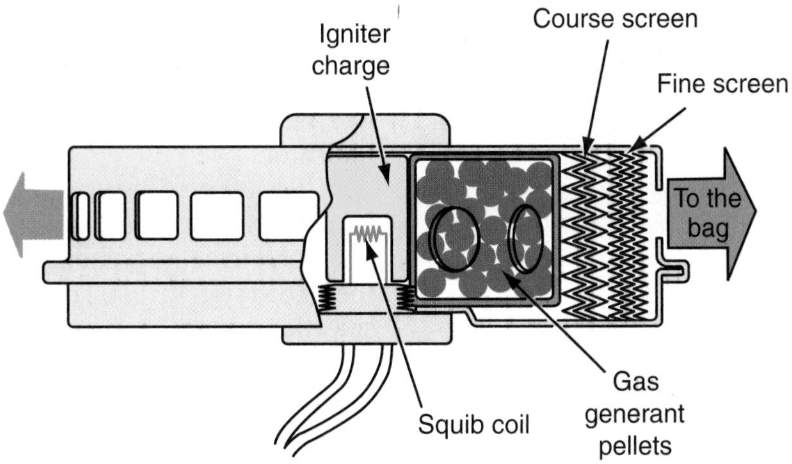

Figure 24-26 The action of an air bag module releasing nitrogen gas to an air bag.

the squib, an electrical arc is formed between two pins. The spark ignites a canister of gas and causes a rapid expansion of the gas, which deploys, or inflates, the air bag.

The inflation assembly is composed of a gas generator (called a generant) containing sodium azide and copper oxide or potassium nitrate propellant. The ZPP ignites the propellant charge. During ignition, large quantities of hot, expanding nitrogen gas are produced very quickly and quickly inflate the air bag. As the nitrogen moves into the air bag, it is filtered to remove sodium hydroxide dust formed during the chemical reaction.

CAUTION!

Wear gloves and eye protection when handling a deployed air bag module. Sodium hydroxide residue may remain on the bag and cause a skin irritation.

Not all air bags use nitrogen gas to inflate the bag; some use a solid propellant and compressed argon gas **(Figure 24–27)**. Argon has a stable structure, cools more quickly, and is inert as well as nontoxic. Argon is commonly used for passenger-side and side protection air bags.

A mounting plate and retainer ring attach the air bag assembly to the inflator. They also keep the entire air bag module connected to the steering wheel.

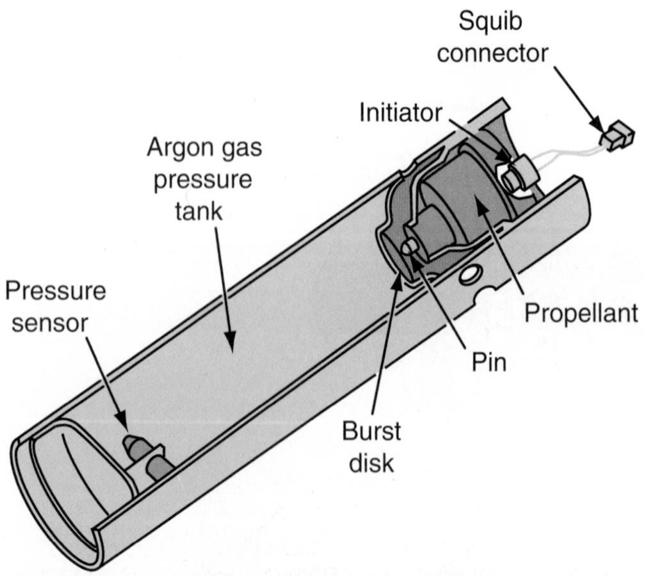

Figure 24–27 An inflator module that uses argon gas.

The bag itself is made of a thin, nylon fabric that is folded into the steering wheel, dash, seat, or door. The powdery substance released from the air bag when it is deployed is regular cornstarch or talcum powder. These powders are used by the air bag manufacturers to keep air bags lubricated and pliable while they are in storage. The entire module must be replaced as one unit when repair of the air bag system is required.

DIAGNOSIS

Before diagnosing the system, perform a system check by observing the air bag warning light and comparing your findings with those described in the service manual for the vehicle. To check the status of the air bag system, make sure the ignition has been off for at least 2 seconds. Then turn the ignition on. The SRS warning lamp should turn on and remain on for about 6 seconds. During this time, the system is performing a preliminary check of the system, including the pretensioners.

If the system detects a problem, the SRS warning light will remain on. If the lamp flashes or turns off and then turns on again, low source voltage may be indicated. A system problem is also indicated when the lamp does not come on when the ignition is initially turned on.

If any of these conditions are present, the system needs to be checked. A thorough visual inspection of sensor integrity is the best place to start when diagnosing a system that is disarmed because of a fault. Damage from a collision or mishandling during a nonrelated repair can set up a fault area, which will disarm the air bag system.

The passenger air bag indicator should also be observed. Its display is an indication of the condition of the passenger weight sensor as well as the air bag's modules. Always refer to the service information to interpret the indicators. Normal displays will vary with what is detected in the passenger seat. If a problem is detected, the SRS warning lamp will be lit, as will the passenger air bag "OFF" indicator.

If the action of the warning lamp indicates a problem, the system should be checked for DTCs. If there is a problem in the passenger side, check the DTCs in the air bag system first. Then check the occupant classification system.

Retrieving Trouble Codes

If the system detects a problem in the SRS system, the malfunction data will be stored in memory and the warning indicator will be lit. Normally two types of faults are stored. Active DTCs will turn the air bag warning lamp on, whereas stored codes are intermittent problems and probably will not turn on the warning lamp.

SRS problems are difficult to verify; therefore, DTCs are extremely important for troubleshooting the system. Most systems have two- and five-digit DTCs. The two-digit codes are flash codes displayed with the SRS warning indicator. The five-digit codes are displayed on a scan tool. It is important to note that when the negative battery cable is disconnected, the system's memory is erased. Therefore, DTCs should be retrieved before disconnecting the battery.

Flash Codes On vehicles that display codes with the warning light or on the digital instrument panel, make sure you follow the procedure prescribed by the manufacturer to retrieve the codes. Normally a jumper wire is connected across two terminals in the DLC with the ignition switch on. Make sure the wire is connected correctly and does not contact other pins in the connector. Once the jumper is in place, observe the action of the SRS warning lamp. Count the blinks and refer to the manufacturer's code table to interpret the code. If there is more than one stored DTC, the second code will flash shortly after the first code is displayed. In most cases, the codes will be erased when the ignition is turned off.

Scan Tool DTC Retrieval To retrieve codes, connect the scan tool **(Figure 24–28)** to the DLC and turn the ignition on. Follow the instructions for the scan tool to retrieve air bag information. Record all stored and active codes. Diagnose the cause of the codes in order, from the lowest number to the highest. Stored codes can be erased with the scan tool but active codes will only be erased when the problem is corrected.

CAUTION!

Testing individual parts of the system must be done with care. Not following the correct procedure or using the wrong tools can cause an air bag to deploy. This is not only dangerous but it is also very expensive. Never attempt to check the resistance of an air bag module.

Once the codes are retrieved, refer to the manufacturer's information to identify the steps for isolating and correcting the problem.

Figure 24-28 An OBD-II scan tool that checks the air bag, antilock brake, and engine control systems. *Courtesy of SPX Service Solutions*

Air Bag Simulator To safely test SRS components, the use of an air bag simulator **(Figure 24–29)** is recommended. This simulator is installed in place of the air bag. The simulator can be adjusted to provide the normal electrical load of the air bag, thereby allowing

Figure 24-29 An air bag simulator is used as a substitute for the actual air bag during testing. *Courtesy of SPX Service Solutions*

accurate testing of the circuits without the fear of accidental air bag deployment.

SERVICING THE AIR BAG SYSTEM

Whenever working on or around air bag systems, it is important to follow all safety warnings **(Figure 24–30)**. Examples of these warnings follow:

1. When performing any work on the inside of a vehicle, make sure you are aware of the location of all air bags and exercise caution when working in those areas.
2. Wear safety glasses when servicing the air bag system.
3. Wait at least 30 minutes after disconnecting the battery before beginning any service on or around the air bag system. The reserve energy module stores enough power to deploy the air bag after battery voltage is lost.
4. Always follow service procedures exactly as given by the manufacturer. Failure to do this

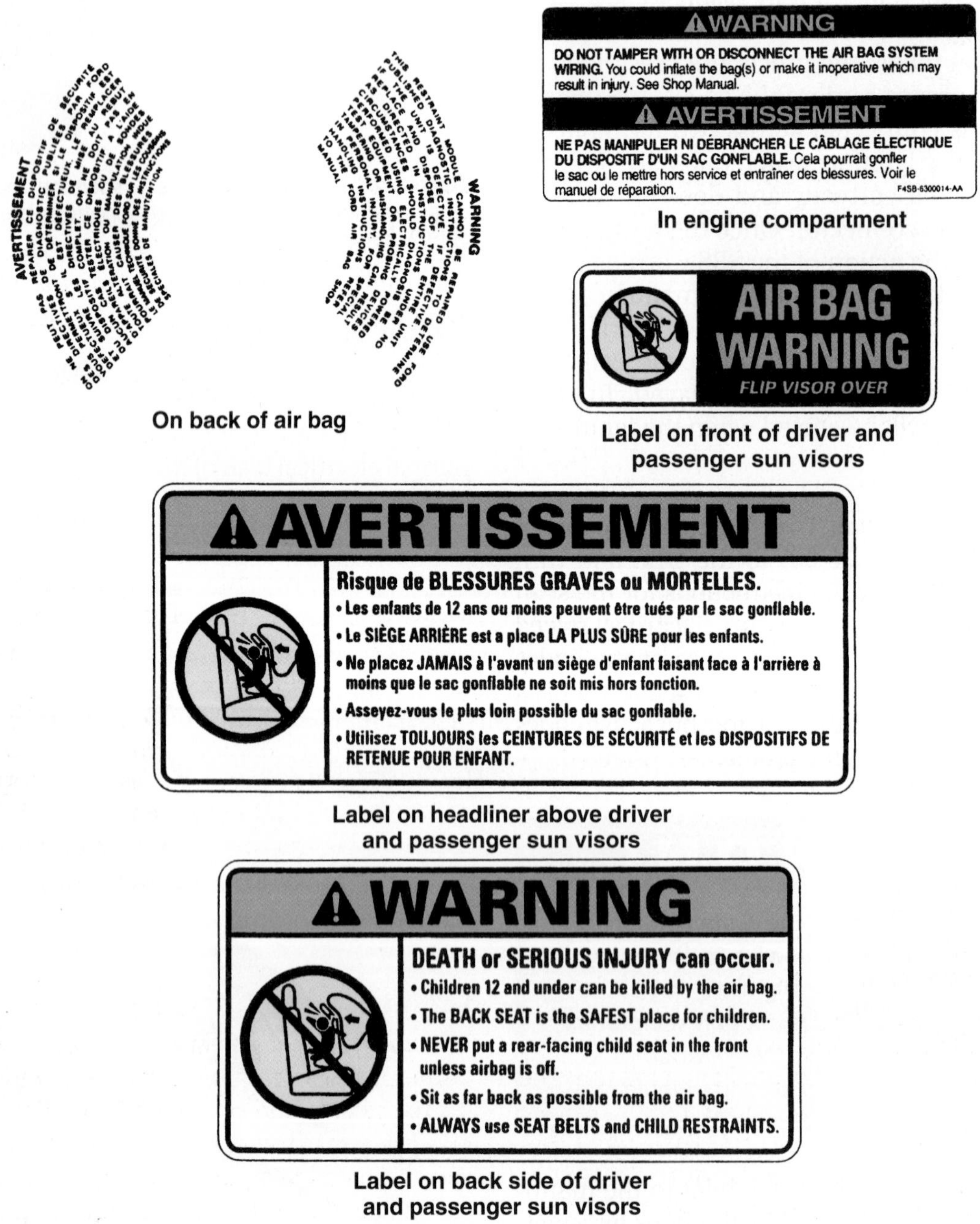

Figure 24–30 Various air bag warning labels. *Courtesy of Ford Motor Company*

can cause the SRS to deploy. Also, if the system is not serviced properly, it may not work when it needs to.

5. Never disassemble or attempt to repair any parts in order to reuse them; always replace them with new ones.
6. Handle all air bag sensors with care. Do not strike or jar a sensor in such a manner that deployment may occur.
7. When carrying a live air bag module, face the trim and bag away from your body.
8. Do not carry the module by its wires or connector.
9. When placing a live module on a bench, face the trim and air bag up.
10. Deployed air bags may have a powdery residue on them. Sodium hydroxide is produced by the deployment reaction and is converted to sodium carbonate when it comes into contact with atmospheric moisture. It is unlikely that sodium hydroxide will still be present. However, wear safety glasses and gloves when handling a deployed air bag. Wash your hands immediately after handling the bag.
11. A live air bag module must be deployed before disposal. Because the deployment of an air bag is through an explosive process, improper disposal may result in injury and fines. A deployed air bag should be disposed of according to EPA and manufacturer procedures.
12. Do not use a battery- or AC-powered voltmeter, ohmmeter, or any other type of test equipment not specified in the service manual to test the air bag module. Never use a testlight to probe for voltage.
13. After work on the SRS is completed, perform the SRS warning light check.

Disarming Mechanical Systems

Because mechanical air bag systems are not tied to the electrical system, they cannot be deactivated by disconnecting the battery. The parts responsible for triggering the air bags and seat belt pretensioners are marked with "Danger Zone" in red. In some cases, the air bags cannot be deactivated and care must be taken not to jar the vehicle or close a door with an obstacle between the door and seat. To disarm other systems, certain cables need to be disconnected or cut. For example, on Volvos with side air bags, it is a black ribbed cable located between the bottom and back cushions of the seat. Always check the service manual before working on or near these systems.

CAUTION!

A two-stage air bag may appear to be fully deployed when only its first stage has deployed. Care must be taken to make sure that two-stage air bag have been fully deployed before handling them. Always assume that any deployed two-stage air bag has an active stage two. Improper handling or servicing can activate the inflator module and cause personal injury. Always follow the manufacturer's recommended handling procedures.

Service Guidelines

An air bag module is serviced as a complete assembly. Technicians repairing these systems are also advised to service crash sensors, mercury switches, and any other related components in assembly groupings. A damaged crash sensor should be replaced. It is a good idea to replace the entire set if a failure or degradation of any single sensor is found.

Photo Sequence 22 covers a typical procedure for replacing an air bag module.

The steering column clockspring should be maintained in its correct index position at all times. Failure to do so can cause damage to the enclosure, wiring, or module. Any of these situations can cause the air bag system to default into a nonoperative mode. The clockspring should be replaced any time it has been removed.

Before returning the vehicle to the customer after service, make sure the sensors are firmly fastened to their mounting fixtures, with their arrows facing forward. Be certain all the fuses are correctly rated and replaced. Make sure a final check is made for codes using the approved scan tool. Carefully recheck the wire and harness routing before releasing the car.

OTHER PROTECTION SYSTEMS

To make vehicles safe and to protect the occupants inside the vehicle, manufacturers include many different systems and options. What follows is a quick look at a few of these. By no means is this discussion conclusive; there are many things about how a vehicle is made that influence the protection and safety it offers. Basically, cars that offer good protection are

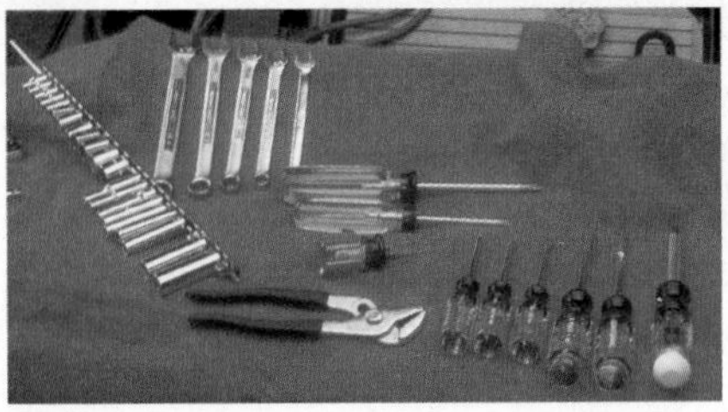

P22-1 *Tools required to remove the air bag module: safety glasses, seat covers, screwdriver set, Torx driver set, battery terminal pullers, battery pliers, assorted wrenches, ratchet and socket set, and service manual.*

P22-2 *Place the seat and fender covers on the vehicle.*

P22-3 *Place the front wheels in the straight ahead position and turn the ignition switch to the LOCK position.*

P22-4 *Disconnect the negative battery cable.*

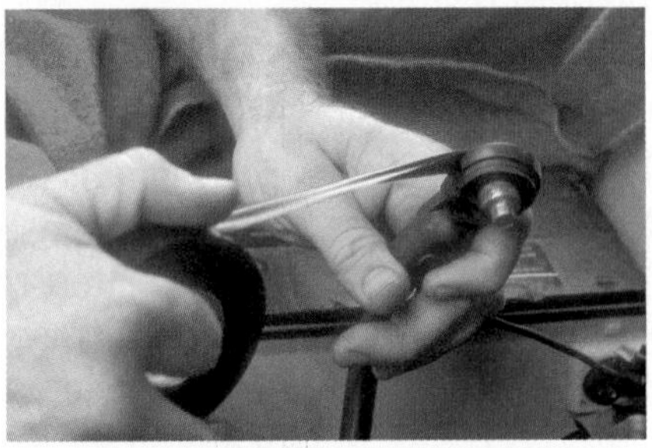

P22-5 *Tape the cable terminal to prevent accidental connection with the battery post. Note: A piece of rubber hose can be substituted for the tape.*

P22-6 *Remove the SIR fuse from the fuse box. Wait 30 minutes to allow the reserve energy to dissipate.*

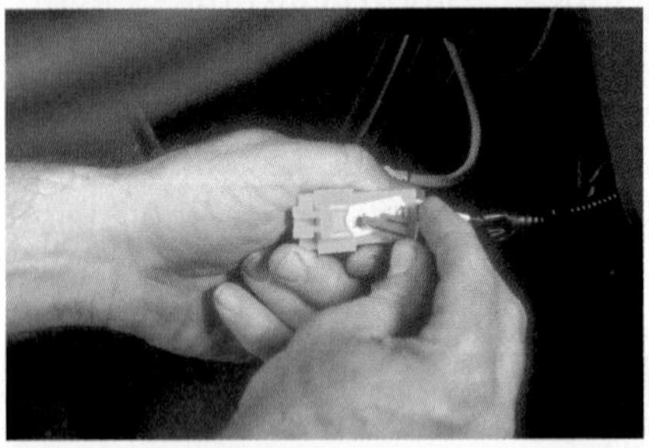

P22-7 *Remove the connector position assurance (CPA) from the yellow electrical connector at the base of the steering column.*

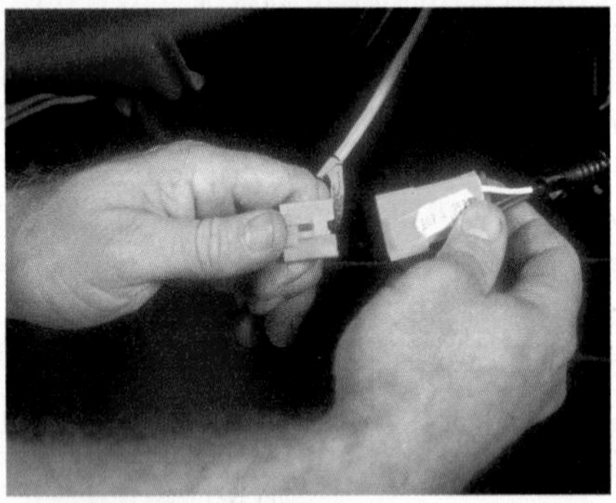

P22-8 *Disconnect the yellow two-way electrical connector.*

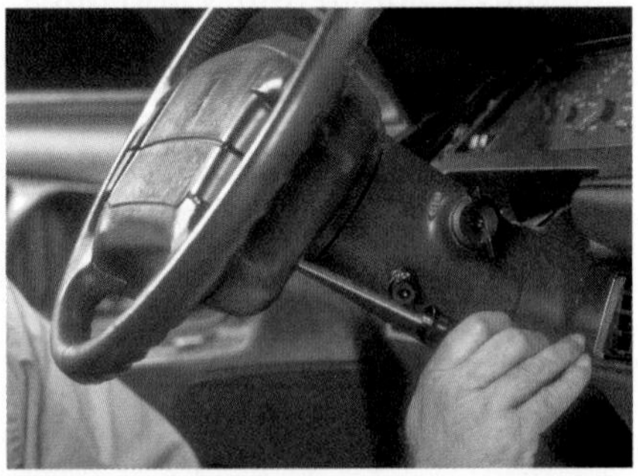

P22-9 *Remove the four bolts that secure the module from the rear of the steering wheel.*

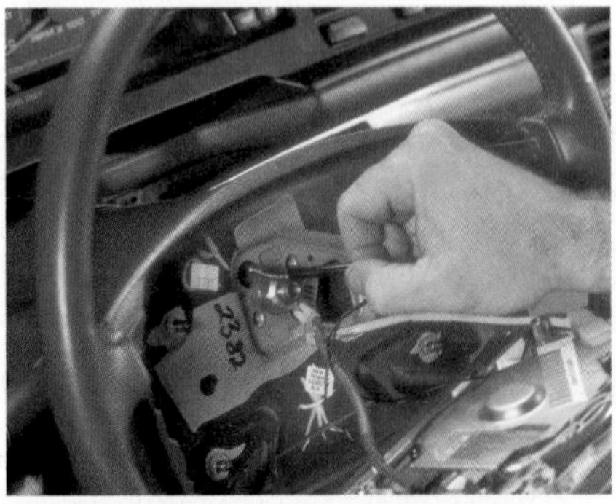

P22-10 *Rotate the horn lead ¼ turn and disconnect.*

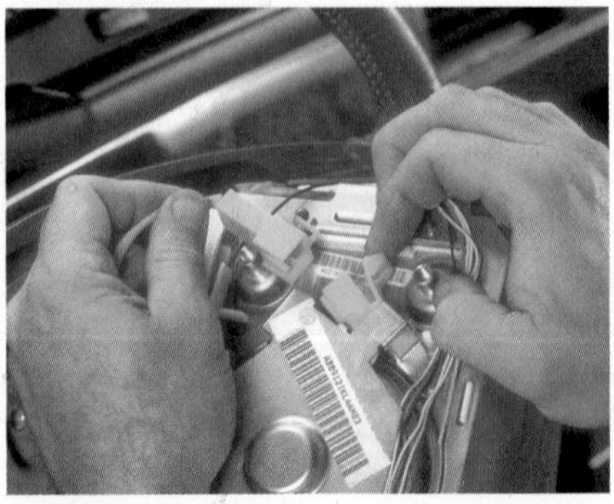

P22-11 *Disconnect the electrical connectors.*

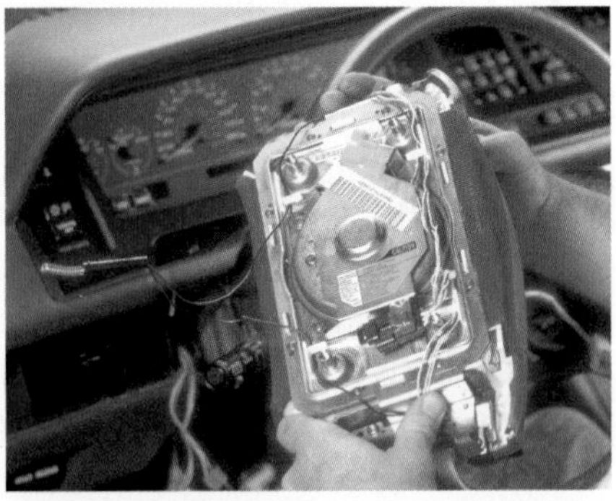

P22-12 *Remove the module.*

those that are constructed to maintain integrity when impacted on. This construction includes side door beams, crumple zones, and reinforced areas of the frame.

Crumple zones are areas of the body that will bend or break away to protect the passengers inside the vehicle. These crumple zones absorb or take on the impact while keeping the passenger compartment undisturbed.

Headrests

Nearly two-thirds of all injuries from collisions are soft-tissue related, commonly referred to as whiplash. Good head restraints and proper adjustment help prevent these injuries. A properly adjusted headrest stops the head and neck from extending backward on impact. Nine out of ten people do not adjust the headrest for their height and comfort. A headrest should be positioned at least to the top of their ear and less than 4 inches (10 cm) from the back of their head.

New systems that automatically adjust the headrest have been developed. These headrest systems move the headrest up and forward when the vehicle is hit from behind. If the vehicle was in a collision, the action of the headrest should be checked. This is done by measuring the amount the headrest can be moved **(Figure 24–31)**.

Event Data Recorder (EDR)

Some adaptive air bag systems have an event data recorder (EDR) that records data from a crash or a near crash. The recorder is normally inside the air bag sensor assembly. It records air bag system diagnostic data, air bag deployment data, seat belt status, engine speed, throttle and brake pedal data, position of the transmission selector, and position of the driver's seat. This information is used for safety improvements only.

Precollision System

A precollision system predicts possible collisions with an obstacle, based on information it receives from various sensors. The system uses a millimeter wave radar sensor to predict a possible collision and retracts the seat belts before the collision. The inputs also are used to control the brake system. Millimeter wave radar uses an extremely high frequency with an extremely short wavelength. Millimeter wave radar does a good job at recognizing obstacles.

The precollision system removes the slack in the front seat belts and warns the driver of an impending collision. The sensor only detects metal objects and will not respond to people, bicycles, trees, and animals. It also will not respond if the seat belts are not buckled.

Rollover Protection

Some convertibles have a built-in roll bar to protect the passengers in case of a rollover. These units are permanent structures of the vehicle. Others have automatic systems that provide for this. They deploy a roll bar from behind the headrests when the vehicle experiences extreme tilting, when the wheels lose contact with the ground, or during a serious accident **(Figure 24–32)**.

Vehicles with rollover protection use sensors to monitor the speed of the rollover and inflate the air bags accordingly. During a rollover, the air bags stay inflated for a longer than normal time to keep the occupants safe until the vehicle comes to a rest. These systems also take the slack out of the seat belts, shut off the fuel pump, and disconnect the battery when a rollover is sensed.

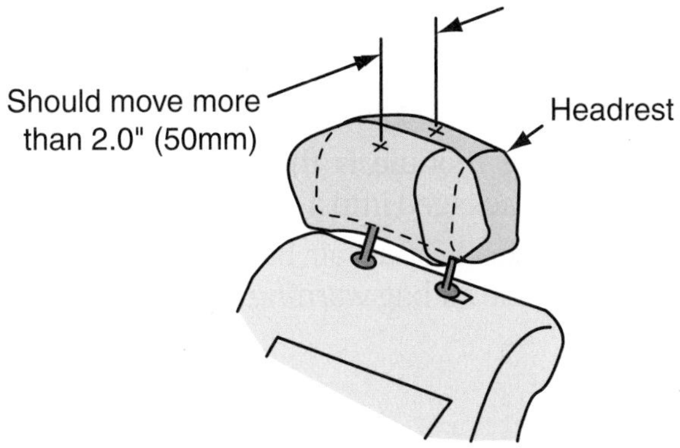

Figure 24-31 Checking an active headrest.

Figure 24-32 This roll bar pops out when the system anticipates a potential rollover. *Courtesy of Chrysler LLC*

Hybrid vehicles also have a rollover protection feature. This system isolates the high-voltage circuits whenever the air bags are deployed.

Four-Point Seat Belts

Although not in use at this time, Ford Motor Company and other companies are studying the use of four-point seat belts. These are currently not legal but may be in 2010. This belt system has two shoulder straps integrated into the frame of the seat. The shoulder straps and the two lap belts buckle together at the center of the passenger's waist. The design allows the forces from a crash to be evenly distributed across the body.

Air Bag Seat Belts

An inflatable seat belt for rear passengers can control the movement of a passenger's head and neck during a crash. This inflatable belt spreads the crash forces over more of the body than a standard belt. The shoulder strap contains a folded, cylindrical air bag that is deployed during a collision.

CASE STUDY

A late-model car with air bags was brought to the shop because its air bag warning lamp was always on. The customer was afraid the air bags would deploy while he was driving. The service writer explained what was required for the air bag to inflate. The customer appreciated the explanation but was still concerned about the warning light. The car was given to a technician for diagnostics. She verified the complaint and then conducted a visual inspection of all of the air bag circuits. Finding nothing noticeably wrong, she connected a scan tool to retrieve trouble codes. The DTC indicated an open driver's side air bag squib circuit. Following the procedure outlined in the vehicle's service manual, she removed the air bag module and tested the clockspring circuit. This circuit was fine so she replaced the air bag module. She knew this would take care of the problem, simply because everything before the module was good. The service manual had firm cautions about not testing the resistance of the squib because that could cause accidental deployment. Therefore, she did not attempt to test it. After the new module was installed, she ran a diagnostic check to make sure the problem was solved. The warning lamp was off and the customer was satisfied.

KEY TERMS

Active restraint system
Adaptive air bags
Air bag sensing diagnostic monitor (ASDM)
Clockspring
Inertia lock retractors
Passive restraint system
Pretensioner
Second-generation air bags
Supplemental inflatable restraint (SIR)
Smart air bags
Supplemental restraint system (SRS)
Squib
Zeronic potassium perchlorate (ZPP)

SUMMARY

- All new vehicles built or sold in the United States must have one or both types of passive restraints: seat belts or air bags.
- Restraint systems are either active or passive.
- When servicing seat belts, inspect the webbing, buckles, retractors, and anchorage.
- An air bag is inflated or deployed by rapid expansion of a gas fired by igniter or squib.
- Smart air bags have two possible stages of deployment in an attempt to match the severity of the impact and/or the size and weight of the seat's occupant.
- The electrical circuit of an air bag system includes impact sensors and an electronic control module.
- The air bag module is the air bag and inflator assembly packaged into a single unit.
- A system check of an air bag system consists of observing the air bag warning light.

- The control module will store trouble codes that can be retrieved by either a scan tool or flash codes.
- Before doing any work on an air bag system, disconnect the battery.
- Care must be taken when removing a live (not deployed) air bag. Be sure the bag and trim cover are pointed away from you.

REVIEW QUESTIONS

1. What is the difference between a passive and an active restraint system?
2. *True or False?* Second-generation air bags rely on compressed air to deploy the bag.
3. What air bag system device contains the ZPP?
4. What is a crumble zone?
5. *True or False?* The air bag diagnostic monitor supplies back-up power to the air bag module in the event that the battery or cables are damaged during the accident.
6. *True or False?* The powdery substance released from the air bag when it is deployed is sodium hydroxide, which can cause skin irritation.
7. *True or False?* When servicing the steering wheel or column, disconnect and remove the clockspring to make sure it is in good shape before reusing it.
8. Which of the following statements about the SRS warning lamp is *not* true?
   a. When the ignition is first turned on, the indicator should turn on and remain on for approximately 6 seconds.
   b. If the system seems to be able to work correctly, the lamp will stay lit as the engine runs.
   c. If the lamp flashes or turns off and then turns on again, low source voltage may be indicated.
   d. A system problem is indicated when the lamp does not turn on when the ignition is initially turned on.
9. Describe the two basic types of passenger weight detection sensors used in most vehicles.
10. What type of seat belt operates automatically with no action required by the vehicle's occupant?
    a. passive restraint
    b. retractor
    c. active restraint
    d. anchorage
11. Which of the following is false?
    a. The air bag igniter assembly is a spark plug-type device with two pins that current must jump across.
    b. The air bag igniter assembly creates a spark that ignites a canister of gas generating zerconic potassium perchlorate.
    c. The air bag igniter assembly is housed in the module assembly.
    d. None of the above.
12. List at least five service precautions that must be adhered to when working on or near air bag systems.
13. Which of these statements about air bag sensors is *not* true?
    a. Roller-type sensors rely on a roller held in place with a magnet. The circuit is closed when the roller moves and strikes the spring contact.
    b. Mass-type sensors rely on a ball and a magnet. The circuit is complete when the ball makes contacts with the electrical contacts in the sensor.
    c. An accelerometer contains a piezoelectric element that generates an analog voltage in relation to the severity of the deceleration forces.
    d. An accelerometer also senses the direction of an impact force.
14. Which of the following statements about two-stage air bags is *not* true?
    a. When low-pressure air bag deployment is desired, only one squib is fired.
    b. When the vehicle is in a severe collision and the occupant needs maximum protection, the squib related to the larger container of gas fires.
    c. For rapid deployment, both squibs fire at the same time.
    d. To phase in full deployment, one squib is fired, then a few milliseconds later the other is.
15. Which of the following statements about seat belt retractors is *not* true?
    a. They stow away the seat belts when they are not being used.

b. They allow freedom of movement for the occupant of the seat.

c. They can tighten up and pull the occupant back during a crash.

d. On some vehicles, they work in concert with the air bag system.

ASE-STYLE REVIEW QUESTIONS

1. Technician A says that compressed argon gas is often used to deploy passenger-side and side impact air bags. Technician B says that compressed argon gas is used to deploy some driver's side air bags. Who is correct?

   a. Technician A c. Both A and B
   b. Technician B d. Neither A nor B

2. Technician A says that seat belt webbing should be replaced if it is bowed. Technician B says that webbing does not need to be replaced if the color has merely faded due to exposure to the sun. Who is correct?

   a. Technician A c. Both A and B
   b. Technician B d. Neither A nor B

3. Technician A says that the module assembly will disarm the air bag system if certain faults occur. Technician B says that at least two sensors must signal the air bag in order to trigger it. Who is correct?

   a. Technician A c. Both A and B
   b. Technician B d. Neither A nor B

4. While testing seat belt retractors: Technician A grasps a vehicle-sensitive-type belt and jerks it. Technician B grasps a webbing-sensitive-type belt and jerks it. Who is correct?

   a. Technician A c. Both A and B
   b. Technician B d. Neither A nor B

5. Technician A checks the clockspring electrical connections for signs of damage when replacing a deployed air bag module. Technician B back probes an air bag system with a multimeter to determine if the system is in good working order. Who is correct?

   a. Technician A c. Both A and B
   b. Technician B d. Neither A nor B

6. Technician A removes the ground battery cable before servicing any component in the air bag system. Technician B wears safety glasses and protective gloves when handling inflated air bags. Who is correct?

   a. Technician A c. Both A and B
   b. Technician B d. Neither A nor B

7. Technician A says that ignition of the air bag only occurs when an outside sensor and an inside sensor are closed. Technician B says that the safing sensor determines if the collision is severe enough to inflate the air bag. Who is correct?

   a. Technician A c. Both A and B
   b. Technician B d. Neither A nor B

8. Technician A says that the steering column clockspring should be maintained in its correct index position at all times. Technician B says that failure to keep the clockspring straight ahead or in the neutral position relative to the steering wheel can cause damage to the enclosure, wiring, or module. Who is correct?

   a. Technician A c. Both A and B
   b. Technician B d. Neither A nor B

9. While diagnosing a seat belt that will not buckle: Technician A looks inside the buckle and disassembles it and removes any obstructions that may prevent it from latching onto the belt latch. Technician B replaces the buckle if an obstruction cannot be easily removed from the buckle. Who is correct?

   a. Technician A c. Both A and B
   b. Technician B d. Neither A nor B

10. While diagnosing the cause of a constant SRS warning light: Technician A follows the service procedures and connects a jumper wire across designated pins in the DLC then records the flashed codes. Technician B connects a scan tool to the DLC and records the displayed codes. Who is correct?

    a. Technician A c. Both A and B
    b. Technician B d. Neither A nor B

ENGINE PERFORMANCE

Section 4

ENGINE PERFORMANCE SYSTEMS

CHAPTER 25

OBJECTIVES

■ State the purpose of the major engine performance systems/components. ■ Explain what is meant by open loop and closed loop. ■ Explain the reasons for OBD-II. ■ Explain the requirements to illuminate the malfunction indicator light in an OBD-II system. ■ Briefly describe the monitored systems in an OBD-II system. ■ Describe an OBD-II warm-up cycle. ■ Explain trip and drive cycle in an OBD-II system. ■ Describe how engine misfire is detected in an OBD-II system. ■ Describe the differences between an A misfire and a B misfire. ■ Describe the purpose of having two oxygen sensors in an exhaust system. ■ Briefly describe what the comprehensive component monitor looks at. ■ Retrieve and record stored diagnostic trouble codes; clear codes. ■ Diagnose the causes of emissions or driveability concerns resulting from malfunctions in the computerized engine control system with stored diagnostic trouble codes. ■ Diagnose emissions or driveability concerns resulting from malfunctions in the computerized engine control system with no stored diagnostic trouble codes; determine necessary action. ■ Obtain and interpret scan tool data.

Engine performance systems are those responsible for how an engine runs. How well an engine runs depends on the combustion process. Today's systems are designed to achieve as close to complete combustion as possible. Basically if all of the fuel that enters an engine's cylinder is burned, combustion is complete.

The requirements for complete combustion are simple. However, achieving these is not. Complete combustion will occur when the correct amount of air is mixed, in a sealed container, with the correct amount of fuel and shocked by the correct amount of heat at the correct time. So to achieve complete combustion, the engine's cylinders must be sealed during the compression and power strokes, allow air to freely enter the cylinder during the intake stroke, and allow gases to freely leave during the exhaust stroke. The amount of air and fuel must be precisely controlled, as should the spark for ignition. Because the engine runs at different speeds, loads, and temperatures, these requirements are very difficult to meet.

Emission control devices are added to the vehicle because complete combustion at all times has not been achieved. These devices reduce the amount of bad vehicle emissions. They also affect the operation of the engine and are, therefore, an engine performance system.

IGNITION SYSTEMS

For complete combustion, the ignition system must supply properly timed, high-voltage surges across each pair of spark plug electrodes **(Figure 25–1)** at the proper time under all engine operating conditions. This is quite a task: Consider a six-cylinder engine running at 4,000 rpm; the ignition system must supply 12,000 sparks per minute because it must fire

Figure 25-1 An ignition system has the sole purpose of providing the spark to start combustion. *Courtesy of Honeywell International Inc.*

Figure 25-2 An ignition coil for an electronic (distributorless) ignition system.

three spark plugs per revolution. These plug firings must also occur at the correct time and generate the correct amount of heat. If the ignition system fails to do these things, fuel economy, engine performance, and emission levels will be adversely affected. There are basically two types of ignition systems: **distributor ignition (DI)** and **electronic ignition (EI)** (**Figure 25–2**) or distributorless ignition systems (DIS).

Purpose of the Ignition System

For each cylinder, the ignition system has three primary jobs:

- It must generate an electrical spark with enough heat to ignite the air-fuel mixture in the combustion chamber.
- It must maintain that spark long enough to allow total combustion of the fuel in the chamber.
- It must deliver the spark to each cylinder to allow combustion to begin at the right time during the compression stroke.

In order for an engine to produce its maximum amount of power, the maximum pressure created by combustion should be present when the piston is at 10 to 23 degrees ATDC. Because the combustion process requires a short period to complete, normally measured in thousandths of a second, combustion must begin before the piston is on its power stroke. Therefore, the delivery of the spark must be timed to arrive at some point before the piston reaches TDC.

Determining how much before TDC is complicated. The speed of the piston moving from its compression stroke to the power stroke changes, whereas the time required for combustion stays the same. This means the spark should be delivered earlier as the engine's speed increases (**Figure 25–3**). However, when the engine is doing much work, the load on the crankshaft tends to slow down the acceleration of the piston, and spark delivery should be somewhat delayed.

When the spark should be delivered also depends on several other factors that can affect combustion times. Higher compression ratios tend to speed up combustion. Higher octane fuels ignite less easily and require more burning time. Increased vaporization and turbulence tend to decrease combustion times. Other factors, including intake air temperature, humidity, and barometric pressure, also affect combustion. Because of all of these, delivering the spark at the right time is a difficult task.

Engine Speed At higher engine speeds, the crankshaft rotates through more degrees in a given period. If combustion is to be completed by 10 degrees ATDC, ignition timing must begin sooner or be advanced. When the engine is cranked to start it, the ignition timing is retarded so the engine can start while it is turning over slowly.

Engine Load The load on an engine is related to the work it must do. Driving up hills or pulling extra weight increases engine load. Under load there is resistance on the crankshaft; therefore, the pistons have a harder time moving through their strokes. This is evident by the low measured vacuum during heavy loads.

Under light loads a high vacuum exists in the intake manifold. The density of the air-fuel mixture

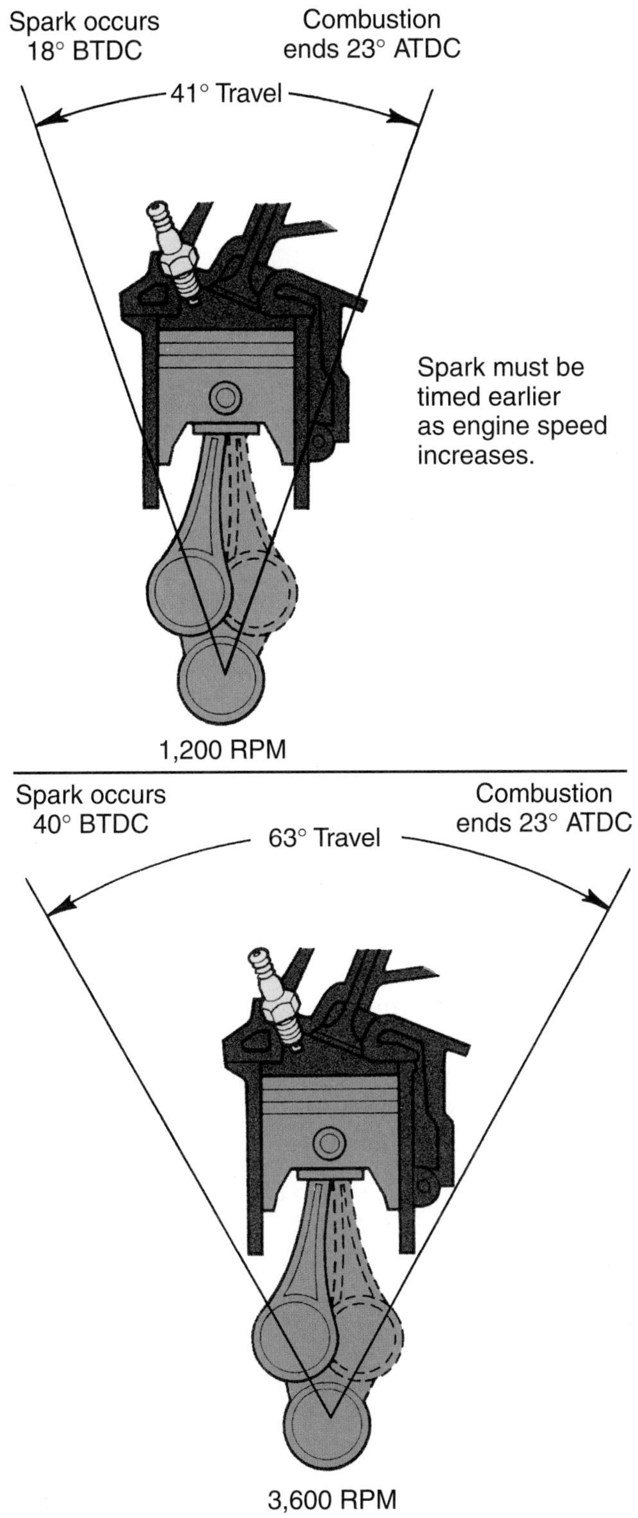

Figure 25-3 Ignition timing must begin earlier as engine speed increases. *Courtesy of Ford Motor Company*

drawn into the cylinders is low. On compression this thin mixture produces less combustion pressure and combustion time is slow. To complete combustion by 10 degrees ATDC, ignition timing must be advanced.

Under heavy loads, when the throttle plate is near fully opened, the vacuum in the cylinders is low and a very dense mass of air and fuel is drawn into the cylinders. High combustion pressure and rapid burning occur. In this case, the ignition timing must be retarded to prevent the combustion process from ending before 10 degrees ATDC.

Ignition Timing

Ignition timing refers to the precise time spark occurs and is specified by referring to the position of the number 1 piston in relation to crankshaft rotation. Early engines had ignition timing reference marks located on a pulley or flywheel to indicate the position of the number 1 piston. This was used to set initial ignition timing. For those engines, the manufacturers typically list an initial or **base ignition timing** specification.

Firing Order

The purpose of an ignition system extends beyond supplying the correct amount of heat at the correct time; it must do this for all of the engine's cylinders according to that engine's firing order.

Each cylinder must produce power once every 720 degrees of crankshaft rotation. Therefore, the ignition system must provide a spark at the right time so that each cylinder can have a power stroke at its own appropriate time. To do this, the ignition system must monitor the rotation of the crankshaft and the relative position of each piston to determine which piston should be delivered the spark. The spark for all cylinders must be delivered at the right time. How the ignition system does this depends on the design of the system.

Computer-Controlled Systems

With computerized ignition systems, inputs usually consist of engine temperature, engine speed, and manifold vacuum. There may be other sensors for throttle position, incoming air temperature, or engine knocking (detonation). The computer processes these inputs and advances or retards spark timing as required. This causes changes in engine operation, which sends new messages to the computer. The computer can constantly adjust timing for maximum efficiency **(Figure 25–4)**.

The advantage of the electronic spark control system is threefold. It compensates for changes in engine (and sometimes outside air) temperature. It makes changes at a rate many times faster than older systems. And, finally, it has a feedback mechanism in which sensor readings allow it to constantly compensate for changing conditions.

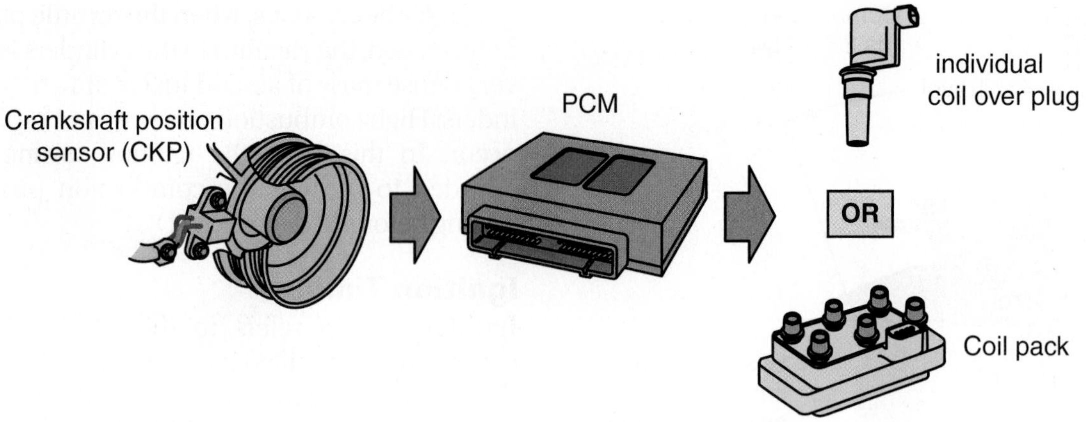

Figure 25-4 Basic layout for a computer-controlled ignition system.

FUEL SYSTEM

The fuel delivery system has the important role of delivering fuel to the fuel injection system. The fuel must also be delivered in the right quantities and at the right pressure. The fuel must also be clean when it is delivered.

A typical fuel delivery system includes a fuel tank, fuel lines, fuel filters, and a pump **(Figure 25–5)**. The system works by using a pump to draw fuel from the fuel tank and passing it under pressure through fuel lines and filters to the fuel injection system. The filter removes dirt and other harmful impurities from the fuel. A fuel line pressure regulator maintains a constant high fuel pressure. This pressure generates the spraying force needed to inject the fuel. Excess fuel not required by the engine returns to the fuel tank through a fuel return line.

Fuel Injection

Electronic fuel injection (EFI) has proven to be the most precise, reliable, and cost-effective method of delivering fuel to the combustion chambers of today's engines. EFI systems are computer controlled and designed to provide the correct air-fuel ratio for all engine loads, speeds, and temperature conditions.

Although fuel injection technology has been around since the 1920s, it was not until the 1980s that manufacturers began to replace carburetors with **electronic fuel injection (EFI)** systems. Many of the early EFI systems were **throttle body injection (TBI)** systems in which the fuel was injected above the throttle plates. A similar system, **central port injection (CPI)**, has the injector assembly located in the lower half of the intake manifold. TBI systems have been replaced by **port fuel injection (PFI)**, which has

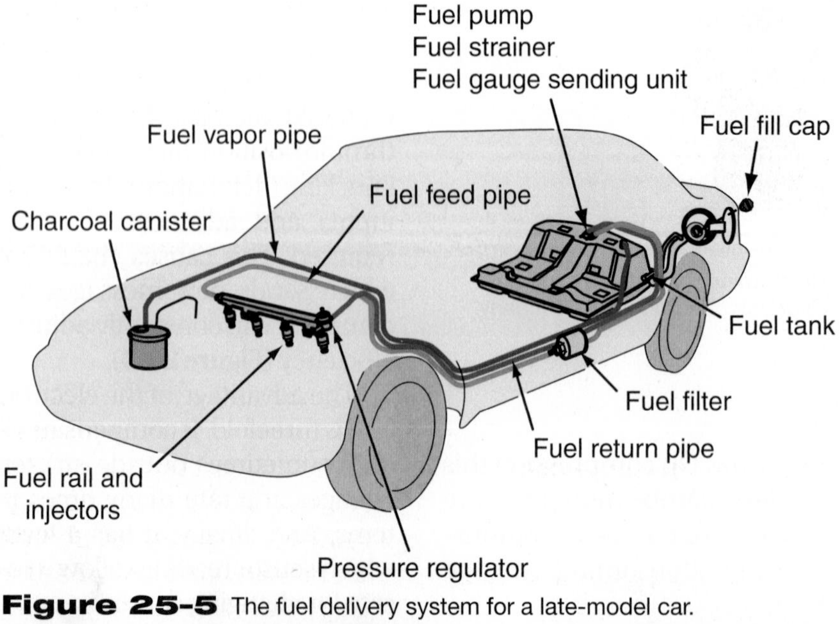

Figure 25-5 The fuel delivery system for a late-model car.

injectors located in the intake ports of the cylinders. All new cars have been equipped with an EFI system since the 1995 model year to fulfill OBD-II requirements. Recently some engines have been equipped with **gasoline direct-injection (GDI)**. In these systems, the fuel is injected directly into the cylinders. Direct injection has been used for years with diesel fuels but has not been successfully used on gasoline engines until lately.

TBI systems have a throttle body assembly mounted on the intake manifold in the position usually occupied by a carburetor. The throttle body assembly usually contains one or two injectors.

On PFI systems, fuel injectors are mounted at the back of each intake valve **(Figure 25–6)**. They use a throttle plate assembly to control intake air, but fuel never passes by the plate.

Most EFI systems only inject fuel during the engine's intake cycle. The engine's fuel needs are measured by intake airflow past a sensor or by intake manifold pressure (vacuum). The airflow or manifold vacuum sensor converts its reading to an electrical signal and sends it to the engine control computer. The computer processes this signal (and others) and calculates the fuel needs of the engine. The computer then sends an electrical signal to the fuel injector or injectors. This signal determines the amount of time the injector opens and sprays fuel. This interval is known as the injector pulse width.

When determining the amount of fuel required at any given time, the PCM also looks at throttle position, engine speed, crankshaft position, engine temperature, inlet air temperature, and oxygen in the exhaust inputs.

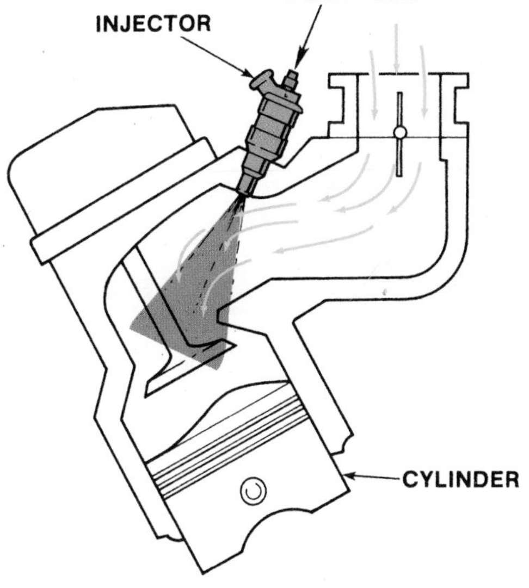

Figure 25–6 In a port injection system, air and fuel are mixed right outside the combustion chamber.

AIR INDUCTION SYSTEM

An internal combustion engine needs air to run. This air supply is drawn into the engine by the vacuum created during the pistons' intake stroke. Controlling the flow of air is the job of the **air induction system**.

Prior to the introduction of emission control devices, the induction system was quite simple. It consisted of an air cleaner housing mounted on top of the engine with a filter inside the housing. Its function was to filter dust and grit from the air being drawn into the engine.

Modern air induction systems filter the air and do much more. The introduction of emission standards and fuel economy standards encouraged the development of intake air temperature controls. The air intake system on a modern fuel-injected engine is complicated **(Figure 25–7)**. Ducts channel cool air from outside the engine compartment to the throttle plate assembly. The air filter has been moved to a position below the top of the engine to allow for aerodynamic body designs. Electronic meters measure airflow, temperature, and density.

Many manufacturers equip their engines with turbo- **(Figure 25–8)** or superchargers. These increase engine efficiency by forcing more air into the cylinders.

Air-Fuel Mixtures

The amount of air mixed with the fuel is called the air-fuel ratio. The ideal air-fuel ratio for most operating conditions of a gasoline engine is approximately 14.7 pounds of air mixed with 1 pound of gasoline. This provides an ideal ratio of 14.7:1. Because air is so much lighter than gasoline, it takes nearly 10,000 gallons of air mixed with 1 gallon of gasoline to achieve an air-fuel ratio of 14.7:1. This is why proper air delivery is as important as fuel delivery.

When the mixture has more air than the ideal ratio calls for, the mixture is said to be lean. Ratios of 15 to 16:1 provide the best fuel economy from gasoline engines. Mixtures that have a ratio below 14.7:1 are considered rich mixtures. Rich mixtures (12 to 13:1) provide more power production from the engine but greater fuel consumption.

EMISSION CONTROL SYSTEMS

Emission controls have one purpose and that is to reduce the amount of pollutants and environmentally damaging substances released by vehicles. The consequences of the pollutants are grievous **(Figure 25–9)**. The air we breathe and water we drink have become contaminated with chemicals that adversely affect our health. It took many years for the public

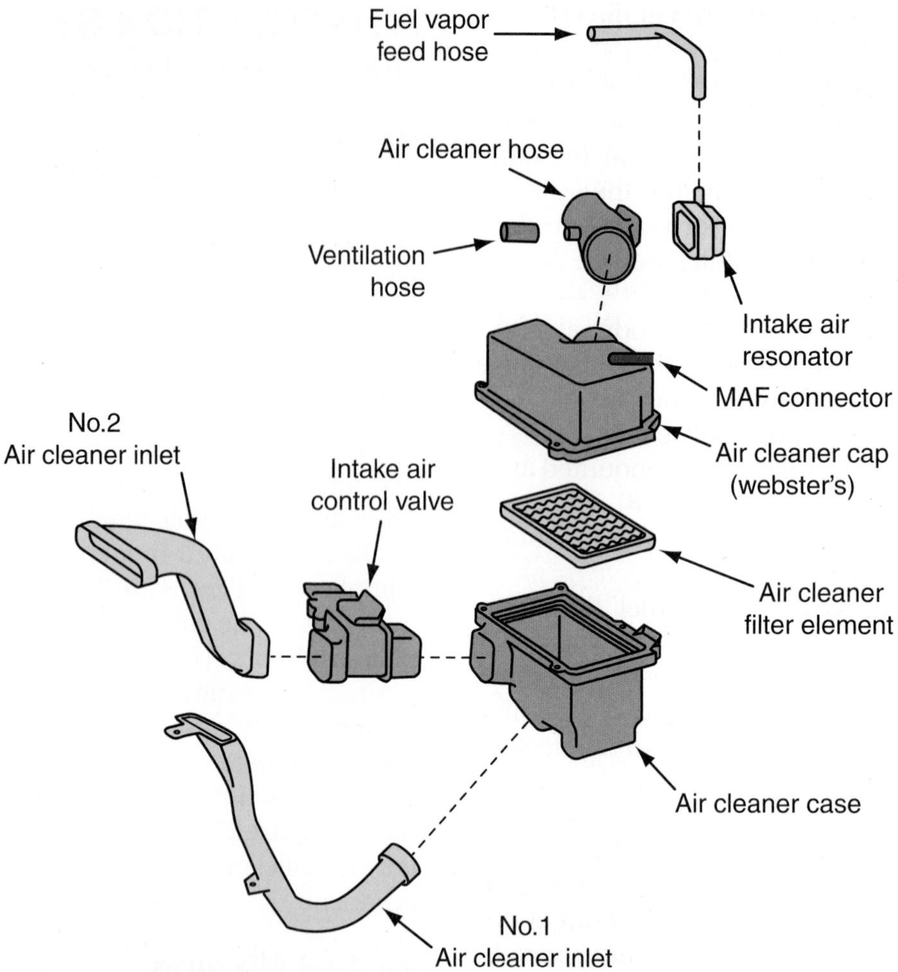

Figure 25-7 An air induction system for a late-model SUV.

Figure 25-8 A turbocharger. The red represents the flow of exhaust and the blue shows the intake air that is pressurized.

Figure 25-9 Dirty exhaust is bad for everyone.

and industry to address the problem of these pollutants. Not until smog became an issue did anyone in power really care and do something about these pollutants.

Smog not only appears as dirty air, it is also an irritant to the eyes, nose, and throat. The things necessary to form photochemical smog are hydrocarbons (HC) and oxides of nitrogen (NO_x) exposed to sunlight in stagnant air. HC in the air that reacts with the NO_x causes these two chemicals to react and form photochemical smog.

There are three main automotive pollutants: HC, carbon monoxide (CO), and NO_x. Particulate (soot) emissions are also present in diesel engine exhaust. HC emissions are caused largely by unburned fuel from the combustion chambers. HC emissions can also originate from evaporative sources such as the gasoline tank. CO emissions are a by-product of the combustion process, resulting from incorrect air-fuel mixtures. NO_x emissions are caused by nitrogen and oxygen uniting at cylinder temperatures above 2,300°F (1,261°C). Current concerns include carbon dioxide (CO_2), which is said to contribute to global warming.

Computer-Controlled Systems

The EGR valve, air pump, and evaporative emissions canister are controlled by the PCM. The computer keeps the level of the three major pollutants (CO, HC, and NO_x) at acceptably low levels. Other emission control devices may also be wholly or partly controlled by the computer. The control of the air, fuel, and ignition systems also contributes significantly to the control of emissions.

ENGINE CONTROL SYSTEMS

As manufacturers come closer to achieving complete combustion, engines are able to produce more power, use less fuel, and emit fewer pollutants. This has been made possible by technological advances, primarily electronics.

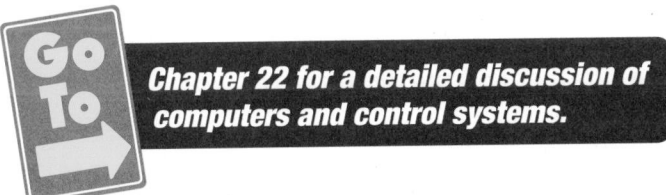

Go To ***Chapter 22 for a detailed discussion of computers and control systems.***

The computer is an engine control system that functions like other computers. It receives inputs, processes information, and commands an output. The primary computer in an engine control system is the **engine control module** (ECM) or the **powertrain control module** (PCM).

SHOP TALK

A PCM controls more than the engine and it often performs the functions of the ECM as well. In this text, the term *PCM* is used unless something is specifically part of an ECM.

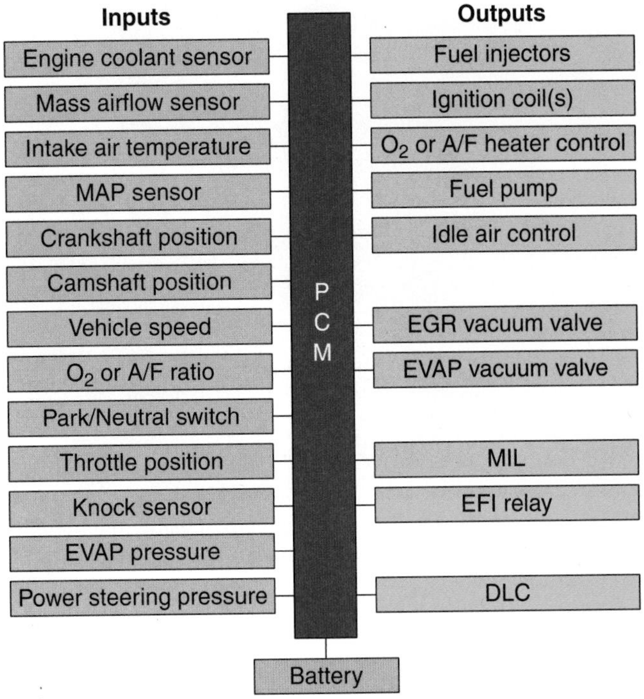

Figure 25–10 A basic look at an engine control system.

The engine control system relies on sensors that convert engine operating conditions such as temperature, engine and vehicle speeds, throttle position, and other conditions into electrical signals that are constantly monitored by the PCM **(Figure 25–10)**. The PCM also senses some conditions through electrical connections. These include voltage changes at various components.

The PCM sorts the input signals and compares them to parameters programmed in it. This is the processing role of the computer. Based on the comparison, the computer may command a change in the operation of a component or system. The PCM also monitors the activity of the system and can detect any problems that occur. At that point, it will set a DTC and store other diagnostic information. The PCM may also store vehicle information, such as the VIN, calibration identification (CAL ID), and calibration verification. These are used to ensure all calibration settings match the vehicle.

Based on the input information and the programs, the PCM decides the best operating parameters and sends out commands to various outputs or actuators. These commands are first sent to output drivers that cause an output device to turn on or off. These outputs include solenoids, relays, lights, motors, clutches, and heaters.

The PCM is linked with several other control modules. They share information and, in some cases, one control module controls another. The shared

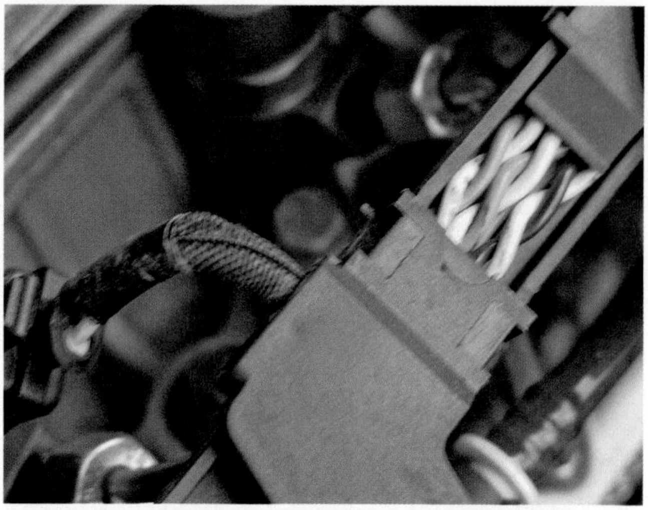

Figure 25-11 The twisted pairs of wires that serve as data buses in a multiplexed system.

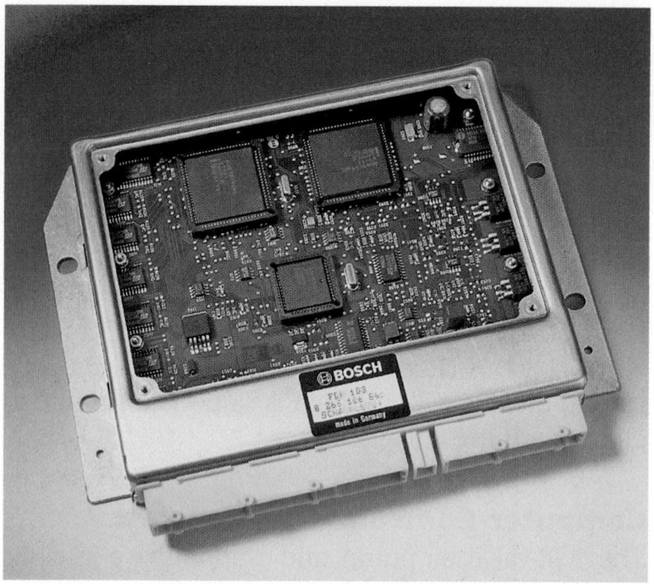

Figure 25-12 A PCM. *Courtesy of Chrysler LLC*

information is present on the CAN data bus in most vehicles (**Figure 25–11**).

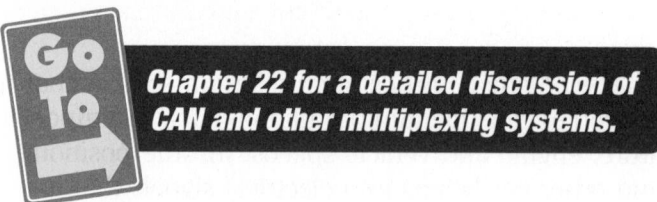

Go To Chapter 22 for a detailed discussion of CAN and other multiplexing systems.

System Components

The sensors, actuators, and computer communicate through the use of electronic and multiplexed circuits. For example, when the incoming voltage signal from the coolant sensor tells the PCM that the engine is getting hot, the PCM sends out a command to turn on the electric cooling fan. The PCM does this by grounding the relay circuit that controls the electric cooling fan. When the relay clicks on, the electric cooling fan starts to spin and cools the engine. The information may also be used to alter the air-fuel ratio and ignition timing.

In a PCM (**Figure 25–12**), RAM is used to store data collected by the sensors, the results of calculations, and other information that is constantly changing during engine operation. Information in volatile RAM is erased when the ignition is turned off or when the power is disconnected. Nonvolatile RAM does not lose its data if its power source is disconnected.

The computer's permanent memory is stored in ROM or PROM and is not erased when the power source is disconnected. ROM and PROM are used to store computer-controled system strategy and look-up tables. PROM normally contains the specific information about the vehicle.

The look-up tables (sometimes called maps) contain calibrations and specifications. Look-up tables indicate how an engine should perform. For example, information (a reading of 20 in. Hg) is received from the manifold absolute pressure (MAP) sensor. This information and the information from the engine speed sensor are compared to a table for spark advance. This table tells the computer what the spark advance should be for that throttle position and engine speed (**Figure 25–13**). The computer then modifies the spark advance.

When making decisions, the PCM is constantly referring to three sources of information: the look-up tables, system strategy, and the input from sensors. The computer makes informed decisions by comparing information from these sources.

Computer Logic

In order to control an engine system, the computer makes a series of decisions. Decisions are made in a step-by-step fashion until a conclusion is reached. Generally, the first decision is to determine the engine mode. For example, to control air-fuel mixture the computer first determines whether the engine is cranking, idling, cruising, or accelerating. Then the computer can choose the best system strategy for the present engine mode. In a typical example, sensor input indicates that the engine is warm, rpm is high, manifold absolute pressure is high, and the throttle plate is wide open. The computer determines that the vehicle is under heavy acceleration or has a wide-open throttle. Next, the computer determines the goal to be reached. For example, with heavy acceleration, the goal is to create a rich air-fuel mixture. At wide-open throttle, with high manifold absolute

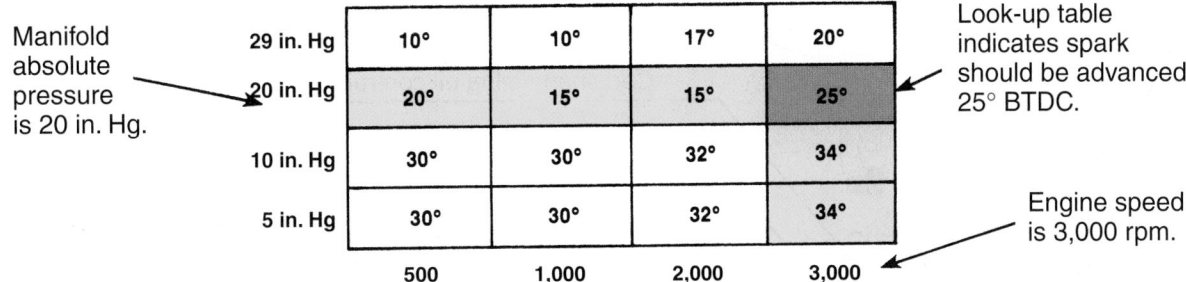

Manifold absolute pressure is 20 in. Hg.

Look-up table indicates spark should be advanced 25° BTDC.

Engine speed is 3,000 rpm.

29 in. Hg	10°	10°	17°	20°
20 in. Hg	20°	15°	15°	25°
10 in. Hg	30°	30°	32°	34°
5 in. Hg	30°	30°	32°	34°
	500	1,000	2,000	3,000

Figure 25-13 Example of a base spark advance look-up table. *Courtesy of Ford Motor Company*

pressure and coolant temperature of 170°F, the table indicates that the air-fuel ratio should be 13:1, that is, 13 pounds of air for every 1 pound of fuel. An air-fuel ratio of 13:1 creates the rich air-fuel mixture needed for heavy acceleration.

In a final series of decisions, the computer determines how the goal can be achieved. In our example, a rich air-fuel mixture is achieved by increasing fuel injector pulse width. The injector nozzle remains open longer and more fuel is drawn into the cylinder, providing the additional power needed.

Additional Engine Controls

The PCM for some engines controls other systems that reduce fuel consumption and emissions.

Variable Valve Timing Many engines use camshaft advance drums (phasers) and other devices to alter camshaft timing. Most of these use oil pressure to move the vanes inside the phaser. The pressurized oil is controlled by a solenoid, which is controlled by the PCM. Like all control systems, the PCM receives several inputs and sends out commands to alter the timing. Some engines also have variable valve lift along with variable timing; these systems are also controlled by oil pressure that is applied by commands of the PCM (**Figure 25–14**).

Cylinder Deactivation A few engines and perhaps more in the future are equipped with a feature that should increase the fuel economy of larger engines without sacrificing an engine's ability to do work when operating under a light load. This system is called "Displacement on Demand" by General Motors and has various other names by other manufacturers. This is an advanced valve train system that selectively turns off half of the engine's cylinders when there is light load on the engine. The engine continues to be even firing on the remaining cylinders. The system provides for smooth transition as the cylinders are deactivated and activated because the deactivation occurs within one camshaft revolution, which is about 120 ms at 1,000 rpm or 40 ms at 3,000 rpm.

The PCM processes data from over twenty sensors to govern the system. It calculates the current torque from all of the cylinders and projects how hard half of them would need to work to match that output. It then deactivates half of the intake and exhaust lifters (**Figure 25–15**). When deactivated, these hydraulic lifters shorten to keep the valves closed. The control of the lifters is done by sending hydraulic pressure to the central part of the lifter, where it works normally, or to the top, where it is diverted and the lifter does not pump up.

At the same time the lifters are deactivated, the PCM turns off the injectors and adjusts the mixture to the other cylinders. It also advances the ignition timing for the cylinders still working and opens the throttle to allow more to reach those cylinders. These actions allow the engine to produce more torque.

Honda hybrids use a system that can close the intake and exhaust valves on all four cylinders during deceleration. This system is called the cylinder idling system and increases the amount of energy captured during regenerative braking.

With the valves closed, the pistons in those cylinders move freely. This, in turn, reduces the amount of engine braking or resistance that takes place during deceleration. By reducing the amount of engine's mechanical resistance during deceleration, the motor/generator can provide maximum resistance by producing more electricity.

The cylinder idling system reverses the role of Honda's variable valve timing system (VTEC). When a VTEC engine is running at high speeds, the PCM directs high oil pressure to lock together a pair of rocker arms at each cylinder to increase horsepower. The movement of a hydraulic piston makes the connection between the two rocker arms, each riding on a differently shaped camshaft lobe and each working a separate valve. As a result, both valves open to improve engine breathing. In the cylinder idling system, low engine speeds and engine oil pressure deactivate the rocker arms. The system also has two rocker arms per valve: a valve lift rocker arm that follows the lobe of the camshaft and a cylinder idle rocker arm

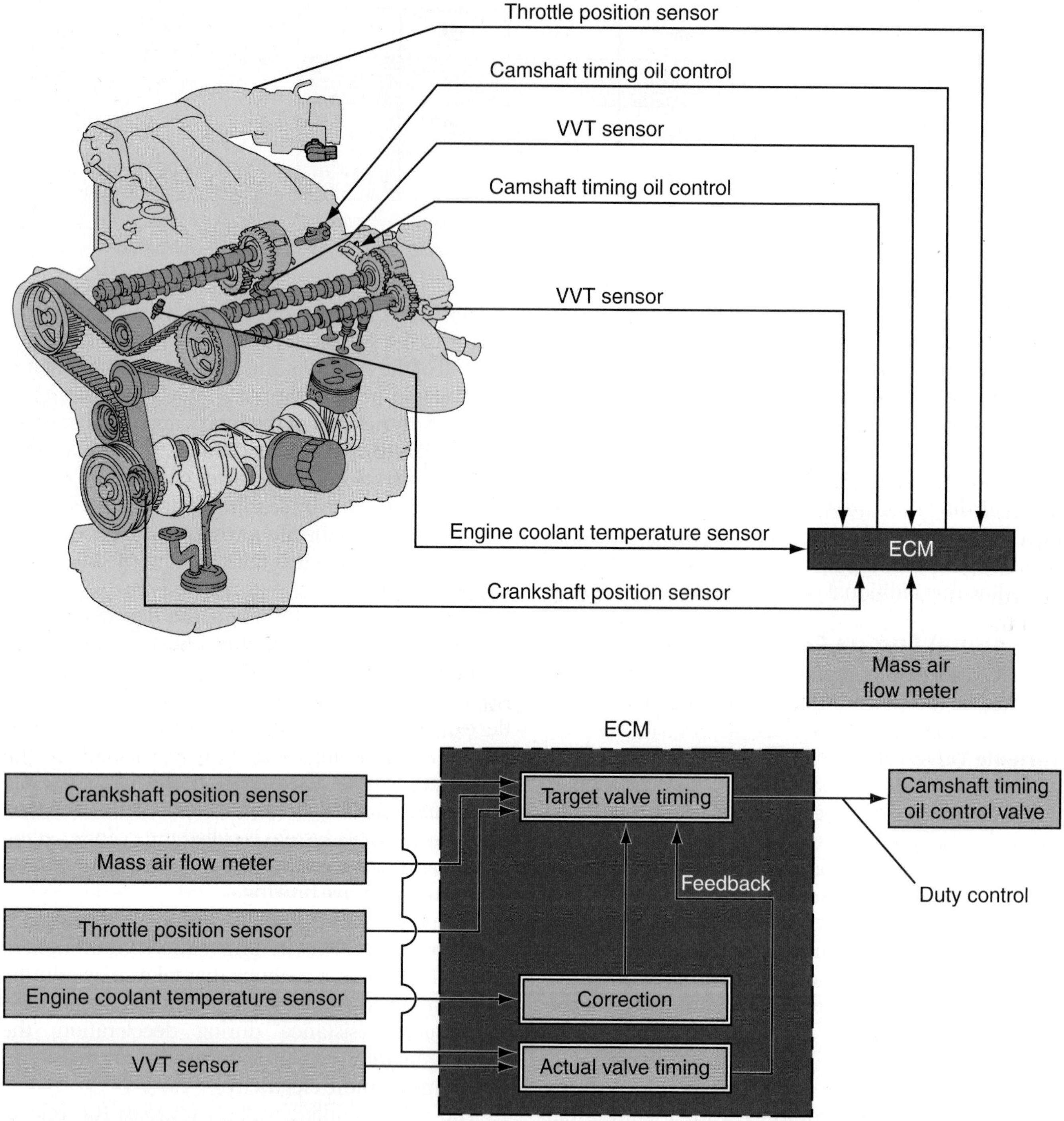

Figure 25–14 Toyota's variable valve timing (VVT-i) system. *Courtesy of Toyota Motor Sales, U.S.A., Inc.*

that actually opens the valve. The two rocker arms are connected by a piston. When a cylinder is deactivated or idled, hydraulic pressure moves the piston, and the connection between the two rocker arms is gone (**Figure 25–16**). This allows the valve lift rocker arm to move along the camshaft lobe without opening the valve.

Electronic Throttle Control Many engines have no mechanical connection between the throttle pedal and the throttle plates. This connection is made by wire. The PCM controls throttle opening based on several inputs, including a throttle pedal sensor (**Figure 25–17**). Electronic throttle control is used because it improves driveability and idle speed stability and eliminates cruise control actuators.

Variable Intake Manifolds In an attempt to match airflow into the engine with engine speed and other variables, some engines have intake manifolds with

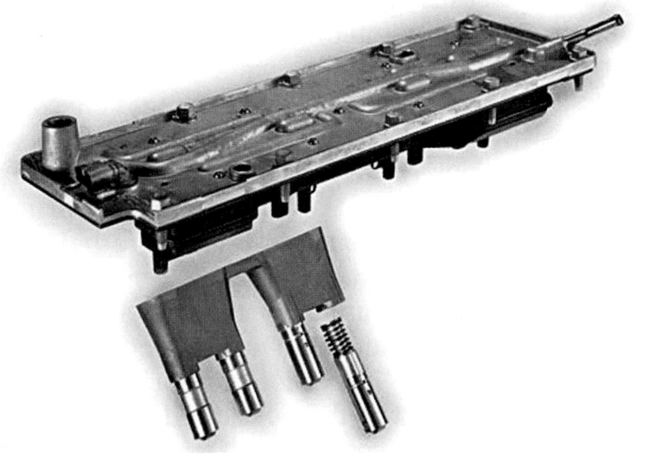

Figure 25-15 Parts of the valve train assembly used for cylinder deactivation. *Courtesy of Delphi Corporation*

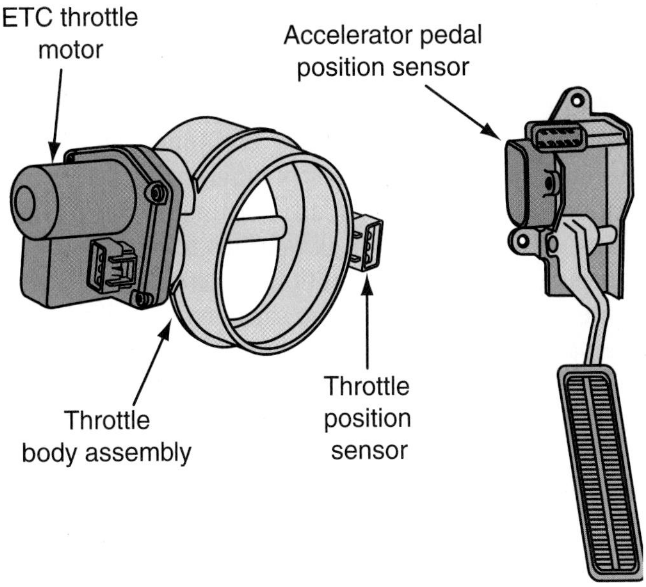

Figure 25-17 An electronic throttle control system.

two or more runners for each cylinder. The PCM opens and closes runners of different lengths to match the engine's needs. The length of the runner determines the amount of air that can be moved into the cylinders. It also affects the turbulence of that air.

Control of Nonengine Functions

Some devices that are not directly connected to the engine are also controlled by the PCM to ensure maximum efficiency. For example, air conditioner compressor clutches can be turned on or off, depending on various conditions. One common control procedure turns off the compressor when the throttle is fully opened. This allows maximum engine acceleration by eliminating the load of the compressor.

On some vehicles, the torque converter lock-up clutch is applied and released by a signal from the computer. The clutch is applied by transmission hydraulic pressure, which is controlled by electrical solenoids that are in turn controlled by the computer.

In most cases, the PCM works with other control modules to control a system. Examples of these are antilock brake systems, traction and stability control systems, and other accessories.

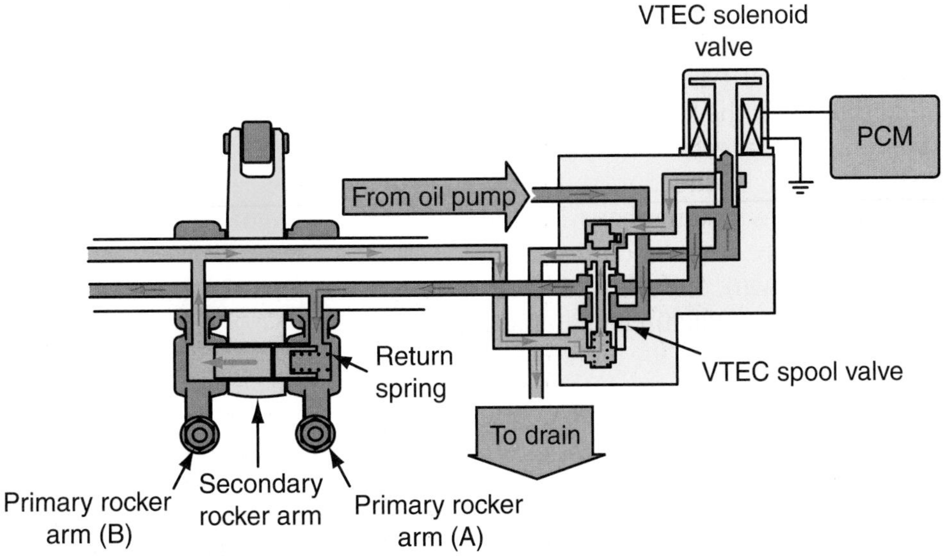

Figure 25-16 A look at Honda's VTEC system. The PCM controls oil pressure to move a piston that locks the primary rocker arms together. Their movement then follows the secondary rocker arm, which is riding on the camshaft.

ON-BOARD DIAGNOSTIC SYSTEMS

Because all manufacturers have continually updated, expanded, and improved their computerized control systems, there are hundreds of different domestic and import systems on the road. Fortunately for technicians, OBD-II called for all vehicles to use the same terms, acronyms, and definitions to describe their components. They also have the same type of diagnostic connector, the same basic test sequences, and display the same trouble codes. OBD-II began in 1996 and has been on all vehicles sold in North America since 1997.

The primary goal of OBD systems is to reduce vehicle emissions and reduce the possibility of future emission increases by detecting and reporting system malfunctions.

OBD-I (On-Board Diagnostic System, Generation 1)

OBD-I systems were first used in 1988. The ECM was capable of monitoring critical emission-related parts and systems and illuminate a malfunction indicator if a defect was found. The **malfunction indicator lamp (MIL)** was in the instrument panel. Most OBD-I systems used flash codes to display DTCs. The codes were displayed with the MIL. Often, the codes were displayed by jumping across terminals at a **diagnostic data link connector (DLC)**. Manufacturers provided lists of what the codes represented, along with step-by-step diagnostic procedures for identifying the exact fault.

Typically, the DTCs represented problems with the sensors, fuel metering system, and the operation of the EGR valve. If any of these were open, shorted, had high resistance, or were operating outside a normal range, a code was set. The MIL not only helped with diagnostics, but also alerted the driver that there was a problem. The MIL would turn off when the condition returned to normal; however, the DTC remained in memory until it was erased by a technician.

OBD-I was a step in the right direction, but it had several faults. It monitored few systems, had a limited number of DTCs (these were not standardized, so each manufacturer had its own), and allowed a limited use of serial data; most manufacturers required a specific scan tool and procedure, and the names used to describe a component varied across the manufacturers and their model vehicles.

OBD-II (On-Board Diagnostic System, Generation 2)

OBD-II was established to overcome some of the weaknesses of OBD-I. This was possible because of the advances made in computer technology and was necessary because of stricter emissions standards. OBD-I systems monitored only a few emission-related parts and were not set to maintain a specific level of emissions. OBD-II was developed to be a more comprehensive monitoring system and to allow more accurate diagnosis by technicians.

Studies estimate that approximately 50% of the total emissions from late-model vehicles are the result of emission-related problems. OBD-II systems are designed to ensure that vehicles remain as clean as possible over their entire life. During an emissions or "smog" check, an inspection computer can be plugged into the DLC of the vehicle and read the data from the vehicle's computer. If emission-related DTCs are present, the vehicle will fail the test.

OBD-II added monitor functions for such things as catalyst efficiency, engine misfire detection, evaporative system, secondary air system, and EGR system flow rate. These monitors detect problems that would affect emissions levels. Also, a serial data stream of twenty basic data parameters and common DTCs was adopted.

OBD-II systems monitor the effectiveness of the major emission control systems and anything else that may affect emissions and will illuminate the MIL when a problem is detected. During the monitoring functions, every part that can affect emission performance is checked by a diagnostic routine to verify that it is functioning properly. OBD-II systems must illuminate the MIL **(Figure 25–18)** if the vehicle's conditions allow emissions to exceed 1.5 times the allowable standard for that model year based on a federal test procedure (FTP). When a component or strategy failure would allow emissions to exceed this level and the fault was detected during two consecutive trips,

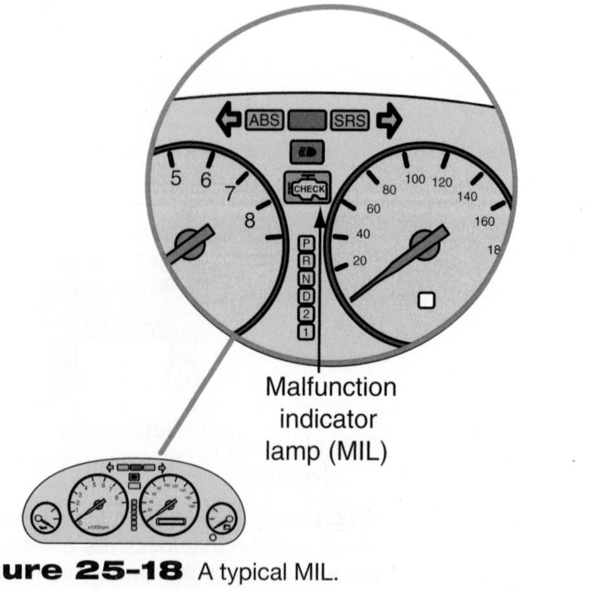

Figure 25-18 A typical MIL.

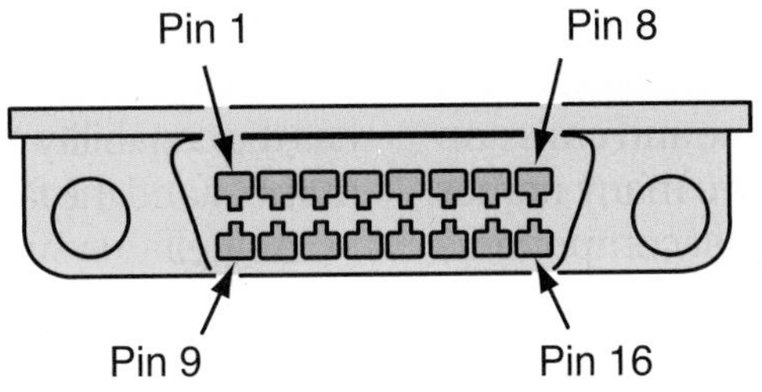

Pin 1: Manufacturer discretionary	Pin 9: Manufacturer discretionary
Pin 2: J1850 bus positive	Pin 10: J1850 bus negative
Pin 3: Manufacturer discretionary	Pin 11: Manufacturer discretionary
Pin 4: Chassis ground	Pin 12: Manufacturer discretionary
Pin 5: Signal ground	Pin 13: Manufacturer discretionary
Pin 6: Manufacturer discretionary	Pin 14: Manufacturer discretionary
Pin 7: ISO 1941-2 "K" line	Pin 15: ISO 9141-2 "L" line
Pin 8: Manufacturer discretionary	Pin 16: Battery power

Figure 25-19 A standard OBD-II DLC with pin designations.

the MIL would illuminate to inform the driver of a problem and a DTC would be stored in the PCM.

Besides increasing the capability of the PCM, additional hardware is required to monitor and maintain emissions performance. Examples of these are the addition of a heated oxygen sensor down the exhaust stream from the catalytic converter, upgrade of specific connectors, components designed to last the mandated 80,000 miles or 8 years, more precise crankshaft or camshaft position sensors, and a new standardized 16-pin DLC.

Rather than use a fixed, unalterable PROM, OBD-II PCMs have an EEPROM to store a large amount of information. The EEPROM stores data without the need for a continuing source of electrical power. It is an integrated circuit that contains the program used by the PCM to provide powertrain control. It is possible to erase and reprogram the EEPROM without removing it from the computer. When a modification to the PCM operating strategy is required, the EEPROM may be reprogrammed through the DLC using computer software.

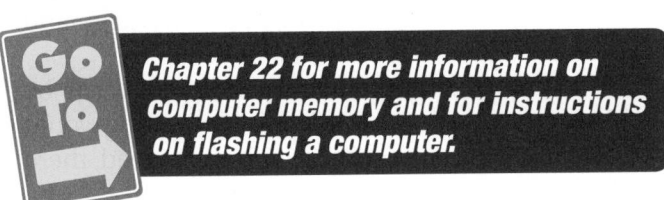

Go To ***Chapter 22 for more information on computer memory and for instructions on flashing a computer.***

For example, if the vehicle calibrations are updated for a specific car model sold in California, a computer may be used to erase the EEPROM. After the erasing procedure, the EEPROM is reprogrammed with the updated information. A CD-ROM disk is used to complete the reprogramming procedure. Manufacturers periodically send authorized service facilities the disks required for current updating of the EEPROMs. PCM recalibrations must be directed by a service bulletin or recall letter.

Data Link Connector OBD-II standards require the DLC to be easily accessible while sitting in the driver's seat (**Figure 25–19**). The DLC cannot be hidden behind panels and must be accessible without tools. The connector pins are arranged in two rows and are numbered consecutively. Seven of the sixteen pins have been assigned by the OBD-II standard. They are used for the same information, regardless of the vehicle's make, model, and year. The remaining nine pins can be used by the individual manufacturers to meet their needs and desires.

The connector is "D"-shaped and has guide keys that allow the scan tool to only be installed one way. Using a standard connector and designated pins allows data retrieval with any scan tool designed for OBD-II. Some vehicles meet OBD-II standards by providing the designated DLC along with their own connector for their own scan tool. Often a vehicle will have more than one DLC, each with its own purpose. Due to OBD standards, the OBD DLC will always be located within a foot, to the right or left, of the steering column.

SHOP TALK

When a vehicle has a 16-pin DLC, this does not necessarily mean that the vehicle is equipped with OBD-II.

OBD-II Terms All vehicle manufacturers must use the same names and acronyms for all electric and electronic systems related to the engine and emission control systems. Previously, there were many names for the same component. Now all similar components will be referred to with the same name. Beginning with the 1993 model year, all service information has been required to use the new terms. This new terminology is commonly called J1930 terminology because they conform to the SAE standard J1930.

OBD-II for Light-Duty Diesels The OBD-II systems are also mandated for all diesel engine vehicles that weigh 14,000 pounds or less. These systems are very similar to those found in gasoline engines. The exceptions to this are the exclusion of the systems that are unique to gasoline engines and the inclusion of those systems unique to a diesel engine.

OBD-III (On-Board Diagnostic System, Generation 3)

Although not implemented at the printing of this book, there is a possibility that soon OBD-III will be a fact in California and elsewhere. OBD-III is aimed at minimizing the delay between the detection of an emissions failure by the OBD-II system and the actual repair of the vehicle. It has been said that the check engine light is a poor motivator for prompt repair and many repairs to the emissions-related parts of vehicles are being delayed until the mandatory emissions inspection is approaching. In other words, vehicles are running around with problems that increase emissions and some owners are doing nothing about it.

OBD-III mandates include a way to automatically read the OBD-II codes off a vehicle. The systems may use cell phone technology and/or remote stations that read the information on a vehicle's computer without physical hookup via radio or satellite.

A cell phone-like device would be plugged into a car's computer. If the computer finds a problem, it would automatically call the Division of Motor Vehicles and report it. The driver would receive notification that the car needs to be fixed.

OBD-III may have communications capabilities enabling information to be passed from the vehicle to remote stations in real time at random, perhaps without the driver's knowledge. If the system's sensors detect a problem, not only would the MIL illuminate, the associated data would also be received by roadside receivers or satellites that are randomly monitoring vehicles. These stations may automatically receive reports on an emissions problem as soon as it occurs.

Currently, roadside readers are capable of reading eight lanes of bumper-to-bumper traffic at 100 mph and can be used from a fixed location with portable units or a mobile unit. If a fault is detected by the reader unit, it has the capability of sending the vehicle's identification number and the fault codes to a central location.

A satellite system can be used with a cellular phone hookup or with location monitoring technology. The vehicle would receive an alert through either of these and would send the vehicle's location, VIN, the date and time, and all OBD data.

Another possibility is to have owners drive through specially designed stations where a vehicle's transponder would transmit information about the vehicle, including OBD-II information. The computers in the station would record the data. If the data showed a problem with the vehicle's emission system, the owner of the vehicle would get a notice to have it fixed.

Another possibility, and a very likely one, is to equip each vehicle with a button that is depressed by the driver once a year to establish communications with a central point and send data to it. Owners with unsatisfactory reporting data would be contacted to bring their vehicles into a shop for more detailed inspection. If the OBD system reports no emissions problems, the owner is allowed to renew the vehicle's registration and the vehicle will not need to be tested prior to that. However, if the owner does not depress the button or if the central computer receives signals indicating emissions problems, the owner will receive a notice informing him or her that the vehicle must be brought in for emissions testing.

OBD-III programs can be incorporated into current inspection and maintenance program. OBD-III might also be used to generate out-of-cycle inspections. Once a fault is detected, a notice could be mailed to the owner requiring an out-of-cycle inspection within a certain number of days or at the next registration or at resale, or the owner will be fined.

This type of vehicle monitoring has raised some fear in car owners. They feel that the government will be able to know too much about their driving habits and driving routes, and they want their privacy protected. For this reason, the final design and method for OBD-III have not been decided.

Government officials answer these concerns by looking at the positives of OBD-III. Monitoring would not be continuous under the program; instead, there would be a specific number of random checks of a vehicle per year. There would be more accurate detection of high emitters and a more comprehensive inspection of these vehicles. Also, there would be quite a cost savings and increased convenience for consumers because their vehicles would not need to go to an inspection testing station on a regular basis if the central computer received nothing but good news.

Currently, GM's OnStar systems can determine if the MIL is illuminated. The system can also remotely read DTCs. However, the communications specialists at the OnStar center are not allowed to tell the owner what is specifically wrong with the vehicle. They are only authorized to describe the urgency of the problem for the driver.

SYSTEM OPERATION

The PCM will operate in different modes based on the conditions. These modes are often referred to as control loops. The PCM does not always process all information it receives. It is programmed to ignore or modify some inputs according to the current operating conditions.

Closed-Loop Mode

During the **closed-loop** mode, the PCM receives and processes all information available. Sensor inputs are sent to the PCM; the PCM compares those values to its programs, then sends commands to the output devices. The output devices adjust timing, air-fuel ratio, and emission control operation. The resulting engine operation will result in new inputs from the sensors. Those new messages are sent to the PCM, where the effectiveness of the change can be processed and corrections made as necessary. This continuous cycle of information is called a closed loop.

Closed control loops are often referred to as feedback systems. This means that the sensors provide constant information, or feedback, on what is taking place in the engine. This allows the PCM to constantly monitor, process, and send out new output commands.

Open-Loop Mode

When the engine is cold, most electronic engine controls go into **open-loop** mode. In this mode, the control loop is not a complete cycle because the computer does not react to feedback information. Instead, the computer makes decisions based on preprogrammed information that allows it to make basic ignition or air-fuel settings and to disregard sensor inputs. The open-loop mode is activated when a signal from the temperature sensor indicates that the engine temperature is too low for gasoline to properly vaporize and burn in the cylinders. Systems with unheated oxygen sensors may also go into the open-loop mode while idling, or at any time that the oxygen sensor cools off enough to stop sending a good signal, and at wide-open throttle.

SHOP TALK

Most late-model engines have a heated oxygen sensor that reduces the time a PCM will be in open loop. If the system can go to closed loop just 1 minute sooner, the amount of pollutants released will be cut nearly in half.

Fail-Safe or Limp-In Mode

Most computer systems also have what is known as the fail-safe or limp-in mode. The limp-in mode is nothing more than the computer's attempt to take control of vehicle operation when input from one of its critical sensors has been lost or is well out of its normal range. To be more specific, if the computer sees a problem with the signal from a sensor, it either works with fixed values in place of the failed sensor input, or, depending on which input was lost, it can also generate a modified value by combining two or more related sensor inputs.

To illustrate this, assume that the MAP sensor stops working. Instead of an actual MAP measurement, the computer compensates by creating an artificial MAP signal from the combination of throttle position input and engine speed data. Although this may not result in the most efficient operation, considering the alternatives, a modified MAP signal is better than no MAP signal at all. This allows the engine to run until the driver can reach a service location.

Adaptive Strategy

A system's adaptive strategy is based on a plan for the timing and control of computer-controlled systems during different operating conditions. If a computer has adaptive strategy capabilities, it can actually learn from past experience. For example, the normal voltage signals from the TP sensor to the PCM range from 0.6 to 4.5 volts. If a 0.2-volt signal is received, the PCM may regard this signal as the result of a worn TP sensor and assign this lower voltage to the normal low-voltage signal. The PCM will add 0.4 volt to the 0.2 volt it received. All future signals from the various throttle positions will also have 0.4 volt added to the signal. Doing this calculation adjusts for the worn TP sensor and ensures that the engine will operate normally. If the input from a sensor is erratic or considerably out of range, the PCM may totally ignore the input.

Most adaptive strategies have two parts: short term and long term. Short-term strategies are those immediately enacted by the computer to overcome a change in operation. These changes are temporary. Long-term strategies are based on the feedback about the short-term strategies. These changes are more permanent.

OBD-II MONITORING CAPABILITIES

OBD-II monitors the performance of emission and other related systems. The purpose of these monitors is to detect failing systems and not wait until they fail before illuminating the MIL. OBD-II systems will perform certain tests on various subsystems of the engine management system. If one or more monitored systems are found to have a malfunction, the MIL will illuminate to alert the driver of a problem. Some monitors run continuously, whereas others will run only when certain operating conditions are present during the drive cycle. These conditions are called the **enable criteria**.

Drive Cycle The OBD-II **drive cycle** is operating conditions that must exist before the self-diagnosis can take place. A drive cycle **(Figure 25–20)** includes an engine start and operation that brings the vehicle into closed loop and includes whatever specific operating conditions are necessary either to initiate and complete a specific monitoring sequence or to verify a symptom or repair.

In a complete drive cycle, all monitors must be completed followed by the catalyst monitor. The catalyst monitor must be completed after the other five monitors are completed in a trip. It is possible that not all monitors will be completed during a drive cycle. Each monitor has its own enable criteria that may not be met during a drive cycle. Also, a test may be interrupted by a fault in another system.

OBD-II Trip A **trip** is a drive cycle that includes all of the conditions (enable criteria) required for a monitor to run. To run all monitors, the vehicle must be driven at different speeds and conditions. During diagnosis, it may be necessary to complete a trip for a monitor to verify the problem or the repair. Depending on the monitor, the system tests the component or system once per trip.

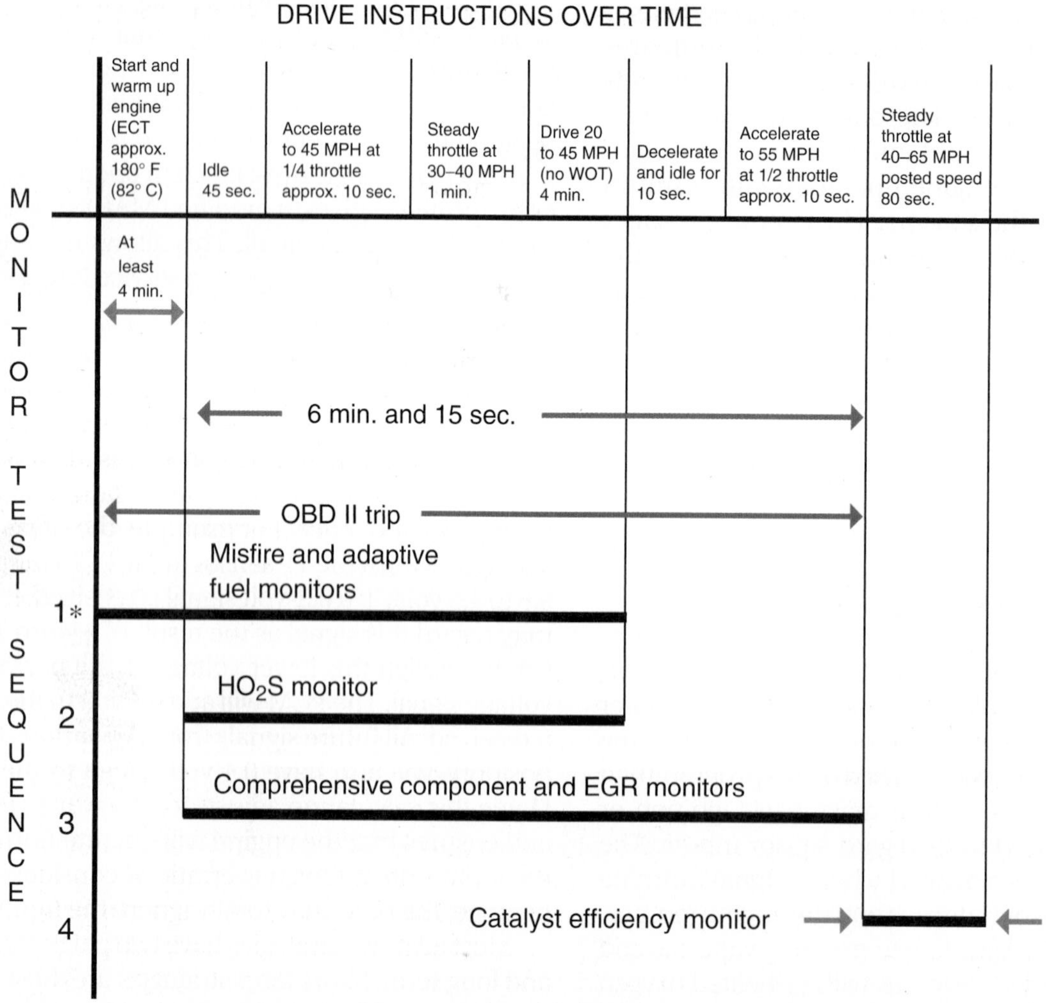

* Since the misfire, adaptive fuel, EGR (requiring idles and accelerations), and comprehensive monitors are continuously checked by the OBD-II system, the test sequence may vary on each vehicle due to outside ambient temperature, engine/vehicle performance temperature, and driving conditions.

Figure 25-20 OBD trip cycle. *Courtesy of Ford Motor Company*

Warm-Up Cycle OBD-II standards define a warm-up cycle as the period from when the engine is started until the engine temperature has increased by at least 60°F (16°C) and has reached at least 160°F (88°C).

Catalyst Efficiency Monitor

OBD-II vehicles use a minimum of two oxygen sensors. One of these is used for feedback to the PCM for fuel control, and the other, located at the rear of the catalytic converter, gives an indication of the efficiency of the converter. The downstream O_2 sensor is sometimes called the **"catalyst monitor sensor" (CMS)**. The catalyst efficiency monitor compares the signals between the two O_2 sensors to determine how well the catalyst is working.

One heated O_2 sensor (HO_2S) is mounted in the exhaust manifold and the additional HO_2S is mounted downstream from the catalytic converter. The HO_2S are identified by their position and location relative to the converters. S1 means the O_2 sensor is upstream or before the catalytic converter, and S2 is downstream or after the catalytic converter. On V-type engines, the additional designation B1 means the sensor is installed in the bank for cylinder No. 1. B2 indicates it is for the other bank of cylinders **(Figure 25–21)**.

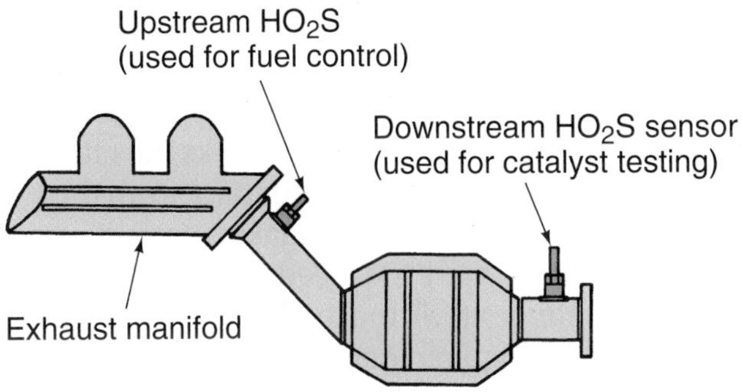

Figure 25-21 OBD-II system with heated oxygen sensors in the exhaust manifold and after the catalytic converter. *Courtesy of Ford Motor Company*

The downstream HO_2S are designed to prevent the collection of condensation on their ceramic material. The internal heater is not turned on until the ECT sensor indicates a warmed-up engine. This action prevents cracking of the ceramic. Gold-plated pins and sockets are used in the HO_2S, and the downstream and upstream sensors have different wiring harness connectors.

A catalytic converter stores oxygen during lean engine operation and gives up this stored oxygen during rich operation to burn up excessive HCs. Catalytic converter efficiency is measured by monitoring the O_2 storage capacity of the converter during closed loop operation.

When the catalytic converter is storing oxygen properly, the downstream HO_2S provide fewer cross counts (low-frequency) voltage signals. If the catalytic converter is not storing O_2 properly, the cross counts of the voltage signal increase on the downstream HO_2S. When the signals from the downstream HO_2S approach the cross counts of the upstream sensors **(Figure 25–22)**, a DTC is set in the PCM memory. If the fault occurs on three drive cycles, the MIL light is illuminated.

Misfire Monitor

If a cylinder misfires, HC is exhausted from the cylinder and enters the catalytic converter. A misfire means a lack of combustion in at least one cylinder for at least one combustion event. A misfire pumps unburned fuels through the exhaust. Although the converter can handle an occasional sample of raw fuel, too much fuel to the converter can overheat and destroy it. The honeycomb material in the converter may melt into a solid mass. If this happens, the converter is no longer efficient in reducing emissions.

Cylinder misfire monitoring requires measuring the contribution of each cylinder to total engine power. The misfire monitoring system uses a highly

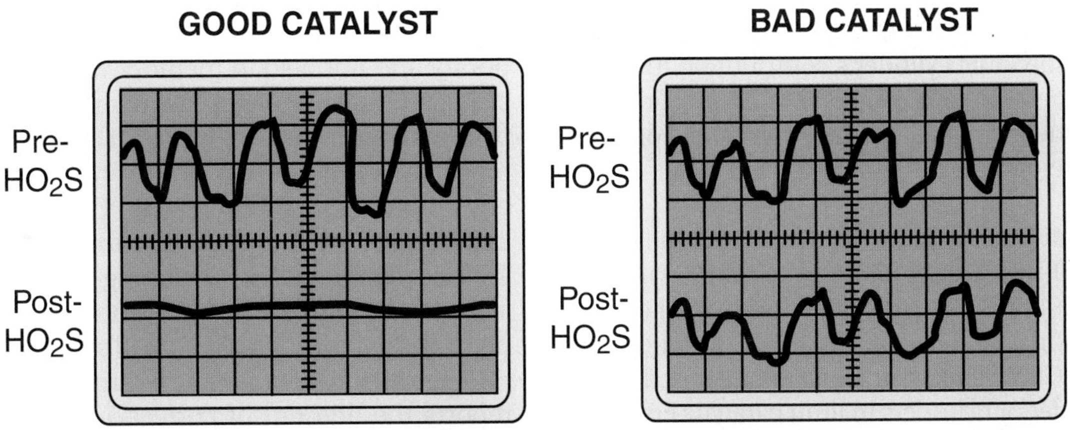

Figure 25-22 Oxygen sensor signal for a good and bad catalytic converter.

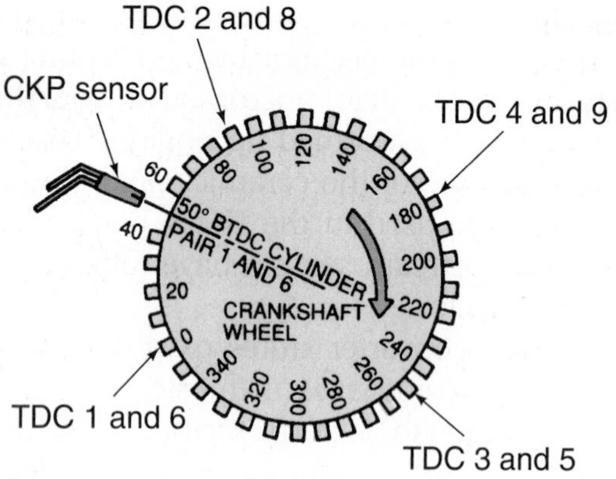

Figure 25-23 High data rate crankshaft sensor used for misfire detection. *Courtesy of Ford Motor Company*

accurate crankshaft angle measurement to measure the crankshaft acceleration each time a cylinder fires. A high data rate crankshaft sensor is required for this **(Figure 25–23)**. If a cylinder is contributing normal power, a specific crankshaft acceleration time occurs. When a cylinder misfires, the cylinder does not contribute to engine power, and crankshaft acceleration for that cylinder is slowed. This monitor runs continuously while the engine is running.

Most OBD-II systems allow a random misfire rate of about 2% before a misfire is flagged as a fault. It is important to note that this monitor only looks at the crankshaft's speed during a cylinder's firing stroke. It cannot determine if the problem is fuel-, ignition-, or mechanically related. Misfires are categorized as type A, B, or C. Type A could cause immediate catalyst damage. Type B could cause emissions of 1.5 times the design standard, and type C could cause an I/M failure. When there is a type A misfire, the MIL will flash. If there is a type B misfire, the MIL will turn on but will not flash. Type C misfires will typically not cause the MIL to light or flash.

The misfire monitoring sequence includes an adaptive feature that compensates for variations in engine operation caused by manufacturing tolerances and component wear. It also has the capability to allow vibration at different engine speeds and loads. When an individual cylinder's contribution to engine speed falls below a certain threshold, the misfire monitoring sequence calculates the vibration, tolerance, and load factors before setting a misfire code.

Type A Misfires The monitor checks for type A misfires in 200-rpm increments. If a cylinder misfires between 2% and 20% of the time, the monitor considers the misfiring to be excessive. This condition may cause the PCM to shut off the fuel injectors at the misfiring cylinder or cylinders to limit catalytic converter heat. When the engine is operating under heavy load, the PCM will not turn off the injectors. If a type A misfire is not corrected, the catalytic converter can be damaged.

If the misfire monitor detects a type A cylinder misfire and the PCM does not shut off the injector or injectors, the MIL light begins flashing. When the misfire monitor detects a type A cylinder misfire, and the PCM shuts off an injector or injectors, the MIL light is illuminated continually.

Type B Misfires To detect a type B cylinder misfire, the misfire monitor checks cylinder misfiring over a 1,000-rpm period. If cylinder misfiring exceeds 2% to 3% during this period, the monitor considers the misfiring to be excessive. This amount of cylinder misfiring may not overheat the catalytic converter, but it may cause excessive emission levels. When a type B misfire is detected, a pending DTC is set in the PCM memory. If this fault is detected on a second consecutive drive cycle, the MIL light is illuminated.

Type C Misfires This type of misfire can cause a vehicle to fail an emissions test but will normally not damage the catalytic converter or raise the emissions levels beyond 1.5 times the FTP.

Fuel System Monitoring

The PCM continuously monitors conditions and adjusts the amount of fuel delivered by the injectors. Most of these changes are within a particular range around a base setting. The PCM checks the effectiveness of these changes through feedback from the oxygen sensor. The PCM then makes corrections as necessary. When the PCM needs to make a change outside that range, it will make a fuel trim adjustment. Fuel trim is the required new setting compared to the basic injector pulse width. Remember, the amount of fuel injected into the engine is controlled by the amount of time the injector is turned on.

The system allows for short-term and long-term fuel trim. **Short-term fuel trim** makes minor adjustments to the pulse width. These adjustments are temporary and not held in memory when the ignition is switched off. **Long-term fuel trim** is set by the effectiveness of the short-term trim. If the short-term trim meets the requirements of the engine during different operating conditions, the PCM may use those adjustments as the new base for injection timing. Those adjustments become long-term trim and are stored in memory. The PCM stores fuel trim data in blocks based on engine load and speed.

The fuel trim monitor is a continuous monitor and reports the amount of correction made with short term and long term. By monitoring the total amount of fuel trim (short term plus long term), the PCM is checking its ability to control the air-fuel ratio. If the PCM must move the short-term and long-term fuel trims to the lean or rich limits, a DTC is set on the next trip if the condition is still necessary.

Heated Oxygen Sensor Monitor

The system also monitors lean-to-rich and rich-to-lean time responses of the oxygen sensor. This test can pick up a lazy O_2 sensor that cannot switch fast enough to keep proper control of the air-fuel mixture. The time it takes the heated O_2 sensors to send a clean signal to the PCM is also monitored. This gives an indication of how the heater circuit is working.

All of the system's HO_2S are monitored once per drive cycle, but the heated oxygen sensor monitor provides separate tests for the upstream and downstream sensors. The heated oxygen sensor monitor checks the voltage signal frequency of the upstream HO_2S. At certain times, the heated oxygen sensor monitor varies the fuel delivery and checks for HO_2S response. A slow response in the sensor voltage signal frequency indicates a faulty sensor. The sensor signal is also monitored for excessive voltage.

The heated oxygen sensor monitor also checks the frequency of the rear HO_2S signals and checks these signals for excessively high voltage. If the monitor does not detect signal voltage frequency within a specific range, the rear HO_2S are considered faulty. The heated oxygen sensor monitor will command the PCM to vary the air-fuel ratio to check the response of the rear HO_2S.

EGR System Monitoring

The EGR monitors use different strategies to determine if the system is operating properly. Some monitor the temperature within the EGR passages. A high temperature indicates that the valve is open and exhaust gases are moving through the passages. Other systems look at the MAP signal, energize the EGR valve, and look for corresponding change in vacuum levels. As the valve is opening, there should be a drop in vacuum. Some systems open the EGR during coast down and monitor the change in STFT.

The EGR monitor looks at the operation of the EGR valve and the flow rates of the system. It also looks for shorts or opens in the circuit. If a fault is detected in any of the EGR monitor tests, a DTC is set in the PCM memory. If the fault occurs during two drive cycles, the MIL light is illuminated. The EGR monitor operates once per OBD-II trip.

There are many different EGR systems used today. The pressure feedback EGR, linear EGR, and delta pressure feedback EGR systems are the most common. If the system uses a delta pressure feedback EGR (DPFE), there is an orifice located under the EGR valve. Small exhaust pressure hoses are connected from each side of this orifice to the DPFE sensor. During the EGR monitor, the PCM first checks the DPFE signal. If this sensor signal is within the normal range, the monitor proceeds with the tests.

The PCM checks EGR flow by checking the DPFE signal against an expected DPFE value for the operating conditions at steady throttle within a specific rpm range.

With the EGR valve closed, the PCM checks for pressure difference at the two pressure hoses connected to the DPFE sensor. When the EGR valve is closed and there is no EGR flow, the pressure should be the same at both pipes. If the pressure is different at these two hoses, the EGR valve is stuck open.

The PCM commands the EGR valve to open and then checks the pressure at the two exhaust hoses connected to the DPFE sensor. With the EGR valve open and EGR flow through the orifice, there should be higher pressure at the upstream hose than at the downstream hose **(Figure 25–24)**.

Evaporative (EVAP) Emission System Monitor

In addition to the various components and failures that could affect the tailpipe emissions on a vehicle, OBD-II monitors the fuel evaporative systems. The system tests the ability of the fuel tank to hold pressure and the purge system's ability to vent the gas fumes from the charcoal canister when commanded to do so by the PCM.

Chrysler and others often use a leak detection pump (LDP) to detect leaks in the EVAP system **(Figure 25–25)**. When specific operating conditions are met, the PCM powers the pump to test the EVAP system. The pump pressurizes the system. As pressure builds, the cycling rate of the pump decreases. If the system does not leak, pressure will continue to build until the pump shuts off. If there is a leak, pressure will not build up and the pump will continue to run. The pump continues to run until the PCM determines it has run a complete test cycle and then sets a DTC. If there are no leaks, the PCM will run the purge monitor. Because of the pressure in the system from the pump, the cycle rate should be high. If no leaks are present and the purge cycle is high, the system passed the test.

In most systems, the EVAP monitor has two parts: the system integrity test that checks for leaks in the tank, gas cap, fuel lines and hoses, canister, and vapor lines, and the purge flow test that checks for blockage

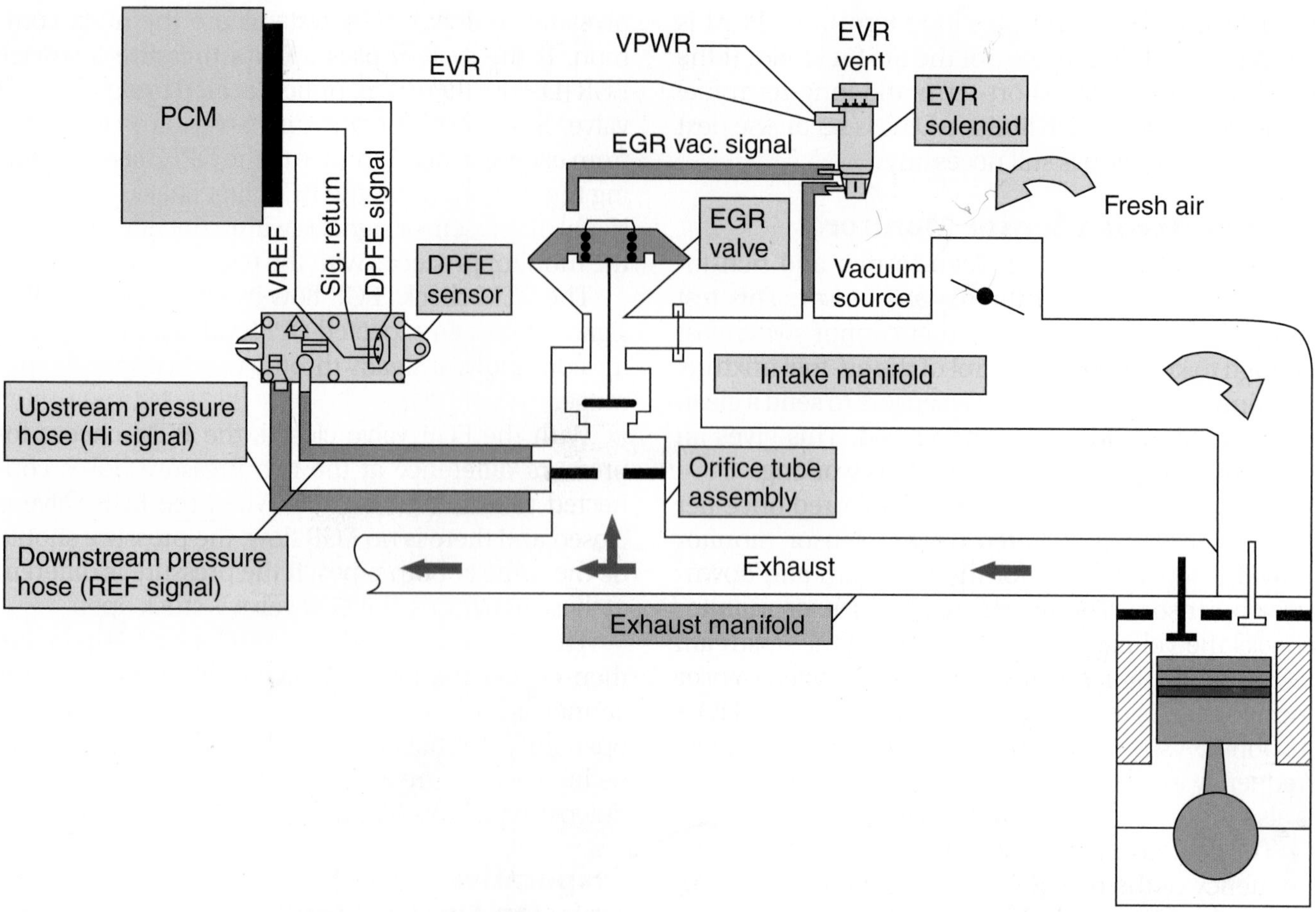

Figure 25-24 An EGR system with a delta pressure feedback sensor. *Courtesy of Ford Motor Company*

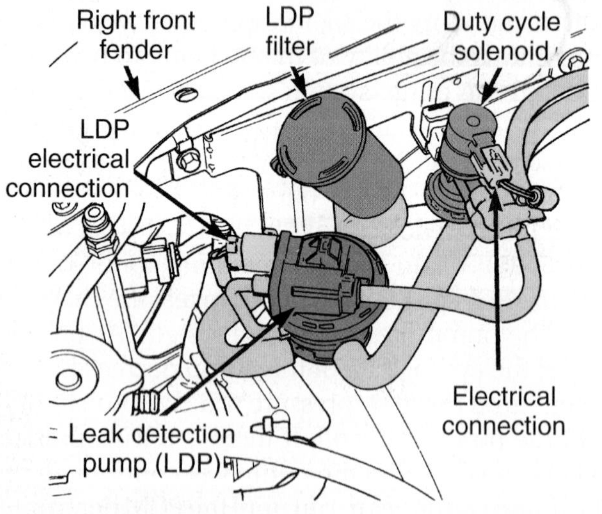

Figure 25-25 Typical location of the LDP and EVAP canister purge solenoid. *Courtesy of Chrysler LLC*

in the vapor lines and purge solenoid. During the integrity test, the PCM closes the canister vent solenoid and pulses the purge solenoid until the tank sensor reads a negative pressure of 20 inches of water. The PCM then closes the purge solenoid and measures the time it takes for the vacuum in the tank to decay. The EVAP purge volume test opens the canister vent solenoid and then varies the duty cycle of the purge solenoid until a measured change in STFToccurs. The EVAP monitor will be suspended whenever the fuel tank is less than one-quarter full or more than three-quarters full.

A few EVAP systems have a purge flow sensor (PFS) connected to the vacuum hose between the canister purge solenoid and the intake manifold **(Figure 25–26)**. The PCM monitors the PFS signal once per drive cycle to determine if there is vapor flow or no vapor flow through the solenoid to the intake manifold.

Enhanced EVAP Systems All new vehicles have an enhanced evaporative system monitor since 2003. This system detects leaks and restrictions in the EVAP system. This monitor first checks the integrity of the EVAP system. This test is only run when certain enable criteria are met. Once the integrity test is run, the monitor will conduct a vacuum pull-down test. During this test, the PCM commands the EVAP canister vent to close. It then opens the vapor management valve, which allows intake vacuum to draw a small amount of vacuum on the system. If the system has a leak, vacuum will not be held in the system and the

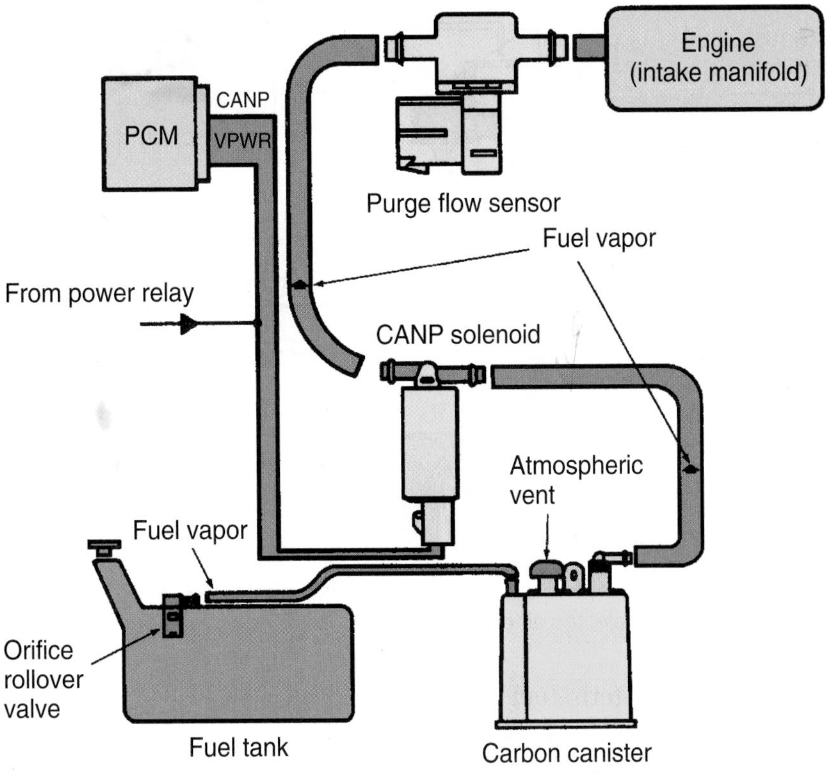

Figure 25-26 An EVAP system with a purge flow sensor. *Courtesy of Ford Motor Company*

monitor will set a DTC. If there are no leaks, the monitor will close the vapor management valve and monitor the system's ability to hold a vacuum. The final check is the vapor generation test. The PCM releases the vacuum in the system and then closes the EVAP system. The monitor looks for pressure changes in the system. An increase in pressure indicates excessive vapor generation.

A specially designed fuel tank filler cap is used on these systems. In these systems, an evaporative system leak or a missing fuel tank cap will cause the MIL is turn on. Also, if the fuel filler cap is not on tight enough, the OBD system can detect leaking vapor and the MIL will light up. If the filler cap is then tightened, the indicator will generally go out after a short period.

Secondary Air Injection (AIR) System Monitor

The AIR system operation can be verified by turning the system on to inject air upstream of the O_2 sensor while monitoring its signal. Many designs inject air into the exhaust manifold when the engine is in open loop and switch the air to the converter when it is in closed loop. If the air is diverted to the exhaust manifold during closed loop, the O_2 sensor thinks the mixture is lean and the signal should drop.

SHOP TALK

Many engines do not need to have a secondary air injection system; therefore this monitor does not run.

On some vehicles, the AIR system is monitored with passive and active tests. During the passive test, the voltage of the precatalyst HO_2S is monitored from startup to closed-loop operation. The AIR pump is normally on during this time. Once the HO_2S is warm enough to produce a voltage signal, the voltage should be low if the AIR pump is delivering air to the exhaust manifold. The secondary AIR monitor will indicate a pass if the HO_2S voltage is low at this time. The passive test also looks for a higher HO_2S voltage when the AIR flow to the exhaust manifold is turned off by the PCM. When the AIR system passes the passive test, no further testing is done. If the AIR system fails the passive test or if the test is inconclusive, the AIR monitor in the PCM proceeds with the active test.

During the active test, the PCM cycles the AIR flow to the exhaust manifold on and off during closed-loop operation and monitors the precatalyst HO_2S voltage and the short-term fuel trim value. When the AIR flow to the exhaust manifold is turned on, the sensor's

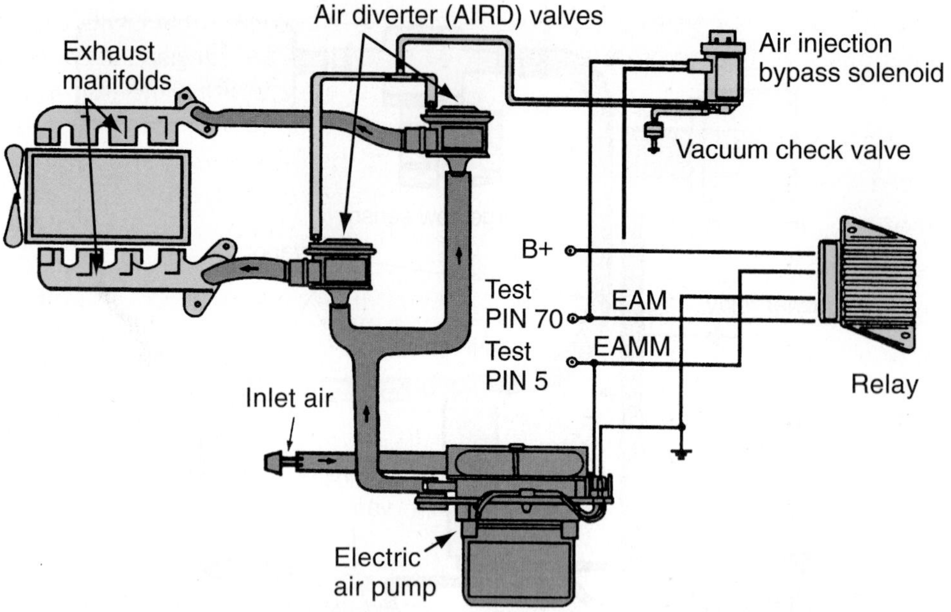

Figure 25-27 An electric air pump system. *Courtesy of Ford Motor Company*

voltage should decrease and the short-term fuel trim should indicate a richer condition. The secondary AIR system monitor illuminates the MIL and stores a DTC in the PCM's memory if the AIR system fails the active test on two consecutive trips.

Some vehicles have an electric air pump system. In this system, the air pump is controlled by a solid-state relay. The relay is operated by a signal from the PCM. An air-injection by-pass solenoid is also operated by the PCM. This solenoid supplies vacuum to dual air diverter valves **(Figure 25–27)**.

The PCM monitors the relay and air pump to determine if secondary air is present. This monitor functions once per drive cycle. When a malfunction occurs in the air pump system on two consecutive drive cycles, a DTC is stored and the MIL is turned on. If the malfunction corrects itself, the MIL is turned off after three consecutive drive cycles in which the fault is not present.

Thermostat Monitor

Present on all 2002 and newer vehicles, the thermostat monitor checks the engine and its cooling system for defects that would affect engine temperature and prevent the engine from reaching normal operating temperature. The goal of this monitor is to identify anything that may stop the PCM from going into closed loop. The monitor checks the time it takes for the cylinder head to reach a specific temperature. If the required temperature is not reached within the desired time, a malfunction is indicated and the monitor will set a pending code. The pending code becomes a real DTC if the malfunction is detected on two consecutive drive cycles. The MIL will be illuminated at that time.

PCV Monitor

The PCV valve removes unwanted vapors from the crankcase. Most of these vapors are HC. The system uses engine vacuum to draw the vapors out and into the intake. If there is a vacuum leak, the system will not work and HC can enter the atmosphere. Also, if there is a leak the engine may stall or not start. The leak causes the engine to run lean, especially at idle. The PCV monitor looks at HO_2S signals for consistent lean readings and the lack of switching from rich to lean. Abnormal signals will cause the monitor to illuminate the MIL after two consecutive driving cycles and will store one or more of the DTCs.

Comprehensive Component Monitor

The comprehensive component monitor (CCM) is a continuous monitor. It looks at the inputs and outputs that would affect emission levels. The system looks at any electronic input that could affect emissions. The strategy is to look for opens and shorts or input signal values that are out of the normal range. It also looks to see if the actuators have their intended effect on the system and to monitor other abnormalities.

The CCM uses several strategies to monitor inputs and outputs. One strategy for monitoring inputs involves checking inputs or electrical defects and out-of-range values by checking the input signals at the analog digital converter. This monitor also checks the circuits for the various outputs. These circuit checks look for continuity and out-of-range values. If an open circuit is detected, a DTC will be set.

The CCM also checks frequency signal inputs by performing rationality checks. During a rationality check, the monitor uses other sensor readings and calculations to determine if a sensor reading is proper for the present conditions. The following get rationality checks:

- Crankshaft position sensor (CKP)
- Output shaft speed sensor (OSS)
- Camshaft position sensor (CMP)
- Vehicle speed sensor (VSS)

The functional test of the CCM checks most of the outputs by monitoring the voltage of each output solenoid, relay, or actuator at the output driver in the PCM. If the output device is off, this voltage should be high. This voltage is pulled low when the output is turned on.

OBD-II SELF-DIAGNOSTICS

It is important to remember that although the diagnostic capabilities of OBD-II are great, the system does not check everything and is not capable of finding the cause of all driveability problems. Not all problems will activate the MIL or store a DTC. A DTC only indicates that a problem exists somewhere in the circuit of the sensor or output. A technician's job is to find the problem. Often DTCs are set that indicate an out-of-range reading by a sensor. This does not mean the sensor is bad. The abnormal signals can be caused by problems in the air, fuel, ignition, emission control, and other systems. Retrieving the information stored in the PCM is a starting point for diagnosis.

When no DTCs are present but there is a driveability problem, or when determining the true cause of the DTC, the basic engine should be checked before moving to the support systems. Many engine control systems offer a running compression test. This test runs much like the misfire monitor. The speed of the crankshaft between cylinder firings is measured and compared. If one or more cylinders accelerate slower on the power stroke, there may be a compression problem. This test should be followed up with the other standard engine mechanical tests.

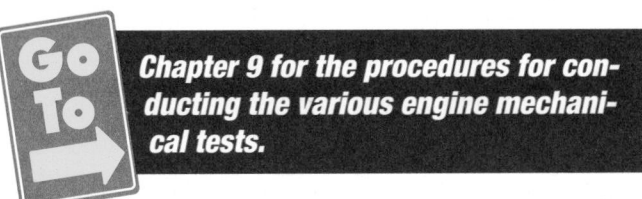

Go To ***Chapter 9 for the procedures for conducting the various engine mechanical tests.***

MIL

According to OBD-II regulations, the PCM must illuminate the MIL when it detects a problem that affects emissions. Also, when a malfunction is detected, a DTC must be set. Depending on the problem, the MIL will either light and stay on or it will blink. The action of the MIL depends on the monitor and the problem. For example, a misfire that would allow emissions to exceed regulations but not damage the catalyst will light the MIL. If the misfire will raise the catalytic converter's temperature enough to destroy it, the MIL will blink.

The MIL will be turned off if the misfire did not occur during three consecutive drive cycles. This requirement is true for most monitors. But an appropriate DTC will be recorded, as will freeze frame data. If the same fault is not detected during 40 warm-up cycles, the DTC and freeze frame data are erased from active memory. However, the DTC and freeze frame data will be stored as a history code with freeze frame data until it is cleared.

Diagnostic Trouble Codes

OBD-II codes are standardized, which means that most DTCs mean the same thing regardless of the vehicle. However, vehicle and scan tool manufacturers can have additional DTCs and add more data streams, report modes, and diagnostic tests. DTCs are designed to indicate the circuit and the system where a fault has been detected.

An OBD-II DTC is a five-character code with both letters and numbers **(Figure 25–28)**. This is called the alphanumeric system.

The first character of the code is a letter. This defines the system where the code was set. Currently there are four possible first character codes:

First digit - Letter indicates component group area P = Powertrain B = Body C = Chassis U = Network communications
Second digit 0 = SAE or OBD mandated 1 = Manufacturer specific
Third digit - Subgroup, powertrain subgroups: 0 = Total system 1 = Fuel and air metering 2 = Fuel and air metering 3 = Ignition system or misfire 4 = Auxiliary emission controls 5 = Idle speed control 6 = PCM and auxiliary inputs 7 = Transmission 8 = Transmission
Fourth and Fifth Digits - Defines the area or component and basic problem

Figure 25-28 Interpreting OBD-II DTCs.

"B" for body, "C" for chassis, "P" for powertrain, and "U" for undefined. The U-codes are designated for future use.

The second character is a number. This defines the code as being a mandated code or a special manufacturer code. A "0" code means that the fault is defined or mandated by OBD-II. A "1" code means the code is manufacturer specific. Codes of "2" or "3" are designated for future use.

The third through fifth characters are numbers. These describe the fault. The third character tells indicates where the fault occurred. The remaining two characters describe the exact condition that set the code. The numbers are organized so that the various codes related to a particular sensor or system are grouped together.

Not all DTCs will cause the MIL to light; this depends on the monitor and the problem. DTCs that will not affect emissions will never illuminate the MIL. Basically there are three types of DTCs. An active or current DTC represents a fault that was detected and occurred during two trips. When a two-trip fault is detected for the first time, a DTC is stored as a **pending code**. A pending DTC is a code representing a fault that has occurred but that has not occurred enough times to illuminate the MIL. There are some DTCs that will set in one trip; again this depends on the monitor.

Freeze Frame Data

One of the mandated capabilities of OBD-II is the **"freeze frame"** or snapshot feature. Although the regulations mandate just emission-related DTCs, manufacturers can choose to include this feature for other systems. With this feature, the PCM takes a snapshot of the activity of the various inputs and outputs at the time the PCM illuminated the MIL. The PCM uses this data for identification and comparison of similar operating conditions if the same problem occurs in the future. This feature is also valuable to technicians, especially when trying to identify the cause of an intermittent problem. The action of sensors and actuators when the code was set can be reviewed. This can be a great help in identifying the cause of a problem. The information held in freeze frame are actual values; they have not been altered by the adaptive strategy of the PCM (**Figure 25–29**).

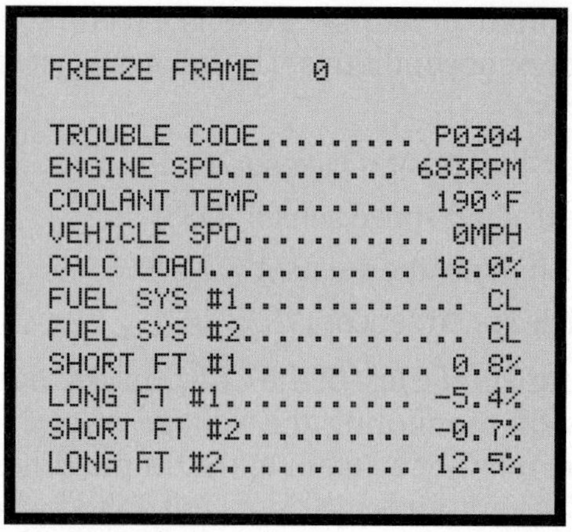

Figure 25-29 An example of the data captured in a freeze frame or snapshot for a DTC. *Courtesy of Toyota Motor Sales, U.S.A., Inc.*

Once a DTC and the related freeze frame data are stored in memory, they will stay there even if other emission-related DTCs are set. This data can only be removed with a scan tool. When a scan tool is used to erase a DTC, it automatically erases all associated freeze frame data. The data is stored by priority; information related to misfire and fuel control have priority over other DTCs. Fuel system misfires will overwrite any other type of data except for other fuel system misfire data.

Test Modes

All OBD-II systems have the same basic test modes and all are accessible with an OBD-II scan tool. Always refer to the manufacturer's information when using these test modes for diagnosis. Photo Sequence 23 shows the procedure for preparing a Snap-on Modis to read OBD-II data.

Mode 1 is the **parameter identification (PID)** mode. This mode allows access to current emission-related data values of inputs and outputs, calculated values, and system status information. Some PID values are manufacturer specific; others are common to all vehicles. This information is referred to as **serial data** (**Figure 25–30**).

Mode 2 is the freeze frame data access mode. This mode permits access to emission-related data values from specific generic PIDs. The number of freeze frames that can be stored is limited. The freeze frame information updates if the condition recurs.

Mode 3 permits scan tools to obtain stored DTCs. The information is transmitted from the PCM to the scan tool following a Mode 3 request. The DTC or its descriptive text, or both, can be displayed on the scan tool.

Mode 4 is the PCM reset mode. It allows the scan tool to clear all emission-related diagnostic information from its memory. When this mode is activated, all DTCs, freeze frame data, DTC history, monitoring test results, status of monitoring test results, and on-board test results are cleared and reset.

Mode 5 is the oxygen sensor monitoring test. This mode gives the actual oxygen sensor outputs

```
CURRENT DATA

ENGINE SPD......... 2260RPM
COOLANT TEMP......... 190°F
VEHICLE SPD.......... 60MPH
IGN ADVANCE.......... 38.0°
CALC LOAD............ 37.2%
MAF.............. 1.21b/min
THROTTLE POS......... 10.1%
INTAKE AIR............ 93°F
FUEL SYS #1........ OLDRIVE
FUEL SYS #2........ OLDRIVE
SHORT FT #1........... 0.0%
LONG FT #1........... -1.5%
SHORT FT #2........... 0.0%
LONG FT #2........... -1.5%
O2S B1 S1.......... 0.705V
O2FT B1 S1............ 0.0%
O2S B1 S2.......... 0.120V
O2FT B1 S2.......... UNUSED
O2S B2 S1.......... 0.660V
O2FT B2 S1............ 0.0%
MIL..................... ON
# CODES.................. 1
MISFIRE MON.......... AVAIL
FUEL SYS MON......... AVAIL
COMP MON............. AVAIL
CAT EVAL............. COMPL
HTD CAT EVAL........... N/A
EVAP EVAL........... INCMPL
2nd AIR EVAL........... N/A
A/C EVAL............... N/A
O2S EVAL............ INCMPL
O2S HTR EVAL........ INCMPL
EGR EVAL............. COMPL
OBD CERT............ OBD II
```

Figure 25-30 An example of what is available when a technician looks at the serial data on a scan tool. *Courtesy of Toyota Motor Sales, U.S.A., Inc.*

```
CONTINUOUS TESTS
ECU:  $10 (Engine)
Number of Tests: 1

P0401
EGR Flow Insufficient
Detected
```

```
PENDING CODES
ECU:   Engine
Number of DTDs: 1

P0401
EGR Flow Insufficient
Detected
```

Figure 25-31 In test Mode 7, the DTCs reported in Continuous Tests and Pending Codes are pending DTCs. If the conditions are detected again, the DTCs will be stored as active DTCs. *Courtesy of Toyota Motor Sales, U.S.A., Inc.*

during the test cycle. These are stored values, not current values that are retrieved in Mode 1. This information is used to determine the effectiveness of the catalytic converter.

Mode 6 is the output state mode (OTM) and can be used to identify potential problems in the non-continuous monitored systems.

Mode 7 reports the test results for the continuous monitoring systems (**Figure 25–31**).

Mode 8 is the request for control of an on-board system test or component mode. It allows the technician to control the PCM, through the scan tool, to test a system. In some cases, the scan tool will only set the conditions for conducting a test and not actually conduct the test. An example of this is the EVAP leak test, which is done with other test equipment.

Mode 9 is the request for vehicle information mode. This mode reports the vehicle's identification number, calibration identification, and calibration verification. This information can be used to see if the most recent calibrations have been programmed into the PCM.

BASIC DIAGNOSIS OF ELECTRONIC ENGINE CONTROL SYSTEMS

Diagnosing a computer-controlled system is much more than accessing DTCs. You need to know what to test, when to test it, and how to test it. Because the capabilities of the engine control computer have evolved from simple to complex, it is important to know the capabilities of the system you are working with before attempting to diagnose a problem. Refer to the service manual for this information. After you understand the system and its capabilities, begin your diagnosis using your knowledge and logic.

The importance of logical troubleshooting cannot be overemphasized. The ability to diagnose a problem (to find its cause and its solution) is what separates an automotive technician from a parts changer.

PHOTO SEQUENCE 23 Preparing a Snap-on Modis to Read OBD II Data

Reproduced under license from Snap-on Incorporated. All of the marks are marks of their owners.

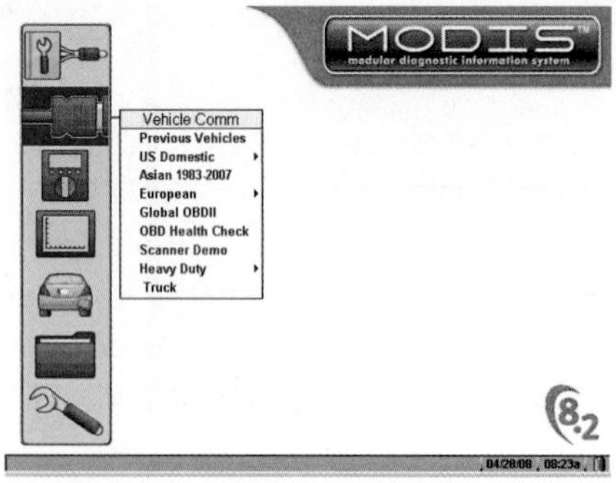

P23-1 *Once the Modis is connected to the vehicle, select the category you wish to check. Global OBD-II is for all nonmanufacturer specific data.*

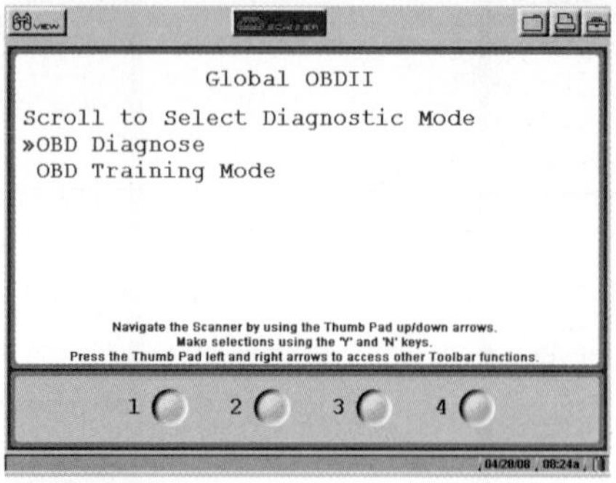

P23-2 *After the tool enters into the generic OBD-II mode, select Diagnose to check the system.*

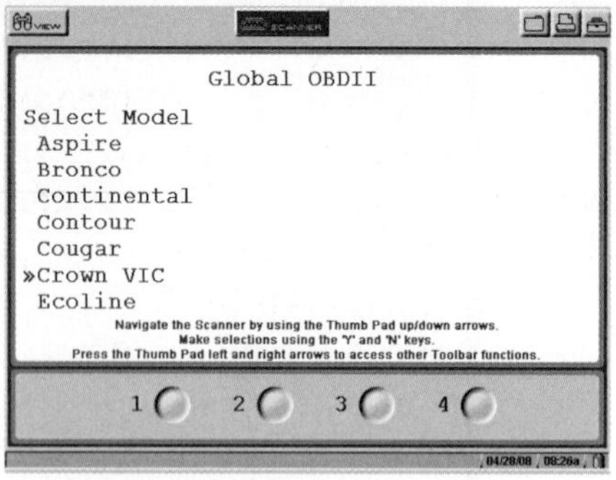

P23-3 *Select the make of the vehicle and then the model.*

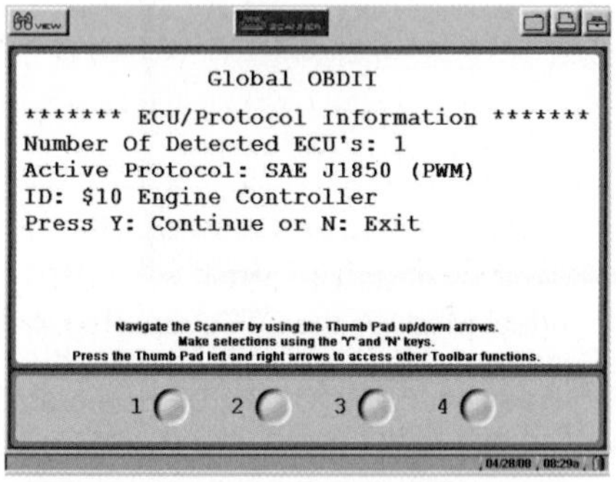

P23-4 *The tool will display basic information about the control system.*

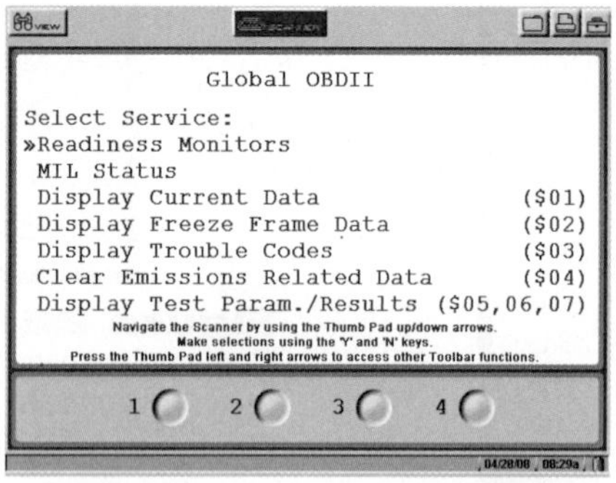

P23-5 *The display will ask what you want to do.*

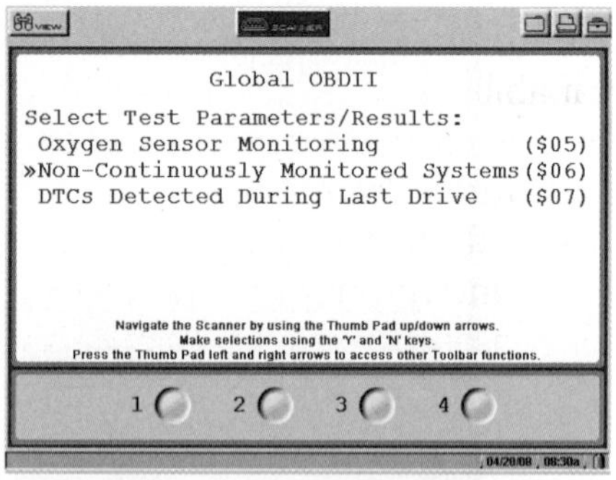

P23-6 *Mode 6 and other test parameters can be selected.*

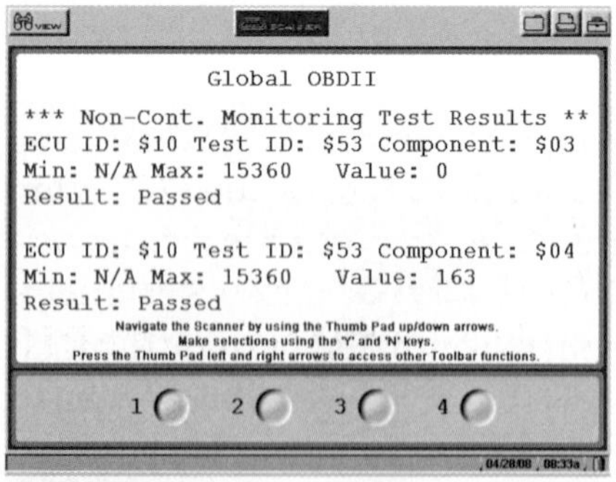

P23-7 *Test results for the selected tests will appear on the tool.*

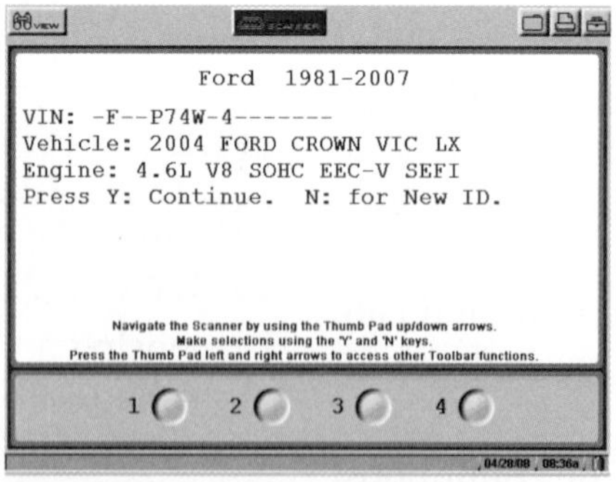

P23-8 *Mode 6 data can be retrieved by entering specific vehicle information, such as the VIN.*

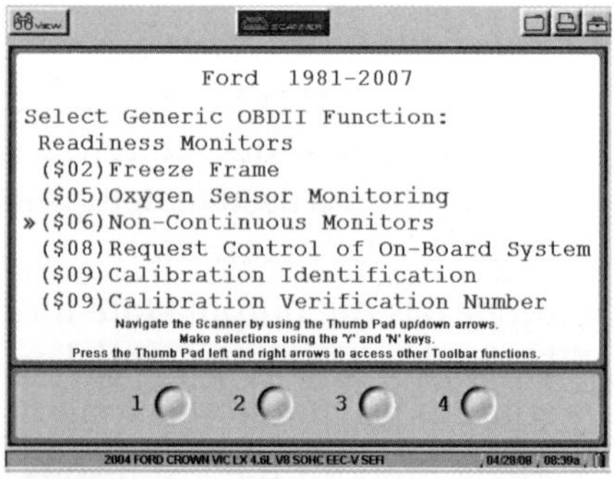

P23-9 *Data from the noncontinuous monitors (Mode 6) will be displayed.*

Logical Diagnosis

When faced with an abnormal engine condition, the best automotive technicians compare clues (such as meter readings, oscilloscope readings, visible problems) with their knowledge of proper conditions and discover a logical reason for the way the engine is performing. Logical diagnosis means following a simple basic procedure. Start with the most likely cause and work to the most unlikely. In other words, check out the easiest, most obvious solutions first before proceeding to the less likely, and more difficult, solutions. Do not guess at the problem or jump to a conclusion before considering all of the factors.

The logical approach has a special application to troubleshooting electronic engine controls. Always check all traditional nonelectronic engine control possibilities before attempting to diagnose the electronic engine control system. For example, low battery voltage might result in faulty sensor readings.

Repair Information

All late-model engine controls have self-diagnosis capabilities. A malfunction in any sensor, output device, or in the computer itself is stored as a DTC. DTCs are retrieved and the indicated problem areas checked further. Correct diagnosis depends on correctly interpreting all data collected and performing all subsequent tests properly. The following service information will help:

- Instructions for retrieving DTCs and obtaining freeze frames
- Instructions for diagnosing a no communication problem between the system and the scan tool
- Current technical bulletins for the vehicle
- A fail-safe chart that shows what strategy is taken when certain DTCs are set
- The operating manual for the scan tool
- A DTC chart with the codes and possible problem areas
- A parts locator for the system
- An electrical wiring diagram for the system
- Identification of the various terminals of the PCM
- A DTC troubleshooting guide
- Component testing sequences

DIAGNOSING OBD-II SYSTEMS

At least one drive cycle is required for the monitors to run. All OBD-II scan tools include a readiness function that shows all of the monitoring sequences and the status of each: complete or incomplete. Incomplete can mean the monitor did not complete, judgment is withheld pending further testing, the monitor did not operate, or the monitor operated and recorded a failure. The "Readiness Test" and "Monitor Status" screens on most scan tools contain identical information (**Figure 25–32**).

Readiness monitor tools are available (**Figure 25–33**). These plug into the DLC and alert the technician when a drive cycle is completed. These can save much time while trying to gather data before and after a repair. They are especially handy after a repair has been made. The vehicle is taken for a test drive and when the readiness monitor flashes and/or beeps, the vehicle can be brought back into the shop and checked for DTCs.

Troubleshooting OBD-II Systems

The following steps provide a general outline for troubleshooting OBD-II systems. There are slight variations in different years and with different models. Always refer to the manufacturer's information before beginning your diagnosis. Troubleshooting OBD-II systems involves a series of steps as listed in **Figure 25–34**. Here is a brief discussion of each step:

1. *Interview the Customer.* Gather as much information as possible from the customer. Ask the customer to describe the driving conditions present when the problem appears. This should include weather, traffic, and speed.
2. *Check the MIL.* The MIL should turn on when the ignition is turned on and the engine is not running. When the engine is started, the MIL should go off. If either of these does not occur, troubleshoot the lamp system before continuing.
3. *Connect the Scan Tool.* Make sure the tool is OBD-II compliant.
4. *Check DTC(s) and Freeze Frame Data.* When using a scan tool, an asterisk (*) next to the DTC often indicates there is stored freeze frame data associated with that DTC. During diagnostics, it is helpful to know the conditions present when the DTC was set, such as: Was the vehicle running or stopped, what was the engine's temperature, and was the air-fuel ratio rich or lean (**Figure 25–35**)? Print or record all DTC and related information. If there was a no or poor communication DTC, solve that problem before continuing.

Go To **Chapter 22 for CAN communication problem diagnostic procedures.**

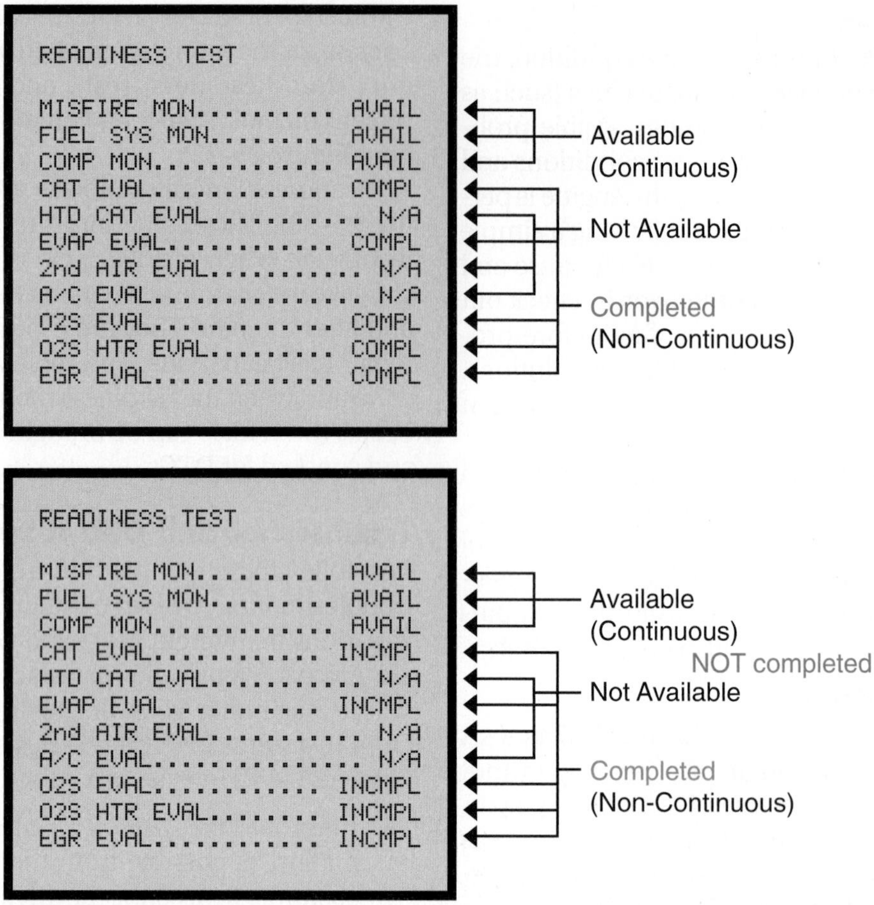

Figure 25-32 These scan tool displays show which monitors have been completed, are available, are not available, and have been completed. *Courtesy of Toyota Motor Sales, U.S.A., Inc.*

Figure 25-33 A "Ready Scan" can be used to determine when a true drive cycle has been completed prior to diagnosis. The tool has a green LED plus sound that alerts a technician when a monitor has been completed. *Courtesy of SPX Service Solutions*

5. *Check Service History and Service Publications.* There may be a TSB or other service alert that may have the necessary repair information. Check these and follow the procedures outlined in them before continuing. Service history can give clues about the cause of the problem because the problem may be related to a recent repair.
6. *Clear DTC and Freeze Frame Data.* Using the scan tool, clear all DTCs and freeze frame data **(Figure 25–36)**. Make sure you do this only after you have recorded all retrieved data.
7. *Visual Inspection.* Take a quick look at the very basics. Check all wires to make sure they are firmly connected and not damaged. Try not to wiggle the wires while doing this; a wiggle may correct an intermittent problem and it may be hard to find later. Conduct a visual inspection on the battery and fuel level. Correct any problems if necessary. If the engine does not start, move down to step 9.
8. *Check DTCs.* Repeat step 4. If there are no DTCs, check the status of the monitors' readiness and

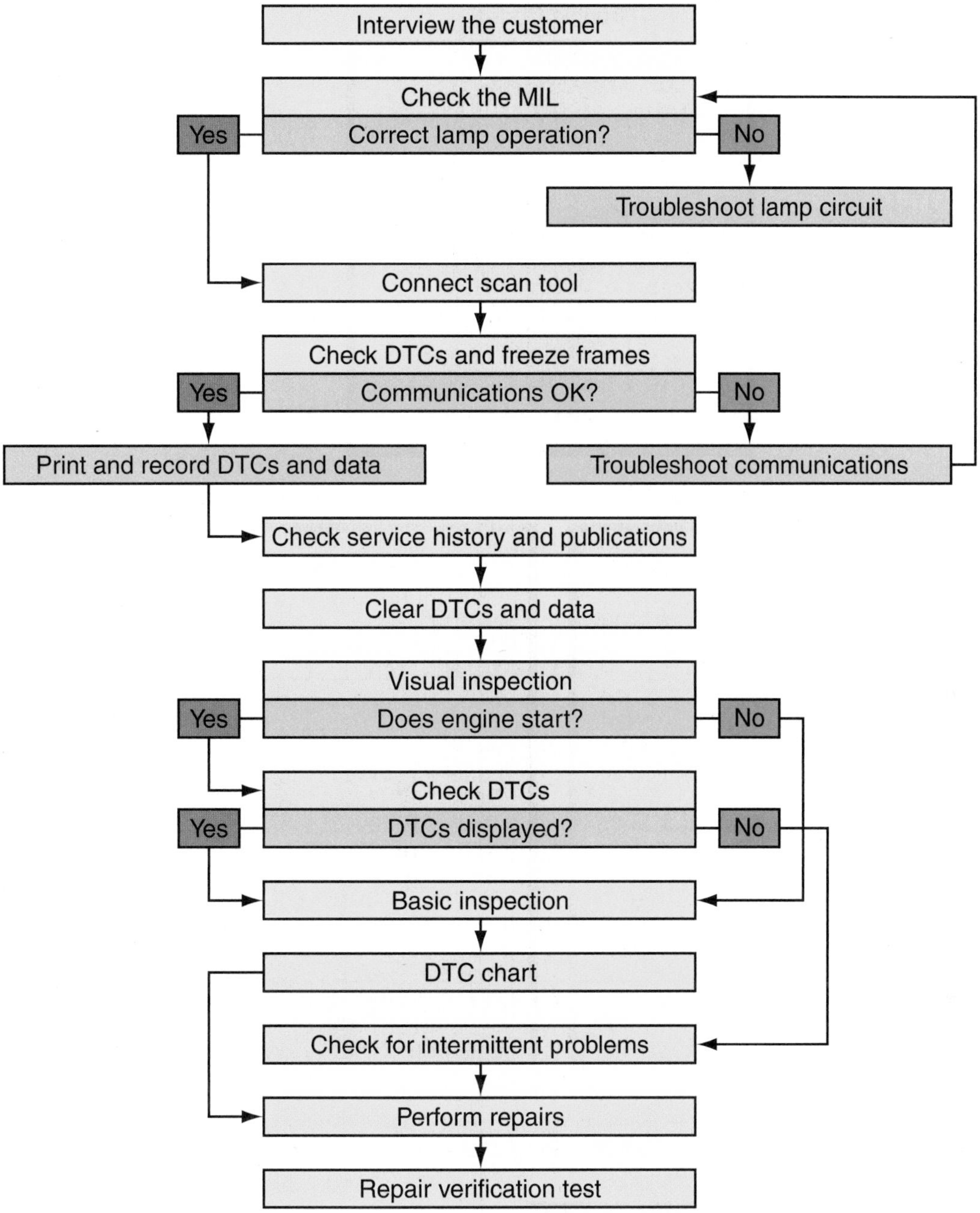

Figure 25-34 The steps that should be followed when troubleshooting OBD-II systems.

the pending codes on the scan tool. Do what is necessary to complete the necessary drive cycles before continuing. For many DTCs, the PCM will enter into the fail-safe mode. This means the PCM has substituted a value to allow the engine to run. Refer to the service information to determine if any DTCs indicate the fail-safe mode; if so, follow the appropriate diagnostic steps. If there is a DTC, proceed to step 10. If no DTCs are present, go to step 11.

9. *Basic Inspection.* When the DTC is not confirmed in the previous check, diagnosis of the engine's support system should be done. Use the problem symptoms chart in the service manual to diagnose when a no code is displayed but the problem is still occurring.
10. *DTC Chart.* Use the DTC chart to determine what was detected, the probable problem areas, and how to diagnose that DTC (**Figure 25–37**).
11. *Check for Intermittent Problems.* If the cause of the problem has not yet been determined, proceed to check for an intermittent problem.
12. *Perform Repairs.* Once the cause of the concern has been identified, perform all required services.
13. *Repair Verification Test.* After making a repair, check your work by repeating steps 2, 3, and 4.

Trouble Codes

```
ECU: $10 (Engine)
Number of DTCs: 3

* P0304
  Cylinder 4 Misfire
  Detected

  P0100
  Mass or Volume Air
  Flow Circuit Malfunction

* P0110
  Intake Air Temperature
  Circuit Malfunction
```

Priority Freeze Frame

```
TROUBLE CODE......... P0304
CALC LOAD.............. 18%
ENGINE SPD.......... 683RPM
COOLANT TEMP....... 190.4°F
INTAKE TEMP........ 125.6°F
CTP SW.................. ON
VEHICLE SPD........... 0MPH
SHORT FT #1........... 0.7%
LONG FT #1........... -5.5%
SHORT FT #2.......... -0.9%
LONG FT #2........... 12.4%
FUEL SYS #1............. CL
FUEL SYS #2............. CL
FC IDL................. OFF
STARTER SIG............ OFF
A/C SIG................ OFF
PNP SW [NSW]............ ON
ELECT LOAD SIG......... OFF
STOP LIGHT SW.......... OFF
ENG RUN TIME............ 80
```

Non-Priority Freeze Frame

```
TROUBLE CODE......... P0110
CALC LOAD............... 0%
ENGINE SPD.......... 662RPM
COOLANT TEMP....... 192.2°F
INTAKE TEMP........ -40.0°F
CTP SW.................. ON
VEHICLE SPD........... 0MPH
SHORT FT #1........... 1.5%
LONG FT #1........... -5.5%
SHORT FT #2........... 1.5%
LONG FT #2........... 12.4%
FUEL SYS #1............. CL
FUEL SYS #2............. CL
FC IDL................. OFF
STARTER SIG............ OFF
A/C SIG................. ON
PNP SW [NSW]............ ON
ELECT LOAD SIG......... OFF
STOP LIGHT SW.......... OFF
ENG RUN TIME............. 0
```

Figure 25-35 The scan tool will display stored DTCs and the freeze frame data for some of them. *Courtesy of Toyota Motor Sales, U.S.A., Inc.*

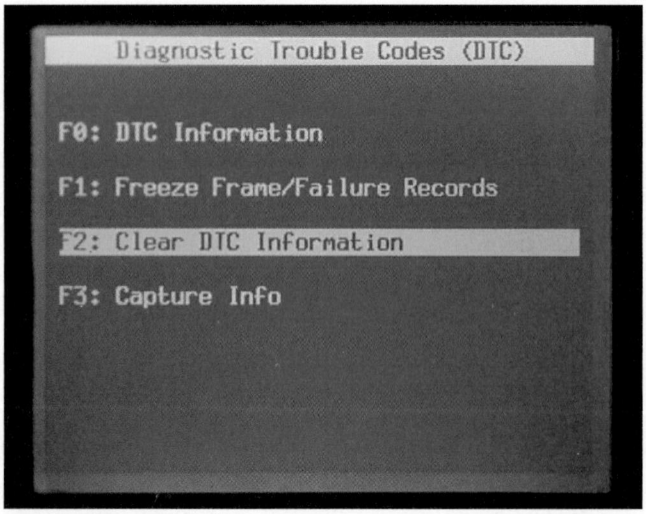

Figure 25-36 Activate the initial retrieval of DTCs and freeze frame data, clear the DTC information, and run the vehicle to see if a DTC is reset.

Intermittent Faults

An intermittent fault is a fault that is not always present. It may not activate the MIL or cause a DTC to be set. Therefore, intermittent problems can be difficult to diagnose. By studying the system and the relationship of each component to another, you should be able to create a list of possible causes for the intermittent problem. To help identify the cause, follow these steps:

1. Observe the history DTCs, DTC modes, and freeze frame data.
2. Call technical assistance for possible solutions. Combine your knowledge of the system with the service information that is available.
3. Evaluate the symptoms and conditions described by the customer.

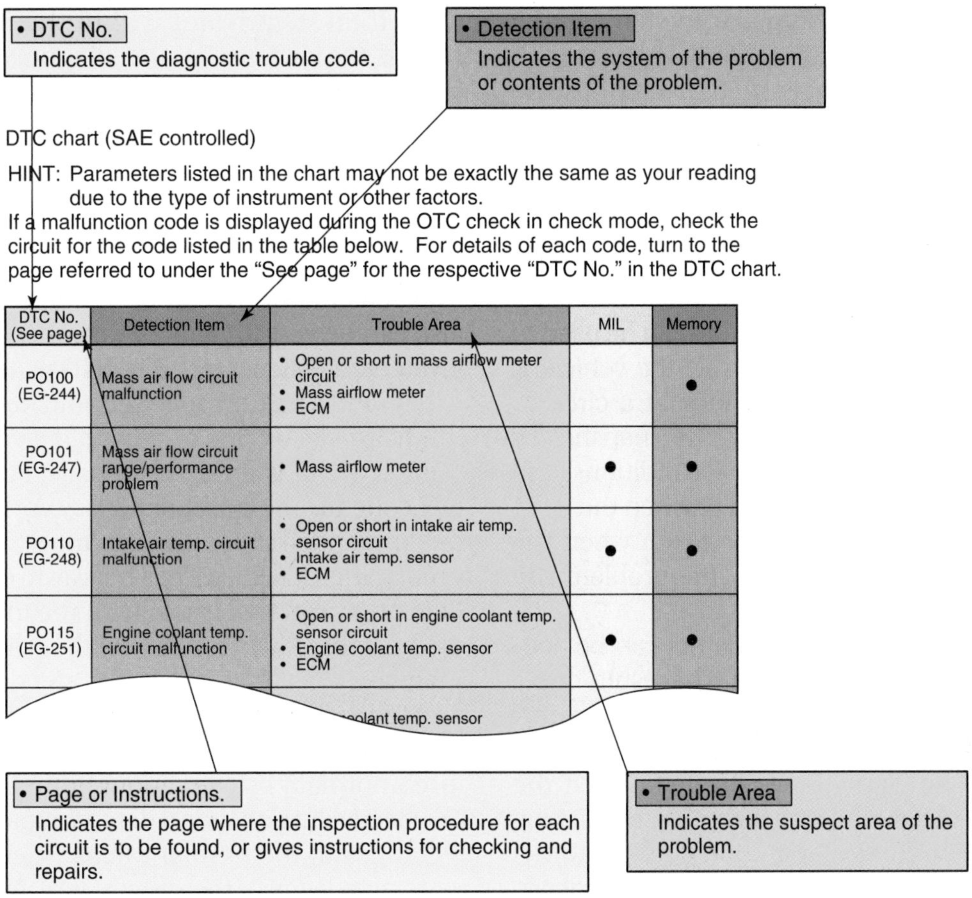
• DTC No.
Indicates the diagnostic trouble code.
• Detection Item
Indicates the system of the problem or contents of the problem.
DTC chart (SAE controlled)
HINT: Parameters listed in the chart may not be exactly the same as your reading due to the type of instrument or other factors.
If a malfunction code is displayed during the OTC check in check mode, check the circuit for the code listed in the table below. For details of each code, turn to the page referred to under the "See page" for the respective "DTC No." in the DTC chart.
DTC No. (See page)
Detection Item
Trouble Area
MIL
Memory
PO100 (EG-244)
Mass air flow circuit malfunction
• Open or short in mass airflow meter circuit
• Mass airflow meter
• ECM
PO101 (EG-247)
Mass air flow circuit range/performance problem
• Mass airflow meter
PO110 (EG-248)
Intake air temp. circuit malfunction
• Open or short in intake air temp. sensor circuit
• Intake air temp. sensor
• ECM
PO115 (EG-251)
Engine coolant temp. circuit malfunction
• Open or short in engine coolant temp. sensor circuit
• Engine coolant temp. sensor
• ECM
...olant temp. sensor
• Page or Instructions.
Indicates the page where the inspection procedure for each circuit is to be found, or gives instructions for checking and repairs.
• Trouble Area
Indicates the suspect area of the problem.

Figure 25-37 An explanation of a typical DTC chart. *Courtesy of Toyota Motor Sales, U.S.A., Inc.*

4. Use a check sheet to identify the circuit or electrical system component that may have the problem.
5. Follow the suggestions for intermittent diagnosis found in service material.
6. Visually inspect the suspected circuit or system.
7. Use the data capturing capabilities of the scan tool.
8. Test the circuit's wiring for shorts, opens, and high resistance. This should be done with a DMM and in the typical manner, unless instructed differently in the service manual.

Most intermittent problems are caused by faulty electrical connections or wiring. Refer to a wiring diagram for each of the suspected circuits or components. This will help identify all of the connections and components in that circuit. The entire electrical system of the suspected circuit should be carefully and thoroughly inspected. Check for burnt or damaged wire insulation, damaged terminals at the connectors, corrosion at the connectors, loose connectors, wire terminals loose in the connector, and disconnected or loose ground wire or straps.

To locate the source of the problem, a voltmeter can be connected to the suspected circuit and the wiring harness wiggled **(Figure 25–38)**. As a guideline for what voltage should be expected in

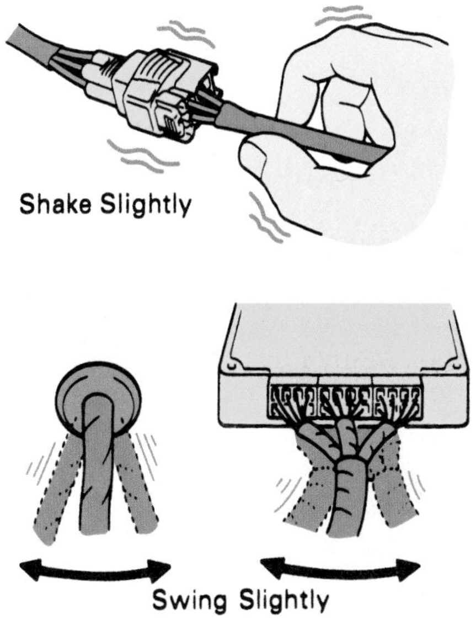
Shake Slightly
Swing Slightly

Figure 25-38 The wiggle test can be used to locate intermittent problems. *Courtesy of Toyota Motor Sales, U.S.A., Inc.*

the circuit, refer to the reference value table in the service manual. If the voltage reading changes with the wiggles, the problem is in that circuit. The vehicle can also be taken for a test drive with the voltmeter connected. If the voltmeter readings become abnormal with changing operating conditions, the circuit being observed probably has the problem.

The vehicle can also be taken for a test drive with the scan tool connected. The scan tool can be used to monitor the activity of a circuit while the vehicle is being driven. This allows you to look at a circuit's response to changing conditions. The snapshot or freeze frame feature stores engine conditions and operating parameters at command or when the PCM set a DTC. If the snapshot can be taken when the intermittent problem is occurring, the problem will be easier to diagnose.

With an OBD-II scan tool, actuators can be activated and their functionality tested. The results of the change in operation can be monitored. Also, the outputs can be monitored as they respond to input changes. When an actuator is activated, watch the response on the scan tool. Also listen for the clicking of the relay that controls that output. If no clicking is heard, measure the voltage at the relay's control circuit; there should be a change of more than 4 volts when it is activated.

To monitor how the PCM and an output respond to a change in sensor signals, use the scan tool. Select the mode that relates to the suspected circuit and view, then record, the scan data for that circuit. Compare the reading to specifications. Then create a condition that would cause the related inputs to change. Observe the data to see if the change was appropriate.

Serial Data

PIDs are codes used to request data from a PCM. A scan tool is used for the request and receipt of the serial data. To do this, the desired PID **(Figure 25–39)** is entered into the scan tool. The device connected to the CAN bus that is responsible for this PID reports the value for that PID back to the bus and is read on the scan tool. Many PIDs are standard for all OBD-II systems; however, not all vehicles will support all PIDs and there are manufacturer-defined PIDs that are not part of the OBD-II standard.

Manufacturers list and define the various applicable PIDs in their service information. This information should be compared to the observed data. If an item is not within the normal values, record the difference and diagnose that particular item.

Using Mode 6

Mode 6 allows access to the results of various monitor diagnostic tests. The test values are stored at the time each monitor is completed. The information available in Mode 6 can be extremely helpful, but it also can be very difficult to decipher. Mode 6 data is given with $ signs, test IDs (TIDs), and content IDs (CIDs) and a mixture of letters and numbers. Most of this has no face value because it is given in "hexadecimals," which is foreign language. For example, the hex code for the number 10 is $0A. This is why it is recommended that only scan tools capable of interpreting this data be used. The way to observe Mode 6 data on a scan tool varies with the make and model of the tool. Always refer to the tool's instruction manual.

Mode 6 data can help identify the cause of problems when a DTC has not been retrieved. Manufacturers normally list the various normal Mode 6 values. To effectively use Mode 6 data, compare the captured reading with the normal values. Then use logic and your knowledge of the system, part, and electricity to determine what is abnormal and what could be causing that. Some manufacturers recommend using Mode 6 to diagnose specific systems. Using this data is also very handy in diagnosing any system of the control circuit.

Repairing the System

After identifying the cause of the problem, repairs should be made. When servicing or repairing OBD-II circuits, the following guidelines are important:

- Do not connect aftermarket accessories into an OBD-II circuit.
- Do not move or alter grounds from their original locations.
- Always replace a relay with an exact replacement. Damaged relays should be thrown away, not repaired.
- Make sure all connector locks are in good condition and are in place.
- After repairing connectors or connector terminals, make sure the terminals are properly retained and the connector is sealed.

Freeze Frame	Acronym	Description	Measurement Units
X	AAT	Ambient Air Temperature	Degrees
X	AIR	Secondary Air Status	On/Off
X	APP_D	Accelerator Pedal Position D	%
X	APP_E	Accelerator Pedal Position E	%
X	APP_F	Accelerator Pedal Position F	%
X	CATEMP11	Catalyst Temperature Bank 1, Sensor 1	Degrees
X	CATEMP12	Catalyst Temperature Bank 1, Sensor 2	Degrees
X	CATEMP21	Catalyst Temperature Bank 2, Sensor 1	Degrees
X	CATEMP22	Catalyst Temperature Bank 2, Sensor 2	Degrees
	IAT	Intake Air Temperature	Degrees
X	LOAD	Calculated Engine Load	%
X	LOAD_ABS	Absolute Load Value	%
X	LONGFT1	Current Bank 1 Fuel Trim Adjustment (kamref1) from Stoichiometry, Which Is Considered Long Term	%
X	LONGGFT2	Current Bank 2 Fuel Trim Adjustment (kamref2) from Stoichiometry, Which Is Considered Long Term	%
X	MAF	Mass Air Flow Rate	gm/s-lb/min
	MIL_DIST	Distance Traveled with MIL on	Kilometer
X	O2S11	Bank 1 Upstream Oxygen Sensor (11)	Volts
X	O2S12	Bank 1 Downstream Oxygen Sensor (12)	Volts
X	O2S13	Bank 1 Downstream Oxygen Sensor (13)	Volts
X	O2S21	Bank 2 Upstream Oxygen Sensor (21)	Volts
X	O2S22	Bank 2 Downstream Oxygen Sensor (22)	Volts
X	O2S23	Bank 2 Downstream Oxygen Sensor (23)	Volts
X	SHRTFT1	Current Bank Fuel Trim Adjustment (lambse1) from Stoichiometry, Which Is Considered Short Term	%
X	SHRTFT2	Current Bank 2 Fuel Trim Adjustment (lambse1) from Stoichiometry, Which Is Considered Short Term	%
X	SPARKADV	Spark Advance Requested	Degrees
X	SPARK_ACT	Spark Advance Actual	Degrees
	TAC_PCT	Commanded Throttle Actuator	%
X	TP	Throttle Position	%
X	VSS	Vehicle Speed Sensor	km/h-mph

Figure 25-39 A sample list of generic PIDs.

P24-1 *Be sure the engine is at normal operating temperature and the ignition switch is off.*

P24-2 *Install the proper module for the vehicle and system into the scan tool.*

P24-3 *Connect the scan tool power leads to the battery or cigar lighter (depending on the design of the scan tool).*

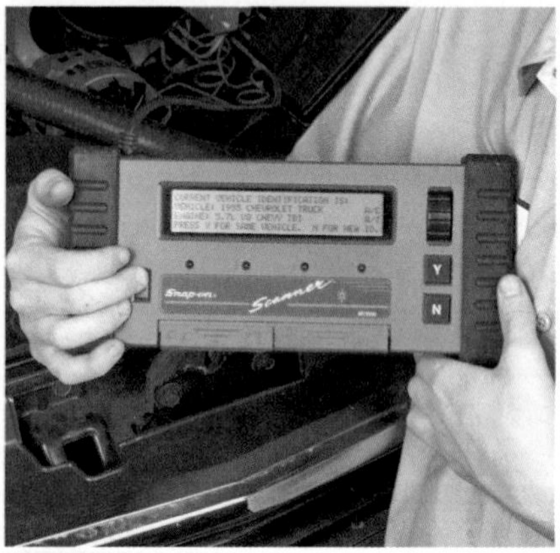

P24-4 *Enter the vehicle's model year and VIN code into the scan tool.*

P24-5 *Select the proper scan tool adapter for the vehicle's DLC.*

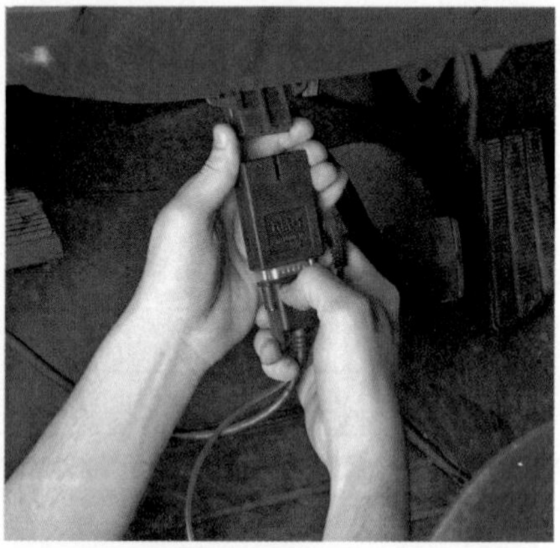

P24-6 *Connect the scan tool to the DLC.*

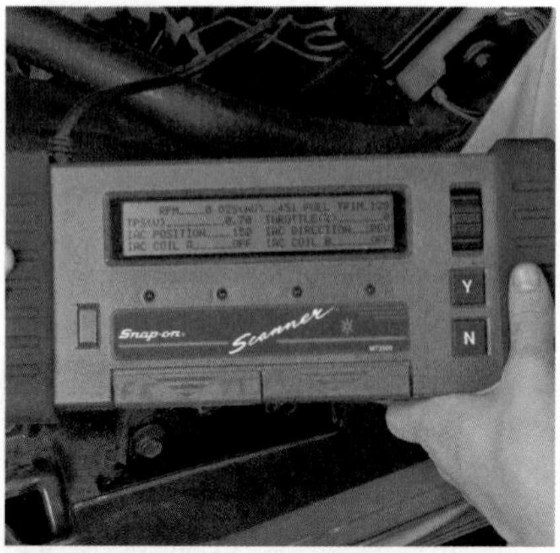

P24-7 *Retrieve the DTCs with the scan tool. Interpret the codes by using the service manual.*

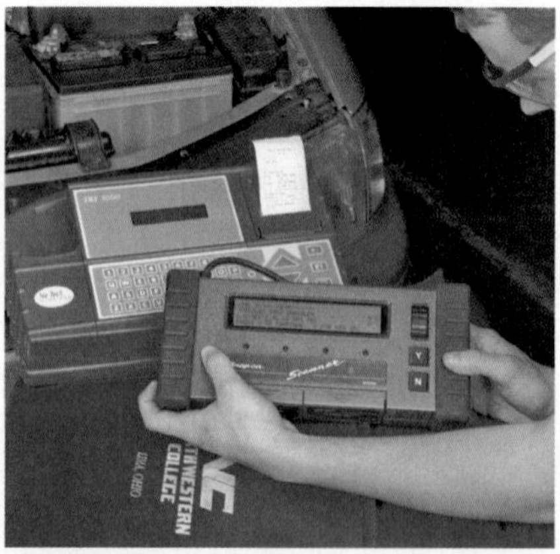

P24-8 *Start the engine and obtain the input sensor and output actuator data on the scan tool. If a printer for the tool is available, print out the data report.*

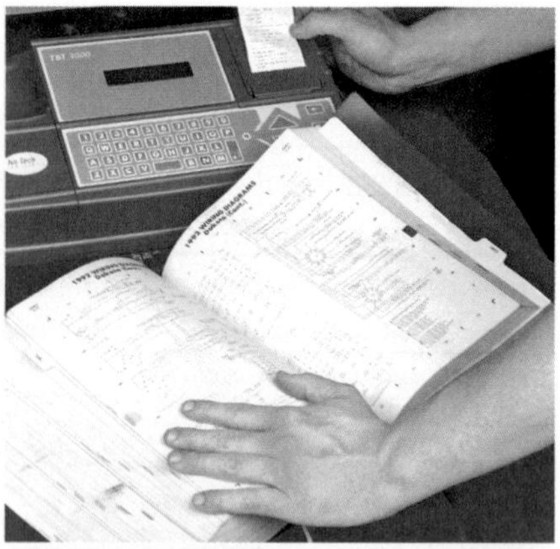

P24-9 *Compare the input sensor and actuator data to the specifications given in the service manual. Mark all data that are not within specifications.*

- When installing a fastener for an electrical ground, be sure to tighten it to the specified torque.

After repairs, the system should be rechecked to verify that the repair took care of the problem. This may involve road testing the vehicle in order to verify that the complaint has been resolved.

DIAGNOSING OBD-I SYSTEMS

OBD-I systems have limited self-diagnostic capabilities. By entering into a self-test mode, the system is able to evaluate its condition. If problems are found, they may be identified as either hard faults (on-demand) or intermittent failures. Each type of fault or failure is assigned a numerical trouble code that is stored in memory.

A hard fault is a problem found in the system at the time of the self-test. An intermittent fault, on the other hand, indicates that a malfunction occurred (for example, a poor connection causing an intermittent open or short) but was not present during the self-test. Nonvolatile RAM allows intermittent faults to be stored for up to a specific number of ignition key on/off cycles. If the trouble does not reoccur during that period, it is erased from the memory.

There are various ways to access the trouble codes stored in the computer. Most manufacturers have specific equipment designed to monitor and test the system's components. Aftermarket companies also manufacture scan tools that have the capability to read and record the system's input and output signals.

Visual Inspection

Begin diagnostics with a visual check of the engine and its systems; include the following:

1. Inspect the air filter and related hardware around the filter.
2. Inspect the entire PCV system.
3. Check to make sure the EVAP canister is not saturated or flooded.
4. Check the charging system for loose, damaged, and corroded wires or connections.
5. Check the condition of the battery and its terminals and cables.
6. Make sure all vacuum hoses are connected and are not pinched or cut.
7. Check all sensors and actuators for signs of physical damage.

Unlocking Trouble Codes

Although the parts in any computerized system are amazingly reliable, they do occasionally fail. Diagnostic charts in service manuals help you through troubleshooting procedures. Start at the top and follow the sequence down.

The MIL should not be illuminated after the engine has started. If it is, the computer has found a problem. OBD-I diagnostic procedures and DTCs differ with vehicle make and year. Each time the key is turned to the on position, the system does a self-check. The self-check makes sure that all of the bulbs, fuses, and electronic modules are working. If the self-test finds a problem, it might store a code for later servicing. It may also instruct the computer to turn on the MIL to show that service is needed.

Retrieving Trouble Codes

Prior to OBD-II, each car manufacturer required a different method for retrieving DTCs. In fact, there were different procedures for different models and engines made by the same manufacturer. The diagnostic connectors looked different and were located in different places. Also, the diagnostic codes represented different things. Although all of the computer systems did basically the same thing, diagnostic methods were as different as day and night. Always check the service information for the vehicle being serviced before attempting to retrieve DTCs.

To retrieve DTCs, some manufacturers require that certain pins at the DLC be connected, particular connectors disconnected, or a specific key-on and key-off sequence be followed. In these cases, the DTCs appear as flash codes displayed with the MIL or on a voltmeter. In most cases a scan tool can be used to view the DTCs.

Using a Scan Tool

Scan tools are available to diagnose nearly all engine control systems. When using a scan tool on these early systems, make sure it is compatible with the system. Often, a manufacturer specific scan tool is required.

Photo Sequence 24 (see page 758) shows a typical procedure for using a scan tool on an OBD-I system. The use of a scan tool varies with the make of the tool, but most require an initial entry of vehicle information, including the VIN.

After the tool has been programmed, the technician selects the desired test sequence on the scan tool. These selections also vary with the type of scan tool and the vehicle being tested.

CASE STUDY

A customer brought in his OBD-II equipped car into the dealership complaining of poor fuel economy. The customer stated that the gas mileage has been declining since the car was new. This is not normal; gas mileage normally improves slightly as the engine is broken in. The customer had no other complaints. Because this is a very difficult problem to verify, the technician began the diagnosis process with a visual inspection and found nothing out of the ordinary. The car's MIL was not lit.

He then connected a scan tool to the DLC and reviewed the data. Comparing the input data being displayed to the normal range of values listed in the service manual, he discovered that the upstream O_2 sensor was biased rich. That meant the PCM was seeing a lean condition and adding fuel to correct this problem. To verify this, the technician watched the fuel trim. Sure enough, the long-term fuel trim had moved to add more fuel. Normally this is caused by a vacuum leak or restricted fuel injectors. The latter probable cause seemed unlikely because the O_2 sensor showed that added fuel was being delivered. Therefore, the technician began to look for a possible cause of a vacuum leak.

The process continued for quite some time as he checked all of the vacuum hoses and components. Nothing appeared to be leaking. As he leaned over to check something in the back of the engine, he heard a slight "pffft." The noise had somewhat of a rhythm to it and he focused his attention to it. It did not sound like a vacuum leak. As he increased the engine's speed, the noise pulses became closer together and soon became a constant noise. The noise appeared to be coming from the lower part of the engine.

Using a stethoscope, he was able to identify the source of the noise. It appeared that the gasket joining the exhaust manifold to the exhaust pipe was not seated properly. To verify this, he raised the car on a hoist and took a look. He found that the retaining bolts were loose. He took a quick look at the gasket and found it to be in reasonable shape, then tightened the bolts to specifications.

Not sure that the noise or the exhaust leak was related to the problem but suspecting that it could be, he connected a lab scope to the oxygen sensor and watched its activity. The sensor's signal no longer showed a bias. It appears that the leak in the exhaust was pulling air into the exhaust between each pulse. This was adding oxygen to the exhaust stream, causing the computer to think the mixture was lean.

KEY TERMS

Air induction system
Base ignition timing
Catalyst monitor sensor (CMS)
Central port injection (CPI)
Closed loop
Diagnostic data link connector (DLC)
Distributor ignition (DI)
Drive cycle
Electronic fuel injection (EFI)
Electronic ignition (EI)
Enable criteria
Engine control module (ECM)
Freeze frame
Gasoline direct-injection (GDI)
Ignition timing
Long-term fuel trim
Malfunction indicator lamp (MIL)
Open loop
Parameter identification (PID)
Pending code
Port fuel injection (PFI)
Powertrain control module (PCM)
Serial data
Short-term fuel trim
Throttle body injection (TBI)
Trip

SUMMARY

- To have complete combustion, there must be the correct amount of fuel mixed, in a sealed container, with the correct amount of air and this must be shocked by the correct amount of heat at the right time.
- The ignition system is responsible for delivering the spark that causes combustion.
- The fuel system must deliver the fuel from the fuel tank to the fuel injectors, which spray fuel into the cylinders.

- The air induction system delivers the air to the cylinders.
- Emission control devices are added to engines because an engine does not experience complete combustion during all operating conditions.
- Engine control systems operate in a loop—input, process, and control.
- An engine control module will operate in open or closed loop. In closed loop, the computer processes all inputs.
- Most engine control systems have self-diagnostic capabilities. By entering this mode, the computer is able to evaluate the entire control system, including itself.
- According to OBD-II standards, all vehicles have a universal DLC with a standard location, a standard list of DTCs, a standard communication protocol, common use of scan tools on all vehicle makes and models, common diagnostic test modes, the ability to record and store a snapshot of the operating conditions that existed when a fault occurred, and a standard glossary of terms, acronyms, and definitions that must be used for all components in electronic control systems.
- OBD-III adds a telecommunication system to OBD-II.
- An OBD-II system has many monitors to check system operation, and the MIL light is illuminated if vehicle emissions exceed 1.5 times the allowable standard for that model year.
- Monitors included in OBD-II are: catalyst efficiency, engine misfire, fuel system, heated exhaust gas oxygen sensor, EGR, EVAP, secondary air injection, thermostat, and comprehensive component monitors.
- OBD-II vehicles use a minimum of two oxygen sensors. One of these is used for feedback to the PCM for fuel control and the other gives an indication of the efficiency of the converter.
- An OBD-II drive cycle includes whatever specific operating conditions are necessary either to initiate and complete a specific monitoring sequence or to verify a symptom or a repair.
- OBD-II's short-term fuel trim and long-term fuel trim strategies monitor the oxygen sensor signals and use this information to make adjustments to the fuel control calculations. The adaptive fuel control strategy allows for changes in the amount of fuel delivered to the cylinders according to operating conditions.
- An OBD-II system monitors the entire emissions system, switches on a MIL if something goes wrong, and stores a fault code in the PCM when it detects a problem.
- OBD-II regulations require that the PCM monitor and perform some continuous tests on the engine control system and components. Some OBD-II tests are completed at random, at specific intervals, or in response to a detected fault.
- OBD-II systems note the deterioration of certain components before they fail, which allows owners to bring in their vehicles at their convenience and before it is too late.
- The MIL informs the driver that a fault that affects the vehicle's emission levels has occurred. After making the repair, technicians may need to take the vehicle for three trips to ensure that the MIL does not illuminate again.
- Most intermittent problems are caused by faulty electrical connections or wiring.
- OBD-I systems have limited and unique self-diagnostic routines.

REVIEW QUESTIONS

1. Describe the difference between an open- and a closed-loop operation.
2. Explain the use and importance of system strategy and look-up tables in the computerized control system.
3. Describe an OBD-II warm-up cycle.
4. Explain the trip and drive cycle in an OBD-II system.
5. Describe how engine misfire is detected in an OBD-II system.
6. Describe the purpose of having two oxygen sensors in an exhaust system.
7. Briefly describe the monitors in an OBD-II system.
8. Type B engine misfires are excessive if the misfiring exceeds ______ to ______% in a ______ rpm period.
9. The ______ monitor system checks the action of the canister purge system.
10. The ______ monitor system has a(n) ______ and ______ test to check the efficiency of the air injection system.

11. The fuel monitor checks _____ _____ fuel trim and _____ _____ fuel trim.
12. Which of the following is the least likely cause for a misfire monitor failure on an OBD-II system?
    a. defective aspirator valve
    b. EVAP system faults
    c. EGR system faults
    d. restricted exhaust
13. Which of the following statements is *not* true?
    a. DTCs that will not affect emissions will never illuminate the MIL.
    b. An active or current DTC represents a fault that was detected and occurred during two trips.
    c. When a two-trip fault is detected for the second time, a DTC is stored as a pending code.
    d. There are some DTCs that will set in one trip.
14. A computer is capable of doing all of the following except _____.
    a. receive input data
    b. process input data according to a program and monitor output action
    c. control the vehicle's operating conditions
    d. store data and information
15. Which of the following memory circuits is used to store trouble codes and other temporary information?
    a. read-only memory
    b. programmed read-only memory
    c. random access memory
    d. all of the above

ASE-STYLE REVIEW QUESTIONS

1. Technician A says that the oxygen sensor provides the major input during the open-loop mode. Technician B says that the coolant temperature sensor controls open-loop mode operation. Who is correct?
    a. Technician A
    b. Technician B
    c. Both A and B
    d. Neither A nor B
2. While discussing OBD-II systems: Technician A says that the PCM illuminates the MIL if a defect causes emission levels to exceed 2.5 times the emission standards for that model year vehicle. Technician B says that if a misfire condition threatens engine or catalyst damage, the PCM flashes the MIL. Who is correct?
    a. Technician A
    b. Technician B
    c. Both A and B
    d. Neither A nor B
3. While discussing the catalyst efficiency monitor: Technician A says that if the catalytic converter is not reducing emissions properly, the voltage frequency increases on the downstream HO_2S. Technician B says that if a fault occurs in the catalyst monitor system on three drive cycles, the MIL will be illuminated. Who is correct?
    a. Technician A
    b. Technician B
    c. Both A and B
    d. Neither A nor B
4. While discussing monitoring systems: Technician A says that the fuel system monitor checks the short-term and long-term fuel trims. Technician B says that the heated oxygen sensor monitoring system checks lean-to-rich and rich-to-lean response times. Who is correct?
    a. Technician A
    b. Technician B
    c. Both A and B
    d. Neither A nor B
5. While discussing the comprehensive monitoring system: Technician A says that it tests various input circuits. Technician B says that it tests various output circuits. Who is correct?
    a. Technician A
    b. Technician B
    c. Both A and B
    d. Neither A nor B
6. While discussing the MIL on OBD-II systems: Technician A says that the MIL will flash if the PCM detects a fault that would damage the catalytic converter. Technician B says that whenever the PCM has detected a fault, it will turn on the MIL. Who is correct?
    a. Technician A
    b. Technician B
    c. Both A and B
    d. Neither A nor B
7. While discussing diagnostic procedures: Technician A says that after preliminary system checks have been made, the DTCs should be cleared from the memory of the PCM. Technician B says that serial data can be helpful when there are no

DTCs but there is a fault in the system. Who is correct?

a. Technician A
b. Technician B
c. Both A and B
d. Neither A nor B

8. Technician A says that the enable criteria are the criteria that must be met before the PCM completes a monitor test. Technician B says that a drive cycle includes operating the vehicle under specific conditions so that a monitor test can be completed. Who is correct?

a. Technician A
b. Technician B
c. Both A and B
d. Neither A nor B

9. While discussing PCM monitor tests: Technician A says that the monitor tests are prioritized. Technician B says that the monitor tests are run when the scan tool activates them. Who is correct?

a. Technician A
b. Technician B
c. Both A and B
d. Neither A nor B

10. While discussing the misfire monitor: Technician A says that while detecting type A misfires, the monitor checks cylinder misfiring over a 500-rpm period. Technician B says that while detecting type B misfires, the monitor checks cylinder misfires over a 1,000-rpm period. Who is correct?

a. Technician A
b. Technician B
c. Both A and B
d. Neither A nor B

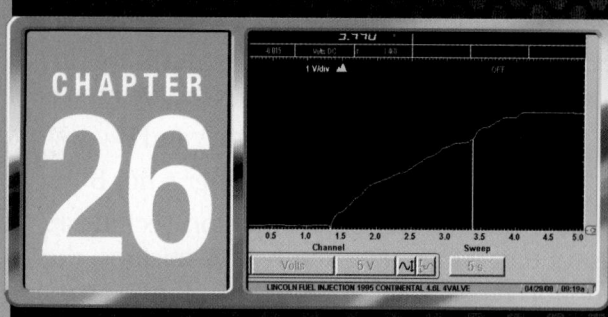

CHAPTER 26 DETAILED DIAGNOSIS AND SENSORS

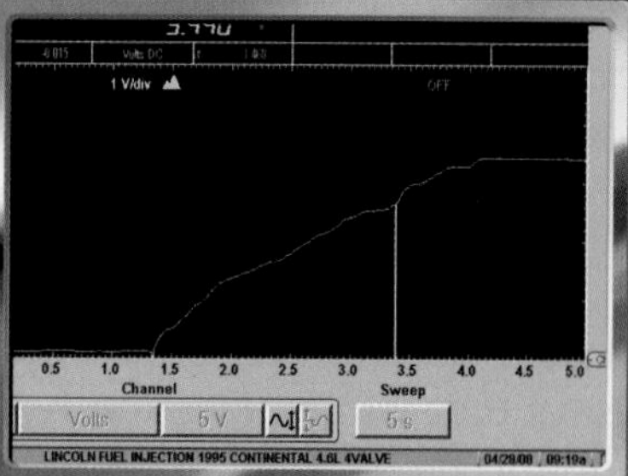

OBJECTIVES

■ **Perform a scan tester diagnosis on various vehicles.** ■ **Conduct preliminary checks on an OBD-II system.** ■ **Use a symptom chart to set up a strategic approach to troubleshooting a problem.** ■ **Monitor the activity of OBD-II system components.** ■ **Diagnose computer voltage supply and ground wires.** ■ **Test and diagnose switch-type input sensors.** ■ **Test and diagnose variable resistance-type input sensors.** ■ **Test and diagnose generating-type input sensors.** ■ **Test and diagnose output devices (actuators).** ■ **Perform active tests of actuators using a scan tool.**

On-board diagnostic systems will lead a technician to the area of a driveability or emissions problem. Many different input sensors are involved in the overall driveability of a vehicle. This includes diesel- and gasoline-powered engines. Because of the use of computer networks, the inputs from the various sensors play an important part in the operation of all engine performance systems. Because the signals from these sensors are shared by many control modules, one sensor cannot be designated as affecting only one system. Often, a DTC will be set that reflects a problem affecting more than one system. These problems are typically caused by a faulty sensor or sensor circuit. This chapter looks at the most common sensors and how to test each.

USING SCAN TOOL DATA

The engine control module (ECM) or PCM constantly monitors information from various switches, sensors, and other control modules. It controls the operation of systems that affect vehicle performance and emission levels and monitors emission-related systems for deterioration. OBD-II monitors set a DTC when the performance of a system can cause elevated emissions. The ECM also alerts the driver of emission-related concerns by illuminating the MIL. At the same time, a DTC is set defining the area that caused the MIL to be lit. The DTC is retrieved with a scan tool **(Figure 26–1)**.

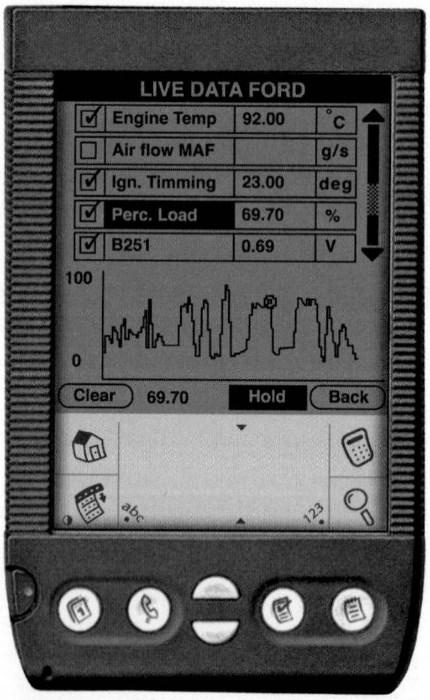

Figure 26–1 A PDA equipped to work as a scan tool and lab scope. *Courtesy of SPX Service Solutions*

Visual Inspection

Diagnosis should begin with a thorough underhood inspection. Often the cause of a driveability problem can be discovered by doing the following:

- Check all vacuum hoses. Make sure they are connected and are not pinched, cut, or cracked.
- Check the condition and tension of the generator drive belt.

Figure 26-2 Check all of the wires in the engine compartment. Look for damaged or burned wires.

- Check all of the wires in the engine compartment. Look for loose and corroded connections and damaged **(Figure 26–2)** or burned wires.
- Inspect the PCM, wiring harness, sensors, and outputs.
- Check the level and condition of the engine's coolant.
- Check the level and condition of the transmission fluid.
- Check the air filter. Also check the air intake system for restrictions and leaks.

CAUTION!

Before running the engine to diagnose a problem, make sure the parking brake is applied and the gear selector is placed firmly in the Park position on automatic transmission vehicles or in the neutral position on manual transmission vehicles. Also, block the drive wheels.

Service Information

After retrieving the DTCs, find their descriptions in the service information. Normally the descriptions are followed by additional information to help diagnosis. As can be seen in **Figure 26–3**, there is more than one possible cause of the problem. One is the sensor itself and the other two concern the sensor's circuit. Detailed testing will identify the exact cause.

Notice that the description also leads to pinpoint tests. These are designed to guide the technician through a step-by-step procedure. To be effective, each step should be performed, in the order given, until the problem has been identified.

Make sure to check all available service information related to the DTCs. There may be a TSB related to the code and following those procedures may fix the problem. Also, make sure the ECM/PCM is programmed with the most current software.

Monitor Failures

All OBD-II scan tools have a readiness function showing all of the monitoring sequences and the status of each: complete or incomplete. If vehicle travel time, operating conditions, or other parameters were insufficient for a monitoring sequence to complete a test, the scanner will indicate which monitoring sequence is not yet complete.

The specific set of driving conditions that will set the requirements for all OBD-II monitoring sequences follows:

1. Start the engine. Do not turn off the engine for the remainder of the drive cycle.
2. Drive the vehicle for at least 4 minutes.
3. Continue to drive until the engine's temperature reaches 180° or more.
4. Idle the engine for 45 seconds.

P0117 – Engine Coolant Temperature (ECT) Sensor 1 Circuit Low			
Description:	Indicates the sensor signal is less than the self-test minimum. The ECT sensor minimum is 0.2 volt or 121°C (250°F).		
Possible Causes:	• Grounded circuit in the harness • Damaged sensor • Incorrect harness connection		
Diagnostic Aids:	A concern is present if an ECT PID reading less than 0.2 volt with the key ON engine OFF or during any engine operating mode.		
Application	**Key On Engine Off**	**Key On Engine Running**	**Continuous Memory**
All		Go to Pinpoint Test 23L	

Figure 26-3 A description of a DTC, as given in typical service information, with additional information to help diagnosis.

5. Open the throttle to about 25% to accelerate from a standstill to 45 mph in about 10 seconds.
6. Drive between 30 and 40 mph with a steady throttle for at least 1 minute.
7. Drive for at least 4 minutes at speeds between 20 and 45 mph. If during this phase you must slow down below 20 mph, repeat this phase of the drive cycle. During this time, do not fully open the throttle.
8. Decelerate and idle for at least 10 seconds.
9. Accelerate to 55 mph with about half throttle.
10. Cruise at a constant speed of 40 to 65 mph for at least 80 seconds with a steady throttle.
11. Decelerate and allow the engine to idle.
12. Using the scan tool, check for diagnostic system readiness and retrieve any DTCs.

When most monitor tests are run and a system or component fails a test, a pending code is set. When the fault is detected a second time, a DTC is set and the MIL is lit **(Figure 26–4)**. It is possible that a DTC for a monitored circuit may not be entered into memory even though a malfunction has occurred. This may happen when the monitoring criteria have not been met.

The PCM's output signals are constantly being monitored by the diagnostic management software.

Monitor	Failure can be caused by:
Catalyst Monitor	Fuel Contaminants Leaking Exhaust Engine Mechanical Problems Defective Upstream or Downstream Oxygen Sensor Circuits Defective PCM
Fuel System Monitor	A Defective Fuel Pump Abnormal Signal from the Upstream HO2S Engine Temperature Sensor Faults Malfunctioning Catalytic Converter MAP or MAF Related Faults Cooling System Faults EGR System Faults Fuel Injection System Faults Ignition System Faults Vacuum Leaks Worn Engine Parts
EGR Monitor	Faulty EGR Valve Faulty EGR Passages or Tubes Loose or Damaged EGR Solenoid Wiring and/or Connectors Damaged DPFE or EGR VP Sensor Disconnected or Loose Electrical Connectors to the DPFE or EGR VP Sensors Disconnected, Damaged, or Misrouted EGR Vacuum Hoses
EVAP Monitor	Disconnected, Damaged, or Loose Purge Solenoid Connectors and/or Wiring Leaking Hoses, Tubes, or Connectors in the EVAP System Vacuum and/or Vent Hoses to the Solenoid and Charcoal Canister Are Misrouted Plugged Hoses from the Purge Solenoid to the Charcoal Canister Loose or Damaged Connectors at the Purge Solenoid Fuel Tank Cap Not Tightened Properly or It Is Missing
Misfire Monitor	Fuel Level Too Low During Drive Cycle Dirty or Defective Fuel Injectors Contaminated Fuel Defective Fuel Pump Restricted Fuel Filter

Figure 26-4 Possible causes of OBD-II monitor failures.

	EGR System Faults EVAP System Faults Restricted Exhaust System Faulty Secondary Ignition Circuit Damaged, Loose, or Resistant PCM Power and/or Ground Circuits.
Oxygen Sensor Monitor	Malfunctioning Upstream and/or Downstream Oxygen Sensor Malfunctioning Heater for the Upstream or Downstream Oxygen Sensor A Faulty PCM Defective Wiring to and/or from the Sensors
AIR System Monitor	Faulty Secondary AIR Solenoid and/or Relay Damaged, Loose, or Disconnected Wiring to the Secondary Air Solenoid and/or Relay Circuit Defective Aspirator Valve Disconnected or Damaged AIR Hoses And/or Tubes A Defective Electric or Mechanical Air Pump Air Pump Drive Belt Missing Faulty AIR Check Valve

Figure 26-4 *(Continued)*

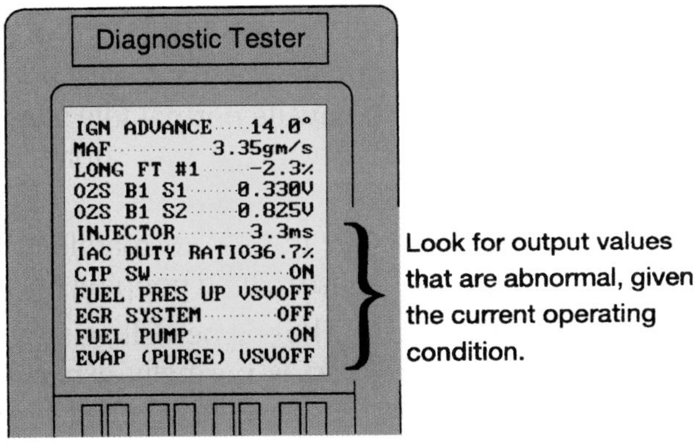

Figure 26-5 Use serial data to identify abnormal conditions. *Courtesy of Toyota Motor Sales, U.S.A., Inc.*

These signals are observed to detect opens and shorts in the circuits. Some of the output devices are also monitored for their effectiveness and functionality **(Figure 26–5)**. The PCM is programmed to expect certain things to happen when it commands a device to do something. Perhaps the simplest example of this is idle speed. The PCM has control over the engine's idle speed. This is accomplished in many different ways; one of these is an idle speed (IAC) motor. The PCM sends voltage signals to the motor, which changes the idle speed. When the PCM commands the motor to extend, the engine's speed should decrease. The PCM watches engine speed and continues to adjust the motor until the desired speed is reached. If the PCM commands the motor to extend and the engine's speed does not decrease, the PCM knows that the motor is not doing what it is supposed to be doing. As a result, the PCM sets a DTC and illuminates the MIL.

Freeze Frame Data

Freeze frame data associated with a DTC should also be retrieved. This data contains values from specific generic PIDs that provide information about the operating conditions present when the code was set. It can also be used to identify components that should be tested.

Misfire freeze frame (MFF) data contains some unique PIDs. MFF data is not part of freeze frame data that is stored with a DTC. It is used only for identifying the cause of a misfire. Generic freeze frame data is also captured when a misfire occurs. However, that data will not represent what happened at the time of the misfire; rather, it will show what happened after the misfire. MFF data is captured at the time of the highest rate of misfire and not when the DTC is set.

SHOP TALK

Many electronic control systems have adaptive strategies that allow the engine to run when one or more inputs fail. The ECM/PCM relies on a predetermined value when it senses an out-of-limit value from a sensor. When looking at data, the input signal may not change with changes in conditions, and this can be misleading. Check for an error message when retrieving DTCs.

Mode 6 Data

The data available in Mode 6 can also be used to identify the cause of a concern. Mode 6 allows access to the results of the various monitor diagnostic tests. These values are stored when a specific monitor has completed a test. Mode 6 data is very valuable because it may show near out-of-range values. These values will not trigger a DTC but can affect the operation of a system or component.

SYMPTOM-BASED DIAGNOSIS

At times, no DTCs are set by the computer but a driveability problem exists. This is when a technician must look at the various engine systems to discover the cause of the concern. What system or part to test is dictated by a description of the problem or its symptoms. Before diagnosing a problem based on its symptoms, make sure that:

- The ECM and MIL are operating correctly.
- There are no stored DTCs.
- All data observed on the scan tool is within normal ranges.
- There is communication between the scan tool and the control system.
- There are no TSBs available for the current symptom.
- All of the grounds for the ECM/PCM are sound.
- The air filter is not restricted.
- All vehicle modifications are identified.
- The vehicle's tires are properly inflated and are the correct size.

Common Symptoms

Repair manuals typically have a section dedicated to symptom-based diagnosis. Although a customer may describe a problem in nontechnical terms, you should summarize the concern to match one or more the various symptoms listed by the manufacturer. Here are some common driveability symptoms and a brief description of each:

- *Hard Start/Long Crank Hard Start:* Engine cranks OK, but it does not start unless it is cranked for a long time. Once running, it may run normally or immediately stall.
- *No Crank:* The starting system does not turn the engine over.
- *No Start (Engine Cranks):* The engine turns over normally but does not start even after a prolonged cranking.

WARNING!

Avoid long periods of engine cranking; raw fuel may load the exhaust system and damage the catalytic converter after the engine starts.

- *Slow Return to Idle:* When the accelerator pedal is released it takes some time for the engine to return to its normal idle speed.
- *Fast Idle:* The engine idles at a higher-than-normal speed or does not return to the normal idle speed when the throttle is released.
- *Runs On (Diesels):* A fast idle can cause the engine to attempt to run after the ignition is turned off. This is called dieseling because combustion is caused by the heat inside the combustion chamber.
- *Rough or Unstable Idle and Stalling:* The engine shakes while idling or changes idle speed constantly. This concern may cause the engine to stall.
- *Low/Slow Idle or Stalls/Quits during Deceleration:* Engine speed drops below its normal idle speed. This can cause stalling when the accelerator pedal is released.
- *Backfire:* Fuel ignites in the intake manifold or exhaust system. This causes a loud popping noise.
- *Lack of or Loss of Power:* The engine is sluggish and provides less power than is normally expected. The engine seems not to increase in speed when the accelerator pedal is depressed.
- *Cuts Out, Misses:* A steady pulsation or jerking at low engine speeds, especially during heavy engine loads. The exhaust may have a spitting sound at idle or low speeds.
- *Hesitation or Stumble:* A momentary lack of response to the accelerator pedal. This concern can occur at any vehicle speed but is more noticeable during acceleration from a stop.
- *Surges:* The power output from the engine seems to change while operating with a steady throttle or while cruising.
- *Detonation/Spark Knock:* The engine makes sharp metallic knocking sounds that is usually worse during acceleration.
- *Poor Fuel Economy:* Fuel economy is noticeably lower than expected or lower than it was before.

Figure 26–6 is a symptom chart for common concerns that have not set a DTC and the component or system that could cause the problem. Each potential

Concern	Component/System
Hard Start/Long Crank	Battery Condition and Current Draw Starting System Fuel/Ignition Intake Air System MAF Sensor Exhaust System PCV System EVAP System
No Crank	Anti-Theft Devices Base Engine Starting system
No Start (Engine Cranks)	Anti-Theft Devices Fuel/Ignition Intake Air System Exhaust System Restrictions Base Engine
Slow Return to Idle	Vacuum Leaks Throttle Body PCV System Intake Air System Leaks
Fast Idle or Runs On (Diesels)	Base Engine Check for Air Leaks Abnormal Engine Temperature TP Sensor Circuits Intake Air System Leaks
Low/Slow Idle or Stalls/Quits during Deceleration	Verify the Fuel Filler Cap is tightened. Automatic Transmission Fuel Delivery System Intake Air System Charging System Base Engine
Backfires	Secondary Ignition Fuel Delivery System Base Engine Exhaust System
Lack of or Loss of Power	Automatic Transmission Fluid Throttle Linkage Air Cleaner Element Check PIDS: LONGFT1/LONGFT2 and VPWR Fuel Delivery System Secondary Ignition MAF Sensor Exhaust System Variable Camshaft Timing System Accelerator Pedal Position Sensor Base Engine Brake System Drag or Binding

(Continued)

Figure 26-6 A no diagnostic trouble codes (DTCs) present symptom chart.

	Automatic Transmission Supercharger or Turbocharger Bypass System
Spark Knock	Abnormal Engine Temperature Coolant Level and Concentration MAF Sensor Base Engine Fuel Delivery System Secondary Ignition System PCV System Contaminated Engine Oil
Poor Fuel Economy	Check PIDs: LONGFT1/LONGFT2 and VPWR Abnormal Engine Temperature Ignition/Fuel System Exhaust System Variable Camshaft Timing System Transmission Fluid Level Automatic Transmission PCV System Incorrect PCM Programming Brake Drag Base Engine Concerns Incorrect PCV Valve Contaminated MAF Sensor Intake Air System

Figure 26-6 *(Continued)*

problem area should be checked. It is important to realize that some problems may cause more than one symptom.

BASIC TESTING

Diagnosis of electronic engine control systems includes much more than retrieving DTCs. Individual components and their circuits must be inspected and tested. The operation of some components can be monitored with the scan tool; however, additional tests are normally necessary. These tests include:

- *Ohmmeter Checks.* Most sensors and output devices can be checked with an ohmmeter. For example, an ohmmeter can be used to check a temperature sensor. Normally, the ohmmeter reading is low on a cold engine and high or infinity on a hot engine if the sensor is a PTC. If the sensor is an NTC, the opposite readings would be expected. Output devices such as coils or motors can also be checked with an ohmmeter.
- *Voltmeter Checks.* Many sensors, output devices, and their wiring can be diagnosed by checking the voltage to them, and in some cases, from them. Even some oxygen sensors can be checked in this manner.
- *Lab Scope Checks.* The activity of sensors and actuators can be monitored with a lab scope or a graphing multimeter. By watching their activity, the technician is doing more than testing them. Often problems elsewhere in the system will cause a device to behave abnormally. These situations are identified by the trace on a scope and by the technician's understanding of a scope and the device being monitored.

In some cases, a final check can be made only by substitution. Substitution is not an allowable diagnostic method under the mandates of OBD-II, nor is it the most desirable way to diagnose problems. However, sometimes it is the only way to verify the cause of a problem. When substituting, replace the suspected part with a known good unit and recheck the system. If the system now operates normally, the original part is defective.

Testing Sensors

To monitor engine conditions, the computer uses a variety of sensors. All sensors perform the same basic function. They detect a mechanical condition (movement or position), chemical state, or temperature

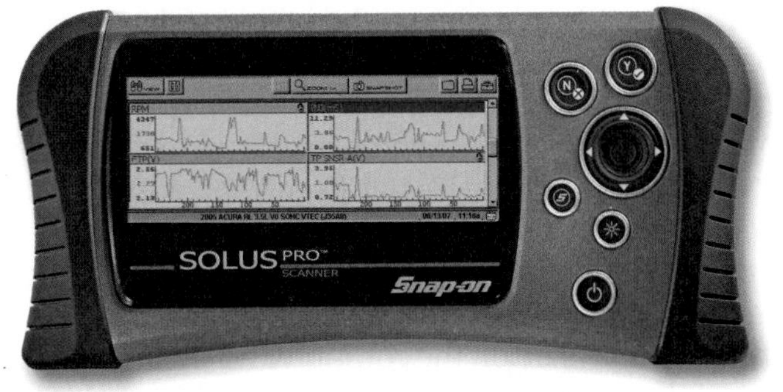

Figure 26-7 Many different types of lab scopes are available to diagnose electronic control systems; make sure to follow the operating procedures given by the manufacturer. *Reproduced under license from Snap-on Incorporated. All of the marks are marks of their owners.*

condition and change it into an electrical signal that can be used by the PCM to make decisions.

If a DTC directs you to a faulty sensor or sensor circuit, or if you suspect a faulty sensor, it should be tested. Always follow the procedures outlined by the manufacturer when testing sensors and other electronic components. Also, make sure you have the correct specifications for each part tested. Sensors are tested with a DMM, scan tool, and/or lab scope or GMM.

Because the controls are different on the various types of automotive lab scopes (**Figure 26–7**) and GMMs, make sure you follow the instructions of the scope's manufacturer. If the scope is set wrong, the scope will not break. It just will not show you what you want to be shown. To help with understanding how to set the controls on a scope, keep the following things in mind. The vertical voltage scale must be adjusted in relation to the expected voltage signal. The horizontal time base or milliseconds per division must be adjusted so the waveform appears properly on the screen. Many waveforms can be clearly displayed when the horizontal time base is adjusted correctly.

The trigger is the signal that tells the lab scope to start drawing a waveform. A marker indicates the trigger line on the screen, and minor adjustments of the trigger line may be necessary to position the waveform in the desired vertical position. Trigger slope indicates the direction in which the voltage signal is moving when it crosses the trigger line. A positive trigger slope means the voltage signal is moving upward as it crosses the trigger line, whereas a negative trigger slope indicates that the voltage signal is moving downward when it crosses the trigger line.

Software packages, often programmed in a lab scope or GMM, are available to help you properly interpret scope patterns and set up a lab scope. These also contain an extensive waveform library that you can refer to and find what the normal waveform of a particular device should look like. The library also contains the waveforms caused by common problems. You can also add to the library by transferring waveforms to a PC from the lab scope. After the waveforms have been transferred, notes can be added to the file. The software may also include the theory of operation, scope setup information, and diagnostic procedures for common inputs and outputs.

There are many different types of sensors; their design depends on what they are monitoring. Some sensors are simple on-off switches. Others are some form of variable resistor that changes resistance according to temperature changes. Some sensors are voltage or frequency generators, whereas others send varying signals according to the rotational speed of a device. Knowing what they are measuring and how they respond to changes are the keys to being able to accurately test an input sensor.

Some inputs to the PCM come from another control module or are simply a connection from a device. Examples of this are the battery voltage input and the heated windshield module. The battery's voltage is available on the data bus and many control modules need this information. There is no sensor involved, just a connection from the battery to the bus. The heated windshield module tells the computer when the heated windshield system is operating. This helps the PCM to accurately determine engine load and control idle speed.

DIAGNOSIS OF COMPUTER VOLTAGE SUPPLY AND GROUND WIRES

SHOP TALK

Never replace a computer unless the ground wires and voltage supply wires are proven to be in satisfactory condition.

All PCMs (OBD-II and earlier designs) cannot operate properly unless they have good ground connections and the correct voltage at the required terminals. A wiring diagram for the vehicle being tested must be used for these tests. Backprobe the battery terminal at the PCM and connect a digital voltmeter from this terminal to ground (**Figure 26–8**). Always use a good engine ground.

The voltage at this terminal should be 12 volts with the ignition switch off. If 12 volts are not available at this terminal, check the computer fuse and related circuit. Turn on the ignition switch and connect the

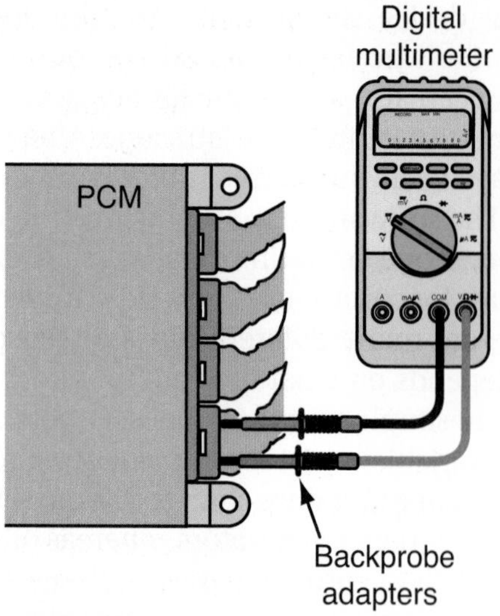

Figure 26-8 Using a digital voltmeter to check the PCM's circuit.

red voltmeter lead to the other battery terminals at the PCM with the black lead still grounded. The voltage measured at these terminals should also be 12 volts with the ignition switch on. When the specified voltage is not available, test the voltage supply wires to these terminals. These terminals may be connected through fuses, fuse links, or relays.

Ground Circuits

Ground wires usually extend from the computer to a ground connection on the engine or battery. With the ignition switch on, connect a digital voltmeter from the battery ground to the computer ground. The voltage drop across the ground wires should be 30 millivolts or less. If the voltage reading is greater than that or more than that specified by the manufacturer, repair the ground wires or connection.

Not only should the computer ground be checked, but so should the ground (and positive) connection at the battery. Checking the condition of the battery and its cables should always be part of the initial visual inspection before beginning diagnosis of an engine control system.

A voltage drop test is a quick way of checking the condition of any wire. To do this, connect a voltmeter across the wire or device being tested. Place the positive lead on the most positive side of the circuit. Then turn on the circuit. Ideally there should be a 0-volt reading across any wire unless it is a resistance wire, which is designed to drop voltage; even then, check the drop against specifications to see if it is dropping too much.

A good ground is especially critical for all reference voltage sensors. The problem here is not obvious until it is thought about. A bad ground will cause the reference voltage (normally 5 volts) to be higher than normal. Normally in a circuit, the added resistance of a bad ground would cause less voltage at a load. Because of the way reference voltage sensors are wired, the opposite is true. If the reference voltage to a sensor is too high, the output signal from the sensor to the computer will also be too high. As a result, the computer will be making decisions based on the wrong information. If the output signal is within the normal range for that sensor, the computer will not notice the wrong information and will not set a DTC.

To show how the reference voltage increases with a bad ground, let us look at a voltage divider circuit. This circuit is designed to provide a 5-volt reference signal off the tap. A vehicle's computer feeds the regulated 12 volts to a similar circuit to ensure that the reference voltage to the sensors is very close to 5 volts. The voltage divider circuit consists of two resistors connected in series with a total resistance of 12 ohms. The reference voltage tap is between the two resistors. The first resistor drops 7 volts (**Figure 26–9**), which leaves 5 volts for the second resistor and for the reference voltage tap. This 5-volt reference signal will be always available at the tap, as long as 12 volts are available for the circuit.

If the circuit has a poor ground, one that has resistance, the voltage drop across the first resistor will be decreased. This will cause the reference voltage to increase. In **Figure 26–10**, to simulate a bad

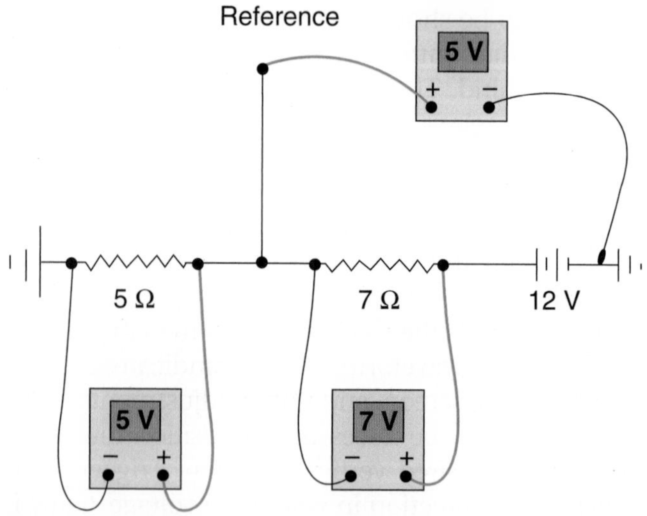

Figure 26-9 A voltage divider circuit with voltage values.

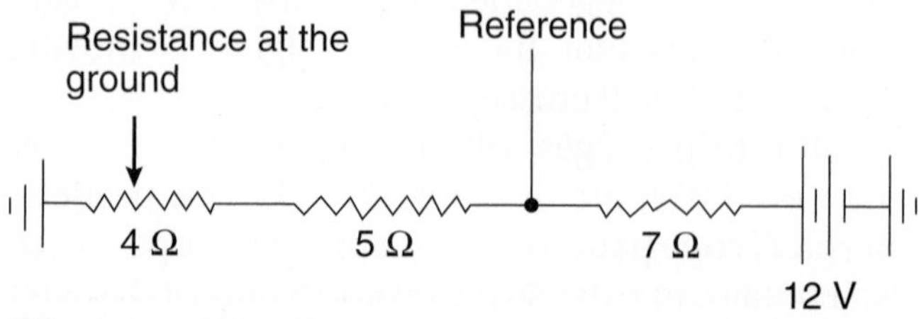

Figure 26-10 Voltage divider circuit with a bad ground.

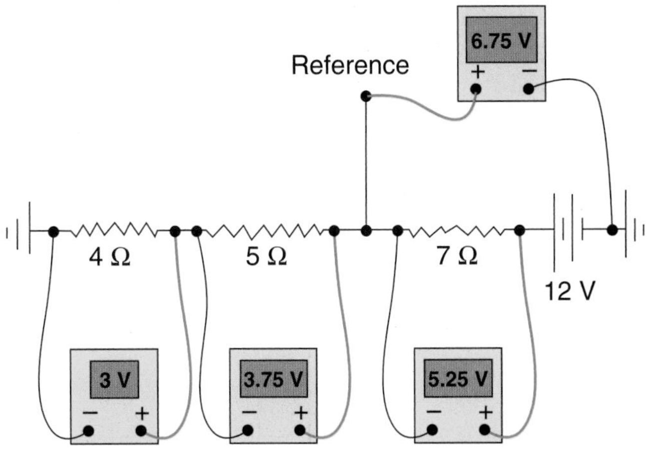

Figure 26-11 Figure 26–10 with voltage readings.

ground, a 4-ohm resistor was added into the circuit at the ground connection at the battery. This increases the total resistance of the circuit to 16 ohms and decreases the current flowing throughout the circuit. With less current flow through the circuit, the voltage drop across the first resistor decreases to 5.25 volts **(Figure 26–11)**. This means the voltage available at the tap will be higher than 5 volts; it will be 6.75 volts.

Electrical Noise Poor grounds can also allow EMI or noise to be present on the reference voltage signal. This noise causes small changes in the voltage going to the sensor. Therefore, the output signal from the sensor will also have these voltage changes. The computer will try to respond to these small rapid changes, which can cause a driveability problem. The best way to check for noise is to use a lab scope.

Connect the lab scope between the 5-volt reference signal into the sensor and the ground. The trace on the scope should be flat **(Figure 26–12)**. If noise is present, move the scope's negative probe to a known good ground. If the noise disappears, the sensor's ground circuit is bad or has resistance. If the noise is still present, the voltage feed circuit is bad or there is EMI in the circuit from another source, such as the AC generator. Find and repair the cause of the noise.

Circuit noise may be present at the positive side or negative side of a circuit. It may also be evident by a flickering MIL, a popping noise on the radio, or by an intermittent engine miss. Noise can cause a variety of problems in any electrical circuit. The most common sources of noise are electric motors, relays and solenoids, AC generators, ignition systems, switches, and A/C compressor clutches. Typically, noise is the result of an electrical device being turned on and off. Sometimes the source of the noise is a defective suppression device. Manufacturers include these devices to minimize or eliminate electrical noise. Some of the commonly used noise suppression devices are resistor-type

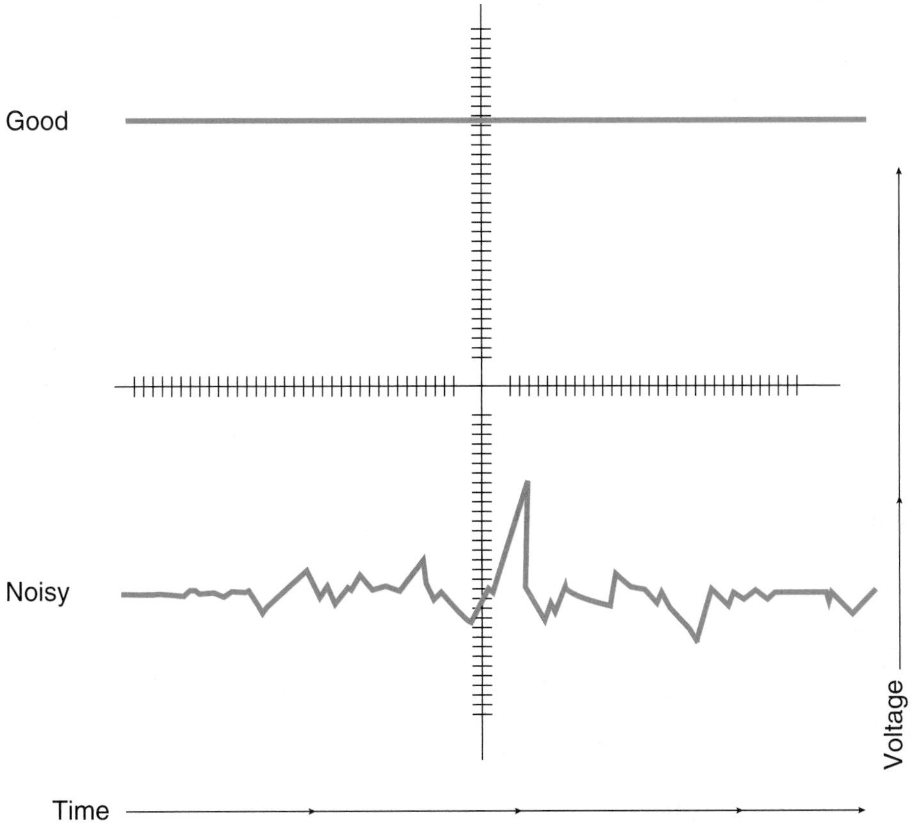

Figure 26-12 *(Top)* A good voltage signal. *(Bottom)* A voltage signal with noise.

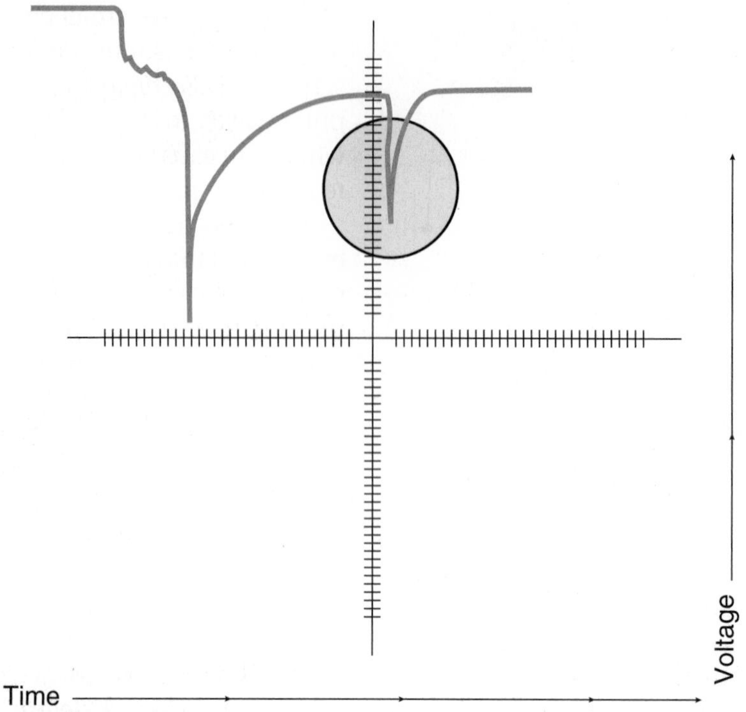

Figure 26-13 A trace of an electromagnetic clutch with a bad clamping diode.

secondary cables and spark plugs, shielded cables, capacitors, diodes, and resistors. Capacitors or chokes are used to control noise from a motor or generator. If the source of the noise is not a poor ground or a defective component, check the suppression devices.

Clamping Diodes Clamping diodes are placed in parallel to coil windings to limit high-voltage spikes. These voltage spikes are induced by the collapsing of the magnetic field around a winding in a solenoid, relay, or electromagnetic clutch. The field collapses when current flow to the winding is stopped. The diode prevents the voltage from reaching the computer and other sensitive electronic parts. When the diode fails to suppress the voltage spikes, the transistors inside the computer can be destroyed. If the diode is bad, a negative spike will appear in a voltage trace **(Figure 26–13)**. Resistors are also used to suppress voltage spikes. They do not eliminate the spikes; rather, they limit the intensity of the spikes. If a voltage trace has a large spike and the circuit is fitted with a resistor to limit noise, the resistor may be bad.

SWITCHES

Switches are turned on and off through an action of a device or by the actions of the driver. Switches are either normally open or normally closed. Switches send a digital signal to the PCM; they are either open or closed. Some switches are provided with a reference voltage of 5 or 12 volts by the PCM. An example of this is a neutral drive/neutral gear switch (NDS). This switch lets the PCM know when the transmission has been shifted into a gear. If the transmission is in park or neutral, the switch is closed. It sends a voltage signal of 1 volt or less to the PCM. When the transmission is placed into a gear, the switch opens and sends a signal above 5 volts to the PCM.

Some switches control the ground side of the circuit. These circuits contain a fixed resistor connected in series with the switch. When the switch is closed, the voltage signal to the PCM is low or zero. When the switch is open, there is a high-voltage signal. Common grounding switches include:

- Idle tracking switch
- Power steering pressure switch
- Overdrive switch
- Clutch pedal position switch

Supply or power side switches are the most commonly used and work in the opposite way. They send a 5- or 12- volt signal to the PCM when they are closed. When the switch is open, there is no voltage at the PCM. Common supply side switches include:

- Ignition switch
- Park/neutral switch
- Air-conditioning (A/C) demand sensor
- Brake switch
- High gear switch

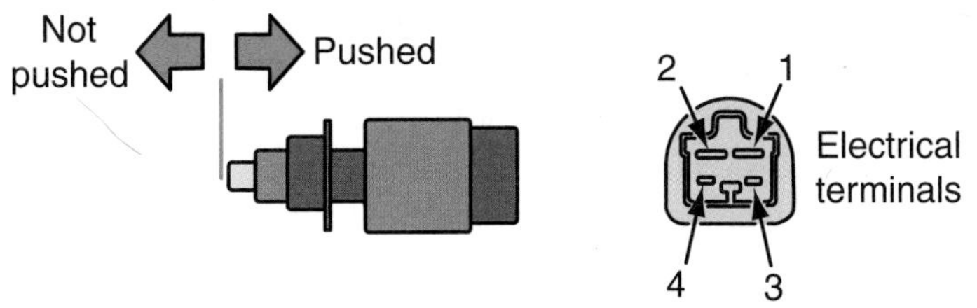

Tester connection	Switch condition	Specified value
1–2	Switch pin released	Below 1Ω
	Switch pin pushed in	10kΩ or higher
3–4	Switch pin released	10kΩ or higher
	Switch pin pushed in	Below 1Ω

Figure 26-14 Check the service information for the proper testing points for a switch.

Testing Switches

A switch can be easily tested with an ohmmeter. Disconnect the connector at the switch. Refer to the wiring diagram to identify the terminals at the switch if there are more than two **(Figure 26–14)**. Connect the ohmmeter across the switch's terminals. Perform whatever action is necessary to open and close the switch. When the switch is open, the ohmmeter should have an infinite reading. When the switch is closed, there should zero resistance. If the switch reacts in any other way, the switch is bad and should be replaced.

Switches can also be checked with a voltmeter. The signal to the PCM from the supply side switches should be 0 volts with the switch open and supply voltage when the switch is close. This indicates to the ECM that a change has taken place. Using a voltmeter is preferred because it tests the circuit as well as the switch. If less than supply voltage is present with the switch closed, there is unwanted resistance in the circuit. Again, expect the opposite readings on a ground side switch.

Some switches are adjustable and must be set so they close and open at the correct time. An example of this is the clutch switch. This switch is used to inform the computer when there is no load (clutch pedal depressed) on the engine. The switch is also connected into the starting circuit. The switch prevents starting of the engine unless the clutch pedal is fully depressed. The switch is normally open when the clutch pedal is released. When the clutch pedal is depressed, the switch closes and completes the circuit between the ignition switch and the starter solenoid. It also sends a signal of no-load to the PCM.

Most grounding switches react to some mechanical action to open or close. There are some, however, that respond to changes of conditions. These may respond to changes in pressure or temperature. An example of this type of switch is the power steering pressure switch. This switch informs the PCM when power steering pressures reach a particular point. When the power steering pressure exceeds that point, the PCM knows there is an additional load on the engine and will increase idle speed.

To test this type of switch, monitor its activity with a DMM or lab scope. With the engine running at idle speed, turn the steering wheel to its maximum position on one side. The voltage signal should drop as soon as the pressure in the power steering unit has reached a high level. If the voltage does not drop, either the power steering assembly is incapable of producing high pressures or the switch is bad.

Temperature responding switches operate in the same way. When a particular temperature is reached, the switch opens. This type of switch is best measured by removing it and submerging it in heated water. Watching the ohmmeter as the temperature increases, a good temperature responding switch will open (have an infinite reading) when the water temperature reaches the specified amount. If the switch fails this test, it should be replaced.

TEMPERATURE SENSORS

The PCM changes the operation of many components and systems based on temperature. Nearly all temperature sensors are NTC thermistors and operate in the same way. Their resistance changes with a change in temperature. The PCM supplies a reference voltage of 5 volts to the sensor. That voltage is changed by the change of the resistor's resistance and is fed back through a ground wire to the PCM. Based on the return voltage, the PCM calculates the exact temperature. When the sensor is cold, its resistance is high, and the return voltage signal is also high. As the sensor warms up, its resistance drops and so does the voltage signal.

Engine Coolant Temperature (ECT) Sensor

The **engine coolant temperature (ECT) sensor** is a thermistor. By measuring ECT, the PCM knows the average temperature of the engine. Temperature is used to regulate many engine functions, such as the fuel injection system, ignition timing, variable valve timing, transmission shifting, EGR, and canister purge, and controlling the open- and closed-loop operational modes of the system. The ECT sensor is normally located in an engine coolant passage just before the thermostat. On cars built prior to OBD-II, a coolant switch may be used. This type of sensor may be designed to remain closed within a certain temperature range or to open only when the engine is warm.

A faulty ECT sensor or sensor circuit can cause a variety of problems. The most common is the failure to switch to the closed-loop mode once the engine is warm. ECT sensor problems are often caused by wiring faults or loose or corroded connections rather than the sensor itself. Many testers are able to show where to place the probes of the tester to check things like the ECT **(Figure 26–15)** A defective ECT sensor or circuit may cause the following problems:

1. Hard engine starting
2. Rich or lean air-fuel ratio
3. Improper operation of emission devices
4. Reduced fuel economy

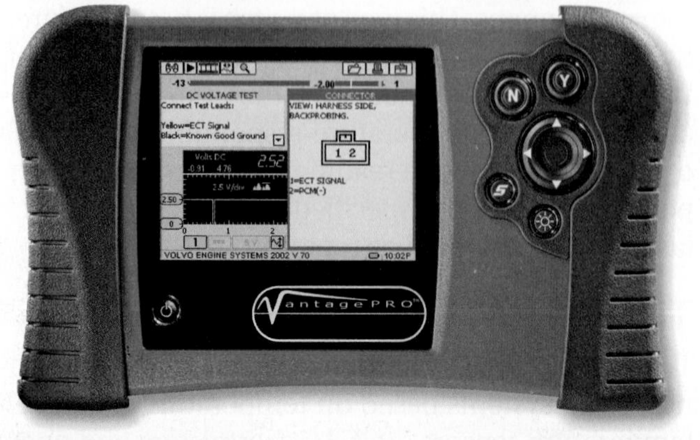

Figure 26-15 An engine coolant temperature (ECT) sensor. *Reproduced under license from Snap-on Incorporated. All of the marks are marks of their owners.*

5. Improper converter clutch lockup
6. Hesitation on acceleration
7. Engine stalling
8. Transmission will not shift into high gear or will shift late.

Intake Air Temperature (IAT) Sensor

The **intake air temperature (IAT) sensor** is also called an air charge temperature sensor. Its resistance decreases as the incoming air temperature increases and increases as incoming air temperature decreases **(Figure 26–16)**. The PCM uses the air temperature information as a correction factor in the calculation of fuel, spark, and airflow. For example, the PCM uses this input to help calculate fuel delivery. Because cold intake air is denser, a richer air-fuel ratio is required.

On engines equipped with a MAP sensor, the IAT is installed in an intake air passage. On other engines, the IAT **(Figure 26–17)** is normally an

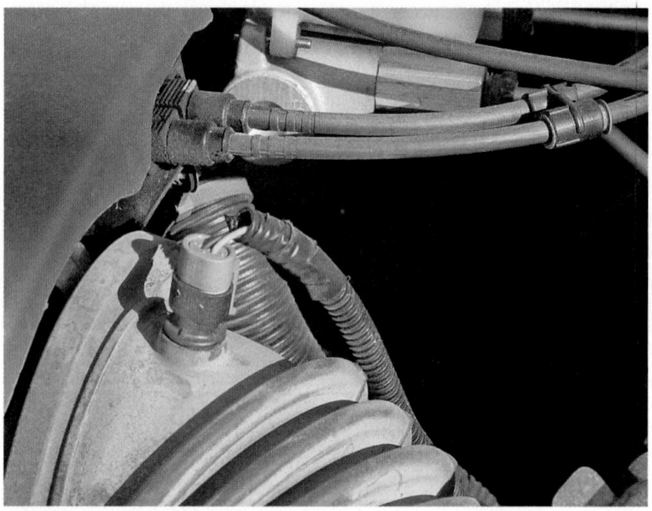

Figure 26-16 An air temperature sensor.

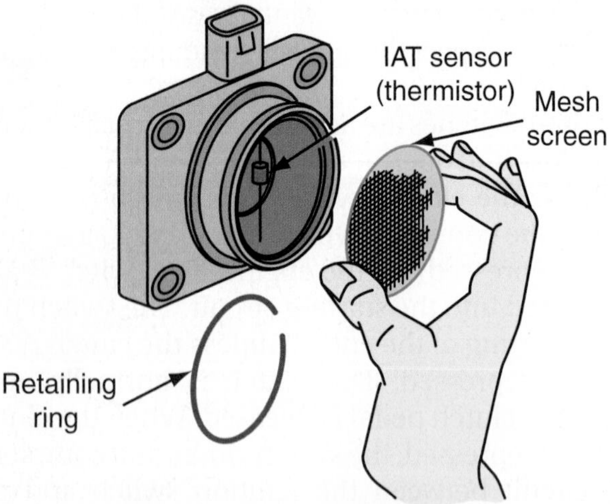

Figure 26-17 A typical intake air temperature (IAT) sensor incorporated into a MAF sensor.

integral part of the mass airflow (MAF) sensor. Most control systems compare the inputs from the IAT and the ECT to determine if the engine is attempting a cold start.

A defective IAT sensor may cause the following problems:

1. Rich or lean air-fuel ratio
2. Hard engine starting
3. Engine stalling or surging
4. Acceleration stumbles
5. Excessive fuel consumption

Other Temperature Sensors

Many other temperature sensors may be used on engines. Their application depends on the engine and the control system. Some turbo- or supercharged engines have two IAT sensors: one before the charger and one after. This is done to monitor the change in temperature as the air is forced into the cylinders.

Some engines have a cylinder head temperature (CHT) sensor installed in the cylinder head to measure its temperature. The primary function of this is to detect engine overheating. When high metal temperatures are reported to the PCM, it will enter into its fail-safe cooling strategy mode.

Other common temperature sensors include an engine oil temperature (EOT) sensor, fuel rail pressure temperature (FRPT) sensor, and EGR temperature sensor. Also, many vehicles built prior to OBD-II use a temperature sensor to directly control the electric radiator fans.

Testing

Temperature sensor circuits should be tested for opens, shorts, and high resistance. Often if one of these problems exist, a DTC will be set. Scan tool data can also give an indication of the condition of the sensor and related circuits. If the observed temperature is the coldest possible value, the circuit is open. If the temperature is the highest possible, the circuit has a short. Most service manuals give a procedure for checking for opens and shorts. A jumper wire is inserted across specific terminals of the circuit and the data is observed. This should cause the readings to go high or hot. If the connector to the sensor is disconnected, the readings should drop to cold. On some vehicles, this test will not work because the PCM will react by using a PID value. High-resistance problems will cause the PCM to respond to a lower temperature than the actual temperature. This can be verified by using a good thermometer (infrared is best) to measure the temperature and compare it to the readings

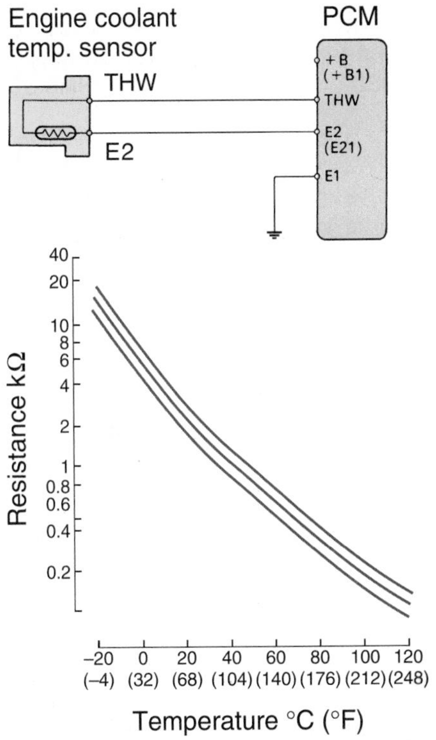

Figure 26–18 Specifications for an ECT sensor. *Courtesy of Toyota Motor Sales, U.S.A., Inc.*

on the scan tool. There will be some difference, but it should be minor if the sensor circuit is working properly. Unwanted resistance in the circuit can cause poor engine performance, poor fuel economy, and engine overheating.

Temperature sensors can be tested by removing them and placing them in a container of water with an ohmmeter connected across the sensor terminals. A thermometer is also placed in the water. When the water is heated, the sensor should have the specified resistance at any temperature **(Figure 26–18)**. Replace the sensor if it does not have the specified resistance. Manufacturers give a temperature and resistance chart for each of the temperature sensors.

WARNING!

Never apply an open flame or a heat gun to an ECT or IAT sensor for test purposes. This action will damage the sensor.

The wiring to the sensor can also be checked with an ohmmeter. With the wiring connectors disconnected from the sensor and the computer, connect an ohmmeter from each sensor terminal to the computer terminal to which the wire is connected. Both sensor wires should indicate less resistance than

COLD CURVE 10,000-OHM RESISTOR USED		HOT CURVE CALCULATED RESISTANCE OF 909 OHMS USED	
−20°F	4.70 V	110°F	4.20 V
−10°F	4.57 V	120°F	4.00 V
0°F	4.45 V	130°F	3.77 V
10°F	4.30 V	140°F	3.60 V
20°F	4.10 V	150°F	3.40 V
30°F	3.90 V	160°F	3.20 V
40°F	3.60 V	170°F	3.02 V
50°F	3.30 V	180°F	2.80 V
60°F	3.00 V	190°F	2.60 V
70°F	2.75 V	200°F	2.40 V
80°F	2.44 V	210°F	2.20 V
90°F	2.15 V	220°F	2.00 V
100°F	1.83 V	230°F	1.80 V
110°F	1.57 V	240°F	1.62 V
120°F	1.25 V	250°F	1.45 V

Figure 26-19 Voltage drop specifications for an ECT sensor. *Courtesy of Chrysler LLC*

specified by the manufacturer. If the wires have high resistance, the wires or wiring connectors must be repaired.

WARNING!

Before disconnecting any computer system component, be sure that the ignition switch is turned off. Disconnecting components may cause high induced voltages and computer damage.

With the sensor installed in the engine, the sensor terminals may be backprobed to connect a digital voltmeter to them. The sensor should provide the specified voltage drop at any coolant temperature **(Figure 26–19)**. To record accurate readings, make sure the meter leads have a good connection.

PRESSURE SENSORS

Most pressure sensors are piezoresistive sensors. A silicon chip in the sensor flexes with changes in pressures. The amount of flex dictates the voltage signal sent out from the sensor. One side of the chip is exposed to a reference pressure, which is either a perfect vacuum or a calibrated pressure. The other side is the pressure that will be measured. As the chip flexes in response to pressure, its resistance changes. This changes the voltage signal sent to the PCM. The PCM looks at the change and calculates the pressure change.

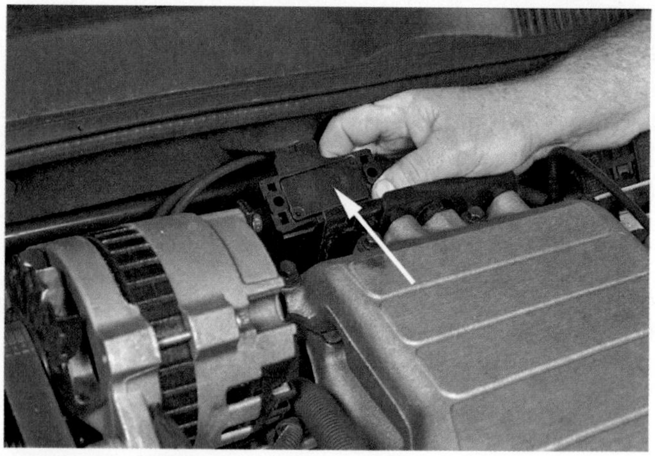

Figure 26-20 A MAP sensor.

Manifold Absolute Pressure (MAP) Sensor

A **manifold absolute pressure (MAP) sensor** senses air pressure or vacuum in the intake manifold. The sensor measures manifold air pressure against an absolute pressure. The MAP sensor uses a perfect vacuum as a reference pressure. The MAP sensor **(Figure 26–20)** measures changes in the intake manifold pressure that result from changes in engine load and speed. The PCM sends a voltage reference signal to the MAP sensor. As the pressure changes, the sensor's resistance also changes. The control module determines manifold pressure by monitoring the sensor output voltage.

The PCM uses the MAP signals to calculate how much fuel to inject in the cylinders and when to ignite the cylinders. A defective MAP sensor may cause a rich or lean air-fuel ratio, excessive fuel consumption, and engine surging. The MAP signal may also be used to regulate the EGR.

High manifold pressure (low vacuum) resulting from full throttle operation requires more fuel. Low pressure or high vacuum requires less fuel. At closed throttle, the engine produces a low MAP value. Wide-open throttle produces a high value. The highest value results when manifold pressure is the same as the pressure outside the manifold and 100% of the outside air is being measured. The use of this sensor also allows the control module to automatically adjust for different altitudes. A PCM with a MAP sensor relies on an IAT sensor to calculate intake air density.

Many EFI systems with MAF sensors do not have MAP sensors. However, there are a few engines with both of these sensors. These use the MAP mainly as a backup if the MAF fails. When the EFI system has a MAF, the computer calculates the intake airflow from the MAF and rpm inputs.

The MAP is the second most important sensor in the fuel management system (the CKP is more important). The basic injector pulse width is set according to the MAP signal. A defective MAP can cause a number of problems, including a no-start condition, reduced power output, and heavy engine surging.

Testing a MAP

A defective MAP sensor may cause a rich or lean air-fuel ratio, excessive fuel consumption, and engine surging. The sensor is mounted on the intake manifold or someplace high in the engine compartment. A hose supplies the sensor with engine vacuum. Inspect the sensor, its electrical connectors, and the vacuum hose. The hose should be checked for cracks, kinks, and proper fit.

The PCM supplies a 5-volt reference signal to the sensor. Begin your diagnosis of the MAP circuit by measuring that voltage. With the ignition switch on, backprobe the reference wire and measure the voltage. If the reference wire does not have the specified voltage, check the reference voltage at the PCM. If the voltage is within specifications at the PCM but low at the sensor, repair the wire. When this voltage is low at the PCM, check the voltage supply wires and ground wires for the PCM. If the wires are good, replace the computer.

A MAP sensor can be monitored with a scan tool through specific PIDs (**Figure 26–21**). When using a scan tool, make sure to use the correct specifications and follow the subsequent tests given in the service information.

With the ignition switch on, connect the voltmeter from the sensor ground wire to the battery ground. If the voltage drop across this circuit exceeds specifications, repair the ground wire from the sensor to the computer.

With the ignition on, backprobe the MAP sensor signal wire and measure the voltage. The voltage reading indicates the barometric pressure (BARO) signal from the MAP sensor to the PCM. Many MAP sensors send a barometric pressure signal to the computer each time the ignition switch is turned on and each time the throttle is in the wide-open position. If the BARO signal does not equal the MAP signal with the ignition on and the engine off, replace the MAP sensor.

To check the voltage signal of a MAP, turn the ignition switch on and connect a voltmeter to the MAP sensor signal wire. Connect a vacuum pump to the MAP sensor vacuum connection and apply vacuum to the sensor. Manufacturers list the expected voltage drop at different vacuum levels (**Figure 26–22**). If the MAP sensor voltage is not within specifications at any vacuum, replace the sensor.

SHOP TALK

MAP sensors have a much different calibration on turbocharged engines than on nonturbocharged engines. Be sure you are using the proper specifications for the sensor being tested.

To check a MAP sensor with a lab scope, connect the scope to the MAP output and a good ground. When the engine is accelerated and returned to idle, the output voltage should increase and decrease (**Figure 26–23**). If the engine is accelerated and the MAP sensor voltage does not rise and fall, or if the signal is erratic, the sensor or sensor wires are defective.

Sensors/Inputs	PCM Pin	Measured/PID Values				Units Measured/PID
		KOEO	Hot Idle	48 KM/H (30 MPH)	89 KM/H (55 MPH)	
MAP	E62	4*	1–1.4*	1.8–2.1*	1.9–2.3*	DCV

* Value may vary 20% depending on altitude, operating conditions, weather, and other factors.

Figure 26-21 The PIDs for a typical MAP sensor.

	Applied Vacuum				
in. Hg	4	8	12	16	20
kPa	14	27	40	54	68
VOLTAGE DROP	0.3 – 0.5	0.7 – 0.9	1.1 – 1.3	1.5 – 1.7	1.9 – 2.1

Figure 26-22 An example of how the voltage drop across a MAP changes with the amount of vacuum applied to the sensor.

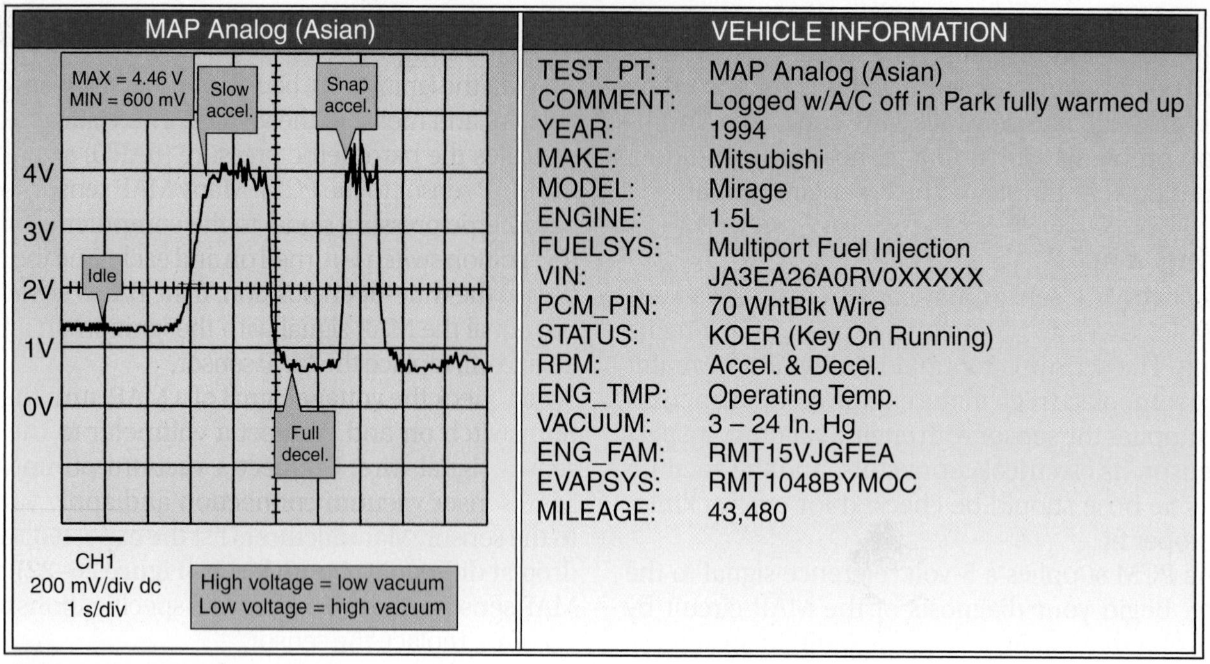

Figure 26-23 Trace of a normal MAP sensor. *Courtesy of Progressive Diagnostics—WaveFile AutoPro*

Some MAP sensors produce a digital voltage signal of varying frequency; begin diagnosis by checking the voltage reference wire and the ground wire. Continue testing by following these steps:

PROCEDURE

1. Turn off the ignition switch, and disconnect the wiring connector from the MAP sensor.
2. Connect the connector on the MAP sensor tester to the MAP sensor.
3. Connect the MAP sensor tester battery leads to a 12-volt battery.
4. Connect a pair of digital voltmeter leads to the MAP tester signal wire and ground.
5. Turn on the ignition switch and observe the barometric pressure voltage signal on the meter. If this voltage signal does not equal specifications, replace the sensor.
6. Supply the specified vacuum to the MAP sensor with a vacuum pump.
7. Observe the voltmeter reading at each specified vacuum. If the MAP sensor voltage signal does not equal the specifications at any vacuum, replace the sensor.

Photo Sequence 25 shows a typical procedure for testing a Ford MAP sensor, which has a varying frequency output. The sequence uses a MAP sensor tester and voltmeter. If a MAP tester is not available, the sensor can be checked with a DMM that measures frequency. Set the meter to read 100 to 200 hertz and connect it to the MAP sensor. Measure the voltage, duty cycle, and frequency at the sensor with no vacuum applied. Then apply about 18 in. Hg of vacuum to the MAP. Observe and record the same readings. A good MAP will have about the same amount of voltage and duty cycle with or without the vacuum. However, the frequency should decrease **(Figure 26–24)**. Normally, a frequency of about 155 hertz is expected at sea level with no vacuum applied to the MAP. When vacuum is applied, the frequency should decrease to around 95 hertz.

A lab scope can be used to check a MAP sensor. The upper horizontal line of the trace should be at 5 volts, and the lower horizontal line should be close to zero **(Figure 26–25)**. Check the waveform for unusual movements of the trace. If the waveform is anything but normal, replace the sensor.

Vapor Pressure Sensor (VPS)

The **vapor pressure sensor (VPS)** measures the vapor pressure in the evaporative emission control system. This sensor is capable of responding to slight pressure changes. The sensor has two chambers divided by a silicon chip. One chamber, called the reference chamber, is exposed to atmospheric pressure. The other chamber is exposed to vapor pressure. Changes in vapor pressure cause the chip to flex, which causes the signal to the PCM to change **(Figure 26–26)**. In most cases, the sensor receives a 5-volt reference voltage. The voltage signal represents the difference between vapor pressure and atmospheric pressure.

P25-1 *Remove the MAP sensor's electrical connector and vacuum hose.*

P25-2 *Connect the appropriate connector of the MAP sensor tester to the MAP sensor.*

P25-3 *Connect the remaining tester connector to the MAP sensor's electrical connector.*

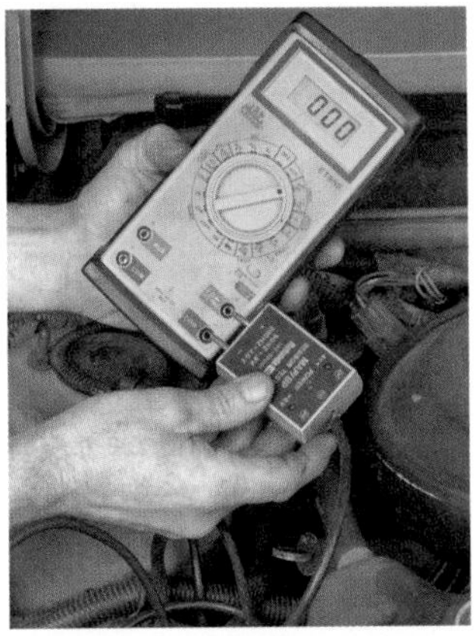

P25-4 *Insert the voltage terminals of the MAP sensor tester into the test lead terminals on a DMM; make sure the polarity is correct.*

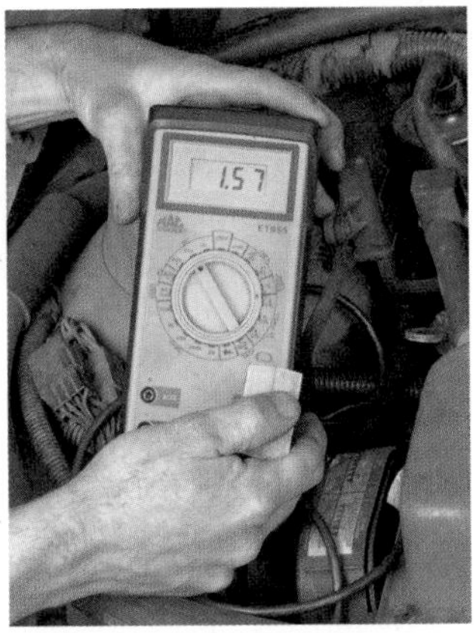

P25-5 *Observe the MAP sensor barometric pressure voltage reading on the voltmeter. Compare this reading to the specifications.*

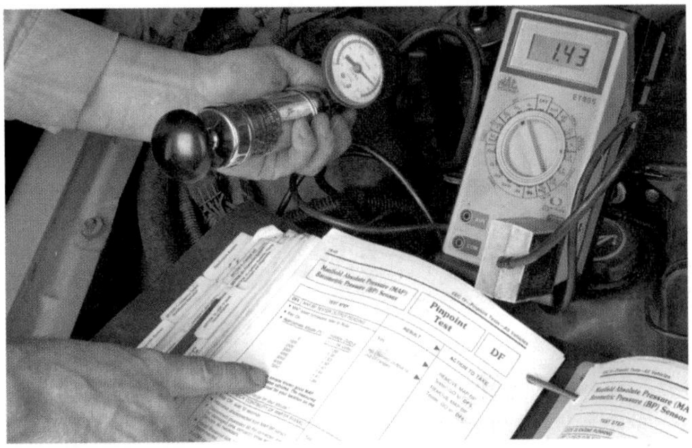

P25-6 *Connect a hand-operated vacuum pump to the MAP sensor and apply 5 in. Hg to the sensor. Observe the sensor's voltage signal on the voltmeter. Compare the reading to specifications.*

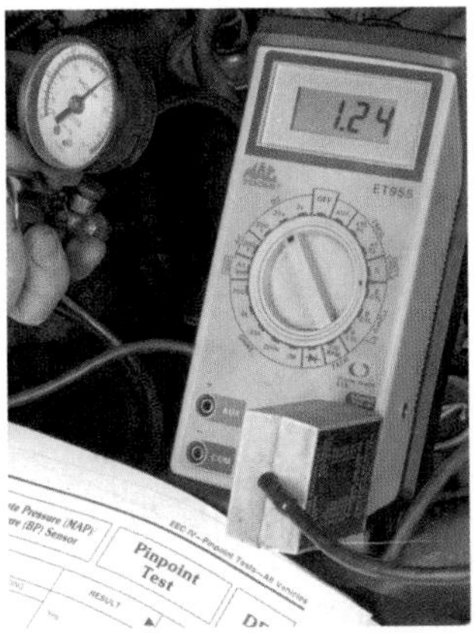

P25-7 *Apply 10 in. Hg to the MAP sensor. Observe the voltage reading now and compare this to specifications.*

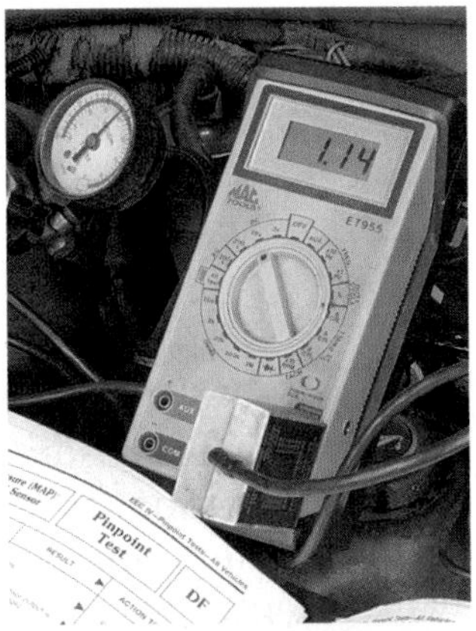

P25-8 *Increase the vacuum to the MAP sensor so that it now has 15 in. Hg. Observe the voltage reading now and compare this to specifications.*

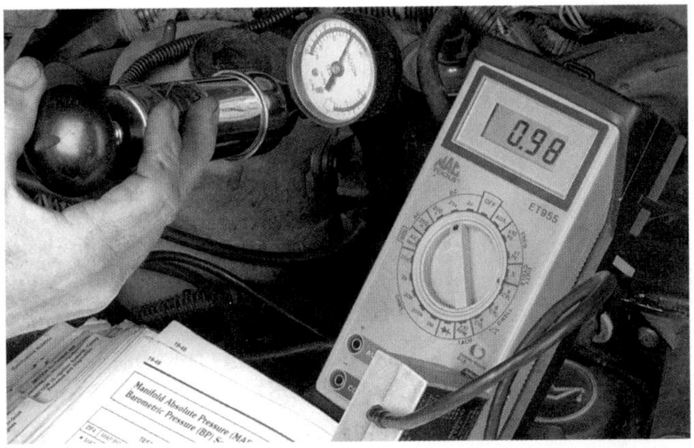

P25-9 *Increase the vacuum to 20 in. Hg. Observe the voltage reading now and compare this to specifications. If the voltage signals from any of the tests do not match specifications, the MAP sensor needs to be replaced. Disconnect the tester and reconnect the electrical connector and vacuum hose to the MAP after you have completed your testing.*

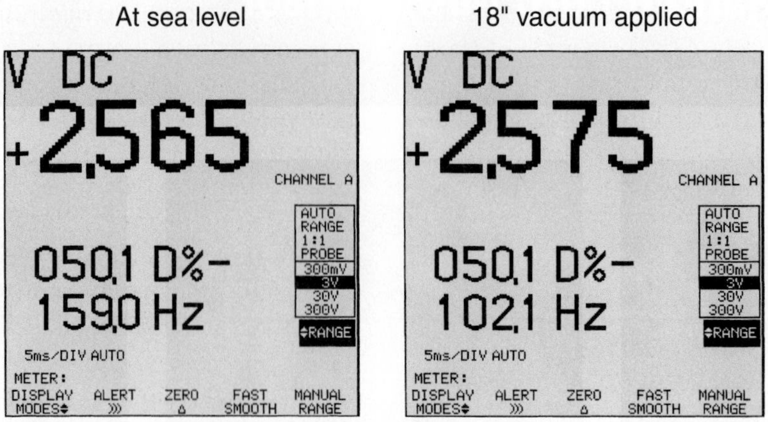

Figure 26-24 Reaction of a good Ford MAP sensor with vacuum applied. *Reproduced with permission of Fluke*

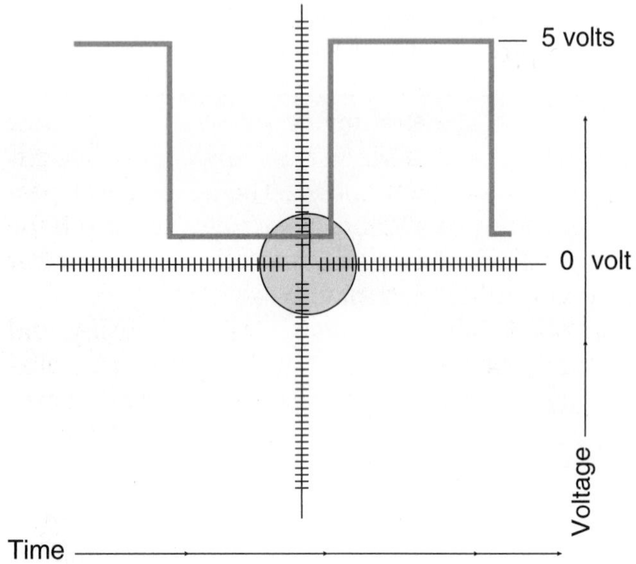

Figure 26-25 A good Ford MAP sensor signal.

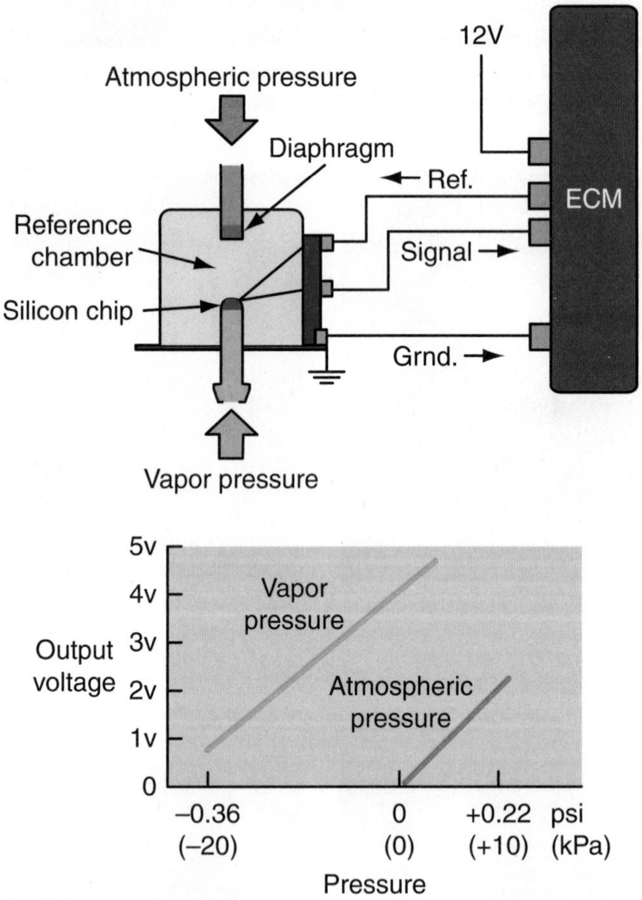

Figure 26-26 The operation of a vapor pressure sensor.

The return or signal voltage will be high when the vapor pressure is high.

The sensor may be on the fuel pump or in a remote location. When the VPS is remotely mounted, a hose connects the sensor to a vapor pressure port. In some cases, the sensor has an additional hose that supplies atmospheric pressure to the VPS. If these hoses are reversed, the PCM will see high vapor pressure and set a DTC. All hoses should be checked for leaks, kinks, and secure connections. The reference voltage should also be checked. The operation of the sensor is checked in the same way as a MAP, except pressure, not vacuum, is applied to the sensor. Refer to the specifications for the amount of testing pressure to apply and the subsequent voltage signal from the sensor.

Other Pressure Sensors

Other pressure sensors are used on some engines. Their application depends on the engine and the control system. Many of these are related to the EGR system. The most commonly used is the feedback pressure EGR sensor. This sensor's voltage signal tells the PCM how much the EGR valve is open. The PCM uses this input to control the vacuum to the EGR valve and control air-fuel ratios and ignition timing.

Some engines have a fuel rail pressure sensor. This sensor is found on returnless fuel systems and measures fuel pressure near the fuel injectors. The PCM uses this input to adjust fuel injector pulse width. Turbo- or supercharged engines have a pressure sensor that monitors the amount of boost.

MASS AIRFLOW (MAF) SENSORS

The **mass airflow (MAF) sensor** measures the flow of air entering the engine **(Figure 26–27)**. This measurement of intake air volume is used to calculate engine load (throttle opening and air volume). It is similar to the relationship of engine load to MAP or vacuum sensor signal. Engine load inputs are used to control the fuel injection and ignition systems, as well as shift timing in automatic transmissions. The airflow sensor is placed between the air cleaner and throttle plate assembly or inside the air cleaner assembly.

There are different types of MAF sensors. The most commonly used design is the hot-wire MAF. In a hot-wire-type MAF **(Figure 26–28)**, a wire, called the hot wire, is positioned so the intake air flows over it. The sensor also has a thermistor, sometimes referred to as the cold wire, located beside the hot wire that measures intake air temperature. The sensor also contains a control module. Current from the PCM keeps the hot wire at a constant temperature above ambient temperature, normally 392°F (200°C). Airflow past the hot wire causes it to lose heat and the PCM responds by sending more current to the wire. The increased current flow keeps the hot wire at its desired temperature. The current required to maintain the hot wire's temperature is proportional to mass airflow. The sensor measures the current and sends a voltage signal to the PCM. The PCM interprets the signal to determine mass airflow. Most MAF sensors have an integrated IAT sensor.

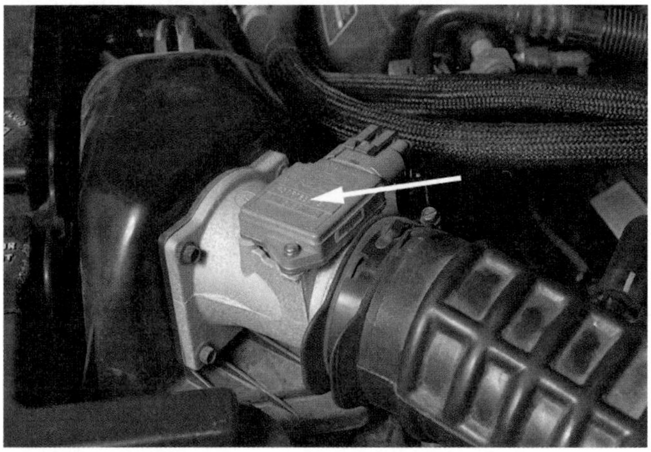

Figure 26-27 A mass airflow sensor.

Testing a MAF Sensor

The test procedure for hot-wire MAF sensors varies with the vehicle make and year. Always follow the test procedure in the appropriate service manual. Most often diagnosis of a MAF sensor involves visual, circuit, and component checks. The MAF sensor passage must be free of debris to operate properly. If the passage is plugged, the engine will usually start but run poorly or stall and may not set a DTC.

Check the air inlet system (air filter, housing, and ductwork) for obstructions, blockage, proper installation, and sealing. Check the screen of the MAF sensor for dirt and other contaminants. Check the throttle plate bore for dirt buildup.

SHOP TALK

DTC P0103 (Mass or Volume Airflow Circuit High) can be set because of debris blocking the MAF sensor screen.

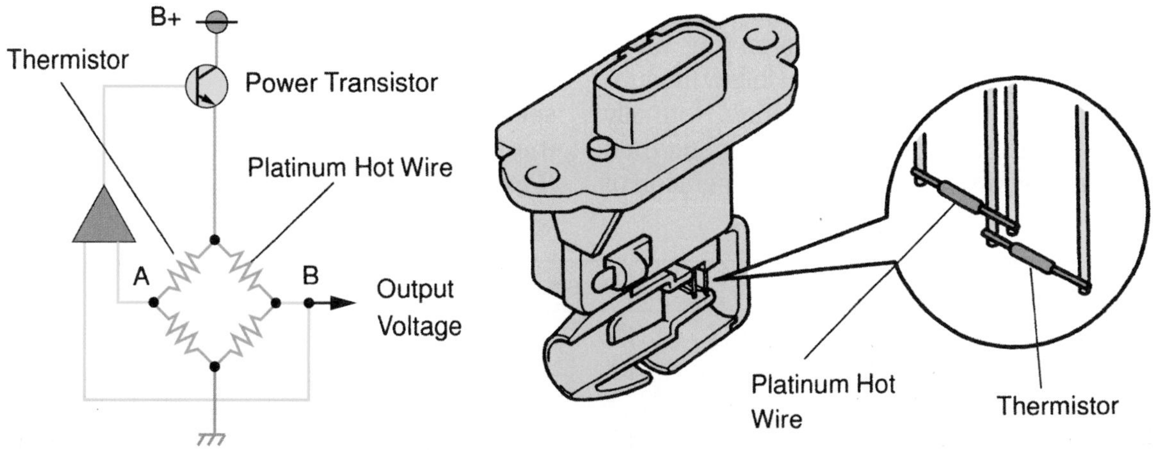

Figure 26-28 A hot-wire MAF. *Courtesy of Toyota Motor Sales, U.S.A., Inc.*

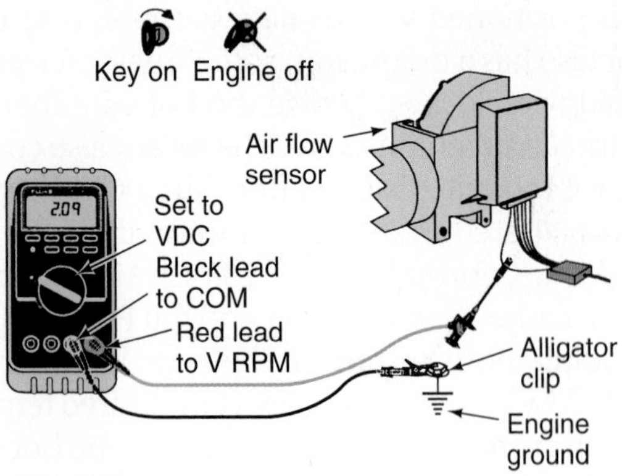

Figure 26-29 A voltmeter connected to measure the signal from a MAF sensor. *Reproduced with permission of Fluke*

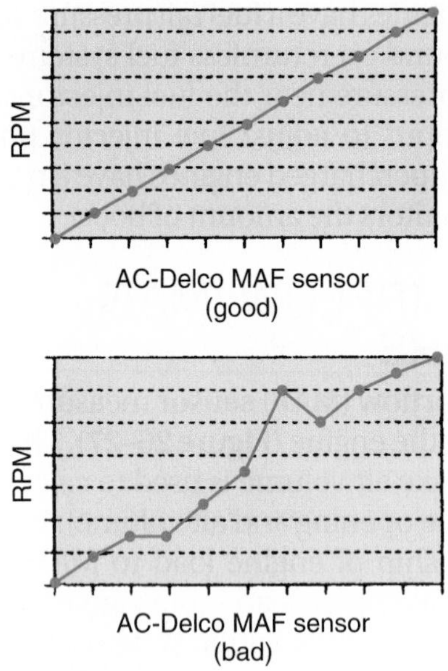

Figure 26-30 Satisfactory and unsatisfactory MAF sensor frequency readings. *Reproduced with permission of Fluke*

Make sure the electrical connections to the MAF are sound. Check the reference voltage to the sensor and the ground circuit. To check the MAF sensor's voltage signal and frequency, connect a voltmeter to the MAF voltage signal wire and a good ground **(Figure 26–29)**. Start the engine and observe the voltmeter reading. On some MAF sensors, this reading should be 2.5 volts. Lightly tap the MAF sensor housing with a screwdriver handle and watch the voltmeter pointer. If the pointer fluctuates or the engine misfires, replace the MAF sensor. Loose internal connections will cause erratic voltage signals and engine misfiring and surging. Most MAF sensors can be checked by supplying power and a ground to the correct sensor terminals, connecting a voltmeter to the signal wire, and blowing air through the sensor.

Set the DMM so that it can read the frequency of DC voltage. With it still connected to the signal wire and ground, the meter should read about 30 hertz (Hz) with the engine idling. Now increase the engine speed and record the meter reading at various speeds. Graph the frequency readings. The MAF sensor frequency should increase smoothly and gradually in relation to engine speed. If the MAF sensor frequency reading is erratic, replace the sensor **(Figure 26–30)**.

A MAF sensor can be monitored with a scan tool through specific PIDs. When using a scan tool, make sure to use the correct specifications and follow the subsequent tests given in the service information. Normally the engine is run at 1,500 rpm for 5 seconds and then allowed to idle. The MAF return signal and operation of the sensor can be observed **(Figure 26–31)**. Often the manufacturer recommends that the sensor be observed at different speeds.

While diagnosing some MAF sensors with a scan tool, the test may display grams per second. This mode provides an accurate test of the MAF sensor. The grams-per-second reading should normally be 4 to 7 with the engine idling. This reading should gradually increase as the engine speed increases. When the engine speed is constant, the grams-per-second reading should remain constant. If the grams-per-second reading is erratic at a constant engine speed or if this reading varies when the sensor is tapped lightly, the sensor is defective. A MAF sensor DTC may not be present with an erratic grams-per-second reading, but the erratic reading indicates a defective sensor.

Frequency varying types of MAF sensors can be tested with a lab scope. The waveform should appear as a series of square waves **(Figure 26–32)**. When the engine speed and intake air flow increase, the frequency of the MAF sensor signals should increase smoothly and proportionately to the change in engine speed. If the MAF or connecting wires are defective, the trace will show an erratic change in frequency **(Figure 26–33)**.

Sensors/Inputs	PCM Pin	Measured/PID Values				Units Measured/PID
		KOEO	Hot Idle	48 KM/H (30 MPH)	89 KM/H (55 MPH)	
MAF	E25	0	0.6	1.1	1.3	DCV

Figure 26-31 The PIDs for a typical MAF sensor.

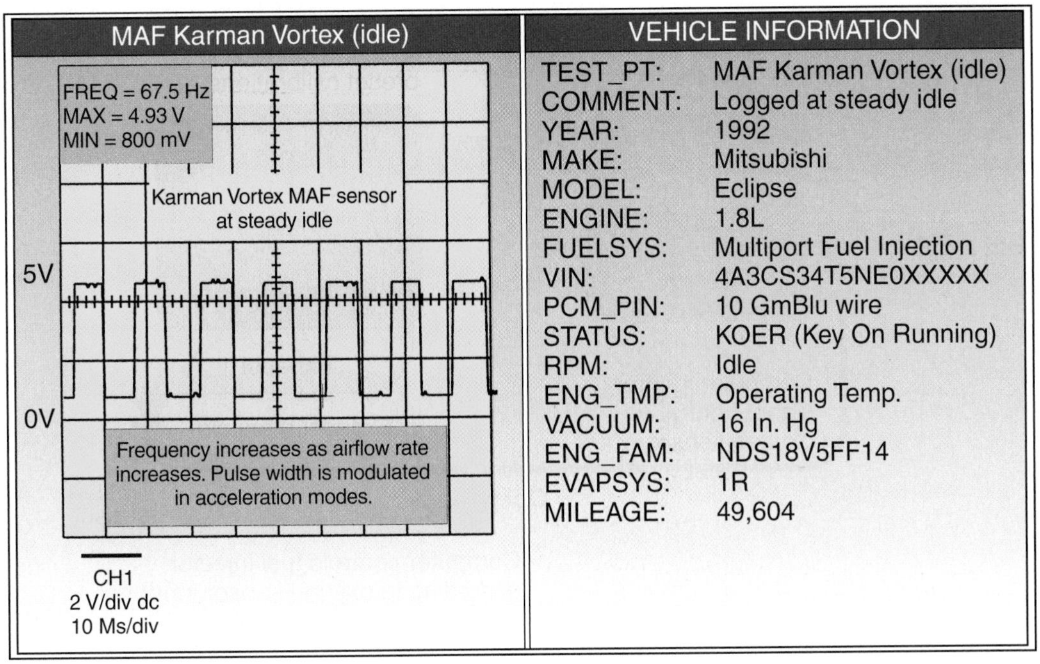

Figure 26-32 A normal trace for a frequency varying MAF sensor. *Courtesy of Progressive Diagnostics—WaveFile AutoPro*

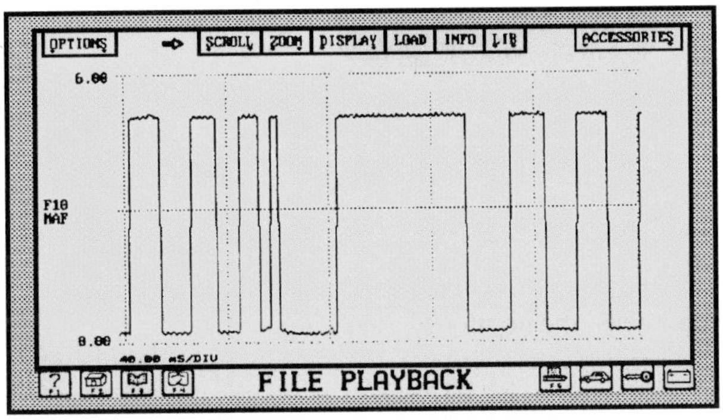

Figure 26-33 The trace of a defective frequency varying MAF sensor. *Courtesy of EDGE Diagnostics Systems*

SHOP TALK

Often O_2S are called "lambda" sensors. The term *lambda* is used to refer to "air-fuel ratio" and "normal." Technically it refers to normal and is represented by the Greek letter λ. It is best to think of lambda as meaning a reference point for normal or ideal air-fuel mixture. This mixture is typically called stoichiometric. A lambda sensor measures the variance from stoichiometric, which is about the same thing an O_2S does.

OXYGEN SENSORS (O_2S)

The exhaust gas oxygen sensor (O_2S) **(Figure 26–34)**, or air-fuel ratio sensor, is the key sensor in the closed-loop mode **(Figure 26–35)**. The O_2S is threaded into the exhaust manifold or into the exhaust pipe near the engine. The PCM uses an O_2S to ensure that the air-fuel ratio is correct for the catalytic converter. The PCM adjusts the amount of fuel injected into the cylinders based on the O_2S signal.

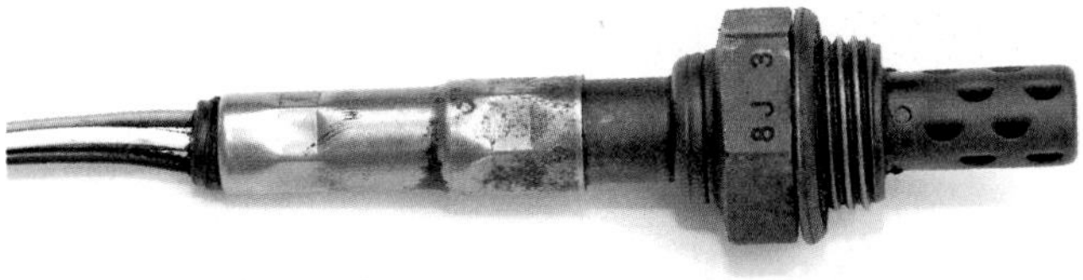

Figure 26-34 A typical oxygen sensor.

OBD-II standards require an O_2S before and after the catalytic converter **(Figure 26–36)**. The O_2S before the converter is used for air-fuel ratio adjustments. This sensor is referred to as Sensor 1. On V-type engines, one sensor is referred to as Bank 1 Sensor 1 and the other as Bank 2 Sensor 1. The O_2S after the catalytic converter is used to determine catalytic converter efficiency. This sensor is referred to as Sensor 2. With two catalytic converters, one sensor is Bank 1 Sensor 2 and the other is Bank 2 Sensor 2. Often Sensor 2 is called the catalyst monitor sensor. Some engines have more than two O_2S in an exhaust bank. The sensor with the highest number designation is the catalyst monitor sensor.

O_2S generate a voltage signal based on the amount of oxygen in the exhaust gas. They compare the oxygen content in the exhaust gas with the oxygen content of the outside air. As the difference between the exhaust

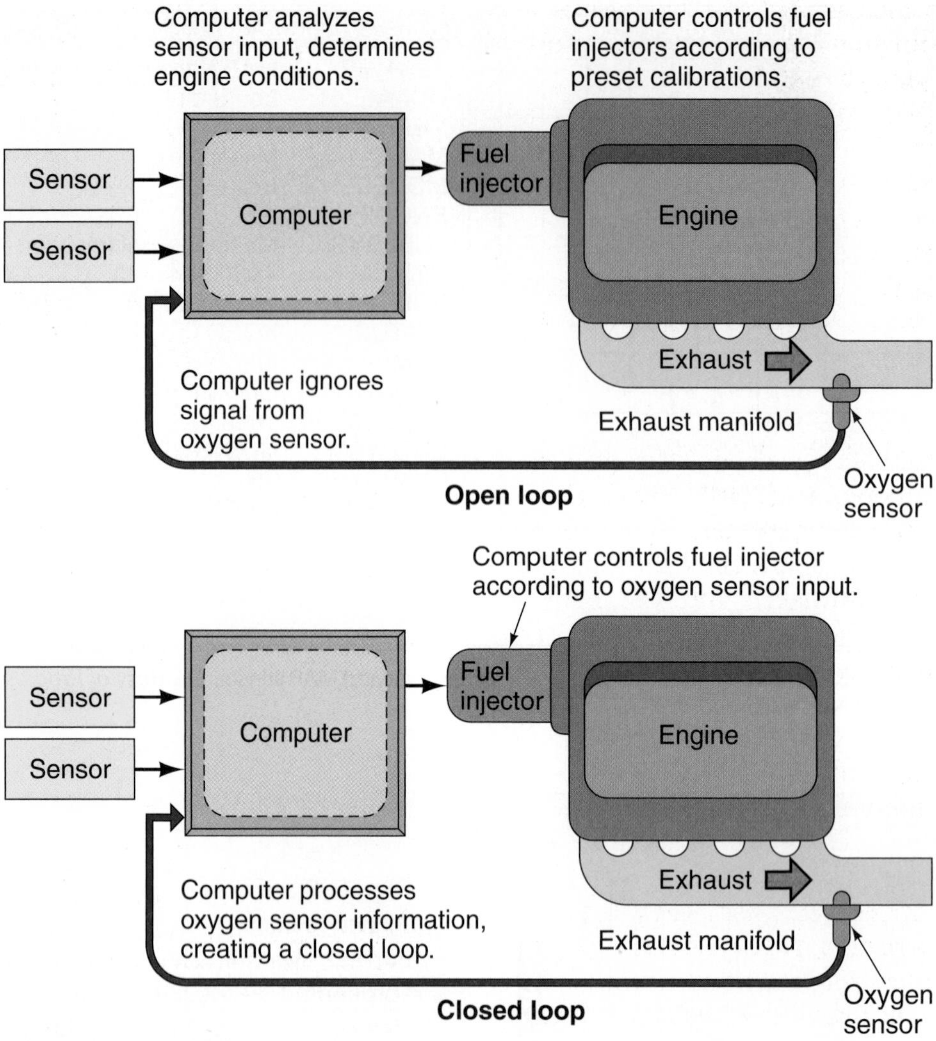

Figure 26-35 The O_2S signal is disregarded when the control system is in open loop. *Courtesy of Ford Motor Company*

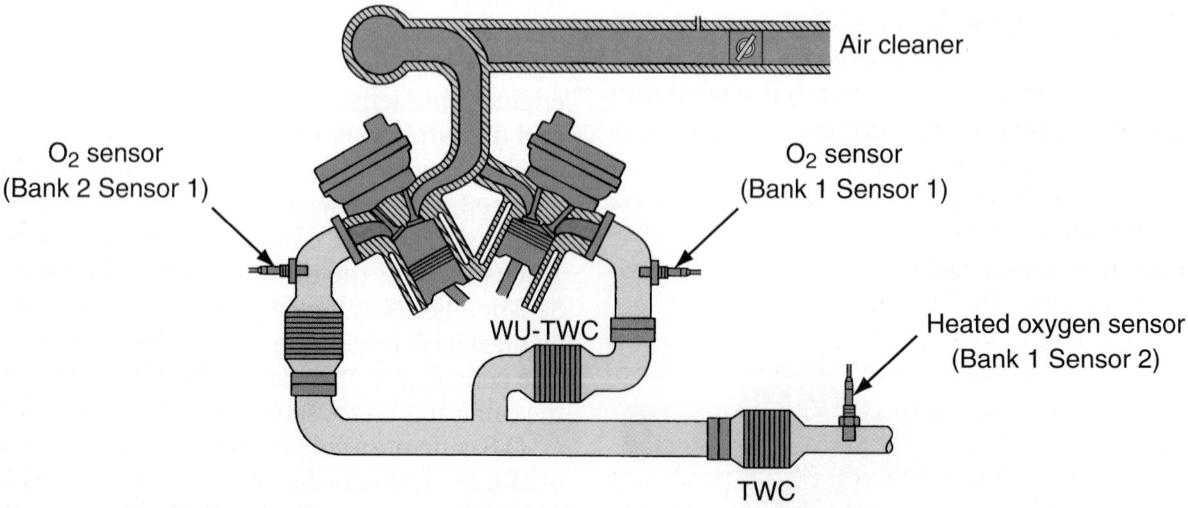

Figure 26-36 The pre- and postcatalytic oxygen sensors in an OBD-II system. *Courtesy of Toyota Motor Sales, U.S.A., Inc.*

and atmosphere increases, the voltage signal also increases. Lean mixtures release large amounts of oxygen and the voltage signal from the O_2S is low. Rich mixtures release lower amounts of oxygen and the resulting O_2 voltage signal is high. Sensor output ranges from 0.1 volt (lean) to 0.9 volt (rich). A perfectly balanced, stoichiometric, air-fuel mixture of 14.7:1 produces an output of around 0.5 volt. When the sensor reading is lean, the computer enriches the air-fuel mixture to the engine. When the sensor reading is rich, the computer leans the air-fuel mixture.

Heated Oxygen Sensors

To generate an accurate signal, an O_2S must operate at a minimum temperature of 750°F (400°C). Current O_2S have a built-in heating element to quickly heat them and keep them hot at idle and light load conditions. The heater is controlled by the PCM. Early O_2S did not have a heater and required some time for the exhaust to warm them. This resulted in extended periods of open-loop operation.

Heated oxygen sensors (HO_2S) have three or four wires connected to them. The additional wires provide voltage for the internal heater in the sensors. HO_2S are sometimes referred to as heated exhaust gas sensors (HEGOs). The heater is not on all of the time. The PCM opens and closes, duty cycles, the ground for the heater circuit as needed. The cycling of current to the heater protects the ceramic material of the heater from being overheated, which would cause it to break.

Zirconium Dioxide Oxygen (ZrO_2) Sensors

Zirconium dioxide oxygen (ZrO_2) sensors are the most commonly used O_2S, although they are being replaced with air-fuel ratio sensors on current vehicles. These have a zirconia (zirconium dioxide) element, platinum electrodes **(Figure 26–37)**, and a heater. The zirconia element has one side exposed to the exhaust stream, and the other side open to the atmosphere through the sensor's wires. Each side has a platinum electrode attached to the zirconium dioxide element. The platinum electrodes conduct the voltage generated. Contamination or corrosion of the platinum electrodes or zirconia elements will reduce the voltage signal output. This type of O_2S is sometimes referred to as a narrow range sensor because it cannot detect the small changes in oxygen content produced by minor changes of the air-fuel mixture.

Titanium Dioxide (TiO_2) Sensors Titanium dioxide oxygen (TiO_2) sensors are found on a few vehicles. These sensors do not generate a voltage signal. Instead they act like a variable resistor, altering a 5-volt reference signal supplied by the control module. Titanium sensors send a low-voltage signal (below 2.5 volts) with low oxygen content and a high-voltage signal (above 2.5 volts) with high oxygen content. Variable-resistance O_2S do not need an outside air reference. This eliminates the need for internal venting to the outside.

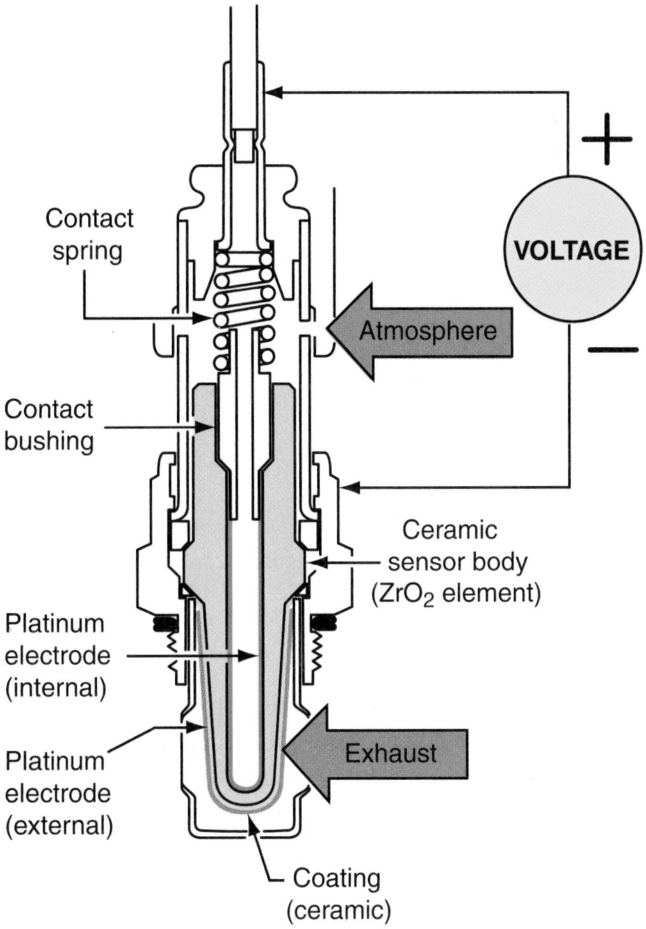

Figure 26-37 The voltage signal from an O_2S results from the difference in voltage at the two platinum plates inside the sensor.

Air-Fuel Ratio (A/F) Sensor

Although an **air-fuel ratio (A/F) sensor** looks like a conventional O_2S, internally it is different. It also operates differently; its voltage output increases as the mixture becomes leaner. The sensor does not directly produce a voltage signal; rather, it changes current. A detection circuit in the PCM monitors the change and strength of the current flow from the A/F sensor and generates a voltage signal proportional to the exhaust oxygen content **(Figure 26–38)**. Based on the voltage signal, the PCM is able to calculate the air-fuel ratio over a wide range of conditions and quickly adjust the amount of fuel required to maintain a stoichiometric ratio. The wide range of operation allows the sensor to measure very lean

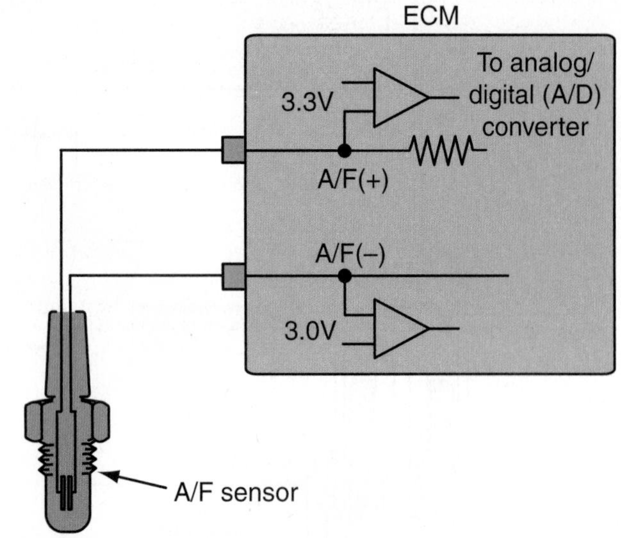

Figure 26-38 The detection circuit for an A/F sensor.

conditions. When the engine is running in lean, the oxygen content of the exhaust is higher than a normal O_2S is capable of measuring. An A/F sensor also operates at a much higher temperature, (1,200°F, 650°C), than a conventional O_2S.

The detection circuit is always measuring the direction and how much current is being produced. When the mixture is at a stoichiometric ratio, no current is generated by the sensor. The voltage signal from the detection circuit is 3.3 volts **(Figure 26–39)**. When the fuel mixture is rich (low exhaust oxygen content), the A/F produces a negative current flow and the detection circuit will produce a voltage below 3.3 volts. A lean mixture, which has more oxygen in the exhaust, produces a positive current flow, and the detection circuit will produce a voltage signal above 3.3 volts.

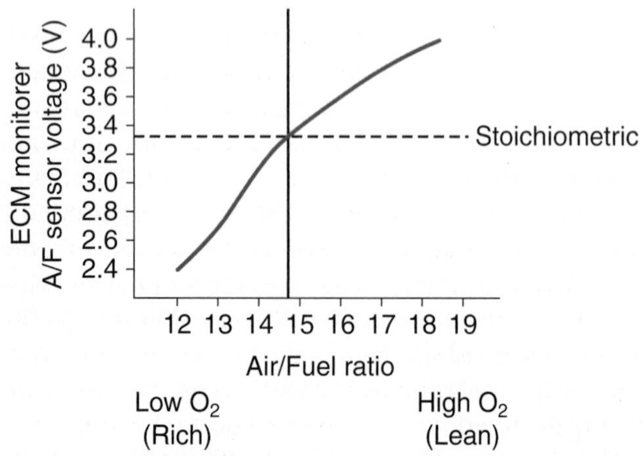

Figure 26-39 A chart showing the voltage outputs from an A/F sensor at different air-fuel ratios.

Checking Oxygen Sensors and Circuits

SHOP TALK

The engine must be at normal operating temperature before the O_2S is tested. Always refer to the specifications and testing procedures supplied by the manufacturer.

Keep in mind that several things can cause an O_2S to appear bad other than a faulty sensor. Common causes of abnormal O_2S operation include incorrect fuel pressure, a malfunctioning AIR system, an EGR leak, a leaking injector, a vacuum leak, and a contaminated MAF sensor. It is important to determine if it is the O_2S itself or some other factor that is causing the O_2S to behave abnormally.

The accuracy of the O_2S reading can be affected by air leaks in the intake or exhaust manifold. A misfiring spark plug that allows unburned oxygen to pass into the exhaust also causes the sensor to give a false lean reading. An O_2S can be contaminated and become lazy. *Lazy* is a term used to describe a common symptom; it takes a longer-than-normal time for the sensor to switch from lean to rich or rich to lean. O_2S and their heaters should also be checked for excessive resistance, opens, and shorts to ground.

Identifying the Cause of O_2S Contamination Many things can cause an O_2S to become contaminated. Before simply replacing a contaminated sensor, find out why and how it was contaminated. Begin by examining the engine for leaks; oil, coolant, and other liquids can plug the pores of the sensor and cause it to respond slowly and inaccurately to the amount of oxygen in the outside air or in the exhaust. If no leaks are evident, check the vehicle's service history. It is possible that recent problems that may or may not have been corrected are the cause of the contamination. For example, if the engine had some service done to it, RTV that was not designed for use around O_2S may have been used.

You may also discover the cause by removing the sensor. The color and smell of the sensor may indicate the problem. If the sensor has a sweet smell, it is undoubtedly contaminated by engine coolant. If it smells burnt, there is a good chance that oil has melted onto the sensor. Silicone and engine coolant will leave white deposits on the sensor. Brown coloring may indicate oil contamination, and black means it was contaminated by a rich air-fuel mixture.

Testing with a Scan Tool The OBD-II O_2S monitor checks for sensor circuit faults, slow response rate,

Definition	(Bank 1 Sensor 1)	(Bank 1 Sensor 2)	(Bank 2 Sensor 1)	(Bank 2 Sensor 2)	(Bank 1 Sensor 3)
HEATER RESISTANCE	P0053	P0054	P0059	P0060	P0055
HIGH VOLTAGE	P0132	P0138	P0152	P0158	P0144
SLOW RESPONSE	P0133	P0139	P0153	P0159	
HEATER CIRCUIT	P0135	P0141	P0155	P0161	P0147
SIGNALS SWAPPED	P0040 (Bank 1 Sensor 1/Bank 2 Sensor 1)		P0041 (Bank 1 Sensor 2/Bank 2 Sensor 2)		

Figure 26-40 A bad oxygen sensor can cause a number of problems and set a variety of DTCs.

and malfunctions in the sensor's heater circuit. There is a separate DTC for each condition for each sensor **(Figure 26–40)**. However, the catalyst monitor sensor (Sensor 3) is not monitored for response rate. In most cases, the O_2S is monitored by the PCM once per trip. The PCM tests the sensor by looking at the return signal after the air-fuel ratio has been changed. The faster the sensor responds, the better the sensor. Diagnostic Mode 5 reports the results of this monitor test.

The PCM will store a code when the sensor's output is not within the desired range. The normal range is between 0 and 1 volt and the sensor should constantly toggle from close to 0.2 to 0.8 volt then back to 0.2. If the range that the sensor toggles in is within the specifications, the computer will think everything is normal and respond accordingly. This, however, does not mean the sensor is working properly.

Watching the scan tool while the engine is running, the O_2S voltage should move to nearly 1 volt than drop back to close to 0 volts. Immediately after it drops, the voltage signal should move back up. This immediate cycling is an important function of an O_2S. If the response is slow, the sensor is lazy and should be replaced. With the engine at about 2,500 rpm, the O_2S should cycle from high to low ten to forty times in 10 seconds. The toggling is the result of the computer constantly correcting the air-fuel ratio in response to the feedback from the O_2S. When the O_2S reads lean, the computer will richen the mixture. When the O_2S reads rich, the computer will lean the mixture. When the computer does this, it is in control of the air-fuel mixture.

Testing with a DMM There are a number of ways an O_2S can be checked. Photo Sequence 26 covers the use of the min/max function on a DMM. If the voltage readings from the min/max test are not what they should be, see how the sensor reacts to lean conditions by causing a vacuum leak. Then check its response to a rich mixture by introducing some propane into the intake. If the sensor responds as it should to these changes, the cause for the previous abnormal readings is probably not the sensor.

WARNING!

An O_2S must be tested with a digital voltmeter. If an analog meter is used, the sensor may be damaged.

Connect the voltmeter between the O_2S wire and ground. Backprobe the connector near the O_2S to connect the voltmeter to the sensor signal wire. If possible, avoid probing through the insulation to connect a meter to the wire. With the engine idling, the sensor voltage should be cycling from low voltage to high voltage. The signal from most O_2S varies between 0 and 1 volt.

If the voltage is continually high, the air-fuel ratio may be rich or the sensor may be contaminated by RTV sealant, antifreeze, or lead from leaded gasoline. If the O_2S voltage is continually low, the air-fuel ratio may be lean, the sensor may be defective, or the wire between the sensor and the computer may have a high-resistance problem. If the O_2S voltage signal remains in a midrange position, the computer may be in open loop or the sensor may be defective.

The sensor can also be tested after it is removed from the exhaust manifold. Connect the voltmeter between the sensor wire and the case of the sensor. Using a propane torch, heat the sensor element. The propane flame keeps the oxygen in the air away from the sensor element, causing the sensor to produce voltage. The voltage of the sensor element should be nearly 1 volt while the sensor is in the flame. The voltage should drop to zero immediately when the flame is removed from the sensor. If the sensor does not produce the specified voltage or if the sensor does not quickly respond to the change, it should be replaced.

If a defect in the O_2S signal wire is suspected, backprobe it at the computer and connect a digital

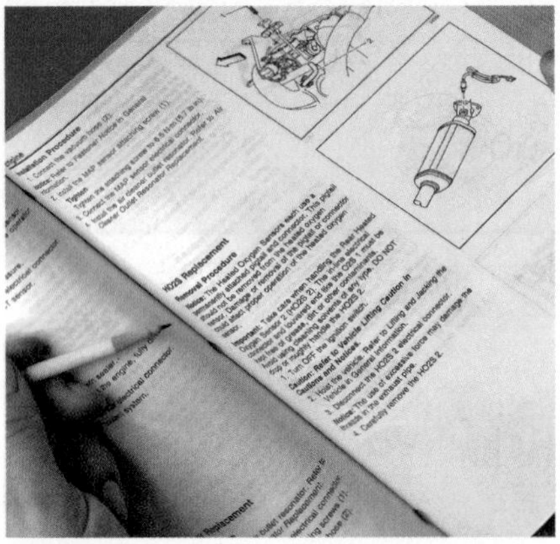

P26-1 *Locate the oxygen sensor in a wiring diagram for the vehicle and identify what part of the sensor each wire is connected to.*

P26-2 *Connect the positive lead of the meter to the power wire for the sensor's heater. Connect the meter's negative lead to a good ground.*

P26-3 *Place the meter where you can see it from the driver's seat.*

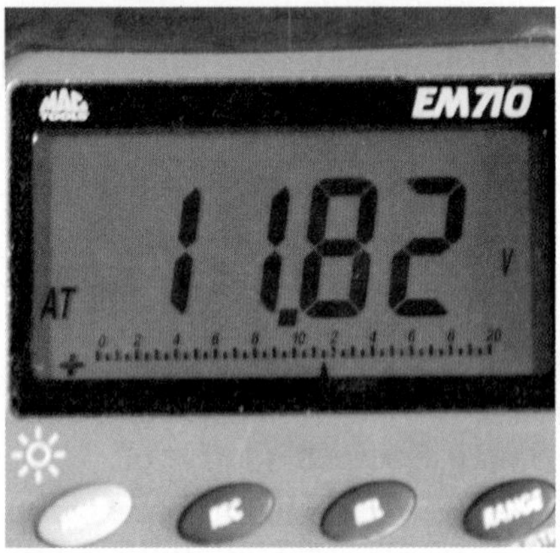

P26-4 *Start the engine and observe the voltage reading as the engine initially starts.*

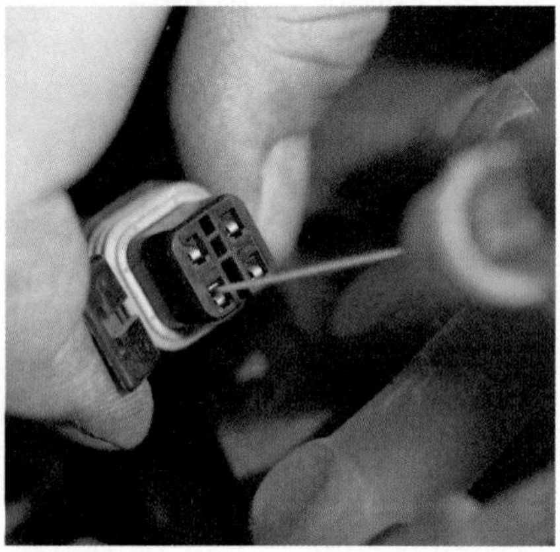

P26-5 *Turn off the engine and move the positive meter lead to the sensor's signal wire. Keep the negative lead grounded.*

P26-6 *Restart the engine and allow it to reach normal operating temperature. Look at the meter to make sure the sensor's signal is toggling from low to high voltage.*

P26-7 *Press the Min/Max button on the meter and observe the voltage. This reading will be the minimum voltage and should be about 0.1 volt.*

P26-8 *Press the Min/Max button again to observe the maximum voltage reading. This should be about 0.9 volt.*

P26-9 *Press the Min/Max button again to read the average voltage. This reading should be about 0.45 volt. Repeat this test at different speeds to get a good look at how well the O_2 sensor responds.*

voltmeter from the signal wire to ground with the engine idling. The difference between the voltage readings at the sensor and at the computer should not exceed the vehicle manufacturer's specifications. A typical specification for voltage drop across the average sensor wire is 0.02 volt.

Now check the sensor's ground. With the engine idling, connect the voltmeter from the sensor case to the sensor ground wire on the computer. Typically the maximum allowable voltage drop across the sensor ground circuit is 0.02 volt. If the voltage drop across the sensor ground exceeds specifications, repair the ground wire or the sensor ground in the exhaust manifold.

With the O_2S wire disconnected, connect an ohmmeter across the heater terminals in the sensor connector. If the heater does not have the specified resistance, replace the sensor.

Most engines are fitted with heated O_2S. A PTC thermistor inside the O_2S heats up as current passes through it. The PCM turns on the circuit based on ECT and engine load (determined from the MAF or MAP sensor signal). The higher the temperature of the heater, the greater its resistance. If the heater is not working, the sensor warm-up time is extended and the computer stays in open loop longer. In this mode, the computer supplies a richer air-fuel ratio. As a result, the engine's emissions are high and its fuel economy is reduced.

To test the heater circuit, disconnect the O_2S connector and connect a voltmeter between the heater voltage supply wire and ground **(Figure 26–41)**. With the ignition switch on, 12 volts should be supplied on this wire. If the voltage is less than 12 volts, repair the fuse in this voltage supply wire or the wire itself.

Testing with a Lab Scope A lab scope or GMM allows a look at the operation of the O_2S while it responds to changes in the air-fuel mixture. Connect the lab scope to the sensor's signal wire and a good ground **(Figure 26–42)**. Set the scope to display the trace at 200 millivolts per division and 500 milliseconds per division.

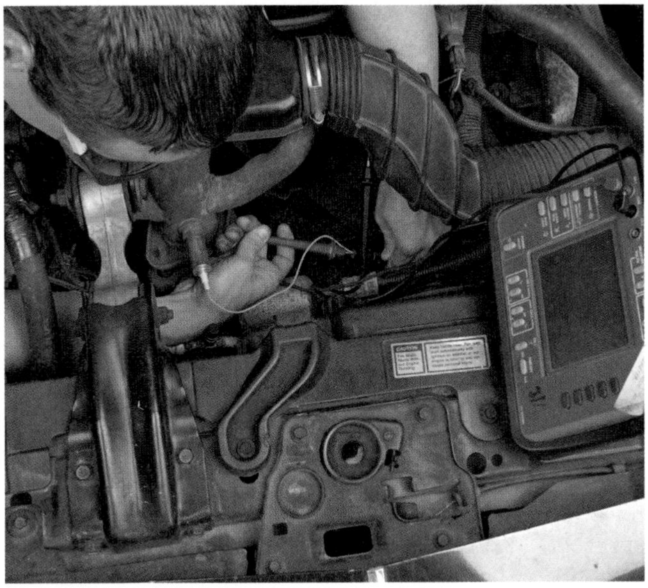

Figure 26-42 The correct way to connect a lab scope to an oxygen sensor.

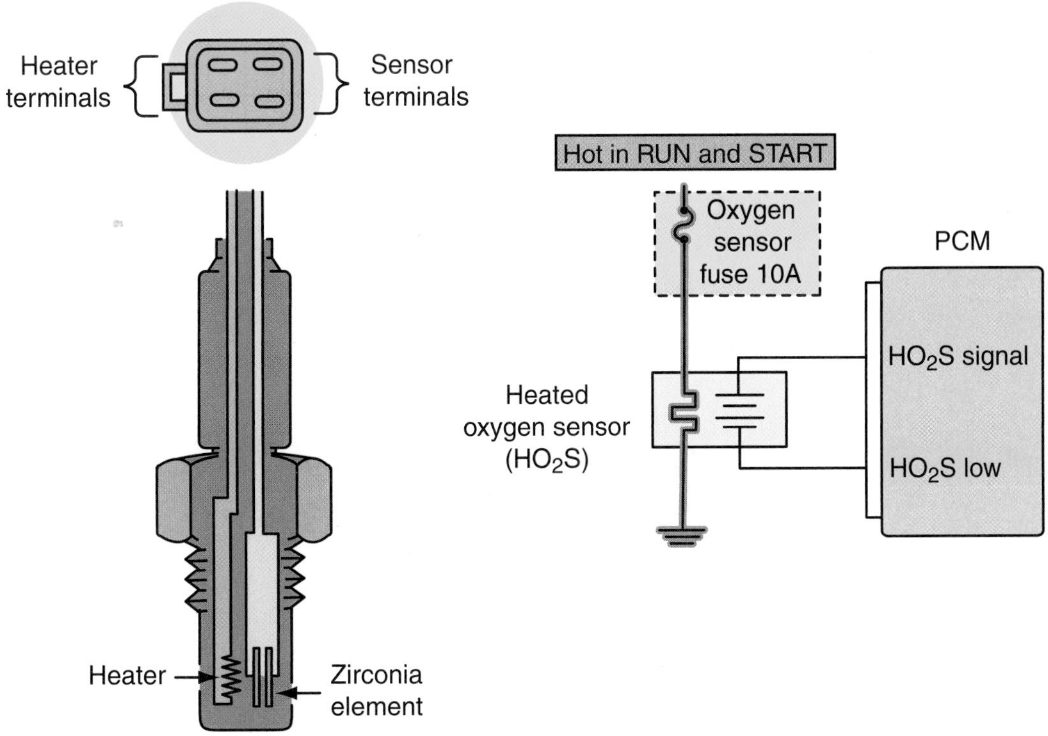

Figure 26-41 Using a wiring diagram to identify the terminals on a heated oxygen sensor.

SHOP TALK

ZrO_2 and TiO_2 sensors produce an analog signal. A/F ratio sensors only produce a signal after their current output has been sent to the PCM; make sure you refer to the manufacturer's information before making judgment on an A/F sensor.

The O_2S can be biased rich or lean, or not work at all, or work too slowly. Begin your testing by allowing the engine and O_2S to warm up. Watch the waveforms (**Figure 26–43**). If the sensor's voltage toggles between 0 and 500 millivolts, it is toggling within its normal range but it is not operating normally. It is biased low or lean. As a result, the computer will be constantly adding fuel to try to reach the upper limit of the sensor. Something is causing the sensor to be biased lean. If the toggling only occurs at the higher limits of the voltage range, the sensor is biased rich.

SHOP TALK

Some PCMs are programmed with a bias for the O_2S. This means the voltage output of the sensor does not exactly reflect how the PCM interprets the signals from the sensor. Normally, the bias is added to the actual reading. Make sure to refer to the service information to identify any voltage biases that may occur. If you do not do this, you will end up with a faulty diagnosis.

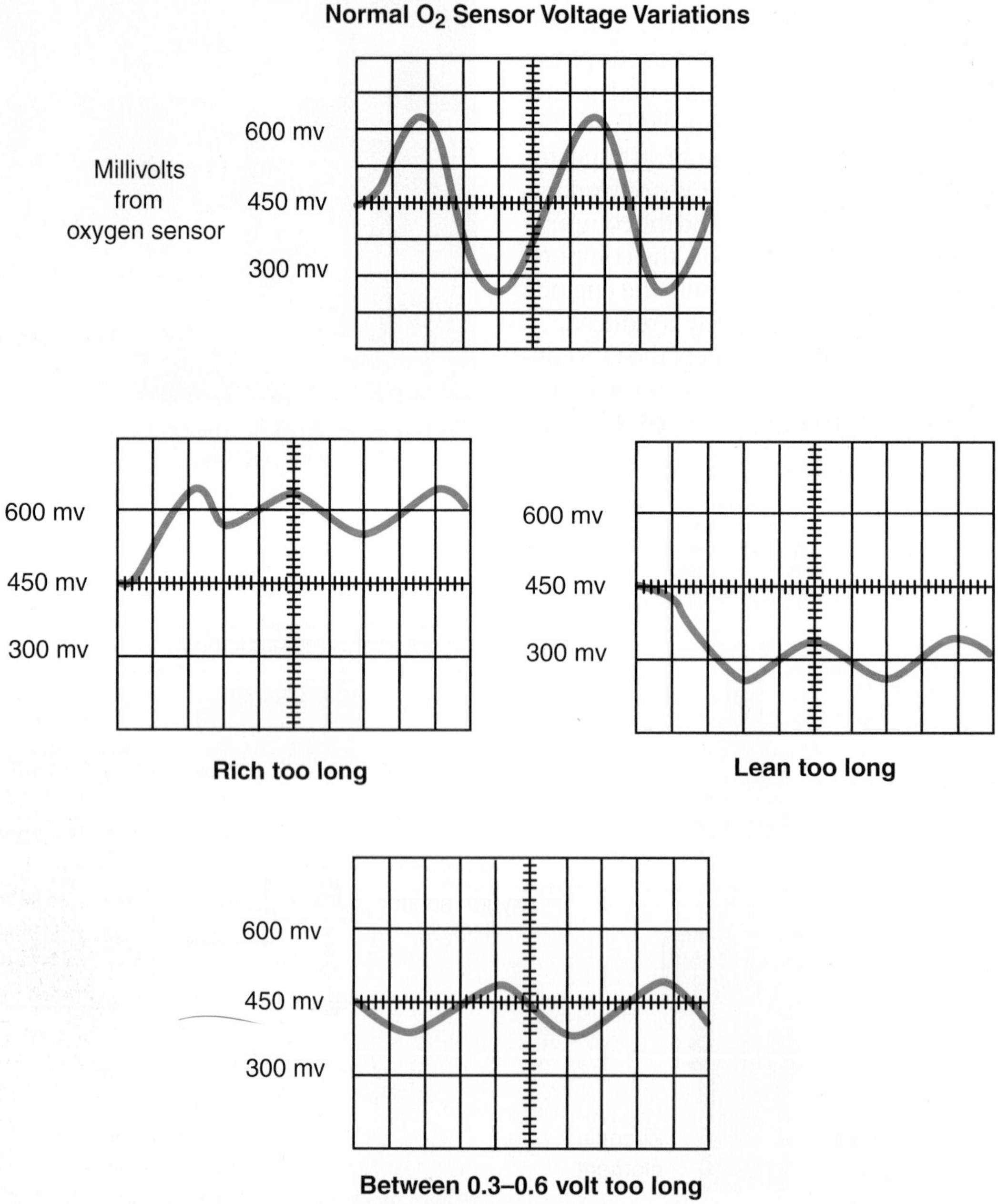

Figure 26-43 Normal and abnormal O_2S waveforms.

When the mixture is rich, combustion has a better chance of being complete. Therefore, the oxygen levels in the exhaust decrease. The O_2S output will respond to the low O_2 with a high-voltage signal. Remember that the PCM will always try to do the opposite of what it receives from the O_2S. When the O_2 shows lean, the PCM goes rich, and vice versa. When a lean exhaust signal is not caused by an air-fuel problem, the PCM does not know what the true cause is and will richen the mixture in response to the signal. This may make the engine run worse than it did.

The signals from the front (upstream) and rear (downstream) O_2S should be compared. This will not only help in determining the effectiveness of the catalytic converter, but it will also help determine the condition of each sensor. Use a two-channel scope or GMM **(Figure 26–44)**. Connect the meter to the signal wire in both harnesses. Start the engine and allow the sensors to warm up. Then raise the engine's speed to 2,000 to 2,500 rpm and observe the waveforms **(Figure 26–45)**. In the waveform, it is evident that the upstream sensor is toggling correctly, and the upstream sensor shows the catalyst is working properly.

The voltage signal from an upstream O_2S should have seven cross counts within 5 seconds with the engine running without a load at 2,500 rpm. The downstream O_2S should have less cross counts and have a lower amplitude than the upstream sensor. O_2 signal **cross counts** **(Figure 26–46)** are the number of times the O_2 voltage signal changes above or below 0.45 volt in a second. If there are not enough cross counts, the sensor is contaminated or lazy. It should be replaced. The downstream O_2S signal does not toggle when the converter is warmed up and the engine is idling. This is because the converter is using all of the oxygen that is in the exhaust.

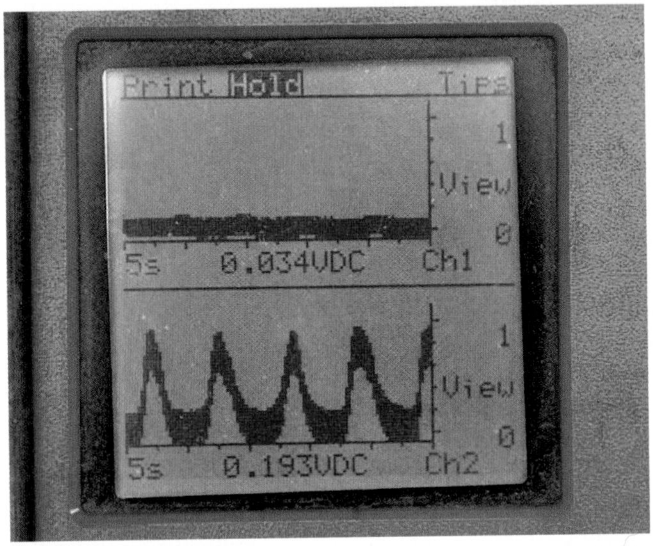

Figure 26-45 A comparison of the upstream (Ch2) and the downstream (Ch1) O_2 sensors.

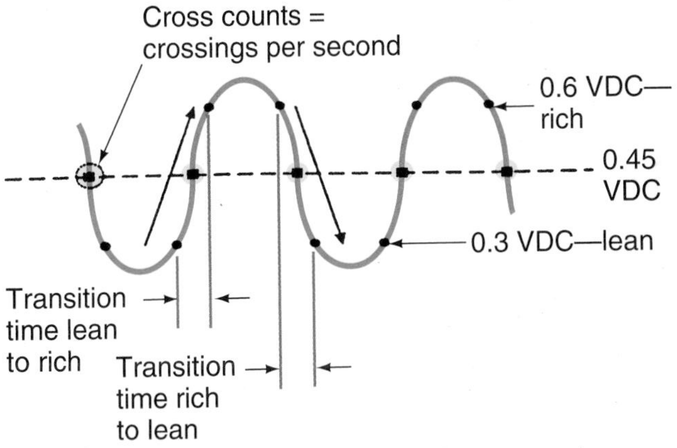

Figure 26-46 O_2 sensor signal cross counts.
Courtesy of SPX Service Solutions

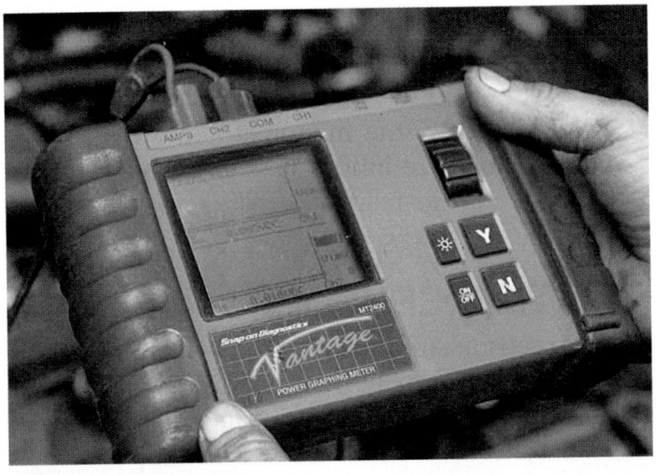

Figure 26-44 A two-channel GMM.

Another check of the responsiveness of an O_2S involves its reaction to an overly rich and lean mixtures. Insert the hose of a propane enrichment tool into the power brake booster vacuum hose or simply install it into the nozzle of the air cleaner assembly. This will drive the mixture rich. Most good O_2S will produce almost 1 volt when driven full rich. The typical specification is at least 800 millivolts. If the voltage does not go high, the O_2S is bad and should be replaced. Now, remove the propane bottle and cause a vacuum leak by pulling off an intake vacuum hose. Watch the scope to see how the O_2S reacts. It should drop to under 175 millivolts. If it does not, replace the sensor. These tests check the O_2S, not the system; therefore, they are reliable O_2S checks.

Testing Air-Fuel Ratio (A/F) Sensors

A/F sensors cannot be tested in the same way as an O_2S. The A/F sensor has two signal wires. The PCM supplies 3 volts to one wire and 3.3 volts to the other. If a voltmeter is connected across these two wires, the difference in potential will be measured. This reading will have no meaning because the 0.3-volt difference will always be there regardless of the oxygen content. This is because the sensor changes current, not voltage. The heater circuit can be checked with a voltmeter, and the heater can be checked with an ohmmeter.

To observe the changes in voltage in relationship to changes in oxygen content, use a scan tool that has the correct software. Some scan tools are not able to read the data from the sensor detection circuit, and a PID for the A/F sensor will not be available. Some scan tools will convert the range of the voltage signals to 0 to 1 volt. This is done by dividing the output from the detection circuit by 5. To calculate the actual voltage signal, multiply the measured voltage by 5. A reading of 0.66 volt equals 3.3 volts, which is the amount that should be present when the mixture is stoichiometric. Remember, the voltage from an A/F sensor will increase as the mixture goes lean and decrease with a rich mixture.

Unless there is a large change in the air-fuel ratio, the voltage readings will toggle very little. The air-fuel ratio is tightly controlled by the PCM and only minor adjustments are normally made. However, if the output voltage of the A/F sensor stays at 3.30 V regardless of conditions, the sensor circuit may be open. If voltage from the sensor stays at 2.8 V or less or 3.8 V or more, the sensor circuit may be shorted.

The action of the sensor can also be observed with an ammeter. Place the ammeter in series with the 3.3-volt signal wire. Separate the connector to the sensor. Use jumper wires to connect the terminals for the heater and the 3-volt signal wire. Then connect the positive lead of the meter to the 3.3-volt signal wire terminal on the PCM side of the connector and the negative lead to the terminal in the other half of the connector. Run the engine and observe the meter. When the mixture is at stoichiometric, there should be 0 amps. When the mixture is rich, there should be negative current flow. And when there is a lean mixture, the current should move to positive.

The A/F sensor monitor is similar to the O_2S monitor but has different operating parameters. The monitor checks for sensor circuit malfunctions and slow response rate and for problems in the sensor's heater circuit. If a fault is found, a DTC that identifies the sensor and type of fault will be set. The PCM tests the performance of A/F sensors by measuring the signal response as the amount of fuel injected into the cylinders is changed. A good sensor will respond very quickly. The results of the monitor test are not reported in Mode 5. Mode 6 is used to determine if the A/F sensors passed or failed the test.

HO_2S and A/F Sensor Repair

If the HO_2S wiring, connector, or terminal is damaged, the entire O_2S assembly should be replaced. Do not attempt to repair the assembly. In order for this sensor to work properly, it must have a clean air reference. The sensor receives this reference from the air that is present around the sensor's signal and heater wires. Any attempt to repair the wires, connectors, or terminals could result in the obstruction of the air reference and degraded O_2S performance.

Additional guidelines for servicing an HO_2S follow:

- Do not apply contact cleaner or other materials to the sensor or wiring harness connectors. These materials may get into the sensor, causing poor performance.
- Ensure that the sensor pigtail and harness wires are not damaged in such a way that the wires inside are exposed. This could provide a path for foreign materials to enter the sensor and cause performance problems.
- Ensure that neither the sensor nor the wires are bent sharply or kinked. Sharp bends, kinks, and so on could block the reference air path through the lead wire.
- Do not remove or defeat the O_2S ground wire. Vehicles that utilize the ground wired sensor may rely on this ground as the only ground contact to the sensor. Removal of the ground wire will cause poor engine performance.
- To prevent damage due to water intrusion, be sure that the sensor's seal remains intact on the wiring harness.
- Use a socket or tool designed to remove the sensor; failure to do this may result in damage to the sensor and/or wiring.
- If suggested by the manufacturer, apply a light coat of antiseize lubricant to the threads of the sensor before installing it.

POSITION SENSORS

Position sensors are used to monitor the position of something from totally closed to totally open positions. Position sensors are basically potentiometers and all operate in the same way. A wiper arm is connected to a moving part on one end and the other end is in contact with a resistor, the resistor is supplied with a reference voltage. As the part moves, so does the wiper arm. As the wiper arm moves on the resistor, the available voltage at the point where the wiper arm contacts the resistor is the signal voltage sent to the PCM. The PCM then interprets the part's position according to the voltage.

Throttle Position (TP) Sensor

Throttle position (TP) sensors (Figure 26–47) send a signal to the PCM regarding the rate of throttle opening and the relative throttle position. The wiper arm in the sensor is rotated by the throttle shaft **(Figure 26–48)**. As the throttle shaft moves, the wiper arm moves to a new location on the resistor. The return voltage signal tells the PCM how much the throttle plates are open. As the signal tells the PCM that the throttle is opening, the PCM enriches the air-fuel mixture to maintain the proper air-fuel ratio. The TP sensor is mounted on the throttle body. A separate idle contact switch or wide-open throttle (WOT) switch may also be used to signal when the throttle is in those positions.

A basic TP sensor has three wires. One wire carries the 5-volt reference signal, another serves as the ground for the resistor, and the third is the signal wire. When the throttle plates are closed, the signal voltage will be around 0.6 to 0.9 volt. As the throttle opens, there is less resistance between the beginning of the resistor and the place of wiper arm contact. Therefore, the voltage signal increases. At WOT the signal will be approximately 3.5 to 4.7 volts. Often the terminals in the connector for the sensor are gold plated. The plating makes the connector more durable and corrosion resistant.

TP Sensors for Electronic Throttle Control The TP sensor for electronic throttle systems has two wiper arms and two resistors in a single housing **(Figure 26–49)**. Therefore, they have two signal wires. This is done to ensure accurate throttle plate position in case one sensor fails. Some of these TP sensors have different voltage signals on the signal wire. However, they work in the same way. One of the signals starts at a higher voltage and has a different change rate. Other designs of this sensor have a signal that decreases with throttle opening and the other increases with the opening. In either case, the PCM uses both signals to determine throttle opening and depends on one if the other sends an out-of-range signal.

Testing a TP Sensor

The most common symptoms of a bad or misadjusted TP sensor are engine stalling, improper idle speed, and hesitation or stumble during acceleration. The fuel mixture leans out because the computer does not receive the right signal telling it to add fuel as the throttle opens. Eventually, the O_2S senses the problem and adjusts the mixture, but not before the engine stumbles.

The initial setting of the sensor is critical. The voltage signal that the computer receives is referenced to this setting. Many service manuals list the initial TP

Figure 26–47 A throttle position (TP) sensor.

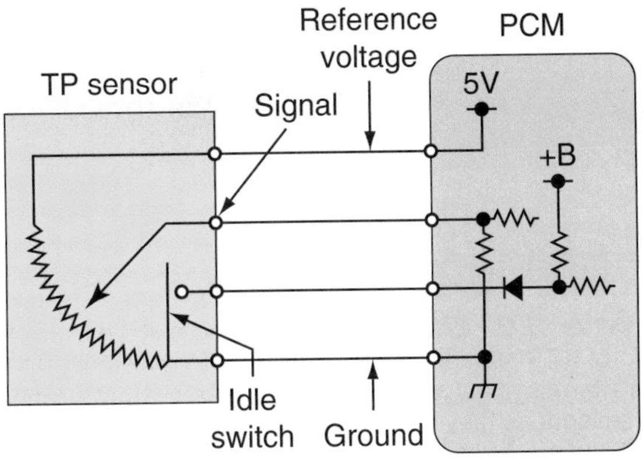

Figure 26–48 Basic circuit for a TP sensor.

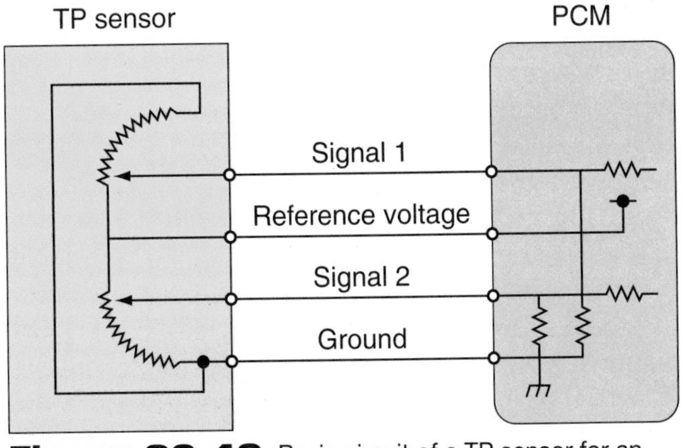

Figure 26–49 Basic circuit of a TP sensor for an electronic throttle system.

sensor setting to the nearest 0.01 volt, a clear indication of the importance of this setting.

With the ignition switch on, connect a voltmeter between the reference wire to ground. Normally, the voltage reading should be 5 volts. If the reference wire is not supplying the specified voltage, check the voltage on this wire at the computer terminal. If the voltage is within specifications at the computer but low at the sensor, repair the reference wire. When this voltage is low at the computer, check the voltage supply wires and ground wires on the computer. If these wires are satisfactory, replace the computer.

SHOP TALK

When the throttle is opened gradually to check the TP sensor voltage signal, tap the sensor lightly and watch for fluctuations on the voltmeter pointer, indicating a defective sensor.

With the ignition switch on, connect a voltmeter from the sensor signal wire to ground. Slowly open the throttle and observe the voltmeter (**Figure 26–50**). The voltmeter reading should increase smoothly and gradually. Typical TP sensor voltage readings would be 0.5 to 1 volt with the throttle in the idle position, and 4 to 5 volts at WOT. If the TP sensor does not have the specified voltage or if the voltage signal is erratic, replace the sensor.

A TP sensor can also be checked with an ohmmeter. Most often, the total resistance of the sensor is given in the specifications. If the sensor does not meet these, it should be replaced.

Adjustment of the TP sensor can be made on some engines. Incorrect TP sensor adjustment may cause inaccurate idle speed, engine stalling, and acceleration stumbles. Follow these steps to adjust a typical TP sensor:

1. Backprobe the TP sensor signal wire and connect a voltmeter from this wire to ground.
2. Turn on the ignition switch and observe the voltmeter reading with the throttle in the idle position.
3. If the TP sensor does not provide the specified voltage, loosen the TP sensor mounting bolts and rotate the sensor housing until the specified voltage is indicated on the voltmeter (**Figure 26–51**).
4. Hold the sensor in this position and tighten the mounting bolts to the specified torque.

TP sensors can also be tested with a lab scope. Connect the scope to the sensor's output and a good ground and watch the trace as the throttle is opened and closed. The resulting trace should look smooth and clean without any sharp breaks or spikes in the signal (**Figure 26–52**). A bad sensor

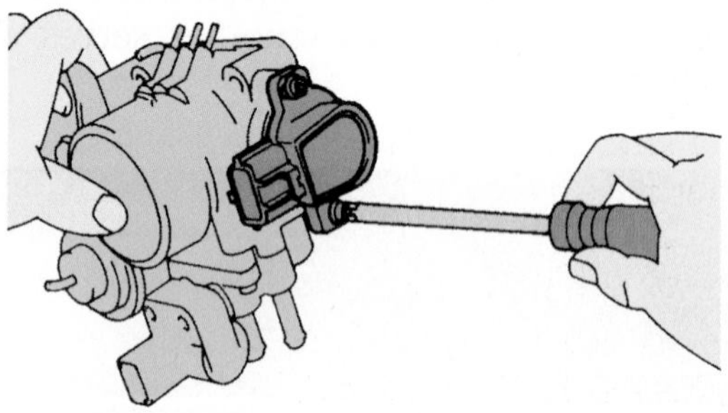

Figure 26-51 Loosen the TP sensor's mounting screws to adjust the sensor. *Courtesy of Toyota Motor Sales, U.S.A., Inc.*

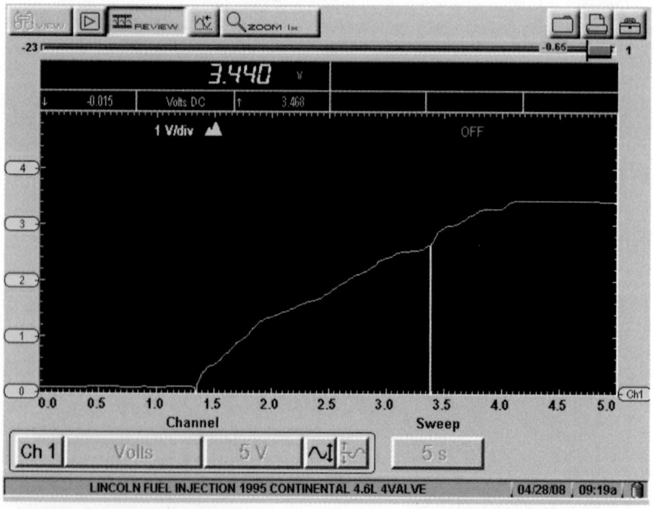

Figure 26-52 This is an analog pattern from the sweep of a TP sensor from its minimum to maximum range. This pattern is also typical of the sweep of ECT, MAP, and MAF sensors as they move through the total range. *Reproduced under license from Snap-on Incorporated. All of the marks are marks of their owners.*

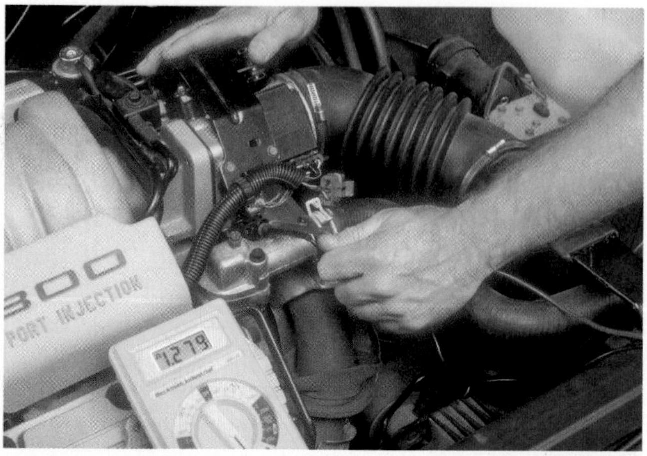

Figure 26-50 Watch the voltmeter as the throttle is opened slowly.

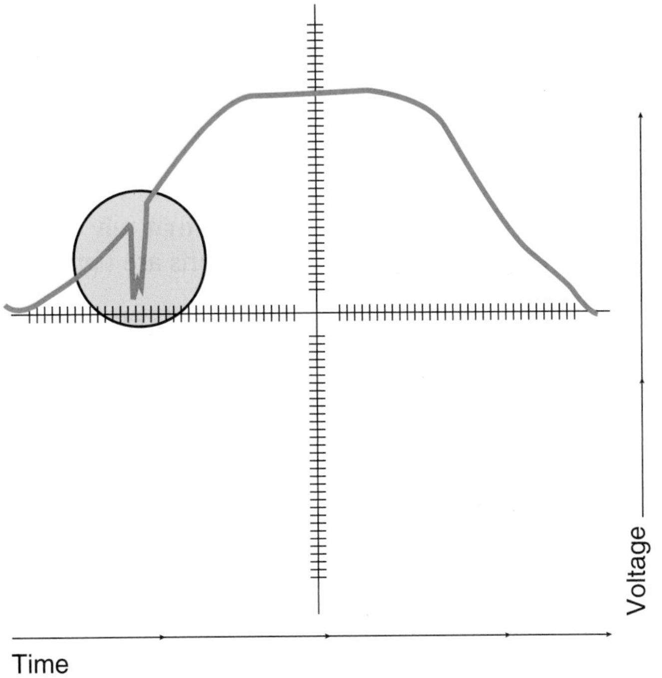

Figure 26-53 The waveform of a defective TP sensor. Notice the glitch while the throttle opens.

will typically have a glitch (a downward spike) somewhere in the trace **(Figure 26–53)** or will not have a smooth transition from high to low. These glitches are an indication of an open or short in the sensor.

The action of a TP sensor can also be monitored on a scan tool. Compare the position, expressed in a percentage, to the voltage specifications for that throttle position.

Be careful; some TP sensors have four wires. The additional wire is connected to an idle switch. Normally, when the switch is closed there will be 0 volts and battery voltage when the switch is open. Check the wiring diagram before measuring voltage here and before deciding if the switch and circuit are good.

EGR Valve Position Sensor

Manufacturers use a variety of sensors or switches to determine when and how far the EGR valve is open. This information is used to adjust the air-fuel mixture and EGR flow rates. The exhaust gases introduced by the EGR valve into the intake manifold reduce the available oxygen and thus less fuel is needed in order to maintain low HC levels in the exhaust. Most EGR valve position sensors are linear potentiometers mounted on top of the EGR valve and detect the height of the EGR valve. When the EGR valve opens, the potentiometer stem moves upward and a higher voltage signal is sent to the PCM. These sensors work in the same way as a TP sensor.

Most EGR valve position sensors have a 5-volt reference wire, voltage signal wire, and ground wire. To test one, measure the voltage at the signal wire with the ignition on and the engine not running. The meter should read about 0.8 volt. Then connect a vacuum pump to the EGR valve and slowly increase the vacuum to about 20 in. Hg. The voltage signal should smoothly increase to 4.5 V at 20 in. Hg. If the signal voltage does not reach the specified voltage, replace the sensor.

These sensors can also be checked with a lab scope or GMM. Watch the rise of the waveform as vacuum is applied. Also look for any glitches in the waveform. These are hard to see on a voltmeter unless they are real severe. The trace should be clean and smooth.

Accelerator Pedal Position (APP) Sensor

The APP sensor is used with electronic throttle control systems. It converts accelerator pedal movement into electrical signals. Like the TP sensor for the electronic throttle system, the APP is based on two potentiometers. The PCM uses the signals from this sensor to determine power or torque demand. In turn, the PCM opens or closes the throttle plate and adjusts the amount of fuel that is injected into the cylinders.

An APP is tested in the same way as other variable resistor sensors.

SPEED SENSORS

Speed sensors measure the rotational speed of something. The PCM uses these signals in a number of ways, depending on the system. Speed sensors are either Hall-effect switches or magnetic pulse generators. Identifying the type of sensor used in a particular application dictates how the sensor should be tested.

Vehicle Speed Sensor (VSS)

The most common **vehicle speed sensor** (VSS) is a magnetic pulse generator (variable reluctance) sensor **(Figure 26–54)**. However, some use a Hall-effect switch. A VSS generates a waveform at a frequency that is proportional to the vehicle's speed. When the vehicle is moving at a low speed, the sensor produces a low-frequency signal. As vehicle speed increases, so does the frequency of the signal. The PCM uses the VSS signal to help control the fuel injection system, ignition system, cruise control, EGR flow, canister purge, transmission shift timing, variable steering, and torque converter clutch lock-up timing. The signal is also used to initiate diagnostic routines. The

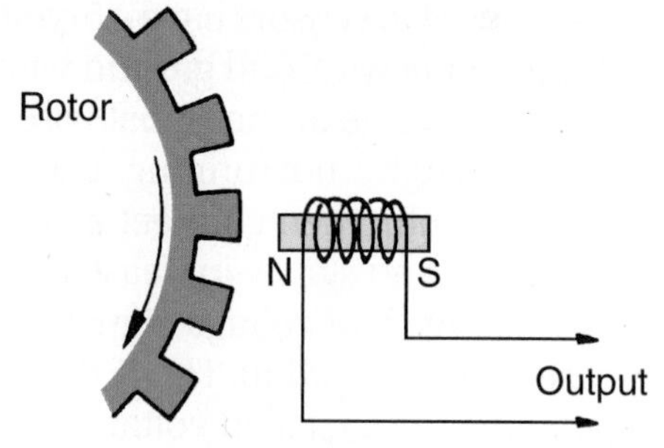

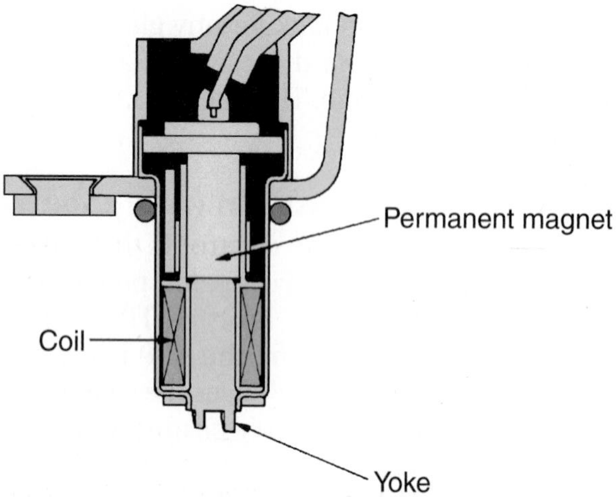

Figure 26-54 A variable speed sensor (VSS).

VSS is also used on some vehicles to limit the vehicle's speed. When a predetermined speed is reached, the PCM limits fuel delivery.

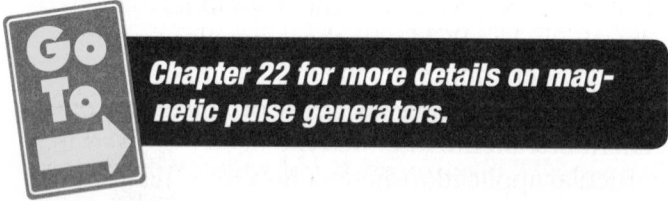

Go To *Chapter 22 for more details on magnetic pulse generators.*

The VSS is mounted on the transmission or transaxle case in the speedometer cable opening where it can measure the rotational speed of the output shaft. On earlier models, the VSS was connected to the speedometer cable. The rotation of a trigger wheel on the output shaft creates a pulsating voltage in the sensor. The frequency of the voltage increases with an increase in speed.

Troubleshooting a VSS

A defective VSS may cause different problems depending on the control systems. If the PCM does not receive a VSS signal, it will set a DTC; it may also set a code if the signal does not correlate with other inputs. A defective VSS circuit can cause increased fuel consumption, poor idle, improper converter clutch lockup, improper cruise control operation, and inaccurate speedometer operation.

A VSS should be carefully inspected before it is tested. Check the wiring harness and connectors at the sensor and the control modules that rely on the VSS signal. Make sure the connections are tight and not damaged.

The voltage in the VSS circuit should be checked for evidence of an open, short, or high resistance. Most of these measurements can be taken at the PCM connector; refer to the service manual to identify the exact measuring points.

The VSS can be tested with a scan tool or lab scope **(Figure 26–55)**; however, make sure to follow the manufacturers' test procedures when doing this. The action of a VSS is best observed while the vehicle is moving. Connect a lab scope, GMM, or scan tool and operate the vehicle at the reference speeds given in the service information. Compare the measurements to specifications. If the measurements meet the specifications, the VSS is working properly and any VSS-related problem is probably caused by the PCM. If the measurements are outside the specifications and the wiring is sound, the sensor is bad.

A VSS can also be checked with the vehicle on a hoist. The vehicle should be positioned so the drive wheels are free to rotate. Backprobe the VSS output wire and connect the voltmeter leads from this wire to ground. Select the 20 V AC scale on the voltmeter. Then start the engine.

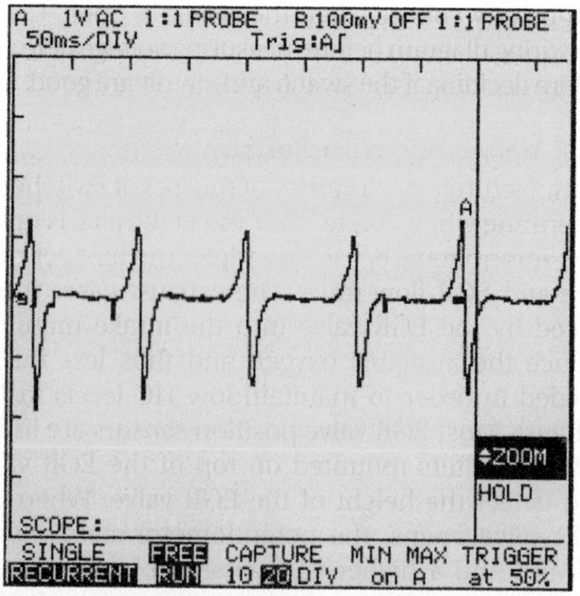

Figure 26-55 The waveform for a good VSS (PM generator). *Reproduced with permission of Fluke*

Place the transmission into a forward gear. If the VSS voltage signal is not 0.5 volt or more, replace the sensor. If the VSS signal is correct, backprobe the VSS terminal at the PCM and measure the voltage with the drive wheels rotating. If 0.5 volt is available at this terminal, the trouble may be in the PCM.

When 0.5 volt is not available at this terminal, turn the ignition switch off and disconnect the wire from the VSS to the PCM. Connect the ohmmeter leads across the wire. The meter should read 0 ohms. Repeat the test with the ohmmeter leads connected to the VSS ground terminal and the PCM ground terminal. This wire should also have 0 ohms resistance. If the resistance in these wires is more than specified, repair the wires.

Speed sensors can also be checked with an ohmmeter. Most manufacturers list a resistance specification. The resistance of the sensor is measured across the sensor's terminals. The typical range for a good sensor is 800 to 1,400 ohms of resistance.

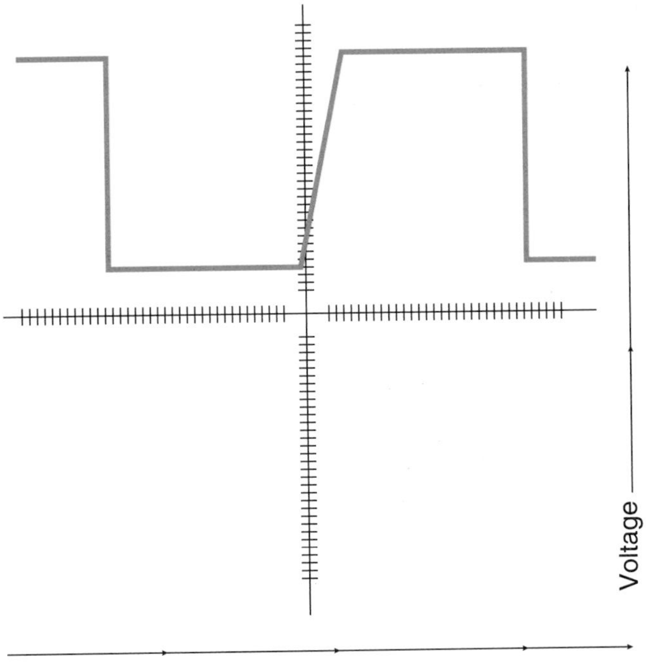

Figure 26-56 A Hall-effect switch with a bad transistor.

Hall-Effect Sensors To test a Hall-effect switch, disconnect its wiring harness. Connect a voltage source of the correct low-voltage level across the positive and negative terminals of the Hall layer. Then connect a voltmeter across the negative and signal voltage terminals. Insert a metal feeler gauge between the Hall layer and the magnet. Make sure the feeler gauge is touching the Hall element. If the sensor is operating properly, the meter will read close to battery voltage. When the feeler gauge blade is removed, the voltage should decrease. On some units, the voltage will drop to near zero. Check the service manual to see what voltage you should observe when inserting and removing the feeler gauge.

When observing a Hall-effect sensor on a lab scope, pay attention to the downward and upward pulses. These should be straight. If they appear at an angle **(Figure 26–56)**, this indicates the transistor is faulty, causing the voltage to rise slowly. The waveform should be a clean and flat square wave. Any change from a normal trace means the sensor should be replaced.

Other Speed Sensors

Speed sensors are used in several other places. Most commonly they are used to monitor the speed of various shafts within an automatic transmission. They provide the PCM signals that may indicate slipping inside the transmission. Some vehicles also have a speed sensor built into engine's cooling fan clutch.

POSITION/SPEED SENSORS

Position/speed sensors tell the PCM the speed, change of speed, and position of a component. The two most commonly used are the camshaft position sensor and crankshaft position sensor. The inputs from these sensors are used to control ignition timing, fuel injection delivery, and variable valve timing. The sensor does this using a Hall-effect switch or magnetic pulse generator; both designs generate a voltage signal.

These sensors are most often magnetic pulse generators. The sensor is mounted close to a pulse wheel or rotor that has notches or teeth. As each tooth moves past the sensor, an AC voltage pulse is induced in the sensor's coil. As the rotor moves faster, more pulses are produced. Speed is calculated by the frequency of the signal, which is the number of pulses in a second. To provide a clean strong signal, the distance between the sensor and the rotor has a specification. If the sensor is too far away from the rotor, the signal will be weak. The ECM determines the speed that the component is revolving based on the number of pulses.

Magnetic pulse generators create an AC voltage signal and their wiring harness is normally a twisted and/or shielded pair of wires.

Crankshaft Position (CKP) Sensor

The rotor for a **crankshaft position (CKP) sensor** has several teeth (the number varies with application)

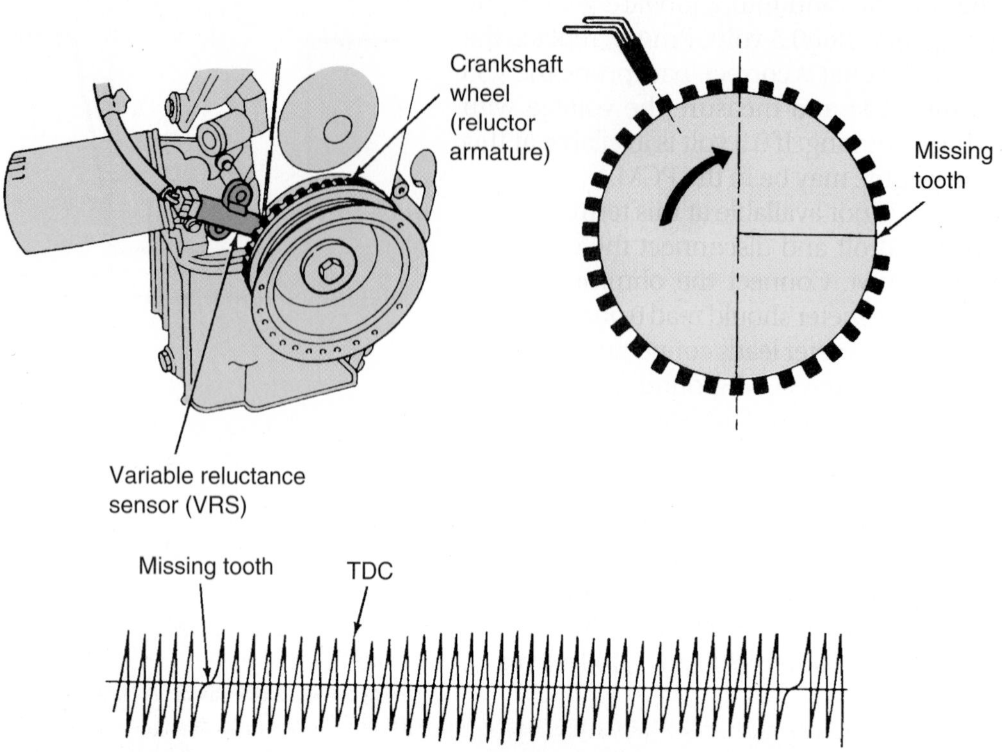

Figure 26-57 Sensor activity to monitor crankshaft speed and position, as well as the location of the number one piston. *Courtesy of Ford Motor Company*

equally spaced around the outside of the rotor. One or more teeth are missing at fixed locations. These missing teeth provide a reference point for the PCM to determine crankshaft position (**Figure 26–57**). For example, the pulse wheel may have a total of thirty-five teeth spaced 10 degrees apart and an empty space where the thirty-sixth tooth would have been. The thirty-five teeth are used to monitor crankshaft speed; the gap is used to identify which pair of cylinders is approaching TDC. The input from the camshaft position sensor signals is used in order to determine which of these two cylinders is on its firing stroke and which is on the exhaust stroke.

Input from the CKP (**Figure 26–58**) is critical to the operation of the ignition system. This input is also used by the PCM to determine if a misfire has occurred. This is done by looking at the time intervals between the teeth. If a misfire occurs, the transition from one tooth to another will be slower than normal.

Checking a CKP Like all electrical devices, the sensor and its wiring should be checked for corrosion and damage. If the PCM detects abnormal signals it will set a DTC. The CKP is checked during several monitor tests. On engines with both a CKP and camshaft position sensor, the PCM will compare those two inputs and set a code if they do correlate with each other.

Figure 26-58 A crankshaft position (CKP) sensor.

The operation of the sensor can be monitored with a scan tool and lab scope. The waveforms from most CKP sensors will have a number of equally spaced pulses and one double pulse or sync signal, as shown in **Figure 26–59**. The number of evenly spaced pulses equals the number of cylinders that the engine has. A Hall-effect sensor generates a digital signal. Carefully examine the trace. Any glitches indicate a problem with the sensor or sensor circuit.

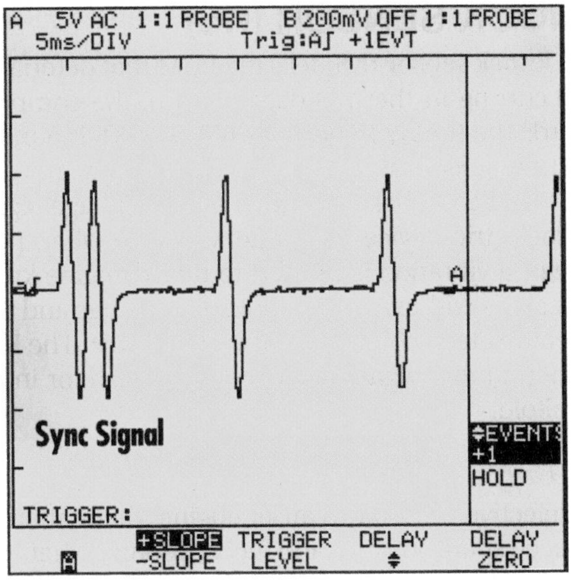

Figure 26-59 The trace of a good crankshaft position sensor. *Reproduced with permission of Fluke*

Replacing a CKP Although the exact procedure for replacing a CKP varies with manufacturer and engine, there are some common steps. The gap between the sensor and the rotor must be set correctly (**Figure 26–60**). This is done in several ways, depending on application. Often this is done with a special alignment tool or spacer that comes with the replacement sensor. If the sensor has an O-ring, make sure to apply a light coat of oil onto the seal. Failure to do this will prevent the sensor from being fully seated, which will affect its output signals and allow for oil leakage.

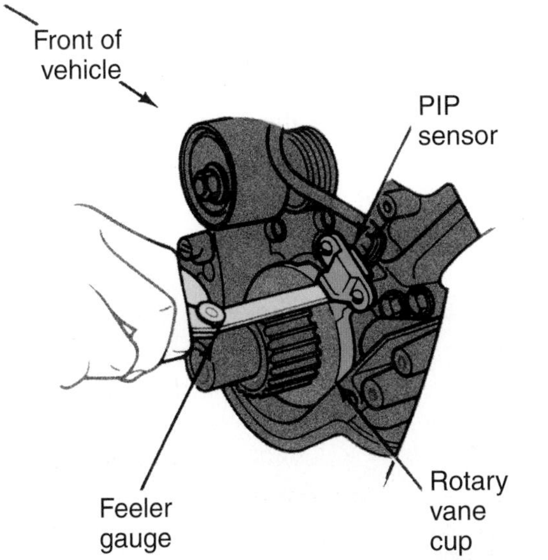

Figure 26-60 The gap between the CKP (PIP) sensor and the rotor is critical. *Courtesy of Ford Motor Company*

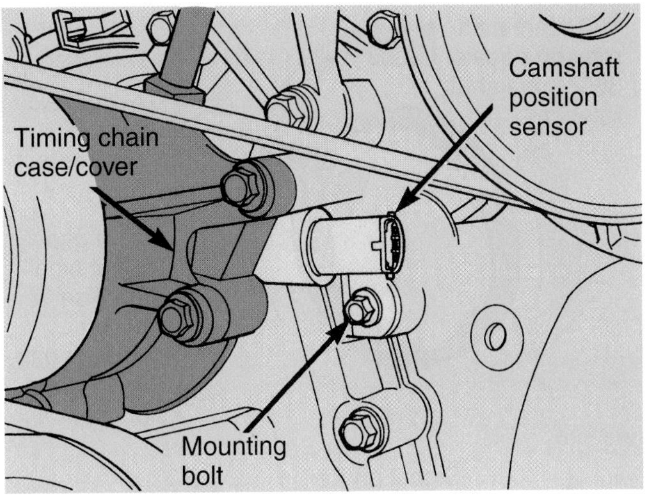

Figure 26-61 A camshaft sensor mounted to the timing chain cover. *Courtesy of Chrysler LLC*

Camshaft Position (CMP) Sensor

The **camshaft position (CMP) sensor** monitors the position of the camshaft (**Figure 26–61**). The CMP's output is used with the CKP to determine when cylinder number 1 is on its compression stroke. This information is used for control of the fuel injection, direct ignition, and variable valve timing systems. Engines with variable valve timing typically have two CMPs, one for each engine bank. On some engines, a bad CMP will prevent the misfire monitor from completing.

CMP sensors can be magnetic pulse generators (**Figure 26–62**) or Hall-effect sensors (**Figure 26–63**). The type of sensor is identified by the number of wires connected to the sensor: A magnetic pulse sensor has two wires and the Hall-effect sensor has three.

CMP Sensor Service Most diagnostic procedures for the CKP apply to the CMP sensor. There are many DTCs related to the CMP. If one of these is set, follow the testing procedures given by the manufacturer.

Figure 26-62 A camshaft sensor and notched cam gear. *Courtesy of Chrysler LLC*

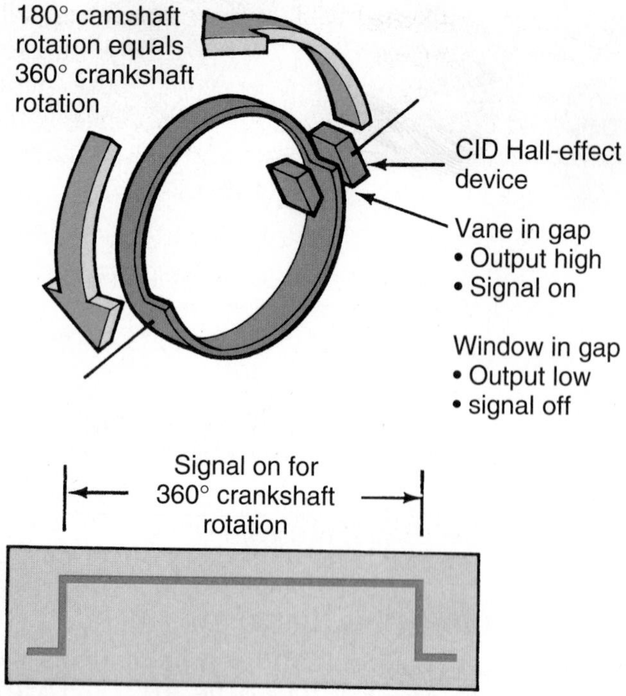

Figure 26-63 A Hall-effect camshaft sensor and its resultant signal. *Courtesy of Ford Motor Company*

These include a thorough inspection of the sensor and its wiring. When observing a CMP on a lab scope, make sure you know what type the sensor is. Magnetic pulse generators create an analog signal, whereas Hall-effect sensors provide a digital signal. Magnetic pulse generators can also be checked with an ohmmeter (**Figure 26–64**). The procedures may also include a verification of camshaft timing. Make sure to lubricate the O-ring when installing a new sensor and tighten the retaining bolt to specifications.

Go To Chapter 13 for the procedure for verifying and adjusting camshaft timing.

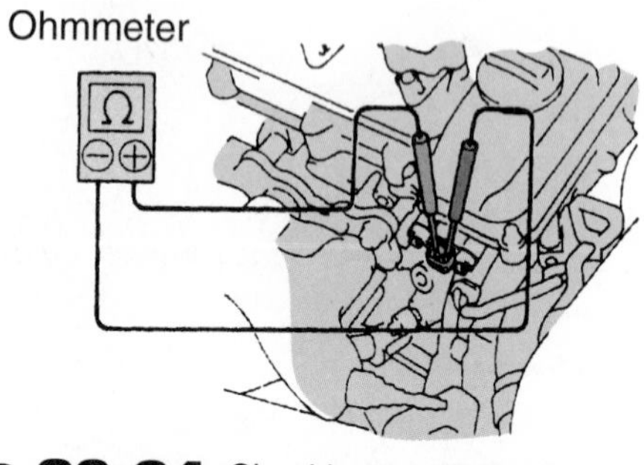

Figure 26-64 Checking a camshaft sensor with an ohmmeter. *Courtesy of Toyota Motor Sales, U.S.A., Inc.*

KNOCK SENSOR (KS)

The **knock sensor (KS)** tells the PCM that detonation is occurring in the cylinders. In turn, the computer retards the timing (**Figure 26–65**). The KS is a piezoelectric device that works like a microphone and converts engine knock vibrations into a voltage signal. Piezoelectric devices generate a voltage when pressure or a vibration is applied to them. Engine knock typically is within a specific frequency range and a KS is set to detect vibrations within that range. The KS is located in the engine block, cylinder head, or intake manifold.

Testing a KS

A defective KS may cause engine detonation or reduced spark advance and fuel economy. When a KS is removed and replaced, the sensor must be tightened to its specified torque. The procedure for checking a KS varies, depending on the vehicle make and year. The KS is checked with several monitors by the PCM. The PCM checks the input from the sensor for abnormal readings that may be caused by an open, short, or high resistance. Always follow the vehicle manufacturer's recommended test procedure and specifications.

A KS can be checked by observing it on a scope while tapping the engine at a point close to the sensor. The waveform should react to the tapping. This test may not work on all KS. The sensor will not respond to the tapping if it is not synchronized to the CKP signal, which is the normal situation on some engines.

A KS and its circuit can also be checked with a voltmeter. Check the voltage input at the sensor and the power feed from the PCM. If the specified voltage is not available to the sensor, backprobe the KS wire at the PCM and measure the voltage. If the voltage is satisfactory at this terminal, repair the KS wire. If the voltage is not within specifications at the PCM, replace the PCM.

Figure 26-65 A knock sensor (KS).

A scan tool can also be used to diagnose a KS. During a road test, open the throttle quickly and observe the scan tool. There should be at least one count when the throttle is opened.

COMPUTER OUTPUTS AND ACTUATORS

Once the PCM's programming instructs that a correction or adjustment must be made in the controlled system, an output signal is sent to a control device or actuator. These actuators—solenoids, switches, relays, or motors—physically act or carry out the command sent by the PCM.

Actuators are electromechanical devices that convert an electrical current into mechanical action. This mechanical action can then be used to open and close valves, control vacuum to other components, or open and close switches. When the PCM receives an input signal indicating a change in one or more of the operating conditions, the PCM determines the best strategy for handling the conditions. The PCM then controls a set of actuators to achieve a desired effect or strategy goal. In order for the computer to control an actuator, it must rely on a component called an output driver.

The circuit driver usually applies the ground circuit of the actuator **(Figure 26–66)**. The ground can be applied steadily if the actuator must be activated for a selected amount of time, or the ground can be pulsed to activate the actuator in pulses. Output drivers are transistors or groups of transistors that control the actuators. These drivers operate by the digital commands from the PCM. If an actuator cannot be controlled digitally, the output signal must pass through an A/D converter before flowing to the actuator. The major actuators in a computer-controlled engine include the following components:

- *Air Management Solenoids.* Secondary air bypass and diverter solenoids control the flow of air from the air pump to either the exhaust manifold (open loop) or the catalytic converter (closed loop).
- *Evaporative Emission (EVAP) Canister Purge Valve.* This valve is controlled by a solenoid. The valve controls when stored fuel vapors in the canister are drawn into the engine and burned. The computer only activates this solenoid valve when the engine is warm and above idle speed.
- *EGR Flow Solenoids.* EGR flow may be controlled by electronically controlled vacuum solenoids. The solenoid valves are supply manifold vacuum to the EGR valve when EGR is required or may vent vacuum when EGR is not required.
- *Fuel Injectors.* These solenoid valves deliver the fuel spray in fuel-injected systems.
- *Idle Speed Controls.* These actuators are small electric motors. On carbureted engines, this idle speed motor is mounted on the throttle linkage. On fuel-injected systems a stepper motor may be used to control the amount of air bypassing the throttle plate.

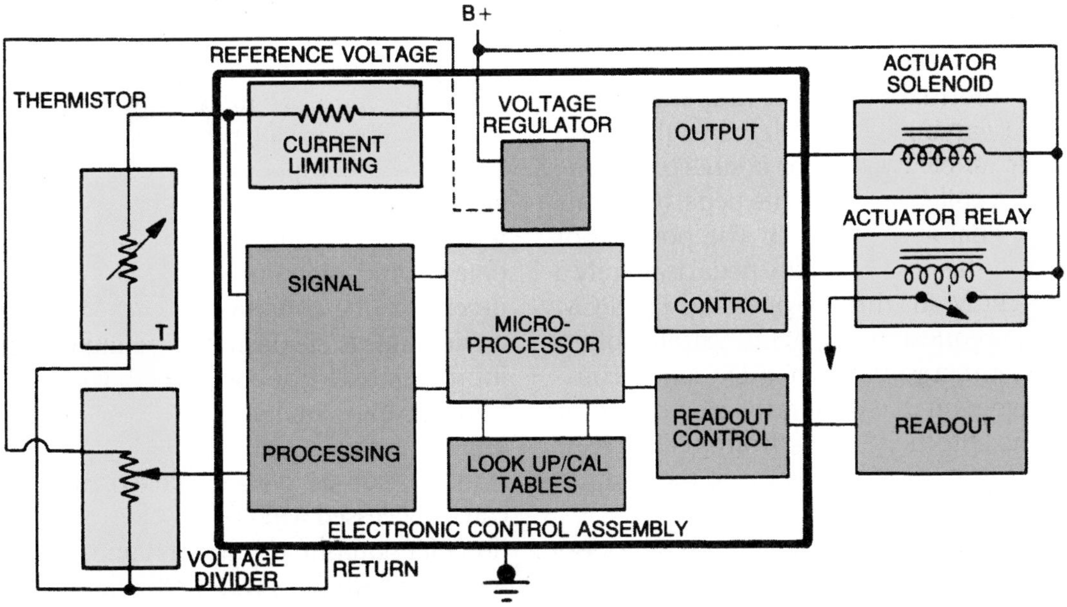

Figure 26-66 Output drivers in the computer usually supply a ground for the actuator solenoids and relays. *Courtesy of Ford Motor Company*

- *Ignition Module.* This is actually an electronic switching device triggered by a signal from the control computer. The ignition module may be a separate unit or may be part of the PCM.
- *Mixture Control (MC) Solenoids.* On some carbureted engines, an electrical solenoid is used to operate the metering rods and idle air bleed valve in the carburetor.
- *Motors and Lights.* Using electrical relays, the computer is used to trigger the operation of electric motors, such as the fuel pump, or various warning light or display circuits.
- *Other Solenoids.* Computer-controlled solenoids may also be used in the operation of cruise control systems, torque converter lock-up clutches, automatic transmission shift mechanisms, and many other systems where mechanical action is needed.

Electronic Throttle Control

Like modern aircraft, the acceleration on some late-model vehicles works on the "drive-by-wire" principle, which is typically called **electronic throttle control (ETC)**. ETC interprets gas pedal movement by the driver and allows for precise throttle control, which helps to improve fuel economy and performance while reducing emissions.

Instead of a throttle cable and mechanical linkage to the throttle body, the connection is made through wires **(Figure 26–67)**. Although these systems are electronically controlled and operated, some still have a mechanical backup system or resort to partial throttle if something goes wrong with the electronic system.

One or two position sensors are attached to the accelerator pedal assembly, sending position and rate of change information to the PCM. The pedal's sensor sends a varying voltage signal to the PCM, which controls an electric motor connected to the throttle plates. A coiled spring in the pedal assembly gives the gas pedal a normal feel. The position or rate of change in the position of the pedal is merely a request to the PCM for throttle opening. The PCM processes this request along with various other inputs and its programming. It then sends commands to a driver unit that powers the electric motor attached to the throttle. Signals from a TP sensor allow the PCM to track the position of the throttle plate.

Electronic throttles are easily adaptable to support cruise control and traction control systems. In the latter, if the wheels spin, the system can close the throttle until wheel spin is no longer detected. Throttle control is also integrated into automatic shifting. With electronic control, the throttle can be closed slightly to reduce engine output during a shift, providing smoother gear changes.

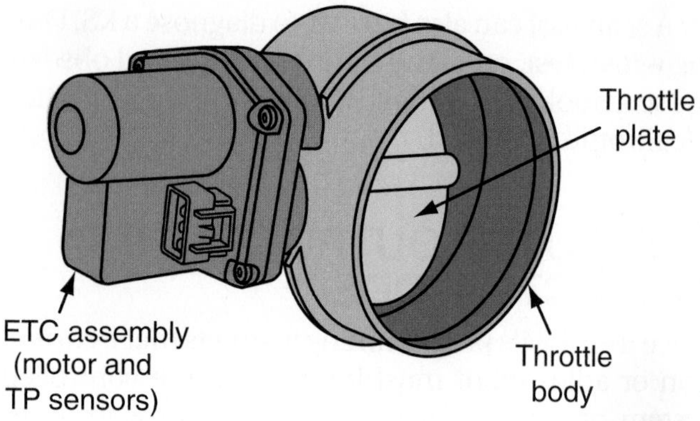

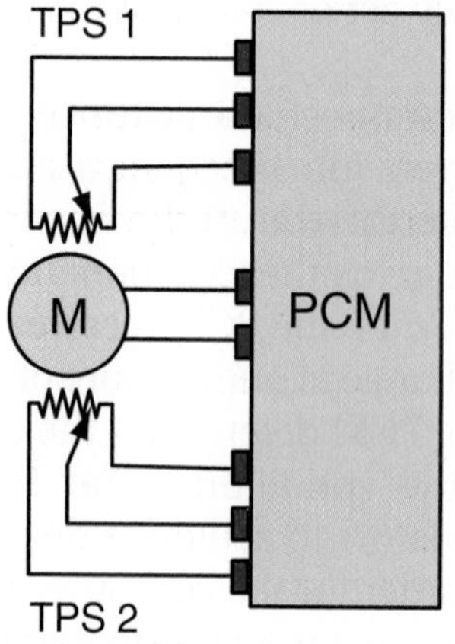

Figure 26-67 A cutaway of an electronic throttle control assembly.

TESTING ACTUATORS

Most systems allow for testing of the actuator through a scan tool. Actuators that are duty cycled by the computer are more accurately diagnosed through this method. Prior to diagnosing an actuator, make sure the engine's compression, ignition system, and intake system are in good condition. Serial data can be used to diagnose outputs using a scanner. The displayed data should be compared against specifications to determine the condition of any actuator. Also, when an actuator is suspected to be faulty, make sure the inputs related to the control of that actuator are within normal range. Faulty inputs will cause an actuator to appear faulty.

Many control systems have operating modes that can be accessed with a scan tool to control the operation of an output. Common names for this mode are the output state control (OSC) and output test mode (OTM). In this mode, an actuator can be enabled or disabled; the duty cycle or the movement of the actuator can be increased or decreased. While the actuator is being controlled, related PIDs are observed as an indication of how the system reacted to the changes. The actuators that can be controlled by this mode vary. Always refer to the service information to determine what can be checked and how it should be checked.

Testing with a DMM

If the actuator is tested by other means than a scanner, always follow the manufacturer's recommended procedures. Because many actuators operate with 5 to 7 volts, never connect a jumper wire from a 12-volt source unless directed to do so by the appropriate service procedure. Some actuators are easily tested with a voltmeter by testing for input voltage to the actuator. If there is the correct amount of input voltage, check the condition of the ground. If both of these are good, then the actuator is faulty. If an ohmmeter needs to be used to measure the resistance of an actuator, disconnect it from the circuit first.

When checking anything with an ohmmeter, logic can dictate good and bad readings. If the meter reads infinite, this means there is an open. Based on what you are measuring across, an open could be good or bad. The same is true for very low resistance readings. Across some things, this would indicate a short. For example, you do not want an infinite reading across the windings of a solenoid. You want low resistance. However, you want an infinite reading from one winding terminal to the case of the solenoid. If you have low resistance, the winding is shorted to the case.

Testing Actuators with a Lab Scope

Most computer-controlled circuits are ground-controlled circuits. The PCM energizes the actuator by providing the ground. On a scope trace, the on-time pulse is the downward pulse. On positive-feed circuits, where the computer is supplying the voltage to turn a circuit on, the on-time pulse is the upward pulse. One complete cycle is measured from one on-time pulse to the beginning of the next on-time pulse.

Actuators are electromechanical devices, meaning they are electrical devices that cause some mechanical action. When actuators are faulty, it is because they are electrically faulty or mechanically faulty. By observing the action of an actuator on a lab scope, you will be able to watch its electrical activity. Normally if there is a mechanical fault, this will affect its electrical activity as well. Therefore, you get a good sense of the actuator's condition by watching it on a lab scope.

To test an actuator, you need to know what it basically is. Most actuators are solenoids. The computer controls the action of the solenoid by controlling the pulse width of the control signal. You can see the turning on and off of the solenoid **(Figure 26–68)** by watching the control signal. The voltage spikes are caused by the discharge of the coil in the solenoid.

Some actuators are controlled pulse-width modulated signals **(Figure 26–69)**. These signals show a changing pulse width. These devices are controlled by varying the pulse width, signal frequency, and voltage levels.

Both waveforms should be checked for amplitude, time, and shape. You should also observe changes to the pulse width as operating conditions

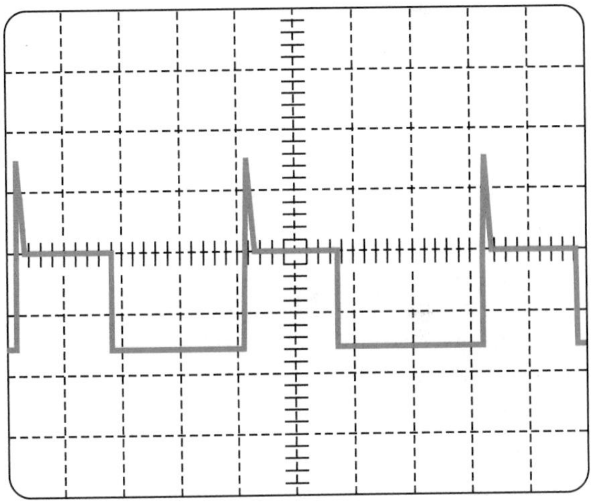

Figure 26-68 A typical solenoid control signal.

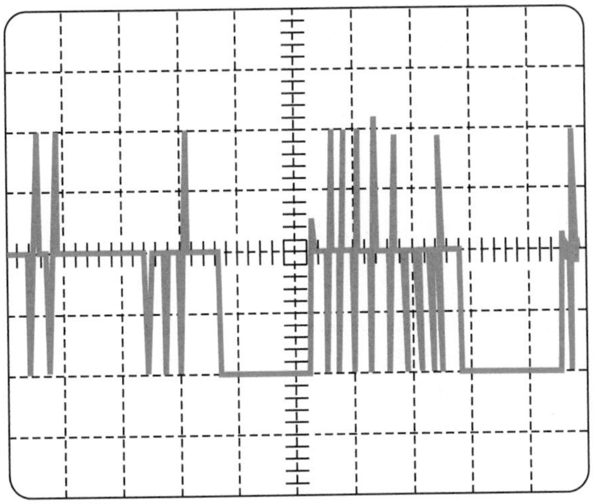

Figure 26-69 A typical pulse-width modulated solenoid control signal.

change. A bad waveform will have noise, glitches, or rounded corners. You should be able to see evidence that the actuator immediately turns off and on according to the commands of the computer.

A fuel injector is actually a solenoid. The PCM's signals to an injector vary in frequency and pulse width. Frequency varies with engine speed, and the pulse width varies with fuel control. Increasing an injector's on time increases the amount of fuel delivered to the cylinders. The trace of a normally operating fuel injector is shown in **Figure 26–70**.

Repairing the System

After isolating the source of the problem, the repairs should be made. The system should then be rechecked to verify that the repair took care of the problem. This may involve road testing the vehicle in order to verify that the complaint has been resolved.

When servicing or repairing OBD-II circuits, the following guidelines are important:

- Do not connect aftermarket accessories into an OBD-II circuit.
- Do not move or alter grounds from their original locations.
- Always replace a relay in an OBD-II circuit with an exact replacement. Damaged relays should be thrown away, not repaired.
- Make sure all connector locks are in good condition and are in place.
- After repairing connectors or connector terminals, make sure the terminals are properly retained and the connector is sealed.
- When installing a fastener for an electrical ground, be sure to tighten it to the specified torque.

Verification of repair is more comprehensive for vehicles with OBD-II system diagnostics than earlier vehicles. Following a repair, the technician should perform the following steps:

1. Review the fail records and the freeze frame data for the DTC that was diagnosed. Record the fail records or freeze frame data.
2. Use the scan tool's clear DTCs or clear information functions to erase the DTCs.
3. Operate the vehicle within the conditions noted in the fail records or the freeze frame data.
4. Monitor the status information for the specific DTC until the diagnostic test associated with that DTC runs.

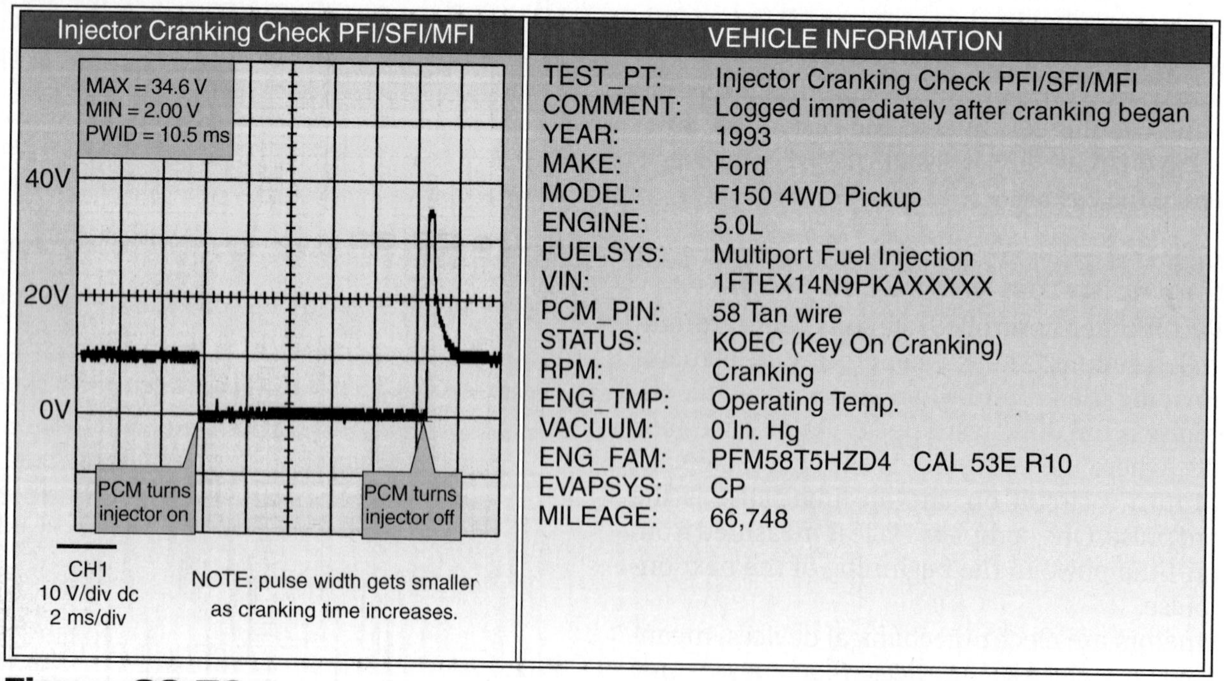

Figure 26-70 The trace of a normally operating fuel injector. *Courtesy of Progressive Diagnostics—WaveFile AutoPro*

CASE STUDY

A customer brought a late-model Dodge into the shop. The concern was that sometimes the engine idled rough and stalled and occasionally seemed like it was missing or surging. The customer also stated that this occurred any time, not in any particular driving condition.

The technician took the car out for a road test and did not witness any surging. However, once she pulled it back into the shop, the engine idled very poorly. She turned off the engine and did a quick underhood inspection. She then connected a scan tool to retrieve any DTCs. She found that code P0123 was set. This code means the PCM has detected voltage signals from the TP sensor that were too high.

Applying her knowledge of Ohm's law to the problem, the technician determined that the problem was caused by a decrease in circuit resistance or the reference voltage from the PCM was too high. She checked the reference voltage at the TP sensor and the PCM and found that both are exactly 5 volts. She then did a more thorough inspection of the sensor wires but found nothing. Then she cleared the codes and ran the engine. At that time the engine had a smooth idle. She wiggled the wiring harness from the PCM to the sensor. The idle quality did not change. Then she turned off the engine and retrieved the DTCs. No code was pulled.

She had confirmed what she suspected before; this was an intermittent problem. She turned off the engine and again wiggled the wiring. The engine was started again and this time the idle was poor. She pulled the codes and found that the same DTC was set. She knew that she was close to identifying the cause of the problem. A careful inspection of the harness and wires was conducted. No loose connections were found and there were no signs of burnt, cut, or chafed wires. No problem was found.

Feeling a bit discouraged, she once again inspected the harness. This time she noticed that as she pulled on the harness to inspect it, the TP sensor moved. She reached up to the sensor to find that it was very loose. She then tightened the mounting bolts and started the engine. It idled smoothly. The DTCs were again cleared and the vehicle taken out on a road test. On return to the shop, the engine idle was still smooth. Again codes were pulled but none were set. She determined that the loose sensor was the cause of the intermittent problem, and she was correct. Code P0123 can be set by a faulty PCM, a faulty TP sensor, a poorly mounted TP sensor, or a wire-to-wire short or short-to-ground in the TP sensor circuit.

KEY TERMS

Air-fuel ratio (A/F) sensor
Camshaft position (CMP) sensor
Crankshaft position (CKP) sensor
Cross counts
Electronic throttle control (ETC)
Engine coolant temperature (ECT) sensor
Heated oxygen sensor (HO_2S)
Intake air temperature (IAT) sensor
Knock sensor (KS)
Manifold absolute pressure (MAP) sensor
Mass airflow (MAF) sensor
Throttle position (TP) sensor
Vapor pressure sensor (VPS)
Vehicle speed sensor (VSS)

SUMMARY

- Troubleshooting electronic engine control systems involves much more than retrieving trouble codes.
- The following can be used to check individual system components: visual checks, ohmmeter checks, voltmeter checks, and lab scope checks.
- Service bulletin information is absolutely essential when diagnosing engine control system problems.
- All PCMs (OBD-II and earlier designs) cannot operate properly unless they have good ground connections and the correct voltage at the required terminals.
- A voltage drop test is a quick way of checking the condition of any electrical conductor or terminal connector.

- Poor grounds can also allow EMI or noise to be present on the reference voltage signal. The best way to check for noise is to use a lab scope.
- It is important to understand how a sensor works and what it measures before testing it.
- Sensors measure temperature, chemical characteristics, pressure, speed, position, and sound.
- Most sensors can be checked with a voltmeter, ohmmeter, scan tool, lab scope, and GMM.
- Most computer-controlled actuators are electromechanical devices that convert the output commands from the computer into mechanical action. These actuators are used to open and close switches, control vacuum flow to other components, and operate valves depending on the system's requirements.
- Most systems allow for testing of the actuator through a scan tool.
- When checking anything with an ohmmeter, logic can dictate good and bad readings. If the meter reads infinite, this means there is an open. Based on what you are measuring across, an open could be good or bad. The same is true for very low-resistance readings. Across some things, this would indicate a short.
- Actuators can be accurately tested with a lab scope.

REVIEW QUESTIONS

1. The PCM uses the VSS signal to help control many systems. List five of them.
2. List the four ways that individual components can be checked.
3. *True or False?* A bad ground can cause an increase in the reference voltage to a sensor.
4. *True or False?* An A/F ratio sensor can be tested in the same way as an oxygen sensor.
5. A typical normal oxygen sensor signal will toggle between ______ and ______ volts.
6. Explain how a bad circuit ground can affect a sensor's reference voltage.
7. When an engine is running lean, the voltage signal from the oxygen sensor will be ______ (low or high).
8. Describe the procedure for testing a MAP sensor that produces a digital voltage signal of varying frequency.
9. Many control systems have operating modes that can be accessed with a scan tool to control the operation of an output. What are the names of the two most common modes for controlling outputs?
10. Which of the following statements about zirconium oxygen sensors is *not* true?
    a. The normal operating range for an O_2S is between 0 and 1 volt.
    b. If the sensor's voltage toggles between 0 volts and 500 millivolts, it is operating normally.
    c. The voltage signal from an upstream O_2S should have seven cross counts within 5 seconds with the engine running without a load at 2,500 rpm.
    d. Engine control systems monitor the activity of the O_2S and store a code when the sensor's output is not within the desired range.
11. Describe the typical procedure for adjusting a TP sensor.
12. Why does the rotor for a CKP have one or more missing teeth?
13. A system may fail the catalyst monitor test for all of these reasons, *except* ______.
    a. defective upstream or downstream oxygen sensor circuits
    b. a leaking exhaust
    c. fuel contaminants
    d. ignition system faults
14. A defective IAT sensor or circuit may cause the following problems, *except* ______.
    a. hard engine starting
    b. rich or lean air-fuel ratio
    c. improper converter clutch lockup
    d. reduced fuel economy
15. Which of the following is the least likely cause of a no-start condition?
    a. faulty antitheft devices
    b. KS
    c. fuel or ignition system faults
    d. faulty intake air system

ASE-STYLE REVIEW QUESTIONS

1. Technician A says that an oxygen sensor can be a voltage-producing sensor. Technician B says that an oxygen sensor is a thermistor sensor. Who is correct?
    a. Technician A
    b. Technician B
    c. Both A and B
    d. Neither A nor B

2. While discussing zirconium O_2S diagnosis: Technician A says that the voltage signal on a satisfactory O_2S should always be cycling between 0.5 V and 1 V. Technician B says that a contaminated O_2S provides a continually low voltage signal. Who is correct?

   a. Technician A c. Both A and B
   b. Technician B d. Neither A nor B

3. While discussing ECT sensor diagnosis: Technician A says that a defective ECT sensor may cause hard cold engine starting. Technician B says that a defective ECT sensor may cause improper operation of emission control devices. Who is correct?

   a. Technician A c. Both A and B
   b. Technician B d. Neither A nor B

4. While discussing vehicle speed sensor tests: Technician A says to use an ohmmeter to test the resistance of the coil. Technician B says that the voltage generated by the sensor can be measured by connecting a voltmeter across the sensor's terminals. Who is correct?

   a. Technician A c. Both A and B
   b. Technician B d. Neither A nor B

5. While discussing TP sensor diagnosis: Technician A says that a four-wire TP sensor contains an idle switch. Technician B says that in some applications, the TP sensor mounting bolts may be loosened and the TP sensor housing rotated to adjust the voltage signal with the throttle in the idle position. Who is correct?

   a. Technician A c. Both A and B
   b. Technician B d. Neither A nor B

6. While diagnosing the cause of a hard starting condition: Technician A checks the ECT sensor. Technician B checks the MAP sensor. Who is correct?

   a. Technician A c. Both A and B
   b. Technician B d. Neither A nor B

7. While discussing testing OBD-II system components: Technician A says that a testlight can be used on 12-volt circuits. Technician B says that a digital voltmeter can be used on the circuits. Who is correct?

   a. Technician A c. Both A and B
   b. Technician B d. Neither A nor B

8. When testing a frequency varying MAF: Technician A says that the frequency should increase with an increase in engine speed. Technician B says that when observed on a lab scope, the signal from the sensor should be a square wave. Who is correct?

   a. Technician A c. Both A and B
   b. Technician B d. Neither A nor B

9. Technician A says that the oxygen sensor provides the major input during the open loop mode. Technician B says that the coolant temperature sensor controls open loop mode operation. Who is correct?

   a. Technician A c. Both A and B
   b. Technician B d. Neither A nor B

10. While discussing the MIL on OBD-II systems: Technician A says that the MIL will flash if the PCM detects a fault that would damage the catalytic converter. Technician B says that whenever the PCM has detected a fault, it will turn on the MIL. Who is correct?

    a. Technician A c. Both A and B
    b. Technician B d. Neither A nor B

CHAPTER

27

IGNITION SYSTEMS

OBJECTIVES

■ Name and describe the three basic types of ignition systems. ■ Name the two major electrical circuits used in all ignition systems and their common components. ■ Describe the operation of ignition coils, spark plugs, and ignition cables. ■ Explain how high voltage is induced in the coil secondary winding. ■ Describe the various types of spark timing systems, including electronic switching systems and their related engine position sensors. ■ Explain the basic operation of a computer-controlled ignition system. ■ Describe the operation of a distributor-based ignition system. ■ Describe the operation of a distributorless ignition system.

The ignition system in today's vehicles is an integral part of the electronic engine control system. The engine control module (ECM) controls all functions of the ignition system and constantly corrects the spark timing. The desired ignition timing is calculated by the PCM according to inputs from a variety of sensors. These inputs allow the PCM to know the current operating conditions. The PCM matches those conditions to its programming and controls ignition timing accordingly. It is important to remember that there has always been a need for engine speed- and load-based timing adjustments. Electronic systems are very efficient at making these adjustments. Many of the inputs used for ignition system control are also used to control other systems, such as fuel injection. These inputs are available on the CAN buses **(Figure 27–1)**.

There are three basic ignition system designs: distributor-based (DI) systems, **distributorless ignition systems (DLI)**, and **direct ignition systems (DIS)**. The latter two designs are designated as electronic ignition (EI) systems by the SAE. DIS is the commonly used design on today's engines.

BASIC CIRCUITRY

All ignition systems consist of two interconnected electrical circuits: a **primary** (low-voltage) **circuit** and a **secondary** (high-voltage) **circuit** **(Figure 27–2)**.

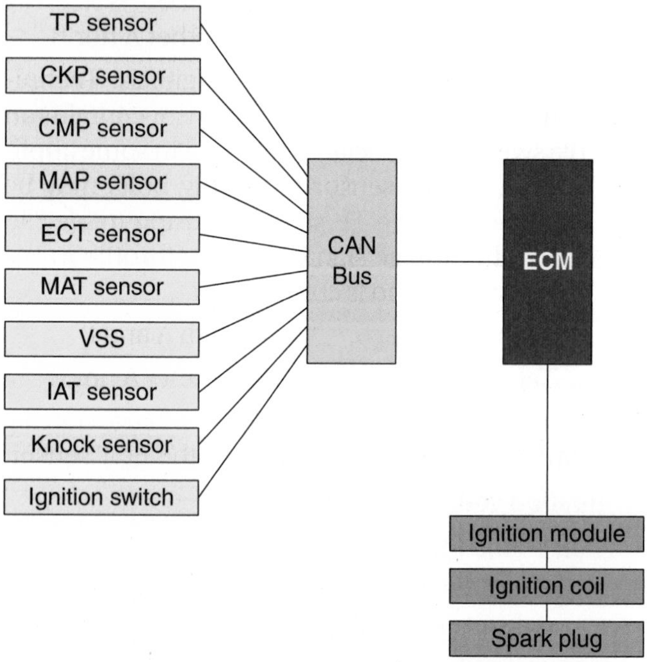

Figure 27–1 Many of the inputs used for ignition system control are also used to control other systems and are available on the CAN buses.

Depending on the exact type of ignition system, components in the primary circuit include the following:

- Battery
- Ignition switch
- Ballast resistor or resistance wire (some systems)
- Starting bypass (some systems)

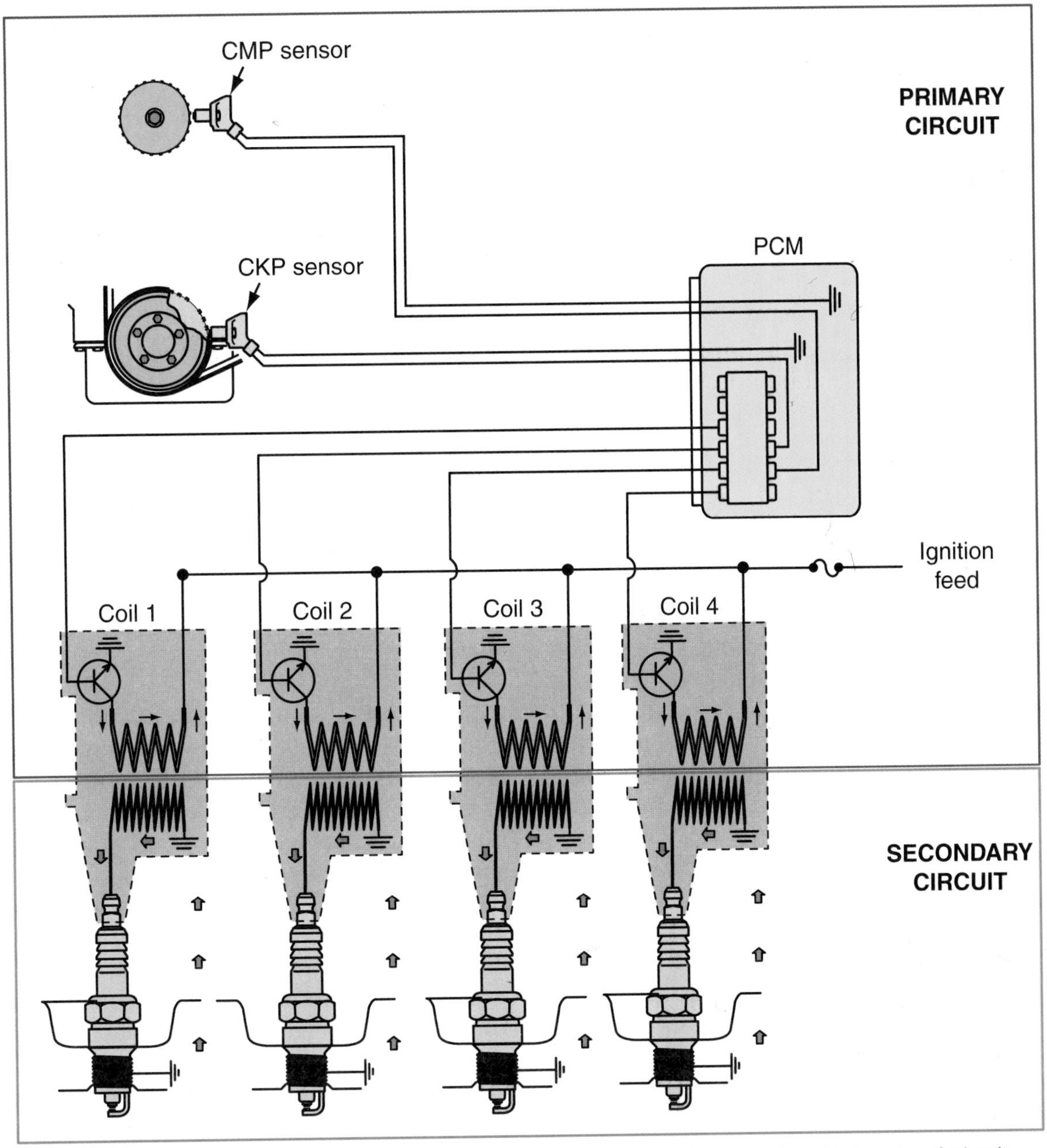

Figure 27-2 Ignition systems have a primary (low-voltage) and a secondary (high-voltage) circuit.

- Ignition coil primary winding
- Triggering device
- Switching device or control module (igniter)

The secondary circuit includes these components.

- Ignition coil secondary winding
- Distributor cap and rotor (some systems)
- High-voltage cables (some systems)
- Spark plugs

Primary Circuit Operation

When the ignition switch is on, current from the battery flows through the ignition switch and primary circuit resistor to the primary winding of the ignition coil. From there it passes through some type of switching device and back to ground. The current flow through the ignition coil's primary winding creates a magnetic field. As the current continues to flow, the magnetic field gets stronger. When the triggering device signals to the switching unit that the piston is approaching TDC on the compression stroke, current flow is stopped. This causes the magnetic field around the primary winding to collapse across the secondary winding. The movement of the magnetic field across the winding induces a high voltage in the secondary winding. The action of the secondary circuit begins at this point.

Some older ignition systems had a **ballast resistor** or resistance wire connected between the ignition switch and the positive terminal of the coil. This resistor limited the voltage and current to the coil. Today, ignition systems do not use a resistor, and voltage to the coil is controlled by the PCM.

Secondary Circuit Operation

The secondary circuit carries high voltage to the spark plugs. The exact manner in which the secondary circuit delivers these high-voltage surges depends on the system. Until 1984 all ignition systems used some type of distributor to accomplish this job. However, in an effort to reduce emissions, improve fuel economy, and boost component reliability, most auto manufacturers are now using distributorless or electronic ignition (EI) systems.

DI Systems In a distributor ignition system, the high voltage from the secondary winding is delivered to the distributor by an ignition cable. The cable connects the coil with a terminal in the center of the distributor cap. The distributor then distributes the high voltage to the individual spark plugs through a set of ignition cables **(Figure 27–3)**. The cables are arranged in the distributor cap according to the firing order of the engine. A rotor driven by the distributor shaft rotates and completes the electrical path from the secondary winding of the coil to the individual spark plugs. The distributor delivers the spark to match the compression stroke of the piston. The distributor assembly may also have the capability of advancing or retarding ignition timing.

The distributor cap is mounted on top of the distributor assembly and an alignment notch in the cap fits over a matching lug on the housing. Therefore, the cap can only be installed in one position, which ensures the correct firing sequence.

The rotor is positioned on top of the distributor shaft, and a projection inside the rotor fits into a slot in the shaft. This allows the rotor to be installed in only one position. A metal strip on the top of the rotor makes contact with the center distributor cap terminal,

Figure 27-3 A typical distributor.

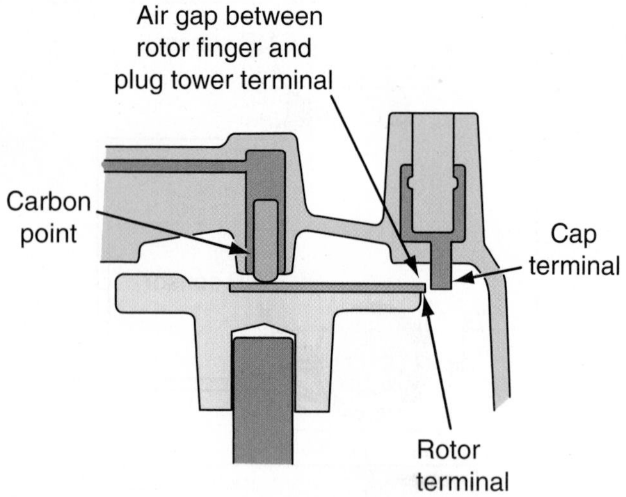

Figure 27-4 The relationship of a rotor and distributor cap.

and the outer end of the strip rotates past the cap terminals **(Figure 27–4)**. This action completes the circuit between the ignition coil and the individual spark plugs according to the firing order.

EI Systems EI systems have no distributor; spark distribution is controlled by an electronic control unit and/or the vehicle's computer **(Figure 27–5)**. Instead of a single ignition coil for all cylinders, each cylinder may have its own ignition coil, or two cylinders may share one coil. The coils are wired directly to the spark plug they control. An ignition control module, tied into the vehicle's computer control system, controls the firing order and the spark timing and advance.

The energy produced by the secondary winding is voltage. This voltage is used to establish a complete circuit so current can flow. The excess energy is used to maintain the current flow across the spark plug's gap. Distributorless ignition systems are capable of producing much higher energy than conventional ignition systems.

Since DI and EI systems are firing spark plugs with approximately the same air gap across the electrodes, the voltage required to start firing the spark plugs in both systems is similar. If the additional energy in the EI systems is not released in the form of voltage, it will be released in the form of current flow. This results in higher firing current and longer spark plug firing times. The average firing time across the spark plug electrodes in an EI system is 1.5 milliseconds compared to approximately 1 millisecond in a DI system. This extra time may seem insignificant, but it is very important. Current emission standards demand leaner air-fuel ratios, and this additional spark duration on EI systems helps to prevent cylinder misfiring with leaner air-fuel ratios. For this reason, car manufacturers have equipped their engines with EI systems.

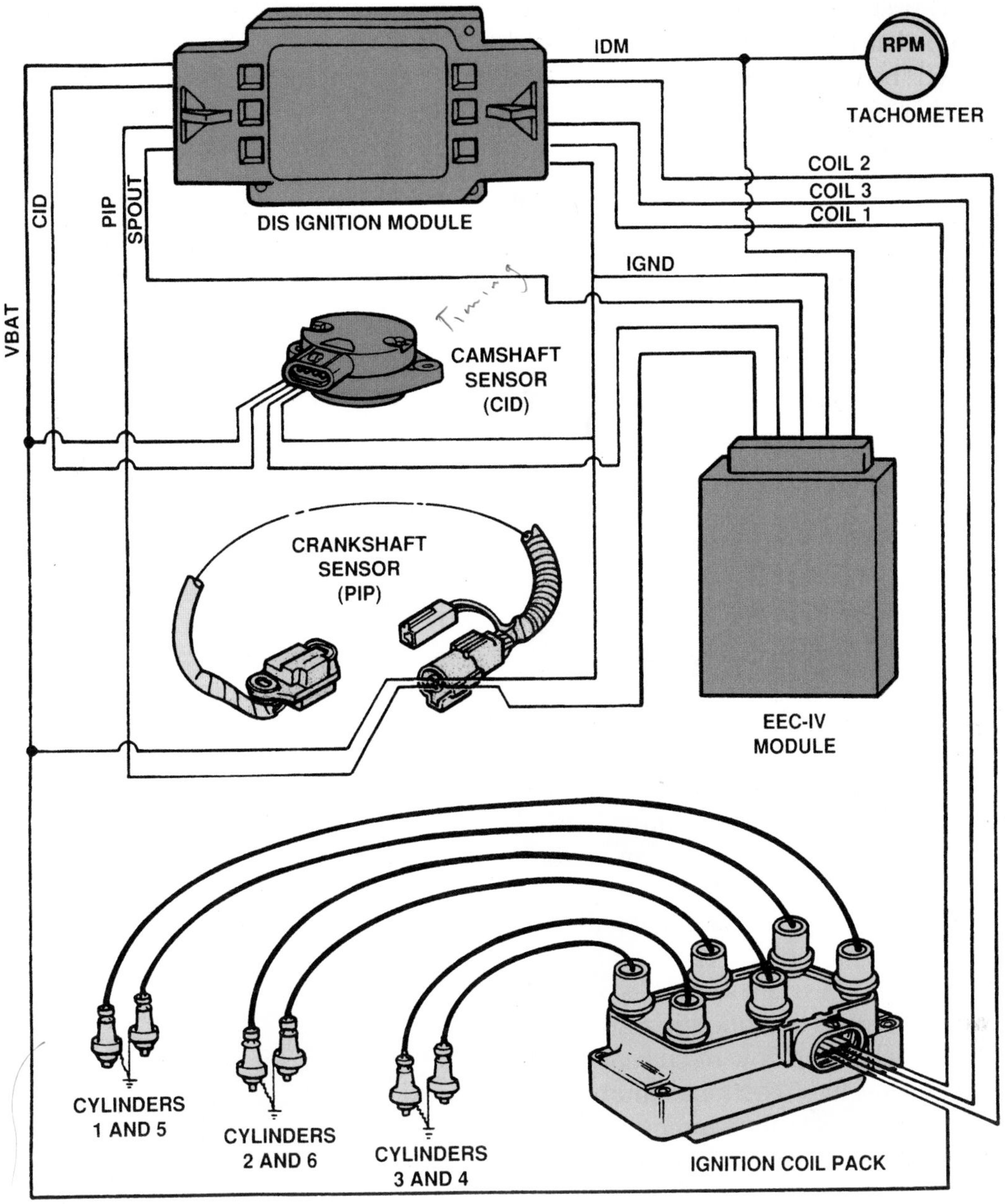

Figure 27-5 An electronic ignition system for a six-cylinder engine. *Courtesy of Ford Motor Company*

IGNITION COMPONENTS

All ignition systems share a number of common components. Some, such as the battery and ignition switch, perform simple functions. The battery supplies low-voltage current to the ignition primary circuit. The current flows when the ignition switch is in the start or run position. Full-battery voltage is always present at the ignition switch, as if it were directly connected to the battery.

Ignition Coils

To generate a spark to begin combustion, the ignition system must deliver high voltage to the spark plugs. Because the amount of voltage required to bridge the gap of the spark plug varies with the operating conditions, most late-model vehicles can easily supply 30,000 to 60,000 volts to force a spark across the air gap. Since the battery delivers 12 volts, a method of stepping up the voltage must be used. Multiplying battery voltage is the job of a coil.

The ignition coil is a **pulse transformer** that transforms battery voltage into short bursts of high voltage. As explained previously, when a magnetic field moves across a wire, voltage is induced in the wire.

If a wire is bent into loops forming a coil and a magnetic field is passed through the coil, an equal amount of voltage is generated in each loop of wire. The more loops of wire in the coil, the greater the total voltage induced. If the speed of the magnetic field is doubled, the voltage output doubles.

An ignition coil uses these principles and has two coils of wire wrapped around an iron core. An iron or steel core is used because it has low **inductive reluctance**. In other words, iron freely expands or strengthens the magnetic field around the windings. The first, or primary, coil is normally composed of 100 to 200 turns of 20-gauge wire. This coil of wire conducts battery current. When a current is passing through the primary coil, it magnetizes the iron core. The strength of the magnet depends directly on the number of wire loops and the amount of current flowing through those loops. The secondary coil of wires may consist of 15,000 to 25,000, or more, turns of very fine copper wire.

Because of the effects of counter EMF on the current flowing through the primary winding, it takes some time for the coil to become fully magnetized or saturated. Therefore, current flows in the primary winding for some time between firings of the spark plugs. The period of time during which there is primary current flow is often called **dwell**. The length of the dwell period is important.

When current flows through a conductor, it will immediately reach its maximum value as allowed by the resistance in the circuit. If a conductor is wound into a coil, maximum current will not be immediately achieved. As the magnetic field begins to form as the current begins to flow, the magnetic lines of force of one part of the winding pass over another part of the winding **(Figure 27–6)**. This tends to cause an opposition to current flow. This occurrence is called **reactance**. Reactance causes a temporary resistance to current flow and delays the flow of current from reaching its maximum value. When maximum current flow is present in a winding, the winding is said to be saturated and the strength of its magnetic field will also be at a maximum.

Saturation can only occur if the dwell period is long enough to allow for maximum current flow through the primary windings. A less-than-saturated coil will not be able to produce the voltage it was designed to produce. If the energy from the coil is too low, the spark plugs may not fire long enough or may not fire at all. If the current is applied longer than needed to fully saturate the winding, the coil will overheat.

A typical coil requires 2 to 6 milliseconds to become saturated. The actual required time depends on the resistance of the coil's primary winding and the voltage applied to it. Some early systems electronically limit the primary current flow at low speeds to prevent the coil from overheating. When the engine reaches higher speeds, the current limitation feature is disabled.

When the primary coil circuit is suddenly opened, the magnetic field instantly collapses. The sudden collapsing of the magnetic field produces a very high voltage in the secondary windings. This high voltage is used to push current across the gap of the spark plug.

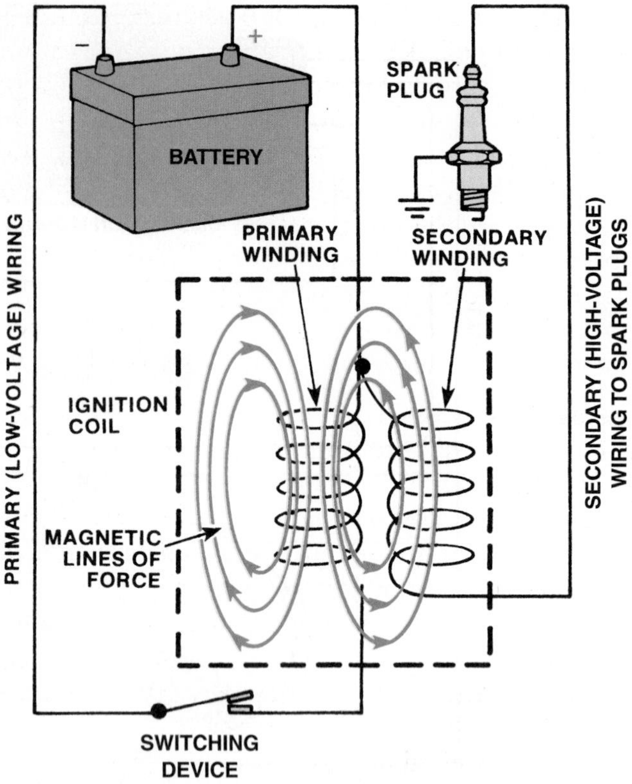

Figure 27–6 Current passing through the coil's primary winding creates magnetic lines of force that cut across and induce voltage in the secondary windings.

Ignition Coil Construction Older engines were equipped with ignition coils that were contained in a metal housing filled with oil to help cool the windings. Today's coils are air cooled. This is now possible because an individual coil is not responsible for providing the firing voltage for all spark plugs. Today's coils fire just one or two plugs. Many different coil designs are found on today's vehicles **(Figure 27–7)**. The actual design depends on the ignition system and the application.

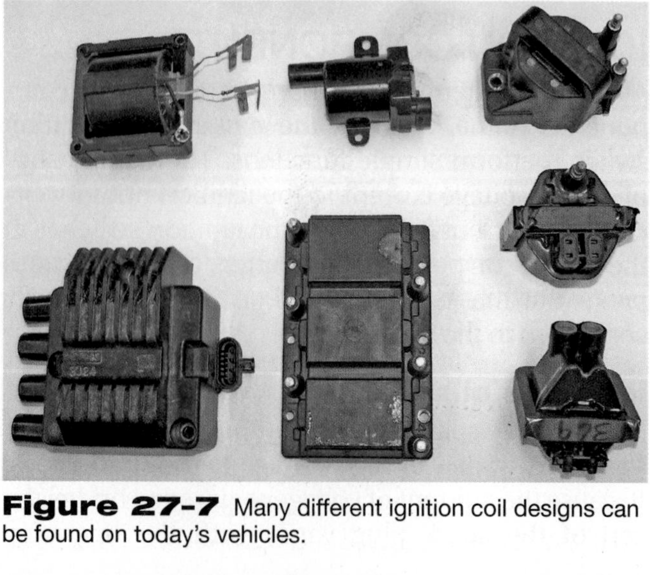

Figure 27–7 Many different ignition coil designs can be found on today's vehicles.

A laminated soft iron core is positioned in the center of each coil. The secondary winding is wound around the core and the primary winding is wound around the secondary winding. The two ends of the primary winding are on the outside of the coil housing and are labeled as positive and negative. One end of the secondary winding is internally connected to the positive terminal of the primary winding; the other end is connected to the spark plug circuit. The wires used to make up the windings are covered with insulation to prevent the wires from shorting to each other.

Secondary Voltage The typical amount of secondary coil voltage required to jump the spark plug gap is 10,000 volts. Most coils have at least 25,000 volts available from the secondary. The difference between the required voltage and the maximum available voltage is referred to as secondary reserve voltage. This reserve voltage is necessary to compensate for high cylinder pressures and increased secondary resistances as the spark plug gap increases through use. The maximum available voltage must always exceed the required firing voltage or ignition misfire will occur. If there is an insufficient amount of voltage available to push current across the gap, the spark plug will not fire.

In most ignition systems with a distributor, only one ignition coil is used. The high voltage of the secondary winding is directed, by the distributor, to the various spark plugs in the system. Therefore, there is one secondary circuit with a continually changing path.

While distributor systems have a single secondary circuit with a continually changing path, distributorless systems have several secondary circuits, each with an unchanging path.

SPARK PLUGS

Spark plugs provide the crucial **air gap** across which the high voltage from the coil causes an arc or spark. The main parts of a spark plug are a steel shell; a ceramic core or insulator, which acts as a heat conductor; and a pair of electrodes, one insulated in the core and the other grounded on the shell. The shell holds the ceramic core and electrodes in a gas-tight assembly and has threads for plug installation in the engine **(Figure 27–8)**. The insulator material may be alumina silicate or a black-glazed, zirconia-enhanced ceramic insulator to provide for increased durability and strength. The shell may be coated with corrosion resistance material and/or materials that prevent the threads from seizing to the cylinder head.

A terminal post on top of the center electrode is the connecting point for the spark plug cable. Current flows through the center of the plug and arcs from the tip of the center electrode to the ground electrode. The

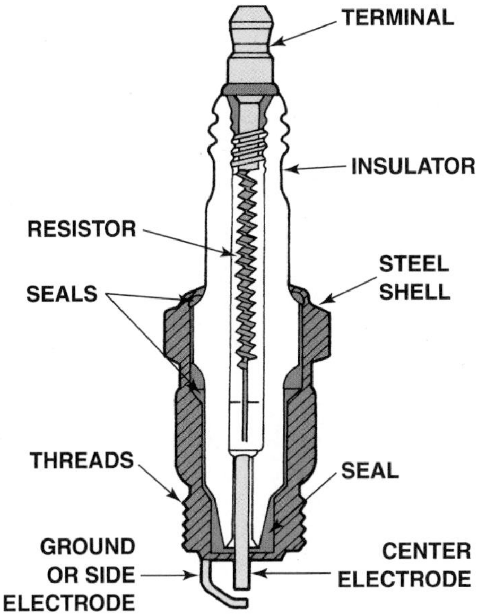

Figure 27-8 Components of a typical spark plug.

center electrode is surrounded by the ceramic insulator and is sealed to the insulator with copper and glass seals. These seals prevent combustion gases from leaking out of the cylinder. Ribs on the insulator increase the distance between the terminal and the shell to help prevent electric arcing on the outside of the insulator. The steel spark plug shell is crimped over the insulation, and a ground electrode, on the lower end of the shell, is positioned directly below the center electrode. There is an air gap between these two electrodes.

Spark plugs come in many different sizes and designs to accommodate different engines **(Figure 27–9)**.

Size Automotive spark plugs are available with a thread diameter of 12 mm, 14 mm, and 18 mm. The 18-mm spark plugs are mostly found on older engines and have a tapered seat that seals, when tightened properly, into a tapered seat in the cylinder head. The 12-mm and 14-mm plugs can have a tapered seat or a flat seat that relies on a thin steel gasket to seal in its bore in the cylinder head. All spark plugs have a hex-shaped outer shell that accommodates a socket wrench for installation and removal. A 12-mm plug has a 5⁄8- or 11⁄16-inch (16 or 18 mm) hex, a 14-mm plug with a tapered seat has a 5⁄8-inch (16 mm) hex, and 14-mm gasketed and 18-mm plugs have a 13⁄16-inch (20.6 mm) hex on the shell. A tapered plug should never be used in an engine designed to use a gasketted plug, or vice versa.

Reach One important design characteristic of spark plugs is the **reach (Figure 27–10)**. This refers to the length of the shell from the contact surface at the seat to the bottom of the shell, including both threaded and nonthreaded sections. Reach is crucial because the plug's air gap must be properly placed in the combustion chamber to produce the correct amount of heat.

THREAD SIZE AND HEX. SIZE

Letter	Thread Size	Hex. Size	Description
L	18mm	22.0mm	
M	18mm	25.4mm	
MA	18mm	20.6mm	Taper seat
J	14mm	20.6mm	Extended electrodes
P	14mm	20.6mm	Platinum tipped electrodes
PQ	14mm	16.0mm	Platinum tipped electrodes
Q	14mm	16.0mm	
QJ	14mm	16.0mm	Extended electrode
K	14mm	16.0mm	ISO
KJ	14mm	16.0mm	ISO
PK	14mm	16.0mm	Platinum tipped electrodes
SK	14mm	16.0mm	Iridium tipped electrode
S	14mm	20.6mm	Special surface gap for Mazda R.E./thread reach 21.5mm
SF	14mm	20.6mm	Surface gap
T	14mm	20.6mm	Taper seat
W	14mm	20.6mm	
X	12mm	18.0mm	
XU	12mm	16.0mm	
U	10mm	16.0mm	
Y	8mm	13.0mm	

THREAD REACH

Letter	Reach	Description
E (Flat seat)	19.0 (3/4") or 20.0mm	W16EXR-U W25EBR
E (Taper seat)	.708"	T16EPR-U
F	12.7mm (1/2")	W20FP-U
FE	19.0mm (3/4") Half thread	U24FER-9
G	21.8mm	X27GPR-U
L	11.2mm (7/16")	W14L
(None):		
18mm Thread (Flat seat)	12.0mm	M24S, L14-U
14mm Thread (Flat seat)	9.5mm (3/8")	W20S-U, W9PR-U
18mm Thread (Taper seat)	.480"	MA16PR-U
14mm Thread (Taper seat)	.460" or .325"	T16PR-U T20M-U

SPECIAL GAP CONFIGURATION

Letter	Description	Example
GL	Platinum center electrode	X22EPR-GL
L	Special type for Honda CVCC and extra project type for moped	W20ESR-L11 W14FP-UL
S	Semi-surface gap	W20EPR-S11
U	U-grooved ground electrode	W16EX-U
US	Star center electrode with U-groove	W14-US
V	Thin center electrode	W24ES-V
Z	Thin platinum center electrode with tapered ground electrode	W24ES-ZU
C	Cut-back ground electrode	W27EMR-C
P	Platinum tipped plug for DIS	PQ20R-PB

SPECIAL DESIGN

Letter	Description	Example
A	Dual ground electrodes for Mazda R.E.	W22EA
A	Electrode projection (7.0mm)	QJ16AR-U
B	Triple ground electrodes	W20EPB
B	Electrode projection (9.5mm)	J16BR-U
C	Electrode projection (5.0mm)	QJ20CR11
D	4-ground electrodes for Mazda R.E.	W27EDR14
H	Electrode projection (8.5mm)	QJ16HR-U
K	Special type for Honda CVCC	W16EKR-S11
LM	Special type for lawnmowers	W14LM-U
M	Compact type	W20M-U
N	Racing type (nickel ground electrode)	W27EN
Pt	Racing type (platinum ground electrode)	W27Ept
P	Projected insulator nose	W16EP-U
S	Regular type	W24ES-U
T	Dual ground electrodes for Toyota T.G.P.	W20ET-S
X	Extra projected insulator nose	W16EX-U

Figure 27-9 The different designs of spark plugs have unique part numbers that can be interpreted by charts given by the manufacturer. *Courtesy of DENSO corporation.*

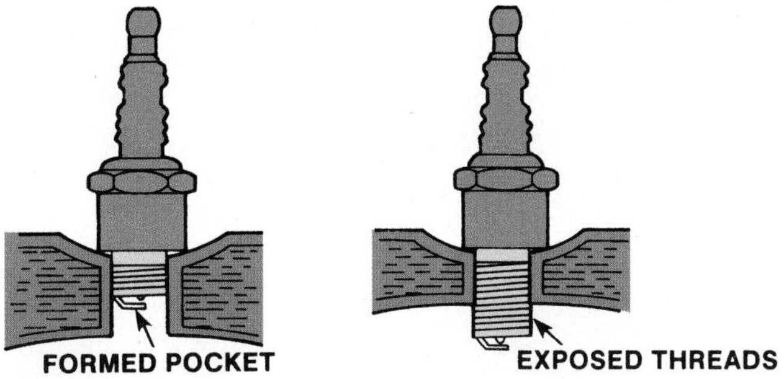

Figure 27-10 Spark plug reach: long versus short.

When a plug's reach is too short, its electrodes are in a pocket and the arc is not able to adequately ignite the mixture. If the reach is too long, the exposed plug threads can get so hot they will ignite the air-fuel mixture at the wrong time and cause preignition. *Preignition* is a term used to describe abnormal combustion, which is caused by something other than the heat of the spark.

Heat Range When the engine is running, most of the plug's heat is concentrated on the center electrode. Heat is quickly dissipated from the ground electrode because it is attached to the shell, which is threaded into the cylinder head. Coolant circulating in the head absorbs the heat and moves it through the cooling system. The heat path for the center electrode is through the insulator into the shell and then to the cylinder head. The **heat range** of a spark plug is determined by the length of the insulator before it contacts the shell. In a cold-range spark plug, there is a short distance for the heat to travel up the insulator to the shell. The short heat path means the electrode and insulator will maintain little heat between firings (**Figure 27–11**).

In a hot spark plug, the heat travels farther up the insulator before it reaches the shell. This provides a longer heat path and the plug retains more heat. A spark plug needs to retain enough heat to clean itself between firings, but not so much that it damages itself or causes preignition of the air-fuel mixture in the cylinder.

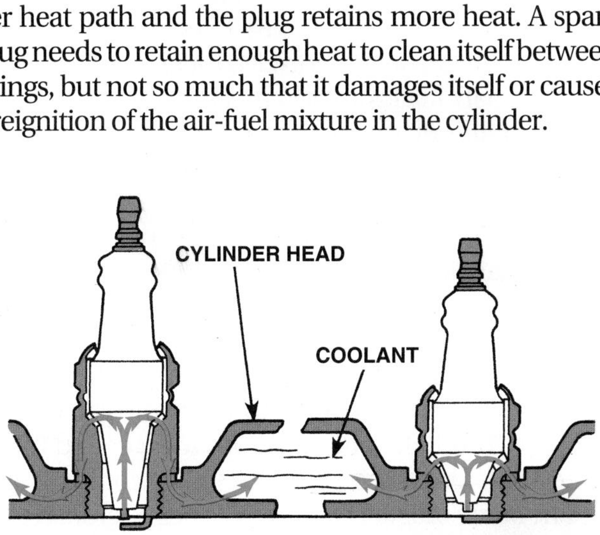

Figure 27-11 Spark plug heat range: hot versus cold.

The heat range is indicated by a code within the plug number imprinted on the side of the spark plug, usually on the porcelain insulator.

Resistor Plugs Most automotive spark plugs have a resistor (normally about 5 K ohms) between the top terminal and the center electrode. The resistance increases firing voltage. Some spark plugs use a semiconductor material to provide for this resistance. The resistor also reduces RFI, which can interfere with, or damage, radios, computers, and other electronic accessories, such as GPS systems. If an engine was originally equipped with resistor plugs, resistor plugs should be installed when the originals are replaced.

WARNING!

Using a nonresistor plug on some engines may cause erratic idle, high-speed misfire, engine run-on, power loss, and abnormal combustion.

Spark Plug Gaps The correct spark plug air gap is essential for achieving optimum engine performance and long plug life. A gap that is too wide requires higher voltage to jump the gap. If the required voltage is greater than what is available, the result is **misfiring**. Misfiring results from the inability of the ignition to jump the gap or maintain the spark. A gap that is too narrow requires lower voltages and can lead to rough idle and prematurely burned electrodes, due to higher current flow.

Electrodes The materials used in the construction of a spark plug's electrodes determine the longevity, power, and efficiency of the plug. The construction and shape of the tips of the electrodes are also important.

The electrodes of a standard spark plug are made with copper and some use a copper-nickel alloy. Copper is a good electrical conductor and offers some resistance to corrosion. Copper melts at 1,981°F so it is more than suitable for use in an internal combustion engine.

Platinum electrodes are used to extend the life of a plug (**Figure 27–12**). Platinum melts at 3,200°F (1,760°C) and is highly resistant to corrosion. Although platinum is an extremely durable material, it is an expensive precious metal; therefore, platinum spark plugs cost more than copper plugs. Also, platinum is not as good a conductor as copper. Spark plugs are available with only the center electrode made of platinum (called single-platinum) and with the center and ground electrodes made of platinum (called double-platinum). Some platinum plugs have a very small center electrode combined with a sharp pointed

Figure 27-12 A platinum-tipped spark plug. *Courtesy of Bob Freudenberger*

Figure 27-13 The spark plug has a small diameter iridium center electrode and a grooved ground electrode. *Courtesy of DENSO Corporation*

ground electrode designed for better performance. Platinum plugs with two ground electrodes are used in waste spark systems.

Until recently, platinum was considered the best material to use for electrodes because of its durability. However, iridium is six times harder, eight times stronger, and has a melting point 1,200° higher than platinum. Iridium is a precious, silver-white metal and one of the densest materials found on earth. A few spark plugs use an iridium alloy as the primary metal complimented by rhodium to increase oxidation wear resistance. This iridium alloy is so durable that it allows for an extremely small center electrode. A typical copper/nickel plug has a 2.5 mm diameter center electrode and a platinum plug has a 1.1 mm diameter. An iridium plug can have a diameter as small as 0.4 mm **(Figure 27–13)**, which means the firing voltage requirements are decreased. Iridium is also used as an alloying material for platinum.

Another rare and hard material used to make electrodes is yttrium. Yttrium has a silvery-metallic luster and has a melting point of 2,773°F (1,523°C). Yttrium is fairly stable in air but oxidizes readily when heated. (Moon rocks contain yttrium.) Yttrium produces a highly adhesive oxide layer that makes the spark plug very durable and reliable, thereby extending its service life.

Electrode Designs Spark plugs are available with many different shapes and numbers of electrodes. When trying to ascertain the advantages of each design, remember the spark is caused by electrons moving across an air gap. The electrons will always jump in the direction of the least electrical resistance. Therefore, if there are

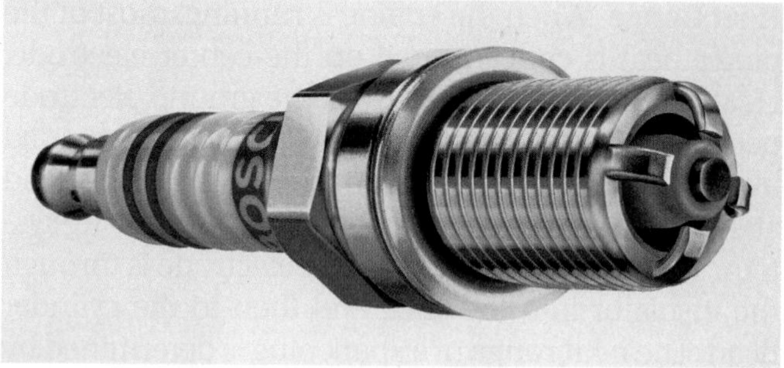

Figure 27-14 A spark plug with four ground electrodes. *Courtesy of Robert Bosch GmbH, www.bosch-presse.de*

four ground electrodes to choose from, the electrons will jump to the closest. Also, keep in mind that the quality and pressure of the air in the air gap influences the resistance of the air gap. Again, the electrons will jump across the path of least resistance. Therefore, spark plugs with four ground electrodes do not typically supply a spark to all four electrodes **(Figure 27–14)**.

The shape of the ground electrode may also be altered. A flat, conventional electrode tends to crush the spark, and the overall volume of the flame front is smaller. A tapered ground electrode increases flame front expansion and reduces the heat lost to the electrode.

Some ground electrodes have a U-groove machined into the side that faces the center electrode. The U-groove allows the flame front to fill the gap formed by the U. This ball of fire develops a larger and hotter flame front, leading to a more complete combustion.

One brand of spark plug has a V-shaped ground electrode. This style of electrode does not block the flame front and allows it to travel upward through the V notch into the combustion chamber. These spark plugs may be equipped with three separate points of platinum, one at each end of the V and the other at the center electrode.

There are also different center electrode designs. These variations are based on the diameter and shape of the electrode. A small diameter center electrode requires less firing voltage and tends to have a longer service life. Some center electrodes are tapered.

Some center electrodes have a V groove machined in them to force the spark to the outer edge of the ground electrode, placing it closer to the air-fuel mixture. This allows for quicker ignition of the air-fuel mixture. V-grooved center electrodes also require lower firing voltages.

On some spark plugs, the center electrode does not extend from the insulator and the spark is generated across the end of the plug. With this design, the ground electrode does not block the flame front. This arrangement is called a surface gap and is intended to prevent carbon fouling, timing drift, and misfiring.

Ignition Cables

Spark plug cables, or ignition cables, make up the secondary wiring. These cables carry the high voltage from the distributor or the multiple coils to the spark plugs. The cables are not solid wire; instead they contain carbon fiber cores that act as resistors in the secondary circuit (**Figure 27–15**). They cut down on radio and television interference, increase firing voltages, and reduce spark plug wear by decreasing current. Insulated boots on the ends of the cables strengthen the connections as well as prevent dust and water infiltration and voltage loss.

Some ignition cables are called *variable pitch* resistor cables. These cables rely on tightly wound and loosely wound copper wire around a layer of ferrite magnetic material wrapped over a fiberglass strand core. This construction creates the necessary resistance with a fraction of the impedance found in solid carbon core-type wire sets.

Some engines have spark plug cable heat shields (**Figure 24–16**) pressed into the cylinder head. These shields surround each spark plug boot and spark plug: They protect the spark plug boot from damage due to the extreme heat generated by the nearby exhaust manifold.

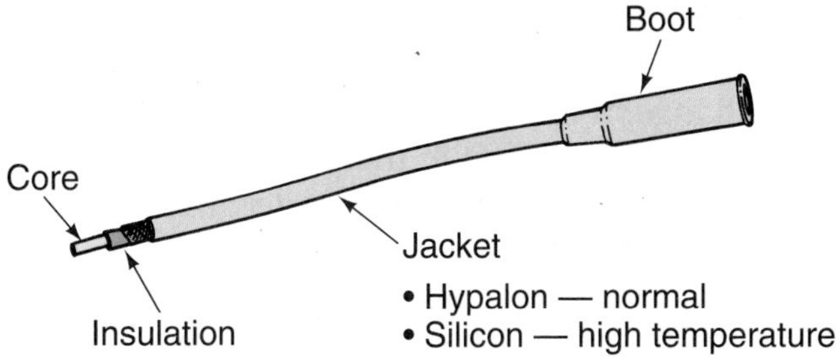

Figure 27–15 Spark plug cable construction. *Courtesy of Chrysler LLC*

Figure 27–16 Spark plug boot heat shields.

TRIGGERING AND SWITCHING DEVICES

Triggering and switching devices are used to ensure the spark occurs at the correct time. A triggering device is simply a device that monitors the movement of the engine's pistons. A switching device is what controls current flow through the primary winding. When the triggering device sends a signal to the switching device that the piston of a particular cylinder is on the compression stroke, the switching device stops current flow to the primary winding. This interruption of current flow happens when the PCM decides it is best to fire the spark plug.

Electronic switching components are normally located in an ignition control module, which may be part of the vehicle's PCM. On older vehicles, the ignition module may be built into the distributor or mounted in the engine compartment.

The ignition module advances or retards the ignition timing in response to engine conditions. Early systems had little control of timing and used mechanical or vacuum devices to alter timing. Today's computer-controlled systems have full control and can adjust ignition timing in response to the input signals from a variety of sensors and the programs in the computer.

Most electronically controlled systems use an NPN transistor to control the primary ignition circuit, which ultimately controls the firing of the spark plugs. The transistor's emitter is connected to ground. The collector is connected to the negative terminal of the coil. When the triggering device supplies a small current to the base of the transistor, current flows through the primary winding of the coil. When the current to the base is interrupted, the current to the coil is also interrupted. An example of how this works is shown in **Figure 27–17**, which is a simplified diagram of an electronic ignition system.

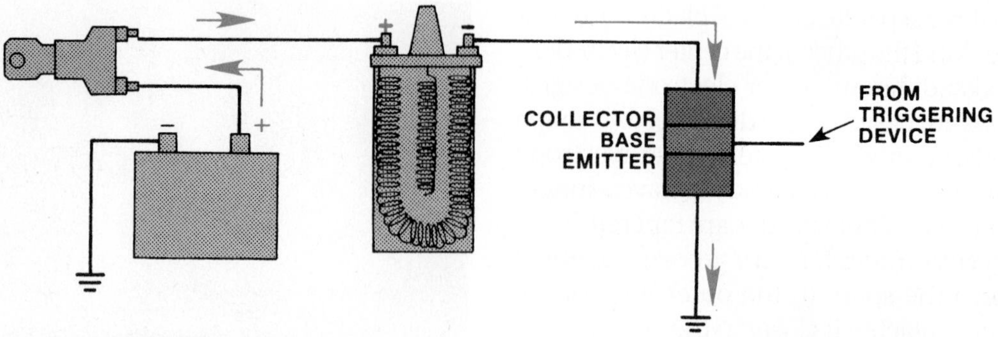

Figure 27-17 When the triggering device supplies a small amount of current to the transistor's base, the primary coil circuit is closed and current flows.

ENGINE POSITION SENSORS

The time when the primary circuit must be opened and closed is related to the position of the pistons and the crankshaft. Therefore, the position of the crankshaft is used to control the flow of current to the base of the switching transistor.

A number of different types of sensors are used to monitor the position of the crankshaft and control the flow of current to the base of the transistor. These engine position sensors and generators serve as triggering devices and include magnetic pulse generators, metal detection sensors, Hall-effect sensors, and photoelectric (optical) sensors.

These sensors can be located inside the distributor or mounted on the outside of the engine to monitor crankshaft position (CKP). In many cases, the input from a CKP is supplemented by inputs from a CMP. On nearly all late-model engines, the CKP and CMP are magnetic pulse generators or Hall-effect switches. When the triggering devices are in the distributor, they can be any of the following types of sensors.

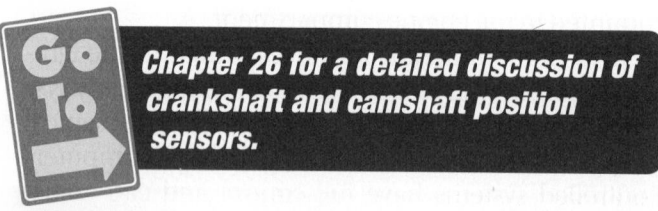

Go To Chapter 26 for a detailed discussion of crankshaft and camshaft position sensors.

Magnetic Pulse Generator

Basically, a magnetic pulse generator or inductance sensor consists of two parts: a trigger wheel and a pickup coil. The trigger wheel may also be called a reluctor, pulse ring, armature, or timing core. The pickup coil, which consists of a length of wire wound around permanent magnet, may also be called a stator, sensor, or pole piece. Depending on the type of ignition system used, the timing disc may be mounted on the distributor shaft, at the rear of the crankshaft, or behind the crankshaft vibration damper (**Figure 27–18**).

Figure 27-18 Location of a typical crankshaft position (CKP) sensor.

The magnetic pulse or PM generator operates on the principles of electromagnetism. A voltage is induced in a conductor when a magnetic field passes over the conductor or when the conductor moves over a magnetic field. The magnetic field is provided by a magnet in the pickup unit, and the rotating trigger wheel provides the required movement through the magnetic field to induce voltage.

As the trigger wheel rotates past the pickup coil, a weak AC signal is induced in the pickup coil. This signal is sent to the ignition module. In early ignition systems, the change in polarity was used as a signal to prepare the ignition coil for another spark plug firing.

When a tooth is aligned to the pickup coil, the magnetic field is not expanding or contracting. There is no change in the magnetic field and at that point zero voltage is induced in the pickup coil. The zero voltage signal from the coil is called the timing or "sync" pulse and is used by the PCM as the basis for timing the events in the ignition system. The timing pulses correspond with the position of each piston within their cylinder.

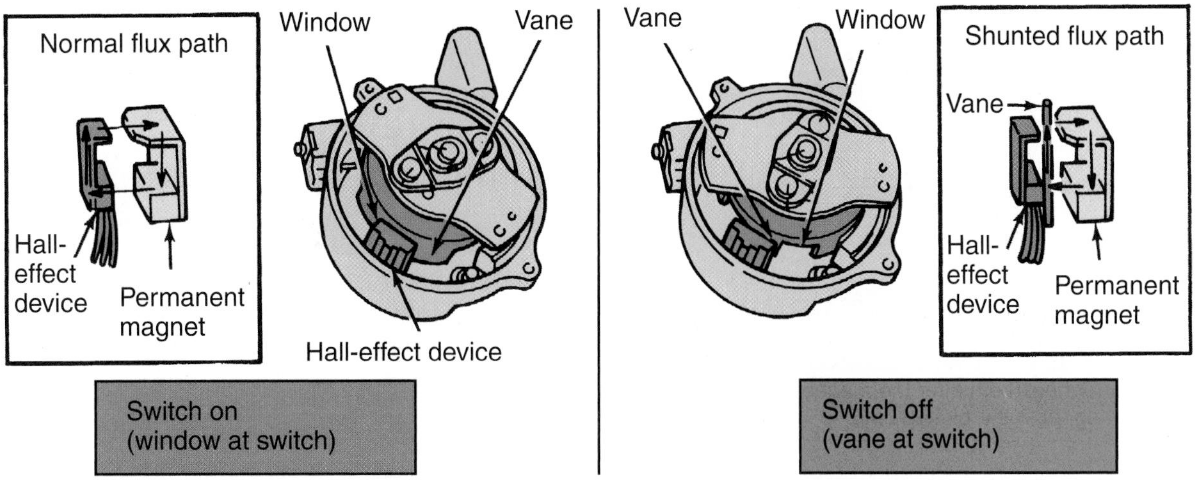

Figure 27-19 Operation of a Hall-effect switch. *Courtesy of Ford Motor Company*

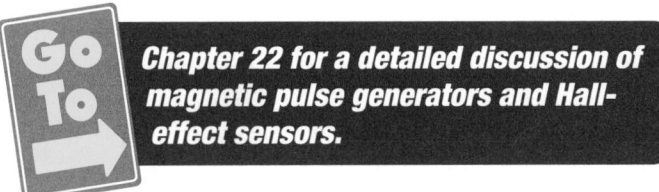

Go To ***Chapter 22 for a detailed discussion of magnetic pulse generators and Hall-effect sensors.***

Hall-Effect Sensor

The Hall-effect sensor or switch is the most commonly used engine position (CKP) sensor. A Hall-effect sensor produces an accurate voltage signal throughout the entire speed range of an engine. It also produces a square wave signal that is more compatible with computers. In an ignition system, the shutter blades are mounted on the distributor shaft **(Figure 27–19)**, flywheel, crankshaft pulley, or cam gear so the sensor can generate a position signal as the crankshaft rotates. A Hall-effect sensor may be normally on or off depending on the system and its circuitry. When a normally off sensor is used, there is maximum voltage output from the sensor when the magnetic field is blocked by the shutter. The opposite is true for normally on sensors. They have a voltage output when the magnetic field is not blocked.

A typical Hall-effect sensor has three wires connected to it. One wire is the reference voltage wire. The PCM supplies a reference voltage of 5 to 12 volts, depending on the system. The second wire delivers the output signal from the sensor to the PCM, and the third wire provides a ground for the sensor.

The signal from a Hall-effect CKP is also used to match fuel injector timing with the engine's firing order on engines equipped with sequential fuel injection. Hall-effect switches are also used as camshaft position (CMP) sensors. When the engine is being started, the PCM receives a signal from the CKP, but the spark plugs will not fire until the PCM receives a reference pulse from the CMP. After the engine starts, the PCM no longer relies on the CMP for ignition sequencing. However, if the CMP is bad, the engine will not restart. If the CKP goes bad, the engine will typically not start or run.

Photoelectric Sensor

Some early distributor ignition systems relied on photoelectric sensors **(Figure 27–20)** to monitor engine position. They consisted of an LED, a light-sensitive phototransistor (photo cell), and a slotted disc called an interrupter. As the interrupter rotated between the LED and the photo cell, pulsating voltage was generated in the photo cell. This voltage was passed onto the ignition module and was used as the basis for all ignition timing.

Figure 27-20 A distributor with an optical-type pickup.

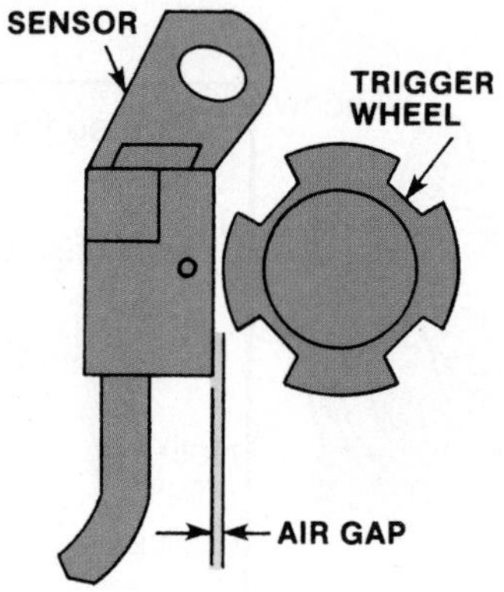

Figure 27-21 In a metal detecting sensor, the revolving trigger wheel teeth alter the magnetic field produced by the electromagnet in the pickup coil.

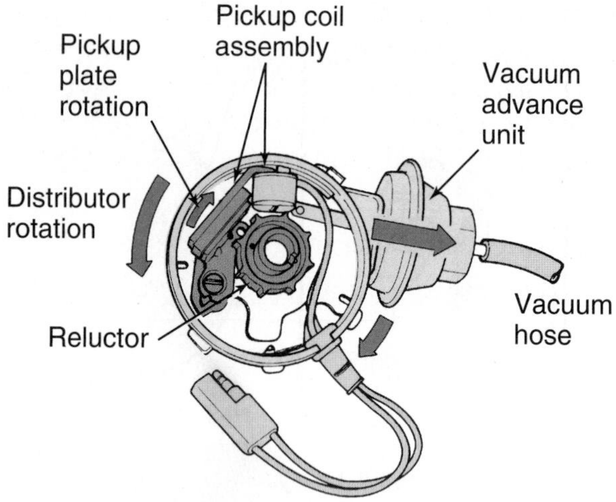

Figure 27-22 Typical vacuum advance unit operation. *Courtesy of Chrysler LLC*

Metal Detection Sensors

Metal detection sensors are found on early electronic ignition systems. They work much like a magnetic pulse generator with one major difference.

A trigger wheel is pressed over the distributor shaft and a pickup coil detects the passing of the trigger teeth as the distributor shaft rotates. However, unlike a magnetic pulse generator, the pickup coil of a metal detection sensor does not have a permanent magnet. Instead, the pickup coil is an electromagnet. A low level of current is supplied to the coil by an electronic control unit, inducing a weak magnetic field around the coil. As the reluctor on the distributor shaft rotates, the trigger teeth pass very close to the coil (**Figure 27–21**). As the teeth pass in and out of the coil's magnetic field, the magnetic field builds and collapses, producing a corresponding change in the coil's voltage. The voltage changes are monitored by the control unit to determine crankshaft position.

Timing Retard and Advance

One of the most important duties of an ignition system is to provide the spark at the correct time. On late-model engines, this is the job of the ignition module. On earlier ignition systems, this was accomplished at the distributor through mechanical and vacuum-responsive devices.

Mechanical devices used weights and springs attached to the distributor shaft that moved with engine speed to advance timing. When engine speed increased, the weights moved out and moved the mounting plate for the triggering unit. The shifting of the triggering unit caused it to send a signal to the switching unit earlier, causing an advance in timing. When engine speed decreased, the weights were pulled by the springs and the timing retards. These units responded solely to engine speed.

Vacuum Advance Units that changed ignition timing in response to engine load were also fitted to distributors (**Figure 27–22**). A vacuum advance unit, comprised of a spring-loaded diaphragm, was attached to the triggering device plate. Vacuum was applied to one side of the diaphragm and atmospheric pressure was applied to the other side. Any increase in vacuum allowed atmospheric pressure to move the diaphragm, which caused a movement of the triggering device's mounting plate. The more vacuum that was present on one side, the more the atmospheric pressure could move the diaphragm and advance the timing. A spring was used to return the plate toward its rest position when vacuum decreased. At full throttle, there is no vacuum advance. Maximum vacuum advance is only at cruising speeds with no load.

DISTRIBUTOR IGNITION SYSTEM OPERATION

The primary circuit of a DI system is controlled by a triggering device and a switching device located inside the distributor or external to it. Although these systems are no longer used by auto makers, there are many of them still on the road and they need service.

Distributor

The reluctor, or trigger wheel, and distributor shaft assembly rotate on bushings in the aluminum distributor housing. A roll pin extends through a retainer and the distributor shaft to hold the shaft in place in the distributor. Another roll pin is used to fasten the drive gear to the lower end of the shaft. This drive gear typically meshes with a drive gear on the engine's camshaft. The gear size is designed to drive the distributor shaft at the same speed as the camshaft, which rotates at one-half the speed of the crankshaft.

Through the years there have been many different designs of DI systems. All operate in the basically the same way but are configured differently. The systems described in this section represent the different designs used by manufacturers. These designs are based on the location of the electronic control module (unit) (ECU) and/or the type of triggering device used.

- DI systems with internal ignition module
- DI systems with external and remote ignition module
- DI systems with the ignition modules mounted on the distributor

Computer-Controlled DI Systems After the manufacturers eliminated the mechanical and vacuum advance mechanisms on their distributors, the ECM or PCM controlled ignition timing. This allowed for more precise control of ignition timing and provided improved combustion. The PCM adjusted the ignition timing according to engine speed, engine load, coolant temperature, throttle position, and intake manifold pressure. These systems varied with application and used a variety of triggering devices.

ELECTRONIC IGNITION SYSTEMS

Very few newer engines are equipped with a distributor; rather they have electronic ignitions. In the past, the term *electronic ignition* was designated to those ignition systems that used electronic controls. Today, electronic ignitions are those that do not use a distributor. There are two types of EI systems used on today's engines: waste spark **(Figure 27–23)** and coil-over-cylinder **(Figure 27–24)** systems. In both cases,

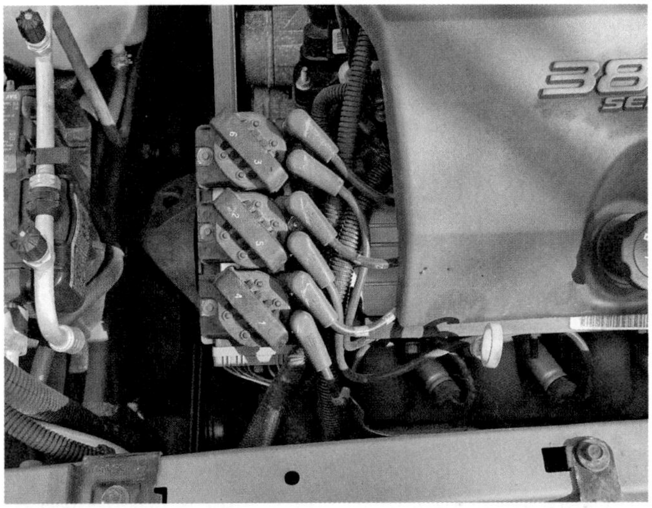

Figure 27-23 A coil pack for a double-ended or waste spark ignition system.

Figure 27-24 A coil per cylinder ignition system.

an ignition module, controlled by the PCM, controls the firing order and ignition timing. A crank sensor is used to trigger the ignition system.

There are many advantages of a distributorless ignition system over one that uses a distributor. Here are some of the more important ones:

- Elimination of a rotor and its subsequent resistance.
- No moving parts and therefore requires little maintenance.
- It is possible to control the ignition of individual cylinders to meet specific needs.
- Flexibility in mounting location. This is important because of today's smaller engine compartments.
- Reduced radio frequency interference because there is no rotor to cap gap.
- Elimination of a common cause of ignition misfire, the buildup of water and ozone/nitric acid in the distributor cap.
- Elimination of mechanical timing adjustments.
- Places no mechanical load on the engine in order to operate.
- Increased available time for coil saturation.
- Increased time between firings, which allows the coil to cool more.

Double-Ended Coil or Waste Spark Systems

Double-ended or waste spark ignition systems use one ignition coil for two spark plugs **(Figure 27–25)**. Both ends of the coil's secondary side are directly connected to a spark plug, which means that two plugs are ignited at the same time; one is fired on the compression stroke of one cylinder and the other is fired on the exhaust stroke of another.

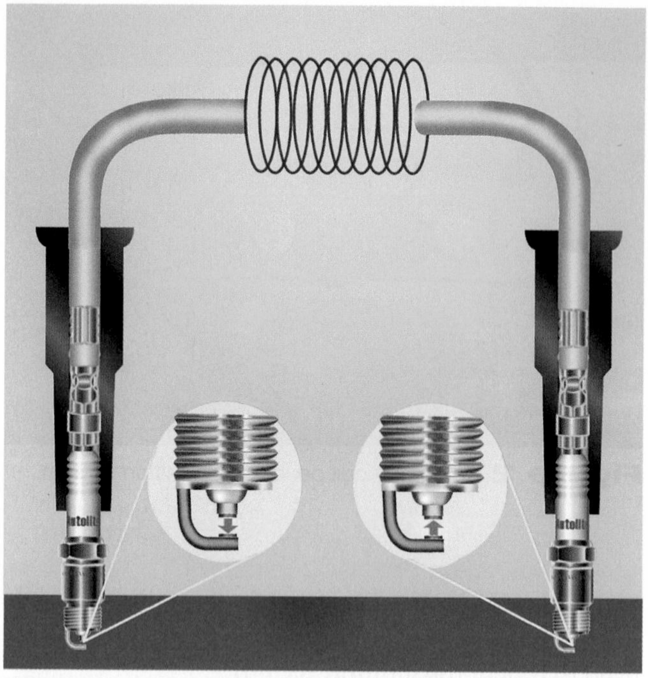

Figure 27-25 An EI system with a double-ended coil. *Courtesy of Honeywell International Inc.*

A four-cylinder engine has two ignition coils, a six-cylinder has three, and an eight-cylinder has four. The computer, ignition module, and various sensors combine to control spark timing.

The computer collects and processes information to determine the ideal amount of spark advance for the operating conditions. The ignition module uses crank/cam sensor data to control the timing of the primary circuit in the coils **(Figure 27–26)**. Remember that there is more than one coil in a distributorless ignition system. The ignition module synchronizes the coils' firing sequence in relation to crankshaft position and firing order of the engine. Therefore, the ignition module takes the place of the distributor.

Primary current is controlled by transistors in the control module. There is one switching transistor for each ignition coil in the system. The transistors complete the ground circuit for the primary, thereby allowing for a dwell period. When primary current flow is interrupted, secondary voltage is induced in the coil and the coil's spark plug(s) fire. The timing and sequencing of ignition coil action is determined by the control module and input from a triggering device.

The control module is also responsible for limiting the dwell time. In EI systems there is time between plug firings to saturate the coil. Achieving maximum current flow through the coil is great if the system needs the high voltage that may be available. However, if the high voltage is not needed, the high current is not needed and the heat it produces is not desired. Therefore, the control module is programmed to only allow total coil saturation when the very high voltage is needed or the need for it is anticipated.

The ignition module also adjusts spark timing below 400 rpm (for starting) and when the vehicle's control computer by-pass circuit becomes open or grounded. Depending on the exact EI system, the ignition coils can be serviced as a complete unit or separately. The coil assembly is typically called a **coil pack** and is comprised of two or more individual coils.

Waste Spark Double-ended coil systems are based on the **waste spark** method of spark distribution. Both ends of the ignition coil's secondary winding are connected to a spark plug. Therefore, one coil is connected in series with two spark plugs. The two spark plugs belong to cylinders whose pistons rise and fall together. With this arrangement, one cylinder of each pair is on its compression stroke and the other is on its exhaust stroke when the spark plugs are fired. Typically, cylinder pairings are:

- Four-cylinder engines: 1 & 2 and 3 & 4
- V6 engines: 1 & 4, 2 & 5, and 3 & 6
- Inline six cylinders: 1 & 6, 2 & 5, and 4 & 3
- V8 engines: 1 & 4, 3 & 8, 6 & 7, and 2 & 5 or 1 & 6, 8 & 5, 4 & 7, and 2 & 3
  (The pairings on V8s will vary as manufacturers vary how they number the cylinders.)

Due to the way the secondary coils are wired, when the induced voltage cuts across the primary and secondary windings of the coil, one plug fires in the normal direction—positive center electrode to negative side electrode—and the other plug fires just the reverse side to center electrode **(Figure 27–27)**. Both plugs fire simultaneously, completing the series circuit. Each plug always fires the same way on both the exhaust and compression strokes.

The coil is able to overcome the increased voltage requirements caused by reversed polarity and still fire two plugs simultaneously because each coil is capable of producing up to 100,000 volts. There is very little resistance across the plug gap on exhaust, so the plug requires very little voltage to fire, thereby providing its mate (the plug that is on compression) with plenty of available voltage.

Some EI systems use the waste spark method of firing but only have one secondary wire coming off each ignition coil. In these systems **(Figure 27–28)**, one spark plug is connected directly to the ignition coil and the companion spark plug is connected to the coil by a high-tension cable.

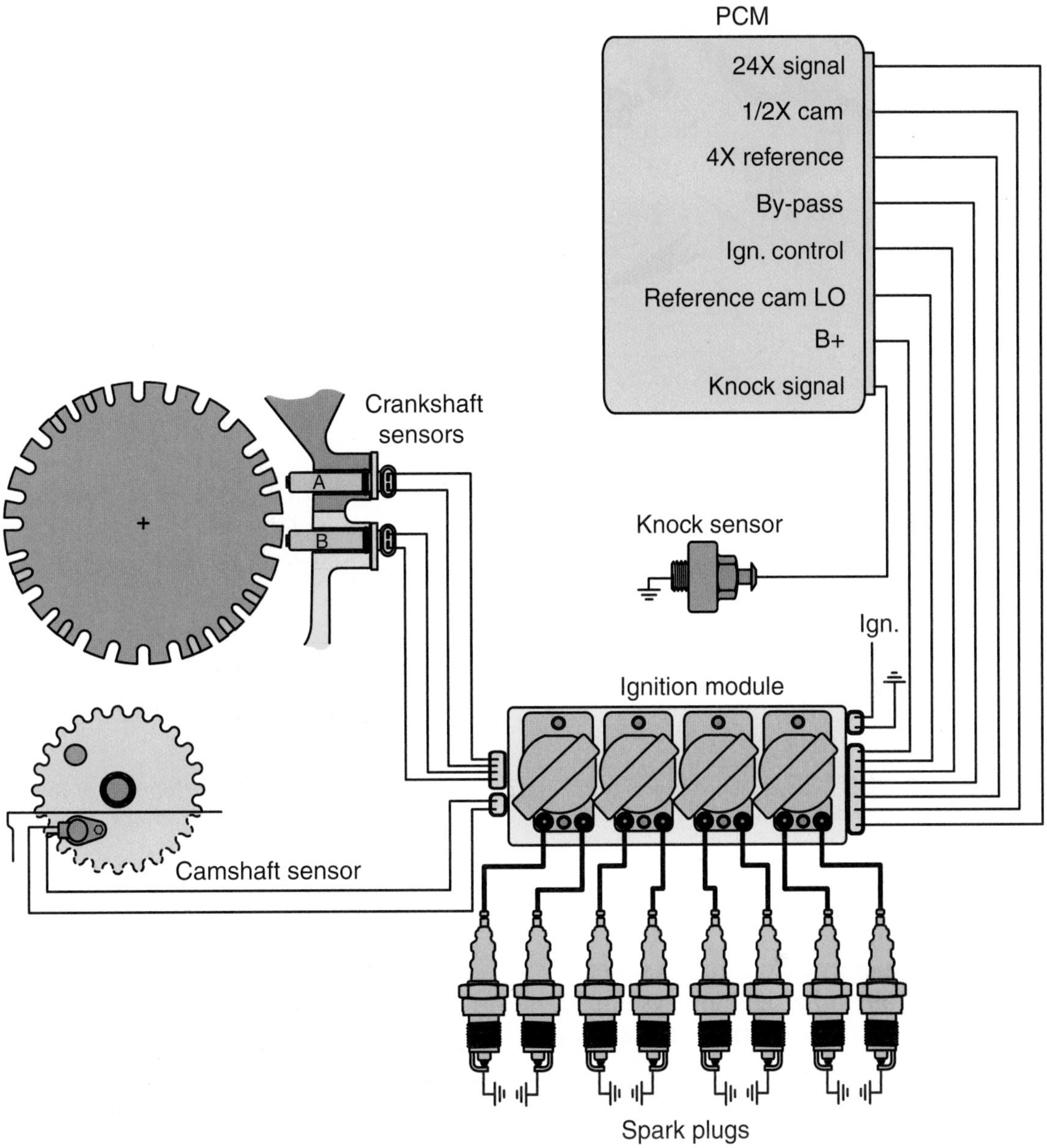

Figure 27-26 An EI system with two crankshaft position sensors and one camshaft position sensor.

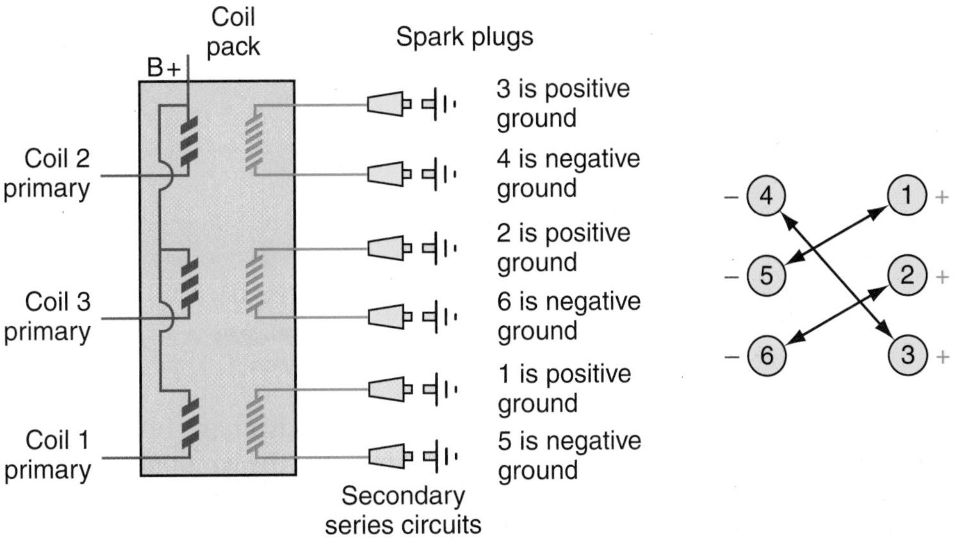

Figure 27-27 Polarity of spark plugs in an EI system. *Courtesy of Ford Motor Company*

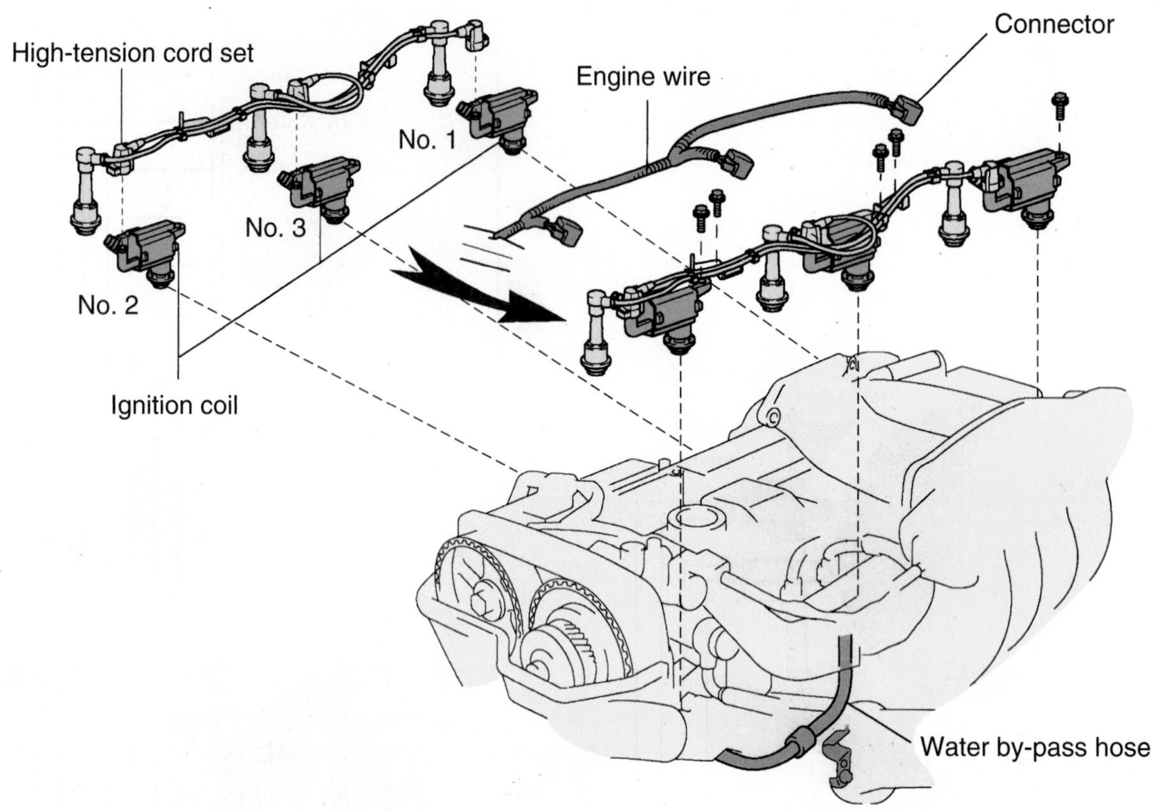

Figure 27-28 A six-cylinder engine with three coils and three spark plug wires. *Courtesy of Toyota Motor Sales, U.S.A., Inc.*

Coil-per-Cylinder Ignition

The operation of a coil-per-cylinder ignition system is basically the same as any other ignition system. By definition, these systems have an individual coil for each spark plug. There are two different designs of coil-per-cylinder systems used today: the **coil-over-plug (COP)** and the separate coil. COP systems rely on a single assembly of an ignition coil and spark plug **(Figure 27–29)**. In these systems, the spark plug is directly attached to the coil and there is no spark plug wire.

The separate coil system is often called a coil-by-plug or coil-near-plug ignition system **(Figure 27–30)**. These systems have individual coils mounted near the plugs and use a short secondary plug wire to connect the coil to the plug. These systems are used when the location of the spark plug does not allow enough room to mount individual coils over the plugs, or when the plugs are too close to the exhaust manifold.

Having one coil for each spark plug allows for more time between each firing, which increases the life of the coil by allowing it to cool. In addition, it also allows for more saturation time, which increases the coil's voltage output at high engine speeds. The increased output makes the coils more effective with lean fuel mixtures, which require higher firing voltages.

Figure 27-29 A coil-on-plug assembly. *Courtesy of Visteon Corporation™*

Another advantage of using the coil-per-cylinder system is that the ignition timing at each cylinder can be individually changed for maximum performance and to respond to knock sensor signals. Other advantages of a coil-per-cylinder system are that all of the

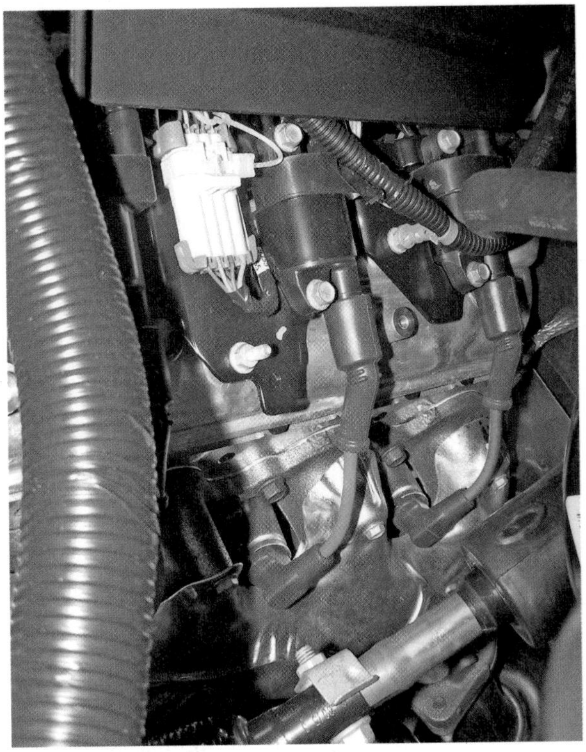

Figure 27-30 A coil-near-plug system.

Figure 27-31 In a COP system, the coil is mounted directly above the spark plug.

engine's spark plugs fire in the same direction and coil failure will affect only one cylinder.

In a typical coil-per-cylinder system, a crankshaft position sensor provides a basic timing signal. This signal is sent to the PCM. The PCM is programmed with the firing order for the engine and determines which ignition coil should be turned on or off. Some engines require an additional timing signal from the camshaft position sensor. On some systems, there is also a coil capacitor for each bank of coils for radio noise suppression.

Coil-over-Plug (COP) Ignition The true difference between COP and other ignition systems is that each coil is mounted directly atop the spark plug **(Figure 27–31)**, so the voltage from the coil goes directly to the plug's electrodes without passing through a plug wire. This means there are no secondary wires to come loose, burn, leak current, break down, or replace. Eliminating plug wires also reduces radio frequency interference (RFI) and electromagnetic interference (EMI) that can interfere with computer systems. However, the absence of plug wires also means that the coils need to be removed and reconnected with adapters or plug wires to test for spark, connect a pickup for an ignition scope, or perform a manual cylinder power balance test.

Twin Spark Plug Systems

Most engines have one spark plug per cylinder, but some have two. One spark plug is normally located on the intake side of the combustion chamber and the other is at the exhaust side. When ignition takes place in two locations within the combustion chamber, more efficient combustion and cleaner emissions are possible. Two coil packs are used, one for the intake side and the other for the exhaust side. These systems are called **dual** or **twin plug** systems **(Figure 27–32)**.

Some engines fire only one plug per cylinder during starting. The additional plug fires once the engine is running. During dual plug operation, the two coil packs are synchronized so the two plugs of each cylinder fire at the same time. Therefore, in a waste spark system, four spark plugs are fired at a time: two during the compression stroke of a cylinder and two during the exhaust stroke of another cylinder.

EI SYSTEM OPERATION

From a general operating standpoint, most electronic ignition systems are similar. One difference in design is the number of ignition coils. COP systems have the same number of coils as the engine has cylinders. Waste spark systems have half the number of coils as there are cylinders. Perhaps the biggest difference in system operation is based on the use of CKP and CMP sensors.

All systems have a CKP to monitor crankshaft position and engine speed. Some also monitor the relative position of each cylinder. Not all systems have a CMP; some have more than one. The signals from a CMP sensor are used for cylinder identification and

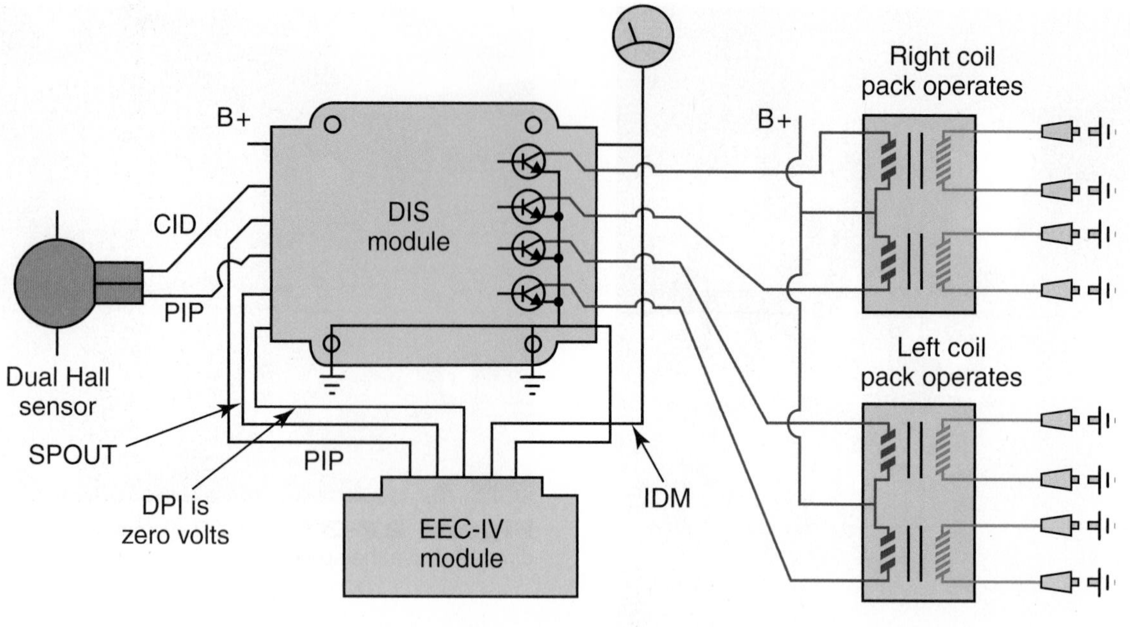

Figure 27-32 A dual plug system for a four-cylinder engine. *Courtesy of Ford Motor Company*

for verifying the correlation between the position of the crankshaft and the camshafts. The design of the trigger wheels or rotors for these two sensors also varies. The design is primarily based on whether the sensor is a magnetic pulse (variable reluctance) or Hall-effect sensor. Both can be used for either sensor. Inputs from these sensors are critical to the operation of the fuel injection and ignition systems.

The layout and operation of these sensors are designed to provide fast engine starts and synchronization of the fuel injection and ignition systems with the position of the engine's individual pistons.

Hall-Effect Sensors

Many Hall-effect sensors rely on pulleys or harmonic balancers with interrupter rings or shutters. In many cases, the crankshaft pulley has half as many windows as the engine has cylinders. As the crankshaft rotates and the interrupter passes in and out of the Hall-effect switch, the switch turns the module reference voltage on and off. The signals are identical and the control module cannot distinguish which of these signals to assign to a particular coil. The signal from the cam sensor gives the module the information it needs to synchronize the crankshaft sensor signals with the position of the number one cylinder. From there the module can energize the coils according to the firing order of the engine. Once the engine has started, the camshaft signal serves no purpose.

Other systems use a dual crankshaft sensor located behind the crankshaft pulley. The pulley has two sets of two interrupter rings **(Figure 27–33)** that rotate through the Hall-effect switches at the dual crankshaft sensor. The inner ring with equally spaced blades rotates through the inner Hall-effect switch, whereas the outer ring with one opening rotates

Figure 27-33 This crankshaft pulley for a six-cylinder engine has two sets of interrupter rings.

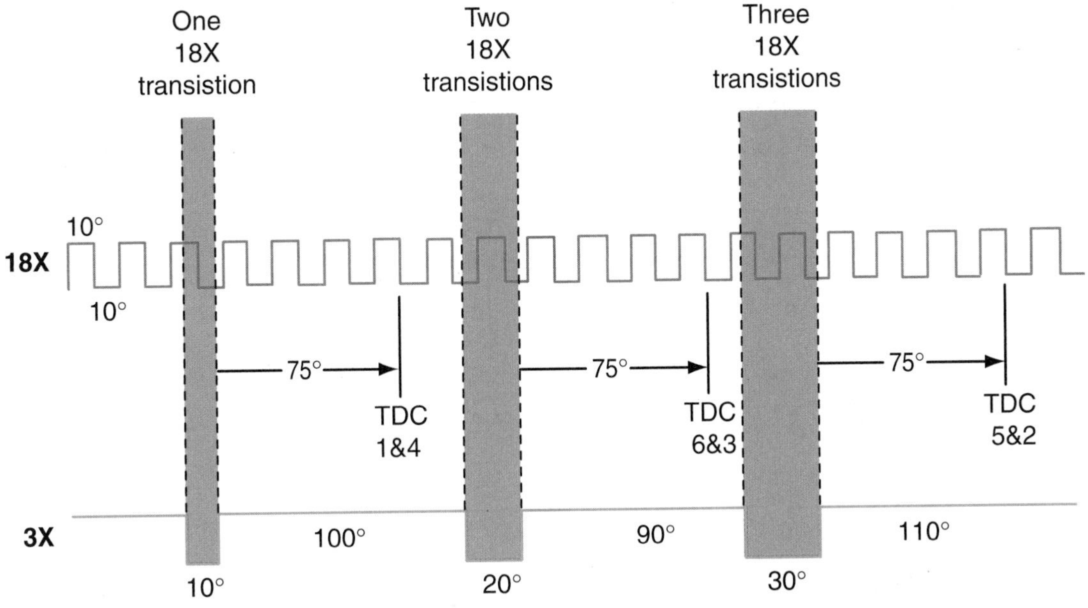

Figure 27-34 3X and 18X crankshaft signals.

through the outer Hall effect. The inner sensor provides leading edge signals and the outer sensor produces one leading edge during one complete revolution of the crankshaft. The outer sensor is the SYNC sensor. The signal from this sensor informs the ignition module regarding crankshaft position. The SYNC signal occurs once per crankshaft revolution.

Another example of a dual CKP sensor has an inner ring on the crankshaft pulley with three blades of unequal lengths with unequal spaces between the blades. On the outer ring there are eighteen blades of equal length with equal spaces between the blades. The signal from the inner sensor is referred to as the 3X signal, whereas the signal from the outer sensor is called the 18X signal. The ignition module knows which coil to fire from the number of 18X signals received during each 3X window rotation (**Figure 27–34**). For example, when two 18X signals are received, the coil module is programmed to sequence coil 3-6 next in the firing sequence. Within 120 degrees of crankshaft rotation, the coil module can identify which coil to sequence and thus start firing the spark plugs. Once the engine is running, the system uses the 18X signal for crankshaft position and speed information. The cam sensor signal is used for injector sequencing, but it is not required for coil sequencing.

Many CMP sensors are Hall-effect sensors and produce a square wave signal (**Figure 27–35**). The sensor may respond to a single slot on a camshaft pulley or the pulley will have several. One design has

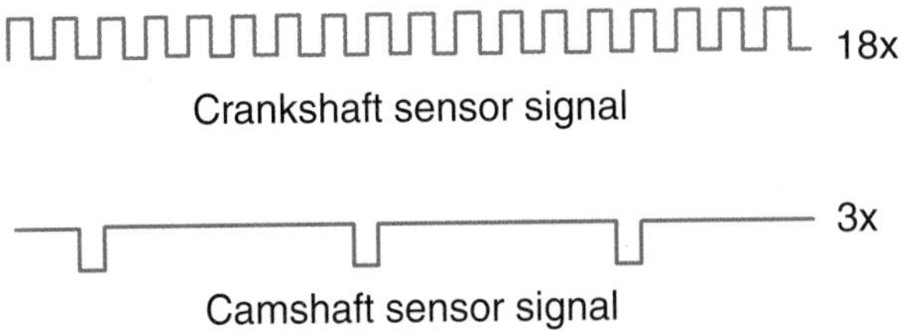

Figure 27-35 The relationship between crankshaft and camshaft signals on a GM 3.8L SFI engine.

four narrow and wide slots on the trigger wheel. The PCM uses the narrow and wide signal patterns to identify camshaft position, or which cylinder is on its compression stroke and which is on the exhaust. The PCM can then calculate the correct timing and sequencing for the spark plugs and fuel injectors.

Magnetic Pulse Generators

Many late-model engines use magnetic pulse generators as CKP sensors. The trigger wheel, also called the reluctor, can be located behind the crankshaft pulley, inside the engine in the middle of the crankshaft, or at the flywheel. Again the design of the trigger wheel depends on the application and operation of the system.

In more basic systems, the trigger wheel is located behind the crankshaft pulley. If the engine is a six cylinder, there will be seven slots in the reluctor, six of which are spaced exactly 60 degrees apart and the seventh notch is located 10 degrees from the number six notch and is used to synchronize the coil

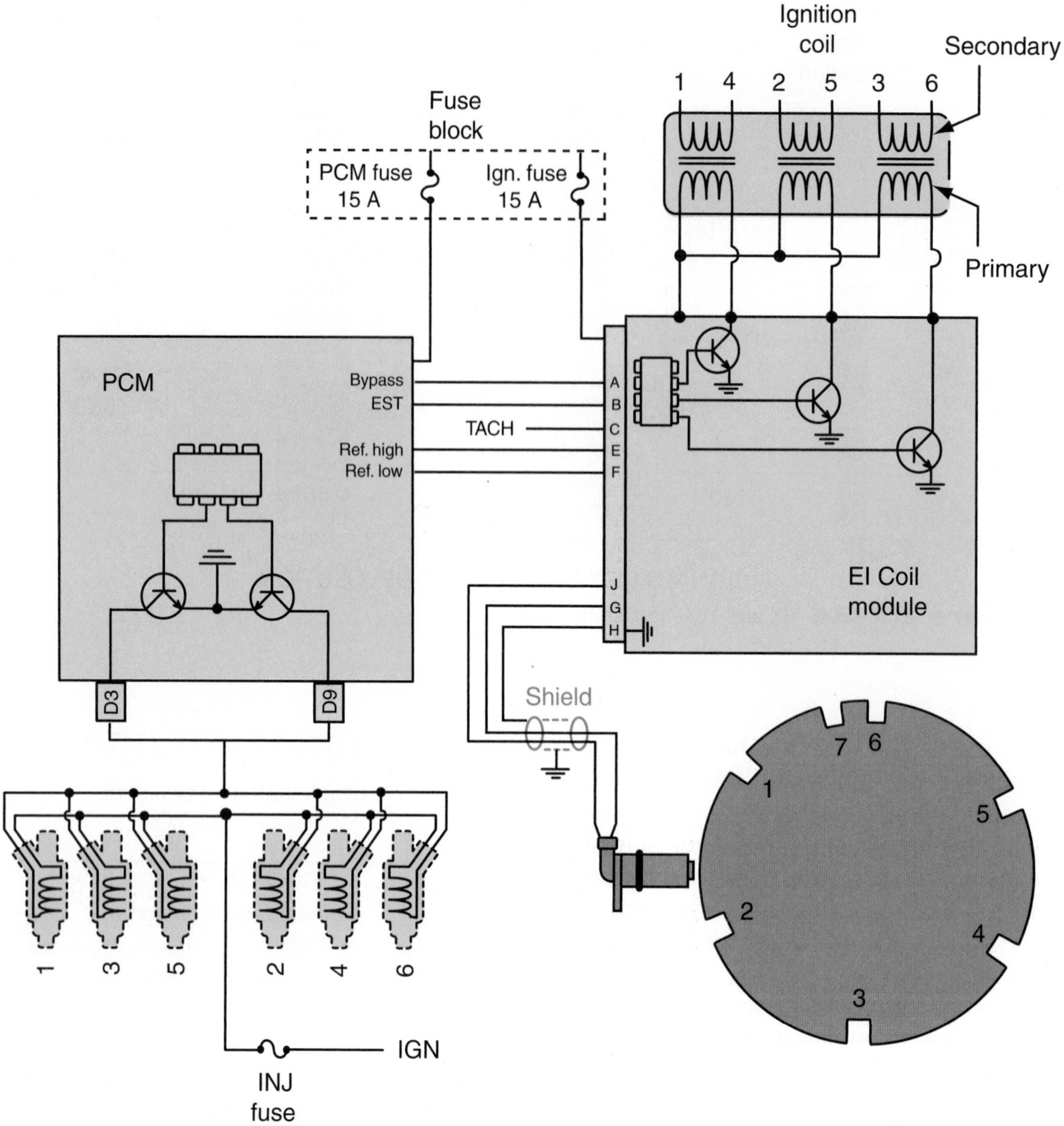

Figure 27–36 Schematic of an EI system with a magnetic pulse generator-type crankshaft sensor. Note the notches on the crankshaft timing wheel.

firing sequence in relation to crankshaft position **(Figure 27–36)**. The same triggering wheel can be and is used on four-cylinder engines. The computer only needs to be programmed to interpret the signals differently than for a six-cylinder engine.

The CKP sensor generates a small AC voltage each time one of the machined slots passes by. By counting the time between pulses, the ignition module picks out the unevenly spaced seventh slot and starts the calculation of the ignition coil sequencing. Similar systems are used with more slots or teeth machined into the reluctor. There is always at least one gap for cylinder identification purposes.

In many cases, the gap is used to identify the position of the crankshaft during engine cranking. A CMP sensor is used to determine what stroke the cylinders are on.

A unique system is found in Cadillac's Northstar engine. The engine has two CKP sensors **(Figure 27–37)**. A reluctor with twenty-four evenly spaced gaps and eight unevenly spaced gaps is cast onto the center of the crankshaft. When the reluctor rotates past the sensors, each sensor produces thirty-two high- and low-voltage signals per crankshaft revolution. The "A" sensor is positioned in the upper crankcase, and the "B" sensor is positioned in the lower crankcase. Because the A sensor is above the B sensor, the signal from the A sensor occurs 27 degrees before the B sensor signal.

The signals from the two sensors are sent to the ignition module. This module counts the number of

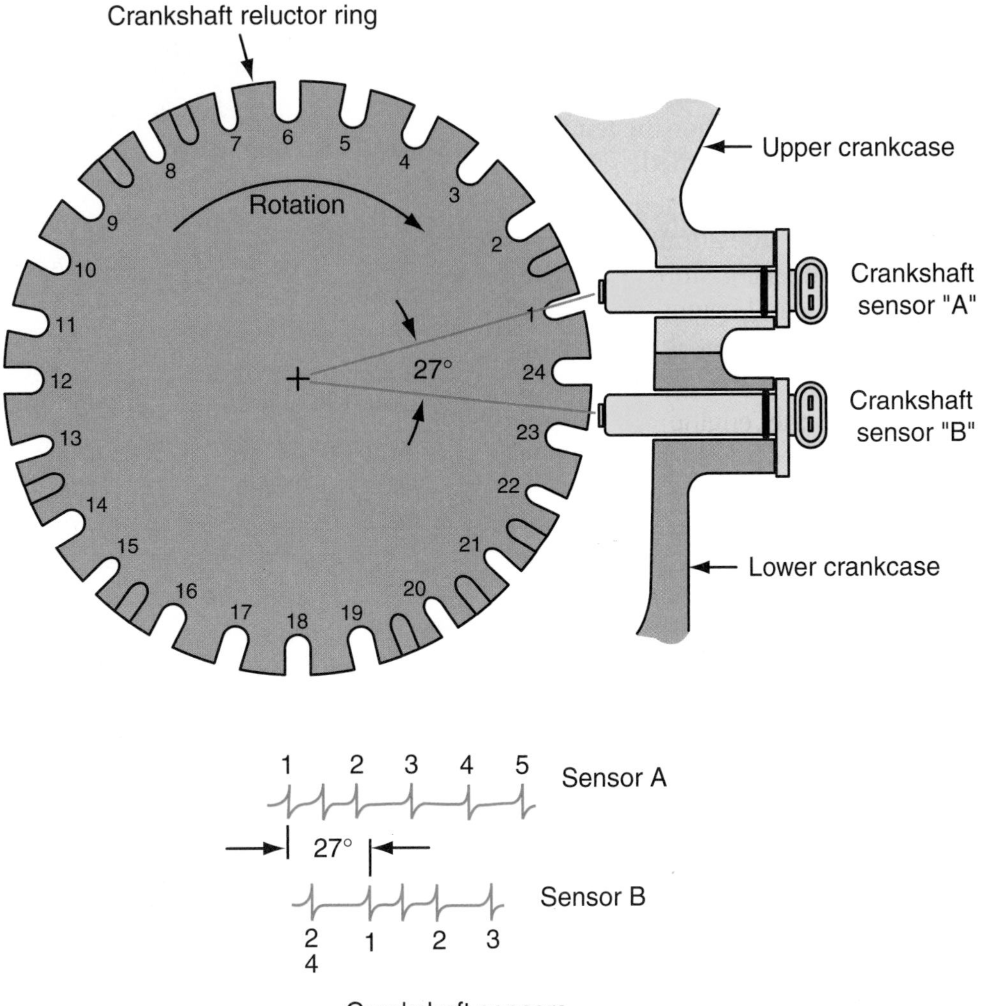

Figure 27-37 A and B crankshaft sensors in a Northstar engine.

signals from one sensor that are between the other sensor's signals to sequence the ignition coils properly. The CMP sensor produces one high- and one low-voltage signal every camshaft revolution, or every two crankshaft revolutions. The PCM uses the camshaft position sensor signal to sequence the injectors properly.

Misfire Detection

A high data rate CKP sensor is used to detect engine misfires, and the CMP is used to identify which cylinder is misfiring. Misfires are detected by variations in crankshaft rotational speed for each cylinder. An interesting feature of most misfire monitors is the ability of the PCM to distinguish an actual misfire from other things that may cause the engine's speed to fluctuate. Driving on a rough road can cause the vehicle's wheels to change rotational speed; this in turn will affect the rotational speed of the crankshaft. To determine whether the engine has misfired or the vehicle is merely driving on a poor surface, the PCM receives wheel speed data from the antilock brake system. A rough road will cause variances in wheel speed and the PCM looks at that data before concluding a misfire occurred.

Basic Timing

The PCM totally controls ignition timing and ignition timing is not adjustable. When the engine is cranked for starting, the PCM sets the timing at a fixed value. This value is used until the engine is running at a predetermined speed. Once that speed is met, the PCM looks at several inputs, including engine speed, load, throttle position, and engine coolant temperature, and makes adjustments accordingly. The PCM continues to rely on those inputs and its programmed strategy throughout operation. All PCMs have limits as to how far the timing can be retarded and advanced.

Timing Corrections The PCM adjusts ignition timing according to its programming and sensor inputs. There are times when the timing is adjusted or corrected to compensate for slight changes in the operating conditions or abnormal occurrences.

- *Temperature:* Ignition timing is advanced when the coolant temperature is low. When the temperature is very high, the timing is retarded.
- *Engine Knock:* When a knock is detected, the PCM retards the timing in fixed steps until the knock disappears. When the knocking stops, the PCM stops retarding the timing and begins to advance the timing in fixed steps unless the knocking reoccurs.
- *Stabilizing Idle:* When the engine idle speed moves away from the desired idle speed, the PCM will adjust the timing to stabilize the engine speed. It is important to know that ignition timing changes are made only to correct minor idle problems. If the engine's speed is above the desired speed, the timing is retarded and when it is too low, the timing is advanced.
- *EGR Operation:* When the EGR valve opens, the timing is advanced. The amount of advance depends on intake air volume and engine speed.
- *Transition Correction:* When the vehicle is accelerated immediately after deceleration, the timing is temporarily advanced or retarded to smoothen the transition.
- *Torque Control:* To provide smooth shifting of an automatic transmission, the PCM will temporarily retard the ignition timing to reduce the engine's torque when the transmission is beginning to change gears.
- *Traction Control Correction:* When excessive wheel slippage occurs, the PCM will retard the timing to reduce the torque output from the engine. Once the slippage has been corrected, timing returns to normal.

CAUTION!

Since EI systems have considerably higher maximum secondary voltage compared to distributor-type ignition systems, greater electrical shocks are obtained from EI systems. Although such shocks may not be directly harmful to the human body, they may cause you to jump or react suddenly, which could result in personal injury. For example, when you jump suddenly as a result of an EI electrical shock, you may hit your head on the vehicle hood or push your hand into a rotating cooling fan.

CASE STUDY

A customer had her late-model Ford product towed into the shop because it would not start. The car was equipped with a distributorless ignition system. The technician attempted to verify the complaint but found the battery totally dead—probably caused by continued attempts to start the engine by the owner and the tow truck operator. After several hours of slow charging the battery, the technician attempted to start the engine. It seemed to crank normally, but it, indeed, would not start.

Knowing the engine needed air, fuel, and spark to start, he began a basic check of those systems. The air intake seemed to be clear, but there was no spark at the test spark plug he inserted into one of the spark plug circuits. Before continuing with any tests on the ignition system, he checked the fuel system. Beginning with the voltage to the TP sensor, this measurement was normal. He then checked the fuel pressure of the system; this too was found to be normal. He assumed at this point that fuel was not the problem.

He then connected a scan tool to the system to watch cranking rpm and to see if the running or relative compression was normal. Again all seemed normal. Then he retrieved the DTCs and found one—P0350, "Ignition Coil Primary/Secondary Circuit." This DTC is normally set when the PCM does not receive a valid ignition diagnostic monitor pulse signal from the ignition module. There are many causes for this, including an open coil driver circuit, a short to ground in the coil driver circuit, a short to power in the coil driver, a faulty coil assembly, or an open or short in the ignition start/run circuit. Because of the many possible causes, he follows the pinpoint tests given in the service information for this code. However, he knows that because the engine does not start, the problem is something that is common to all cylinders. Therefore, it is unlikely that the cause would be a bad driver or driver circuit. The next step in the pinpoint testing routine verifies his conclusion. He checks for voltage from the ignition switch to the coil pack and finds none. By carefully inspecting the circuit, he finds a burnt connector. The connector apparently had made contact with the exhaust manifold. He makes the repair and the engine starts.

KEY TERMS

Air gap
Ballast resistor
Coil-over-plug (COP)
Coil pack
Direct ignition system (DIS)
Distributorless ignition system (DLI)
Dual plug
Dwell
Heat range
Inductive reluctance
Misfiring
Primary circuit
Pulse transformer
Reach
Reactance
Secondary circuit
Twin plug
Waste spark

SUMMARY

- The ignition system supplies high voltage to the spark plugs to ignite the air-fuel mixture in the combustion chambers.
- The ignition system has two interconnected electrical circuits: a primary circuit and a secondary circuit.
- The primary circuit supplies low voltage to the primary winding of the ignition coil. This creates a magnetic field in the coil.
- A switching device interrupts primary current flow, collapsing the magnetic field and creating a high-voltage surge in the ignition coil secondary winding.
- The switching device used in electronically controlled systems is an NPN transistor.
- The secondary circuit carries high-voltage surges to the spark plugs. On some systems, the circuit runs from the ignition coil, through a distributor, to the spark plugs.
- The distributor may house the switching device plus timing advance mechanisms. Some systems locate the switching device outside the distributor housing.
- Ignition timing is directly related to the position of the crankshaft. Magnetic pulse generators and Hall-effect sensors are the most widely used engine position sensors. They generate an electrical signal at certain times during crankshaft rotation. This signal triggers the electronic switching device to control ignition timing.
- Distributors are seldom found on today's engines. Nearly all of today's engines are equipped with an EI system for which there are primarily two different designs: double-ended coil and coil-per-cylinder.
- In computer-controlled ignitions, the computer receives input from numerous sensors. Based on this data, the computer determines the optimum firing time and signals an ignition module to activate the secondary circuit at the precise time needed.
- In some systems, the camshaft sensor signal informs the computer when to sequence the coils and fuel injectors.
- The crankshaft sensor signal provides engine speed and crankshaft position information to the computer.
- Some EI systems have a combined crankshaft and SYNC sensor at the front of the crankshaft.

REVIEW QUESTIONS

1. Explain how voltage is induced in the distributor pickup coil as the reluctor high point approaches alignment with the pickup coil.
2. Explain why dwell time is important to ignition system operation.
3. Name the engine operating conditions that most affect ignition timing requirements.
4. Explain how the plugs fire in a two-plug-per-coil EI system.
5. Explain the components and operation of a magnetic pulse generator.
6. What happens when the low-voltage current flow in the coil primary winding is interrupted by the switching device?
   a. The magnetic field collapses.
   b. A high-voltage is induced in the coil secondary winding.
   c. Both a and b.
   d. Neither a nor b.
7. *True or False?* A spark plug with two ground electrodes will provide two separate sparks when fired.
8. Why is high voltage needed to establish a spark across the gap of a spark plug?
9. List the advantages of having one ignition coil per cylinder.

10. Which of the following is a function of all ignition systems?
    a. to generate sufficient voltage to force a spark across the spark plug gap
    b. to time the arrival of the spark to coincide with the movement of the engine's pistons
    c. to vary the spark arrival time based on varying operating conditions
    d. all of the above
11. Reach, heat range, and air gap are all characteristics that affect the performance of which ignition system component?
    a. ignition coils
    b. ignition cables
    c. spark plugs
    d. breaker points
12. *True or False?* The spark plug wire is eliminated in all coil-per-cylinder ignition systems to reduce maintenance and the chances of EMI and RFI.
13. Modern ignition cables contain fiber cores that act as a ______ in the secondary circuit to cut down on radio and television interference and reduce spark plug wear.
14. The magnetic field surrounding the coil in a magnetic pulse generator moves when the ______.
    a. reluctor tooth approaches the coil
    b. reluctor tooth begins to move away from the pickup coil pole
    c. reluctor is aligned with the pickup coil pole
    d. both a and b
15. Which of the following electronic switching devices has a reluctor with wide shutters rather than teeth?
    a. magnetic pulse generator
    b. metal detection sensor
    c. Hall-effect sensor
    d. all of the above

ASE-STYLE REVIEW QUESTIONS

1. Technician A says that a magnetic pulse generator is equipped with a permanent magnet. Technician B says that a Hall-effect switch is equipped with a permanent magnet. Who is correct?
    a. Technician A   c. Both A and B
    b. Technician B   d. Neither A nor B
2. While discussing ignition systems: Technician A says that an ignition system must supply high-voltage surges to the spark plugs. Technician B says that the system must maintain the spark long enough to burn all of the air-fuel mixture in the cylinder. Who is correct?
    a. Technician A   c. Both A and B
    b. Technician B   d. Neither A nor B
3. While discussing ignition timing requirements: Technician A says that more advanced timing is desired when the engine is under a heavy load. Technician B says that more advanced timing is desired when the engine is running at high engine speeds. Who is correct?
    a. Technician A   c. Both A and B
    b. Technician B   d. Neither A nor B
4. While discussing secondary voltage: Technician A says that the normal required secondary voltage is higher at idle speed than at wide-open throttle conditions. Technician B says that the maximum available secondary voltage must always exceed the normally required secondary voltage. Who is correct?
    a. Technician A   c. Both A and B
    b. Technician B   d. Neither A nor B
5. Technician A says that an ignition system must generate sufficient voltage to force a spark across the spark plug gap. Technician B says that the ignition system must time the arrival of the spark to coincide with the movement of the engine's pistons and vary it according to the operating conditions of the engine. Who is correct?
    a. Technician A   c. Both A and B
    b. Technician B   d. Neither A nor B
6. In EI systems using one ignition coil for every two cylinders: Technician A says that two plugs fire at the same time with one wasting the spark on the exhaust stroke. Technician B says that one plug fires in the normal direction (center to side electrode) and the other in reversed polarity (side to center). Who is correct?
    a. Technician A   c. Both A and B
    b. Technician B   d. Neither A nor B
7. While discussing EI systems: Technician A says that all systems rely on CKP and CMP signals to synchronize the ignition system with piston movement. Technician B says that CKP and CMP signals are used to synchronize the operation of

the ignition and fuel injection systems with piston position. Who is correct?

a. Technician A
b. Technician B
c. Both A and B
d. Neither A nor B

8. While discussing EI systems: Technician A says that when the engine is being cranked, the PCM relies only on the signals from the ECT to determine initial ignition timing. Technician B says that when an engine is initially started and is running at a predetermined speed, the PCM looks at several inputs, including engine speed, load, throttle position, and engine coolant temperature, and makes adjustments accordingly. Who is correct?

    a. Technician A
    b. Technician B
    c. Both A and B
    d. Neither A nor B

9. While discussing ignition timing control in EI systems: Technician A says that when the engine coolant temperature is low, the timing is retarded until normal operating temperature is reached. Technician B says that the PCM will retard the timing during EGR operation. Who is correct?

    a. Technician A
    b. Technician B
    c. Both A and B
    d. Neither A nor B

10. While discussing spark plugs: Technician A says that too narrow of a spark plug gap can cause a rough idle and premature wear of the electrodes. Technician B says that a spark plug with the wrong reach can cause preignition. Who is correct?

    a. Technician A
    b. Technician B
    c. Both A and B
    d. Neither A nor B

CHAPTER 28

IGNITION SYSTEM DIAGNOSIS AND SERVICE

OBJECTIVES

■ Perform a no-start diagnosis and determine the cause of the condition. ■ Determine the cause of an engine misfire. ■ Perform a visual inspection of ignition system components, primary wiring, and secondary wiring to locate obvious trouble areas. ■ Describe what an oscilloscope is, its scales and operating modes, and how it is used in ignition system troubleshooting. ■ Test the components of the primary and secondary ignition circuits. ■ Test individual ignition components using test equipment such as a voltmeter, ohmmeter, and testlight. ■ Service and install spark plugs. ■ Describe the effects of incorrect ignition timing. ■ Check and set (when possible) ignition timing. ■ Diagnose engine misfiring on EI-equipped engines.

This chapter concentrates on testing ignition systems and their individual components. It must be stressed, however, that there are many variations in the ignition systems used by auto manufacturers. The tests covered in this chapter are those generally used as basic troubleshooting procedures. Exact test procedures and the ideal troubleshooting sequence will vary among vehicle makers and individual models. Always consult the vehicle's service manual when performing ignition system service.

Two important precautions should be taken during all ignition system tests:

1. Turn the ignition switch off before disconnecting any system wiring.
2. Do not touch any exposed connections while the engine is cranking or running.

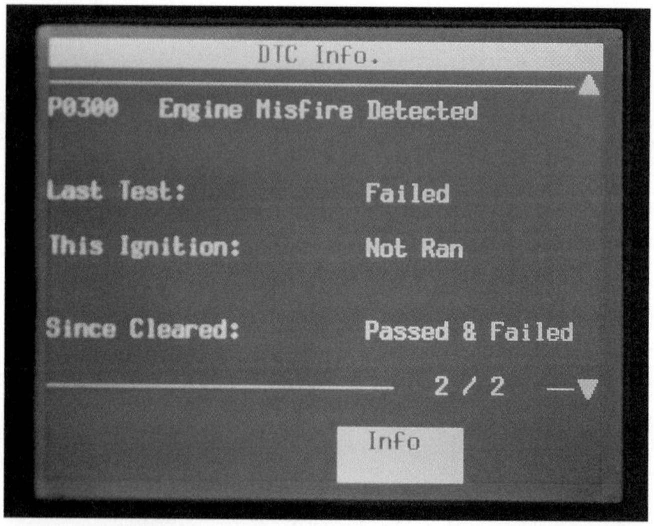

Figure 28–1 The PCM will store a DTC when it detects a misfire.

MISFIRES

When something prevents complete combustion, the result is a misfire or incomplete combustion. Misfires can cause lack of power, poor gas mileage, excessive exhaust emissions, and a rough running engine. Misfires (**Figure 28–1**) are not always caused by the ignition system; other systems also can cause them. A spark plug misfires when it has a weak spark or does not fire at all. Misfires can be caused by a fouled spark plug, a bad coil, problems in the primary or secondary ignition circuit, or an incorrect plug gap.

Abnormal Combustion

Incomplete combustion is not the only abnormal condition engines may experience; they may also experience detonation. Detonation is usually caused by excessively advanced ignition timing, engine overheating, excessively lean mixtures, or the use of low-octane gasoline. Detonation can cause physical damage to the pistons, valves, bearings, and spark plugs.

Preignition can cause pinging or spark knocking. Any hot spot within the combustion chamber can cause preignition. Common causes of preignition are incandescent carbon deposits in the combustion chamber, a faulty cooling system, too hot of a spark plug, poor engine lubrication, and cross firing. Preignition usually leads to detonation; preignition and detonation are two separate events.

GENERAL IGNITION SYSTEM DIAGNOSIS

The ignition system should be tested whenever you know or suspect there is no spark, not enough spark, or when the spark is not being delivered at the correct time to the cylinders.

Common vs. Noncommon Problems

In most cases, all ignition problems can be divided into two types: common and noncommon. Common problems are those that affect all cylinders, and noncommon problems are those that affect one or more cylinders but not all. Common ignition components include the parts of the primary circuit and the secondary circuit up to the distributor's rotor in DI systems. Noncommon parts are the individual spark plug terminals inside the distributor cap, spark plug wires, and the spark plugs. With EI systems, the individual coils are noncommon parts.

The best indicator of a noncommon problem is the reading on a vacuum gauge (**Figure 28–2**). If a vacuum gauge is connected to a four-cylinder engine at idle and the needle of the gauge is within the normal vacuum range for three-fourths of the time and drops one-fourth of the time, this indicates that three of the cylinders are working normally while the fourth is not. The cause of the problem is noncommon. If that cylinder is sealed and all cylinders are receiving the correct amount of air and fuel, the problem must be in the ignition system. The problem is in the noncommon parts of the ignition system.

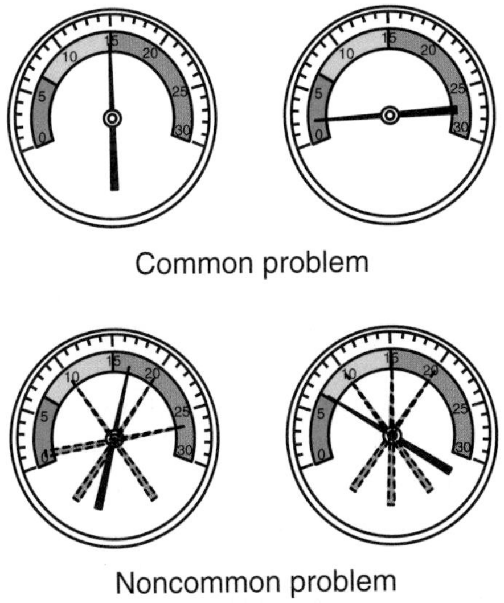

Figure 28-2 The reaction of a vacuum gauge when there is a common or noncommon problem.

Determining if the ignition problem is common or noncommon is a good way to start troubleshooting the ignition system. By dividing the ignition system into common and noncommon parts, you will test only those parts that could cause the problem.

Generally when an engine runs unevenly, the cause is a noncommon problem. If the engine does not start, the problem is probably a common one. EI systems, especially coil-per-cylinder systems, make troubleshooting a little easier. The PCM or ignition module may be the only part that is common to all cylinders. The coils and all of the secondary circuit are common to only one or two cylinders. For example, if a coil in a waste fire system is bad, two cylinders will be affected and not the entire engine.

VISUAL INSPECTION OF IGNITION SYSTEMS

Begin all diagnosis by gathering as much information as possible from the customer. Then conduct a careful visual inspection. The system should be checked for obvious problems. Although no-start problems and incorrect ignition timing are caused by the primary circuit, the secondary circuit can be the cause of driveability problems and should be carefully checked. In addition to the ignition system, inspect all related electrical connectors or fuses, vacuum lines, air intake system, and cooling system. Also check available service information that may relate to the symptoms.

Symptoms commonly caused by ignition system problems include (keep in mind that the ignition system is not the only thing that can cause these):

- *Hard starting*—The engine requires an excessive amount of time to start.
- *Rough idle*—The engine idles poorly and may stall.
- *Engine stalling*—The engine quits unexpectedly. It may occur right after engine startup, while idling, or during deceleration.
- *Hesitation*—The engine does not immediately respond to opening of the throttle.

- *Stumble*—The engine temporarily loses power during acceleration.
- *Poor acceleration*—The vehicle accelerates slower than expected.
- *Surge*—The engine's speed fluctuates with a constant throttle during idle, steady cruise, acceleration, or deceleration.
- *Bucking*—The vehicle jerks shortly after acceleration or deceleration.
- *Knocking (pinging)*—The engine makes a sharp metallic noise during acceleration.
- *Backfire and afterfire*—Backfire is a loud pop coming from the intake system, usually during rapid throttle opening. Afterfire is a popping that occurs in the exhaust system, usually during quick deceleration.

Scan Tools

Today's ignition systems are part of the engine control system. Part of the visual inspection should include a check of the MIL. If it is operating correctly and a fault is emissions related, the lamp will remain on after the engine has started. Also check all TSBs related to the vehicle. Look for bulletins that recommend reflashing the PCM. Sometimes the EEPROM needs to be reprogrammed due to changes made in the strategy or calibrations after the vehicle was produced.

Go To Chapter 22 for flashing procedures.

DTCs and scan tool data should be retrieved during the initial diagnostic routine **(Figure 28–3)**. This includes KOEO, KOER, and continuous self-test DTCs. Also make sure to record any and all freeze frame data associated with the DTCs. Use the pinpoint tests to identify what set the code. Often these will lead to the exact cause of the problem; other times you will need to do further testing. If more than one DTC was set, refer to the vehicle's wiring diagram to identify parts and circuits that may be common to those codes. If the scan tool is unable to communicate with the system, follow the tests prescribed by the manufacturer to correct that problem. If DTCs are retrieved, correct the problems that are causing the trouble codes before moving on. Fixing the cause of the codes may correct the ignition problem.

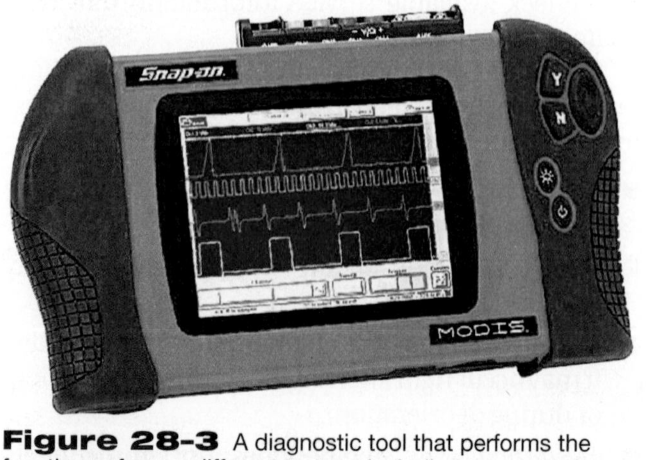

Figure 28-3 A diagnostic tool that performs the functions of many different testers, including a scan tool. *Reproduced under license from Snap-on Incorporated. All of the marks are marks of their owners.*

If no DTCs were retrieved, look at the system's serial data. First identify the appropriate PIDs. The PID test mode allows access to PCM information, including input signals, outputs, calculated values, and the status of the system and monitors. While observing serial data, look at the inputs. Identify any signals that are outside the normal range. Do the same with the outputs. Mode 6 data can also be helpful. This mode displays the test values stored at the time a particular monitor was completed **(Figure 28–4)**.

SHOP TALK

Keep in mind that the PCM may assume that the input signals are correct while it controls an output device. This means incorrect inputs can cause an output to appear out of range because the PCM is driving it outside the normal range. Check the input signals before checking the outputs.

If no codes were retrieved and all appears to be normal on the data stream, diagnosis should be based on symptoms and detailed testing of the ignition system.

Primary Circuit

Primary ignition system wiring should be checked for tight connections. Electronic circuits operate on very low voltage. Voltage drops caused by corrosion or dirt can cause running problems. Missing or broken tab locks on wire terminals are often the cause of intermittent ignition problems due to vibration or thermal related failure.

Test the integrity of a suspect connection by tapping, tugging, and wiggling the wires while the engine is running. Be gentle. The object is to re-create an ignition interruption, not to cause permanent circuit damage. With the engine off, separate the suspect connectors and check them for dirt and corrosion. Clean the connectors according to the manufacturer's recommendations.

J1979 MISFIRE MODE $06 DATA			
Monitor ID	**Test ID**	**Description for CAN**	**Increments**
A1	$80	Total engine misfire and catalyst damage misfire rate	%
A1	$81	Total engine misfire and emission threshold misfire rate	%
A1	$82	Highest catalyst damage misfire and catalyst damage threshold misfire rate	%
A1	$83	Highest emission threshold misfire and emission threshold misfire rate	
A1	$84	Inferred catalyst mid-bed temperature	°C
A2-AD	$0B	Misfire counts for last 10 drive cycles	events
A2-AD	$0C	Misfire counts for last/current drive cycle	events
A2-AD	$80	Cylinder "X" misfire rate and catalyst damage misfire rate	%
A2-AD	$81	Cylinder "X" misfire rate and emission threshold misfire rate	%

Figure 28-4 Mode 6 data from the misfire monitor.

Do not overlook the ignition switch as a source of intermittent ignition problems. A loose mounting rivet or poor connection can result in erratic spark output. To check the switch, gently wiggle the ignition key and connecting wires with the engine running. If the ignition cuts out or dies, the problem is located.

Carefully inspect the wires and belts for the charging system. Also, check the charging voltage at the battery. The efficiency of an ignition system depends on the voltage it receives. If battery or charging system voltage is low, the input to the primary side of the coil will also be low.

Moisture can cause a short to ground or reduce the amount of voltage available to the spark plugs. This can cause poor performance or a no-start condition. Carefully check the ignition system for signs of moisture. **Figure 28–5** shows the common places where moisture may be present in a DI system.

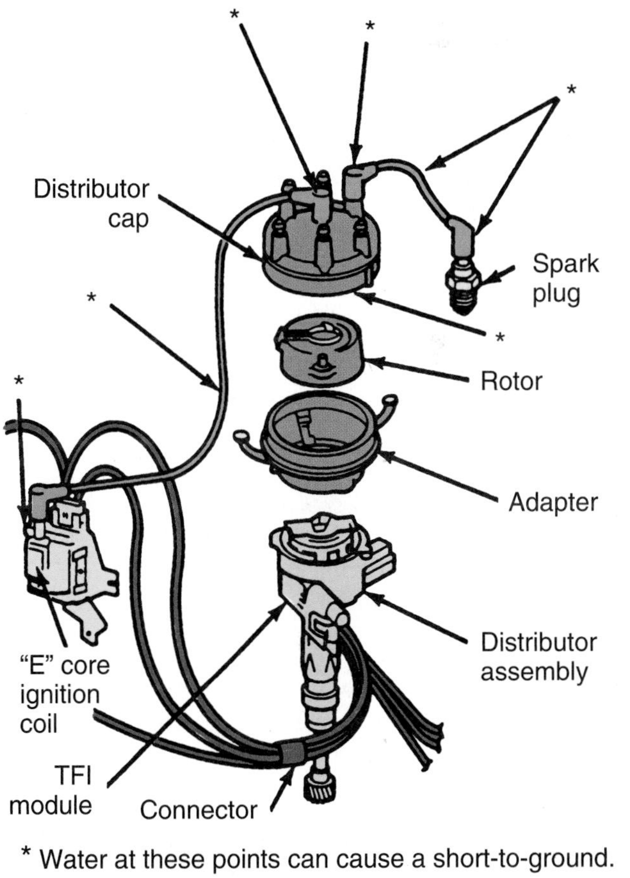

Figure 28-5 Places to check for moisture in an ignition system. *Courtesy of Ford Motor Company*

Ground Circuits

Ground straps are often neglected, or worse, left disconnected after routine service. With the increased use of plastics in today's vehicles, ground straps may mistakenly be reconnected to a nonmetallic surface. The result of any of these problems is that the current that was to flow through the disconnected or improperly grounded strap is forced to find an alternate path to ground. Sometimes the current attempts to back up through another circuit. This may cause the circuit to operate erratically or fail altogether. The current may also be forced through other components, such as wheel bearings or shift and clutch cables that are not meant to handle current flow, causing them to wear prematurely or become seized in their housing.

Examples of bad ground-circuit-induced ignition failures include burned ignition modules resulting from missing or loose coil ground straps and intermittent ignition operation resulting from a poor ground at the control module. Poor ground can be identified by conducting voltage drop tests and by monitoring the circuit with a lab scope.

When conducting a voltage drop test, remember that the circuit must be turned on and have current flowing through it. If the circuit is tested without current flow, the circuit will show zero voltage drop, which would indicate that it is good regardless of the amount of resistance present.

The same is also true when checking a ground with the lab scope. Make sure the circuit is on. If the ground is good, the trace on the scope should be at 0 volts and be flat. If the ground is bad, some voltage will be indicated and the trace will not be flat (**Figure 28–6**).

Often a bad sensor ground will cause the same symptoms as a faulty sensor. Before condemning a sensor, check its ground with a lab scope. **Figure 28–7** shows the output of a good Hall-effect switch with a bad ground.

SERVICE TIP

When checking the ignition system with a lab scope, gently tap and wiggle the components while observing the trace. This may indicate the source of an intermittent problem.

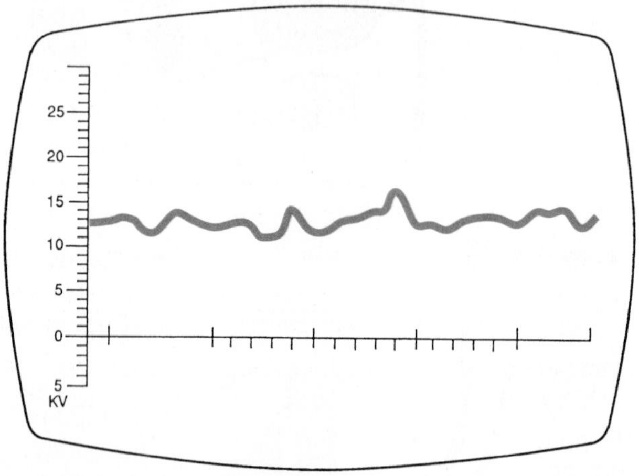

Figure 28-6 A voltage signal caused by a poor ignition module ground. *Courtesy of SPX Service Solutions*

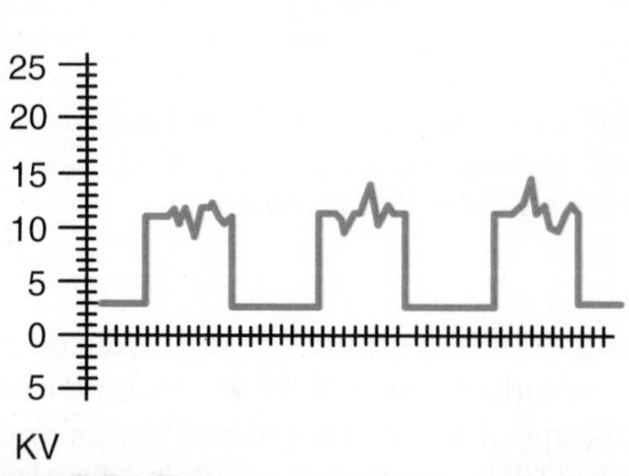

Figure 28-7 A voltage trace with ignition system noise due to a bad ground. Notice that the trace does not reach zero.

Electromagnetic Interference

Electromagnetic interference (EMI) can cause problems with the vehicle's computer. EMI is produced when electromagnetic radio waves of sufficient amplitude escape from a wire or conductor. Unfortunately, an automobile's spark plug wires, ignition coil, and AC generator coils all possess the ability to generate these radio waves. EMI can alter signals from sensors and to actuators. The result may be an intermittent driveability problem that may appear to be caused by many different systems.

To minimize the effects of EMI, check to make sure that sensor wires running to the computer are routed away from potential EMI sources. Rerouting a wire by no more than an inch or two may keep EMI from falsely triggering or interfering with computer operation.

Connecting a lab scope to voltage and ground wires can identify EMI problems. Common problems such as poor spark plug wire insulation will allow EMI.

Sensors

A voltage pulse from a crankshaft position sensor (**Figure 28–8**) activates the transistor in the control module. In most ignition systems, this sensor is either a magnetic pulse generator or Hall-effect sensor. These sensors are mounted either on the distributor shaft or the crankshaft.

The reluctor or pole piece of a magnetic pulse generator is replaced only if it is broken or cracked. The pickup coil wire leads can become grounded if their insulation wears off as the breaker plate moves with the vacuum advance unit (**Figure 28–9**). Inspect these leads carefully. Position these wires so that they do not rub the breaker plate as it moves.

Under unusual circumstances, the nonmagnetic reluctor can become magnetized and upset the pickup coil's voltage signal to the control module. Use a steel feeler gauge to check for signs of magnetic

Figure 28-8 The wiring to the crankshaft sensor should be carefully inspected.

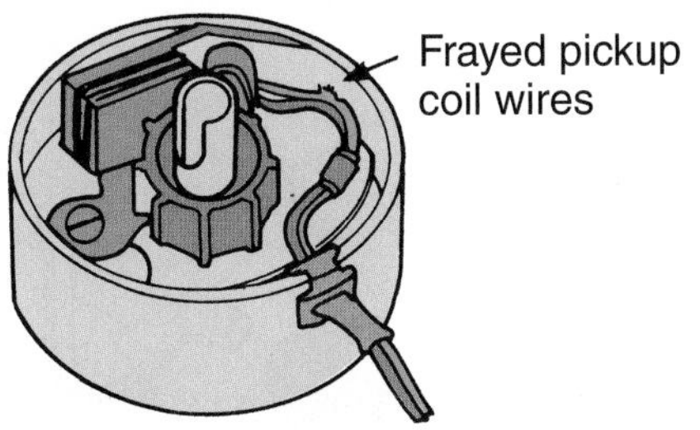

Figure 28-9 Inspect pickup coil wiring for damage.

attraction and replace the reluctor if the test is positive. On some systems, the gap between the pickup and the reluctor must be checked and adjusted to manufacturer's specifications. To do this, use a properly sized nonmagnetic feeler gauge to check the air gap between the coil and reluctor. Adjust the gap if it is out of specification.

Hall-effect sensor problems are similar to those of magnetic pulse generators. This sensor produces a voltage when it is exposed to a magnetic field. The Hall-effect assembly is made up of a permanent magnet located a short distance away from the sensor. Attached to the distributor shaft is a shutter wheel. When the shutter is between the sensor and the magnet, the magnetic field is interrupted and voltage immediately drops to zero. This drop in voltage is the signal to the ignition module. When the shutter leaves the gap between the magnet and the sensor, the sensor produces voltage again.

Control Modules

Electronic ignitions use transistors as switches. These transistors are contained inside the control module that can be mounted to or in the distributor, remotely mounted to a surface inside the engine compartment, mounted below the ignition coil pack **(Figure 28–10)**, or be integral to the PCM. Control modules should be tightly mounted to clean surfaces. A loose mount can cause intermittent misfires or no-start conditions. Often the module is grounded through its mounting. A loose mounting also can cause heat buildup that can damage and destroy the transistors and other electronic components inside the module. Some manufacturers recommend the use of special heat-conductive silicone grease between the control unit and its mounting. This helps conduct heat away from the module, reducing the chance of heat-related failure. During the visual inspection, check all electrical connections to the module. They must be clean and tight.

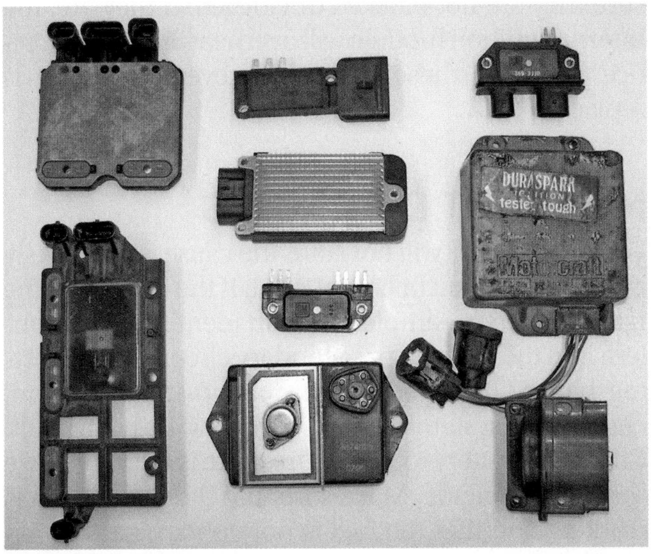

Figure 28-10 When the ignition module is not part of the PCM, it can have many different shapes and designs.

Secondary Circuit

Spark plug (ignition) and coil cables should be pushed tightly into the distributor cap and coil and onto spark plugs. Inspect all secondary cables for cracks and worn insulation, which cause high-voltage leaks. Inspect all of the boots on the ends of the secondary wires for cracks and hard, brittle conditions. Replace the wires and boots if they show evidence of these conditions. Most manufacturers recommend spark plug wire replacement only in complete sets.

The secondary coil cable should also be inspected **(Figure 28–11)**. When checking this cable, check the ignition coil. The coil should be inspected for cracks or any evidence of leakage in the coil tower.

Secondary cables must be connected according to the firing order. Refer to the manufacturer's service manual to determine the correct firing order and cylinder numbering.

White or grayish powdery deposits on secondary cables at the point where they cross or near metal parts indicate that the cables' insulation is faulty. The deposits occur because the high voltage in the cable has burned the dust collected on the cable. Such faulty insulation may produce a spark that sometimes can be heard and seen in the dark. An occasional glow

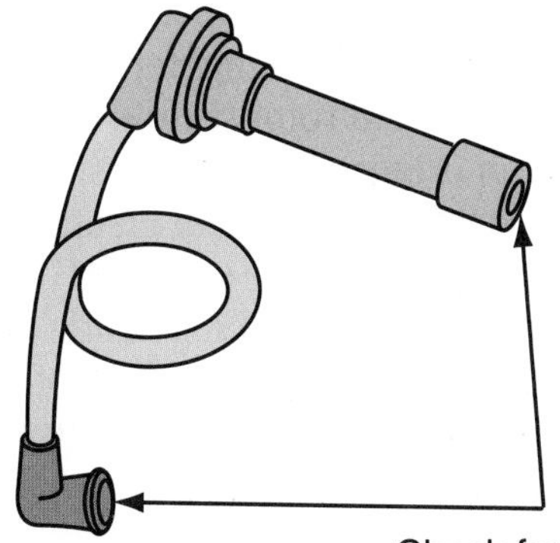

Figure 28-11 Carefully inspect the secondary cables.

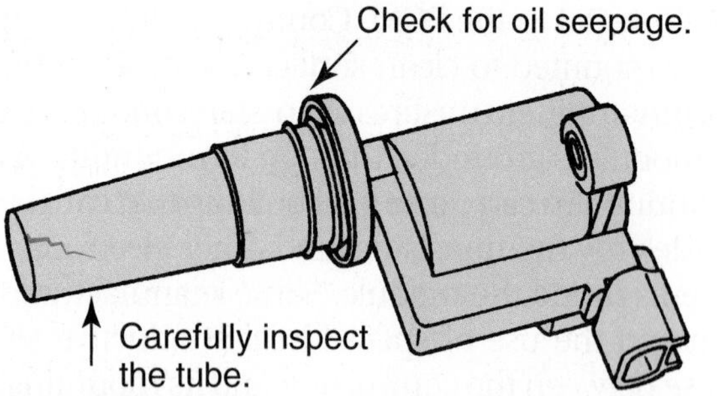

Figure 28-12 The plastic assembly for COP assembly should be carefully inspected.

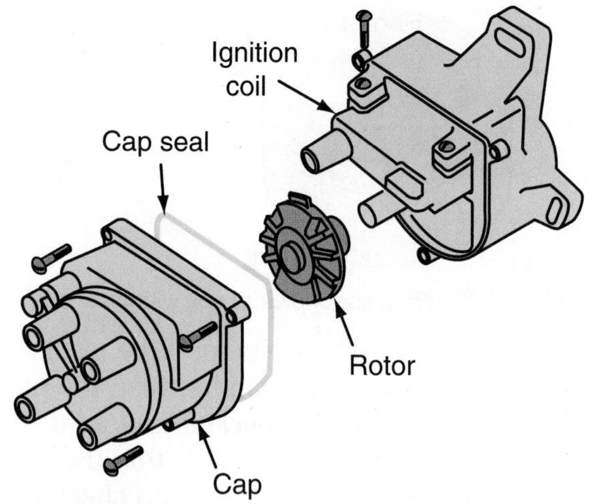

Figure 28-13 Inspect the distributor cap and rotor.

around the spark plug cables, known as a **corona effect**, is not harmful but indicates that the cable should be replaced.

Spark plug cables from consecutively firing cylinders should cross rather than run parallel to one another. Spark plug cables running parallel to one another can induce firing voltages in one another and cause the spark plugs to fire at the wrong time.

On distributorless or electronic ignition (EI) systems, visually inspect the secondary wiring connections at the individual coil modules. Make sure all of the spark plug wires are securely fastened to the coil and the spark plug. If a plug wire is loose, inspect the terminal for signs of burning. The coils should be inspected for cracks or any evidence of leakage in the coil tower. Check for evidence of terminal resistance. A loose or damaged wire or bad plug can lead to carbon tracking of the coil. If this condition exists, the coil must be replaced.

On COP systems, carefully check the tubes that fit around the terminal of the spark plugs (**Figure 28–12**). If the tube is cracked, voltage can leak out, jump to the cylinder head, and cause a misfire. Also, make sure the coil assembly fits snugly over the spark plug and is securely mounted.

Distributor Cap and Rotor

The distributor cap should be properly seated on its base. All clips or screws should be tightened securely.

The distributor cap and rotor should be removed for visual inspection (**Figure 28–13**). Physical or electrical damage is easily recognizable. Electrical damage from high voltage can include corroded or burned metal terminals and **carbon tracking** inside distributor caps. Carbon tracking is the formation of a line of carbonized dust between distributor cap terminals or between a terminal and the distributor housing. Carbon tracking indicates that high-voltage electricity has found a low-resistance conductive path over or through the plastic. The result is a misfire or a cylinder that fires at the wrong time. Check the outer cap towers and metal terminals for defects. Cracked plastic requires replacement of the unit. Damaged or carbon-tracked distributor caps or rotors should be replaced.

The rotor should be inspected carefully for discoloration and other damage. Inspect the top and bottom of the rotor carefully for grayish, whitish, or rainbow-hued spots. Such discoloration indicates that the rotor has lost its insulating qualities. High voltage is being conducted to ground through the plastic.

If the distributor cap or rotor has a mild buildup of dirt or corrosion, it should be cleaned. If it cannot be cleaned up, it should be replaced. Small round brushes are available to clean cap terminals. Wipe the cap and rotor with a clean shop towel, but avoid cleaning them in solvent or blowing them off with compressed air, which may contain moisture and may result in high-voltage leaks.

Check the distributor cap and housing vents. Make sure they are not blocked or clogged. If they are, the internal ignition module will overheat. It is good practice to check these vents whenever a module is replaced.

NO-START DIAGNOSIS

When an engine will not start, the cause is most likely a common circuit or component. If the cause is in the ignition system, simple tests can identify if the problem is in the primary or secondary circuit. Before testing the ignition system because of a no-start condition, check other systems to make sure they are not at fault. Check the battery to make sure there is ample voltage to start the engine. Make sure that the fuel system is working and that the fuel is not contaminated. Also check for severe vacuum leaks, low compression, and a broken cam belt or chain.

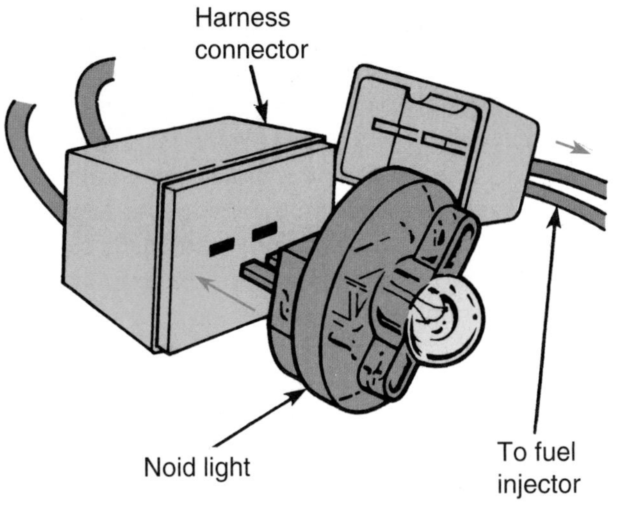

Figure 28-14 Use a noid light at the injectors when there is a no-start condition. If the injectors pulse while the engine is cranked, the triggering unit for the primary ignition circuit should be okay.

SHOP TALK

While checking the operation of the fuel injection system when there is a no-start condition, use a noid light at the injectors (Figure 28–14). If the injectors pulse while the engine is cranked, the triggering unit for the primary ignition circuit should be okay. The injection system uses the same signals to pulse the injectors.

If the problem is caused by an ignition fault, follow this procedure to determine the exact cause of the problem. Often manufacturers include a detailed troubleshooting tree in their service manuals to help identify the cause of the no-start condition.

PROCEDURE

Basic No-Start Diagnosis

STEP 1 Connect a test spark plug to the spark plug wire and ground the spark plug case.

STEP 2 Crank the engine and observe the spark plug. If there is a bright, snapping, blue spark, the ignition is working properly.

STEP 3 If the test spark plug does not fire, check for coil output at the coil terminal.

STEP 4 If there is no spark, connect a testlight or DMM from the negative side of the coil to ground. Turn on the ignition switch. In most cases, the testlight should light. If the testlight is "off," there is an open circuit in the coil primary winding or in the circuit from the ignition switch to the coil battery terminal.

STEP 5 With the testlight or DMM still connected, crank the engine. If the light flashes, the primary circuit is okay and the problem is a bad coil.

STEP 6 If the light does not flash, check the voltage from the ignition switch to the positive side of the coil. If there is no voltage, the problem is in that circuit or the switch. If there is voltage at the positive side of the coil, the problem is the pickup unit or the control module.

STEP 7 Keep in mind that on some vehicles, the PCM will not send power to the coil until it receives a CKP signal. A magnetic pulse generator can be checked with an ohmmeter, DMM, or scope. A Hall-effect sensor should be checked with a DMM or scope. Compare your findings to specifications.

STEP 8 If the pickup unit is good, suspect the ignition module. Make sure all wiring to and from the module are good.

SHOP TALK

When using a test spark plug, make sure you use the correct one for the system. There are two different types: low-voltage and high-voltage test plugs. Low-voltage plugs will fire if around 25 kV are applied to them. High-voltage plugs need 35 kV. If you use a high-voltage plug on a low-voltage system, it may not spark, leading you to believe that there is an ignition problem when there may not be. To determine which test plug to use, look at the specifications of the ignition system.

No-Start Diagnosis of EI Systems

When an engine with an EI system has a no-start problem, begin diagnosis with a visual inspection of the ignition system. Check for good primary connections. Inspect the coils and all related wiring. Check the CKP and CMP sensors and their wiring for damage. If these sensors fail or if there is resistance in the connections, the engine may not start.

There are many ignition-related DTCs that may be set by the PCM when there is a no-start condition. Always retrieve the codes and follow their pinpoint tests when diagnosing an EI system. Also, by monitoring serial data, the cause of the problem may be quickly identified. If the PCM set no codes, follow this procedure to identify the cause of a no-start problem.

Keep in mind that it is very unlikely that a no-start concern is caused by a noncommon circuit or component, such as an ignition coil.

PROCEDURE

No-Start Diagnosis for EI Systems

STEP 1 Connect a test spark plug **(Figure 28–15)** to the spark plug wire and ground the spark plug case.

Figure 28-15 A test spark plug for high-voltage ignition systems. *Reproduced under license from Snap-on Incorporated. All of the marks are marks of their owners.*

STEP 2 Crank the engine and observe the spark plug. If there is a bright, snapping, blue spark, the ignition is working properly.

STEP 3 If the spark is weak or there is no spark, check the wiring to and from the PCM.

STEP 4 If the power and ground circuits for the PCM are okay, connect a voltmeter from the input (battery) terminal on each coil pack to ground. With the ignition switch on, the voltmeter should read 12 volts. If the voltage is less than that, check the system's wiring diagram to determine what is included in the coil's power feed circuit.

STEP 5 If battery voltage is present, check the voltage drop across each of the components and wires to identify the location of an open or high resistance.

STEP 6 If none is found, check the crankshaft and camshaft position sensors. Both of these sensor circuits can be checked with a voltmeter, ohmmeter, or DSO. If the sensors are receiving the correct amount of voltage and have good low-resistance ground circuits, their output should be a digital or a pulsing voltage signal while the engine is cranking. Compare their resistance readings to specifications. If any readings are abnormal, the circuit needs to be repaired or the sensor needs to be replaced.

SHOP TALK

When using a DMM to check a crankshaft or camshaft sensor, crank the engine for a very short time and observe the meter. The reading should cycle from around 0 volts to 9 to 12 volts. Because digital meters do not react instantly, it is difficult to see the changes if the engine is cranked continually.

DIAGNOSING WITH AN ENGINE ANALYZER

It is impossible to accurately troubleshoot any ignition system without performing various electrical tests. An engine analyzer houses most of the necessary test equipment to do a complete engine performance analysis. Using an engine analyzer is a good way to determine if the driveability problem is caused by the ignition system. Also, the analyzer can be used to check individual ignition parts and circuits.

On some engine analyzers, a service test screen that lists all the tests that may be performed appears. Any individual test may be selected. For example, you may want to use the multimeter to measure the resistance of the coil windings. Move the cursor to select MULTIMETER and the meter readings will be displayed on the screen. Now the multimeter leads may be connected to measure volts, amperes, or ohms.

If COMPREHENSIVE TESTS is selected, the analyzer automatically performs a complete test sequence. When RUN CUSTOM TESTS is selected, you can design your own test sequence. The test capabilities of the analyzer depend on the manufacturer and model. Always follow the procedures given in the operating manual.

Cylinder Performance Test

During the cylinder performance test (also called the power balance test), the analyzer momentarily stops the ignition system from firing one cylinder at a time. During this brief time, the rpm drop is recorded. When a cylinder is not contributing to engine power due to low compression or some other problem, there will be very little rpm drop when that cylinder stops firing. During the test, some analyzers record the actual rpm drop, whereas others show the percentage of rpm drop. Many analyzers also record the amount of hydrocarbons (HC) emitted when each cylinder stops firing. If a cylinder was misfiring prior to the test, the cylinder will have high HC emissions. Therefore, when this cylinder stops firing during the test, there will not be much

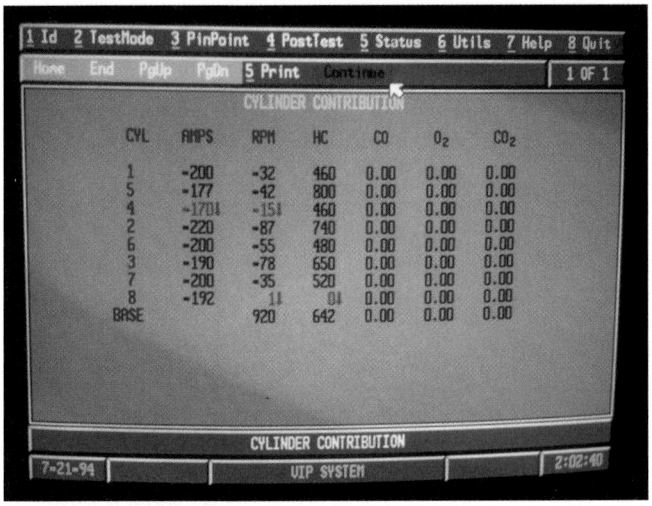

Figure 28–16 Cylinder performance test results showing one bad cylinder.

of a change in HC emissions. A cylinder with low compression or a problem that causes incomplete combustion will not have much rpm drop or HC change during the cylinder performance test **(Figure 28–16)**.

Dynamic Compression Test On some analyzers, the cranking amperes are displayed for each cylinder. When a cylinder has low compression, the starter draw will be reduced for that cylinder. Other analyzers assign a value of 100% to the cylinder with the highest cranking amperes, and then provide a percentage reading for each of the other cylinders in relation to the cylinder with the 100% reading. The results of this test are similar to a power balance test.

Battery, Starting, and Charging System Tests

During a battery test, the analyzer places a load on the battery and may record the battery's open circuit voltage, load voltage, and available cold cranking amperes. While performing the starter, or cranking, test, some analyzers record cranking volts, cranking current, cranking vacuum, cranking speed, cranking dwell, coil input volts, and HC.

Charging tests include regulator voltage, generator current, and generator waveforms.

Emission Level Analysis

The emission level tests use an exhaust gas analyzer. The exhaust gas analysis screen may include HC, CO, O_2, CO_2, NO_x, engine temperature, exhaust temperature, vacuum, and engine rpm.

Engine Computer System Diagnosis

Many engine analyzers will diagnose engine computer systems in much the same way as a scan tester. The analyzer performs tests on common engine control systems.

Ignition Performance Tests

Ignition performance tests on most engine analyzers include primary circuit tests, secondary kilovolt (kV) tests, acceleration test, scope patterns, and cylinder miss recall. Some secondary kV tests include a snap kV test in which the analyzer directs the technician to accelerate the engine suddenly. When this action is taken, the firing kV should increase evenly on each cylinder. Some analyzers display the burn time for each cylinder with the secondary kV tests. The burn time is measured in milliseconds (ms).

The secondary kV display from an EI system includes average kV for each cylinder on the compression stroke and average kV for the matching cylinder that fires at the same time on the exhaust stroke. The burn time is also included on the secondary kV display from an EI system **(Figure 28–17)**.

Some analyzers are capable of freezing scope patterns and storing them in memory. These can be later recalled and reviewed to identify intermittent cylinder misfires.

Scope Patterns

An oscilloscope or "scope" converts the electrical activity of the ignition system into a visual image showing voltage changes over a given period. This information is displayed on a screen in the form of a continuous voltage line called a pattern or trace **(Figure 28–18)**. By studying the pattern, a technician can see what the ignition system is doing.

Always follow the instructions for the specific analyzer when connecting the test leads and operating the scope. Scopes typically have at least four leads for distributor ignitions: a primary pickup that connects to the negative terminal of the ignition coil, a ground lead that connects to a good ground, a secondary pickup that clamps around the coil's high-tension

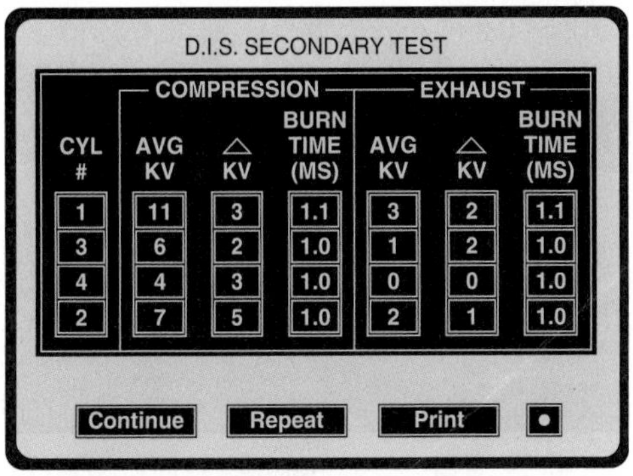

Figure 28–17 Secondary kV display on an EI system. *Courtesy of SPX Service Solutions*

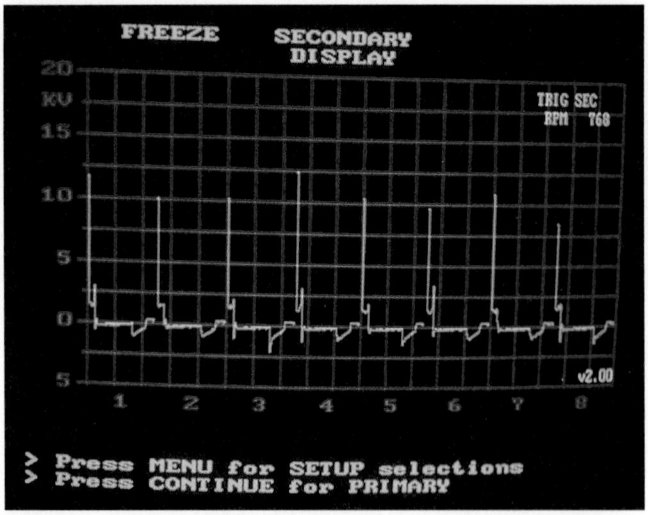

Figure 28-18 The secondary pattern for an eight-cylinder engine.

wire, and a trigger pickup that clamps around the spark plug wire of the number 1 cylinder.

Connecting the scope to a DLS system requires adapters **(Figure 28–19)** or additional test leads. The leads are connected to the individual spark plug wires **(Figure 28–20)**. On some scopes, the companion cylinders of a waste spark system are viewed at the same time. Other scopes display all of the cylinders, allowing technicians to compare the activity of the cylinders.

Figure 28-19 Leads for connecting to an EI system. *Reproduced under license from Snap-on Incorporated. All of the marks are marks of their owners.*

Figure 28-20 The tester's leads are connected to the individual spark plug wires. *Reproduced under license from Snap-on Incorporated. All of the marks are marks of their owners.*

Adapters must also be used to monitor the secondary circuit on COP systems **(Figure 28–21)**. These adapters also allow for cylinder-to-cylinder comparisons. Most COP systems can be tested with a low-amperage probe through the ignition primary circuit.

Scales A typical scope screen has two vertical voltage scales: one on the left and one on the right. Typically, the scale on the left is divided into increments of 1 kV (1,000 volts) and ranges from 0 to 25 kV. This scale is useful for testing secondary voltage. It can also be used to measure primary voltage by interpreting the scale in volts rather than kilovolts. The scale on the right side is divided into increments of 2 kV and has a range of 0 to 50 kV. This scale is also used for testing secondary voltage. This scale can also be used to measure primary voltage in the 0- to 500-volt range.

Figure 28-21 A COP test adapter. *Reproduced under license from Snap-on Incorporated. All of the marks are marks of their owners.*

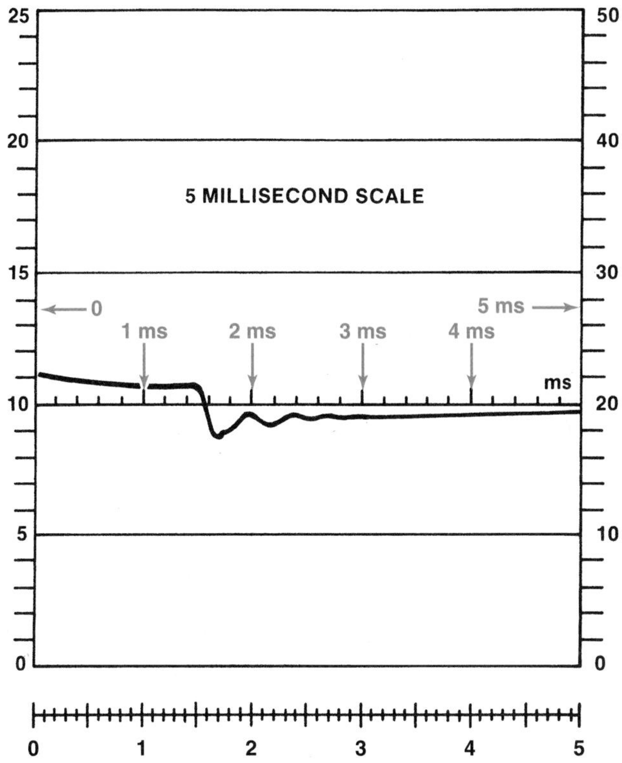

Figure 28-22 A 5-millisecond pattern showing spark duration.

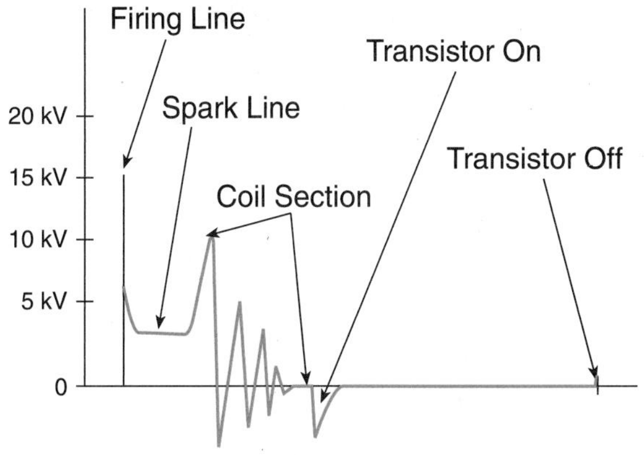

Figure 28-23 A typical secondary pattern.

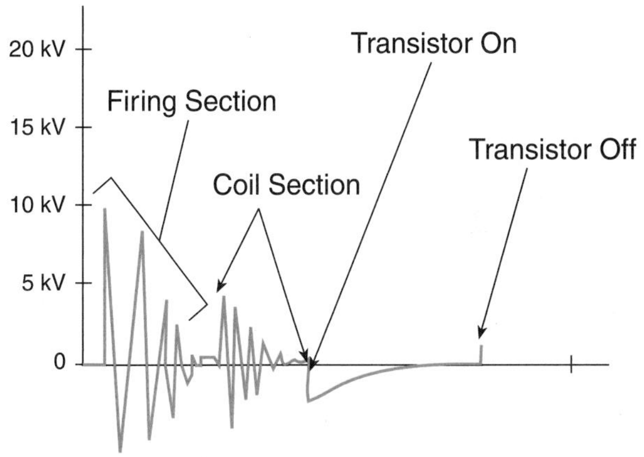

Figure 28-24 A typical primary pattern.

The screen also has a horizontal time scale located at the bottom. The time may be expressed as percent of dwell or in milliseconds. The percent of dwell scale is divided into increments of 2 percentage points and ranges from 0% to 100%. This represents one complete ignition cycle. The millisecond scale is typically broken down into units of 0 to 5 ms or 0 to 25 ms. The 5 ms scale is often used to measure the duration of the spark **(Figure 28–22)**. The complete firing pattern can normally be displayed in the 25-millisecond mode. A scope displays changes in voltage over time from left to right, similar to reading a book.

Understanding Single Cylinder Patterns A typical ignition waveform can represent the secondary or primary circuit. A typical secondary pattern is shown in **Figure 28–23**. A typical primary circuit pattern is shown in **Figure 28–24**. The primary pattern is used when secondary circuit connections are not possible or to observe cylinder timing problems. The main sections of a secondary waveform include:

- *Firing Line.* The **firing line** appears on the left side of the screen. The height of the firing line represents the voltage needed to overcome the resistance in that secondary circuit and to initiate a spark across the gap of the spark plug. Typically, around 10,000 volts are required. Keep in mind that cylinder conditions have an effect on this resistance. Leaner air-fuel mixtures increase the resistance and increase the required **firing voltage**.
- *Spark Line.* Once the resistance in the secondary is overcome, the spark jumps the plug gap, establishing current flow. The time the spark actually lasts is represented by the **spark line**. The spark line begins at the firing line and continues until the voltage from the coil drops below the level needed to keep current flowing across the gap.
- *Intermediate Section.* After the spark line is the **intermediate section** or coil-condenser zone. It shows the remaining coil voltage as it dissipates or drops to zero. Remember, once the spark has ended, there is still voltage in the ignition coil. This voltage must leave the coil before the coil can be prepared for another cylinder firing. The voltage moves back and forth within the primary circuit until it drops to zero. Notice that the voltage traces in this section steadily drop in height until the coil's voltage is zero.

- *Dwell Section.* The next section of the waveform begins with the primary circuit current "on" signal. It appears as a slight downward turn followed by several small oscillations. The slight downward curve occurs just as current begins to flow through the coil's primary winding. The oscillations that follow indicate the beginning of the magnetic field buildup in the coil. This curve marks the beginning of a period known as the dwell section of zone. The end of the dwell zone occurs when the primary current is turned off by the switching device. The trace turns sharply upward at the end of the dwell zone. Turning off primary current flow collapses the magnetic field around the coil and generates another high-voltage surge for the next cylinder in the firing order. Remember, the primary current off signal is the same as the firing line for the next cylinder. The length of the dwell section represents the amount of time that current is flowing through the primary.

Most scope patterns look more or less like the one just described. The patterns produced by some systems may have fewer oscillations in the intermediate section. Patterns may also vary slightly in the dwell section. The length of this section depends on when the control module turns the transistor on and off.

Older DI systems used a fixed dwell period. The number of dwell degrees remained the same during all engine speeds. So if the engine has 30 degrees of dwell at idle, it should have 30 degrees of dwell at 2,000 rpm. This is not saying that the actual amount of time has remained the same. A fixed dwell of 30 degrees at 2,000 rpm gives the ignition coil only ¼ the saturation time, in milliseconds, that it has at 500 rpm.

Most control modules provide a variable dwell; dwell time changes with engine speed. At idle and low rpm speeds, a short dwell provides enough time for complete ignition coil saturation **(Figure 28–25A)**. The current on and current off signal appears very close to each other, usually less than 20 degrees. As engine speed increases, the control module lengthens the dwell degrees **(Figure 28–25B)**. This, of course, increases the available time for coil saturation.

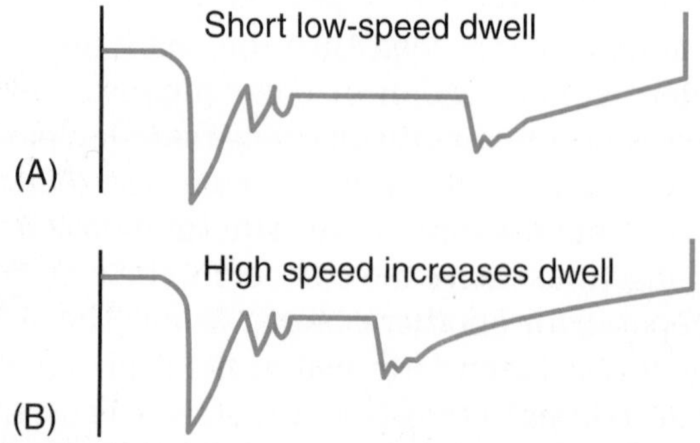

Figure 28-25 An example of a variable dwell ignition system.

Many late-model systems are **current limiting**. These systems saturate the ignition coil quickly by passing high current through the primary winding for a fraction of a second. Once the coil is saturated, the need for high current is eliminated, and a small amount of current flows to keep the coil saturated.

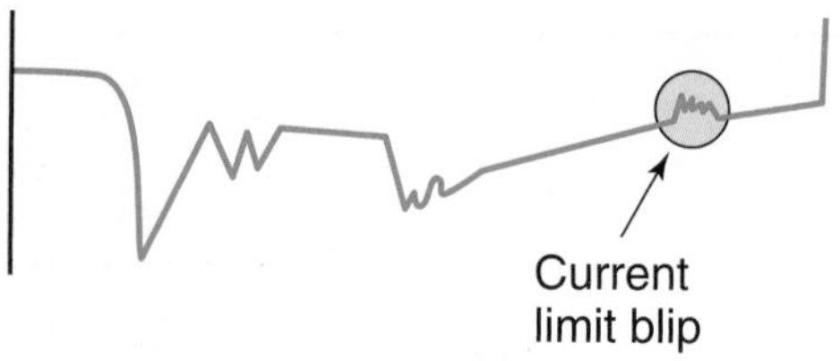

Figure 28-26 A pattern showing an ignition system limiting current during dwell.

The point at which the control module cuts back from high to low current appears as a small blip or oscillation during the dwell section of the pattern **(Figure 28–26)**. At high engine speeds, this blip may be missing. In an attempt to keep the coil saturated, the module may not stop sending high current to the coil. The blip may also not appear if a replacement module does not have a current-limiting circuit, the primary winding has excessive resistance, or the coil is otherwise faulty. Further testing of the coil is needed to pinpoint the cause of the missing blip.

Pattern Display Modes The scope can display waveforms in several ways. When the display pattern is selected, the scope displays all the cylinders in a row from left to right as shown in **Figure 28–27**. The

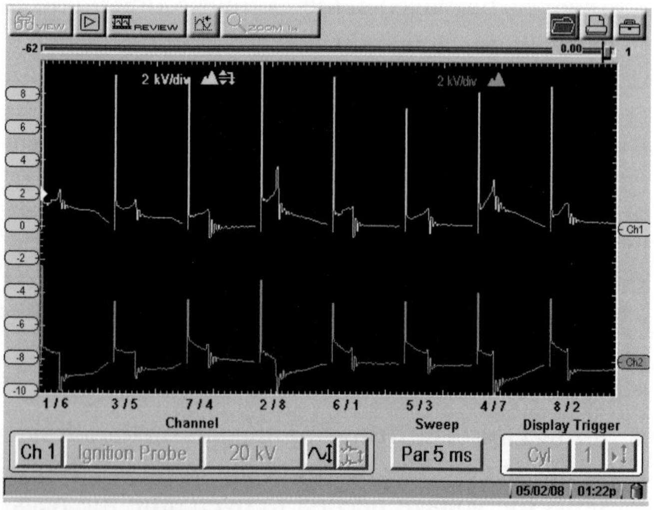

Figure 28-27 Typical parade (display) patterns for the primary and secondary circuits. *Reproduced under license from Snap-on Incorporated. All of the marks are marks of their owners.*

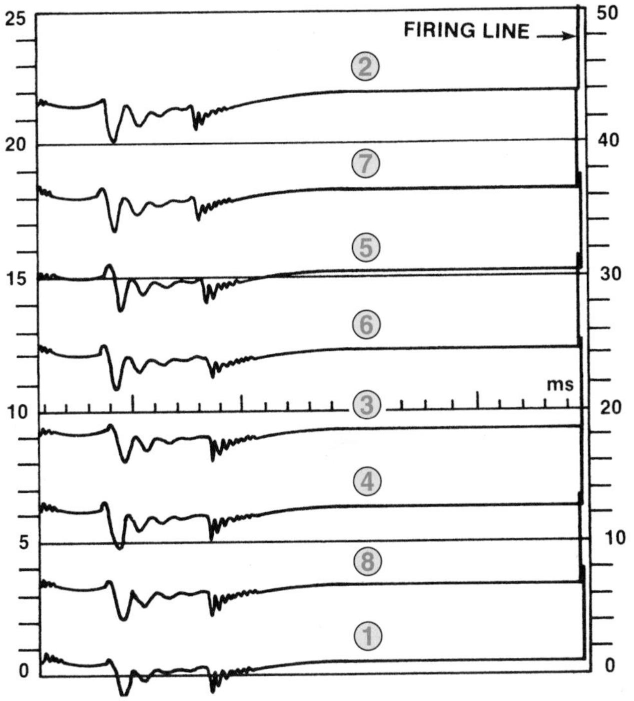

Figure 28-28 A secondary raster pattern.

cylinders are arranged according to the engine's firing order. The pattern begins with the spark line of cylinder #1 and ends with the firing line for cylinder #1. This display pattern is commonly used to compare the voltage peaks for each cylinder.

Another display mode is the **raster pattern** **(Figure 28–28)**. A raster pattern stacks the waveforms of the cylinders one above the other. Cylinder #1 is displayed at the bottom of the screen and the rest of the cylinders are arranged above it according to the engine's firing order. In a raster pattern, the waveform for each cylinder begins with the spark line and ends with the firing line. This allows for a much closer inspection of the voltage and time trends than is possible with the display pattern.

A **superimposed pattern** displays all of the patterns one on top of the other. Like the raster pattern, the superimposed voltage patterns are displayed the full width of the screen, beginning with the spark line and ending with the firing line. A superimposed pattern is used to identify variations of one cylinder's pattern from the others.

SHOP TALK

Often a DSO is only capable of displaying one cylinder at a time. To look at the rest of the signals, the input pickup must be moved to the other cylinders one at a time. This means it is not possible to view a parade, raster, superimposed pattern.

Spark Plug Firing Voltage In a secondary pattern, the firing line is the highest line in the pattern. The firing line is affected by anything that adds resistance to the secondary circuit. This includes the condition of the spark plugs or the secondary circuit, engine temperature, fuel mixture, and compression pressures. The normal height of a firing line, with the engine at idle, should be between 7 and 13 kV with no more than a 3 kV variation between cylinders. If one or more firing lines are too low or high, the cause is something that is common to only those cylinders. If the firing lines are all too high or low, the problem is something that is common to all cylinders. **Table 28-1** covers most of the things that could cause abnormal firing lines.

High resistance in the secondary circuit also produces a spark line that is higher in voltage and has a steep slope with shorter firing durations. A good spark line should be relatively flat and measure 2 to 4 kV in height.

The voltage required to fire a spark plug increases when the engine is under load. The voltage increase is moderate and uniform if the spark plugs are in good condition and properly gapped. To test the spark plugs under load, note the height of the firing lines at idle speed. Then quickly open and release the throttle (snap accelerate) and note the rise in the firing lines while checking the voltages for uniformity. A normal rise should be between 3 and 4 kV. If the rise is not equal on all cylinders or if the rise is too low or too high, the spark plugs are probably faulty. Also, watch the decrease in their height as the throttle is closed. If the cylinders do not uniformly drop, there is high resistance in the circuits that were different.

The firing lines should also be checked while the engine is cranking and the plug wires are disconnected from the plugs. Insert a test plug into the plug wire. This will give an indication of the condition of the coil(s) and the secondary insulation. Follow the specified procedures to prevent the engine from starting. Then crank the engine and observe the firing lines. They should reach at least 30 kV; less indicates a possible bad coil. If some of the firing lines are low, the secondary insulation is breaking down or the coil for those cylinders is bad.

Spark Duration The amount of time that the spark plug is actually firing is important. This is called spark duration. **Spark duration** is represented by the length of the spark line and is measured in milliseconds. Most engines have a spark duration of approximately 1.5 milliseconds. Short spark durations cannot provide complete combustion and can cause increased emissions levels and a loss of power. If the spark duration is too long, the spark plug electrodes might wear prematurely. When the ignition pattern has a

TABLE 28-1 FIRING LINE DIAGNOSIS

Condition	Probable Cause	Remedy
Firing voltage lines the same, but abnormally high	1. Retarded ignition timing 2. Fuel mixture too lean 3. High resistance in coil wire 4. Corrosion in coil tower terminal 5. Corrosion in distributor coil terminal	1. Rest ignition timing. 2. Readjust carburetor or check for vacuum leak. 3. Replace coil wire. 4. Clean or replace coil. 5. Clean or replace distributor cap.
Firing voltage lines the same, but abnormally low	1. Fuel mixture too rich 2. Breaks in coil wire causing arcing 3. Cracked coil tower causing arcing 4. Low coil output 5. Low engine compression	1. Readjust carburetor or check for plugged air filter. 2. Replace coil wire. 3. Replace coil. 4. Replace coil. 5. Determine cause and repair.
One or more, but not all, firing voltage lines higher than the others	1. Idle mixture not balanced 2. EGR valve stuck open 3. High resistance in spark plug wire 4. Cracked or broken spark plug insulator 5. Intake vacuum leak 6. Defective spark plugs 7. Corroded spark plug terminals	1. Readjust idle mixture. 2. Inspect or replace EGR valve. 3. Replace spark plug wires. 4. Replace spark plugs. 5. Repair leak. 6. Replace spark plugs. 7. Replace spark plugs.
One or more, but not all, firing voltage lines lower	1. Curb idle mixture not balanced 2. Breaks in plug wires causing arcing 3. Cracked coil tower causing arcing 4. Low compression 5. Defective or fouled spark plugs	1. Readjust idle mixture. 2. Replace spark plug wires. 3. Replace coil. 4. Determine cause and repari. 5. Replace spark plugs.
Cylinders not firing	1. Cracked distributor cap terminals 2. Shorted spark plug wire 3. Merchancial problem in engine 4. Defective spark plugs 5. Spark plugs fouled	1. Replace distributor cap. 2. Determine cause of short and replace wire. 3. Determine problem and correct. 4. Replace spark plugs. 5. Replace spark plugs.

long spark line, it normally follows a short firing line, which may indicate a fouled spark plug, low compression, or a spark plug with a narrow gap.

The spark line should be at a right angle to the firing line. It should also be relatively flat but show some small ripples. These ripples result from the turbulence inside the combustion chamber. The slope of the spark line gives a picture of air-fuel ratio.

SHOP TALK

The spark line may appear as if the secondary path is complete, but the spark plug may not be firing. This can be caused by problems that allow the voltage to jump before it reaches the spark plug. If the spark line is totally flat, is longer than 2 ms, and is less than 500 volts, there is a good chance that the plug is not firing.

Coil Condition The coil/condenser section shows the voltage reserve in the coil. The reserve can be determined by looking at the height of the oscillations. The heights should uniformly decrease to a zero value. If the waveform does not have normal oscillations in its intermediate section, check for a possible short in the coil by testing the resistance of the primary and secondary windings.

The available voltage output from a coil can also be checked with the coil output test. Following are steps to safely perform a coil output test:

1. Install a test plug in the coil wire or a plug wire if there is no coil wire.
2. Set the scope on display with a voltage range of 50 kV.
3. Crank the engine and note the height of the firing line. The firing line should exceed 35 kV. Lower-

than-specified voltages may indicate lower-than-normal available voltage in the primary circuit. The control module may have developed high internal resistance. The coil or coil cable may also be faulty.

Primary Circuit Checks The primary ignition pattern shows the action of the primary circuit. To be able to spot abnormal sections of a primary waveform, you must know what causes each change of voltage and time in a normal primary waveform. Although the true cycle of the primary circuit begins and ends when the switching transistor is turned on, the displayed pattern begins right after the transistor is turned off. At this moment in time, the magnetic field around the windings collapses and a spark plug is fired.

SHOP TALK

Always carefully check the primary circuit; a 1-volt loss in the primary circuit can reduce secondary output by up to 10,000 volts.

Looking at the primary pattern shown in **Figure 28–29**, the trace at the left represents the collapsing of the primary winding after the transistor turns off and primary current flow is interrupted. The height of these oscillations depends on the current that was flowing through the winding right before it was stopped. The amount of current flow depends on the time it was able to flow, the voltage applied to the winding, and the resistance of the winding. High primary circuit resistance will reduce the maximum amount of current that can flow through the winding. Reduced current flow through the winding will reduce the amount of voltage that can be induced when the field collapses.

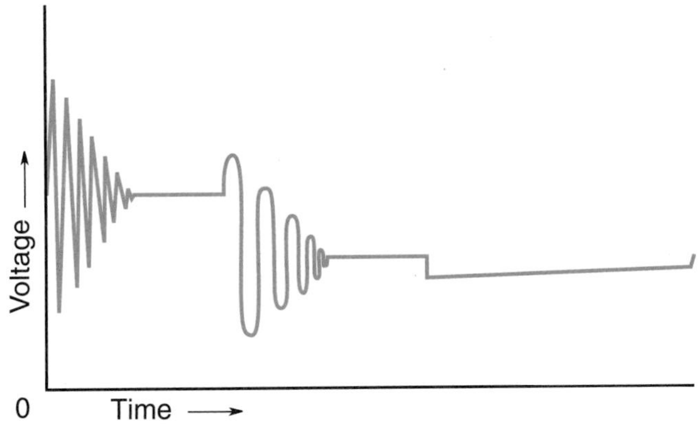

Figure 28-29 A typical primary pattern.

During the collapsing of the primary winding, the spark plug is firing. The primary circuit's trace shows sharp oscillations of decreasing voltages. The overall shape of this group of oscillations should be conical and should last until the spark plug stops firing.

After the firing of the plug, some electrical energy remains in the coil. This energy must be released prior to the next dwell cycle. The next set of oscillations shows the dissipation of this voltage. These oscillations should be smooth and become gradually smaller until the 0-volt line is reached. At that point there is no voltage left and the coil is ready for the next dwell cycle.

Immediately following this dissipation of coil energy is the transistor "on" signal. This is when current begins to flow through the primary circuit. It is the beginning of dwell. When the transistor turns on, there should be a clean and sharp turn in the trace. A clean change indicates that the circuit was instantly turned on. If there is any sloping or noise at this part of the signal, something is preventing the circuit from being instantly turned on. When looking at a superimposed primary pattern, any variation between cylinders will show up as a blurred or noisy transistor "on" signal **(Figure 28–30)**.

If there are erratic voltage spikes at the transistor "on" signal, the ignition module may be faulty, the distributor shaft bushings may be worn, or the armature is not securely fit to the distributor shaft. The problem is preventing dwell from beginning smoothly and causing the engine to have a rough idle, intermittent miss, and/or higher-than-normal HC emission levels.

During dwell, the trace should be relatively flat. However, many ignition systems have features that change current flow during dwell. These features are designed to allow complete coil saturation only when

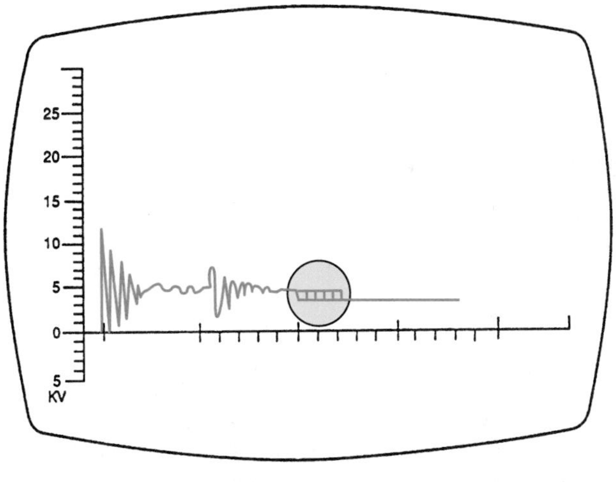

Figure 28-30 A superimposed primary pattern showing cylinders with different transistor "on" times. *Courtesy of SPX Service Solutions*

that is needed. By reducing the amount of current, the amount of voltage induced in the secondary is also reduced.

Stress Testing Components

Often, an intermittent ignition problem only occurs under certain conditions such as extremes in heat or cold, or during rainy or humid weather. Careful questioning of the customer should lead to determining if the problem is stress condition related. Does the problem occur on cold mornings? Does it occur when the engine is fully warmed up? Is it a rainy day problem? If the answer to any of these questions is positive, you can reproduce the same conditions in the shop during stress testing.

CAUTION!

When using cool-down sprays, always wear eye protection and avoid spraying your skin or clothing. Also keep away from the belts and fans. Use extreme caution.

Cold Testing With the scope on raster, cool major ignition components such as the control module, pickup coil, and major connections one at a time using a liquid cool-down agent. After cooling a component, watch the pattern for any signs of malfunction, particularly in the dwell zone. If there is no sign of malfunction, cool down the next component after the first has warmed to normal operating temperature. Cooling (or heating) more than one component at a time provides inconclusive results.

Heat Testing To heat stress components, use a heat gun or hair dryer to direct hot air into the component. Heat guns intended for stripping paint and other household jobs can become extremely hot and melt plastic, wire insulation, and other materials. Use a moderate setting and proceed cautiously. Look for changes in the dwell section of the trace, particularly in the variable dwell or current-limiting areas. If connections appear to be the problem, disconnect them, clean the terminals, and coat them with dielectric compound to seal out dirt and moisture.

Moisture Testing A wet stress test is performed by lightly spraying the components, coil and ignition cables, and connections with water. Do not flood the area; a light mist does the job. A scope set on raster or display helps pinpoint problems, but it is often possible to hear and feel the miss or stutter without the use of a scope. As with heat and cold testing, do not spray down more than one area at a time or results could be misleading. If you suspect a poor connection, clean and seal it, then retest it.

DIAGNOSING WITH A DSO OR GMM

Commonly used diagnostic tools are modular units that can serve as many different tools, including a scan tool, lab scope, DMM, and GMM. They may also contain libraries of information that serve as references for diagnostics, such as the basic operation of a component or system, test procedures, identification of pins in a connector, location of the component or connector, specifications, and diagnostic tips. Some even have a database of known-good waveforms in their memory, which again help in the diagnosis of a problem. All of these features make diagnosis simpler.

Each model of these tools has many unique operating procedures and the tool's user manual must be used to have accurate results. What follows are some general guidelines for using these tools to diagnose an ignition system.

After vehicle identification information is entered, the test mode is selected. This mode allows you to test individual components. The tool will display how to properly connect the tool to the component and, in most cases, display the expected reading for those tests. There will also be the option of displaying the test results in a variety of ways: digital, digital graphs or waveforms, analog graphs or waveforms, or a combination of these.

Using the DSO or GMM

The graphing meter function is much like using a scope, except that it also displays digital MIN/MAX readings with the trace. Many have a four-channel capability, which means you can look at the signals from four different sensors and/or outputs at the same time. This is a great help during diagnosis. This mode also allows a technician to observe any glitches or noise that may affect operation.

To aid in diagnosing intermittent problems, screens can be frozen, saved, and printed for review later. The tool can typically be connected to the vehicle, and then the vehicle can be taken for a road test. Intermittent problems can be observed and the data around those problems can be stored in the meter's memory for review after the road test.

In most cases, additional leads must be attached to the lab scope or GMM to read additional channels. These are color-coded and the color of the display

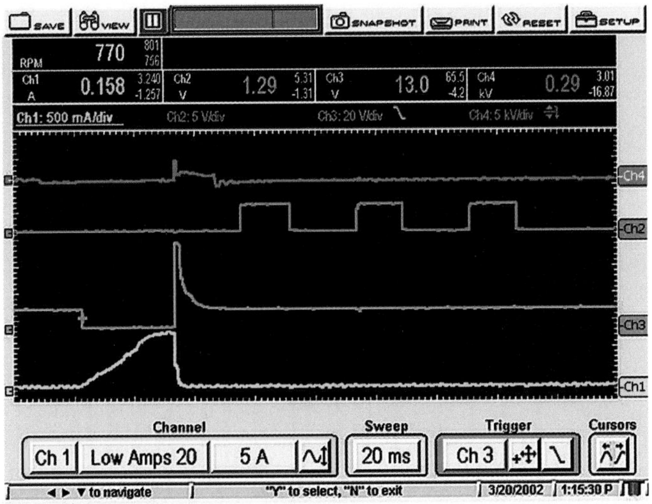

Figure 28-31 Waveforms on four different channels: CH1 in yellow, CH2 in blue, CH3 in green, and CH4 in red. *Reproduced under license from Snap-on Incorporated. All of the marks are marks of their owners.*

matches the color of the lead. That lead is attached to the point where the desired signal can be monitored. A ground lead must also be attached. The color-coding throughout allows for easy identification of what is being monitored **(Figure 28–31)**. The lab scope may allow a look at DC volts, low amps, ignition secondary voltages, vacuum, and pressure.

The readings or waveforms for the component or system monitored can be looked at individually or at the same time. In most cases, channel 1 will always be displayed. If the tool is capable of two or more channels, these can be selected individually. However, on many GMMs, channels 1 and 2 are always displayed and the additional channels must be selected.

Certain settings and controls are critical to recording a worthwhile waveform, including:

- *Scale*—Used to measure the events of one or more waveforms.
- *Filter*—Minimizes and cleans up waveforms from unwanted noise in order to look at voltages.
- *Threshold*—Changes the reference point for the waveform and is only used when measuring frequency, duty cycle, dwell, and pulse width. This may be manually or automatically set.
- *Peak Detect*—Used to capture spikes and glitches in signals. When it is off, the tool collects just enough data to form a waveform. When it is on, the tool collects more data than is needed. This is why it allows you to capture a glitch or noise.
- *Time Scale (Sweep)*—Sets the amount of time data will be displayed; more data can be observed with a faster sweep. Also, the usefulness of peak detect is increased with a longer sweep. However, long sweeps have a slow sample rate and when observing ignition systems, the firing line may not appear.
- *Voltage Scale (Sensitivity)*—Adjusts how sensitive the scope will be to voltage changes. The lower the setting, the more sensitive the ignition scope will be to detecting cylinder firing.
- *Trigger*—Sets the criteria that should be recognized to start the display of data.
- *Trigger Slope*—Sets the direction that the waveform must be moving toward to start the waveform.
- *Trigger Level*—Used to place the trigger point along the horizontal or vertical axis of the display.

Testing the Ignition System

SHOP TALK

It is important that the correct adapters and leads be used when using most testers. These are designed for specific purposes and if the wrong ones are used, inaccurate measurements will result. For example, there are many different COP adapters because there are many different designs of COP systems (Figure 28–32). Also, adapters allow for connection to a waste spark system that allows a view of the total system.

Prior to testing an ignition system, the diagnostic tool must be programmed to match the type of ignition system found on the vehicle; this includes ignition type, number of cylinders, firing order, polarity of the spark plugs, and the source of the engine speed reference. Each type of ignition system also

Figure 28-32 Examples of the various adapters and connectors needed to connect a lab scope or GMM to the different ignition systems.

requires different adapters and different setup procedures.

All ignition scopes must have a means to monitor the secondary and a way to monitor the switching of the primary circuit. The primary is monitored with a reference or engine speed pickup. This pickup is placed around the number 1 spark plug wire on all systems. On DLS systems, the tool must be programmed with the number of cylinders, cylinder firing order, and plug polarities. On coil-per-cylinder systems, the tool must know the number of cylinders and the firing order.

It may take a few seconds for the tool to synchronize itself with the ignition system. Also, on some tools connected to waste spark systems, the firing of the plugs that are on the compression stroke will only appear on channel 1 and those that occur on the exhaust stroke will appear on channel 2.

All events of the ignition system will appear in the waveform, as they were in the waveform shown on an ignition scope. This means the interpretation of the patterns is the same regardless of what is displaying it.

The primary difference between an ignition scope and a DSO or GMM is that the latter can display the min/max values for firing kV, spark kV, and spark duration for each cylinder. On an ignition scope, those values must be visually observed.

Both the power and waste spark of an EI system can be observed and compared for each coil. Photo Sequence 27 shows the procedure for connecting a scope to a DLS and how to interpret the results. Keep in mind that if one plug in a DLS system is fouled, it will affect the other plug in the circuit. Observing the activity of both will identify the problem plug.

An Undetected Cylinder When the firing of a cylinder is not detected by the tool, there will be a void in the ignition system's waveform. This can be caused by an ignition problem or improper setup of the tool. To determine if it is the latter, lower the sensitivity setting. If the cylinder still is not displayed, it is fair to assume that the cylinder is not firing.

IGNITION TIMING

The primary circuit controls the secondary circuit. Therefore, it controls ignition timing. Most primary circuit problems in a computer-controlled ignition system result in starting problems or poor performance due to incorrect timing.

If engine performance is poor, the cause of the problem can be many things. There can be a problem with the engine such as poor compression, incorrect valve timing, overheating, and so on. The air-fuel mixture or the ignition timing is incorrect. When the ignition timing is not correct, many tests will point to the problem. Incorrect ignition timing will cause incomplete combustion at one or all engine speeds. Incomplete combustion will cause excessive O_2 in the exhaust. This will cause the PCM to try to correct the apparent lean mixture (**Figure 28–33**). Incorrect timing is not a lean condition, but the PCM cannot tell that the timing is wrong. It only knows there is too much O_2 in the exhaust. Under this condition the waveform from the O_2 sensor will be lean biased.

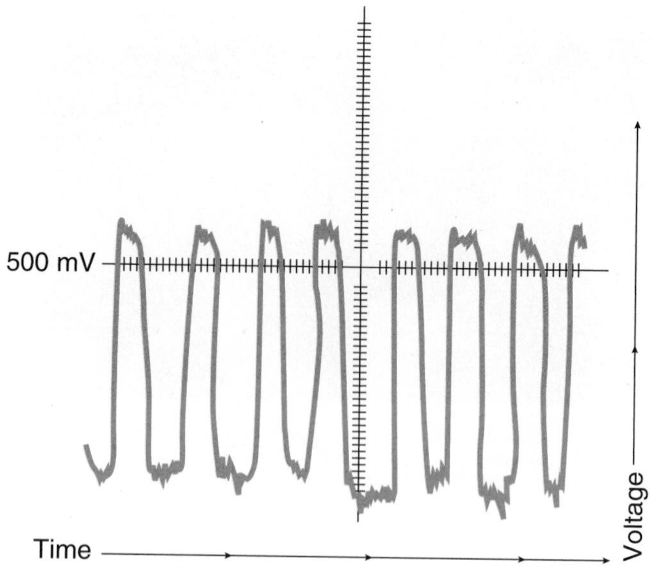

Figure 28-33 A lean-biased O_2 sensor.

SHOP TALK

Observing the waveform of an O_2 sensor can help in the diagnosis of engine performance problems. **Figure 28–34** shows how ignition problems affect the signal from the O_2 sensor. Keep in mind that during complete combustion, nearly all of the O_2 in the combustion chamber is combined with the fuel. This means there will be little O_2 in the exhaust. As combustion becomes more incomplete, the levels of O_2 increase.

Excessive O_2 in the exhaust will also show up on an exhaust gas analyzer. With incorrect timing you should also see higher-than-normal amounts of HC. Keep in mind that it takes approximately 7 seconds for the exhaust to be analyzed. If you slowly accelerate the engine and see the HC and O_2 levels on the exhaust gas analyzer rise, the condition that existed

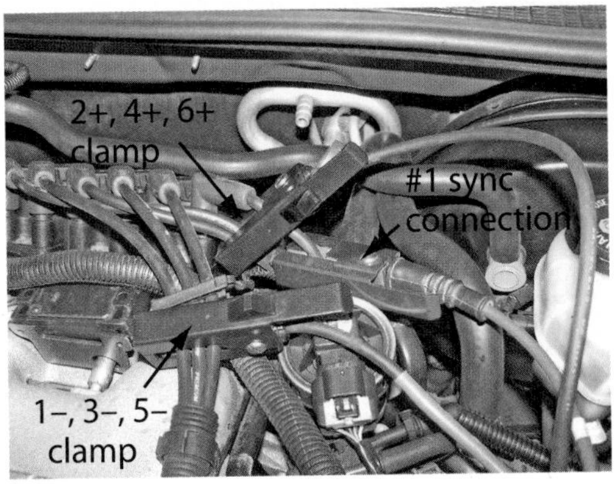

P27-1 *To observe the activity of an EI system on a scope, special adapters are required.*

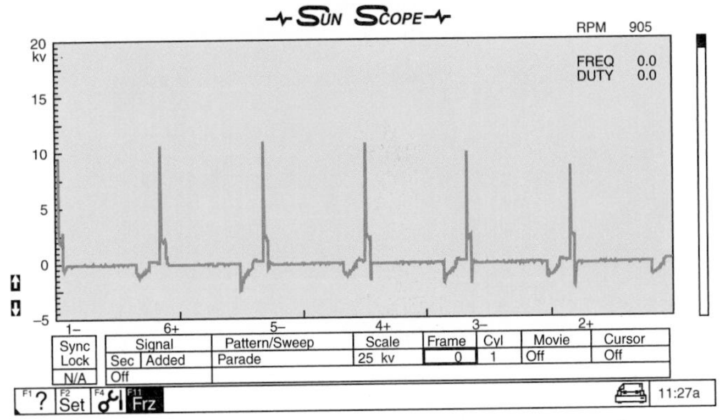

P27-2 *A secondary pattern for the system. Note the waveform for three of the cylinders (6+, 4+, and 2+) fire positively and the remaining three fire negatively (1-, 3-, and 5-).*

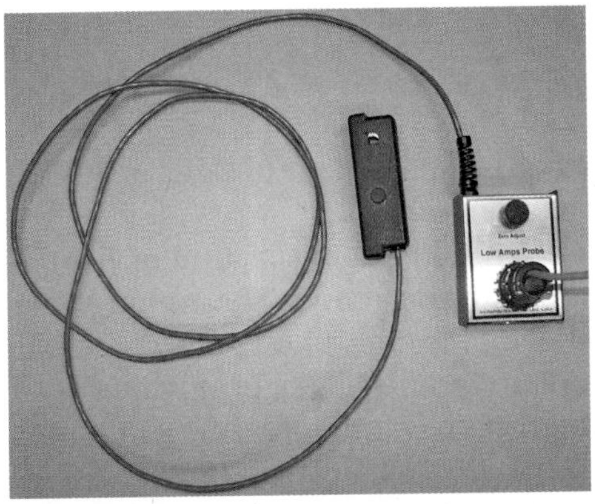

P27-3 *To look at the primary circuit, connect a low-amp probe to the scope. Make sure it is compatible with the scope.*

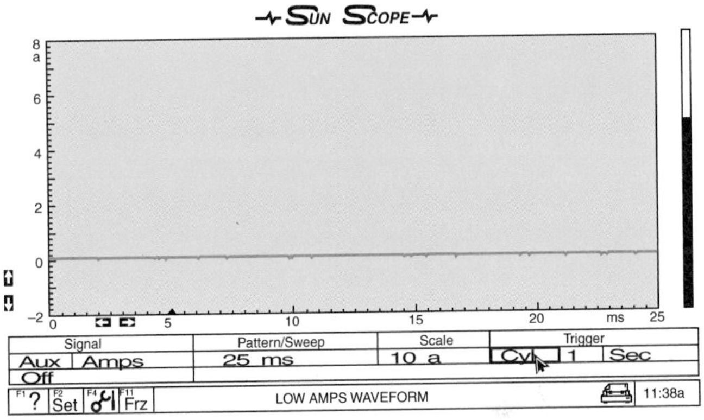

P27-4 *The baseline on the scope for amperage must be set to zero before the probe is connected to the system.*

P27-5 *Install the probe around the ignition feed at the ignition module.*

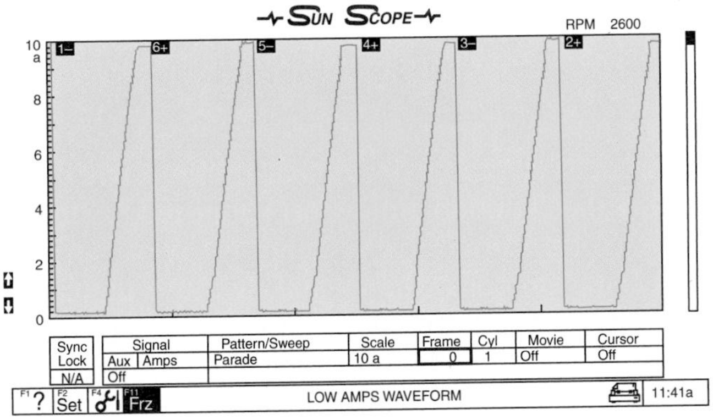

P27-6 *Once the amp probe is synchronized to the firing order, primary patterns for all cylinders will appear. The slope on the front edge of the waveform shows the time it took for coil saturation. The flat tops of the waveform result from the module limiting the current to the coil.*

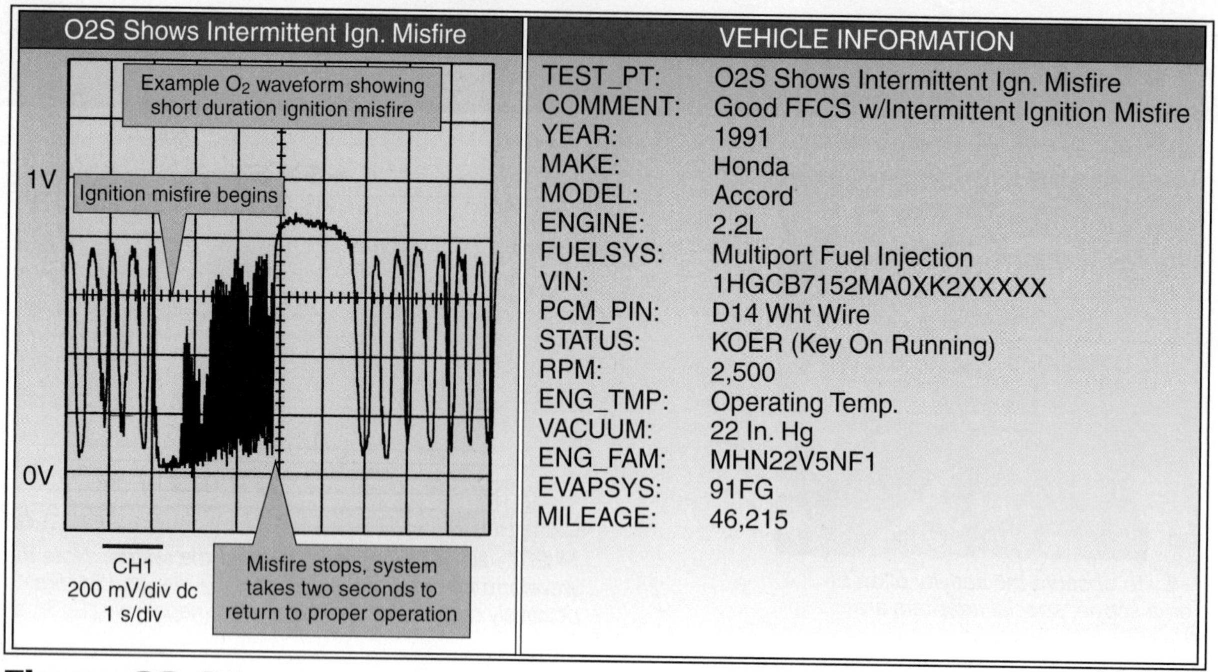

Figure 28-34 An O_2 sensor signal caused by an ignition problem. *Courtesy of Progressive Diagnostics—wavelife AutoPro*

7 seconds earlier was the cause of the rise in emissions levels. To make this easier to track, make sure you hold the engine at each test speed for at least 7 seconds. This way you will be able to observe the rise (or fall) of the emissions levels at that particular speed.

Incorrect ignition timing will also affect manifold vacuum readings and ignition system waveforms on a scope. When anything indicates a problem with the primary ignition circuit, the suspected parts should be tested. Symptoms of overly advanced timing include pinging or engine knock. Insufficient advance or retarded timing at higher engine speeds could cause hesitation and poor fuel economy.

Setting Ignition Timing

Only engines equipped with a distributor may need to have their ignition timing set or adjusted. On these systems, the correct base timing is critical for the proper operation of the engine. Because the computer bases its control to the base timing setting, all other ignition timing settings will also be wrong. On DI systems, the base timing is adjustable. On others, if the base timing is wrong, the ignition module, distributor, or PCM may need to be replaced.

Each ignition system has its own set of procedures to check ignition timing; always refer to the vehicle's emissions underhood label or the appropriate service manual before proceeding. These give the correct procedure for disabling the computer's control of the timing, the correct timing specifications, and the conditions that must be present when checking or adjusting base ignition timing.

To check the ignition timing, a timing light is aimed at the ignition timing marks. The timing marks are usually located on the crankshaft pulley or on the flywheel. A stationary pointer, line, or notch is positioned above the rotating timing marks. The timing marks are lines on the crankshaft pulley **(Figure 28–35)** or flywheel that represent various positions of the piston as it relates to TDC. When piston 1 is at TDC, the timing line or notch will line up with the zero reference mark on the timing plate. Usually an engine is timed so that the number 1 spark plug fires several degrees BTDC. The timing light flashes every time the number 1 spark plug fires. When pointed at the timing marks, the strobe of the light will freeze the spinning timing marks as it passes the timing scale. The ignition timing is checked by observing the degrees of crankshaft rotation (BTDC or ATDC) when the spark plug fires.

Spark plug gap and idle speed must be correct before checking or setting ignition timing. Also, the engine must be at operating temperature. After you have a base timing reading, compare it to the specifications. As an example, if the specification calls for 10 degrees before TDC and your reading was 3 degrees before TDC, the timing is retarded 7 degrees. This means the timing must be advanced by 7 degrees. To do this, rotate the distributor until the timing marks align at 10 degrees. Then retighten the distributor holddown bolt.

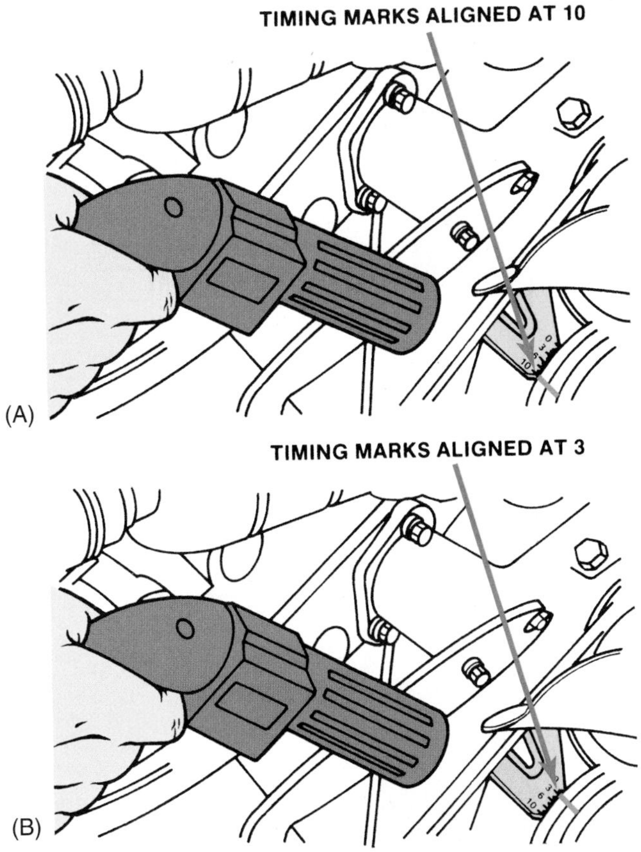

Figure 28-35 (A) Timing marks illuminated by a timing light at 10 degrees BTDC and (B) timing marks at 3 degrees BTDC.

BASIC PRIMARY CIRCUIT COMPONENTS

The primary circuit has the responsibility of controlling the action of the secondary and , in many cases, is the cause of starting problems. The components in the primary ignition circuit vary with manufacturer and design. It is important to identify the components of the circuit and the appropriate test methods. Always work systematically through a circuit, testing each wire, connector, and component. Do not jump from one component to another. It is possible that the component inadvertently overlooked is the one causing the trouble. Always compare the readings with specifications given by the manufacturer.

Ignition Switch

An ignition switch supplies voltage to the ignition control module and/or the ignition coil. Often an ignition system has two wires connected to the run terminal of the ignition switch. One is connected to the module. The other is connected to the primary resistor and coil. The start terminal of the switch is also wired to the module.

You can check for voltage using either a 12-volt test light or a DMM. To use a testlight, turn the ignition key off and disconnect the wire connector at the module. Also, disconnect the S terminal of the starter solenoid to prevent the engine from cranking when the ignition is in the run position. Turn the key to the run position and probe the red wire connection to check for voltage. Also check for voltage at the battery terminal of the ignition coil using the testlight.

Next, turn the key to the start position and check for voltage at the white wire connector at the module and the battery terminal of the ignition coil. If voltage is present, the switch and its circuit are okay.

To make the same test using a DMM, turn the ignition switch to the off position and back-probe, with the meter's positive lead, the power feed wire at the module. Connect the meter's negative to a good ground at the distributor base. Turn the ignition to the run or start position as needed, and measure the voltage. The reading should be at least 90% of battery voltage.

Ignition Coil Resistance

Ignition coils, like all parts that contain electrical windings, can be checked with an ohmmeter. In an ignition coil there are two separate windings and each has a different resistance value. This is due to the wire size and the number of windings. Always refer to the specifications prior to testing a coil. If a measurement is not within specifications, the coil or coil assembly should be replaced.

To check the primary windings, set the ohmmeter to the auto-range mode and connect the meter across to the primary coil (BAT and TACH) terminals **(Figure 28–36)**. An infinite ohmmeter reading indicates an open winding. Higher-than-normal readings indicate the presence of excessive resistance. If the measurement is less than the specified resistance, the windings are shorted.

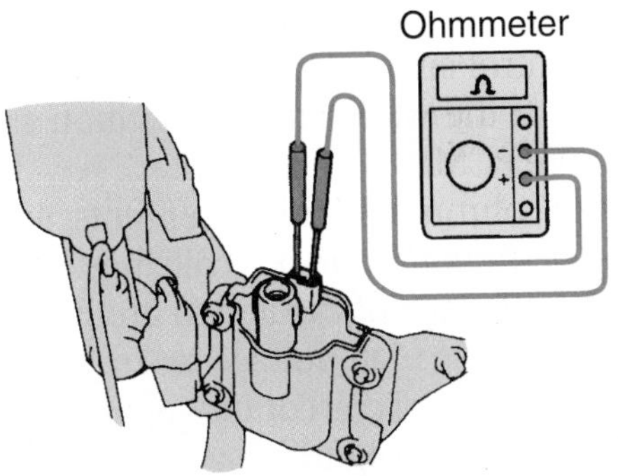

Figure 28-36 An ohmmeter connected to primary coil terminals. *Courtesy of Toyota Motor Sales, U.S.A., Inc.*

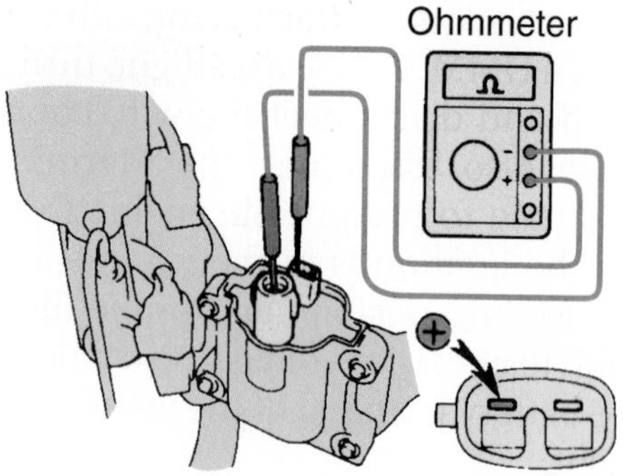

Figure 28–37 An ohmmeter connected from one primary terminal to the coil tower to test secondary winding. *Courtesy of Toyota Motor Sales, U.S.A., Inc.*

To check the secondary winding, connect the meter between the coil's secondary terminal and the positive (BAT) terminal of the coil **(Figure 28–37)**. A meter reading below the specified resistance indicates a shorted secondary winding. An infinite meter reading indicates that the winding is open. Higher-than-normal readings indicate the presence of excessive resistance.

In some coils, the secondary winding is connected to the coil's frame. When testing the secondary winding in these coils, the ohmmeter must be connected from the secondary coil terminal to the coil frame or to the ground wire terminal extending from the coil frame.

The secondary windings of a waste spark ignition coil are not checked in the same way as other coils. Each coil has two secondary terminals. The coil is checked by connecting the meter across the two secondary terminals **(Figure 28–38)**. As with other coils, compare the readings to specifications. COP coils are checked in the same way as other coils.

If the ignition coils check out with an ohmmeter but still do not spark, check the voltage from the ignition relay. There should be 12 volts. If this voltage is not present, check for battery voltage to the relay. If this voltage is there, check the relay control wires from the PCM. If the circuit is good, the relay should be replaced. If no voltage is available to the relay, check the relay control circuit from the PCM and the ignition switch. If the wires and the ignition switch are good, test the PCM.

Although ohmmeter measurements are a good indication of the condition of a coil, they do not check for defects such as poor insulation around the windings, which causes high-voltage leaks. Therefore, an accurate indication of coil condition is the coil voltage output test with a test spark plug connected from the coil secondary wire to ground as explained in the no-start diagnosis.

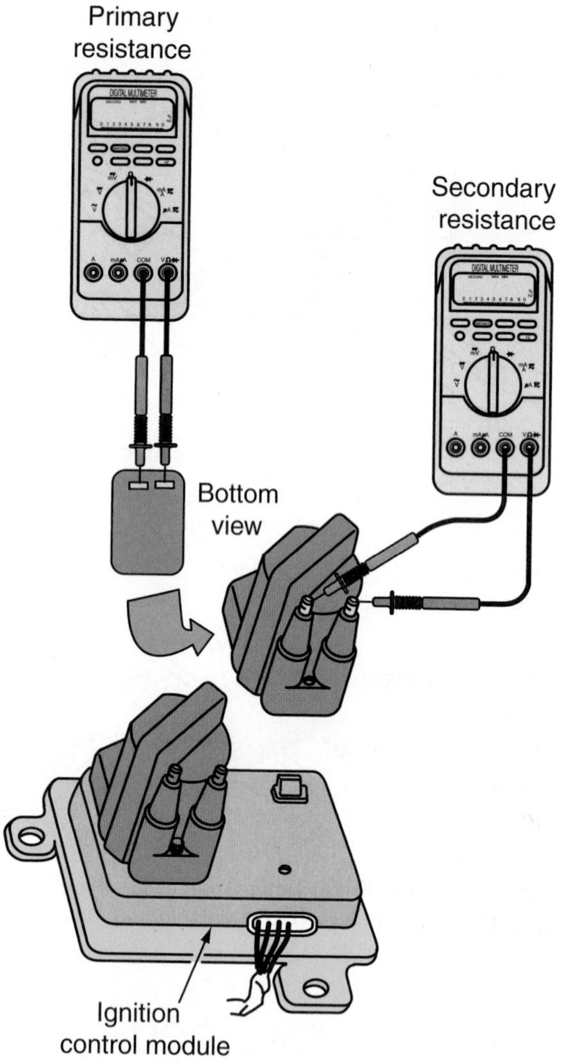

Figure 28–38 Meter connections for testing the resistance of a double-ended ignition coil.

High-Voltage Diodes Some secondary ignition coil windings contain a high-voltage diode. Normally when the coil has one, the manufacturer does not recommend testing the resistance of the secondary winding. The coil's output can be measured but this test should only be done after the primary windings have been checked.

Crankshaft/Camshaft Sensors

Although ignition timing on most EI systems is not adjustable, the air gaps at the crankshaft and camshaft sensors will affect the operation of the ignition system. On some engines, this gap is adjustable and on others it is an indication that the sensor should be replaced. If there is no provision for adjusting the gap and the gap is incorrect, the sensor should be replaced.

When checking the gap, make sure there are no signs of damage to the rotating vane assembly. Measure the gap with a nonmagnetic feeler gauge. Compare your findings to specifications. If the gap is not

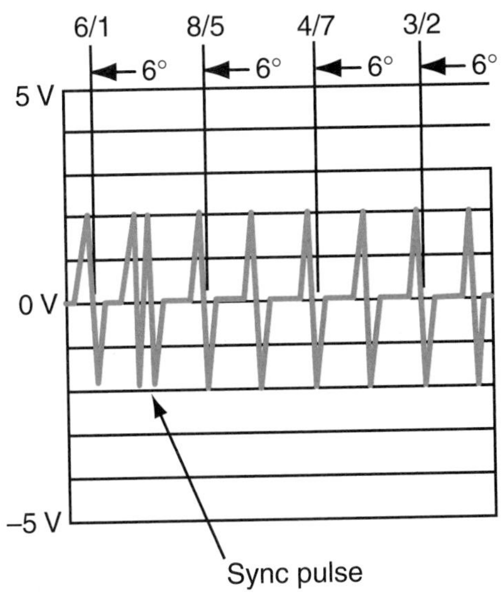

Figure 28-39 A waveform for a nine-slot trigger wheel for the crankshaft sensor.

correct, the sensor should be replaced or the gap adjusted.

The sensors can also be checked with an ohmmeter. Connect the negative lead to the ground terminal at the sensor and measure the resistance between that terminal and the others one at a time. The resistance should be within the specified range; if not replace the sensor.

Also, the action of the sensors can be monitored with an AC voltmeter or a lab scope connected across the sensor terminals to check the sensor signal while the engine is cranking. Meter readings below the specified value indicate a shorted sensor winding, whereas infinite meter readings prove that the sensor winding is open. On a lab scope, the pattern should be smooth, and all peaks, with the exception of the sync pulse, should be at the same height **(Figure 28–39)**.

SHOP TALK

While observing the wave patterns of a CKP, remember that the pattern can be altered by engine misfiring (Figure 28–40).

On some engines, the crankshaft sensor is mounted inside the engine. These sensors are continuously splashed with engine oil. Often these sensors fail because engine oil enters the sensor and shorts out the winding.

A quick check of the condition of the permanent magnet in a sensor can be made by placing a flat steel tool (such as the blade of a feeler gauge set) near the sensor. It should be attracted to the sensor if the sensor's magnet is okay.

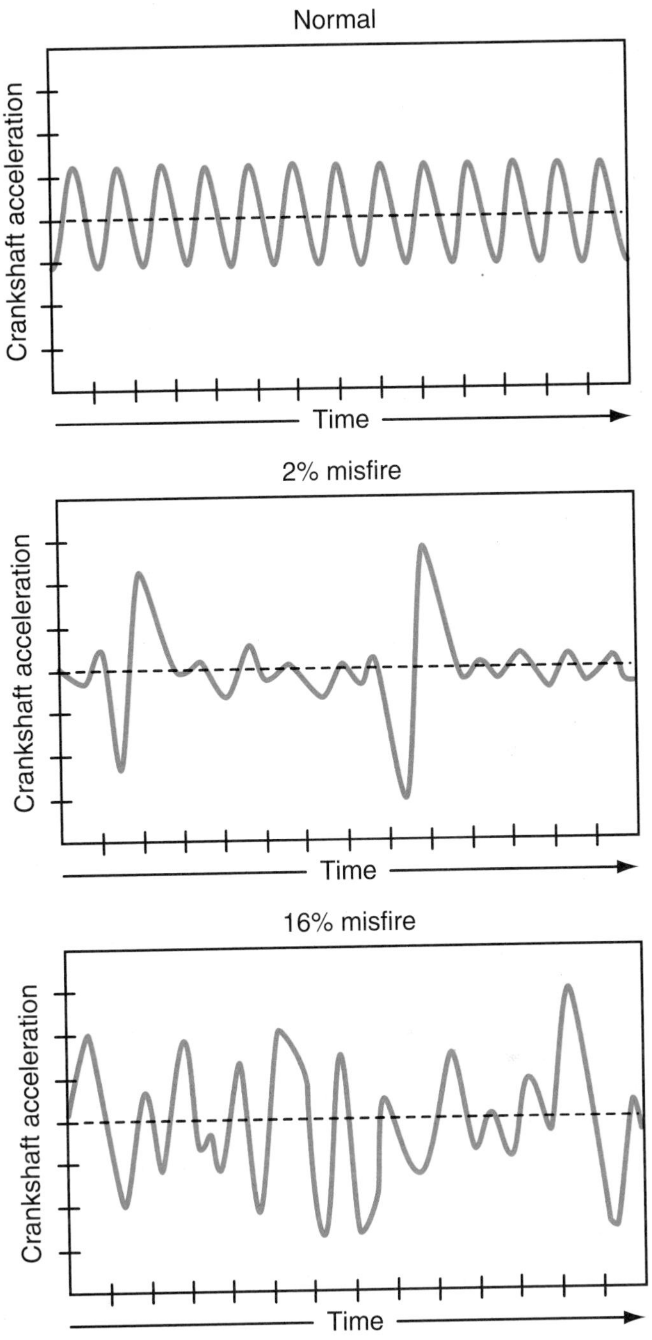

Figure 28-40 The reaction of a CKP's waveform to misfires.

If a crankshaft or camshaft sensor needs to be replaced, always clean the sensor tip and install a new spacer (if so equipped) on the sensor's tip. New sensors typically have a spacer already installed on the sensor **(Figure 28–41)**. Install the sensor until the spacer lightly touches the sensor ring and tighten the sensor mounting bolt.

Pickup Coils

It is important that all wires and connectors between the trigger sensor and ignition module and module to the PCM be visually checked as well as checked for excessive resistance with an ohmmeter.

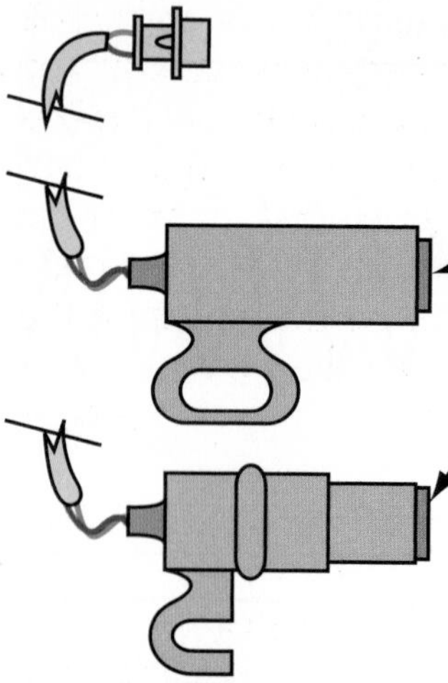
Spacer #5252229
MUST be in place
before installation

Figure 28-41 Some new CKP sensors rely on a paper spacer to set the sensor's gap to specifications.

To test magnetic pulse and metal detection pickup coils inside a distributor or used as a CKP or CMP with an ohmmeter, connect the ohmmeter from one of the pickup leads to ground to test for a short to ground. If there is a short to ground, the ohmmeter will have a reading of less than specifications. Most pickup coils have 150 to 900 ohms resistance, but always refer to the manufacturer's specifications. If the pickup coil is open, the ohmmeter will display an infinite reading. When the pickup coil is shorted, the ohmmeter will display a reading lower than specifications.

While the ohmmeter leads are connected, pull on the pickup leads and watch for an erratic reading, indicating an intermittent open in the pickup leads.

Often the gap between the pickup coil and reluctor is adjustable. The gap should be measured with a nonmagnetic feeler gauge placed between the reluctor high points and the pickup coil. If adjustment is required, loosen the pickup and move it until the specified air gap is obtained. Retighten the pickup coil bolts to the specified torque. Some pickup coils are riveted to the pickup plate. A pickup gap adjustment is not required for these pickup coils.

Voltmeter Checks If the resistance is within specifications, the circuit should be checked with a voltmeter. Turn the ignition on and connect the voltmeter across the voltage input wire and ground. Compare the reading to specifications.

When the voltmeter is connected from the reference wire or the sync wire to an engine ground, the voltmeter should cycle from nearly 0 to 5 volts while the engine is cranking. If the pickup signal is not within specifications, the pickup is defective.

Hall-Effect Sensors

Prior to testing a Hall-effect pickup, the ohmmeter should be connected across each of the wires between the pickup and the computer with the ignition switch off. A computer terminal and pickup coil wiring diagram is essential for these tests. Satisfactory wires have nearly 0 ohm resistance, while higher or infinite readings indicate defective wires.

With the voltmeter hooked up, insert a steel feeler gauge or knife blade between the Hall layer and magnet. If the sensor is good, the voltmeter should read within 0.5 volt of battery voltage when the feeler gauge or knife blade is inserted and touching the magnet. When the feeler gauge or blade is removed, the voltage should read less than 0.5 volt.

In the following tests, the distributor connector is connected to the unit and the connector backprobed. With the ignition switch on, a voltmeter should be connected from the voltage input wire to ground. The specified voltage should appear on the meter. The ground wire should be tested with the ignition switch on and a voltmeter connected from the ground wire to a ground connection near the distributor. With this meter connection, the meter indicates the voltage drop across the ground wire, which should not exceed 0.2 volt.

Connect a digital voltmeter from the pickup signal wire to ground. If the voltmeter reading does not fluctuate while cranking the engine, the pickup is defective. However, if the voltmeter reading fluctuates from nearly 0 volt to between 9 and 12 volts, that indicates a satisfactory pickup. During this test, the voltmeter reading may not be accurate because of the short duration of the voltage signal. If the Hall-effect pickup signal is satisfactory and the 12-volt test lamp did not flutter during the no-start test, the ignition module is probably defective.

Using a Logic Probe

The primary can also be checked with a logic probe. There are three lights on a logic probe. The red light illuminates when the probe senses more than 10 volts. When there is a good ground, the green light turns on. The yellow light flashes whenever the voltage changes. This light is used to monitor a pulsing signal, such as one produced by a digital sensor like a Hall-effect switch.

To check the primary circuit with a logic probe, turn the ignition on. Touch the probe to both (positive and negative) primary terminals at the coil. The red light should come on at both terminals (**Figure 28–42**), indicating that at least 10 volts are available to the coil and that there is continuity through the coil. If the red light does not come on

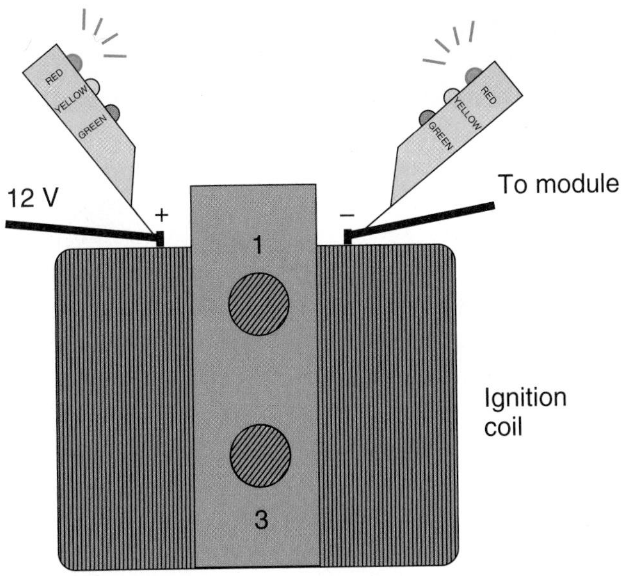

Figure 28-42 The red light on a logic probe should turn on when touched to both sides of the primary winding.

when the positive side of the coil is probed, check the power feed circuit to the coil. If the light comes on at the positive terminal but not at the negative, the coil has excessive resistance or is open.

Now move the probe to the negative terminal of the coil and crank the engine. The red and green lights should alternately flash, indicating that over 10 volts are available to the coil while cranking and that the circuit is switching to ground. If the lights do not come on, check the ignition power feed circuit from the starter. If the red light comes on but the green light does not, check the crankshaft or camshaft sensor. If these are working properly, the ignition module is probably defective.

A Hall-effect switch is also easily checked with a logic probe. If the switch has three wires, probe the outer two wires with the ignition on **(Figure 28–43)**. The red light should come on when one of the wires is probed, and the green light should come on when the other wire is probed. If the red light does not turn on at either wire, check the power feed circuit to the sensor. If the green light does not come on, check the sensor's ground circuit.

Figure 28-43 One end terminal of a Hall-effect sensor connector should cause the red light to come on; the other end terminal should cause the green light to come on.

Backprobe the center wire and crank the engine. All three lights should flash as the engine is cranked. The red light will come on when the sensor's output is above 10 volts. As this signal drops below 4 volts, the green light should come on. The yellow light will flash each time the voltage changes from high to low. If the logic probe's lights do not respond in this way, check the wiring at the sensor. If the wiring is okay, replace the sensor.

Using a Lab Scope

If the crankshaft sensor is a Hall-effect switch, square waves should be seen on the scope. All pulses should normally be identical in spacing, shape, and amplitude. By using a dual-trace lab scope, the relationship between the crankshaft sensor and the ignition module can be observed. During starting, the module will provide a fixed amount of timing advance according to its program and the cranking speed of the engine. By observing the crankshaft sensor output and the ignition module, this advance can be observed **(Figure 28–44)**. The engine will not start if the ignition module does not provide for a fixed amount of timing advance.

Knock Sensors

Most systems have knock sensors that retard timing when the engine is experiencing ping or knocking. A quick check of a knock sensor is made by watching ignition timing while the engine is running and the engine block is tapped with the handle of a screwdriver. The noise should cause a change in timing. Photo Sequence 28 shows this procedure.

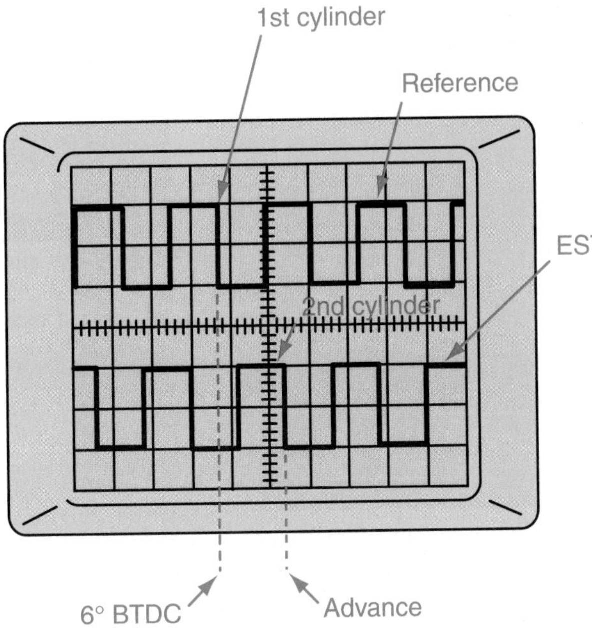

Figure 28-44 Electronic spark timing (EST) and crankshaft sensor signals compared on a dual-trace scope.

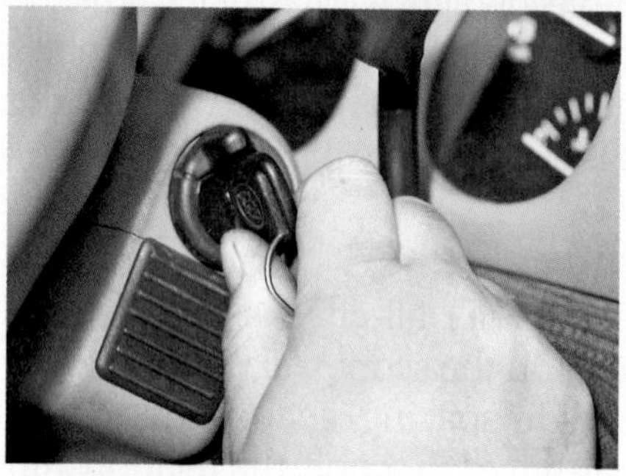

P28-1 *Make sure the engine is at normal operating condition and the ignition switch is off.*

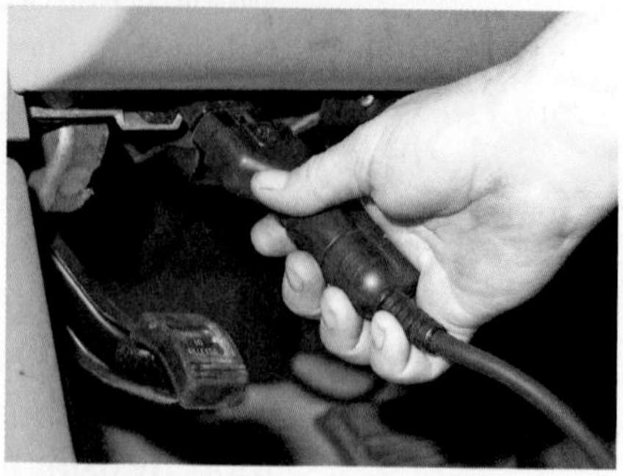

P28-2 *Connect the scan tool with the correct adapter to the DLC.*

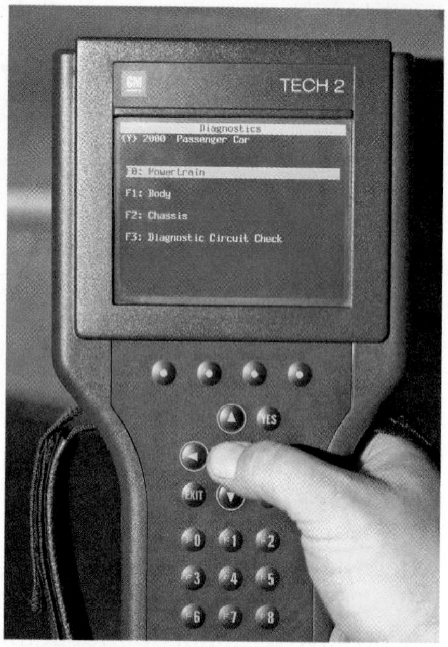

P28-3 *Program the scan tool for the vehicle and system being tested.*

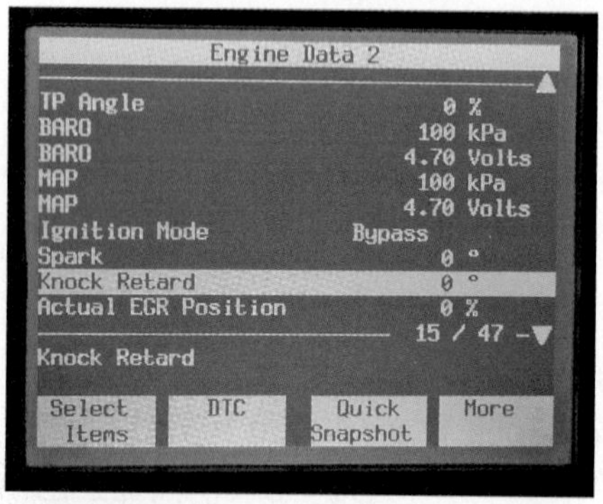

P28-4 *Then select "knock sensor" on the scan tool.*

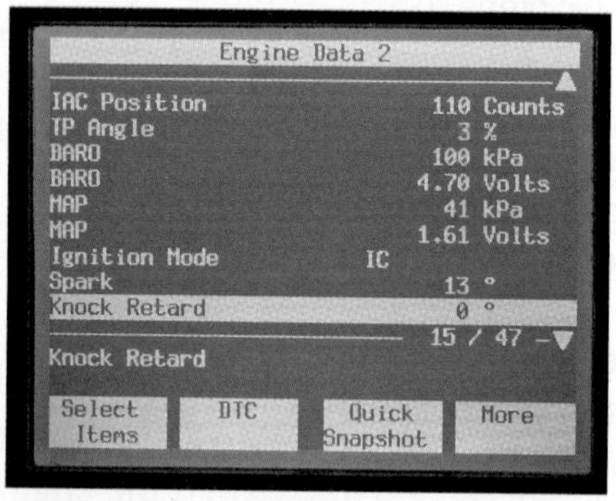

P28-5 *Look at the knock sensor signal with the engine running at about 1,500 rpm. There should be no signal from the knock sensor.*

P28-6 *Tap on the right exhaust manifold above the knock sensor with a hammer to simulate a knock. Watch the scan tool.*

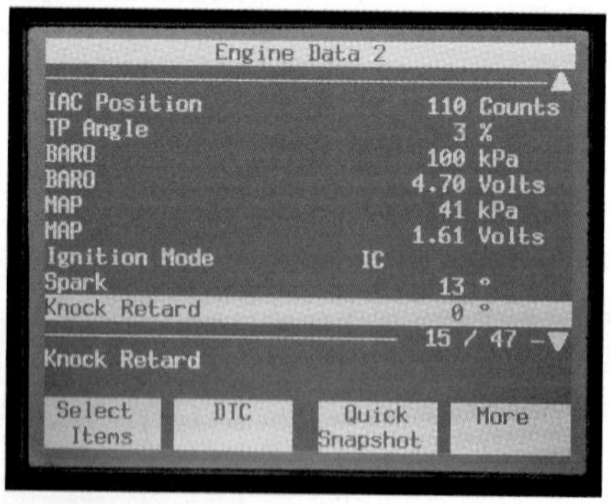

P28-7 *The knock sensor should now show a change in ignition timing. If it does, the knock sensor and circuit are working.*

Control Module

The most effective way to test a control module is to use an ignition module tester. This tester evaluates and determines if the module is operating within a given set of parameters. It does this by simulating normal operating conditions while looking for faults in the module's components. If a module tester is not available, check out all other system components before condemning the control module.

DISTRIBUTOR SERVICE

Typically the distributor will need to be removed to service it. Before removing it, grasp the distributor shaft and move it from side to side. If any movement is detected, remove the distributor and check the bushing. Lightly clamp the distributor in a soft-jaw vise. Clamp a dial indicator on the top of the distributor housing. Position the plunger of the indicator so that it rests on the distributor shaft. When the shaft is pushed horizontally, watch the movement of the indicator. Compare this to the specifications given in the service manual. If the movement exceeds the allowed amount, the distributor bushings and/or shaft are worn. Many manufacturers recommend complete distributor replacement rather than bushing or shaft replacement.

A typical procedure for removing a distributor follows:

PROCEDURE

To remove a distributor:

1. Disconnect the electrical connector and the vacuum advance hose, if the distributor has them.
2. Remove the distributor cap and note the position of the rotor. On some vehicles, it may be necessary to remove the spark plug wires from the cap prior to cap removal.
3. Note the position of the vacuum advance, then remove the distributor holddown bolt and clamp.
4. Pull the distributor from the engine. Most distributors will need to be twisted as they are pulled out of their bore. Note the direction of rotation.
5. Once the distributor is removed, install a shop towel in the distributor opening to keep foreign material out of the engine block.

A typical procedure for installing and timing the distributor follows:

PROCEDURE

The following procedure may be followed to install the distributor and time it to the engine:

1. Lubricate the O-ring on the distributor shaft.
2. Position the rotor so that it is aligned with the mark made to the distributor housing prior to removal.
3. Align the distributor to the mark made on the engine block during removal.
4. Lower the distributor into the engine block; make sure the distributor drive is fully seated. Distributors equipped with a helical drive gear will rotate as the distributor is being installed, causing the distributor to move away from the reference marks. Pay attention to how much the rotor moves, then remove the distributor and move the rotor backward the same amount. This should allow the shaft to rotate while the distributor is being installed and still be aligned with the reference marks.
5. Make sure the distributor housing is fully seated against the engine block; sometimes it may be necessary to wiggle or rock the distributor to seat it fully into the drive gear. Distributors with drive lugs must be mated with the drive grooves in the camshaft. Both are offset to eliminate the possibility of installing the distributor 180 degrees out of time.
6. Rotate the distributor a small amount so the timer core teeth and pickup teeth are aligned.
7. Install the distributor holddown clamp and bolt **(Figure 28–45)**, and leave the bolt slightly loose.
8. Install the spark plug wires in the direction of distributor shaft rotation and in the correct cylinder firing order.
9. Connect the wiring for the distributor. The vacuum advance hose is usually left disconnected until the timing is set with the engine running.

SECONDARY CIRCUIT TESTS AND SERVICE

A coil secondary winding and spark plug are found in all ignition systems. DI, distributorless, and coil-near-cylinder systems also have spark plug wires. DI systems also have a distributor cap and rotor and sometimes a lead for the ignition coil.

Distributor Cap and Rotor

Because there is an air gap between the rotor and the distributor cap, electrical arcing takes place between the two. This causes a deterioration of the rotor tip

Figure 28-45 The holddown clamp and bolt for a distributor.

Figure 28-46 Before installing the spark plug wires to the distributor cap, identify the terminal for cylinder number 1 and then follow the engine's firing order.

and the distributor cap terminals. This added resistance could cause cylinder misfires.

When installing a new cap or rotor, it is always wise to replace both at the same time. Make sure both are fully seated before attempting to start the engine. Also remember that the spark plug wires need to be arranged on the cap according to the firing order **(Figure 28–46)**.

Waste Spark Systems

Standard test procedures using an oscilloscope, an ohmmeter, and a timing light can be used to diagnose problems in distributorless ignition systems **(Figure 28–47)**. Keep in mind, however, that problems involving one cylinder may also occur in its companion cylinder that fires off the same coil. Follow the testing procedures given in the service manual. Specific DTCs are designed to help troubleshoot ignition problems in these systems. The diagnostic procedure for EI systems varies depending on the vehicle make and model year.

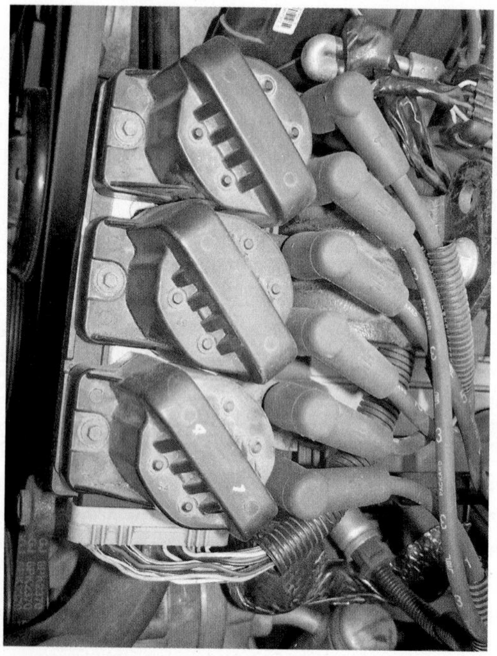

Figure 28-47 A distributorless ignition system.

There is a separate primary circuit for each coil. If one coil does not work properly, it may be caused by something common or not common to the other coils. Regardless of the system's design, there are common components in all electronic ignitions: an ignition module, crankshaft and/or camshaft sensors, ignition coils, a secondary circuit, and spark plugs. Most of these components are common to only the spark plug or spark plugs to which they are connected. The components that are common to all cylinders (such as the camshaft sensor) will, most often, be the cause of no-start problems. The other components will cause misfire problems.

Testing the secondary circuit of a DLS system is just like testing the secondary of any other type of ignition system. The spark plug wires and spark plugs should be tested to ensure that they have the appropriate amount of resistance. Because the resistance in the secondary dictates the amount of voltage the spark plug will fire with, it is important that secondary resistance be within the desired range.

A quick way to examine the secondary circuit is with an ignition scope. On the scope, the firing line and the spark plug indicate the resistance of the secondary. While observing the firing line, remember that the height of the line increases with an increase in resistance. The length of the spark line decreases as the firing line goes higher. This means that high resistance will cause excessively high firing voltages and reduced spark times.

When checking a DLS system with a scope, remember that in waste spark systems half of the plugs fire with reverse polarity. This means half of the firing lines will be higher than the other firing lines. Normally, reverse firing requires 30% more voltage than normal firing.

Excessive resistance is not the only condition that will affect the firing of a spark plug. Spark plug wires and the spark plug itself can allow the high voltage to leak and establish current through another metal object instead of the electrodes of the spark plug. When this happens, the spark plug does not fire and combustion does not take place within the cylinder.

Also keep in mind that the secondary circuit is completed through the metal of the engine. If the spark plugs are not properly torqued into the cylinder heads, the threads of the spark plug may not make good contact and the circuit may offer resistance. Always tighten spark plugs to their specified torque. Make sure not to cross-thread them.

Most manufacturers recommend the use of an antiseize compound on spark plug threads. This compound must be applied in the correct amounts and at the correct place. Too little compound will cause gaps in the contact between the spark plug threads and the spark plug bores. Too much may allow the spark to jump to a buildup rather than the spark plug electrode.

Coil-over-Plug Systems

Remember, an individual coil problem will cause misfiring in only one cylinder. Ignition coils are tested with an ohmmeter in much the same way as other ignition coils. If the resistance is out of specifications, the coil should be replaced. Intermittent coil problems can be caused by corrosion at the electrical connectors to the coil. The action of the

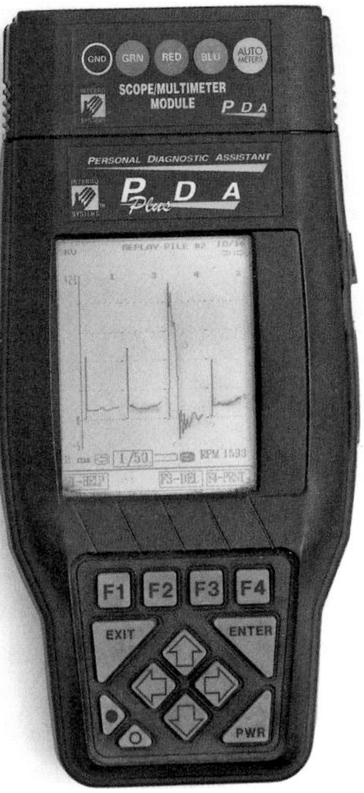

Figure 28–48 The scope pattern for a COP system with one faulty cylinder.

secondary can also be monitored on a lab scope (**Figure 28–48**).

Codes retrieved from the PCM will indicate whether the misfire is a general one or an individual cylinder. Because a COP ignition problem will affect only one cylinder, the general misfire code (P0300) is probably caused by a fuel delivery problem or a vacuum leak.

DTCs that indicate a misfire at an individual cylinder (P0301, P0302, P0303, etc.) are typically caused by a dirty or defective fuel injector, a fouled spark plug, a bad coil, or an engine mechanical problem. If the misfire is caused by a fuel injector problem, a fuel injector code (P0201, P0202, P0203, etc.) will also be retrieved; these identify the affected cylinder.

If a crankshaft position sensor is bad, there will no timing reference and this can prevent the engine from starting or have a hard time starting.

Each ignition coil has a driver circuit in the PCM that controls primary current flow. If there is a bad driver circuit, that spark plug will not fire. Also, keep in mind that an engine may start with a faulty camshaft sensor but it may only run in the fail-safe or limp-in mode because the fuel injectors cannot be synchronized without the camshaft signal.

Coil-near-plug systems can be checked with a scope or graphing meter in the same way as other ignition systems. The pickup for secondary signals is installed over the spark plug wire. However, COP

Figure 28-49 A coil-near-plug ignition system.

systems require special adapters to connect the scope or analyzer to the ignition system. Some low-amperage current probes can also be used to monitor the activity of an individual coil in a COP system. Also, make sure to check the spark plug wires **(Figure 28–49)**.

Spark Plugs

All of the ignition system is designed to do no more than supply the voltage necessary to cause a spark across the gap of a spark plug. This simple event is what starts the combustion process. Needless to say, a healthy spark plug is extremely important to the combustion process. Spark plug replacement is part of the preventive maintenance program for all vehicles. The recommended replacement interval depends on a number of factors but ranges from 20,000 to 100,000 miles (32,000 to 160,000 km).

Removal of an engine's spark plugs is pretty straightforward. Remove the cables from each plug, being careful not to pull on the cables. Instead, grasp the boot **(Figure 28–50)** and gently twist it off. (To save time and avoid confusion later, use masking tape

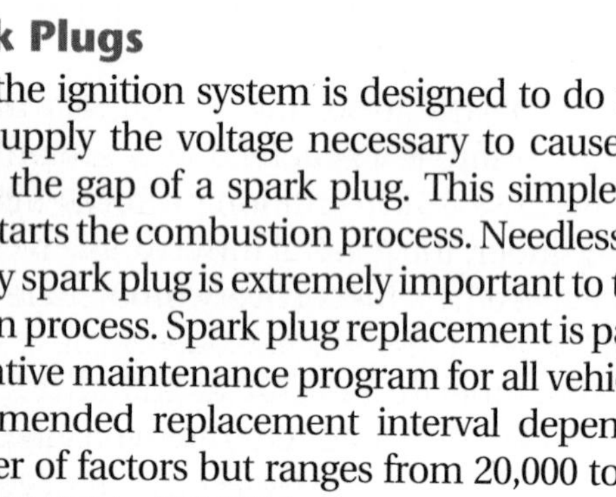

Figure 28-50 Grab hold of the boot and twist while pulling it off a spark plug. *Courtesy of Toyota Motor Sales, U.S.A., Inc.*

Figure 28-51 Use a spark plug socket to remove the plugs.

to mark each of the cables with the number of the plug it attaches to.)

Using a spark plug socket and ratchet, loosen each plug a couple of turns. A spark plug socket should be used because it has an internal rubber bushing to prevent plug insulator breakage. Spark plug sockets can have either a ⅜- or ½-inch drive, and most have an external hex so that they can be turned using an open-end or box wrench.

Once the plugs are loose, use compressed air to blow dirt away from the base of the plugs. Then remove the plugs, making sure their gaskets have also been removed (if applicable). When the spark plugs are removed **(Figure 28–51)**, they should be set in order so the spark plug from each cylinder can be examined.

Check the threads in the cylinder head for damage. Normally you can do this by feel as you remove a spark plug. If the plug does not turn out smoothly after it is loose, the threads may be damaged. Often the threads can be cleaned up with a spark plug thread chaser. Also, check the threads on the spark plug. Look for damage or metal embedded in the threads, as these are sure signs of problems. If the cylinder head is aluminum, it may be necessary to install a threaded insert into the spark plug bore.

Inspecting Spark Plugs

Once the spark plugs have been removed, it is important to "read" them. In other words, inspect them closely, noting in particular any deposits on the plugs and the degree of electrode erosion. A normal-firing spark plug will have a minimum amount of deposits on it and will be colored light tan or gray

Figure 28-52 Normal spark plug. *Courtesy of Federal-Mogul Corporation*

Figure 28-53 A worn spark plug. *Courtesy of Federal-Mogul Corporation*

(**Figure 28–52**). However, there should be no evidence of electrode burning, and the increase of the air gap should be no more than 0.001 inch (0.0254 mm) for every 10,000 miles (16,000 km) of engine operation. A plug that exceeds this wear should be replaced and the cause of excessive wear corrected (**Figure 28–53**). Worn or dirty spark plugs may work fine at idle or low speeds, but they frequently fail during heavy loads or higher engine speeds.

It is possible to diagnose a variety of engine conditions by examining the electrodes of the spark plugs. Ideally, all of the plugs from an engine should look alike. Whenever plugs from different cylinders look different, a problem exists in those cylinders. The following are examples of plug problems and how they should be dealt with.

Cold Fouling This condition is the result of an excessively rich air-fuel mixture. It is characterized by a layer of dry, fluffy, black carbon deposits on the tip of the plug (**Figure 28–54**). **Cold fouling** is caused by a rich air-fuel mixture or an ignition fault that causes the spark plug not to fire. If only one or two of the plugs show evidence of cold fouling, sticking valves are the likely cause. The plug can be used again, provided its electrodes are filed and the air gap is reset.

Figure 28-54 A cold- or carbon-fouled spark plug. *Courtesy of Federal-Mogul Corporation*

Correct the cause of the problem before reinstalling or replacing the plugs.

SHOP TALK

If cold fouling is present on a vehicle that operates a great deal of the time at idle and low speeds, plug life can be lengthened by using hotter spark plugs.

Wet Fouling When the tip of the plug is practically drowned in excess oil, this condition is known as **wet fouling** (**Figure 28–55**). In an overhead valve engine, the oil may be entering the combustion chamber by flowing past worn valve guides or valve guide seals. If the vehicle has an automatic transmission, a likely cause of wet-fouled plugs is a defective vacuum modulator that is allowing transmission fluid to enter the chamber. On high-mileage engines, check for worn rings or excessive cylinder wear. The best solution is to correct the problem and replace the plugs with the specified type.

Splash Fouling **Splash fouling** occurs immediately following an overdue tune-up. Deposits in the

Figure 28-55 A wet- or oil-fouled spark plug. *Courtesy of Federal-Mogul Corporation*

Figure 28-56 A splash-fouled spark plug. *Courtesy of Federal-Mogul Corporation*

Figure 28-58 This spark plug shows signs of overheating. *Courtesy of Federal-Mogul Corporation*

combustion chamber, accumulated over a period of time due to misfiring, suddenly loosen when the temperature in the chamber returns to normal. During high-speed driving, these deposits can stick to the hot insulator and electrode surfaces of the plug (**Figure 28–56**). These deposits can actually bridge across the gap, stopping the plug from sparking. Normally splash-fouled plugs can be cleaned and reused.

Gap Bridging A plug with a bridged gap (**Figure 28–57**) is not frequently seen in automobile engines. It occurs when flying carbon deposits within the combustion chamber accumulate over a long period of stop-and-go driving. When the engine is suddenly placed under a hard load, the deposits melt and bridge the gap, causing misfire. This condition is best corrected by replacing the plug.

Glazing Under high-speed conditions, the combustion chamber deposits can form a shiny, yellow glaze over the insulator. When it gets hot enough, the glaze acts as an electrical conductor, causing the current to follow the deposits and short out the plug. **Glazing** can be prevented by avoiding sudden wide-open throttle acceleration after sustained periods of low-speed or idle operation. Because it is virtually impossible to remove glazed deposits, glazed plugs should be replaced.

Overheating This condition is characterized by white or light-gray blistering of the insulator. There may also be considerable electrode gap wear (**Figure 28–58**). Overheating can result from using too hot a plug, over-advanced ignition timing, detonation, a malfunction in the cooling system, an overly lean air-fuel mixture, using fuel too low in octane, an improperly installed plug, or a heat-riser valve that is stuck closed. Overheated plugs must be replaced.

Turbulence Burning When turbulence burning occurs, the insulator on one side of the plugs wears away as the result of normal turbulence in the combustion chamber. As long as the plug life is normal, this condition is of little consequence. However, if the spark plug shows premature wear, overheating can be the problem.

Preignition Damage is caused by excessive engine temperatures. Preignition damage is characterized by melting of the electrodes or chipping of the electrode tips (**Figure 28–59**). When this problem occurs, look for the general causes of engine overheating, including over-advanced ignition timing, a burned

Figure 28-57 There is no longer a gap on this spark plug. *Courtesy of Federal-Mogul Corporation*

Figure 28-59 A spark plug with preignition damage. *Courtesy of Federal-Mogul Corporation*

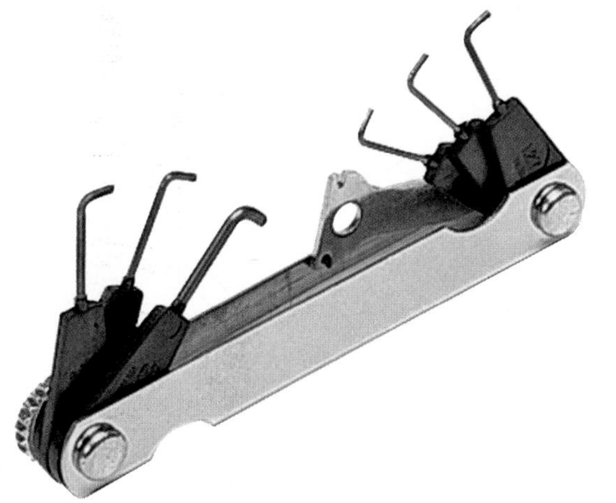

Figure 28-60 A round wire-type feeler gauge set. *Reproduced under license from Snap-on Incorporated. All of the marks are marks of their owners.*

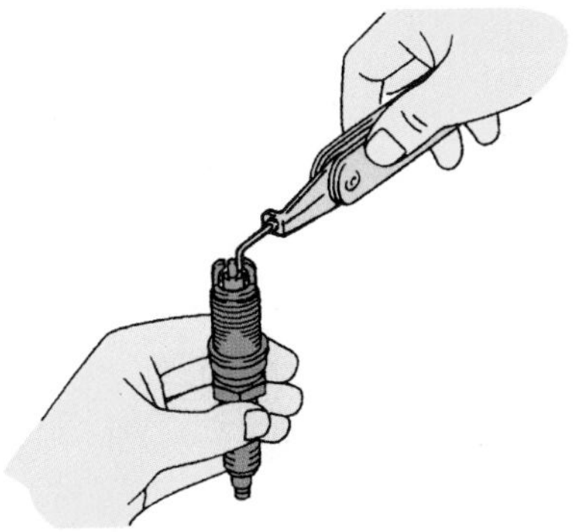

Figure 28-61 The gap between the center electrode and both ground electrodes should be checked and adjusted to specifications. *Courtesy of Toyota Motor Sales, U.S.A., Inc.*

head gasket, and using fuel too low in octane. Other possibilities include loose plugs or using plugs of the improper heat range. Do not attempt to reuse plugs with preignition damage.

Regapping Spark Plugs

Both new and used spark plugs should have their air gaps set to manufacturer's specifications. Always use round wire gauges when checking and setting the gap **(Figure 28–60)**.

After the gap has been adjusted, make sure that the ground electrode is as horizontal as it can be.

Always check the air gap of a new spark plug before installing it. Never assume the gap is correct just because the plug is new. Do not try to reduce a plug's air gap by tapping the side electrode on a bench. Use a spark plug gapping tool to bend the ground electrode to its correct height. When doing this, be careful not to contact or put pressure on the center electrode. This is especially critical with fine wire platinum and iridium plugs. Also while bending the ground electrode, try to keep it in alignment with the center electrode.

Some engines are equipped with spark plugs that have more than one ground electrode. The gap between the center electrode and each ground electrode should be checked **(Figure 28–61)**. If the gap between the center electrode and one of the ground electrodes is less than that of the others, spark will occur only at the smallest gap. This is also true of V-shaped ground electrodes. If one leg of the vee is closer to the center electrode than the other, the spark will always occur across the shortest distance.

The gap of spark plugs with a surface gap and of some with more than one ground electrode cannot be adjusted with conventional tools, and most manufacturers recommend that the gap be left alone.

PROCEDURE

Spark Plug Installation

STEP 1. Wipe dirt and grease from the plug seats with a clean cloth.

STEP 2. Verify that the replacement spark plugs are the correct ones for the engine by matching the part number to its application **(Figure 28–62)**. Never assume that the plugs that were removed from the engine are the correct type.

STEP 3. Adjust the air gap, as needed.

STEP 4. Check the service manual to see if antiseize compound should be applied to the plug's threads **(Figure 28–63)**

STEP 5. Install the plugs and tighten them with your hand. If the plugs cannot be installed easily by hand, the threads in the cylinder head may need to be cleaned with a thread-chasing tap. Be especially careful not to cross-thread the plugs when working with aluminum heads.

STEP 6. Tighten the plugs with a torque wrench, following the vehicle manufacturer's specifications or the values shown in **Figure 28–64**.

Secondary Ignition Wires

Inspect all the spark plug wires and the secondary coil wire for cracks and worn insulation, which cause high-voltage leaks. Inspect all the boots on the ends of the plug wires and coil secondary wire for cracks and hard, brittle conditions.

Make sure that the secondary cables are secured tightly to the spark plugs and the ignition coil(s). Also check the cables for any damage or signs of arcing. Replace the wires and boots if they show evidence of these conditions. Most manufacturers recommend that spark plug wires be replaced as a complete set.

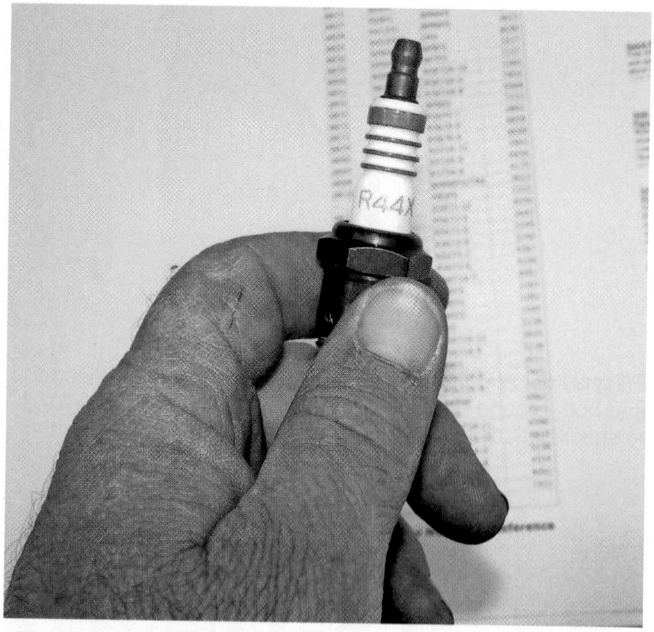

Figure 28-62 Make sure that the replacement spark plugs are the correct ones for the application.

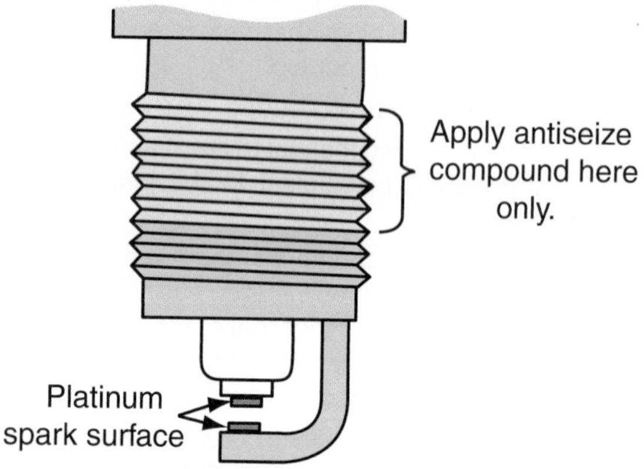
Apply antiseize
compound here
only.
Platinum
spark surface

Figure 28-63 Proper placement of antiseize compound on the threads of a spark plug.

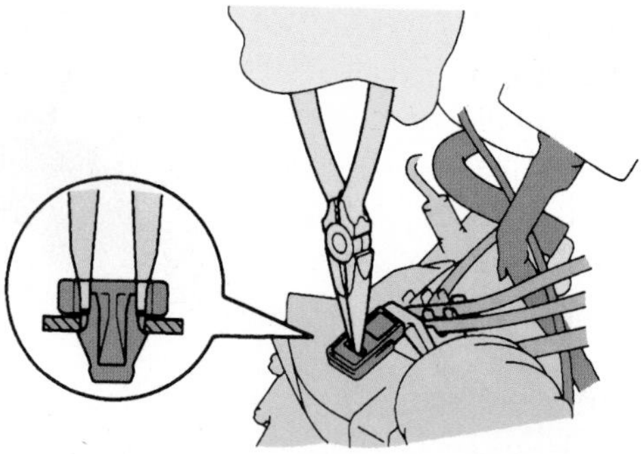

Figure 28-65 Use needle-nose pliers to remove the wires from the spark plug cable protector. *Courtesy of Toyota Motor Sales, U.S.A., Inc.*

SHOP TALK

Many ignition systems use locking tabs to secure the spark plug cable to the ignition coil. To remove the cable from the coil, squeeze the locking tabs with needle-nose pliers **(Figure 28–65)** or use a screwdriver to lift up the locking tab. Then disconnect the cables from the spark plugs by firmly holding the boot and, with a twist-pull effort, remove the cable from the plug. When reconnecting the cable to the coil, make sure that the locking tabs are in place by pressing down on the center of the cable terminal. Also, make sure that the routing of the cables is correct.

Resistance Checks The resistance of secondary cables can be checked with an ohmmeter. Do this by removing the cable and measuring across the cable. On DI systems, the spark plug wires may be left in the distributor cap when measuring resistance. This will also check the cap-to-cable connections. Set the ohmmeter to the X1,000 scale, and connect the ohmmeter leads from the end of a spark plug wire to the appropriate terminal inside the distributor cap.

Spark Plug Type	Thread Diameter	Cast-Iron Head (LB-FT.)	Aluminum Head (LB-FT.)
Flat seat w/gasket	18 mm	25–33	25–33
Conical seat/no gasket	18 mm	14–22	14–22
Flat seat w/gasket	14 mm	18–25	18–22
Flat seat w/gasket	12 mm	10–18	10–15
Conical seat/no gasket	14 mm	10–18	7–15

Figure 28-64 Typical spark plug torque specifications.

If the ohmmeter reading is more than specified by the manufacturer, remove the wire from the cap and check the wire alone. If the wire has more resistance than specified, replace the wire. When spark plug wire resistance is satisfactory, check the cap terminal for corrosion. Repeat the ohmmeter tests on each spark plug wire and the coil secondary wire.

Replacing Spark Plug Wires

When spark plug wires are being installed, make sure they are routed properly as indicated in the vehicle's service manual. When removing the spark plug wires from a spark plug, grasp the spark plug boot tightly, and twist while pulling the cable from the end of the plug. When installing a spark plug wire, make sure the boot is firmly seated around the top of the plug, then squeeze the boot to expel any air that may be trapped inside.

Two spark plug wires should not be placed side by side for a long span if these wires fire one after the other in the cylinder firing order. When two spark plug wires that fire one after the other are placed side by side for a long span, the magnetic field from the wire that is firing builds up and collapses across the other wire. This magnetic collapse may induce enough voltage to fire the other spark plug and wire when the piston in this cylinder is approaching TDC on the compression stroke. This action may cause detonation and reduced engine power.

Also make sure that the wires are secure in their looms and that the looms are properly placed.

CASE STUDY

An irate customer brought his late-model Chevrolet back to the service department, complaining that the car ran worse than it did before he had it serviced the previous day. The previous complaint was extreme detonation. After interviewing the customer, the service writer found that the engine no longer spark knocked but was very sluggish. According to the repair order, a new knock sensor was installed.

The technician who had done the previous work was assigned this repair. When the engine was started, the MIL was on. The technician immediately connected a scan tool and found a service code of P0325, which indicated a problem with the knock sensor circuit. This was the same code that led him to replace the knock sensor earlier. Since he could not remember erasing the codes after installing the new sensor, he erased the codes. He assumed this was the problem and that the PCM was compensating for a bad knock sensor when in fact it was new and good. Fortunately, the technician took the car for a test drive to verify the repair. Almost immediately after starting the test drive, the MIL was lit again.

Back in the shop, he reconnected the scan tool and code P0325 was present again. A bit bewildered and frustrated, he resorted to looking in the service manual for guidance. He followed the troubleshooting sequence for the code, only to find that the wire from the sensor to the electronic spark control module was disconnected. Apparently he did not connect the wire to the sensor after he replaced the sensor. He connected the wire and visually inspected all hoses and wires around the area where he had been working. He then erased the code and took the car on another test drive. This time the MIL did not come on, and the car ran quite well. The disconnected wire prevented the control module from advancing the timing as needed and, therefore, the engine ran poorly. This experience taught the technician many lessons, one of which was that a trouble code identifies a problem circuit, not necessarily a problem component.

KEY TERMS

Carbon tracking
Cold fouling
Corona effect
Current limiting
Display pattern
Firing line
Glazing
Intermediate section
Kilovolt (kV)
Raster pattern
Spark duration
Spark line
Splash fouling
Superimposed pattern
Wet fouling

SUMMARY

- Secure wiring and connections are important to ignition systems. Loose connections, corrosion, and dirt can adversely affect performance.
- Wires, connections, and ignition components can be tested for intermittent failure by wiggling them or stress testing by applying heat, cold, or moisture.

- A scope provides a visual representation of voltage changes over time.
- Waveforms can be viewed in different modes and scales on a scope. Secondary and primary ignition circuits can be viewed.
- Ignition patterns can be broken down into three main sections or zones: firing section, intermediate section, and dwell section.
- The firing line and spark line display firing voltage and spark duration.
- The intermediate section shows coil voltage dissipation.
- The dwell section shows the activation of primary coil current flow and primary coil current switch off. The primary current off signal is also the firing line for the next cylinder in the firing order.
- Current limiting ignition systems saturate the ignition coil very quickly with high current flow and then cut back or limit current flow to maintain saturation. This system extends coil life.
- Precautions must always be taken to avoid open circuits during ignition system testing. A special test plug is used to limit coil output during testing. Always use the correct test plug for the system.
- Firing voltages are normally between 7 and 13 kV with no more than 3 kV variation between cylinders.
- High secondary circuit resistance produces a higher-than-normal firing line and shorter spark lines.
- Individual ignition components are commonly tested for excessive internal resistance using an ohmmeter. A voltmeter or scope can also be used to monitor their operating voltages.
- Proper spark plug gapping and installation are important to ignition system operation. Spark plug condition, such as cold fouling, wet fouling, and glazing, is often a good indication of other problems.
- Standard test procedures using an oscilloscope, GMM, and/or DMM can be used to diagnose problems in EI and DI systems.
- Often, if a crank or cam sensor fails, the engine will not start. These sensor circuits can be checked with a voltmeter. If the sensors are receiving the correct amount of voltage and have good low-resistance connections, their output should be a square wave or a pulsing analog signal while the engine is cranking.
- The resistance of COP ignition coils can be checked in the same way as conventional coils; however, different meter connections are required to test waste spark coils.

REVIEW QUESTIONS

1. Name the three types of stress testing used to test for intermittent ignition component problems and list the procedures for conducting each type of test.
2. Why is the procedure for checking the resistance of a waste spark ignition coil different from the procedures for checking other types of ignition coils?
3. Name the three types of trace pattern display modes used on an oscilloscope and give examples of when each mode is most useful.
4. List the common types of spark plug fouling and the typical problems each type of fouling indicates.
5. List at least two methods of checking the operation of Hall-effect sensors.
6. What happens if one of the ground electrodes of a spark plug with two or more electrodes is closer to the center electrode than the other?
7. *True or False?* Input signals should always be checked before checking output signals.
8. Leaner air-fuel mixtures ______.
   a. decrease the electrical resistance inside the cylinder and decrease the required firing voltage
   b. increase the electrical resistance inside the cylinder and increase the required firing voltage
   c. increase the electrical resistance inside the cylinder and decrease the required firing voltage
   d. have no measurable effect on cylinder resistance
9. Describe the basic procedure for finding the cause of a no-start problem on an engine equipped with an EI system.
10. What does the spark line represent?
11. *True or False?* On some engines, if the gap between the crankshaft sensor and its trigger wheel is outside specifications, the sensor should be replaced.
12. What is the typical procedure for checking the resistance of the primary winding in an ignition coil?
13. While checking a pickup coil with an ohmmeter, a higher-than-normal reading indicates that the pickup unit is ______.
   a. shorted
   b. open
   c. has high resistance
   d. none of the above

14. A CKP sensor can be checked with all of the following *except* a(n) ______.
    a. logic probe
    b. voltmeter
    c. ammeter
    d. lab scope
15. Which of the following will *not* cause one or more, but not all, firing lines to be higher than normal?
    a. high resistance in the spark plug wire
    b. retarded ignition timing
    c. faulty fuel injector
    d. defective spark plug

ASE-STYLE REVIEW QUESTIONS

1. The firing lines on an oscilloscope pattern are all abnormally low. Technician A says that the problem is probably low coil output. Technician B says that the problem could be an overly rich air-fuel mixture. Who is correct?
    a. Technician A
    b. Technician B
    c. Both A and B
    d. Neither A nor B
2. While testing the coils in an EI system: Technician A says that an infinite reading means that the winding has zero resistance and is shorted. Technician B says that the primary windings in each coil should be checked for shorts to ground. Who is correct?
    a. Technician A
    b. Technician B
    c. Both A and B
    d. Neither A nor B
3. While discussing no-start diagnosis with a test spark plug: Technician A says that if a test light flutters at the coil tach terminal but the test spark plug does not fire when connected from the coil secondary wire to ground with the engine cranking, the ignition coil is defective. Technician B says that if a test spark plug connected to a terminal of a waste spark ignition coil fires when the engine is cranking but the engine does not run, a bad PCM is indicated. Who is correct?
    a. Technician A
    b. Technician B
    c. Both A and B
    d. Neither A nor B
4. Technician A says that EMI can affect sensor signals. Technician B says that EMI can cause intermittent driveability problems. Who is correct?
    a. Technician A
    b. Technician B
    c. Both A and B
    d. Neither A nor B
5. While discussing waste spark EI systems: Technician A says that a spark plug with too wide of a gap can affect the firing of its companion spark plug. Technician B says that improper spark plug torque can cause an engine misfire. Who is correct?
    a. Technician A
    b. Technician B
    c. Both A and B
    d. Neither A nor B
6. While discussing how to test a crankshaft position sensor: Technician A says that logic probe can be used. Technician B says that a DMM can be used. Who is correct?
    a. Technician A
    b. Technician B
    c. Both A and B
    d. Neither A nor B
7. While discussing the possible causes for a no-start condition on an EI-equipped engine: Technician A says that a shorted crankshaft sensor may prevent the engine from starting. Technician B says that a shorted spark plug may stop the engine from starting. Who is correct?
    a. Technician A
    b. Technician B
    c. Both A and B
    d. Neither A nor B
8. While discussing the diagnosis of an electronic ignition (EI) system in which the crankshaft and camshaft sensor tests are satisfactory but a test spark plug connected from the spark plug wires to ground does not fire: Technician A says that the coil assembly may be defective. Technician B says that the voltage supply wire to the coil assembly may be open. Who is correct?
    a. Technician A
    b. Technician B
    c. Both A and B
    d. Neither A nor B
9. While discussing EI service and diagnosis: Technician A says that the crankshaft sensor may be rotated to adjust the basic ignition timing. Technician B says that the crankshaft sensor may be moved to adjust the clearance between the sensor and the rotating blades on some EI systems. Who is correct?
    a. Technician A
    b. Technician B
    c. Both A and B
    d. Neither A nor B
10. While discussing engine misfire diagnosis: Technician A says that a defective EI coil may cause cylinder misfiring. Technician B says that the engine compression should be verified first if the engine is misfiring continually. Who is correct?
    a. Technician A
    b. Technician B
    c. Both A and B
    d. Neither A nor B

CHAPTER 29

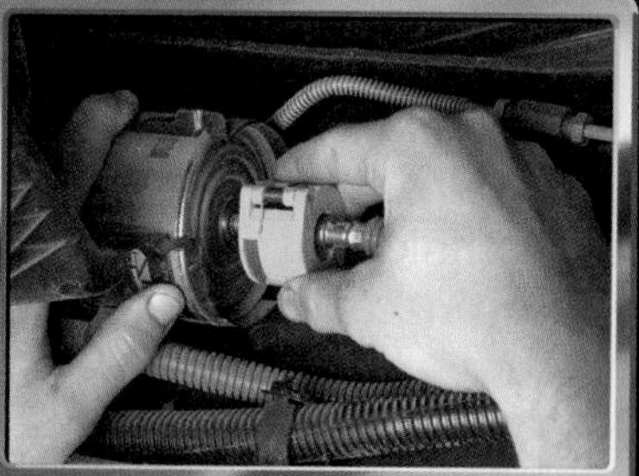

FUEL DELIVERY SYSTEMS

OBJECTIVES

■ Describe the components of a fuel delivery system and the purpose of each. ■ Conduct a visual inspection of a fuel system. ■ Relieve fuel system pressure. ■ Inspect and service fuel tanks. ■ Inspect and service fuel lines and tubing. ■ Describe the different fuel filter designs and mountings. ■ Remove and replace fuel filters. ■ Explain how common electric fuel pump circuits work. ■ Conduct a pressure and volume output test on a mechanical and electric fuel pump. ■ Service and test electric fuel pumps.

To have an efficient-running engine there must be the correct amount of fuel. To provide this, fuel must be stored, pumped out of storage, piped to the engine, filtered, and delivered to the fuel injectors.

The fuel system in today's vehicles is designed to prevent fuel vapors from entering the atmosphere. They are normally called returnless on-demand systems. In older systems, a fuel pump delivered fuel under pressure to the fuel injectors. A pressure regulator at the injectors controlled the fuel pressure by sending excess fuel back to the fuel tank. This is a return fuel system **(Figure 29–1)**.

In a returnless system **(Figure 29–2)**, the pressure regulator is in the fuel tank and excess fuel is released to the tank. There is no need for a return line. In a return system, the fuel sent back to the tank has been heated by underhood temperatures. The introduction of the warm fuel to the tank causes the fuel to evaporate. Fuel pressure and volume are controlled

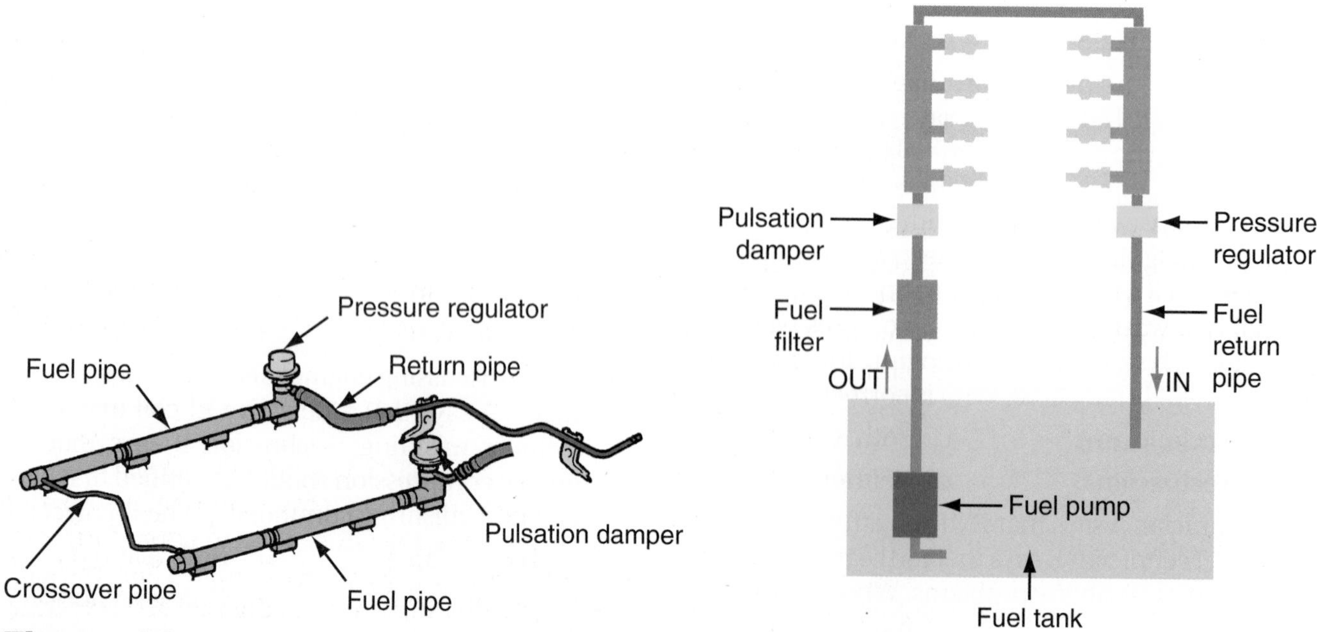

Figure 29-1 A return fuel delivery system. *Courtesy of Toyota Motor Sales, U.S.A., Inc.*

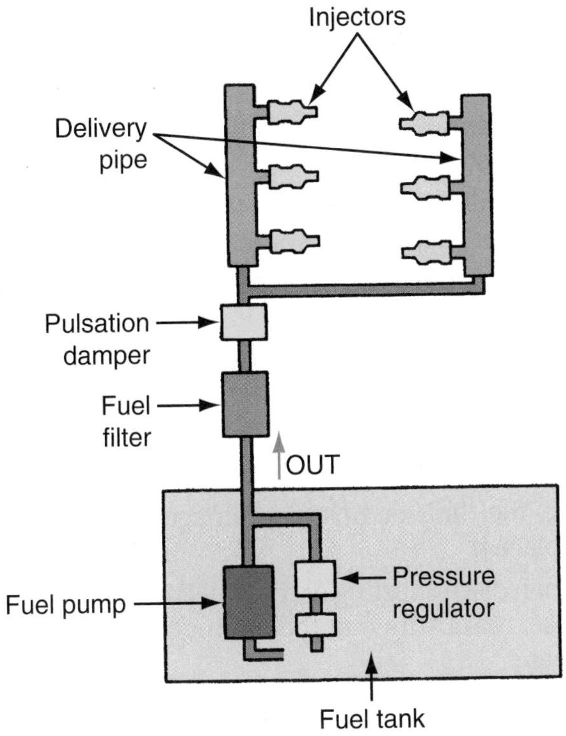

Figure 29-2 A returnless fuel delivery system.
Courtesy of Toyota Motor Sales, U.S.A., Inc.

by the PCM according to the existing operating conditions.

BASIC FUEL SYSTEM DIAGNOSIS

The fuel system should be checked whenever there is evidence of a fuel leak, fuel smell, or evidence of inadequate fuel supply. The fuel system should be checked whenever basic tests suggest that there is too little or too much fuel being delivered to the cylinders. Lean mixtures are often caused by insufficient amounts of fuel being drawn out of the fuel tank. Lean mixtures can cause bad results in many different diagnostic tests, including high HC, O_2, and NO_x readings on an exhaust analyzer and high firing lines on a scope.

When no fuel is delivered to the engine, the engine will not start. On carburetor and throttle-body-injected engines, it is easy to determine if fuel is being delivered. Simply look down the throttle body. If the surfaces are wet or you see fuel being sprayed while cranking the engine, fuel is there. With port injection, it is a little more difficult. Connect a fuel pressure gauge to the fuel line or rail and observe the fuel pressure while cranking the engine. Testing fuel pressure is described later in this chapter. However, if there is no fuel pressure while cranking, there is no fuel being delivered to the engine.

CAUTION!

Gasoline is very volatile and flammable. Never expose it to open flame or extreme heat. Disconnect the negative battery cable before doing anything that may release gas vapors. Use containers to catch gasoline and rags to wipe up any spills. Use a flashlight or an enclosed fluorescent tube or LED lamp designed for safe use around fuels. When working with a fuel system, always have a Class B fire extinguisher nearby.

There are many components in the fuel system. These can be grouped into two categories: fuel delivery and the fuel injection system. Diagnosis and basic service to the fuel delivery system are covered in this chapter. All tests given in this chapter assume that the fuel is good and not severely contaminated.

Contaminated Fuel

Obviously, water does not burn. Therefore, water in the fuel tank can cause a driveability problem. If there is water in the fuel, drain the fuel tank, replace the fuel filter, and refill the tank with fresh gasoline. Also, if any fuel has sat for a while, it becomes less volatile or stale. When fuel is stored a long time and exposed to air and heat, the fuel begins to break down and evaporate, leaving behind large molecules of carbon and gum. The separation of these materials from the fuel lowers its volatility. Also, the molecules can collect in and restrict the fuel lines and injectors. If the molecules are injected into the cylinders, they will not burn and they can cause abrasion in the cylinders. If a fuel smells sour and has been stored for quite some time, the fuel is stale. It may be usable; however, it should be mixed with as much fresh fuel as possible.

There are many products available to partially revitalize the fuel; however, if the fuel is so stale that an engine will not run on it, drain it and refill the tank with fresh gasoline. If fuel will be stored for a while, a fuel stabilizer should be added to it before it is stored.

GUIDELINES FOR SAFELY WORKING ON FUEL SYSTEMS

Many things need to be considered before working on a fuel system. Fuel in vapor and liquid form presents many potential hazards. Fuel plus heat presents even more! Also, dispose of all drained fuel according to local regulations.

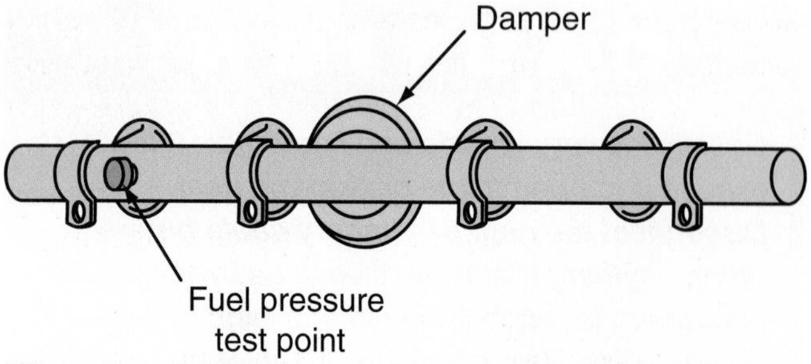

Figure 29-3 The typical location of the fuel pressure test port.

Before loosening or disconnecting fuel lines, all pressure in the system must be released. Fuel injection systems operate at high fuel pressures and are designed to hold most of that pressure when the engine is not running. This residual pressure allows for fast engine starting. When a fuel line that has pressurized fuel in it is loosened, the fuel will spray uncontrollably as soon as it can. The fuel can spray on something hot and cause a fire or spray into your eyes and cause a serious injury.

Most fuel injection fuel rails have a fuel pressure test port (often referred to as the Schrader valve) on the fuel rail (**Figure 29–3**). The pressure can be relieved at this port. Begin by disconnecting the negative battery cable. Then loosen the fuel tank filler cap to relieve any vapor pressure built up in the tank. Wrap a shop towel around the fuel pressure test port on the fuel rail and remove the dust cap from this valve. Connect a fuel pressure gauge to the fuel pressure test port on the fuel rail. Install a bleed hose onto the gauge and put the free end into an approved gasoline container. Then open the gauge bleed valve to relieve fuel pressure from the system into the gasoline container. Be sure that all the fuel in the bleed hose is drained into the gasoline container.

If the system does not have a test port, the pressure can be relieved by loosening the fuel tank filler cap to relieve any tank vapor pressure. Then remove the fuel pump fuse. Start and run the engine until the fuel in the lines is used up and the engine stops. Crank the engine with the starter for about 3 seconds to relieve any remaining fuel pressure.

Additional safety guidelines include:

- Always wear eye protection and follow all other safety rules to prevent injury to yourself or others when servicing fuel systems.
- When working on a fuel system in the engine compartment, disconnect the negative cable of the battery. An electrical spark may cause a fire or explosion.
- Slowly remove the fuel filler cap. If the cap is venting vapor or if you hear a hissing sound, wait until it stops before completely removing the cap.
- Do not smoke when working on or near any fuel-related component.
- Do not allow heat or flames to be near while working on or near the fuel system.
- Remove all electronic devices, such as cell phones, pagers, and audio equipment, from your clothing when working on or near the fuel system.
- Handle and store all fuels with the utmost caution.
- Clean all fuel spills immediately; spilled fuel may be ignited by hot components.
- If a fuel line or hose is damaged in any way, replace it.
- When disconnecting or reconnecting a fuel line or hose, make sure that the mating parts are totally clean.
- After disconnecting a fuel line or hose, plug both ends to prevent dirt from entering.
- When disconnecting and reconnecting a fuel line or hose, only use the tools designed for that connection. Using the wrong tool can cause a poor connection that can result in a fuel leak.
- Use fuel line pinch-off tools or rubber fuel lines to minimize fuel spillage when changing fuel filters and pumps.

FUEL TANKS

Fuel tanks include devices that prevent vapors from leaving the tank. For example, to contain vapors and allow for expansion, contraction, and overflow that result from changes in the temperature, the fuel tank has a separate air chamber dome at the top. All fuel tank designs provide some control of fuel height when the tank is filled. Frequently, this control is achieved by using vent lines with the filler tube or tank (**Figure 29–4**). These fuel height controls allow only 90% of the tank to be filled. The remaining 10% is for expansion during hot weather. Some fuel tanks have an overfill limiting valve to prevent overfilling of the tank.

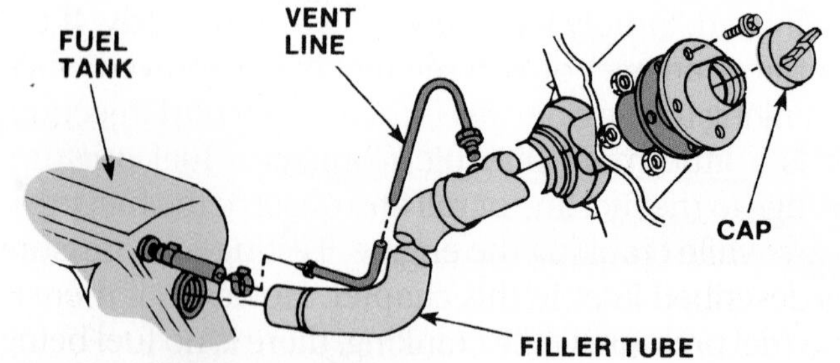

Figure 29-4 Vent lines within the fuel tank filler tube control the fuel level.

Fuel tanks are constructed of pressed corrosion-resistant steel, aluminum, or molded polyethylene plastic. Aluminum and molded plastic tanks are the most commonly used.

Most tanks have slosh baffles or surge plates to prevent fuel from splashing around inside the tank. In addition to slowing down fuel movement, the plates tend to keep the fuel pickup and sending unit for the fuel gauge immersed in the fuel during hard braking and acceleration. The plates or baffles also have holes or slots in them to permit the fuel to move from one end of the tank to the other. With few exceptions, the fuel tank is located in the rear of the vehicle.

A fuel tank has an inlet filler tube and cap. The location of the filler tube depends on the tank design. All current filler tubes have a built-in restrictor that prevents the entry of the larger leaded fuel delivery nozzle at gas pumps. The filler tube can be a rigid one-piece tube soldered to the tank or can be made of multiple pieces.

Some form of liquid vapor separator is incorporated into nearly every fuel tank. This separator stops liquid fuel or bubbles from reaching the vapor storage canister. It can be located inside the tank, on the tank, in the fuel vent lines, or near the fuel pump. Check the service manual for the exact location of the liquid vapor separator and the routing of the hoses to it.

Inside the fuel tank, there is also a sending unit that includes a pickup tube and float-operated fuel gauge sender unit. Most current fuel pumps are installed inside the tank and the pickup and fuel gauge sensor are part of that assembly **(Figure 29–5)**. A fuel strainer attaches to the pickup tube. The fuel strainer, sometimes referred to as a sock, is made of woven plastic. The strainer serves as a filter, stopping any rust or dirt that may be in the fuel from entering into the fuel pump. The fuel tank also has vent valves that are connected via hoses to a charcoal canister that collects HC emissions when the engine is running.

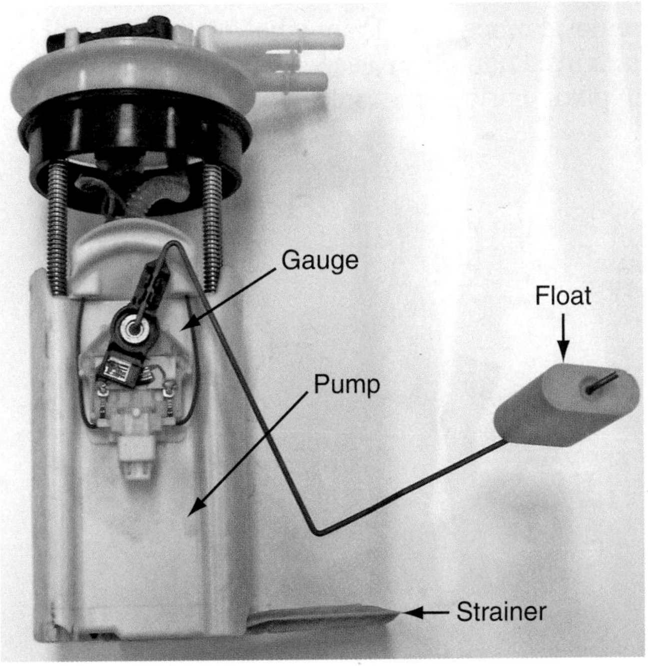

Figure 29-5 A combination electric fuel pump and sending unit.

Inspection

Fuel tanks should be inspected for leaks; road damage; corrosion and rust on metal tanks; loose, damaged, or defective seams; loose mounting bolts; and damaged mounting straps. Leaks in the fuel tank, lines, or filter may cause a gasoline odor in and around the vehicle, especially during low-speed driving and idling.

A weak seam, rust, or road damage can cause leaks in the metal fuel tank. The best method of permanently solving this problem is to replace the tank. Another method is to remove the tank and steam clean or boil it in a caustic solution to remove the gasoline residue. After this has been done, the leak can be soldered or brazed by an appropriately equipped specialty shop.

Holes in a plastic tank can sometimes be repaired by using a special tank repair kit. Be sure to follow manufacturer's instructions when doing the repair.

When a fuel tank is leaking dirty water or has water in it, the tank must be cleaned, repaired, or replaced.

SERVICE TIP

When a fuel tank must be removed, if possible, ask the customer to bring the vehicle to the shop with a minimal amount of fuel in the tank. ■

Fuel Tank Draining

WARNING!

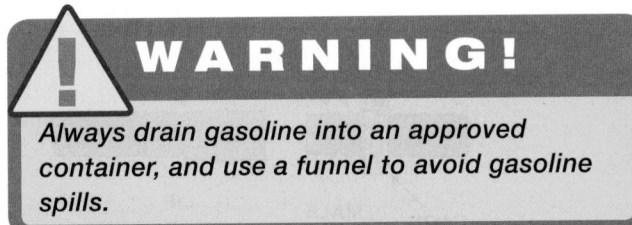

Always drain gasoline into an approved container, and use a funnel to avoid gasoline spills.

The fuel tank must be drained prior to tank removal. Begin by removing the negative cable from the battery. Then raise the vehicle on a hoist. Make sure you have an approved gasoline container and are prepared to catch all of the fuel before proceeding. If the tank has a drain bolt, remove it to drain the fuel. If the fuel tank does not have a drain bolt, locate the fuel tank drainpipe or filler pipe. Using the proper adapter, connect the intake hose from a hand-operated or air-operated pump to the pipe. Insert the discharge hose from the hand-operated or air-operated pump into an approved gasoline container, and operate the pump until all the fuel is removed from the tank.

CAUTION!

Abide by federal and state laws for the disposal of contaminated fuels. Be sure to wear eye protection when working under the vehicle.

Fuel Tank Service

In most cases, the fuel tank must be removed for servicing. The procedure for removing a fuel tank varies depending on the vehicle make and year. Always follow the procedure in the vehicle manufacturer's service manual. What follows is a typical procedure:

PROCEDURE

STEP 1 Disconnect the negative terminal from the battery.

STEP 2 Relieve the fuel system pressure and drain the fuel tank.

STEP 3 Raise the vehicle on a hoist or lift the vehicle with a floor jack and lower the chassis onto jack stands.

STEP 4 Use compressed air to blow dirt from the fuel line fittings and wiring connectors.

STEP 5 Remove the fuel tank wiring harness connector from the body harness connector.

STEP 6 Remove the ground wire retaining screw from the chassis if used.

STEP 7 Disconnect the fuel lines from the fuel tank. If these lines have quick-disconnect fittings, follow the manufacturer's recommended removal procedure in the service manual. Some quick-disconnect fittings are hand releasable, and others require the use of a special tool **(Figure 29–6)**.

STEP 8 Wipe the filler pipe and vent pipe hose connections with a shop towel, and then disconnect the hoses from the filler pipe and vent pipe to the fuel tank.

STEP 9 Unfasten the filler from the tank. If it is a rigid one-piece tube, remove the screws around the outside of the filler neck near the filler cap. If it is a three-piece unit, remove the neoprene hoses after the clamp has been loosened.

STEP 10 Loosen the bolts holding the fuel tank straps to the vehicle **(Figure 29–7)** until they are about two threads from the end.

STEP 11 Holding the tank securely against the underchassis with one hand, remove the strap bolts and lower the tank to the ground. When lowering the tank, make sure all wires and tubes are unhooked. Be careful as small amounts of fuel might still be in the tank.

WARNING!

Do not heat the bolts on the fuel tank straps in order to loosen them. The heat could ignite the fuel fumes.

To reinstall a repaired or new fuel tank, reverse the removal procedure. Be sure that all the rubber or felt tank insulators are in place. Then, with the tank straps in place, position the tank. Loosely fit the tank straps

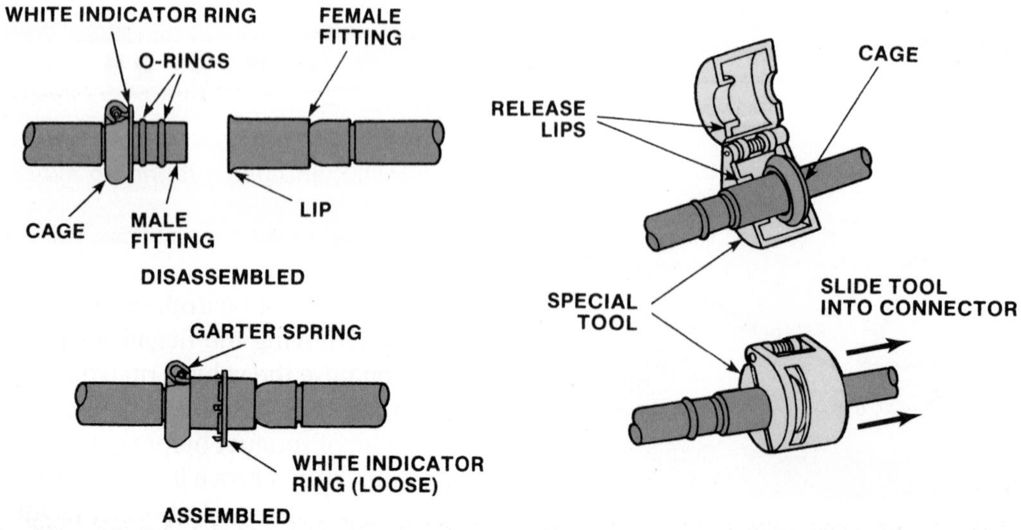

Figure 29-6 Some quick-connect fittings require the use of a special tool to separate them.

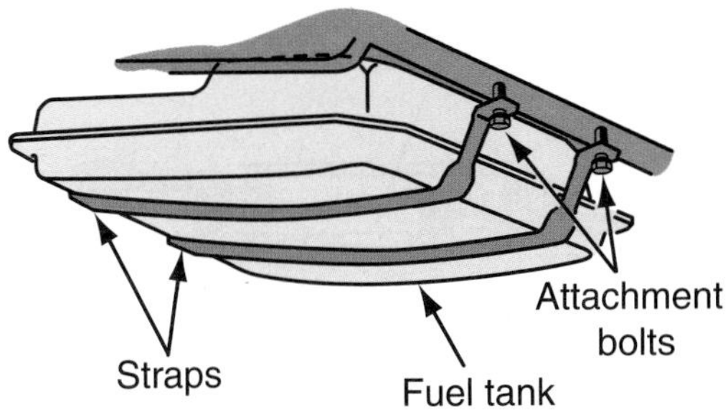

Figure 29-7 Front and rear fuel tank strap mounting bolts.

around the tank, but do not tighten them. Make sure that the hoses, wires, and vent tubes are connected properly **(Figure 29–8)**. Check the filler neck for alignment and for insertion into the tank. Tighten the strap bolts and secure the tank to the car. Install all of the tank accessories (vent line, sending unit wires, ground wire, and filler tube). Fill the tank with fuel and check it for leaks, especially around the filler neck and the pickup assembly. Reconnect the battery and check the fuel gauge for proper operation.

FILLER CAPS

Filler tube caps (commonly called gas or fuel caps) seal the fuel tank while allowing refilling of the tank. Filler caps are nonventing and have some type of pressure-vacuum relief valve arrangement **(Figure 29–9)**. Under normal conditions, the valve is closed. When extreme pressure or vacuum is present, the relief valve opens to prevent the tank from collapsing or balooning. Once the pressure or vacuum is relieved, the valve closes.

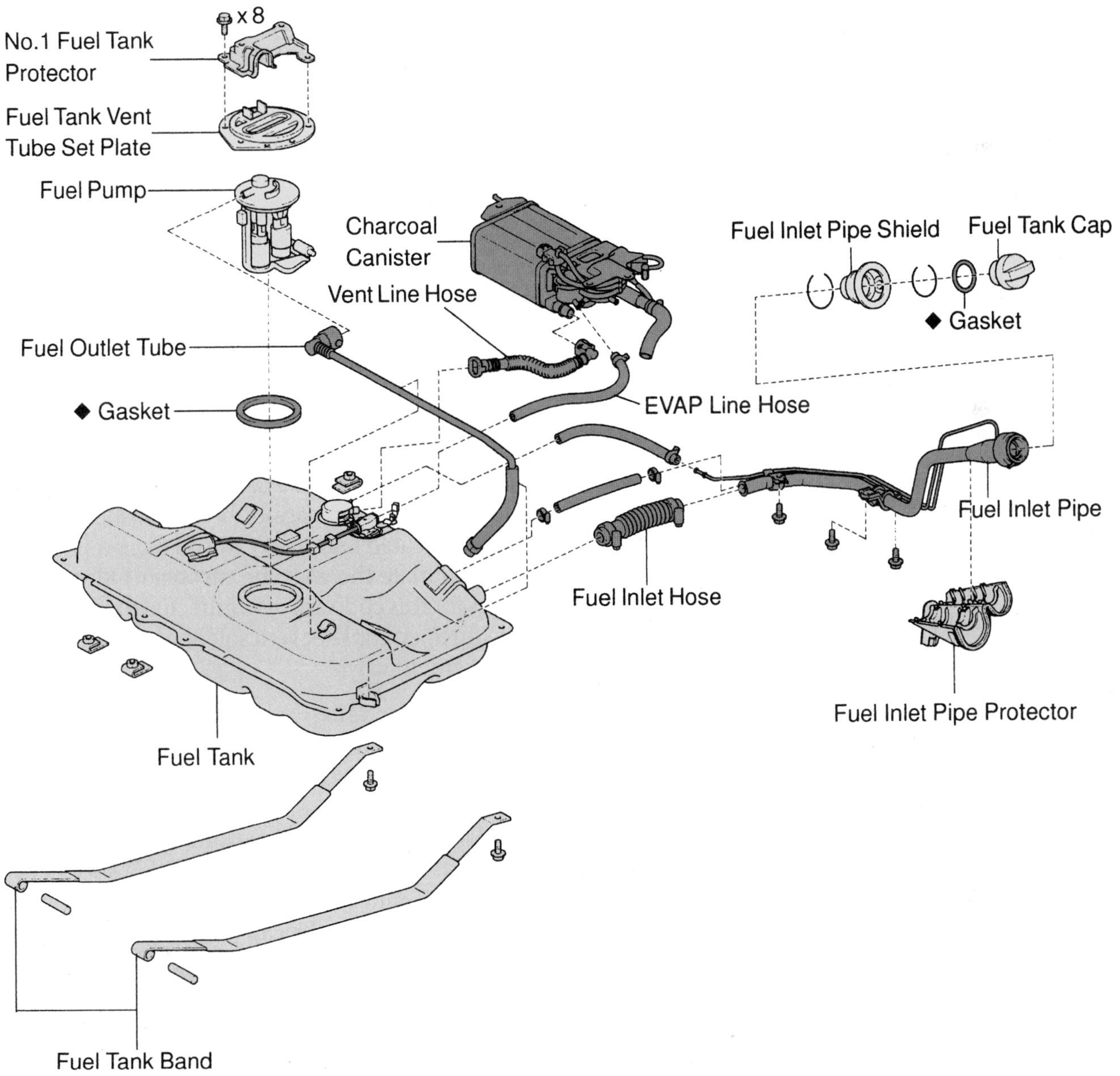

Figure 29-8 The hoses, wires, and tubes normally connected to a fuel tank. *Courtesy of Toyota Motor Sales, U.S.A., Inc.*

Figure 29-9 A cutaway of a pressure-vacuum gasoline filler cap. *Courtesy of Stant Corporation, www.stant.com*

Figure 29-10 The gas cap for current vehicles is tethered to the vehicle and threaded into the filler tube.

The filler cap of late-model vehicles is tethered to the vehicle **(Figure 29–10)**. The cap is threaded into the upper end of the filler pipe. The threaded area on the cap is designed to allow any remaining tank pressure to escape during cap removal. The cap and filler neck are designed to prevent overtightening. To install the cap, turn it clockwise until a clicking noise is heard. This indicates that the cap is properly tightened and fully seated. A fuel filler cap that is not fully seated may cause a malfunction in the emission system.

WARNING!

When a tank filler cap is replaced, the replacement cap must be exactly the same as the original cap or a malfunction may occur in the filling and venting system, resulting in higher emission levels and the escape of dangerous HCs.

OBD-II Monitor

Late-model vehicles that meet enhanced evaporative requirements have a vacuum-based evaporative system integrity check. If the gas cap is loose or missing, the ECM/PCM will detect an evaporative system leak and will illuminate a warning message. On some vehicles the "Check fuel cap" message will appear each time the engine is started until the system turns the message off. The message may turn off after the cap is replaced or tightened until at least one click is heard. If the message does not turn off, there be a leak in the system or the circuit for the message is faulty.

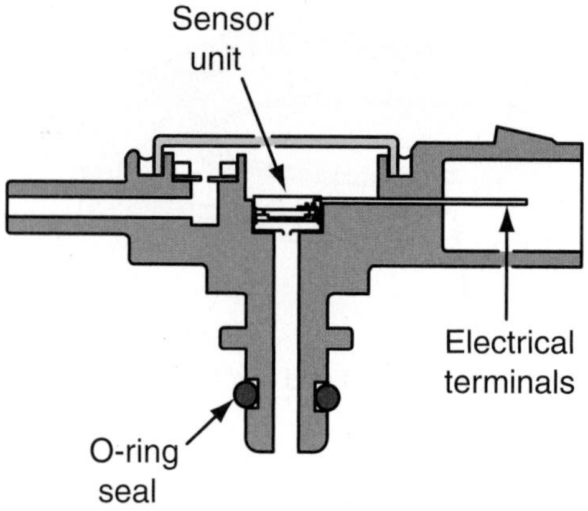

Figure 29-11 A fuel tank pressure (FTP) sensor.

A fuel tank pressure (FTP) sensor is a transducer that converts the absolute pressure in the fuel tank into an input for the PCM **(Figure 29–11)**. The integrity check is done by creating a vacuum in the tank and measuring how well it holds the vacuum. If the gas cap is off or loose, the tank will not hold a vacuum. Before a vacuum is formed in the tank, the canister vent solenoid is closed to seal the entire evaporative system. Then the vapor management valve creates a slight negative pressure in the tank. If the desired amount of vacuum cannot be established, a system leak is indicated and the PCM will store a P0455 DTC and illuminate the warning message. Other possible causes for this code are disconnected or kinked vapor lines, an open canister vent solenoid, or a closed vapor management valve.

Fuel Cap Testing

A gas cap should be checked whenever the PCM detects a leak in the evaporative system and the cap is securely fastened. Also, some states, such as California, mandate a gas cap check as part of the annual emissions tests. A gas cap is checked with a special tester. The cap is connected to the tester by an adapter specifically designed for the cap **(Figure 29–12)**. The tester then applies pressure to the cap and monitors its ability to hold the pressure. The readout on the tester simply says PASS or FAIL. A cap that has failed should be replaced. However, it is important that the correct adapter was used for the cap. If the wrong adapter was used, the cap will fail the test even if it is good.

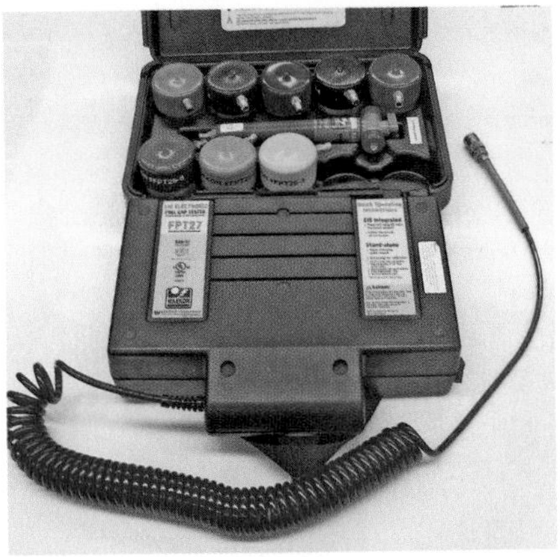

Figure 29-12 A fuel cap tester with various adapters.

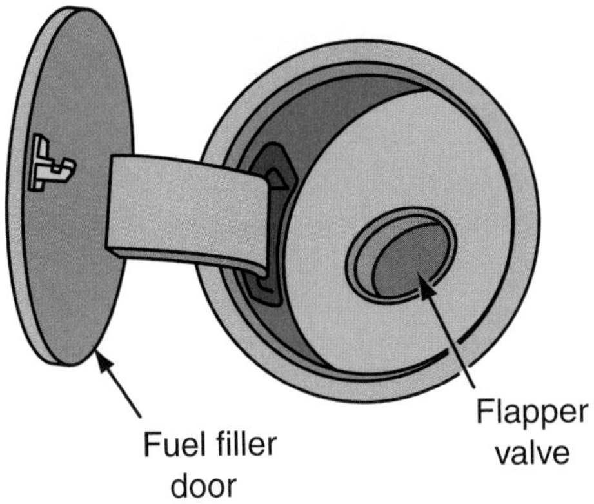

Figure 29-13 A capless fuel filler tube with its flapper valve.

Capless Fuel System

Ford Motor Company's Ford GT was the first modern production car to meet all emissions standards without a gas cap **(Figure 29–13)**. This technology is used on many 2008 and newer Ford vehicles. It is a very simple design. A spring-loaded flapper valve is positioned at the opening for the filler neck. This valve tightly seals the tank until a fuel nozzle is inserted into the opening. The nozzle opens the valve and allows for the refilling of fuel. As soon as the nozzle is removed, the valve is shut by the springs. A capless fuel system reduces the time the fuel vapors can escape during refueling. It also makes it more convenient for the consumer, because there is no cap to tighten. The filler door, which is outside the filler tube, helps to seal in the fuel and fuel vapors.

FUEL LINES AND FITTINGS

The fuel lines **(Figure 29–14)** carry fuel from the tank to the fuel filter and fuel injection assembly. Fuel lines can be made of either metal tubing or flexible nylon or synthetic rubber hose. The latter must be able to resist gasoline. The hoses must also be nonpermeable, so gas and gas vapors cannot evaporate through the hose. Ordinary rubber hose, such as that used for vacuum lines, deteriorates when exposed to gasoline. Only hoses made for fuel systems should be used. Similarly, vapor vent lines must be made of material that resists attack by fuel vapors.

Fuel supply lines from the tank to the injectors are routed to follow the frame along the underchassis of the vehicle. Generally, rigid lines are used extending from near the tank to a point near the fuel pump and fuel filter. To prevent ruptures during a rear impact, the gaps between the frame and tank or fuel pump are joined by short lengths of flexible hose.

Many fuel tanks have vent hoses to allow air in the fuel tank to escape when the tank is being filled with fuel. Vent hoses are usually installed alongside the filler neck. Replacement vent hoses are usually marked with the designation **EVAP** to indicate their intended use. The inside diameter of a fuel delivery hose is generally larger (5/16 to 3/8 inch [7.94 to 9.35 mm]) than that of a fuel return hos (1/4 inch [6.35 mm]).

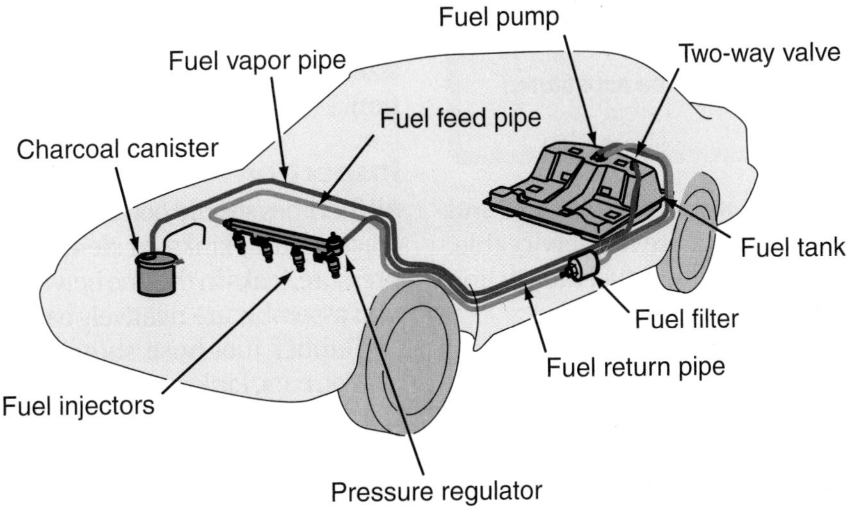

Figure 29-14 A typical layout of the fuel lines on a late-model car.

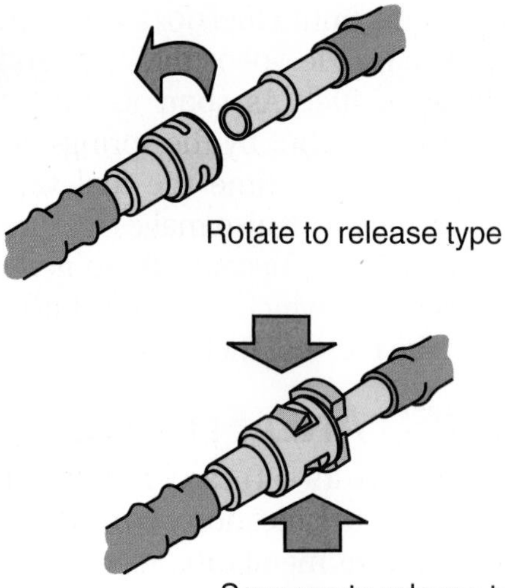

Figure 29-15 Quick-disconnect hand releasable fuel line fittings.

To control the rate of vapor flow from the fuel tank to the vapor storage tank, a plastic or metal restrictor may be placed in either the end of the vent pipe or in the vapor-vent hose itself. When the latter hose must be replaced, the restrictor must be removed from the old vent hose and installed in the new one.

Fittings

Sections of fuel line are assembled together by fittings. Some of these fittings are a threaded-type fitting, while most are a quick-release design. Many fuel lines have quick-disconnect fittings with a unique female socket and a compatible male connector. These quick-disconnect fittings are sealed by an O-ring inside the female connector. Some of these quick-disconnect fittings have hand-releasable locking tabs **(Figure 29–15)**, while others require a special tool to release the fitting **(Figure 29–16)**.

WARNING!

Other types of O-rings should not be substituted for a Viton O-ring.

The interior components, such as the O-rings and spacers, of quick-connect fittings are not serviceable. If the fitting is damaged, the complete fuel tube or line must be replaced.

Some fuel lines have threaded fittings with an O-ring seal to prevent fuel leaks. These O-ring seals are usually made from Viton, which resists deterioration from gasoline. On some other fuel lines, the fuel hose is clamped to the steel line and the hose and clamp must be properly positioned on the steel line **(Figure 29–17)**.

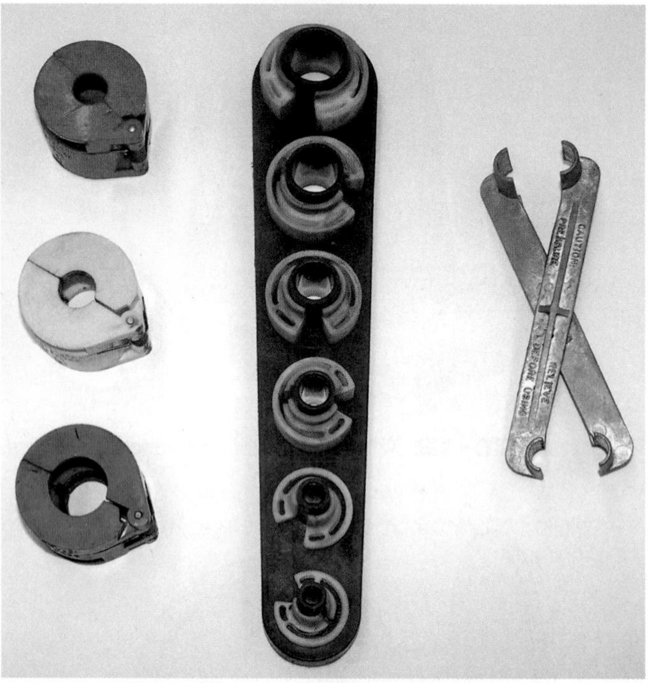

Figure 29-16 An assortment of quick-disconnect tools.

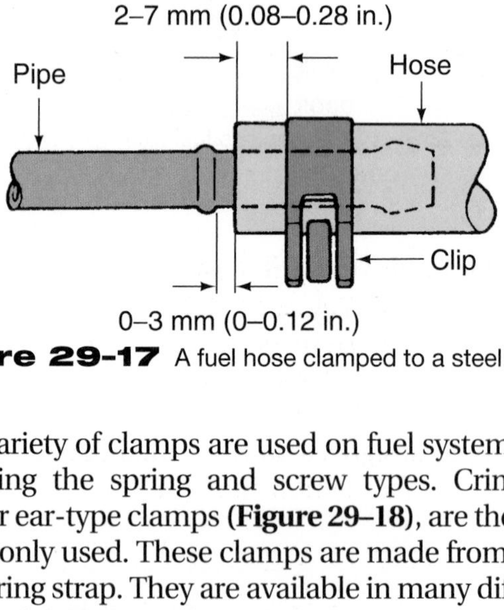

Figure 29-17 A fuel hose clamped to a steel tubing.

A variety of clamps are used on fuel system lines, including the spring and screw types. Crimp, or Oetiker ear-type clamps **(Figure 29–18)**, are the most commonly used. These clamps are made from a single, spring strap. They are available in many different sizes and designs, each made for a particular connection. They are tightened with a special crimping tool.

Inspection

All fuel lines should occasionally be inspected for holes, cracks, leaks, kinks, or dents. Since the fuel is under pressure, leaks in the line between the pump and injection assembly are relatively easy to recognize.

Rubber fuel hose should be inspected for leaks, cracks, cuts, kinks, oil soaking, and soft spots or deterioration. If any of these conditions is found, the fuel hose should be replaced. When rubber fuel hose is installed, the hose should be installed to the proper depth on the metal fitting or line.

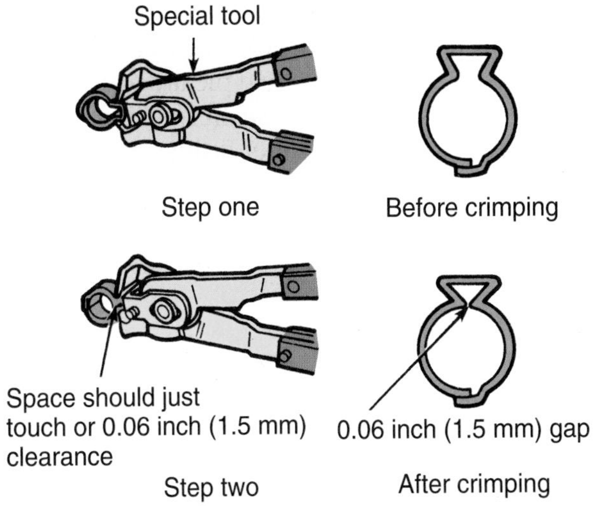

Figure 29-18 A special tool is required to tighten crimp clamps.

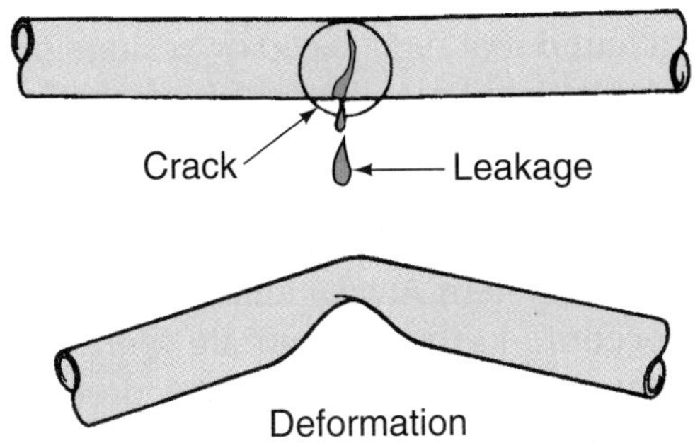

Figure 29-19 Steel tubing should be inspected for leaks, kinks, and deformation. *Courtesy of Toyota Motor Sales, U.S.A., Inc.*

Steel tubing should be inspected for leaks, kinks, and deformation (**Figure 29–19**). Tubing should also be checked for loose connections and proper clamping to the chassis. If the tubing's threaded connections are loose, they must be tightened to the specified torque. Some threaded fuel line fittings contain an O-ring. If the fitting is removed, the O-ring should be replaced.

Nylon fuel pipes should be inspected for leaks, nicks, scratches and cuts, kinks, melting, and loose fittings. If these fuel pipes are damaged in any way, they must be replaced. Nylon fuel pipes must be secured to the chassis at regular intervals to prevent fuel pipe wear and vibration.

WARNING!

Always cover a nylon fuel pipe with a wet shop towel before using a torch or other source of heat near the line. Failure to observe this precaution may result in fuel leaks, personal injury, and property damage.

WARNING!

If a vehicle has nylon fuel pipes, do not expose the vehicle to temperatures above 194°F (90°C) for any extended period to avoid damage to the pipes.

Line Replacement

When a damaged fuel line is found, replace it with one of similar construction—steel tubing with steel, and flexible tubing with nylon or synthetic rubber. When installing flexible tubing, always use new clamps. The old ones lose some of their tension when they are removed and do not provide an effective seal when used on the new line.

CAUTION!

Do not substitute aluminum or copper tubing for steel tubing. Never use a hose within 4 inches of any hot engine or exhaust system component.

Any damaged or leaking fuel line must be replaced. To fabricate a new fuel line, select the correct tube and fitting dimension and start with a length that is slightly longer than the old line. With the old line as a reference, use a tubing bender to form the same bends in the new line as those that exist in the old. Although steel tubing can be bent by hand to obtain a gentle curve, any attempt to bend a tight curve by hand usually kinks the tubing. To avoid kinking, always use a bending tool like those shown in **Figure 29–20**.

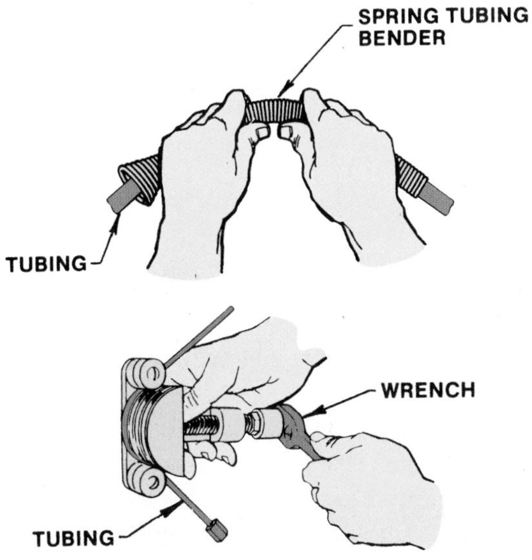

Figure 29-20 Two types of bending tools for steel tubing.

Nylon fuel pipes provide a certain amount of flexibility and can be formed around gradual curves under the vehicle. Do not force a nylon fuel pipe into a sharp bend, because doing so may kink the pipe and restrict the flow of fuel. When nylon fuel pipes are exposed to gasoline, they may become stiffer, making them more susceptible to kinking. Be careful not to nick or scratch nylon fuel pipes.

WARNING!

When connecting threaded fittings, make sure that the threads are aligned before tightening the connection. If the fittings are cross-threaded, fuel leaks will inevitably result. Also, always tighten fuel line fittings to the specified torque. If a fitting leaks, check the O-rings and seals.

FUEL FILTERS

Automobiles and light trucks usually have an in-tank strainer and a gasoline filter. The strainer, located in the gasoline tank, is made of a finely woven fabric. The purpose of this strainer is to prevent large contaminant particles from entering the fuel system where they could cause excessive fuel pump wear or plug fuel metering devices. It also helps to prevent passage of any water that might be present in the tank. Servicing of the fuel tank strainer is seldom required; however, if the gasoline usually used contains large amounts of alcohol, the strainer will need to be replaced often.

A fuel filter is connected in the fuel line between the fuel tank and the engine. Many of these filters are mounted under the vehicle **(Figure 29–21)**, and others are mounted in the engine compartment. Most fuel filters contain a pleated paper element mounted in the filter housing, which may be made from metal or plastic. Paper filter elements are efficient at removing and trapping small particles, as well as large-size contaminants. Fuel filters are typically contained in a metal case, but some have a plastic housing. On many fuel filters, the inlet and outlet fittings are identified, and the filter must be installed properly. An arrow on some filter housings indicates the direction of fuel flow through the filter.

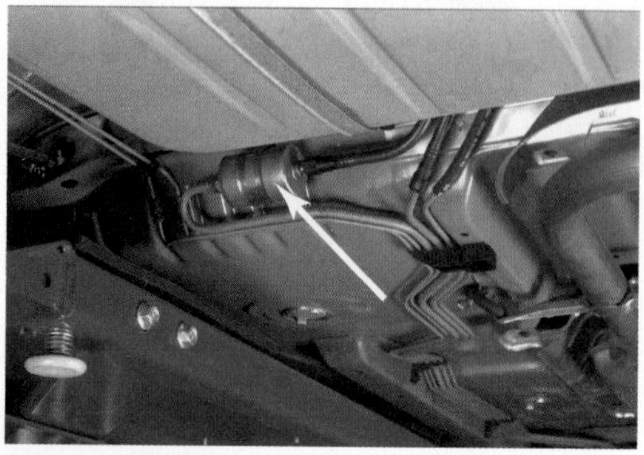

Figure 29-21 An inline fuel filter mounted under a vehicle.

Servicing Filters

Fuel filters **(Figure 29–22)** and elements are serviced by replacement only. Some vehicle manufacturers recommend fuel filter replacement at 30,000 miles (48,000 km). Always replace the fuel filter at the vehicle manufacturer's recommended mileage. If dirty or contaminated fuel is placed in the fuel tank, the filter may require replacing before the recommended mileage. A plugged fuel filter may cause the engine to surge and cut out at high speed or hesitate on acceleration. A restricted fuel filter causes low fuel pump volume.

The fuel filter replacement procedure varies depending on the make and year of the vehicle and the type of fuel system. Always follow the filter replacement procedure in the appropriate service manual. Photo Sequence 29 shows a typical procedure for relieving fuel pressure and removing a fuel filter.

To install a new filter, begin by wiping the male tube ends of the new filter with a clean shop towel. Apply a few drops of clean engine oil to the male tube ends on the filter. Check the quick connectors to be sure the large collar on each connector has rotated back to the original position. The springs must be visible on the inside diameter of each quick connector. Then install the filter, in the proper direction, and leave the mounting bolt slightly loose. Install the outlet connector onto the filter outlet tube and press the

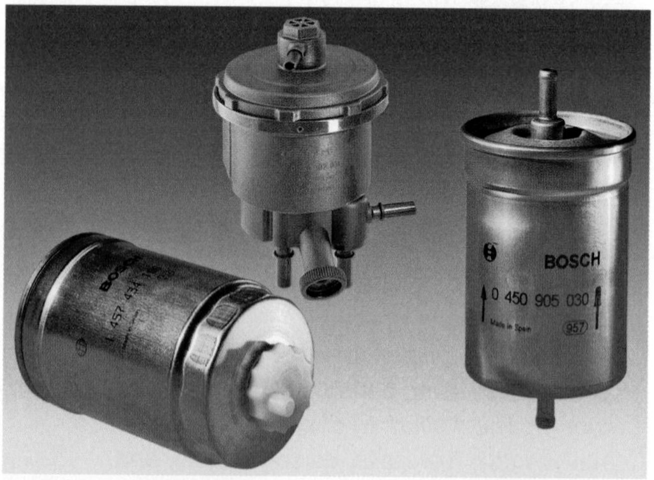

Figure 29-22 An assortment of fuel filters.
Courtesy of Robert Bosch GmbH, www.bosch-presse.de

P29-1 *Disconnect the negative cable at the battery.*

P29-2 *Loosen the fuel tank filler cap to relieve any fuel tank vapor pressure.*

P29-3 *Wrap a shop towel around the Schrader valve on the fuel rail and remove the dust cap from the valve.*

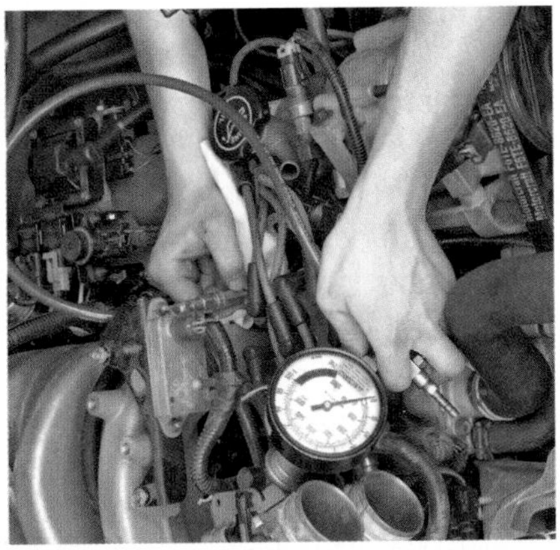

P29-4 *Connect the fuel pressure gauge to the Schrader valve.*

P29-5 *Install the free end of the gauge bleed hose into an approved gasoline container, and open the gauge bleed valve to relieve the fuel pressure.*

P29-6 *Place the vehicle on the hoist and position the lift arms according to manufacturer's recommendations. Then raise the vehicle.*

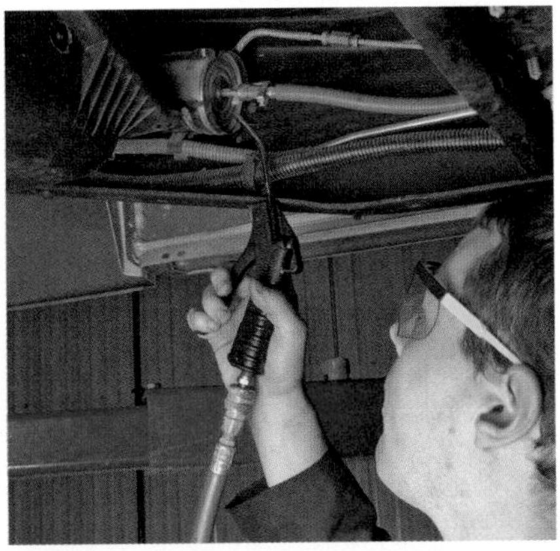

P29-7 *Flush the fuel filter line connectors with water, and use compressed air to blow debris off and away from the connectors.*

P29-8 *Follow the recommended procedures for disconnecting the fuel inlet connector.*

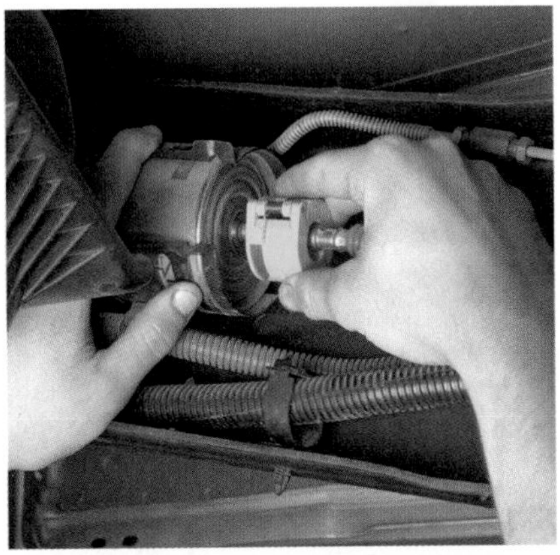

P29-9 *Follow the recommended procedures for disconnecting the fuel outlet connector. Then remove the fuel filter.*

connector firmly in place until the spring snaps into position. Grasp the fuel line and try to pull this line from the filter to be sure the quick connector is locked in place. Then do the same with the inlet connector. Now tighten the filter-retaining bolt to the specified torque. Once everything is connected, lower the vehicle, start the engine, and check for fuel leaks at the filter.

FUEL PUMPS

A fuel pump draws fuel from the fuel tank and pushes it through fuel lines to the engine's injection system. All current vehicles use an electric fuel pump. Older vehicles with carburetors had mechanical pumps; these will not be discussed.

An electric fuel pump can be located inside or outside the fuel tank. Diaphragm, plunger, or bellows types are found on non-EFI systems and are referred to as demand pumps. When the ignition is turned on, the pump starts to run and shuts off automatically when the fuel line is pressurized. When there is a demand for more fuel, the pump turns on again. **Figure 29–23** shows a typical wiring diagram for an electric fuel pump.

An in-tank electric pump is usually the rotary type. Some vehicles have an in-tank pump and a second pump mounted under the vehicle. An in-tank fuel pump has a small DC electric motor with an impeller mounted on the end of the motor's shaft. A pump cover, with inlet and discharge ports, is mounted over the impeller. When the armature and impeller rotate, fuel is moved from the tank to the inlet port, and the impeller forces the fuel around the impeller cover and out the discharge port (**Figure 29–24**).

Fuel moves from the discharge port through the inside of the motor and out the check valve and outlet connection, which is connected via the fuel line to the fuel filter and underhood fuel system components. A pressure relief valve near the check valve opens if the fuel supply line is restricted and pump pressure becomes very high. When the relief valve opens, fuel is returned through this valve to the pump inlet. Each time the engine is shut off, the check valve prevents fuel from draining out of the fuel system and back into the fuel tank.

Fuel pumps are mounted inside the tank to reduce noise, keep them cool, and keep the entire fuel line pressurized to prevent premature fuel evaporation. Although it is dangerous to have a spark near gasoline and there is a great potential for sparks between an electric motor's armature and brushes, the in-tank fuel pump is safe because there is no oxygen to support combustion in the tank.

Diesel Engines Fuel injection is used on all diesel engines. Older diesel engines had a distributor-type injection pump driven and regulated by the engine. The pump supplied fuel to injectors according to the engine's firing order. Newer diesel engines are equipped with common rail or direct injection (DI) systems. In these systems, an engine-driven fuel pump delivers fuel to the injectors at a very high pressure, about 26,000 psi (180 Mpa or 1,800 bar). In a common rail system, the computer controls the individual injectors that are fed fuel by the common rail.

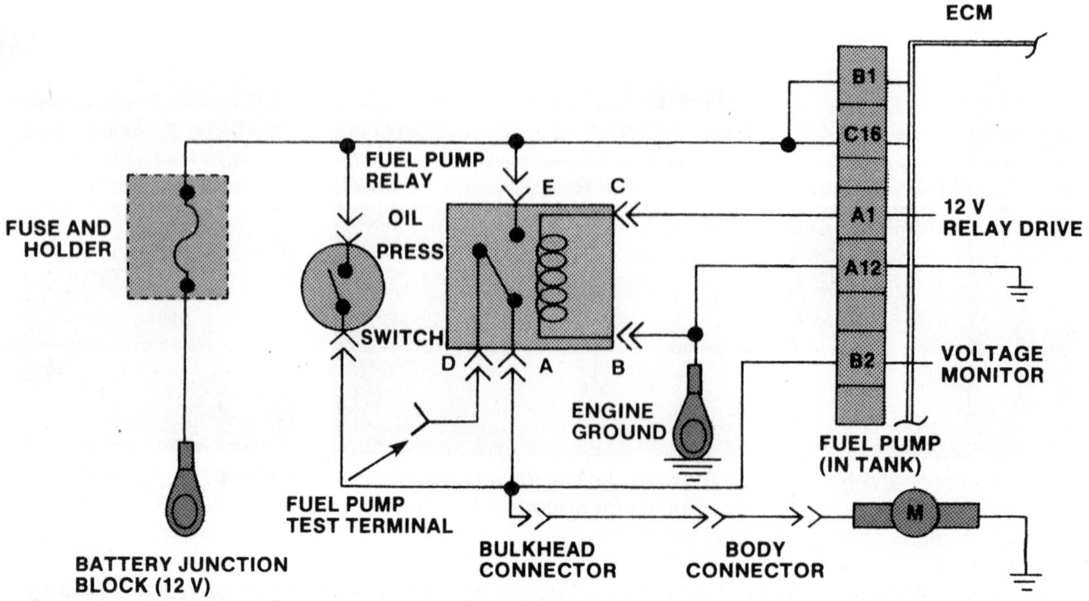

Figure 29-23 A typical wiring diagram for an electric fuel pump.

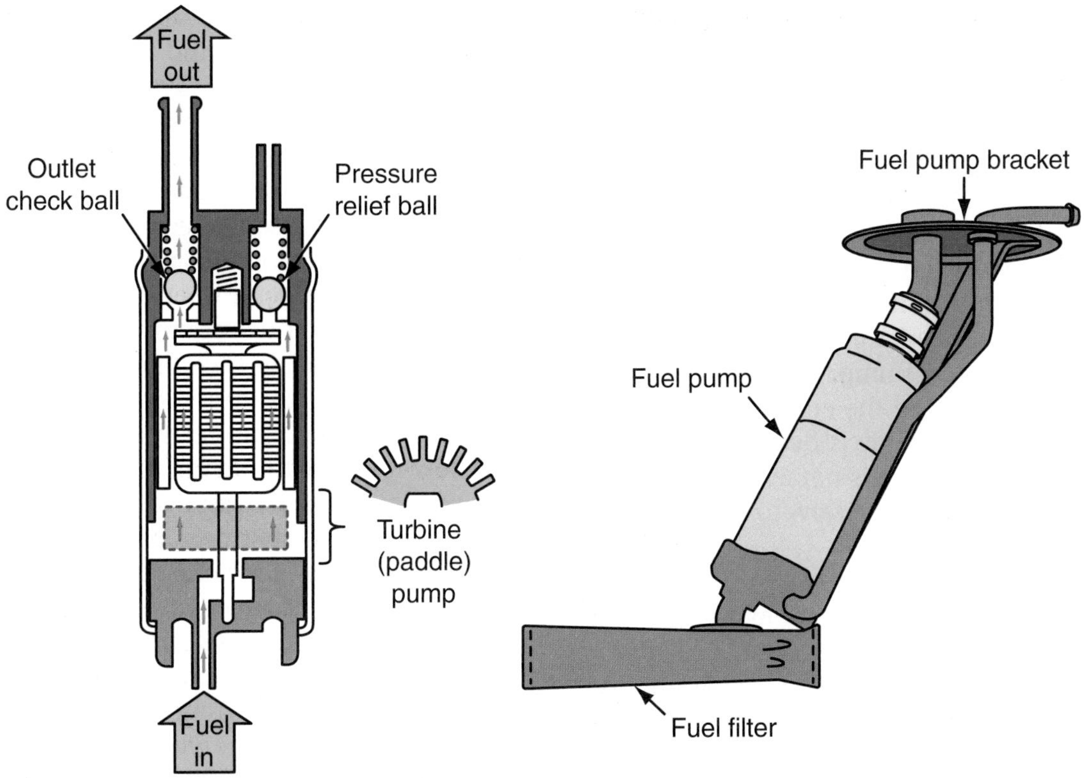

Figure 29-24 An electric fuel pump.

Fuel Pump Circuits

Electric fuel pump circuits vary depending on the vehicle make and year. All fuel pumps on late-model vehicles are controlled by the PCM according to inputs received on the CAN bus and other specific sensors. Fuel pumps are controlled by the PCM or through a designated electronic control unit tied into the PCM **(Figure 29–25)**. In some cases, the output of the pump is controlled by the PCM through pulse-width modulation. In these systems, the pump's output during closed loop is monitored by a fuel rail pressure sensor.

In a GM fuel pump circuit (other manufacturers use similar systems), the PCM supplies voltage to the winding of the fuel pump relay when the ignition switch is turned on. This action closes the relay contacts and voltage is supplied through the relay to the fuel pump. The fuel pump remains on while the engine is cranking or running. If the ignition switch is on for 2 seconds and the engine is not cranked, the PCM turns off the voltage to the fuel pump relay to stop the pump.

The PCM also shuts off the fuel pump when the following occur:

- The vehicle experiences long, high-speed, closed throttle coast down. Fuel is shut off to prevent damage to the catalytic converters, reduce emissions, and increase the effects of engine braking.

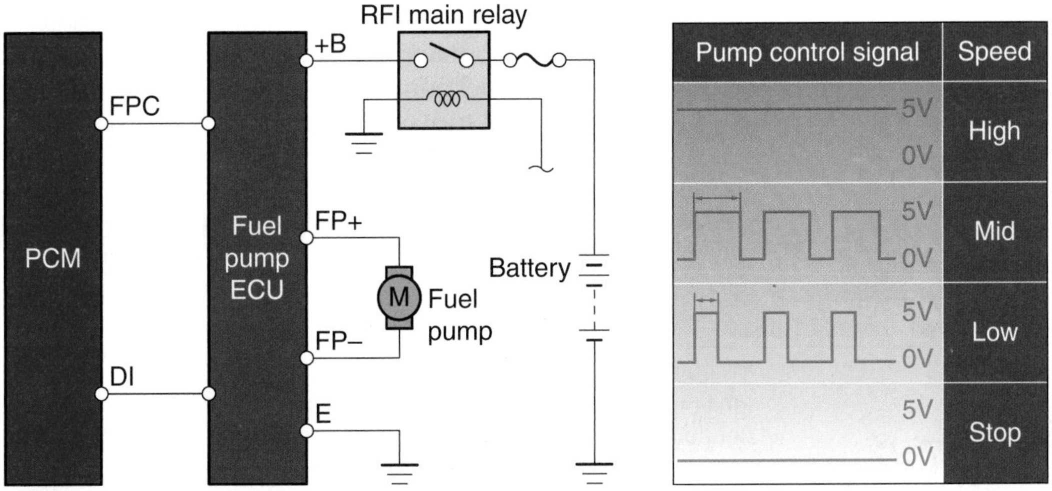

Figure 29-25 The basic circuit for a multiple-speed pulse-modulated fuel pump. *Courtesy of Toyota Motor Sales, U.S.A., Inc.*

- The engine speed exceeds a predetermined limit.
- The speed of the vehicle exceeds the speed rating of the tires.
- The vehicle has been in a collision.
- The air bag has deployed.
- A fuel line has ruptured.

An oil pressure switch is connected parallel to the fuel pump relay points. If the relay becomes defective, voltage is supplied through the oil pressure switch points to the fuel pump. This action keeps the fuel pump operating and the engine running, even though the fuel pump relay is defective. When the engine is cold, oil pressure is not available immediately, and the engine may be slow to start if the fuel pump relay is defective.

The fuel pump relay in Chrysler EFI systems is referred to as an automatic shutdown (ASD) relay. With the ignition switch turned on, the PCM grounds the windings of the relay and the relay points close. These supply voltage to the fuel pump, positive primary coil terminal, oxygen sensor heater, and the fuel injectors in some systems (**Figure 29–26**).

On Chrysler products with a power module and logic module, the engine must be cranked before the power module grounds the ASD relay winding. The later model PCM grounds the ASD relay winding when the ignition switch is turned on, and the relay remains closed while the engine is cranking or running. If the ignition switch is on for one-half second and the engine is not cranked, the PCM opens the circuit from the ASD relay winding to ground. Under this condition, the ASD relay points open and voltage is no longer supplied to the fuel pump, positive primary coil terminal, injectors, and oxygen sensor heater.

Later model Chrysler fuel pump circuits have a separate ASD relay and a fuel pump relay. In these circuits, the fuel pump relay supplies voltage to the fuel pump, and the ASD relay powers the positive primary coil terminal, injectors, and oxygen sensor heater. The ASD relay and the fuel pump relay operate the same as the previous ASD relay. The PCM grounds both relay windings through the same wire.

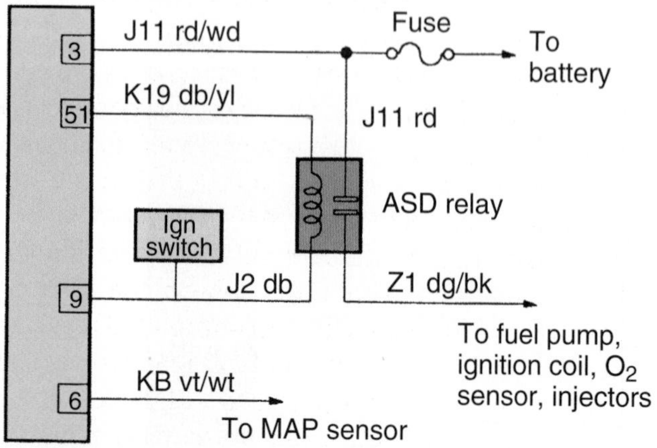

Figure 29-26 A Chrysler fuel pump circuit with an ASD relay. *Courtesy of Chrysler LLC*

On many Toyota vehicles, the fuel pump relay is called a circuit opening relay. This relay has dual windings and it is mounted on the firewall. One of the windings of the circuit opening relay is connected between the starter relay points and ground, and the second relay winding is connected from the battery positive terminal to the PCM. When the engine is cranking and the starter relay points are closed, current flows through the starter relay points and the circuit opening relay winding to ground. This current flow creates a magnetic field around the circuit opening relay winding that closes the relay points. When these points close, current flows through the points to the fuel pump (**Figure 29–27**).

Once the engine starts, the starter relay is no longer energized, and current stops flowing through these relay points and the winding of the circuit opening relay. However, the PCM grounds the other winding of the circuit opening relay when the engine starts. This action keeps the relay points closed while the engine is running.

Rollover Protection

Electric fuel pump circuits include some sort of rollover protection. Typically this includes the installation of an **inertia switch** that shuts off the fuel pump if the vehicle in involved in a collision or rolls over. A typical inertia switch (**Figure 29–28**) consists of a permanent magnet, a steel ball inside a conical ramp, a target plate, and a set of electrical contacts. The magnet holds the steel ball at the bottom of the ramp. In the event of a collision, the inertia of the ball causes it to break away from the magnetic field and roll up the ramp. When it strikes the target plate, the electrical contacts open and the circuit between the PCM and fuel pump control unit opens, causing the fuel pump to turn off. The switch has a reset button that must be depressed for at least 1 second before the pump will operate again.

If the ignition switch is on and the fuel line is broken during an accident, the PCM turns off the fuel pump relay to prevent the fuel pump from pumping gasoline from the ruptured fuel line.

Passive Restraint Systems Most passive restraint systems will send a signal to the PCM when an air bag is deployed. The PCM, in turn, shuts down the power to the fuel pump. In most cases, it is the center air bag sensor assembly that sends the deployment signal to

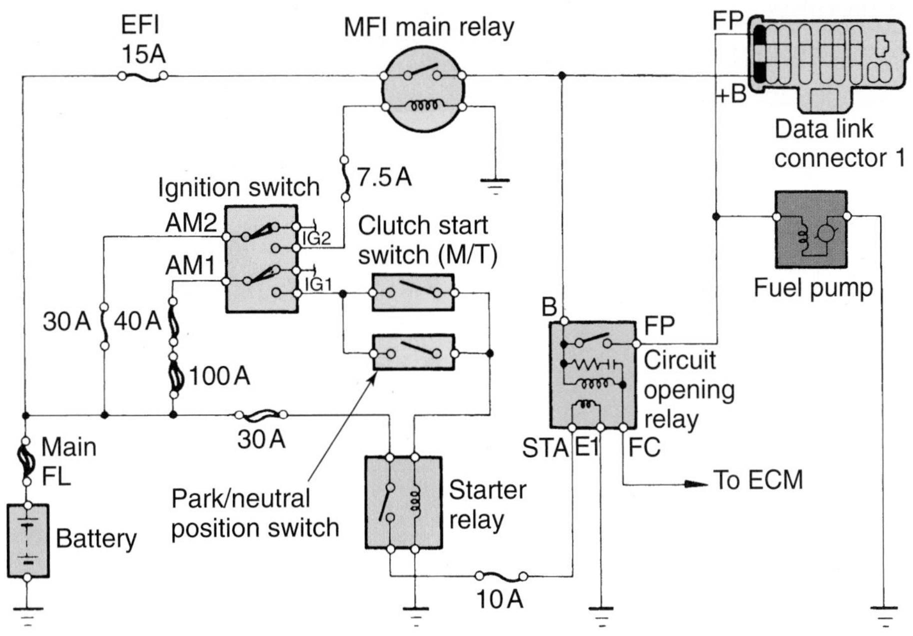

Figure 29-27 A wiring diagram for the circuit opening relay.

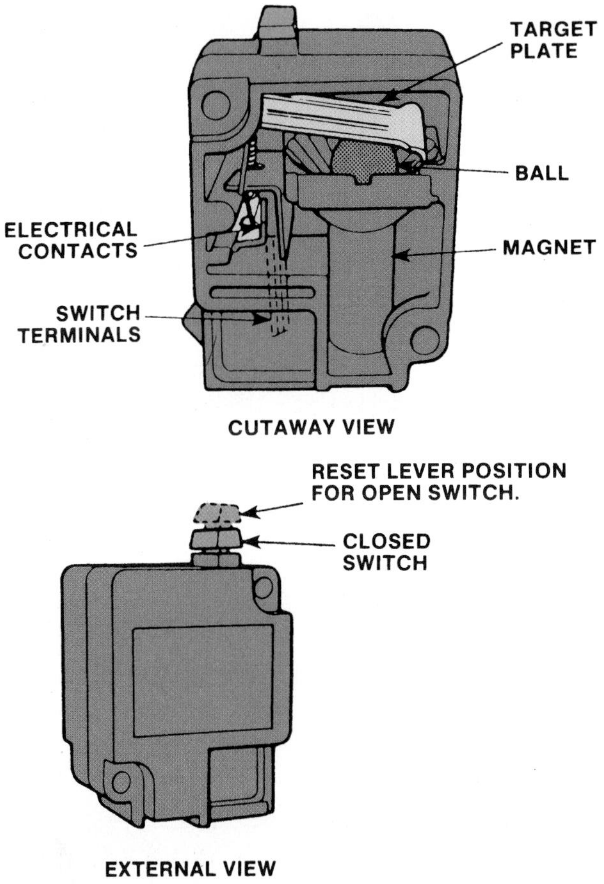

Figure 29-28 Details of a Ford inertia switch.

the ECM through the CAN. This air bag sensor assembly contains a deceleration sensor, a safing sensor, a drive circuit, a diagnosis circuit, and an ignition control circuit. The main sensor for the deployment of an air bag and the de-powering of the fuel pump is the deceleration sensor.

The circuit constantly monitors its own operation and readiness; if it detects a malfunction, the SRS warning light will illuminate and a DTC will be stored.

Troubleshooting

The fuel pressure test is the commonly used test for the fuel pump and related parts. Before connecting this test, carefully inspect the system for leaks and repair them before continuing. Then relieve the pressure in the system. When doing this, make sure to collect all spilled fuel. Scan tools can be used to shut down the fuel pump on many systems. Also, the PCM may set a DTC as a result of shutting off the fuel pump, make sure to clear that code after testing.

Fuel pressure is read with a fuel pressure gauge. The proper procedure for testing fuel pump pressure is shown in Photo Sequence 30. These photos outline the steps to follow while performing the test on an engine with fuel injection. To conduct this test on specific fuel injection systems, refer to the service manual for instructions. Most domestic systems have a **Schrader valve** on the fuel rail, which can be used to connect the fuel pressure gauge. If the system does not have a Schrader valve, a tee should be installed in the fuel supply line to connect the gauge **(Figure 29–29)**.

On some engines, the fuel rail is fitted with a fuel pulsation damper. The point where the damper attaches to the fuel rail is the recommended place for connecting the pressure gauge. To connect the gauge, place a rag over the damper unit and loosen it one turn with a wrench. After all pressure is released, remove the damper unit and connect the gauge into the damper's fitting **(Figure 29–30)**.

PHOTO SEQUENCE 30

Checking Fuel Pressure on a PFI System

P30-1 *Many problems on today's cars can be caused by incorrect fuel pressure. Therefore, checking fuel pressure is an important step in diagnosing driveability problems.*

P30-2 *Prior to testing the fuel pump, a careful visual inspection of the injectors, fuel rail, and fuel lines and hoses is necessary. Any sign of a fuel leak should be noted and the cause corrected immediately.*

P30-3 *The supply line into the fuel rail is a likely point of leakage. Check the area around the fitting to make sure no leaks have occurred.*

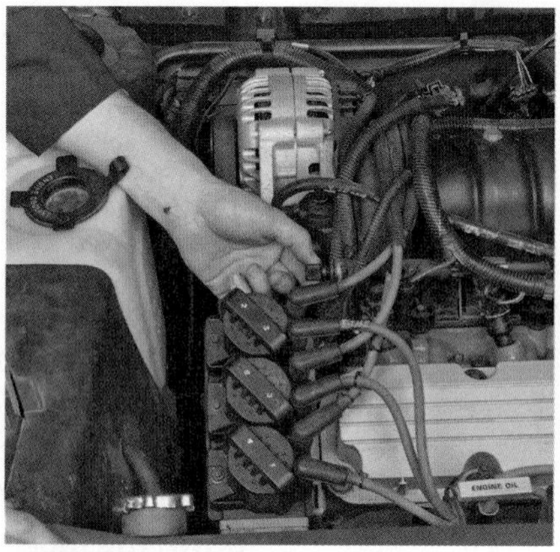

P30-4 *Most fuel rails are equipped with a test fitting that can be used to relieve pressure and to test pressure.*

P30-5 *To test fuel pressure, connect the appropriate pressure gauge to the fuel rail test fitting (Schrader valve).*

P30-6 *Connect a hand-held vacuum pump to the fuel pressure regulator.*

P30-7 *Turn the ignition switch to the run position and observe the fuel pressure gauge. Compare the reading to specifications. A reading lower than normal indicates a faulty fuel pump or fuel delivery system.*

P30-8 *To test the fuel pressure regulator, create a vacuum at the regulator with the vacuum pump. Fuel pressure should decrease as vacuum increases. If pressure remains the same, the regulator is faulty.*

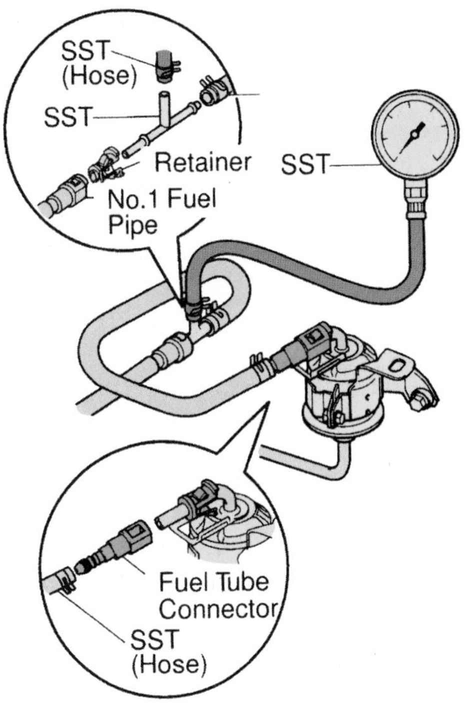

Figure 29-29 On fuel systems that do not have a Schrader valve, it may be necessary to fit a tee in the fuel line to connect the fuel pressure gauge. *Courtesy of Toyota Motor Sales, U.S.A., Inc.*

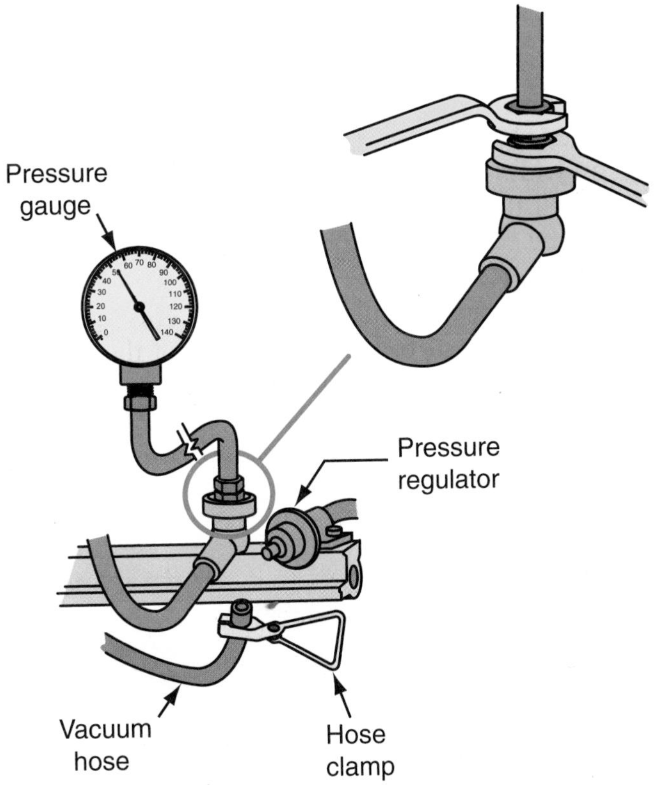

Figure 29-30 Connecting a fuel pressure gauge to the fuel pulsation damper fitting.

Often the specifications for fuel pressure are for key on engine off conditions. This means there will be no signal from the CKP, which means the fuel pump will not run for very long with the key on. Many systems will energize the fuel pump for only a few seconds prior to cranking. Unless the pump is running, it is impossible to measure fuel pressure. On these engines, you must control the operation of the pump with a scan tool.

The action of the fuel pump can also be checked by controlling the vacuum to the fuel pressure regulator. With 20 in. Hg applied to the regulator, the fuel pressure should drop. When there is no vacuum to the regulator, the fuel pressure should rise. This check also verifies that the regulator is working properly.

WARNING!

When testing a fuel system, do not let fuel contact any electrical wiring. Even the smallest spark could ignite the fuel.

Some manufacturers recommend that the pressure be measured while the engine is idling. Always make sure that you are using the correct conditions and specifications. Pressure will be slightly lower when the engine is running. Typically the pressure will rise and fall because the injectors are opening and closing, causing the gauge to fluctuate slightly. Meter fluctuations can, however, indicate air in the system.

Remember, if the fuel pressure is outside specifications, driveability problems can result. Excessive pressure causes a rich air-fuel mixture and insufficient pressure results in a leaner-than-normal mixture.

Interpreting the Results High fuel pressure readings usually indicate a faulty pressure regulator or an obstructed return line. To identify the cause, disconnect the fuel return line at the tank. Use a length of hose to route the returning fuel into a container. Start the engine and note the pressure reading at the engine. If fuel pressure is now within specifications, check for an obstruction in the return system at the tank. The fuel reservoir check valve or aspirator jet might be clogged.

If the fuel pressure still reads high with the return line disconnected, note the volume of fuel flowing through the line. Little or no fuel flow can indicate a plugged return line. Shut off the engine and connect a length of hose directly to the fuel pressure regulator return port to bypass the return hose. Restart the engine and again check the pressure. If bypassing the return line brings the readings back within specifications, a plugged return line is the culprit.

If pressure is still high, apply vacuum to the pressure regulator. If there is still no change, replace the pressure regulator. If applying vacuum to the regulator lowers fuel pressure, the vacuum hose to the regulator might be plugged, leaking, or misrouted.

Low fuel pressure, on the other hand, can be due to a clogged fuel filter, restricted fuel line, weak pump, leaky pump check valve, defective fuel pressure regulator, or dirty filter sock in the tank. It is possible to rule out filter and line restrictions by checking the pressure at the pump outlet. A higher reading at the pump outlet (at least 5 psi) means there is a restriction in the filter or line. If the reading at the pump outlet is unchanged, then the pump either is weak or is having trouble picking up fuel (clogged filter sock in the tank). Either way it is necessary to get inside the fuel tank. If the filter sock is gummed up with dirt or debris, it is also wise to clean out the tank when the filter sock is cleaned or replaced.

Another possible source of trouble is the pump's check valve. Some pumps have one, whereas others have two (positive displacement roller vane pumps). The check valve prevents fuel movement through the pump when the pump is off so residual pressure remains at the injectors. This can be checked by watching the fuel pressure gauge after the engine is shut off.

Residual Pressure Often the cause of poor starting is the lack of residual pressure in the system. This can be checked by looking at the pressure after the engine has been run and then turned off. The system should hold about the same pressure, for about 5 minutes, as it did during the pressure test. If the pressure drops off quickly after the engine and ignition are turned off, there is a leakage problem in an injector, fuel pump, connectors, hoses, or pressure regulator.

Fuel Volume Test If the fuel pressure is within specifications, you cannot conclude that the fuel delivery system is fine. Fuel volume or the pump's capacity to cause fuel flow is also important and should be tested according to the procedures outlined in the service manual. This test measures the flow rate of the pump and can help isolate fuel system restrictions or weak pumps. The test is conducted by collecting the fuel dispensed during a period of time, which is normally 30 seconds. Disconnect the return line on returnable systems or the supply line on returnless systems and connect a hose to end of the line. Put the other end of the hose into a graduated container (**Figure 29–31**). Turn on the fuel pump for about thirty seconds and measure the amount of fuel in the container. Normally the desired amount is one pint (0.47 liters). The flow of fuel into the container should be smooth and continuous with no signs of air bubbles. Results other than these indicate a bad pump or restrictions in the delivery system.

Figure 29–31 The setup for checking the fuel pump volume.

After checking the fuel pressure and volume, remove the pressure gauge and all adapters and hoses installed for the tests. Reinstall and tighten the fuel filler cap. Then turn the ignition on and check for fuel leaks.

Using a Lab Scope Monitoring the voltage to the fuel pump with a lab scope or GMM can give an idea of how well the pump is working. The voltage traces should be consistent throughout the time period set on the scope. The pattern is made up of a series of small humps. Each of the humps represents one commutator segment in the motor. Each of the humps should look the same. If there are variances in the pattern, a worn pump is indicated (**Figure 29–32**). The voltage traces show the voltage needed to establish current flow in the pump's windings.

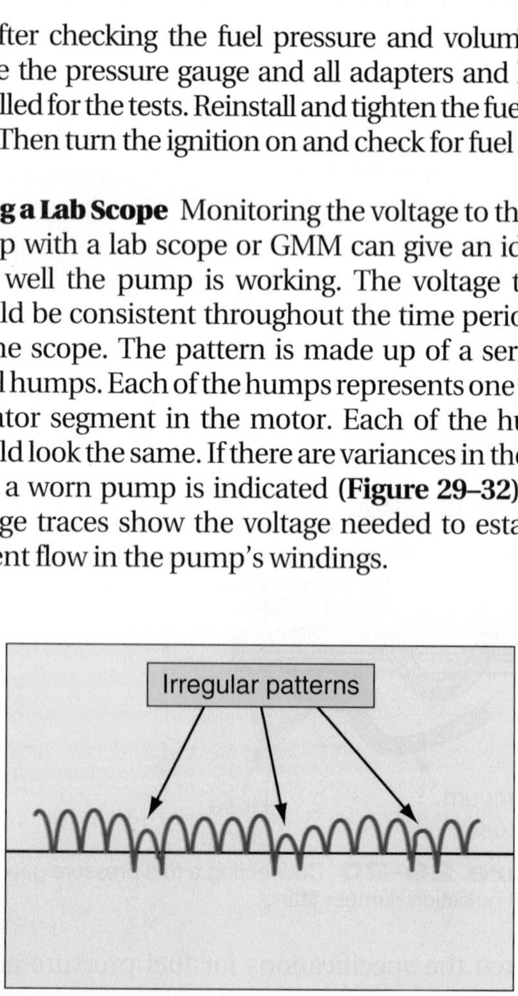

Figure 29–32 Monitoring a fuel pump with a lab scope or GMM can show a worn pump, as evidenced by irregular voltage spikes.

Min/Max Fuel Pressure	Normal Draw
10–15 psi (70–105 kPa)	2–4 amps
35–60 psi (240–415 kPa)	4–9 amps

Figure 29-33 A chart showing the typical current draw of low- and high-pressure fuel pumps.

Using a DMM A pressure transducer can be used to connect a DMM (or lab scope) to the fuel system to measure fuel pressure. This is a safe way to monitor fuel pressure during a road test. The transducer is connected in the same way as a pressure gauge. Check the DMM's manual to determine how to interpret the readings.

A DMM with a current probe can be used to monitor the current flow to the fuel pump. Looking at the current can give a good picture of the condition of the pump and the overall fuel delivery system. The normal amount of current draw for a fuel pump may be listed in the vehicle's specifications. If it is not, use the values in **Figure 29–33** as a guideline.

In motor circuits, excessive current means the armature is rotating slower than normal. Therefore, if the pump is drawing too much current, it is working harder than it should. This can be caused by a restricted fuel filter or return line.

The opposite is true; if the current levels are low, the motor is spinning too fast. This can be caused by very low fuel levels in the tank. Current can also be too low due to high resistance in the circuit. Check the connectors from the relay and to the pump. Also, check the ground circuit for the pump. If the circuit is fine, check the pump.

SHOP TALK

All electrical tests assume that the battery is fully charged. Often when there is a fuel pump problem, the engine has been cranked often and long. This lowers the battery's voltage and will affect all tests on the pump and its circuit.

No-Start Diagnosis

When an engine fails to start because there is no fuel delivery, the first check is the fuel gauge. A gauge that reads higher than a half tank probably means there is fuel in the tank, but not always. A defective sending unit or miscalibrated gauge might be giving a false indication. Sticking a wire or dowel rod down the fuel tank filler pipe tells whether there really is fuel in the tank. If the gauge is faulty, repair or replace it.

Listen for pump noise. When the key is turned on, the pump should hum for a couple of seconds to build system pressure. The pump may be energized through an oil-pressure switch (the purpose of which is to shut off the flow of fuel in case of an accident that stalls the engine). On most late-model cars with computerized engine controls, the computer energizes a pump relay **(Figure 29–34)** when it receives a cranking signal from the distributor pickup or crankshaft sensor. An oil-pressure switch might still be included in the circuitry for safety purposes and to serve as a backup in case the relay or computer signal fails. Failure of the pump relay or computer driver signal can cause slow starting because the fuel pump does not come on until the engine cranks long enough to build up sufficient oil pressure to trip the oil-pressure switch.

The pump might be good, but if it does not receive voltage and have a good ground, it does not run. To check the ground, connect a testlight across the ground and feed wires at the pump to check for voltage, or use a voltmeter to read actual voltage and an ohmmeter to check ground resistance. The latter is the better test technique because a poor ground connection or low voltage can reduce pump operating speed and output. If the electrical circuit checks out but the pump does not run, the pump is probably bad and should be replaced.

No voltage at the pump terminal when the key is on and the engine is cranking indicates a faulty oil-pressure switch, pump relay, relay drive circuit in the computer, or a wiring problem. Check the pump fuse to see if it is blown. Replacing the fuse might restore power to the pump, but until you have found out what caused the fuse to blow, the problem is not solved. The most likely cause of a blown fuse would be a short in the wiring between the relay and pump, or a short inside the oil-pressure switch or relay, or a bad fuel pump.

A faulty oil-pressure switch can be checked by bypassing it with a jumper wire. If doing this restores power to the pump and the engine starts, replace the switch. If an oil-pressure switch or relay sticks in the closed position, the pump can run continuously whether the key is on or off, depending on how the circuit is wired.

To check a pump relay, use a test light to check across the relays and ground terminals to tell if the relay is getting battery voltage and ground. Next, turn off the ignition, wait about 10 seconds, then turn it on. The relay should click and you should see battery voltage at the relay's pump terminal. If nothing happens, repeat the test, checking for voltage at the relay terminal wired to the computer. The presence of a voltage signal here means the computer is doing its job but the relay is failing to close and should be replaced. No voltage signal from the computer indicates an open in that wiring circuit or a fault in the computer itself.

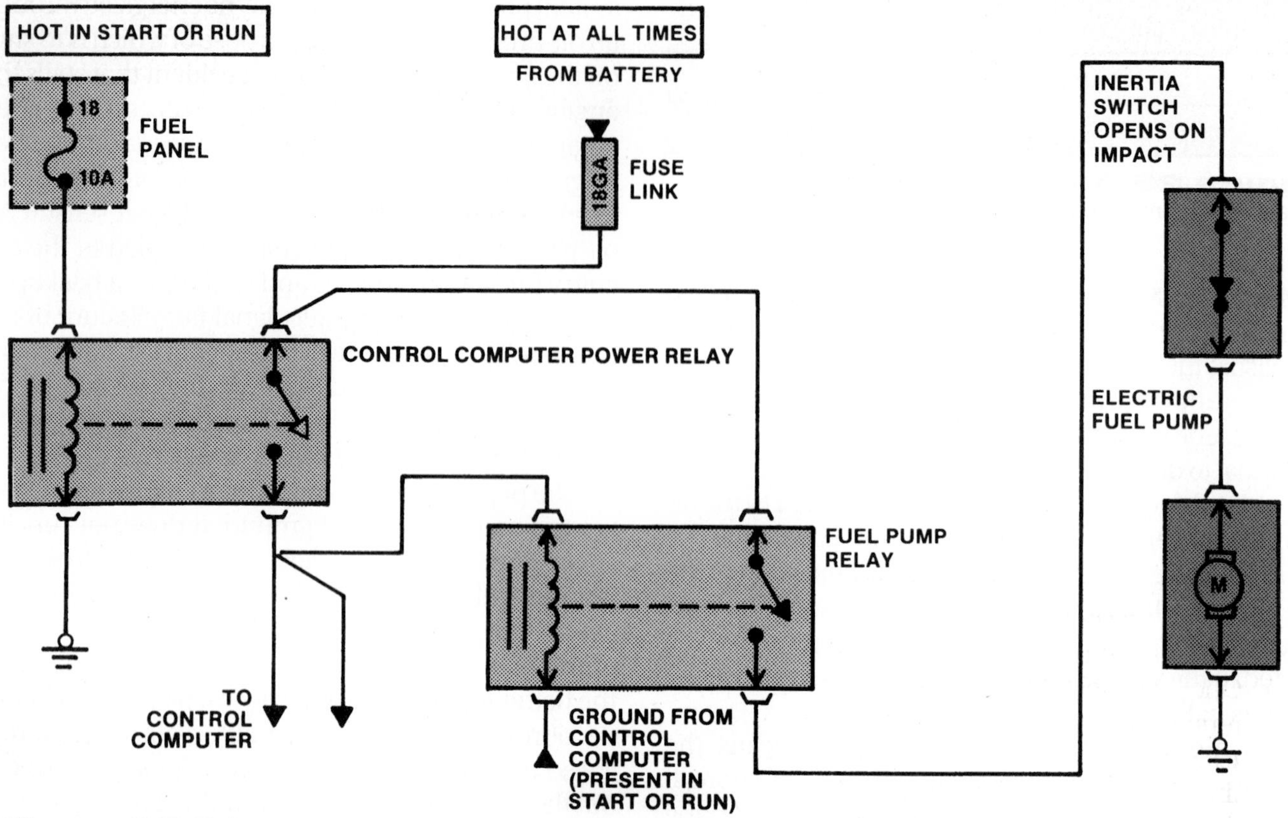

Figure 29-34 An electrical fuel delivery system wiring diagram.

Replacement

When replacing an electric pump, be sure that the new or rebuilt replacement unit meets the minimum requirements of pressure and volume for that particular vehicle. This information can be found in the service manual. If the fuel pump is mounted in the fuel tank, the procedure for replacement is different than if the unit is external to the tank.

External Fuel Pump Before removing the fuel pump, disconnect the negative battery cable. Then disconnect the electrical connectors on the fuel pump. Label the wires to aid in connecting it to the new pump. Reversing polarity on most pumps destroys the unit.

Now disconnect the fuel lines at the pump. These lines should also be labeled so they are installed correctly on the new pump.

Loosen and remove the bolts holding the pump in place. Remove the pump by pushing the pump up until the bottom is clear of the bracket. Swing the pump out to the side and pull it down to free it from the rubber fuel line coupler. The rubber sound insulator between the bottom of the pump and bracket and the rubber coupler on the fuel line are normally discarded because new ones are included with the replacement pump. Some pumps have a rubber jacket around them to quiet the pump. If this is the case, slip off the jacket and put it on the new pump.

CAUTION!

Avoid the temptation to test the new pump before reinstalling the fuel tank by energizing it with a couple of jumper wires. Running the pump dry can damage it because the pump relies on fuel for lubrication and cooling.

Compare the replacement pump with the old one. If necessary, transfer any fuel line fittings from the old pump to the new one. When inserting the new pump back into its bracket, be careful not to bend the bracket. Make sure the rubber sound insulator under the bottom of the pump is in place. Install a new filter sock (if so equipped) on the pump inlet and reconnect the pump wires. Be absolutely certain you have correct polarity.

If the fuel was removed from the tank, replace it. Make sure all electrical connections are reconnected and that all fuel lines and hoses are properly fastened and tightened. Then reconnect the ground terminal at the battery. Start the engine and check all connections for fuel leaks.

CAUTION!

Never turn on the ignition switch or crank the engine with a fuel line disconnected. This action will result in gasoline discharge from the disconnected line, which may result in a fire, causing personal injury and/or property damage.

Internal Fuel Pump On many vehicles, the fuel tank must be removed to replace the fuel pump and/or fuel gauge sending unit. On other vehicles, the unit can be serviced through an opening in the vehicle's trunk or under the rear seat **(Figure 29–35)**. Some vehicles have a separate fuel pump and gauge sending unit, while others have both contained in a single unit. Once the fuel tank is out of the vehicle, if necessary, remove the unit from the tank.

These units are often held in the tank by either a retaining ring or screws. The easiest way to remove a retaining ring is to use a special tool designed for this purpose. This tool fits over the metal tabs on the retaining ring, and after about a quarter turn, the ring comes loose and the unit can be removed. If the special tool is not available, a brass drift punch and ball-peen hammer usually can do the job.

When removing the unit from the tank, be very careful not to damage the float arm, the float, or the fuel gauge sender. Check the unit carefully for any damaged components. Shake the float and if fuel can be heard inside, replace it. Make sure the float arm is not bent. It is usually wise to replace the sock and O-ring before replacing the unit. Check the fuel gauge and sender unit as described in the service manual. When reinstalling the pickup pipe-sending unit, be very careful not to damage any of the components.

Once the unit is removed, check the filter on the fuel pump inlet. If the filter is contaminated or damaged, replace the filter. Inspect the fuel pump inlet for dirt and debris. Replace the fuel pump if these foreign particles are found in the pump inlet.

If the pump inlet filter is contaminated, flush the tank with hot water for at least 5 minutes. Dump all the water out of the tank through the pump opening in the tank. Shake the tank to be sure all the water is removed. Allow the tank to sit and air dry before reinstalling it or adding fuel to it. Remember, gasoline fumes are extremely ignitable, so keep all open flames and sparks away from the tank while it is drying.

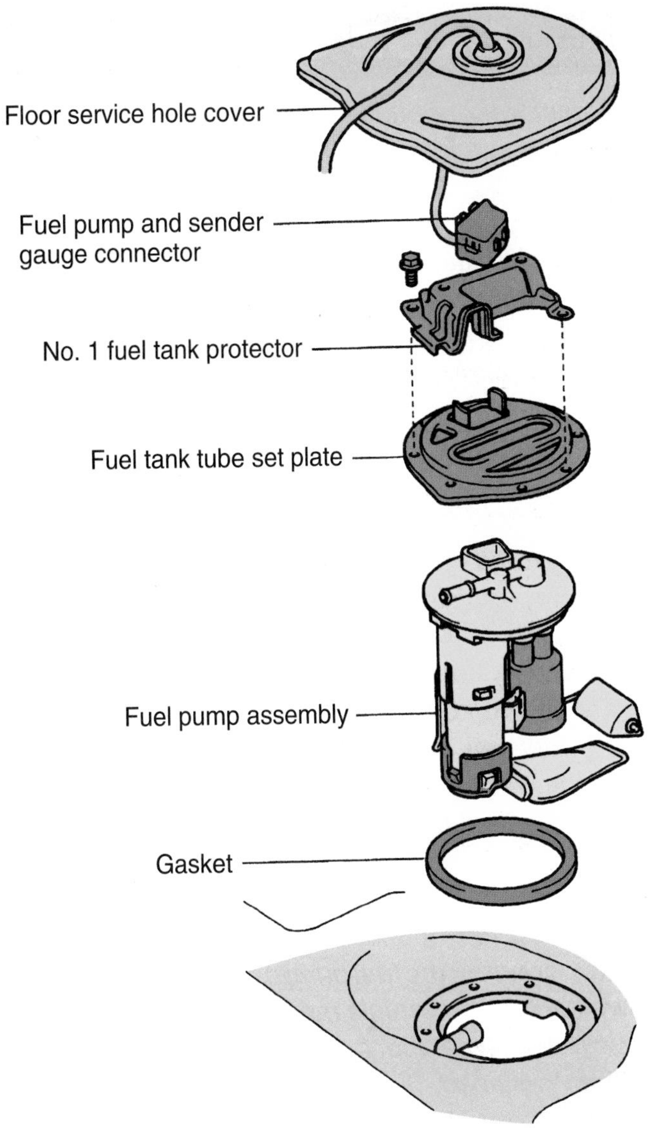

Figure 29–35 Some fuel pumps can be serviced through an access hole in the car's floor. *Courtesy of Toyota Motor Sales, U.S.A., Inc.*

Check all fuel hoses and tubing on the fuel pump assembly. Replace fuel hoses that are cracked, deteriorated, or kinked. When fuel tubing on the pump assembly is damaged, replace the tubing or the pump.

Make sure the sound insulator sleeve is in place on the electric fuel pump, and check the position of the sound insulator on the bottom of the pump.

Clean the pump and sending unit mounting area in the fuel tank with a shop towel, and install a new gasket or O-ring on the pump and sending unit. Install the fuel pump and gauge sending unit assembly in the fuel tank and secure this assembly in the tank using the vehicle manufacturer's recommended procedure.

CASE STUDY

A customer complained about severe engine surging in a late-model Ford. When the technician lifted the hood, he noticed many fuel system and ignition system components had been replaced recently. The customer was asked about previous work done on the vehicle, and he indicated this problem had existed for some time. Several shops had worked on the vehicle, but the problem still persisted. The fuel pump and filter were replaced, as were some ignition coils and spark plugs.

The technician road tested the vehicle and found it did have a severe surging problem at freeway cruising speeds. From past experience, the technician thought the surging was caused by lack of fuel supply. The technician decided to connect a fuel pressure gauge at the fuel rail of the injection assembly. The gauge was securely taped to one of the windshield wiper blades so the gauge could be observed from the passenger compartment. The technician drove the vehicle on a second road test and found when the surging problem occurred, the fuel pump pressure dropped well below the vehicle manufacturer's specifications. Since the fuel pump and filter had been replaced, the technician concluded the problem must be in the fuel line or tank.

The technician returned to the shop and raised the vehicle on a hoist. The steel fuel tubing appeared to be in satisfactory condition. However, a short piece of rubber fuel hose between the steel fuel tubing and the fuel line entering the fuel tank was flattened and soft in the center. This fuel hose was replaced and rerouted to avoid kinking. Another road test proved the surging problem was eliminated.

The flattened fuel hose restricted fuel flow, and at higher speeds the increased vacuum from the fuel pump made the flattened condition worse, which restricted the fuel flow and caused the severe surging problem.

KEY TERMS

EVAP
Inertia switch
Schrader valve

SUMMARY

- A typical fuel delivery system includes a fuel tank, fuel lines, a fuel filter, a fuel pressure regulator, and a fuel pump.
- The fuel system should be checked whenever there is evidence of a fuel leak or smell.
- The fuel system should also be checked whenever basic tests suggest that there is too little or too much fuel being delivered to the cylinders.
- Fuel delivery problems typically cause no-start or loss of power problems.
- Because electronic fuel injection systems have a residual fuel pressure, this pressure must be relieved before disconnecting any fuel system component.
- Fuel tanks have devices that prevent fuel vapors from leaving the tank. All fuel tanks have a filler tube and a nonvented cap.
- The fuel tank should be inspected for leaks; road damage; corrosion; rust; loose, damaged, or defective seams; loose mounting bolts; and damaged mounting straps.
- Leaks in the metal fuel tank can be caused by a weak seam, rust, or road damage. The best way to permanently solve these problems is to replace the tank.
- In-tank fuel pumps and fuel level gauge sending units are held in the tank by either a retaining ring or screws.
- Fuel lines are made of seamless, double-wall metal tubing, flexible nylon, or synthetic rubber hose. The latter must be able to resist gasoline and be nonpermeable to fuel and fuel vapors. Rubber hose, such as that used for vacuum lines, deteriorates when exposed to gasoline. Only hoses made for fuel systems should be used for replacement.
- Vapor vent lines must be made of material that resists attack by fuel vapors. Replacement vent hoses are usually marked with the designation "EVAP" to indicate their intended use.
- All fuel lines should occasionally be inspected for holes, cracks, leaks, kinks, or dents.
- Automobiles and light trucks have an in-tank strainer and a fuel filter. The strainer, located in the gasoline tank, is made of a finely woven fabric. The strainer prevents large contaminant particles from entering the fuel system where they could cause excessive fuel pump wear.

- To determine if the fuel pump is in satisfactory operating condition, tests for both fuel pump pressure and fuel pump capacity should be performed.
- High fuel pressure readings normally indicate a faulty pressure regulator or an obstructed return line.
- Low pressure can be caused by a clogged fuel filter, restricted fuel line, weak pump, leaky pump check valve, defective fuel pressure regulator, or dirty filter sock in the tank.
- An inertia switch in the fuel pump circuit opens the fuel pump circuit immediately if the vehicle is involved in a collision.
- SRSs automatically shut off the fuel pump when an air bag is deployed.

REVIEW QUESTIONS

1. Fuel pump ______ is a statement of the volume of the flow of the pump.
2. Explain the purpose of the relief valve and one-way check valve in an electric fuel pump.
3. Most fuel tank filler caps contain a pressure valve and a ______.
   a. vapor separator
   b. vacuum relief valve
   c. one-way check valve
   d. surge plate
4. What type of fire extinguisher should you have close by when you are working on fuel system components?
5. What is the first thing that should be disconnected when removing a fuel tank?
6. Why is a plastic or metal restrictor placed in either the end of the vent pipe or in the vapor-vent hose on some vehicles?
7. Low fuel pump pressure causes a ______ mixture and excessive pressure causes a ______ mixture.
8. *True or False?* Fuel pressure typically is at its highest level when vacuum is applied to the pressure regulator.
9. If fuel pump pressure or volume is less than specified, which of the following is *not* a likely cause of the problem?
   a. a restricted fuel filter
   b. a faulty fuel pump
   c. restricted fuel return lines
   d. restricted fuel supply lines
10. Which of the following statements about a fuel volume test is *not* true?
    a. This test measures the flow rate of the pump and can help isolate fuel system restrictions or weak pumps.
    b. The test is conducted by collecting the fuel dispensed by the certain during a period of time, normally 5 seconds.
    c. The flow of fuel into the container should be smooth and continuous with no signs of air bubbles.
    d. Poor results may indicate a bad pump or restrictions in the delivery system.
11. Low fuel pressure can be caused by all of the following *except* ______.
    a. a clogged fuel filter
    b. a restricted fuel line
    c. a fuel pump that draws too high amperage
    d. a dirty filter sock in the tank
12. If a fuel system loses pressure immediately after the ignition is turned off, the possible problems are all of the following *except* ______.
    a. a leaking injector
    b. restricted fuel lines
    c. leaking connectors or hoses
    d. a faulty pressure regulator
13. List at least five safety precautions that should be adhered to when working with fuel systems.
14. Describe the operation of the evaporative system integrity check.
15. *True or False?* On today's vehicles, an electric fuel pump runs only when the engine is running.

ASE-STYLE REVIEW QUESTIONS

1. Technician A says that excessively high pressure from an electric fuel pump can be caused by a faulty pressure regulator. Technician B says that the typical problem may be an obstructed return line. Who is correct?
   a. Technician A
   b. Technician B
   c. Both A and B
   d. Neither A nor B
2. Technician A replaces a damaged steel fuel line with one made of synthetic rubber. Technician B

replaces a damaged steel fuel line with one made of steel. Who is correct?

a. Technician A
b. Technician B
c. Both A and B
d. Neither A nor B

3. While discussing electric fuel pumps: Technician A says that some electric fuel pumps are combined in one unit with the gauge sending unit. Technician B says that on an engine with an electric fuel pump, low engine oil pressure may cause the engine to stop running. Who is correct?

a. Technician A only
b. Technician B only
c. Both A and B
d. Neither A nor B

4. To relieve fuel pressure on an EFI car: Technician A connects a pressure gauge to the fuel rail. Technician B disables the fuel pump and runs the car until it dies. Who is correct?

a. Technician A
b. Technician B
c. Both A and B
d. Neither A nor B

5. While discussing quick-disconnect fuel line fittings: Technician A says that some quick-disconnect fittings may be disconnected with a pair of snapring pliers. Technician B says that some quick-disconnect fittings are hand releasable. Who is correct?

a. Technician A
b. Technician B
c. Both A and B
d. Neither A nor B

6. While discussing quick-disconnect fuel line fittings: Technician A says that on some hand-releasable, quick-disconnect fittings, the fitting may be removed by pulling on the fuel line. Technician B says that some hand-releasable, quick-disconnect fittings may be disconnected by twisting the large connector collar in both directions and pulling on the connector. Who is correct?

a. Technician A
b. Technician B
c. Both A and B
d. Neither A nor B

7. While discussing the various fuel filters used in current automobiles: Technician A says that all systems have a vapor separator that prevents gaseous fuel from being sent to the fuel pump. Technician B says that fuel systems have a strainer located in the fuel tank that prevents large contaminants from entering the fuel system. Who is correct?

a. Technician A
b. Technician B
c. Both A and B
d. Neither A nor B

8. While discussing fuel tank filler pipes and caps: Technician A says that the threaded filler cap should be tightened until it clicks. Technician B says that the vent pipe is connected from the top of the filler pipe to the bottom of the fuel tank. Who is correct?

a. Technician A only
b. Technician B only
c. Both A and B
d. Neither A nor B

9. While discussing electric fuel pumps: Technician A says that the one-way check valve prevents fuel flow from the underhood fuel system components into the fuel pump and tank when the engine is shut off. Technician B says that the one-way check valve prevents fuel flow from the pump to the fuel filter and fuel system if the engine stalls and the ignition switch is on. Who is correct?

a. Technician A only
b. Technician B only
c. Both A and B
d. Neither A nor B

10. While discussing electric fuel pumps: Technician A says that some electric fuel pumps in EFI systems are computer controlled. Technician B says that fuel pump pressure is determined by measuring the amount of fuel that the pump will deliver in a specific length of time. Who is correct?

a. Technician A
b. Technician B
c. Both A and B
d. Neither A nor B

ELECTRONIC FUEL INJECTION

CHAPTER 30

OBJECTIVES

■ Explain the differences in the point of injection in throttle body or port injection systems. ■ Describe the difference between a sequential fuel injection (SFI) system and a multiport fuel injection (MFI) system. ■ Explain the design and function of major EFI components. ■ Describe the inputs used by the computer to control the idle air control and idle air control by-pass air motors. ■ Describe how the computer supplies the correct air-fuel ratio on a throttle body injection (TBI) system. ■ Explain how the clear flood mode operates on a TBI system. ■ Explain why manifold vacuum is connected to the pressure regulator in an MFI system. ■ Describe the operation of the pressure regulator in a returnless EFI system. ■ Describe the operation of the central injector and poppet nozzles in a central port injection (CMFI) system. ■ Describe the operation of direct gasoline injection systems. ■ Describe the operation of the injection systems used in light- and medium-duty diesel engines.

This chapter discusses the components of electronic fuel injection (EFI) systems and explains how the various designs of EFI work. EFI systems are computer controlled and designed to provide the correct air-fuel ratio for all engine loads, speeds, and temperature conditions. The computer monitors the operating conditions and attempts to provide the engine with the ideal air-fuel ratio. The ideal fuel ratio is often called the **stoichiometric ratio**. A stoichiometric mixture is one that has the air-to-fuel ratio necessary for complete combustion of the fuel **(Figure 30–1)**. This means all of the fuel and the oxygen in the air are completely consumed during combustion. Different fuels have a different stoichiometric ratio **(Figure 30–2)**. The stoichiometric mixture for gasoline is 14.7:1.

A stoichiometric mixture is theoretically the best combination of fuel and air to provide for total combustion. The ratio, however, is based on an ideal environment for combustion. This environment rarely exists and the injection system's controls vary the ratio in response to the inefficiencies. It also responds to changes in conditions that affect the combustion process. The air-fuel ratio also changes for starting, maximum power, and maximum fuel economy. The stoichiometric ratio allows the catalytic converters to work more efficiently.

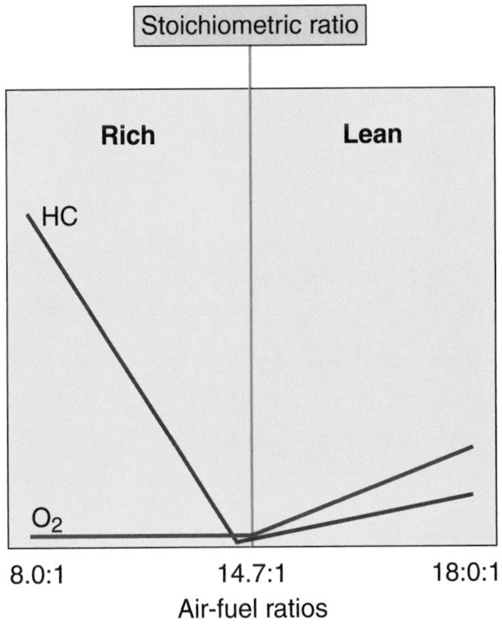

Figure 30–1 A graph showing how the amount of raw gasoline (HC) and air (O_2) released in the exhaust changes with a change in the air-fuel ratio.

Fuel Type	Stoichiometric A/F Ratio
Natural gas	17.2:1
Typical gasoline*	14.7:1
Diesel	14.6:1
Ethanol	9.0:1
Methanol	6.4:1

* Gasoline types vary according to the percentage of alcohol added.

Figure 30-2 Different fuels have different stoichiometric air-fuel ratios.

BASIC EFI

In an EFI system (**Figure 30–3**), the computer must know the amount of air entering the engine so it can supply the correct amount of fuel for that amount of air. In systems with a MAP sensor, the computer calculates the amount of intake air based on MAP and rpm input signals. This type of EFI system is referred to as a **speed density** system, because the computer calculates the air intake flow according to engine speed and intake manifold vacuum. Because air density changes with air temperature, an intake air temperature sensor is also used.

Today the most commonly used EFI system is the mass airflow (MAF) system. This system relies on a MAF sensor that directly measures the amount of intake air. The most common type of MAF sensor is the hot wire design. MAF systems are very responsive to changes in operating conditions because they actually measure, rather than compute, airflow.

During closed loop, EFI systems rely on the input from a variety of sensors before adjusting the air-fuel ratio. Based on all of the inputs, the PCM is able to determine the current operating conditions of the engine such as: starting, idle, acceleration, cruise, deceleration, and operating temperature. The PCM gathers the inputs and refers to the look-up tables in its memory to determine the ideal air-fuel ratio for the current conditions. During open loop, fuel is delivered according to predetermined parameters held in the PCM's memory

When conditions, such as starting or wide-open throttle, demand that the signals from the oxygen sensor be ignored, the system operates in open loop. Pre-OBD-II systems may also go into the open-loop mode while idling, or at any time that the oxygen sensor cools off enough to stop sending a good signal, and at wide-open throttle.

Fuel Injectors

Fuel injectors are electromechanical devices that meter and atomize fuel so it can be sprayed into the intake manifold. Fuel injectors resemble a spark plug in size and shape. O-rings are used to seal the injector at the intake manifold, throttle body, and/or fuel rail mounting positions. These O-rings provide thermal

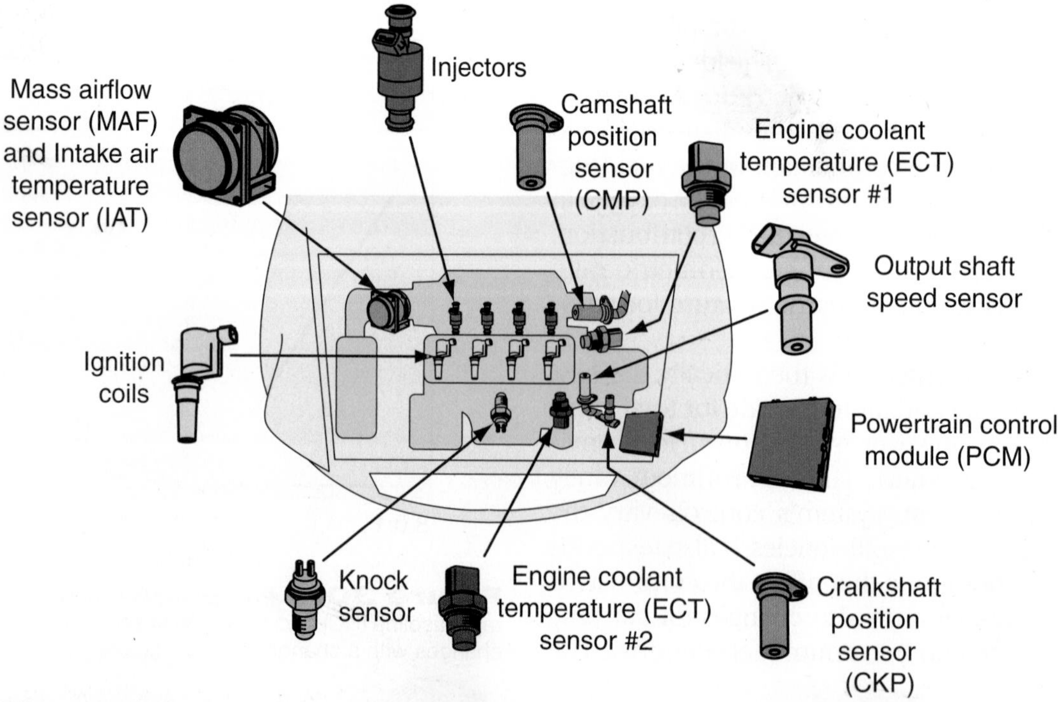

Figure 30-3 A typical electronic fuel injection system.

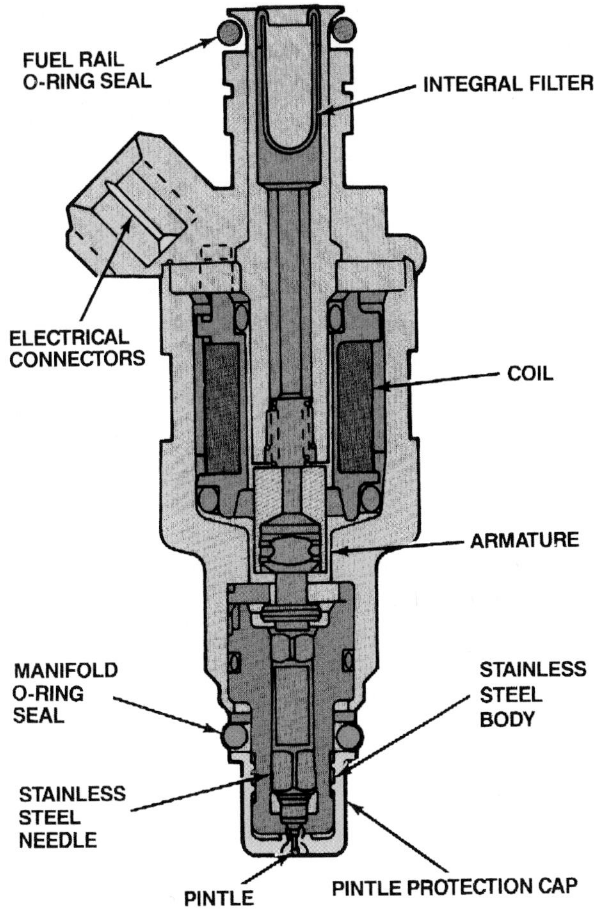

Figure 30-4 A typical fuel injector used in multiport fuel injection systems. *Courtesy of Ford Motor Company*

insulation to prevent the formation of vapor bubbles and promote good hot start characteristics. They also dampen potentially damaging vibration. When the injector is electrically energized, a fine mist of fuel sprays from the injector tip.

Most injectors consist of a solenoid, a needle valve, and a nozzle **(Figure 30–4)**. The solenoid is attached to the nozzle valve. The PCM controls the injector by controlling its ground circuit through a driver circuit. When the solenoid winding is energized, it creates a magnetic field that draws the armature back and pulls the needle valve from its seat. Fuel then sprays out of the nozzle. When the solenoid is de-energized, the magnetic field collapses and a helical spring forces the needle valve back on its seat, shutting off fuel flow.

Another injector design uses a ball valve and valve seat. In this case, the magnetic field created by the solenoid coil pulls a plunger upward, lifting the ball valve from its seat. A spring is used to return the valve to its seated or closed position.

Each fuel injector has a two-wire connector. One wire supplies voltage to the injector. This wire may connect directly to the fuse panel or to the PCM, which, in turn, connects to the fuse panel. In a few systems, a resistor under the hood or in the PCM is used to reduce the 12-volt battery supply voltage to 3 volts or less. The second wire is a ground wire. This ground wire is connected to the driver circuit inside the PCM.

The amount of fuel released by an injector depends on fuel pressure and the length of time the injector is energized. Fuel pressure is mainly controlled by a pressure regulator, and the injector's pulse width is controlled by the PCM. Typical pulse widths range from 1 to 10 milliseconds at full load. The PCM controls the pulse width according to various input sensor signals, operating conditions, and its programming. The primary inputs are related to engine load and engine coolant temperature. Cold starting requires the longest pulse width.

Different engines require different injectors. Injectors are designed to pass a specified amount of fuel when opened. In addition, the number of holes at the tip of the injector varies with engines and model years. Fuel injectors can be top fuel feeding, side feeding, or bottom feeding **(Figure 30–5)**. Top- and side-feed injectors are primarily used in port injection systems that operate using high fuel system pressures. Bottom-feed injectors are used in throttle body systems. Bottom-feed injectors are able to use fuel pressures as low as 10 psi.

There have been some problems with deposits on injector tips. Because small quantities of gum are present in gasoline, injector deposits usually occur when this gum bakes onto the injector tips after a hot engine is shut off. Most manufacturers use fuel injectors designed to reduce the chance of deposit buildup at the tips. Also, oil companies have added a detergent to their gasoline to help prevent injector tip deposits.

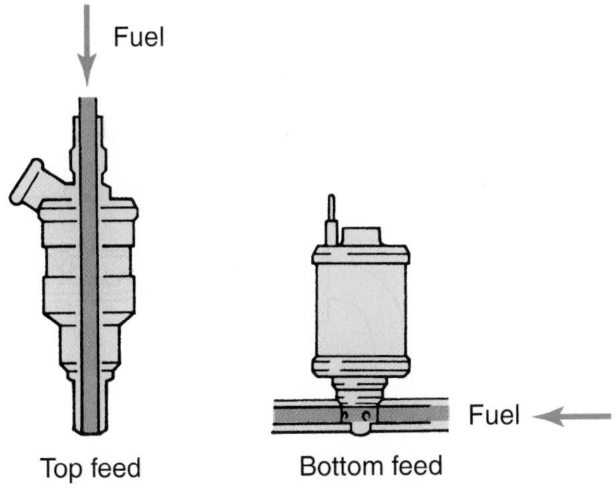

Figure 30-5 Examples of top feed and bottom feed injectors.

Performance TIP *Increasing the amount of fuel injected when the engine is in open loop can boost an EFI engine's performance. This normally can be done by installing larger injectors, increasing the fuel pressure, and lengthening the injectors' pulse width. The latter is done by reprogramming the PCM. These changes will have little effect on the engine when it is in closed loop. Of course, there will be no increase in power if there is not enough air for the increased fuel.*

Idle Speed Control

Idle speed control is a function of the PCM. Based on operating conditions and inputs from various sensors, the PCM regulates the idle speed. In throttle body and port EFI systems, engine idle speed is controlled by bypassing a certain amount of airflow past the throttle valve in the throttle body housing. Two types of air by-pass systems are used: auxiliary air valves and **idle air control (IAC)** valves. IAC valve systems are more common **(Figure 30–6)**. Most TBI units are fitted with an idle speed motor.

The IAC system is a stepper motor or actuator that positions the IAC valve in the air by-pass channel around the throttle valve. The IAC valve is part of the throttle body casting. The PCM calculates the amount of air needed for smooth idling based on input data, such as coolant temperature, engine load, engine speed, and battery voltage.

If engine speed is lower than desired, the PCM activates the motor to retract the IAC valve. This opens the channel and diverts more air around the throttle valve. If engine speed is higher than desired, the valve is extended and the by-pass channel is made smaller. Air supply to the engine is reduced and engine speed falls.

During cold starts idle speed can be as high as 2,100 rpm to quickly raise the temperature of the catalytic converter. The PCM maintains cold idle speed for approximately 40 to 50 seconds even if the driver attempts to alter it by kicking the accelerator. After this preprogrammed time, depressing the accelerator pedal sends a signal to the PCM to reduce idle speed.

Some engines are equipped with an auxiliary air valve to aid in the control of idle speed. Unlike the IAC valve, the auxiliary air valve is not controlled by the PCM. Like the IAC system, however, the auxiliary air valve provides additional air during cold engine idling.

Inputs

The ability of the fuel injection system to control the air-fuel ratio depends on its ability to properly time the injector pulses with the compression stroke of each cylinder and its ability to vary the injector "on" time, according to changing engine demands. Both tasks require the use of sensors that monitor the operating conditions of the engine. The PCM receives these signals from the CAN bus and inputs sent directly to the computer **(Figure 30–7)**.

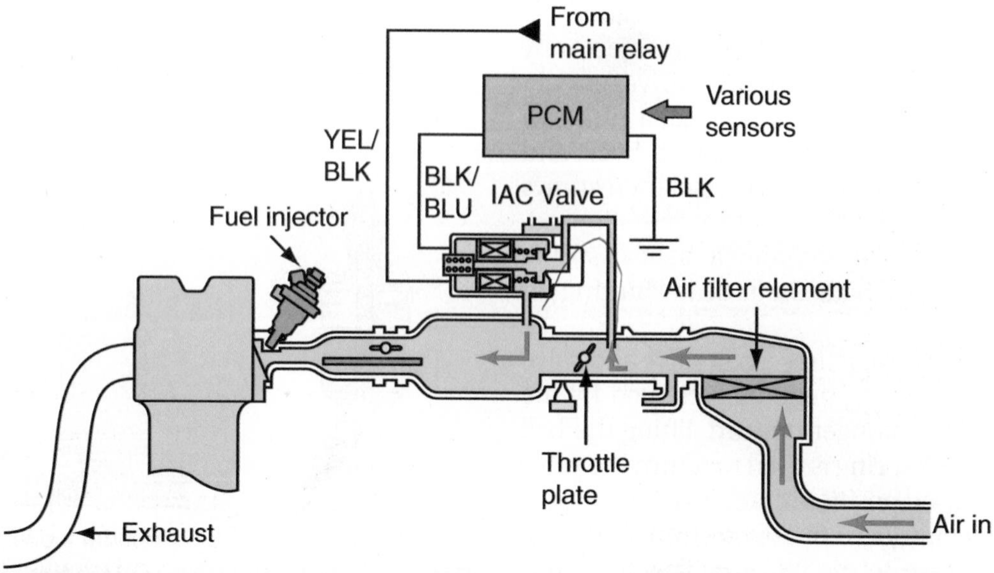

Figure 30-6 An idle air control system.

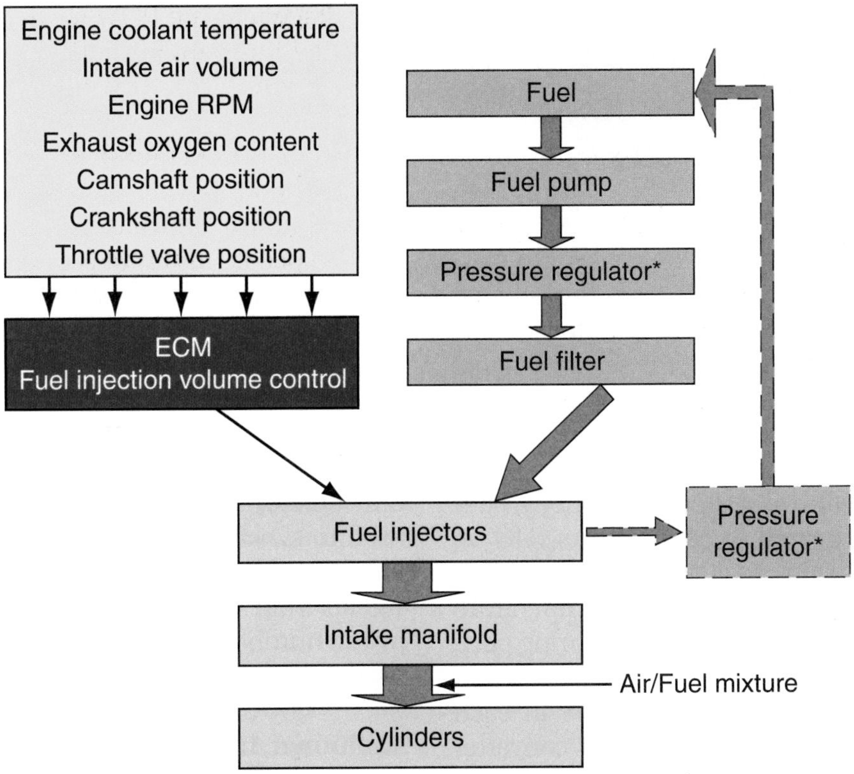

Figure 30-7 A basic electronic fuel injection system. *Courtesy of Toyota Motor Sales, U.S.A., Inc.*

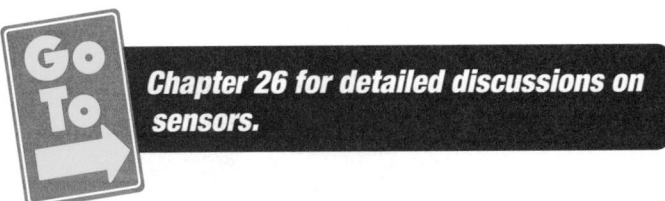

Go To *Chapter 26 for detailed discussions on sensors.*

MAF Sensor The mass airflow (MAF) sensor converts air flowing past a heated sensing element into an electronic signal **(Figure 30–8)**. The strength of this signal is determined by the energy needed to keep the element at a constant temperature above the incoming ambient air temperature. As the volume and density (mass) of airflow across the heated element change, the temperature of the element is affected and the current flow to the element is adjusted to maintain the desired temperature. The varying current parallels the particular characteristics of the incoming air (hot, dry, cold, humid, high/low pressure). The PCM monitors the changes in current to determine air mass and to calculate fuel requirements.

Figure 30-8 A mass airflow (MAF) sensor.

MAP Sensor The manifold absolute pressure (MAP) sensor measures changes in the intake manifold pressure that result from changes in engine load and speed. At closed throttle, the engine produces a low MAP value. A wide-open throttle produces a high value. MAP output is the opposite of what is measured on a vacuum gauge. The use of this sensor also allows the control computer to adjust automatically for different altitudes. The MAP signal may also be used to inform the PCM when the EGR valve is open during the EGR monitor test.

Figure 30-9 An oxygen sensor.

Oxygen Sensors (O_2S) The signals from the exhaust gas oxygen sensor (O_2S) are used to monitor the air-fuel mixture **(Figure 30–9)**. When the sensor's signal indicates a lean mixture, the computer enriches the air-fuel mixture to the engine. When the sensor reading is rich, the computer leans the air-fuel mixture. Because an O_2S must be hot to operate properly, late-model engines use heated oxygen sensors (HO_2S). These sensors have an internal heating element that allows the sensor to reach operating temperature quickly and to maintain its temperature during periods of idling or low engine load.

OBD-II systems have two or more O_2S in each exhaust system, one before the catalytic converter (upstream) and one after it (downstream). The signals from the upstream sensor readings are used to monitor the air-fuel ratio and the downstream sensors are used to monitor the effectiveness of the catalytic converter.

Many engines are fitted with air-fuel ratio sensors in the upstream position. These sensors can react to very minor changes in the air-fuel ratio. This allows the PCM to have precise control of the fuel injection system.

IAT Sensor Cold air is denser than warm air. Cold, dense air can burn more fuel than the same volume of warm air because it contains more oxygen. The intake air temperature (IAT) sensor measures air temperature and sends an electronic signal to the PCM. The computer uses this input along with the air volume input in determining the amount of oxygen entering the engine.

ECT Sensor The engine coolant temperature (ECT) sensor signals the PCM when the engine needs cold enrichment, as it does during warmup. This adds to the base pulse width but decreases to zero as the engine warms up. The PCM may also order a richer mixture when the engine is overheating. A rich mixture burns cooler.

TP Sensor The throttle position (TP) sensor allows the PCM to monitor throttle position. The signals from this sensor are used to clarify load and operating conditions. This potentiometer reacts directly to the movement of the throttle plate. A sudden increase in TP voltage tells the PCM to momentarily enrich the mixture to prevent hesitation and stumbling during acceleration.

CKP Sensor The crankshaft position (CKP) sensor is used to monitor engine speed. This signal advises the PCM to adjust the pulse width of the injectors for engine speed. This input is the most important input in the fuel management system. It is used to synchronize the injectors with events in the cylinders. The signals from the CKP are often used with the signals from the CMP to determine which cylinders are on the compression stroke.

CMP Sensor The camshaft position (CMP) sensor is used to synchronize the firing of the injectors with the individual cylinders in the engine. By using the signals from the CMP, the PCM can determine that piston number 1 is on the compression stroke. This is used for fuel injection timing.

Additional Input Information Sensors Additional sensors are also used to provide the following information on engine conditions:

- Vehicle speed
- Air conditioner operation
- Gearshift lever position
- Battery voltage
- EGR valve position

Operational Modes

All fuel injection systems operate in response to inputs. However, the PCM's programming allows it to define the conditions and establish a summary of those conditions. The PCM then controls the delivery of fuel according to that mode of operation. Different EFI systems have different operational modes, but most have starting, run, clear flood, acceleration, and deceleration.

Starting Mode When the ignition switch is initially moved to the start position, the PCM turns on the fuel pump for about 2 seconds. When the PCM receives a good signal from the CKP sensor, it energizes the fuel pump to allow for starting. If a CKP signal is not present, the fuel pump is shut off. With a CKP signal, the PCM controls injector timing and bases the pulse width of the injectors entirely on the engine's coolant temperature and load. Once the engine is cranking, the PCM sets the injectors' pulse width according to inputs from the MAF, IAT, ECT, and TP sensors. In some cases, as the engine is cranking, the injectors may prime the cylinders with a

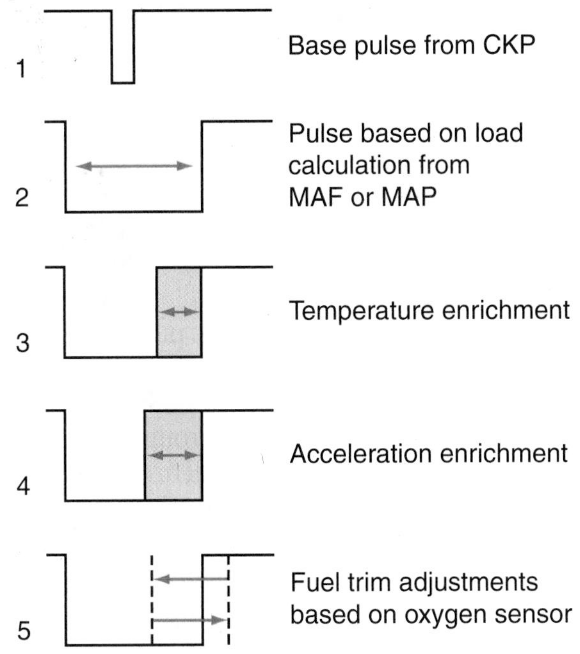

Figure 30-10 A graphical representation of low pulse width is calculated.

spray of fuel to help get the engine started. The system stays in starting mode until the engine is rotating at a predetermined speed.

Run Mode Once the engine has started and is running above a predetermined speed, the system will operate in open loop. In open loop, the PCM sets injector pulse width according to MAF, IAT, ECT, and TP sensor signals. The system stays in open loop until the PCM receives good signals from the O_2S and a predetermined engine temperature has been reached. Once these conditions have been met, the system moves into closed loop. In closed loop, the PCM adjusts the pulse width according to inputs from a variety of sensors, but primarily the O_2S (**Figure 30–10**).

Clear Flood Mode At times the engine will not start because it has received too much fuel; this is called flooding. When an engine floods, the excess fuel must be pumped out of the cylinders. This is done by fully depressing the accelerator pedal and cranking the engine. The **clear flood mode** is not an automatic process; it is initiated by depressing and holding the accelerator pedal down. When the PCM detects a wide-open throttle, it will go into the acceleration enrichment mode for 3 seconds. If the throttle is held open and the engine's speed is below a predetermined rpm, the system will return to the start mode. In some cases, the PCM will completely turn off the injectors if engine cranking continues for a long period.

Acceleration Mode Based on signals from the TP and MAF sensors, the PCM can tell when the vehicle is being accelerated. To compensate for the sudden rush of intake air as the throttle is opened, the PCM increases the injectors' pulse width. The pulse width change is calculated by the PCM according to inputs from the CKP, MAP, ECT, MAF, and TP sensors. Once the PCM determines that the vehicle is no longer accelerating, the EFI system is returned to the run mode.

Deceleration Mode Inputs from the MAF and TP sensors are also used by the PCM to detect deceleration. During deceleration, the PCM reduces injector pulse width. Some systems will totally shut off the fuel when the vehicle is rapidly decelerating. Some vehicles will totally shut down the fuel system for a brief period during deceleration.

Fuel Trim

OBD-II systems constantly monitor the signals from the oxygen or air-fuel ratio sensors while it is operating in closed loop. Based on these inputs and programming instructions, the PCM alters the injectors' pulse width to provide the best possible combination of driveability, fuel economy, and emission control. The adjustments made to the base (programmed) pulse width for operating conditions is called fuel trim. Fuel trim can be monitored for diagnostic purposes. The base pulse width is given a value of 0% and all changes are expressed as a negative or positive value. A positive fuel trim value means the PCM detects a lean mixture and is increasing the pulse width to add more fuel to the mixture. A negative fuel trim value means the PCM is reducing the amount of fuel by decreasing the pulse width to compensate for a rich condition.

The system allows for short-term and long-term fuel trim. Short-term fuel trim (STFT) represents changes made immediately in response to HO_2S signals. Long-term fuel trim (LTFT) represents changes made to set a new base pulse width. LTFT responds to the trends of the STFT. STFT is erased when the ignition is turned off. LTFT remains in the PCM's memory.

THROTTLE BODY INJECTION (TBI)

For many auto manufacturers, TBI served as a stepping stone from carburetors to more advanced fuel injection systems. TBI units were used on many engines during the 1980s and 1990s. The throttle body unit is mounted directly to the intake manifold. The injector(s) spray fuel down into a throttle body chamber leading to the intake manifold. The intake manifold feeds the air-fuel mixture to all cylinders.

Four-cylinder engines have a single throttle body assembly with one injector and throttle plate, whereas

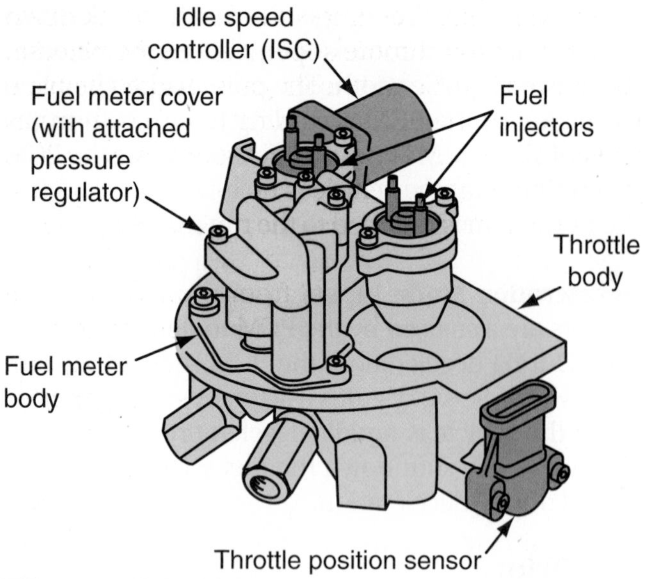

Figure 30-11 Dual throttle body assembly used on a six- or an eight-cylinder engine.

V6 and V8 engines are usually equipped with dual injectors and two throttle plates on a common throttle shaft **(Figure 30–11)**.

The throttle body assembly contains a pressure regulator, injector or injectors, TP sensor, idle speed control motor, and a throttle shaft and linkage assembly. The throttle body casting has ports that can be located above, below, or at the throttle plate depending on the design. These ports provide vacuum signals for the MAP sensor and for emission control devices, such as the EGR valve, the EVAP system, and so on.

The fuel pressure regulator is similar to a diaphragm-operated relief valve. Fuel pressure is on one side of the diaphragm and atmospheric pressure is on the other side. The regulator is designed to provide a constant pressure to the fuel injector throughout the range of engine loads and speeds. If regulator pressure is too high, a strong fuel odor is emitted and the engine runs too rich. On the other hand, regulator pressure that is too low results in poor engine performance, or detonation can take place due to the lean mixture.

CAUTION!

Always relieve the fuel pressure before disconnecting a fuel system component to avoid gasoline spills that may cause a fire, resulting in personal injury and/or property damage. Also, always wear safety glasses when there is a chance of fuel spray.

Throttle body systems are not as efficient as port injection systems. Fuel is not distributed equally to all cylinders, the air and fuel from the TBI unit pass through the intake manifold, and the length and shape of the manifold's runners affect distribution. There is also a potential problem of fuel condensing and forming puddles in the manifold when the manifold is cold.

Fuel Injectors

In a TBI system, the fuel injector is pulsed on and off by the PCM. Surrounding the inlet of the injector is a fine screen filter to remove any remaining contaminants in the fuel before it flows through the injector. When the injector is energized, fuel under pressure is injected at the directly onto the throttle shaft.

The type of fuel injector used on an engine depends on the system. Most use ball or needle valves. The valves are normally closed and open when commanded by the PCM. The injector is typically controlled through its ground circuit.

Pressure Regulation

Each injector is sealed into the throttle body with O-ring seals, which prevent fuel leakage around the injector. Fuel is supplied to the injector through a passage from the pressure regulator. The regulator has a diaphragm and valve assembly. The valve is held closed by a spring. At a specific fuel pressure, the regulator diaphragm is forced upward to open the valve, and some fuel is returned to the fuel tank to reduce the pressure.

After the pressure has decreased, the spring closes the regulator valve. This action causes the fuel pressure to increase and reopen the pressure regulator valve. The constant opening and closing of the valve maintains the pressure at the desired level. In most TBI systems, the fuel pressure is maintained at 10 to 25 psi (70 to 172 kPa).

To prevent fuel boiling in the TBI assembly, the pressure is kept high. When the pressure on a liquid is increased, the boiling point of the liquid is raised proportionally. If fuel boils in the TBI assembly, vapor and fuel will be discharged from the injectors. Vapor discharge from the injectors creates a lean air-fuel ratio, which results in lack of engine power and acceleration stumbles.

Injector Internal Design and Electrical Connections

The injector's valve is held closed by a spring. In this position, the valve closes the metering orifices in the end of the injector. Openings in the sides of the injector allow fuel to enter the cavity surrounding the injector tip. The tip of the injector may contain up to six metering orifices, but some injectors have a single metering orifice. Injector design varies depending on the manufacturer.

Each injector has two terminals, and an internal coil is connected across these terminals. The stem of the valve is positioned in the center of the injector's coil. When the ignition switch is turned on, 12 volts are supplied to one of the injector terminals, and the other terminal is connected to the computer. When the computer grounds this terminal, current flows through the injector coil to ground in the computer. The windings are now energized and the plunger is pulled up. The injector then sprays fuel out of the injector orifices into the air stream above the throttle.

In most TBI systems, the computer grounds an injector each time a signal is received from the distributor pickup. This type of TBI system may be referred to as a synchronized system because the injector pulses are synchronized with the pickup signals. In a dual injector throttle body assembly, the computer grounds the injectors alternately under most operating conditions.

PORT FUEL INJECTION (PFI)

PFI systems use at least one injector at each cylinder. They are mounted in the intake manifold near the cylinder head where they can inject a fine, atomized fuel mist as close as possible to the intake valve (**Figure 30–12**). Delivering the fuel mist right outside the combustion chamber allows the fuel to break down and vaporize a little more before it enters the cylinder. Some port injection systems have a director plate right under the nozzle of the injector. This plate has several small holes that break up the fuel as it is sprayed through the plate.

Through the years, many different PFI systems have been used and although they have things in common, they do not fire the injectors in the same way. PFI systems can be divided into two basic categories: **multiport injection (MPI)** systems and **sequential fuel injection (SFI)** systems; each is defined by injector control.

MPI Systems

Due to OBD-II regulations, MPI systems are no longer used; rather, all engines are now fitted with SFI systems. In MPI systems, the injectors were grouped together and the injectors in each group fired at the same time. Some MPI systems fired all of the injectors simultaneously. This system offered easy programming and relatively fast adjustments to the air-fuel mixture. The injectors were connected in parallel so they shared one driver circuit and the PCM sent out one signal for all injectors. The amount of fuel required for each four-stroke cycle was divided in half and delivered in two injections, one for every 360 degrees of crankshaft rotation. The fact that the intake charge

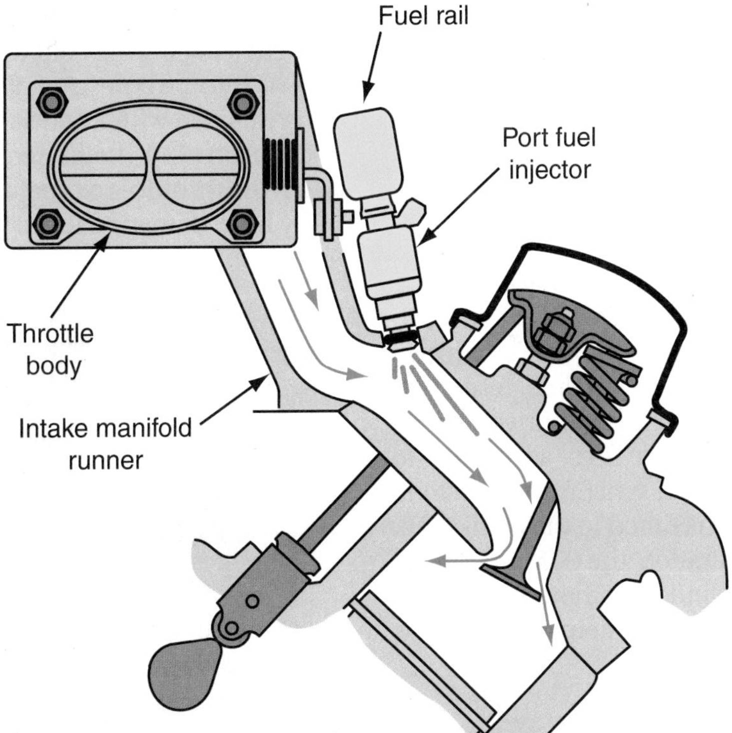

Figure 30-12 Port injection sprays fuel into the intake port and fills the port with fuel vapor before the valve opens.

had to wait in the manifold for varying periods was the system's major drawback.

In grouped systems, there is a driver and ground circuit for each group. When the injectors are split into two equal groups, the groups are fired alternatively, with one group firing each engine revolution.

Because only two injectors can be fired relatively close to the time when the intake valve is about to open, the fuel charge for the remaining cylinders must stand in the intake manifold for varying periods. These periods are very short; therefore, the standing of fuel in the intake manifold was not that great a disadvantage of MPI systems. At idle speeds this wait was about 150 milliseconds, and at higher speeds the time was much less.

A commonly used MPI system was the speed density design. Sensors in a typical speed density system included a TP sensor, O_2S, ECT, and KS, which worked with the ignition and injection system. Because speed density systems had no means to directly measure intake airflow, the injectors' pulse width was controlled by the PCM's programming and minor adjustments were not possible. Speed density systems were eventually replaced with MAF systems. MAF systems were more responsive to intake air conditions.

Sequential Fuel Injection

SFI systems use a MAF or MAP sensor as well as a variety of other sensors. These systems control each injector separately so that it fires just before the intake valve opens. This means that the mixture is never static in the intake manifold and adjustments to the mixture can be made instantaneously between the firing of one injector and the next. Sequential firing is the most accurate and desirable method of regulating port injection.

To meet OBD-II regulations, SFI systems are capable of turning off the injector at a misfiring cylinder.

In SFI systems, each injector is connected individually to the computer, and the computer completes the ground for each injector one at a time. A sequential system has one injector per injector driver.

When the injection system fires according to crankshaft speed, it is called a synchronous system. The action of the injectors is timed to the crankshaft. During starting and acceleration, the PCM may inject extra fuel into all of the cylinders without referring to inputs from the CKP. When the system does this, it is called asynchronous injection.

Throttle Body

The throttle body in a PFI system controls the amount of air that enters the engine as well as the amount of vacuum in the intake manifold. A typical throttle body assembly is comprised of an IAC valve assembly, an idle air orifice, single or double bores with throttle plates, and a TP sensor **(Figure 30–13)**.

Figure 30–13 A TP sensor mounted to a throttle body.

The throttle body housing assembly is a single-piece aluminum or plastic casting. The throttle bore and throttle plates control the amount of intake air that enters the engine. The throttle shaft and plate(s) are by the accelerator pedal, via a linkage comprised of a cam and cable. The TP sensor monitors the movement and position of the plates and sends a signal to the PCM. On some systems, a small amount of coolant is routed through a passage in the throttle body to prevent icing during cold weather.

The throttle bores and plates in many throttle bodies are coated with a special sealant. This sealant has two purposes: It helps seal the plates to the bores when the throttle is closed, and it protects the throttle body from damage caused by contaminants, such as sludge, that may build up in the intake manifold.

SHOP TALK

Throttle body assemblies that have this protective coating should not be cleaned, because any cleaning may remove the sealant. Most coated throttle bodies have a warning sticker affixed to them for identification.

The MAF sensor measures the amount of intake air as well as the air from the idle air orifice and the positive crankcase ventilation (PCV) system. This allows the PCM to know how much air is entering the intake manifold. Some throttle bodies also have a fresh air outlet above the throttle plates for the PCV and IAC systems.

When the engine is at idle and the throttle plates are closed, a small amount of air enters the engine through the idle air orifice. The IAC system allows additional air to enter if the PCM commands it to do so. Idle speed is totally controlled by the PCM and there are no provisions for adjusting it at the throttle body. Some throttle bodies have a throttle stop screw that prevents the throttle plate from seizing in its bore when it is quickly closed.

Throttle bodies also have several vacuum taps or outlets located below the throttle plates. These are used to monitor engine vacuum and provide vacuum to the PCV valve, exhaust gas recirculation (EGR) valve, evaporative emission (EVAP) system, and A/C controls.

Fuel Delivery

The fuel injectors are separately controlled and are mounted to the intake manifold and supplied pressurized fuel through fuel lines. These fuel lines run to each cylinder from a fuel manifold, usually referred to as a **fuel rail** (**Figure 30–14**). The fuel rail also has a fuel pressure test port, and on some engines, a fuel rail pressure (FRP) sensor. The rail is supplied fuel by the fuel pump and distributes the same amount of fuel at the same pressure to each cylinder. Because the fuel is dispensed at each cylinder and not the intake manifold, there is little or no fuel to wet the manifold walls; therefore, there is no need for manifold heat or early fuel evaporation system.

The fuel rail assembly on a PFI system of V6 and V8 engines usually consists of a left- and right-hand rail assembly. The two rails can be connected by crossover and return fuel tubes. A typical fuel rail is shown in **Figure 30–15**.

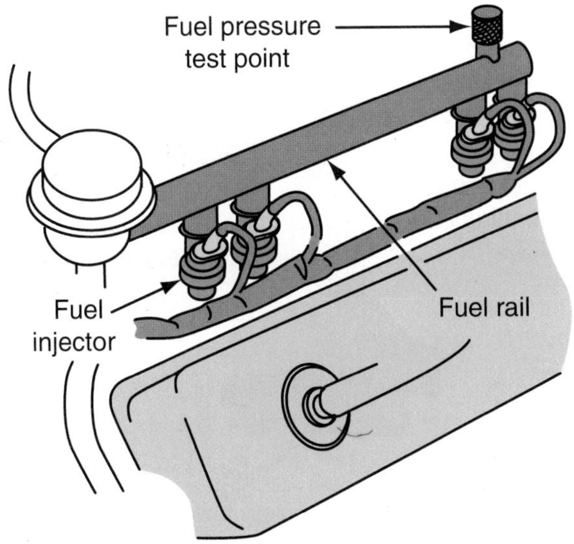

Figure 30–14 A typical fuel rail.

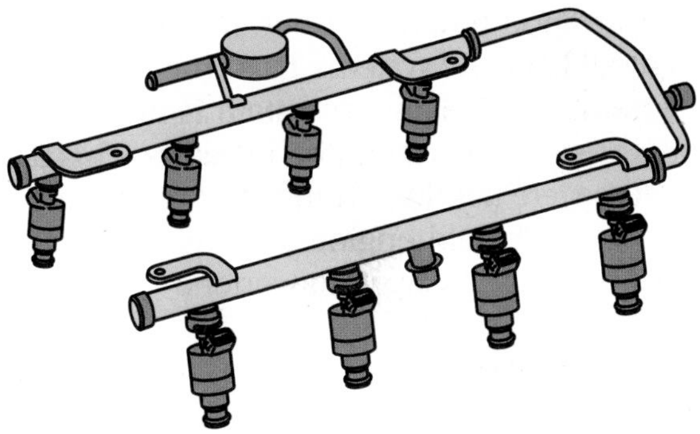

Figure 30–15 A one-piece fuel rail for a V8 engine.

Some engines are now equipped with variable induction intake manifolds that have separate runners for low and high speeds. This technology is only possible with port injection.

Other engines incorporate a fuel pressure sensor that monitors the pressure at the fuel injectors. The PCM controls fuel pressure based on this input. The sensor receives a reference voltage of 5 volts, and changes in fuel rail pressure are reflected by the signal back to the PCM. Basically, when fuel pressure is high, the signal has high voltage and when the pressure is low, the signal is low.

Pulsation Damper Some engines have a pulsation damper on their fuel rails. These dampers reduce the pressure pulsations caused by the rapid opening and closing of the fuel injectors. Without control over these fluctuations, the fuel pressure at each injector can be affected. The damper works to control the volume of fuel in the fuel rail. When the pressure in the rail quickly drops, the damper temporarily reduces the volume of fuel in the rail to prevent the fuel pressure from becoming too low. The opposite occurs when the pressure quickly increases. The damper also reduces fuel noise and maintains pressure during engine cool down.

Injector Control

The computer is programmed to ground the injectors well ahead of the actual intake valve openings so the intake ports are filled with fuel vapor before the intake valves open. In both SFI and MFI systems, the computer supplies the correct injector pulse width to provide a stoichiometric air-fuel ratio. The computer increases the injector pulse width to provide air-fuel ratio enrichment while starting a cold engine.

SFI requires inputs from the CKP and CMP sensors in order to determine when to fire each injector. On many systems, once the position of piston number 1 from the CKP is received by the PCM, signals from the CMP are used to synchronize the injectors to the engine's firing order.

A clear flood mode is also available in the computer in MFI and SFI systems. On some TBI, MFI, and SFI systems, if the ignition system is not firing, the computer stops operating the injectors. This action prevents severe flooding from long cranking periods while starting a cold engine and prevents raw fuel from entering the catalytic converters.

Pressure Regulators

PFI systems require an additional control not required by TBI units. In a TBI, the injectors are mounted above the throttle plates and are not affected by fluctuations in manifold vacuum; PFI have their tips located in the manifold where constant changes in vacuum would affect the amount of fuel injected. To compensate for these fluctuations, port injection systems have a fuel pressure regulator that senses manifold vacuum and continually adjusts the fuel pressure to maintain a constant pressure drop across the injector. The pressure regulators respond to changes in manifold pressure due to engine load.

PFI systems have a fuel pressure regulator located on or near the fuel rail **(Figure 30–16)**. A diaphragm and valve assembly is positioned in the center of the regulator, and a diaphragm spring seats the valve on the fuel outlet. This is a "return" fuel system.

When fuel pressure reaches the regulator setting, the diaphragm moves against the spring tension and the valve opens. This allows fuel to flow through the return line to the fuel tank. The fuel pressure drops slightly when the regulator valve opens and builds when the spring closes the valve.

A vacuum hose is connected from the intake manifold to the vacuum inlet on the pressure regulator. This hose supplies vacuum to the diaphragm. The vacuum works against fuel pressure to move the diaphragm and open the valve. When the engine is running at idle speed, high manifold vacuum is applied to the pressure regulator. Under this condition, fuel pressure opens the regulator valve.

If the engine is operating at wide-open throttle (WOT), no vacuum (high manifold pressure) is applied to the pressure regulator. This high pressure prevents the regulator valve from opening. This is good because the injectors are discharging fuel into a higher pressure when compared to idle speed conditions. If the fuel pressure remained constant at idle and WOT conditions, the injectors would discharge less fuel into the higher pressure in the intake manifold at WOT. The increase in fuel pressure supplied by the pressure regulator at WOT maintains the same pressure drop across the injectors at idle speed and WOT. When this same pressure drop is maintained, the change in pressure at the injector does not affect the amount of fuel discharged by the injectors.

Returnless Systems Nearly all late-model injection systems have a fuel pressure regulator as part of the fuel sender and pump assembly in the fuel tank **(Figure 30–17)** and are called returnless systems. The term is applied because the regulator sits inside the fuel tank and there is no designated return fuel line and its action is not related to engine vacuum. With a returnless system, only the fuel needed by the engine is filtered, thus allowing the use of a smaller fuel filter. When the engine is at idle, the fuel pressure should be between 55 and 60 psi (80 and 410 kPa). If the pressure regulator supplies a fuel pressure that is too low or too high, a driveability problem can result. The system also has a switch that shuts off the fuel when a collision has occurred.

Figure 30–16 A typical fuel pressure regulator for a port injection system.

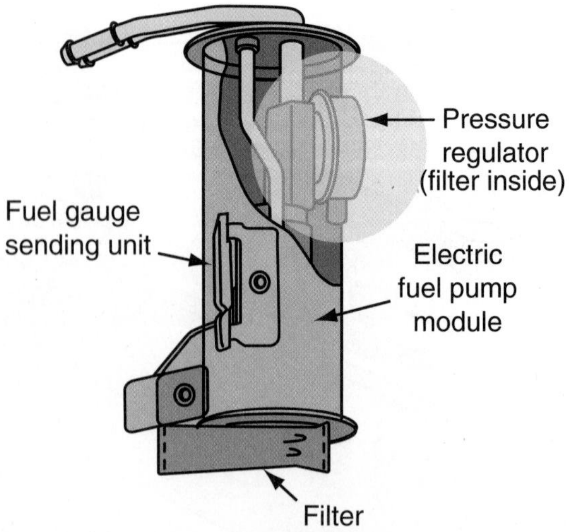

Figure 30–17 A returnless fuel system with the pressure regulator and filter mounted in the fuel tank with the fuel pump and fuel gauge sending unit.

Throttle-by-Wire Systems

Throttle-by-wire systems are becoming more common. These systems eliminate the need for a throttle cable and linkage. They also provide many other benefits over traditional throttle controls, such as improved driveability, increased fuel efficiency, and decreased emissions. Throttle-by-wire systems are used by many manufacturers; each calls their system by a different name, the most common being electronic throttle body (ETB) and electronic throttle control (ETC). There are differences in the programming and hardware used.

Based on inputs from accelerator pedal sensors and its programming, the PCM electronically controls the position of the throttle plate **(Figure 30–18)**. A DC motor that is controlled by the PCM is used to move the plate. The motor may be an integral part of the throttle body or a separate unit mounted to the throttle body.

As a safety feature, two springs are attached to the throttle plate shaft. One, which is the stronger of the two, is used to open the throttle plate slightly if the PCM loses control of engine speed. This allows for limp home operation. The other spring is used to close the throttle. To prevent the throttle plate from closing too much and possibly binding its bore, a hard stop is on the throttle body assembly to limit the closure of the throttle plate.

Driver input to the PCM is delivered by accelerator pedal sensors. A system may have more than one pedal sensor or may use a sensor that sends out more than one signal. In either case, the PCM will receive multiple signals regarding accelerator pedal position. This redundancy is important because it allows the PCM to closely monitor the action of the system. The multiple signals ensure that the PCM is receiving correct information.

The TP sensor also provides redundant signals. The multiple signals ensure that the PCM knows where the throttle plate is positioned at all times. These TP sensors are recognizable because they have four electrical leads.

Redundancy also occurs in the processing of the inputs and the controlling of the throttle motor. The throttle system is monitored by two separate processors inside the PCM. Both are looking at the same things and if one determines that something is wrong with the other, it will override the commands of that processor.

In addition to eliminating the throttle linkage, electronic throttle systems also eliminate the need for an IAC valve and idle air orifice. These are not needed because the PCM can control throttle plate opening to meet the air needs of the engine when it is idling.

ETCs are also used with variable valve timing. Because the engine's power changes with a change in valve timing, the desired position of the throttle for those conditions also changes. The PCM can instantly change the throttle position to eliminate any noticeable change in engine power.

ETC also works with electronic-controlled automatic transmissions. The PCM can alter throttle position when the transmission is shifting gears to improve shift quality. This allows the transmission to make quicker gear changes smoothly.

Other advantages include the elimination of cruise control actuators while providing improved speed

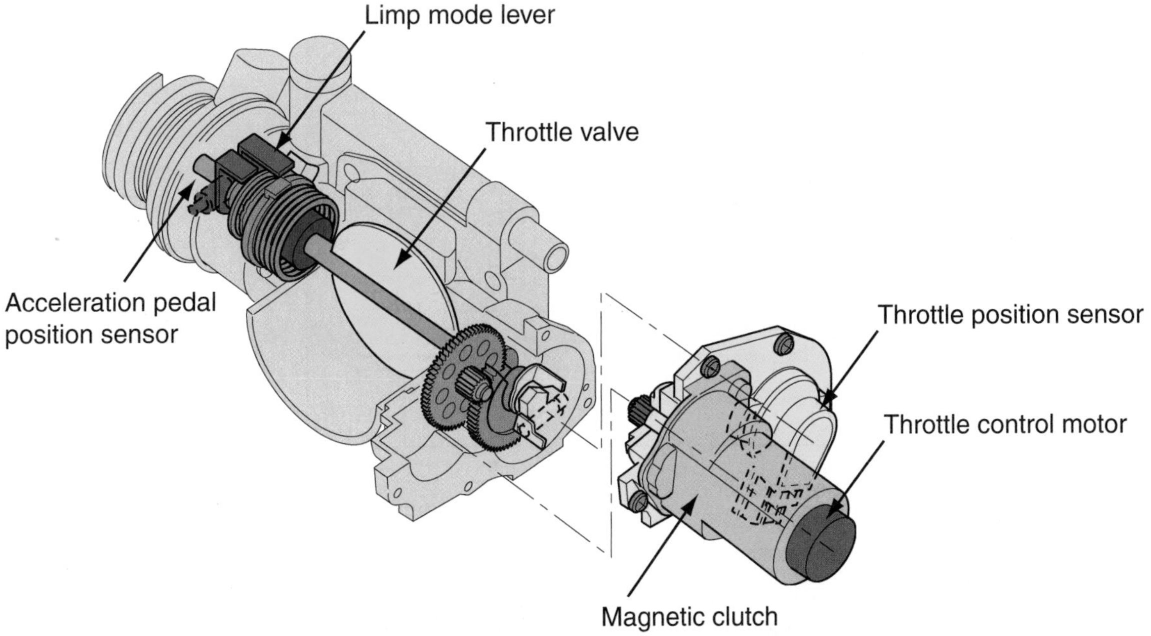

Figure 30–18 An electronic throttle body (ETB) assembly. *Courtesy of Toyota Motor Sales, U.S.A., Inc.*

control. ETC also allows the PCM to be more effective when limiting engine and vehicle speeds. ETC is also used in automatic traction control systems.

CENTRAL MULTIPORT FUEL INJECTION (CMFI)

In a **central port (CPI)** or **central multiport injection (CMFI)** system, a central injector assembly is mounted in the lower half of the intake manifold. The CMFI system uses one injector to control the fuel flow to individual **poppet nozzles** (**Figure 30–19**). The CMFI injector assembly consists of a fuel metering body, a pressure regulator, one fuel injector, poppet nozzles with nylon fuel tubes, and a gasket seal. The injector distributes metered fuel through a distribution gasket that seals the injector to the lines connected to the nozzles. To meet OBD-II regulations, shut off the solenoids that were installed at the poppet injectors to allow the injectors to be shut off at misfiring cylinders.

Pressure Regulator

The pressure regulator is mounted with the central injector. The regulator is inside the intake manifold and intake vacuum is supplied through an opening in the regulator cover. Normally, a regulator spring pushes downward on a diaphragm and closes the valve. Fuel pressure from the fuel pump pushes the diaphragm upward. When the pressure exceeds that of the spring, the valve opens. Pressure is decreased as fuel flows through the valve and a return line to the fuel tank (**Figure 30–20**).

The pressure regulator maintains fuel pressure at 54 to 64 psi (370 to 440 kPa), which is higher than many PFI systems. Higher pressure is required in the CMFI system to prevent fuel vaporization from the extra heat encountered with the CMFI assembly, poppet nozzles, and lines mounted inside the intake manifold.

Injector Design and Operation

The armature is placed under the injector winding in the central injector. The lower side of this armature acts as a valve that covers the outlet ports to the nylon tubes and poppet nozzles. A supply of fuel at a constant pressure surrounds the armature while the ignition switch is on. Each time the PCM grounds the injector, the armature is lifted up, which opens the injector ports. Fuel is then forced to the poppet nozzles (**Figure 30–21**).

The PCM controls the amount of fuel delivered by the central injector by controlling its pulse width. The injector winding has low resistance, and the PCM operates the injector with a peak-and-hold current. When the PCM grounds the injector winding, the current flow in this circuit increases rapidly to 4 amperes. When the current flow reaches this value, a current-

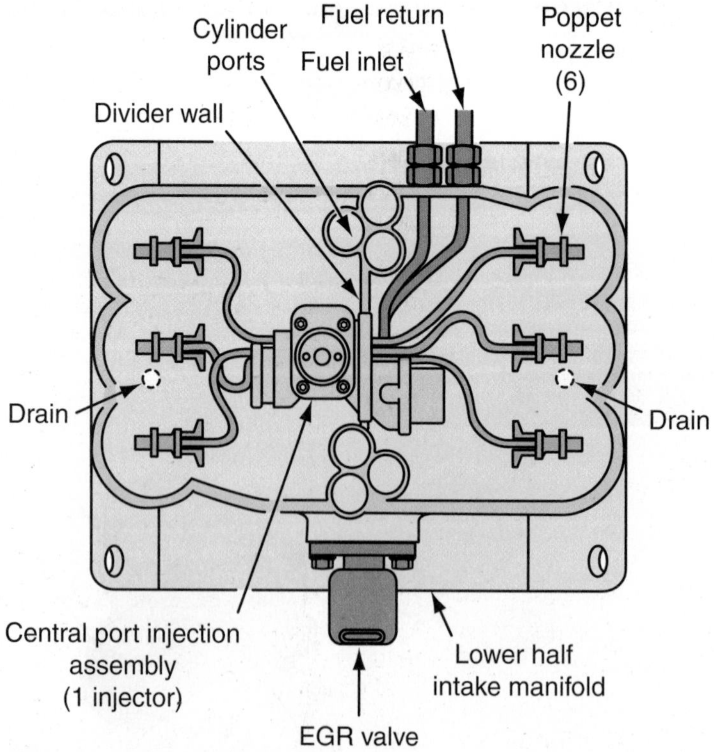

Figure 30-19 Central multiport fuel injection components in the lower half of the intake manifold.

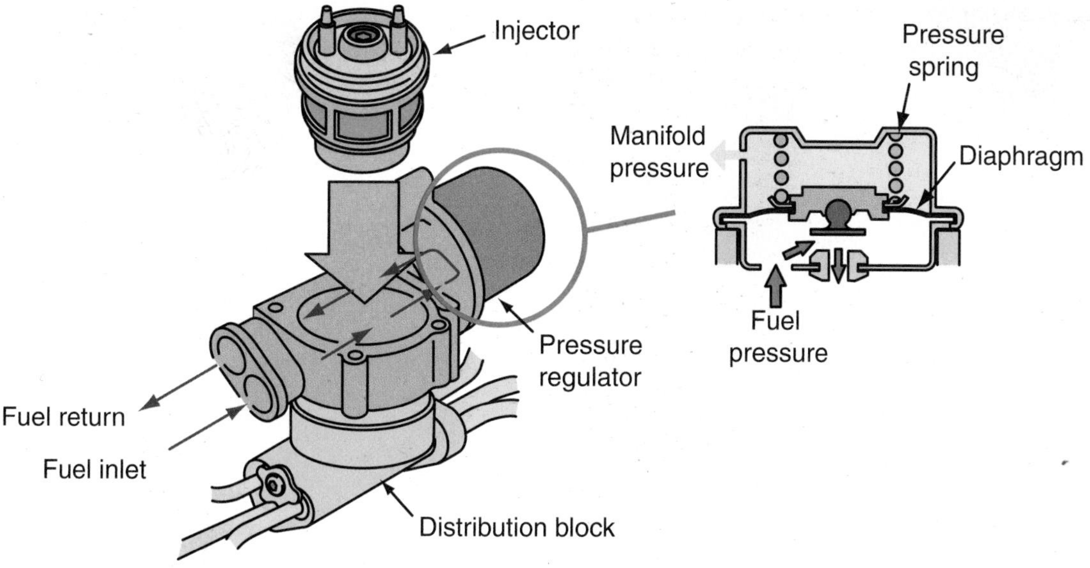

Figure 30-20 A pressure regulator for a CMFI system.

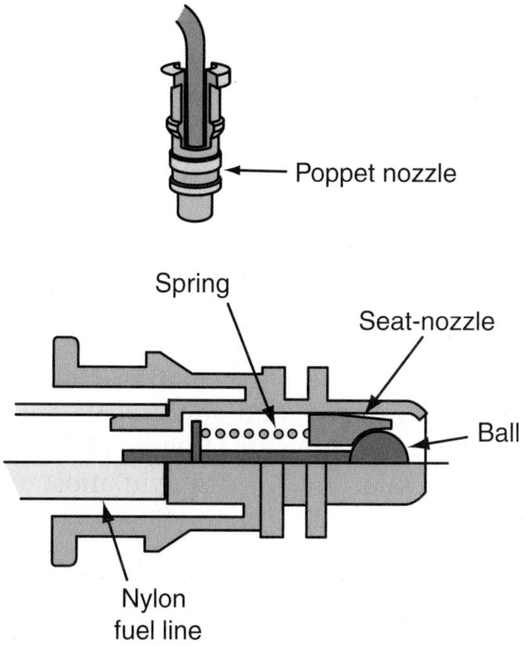

Figure 30-21 The internal design of a poppet nozzle.

limiting circuit in the PCM limits the current flow to 1 ampere for the remainder of the injector pulse width. The peak-and-hold function provides faster injector opening and closing.

Poppet Nozzles

The poppet nozzles are snapped into openings in the lower half of the intake manifold, and the tip of each nozzle directs fuel into an individual intake port. Each poppet nozzle contains a check ball and seat at the tip of the nozzle. A spring holds the check ball in the closed position. When fuel pressure is applied to the nozzles, the pressure forces the check ball to open against spring pressure. The poppet nozzles open when the fuel pressure exceeds 37 to 43 psi (254 to 296 kPa), and the fuel sprays from these nozzles into the intake ports.

When fuel pressure drops below this value, the nozzles close. Under this condition, approximately 40 psi (276 kPa) fuel pressure remains in the nylon lines and poppet nozzles. This pressure prevents fuel vaporization in the lines and nozzles during hot engine operation or hot soak periods. If a leak occurs in a line or other CMFI component, fuel drains from the bottom of the intake manifold through two drain holes to the center cylinder intake ports.

GASOLINE DIRECT-INJECTION SYSTEMS

Direct injection has been around for many years on diesel engines. Until recently, this type of injection system has been seldom used with gasoline. With direct injection, highly pressurized fuel is sprayed directly into the cylinders. This type of injection is used by many auto manufacturers and has many different names **(Figure 30–22)**. The most commonly used in gasoline direct injection (GDI); GDI, however, is a registered trademark of Mitsubishi Motors and can only be used by that manufacturer. Therefore, the systems are called gasoline direct injection (GDi), fuel stratified injection (FSI), high precision injection, direct injection (DI), or direct injection spark ignition (DISI). The fuel is highly pressurized when it is sprayed into the cylinders. Under this pressure, the fuel arrives as very fine droplets.

Injectors

With gasoline direct injection, specially designed injectors deliver the fuel into the high pressures and temperatures in the cylinders **(Figure 30–23)**. To

Figure 30-22 A turbocharged direct-injection-equipped engine from Volkswagen.

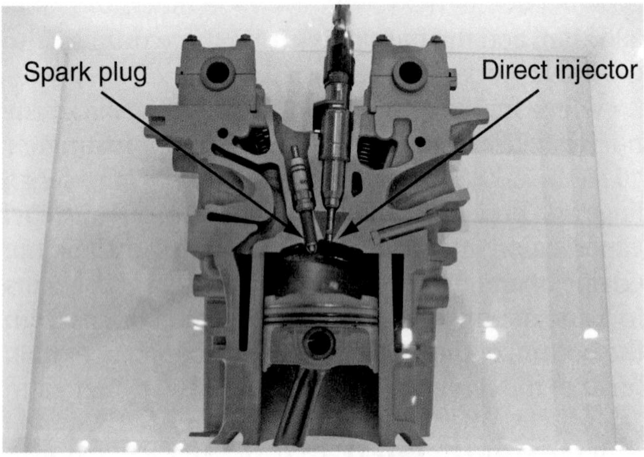

Figure 30-23 The direct injector is placed in the combustion chamber so its spray of fuel is aimed at the spark plug.

Figure 30-24 An injector for gasoline direct injection. *Courtesy of Delphi Corporation*

prevent the heat from igniting the fuel in the injector, the injectors are designed to completely seal after the fuel is sprayed. The injectors must also be able to spray the fuel at a much higher pressure than what is in the cylinder. If this did not happen, the fuel would not enter the cylinder; instead the cylinder's pressure would enter the injector. Remember, a high pressure always moves to a point of lower pressure. The injectors need more electrical power to work with the high pressure; therefore, the PCM has separate high voltage and driver circuits for each injector. The injectors must also be resistant to deposit formations and provide a highly atomized and directed spray of fuel. The fuel injectors are normally made with a small extended tip **(Figure 30–24)**. This allows the injector to quickly move the heat, which is absorbed during combustion, to the cooling jackets in the cylinder head.

The PCM controls the pulse width and timing of each injector and allows the system to operate in very distinctly different modes. Fuel is injected before or after the intake valve is closed, depending on the operational mode. The pulse width also changes with the operational mode, and adjustments are made according to inputs from the MAF and IAT sensors.

Solenoid injectors are used in most direct-injection systems. However, some are piezoelectrically actuated injectors. Piezoelectric injectors rely on stacked crystals. When voltage is applied to the crystals, they change size. The pintle of the injector is attached to the crystals and when the size of the crystals changes, the pintle moves. Piezoelectric injectors have a much faster response time than solenoid injectors.

With GDI, fuel can be injected at any time, not only when the intake valve is open. Also, the injectors can pulse twice during the transition from the compression stroke to combustion. The two pulses promote complete combustion when the PCM senses that operating conditions may prevent a complete burning of the fuel.

High-Pressure Fuel Pump

Gasoline is moved from the fuel tank to the engine in a conventional way, which is by an in-tank electric pump. The fuel is delivered to a mechanical, high-pressure pump driven by an eccentric on the end of a

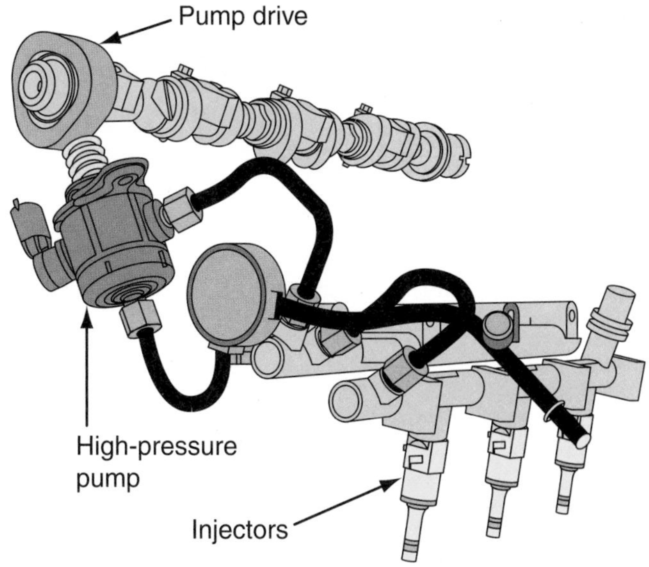

Figure 30-25 The high-pressure fuel pump for a GDI system is driven by a camshaft.

camshaft **(Figure 30–25)**. This pump supplies fuel to a variable-pressure fuel rail. The individual injectors are attached to the rail. A GDI system can operate with pressures of 435 to 1,885 psi (33 to 130 bar). The pump is a volume-controlled, high-pressure pump. It only moves the required amount of fuel to the injectors **(Figure 30–26)**.

The PCM regulates the pressure in the fuel rail based on signals from the inputs from the fuel rail pressure regulator that is located on the fuel rail or is part of the high-pressure pump. The pressure is regulated by controlling the amount of fuel that enters the high-pressure pump or by changing the effective pumping stroke of the pump. Controlling the pump inlet is the most common. The PCM controls the power and ground circuits of the regulator. When the regulator is de-energized, the inlet valve is held open by spring pressure. When it is energized, it closes the valve. Through pulse width modulation, the pressure from the pump can be maintained at a value that is best for the operating conditions.

The PCM uses CKP and CMP signals to synchronize the action of the regulator with position and movement of the eccentric on the camshaft.

To protect the system from excessive pressures, there are pressure relief valves in the pump or fuel rail. When pressure reaches a predetermined value, some fuel leaves the relief valve and returns to the fuel tank.

Operational Modes

Most direct-injection systems can operate in three distinct modes: lean burn, stoichiometric, and full power. Each of these modes has different air-fuel ratios, injection timing, and pump pressures. The PCM chooses the mode based on operating conditions. The lean mode relies on a stratified charge for combustion, and the stoichiometric and full-power modes rely on a homogeneous mixture, which means that the air and fuel are well mixed **(Figure 30–27)**. The PCM also must be able to smoothly transition the move from one to another. The systems also have a limp-home mode if the PCM detects a problem with the system. During this time, the PCM will use a fuel strategy that allows the engine to run well enough to drive the vehicle in for service.

It is important to note that not all direct-injection systems have a lean burn mode. These systems use this technology for power gains and emission reduction, not to minimize fuel consumption.

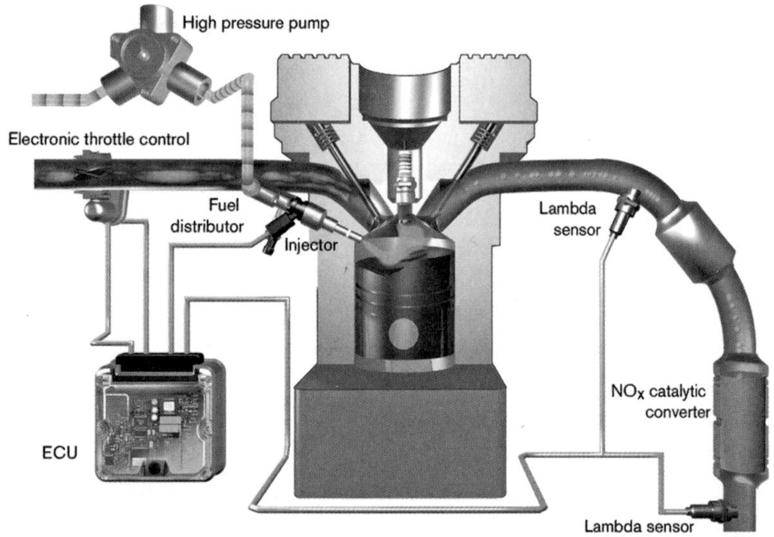

Figure 30-26 The component layout for a GDI system. *Courtesy of Robert Bosch GmbH, www.bosch-presse.de*

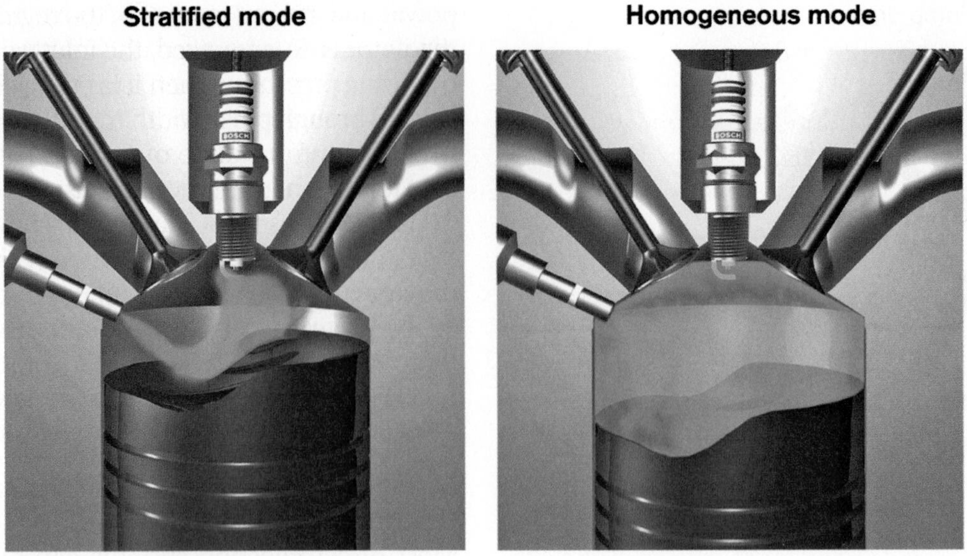

Figure 30-27 Two of the basic operation modes of a GDI system. *Courtesy of Robert Bosch GmbH, www.bosch-presse.de*

Lean Burn Mode Direct-injected engines are able to run at very lean mixtures, with air-fuel ratios as high as 60:1. This benefit is why GDI engines are capable of drastically reducing fuel consumption. The engines run in the lean mode when the vehicle is cruising with a very light load. In addition to a reduction in fuel consumption, exhaust emissions are also lowered. The lean mixtures are possible because the system allows for a stratified charge.

The placement of the injector tip in the combustion chamber is an important feature of a direct-injection system, especially when it is operating in the lean mode. A small amount of fuel is sprayed near the spark plug when the piston has nearly completed its compression stroke but before ignition occurs. There is enough fuel around the spark plug to cause combustion. This local area of combustion is the **stratified charge**. The area surrounding that small area has little or no fuel and is mostly air or recirculated exhaust gas. The air surrounding the stratified charge forms an insulating cushion to keep fuel away from the cylinder walls.

Stoichiometric Mode During medium-load operation, the system operates in the stoichiometric mode. The air-fuel ratio is near stoichiometric and the fuel is injected during the intake stroke.

Full-Power Mode The full-power mode is used during heavy loads and hard acceleration. The air-fuel mixture is slightly richer than stoichiometric and the fuel is again injected during the intake stroke.

Compression Ratios

The ability of a direct injection to change air-fuel ratios and injection timing over a wide range also allows it to eliminate most conditions that would cause engine knock. This means GDI engines can operate at higher compression ratios without requiring the use of high-octane gasoline. The benefit of high compression is simply higher compression extracts more energy from each droplet of fuel. Therefore, running higher compression ratios provides increased engine horsepower and torque without consuming more fuel.

Higher compression ratios are also possible because the small droplets of fuel injected directly into the cylinder tend to cool the mixture in the cylinder. This cooling makes the mixture denser, means more power can be produced, and makes the mixture less likely to detonate. Because regardless of the operational mode the fuel is sprayed around the spark plug, the mixture burns quickly. This means there is less need for spark advance; this also decreases the chances of detonation.

Advantages of GDI

When compared to other injection systems, direct injection has the following advantages:

- Increases fuel efficiency
- Provides higher power output
- Increases the engine's volumetric efficiency
- Lowers engine thermal losses
- Decreases emissions (NO_x may increase if combustion temperatures go too high.)
- Allows the engine to have high compression without the need to use high-octane fuel
- Reduces most of the turbo lag when used with a turbocharger

GDI Plus SFI

Some engines from Toyota and Lexus use a combination of direct and indirect injection. Each cylinder has two injectors: one at the intake port and the other directly in the cylinder **(Figure 30–28)**. Both sets of injectors work in the same way as they would in a normal setup. However, the PCM shuts down the port injection when it is not needed. When the engine is cold or running at low to middle speeds with a light load, both injection systems operate. During high engine speeds and heavy loads, only direct injection is used. During cold starts, the port and direct injectors work together to create a lean stratified charge. Fuel is initially injected into the intake port during the exhaust stroke. Fuel is then injected from the direct injector near the end of the piston's compression stroke. This creates a stratified charge. This charge burns rapidly and therefore quickly heats the chamber; this in turn quickly warms up the catalytic converter.

The PCM controls the injection volume and timing of each injector according to engine load, intake airflow, temperature, and other inputs. This system takes advantage of the benefits of both types of injection systems. A port injection system is more efficient at slight throttle openings, whereas a direct-injection system is best with engine speed and load. The goal of

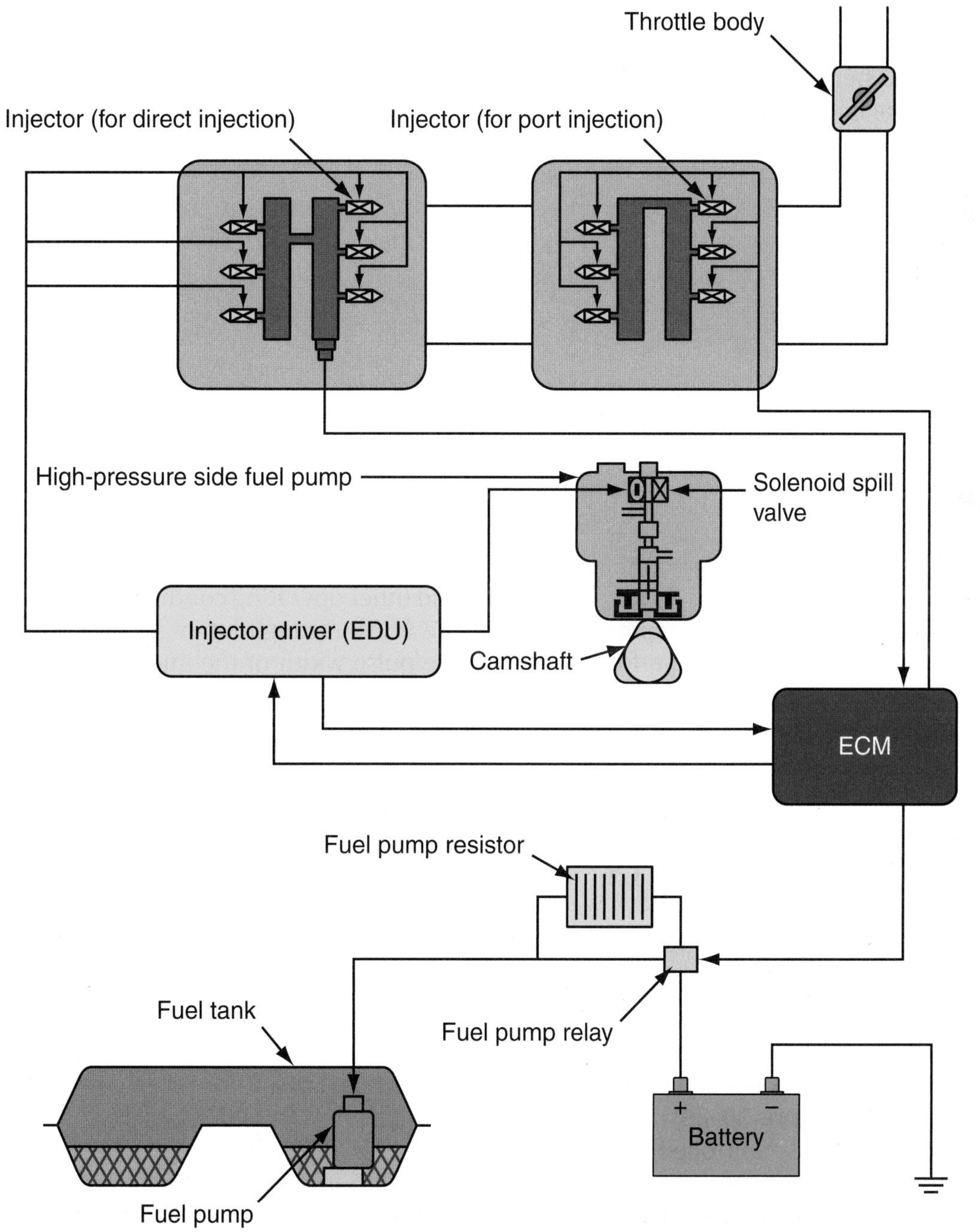

Figure 30–28 Toyota's dual system.

the system is improved performance and decreased fuel consumption and emissions during all operating conditions.

Fuel is delivered to both injection systems by the low-pressure fuel pump in the tank. The fuel for the direct-injection system is sent to a high-pressure pump driven by the exhaust camshaft. At the end of the camshaft there is a three-sided lobe that rides on the plunger of the high-pressure pump. The plunger is moved up and down, or stroked, three times for every one complete revolution of the lobe.

The injectors in the direct system have a two slit orifice at the nozzle. This nozzle is designed to provide a fuel spray that fans out toward the spark plug. The injectors are pulsed by an electronic driver unit (EDU) that sends a high-voltage signal to the injectors. The EDU is controlled by the PCM by pulse width modulation. Therefore, the PCM is able to control the timing of the injectors.

The PCM also controls the pressure in the direct-injection system. In doing this, the PCM also is controlling the amount of fuel that is injected into the cylinders. A fuel pressure sensor on the fuel rail monitors the pressure and sends a signal to the PCM. The PCM calculates the required pressure for the current conditions and orders the EDU to alter the action of a spill control valve, if necessary.

The spill valve is located at the inlet passage of the high-pressure pump and is used to control the pump discharge pressure. It is electrically opened and closed by the EDU. The pressure is regulated by controlling the amount of time the valve is closed.

Fuel is drawn in by the pump when the valve is open and the pump's plunger is moved downward by its spring. The EDU will then close the spill valve and the pump's plunger will move up. The force of the lobe on the plunger will put pressure on the fuel inside the pump. Once the fuel pressure is strong enough to open a check valve at the outlet of the pump, fuel will flow through the fuel rail to the injectors.

LIGHT- AND MEDIUM-DUTY DIESEL FUEL INJECTION

Most late-model diesel injection systems are very similar to GDI systems. Actually, it is more accurate to say that the GDI systems are very similar to today's light- and medium-duty diesel injection systems. Nearly all of today's diesel systems have a common rail injection system **(Figure 30–29)** in which the injectors are totally directly controlled by the PCM. However, some diesel engines use a mechanical-type system and only the fuel pressure and emission controls are regulated by the PCM.

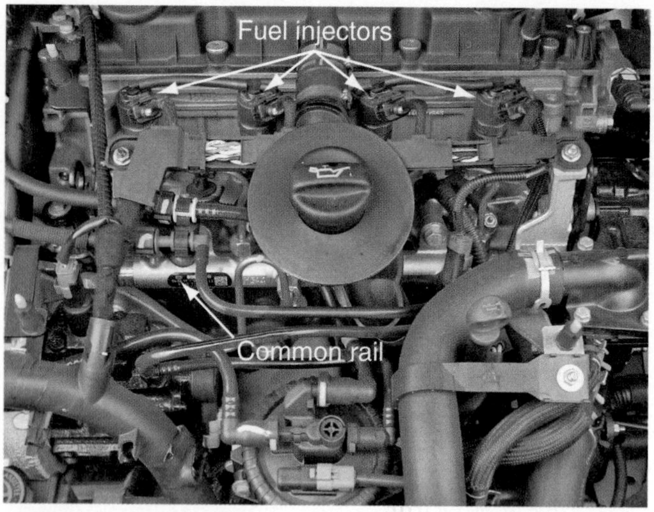

Figure 30–29 A common rail diesel fuel injection system.

To meet OBD standards, diesel control systems use many of the same inputs as a gasoline engine (such as the MAP, IAT, MAF, TP, CKP, and CMP sensors). They also share many of the same diagnostic routines. The only difference, and perhaps the biggest one, is the way the air-fuel mixture is ignited. With compression injection, fuel control is one of the keys to engine efficiency. Most systems control the timing and fuel pressure electronically. Diesel engines run at much higher compression ratios than gasoline engines; therefore, the injectors need to release fuel at a much higher pressure.

In a typical system, the PCM calculates engine speed using inputs from the CKP. The PCM then determines the ideal fuel requirements for that speed and other operating conditions. In response to these, the PCM, through a fuel injection module, controls the pulse width of the injectors and pressure in the fuel rail. A mechanically driven pump supplies high pressure to the fuel rail. The fuel rail is fitted with a pressure regulator that is controlled by commands from the PCM. The pressure in the fuel rail can range from 2,000 psi (138 bar) to 25,000 psi (1,724 bar). Often, fuel is routed through the fuel injection module to cool it before it is delivered to the injectors.

The design of the injectors varies with application; some are solenoid actuated and others use piezoelectric injectors. Some high-pressure common rail fuel systems are capable of injecting fuel five times during one compression stroke. This is done to keep the fuel under high pressure, reduce combustion noise, and provide lower emissions. High voltage is required to operate most solenoid injectors, normally around 90 volts. Some, however, require only battery voltage.

Diesel engines produce 20% to 30% less carbon dioxide (CO_2) emissions, lower carbon monoxide

(CO) emissions, better fuel economy, and more low-speed torque than a comparably sized gasoline engine. With precise control of fuel delivery through electronics, the use of exhaust-stream after treatment, and an oxidation catalytic converter; oxides of nitrogen (NO_x) and particulate matter (PM) are also drastically reduced.

Electromechanical Systems

A few diesels have mechanically actuated injectors that are fitted with a solenoid to control fuel pressure. In these units, the high-pressure pump and injector are in a single assembly and there is one these for each cylinder. The pumps are driven by a special lobe on the camshaft; this third lobe sits next to the intake and exhaust lobes for each cylinder. Because the injectors are mechanically driven, their timing is fixed. Fuel control is accomplished by regulating fuel pressure. To do this, the PCM controls the solenoids at the injectors.

CASE STUDY

A customer phoned to say that he was having his SFI Cadillac towed to the shop because the engine had stopped and would not restart. Before the technician started working on the car, he routinely checked the oil and coolant condition. The engine oil dipstick indicated that the crankcase was severely overfilled with oil and the oil had a strong smell of gasoline. The technician checked the ignition with a test spark plug and found the ignition system to be firing normally. Of course, the technician thought that the no-start problem must be caused by the fuel system and the most likely causes would be the fuel pump or filter.

The technician removed the air cleaner hose from the throttle body and removed the air filter to perform a routine check of the filter and the throttle body. The throttle body showed evidence of gasoline lying at the lower edge of the throttle bore. The technician asked a coworker to crank the engine while he looked in the throttle body. While cranking the engine, gasoline flowed into the throttle body below the throttle plate. The technician thought this situation was impossible. An SFI system does not inject fuel into the throttle body.

He began thinking about how fuel could be getting into the throttle body and reasoned that the fuel had to be coming through one of the vacuum hoses. Next, he thought about which vacuum hose could be the source of this fuel. He remembered that the pressure regulator vacuum hose is connected to the intake manifold. He then removed the vacuum hose from the fuel pressure regulator and placed the end of the hose into a container. When the engine was cranked, fuel squirted out of the hose. This meant that the fuel pressure regulator diaphragm was leaking and allowing fuel to be drawn into the manifold.

A new fuel pressure regulator was installed and the spark plugs were cleaned. Then the engine's oil was changed with the filter. After all of this, the engine started and ran normally.

KEY TERMS

Central port injection (CPI)
Central multiport injection (CMFI)
Clear flood mode
Fuel rail
Idle air control (IAC)
Multiport injection (MPI)
Poppet nozzles
Sequential fuel injection (SFI)
Speed density
Stoichiometric ratio
Stratified charge

SUMMARY

- EFI systems are computer controlled and designed to provide the correct air-fuel ratio for all engine loads, speeds, and temperature conditions. The ideal fuel ratio is called the stoichiometric ratio, which means the air-to-fuel ratio can allow for complete combustion.
- EFI systems rely on inputs from various sensors; these include airflow, air temperature, mass airflow, manifold absolute pressure, exhaust oxygen context, coolant temperature, and throttle position sensors.
- The volume airflow sensor and mass airflow sensors determine the amount of air entering the engine. The MAP sensor measures changes in the intake manifold pressure that result from changes in engine load and speed.

- The heart of the fuel injection system is the electronic control unit. The PCM receives signals from all the system sensors, processes them, and controls the fuel injectors.
- In an EFI system, the computer supplies the proper air-fuel ratio by controlling injector pulse width.
- Port injection systems use one of four firing systems: grounded single fire, grouped double fire, simultaneous double fire, or sequential fire. All vehicles built after 1996 use SFI.
- Two types of fuel injectors are commonly used: top feed and bottom feed. Top-feed injectors are used in port injection systems. Bottom-feed injectors are used in throttle body injection systems.
- In any EFI system, the fuel pressure must be high enough to prevent fuel boiling.
- Most computers provide a clear flood mode if a cold engine becomes flooded. Pressing the gas pedal to the floor while cranking the engine activates this mode.
- In an SFI system, each injector has an individual ground wire connected to the computer.
- The pressure regulator maintains the specified fuel system pressure and returns excess fuel to the fuel tank.
- In a returnless fuel system, the pressure regulator and filter assembly is mounted with the fuel pump and gauge sending unit assembly on top of the fuel tank. This pressure regulator returns fuel directly into the fuel tank.
- A central multiport injection system has one central injector and a poppet nozzle in each intake port. The central injector is operated by the PCM, and the poppet nozzles are operated by fuel pressure.
- GDI systems inject gasoline directly into the combustion chamber and produces a stratified air-fuel charge that allows for complete combustion with lean air-fuel ratios.
- GDI systems use special injectors that spray the gasoline at very high pressures and seal extremely well when they are not open.
- Most current light- and medium-duty diesel engines use electronically controlled common rail injection systems.

REVIEW QUESTIONS

1. Explain the major differences between throttle body fuel injection and port fuel injection systems.
2. What is meant by sequential firing of fuel injectors?
3. Describe the purpose of an ECT signal on an EFI system.
4. Explain how the computer controls the air-fuel ratio on an EFI system.
5. *True or False?* Diesel engines require higher fuel pressures than comparably sized gasoline engines.
6. Describe the purpose of a manifold absolute pressure (MAP) sensor.
7. Explain the basic operation of a CMFI system.
8. *True or False?* In GDI systems, the PCM controls fuel pressure through pulse width modulation of the electric high-pressure fuel pump.
9. When the throttle plates are closed in a port injection system, where does the air needed for idle speeds come from?
10. If the injector pulse width is increased on TBI, MFI, or SFI systems, the air-fuel ratio becomes ________.
11. When the engine is idling, the pressure regulator provides ________ fuel pressure compared to the fuel pressure at wide-open throttle.
12. Why are special fuel injectors needed for GDI systems?
13. How does a GDI system provide for a stratified charge in the combustion chamber?
14. How does a piezoelectric injector work?
15. Which of the following statements about fuel trim is *not* true?
    a. A positive fuel trim value means that the PCM detects a lean mixture and is decreasing the pulse width to add more fuel to the mixture.
    b. Short-term fuel trim represents changes made immediately in response to HO_2S signals.
    c. Long-term fuel trim represents changes made to set a new base pulse width.
    d. Long-term fuel trim responds to the trends of the short-term fuel trim.

ASE-STYLE REVIEW QUESTIONS

1. While discussing EFI systems: Technician A says that the PCM provides the proper air-fuel ratio by controlling the fuel pressure. Technician B says that the PCM provides the proper air-fuel

ratio by controlling injector pulse width. Who is correct?

a. Technician A
b. Technician B
c. Both A and B
d. Neither A nor B

2. While discussing electronic fuel injection principles: Technician A says that the PCM always adjusts the air-fuel ratio in response to the O_2S signals. Technician B says that cold air is dense, which is why the PCM enrichens the mixture when the engine is cold. Who is correct?

a. Technician A
b. Technician B
c. Both A and B
d. Neither A nor B

3. While discussing returnless fuel systems: Technician A says that in a returnless fuel system the pressure regulator is mounted on the fuel rail. Technician B says that in this type of fuel system the pump and pressure regulator are combined in one unit. Who is correct?

a. Technician A
b. Technician B
c. Both A and B
d. Neither A nor B

4. While discussing fuel boiling in the fuel rail: Technician A says that fuel boiling in the fuel rail causes a lean air-fuel ratio. Technician B says that the computer will compensate for the improper air-fuel ratio caused by fuel boiling in the fuel rail. Who is correct?

a. Technician A
b. Technician B
c. Both A and B
d. Neither A nor B

5. While discussing throttle-by-wire systems: Technician A says that a spring on the throttle shaft provides the limp-home feature. Technician B says that most systems have redundant processors and the TP and accelerator pedal sensors send redundant input signals. Who is correct?

a. Technician A
b. Technician B
c. Both A and B
d. Neither A nor B

6. While discussing the compression ratios used with engines equipped with a GDI system: Technician A says that higher ratios can be used because ignition timing is normally very advanced. Technician B says that higher ratios are possible because the small droplets of fuel injected directly into the cylinder tend to cool the mixture in the cylinder. Who is correct?

a. Technician A
b. Technician B
c. Both A and B
d. Neither A nor B

7. While discussing air-fuel ratios: Technician A says that a stoichiometric mixture is theoretically the best combination of fuel and air to provide for total combustion. Technician B says that a stoichiometric mixture always results in complete combustion. Who is correct?

a. Technician A
b. Technician B
c. Both A and B
d. Neither A nor B

8. Technician A says that air-fuel ratios can be regulated by controlling the "on time" of the injectors. Technician B says that the air-fuel ratio can be regulated by controlling the pressure applied to the individual injectors. Who is correct?

a. Technician A
b. Technician B
c. Both A and B
d. Neither A nor B

9. While discussing GDI systems: Technician A says that this type of system allows engines to run at very lean air-fuel ratios and at higher-than-normal compression ratios. Technician B says that GDI systems eliminate the need to have an EGR valve on the engine. Who is correct?

a. Technician A
b. Technician B
c. Both A and B
d. Neither A nor B

10. While discussing SFI systems: Technician A says that when the injection system fires according to crankshaft position and speed, it is operating in the synchronous mode. Technician B says that one of the advantages of SFI systems is that they are always in the synchronous mode. Who is correct?

a. Technician A
b. Technician B
c. Both A and B
d. Neither A nor B

CHAPTER 31

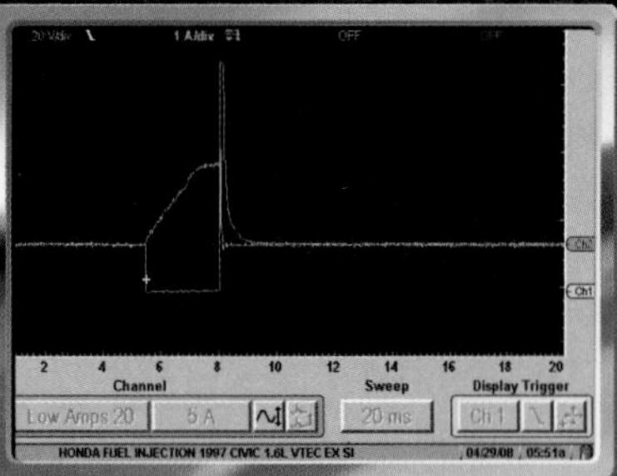

FUEL INJECTION SYSTEM DIAGNOSIS AND SERVICE

OBJECTIVES

■ **Perform a preliminary diagnostic procedure on a fuel injection system.** ■ **Remove, clean, inspect, and install throttle body assemblies.** ■ **Explain the results of incorrect fuel pressure in a TBI, MFI, or SFI system.** ■ **Perform an injector balance test and determine the injector condition.** ■ **Clean injectors on an MFI or SFI system.** ■ **Perform an injector sound, ohmmeter, noid light, and scope test.** ■ **Perform an injector flow test and determine injector condition.** ■ **Perform an injector leakage test.** ■ **Remove and replace the fuel rail, injectors, and pressure regulator.** ■ **Check the components of a GDI system.** ■ **Diagnose causes of improper idle speed on vehicles with fuel injection.**

Troubleshooting fuel injection systems requires systematic step-by-step test procedures. With so many interrelated components and sensors controlling fuel injection performance **(Figure 31–1)**, a hit-or-miss approach to diagnosing problems can quickly become frustrating, time-consuming, and costly.

Fuel injection systems are integrated into engine control systems **(Figure 31–2)**. The self-test modes of these systems are designed to help in engine diagnosis. Unfortunately, when a problem upsets the smooth operation of the engine, many service technicians automatically assume that the computer (PCM) is at fault. But in the vast majority of cases, complaints about driveability, performance, fuel mileage, roughness, or hard starting or no-starting are due to something other than the computer itself (although many problems are caused by sensor malfunctions that can be traced using the self-test mode).

Before condemning sensors as defective, remember that weak or poorly operating engine components can often affect sensor readings and result in poor performance. For example, a sloppy timing chain or bad rings or valves reduce vacuum and cylinder pressure, resulting in a lower exhaust temperature. This can affect the operation of a perfectly good oxygen or lambda sensor, which must heat up to approximately 600°F (315°C) before functioning in its closed loop mode.

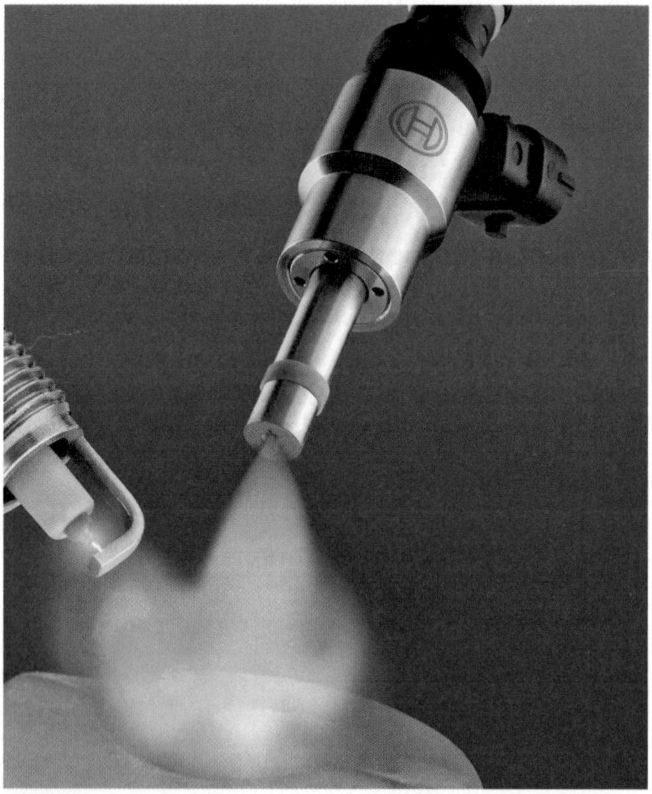

Figure 31–1 The action of a fuel injector and spark plug to start combustion. *Courtesy of Robert Bosch GmbH, www.bosch-presse.de*

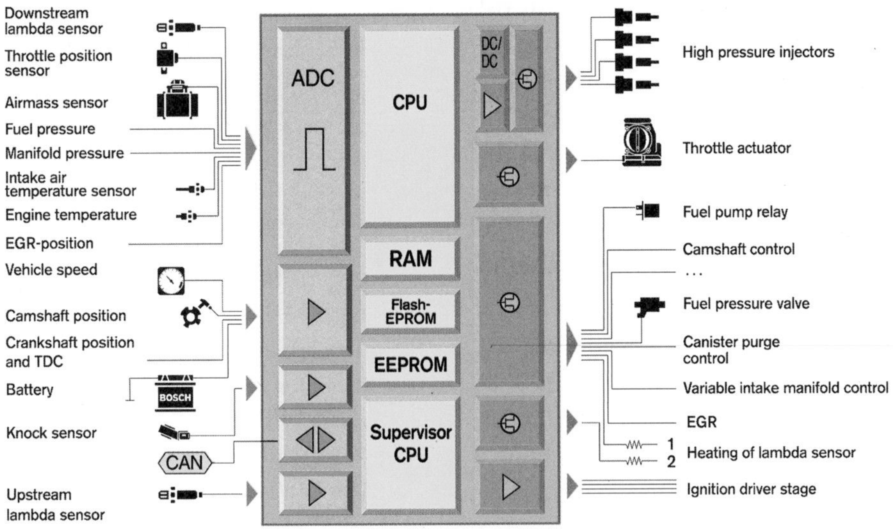

Figure 31-2 The layout of a late-model EFI system. *Courtesy of Robert Bosch GmbH, www.bosch-presse.de*

PRELIMINARY CHECKS

Before conducting any tests on the injection and engine control systems, be certain of the following:

- The battery is fully charged, with clean terminals and connections.
- The charging and starting systems are operating properly.
- All fuses and fusible links are intact.
- All electrical harnesses are routed properly and their connectors and terminals are free of corrosion and tight (**Figure 31–3**).
- The MIL is working properly.
- All parts of the air induction system are in good condition and they do not leak. Also, the air filter should be clean.
- All vacuum lines are in good condition, properly routed, and attached.
- The PCV system is working properly.
- All emission control devices are in place, connected, and operating properly.
- The level and condition of the engine's coolant is good.
- The ECT circuit is in good condition.
- The engine is in good mechanical condition.
- The gasoline in the tank is of good quality and has not been substantially cut with alcohol or contaminated with water.
- The exhaust system is not leaking.

Figure 31-3 All underhood wiring should be carefully inspected.

Service Precautions

These precautions must be observed when servicing electronic fuel injection systems:

- Always relieve the fuel pressure before disconnecting any component in the fuel system.
- Never turn the ignition switch on while any component of the fuel system is disconnected.

- Use only the test equipment recommended by the vehicle manufacturer.
- When arc welding is necessary, disconnect both battery cables before welding. Also place the ground for the welder as close as possible to the welding electrode or wire to minimize the electrical risk to the modules on the vehicle.
- When disconnecting the battery, always disconnect the negative cable first.
- Use a grounding strap to prevent static electric discharges when handling computers, modules, and computer chips.
- Isolate or disable all high-voltage systems before testing or servicing the fuel system.

BASIC EFI SYSTEM CHECKS

Fuel injection problems normally cause no-start or poor driveability problems. The causes of these problems can involve many different sensors and other systems. Of course, the fuel delivery system and the injection system can also cause problems. The EFI system must have good communications with the CAN.

Keep in mind that the PCM constantly adjusts the activity of the fuel injectors to meet the current operating conditions. Therefore, abnormal conditions can cause abnormal operation of the EFI system.

Diagnosis should begin with a check of the MIL; if it stays on or is flashing, the PCM has detected a problem and has set a DTC. The absence of a DTC does not mean that the system is operating normally. A DTC is set only when the PCM sees values that are outside a range or would affect emissions.

Connect a scan tool to the vehicle and retrieve any DTCs. Use the manufacturer's recommended test sequences for identifying the cause of the DTC. Many fuel injection problems set a fuel trim-related DTC. Fuel trim should be observed whether or not a DTC was set.

Fuel Trim

The fuel injection system can provide limited corrections for changes in conditions. The fuel system monitor tracks the amount of these corrections. The corrections to base or programmed injection pulse width and timing are called fuel trim and these can be short-term (STFT) and long-term fuel trims (LTFT). The fuel trim monitor is a continuous monitor that looks at the sum of the STFT and LTFT to assess its ability to control the air-fuel ratio. If the PCM detects that it must make an extreme change in fuel strategy, it will set a DTC.

> **SHOP TALK**
>
> The STFT and LTFT should be looked at before beginning any other diagnosis of a driveability problem. Fuel trim allows you to look at what the PCM is doing to control fuel delivery.

Keep in mind that the actual fuel trim will be the opposite of the DTC. If a system is too lean, DTCs P0171 or P0174 are set because the system is making the mixture richer **(Figure 31–4)**. Likewise, DTCs P0172 and P0175 are set if the system is too rich and the system must reduce the injectors' pulse width to lean the mixture. Also keep in mind that a fuel system DTC does not mean that the fuel system is at fault. Rather, there is a condition that is forcing the PCM to put fuel trim outside a desired range.

> **SHOP TALK**
>
> Prior to OBD-II, fuel correction data was given different names. For example, what is now called STFT was called "block integrator" by GM and it called LTFT "block learn." When working on vehicles with OBD-I systems, make sure you refer to the service information to identify what you are looking at.

Signals from the air-fuel or oxygen sensors cause the PCM to make adjustments to the fuel trim. STFT

TROUBLE CODE.........	P0171
CALC LOAD.............	45%
MAP................	46KPa-a
ENGINE SPD..........	647rpm
COOLANT TEMP.......	197.6°F
INTAKE AIR..........	77.0°F
CTP SW.................	ON
VEHICLE SPD...........	0MPH
SHORT FT #1...........	0.7%
LONG FT #1...........	35.8%
FUEL SYS #1.............	CL
FUEL SYS #2.........	UNUSED
FC IDL.................	OFF
STARTER SIG...........	OFF
A/C SIG................	OFF
PNP SW [NSW]..........	OFF
ELECT LOAD SIG.........	OFF
STOP LIGHT SW..........	ON
PS OIL PRESS SW........	OFF
PS SIGNAL..............	ON
ENG RUN TIME...........	92

Figure 31-4 In this screen, DTC P0171 (system too lean) was set when the long-term fuel trim exceeded 35%. *Courtesy of Toyota Motor Sales, U.S.A., Inc.*

adjustments are temporary and are not stored in the PCM's memory. LTFT, however, are stored and influences the base timing and duration of the injectors.

While observing fuel trim on a scan tool, zero is the midpoint of the fuel strategy when the PCM is in closed loop. A change in fuel trim is presented as a percentage: Numbers without a minus sign mean fuel is being added, and numbers with a minus sign mean fuel is being subtracted. The constant change or crossing above and below the zero line indicates proper system operation. If the STFT readings are constantly on either side of the zero line, the engine is not operating efficiently.

Some intermittent fuel-related driveability problems may be diagnosed by recording the computer's data stream. Before you jump into the data stream, look at the STFT and LTFT.

During open loop, the PCM changes pulse width without feedback from the O_2 sensor, and the short-term adaptive memory value is 1. The number 1 represents a 0% change. Once the engine warms up, the PCM moves into closed loop and begins to recognize the signals from the O_2 sensor. The system remains in closed loop until the throttle is fully opened or the engine's temperature drops below a specified limit. In both of these cases, the system goes into open loop.

When the system is in open loop, the injectors operate at a fixed base pulse width. During closed-loop operation, the pulse width is either lengthened or shortened according to the inputs from sensors. As the voltage from the O_2 sensor increases in response to a rich mixture, the STFT decreases, which means the pulse width is shortened. Decreases in STFT are indicated on a scan tool as a number below 1. For example, the short-term adaptive value of 0.75 means the pulse width was shortened by 25% and the percent of change on the scan tool will be −25. A short-term adaptive value of 1.25 means the pulse width was lengthened by 25%. The latter will be displayed as +25.

STFT will normally toggle in response to the O_2 sensor **(Figure 31–5)**. When the O_2 sensor has a rich signal, the STFT goes lean to adjust the air-fuel ratio. The O_2 sensor will then have a lean signal and the STFT will go rich. This cycling is continuous and normal.

Condition	Exhaust Oxygen Content	Fuel Trim Correction	Result
Lean	High	FT % increases	Adds fuel
Rich	Low	FT % decreases	Subtracts fuel

Figure 31-5 STFT will normally toggle in response to the oxygen content in the exhaust.

Once the engine reaches a specified temperature (normally 180°F [82°C]), the PCM begins to update the LTFT. This setting is based on engine speed and the short fuel trim. If the STFT moves 3% and stays there for a period of time, the PCM adjusts the LTFT. The LTFT works to bring the STFT close to a 0% correction.

If a lean condition caused by a vacuum leak, restricted injectors, dirty MAF, clogged fuel filter, or bad fuel pump exists, the LTFT will have a + number. If the injectors leak or the fuel pressure regulator is faulty, there will be a rich condition. The rich condition will be evident by – LTFT numbers. If the engine's condition is too far toward either the lean or rich side, the LTFT will not compensate and a DTC will be set.

To diagnose the fuel control system, observe the data stream on the scan tool, especially the STFT. With the engine running, pull off a large vacuum hose and watch for the STFT value to rise above zero to as high as 33%. Now look at the LTFT value; it should have risen above zero as it learned the engine's condition. A smaller vacuum leak must be made to watch the STFT return to zero as the result of LTFT compensation.

Because V-type engines have two sets of O_2 sensors, a pair of STFT and LTFT will appear on the scan tool. Use them to isolate the side of the engine that is running rich or lean by simply observing the codes. A high LTFT on one bank indicates there is a problem in that bank, possibly a vacuum leak or EGR problem.

Other systems can cause the fuel trim to be out of range. They include the following:

- Air induction system
- PCV system
- Engine coolant temperature sensor
- Ignition system
- Exhaust system
- O_2 or A/F sensor (check these after all other systems have been determined to be working properly)

Go To **Chapter 26 for basic diagnostics of computer systems and detailed testing of sensors.**

Oxygen Sensor Diagnosis

Carefully check the wiring and connectors to the O_2 sensors for damage and evidence of unwanted resistance. Also, check for intake and exhaust system leaks. These conditions would tend to create more oxygen in the exhaust and the PCM will respond by adding fuel to themixture. A misfiring spark plug allows unburned fuel and oxygen in the exhaust, which also causes the sensor to give a false lean reading.

Before testing an O_2 sensor, refer to the correct wiring diagram to identify the terminals at the sensor. Most late-model engines use heated oxygen sensors (HO_2S). These sensors have an internal heater that helps to stabilize the output signals. Most HO_2S have four wires connected to them. Two are for the heater and the other two are for the sensor.

SHOP TALK

On some older engines, O_2 sensor heater problems (and faulty O_2 sensor signals) can result from a poorly grounded exhaust system. A poor ground adds resistance to the circuit, which prevents the heater from heating the sensor sufficiently. This can cause faulty sensor output signals. Often the problem can be corrected by tightening the exhaust manifold-to-engine bolts.

An O_2 sensor can be checked with a voltmeter. Connect it between the O_2 sensor wire and ground. The sensor's voltage should cycle from low to high. The signal from most O_2 sensors varies between 0 and 1 volt. If the voltage is continually high, the air-fuel ratio may be rich or the sensor may be contaminated. When the O_2 sensor voltage is continually low, the air-fuel ratio may be lean, the sensor may be defective, or the wire between the sensor and the PCM may have a high resistance. If the O_2 sensor signal remains in a mid-range position, the computer may be in open loop or the sensor may be defective.

If the O_2 sensor signal sits at or is close to zero, unplug the sensor. If the voltage increases while the sensor is being unplugged, the sensor is probably shorted to ground. If O_2 sensor voltage sits at or is close to 1 volt, check the wiring at the sensor to see if the heater power feed wire or connector is shorted to the sensor's output signal wire.

The activity of an O_2 sensor is best monitored with a lab scope. The scope is connected to the sensor in the same way as a voltmeter. The switching of the sensor should be seen as the sensor signal goes to lean to rich to lean continuously (**Figure 31–6**). If the pattern toggles at high voltages, there is a rich condition (**Figure 31–7**). With a lean mixture, the pattern will toggle at low voltages.

The activity of the sensor can also be monitored on a scan tool. When watching the voltage while the engine is running, it should move to nearly 1 volt then drop back to close to 0 volts. Immediately after it drops, the voltage should move back up. This immediate cycling is an important function of an O_2 sensor. If the response is slow, the sensor is lazy and should be replaced. With the engine at about 2,500 rpm, the O_2 sensor should cycle ten to forty times in 10 seconds. When testing the O_2 sensor, make sure the sensor is heated and the system is in closed loop. Some OBD-II systems have a very low data transfer rate, which will cause the results of this test to be inaccurate. Check the service information before coming to conclusions.

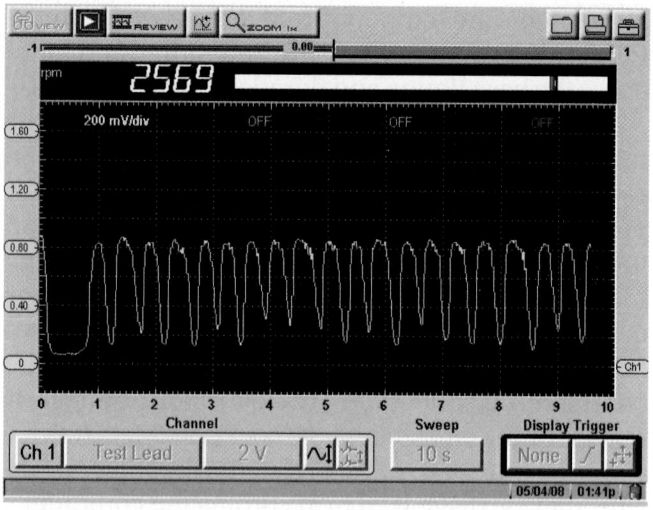

Figure 31-6 A good O_2 sensor waveform. *Reproduced under license from Snap-on Incorporated. All of the marks are marks of their owners.*

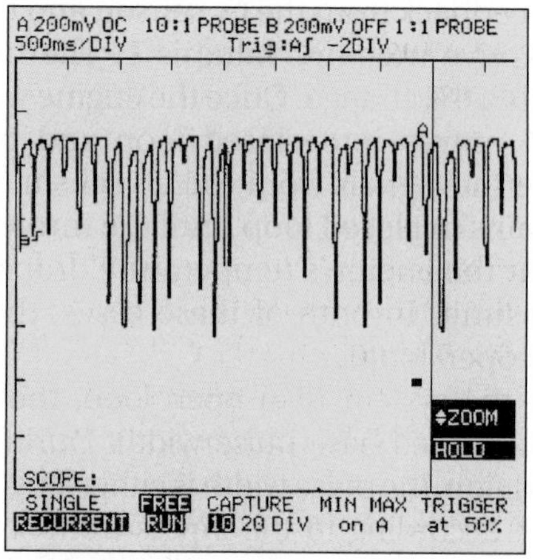

Figure 31-7 An O_2 sensor signal caused by a leaking injector. *Reproduced with permission of Fluke*

A/F Sensors

The signals from air-fuel ratio sensors are different from those of a conventional O_2 sensor. Therefore, testing and evaluating these sensors are different. When some of these sensors are looked at with an OBD-II generic scan tool, the voltage range will appear the same as a conventional O_2 sensor, 0 to 1 volt.

Exhaust Oxygen Content	Direction of Current Flow	Voltage Signal	A/F Mixture Is:
Low	Negative	Below 3.3	Rich
Stoichiometric	Zero	3.3	14.7:1
High	Positive	Above 3.3	Lean

Figure 31-8 A table showing the action of an air-fuel ratio sensor.

However, because an A/F sensor operates differently, the normal toggling of the voltage does not appear. The voltage tends to stay around 0.66 volt. When a lean condition is created with a vacuum leak, the voltage will increase to about 0.8 volt and return to 0.66 volt when the vacuum line is reconnected. A normal O_2 sensor would have dropped to a low voltage and risen to a higher voltage after the vacuum line was reconnected. If a technician does not know that the engine has an A/F sensor, an assumption would be made that the sensor or PCM is bad.

Scan tools designed for A/F sensors will display the voltage signals within a range of 2.4 to 4 volts. The 0.66 volt on the generic scan tool would read as 3.2 volts on the enhanced scan tool. The voltage increases with a lean mixture and drops with a rich mixture (**Figure 31–8**). This is opposite of a conventional O_2 sensor.

An A/F sensor can be tested by manually causing a lean or rich mixture. The voltage signal should quickly respond to those changes. If the voltage does not change or changes very slowly, check the sensor's heater circuit. A/F sensors need to be hot before they will work correctly. The PCM controls the heater circuit by duty cycling the heater through the ground-side of the heater. The power feed to the heater should have battery voltage. If it does not, diagnose that circuit. The circuit can also be checked with a current probe or lab scope. When looking at the heater circuit's current flow, changes in amperage should be evident. Pay attention to the peak amperage; normally this should not exceed 6 amps. Refer to the service information for specific values. In many cases, Mode $06 will provide heater current values.

Air Induction System Checks

In a fuel injection system (particularly designs that rely on airflow meters or mass airflow sensors), all the air entering the engine must be accounted for by the air measuring device. If it is not, the air-fuel ratio becomes overly lean. For this reason, cracks or tears in the plumbing between the airflow sensor and

Figure 31-9 Carefully inspect the ductwork and hoses of the air induction system.

throttle body are potential air leak sources that can affect the air-fuel ratio.

During a visual inspection of the air control system, pay close attention to these areas, looking for cracked or deteriorated ductwork (**Figure 31–9**). Also make sure that all induction hose clamps are tight and properly sealed. Look for possible air leaks in the crankcase, for example, around the dipstick tube and oil filter cap. Any extra air entering the intake manifold through the PCV system is also not measured and can upset the delicately balanced air-fuel mixture at idle.

It is important to note that vacuum leaks may not affect the operation of engines fitted with a speed density fuel injection system. This does not mean that vacuum leaks are okay, it just means that the operating system may be capable of adjustments that allow the engine to run well in spite of the vacuum leak. This is true for vacuum leaks that are common to all cylinders. If the vacuum leak affects one or two cylinders, the computer cannot compensate for the unmetered air and those two cylinders will not operate efficiently.

Airflow Sensors

When diagnosing poor fuel economy, erratic performance, hesitation, or hard-starting problems, make the following checks to determine if the airflow sensor is at fault.

Mass Airflow Sensors MAF sensors measure intake air before it reaches the throttle plate. If any air bypasses the sensor and enters the combustion

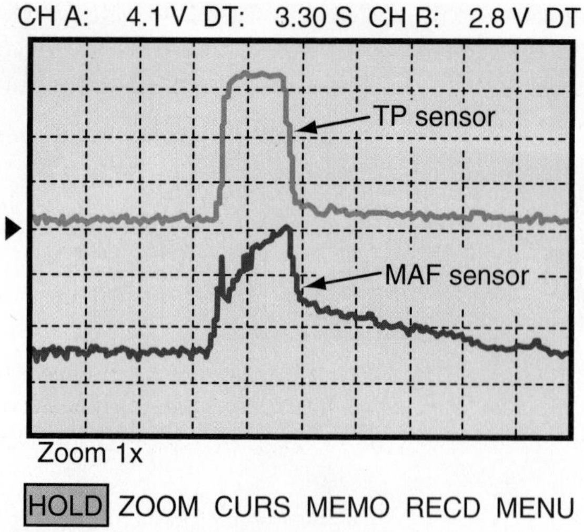

Figure 31–10 Waveforms of TP and MAF sensors showing how they simultaneously change with acceleration and deceleration.

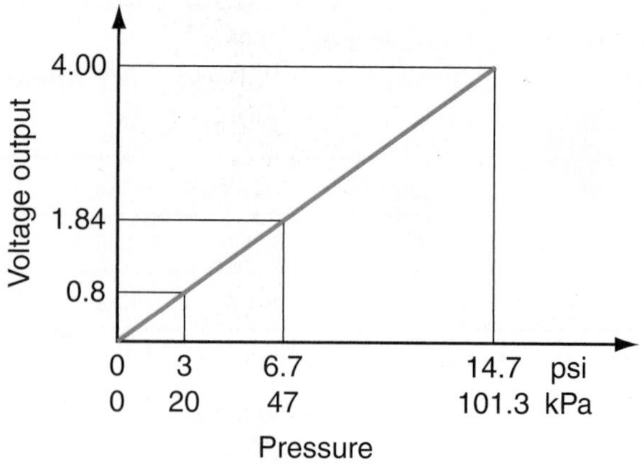

Figure 31–11 A chart showing how a typical MAP sensor's signal changes with a change in pressure.

chambers without being measured, the engine will run lean. A vacuum leak or a leak between the MAF and throttle plate will also reduce fuel delivery. The resulting lean mixture will cause the PCM to store DTCs indicating excessive fuel corrections and/or lean misfires.

The action of a MAF can be observed on a lab scope **(Figure 31–10)**. As the throttle is opened, the MAF signal should increase due to the increase in airflow.

If a MAF is suspected as causing a lean condition, remove it and inspect the hot wire for signs of debris. This wire can be cleaned by gently wiping it with a camel hair artist brush soaked in a carburetor or similar cleaner. Never soak or immerse the MAF sensor in any cleaner. If the wire is dirty, check the "burn-off" relay. Some manufacturers recommend that the sensor be replaced if the wire is dirty.

SHOP TALK

To quickly diagnose an intermittent failure of GM's MAF, start the engine, let it idle, and lightly tap the sensor with a plastic mallet or screwdriver handle. If the engine stalls, runs worse, or the idle quality improves, the MAF sensor is probably defective and should be replaced. Similarly, if the engine does not start or idles poorly, unplug the MAF. If the engine starts or runs better with the sensor unplugged, the sensor should be replaced.

Manifold Absolute Pressure (MAP) Sensors Speed density systems rely on MAP signals. Increased manifold pressure (lower vacuum) causes the injectors' pulse width to increase regardless of the cause of the pressure change (open throttle, vacuum leak, recirculated exhaust gas, or exhaust pressure).

MAP sensors respond in the opposite way as MAFs and VAFs. Vacuum leaks in speed density systems can be verified by watching the IAC counts on a scan tool. Low vacuum will result in low IAC counts. In response, the PCM will increase the amount of fuel, which will cause the idle speed to increase. To correct this, the PCM will attempt to reduce the idle speed by closing the IAC. MAP sensors respond directly to changes in pressure **(Figure 31–11)**. Special MAP transducers are available to allow a DMM to monitor MAP signals **(Figure 31–12)**.

Throttle Body

The throttle body **(Figure 31–13)** allows the driver to control the amount of air that enters the engine, thereby controlling the speed of the engine. Each type of throttle body assembly is designed to allow a certain amount of air to pass through it at a particular

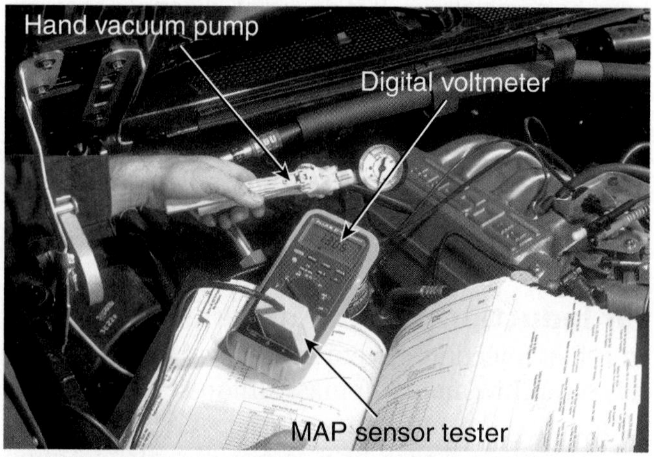

Figure 31–12 This adapter changes the frequency signal from a MAP to an analog signal that can be monitored with a DMM.

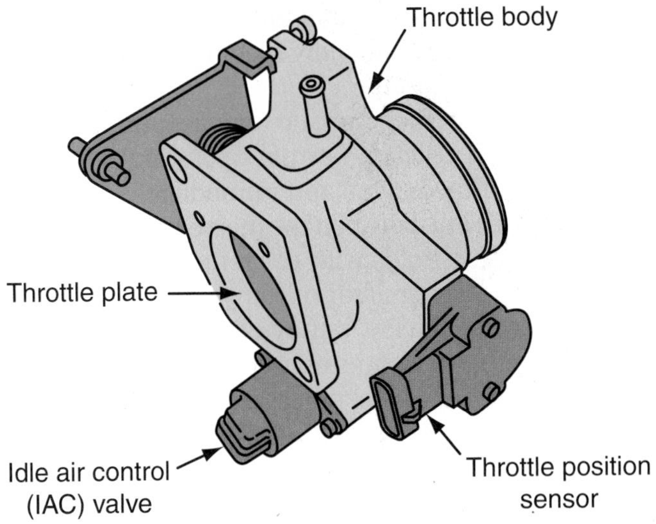

Figure 31-13 A throttle body assembly.

amount of throttle opening. If anything accumulates on the throttle plates or in the throttle bore, the amount of air that can pass through is reduced. This normally causes an idle problem.

These deposits can be cleaned off the throttle assembly and the airflow through them restored. Begin by removing the air duct from the throttle assembly; that will give access to the plate and bore. The deposits can be cleaned with a spray cleaner or wiped off with a cloth. If either of these cleaning methods does not remove the deposits, the throttle body should be removed, disassembled, and placed in an approved cleaning solution.

A pressurized can of throttle body cleaner may be used to spray around the throttle area without removing and disassembling the throttle body. The throttle assembly can also be cleaned by soaking a cloth in carburetor solvent and wiping the bore and throttle plate to remove light to moderate amounts of carbon residue. Also, clean the backside of the throttle plate. Then, remove the idle air control valve from the throttle body (if so equipped) and clean any carbon deposits from the pintle tip and the IAC air passage.

If the intake manifold setup has diaphragms or solenoids that control the selection of manifold runners, make sure you allow some of the cleaning solution to be drawn in by the engine while it is running in order to clean the valves, sometimes called *butterflies,* controlled by the solenoids or diaphragms. Dirty switchover valves can cause hard starting, poor performance, increased oil consumption, and DTCs.

Throttle Body Inspection Throttle body inspection and service procedures vary widely depending on the year and make of the vehicle. However, some components such as the TP sensor are found on nearly all throttle bodies. Since throttle bodies have some common components, inspection procedures often involve checking common components.

PROCEDURE

Throttle Body Inspection

STEP 1 Check for smooth movement of the throttle linkage from idle position to the wide-open position.

STEP 2 Check the throttle linkage and cable for wear and looseness.

STEP 3 Check the vacuum at each vacuum port on the throttle body while the engine is idling and while it is running at a higher speed.

STEP 4 Apply vacuum to the throttle opener. Then, disconnect the TP sensor connector, and test the TP sensor with an ohmmeter connected across the appropriate terminals.

STEP 5 Loosen the two TP sensor mounting screws and rotate the TP sensor as required to obtain the specified ohmmeter readings. Then retighten the mounting screws. If the TP sensor cannot be adjusted to obtain the proper ohmmeter readings, replace the TP sensor.

STEP 6 Operate the engine until it reaches normal operating temperature, and check the idle speed on a tachometer. The idle speed should be 700 to 800 rpm.

STEP 7 Disconnect and plug the vacuum hose from the throttle opener. Maintain 2,500 engine rpm.

STEP 8 Release the throttle valve and observe the tachometer reading. When the throttle linkage strikes the throttle opener stem, the engine speed should be between 1,300 and 1,500 rpm. Adjust the throttle opener, as necessary, and reconnect the throttle opener vacuum hose.

Throttle Body Removal and Cleaning Whenever it is necessary to remove the throttle body assembly for replacement or cleaning, make sure you follow the procedures outlined by the manufacturer. Also, begin by connecting a 12-volt power supply (if available) to the cigarette lighter socket and disconnect the negative battery cable. If the vehicle is equipped with an air bag, wait 1 minute. Then remove the throttle body according to recommendations. Once the assembly has been removed, remove all nonmetallic parts such as the TP sensor, IAC valve, throttle opener, and the throttle body gasket from the throttle body. Now it is safe to clean the throttle body assembly in the recommended throttle body cleaner and blow dry with compressed air. Blow out all passages in the throttle body assembly.

Before reinstalling the throttle body assembly, make sure all metal mating surfaces are clean and free from metal burrs and scratches. Make sure you have new gaskets and seals for all sealing surfaces before you begin to reinstall the assembly. After everything that was disconnected is reconnected, reconnect the negative battery cable and disconnect the 12-volt power supply.

Fuel System Checks

CAUTION!

Dispose of fuel-soaked rags or towels by placing them in a fireproof container.

Go To **Chapter 29 for a detailed discussion on fuel delivery systems and how to test and service them.**

Checking for fuel delivery is a simple task on throttle body systems. Remove the air cleaner, crank the engine, and the injector should be spraying some fuel. It is impossible to visually check the spray pattern and volume of port system injectors. However, performing simple fuel pressure and fuel volume tests will give an accurate evaluation of the fuel delivery system. Keep in mind that fuel pressure affects the output of a fuel injector: If an injector has the same pulse rate but receives low pressure there is less fuel; if the pressure is high, the amount of fuel increased.

Low fuel pressure can cause a no-start or poor-run problem. It can be caused by a clogged fuel filter, a faulty pressure regulator, or a restricted fuel line anywhere from the fuel tank to the fuel filter connection.

High fuel pressure readings will result in a rich-running engine. A restricted fuel return line to the tank or a bad fuel regulator may be the problem. To isolate the cause of high pressure, relieve system pressure and connect a tap hose to the fuel return line. Direct the hose into a container and energize the fuel pump. If fuel pressure is now within specifications, the fuel return line is blocked. If pressure is still high, the pressure regulator is faulty.

If fuel pressure is within specs but the pressure slowly bleeds down, there may be a leak in the fuel pressure regulator, the fuel pump check valve, or the injectors themselves. Hard starting is a common symptom of internal system leaks. Also, fuel starvation and lean conditions can occur when the injectors drain the fuel rail faster than the fuel pump can fill it. This could be caused by low fuel pressure or inadequate delivery volume.

Connect the fuel pressure gauge and run the engine until it reaches normal operating temperature. Turn off the engine and immediately pinch off the fuel return line between the fuel regulator and the tank. If pressure holds with the return line pinched, the leak is in the regulator or return line. If the pressure still drops quickly, remove the clamp from the line. Now run the fuel pump to restore normal pressure then immediately pinch off the supply hose between the fuel pump and the inlet on the fuel rail. If the system now maintains pressure, it is probable that the fuel pump check valve is leaking. If the pressure still drops, there is an external leak in the rail or the injectors are leaking.

Returnless Systems Nearly all late-model vehicles use a returnless fuel system. In these systems, the pressure regulator is part of the fuel pump assembly in the fuel tank. They may also have a pressure sensor on the fuel rail and modulate the pulse width of the fuel pump to control pressure. The pulse width is controlled by the PCM. High pressures in these systems can be caused by a faulty pressure regulator, pressure sensor, fuel pump, PCM, or electrical circuits.

Injector Checks

A fuel injector is nothing more than a solenoid-actuated fuel valve. Its operation is quite basic in that as long as it is held open and the fuel pressure remains steady, it delivers fuel until it is told to stop.

Because all fuel injectors operate in a similar manner, fuel injector problems tend to exhibit the same failure characteristics. The main difference is that, in a TBI design, generally all cylinders will suffer if an injector malfunctions, whereas in port systems the loss of one injector will only affect one cylinder.

An injector that does not open causes hard starts on port-type systems and an obvious no-start on single-point TBI designs. An injector that is stuck partially open causes loss of fuel pressure (most noticeably after the engine is stopped and restarted within a short time period) and flooding due to raw fuel dribbling into the engine. Buildups of gum and other deposits on the tip of an injector can reduce the amount of fuel sprayed by the injector or they can prevent the injector from totally sealing, allowing it to leak. Since injectors on MFI and SFI systems are subject to more heat than TBI injectors, port injectors have more problems with tip deposits.

Because an injector adds the fuel part to the air-fuel mixture, it is obvious that any defect in the fuel injection system will cause the mixture to go rich or lean. If the mixture is too rich and the PCM is in control of the air-fuel ratio, a common cause is that one or more injectors are leaking. An easy way to verify this on port-injected engines is to use an exhaust gas analyzer.

With the engine warmed up, but not running, remove the air duct from the airflow sensor. Then insert the gas analyzer's probe into the intake plenum area. Be careful not to damage the airflow sensor or throttle plates while doing this. Look at the HC readings on the analyzer. They should be low and drop as time passes. If an injector is leaking, the HC reading will be high and will not drop. This test does not locate the bad injector, but does verify that one or more are leaking.

Another cause of a rich mixture is a leaking fuel pressure regulator. If the diaphragm of the regulator is ruptured, fuel will move into the intake manifold through the diaphragm, causing a rich mixture. The regulator can be checked by using two simple tests. After the engine has been run, disconnect the vacuum line to the fuel pressure regulator **(Figure 31–14)**. If there are signs of fuel inside the hose or if fuel comes out of the hose, the regulator's diaphragm is leaking. The regulator can also be tested with a hand-operated vacuum pump. Apply 5 in. Hg (127 mm Hg) to the regulator. A good regulator diaphragm will hold that vacuum.

Checking Voltage Signals When an injector is suspected as the cause of a lean problem, the first step is to determine if the injector is receiving a signal (from the PCM) to fire.

WARNING!

When performing this test, make sure to keep off the accelerator pedal. On some models, fully depressing the accelerator pedal activates the clear flood mode, in which the voltage signal to the injectors is automatically cut off. Technicians unaware of this waste time tracing a phantom problem.

Once the injector's electrical connector has been removed, check for voltage at the injector using a high impedance test light or a convenient noid light that plugs into the connector. After making the test connections, crank the engine. The noid light flashes if the computer is cycling the injector on and off. If the light is not flashing, the computer or connecting wires are defective. If sufficient voltage is present after checking each injector, check the electrical integrity of the injectors themselves.

An ohmmeter can be used to test the electrical soundness of an injector. Connect the ohmmeter across the injector terminals **(Figure 31–15)** after the wires to the injector have been disconnected. If the meter reading is infinite, the injector winding is open. If the meter shows more resistance than the specifications call for, there is high resistance in the winding. A reading that is lower than the specifications indicates that the winding is shorted. If the injector is even a little bit out of specifications, it must be replaced.

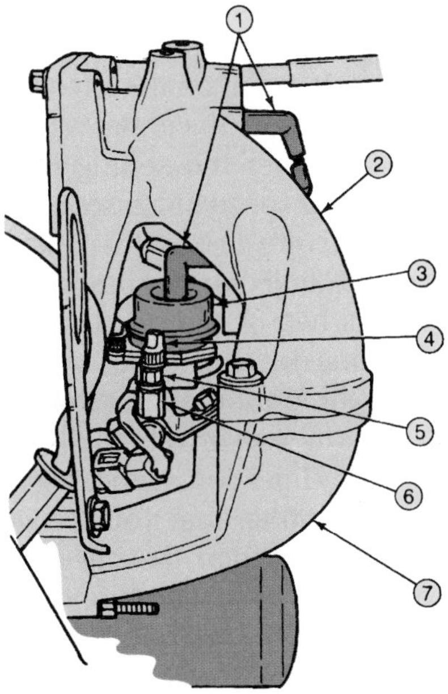

Item	Description
1	Fuel pressure regulator vacuum tube
2	Upper intake manifold
3	Fuel pressure regulator
4	Fuel pressure relief valve cap
5	Fuel pressure relief valve
6	Fuel injection supply manifold
7	Lower intake manifold

Figure 31–14 The typical location of a fuel pressure regulator on an inline EFI engine.
Courtesy of Ford Motor Company

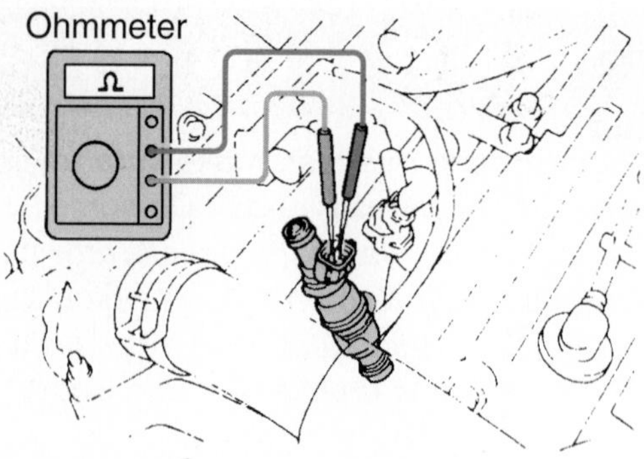

Figure 31-15 Checking an injector with an ohmmeter. *Courtesy of Toyota Motor Sales, U.S.A., Inc.*

Injector Balance Test If the injectors are electrically sound, perform an injector pressure balance test. This test will help isolate a clogged or dirty injector. Photo Sequence 31 shows a typical procedure for testing injector balance. An electronic injector pulse tester is used for this test. As each injector is energized, a fuel pressure gauge is observed to monitor the drop in fuel pressure. The tester is designed to safely pulse each injector for a controlled length of time. The tester is connected to one injector at a time. The ignition is turned on until a maximum reading is on the pressure gauge. That reading is recorded and the ignition turned off. With the tester, activate the injector and record the pressure reading after the needle has stopped pulsing. This same test is performed on each injector.

The difference between the maximum and minimum reading is the pressure drop. Ideally, each injector should drop the same amount when opened. Typically a variation of 1.5 to 3 psi (20 kPa) after each injector is energized is cause for concern. If there is no pressure drop or a low pressure drop, suspect a restricted injector orifice or tip. A higher-than-average pressure drop indicates a rich condition. When an injector plunger is sticking in the open position, the fuel pressure drop is excessive. If there are inconsistent readings, the nonconforming injectors either have to be cleaned or replaced.

If the injector's orifice is dirty or otherwise restricted, there will not be much pressure decrease when the injector is energized. Stumbles during acceleration, engine stalling, and erratic idle are all caused by restricted injector orifices.

If an excessive amount of pressure drop is observed, it is likely that an injector's plunger is sticking open. A sticking injector may result in a rich air-fuel mixture.

Injector Sound Test If the injector's electrical leads are difficult to access, an injector power balance test is hard to perform. As an alternative, start the engine and use a technician's stethoscope to listen for correct injector operation. A good injector makes a rhythmic clicking sound as the solenoid is energized and de-energized several times each second. If a clunk-clunk instead of a steady click-click is heard, chances are the problem injector has been found. Cleaning or replacement is in order. If an injector does not produce any clicking noise, the injector, connecting wires, or PCM may be defective. When the injector clicking noise is erratic, the injector plunger may be sticking. If there is no injector clicking noise, proceed with the injector resistance test and noid light test to locate the cause of the problem. If a stethoscope is not handy, use a thin steel rod, wooden dowel, or fingers to feel for a steady on/off pulsing of the injector solenoid.

Oscilloscope Checks An oscilloscope can be used to monitor the injector's pulse width and duty cycle when an injector-related problem is suspected. The pulse width is the time in milliseconds that the injector is energized. The duty cycle is the percentage of on-time to total cycle time.

To check the injector's firing voltage on the scope, a typical hookup involves connecting the scope's positive lead to the injector supply wire and the scope's negativelead to an engine ground.

Fuel injection signals vary in frequency and pulse width. The pulse width is controlled by the PCM, which varies it to control the air-fuel ratio. The frequency varies depending on engine speed. The higher the speed, the more pulses per second there are. Most often the injector's ground circuit is completed by a driver circuit in the PCM. All of these factors are important to remember when setting a lab scope to look at fuel injector activity. Set the scope to read 12 volts, then set the sweep and trigger to allow you to clearly see the on signal on the left and the off signal on the right. Make sure the entire waveform is clearly seen. Also remember the setting may need to be changed as engine speed increases or decreases.

Fuel injectors are fired either individually or in groups. When the injectors are fired in groups, a driver circuit controls two or more injectors. On some V-type engines, one driver fires the injectors on one side of the engine, while another fires the other side. Each fuel injector has its own driver transistor in sequential and throttle body injection. It is extremely important, while troubleshooting, that you recognize how the injectors are fired. When the injectors are fired in groups, there can be a common or noncommon cause of the problem. For example, a defective driver circuit in the PCM would affect all of the injectors in a group, not just one. Conversely, if one injector in the group is not firing, the problem cannot be the driver.

P31-1 *Connect the fuel pressure gauge to the Schrader valve on the fuel rail, and then relieve the pressure in the system.*

P31-2 *Disconnect the number 1 injector and connect the injector pulse tester to the injector's terminals.*

P31-3 *Connect the injector pulse tester's power supply leads to the battery.*

P31-4 *Cycle the ignition switch several times until the system pressure is at the specified level.*

P31-5 *Push the injector pulse tester switch and record the pressure on the pressure gauge. Subtract this reading from the measured system pressure. The answer is the pressure drop across that injector.*

P31-6 *Move the injector tester to the number 2 injector and cycle the ignition switch several times to restore system fuel pressure.*

P31-7 *Depress the injector pulse tester's switch and observe the fuel pressure. Again the difference between the system pressure and the pressure when an injector is activated is the pressure drop across the injector.*

P31-8 *Move the injector tester's leads to the number 3 injector and cycle the ignition switch to restore system pressure.*

P31-9 *Depress the switch on the tester to activate that injector and record the pressure drop. Continue the procedure for all injectors, then compare the results of each to specifications and to each other.*

To read the injector waveform on group fuel injection systems, the scope must be connected to one injector harness for each group. Since all of the injectors in the group share the same circuit, a problem in one will affect the entire waveform for the group. The only way to isolate an injector electrical problem is to disconnect the injectors, one at a time. If the waveform improves when an injector is disconnected, that injector has a problem. If the waveform never cleans up, the problem is in the driver circuit or the wiring harness.

In sequential fuel injection systems each injector has its own driver circuit and wiring. To check an individual injector, the scope must be connected to that injector. This is great for locating a faulty injector. If the scope has a memory feature, a good injector waveform can be stored and recalled for comparison to the suspected bad fuel injector pattern. To determine if a problem is the injector itself or the PCM and/or wiring, simply swap the injector wires from an injector that had a good waveform to the suspect injector. If the waveform cleans up, the wiring harness or the PCM is the cause of the problem. If the waveform is still not normal, the injector is to blame.

There are three different types of fuel injector circuits. In the conventional circuit, the driver constantly applies voltage to the injector. The circuit is turned on when a ground is provided. The waveform for this type of injector circuit is shown in **Figure 31–16**. Notice that there is a single voltage spike at the point where the injector is turned off. The total on-time of the injector is measured from the point where the trace drops (on the left) to the point where it rises up (next to the voltage spike).

Peak and hold injector circuits use two driver circuits to control injector action. Both driver circuits complete the circuit to open the injector. This allows for high current at the injector, which forces the injector to open quickly. After the injector is open, one of the circuits turns off. The second circuit remains on to hold the injector open. This is the circuit that controls the pulse width of the injector. This circuit also contains a resistor to limit current flow during on-time. When this circuit turns off, the injector closes. When looking at the waveform for this type of circuit (**Figure 31–17**), there will be two voltage spikes. One is produced when each circuit opens. To measure the on-time of this type injector, measure from the drop on the left to the point where the second voltage spike is starting to move upward.

A **pulse-modulated injector** circuit uses high current to open the injector. Again this allows for quick injector firing. Once the injector is open, the circuit ground is pulsed on and off to allow for a long on-time without allowing high current flow through the circuit. To measure the pulse width of this type of injector (**Figure 31–18**), measure from the drop on the left to the beginning of the large voltage spike, which should be at the end of the pulses.

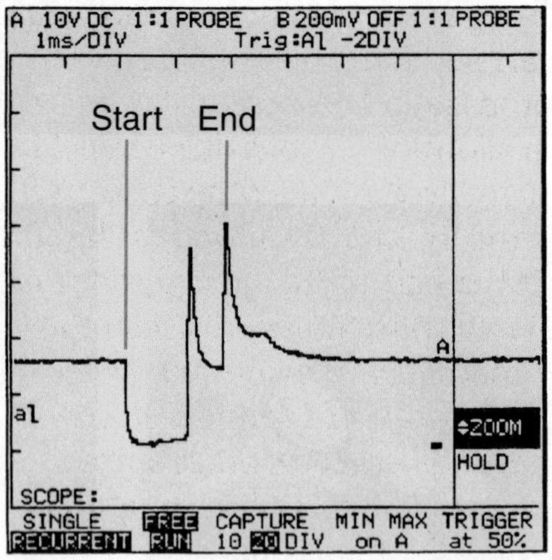

Figure 31–17 The waveform for a peak and hold fuel injector driver circuit. *Reproduced with permission of Fluke*

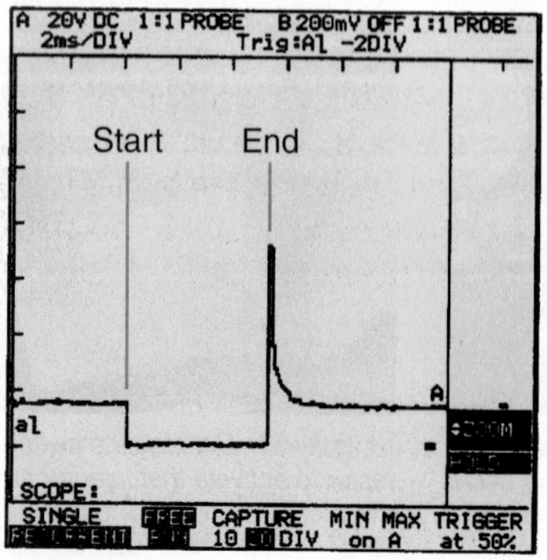

Figure 31–16 The waveform for a conventional fuel injector driver circuit. *Reproduced with permission of Fluke*

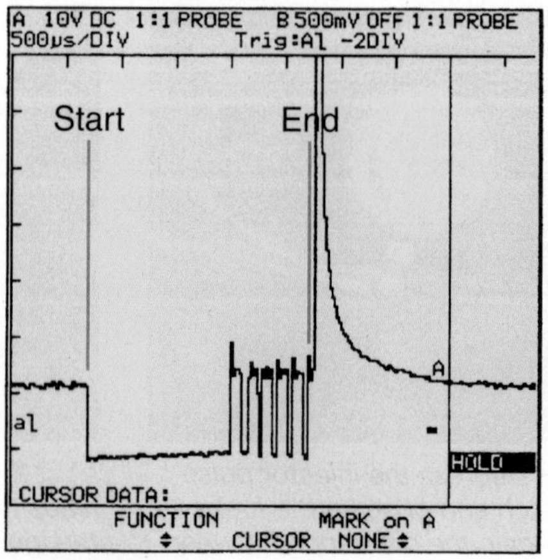

Figure 31–18 The waveform for a pulse-modulated fuel injector driver circuit. *Reproduced with permission of Fluke*

For all types of injectors, the waveform should have a clean, sudden drop in voltage when it is turned on. This drop should be close to 0 volts. Typically, the maximum allowable voltage during the injector's on-time is 600 millivolts. If the drop is not perfectly vertical, either the injector is shorted or the driver circuit in the PCM is bad. If the voltage does not drop to below 600 millivolts, there is resistance in the ground circuit or the injector is shorted. When comparing one injector's waveform to another, check the height of the voltage spikes. The voltage spike of all injectors in the same engine should have approximately the same height. If there is a variance, the power feed wire to the injector with the variance or the PCM's driver for that injector is faulty.

While checking the injectors with a lab scope, make sure the injectors are firing at the correct time. To do this, use a dual trace scope and monitor the ignition reference signal and a fuel injector signal at the same time. The two signals should have some sort of rhythm between them. For example, there can be one injector firing for every four ignition reference signals **(Figure 31–19)**. This rhythm is dependent upon several things; however, it does not matter what the rhythm is; it only matters that the rhythm is constant. If the injector's waveform is fine but the rhythm varies, the ignition reference sensor circuit is faulty and not allowing the injector to fire at the correct time. If the ignition signal is lost because of a faulty sensor, the injection system will also shut down. If the injector circuit and the ignition reference circuit shut down at the same time, the cause of the problem is probably the ignition reference sensor. If the injector circuit shuts off before the ignition circuit, the problem is the injector circuit or the PCM.

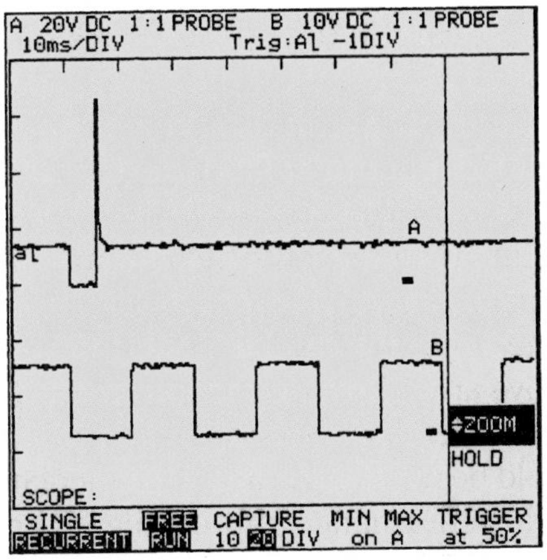

Figure 31–19 On a dual trace scope, you can compare the ignition reference signal to the injector's ON signal. *Reproduced with permission of Fluke*

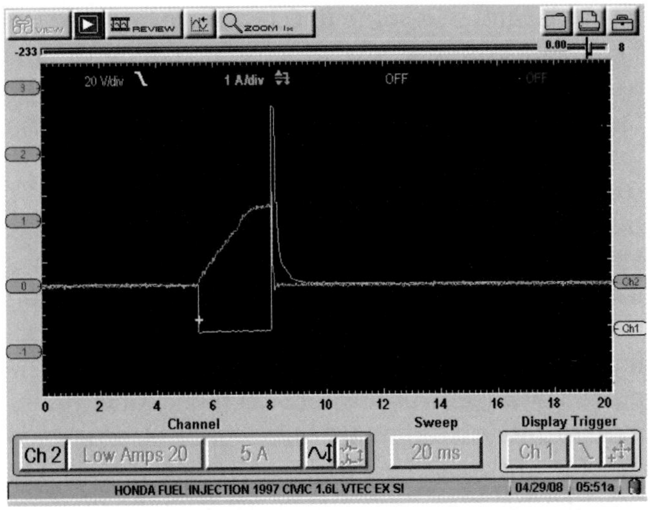

Figure 31–20 Comparing injector current ramping (green) to the voltage signals (yellow). *Reproduced under license from Snap-on Incorporated. All of the marks are marks of their owners.*

Current Ramping When using a DSO or GMM, the ramping current should be compared to the voltage **(Figure 31–20)**. When the injector is grounded, current flows to it. The current trace should show a steady build until the injector fires. At that point, the current should drop sharply. When one looks at the build of current, any unwanted resistance in the injector's coil will show up as periodic drops during its ramp. Although the injector may be firing in spite of the resistance, the injector is bad or will be soon.

It is important to understand the stages of voltage at the injector to understand what the current should be doing. Before the injector fires, there is system voltage to it but no current flow. Once the PCM grounds the injector, current begins to flow. When there is enough current to turn on the injector, the injector opens. At this point, if the voltage does not drop close to zero, there could be ground problems or a weak driver circuit. Also, the current may have a rough time building. After the injector is fired, current stops and the voltage should move high. Immediately after the injector is turned off, the magnetic field in the coil collapses, causing another voltage spike. If the magnetic field in the coil is weak, this secondary spike will not be very high. This can be caused by high resistance or a short in the injector's windings. It can also be caused by low current to the injector.

INJECTOR SERVICE

Because a single injector can cost up to several hundred dollars, arbitrarily replacing injectors when they are not functioning properly, especially on multiport systems, can be an expensive proposition. If injectors are electrically defective, replacement is the only alternative. However, if the injector balance test indicated that some injectors were restricted or if the vehicle is exhibiting rough idle, stalling, or slow or uneven acceleration, the injectors may just be dirty and require a good cleaning.

Injector Cleaning

Before discussing the typical cleaning systems available and how they are used, several cleaning precautions are in order. First, never soak an injector in cleaning solvent. Not only is this an ineffective way to clean injectors, but it most likely will destroy the injector in the process. Also, never use a wire brush, pipe cleaner, toothpick, or other cleaning utensil to unblock a plugged injector. The metering holes in injectors are drilled to precise tolerances. Scraping or reaming the opening may result in a clean injector but it may also be one that is no longer an accurate fuel-metering device.

The basic premise of all injection cleaning systems is similar in that some type of cleaning chemical is run through the injector in an attempt to dissolve deposits that have formed on the injector's tip. The methods of applying the cleaner can range from single-shot, premixed, pressurized spray cans to self-mix, self-pressurized chemical tanks resembling bug sprayers. The premixed, pressurized spray can systems are fairly simple and straightforward to use since the technician does not need to mix, measure, or otherwise handle the cleaning agent.

Automotive parts stores usually sell pressurized containers of injector cleaner with a hose for Schrader valve attachment. During the cleaning process, the engine runs on the pressurized container of propane and injector cleaner. Fuel pump operation must be stopped to prevent the pump from forcing fuel up to the fuel rail. Disconnect the wires from the in-tank fuel pump or the fuel pump relay to disable the fuel pump. Plug the fuel return line from the fuel rail to the tank. Connect a can of injector cleaner to the Schrader valve on the fuel rail and run the engine for about 20 minutes on the injector solution.

Other systems require the technician to assume the role of chemist and mix up a desired batch of cleaning solution for each application. The chemical solution then is placed in a holding container and pressurized by hand pump or shop air to a specified operating pressure. The injector cleaning solution is poured into a canister on some injector cleaners and shop air supply is used to pressurize the canister to the specified pressure. The injector cleaning solution contains unleaded fuel mixed with injector cleaner.

The container hose is connected to the Schrader valve on the fuel rail **(Figure 31–21)**. Disable the fuel pump according to the car manufacturer's instructions (for example, pull the fuel pump fuse, disconnect a lead at pump). Clamp off the fuel pump return line at the flex connection to prevent the cleaner from seeping into the fuel tank. Set and connect the cleaning system so it can circulate the cleaning solution through the fuel rail with the engine off. To do this, adjust the machine's delivery pressure to a pressure higher than normal delivery pressure. To flush the entire fuel rail, including the injector inlet screens and regulator, adjust the machine's delivery pressure higher than the fuel pressure regulator's normal pressure setting.

Figure 31–21 An injector cleaner connected to a fuel rail. *Courtesy of SPX Service Solutions*

Readjust the machine to a pressure slightly lower than the normal regulated pressure and then open the cleaner's control valve one-half turn or so to prime the injectors, and then start the engine. If available, set and adjust the cleaner's pressure gauge to approximately 5 psi (34.47 kPa) below the operating pressure of the injection system and let the engine run at 1,000 rpm for 10 to 15 minutes or until the cleaning mix has run out. If the engine stalls during cleaning, simply restart it. Run the engine until the recommended amount of fluid is exhausted and the engine stalls. Shut off the ignition, remove the cleaning setup, and reconnect the fuel pump.

After removing the clamping devices from around the fuel lines, start the car. Let it idle for 5 minutes or so to remove any leftover cleaner from the fuel lines. In the more severely clogged cases, the idle improvement should be noticeable almost immediately. With more subtle performance improvements, an injector balance test verifies the cleaning results. Once the injectors are clean, recommend the use of an in-tank cleaning additive or a detergent-laced fuel.

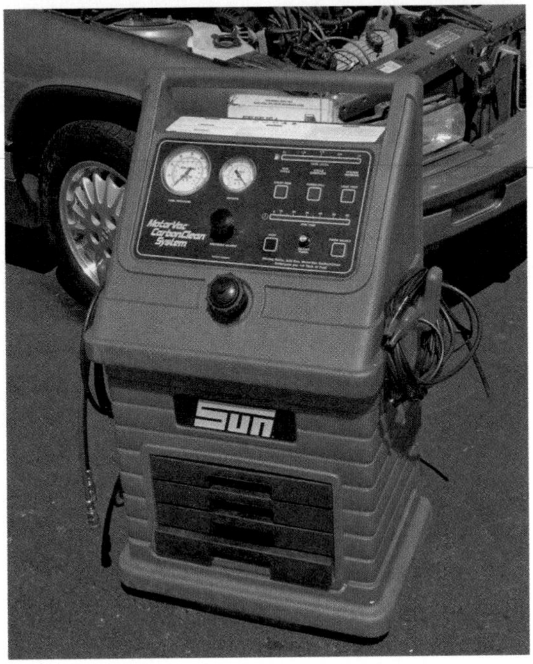

Figure 31-22 A fuel injector cleaner cart.

The more advanced units feature electrically operated pumps neatly packaged in roll-around cabinets that are quite similar in design to an A/C charging station **(Figure 31–22)**.

After the injectors are cleaned or replaced, rough engine idle may still be present. This problem occurs because the adaptive memory in the computer has learned previously about the restricted injectors. If the injectors were supplying a lean air-fuel ratio, the computer increased the pulse width to try to bring the air-fuel ratio back to stoichiometric. With the cleaned or replaced injectors, the adaptive computer memory is still supplying the increased pulse width. This action makes the air-fuel ratio too rich now that the restricted injector problem does not exist. With the engine at normal operating temperature, drive the vehicle for at least 5 minutes to allow the adaptive computer memory to learn about the cleaned or replaced injectors. Afterward, the computer should supply the correct injector pulse width and the engine should run smoothly. This same problem may occur when any defective computer system component is replaced.

FUEL RAIL, INJECTOR, AND REGULATOR SERVICE

There are service operations that will require removing the fuel injection fuel rail, pressure regulator, and/or injectors. Most of these are not related to fuel system repair. However, when it is necessary to remove and refit them, it is important that it be done carefully and according to the manufacturer's recommended procedures.

Injector Replacement

Photo Sequence 32 outlines a typical procedure for removing and installing an injector. Consult the vehicle's service manual for instructions on removing and installing injectors. Before installing the new one, always check to make sure the sealing O-ring is in place. Also, prior to installation, lightly lubricate the sealing ring with engine oil or automatic transmission fluid (avoid using silicone grease, which tends to destroy an O_2S and clog the injectors) to prevent seal distortion or damage.

WARNING!

Cap injector openings in the intake manifold to prevent the entry of dirt and other particles. Also, after the injectors and pressure regulator are removed from the fuel rail, cap all fuel rail openings to keep dirt out of the fuel rail.

Fuel Rail, Injector, and Pressure Regulator Removal

The procedure for removing and replacing the fuel rail, injectors, and pressure regulator varies depending on the vehicle. On some applications, certain components must be removed to gain access to these components. The system must be relieved of any and all pressure before the fuel lines are opened to remove any of the components. Before removing the fuel rail, wipe off any dirt from the fuel rail with a shop towel. Then loosen the fuel line clamps on the fuel rail, if so equipped. If these lines have quick-disconnect fittings, grasp the larger collar on the connector and twist in either direction while pulling on the line to remove the fuel supply and return lines **(Figure 31–23)**. Now, remove the vacuum line from the pressure regulator and disconnect the electrical connectors from the injectors. The fuel rail is now ready to be removed. On some engines, the fuel rail is held in place by bolts; they need to be removed before pulling the fuel rail free. When pulling the fuel rail away from the engine, pull with equal force on each side of the fuel rail to remove the rail and injectors.

WARNING!

Do not use compressed air to flush or clean the fuel rail. Compressed air contains water, which may contaminate the fuel rail.

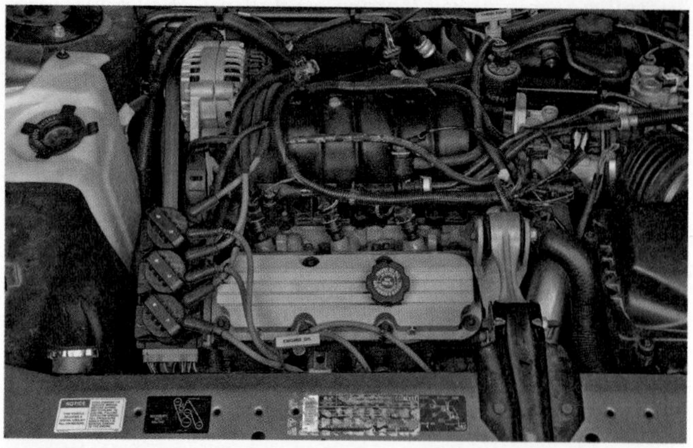

P32-1 *Often an individual injector needs to be replaced. Random disassembly of the components and improper procedures can result in damage to one of the various systems located near the injectors.*

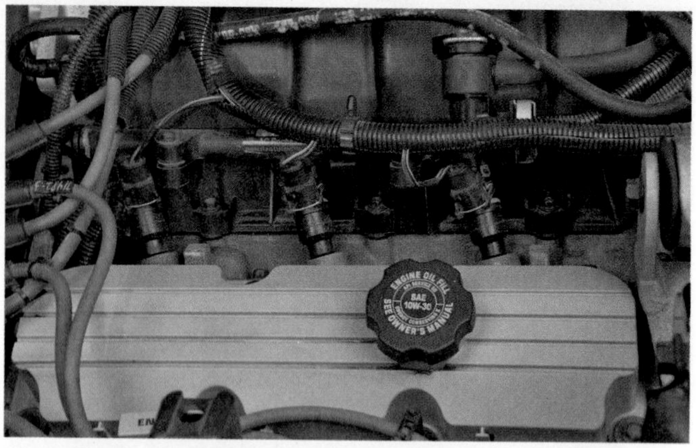

P32-2 *The injectors are normally attached directly to a fuel rail and inserted into the intake manifold or cylinder head. They must be positively sealed because high-pressure fuel leaks can cause a serious safety hazard.*

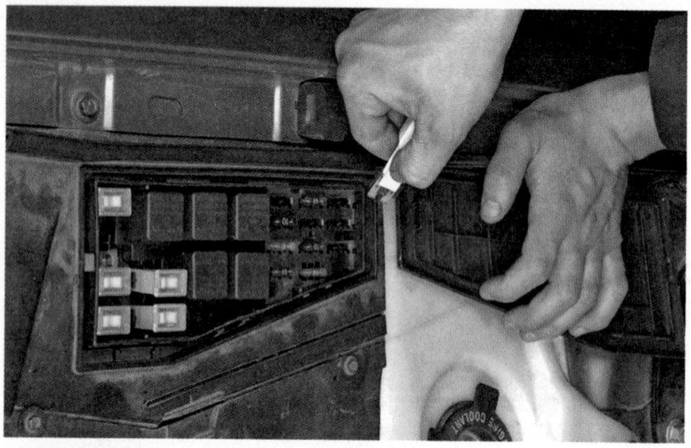

P32-3 *Prior to loosening any fitting in the fuel system, the fuel pump fuse should be removed.*

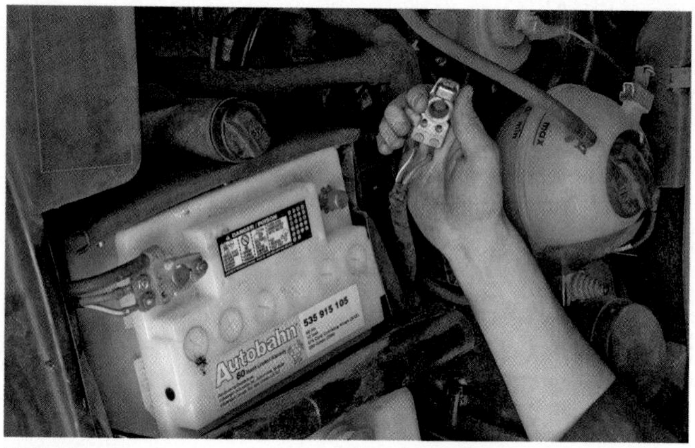

P32-4 *As an extra precaution, many technicians disconnect the negative cable at the battery.*

P32-5 *To remove an injector, the fuel rail must be able to move away from the engine. The rail-holding brackets should be unbolted and the vacuum line to the pressure regulator disconnected.*

P32-6 *Disconnect the wiring harness to the injectors by depressing the center of the attaching wire clip.*

P32-7 *The injectors are held to the fuel rail by a clip that fits over the top of the injector. O-rings at the top and at the bottom of the injector seal the injector.*

P32-8 *Pull up on the fuel rail assembly. The bottoms of the injectors will pull out of the manifold while the tops are secured to the rail by clips.*

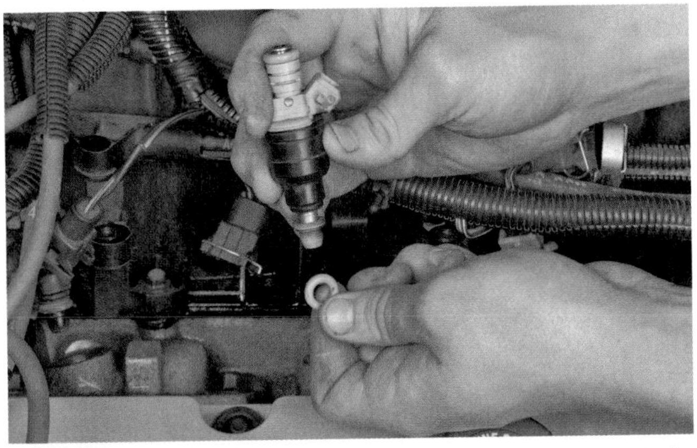

P32-9 *Remove the clip from the top of the injector and remove the injector unit. Install new O-rings onto the new injector. Be careful not to damage the seals while installing them, and make sure they are in their proper locations.*

P32-10 *Install the injector into the fuel rail and set the rail assembly into place.*

P32-11 *Tighten the fuel rail holddown bolts according to manufacturer's specifications.*

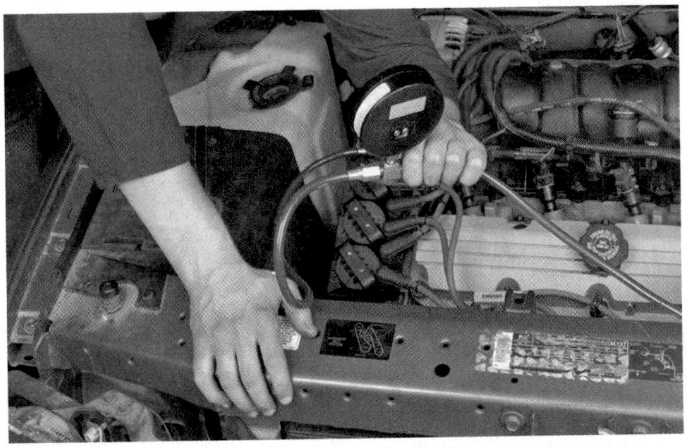

P32-12 *Reconnect all parts that were disconnected. Install the fuel pump fuse and reconnect the battery. Turn the ignition switch to the run position and check the entire system for leaks. After a visual inspection has been completed, conduct a fuel pressure test on the system.*

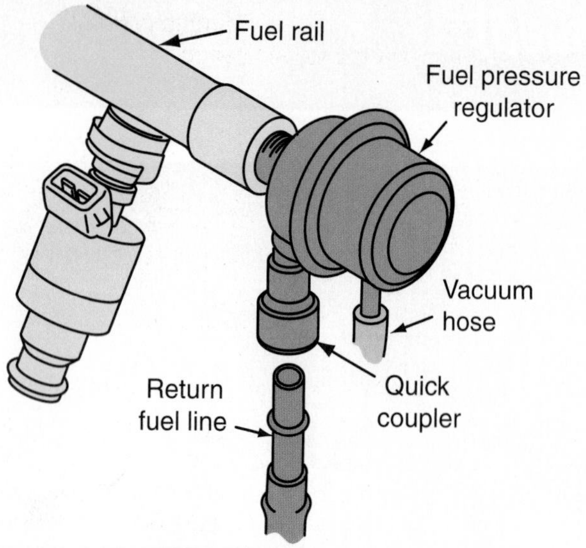

Figure 31-23 Fuel supply and return lines connected to the fuel rail.

Figure 31-24 A GDI injector.

Prior to removing the injectors and pressure regulator, the fuel rail should be cleaned with a spray-type engine cleaner. Normally the approved cleaners are listed in the service manual. After the rail is cleaned the injectors can be pulled from the fuel rail. Use snapring pliers to remove the snapring from the pressure regulator cavity. Note the original direction of the vacuum fitting on the pressure regulator and pull the pressure regulator from the fuel rail. Clean all components with a clean shop towel. Be careful not to damage fuel rail openings and injector tips. Check all injector and pressure regulator openings in the fuel rail for metal burrs and damage.

WARNING!

Do not immerse the fuel rail, injectors, or pressure regulator in any type of cleaning solvent. This action may damage and contaminate these components.

When reassembling the fuel rail with the injectors and pressure regulator, make sure all O-rings are replaced and lightly coated with engine oil. Assemble the fuel rail in the reverse order as that used for disassembly. After the rail and injectors are in place and everything connected to them, reconnect the negative battery terminal and disconnect the 12-volt power supply from the cigarette lighter. Then start the engine and check for fuel leaks at the rail and be sure the engine operation is normal.

Special GDI Checks

The injectors in a GDI system **(Figure 31–24)** are best checked with an ohmmeter. Resistance checks can identify if the injector has an open or a short. If an injector does not have the specified resistance, it should be replaced. Of course, there are also designated DTCs that may lead you to suspect a problem with an injector. Those DTCs are related to particular injectors, unless the malfunction is something that affects many of them.

Because GDI operates under very high pressures, a typical fuel volume check should not be done. However, the pressure of the fuel can be checked as well as the condition of the high-pressure pump. Fuel pressure is best tested by controlling the activity of the pump with a scan tool. With the scan tool and fuel pressure gauge connected, start the engine. Select the mode on the tool that allows for control of the fuel pressure (this mode will vary with the type of tool). Attempt to bring the fuel pressure up to the specified pressure. While changing the pressure, the reading should fluctuate. Also listen to the sound of the pump. It too should cycle. If any of these conditions do not exist, check the pump, its electrical circuits, pressure sensor, and the PCM.

The fuel injector **(Figure 31–25)** and pump control can be checked with an ohmmeter **(Figure 31–26)**. Connect the meter across the terminals at the connector. Compare your readings to specifications. If the reading does not match specs, replace the injector or pump. The fuel pressure sensor can also be checked with a voltmeter. With the engine running and a fuel pressure gauge connected, backprobe the sensor's output signal and compare the pressure gauge reading and voltage to specifications.

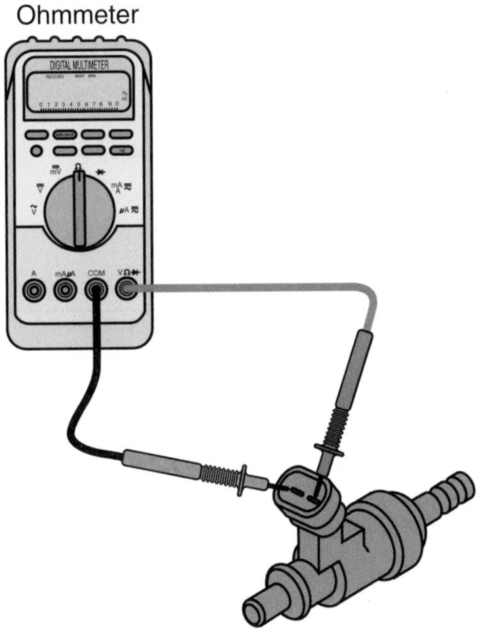

Figure 31–25 A GDI injector can be checked with an ohmmeter.

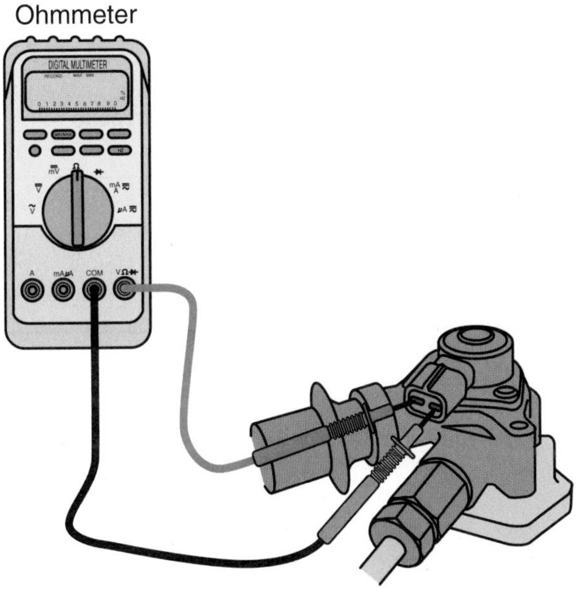

Figure 31–26 The control unit for the high-pressure fuel pump in a GDI system can be checked with an ohmmeter.

ELECTRONIC THROTTLE CONTROLS

An electronic throttle control (ETC) system normally includes a throttle actuator or motor, TP sensors, accelerator pedal position (APP) sensors, an electronic throttle control module, and a relay. The control module may be part of the PCM or may be a separate unit. The actuator responds to commands from the PCM **(Figure 31–27)**.

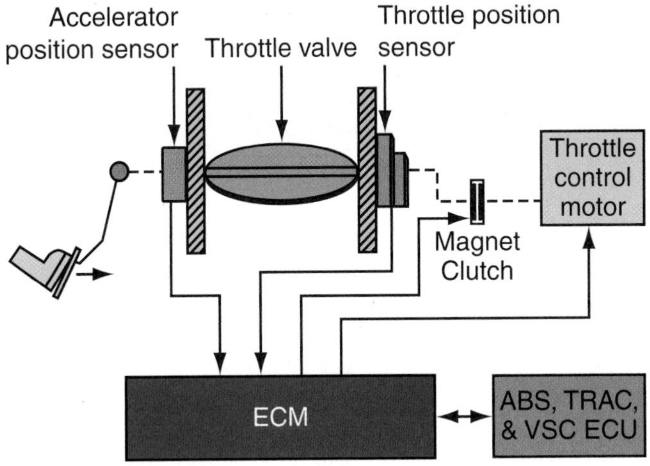

Figure 31–27 A simplistic look at an electronic throttle control system. *Courtesy of Toyota Motor Sales, U.S.A., Inc.*

The ETC system relies on redundant inputs and processors. Two accelerator pedal position signals are sent to the PCM, each having a different voltage range. The PCM processes these signals along with other inputs to set the throttle plate at the required position. The throttle plate is controlled by a DC motor that is controlled by the PCM. The PCM has two driver circuits for the motor; one causes the plate to open and the other closes it. The exact position of the throttle plate is monitored by two TP sensors; the signal range of these also differ. It also has redundant monitors that track the system's effectiveness. When the PCM detects a fault, the monitor will set a DTC and/or put the system into a limp-home mode **(Figure 31–28)**.

Diagnostic Monitors

The PCM monitors the voltage levels of the APP sensors, TP sensors, and the throttle motor circuit. It also monitors the return rate of the two return springs. The APP sensors are potentiometers mounted at the accelerator pedal. Depending on the vehicle, there can be two separate APP sensors or one sensor capable of sending two separate signals. One signal is used to define the position of the pedal. The signal varies between 0 and 5 volts according to the pedal's position. The other signal is used to monitor the sensor itself. The PCM compares the two signals and if there is a difference in the reported pedal position, the PCM determines that the sensor is bad and will illuminate the MIL and set a DTC.

The TP sensor (there may be two separate sensors or one capable of providing two separate signals) is mounted to the throttle body. This sensor sends a signal noting the position of the throttle. The PCM uses one of the signals as feedback. This is done by adding the TP voltage to the APP voltage. This sum must equal 5 volts. If the sum is greater or less than this, the PCM will illuminate the MIL and set a DTC.

DTC	Description	Probable Causes	MIL
P0120	Throttle / Pedal Position Sensor / Switch "A" Circuit Malfunction	1. TP sensor 2. ECM	ON
P0121	Throttle / Pedal Position Sensor / Switch "A" Circuit Range / Performance Problem	1. TP sensor	ON
P0122	Throttle / Pedal Position Sensor / Switch "A" Circuit Low Input	1. TP sensor 2. Short in APP circuit 3. Open in TP circuit 4. ECM	ON
P0123	Throttle / Pedal Position Sensor / Switch "A" Circuit High Input	1. TP sensor 2. Open in APP circuit 3. Open in ETC circuit 4. Short between TP and APP circuits 5. ECM	ON
P0220	Throttle / Pedal Position Sensor / Switch "B" Circuit	1. TP sensor 2. ECM	ON
P0222	Throttle / Pedal Position Sensor / Switch "B" Circuit Low Input	1. TP sensor 2. Short in APP circuit 3. Open in TP circuit 4. ECM	ON
P0223	Throttle / Pedal Position Sensor / Switch "B" Circuit High Input	1. TP sensor 2. Open in APP circuit 3. Open in ETC circuit 4. Short between TP and APP circuits 5. ECM	ON
P0505	Idle Control System Malfunction	1. ETC 2. Air induction system 3. PCV hose connection 4. ECM	ON
P050A	Cold Start Idle Air Control System Performance	1. Throttle body assembly 2. MAF sensor 3. Air induction system 4. PCV hose connections 5. VVT system 6. Air cleaner element 7. ECM	ON
P050B	Cold Start Ignition Timing Performance	1. Throttle body assembly 2. MAF sensor 3. Intake system 4. PCV hose connection 5. VVT system 6. Air filter element 7. ECM	ON
P060A	Internal Control Module Monitoring Processor Performance	ECM	ON
P060D	Internal Control Module Accelerator Pedal Position Performance	ECM	ON
P060E	Internal Control Module Throttle Position Performance	ECM	ON
P0657	Actuator Supply Voltage Circuit / Open	ECM	ON

Figure 31-28 Some examples of generic DTCs related to the electronic throttle control system.

The PCM also looks at the voltages to the control unit and the actuator assembly, in addition to the amount of current required to move the throttle plate. If it detects lower-than-normal voltage to the control unit or actuator, it will disable the throttle control system and set it into the limp-home mode. It will also illuminate the MIL and set the DTC. The PCM continuously monitors the current flow through the actuator. If the current is too high or low, the PCM recognizes this as a problem. The PCM also looks at the position of the throttle and the current flowing to it. If the position does not change when it is commanded to do so, the MIL will be illuminated and a DTC set.

Fail-Safe Mode When DTCs relating to the ETC system are set, the PCM will enter into a fail-safe of limp-home mode. During this time the actuator is no longer controlled by the PCM. However, the throttle plate is held slightly open by a spring. This limited throttle opening restricts the ability of the engine to provide power. If the accelerator pedal is depressed, the PCM will control engine output by controlling the fuel injection and ignition systems. This will allow the vehicle to be driven slowly.

Idle Speed

Although there is no adjustment for idle speed, idle speed checks can give an indication of the condition of the electronic throttle system. The best way to monitor idle speed is with a scan tool. Before checking the speed, make sure that the MIL is not illuminated and that there are no DTCs set. Also make sure that the ignition, air induction, and PCV systems are okay.

Set the parking brake and disconnect the connector to the evaporative emission (EVAP) canister purge valve. Connect the scan tool to the DLC and make sure there is communication between the tool and the vehicle. Start the engine and make sure all accessories are turned off. Increase the speed to about 3,000 rpm and hold it there until the cooling fan turns on. Allow the engine to idle; check the idle speed and compare it to specs. Then turn on some heavy load accessories, such as MAX air conditioning with the blower on high and the high-beam headlights. Check the idle speed. If the idle speed does not match specifications, conduct the idle learn procedure. If the speed is still wrong, further diagnostics of the system and fuel injection system are necessary. Once the idle speed matches the specifications, reconnect the EVAP canister purge valve.

Idle Learn All throttle positions reported to the PCM reflect a change from a base reading. It is extremely important that the PCM knows that base. Manufacturers have prescribed procedures for teaching the PCM this base. Those procedures must be followed exactly as stated. The learn or relearn procedure should be completed anytime the PCM has been replaced or updated, and after the throttle body has been cleaned or replaced. Often no-DTC idle speed problems are solved by completing the idle learn or relearn procedure. It is important to note that the system does not need to relearn if the battery has been disconnected. The system will automatically go through the procedure when the engine is restarted.

Idle learn is normally done by running the engine at idle for a prescribed time after it has reached normal operating temperature and the PCM has been reset by a scan tool. The idle learn process can be monitored and verified with a scan tool.

General Diagnostics

There are several DTCs assigned to problems with the ETC system. All of the major components have several DTCs assigned to them. There are also pinpoint tests for each component. Most of the components can be checked with a voltmeter, ohmmeter, or scope. The parts should be inspected prior to testing. This is especially true of the throttle body assembly. It should be checked for any buildup of dirt on the plate and in the bores. If needed, clean the assembly. Also make sure that the plate moves freely in the bore.

Ohmmeter checks are made across specific terminals at the throttle body. On most units, the connection is made across the positive (M+) feed to the motor and the negative (M−) terminals. This setup measures the resistance of the windings in the actuator's motor. Some throttle bodies have the TP sensor built into the assembly. In these cases, the TP sensor is checked at different terminals in the same connector. Always refer to the service information to identify the test terminals and resistance specifications.

A voltmeter can be used to check the operation of the TP and APP sensors. This is done in the same way as when checking typical sensors, except that with ETC systems there are two output signals to monitor. The sensors can also be checked with a lab scope.

The throttle motor can also be checked with a lab scope. Connect the scope to the M+ terminal and to ground. Then observe the activity at the M− terminal and ground (**Figure 31–29**). Remember, the waveform should change with a change in throttle angle.

IDLE SPEED CHECKS

The idle speed of fuel-injected engines without an electronic throttle is regulated by controlling the amount of air that bypasses the airflow sensor or throttle plates. When one of these vehicles has an idling problem, check the linkage and vacuum lines

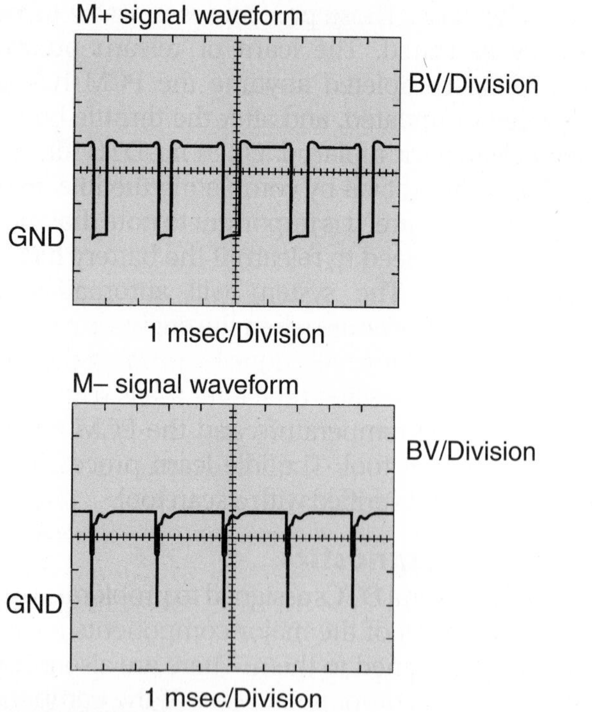

Figure 31-29 Waveforms for both power feeds (M+ and M–) at an electronically controlled throttle body. *Courtesy of Toyota Motor Sales, U.S.A., Inc.*

before going any further with your diagnosis. Although idle speed is not typically adjustable, some engines do have provisions for setting the speed. Always refer to the decal in the engine compartment to identify the conditions that must be present before adjusting the speed. Also, refer to the service information before making any adjustments. The idle speed and quality for most engines are controlled by the PCM through the IAC.

IAC Checks

If the idle speed is not within specifications, the system should be looked at with a scan tool. Make sure there is proper communications between the scan tool and the vehicle. Also make sure that all CAN communications are good.

Many different inputs will affect the performance of the IAC as well as cause other driveability problems. Following are examples:

- With higher-than-normal TP signals at idle, the PCM interprets this as the throttle being opened and will close the IAC to decrease idle speed.
- If the resistance of the ECT is too high, a higher-than-normal voltage signal is sent to the PCM, which interprets this as a cold engine. The PCM will then open the IAC to increase speed.
- If the resistance of the ECT is too low, a lower-than-normal voltage signal is sent to the PCM, which interprets this as a hot engine. The PCM will then open the IAC to increase speed.
- A stuck closed A/C switch will signal to the PCM that the A/C is always on and the PCM will always order an increased idle speed.

SHOP TALK

On most vehicles, the scan tool will indicate the status of input switches as closed or open or high or low. Most input switches provide a high-voltage signal to the PCM when they are open and a low-voltage signal if they are closed.

If the sensors and circuits seem to be fine, check the IAC. The IAC can be monitored with a scan tool. Most will have a test mode that allows for manual control of the IAC. The IAC motor counts are observed. When the IAC is operated in its low range, the PCM will move the IAC in approximately 16 steps **(Figure 31–30)**. When operated in its high range, there should be approximately 112 steps or counts **(Figure 31–31)**. The normal

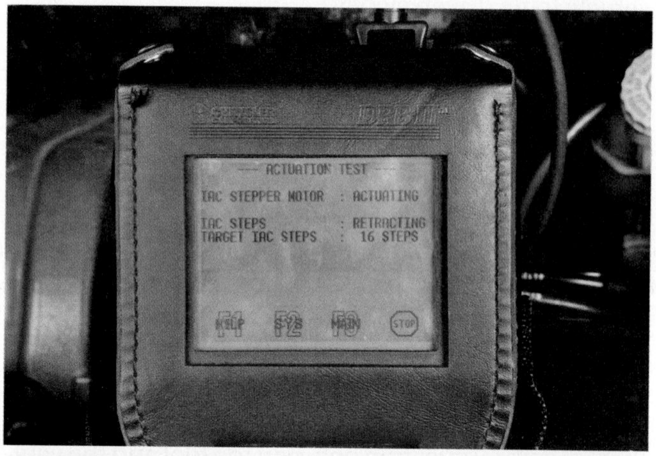

Figure 31-30 When the IAC is operated in its low range, the PCM will move the IAC in approximately 16 steps.

Figure 31-31 When operated in its high range, the IAC should move approximately 112 steps or counts.

range of counts is given in the appropriate service information.

Most technicians look for consistent idle counts on a warm engine to verify that the IAC is not sticking or malfunctioning. During the check, devices such as the A/C system can be operated and the scan tool watched. If the counts change when the A/C is turned on and off, the IAC, connecting wires, and PCM are working fine. When the counts do not change, further diagnosis is required. Always follow the diagnostic procedures given by the manufacturer.

SHOP TALK

If the IAC motor counts are below 12 on the scan tool, look for a vacuum leak. The PCM will run the valve all the way down to compensate for the leak. If no leak is found, then check the circuit between the PCM and the motor. The circuit is likely open between the PCM and the motor. Wiggle the wires on the IAC motor and observe the reading on the scan tool. If the count reading changes while wiggling the wires, you have found the problem.

IAC problems on OBD-II vehicles will set one or more DTCs. Although there are DTCs designated for the IAC system, keep in mind that the operation of the entire engine performance system can affect idle speed and quality. Therefore, all DTCs could be the cause of an idle problem.

It is important to note that the PCM cannot effectively control idle if it does not know when the throttle is closed. This is also important for all other engine speeds. The PCM sets the closed throttle reference according to the lowest TP voltage signal since the engine was started. The PCM does this each time the engine is started. This reference voltage is called the TPREL PID on most systems. When looking at this PID with the throttle closed, the value should be C/T (closed throttle). If any other value is noted, the engine will have a higher-than-normal idle speed. Abnormal readings are typically caused by a bad TP sensor, loose or worn throttle plates, or excessive noise in the TP or related circuits.

Most late-model vehicles have a feature that adjusts the calibration of the IAC according to the wear of system components; this is called idle air trim. The system constantly monitors the engine and its systems and determines the ideal idle speed. The idle speed is based on look-up tables. If the corrections exceed predetermined levels, the PCM will set a DTC. Once the problem is solved and corrected, the system must relearn its base idle trim values. This process is completed with a scan tool and should be done whenever a part of the IAC system is replaced or a repair is made to something that affects idle speed.

Servicing the IAC Motor

On some vehicles, there is a provision to manually move the IAC plunger through the scan tool. When this test mode is selected, the PCM is ordered to extend and retract the IAC motor plunger every 2.8 seconds. When this plunger extends and retracts properly, the motor, connecting wires, and PCM are in normal condition. If the plunger does not extend and retract, further diagnosis is necessary to locate the cause of the problem. On other vehicles, this same test can be conducted with a jumper wire connecting two terminals of the DLC together.

Carbon deposits in the IAC motor air passage in the throttle body or on the IAC motor's pintle can cause erratic idling and engine stalling. Remove the motor from the throttle body, and inspect the throttle body air passage for carbon deposits. If there are heavy carbon deposits, remove the complete throttle body for cleaning. Clean the IAC air passage, motor sealing surface, pintle valve, and pintle seat with throttle body cleaner.

WARNING!

Be careful while cleaning the assembly; the IAC motor can be damaged if throttle body cleaner is allowed to get inside the motor.

The motor can be checked with an ohmmeter. If the ohmmeter readings are not within specifications, replace the motor. If a new IAC motor is installed, make sure that the part number, pintle shape, and diameter are the same as those on the original motor. Measure the distance from the end of the pintle to the shoulder of the motor casting **(Figure 31–32)**. Move the pintle until it is at the specified distance. If the pintle is extended too far, the motor can be damaged during installation.

Install a new gasket or O-ring on the motor. If the motor is sealed with an O-ring, lubricate the ring with transmission fluid. If the motor is threaded into the throttle body, tighten the motor to the specified torque. When the motor is bolted to the throttle body, tighten the mounting bolts to the specified torque.

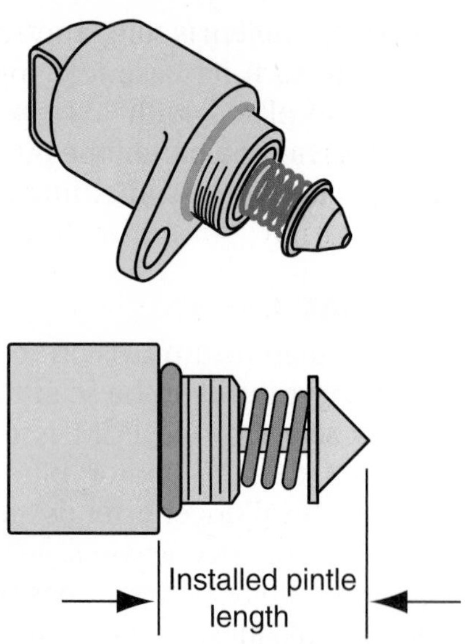

Figure 31-32 Measure the distance the pintle of an IAC motor is extended before installing the motor.

CASE STUDY

A customer brought in a late-model car complaining of poor fuel economy, poor overall performance, and the engine seemed to shake when idling or driven at a constant speed. The technician started the engine and verified that it did have a slight shake. The shake was accompanied by a change in sound. He suspected that the engine was misfiring. Being a recent graduate of a post-secondary automotive program, he had little experience so he followed what he was taught rather than jump to a conclusion.

Knowing that a misfire can be caused by an engine, fuel, air, or ignition problem, he began his diagnosis, planning to eliminate systems that were not causing the misfire. First he conducted a visual inspection and found nothing abnormal, except that the MIL was illuminated but not flashing. This meant the PCM had set a DTC but the conditions present when the code was set were not serious enough to destroy the catalytic converter. He then connected a vacuum gauge to the intake manifold. The results of this may eliminate the mechanical parts of the engine or give him clues as to where to go next. The needle of the gauge fluctuated more than it should, but the overall reading was about 17 in. Hg. Based on this, he concluded that the fluctuation was due to the miss. The overall reading, however, indicated that the engine itself was rather sound.

He then connected a scan tool to the vehicle and looked at the fuel trim PIDs. He remembered that fuel trims of about ±10% indicate possible ignition problems and fuel trims of about ±20% suggest an air-fuel ratio problem. In this case, he found the STFT to be 28%. Now he knew that the misfire was caused by the air or fuel systems. This was an important step because misfires can set a variety of DTCs and a fault in one system can cause a DTC that relates to another system.

He also noticed that the O_2 sensor signals went excessively lean at times that seemed to correlate with the miss. Retrieving the DTC P0303, he could assume that the problem was at cylinder number 3. Knowing that the problem is probably caused by the fuel injection system, he checked the voltage to the number 3 injector. Battery voltage was there. He then checked the pulsing of the injector with a noid light. The injector seemed to be pulsing at the correct times. He concluded that the circuits to the injector were fine and that the PCM was trying to control it.

He therefore assumed that the problem was the injector and planned on replacing it. When he unplugged the electrical connector to the injector, he saw the problem. The electrical contacts for the injector were corroded. The corrosion was not bad enough to stop the injector from working, but its presence reduced the current to the injector. He replaced the injector and cleaned the terminals in the connector to the injector.

To verify the repair, he erased the DTC and started the engine. The engine idled smoothly and ran smoothly during a test drive. Once back in the shop, he noticed that the MIL had not relit but he looked for DTCs with the scan tool. A new DTC was not set and the cause of the problem was probably solved.

KEY TERMS

Peak and hold injector

Pulse-modulated injector

SUMMARY

- Always relieve the fuel pressure before disconnecting any component in the fuel system.
- Always turn off the ignition switch before connecting or disconnecting any system component or test equipment.
- An O_2 sensor can be checked with a voltmeter connected between the sensor wire and ground. The sensor's voltage should be cycling from low voltage to high voltage.
- The signal from most O_2 sensors varies between 0 and 1 volt. If the voltage is continually high, the air-fuel ratio may be rich or the sensor may be contaminated.
- When the O_2 sensor voltage is continually low, the air-fuel ratio may be lean, the sensor may be defective, or the wire between the sensor and the computer may have high resistance.
- If the O_2 sensor voltage signal remains in a midrange position, the computer may be in open loop or the sensor may be defective.
- A/F ratio sensors respond to oxygen in the opposite way as O_2 sensors.
- The activity of the sensor can be monitored on a scan tool or lab scope.
- An injector that does not open causes hard starts on port-type systems and an obvious no-start on single-point TBI designs.
- An injector that is stuck partially open causes a loss of fuel pressure and flooding due to raw fuel dribbling into the engine.
- Buildups of gum and other deposits on the tip of an injector can reduce the amount of fuel sprayed by the injector and can prevent it from totally sealing, allowing it to leak.
- A rich mixture can be caused by a leaking fuel pressure regulator. If the regulator's diaphragm is ruptured, fuel will flow into the intake manifold, causing a rich mixture.
- When an injector is suspected as the cause of a lean problem, determine if the injector is receiving a signal to fire. Check for voltage at the injector using a high-impedance testlight or noid light.
- An ohmmeter can be used to test the electrical soundness of an injector.
- An injector pressure balance test will help isolate a clogged injector.
- An oscilloscope can be used to monitor the injector's pulse width and duty cycle.
- For all types of injectors, the waveform on a scope should have a clean, sudden drop in voltage when it is turned on.
- Never soak an injector in cleaning solvent or use a wire brush, pipe cleaner, toothpick, or other cleaning utensil to unblock a plugged injector.
- In a fuel injection system, idle speed is regulated by controlling the amount of air that is allowed to bypass the airflow sensor or throttle plates. When a car tends to stall or idles too fast, look for obvious problems like binding linkage and vacuum leaks first. Then check the minimum idle checking/setting procedure described on the underhood decal.

REVIEW QUESTIONS

1. List the three things that must occur for an EFI to operate properly.
2. What is indicated by trouble codes?
3. What is the correct procedure for checking an oxygen sensor with a DMM?
4. What is the difference between STFT and LTFT?
5. The PCM checks for a closed throttle plate each time the engine starts. This becomes the base for all throttle settings and can be viewed by looking at the PID with the throttle closed. Which of the following would *not* cause an abnormal reading and a higher-than-normal idle speed?
   a. a bad TP sensor
   b. a bad IAC valve
   c. loose or worn throttle plates
   d. excessive noise in the TP or related circuits
6. Which of the following is a likely cause of a no-start condition if the engine starts when the electrical connector to the MAF is disconnected?
   a. a defective oxygen sensor
   b. a defective PCM
   c. a defective MAP sensor
   d. a defective MAF sensor
7. What problem may result from dirt buildup on an engine's throttle plates?
8. Which of the following would *not* cause a hard-to-start problem on a PFI engine?
   a. an electrically open fuel injector
   b. a defective oxygen sensor
   c. a leaking fuel pressure regulator
   d. dirty injectors
9. What is the correct way to test an injector with an ohmmeter?
10. What is the difference between the pulse width and the duty cycle of an injector?

11. How can you use a dual trace scope to make sure that the injectors are firing at the correct time?
12. When conducting a fuel pressure test on the high-pressure pump in a GDI system, what is indicated by a fluctuating needle on the pressure gauge?
13. What is the purpose of having two accelerator pedal signals in an electronic throttle control system?
14. What is indicated by a negative LFFT value?
15. *True or False?* The signals from an air-fuel ratio sensor are identical to those from a conventional oxygen sensor.

ASE-STYLE REVIEW QUESTIONS

1. While discussing the causes of higher-than-specified idle speeds: Technician A says that an intake manifold vacuum leak may cause a high idle speed. Technician B says that if the TP sensor voltage signal is higher than specified, the idle speed may be higher than normal. Who is correct?
   a. Technician A
   b. Technician B
   c. Both A and B
   d. Neither A nor B
2. While discussing IAC valve diagnosis: Technician A says that on some vehicles, a jumper wire may be connected to specific DLC terminals to check the IAC valve operation. Technician B says that if the scan tester indicates zero IAC valve counts, there may be an open circuit between the PCM and the IAC valve. Who is correct?
   a. Technician A
   b. Technician B
   c. Both A and B
   d. Neither A nor B
3. While discussing IAC motor removal, service, and replacement: Technician A says that throttle body cleaner may be used to clean the IAC motor internal components. Technician B says that on some vehicles, IAC motor damage occurs if the pintle is extended more than specified during installation. Who is correct?
   a. Technician A
   b. Technician B
   c. Both A and B
   d. Neither A nor B
4. While discussing injector testing: Technician A says that a defective injector may cause cylinder misfiring at idle speed. Technician B says that restricted injector tips may result in acceleration stumbles. Who is correct?
   a. Technician A
   b. Technician B
   c. Both A and B
   d. Neither A nor B
5. While discussing airflow sensors: Technician A says that with mass airflow and volume airflow sensor systems, if any air bypasses the sensors, the engine will run lean. Technician B says that vacuum leaks in a speed density system will decrease injector pulse width. Who is correct?
   a. Technician A
   b. Technician B
   c. Both A and B
   d. Neither A nor B
6. While discussing scan tool diagnosis of TBI, MFI, and SFI systems: Technician A says that the scan tool will erase fault codes quickly on many systems. Technician B says that many scan tools will store sensor readings during a road test and then play back the results in a snapshot test mode. Who is correct?
   a. Technician A
   b. Technician B
   c. Both A and B
   d. Neither A nor B
7. While discussing a high idle speed problem: Technician A says that higher-than-normal idle speed may be caused by low electrical system voltage. Technician B says that higher-than-normal idle speed may be caused by a defective coolant temperature sensor. Who is correct?
   a. Technician A
   b. Technician B
   c. Both A and B
   d. Neither A nor B
8. While discussing the causes of a rich air-fuel ratio: Technician A says that a rich air-fuel ratio may be caused by low fuel pump pressure. Technician B says that a rich air-fuel ratio may be caused by a defective coolant temperature sensor. Who is correct?
   a. Technician A
   b. Technician B
   c. Both A and B
   d. Neither A nor B
9. While diagnosing an idle speed problem: Technician A says that higher-than-normal TP signals at idle will cause the IAC to close, which will decrease idle speed. Technician B says that if the resistance of the ECT is too low, a lower-than-normal voltage signal is sent to the PCM, which interprets this as a hot engine and will close the IAC to decrease speed. Who is correct?
   a. Technician A
   b. Technician B
   c. Both A and B
   d. Neither A nor B
10. While looking at fuel trim values: Technician A says that restricted or dirty injectors may be evident by a negative LTFT. Technician B says that a faulty fuel pressure regulator may result in a negative LTFT. Who is correct?
    a. Technician A
    b. Technician B
    c. Both A and B
    d. Neither A nor B

INTAKE AND EXHAUST SYSTEMS

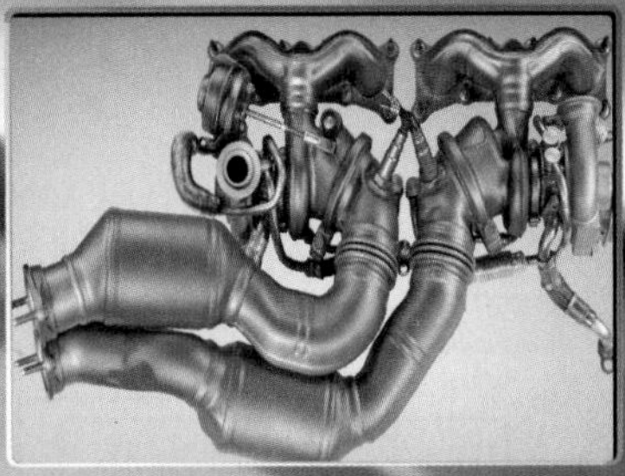

CHAPTER 32

OBJECTIVES

■ Describe how the engine creates vacuum and how vacuum is used to operate and control many automotive devices. ■ Explain the operation of the components in the air induction system, including ductwork, air cleaner/filters, and intake manifolds. ■ Inspect and troubleshoot vacuum and air induction systems. ■ Explain the purpose and operation of a turbocharger. ■ Inspect a turbocharger, and describe some common turbocharger problems. ■ Explain supercharger operation, and identify common supercharger problems. ■ Explain the operation of exhaust system components, including the exhaust manifold, gaskets, exhaust pipe and seal, catalytic converter, muffler, resonator, tailpipe, and clamps, brackets, and hangers. ■ Properly perform an exhaust system inspection, and service and replace exhaust system components.

An internal combustion engine needs air for combustion. It also needs to allow spent gases to leave the cylinder after combustion. The focus of this chapter is on the intake and exhaust systems. Both of these are often overlooked but are very important.

The reason air enters a cylinder is simply a basic law of physics—high pressure always moves toward an area of low pressure. Therefore, outside air moves into the cylinders because of the vacuum formed on the intake stroke.

VACUUM SYSTEMS

The vacuum in the intake manifold not only draws air into the cylinders; it also is used to operate or control many systems, such as emission controls, power brake boosters, parking brake releases, heater/air conditioners, and cruise controls **(Figure 32–1)**. Vacuum is applied to these systems through a network of hoses, tubes, and control valves.

Vacuum Basics

The term *vacuum* refers to any pressure that is lower than the earth's atmospheric pressure at any given altitude. The higher the altitude, the lower the atmospheric pressure.

Go To ***Chapter 8 for a detailed explanation of atmospheric pressure and vacuum.***

Vacuum is measured in relation to atmospheric pressure. At sea level, atmospheric pressure is 14.7 psi and will appear as zero on most pressure gauges. This does not mean that there is no pressure; rather, it means the gauge is designed to read pressures greater than atmospheric pressure. Measurements taken on this type of gauge are given in pounds per square inch and should be referred to as psig (pounds per square inch gauge). Some gauges read in bar, kP, or inches of mercury. Gauges and other measuring devices that include atmospheric pressure in their readings also display their measurements in psi. However, the measurements on these gauges should be referred to as psia (pounds per square inch absolute). There is a big difference between 12 psia and 12 psig: 12 psia is less than atmospheric pressure and therefore would represent a vacuum, whereas 12 psig would be approximately 26.7 psia. Vacuum, therefore, is any pressure less than 0 psig or 14.7 psia **(Figure 32–2)**. Normally a

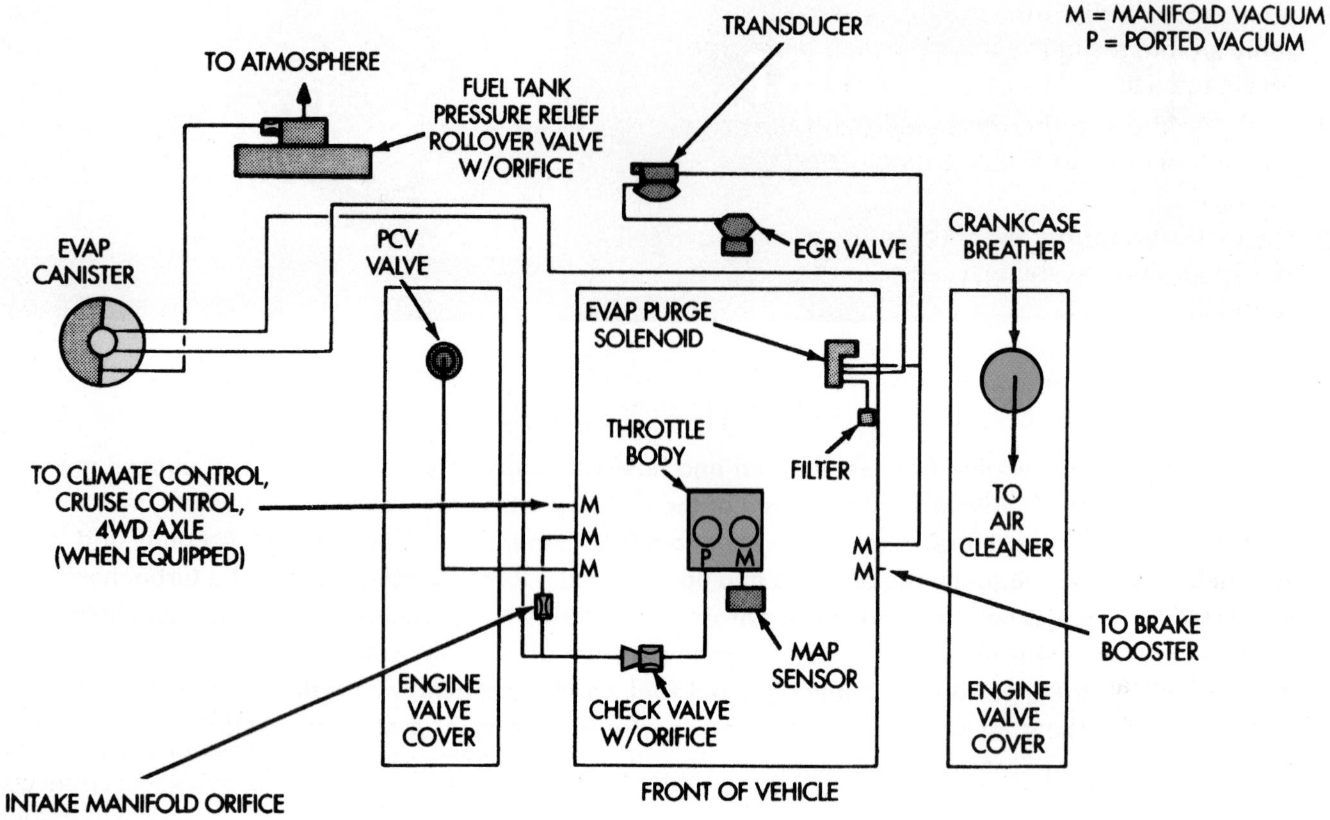

Figure 32-1 Typical vacuum devices and controls. *Courtesy of Chrysler LLC*

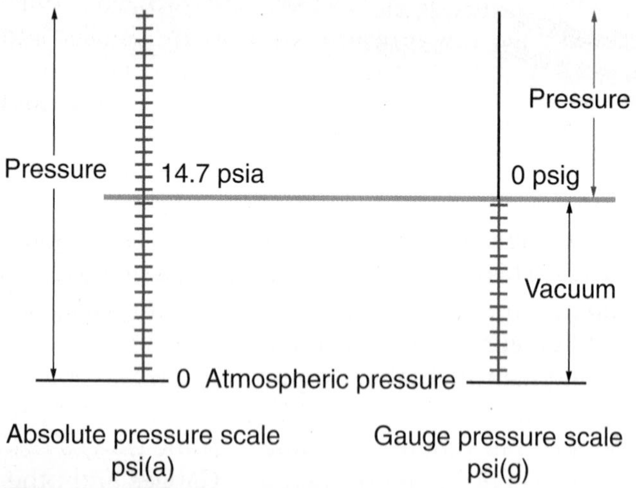

Figure 32-2 A comparison of psia and psig.

measure of vacuum is given in inches of mercury (in. Hg). Vacuum may also be expressed in units of kilopascals and bar. Normal atmospheric pressure at sea level is about 1 bar or 100 kilopascals.

Engine vacuum is created by the downward movement of the piston during the intake stroke. With the intake valve open and the piston moving downward, a partial vacuum is created within the cylinder and intake manifold. The air passing the intake valve does not move fast enough to fill the cylinder, thereby causing the lower pressure. This partial vacuum is continuous in a multicylinder engine, because at least one cylinder is always at some stage of its intake stroke.

Diagnosis and Troubleshooting

Vacuum problems can produce or contribute to the following driveability symptoms:

- Stalling
- Poor starting
- Backfiring
- Rough idle
- Poor acceleration
- Rich or lean stumble
- Overheating
- Detonation, or knock or pinging
- Rotten eggs exhaust odor
- Poor fuel economy

Whenever there is a driveability problem, do the following:

- Inspect vacuum hoses for improper routing or disconnections.

- Look for cut or disconnected hoses that will allow more air into the intake manifold than the engine is calibrated for.
- Look for kinks in the lines and hoses that can cut off vacuum to a component, thereby disabling it.
- Check for vacuum hose routing and wear near hot spots, such as the exhaust manifold or EGR tubes.
- Make sure there is no evidence of oil or transmission fluid in vacuum hose connections. (Valves can become contaminated by oil getting inside.)
- Inspect vacuum system devices for damage (dents in cans; by-pass valves; broken nipples on valves; broken "tees" in vacuum lines, and so on).

Any defective hoses should be replaced one at a time to avoid misrouting. OEM vacuum lines may be installed in a harness consisting of $\frac{1}{8}$-inch or larger outer diameter and $\frac{1}{16}$-inch inner diameter nylon hose with bonded nylon or rubber connectors. Occasionally, a rubber hose might be connected to the harness. The nylon connectors have rubber inserts to provide a seal between the nylon connector and the connection (nipple). In recent years, many manufacturers have been using ganged steel vacuum lines.

Vacuum Test Equipment

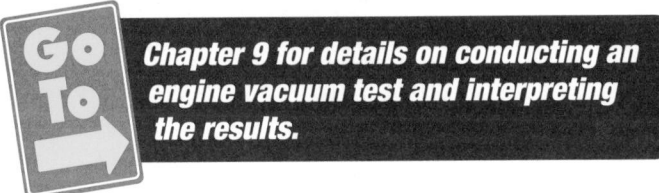

Go To ***Chapter 9 for details on conducting an engine vacuum test and interpreting the results.***

With a vacuum gauge connected to the intake manifold and the engine warm and idling, watch the action of the gauge's needle. A healthy engine will give a steady, constant vacuum reading between 17 and 22 in. Hg. Some four- and six-cylinder engines, however, may have a normal reading of 15 in. Hg, and many high-performance engines will have an unsteady but consistent reading.

An emissions vacuum schematic is given on an under-hood decal. This shows the vacuum hose routings and vacuum source for all emissions-related equipment. The vacuum schematic in **Figure 32–3** shows the relationship and position of components as they are mounted on the engine. It is important to remember that these schematics only show the vacuum-controlled parts of the emission system. The location and hose routing for other vacuum devices can be found in the service information.

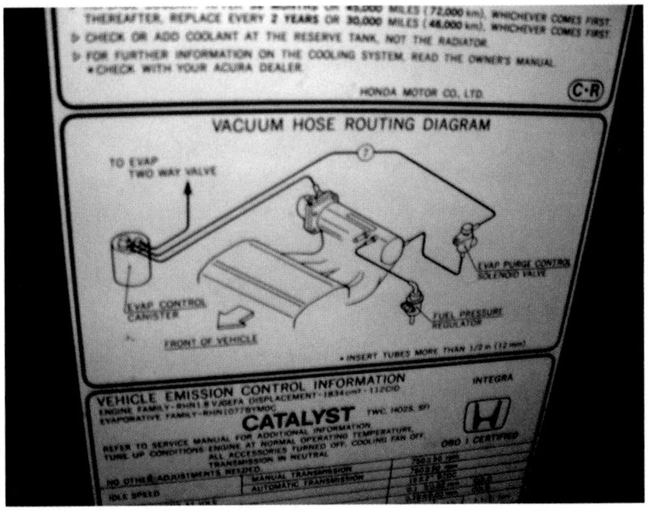

Figure 32-3 An under-hood vacuum hose routing diagram.

AIR INDUCTION SYSTEM

The air induction system directs outside air to the engine's cylinders. The induction system is comprised of ductwork that channels outside air to an air cleaner that removes dirt from the air, ductwork that connects the filter to the throttle body, and an intake manifold that distributes the air to the engine's cylinders (**Figure 32–4**). Within the induction system are sensors that measure intake air temperature and airflow.

An inspection of the air induction system should be part of diagnosing a driveability problem. Make sure that the intake ductwork is properly installed and that all connections are airtight—especially those between an airflow sensor or remote air cleaner and the throttle body.

Air Cleaner/Filter

The primary purpose of the air filter is to prevent airborne contaminants and abrasives from entering the cylinders. These contaminants can cause serious

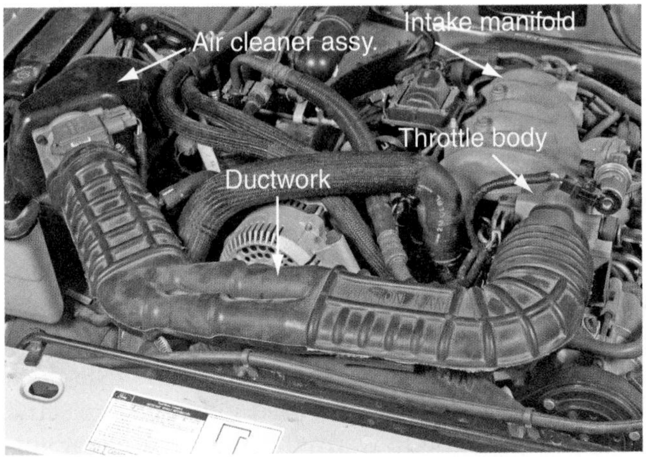

Figure 32-4 A typical late-model air induction system.

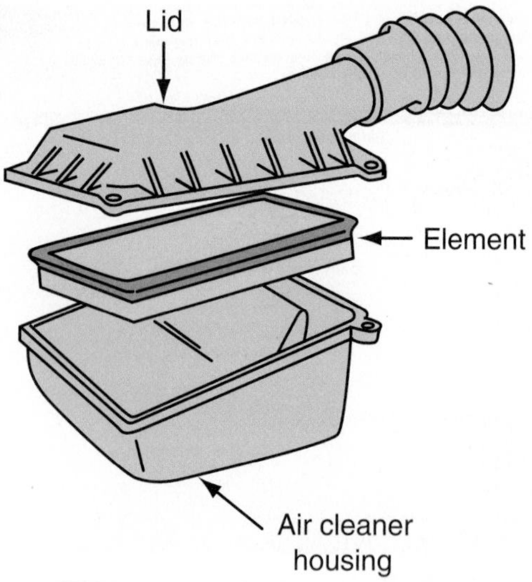

Figure 32-5 A typical flat air filter element.

damage and appreciably shorten engine life. Therefore, all intake air should pass through the filter before entering the engine.

The air filter is inside a sealed air cleaner assembly. This assembly is also used to direct the airflow and reduce the noise caused by the movement of intake air **(Figure 32–5)**. The air cleaner also provides filtered air to the PCV system and provides engine compartment fire protection in the event of backfire.

If the air filter becomes very dirty, the dirt can block the flow of air into the engine. Restricted airflow to the engine can cause poor fuel economy, poor performance, and high emissions.

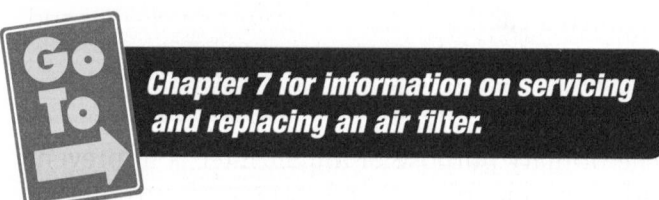

Go To ***Chapter 7 for information on servicing and replacing an air filter.***

INTAKE MANIFOLDS

An intake manifold distributes the clean air or air-fuel ratio as evenly as possible to each cylinder of the engine.

Older engines had cast-iron intake manifolds. The manifold delivered air and fuel to the cylinders and had short runners. These manifolds were either wet or dry. Wet manifolds had coolant passages cast directly in them. Dry manifolds did not have coolant passages but had exhaust passages through the manifold to heat the floor of the manifold. This helped to vaporize the fuel before it arrived in the cylinders. Other dry manifold designs used some sort of electric heater unit or grid to warm the bottom of the manifold. Heating the floor of the manifold stopped fuel from condensing in the plenum area. Good fuel vaporization and the prevention of condensation allowed for delivery of a more uniform air-fuel mixture to the individual cylinders.

Intake manifolds also serve as the mounting point for many intake-related accessories (such as the fuel injectors, fuel rail, and throttle body) and sensors **(Figure 32–6)**. Some include a provision for mounting the thermostat and thermostat housing. In addition, connections to the intake manifold provide a vacuum source for the exhaust gas recirculation (EGR) system, power brakes, and/or heater and air-conditioning airflow control doors.

Intake manifolds for today's engines are typically made of die-cast aluminum **(Figure 32–7)** or plastic **(Figure 32–8)**. These materials are used to reduce weight. Because these intake manifolds only deliver air to the cylinders, fuel vaporization and condensation are not of any concern. The primary goal of their design is the capability to deliver equal amounts of air to each cylinder.

Design Variations

Basic manifold design varies with the different types of engine. For example, the intake manifold for a four-cylinder engine has either four runners or two runners that break into four near the cylinder head. Inline six-cylinder engines have six runners or three that branch off into six near the cylinder head. On V-type engines, there are individual runners for each cylinder.

An intake manifold has two basic components: a plenum area and runners. As air first enters the intake manifold, it moves into the plenum. The air then moves from the plenum through the runners to the cylinders. The size and shape of the plenum and runners are designed for a specific engine and application.

The plenum serves as a reservoir for the air and is used to distribute the intake charge evenly and to enhance engine breathing. The shape of the runners is different for PFI and GDI engines than other engines. An intake manifold that delivers both air and fuel is designed to cause turbulence so that the air and fuel mix while they are being delivered to the cylinders. When the intake manifold only delivers air, there is no need for turbulence and the runners provide a smooth direct flow of the air. Air-only runners have smooth finishes and a minimum number of bends. The length of the runners is designed to achieve the best performance during a particular range of engine speeds.

An engine is most volumetrically efficient when a maximum amount of air enters the cylinders. Peak engine torque occurs when the engine is most efficient. Generally, an engine designed for maximum

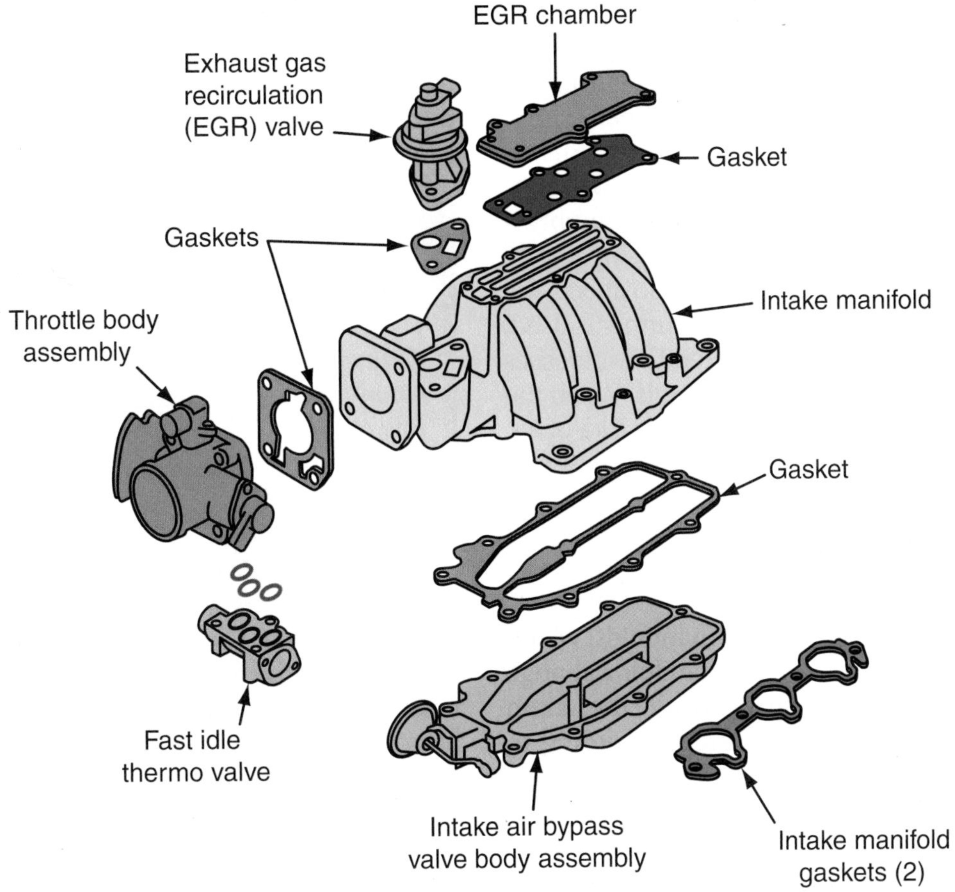

Figure 32-6 A late-model intake manifold for a V6 engine.

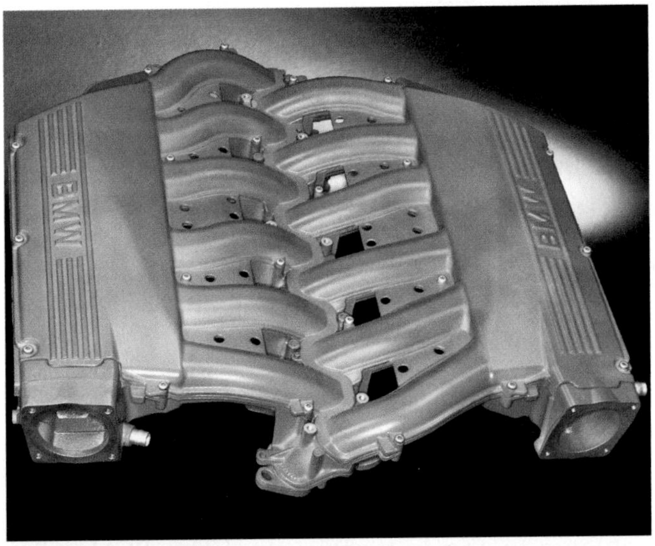

Figure 32-7 A die-cast aluminum intake manifold. *Courtesy of BMW of North America, LLC*

Figure 32-8 A plastic intake manifold for an inline five-cylinder engine. *Courtesy of BMW of North America, LLC*

torque and horsepower at high speeds has shorter runners than one that provides high torque at lower speeds. The length of the runners changes when the engine develops its peak torque.

One of the things that most do not think about is the behavior of air when it enters a runner. A simple thought is the air arrives and stays there until the intake valve opens. It then flows past the valve into the cylinder. Actually, the air is moving at a pretty good speed when it is pulled into the cylinder and must come to a halt when the valve closes. The air does not sit there until the valve opens again; rather, it bounces off the closed valve and heads toward the plenum area. When the air reaches the plenum, it meets a rush of incoming air and bounces back toward the intake valve. The air, however, bounces back quicker than it did when it left the intake valve. This is due to a push given by the intake air.

A runner is designed to take this bouncing air and send it back to the intake valve in time for the next opening. This timing determines the length of the runner and results in a stronger intake charge because the air is under pressure. In most manifolds, this air wave bounces several times before the intake valve opens again. The bouncing effect and resultant pressure of the air is called acoustic supercharging.

The inside diameter of the runners also affects the delivery of air. When a small diameter runner is used, the air will move into the cylinders faster. This increases volumetric efficiency at low engine speeds. When the engine is running at higher speeds, it needs a lot of air. Small diameter runners would restrict the airflow and hurt the engine's efficiency. Therefore, larger diameter runners are needed at high engine speeds.

Variable Intake Manifolds

Many engines have variable intake manifolds, which are controlled by the PCM. These manifolds change the size of the plenum area and/or change the length and effective diameter of the runners according to engine speed and load. The use of these manifolds allows the engine to experience high volumetric efficiency with more than one range of engine speeds.

The operation and design of these manifolds vary with manufacturer and engine. Systems that alter the plenum area have two small plenums. Depending on the system, only one of the plenums is used during low speeds. In other systems, the plenums are divided and are used for specific cylinders. In both cases, when the engine reaches a particular speed, the plenums are opened and work together to create a larger plenum area. These systems are commonly referred to as **intake manifold tuning (IMT)** systems.

IMT systems have a motor connected to a butterfly valve in the center of the manifold **(Figure 32–9)**. The valve is closed during low speeds, keeping the two plenum areas separated. When commanded by the PCM, the valve opens and allows the two plenums to become one large plenum. The PCM receives feedback on the position of the valve through a position sensor on the motor.

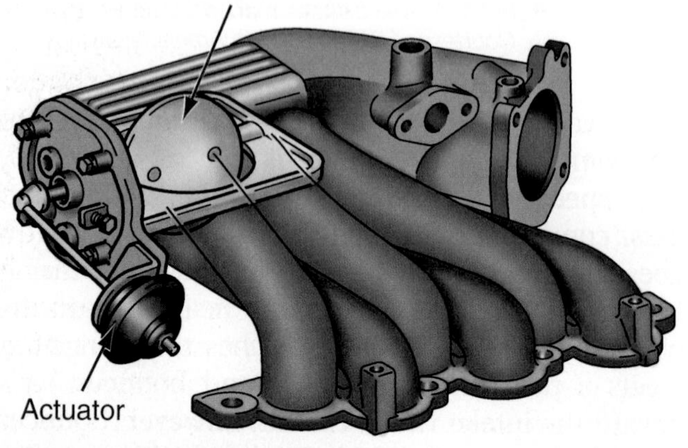

Figure 32-9 An IMT control valve. *Courtesy of Toyota Motor Sales, U.S.A., Inc.*

> **WARNING!**
>
> *The butterfly valve is moved with great force. Always keep your fingers away from the valve when the system is energized. Failure to do this may result in a serious injury.*

The most common variable intake manifold designs change the path of air between long and short runners or between small-diameter and large-diameter runners according to engine speed. These systems are typically called **intake manifold runner control (IMRC)** systems. Intake air passes through long- or small-diameter runners at low speeds and is routed through short- or large-diameter runners at high speeds. The switching of runners is mostly done by controlling a butterfly valve that opens and closes the short- or large-diameter runners. Ultimately, the overall volume of air is controlled by the throttle plate; the butterfly valve in the manifold merely controls the routing of the air.

Changing the runners for different speeds allows for the benefits of acoustic supercharging as well as providing increased airflow at high speeds. It is important to realize that too much airflow at low speeds can actually hurt engine performance because the engine does not need it and the resultant air waves are hard to time to intake valve opening. All IMRC systems must have a feedback system, according to OBD-II standards. If the IMRC system is not working correctly, a DTC is set.

The butterfly valve in IMRC systems is vacuum or electrically controlled. In vacuum systems, a vacuum actuator is mounted on the manifold **(Figure 32–10)**. Vacuum to the actuator is controlled by a PCM-regulated solenoid. Linkage connects the actuator to the butterfly valve. The PCM relies mainly on inputs from the TP, ECT, and CKP sensors to determine when to open or close the butterfly valve. At low speeds, the solenoid is energized. This allows manifold vacuum to hold the valve closed. Once engine speed and other conditions are met, the solenoid is turned off and springs on the butterfly valve force the valve open.

In electrical systems, a motor is used to move the butterfly valve. There can be one valve per cylinder or

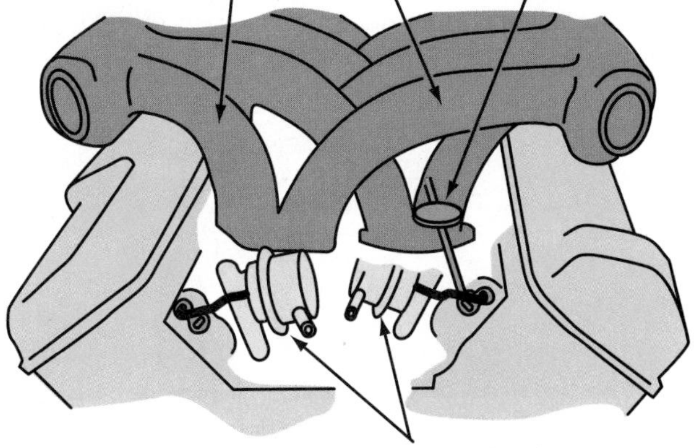

Figure 32–10 The vacuum controls and butterflies used to switch between the long and short intake runners.

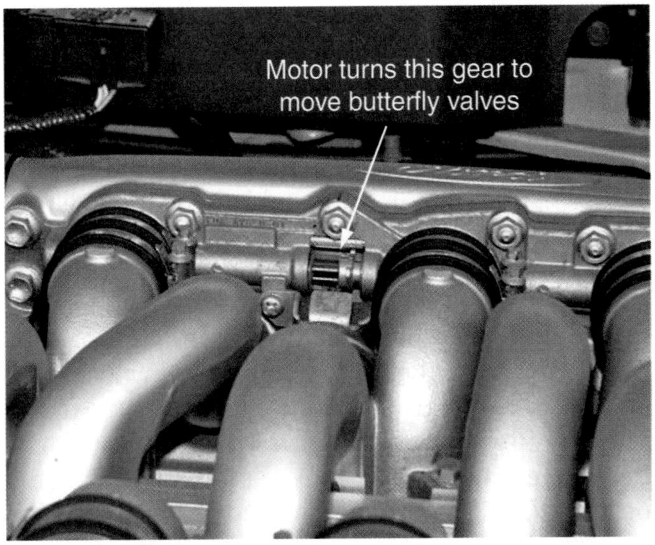

Figure 32–11 An electric motor meshes with this gear to rotate the shaft for the butterfly valves in this variable intake manifold.

one valve per bank of runners **(Figure 32–11)**. The valve can be normally open and remain open, or it is closed by the solenoid when the valve for the other runner is opened. Often there is a butterfly valve common to the high-speed runners. This valve is located in the plenum area and switches the routing of the air inside the plenum. The action of the motor is controlled by the PCM and can be duty cycled to be able to respond to current engine speeds.

Servicing an Intake Manifold

There are few reasons why an intake manifold would need to be replaced or repaired. If the manifold is cracked or the sealing surfaces are severely damaged, it should be replaced. When the manifold is removed or there is evidence of a leak between the manifold and the cylinder head, the sealing surfaces should be checked for flatness. This check should include coolant and oil passages. Minor imperfections on the surface can be filed smooth; however, never try to repair any serious damage.

A leaking manifold gasket is one of the most common causes of internal coolant leaks. A bad gasket can also cause vacuum leaks. When installing an intake manifold, use new gaskets and seals. Make sure that the manifold and gasket are aligned properly. On V-type engines, the use of guide bolts helps to ensure proper alignment. Make sure that all of the attaching bolts are tightened to specifications and in the correct order.

Some engines have two-piece intake manifolds. The two parts are sealed together with gaskets and seals. These should be replaced when the manifold is separated, and bolts should be tightened to specifications.

FORCED INDUCTION SYSTEMS

Engines cannot produce the amount of power they are capable of at high speeds because they do not receive enough air. This is the reason why many "race cars" have hood scoops. Hood scoops deliver cool air under pressure to the intake manifold and provide an open source for the air. With today's body styles, hood scoops are not particularly desired because they increase air drag. Therefore, other methods are used to increase the volume of intake air and compression. Variable intake manifolds and valve timing certainly help with this.

Keep in mind that the power generated by the internal combustion engine is directly related to the amount of air that is compressed in the cylinders. In other words, the greater the compression (within reason), the greater the output of the engine. Two approaches can be used to increase an engine's effective compression. One is to modify the engine to increase its compression ratio. This can be done in many ways, including the use of domes or high-top pistons, altered crankshaft strokes, or changes in the shape and structure of the combustion chamber.

Another, less expensive way to increase compression (and engine power) without physically changing the shape of the combustion chamber is to simply increase the intake charge. By pressurizing the intake mixture before it enters the cylinder, more air and fuel molecules can be packed into the combustion chamber. The two ways to artificially increase the amount of airflow into the engine are known as turbocharging and supercharging.

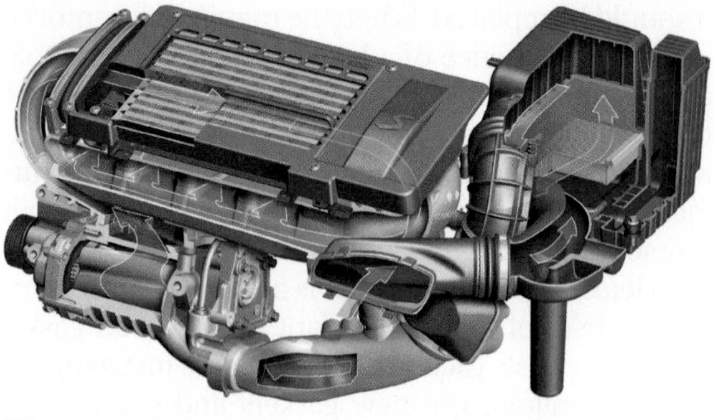

Figure 32–12 Routing of the boosted air in and out of an intercooler. *Courtesy of BMW of North America, LLC*

Figure 32–13 The intercooler for this engine is above the engine.

Both of these systems force more air into the intake manifold by compressing the air before it reaches the manifold. Turbocharging uses exhaust gases, the heat of which is normally wasted, and supercharging relies on the rotation of the engine to compress the air. Both systems offer benefits but have some limitations. The biggest disadvantage of using either system is related to the compression of air. When air is compressed, its heat increases. High-temperature air is less dense, meaning there is less air or oxygen in the air. Most turbocharged or supercharged systems use an intercooler to increase the density of the air.

Intercoolers

The **intercooler** cools the turbocharged or supercharged air before it reaches the combustion chamber **(Figure 32–12)**. When intake air is compressed, its temperature increases greatly. As a result, air density is reduced and the cylinders receive less fresh air than what could be from the boosted intake air. Also, the increased temperature of the air increases the chances of engine knock. To offset these consequences, many turbocharger and supercharger systems are fitted with an intercooler or charge air cooler. The removal of heat from the pressurized air going into the intercooler increases the density of the air, which improves efficiency, engine horsepower, and torque.

Intercoolers are really radiators for the intake air. The heat from the compressed air that passes though it is removed and dissipated to the atmosphere. An intercooler system typically consists of an added radiator in the grille area or above the engine **(Figure 32–13)**, a coolant reservoir (separate from the reservoir of the engine's cooling system), a pump, and hoses and tubes to connect the components **(Figure 32–14)**. An intercooler is always located after the turbocharger or supercharger and before the intake manifold. As the heated air flows through the intercooler, heat is transferred to the coolant circulating through the intercooler. The coolant is cooled by the air passing through the intercooler. The amount of coolant moving through the intercooler is normally controlled by the PCM. Therefore, the PCM controls the temperature of the incoming air.

TURBOCHARGERS

Turbochargers are used to increase engine power by compressing intake air before it enters the engine. Turbochargers are air pumps driven by the engine's exhaust stream. The heat and pressure of the exhaust gases spin the turbine blades (hence the name turbocharger) of the pump. The turbine wheel is connected to a compressor wheel. As the turbine spins, so does the compressor wheel. The compressor wheel spins at very high speeds and compresses the intake air. The compressed air is then sent to the cylinders. Because exhaust gas is a waste product, the energy developed by the turbine is said to be free because it theoretically does not use any of the engine's power it helps to produce.

Turbochargers are used on both diesel and gasoline engines. The main advantage of their use is that they allow for an increase of power without a substantial decrease in fuel economy. This is because they boost the engine's power output only when extra power is needed. A small engine, in most cases, can be used to provide low fuel consumption and emissions levels. When increased power is needed, the turbocharger is activated.

Construction

A turbocharger is normally located close to the exhaust manifold. An exhaust pipe runs between the exhaust manifold and the turbine housing to carry the exhaust flow to the turbine wheel **(Figure 32–15)**. Another pipe connects the compressor housing intake to the throttle plate assembly or intake manifold.

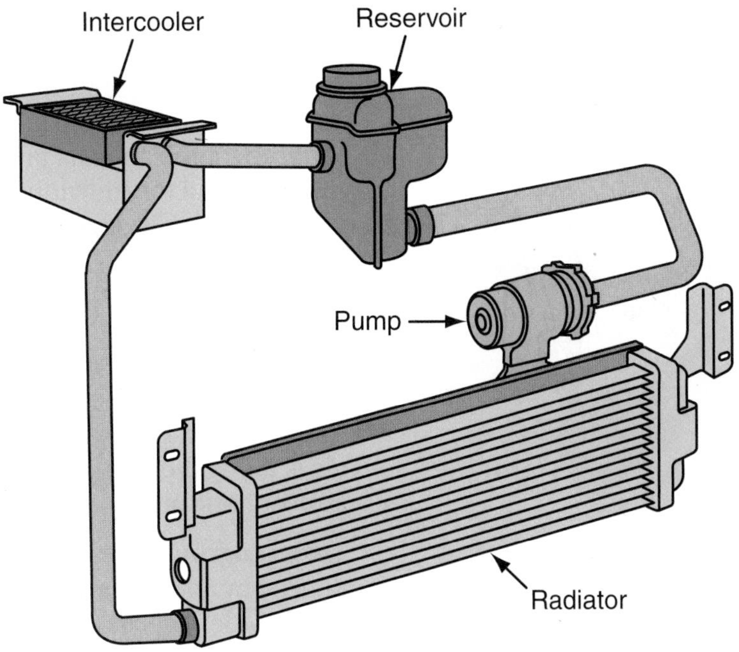

Figure 32-14 The main parts of a water-to-air intercooler.

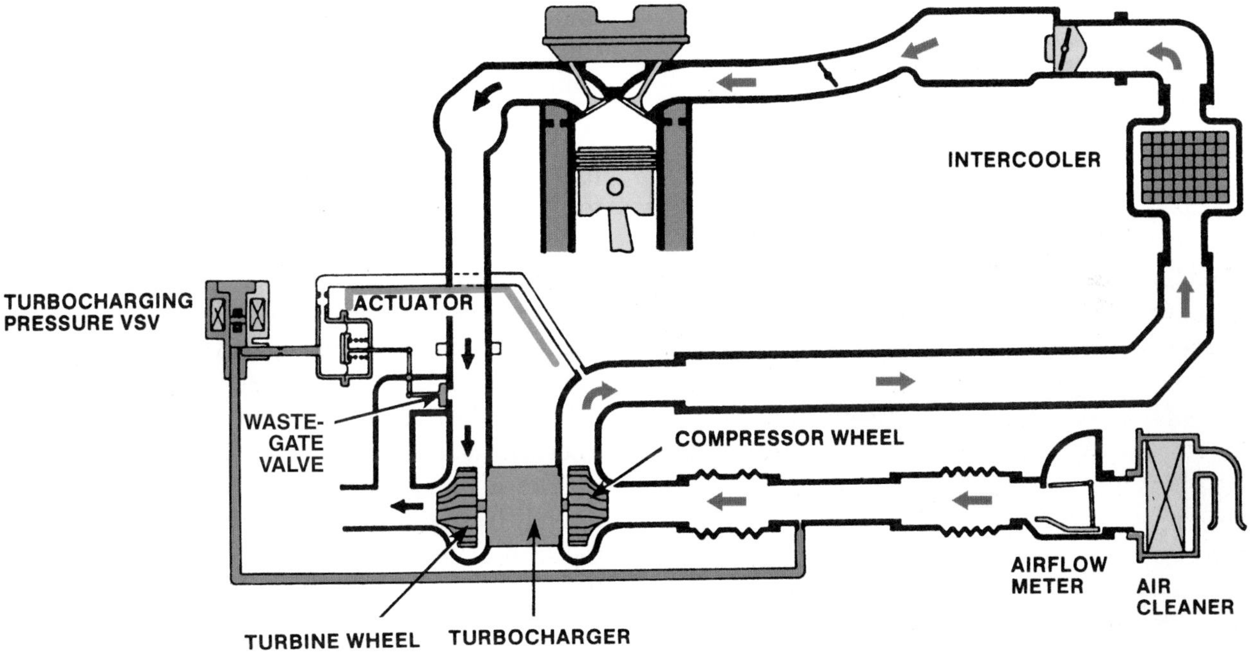

Figure 32-15 Exhaust gas and airflow in a typical turbocharger system. *Courtesy of Toyota Motor Sales, U.S.A., Inc.*

A typical turbocharger, usually called a turbo, has the following components **(Figure 32–16)**:

- Turbine or hot wheel
- Shaft
- Compressor or cold wheel
- Center housing and rotating assembly (CHRA)
- Wastegate valve
- Actuator

Inside the turbocharger, the turbine wheel (hot wheel) and the compressor wheel (cold wheel) are mounted on the same shaft. The CHRA houses the shaft, shaft bearings, turbine seal assembly, and compressor seal assembly. Each wheel is encased in its own spiral-shaped enclosure in the housing that serves to control and direct the flow of exhaust and intake air. Because the turbine wheel is in the exhaust path, it gets very hot. It also spins at very high speeds; therefore, it is normally made of a heat-resistant cast iron. A turbocharger is equipped with a wastegate

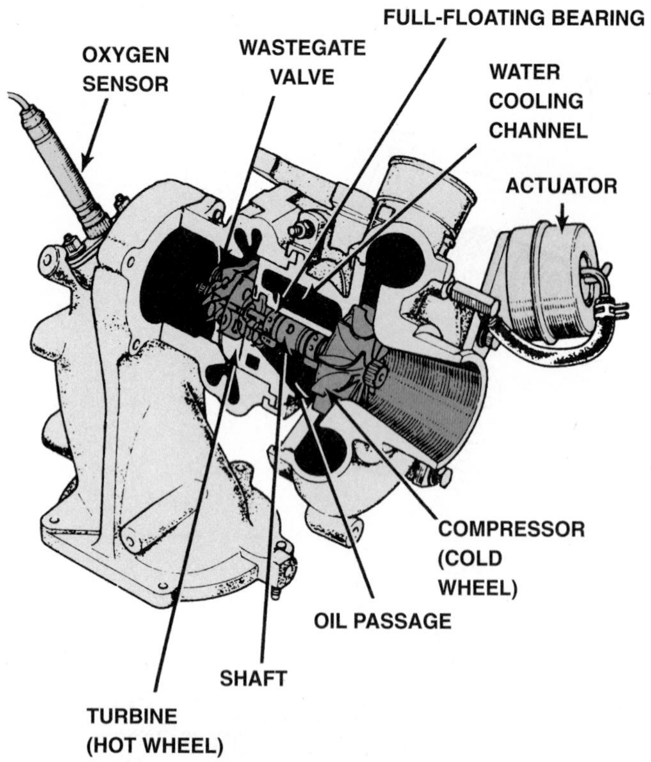

Figure 32-16 A cross section of a turbocharger shows the turbine wheel, the compressor wheel, and their connecting shaft. *Courtesy of Toyota Motor Sales, U.S.A., Inc.*

Figure 32-17 A wastegate.

valve to control the pressure of the air delivered to the cylinders **(Figure 32–17)**.

The turbocharger typically begins to compress the intake air when the engine's speed is above 2,000 rpm. The force of the exhaust flow is directed against the side of the turbine wheel. As the hot gases hit the turbine wheel, causing it to spin, the specially curved turbine fins direct the exhaust gases toward the center of the housing where they exit. This action creates a flow called a vortex. Once the turbine starts to spin, the compressor wheel (shaped like a turbine wheel in reverse) also starts to spin. Intake air is drawn into the housing and is caught by the whirling blades of the compressor and thrown outward by centrifugal force. From there the air exits under pressure to the intake manifold and the individual cylinders.

Air is typically drawn into the cylinders by the difference in pressure between the atmosphere and engine vacuum. A turbocharger, however, is capable of pressurizing the intake charge above normal atmospheric pressure. *Turbo boost* is the term used to describe the positive pressure increase created by a turbocharger. For example, 10 psi of boost means the air is being fed into the engine at 24.7 psi (14.7 psi atmospheric plus 10 pounds of boost).

Turbo Lag Increases in horsepower are normally evidenced by an engine's response to a quick opening of the throttle. The lack of immediate throttle response is felt with some turbocharged engines. This delay or **turbo lag** occurs because exhaust gas requires a little time to build enough energy to spin the wheels fast enough to respond to the engine's speed. This causes the power from the engine to temporarily lag behind what is needed for the conditions.

Wastegate Valve

If the pressure of the air from a turbocharger becomes too high, knocking occurs, engine output decreases, and the pressure created by the combustion of the air-fuel mixture can become so great that the engine may self-destruct. To prevent this, turbochargers have a **wastegate** valve. The wastegate valve is part of the turbine housing. It allows a certain amount of exhaust gas to bypass the turbine when the boost pressure exceeds a certain value. This action reduces the pressure.

The action of the wastegate can be controlled directly by manifold pressure or by the PCM according to manifold pressure. Most late-model systems have PCM-controlled wastegates. In non-PCM systems, an actuator that senses the air pressure in the induction system opens the wastegate when the pressure becomes too high. This action decreases the amount of exhaust that reaches the turbine. This, in turn, reduces turbine and compressor wheel speed and decreases the output pressure from the turbocharger. When the pressure in the intake manifold is not great enough to override the spring in the actuator, the wastegate is closed and all of the exhaust flows past the turbine; therefore, the turbine spins accordingly. When the boost pressure overcomes the tension of the actuator's spring, the actuator opens the wastegate valve and some exhaust gas is diverted around the turbine wheel. As a result, turbine speed is controlled, as is boost pressure.

On late-model engines, the wastegate is controlled by the PCM that directly controls a solenoid that controls vacuum to the wastegate. When vacuum is introduced to the wastegate, it opens to allow exhaust gases to bypass the turbine. The action of the solenoid is controlled by the PCM according to various inputs. The PCM also adjusts ignition timing and air-fuel mixtures according to turbocharger output. Retarding spark timing is an often-used method of controlling detonation on turbocharged engines. Unfortunately, any time the ignition is retarded, power is lost, fuel economy suffers, and the engine tends to run hotter. Most systems use knock-sensor signals to retard timing only when detonation is detected. These sensors are also used to limit boost pressure according to the octane rating of the fuel being used. This maximizes engine performance and reduces the chances of engine knocking during all conditions regardless of the fuel's octane rating.

When the PCM detects excessive manifold pressure, it opens the wastegate and also enriches the mixture. This rich mixture reduces combustion temperature, which helps to cool the turbocharger and combustion chamber.

Turbo Lubrication and Cooling

Keeping a turbocharger cool and well lubricated is essential to the unit's durability. The turbine wheel faces very high temperatures. The temperature of the exhaust from a gasoline engine can surpass 1,800°F (982°C). The exhaust from a diesel engine is cooler and ranges from 1,400° to 1,500°F (768° to 816°C). This heat can destroy the turbocharger if it is not controlled.

The turbine and compressor wheels rotate at very high speeds (up to 100,000 rpm). Their shaft is mounted on full-floating bearings designed to keep the shaft secure while it is rotating. The bearings are lubricated by engine oil and rotate freely on the shaft and in the housing. Engine oil is delivered to the turbocharger through an oil inlet pipe and then circulated to the bearings. The bearings have outer seals to prevent oil leakage. The oil passes over the bearings and returns to the engine's oil pan through an oil outlet pipe. The circulation of the oil cools the turbocharger.

A turbocharger may be cooled by engine coolant that circulates through coolant channels built into the housing. Coolant is sent from the engine's thermostat housing through a coolant inlet pipe to the turbo housing. After the coolant has circulated through the housing, it is returned to the water pump through the coolant outlet pipe.

Various Turbocharger Designs

In an effort to increase the efficiency of turbocharged engines, manufacturers have developed various designs of turbochargers and their control systems. Common alternative designs are called **variable nozzle turbine (VNT)** and variable geometry turbochargers. In these, the cross-sectional area through which the exhaust flows is variable. This area is adjusted via movable vanes that change their angles according to turbine speed **(Figure 32–18)**. At lower engine speeds the vanes restrict exhaust flow, thereby increasing boost pressure. At higher engine speeds, the vanes open wider and exhaust backpressure decreases. This allows the turbocharger to provide more boost at lower engine speeds without producing too much boost at higher speeds. It is claimed that the use of a variable turbocharger can reduce a gasoline engine's fuel consumption by 20%.

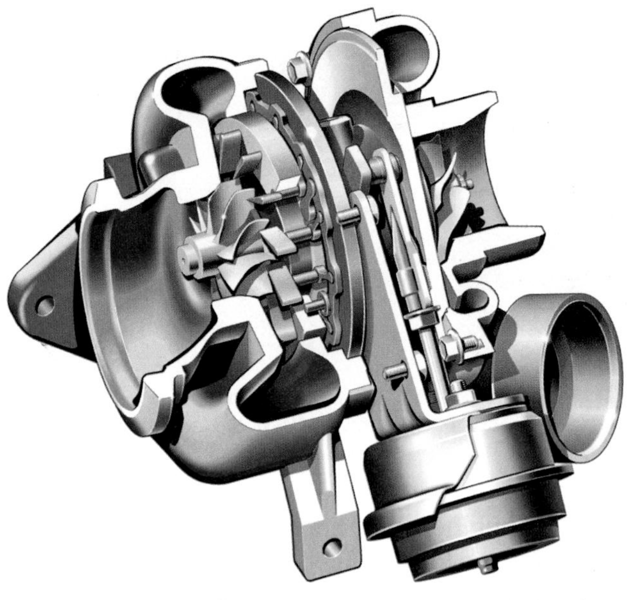

Figure 32–18 A variable turbocharger. Notice the movable turbine blades. *Courtesy of Chrysler LLC*

Variable turbos do not have a wastegate. They provide a higher boost at lower engine speeds and are more responsive to changes in engine load. They also help reduce the effects of turbo lag.

Twin Turbochargers Some engines have two turbochargers. The action of the two depends on the application. Some engines have a turbo for one-half the cylinders and another for the other half **(Figure 32–19)**. Some V-type engines have a turbocharger for each bank of cylinders. The turbos use the exhaust from specific cylinders to compress the air for those same cylinders.

Other engines have two different-sized turbos. Each turbo is designed for specific conditions. Normally, the smaller of the two spools (spins) up to speed very quickly. This reduces turbo lag. The larger one is slower to get up to speed but adds the boost at higher engine speeds. This is a two-stage design: one

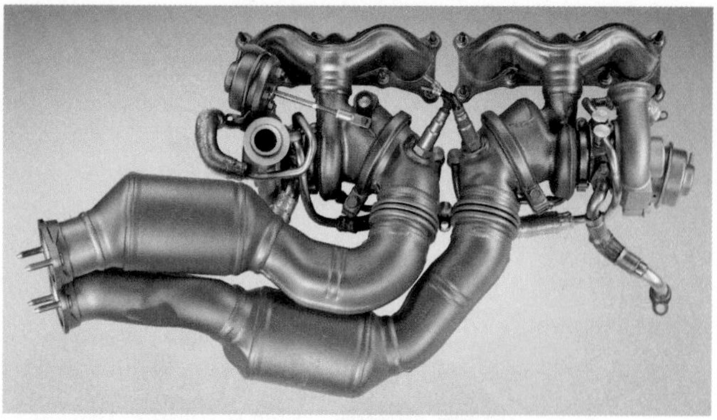

Figure 32-19 A twin turbocharger setup on an inline six-cylinder engine. *Courtesy of BMW of North America, LLC*

for lower engine speeds and immediate increase of speed and one for sustained power.

In a typical twin turbocharger system, the function of the two turbochargers is controlled by the operating mode of the larger turbocharger. Its operation is controlled by control valves that regulate exhaust gases to it and the amount of air from it.

During low engine speeds and loads, boost is only provided by the smaller unit. The control valves of the other turbo have it disabled. When boost pressure from the first turbo reaches a predetermined level, exhaust gas is allowed to flow to the second turbo. At this time it is spinning but not providing any boost. This is a preparation step; the second turbine is spinning before it is needed. Once the load or speed conditions demand more power, a valve opens and allows the boost pressure from the second turbo to enter the intake manifold. At this time, boost from both turbos is sent to the intake manifold. In some systems, the higher pressure from the larger turbo causes the wastegate valve, which is an integral part of the smaller turbo, to open and control the maximum boost.

When the engine moves from high to low speed, a control valve stops the flow from the second turbo to the intake. Another control valve then blocks off exhaust flow to the turbo. These actions prevent high boost during deceleration.

Turbocharger Inspection

USING SERVICE INFORMATION

General service procedures for turbocharger systems are normally in a separate section of service manuals. Individual inputs and outputs that are part of the electronic control system are covered in the engine control or performance section of the manual.

To inspect a turbocharger, start the engine and listen to the sound that the turbo system makes. As you become more familiar with this characteristic sound, it will be easier to identify an air leak between the compressor outlet and engine or an exhaust leak between engine and turbo by the presence of a higher pitched sound. If the turbo sound cycles or changes in intensity, the likely causes are a plugged air cleaner or loose material in the compressor inlet ducts or dirt buildup on the compressor wheel and housing.

After listening, check the air cleaner and remove the ducting from the air cleaner to turbo and look for dirt buildup or damage from foreign objects. Check for loose clamps on the compressor outlet connections and check the engine intake system for loose bolts or leaking gaskets. Then disconnect the exhaust pipe and look for restrictions or loose material. Examine the exhaust system for cracks, loose nuts, or blown gaskets. Rotate the turbo shaft assembly. Does it rotate freely? Are there signs of rubbing or wheel impact damage?

Visually inspect all hoses, gaskets, and tubing for proper fit, damage, and wear. Check the low-pressure, or air cleaner, side of the intake system for vacuum leaks.

Pressure Testing The performance of a turbocharger can be checked with a pressure gauge. The gauge is connected to the intake manifold and the amount of boost observed. This is best done on a road test. This check can also verify the action of the wastegate and its controls. Manifold pressure can be monitored with a scan tool as well.

SHOP TALK

When oil leakage is noted at the turbine end of the turbocharger, check the turbocharger oil drain tube and the engine crankcase breathers for restrictions. When sludged engine oil is found, the engine's oil and oil filter must be changed.

On the pressure side of the system you can check for leaks by using soapy water. After applying the soap mixture, look for bubbles to pinpoint the source of the leak.

Exhaust leaks upstream to the turbocharger housing will also affect turbo operation. If exhaust gases escape prior to entering the housing, the reduced temperature and pressure will cause a proportionate reduction in boost and a loss of power. If the wastegate does not appear to be operating properly (too much or too little boost), check to make sure that the connecting linkage is operating smoothly and not binding. Also make sure that the pressure sensing hose is clear and properly connected.

SHOP TALK

Because of the needed balance and sealing of the units, automotive technicians do not rebuild turbochargers. Most repair shops do not have the required equipment; therefore, all repairs to turbochargers are done at specialty shops.

Wastegates Wastegate problems can usually be traced to carbon buildup, which keeps the unit from closing or causes it to bind. A defective diaphragm or leaking vacuum hose can result in an inoperative wastegate. Before condemning the wastegate, check the ignition timing, the spark-retard system, vacuum hoses, knock sensor, oxygen sensor, and computer.

CAUTION!

When removing carbon deposits from turbine and wastegate parts, never use a hard metal tool or sandpaper. Remember that any gouges or scratches on these parts can cause severe vibration or damage to the turbocharger. To clean these parts, use a soft brush and a solvent.

Common Turbocharger Problems With proper care and servicing, a turbocharger will provide years of reliable service. Most turbocharger failures are caused by lack of lubrication, ingestion of foreign objects, or contamination of the lubricant **(Figure 32–20)**.

Replacing a Turbocharger

If the turbocharger is faulty, it should be replaced with a new or rebuilt unit. Always follow the procedure given in the service information. Before removing the turbocharger, plug the intake and exhaust ports and the oil inlet to prevent the entry of dirt or other foreign material. While replacing it, check for an accumulation of sludge in the oil pipes and, if necessary, clean or replace the oil pipes. Do not drop or bang the new unit against anything or grasp it by easily deformed parts.

When installing a new turbocharger, put 20 cc (0.68 oz.) of oil into the turbocharger oil inlet and turn the compressor wheel by hand several times to spread oil to the bearings.

After installing the new turbocharger, or after starting an engine that has been unused, there can be a considerable lag after engine startup before the oil pressure is sufficient to deliver oil to the turbo's bearings. To prevent this problem, follow these simple steps:

- Make certain that the oil inlet and drain lines are clean before connecting them.
- Be sure the engine oil is clean and at the proper level.
- Fill the oil filter with clean oil.
- Leave the oil drain line disconnected at the turbo and crank the engine without starting it until oil flows out of the turbo drain port.
- Disconnect the fuel supply lines and crank the engine for 30 seconds to distribute oil throughout the engine.
- Allow the engine to idle for 60 seconds.
- Connect the drain line, start the engine, and operate it at low idle for a few minutes before running it at higher speeds.

Figure 32-20 Damaged turbocharger wheels.

Maintenance

Turbochargers require no maintenance; however, the engine's oil and filter must be changed on a regular basis. The units operate at high speeds and high temperatures. Poor lubrication will cause the turbo to self-destruct. Also, the high heat breaks down the oil and, therefore, its service life is shorter than it would be in a conventional engine. Manufacturers typically recommend that a specific type of oil be used in these engines. After the oil and filter have been changed, the engine should run at idle for at least 30 seconds. Doing this allows oil to circulate through the turbo.

In addition to frequent oil changes, the air cleaner and filter assembly needs to be maintained. This includes making sure that there are no air leaks in the system. The slightest amount of dirt entering the turbo can damage its turbine and compressor wheels. PCV valves and filters also need to be maintained on a regular basis.

Turbo Startup and Shutdown

The number one cause of turbo failure is poor lubrication. Because its bearings are not well lubricated immediately after the engine is started, allow the engine to idle for some time before putting it under a load. If the engine has not been run for a day or more, start the engine and allow it to idle for 3 to 5 minutes to prevent oil starvation to the turbo. Engine lube systems have a tendency to bleed down. When the engine has sat for a long time, it is wise to crank the engine without starting it until a steady oil pressure reading is observed. This is called priming the lubricating system. A turbocharged engine should never be operated under load if the engine has less than 30 psi of oil pressure. The same starting procedure should be followed in cold weather. The thick engine oil will take a longer period to flow. Low oil pressure and slow oil delivery during engine starting can destroy the bearings in a turbocharger.

When the engine has been operated at high speeds or heavy loads, the turbine wheel has been exposed to very hot exhaust gases. This heat is transferred to the turbine's shaft and the compressor wheels. If the engine is turned off immediately after high-speed driving, oil and coolant to the turbo unit will immediately stop. The shaft, however, will still spin due to its inertia. Poor lubrication at this point will destroy the bearings and shaft. Therefore, the engine should idle for 20 to 120 seconds after a hard run to allow the shaft to cool. When the engine is idling, exhaust temperatures drop to 573° to 752°F (300° to 400°C). The cooler exhaust gases help cool the shaft and its lubricating oil.

SUPERCHARGERS

SHOP TALK

Supercharging fascinated auto engineers well before they decided to steer with a wheel instead of a tiller. The 1906 American Chadwick had a supercharger. Since then many engines have been fitted with superchargers. Supercharged Dusenbergs, Hispano-Suizas, and Mercedes-Benzes were giants among luxury-car marques, as well as winners on the race tracks in the 1920s and 1930s. Then, after World War II, supercharging started to fade, although both Ford and American Motors sold supercharged passenger cars into the late 1950s. However, after being displaced first by larger V8 engines then by turbochargers, the supercharger started to make a comeback with the 1989 models. Today some engines have a supercharger as an alternative to the turbo **(Figure 32–21)**.

Figure 32-21 A supercharger mounted to a V6 engine.

Superchargers are positive displacement air pumps driven directly by the engine's crankshaft via a V-ribbed belt. They improve horsepower and torque by increasing air pressure and density in the intake manifold. The pressure boost is proportional to engine speed.

A typical supercharger is normally made up of a magnetic clutch, two rotors, two shafts, two rotor gears, housing, and a rear plate and cover **(Figure 32–22)**. Some superchargers have four rotor assemblies. The drive belt connects the engine to the magnetic clutch connected to one rotor shaft. Gears connect the two aluminum rotor shafts and drive them in opposite directions. The rotors have three helical lobes. The rotors are press-fit onto the rotor shafts and then held in position by pins and serrations. The rotors turn

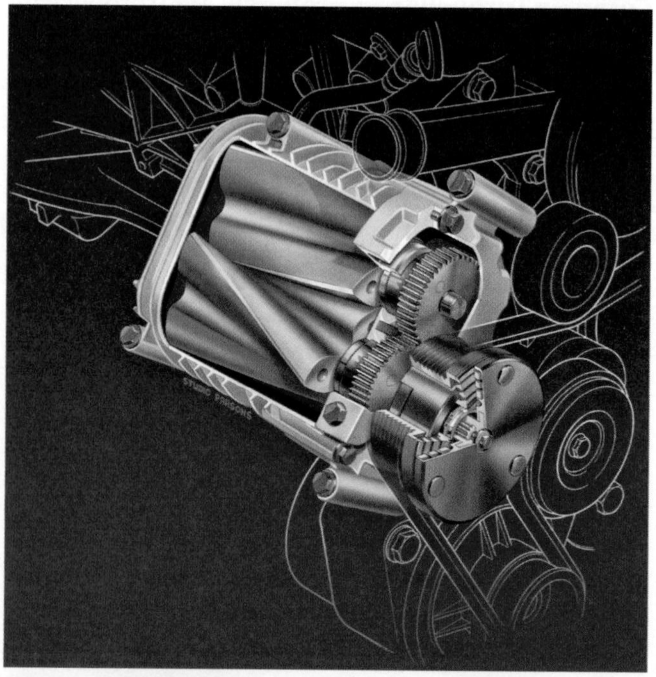

Figure 32-22 A drawing showing the drive setup for a supercharger. *Courtesy of Chrysler LLC*

within a sealed housing and pressurize the air as they rotate. The rotors' shafts are supported by ball or needle bearings in the rear plate and are lubricated by oil in the supercharger unit.

The magnetic clutch allows the PCM to engage and disengage the drive power to the supercharger. The PCM also can control the amount of fuel injected and the ignition timing during boost. When the clutch is energized, the engine drives the supercharger. The PCM de-energizes the clutch when the engine is running under a light load. The clutch is comprised of a stator, pulley, pressure plate, and hub.

The clutch pulley rotates with the drive belt and drives the stator. When the stator is energized, the pressure plate is pressed against the clutch pulley, locking them together. The clutch hub is splined to the rotor's shaft and the rotors rotate as a unit. To reduce the shock caused by the quick engagement of the clutch, most clutch assemblies have a rubber damper between the boss of the clutch hub and the pressure plate.

Normally, there is a 0.02 in. (0.5 mm) clearance between the clutch hub and clutch pulley. If this clearance increases because of wear, noise will result. The clearance can be corrected by changing the thickness of the adjusting shim.

To handle the higher operating temperatures imposed by supercharging, the engine is typically fitted with an engine oil cooler. This water-to-oil cooler is generally mounted between the engine front cover and oil filter.

Supercharger Operation

Figure 32–23 illustrates the flow of the air through a supercharger. The air comes in through the remote-mounted air cleaner and the mass airflow meter. It then moves through the throttle plate assembly and passes through the supercharger inlet plenum assembly, which is bolted to the back of the supercharger.

The air enters the supercharger and is pressurized by the spinning rotors. It then exits through the top of the supercharger by way of the air outlet adapter. As the air is compressed, its temperature increases. Because cooler, denser air is desired for increased power, the heated air is routed through an intercooler. An intercooler can decrease the temperature of the air by 150°F (66°C).

This cooled air then passes through to the intake manifold adapter assembly, which is bolted to the rear of the intake manifold. When the intake valves open, the air is forced into the combustion chambers where it is mixed with fuel delivered by the fuel injectors.

Supercharger By-Pass System

Unlike a turbocharger, the supercharger does not require a wastegate to limit boost and prevent a potentially damaging overboost condition. Because the speed of the supercharger is directly linked to the engine speed, its pumping power is limited by the rpm of the engine itself rather than revolutions produced by exhaust gases. Supercharger boost is therefore directly controlled through the opening and closing of the throttle or a by-pass system that controls the air leaving the supercharger. The by-pass system may be electrically or vacuum controlled.

The by-pass circuit allows the supercharger to idle when the extra power is not needed. The by-pass routes any excess air in the intake manifold back through the supercharger inlet plenum assembly, allowing the engine to run, in effect, normally aspirated. This eliminates any boost from the supercharger. The by-pass system reduces air handling losses when boost is not needed and this results in better fuel economy.

Some systems use a PCM-controlled stepper motor to control the amount of air that bypasses the supercharger. The PCM determines the required boost pressure based on current engine conditions and controls the operation of the magnetic clutch and by-pass valve.

Other systems have a vacuum motor that regulates the amount of air to be bypassed. As the power demands from the engine increase, a vacuum motor controls a butterfly valve that routes more or less air to the intake manifold, thereby changing the boost. When this bypass is completely closed, boost can reach about 12 psi. When the actuator is open, during high-vacuum engine conditions, the air bypasses the supercharger. As the throttle is opened and engine vacuum decreases, the actuator closes and allows more air into the supercharger.

Supercharger Designs

Although there have been a number of supercharger designs on the market over the years, the most popular is the **Roots** type. This design uses a pair of three lobed rotor vanes driven by the crankshaft. The lobes force air into the intake manifold.

The key to the supercharger's operation, of course, is primarily the design of the rotors. Some Roots-type superchargers use straight-lobe rotors that result in uneven pressure pulses and, consequently, relatively high noise levels. Therefore, the supercharger used with most of today's engines uses a helical design for the two rotors. The helical design evens out the pressure pulses in the blower and reduces noise. It was found that a 60-degree helical twist works best for equalizing the inlet and outlet volumes.

Another benefit of the helical rotor design is it reduces carry-back volumes—air that is carried back to the inlet side of the supercharger because of the

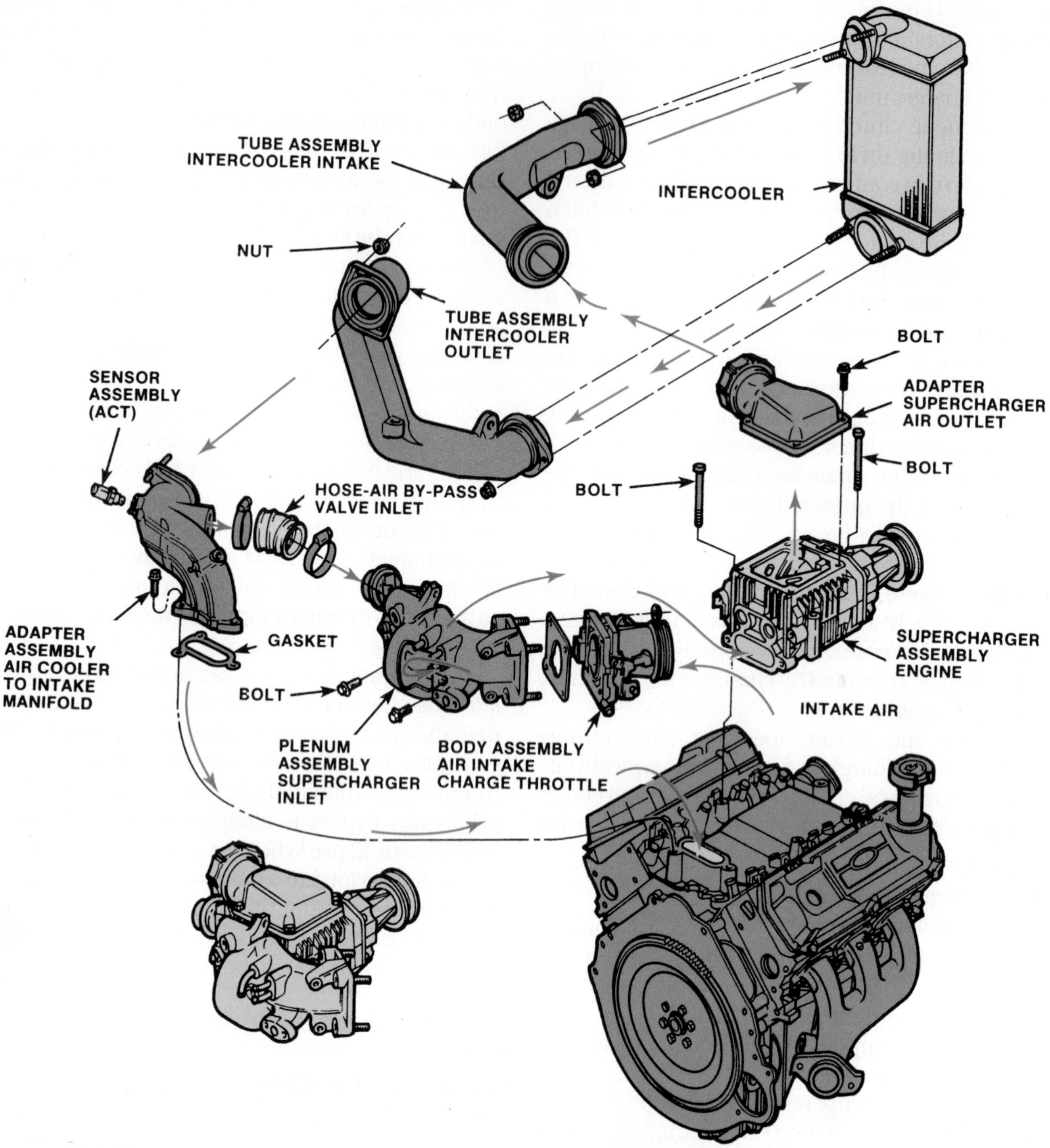

Figure 32-23 Airflow through a supercharger assembly into the engine.

unavoidable spaces between the meshing rotors—which represents a loss of efficiency.

Another popular supercharger design, especially in Europe, is the **G-Lader** (spiral) supercharger, which is based on a 1905 French design. Spiral ramps in both sides of the rotor intermesh with similar ramps in the housing. Unlike most superchargers, the rotor of the G-Lader does not spin on its axis; rather it moves around an eccentric shaft. This motion draws in air, squeezes it inward through the spiral, which compresses it, then forces it through ducts in the center into the engine. Airflow is essentially constant, so intake noise is lower than that of a Roots blower. Because there is only a slight wiping motion between the spiral and housing, wear is minimal.

Supercharger Problems

Many of the problems and their remedies given for turbochargers hold good for superchargers. There are also problems associated specifically with the supercharger. Refer to the service manual for the symptoms of supercharger failure and a summary of the causes and the recommended repairs.

USING SERVICE INFORMATION

Service procedures for supercharger systems are normally in a separate unit of a service manual.

Maintenance

A supercharger normally has its own oil supply and requires special oil for lubrication. Its oil level must be checked and corrected periodically. Also, the unit ought to be inspected regularly for oil leaks. Like a turbocharger, any dirt in the intake air can destroy the unit. The induction system must be checked for leaks and the filter changed on a regular basis.

Supercharger + Turbocharger

A few European vehicles are equipped with a gasoline engine that has a supercharger and a turbocharger. These systems are called "twincharger" systems. The supercharger is used to boost power at low speeds and the turbocharger boosts power at high speeds. These engines have direct injection and can provide as much power as larger engines with low fuel consumption and emissions.

The supercharger is belt driven and spins at five times the speed of the crankshaft. A magnetic clutch engages the supercharger when it is needed. The supercharger is connected in series with the turbocharger. A control flap facilitates intake air to both the turbocharger and supercharger. The maximum boost of the twincharger system is about 36 psi (2.5 bar) at 1,500 rpm. This provides for an increase in torque at low speeds and a substantial increase in horsepower at high speeds. This also results in a decrease in fuel consumption and emissions.

The supercharger provides boost from just over 2,400 rpm and then the turbo kicks in and provides all of the boost. With this system there is no turbo lag.

EXHAUST SYSTEM COMPONENTS

The various components of the typical exhaust system include the following:

- Exhaust manifold
- Exhaust pipe and seal
- Catalytic converter
- Muffler
- Resonator
- Tailpipe
- Heat shields
- Clamps, brackets, and hangers
- Exhaust gas oxygen sensors

All the parts of the system are designed to conform to the available space of the vehicle's undercarriage and yet be a safe distance above the road.

CAUTION!

When inspecting or working on the exhaust system, remember that its components get very hot when the engine is running. Contact with them could cause a severe burn. Also, always wear safety glasses or goggles when working under a vehicle.

Exhaust Manifold

The exhaust manifold **(Figure 32–24)** collects the burnt gases as they are expelled from the cylinders and directs them to the exhaust pipe. Exhaust manifolds for most vehicles are made of cast or nodular iron. Many newer vehicles have stamped, heavy-gauge sheet metal or stainless steel units.

Inline engines have one exhaust manifold. V-type engines have an exhaust manifold on each side of the engine. An exhaust manifold will have either three, four, or six passages, depending on the type of engine. These passages blend into a single passage at the other end, which connects to an exhaust pipe. From that point, the flow of exhaust gases continues to the catalytic converter, muffler, and tail pipe, then exits at the rear of the car.

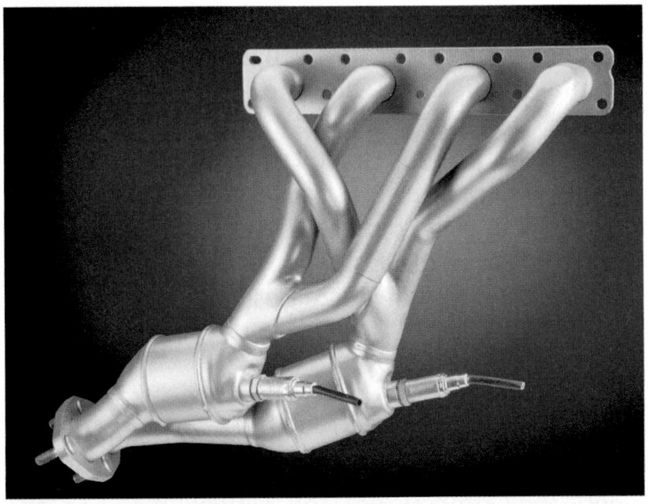

Figure 32-24 This exhaust manifold has two separate mini- or warm-up catalytic converters. *Courtesy of BMW of North America, LLC*

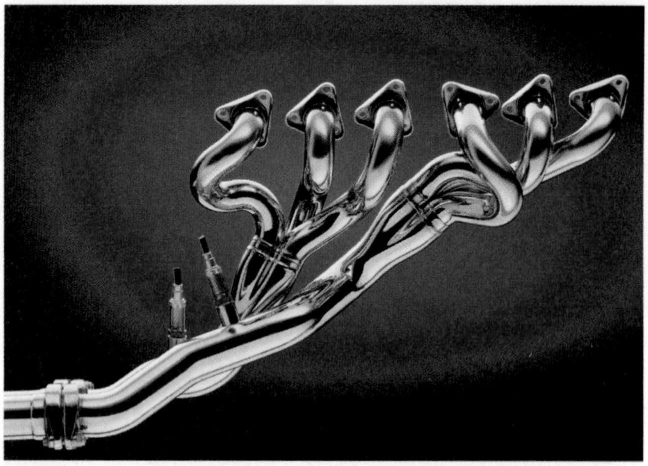

Figure 32-25 A tuned exhaust manifold, called a header. *Courtesy of BMW of North America, LLC*

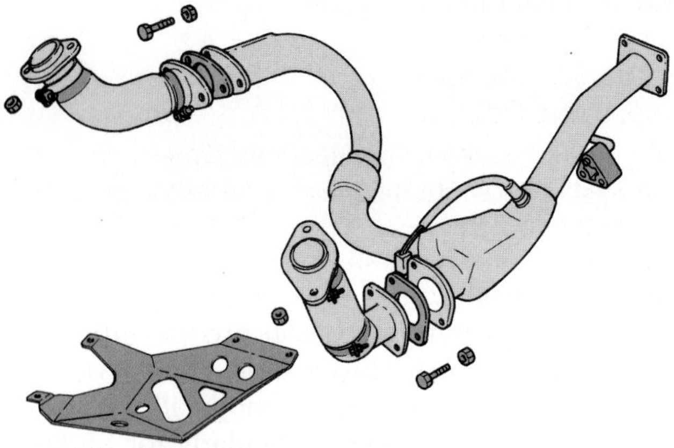

Figure 32-26 The front exhaust pipe assembly for a V6 engine. *Courtesy of American Isuzu Motors Inc.*

V-type engines may be equipped with a dual exhaust system that consists of two almost identical, but individual systems in the same vehicle.

Exhaust systems are designed for particular engine-chassis combinations. Exhaust system length, pipe size, and silencer size are used to tune the flow of gases within the exhaust system. Proper tuning of the exhaust manifold tubes can actually create a partial vacuum that helps draw exhaust gases out of the cylinder, improving volumetric efficiency. Separate, tuned exhaust headers **(Figure 32–25)** can also improve efficiency by preventing the exhaust flow of one cylinder from interfering with the exhaust flow of another cylinder. Cylinders next to one another may release exhaust gas at about the same time. When this happens, the pressure of the exhaust gas from one cylinder can interfere with the flow from the other cylinder. With separate headers, the cylinders are isolated from one another, interference is eliminated, and the engine breathes better.

Perhaps the largest performance gain from using an exhaust header is that it increases the engine's volumetric efficiency. A low pressure is present in the exhaust each time a pulse of exhaust ends. A header uses this low pressure to pull exhaust gases out of the cylinder when the exhaust valve opens. The low pressure also helps draw more air into the cylinder during valve overlap. Enhancing exhaust flow and increasing intake flow increases the efficiency of the engine.

Exhaust Pipe and Seal

The exhaust pipe is metal pipe—either aluminized steel, stainless steel, or zinc-plated heavy-gauge steel—that runs under the vehicle between the exhaust manifold and the catalytic converter **(Figure 32–26)**.

SHOP TALK

The exhaust manifold gasket seals the joint between the head and exhaust manifold. Many new engines are assembled without exhaust manifold gaskets. This is possible because new manifolds are flat and fit tightly against the head without leaks. Exhaust manifolds go through many heating/cooling cycles. This causes stress and some corrosion in the exhaust manifold. Removing the manifold usually distorts the manifold slightly so that it is no longer flat enough to seal without a gasket. Exhaust manifold gaskets are normally used to eliminate leaks when exhaust manifolds are reinstalled.

CATALYTIC CONVERTERS

A **catalytic converter** **(Figure 32–27)** is part of the exhaust system and a very important part of the emission control system. The catalytic converter is located

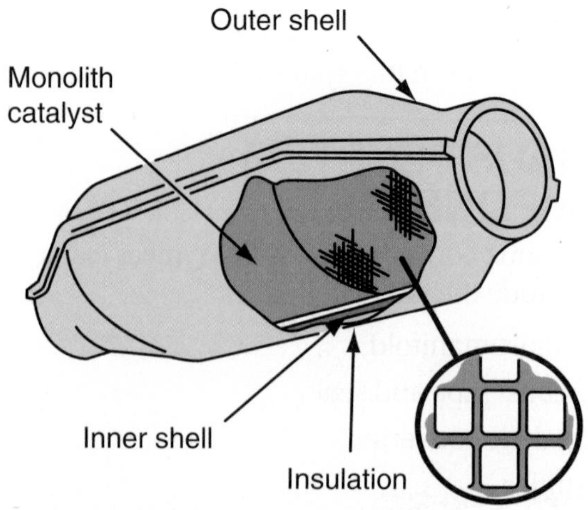

Figure 32-27 A catalytic converter.

ahead of the muffler in the exhaust system. The extreme heat in the converter oxidizes the exhaust that flows out of the engine. Because it is part of both systems, it has a role in both. As an emission control device, it is responsible for converting undesirable exhaust gases into harmless gases. As part of the exhaust system, it helps reduce the noise level of the exhaust. A catalytic converter contains a ceramic element coated with a catalyst. A catalyst is a substance that causes a chemical reaction in other elements without actually becoming part of the chemical change and without being used up or consumed in the process.

Go To **Chapter 33 for a detailed discussion on catalytic converters.**

Catalytic converters may be pellet type (used in older systems) or monolithic type. Exhaust gases pass over this bed of catalyst material. In a monolithic-type converter, the exhaust gases pass through a honeycomb ceramic block. The converter beads or ceramic block are coated with a thin coating of cerium, platinum, palladium, and/or rhodium and are held in a stainless steel container. These elements are used alone or in combination with each other to change the undesirable emissions into harmless compounds.

Since the late 1980s, vehicles have had a three-way converter (TWC) that treats all three controlled emission gases. It oxidizes HC and CO by adding oxygen and reduces NO_x by removing oxygen from the nitrogen oxides. Diesel engines have a particulate oxidizer catalytic converter **(Figure 32–28)** that collects and cleans the particulates from diesel fuel that would normally be emitted from a diesel engine as black smoke.

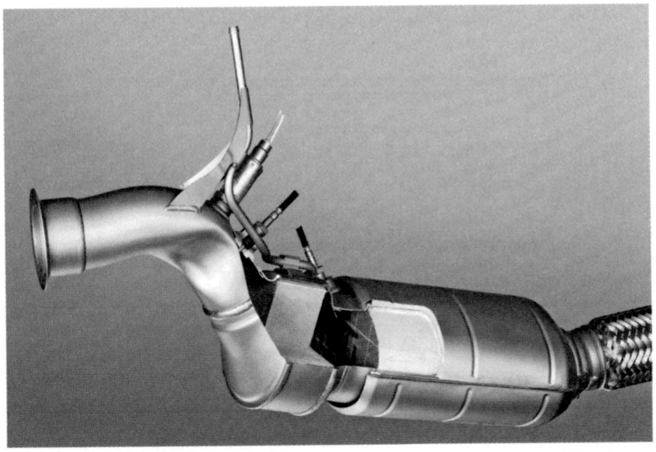

Figure 32-28 A particulate filter for a diesel engine. *Courtesy of BMW of North America, LLC*

All late-model engines have a mini-catalytic converter that is either built in the exhaust manifold or located next to it. These converters are used to clean the exhaust during engine warmup and are commonly called warm-up converters.

Many older catalytic converters had an air hose connected from the AIR system to the oxidizing catalyst. This air helped the converter work by making extra oxygen available. Fresh air added to the exhaust at the wrong time could overheat the converter and produce NO_x, something the converter is trying to destroy.

OBD-II regulations call for a monitoring system that looks at the effectiveness of the converter. This system uses two oxygen sensors, one before the catalyst and one after **(Figure 32–29)**. If the sensors' outputs are the same, the converter is not working properly and the MIL will light and a DTC will be set.

Converter Problems

The converter is normally a trouble-free emission control device; however, it can go bad or become plugged. Often these problems are caused by overheating the converter. When raw fuel enters the exhaust because of an engine misfiring, the temperature of the converter quickly increases. The heat can melt the catalyst materials inside the converter, causing a major restriction to the flow of exhaust.

A plugged converter or any exhaust restriction can cause loss of power at high speeds, stalling after starting (if totally blocked), a drop in engine vacuum as engine rpm increases, or sometimes popping or backfiring at the carburetor.

The best way to determine if a catalytic converter is working is to check the quality of the exhaust. This is done with a five-gas exhaust analyzer. The results of this test should show low emission levels if the converter is working properly.

Another way to test a converter is to use a handheld digital infrared thermometer **(Figure 32–30)** or **pyrometer**, which is an electronic device that measures heat. When placing the pyrometer close to the exhaust pipe just ahead of and just behind the converter, there should be an increase in temperature as the exhaust gases pass through the converter. If the outlet temperature is the same or lower, nothing is happening inside the converter. This means the converter should be replaced. If there is only a slight difference in temperature, check the activity of the oxygen sensor (O_2S) before condemning the converter. The efficiency of today's converters depends on the normal swings of rich and lean mixtures. A biased O_2S can affect converter activity. If the O_2S is working fine, the converter should be

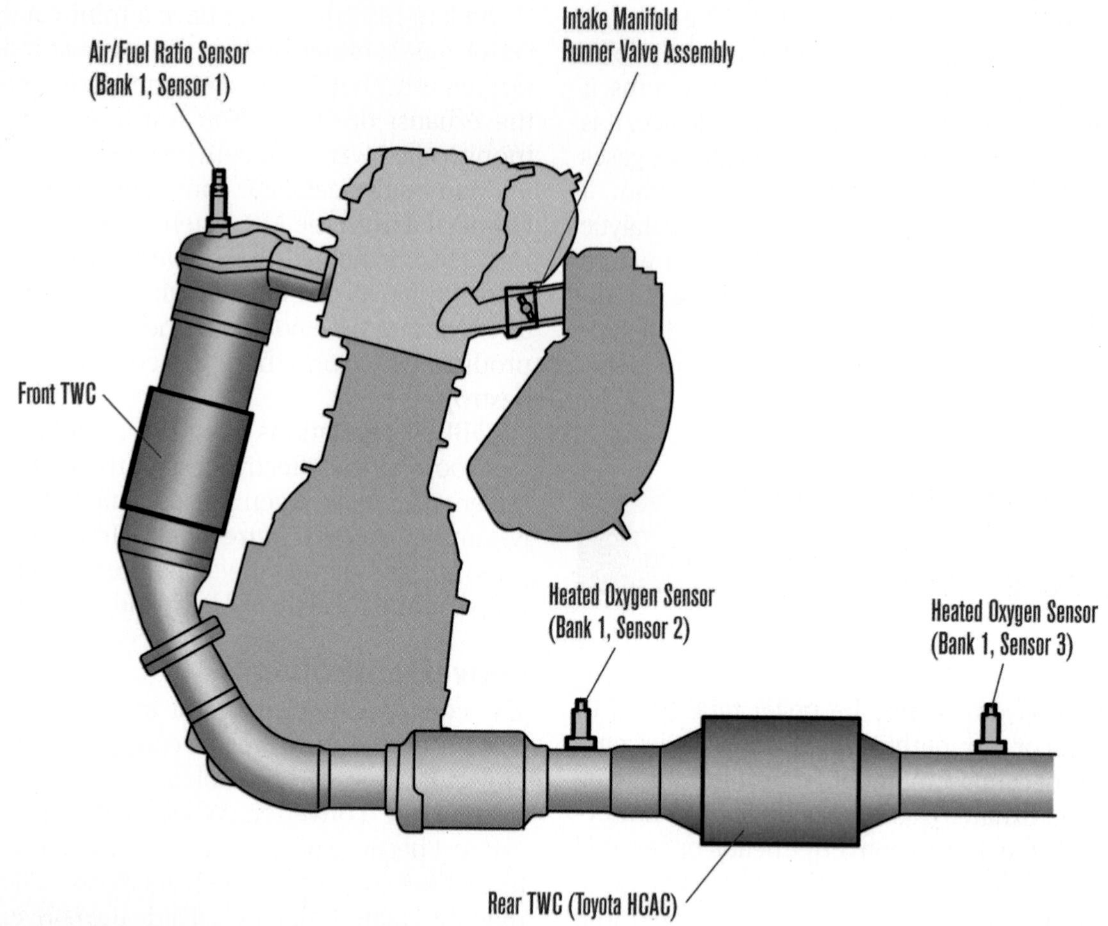

Figure 32-29 The basic configuration of an OBD-II exhaust manifold and pipe with its oxygen sensors and catalytic converter. *Courtesy of Toyota Motor Sales, U.S.A., Inc.*

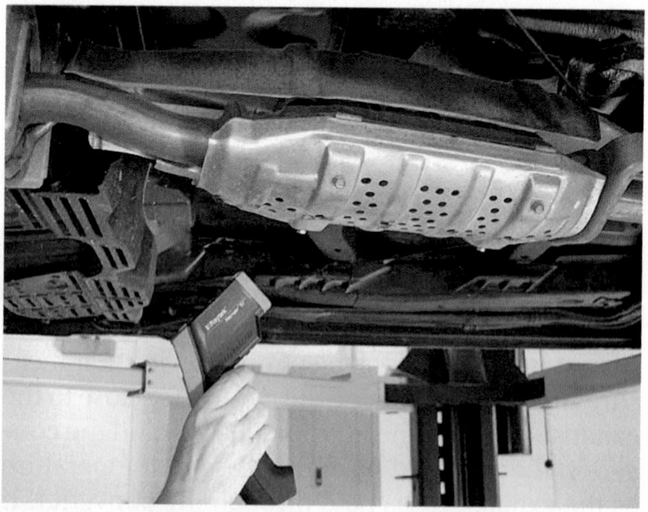

Figure 32-30 An infrared thermometer can be used to check the effectiveness of a catalytic converter.

replaced. Further testing of a catalytic converter is included in Chapter 34.

Mufflers

The **muffler** is a cylindrical or oval-shaped component, generally about 2 feet (0.6 meter) long, mounted in the exhaust system about midway or toward the rear of the-car. Inside the muffler is a series of baffles, chambers, tubes, and holes to break up, cancel out, or silence the pressure pulsations that occur each time an exhaust valve opens.

Two types of mufflers are frequently used on passenger vehicles **(Figure 32–31)**. Reverse-flow mufflers change the direction of the exhaust gas flow through the inside of the unit. This is the most common type of automotive muffler. Straight-through mufflers permit

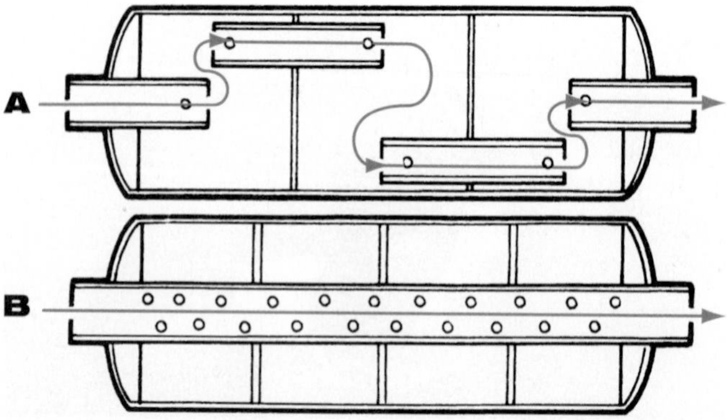

Figure 32-31 (A) A reverse-flow muffler; (B) a straight-through muffler.

exhaust gases to pass through a single tube. The tube has perforations that tend to break up pressure pulsations. They are not as quiet as the reverse-flow type.

In recent years there have been several important changes in the design of mufflers. Most of these changes have been centered at reducing weight and emissions, improving fuel economy, and simplifying assembly. These changes include the following:

New Materials More and more mufflers are being made of aluminized and stainless steel. Using these materials reduces the weight of the units and extends their lives.

Double-Wall Design Retarded engine ignition timing that is used on many small cars tends to make the exhaust pulses sharper. Many cars use a double-wall exhaust pipe to better contain the sound and reduce pipe ring.

Rear-Mounted Muffler More and more often, the only space left under the car for the muffler is at the very rear. This means the muffler runs cooler than before and is more easily damaged by condensation in the exhaust system. This moisture, combined with nitrogen and sulfur oxides in the exhaust gas, forms acids that rot the muffler from the inside out. Many mufflers are being produced with drain holes drilled into them.

Backpressure Even a well-designed muffler produces some **backpressure** in the system. Backpressure reduces an engine's volumetric efficiency, or ability to "breathe." Excessive backpressure caused by defects in a muffler or other exhaust system part can slow or stop the engine. However, a small amount of backpressure can be used intentionally to allow a slower passage of exhaust gases through the catalytic converter. This slower passage results in more complete conversion to less harmful gases. Also, no backpressure may allow intake gases to enter the exhaust.

Resonator

On some older vehicles, there is an additional muffler, known as a **resonator** or silencer. This unit is designed to further reduce or change the sound level of the exhaust. It is located toward the end of the system and generally looks like a smaller, rounder version of a muffler.

Tailpipe

The **tailpipe** is the last pipe in the exhaust system. It releases the exhaust fumes into the atmosphere beyond the back end of the car.

Heat Shields

Heat shields are used to protect other parts from the heat of the exhaust system and the catalytic converter (**Figure 32–32**). They are usually made of pressed or perforated sheet metal. Heat shields trap the heat in the exhaust system, which has a direct effect on maintaining exhaust gas velocity.

Heat shields on a catalytic converter also prevent the heat from a catalytic converter connected to a misfiring engine from setting grass or other materials on fire while it is parked. There is enough heat from an overheated converter to melt asphalt.

Flex Joints The exhaust systems on FWD vehicles have a flex joint somewhere in the front exhaust pipe. This joint allows the engine to move or roll without moving the exhaust system with it. The joint prevents the exhaust from hitting the vehicle's underbody as well as prevents the pipe from cracking due to the stress put on it.

Clamps, Brackets, and Hangers

Clamps, brackets, and hangers are used to properly join and support the various parts of the exhaust system. These parts also help to isolate exhaust noise by preventing its transfer through the frame (**Figure 32–33**) or body to the passenger compartment. Clamps help to secure exhaust system parts to one another. The pipes are formed in such a way that one slips inside the other. This design makes a close fit. A U-type clamp usually holds this connection tight. Another important job of clamps and brackets is to hold pipes to the bottom of the vehicle. Clamps and brackets must be designed to allow the exhaust system to vibrate without transferring the vibrations through the car.

There are many different types of flexible hangers available, each designed for a particular application. Some exhaust systems are supported by doughnut-shaped rubber rings between hooks on the exhaust component and on the frame or car body. Others are supported at the exhaust pipe and tailpipe connections by a combination of metal and reinforced fabric hanger. Both the doughnuts and the reinforced fabric allow the exhaust system to vibrate without breakage that could be caused by direct physical connection to the vehicle's frame.

Some exhaust systems are a single unit in which the pieces are welded together by the factory. By welding instead of clamping the assembly together, car makers save the weight of overlapping joints as well as that of clamps.

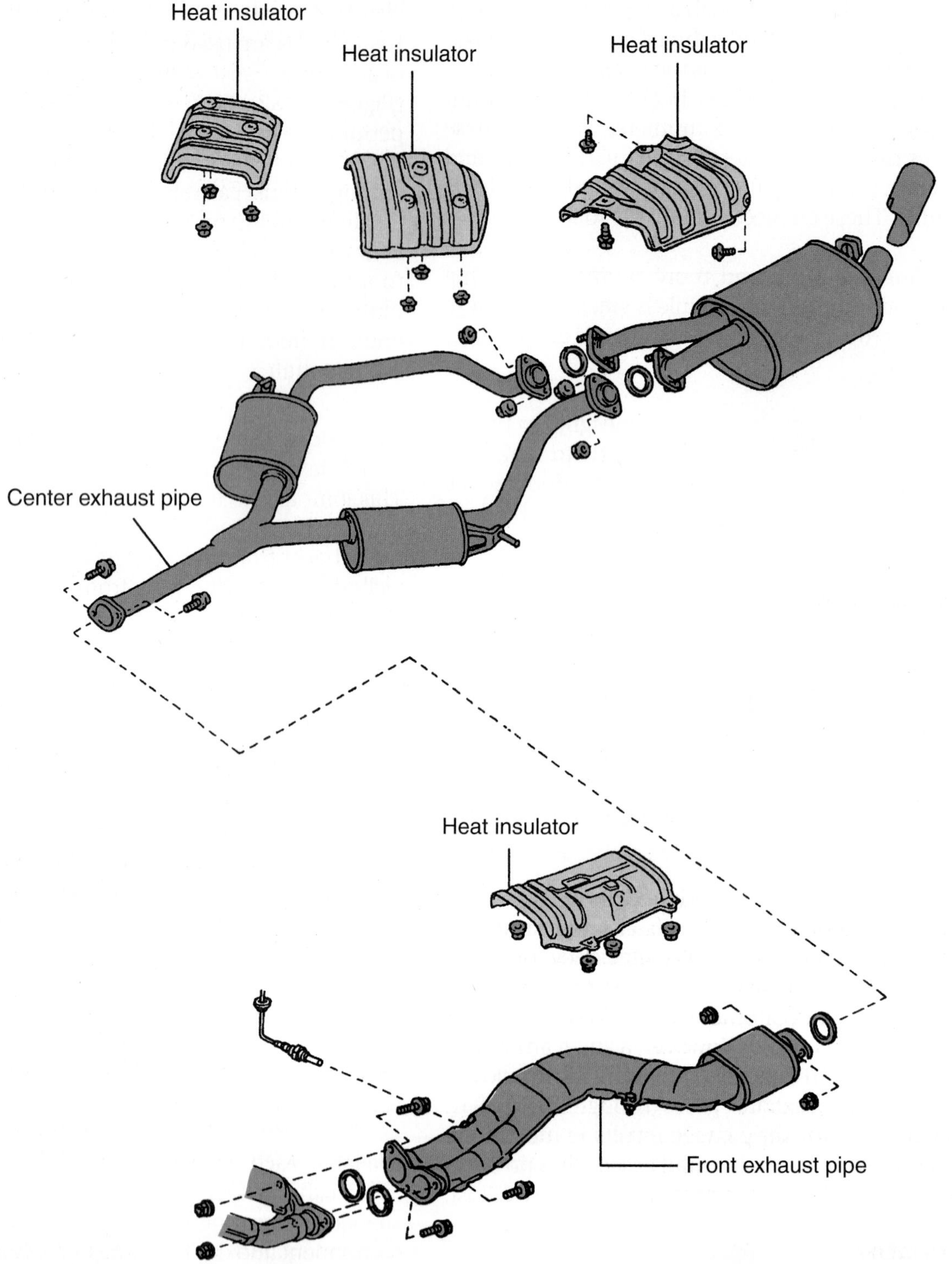

Figure 32-32 The typical location of heat shields in an exhaust system. *Courtesy of Toyota Motor Sales, U.S.A., Inc.*

EXHAUST SYSTEM SERVICE

Exhaust system components are subject to physical and chemical damage. Any physical damage to an exhaust system part that causes a partially restricted or blocked exhaust system usually results in loss of power or backfire up through the throttle plate(s). In addition to improper engine operation, a blocked or restricted exhaust system causes increased noise and air pollution. Leaks in the exhaust system caused by either physical or chemical (rust) damage could result in illness, asphyxiation, or even death.

Exhaust System Inspection

Most parts of the exhaust system, particularly the exhaust pipe, muffler, and tailpipe, are subject to rust,

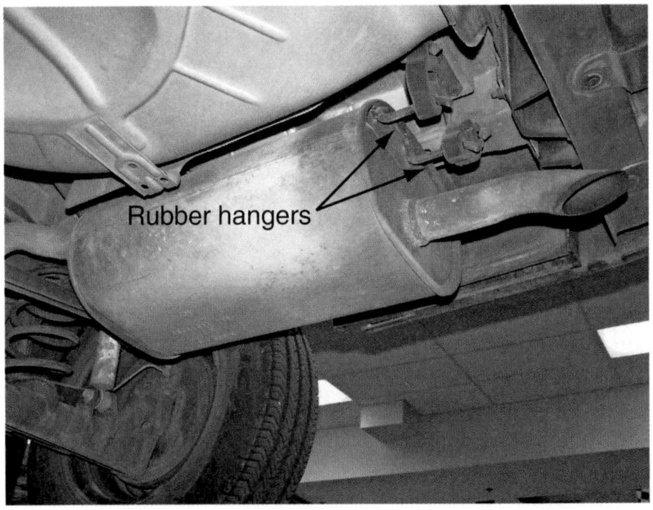

Figure 32-33 Rubber hangers are used to keep the exhaust system in place without allowing it to contact this pickup's frame.

corrosion, and cracking. Broken or loose clamps and hangers can allow parts to separate or hit the road as the car moves.

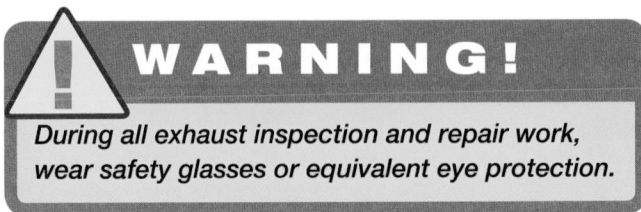

> **WARNING!**
>
> *During all exhaust inspection and repair work, wear safety glasses or equivalent eye protection.*

All inspections should include listening for abnormal noises. The inspection should also include a careful look at the system while the vehicle is raised on a hoist. **Table 32–1** lists some of the common exhaust systems problems and their likely causes.

Before beginning work on the system, be sure that it is cool to the touch. Some technicians disconnect the battery ground to avoid short-circuiting the electrical system. Finally, check the system for critical clearance points so the system does not contact the chassis while the vehicle is driven.

Exhaust Restriction Test Often leaks and rattles are the only things looked for in an exhaust system. The exhaust system should also be tested for blockage and restrictions. Collapsed pipes or clogged converters and/or mufflers can cause these blockages.

There are many ways to check for a restricted exhaust. The sound of the exhaust can indicate a restriction. With a restriction, the exhaust will wheeze as it struggles to exit the exhaust system. Although this is not the most effective way to determine if there is a restriction, it is a good start.

Checking the pressure built up in the exhaust is the best way to determine if the system is blocked. This is done by installing a pressure gauge in the O_2S bore in the exhaust pipe. At idle, the gauge should read less than 1.5 psi (10 kPa) and should be less than 2.5 psi (17 kPa) at 2,500 rpm.

Some technicians check for a restricted exhaust with a vacuum gauge. With the gauge connected to the intake manifold, the engine's speed is raised and held. The vacuum reading should rise and either hold there or increase slightly. If the vacuum decreases, there is an exhaust restriction.

Exhaust Leaks Exhaust leaks are often identified by sound, although very small leaks can be difficult to locate. One of the most effective ways to identify the source of a leak in the system is the use of a smoke machine. When smoke is introduced to the exhaust system, a trace of smoke will identify the source of the leak.

Replacing Exhaust System Components

Before beginning work on an exhaust system, make sure it is cool to the touch. Some technicians disconnect the battery's negative cable before starting to work to avoid short-circuiting the electrical system. Soak all rusted bolts, nuts, and other removable parts with a good penetrating oil. Finally, check the system for critical clearance points so they can be maintained when new components are installed.

Most exhaust work involves the replacement of parts. When replacing exhaust parts, make sure the new parts are exact replacements for the original parts. Doing this will ensure proper fit and alignment as well as ensure acceptable noise levels.

Exhaust system component replacement might require the use of special tools **(Figure 32–34)** and welding equipment.

Exhaust Manifold Servicing As mentioned, the manifold itself rarely causes any problems. On occasion, an exhaust manifold will warp because of excess heat. A straightedge and feeler gauge can be used to check the machined surface of the manifold.

Another problem—also the result of high temperatures generated by the engine—is a cracked manifold. This usually occurs after the car passes through a large puddle and cold water splashes on the manifold's hot surface. If the manifold is warped beyond manufacturer's specifications or is cracked, it must be replaced.

TABLE 32-1 EXHAUST SYSTEM SYMPTOM CHART

Condition	Possible Causes
Rattle, squeaks, or buzzing	Loose or damaged heat shield Loose or damaged exhaust isolators or isolator hanger brackets Loose or damaged catalytic converter or muffler Exhaust contacting the chassis
Droning or clunking	Loose or damaged exhaust isolators Exhaust contacting the chassis
Whistling, humming, or ticking noise that tends to change as the engine warms	Broken, loose, or missing exhaust manifold fasteners or gaskets Loose heated oxygen or catalyst monitor sensor Exhaust system leak
High-frequency hissing or rushing noise	Exhaust system leak
Pinging noise that occurs when the exhaust system is hot and the engine is turned off	Cool-down pinging is normal. It is caused by the exhaust system expanding and contracting during heating and cooling.
Vibration at idle and at low speeds; may be accompanied by clunking or buzzing	Loose or damage exhaust isolators or isolator hanger brackets Exhaust damper broken or out of position Exhaust contacting the chassis
Vehicle has low or no power	Exhaust pipe pinched or crushed Damaged catalytic converter Loose obstruction in the exhaust system Restricted exhaust (possibly ice in the muffler or tailpipe)
Burning smell	Foreign material in the exhaust system Missing heat shields
Sulfur or rotten egg smell from exhaust	Rich fuel conditions Misfire conditions Excessive sulfur content in fuel
Visible rust on the surface of the exhaust pipes	If there are no perforations, the condition is normal.

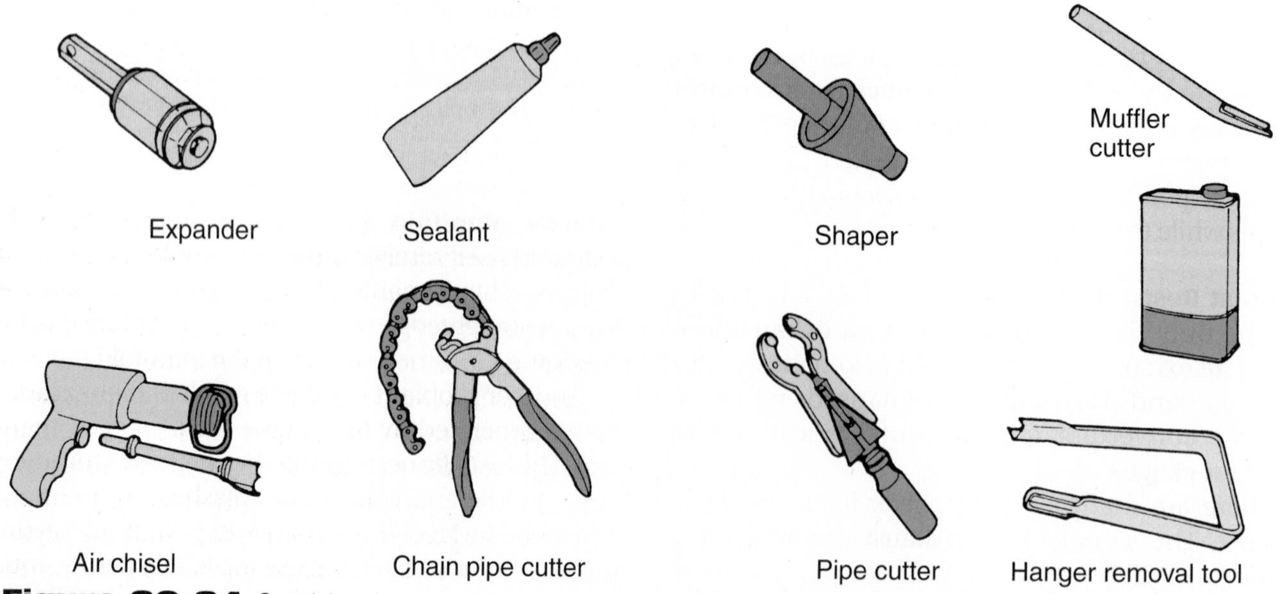

Figure 32-34 Special tools required for exhaust work.

Replacing Leaking Gaskets and Seals The most likely spot to find leaking gaskets and seals is between the exhaust manifold and the exhaust pipe **(Figure 32–35)**.

Most often exhaust bolts are quite rusted and can be difficult to loosen. This is why it is wise to soak the bolts and nuts with penetrating fluid before attempting to disassemble the system **(Figure 32–36)**.

When installing exhaust gaskets, carefully follow the recommendations on the gasket package label and instruction forms. Read through all installation steps before beginning. Take note of any of the original equipment manufacturer's recommendations in service manuals that could affect engine sealing. Manifolds warp more easily if an attempt is made to remove them while they are still hot. Remember, heat expands metal, making assembly bolts more difficult to remove and easier to break.

Figure 32–35 Leaking gaskets and seals are often found between the exhaust manifold and pipe.

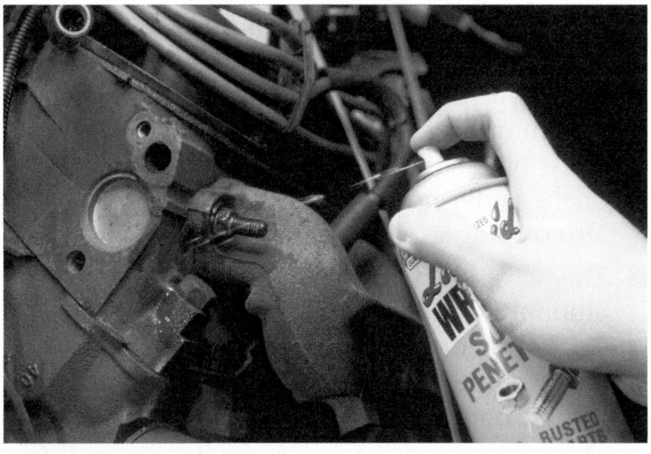

Figure 32–36 Before attempting to disassemble the exhaust system, spray all nuts and bolts with penetrating oil.

To replace an exhaust manifold gasket, follow the torque sequence in reverse to loosen each bolt. Repeat the process to remove the bolts. Doing this minimizes the chance that components will warp.

Any debris left on the sealing surfaces increases the chance of leaks. A good gasket remover will quickly soften the old gasket debris and adhesive for quick removal. Carefully remove the softened pieces with a scraper and a wire brush. Be sure to use a nonmetallic scraper when attempting to remove gasket material from aluminum surfaces.

Inspect the manifold for irregularities that might cause leaks, such as gouges, scratches, or cracks. Replace it if it is cracked or badly warped. File down any imperfections to ensure proper sealing of the manifold.

Retap and redie all threaded bores. This ensures tight, balanced clamping forces on the gasket. The most common cause of manifold warpage is incorrectly torqued manifold studs and nuts. Often when an exhaust manifold is removed, the studs for the manifold unthread with the nuts. Before reinstalling the manifold, separate the nuts from the studs and thoroughly clean both. In many cases it is best to replace the studs and nuts. Make sure that the studs and nuts are exact replacements for the original parts. Lubricate all threads with a high-temperature antiseize lubricant.

Apply a small amount of contact adhesive onto the mounting surface to hold the gasket in place. Align the gasket before the adhesive dries. Allow the adhesive to completely dry before installing the manifold.

Install the bolts or nuts finger-tight. Tighten them in three steps—one-half, three-quarters, and full torque. Always follow the specifications given for that engine. Normally the nuts should be tightened in a particular sequence. This sequence typically begins at the center and continues outward in an X pattern.

Replacing Exhaust Pipes In most cases, the exhaust system is replaced as a unit. Doing this ensures a proper fit and saves much time. However, there are times when only a section or component needs to be replaced. When doing this take care not to damage any surrounding parts. To replace an exhaust pipe, support the converter to keep it from falling and remove the O_2S. Remove any hangers or clamps holding the exhaust pipe to the frame. Unbolt the flange holding the exhaust pipe to the exhaust manifold. Disconnect the pipe from the converter and pull the front exhaust pipe loose and remove it. If the pipe is sealed with a gasket, replace it when installing the new pipe.

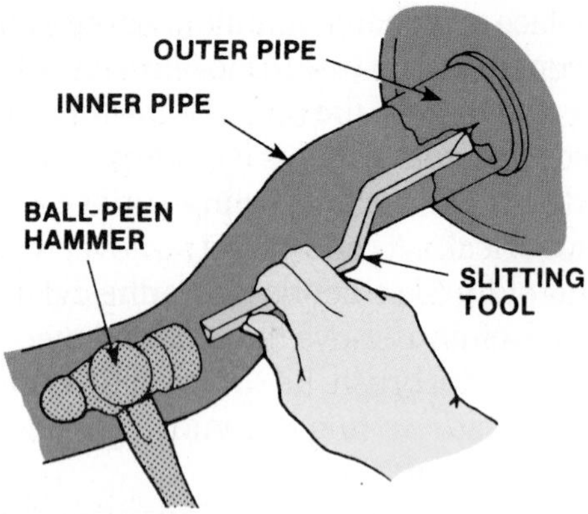

Figure 32-37 Removing a rusted-on muffler.

CAUTION!

Be sure to wear safety goggles to protect your eyes and work gloves to prevent cutting your hands on the rusted parts.

When trying to replace a part in the exhaust system, you may run into parts that are rusted together. This is especially a problem when a pipe slips into another pipe or the muffler. If you are trying to reuse one of the parts, you should carefully use a cold chisel or slitting tool **(Figure 32–37)** on the outer pipe of the rusted union. You must be careful when doing this, because you can easily damage the inner pipe. It must be perfectly round to form a seal with a new pipe.

CASE STUDY

A customer brought in his late-model Honda to have its charging system checked. The charging gauge on the dash consistently showed a low reading. The technician checked it out and found that the generator had a low output. The generator was replaced and appeared to be working fine. Two weeks later the customer returned the car with the same complaint. Again the output was checked and was found to be low. The assumption was made that the replacement generator must be defective so another was installed. In less than a week, the customer came back with the same complaint. This time the technician did more checking and found what was evidently causing the problem. It was right there! The owner had a supercharger installed on the engine. The drive for the supercharger was the generator. Whenever the engine was put under load, the drag of the supercharger would pull the generator up. The cause of the repeated generator failure was the tension of the drive belt on the generator's rotor. This quickly wore the shaft's bearings. The owner was advised of the problem and after much discussion was convinced to replace the supercharger with a turbocharger. Once this was done, there were no more charging system problems.

KEY TERMS

Backpressure
Catalytic converter
G-Lader
Intake manifold runner control (IMRC)
Intake manifold tuning (IMT)
Intercooler
Muffler
Pyrometer
Resonator
Roots
Supercharger
Tailpipe
Turbocharger
Turbo lag
Variable nozzle turbine (VNT)
Wastegate

SUMMARY

- The reason air enters the engine's cylinders is a basic law of physics—high pressure always moves toward an area of low pressure. Therefore, outside air moves into the cylinders because of the vacuum formed on the intake stroke.
- The vacuum in the intake manifold operates many systems such as emission controls, brake boosters, heater/air conditioners, cruise controls, and more. Vacuum is applied through an elaborate system of hoses, tubes, and relays. A diagram of emission system vacuum hose routing is located on the under-hood decal. Loss of vacuum can create many driveability problems.
- The air induction system allows a controlled amount of clean, filtered air to enter the engine. Cool air is drawn in through a fresh air tube. It passes through an air cleaner before entering the throttle body.
- The intake manifold distributes the air or air-fuel mixture as evenly as possible to each cylinder.

Intake manifolds are made of cast iron, aluminum, or plastic.

- Two approaches can be used to increase an engine's volumetric efficiency. One is to modify the internal configuration of the engine to increase the compression ratio. Another way is to increase the quantity of the intake charge; this is done by turbocharging or supercharging.
- Some turbocharged and supercharged engines are equipped with an intercooler that is designed to cool the compressed air from the turbocharger or supercharger.
- A turbocharger does not require a mechanical connection between the engine and pressurizing pump to compress the intake gases. It relies on the rapid expansion of hot exhaust gases exiting the cylinders to spin turbine blades.
- A typical turbocharger consists of a turbine, shaft, compressor, and housings. A wastegate manages turbo output by controlling the amount of exhaust gas that is allowed to enter the turbine housing.
- Turbochargers are lubricated by pressurized and filtered engine oil that is line fed to the unit's oil inlet.
- Reducing the diameter of the turbine and compressor wheels reduces turbo lag but limits the amount of boost. Using two small-diameter turbos provides good boost and very little turbo lag.
- If the turbo sound cycles or changes in intensity, the likely causes are a plugged air cleaner or loose material in the compressor inlet ducts or dirt buildup on the compressor wheel and housing. Most turbocharger failures are caused by one of the following reasons: lack of lubricant, ingestion of foreign objects, or contamination of lubricant. Turbo lag occurs when the turbocharger is unable to meet the immediate demands of the engine.
- Superchargers are air pumps connected directly to the crankshaft by a belt. The positive connection yields instant response and pumps air into the engine in direct relationship to crankshaft speed.
- The most popular supercharger design is the Roots type. Another supercharger design is the G-Lader supercharger.
- A vehicle's exhaust system carries away gases from the passenger compartment, cleans the exhaust emissions, and muffles the sound of the engine. Its components include the exhaust manifold, exhaust pipe, catalytic converter, muffler, resonator, tailpipe, heat shields, clamps, brackets, and hangers.
- The exhaust manifold is a bank of pipes that collects the burned gases as they are expelled from the cylinders and directs them to the exhaust pipe. Engines with all the cylinders in a row have one exhaust manifold. V-type engines have an exhaust manifold on each side of the engine. The exhaust pipe runs between the exhaust manifold and the catalytic converter.
- The catalytic converter uses the catalysts to change CO, HC, and NO_x into water vapor, CO_2, N, and O_2.
- The muffler consists of a series of baffles, chambers, tubes, and holes to break up, cancel out, and silence pressure pulsations. Two types commonly used are the reverse-flow and the straight-through mufflers.
- Some vehicles have an additional muffler called a resonator to further reduce the sound level of the exhaust. The tailpipe is the end of the pipeline carrying exhaust fumes to the atmosphere beyond the back end of the car. Heat shields protect vehicle parts from exhaust system heat. Clamps, brackets, and hangers join and support exhaust system components.
- Exhaust system components are subject to both physical and chemical damage. The exhaust can be checked by listening for leaks and by visual inspection. Most exhaust system servicing involves the replacement of parts.

REVIEW QUESTIONS

1. What is the difference between psia and psig readings?
2. What can be used to check for leaks on the pressure side of a turbocharger system?
3. A mini-converter is used ______.
   a. on small vehicles where a normal converter will not fit properly
   b. on engines that use leaded fuels
   c. in conjunction with EGR systems to supply clean exhaust for the cylinders
   d. to reduce emissions during engine warmup
4. *True or False?* Engine misfires can cause a catalytic converter to overheat.
5. What is acoustic supercharging?
6. Describe the basic operation of an intake manifold runner control (IMRC) system.
7. What is the purpose of an intercooler?
8. *True or False?* The wastegate on variable turbochargers is controlled by the PCM.
9. Why is a wastegate not needed on a supercharger?

10. A restricted exhaust system can cause ______.
    a. stalling
    b. loss of power
    c. backfiring
    d. all of the above
11. Which of the following is *not* characteristic of a turbocharger?
    a. It is used to increase engine power by compressing the air that goes into the combustion chambers.
    b. It usually is located close to the exhaust manifold.
    c. It utilizes an exhaust-driven turbine wheel.
    d. It requires a mechanical connection between the engine and the pressurizing pump to compress the intake gases.
12. Ten psi of turbo boost means that air is being fed into the engine at ______ when the engine is operating at sea level.
    a. 4.7 psia
    b. 10 psia
    c. 14.7 psia
    d. 24.7 psia
13. What manages turbo output by controlling the amount of exhaust gas entering the turbine housing?
    a. wastegate
    b. turbine seal assembly
    c. hot wheel
    d. cold wheel
14. What is the first step in turbocharger inspection?
    a. Check the air cleaner for a dirty element.
    b. Open the turbine housing at both ends.
    c. Start the engine and listen to the system.
    d. Remove the ducting from the air cleaner to the turbo and examine the area.
15. Which of the following statements concerning superchargers is incorrect?
    a. Superchargers must overcome inertia and spin up to speed as the flow of exhaust gas increases.
    b. Superchargers do not require a wastegate to limit boost.
    c. A bypass is designed into the system to allow the supercharger to idle along when extra power is not needed.
    d. Superchargers improve horsepower and torque.

ASE-STYLE REVIEW QUESTIONS

1. Technician A says that a turbocharger has its own self-contained lubrication system. Technician B says that a turbocharger should not be operated when the engine's oil pressure is lower than 30 psi. Who is correct?
   a. Technician A c. Both A and B
   b. Technician B d. Neither A nor B
2. Technician A makes sure that the exhaust system is cool to the touch before working on it. Technician B disconnects the battery's negative cable before starting to work. Who is correct?
   a. Technician A c. Both A and B
   b. Technician B d. Neither A nor B
3. While discussing MAF-equipped engines: Technician A says that disconnected vacuum hoses admit more air into the intake manifold than the engine is calibrated for. Technician B says that the most common result of a vacuum leak is a rough-running engine due to a richer air-fuel mixture. Who is correct?
   a. Technician A c. Both A and B
   b. Technician B d. Neither A nor B
4. Technician A uses sandpaper to remove carbon deposits from turbocharger wastegate parts. Technician B scrapes off heavy deposits before attempting to clean the unit. Who is correct?
   a. Technician A c. Both A and B
   b. Technician B d. Neither A nor B
5. Technician A says that a vacuum leak results in less air entering the engine, which causes a richer air-fuel mixture. Technician B says that a vacuum leak anywhere on an MAF-equipped engine can cause the engine to run poorly. Who is correct?
   a. Technician A c. Both A and B
   b. Technician B d. Neither A nor B
6. Technician A says that vacuum hose routing for the entire vehicle is illustrated on the under-hood decal. Technician B says that this decal illustrates emission system vacuum hose routing. Who is correct?
   a. Technician A c. Both A and B
   b. Technician B d. Neither A nor B
7. Before replacing any exhaust system component: Technician A soaks all old connections with penetrating oil. Technician B checks the old

system's routing for critical clearance points. Who is correct?

a. Technician A
b. Technician B
c. Both A and B
d. Neither A nor B

8. A vehicle's intake manifold is warped beyond the manufacturer's specifications: Technician A replaces it. Technician B cuts the surface flat. Who is correct?

a. Technician A
b. Technician B
c. Both A and B
d. Neither A nor B

9. Technician A says that a low vacuum reading can be caused by incorrect ignition timing. Technician B says that an engine with low compression will have a low vacuum reading. Who is correct?

a. Technician A
b. Technician B
c. Both A and B
d. Neither A nor B

10. Technician A says that the exhaust manifold gasket seals the joint between the exhaust manifold and the exhaust pipe. Technician B says that a resonator helps to reduce exhaust noise. Who is correct?

a. Technician A
b. Technician B
c. Both A and B
d. Neither A nor B

CHAPTER 33

EMISSION CONTROL SYSTEMS

OBJECTIVES

■ Explain why hydrocarbon (HC) emissions are released from an engine's exhaust. ■ Explain how carbon monoxide (CO) emissions are formed in the combustion chamber. ■ Describe how oxides of nitrogen (NO_x) are formed in the combustion chamber. ■ Describe how carbon dioxide (CO_2) is formed in the combustion chamber. ■ Describe oxygen (O_2) emissions in relation to air-fuel ratio. ■ Describe the operation of an evaporative control system during the canister purge and nonpurge modes. ■ Explain the purpose of the positive crankcase ventilation system. ■ Describe the operation of the detonation sensor and electronic spark control module. ■ Describe the purpose and operation of an exhaust gas recirculation (EGR) valve. ■ Define the purpose of a catalytic converter. ■ Describe the operation of a secondary air injection system. ■ Describe the emission controls commonly found on today's light-duty diesel engines.

The emission controls on today's cars and trucks are an integral part of the engine and electronic engine control system. Perhaps it is better to say that the electronic control systems are really emission control systems. The drive to have cleaner and more fuel-efficient vehicles has led to many of the control systems now in place. These systems have also contributed to significant increases in power and reliability and improved driveability.

POLLUTANTS

Automotive emissions that are of the most concern to environmentalists, engineers, and technicians are hydrocarbon (HC), carbon monoxide (CO), oxides of nitrogen (NO_x), and oxygen (O_2). These gases contribute to air and water pollution and climate change. The exception to this is O_2, which is not a pollutant. O_2 emissions are, however, an indication of the completeness of the combustion process. The O_2 content in an engine's exhaust also is monitored during emissions inspections to detect holes in the exhaust pipes or broken pipes. Both of these would dilute the exhaust sample.

An exhaust gas that is not monitored in most emissions testing is sulfur dioxide (SO_2). This is a colorless gas that has a rotten egg smell. It is caused by large amounts of sulfur content in gasoline and is produced by the action of the catalytic converter. SO_2 can cause heart problems, asthma, and other respiratory problems.

HC, NO_x, and CO emissions are caused by many different things; the most important are temperatures at different spots within the combustion chamber and the air-fuel ratio. It is interesting to note that the conditions for minimizing HC emissions are the same as those that cause high NO_x emissions (**Figure 33–1**).

The allowable amounts of automobile emissions are regulated by the government. Through the years, the maximum allowable amount emitted by an automobile has decreased (**Figure 33–2**). Before you can understand the purpose of each emission control device and how to diagnose and service them, you must have an understanding of why these gases are emitted by an automobile.

All vehicles for the past 40 or more years are equipped with devices that reduce the levels of emissions released by the exhaust system. These are commonly referred to as "tailpipe" emissions. The following discussion of the various gases addresses their formation before they are reduced by the various emission control devices.

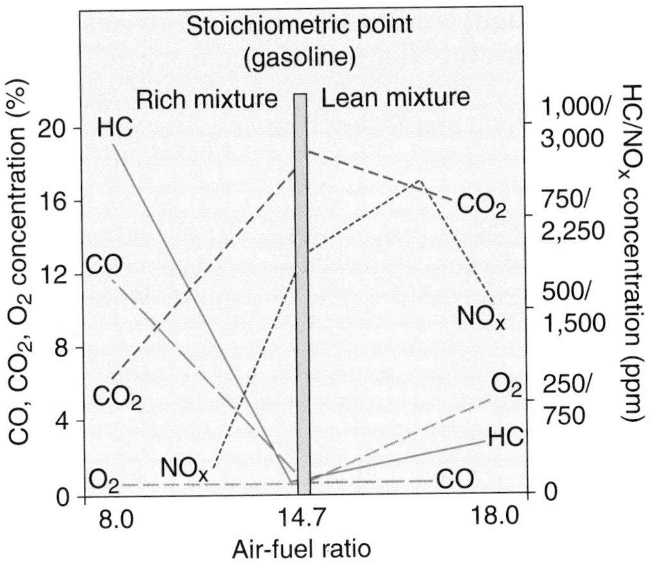

Figure 33-1 This chart shows how difficult it is to control exhaust emissions. Notice that NO_x and CO_2 are high when the HCs are low.

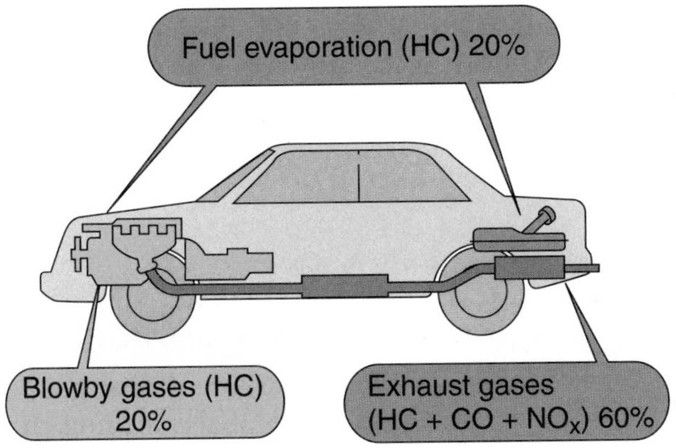

Figure 33-2 The main sources of automotive emissions.

Hydrocarbons

Hydrocarbon (HC) emissions are caused by incomplete combustion, and these emissions are actually molecules of unburned or partially burned fuel. Even an engine in good condition with satisfactory ignition and fuel systems will release some HC. Evaporative emissions from the fuel storage and delivery system are also a source of HC emissions.

HC emissions from the combustion process result when the following occur:

- The air-fuel mixture is compressed into the small sheltered areas of the combustion chamber, where the flame front cannot ignite the fuel. These include the areas formed by the top piston ring and the cylinder wall, any crevices formed by the head gasket, spark plug threads, and the valve seat.
- Fuel is absorbed by the oil on the walls of the cylinder.
- Fuel is absorbed by and/or is contained within carbon deposits in the combustion chamber.
- When the flame front approaches the cooler cylinder wall, the flame front quenches, leaving some unburned HCs.
- Some of the air-fuel mixture is left unburned because the flame front stops before igniting all of the mixture (misfire).
- The fuel does not adequately mix with the air and ignite prior to the end of combustion.
- The closed exhaust valve allows some fuel to leak during the intake, compression, and power strokes.
- There is a misfire caused by an ignition system fault.

An excessively lean air-fuel ratio results in cylinder misfiring and high HC emissions. A very rich air-fuel ratio also causes higher-than-normal HC emissions. At the stoichiometric air-fuel ratio, HC emissions are low.

Carbon Monoxide

Carbon monoxide (CO) is a by-product of combustion. CO is a poisonous chemical compound of carbon and oxygen. CO is a colorless, odorless, and highly poisonous gas that can cause dizziness, headaches, impaired thinking, and death by O_2 starvation. It forms in the engine when there is not enough O_2 to combine with the carbon during combustion. When there is enough O_2 in the mixture, carbon dioxide (CO_2) is formed. CO_2 is not a pollutant and is the gas used by plants to manufacture oxygen. CO is primarily found in the exhaust but can also be in the crankcase.

CO emissions are caused by a lack of air or too much fuel in the air-fuel mixture. CO will not occur if combustion does not take place in the cylinders; therefore, the presence of CO means combustion is taking place. As the air-fuel ratio becomes richer, the CO levels increase **(Figure 33–3)**. At the stoichiometric air-fuel ratio, the CO emissions are very low. If the air-fuel ratio is leaner than stoichiometric, the CO emissions remain very low. Therefore, CO emissions are a good indicator of a rich air-fuel ratio, but they are not an accurate indication of a lean air-fuel ratio.

Nitrogen Oxides

Nitrogen oxides (NO_x) are the various compounds of nitrogen and O_2 formed during the combustion

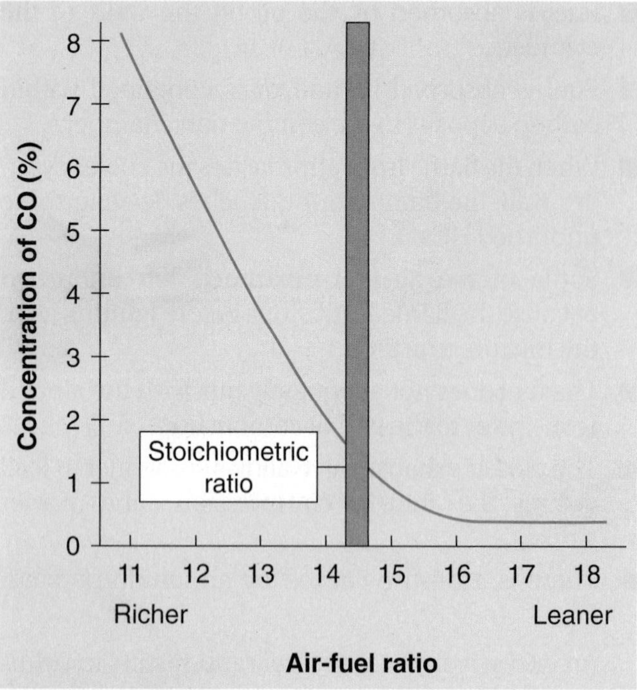

Figure 33-3 The amount of CO in the exhaust varies with the air-fuel mixture.

process. Both nitrogen and O_2 are present in the air used for combustion. Exposure to NO_x can cause respiratory problems, such as lung irritation, bronchitis, or pneumonia. Photochemical smog results when HCs and NO_x are combined with sunlight. Smog appears as brownish ground-level haze. Smog can cause many health problems, including chest pains, shortness of breath, coughing and wheezing, and eye irritation. When NO_x in the atmosphere mixes with rain (H_2O), nitric acid (HNO_3) or acid rain is created.

The formation of NO_x is the result of high combustion temperatures and pressures. When combustion temperatures reach more than 2,300°F (1,261°C), the N and the O_2 in the air begin to combine and form NO_x.

SHOP TALK

It is often said that NO_x forms at 2,500°F (1,371°C); studies have shown that the formation of NO_x actually begins around 2,300°F (1,261°C). The actual temperature is not really important; what is important is the thought that keeping combustion temperatures low keeps NO_x low.

Because outside air is 78% N_2, the gas cannot be prevented from entering the engine. Therefore, the only way to control NO_x is to prevent N_2 from joining with oxygen during the combustion process. This is done by controlling the temperature of the combustion process. This, however, can be very tricky. When the mixture is slightly rich, the combustion temperature drops and there is less chance for the production of NO_x. However, a rich mixture will also lead to an increase in CO and HC emissions. Likewise, when the mixture is slightly lean there is less of a chance for CO and HC emissions, but there is a greater chance for the production of NO_x because combustion temperature increases.

The "x" in NO_x stands for the proportion of oxygen mixed with a nitrogen atom. The "x" is a variable, which means it could be the number 1, 2, 3, and so on, therefore, the term *NO_x* refers to many different oxides of nitrogen (NO, NO_2, NO_3, etc.). NO_x emissions from an engine are mostly nitric oxide (NO), with less than 1% of the total NO_x being nitrogen dioxide (NO_2). NO is unhealthy and contributes to the greenhouse effect. NO_2 is a very toxic gas and contributes to the formation of smog, ozone, and acid rain.

It is important to note that diesel engines (because combustion temperatures of the fuel are low) produce more NO_2 than gasoline engines. About one-third of the nitrogen converted in a diesel engine becomes NO_2.

Carbon Dioxide

Carbon dioxide (CO_2) is not a pollutant; however, it has been linked to another environmental concern—the "greenhouse effect." CO_2 is a **greenhouse gas** and may be the major cause of global warming. It is claimed that 14% of all CO_2 emissions in North America are from the exhaust of automobiles and 27% of the CO_2 is caused by all transportation.

From an efficiency standpoint, carbon dioxide is an ideal by-product of combustion. Therefore, large amounts of CO_2 in the exhaust are desired. As the air-fuel ratio goes from 9:1 to 14.7:1, the CO_2 levels gradually increase from approximately 6% to 13.5%. CO_2 levels are highest when the air-fuel ratio is slightly leaner than stoichiometric. The production of CO_2 is directly related to the amount of fuel consumed; therefore, more fuel-efficient engines produce less. Each gallon of gasoline produces about 19.4 pounds (8.8 kg) of CO_2.

To put this in perspective:

- A vehicle that gets 10 mpg will emit 1.94 lb of CO_2 per mile, or 11.6 tons a year if driven 12,000 miles per year.
- A vehicle that gets 20 mpg will emit 0.97 lb of CO_2 per mile, or 6 tons a year if driven 12,000 miles per year.

- A vehicle that gets 30 mpg will emit 0.65 lb of CO_2 per mile, or 3.9 tons a year if driven 12,000 miles per year.

To reduce CO_2 levels in the exhaust, engineers are working hard to decrease fuel consumption. This, however, is difficult, because many of the methods used to increase fuel efficiency, such as lean mixtures, increase CO_2 and other emission levels. The concern for CO_2 emissions is one of the primary reasons for the continued exploration of alternative fuel and power sources for automobiles. It is also one of the main reasons the government has raised the CAFÉ mark to 35 mpg.

Other than being emitted from an engine's exhaust or the burning of fossil fuels, CO_2 is also emitted by nature. There are many natural sources of CO_2 such as the oceans and decaying plantlife. Plants also use CO_2 in their photosynthesis process, which results in a release of oxygen to the atmosphere. The current concern on global warming suggests that there is more carbon dioxide in the atmosphere than can be used by plantlife. In fact, the more CO_2 molecules there are in the atmosphere, the warmer the earth gets.

To bring the amount of CO_2 in the atmosphere to an acceptable level, we need more plantlife and we need to reduce the amount that is released to the atmosphere. A target for reducing CO_2 is the automobile. There are approximately 700 million automobiles on the world's roads. They produce about 2.8 billion tons of CO_2 annually, which is close to 20% of the world's total CO_2 emissions from burning fossil fuels (**Figure 33–4**). The quantity of CO_2 emissions from vehicles is measured in grams per mile (g/mi) or grams per kilometer (g/km).

Other substances are also considered greenhouse gases. These include water vapor, methane, and nitrous oxide. Greenhouse gases allow the light from the sun to freely enter the atmosphere. When the sunlight reaches the earth's surface, some of it is absorbed and warms the earth. The rest of the sunlight is reflected back to the atmosphere as heat. Greenhouse gases absorb and trap the heat. This process is considered to be the cause of global warming.

There are currently no standards for CO_2 emissions in North America; however, the European Union has standards that all new vehicles must meet. California and other states also are considering a CO_2 standard. Whether there is a CO_2 standard or not, the government is imposing new CAFÉ standards that will, in effect, reduce CO_2 emissions.

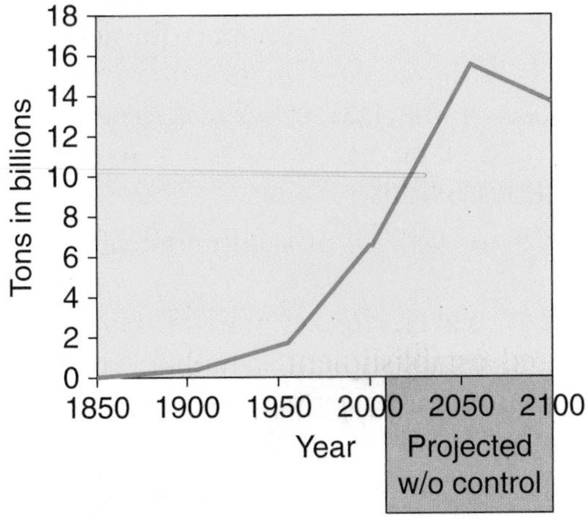

Figure 33-4 Total worldwide carbon emissions from burning fossil fuels.

CO_2 Reduction at Factories Not only are manufacturers trying to reduce CO_2 levels emitted by the exhaust, they are also working to reduce the amount of greenhouse gases from their factories. GM, for example, announced a goal of reducing CO_2 emissions from its facilities by 40% by the year 2010. They are well on their way. The strategies to this include using less energy, reducing waste, increasing the use of renewable resources, and increasing the efficiency of the entire manufacturing process.

Oxygen

Oxygen (O_2) is not a pollutant; therefore, its presence in the exhaust does not pose any threat to our environment. However, too much oxygen in the exhaust does indicate that an improper air-fuel mixture or poor combustion has occurred in the engine.

If the air-fuel ratio is rich, all the oxygen in the air is mixed with fuel, and the O_2 levels in the exhaust are very low. When the air-fuel ratio is lean, there is not enough fuel to mix with all the air entering the engine, and O_2 levels in the exhaust are higher. Therefore, O_2 levels are a good indicator of a lean air-fuel ratio, and they are not affected by catalytic converter operation.

Water (H_2O)

H_2O should be emitted from a vehicle's exhaust. It is the result of oxidation, whereby HC is oxidized by the converter to form CO_2 and H_2O. The amount of water emitted depends on a number of things, including the effectiveness of the converter and the amount of HC and O_2 in the exhaust before it passes through the converter. Normally during cold engine startup, steam is emitted from the exhaust. The steam results from the heating of the water present in the converter and exhaust system. Much of this water is formed by condensation due to the cooling of the exhaust system.

Diesel Emissions

Diesel engines are the most efficient of all internal combustion engines. They have low fuel consumption rates and produce low amounts of greenhouse gases. However, there are exhaust emissions that are of concern. Some of these are currently regulated and others will be in the future. A typical diesel engine emits:

- Carbon (soot) and various carbon-based compounds
- NO_X
- Water
- Carbon monoxide
- Sulfur dioxide
- Various hydrocarbon-based compounds

Soot is the most obvious emission and is often referred to as diesel particulates. These particulates are mostly comprised of carbon-based substances that tend to absorb the other contaminants in a diesel engine's exhaust. Many factors influence the amount of soot released, including the fuel used and the engine design. Particulate emissions are the main obstacle for using diesel engines in passenger cars and light- and medium-duty trucks. California and other states have set a standard for particulate emissions. These standards will become more stringent as diesel technology advances. These emissions are measured in grams per mile or kilometer.

EMISSION CONTROL DEVICES

According to a document based on a study by the Environmental Protection Agency (EPA), passenger cars are responsible for 17.8% of the total hydrocarbon emissions, 30.9% of the total carbon monoxide emissions, and 11.1% of the oxides of nitrogen emissions. After more than 40 years of emission regulations, these figures remain staggering! Imagine what these figures would be if automotive emissions had remained unregulated during the past 40 years!

Emission standards have been one of the driving forces behind many of the technological changes in the automotive industry. Catalytic converters and other emission systems were installed to meet emission standards. These standards have become progressively more stringent through the years, and the engine designs and devices required to meet the standards has become very complex.

Legislative History

SHOP TALK

Through the years, vehicles have been built specifically for the State of California. The California Air Resources Board (CARB) has led the way for many emissions laws and standards. Many are specific to the state, although most of these have eventually been implemented by other states, provinces, and countries. Check the labeling on the vehicle to see if the vehicle has special equipment or calibrations designed for use in California and not found in vehicles for other states.

The driving force behind the development of emission control devices has been the various federal clean air acts put into effect over the years. The following are some key acts that have shaped the design of today's vehicles:

- The Clean Air Act of 1963 identified the automobile as a major contributor to air pollution and was responsible for as much as 40% of all emissions.
- The Clean Air Act Amendments of 1965 required auto manufacturers to install emissions control devices on all passenger cars and light trucks by the 1968 model year.
- The Clean Air Act Amendments of 1970 established nationwide air quality standards and linked federal building and highway funds to meeting those standards. Areas that did not meet the standards were required to institute a plan to correct the problem or lose the funds.
- The Clean Air Act Amendments of 1977 mandated that areas that did not meet air quality standards must establish and enforce basic inspection and maintenance (basic I/M) programs for all passenger cars and light trucks. The purpose of the I/M programs was to test and repair the effectiveness of all systems that affected vehicle emissions.
- The Clean Air Act Amendments of 1990 again reduced the allowable amount of exhaust emissions. However, the key part of this act was the required establishment of enhanced inspection and maintenance (enhanced I/M) programs in areas that did not meet air quality standards.
- The Energy Independence and Security Act of 2007 required a 40% increase in vehicle fuel economy by the year 2020. It also established

mandates for increased use of renewable fuels. This act also focused on the reduction of CO_2 emissions, although no standard was set.

Inspection and Maintenance Programs

Emission standards set the maximum allowable amounts of emissions from a new automobile. These standards have become stricter through time. Amendments to the acts called for emission testing of vehicles on the road. These periodic tests are part of the inspection and maintenance (I/M) program. The purpose of these inspections is simply to identify those vehicles that have been tampered with or have not received good maintenance. Studies have shown that 20% of the vehicles on the road are not being properly maintained, and those vehicles account for more than 90% of the emissions from automobiles. I/M programs are designed to identify these vehicles and make the necessary repairs to allow those vehicles to have acceptable amounts of emissions.

The first Clean Air Act set emission standards for new cars. To make sure the new vehicles met these standards, the federal test procedure (FTP) was instituted. The test is performed on a random sample of preproduction vehicles. These vehicles are used to represent the vehicles for the next model year. Their emissions are carefully checked and compared to the standards established for that model year. If the emission levels meet or are lower than the standards, the vehicle is then certified.

The FTP uses an inertia weight dynamometer (dyno) that allows the vehicle to be driven under varying loads. The dyno is capable of simulating actual driving conditions that the test vehicle would encounter by changing the load applied to the drive wheels. Emission levels are measured with a constant volume sampling (CVS) system that measures the mass of HC, CO, CO_2, and NO_x emitted from the vehicle in grams per mile. These exhaust analyzers are much more precise, complex, and expensive than the exhaust analyzer found in most shops.

The act also prompted Californians to create CARB. CARB's purpose was to implement strict air standards, which later became federal standards. One of the approaches made by CARB to clean the air was to start periodic motor vehicle inspection (PMVI) programs. This inspection included a tailpipe emissions test and an underhood inspection. The tailpipe test certifies that the vehicle's exhaust emissions (HC and CO) are within the limits of the law. The emissions were measured in parts per million (ppm) and were taken with the engine at idle. The allowable emissions are three to four times higher than those required for a new vehicle. This provides some tolerance for engine and system wear. The underhood and/or vehicle inspection verifies that the emission control systems have not been tampered with or disconnected. Today, California is not the only state that requires annual emissions testing. Many states have incorporated an emissions test with their annual vehicle registration procedures.

During the 1980s, this basic I/M program was changed to include tests conducted under a load with a dyno. This test was called the accelerated simulated mode (ASM). This test measured CO, HC, and NO_x emissions.

After the implementation of the Clean Air Act of 1990, more precise testing was instituted. The result is called the I/M 240 test. Many states have implemented an I/M 240 or similar program. The I/M 240 (or enhanced I/M) tests vehicle emissions while it is operating under a variety of load conditions and speeds. While on the dyno, the vehicle is operated for up to 240 seconds and under different load conditions **(Figure 33–5)**. The test drive on the dyno simulates both in-traffic and highway driving and stopping and includes the same conditions as the FTP. However, the I/M 240 test is only a small portion of a complete FTP.

The I/M 240 consists of three separate tests: (1) a transient, mass emission tailpipe test, (2) an evaporative system purge flow test, and (3) an evaporative system pressure test. The test results are given in the same increments as the FTP (grams per mile) and therefore can be directly related to the FTP standards. The test also measures HC, CO, CO_2, and NO_x emissions during the I/M 240 drive cycle.

Basically the same equipment used for the FTP is used for an I/M 240 test. A variable inertia weight dynamometer is used because it can be adjusted to

Figure 33-5 Emission levels are commonly checked with the vehicle running on a dyno. *Courtesy of Robert Bosch GmbH, www.bosch-presse.de*

match the weight of the vehicle and allows the vehicle to be driven under a variety of loads. Emission levels are determined by collecting the exhaust in a CVS and analyzing its contents through the use of very sophisticated equipment. This equipment is also similar to those used during the FTP; they include the following:

- Nondispersive infrared (NDIR) tester for measuring CO and CO_2
- Flame ionization detector (FID) for measuring HC
- Chemiluminescence detector for measuring NO_x

HC emissions from a vehicle's EVAP can be higher than the HC emissions in an engine's exhaust. Therefore, monitoring this system is an important part of the I/M 240 test. The I/M 240 test includes a visual inspection of the EVAP system and a purge volume and fuel tank pressure test. The purge test measures the flow of fuel vapors into the engine's intake during the test's drive cycle. The pressure test is used to check for leaks that would allow vapors to be released into the atmosphere.

Vehicle Emission Control Information (VECI)

All vehicles have a **vehicle emission control information (VECI)** decal that gives specific emission control information for that vehicle and engine **(Figure 33–6)**. These decals are normally located on the underside of the hood or on the radiator support frame. The information contained on the VECI is important when conducting an I/M test and when diagnosing or repairing an emissions-related problem. Most of the information contained on the decal is expressed by acronyms; **Figure 33–7** gives some examples of these.

Engine/Evaporative Emission System Information Starting in 1994, all manufacturers must use a standardized system to identify their individual engine and EVAP system families. These names must be

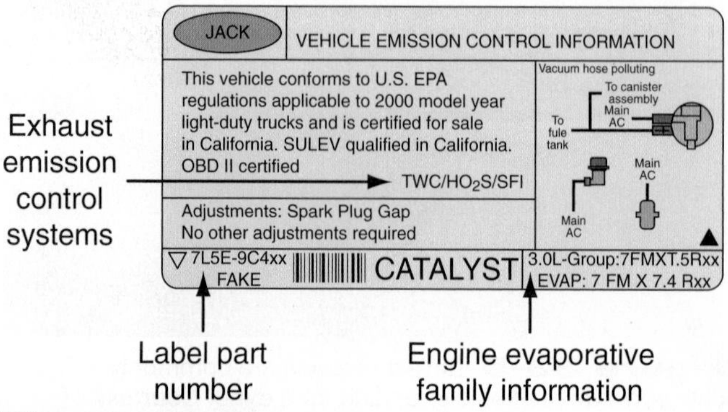

Figure 33-6 An example of a VECI.

Acronym	Definition
CARB	California Air Resource Board
CI	Cylinder Injection
EPA	Environmental Protection Agency
EVAP	Evaporative Emissions
GVW	Gross Vehicle Weight
GVWR	Gross Vehicle Weight Rating, curb weight plus payload
HO_2S	Heated Oxygen Sensor
ILEV	Inherently Low Emission Vehicle
LDDT	Light Duty Diesel Truck categories
LEV	Low Emission Vehicle
LVW	Loaded Vehicle Weight, curb weight plus 300 lb (136.08 Kg)
MDV	Medium Duty Vehicle
MHDDE	Medium Heavy Duty Diesel Engine
MPI	Multi-Port Injection
MY	Model Year
NCP	Non-Compliance Penalty
OBD	On Board Diagnostic
ORVR	On-Board Refueling Vapor Recovery
PC	Passenger Car
PZEV	Partially Zero Emission Vehicle
SFI	Sequential Fuel Injection
SULEV	Super Ultra Low Emission Vehicle
TWC	Three Way Catalyst
ULEV	Ultra Low Emission Vehicle
ZEV	Zero Emission Vehicle

Figure 33-7 Vehicle emission control information (VECI) acronym definitions.

twelve characters long and are shown in a box on the VECI. The first twelve-character ID contains the size of the engine and its family group. On the second line is another twelve-character ID. This identifies the family name of the EVAP system. Both of these names are specific to that vehicle.

Base Engine Calibration Information Important engine (powertrain) calibration information is normally given in the lower right-hand corner of the vehicle's certification label. The vehicle certification label is typically affixed on the left front door or door post. Base engine calibration information is limited to a maximum of five characters per line and no more than two lines. This coding is used during diagnostics and service. The certification label also contains a coded description of the vehicle.

Classifications of Emission Control Devices

All emission control systems fall into one of three classifications: evaporative control systems, precombustion, and postcombustion.

The **evaporative control (EVAP)** system is a sealed system. It traps the fuel vapors (HC) that would normally escape from the fuel delivery system into the air.

Most pollution control systems used today prevent emissions from being created in the engine, either during or before the combustion cycle. The common **precombustion control** systems are the PCV, engine modifications, spark control, and exhaust gas recirculation (EGR) systems.

Postcombustion control systems clean up the exhaust gases after the fuel has been burned. Secondary air or air injector systems put fresh air into the exhaust to reduce HC and CO to harmless water vapor and carbon dioxide by a chemical (thermal) reaction with oxygen in the air. Catalytic converters are the most effective postcombustion emission control; they reduce NO_x, HC, and CO emissions.

EVAPORATIVE EMISSION CONTROL SYSTEMS

Fuel vapors from the gasoline tank and the carburetor float bowl were brought under control with the introduction of EVAP systems. These systems were first installed in 1970 model cars sold in California and in most domestic-made cars beginning with 1971 models. Through the years, EVAP emissions have been closely monitored and the control systems modified to minimize the chances of vapors entering the atmosphere. EVAP emissions are limited by law, and the current limit in the United States is 2 grams of HC per hour.

Current systems are computer controlled and are monitored by the OBD-II system. Most current EVAP systems include the following components:

- A domed fuel tank in which its upper portion is raised. Fuel vapors rise to this upper portion and collect.
- Canister vent solenoid valve
- A special filler design to limit the amount of fuel that can be put in the tank
- Fuel tank vacuum or pressure sensor
- Fuel lines
- Vapor lines
- Fuel tank cap
- A vapor separator in the top of the fuel tank **(Figure 33–8)**
- Charcoal (EVAP) canister
- Purge lines
- Purge solenoid valve (vapor management valve)

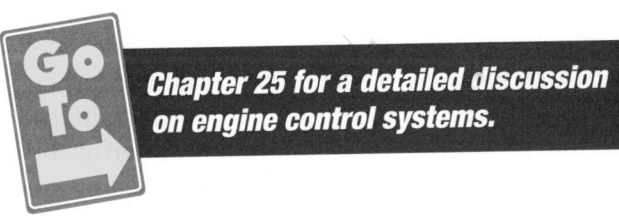

Go To ***Chapter 25 for a detailed discussion on engine control systems.***

Fuel tanks are sealed units and are designed to prevent vapors that result from the evaporation of the gasoline from entering the atmosphere. Also, special

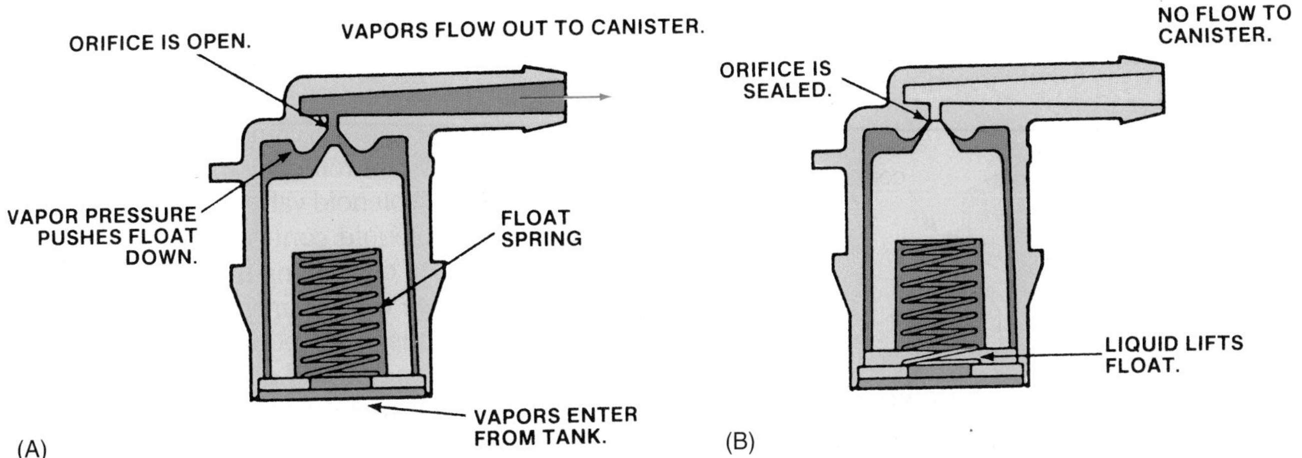
ORIFICE IS OPEN.
VAPORS FLOW OUT TO CANISTER.
VAPOR PRESSURE PUSHES FLOAT DOWN.
FLOAT SPRING
VAPORS ENTER FROM TANK.
(A)
NO FLOW TO CANISTER.
ORIFICE IS SEALED.
LIQUID LIFTS FLOAT.
(B)

Figure 33-8 (A) Normal operation of a vapor separator, (B) with liquid in the separator.

devices are used to reduce the amount of gas vapors that escape from the fuel tank while the vehicle is being refueled. When the fuel cap is removed on all late-model vehicles, the filler neck is sealed with a hinged and spring-loaded flap. The size of the flap is large enough to allow an unleaded fuel nozzle to enter but too small for a leaded nozzle. Gasoline pump nozzles have also been modified to prevent fuel vapors from entering the atmosphere during refueling. Some nozzles are equipped with a rubber boot that seals the nozzle to the filler neck. Other nozzles are designed to create a vacuum that draws in vapors as the liquid fuel is being pumped into the tank. Most late-model vehicles have an on-board refueling vapor recovery (ORVR) system, and the special nozzles are redundant and not really needed.

The EVAP system moves the built-up vapors to a canister where they are stored until the system purges the canister and allows the vapors to enter the intake manifold. The EVAP system also allows some atmospheric pressure to enter the tank. This prevents the buildup of vacuum in the tank that could cause the tank to collapse.

Fuel vapors inside the fuel tank are vented at the top of the tank through the vapor separator. The separator collects droplets of liquid fuel and directs them back into the tank. The vapors leave the separator and move to the canister through the vapor line.

The **charcoal** (carbon) **canister** **(Figure 33–9)** is normally located in the engine compartment. Fuel vapors from the gas tank are routed to and absorbed onto the surfaces of the canister's charcoal granules. When the vehicle is restarted, vapors are drawn by the vacuum into the intake manifold to be burned by the engine. Canister purging varies widely with vehicle make and model. On new vehicles, the PCM controls when the canister will be purged. In some instances a fixed restriction allows constant purging whenever there is manifold vacuum. In others, a staged valve provides purging only at speeds above idle.

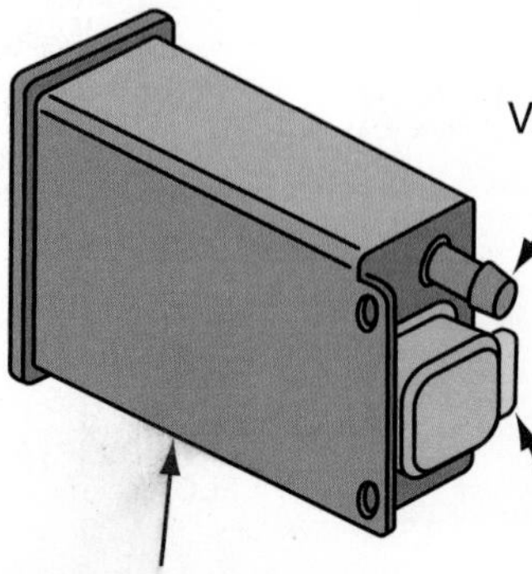

Figure 33-9 A charcoal canister.

The **canister purge valve** is normally closed. It opens the inlet to the purge outlet when vacuum is applied. Some units incorporate a thermal delay valve so the canister is not purged until the engine reaches operating temperature. Purging at idle or with a cold engine creates other problems, such as rough running and increased emissions because of the additional vapor added to the intake manifold.

The canister contains a liquid fuel trap that collects any liquid fuel entering the canister. Condensed fuel vapor forms liquid fuel. This liquid is returned from the canister to the tank when a vacuum is present in the tank. This liquid fuel trap prevents liquid fuel from contaminating the charcoal in the canister.

WARNING!

Gasoline vapors are extremely explosive! Do not smoke or allow sources of ignition near any component of the EVAP system. Explosion of gasoline vapors may result in property damage or personal injury.

Early EVAP systems were not controlled by the PCM but instead relied on a ported vacuum purge port, a vacuum check valve, and a thermo vacuum valve (TVV). The latter was used to prevent purging when the engine was cold. The canister was purged whenever the throttle plate was open enough to expose the vacuum port to engine vacuum. The vacuum then opened the check valve to allow vapors to move to the intake manifold. The check valve also served to keep the system sealed when the engine was not running. The amount of purged vapors was controlled by a fixed orifice. This meant that any time the controlling port had vacuum, the same amount of fuel vapors was sent to the engine, regardless of engine load or speed. This resulted in driveability problems during some condition because the air-fuel mixture became too rich.

To gain more control of canister purging, the EVAP operation is now controlled by the PCM. These systems use a purge solenoid valve that is duty cycled by the PCM. The solenoid controls the vacuum to the canister, therefore controlling the amount of vapors purged. The purge valve is only open and controlled when the system is in closed loop, and then it only opens when the engine and conditions can respond to the extra enrichment of the vapors.

PCM-controlled EVAP systems allow for precise control of purge flow and vapor volume. Because the system responds to current engine and operating conditions, the purging of the vapors does not affect driveability.

Enhanced Evaporative Emission (EVAP) System

To meet OBD-II regulations, late-model vehicles have an enhanced EVAP system. Enhanced EVAP systems operate in much the same way as previous PCM-controlled EVAP systems, but they also conduct tests that can detect small, 0.020-inch (0.5 mm) system leaks and monitor canister purge flow. The tests are only run when certain conditions are present. These conditions vary with make, model, and engine type.

These systems have a fuel tank pressure (FTP) sensor and EVAP canister vent (CV) solenoid in addition to typical EVAP components **(Figure 33–10)**. Some systems are also equipped with a pump often called the leak detection pump (LDP). The CV solenoid seals the charcoal canister from the atmosphere when the system is conducting the leak check monitor. The FTP sensor measures pressure or vacuum in the fuel tank and compares it to atmospheric pressure. This input is also used to check for pinched or restricted vapor lines **(Figure 33–11)**. A vacuum is created in the system during the leak test; the FTP sensor measures the vacuum at the beginning of the test and again after a fixed period. If vacuum cannot be built into the system, or if the system will not hold a vacuum, a leak is evident. A signal fuel level input (FLI) sensor is used during the leak check to determine how much fuel is in the tank. This determines how long it should take to build a vacuum in the tank. The FLI input is also used to determine if the EVAP and other monitors can run. Too low or too high of a fuel level will not allow the monitors to run.

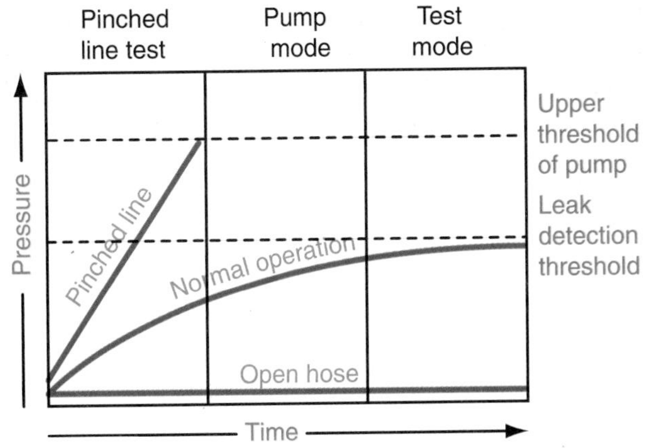

Figure 33–11 The rate of pressure buildup determines if the system has pinched lines or a leak, or if it is operating normally.

On systems with an LDP, the PCM turns on the pump when it is checking the system. The pump

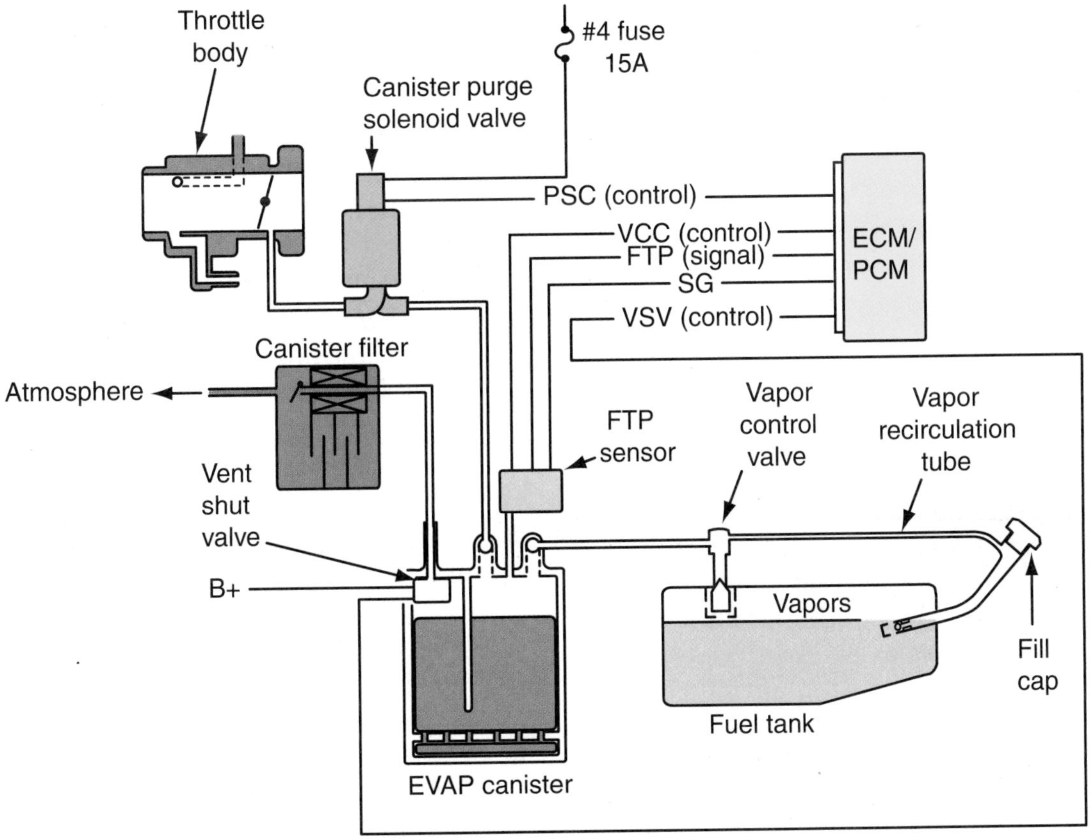

Figure 33–10 A diagram of an enhanced EVAP system.

pressurizes the system. As pressure builds, the cycling rate of the pump decreases. If there is no leak in the system, the pressure will build until the pump shuts off. If there is a leak, pressure will not build up and will not shut down the pump. The pump will continue to run until the PCM determines it has run a complete test cycle.

If the PCM senses that there is no leak in the system, it will run the purge monitor. This test is completed by calculating or measuring the amount of vapors that are being purged. In many systems, the PCM calculates the purge flow based on MAF or MAP sensor data and compares it to FTP or fuel trim data. When fuel trim is used, the purge monitor determines if the purge system is functioning properly by applying a long duty cycle to the purge solenoid. The subsequent change in STFT is monitored. If the system is working correctly as the duty cycle of the solenoid is increased, the STFT should change proportionately. This is the most commonly used method for checking the operation of the purge system.

Other systems use a purge flow sensor connected to a vacuum hose between the purge solenoid and the intake manifold. The PCM monitors the signal from the sensor once per drive cycle to determine if there is vapor flow or no vapor flow through the solenoid to the intake manifold. Other EVAP systems have a vapor management valve connected in the vacuum hose between the canister and the intake manifold. The vapor management valve is a normally closed valve. The PCM operates the valve to control vapor flow from the canister to the intake manifold. The PCM also monitors the valve's operation to determine if the EVAP system is purging vapors properly. If no leaks are present and the purge cycle is correct, the system has passed the test.

SHOP TALK

It may be impossible to refuel a late-model vehicle when the engine is running. This is normal! If the PCM is conducting a check of the EVAP system, the vent valve will be closed and the resultant pressure in the tank will stop fuel flow from the gas pump's nozzle. This means it will be impossible to fill the tank.

If the gas cap is off or loose during the leak test, the PCM will detect this large leak and the warning lamp will illuminate. With this leak, the system will not continue its leak test.

PRECOMBUSTION SYSTEMS

Systems designed to prevent or limit the amount of pollutants produced by an engine are called precombustion emission control devices. Although there are specific systems and engine designs that are classified this way, anything that makes an engine more efficient can be categorized as a precombustion emission control device.

Engine Design Changes

The basic engine has been modified through the years to increase its overall efficiency. Many of the changes have occurred inside the engine. Others involve the fuel and ignition systems. The result of these changes is improved performance and driveability and a decrease in exhaust emissions. Following is a summary of some of those changes.

- *Better sealing pistons.* Blowby gases are reduced through the use of better sealing piston rings and improved cylinder wall surfaces. Many engines are also fitted with low-friction piston rings. This increases fuel economy and engine power.
- *Combustion chamber designs.* The primary goal in designing combustion chambers is the reduction or elimination of the quench area. Another trend in combustion chamber design is locating the spark plug closer to the center of the chamber. Manufacturers have also worked with designs that cause controlled turbulence in the chamber. This turbulence improves the mixing of the fuel with the air, which improves combustion.
- *Lower compression.* By keeping the compression ratio low, combustion temperatures can be kept below the point where NO_x is formed. However, new developments have allowed the use of higher compression ratios on some high-performance engines.
- *Decreased friction.* Overcoming the friction of the engine's moving parts reduces the power and energy lost during engine operation. Improved engine oils, new component materials, and weight reductions have had the biggest impact on reducing friction.
- *Intake manifold designs.* Thanks to the wide and successful use of port fuel injection, intake manifold are designed to distribute equal amounts of air to each cylinder. The use of plastic intake manifolds has allowed for smoother runners and better heat control of the air.
- *Improved cooling systems.* High engine temperatures reduce HC and CO emissions. However, they also make the formation of NO_x harder to control. Engine cooling systems are designed to run at high

temperatures but are prevented from getting too hot, thereby limiting the production of NO_x. Today's engine control systems incorporate many features that change the air-fuel mixture, ignition timing, and idle speed to control the engine's temperature.

- *Spark control systems.* Spark control systems have been in use since the earliest gasoline engines. It was discovered that the proper timing of the ignition spark helped to reduce exhaust emissions and develop more power output. Incorrect timing affects the combustion process. Incomplete combustion results in HC emissions. High CO emissions can result from incorrect ignition timing. Advanced timing can also increase the production of NO_x. When timing is too far advanced, combustion temperatures rise. For every 1° of overadvance, the temperature increases by 125°F (51.57°C). Spark control on today's engines is handled by the PCM. Through input signals from various sensors, the PCM adjusts ignition timing for optimal performance with minimal emissions levels.

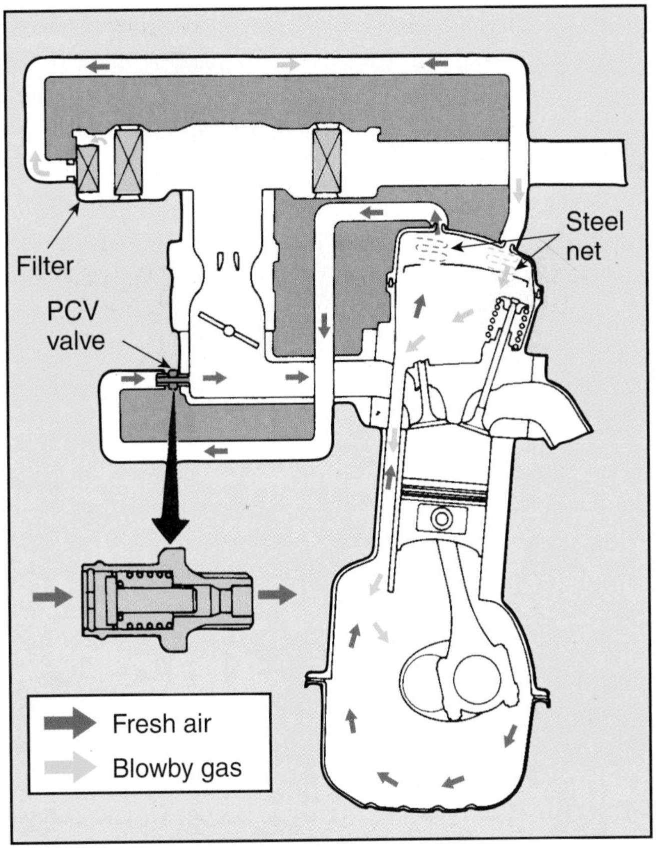

Figure 33-12 A typical PCV system. *Copyright, Nissan (2008). Nissan and the Nissan logo are registered trademarks of Nissan.*

PCV Systems

In late 1959, California established the first standards for automotive emissions. In 1967, the federal Clean Air Act was amended to provide standards that applied to automobiles. The first controlled automotive emission was crankcase vapors. During combustion some unburned fuel and other products of combustion leak past the piston rings and move into the crankcase. This leakage is called blowby. Blowby gases are largely HC gases which are PCV systems route the gases, which are mixed with outside air, into the engine's intake. From there the gases are drawn into the cylinders and burned. PCV systems were installed on all cars beginning with the 1963 models.

Blowby must be removed from the engine before it condenses in the crankcase and reacts with the oil to form sludge. Sludge, if allowed to circulate with engine oil, corrodes and accelerates the wear of pistons, piston rings, valves, bearings, and other internal parts of the engine. Blowby also carries some unburned fuel into the crankcase. If not removed, the unburned fuel dilutes the engine's oil. When oil is diluted, it does not lubricate the engine properly, which causes excessive wear.

Blowby gases must also be removed from the crankcase to prevent premature oil leaks. Because these gases enter the crankcase by the pressure created during combustion, they pressurize the crankcase. The gases exert pressure on the oil pan gasket and crankshaft seals. If the pressure is not relieved, oil is eventually forced out of these seals.

Operation The PCV system uses engine vacuum to draw fresh air through the crankcase. This fresh air enters through the air filter or through a separate PCV breather filter.

When the engine is running, intake manifold vacuum is supplied to the PCV valve. This vacuum moves air through the clean air hose into the rocker arm or camshaft cover. From there, the air flows through openings in the cylinder head into the crankcase where it mixes with blowby gases. The mixture of blowby gases and fresh air flows up through cylinder head to the PCV valve. Vacuum draws the blowby gases through the PCV valve into the intake manifold **(Figure 33–12)**. The blowby gases mix with the intake air and enter the combustion chambers where they are burned.

PCV Valve The PCV valve **(Figure 33–13)** is usually mounted in a rubber grommet. A hose connects the valve to the intake manifold. A clean air hose is connected from the air filter to the opposite rocker arm cover. A filter is positioned at end of the clean air hose. On some systems, the PCV valve is mounted in a vent module, and the clean air filter is located in this module.

A PCV valve contains a tapered valve. When the engine is not running, a spring keeps the valve seated

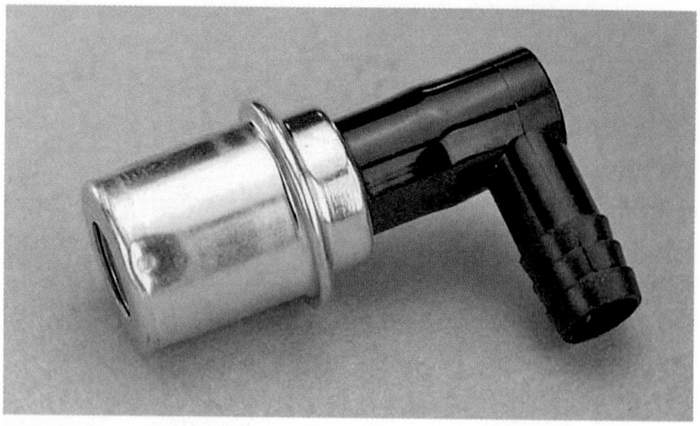

Figure 33–13 A PCV valve.

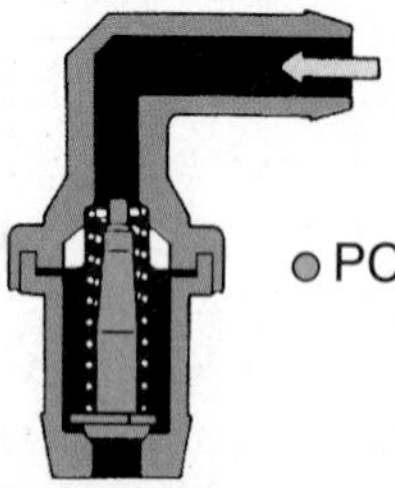

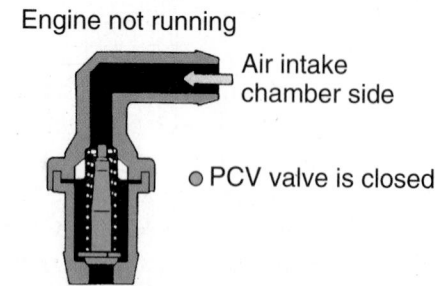

Figure 33–14 The PCV valve position with the engine not running. *Courtesy of Toyota Motor Sales, U.S.A., Inc.*

against the valve housing **(Figure 33–14)**. During idle or deceleration, high intake manifold vacuum moves the valve upward against the spring tension. Under this condition, the blowby gases flow through a small opening in the valve. Because the engine is not under heavy load, the amount of blowby gas is minimal and the small valve opening is all that is needed to move the blowby gases out of the crankcase.

Manifold vacuum drops off during part-throttle operation. As the vacuum signal to the PCV valve decreases, a spring moves the tapered valve downward to increase the opening **(Figure 33–15)**. Because engine load is higher at part-throttle operation than at idle, blowby gases are increased. The larger opening of the valve allows all the blowby gases to be drawn into the intake manifold.

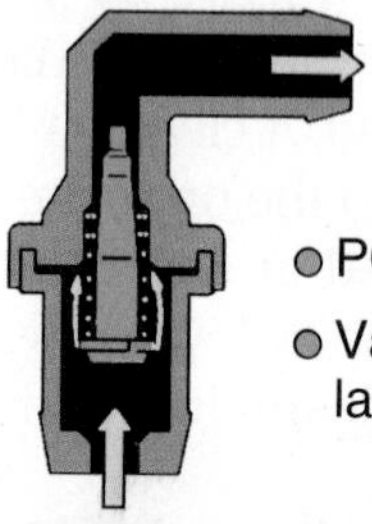

Figure 33–15 The PCV valve position during part-throttle operation. *Courtesy of Toyota Motor Sales, U.S.A., Inc.*

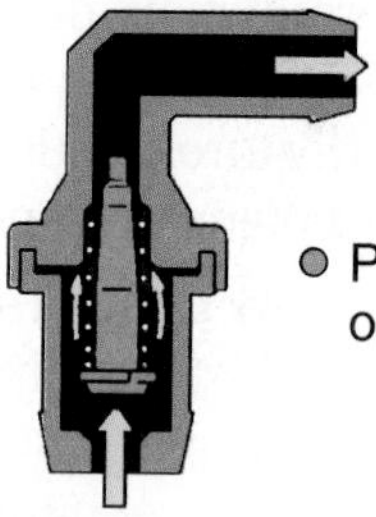

Figure 33–16 The PCV valve position during hard acceleration or heavy load. *Courtesy of Toyota Motor Sales, U.S.A., Inc.*

When the engine is operating under heavy load and at wide-open throttle, the decrease in intake manifold vacuum allows the spring to move the tapered valve further down in the PCV valve **(Figure 33–16)**. This provides a larger opening through the valve. Because higher engine load results in more blowby gases, the larger PCV valve opening is necessary to allow these gases to flow through the valve into the intake manifold.

When worn rings or scored cylinders allow excessive blowby gases into the crankcase, the PCV valve opening may not be large enough to allow these gases to flow into the intake manifold. Under this condition, the blowby gases create a pressure in the crankcase, and some of these gases are forced through the clean air hose and filter into the air cleaner. When this action occurs, there is oil in the PCV filter and air cleaner. This same action occurs if the PCV valve is restricted or plugged.

When there is high crankcase pressure and the engine is under heavy load, the PCV gases can experience reverse flow. The high pressure and high concentration of gases accompanied with very low engine vacuum allow the gases to move out of the intake manifold. The vacuum is too weak to draw in the gases. This results in an accumulation of oil in the throttle body. Therefore, oil buildup inside the throttle body can be an indication of high crankcase pressure.

If the PCV valve sticks in the wide-open position, excessive airflow through the valve causes rough idle operation. If a backfire occurs in the intake manifold, the tapered valve is seated in the PCV valve as if the engine was not running. This action prevents the backfire from entering the engine, where it could cause an explosion.

Fixed Orifice Tube PCV System Some engines are equipped with a PCV system that does not use a PCV valve. Rather, the blowby gases are routed into the intake manifold through a fixed orifice tube **(Figure 33–17)**. The system works the same as if it had a valve, except that the system is regulated only by the vacuum at the orifice. The size of the orifice limits the amount of blowby flow into the intake.

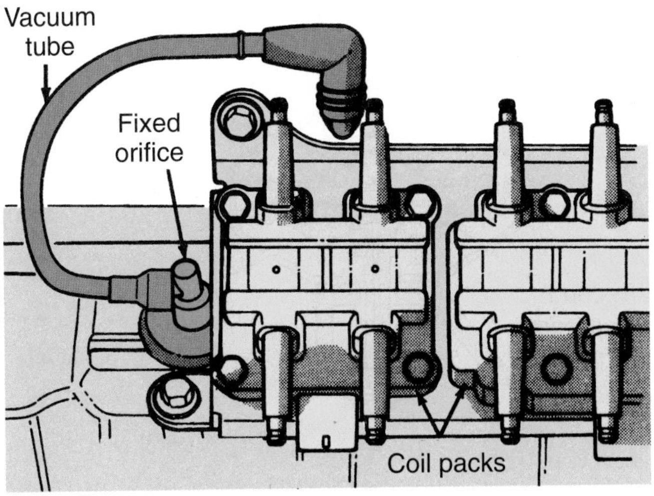

Figure 33-17 A fixed orifice tube-type PCV system. *Courtesy of Chrysler LLC*

Heated PCV Systems Crankcase vapors contain some moisture, which means the water in the vapors can freeze when the engine has sat in cold weather. This can cause the PCV system not to work until the ice melts, which may be a while. When the PCV system is not working, excess pressures can build up in the crankcase and force blowby gases out. To prevent this from occurring, some engines are equipped with heated PCV systems.

Heated systems have a heated PCV valve or heated PCV tube. The valves can be coolant heated **(Figure 33–18)** or electrically heated. Coolant heated valves have passages that allows the coolant to flow around the valve. Some electrically heated valves are controlled by the PCM, whereas others use a thermistor in the heater's wiring harness. Heated tubes rely on electrical heaters that are either PCM controlled or use a thermistor to control the heat.

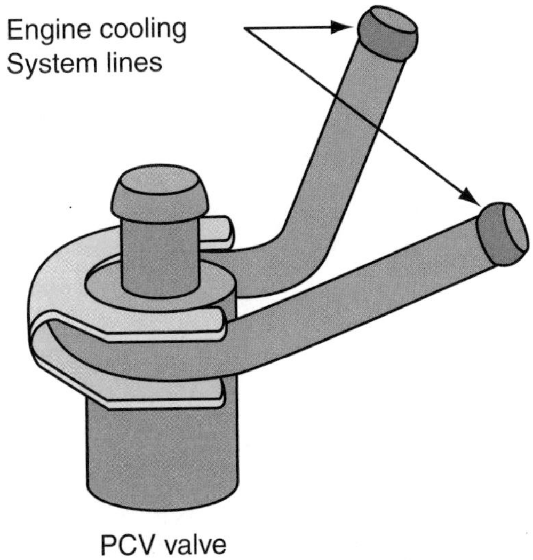

Figure 33-18 A coolant heated PCV valve.

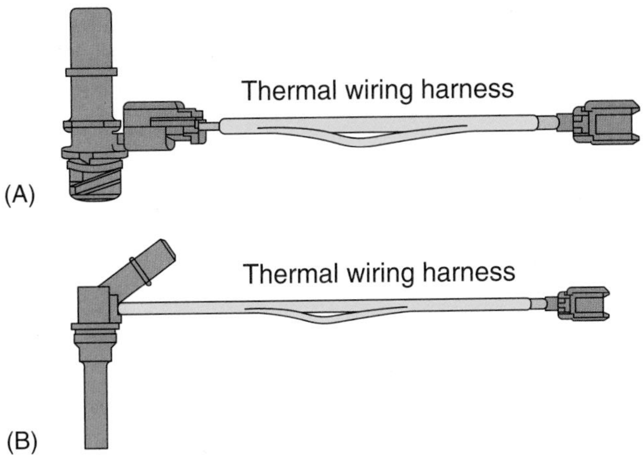

Figure 33-19 (A) An electrically heated PCV valve, and (B) an electrically heated PCV tube.

Electrically heated tubes or valves have a heating element as part of the valve, the connection between the valve and the PCV tube, or in the tube **(Figure 33–19)**. When the heating element is controlled by a thermistor, voltage is applied to the element when it is cold. Once warm, the resistance is so high that the voltage and amperage is too low to energize the element.

When the PCM controls the heating element, the heater is directly controlled by the PCM. The PCM uses IAT signals to determine when to energize the heating element.

PCV Monitor Vehicles that have OBD-II PCV monitoring capabilities use special PCV valves. These valves are designed so they create a total seal when installed. Most use a cam-lock thread design that requires one-quarter to one-half turn to lock them in place **(Figure 33–20)**. The locking mechanism is designed to eliminate the chance of the valve accidentally becoming loose in its grommet.

EGR Systems

Most vehicle manufacturers started to provide emission control systems that reduce NO_x as early as 1970. The EGR system releases a sample of exhaust gases

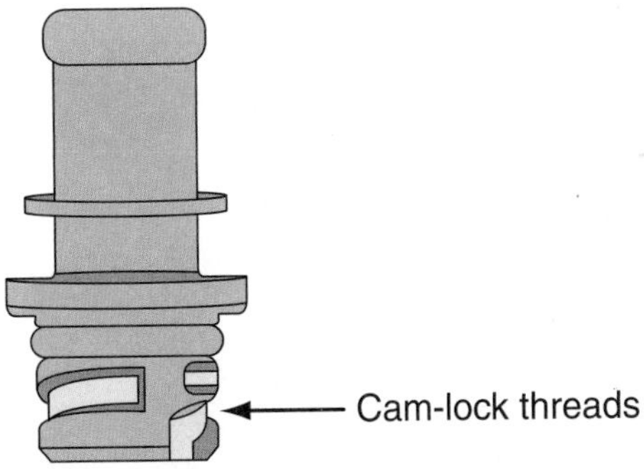

Figure 33-20 A PCV valve with cam-lock threads that prevent it from accidentally becoming loose.

into the intake's air-fuel mixture. This lowers the peak temperature of combustion and therefore reduces the chances of NO_x being formed. The recirculated exhaust gas dilutes the air-fuel mixture. Because exhaust gas does not burn, this lowers the combustion temperature and reduces NO_x emissions. At lower combustion temperatures, the nitrogen in the incoming air is simply carried out with the exhaust gases.

Driveability problems can result from having too much recirculated exhaust gas in the combustion chamber. This is especially true when the there is a high demand for engine power. Also, poor control of EGR flow can cause starting and idling problems. This is why EGR flow is disabled during cold starting, at idle, and at throttle openings of more than 50%. There is maximum EGR flow only when the vehicle is at a cruising speed with a very light load.

Many late-model engines do not have an EGR system. Rather, they rely on variable valve timing to prevent all of the exhaust gases from leaving the cylinder during the exhaust stroke during some operating conditions. The retention of the exhaust serves the same purpose as the EGR system. Other engines without an EGR system use other technologies to reduce combustion temperatures.

OBD-II systems monitor the EGR system to determine if the system is operating properly. These monitors use a variety of sensors and methods. If a fault is detected in any of the EGR monitor tests, a DTC is set. If the fault occurs during two drive cycles, the MIL is illuminated. The EGR monitor operates once per OBD-II trip.

EGR Valve Many older engines are equipped with a vacuum-operated EGR valve **(Figure 33–21)** to regulate the flow of exhaust gas into the intake manifold. Most late-model engines have an electrically controlled EGR valve. Typically, the EGR valve is mounted to the intake manifold.

Figure 33-21 An EGR valve.

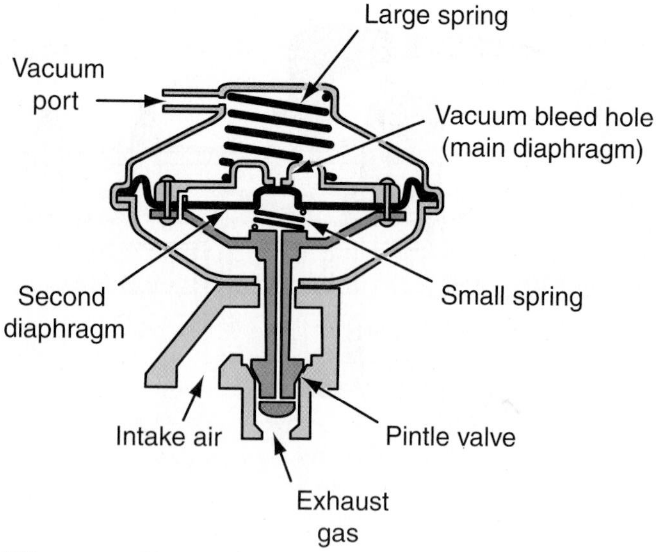

Figure 33-22 A typical design of a vacuum-controlled EGR valve.

Figure 33–22 illustrates the basic design of a vacuum-controlled EGR valve. The EGR valve is a flow control valve. A small exhaust crossover passage in the intake manifold admits exhaust gases to the inlet port of the EGR valve. Opening the EGR valve allows exhaust gases to flow through the valve **(Figure 33–23)**. Here the exhaust gas mixes with the intake air or air-fuel mixture in the intake manifold. This dilutes the mixture so combustion temperatures are minimized.

On some engines, the exhaust gas from the EGR system is distributed through passages in the cylinder heads and distribution plates to each intake port. The distribution plates are positioned between the cylinder

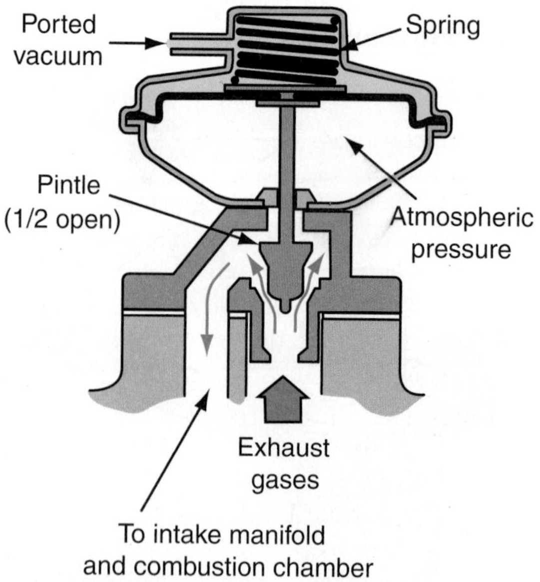

Figure 33-23 When the EGR is open a small amount of exhaust gas recirculates from the exhaust manifold to the intake manifold.

heads and the intake manifold. Because the exhaust gas from the EGR system is distributed equally to each cylinder, smoother engine operation results.

To regulate the amount of EGR flow, EGR valves have a fixed orifice or a tube with a narrow inside diameter. On some engines, the gasket for the EGR valve provides the orifice.

Vacuum EGR Valve Controls Vacuum is used to control the operation of EGR valves on older engines. Many different vacuum controls have been used. Ideally the EGR system should operate when the engine reaches operating temperature and/or when the engine is operating under conditions other than idle or wide-open throttle (WOT). The following are various controls that relate directly to vacuum-controlled EGR systems.

- The **thermal vacuum switch (TVS)** senses the air temperature. When the engine reaches operating temperature, the TVS opens to supply vacuum to the EGR valve.
- The **ported vacuum switch (PVS)** senses coolant temperature. The PVS cuts off vacuum when the engine is cold and allows vacuum to the EGR valve when the engine is warm.
- Some engines have an **EGR delay timer control** system, which prevents EGR operation for a predetermined amount of time after warm engine startup.
- Some applications have a WOT valve to cut off EGR flow at WOT.

Backpressure EGR Many engines have a **backpressure transducer** that modulates, or changes, the amount the EGR valve opens. It controls the amount of air bleed in the EGR vacuum line according to the level of exhaust gas pressure, which is dependent on engine speed and load. The backpressure transducer may be a separate unit or incorporated into the EGR valve. There are two basic types of backpressure EGR systems.

A positive backpressure EGR valve has a bleed port and valve positioned at the center of a diaphragm. A spring holds the bleed valve open. An exhaust passage connects the lower end of the valve through the stem to the bleed valve. When the engine is running, exhaust pressure is applied to the bleed valve. At low engine speeds, exhaust pressure is not high enough to close the bleed valve. Because the vacuum supplied to the diaphragm is bled off, the valve remains closed. As engine and vehicle speed increase, the exhaust pressure also increases. Eventually exhaust pressure closes the bleed port and vacuum lifts the diaphragm and opens the valve.

In a negative backpressure EGR valve, the bleed port is normally closed. An exhaust passage connects the lower end of the tapered valve through the stem to the bleed valve. When the engine is running at lower speeds, there is a high-pressure pulse in the exhaust system. However, between these high-pressure pulses, there are low-pressure pulses. As the engine speed increases, the high-pressure pulses become closer together. The negative exhaust pressure pulses decrease and the bleed valve closes and opens the EGR valve.

PCM-Controlled EGR Valves PCM-controlled EGR valves are typically vacuum or electrically operated. When vacuum operated, the system looks at the pressure drop across the metering orifice in the exhaust feed tube or the valve as the valve opens and closes. At the orifice is a differential pressure feedback EGR sensor that sends a signal to the PCM **(Figure 33–24)**.

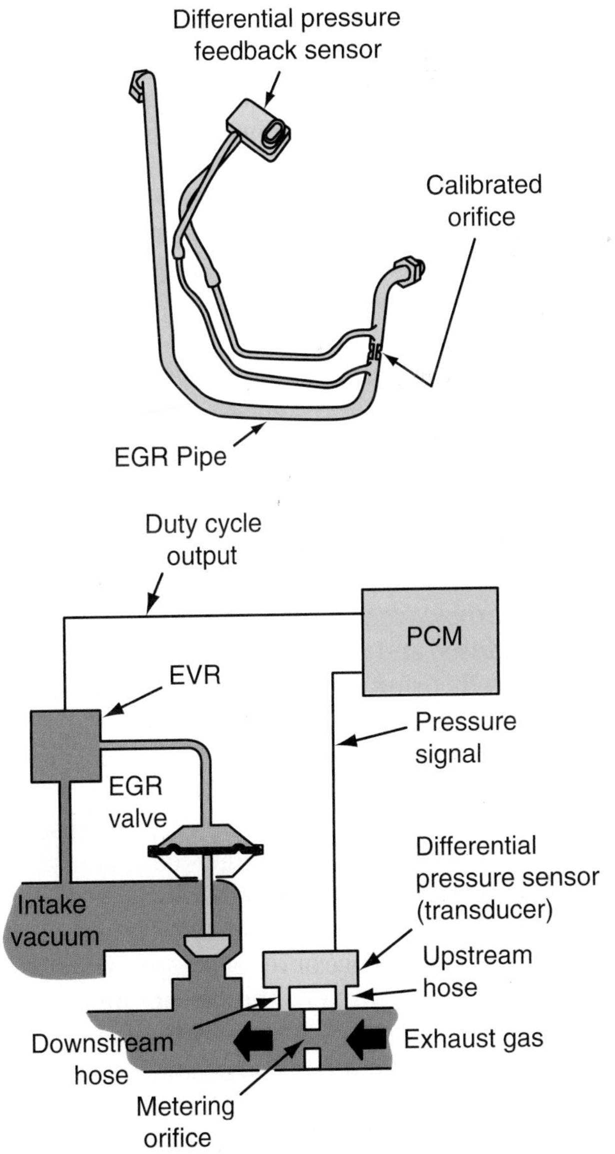

Figure 33-24 An EGR system with a DPFE sensor.

This is called the **differential pressure feedback EGR (DPFE)** system.

In this system, the PCM calculates the desired amount of EGR flow according to the current operating conditions. The PCM looks at the inputs from many sensors before determining this value. It then calculates the necessary pressure drop across the orifice to obtain this flow. Once the value is determined, the PCM sends commands to the EGR vacuum regulator solenoid. The solenoid is duty cycled by the PCM. As the duty cycle increases, more vacuum is sent to the valve and it remains open for longer periods.

As exhaust gases pass through the valve, they must also pass through the orifice. The DPFE sensor measures the pressure drop across the orifice and sends a feedback signal to the PCM. Based on this signal, the PCM can make corrections to the operation of the EGR valve. Normally the voltage signal from the DPFE sensor is 0 to 5 volts and the voltage is directly proportional to the pressure drop.

Some EGR valves have an exhaust gas temperature sensor. This sensor contains an NTC thermistor; an increase in exhaust temperature decreases the sensor's resistance. Two wires are connected from the temperature sensor to the PCM. The PCM senses the voltage drop across this sensor. Cool exhaust temperature and higher sensor resistance cause a high-voltage signal to the PCM, whereas hot exhaust temperature and low sensor resistance result in a low-voltage signal.

Electric Exhaust Gas Recirculation (EEGR) System The EEGR system allows for precise control of NO_x production without relying on engine vacuum. The EEGR system uses an electric motor in the EGR valve. Normally the EEGR valve is water or air cooled. The PCM controls the stepper motor that controls the position of the EGR valve's pintle valve. The position of the valve determines the rate of EGR flow. A spring keeps the valve closed and must be overcome by the force of the motor. By using an electric motor to control the valve, the system has no need for a vacuum diaphragm, vacuum regulator solenoid, orifice or orifice tube, or DPFE sensor.

The PCM receives signals from various sensors to determine the current operating conditions. The PCM then calculates the desired amount of EGR for those conditions. Then the PCM commands the motor to move (advance or retract) a specific number of discrete steps. Normally, the stepper motor has a fixed number of possible steps, each relating to the position of the pintle valve. The position of the pintle determines the EGR flow.

Other electric EGR valves rely on solenoids that control the amount of EGR flow. A **digital EGR valve** contains up to three electric solenoids operated directly by the PCM (**Figure 33–25**). Each solenoid contains a movable plunger with a tapered tip that seats in an orifice. When a solenoid is energized, its plunger is lifted and exhaust gas is allowed to recirculate through the orifice into the intake manifold. Each of the solenoids and orifices has a different size. The PCM can operate one, two, or three solenoids to supply the amount of exhaust recirculation required to provide optimum control of NO_x emissions.

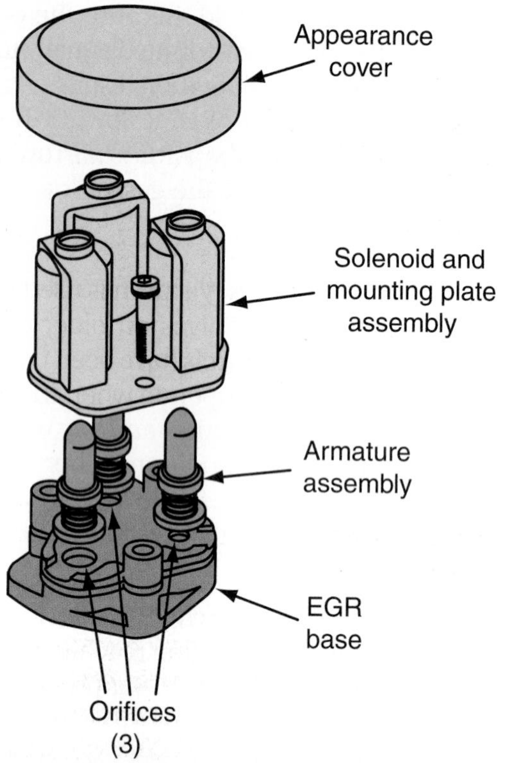

Figure 33-25 A digital EGR with three solenoids.

A **linear EGR valve** has a single solenoid or stepper motor operated by the PCM. A tapered pintle is positioned on the end of the solenoid's plunger. When the solenoid is energized, the plunger and tapered valve are lifted and exhaust gas is allowed to recirculate into the intake manifold (**Figure 33–26**). The EGR valve contains an EGR valve position sensor, which is a linear potentiometer. The signal from this sensor varies from approximately 1 V with the EGR valve closed to 4.5 V with the valve wide open. The PCM controls the EGR solenoid winding through pulse width modulation to provide accurate control of the plunger and EGR flow. A sensor sends feedback to the PCM to let it know that the commanded valve position was achieved.

Intake Heat Control Systems

HC and CO exhaust emissions are highest when the engine is cold. The introduction of warm combustion air improves the vaporization of the fuel in the

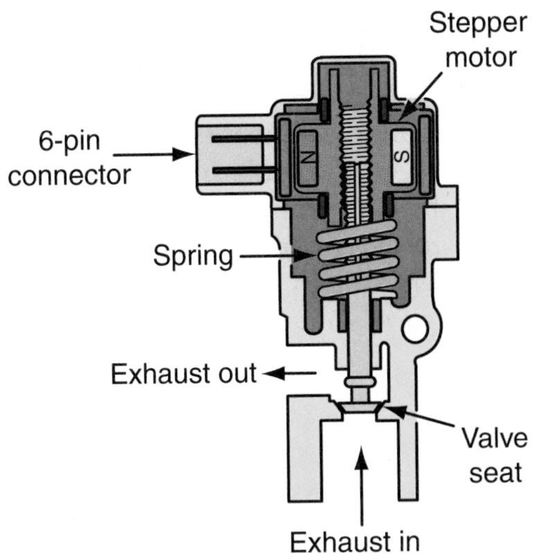

Figure 33-26 Basic construction of an electronically operated EGR valve with a stepper motor.

fuel injector throttle body or intake manifold. On older engines equipped with a throttle body injection (TBI) unit, the fuel is delivered above the throttles, and the intake manifold is filled with a mixture of air and gasoline vapor. Some intake manifold heating is required to prevent fuel condensation, especially when the intake manifold is cool or cold. Therefore, these engines have intake manifold heat control devices such as heated air inlet systems, manifold heat control valves, and **early fuel evaporation (EFE)** heaters.

A heated air inlet control may be used on engines with TBI. This system controls the temperature of the air on its way to the throttle body. Another system used is an exhaust manifold heat control valve that routes exhaust gases to warm the intake manifold when the engine is cold. This heats the air-fuel mixture in the intake manifold. These control valves can be either vacuum or thermostatically operated.

Some intake heat systems are computer controlled. These systems use an EFE heater. The EFE heater is a resistance grid that heats the mixture as it passes from the throttle body to the manifold. The engine coolant temperature sensor sends a signal to the PCM in relation to coolant temperature. At a preset temperature, the PCM grounds the mixture heater relay winding, which closes the relay's contacts. When the coolant temperature reaches a preset point, the PCM opens the ground circuit and the relay contacts open and shut off the current flow to the heater.

Port injection engines do not need heat risers or EFE heaters because the intake manifold delivers only air to the cylinders. Fuel is discharged into the intake ports near the intake valves or directly into the cylinders. Therefore, there is no need to warm the fuel.

POSTCOMBUSTION SYSTEMS

Postcombustion emission control devices clean up the exhaust after the fuel has been burned but before the gases exit the vehicle's tailpipe. An excellent example of this is the catalytic converter. A converter is one of the most effective emission control devices on a vehicle for reducing HC, CO, and NO_x.

Another post combustion system is the secondary air or air injection system. This system forces fresh air into the exhaust stream to cause a secondary combustion and reduce HC and CO emissions.

Catalytic Converters

One of the most important developments for lowering emission levels has been the availability and use of unleaded gasoline. Since 1971, engines have been designed to operate on unleaded fuels. Removing lead from gasoline eliminates lead particles in the exhaust. It also increases spark plug life, which is important for decreasing emissions. Also, the use of unleaded fuel avoids the formation of lead deposits in the combustion chambers that tend to increase HC emissions. Unleaded fuels also led to the use of catalytic converters, which provides a way to oxidize CO and HC emissions in an engine's exhaust.

Beginning with the 1975 model year, passenger cars and light trucks have been equipped with converters. A catalytic converter is positioned within the exhaust system and converts various emissions into less harmful gases. Today's catalytic converters are extrememly effective in reducing the amount of HC, CO, and NO_x emitted from a vehicle's tailpipe.

Most current vehicles have two converters in each exhaust stream. If the engine has more than one exhaust manifold, there is an exhaust stream from each and each of those streams has two converters. The first converter is located close to the exhaust manifold. Because the effectiveness of a converter depends on its temperature, placing a converter close to the manifold allows it to warm up quickly. The converters are also small, which helps them to heat up quickly. These converters are called light-up or warm-up converters, or precats (**Figure 33–27**). Their primary purpose is to reduce emissions while the main converters are warming up.

The main converter is located behind the precat. On some vehicles, it may be connected to two precats by a Y-pipe. Depending on the engine, a vehicle can have one or two main converters plus up to four precats.

The effectiveness of the converter is measured by the catalyst monitor of OBD-II systems. The monitor relies on heated oxygen sensors to measure the converter's effectiveness. The location and number of

Figure 33-27 A precat.

HO_2S found in an exhaust stream vary with vehicle design and the emission certification level (LEV, ULEV, PZEV, etc.) of the vehicle **(Figure 33–28)**. Most vehicles have two HO_2S in each exhaust stream. In each stream there is an HO_2S in the front exhaust pipe before the catalyst. The front sensors (HO_2S11/HO_2S21) are used for fuel control. An additional HO_2S is located after the catalyst and is used to monitor catalyst efficiency.

Many PZEVs have three HO_2S in each exhaust stream. The first HO_2S is located near the exhaust manifold and is used for fuel control. The second is in the center of the converter and monitors the amount of oxygen available in the converter. The third is after the converter and is used for long-term fuel trim control and for monitoring the effectiveness of the converter.

The converter is designed to respond to ever-changing exhaust quality. The amount and type of undesired gases change with operating conditions and driving modes **(Figure 33–29)**.

A catalytic converter is basically a housing shaped like a muffler that contains two or more ceramic elements coated with a catalyst. The catalysts are responsible for the chemical changes that occur in the converter. A catalyst is something that causes a chemical reaction without being part of the reaction. As the exhaust gases pass over the catalyst, most of the harmful gases are changed to harmless gases. Internally, the ceramic elements are designed to expose the exhaust gases to as much surface area as possible. Ceramic materials are coated with the catalyst material to minimize the amount of catalyst material necessary. Most catalyst materials are precious metals that are quite expensive.

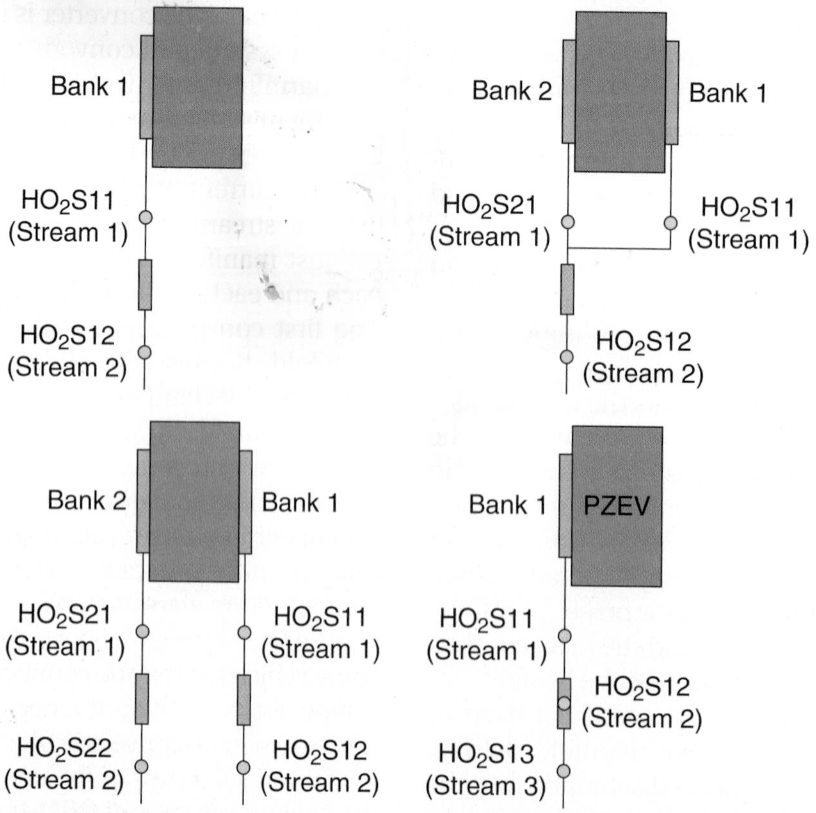

Figure 33-28 The location and number of HO_2S found in an exhaust stream varies with vehicle design and the emission certification level of the vehicle.

Emission	% of Exhaust During:			
	Idle	Acceleration	Cruise	Deceleration
CO	5.2%	5.2%	0.8%	4.2%
HC	0.08%	0.04%	0.03%	0.4%
NO_x	0.003%	0.3%	0.15%	0.006%

Legend:

HIGHEST	
LOWEST	

Figure 33-29 The approximate emission amounts during different driving modes.

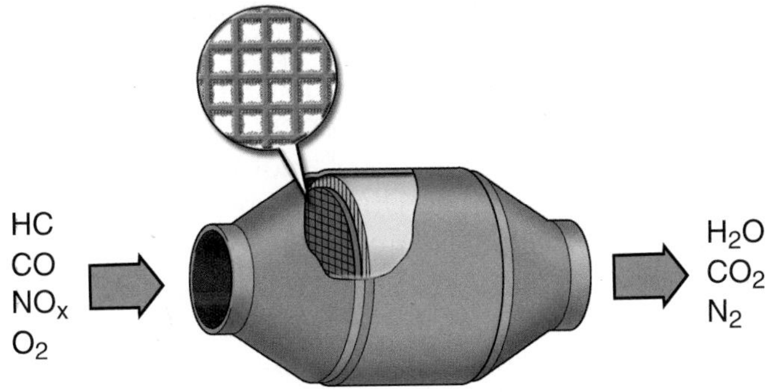

Figure 33-30 A honeycomb monolith-type catalytic converter. *Courtesy of Toyota Motor Sales, U.S.A., Inc.*

The catalyst coated ceramic elements have a **honeycomb monolith** design or are **ceramic beads**. Nearly all converters used on today's vehicles have the honeycomb structure **(Figure 33–30)**. Early converters were made with either design. The beads, or pellets, have a porous surface and are approximately ⅛ inch (3 mm) in diameter. In spite of their small size, their construction provides up to 4 sq. in. (10 sq. mm) of pore surface area. The exhaust gases move to this area and are changed by the catalyst materials in the pores.

The honeycomb monolith design looks like a honeycomb and each opening has 1 to 2,000 pores that are about 0.04 inch (1 mm) in size separated by thin walls. This allows an extremely large area for the gases to adhere and react to.

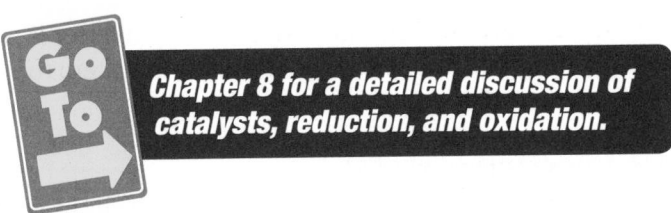

Go To *Chapter 8 for a detailed discussion of catalysts, reduction, and oxidation.*

Prior to OBD-II, catalytic converters contained two different types of catalysts: a **reduction catalyst** and an **oxidation catalyst**. The two separate catalysts created a dual-bed converter. Exhaust gases passed over the first or reduction bed where NO_x emissions were eliminated. Then the exhaust passed to the second where they were oxidized to eliminate CO and HC emissions.

During reduction, as NO_x gases pass over the catalyst, the N atoms are pulled from the NO_x molecules and combined with other N atoms to form N_2, which passes through the converter. The released O_2 atoms react with the CO in the exhaust stream and form CO_2 or pass through to the second bed. The result of NO_x reduction is pure N_2 plus O_2 or CO_2.

During the oxidation phase inside the converter, HC and CO molecules experience a second combustion. This occurs because of the presence of O_2 and the temperature of the converter. The result of this combustion or oxidation process is water vapor (H_2O) and CO_2.

All late-model vehicles use a **three-way converter (TWC)** that decreases HC, CO, and NO_x emissions **(Figure 33–31)**. The catalyst used in the reduction bed is either platinum or rhodium. When NO_x is exposed to hot rhodium (Rh), it breaks down into O_2 and N_2 molecules. Some of the free O_2 molecules combine with CO molecules, and the resultant gases are O_2, CO_2, and N_2, which move to the oxidation

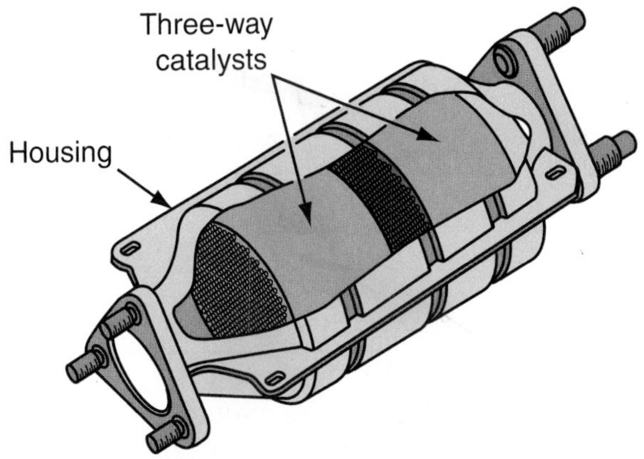

Figure 33-31 A typical three-way catalytic converter.

Emission	Process	Action	Result
NO_x	*REDUCTION*	2NO+2CO	N_2+$2CO_2$
HC	*OXIDATION*	HC+O_2	CO_2+H_2O
CO	*OXIDATION*	2CO+O_2	$2CO_2$

Figure 33-32 The action of a dual-bed TWC.

catalyst. The oxidation catalyst, normally platinum (Pt) and palladium (Pd), combines CO and HC with the O_2 released by the reduction catalyst or in the exhaust to form CO_2 and H_2O **(Figure 33–32)**.

The presence of O_2 is important to the reduction and oxidizing processes. Early TWCs relied on fresh air injected by the secondary air system between the two catalysts. This air intake was controlled by the secondary AIR system. Other converters had a layer of cerium in the center section of the converter. The element cerium has the ability to store O_2. Late-model converters rely on the O_2 content in the exhaust. The amount of O_2 in the exhaust bounces up and down as the PCM makes slight changes to the air-fuel mixture during closed-loop operation.

As the PCM adjusts the air-fuel ratio around the desired stoichiometric ratio, it constantly toggles the mixture between slightly lean to slightly rich. This action provides the necessary O_2 and CO for the TWC. High CO content is necessary for reducing NO_x emissions, whereas high O_2 is required for the oxidation of CO and HC. When the mixture is lightly rich, more CO and less O_2 are present in the exhaust. When the mixture is slightly lean, CO content decreases and O_2 content increases **(Figure 33–33)**. The converter stores some of the O_2 to allow for better oxidation during the rich cycle.

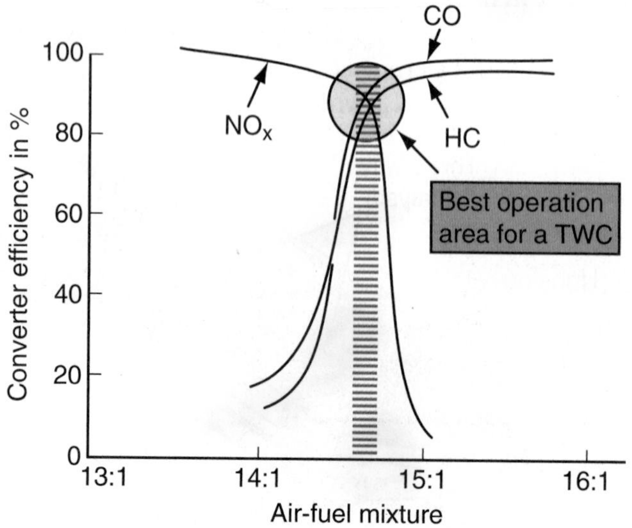

Figure 33-33 The efficiency of a catalytic converter is at its highest level when there is a stoichiometric mixture.

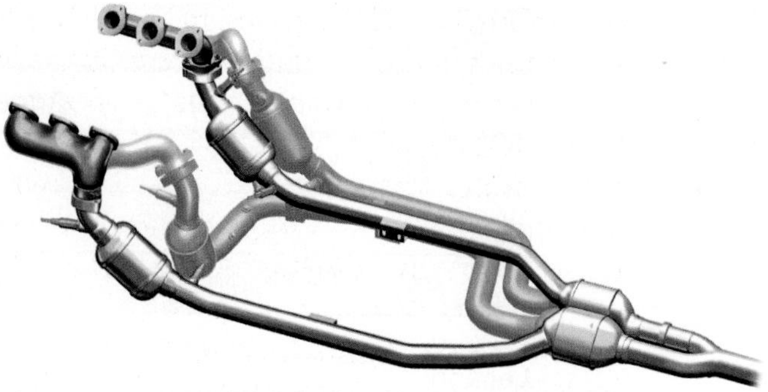

Figure 33-34 Two variations of the same exhaust system, each with a precat right after the exhaust manifold. *Courtesy of Chrysler LLC*

The efficiency of a catalytic converter is affected by its temperature. The temperature of the exhaust gases heat up the converter. The normal operating temperature for most converters is about 900°F (500°C). As the temperature of a converter increases, its efficiency increases. During converter warmup, the point at which the converter is operating at more than 50% efficiency is called catalyst light off. This normally occurs at 475°F to 575°F (246°C to 302°C). It takes a while for the converter to reach this temperature, especially when it is mounted away from the engine and under the vehicle. During this warm-up time, exhaust emissions levels are high. To provide cleaner exhaust after a cold start, precats are used **(Figure 33–34)**.

Other Converter Designs Engines that run on very lean mixtures produce high amounts of NO_x. To reduce these emissions, many of these vehicles have an additional catalytic converter, called a storage or adsorber converter. After the exhaust gases leave a three-way converter, they flow through a special NO_x storage converter. This converter is coated with barium and extracts the nitrogen oxides from the exhaust and stores them until its nitrogen oxide sensor senses that the storage converter is filled. At that time, the sensor sends a signal to the PCM and the system starts to deliver a richer air-fuel mixture. As this richer exhaust flows through the storage converter, it regenerates the converter and the nitrogen oxides are converted into harmless nitrogen. When the converter is free of nitrogen oxide, the sensor signals the system to run lean mixtures again.

Air Injection Systems

An **air injection reactor (AIR)** system was built into cars and light trucks sold in California in 1966 and used in all automobiles for several years. The AIR system reduced the amounts of HC and CO in the exhaust by injecting fresh air into the exhaust manifolds. The air caused combustion of the gases in the exhaust manifolds and pipes, thereby reducing the amount of the gases emitted from the tailpipe. O_2 in the air combines with the HC and CO to oxidize them and produce harmless water vapor and CO_2. The air was delivered by an air pump or through a pulse-air system that relied on exhaust pulses that created a vacuum and drew in outside air.

As manufactuirers gained more control over emissions through engine design and advanced emission control systems, the purpose of the AIR system changed. It became known as the secondary air injection system and was modified to allow catalytic converters to operate more efficeintly. The system injected air into the catalytic converter. This helped the converter oxidize and reduce the gases entering into it. AIR systems are not commonly used today. Improved combustion and better catalytic converters have eliminated their need.

A typical system with an air pump is shown in **Figure 33–35**. An **air pump (Figure 33–36)**, sometimes referred to as a smog pump, produces pressurized air that is sent to the exhaust manifold, or exhaust ports, and to the catalytic converter. The air pump is driven by a belt from the crankshaft or an electric motor. An **air control valve** (or **air-switching valve**) is a vacuum-operated valve that routes the air from the pump either to the exhaust manifold or to the catalytic converter. A thermal vacuum switch controls the vacuum to the air control valve. When the coolant is cold, it signals the valve to direct air to the exhaust manifold. When the engine is warm, the extra air in the manifold affects EGR operation, so the air control valve directs the air to the converter, where it aids the converter in oxidizing emissions.

Figure 33-36 An air injection pump.

An **air by-pass (AIRB) valve** (or **diverter [AIRD] valve**) located between the air pump and the air control valve diverts, or detours, air during deceleration. Excess air in an exhaust rich with fuel can produce a backfire or explosion in a muffler. A vacuum signal operates the by-pass valve during deceleration and the compressed air is diverted to the atmosphere. One-way check valves allow air into the exhaust but

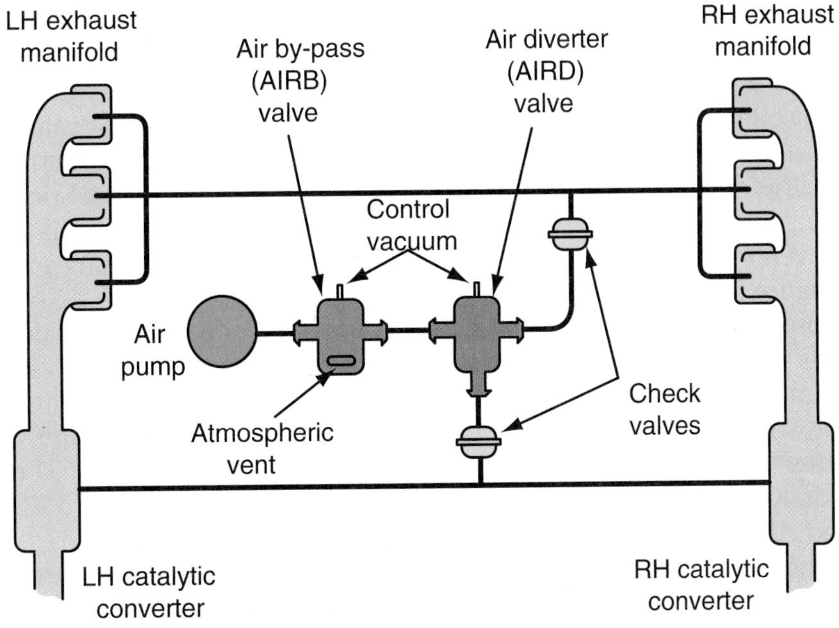

Figure 33-35 A typical pump-type air injection system.

prevent exhaust from entering the pump in the event the drive belt breaks. Their location in the system is behind the air control valve and before the exhaust manifold and catalytic converter.

Pulse (Nonpump) Systems Some early systems did not have an air pump; rather, they relied on the natural exhaust pressure pulses to pull air from the air cleaner into the exhaust manifolds and/or the catalytic converter. A manifold pipe is installed in the exhaust manifold for each cylinder of the engine. The inner end of these pipes is positioned near the exhaust port. The outer ends of the manifold pipes are connected to a metal container, and a one-way check valve is mounted between the outer end of each pipe and the metal container.

At lower engine speeds, each negative pressure pulse in the exhaust manifold moves air from the air cleaner through the metal container, check valve, and manifold pipe into an exhaust port. High-pressure pulses in the exhaust manifold close the one-way check valves and prevent exhaust from entering the system.

The incoming O_2 reduces the HC and CO content of the exhaust gases by continuing the combustion of unburned gases in the same way as systems with an air pump.

Electronic Secondary Air Systems The role and use of the secondary AIR system has decreased through the years. This is because there are fewer HC and CO emissions in the exhaust of a typical engine. When engines are fitted with a secondary AIR system, they are monitored by the OBD-II system and are solely used to supply O_2 to the catalytic converter. The air from the system not only helps clean up the emissions by causing combustion, it also serves to heat the converter so it can work more effectively.

The typical electronic secondary air system, like the conventional air injection system, consists of an air pump connected to a secondary air by-pass valve, which directs the air either to the atmosphere or to the catalytic converter.

The air pump is driven by an electric motor controlled by the PCM (**Figure 33–37**). Intake air passes through a centrifugal filter fan at the front of the pump where foreign materials are separated from the air by centrifugal force. In some systems, air flows from the pump to an AIRB valve, which directs the air either to the atmosphere or to the AIRD valve. The AIRD valve directs the air to the catalytic converter.

Both the AIRB and AIRD valves have solenoids that are controlled by the PCM. When either solenoid is energized, vacuum is applied to the AIRB valve and secondary air is vented to the atmosphere. When no vacuum is applied to the AIRD valve, secondary air (if present) is directed to the catalytic converter.

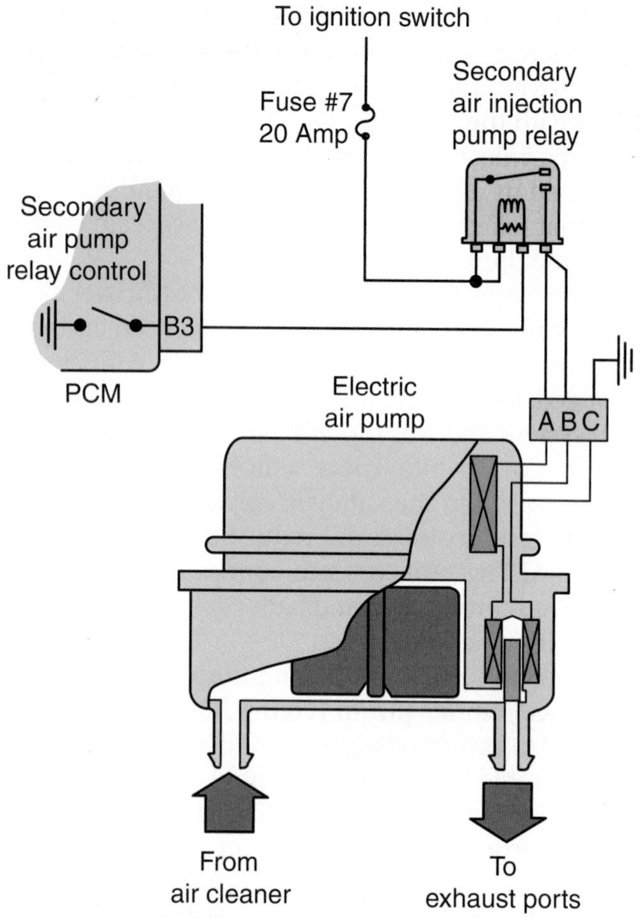

Figure 33–37 An electric air pump circuit.

There are two check valves in the secondary air system. Secondary air must flow through a check valve before it reaches the catalytic converter. These check valves prevent the backflow of exhaust gases into the pump in the event of an exhaust backfire or if the pump fails.

In the by-pass mode, vacuum is not applied either to the AIRB or the AIRD valve and secondary air is vented to the atmosphere. Secondary air may be vented or bypassed due to a fuel-rich condition or during deceleration. Secondary air is also typically bypassed during cold engine cranking and cold idle conditions.

Upstream Mode During the downstream mode, secondary air is routed through the AIRB valve and the AIRD valve to the exhaust manifold. The upstream mode is actuated when the PCM senses a warm crank/startup condition. The secondary airflow remains upstream for 1 to 3 minutes after startup to help control emissions.

The air-fuel mixture at startup is typically very rich. This rich mixture results in unburned HC and CO in the exhaust after combustion. By switching to the upstream mode, the hot HC and CO mix with the incoming secondary air and are burned up.

This reburning of HC and CO causes the exhaust gases to get hotter, which in turn heats up the O_2 sensor. This allows the electronic engine control system to switch to the closed-loop operation sooner because the O_2 sensor is ready to function sooner. It should be noted that the upstream mode increases the O_2 level in the exhaust gases. As a result, the voltage signal from the O_2 sensor is at a continuous low level. The computer interprets this signal as a continuous lean condition. This means the upstream mode results in inaccurate exhaust gas oxygen measurements. To solve this dilemma, the computer automatically switches to the open-loop fuel control whenever the upstream mode is activated. It ignores the O_2 sensor input.

Downstream Mode During the downstream mode, secondary air is routed through the AIRB valve and the AIRD valve to the catalytic converter. The fresh secondary air allows the converter to reduce the NO_x emissions. The secondary air system operates in this mode during most conditions. The PCM is in closed loop when the secondary air system is diverted downstream.

DIESEL EMISSION CONTROLS

In the early 1980s, diesel-powered cars were offered by some manufacturers in North America but that trend ended quickly because they were noisy, unreliable, and dirty. However, with technological advances and the availability of low sulfur diesel fuel, diesel engines in cars and light trucks will become more common soon. Diesel engines are capable of providing much better fuel economy than gasoline engines of the same size. They also provide high amounts of torque at low engine speeds.

Go To ***Chapter 9 for a basic description of how diesel engines operate.***

Today's diesel engines are very durable. This is because they tend to be overbuilt to withstand very high-compression pressures and the shock loading from the detonation of the air-fuel mixture. The downside is that diesel engines tend to be heavier than comparably sized gasoline engines. All reciprocating parts must be built stronger and, therefore, are heavier. Also, to produce the horsepower needed for an application, diesel engines must have a higher displacement than would be used in a gasoline engine. In spite of the increased size, diesel engines consume less fuel. To counter the increased weight and displacement, many diesels are fitted with turbochargers.

Diesel engines, especially those equipped with turbochargers, produce a substantial amount of torque at low engine speeds. One reason for this is the high-compression ratios required to ignite the fuel. Also, when the fuel is injected into the highly compressed air, combustion occurs more violently and lasts longer than in a spark ignition engine. This advantage, however, is what causes the knocking noise that many find annoying. Diesel engines also have a much longer stroke than a gasoline engine. The longer stroke and long burning time produce high-torque outputs. The turbocharger used on a diesel engine can provide up to 30 psi of boost, which is much more than a gasoline engine can withstand. This boost significantly increases the engine's output. Diesel engines also waste less heat, which means that more of the energy of the fuel is used to power the vehicle.

The use of cleaner fuels and new technologies, such as PCM-controlled fuel injection systems, allows today's diesel engines to have emission levels that match those of gasoline engines. Clean diesel exhaust is further possible because the vehicles can be fitted with EGR and PCV valves, catalytic converters, and particulate filters. Technology has also allowed diesel engines to run quietly.

Diesel engines obtained their reputation for being dirty because they emit soot or particulate matter (PM). PM is mostly unburned but solid HC molecules. The use of a particulate filter along with new fuel injection systems and redesigned combustion chambers have brought soot emissions under control. NO_x emissions are also high from a diesel engine and various systems are being used to control these emissions. However, diesel engines emit less HC, CO, and CO_2 than spark ignition gasoline engines.

Today's diesel engines are fitted with OBD systems similar to those found on gasoline engines. These systems monitor the effectiveness of various emission-related devices, such as the fuel system, EGR system, catalyst, NO_x adsorber/trap, PM trap, and PCV system. They also have misfire and comprehensive component monitors. **Figure 33–38** shows how the emission standards for diesel engines have changed through the years.

Low-Sulfur Fuel

Eighty percent of all diesel fuel available for sale must be ultra-low-sulfur fuel. Legislation has required fuel suppliers to remove nearly all sulfur from most of the diesel fuel they produce for

Year	CO_2	NO_x	HC	PM
1988	15.5	6.0	1.3	0.60
1991	15.5	5.0	1.3	0.25
1994	15.5	5.0	1.3	0.10
1998	15.5	4.0	1.3	0.10
2004	15.5	**2.4***	–	0.10
2007#	15.5	0.2	0.14	0.01

* This value is not just a NO_x standard; it is for nonmethane hydrocarbons (NMHC) + NO_x, which is why there is no standard for HC.

The PM standard went into full effect in the 2007 model year. The NO_x and HC standards have been phased in from that time and must be met by 2010.

Figure 33-38 Emission standards for diesel engines (given in grams per brake horsepower per hour).

on-highway use. Previously up to 500 ppm were allowed. The new limit is 15 ppm. The use of this fuel not only reduces sulfur-related emissions, but it also allows diesel engines to be equipped with typically sulfur-intolerant exhaust emission controls, such as particulate filters and NO_x catalysts. This means diesel exhaust now has the capability to be as clean as that from a gasoline engine.

Diesel fuel is a light fuel oil and is denser than gasoline. This means diesel fuel contains more energy per liter or gallon. Diesel fuel is also a better lubricant than gasoline; therefore, it has less effect on the desired oil film on piston rings and cylinder walls. This explains why diesel engines tend to be more durable than gasoline engines.

Diesel engines, especially older ones without electronic fuel injection systems, can run on a wide variety of other fuels. The most widely used alternate fuel is vegetable oil made from plants; these fuels are often referred to as biodiesel fuels. Normally, diesel engines can run biodiesel fuel without being modified.

Diesel Fuel Injection

Go To *Chapter 30 for a discussion and explanation of common fuel injection systems for diesel engines.*

Like gasoline engines, the overall efficiency of a diesel engine depends on fuel delivery. With diesel engines, the fuel injection system is definitely an emission control system. The timing of the start of injection is controlled by the PCM. When injection begins at precisely the correct time, emissions are minimized and fuel economy maximized. When injection begins too soon, high temperatures and pressures are formed in the cylinders. This causes an increase in NO_x emissions. When injection begins late, incomplete combustion will occur and HC, CO, and PM emissions increase.

The amount of fuel injected into a cylinder is controlled by the pulse width of the injectors and the pressure of the fuel delivered. The PCM calculates the desired amount of fuel based on inputs from a variety of sensors, including the CKP, CMP, BP, MAF, and accelerator pedal position (PPS) sensors.

Most current light-duty diesel engines use a common rail injection system. These systems use high pressure and this pressure is provided equally to all injectors. This allows the PCM to precisely control and monitor the amount of fuel injected into the cyclinders. The introduction of highly pressurized fuel into the cylinders provides improved atomization of the fuel. This, along with precise timing, makes it possible for the engines to run cleaner and quieter. Common rail systems use solenoid-operated or piezo-inline injectors **(Figure 33–39)**.

Figure 33-39 A solenoid-operated injector (left) and a piezo-inline injector (right); both are commonly used in diesel common rail systems. *Courtesy of Robert Bosch GmbH, www.bosch-presse.de*

The use of piezoelectric injectors increases the PCM's ability to control injection timing. New common rail designs have injectors that are activated by hundreds of thin piezo crystal wafers. Piezo crystals expand quickly when a current is applied to them. The ability of these injectors to quickly respond also allows the PCM to control several injector firings in a single combustion stroke. Some systems fire the injectors five times per stroke.

Many systems fire the injector three times. To reduce noise, the PCM fires the injectors to allow a small amount of fuel to enter the cylinder a few ten-thousandths of a second before firing the injector for combustion. This small spray of fuel begins the combustion process and is called the pilot injection. The pilot injection decreases the harshness of the combustion that occurs when the main injection takes place. A third injection of fuel takes place at the end of the combustion stroke to lower the temperature inside the cylinder.

Glow Plugs

Diesel engines can be very hard to start when they are cold. This is because the cold metal of the engine absorbs the heat generated by the compression of the intake air. This means the air is not hot enough to ignite the fuel. Most diesel engines use small electric heaters called glow plugs inside the cylinder to heat the intake air and help ignite fuel. Other engines have heaters built into the intake manifold to warm intake air until the engine reaches operating temperature. The PCM controls and monitors the operation of the glow plugs.

Typically, when the engine's coolant temperature is below 48°F (9°C), the glow plugs are energized. During this time, the driver is told to wait before attempting to the start the engine. Once the warning lamp turns off, the engine is ready to start. The glow plugs remain energized for a period of time to reduce emissions and improve driveability while the engine is warming up. Some engines have heaters in their coolant passages to allow the engine to warm up faster.

PCV System

Diesel engines emit crankcase gases just like gasoline engines. However, control of these gases is much different in diesel engines because they produce very little vacuum and, therefore, conventional PCV systems do not work. Vacuum is produced in a diesel engine in the opposite way as a gasoline engine—little to no vacuum at idle with an increase as engine speed increases. Therefore, a conventional PCV system cannot work on a diesel engine. Many diesel engines release these gases to the atmosphere through a crankcase breather or downdraft tube just like early gasoline engines. Both are a source of HC and PM emissions and are undesirable. Some systems have a PM filter that reduces those emissions but do not allow the engine to meet emission standards.

The industry has taken many different steps to control these emissions. One of these steps is the installation of a multistage filter system that is designed to collect, coalesce, and return the emitted crankcase oils to the oil sump. Another method is used on engines with a turbocharger. On these systems, intake air is drawn through an air filter and into the MAF. After the MAF, a hose connected from the valve cover draws in crankcase fumes into the intake air. Because of the low vacuum, the amount of air is very low so the system also relies on the vacuum produced by the intake for the turbocharger. The movement of air into the turbo creates a vacuum that draws the crankcase gases into the intake track.

Crankcase Depression Regulator (CDR) The **crankcase depression regulator (CDR)** valve is very similar to the PCV valve used in gasoline engines. It directs crankcase vapors back into the combustion chambers but is designed to work at very low levels of vacuum. It also maintains crankcase pressures to prevent oil consumption and oil leaks due to excessive buildup of pressure in the crankcase. A CDR contains a large silicone rubber/synthetic diaphragm and return spring. When vacuum is introduced to the valve, the valve opens against spring tension and allows crankcase gases to flow into the intake. CDR valves are used on both turbo and nonturbo engines but must be calibrated for the application.

EGR Systems

Normally, the efficiency of diesel engines results in high amounts of NO_x emissions. Therefore, EGR systems are integrated into the latest diesel engines. The EGR systems are quite the same as those used on gasoline engines, which means a sample of exhaust is introduced into the combustion chambers to reduce combustion temperatures. One of the main differences is that most manufacturers cool the incoming EGR gases before introducing them into the cylinders. This reduces the temperature of combustion and therefore reduces the amount of NO_x emitted by the exhaust (**Figure 33–40**). Most systems with EGR coolers use engine coolant that passes through a separate circuit to cool the recirculated exhaust gases.

The PCM operates and monitors the EGR system. EGR flow is controlled by the PCM through a digital EGR valve. EGR flow will only occur when the engine is at a predetermined level and other conditions are

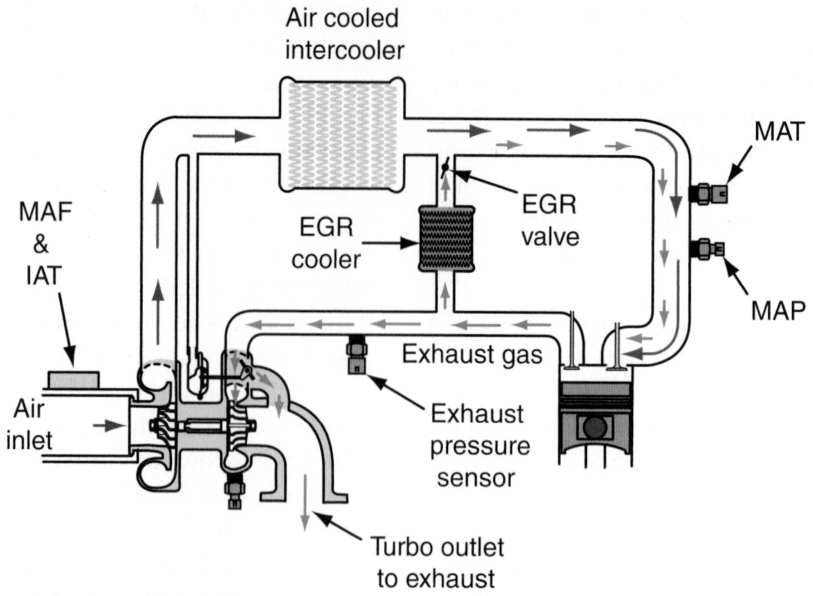

Figure 33-40 Intake airflow on a turbocharged diesel engine with an EGR cooler.

met. The PCM's EGR monitor consists of a series of electrical and functional tests that monitor the operation of the EGR system.

Catalytic Converters

The most common catalytic converter for diesel engines is an oxidation catalyst. This converter uses the O_2 in the exhaust to oxidize CO to form CO_2 and oxidize HC to form H_2O and CO_2. The oxidation catalyst is typically Pt. The converter also reduces the amount of soot. Some engines are fitted with diesel particulate filters that capture the remaining soot.

To control NO_x emissions, engines may also have a NO_x adsorber catalyst built into the oxidation converter or as a separate unit. This converter is comprised of an alkaline metal (typically barium) to store NO_x and a NO_x reduction catalyst (typically Rh).

Particulate Filter

Limiting the amount of PM or soot from the exhaust of a diesel engine is a top priority. Research has suggested that long-term exposure to diesel soot can cause cancer. Soot can be controlled by running the engine on a lean mixture and using a high-pressure common rail injection system. This brings soot emissions down significantly, as does the oxidizing catalytic converter. However, to meet EPA emission standards for PM emissions, late-model vehicles also have a **particulate filter**.

Particulate filters are placed into the exhaust system after the catalytic converter. Sometimes, the PM filter is part of the converter assembly **(Figure 33–41)**. Particulate filters are designed to trap PM. Early PM filters needed to be cleaned as part of a preventive maintenance program. Newer designs periodically

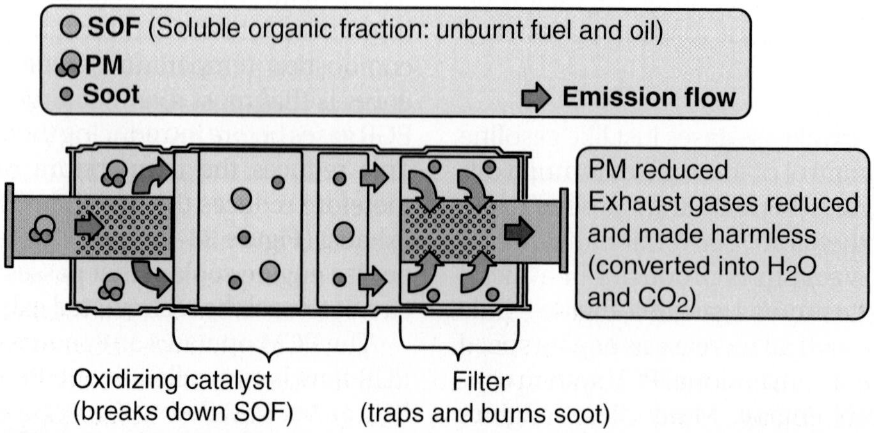

Figure 33-41 A continuous-regenerative diesel particulate catalyst assembly.

burn off the collected PM and are designed to last the life of the vehicle without any special maintenance. The use of low-sulfur fuel has allowed for the installation of these new filters. High-sulfur fuel creates large amounts of ash buildup in the PM filter. Even with low-sulfur fuels, ash from the fuel and the burn-off of the collected soot will accumulate in the filter.

Customer Care

Customers with late-model diesel-powered vehicles with particulate filters need to be made aware that the use of regular diesel fuel in place of low-sulfur diesel fuel will destroy the particulate filter. Regular diesel fuel has high-sulfur content and is intended to be used in farm and construction equipment only.

The filter is fitted with a sensor that measures exhaust backpressure. As ash builds up in the filter, exhaust backpressure increases. When the sensor informs the PCM that backpressure has reached a specified level, the PCM will order the fuel injection to spray an additional amount of fuel. The extra fuel will cause the oxidation catalytic converter to heat up and this heat will burn off the ash in the PM filter. The period of time the PCM adds fuel to clean the filter is called the regeneration cycle.

SHOP TALK

If the owner of a diesel-powered vehicle wants to install chrome tips on the tailpipes, there may not be the correct size available. The size and shape of the pipe is designed to draw cold air in to cool the exhaust while the particulate filter is in the self-cleaning mode. Also, make sure that the owner knows to keep the ends of the pipe clean. Any buildup of mud or other debris will block the intake of air and this can destroy the particulate filter.

Selective Catalytic Reduction (SCR) Systems

To reduce NO_x emissions from a diesel vehicle, two separate technologies are used: selective catalytic reduction (SCR) and NO_x traps or adsorbers. Conventional reduction catalytic converters do not work well on diesel engines. This is because the O_2 content in the exhaust is high and their converters operate at a low temperature. This means a reduction converter would not be very effective.

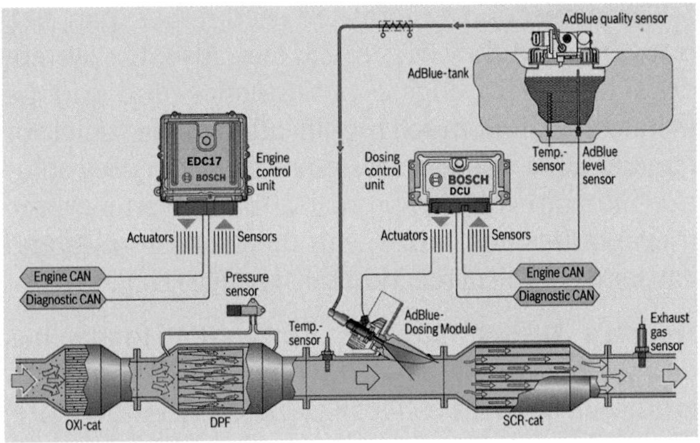

Figure 33-42 The basic setup for multiple catalytic converters (oxidation, PM, and SCR), along with the various electronic controls. *Courtesy of Robert Bosch GmbH, www .bosch-presse.de*

To control NO_x emissions, many of the new diesel vehicles will have **selective catalytic reduction (SCR)** systems. SCR is a process where a reductant is injected into the exhaust stream and then absorbed onto a catalyst **(Figure 33–42)**. The reductant removes O_2 from a substance and combines another substance with the O_2 to form another compound. In this case, O_2 is separated from the NO_x and is combined with hydrogen to form water.

The common reductants used in SCR systems are ammonia and urea water solutions. When the tanks are empty the emission levels will not be satisfactory. The reductant is injected in the exhaust stream over a catalyst. These special catalytic converters work well only when they are within a specific temperature range. The engine's control unit is programmed to keep the temperature of the exhaust within that range. Also, the amount of reductant sprayed into the exhaust must be proportioned to the amount of exhaust flow. This is also something controlled by the engine's control module.

The tanks that hold the reductant must be able to be refilled. This is a concern of the EPA and different ways of enforcing drivers to refill the reductant tank are being tried. Some systems have a warning lamp that informs the driver of a low tank; other vehicles have a warning lamp, and the PCM will put the vehicle into a limp-home mode when the tank is very low or empty. Some vehicles will not start if the tank is empty.

Urea The common reductant used is SCR systems is an ammonia-like substance called urea. **Urea** is an organic compound made of carbon, nitrogen, oxygen, and hydrogen. It is found in the urine of mammals and amphibians. Urea helps to eliminate more than 90% of the nitrogen oxides in the exhaust gases. Many European manufacturers call their urea injection systems the Blutec emissions treatment system.

Urea injection systems are used rather than NO_x traps because they cost much less. Also, the system does not affect engine performance and can be installed without much modification to the vehicle or engine. Urea injection systems squirt the urea solution into the exhaust pipe, before the converter, where it evaporates and mixes with the exhaust gases and causes a chemical reaction that reduces NO_x.

Honda's Diesel Catalytic Converter Honda has developed a special catalytic converter that produces ammonia from the exhaust gases. There is, therefore, no need to inject ammonia into the exhaust stream and there is no need to maintain a supply of it in tanks. The engine is fitted with a lean NO_x (LNT) catalyst, also referred to as the NO_x absorber or de NO_x catalyst. The converter has two layers. One layer absorbs NO_x from the exhaust gas and converts some of the NO_x into ammonia. The other layer absorbs the ammonia. The basic operation of the converter is controlled by the PCM. The PCM toggles the air-fuel ratio between rich and lean. When the mixture is lean, the exhaust has large amounts of O_2 and NO_x. When the mixture goes rich, it contains large amounts of hydrogen. The rate of the toggling (or cycling) of the mixture depends primarily on vehicle speed.

NO_x is absorbed by the Pt in the lower layer of the catalytic converter during normal (lean) operation.

Then periodically, for a few seconds, the PCM switches to a rich mixture to remove the stored NO_x so the catalyst can begin to trap NO_x again. When the system switches to a slightly rich mixture, there is more hydrogen in the exhaust. The LNT generates its own ammonia (NH_3) by causing a reaction between the stored NO_x and the hydrogen. The upper layer of the converter temporarily stores the ammonia until the system switches back to a lean mixture. At that time, the ammonia reacts with the NO_x in the exhaust gas and reduces it to N_2.

CASE STUDY

A customer complains of severe detonation. The technician checks all the EGR vacuum lines and finds them intact and properly routed.

The technician pushes the EGR valve stem against spring pressure; it moves freely and returns fully. With the engine at normal operating temperature, a helper opens the throttle enough to reach 2,500 rpm while the technician watches the EGR valve stem. When the throttle is moved, the valve stem does not retract. The technician removes the hose to the EGR valve and feels no vacuum when the engine is revved. Next, the technician pulls off the source vacuum line from the thermostatic vacuum switch and feels for vacuum. Because vacuum is present, but none is getting to the EGR valve with the engine warm, the technician replaces the thermostatic vacuum switch. The problem should be solved.

KEY TERMS

Air bypass (AIRB)
Air control valve
Air diverter (AIRD)
Air injection reactor (AIR)
Air-switching valve
Backpressure transducer
Canister purge valve
Carbon dioxide (CO_2)
Ceramic beads
Charcoal canister
Crankcase depression regulator (CDR)
Differential pressure feedback EGR (DPFE)
Digital EGR valve
Early fuel evaporation (EFE)
EGR delay timer control
Evaporative control (EVAP)
Greenhouse gas
Honeycomb monolith
Linear EGR valve
Oxidation catalyst
Particulate filter
Ported vacuum switch (PVS)
Postcombustion control
Precombustion control
Reduction catalyst
Selective catalytic reduction (SCR)
Thermal vacuum switch (TVS)
Three-way converter (TWC)
Urea
Vehicle emission control information (VECI)

SUMMARY

- Unburned hydrocarbons, carbon monoxide, and oxides of nitrogen are three types of emissions being controlled in gasoline engines. HC emissions are unburned gasoline released by the engine because of incomplete combustion. CO emissions are a by-product of combustion and are caused by a rich air-fuel ratio. Oxides of nitrogen (NO_x) are formed when combustion temperatures reach more than 2,300°F (1,261°C).
- CO_2 is a product of combustion and not a pollutant; however, it is considered a greenhouse gas.
- An evaporative (EVAP) emission system stores vapors from the fuel tank in a charcoal canister

until certain engine operating conditions are present. When those conditions are present, fuel vapors are purged from the charcoal canister into the intake manifold.

- Precombustion control systems prevent emissions from being created in the engine, either during or before the combustion cycle. Postcombustion control systems clean up exhaust gases after the fuel has been burned. The evaporative control system traps fuel vapors that would normally escape from the fuel system into the air.
- The PCV system removes blowby gases from the crankcase and recirculates them to the engine intake.
- With the engine running at idle speed, the high intake manifold vacuum moves the PCV valve toward the closed position.
- During part-throttle operation, the intake manifold vacuum decreases and the PCV valve spring moves the valve toward the open position. As the throttle approaches the wide-open position, intake manifold vacuum decreases and the spring moves the PCV valve further toward the open position.
- A digital EGR valve has up to three electric solenoids operated by the PCM.
- A linear EGR valve contains an electric solenoid that is operated by the PCM with a pulse width modulation (PWM) signal.
- A pressure feedback electronic (PFE) sensor sends a voltage signal to the PCM in relation to the exhaust pressure under the EGR valve.
- Early secondary air injection systems pumped air into the exhaust ports during engine warmup and delivered air to the catalytic converters with the engine at normal operating temperatures. Newer ones move air into the catalytic converter to help in the oxidation process.
- Today's vehicles have a three-way catalytic converter that reduces NO_x and oxidizes HC and CO.
- Modern diesel engines are equipped with typical emission controls but also may have an SCR system and/or a particulate filter.

REVIEW QUESTIONS

1. Explain why a small PCV valve opening is adequate at idle speed.
2. At what temperature do nitrogen atoms combine with oxygen atoms to form NO_x?
3. Describe the operation of a digital EGR valve.
4. Name the three types of emissions being controlled in gasoline engines.
5. The PCV system prevents ______ ______ from escaping to the atmosphere.
6. In a negative backpressure EGR valve, if the exhaust pressure passage in the stem is plugged, the bleed valve remains ______.
7. Describe how a selective catalytic reduction system works.
8. HC emissions may come from the tailpipe or ______ sources.
9. What types of catalyst are typically used to reduce NO_x and to oxidize HC and CO?
10. Why is a PCV system critical to an engine's durability?
11. Which of the following systems is designed to reduce NO_x and has little or no effect on overall engine performance?
    a. PCV
    b. EGR
    c. AIR
    d. EVAP
12. What is a catalyst?
13. Rather then rely on the AIR system for extra oxygen in a TWC, what do most late model vehicles do to support the oxidation process?
14. Which of the following statements about carbon dioxide is *not* true?
    a. CO_2 levels are highest when the air-fuel ratio is slightly leaner than stoichiometric.
    b. The production of CO_2 is directly related to the amount of fuel consumed; therefore, more fuel-efficient engines produce less CO_2.
    c. To reduce CO_2 levels in the exhaust, engineers are working hard to decrease fuel consumption to keep CO_2 low.
    d. The concern for CO_2 emissions is one of the primary reasons for the delayed use of alternative fuels for automobiles.
15. Why do the EGR systems on many late-model diesel vehicles include an EGR cooler?

ASE-STYLE REVIEW QUESTIONS

1. While discussing PCV valve operation: Technician A says that the PCV valve opening is decreased at part-throttle operation compared

to idle operation. Technician B says that the PCV valve opening is decreased at wide-open throttle compared to part throttle. Who is correct?

a. Technician A only	c. Both A and B
b. Technician B only	d. Neither A nor B

2. While discussing EGR systems with a pressure feedback electronic (PFE) sensor: Technician A says that the PFE sensor sends a signal to the PCM in relation to intake manifold pressure. Technician B says that the PCM corrects the EGR flow if the actual flow does not match the requested flow. Who is correct?

a. Technician A only	c. Both A and B
b. Technician B only	d. Neither A nor B

3. While discussing evaporative (EVAP) systems: Technician A says that the coolant temperature has to be above a preset value before the PCM will operate the canister purge solenoid. Technician B says that the vehicle speed has to be above a preset value before the PCM will operate the canister purge solenoid. Who is correct?

a. Technician A only	c. Both A and B
b. Technician B only	d. Neither A nor B

4. Technician A says that the EGR vent solenoid is normally open. Technician B says that the EGR control solenoid is normally open. Who is correct?

a. Technician A only	c. Both A and B
b. Technician B only	d. Neither A nor B

5. While discussing SCR systems: Technician A says that they are used in place of a conventional catalytic converter to reduce NO_x, CO, and HC. Technician B says that the most common reductant used is SCR systems is an ammonia-like substance called urea. Who is correct?

a. Technician A only	c. Both A and B
b. Technician B only	d. Neither A nor B

6. While discussing automotive emissions: Technician A says that oxygen emissions are monitored because they are an indication of the completeness of the combustion process. Technician B says that the conditions for minimizing HC emissions are the same as those that cause high NO_x emissions. Who is correct?

a. Technician A only	c. Both A and B
b. Technician B only	d. Neither A nor B

7. Technician A says that during oxidation, as NO_x gases pass over the catalyst, the N atoms are pulled from the NO_x molecules and are absorbed by the catalyst. The released oxygen atoms then flow through the converter to the second bed. The result of NO_x oxidation is pure nitrogen plus oxygen or carbon dioxide. Technician B says that during the reduction phase inside the converter, hydrocarbon and carbon monoxide molecules experience a second combustion. The result of this combustion or oxidation process is water vapor and carbon dioxide. Who is correct?

a. Technician A only	c. Both A and B
b. Technician B only	d. Neither A nor B

8. While discussing PCV systems without a PCV valve: Technician A says that blowby gases are routed into the intake manifold through a fixed orifice tube. The system works the same as if it had a valve, except that the system is regulated only by the vacuum at the orifice and the size of the orifice. Technician B says that systems without a PCV valve have other devices that eliminate the need for a PCV system. Who is correct?

a. Technician A only	c. Both A and B
b. Technician B only	d. Neither A nor B

9. While diagnosing the cause of high HC and O_2 emissions: Technician A says that the HC may be elevated because of an ignition problem. Technician B says that the O_2 may be elevated because of an ignition problem. Who is correct?

a. Technician A only	c. Both A and B
b. Technician B only	d. Neither A nor B

10. While discussing electric exhaust gas recirculation (EEGR) systems: Technician A says that some are air cooled. Technician B says that some are cooled by engine coolant. Who is correct?

a. Technician A only	c. Both A and B
b. Technician B only	d. Neither A nor B

EMISSION CONTROL DIAGNOSIS AND SERVICE

CHAPTER 34

OBJECTIVES

■ Use DTCs to initially diagnose emissions problems. ■ Briefly describe the emissions-related monitoring capabilities of an OBD-II system. ■ Describe the reasons why certain gases are formed during combustion. ■ Describe the inspection and replacement of PCV system parts. ■ Diagnose engine performance problems caused by improper EGR operation. ■ Diagnose and service the various types of EGR valves. ■ Diagnose EGR vacuum regulator (EVR) solenoids. ■ Diagnose and service the various intake heat control systems. ■ Check the efficiency of a catalytic converter. ■ Diagnose and service secondary air injection systems. ■ Diagnose and service evaporative (EVAP) systems.

The quality of an engine's exhaust depends on two things: One is the effectiveness of the emission control devices; the other is the efficiency of the engine. A totally efficient engine changes all of the energy in the fuel into heat energy. This heat energy is the power produced by the engine. To run, an engine must receive fuel, air, and heat.

In order for an engine to run efficiently, it must have fuel mixed with the correct amount of air. This mixture must be shocked by the correct amount of heat (spark) at the correct time. All of this must happen in a sealed container or cylinder. When these conditions are met, a great amount of heat energy is produced and the fuel and air combine to form water and carbon dioxide.

Because it is nearly impossible for an engine to receive the correct amounts of everything, a good running engine will emit some amounts of pollutants. It is the job of the emission control devices to clean them up. It is important to remember that the primary purpose of OBD systems is to reduce emissions; therefore, the circuits and controls in those systems must work properly in order to ensure low emissions.

The three emissions controlled in gasoline and diesel engines are unburned hydrocarbons (HC), carbon monoxide (CO), and oxides of nitrogen (NO_x). Federal laws require new cars and light trucks to meet specific emissions levels. State governments have also passed laws requiring that car owners maintain their vehicles so that the emissions remain below an acceptable level. Most states and provinces require an annual emissions inspection to meet that goal. The most common is the I/M 240 test. The results of this test can be a great tool for solving emissions and driveability problems.

I/M 240 TEST

An I/M 240 test checks the emission levels of a vehicle while it is operating under a variety of conditions and loads. This test gives a true look at the exhaust quality of a vehicle as it is working. The testing sequence normally begins with a leakage test **(Figure 34–1)** and a functional test **(Figure 34–2)** of the EVAP system and a complete visual inspection of the emission control system. Many states have plans to phase out I/M 240 testing and rely totally on OBD-II tests.

During the test, the vehicle is loaded to simulate a short drive on city streets then a longer drive on a highway. The complete test cycle includes acceleration, deceleration, and cruising. During the test, the vehicle's exhaust is collected by a CVS system that makes sure a constant volume of ambient air and exhaust pass through the exhaust analyzer. The CVS exhaust hose covers the entire exhaust pipe; therefore, it collects all of the exhaust. The hose contains a mixing tee that draws in outside air to maintain a constant volume to the gas analyzer. This is important for the calculation of mass exhaust emissions.

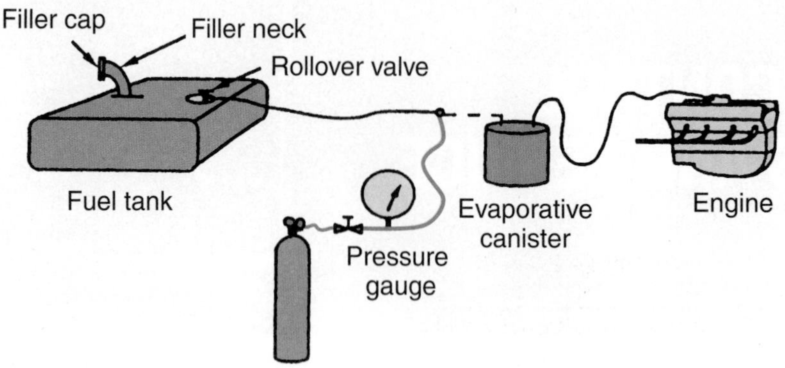

Figure 34-1 A typical setup for conducting a leakage test on an evaporative emission control system. *Courtesy of the U.S. EPA*

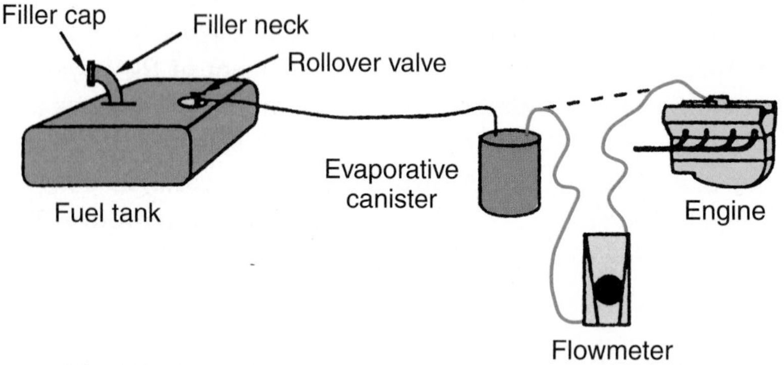

Figure 34-2 A typical setup for conducting a functional test of the evaporative emission control system. *Courtesy of the U.S. EPA*

The test is conducted on a chassis dynamometer **(Figure 34–3)**, which loads the drive wheels to simulate real-world conditions. Testing a vehicle while it is driven through various operating conditions on a chassis dynamometer is called transient testing, and testing a vehicle at a constant load on a dyno is called steady state testing.

At most I/M testing stations when a vehicle enters an inspection lane, a lane inspector greets the driver. The inspector receives the needed information from the driver, and then directs the driver to the customer waiting area. The inspector moves the vehicle to be visually inspected. Upon completion of the inspection, the inspector enters all pertinent information into the computer. After this information has been entered, the inspector moves the car onto the inertia simulation dynamometer.

Once on the dyno, the inspector raises the hood of the vehicle and moves a large cooling fan in front of the car. The engine is allowed to idle for at least 10 seconds, and then testing begins. The transient driving cycle is 240 seconds and includes periods of idle, cruise, and varying accelerations and decelerations. During the drive cycle, the car's exhaust is collected by the CVS system. The inspector driving the car must follow an electronic, visual depiction of the speed, time, acceleration, and load relationship of the transient driving cycle. For cars equipped with a manual transmission, the inspector is prompted by the computer to shift gears according to a predetermined schedule.

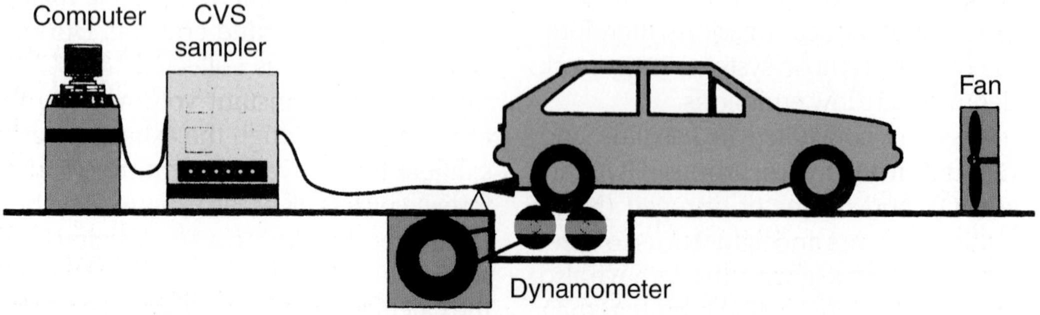

Figure 34-3 Components of a typical I/M 240 test station.

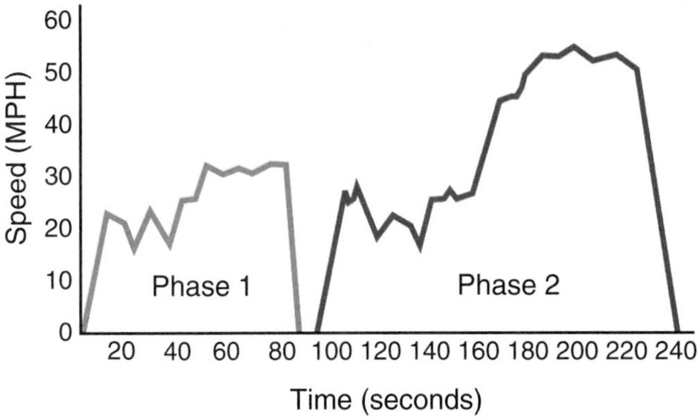

Figure 34–4 The drive trace for an I/M 240 test.

The I/M 240 drive cycle is displayed by a trace. The trace is based on road speed versus time. The test is comprised of two phases—Phase 1 and Phase 2—and combining the results of these two phases results in the test composite **(Figure 34–4)**. Phase 1 is the first 95 seconds of the drive cycle. The car travels 56 hundredths of a mile. This phase represents driving on a flat highway at a maximum speed of 32.4 miles per hour under light to moderate loads. Phase 2 is from the 96th second of the test to the end, or the 240th second. The vehicle travels 1.397 miles on a flat highway at a maximum speed of 56.7 miles per hour, including a hard acceleration to highway speeds.

The end result of the I/M 240 test is the measurement of the pollutants emitted by the vehicle during normal, on-the-road driving. After the test, the customer receives an inspection report. If the car failed the emissions test, it must be fixed.

The I/M 240 test measures HC, CO, NO_x, and CO_2. The job of a technician is to determine what loads and engine speeds the vehicle failed. The drive cycle is actually six operating modes. These modes need to be duplicated to identify why the vehicle failed and if the problem has been fixed:

Mode one—idle, no load at 0 miles per hour

Mode two—acceleration from 0 to 35 miles per hour

Mode three—acceleration from 35 to 55 miles per hour

Mode four—a steady cruise at 35 miles per hour

Mode five—a steady high cruise at 55 miles per hour

Mode six—decelerations from 35 miles per hour to 0 and from 55 miles per hour to 0.

Chassis Dynamometer

Chassis dynamometers are used during the I/M 240 test and can be valuable when diagnosing other driveability problems, including finding the cause of low power, overheating, and speedometer accuracy. A chassis dyno, or **dynamometer**, is designed to simulate the various road conditions in which a vehicle is driven. With a chassis dyno, a vehicle can be driven on rollers **(Figure 34–5)**. This allows the vehicle to be driven through the test conditions while it is stationary.

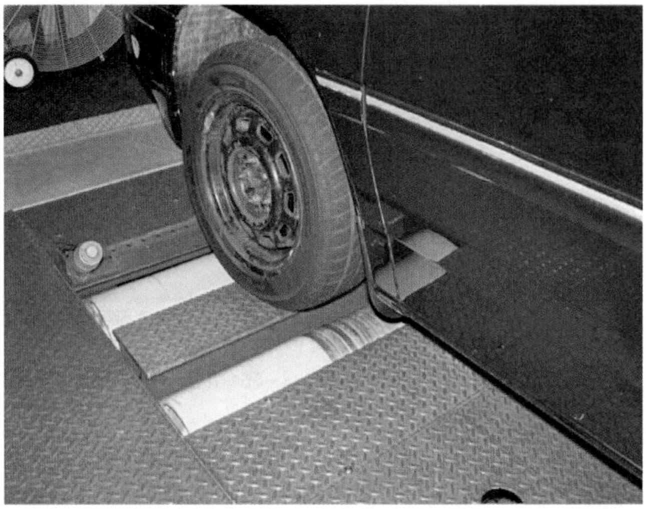

Figure 34–5 The wheels of this vehicle are ready to be dropped into the rollers of a chassis dyno.

Most chassis dynos are of the absorption type, which means they can simulate a variety of loads on the vehicle. This means the dyno is able to absorb the power exerted by the drive wheels and transfer that power to the air or water circulating through the dyno system. A dyno also has a program that measures torque and speed.

There are several types of absorption dynos available: waterbrake, eddy current, DC, and AC motors. A chassis dyno must be able to operate at any speed and load the vehicle to any level of torque that the test requires. After a vehicle is positioned on the dyno, basic information about the vehicle needs to be programmed into the dyno's computer. The computer will then adjust the resistance of the rollers accordingly to simulate various road conditions.

During I/M testing, a heated transfer hose is inserted into the vehicle's tailpipe to collect the exhaust gases and direct them into the constant volume sampler. Then the vehicle is driven at a variety of speeds and loads.

It is important to realize that OBD-II systems do not return all systems back to their normal state after repairs. The repaired vehicles should be run through a drive cycle to make sure that the repairs have been made correctly, and that all vehicle parameters are within acceptable limits.

It should be also noted that all emission testing programs will fail a vehicle if the check engine light is illuminated or if the readiness code is not at its normal or ready state.

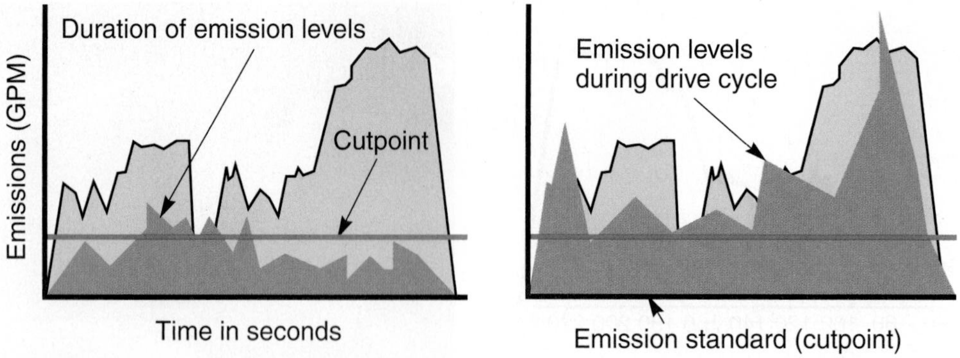

Figure 34-6 (Left) The emissions level is acceptable. (Right) The emissions level is not acceptable.

Other I/M Testing Programs

Many state I/M programs only measure the emissions from a vehicle while it is idling. The test is conducted with a certified exhaust gas analyzer. The measurements of the exhaust sample are then compared to standards dictated by the state according to the production year of the vehicle.

Some states also include a preconditioning mode at which the engine is run at a high idle (approximately 2,500 rpm), or the vehicle is run at 30 mph on a dynamometer for 20 to 30 seconds prior to taking the idle tests. This preconditioning mode heats up the catalytic converter, allowing it to work at its best. Some programs include the measurement of the exhaust gases during a low constant load on the dyno or during a constant high idle. These measurements are taken in addition to the idle tests. A visual inspection and functional test of the emission control devices is part of some I/M programs. If the vehicle has tampered, nonfunctional, or missing emission control devices, the vehicle will fail the inspection.

CARB developed a test that incorporates steady state and transient testing. This test is called the **acceleration simulation mode (ASM)** test. The ASM test includes a high-load steady state phase and a 90-second transient test. This test is an economical alternative to the I/M 240 test, which requires a dyno with a computer-controlled power absorption unit. The ASM can be conducted with a chassis dyno and a five-gas analyzer.

Another alternative to the I/M 240 test is the **repair grade (RG-240)** test. This program uses a chassis dynamometer, constant volume sampling, and a five-gas analyzer. It is very similar to the I/M 240 test, but it is more economical. The primary difference between the I/M 240 and the RG-240 is the chassis dyno. The dyno for the RG-240 is less complicated but nearly matches the load simulation of the I/M 240 dyno.

Interpreting the Results of an I/M Test

The report from an I/M 240 test shows the amount of gases emitted during the different speeds and loads of the test. The report also gives the average output for each of the gases and shows the cutpoint for the various gases. The **cutpoint** is the maximum allowable amount of each gas. Nearly all vehicles will have some speeds and conditions where the emissions levels are above the cutpoint. A vehicle that fails the test will have many areas of high pollutant output.

To use the report as a diagnostic tool, pay attention to all of the gases and to the loads and speeds at which the vehicle went over the cutpoint **(Figure 34–6)**. Think about what system or systems are responding to the load or speed. Pay attention not only to the gases that are above the cutpoint, but also to those that are below. Also, think about the relationship of the gases at specific speeds and loads. For example, if the HC readings are above the cutpoint at a particular speed and the NO_x readings are slightly below the cutpoint at the same speed, fixing the HC problem will probably cause the NO_x to increase above the cutpoint. As combustion is improved, the chances of forming NO_x also increases. Therefore, it wise to consider the possible causes of the almost too high NO_x, in addition to the high HC. Consider all of the correlations between the gases when diagnosing a problem.

TESTING EMISSIONS

Testing the quality of the exhaust is both a procedure for testing emission levels and a diagnostic routine. An exhaust analyzer is one of the valuable diagnostic tools for driveability problems **(Figure 34–7)**. By looking at the quality of an engine's exhaust, a technician is able to look at the effects of the combustion process. Any defect can cause a change in exhaust quality. The amount and type of change serves as the basis of diagnostic work.

Exhaust Analyzer

Early emission analyzers measured the amount of HCs and COs in the exhaust. HCs in the exhaust are raw, unburned fuel. HC emissions indicate that

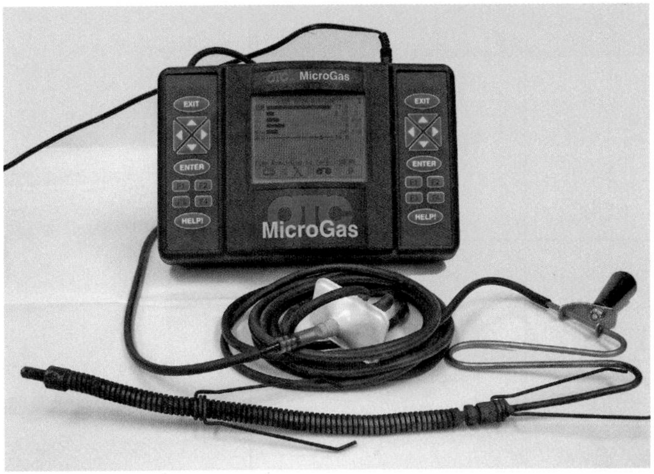

Figure 34–7 A hand-held exhaust gas analyzer.

complete combustion is not occurring in the engine. Emissions analyzers measure HC in **parts per million (ppm)** or grams per mile (g/mi). CO is an odorless, toxic gas that is the product of combustion and is typically caused by a lack of air or excessive fuel. CO is typically measured as a percent of the total exhaust.

SHOP TALK

The exhaust analyzers used in I/M 240 and similar tests do not measure the emissions levels in ppm or a percentage. Rather, they measure the gases in grams per mile (g/mi). This is because an I/M test is dynamic; during the test the vehicle is operated at a variety of speeds and loads. There is no direct equivalency between the two measuring systems.

An exhaust analyzer has a long sample hose with a probe in the end of the hose. Before measuring the exhaust gases, disable the air injection system, if the engine is equipped with one. Start the engine. Once the engine is warmed up, insert the probe in the tailpipe **(Figure 34–8)**. When the analyzer is turned on, an internal pump moves an exhaust sample from the tailpipe through the sample hose and the analyzer. A water trap and filter in the hose removes moisture and carbon particles.

The pump forces the exhaust sample through a sample cell in the analyzer. The exhaust sample is then vented to the atmosphere. In the sample cell, the analyzer determines the quantities of HC and CO, if the analyzer is a two-gas analyzer, or HC, CO, carbon dioxide (CO_2), and oxygen (O_2), if it is a four-gas analyzer. Most newer analyzers also measure NO_x and are called five-gas analyzers **(Figure 34–9)**.

Figure 34–8 The analyzer's probe is inserted into the vehicle's tailpipe.

Figure 34–9 A five-gas exhaust analyzer.

By measuring NO_x, CO_2, and O_2, in addition to HC and CO, a technician gets a better look at the efficiency of the engine **(Figure 34–10)**. Ideally, the combustion process will combine fuel (HC) and O_2 to form water and CO_2. Although most of the air brought into the engine is nitrogen, this gas should not become part of the combustion process and should pass out of the exhaust as nitrogen.

Maximum limits for the measured gases are set by regulations according to the model year of the vehicle. These limits also vary by state or locale. It is always desirable to have low amounts of four of

	IDLE	2500 RPM	PROBABLE CAUSE
HC ppm CO% CO_2% O_2% NO_x ppm	0–150 1–15 10–12 0.5–2.0 100–300	0–75 0–0.8 11–13 0.5–1.25 200–1,000	Normal reading
HC ppm CO% CO_2% O_2% NO_x ppm	0–150 3.0+ 8–10 0–0.5 0–200	0–75 3.0+ 9–11 0–0.5 100–500	Rich mixture
HC ppm CO% CO_2% O_2% NO_x ppm	0–150 0–1.0 8–10 1.5–3.0 300–1,000	0–75 0–0,25 11 1.0–2.0 1,000+	Lean mixture
HC ppm CO% CO_2% O_2% NO_x ppm	50–850 0–0.3 5–9 4–9 300–1,000	50–750 0–0.3 6–10 2–7 1,000+	Lean misfire
HC ppm CO% CO_2% O_2% NO_x ppm	50–850 0.1–1.5 6–8 4–12 0–200	50–750 0–0.8 8–10 4–12 100–500	Misfire

Figure 34-10 The readings of the chemicals in the exhaust can lead the technician to the cause of a driveability problem.

the five measured gases at all engine speeds. The only gas that is desired in high percentages is CO_2. Normally, the desired amount of CO_2 is 13.4% or more. The other gases should be kept to the lowest levels possible. Typically, CO should be 0.5% or less, O_2 should be about 0.1–1%, HC should be about 100 ppm or less, and NO_x should be 200 ppm or less.

Many of the emission control devices that have been added to vehicles over the past 30 years have decreased the amount of HC and CO in the exhaust. This is especially true of catalytic converters. These devices alter the contents of the exhaust. Therefore, checking the HC and CO contents in the exhaust may not be a true indication of the operation of an engine. However, a look at the other measured gases, those not altered by the system, can give a good picture of the engine and its systems.

Interpreting the Results

When using the exhaust analyzer as a diagnostic tool, it is important to realize that the severity of the problem dictates how much higher than normal the readings will be.

The levels of HC and CO in the exhaust are a direct indication of engine performance. Unburned HCs are particles, usually vapors, of gasoline that have not been burned during combustion. They are present in the exhaust and in crankcase vapors. Of course, any raw gas that evaporates out of the fuel system is classified as HC. HCs in the exhaust are a sign of incomplete combustion; the better the combustion, the less HCs in the exhaust. Today's engines are designed to be efficient and therefore release lower amounts of HCs in the exhaust. However, it is only possible to have complete combustion in a laboratory; therefore, all engines will release some HCs. This is because there is always some fuel in the combustion chamber that the flame front cannot reach.

CO forms when there is not enough O_2 to combine with the carbon during combustion. CO is a by-product of combustion. If combustion does not take place, CO will be low. When the engine receives enough O_2 in the mixture, CO_2 is formed. Lower levels of CO are to be expected when HC readings are high. When misfiring occurs, there is less total combustion in the cylinders. Therefore, CO will decrease slightly.

High NO_x readings indicate high combustion temperatures. CO_2 is a desired element in the exhaust stream and is only present when there is combustion; thus the more CO_2 in the exhaust stream, the better.

O_2 is used to oxidize CO and HC into water and CO_2. Ideally, there should be very low amounts of O_2 in the exhaust. CO and O_2 are inversely related: When one goes up, the other goes down.

SHOP TALK

Because increases in CO tend to lower NO_x emissions, it is possible that NO_x readings will increase after you have corrected the cause of the high CO readings.

General Guidelines When using gas measurements to diagnose a driveability problem, keep the following mind:

- HC is the result of incomplete combustion.
- High CO indicates an excessively rich air-fuel mixture caused by a restriction in the air intake system or too much fuel being delivered to the cylinders.
- Low CO does not indicate a lean mixture; rather, as the mixture becomes more lean, HC increases.
- When CO goes up, O_2 goes down.
- O_2 is used by the catalytic converter and may be low.
- O_2 readings are higher on vehicles with properly operating air injection systems.
- If there is a high O_2 reading without a low CO reading, an air leak in the exhaust system may be indicated.
- As the mixture goes lean, O_2 levels increase.
- As the mixture goes rich, O_2 moves and stays low.
- If there is high HC and low CO in the exhaust because of a lean misfire, the amount of O_2 will be high.
- NO_x goes high when there is a lean mixture or when the engine gets hot.
- NO_x is highest when CO and HC are lowest.
- Minimal NO_x is produced while an engine idles.
- At stoichiometric (lambda—λ), HC and CO will be at their lowest levels and CO_2 will be at its highest.
- If there is an extremely rich air-fuel mixture, there will be high HC and CO and low NO_x and O_2 readings.
- A deficiency of O_2 reduces NO_x.

For specific causes of undesirable amounts of exhaust gases, see **Figure 34–11**.

Problems with emission control devices will not only cause an increase in emission levels but can cause driveability problems. A no-start or hard-to-start problem can be caused by the obvious things, plus a disconnected or damaged vacuum hose, a malfunction of the EVAP canister purge, or a stuck open EGR valve. If the engine runs rough or stalls, the problem can be a problem with the EVAP canister purge system, a disconnected or damaged vacuum hose, a stuck open EGR valve, or a faulty PCV system. Excessive oil consumption can be caused by bad piston rings, poor valve oil seals, or a plugged PCV hose. Also, poor fuel economy can be caused by a number of things, including a problem with the EGR system.

BASIC INSPECTION

Before diagnosing the cause of excessive emissions, make sure you inspect the entire vehicle for obvious problems. The results of the inspection may help pinpoint the cause of the problem.

- Check the battery's voltage.
- Inspect/clean the air filter for dirt or contamination.
- Disconnect the air inlet line hose from the charcoal canister. Then check that air can flow freely into the air inlet. If air cannot flow freely, repair or replace it.
- Check and adjust the idle speed.
- Make sure that base timing is set to specifications.
- Check the fuel pump pressure.
- Check the engine's mechanical condition.
- Check all accessible electrical connections and vacuum and air induction hoses and ducts.
- Check the condition of the PCM main grounds.
- Check all EVAP, EGR, and PCV hoses, connections, and seals for signs of leakage and damage.
- Inspect for unwanted fuel entering the intake manifold from the EVAP system, fuel pressure regulator diaphragm, or PCV system.
- Check the air tightness of the fuel tank and filler pipe.

Part of the inspection should also include a look at the MIL. If the MIL is lit, a DTC is stored, and this can lead to the cause of excessive emissions. Keep in mind that the primary purpose of an OBD system is to limit the amount of pollutants emitted by the engine. This is why retrieving DTCs from the control system is the best beginning for diagnosing emission problems. OBD-II systems include several system monitors that relate to exhaust emission.

OBD-II Monitors

The OBD-II system monitors the emission control systems and components that can affect tailpipe or

Condition	Possible Condition/Problem
Excessive HC emissions	Ignition system misfiring—a fouled spark plug or defective spark plug wire Incorrect ignition timing Excessively lean air/fuel ratio Dirty fuel injector Low cylinder compression Excessive EGR dilution Defective valves, guides, or lifters Vacuum leaks Defective system input sensor
Excessive CO emissions	Rich air-fuel mixtures Plugged PCV valve or hose Dirty air filter Leaking fuel injectors Higher-than-normal fuel pressures Ruptured diaphragm in the fuel pressure regulator Defective system input sensor
Excessive HC and CO emissions	Plugged PCV system Excessively rich air-fuel ratio AIR pump inoperative or disconnected Engine oil diluted with gasoline
Lower-than-normal O_2 emissions	Rich air-fuel mixtures Dirty air filter Faulty injectors Higher-than-normal fuel pressures Defective system input sensor Restricted PCV system Charcoal canister purging at idle and low speeds
Lower-than-normal CO_2 emissions	Leaking exhaust system Rich air-fuel mixture
Higher-than-normal O_2 emissions	An engine misfire Lean air-fuel mixtures Vacuum leaks Lower-than-specified fuel pressures Defective fuel injectors Defective system input sensor
Higher-than-normal NO_x emissions	An overheated engine Carbon deposits on the intake valves Lean air-fuel mixtures Vacuum leaks Overadvanced ignition timing Defective EGR system Ineffective reduction catalytic converter
Higher-than-normal NO_x and HC emissions	Poor flow through the radiator A partially closed thermostat Bad water pump An inactive EGR system

Figure 34-11 Possible conditions causing various emission levels.

Higher-than-normal HC, O_2, and NO_x emissions	A false signal from the O_2 sensor An out-of-calibration MAP sensor Plugged injector Low fuel pressure Vacuum leak Engine knock Overadvanced ignition timing
Higher-than-normal HC, CO, NO_x. emissions	Carbon buildup on top of the piston or cylinder walls A worn or slipping timing belt

Figure 34–11 *(Continued)*

evaporative emissions. Through this monitoring process, problems are detected before emissions are 1.5 times greater than applicable emission standards. Each monitor requires one or more diagnostic tests. The monitors run continuously or noncontinuously and only when certain enabling criteria have been met. If a system or component fails these tests, a DTC is stored and the MIL is illuminated within two driving cycles.

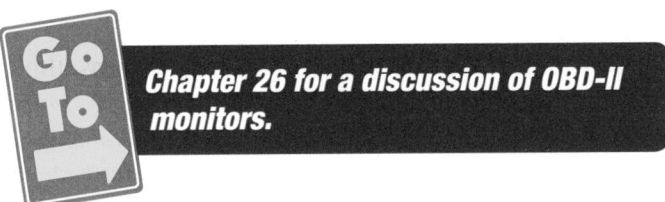

Go To ***Chapter 26 for a discussion of OBD-II monitors.***

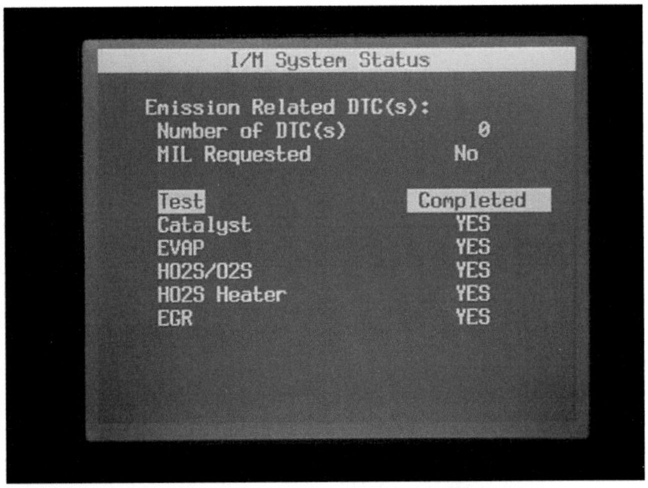

Figure 34–12 An I/M status screen on a scan tool showing the monitors have been completed.

A pending DTC is stored in the PCM when a concern is initially detected. Pending DTCs are displayed as long as the concern is present. Note that OBD regulations require a complete concern-free monitoring cycle to occur before erasing a pending DTC. This means that a pending DTC is erased on the next power-up after a concern-free monitoring cycle. However, if the concern is still present after two consecutive drive cycles, the MIL is illuminated. Once the MIL is illuminated, three consecutive drive cycles without a concern detected are required to turn off the MIL. The DTC is erased after forty engine warm-up cycles once the MIL is extinguished.

A part of many mandatory I/M inspections is taking a look at the OBD system tests. This is done by observing the system status display on a scan tool. If the diagnostic tests for a particular monitor have been completed, the scan tool will indicate this with a YES **(Figure 34–12)**. A system monitor is complete when either all of the tests comprising the monitor have been run and had satisfactory results, or when DTCs related to the monitor have turned on the MIL. If for any reason the tests are not completed, the system status display will indicate NO under the completed column. If the system has set a DTC that is associated with one of the regulated systems, the required tests may not be able to run. If there are no DTCs that would prevent the vehicle from completing a particular test, the vehicle should be operated according to the recommended drive cycle so that all enable criteria are met.

Not all vehicles have the same emission control systems. For example, a vehicle may not be equipped with an AIR or EGR. Therefore, the status of these monitors will not be shown. Also, the status of some monitors, such as the misfire and comprehensive component monitors, may not be listed. These monitors run continuously and, therefore, their status is not required by OBD-II.

Often, the procedures for servicing an emission-related system begin with the clearing of all DTCs. Doing this will also reset the status indicators for the monitors. Once repairs are completed, the vehicle should be operated according to the drive cycle recommended by the manufacturer. This will allow the monitors to complete their tests.

PROCEDURE

A typical procedure for allowing the I/M system status tests to complete follows:

STEP 1 Observe the DTCs with a scan tool.

If there is a DTC that would prevent the system status tests from completing, diagnose it before continuing.

STEP 2 Check the available service information to see if there are software updates for the vehicle.

STEP 3 Check the scan tool for the current status of the monitors. If the EVAP monitor has not been completed, diagnose and repair the system before continuing.

STEP 4 Turn off all accessories.

STEP 5 Open the hood, set the parking brake, and place the transmission in park for automatic transmissions or neutral for manual transmissions.

STEP 6 Start the engine and allow it to idle for about 2 minutes.

STEP 7 Close the hood, release the parking brake, and take the vehicle on a test drive.

STEP 8 Lightly accelerate to 45 to 50 mph (72 to 80 km/h) and maintain this speed for about 5 minutes after the engine reaches normal operating temperature.

STEP 9 Lightly accelerate to 55 mph (90 km/h) and maintain this speed for about 2 minutes.

STEP 10 Let off the throttle and allow the vehicle to decelerate for about 10 seconds.

STEP 11 Stop the vehicle and allow it to idle for about 2 minutes.

STEP 12 Turn off the ignition and allow the vehicle to sit undisturbed for about 45 minutes.

STEP 13 Check the status of the monitors with a scan tool.

STEP 14 If all tests were completed, retrieve all hard and pending DTCs.

STEP 15 If the tests were not completed, repeat the procedure.

EVAPORATIVE EMISSION CONTROL SYSTEM DIAGNOSIS AND SERVICE

WARNING!

If gasoline odor is present in or around a vehicle, check the EVAP system for cracked or disconnected hoses, and check the fuel system for leaks. Gasoline leaks or vapors can cause an explosion, resulting in personal injury and/or property damage. The cause of fuel leaks or fuel vapor leaks should be repaired immediately.

Most I/M inspections begin with a check of the EVAP system. This includes checking the results of the EVAP monitor's diagnostic tests. The EVAP system is monitored for leaks and its ability to move fuel vapor when commanded by the PCM to do so. Problems that are detected will set a DTC. Normally this is the only way a driver knows there is an EVAP problem. At times, the owner may complain of a gasoline odor in or around the vehicle. The common tools for troubleshooting an EVAP system are a diagnostic tester and pressure kit. The exact procedures and tools will vary with each application.

EVAP Monitors

Many different EVAP systems and monitoring systems have been used by the manufacturers. It is important to correctly identify the system used on a particular vehicle before doing any diagnostic procedures or service on the vehicle. A system leak can generate multiple DTCs depending on the component and location of the leak. The monitor also checks the electrical integrity of the system and the vapor purge rate **(Figure 34–13)**. All EVAP monitor DTCs require two trips.

To monitor the vapor purge in most systems, the PCM changes the duty cycle of the purge solenoid while monitoring the STFT. A problem is indicated by uncorrelated changes in the STFT. Other systems use a pressure sensor to monitor the pressure in the system and the purge side of the charcoal canister. Normally, the changes in the pressures between the two are very small. A difference in pressure should be evident when the PCM commands the purge solenoid to open. If there is no difference in pressure, the PCM determines that the purge system is not working properly and will set a DTC. Enhanced EVAP systems may also calculate purge flow rate by looking at the inputs from the MAF sensor or by comparing MAP sensor readings with the input from the tank pressure sensor.

Early EVAP systems were designed to detect leaks that were 0.040 inch (1 mm) and greater. These systems are often called nonintrusive systems. Starting with the 2000 model year, a new EVAP monitor system has been implemented. This system is designed to detect leaks of 0.020 inch (0.5 mm). These are called intrusive systems.

There are two different methods of leak detection used by enhanced ODB systems: system pressure and vacuum. Vacuum testing requires two solenoid valves and a fuel tank absolute-pressure sensor. The vacuum comes from the engine or an on-board vacuum pump (commonly referred to as the leak detection pump or LDP). Using the pressure method of checking the system tends to be more precise but entails the use of

EVAP Code	Description
P0441	Incorrect or uncommanded purge flow
P0442	Small leak
P0443	Purge solenoid electrical fault
P0446	Blocked cannister vent
P0449	Cannister vent solenoid electrical fault
P0452	Tank pressure sensor voltage low
P0453	Tank pressure sensor voltage high
P0454	Tank pressure sensor voltage noisy
P0455	Large leak
P0456	Very small leak
P0457	Gross leak
P0460	Fuel level circuit
P0461	Fuel level sensor stuck/noisy
P0462	Fuel level sensor voltage low
P0463	Fuel level sensor voltage high
P0464	Fuel level sensor voltage noisy at idle
P1443 (Ford)	Gross leak, no flow
P1450 (Ford)	Excessive vacuum
P1451 (Ford)	Vent valve circuit fault
P1486 (Chrysler)	Pinched hose
P1494 (Chrysler)	LDP fault
P1495 (Chrysler)	LDP solenoid circuit

Figure 34–13 Common EVAP codes.

more components. The method used by the manufacturers varies with vehicle made and model year.

Scan Tool Being a system controlled and monitored by the OBD-II, the MIL will illuminate to alert the driver that there is a problem if a failure is detected. When a defect occurs in the canister purge solenoid and related circuit, a DTC is normally set. If a DTC related to the EVAP system is set, always correct the cause of the code before doing any further diagnosis of the EVAP system.

When the scan tool is set to the appropriate mode, it will indicate whether the purge solenoid is on or off. With the engine idling, the purge solenoid should be off. Leave the scan tool connected and road test the vehicle. Be sure all the conditions required to energize the purge solenoid are present, and observe this solenoid status on the scan tool. The tester should indicate when the purge solenoid is on. If the purge solenoid is not on under the necessary conditions, check the power supply wire to the solenoid, solenoid winding, and wire from the solenoid to the PCM.

Diagnosis

If the EVAP system is purging vapors from the charcoal canister when the engine is idling or operating at very low speed, rough engine operation will occur, especially at higher ambient temperatures. Cracked hoses or a canister saturated with gasoline may allow gasoline vapors to escape to the atmosphere, resulting in gasoline odor in and around the vehicle.

The entire EVAP system should be carefully inspected (**Figure 34–14**). All of the hoses in the system should be checked for leaks, restrictions, and loose connections. Repair or replace them as necessary. Also check the canister for cracks or damage; if they are found, replace the charcoal canister assembly.

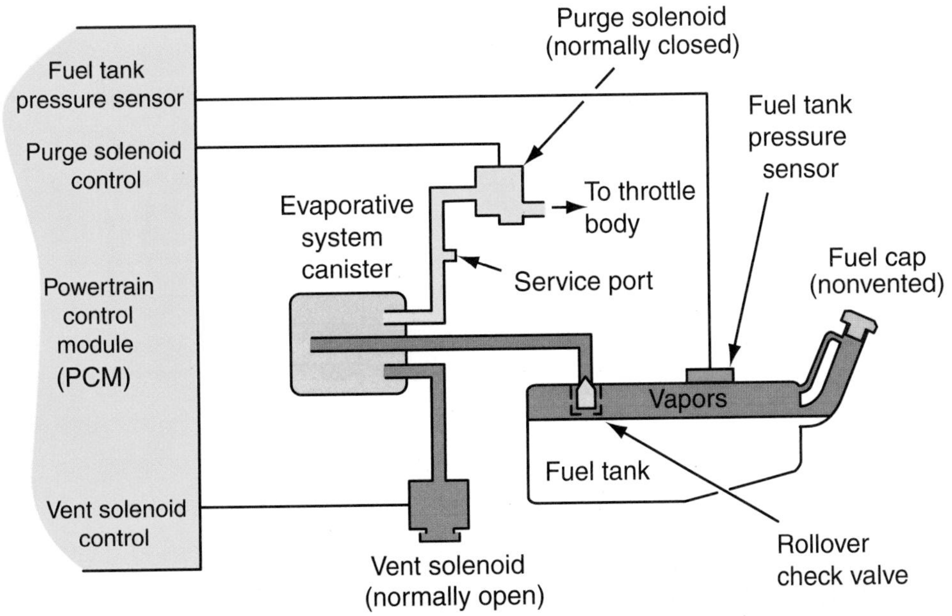

Figure 34–14 Check all hoses and electrical connectors and wires in the EVAP system.

Figure 34-15 The charcoal canister for a typical EVAP system.

The operation of the charcoal canister **(Figure 34–15)** can be quickly checked on most vehicles. Close the purge port, gently blow into its vent port, and check that air flows from the port. If this does not happen, replace the charcoal canister assembly. Close the vent port and gently blow into its air inlet port. check that air flows from the purge port. If this does not happen, replace the charcoal canister assembly. With the purge port and air inlet port closed, apply low air pressure (about 2.8 psi [19.6 kPa]) to the vent port. Make sure that the canister can hold that pressure for at least 1 minute. If this does not happen, replace the charcoal canister assembly.

Also make certain that the canister filter is not completely saturated. Remember that a saturated charcoal filter can cause symptoms that can be mistaken for fuel system problems. Rough idle, flooding, and other conditions can indicate a canister problem. A canister filled with liquid or water causes backpressure in the fuel tank. It can also cause richness and flooding symptoms during purge or startup.

To test for saturation, unplug the canister momentarily during a diagnosis procedure and observe the engine's operation. If the canister is saturated, either it or the filter must be replaced depending on its design; that is, some models have a replaceable filter, whereas others do not.

A vacuum leak in any of the evaporative emission components or hoses can cause starting and performance problems, as can any engine vacuum leak. It can also cause complaints of fuel odor. Incorrect connection of the components can cause rich stumble or lack of purging (resulting in fuel odor).

The canister purge solenoid **(Figure 34–16)** winding may be checked with an ohmmeter. With the tank pressure control valve removed, try to blow air through the valve with your mouth from the tank side of the valve. Some restriction to airflow should be felt until the air pressure opens the valve. Connect a vacuum hand pump to the vacuum fitting on the valve and apply 10 in. Hg to the valve. Now try to blow air through the valve from the tank side. Under this condition, there should be no restriction to airflow. If the tank pressure control valve does not operate properly, replace the valve.

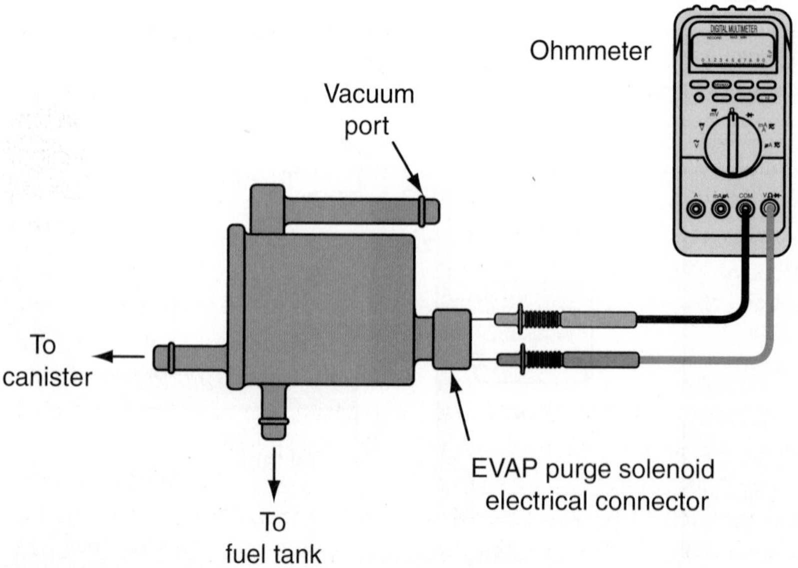

Figure 34-16 Checking an EVAP purge solenoid.

The electrical connections in the EVAP system should be checked for looseness, corroded terminals, and worn insulation. Measure the voltage drop across the solenoid and compare your readings to specifications. If the readings are outside specifications, replace the charcoal canister assembly.

SHOP TALK

It is recommended to never wiggle hoses or tighten fittings and caps during a visual inspection until the system has been pressurized for further testing.

Purge Test

Certain conditions will increase the amount of fuel vapors stored in the charcoal canister and therefore should be avoided prior to I/M tests. These include prolonged periods of idling or prolonged periods of sitting in the sun on a hot day. Most vehicles should be driven at highway speeds for about 5 minutes to allow them to complete their normal purge cycle before I/M testing. Excessive fuel vapors can cause an increase in CO during testing. To determine if the canister is the cause of high CO, isolate the EVAP system and retest the emission levels. The EVAP can be isolated by disconnecting the purge hose from the throttle body or intake manifold. If the EVAP system appears to be the cause of high emissions, the system needs to be carefully checked.

Before checking the purge rate of the EVAP system, allow the engine to run until it reaches normal operating temperature. Connect the purge flow tester's flow in series with the engine and evaporative canister. Zero the gauge of the tester with the engine off, and then start the engine. With the engine at idle, turn on the tester and record the purge flow rate and accumulated purge volume. Gradually increase engine speed to about 2,500 rpm and record the purge flow. A good working system will have at least 1 liter of flow within a few seconds. Most vehicles will have a flow of 25 liters during the time period required for an I/M 240 test. If the system does not establish at least 1 liter of flow during the 240 seconds, the system needs to be carefully diagnosed, repaired, and retested.

Leak Tests

The operation of a pressure test is often called a pressure decay test. It checks for leaks that allow fuel vapors to escape into the atmosphere. EVAP leak testers pressurize the system with a very low pressure (14 inches of water or about 0.5 psi). The tester typically uses pressurized nitrogen to fill the system.

Figure 34-17 An EVAP service port.

Nearly all OEMs recommend nitrogen rather than compressed air. In fact many are concerned about introducing oxygen (outside air has much oxygen in it) into the system because it can cause an explosion.

Before using an EVAP leak tester, make sure that the fuel tank is at least half full. The tester kit will include a pressure guideline chart. The chart contains various size fuel tanks and different fuel levels. Interpreting the chart will let you know how much pressure should be applied to the system. The tester is normally connected to the EVAP service port **(Figure 34–17)**. If the vehicle does not have the service port, use the correct adapter for the vehicle's filler neck. Remove the fuel cap and inspect the filler neck. If the neck is rusted or damaged, this could be a source of leakage. Connect and tighten the adapter to the filler neck. Then connect the tester to the adapter. Remember to check the gas cap because it will be bypassed when leak testing is done at the filler neck.

Before an OBD-II system can be tested with pressurized N_2, a scan tool must be used to set the system for an EVAP test. Doing this will close the canister's vent. Ground the tester **(Figure 34–18)** to the vehicle before turning it on, which is typically done by connecting a jumper wire from the ground screw on the tester to a bolt at the filler neck. Doing this eliminates the chance of **electrostatic discharge (ESD)**. ESD can ignite the fuel vapors. Then pinch off the vent hose at the carbon canister on non-OBD-II systems.

Adjust the output pressure from the tester so it agrees with the chart mentioned previously. Once the

Figure 34-18 An EVAP leak tester. *Reproduced under license from Snap-on Incorporated. All of the marks are marks of their owners.*

tester applies pressure to the system, its gauge will show how much pressure the system is able to hold **(Figure 34–19)**. The pressure will initially fall and rise before it stabilizes. If the system remains above 8 inches of water after 2 minutes, the vehicle has passed the test. Obviously if there is a leak, the system will not be able to hold pressure. If the pressure drops dramatically, listen for leaks at the fuel cap, tank, connections, valves, and hoses.

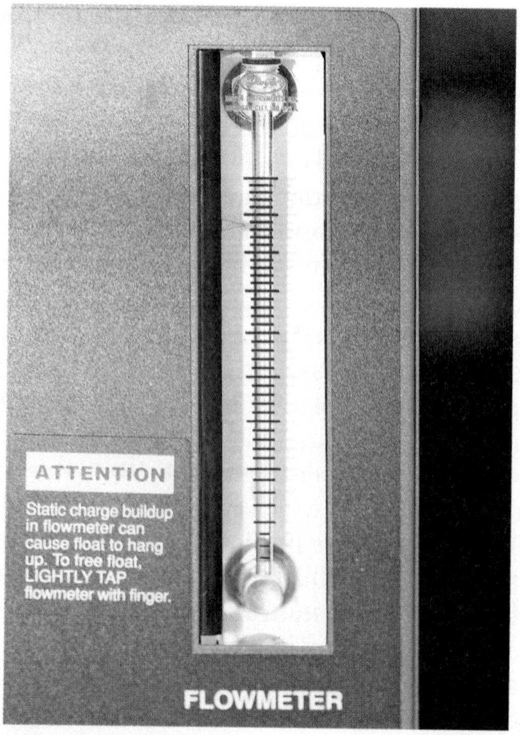

Figure 34-19 The flowmeter for an EVAP leak tester.

The source of the leak may be found by spraying the suspected areas with soapy water and looking for bubbles. Leaks can also be found by using an ultrasonic leak detector that will detect the sound made by the leaking vapors. An additional way to find the leak is through the use of an exhaust analyzer. This method can be a bit frustrating because it will only respond to fuel vapors. Because the nitrogen in the system will push the vapors out, the probe of the analyzer must be placed at the suspected area prior to or immediately after the system is pressurized. Obviously, the analyzer will read high HCs at the point of the leak. Once the source of the leak is found, make the necessary repairs and retest the system.

Using Dye To help identify the source of a leak, ultraviolet dye can be added to the system. The dye will identify the source of a leak when the system is inspected with a UV lamp. The dye is approved by many OEMs and will not harm the system, catalytic converters, or the H_2OS.

Smoke Test A very popular way to identify leaks is with a smoke machine. Nearly all OEMs recommend this method. The smoke machine vaporizes a specially formulated, highly refined mineral oil-based fluid. The machine then introduces the resultant smoky vapor into the EVAP system.

Pressurized nitrogen pushes the smoke through the system. The source of a leak is identified by the escaping smoke **(Figure 34–20)**. As when pressure testing the system, the system must be sealed to get accurate results. This means the canister vent port must be blocked off. On some vehicles the solenoid can be closed through commands input with a scan tool.

Figure 34-20 During a smoke test, smoke will be seen at all points of leakage. Be sure to inspect the entire system.

Fuel Cap Tester Most fuel cap testers come with a variety of fuel cap adapters. Always use the one that is appropriate for the vehicle being tested. Tighten the cap to the adapter and connect the cap and adapter to the tester. Turn on the tester. The tester will create a pressure on the cap and monitor the cap's ability to hold the pressure. In most cases, the tester will illuminate lights, indicating that the cap is good or bad.

PCV SYSTEM DIAGNOSIS AND SERVICE

No adjustments can be made to the PCV system. Service of the system involves careful inspection, functional tests, and replacement of faulty parts. Some engines use a fixed orifice tube in place of a valve. These should be cleaned periodically with a pipe cleaner soaked in carburetor cleaner. Although there is no PCV valve, this type of system is diagnosed in the same way as those systems with a valve. When replacing a PCV valve, match the part number on the valve with the vehicle maker's specifications for the proper valve. If the valve cannot be identified, refer to the part number listed in the manufacturer's service manual. Newer PCV valves have locking devices that prevent them from becoming loose or falling out. Make sure that the lock is fully engaged when installing the valve.

Consequences of a Faulty PCV System

If the PCV valve is stuck open, excessive airflow through the valve causes a lean air-fuel ratio and possible rough idle or engine stalling. When the PCV valve or hose is restricted, excessive crankcase pressure forces blowby gases through the clean air hose and filter into the air cleaner. Worn rings or cylinders cause excessive blowby gases and increased crankcase pressure, which forces blowby gases through the hose and filter into the air cleaner. A restricted PCV valve or hose may also cause an accumulation of moisture and sludge in the engine and engine oil.

Leaks at the engine gaskets not only will cause oil leaks, but will allow blowby gases to escape into the atmosphere. The PCV system will also draw unfiltered air through these leaks. This can result in premature wear of engine parts, especially when the vehicle is operated in dusty conditions. Check the engine for signs of oil leaks. Be sure the oil filler cap and dipstick fit and seal properly. If these do not seal, they can be the source of false air and cause a change in LTFT on MAF systems.

Visual Inspection

The PCV valve can be located in several places. The most common location is in a rubber grommet in the valve cover. It can be installed in the middle of the hose connections, as well as directly in the intake manifold.

Once the PCV valve is located, make sure that all the PCV hoses are properly connected and have no breaks or cracks. Remove the air cleaner and inspect the air and crankcase filters. Crankcase blowby can clog these with oil. Clean or replace such filters. Oil in the air cleaner assembly indicates that the PCV valve or hoses are plugged. Make sure you check these and replace the valve and clean the hoses and air cleaner assembly. When the PCV valve and hose are in satisfactory condition and there was oil in the air cleaner assembly, perform a cylinder compression test to check for worn cylinders and piston rings.

Functional Checks of the PCV System

Before beginning the functional checks, double-check the PCV valve part number to make certain that the correct valve is installed. If the correct valve is being used, continue by disconnecting the PCV valve from the valve cover, intake manifold, or hose. Start the engine and let it run at idle. If the PCV valve is not clogged, a hissing is heard as air passes through the valve. Place a finger over the end of the valve to check for vacuum **(Figure 34–21)**. If there is little or no vacuum at the valve, check for a plugged or restricted hose. Replace any plugged or deteriorated hoses. Turn off the engine and remove the PCV valve. Shake the valve and listen for the rattle of the check needle inside the valve. If the valve does not rattle, replace it.

Some vehicle manufacturers recommend that the valve be checked by removing it from the valve cover and hose. Connect a hose to the inlet side of

Figure 34-21 With the engine at idle, vacuum should be felt at the PCV valve.

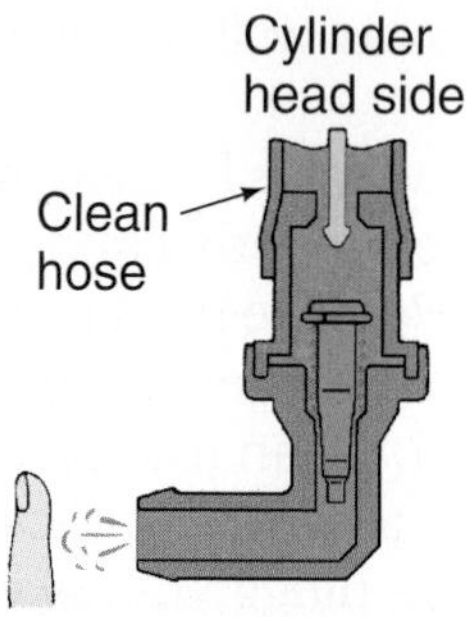

Figure 34-22 Blowing through the inlet of a PCV valve, air should freely flow through. *Courtesy of Toyota Motor Sales, U.S.A., Inc.*

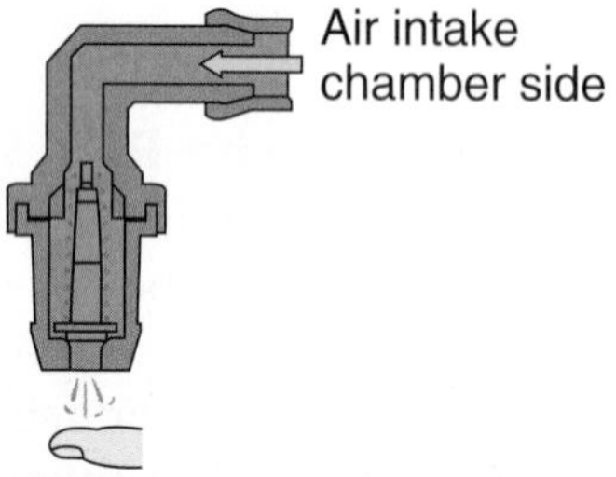

Figure 34-23 Blowing through the outlet of a PCV valve, air should barely flow through. *Courtesy of Toyota Motor Sales, U.S.A., Inc.*

the PCV valve, and blow air through the valve with your mouth while holding your finger near the valve outlet **(Figure 34–22)**. Air should pass freely through the valve. If air does not pass freely through the valve, replace the valve. Move the hose to the outlet side of the PCV valve and try to blow back through the valve **(Figure 34–23)**. It should be difficult to blow air through the PCV valve in this direction. If air passes easily through the valve, replace the valve.

CAUTION!

Do not attempt to suck through a PCV valve with your mouth. Sludge and other deposits inside the valve are harmful to the human body.

Another simple check of the PCV valve can be made by pinching the hose between the valve and the intake manifold **(Figure 34–24)** with the engine at idle. You should hear a clicking sound from the valve when the hose is pinched and unpinched. A sound scope or stethoscope will help hear the clicking. If no clicking sound is heard, check the PCV valve grommet for cracks or damage. If the grommet is okay, replace the PCV valve.

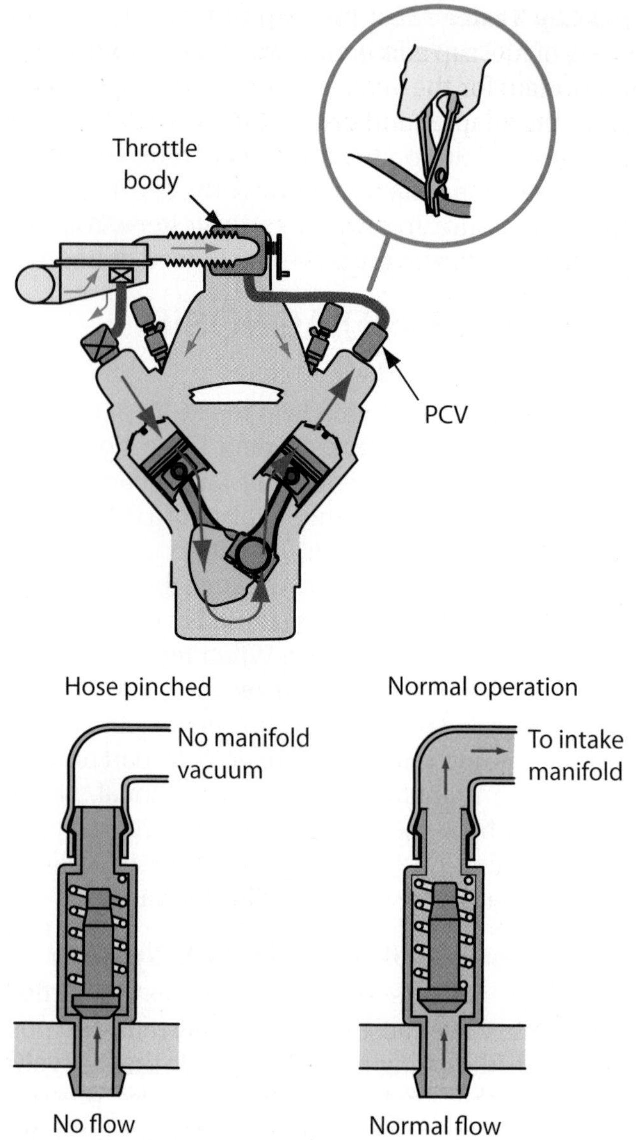

Figure 34-24 When the PCV hose is pinched, the valve should click.

The PCV system can be checked with an exhaust analyzer. Check and record the CO and O_2 measurements with the engine at idle. Then pull the PCV valve out of the engine and allow it to draw in air from under the hood. Check the CO and O_2 readings. The CO should decrease and the O_2 increase. If the readings did not change, clean the PCV system and replace the valve. If the CO decreased by 1% or more, the engine's oil may be diluted with raw fuel. Locate the cause of the fuel leakage. Then change the engine's oil and filter. Now cover the open end of the valve and observe the CO and O_2 readings after analyzer stabilizes. The CO should increase and the O_2 decrease. If readings were the same when the valve was open to underhood air or the readings changed very little, check the valve or the hose for restrictions. Clean the system and replace the valve.

Remember that proper operation of the PCV system depends on a sealed engine. The crankcase is sealed by the dipstick, valve cover, gaskets, and sealed filler cap. If oil sludging or dilution is found and the PCV system is functioning properly, check the engine for oil leaks and correct them to ensure that the PCV system can function as intended. Also, be aware of the fact that a very worn engine may have more blowby than the PCV system can handle. If there are symptoms that indicate the PCV system is plugged (oil in air cleaner, saturated crankcase filter, etc.) but no restrictions are found, check the wear of the engine.

Diesel Crankcase Ventilation Systems

Until 2007, most diesel engines removed crankcase pressures and blowby with a downdraft tube. This system is very similar to those used on gasoline engines until the late 50s. Although the systems had a filter to collect some particulate matter, most was vented to the atmosphere.

Since then diesel engines are required to control crankcase emissions. The most common systems have a filter that collects the blowby gases and a system to return them to the engine's intake or crankcase. These systems are designed to self-clean but problems in this system can occur. Therefore, it is important to inspect and clean the system as necessary as well as check the engine for excess crankcase pressures. When pressure builds up because the valve is stuck closed, crankcase pressure can force oil past some gaskets and seals. If the valve is stuck open, oil from the crankcase will be drawn into the engine and burned as fuel; however, it is heavier and thicker than diesel fuel and will cause excessive heat in the cylinders. If there oil or oil mist is present in the intake, check the valve.

Measuring Crankcase Pressure

To measure crankcase pressure, bring the engine up to normal operating temperature. Then connect a water manometer to the dipstick tube after the dipstick has been removed. With the engine at idle, observe the reading on the meter. If the reading is greater than 1 inch H_2O check the crankcase valve. If the valve is clean and working properly, check the compression and leakage of the cylinders. If the crankcase pressure is less than 1 inch H_2O, raise the engine to above 2,000 rpm. If the pressure drops to a negative value of 4 to 5 inches H_2O, the system is okay. If the drops are more than that, replace the valve. Keep in mind that a restriction in the air intake system will increase inlet vacuum in turbocharged engines and the intake vacuum on nonturbocharged engines.

SHOP TALK

If a water manometer is not available, you can make one with clear plastic tubing that will fit over the dipstick tube. Form a 3-inch "U" bend in the hose with at least 12 inches on each side of the "U." Attach the hose to a board with large staples spaced at 1-inch intervals. These intervals will be used for measurement. Fill the tube with water so that approximately one-third of the tube has water in it. On the wood, mark the level of the water. Connect one end of the tube to the dipstick tube and start the engine. Measure the amount the water moved; this is the pressure inside the crankcase as measured in inches of H_2O.

EGR SYSTEM DIAGNOSIS AND SERVICE

Manufacturers calibrate the amount of EGR flow for each engine. Too much or too little can cause performance problems by changing the engine breathing characteristics. Also, with too little EGR flow, the engine can overheat, detonate, and emit excessive amounts of NO_x. When any of these problems exist and it seems likely that the EGR system is at fault, check the system. Typical problems that show up in ported EGR systems follow:

- *Rough idle* can be caused by a stuck open EGR valve, a PVS that fails to open, dirt on the valve seat, or loose mounting bolts (this also causes a vacuum leak and a hissing noise).
- *No-start, surging, or stalling* can be caused by an open EGR valve.
- *Detonation (spark knock)* can be caused by any condition that prevents proper EGR gas flow, such as a valve stuck closed, leaking valve diaphragm, restrictions in flow passages, disconnected EGR, or a problem in the vacuum source.
- *Excessive NO_x emissions* can be caused by any condition that prevents the EGR from allowing the correct amount of exhaust gases into the cylinder or anything that allows combustion temperatures.
- *Poor fuel economy* is typically caused by the EGR system if it relates to detonation or other symptoms of restricted or zero EGR flow.

Scan Tool

On OBD-II systems, the EGR system monitor is designed to test the integrity and flow characteristics of the EGR system. The monitor is activated during EGR operation and after certain engine conditions are met. Input from the ECT, IAT, TP, and CKP sensors are required to activate the monitor. Once activated, a typical EGR monitor carries out these tests:

- The differential pressure feedback EGR sensor and circuit are continuously tested for opens and shorts.
- The EGR vacuum regulator solenoid is continuously tested for opens and shorts. The monitor compares circuit voltage to what should exist when the EGR valve is in its commanded state.
- Non-DPFE systems monitor STFT while opening the EGR value; the STFT should decrease.
- The EGR flow rate test is checked by comparing the actual differential pressure feedback EGR circuit voltage to the desired EGR flow voltage for the current operating conditions to determine if the EGR flow rate is acceptable.
- The hoses connected to the differential pressure feedback EGR sensor hoses are tested once per drive cycle. They are checked for restrictions and opens when the EGR valve is closed and during acceleration. The monitor checks the voltage from the differential pressure feedback EGR sensor. A reading not within the normal range for a closed valve indicates a problem with the hose.
- Checking for a stuck open EGR valve or some EGR flow at idle is continuously done at idle. The monitor compares the differential pressure feedback EGR circuit voltage to the differential pressure feedback EGR circuit voltage stored during key on engine off operation to determine if there is EGR flow.
- The MIL is illuminated after one of these tests fails on two consecutive drive cycles.

Different methods of EGR flow monitoring are used; these include temperature sensors, manifold pressure changes, fuel trim changes, and differential pressure measurement. Using temperature sensors, an EGR temperature sensor is installed in the EGR passageway. During normal EGR flow, the temperature of the EGR temperature sensor will rise at least 95°F (35°C) above ambient air temperature. When the EGR valve is open, the ECM compares EGR temperature to intake air temperature. If the temperature does not rise a specified amount over ambient temperature, the ECM assumes there is a problem in the system, and this information is stored in the ECM.

MAP systems base their calculations on the assumption that low intake manifold vacuum means the engine is under a heavier load and greater airflow. Too much EGR flow will increase manifold pressure and the PCM will interpret this as an increase in airflow and will increase the amount of fuel delivered to the cylinders. The system compares MAP readings with the reading from the H_2OS. The PCM will also try to compensate for this by correcting the fuel trim. Excessive EGR flow will cause negative fuel trims, and low EGR flow will cause higher-than-normal fuel trim readings. If fuel trim readings seem to be out of line, EGR flow may be the cause.

When the PCM detects a problem in the system, it will set a DTC. It is important to note that the conditions required to set a DTC vary with manufacturer and model. It is also important to realize that inputs from the engine control system EGR operation. Therefore, all engine-related DTCs should be dealt with before moving into detailed diagnostics of the EGR system. Some DTCs pertain solely to the EGR valve and its solenoids (**Figure 34–25**).

EGR System Troubleshooting

Before attempting to troubleshoot or repair a suspected EGR system on a vehicle, make sure that the

DTC	Description
P0106	MAP sensor rationality error
P0107	MAP sensor voltage low
P0108	MAP sensor voltage high
P0109	MAP sensor intermittent (non-MIL) problem
P0400	EGR system leak detected
P0401	Insufficient EGR flow
P0402	DPFE EGR stuck open
P0403	EVR circuit open or shorted
P0404	EGR control circuit range/performance problem
P0405	EGR valve position sensor circuit low
P0406	EGR valve position sensor circuit high
P1400 (Ford)	DPFE circuit low
P1401 (Ford)	DPFE circuit high
P1405 (Ford)	Upstream hose off or plugged
P1406 (Ford)	Downstream hose off or plugged
P1409 (Ford)	EVR circuit open or shorted
P2413 (Honda)	EGR system malfunction
P2457 (Ford)	Insufficient EGR cooler performance

Figure 34-25 Samples of EGR-related DTCs.

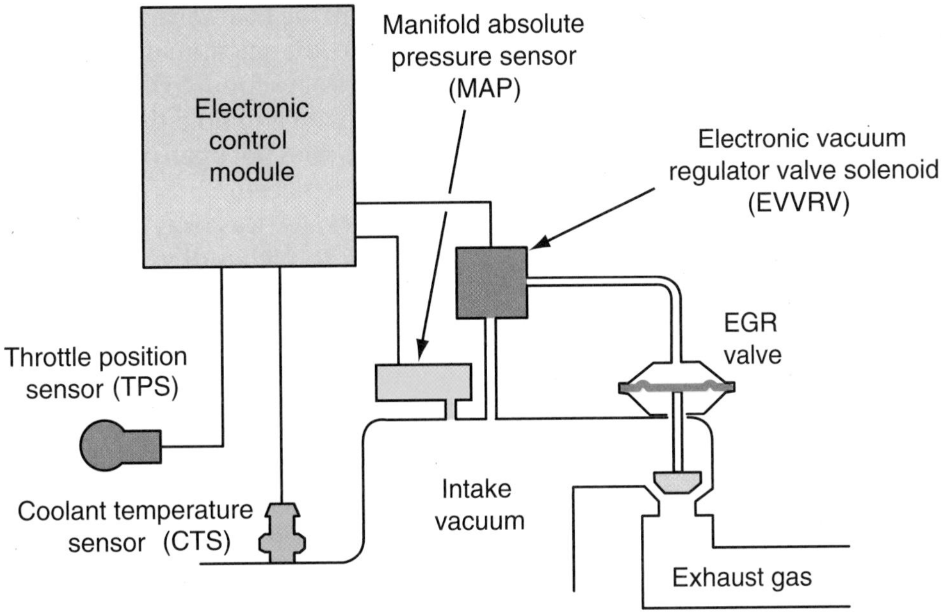

Figure 34-26 The circuit of a typical computer-controlled EGR system.

engine is mechanically sound, that the injection system is operating properly, and that the spark control system is working properly.

Most often an electronically controlled EGR valve functions **(Figure 34–26)** in the same way as a vacuum-operated valve. Apart from the electronic control, the system can have all of the problems of any EGR system. Those that are totally electronic and do not use a vacuum signal can have the same problems as others, with the exception of vacuum leaks and other vacuum-related problems. Sticking valves, obstructions, and loss of vacuum produce the same symptoms as on non–electronic-controlled systems. If an electronic control component is not functioning, the condition is usually recognized by the PCM. The solenoids, or the EVR, should normally cycle on and off frequently when EGR flow is being controlled (warm engine and cruise rpm). If they do not, a problem in the electronic control system or the solenoids is indicated. Generally, an electronic control failure results in low or zero EGR flow and might cause symptoms like overheating, detonation, and power loss.

EGR solenoids are used with all types of EGR valves, especially backpressure-type valves. The PCM uses a solenoid to regulate ported or manifold vacuum to the EGR valve.

The solenoid is actually a vacuum switch. The PCM controls the switch through pulse width modulation. No vacuum is sent to the valve unless the PCM allows it. The EGR solenoid has two or more vacuum lines and an electrical connector. The solenoid also has an air bleed and sometimes an air filter. Vacuum is bled off through the filter vent. If the filter becomes clogged, the EGR valve will open too much and cause a driveability problem.

Before attempting any testing of the EGR system, visually inspect the condition of all vacuum hoses for kinks, bends, cracks, and flexibility. Replace defective hoses as required. Check vacuum hose routing against the underhood decal or the service manual; correct any misrouted hoses. If the system is fitted with an EVP sensor, the wires routed to it should also be checked.

If the EGR valve remains open at idle and low engine speed, the idle operation is rough and surging occurs at low speed. When this problem is present, the engine may hesitate on low-speed acceleration or stall after deceleration or after a cold start. If the EGR valve does not open, engine detonation occurs. When a defect occurs in the EGR system, a DTC is usually set in the PCM memory.

EGR Valves and Systems Testing

On many older engines, a single diaphragm EGR valve can be checked with a hand-operated vacuum pump **(Figure 34–27)**. Before conducting this test, make sure that the engine produces enough vacuum to properly operate the valve. This is done by connecting a vacuum gauge to the engine's intake manifold. Then start the engine and gradually increase speed to 2,000 rpm with the transmission in neutral. The reading should be above 16 inches of mercury. If not, there could be a vacuum leak or exhaust restriction. Before continuing to test the EGR, check the MAP and/or correct the problem of low vacuum.

To check the valve with a vacuum pump, remove the vacuum supply hose from the EGR valve port.

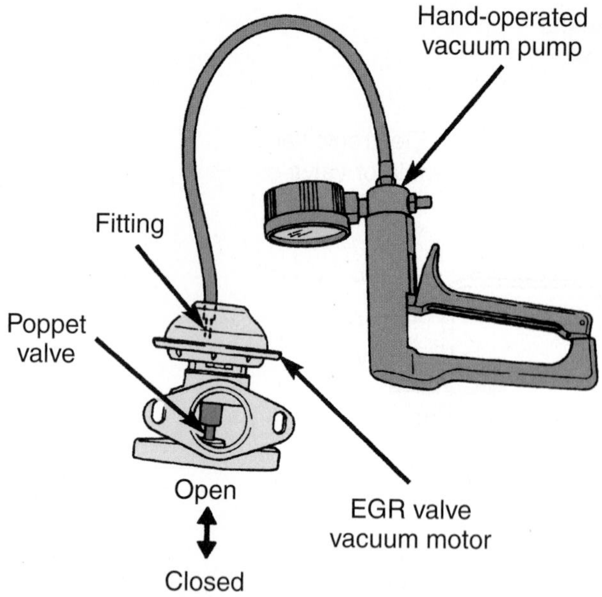

Figure 34-27 Watch the action of the valve when vacuum to it is applied and released. *Courtesy of Chrysler LLC*

Connect the vacuum pump to the port and supply 18 inches of vacuum. Observe the EGR diaphragm movement. When the vacuum is applied, the diaphragm should move. If the valve diaphragm did not move or did not hold the vacuum, replace the valve.

With the engine at normal operating temperature, observe the engine's idle speed. If necessary, adjust the idle speed to the emission decal specification. Slowly apply 5 to 10 inches of vacuum to the EGR valve. The idle speed should drop more than 100 rpm (the engine may stall) and then return to normal again when the vacuum is removed. If the idle speed does not respond in this manner, remove the valve and check for carbon in the passages under the valve. Clean the passages as required or replace the EGR valve. Carbon may be cleaned from the lower end of the EGR valve with a wire brush, but do not immerse the valve in solvent and do not sandblast the valve. Also, make sure that the vacuum hoses are in good condition and properly routed.

Diagnosis of a Negative Backpressure EGR Valve

A negative backpressure EGR valve is identified by the letter "N" stamped on it. The valve is opened by a combination of engine vacuum and the negative exhaust system pulses that occur as each exhaust valve closes. As soon as the valve opens, backpressure is reduced slightly, which opens a vacuum bleed and the valve quickly closes. This causes the opening of the valve to modulate according to negative exhaust system pulses.

With the engine at normal operating temperature and the ignition switch off, disconnect the vacuum hose from the EGR valve and connect a hand vacuum pump to the vacuum fitting on the valve. Supply 18 inches of vacuum to the EGR valve. The EGR valve should open and hold the vacuum for 20 seconds. If the valve does not open or cannot hold the vacuum, it must be replaced.

If the valve was okay in the first test, continue by applying 18 inches of vacuum to the valve and start the engine. The vacuum should drop to zero, and the valve should close. If the valve does not react this way, replace it.

Diagnosis of a Positive Backpressure EGR Valve

A positive backpressure EGR valve can be identified by the letter "P" stamped next to the part number and date code. It has a thicker-than-normal pintle shaft because it is hollow. The hollow design allows exhaust gases to flow into the shaft and push up on it. With positive backpressure from the exhaust system, the shaft rises and seals the control valve. Once the control valve is closed, it allows applied vacuum to pull up on the diaphragm. With low backpressure, the valve will not hold vacuum and the vacuum is bled to the atmosphere. As engine load increases, so does engine backpressure, which causes the control valve inside the EGR to trap vacuum and open up. To test this valve, bring the engine up to 2,000 rpm to create backpressure, then apply vacuum. EGR should open and cause a 100 rpm drop or more. Positive backpressure EGR valves are used in simple vacuum-controlled systems as well as more complex pulse width modulated applications.

Diagnosis of a Digital EGR Valve

Digital EGR valves are only found on GM products. They are totally electronically controlled units. They have two or three solenoids, and part of the valve is always open. Use a scan tool to check a digital EGR valve. Start the engine and allow it to run at idle speed. Select the EGR control on the scan tool, and then energize the solenoids one at a time. Engine speed should drop slightly as each EGR solenoid is energized.

If the EGR valve does not respond correctly, make sure 12 volts are applied to the EGR valve. Then check the resistance of the valve. Connect an ohmmeter across the electrical terminals on the valve (**Figure 34–28**); the windings can be checked for opens, shorts, and excessive resistance. If any resistance reading is not within specs, the valve should be replaced. Visually check all of the wires between the EGR valve and the PCM. Also make sure that the EGR passages are not restricted or plugged. To do this, you will need to remove the valve.

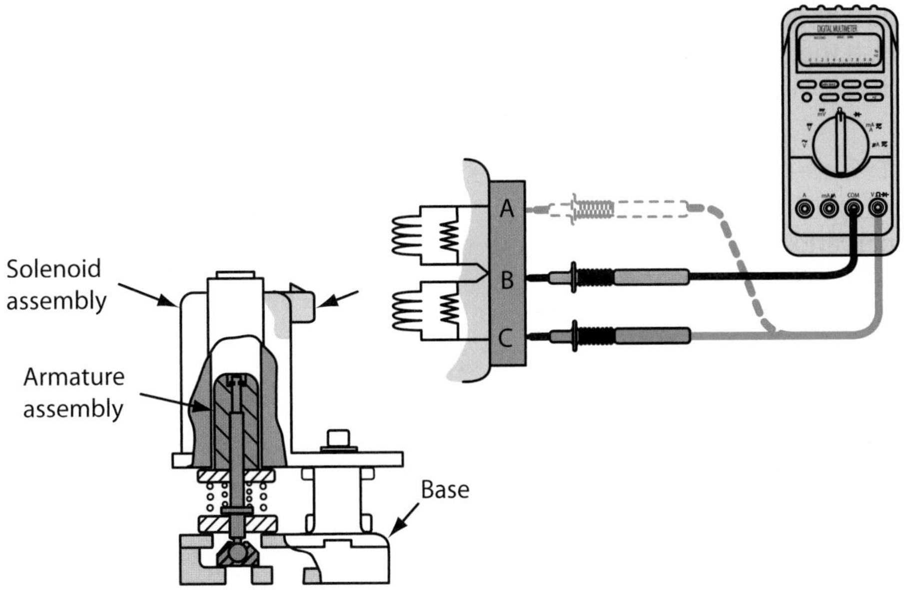

Figure 34-28 Ohmmeter connections for checking a digital EGR valve.

Linear EGR Valve Diagnosis

The correct procedure for diagnosing a linear EGR valve (**Figure 34–29**) will vary, depending on the vehicle make and model year. Always follow the recommended procedure in the vehicle manufacturer's service manual. A scan tool may be used to diagnose a linear EGR valve. The engine should be at normal operating temperature. Because the linear EGR valve has an EVP sensor, the actual pintle position may be checked on the scan tool. The pintle position should not exceed 3% at idle speed. The scan tool may be operated to command a specific pintle position, such as 75%, and this commanded position should be achieved within 2 seconds. With the engine idling, select various pintle positions and check the actual pintle position. The pintle position should always be within 10% of the commanded position.

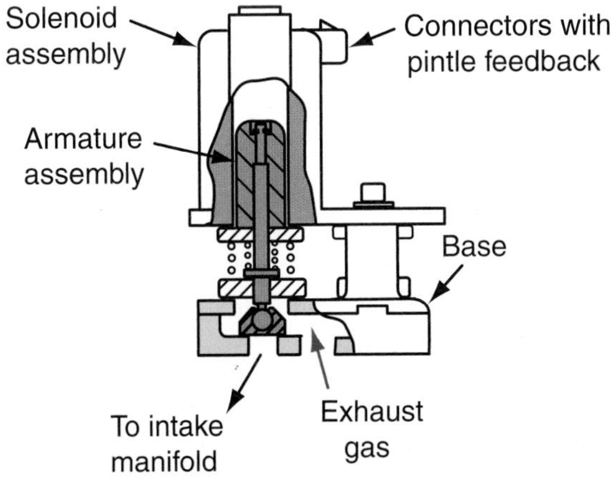

Figure 34-29 A linear EGR valve relies on a solenoid to move the pintle.

If a linear EGR valve does not operate properly, check the fuse in the supply wire to the EGR valve. Also check for open circuits, grounds, and shorts in the wires connected from the EGR valve to the PCM (**Figure 34–30**). Verify that the EVP sensor is receiving a 5-volt reference signal and verify that the ground circuit is good. If these are okay, remove the valve with the wiring harness still connected to it. Then connect a DMM across the pintle position wire at the EGR valve to ground and manually push the pintle upward.

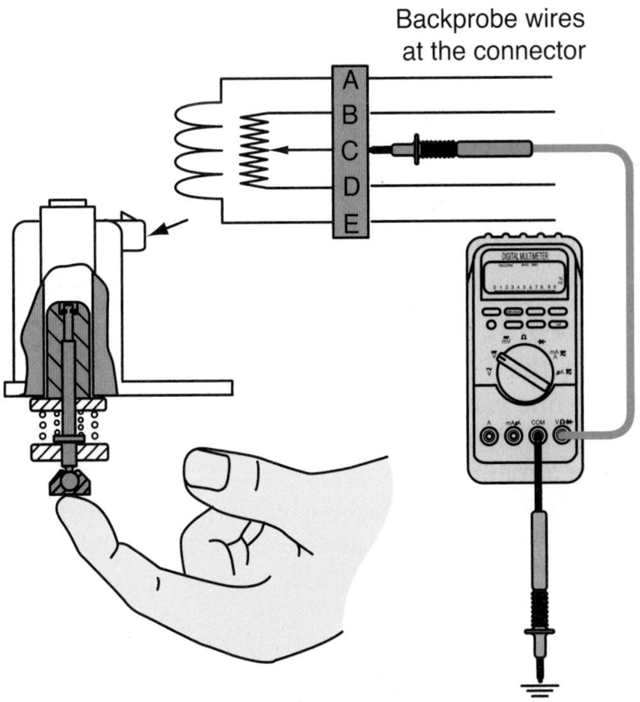

Figure 34-30 To check a linear EGR valve, check the voltage at the various pins of its connector.

The voltmeter reading should change from approximately 1 to 4.5 volts. If the EGR valve did not operate properly, it should be replaced.

Checking EGR Efficiency

Although most testing of EGR valves involves the valve's ability to open and close at the correct time, we are not really testing what the valve was designed to do—control NO_x emissions. EGR systems should be tested to see if they are doing what they were designed to do.

Many technicians wrongly conclude by thinking that an EGR valve is working properly if the engine stalls or idles very rough when the EGR valve is opened. Actually this test just shows that the valve was closed and that it will open. A good EGR valve opens and closes, but it also allows the correct amount of exhaust gas to enter the cylinders. EGR valves are normally closed at idle and open at approximately 2,000 rpm. This is where the EGR system should be checked.

To check an EGR system, use a five-gas exhaust analyzer. Allow the engine to warm up, and then raise the engine speed to around 2,000 rpm. Watch the NO_x readings on the analyzer. In most cases, NO_x should be below 1,000 ppm. It is normal to have some temporary increases over 1,000 ppm; however, the reading should be generally less than 1,000. If the NO_x is above 1,000, the EGR system is not doing its job. The exhaust passage in the valve is probably clogged with carbon.

If only a small amount of exhaust gas is entering the cylinder, NO_x will still be formed. A restricted exhaust passage of only $\frac{1}{8}$ inch will still cause the engine to run rough or stall at idle, but it is not enough to control combustion chamber temperatures at higher engine speeds. Never assume that the EGR passages are okay just because the engine stalls at idle when the EGR is fully opened.

Electronic EGR Controls

When the EGR valve checks out and everything looks fine visually but a problem with the EGR system is evident, the EGR controls should be tested. Often a malfunctioning electronic control will trigger a DTC. Service manuals give the specific directions for testing these controls; always follow them. The tests that follows are given as examples of those test procedures.

Some EGR valves are electronic/mechanical EGR valves. These valves have different names depending on the application. These types of valves operate in the same way as a single diaphragm EGR valve. However, they have a position sensor above the EGR diaphragm. This tells the PCM how far the valve is open. The valve position sensor is a potentiometer and can wear. The sensor can be checked with a DMM or lab scope. The pattern should show a clean sweep as the valve is opened and closed. The EGR monitor system watches the output from the sensor and if the sensor's voltage reading is too high or low, a DTC will be set.

EGR Vacuum Regulator (EVR) Tests

Connect a pair of ohmmeter leads to the EVR terminals to check the winding for open circuits and shorts. An infinite ohmmeter reading indicates an open circuit, whereas a lower-than-specified reading means the winding is shorted. Then connect the ohmmeter leads from one of the EVR solenoid terminals to the solenoid case. You should get an infinite reading; a low ohmmeter reading means the winding is shorted to the case.

SHOP TALK

The same quad driver in a PCM may operate several outputs. On General Motors' computers, quad drivers sense high current flow. If a solenoid winding is shorted and the quad driver senses high current flow, the quad driver shuts down all the outputs it controls. This prevents damage caused by the high current flow. When the PCM does not operate an output or outputs, always check the resistance of the output's solenoid windings before replacing the PCM. A lower-than-specified resistance in a solenoid winding indicates a shorted condition, and this problem may explain why the PCM quad driver is not operating the outputs.

A scan tool may be used to diagnose the operation of an EVR solenoid. The procedure is shown in Photo Sequence 33. This procedure is given as an example; always follow the procedures given in the service manual.

SHOP TALK

In some EGR systems, the PCM energizes the EVR solenoid at idle and low speeds. Under this condition, the solenoid shuts off vacuum to the EGR valve. When the proper input signals are available, the PCM de-energizes the EVR solenoid and allows vacuum to the EGR valve.

P33-1 *Disconnect the connector to the EGR solenoid and connect the leads of an ohmmeter to the solenoid's terminals.*

P33-2 *Compare your readings to the specifications for the solenoid.*

P33-3 *Connect the ohmmeter leads to one of the solenoid's terminals and to ground. An infinite reading means the solenoid is not shorted to ground.*

P33-4 *Reconnect the wiring to the connector and run the engine to bring it to normal operating temperature. While the engine is running, prepare the scan tool for the vehicle.*

P33-5 *Turn off the engine and connect the power cable of the scan tool to the vehicle.*

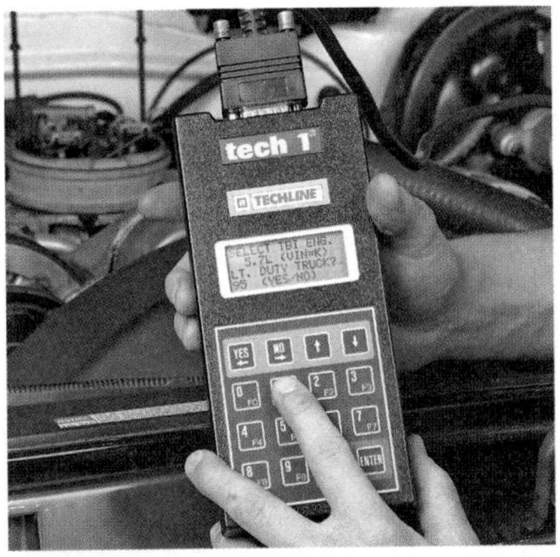

P33-6 *Enter the necessary information into the scan tool.*

P33-7 *Connect the scan tool to the DLC.*

P33-8 *Start the engine and obtain the EGR data on the scanner. The EGR valve should be off and remain off while the engine is idling.*

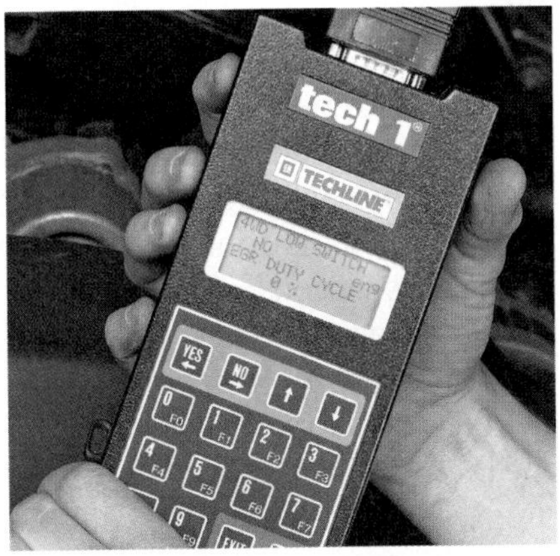

P33-9 *Take the vehicle for a test drive with the scan tool still connected. The EGR solenoid should cycle to ON once the vehicle is at a cruising speed. If it does not, check the solenoid and associated circuits.*

Exhaust Gas Temperature Sensor Diagnosis

To test an exhaust gas temperature sensor, remove it and place it in a container of oil. Place a thermometer in the oil and heat the container. Connect the ohmmeter leads to the exhaust gas temperature sensor terminals. The exhaust gas temperature sensor should have the specified resistance at various temperatures.

Diesel Engines

Recently, diesel engines have been equipped with EGR valves. These systems release a sample of exhaust gases into the intake of the turbocharger or the intake manifold. When the exhaust gases pass through the intercooler, the temperature is decreased, which lowers the chances of NO_x formation.

There are basically two types of EGR systems used on diesel engines:

- High-pressure EGR captures the exhaust gas prior to the turbocharger and redirects it back into the intake air. Sometimes, the system will have a catalyst in the high-pressure EGR loops to reduce PM levels that are recirculated back through the combustion process.
- A low-pressure EGR collects the exhaust after the turbocharger and a diesel particulate filter and returns it to the intercooler. Diesel PM filters are always used with a low-pressure EGR system to make sure large amounts of particulate matter are not recirculated to the engine, which would result in accelerated wear in the engine and turbocharger.

CATALYTIC CONVERTER DIAGNOSIS

The catalytic converter monitor looks at a converter's ability to store O_2. O_2 storage is only one function of a converter but is a good indication of how efficient the converter is. As the catalyst efficiency declines due to thermal and chemical deterioration, its ability to store O_2 also declines. Therefore, OBD-II systems compare the O_2 content in the exhaust before and after the converter. This is done by monitoring the signals from O_2 sensors placed before and after the converter.

The catalyst monitor will run after the HO_2S monitor has been completed, when there are no DTCs stored by the secondary AIR and EVAP systems. Inputs from the ECT, IAT, MAF, CKP, TP, and vehicle speed sensors are required to enable the catalyst efficiency monitor. After the engine has warmed up and the necessary inputs are available, the PCM will calculate whether the converter has warmed up or not. If it is warm, the monitor will run. Converter efficiency is determined by comparing the precatalyst HO_2S or A/F sensor signal with the signal from the postcatalyst HO_2S. The PCM looks at the signal differences between the two sensors to measure converter efficiency (**Figure 34–31**).

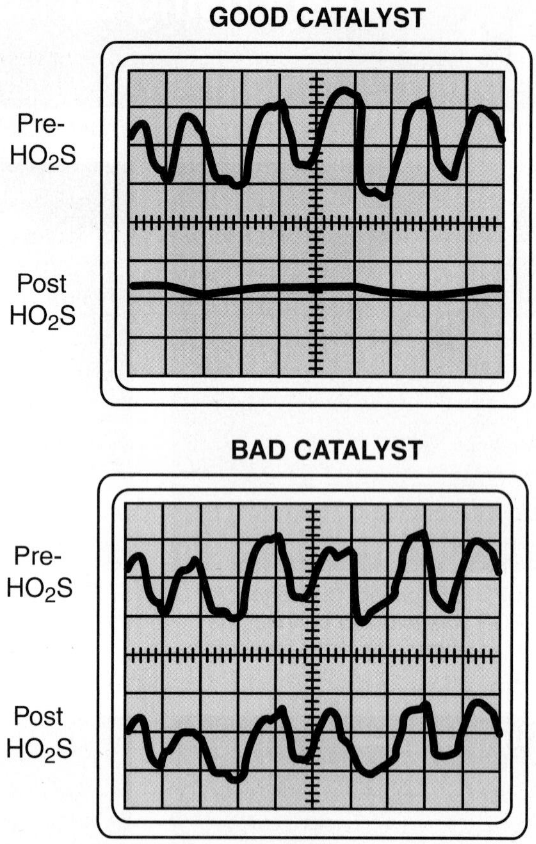

Figure 34–31 A comparison of the HO_2S signals for a good and a bad catalytic converter.

During normal operation, the front HO_2S switches more often and with a greater amplitude than the rear HO_2S. The rear HO_2S also has a shorter signal. The monitor compares the cross counts of each sensor as well as the signal length. When the converter has lost some of its ability to store O_2, the postcatalyst or downstream HO_2S signal begins to switch more rapidly with increasing amplitude and signal length. It starts to look like the signal from the precatalyst or upstream HO_2S.

When the signals become alike and stay that way through a number of drive cycles, the PCM will set a DTC and illuminate the MIL. The activity of the HO_2S can be monitored on a scan tool or lab scope. Doing this will help determine if either the sensors or the converter is bad. A converter-related DTC does not always indicate that the converter is bad. These DTCs can be set for a number of other reasons, such as:

- A small leak in the secondary AIR system
- A slight misfire that is causing extra O_2 to enter the exhaust stream
- An exhaust leak downstream of the front HO_2S

Converter Diagnosis

Typically, catalytic converters fail because of deterioration of the catalyst material or because of physical damage. A converter should be checked for cracks and dents. It is also possible that the internal components of the converter are damaged or broken. A quick test of internal damage is done with a rubber mallet. The converter is smacked with the mallet. If the converter rattles, it needs to be replaced and there is no need to do other testing. A rattle indicates loose catalyst substrate, which will soon rattle into small pieces. This test is not used to determine if the catalyst is good.

Converters often fail because the catalyst material becomes coated with foreign materials. This normally is the result of contaminated fuel, sealants, or coolant entering the exhaust stream. A buildup of this material reduces the catalysts' ability to reduce NO_x and oxidize HC and CO.

An overheated converter can become plugged and restrict exhaust flow. The typical cause of an overheated converter is engine misfiring. A plugged converter or any exhaust restriction can cause loss of power at high speeds, stalling after starting (if totally blocked), or sometimes popping or backfiring at the intake manifold.

A vacuum gauge can be used to watch engine vacuum while the engine is accelerated. Another way to check for a restricted exhaust or catalyst is to insert a pressure gauge in the exhaust manifold's bore for the O_2 sensor **(Figure 34–32)**. With the gauge in place, hold the engine's speed at 2,000 rpm and watch the gauge. The desired pressure reading will be less than 1.25 psi. A very bad restriction will give a reading of over 2.75 psi.

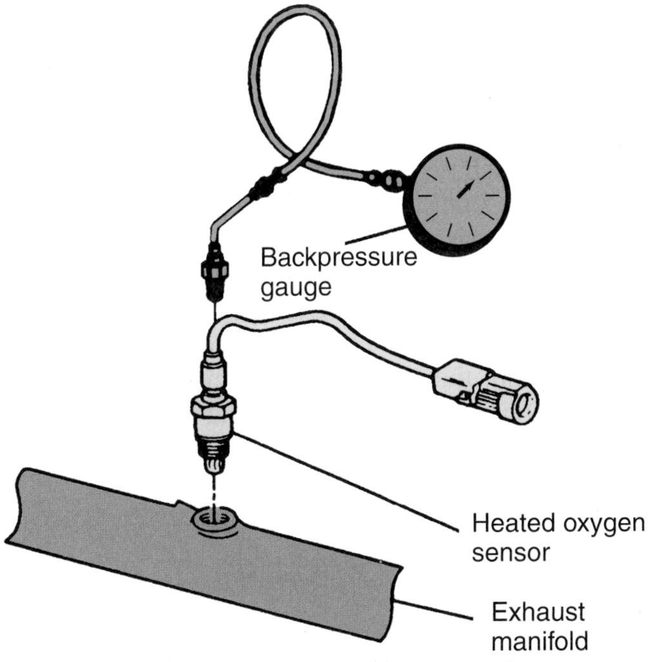

Figure 34-32 To measure exhaust system backpressure, insert a pressure gauge into the oxygen sensor's bore in the exhaust.

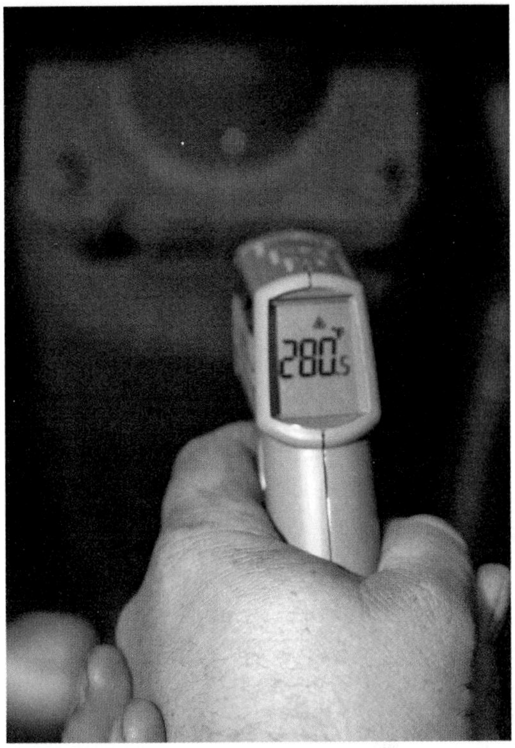

Figure 34-33 When doing the delta temperature test, check the temperature of the converter.

Delta Temperature Test The converter should be checked for its ability to convert CO and HC into CO_2 and water. This can be done by conducting a delta temperature test. This test is conducted with a handheld digital pyrometer. Bring the engine to normal operating temperature and run it at 2,500 rpm for 2 minutes. Hold the engine at that speed and place the pyrometer near the inlet of the converter **(Figure 34–33)** and then just behind the converter **(Figure 34–34)**; record the temperatures. There should be an increase of at least 50°F (10°C) or 8% above the inlet temperature reading as the exhaust gases pass through the converter. If the outlet temperature is the same or lower, nothing is happening inside the converter. To do its job efficiently, the converter needs a steady supply of O_2 from the air pump. A bad pump, a faulty diverter valve or control valve, leaky air connections, or a faulty computer control over the air injection system could be preventing the needed O_2 from reaching the converter. If the converter fails this test, check those systems.

Oxygen Storage Test The O_2 storage test is based on the fact that a good converter stores O_2. Begin by warming up a four- or five-gas analyzer. Disable the air injection system **(Figure 34–35)**. Once the converter

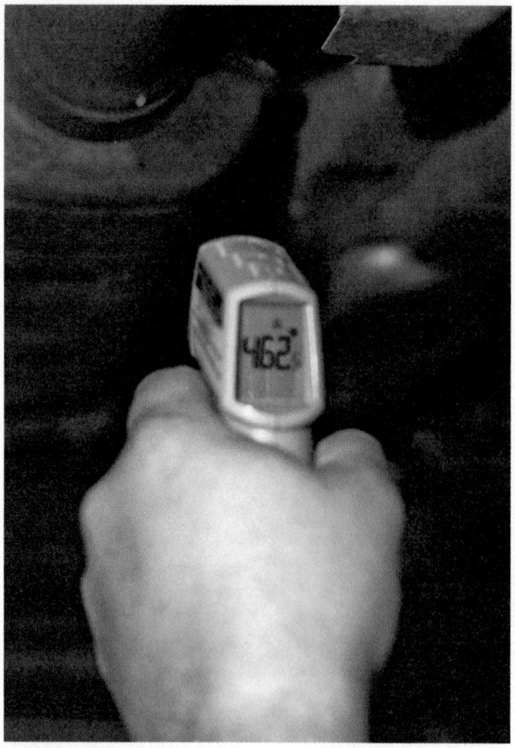

Figure 34-34 Compare the outlet temperature with the inlet to see how efficient the converter is.

Figure 34-35 Before conducting an oxygen storage test, disable the air injection system.

Figure 34-36 While checking converter efficiency, you may need to add some propane into the intake to enrich the mixture.

is warmed up, insert the analyzer's probe into the tailpipe, and hold the engine at 2,000 rpm. Watch the O_2 readings. Once the numbers stop dropping, check the O_2 level on the gas analyzer. The O_2 readings should be about 0.0%. This shows that the converter is using the available O_2. Immediately after the O_2 drops, quickly snap the throttle and watch the O_2 reading just as the CO begins to increase. If the O_2 now exceeds 1.2%, the converter is failing this test. If the O_2 readings never reached zero, the test may need to be repeated after adding some propane through the air intake until all of the O_2 stored in the converter is depleted.

Checking Converter Efficiency This converter test uses a principle that checks the converter's efficiency. Before beginning this test, make sure that the converter is warmed up and there are no ignition problems, vacuum leaks, or fuel restrictions. Disable the air injection system and the disconnect the HO_2S. Calibrate a four- or five-gas analyzer and insert its probe into the tailpipe. With a propane enrichment tool (**Figure 34–36**), richen the air-fuel mixture until the CO reading is about 2%. Then reconnect the air injection system. Observe the HC, CO, and O_2 readings. If the converter is working correctly, the O_2 should increase and HC and CO should decrease when the air injection system is reconnected. If the O_2 increased but the CO and HC did not change much, or if the O_2 is higher than the CO and the CO is greater than 0.5%, the converter is faulty. If the O_2 is lower than the CO, the converter is not oxidizing HC and CO.

AIR SYSTEM DIAGNOSIS AND SERVICE

Not all engines are equipped with an air injection system; only those that need them to meet emissions standards have them. Therefore, air injection systems are vital to proper emissions on engines equipped with them. Each system has its own test procedure; always follow the manufacturer's recommendations for testing.

Most AIR systems are computer controlled and rely on solenoids to control the direction of airflow to the exhaust manifold or to the converter. When the system is in closed loop, the air from the air injection system must be directed away from the O_2 sensor. Some systems have switching valves that allow a small amount of air to flow past the O_2 sensor. The computer knows how much and adjusts the O_2 input accordingly. Sometimes the amount of air that can move through a closed switching valve is marked on its housing. The pump has at least two hoses. The inlet hose is the larger of the two and connects to the air filter assembly or a small dedicated air filter. The output hose carries output air through the valve and into the exhaust.

Figure 34-37 An electric air pump.

Secondary AIR Monitor

The operation of the secondary air injection (AIR) system is checked by an OBD-II monitor. The monitor looks at the complete electrical circuit for the AIR system, especially the electric pump (if so equipped) and pump relay. It checks it for shorts, opens, and high resistance. It checks the ability of the system to inject air into the exhaust. It does this through input from the MAF. It compares the MAF signals when the system is off and when it is energized. The condition of the pump and hoses is also checked at this time.

The monitor runs when the AIR system is operating but when the enable criteria are met. Most AIR systems will set DTCs in the PCM if there is a fault in the solenoids and related wiring. In some AIR systems, DTCs are set in the PCM memory if the airflow from the pump is continually upstream or downstream. Always use a scan tool to check for any DTCs related to the AIR system, and correct the causes of these codes before proceeding with further system diagnosis. When a fault is detected, a DTC is set and the MIL will be lit if the fault is detected during two consecutive drive cycles.

Some late-model AIR systems use an electric air pump controlled by the PCM **(Figure 34–37)**. These systems have an AIR solenoid and solenoid relay. When the PCM provides a ground for the relay, battery voltage is applied to the solenoid and the pump. Typically, DTCs will be set if one of the components fails or if the hoses or check valves leak. A quick check of the system can be made with a scan tool.

Set the scan tool to watch the voltage at the O_2 sensor(s). Start the engine and allow it to idle. Once the engine has reached normal operating temperature, enable the AIR system and check the HO_2S voltages. If the voltages are low, the AIR pump, solenoid, and shutoff valve are working properly. If the voltages are not low, each component of the system needs to be checked and tested. For testing purposes, most bidirectional scan tools can turn the AIR pump on for testing purposes.

Secondary AIR System Service and Diagnosis

The first step in diagnosing a secondary air injection system is to check all vacuum hoses and electrical connections in the system. Most belt-driven AIR pumps have a centrifugal filter behind the pulley to keep dirt out of the pump. Air flows through this filter into the pump. The pulley and filter are bolted to the pump shaft and are serviced separately **(Figure 34–38)**. If the pulley or filter is bent, worn, or damaged, it should be replaced. Also check the AIR pump's belt for condition and tension and correct it as necessary. The pump assembly is usually not serviced.

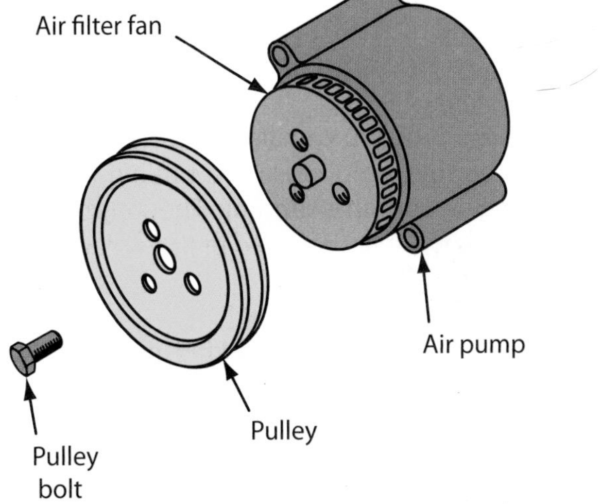

Figure 34-38 An AIR pump pulley and filter assembly.

In some AIR systems, pressure relief valves are mounted in the AIRB and AIRD valves. Other AIR systems have a pressure relief valve in the pump. If the pressure relief valve is stuck open, airflow from the pump is continually exhausted through this valve, which causes high tailpipe emissions.

If the hoses in the AIR system show evidence of burning, the one-way check valves are leaking, which allows exhaust to enter the system. Leaking air manifolds and pipes result in exhaust leaks and excessive noise.

If the AIR system does not pump air into the exhaust ports during engine warmup, HC emissions are high during this mode, and the HO_2S, or sensors, takes longer to reach normal operating temperature. Under this condition, the PCM remains in open loop longer. Because the air-fuel ratio is richer in open loop, fuel economy is reduced.

When the AIR system pumps air into the exhaust ports with the engine at normal operating temperature, the additional air in the exhaust stream causes lean signals from the HO_2S or sensors. The PCM responds to these lean signals by providing a richer air-fuel ratio. This increases fuel consumption. A vehicle can definitely fail an emission test because of air flowing past the HO_2S when it should not be. If the HO_2S is always sending a lean signal back to the computer, check the air injection system.

Noise Diagnosis

Leaks in the AIR system can cause a noise. It may sound like an exhaust leak or a hissing, depending on where the leak is. To verify that the system is leaking, disconnect the pressure hose from the switching or combination valve. Plug the end of the hose and run the pump. Normally the sound will be amplified and can be found. The pump itself can be the source of a leak. This typically results in a whistling noise when the pump is running. At times, the source of the leak can be found by feeling around the pump as it runs. A common source of leakage is a bad or loose seal at the pump shaft.

A pump problem can also be the cause of a noise. One common pump noise is a rattling that is heard only when the pump is running. The common cause of this noise is worn or damaged pump isolator mounts.

System Efficiency Test

When the AIR system is working properly, HC, CO and CO_2 are decreased and O_2 is increased. Run the engine at about 1,500 rpm with the secondary AIR system on (enabled). Using an exhaust gas analyzer, measure and record the emission levels. Next, disable the secondary AIR system and continue to allow the engine to idle. Again, measure and record the emission levels in the exhaust gases. The O_2 readings should be at least 4% less than they were when the AIR system was enabled. Less than that indicates an AIR problem.

AIR Component Diagnosis

Not all AIR systems have the same components. The following are some of the more common parts used in today's AIR systems.

AIRB Solenoid and Valve When the engine is started, listen for air being exhausted from the AIRB valve for a short period. If this air is not exhausted, remove the vacuum hose from the AIRB and start the engine. If air is now exhausted from the AIRB valve, check the AIRB solenoid and connecting wires. When air is still not exhausted from the AIRB valve, check the air supply from the pump to the valve. If the air supply is available, replace the AIRB valve.

During engine warmup, remove the hose from the AIRD valve to the exhaust ports and check for airflow from this hose. If airflow is present, the system is operating normally in this mode. When air is not flowing from this hose, remove the vacuum hose from the AIRD valve and connect a vacuum gauge to this hose. If vacuum is above 12 in. Hg, replace the AIRD valve. When the vacuum is zero, check the vacuum hoses, the AIRD solenoid, and connecting wires.

SHOP TALK

With the engine at normal operating temperature, the AIR system sometimes moves back into the upstream mode with the engine idling. It may be necessary to increase the engine speed to maintain the downstream mode.

With the engine at normal operating temperature, disconnect the air hose between the AIRD valve and the catalytic converters and check for airflow from this hose. When airflow is present, system operation in the downstream mode is normal. If there is no airflow from this hose, disconnect the vacuum hose from the AIRD valve and connect a vacuum gauge to the hose. When the vacuum gauge indicates zero vacuum, replace the AIRD valve. If some vacuum is indicated on the gauge, check the hose, the AIRD solenoid, and connecting wires.

Combination Valve This valve is typically part of the AIR monitoring system. It can be quickly checked with a hand-held vacuum pump. Apply vacuum to the valve. It should open. If it does not, replace it. If it does open, check the valve's vacuum source for adequate vacuum and the solenoids that control vacuum to it.

Check Valve All of the types of air injection systems have at least one thing in common—a one-way check valve. The valve opens to let air in but closes to keep exhaust from leaking out. The check valve can be checked with an exhaust gas analyzer. Start the engine and hold the probe of the exhaust gas analyzer near the check valve port. If any amount of CO or CO_2 is read, the valve leaks. If this valve is leaking, hot exhaust is also leaking, which could ruin the other components in the air injection system.

CASE STUDY

A late-model GM vehicle was brought into the shop because the "Service Engine Soon" lamp was on constantly. A scan tool was used to retrieve the set DTCs, and a P0442 code had been set by the system. This code indicated that there is a small EVAP leak. The technician checked the gas cap to make sure it was tight; it was. He visually inspected the system and found no apparent signs of leakage. He then connected an EVAP pressure tester and pressurized the system. No leaks were evident. Knowing the code was set because of a leak, he was hesitant to assume there was no leak. In addition, the code said it was a small leak, which might be hard to find. He then added smoke to the system, hoping that the small leak would appear. Again, no leak was evident. It then occurred to the technician that the problem was an intermittent one, so he started to move the EVAP hoses to check for looseness. Sure enough, when the canister inlet hose was moved, the meter on the pressure tester dropped and smoke started to come out of the connection of the hose to the canister. After looking at the connection, the technician discovered that the hose clamp was loose. He tightened it, then retested the system for leaks and none were found. To be safe, he erased the DTC and took the vehicle out on the road. He completed what he believed to be several drive cycles. The DTC was not reset; therefore, the problem was solved.

KEY TERMS

Acceleration simulation mode (ASM)
Cutpoint
Dynamometer
Electrostatic discharge (ESD)
Parts per million (ppm)
Repair grade (RG-240)

SUMMARY

- The quality of an engine's exhaust depends on the effectiveness of the emission control devices and the efficiency of the engine.
- The three emissions controlled in gasoline and diesel engines are unburned hydrocarbons (HC), carbon monoxide (CO), and oxides of nitrogen (NO_x).
- Most states and provinces require an annual emission inspection, the most common of which is the I/M 240 test.
- Chassis dynamometers are used during the I/M 240 test and can be valuable when diagnosing other driveability problems, including finding the cause of low power, overheating, and speedometer accuracy.
- The report from an I/M 240 test shows the amount of gases emitted during the different speeds and loads of the test. These can be valuable when diagnosing an emission or engine problem.
- An exhaust analyzer is used to look at the quality of an engine's exhaust, which can indicate the quality of the combustion process taking place in the engine.
- Unburned hydrocarbons are particles of gasoline that have not been burned during combustion. They are present in the exhaust and in crankcase vapors.
- Carbon monoxide forms when there is not enough oxygen to combine with the carbon during combustion.
- High NO_x readings indicate high combustion temperatures.
- CO_2 is a desired element in the exhaust stream and is only present when there is combustion.
- O_2 is used to oxidize CO and HC into water and CO_2 and, therefore, very low amounts of O_2 in the exhaust are desirable.
- The OBD-II system monitors the emission control systems and components that can affect tailpipe or evaporative emissions.

- EVAP systems can be tested with a scan tool, DMM, hand-held vacuum pump, pressure gauge, and leak tester.
- PCV systems are most commonly checked visually or checked with an exhaust gas analyzer.
- EGR systems can be checked visually or with a scan tool, exhaust analyzer, hand-held vacuum pump, and DMM.
- The most common ways to check the efficiency and operation of a catalytic converter are to monitor the pre- and postconverter HO_2S and retrieve DTCs, or conduct the delta temperature, oxygen storage, and efficiency tests.
- Secondary AIR systems are typically checked with a scan tool or exhaust analyzer.

REVIEW QUESTIONS

1. A lean air-fuel ratio causes HC emissions to ______.
2. What will result from too little EGR flow? And what can cause a reduction in the flow?
3. What can result from a charcoal canister that is filled with liquid or water?
4. What happens if a PCV valve is stuck in the open position?
5. When testing a negative and a positive backpressure EGR valve with a vacuum gauge, what are the differences in the expected results?
6. How do you test the efficiency of a secondary AIR system?
7. *True or False?* No-start, surging, or stalling can be caused by an EGR valve that does not open.
8. List five common causes for high HC emissions.
9. Describe carbon monoxide (CO) emissions in relation to air-fuel ratio.
10. A dirty catalytic converter can cause all of the following *except* ______.
    a. stalling after the engine starts
    b. decreased HC emissions from the tailpipe
    c. a drop in engine vacuum
    d. decreased production of NO_x in the cylinders
11. Which of the following statements about EVAP systems is *not* true?
    a. If the system is purging vapors from the charcoal canister when the engine is running at high speeds, rough engine operation will occur.
    b. Cracked hoses or a canister saturated with gasoline may allow gasoline vapors to escape to the atmosphere, resulting in gasoline odor in and around the vehicle.
    c. Rough idle, flooding, and other similar conditions can indicate a saturated canister.
    d. A vacuum leak in the system can cause starting and performance problems.
12. As a catalytic converter begins to deteriorate, the signal from the postcatalyst HO_2S becomes ______ the signal of the precatalyst HO_2S.
    a. shorter than
    b. more like
    c. larger than
    d. flatter than
13. How much pressure does a typical EVAP pressure tester apply to the system during testing?
    a. 14 in. Hg
    b. 14 psi
    c. 1 in. H_2O
    d. 0.5 psi
14. Which of the following exhaust gases is typically not measured during an I/M 240 test?
    a. HC
    b. O_2
    c. NO_x
    d. CO
15. Which of following AIR faults is the least likely to prevent the required air from reaching a catalytic converter?
    a. a bad air pump
    b. an open switching or control valve
    c. out-of-range input signals
    d. leaking air connections

ASE-STYLE REVIEW QUESTIONS

1. While discussing the proper way to test a catalytic converter: Technician A says that a pressure gauge can be inserted into the oxygen sensor bore and the backpressure caused by the converter measured. Technician B says that restrictions in the converter can be checked with a vacuum gauge while the engine is being quickly accelerated. Who is correct?
    a. Technician A
    b. Technician B
    c. Both A and B
    d. Neither A nor B

2. While discussing tailpipe emissions and cylinder misfiring: Technician A says that cylinder misfiring causes a significant increase in HC emissions. Technician B says that cylinder misfiring results in a large increase in CO emissions. Who is correct?
   a. Technician A
   b. Technician B
   c. Both A and B
   d. Neither A nor B
3. While discussing catalytic converter diagnosis: Technician A says that a delta temperature test should be conducted. Technician B says that a good converter is evident by low amounts of CO_2 in the exhaust. Who is correct?
   a. Technician A
   b. Technician B
   c. Both A and B
   d. Neither A nor B
4. While discussing EGR valve diagnosis: Technician A says that if the EGR valve does not open, the engine may hesitate on acceleration. Technician B says that if the EGR valve does not open, the engine may detonate on acceleration. Who is correct?
   a. Technician A
   b. Technician B
   c. Both A and B
   d. Neither A nor B
5. While discussing EGR valve diagnosis: Technician A says that a defective throttle position sensor may affect the EGR valve operation. Technician B says that a defective engine coolant temperature (ECT) sensor may affect the EGR valve operation. Who is correct?
   a. Technician A
   b. Technician B
   c. Both A and B
   d. Neither A nor B
6. When discussing the diagnosis of a positive backpressure EGR valve: Technician A says that with the engine running at idle speed, if a hand pump is used to supply vacuum to the EGR valve, the valve should open at 12 in. Hg of vacuum. Technician B says that with the engine not running, any vacuum supplied to the EGR valve should be bled off and the valve's diaphragm should not move. Who is correct?
   a. Technician A
   b. Technician B
   c. Both A and B
   d. Neither A nor B
7. While diagnosing a PCV problem, Technician A says that a stuck open PCV valve will cause a richer than normal air-fuel mixture. Technician B says that oil in the air cleaner assembly can be caused by worn piston rings. Who is correct?
   a. Technician A
   b. Technician B
   c. Both A and B
   d. Neither A nor B
8. Technician A says that the AIR by pass valve directs secondary air either to the exhaust manifold or to the catalytic converter. Technician B says that secondary air may be vented during deceleration. Who is correct?
   a. Technician A
   b. Technician B
   c. Both A and B
   d. Neither A nor B
9. While discussing PCV system diagnosis: Technician A says that a defective PCV valve may cause rough idle operation. Technician B says that satisfactory PCV system operation depends on a properly sealed engine. Who is correct?
   a. Technician A
   b. Technician B
   c. Both A and B
   d. Neither A nor B
10. While discussing EVAP testing: Technician A says that most OEMs recommend that clean, high-pressure shop air be used to pressurize the system to test for leaks. Technician B says that nitrogen is typically used to drive the smoke through the system when checking the system with a smoke machine. Who is correct?
    a. Technician A
    b. Technician B
    c. Both A and B
    d. Neither A nor B

CHAPTER 35

FUELS AND OTHER ENERGY SOURCES

OBJECTIVES

■ Describe the basic composition of gasoline. ■ Explain why materials are added to gasoline to make it more efficient. ■ Name the common substances used as oxygenates in gasoline and explain what they do. ■ Describe how the quality of a fuel can be tested. ■ Explain the advantages and disadvantages of the various alternative fuels. ■ Explain the differences between the different types of hybrid vehicles. ■ Explain how a fuel cell works.

This chapter takes a look at the fuels and other energy sources used to propel a vehicle. Although there are several types of fuels for automotive use, gasoline is the most commonly used and most readily available. However, there is much interest in finding suitable alternatives to gasoline, including the use of electricity. Much research is being done on electric, hybrid electric, and fuel cell vehicles. These are discussed along with the various automotive fuels.

Regardless of the type of fuel used for combustion, efficiency depends on having the correct amount of air mixed with the correct amount of fuel. The ideal air-fuel or stoichiometric ratio for a gasoline engine is approximately 14.7 pounds of air mixed with 1 pound of gasoline. This provides a ratio of 14.7:1. Different fuels have different stoichiometric ratios. Because air is so much lighter than gasoline, it takes nearly 10,000 gallons of air mixed with 1 gallon of gasoline to achieve this air-fuel ratio. Lean ratios of 15 to 16:1 provide the best fuel economy. Rich mixtures have a ratio below 14.7:1 and provide more power from the engine but greater fuel consumption (**Figure 35–1**).

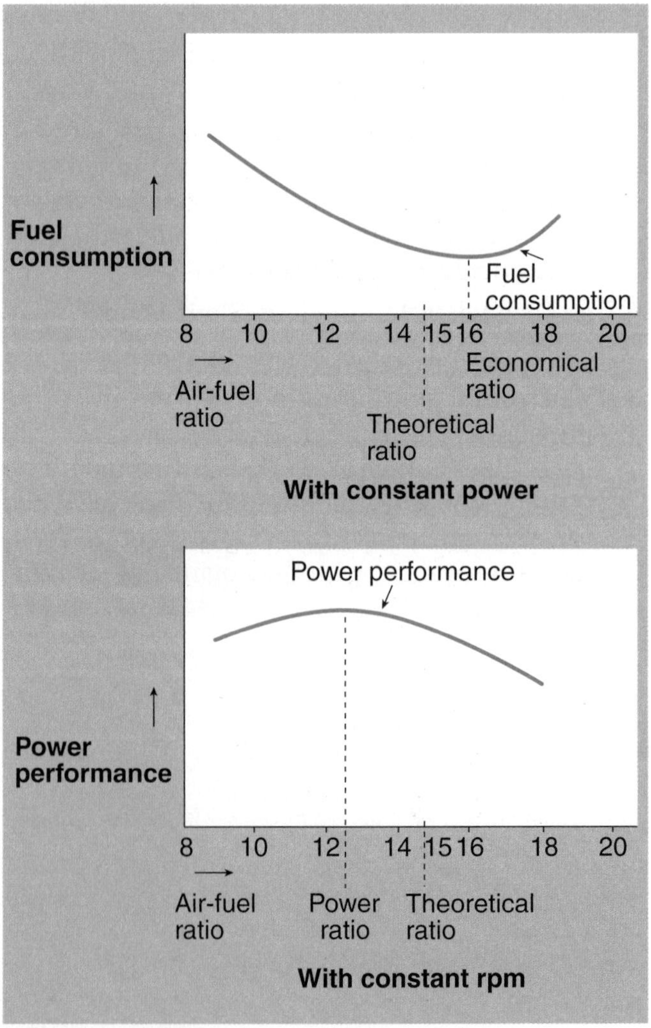

Figure 35-1 Fuel consumption and performance at various air-fuel ratios.

CRUDE OIL

Crude oil is also called **petroleum**, which means oil from the earth. The name fits; crude oil is drawn out of oil reservoirs and sands below the earth's surface. The oil extracted from the earth is called crude because it has yet to be processed or refined. Crude oil is commonly referred to as a fossil fuel because it is naturally produced by the decaying of plants and animals that lived a long time ago and were covered by dirt for many years. Crude oil is a liquid that varies in appearance. Normally, it has a dark brown or black color, but it can also be yellow or greenish.

Although the composition of crude oil varies, it typically is:

- 84% carbon
- 14% hydrogen
- 1% to 3% sulfur, in the form of hydrogen sulfide, sulfides, disulfides, and elemental sulfur
- Less than 1% nitrogen
- Less than 1% oxygen
- Less than 1% metals, normally nickel, iron, vanadium, copper, and arsenic
- Less than 1% salts, in the form of sodium chloride, magnesium chloride, and calcium chloride

The high concentration of carbon and hydrogen is why products produced from crude oil are called hydrocarbon fuels or compounds.

Petroleum Products

Most of the petroleum extracted from the earth is processed into hydrocarbon fuels, such as asphalt, wax, gasoline, diesel fuel, kerosene jet fuel, heating and other fuel oils, lubricating oils and greases, liquefied petroleum gas, and natural gas. About 16% of the crude oil is processed to help make a variety of products, such as polymers, plastics, detergents, deodorants, and medicines.

Hydrocarbons The hydrocarbons (HCs) in crude oil have many different lengths and structures. Therefore, the only thing the different hydrocarbons have in common is that they contain carbon and hydrogen. The number of carbon atoms in an HC molecule defines its length. Sometimes that number, when combined with the number of hydrogen atoms, is called a chain. The HC with the shortest chain is methane (CH_4), which is a very light gas. Longer chains with five or more carbons are liquids or solids. Asphalt has thirty-five or more carbon atoms. HCs contain a great amount of energy, which is why they have been used as a source of energy for many years.

Each of the different HCs has a distinct purpose. Therefore, each of the HCs must be separated from crude oil in order to be useful. This separation process is called refining. After refining, one barrel (42 gallons or 159 liters) of crude oil will produce 20 gallons (75.7 L) of gasoline, 7 gallons (26.5 L) of diesel fuel, and smaller amounts of various other petroleum products.

Refining

Oil refining separates the HCs into useful substances. A refinery is the place where the separation occurs. The easiest and most common way to separate the various HCs (called fractions) is through a process called **fractional distillation** **(Figure 35–2)**. The basis of this method is simply that the different HC chain lengths have progressively higher boiling points. This means that the shorter chains require less heat to vaporize. Here are some examples (keep in mind that there are types of each of the following based on its quality):

- Propane will boil at less than 104°F (40°C).
- Gasoline will boil at 104° to 401°F (40° to 205°C).
- Jet fuel will boil at 350° to 617°F (175° to 325°C).
- Diesel fuel will boil at 482° to 662°F (250° to 350°C).
- Lubricating oil will boil at 572° to 700°F (300° to 370°C).
- Asphalt will boil at temperatures greater than 1,112°F (600°C).

During fractional distillation, crude oil is heated with high-pressure steam to about 1,112°F (600°C). This causes all of the crude oil to boil, forming vapor. The vapor moves into the fractional distillation column that contains many trays or plates. The trays have holes to allow the vapor to pass through. The heated oil vapors heat the column, causing the bottom to be very hot and the top cool. As the vapor moves up the column, it cools. The vapor condenses or becomes a liquid when it reaches the point in the column where the temperature is equal to the fractions' boiling temperature. Therefore, the fractions with the lowest boiling point will condense at the highest level within the column and those with high boiling points will condense at lower levels. The various trays collect the condensation and pass the liquid out of the column.

Very few of the fractions that leave the column are ready to be used. They must be treated and cleaned to remove impurities. Also, refineries combine the various fractions to make a desired product. For example, different octane ratings of gasoline are possible by mixing different fractions. Some fractions are chemically altered so they can be used for their specific application, and others are chemically processed to produce other fractions. The finished products are stored until they can be delivered to their markets.

Chemical Processing Some fractions are processed chemically to produce a different type of HC. For example, chemical processing can break down longer chains into shorter ones. Doing this allows the

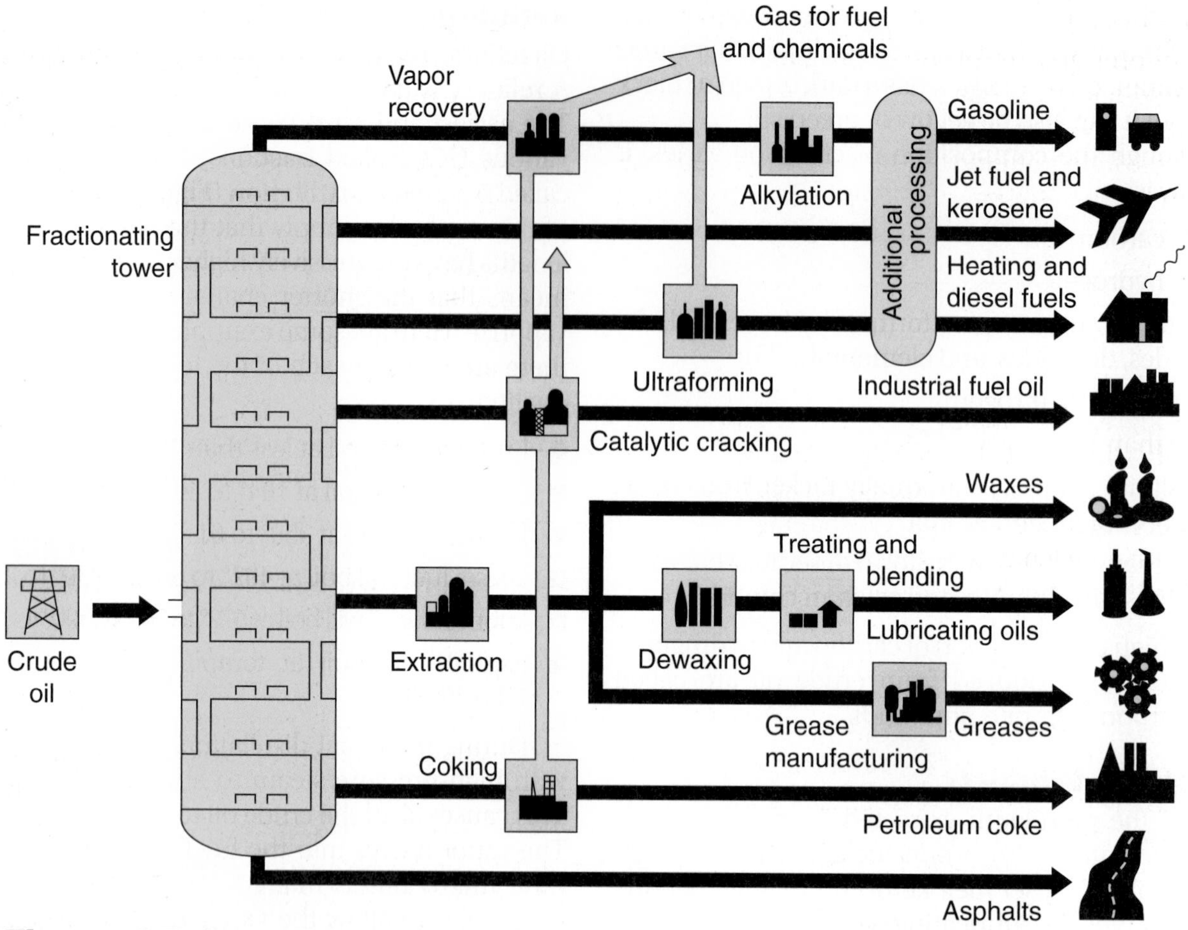

Figure 35-2 The refining process for crude oil. *Courtesy of the American Petroleum Institute*

refineries to alter some of the HCs to meet market demands. Through processing, if the demand for gasoline is high, diesel fuel can be altered to become gasoline.

The process of breaking down a large HC into a small one is called cracking. During cracking, large HCs are introduced to high temperature and sometimes high-pressure conditions, forcing the HCs to break apart. Catalysts are often used to speed up the cracking process.

At times, small HCs are combined to make larger ones. This process is called unification and relies on a catalyst to force the HCs to combine.

Also, the structure of HC molecules can be rearranged to provide a different HC. This is commonly used to produce octane boosters for gasoline.

Cleaning and Blending The fractions captured from fractional distillation and chemical processing are treated in a variety of ways to remove all impurities. Some of the techniques used by refineries include passing the fractions through sulfuric acid to remove unsaturated HCs, a drying column to remove water, and hydrogen-sulfide scrubbers to remove sulfur.

After they are cleaned, the base fraction is blended with small amounts of other fractions to make various products, such as various grades of gasoline and lubricating oil and greases.

Concerns

Fossil fuels are used for many purposes; in fact, due to their high-energy density and relative abundance, they are the world's most important energy source. However, their use comes with costs. Although it appears that there is plenty of oil available today, we may run out in the future.

It is estimated that there is approximately 3.74 trillion barrels (440 km^3) of oil reserves, including oil sands, available. This seems like a lot; however, the current level of oil consumption is about 84 million barrels (3.6 km^3) per year. This means the oil from known oil reserves will be gone by 2039.

Another concern is that when HCs are burned, they release CO_2. This is a growing concern because CO_2 has been linked to global warming. Although there is much emphasis on reducing CO_2 emissions, it is important to realize that of the total amount of CO_2 emissions worldwide, less than 4% is man-made. The rest is from nature; things like breathing and plant and animal decay contribute greatly to the buildup of CO_2 in the atmosphere.

Burning fossil fuels for transportation contributes to about one-quarter of the man-made CO_2 emissions. Reducing this has been the goal of manufacturers and governments.

The declining amount of oil in reserves and the concern for the environment are the leading factors in the development and use of alternate fuels and energy sources.

GASOLINE

Gasoline is a complex mixture of approximately 300 various ingredients, mainly HCs. The chemical symbol for this liquid is C_8H_{15}, which indicates that each molecule of gasoline contains eight carbon atoms and fifteen hydrogen atoms. Gasoline is a colorless liquid with excellent vaporization capabilities.

Oil refiners must meet gasoline standards set by the American Society for Testing and Materials (ASTM), the EPA, some state requirements, and their own company standards. Many of the performance characteristics of gasoline can be controlled during refining and blending. Many additives are blended into gasoline before it is available to the public (**Figure 35–3**). The major factors affecting fuel performance are antiknock quality, volatility, sulfur content, and deposit control.

Purpose	Additive
Octane enhancer	Methyl *t*-butyl ether (MTBE) *t*-butyl alcohol (TBA) Ethanol Methanol
Antioxidant	Butylated methyl, ethyl and dimethyl phenols Various other phenols and amines
Metal deactivator	Disalicylidene-*N*-methyldipropylene-triamine *N,N'*-disalicylidene-1,2-Ethanediamine Other related amines
Ignition controller	Tri-o-cresylphosphate (TOCP)
Icing inhibitor	Isopropyl alcohol
Detergent/dispersant	Various phosphates, amines, phenols, alcohols, and carboxylic acids
Corrosion inhibitor	Carboxylic, phosphoric, and sulfonic acids

Figure 35-3 The various additives blended with gasoline for today's vehicles.

Antiknock Quality

An **octane** number or rating was developed by the petroleum industry so the **antiknock** quality of a gasoline could be rated. The octane number is a measure of the fuel's tendency not to experience detonation in the engine. The higher the octane rating, the less the engine has of a tendency to knock.

Two methods are used for determining the octane number of gasoline: the **motor octane number (MON)** method and the **research octane number (RON)** method. Both use a laboratory single-cylinder engine equipped with a variable head and knock meter to measure knock intensity. A test sample of the fuel is used in the engine as the engine's compression ratio and air-fuel mixture are adjusted to develop a specific knock intensity. There are two primary standard reference fuels: **isooctane** and **heptane.** Isooctane does not knock in an engine but is not used in gasoline because of its expense. Heptane knocks severely in an engine. Isooctane has an octane number of 100. Heptane has an octane number of zero.

A fuel of unknown octane value is run in the test engine equipped with a variable compression cylinder head and a knock meter. The severity of knock is measured. Various proportions of isooctane and heptane are run in the engine to duplicate the severity of the engine knock when the test fuel was run. When the knock caused by the isooctane and heptane mixture matches that caused by the fuel being tested, the octane number is established by the percentage of isooctane in the mixture. For example, if 85% isooctane and 15% heptane produced the same knock severity as the tested fuel, that fuel would be rated as having an octane rating of 85.

The octane rating required by law and the one displayed on gasoline pumps is the **Antiknock Index (AKI)**. It is the average of RON and MON. The AKI is stated as (R+M)/2.

By itself, the octane rating has nothing to do with fuel economy or engine efficiency. Only when an engine is designed to require the use of higher octane gasoline can the value of the fuel be obtained. Most modern engines are designed to operate efficiently with regular grade gasoline and do not require high-octane gasoline. One of the things to remember about high-octane fuel is that it burns slower than low-octane gasoline. This is why it is less likely to cause detonation. Most engine control systems have a sensor to detect if a knock is occurring so the PCM can retard the ignition timing to prevent detonation.

There is less heat energy in high-octane gasoline and, therefore, less power is generated during combustion. Using a higher-octane gasoline than required will not produce more power unless the engine has a knock sensor that allows the system to adjust ignition timing to take advantage of the higher octane. Higher-octane gasoline is used in high-performance engines because they have high compression ratios, which provide greater power output.

Volatility

Gasoline is very volatile. It readily evaporates, so its vapor adequately mixes with air for combustion. Only vaporized fuel supports combustion. The volatility of gasoline affects the following performance characteristics or driving conditions:

- *Cold Starting and Warmup.* A fuel can cause hard starting, hesitation, and stumbling during warmup if it does not readily vaporize. A fuel that vaporizes too easily in hot weather can form vapor in the fuel delivery system, causing **vapor lock** or a loss of performance. If gasoline vaporizes while it is in a fuel line, it can stop the flow of gasoline. Rather than flow through the line, the pressurized fuel will compress the vapor, not move it. Vapor lock can cause a variety of driveability problems.
- *Altitude.* Gasoline vaporizes more easily at high altitudes, so volatility is controlled in blending according to the elevation of the location where the fuel is sold.
- *Crankcase Oil Dilution.* A fuel must vaporize to prevent diluting the crankcase oil with liquid fuel or break down the oil film on the cylinder walls, causing scuffing or scoring. The liquid eventually enters the oil in the crankcase, forming an accumulation of sludge, gum, and varnish as well as affecting the lubrication properties of the oil.

Gasoline blended for summer (hot weather) use is less volatile (does not burn as easily) than winter gasoline. Also, in high altitude areas, fuels must be blended to have higher volatility.

There are three methods of measuring the volatility of a fuel. The most common is the **Reid vapor pressure (RVP)** test. The RVP test is performed by placing a sample of gasoline into a sealed metal container that has a pressure measuring device attached to it. The container is submerged in heated (100°F or 38°C) water. As the fuel is heated, it vaporizes. Remember, the more volatile a fuel is, the easier it will vaporize. As fuel vaporizes, it creates vapor pressure within the container. Fuels that are more volatile will create more pressure. Vapor pressure is measured in psi.

Sulfur Content

Gasoline can contain some of the sulfur present in the crude oil. Sulfur content is reduced at the refinery to limit the amount of corrosion it can cause in the engine and exhaust system. When the hydrogen in the HCs of the fuel is burned, one of the by-products of combustion is water. Water leaves the combustion chamber as steam but can condense back to water when passing through a cool exhaust system. When the engine is shut off and cools, steam condenses back to a liquid and forms water droplets. Steam present in crankcase blowby also condenses to water.

When the sulfur in the fuel is burned, it combines with oxygen to form **sulfur dioxide**. This compound can combine with water to form sulfuric acid, which is a highly corrosive compound. This type of corrosion is the leading cause of exhaust valve pitting and exhaust system deterioration. Sulfuric acid also attacks the lead linings of the main and rod bearings. This is one reason engine oil needs to be changed regularly. With catalytic converters, the sulfur dioxide can cause the obnoxious odor of rotten eggs during engine warmup. To reduce corrosion caused by sulfuric acid, the sulfur content in gasoline is limited to less than 0.01%.

Deposit Control

Several additives are put into gasoline to control harmful deposits, including gum or oxidation inhibitors, detergents, metal deactivators, and rust inhibitors.

BASIC GASOLINE ADDITIVES

At one time, all a gasoline-producing company had to do to produce its product was to pump the crude from the ground, run it through the refinery to separate it, dump in a couple of grams of lead per gallon, and deliver the finished product to the service station. Of course, automobiles were much simpler then and what they burned was not very critical. As long as gasoline vaporized easily and did not cause the low-compression engines to knock, everything was fine.

Lead compounds, such as **tetraethyl lead (TEL)** and **tetramethyl lead (TML)**, were added to gasoline to increase its octane rating. However, since the mid-1970s, vehicles have been designed to run on unleaded gasoline only. Leaded fuels are no longer available as automotive fuels. Because of the poisoning effect lead has on a catalytic converter; today's gasolines are limited to a lead content of 0.06 gram per gallon. To achieve the desired octane rating, **methylcyclopentienyl manganese tricarbonyl (MMT)** is added to gasoline.

Anti-Icing or Deicer

Isopropyl alcohol is added seasonally to gasoline as an anti-icing agent to prevent fuel line freeze-up in cold weather.

Metal Deactivators and Rust Inhibitors

These additives are used to inhibit reactions between the fuel and the metals in the fuel system that can form abrasive and filter-plugging substances.

Gum or Oxidation Inhibitors

Some gasolines contain aromatic amines and phenols to prevent the formation of gum and varnish. During storage, harmful gum deposits can form due to the reaction between some gasoline molecules with each other and with oxygen. Oxidation inhibitors are added to promote gasoline stability. They help control gum, deposit formation, and staleness.

Gum content is influenced by the age of the gasoline and its exposure to oxygen and certain metals such as copper. If gasoline is allowed to evaporate, its residue can form gum and varnish.

Detergents

Detergent additives are designed to do only what their name implies—clean certain critical parts inside the engine. They do not affect octane.

Performance TIP *Adding nitrous oxide to the air-fuel mixture is not something done by oil refineries. Rather it is commonly done by those seeking more instantaneous power from their engines. Nitrous oxide is injected as a dense liquid. When nitrous oxide is heated, it breaks down into nitrogen and oxygen. This provides more oxygen inside the cylinder when the fuel ignites. Because there is more oxygen, more fuel can be injected into the cylinder. The engine therefore produces more power. Nitrous oxide also improves engine performance by cooling the gases in the cylinder, thereby making the air denser. Nitrous oxide is injected into the engine's intake when the driver pushes a button to activate the system. Nitrous kits, which include nearly all that is needed to add the system to an engine, are available for many engines. The nitrous tanks typically store enough nitrous for 3 to 5 minutes of operation.*

OXYGENATES

Oxygenates are compounds, such as alcohols and ethers, that contain oxygen. By carrying oxygen, the fuel tends to lean the mixture. Oxygenates improve combustion efficiency, thereby reducing emissions. Many oxygenates also serve as excellent octane enhancers when blended in gasoline (**Figure 35–4**). Oxygenated fuels tend to have lower CO emissions.

It should be noted that the use of oxygenated gasoline may cause a slight decrease in fuel economy in late-model vehicles. This is due to the HO_2S detecting extra oxygen and the PCM responding to this by richening the mixture. Also, the alcohol used in oxygenated gasoline may cause swelling of an older vehicle's fuel hoses, gaskets, and O-rings.

Oxygenates added to gasoline produce what is referred to as **reformulated gasoline (RFG)**. RFG is also called "cleaner-burning" gasoline and costs slightly more than normal gasoline. RFG can be used in most engines with no modifications.

Ethanol

By far the most widely used gasoline oxygenate additive is **ethanol** (ethyl alcohol), or grain alcohol. Ethanol is a noncorrosive and relatively nontoxic alcohol made from renewable biological sources. Blending 10% ethanol into gasoline results in an increase of 2.5 to 3 octane points. With ethanol-blended gasoline, air toxics are about 50% less.

In addition to octane enhancement, ethanol blending keeps the fuel injectors cleaner and less subject to corrosion due to the detergent additives found in most ethanol. Ethanol can also loosen contaminants and residues that may have gathered in the vehicle's fuel system.

All alcohols have the ability to absorb the water in the fuel system from condensation. This reduces the chances of fuel line freeze-up during cold weather. Ethanol also decreases CO emissions due to the higher oxygen content of the fuel.

Ethanol blends are approved by all auto manufacturers because of their clean air benefits. Older engines with nonhardened valve seats may need a lead

	Ethanol	MTBE	ETBE	TAME
Chemical formula	CH_3CH_2OH	$CH_3OC(CH_3)_3$	$CH_3CH_2OC(CH_3)_3$	$(CH)_3CCH_2OCH_3$
Octane, (R+M)/2	115	110	111	105
Oxygen content, % by weight	34.73	18.15	15.66	15.66
Blending vapor pressure, RVP	18	8	4	1.5

Figure 35-4 The typical properties of the common oxygenates.

substitute added to gasoline or ethanol blends to prevent premature valve seat wear. The chance of valve burning is decreased when ethanol is used because ethanol burns cooler than gasoline.

Methanol

Methanol is the lightest and simplest of the alcohols and is also known as wood alcohol. It can be distilled from coal or renewable sources, but most of what is used today is derived from natural gas.

Many automakers continue to warn motorists about using a fuel that contains more than 10% methanol and cosolvents by volume. Methanol is recognized as being far more corrosive to fuel system components than ethanol, and it is this corrosion that has automakers concerned. Methanol is also highly toxic and there are safety concerns with ingestion, eye or skin contact, and inhalation.

Methanol can be used directly as an automotive fuel but the engine must be modified for its use. It can also be used in flexible fuel vehicles as M85, which is 85% methanol. However, this is not very common because manufacturers are no longer supplying methanol-powered vehicles. In the future, methanol could be the fuel of choice for providing hydrogen to power fuel cell vehicles.

MTBE

In the past, methyl tertiary butyl ether (MTBE) was used as an octane enhancer because of its excellent compatibility with gasoline. Methanol can be used to make MTBE. However, MTBE production and its use have declined because it was found to contaminate groundwater. As of 2004, MTBE is no longer used in gasoline and has been replaced by ethanol and other oxygenates such as tertiary amyl methyl ether (TAME) and ethyl tertiary butyl ether (ETBE).

Aromatic Hydrocarbons

These are petroleum-derived compounds, including benzene, xylene, and toluene, that are being used as octane boosters.

GASOLINE QUALITY TESTING

Two tests can be done to test the quality of gasoline: the Reid vapor pressure test and the alcohol content test.

Testing the RVP of Gasoline

RVP is a measure of the volatility of gasoline. Fuels that are more volatile vaporize more easily, creating more pressure. Increasing the RVP of a gasoline permits the engine to start easier in cold weather. The RVP of winter blend gasoline is about 9.0 psi. Summer grade is typically around 7.0 psi.

A special fuel vapor pressure tester is needed to test the RVP of gasoline. Make sure the gasoline that is being tested is cool. Then put a sample in the tester's container and secure seal the container as soon as the gasoline is in it. Put hot water in another container and put the container holding the fuel into it. Make sure that most of the container holding the fuel is covered by water. Connect the pressure gauge assembly to the container holding the gasoline. Put a thermometer in the water. When the water temperature is 105°F (40°C) for at least 2 minutes, take your pressure reading and compare it to specifications.

Alcohol in Fuel Test

Pump gasoline may contain a small amount of alcohol, normally up to 10%. If the amount is greater than that, problems may result, such as fuel system corrosion, fuel filter plugging, deterioration of rubber fuel system components, and a lean air-fuel ratio. These fuel system problems caused by excessive alcohol in the fuel may cause driveability complaints such as lack of power, acceleration stumbles, engine stalling, and no-start. If the correct amount of fuel is being delivered to the engine and there is evidence of a lean mixture, check for air leaks in the intake then check the gasoline's alcohol content.

There are many different ways to check the percentage of alcohol in gasoline. Some are more exact than others and some require complex instruments.

PROCEDURE

To check the amount of alcohol in a sample of gasoline:

STEP 1 Obtain a 100-milliliter (mL) cylinder graduated in 1 mL divisions.

STEP 2 Fill the cylinder to the 90 mL mark with gasoline.

STEP 3 Add 10 mL of water to the cylinder so it is filled to the 100 mL mark.

STEP 4 Install a stopper in the cylinder, and shake it vigorously for 10 to 15 seconds.

STEP 5 Carefully loosen the stopper to relieve any pressure.

STEP 6 Install the stopper and shake vigorously for another 10 to 15 seconds.

STEP 7 Carefully loosen the stopper to relieve any pressure.

STEP 8 Place the cylinder on a level surface for 5 minutes to allow liquid separation.

STEP 9 Observe the liquid. Any alcohol in the fuel is absorbed by the water and settles to the bottom. If the water content in the bottom of the cylinder exceeds 10 mL, there is alcohol in the fuel. For example, if the water content is now 15 mL, there was 5% alcohol in the fuel. NOTE: Because this procedure does not extract 100% of the alcohol from the fuel, the percentage of alcohol in the fuel may be higher than indicated.

ALTERNATIVES TO GASOLINE

The concerns of burning fossil fuels and the rapid decline of their reserves have led to a near frantic search for alternative fuels. While determining the viability of an alternative fuel, many things are considered, including emissions, cost, availability, fuel consumption, safety, engine life, fueling facilities, weight and space requirements for fuel tanks, and the range of a fully fueled vehicle. By using alternative fuels, we not only can reduce our reliance on oil, but we can reduce emissions and the effects an automobile's exhaust has on global warming.

Much attention has been paid to renewable fuel sources. **Renewable fuels** are those derived from nonfossil sources and produced from plant or animal products or wastes (biomass). Biomass fuels, such as biodiesel and ethanol, can be burned in internal combustion engines. Biomass fuels tend to be carbon neutral, which means that during combustion, they release the same amount of CO_2 that was absorbed from the atmosphere when the plant or animal was living. Combustion does not cause an increase in CO_2 emissions.

Ethanol and methanol are used as oxygenates for blending with gasoline. They can also be used as the primary energy source for internal combustion engines. However, because ethanol is made from renewable sources it is the most commonly used.

Ethanol

Ethanol is a high-quality, low-cost, high-octane fuel (rated at 115) that burns cleaner than gasoline. The use of ethanol as a fuel is not new. Ford's Model T was designed to run on ethyl alcohol (ethanol). Ethanol (CH_3CH_2OH), commonly called grain alcohol, is a renewable fuel made from nearly anything that contains carbon (**Figure 35–5**). It is most commonly produced by fermenting and distilling corn, cornstalks, sugar cane, other grains, or biomass waste.

Because ethanol is an alcohol, it can absorb the moisture that may be present in a fuel system. The absorbed water is simply passed with the fuel and burned by the engine. However, if the moisture content in the fuel becomes too high, the water will separate from the fuel and drop to the bottom of the fuel tank. If this is suspected, remove all fuel and water from the tank and refill it with clean ethanol-blended fuel.

For automotive use, ethanol is blended with gasoline. The common blends are an E10 blend, which is 10% ethanol and 90% gasoline, and E85, which is 85% ethanol. Most gasoline-powered vehicles in North America can run on blends of up to 10% ethanol, and some are equipped to run on E85.

The use of E85 has many advantages over the use of traditional gasoline:

- It is produced in the United States and can reduce our reliance on foreign oil.
- Vehicles do not need many modifications to use it.
- Its emissions are cleaner than those of a gasoline engine.
- CO_2 emissions are much lower.
- Ethanol-blended fuel keeps the fuel system clean because it does not leave varnish or gummy deposits.

However, the infrastructure for E85 is weak. There are very few fuel filling stations that offer E85. The

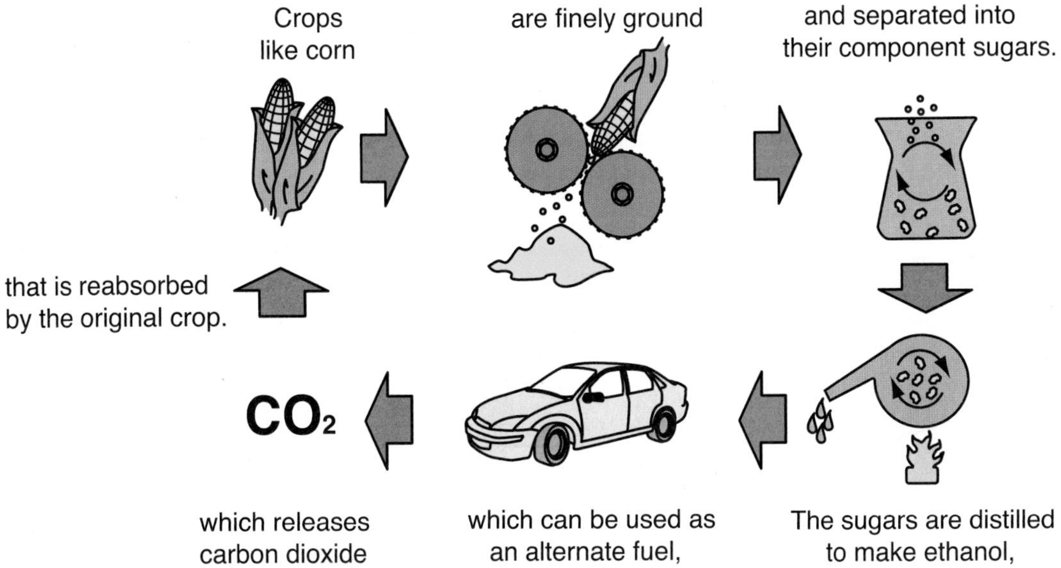
Crops like corn
are finely ground
and separated into their component sugars.
The sugars are distilled to make ethanol,
which can be used as an alternate fuel,
which releases carbon dioxide
CO2
that is reabsorbed by the original crop.

Figure 35-5 The carbon cycle of ethanol.

amount of energy it takes to produce E85 is more than the energy it provides. E85 also has about 25% less energy than gasoline; therefore, fuel economy will decrease by about that much in a typical vehicle.

Methanol Methanol (CH_3OH) is a clean-burning alcohol fuel that is most often made from natural gas but can also be produced from coal and biomass. Because North America has an abundance of these materials, the use of methanol can decrease the dependence on foreign oils. Methanol use as a fuel has declined through the years but may become the fuel for fuel cell vehicles. It has been the fuel of choice for Indianapolis-type race cars since 1965. However, starting in 2007, the Indy Racing League (IRL) cars switched to pure ethanol for all races.

Currently, these alcohols are mixed with 15% gasoline, creating M85 and E85. The small amount of gasoline improves the cold-starting ability of the alcohols.

Flex Fuel Vehicles

Many vehicles are designed to use something other than gasoline as fuel. These are commonly referred to as dedicated vehicles or bi- or multiple-fuel vehicles. Dedicated vehicles are those designed to use one particular type of alternative fuel, such as diesel fuel. Bi- and multiple-fuel vehicles are designed to use more than one fuel. Bi-fuel vehicles have two separate tanks and can operate on one fuel or the other, such as natural gas or unleaded gasoline. A bi-fuel vehicle has two separate fuel systems; the two fuels are mixed inside the engine.

Flexible fuel vehicles (FFV) can run on ethanol or gasoline, or a mixture of the two **(Figure 35–6)**. The alcohol fuel and gasoline are stored in the same tank. This gives the driver flexibility and convenience when refilling the fuel tank. Many vehicles are fitted with systems that allow the use of multiple fuels. These include vehicles from Chrysler, Ford Motor Co., General Motors, and Nissan. Flex fuel vehicles have a clover leaf symbol (internally or externally) that shows they can use multiple fuels.

Most of these vehicles use a sensor that detects the blend of fuel in the tank, and the PCM adjusts the operating parameters accordingly. Others have a virtual fuel sensor that relies on inputs from the HO_2S for oxygen readings. These systems adjust the air-fuel mixture according to the oxygen readings of the different fuel compositions that may be in the fuel tank.

Propane/LPG

Propane, also referred to as **liquefied petroleum gas** or **LP gas**, is used by many fleets around the world in taxis, police cars, school buses, and trucks. LP gas is similar to gasoline chemically. It is called liquid petroleum because it is stored as a liquid in a pressurized bottle. The pressure increases the boiling point of the liquid and prevents it from vaporizing. LP gas burns clean because it vaporizes at atmospheric temperatures and pressures. This means it emits less HCs, CO_2, and CO. Propane is a clean-burning fuel that provides a driving range closer to gasoline than other alternative fuels.

Propane allows for quick starting, even in the coldest of climates. It also has a higher octane rating than gasoline. However, there is a reduction of engine power output (about 5%) because it is difficult to fill the cylinders with the gas. Propane is a dry fuel that enters the engine as vapor. Gasoline, on the other hand, enters the engine as tiny droplets of liquid. LP gas is a good alternative to gasoline but it is a fossil fuel and therefore is not a favored alternative fuel for the future.

LP gas vehicles have special tanks or cylinders to store the gas **(Figure 35–7)**. However, the gas is stored at about 200 pounds per square inch. Under this pressure, the gas turns into a liquid and is stored as a

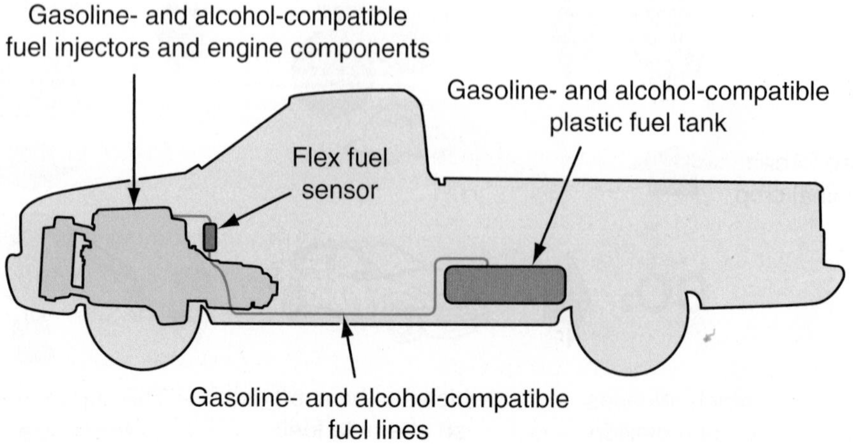

Figure 35-6 A flexible fuel vehicle.

Figure 35-7 The layout of a propane car.

liquid. When the liquid propane is drawn from the tank, it warms and changes back to a gas before it is burned in the engine. The propane fuel system is a completely closed system.

Compressed Natural Gas

Natural gas, **compressed natural gas (CNG)** and liquefied natural gas (LNG), is a very clean-burning fuel. There is an abundant supply of natural gas. It burns cleaner and it is less expensive than gasoline. Combustion with CNG also results in 25% less CO_2 emissions. These factors make the use of natural gas an attractive alternative fuel, especially to companies with fleets of vehicles. However, it is reliant on fossil oil and therefore does not address the issue of continued reliance on fossil fuel. Typically, CNG is used in light- and medium-duty vehicles, whereas LNG is used in transit buses, train locomotives, and long-haul semi-trucks.

CNG must be safely stored in cylinders at pressures of 2,400, 3,000, or 3,600 pounds per square inch. This is the biggest disadvantage of using CNG as a fuel. The space occupied by these cylinders takes away luggage and, sometimes, passenger space. As a result, CNG vehicles have a shorter driving range than comparable gasoline vehicles. Bi-fuel vehicles are equipped to store both CNG and gasoline and will run on either.

Natural gas turns into a liquid when it is cooled to −263.2°F (−164°C). Because it is a liquid, a supply of LNG takes up less room in the vehicle than does CNG. Therefore, the driving range of an LNG vehicle is longer than a comparable CNG vehicle. However, the fuel must be dispensed and stored at extremely cold temperatures. This requires refrigeration units that also take up space. This is why LNG is not a practical fuel for personal use and is only used in heavy-duty applications.

The chief disadvantage of CNG at present is its limited distribution network. Fuel facilities are needed in greater numbers than are currently in existence due to the relatively shorter range of CNG vehicles. The space taken by the CNG cylinders and their weight, about 300 pounds, also would be considered disadvantages in most applications. The basic components of a **natural gas vehicle (NGV)** are shown in **Figure 35–8**.

CNG is injected into the cylinders by a high-pressure injector at the instant ignition occurs. The high pressures and temperature in the cylinder speeds the pre-flame reaction to start the ignition of the fuel injected into the cylinder. The compressed, hot air in the cylinder causes the fuel to ignite quickly.

P-Series Fuels P-series is a new fuel classified as an alternative fuel. It is a blend of natural gas liquids, ethanol, and biomass-derived cosolvents. **P-series fuels** are clear, colorless, 89–93 octane, liquid blends that are formulated to be used alone or freely mixed, in any proportion, with gasoline. Like gasoline, low vapor pressure formulations are produced to prevent excessive evaporation during summer, and high vapor pressure formulations are used for easy starting in cold weather.

Each gallon of P-series fuel emits approximately 50% less CO_2, 35% less HCs, and 15% less CO than gasoline. It also has 40% less ozone-forming potential. Other benefits of P-series fuels include:

- They could be 96% derived from domestic sources.
- More than 60% of the energy content in the fuel is derived from renewable sources.
- They could reduce fossil energy use by 49% to 57% and petroleum use by 80% compared to gasoline.
- Greenhouse emissions are 45% to 50% below those of reformulated gasoline.

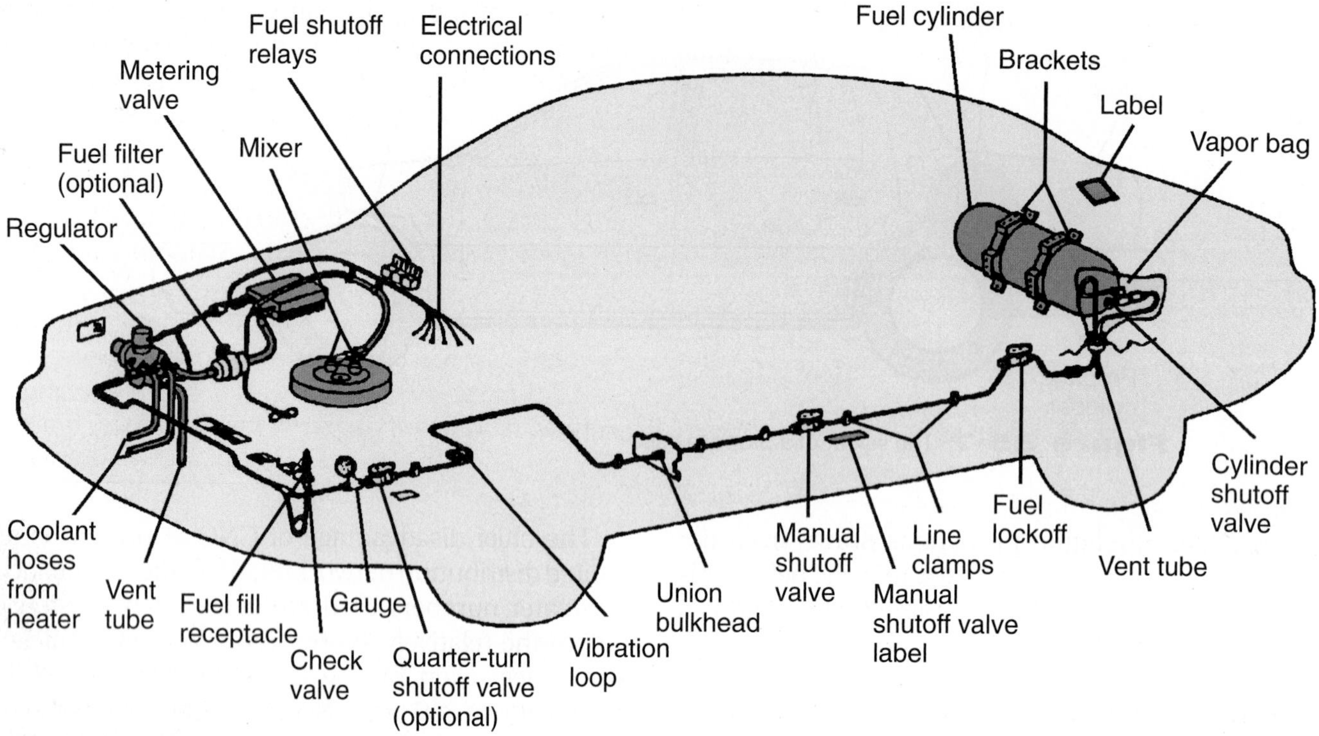

Figure 35-8 NCV system components. *Courtesy of Teleflex GFI Control Systems Inc.*

Hydrogen

Hydrogen is cited by some as the fuel of the future because it is full of energy due to its atomic structure and abundance. It is the simplest and lightest of all elements and has one proton and one electron **(Figure 35–9)**. Hydrogen is a colorless and odorless gas. It is one of the most abundant elements on earth. However, it is only found in compound form. The combination of hydrogen and oxygen forms water. Fossil fuels are combinations of carbon and hydrogen, or HCs.

Hydrogen is extracted from various substances through a process that pulls hydrogen out of its bond with another element or elements. Hydrogen is commonly extracted from water, fossil fuels, coal, and biomass. The two most common ways that hydrogen is produced are steam reforming and electrolysis. Currently it costs much more to produce hydrogen than it does to produce other fuels such as gasoline. This, again, is an obstacle and the focus of much research.

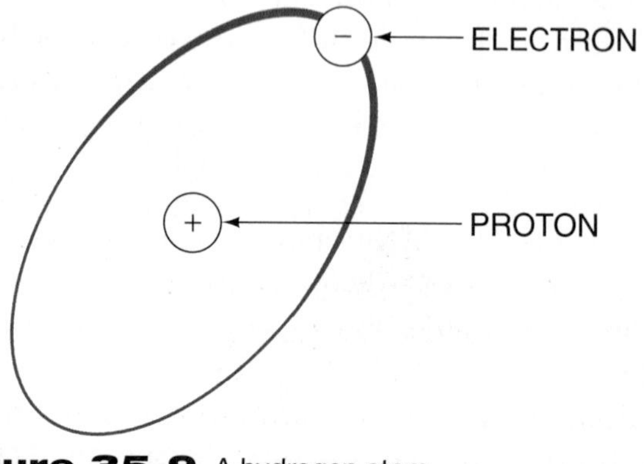

Figure 35-9 A hydrogen atom.

Steam reforming is the most common method used to produce hydrogen. This process uses high-temperature steam to extract hydrogen from natural gas or methane. During electrolysis, electrical current is passed through water. The water then separates into hydrogen and oxygen. The hydrogen atoms collect at a negatively charged cathode, and the oxygen atoms collect at the positively charged anode.

In-vehicle **reformers** can be used to extract hydrogen from a fossil fuel. Reformers make the vehicle more practical because the fuel supply is easily replenished. However, reformers do have some emissions issues, are costly, and take up valuable vehicle space. There is also an issue of the purity of the fuels that will be reformed.

Hydrogen Fuel To demonstrate the energy in hydrogen, there are hydrogen bombs. Some manufacturers are experimenting with burning hydrogen in internal combustion engines. There is also research being done with adding hydrogen to other fuels. In both of these cases, exhaust emissions are reduced without a great decrease in power output; in some cases the power actually increased.

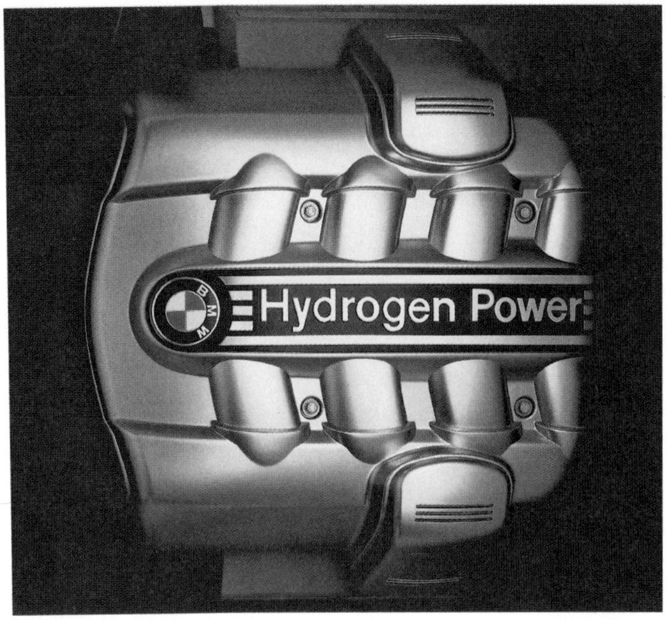

Figure 35-10 A hydrogen-powered internal combustion engine. *Courtesy of BMW of North America, LLC*

Three major auto manufacturers have developed and tested hydrogen-fueled internal combustion engines; actually these vehicle have bi-fuel capabilities. BMW's bi-fueled V12 engine uses liquefied hydrogen or gasoline as its fuel **(Figure 35–10)**. When running on hydrogen, the engine emits zero CO_2 emissions. To store the liquefied hydrogen, the storage tank is kept at a constant temperature of −423°F (−253°C). At this temperature, the liquid hydrogen has the highest possible energy density.

Ford and Mazda have also developed vehicles with hydrogen power. Mazda is using its rotary engine, which it claims is ideal for using hydrogen fuel. The concept vehicles from both manufacturers are also bi-fuel vehicles. Ford has converted a V10 and a 2.3-liter, I-4 engine to run on hydrogen. Engine modifications include a higher compression ratio, special fuel injectors, and a modified electronic control system. When running on hydrogen, the engine is more than 10% more efficient than when it runs on gasoline and emissions levels are very close to zero. Because the fuel contains no carbon, there are no carbon-related emissions (CO, HC, or CO_2).

Typically, an engine running on hydrogen produces less power than a same-sized gasoline-powered engine. Ford added a supercharger with an intercooler to the engines to compensate for the loss of power.

Infrastructure and Storage Other than manufacturing costs, the biggest challenge for hydrogen-powered vehicles is the lack of an infrastructure. Vehicles need to be able to be refueled quickly and conveniently.

Hydrogen contains more energy per weight than any other fuel, but it contains much less energy by volume. This makes storing enough hydrogen for an acceptable driving range very difficult. Naturally, you can store more in a larger container but that container would take up more space and add considerable weight to the vehicle.

Hydrogen is normally stored as a liquid or as a compressed gas. When stored as a liquid, it must be kept very cold. Keeping it that cold adds weight and complexity to the storage system. The tanks required for compressed hydrogen need to be very strong, and that translates to weight. Also, higher pressures mean more hydrogen can be packed into the tank but the tank must be made stronger before the pressure can be increased.

DIESEL FUEL

Diesel fuel is designed to be used by diesel engines and is therefore not an alternative fuel for gasoline engines. Diesel fuel is a fossil fuel but it has different properties and characteristics than gasoline. Diesel fuel is heavier and has more carbon atoms, and it has about 15% more energy density. Diesel engines provide more torque than a gasoline engine of the same size and consume less fuel per mile. Because diesel engines burn less fuel, they also emit less CO_2 **(Figure 35–11)**.

The small, high-speed diesel engines found in automobiles require a high-quality and highly volatile fuel because they cannot tolerate excessive carbon deposits that result from low-volatile fuels. Large, low-speed industrial diesels are relatively unaffected by carbon deposits and can run on low-quality fuel.

The diesel fuel's ignition quality is measured by a **cetane** rating. Much like the octane number, cetane is measured in a single-cylinder test engine with a variable compression ratio. The diesel fuel is compared to cetane, which is a colorless, liquid HC that has excellent ignition qualities. Cetane is rated at 100. The higher the cetane number, the shorter the ignition lag time (delay time) from the point the fuel enters the combustion chamber until it ignites.

In fuels that are readily available, the cetane number ranges from 40 to 55 with values of 40 to 50 being most common. These cetane values are satisfactory for medium-speed engines whose rated speeds are from 500 to 1,200 rpm and for high-speed engines rated over 1,200 rpm. Low-speed engines rated below 500 rpm can use fuels in the 30+ cetane range. The cetane number can be improved with the addition of certain compounds such as ethyl nitrate, acetone peroxide, and amyl nitrate.

Minimum quality standards for diesel fuel grades have been set by the American Society for Testing

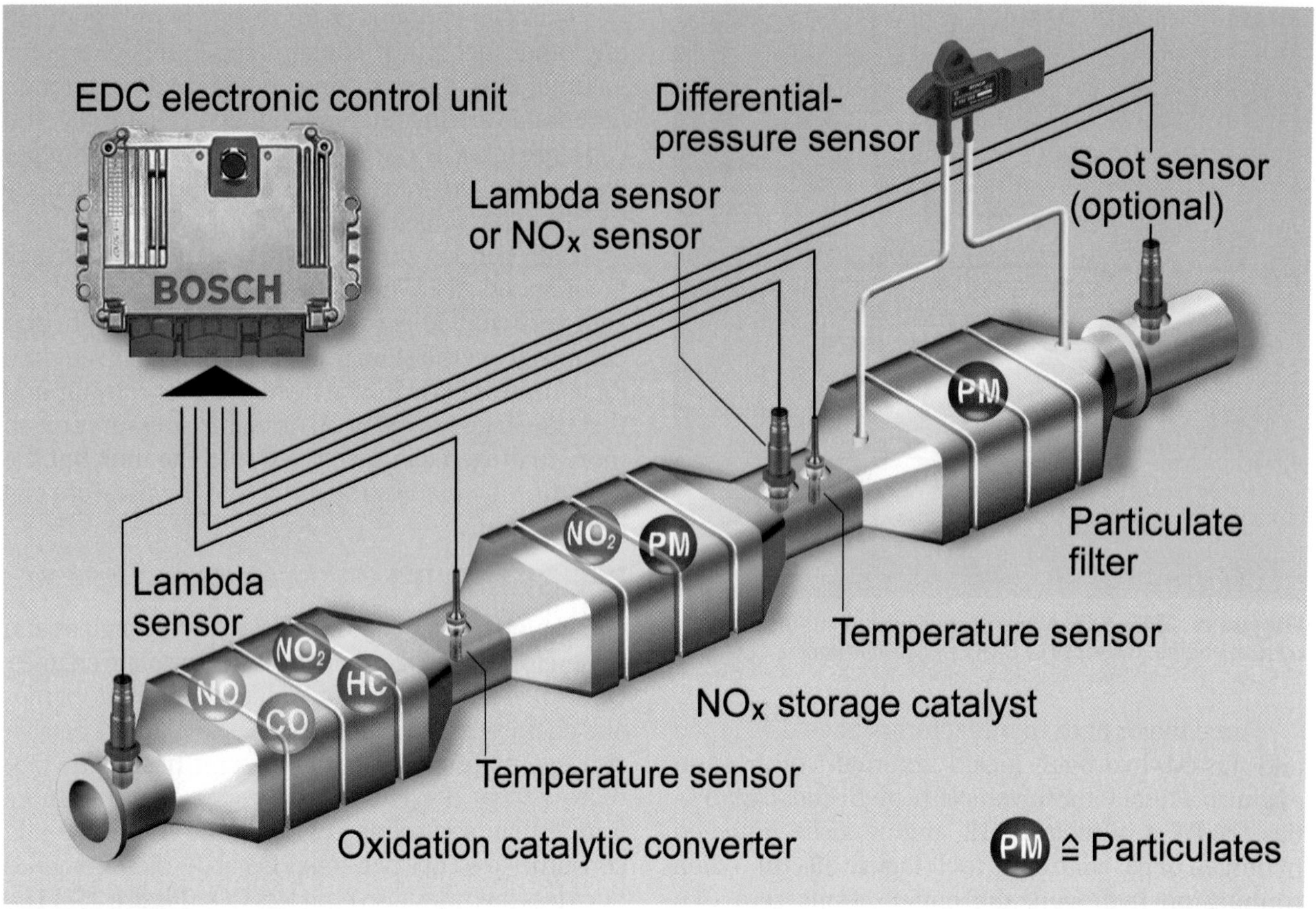

Figure 35-11 The components required to clean the exhaust from a diesel engine. *Courtesy of Robert Bosch GmbH, www.bosch-presse.de*

Materials. Two grades of diesel fuels, Number 1 and Number 2, are available. Number 2 diesel fuel is the most popular and widely distributed. Number 1 diesel fuel is less dense than Number 2, with lower heat content. Number 1 diesel fuel is blended with Number 2 to improve starting in cold weather. In the winter, diesel fuel is likely to be a mixture of Number 1 and 2 fuels. In moderately cold climates, the blend may be 90% Number 2 to 10% Number 1. In very cold climates, the ratio may be as high as 50:50. Diesel fuel economy can be expected to drop off during the winter months due to the use of Number 1 diesel in the fuel blend. The drawbacks with normal diesel are:

- Diesel fuel is a high-sulfur content fuel.
- Diesel engines emit particulates, commonly called soot.
- Diesels have high combustion temperatures, which result in excessive nitrogen oxide emissions.
- Diesel fuel is also prone to "waxing" in cold weather, a term for the solidification of diesel oil into a crystalline state.

Low-Sulfur Diesel Fuel Recently, a cleaner or low-sulfur diesel fuel has become available through legal mandates. The intent of the law was to significantly reduce the amount of sulfur in the fuel. Previously the standards allowed diesel fuel to contain up to 500 ppm of sulfur. The new clean diesel fuel is limited to 15 ppm. This fuel will allow diesels to emit less NO_x, soot, and other unwanted sulfur compounds.

Biodiesel Fuels

Fuels derived from renewable biological sources for use in diesel engines are known as **biodiesel fuels**. Animal fats, recycled restaurant grease, and vegetable oils derived from crops such as soybeans, canola, corn, and sunflowers are used in the production of biodiesel fuel. Biodiesel fuel can be used directly or be blended with diesel fuel. Pure biodiesel is biodegradable, nontoxic, and free of sulfur and aromatics.

One of the most common biodiesel fuels is vegetable oil from a wide variety of plants. Many diesel engines can be run on vegetable oil without modification, and others only need slight modifications. It is important to note that the first diesel engine ran on peanut oil, rather than fossil fuel. The only true limit as to what type of fuel a diesel engine can use is the fuel's ability to melt with temperature increases and flow through the fuel lines while lubricating the injector pump and injectors.

Not all of the biodiesel fuel produced is used as a stand-alone fuel. Often biodiesel fuel is mixed with petroleum-based diesel fuel. The two most common are B5, which is 95% petroleum-based and 5% vegetable- or animal fat-based fuel, and B20, which contains 20% biodiesel blended into regular diesel fuel.

The use of biodiesel fuel has many advantages:

- It can help reduce the nation's dependency on imported oil.
- Because it is carbon neutral, it can reduce CO_2 emissions by 20% to 60%.
- HC emissions can be reduced by 50%.
- Sulfur emissions can essentially be eliminated.
- CO emissions can be reduced by an average of 48%.
- PM emissions can be reduced by 47%.
- It enables diesel engines to run smoother and quieter and have a longer life.

However, due to production costs, biodiesel is more expensive than petroleum-based diesel fuel. There are also problems with the infrastructure; biodiesel is not readily available. Also, there is an increase in the amount of NO_x in the exhaust when biodiesel is used. Biodiesel cannot be used in vehicles manufactured before 1992 without modifying the fuel systems of those vehicles. For most late-model diesel engines, little, if any, modifications need to be done.

ELECTRIC VEHICLES

Electric vehicles are common in manufacturing, shipping, and other industrial plants, where the exhaust of an internal combustion engine could cause illness or discomfort to the workers in the area. They are used on golf courses, where the quiet operation adds to the relaxing atmosphere. They are also commonly used in the downtown areas of large cities and large campuses where peace, quiet, and fresh air are a priority.

Figure 35-12 A pair of current electric-powered sports cars. *Courtesy of Tesla Motors*

Nearly all of the major automobile manufacturers have researched and developed electric vehicles; however, currently none of them has any available. Some smaller manufacturers are offering electric vehicles at this time **(Figure 35–12)**.

Electric vehicles (EVs) use one or more electric motors powered by electricity stored in batteries. The source of the electricity can also be fuel cells, or photovoltaic (PV) or solar cells, that convert the sun's energy into electricity. The electricity for the motors is stored in a battery that must be recharged from an external electrical power source. The basic components of an EV are shown in **Figure 35–13**.

The drivetrain of an electric drive vehicle is much more efficient than the drivetrain of an internal combustion engine. EVs are zero-emission vehicles because they do not directly pollute the air. The only pollution associated with them is the result of creating the electricity to charge their batteries. Even considering those emissions, EVs are more than 99% cleaner than the cleanest gasoline vehicle.

The biggest disadvantage of an EV is the limited driving range. Most EVs made by the OEMs have less than a 120-mile range before the batteries need to be recharged. Much research is being done to extend the range and to decrease the required recharging times. The batteries are recharged by plugging them into a

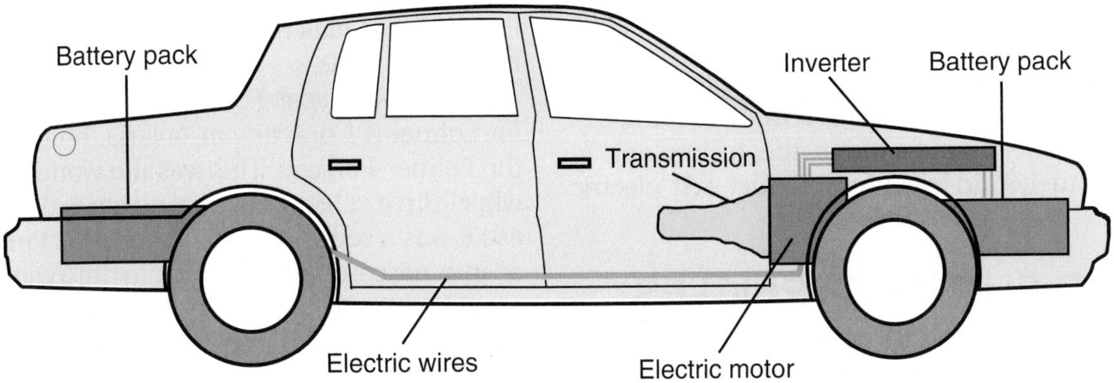

Figure 35-13 An electric car.

Figure 35-14 Recharging an electric vehicle with an inductive charge. *Copyright, Nissan (2008). Nissan and the Nissan logo are registered trademarks of Nissan.*

recharging outlet at home **(Figure 35–14)** or other locations. The required recharging time varies with the type of charger, the size and type of battery, and other factors. Normal recharge time is 4 to 8 hours.

To extend the driving range, many systems are being used in EVs. The most common is regenerative braking. This, coupled with the use of highly efficient accessories (such as a heat pump for passenger heating and cooling), have extended the range. However, the range is not great enough to meet the needs of a typical driver. The wide use of EVs in the future depends on the development of batteries that can extend the range of operation. Some of the technologies being used or evaluated include lead-acid, nickel cadmium, nickel iron, nickel zinc, nickel metal hydride, sodium nickel chloride, zinc bromine, sodium sulfur, lithium, zinc air, and aluminum air batteries. The new batteries must also be able to be charged in a much shorter time.

Although EVs are currently not a viable transportation option, the lessons learned from building EVs have given the manufacturers the technology to move on to hybrid electric and fuel cell electric vehicles.

HYBRID ELECTRIC VEHICLES

Hybrid electric vehicles (HEVs) are or will soon be available from all of the major automobile manufacturers. Any vehicle that combines two or more

Figure 35-15 A Toyota Prius. *Courtesy of Toyota Motor Sales, U.S.A., Inc.*

sources of power is called a hybrid. Most of the technology used in these vehicles is discussed in the appropriate sections of this book, according to the system. Current HEVs have an internal combustion engine and one or more electric motors. How these are used depends on the design of the system. Toyota introduced the first mass-produced hybrid vehicle in 1997. The hybrid, the Prius, was only available in Japan until the 2000 model year, when it is was brought to North America **(Figure 35–15)**. Since then many different models have been available.

Series Hybrids

Some HEV designs are close to being an EV in that the engine is used to drive a generator that charges the battery. The electric motor powers the vehicle. The engine is there only to extend the vehicle's driving range. The engine may be a gasoline or diesel engine. A computer controls the operation of the engine depending on the power needs of the battery. The generator also works as a starter motor. When the computer senses that system voltage is low, the engine quickly starts and drives the generator. An electric motor can only operate at maximum power when it receives full voltage; the batteries, generator, and engine provide for this. Currently, series hybrids are only in development; GM is working on one (called the Chevy Volt) that may be released in 2010.

In 1898, Dr. Jacob Ferdinand Porsche, an engineer for Lohner & Company in Austria, built his first car, the Lohner-Porsche. This was the world's first front-wheel-drive vehicle. His second car is of more interest; it was a series-hybrid vehicle. This Porsche used an internal combustion engine to run a generator that provided the power to electric motors that powered individual wheels through the hub of the wheels. This vehicle was used as a military supply carrier during World War I.

SHOP TALK

Do not get confused between the motor and the engine in a hybrid vehicle. A vehicle's engine has been, and still is, often called a motor. It is not a motor in spite of the fact that we put motor oil in it. The real name for motor oil should be engine oil. By definition, a motor is a machine that converts electrical energy into mechanical energy. An engine converts chemical energy into mechanical energy.

Parallel Hybrids

The most common hybrid vehicles rely on power from the electric motor or engine and, in some cases, from both **(Figure 35–16)**. When the vehicle moves from a stop and has a light load, the motor moves the vehicle. Power for the electric motor comes from stored electricity in the battery pack. During normal driving conditions, the engine is the main power source. Engine power is also used to rotate a generator that recharges the storage batteries. The motor is run to add power to the powertrain. A computer controls the operation of the motor depending on the power needs of the vehicle. During full throttle or heavy load operation, the motor is turned on to increase the output of the powertrain. The electric motor can also act as a generator. During deceleration, the motor works as a generator to charge the batteries and to help slow down the vehicle.

These systems are commonly referred to as parallel systems, but because the engine is used sometimes to drive the generator, it is more correct to say that these are series-parallel systems.

Hybrid configurations are further defined by the role of the electric motor. A **full hybrid** is a vehicle that can run on just the engine, just the batteries, or a combination of the two. The Prius and the Escape are examples of full hybrids. An **assist hybrid** cannot be powered only by the electric motor. The electric motor helps or assists the engine to overcome increased load. At other times, the vehicle is powered by the engine. A micro or **mild hybrid** is a vehicle equipped with stop-start technology combined with regenerative braking. The electric motor/generator may offer some additional power to the drivetrain. There are two additional classifications: the performance hybrid (some call this "muscle hybrid"), which is a full hybrid designed for improved acceleration

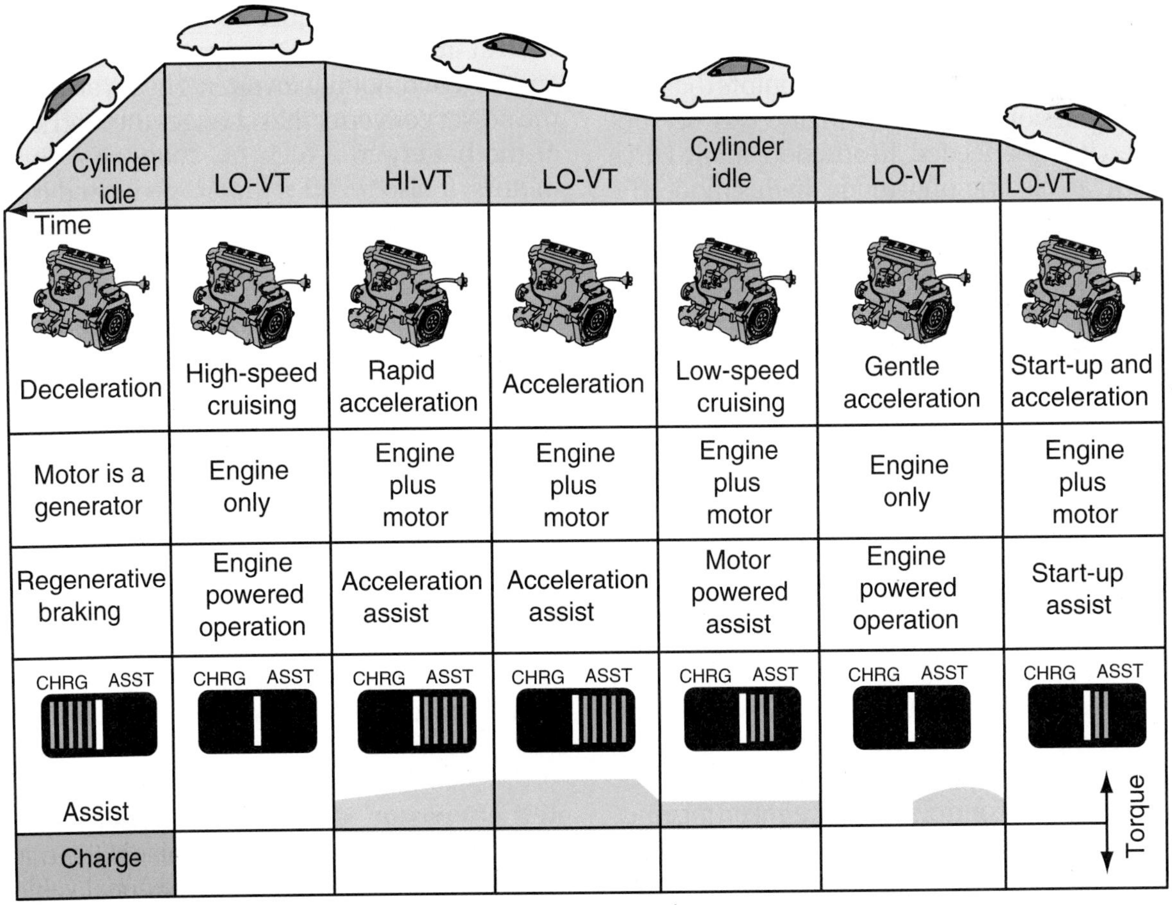

Cylinder idle
LO-VT
HI-VT
LO-VT
Cylinder idle
LO-VT
LO-VT
Time
Deceleration
High-speed cruising
Rapid acceleration
Acceleration
Low-speed cruising
Gentle acceleration
Start-up and acceleration
Motor is a generator
Engine only
Engine plus motor
Engine plus motor
Engine plus motor
Engine only
Engine plus motor
Regenerative braking
Engine powered operation
Acceleration assist
Acceleration assist
Motor powered assist
Engine powered operation
Start-up assist
CHRG ASST
Assist
Charge
Torque

Figure 35–16 The modes of operation for Honda's hybrid vehicles.

rather than fuel economy, and a plug-in hybrid, which is a full hybrid that can use an external electrical source to charge the batteries, thereby extending the electric-only driving range.

Advantages

Hybrids consume significantly less fuel than vehicles powered by gasoline alone. Therefore, they can reduce the country's dependence on fossil fuels and foreign oil. There is a substantial difference in fuel consumption of a hybrid and a conventional gasoline-powered vehicle. HEVs have a higher initial cost than comparable conventional vehicles and tend to be heavier. The added weight decreases fuel economy, which is why some larger hybrid SUVs see little gain in fuel mileage. The additional weight results from the addition of large battery packs and electric motor/generators. These same items contribute to the higher cost as well. However, the cost of some of these items will decrease as the volume of production increases. Currently, an HEV costs nearly $4,000 more to produce.

Hybrids can have more than 90% fewer emissions than the cleanest conventional vehicles. HEVs also produce significantly lower total fuel cycle ("well to wheel") emissions when compared to equivalently sized conventional vehicles. HEVs emit one-third to one-half less CO_2 than conventional vehicles.

The low emissions result from the use of smaller and more efficient internal combustion engines. The engine's power is boosted by electric motors that produce zero emissions. Also, the engine can be shut down when it is not needed. In addition, many HEVs can move in an electric-only mode. In this mode, the vehicle has no emissions. Clean electricity is also used to power many accessories and other equipment that typically are driven by the engine. This means the engine has less work to do and therefore will use less fuel and emit fewer pollutants.

Hybrids will never be zero-emission vehicles, because they rely on an engine for much of its power. However, most are rated as being close to zero-emission vehicles.

Technology

Today's hybrids are rolling examples of modern technology. The switching between the electric motor and gasoline engine is controlled by computers, as are other features of vehicle. The control systems are extremely complex. They have very fast processing speeds and real-time operating systems. The individual computers in the control system are linked together and communicate with each other by high-speed "communication buses," controller area network (CAN) communications take place between many control systems **(Figure 35–17)**: the electric motor controller, engine controller, battery management system, brake system controller, transmission controller, and electrical grid controller, and some systems have 12- or 42-volt components that also must be controlled. The basic components of a hybrid vehicle include batteries, fuel tank, transmission, electric motor, inverter, and internal combustion engine.

Batteries The voltage of the battery packs **(Figure 35–18)** in a hybrid vehicle depends on the system and the manufacturer.The voltage range is from 42 to 330 volts. Most battery packs are comprised of several small batteries connected together to provide the required voltage. Most hybrids also have an additional 12-volt battery to power conventional electrical items, such as lighting, wipers, sound systems, etc.

Go To ***Chapter 17 for a discussion on the different types of batteries used in hybrid vehicles.***

Electronics Most of the advanced electronics for a hybrid system are contained in a single, water- or air-cooled assembly **(Figure 35–19)**. This unit may contain an inverter, DC-DC converter, boost converter, and air-conditioning inverter. The inverter is an electric power converter that changes the high DC voltage of the battery to a high AC voltage for the electric motors. It also rectifies the AC generated by generators to recharge the high-voltage battery pack.

The DC-DC converter reduces some of the high DC voltage from the battery pack to 12-volt DC in order to charge the separate 12-volt battery. The voltage of the auxiliary battery is monitored by the converter, which attempts to keep it at a constant level. A boost converter can be used to boost the voltage to the motor up to 500 volts. This increased voltage increases the power output of the motor. The air-conditioning inverter alters battery voltage so that it can be used to power the air-conditioning compressor.

Layouts

Many different layouts and systems are used in today's hybrid vehicles. What follows is a basic discussion on these setups.

Belt Alternator Starter A **belt alternator starter (BAS)** system **(Figure 35–20)** replaces the traditional starter and generator in a conventional vehicle. The unit is located where the generator would normally

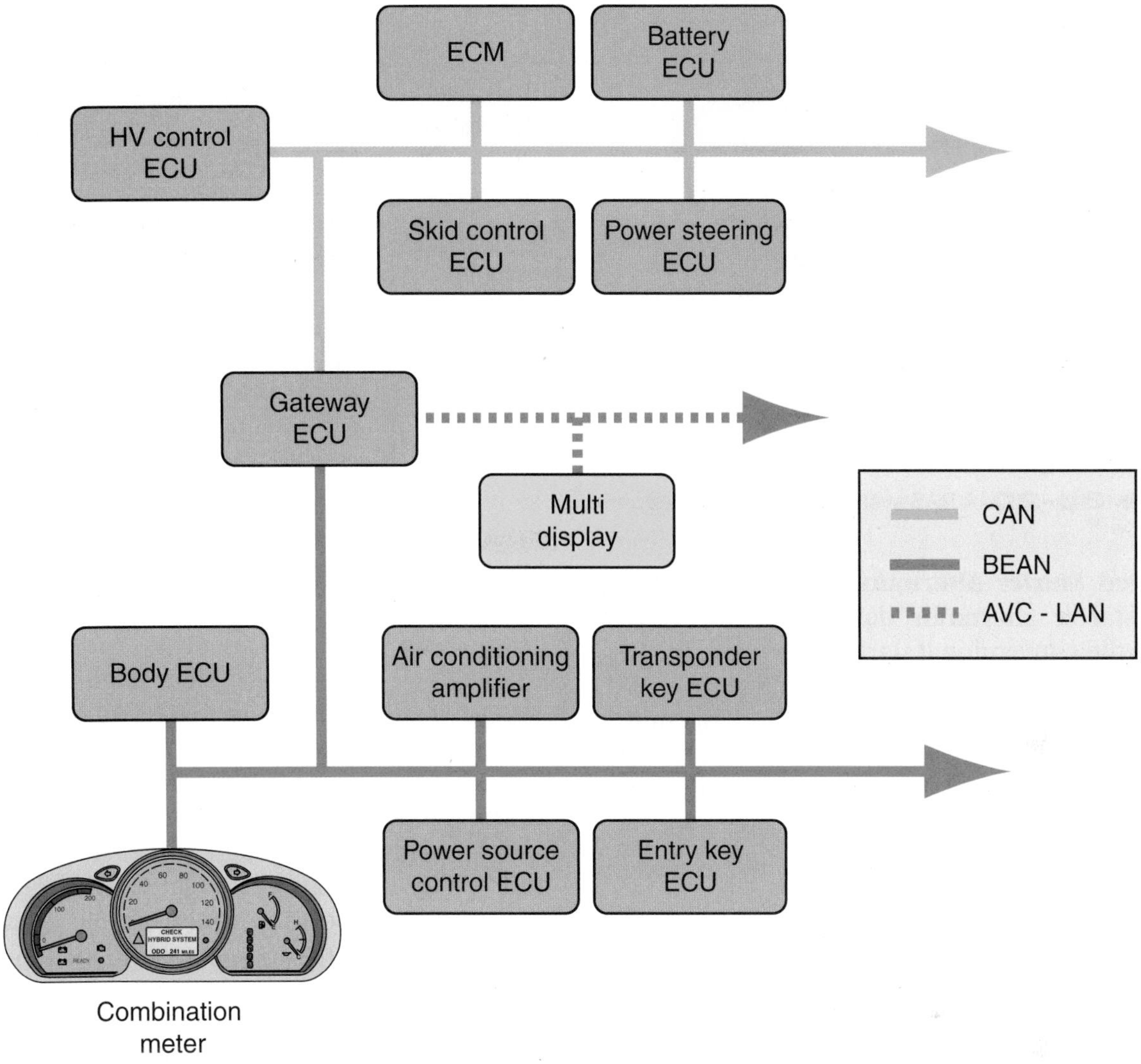

Figure 35-17 The individual computers in the control system are linked together and communicate with each other by high-speed "communication buses" known as controller area network (CAN).

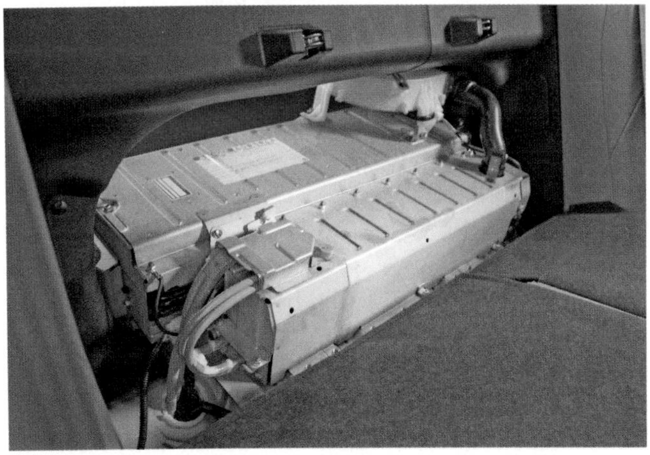

Figure 35-18 The battery pack in a Toyota Camry hybrid. *Courtesy of Toyota Motor Sales, U.S.A., Inc.*

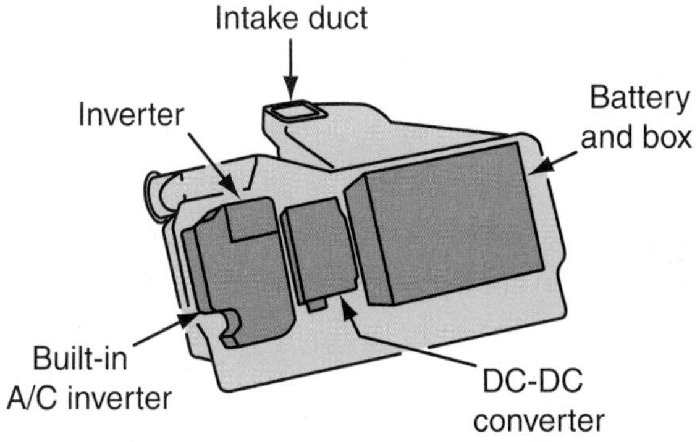

Figure 35-19 Honda's hybrid intelligent power unit.

be and is connected to the engine's crankshaft by a drive belt. This unit, an electric motor, serves as the starting motor and generator. When the engine is running, a drive belt spins the rotor of the motor and the motor acts as a generator to charge the batteries. To start the engine, the motor's rotor spins and moves the drive belt, which in turn cranks the engine. These systems have the capability of providing stop-start, regenerative braking, and high-voltage generation. Some also offer a small amount of engine assist. They are typically connected to a 42-volt power source.

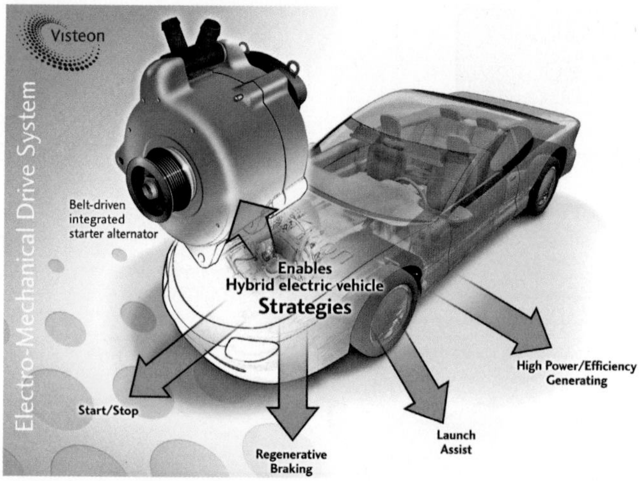

Figure 35-20 A BAS system. *Courtesy of Visteon Corporation™*

Integrated Starter Alternator Damper The **integrated starter alternator damper (ISAD)** system replaces the conventional starter, generator, and flywheel with an electronically controlled compact AC asynchronous induction electric motor. This unit is typically housed in the transmission's bell housing between the engine and the transmission, although it can be mounted to the side of the transmission.

The electricity generated by the system is used to recharge the 12- and 42-volt battery packs; both of these are used to power the various vehicle systems. The ISAD system is very similar to Honda's integrated motor assist (IMA) system. Honda introduced the basic concept of placing an electric motor between the engine and transmission and many variations of its design have been developed. Honda calls its system the **integrated motor assist (IMA)** hybrid system **(Figure 35–21)**.

Figure 35-21 The integrated motor assist (IMA) is placed between the engine and transaxle. *Courtesy of American Honda Motor Company*

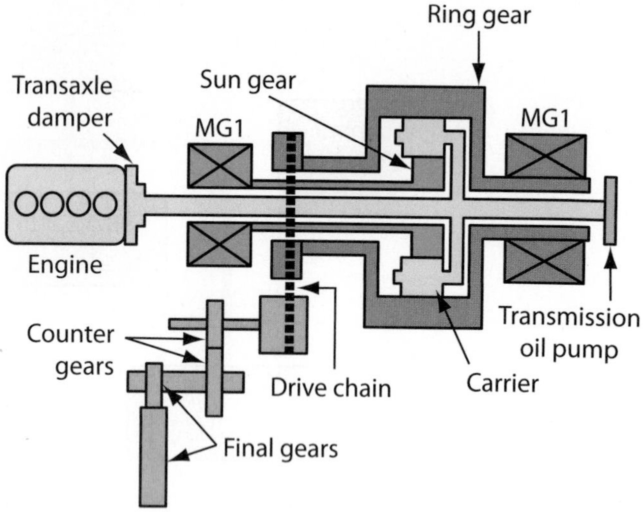

Figure 35-22 The layout of the main components connected to Toyota's power-splitting device.

Power-Split System Currently, Ford, Nissan, and Toyota use a power-split system **(Figure 35–22)**. This system is the basis for many series-parallel hybrid vehicles. This system is capable of instantaneously switching from one power source to another or combining the two. The unit functions as a continuously variable transaxle, although it does not use the belts and pulleys normally associated with CVTs. This transaxle also does not have a torque converter or clutch. Rather, a damper is used to cushion engine vibration and the power surges that result from the sudden engagement of power to the transaxle.

The power-split unit is basically comprised of a planetary gear set and two electric motors. When used with high-output engines, the power-split unit also has an additional reduction planetary gear set.

Go To **Chapter 40 for a detailed discussion of the power-split unit and other hybrid transmissions.**

Electric 4WD Some 4WD hybrids use an electric motor, differential, and rear transaxle housing to drive the rear wheels. This unit is not mechanically tied to the front drive axles. Rather, its action is controlled by electronics **(Figure 35–23)**. This allows the system to be capable of responding to operating conditions by varying the distribution of torque between the front and rear axles.

Two-Mode Transmission GM, BMW, and Chrysler worked together to develop a two-mode full hybrid system that can be used with gasoline or diesel

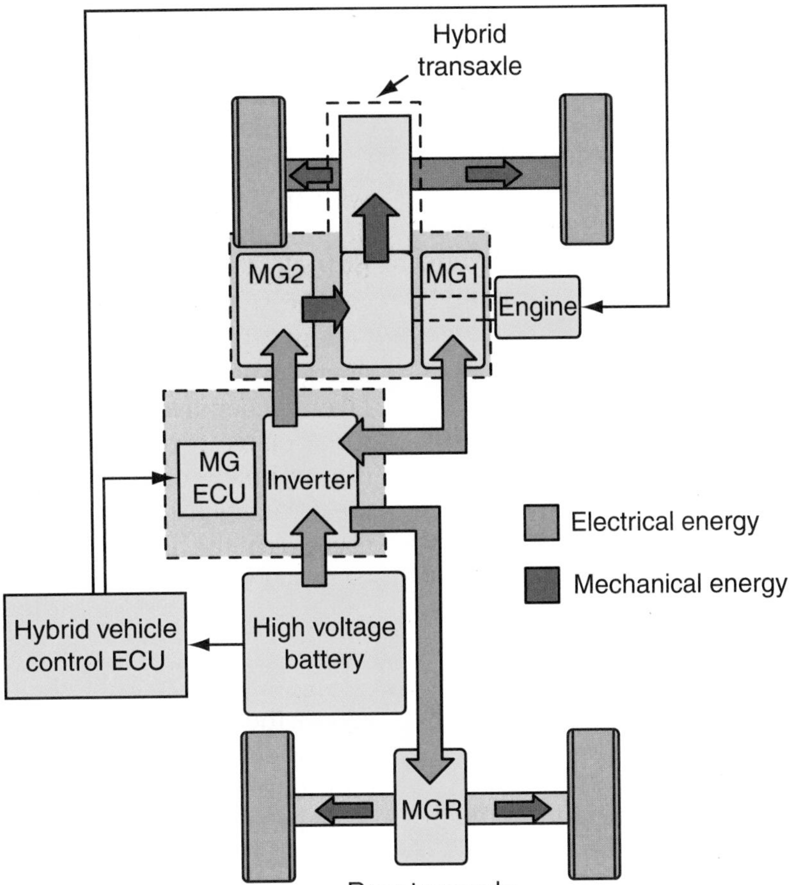

Figure 35-23 The rear axle is driven by an electric motor (MGR) and is not mechanically linked to the front drive axles.

engines. The entire drive system fits into a standard transmission housing and is basically three simple planetary gear sets coupled to two electric motors, which are electronically controlled. The system has two distinct modes of operation. It operates in the first mode during low-speed and low-load conditions and the second mode is used while cruising at highway speeds.

Two compact AC motors are connected to the transmission's gear sets **(Figure 35–24)**. The result is a continuously variable transmission that is based totally on planetary gears. The gears work to increase the torque output of the motors. This enables the system to rely on a relatively low voltage, which in turn means that the inverter, converter, and controller can be made lighter and smaller. The NiMH battery pack has a nominal voltage of 300 volts and is contained in a housing equipped with a cooling circuit. Like other hybrids, keeping the battery pack within a specified range is a top priority.

Audi, VW, and Porsche are all developing hybrid vehicles. Some are seeking fuel economy, whereas others are striving for increased performance. The systems they are working on depend on their goals, but it is fair to say that if overall performance is sought, they will use electric motors to supplement the engine and/or four-wheel-drive system. If economy is sought, they will use electric motors to propel the vehicles when the engine is not

Figure 35-24 GM, Chrysler LLC, and BMW of North America, LLC, two-mode hybrid electric motors and planetary gear sets. *Courtesy of Chrysler LLC*

needed. For example, Porsche's system for its Cayenne hybrid has a strong enough electric motor that will allow full electric operation while the vehicle is traveling at highway speeds. This is the time when low horsepower is needed to maintain a cruising speed.

Keep in mind that BMW is one of the partners with GM and Chrysler in developing the two-mode hybrid system. When BMW releases a hybrid, it will undoubtedly be based on that system.

Plug-in Hybrids

Plug-in hybrid electric vehicles (PHEVs) are full hybrids with larger batteries and the ability to recharge from an electric power grid. They are equipped with a power socket that allows the batteries to be recharged when the engine is not running **(Figure 35–25)**. The socket can be plugged into a normal 120-volt outlet. Charged overnight, PHEVs can drive up to 60 miles without the engine ever turning on. When the batteries run low, the engine starts and powers the vehicle and the generator to charge the batteries.

The biggest advantage of plug-in hybrids is they can be driven in an electric-only mode for a much greater distance. During that time, the vehicle consumes no fuel. Under normal conditions, a plug-in hybrid can be twice as fuel efficient as a regular hybrid. A fully charged PHEV will produce half the emissions of a normal HEV. This is simply due to the fact that there are no emissions when the engine is not running.

The manufacturing costs of a PHEV are about 20% higher than a regular HEV. The increase in cost is mainly due to the price of the larger batteries. No current hybrid needs to be plugged in to recharge the batteries. However, there is much research being done on "plug-in" hybrids.

Figure 35-25 A plug-in hybrid being charged with an external power source. *Courtesy of Toyota Motor Sales, U.S.A., Inc.*

Hydraulic Hybrids

Hydraulic hybrids function in the same way as HEVs, except energy for the alternative power source is stored in tanks of hydraulic fluid under pressure rather than in batteries. Also, rather than being fitted with an electric motor, these vehicles have a hydraulic propulsion system, which can power the vehicle by itself.

The stored hydraulic energy propulsion (SHEP) system captures energy during braking and uses that energy when the vehicle is accelerating from a stop. Like regenerative braking systems, the system captures a large percentage of the energy normally lost during braking. That energy is stored in hydraulic tanks attached to the vehicle's chassis. The system also aids in the halting of the vehicle. When the brake pedal is depressed, the control unit opens solenoids that send fluid from the low-pressure tank to the pump at the drive shaft. As the drive shaft turns, it turns the pump, and fluid pressure increases. This causes the pump and drive shaft to slow down. The fluid under pressure now moves to the high-pressure tank for storage.

During acceleration, the system's computer instructs the pump to send the stored high-pressure fluid back to the drive shaft and to the low-pressure tank. At this point, the vehicle moves without power from the engine and without burning any fuel. Once the computer senses that the energy stored in the tanks has been used, the engine will start and take over the operation of the vehicle. The energy is restored in the tanks during the next brake application.

Hydraulic hybrid technology is being developed for use in heavy vehicles like buses, trucks, and military vehicles. It is projected that hydraulic hybrids will use 60% to 70% less fuel than conventional delivery vehicles. The trucks will also have lower emissions because the engine is not used during acceleration.

Hybrid Service

Hybrid systems rely on very high voltages. All of the high-voltage cables are covered in orange sleeves for easy identification, and you should always follow the procedures for disarming the high-voltage system before performing any service on or near the high-voltage circuits. This is important for all services, not just electrical. Air-conditioning, engine, transmission, and body work can require services completed around and/or with high-voltage systems. If there is any doubt as to whether something has high voltage or not, or if the circuit is sufficiently isolated, test it before touching anything.

Go To ***Chapter 17 for the procedures for isolating the high-voltage system in common hybrid vehicles.***

Maintenance and Service Hybrid vehicles are maintained and serviced in the same way as conventional vehicles, except for the hybrid components. The latter includes the high-voltage battery pack and circuits, which must be respected when doing any service on the vehicles. The manufacturers list the recommended service intervals in their service and owner's manuals. Nearly all of the items are typical of a conventional vehicle. Care needs to be taken to avoid anything orange while carrying out the maintenance procedures.

Go To ***Chapter 7 for guidelines on performing preventive maintenance on a hybrid vehicle.***

For the most part, actual service to the hybrid system is not something that is done by technicians unless they are certified to do so by the automobile manufacturer. Diagnosing the systems varies with the manufacturer, although certain procedures apply to all. Keep in mind that a hybrid has nearly all of the basic systems as a conventional vehicle and they are diagnosed and serviced in the same way.

One of the things to pay attention to, both from the owner's or service manuals and driving experience, is the stop-start feature. You need to know when the engine will normally shut down and restart. Without this knowledge, or the knowledge of how to prevent this, the engine may start on its own when you are working under the hood. Needless to say, this can create a safety hazard. Unless the system is totally shut down, the engine may start at any time when its control system senses that the battery needs to be recharged.

The computer-controlled systems are extremely complex, especially in assist and full hybrids, and are very sensitive to voltage changes. This is why the manufacturers recommend a thorough inspection of the auxiliary battery and connections every 6 months.

Diagnostics A hybrid vehicle has unique systems that require special procedures and test equipment. It is imperative that you have good information before attempting to diagnose these vehicles. Also make sure you follow all test procedures precisely as they are given.

Hybrids present unique considerations when they have a driveability problem. The problem can be caused by the hybrid system, engine, or transmission. Determining which system is at fault can be difficult. On some hybrids, it is possible to shut down the hybrid system and drive the vehicle solely by engine power. On others, such as the Toyota and Ford hybrids, this is not possible. If electric power can be shut off and the vehicle still drives poorly, the problem is the engine or transmission. If it is not possible to shut down either power source, your diagnosis must be based on the symptom and information retrieved with a scan tool.

SHOP TALK

It should be remembered that the hybrid system is looked at as another emission control device by OBD-II, and all information from the computers will relate to emissions first.

Diagnosis of a hybrid vehicle should follow a logical approach. The first step is gathering as much information as possible from the customer about the concern. This is followed by a thorough inspection of the vehicle. A road test is then taken to verify the problem as well as to define the problem. After having a good understanding of what is and is not working properly, all service bulletins that relate to the problem should be read and the appropriate procedures followed. If the cause is still unknown, specific tests should be conducted to pinpoint it. Once the cause is identified, repairs should be made and then the repair should be verified.

All warning lamps in the instrument panel should be checked **(Figure 35–26)**. If any of these remain on after the engine is started, the cause should be identified and corrected before continuing with diagnosis. Lastly, a scan tool should be used to retrieve any fault codes held in the computer's memory. In many cases, a manufacturer-specific scan tool is required to test hybrids. Aftermarket scan tools may be able to retrieve codes and display some data, but they may have limited capabilities. Also, follow the prescribed sequence for retrieving and responding to all diagnostic trouble codes (DTCs).

Test Equipment To test high-voltage systems you need a Category 3 (CAT III) digital volt ohmmeter and, of course, a good pair of insulating gloves. Although the high-voltage system can be isolated from the rest

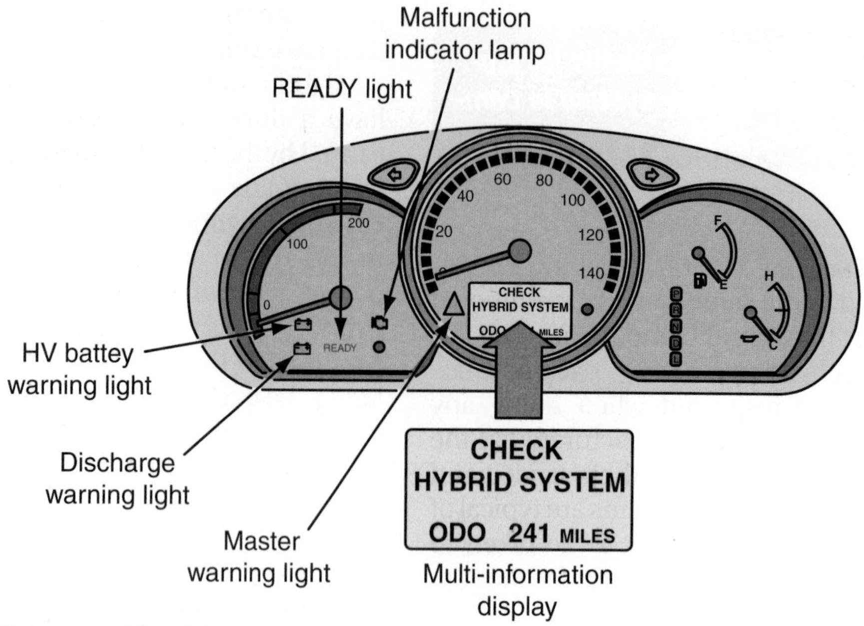

Figure 35-26 An example of some of the warning lights on a hybrid vehicle.

of the vehicle, high voltage is still at and around the battery pack. When checking for the presence of high voltage, make sure to check the inverter assembly. This is typically where the ultracapacitors are and until they are discharged, they are lethal.

Another tool that will save much time and effort during diagnosis is an **insulation resistance tester**. These meters can check for voltage leakage from the insulation of the high-voltage cables. Obviously, no leakage is desired and any leakage can cause a safety hazard as well as damage to the vehicle. Minor leakage can also cause hybrid system-related driveability problems. This meter should be a CAT III meter for the same reasons as the DVOM. In fact, these meters often have the capability of checking the resistance and voltage of the circuits like a DVOM.

To measure insulation resistance, system voltage is selected at the meter and the probes placed at their test position. The meter will display the voltage it detects. Normally, resistance readings are taken with the circuit de-energized, and this is true for resistance checks with this meter unless you are checking the effectiveness of the cable or wire insulation. In this case, the meter is measuring the insulation's effectiveness and not its resistance.

WARNING!

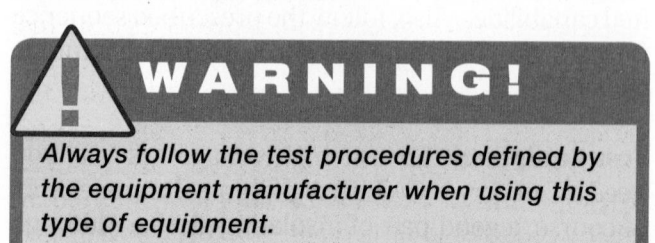

Always follow the test procedures defined by the equipment manufacturer when using this type of equipment.

The probes for the meters should have safety ridges or finger positioners. These help prevent physical contact between your fingertips and the meter's test leads.

FUEL CELL ELECTRIC VEHICLES

Fuel cell electric vehicles (FCEVs) are the result of the many years of research and development on electric and hybrid vehicles. All electric-drive vehicles share many of the same technologies but differ greatly in the source of energy used to power the electric motors that are used to move the vehicle **(Figure 35–27)**. Nearly all manufacturers are experimenting with fuel cells. A Honda spokesman predicted that fuel cell vehicles could have a market share of 5% by 2020. A Toyota spokesman said that the introduction of a fuel cell vehicle for consumers will not happen before 2010. This is the same year that GM has said it plans to have a production-ready fuel cell vehicle. GM also plans to sell one million FCEVs by the year 2020. We do not know when FCEVs will be a common sight on the road, but we do know that many companies are working to get them there.

This technology is not new, nor is it unproven; the National Aeronautics and Space Administration (NASA) has been using this technology in its spacecraft for years. Fuel cells provide the energy for the various electronic devices on-board the spacecraft.

FCEVs have electric motors, but the energy source for those motors is not necessarily batteries. The true source of electrical power is a fuel cell. The energy generated by a fuel cell can directly power the traction motor of the vehicle and/or be stored in a battery or

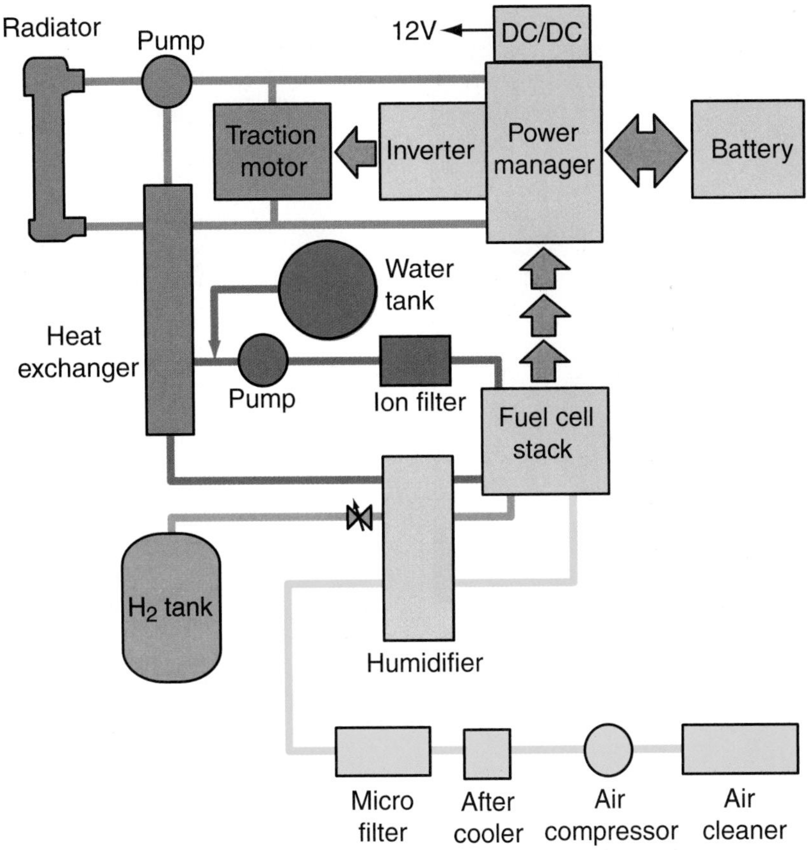

Figure 35-27 The basic layout for a fuel cell vehicle.

ultracapacitor. If there is no storage (battery or capacitor) in the system, regenerative braking does not exist. However, when equipped with a battery or capacitor, regenerative braking is used and the energy stored in either one can provide power boosts for the vehicle. An external energy source is not required to refill the electrical storage unit; however, the fuel used in the fuel cell must be refilled. An FCEV with a storage battery or capacitor is considered a fuel cell hybrid.

Fuel cell vehicles use hydrogen as their fuel or energy source. The supply of hydrogen can be stored in tanks in the vehicle or can be provided by a reformer that extracts hydrogen from another fuel, such as gasoline, methanol, or natural gas. Pure water and heat are the only emissions from a fuel cell.

A fuel cell vehicle is much like a battery-operated EV. It operates like one and has many of the same characteristics as one: Electricity powers an electric motor to drive the vehicle, the vehicle operates very quietly, and the output of CO_2 and other harmful emissions is zero. The main powertrain components in a typical fuel cell vehicle are:

- Fuel cell stack—An electrical generation device made up of several individual fuel cells
- High-pressure hydrogen supply system or reformer with a fuel tank
- Air supply system—An air pump to supply the fuel cells with air
- Humidification system—Recycles water vapor generated in the FC stack to humidify the hydrogen and air
- Fuel cell cooling system
- Storage battery or ultracapacitor
- Traction motor and transmission
- Control module and related inputs and outputs—Includes a DC/DC converter

Fuel Cells

A fuel cell **(Figure 35–28)** produces electricity through an electrochemical reaction that combines hydrogen and oxygen to form water. The basic principle is the opposite of electrolysis. **Electrolysis** is the process of separating a water molecule into oxygen and hydrogen atoms by passing a current through an electrolyte placed between two electrodes **(Figure 35–29)**. In a fuel cell there is no combustion; the reaction is purely chemical. Catalysts are used to combine the fuel (hydrogen) with oxygen. The reaction releases electrons or electrical energy. Fuel cells have no moving parts and can continue to work until the fuel supply is depleted. In other words, the driving range of a fuel cell vehicle is largely dependent on the amount of fuel it can carry.

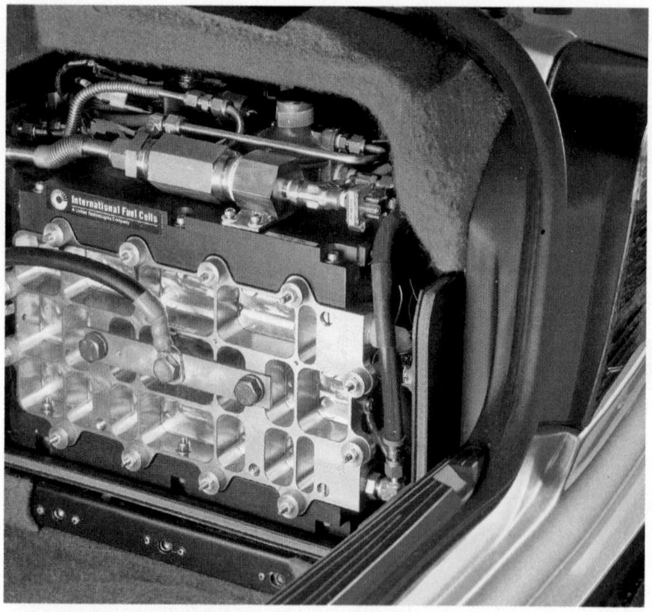

Figure 35-28 A trunk-mounted fuel cell. *Courtesy of BMW of North America, LLC*

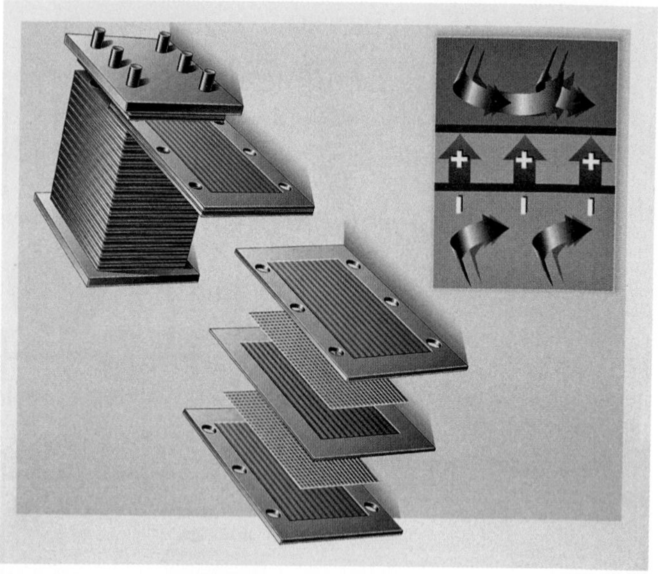

Figure 35-30 A fuel cell stack. *Courtesy of Chrysler LLC*

A single fuel cell produces very low voltage, normally less than 1 volt. To provide the amount of power needed to propel a vehicle, several hundred fuel cells are connected in series. This assembly is the **fuel cell stack** (**Figure 35–30**), which is called this because the cells are layered or stacked next to each other. Each fuel cell produces electricity and the combined output of the cells is used to power the vehicle.

A fuel cell has two electrodes coated with a catalyst. The electrodes are separated from each other by an electrolyte and from the case by separators. One of the electrodes has a positive polarity, the anode, and the other is negative and is the cathode. The electrolyte is most often a polymer membrane, called the proton or ion exchange membrane. Polymers can be very resistant to chemicals and can serve as an electrical insulator. The polymer membrane in a fuel cell does both. The catalyst, normally platinum, on the electrodes causes the chemical reaction in the fuel cell. A fuel cell consumes only hydrogen and oxygen. The oxygen is delivered to the cell by an air compressor that draws air in from outside the fuel cell. Hydrogen is fed into the fuel cell from a pressurized tank or a reformer. In a direct hydrogen fuel cell vehicle, there are zero

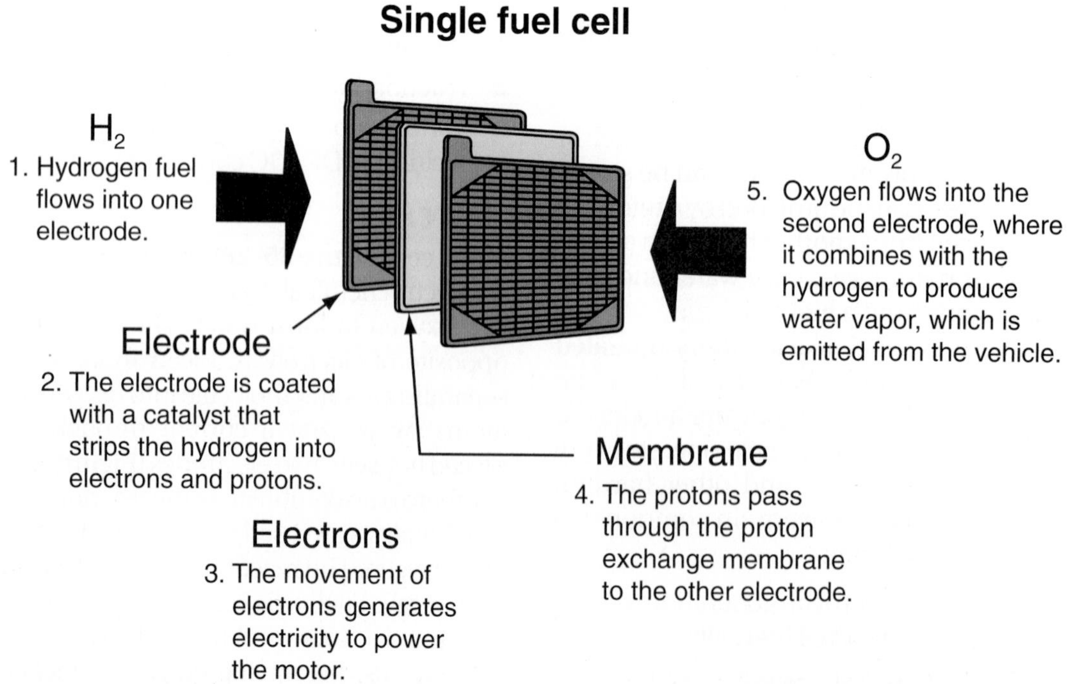

Single fuel cell
H_2
1. Hydrogen fuel flows into one electrode.
O_2
5. Oxygen flows into the second electrode, where it combines with the hydrogen to produce water vapor, which is emitted from the vehicle.
Electrode
2. The electrode is coated with a catalyst that strips the hydrogen into electrons and protons.
Membrane
4. The protons pass through the proton exchange membrane to the other electrode.
Electrons
3. The movement of electrons generates electricity to power the motor.

Figure 35-29 The basic operation of a fuel cell.

emissions. However, if the hydrogen is extracted from a hydrocarbon-based fuel (typically methanol) by a reformer, there will be some vehicle emissions.

When hydrogen is delivered to the anode, the catalyst causes the hydrogen atoms to separate into electrons and protons. Electrons always move to something more positive but cannot pass through the membrane. Therefore, their only path to the positive side of the fuel cell is through an external circuit. The movement of the electrons through that circuit results in direct current flow. It is this current flow that powers the vehicle's electric propulsion motors.

Oxygen enters the other side of the fuel cell and reacts to the catalyst on the cathode. This reaction splits the oxygen molecules into oxygen ions. The protons (hydrogen ions) that were released from the hydrogen at the anode move toward the oxygen ions. The membrane that separates the two electrodes will only allow protons to pass through and they do. At the cathode, the hydrogen ions bond with the oxygen ions to form water.

To function, the ion exchange membrane must be kept moist. Therefore, some of the water produced by the fuel cell is used to humidify the incoming hydrogen and oxygen. The remaining water is emitted as exhaust from the fuel cell. Some heat is also emitted by the fuel cell. The heat is either released to the outside air or captured and used to heat the fuel cell; it also can be used to heat the passenger compartment.

Types of Fuel Cells

The fuel cell described previously is currently the most commonly used fuel cell in concept vehicles, the proton exchange membrane (PEM). There are many other different designs; some are totally impractical for use in an automobile, whereas others show promise. Most fuel cell designs vary by size, weight, fuel, cost, and operating temperature. However, *all* have two electrodes and an electrolyte between them. The following are the various types currently being researched.

Proton Exchange Membrane (PEM) Fuel Cell The PEM, or derivatives of it, is a favored design because it allows for adjustable outputs, which is necessary for driving. The speed of the vehicle can be controlled by controlling the output of the fuel cell. Although it is quite compact, it is capable of providing high outputs. It also operates at a low temperature of 176°F (80°C). However, it is expensive to manufacture. Much of the cost is because the catalysts are platinum based.

Solid Oxide Fuel Cell (SOFC) The SOFC may be the first design to be used in a mass-produced automobile. However, it will not be used to power a traction motor. Rather, it may be used to replace the belt-driven generator (alternator) on internal combustion engines. The SOFC can provide high power levels, which means more accessories can be electrically driven. These cells have a ceramic anode, ceramic cathode, and a solid electrolyte (**Figure 35–31**). The cells operate at very high temperatures. The high operating temperature is the reason these fuel cells may not be used to power a vehicle. When this high heat is generated, it must be released. Releasing a large quantity of high heat can cause many problems

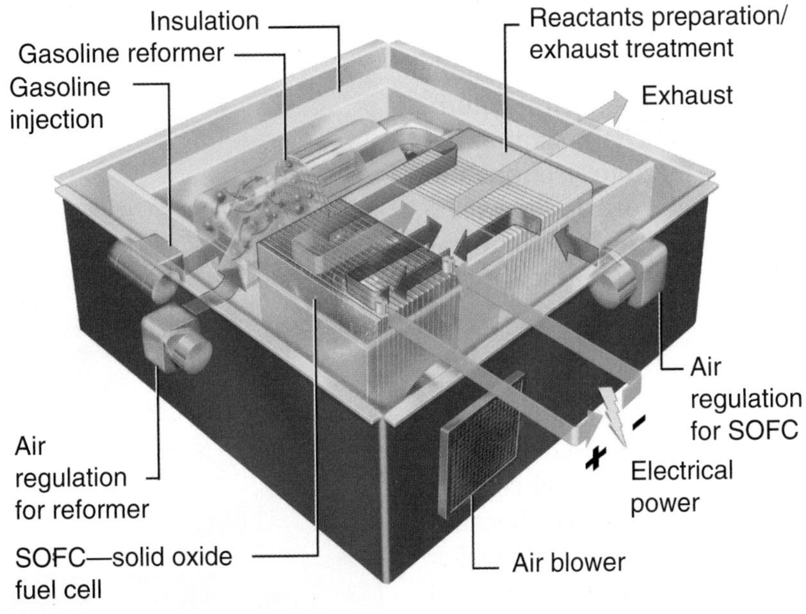

Figure 35-31 An SOFC fuel cell with a gasoline reformer. *Courtesy of BMW of North America, LLC*

in other automotive systems. When used to replace the alternator, the quantity of heat is far less and it can be easily moved away from the vehicle.

Molten Carbonate Fuel Cell (MCFCs) The MCFC is unlikely to ever be used in an automobile; it is best suited as a power generator to supply factories and perhaps cities. It has the ability to generate electricity from coal-based fuels or natural gas. This fuel cell uses a liquefied carbonate salt as its electrolyte. This fuel cell operates at high temperatures but its output is strongly dependent on operating temperatures, because a drop of just 100°F (50°C) will drop its output by as much as 15%. It also needs an intake of CO_2, which is one of its by-products.

Direct Methanol Fuel Cell (DMFC) The DMFC is a type of PEM fuel cell. Liquid methanol, rather than hydrogen, is the fuel oxidized at the anode and the oxygen from the outside is reduced at the cathode. A mixture of methanol and water is delivered directly into this modified PEM cell and releases the hydrogen needed by the fuel cell. This cell uses a thin membrane lightly covered on both sides with a layer of a platinum-based catalyst. It is not as efficient as the PEM fuel cell and the response of this fuel cell is slower than a PEM. Also, the DMFC has emissions that are not present with other fuel cell designs. As the hydrogen is removed from the methanol, carbon is released. The carbon and hydrogen combine with the oxygen at the outlet of the cell to form CO_2 and water.

Phosphoric Acid Fuel Cell The phosphoric acid fuel cell is the most commercially used fuel cell and can operate at 37% to 42% efficiency. This fuel cell uses liquid phosphoric acid as the electrolyte with electrodes made of carbon paper coated with a platinum catalyst. It operates at a relatively high temperature. The operation of the fuel cell is much the same as the others; a catalyst separates the fuel into electrons and protons. It is unlikely that this fuel cell design will be used in an automobile because of the high temperatures, the reluctance to have varying outputs, and its cost.

Alkaline Fuel Cell (AFC) The AFC is the one used primarily by NASA. It is expensive but highly efficient. In a spacecraft, the water (its exhaust) is used as drinking water for the space travelers. This fuel cell will undoubtedly never be used in automobiles because of its cost. They are also very sensitive to CO_2, which means they do best where all CO_2 can be removed from the incoming supply of air. This fuel cell operates in the same way as a PEM.

Obstacles for Fuel Cell Vehicles

Fuel cells provide clean energy and are quite efficient. These are the main reasons for considering their use in future automobiles **(Figure 35–32)**. The lack of a hydrogen infrastructure is one of their biggest obstacles. This is an obstacle of practicality and consumer acceptance, not of engineering. In order for a fuel cell vehicle to be practical, its fuel must be readily available.

Other obstacles that need to be conquered before fuel cell technology is incorporated in automobiles include the following

- FCEVs must be able to store hydrogen. Storing hydrogen in pressure tanks may be the answer, but high pressures are required and high-pressure tanks are very expensive. A typical fuel cell vehicle stores hydrogen at 5,000 psi (352 kg/cm) and has a driving range of about 150 miles (241 km). Doubling the pressure would nearly double the driving range. To double the pressure, stronger tanks are required, which adds to the cost of the tanks. To avoid the need to store hydrogen, reformers can be used. However, a reformer has undesirable emissions, such as CO_2. Its use also does not reduce our dependence on fossil fuels. Reformers are slow and require long run times before they can provide enough hydrogen to move a vehicle a few feet.

Figure 35-32 The engine compartment of a fuel cell car. *Courtesy of Chrysler LLC*

- The cost of a fuel cell and its supporting systems is extremely high. It is estimated that the cost of one fuel cell vehicle is $1 million. Obviously, the cost will come down as more FCEVs are built.
- An issue that may seem to be an odd concern is noise. FCEVs are too quiet. If the vehicles can be seen but not heard, their approach to the rear of pedestrians and bicyclists can present dangerous situations.
- To control the output of a fuel cell and therefore the speed of the vehicle, advanced electronics are necessary.
- The traction motors used in an FCEV must be very efficient and able to respond to changing driving conditions.
- Most fuel cells take some time to start, especially when they are cold. Freezing temperatures can kill a fuel cell. Ice on the membrane can destroy it or at least stop the fuel cell from working. An exhaust system plugged with ice will shut down a fuel cell. All water must be able to be removed from the fuel cell after it has been shut down. This requires energy from a storage device.
- Heat must be carefully controlled. Fuel cells become heated while they operate and operate best within a particular temperature range. That range depends on the type of fuel cell. PEM cells operate best at a lower temperature than conventional gasoline engines. This means the cooling system must be more efficient with larger and/or more radiators than those used in conventional vehicles. This becomes more of a challenge when one considers that the electronics and traction motors must also be kept cool, requiring an additional cooling system because they operate at a different temperature range than the fuel cell stack. An additional cooling problem exists when the vehicle is equipped with a high-voltage battery pack and/or ultracapacitors.
- Another heat-related problem is generated by the air compressor that feeds outside air into the fuel cell. As air is compressed, its temperature increases. Because the fuel cell works best with a specific temperature range, the compressed air can heat up the cell beyond that range. To eliminate this, intercoolers must be added to the air compressor system. These, again, occupy space. There is also the problem of filtering the incoming air. Ideally, the incoming air would be free of all dirt and other contaminants. A filtering system occupies space and has an impact on the overall layout and design of the vehicle.

CASE STUDY

A customer brought her late-model Prius into the shop. She felt everything was fine with the car but had a concern. She noticed while washing it that the front brake rotors were very rusty. Her concern was simply that they looked like they would not work properly. The technician took a look and verified the concern. He then pulled the wheels and checked the brakes because rusted rotors could be caused by the brakes not applying. He found nothing wrong with the brake system, so he removed the rotors and lightly machined them to bring back a shiny smooth finish. When he presented it to the customer, she was happy. However, 2 months later the customer brought the car back with the same concern. Quite puzzled, the technician again checked the brakes and found nothing wrong. He began to do some research and asked other technicians about the cause of this. Someone mentioned regenerative braking, which he knew little about. After doing quite a bit of thinking and studying, he realized that the rust was being formed on the discs because the brakes were not being used much. Regenerative braking slows the car down and the brakes only bring it to a halt when there is a hard stop. He then cleaned off the rotors again, and explained the cause of concern to the customer. She accepted it, especially when he told her to make a few hard stops every once in a while so the rotors will not collect rust.

KEY TERMS

Antiknock
Antiknock Index (AKI)
Assist hybrid
Belt alternator starter (BAS)
Biodiesel fuels
Cetane
Compressed natural gas (CNG)
Electrolysis
Ethanol
Flexible fuel vehicles (FFV)
Fractional distillation
Fuel cell electric vehicle (FCEV)
Fuel cell stack
Full hybrid
Heptane
Hydraulic hybrid

Hydrogen
Insulation resistance tester
Integrated motor assist (IMA)
Integrated starter alternator damper (ISAD)
Isooctane
Liquefied petroleum gas (LP gas)
Methanol
Methylcyclopentienyl manganese tricarbonyl (MMT)
Mild hybrid
Motor octane number (MON)
Natural gas vehicle (NGV)
Octane
Oxygenates
Petroleum
Plug-in hybrid electric vehicle (PHEV)
P-series fuels
Reformer
Reformulated gasoline (RFG)
Reid vapor pressure (RVP)
Renewable fuels
Research octane number (RON)
Sulfur dioxide
Tetraethyl lead (TEL)
Tetramethyl lead (TML)
Vapor lock

SUMMARY

- Crude oil is also called petroleum. It is drawn out of oil reservoirs and sands below the earth's surface.
- The hydrocarbons in crude oil have many different lengths and structures.
- The easiest and most common way to separate the various hydrocarbons (called fractions) is through a process called fractional distillation.
- Gasoline is a complex mixture of approximately 300 various ingredients, mainly hydrocarbons.
- Gasoline's octane number or rating gives the antiknock quality of a gasoline.
- Octane ratings are measured by the motor octane number (MON) method and the research octane number (RON) method. The octane rating required by law is the Antiknock Index (AKI). It is stated as (R+M)/2.
- The most common method for measuring the volatility of a fuel is the Reid vapor pressure (RVP) test.
- Several additives are put into gasoline to control harmful deposits, including gum or oxidation inhibitors, detergents, metal deactivators, and rust inhibitors.
- Oxygenates are compounds such as alcohols and ethers that are added to fuel to improve combustion efficiency and enhance octane ratings.
- The most widely used gasoline oxygenate additive is ethanol.
- Renewable fuels are those derived from nonfossil sources and produced from plant or animal products or wastes (biomass).
- Flexible fuel vehicles (FFV) can run on ethanol or gasoline or a mixture of the two. The alcohol fuel and gasoline are stored in the same tank.
- Hydrogen is full of energy. Due to its atomic structure and abundance it may be the fuel of the future.
- Diesel fuel is heavier and has more carbon atoms; it has about 15% more energy density than gasoline.
- Diesel fuel's ignition quality is measured by a cetane rating.
- Fuels derived from renewable biological sources for use in diesel engines are known as biodiesel fuels. Animal fats, recycled restaurant grease, and vegetable oils derived from crops such as soybeans, canola, corn, and sunflowers are used in the production of biodiesel fuel.
- Electric vehicles (EVs) use one or more electric motors that are powered by electricity stored in batteries.
- Hybrid electric vehicles (HEVs) have an internal combustion engine and one or more electric motors.
- There many types of hybrid vehicles, the most common of which are the full hybrid, assist hybrid, and mild hybrid.
- All hybrids are fitted with a high-voltage battery pack that must be isolated from the rest of the vehicle before service is performed.
- The basic layout for a hybrid contains one of the following systems: belt alternator starter, integrated starter alternator damper, power-split transaxle, or two-mode transmission.
- Hybrid vehicles are maintained and serviced in the same way as conventional vehicles, except for the hybrid components.
- The main powertrain components in a typical fuel cell vehicle include a fuel cell stack, high-pressure hydrogen supply or reformer with a fuel tank, air supply system, humidification system, fuel cell cooling system, storage battery or ultracapacitor, traction motor and transmission, and control module and related inputs and outputs.
- A fuel cell produces electricity through a process that works in the opposite way as electrolysis in which hydrogen and oxygen are combined to form water.

REVIEW QUESTIONS

1. What type of hybrid vehicle only uses the engine to run a generator to charge the battery and/or supply power for the motor?
2. *True or False?* The diesel fuel's antiknock quality is measured by the cetane rating.
3. Briefly explain how a fuel cell works.
4. What is a hybrid vehicle?
5. What does the Reid vapor pressure (RVP) test measure?
6. List three alternative fuels for an internal combustion engine and briefly explain where they come from.
7. Explain why hybrid vehicles will never be considered zero-emission vehicles.
8. Explain why a hybrid vehicle is more efficient than a vehicle powered only by an internal combustion engine.
9. Explain the differences among a full hybrid, an assist hybrid, and a mild hybrid.
10. Describe the basic operation of a hydraulic hybrid.
11. Describe the primary advantage of plug-in hybrid vehicles.
12. What is the basic fuel for a fuel cell?
    a. hydrogen
    b. methanol
    c. electricity
    d. gasoline
13. The component that changes the molecular structure of hydrocarbons into hydrogen-rich gas is called a ______.
    a. pressure container
    b. fuel cell
    c. fuel injector
    d. reformer
14. Which of the following statements about hydrogen is *not* true?
    a. Hydrogen displaces air, so any release in an enclosed space could cause asphyxiation.
    b. Hydrogen must be stored as a compressed gas.
    c. Hydrogen is nontoxic.
    d. Hydrogen is highly flammable and there is risk for an explosion.
15. Which of the following chemicals is commonly added to gasoline to increase its octane rating?
    a. isooctane
    b. heptane
    c. sulfur
    d. ethanol

ASE-STYLE REVIEW QUESTIONS

1. Technician A says that the use of methanol in internal combustion engines has declined over the years. Technician B says that the use of MTBE as a gasoline additive has declined over the years. Who is correct?
    a. Technician A
    b. Technician B
    c. Both A and B
    d. Neither A nor B
2. Technician A says that reformulated gasoline produces a leaner air-fuel mixture. Technician B says that RFG generates more carbon dioxide. Who is correct?
    a. Technician A
    b. Technician B
    c. Both A and B
    d. Neither A nor B
3. Technician A says that dedicated vehicles are those designed to use one particular type of fuel. Technician B says that bi-fuel vehicles can operate solely on an alcohol-based fuel or unleaded gasoline, or a mixture of the two, which gives the driver flexibility and convenience when refilling the fuel tank. Who is correct?
    a. Technician A
    b. Technician B
    c. Both A and B
    d. Neither A nor B
4. Technician A says that assist hybrids are those that use an electric motor to assist the internal combustion engine but do not have the ability to move the vehicle solely by battery power. Technician B says that all hybrids that do not have the ability to use the electric motor in propulsion are mild hybrids. Who is correct?
    a. Technician A
    b. Technician B
    c. Both A and B
    d. Neither A nor B
5. Technician A says that a fuel cell produces electricity through an electrochemical reaction that combines hydrogen and oxygen to form water. Technician B says that in a fuel cell, catalysts are used to ignite the hydrogen, causing a release of electrons or electrical energy. Who is correct?
    a. Technician A
    b. Technician B
    c. Both A and B
    d. Neither A nor B

6. While discussing working on hybrid vehicles: Technician A says to make sure that the high-voltage system is shut down and isolated from the vehicle before working near or with any high-voltage component. Technician B says that when working on or near the high-voltage system, even when it is de-powered, always use insulated tools. Who is correct?

   a. Technician A
   b. Technician B
   c. Both A and B
   d. Neither A nor B

7. While discussing the DC-DC converter used in most hybrids: Technician A says that it changes some of the DC voltage from the battery module to an AC voltage for the electric motors. Technician B says that it provides the power to operate the car's 12-volt electrical system and accessories. Who is correct?

   a. Technician A
   b. Technician B
   c. Both A and B
   d. Neither A nor B

8. Technician A says that all FCEVs have regenerative braking. Technician B says that all FCEVs have an auxiliary 12-volt battery. Who is correct?

   a. Technician A
   b. Technician B
   c. Both A and B
   d. Neither A nor B

9. Technician A says that gasoline can be used as a source of hydrogen for a fuel cell. Technician B says that carbon dioxide can be used as a source of hydrogen for a fuel cell. Who is correct?

   a. Technician A
   b. Technician B
   c. Both A and B
   d. Neither A nor B

10. While discussing hydrogen: Technician A says that a hydrogen atom has one proton and two electrons. Technician B says that hydrogen is produced in a fuel cell when water is broken down into its basic elements. Who is correct?

    a. Technician A
    b. Technician B
    c. Both A and B
    d. Neither A nor B

MANUAL TRANSMISSIONS AND TRANSAXLES

Section 5

CHAPTER 36

CLUTCHES

OBJECTIVES

■ Describe the various clutch components and their functions. ■ Name and explain the advantages of the different types of pressure plate assemblies. ■ Name the different types of clutch linkages. ■ List the safety precautions that should be followed during clutch servicing. ■ Explain how to perform basic clutch maintenance. ■ Name the six most common problems that occur with clutches. ■ Explain the basics of servicing a clutch assembly.

The clutch is located between the transmission and engine where it provides a mechanical coupling between the engine's flywheel and the transmission's input shaft. The driver operates the clutch through a linkage that extends from the passenger compartment to the **bell housing** (also called the clutch housing) between the engine and the transmission.

All manual transmissions require a clutch to engage or disengage the transmission. If the vehicle had no clutch and the engine was always connected to the transmission, the engine would stop every time the vehicle was brought to a stop. The clutch allows the engine to idle while the vehicle is stopped. It also allows for easy shifting between gears. (Of course, all of this applies to manual transaxles as well.)

The clutch engages the transmission gradually by allowing a certain amount of slippage between the transmission's input shaft and the flywheel. **Figure 36–1** shows the components needed to do this: the flywheel, clutch disc, pressure plate assembly, clutch release bearing (or throwout bearing), and the clutch fork.

OPERATION

The basic principle of clutch operation is shown in **Figure 36–2**. The pressure plate and flywheel are the drive or input members of the assembly. The **clutch disc**, also called the **friction disc**, is the driven or output member and is connected to the transmission's input shaft. As long as the clutch is disengaged (clutch pedal depressed), the drive members turn independently of the driven member, and the engine is disconnected from the transmission. However, when the clutch is engaged (clutch pedal released), the pressure plate moves toward the flywheel and the clutch disc is squeezed between the two revolving drive members and forced to turn at the same speed.

WARNING!

Use the appropriate cleaning liquid and equipment before and during disassembling a clutch assembly. Some clutch discs were made with asbestos. Inhalation of asbestos can cause serious illnesses. Assume that all clutch discs have asbestos and follow the procedures for containing it.

Flywheel

The flywheel, an important part of the engine, is also the main driving member of the clutch. It is normally

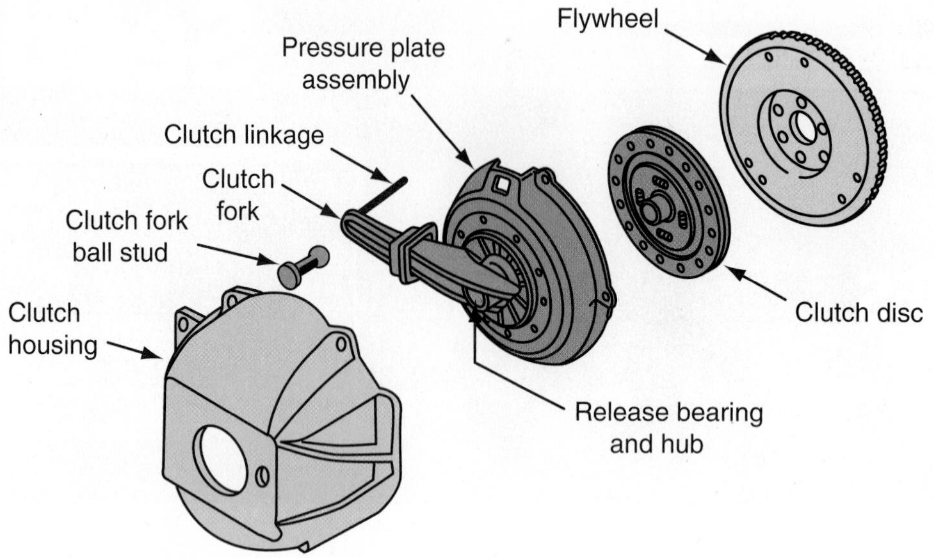

Figure 36-1 Major parts of a clutch assembly.

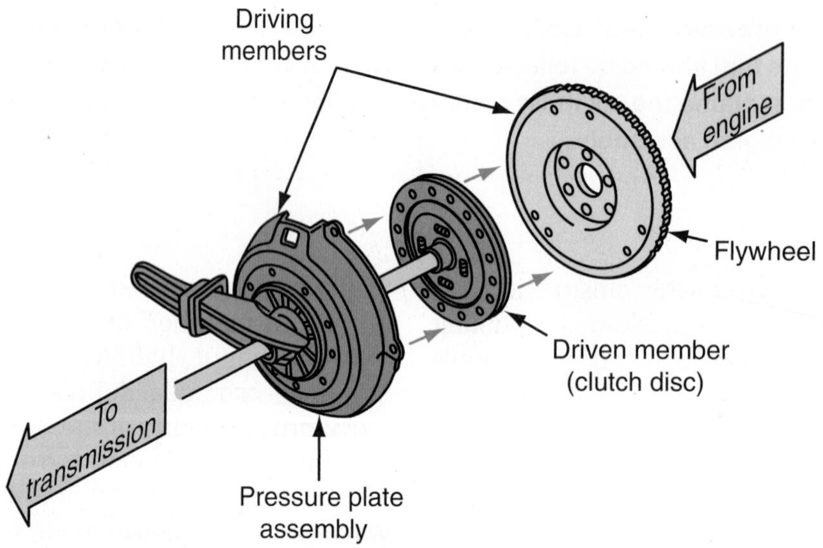

Figure 36-2 When the clutch is engaged, the driven member is squeezed between the two driving members. The transmission is connected to the driven member.

made of nodular or gray cast iron, which has a high graphite content to lubricate the engagement of the clutch. Welded to or pressed onto the outside diameter of the flywheel is the starter ring gear. The large diameter of the flywheel allows for an excellent gear ratio of the starter drive to ring gear, which provides for ample engine rotation during starting. The rear surface of the flywheel is a friction surface machined very flat to ensure smooth clutch engagement. The flywheel also provides some absorption of torsional vibration of the crankshaft. It further provides the inertia to rotate the crankshaft through the four strokes.

The flywheel has two sets of bolt holes drilled into it. The inner set is used to fasten the flywheel to the crankshaft, and the outer set provides a mounting plate for the pressure plate assembly. A bore in the center of the flywheel and crankshaft holds the **pilot bushing** or bearing, which supports the front end of the transmission input shaft and maintains alignment with the engine's crankshaft. Often a ball or roller needle bearing is used instead of a pilot bushing. Some transaxles have a short, self-centering input shaft that does not require a pilot bushing or bearing.

Dual-Mass Flywheel A few cars and light trucks use a dual-mass flywheel. These flywheels are used to reduce vibrations transmitted through the transmission, provide for smoother shifting, and reduce gear noise. Dual-mass flywheels can reduce the oscillations of the crankshaft before they move through the

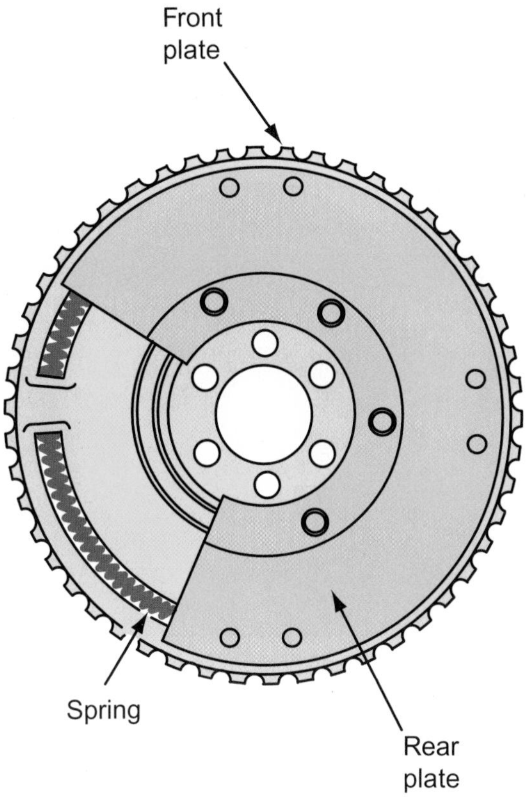

Figure 36–3 A dual-mass flywheel.

Figure 36–4 A clutch disc.

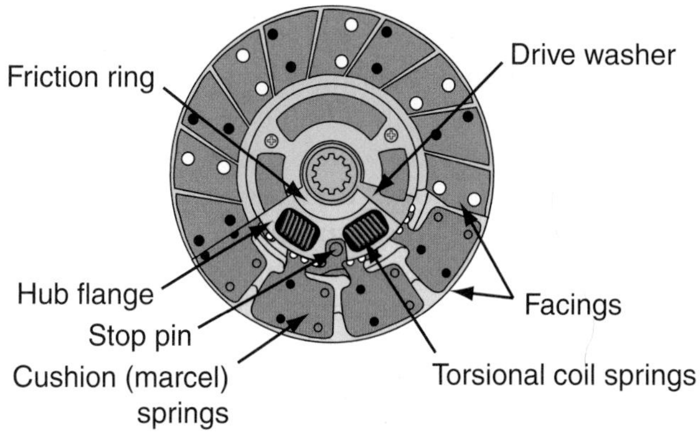

Figure 36–5 The major parts of a clutch disc.

transmission **(Figure 36–3)**. The flywheel consists of two rotating plates connected by a spring and damper system. The forwardmost portion of the flywheel is bolted to the end of the crankshaft and smooths out the crankshaft's oscillations. The pressure plate of the clutch is bolted to the rearward portion of the flywheel. Engine torque moves from the front plate through the damper and spring assembly to the rear plate before it enters the transmission.

Some have a torque-limiting feature that prevents damage to the transmission during peak torque loads. The rotation of the two flywheel plates can differ by as much as 360 degrees. This allows the forward plate to absorb torque spikes and not pass them along through the transmission.

Clutch Disc

The clutch disc **(Figure 36–4)** is splined to the transmission's input shaft and receives the driving motion from the flywheel and pressure plate assembly and transmits that motion to the transmission input shaft. The parts of a clutch disc are shown in **Figure 36–5**.

There are two types of friction facings. Molded friction facings are preferred because they withstand greater pressure plate loading force without damage. Woven friction facings are used when additional cushioning action is needed for clutch engagement. Until recently, the material that was molded or woven into facings was predominantly asbestos. Now, because of the hazards associated with asbestos, other materials such as paper-base and ceramics are being used instead. Particles of cotton, brass, rope, and wire are added to prolong the life of the clutch disc and provide torsional strength.

Grooves are cut across the face of the friction facings. This promotes clean disengagement of the driven disc from the flywheel and pressure plate; it also promotes better cooling. The facings are riveted to wave springs, also called **cushioning springs**, which cause the contact pressure on the facings to rise gradually as the springs flatten out when the clutch is engaged. These springs reduce chatter when the clutch is engaged and also reduce the chance of the clutch disc sticking to the flywheel and pressure plate surfaces when the clutch is disengaged. The wave springs and friction facings are fastened to the steel disc.

The clutch disc is designed to absorb such things as crankshaft vibration, abrupt clutch engagement, and driveline shock. Torsional coil springs allow the

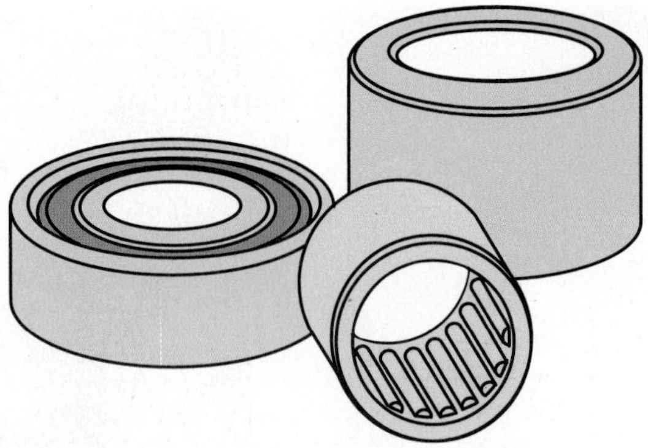

Figure 36-6 Different designs of pilot bushings and bearings used in today's clutch assemblies.

disc to rotate slightly in relation to the pressure plate while they absorb the torque forces. The number and tension of these springs is determined by engine torque and vehicle weight. Stop pins limit this torsional movement to approximately ⅜ inch.

Pilot Bushing/Bearing

The pilot bushing or bearing **(Figure 36–6)** is sometimes used to support the outer end of the transmission's input shaft. This shaft is splined to the clutch disc and transmits power from the engine (when the clutch is engaged) to the transmission. The transmission end of the input shaft is supported by a large bearing in the transmission case. Because the input shaft extends unsupported from the transmission, a pilot bushing is used to keep it in position. By supporting the shaft, the pilot bushing keeps the clutch disc centered in the pressure plate.

Pressure Plate Assembly

The purpose of the pressure plate assembly **(Figure 36–7)** is twofold. First, it must squeeze the clutch disc onto the flywheel with sufficient force to transmit engine torque efficiently. Second, it must move away from the clutch disc so the clutch disc can stop rotating, even though the flywheel and pressure plate continue to rotate.

Basically, there are two types of pressure plate assemblies: those with coil springs and those with a **diaphragm** spring. Both types have a stamped steel cover that bolts to the flywheel and acts as a housing to hold the parts together. In both, there is also the pressure plate, which is a heavy, flat ring made of nodular or gray cast iron. The assemblies differ in the manner in which they move the pressure plate toward and away from the clutch disc.

Figure 36-7 A diaphragm spring clutch pressure plate.

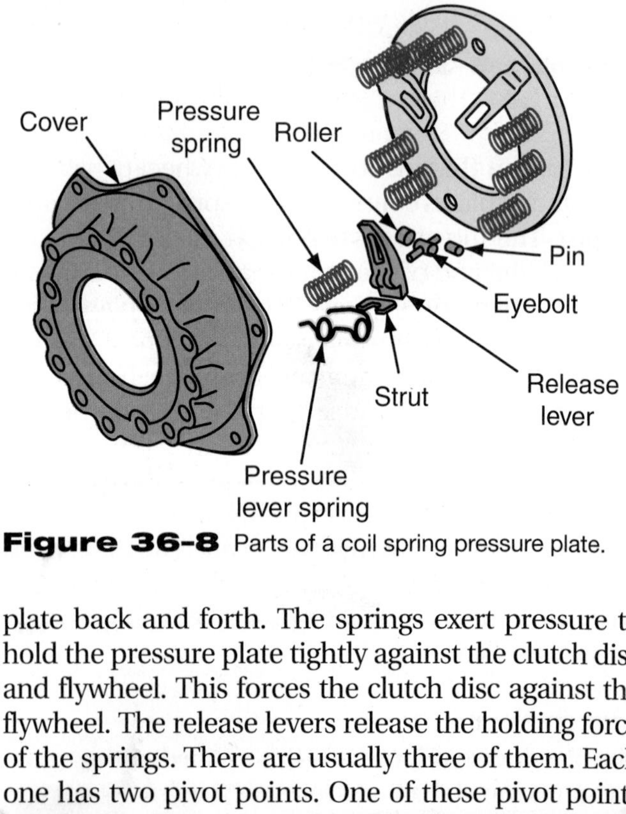

Figure 36-8 Parts of a coil spring pressure plate.

Coil Spring Pressure Plate Assembly A coil spring pressure plate assembly, shown in **Figure 36–8**, uses coil springs and release levers to move the pressure plate back and forth. The springs exert pressure to hold the pressure plate tightly against the clutch disc and flywheel. This forces the clutch disc against the flywheel. The release levers release the holding force of the springs. There are usually three of them. Each one has two pivot points. One of these pivot points attaches the lever to a pedestal cast into the pressure plate and the other to a release lever yoke/keybolt bolted to the cover. The levers pivot on the pedestals and release lever yokes to move the pressure plate through its engagement and disengagement operations. To disengage the clutch, the release bearing pushes the inner ends of the release levers forward toward the flywheel. The release levers are class one levers, which means the fulcrum is between the effort and the load. Each end of the lever moves in the opposite direction. When force pushes one end of the lever down, the other end moves up. In a coil spring pressure plate, the release lever yokes act as fulcrums for the levers, and the outer ends of the release levers move backward, pulling the pressure plate away from the clutch disc. This compresses the springs and releases the clamping of the clutch disc between the pressure plate and flywheel.

When the clutch is engaged, the release bearing moves away from the pressure plate. This allows the pressure plate springs to push the pressure plate and clutch disc against the flywheel, allowing power transfer from the engine to the transmission.

Some of the advantages and disadvantages of a coil spring pressure plate include the following:

- Clamping pressure is increased or decreased by changing the number of springs and their tension.
- They require more pedal effort than diaphragm pressure plates.
- As the clutch disc wears, the coil springs expand and their clamping force is reduced.
- Because of these disadvantages, passenger cars and light trucks are now almost exclusively equipped with diaphragm spring clutches because of the disadvantages of coil spring pressure plates.

Diaphragm Spring Pressure Plate Assembly The diaphragm spring pressure plate assembly relies on a cone-shaped diaphragm spring between the pressure plate and the pressure plate cover to move the pressure plate back and forth. The diaphragm spring (sometimes called a Belleville spring) is a single, thin sheet of metal that works in the same manner as the bottom of an oil can. The metal yields when pressure is applied to it. When the pressure is removed, the metal springs back to its original shape. The center portion of the diaphragm spring is slit into numerous fingers that act as release levers **(Figure 36–9)**.

During clutch disengagement, these fingers are moved forward by the release bearing. The diaphragm spring pivots over the fulcrum ring (also called the pivot ring), and its outer rim moves away from the flywheel. The retracting springs pull the pressure plate away from the driven disc and disengage the clutch.

When the clutch is engaged, the release bearing and the fingers of the diaphragm spring move toward the transmission. As the diaphragm pivots over the pivot ring, its outer rim forces the pressure plate against the clutch disc so the clutch is engaged to the flywheel.

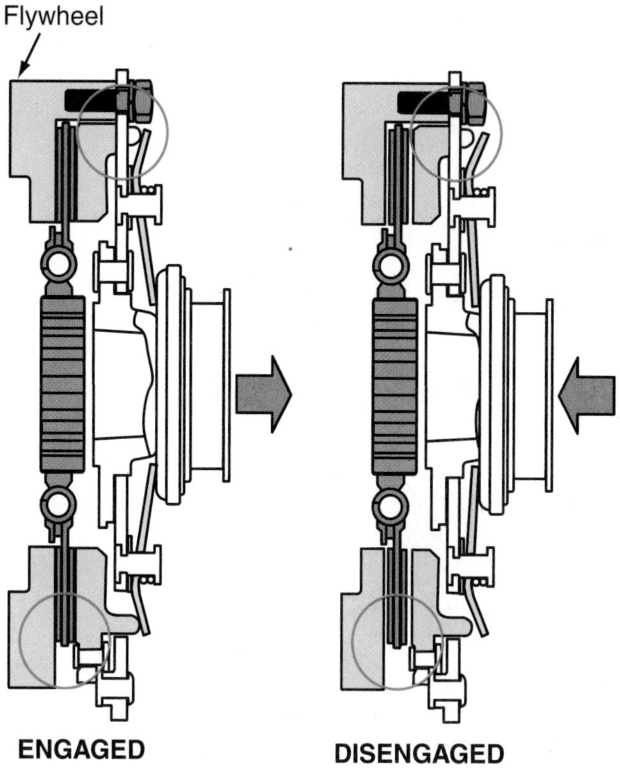

Figure 36–9 The action of a diaphragm spring-type pressure plate assembly.

Diaphragm spring pressure plate assemblies have the following advantages over other types of assemblies:

- Compactness.
- Less weight.
- Fewer moving parts to wear out.
- Little pedal effort is required from the operator.
- They provide a balanced force around the pressure plate so rotational unbalance is reduced.
- Clutch disc slippage is less likely to occur. Mileage builds because the force holding the clutch disc to the flywheel does not change throughout its service life.

Clutch Release Bearing

The **clutch release bearing**, also called a **throwout bearing**, is usually a sealed, prelubricated ball bearing **(Figure 36–10)**. Its function is to smoothly and quietly move the pressure plate release levers or diaphragm spring through the engagement and disengagement process.

The release bearing is mounted on a hub, which slides on a hollow shaft at the front of the transmission housing. This hollow shaft, shown in **Figure 36–11**, is part of the transmission bearing retainer.

To disengage the clutch, the release bearing is moved on its shaft by the **clutch fork**. As the release

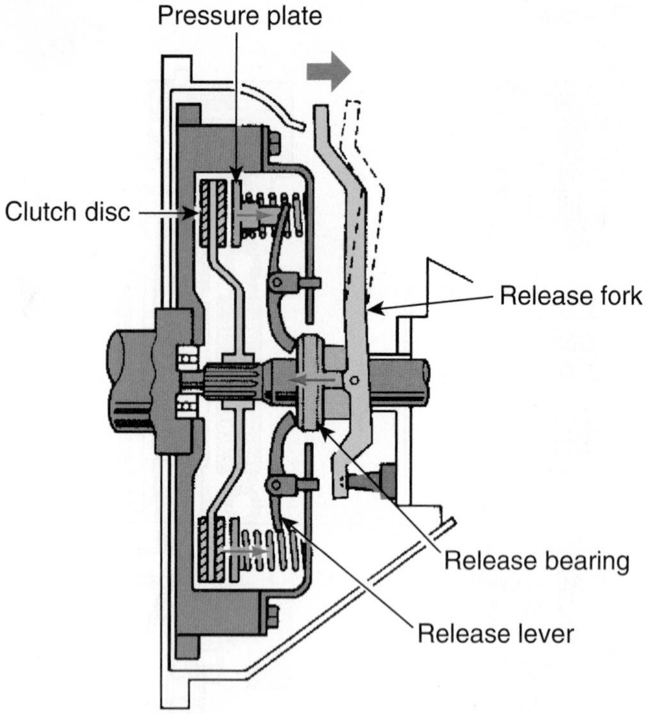

Figure 36-10 The action of the release fork and bearing on the pressure plate when the clutch pedal is depressed.

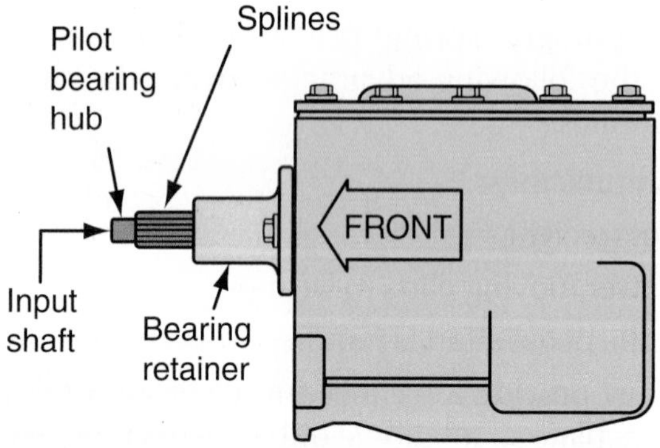

Figure 36-11 The release bearing slides on the hollow shaft of the transmission's front bearing retainer.

bearing contacts the release levers or diaphragm spring of the pressure plate assembly, it begins to rotate with the rotating pressure plate assembly. As the release bearing continues forward, the clutch disc is disengaged from the pressure plate and flywheel.

To engage the clutch, the release bearing slides to the rear of the shaft. The pressure plate moves forward and traps the clutch disc against the flywheel to transmit engine torque to the transmission input shaft. Once the clutch is fully engaged, the release bearing is normally stationary.

Rotating Release Bearing Self-adjusting clutch linkages, used on many vehicles, apply just enough tension to the clutch control cable to keep a constant light pressure against the release bearing. As a result, the release bearing is kept in contact with the release levers or diaphragm spring of the rotating pressure plate assembly. The release bearing rotates with the pressure plate.

Clutch Fork

The clutch fork is a forked lever that pivots on a support shaft or ball stud located in an opening in the bell housing. The forked end slides over the hub of the release bearing and the small end protrudes from the bell housing and connects to the clutch linkage and clutch pedal. The clutch fork moves the release bearing and hub back and forth during engagement and disengagement.

Clutch Linkage

The clutch linkage is a series of parts that connects the clutch pedal to the clutch fork. It is through the clutch linkage that the operator controls the engagement and disengagement of the clutch assembly smoothly and with little effort.

Shaft and Lever Linkage Found on older vehicles, the shaft and lever clutch linkage has many parts and pivot points and transfers the movement of the clutch pedal to the release bearing via shafts, levers, and bell cranks. On some vehicles, the pivot points were equipped with grease fittings; others had low-friction plastic grommets and bushings at their pivot points.

A typical shaft and lever clutch control assembly includes a release lever and rod, an equalizer or cross shaft, a pedal-to-equalizer rod, an assist or over-center spring, and the pedal assembly. Depressing the pedal moves the equalizer, which, in turn, moves the release rod. When the pedal is released, the assist spring returns the linkage to its normal position and removes the pressure on the release rod. This action causes the release bearing to move away from the pressure plate.

Cable Linkage A cable linkage can perform the same controlling action as the shaft and lever linkage but with fewer parts. The clutch cable system does not take up much room. It also has the advantage of flexible installation so it can be routed around the power brake and steering units. These advantages help to make it the most commonly used clutch linkage.

The clutch cable (**Figure 36–12**) is made of braided wire. The upper end is connected to the top of the clutch pedal arm, and the lower end is fastened to the clutch fork. It is designed with a flexible outer housing that is fastened at the fire wall and the clutch housing.

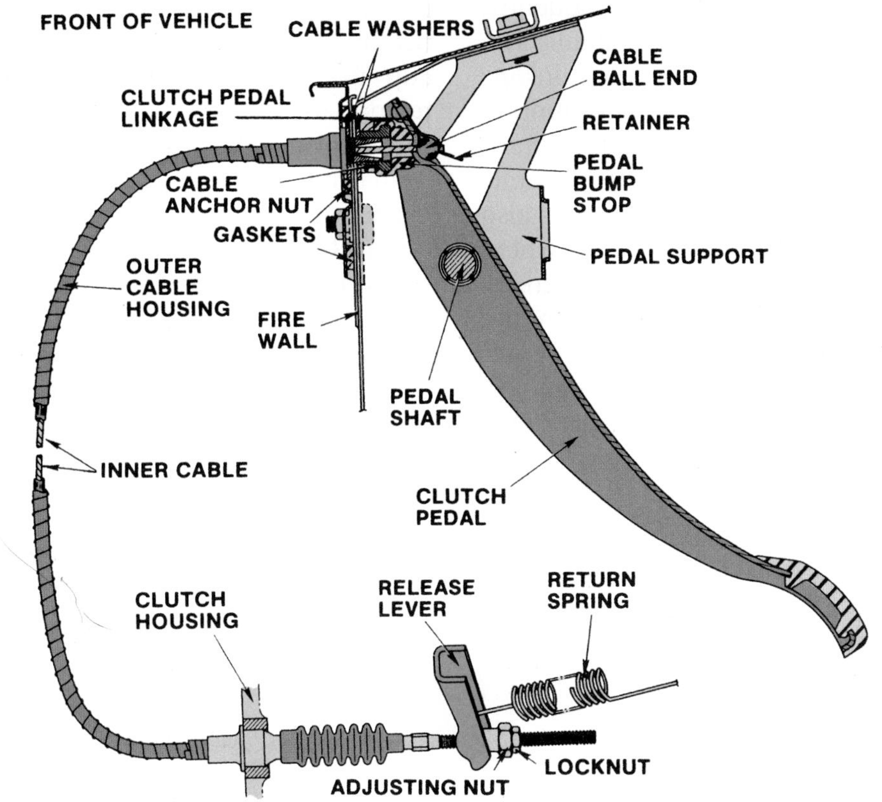

Figure 36-12 A typical clutch cable system.

When the clutch pedal is pushed to the disengaged position, it pivots on the pedal shaft and pulls the inner cable through the outer housing. This action moves the clutch fork to disengage the clutch. The pressure plate springs and springs on the clutch pedal provide the force to move the cable back when the clutch pedal is released.

Self-Adjusting Clutch Self-adjusting clutch mechanisms monitor clutch pedal play and automatically adjust it when necessary.

Usually the self-adjusting clutch mechanism is a ratcheting mechanism located at the top of the clutch pedal behind the dash panel (**Figure 36–13**). The ratchet is designed with a pawl and toothed segment, and a pawl tension spring is used to keep the pawl in contact with the toothed segment. The pawl allows the toothed segment to move in only one direction in relation to the pawl.

The clutch cable is guided around and fastened to the toothed segment, which is free to rotate in one direction (backwards) independently of the clutch pedal. The tension spring pulls the toothed segment backwards.

When the clutch cable develops slack due to stretching and clutch disc wear, the cable is adjusted automatically when the clutch is released. The tension spring pulls the toothed segment backwards and allows the pawl to ride over to the next tooth. This effectively shortens the cable. Actually, the cable is not really shortened; but the slack has been reeled in by the repositioning of the toothed segment. This self-adjusting action takes place automatically during the clutch's operational life.

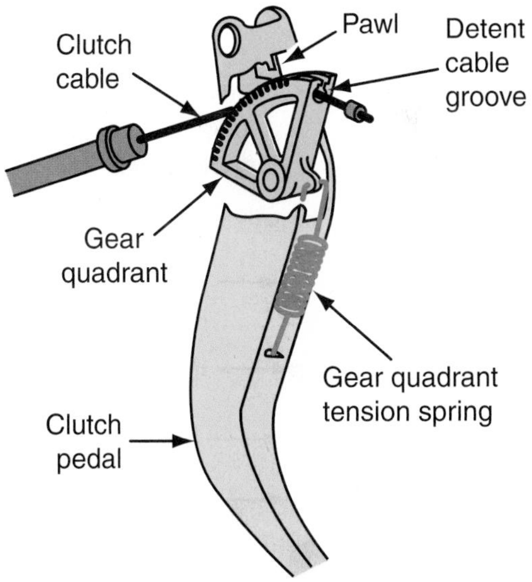

Figure 36-13 A typical clutch cable self-adjusting mechanism.

Hydraulic-Operated Clutch Linkage Frequently, the clutch assembly is controlled by a hydraulic system (**Figure 36–14**). In the hydraulic clutch linkage system, hydraulic (liquid) pressure transmits motion from one sealed cylinder to another through a hydraulic line. Like the cable linkage assembly, the hydraulic linkage is compact and flexible. Cable linkages also allow engineers to place the release fork anywhere that gives them more flexibility in body design. In addition, the hydraulic pressure developed by the master cylinder decreases required pedal effort and provides a precise method of controlling clutch operation. Brake fluid is commonly used as the hydraulic fluid in hydraulic clutch systems.

A hydraulic clutch master cylinder is shown in **Figure 36–15**. Its pushrod moves the piston and primary cup to create hydraulic pressure. The snapring restricts the travel of the piston. The secondary cup at the snapring end of the piston stops hydraulic fluid from dripping into the passenger compartment. The piston return spring holds the primary cup and piston in the fully released position. Hydraulic fluid is stored in the reservoir on top of the master cylinder housing.

The slave cylinder body may have a bleeder valve to bleed air from the hydraulic system for efficient clutch linkage operation. The cylinder body is threaded for a tube and fitting at the fluid entry port.

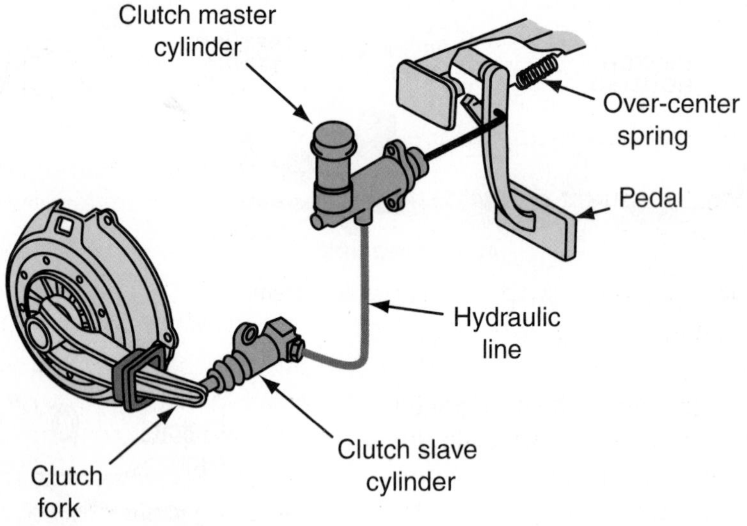

Figure 36–14 A typical hydraulic clutch linkage arrangement.

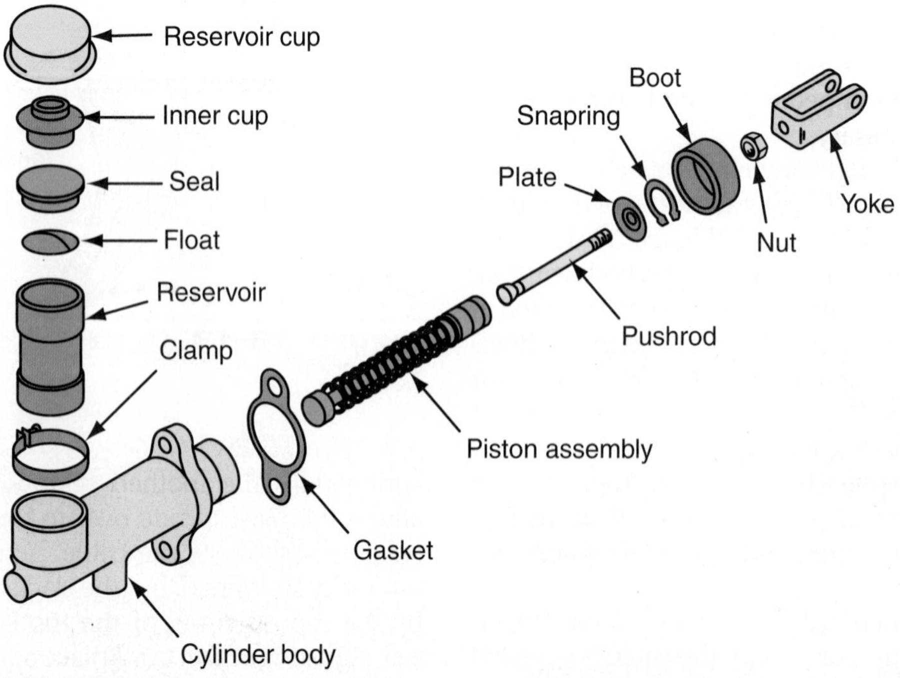

Figure 36–15 Parts of a hydraulic clutch master cylinder.

Rubber seal rings are used to seal the hydraulic pressure between the piston and the slave cylinder walls. The piston retaining ring is used to restrict piston travel to a certain distance. Piston travel is transmitted by a pushrod to the clutch fork. The pushrod boot keeps contaminants out of the slave cylinder.

When the clutch pedal is depressed, the movement of the piston and primary cup develops hydraulic pressure that is displaced from the master cylinder, through a tube, into the slave cylinder. The slave cylinder piston movement is transmitted to the clutch fork, which disengages the clutch.

When the clutch pedal is released, the primary cup and piston are forced back to the disengaged position by the master cylinder piston return spring. External springs move the slave cylinder pushrod and piston back to the engaged position. Fluid pressure returns through the hydraulic tubing to the master cylinder assembly. There is no residual pressure in the system when the clutch assembly is in the engaged position.

Internal Slave Cylinders An internal concentric slave cylinder is found on some cars and light trucks. These units are actually a combination of the slave cylinder and the clutch release bearing **(Figure 36–16)**. By having the slave cylinder directly behind the release bearing, the movement of the release bearing is linear. In other clutch linkage designs, the release bearing moves through an arc as it engages and disengages the clutch.

An internal slave cylinder is a doughnut-shaped unit that mounts to the front of the transmission, and the transmission's input shaft passes through it. The slave cylinder is either bolted to the transmission's front bearing cover or is held by a pressed pin. In many cases, the transmission must be removed to properly diagnose and service this type of slave cylinder.

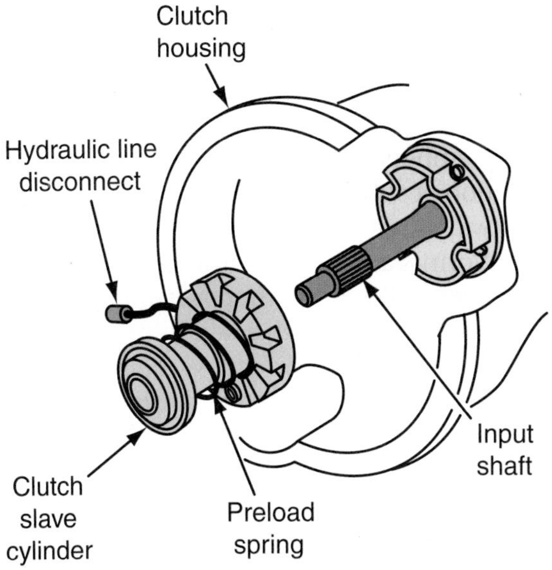

Figure 36-16 A concentric internal clutch slave cylinder.

CLUTCH SERVICE SAFETY PRECAUTIONS

When servicing the clutch, exercise the following precautions:

- Always wear eye protection when working underneath a vehicle.
- Remove asbestos dust only with a special, approved vacuum collection system or an approved liquid cleaning system.
- Never use compressed air or a brush to clean off asbestos dust.
- Follow all federal, state, and local laws when disposing of collected asbestos dust or liquid containing asbestos dust.
- Never work under a vehicle that is not raised on a hoist or supported by safety or jack stands.
- Use jack stands and special jacks to support the engine and transmission.
- Have a helper assist in removing the transmission.
- Be sure the work area is properly ventilated, or attach a ventilating hose to the vehicle's exhaust system when an engine is to be run indoors.
- Do not allow anyone to stand in front of or behind the automobile while the engine is running.
- Set the emergency brake securely and place the gearshift in neutral when running the engine of a stationary vehicle.
- Avoid touching hot engine and exhaust system parts. Whenever possible, let the vehicle cool down before beginning to work on it.

CLUTCH MAINTENANCE

Clutches may require checking and adjustment of linkage at regular intervals. Vehicles with external clutch linkage require periodic lubrication. These maintenance procedures are explained in this section.

USING SERVICE INFORMATION

Service manuals include adjustment procedures and instructions for clutch removal, inspection, installation, and troubleshooting. They may also offer information to aid in clutch release bearing analysis.

Clutch Linkage Adjustment

Except for systems with self-adjusting mechanisms, the release bearing should not touch the pressure plate release levers when the clutch is engaged (pedal up). Clearance between these parts prevents premature clutch plate, pressure plate, and release bearing wear. As the clutch disc wears and becomes thinner, this clearance is reduced.

Clearance can be ensured by adjusting the clutch linkage so the pedal has a specified amount of play, or **free travel**. Free travel is the distance a clutch pedal moves when depressed before the release bearing contacts the clutch release lever or diaphragm spring of the pressure plate.

To check pedal play, use a tape measure or ruler. Place the tape measure or ruler beside the clutch pedal and the end against the floor of the vehicle and note the reading (**Figure 36–17**). Then depress the clutch pedal just enough to take up the pedal play and note the reading again. The difference between the two readings is the amount of pedal play.

Adjustment should be performed when pedal play is not correct or when the clutch does not engage or disengage properly. To adjust clutch pedal play, refer to the manufacturer's service manual for the correct procedure and adjustment point locations. Often pedal play can be increased or decreased by turning a threaded fastener located either under the dash at the clutch pedal or where the linkage attaches to the clutch fork.

SHOP TALK

Normally, clutch condition will dictate the amount of clutch pedal free play there is. However, there should always be free play with hydraulic clutch systems.

Clean the linkage with a shop towel and solvent, if necessary, before checking it and replacing any damaged or missing parts or cables. Check hydraulic-operated linkage systems for leaks at the clutch master cylinder, hydraulic hose, and slave cylinder. Then adjust the linkage to provide the manufacturer's specified clutch pedal play.

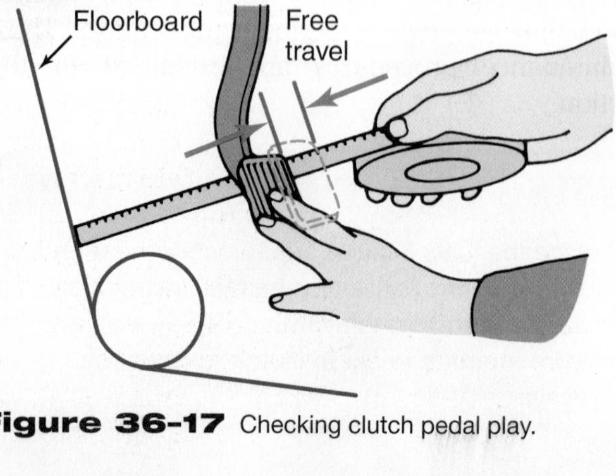

Figure 36-17 Checking clutch pedal play.

External Clutch Linkage Lubrication

External clutch linkage should be lubricated at regular intervals, such as during a chassis lubrication. Refer to the vehicle's service manual to determine the proper lubricant. Many clutch linkages use the same chassis grease used for suspension parts and U-joints. Lubricate all the sliding surfaces and pivot points in the clutch linkage (**Figure 36–18**). The linkage should move freely after lubrication.

On vehicles with hydraulic clutch linkage, check the clutch master cylinder reservoir fluid level. It should be approximately $\frac{1}{4}$ inch (6.35 mm) from the top of the reservoir. If it must be refilled, use approved brake fluid. Also, because the clutch master cylinder does not consume fluid, check for leaks in the master cylinder, connecting flexible line, and slave cylinder, if the fluid is low. On some vehicles, the brake fluid reservoir is shared with the hydraulic clutch.

CLUTCH PROBLEM DIAGNOSIS

Check and attempt to adjust the clutch pedal play before attempting to diagnose any clutch problems. If the friction lining of the clutch is worn too thin (**Figure 36–19**), the clutch cannot be adjusted successfully. The most common clutch problems are described here.

Slippage

Clutch slippage is a condition in which the engine overspeeds without generating any increase in torque to the driving wheels. It occurs when the clutch disc is not gripped firmly between the flywheel and the pressure plate. Instead, the clutch disc slips between these driving members. Slippage can occur during initial acceleration or subsequent shifts.

One way to check for slippage is by driving the vehicle. Normal acceleration from a stop and several gear changes indicate whether the clutch is slipping.

Slippage also can be checked in the shop. Check the service manual for correct procedures. A general procedure for checking clutch slippage follows. Be sure to follow the safety precautions stated earlier.

With the parking brake on, disengage the clutch. Shift the transmission into third gear, and increase the engine speed to about 2,000 rpm. Slowly release the clutch pedal until the clutch engages. The engine should stall immediately.

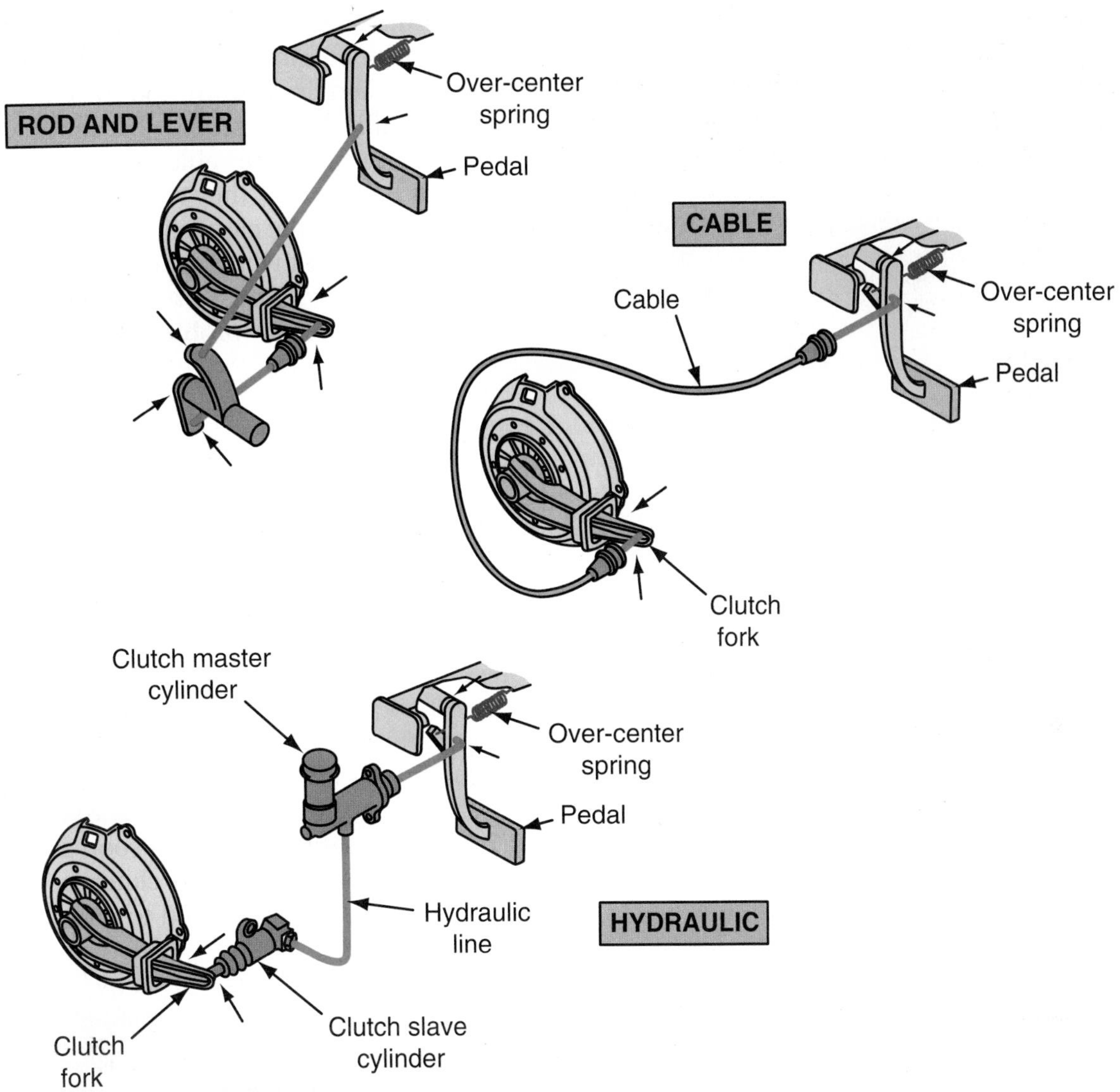

Figure 36–18 Clutch linkage lubrication points.

If it does not stall within a few seconds, the clutch is slipping. Safely raise the vehicle and check the clutch linkage for binding and broken or bent parts. If no linkage problems are found, the transmission and the clutch assembly must be removed so the clutch parts can be examined.

SHOP TALK

Severe or prolonged clutch slippage causes grooving or heat damage to the pressure plate and/or flywheel.

Clutch slippage can be caused by an oil-soaked **(Figure 36–20)** or worn disc facing, warped pressure plate, weak or broken diaphragm spring, or the release bearing contacting and applying pressure to the release levers.

Drag and Binding

If the clutch disc is not completely released when the clutch pedal is fully depressed, clutch drag occurs. Clutch drag causes gear clash, especially when shifting into reverse. It can also cause hard starting and vehicle movement during starting because the engine attempts to turn the transmission input shaft.

To check for clutch drag, start the engine, depress the clutch pedal completely, and shift the transmission into first gear. Do not release the clutch. Then shift the transmission into neutral and wait 5 seconds before attempting to shift smoothly into reverse.

It should take no more than 5 seconds for the clutch disc, input shaft, and transmission gears to come to a complete stop after disengagement. This period, called the clutch spindown time, is normal and should not be mistaken for clutch drag.

If the shift into reverse causes gear clash, raise the vehicle safely and check the clutch linkage for

Figure 36-19 A severely worn clutch disc. *Courtesy of LuK Automotive Systems*

Figure 36-20 Grease and oil on the hub of this disc is an indication that the disc may be contaminated with oil. *Courtesy of LuK Automotive Systems*

binding, broken, or bent parts. If no problems are found in the linkage, the transmission and clutch assembly must be removed so that the clutch parts can be examined.

Clutch drag can occur as a result of a warped disc or pressure plate, a loose disc facing, a defective release lever, or incorrect clutch pedal adjustment that results in excessive pedal play. A binding or seized pilot bushing or bearing can also cause clutch drag. A cracked or broken fire wall around the clutch master cylinder or cable guide can also cause the clutch to bind or drag. Because the cable or master cylinder is mounted to the fire wall, they will be able to move by the force exerted on the clutch linkage.

Binding can result when the splines in the clutch disc hub or on the transmission input shaft are damaged or when there are problems with the release levers.

Chatter

A shaking or shuddering that is felt in the vehicle as the clutch is engaged is known as clutch **chatter (Figure 36–21)**. It usually occurs when the pressure plate first contacts the clutch disc and stops when the clutch is fully engaged.

To check for clutch chatter, start the engine, depress the clutch completely, and shift the transmission into first gear. Increase engine speed to about 1,500 rpm, then slowly release the clutch pedal and check for chatter as the pedal begins to engage. Do not release the pedal completely, or the vehicle might jump and cause serious injury. As soon as the clutch is partially engaged, depress the clutch pedal immediately and reduce engine speed to prevent damage to the clutch parts.

Usually clutch chatter is caused by liquid leaking onto the clutch and contaminating its friction

Figure 36-21 The marks on the surface of this pressure plate are caused by clutch chatter. *Courtesy of LuK Automotive Systems*

surfaces. This results in a mirrorlike shine on the pressure plate or a glazed clutch facing. Oil and clutch hydraulic fluid leaks can occur at the engine rear main bearing seal transmission input shaft seal, clutch slave cylinder, and hydraulic line. Other causes of clutch chatter include broken engine or transmission mounts, loose bell housing bolts, and damaged clutch linkage.

During disassembly, check for a warped pressure plate or flywheel, hot spots on the flywheel, a burned or glazed disc facing, and worn input shaft splines. If the chattering is caused by an oil-soaked clutch disc and no other parts are damaged, then the disc alone needs to be replaced. However, the cause of the oil leak must also be found and corrected.

Clutch chatter can also be caused by broken or weak torsional coil springs in the clutch disc and by the failure to resurface the flywheel when a new clutch disc and/or pressure plate is installed. It is highly recommended that the flywheel be resurfaced every time a new clutch disc or pressure plate is installed.

Figure 36-22 The shiny areas on this pressure plate shows that the release bearing is not evenly contacting the pressure plate. *Courtesy of LuK Automotive Systems*

Pedal Pulsation

Pedal pulsation is a rapid up-and-down movement of the clutch pedal as the clutch disengages or engages. This pedal movement usually is minor, but it can be felt through the clutch pedal. It is not accompanied by any noise. Pulsation begins when the release bearing makes contact with the release levers.

To check for pedal pulsation, start the engine, depress the clutch pedal slowly until the clutch just begins to disengage, and then stop briefly. Resume depressing the clutch pedal slowly until the pedal is depressed to a full stop.

On many vehicles, minor pulsation is considered normal. If pulsation is excessive, the clutch must be removed and disassembled for inspection.

Pedal pulsations can result from the misalignment of parts. Check for a misaligned bell housing or a bent flywheel. Inspect the clutch disc and pressure plate for warpage. Broken, bent, or warped release levers also create misalignment (**Figure 36–22**).

> ***Customer Care***
>
> **If you repair a vehicle with clutch slippage, tactfully inform the customer about the different poor driving habits that can cause this problem. These habits include riding the clutch pedal and holding the vehicle on an incline by using the clutch as a brake.**

Vibration

Clutch vibrations, unlike pedal pulsations, can be felt throughout the vehicle, and they occur at any clutch pedal position. These vibrations usually occur at normal engine operating speeds (more than 1,500 rpm).

There are several possible sources of vibration that should be checked before disassembling the clutch to inspect it. Check the engine mounts and the crankshaft damper pulley. Look for any indication that engine parts are rubbing against the body or frame.

Accessories can also be a source of vibration. To check them, remove the drive belts one at a time. Set the transmission in neutral, and securely set the emergency brake. Start the engine and check for vibrations. Do not run the engine for more than 1 minute with the belts removed.

If the source of vibration is not discovered through these checks, the clutch parts should be examined. Be sure to check for loose flywheel bolts, excessive flywheel runout, and pressure plate cover balance problems.

Noises

Many clutch noises come from bushings and bearings. Pilot bushing noises are squealing, howling, or trumpeting sounds that are most noticeable in cold weather. These bushing noises usually occur when the pedal is being depressed and the transmission is in neutral. Release bearing noise is a whirring, grating, or grinding sound that occurs when the clutch pedal is depressed and stops when the pedal is fully released. It is most noticeable when the transmission is in neutral, but it also can be heard when the transmission is in gear.

Hydraulic-Operated Clutch Diagnosis

Diagnostics of a hydraulic clutch system should begin with an inspection of the fluid. Check the fluid and reservoir for dirt and contamination. Foreign matter in the fluid will destroy the seals and wear grooves in the master and slave cylinders' bores.

A soft clutch pedal, excessive pedal travel, or a clutch that fails to release when the pedal is depressed can be caused by low fluid in the reservoir. To correct this problem, refill the reservoir to the correct level then bleed the system. This problem can also be caused by a faulty or damaged primary or secondary seal in the master cylinder. A leaking secondary seal will be evident by external leaks, whereas a primary seal leak will be internal. To correct either of these problems, replace or rebuild the master cylinder then refill and bleed the system. A leaking slave cylinder should be replaced and the system refilled with fluid and then bled.

If there is an extremely hard pedal, check the pedal mechanism and release fork for binding. If there is evidence of binding, repair and lubricate the assembly to ensure free movement. A hard pedal can also be caused by a blocked compensation port in the master cylinder. The port may be blocked by improper pushrod adjustments or because the piston is binding in the master cylinder bore. If the piston is binding, the master cylinder should be replaced or rebuilt and the hydraulic system flushed, refilled, and bled. This problem may be also caused by swollen cup seals or contamination in the master or slave cylinders. If this is the problem, the master or slave cylinder should be replaced and the system flushed, refilled, and bled. Restricted hydraulic lines can also cause a hard pedal.

Restricted lines can also prevent the clutch from being totally engaged. The residual pressure will keep the release bearing in contact with the pressure plate. This problem will also cause wear on the pressure plate fingers and the release bearing. Restricted lines should be replaced and the system flushed to remove all debris.

If the clutch does not fully engage when the pedal is released, check the pedal and release assemblies for binding or improper adjustment. A swollen primary cup in the master cylinder can also cause this problem. Swollen cups are caused by fluid contamination. This is typically the result of ATF being added to the fluid reservoir instead of DOT-3 brake fluid, which is the most commonly used fluid in a hydraulic-operated clutch system. If this is the case, the master and slave cylinders should be replaced and the system flushed.

CLUTCH SERVICE

A prerequisite for removing and replacing the clutch in a vehicle is removing the driveline or drive shafts and transmission or transaxle.

Removing the Clutch

After raising the vehicle on a hoist, clean excessive dirt, grease, or debris from around the clutch and transmission. Then disconnect and remove the clutch linkage. Cable systems need to be disconnected at the transmission.

On rear-wheel-drive vehicles, remove the driveline and transmission because the engine is somewhat supported by the transmission mounts. It may be necessary to support the engine with a tall jack stand. In some cases, the bell housing is removed with the transmission. In other cases, it is removed after the transmission is removed.

On front-wheel-drive vehicles with transaxles, any parts that interfere with transaxle removal must be removed first. These parts might include drive axles, parts of the engine, brake and suspension system, or body parts. Check the service manual for specific instructions.

The clutch assembly is accessible after the bell housing has been removed. Use an approved vacuum collection system or an approved liquid cleaning system to remove asbestos dust and dirt from the clutch assembly.

Photo Sequence 34 outlines the typical procedure for replacing a clutch disc and pressure plate. Always refer to the manufacturer's recommendations for bolt torque specifications prior to reinstalling the assembly. Compare the new parts with the old to make sure they are the correct replacements. This includes checking the splines of the disc with the input shaft.

While working on a clutch assembly, follow these guidelines:

- Check the bell housing, flywheel, and pressure plate for signs of oil leakage. Identify and repair the cause before installing new parts.
- Make sure that the mating surfaces of the engine block and bell housing are clean. The smallest amount of dirt can cause misalignment, which can cause premature wear of transmission shafts and bearings.
- Check the engine-to-bell housing dowels and dowel bores. Replace or repair any damaged parts.

P34-1 *The removal and replacement of a clutch assembly can be completed while the engine is in or out of the car. The clutch assembly is mounted to the flywheel that is mounted to the rear of the crankshaft.*

P34-2 *Before disassembling the clutch, make sure alignment marks are present on the pressure plate and flywheel.*

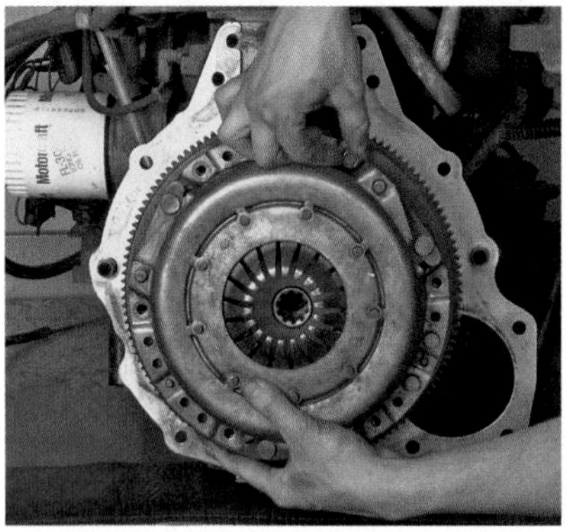

P34-3 *The attaching bolts should be loosened before removing any of the bolts. With the bolts loosened, support the assembly with one hand while using the other to remove the bolts. The clutch disc will be free to fall as the pressure plate is separated from the flywheel. Keep it intact with the pressure plate.*

P34-4 *The surface of the pressure plate should be inspected for signs of burning, grooving, warpage, and cracks. Any faults normally indicate that the plate should be replaced.*

P34-5 *The surface of the flywheel should also be carefully inspected. Normally the flywheel surface can be resurfaced to remove any defects. Also check the runout of the flywheel. The pilot bushing or bearing should also be inspected.*

P34-6 *The new clutch disc is placed into the pressure plate as the pressure plate is moved into its proper location. Make sure the disc is facing the correct direction. Most are marked to indicate which side should be seated against the flywheel surface.*

P34-7 *Install the pressure plate according to the alignment marks made during disassembly. Then install the clutch alignment tool through the hub of the disc and the pilot bearing to center the disc on the flywheel.*

P34-8 *Install the attaching bolts, but do not tighten.*

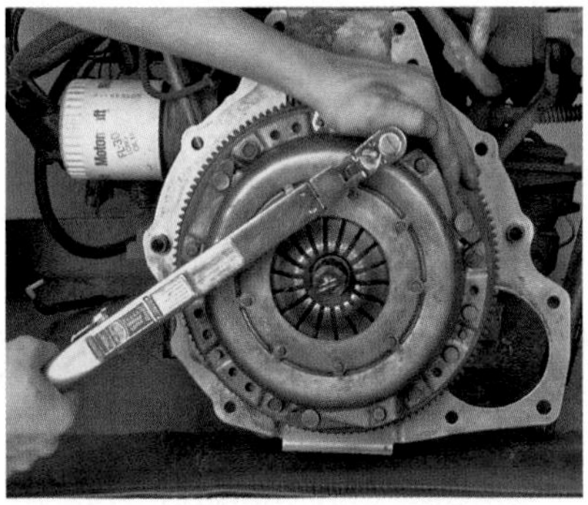

P34-9 *With the disc aligned, tighten the attaching bolts according to the procedures outlined in the service information and check the release finger/lever height after tightening the bolts. In most cases, the flywheel will need to be held in order to tighten the bolts.*

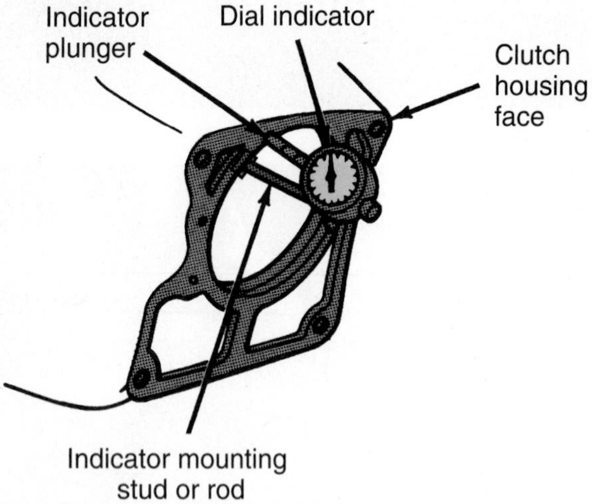

Figure 36-23 Use a dial indicator to check the runout of the clutch housing mounting surfaces. *Courtesy of Chrysler LLC*

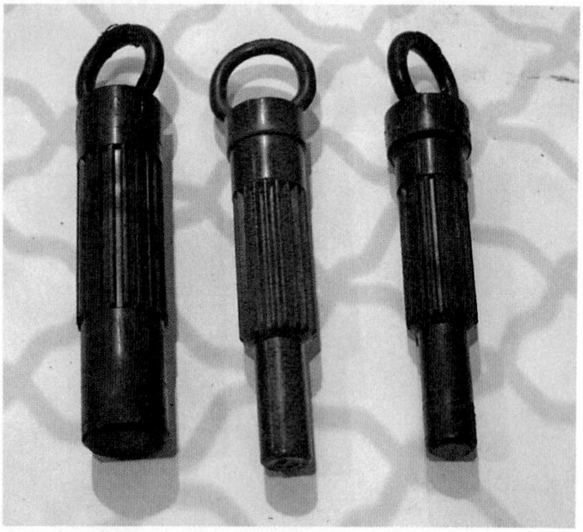

Figure 36-25 An assortment of clutch alignment tools.

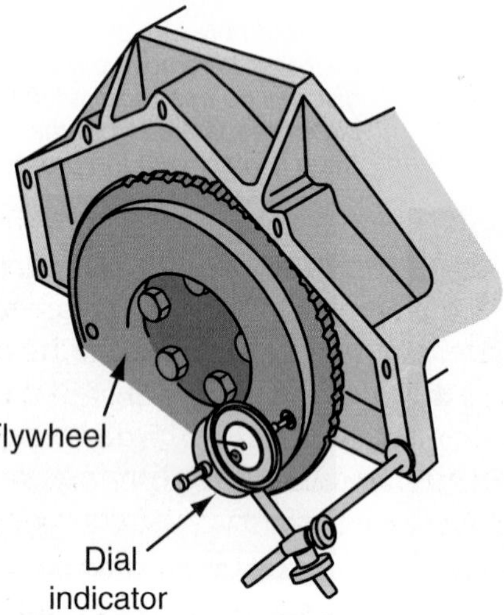

Figure 36-24 Measure the runout of the flywheel with a dial indicator.

- Check the mounting surfaces of the bell housing for damage and runout **(Figure 36–23)**.
- Check the flywheel for signs of burning or excessive wear. Check the runout of the flywheel with a dial indicator **(Figure 36–24)**.
- Check the teeth on the flywheel's ring gear; if there is damage, the ring gear or flywheel should be replaced.
- Use a clutch alignment tool during disassembly and reassembly **(Figure 36–25)**. The tool will keep the disc centered on the pressure plate.
- Loosen and tighten the pressure plate bolts according to the prescribed sequence.
- When installing a flywheel, make sure the bolts are tightened to specifications and in the prescribed sequence (normally a star pattern).
- When measuring the lining thickness of a bonded clutch disc, measure the total thickness of the facing or lining. To measure the wear of a riveted lining, measure the material above the rivet heads.
- Keep grease off the frictional surfaces of the clutch disc, flywheel, and pressure plate.
- Check the pressure plate surface for warpage by laying a straightedge across the surface and inserting a feeler gauge between the surface and the straightedge. Compare the measurement against the specifications given in the service manual.
- Check the release levers of the pressure plate for uneven wear or damage.
- Check the release bearing by turning it with your fingers and making sure it rotates freely.
- Check the clutch for damage.
- Check the pilot bushing or bearing for wear. Replace it if necessary **(Figure 36–26)**.
- Lightly lubricate the input shaft and bearing retainer **(Figure 36–27)**.
- Lubricate the clutch fork pivot points, the inside of the release bearing hub, and the linkages.
- After the clutch assembly has been reinstalled, check the clutch pedal free travel.

SHOP TALK

If the vehicle is a hybrid, make sure you disconnect the high-voltage system before beginning any work. Also, if the electric motor is sandwiched between the engine and transmission (Figure 36–28), make sure you follow all procedures. The permanent magnet used in the motor is very strong and requires special tools to remove and install it.

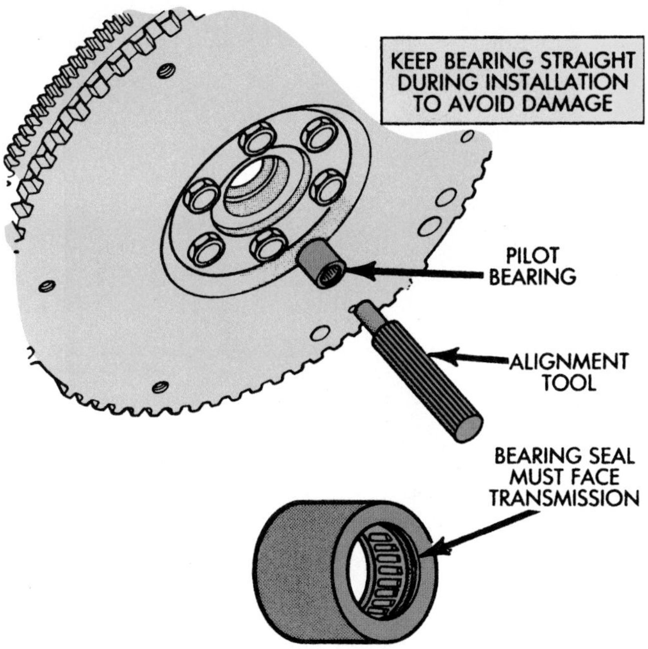

Figure 36-26 A typical method for installing a pilot bearing. *Courtesy of Chrysler LLC*

Figure 36-28 The electric motor for a hybrid vehicle with a clutch assembly attached to it.

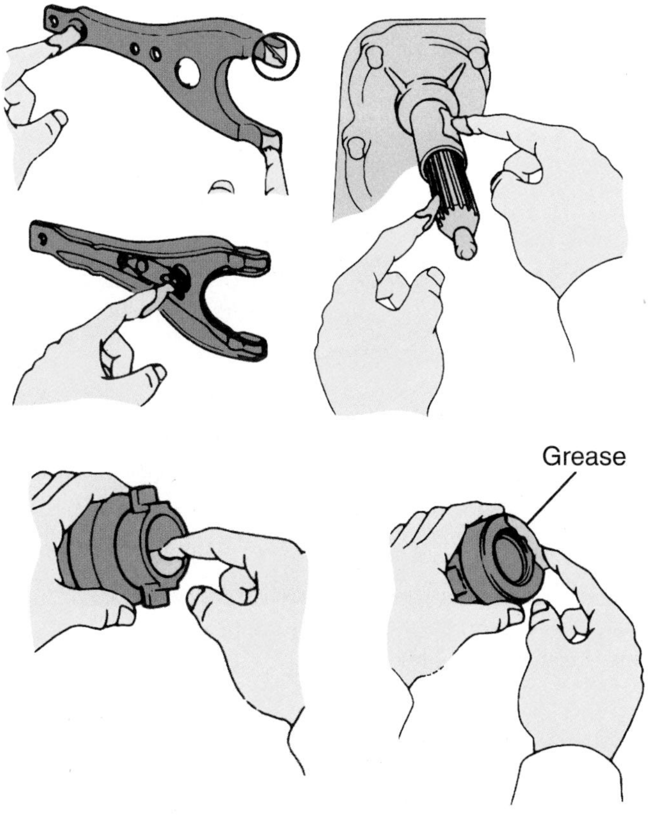

Figure 36-27 All contact points of the clutch release system should be lightly lubricated. Avoid applying too much grease.

Hydraulic-Operated Clutch Linkage Service

The proper fluid level in the reservoir is usually marked on the reservoir. The reservoir is normally mounted to the top of the master cylinder or is part of the master cylinder assembly. The hydraulic system does not consume fluid; therefore, if the fluid is low, check for leaks at the master and slave cylinders and the connecting hydraulic lines. Fill the reservoir only to the fill line to allow the fluid to rise as the clutch disc wears. Overfilling the system will cause slip and premature failure. Air can enter the system through the compensation and bleed ports if the fluid level in the reservoir is too low. The system must be bled to remove the trapped air.

Master cylinder problems are typically external or internal fluid leaks that require that the unit be replaced or rebuilt. Rebuild kits are available for most master cylinders. If a cast-iron master cylinder is rebuilt, the cylinder bore should be honed to remove any imperfections in the bore and new seals used. The bores of aluminum master cylinders should never be honed.

Internal and external leaks are also typical problems for slave cylinders. Seldom are these cylinders rebuilt; rather, they are replaced. Replacing a slave cylinder is rather straightforward on most vehicles. Simply disconnect the hydraulic lines and unbolt the unit. If it appears that the piston of a slave cylinder is seized in its bore, check the movement of the release fork and lever at the clutch before replacing the slave cylinder. Leaks may also result from damaged or corroded hydraulic lines. These lines should be replaced, if damaged, with the same type of tube as was originally installed.

Bleeding the System Whenever the hydraulic system is opened, the entire system should be bled. **Bleeding** may also be necessary if the system has run low on fluid and air is trapped in the lines or cylinders. Bleeding can be accomplished through the use of a power bleeder (the same device used to bleed a brake system), a vacuum bleeder, or the use of a coworker. On most cars, it is impossible to pressurize the system and bleed the hydraulic lines at the same time; therefore, it is important that you have the proper equipment or someone to assist you. The typical procedure for bleeding the system follows:

PROCEDURE

Bleeding the Hydraulic System

STEP 1 Check the entire hydraulic circuit to make sure there are no leaks.

STEP 2 Check the clutch linkage for wear and repair any defects before continuing.

STEP 3 Make sure all mounting points for the master and slave cylinders are solid and do not flex under the pressure of depressing the pedal.

STEP 4 Fill the master cylinder with the approved fluid **(Figure 36–29)**.

STEP 5 Attach one end of a hose to the end of the bleeder screw and the other end into a catch can **(Figure 36–30)**. Loosen the bleed screw at the slave cylinder approximately one-half turn.

STEP 6 Fully depress the clutch pedal, and then move the pedal through three quick and short strokes. Allow the fluid and air to exit the system and then immediately close the bleeder screw.

STEP 7 Release the pedal rapidly.

STEP 8 Recheck the fluid level in the master cylinder.

STEP 9 Repeat steps 3 and 4 until no air is evident in the fluid leaving the bleeder screw.

STEP 10 Close the bleeder screw immediately after the last downward movement of the pedal **(Figure 36–31)**.

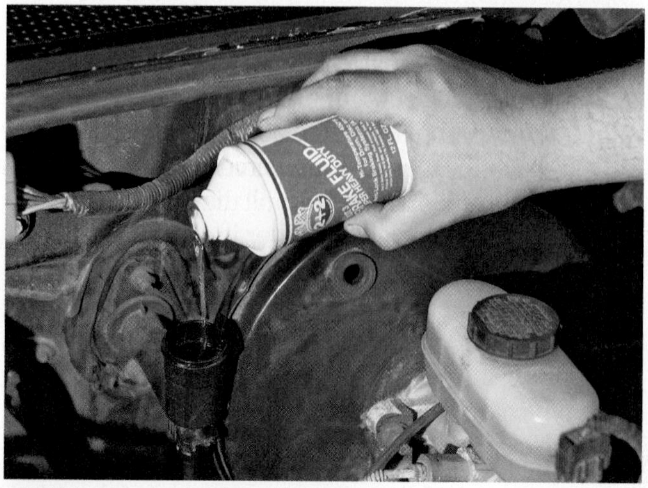

Figure 36-29 Fill the fluid reservoir with the correct fluid.

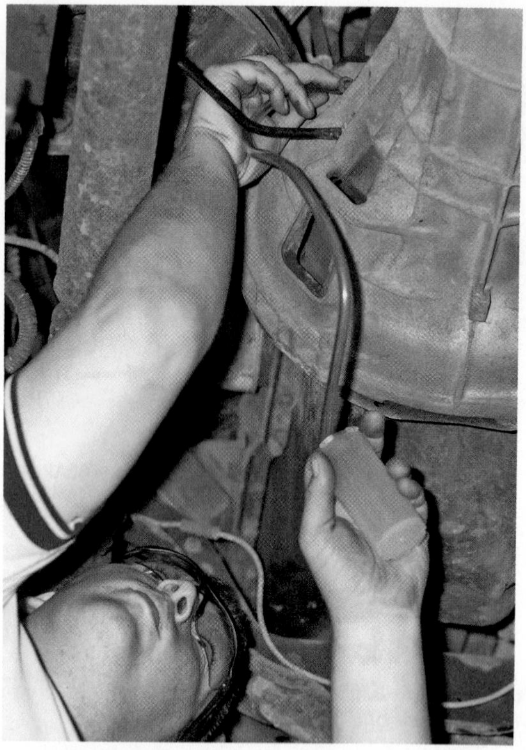

Figure 36-30 Attach a hose between the bleeder screw and a container before beginning to bleed the system.

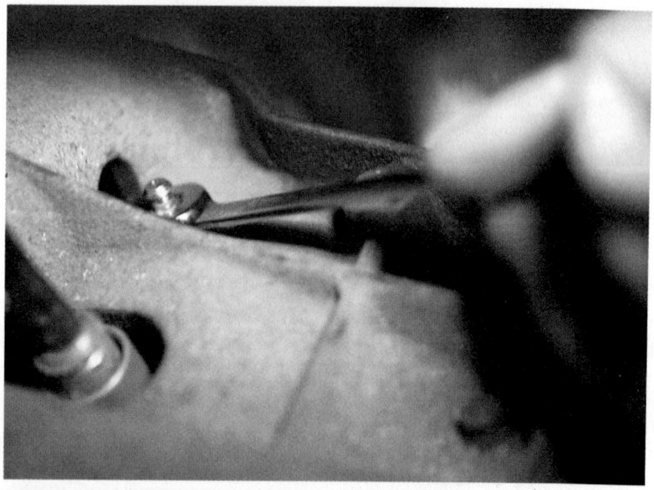

Figure 36-31 Make sure the bleeder screw is tight after the system has been bled.

KEY TERMS

Bell housing
Bleeding
Chatter
Clutch disc
Clutch fork
Clutch release bearing
Clutch slippage
Cushioning springs
Diaphragm spring
Free travel
Friction disc
Pilot bushing
Throwout bearing

CASE STUDY

A customer brings in a car that has a series of driveability problems. It uses an excess amount of fuel and oil, it lacks acceleration, and it seems to lose power while traveling on the highway.

The technician listens to the customer's complaints and asks the right questions. Then he prepares the car for a road test. While doing this, he checks the oil level and finds it to be low. This somewhat verifies that the engine is using excessive amounts of oil, so he removes the spark plugs from the engine to check for signs of oil burning. The spark plugs are clean and show no signs of oil. The technician then checks the engine for leaks hoping to find the cause of the excessive oil loss. The only wet spot is to the rear of the oil pan and the front of the bell housing. A leaking rear main seal is the likely cause of the oil loss. However, this does not explain the other problems of poor fuel mileage and a general lack of power. If the engine had been burning oil, all of the problems could have been related.

He then takes the car on a road test and finds that the clutch is slipping. When the car is in high gear, he steps hard on the accelerator and notices that the engine speed increases while the car's speed stays the same. After some thought, he determines this is the cause of the poor fuel mileage and the lack of power. He further determines that all of the customer's complaints are related to the same problem, a leaking rear main seal. Oil has been leaking out of the engine and onto the clutch disc, causing it to slip. He returns to the shop. He then notifies the customer and gives an estimate for the repairs. The customer authorizes the repair, which includes a new rear main seal, release bearing, pilot bushing, clutch disc, and pressure plate, plus resurfacing the flywheel.

SUMMARY

- The clutch, located between the transmission and the engine, provides a mechanical coupling between the engine flywheel and the transmission's input shaft. All manual transmissions and transaxles require a clutch.
- The flywheel, an important part of the engine, is also the main driving member of the clutch.
- The clutch disc receives the driving motion from the flywheel and pressure plate assembly and transmits that motion to the transmission input shaft.
- The twofold purpose of the pressure plate assembly is to squeeze the clutch disc onto the flywheel and move away from the clutch disc so the disc can stop rotating. There are basically two types of pressure plate assemblies: those with coil springs and those with a diaphragm spring.
- The clutch release bearing, also called a throwout bearing, smoothly and quietly moves the pressure plate release levers or diaphragm spring through the engagement and disengagement processes.
- The clutch fork moves the release bearing and hub back and forth. It is controlled by the clutch pedal and linkage.
- Clutch linkage can be mechanical or hydraulic.
- The self-adjusting clutch is a clutch cable linkage that monitors clutch pedal play and automatically adjusts it when necessary.
- It is important that certain precautions are exercised when servicing the clutch. Clutch maintenance includes linkage adjustment and external clutch linkage lubrication.
- Slippage occurs when the clutch disc is not gripped firmly between the flywheel and the pressure plate. It can be caused by an oil-soaked or worn disc facing, warped pressure plate, weak diaphragm spring, or the release bearing contacting and applying pressure to the release levers.
- Clutch drag occurs if the clutch disc is not completely released when the clutch pedal is fully depressed. It can occur as a result of a warped disc or pressure plate, a loose disc facing, a defective release lever, or incorrect clutch pedal adjustment that results in excessive pedal play.
- Chatter is a shuddering felt in the vehicle when the pressure plate first contacts the clutch disc and it stops when the clutch is fully engaged. Usually chatter is caused when liquid contaminates the friction surfaces.
- Pedal pulsation is a rapid up-and-down movement of the clutch pedal as the clutch disengages or engages. It results from a misalignment of parts.

- Clutch vibrations can be felt throughout the vehicle, and they occur at any clutch pedal position. Sources of clutch vibrations include loose flywheel bolts, excessive flywheel runout, and pressure plate cover balance problems.

REVIEW QUESTIONS

1. Name two types of friction facings.
2. *True or False?* If the fluid reservoir for a hydraulic clutch is low, the entire system should be checked for fluid leaks.
3. What is another name for the diaphragm spring?
4. Name three types of clutch linkages.
5. What is used to measure clutch pedal play?
6. The clutch, or friction, disc is connected to the ________.
   a. engine crankshaft
   b. transmission input shaft
   c. transmission output shaft
   d. transmission countershaft
7. Torsional coil springs in the clutch disc ________.
   a. cushion the driven disc engagement rear to front
   b. are the mechanical force holding the pressure plate against the driven disc and flywheel
   c. absorb the torque forces
   d. are located between the friction rings
8. The pressure plate moves away from the flywheel when the clutch pedal is ________.
9. Which of the following is probably *not* the cause of a vibrating clutch?
   a. excessive crankshaft end play
   b. out-of-balance pressure plate assembly
   c. excessive flywheel runout
   d. loose flywheel bolts
10. When the clutch pedal is released on a hydraulic clutch linkage, the ________.
    a. master cylinder piston is released by spring tension
    b. master cylinder piston is released by hydraulic pressure
    c. slave cylinder is released by hydraulic pressure
    d. slave cylinder piston does not move
11. When the clutch is disengaged, the power flow stops at the ________.
    a. transmission input shaft
    b. driven disc hub
    c. pressure plate and flywheel
    d. torsion springs
12. Insufficient clutch pedal clearance results in ________.
    a. gear clashing while shifting transmission
    b. a noisy front transmission bearing
    c. premature release bearing failure
    d. premature pilot bearing failure
13. Before making a clutch adjustment, it is necessary to ________.
    a. measure clutch pedal free travel
    b. lubricate the clutch linkage
    c. check the hydraulic fluid level
    d. place the transmission in reverse
14. The surface of the pressure plate contacts the ________.
    a. transmission main shaft
    b. throwout bearing
    c. clutch disc
    d. flywheel
15. Which of the following would *not* cause clutch binding?
    a. a warped clutch disc
    b. improper pedal adjustment
    c. an oil-soaked clutch disc
    d. a cracked fire wall

ASE-STYLE REVIEW QUESTIONS

1. Technician A says that clutch slippage is most noticeable in higher gears. Technician B says that clutch slippage is not noticeable in lower gears. Who is correct?
   a. Technician A
   b. Technician B
   c. Both A and B
   d. Neither A nor B
2. While discussing ways to determine if a pilot bushing is faulty: Technician A says that if it operated quietly, it does not need to be replaced. Technician B says that a careful inspection of the transmission's input shaft can determine the condition of the pilot bushing. Who is correct?
   a. Technician A
   b. Technician B
   c. Both A and B
   d. Neither A nor B
3. While discussing the different types of pressure plates: Technician A says that coil spring types are commonly used because they have strong

springs. Technician B says that Belleville-type pressure plates are not commonly used because they require excessive space in the bell housing. Who is correct?

a. Technician A
b. Technician B
c. Both A and B
d. Neither A nor B

4. While discussing the purpose of a clutch pressure plate: Technician A says that the pressure plate assembly squeezes the clutch disc onto the flywheel. Technician B says that the pressure plate moves away from the clutch disc so that the disc can stop rotating. Who is correct?

a. Technician A
b. Technician B
c. Both A and B
d. Neither A nor B

5. Technician A says that an oil-soaked clutch disc can cause clutch chatter. Technician B says that clutch chatter can be caused by loose bell housing bolts. Who is correct?

a. Technician A
b. Technician B
c. Both A and B
d. Neither A nor B

6. While discussing different abnormal clutch noises: Technician A says that pilot bushing noises are most noticeable in cold weather and usually occur when the pedal is being depressed and the transmission is in neutral. Technician B says that release bearing noise is most noticeable when the transmission is in neutral and occurs when the clutch pedal is depressed and stops when the pedal is fully released. Who is correct?

a. Technician A
b. Technician B
c. Both A and B
d. Neither A nor B

7. Technician A says that incorrect clutch adjustment can cause premature clutch disc wear. Technician B says that incorrect clutch adjustment can cause premature clutch release bearing wear. Who is correct?

a. Technician A
b. Technician B
c. Both A and B
d. Neither A nor B

8. While discussing the cause of a pulsating clutch pedal: Technician A says that this can be caused by a low fluid level in the hydraulic system. Technician B says that this can be caused by a warped flywheel. Who is correct?

a. Technician A
b. Technician B
c. Both A and B
d. Neither A nor B

9. Technician A says that if the hydraulic clutch reservoir is overfilled, the clutch may prematurely wear. Technician B says that if the fluid level is too low, the clutch may not disengage. Who is correct?

a. Technician A
b. Technician B
c. Both A and B
d. Neither A nor B

10. Technician A says that all manual transmission vehicles have at least one pilot bearing or bushing for the input shaft. Technician B says that the pressure plate always rotates with the input shaft when the engine is running. Who is correct?

a. Technician A
b. Technician B
c. Both A and B
d. Neither A nor B

CHAPTER 37

MANUAL TRANSMISSIONS AND TRANSAXLES

OBJECTIVES

■ Explain the design characteristics of the gears used in manual transmissions and transaxles. ■ Explain the fundamentals of torque multiplication and overdrive. ■ Describe the purpose, design, and operation of synchronizer assemblies. ■ Describe the purpose, design, and operation of internal and remote gearshift linkages. ■ Explain the operation and power flows produced in typical manual transmissions and transaxles.

The transmission or transaxle (**Figure 37–1**) is a vital link in the powertrain of any modern vehicle. The purpose of the transmission or transaxle is to use gears of various sizes to give the engine a mechanical advantage over the driving wheels. During normal operating conditions, power from the engine is transferred through the engaged clutch to the input shaft of the transmission or transaxle. Gears in the transmission or transaxle housing alter the torque and speed of this power input before passing it on to other components in the drivetrain. Without the mechanical advantage the gearing provides, an engine can generate only limited torque at low speeds. Without sufficient torque, moving a vehicle from a standing start would be impossible.

Figure 37–1 A late-model transaxle.

In any engine, the crankshaft always rotates in the same direction. If the engine transmitted its power directly to the drive axles, the wheels could be driven only in one direction. Instead, the transmission or transaxle provides the gearing needed to reverse direction so the vehicle can be driven backward. There is also a neutral position that stops power from reaching the drive wheels.

TRANSMISSION VERSUS TRANSAXLE

Vehicles are propelled in one of three ways: by the rear wheels, by the front wheels, or by the front and rear wheels. The type of drive system used determines whether a conventional transmission or a transaxle is used.

Vehicles propelled by the rear wheels normally use a transmission. Transmission gearing is located within an aluminum or iron casting called the transmission case assembly. The transmission case assembly is attached to the rear of the engine, which is normally located in the front of the vehicle. A drive shaft links the output shaft of the transmission with the differential and drive axles located in a separate housing at the rear of the vehicle. The differential splits the driveline power and redirects it to the two rear drive axles, which then pass it on to the wheels. For many years, rear-wheel-drive (RWD) systems were the conventional method of propelling a vehicle.

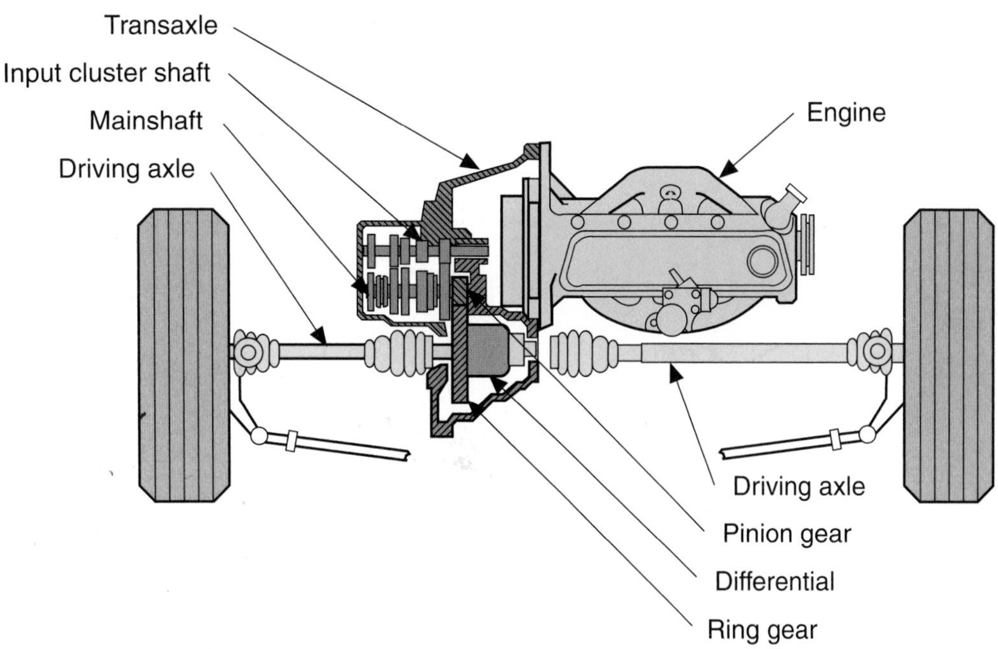

Figure 37-2 The location of typical front-wheel-drive powertrain components.

Front-wheel-drive (FWD) vehicles are propelled by the front wheels. For this reason, they must use a drive design different from that of a RWD vehicle. The transaxle is the special power transfer unit commonly used on FWD vehicles. A transaxle combines the transmission gearing, differential, and drive axle connections into a single case housing located in front of the vehicle **(Figure 37–2)**. This design offers many advantages. One major advantage is the good traction on slippery roads due to the weight of the drivetrain components being directly over the driving axles of the vehicle. It is also more compact and lighter than the transmission of a RWD vehicle. Transverse engine and transaxle configurations also allow for lower hood lines, thereby improving the vehicle's aerodynamics.

Four-wheel-drive vehicles typically use a transmission and transfer case. The transfer case mounts on the side or back of the transmission. A chain or gear drive inside the transfer case receives power from the transmission and transfers it to two separate drive shafts. One drive shaft connects to a differential on the front drive axle. The other drive shaft connects to a differential on the rear drive axle.

Most manual transmissions and transaxles are constant mesh, fully synchronized units. Constant mesh means that whether or not the gear is locked to the output shaft, it is in mesh with its counter gear. All gears rotate in the transmission, except for reverse gear, as long as the clutch is engaged. Fully synchronized means the unit uses a mechanism of rings and clutches to bring rotating shafts and gears to the same speed before shifts occur. This promotes smooth shifting. In a vehicle equipped with a four-speed manual shift transmission or transaxle, all four forward gears are synchronized. Reverse gearing may or may not be synchronized, depending on the type of transmission/transaxle.

Transmission Designs

All automotive transmissions/transaxles are equipped with a varied number of forward speed gears, a neutral gear, and one reverse speed. Transmissions can be divided into groupings based on the number of forward speed gears they have.

Five-speed transmissions and transaxles are now the commonly used units. Some of the early five-speed units were actually four-speeds with an add-on fifth or overdrive gear. Overdrive reduces engine speed at a given vehicle speed, which increases top speed, improves fuel economy, and lowers engine noise. Most late-model five-speed units incorporate a fifth gear in their main assemblies. This is also true of six-speed transmissions and transaxles. The fifth and sixth gears are included in the main assembly and typically provide two overdrive gears. The addition of the two overdrive gears allows the manufacturers to use lower final drive gears for acceleration. The fifth and sixth gears reduce the overall gear ratio and allow for slower engine speeds during highway operation.

Self-Shifting Manual Transmissions

Self-shifting manual transmissions are currently available on some passenger cars and are used in Formula One race cars. These transmissions work like

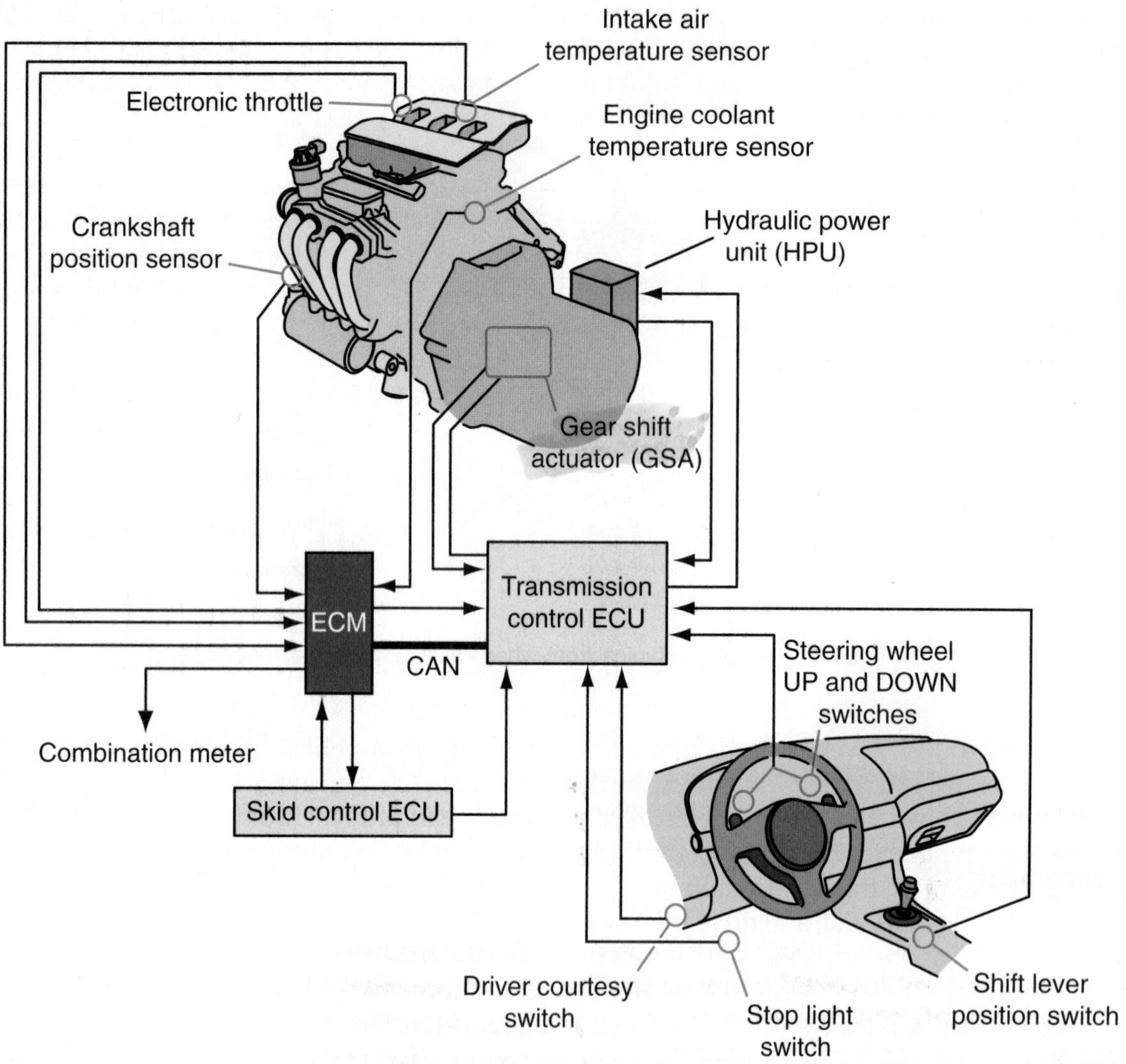

Figure 37-3 The components of a typical sequential manual transmission system.

typical manual transmissions except electronic or hydraulic actuators shift the gears and work the clutch **(Figure 37–3)**. The driver shifts the gears using buttons or paddles on the steering wheel. Some units can also be operated in a fully automatic mode. It is important to realize that these are not automatic transmissions with manual controls!

These transmissions do not use planetary gear sets or torque converters. They are manual transmissions with a computer-controlled actuator connected to the shift forks and a clutch actuator **(Figure 37–4)**. The computer is programmed to shift the transmission automatically at the correct time, in the correct sequence, and to activate the clutch and allow for precise shifting when the driver selects the automatic mode.

The driver can control gear changes by using the shifting mechanism. There is no gearshift linkage or

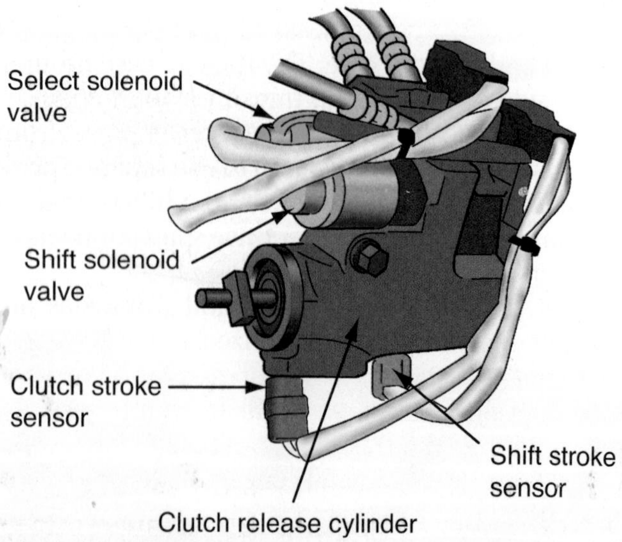

Figure 37-4 The gear shift actuator that engages and disengages the clutch and shifts gears. *Courtesy of Toyota Motor Sales, U.S.A., Inc.*

cable; instead, a sensor at the shifter sends a signal to the computer. The computer, in turn, commands the actuators to engage or disengage the clutch and the gears with very fast response times. Engine torque is controlled during the shift by directly controlling the throttle or ignition/fuel injection system to provide smooth shifts.

Volkswagen/Audi Direct Shift Gearbox (DSG) The Direct Shift Gearbox (DSG) is a six-speed manual transaxle. Two gears can be engaged at the same time through the use of integrated twin multiplate clutches. When the car is moving, one gear is engaged. When the next gearshift point is approached, the next appropriate gear is preselected but its clutch is kept disengaged. The shifting process opens the clutch of the activated gear and closes the other clutch at the same time. The gear change takes place under load so that a permanent flow of power is maintained.

The clutches are hydraulically controlled. Clutch #1 is fixed to the odd-numbered gears plus reverse and clutch #2 controls the even gears. The arrangement hints at being two parallel three-speed transmissions built into one housing. This is how two gears can be engaged at the same time. As an example, when the car is being driven in third gear, fourth gear is already engaged but is not yet active. As soon as the ideal shift point is reached, the clutch for third gear opens while the other closes, activating fourth gear. The opening and closing of the clutches happen at the same time, producing a smooth and quick shift. The entire shift process is completed in a few hundredths of a second.

The most complex component of the DSG is the mechatronic control module located in the transaxle. A heat exchanger is bolted to the housing to keep the computer at desired temperatures. The computer controls the clutches, input and output shafts, cooling, gear selection, and pressures. Five modulation valves, five shift valves, and numerous other valves are used to control the transaxle. In addition to twelve designated sensors, the computer relies on bus data to determine current operating conditions.

BMW Sequential M Gearbox (SMG) Similar to the DSG, BMW's Sequential M Gearbox (SMG) is a self-shifting manual transmission. The SMG is based on a regular six-speed transmission. The SMG does not use a clutch pedal but does have a clutch. Gear shifting is done through paddle shifters on the steering column or with a gearshift lever in the center console. The actual shifting, though, is done by computers, solenoids, hydraulics, and linkages. Shifting characteristics are based on inputs from a variety of sensors **(Figure 37–5)**, driver-selected programs, and the programmed logic in the control module. Basically, the control module interrupts the engine's power for just milliseconds and causes gear changes electrohydraulically while it opens and closes the clutch. The system has a total of eleven settings, six in manual mode and five in the automatic mode. These allow the driver to select the style of driving that is desired.

An additional feature is the *climbing assistant.* This permits pulling away on forward slopes without rolling back. It can be used both in the sequential and automatic modes and for forward and reverse travel.

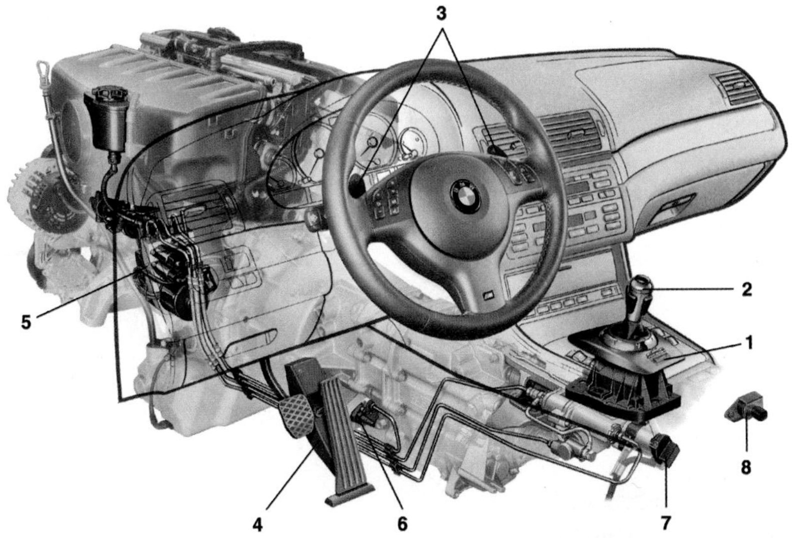

Figure 37-5 BMW's SMG system. Legend: 1 = drivelogic control module; 2 = gearshift; 3 = paddles; 4 = accelerator input; 5 = hydraulic unit; 6 = clutch position sensor; 7 = input shaft speed sensor and transmission oil temperature sensor; 8 = gear selector position sensor. *Courtesy of BMW of North America, LLC*

All the driver needs to do is to press the brake pedal with the car stationary and to pull the rocker switch on the steering wheel for a short period. When the brake is released again, the car is ready to drive away within 2 seconds without first rolling away in an uncontrolled manner.

Another feature, the *acceleration assistant,* offers maximum acceleration. To activate this program, the driver selects the S6 driving program, pushes the selector lever forward, and keeps it in that position while the car is stationary. The driver now fully depresses the accelerator pedal and the optimum engine speed for maximum acceleration is automatically set. When the driver releases the selector lever, the car accelerates as quickly as possible with a minimum amount of wheel slip.

GEARS

The purpose of the gears in a manual transmission or transaxle is to transmit rotating motion. Gears are normally mounted on a shaft, and they transmit rotating motion from one parallel shaft to another **(Figure 37–6)**.

Gears and shafts can interact in one of three ways: the shaft can drive the gear; the gear can drive the shaft; or the gear can be free to turn on the shaft. In this last case, the gear acts as an idler gear.

Sets of gears can be used to multiply torque and decrease speed, increase speed and decrease torque, or transfer torque and leave speed unchanged.

Gear Design

Gear pitch is a very important factor in gear design and operation. Gear pitch refers to the number of teeth per given unit of pitch diameter. A simple way of determining gear pitch is to divide the number of teeth by the pitch diameter of the gear. For example, if a gear has thirty-six teeth and a 6-inch pitch diameter, it has a gear pitch of six **(Figure 37–7)**. The important fact to remember is that gears must have the same pitch to operate together. A five-pitch gear meshes only with another five-pitch gear, a six-pitch only with a six-pitch, and so on.

Spur Gears The **spur gear** is the simplest gear design used in manual transmissions and transaxles. As shown in **Figure 37–8**, spur gear teeth are cut straight across the edge parallel to the gear's shaft. During operation, meshed spur gears have only one tooth in full contact at a time.

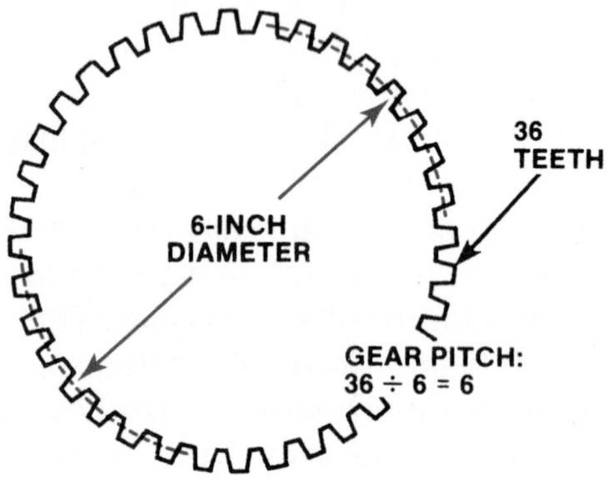

Figure 37-7 Determining gear pitch.

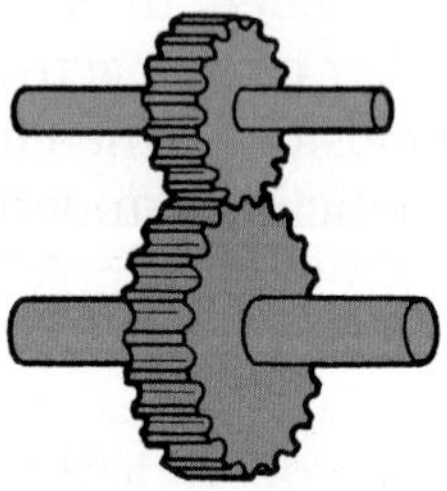

Figure 37-8 Spur gears have teeth cut straight across the gear edge parallel to the shaft.

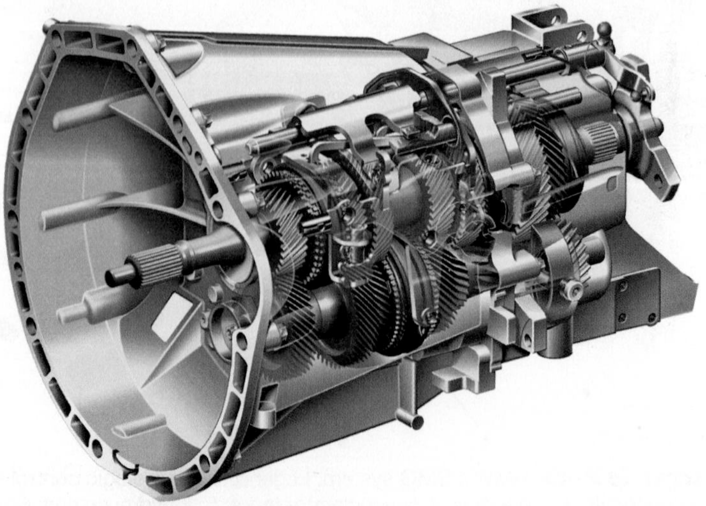

Figure 37-6 The gears in a transmission transmit the rotating power from the engine. *Courtesy of Chrysler LLC*

Its straight tooth design is the spur gear's main advantage. It minimizes the chances of popping out of gear, an important consideration during acceleration/deceleration and reverse operation. For this reason, spur gears are often used for the reverse gear.

The spur gear's major drawback is the clicking noise that occurs as teeth contact one another. At higher speeds, this clicking becomes a constant whine. Quieter gears, such as the helical design, are often used to eliminate this gear whine problem.

SHOP TALK

When a small gear is meshed with a much larger gear, the input gear is often called a pinion or pinion gear, regardless of its tooth design.

Helical Gears A **helical gear** has teeth that are cut at an angle or are spiral to the gear's axis of rotation (**Figure 37–9**). This configuration allows two or more teeth to mesh at the same time, which distributes tooth load and produces a very strong gear. Helical gears also run more quietly than spur gears because they create a wiping action as they engage and disengage the teeth on another gear. One disadvantage is that helical teeth on a gear cause the gear to move fore or aft (axial thrust) on a shaft, depending on the direction of the angle of the gear teeth. This axial thrust must be absorbed by thrust washers and other transmission gears, shafts, or the transmission case.

Helical gears can be either righthanded or lefthanded, depending on the direction the spiral appears to go when the gear is viewed face-on. When mounted on parallel shafts, one helical gear must be righthanded and the other lefthanded. Two gears with the same direction spiral do not mesh in a parallel mounted arrangement.

Spur and helical gears that have teeth cut around their outside diameter edge are called **external gears**. When two external gears are meshed together, one rotates in the opposite direction as the other (**Figure 37–10**). If an external gear is meshed with an internal gear (one that has teeth around its inside diameter), both rotate in the same direction.

Figure 37-9 Helical gears have teeth cut at an angle to the gear's axis of rotation.

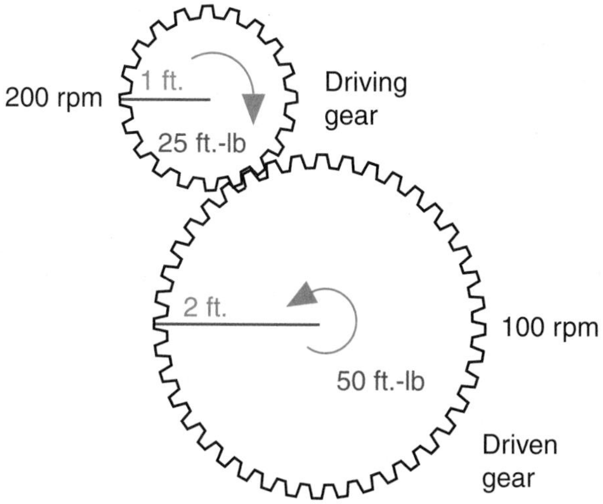

Figure 37-10 Externally meshed gears rotate in opposite directions.

Figure 37-11 An idler gear is used to transfer motion without changing rotational direction.

Idler Gears

An **idler gear** is a gear that is placed between a drive gear and a driven gear. Its purpose is to transfer motion from the drive gear to the driven gear without changing the direction of rotation. It can do this because all three gears have external teeth (**Figure 37–11**).

Idler gears are used in reverse gear trains to change the directional rotation of the output shaft. In all forward gears, the input shaft and the output shaft turn in the same direction. In reverse, the output shaft turns in the opposite direction as the input shaft. This allows the vehicle drive wheels to turn backward.

BASIC GEAR THEORY

Go To Chapter 8 for a detailed explanation on how gears multiple torque.

As gears with different numbers of teeth mesh, each rotates at a different speed and torque. A manual transmission is an assembly of gears and shafts that transmits power from the engine to the drive axle. By moving the shift lever, various gear and speed ratios can be selected. The gears in a transmission are selected to give the driver a choice of both speed and torque. Lower gears allow for lower vehicle speeds but more torque. Higher gears provide less torque but higher vehicle speeds. There is often much confusion about the terms *high* and *low gear ratios.* A gear ratio of 4:1 is lower than a ratio of 2:1. Although numerically the 4 is higher than the 2, the 4:1 gear ratio allows for lower speeds and hence is termed a low gear ratio.

Different gear ratios are necessary because an engine develops relatively little power at low engine speeds. The engine must be turning at a fairly high speed before it can deliver enough power to get the car moving. Through selection of the proper gear ratio, torque applied to the drive wheels can be multiplied.

Transmission Gear Sets

Power is moved through the transmission via four gears (two sets of two gears). Speed and torque are altered in steps. To explain how this works, let us assign numbers to each of the gears. The small gear on the input shaft has twenty teeth. The gear it meshes with has forty. This provides a gear ratio of 2:1. The output of this gear set moves along the shaft of the forty-tooth gear and rotates other gears. The gear involved with first gear has fifteen teeth. This gear rotates with the same speed and with the same torque as the forty-tooth gear. However, the fifteen-tooth gear is meshed with a larger gear with thirty-five teeth. The gear ratio of the fifteen-tooth and the thirty-five-tooth gear set is 2.33:1. However, the ratio of the entire gear set (both sets of two gears) is 4.67:1.

To calculate this gear ratio, divide the driven (output) gear of the first set by the drive (input) gear of the first set. Do the same for the second set of gears, and then multiply the answer from the first by the second. The result is equal to the gear ratio of the entire gear set. The mathematical formula follows:

$$\frac{\text{driven (A)}}{\text{drive (A)}} \times \frac{\text{driven (B)}}{\text{drive (B)}} = \frac{40}{20} \times \frac{35}{15} = 4.67{:}1$$

Most of today's transmissions have at least one overdrive gear. Overdrive gears have ratios of less than 1:1. These ratios are achieved by using a small driving gear meshed with a smaller driven gear. Output speed is increased and torque is reduced. The purpose of overdrive is to promote fuel economy and reduce operating noise while maintaining highway cruising speed.

The driveline's gear ratios are further increased by the gear ratio of the ring and pinion gears in the drive axle assembly. Typical axle ratios are between 2.5 and 4.5:1. The final (overall) drive gear ratio is calculated by multiplying the transmission gear ratio by the final drive ratio. If a transmission is in first gear with a ratio of 3.63:1 and has a final drive ratio of 3.52:1, the overall gear ratio is 12.78:1. If fourth gear has a ratio of 1:1, using the same final drive ratio, the overall gear ratio is 3.52:1. The overall gear ratio is calculated by multiplying the ratio of the first set of gears by the ratio of the second (3.63 times 3.52 equals 12.78).

Reverse Gear Ratios

Reverse gear ratios involve two driving (driver) gears and two driven gears:

- The input gear is driver #1.
- The idler gear is driven #1.
- The idler gear is also driver #2.
- The output gear is driven #2.

If the input gear has twenty teeth, the idler gear has twenty-eight teeth and the output gear has forty-eight teeth. However, since a single idler gear is used, the teeth of it are not used in the calculation of gear ratio. The idler gear merely transfers motion from one gear to another. The calculations for determining reverse gear ratio with a single idler gear follow.

$$\begin{aligned}\text{Reverse gear ratio} &= \frac{\text{driven \#2}}{\text{driver \#1}} \\ &= \frac{48}{20} \\ &= 2.40\end{aligned}$$

If the gear set uses two idler gears (one with twenty-eight teeth and the other with forty teeth), the gear ratio involves three driving gears and three driven gears:

- The input gear is driver #1.
- The #1 idler gear is driven #1.
- The #1 idler gear is also driver #2.
- The #2 idler gear is driven #2.
- The #2 idler gear is also driver #3.
- The output gear is driven #3.

The ratio of this gear set would be calculated as follows:

$$\begin{aligned}\text{Reverse gear ratio} &= \frac{\text{driven \#1} \times \text{driven \#2} \times \text{driven \#3}}{\text{driver \#1} \times \text{driver \#2} \times \text{driver \#3}} \\ &= \frac{28 \times 40 \times 48}{20 \times 28 \times 40} \\ &= \frac{53{,}760}{22{,}400} \\ &= 2.40\end{aligned}$$

As can be seen, idler gears do not affect the gear ratio.

TRANSMISSION/TRANSAXLE DESIGN

The internal components of a transmission or transaxle consist of a parallel set of metal shafts on which meshing gear sets of different ratios are mounted (**Figure 37–12**). By moving the shift lever, gear ratios can be selected to generate different amounts of output torque and speed.

The gears are mounted or fixed to the shafts in a number of ways. They can be internally splined or keyed to a shaft. Gears can also be manufactured as an integral or "clustered" part of the shaft. Gears that must be able to freewheel around the shaft during certain speed ranges are mounted to the shaft using bushings or bearings.

The shafts and gears are contained in a transmission or transaxle case or housing. The components of this housing include the main case body, side or top cover plates, extension housings, and bearing retainers (**Figure 37–13**). The metal components are bolted together with gaskets providing a leak-proof seal at all joints. The case is filled with transmission fluid to provide constant lubrication and cooling for the spinning gears and shafts.

Transmission Features

Although they operate in a similar fashion, the layout, components, and terminology used in transmissions and transaxles are not exactly the same.

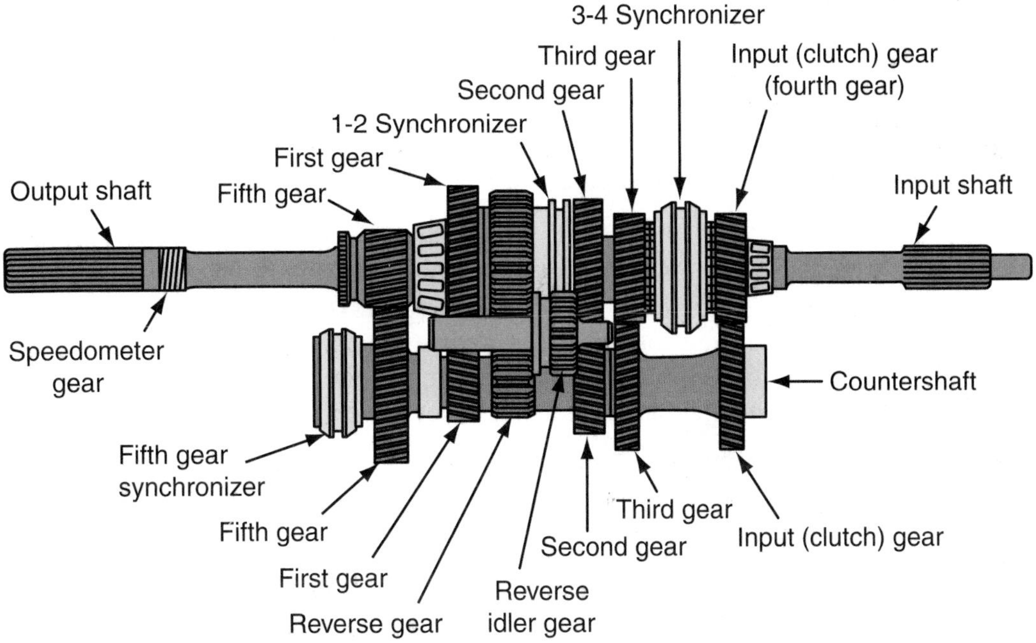

Figure 37–12 The arrangement of the gears and shafts in a typical five-speed transmission.

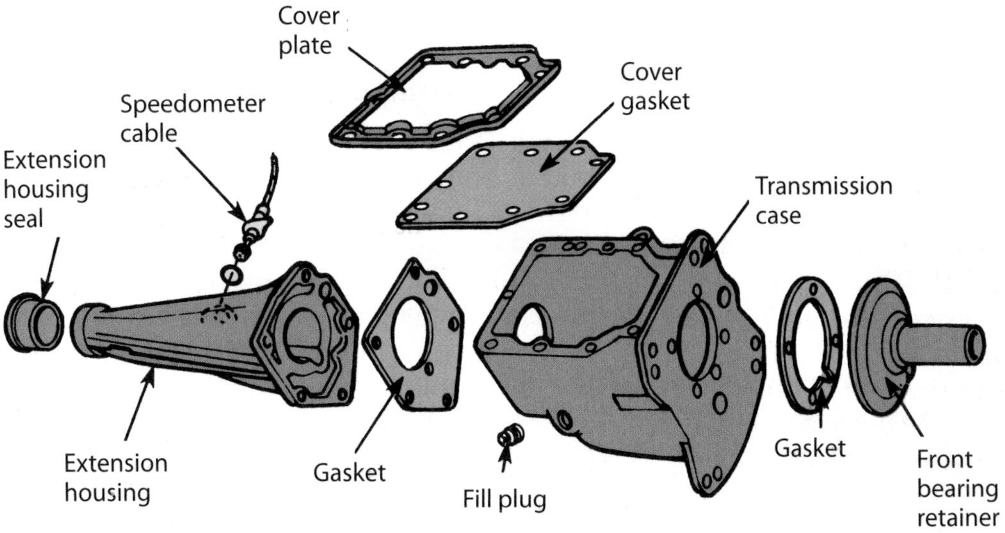

Figure 37–13 Typical manual transmission case components. *Courtesy of Ford Motor Company*

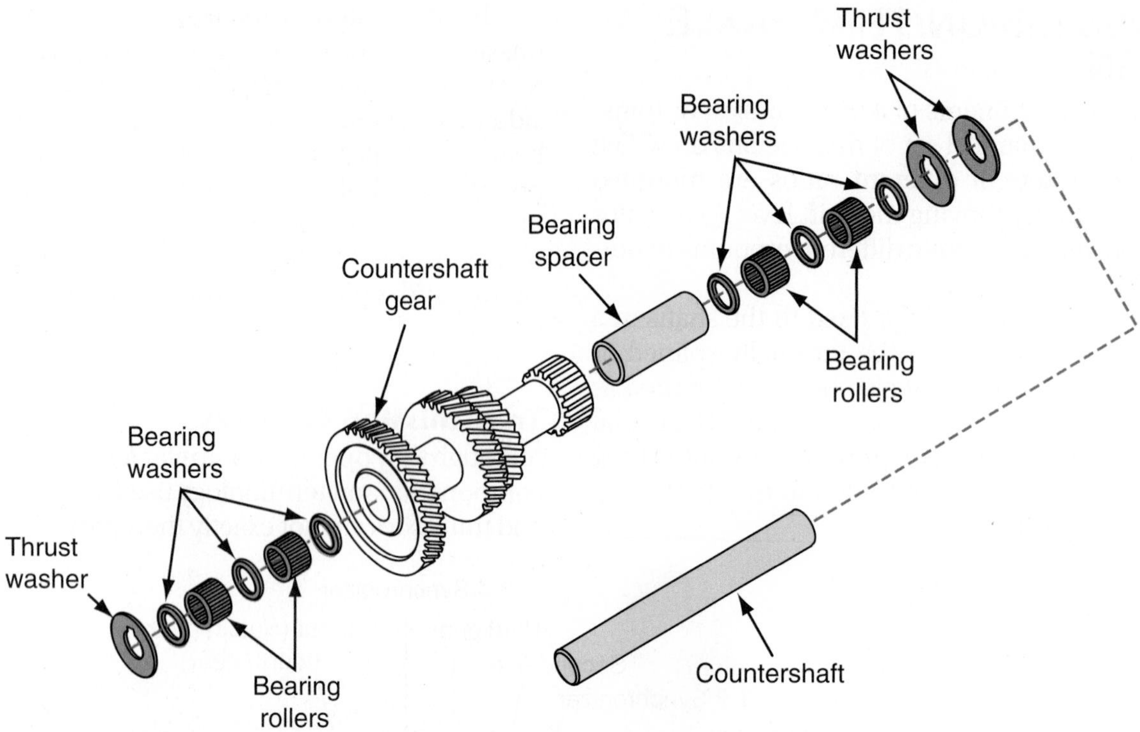

Figure 37-14 A typical countershaft assembly.

Most transmissions have three basic shafts: the mainshaft, input shaft, and **countershaft** (**Figure 37–14**). The speed gears ride on the mainshaft. The mainshaft is also called the output shaft. The input shaft is inline with the mainshaft but is not directly connected to it. This shaft and its clutch gear rotates with the clutch assembly and drives the countergear assembly. The countergears, in turn, cause the speed gears to rotate. The countershaft assembly is often referred to as the cluster gear.

The countershaft is actually several gears machined out of a single piece of steel or iron. The countershaft may also be called the countergear or **cluster gear**. The countergear mounts on roller bearings on the countershaft. The countershaft is pinned in place and does not turn. Thrust washers control the amount of end play of the unit in the transmission case.

The main gears on the main shaft or output shaft transfer rotation from the countergears to the output shaft. The main gears are also called speed gears. They are mounted on the output shaft using roller bearings. Speed gears freewheel around the output shaft until they are locked to it by the engagement of their shift **synchronizer** unit.

Power flows from the transmission **input shaft** to the **clutch gear**. The clutch gear meshes with the large countergear of the countergear cluster. This cluster gear is now rotating. Since the cluster gear is meshed with the speed gears on the mainshaft, the speed gears are also turning.

There can be no power output until one of the speed gears is locked to the mainshaft. This is done by activating a shift fork, which moves its synchronizer to engage the selected speed gear to the mainshaft. Power travels along the countergear until it reaches this selected speed gear. It then passes through this gear back to the mainshaft and out of the transmission to the driveline.

Transaxle Features

Transaxles use many of the design and operating principles found in transmissions. But because the transaxle also contains the differential gearing and drive axle connections, there are major differences in some areas of operation.

A transaxle typically has two separate shafts—an input shaft and an output shaft. The input shaft is the driving shaft. It is normally located above and parallel to the output shaft. Because the input shaft in most transversely mounted transaxles are supported by bearings in the housing, these units do not need a pilot bearing or bushing to support the portion of the input shaft that extends into the clutch assembly. This type of shaft is called a self-centering shaft. The output shaft is the driven shaft. The transaxle's main (speed) gears freewheel around the output shaft unless they are locked to the shaft by their synchronizer assembly. The main speed gears are in constant mesh with the input shaft drive gears. The drive gears turn whenever the input shaft turns.

Item	Description
1.	Mainshaft
2.	Input cluster gear shaft
3.	4th speed gears
4.	3rd speed gears
5.	2nd speed gears
6.	Reverse gears
7.	Reverse idler gears
8.	1st speed gears
9.	5th speed gear drive shaft
10.	5th speed gear
11.	5th gear driveshaft pinion gear
12.	Mainshaft pinion gear
13.	Differential oil seals
14.	CV shafts
15.	Differential pinion gears
16.	Differential side gears
17.	Final drive ring gear
18.	1st/2nd synchronizer
19.	3rd/4th synchronizer
20.	5th synchronizer

Note: Gear #11 is always in mesh with gear #17. This relationship is not shown for clarity reasons.

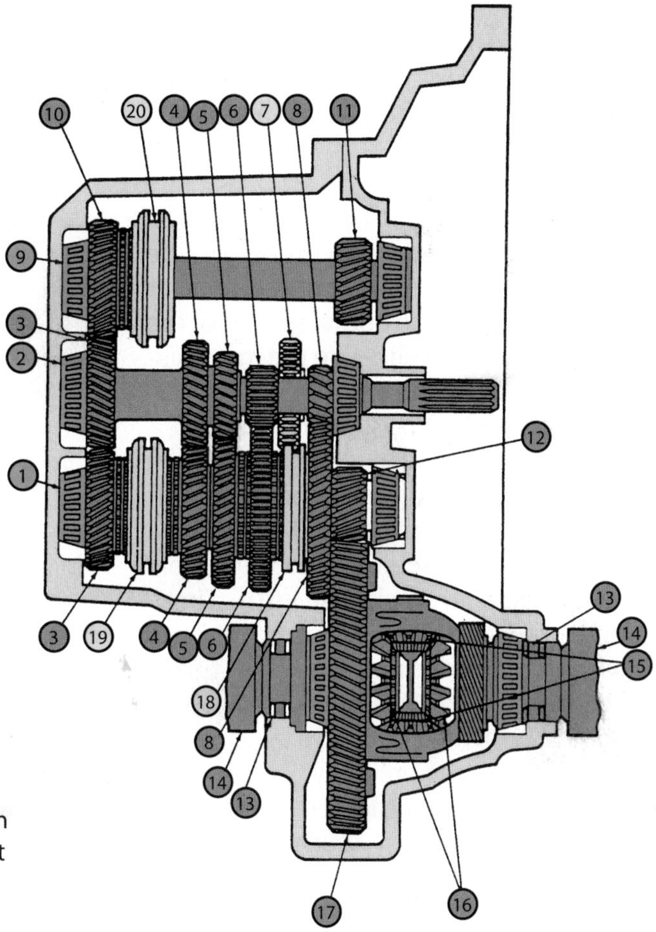

Figure 37-15 A transaxle with three gear shafts. *Courtesy of Ford Motor Company*

The names used to describe transaxle shafts vary between manufacturers. The service manuals of some vehicles refer to the input shaft as the mainshaft and the output as the driven pinion or drive shaft. Others call the input shaft and its gears the input gear cluster and refer to the output shaft as the mainshaft. For clarity, this text uses the terms **input gear cluster** for the input shaft and its drive gears, and **pinion shaft** for the output shaft.

A pinion gear is machined onto the end of the transaxle's pinion shaft. This pinion gear is in constant mesh with the differential ring gear located in the lower portion of the transaxle housing. Because the pinion gear is part of the pinion shaft, it must rotate whenever the pinion shaft turns. With the pinion rotating, engine torque flows through the ring gear and differential gearing to the drive shafts and driving wheels.

Some transaxles have a third shaft designed to offset the power flow on the output shaft. Power is transferred from the output shaft to the third shaft using helical gears and by placing the third shaft in parallel with the output and input shafts. Other transaxles with a third shaft use an offset input shaft that receives the engine's power and transmits it to a mainshaft, which serves as an input shaft. The third shaft is only added to transaxles when an extremely compact transaxle is required (**Figure 37–15**).

SYNCHRONIZERS

The synchronizer performs a number of jobs vital to transmission/transaxle operation. Its main job is to bring components that are rotating at different speeds to one synchronized speed. A synchronizer ensures that the pinion shaft and the speed gear are rotating at the same speed. The second major job of the synchronizer is to actually lock these components together. The end result of these two functions is a clash-free shift. In some transmissions, a synchronizer can have another important job. When spur teeth are cut into the outer sleeve of the synchronizer, the sleeve can act as a reverse gear and assist in producing the correct direction of rotation for reverse operation.

In modern transmissions and transaxles, all forward gears are synchronized. One synchronizer is placed between the first and second gears on the pinion shaft. Another is placed between the third and fourth gears on the **mainshaft**. If the transmission has a fifth gear, it is also equipped with a synchronizer. Reverse gear is not normally fitted with a synchronizer. A synchronizer requires gear rotation to do its job and reverse is selected with the vehicle at a stop.

Synchronizer Design

Figure 37–16 illustrates the most commonly used synchronizer—a **block** or **cone synchronizer**. The synchronizer sleeve surrounds the synchronizer assembly and meshes with the external splines of the clutch hub. The clutch hub is splined to the transmission pinion shaft and is held in position by a snapring. A few transmissions use pin-type synchronizers.

The synchronizer sleeve has a small internal groove and a large external groove in which the shift fork rests. Three slots are equally spaced around the outside of the clutch hub. Inserts fit into these slots and are able to slide freely back and forth. These inserts, sometimes referred to as shifter plates or keys, are designed with a ridge in their outer surface. Insert springs hold the ridge in contact with the synchronizer sleeve internal groove.

The synchronizer sleeve is precisely machined to slide onto the clutch hub smoothly. The sleeve and hub sometimes have alignment marks to ensure proper indexing of their splines when assembling to maintain smooth operation.

Brass, bronze, or powdered iron synchronizing **blocking rings** are positioned at the front and rear of each synchronizer assembly. Some synchronizer assemblies use additional frictional material on the blocking rings to reduce slippage. Each blocking ring has three notches equally spaced to correspond with the three insert keys of the hub. Around the outside of each blocking ring is a set of beveled clutching teeth, called dog teeth, which is used for alignment during the shift sequence. The inside of the blocking ring is shaped like a cone. This coned surface is lined with many sharp grooves.

The cone of the blocking ring makes up only one-half of the total cone clutch. The second or matching

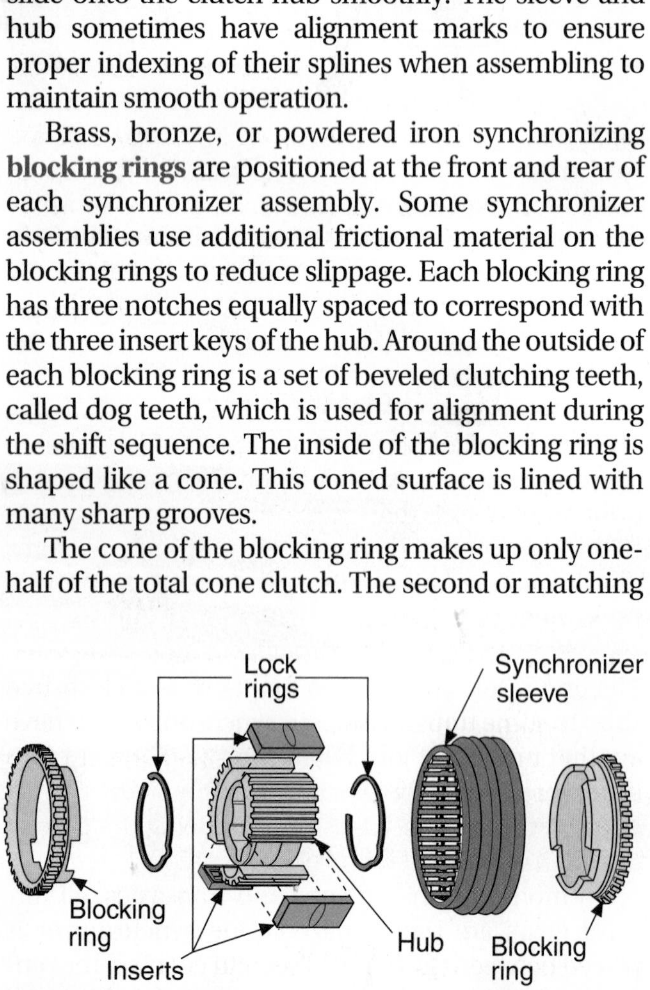

Figure 37–16 An exploded view of a blocking ring-type synchronizer assembly.

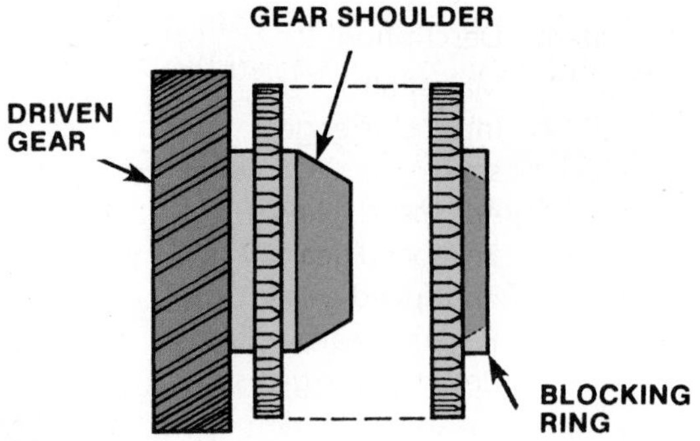

Figure 37–17 Gear shoulder and blocker ring mating surfaces.

half of the cone clutch is part of the speed gear to be synchronized. As shown in **Figure 37–17**, the shoulder of the speed gear is cone shaped to match the blocking ring. The shoulder also contains a ring of beveled clutching teeth designed to align with the clutching teeth on the blocking ring.

Operation

When the transmission is in neutral or reverse, the first/second and third/fourth synchronizers are in their neutral position and are not rotating with the pinion shaft. Gears on the mainshaft are meshed with their countershaft partners and are freewheeling around the pinion shaft at various speeds.

To shift the transmission into first gear, the clutch is disengaged and the gearshift lever is placed in first gear position. This forces the shift fork on the synchronizer sleeve toward the first speed gear on the pinion shaft. As the sleeve moves, the inserts also move because the insert ridges lock the inserts to the internal groove of the sleeve.

The movement of the inserts forces the blocking ring's coned friction surface against the coned surface of the first speed gear shoulder. When the blocking ring and gear shoulder come into contact, the grooves on the blocking ring cone cut through the lubricant film on the first speed gear shoulder and a metal-to-metal contact is made. The contact generates substantial friction and heat. This is one reason bronze or brass blocking rings are used. A nonferrous metal such as bronze or brass minimizes wear on the hardened steel gear shoulder. This frictional coupling is not strong enough to transmit loads for long periods. As the components reach the same speed, the synchronizer sleeve can now slide over the external clutching teeth on the blocking ring and then over the clutching teeth on the first speed gear shoulder. This completes the engagement. Power flow is now from the first speed gear, to the synchronizer sleeve, to the synchronizer clutch hub, to the main output shaft, and out to the driveline.

To disengage the first speed gear from the pinion shaft and shift into second speed gear, the clutch must be disengaged as the shift fork is moved to pull the synchronizer sleeve and disengage it from the first gear. As the transmission is shifted into second gear, the inserts again lock into the internal groove of the sleeve. As the sleeve moves forward, the forward blocking ring is forced by the inserts against the coned friction surface on the second speed gear shoulder. Once again, the grooves on the blocking ring cut through the lubricant on the gear shoulder to generate a frictional coupling that synchronizes the speed gear and shaft speeds. The shift fork can then continue to move the sleeve forward until it slides over the blocking ring and speed gear shoulder clutching teeth, locking them together. Power flow is now from the second speed gear, to the synchronizer sleeve, to the clutch hub, and out through the pinion shaft.

Advanced Synchronizer Designs Many manufacturers are using multiple cone-type synchronizers in their transmissions. These transmissions are fitted with single-cone, double-cone, or triple-cone synchronizers. For example, first and second gears may have triple-cone synchronization, third and fourth may have double-cone, and fifth and sixth may have single-cone.

Double-cone synchronizers **(Figure 37–18)** have friction material on both sides of the synchronizer rings. The extra friction surfaces results in decreased shift effort and greater synchronizer durability. Triple-cone synchronizers provide a third surface of friction material.

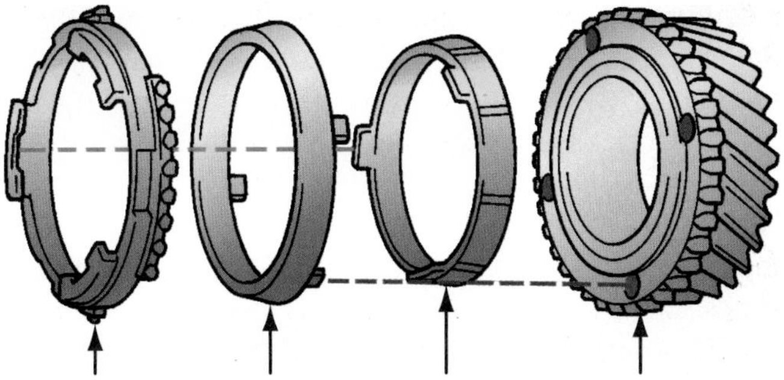

Figure 37–18 A two-cone synchronizer. *Courtesy of Toyota Motor Sales, U.S.A., Inc.*

With multiple-cone synchronizers, the size of the transmission can be reduced. The multiple-cone synchronizers offer a high synchronizer capacity in a smaller package. To obtain the same results in shifting, a single-cone synchronizer would need to have a larger diameter, which would increase the overall size and weight of the transmission.

GEARSHIFT MECHANISMS

Figure 37–19 illustrates a typical transmission shift linkage for a five-speed transmission. As you can see, there are three separate shift rails and forks. Each shift rail/shift fork is used to control the movement of a synchronizer, and each synchronizer is capable of engaging and locking two speed gears to the mainshaft. The shift rails transfer motion from the driver-controlled gearshift lever to the shift forks.

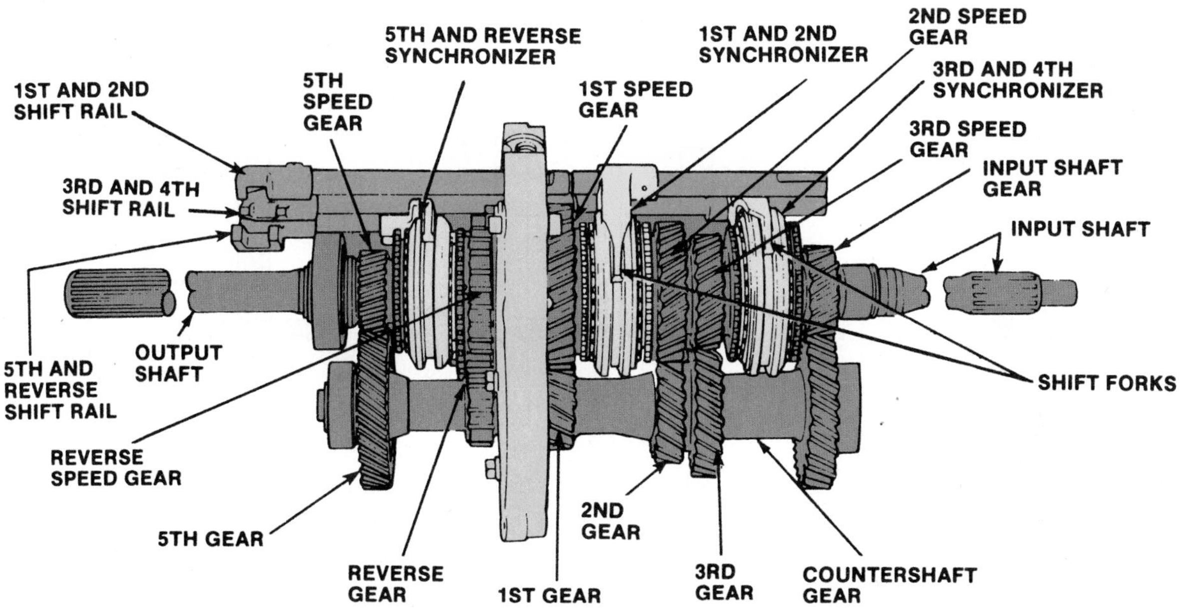

Figure 37–19 An interior view of a five-speed overdrive transmission. Three separate shift rail/shift fork/synchronizer combinations control first/second, third/fourth, and fifth/reverse shifting. *Courtesy of Ford Motor Company*

Figure 37-20 The shift forks, riding on shift rails, fit into the grooves of the synchronizer sleeve.

The **shift forks (Figure 37–20)** are semicircular castings connected to the shift rails with split pins. The shift fork rests in the groove in the synchronizer sleeve and surrounds about one-half of the sleeve circumference.

The gearshift lever is connected to the shift forks by means of a gearshift linkage. Linkage designs vary between manufacturers but can generally be classified as being direct or remote.

Gearshift Linkages

There are two basic designs of gearshift linkages: internal and external. Internal linkages are located at the side or top of the transmission. The control end of the shifter is mounted inside the transmission, as are all of the shift controls. Movement of the shifter moves a **shift rail** and shift fork toward the desired gear. This moves the synchronizer sleeve to lock the selected speed gear to the shaft. This type of linkage is often called a direct linkage, because the shifter is in direct contact with the internal gear shifting mechanisms.

Shift rails are machined with interlock and detent notches. The interlock notches prevent the selection of more than one gear during shifting. When a shift rail is moved by the shifter, interlock pins hold the other shift rails in their neutral position **(Figure 37–21)**. The detent notches and matching spring loaded pins or balls give the driver feedback as to when the shift collar is adequately moved.

As the shift rail moves, a detent ball moves out of its detent notch and drops into the notch for the selected gear. At the same time, an interlock pin moves out of its interlock notch and into the other shift rails.

External linkages function in much the same way, except that rods, external to the transmission, are connected to levers that move the internal shift rails of the transmission **(Figure 37–22)**. Some transaxles are shifted by rods **(Figure 37–23)** or by cable **(Figure 37–24)**.

TRANSMISSION POWER FLOW

The following sections describe the path of power through a typical five-speed transmission.

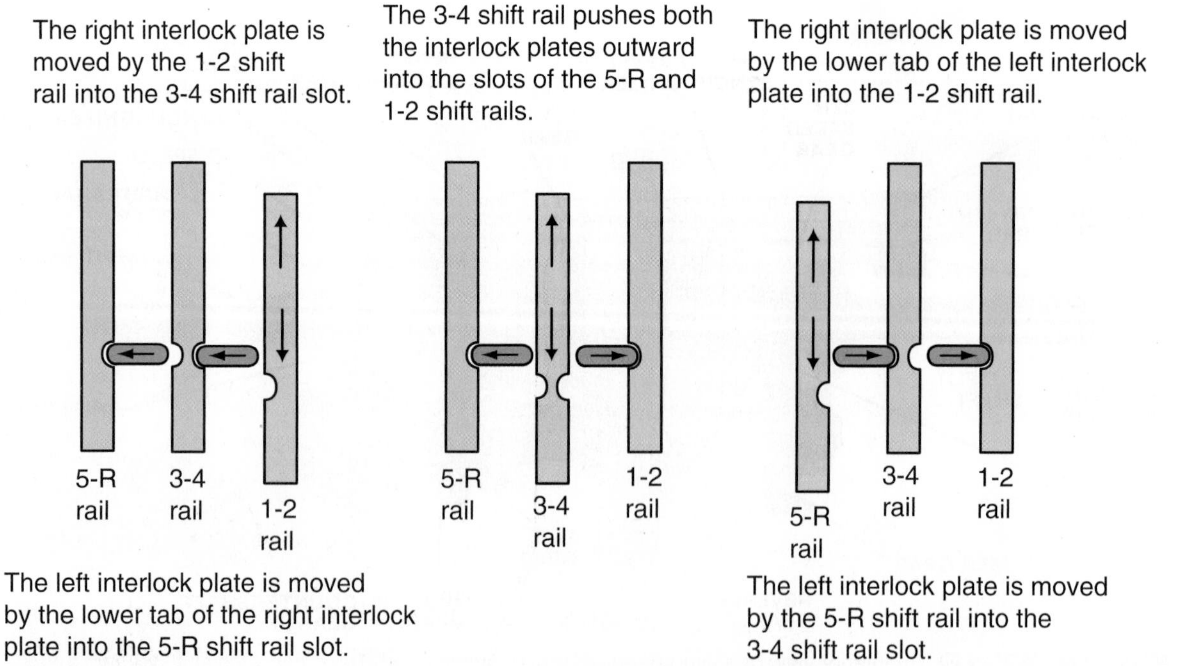
The right interlock plate is moved by the 1-2 shift rail into the 3-4 shift rail slot.
The 3-4 shift rail pushes both the interlock plates outward into the slots of the 5-R and 1-2 shift rails.
The right interlock plate is moved by the lower tab of the left interlock plate into the 1-2 shift rail.
5-R rail
3-4 rail
1-2 rail
5-R rail
3-4 rail
1-2 rail
5-R rail
3-4 rail
1-2 rail
The left interlock plate is moved by the lower tab of the right interlock plate into the 5-R shift rail slot.
The left interlock plate is moved by the 5-R shift rail into the 3-4 shift rail slot.

Figure 37-21 Interlock pins prevent the selection of one or more gears.

Figure 37-22 An external shifter assembly mounted to the transmission.

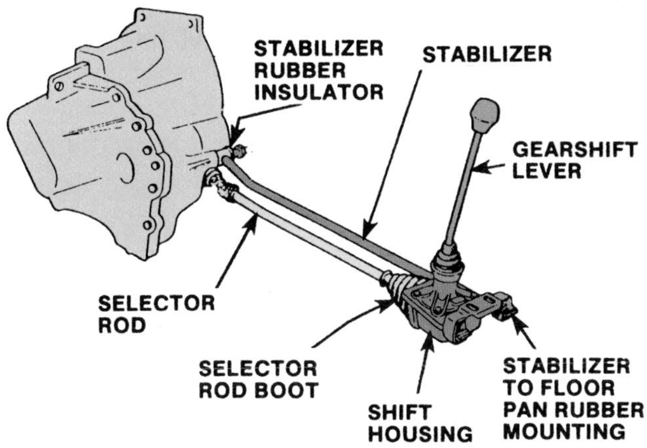

Figure 37-23 A remote gearshift showing linkage, selector rod, and stabilizer (stay bars). *Courtesy of Ford Motor Company*

Neutral

When the gear selector is placed in the neutral position, the input shaft rotates at engine speed. The shaft's clutch gear is in mesh with the countergear, which rotates on the countershaft. The countergears are in mesh with the speed gears. Because none of the speed gears are locked to the mainshaft, the gears spin freely and no torque is applied to the mainshaft.

All gear changes pass through the neutral gear position. When changing gears, one speed gear is disengaged, resulting in neutral, before the chosen gear is engaged. This is important to remember when diagnosing hard-to-shift problems.

First Gear

First gear power flow is illustrated in **Figure 37–25**. Power or torque flows through the input shaft and clutch gear to the countergear. The countergear rotates. The first gear on the cluster drives the first speed gear on the mainshaft. When the driver selects first gear, the first/second synchronizer moves to the rear to engage the first speed gear and lock it to the mainshaft. The first speed gear drives the main (output) shaft, which transfers power to the driveline.

Second Gear

When the shift from first to second gear is made, the shift fork disengages the first/second synchronizer from the first speed gear and moves it until it locks the second speed gear to the mainshaft. Power flow is still

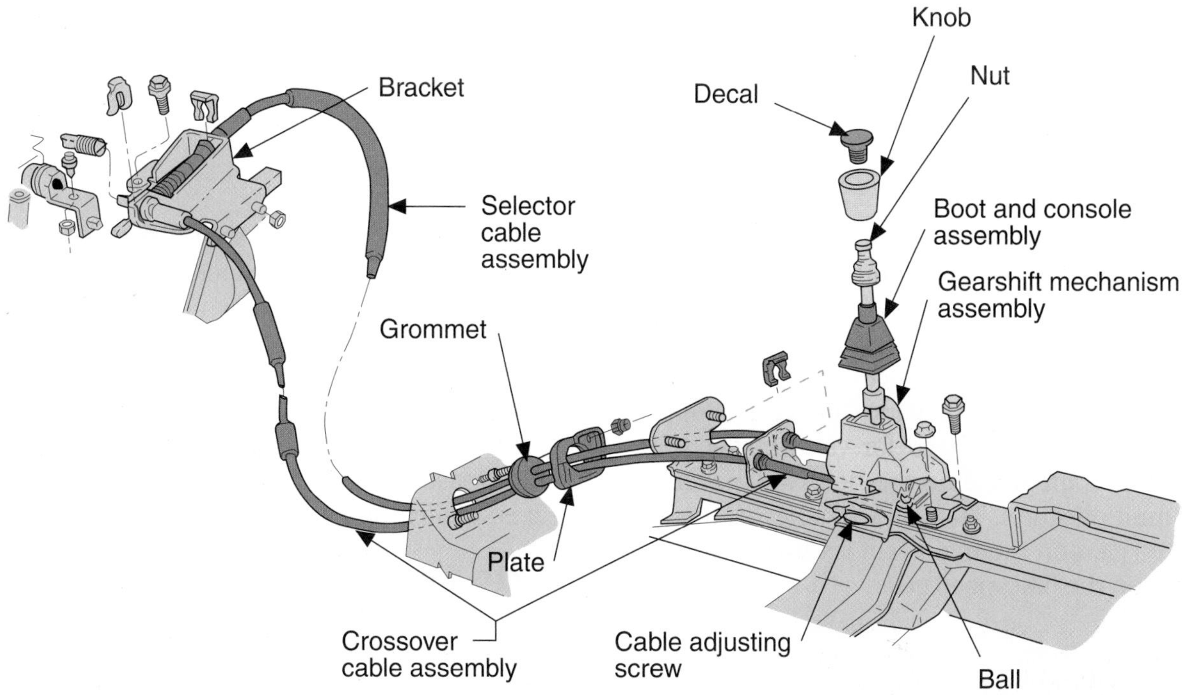

Figure 37-24 A cable-type external gearshift linkage used in a transaxle application. *Courtesy of Chrysler LLC*

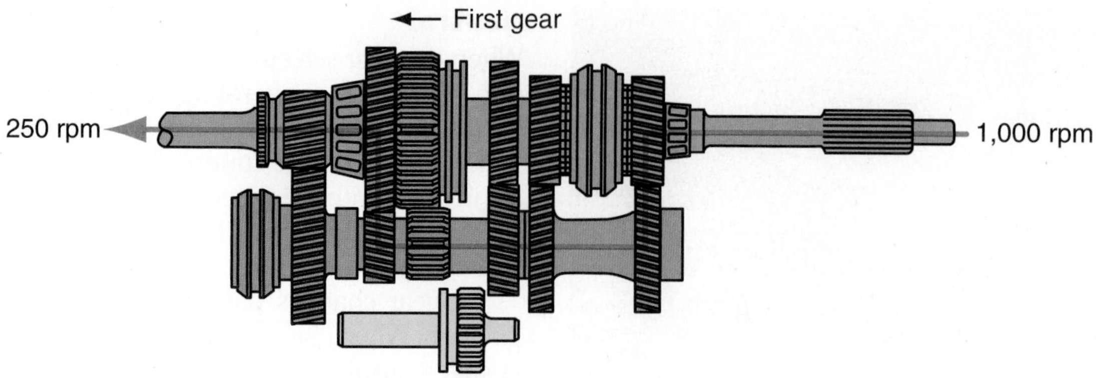

Figure 37-25 Power flow in first gear with a gear ratio of 4:1.

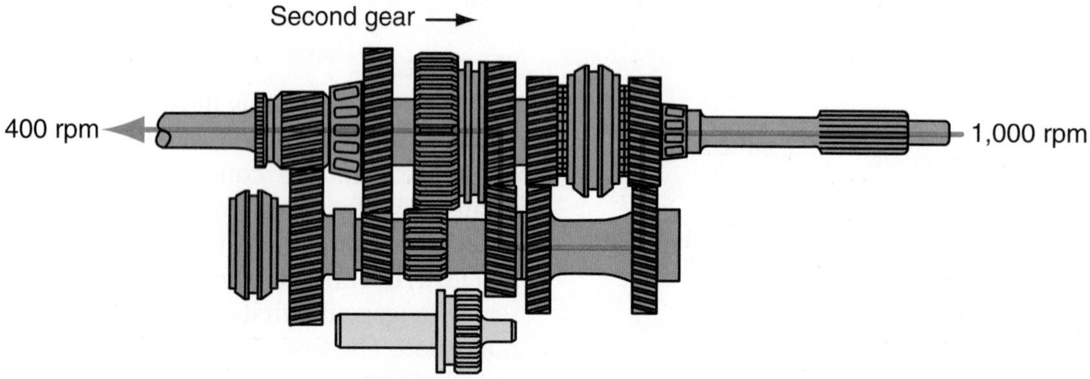

Figure 37-26 Power flow in second gear with a gear ratio of 2.5:1.

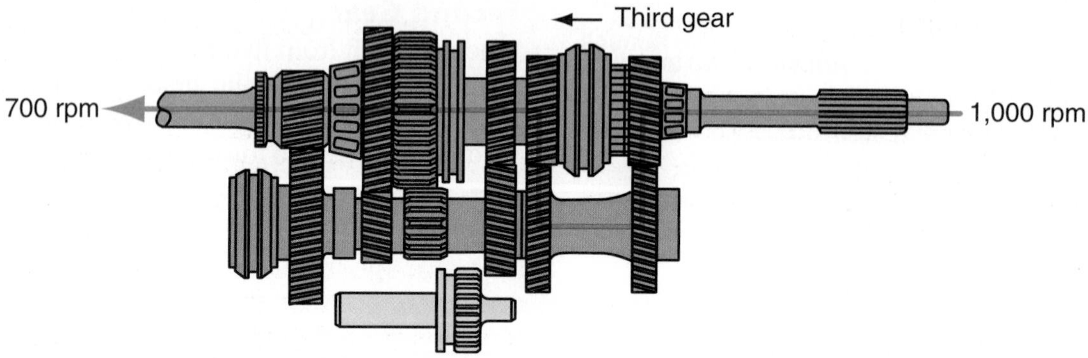

Figure 37-27 Power flow in third gear with a gear ratio of 1.43:1.

through the input shaft and clutch gear to the countergear. However, now the second countergear on the cluster transfers power to the second speed gear locked on the mainshaft. Power flows from the second speed gear through the synchronizer to the mainshaft (output shaft) and driveline **(Figure 37–26)**.

In second gear, the need for vehicle speed and acceleration is greater than the need for maximum torque multiplication. To meet these needs, the second speed gear on the mainshaft is designed slightly smaller than the first speed gear.

Third Gear

When the shift from second to third gear is made, the shift fork returns the first/second synchronizer to its neutral position. A second shift fork slides the third/fourth synchronizer until it locks the third speed gear to the mainshaft. Power flow now goes through the third gear of the countergear to the third speed gear, through the synchronizer to the mainshaft, and driveline **(Figure 37–27)**.

Third gear permits a further decrease in torque and increase in speed. As you can see, the third speed gear is smaller than the second speed gear.

Fourth Gear

In fourth gear, the third/fourth synchronizer is moved to lock the clutch gear on the input shaft to the mainshaft. This means power flow is directly from the input shaft to the mainshaft (output shaft) at a gear

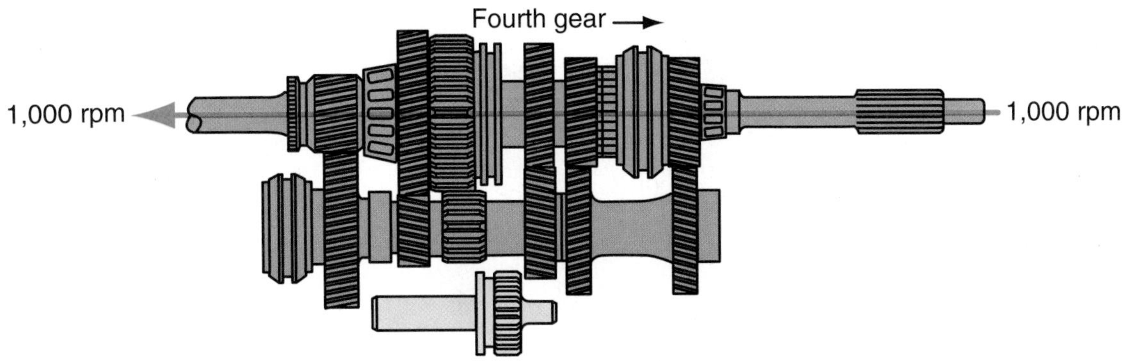

Figure 37-28 Power flow in fourth gear with a gear ratio of 1:1.

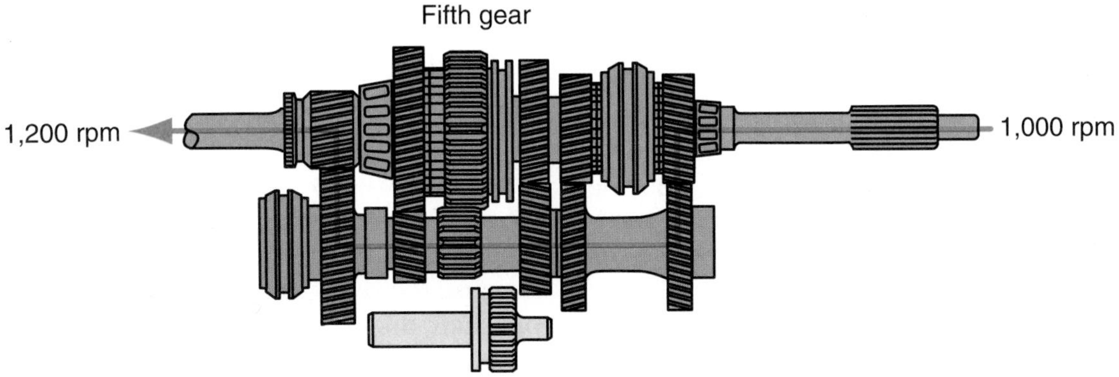

Figure 37-29 Power flow in fifth gear with a gear ratio of 0.83:1.

ratio of 1:1 **(Figure 37–28)**. This ratio results in maximum speed output and no torque multiplication. Fourth gear has no torque multiplication because it is used at cruising speeds to promote maximum fuel economy. The vehicle is normally downshifted to lower gears to take advantage of torque multiplication and acceleration when passing slower vehicles or climbing grades.

Fifth Gear

When fifth gear is selected, the fifth gear synchronizer engages fifth gear to the mainshaft **(Figure 37–29)**. This causes a large gear on the countershaft to drive a smaller gear on the mainshaft, which results in an overdrive condition. Overdrive permits an engine speed reduction at higher vehicle speeds.

Reverse

In reverse gear, it is necessary to reverse the direction of the mainshaft (output shaft). This is done by introducing a reverse idler gear into the power flow path. The idler gear is located between the countershaft reverse gear and the reverse speed gear on the mainshaft. The idler assembly is made of a short drive shaft independently mounted in the transmission case parallel to the countershaft. The idler gear may be mounted near the midpoint of the shaft.

The reverse speed gear is actually the external tooth sleeve of the first/second synchronizer.

When reverse gear is selected, both synchronizers are disengaged and in the neutral position. In the transmission shown in **Figure 37–30**, the shifting linkage moves the reverse idler gear into mesh with the first/second synchronizer sleeve. Power flows through the input shaft and clutch gear to the countershaft. From the countershaft, it passes to the reverse idler gear, where it changes rotational direction. It then passes to the first/second synchronizer sleeve. Rotational direction is again reversed. From the sleeve, power passes to the mainshaft and driveline.

Not all transmissions use speed and idler gears for reverse. For example, reverse gears in most Ford transmissions are helical gears that are in constant mesh with the first gear.

> **SHOP TALK**
>
> In Figure 37–30, the reverse idler gear is drawn below the countershaft. This is not where it really is. It is actually between the first/second synchronizer and the countershaft. To place it there in the drawing would make the illustration unclear.

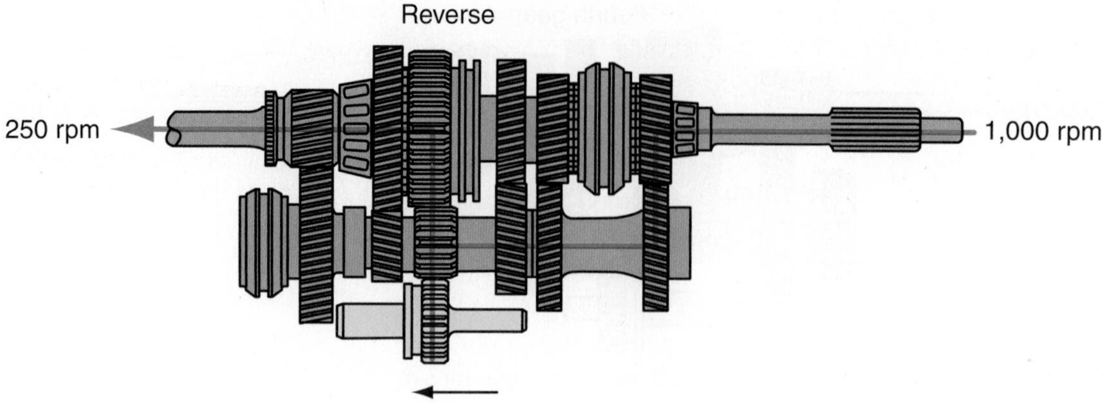

Figure 37-30 Power flow in reverse gear with a gear ratio of 4:1.

TRANSAXLE POWER FLOWS

When studying the power flow patterns in the following section, keep in mind that the views are based on you standing by the right front fender and looking into the engine compartment. This will give an accurate idea of the true rotational direction of the gears and shaft. The transaxles used in these examples are five-speed, two-shaft units **(Figure 37–31)**.

Neutral

When a transaxle is in its "neutral" position, no power is applied to the differential. Because the synchronizer collars are centered between their gear positions, the meshed drive gears are not locked to the output shaft. Therefore, the gears spin freely on the shaft and the output shaft does not rotate.

Forward Gears

When first gear is selected **(Figure 37–32)**, the first/second gear synchronizer engages with first gear. Because the synchronizer hub is splined to the output shaft, first gear on the input shaft drives its mating gear (first gear) on the output shaft. This causes the output shaft to rotate at the ratio of first gear and to drive the differential ring gear at that same ratio.

As the other forward gears are selected, the appropriate shift fork moves to engage the selected

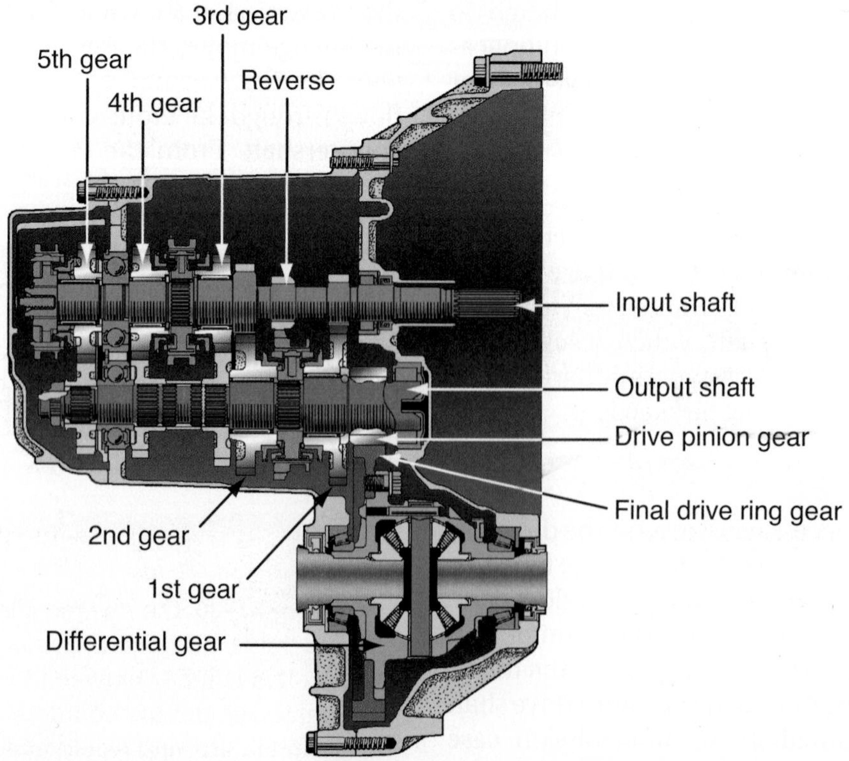

Figure 37-31 A five-speed transaxle with two internal shafts and final drive assembly. *Courtesy of Toyota Motor Sales, U.S.A., Inc.*

synchronizer with the gear. Because the synchronizer's hub is splined to the output shaft, the desired gear on the input shaft drives its mating gear on the output shaft (**Figures 37–33, 37–34, 37–35,** and **37–36**). This causes the output shaft to rotate at the ratio of the selected gear and drive the differential ring gear at that same ratio.

Reverse

When reverse gear is selected on transaxles that use a sliding reverse gear (**Figure 37–37**), the shifting fork forces the gear into mesh with the input and output shafts. The addition of this third gear reverses the normal rotation of first gear and allows the car to change direction.

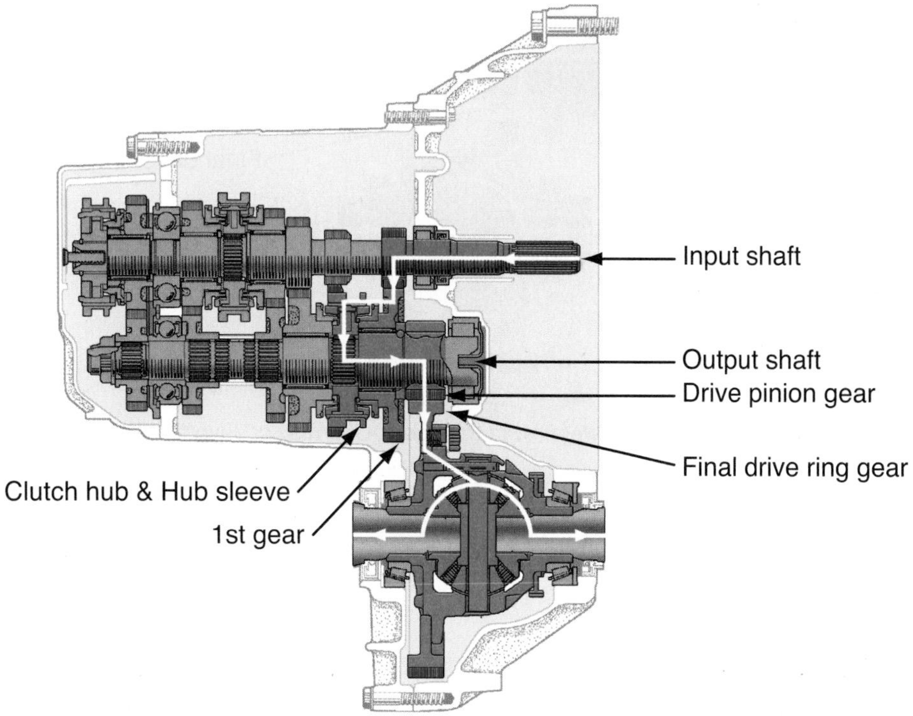

Figure 37-32 Power flow in first gear. *Courtesy of Toyota Motor Sales, U.S.A., Inc.*

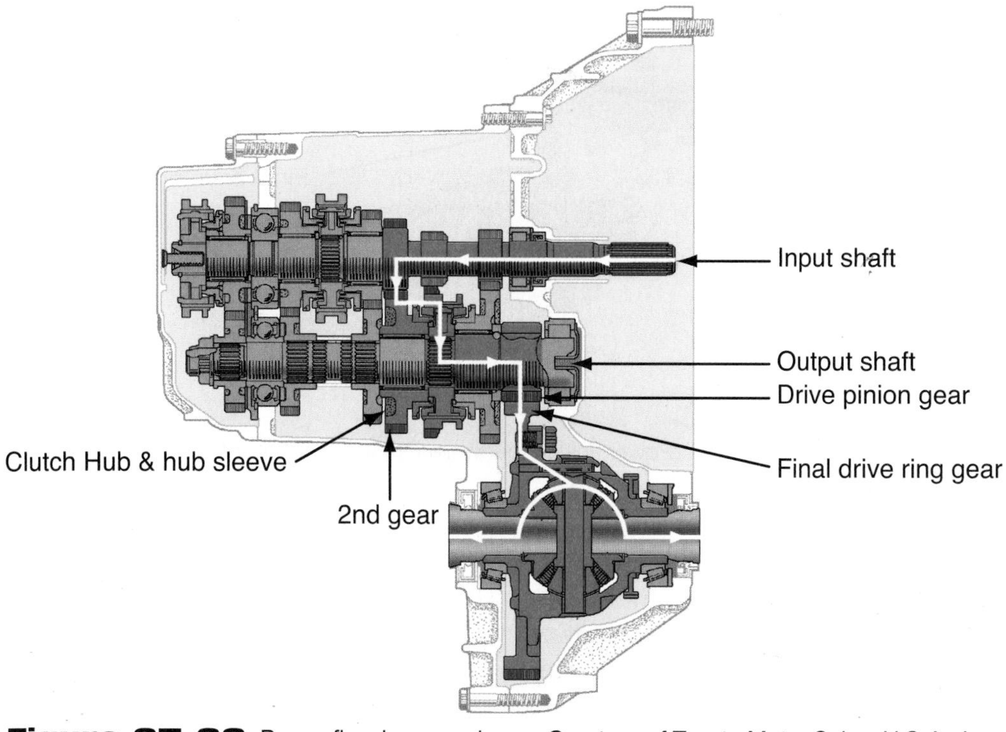

Figure 37-33 Power flow in second gear. *Courtesy of Toyota Motor Sales, U.S.A., Inc.*

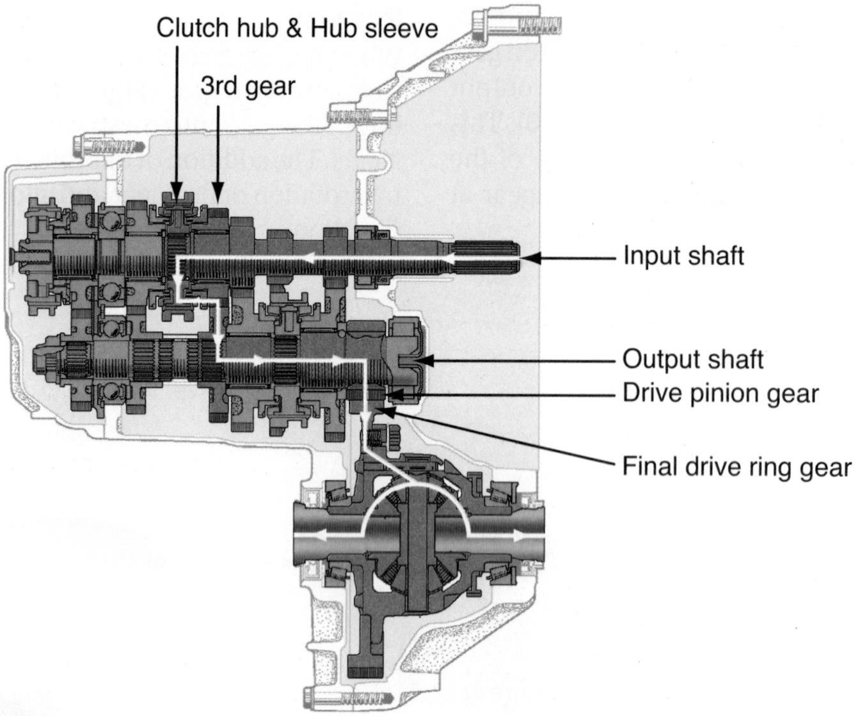

Figure 37-34 Power flow in third gear. *Courtesy of Toyota Motor Sales, U.S.A., Inc.*

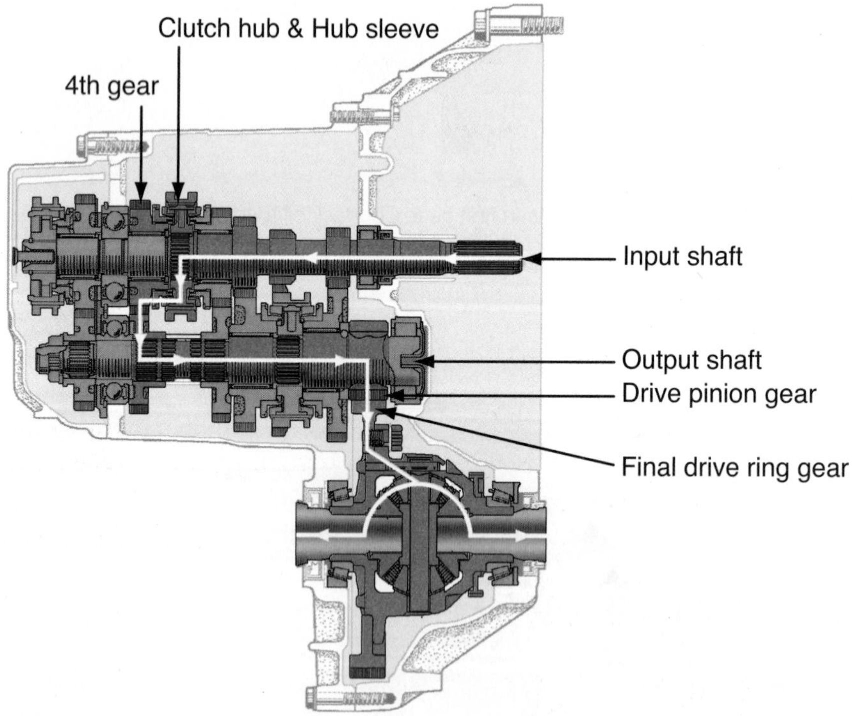

Figure 37-35 Power flow in fourth gear. *Courtesy of Toyota Motor Sales, U.S.A., Inc.*

Differential Action

The final drive ring gear is driven by the transaxle's output shaft. The ring gear then transfers the power to the differential case. The case holds the ring gear with its mating pinion gear. The differential side gears are connected to the drive axles.

One major difference between the differential in a RWD car and the differential in a transaxle is direction of power flow. In a RWD differential, the power flow changes 90 degrees between the drive pinion gear and the ring gear. This change in direction is not needed with most FWD cars. The transverse

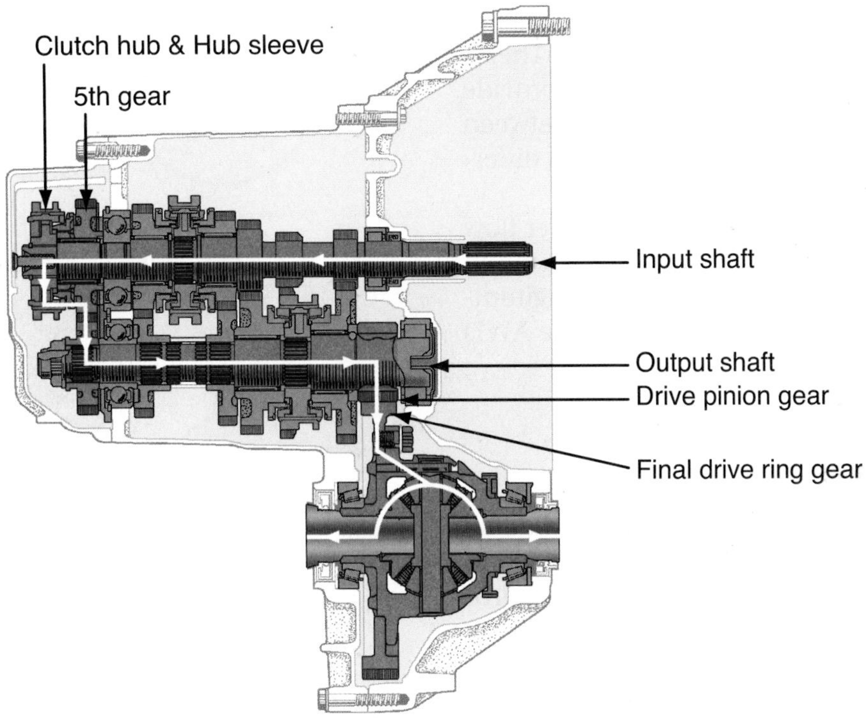

Figure 37-36 Power flow in fifth gear. *Courtesy of Toyota Motor Sales, U.S.A., Inc.*

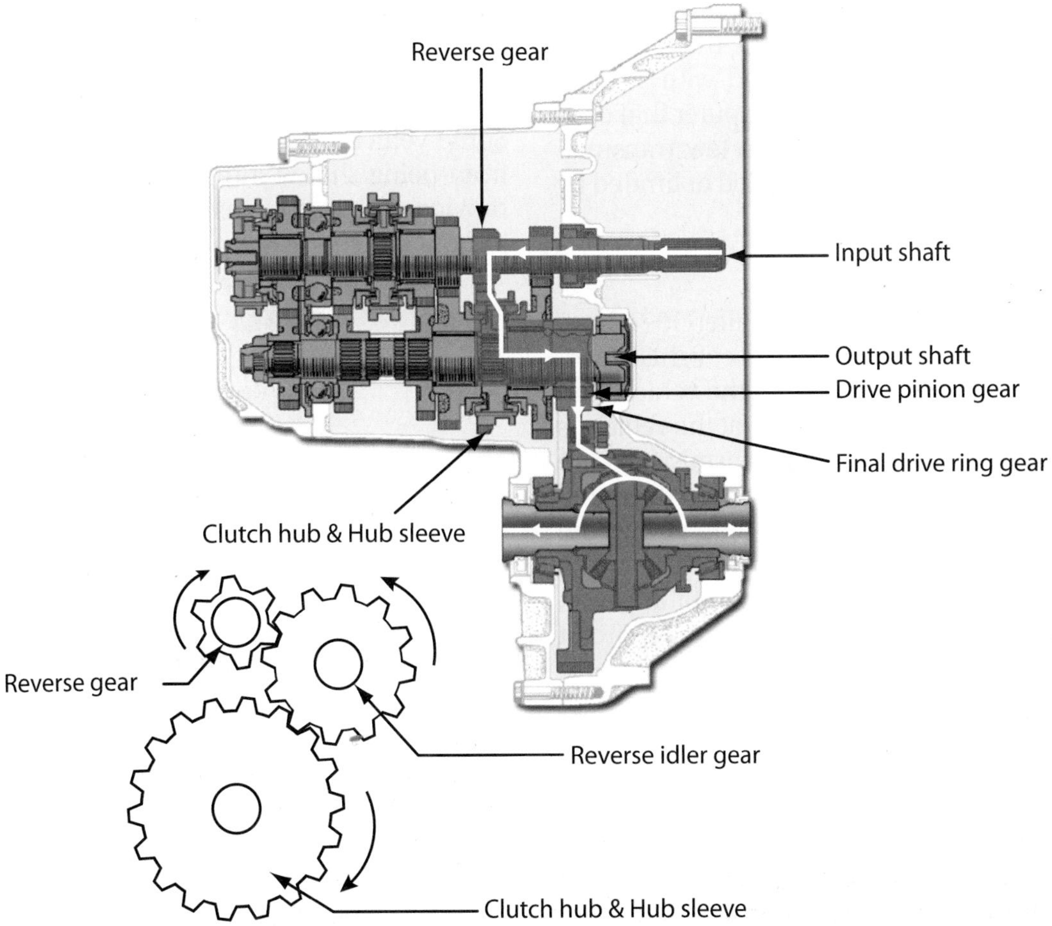

Figure 37-37 Power flow in reverse gear. *Courtesy of Toyota Motor Sales, U.S.A., Inc.*

engine position places the crankshaft so that it already is rotating in the correct direction. Therefore, the purpose of the differential is only to provide torque multiplication and divide the torque between the drive axle shafts so that they can rotate at different speeds.

Some transaxles need the 90-degree power flow change in the differential. These units are used in rear-engine with RWD applications in longitudinally positioned engines with FWD or some AWD vehicles.

FINAL DRIVE GEARS AND OVERALL RATIOS

All vehicles use a differential to provide an additional gear reduction (torque increase) above and beyond what the transmission or transaxle gearing can produce. This is known as the **final drive gear**.

In a transmission-equipped vehicle, the differential gearing is located in the rear axle housing. In a transaxle, however, the final reduction is produced by the final drive gears housed in the transaxle case.

ELECTRICAL SYSTEMS

Although most manual transmissions are not electrically operated or controlled, a few accessories of the car are controlled or linked to the transmission. The transmission may also be fitted with sensors that give vital information to the computer that controls other car systems. There are a few transmissions that have their shifting controlled or limited by electronics.

Reverse Lamp Switch

All vehicles sold in North America after 1971 have been required to have back-up (reverse) lights. Back-up lights illuminate the area behind the vehicle and warn other drivers and pedestrians that the vehicle is moving in reverse. Most manual transmissions are equipped with a separate switch located on the transmission **(Figure 37–38)** but can be mounted to the shift linkage away from the transmission. If the switch is mounted in the transmission, the shifting fork closes the switch and completes the electrical circuit whenever the transmission is shifted into reverse gear. If the switch is mounted on the linkage, the switch is closed directly by the linkage.

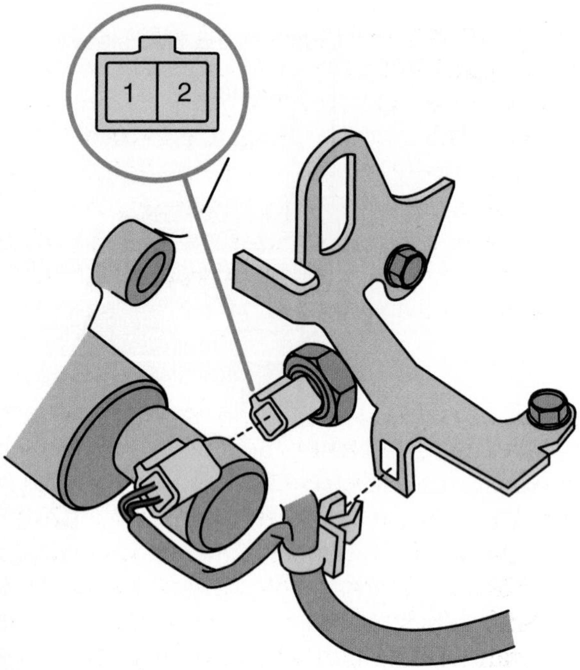

Figure 37-38 A typical back-up light switch for a transaxle.

Vehicle Speed Sensor

Most late-model transmissions and transaxles are fitted with a VSS. This sensor sends an electrical signal to the vehicle's PCM. This signal represents the speed of the transmission's output shaft. The PCM then calculates the speed of the vehicle. This information is used for many systems, such as cruise control, fuel and spark management, and instrumentation. Affected instrumentation includes the speedometer, odometer, and upshift lamp.

Reverse Lockout System

Some vehicles electrically prevent the transmission from being shifted into reverse when the vehicle is moving. These systems are typically called reverse lockout systems **(Figure 37–39)**. Normally, when the

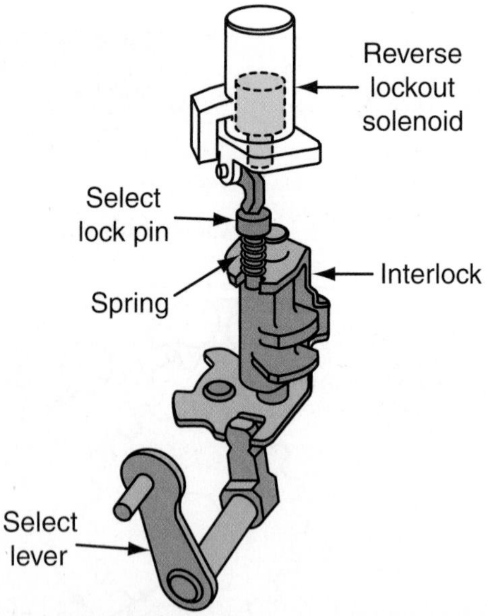

Figure 37-39 A PCM-controlled reverse lockup assembly.

vehicle is moving at a speed of 12 mph (20 km/h) or more, a signal from the PCM energizes the reverse lockout solenoid. The solenoid pushes a cam and lock pin down. This action makes it impossible for the transmission to be shifted into reverse. When the vehicle is sitting still or traveling at lower speeds, the solenoid is not energized. A return spring holds the lock pin away from the interlock assembly and the shifter is free to move into reverse.

Shift Blocking

Some six-speed transmissions have a feature called shift blocking that prevents the driver from shifting into second or third gears from first gear, when the engine's coolant temperature is below a specified degree, the speed of the car is between 12 and 22 mph (20 and 29 km/h), and the throttle is opened less than 35%. Shift blocking helps improve fuel economy. These transmissions are also equipped with reverse lockout, as are some others. This feature prevents the engagement of reverse whenever the vehicle is moving forward.

Shift blocking is controlled by the PCM. A "skip shift" solenoid is used to block off the shift pattern from first gear to second and third gears. The driver moves the gearshift from its up position to a lower position, as if shifting into second gear, but fourth gear is selected.

Customer Care

Just because the technician gets a little dirty in the course of a repair does not mean the vehicle should, too. Treat every car that enters the shop with the utmost care and consideration. Scratches from belt buckles or tools and grease smears on the steering wheel, upholstery, or carpeting are inexcusable, and a sure way of losing business. Always use fender, seat, and floor covers when the job requires them. Check your hands for cleanliness before driving a vehicle or operating the windows and dash controls.

CASE STUDY

A customer brought his 2001 Toyota into the shop and said that the transaxle was jumping out of fifth gear. The technician verified the problem, then proceeded to check the adjustment of the shift linkage. He checked the adjustment and inspected the linkage and found nothing wrong. He proceeded to check the alignment of the transaxle to the engine, and again the cause of the problem was not found.

Suspecting internal damage, the technician removed the transaxle from the car and disassembled it, hoping to find the cause of the problem. Again the cause was not found. The technician then carefully reassembled the transaxle and installed it back in the car. During the road test, he again experienced the problem. He returned the car to the customer and told him that the problem could not be fixed.

Taking this approach is a good way to lose customers and your job. Of course the problem could have been fixed! True, he did check all the right things. However, he probably did not totally disassemble the transaxle to check all the internal parts carefully. A slight bit of wear on the dog teeth of the synchronizer can cause this kind of problem. At times, the wear is difficult to see, especially if the wear is even all around the ring. Most technicians, as a matter of course, will replace the blocking rings whenever a transaxle is taken apart, because they cannot always see the wear. If the technician in this case study had followed this procedure, he probably would not be delivering pizza today.

KEY TERMS

Block synchronizer
Blocking rings
Cluster gear
Clutch gear
Cone synchronizer
Countershaft
External gear
Final drive gear
Gear pitch
Helical gear
Idler gear
Input shaft
Mainshaft
Pinion shaft
Shift forks
Shift rails
Spur gear
Synchronizer

SUMMARY

- A transmission or transaxle uses meshed gears of various sizes to give the engine a mechanical advantage over its driving wheels.

- Transaxles contain the gear train plus the differential gearing needed to produce the final drive gear ratios. Transaxles are commonly used on front-wheel-drive vehicles.
- Transmissions are normally used on rear-wheel-drive vehicles.
- Gears in the transmission/transaxle transmit power and motion from an input shaft to an output shaft. These shafts are mounted parallel to one another.
- Spur gears have straight cut teeth, while helical gears have teeth cut at an angle. Helical gears run without creating gear whine.
- When a small gear drives a larger gear, output speed decreases but torque (power) increases.
- When a large gear drives a smaller gear, output speed increases but torque (power) decreases.
- When two external toothed gears are meshed and turning, the driven gear rotates in the opposite direction of the driving gear.
- Synchronizers bring parts rotating at different speeds to the same speed for smooth clash-free shifting. The synchronizer also locks and unlocks the driven (speed) gears to the transmission/transaxle output shaft.
- Idler gears are used to reverse the rotational direction of the output shaft for operating the vehicle in reverse.
- In typical five-speed transmission shift linkage, there are three separate shift rails and forks. Each shift rail/shift fork is used to control the movement of a synchronizer.
- Gear ratios indicate the number of times the input drive gear is turning for every turn of the output driven gear. Ratios are calculated by dividing the number of teeth on the driven gear by the number of teeth on the drive gear. You can also use the rpm speeds of meshed gears to calculate gear ratios.
- A gear ratio of less than one indicates an overdrive condition. This means the driven gear is turning faster than the drive gear. Speed is high, but output torque is low.
- All vehicles use a gear set in the final drive to provide additional gear reduction (torque increase) above and beyond what the transmission or transaxle gearing can produce.

REVIEW QUESTIONS

1. What determines whether a conventional transmission or a transaxle is used?
2. Name the three drive configurations that a gear set can transfer torque.
3. *True or False?* A reverse idler gear changes the direction of torque flow to the opposite direction of engine rotation.
4. Define the purpose of the final drive gears.
5. Explain the role of shift rails and shift forks in the operation of a transmission or transaxle.
6. *True or False?* In most transmissions and transaxles, there is one synchronizer assembly for each speed gear.
7. The number of gear teeth per unit of measurement of the gear's diameter (such as teeth/inch) is known as gear ______.
8. Which of the following gear ratios indicates an overdrive condition?
   a. 2.15:1
   b. 1:1
   c. 0.85:1
   d. none of the above
9. Which type of gear develops gear whine at high rotational speeds?
   a. spur gear
   b. helical gear
   c. both a and b
   d. neither a nor b
10. When an idler gear is placed between the driving and driven gears, the driven gear ______ ______ ______ ______ ______.
11. The component used to ensure that the mainshaft (output shaft) and main (speed) gear locked to it are rotating at the same speed is known as a ______.
    a. synchronizer
    b. shift detent
    c. shift fork
    d. transfer case
12. In a transaxle, the pinion gear on the pinion shaft meshes with the ______.
    a. reverse idler gear
    b. ring gear
    c. countershaft drive gear
    d. input gear
13. *True or False?* The cone on a synchronizer's blocking ring serves as a cone-clutch assembly during the changing of gears.

14. Which of the following gear ratios provides the highest torque multiplication?
    a. 0.85:1
    b. 2.67:1
    c. 5.23:1
    d. 0.50:1
15. What does a shift blocking system do, and how does it basically work?

ASE-STYLE REVIEW QUESTIONS

1. Technician A says that most transaxles have a countershaft between the main and output shafts. Technician B says that a countershaft is always in motion when the engine is running. Who is correct?
   a. Technician A
   b. Technician B
   c. Both A and B
   d. Neither A nor B
2. While discussing the various types of transmissions found on today's vehicles: Technician A says that one type is the sliding gear transmission. Technician B says that one type is the sliding collar transmission. Who is correct?
   a. Technician A
   b. Technician B
   c. Both A and B
   d. Neither A nor B
3. Technician A says that the countergear or cluster gear is actually several gears machined out of single piece of iron or steel. Technician B says that the countergear is driven by the clutch gear and drives the mainshaft speed gears. Who is correct?
   a. Technician A
   b. Technician B
   c. Both A and B
   d. Neither A nor B
4. Technician A says that reverse lamp switches are activated at the transmission/transaxle by the gearshift lever. Technician B says that reverse lamp switches are activated by a sensor at the input shaft. Who is correct?
   a. Technician A
   b. Technician B
   c. Both A and B
   d. Neither A nor B
5. Technician A says that if a single idler gear is used to obtain reverse gear, its size and number of teeth must be used to calculate gear ratios. Technician B says that an idler gear is often used for reverse gear because it causes the output shaft to rotate in the reverse direction of the input shaft. Who is correct?
   a. Technician A
   b. Technician B
   c. Both A and B
   d. Neither A nor B
6. Technician A says that in a conventional transmission, the speed gears freewheel around the mainshaft until they are locked to it by the appropriate synchronizer. Technician B says that speed gears are an integral part of the countershaft assembly. Who is correct?
   a. Technician A
   b. Technician B
   c. Both A and B
   d. Neither A nor B
7. While discussing the power flow through a five-speed transmission while it is in first gear: Technician A says that power enters on the input shaft, which rotates the countershaft that is engaged with first gear. Technician B says that the first gear synchronizer engages with the clutching teeth of first gear and locks the gear to the main shaft, allowing power to flow from the input gear through the countershaft and to first gear and the main shaft. Who is correct?
   a. Technician A
   b. Technician B
   c. Both A and B
   d. Neither A nor B
8. Technician A says that double-cone synchronizers have friction material on both sides of the synchronizer rings. Technician B says that multiple-cone synchronizers are used with larger engines and require a larger transmission housing. Who is correct?
   a. Technician A
   b. Technician B
   c. Both A and B
   d. Neither A nor B
9. Technician A says that some synchronizer assemblies use frictional material on the blocking rings to reduce slippage. Technician B says that the inside of some blocking rings is shaped like a cone lined with many sharp grooves. Who is correct?
   a. Technician A
   b. Technician B
   c. Both A and B
   d. Neither A nor B
10. While discussing shift mechanisms: Technician A says that the interlock notches hold the transmission in the selected gear. Technician B says that the detent notches prevent the selection of more than one gear during operation. Who is correct?
    a. Technician A
    b. Technician B
    c. Both A and B
    d. Neither A nor B

CHAPTER
38

MANUAL TRANSMISSION/ TRANSAXLE SERVICE

OBJECTIVES

■ **Perform a visual inspection of transmission/transaxle components for signs of damage or wear.** ■ **Check transmission oil level correctly, detect signs of contaminated oil, and change oil as needed.** ■ **Describe the steps taken to remove and install transmissions/transaxles, including the equipment and safety precautions used.** ■ **Identify common transmission problems and their probable causes and solutions.** ■ **Describe the basic steps and precautions taken during transmission/transaxle disassembly, cleaning, inspection, and reassembly procedures.**

When properly operated and maintained, a manual transmission/transaxle normally lasts the life of the vehicle without a major breakdown. All units are designed so the internal parts operate in a bath of oil circulated by the motion of the gears and shafts. Some units also use a pump to circulate oil to critical wear areas that require more lubrication than the natural circulation provides.

Maintaining good internal lubrication is the key to long transmission/transaxle life. If the amount of oil falls below minimum levels, or if the oil becomes too dirty, problems result.

SHOP TALK

Whenever you are diagnosing or repairing a transaxle or transmission, make sure you refer first to the appropriate service manual before you begin.

Prior to beginning any diagnosis, service, or repair work, be sure you know exactly which transmission you are working on. This will ensure that you are following the correct procedures and specifications and are installing the correct parts. Proper identification can be difficult because transmissions cannot be accurately identified by the way they look. The only positive way to identify the exact design of the transmission is by its identification numbers.

Transmission identification numbers are found either as numbers stamped on the case or on a metal tag held by a bolt head. Use a service manual to decipher the identification number. Most identification numbers include the model, gear ratios, manufacturer, and assembly date **(Figure 38–1)**. Whenever you work with a transmission with a metal ID tag, make sure the tag is put back on the transmission so that the next technician will be able to properly identify the transmission.

If the transmission does not have an ID tag, the transmission must be identified by comparing it with those in the vehicle's service manual.

LUBRICANT CHECK

The transmission/transaxle gear oil level should be checked at the intervals specified in the service manual. Normally, these range from every 7,500 to 30,000 miles (12,000 to 48,000 km). For service convenience, many units are now designed with a dipstick and filler tube accessible from beneath the hood. Check the oil with the engine off and the vehicle resting in a level position. If the engine has been running, wait 2 to 3 minutes before checking the gear oil level.

Some vehicles have no dipstick. Instead, the vehicle must be placed on a lift, and the oil level checked through the **fill plug** opening on the side of the unit. Clean the area around the plug before loosening and removing it. Normally, lubricant should be level with, or not more than, ½ inch (12.7 mm) below the fill hole.

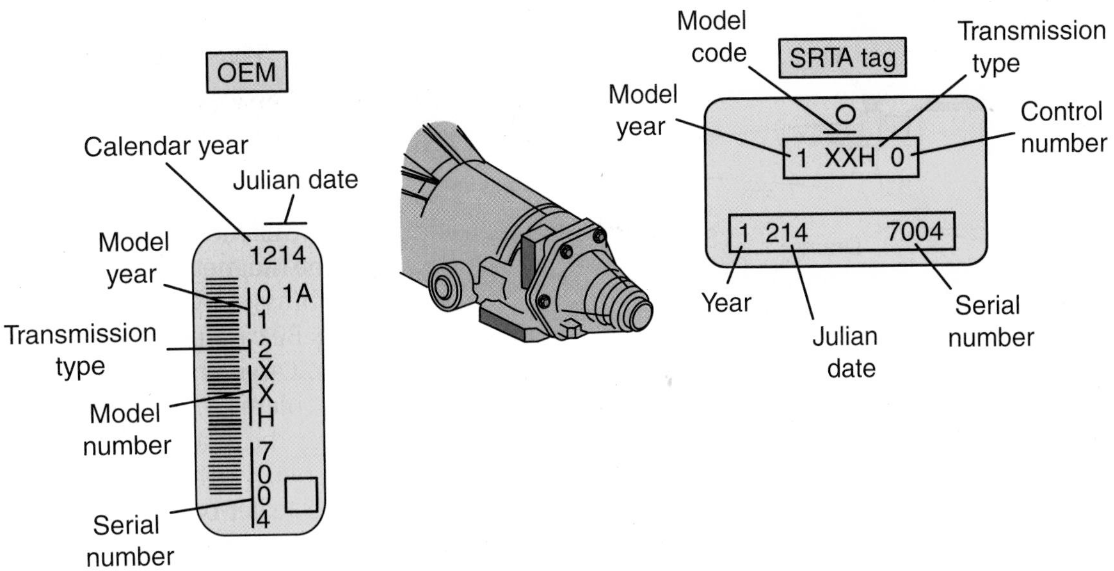

Figure 38-1 Location of and information found on GM transmission ID decals.

Always check the service information for the proper level. Add the proper grade lubricant as needed using a filler pump.

Manual transmission/transaxle lubricants in use today include single- and multiple-viscosity gear oils, engine oils, special hydraulic fluids, and automatic transmission fluid. Always refer to the service manual to determine the correct lubricant and viscosity range for the vehicle and operation conditions (**Figure 38–2**).

Lubricant Leaks

Normally, the location and cause of a transmission fluid leak can be quickly identified by a visual inspection. The following are common causes for fluid leakage:

1. An excessive amount of lubricant in the transmission or transaxle.
2. The use of the wrong type of fluid; it will foam excessively and leave through the vent.
3. A loose or broken input shaft bearing retainer.
4. A damaged input shaft bearing retainer O-ring and/or lip seal.
5. Loose or missing case bolts.
6. Case is cracked or has a porosity problem.
7. A leaking shift lever seal.

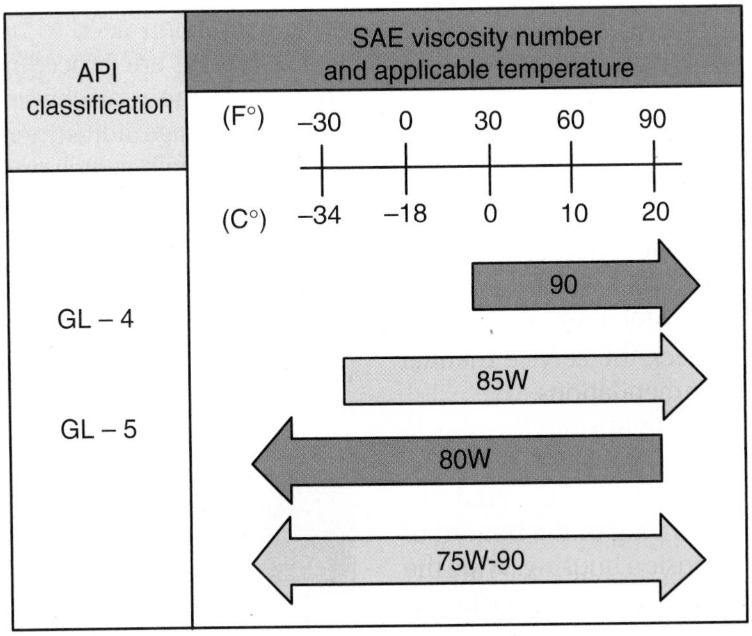

Figure 38-2 Typical transmission/transaxle gear oil classification and viscosity range data.

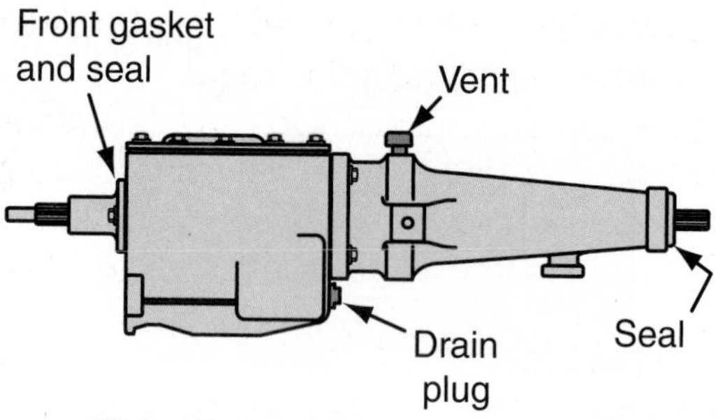

Figure 38-3 Possible sources of fluid leaks.

8. Gaskets or seals are damaged or missing.
9. The drain plug is loose.
10. Bad axle seals.

Fluid leaks from the seal of the extension housing can be corrected with the transmission in the car. Often, the cause for the leakage is a worn extension housing bushing, which supports the sliding yoke. When the drive shaft is installed, the clearance between the sliding yoke and the bushing should be minimal. If the clearance is satisfactory, a new oil seal will correct the leak. If the clearance is excessive, the repair requires that a new seal and a new bushing be installed. If the seal is faulty, the transmission vent should be checked for blockage. If the vent is plugged, the oil will be under high pressure when the transmission is hot, and this pressure can cause seal leakage (**Figure 38–3**).

An oil leak at the speedometer cable can be corrected by replacing the O-ring seal. An oil leak stemming from the mating surfaces of the extension housing and the transmission case may be caused by loose bolts. To correct this problem, tighten the bolts to the specified torque.

Fluid Changes

The manufacturers of most transmissions do not recommend intervals for manual transmission fluid replacement. Older transmissions typically had 20,000-mile fluid change intervals. When the vehicle is operated under severe conditions, such as in high heat or dusty road conditions, the fluid may need to be periodically changed. Check the service manual for the manufacturer's recommendations.

To change the transmission fluid, drive the vehicle to warm the fluid. Then raise the vehicle on a hoist. Make sure the car is level so that all of the fluid can drain out. Locate the oil drain plug in the bottom of the transmission case or extension housing. Wipe the area around the plug and remove it. Catch the fluid in a catch pan positioned below the hole. Let the transmission drain completely. The fluid is normally very thick and it takes some time to drain it all out.

Inspect the drained fluid for gold color metallic and other particles. The gold color particles come from the brass blocking rings of the synchronizers. Metal shavings are typically from the wearing of gears. After the fluid is drained, take a small magnet and insert it into the drain hole. Sweep the magnet around the inside to remove all metal particles. Because brass is not magnetic, the magnet will not attract brass particles. To remove brass shavings, insert a small brush or the end of a rag. Be careful to pull the shavings out, not push them in. Large amounts of metal particles indicate severe problems.

Before refilling the transmission, reinstall the drain plug with a new washer. Some manufacturers recommend that a sealer be used on the plug; check the service manual. Remove the filler plug, which is normally located above the drain plug; Check the service manual to identify the proper type and quantity of fluid for that transmission. Normally the case should be filled until the oil just starts to run out the filler hole or until it is at the bottom of the bore. Reinstall the plug with a new washer. Check the transmission housing vent to make sure it is not blocked with dirt. If the case is not properly vented, the fluid can easily break down and the pressure buildup can cause leaks. Make sure you fill the transmission with the correct type and amount of fluid. Too much or too little fluid can destroy a transmission.

IN-VEHICLE SERVICE

Much service and maintenance work can be done to transmissions while they are in the car. Only when a complete overhaul or clutch service is necessary does the transmission need to be removed from the car. The following are procedures for common service operations: the replacement of a rear oil seal and bushing, linkage adjustments, and replacement of the back-up light switch and the speedometer cable retainer and drive gear.

Rear Oil Seal and Bushing Replacement

Procedures for the replacement of the rear oil seal and bushing on a transmission vary little with each car model. Typically, to replace the rear bushing and seal, follow these steps:

PROCEDURE

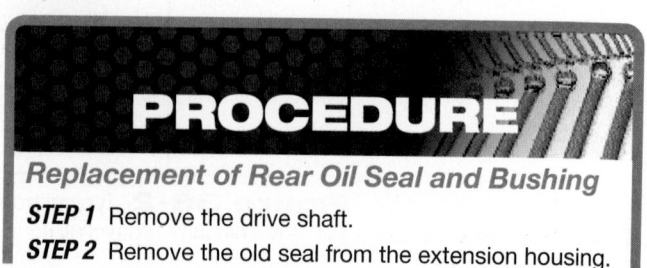

Replacement of Rear Oil Seal and Bushing

STEP 1 Remove the drive shaft.

STEP 2 Remove the old seal from the extension housing.

STEP 3 Insert the appropriate puller tool into the extension housing until it grips the front side of the bushing.

STEP 4 Pull the bushing from the housing **(Figure 38–4)**.

STEP 5 Drive a new bushing into the extension housing.

STEP 6 Lubricate the lip of the seal, then install the new seal in the extension housing **(Figure 38–5)**.

STEP 7 Install the drive shaft.

Linkage Adjustment

Transmissions with internal linkage have no provision for adjustments. However, external linkages, both floor and column mounted, can be adjusted. Linkages are adjusted at the factory, but worn parts may make adjustments necessary. Also, after a transmission has been disassembled, the shift lever and other controls may need adjustment.

Only externally controlled gearshift levers and linkages can be adjusted. To begin the adjustment procedure, raise the car and support it on jack stands. Then follow the procedure given in your service manual.

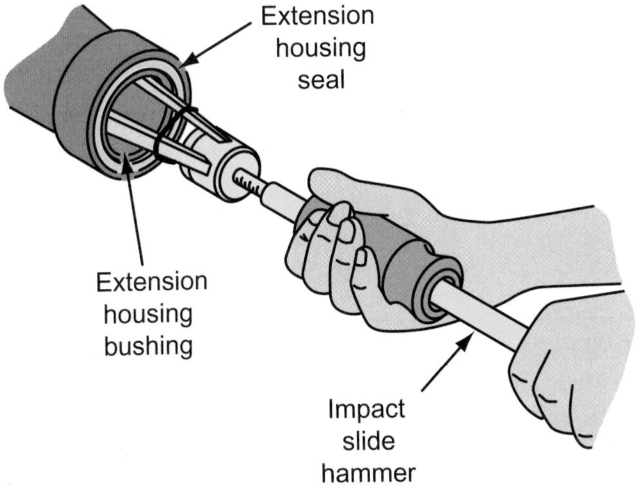

Figure 38–4 Removing the extension housing's seal and bushing.

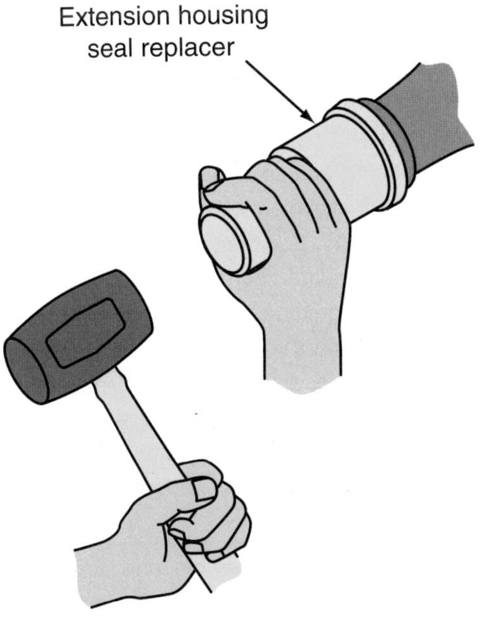

Figure 38–5 Drive the new seal in place with a hammer and seal driver.

Back-Up Light Switch Service

To replace the back-up light switch, disconnect the electrical lead to the switch. Put the transmission into reverse gear and remove the switch. Never shift the transmission until a new switch has been installed. To prevent fluid leaks, wrap the threads of a new back-up light switch with Teflon tape in a clockwise direction before installing it. Tighten the switch to the correct torque and reconnect the electrical wire to it.

Speedometer Drive Gear Service

Begin to remove the speedometer cable retainer and drive gear by cleaning off the top of the speedometer cable retainer. Then remove the holddown screw that keeps the retainer in its bore. Carefully pull up on the speedometer cable, pulling the speedometer retainer and drive gear assembly from its bore. Unscrew the speedometer cable from the retainer.

To reinstall the retainer, lightly grease the O-ring on the retainer and gently tap the retainer and gear assembly into its bore while lining the groove in the retainer with the screw hole in the side of the clutch housing case. Install the holddown screw and tighten it in place.

DIAGNOSING PROBLEMS

Service manuals list the most common problems associated with manual transmissions and transaxles. Proper diagnosis involves locating the exact source of the problem. Many problems that seem transmission/transaxle related may actually be caused by problems in the clutch driveline or differential. Check these areas along with the transmission/transaxle, particularly if you are considering removing the transmission/transaxle for service.

Table 38–1 is a troubleshooting chart for common transmission and transaxle problems.

Remember to begin all diagnostics with an interview of the customer. Then verify the customer's complaint or concern. Also, after repairs are made, make sure you verify the repair.

Visual Inspection

Visually inspect the transmission/transaxle at regular intervals. Perform the following checks:

1. Check for lubricant leaks at all gaskets and seals. The transmission rear seal at the driveline is particularly prone to leakage.

TABLE 38-1 TRANSMISSION/TRANSAXLE TROUBLESHOOTING CHART

Problem	Possible Cause	Remedy
Gear clash when shifting from one gear to another	1. Clutch adjustment incorrect 2. Clutch linkage or cable binding 3. Clutch housing misalignment 4. Lubricant level low or incorrect lubricant 5. Gearshift components or synchronizer blocking rings worn or damaged	1. Adjust the clutch. 2. Lubricate or repair as necessary. 3. Check runout at the rear face of the clutch housing. Correct runout. 4. Drain and refill the transmission/transaxle and check for lubricant leaks if the level was low. Repair as necessary. 5. Remove, disassemble, and inspect the transmission/transaxle. Replace worn or damaged components as necessary.
Clicking noise in any one gear range	1. Damaged teeth on input or intermediate shaft gears (transaxles) or damaged teeth on the countergear, cluster gear assembly, or output shaft gears (transmissions)	1. Remove, disassemble, and inspect the unit. Replace worn or damaged components as necessary.
Does not shift into one gear	1. Gearshift internal linkage or shift rail assembly worn, damaged, or incorrectly assembled 2. Shift rail detent plunger worn, spring broken, or plug loose 3. Gearshift lever worn or damaged 4. Synchronizer sleeves or hubs damaged or worn	1. Remove, disassemble, and inspect the transmission/transaxle cover assembly. Repair or replace components as necessary. 2. Tighten the plug or replace worn or damaged components as necessary. 3. Replace the gearshift lever. 4. Remove, disassemble, and inspect the unit. Replace worn or damaged components.
Locked in one gear—cannot be shifted out of that gear	1. Shift rails worn or broken, shifter fork bent, setscrew loose, center detent plug missing or worn 2. Broken gear teeth on countershaft gear input shaft, or reverse idler gear 3. Gearshift lever broken or worn, shift mechanism in cover incorrectly assembled or broken, worn or damaged gear train components	1. Inspect and replace worn or damaged parts. 2. Inspect and replace the damaged part. 3. Disassemble the transmission/transaxle. Replace the damaged parts of the assembly correctly.
Slips out of gear	1. Clutch housing misaligned 2. Gearshift offset lever nylon insert worn or lever attachment nut loose 3. Gearshift mechanisms, shift forks, shift rail, detent plugs, springs, or shift cover worn or damaged 4. Clutch shaft or roller bearings worn or damaged 5. Gear teeth worn or tapered, synchronizer assemblies worn or damaged, excessive end play caused by worn thrust washers or output shaft gears 6. Pilot bushing worn	1. Check runout at the rear face of the clutch housing. 2. Remove the gearshift lever and check for loose offset lever nut or worn insert. Repair or replace as necessary. 3. Remove, disassemble, and inspect the transmission cover assembly. Replace worn or damaged components as necessary. 4. Replace the clutch shaft or roller bearings as necessary. 5. Remove, disassemble, and inspect the transmission/transaxle. Replace worn or damaged components as necessary. 6. Replace the pilot bushing.
Vehicle moving—rough growling noise isolated in transmission/transaxle and heard in all gears	1. Intermediate shaft front or rear bearings worn or damaged (transaxle) or output shaft rear bearing worn or damaged (transmission)	1. Remove, disassemble, and inspect the transmission/transaxle. Replace damaged components as necessary.

TABLE 38-1 *(Continued)*

Problem	Possible Cause	Remedy
Rough growling noise when engine operating with transmission/ transaxle in neutral	1. Input shaft front or rear bearings worn or damaged (transaxle) or input shaft bearing, countergear, or countershaft bearings worn or damaged (transmission)	1. Remove, disassemble, and inspect the transmission/transaxle. Replace damaged components as necessary.
Vehicle moving—rough growling noise in transmission—noise heard in all gears except direct drive	1. Output shaft pilot roller bearings	1. Remove, disassemble, and inspect the transmission. Replace damaged components as needed.
Transmission/transaxle shifts hard	1. Clutch adjustment incorrect	1. Adjust the clutch.
	2. Clutch linkage binding	2. Lubricate or repair as necessary.
	3. Shift rail binding	3. Check for mispositioned roll pin, loose cover bolts, worn shift rail bores, worn shift rail, distorted oil seal, or extension housing not aligned with the case. Repair as necessary.
	4. Internal bind in transmission/transaxle caused by shift forks, selector plates, or synchronizer assemblies	4. Remove, disassemble, and inspect the unit. Replace worn or damaged components as necessary.
	5. Clutch housing misalignment	5. Check runout at the rear of the clutch housing. Correct runout.
	6. Incorrect lubricant	6. Drain and refill.

2. Check the case body for signs of porosity that show up as leakage or seepage of lubricant.
3. Push up and down on the unit. Watch the transmission mounts to see if the rubber separates from the metal plate. If the case moves up but not down, the mounts require replacement.
4. Move the clutch and shift linkages around and check for loose or missing components. Cable linkages should have no kinks or sharp bends, and all movement should be smooth.
5. Transaxle drive axle boots should be checked for cracks, deformation, or damage.
6. The constant velocity joints on transaxle drive axles should be thoroughly inspected.

Transmission Noise

Most manual transmission/transaxle complaints center around noise in the unit. Once again, be certain the noise is not coming from other components in the drivetrain. Unusual noises may also be a sign of trouble in the engine or transmission mounting system. Improperly aligned engines, improperly torqued mounting bolts, damaged or missing rubber mounts, cracked brackets, or even a stone rattling around inside the engine compartment can create noises that appear to be transmission/transaxle related.

SHOP TALK

If during the test drive you hear a noise you suspect is coming from inside the transmission/ transaxle, bring the vehicle to a stop. Disengage the clutch. If the noise stops with the engine at idle and the clutch disengaged, the noise is probably inside the unit.

Once you have eliminated all other possible sources of noise, concentrate on the transmission/ transaxle unit. Noises from inside the transmission/ transaxle may indicate worn or damaged bearings, gear teeth, or synchronizers. A noise that changes or disappears in different gears can indicate a specific problem area in the transmission.

CAUTION!

When the transmission/transaxle is in gear and the engine is running, the driving wheels and related parts turn. Avoid touching moving parts. Severe physical injury can result from contact with spinning drive axles and wheels.

The type of noise detected may also help indicate the problem.

Rough, Growling Noise This noise can be a sign of several problems in a transaxle or transmission depending on when it occurs. If the noise occurs when the transaxle is in neutral and the engine running, the problem may be the input shaft roller bearings. The input shaft is supported on either end by tapered roller bearings, and these are the only bearings in operation when the transaxle is in neutral. In its early stages, the problem should not cause operational difficulties; but left uncorrected, it grows worse until the bearing race or rolling element fractures. Solving the problem involves transaxle disassembly and bearing replacement.

When the vehicle is moving, both the input and mainshaft (output shaft) are turning in the transaxle. If the noise occurs in forward and reverse gears, but not in neutral, the output or mainshaft bearings are the likely failed component.

If the rough growling noise occurs when the engine is running, the clutch engaged, and the transmission in neutral, the front input shaft bearing is likely at fault. Rough growling when the vehicle is moving in all gears indicates faulty countergear bearings or countershaft-to-cluster assembly needle bearings. If the problem occurs in all gears except direct drive, the bearing at the rear of the transmission input shaft may be at fault. This bearing supports the pilot journal at the front of the transmission output shaft. In all forward gears except direct drive, the input shaft and output shaft turn at two different speeds. In reverse, the two shafts turn in opposite directions. In direct drive, the two shafts are locked together and this bearing does not turn. If the growling noises stop during direct drive operation, the rear input shaft bearing may be at fault. Disassembly, inspection, and replacement of damaged parts is needed.

Clicking or Knocking Noise Normally, helical gears are quiet because the gear teeth are constantly in contact. (When spur cut gear teeth are used for reverse gearing, clicking or a certain amount of **gear whine** is normal, particularly when backing up at faster speeds.)

Clicking or whine in forward gear ranges may indicate worn helical gear teeth. This problem may not require immediate attention.

Chipped or broken teeth are dangerous because the loose parts can cause severe damage in other areas of the transmission/transaxle. Broken parts are usually indicated by a rhythmical knocking sound, even at low speeds. Complete disassembly, inspection, and replacement of damaged parts is the solution to this problem.

Gear Clash

Gear clash is indicated by a grinding noise during shifting. The noise is the result of one gearset remaining partly engaged while another gearset attempts to turn the mainshaft. Gear clash can be caused by incorrect clutch adjustment or binding of clutch or gearshift linkage. Damaged, worn, or defective synchronizer blocking rings can cause gear clash, as can use of an improper gear lubricant.

Hard Shifting

If the shift lever is difficult to move from one gear to another, check the clutch linkage adjustment. Hard shifting may also be caused by damage inside the transmission/transaxle, or by a lubricant that is too thick. Common hard shifting includes badly worn bearings and damaged clutch gears, control rods, shift rails, shift forks, and synchronizers.

Jumping Out of Gear

If the car jumps out of gear into neutral, particularly when decelerating or going down hills, first check the shift lever and internal gearshift linkage. Excessive clearance between gears and the input shaft or badly worn bearings can cause jumping out of gear. Other internal transmission/transaxle parts to inspect are the clutch pilot bearing, gear teeth, shift forks, shift rails, and springs or detents.

Locked in Gear

If a transmission or transaxle locks in one gear and cannot be shifted, check the gearshift lever linkage for misadjustment or damage. Low lubricant level can also cause needle bearings, gears, and synchronizers to seize and lock up the transmission.

If these checks do not resolve the problem, the transmission or transaxle must be removed from the vehicle and disassembled. After disassembly, inspect the internal countergear, clutch shaft, reverse idler, shift rails, shift forks, and springs or detents for damage. Also, check for worn support bearings.

If the problem seems to be in the clutch assembly, make sure the transmission/transaxle is out of gear, set the parking brake, and start the engine. Increase the engine speed to about 1,500 to 2,000 rpm and gradually apply the clutch until the engine torque causes tension at the drive train mounts. Watch the torque reaction of the engine. If the engine's reaction to the torque appears to be excessive, broken or worn drive train mounts may be the cause and not the clutch.

The engine mounts on FWD cars are important to the operation of the clutch and transaxle (**Figure 38–6**). Any engine movement may change the effective length of the shift and clutch control cables and therefore may affect the engagement of the clutch and/or gears. A clutch may slip due to clutch linkage

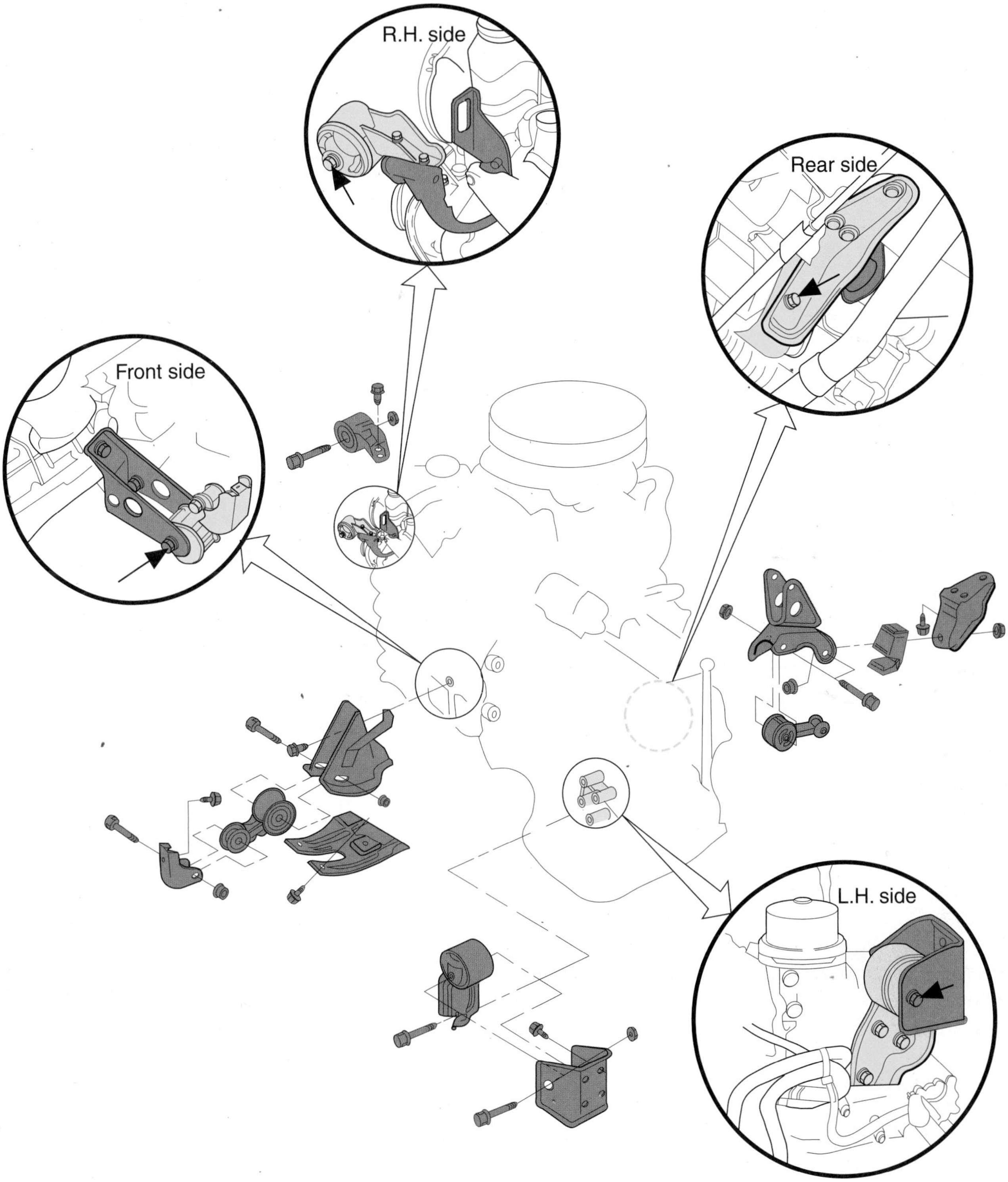

Figure 38-6 Typical engine and transaxle mounts. *Copyright, Nissan (2008). Nissan and the Nissan logo are registered trademarks of Nissan.*

changes as the engine pivots on its mounts. To check the condition of the transaxle mounts, pull up and push down on the transaxle case while watching the mount. If the mount's rubber separates from the metal plate or if the case moves up but not down, replace the mount. If there is movement between the metal plate and its attaching point on the frame, tighten the attaching bolts to an appropriate torque.

If it is necessary to replace the transaxle mount, make sure you follow the procedure for maintaining the alignment of the drive line. Some manufacturers recommend that a holding fixture or special bolt be used to keep the unit in its proper location. A broken clutch cable may be caused by worn mounts and improper cable routing. Inspect all clutch and transaxle linkages and cables for kinks or stretching. Often transaxle problems can be corrected by replacing or repairing the clutch or gearshift cables and linkage.

Shift Linkage

Check the shift linkage for smooth gear changes and full travel. If the linkage cannot move enough to fully engage a gear, the transmission/transaxle will jump out of gear while it is under a load. Some FWD cars have experienced the problem of jumping out of second or fourth gear. This can be caused by improper shifter-to-shifter boot clearances. This prevents the shift forks from moving far enough to fully engage the synchronizer collars to their mating gears. If this is not the cause, check the engine and transmission mounts. Apparent linkage problems can actually be internal transaxle problems.

USING SERVICE INFORMATION

A service manual is absolutely necessary when performing any type of transmission/transaxle disassembly work. Not only does the manual clearly illustrate all components and their disassembly procedure, it also lists many vital specifications, such as shaft and gear thrust (side) clearances, synchronizer ring and cone clearances, and bolt torque values. Special service tools, such as transmission service stands, oil seal presses, bearing replacers, shaft removers, pullers, and installing tools are also illustrated and explained.

TRANSMISSION/TRANSAXLE REMOVAL

Removing the transmission from a RWD vehicle is generally more straightforward than removing one from a FWD model, because there is typically one cross member, one drive shaft, and easy access to the cables, wiring, and bell housing bolts. Transmissions in FWD cars, because of their limited space, can be more difficult to remove as you may need to disassemble or remove large assemblies such as engine cradles, suspension components, brake components, splash shields, or other pieces that would not usually affect RWD transmission removal. The engine may also need to be supported with fixtures while removing the transmission.

RWD Vehicles

The correct procedure for removing a transmission varies with each year, make, and model of vehicle; always refer to the service manual for the correct procedure. Normally the procedure begins with placing the vehicle on a hoist.

Once the vehicle is in position, disconnect the negative battery cable and place it away from the battery. Carefully check under the hood to identify anything that may interfere with transmission removal. Then raise the vehicle and disconnect the parts of the exhaust system that may get in the way. Disconnect all electrical connections and the speedometer cable at the transmission. Make sure you place these away from the transmission so they are not damaged during transmission removal or installation.

Place a drain pan under the transmission and drain the transmission's fluid. Then move the drain pan to the rear of the transmission. Before removing the drive shaft, use chalk to show the alignment of the rear U-joint and the pinion flange (**Figure 38–7**). Then remove the drive shaft.

Disconnect and remove the transmission linkage. It is best to do this by disconnecting as little as possible.

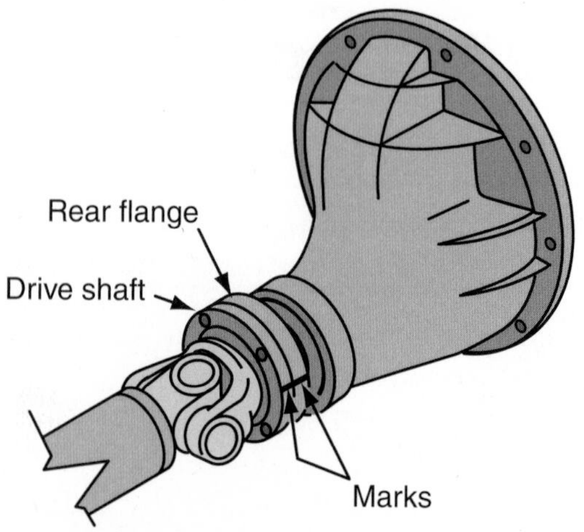

Figure 38-7 To ensure proper balance and phasing of the drive shaft, make alignment marks on the rear flange and the rear yoke.

Place a transmission jack under the transmission and secure the transmission to it. Then loosen and remove the lower bell housing-to-engine block bolts and the cross member at the transmission. After the mount is free from the transmission, lower the transmission slightly so you can easily access the top transmission-to-engine bolts. Loosen and remove the remaining transmission-to-engine bolts.

Slowly and carefully move the transmission away from the engine until the input shaft is out of the clutch assembly. Then slowly lower the transmission. Once the transmission is out of the vehicle, carefully move it to the work area and mount it to a stand or bench.

On some cars, the engine and transmission must be removed as a unit. The assembly is lifted with an engine hoist or lowered underneath the car.

FWD Vehicles

On some vehicles, the recommended procedure may include removing the engine with the transaxle. Always refer to the service manual before proceeding to remove the transaxle. Identify any special tool needs and precautions recommended by the manufacturer. You will waste much time and energy if you do not check the manual first.

Begin removal by placing the vehicle on a lift. Working under the hood, disconnect the battery before loosening any other components. Then disconnect all electrical connectors and the speedometer cable at the transaxle.

Now disconnect the shift linkage or cables and the clutch cable. Identify the transaxle-to-engine bolts that cannot be removed from under the vehicle and remove them. Install the engine support fixture to hold the engine in place while removing the transaxle. Disconnect and remove all items that will interfere with the removal of the transaxle.

Loosen the large nut that retains the outer CV joint, which is splined shaft to the hub. It is recommended that this nut be loosened with the vehicle on the floor and the brakes applied. This will make the job easier and reduces the chance of damaging the CV joints and wheel bearings.

Now raise the vehicle and remove the front wheels. Tap the splined CV joint shaft with a soft-faced hammer to see if it is loose. Most will come loose with a few taps. Some vehicles have an interference fit spline at the hub and you will need a special puller for this type of CV joint. The tool pushes the shaft out and on installation pulls the shaft back into the hub.

The lower ball joint must now be separated from the steering knuckle. The ball joint will either be bolted to the lower control arm or held in the knuckle with a pinch bolt. Once the ball joint is loose, the control arm can be pulled down and the knuckle can be pushed outward to allow the splined CV joint shaft to slide out of the hub. The inboard joint can be pried out or will slide out. Some transaxles have retaining clips that must be removed before the inner joint can be removed. Pull the drive axles out of the transaxle **(Figure 38–8)**.

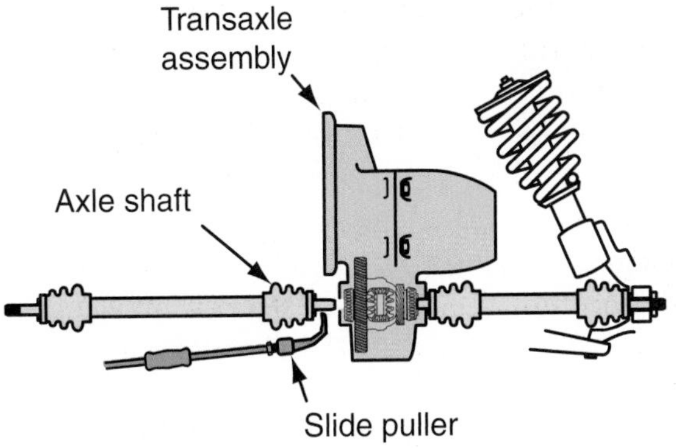

Figure 38-8 The inboard joints are normally pulled out of the transaxle.

While removing the axles, make sure the brake lines and hoses are not stressed. Suspend them with wire to relieve the weight on the hoses and to keep them out of the way.

On some cars, the inner CV joints have flange-type mountings. These must be unbolted for removal of the shafts. In some cases, the flange-mounted drive shafts may be left attached to the wheel and hub assembly and only unbolted at the transmission flange. The free end of the shafts should be supported and placed out of the way.

Now the remaining shift linkages, electrical connections, and speedometer cables should be disconnected. The exhaust system may also need to be lowered or partially removed.

Remove the starter. The starter's wiring may be left connected or you may remove the starter from the vehicle to get it totally out of the way.

With the transmission jack supporting the transmission, remove the transaxle mounts. If the car is equipped with an engine cradle that will separate, remove the half of the cradle that allows for transaxle removal. Then remove all remaining transaxle to engine bolts. Slide the transaxle away from the engine. Make sure the input shaft is out of the clutch assembly before lowering the transmission.

CLEANING AND INSPECTION

Disassembly and overhaul procedures can vary greatly between transmission/transaxle models, so always follow the exact steps outlined in the service manual.

Figure 38-9 Use a dial indicator to measure the end play of the shafts before disassembling the unit.

Clean the transmission/transaxle with a steam cleaner, degreaser, or cleaning solvent. As you begin to disassemble the unit, pay close attention to the condition of its parts. Using a dial indicator, measure and record the endplay of the input and main shafts **(Figure 38–9)**. This information will be needed during the reassembly of the unit for selecting the appropriate selective shims and washers.

SHOP TALK

Before disassembling a transmission, observe the effort it takes to rotate the input shaft through all the forward gears and reverse. Extreme effort in any or all gears may indicate an endplay or preload problem.

Remove the bell housing from the transmission case, extension housing **(Figure 38–10)**, and the side or top cover. The seal and bushing should be removed from the extension housing (tail shaft) prior to cleaning. With the housing and cover removed, the gears, synchronizers, and shafts are exposed and the shift forks can be removed.

Figure 38-10 The extension housing of a transmission needs to be removed before the rest of the unit is disassembled.

SHOP TALK

It is good practice to lay the parts on a clean rag as you remove them, and to keep them in order to aid you during reassembly.

Each transmission design has its own specific service procedures. Photo Sequence 35 guides you through the disassembly of a typical transaxle. Always refer to your service manual prior to overhauling a transmission or transaxle.

In some cases the countershaft must be removed before the input and mainshaft. In other cases, the mainshaft is removed with the extension housing. It may be removed through the shift cover opening. To avoid difficulty in disassembly, follow the recommended sequence. A **gear puller** or hydraulic press is often needed to remove gears and synchronizer assemblies from transmission/transaxle pinion shafts.

Bearing removal and installation procedures require that the force applied to remove or install the bearing should always be placed on the tight bearing race. In some cases, the inner race is tight on the shaft, while in others it is the outer race that is tight in its bore. Removal or installation force should be applied to the tight race. Serious damage to the bearing can result if this practice is not followed.

Use a soft-faced hammer or a brass drift and ball-peen hammer if tapping is required. Never use excessive force or hammering.

During assembly of the transmission, never attempt to force parts into place by tightening the front bearing retainer bolts or extension housing bolts. All parts must be fully in place before tightening any bolts. Check for free rotation and shifting. New gaskets and seals should always be used.

The following are some general cleaning and inspection guidelines that result in quality workmanship and service:

1. Wash all parts, except sealed ball bearings and seals, in solvent. Brush or scrape all dirt from the parts. Remove all traces of the old gasket. Wash the roller bearings in solvent; dry them with a clean cloth. Never use compressed air to spin the bearings.
2. Inspect the front of the transmission case for nicks or burrs that could affect its alignment with

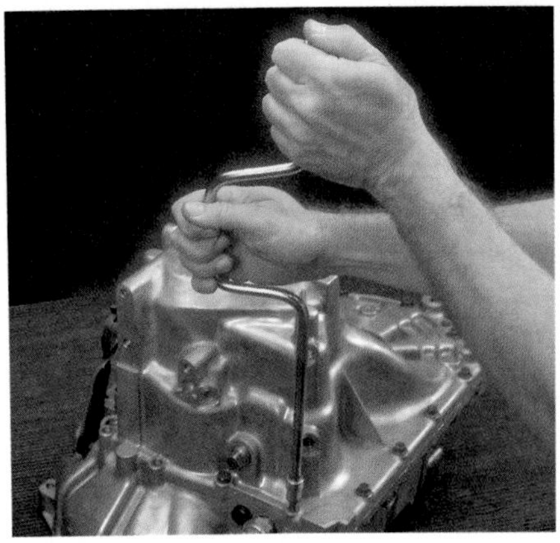

P35-1 *Place the transaxle into a suitable work stand. Remove the reverse idler shaft retaining bolt and detent plunger retaining screw. Then loosen and remove all transaxle case-to-clutch housing attaching bolts.*

P35-2 *Separate the housing from the case. If the housing is difficult to loosen, tap it with a soft mallet.*

P35-3 *Remove the C clip retaining ring from the fifth-gear shift-relay lever pivot pin.*

P35-4 *Remove the fifth-gear shift-relay lever, reverse idler shaft, and reverse idler gear from the case.*

P35-5 *Use a punch to drive the roll pin from the shift-lever shaft.*

P35-6 *Remove the shift-lever shaft by gently pulling on it.*

P35-7 *Remove the reverse kickdown spring assembly.*

P35-8 *Grasp the input and main shafts and lift them as an assembly from the case. Note the position of the shift forks as an aid when reinstalling them.*

P35-9 *Remove the fifth-gear shaft assembly and the fifth-gear shift fork assembly.*

Disassembly of a Typical Transaxle *(continued)*

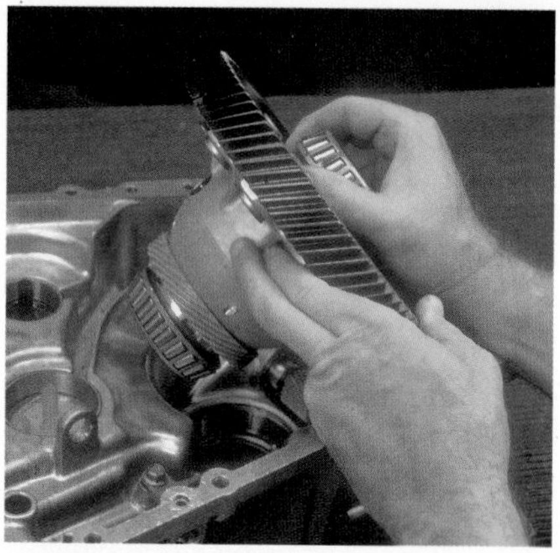

P35-10 *Remove the differential assembly from the case.*

P35-11 *Remove the bolts for the shift-relay lever support bracket, then remove the assembly.*

P35-12 *Carefully separate the shift rail and forks from the gears on the mainshaft.*

P35-13 *Remove the bearing at the fourth-gear end of the mainshaft.*

P35-14 *Slide the fourth gear from the shaft.*

P35-15 *Remove the synchronizer blocking ring from the assembly.*

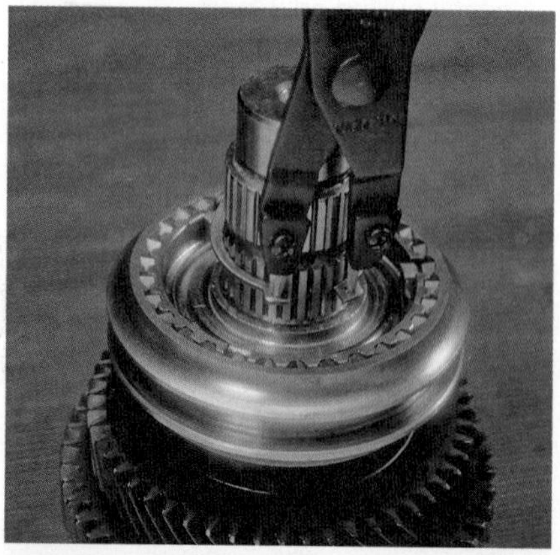

P35-16 *Remove the third- and fourth-gear synchronizer retaining ring.*

P35-17 *Lift the assembly off the shaft. Then remove the remaining gears and synchronizer assembly as a unit.*

P35-18 *Separate the synchronizer's hub, sleeve, and keys, noting their relative positions and scribing their location on the hub and the sleeve prior to separation.*

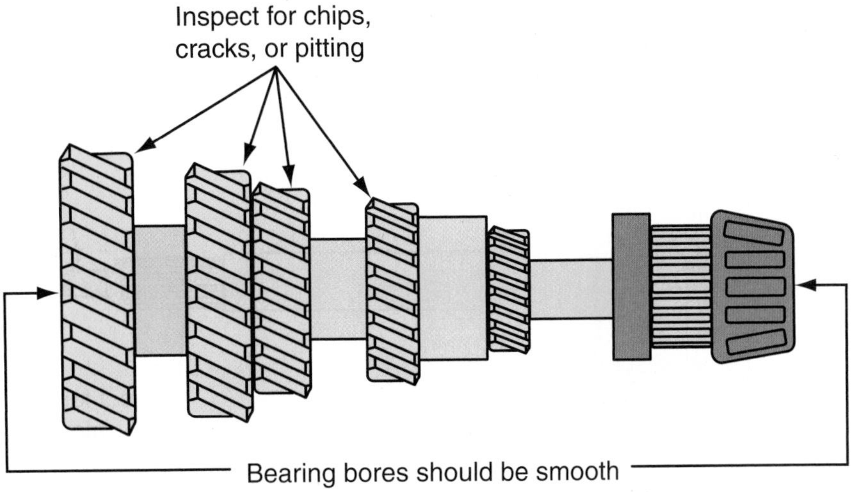
Inspect for chips, cracks, or pitting
Bearing bores should be smooth
Figure 38-11 Carefully inspect the countergear assembly.

the flywheel housing. Remove all nicks and burrs with a fine stone (cast-iron casing) or fine file (aluminum casing).

3. Replace any cover that is bent or distorted. If there are vent holes in the case, make certain they are open.
4. Inspect the seal and bushing in the extension housing. Measure the inside diameter of the bushing and compare that to specifications. Replace them if they are worn or damaged.
5. Inspect the ball bearings by holding the outer ring stationary and rotating the inner ring several times. Inspect the raceway of the inner ring from both sides for pits and spalling. Light particle indentation is acceptable wear, but all other types of wear merit replacement of the bearing assembly. Next, hold the inner ring stationary and rotate the outer ring. Examine the outer ring raceway for wear and replace as needed.
6. Examine the external surfaces of all bearings. Replace the bearings if there are radial cracks on the front and rear faces of the outer or inner rings, cracks on the outside diameter or outer ring, or deformation or cracks in the ball cage.
7. Lubricate the cleaned bearing raceways with a light coat of oil. Hold the bearing by the inner ring in a vertical position. Spin the outer ring several times by hand. If roughness or vibration is felt, or the outer ring stops abruptly, replace the bearing.
8. Replace any roller bearings that are broken, worn, or rough. Inspect their respective races. Replace them as needed.
9. Replace the counter (cluster) gear if its gear teeth are chipped, broken, or excessively worn **(Figure 38–11)**. Replace the countershaft if the shaft is bent, scored, or worn. Also, inspect the bore for the countershaft. If the bore is excessively worn or damaged, the needle bearings will not seat properly against the shaft.
10. Replace the reverse idler gear or sliding gear if its teeth are chipped, worn, or broken. Replace the idler gear shaft if it is bent, worn, or scored.
11. Replace the input shaft and gear if its splines are damaged or if the teeth are chipped, worn, or damaged **(Figure 38–12)**. If the roller bearing surface in the bore of the gear is worn or rough, or if the cone surface is damaged, replace the gear and the gear rollers.
12. Replace all main or speed gears that are chipped, broken, or worn **(Figure 38–13)**.
13. Check the synchronizer sleeves for free movement on their hubs **(Figure 38–14)**. Alignment marks (if present) should be properly indexed.
14. Inspect the synchronizer blocking rings for widened index slots, rounded clutch teeth, and smooth internal surfaces. Remember, the blocking rings must have machined grooves on their internal surfaces to cut through lubricant **(Figure 38–15)**. Units with worn, flat grooves must be replaced. Also, check the clearance between the block ring and speed gear clog teeth against service manual specifications **(Figure 38–16)**.
15. Replace the speedometer drive gear if its teeth are stripped or damaged. Install the correct size replacement gear.
16. Replace the output shaft if there is any sign of wear or runout or if any of the splines are damaged.

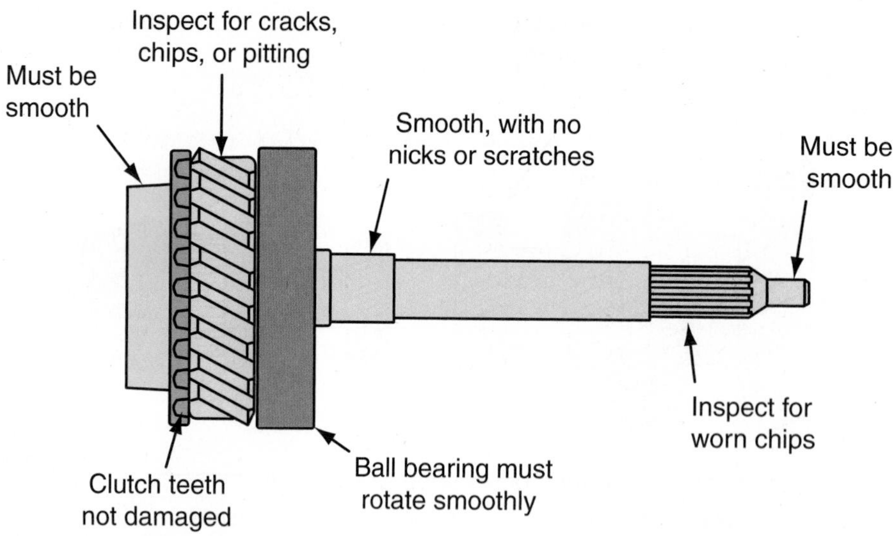
Inspect for cracks, chips, or pitting
Must be smooth
Smooth, with no nicks or scratches
Must be smooth
Inspect for worn chips
Ball bearing must rotate smoothly
Clutch teeth not damaged

Figure 38-12 The input shaft, including the splines, should also be carefully inspected.

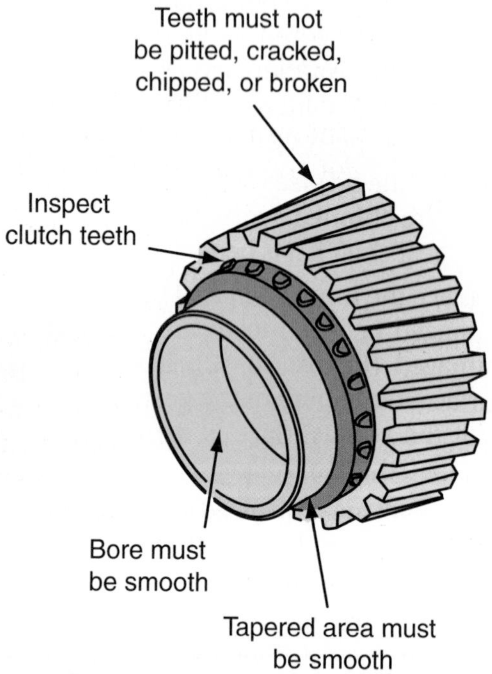
Teeth must not be pitted, cracked, chipped, or broken
Inspect clutch teeth
Bore must be smooth
Tapered area must be smooth

Figure 38-13 Every gear should be checked.

Figure 38-14 The movement of each synchronizer unit should be checked.

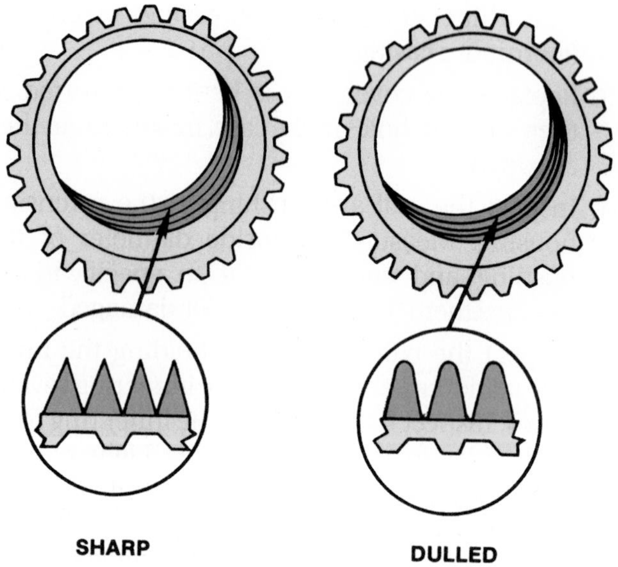
SHARP
DULLED

Figure 38-15 Grooves on the internal surface of the synchronizer blocker ring must be sharp.

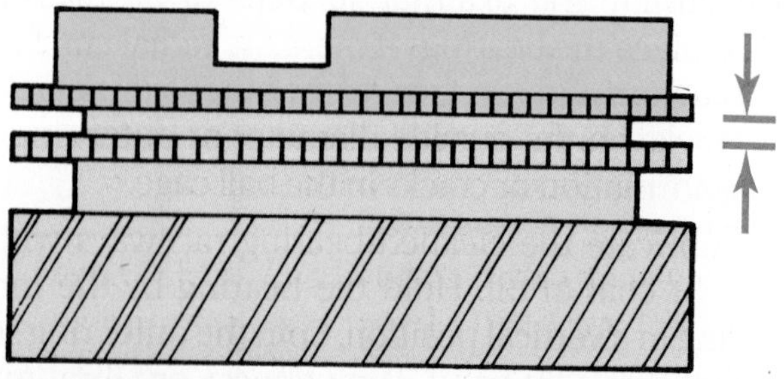

Figure 38-16 The clearance between the synchronizer blocker ring and the gear's clutching teeth must meet specifications.

Aluminum Case Repair

Normally, the case is replaced if it is cracked or damaged. However some manufacturers recommend the use of an epoxy-based sealer on some types of leaks in some

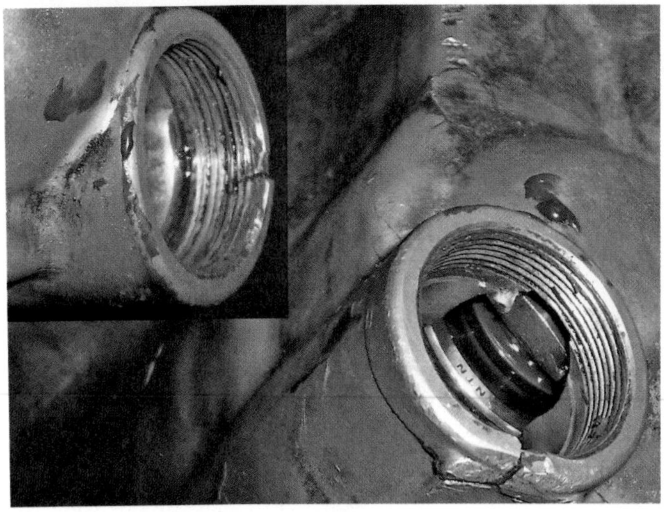

(A)

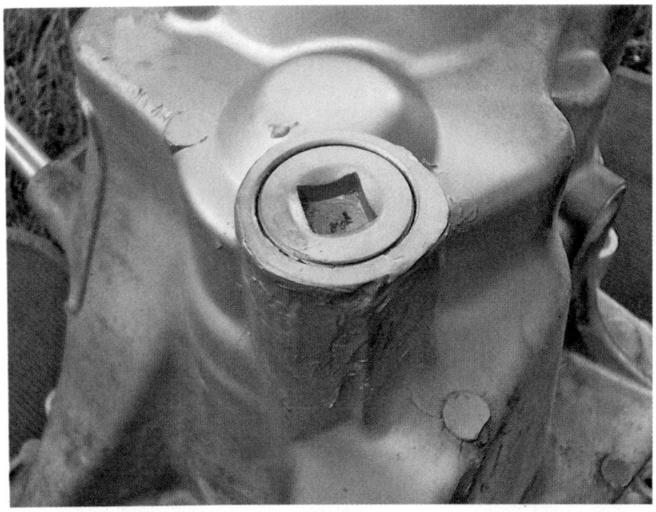

(B)

Figure 38-17 (A) A crack around the filler plug opening. (B) The crack was repaired with epoxy.

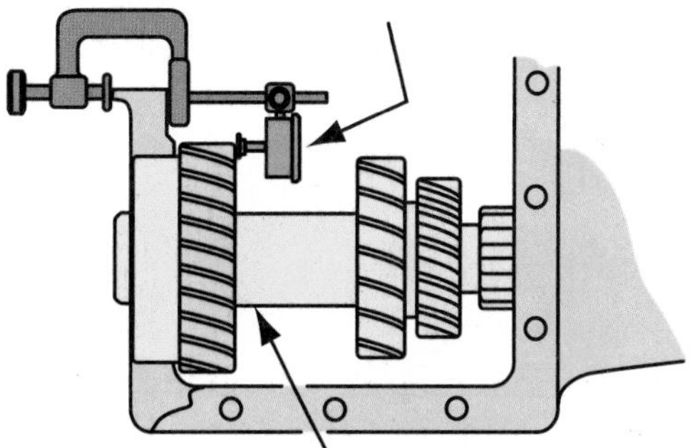

Figure 38-18 A typical setup for checking countershaft end play.

locations on the transmission **(Figure 38–17)**. Refer to the manufacturer's recommendations before attempting to repair a crack or correct for porosity leaks.

If a threaded area in an aluminum housing is damaged, helicoil-type service kits can be used to insert new threads in the bore. Some threads should never be repaired; check the service manual to identify which ones can be repaired.

After all parts are inspected and the defective parts replaced, you can begin to reassemble the transmission/transaxle. While you are doing so, coat all parts with gear lube.

SHOP TALK

If the transmission/transaxle is fitted with paper-type blocking rings, soak them in ATF prior to installing them.

Many late-model transmissions and transaxles have specifications for end play, backlash, and preload; make sure these specifications are met. Follow the procedures given in the service manual for the particular transmission/transaxle being worked on **(Figure 38–18)**. For most transmissions, there are specifications for the end play and preload of the input shaft, the countershaft, and the differential. These are usually set by shims under the bearing caps. Reuse the original shims, if possible.

Specific repair and assembly instructions will vary from transaxle to transaxle and from transmission to transmission. Therefore, before beginning to reassemble the unit, gather the specific information about the unit you are working on. Photo Sequence 36 shows the reassembly of a typical transaxle.

DISASSEMBLY AND REASSEMBLY OF THE DIFFERENTIAL CASE

Although it is a part of the transaxle, the differential is often kept together while making a repair to the transmission part of the transaxle. The differential case normally can be removed as soon as the transaxle case has been separated. It may be the source of the problem and be the only part of the transaxle that needs service. Therefore, the disassembly and reassembly of the differential is set aside from the procedures listed for the transaxle **(Figure 38–19)**.

Begin the disassembly by separating the ring gear from the differential case. Then remove the pinion shaft lock bolt. Remove the pinion shaft, then remove the gears and thrust washers from the case. If the differential side bearings are to be replaced, use a puller to remove the bearings. Use the correct installer for reinstallation of the side bearings.

P36-1 *Lightly oil the parts of the synchronizer.*

P36-2 *Assemble the synchronizer assemblies, being careful to align the index marks made during disassembly.*

P36-3 *Install the synchronizer assemblies onto the mainshaft.*

P36-4 *Install fourth gear and its bearing. The bearing may need to be lightly pressed into position.*

P36-5 *Install and tighten the shift-relay lever support bracket.*

P36-6 *Install the differential assembly into the transaxle case.*

P36-7 *Install the fifth-gear shaft assembly and fork shaft into the case.*

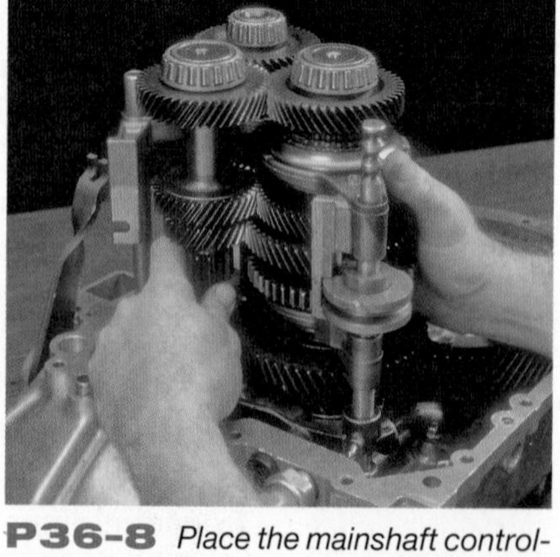

P36-8 *Place the mainshaft control-shaft assembly on the mainshaft so that the shift forks engage in their respective slots in the synchronizer sleeves. Then install the mainshaft assembly.*

P36-9 *Properly position the shift lever, bias and kickdown springs.*

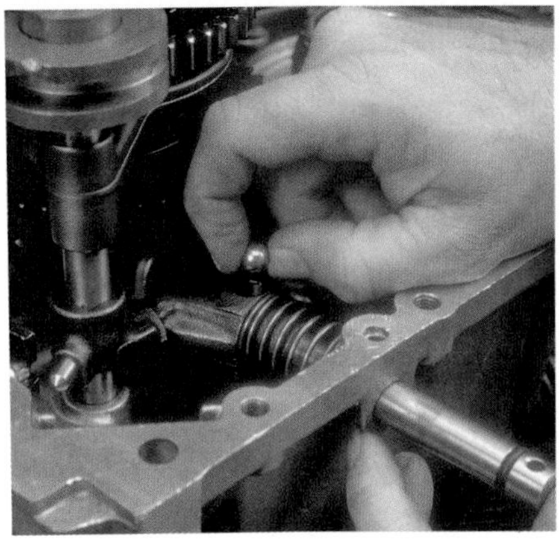

P36-10 *Install the inhibitor spring and ball in the fifth and reverse-inhibitor shaft hole.*

P36-11 *Depress the inhibitor ball and spring using a drift, and slide the shift-lever shaft through the shift lever. Then tap the shaft into its bore in the clutch housing.*

P36-12 *Install the reverse idler gear shaft and gear into the appropriate bore in the case.*

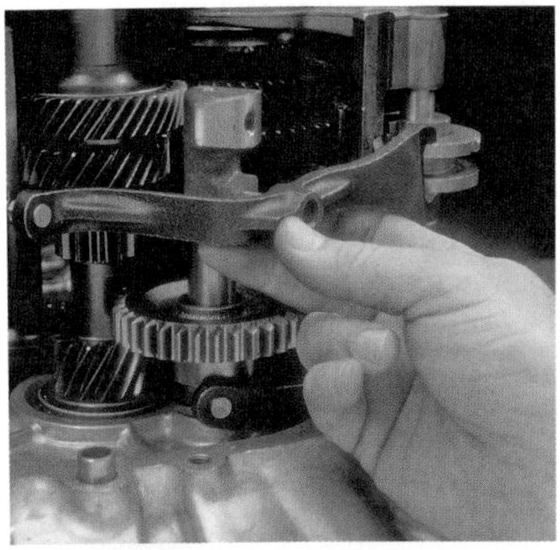

P36-13 *Install the fifth-gear relay shaft and align it with the fifth-gear fork slot and interlock spring.*

P36-14 *Install the retaining clip onto the fifth-gear relay shaft.*

P36-15 *Apply a thin bead of anaerobic sealant on the case's mating surface for the clutch housing.*

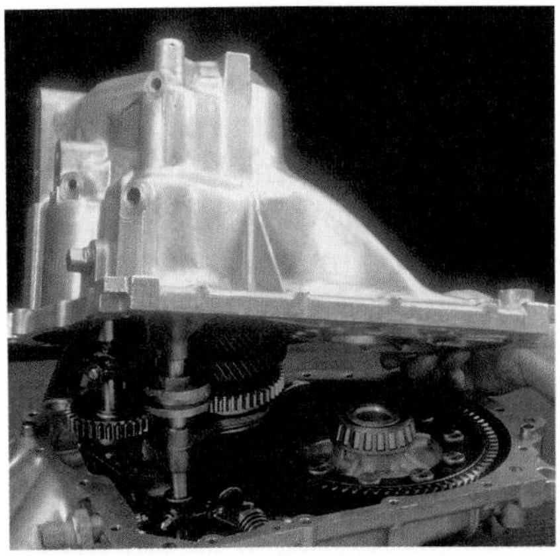

P36-16 *Install the clutch housing to the case. Be careful that the mainshaft, shift-control shaft, and fifth-gear shaft align with the bores in the case.*

P36-17 *After the housing and case are fit snugly together, tighten the attaching bolts to the specified torque and in their proper order.*

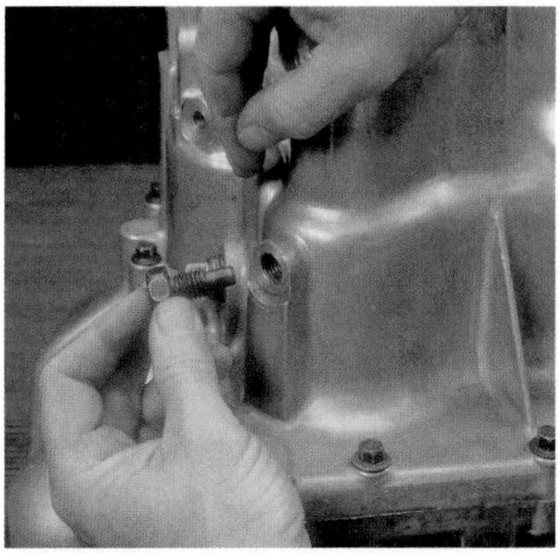

P36-18 *Install and tighten the reverse-idler shaft retaining bolt.*

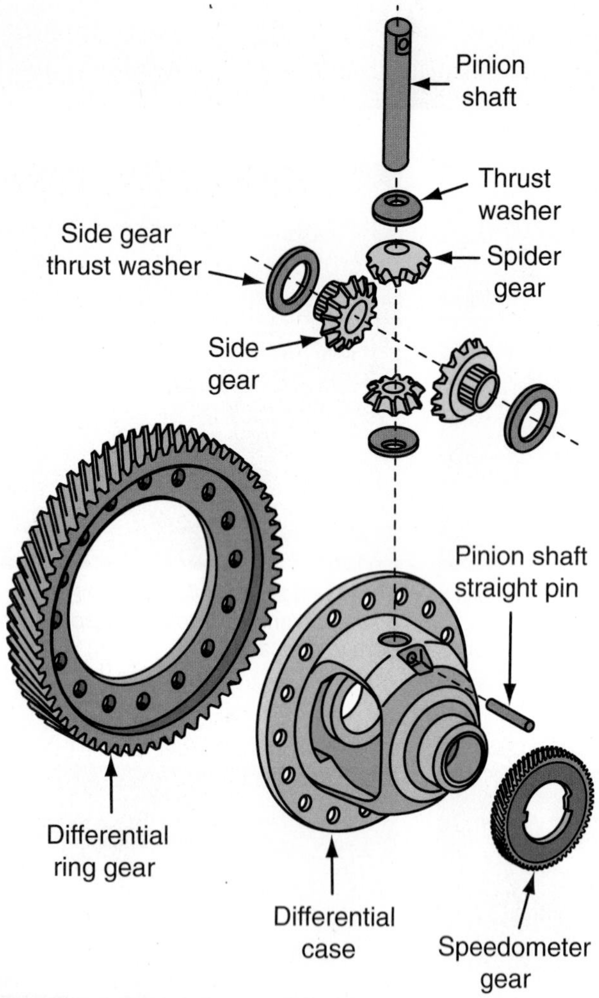

Figure 38-19 The differential assembly for a transaxle.

Clean and inspect all parts. Replace any damaged or worn parts. Install the gears and thrust washers into the case and install the pinion shaft and lock bolt. Tighten the bolt to the specified torque. Attach the ring gear to the differential case and tighten to the specified torque.

Shim Selection

While you are disassembling the differential or transaxle, make sure to keep all shims and bearing races together and identified for reinstallation in their original location. Carefully inspect the bearings for wear and/or damage and determine if a bearing should be replaced. Replacement tapered roller bearings will be available with a nominal thickness service shim. A nominal thickness service shim will handle reshimming the input shaft and output shaft bearings during normal repair.

When it is necessary to replace a bearing, race, or housing, refer to the manufacturer's recommendation for nominal shim thickness. If only other parts of the differential or transaxle are replaced, reuse the original shims. When repairs require the use of a service shim, discard the original shim. Never use the original shim together with the service shim. The shims must be installed only under the bearing cups at the transaxle case end of both the input and output shafts.

REASSEMBLY/REINSTALLATION OF TRANSMISSION/TRANSAXLE

Transmission/transaxle reassembly and reinstallation procedures are basically the reverse of disassembly. Once again, refer to the service manual for any special procedures. New parts are installed as needed, and new gaskets and seals are always used.

Serviceable gears are pressed onto the main shaft using special press equipment. Separate needle bearings should be held in place with petroleum jelly so shafts can be inserted into place. During reassembly, measure shaft end play. Adjust it to specifications with shims, spacers, or snaprings of different thicknesses. All fasteners are tightened to the manufacturer's torque specifications.

Soft-faced mallets can be used to tap shafts and other parts into place. After reassembly, secure the transmission to a transmission jack with safety chains and raise it into place. Before the transmission is reinstalled, inspect and service the clutch as necessary.

Installing the Transmission/Transaxle

SHOP TALK

Always check for free rotation of the transmission in all gears before installing in a vehicle. If the shafts do not rotate freely, identify the cause and correct it.

After the unit is together, install the clutch assembly, and a new throw-out bearing, prior to installing the transmission/transaxle. Generally, installation is the reverse procedure as removal. When installing the transmission, never let the transmission hang by its input shaft. Use a transmission jack to hold the transmission while you guide it into place. The input shaft should be lightly coated with grease prior to installation to aid installation and serve as a lubricant for the pilot bearing. Avoid putting too much grease on the shaft, because the excess may fly off and get on the clutch disc, causing it to slip and/or burn.

Most transmissions are doweled to the engine or bell housing. Use the dowels to locate and support the transmission during installation. Tighten the mounting bolts evenly, making sure nothing is caught between the housings. Then lightly coat the drive shaft's slip joint and carefully insert it into the extension housing to prevent possible damage to the rear oil seal. Reattach and adjust the shift linkage and fill the transmission with the proper fluid.

WARNING!

If the transmission/transaxle does not fit snugly against the engine block or if you cannot move the transmission into place, do not force it. Pull the transmission back and lower it. Inspect the input shaft splines for dirt or damage. Also check the mating surfaces for dirt or obstructions. If you try to force the transmission into place by tightening the bolts, you may break the case.

CASE STUDY

A customer complains that his rear-wheel-drive, manual transmission vehicle is experiencing intermittent operating noise, particularly when the vehicle is "just warming up." On the test drive, the technician notices a low growling noise in the lower gear ranges. The noise disappears at cruising speeds in high (fourth) gear.

The technician returns to the shop and places the vehicle on a lift for further inspection. The driveline, differential, clutch, and wheel bearings all appear to be in good condition. This confirms that the problem is transmission related. The technician suspects a damaged bearing at the rear of the input shaft. This bearing supports the pilot journal at the front of the transmission's output shaft. In all forward gear ranges except direct drive, the input and output shafts turn at different speeds. In reverse gear they turn in opposite directions; but in direct drive (fourth gear in this case), the two shafts are locked together by a synchronizer and turn at the same speed. This relieves the operating pressure placed on the input shaft rear bearing, thus eliminating the growling noise.

The only way this problem can be corrected is to disassemble the transmission and clean, examine, and replace any damaged components. The teardown confirms the technician's diagnosis. The roller bearings are cracked and disintegrating. The pilot journal on the output shaft is also slightly damaged. The shaft is sent to a machine shop where it is undercut and fitted with a press-on steel bushing to return it to the manufacturer's specified diameter. The roller bearings are then replaced and the transmission reassembled. While the transmission is apart, all the components are cleaned and examined closely for damage and wear. The oil-cutting grooves on the first/second synchronizer are dull and flat. Although the customer has not yet experienced jumping out of gear problems in these gear ranges, the technician shows the customer the worn synchronizer and explains the problem it could cause in the near future. He strongly suggests replacement at this time and the customer agrees to this additional work.

KEY TERMS

Fill plug
Gear clash
Gear puller
Gear whine

SUMMARY

- Proper lubrication is vital to long transmission/transaxle life. The transmission gear lubricant must be checked and changed at manufacturer's suggested intervals.
- Metal particles or shavings in the gear lubricant indicate extensive internal wear or damage.
- The first step in diagnosing transmission/transaxle problems is to confirm that the problem exists inside the transmission/transaxle. Clutch and driveline problems may often appear to be transmission/transaxle problems.
- The initial visual inspection should include checks for lubricant leakage at gaskets and seals, transmission mount inspection, clutch and gearshift linkage checks, and drive axle and CV joint inspection.

- Rough growling noise inside the transmission/transaxle housing is an indication of bearing problems.
- A clicking noise may indicate excessive gear tooth wear. Rhythmical knocking is a sign of loose or broken internal components.
- Hard shifting can be caused by shift linkage problems, improper lubricant, or worn internal components, such as bearings, gears, shift forks, or synchronizers.
- Jumping out of gear can be caused by misaligned drivetrain mounts, a worn or poorly adjusted shift linkage, excessive clearance between gears, or badly worn bearings.
- Low lubricant levels, poorly adjusted shift linkages, or damaged internal components can result in transmission/transaxle lockup.
- Always follow service manual recommendations for removing the transmission/transaxle from the vehicle and disassembling it.
- Use recommended bearing pullers, gear pullers, and press equipment to remove and install gears and synchronizers on shafts.
- Clean and inspect all parts carefully, replacing worn or damaged components. Never force components in place during reassembly. Follow all clearance specifications listed in the service manual.
- Always use new snaprings, gaskets, and seals during reassembly.

REVIEW QUESTIONS

1. After draining gear oil from a transaxle, the technician notices that the oil has shiny, metallic particles in it. What does this indicate?
2. List five checks that should be made during the visual inspection of the parts of a transmission/transaxle.
3. List three causes of noise that are not transmission related but may appear to be.
4. What tool is often required to remove gears and synchronizer assemblies from the mainshaft?
5. When removing or installing bearings, where should the force be applied?
6. List five common sources for transmission fluid leaks.
7. *True or False?* Bearing noise increases under load and is usually described as a growl that gets louder with speed.
8. A rough, growling noise that is heard from a transaxle while it is in neutral with the engine running, the vehicle stationary, and the clutch engaged is a likely indication that there is a problem in the ______.
   a. transaxle input shaft bearings
   b. transaxle main (intermediate) shaft bearings
   c. first/second synchronizer assembly
   d. pinion and ring gear interaction
9. A clicking noise during transmission/transaxle operation may be an indication of ______.
   a. worn mainshaft (input shaft) bearings
   b. faulty synchronizer operation
   c. failed oil seals
   d. worn, broken, or chipped gear teeth
10. Worn teeth on a speed gear can cause ______.
    a. gear clash
    b. hard shifting
    c. the transmission to shift into a gear
    d. the gear to jump out
11. Using a lubricant that is thicker than the specified lubricant can lead to ______.
    a. the gear to jump out
    b. hard shifting
    c. gear lockup
    d. gear slippage
12. What should you do to properly identify the type of transmission you are working on?
13. *True or False?* Some manufacturers recommend the use of heavy oil in the transaxle; others may recommend the use of automatic transmission fluid in the transaxle.
14. List at least five items that are typically removed when removing a transaxle.
15. A poorly adjusted shift linkage can cause which of the following problems?
    a. gear clash
    b. hard shifting
    c. the gear to jump out
    d. all of the above

ASE-STYLE REVIEW QUESTIONS

1. Technician A says that the transmission rear seal at the driveline is particularly prone to leakage. When Technician B pushes up and down on the

transmission, he says that the mounts require replacement because the case moves up and down. Who is correct?

a. Technician A
b. Technician B
c. Both A and B
d. Neither A nor B

2. A noise occurs in forward and reverse gears but not in neutral. Technician A says that the input shaft bearing may be bad. Technician B says that the mainshaft bearing may be bad. Who is correct?

a. Technician A
b. Technician B
c. Both A and B
d. Neither A nor B

3. A rough, growling noise occurs when a vehicle with a manual transmission is moving in any gear. Technician A says that the rear input shaft bearing may be at fault. Technician B says that this condition indicates the countergear bearings may be faulty. Who is correct?

a. Technician A
b. Technician B
c. Both A and B
d. Neither A nor B

4. A car jumps out of gear into neutral, particularly when decelerating or going down hills. Technician A checks the shift lever and external gearshift linkage first. Technician B says that the clutch pilot bearing could be the problem. Who is correct?

a. Technician A
b. Technician B
c. Both A and B
d. Neither A nor B

5. While diagnosing a noise from a transmission: Technician A says that the noise is caused by something internal if it is most noticeable during a test drive. Technician B says that the noise is caused by the clutch if it disappears when the clutch is disengaged. Who is correct?

a. Technician A
b. Technician B
c. Both A and B
d. Neither A nor B

6. Technician A says that broken or worn engine and transaxle mounts can cause a transaxle to have shifting problems. Technician B says that poor shift boot alignment can cause a transaxle to jump out of gear. Who is correct?

a. Technician A
b. Technician B
c. Both A and B
d. Neither A nor B

7. While inspecting a transaxle's gears: Technician A says that wear on the back of the gear teeth is normal. Technician B says that it is normal for a reverse idler gear to have small chips on its engagement side. Who is correct?

a. Technician A
b. Technician B
c. Both A and B
d. Neither A nor B

8. During a test drive, a noise that appears to be transmission related disappears when the driver brings the vehicle to a stop and disengages the clutch with the engine at idle. Technician A says that the noise may be caused by an internal transmission problem. Technician B says that the problem may be something interfering with the flywheel or pressure plate. Who is correct?

a. Technician A
b. Technician B
c. Both A and B
d. Neither A nor B

9. While inspecting the synchronizers of a transaxle: Technician A says that if the clutch teeth of the synchronizer are rounded, the synchronizer assembly must be replaced. Technician B says that the movement of the synchronizer sleeve on the shaft should be checked. Who is correct?

a. Technician A
b. Technician B
c. Both A and B
d. Neither A nor B

10. Technician A says that the proper fluid level for a transmission is normally to the bottom of the filler hole. Technician B says that low lubricant levels can cause a transmission to clash during shifting. Who is correct?

a. Technician A
b. Technician B
c. Both A and B
d. Neither A nor B

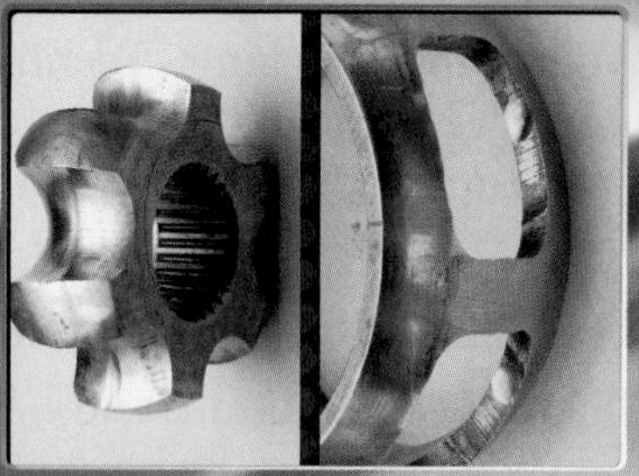

CHAPTER 39

DRIVE AXLES AND DIFFERENTIALS

OBJECTIVES

■ Name and describe the components of a front-wheel-drive axle. ■ Describe the operation of a front-wheel-drive axle. ■ Diagnose problems in CV joints. ■ Perform preventive maintenance on CV joints. ■ Explain the difference between CV joints and universal joints. ■ Name and describe the components of a rear-wheel-drive axle. ■ Describe the operation of a rear-wheel-drive axle. ■ Explain the function and operation of a differential and drive axles. ■ Describe the various differential designs, including complete, integral carrier, removable carrier, and limited slip. ■ Describe the three common types of driving axles. ■ Explain the function of the main driving gears, drive pinion gear, and ring gear. ■ Describe the operation of hunting, nonhunting, and partial nonhunting gears. ■ Describe the different types of axle shafts and axle shaft bearings.

The drive axle assembly transmits torque from the engine and transmission to drive the vehicle's wheels. The drive axle changes the direction of the power flow, multiplies torque, and allows different speeds between the two drive wheels. Drive axles are used for both front-wheel-drive and rear-wheel-drive vehicles.

FRONT-WHEEL-DRIVE (FWD) AXLES

Front-wheel-drive (FWD) axles, also called axle shafts, typically transfer engine torque from the transaxle's differential to the front wheels. One of the most important components of FWD axles is the constant velocity (CV) joint. These joints are used to transfer uniform torque at a constant speed, while operating through a wide range of angles.

On FWD or four-wheel-drive cars, operating angles of as much as 40 degrees are common (**Figure 39–1**). The drive axles must transmit power from the engine to front wheels that must drive, steer, and cope with the severe angles caused by the up-and-down movement of the vehicle's suspension. To accomplish this, these cars must have a compact joint that ensures the driven shaft is rotated at a constant velocity, regardless of angle. CV joints also allow the length of the axle assembly to change as the wheel travels up and down.

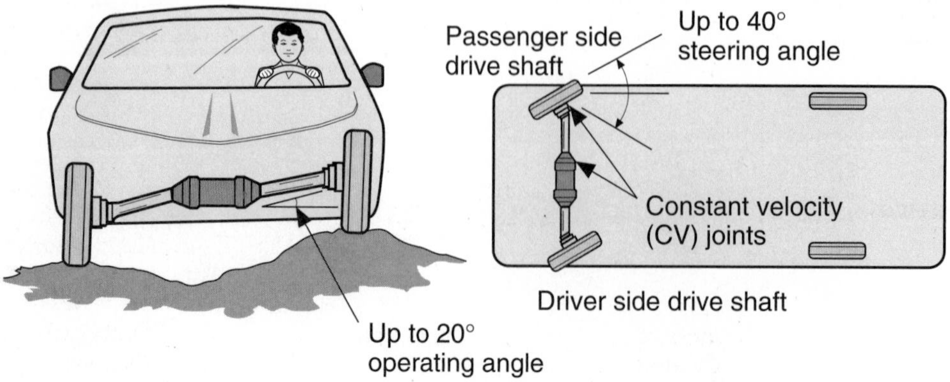

Figure 39–1 FWD drive axle shaft angles.

TYPES OF CV JOINTS

CV joints come in a variety of styles. The different types of joints can be referred to by position (inboard or outboard), by function (fixed or plunge), or by design (ball-type or tripod).

Inboard and Outboard Joints

On FWD vehicles, two CV joints are used on each half shaft (**Figure 39–2**). The joint nearer the transaxle is the inner or **inboard joint**, and the one nearer the wheel is the outer or **outboard joint**. In a RWD vehicle with independent rear suspension, the joint nearer the differential can also be referred to as the inboard joint. The one closer to the wheel is the outboard joint.

Fixed and Plunge Joints

CV joints are either a **fixed joint** (meaning it does not plunge in and out to compensate for changes in length) or a **plunge joint** (one that is capable of in-and-out movement).

In FWD applications, the inboard joint is a plunge joint. This joint allows for a change in the effective length of the axle by allowing it to move in and out on its connection to the transaxle's axle gear. The outboard joint is a fixed joint. The outboard joint must also be able to handle the much greater operating angles required for steering (up to 40 degrees).

In RWD applications with IRS, one joint on each axle shaft can be fixed and the other a plunge or both can be plunge joints. Because the wheels are not used for steering, the operating angles are not as great. Therefore, plunge joints can be used at either or both ends of the axle shafts.

Ball-Type Joints

There are two basic varieties of CV joints: the **ball-type** and **tripod-type joints**. Both types are used as either inboard or outboard joints, and both are available in fixed or plunge designs.

Fixed Ball-Type CV Joints The **Rzeppa joint**, or fixed ball-type joint, consists of an inner ball race, six balls, a cage to position the balls, and an outer housing (**Figure 39–3**). Tracks machined in the inner race and outer housing allow the joint to flex. The inner race and outer housing form a ball-and-socket arrangement. The six balls serve both as bearings between the races and the means of transferring torque from one to the other.

If viewed from the side, the balls within the joint always bisect the angle formed by the shafts on either side of the joint regardless of the operating angle. This reduces the effective operating angle of the joint by half and virtually eliminates all vibration problems. The input speed to the joint is always equal to the output velocity of the joint—thus the description "constant velocity." The cage helps to maintain this alignment by holding the six balls snugly in its windows. If the cage windows become worn or deformed over time, the resulting play between ball and window typically results in a clicking noise when turning. It is important to note that opposing balls in a Rzeppa CV joint always work together as a pair. Heavy wear in the tracks of one ball almost always results in identical wear in the tracks of the opposing ball.

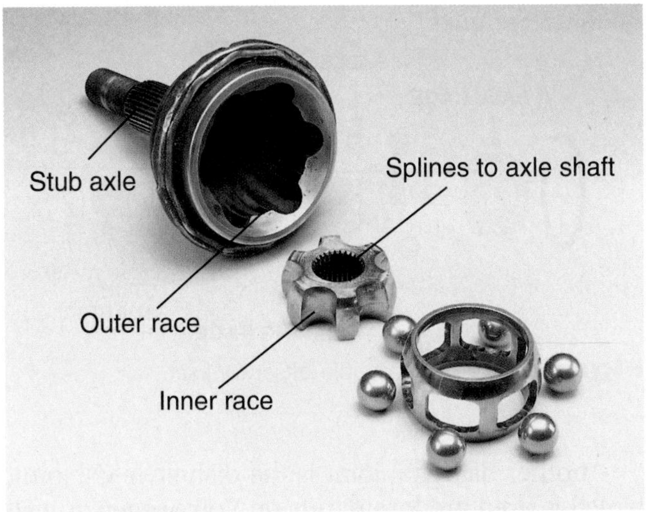

Figure 39-3 A Rzeppa ball-type fixed CV joint. *Courtesy of Federal-Mogul Corporation*

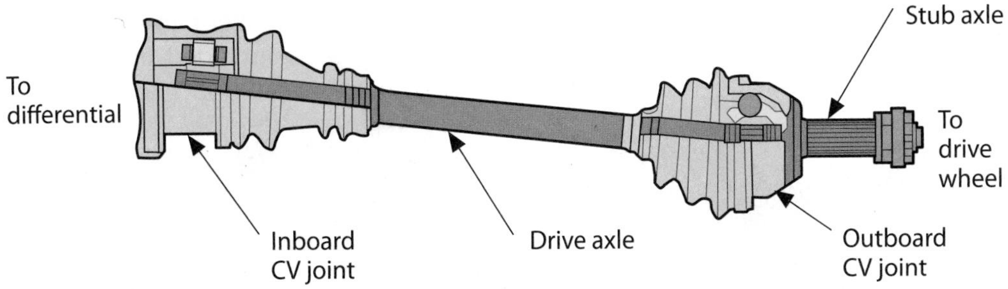

Figure 39-2 A typical FWD drive axle assembly. *Courtesy of Perfect Circle/Dana Corporation*

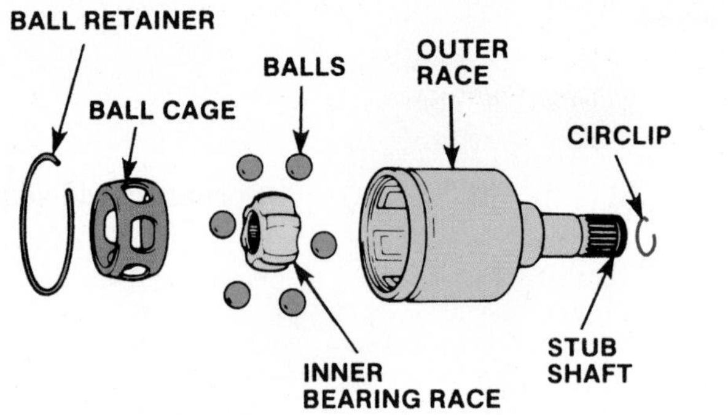

Figure 39-4 A double-offset CV joint.

Another ball-type joint is the dish-style CV joint, which is used predominantly on Volkswagen as well as on many German RWD models. Its design is very similar to the Rzeppa joint.

Plunging Ball-Type Joints There are two basic styles of plunging ball-type joints: the **double-offset** and the **cross groove joints**. This is a more compact design with a flat, doughnut-shaped outer housing and angled grooves.

The double-offset joint **(Figure 39–4)** uses a cylindrical outer housing with straight grooves and is typically used in applications that require higher operating angles (up to 25 degrees) and greater plunge depth (up to 2.4 inches [60 mm]). This type of joint can be found at the inboard position on some FWD half shafts as well as on the propeller shaft of some FWD shafts.

The cross groove joint **(Figure 39–5)** has a much flatter design than any other plunge joint. It is used as the inboard joint on FWD half shafts or at either end of a RWD independent rear suspension axle shaft.

The feature that makes this joint unique is its ability to handle a fair amount of plunge (up to 1.8 inches [46 mm]) in a relatively short distance. The inner and outer races share the plunging motion equally, so less overall depth is needed for a given amount of plunge. The cross groove can handle operating angles up to 22 degrees.

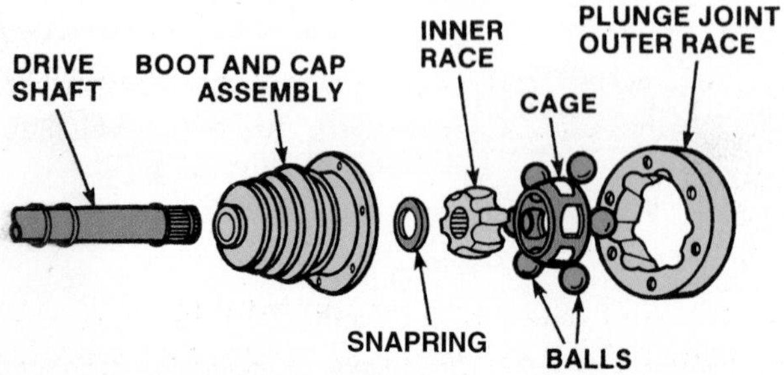

Figure 39-5 A cross-groove joint.

Tripod CV Joints

As with ball-type CV joints, tripod joints come in two varieties: plunge and fixed.

Tripod Plunging Joints Tripod plunging joints **(Figure 39–6)** consist of a central drive part or tripod (also known as a "spider"). This has three trunnions fitted with spherical rollers on needle bearings and an outer housing (sometimes called a "tulip" because of its three-lobed, flowerlike appearance). On some tripod joints, the outer housing is closed, meaning the roller tracks are totally enclosed within it. On others, the tulip is open and the roller tracks are machined out of the housing. Tripod joints are most commonly used as FWD inboard plunge joints.

Fixed Tripod Joints The fixed tripod joint is sometimes used as the outboard joint in FWD applications. In this design, the trunnion is mounted in the outer housing and the three roller bearings turn against an open tulip on the input shaft. A steel locking spider holds the joint together.

The fixed tripod joint has a much greater angular capability. The only major difference from a service standpoint is that the fixed tripod joint cannot be removed from the drive shaft or disassembled because of the way it is manufactured. The complete joint and shaft assembly must be replaced if the joint goes bad.

FRONT-WHEEL-DRIVE APPLICATIONS

FWD half shafts can be solid or tubular, of equal **(Figure 39–7)** or unequal length **(Figure 39–8)**, and come with or without damper weights. Equal-length shafts are used in some vehicles to help reduce torque steer (the tendency to steer to one side as engine power is applied). In these applications, an intermediate shaft is used as a link from the transaxle to one of the half shafts. This intermediate shaft can use an ordinary Cardan universal joint (described later in this chapter) to a yoke at the transaxle. At the outer end is a support bracket and bearing assembly. Looseness in the bearing or bracket can create vibrations. These items should be included in any inspection of the drivetrain components. The small damper weight, called a **torsional damper**, that is sometimes attached to one half shaft serves to dampen harmonic vibrations in the drivetrain and to stabilize the shaft as it spins, not to balance the shaft.

Regardless of the application, outer joints typically wear faster than inner joints because of the increased range of operating angles to which they are

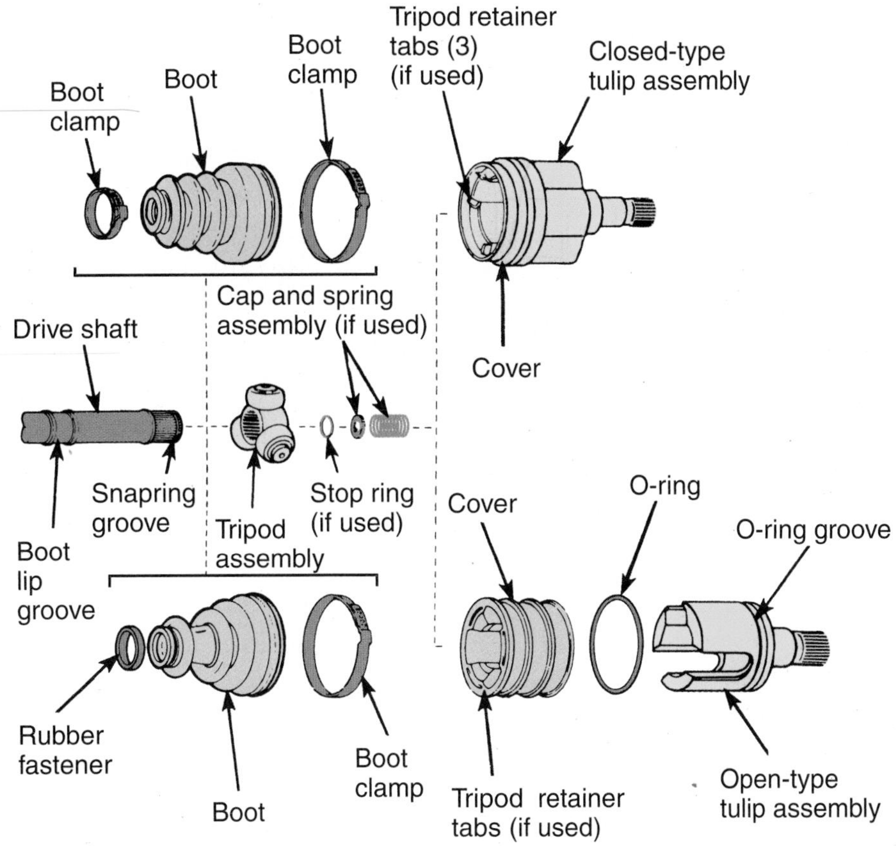

Figure 39-6 Inner tripod plunge-type joints: closed housing and open housing.

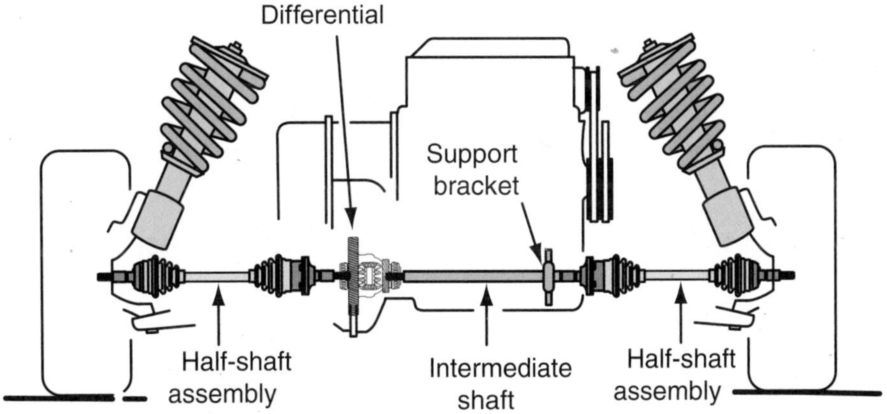

Figure 39-7 Equal-length FWD half shafts with an intermediate shaft.

subjected. Inner joint angles may change only 10 to 20 degrees as the suspension travels through jounce and rebound. Outer joints can undergo changes of up to 40 degrees in addition to jounce and rebound as the wheels are steered. That, combined with more flexing of the outer boots, is why outer joints have a higher failure rate. On average, nine outer CV joints are replaced for every inner CV joint. That does not mean the technician should overlook the inner joints. They wear too. Every time the suspension travels through jounces and rebound, the inner joints must plunge in and out to accommodate the different arcs between the drive shafts and suspension. Tripod inner joints tend to develop unique wear patterns on each of the three rollers and their respective tracks in the housing, which can lead to noise and vibration problems.

Other Applications

CV joints are also found on the front axles of many four-wheel-drive vehicles and on vehicles with rear independent suspension systems **(Figure 39–9)**.

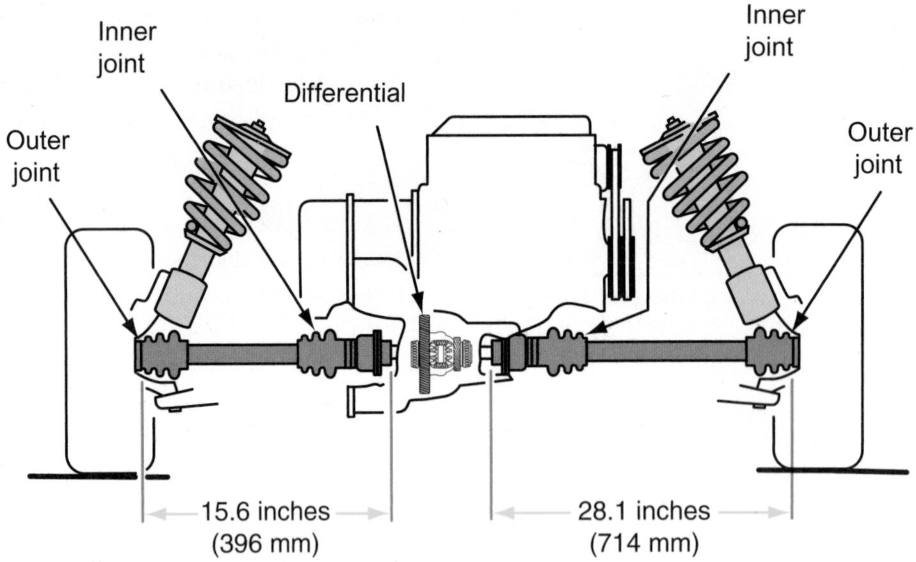

Figure 39-8 Unequal length FWD half shafts.

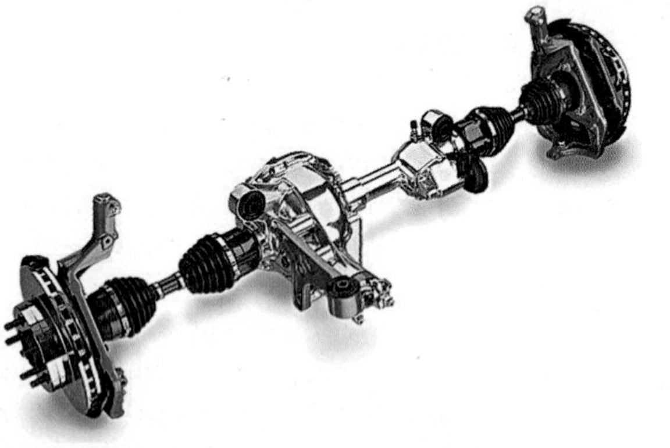

Figure 39-9 A CV joint-equipped rear axle assembly for a vehicle with independent rear suspension. *Courtesy of Dana Corporation*

Their use in these designs offers the same benefits as when they are used for FWD.

CV JOINT SERVICE

With proper service, CV joints can have a long life, despite having to perform extremely difficult jobs in hostile environments. They must endure extreme heat and cold and survive the shock of hitting potholes at high speeds. Fortunately, high-torque loads during low-speed turns and many thousands of high-speed miles normally do not bother the CV joint. It is relatively trouble free unless damage to the boot or joint goes unnoticed.

All CV joints are encased in a protective rubber (neoprene, natural, or silicone) or thermoplastic (Hycrel) boot. The job of the boot is to retain grease and to keep dirt and water out. The importance of the boot cannot be overemphasized because without its protection the joint does not survive. For all practical purposes, a CV joint is lubed for life. Once packed with grease and installed, it requires no further maintenance. A loose or missing boot clamp, or a slit, tear, or a small puncture in the boot itself allows grease to leak out and water or dirt to enter. Consequently, the joint is destroyed.

Although outboard joints tend to wear faster than the inboard ones, the decision as to whether to replace both joints when the half shaft is removed depends on the circumstances. If the vehicle has low miles and joint failure is the result of a defective boot, there is no reason to replace both joints. On a high-mileage vehicle where the bad joint has actually just worn itself out, it might be wise to save the expense and inconvenience of having the half shaft removed twice for CV joint replacement.

Diagnosis and Inspection

Any noise in the engine, drive axle, steering, or suspension is a good reason for a thorough inspection of the vehicle. A road test on a smooth surface is a good place to begin. The test should include driving at average highway speeds, some sharp turns, acceleration, and coasting. Look and listen for the following signs:

- A popping or clicking noise when turning indicates a possible worn or damaged outer joint **(Figure 39–10)**. To help identify the exact cause, put the vehicle in reverse and back up in a circle. If the noise gets louder, the outer joints should be replaced.
- A clunk during accelerating, decelerating, or putting an automatic transaxle into drive can be caused by excessive play in the inner joint on FWD vehicles. A clunking noise when putting an automatic transmission into gear or when

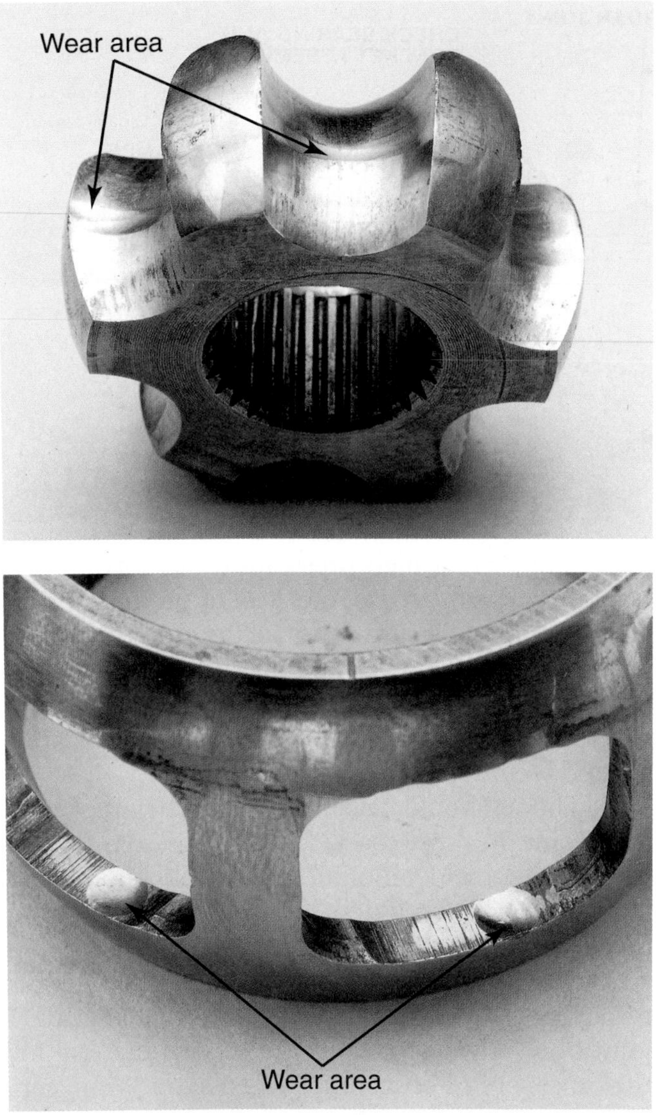

Figure 39-10 A worn cage or race can cause a clicking sound during a turn. *Courtesy of Federal-Mogul Corporation*

starting out from a stop usually indicates excessive play in an inner or outer joint. Be warned, though, that the same kind of noise can also be produced by excessive backlash in the differential gears and transmission. Alternately accelerating and decelerating in reverse while driving straight can reveal worn inner plunge joints. A bad joint clunks or shudders.

- A humming or growling noise is sometimes due to inadequate lubrication of either the inner or outer CV joint. It is more often due to worn or damaged wheel bearings, a bad intermediate shaft bearing on equal-length half-shaft transaxles, or worn shaft bearings within the transmission.
- A shudder or vibration when accelerating is often caused by excessive play in either the inboard or outboard joint but more likely it is the inboard plunge joint. These vibrations can also be caused by a bad intermediate shaft bearing on transaxles with equal-length half shafts. On FWD vehicles with transverse-mounted engines, this kind of vibration can also be caused by loose or deteriorated engine/transaxle mounts. Be sure to inspect the rubber bushings in the engine's upper torque strap to rule out this possibility. A vibration or shudder that increases with speed or comes and goes at a certain speed may be the result of excessive play in an inner or outer joint. A bent axle shaft can cause the same problem. Note, however, that some shudder could also be inherent to the vehicle.
- A cyclic vibration that comes and goes between 45 and 60 mph (72 and 100 km) may lead the technician to think there is a wheel that is out of balance. However, as a rule, an out-of-balance wheel produces a continuous vibration. A more likely cause is a bad inner tripod CV joint. The vibration occurs because one of the three roller tracks has become dimpled or rough. Every time the tripod roller on the bad track hits the rough spot, it creates a little jerk in the driveline, which the driver feels as a cyclic vibration.
- If a noise is heard while driving straight ahead but it ceases while turning, the problem is usually not a defective outer CV joint but a bad front wheel bearing. Turning changes the side load on the bearing, which may make it quieter than before.
- A vibration that increases with speed is rarely due to CV joint problems or FWD half-shaft imbalance. An out-of-balance tire or wheel, an out-of-round tire or wheel, or a bent rim are the most likely causes. It is possible that a bent half shaft, as the result of collision or towing damage, could cause the vibration. A missing damper weight could also be the culprit.

Begin CV joint inspection **(Figure 39–11)** by checking the condition of the boots. Splits, cracks, tears, punctures, or thin spots caused by rubbing call for immediate boot replacement. If the boot appears rotted, this indicates improper greasing or excessive heat, and it should be replaced. Squeeze all boots. If any air escapes, replace the boot.

If the inner boot appears to be collapsed or deformed, venting it (allowing air to enter) might solve the problem. Place a round-tipped rod between the boot and drive shaft. This equalizes the outside and inside air and allows the boot to return to its normal shape.

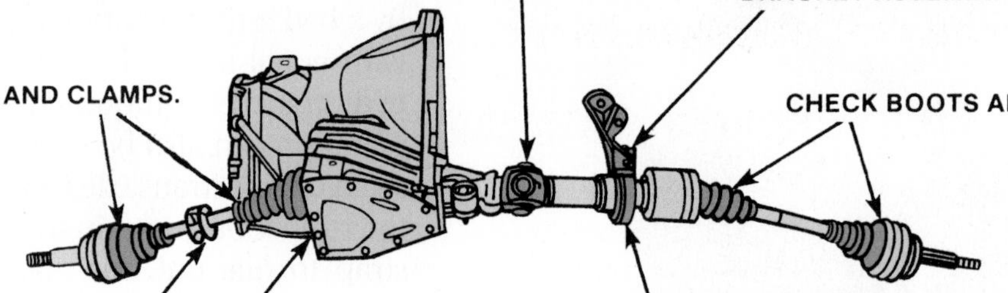
CHECK CARDAN JOINT.
CHECK BEARING AND BRACKET ASSEMBLY.
CHECK BOOTS AND CLAMPS.
CHECK BOOTS AND CLAMPS.
CHECK TORSIONAL DAMPING RING.
CHECK FOR SEAL LEAKS.
CHECK INTERMEDIATE SHAFT BEARING.

Figure 39-11 Inspection points for a FWD vehicle.

Make sure that all boot clamps are tight. Missing or loose clamps should be replaced. If the boot appears loose, slide it back and inspect the grease inside for possible contamination. A milky or foamy appearance indicates water contamination. A gritty feeling when rubbed between the fingers indicates dirt. In most cases, a water- or dirt-contaminated joint should be replaced.

The drive axles should be checked for signs of contact or rubbing against the chassis. Rubbing can be a symptom of a weak or broken spring or engine mount, as well as chassis misalignment. On FWD transaxles with equal-length half shafts, inspect the intermediate shaft U-joint, bearing, and support bracket for looseness by rocking the wheel back and forth and watching for any movement. Oil leakage around the inner CV joints indicates a faulty transaxle shaft seal. To replace the seal, the half shaft must be removed.

Obtaining CV Repair Parts

To repair a drive axle, a complete shaft should be installed. Most aftermarket part suppliers offer a complete line of original equipment drive shafts for FWD vehicles. These shafts come fully assembled and ready for installation. This repair method eliminates the need to tear down and rebuild an old shaft.

If only the CV joints need service, a CV joint service kit should be installed. Joint service kits typically include a CV joint, boot, boot clamps and seals, special grease for lubrication (various joints require different amounts of grease; the correct quantity is packed in each kit), retaining rings, and all other attachment parts.

Part manufacturers also produce a line of complete boot sets for each application, including new clamps and the appropriate type and amount of grease for the joint. CV joints require a special high-temperature, high-pressure grease. Substituting any other type of grease may lead to premature failure of the joint. Be sure to use all the grease supplied in the joint or boot kit. The same rule applies to the clamps. Use only those clamps supplied with the replacement boot. Follow the directions for positioning and securing them.

Old boots should never be reused when replacing a CV joint. In most cases, failure of the old joint is caused by some deterioration of the old boot. Reusing an old boot on a new joint usually leads to the quick destruction of the joint.

Photo Sequence 37 shows the procedure for removing a typical drive axle and replacing a CV joint boot. Always refer to the service manual for the exact service procedure. The diagnosis and service chart shown in **Table 39–1** gives an idea of the types of front-wheel drivetrain problems that can occur.

CV Joint Service Guidelines

The following are some guidelines to follow when servicing CV joints:

- Never jerk or pull on the axle shaft when removing it from a vehicle with tripod inner joints. Doing so may pull the joint apart, allowing the needle bearings to fall out of the roller. Pull on the inner housing, and support the outer end of the shaft until the shaft is completely out.
- Always install new hub nuts and torque them to specifications. This is absolutely necessary to properly preload the wheel bearings. Do not guess. The specifications can vary from 75 to 235 ft.-lb (101 to 318 N-m). Most axle hub nuts are staked in place after they have been tightened **(Figure 39–12)**. Others have a castellated nut that is secured with a cotter pin.

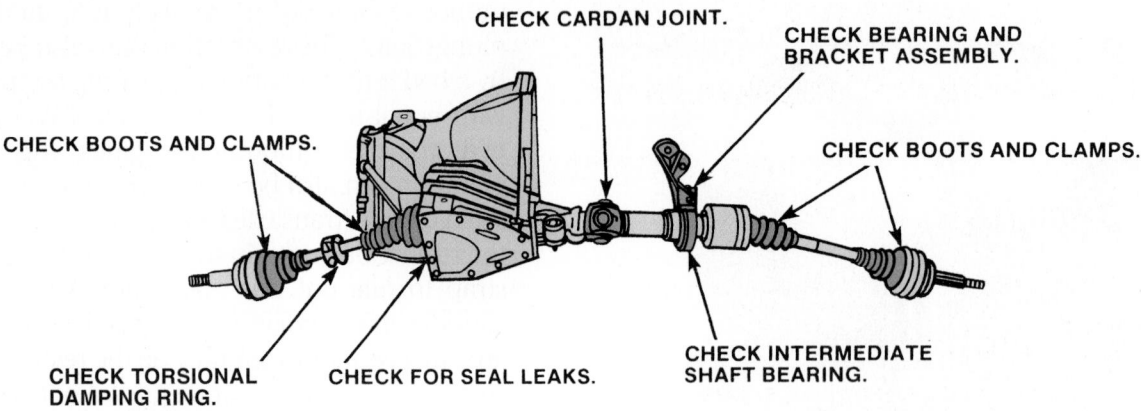

P37-1 *Removing the axle from the car begins with the removal of the wheel cover and wheel hub cover. The hub nut should be loosened before raising the car and removing the wheel.*

P37-2 *After the car is raised and the wheel is removed, the hub nut can be unscrewed from the axle shaft.*

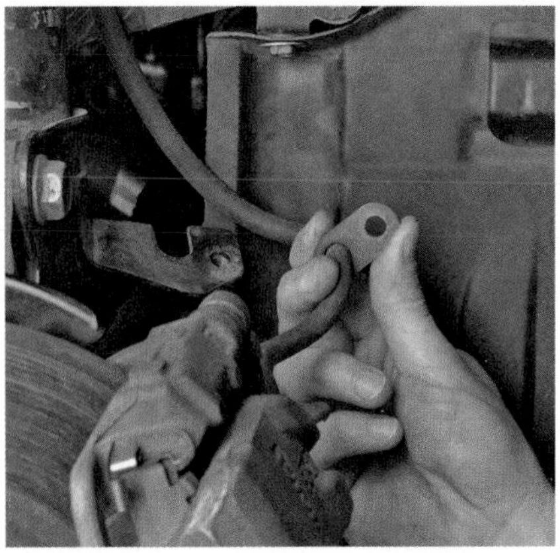

P37-3 *The brake line holding clamp must be loosened from the suspension.*

P37-4 *The ball joint must be separated from the steering knuckle assembly. To do this, first remove the ball joint retaining bolt. Then pry down on the control arm until the ball joint is free.*

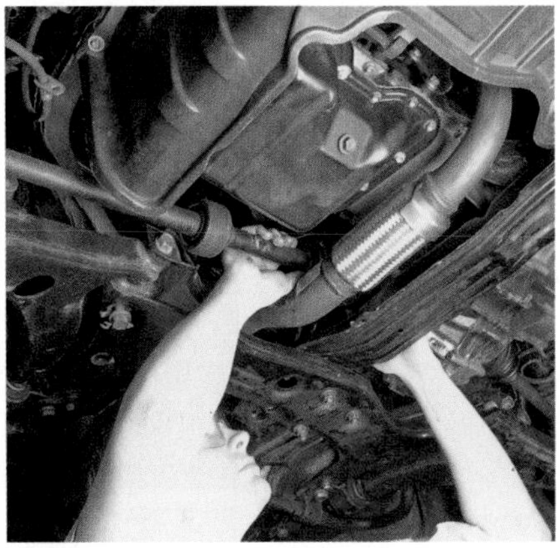

P37-5 *The inboard joint can be pulled free from the transaxle.*

P37-6 *A special tool is normally needed to separate the axle shaft from the hub allowing the axle to be removed from the car. Never hit the end of the axle with a hammer.*

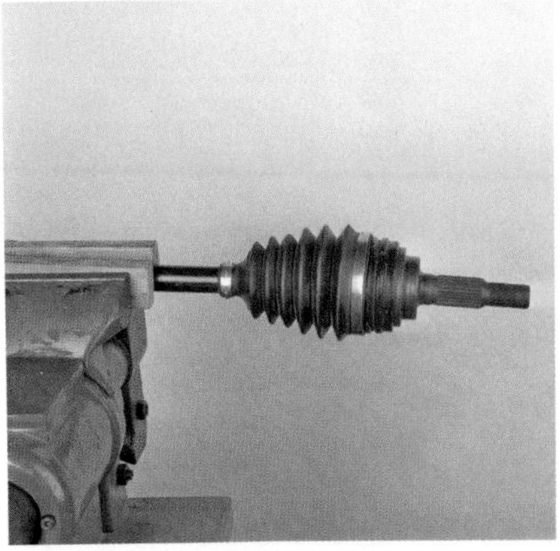

P37-7 *The axle shaft should be mounted in a soft-jawed vise for work on the joint. Pieces of wood on either side of the axle work well to secure the axle without damaging it.*

P37-8 *Begin boot removal by cutting and discarding the boot clamps.*

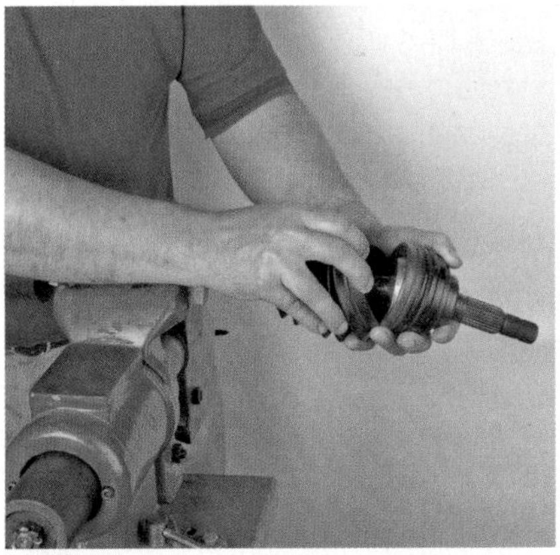

P37-9 *Scribe a mark around the axle to indicate the boot's position on the shaft. Then move the boot off the joint.*

Removing and Replacing a CV Joint Boot (continued)

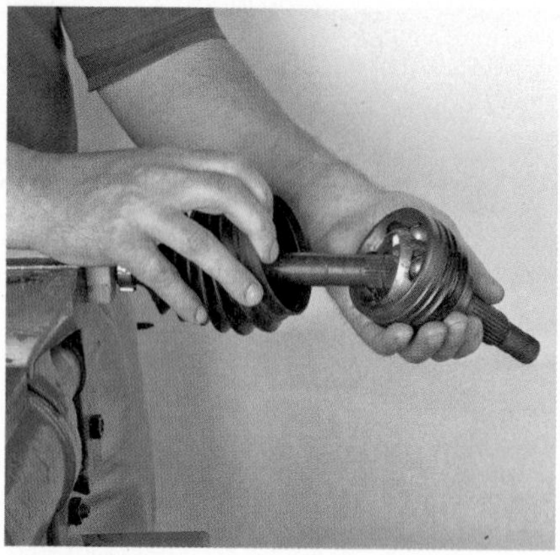

P37-10 *Remove the circlip and separate the joint from the shaft.*

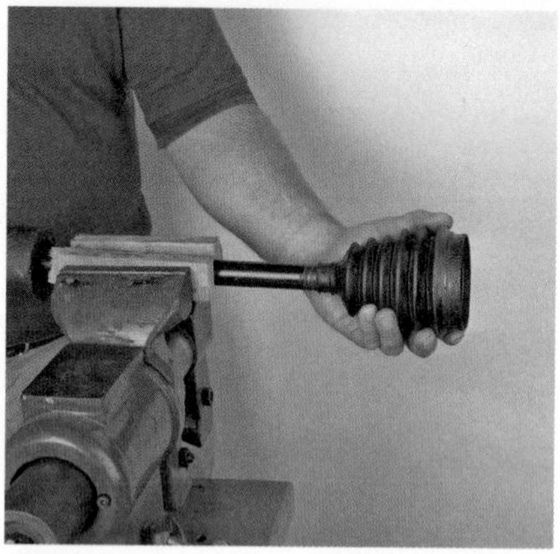

P37-11 *Slide the old boot off the shaft.*

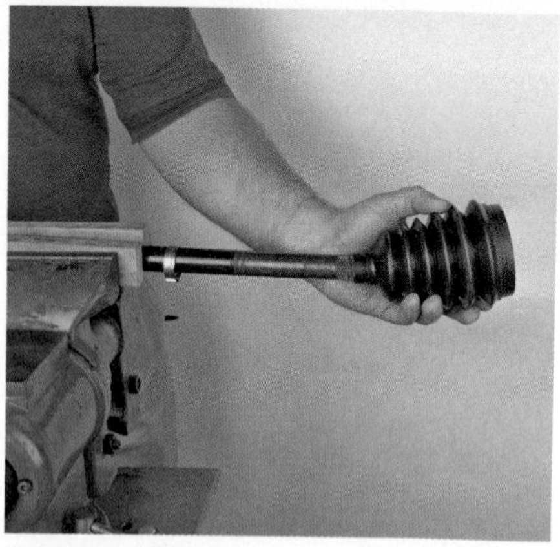

P37-12 *Clean and inspect the joint, then wipe the axle shaft clean and install the new boot onto the shaft.*

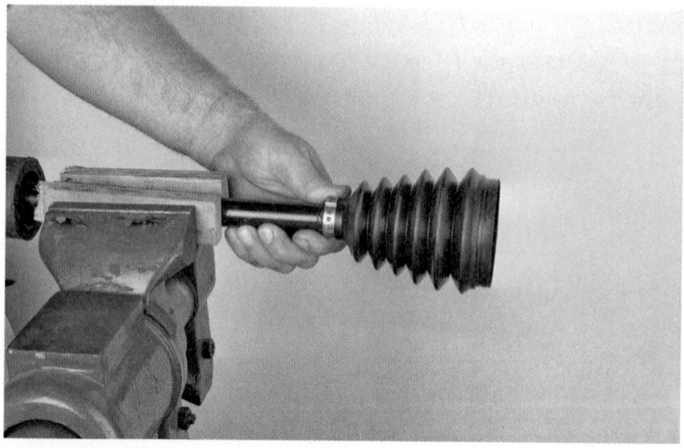

P37-13 *Place the boot into its proper location on the shaft and install a new clamp.*

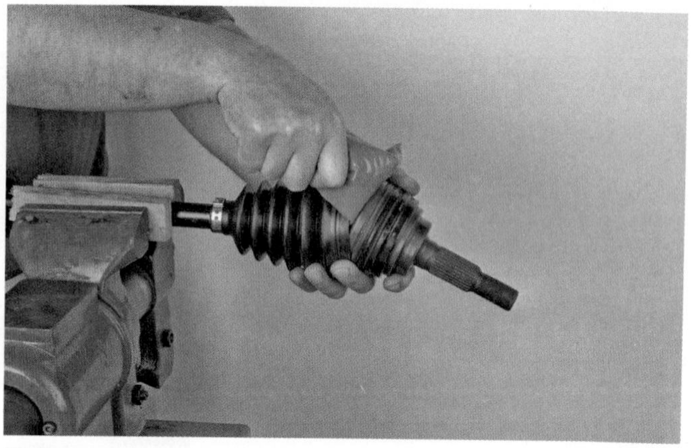

P37-14 *Using a new circlip, reinstall the joint on the shaft. Pack joint grease into the joint and boot. The entire packet of grease that comes with a new boot needs to be forced into the boot and joint.*

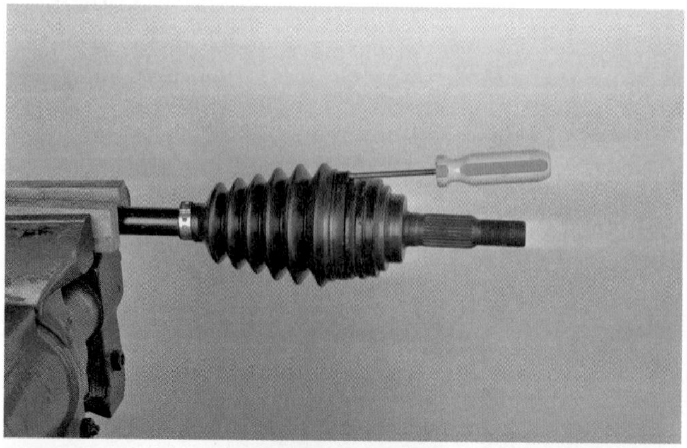

P37-15 *Pull the boot over the joint and into its proper position. Use a dull screwdriver to lift an edge of the boot up to equalize the pressure inside the boot with the outside air.*

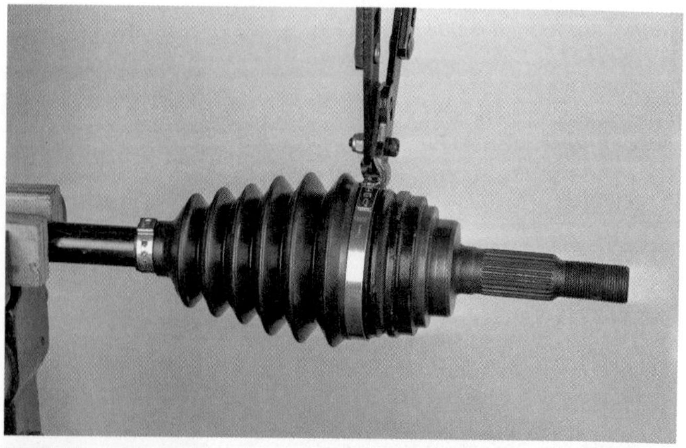

P37-16 *Install the new large boot clamp and reinstall the axle into the car. Torque the hub nut after the wheels have been reinstalled and the car is sitting on the ground. Always follow the correct procedure for the clamp being installed.*

TABLE 39-1 PROBLEM DIAGNOSIS AND SERVICE FOR FWD DRIVELINES

Problem	Possible Cause	Corrective Remedy
Vibrations in steering wheel at highway speeds	Front-wheel balance	Front-wheel unbalance is felt in the steering wheel. Front wheels must be balanced.
Vibrations throughout vehicle	Worn inner CV joints	Worn parts of the inner CV joint are not operating smoothly.
Vibrations throughout vehicle at low speed	Bent axle shaft	Axle shaft does not operate on the center of the axis; thus, vibration develops.
Vibrations during acceleration	Worn or damaged outer or inner CV joints Fatigued front springs	CV joints are not operating smoothly due to damage or wear on parts. Sagged front springs are causing the inner CV joint to operate at too great an angle, causing vibrations.
Grease dripping on ground or sprayed on chassis parts	Ripped or torn CV joint boots	Front-wheel-drive CV joints are immersed in lubricant. If the CV joint boot has a rip or is torn, lubricant leaks out. The condition must be corrected as soon as possible.
Clicking or snapping noise heard when turning curves and corners	Worn or damaged outer CV joint Bent axle shaft	Worn parts are clicking and noisy as loading and unloading on the CV joint takes place. Irregular rotation of the axle shaft is causing a snapping, clicking noise.

Figure 39-12 Most axle hub nuts are staked after they are tightened to lock them in place. *Courtesy of Federal-Mogul Corporation*

- Never use an impact wrench to loosen or tighten axle hub nuts. Doing so may damage the wheel bearings as well as the CV joints.
- On vehicles with antilock brakes, use care to protect the wheel speed sensor and tone ring on the outer CV joint housings. If misaligned or damaged during joint replacement, it can cause wheel speed sensor problems.
- Always recheck the alignment after replacing CV joints. Marking the camber bolts is not enough, because camber can be off as much as three-quarters of a degree due to differences between the size of the camber bolts and their holes.

CV Shaft and Rubber Boot Care Tips

The rubber boots need special care when you are servicing the CV joints, engine, or transaxle. The following tips might save you trouble later:

- Always support the control arm when doing on-the-car balancing of the front wheels to avoid high-speed operation at a steep half-shaft angle. Off-the-car balancing might be a wiser choice.
- Do not use half shafts as lift points for raising a car.
- Use a plastic or metal shield over rubber boots to protect them from accidental tool damage when performing other wheel, brake, suspension, or steering system maintenance.
- Clean only with soap and water.
- Avoid boot contact with gasoline, oil, or degreaser compounds.

REAR-WHEEL DRIVE SHAFTS

A drive shaft must smoothly transfer torque while rotating, changing length, and moving up and down. The different designs of drive shafts all attempt to ensure a vibration-free transfer of the engine's power from the transmission to the differential. This goal is

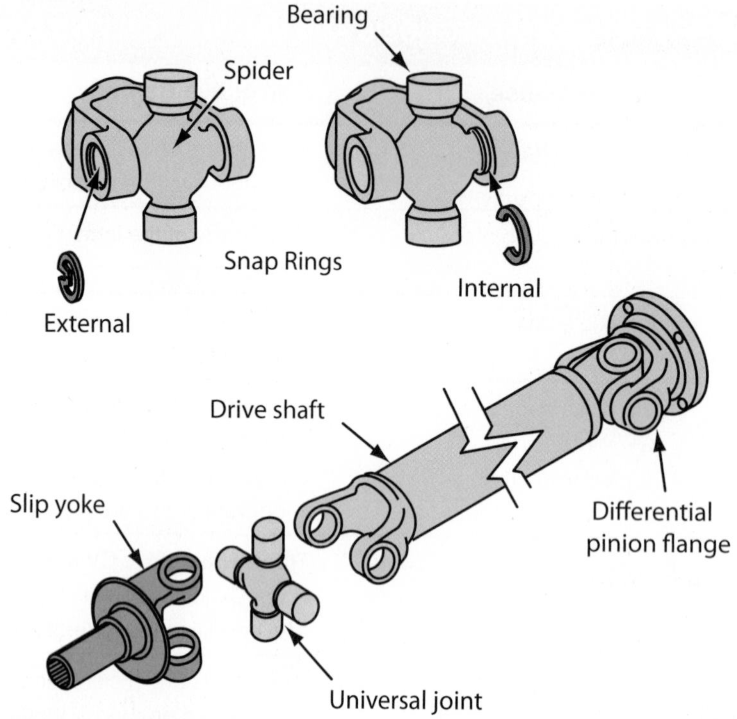

Figure 39-13 A drive shaft assembly with exploded U-joints.

complicated by the fact that the engine and transmission are bolted solidly to the frame of the car, whereas the differential is mounted on springs. As the rear wheels go over bumps in the road or changes in the road's surface, the springs compress or expand, changing the angle of the drive shaft between the transmission and the differential, as well as the distance between the two. To allow for these changes, the Hotchkiss-type drive shaft is fitted with one or more U-joints to permit variations in the angle of the drive, and a slip joint that permits the effective length of the drive shaft to change.

SHOP TALK

When a vehicle is intentionally raised or lowered, the length of the drive shaft should be changed to allow for normal travel of the slip yoke on the output shaft.

Starting at the front or transmission end of a RWD shaft, there is a slip yoke, a universal joint, a drive shaft yoke, and a drive shaft **(Figure 39–13)**. At the rear or differential end, there is another drive shaft yoke and a second universal joint connected to the differential pinion flange.

In addition to these basic components, some drivetrains have a drive shaft support bearing **(Figure 39–14)**. Large vehicles with long drive shafts often use a double U-joint arrangement, called a double

Figure 39-14 A center bearing assembly.

Cardan joint or a CV U-joint, to help minimize driveline vibrations. These vehicles may also have a center bearing. The center bearing allows the length of the shaft to be divided in half. When a center bearing is used, the ends of the drive shaft that are going into the bearing are slip joints.

Slip Yoke

A sliding or **slip yoke** **(Figure 39–15)** is internally splined and its outside diameter is precisely machined to fit into the rear seal. The internal splines slip over the external splines of the output shaft. The slip yoke rotates at output shaft speed and can slide on the splines (hence the name slip yoke). The slip yoke allows the effective length of the drive shaft to change with the movement of the rear suspension and drive axle assembly.

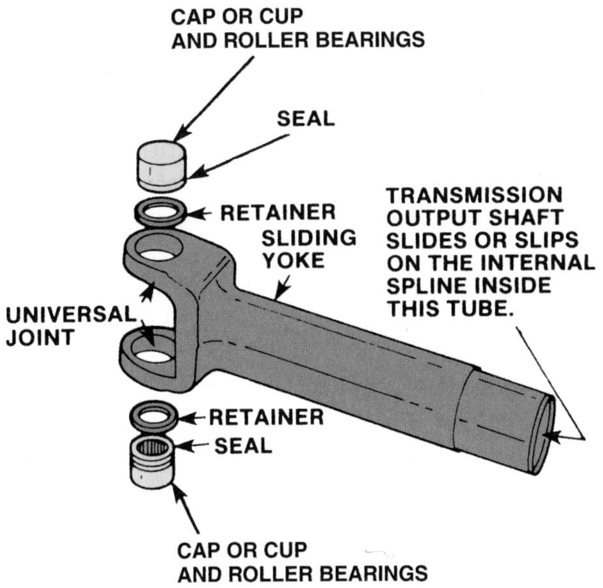

Figure 39-15 A typical slip or sliding yoke.

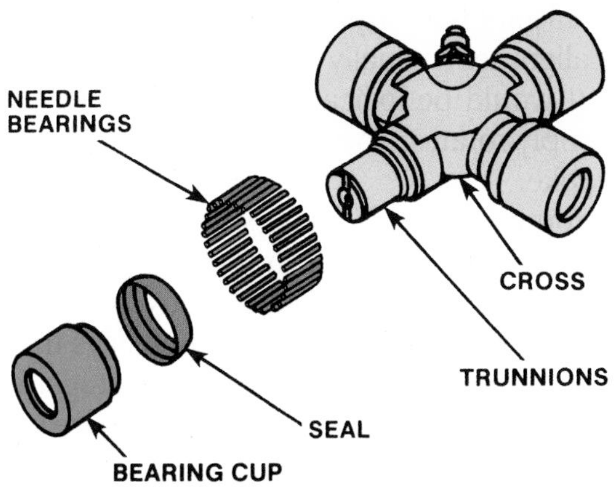

Figure 39-16 A Cardan joint.

Drive Shaft and Yokes

The drive shaft is nothing more than an extension of the transmission output shaft. The drive shaft, which is usually made from seamless steel tubing, transfers engine torque from the transmission to the rear driving axle. The yokes, which are either welded or pressed onto the shaft, provide a means of connecting two or more shafts together. At the present time, a limited number of vehicles are equipped with fiber composite—reinforced fiberglass, graphite, and aluminum—drive shafts. The advantages of using these materials are weight reduction, torsional strength, fatigue resistance, easier and better balancing, and reduced interference from shock loading and torsional problems. Some drive shafts are fitted with a torsional damper to reduce torsional vibrations.

The drive shaft, like any other rigid tube, has a natural vibration frequency. If one end were held tightly, it would vibrate at its own frequency when deflected and released. It reaches its natural frequency at its critical speed. Critical drive shaft speed depends on the diameter of the tube and its length. Diameters are as large as possible and shafts as short as possible to keep the critical speed frequency above the driving speed range. It should be remembered that since the drive shaft generally turns three to four times faster than the tires, proper drive shaft balance is required for vibration-free operation.

OPERATION OF U-JOINTS

The U-joint allows two rotating shafts to operate at a slight angle to each other. A French mathematician named Cardan developed the original joint in the sixteenth century. In 1902, Clarence Spicer modified Cardan's invention for the purpose of transmitting engine torque to an automobile's rear wheels.

The U-joint is basically a double-hinged joint consisting of two Y-shaped yokes, one on the driving or input shaft and the other on the driven or output shaft, plus a cross-shaped unit called the cross (**Figure 39–16**). A U-joint is used to connect the yokes together. The four arms of the cross are fitted with bearings in the ends of the two shaft yokes. The input shaft's yoke causes the cross to rotate, and the two other trunnions of the cross cause the output shaft to rotate. When the two shafts are at an angle to each other, the bearings allow the yokes to swing around on their trunnions with each revolution. This action allows two shafts, at a slight angle to each other, to rotate together.

U-joints allow the drive shaft to transmit power to the rear axle through varying angles that are controlled by the travel of the rear suspension. Because power is transmitted on an angle, U-joints do not rotate at a constant velocity, nor are they vibration free.

Speed Variations (Fluctuations)

Although simple in appearance, the universal joint is more intricate than it seems because its natural action is to speed up and slow down twice in each revolution while operating at an angle. The amount that the speed changes varies according to the steepness of the U-joint's angle.

U-joint operating angle is determined by taking the difference between the transmission installation angle and the drive shaft installation angle. When the universal joint is operating at an angle, the driven yoke speeds up and slows down twice during each drive shaft revolution.

These four speed changes are not normally visible during rotation, but they may be understood more easily after examining the action of a U-joint. A universal joint is a coupling between two shafts not in direct alignment, usually with changing relative positions. It would be logical to assume that the entire unit simply rotates. This is true only for the U-joint's input yoke.

The output yoke's circular path looks like an ellipse because it can be viewed at an angle instead of straight on. This effect can be obtained when a coin is rotated by the fingers. The height of the coin stays the same even though the sides seem to get closer together.

This illusion might seem to be a merely visual effect, but it is more than that. The U-joint rigidly locks the circular action of the input yoke to the elliptical action of the output yoke. The result is similar to what would happen when changing a clock face from a circle to an ellipse.

Like the hands of a clock, the input yoke turns at a constant speed in its true circular path. The output yoke, operating at an angle to the other yoke, completes its path in the same amount of time. However, its speed varies, or is not constant, compared to the input.

Speed fluctuation is more easily visualized when looking at the travel of the yokes by 90-degree quadrants **(Figure 39–17)**. The input yoke rotates at a steady or constant speed through the complete 360-degree turn. The output yoke quadrants alternate between shorter and longer distance travel than the input yoke quadrants. When one point of the output yoke covers the shorter distance in the same amount of time, it must travel at a slower rate. Conversely, when traveling the longer distance (but only 90 degrees) in the same amount of time, it must move faster.

Because the average speed of the output yoke through the four 90-degree quadrants (360 degrees) equals the constant speed of the input yoke during the same revolution, it is possible for the two mating yokes to travel at different speeds. The output yoke is falling behind and catching up constantly. The resulting acceleration and deceleration produces a fluctuating torque and torsional vibrations characteristic of all Cardan U-joints. The steeper the U-joint angle, the greater the fluctuations in speed will be. Conversely, the smaller the angle, the speed will change less.

Phasing of Universal Joints

The torsional vibrations set up by the fluctuations in velocity are transferred down the drive shaft to the next U-joint. At this joint similar speed fluctuation occurs. Since these speed variations take place at equal and opposite angles to the first joint, they cancel out each other. To provide for this canceling effect, drive shafts should have at least two U-joints and their operating angles must be equal to each other. Speed fluctuations can be canceled if the driven yoke has the same point of rotation, or same plane, as the driving yoke. When the yokes are in the same plane, the joints are said to be "in phase."

On a two-piece drive shaft, you may encounter problems if you are not careful. The center driving yoke is splined to the front drive shaft. If the yoke's position on the drive shaft is not indicated in some manner, the yoke could be installed in a position that is out of phase. Manufacturers use different methods of indexing the yoke to the shaft. Some use aligning arrows. Others machine a master spline that is wider than the others. The yoke and shaft cannot be reassembled until the master spline is aligned properly. When there are no indexing marks, you should index

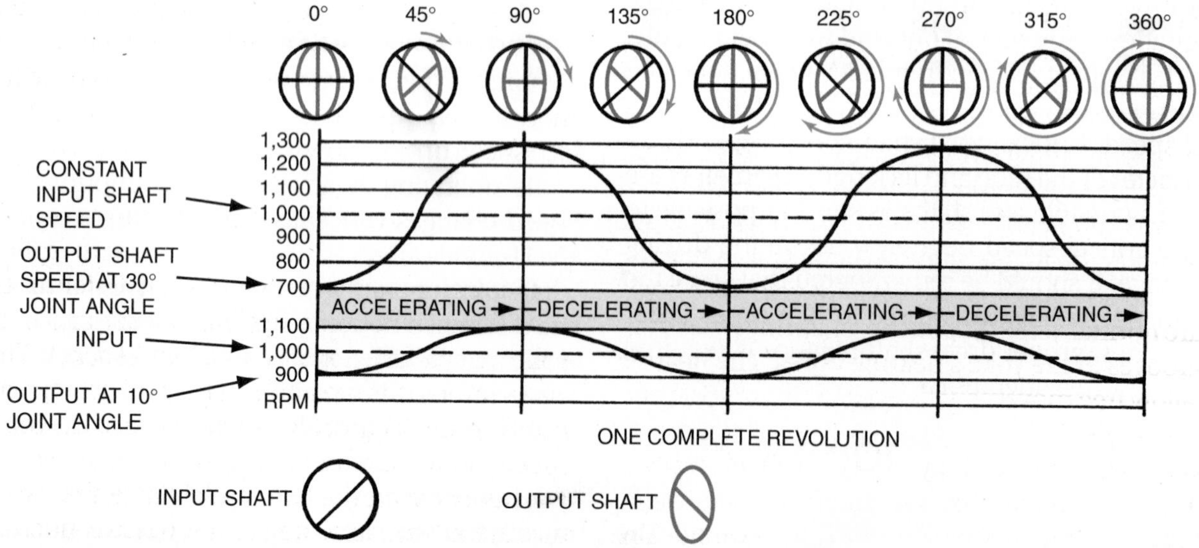

Figure 39-17 A graph showing typical drive shaft yoke speed fluctuations.

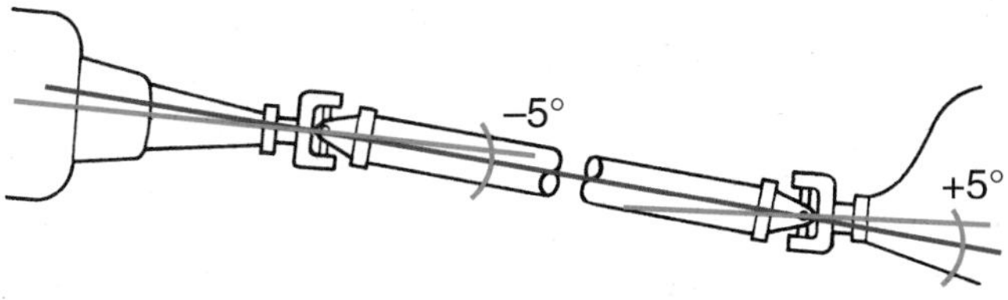

Figure 39-18 When a drive shaft's joints are in phase and have canceling angles, inherent vibrations are reduced.

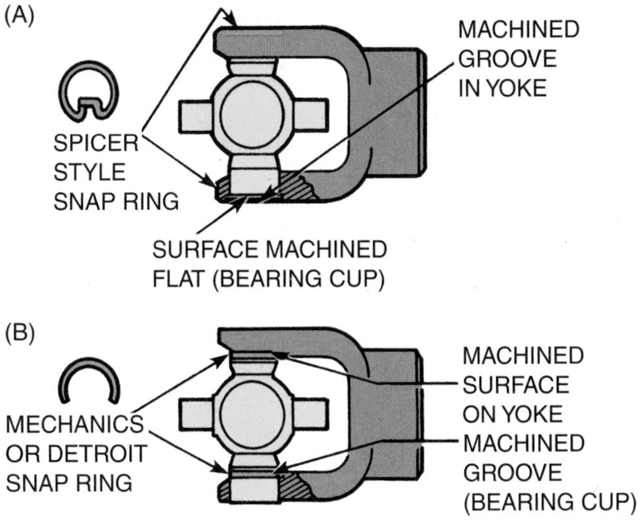

Figure 39-19 (A) A Spicer-style U-joint and (B) a Mechanics or Detroit-style U-joint.

the yoke to the drive shaft before disassembling the U-joint. This saves time and frustration during reassembly.

Canceling Angles

Vibrations can be reduced by using canceling angles **(Figure 39–18)**. Carefully examine the illustration, and note that the operating angle at the front of the drive shaft is offset by the one at the rear of the drive shaft. When the front U-joint accelerates, causing a vibration, the rear universal joint decelerates, causing a vibration. The vibrations created by the two joints oppose and dampen the vibrations from one to the other. The use of canceling angles provides a smoother drive shaft operation.

TYPES OF U-JOINTS

There are three common designs of U-joints: single U-joints retained by either an inside or outside snapring, coupled U-joints, and U-joints held in the yoke by U-bolts or lock plates.

Single Universal Joints

The single Cardan/Spicer U-joint is also known as the cross or four-point joint. These two names aptly describe the single Cardan, since the joint itself forms a cross, with four machined trunnions or points equally spaced around the center of the axis. Needle bearings used to abate friction and provide smoother operation are set in bearing cups. The trunnions of the cross fit into the cup assemblies and the cup assemblies fit snugly into the driving and driven U-joint yokes. U-joint movement takes place between the trunnions, needle bearings, and bearing cups. There should be no movement between the bearing cup and its bore in the universal joint yoke. The bearings are normally held in place by snaprings that drop into grooves in the yoke's bearing bores. The bearing caps allow free movement between the trunnion and yoke. The needle bearing caps may also be pressed into the yokes, bolted to the yokes, or held in place with U-bolts or metal straps.

There are other styles of single U-joints. The method used to retain the bearing caps is the major difference between these designs. The Spicer style **(Figure 39–19A)** uses a snapring that fits into a groove machined in the outer end of the yoke. The bearing cups for this style are machined to accommodate the snapring.

The Mechanics or Detroit/Saginaw style **(Figure 39–19B)** uses an external snapring that fits into a groove machined in the bearing cup on the end closest to the grease seal. When installed, the snapring rests against the machined inside portion of the yoke. On some joints, nylon is injected into the machined grooves to retain the U-joint. When these joints are replaced, new retaining rings are included with the joint and the plastic is not reinjected. Make sure that all of the plastic is removed before installing a new joint.

The Cleveland style is an attempt to combine different joint styles to have more applications from one joint. The bearing cups for this U-joint are machined to accommodate either Spicer or Mechanics style snaprings. If a replacement U-joint comes with both style clips, use the clips that pertain to your application.

Double-Cardan Universal Joint

A **double-Cardan** U-joint is used with split drive shafts and consists of two Cardan U-joints closely connected by a centering ball socket and a center yoke, which functions as a ball and socket. The ball and socket splits the angle of the two shafts between two U-joints **(Figure 39–20)**. Because of the centering socket yoke, the total operating angle is divided equally between the two joints. Since the two joints operate at the same angle, the normal fluctuations that result from the use of a single U-joint are

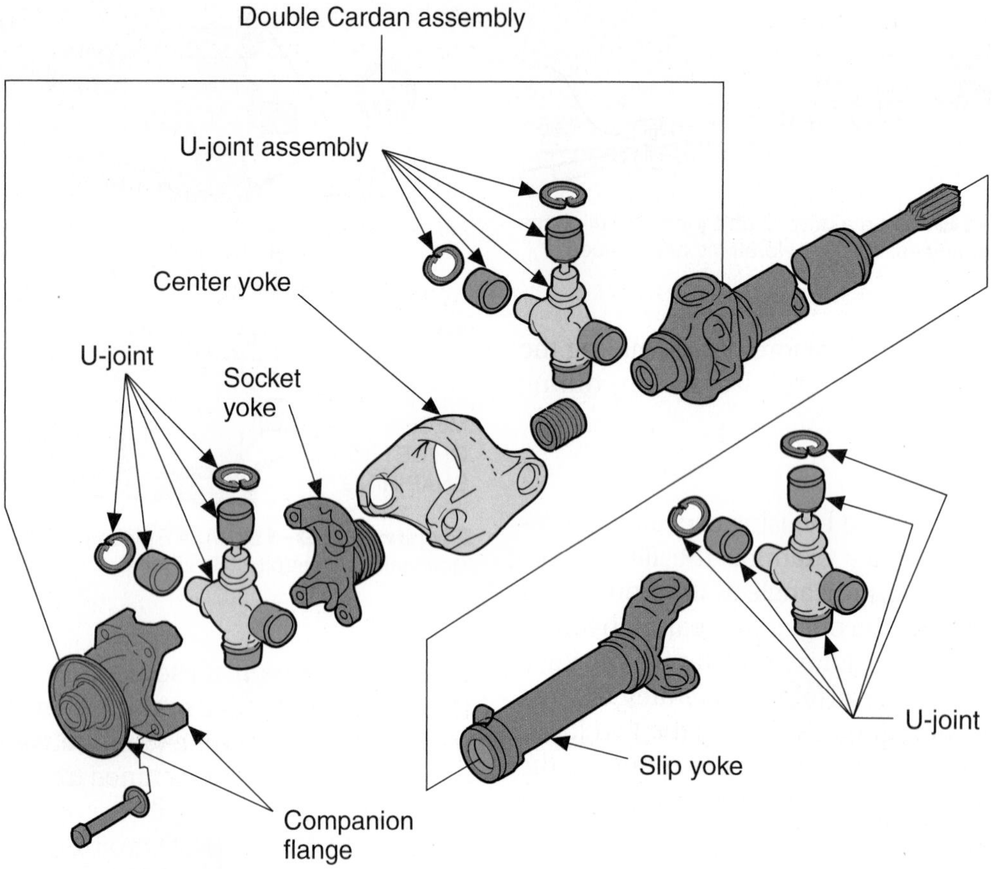

Figure 39-20 A double-Cardan joint. *Courtesy of Ford Motor Company*

canceled out. The acceleration and deceleration of one joint is canceled by the equal and opposite action of the other.

The double-Cardan joint is classified as a CV U-joint. It is most often found in front-engine RWD luxury-type vehicles.

DIAGNOSIS OF DRIVE SHAFT AND U-JOINT PROBLEMS

A failed U-joint or damaged drive shaft can exhibit a variety of symptoms. A clunk that is heard when the transmission is shifted into gear is the most obvious. You can also encounter unusual noise, roughness, or vibration.

To help differentiate a potential drivetrain problem from other common sources of noise or vibration, it is important to note the speed and driving conditions at which the problem occurs. As a general guide, a worn U-joint is most noticeable during acceleration or deceleration and is less speed sensitive than an unbalanced tire (commonly occurring in the 30 to 60 mph [50 to 100 km/h] range) or a bad wheel bearing (more noticeable at higher speeds). Unfortunately, it is often very difficult to accurately pinpoint drivetrain problems with only a road test. Therefore, expand the undercar investigation by putting the vehicle up on the lift, where it is possible to get a good view of what is going on underneath.

SHOP TALK

When diagnosing driveline noise, if a chirping sound that increases with speed is heard, suspect a dry U-joint. The chirping typically occurs with a frequency two to four times faster than the speed of the wheels.

The first problem most likely encountered is an undercar fluid leak. If a lot of lube is escaping from the pinion shaft seal, the drivetrain noise could be caused by a bad pinion bearing. To confirm the problem, start the engine, put the transmission in gear, and listen at the carrier. If the bearing is noisy, it is necessary to make one of those difficult judgment calls. If the bearing sounds fine but the pinion seal is still leaking, suggest an on-the-car seal replacement.

On some vehicles, seal replacement is a simple procedure that involves removing the pinion flange and replacing the seal. Others are a little more complex because the pinion shaft is retained with a nut

Figure 39-21 The tools required to tighten the flange nut.

that must be removed to gain access to the seal. These units require special tools to loosen and tighten the pinion nut **(Figure 39–21)**, which allows for the removal of the flange. Always refer to the service manual for the correct procedure. On many units there is a collapsible spacer behind the pinion nut. Whenever the nut is loosened, a new spacer should be installed before torquing the nut.

If the seal needs to be replaced, check the runout of the flange at the rear axle **(Figure 39–22)**. A damaged seal can be caused by excessive runout. Also inspect the surface of the flange that rides in the seal. During reassembly, make sure the outer surface of the flange is lubricated before pushing it into the seal.

At the other end of the driveline, inspect the transmission's extension housing seal the same way. If it is leaking, the seal itself can be easily replaced. Check the extension housing bushing. That is the most likely reason the seal went bad in the first place. Once the yoke is removed, an internal expanding bearing/bushing puller makes short work of bushing replacement. Before pushing the slip yoke back in after the new seal is installed, make sure the machined surface of the bore is free of scratches, nicks, and grooves that could damage the seal. For that added margin of safety, a little transmission lube or petroleum jelly on the lip of the seal helps the parts slide in easily.

If the seals pass the test, continue driveline examination by inspecting the U-joint's grease seals for signs of rust, leakage, or lubrication breakdown. Also, check for excessive joint movement by firmly grasping and attempting to rotate the coupling yokes back and forth in opposite directions. If any perceptible trunnion-to-bearing movement is felt, the joint should be replaced.

The runout of the drive shaft should also be checked. If there is excessive runout, determine the cause and make the necessary repairs. If the runout is fine, check the phasing of the joints and their angle. To check their operating angle, use an inclinometer. This instrument, when attached to the drive shaft, displays the angle of the drive shaft along any point. Your finding from this test should be compared to specifications. Normally, if the angles are wrong, the rear axle has moved in its mounting.

As a final diagnosis inspection point, check the entire length of the drive shaft for excess undercoating, dents, missing weights, or other damage that could cause an imbalance and result in a vibration. If no damage is found, the drive shaft should be removed and its balance checked by a specialty shop.

When a U-joint is damaged or excessively worn, it must be replaced. Photo Sequence 38 covers the typical procedure for removing a U-joint from a drive shaft. After a replacement joint is obtained, it needs to be installed. Photo Sequence 39 covers the reassembly of a common U-joint.

Figure 39-22 The setup for checking the runout of a companion flange.

DIFFERENTIALS AND DRIVE AXLES

The differential is a geared mechanism located between the driving axles of a vehicle. It rotates the driving axles at different speeds when the vehicle is turning a corner **(Figure 39–23)**. It also allows both axles to turn at the same speed when the vehicle is moving in a straight line. The drive axle assembly directs driveline torque to the vehicle's drive wheels. The gear ratio of the drive axle's ring and pinion gears is used to increase torque. The differential serves to establish a state of balance between the forces between the drive wheels and allows the drive wheels to turn at different speeds when the vehicle changes direction.

P38-1 *Clamp the slip yoke in a vise and support the outer end of the drive shaft.*

P38-2 *Remove the lock rings on the tops of the bearing cups. Make index marks in the yoke so that the joint can be assembled with the correct phasing.*

P38-3 *Select a socket that has an inside diameter large enough for the bearing cup to fit into; usually a 1 1/4-inch socket works.*

P38-4 *Select a second socket that can slide into the shaft's bearing cup bore—usually a 9/16-inch socket.*

P38-5 *Place the large socket against one vise jaw. Position the drive shaft yoke so that the socket is around a bearing cup.*

P38-6 *Position the other socket to the center of the bearing cup opposite to the one in line with the large socket.*

P38-7 *Carefully tighten the vise to press the bearing cup out of the yoke and into the large socket.*

P38-8 *Separate the joint by turning the shaft over in the vise and driving the cross and remaining bearing cup down through the yoke with a brass drift and hammer.*

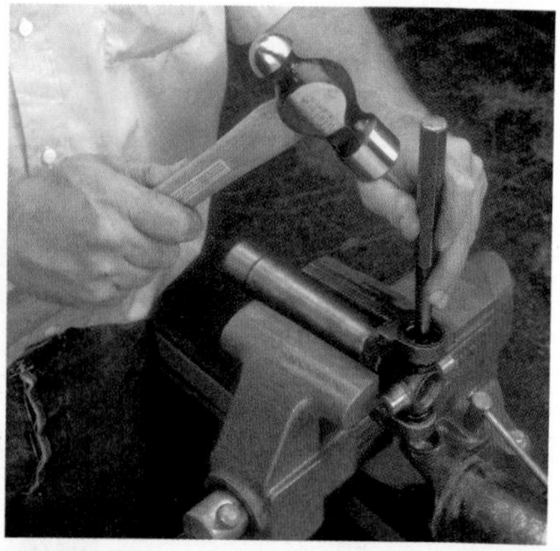

P38-9 *Use a drift and hammer to drive the joint out of the other yokes.*

P39-1 *Clean any dirt from the yoke and the retaining ring grooves.*

P39-2 *Carefully remove the bearing cups from the new U-joint.*

P39-3 *Place the new spider inside the yoke and push it to one side.*

P39-4 *Start one cup into the yoke's ear and over the cross's trunnion.*

P39-5 *Carefully place the assembly in a vise or U-joint bearing press and press the cup partially through the ear.*

P39-6 *Remove the shaft from the vise and push the cross toward the other side of the yoke.*

P39-7 *Start a cup into the yoke's ear and over the trunnion.*

P39-8 *Place the shaft in the vise and tighten the jaws to press the bearing cup into the ear and over the trunnion. Then install the snap rings. Make sure they are seated in their grooves.*

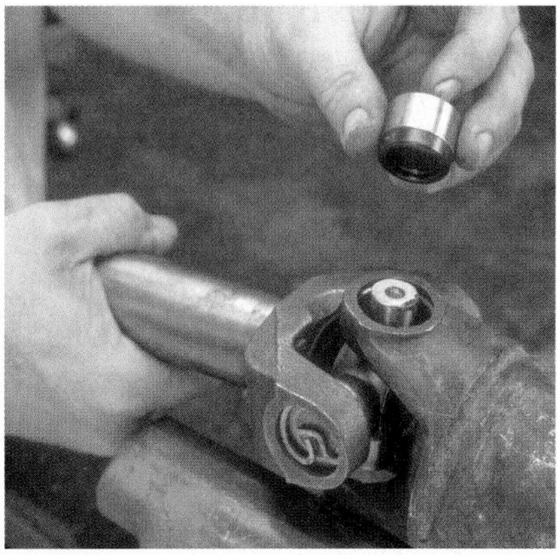

P39-9 *Position the joint's cross in the drive shaft yoke and install the two remaining bearing cups.*

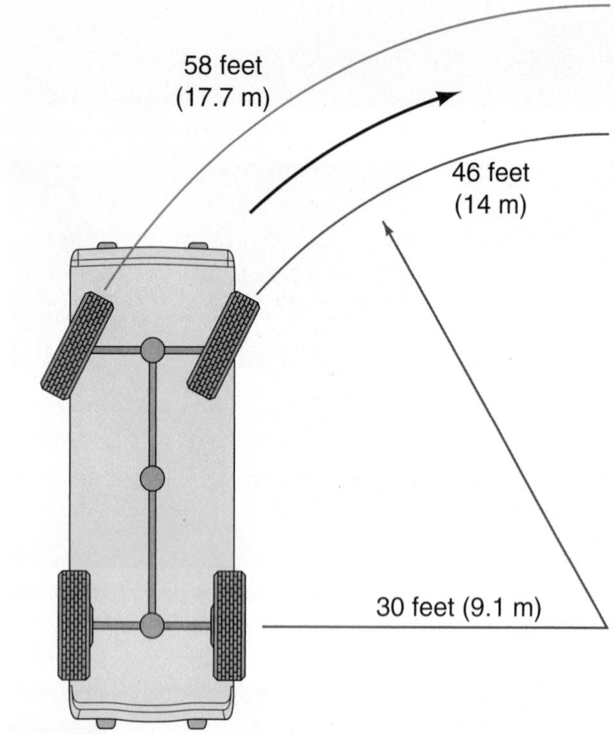

Figure 39-23 Travel of wheels when a vehicle is turning a corner.

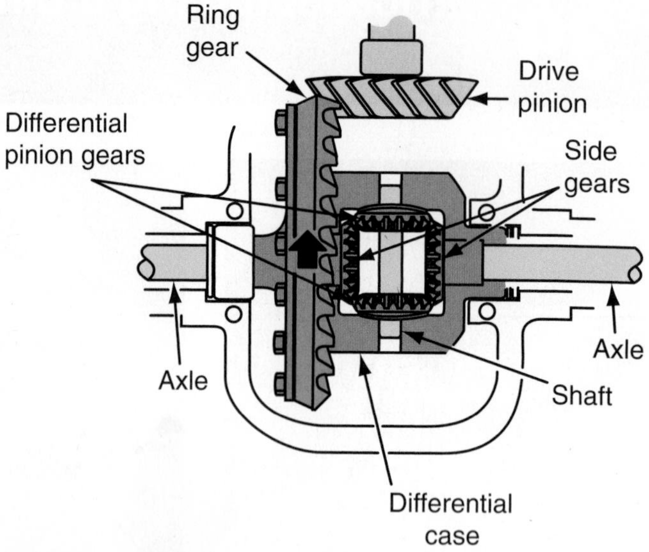

Figure 39-24 The components of a typical final drive unit.

On a FWD car or truck, the differential is normally an integral part of the transaxle assembly located at the front of the vehicle. Transaxle design and operation depends on whether the engine is mounted transversely or longitudinally. With a transversely mounted engine, the crankshaft centerline and drive axle are on the same plane. With a longitudinally mounted power plant, the differential must change the power flow 90 degrees.

On RWD vehicles, the differential is located in the rear axle housing or carrier. The drive shaft connects the transmission with the rear axle gearing. Four-wheel-drive vehicles have differentials on both their front and rear axles.

The differential allows the drive wheels to rotate at different speeds when negotiating a turn or curve in the road and redirects the engine torque from the drive shaft to the rear drive axle shafts. The drive shaft turns in a motion perpendicular to the rotation of the drive wheels. The final drive gears redirect the torque so that the drive axle shafts turn in a motion parallel to the rotation of the drive wheels.

The final drive gears in the drive axle assembly are also sized to provide a gear reduction, or a torque multiplication. Axles with a low (numerically high) gear ratio allow for fast acceleration and good pulling power. Axles with high gear ratios allow the engine to run slower at any given speed, resulting in better fuel conservation.

Components of Final Drives and Differentials

The components of commonly used final drive units are shown in **Figure 39–24**. There are several other basic design arrangements. However, the one most commonly used design has pinion/ring gears and a pinion shaft. In RWD vehicles the gear set is comprised of hypoid gears, whereas FWD units use a planetary gear set or spiral bevel gears. To create a clear picture of the major components of a differential assembly and their required service, RWD units are the focus of the following discussion.

The pinion shaft is mounted in the front of the carrier and is supported by two or three bearings. An overhung pinion gear is supported by two tapered bearings spaced far enough apart to provide the needed leverage to rotate the ring gear and drive axles (**Figure 39–25**). A straddle-mounted pinion gear rests on three bearings: Two tapered bearings on the front support the input shaft and one roller bearing is fitted over a short shaft extending from the rear end of the drive **pinion gear**.

The pinion gear meshes with a **ring gear**. The ring gear is made of hardened steel. The ring gear is bolted to the differential case. When the pinion gear is rotated by the drive shaft, the ring gear rotates and turns the differential case and axle shafts. Most automotive applications have two pinion gears mounted on a straight shaft in the differential case. On heavier trucks, the differential contains four pinion gears mounted on a cross-shaped spider in the differential case. The pinion shafts are mounted in holes in the case (or in matching grooves in the case halves) and are secured by a lock bolt or retaining rings.

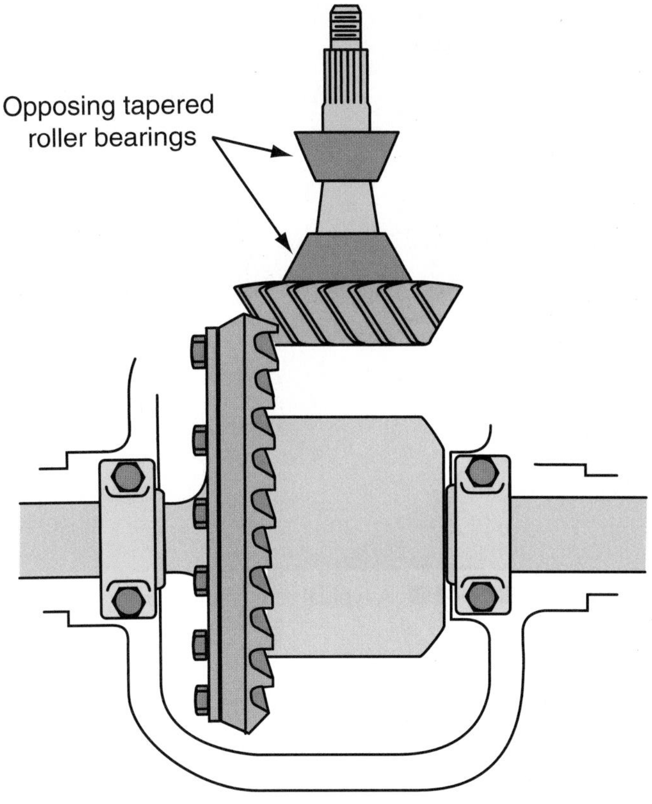

Figure 39-25 The pinion drive is supported by bearings.

Figure 39-27 The flow of oil in a hypoid gear set as it spins. *Courtesy of Dana Corporation*

A differential also contains two side or axle gears. The inside bore of these gears is splined and mates with splines on the ends of the axles. The differential pinion gears and side gears are in constant mesh. The pinion gears are mounted on a pinion gear shaft, which is mounted in the differential casing. As the case turns with the ring gear, the pinion shaft and gears also turn. The pinion gears deliver torque to the side gears.

Hypoid Gear A **hypoid gear** contacts more than one tooth at a time. The hypoid gear also makes contact with a sliding motion. This sliding action provides smooth and quiet operation. With a hypoid gear set, the pinion gear is placed lower in the differential. The drive pinion meshes with the ring gear at a point below its centerline **(Figure 39–26)**.

The sliding effect of two hypoid gears meshing tends to wipe lubricant from the face of the gears, resulting in eventual damage **(Figure 39–27)**. Differentials require the use of extreme pressure-type lubricants. The additives in this type of lubricant allow the lubricant to withstand the wiping action of the gear teeth without separating from the gear face.

Types of Final Drive Gear Sets Ring and pinion gear sets are normally classified as hunting, nonhunting, or partial nonhunting gears. These classifications are based on the number of teeth on the pinion and ring gears. Knowing the type of gear set is important when diagnosing and servicing final drive assemblies.

- *Hunting Gear Set.* When one drive pinion gear tooth contacts every ring gear tooth after several revolutions, this is called a **hunting gear set**. In other words, the drive pinion hunts out each ring gear tooth. A typical hunting gear set may have nine drive pinion teeth and thirty-seven ring gear teeth. The rear-axle ratio for this combination would be 4.11:1.
- *Nonhunting Gear Set.* When one drive pinion gear tooth contacts only certain ring gear teeth, the gear set is a **nonhunting gear set**. A typical

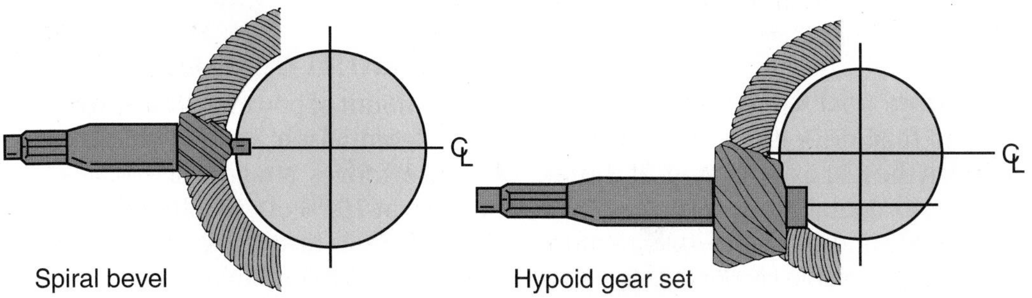

Figure 39-26 In a hypoid gear set, the drive pinion meshes with the ring gear at a point below its centerline.

Paint marking
indicates
position
in which gears
were lapped.

Figure 39-28 Index marks on a ring and pinion gear set. *Courtesy of Ford Motor Company*

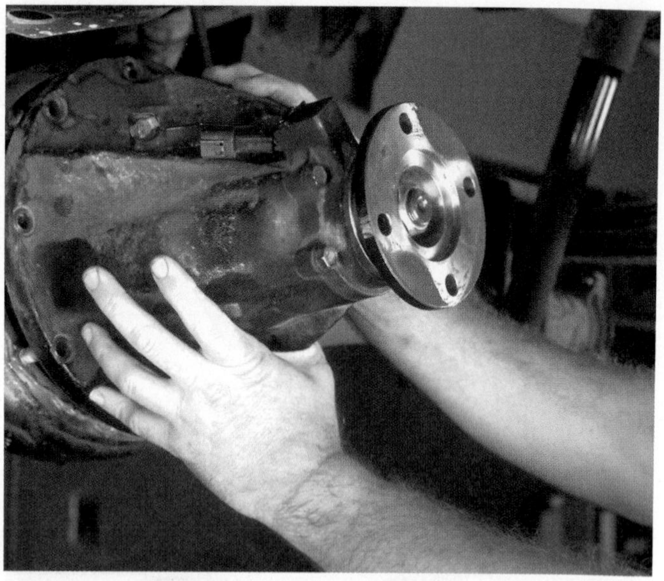

Figure 39-29 A typical removable-carrier axle housing.

nonhunting gear set may have ten drive pinion teeth and thirty ring gear teeth. The rear-axle ratio for this combination would be 3.00:1. For every revolution of the ring gear, each drive pinion tooth would contact the same three teeth of the ring gear. The drive pinion gear teeth do not hunt out all ring gear teeth. Nonhunting gear sets have timing marks that must be aligned **(Figure 39–28)**. These marks allow the gear set to be placed in the same position as they were when the gears were lapped during the manufacturing process. Returning the gears to this position allows for quiet and durable operation.

- *Partial Nonhunting Gear Set.* The difference between nonhunting and **partial nonhunting gear sets** is the number of ring gear teeth that are contacted. In a partial nonhunting gear set, one drive pinion tooth contacts six ring gear teeth instead of three. During the first revolution of the ring gear, one drive pinion tooth contacts three ring gear teeth. During the second revolution of the ring gear, the drive pinion tooth contacts three different ring gear teeth. During every other ring gear revolution, one drive pinion tooth contacts the same ring gear teeth. A typical partial nonhunting gear set may have ten drive pinion teeth and thirty-five ring gear teeth. The rear-axle ratio for this combination would be 3.50:1. These gear sets also have timing marks that must be aligned during service.

Rear Axle Housing and Casing

The differential and final drive gears in a rear-drive vehicle are housed in the rear axle housing, or carrier. The axle housing also contains the two drive axle shafts. Two types of axle housings are found on modern automobiles: the removable carrier and the integral carrier. The removable carrier axle housing is open on the front side. Because it resembles a banjo, it is often called a banjo housing. The backside of the housing is closed to seal out dirt and contaminants and keep in the lubricant. The differential is mounted in a carrier assembly that can be removed as a unit from the axle housing **(Figure 39–29)**. Removable carrier axle housings are most commonly used today on trucks and other heavy-duty vehicles.

The integral housing is most often found on late-model cars and light trucks **(Figure 39–30)**. A cast-iron carrier forms the center of the axle housing. Steel axle tubes are pressed into both sides of the carrier to form the housing. The housing and carrier have a removable rear cover that allows access to the differential assembly. Because the carrier is not removable, the differential components must be removed and serviced separately. For many operations, a case spreader must be used to remove the components. In addition to providing a mounting place for the differential, the axle housing also contains brackets for mounting suspension components such as control arms, leaf springs, and coil springs.

Some vehicles have an ABS speed sensor attached to the carrier housing for rear-wheel lockup prevention during braking.

Differential Operation

The amount of power delivered to each driving wheel by differential is expressed as a percentage. When the vehicle moves straight ahead, each driving wheel rotates at 100% of the differential case speed. When the vehicle is turning, the inside wheel might be getting 90% of the differential case speed. At the same time, the outside wheel might be getting 110% of the differential case speed.

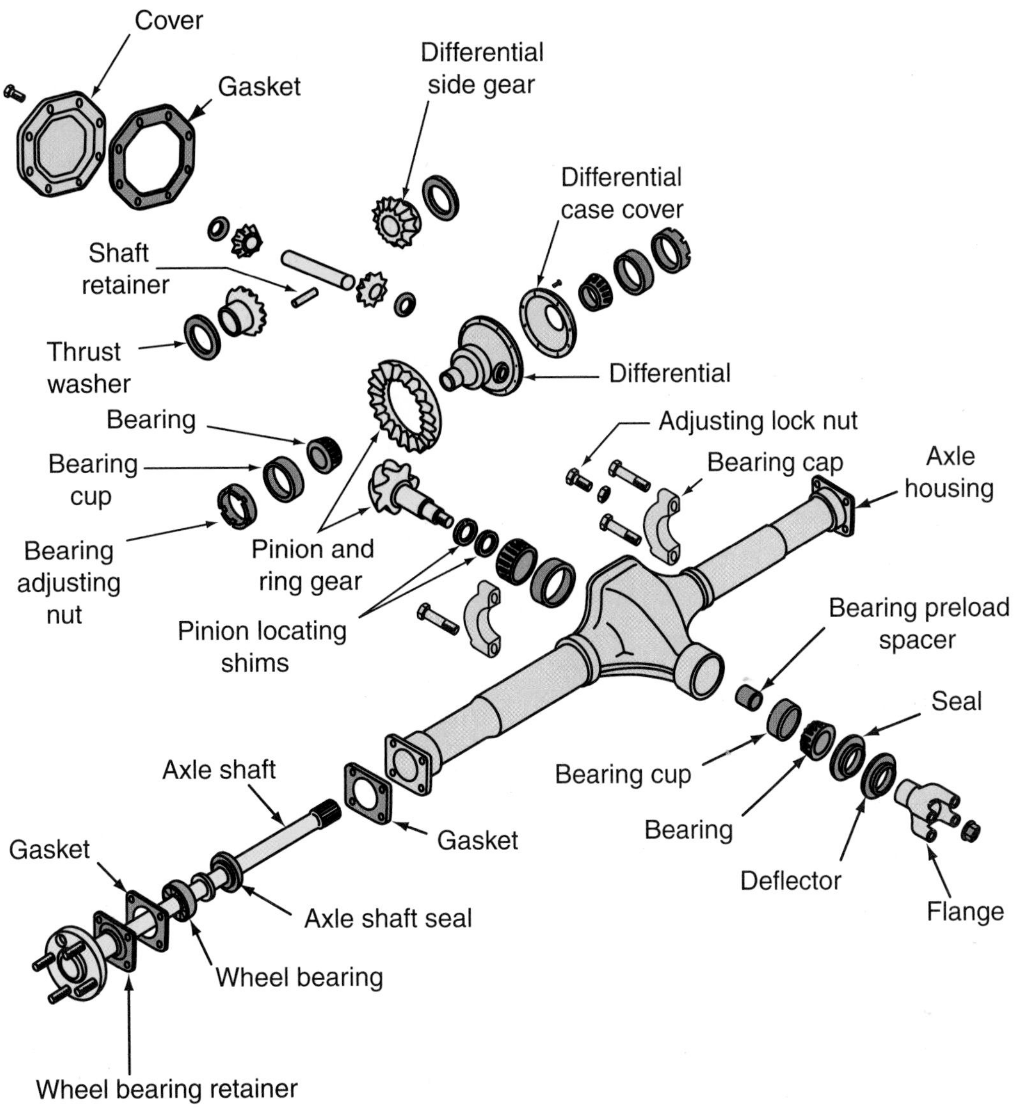

Figure 39-30 An exploded view of an integral-carrier axle housing with a hypoid final drive assembly and semifloating axles.

Power flow through the axle begins at the drive pinion yoke, or companion flange **(Figure 39–31)**. The companion flange accepts torque from the rear U-joint. The companion flange is attached to the drive pinion gear, which transfers torque to the ring gear. As the ring gear turns, it turns the differential case and the pinion shaft. The differential pinion gears transfer torque to the side gears to turn the driving axle shafts. The differential pinion gears determine how much torque goes to each driving axle, depending on the resistance an axle shaft or wheel has while turning. The pinion gears can move with the carrier, and they can rotate on the pinion shaft.

When drive shaft torque is applied to the input shaft and drive pinion, the shaft rotates in a direction that is perpendicular to the vehicle's drive axles. When this rotary motion is transferred to the ring gear, the torque flow changes direction and becomes parallel to the axle shafts and wheels. Because the ring gear is bolted to the differential case, the case must rotate with the ring gear. The pinion gear shaft mounted in the differential case must also rotate with the case and the ring gear. The pinions turn end over end. Gears do not rotate on the pinion shaft when both driving wheels are turning at the same speed. They rotate end over end as the differential case rotates. Because the pinions are meshed with both side gears, the side gears rotate and turn the axle shafts. The ring gear, differential gears, and axle shafts turn together without variation in speed as long as the vehicle is moving in a straight line.

When a vehicle turns into a curve or negotiates a turn, the wheels on the outside of the curve must travel a greater distance than the wheels on the inside of the curve. The outer wheels must then rotate faster than the inside wheels. This would be impossible if the axle shafts were locked solidly to the ring gear. However, the differential allows the outer wheels and

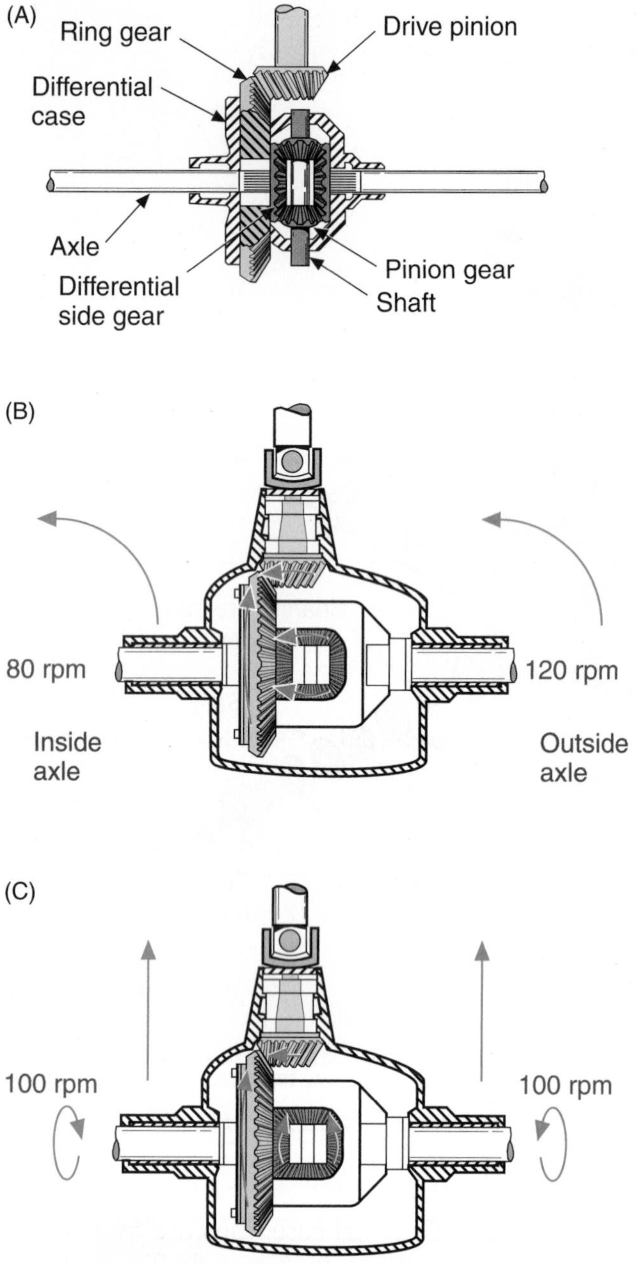

Figure 39–31 (A) Basic differential components; (B) differential action while the vehicle is turning left; (C) differential action while the vehicle is moving straight.

axle shaft to increase in speed and the inner wheels and axle to slow down, thus preventing the skidding and rapid tire wear that would otherwise occur. The differential action also makes the vehicle much easier to control while turning.

For example, when a car makes a sharp right-hand turn, the left-side wheels, axle shaft, and side gear must rotate faster than the right-side wheels, axle shaft, and side gear. The left side of the axle must speed up and the right side must slow down. This is possible because the pinions to which the side gears are meshed are free to rotate on the pinion shaft. The increased speed of the left-side wheels causes the side gear to rotate faster than the differential case. This causes the pinions to rotate and walk around the slowing down side gear. As the pinions turn to allow the left-side gear to increase speed, a reverse action—known as a reverse walking effect—is produced on the right-side gear. It slows down an amount that is inversely proportional to the increase in the left-side gear.

LIMITED-SLIP DIFFERENTIALS

Driveline torque is evenly divided between the two rear drive axle shafts by the differential. As long as the tires grip the road, providing a resistance to turning, the drivetrain forces the vehicle forward. When one tire encounters a slippery spot on the road, it loses traction, resistance to rotation drops, and the wheel begins to spin. Because resistance has dropped, the torque delivered to both drive wheels changes. The wheel with good traction is no longer driven. If the vehicle is stationary in this situation, only the wheel over the slippery spot rotates. When this is occurring, the differential case is driving the differential pinion gears around the stationary side gear.

This situation places stress on the differential gears. When the wheel spins because of traction loss, the speed of some of the differential gears increases greatly, while others remain idle. The amount of heat developed increases rapidly, the lube film breaks down, metal-to-metal contact occurs, and the parts are damaged. If spinout is allowed to continue long enough, the axle could break. The final drive or differential gears can also be damaged from prolonged spinning of one wheel. This is especially true if the spinning wheel suddenly has traction. The shock of the sudden traction can cause severe damage to the drive axle assembly.

To overcome these problems, differential manufacturers have developed the **limited-slip differential (LSD)**. LSDs are manufactured under such names as sure-grip, no-spin, positraction, or equal-lock. Some vehicles use a viscous clutch in their limited-slip drive axles. These units are predominantly used in 4WD vehicles and are discussed in Chapter 43.

Clutch-Based Units

Many LSDs use friction material to transfer the torque applied to a slipping wheel to the one with traction. Those that use a clutch pack **(Figure 39–32)** have two sets (one for each side gear) of clutch plates and friction discs to prevent normal differential action. The friction discs are steel plates with an abrasive coating on both sides. These discs fit over the external splines on the side gears' hub. The clutch plates are also made of steel but have no friction material bonded to them.

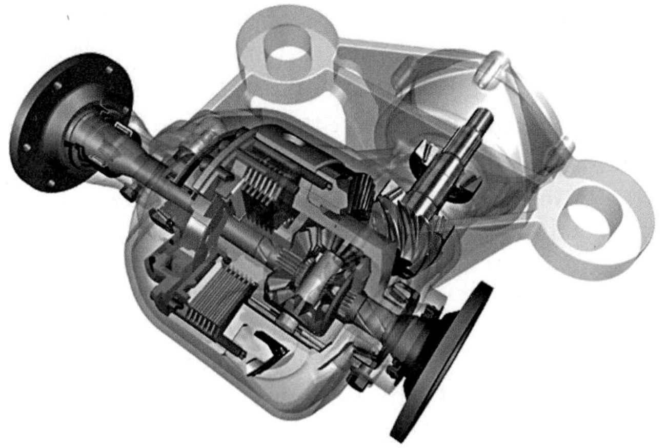

Figure 39-32 A late-model sophisticated LSD with friction clutches. *Courtesy of Dana Corporation*

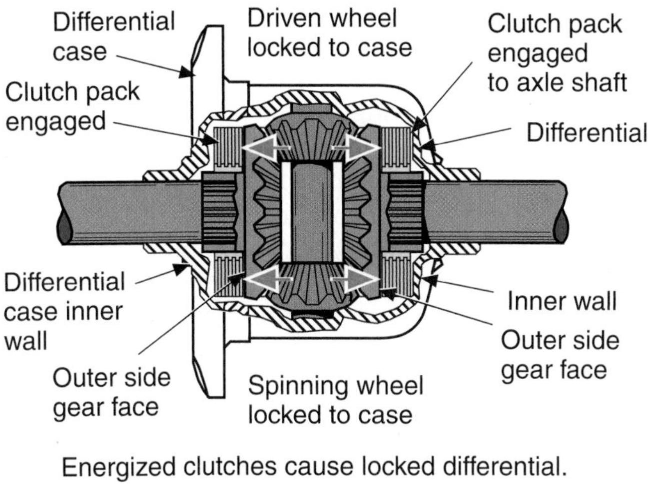

Figure 39-33 Action of the clutches in a limited-slip differential.

The plates are placed between the friction discs and fit into internal splines in the differential case. Pressure is kept on the clutch packs by either an S-shaped spring or coil springs.

As long as the friction discs maintain their grip on the steel plates, the differential side gears are locked to the differential case **(Figure 39–33)**, allowing the case and drive axles to rotate at the same speed and preventing one wheel from spinning faster than the other.

A common LSD uses two cone-shaped parts to lock the side gears to the differential case. The cones are located between the side gears and the case and are splined to the side gear hubs. The exterior surface of the cones is coated with a friction material that grabs the inside surface of the case. Four to six coil springs mounted in thrust plates between the side gears maintain a preload on the cones. When the cones are forced against the case, the axles rotate with the differential case.

The clutch plates and cones are designed to slip when a predetermined amount of torque is applied to

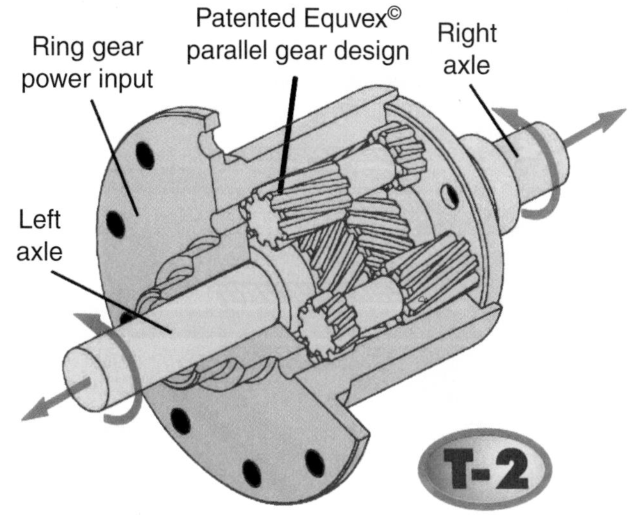

Figure 39-34 A Torsen torque-sensitive LSD. *Courtesy of JTEKT Torsen North America, Inc.*

them, which allows the vehicle to have differential action when it is turning a corner.

Gear-Based Units

Manufacturers are using a wide range of LSD designs other than the typical clutch type. These designs were born out of the need to improve vehicle stability and tire traction. Many are gear-based and are often called torque-bias or torque-sensing (Torsen) units. The basis of these units is a parallel-axis helical gear set **(Figure 39–34)**. The Torsen differential multiplies the torque available from the wheel that is starting to spin or lose traction and sends it to the slower turning wheel with the better traction. This action is initiated by the resistance between the sets of gears in mesh.

Helical-geared LSDs respond very quickly to changes in traction. They also do not bind in turns and do not lose their effectiveness with wear as clutch-based units can.

AXLE SHAFTS

The purpose of an axle shaft is to transfer driving torque from the differential assembly to the vehicle's driving wheels. There are two types of axles: dead and live or drive. A **dead axle** does not drive a vehicle. It merely supports the vehicle load and provides a mounting place for the wheels. The rear axle of a FWD vehicle is a dead axle, as are the axles used on trailers.

A **live axle** is one that drives the vehicle. Drive axles transfer torque from the differential to each driving wheel. Depending on the design, rear axles can also help carry the weight of the vehicle or even act as part of the suspension. Three types of driving axles are commonly used **(Figure 39–35)**: semifloating, three-quarter floating, and full-floating.

All three use axle shafts that are splined to the differential side gears. At the wheel ends, the axles can

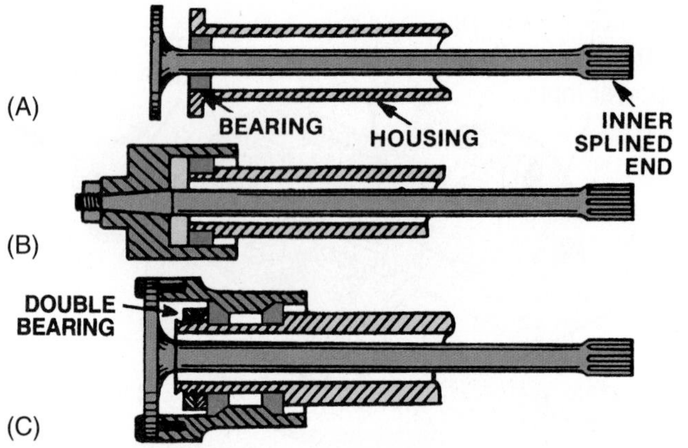

Figure 39-35 The types of rear axle shafts: (A) semifloating; (B) three-quarter floating; and (C) full-floating.

be attached in any one of a number of ways. This attachment defines the type of axle it is and the manner in which the shafts are supported by bearings.

Semifloating Axle Shafts

Semifloating axles help to support the weight of the vehicle. Most RWD vehicles have semifloating axles. The axles are supported by bearings located in the axle housing. An axle shaft bearing supports the vehicle's weight and reduces rotational friction. The inner ends of the axle shafts are splined to the axle side gears. The axle shafts transmit only driving torque and are not acted upon by other forces. Therefore, the axle shafts are said to be floating.

The driving wheels are bolted to the outer ends of the axle shafts. The outer axle bearings are located between the axle shaft and axle housing. This type of axle has a bearing pressed into the end of the axle housing. This bearing supports the axle shaft. The axle shaft is held in place with either a bearing retainer belted to a flange on the end of the axle housing or by a C-shaped washer that fits into grooves machined in the splined end of the shaft. A flange on the wheel end of the shaft is used to attach the wheel.

When semifloating axles are used to drive the vehicle, the axle shafts push on the shaft bearings as they rotate. This places a driving force on the axle housing, springs, and vehicle chassis, moving the vehicle forward. The axle shaft faces the bending stresses associated with turning corners and curves, skidding, and bent or wobbling wheels, as well as the weight of the vehicle. In the semifloating axle arrangement with a C-shaped washer-type retainer, if the axle shaft breaks, the driving wheel comes away from or out of the axle housing.

Three-Quarter Floating Axle

The wheel bearing on a **three-quarter floating axle** is on the outside of the axle housing instead of inside the housing as in the semifloating axle. The wheel hubs are bolted to the end of the axle shaft and are supported by the bearing. In this arrangement, the axle shaft only supports 25% of the vehicle's weight. The weight is transferred through the wheel hub and bearing to the axle housing. Three-quarter floating axles are found on older vehicles and some trucks.

Full-Floating Axle Shafts

Most medium- and heavy-duty vehicles use a **full-floating axle shaft**. This design is similar to the three-quarter floating axle except that two bearings rather than one are used to support the wheel hub. These are slid over the outside of the axle housing and carry all of the stresses caused by torque loading and turning. The wheel hubs are bolted to flanges on the outer end of each axle shaft.

In operation, the axle shaft transmits only the driving torque. The driving torque from the axle shaft rotates the axle flange, wheel hub, and rear driving wheel. The wheel hub forces its bearings against the axle housing to move the vehicle. The stresses caused by turning, skidding, and bent or wobbling wheels are taken by the axle housing through the wheel bearings. If a full-floating axle shaft should break, it can be removed from the axle housing. Because the rear wheels rotate around the rear axle housing the disabled vehicle can be towed to a service area for replacement of the axle shaft.

Independently Suspended Axles

In an independently suspended axle system, the driving axles are usually open instead of being enclosed in an axle housing. The two most common suspended rear driving axles are the DeDion axle system and the swing axle system.

The DeDion axle system resembles a normal driveline. The driving axles look like a drive shaft with U-joints at each end of the axles. A slip joint is attached to the innermost U-joint. The outboard U-joint is connected to the wheel hub, which allows the driving axle to move up and down as it rotates.

On vehicles that use a swing axle, the driving axle shafts can be open or enclosed. An axle fits into the differential by way of a ball-and-socket system. The ball-and-socket system allows the axle to pivot up and down. As the axle pivots, the driving wheel swings up and down. This system best describes the drive axles of a FWD vehicle.

Axle Shaft Bearings

The axle shaft bearing supports the vehicle's weight and reduces rotational friction. In an axle mount, radial and thrust loads are always present on the axle shaft bearing when the vehicle is moving. Radial bearing loads act at 90 degrees to the axle shaft's center of axis. **Radial loading** is always present whether or not the vehicle is moving.

Thrust loading acts on the axle bearing parallel with the center of axis. It is present on the driving wheels, axle shafts, and axle bearings when the vehicle turns corners or curves.

There are three designs of axle shaft bearings used in semifloating axles: ball-type bearing, straight roller bearing, and tapered roller bearing.

The bearing load of primary concern is axle shaft end thrust. When a vehicle moves around a corner, centrifugal force acts on the vehicle body, causing it to lean to the outside of the curve. The vehicle's chassis does not lean because of the tires' contact with the road's surface. As the body leans outward, a thrust load is placed on the axle shaft and axle bearing. Each type of axle shaft bearing handles end thrust differently.

Normally, the way the axles are held in the housing is quite obvious after the rear wheels and brake assemblies have been removed. If the axle shaft is held in by a retainer and three or four bolts, it is not necessary to remove the differential cover to remove the axle. Most ball and tapered roller bearing supported axle shafts are retained in this manner **(Figure 39–36)**. To remove the axle, remove the bolts that hold the retainer to the backing plate, then pull the axle out. Normally, the axle shaft slides out without the aid of a puller. Sometimes a puller is required.

A straight-roller bearing supported axle shaft does not use a retainer to secure it. Rather, a C-shaped washer is used to retain the axle shaft **(Figure 39–37)**. This C-shaped washer is located inside the differential, and the differential cover must be removed to gain access to it. To remove this type of axle, first remove the wheel, brake drum, and differential cover. Then remove the differential pinion shaft retaining bolt and differential pinion shaft. Now push the axle shaft in and remove the C-shaped washer. The axle can now be pulled out of the housing.

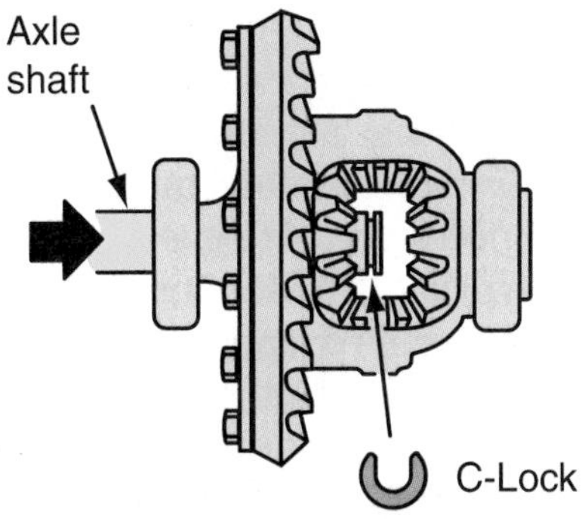

Figure 39-37 The location of C-lock type axle shaft retainers.

Ball bearings are lubricated with grease packed in the bearing at the factory. An inner seal, designed to keep the gear oil from the bearing, rides on the axle shaft just in front of the retaining ring. This type of bearing also has an outer seal to prevent grease from spraying onto the rear brakes. Ball-type axle bearings are pressed on and off the axle shaft. The retainer ring is made of soft metal and is pressed onto the shaft against the wheel bearing. Never use a torch to remove the ring. Rather, drill into it or notch it in several places with a cold chisel to break the seal **(Figure 39–38)**. The ring can then be slid off the shaft easily. Heat should not be used to remove the ring because it can take the temper out of the shaft and thereby weaken

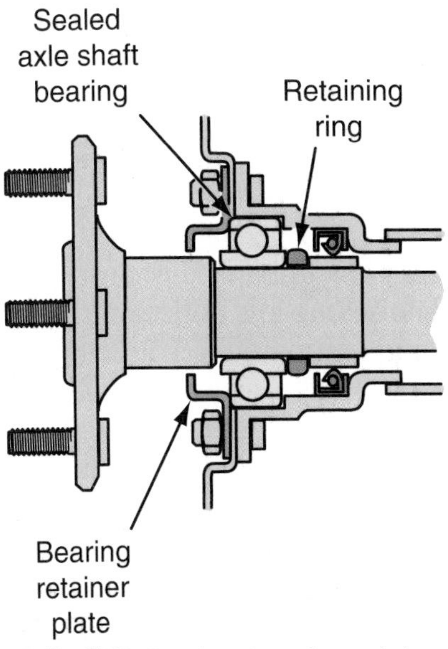

Figure 39-36 The location of an axle bearing retainer.

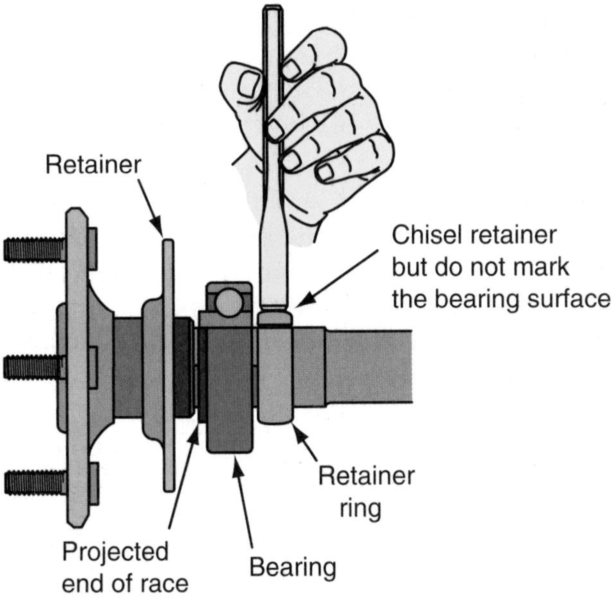

Figure 39-38 Freeing the retainer ring from an axle shaft.

it. Likewise, a torch should never be used to remove a bearing from an axle shaft.

Roller axle bearings are lubricated by the gear oil in the axle housing. Therefore, only a seal to protect the brakes is necessary with these bearings. These bearings are typically pressed into the axle housing and not onto the axle. To remove them, the axle must first be removed and then the bearing pulled out of the housing. With the axle out, inspect the area where it rides on the bearing for pits or scores. If pits or score marks are present, replace the axle.

Tapered-roller axle bearings are not lubricated by gear oil. They are sealed and lubricated with wheel grease. This type of bearing uses two seals and must be pressed on and off the axle shaft using a press. After the bearing is pressed onto the shaft, it must be packed with wheel bearing grease. After packing the bearing, install the axle in the housing. Shaft end play must be checked. Use a dial indicator and adjust the end play to the specifications given in the service manual. If the end play is not within specifications, change the size of the bearing shim.

The installation of new axle shaft seals is recommended whenever the axle shafts have been removed. Some axle seals are identified as being either right or left side. When installing new seals, make sure to install the correct seal in each side. Check the seals or markings of right or left or for color coding.

USING SERVICE INFORMATION

The driveline can create some especially difficult diagnostic problems. The driveline easily picks up vibrations and noises from other parts of the vehicle. A test drive is the best way to begin diagnosis. Most service manuals have a road test checklist that helps with identifying the cause of a noise or vibration.

Diagnosis

The key to diagnosing a drive axle or differential problem is to note what happens during different vehicle speeds and speed changes. Most manufacturers recommend that the vehicle be operated in four distinct modes during the road test.

- *Drive mode.* The vehicle is accelerated; the throttle must be depressed enough to apply sufficient engine torque.
- *Cruise mode.* Vehicle speed is held constant. This means that the throttle must be applied at all times. The speed must be held at a predetermined rpm on a level road.
- *Coast mode.* The throttle is released and the vehicle is allowed to coast down from a specific speed.
- *Float mode.* This is controlled deceleration. The throttle is slowly released. It is important that the brakes be applied during this test mode.

During each of these modes, the stress on the various driveline parts changes. The cause of a problem is identified by thinking about what is under and not under stress during each mode.

CAUTION!

Remember that driving safely is always important! Hard cornering or sudden braking should be avoided. Abusive driving can worsen a problem or create a new one. Carelessness during the test drive can result in an accident. Remember, you are responsible for someone else's vehicle.

SERVICING THE FINAL DRIVE ASSEMBLY

Before removing a final drive unit for service, make sure it needs to be serviced. Typically, problems with the differential and drive axles are first noticed as a leak or noise. As the problem worsens, vibrations or a clunking noise might be felt during certain operating conditions. Diagnosis of the problem should begin with a road test in which the vehicle is taken through the different modes of operation.

Basic Diagnosis

It is common for the source of a noise to be tires, not the final drive unit. To make sure the noises are not caused by tire tread patterns and/or wear, drive the car on various types of road surfaces (asphalt, concrete, and packed dirt). If the noise changes with the road surfaces, it means the tires are the cause of the noise.

Another way to isolate tire noises is to coast at speeds less than 30 mph (48 km/h). If the noise is still heard, the tires are probably the cause. Drive axle and differential noises are less noticeable at these speeds. Accelerate and compare the sounds to those made while coasting. Drive axle and differential noises change. Tire noise remains constant.

Sometimes it is difficult to distinguish between axle-bearing noises and noises coming from the differential. Differential noises often change with the driving mode, whereas axle-bearing noises are usually constant. The sound of the bearing noise usually increases in speed and loudness as vehicle speed increases.

Operational noises are generally caused by bearings or gears that are worn, loose, or damaged. Bearing noises might be a whine or a rumble. A whine is a high-pitched, continuous "whee" sound. A rumble sounds like distant thunder.

Gears can also whine or emit a howl—a very loud, continuous sound. Howling is often caused by low lubricant in the drive axle housing. The meshing teeth scrape metal from each other and can be heard in all gear ranges. If topping up the lubrication level does not alleviate the howling noise, then the drive pinion and ring gear must be replaced.

Disassembly

Although FWD axle final drive units are normally an integral part of the transaxle, most of the procedures for servicing RWD units apply to them. To service a final drive assembly in removable carrier housing, the unit must be removed from the housing. Units in integral carriers are serviced in the housing.

A highly important step in the procedure for disassembling any final drive unit is a careful inspection of each part as it is removed. The bearings should be looked at and felt to determine if there are any defects or evidence of damage.

After the ring and pinion gears have been inspected and before they have been removed from the assembly, check the side play. Using a screwdriver, attempt to move the differential case assembly laterally. Any movement is evidence of side play. Side play normally indicates that as the result of loose bearing cones on the differential case hubs, the differential case must be replaced.

Prior to disassembling the assembly, measure the runout of the ring gear. Excessive runout can be caused by a warped gear, worn differential side bearings, warped differential case, or particles trapped between the gear and case. Runout is checked with a dial indicator mounted on the carrier assembly. The plunger on the indicator should be set at a right angle to the gear. With the dial indicator in position and its dial set to zero, rotate the ring gear and note the highest and lowest readings. The difference between these two readings indicates the total runout of the ring gear. Normally, the maximum permissible runout is 0.003 to 0.004 inches (0.0762 to 0.1016 mm).

To determine if the runout is caused by a damaged differential case, remove the ring gear and measure the runout of the ring gear mounting surface on the differential case. Runout should not exceed 0.004 inch (0.1016 mm). If runout is greater than that, the case should be replaced. If the runout was within specifications, the ring gear is probably warped and should be replaced. A ring gear is never replaced without replacing its mating pinion gear.

Some ring gear assemblies have an excitor ring, used in antilock brake systems. This ring is normally pressed onto the ring gear hub and can be removed after the ring gear is removed. If the ring gear assembly is equipped with an excitor ring, carefully inspect it and replace it if it is damaged.

Prior to disassembling the unit, the drive shaft must be removed. Before disconnecting it from the pinion's companion flange, locate the shaft-to-pinion alignment marks. If they are not evident, make new ones. This avoids assembling the unit with the wrong index, which can result in driveline vibration.

During disassembly, keep the right and left shims, cups, and caps separated. If any of these parts are reused, they must be installed on the same side as they were originally located.

Assembly

When installing a ring gear onto the differential case, make sure the bolt holes are aligned before pressing the gear in place. While pressing the gear, pressure should be evenly applied to the gear. Likewise, when tightening the bolts, always tighten them in steps and to the specified torque. These steps reduce the chances of distorting the gear.

Examine the gears to locate any timing marks on the gear set that indicate where the gears were lapped by the manufacturer. Normally, one tooth of pinion gear is grooved and painted, while the ring gear has a notch between two painted teeth. If the paint marks are not evident, locate the notches. Proper timing of the gears is set by placing the grooved pinion tooth between the two marked ring gear teeth. Some gear sets have no timing marks. These gears are hunting gears and do not need to be timed. Nonhunting and partial nonhunting gears must be timed.

Whenever the ring and pinion gears or the pinion or differential case bearings are replaced, pinion gear depth, pinion bearing preload, and the ring and pinion gear tooth patterns and backlash must be checked and adjusted. This holds true for all types of differentials except most FWD differentials that use helical-cut gears, and taking tooth patterns is not necessary. Nearly all other final drive units use hypoid gears that must be properly adjusted to ensure a quiet operation.

Pinion gear depth is adjusted with shims placed behind the pinion bearing **(Figure 39–39)** or in the housing. The thickness of the drive pinion rear bearing shim controls the depth of the mesh between the pinion and ring gear. To determine and set pinion depth, a special tool is normally used to select the proper pinion shim **(Figure 39–40)**. Always follow the procedures in the service manual when setting up the tool and determining the proper shim.

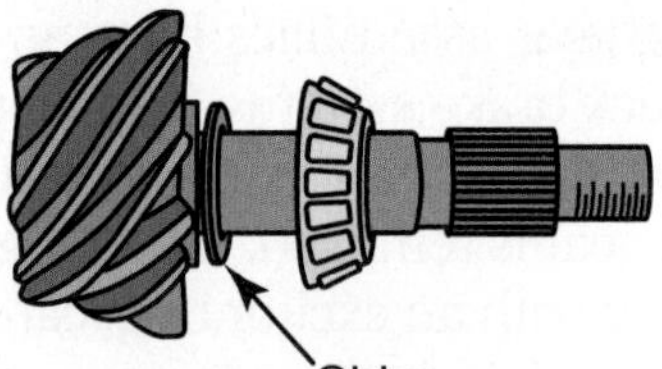

Figure 39-39 The typical placement of a pinion gear depth shim.

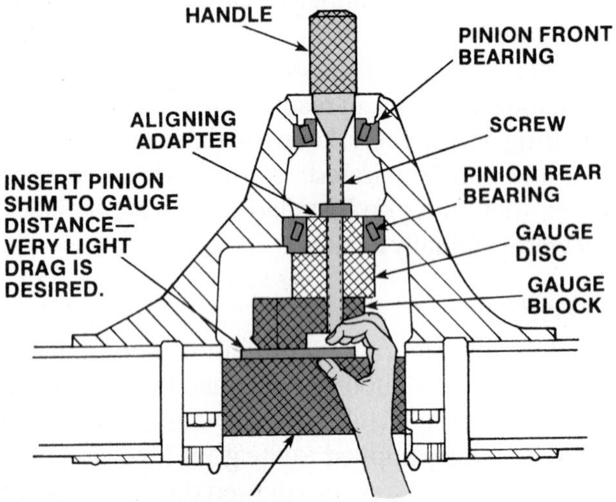

Figure 39-40 A special tool for measuring proper pinion gear depth.

Pinion bearing **preload** is set by tightening the pinion nut until the desired number of inch-pounds is required to turn the shaft. Tightening the nut crushes the collapsible pinion spacer, which maintains the desired preload. Never overtighten and then loosen the pinion nut to reach the desired torque reading. Tightening and loosening the pinion nut damages the **collapsible spacer**. It must then be replaced. For the exact procedures and specifications for bearing preload, refer to the service manual. Incorrect bearing preload can cause differential noise. Some cases use shims to set pinion bearing preload.

It is recommended that a new pinion seal be installed whenever the pinion shaft is removed from the differential. To install a new seal, thoroughly lubricate it and press it in place with an appropriate seal driver.

Backlash of the gear set is adjusted at the same time as the side-bearing preload. Side-bearing preload limits the amount the differential is able to move laterally in the axle housing. Adjusting backlash sets the depth of the mesh between the ring and pinion gear teeth. Both of these are adjusted by shim thickness or by the adjustments made by the side-bearing adjusting nuts. Photo Sequence 40 goes through the typical procedure for measuring and adjusting backlash and side-bearing preload on a gearset that uses shims for adjustment. Photo Sequence 41 covers the same steps for a unit that has adjusting nuts.

A typical procedure for measuring and adjusting backlash and preload involves rocking the ring gear and measuring its movement with a dial indicator. Compare measured backlash with the specifications. Make the necessary adjustments. Then recheck the backlash at four points equally spaced around the ring gear. Normally, backlash should be less than 0.004 inch (0.1016 mm).

The pattern of gear teeth determines how quietly two meshed gears run. The pattern also describes where on the faces of the teeth the two gears mesh. The pattern should be checked during teardown for gear noise diagnosis, after adjusting backlash and side-bearing preload, or after replacing the drive pinion and setting up the pinion bearing preload. The terms commonly used to describe the possible patterns on a ring gear and the necessary corrections are shown in **Figure 39–41**.

To check the gear tooth pattern, paint several ring gear teeth with nondrying Prussian blue, ferric oxide, or red or white lead marking compound (**Figure 39–42**). White marking compound is preferred by many technicians because it tends to be more visible than the others are. Use the pinion gear yoke or companion flange to rotate the ring gear. This will preload the ring gear while it is rotating and will simulate vehicle load. Rotate the ring gear so the painted teeth contact the pinion gear. Move it in both directions enough to get a clearly defined pattern. Examine the pattern on the ring gear and make the necessary corrections.

Most new gear sets purchased today come with a pattern prerolled on the teeth. This pattern provides the quietest operation for that gear set. Never wipe this pattern off or cover it up. When checking the pattern on a new gear set, only coat half of the ring gear

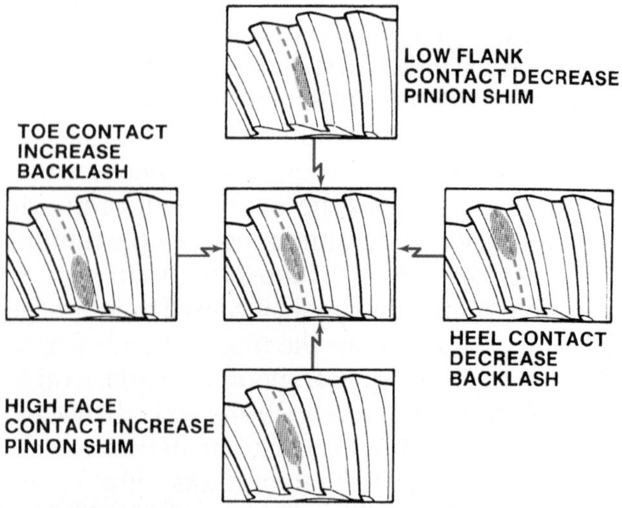

Figure 39-41 Commonly used terms for describing the possible patterns on a ring gear with the recommended corrections. *Courtesy of Ford Motor Company*

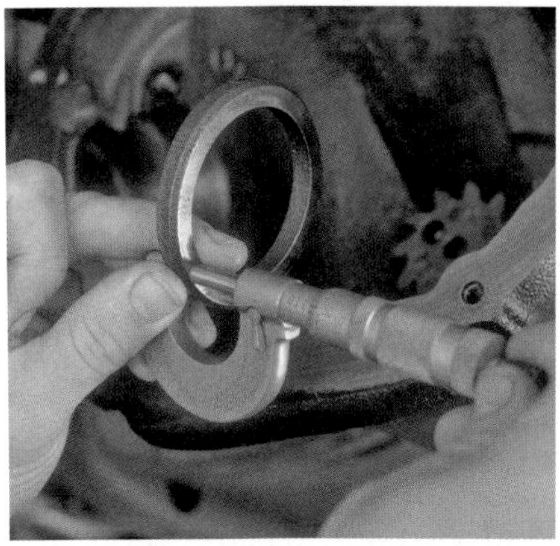

P40-1 *Measure the thickness of the original side bearing preload shims.*

P40-2 *Install the differential case into the housing.*

P40-3 *Install service spacers that are the same thickness as the original preload shims between each bearing cup and the housing.*

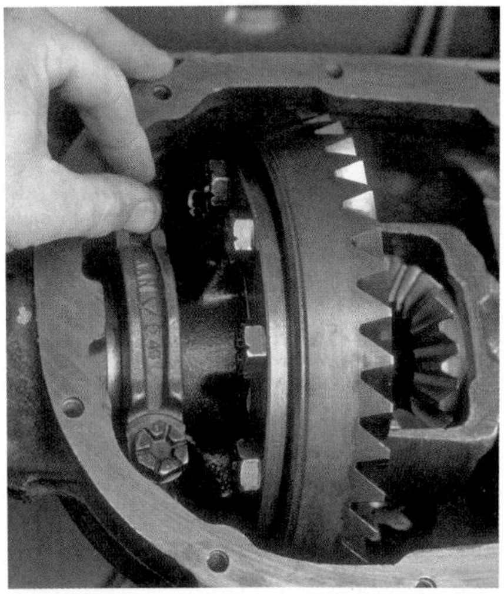

P40-4 *Install the bearing caps and finger tighten the bolts.*

P40-5 *Mount a dial indicator to the housing so that the button of the indicator touches the face of the ring gear. Using two screwdrivers, pry between the shims and the housing. Pry to one side and set the dial indicator to zero, then pry to the opposite side and record the reading.*

P40-6 *Select two shims with a combined thickness to that of the original shims plus the indicator reading, then install them.*

P40-7 *Using the proper tool, drive the shims into position until they are fully seated.*

P40-8 *Install and tighten the bearing caps to specifications.*

P40-9 *Check the backlash and preload of the gear set. Check the backlash by holding the input pinion, rocking the ring gear, and noting the movement on the dial indicator. Adjust the shim pack to allow for the specified backlash. Recheck the backlash at four points equally spaced around the ring gear.*

PHOTO SEQUENCE 41 Measuring and Adjusting Backlash and Side-Bearing Preload on a Final Drive Assembly with Adjusting Nuts

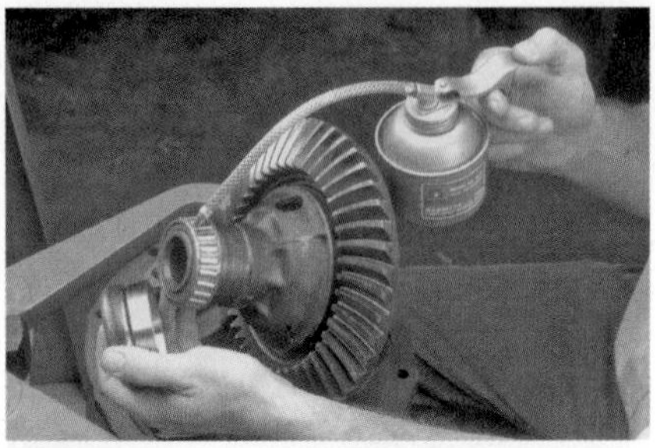

P41-1 *Lubricate the differential bearings, cups, and adjusters.*

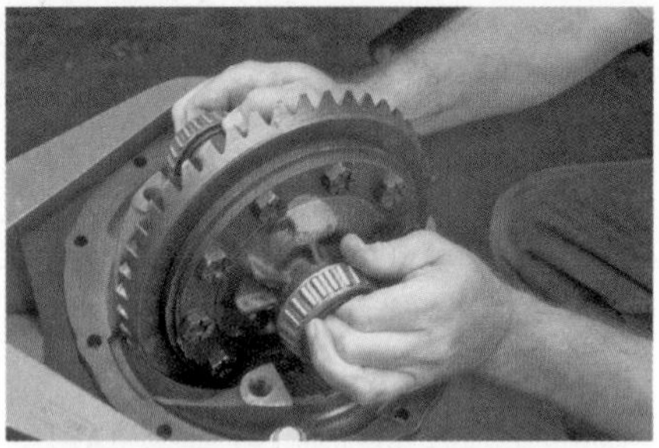

P41-2 *Install the differential case into the housing.*

P41-3 *Install the bearing cups and adjusting nuts onto the differential case.*

P41-4 *Snugly tighten the top bearing cup bolts and finger tighten the lower bolts.*

P41-5 *Turn each adjuster until bearing free play is eliminated with little or no backlash present between the ring and pinion gears.*

P41-6 *Seat the bearings by rotating the pinion several times each time the adjusters are moved.*

P41-7 *Install a dial indicator and position the plunger against the drive side of the ring gear. Set the dial to zero. Using two screwdrivers, pry between the differential case and the housing. Observe the reading.*

P41-8 *Determine how much the preload needs to be adjusted and set the preload by turning the right adjusting nut.*

P41-9 *Check the backlash by rocking the ring gear and noting the movement on the dial indicator.*

P41-10 *Adjust the backlash by turning both adjusting nuts in equal amounts so that the preload adjustment remains unchanged.*

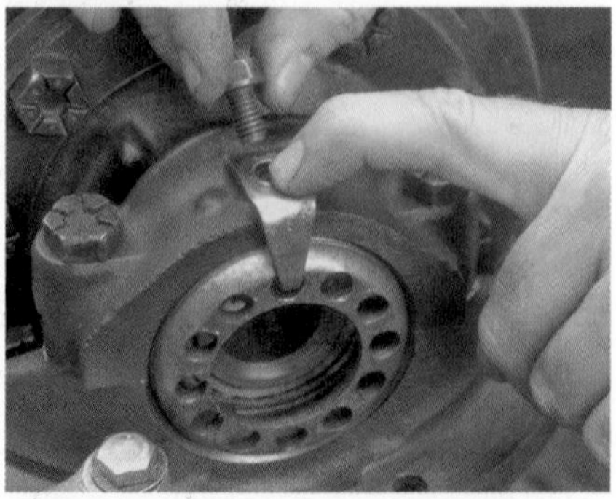

P41-11 *Install the locks on the adjusting nuts.*

P41-12 *Tighten the bearing cap bolts to the specified torque.*

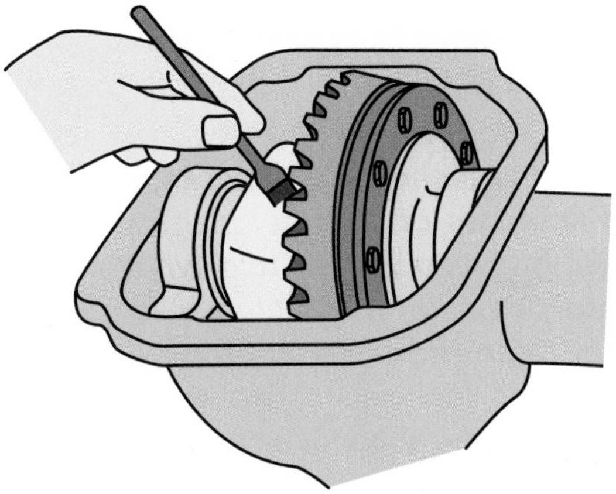

Figure 39-42 To check gear tooth patterns, several teeth of the ring gear are coated with a marking compound and the pinion gear is rotated with the ring gear. The resultant pattern shown on the teeth determines how the gear set ought to be adjusted.

with the marking compound and compare the pattern with the prerolled pattern.

Maintenance

Maintenance includes inspecting the level of and changing the gear lubricant, and lubricating the U-joints if they are equipped with zerk or grease fittings. Most modern U-joints are of the extended life design, meaning they are sealed and require no periodic lubrication. However, it is wise to inspect the joints for hidden grease plugs or fittings.

Proper lubrication is necessary for drive axle durability. Different applications require different gear lubes. The American Petroleum Institute (API) has established a rating system for the various gear lubes available. In general, rear axles use either SAE 80- or 90-weight gear oil for lubrication, meeting API GL-4 or GL-5 specifications. With limited-slip axles, it is very important that the proper gear lube be used. Most often, a special friction modifier fluid should be added to the fluid. If the wrong lubricant is used, damage to the clutch packs and grabbing or chattering on turns will result. If this condition exists, try draining the oil and refilling with the proper gear lube before servicing it.

DIAGNOSING DIFFERENTIAL NOISES

If a whining is heard when turning corners or rounding curves, the problem might be damaged differential pinion gears and pinion shaft. This damage is caused when the inside diameter of the differential pinions and the outside diameter of the differential pinion shaft are scored and damaged. The damage is usually caused by allowing one driving wheel to revolve at high speeds while the opposite wheel remains stationary.

Another gear noise that is common in differentials is the chuckle. A chuckle is a low "heh-heh" sound that occurs when gears are worn to the point where there is excessive clearance between the pinion gear and the ring gear. Chuckle sounds occur most often in the decelerating mode, particularly below 40 mph (65 km/h). As the vehicle decelerates, the chuckle also slows and can be heard all the way to a stop.

A knock or clunk is caused by excessive wear or loose or broken parts. A knock is a repetitious rapping sound that occurs during all phases of driving but is most noticeable during acceleration and deceleration when the gears are loaded.

A clunk is a sharp, loud noise caused by one part hitting another. Unlike a knock, a clunk can be felt as well as heard. Clunks are generally caused by loose parts striking each other.

Limited-slip clutch packs or cones that need servicing might be heard as a chatter or a rapid clicking noise that creates a vibration in the vehicle. Chattering is usually noticed when rounding a corner. A change of differential lubricant and adding friction modifier to the fluid sometimes corrects this problem. After draining the oil, replace it with the manufacturer's suggested friction modifier and lubricant. Road test the vehicle again.

To make sure that the noise heard during the test drive is coming from the differential, stop the vehicle and shift the transmission into neutral. Run the engine at various rpm levels. If the noise is heard during this procedure, it is caused by a problem somewhere other than in the differential.

Vibration Problems

Often the source of vibration is a bent axle or axle flange, or improper mounting of the wheel to the flange. To check the runout of the flange, position a dial indicatior against the outer flange surface of the axle. Apply slight pressure to the center of the axle to remove the endplay in the axle, and then zero the indicator. Slowly rotate the axle one complete revolution and observe the readings on the indicator. The total amount of indicator movement is the total amount of axle flange lateral runout. Compare the measured runout with specifications.

Inspect the wheel studs in the axle flange. If they are broken or bent, they should be replaced. Also check the condition of the threads. If they have minor distortions, run a die over the stud. If the threads are severely damaged, the stud should be replaced. Studs are normally pressed in and out of the flange. Make sure you install the correct size.

CASE STUDY

A customer brings his subcompact, front-wheel-drive car into the shop complaining of recurring noise in the front wheels. The noise is most noticeable when the car is making turns. The owner states he had a similar complaint a few months ago and the shop had replaced an outer CV joint. This corrected the problem until recently. The customer suspects that the same joint went bad again and demands that the shop replace it, free of charge, because it is obvious that the replacement joint was defective.

The service writer records the information from the customer and tells him that he will be notified as soon as the problem is diagnosed. As soon as the customer leaves, the service writer looks up the customer's file and finds that a CV joint had been replaced 2 months ago.

The service writer gives the repair order to the technician, along with the old repair order. The technician begins the diagnostic procedure with a test drive to verify the complaint. From the test drive and a visual inspection, the technician concludes that the same CV joint is faulty. What could cause the joint to fail so soon? Was the replacement joint defective? Was the replacement joint installed incorrectly? Is some other fault causing the joints to fail? No matter what the answer, it seems that the customer will not be charged for this repair. Also, it is likely that the technician will not get paid for the repair.

Upon disassembly of the axle, the technician finds the joint's lubricant to be contaminated with metal shavings and moisture. A thorough inspection of the boot reveals no tears or punctures. While inspecting the boot, it is noticed that the inner end of the boot moves freely on the axle shaft. The technician knows then what had caused the contamination and resulting premature failure of the joint.

When installing the replacement joint and boot, the technician failed to properly tighten the inner boot clamp. This allowed lubricant to leak from and water to enter the boot. A new joint and boot are installed on the axle. The technician verifies the repair by a test drive. The replacement joint took care of the noise. Before releasing the car back to the owner, the technician rechecks the position and tightness of the boot clamps. The customer is called and is told what had happened. Although the customer is not happy about the mistake, he appreciates the honesty of the technician. Two months later he returned to the shop for an oil change. He has been a regular customer ever since.

KEY TERMS

Ball-type joint
Collapsible spacer
Cross groove joint
Dead axle
Double-Cardan joint
Double-offset joint
Fixed joint
Full-floating axle shaft
Hunting gear set
Hypoid gear
Inboard joint
Limited-slip differential (LSD)
Live axle
Nonhunting gear set
Outboard joint
Partial nonhunting gear set
Pinion gear
Plunge joint
Preload
Radial loading
Ring gear
Rzeppa joint
Semifloating axle
Slip yoke
Three-quarter floating axle
Thrust loading
Torsional damper
Tripod-type joint

SUMMARY

- FWD axles generally transfer engine torque from the transaxle to the front wheels.
- Constant velocity (CV) joints provide the necessary transfer of uniform torque and a constant speed while operating through a wide range of angles.
- In FWD drivetrains, two CV joints are used on each half shaft. The different types of joints can be referred to by position (inboard or outboard), by function (fixed or plunge), or by design (ball-type or tripod).
- FWD half shafts can be solid or tubular, of equal or unequal length, and with or without damper weights.
- Most problems with FWD systems are noted by noise and vibration.
- Lubricant is the most important key to a long life for the CV joint.

- A U-joint is a flexible coupling located at each end of the drive shaft between the transmission and the pinion flange on the drive axle assembly.
- A U-joint allows two rotating shafts to operate at a slight angle to each other; this is important to RWD vehicles.
- A failed U-joint or damaged drive shaft can exhibit a variety of symptoms. A clunk that is heard when the transmission is shifted into gear is the most obvious. You can also encounter unusual noise, roughness, or vibrations.
- A differential is a geared mechanism located between the driving axle shafts of a vehicle. Its job is to direct power flow to the driving axle shafts. Differentials are used in all types of power trains.
- The differential performs several functions. It allows the drive wheels to rotate at different speeds when negotiating a turn or curve in the road, and the differential drive gears redirect the engine torque from the drive shaft to the rear-drive axles.
- The final drive and differential of a RWD vehicle is housed in the axle housing, or carrier housing.
- The purpose of the axle shaft is to transfer driving torque from the differential and final drive assembly to the vehicle's driving wheels.
- There are three types of driving axle shafts commonly used: semifloating, three-quarter floating, and full-floating.
- Axle shaft bearings may support the vehicle's weight but always reduce rotational friction.
- Problems with the differential and drive axle shafts are usually first noticed as a leak or noise. As the problem progresses, vibrations or a clunking noise might be felt in various modes of operation.

REVIEW QUESTIONS

1. Name the three ways in which CV joints can be classified.
2. What type of axle housing resembles a banjo?
3. What type of axle merely supports the vehicle load and provides a mounting place for the wheels?
4. What type of floating axle has one wheel bearing per wheel on the outside of the axle housing?
5. How are problems normally first noticed with the differential and drive axles?
6. In front-wheel drivetrains, the CV joint nearer the transaxle is the _______.
   a. inner joint
   b. inboard joint
   c. outboard joint
   d. both a and b
7. A CV joint that is capable of in-and-out movement is a _______.
   a. plunge joint
   b. fixed joint
   c. inboard joint
   d. both a and c
8. The double-offset joint is typically used in applications that require _______.
   a. higher operating angles and greater plunge depth
   b. lower operating angles and lower plunge depth
   c. higher operating angles and lower plunge depth
   d. lower operating angles and greater plunge depth
9. Which type of joint has a flatter design than any other?
   a. double-offset
   b. disc
   c. cross groove
   d. fixed tripod
10. Which of these is the best way to determine which CV joint is faulty?
    a. squeeze test
    b. runout test
    c. visual inspection
    d. road test
11. The single Cardan/Spicer universal joint is also known as the _______.
    a. cross joint
    b. four-point joint
    c. both a and b
    d. neither a nor b
12. The drive shaft component that provides a means of connecting two or more shafts together is the _______.
    a. pinion flange
    b. U-joint
    c. yoke
    d. biscuit
13. Large cars with long drive shafts often use a double-U-joint arrangement called a _______.
    a. Spicer style U-joint
    b. constant velocity U-joint
    c. Cleveland style U-joint

d. none of the above

14. Which type of driving axle supports the weight of the vehicle?
    a. semifloating
    b. three-quarter floating
    c. full-floating
    d. none of the above
15. Which of the following describes the double-Cardan universal joint?
    a. It is most often installed in front-engine rear-wheel-drive luxury automobiles.
    b. It is classified as a constant velocity U-joint.
    c. A centering ball socket is inside the coupling yoke between the two universal joints.
    d. All of the above.

ASE-STYLE REVIEW QUESTIONS

1. Technician A says that a gear tooth pattern identifies ring gear runout. Technician B says that gear patterns are not accurate if there is excessive ring gear runout. Who is correct?
   a. Technician A c. Both A and B
   b. Technician B d. Neither A nor B
2. Technician A says that limited-slip differential clutch packs are designed to slip when the vehicle turns a corner. Technician B says that a special additive is placed in the hypoid gear lubricant to promote clutch pack slippage on corners. Who is correct?
   a. Technician A c. Both A and B
   b. Technician B d. Neither A nor B
3. Technician A says that side-bearing preload limits the amount the differential case is able to move laterally in the axle housing. Technician B says that adjusting backlash sets the depth of the mesh between the ring and pinion gears' teeth. Who is correct?
   a. Technician A c. Both A and B
   b. Technician B d. Neither A nor B
4. Technician A says that a hunting gear set is one in which one drive pinion gear tooth contacts only certain ring gear teeth. Technician B says that a partial nonhunting gear set is one in which one pinion tooth contacts only six ring gear teeth. Who is correct?
   a. Technician A c. Both A and B
   b. Technician B d. Neither A nor B
5. While discussing the possib ing noise during acceleratic that an outer CV joint coul nician B says that insufficier axle may be the cause of t correct?
   a. Technician A c. Both A and B
   b. Technician B d. Neither A nor B
6. While diagnosing the cause of a clicking noise that is heard only when the car is turning: Technician A says that the most probable cause is a worn wheel bearing. Technician B says that the most probable cause is a worn outboard CV joint. Who is correct?
   a. Technician A c. Both A and B
   b. Technician B d. Neither A nor B
7. Technician A says that some axle shafts are retained in the housing by a plate and bolts. Technician B says that a C washer or clip retains some axle shafts in the housing. Who is correct?
   a. Technician A c. Both A and B
   b. Technician B d. Neither A nor B
8. Technician A says that when a car is moving straight ahead, all differential gears rotate as a unit. Technician B says that when a car is turning a corner, the inside differential side gear rotates slowly on the pinion, causing the outside side gear to rotate faster. Who is correct?
   a. Technician A c. Both A and B
   b. Technician B d. Neither A nor B
9. While discussing different types of ring and pinion gear sets: Technician A says that with a nonhunting gear set, each tooth of the pinion will return to the same tooth space on the ring gear each time the pinion rotates. Technician B says that when a hunting gear set rotates, any pinion gear tooth is likely to contact each and every tooth on the ring gear. Who is correct?
   a. Technician A c. Both A and B
   b. Technician B d. Neither A nor B
10. While reviewing the procedure for setting backlash: Technician A says that backlash is adjusted along with side-bearing preload by loosening or tightening side-bearing adjusting nuts. Technician B says that normally, to decrease the amount of backlash, a thin shim is installed on one side of the gear and a thick one on the other side. Who is correct?
    a. Technician A c. Both A and B
    b. Technician B d. Neither A nor B

AUTOMATIC TRANSMISSIONS AND TRANSAXLES

CHAPTER 40

OBJECTIVES

■ Explain the basic design and operation of standard and lockup torque converters. ■ Describe the design and operation of a simple planetary gear set and Simpson geartrain. ■ Name the major types of planetary gear controls used on automatic transmissions and explain their basic operating principles. ■ Describe the construction and operation of common Simpson geartrain-based transmissions and transaxles. ■ Describe the construction and operation of common Ravigneaux geartrain-based transmissions. ■ Describe the construction and operation of transaxles that use planetary gear sets in tandem. ■ Describe the construction and operation of automatic transmissions that use helical gears in constant mesh. ■ Describe the construction and operation of CVTs. ■ Describe the design and operation of the hydraulic controls and valves used in modern transmissions and transaxles. ■ Explain the role of the following components of the transmission control system: pressure regulator valve, throttle valve, governor assembly, manual valve, shift valves, and kickdown valve. ■ Identify the various pressures in the transmission, state their purpose, and tell how they influence the operation of the transmission.

Many rear-wheel-drive (RWD) and four-wheel-drive vehicles are equipped with automatic transmissions **(Figure 40–1)**. Automatic transaxles, which combine an automatic transmission and final drive assembly in a single unit, are used on front-wheel-drive (FWD), all-wheel-drive, and some RWD vehicles **(Figure 40–2)**.

Figure 40-1 A six-speed automatic transmission.
Courtesy of BMW of North America, LLC

An automatic transmission or transaxle selects gear ratios according to engine speed, powertrain load, vehicle speed, and other operating factors. Little effort is needed on the part of the driver, because both upshifts and downshifts occur automatically. A driver-operated clutch is not needed to change gears, and the vehicle can be brought to a stop without shifting to neutral. This is a great convenience, particularly in stop-and-go traffic. The driver can also

Figure 40–2 A cutaway of an automatic transaxle. *Courtesy of Chrysler LLC*

manually select a lower forward gear, reverse, neutral, or park. Depending on the forward range selected, the transmission can provide engine braking during deceleration.

The number of available forward gears in current vehicles varies from four to eight. Some have zero fixed ratios and use a constantly variable design, where the ratio changes according to conditions. Most early automatic transmissions were three- or four-speed units. Today the most common are five- and six-speed units. Most new automatics also feature a lockup torque converter. Some automatic transmissions are fitted with a transfer case that sends torque to additional drive axles to allow for four-wheel or all-wheel drive **(Figure 40–3)**.

Until recently, all automatic transmissions were controlled by hydraulics only. However, most systems now feature computer-controlled operation of the torque converter and transmission. Based on input data supplied by electronic sensors and switches, the computer sets the torque converter's operating mode, controls the transmission's shifting sequence, and in some cases regulates transmission oil pressure.

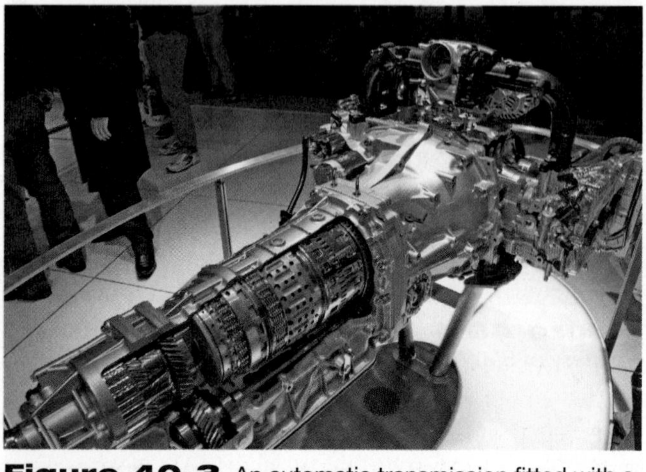

Figure 40–3 An automatic transmission fitted with a transfer case for all-wheel drive.

The most widely used automatic transmissions and transaxles are four-speed units with an overdrive fourth gear. Five- and six-speed transmissions are also used. Many older cars had three speeds and a select group of newer cars have five speeds. Most new automatics also feature a lockup torque converter. Until recently, all automatic transmissions were controlled by hydraulics. However, many systems now feature computer-controlled operation of the torque converter and transmission. Based on input data supplied by electronic sensors and switches, the computer sets the torque converter's operating mode, controls the transmission's shifting sequence, and in some cases regulates transmission oil pressure.

Today's automatic transmissions have four to eight forward speeds. Five- and six-speed units are the most common. Seven- and eight-speed units are mostly found in luxury vehicles. Today's transmissions have at least one overdrive gear to reduce fuel consumption, lower emission levels, and reduce noise while the vehicle is cruising. Today's transmissions also have a lockup torque converter that eliminates loss of power through the torque converter. The torque converter lockup clutch and shifting of the transmission are computer controlled.

TORQUE CONVERTER

Automatic transmissions use a fluid clutch known as a torque converter to transfer engine torque from the engine to the transmission.

The torque converter operates through hydraulic force provided by automatic transmission fluid, often simply called transmission oil. The torque converter changes or multiplies the twisting motion of the engine crankshaft and directs it through the transmission.

The torque converter automatically engages and disengages power from the engine to the transmission in relation to engine rpm. With the engine running at the correct idle speed, there is not enough fluid flow for power transfer through the torque converter. As engine speed is increased, the added fluid flow creates sufficient force to transmit engine power through the torque converter assembly to the transmission.

Design

Nearly all torque converters, or T/Cs, are one-piece, welded units that can only be repaired by specialty shops. They are located between the engine and transmission and are sealed, doughnut-shaped units **(Figure 40–4)** that are always filled with automatic transmission fluid. All of the vital parts of a torque converter are housed within its shell **(Figure 40–5)**.

A flex- or drive plate is used to mount the torque converter to the crankshaft. The flexplate transfers the rotation of the crankshaft to the shell of the torque converter. The flexplate is designed to flex in response to the slight change in torque converter size as pressure builds in it. It is bolted to a flange on the rear of the crankshaft and to mounting pads on the front of the torque converter shell. A heavy flywheel is not needed because the mass of the torque converter and flexplate work like a flywheel to smooth out the intermittent power strokes of the engine. The flexplate or torque converter is surrounded by the ring gear for the starting motor.

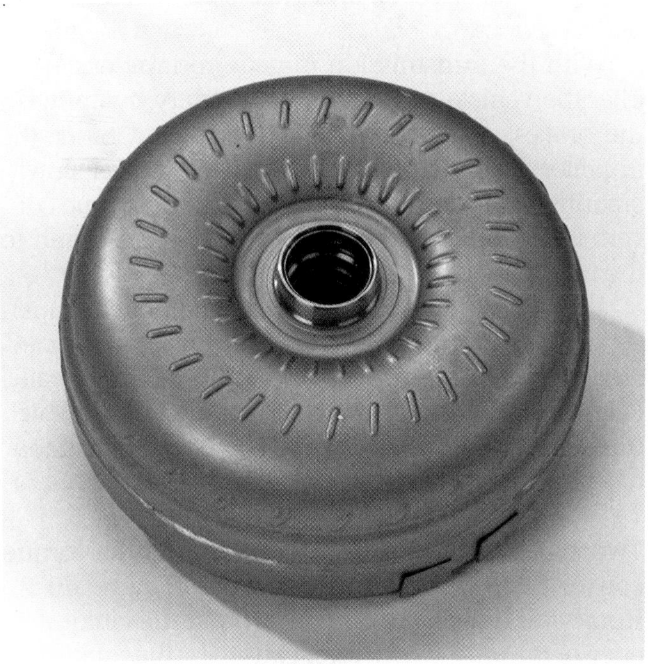

Figure 40-4 A torque converter. *Courtesy of Transtar Industries*

Figure 40-5 A cutaway of a modern torque converter. *Courtesy of BMW of North America, LLC*

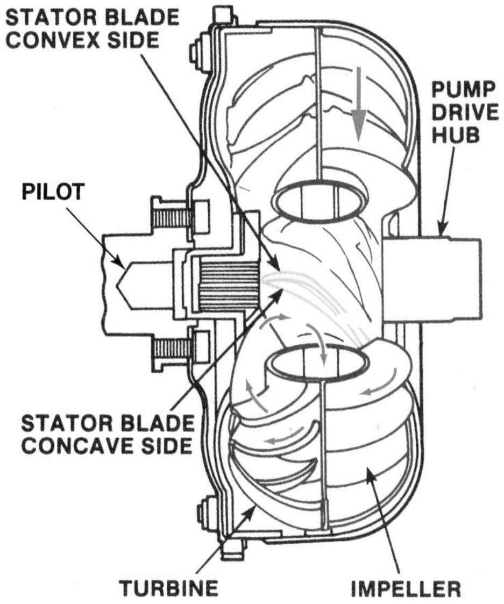

Figure 40-6 A torque converter's major internal parts are its impeller, turbine, and stator. *Courtesy of Ford Motor Company*

Components

A standard torque converter consists of three elements **(Figure 40–6)**: the pump assembly, often called an impeller, the stator assembly, and the turbine.

The **impeller** assembly is the input (drive) member. It receives power from the engine. The **turbine** is the output (driven) member. It is splined to the transmission's turbine shaft. The **stator** assembly is the reaction member or torque multiplier. The stator is supported on a one-way clutch, which operates as an overrunning clutch and permits the stator to rotate freely in one direction and lock up in the opposite direction.

The exterior of the torque converter shell is shaped like two bowls standing on end facing each other. To support the weight of the torque converter, a short stubby shaft projects forward from the front of the torque converter shell and fits into a pocket at the rear of the crankshaft. At the rear of many torque converter shells is a hollow hub with notches or flats at one end, ground 180 degrees apart. This hub is called the pump drive hub. The notches or flats drive the transmission pump assembly. At the front of the transmission within the pump housing is a pump bushing that supports the pump drive hub and provides rear support for the torque converter assembly. Some other transaxles have a separate shaft to drive the pump.

The impeller forms one internal section of the torque converter shell and has numerous curved blades that rotate as a unit with the shell. It turns at engine speed, acting like a pump to start the transmission oil circulating within the torque converter shell.

While the impeller is positioned with its back facing the transmission housing, the turbine is positioned with its back to the engine. The curved blades of the turbine face the impeller assembly.

The turbine blades have a greater curve than the impeller blades, which helps eliminate oil turbulence between the turbine and impeller blades that would slow impeller speed and reduce the converter's efficiency.

The stator is located between the impeller and turbine. It redirects the oil flow from the turbine back into the impeller in the direction of impeller rotation with minimal loss of speed or force. The side of the stator blade with the inward curve is the concave side. The side with an outward curve is the convex side.

Basic Operation

Transmission oil is used as the medium to transfer energy in the T/C. **Figure 40–7A** illustrates the T/C impeller or pump at rest. **Figure 40–7B** shows it being driven. As the pump impeller rotates, centrifugal force throws the oil outward and upward due to the curved shape of the impeller housing.

The faster the impeller rotates, the greater the centrifugal force becomes. In **Figure 40–7B**, the oil is simply flying out of the housing and is not producing any work. To harness some of this energy, the turbine assembly is mounted on top of the impeller (**Figure 40–7C**). Now the oil thrown outward and upward from the impeller strikes the curved vanes of the turbine, causing the turbine to rotate. (There is no direct mechanical link between the impeller and turbine.) An oil pump driven by the converter shell and the engine continually delivers oil under pressure into the T/C through a hollow shaft at the center axis of the rotating torque converter assembly. A seal prevents the loss of fluid from the system.

The turbine shaft is located within this shaft. As mentioned earlier, the turbine shaft is splined to the turbine and transfers power from the torque converter to the transmission's main drive shaft. Oil leaving the turbine is directed out of the torque converter to an external oil cooler and then to the transmission's oil sump or pan.

With the transmission in gear and the engine at idle, the vehicle can be held stationary by applying the brakes. At idle, engine speed is slow. Since the impeller is driven by engine speed, it turns slowly, creating little centrifugal force within the torque converter. Therefore, little or no power is transferred to the transmission.

When the throttle is opened, engine speed, impeller speed, and the amount of centrifugal force generated in the torque converter increase dramatically. Oil is then directed against the turbine blades, which transfer power to the turbine shaft and transmission.

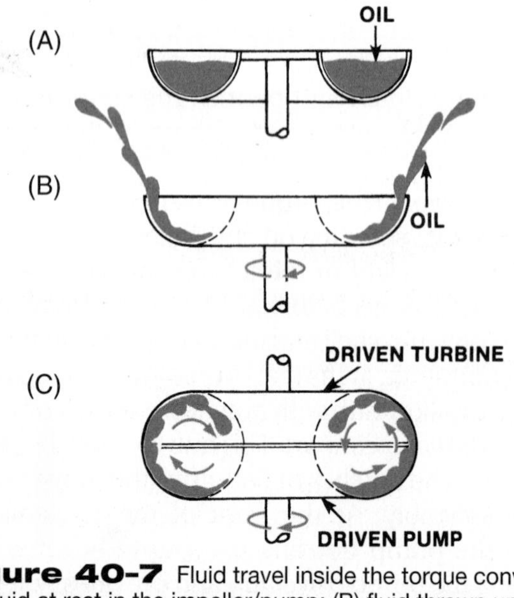

Figure 40-7 Fluid travel inside the torque converter: (A) Fluid at rest in the impeller/pump; (B) fluid thrown up and outward by the spinning pump; and (C) fluid flow harnessed by the turbine and redirected back into the pump.

Types of Oil Flow

Two types of oil flow take place inside the torque converter: rotary and vortex flow (**Figure 40–8**). **Rotary oil flow** is the oil flow around the circumference of the torque converter caused by the rotation of the torque converter on its axis. **Vortex oil flow** is the oil flow occurring from the impeller to the turbine and back to the impeller, at a 90-degree angle from engine rotation.

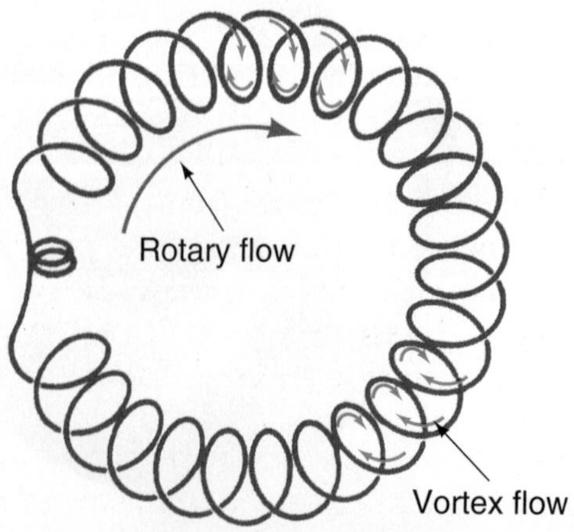

Figure 40-8 The difference between rotary and vortex flow. Note that vortex flow spirals its way around the converter.

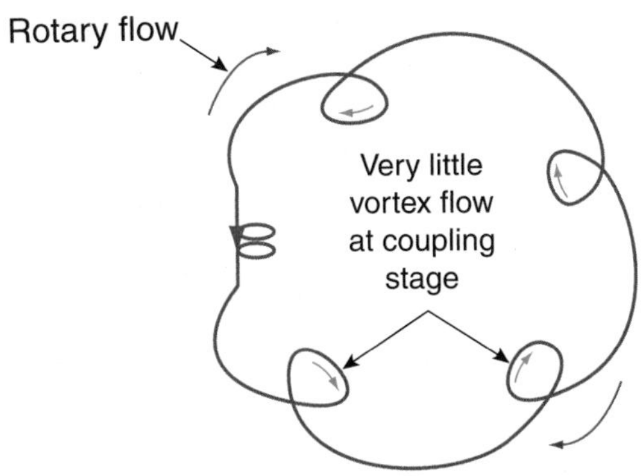

Figure 40–9 Rotary flow is at its greatest at the coupling stage.

Figure 40–9 also shows the oil flow pattern as the speed of the turbine approaches the speed of the impeller. This is known as the **coupling point**. The turbine and the impeller are running at essentially the same speed. They cannot run at exactly the same speed due to slippage between them. The only way they can turn at exactly the same speed is by using a lockup clutch to mechanically tie them together.

The stator mounts through its splined center hub to a mating stator shaft. The stator freewheels when the impeller and turbine reach the coupling stage.

The stator redirects the oil leaving the turbine back to the impeller, which helps the impeller rotate more efficiently **(Figure 40–10)**. Torque converter multiplication can only occur when the impeller is rotating faster than the turbine.

A stator is either a rotating or fixed type. Rotating stators are more efficient at higher speeds because there is less slippage when the impeller and turbine reach the coupling stage.

Overrunning Clutch

An **overrunning clutch** keeps the stator assembly from rotating when driven in one direction and permits overrunning (rotation) when turned in the opposite direction. Rotating stators generally use a roller-type overrunning clutch that allows the stator to freewheel (rotate) when the speed of the turbine and impeller reach the coupling point.

The roller clutch **(Figure 40–11)** is designed with a movable inner race, rollers, accordion (apply) springs, and outer race. Around the inside diameter of the outer race are several cam-shaped pockets. The rollers and accordion springs are located in these pockets.

As the vehicle begins to move, the stator stays in its stationary or locked position because of the difference between the impeller and turbine speeds. This locking mode takes place when the inner race rotates counterclockwise. The accordion springs force the rollers up the ramps of the cam pockets into a wedging contact with the inner and outer races.

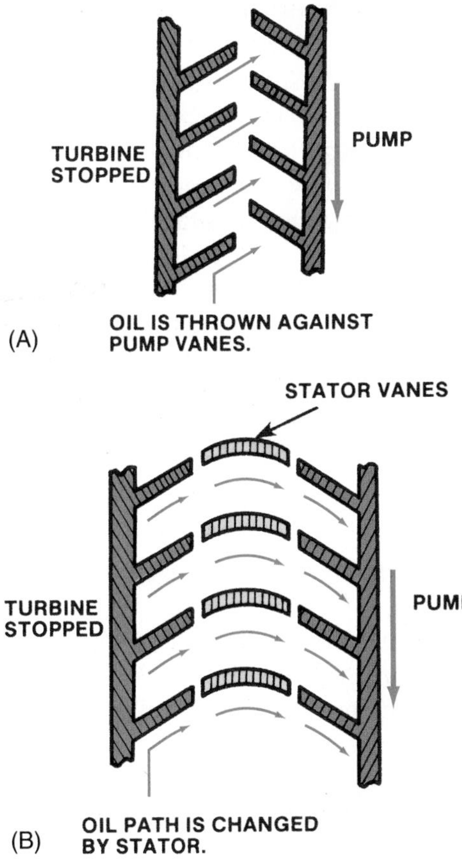

Figure 40–10 (A) Without a stator, fluid leaving the turbine works against the direction in which the impeller or pump is rotating. (B) With a stator in its locked (noncoupling) mode, fluid is directed to help push the impeller in its rotating direction.

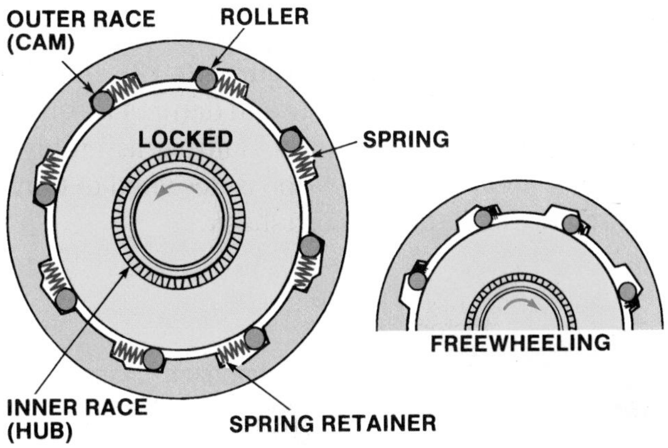

Figure 40–11 A typical roller-type overrunning clutch.

As vehicle road speed increases, turbine speed increases until it approaches impeller speed. Oil exiting the turbine vanes strikes the back face of the stator, causing the stator to rotate in the same direction as the turbine and impeller. At this higher speed, clearance exists between the inner stator race and hub. The rollers at each slot of the stator are pulled around the stator hub. The stator freewheels or turns as a unit.

If the vehicle slows, engine speed also slows along with turbine speed. This decrease in turbine speed allows the oil flow to change direction. It now strikes the front face of the stator vanes, halting the turning stator and attempting to rotate it in the opposite direction.

As this happens, the rollers jam between the inner race and hub, locking the stator in position. In a stationary position, the stator now redirects the oil exiting the turbine so torque is again multiplied.

LOCKUP TORQUE CONVERTER

A lockup torque converter eliminates the 10% slip that takes place between the impeller and turbine at the coupling stage. The engagement of a clutch between the impeller and the turbine assembly greatly improves fuel economy and reduces operational heat and engine speed. The assembly of a lockup torque converter is typically called a **torque converter clutch** (TCC).

Through the years, many different types of TCC systems have been used. The most common design is the electronically controlled lockup piston clutch. Clutch lockup systems can also be fully mechanical, centrifugally controlled, or dependent on a viscous coupling.

Lockup Piston Clutch

The lockup piston clutch has a piston-type clutch located between the front of the turbine and the interior front face of the shell **(Figure 40–12)**. Its main components are a piston plate and damper assembly and a clutch friction plate. The damper assembly is made of several coil springs and is designed to transmit driving torque and absorb shock.

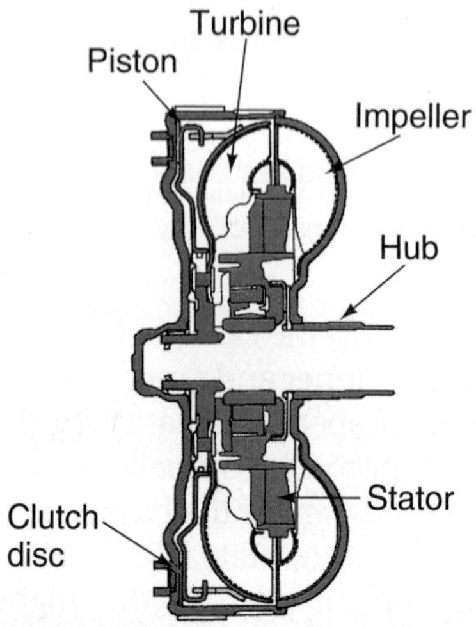

Figure 40-12 A piston-type converter lockup clutch assembly. *Courtesy of Chrysler LLC*

The clutch is controlled by hydraulic valves, which are controlled by the PCM **(Figure 40–13)**. The PCM monitors operating conditions and controls lockup according to those conditions.

To understand how this system works, consider this example. To provide for clutch control, Chrysler adds a three-valve module to its standard transmission valve body. The three valves are the lockup valve, fail-safe valve, and switch valve. The lockup valve actually controls the clutch. The fail-safe valve prevents lockup until the transmission is in third gear. The switch valve directs fluid through the turbine shaft to fill the torque converter.

When the converter is not locked, fluid enters the converter and moves to the front side of the piston, keeping it away from the shell or cover. Fluid flow continues around the piston to the rear side and exits between the neck of the torque converter and the stator support.

During the lockup mode, the switch valve moves and reverses the fluid path. This causes the fluid to move to the rear of the piston, pushing it forward to apply the clutch to the shell and allowing for lockup. Fluid from the front side of the piston exits through the turbine shaft that is now vented at the switch valve.

During acceleration, system fluid pressure increases. If the converter is in its lockup mode, the higher pressure moves the fail-safe valve to block fluid pressure to the lockup valve. Spring tension moves the switch valve, directing fluid pressure to the front side of the piston. The torque converter then returns to its nonlockup mode.

PLANETARY GEARS

Nearly all automatic transmissions rely on planetary gear sets **(Figure 40–14)** to transfer power and multiply engine torque to the drive axle. Compound gear sets combine two simple planetary gear sets so load can be spread over a greater number of teeth for strength and also to obtain the largest number of gear ratios possible in a compact area.

A simple planetary gear set consists of three parts: a sun gear, a carrier with planetary pinions mounted to it, and an internally toothed ring gear or **annulus**. The **sun gear** is located in the center of the assembly **(Figure 40–15)**. It can be either a spur or helical gear design. It meshes with the teeth of the planetary pinion gears. Planetary pinion gears are small gears fitted into a framework called the **planetary carrier**. The planetary carrier can be made of cast iron, aluminum, or steel plate and is designed with a shaft for each of the planetary pinion gears. (For simplicity, planetary pinion gears are called **planetary pinions**.)

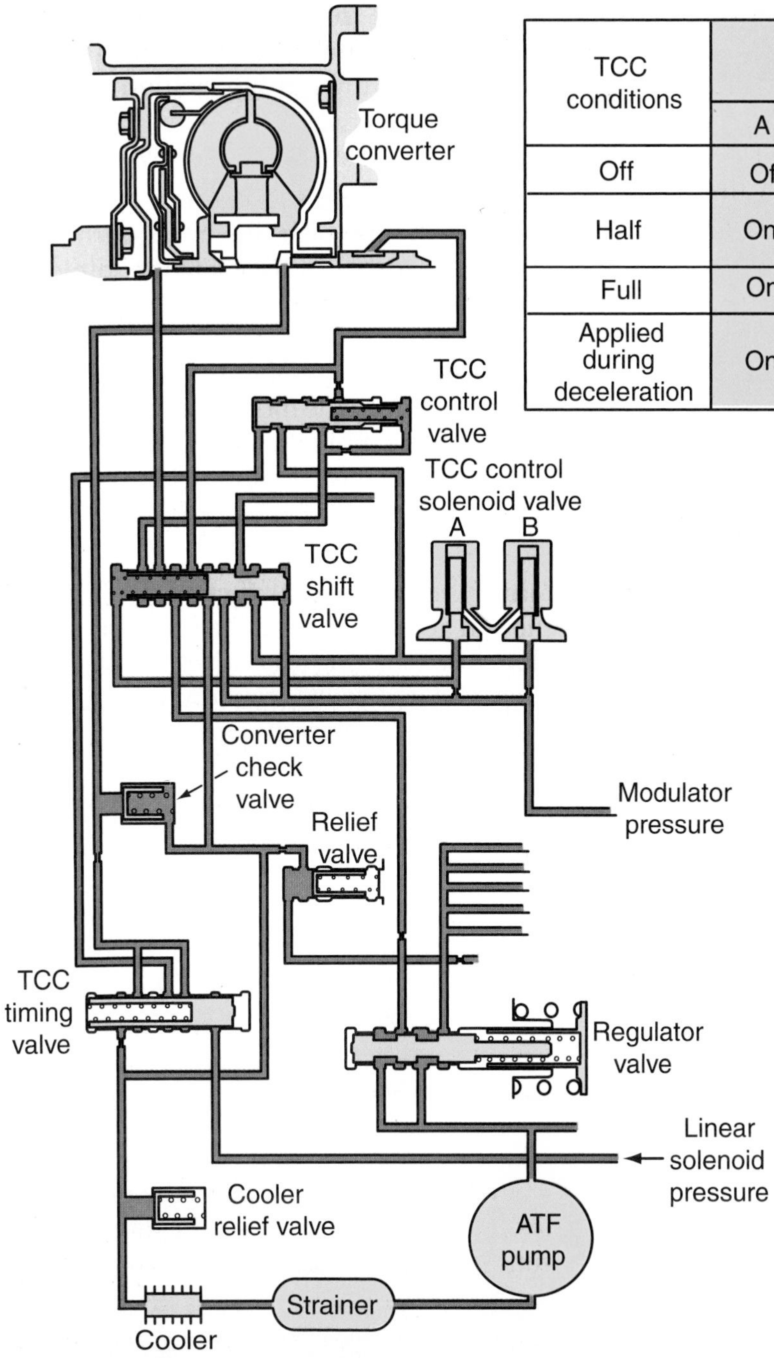

TCC conditions	TCC control solenoid valve		Linear solenoid pressure
	A	B	
Off	Off	Off	High
Half	On	Duty operation Off ←→ On	Low
Full	On	On	High
Applied during deceleration	On	Duty operation Off ←→ On	Low

Figure 40-13 A typical circuit for activating the TCC.

Figure 40-14 A single planetary gear set.

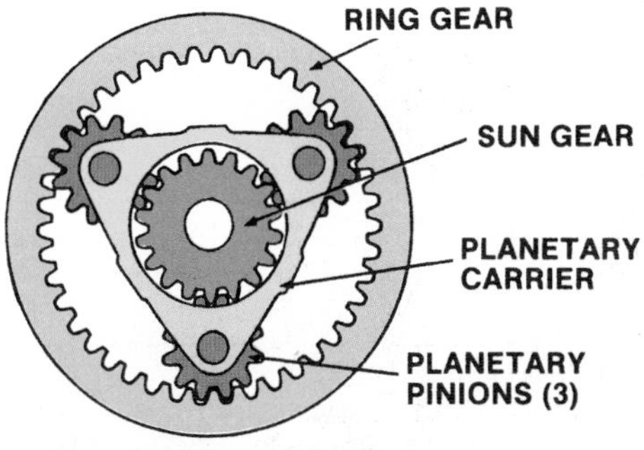

Figure 40-15 Planetary gear configuration is similar to the solar system, with the sun gear surrounded by the planetary pinion gears. The ring gear surrounds the complete gear set.

Planetary pinions rotate on needle bearings positioned between the planetary carrier shaft and the planetary pinions. The carrier and pinions are considered one unit—the midsize gear member.

The planetary pinions surround the sun gear's center axis and they are surrounded by the annulus or ring gear, which is the largest part of the simple gear set. The ring gear acts like a band to hold the entire gear set together and provide great strength to the unit. To help remember the design of a simple planetary gearset, use the solar system as an example. The sun is the center of the solar system with the planets rotating around it; hence, the name planetary gear set.

How Planetary Gears Work

Each member of a planetary gear set can spin (revolve) or be held at rest. Power transfer through a planetary gear set is only possible when one of the members is held at rest, or if two of the members are locked together.

Any one of the three members can be used as the driving or input member. At the same time, another member might be kept from rotating and thus becomes the reaction, held, or stationary member. The third member then becomes the driven or output member. Depending on which member is the driver, which is held, and which is driven, either a torque increase (underdrive) or a speed increase (overdrive) is produced by the planetary gear set. Output direction can also be reversed through various combinations.

Table 40–1 summarizes the basic laws of simple planetary gears. It indicates the resultant speed, torque, and direction of the various combinations available. Also, remember that when an external-to-external gear tooth set is in mesh, there is a change in the direction of rotation at the output. When an external gear tooth is in mesh with an internal gear, the output rotation for both gears is the same.

Maximum Forward Reduction With the ring gear held and the sun gear (the input) turning clockwise, the sun gear rotates the planetary pinions counterclockwise on their shafts. The small sun gear (driving) rotates several times, driving the midsize planetary carrier (the output) one complete revolution, resulting in the most gear reduction or the maximum torque multiplication that can be achieved in one planetary gear set. Input speed is high, but output speed is low.

Minimum Forward Reduction In this combination the sun gear is held and the ring gear (input) rotates clockwise. The ring gear drives the planetary pinions clockwise and walks around the stationary sun gear (held). The planetary pinions drive the planetary carrier (output) in the same direction as the ring gear—forward. This results in more than one turn of the input as compared to one complete revolution of the output. The result is torque multiplication. The planetary gear set is operating in a forward reduction with the large ring gear driving the small planetary carrier. Therefore, the combination produces minimum forward reduction.

Maximum Overdrive With the ring gear held and the planetary carrier (input) rotating clockwise, the three planetary pinion shafts push against the inside diameter of the planetary pinions. The pinions are forced to walk around the inside of the ring gear, driving the sun gear (output) clockwise. In this combination, the

TABLE 40–1 LAWS OF SIMPLE PLANETARY GEAR OPERATION

Sun Gear	Carrier	Ring Gear	Speed	Torque	Direction
1. Input	Output	Held	Maximum reduction	Increase	Same as input
2. Held	Output	Input	Minimum reduction	Increase	Same as input
3. Output	Input	Held	Maximum increase	Reduction	Same as input
4. Held	Input	Output	Minimum increase	Reduction	Same as input
5. Input	Held	Output	Reduction	Increase	Reverse of input
6. Output	Held	Input	Increase	Reduction	Reverse of input
7. When any two members are held together, speed and direction are the same as input. Direct 1:1 drive occurs.					
8. When no member is held or locked together, output cannot occur. The result is a neutral condition.					

midsize planetary carrier is rotating less than one turn and driving the smaller sun gear at a speed greater than the input speed. The result is overdrive with maximum speed increase.

Slow Overdrive In this combination, the sun gear is held and the carrier rotates (input) clockwise. As the carrier rotates, the pinion shafts push against the inside diameter of the pinions and they are forced to walk around the held sun gear. This drives the ring gear (output) faster and the speed increases. The carrier turning less than one turn causes the pinions to drive the ring gear one complete revolution in the same direction as the planetary carrier and a slow overdrive occurs.

Slow Reverse Here the sun gear (input) is driving the ring gear (output) with the planetary carrier held stationary. The planetary pinions, driven by the sun gear, rotate counterclockwise on their shafts. While the sun gear is driving, the planetary pinions are used as idler gears to drive the ring gear counterclockwise. This means the input and output shafts are operating in the opposite or reverse direction to provide a reverse power flow. Since the driving sun gear is small and the driven ring gear is large, the result is slow reverse.

Fast Reverse For fast reverse, the carrier is held, but the sun gear and ring gear reverse roles, with the ring gear (input) now being the driving member and the sun gear (output) driven. As the ring gear rotates counterclockwise, the pinions rotate counterclockwise as well, while the sun gear turns clockwise. In this combination, the input ring gear uses the planetary pinions to drive the output sun gear. The sun gear rotates in reverse to the input ring gear, providing fast reverse.

Direct Drive In the direct drive combination, both the ring gear and the sun gear are input members. They turn clockwise at the same speed. The internal teeth of the clockwise turning ring gear try to rotate the planetary pinions clockwise as well. But the sun gear, which rotates clockwise, tries to drive the planetary pinions counterclockwise. These opposing forces lock the planetary pinions against rotation so the entire planetary gear set rotates as one complete unit, providing direct drive. Whenever two members of the gear set are locked together, direct drive results.

Neutral Operation When no member is held or locked, a neutral condition exists.

SHOP TALK

Following are some tips for remembering the basic operation of a simple planetary gear set:

- When the planetary carrier is the drive (input) member, the gear set provides an overdrive condition. Speed increases, torque decreases.
- When the planetary carrier is the driven (output) member, the gear set provides a forward reduction direction. Speed decreases, torque increases.
- When the planetary carrier is stationary (held), the gear set provides a reverse gear.

COMPOUND PLANETARY GEAR SETS

A limited number of gear ratios are available from a single planetary gear set. Gear sets can be combined to increase the number of available gear ratios. A typical automatic transmission has at least two planetary gear sets (**Figure 40–16**) connected together to provide the various gear ratios needed to efficiently move a vehicle. There are two common designs of compound gear sets: the Simpson gear set, in which two planetary gear sets share a common sun gear, and the Ravingeaux gear set, which has two sun gears, two sets of planet gears, and a common ring gear.

Many transmissions are fitted with additional single planetary gear sets connected in tandem to provide additional forward speed ratios.

Simpson Geartrain

The **Simpson geartrain** is an arrangement of two separate planetary gear sets with a common sun gear, two ring gears, and two planetary pinion carriers. A Simpson geartrain is the most commonly used compound planetary gear set and is used to provide three forward gears. One half of the compound set or one planetary unit is referred to as the front planetary and the other planetary unit is the rear planetary (**Figure 40–17**). The two planetary units do not need to be the same size or have the same number of teeth on their gears. The size and number of gear teeth determine the actual gear ratios obtained by the compound planetary gear assembly.

Gear ratios and direction of rotation are the result of applying torque to one member of either planetary unit, holding at least one member of the gear set, and using another member as the output. For the most part, each automobile manufacturer uses different parts of the planetary assemblies as inputs, outputs, and reaction members. The role of the planetary

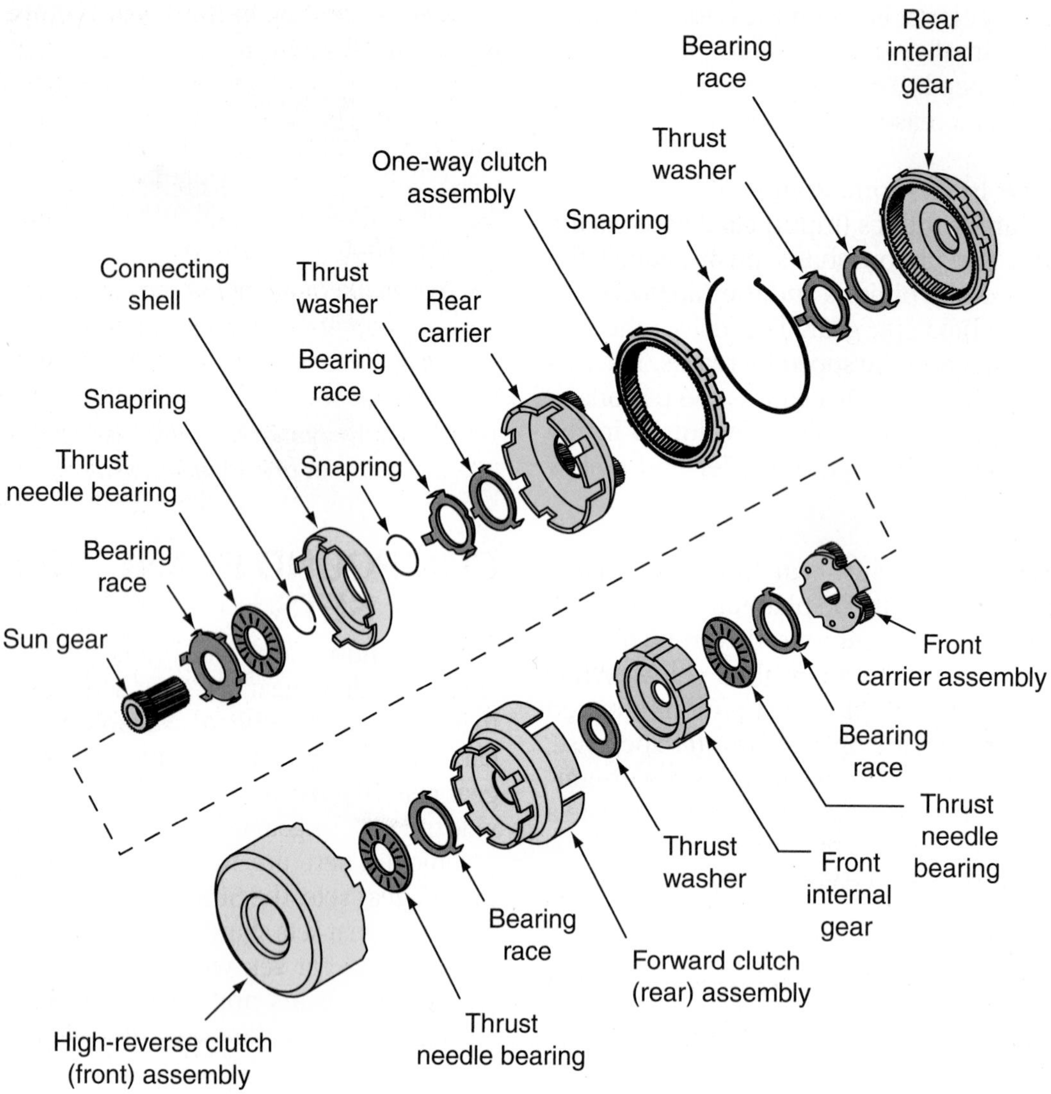

Figure 40-16 A Simpson planetary gear set.

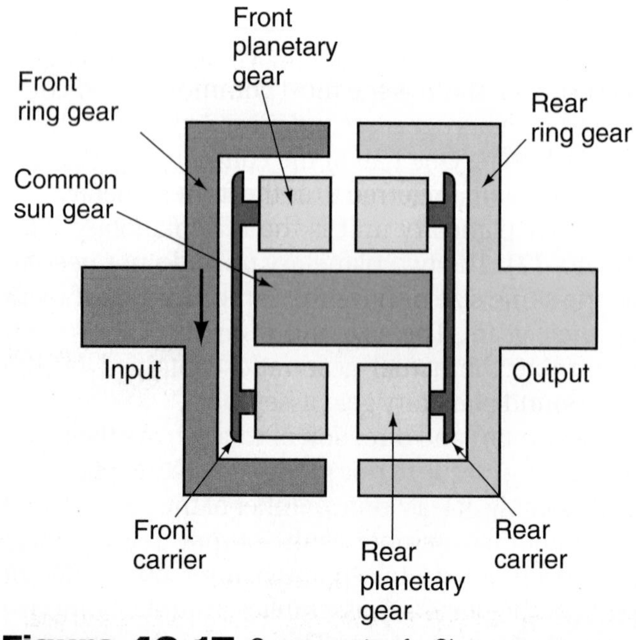

Figure 40-17 Components of a Simpson gear set.

members also varies with the different transmission models from the same manufacturer. There are also many different apply devices used in the various transmission designs.

A Simpson gear set can provide the following gear ranges: neutral, first reduction gear, second reduction gear, direct drive, and reverse. The typical power flow through a Simpson geartrain when it is in neutral has engine torque being delivered to the transmission's input shaft by the torque converter's turbine. No planetary gear set member is locked to the shaft; therefore, engine torque enters the transmission but goes nowhere else.

When the transmission is shifted into first gear (**Figure 40–18**), engine torque is again delivered into the transmission by the input shaft. The input shaft is now locked to the front planetary ring gear that turns clockwise with the shaft. The front ring gear drives the front planet gears, also in a clockwise direction. The front planet gears drive the sun gear

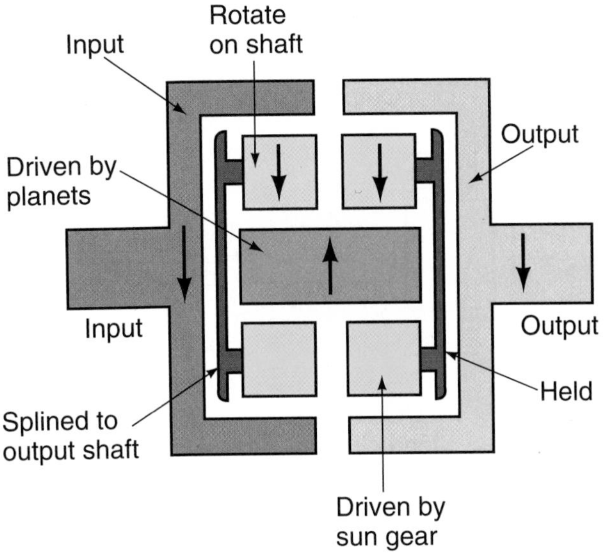

Figure 40-18 Power flow through a Simpson gear set while operating in first gear.

in a counterclockwise direction. The rear planet carrier is locked; therefore, the sun gear spins the rear planet gears in a clockwise direction. These planet gears drive the rear ring gear, which is locked to the output shaft, in a clockwise direction. The result of this power flow is a forward gear reduction.

When the transmission is operating in second gear (**Figure 40–19**), engine torque is again delivered into the transmission by the input shaft. The input shaft is locked to the front planetary ring gear that turns clockwise with the shaft. The front ring gear drives the front planet gears, also in a clockwise direction. The front planet gears walk around the sun gear because it is held. The walking of the planets forces the planet carrier to turn clockwise. Since the carrier is locked to the output shaft, it causes the shaft to rotate in a forward direction with some gear reduction.

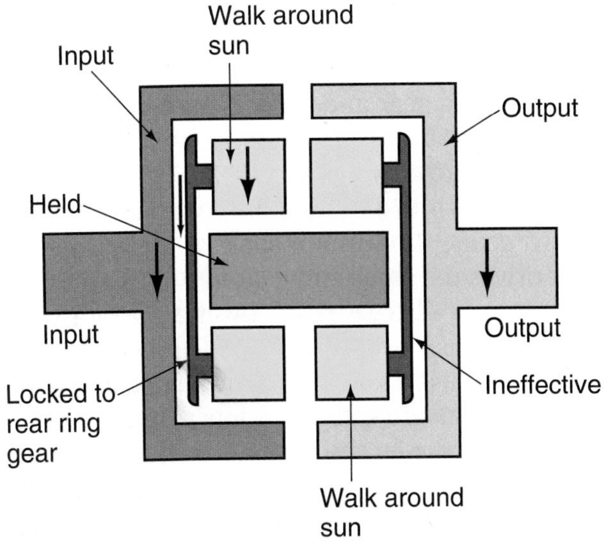

Figure 40-19 Power flow through a Simpson gear set while operating in second gear.

When operating in third gear (**Figure 40–20**), the input is received by the front ring gear, as in the other forward positions. However, the input is also received by the sun gear. Since the sun and ring gear are rotating at the same speed and in the same direction, the front planet carrier is locked between the two and is forced to move with them. Since the front carrier is locked to the output shaft, direct drive results.

To obtain a suitable reverse gear in a Simpson geartrain, there must be a gear reduction, but in the opposite direction as the input torque (**Figure 40–21**). The input is received by the sun gear, as in the third gear position, and rotates in a clockwise direction. The sun gear then drives the rear planet gears in a clockwise direction. The rear planet carrier is held; therefore, the planet gears drive the rear ring gear in a counterclockwise direction. The ring gear is locked to the output shaft that turns at the same speed and direction as the rear ring gear. The result is a reverse gear.

Typically, when the transmission is in neutral or park, no apply devices are engaged, allowing only the

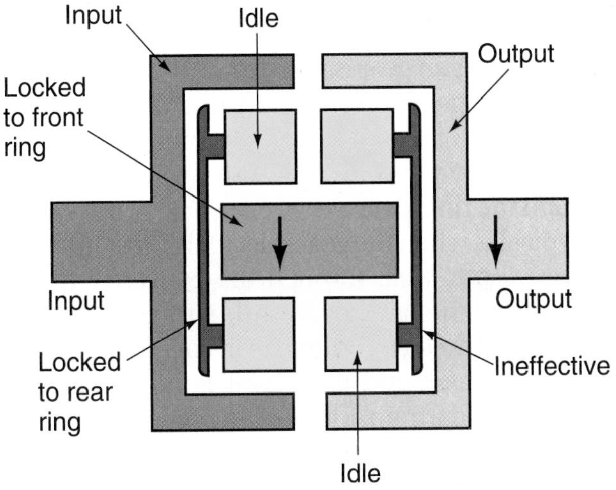

Figure 40-20 Power flow through a Simpson gear set while operating in direct drive.

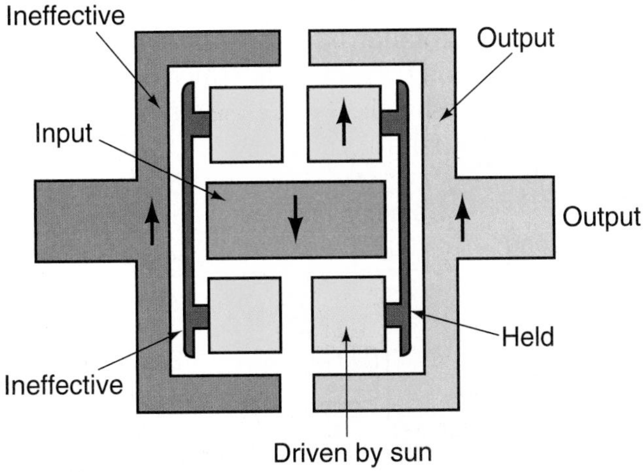

Figure 40-21 Power flow through a Simpson gear set while operating in reverse.

input shaft and the transmission's oil pump to turn with the engine. In park, a pawl is mechanically engaged to a parking gear that is splined to the transmission's output shaft, locking the drive wheels to the transmission's case.

Ravigneaux Geartrain

The **Ravigneaux geartrain**, like the Simpson geartrain, provides forward gears with a reduction, direct drive, overdrive, and a reverse operating range. The Ravigneaux offers some advantages over a Simpson geartrain. It is very compact. It can carry large amounts of torque because of the great amount of tooth contact. It can also have three different output members. However, it has a disadvantage because it is more complex and, therefore, its actions are more difficult to understand.

The Ravigneaux geartrain is designed to use two sun gears: one small and one large **(Figure 40–22)**. They also have two sets of planetary pinion gears: three long pinions and three short pinions. The planetary pinion gears rotate on their own shafts that are fastened to a common planetary carrier. A single ring gear surrounds the complete assembly.

The small sun gear is meshed with the short planetary pinion gears. These short pinions act as idler gears to drive the long planetary pinion gears. The long planetary pinion gears mesh with the large sun gear and the ring gear.

Typically, when the gear selector is in neutral position, engine torque, through the converter turbine shaft, drives the small (forward) sun gear. The sun gear is not locked to another gear, so it freewheels on the pinion gears and the power is not transmitted through the geartrain. There is no power output.

When the transmission is operating in first gear **(Figure 40–23)**, engine torque drives the small sun gear clockwise. The planetary carrier is prevented from rotating counterclockwise; therefore, the sun gear drives the short planet gears counterclockwise. The direction of rotation is reversed as the short planet gears drive the long planet gears, which drive the ring gear and output shaft in a clockwise direction but at a lower speed than the input.

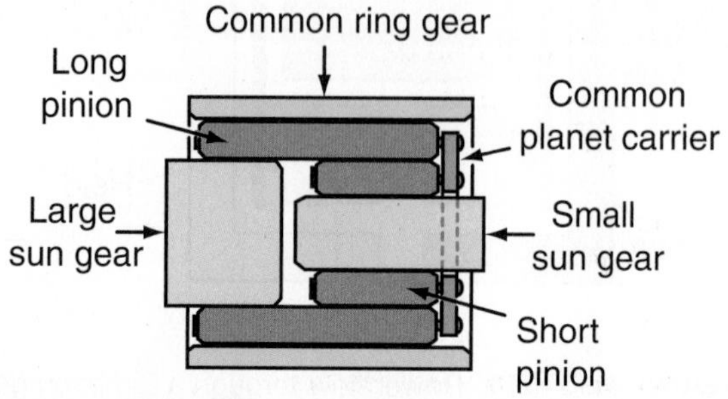

Figure 40-22 The parts of a Ravigneaux gear set.

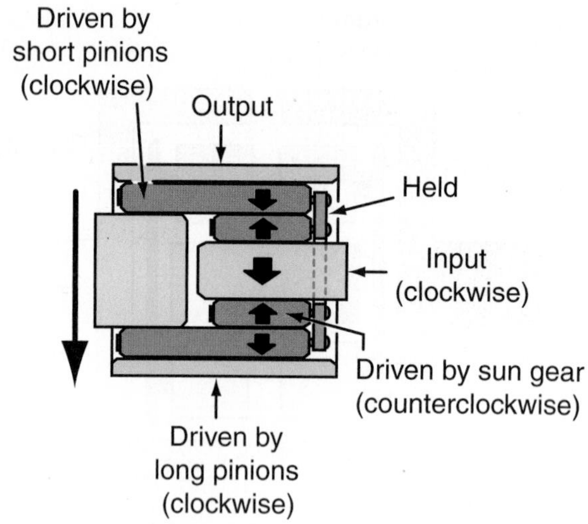

Figure 40-23 Power flow through a Ravigneaux gear set while operating in first gear.

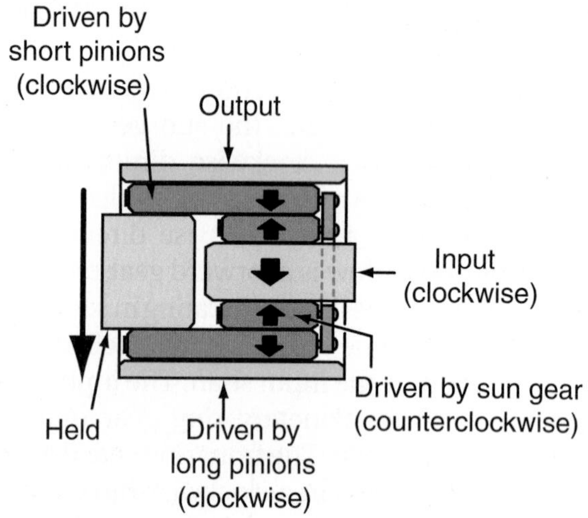

Figure 40-24 Power flow through a Ravigneaux gear set while operating in second gear.

In second gear **(Figure 40–24)** operation, the small sun gear is rotating clockwise and causes the short planet gears to rotate counterclockwise. The direction of rotation is reversed as the short planet gears drive the long planet gears, which walk around the stationary large or reverse sun gear. This walking drives the ring gear and output shaft in a clockwise direction and at a torque reduction.

During third gear **(Figure 40–25)** operation, there are two inputs into the planetary geartrain. As in other forward gears, the turbine shaft of the torque converter drives the small sun gear in a clockwise direction. Input is also received by the planetary gear carrier. Because two members of the geartrain are being driven at the same time, the planetary gear carrier and the forward sun gear rotate as a unit. The long planet gears transfer the torque in a clockwise direction through the gear set to the ring gear and output shaft. This results in direct drive.

To operate in overdrive or fourth gear, input is received only at the planetary carrier in a clockwise

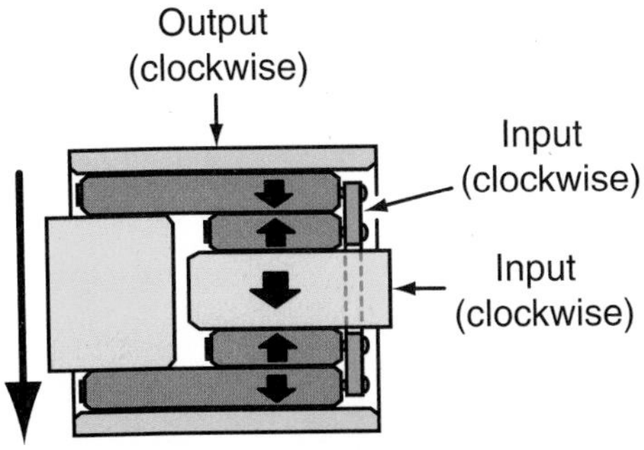

Figure 40-25 Power flow through a Ravigneaux gear set while operating in direct drive.

direction. The long planet gears walk around the stationary reverse sun gear in a clockwise direction and drive the ring gear and output shaft. This results in an overdrive condition.

During reverse gear operation, input is received at the reverse sun gear. The planetary gear carrier is held. The clockwise rotation of the reverse sun gear drives the long planet gears in a counterclockwise direction. The long planets then drive the ring gear and output shaft in a counterclockwise direction with a speed reduction.

Planetary Gear Sets in Tandem

Rather than relying on the use of a compound gear set, some automatic transmissions use two simple planetary units in series (**Figure 40–26**). In this type of arrangement, gear set members are not shared; instead, the holding devices are used to lock different members of the planetary units together.

Although the geartrain is based on two simple planetary gearsets operating in tandem, the combination of the two planetary units does function much like a compound unit. The two tandem units do not share a common member; rather, certain members are locked together or are integral with each other. The front planetary carrier is locked to the rear ring gear and the front ring gear is locked to the rear planetary carrier.

A transaxle may have additional planetary gear sets, including one that is used as the final drive gear set.

Lepelletier System

Some late-model six-, seven-, and eight-speed transmissions use the Lepelletier system. The Lepelletier system connects a simple planetary gear set to a Ravigneaux gear set. This design has been around for many years but was difficult to control. Today's electronic technologies have made it practical. With this design, transmissions can be made with additional forward speeds without an increase in size and weight. In fact, most six-speed transmissions are more compact and are lighter than nearly all four- or five-speed transmissions.

In this arrangement, the ring gear of the simple gear set serves as the input to the gear sets, and the input can be connected to the carrier of the Ravigneaux gear at the same time (**Figure 40–27**). As engine torque passes through different input gears, it drives a variety of combinations of gears in the simple and the Ravigneaux gearsets. These combinations

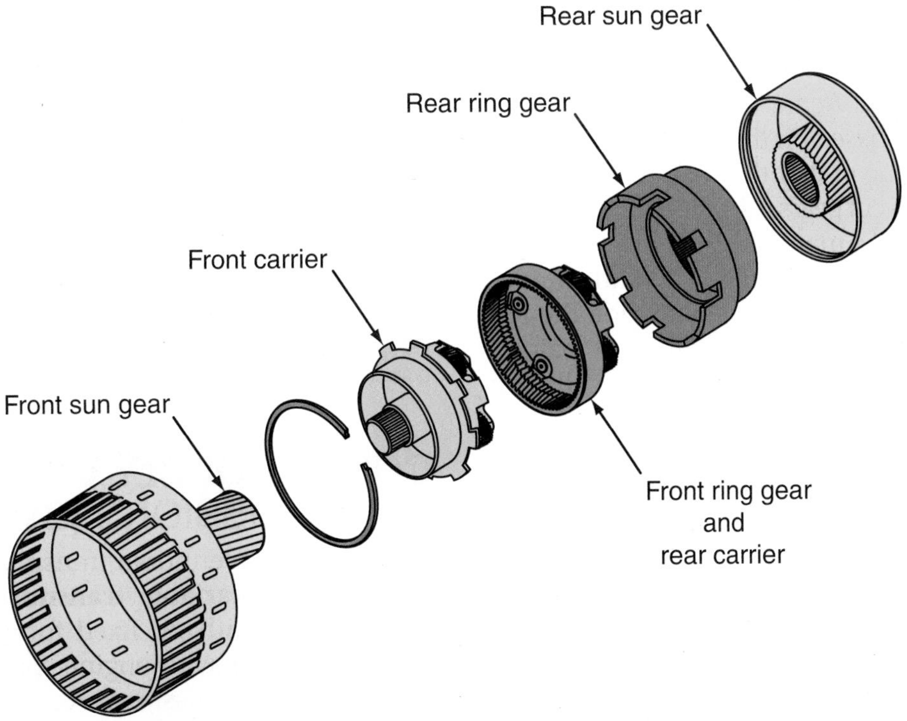

Figure 40-26 Two planetary units with the ring gear of one gear set connected to the planet carrier of the other.

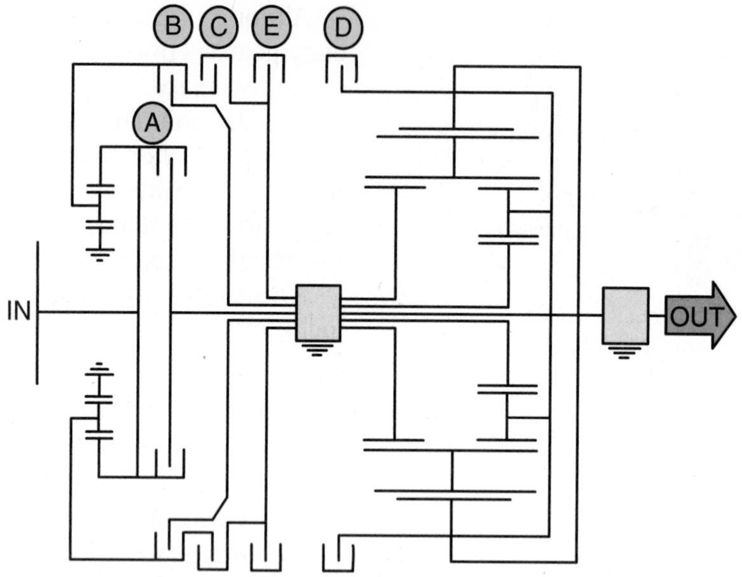

Figure 40-27 A six-speed Lepelletier transmission based on a planetary gear and a Ravigneaux gear and five multiple disc packs (marked A through E).

result in the various forward speed ratios. The ring gear of the Ravigneaux gear set is the output member for the transmission.

In some seven-speed models, the input shaft is always connected to the ring of the simple planetary gear and, in addition, can be connected to the carrier and large sun gear of the Ravigneaux gear. This allows for additional gear combinations. The ring gear of the Ravigneaux gear set still serves as the output.

HONDA'S NONPLANETARY-BASED TRANSMISSION

The Honda nonplanetary-based transaxles are used in many Honda and Acura cars. Saturn automatic transaxles are also based on this design. These transmissions are unique in that they use constant-mesh helical and square-cut gears **(Figure 40–28)** in a manner similar to that of a manual transmission.

These transaxles have a mainshaft and countershaft on which the gears ride. To provide the four forward and one reverse gear, different pairs of gears are locked to the shafts by hydraulically controlled clutches **(Figure 40–29)**. Reverse gear is obtained through the use of a shift fork that slides the reverse gear into position. The power flow through these transaxles is also similar to that of a manual transaxle.

Power flow through these transaxles is similar to that through a manual transaxle. The action of the clutches is much the same as the action of the synchronizer assemblies in a manual transaxle. Honda uses four multiple-disc clutches, a sliding reverse gear, and a one-way clutch to control the gears.

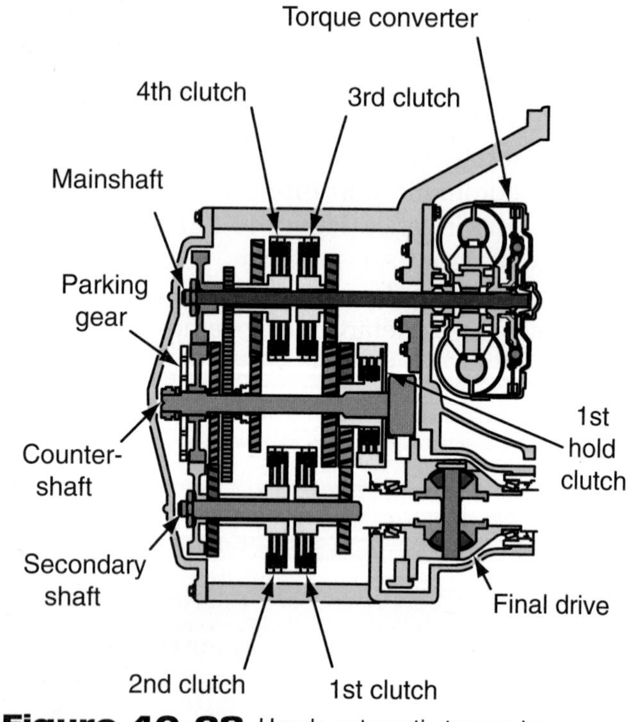

Figure 40-28 Honda automatic transaxles use constant-mesh helical gears instead of planetary gear sets.

CONTINUOUSLY VARIABLE TRANSMISSIONS (CVT)

Another unconventional transmission design, the **continuously variable transmission (CVT)**, is a transmission with no fixed forward speeds. The gear ratio varies with engine speed and temperature. These transmissions are, however, fitted with a one-speed reverse gear. Some CVT transaxles do not have a torque converter; rather, they use a

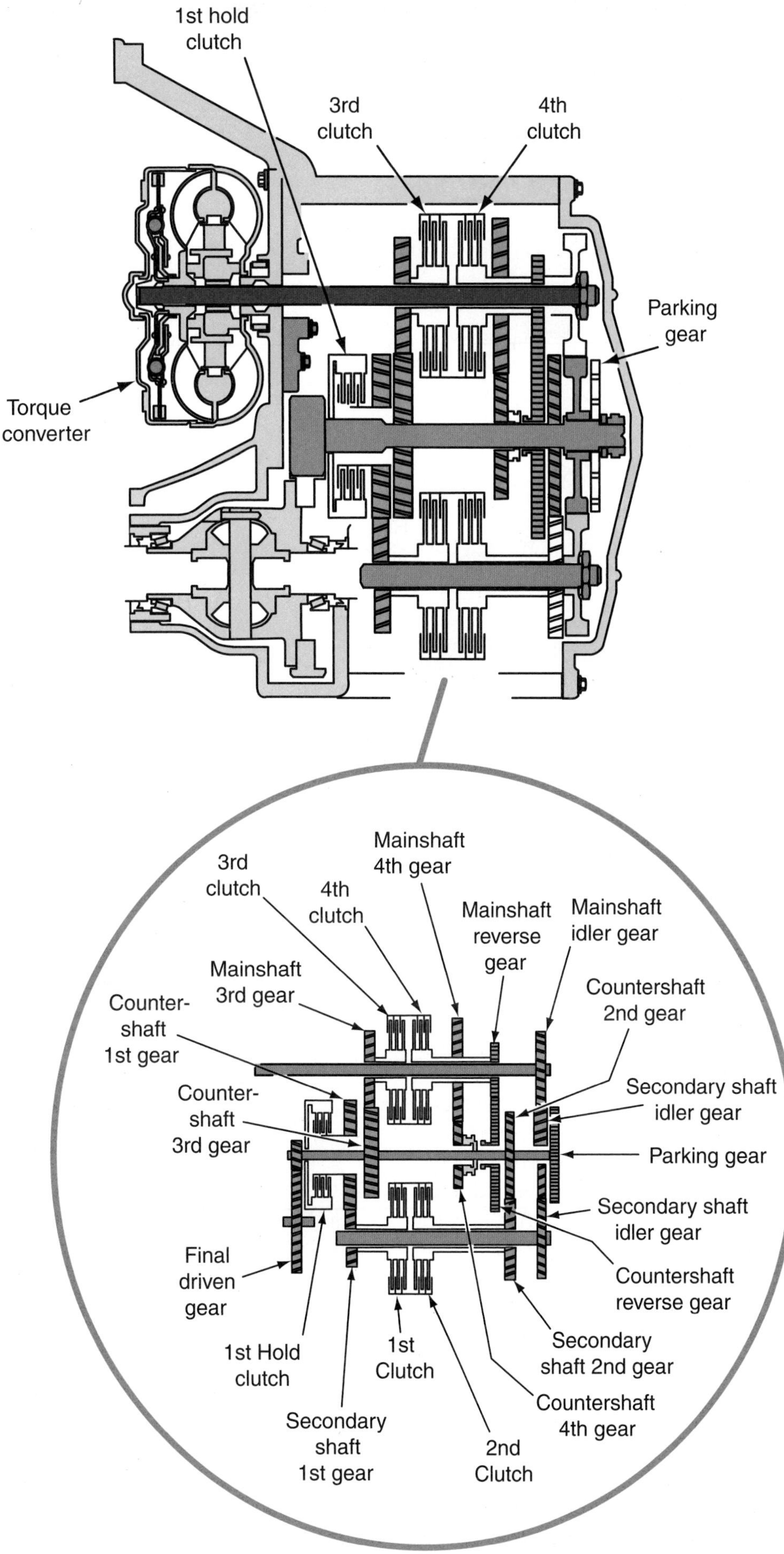

Figure 40-29 Arrangement of gears and reaction devices in a typical nonplanetary gear set transaxle.

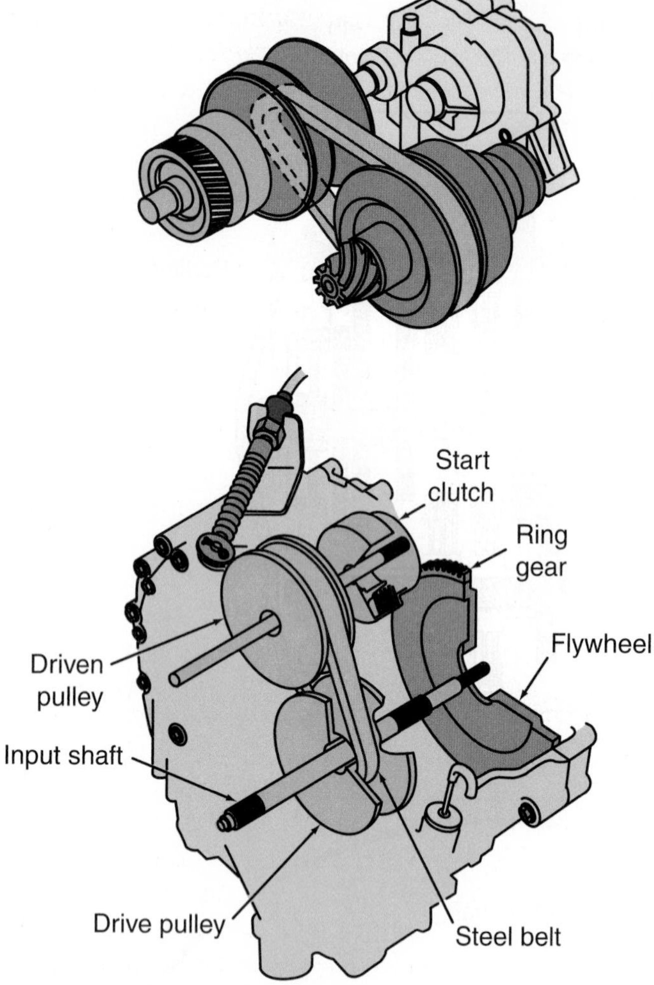

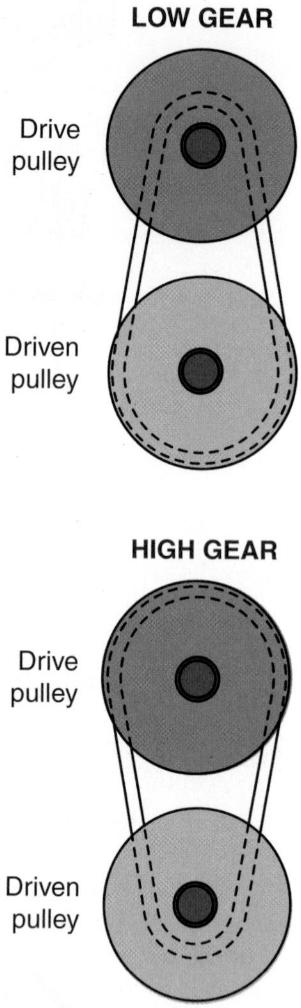

Figure 40-30 Honda's CVT.

manual transmission-type flywheel with a start clutch. Instead of relying on planetary or helical gear sets to provide drive ratios, a CVT uses belts and pulleys **(Figure 40–30)**.

One pulley is the driven member and the other is the drive. Each pulley has a movable face and a fixed face. When the movable face moves, the effective diameter of the pulley changes. The change in effective diameter changes the effective pulley (gear) ratio. A steel belt links the driven and drive pulleys **(Figure 40–31)**.

To achieve a low pulley ratio, high hydraulic pressure works on the movable face of the driven pulley to make it larger. In response to this high pressure, the pressure on the drive pulley is reduced. Since the belt links the two pulleys and proper belt tension is critical, the drive pulley reduces just enough to keep the proper tension on the belt. The increase of pressure at the driven pulley is proportional to the decrease of pressure at the drive pulley. The opposite is true for high pulley ratios. Low pressure causes the driven pulley to decrease in size, whereas high pressure increases the size of the drive pulley.

Figure 40-31 CVTs use pulleys that change size and are connected by a belt. *Copyright, Nissan (2008). Nissan and the Nissan logo are registered trademarks of Nissan.*

Different speed ratios are available any time the vehicle is moving. Because the size of the drive and driven pulleys can vary greatly, vehicle loads and speeds can be changed without changing the engine's speed. With this type of transmission, attempts are made to keep the engine operating at its most efficient speed, thus increasing fuel economy and decreasing emissions.

Many late-model CVTs are equipped with a feature that simulates the activity of a manual shifting automatic transmission. These transmissions have five or six predetermined areas that the pulleys stop in, thereby giving the feel and shift effect of distinct shifts.

Nissan has recently introduced a CVT, called the Extroid CVT, that is based on discs and rollers. This design may be more widely used in the future because it has the capability of withstanding large amounts of engine torque.

Planetary Gear-Based CVTs

Hybrid electric vehicles from Toyota and Ford rely on a planetary gear set to provide for a CVT. The transaxle contains two electric motor/generators, a differential, and a simple planetary gear set. The engine and the motor/generators are connected directly to the planetary gear unit. The planetary gear set is called the power split device because it can transfer power between the engine, motor/generators, drive wheels, and nearly any combination of these. The power split device splits power from the engine to different paths: to drive one of the motor/generators or to drive the car's wheels, or both. The other motor/generator can drive the wheels, assist the engine in driving the wheels, or be driven by the wheels. The speed ratios change in response to the torque applied to the various members of the gear set. In this arrangement, there are basically two sources of torque: the engine and an electric traction motor. Both rotate in the same direction but not at the same speed. Therefore, one can assist the rotation of the other, slow down the rotation of the other, or work together.

Two-Mode Hybrid System

GM, BMW, and Chrysler together developed a two-mode full hybrid system. The **two-mode hybrid system** is another planetary gear-based CVT. The system fits into a standard transmission housing and is basically two planetary gear sets coupled to two electric motors, which are electronically controlled. This combination results in a CVT with motor/generators for hybrid operation. The system has two distinct modes of operation. It operates in the first mode during low speed and low load conditions and in the second mode while cruising at highway speeds.

The two-mode hybrid system can operate solely on electric or engine power or a combination of the two. Electronic controls are used to control the output of the motors and the engine.

Two compact electric motors are connected to the transmission's gear sets. The gears work to increase the torque output of the motors. Typically, when one or both of the motors are not providing propulsion power, they work as generators driven by the engine or by the drive wheels for regenerative braking.

PLANETARY GEAR CONTROLS

Certain parts of the planetary geartrain must be held, while others must be driven to provide the needed torque multiplication and direction for vehicle operation. Planetary gear controls is the general term used to describe transmission bands, servos, and clutches.

Transmission Bands

A **band** is a braking assembly positioned around a stationary or rotating drum or carrier. The band brings a drum to a stop by wrapping itself around the drum and holding it. The band is hydraulically applied by a servo assembly. Connected to the drum is a member of the planetary geartrain. The purpose of a band is to hold a member of the planetary gear set by holding the drum and connecting planetary gear member stationary. Bands provide excellent holding characteristics and require a minimum amount of space within the transmission housing.

When a band closes around a rotating drum, a wedging action takes place to stop the drum from rotating. The wedging action is known as self-energizing action. A typical band is designed to be larger in diameter than the drum it surrounds. This design promotes self-disengagement of the band from the drum when servo apply force is decreased to less than servo release spring tension. A friction material is bonded to the inside diameter of the band.

Typically, if the band will be holding a low-speed drum, the lining material of a band is a semimetallic compound. If the band is designed to hold a high-speed drum, it will have a paper-based lining.

Band lugs are either spot welded or cast as a part of the band assembly. The purpose of the lugs is to connect the band with the servo through the actuating (apply) linkage and the band anchor (reaction) at the opposite end. The band's steel strap is designed with slots or holes to release fluid trapped between the drum and the applying band.

SHOP TALK

A holding planetary control unit is also called a brake or reaction unit because it holds a geartrain member stationary, reacting to rotation.

The bands used in automatic transmissions are rigid, flexible, single wrap, or double wrap types. Steel single wrap bands **(Figure 40–32A)** are used to hold geartrain components driven by high-output engines. Self-energizing action is low because of the rigidity of the band's design. Thinner steel bands are not able to provide a high degree of holding power, but because of the flexibility of design, self-energizing action is stronger and provides more apply force.

The double wrap band is a circular external contracting band normally designed with two or three segments **(Figure 40–32B)**. As the band closes, the segments align themselves around the drum and provide a cushion. The steel body of the double wrap band may be thin or thick steel strapping material. Modern automatic transmissions use thin single or double wrap bands for increased efficiency. Double wrap bands made with heavy thick steel strapping are required for high output engines.

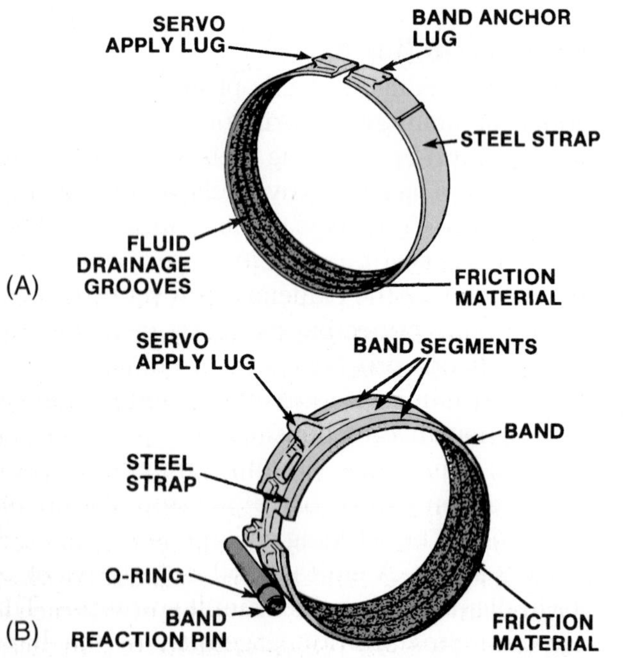

Figure 40-32 (A) Typical single wrap and (B) double wrap transmission band designs. *Courtesy of Chrysler LLC*

Transmission Servos

The **servo** assembly converts hydraulic pressure into a mechanical force that applies a band to hold a drum stationary. Simple and compound servos are used to engage bands in modern transmissions.

Simple Servo In a simple servo **(Figure 40–33)**, the servo piston fits into the servo cylinder and is held in the released position by a coil spring. The piston is sealed with a rubber ring, which keeps fluid pressure confined to the apply side of the servo piston.

To apply a band, fluid pressure is directed to the apply side of the servo piston. The servo piston moves against the return coil spring and develops servo apply force. This force is applied to the band lug through the apply lever and strut. At the opposite end of the band is the anchor strut or end pin and adjustment screw. These hold that end of the band stationary as the band tightens around the rotating drum. The rotating drum comes to a stop and is held stationary by the band.

When servo apply force is released, the return spring forces the servo piston to move in the cylinder. With the servo apply force removed, the band springs free and permits drum rotation. Remember, in most

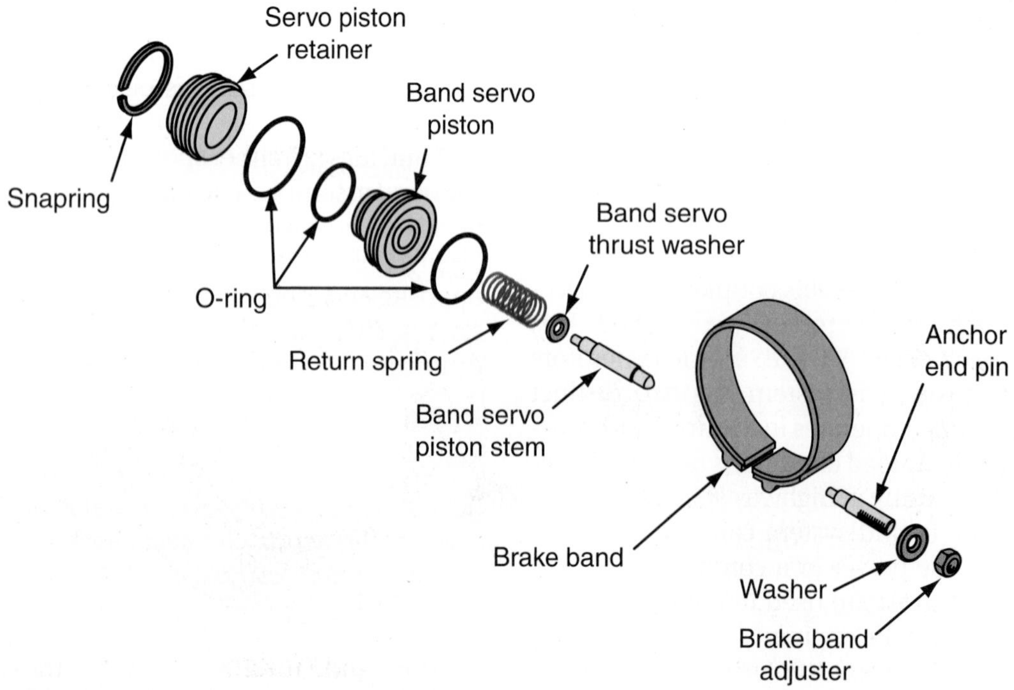

Figure 40-33 A typical band and servo assembly.

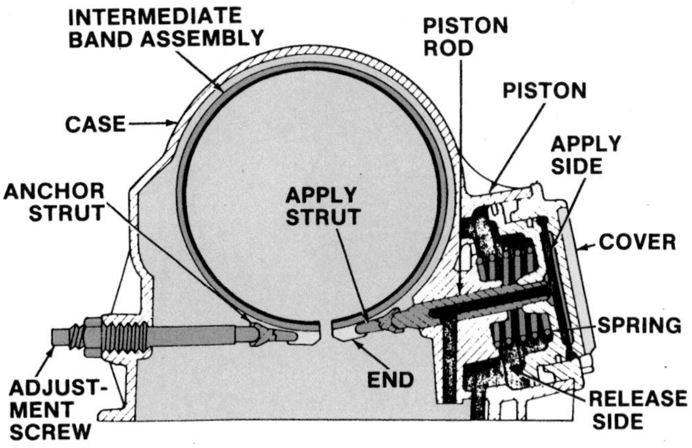

Figure 40–34 A typical compound servo design.
Courtesy of Ford Motor Company

automatic transmissions, hydraulic pressure applies the band and spring pressure releases it.

Compound Servo A compound servo (**Figure 40–34**) has a cylinder that is cast as part of the transmission housing. If the servo is located near the front of the transmission, it uses seal rings capable of withstanding the heat generated by the torque converter and engine.

When the compound servo is applied, fluid pressure flows through the hollow piston pushrod to the apply side of the servo piston. The piston compresses the servo coil spring and forces the pushrod to move one end of the band toward the adjusting screw and anchor. The band tightens around the rotating drum and brings it to a stop. The apply of the compound servo piston is much like the simple servo, but there the similarity ends.

Fluid pressure is applied to the release side of the servo piston when the band is to be released. This provides equal pressure on both sides of the piston and allows the tension of the servo spring to push the piston back up its bore. This action releases the band.

In some transmissions, the servo piston has a larger area on the release side, which causes a more positive release. This design is used to ensure that a band is released before another reaction member is applied.

TRANSMISSION CLUTCHES

In contrast to a band, which can only hold a planetary gear member, transmission clutches are capable of both holding and driving members.

Overrunning Clutches

In an automatic transmission operation, both sprag and roller overrunning clutches are used to hold or drive members of the planetary gear set. These clutches operate mechanically. An overrunning clutch allows rotation in only one direction and operates at all times. One-way overrunning clutches can be either roller-type or sprag-type clutches.

In a roller type (**Figure 40–35**), roller bearings are held in place by springs to separate the inner and outer races of the clutch assembly. One of the races is normally held by the transmission case and is unable to rotate. Around the inside of the outer race are several cam-shaped indentations. The rollers and springs are located in these pockets. Rotation of one race in one direction locks the rollers between the two races, preventing the race from moving. When the race is rotated in the opposite direction, the roller bearings move into the pockets and are not locked and the race is free to rotate.

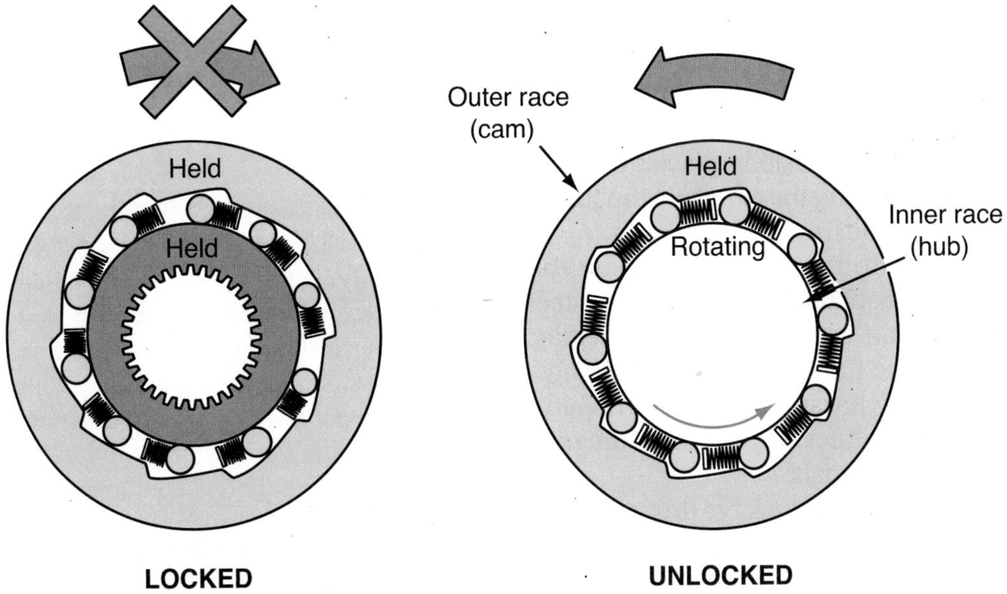

Figure 40–35 The action of a one-way roller clutch.

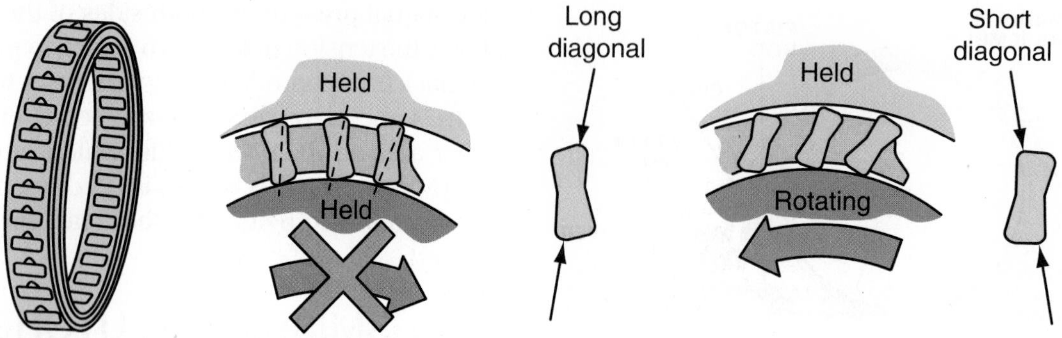

Figure 40-36 The action of a sprag-type one-way clutch.

A one-way sprag clutch **(Figure 40–36)** consists of a hub and drum separated by figure-eight-shaped metal pieces called sprags. The sprags are shaped so that they lock between the races when a race is turned in one direction only. The sprags are longer than the distance between the two races. Springs hold the sprags at the correct angle and maintain the sprags' contact with the races, thereby allowing for instantaneous engagement. When a race rotates in one direction, the sprags lift and allow the races to move independently. When a race is moved in the opposite direction, the sprags straighten and lock the two races together.

Multiple-Disc Clutches

A **multiple-disc clutch** uses a series of flat friction discs to transmit torque or apply braking force. The discs have internal teeth that are sized and shaped to mesh with splines on the clutch assembly hub. Between each pair of friction discs is a flat steel disc. These discs have external teeth that are sized and shaped to mesh with internal splines in the transmission housing or a clutch drum. The combination of friction and steel plates make up the clutch pack.

The clutch pack also has one very thick plate known as the pressure plate. The pressure plate has tabs around the outside diameter to mate with the channels in the clutch drum. It is held in place by a large snapring. Upon engagement, the clutch piston forces the clutch pack against the fixed pressure plate.

The friction discs are sandwiched between the steel plates and pressure plate. Friction discs are steel core plates with friction material bonded to either side. Asbestos was once the universal friction material used but because it is hazardous to human health, cellular paper fibers, graphites, and ceramics are now being used as friction materials.

Clutch packs are enclosed in a large drum-shaped housing that can be either a separate casting **(Figure 40–37)** or part of the existing transmission housing. This drum housing also holds the other clutch components: cylinder, hub, piston, piston return springs, seals, pressure plate, friction plates, and snaprings. The hub is connected to the member of the planetary gear set that will receive the desired braking or transfer force when the clutch is applied or released.

The piston is made of cast aluminum or stamped steel with a seal ring groove around the outside diameter. A rubber seal ring is installed in this groove to retain the fluid pressure that engages the clutch. Piston return springs overcome the reduced fluid pressure in the clutch and move the piston to the disengaged position when clutch action is no longer needed.

Planetary Control Terminology

Table 40–2 is a crossover chart listing the names that the different manufacturers call the same planetary gear control.

This terminology is a sure point of confusion while you are first learning about automatic transmissions. Sometimes manufacturers refer to a clutch or band by the speed gear(s) it controls, others use terms that define what the control does, and others, such as Chrysler, refer to the planetary controls by their location in the transaxle or transmission housing.

SHOP TALK

To make things less confusing, this text follows Chrysler's logic as much as possible. In doing so, there may be some discrepancies between what is in this text and what is in a service manual. Keep in mind that the purpose of this text is to give you a working knowledge of automatic transmissions. The purpose of a service manual is to give you the specific information needed to service or diagnose a specific transmission. Once you have a good understanding of how a transmission works, you will be able to effectively use the information given in a service manual.

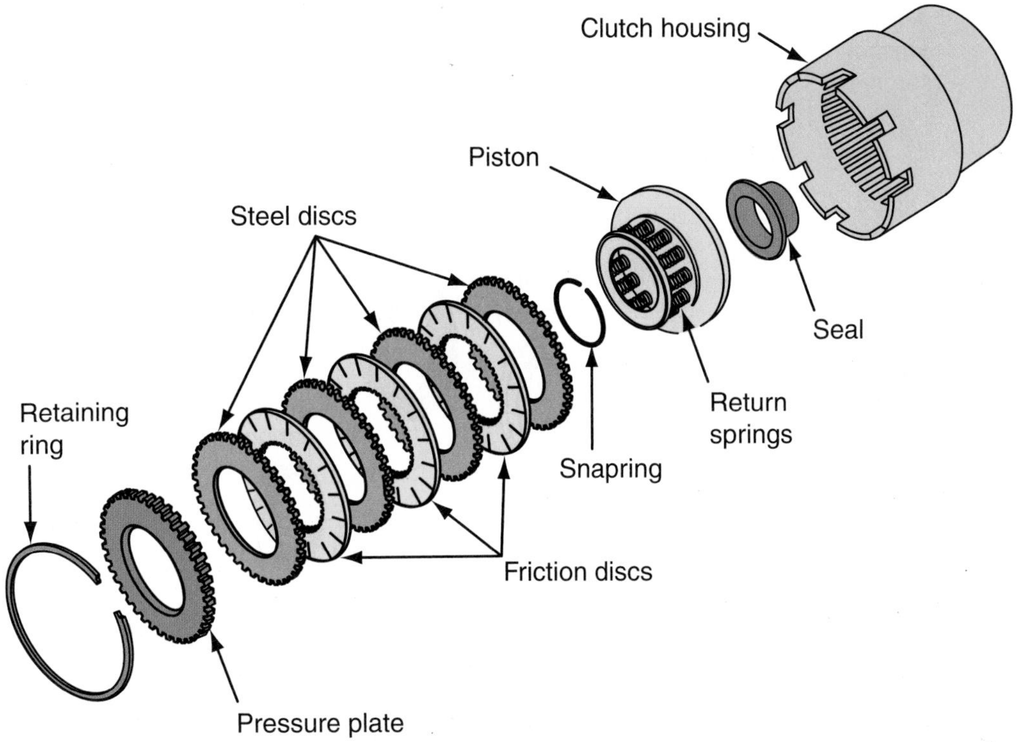

Figure 40-37 A multiple-disc clutch assembly.

TABLE 40-2 PLANETARY CONTROL TERMINOLOGY CROSSOVER CHART

Chrysler	Ford	General Motors
Front clutch	Reverse and high clutch	Direct clutch
Rear clutch	Forward clutch	Forward clutch
Front kickdown band	Intermediate band	Intermediate band
Low and reverse rear band	Low and reverse band or clutch	Low and reverse band or clutch
Overrunning clutch	One-way clutch	Low roller clutch

BEARINGS, BUSHINGS, AND THRUST WASHERS

When a component slides over or rotates around another part, the surfaces that contact each other are called bearing surfaces. A gear rotating on a fixed shaft can have more than one bearing surface; it is supported and held in place by the shaft in a radial direction. Also, the gear tends to move along the shaft in an axial direction as it rotates and is therefore held in place by some other components. The surfaces between the sides of the gear and the other parts are bearing surfaces.

A bearing is a device placed between two bearing surfaces to reduce friction and wear. Most bearings have surfaces that either slide or roll against each other. In automatic transmissions, sliding bearings are used where one or more of the following conditions prevail: low rotating speeds, very large bearing surfaces compared to the surfaces present, and low-use applications. Rolling bearings are used in high-speed applications, high load with relatively small bearing surfaces, and high use.

Transmissions use sliding bearings composed of a relatively soft bronze alloy. Many are made from steel with the bearing surface bonded or fused to the steel. Those that take radial loads are called bushings and those that take axial loads are called thrust washers (**Figure 40–38**). The bearing's surface usually runs against a harder surface such as steel to produce minimum friction and heat wear characteristics.

Bushings are cylindrically shaped and usually held in place by press fit. Since bushings are typically made of a soft metal, they act like a bearing and support many of the transmission's rotating parts (**Figure 40–39**). They are also used to precisely guide the movement of various valves in the transmission's valve body. Bushings can also be used to control fluid flow; some restrict the flow from one part to another, while others are made to direct fluid flow to a particular point or part in the transmission.

Often serving both as a bearing and a spacer, thrust washers are made in various thicknesses. They may have one or more tangs or slots on the inside or outside circumference that mate with the shaft bore to

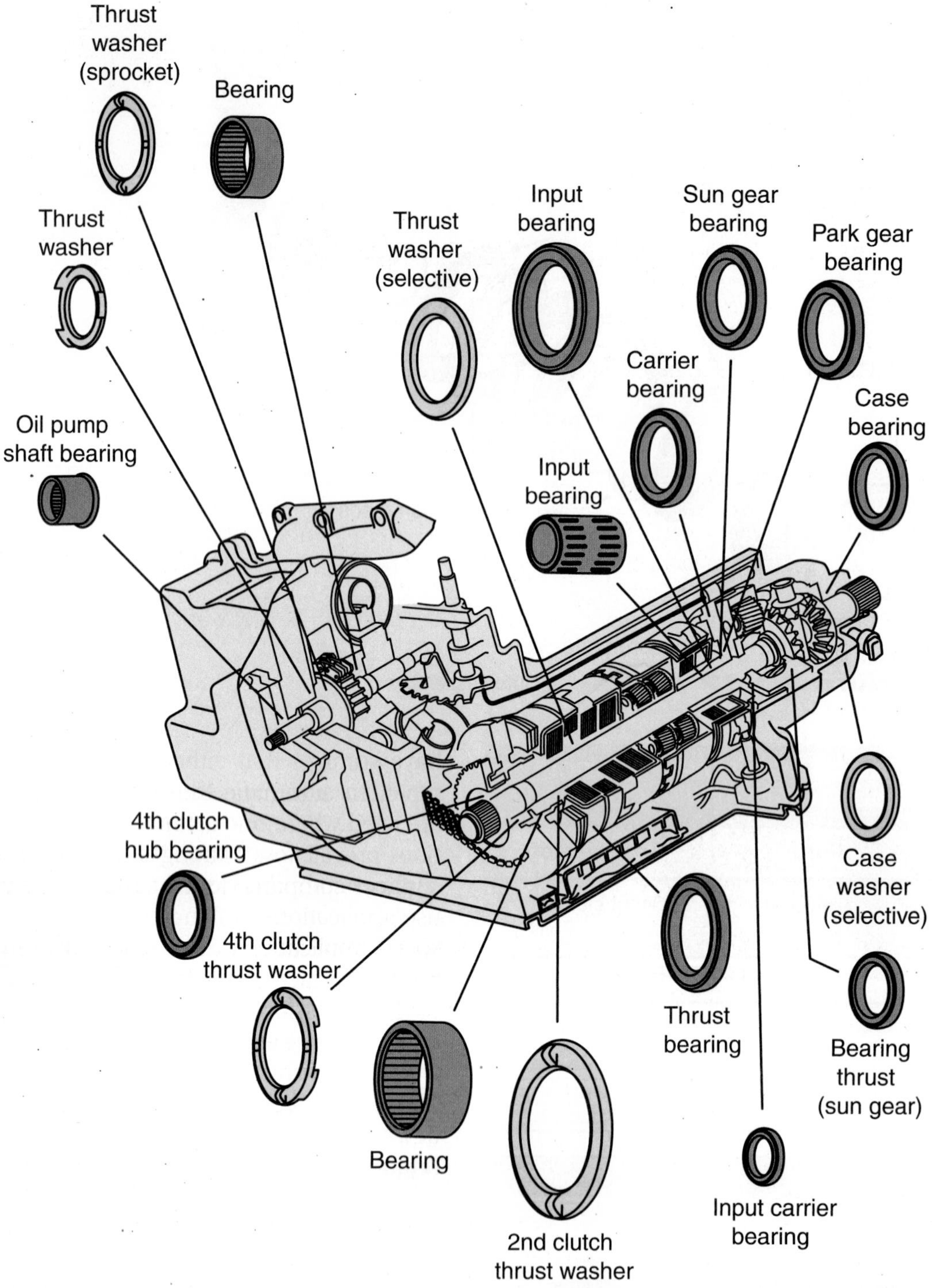

Figure 40-38 The location of various bearings and thrust washers in a typical transaxle.

keep them from turning. Some thrust washers are made of nylon or Teflon, which are used when the load is low. Others are fitted with rollers to reduce friction and wear.

Thrust washers normally control free axial movement or end play. Since some end play is necessary in all transmissions because of heat expansion, proper end play is often accomplished through selective thrust washers. These thrust washers are inserted between various parts of the transmission. Whenever end play is set, it must be set to manufacturer's specifications. Thrust washers work by filling the gap between two objects and become the primary wear item because they are made of softer materials than the parts they protect. Normally, thrust washers are made of copper- or babbit-faced soft steel, bronze, nylon, or plastic.

Torrington bearings **(Figure 40–40)** are thrust washers fitted with roller bearings. These thrust bearings are primarily used to limit end play but also to reduce the friction between two rotating parts. Most

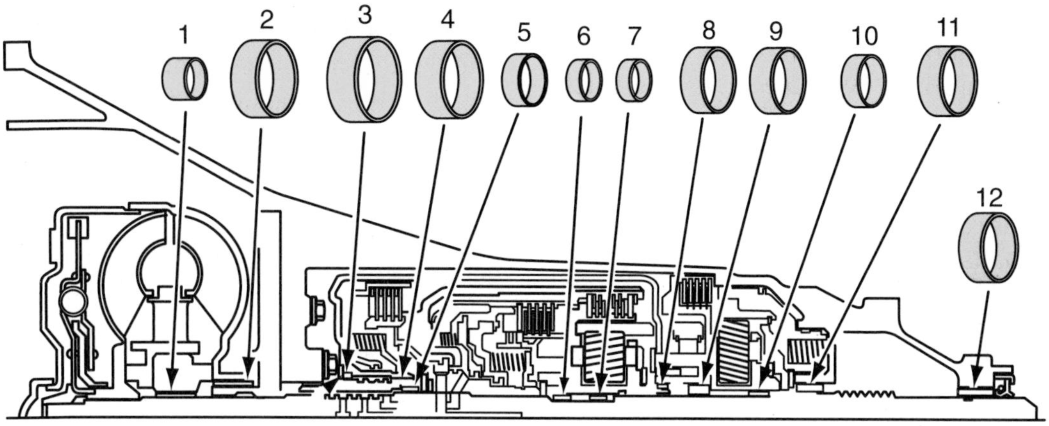

LEGEND

1. Bushing, Stator shaft (front)
2. Bushing, Oil pump body
3. Bushing, Reverse input clutch (front)
4. Bushing, Reverse input clutch (rear)
5. Bushing, Stator shaft (rear)
6. Bushing, Input sun gear (front)
7. Bushing, Input sun gear (rear)
8. Bushing, Reaction carrier shaft (front)
9. Bushing, Reaction gear
10. Bushing, Reaction carrier shaft (rear)
11. Bushing, Case
12. Bushing, Case extension

Figure 40-39 Bushings are used throughout a transmission.

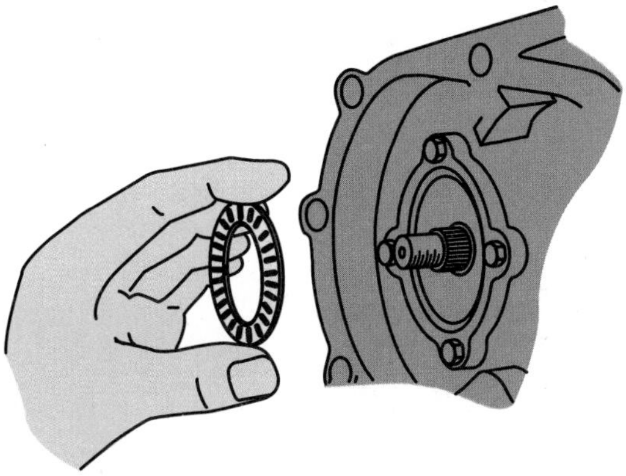

Figure 40-40 A Torrington-type axial thrust bearing.

often, Torrington bearings are used in combination with flat thrust washers to control end play of a shaft or the gap between a gear and its drum.

The bearing surface is greatly reduced through the use of roller bearings. The simplest roller bearing design leaves enough clearance between the bearing surfaces of two sliding or rotating parts to accept some rollers. Each roller's two points of contact between the bearing surfaces are so small that friction is greatly reduced. The bearing surface is more like a line than an area.

If the roller length to diameter is about 5:1 or more, the roller is called a needle and such a bearing is called a needle bearing. Sometimes the needles are loose or they can be held in place by a steel cylinder or by rings at each end. Often the latter are drilled to accept pins at the ends of each needle that act as axles. These small assemblies help save the agony of losing one or more loose needles and the delay caused by searching for them.

Many other roller bearings are designed as assemblies. The assemblies consist of an inner and outer race, the rollers, and a cage. There are roller bearings designed for radial loads and others designed for axial loads.

A tapered roller bearing is designed to accept both radial and axial loads. Its rollers turn on an angle to the bearing assembly's axis rather than parallel to it. The rollers are also slightly tapered to fit the angle of the inner and outer races. The bearing assembly consists of an inner race, the rollers, the cage, and the outer race. Tapered roller bearings are normally used in pairs and are rarely used in automatic transmissions. They are commonly used in final drive units.

The heaviest radial loads in automatic transmissions are carried by either roller or ball bearings. Ball bearings are constructed similarly to roller bearings, except that the races are grooved to accept the balls. The groove radius is slightly larger than the ball radius, which reduces the bearing surface area more than the roller bearing does. A ball bearing can also withstand light axial loads. Lip seals are sometimes built into ball bearings to retain lubricants.

SNAPRINGS

Many different sizes and types of snaprings are used in today's transmissions. External and internal snaprings are used as retaining devices throughout the transmission. Internal snaprings are used to hold servo assemblies and clutch assemblies together. In fact, snaprings are also available in several thicknesses and may be used to adjust the clearance in multiple-disc clutches. Some snaprings for clutch packs are waved to smooth clutch application. External snaprings are used to hold gear and clutch assemblies to their shafts.

GASKETS AND SEALS

The gaskets and seals of an automatic transmission help to contain the fluid within the transmission and prevent the fluid from leaking out of the various hydraulic circuits. Different types of seals are used in automatic transmissions; they can be made of rubber, metal, or Teflon **(Figure 40–41)**. Transmission gaskets are made of rubber, cork, paper, synthetic materials, metal, or plastic.

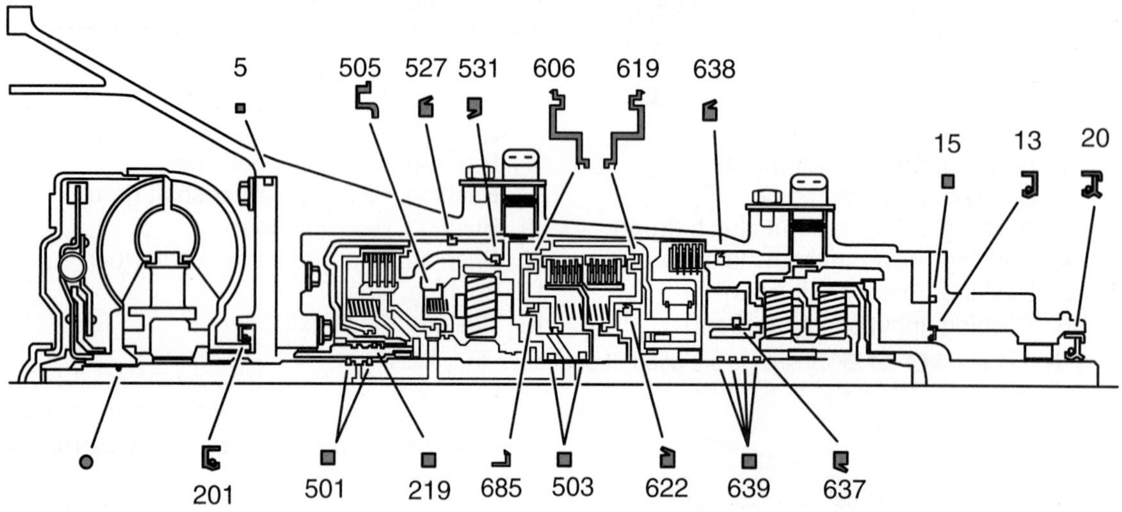

LEGEND

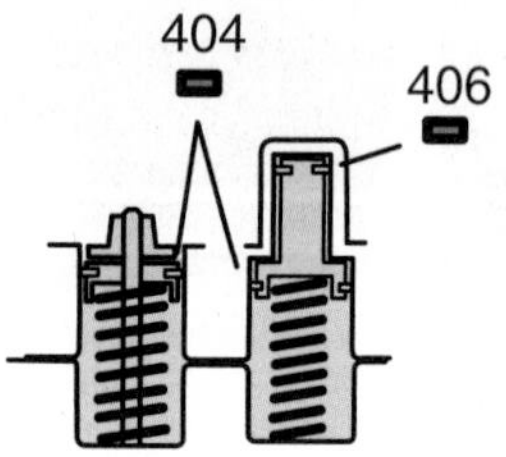

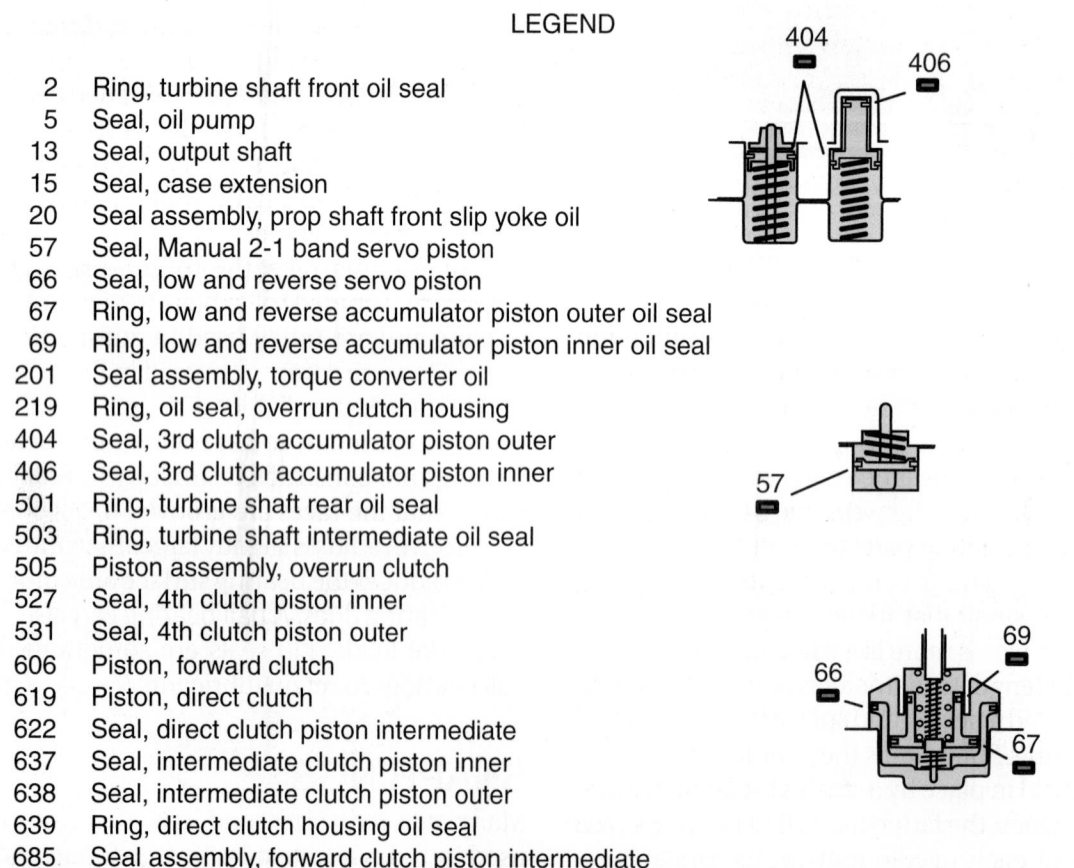

2 Ring, turbine shaft front oil seal
5 Seal, oil pump
13 Seal, output shaft
15 Seal, case extension
20 Seal assembly, prop shaft front slip yoke oil
57 Seal, Manual 2-1 band servo piston
66 Seal, low and reverse servo piston
67 Ring, low and reverse accumulator piston outer oil seal
69 Ring, low and reverse accumulator piston inner oil seal
201 Seal assembly, torque converter oil
219 Ring, oil seal, overrun clutch housing
404 Seal, 3rd clutch accumulator piston outer
406 Seal, 3rd clutch accumulator piston inner
501 Ring, turbine shaft rear oil seal
503 Ring, turbine shaft intermediate oil seal
505 Piston assembly, overrun clutch
527 Seal, 4th clutch piston inner
531 Seal, 4th clutch piston outer
606 Piston, forward clutch
619 Piston, direct clutch
622 Seal, direct clutch piston intermediate
637 Seal, intermediate clutch piston inner
638 Seal, intermediate clutch piston outer
639 Ring, direct clutch housing oil seal
685 Seal assembly, forward clutch piston intermediate

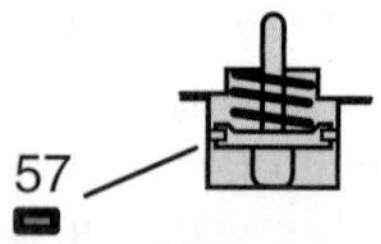

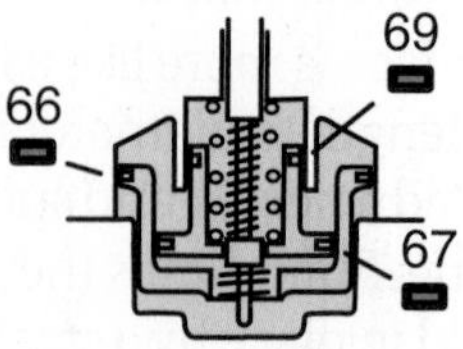

Figure 40-41 The location of various seals and gaskets in a typical transmission.

Gaskets

Gaskets are used to seal two parts together or to provide a passage for fluid flow from one part of the transmission to another. Gaskets are easily divided into two separate groups, hard and soft, depending on their application. Hard gaskets are used whenever the surfaces to be sealed are smooth. This type of gasket is usually made of paper. A common application of a hard gasket is the gasket used to seal the valve body and oil pump against the transmission case. Hard gaskets are also often used to direct fluid flow or to seal off some passages between the valve body and the separator plate.

Gaskets that are used when the sealing surfaces are irregular or in places where the surface may distort when the component is tightened into place are called soft gaskets. A typical location of a soft gasket is the oil pan gasket that seals the oil pan to the transmission case. Oil pan gaskets are typically a composition-type gasket made with rubber and cork. However, some late-model transmissions use an RTV sealant instead of a gasket to seal the oil pan.

Seals

As valves and transmission shafts move within the transmission, it is essential that the fluid and pressure be contained within its bore. Any leakage would decrease the pressure and result in poor transmission operation. Seals are used to prevent leakage around valves, shafts, and other moving parts. Rubber, metal, or Teflon materials are used throughout a transmission to provide for static and dynamic sealing. Both static and dynamic seals can provide for positive and nonpositive sealing. A definition of each of the different basic classifications of seals follows:

- *Static.* A seal used between two parts that do not move in relationship to each other, such as the pan and oil pump-to-case gaskets.
- *Dynamic.* A seal used between two parts that do move in relationship to each other. This movement is either a rotating or reciprocating (up and down) motion. The seal of a clutch piston is an example of this type of seal.
- *Positive.* A seal that prevents all fluid leakage between two parts.
- *Nonpositive.* A seal that allows a controlled amount of fluid leakage. This leakage is typically used to lubricate a moving part.

Three major types of rubber seals are used in automatic transmissions: the O-ring, the lip seal, and the square-cut seal. Rubber seals are made from synthetic rubber rather than natural rubber.

O-rings are round seals with a circular cross section. Normally an O-ring is installed in a groove cut into the inside diameter of one of the parts to be sealed. When the other part is inserted into the bore and through the O-ring, the O-ring is compressed between the inner part and the groove. This pressure distorts the O-ring and forms a tight seal between the two parts.

O-rings can be used as dynamic seals but are most commonly used as static seals. An O-ring can be used as a dynamic seal when the parts have relatively low amounts of axial movement. If there is a considerable amount of axial movement, the O-ring will quickly be damaged as it rolls within its groove. O-rings are never used to seal a shaft or part that has rotational movement.

Lip seals are used to seal parts that have axial or rotational movement. They are round to fit around a shaft but the entire seal does not serve as a seal; rather, the sealing part is a flexible lip. The flexible lip is normally made of synthetic rubber and shaped so that it is flexed when it is installed to apply pressure at the sharp edge of the lip. Lip seals are used around input and output shafts to keep fluid in the housing and dirt out. Some seals are double-lipped.

When the lip is around the outside diameter of the seal, it is used as a piston seal (**Figure 40–42**). Piston seals are designed to seal against high pressures and the seal is positioned so that the lip faces the source of the pressurized fluid. The lip is pressed firmly against the cylinder wall as the fluid pushes against the lip; this forms a tight seal. The lip then relaxes its seal when the pressure on it is reduced or exhausted.

Lip seals are also commonly used as shaft seals. When used to seal a rotating shaft, the lip of the seal is around the inside diameter of the seal and the outer diameter is bonded to the inside of a metal housing.

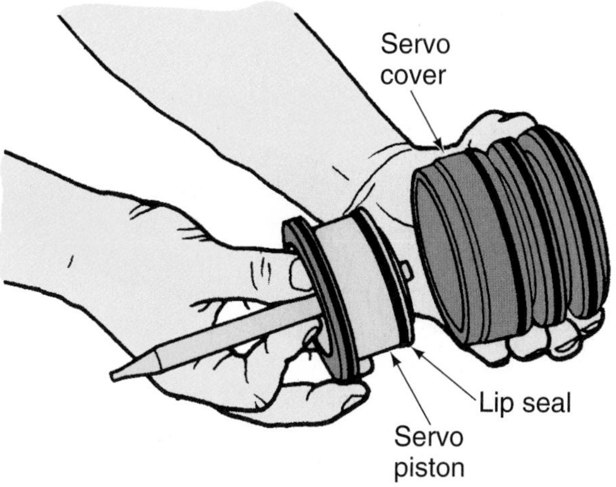

Figure 40-42 Typical application of a lip seal.
Courtesy of Ford Motor Company

The outer metal housing is pressed into a bore. To help maintain good sealing pressure on the rotating shaft, a garter spring is fitted behind the lip. This toroidal spring pushes on the lip to provide for uniform contact on the shaft. Shaft seals are not designed to contain pressurized fluid; rather, they are designed to prevent fluid from leaking over the shaft and out of the housing. The tension of the spring and of the lip is designed to allow an oil film of about 0.0001 inch (0.00254 mm). This oil film serves as a lubricant for the lip. If the tolerances increase, fluid will be able to leak past the shaft and if the tolerances are too small, excessive shaft and seal wear will result.

A **square-cut seal** is similar to an O-ring; however, a square-cut seal can withstand more axial movement than an O-ring can. Square-cut seals have a rectangular or square cross section. They are designed this way to prevent the seal from rolling in its groove when there are large amounts of axial movement. Added sealing comes from the distortion of the seal during axial movement. As the shaft inside the seal moves, the outer edge of the seal moves more than the inner edge causing the diameter of the sealing edge to increase, which creates a tighter seal.

Metal Sealing Rings

There are some parts of the transmission that do not require a positive seal and in which some leakage is acceptable. These components are sealed with ring seals that fit into a groove on a shaft **(Figure 40–43)**. The outside diameter of the ring seals slide against the walls of the bore into which the shaft is inserted. Most ring seals in a transmission are placed near pressurized fluid outlets on rotating shafts to help retain pressure. Ring seals are made of cast iron, nylon, or Teflon.

Three types of metal seals are used in automatic transmissions: **butt-end seals**, open-end seals, and hook-end seals. In appearance, butt-end and open-end seals are much the same; however, when an open-end seal is installed, there is a gap between the ends of the seal. **Hook-end seals (Figure 40–44)** have small hooks at their ends that are locked together during installation to provide better sealing than the open-end or butt-end seals.

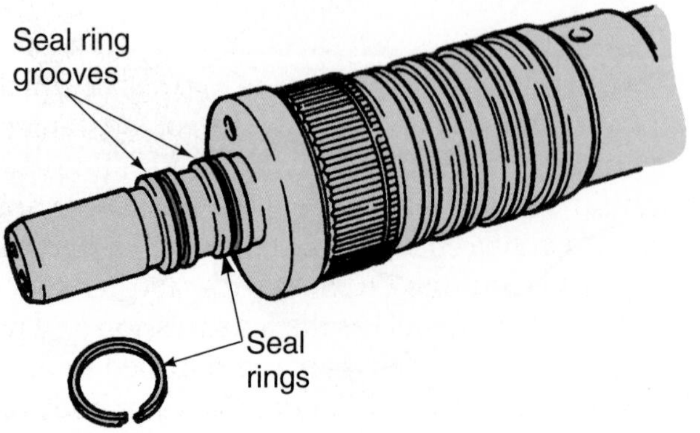

Figure 40-43 Metal sealing rings are fit into grooves on a shaft.

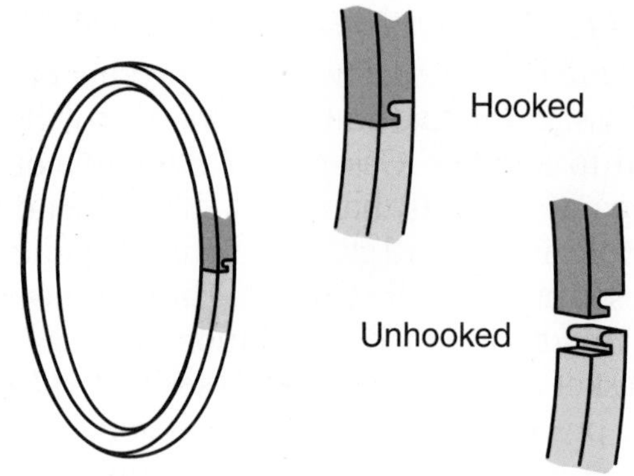

Figure 40-44 Hook-end sealing rings.

Teflon Seals

Some transmissions use Teflon seals instead of metal seals. Teflon provides for a softer sealing surface, which results in less wear on the surface that it rides on and therefore a longer-lasting seal. Teflon seals are similar in appearance to metal seals except for the hook-end type. The ends of locking-end Teflon seals are cut at an angle **(Figure 40–45)** and the locking

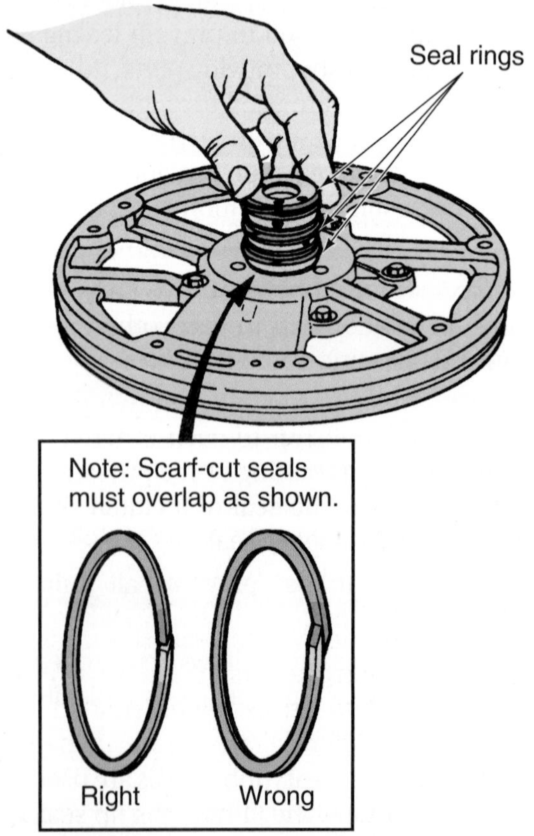

Figure 40-45 Scarf-cut seals. *Courtesy of Ford Motor Company*

hooks are somewhat staggered. These seals are often called scarf-cut seals.

Many late-model transmissions are equipped with solid one-piece Teflon seals. Although the one-piece seal requires some special tools for installation, they provide for a nearly positive seal. These Teflon rings form a better seal than other metal sealing rings.

General Motors uses a different type of synthetic seal on some late-model transmissions. The material used in these seals is Vespel, which is a flexible but highly durable plasticlike material.

FINAL DRIVES AND DIFFERENTIALS

The last set of gears in the drivetrain is the final drive. In most RWD cars, the final drive is located in the rear axle housing. On most FWD cars, the final drive is located within the transaxle. Some FWD cars with longitudinally mounted engines locate the differential and final drive in a separate case that bolts to the transmission. AWD and 4WD vehicles have a final drive unit in the front and rear drive axles.

RWD final drives normally use a hypoid final drive gearset that turns the power flow 90 degrees from the drive shaft to the drive axles. On FWD cars with a transversely mounted engines, the power flow axis is naturally parallel to that of the drive axles; therefore, the power does not need to turn. Simple gear connections can be made to connect the output of the transmission to the final drive.

Final Drive Assemblies

A transaxle's final drive gears provide a way to transmit the transmission's output to the differential section of the transaxle. There are four common configurations used as the final drives on FWD vehicles: helical gear, planetary gear, hypoid gear, and chain drive. The helical, planetary, and chain final drive arrangements are found with transversely mounted engines. Hypoid final drive gear assemblies are normally found in vehicles with a longitudinally placed engine. The hypoid assembly is basically the same unit as would be used on RWD vehicles and is mounted directly to the transmission.

Some transaxles route power from the transmission through two helical-cut gears to a transfer shaft. A helical-cut pinion gear attached to the opposite end of the transfer shaft drives the differential ring gear and carrier. The differential assembly then drives the axles and wheels.

Rather than use helical-cut or spur gears in the final drive assembly, some transaxles use a simple planetary gear set for its final drive. The sun gear of this planetary unit is driven by the final drive sun gear shaft, which is splined to the front carrier and rear ring gear of the transmission's gear set. The final drive sun gear meshes with the final drive planetary pinion gears, which rotate on their shafts in the planetary carrier. The pinion gears mesh with the ring gear, which is splined to the transaxle case. The planetary carrier is part of the differential case, which contains typical differential gearing, two pinion gears, and two side gears.

The ring gear of a planetary final drive assembly has lugs around its outside diameter that fit into grooves machined inside the transaxle housing. These lugs and grooves hold the ring gear stationary. The transmission's output is connected to the planetary gearset's sun gear. In operation, the transmission's output drives the sun gear that, in turn, drives the planetary pinion gears. The pinion gears walk around the inside of the stationary ring gear. The rotating planetary pinion gears drive the planetary carrier and differential case. This combination provides maximum torque multiplication from a simple planetary gearset.

Chain-drive final drive assemblies use a multiple-link chain to connect a drive sprocket, connected to the transmission's output shaft, to a driven sprocket that is connected to the differential's pinion shaft. This design allows for remote positioning of the differential within the transaxle housing. Final drive gear ratios are determined by the size of the driven sprocket compared to the drive sprocket.

HYDRAULIC SYSTEM

A hydraulic system uses a liquid to perform work. In an automatic transmission, this liquid is automatic transmission fluid (ATF). ATF is one of the most complex fluids produced by the petroleum industry for the automobile.

The transmission's pump is the source of all fluid flow in the hydraulic system. It provides a constant supply of fluid under pressure to operate, lubricate, and cool the transmission. **Pressure-regulating valves** change the fluid's pressure to control the shift quality of a transmission and the shift points of the transmission equipped with a governor. **Flow-directing valves** direct the pressurized fluid to the appropriate apply device to cause a change in gear ratios. The hydraulic system also keeps the T/C filled with fluid.

The reservoir for ATF is the transmission's oil pan. Fluid is drawn from the pan and returned to it. The pressure source is the oil pump. The valve body contains control valving to regulate or restrict the pressure and flow of fluid within the transmission. The output devices for the hydraulic system are the servos or clutches operated by hydraulic pressure.

Hydraulic Principles

An automatic transmission uses ATF fluid pressure to control the action of the planetary gearsets. This fluid pressure is regulated and directed to change gears automatically through the use of various pressure regulators and control valves.

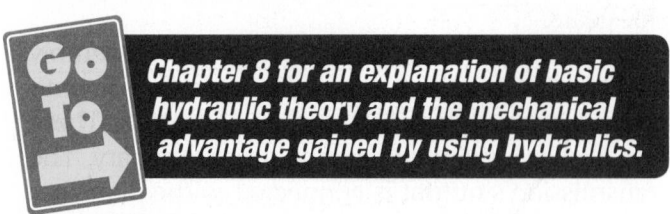

Go To Chapter 8 for an explanation of basic hydraulic theory and the mechanical advantage gained by using hydraulics.

Fluids work well in increasing force because they are perfect conductors of pressure. Fluids cannot be compressed. Therefore, when a piston in a cylinder moves and displaces fluid, that fluid is distributed equally within the circuit.

APPLICATION OF HYDRAULICS IN TRANSMISSIONS

A common hydraulic system within an automatic transmission is the servo assembly, which is used to control the application of a band. The band must tightly hold the drum or planetary carrier it surrounds when it is applied. The holding capacity of the band is determined by the construction of the band and the pressure applied to it. This pressure or holding force is the result of the action of a servo. The servo multiplies the force through hydraulic action.

If a servo has a diameter of 4 inches (102 mm) and has a pressure of 70 psi (4.9 kg/cm^2) applied to it, the apply force of the servo is 880 pounds (399 kg) **(Figure 40–46)**. The force exerted by the

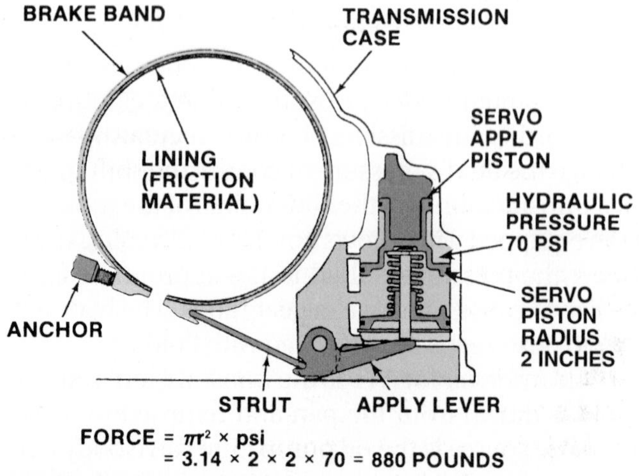

Figure 40–46 Calculating the output force developed by a servo assembly. *Courtesy of Chrysler LLC*

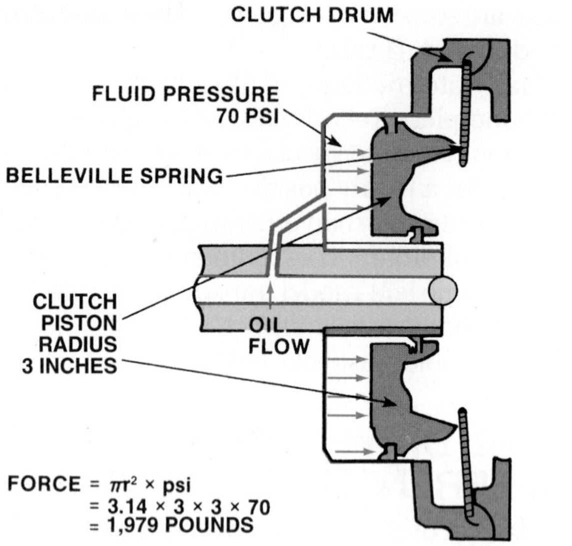

Figure 40–47 Using hydraulics to increase work in a multiple-disc clutch. *Courtesy of Chrysler LLC*

servo is further increased by its lever-type linkage and the self-energizing action of the band. The total force applied by the band stops and holds the rotating drum connected to a planetary gearset member.

A multiple-disc assembly is also used to stop and hold gear set members. This assembly also uses hydraulics to increase its holding force. If the fluid pressure applied to the clutch assembly is 70 psi (4.9 kg/cm^2) and the diameter of the clutch piston is 6 inches (152 mm), the force applying the clutch pack is 1,979 pounds (898 kg). If the clutch assembly uses a **Belleville spring** or piston spring **(Figure 40–47)**, which adds a mechanical advantage of 1.25, the total force available to engage the clutch will be 1,979 pounds (898 kg) multiplied by 1.25, or 2,474 pounds (1,122.5 kg).

Functions of ATF

The ATF circulating through the transmission and torque converter and over the parts of the transmission cools the transmission. The heated fluid typically moves to a transmission fluid cooler, where the heat is removed. As the fluid lubricates and cools the transmission, it also cleans the parts. The dirt is carried by the fluid to a filter, where the dirt is removed.

Another critical job of ATF is its role in shifting gears. ATF moves under pressure throughout the transmission and causes various valves to move. The pressure of the ATF changes with changes in engine speed and load.

ATF is also used to operate the various apply devices (clutches and bands) in the transmission. At the appropriate time, a switching valve opens and sends pressurized fluid to the apply device that engages or disengages a gear. The valving and hydraulic circuits are contained in the valve body.

Reservoir

A fluid reservoir stores fluid and provides a constant source of fluid for the system. In an automatic transmission, the reservoir is the pan, typically located at the bottom of the transmission case. ATF is forced out of the pan by atmospheric pressure and into the pump and is then returned to it after it has circulated through the selected circuits. A transmission dipstick placed within a filler tube is typically used to check the level of the fluid and to add ATF to the transmission. Other transmissions have a side plug on the pan or the transmission to check and replenish fluid level.

Venting

All reservoirs must have an air vent that allows atmospheric pressure to force the fluid into the pump when the pump creates a low pressure at its inlet port. The pans of many automatic transmissions vent through the handle of the dipstick; others rely on a vent in the transmission case. Transmissions must also be vented to allow for the exhaust of built-up air pressure that results from heat and the moving components inside the transmission. The movement of these parts can force air into the ATF, which would not allow it to increase in pressure, cool, or lubricate the transmission properly.

Transmission Coolers

The removal of heat from ATF is extremely important to the durability of the transmission. Excessive heat causes the fluid to break down. Once broken down, ATF no longer lubricates well and has poor resistance to oxidation. Oxidized ATF may damage transmission seals. When a transmission is operated for some time with overheated ATF, varnish is formed inside the transmission. Varnish buildup on valves can cause them to stick or move slowly. The result is poor shifting and glazed or burned friction surfaces. Continued operation can lead to the need for a complete rebuilding of the transmission.

It is important to note that ATF is designed to operate at 175°F (80°C). At this temperature, the fluid should remain effective for 100,000 miles (160,000 km). However, when the operating temperature increases, the useful life of the fluid quickly decreases. A 20°F increase in operating temperature will decrease the life of ATF by one-half!

Transmission housings are fitted with ATF cooler lines **(Figure 40–48A)** that direct the hot fluid from the torque converter to the transmission cooler,

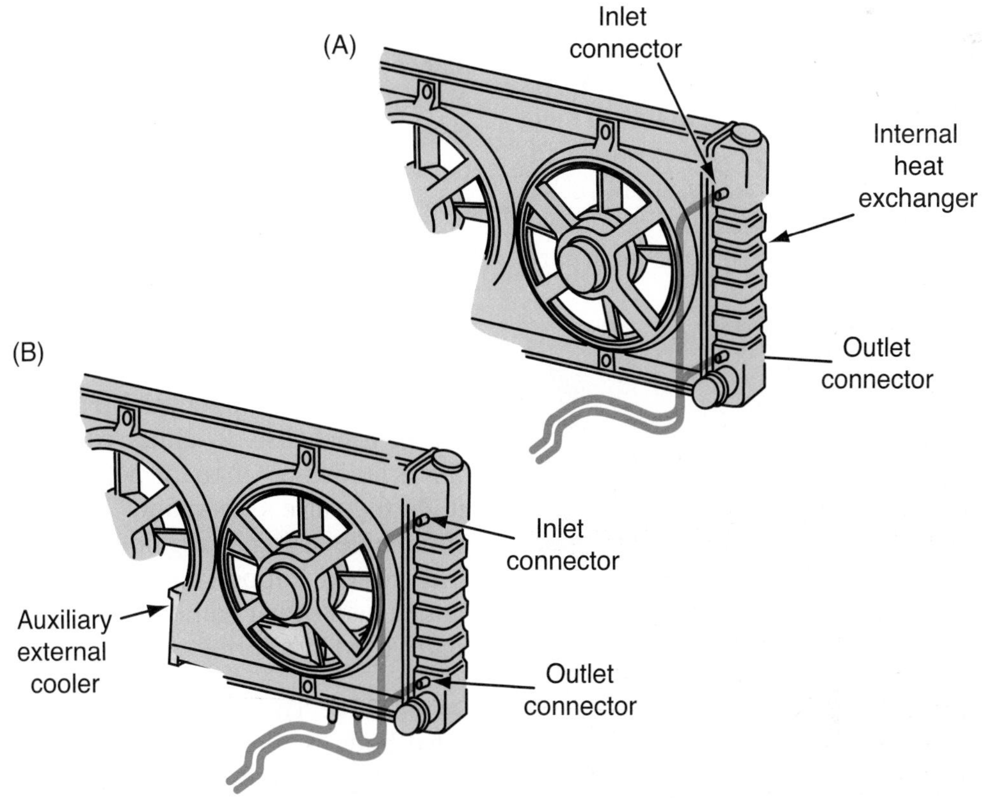

Figure 40-48 (A) A transmission cooler (heat exchanger) located in a radiator. (B) An auxiliary cooler added to the normal cooler circuit.

normally located in the vehicle's radiator. The heat of the fluid is reduced by the cooler and the cool ATF returns to the transmission. In some transmissions, the cooled fluid flows directly to the transmission's bushings, bearings, and gears. Then, the fluid is circulated through the rest of the transmission. The cooled fluid in other transmissions is returned to the oil pan, where it is drawn into the pump and circulated throughout the transmission.

Some vehicles, such as those designed for heavy-duty use, are equipped with an auxiliary fluid cooler **(Figure 40–48B)**, in addition to the one in the radiator. This cooler removes additional amounts of heat from the fluid before it is sent back to the transmission.

Valve Body

For efficient transmission operation, the bands and multiple-disc packs must be released and applied at the proper time. The **valve body** assembly **(Figure 40–49)** is responsible for the control and distribution of pressurized fluid throughout the transmission. This assembly is made of two or three main parts: a valve body, separator plate, and transfer plate. These parts are bolted as a single unit to the transmission housing. The valve body is machined from aluminum or iron and has many precisely machined bores and fluid passages. Various valves are fitted into the bores, and the passages direct fluid to various valves and other parts of the transmission. The separator and transfer plates are designed to seal off some of these passages and to allow fluid to flow through specific passages.

The purpose of a valve body is to sense and respond to engine and vehicle load as well as to meet the needs of the driver. Valve bodies are normally fitted with three different types of valves: spool valves, check ball valves, and poppet valves. The purpose of these valves is to start, to stop, or to use movable parts to regulate and direct the flow of fluid throughout the transmission.

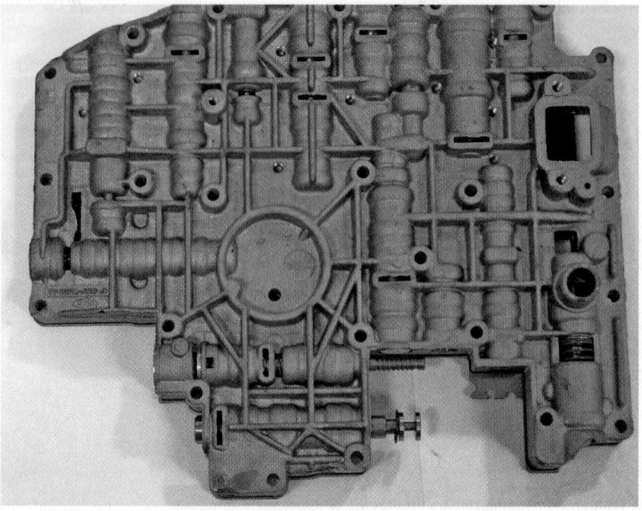

Figure 40–49 A typical valve body.

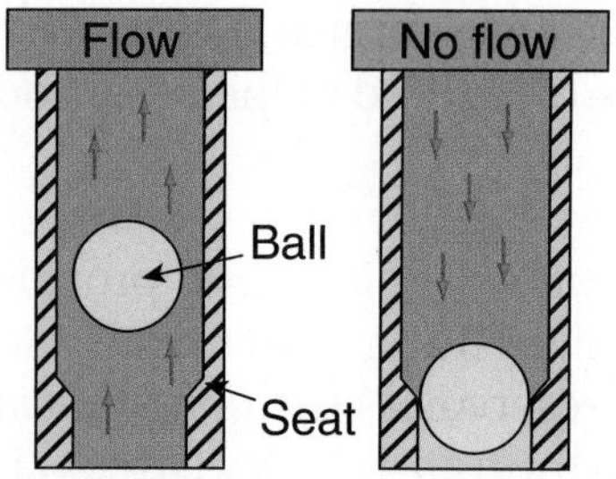

Figure 40–50 The operation of a check ball valve.
Courtesy of Chrysler LLC

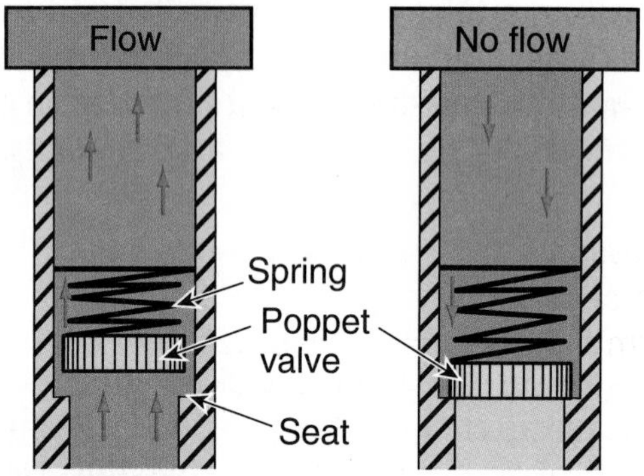

Figure 40–51 Typical poppet valve operation.
Courtesy of Chrysler LLC

Check Ball Valve The **check ball valve** is a ball that operates on a seat located on the valve body. The check ball operates by having a fluid pressure or manually operated linkage force it against the ball seat to block fluid flow **(Figure 40–50)**. Pressure on the opposite side unseats the check ball. Check balls and poppet valves can be normally open, which allows free flow of fluid pressure, or normally closed, which blocks fluid pressure flow.

At times, the check ball has two seats to check and direct fluid flow from two directions, being seated and unseated by pressures from either source.

Poppet Valve A poppet valve **(Figure 40–51)** can be a ball or a flat disc. In either case, the poppet valve blocks fluid flow. Often the poppet valve has a stem to guide the valve's operation. The stem normally fits into a hole acting as a guide to the valve's opening and closing. Poppet valves tend to pop open and closed, hence their name. Normally poppet valves are held closed by a spring.

Spool Valve The most commonly used valve in a valve body is the **spool valve**. A spool valve **(Figure 40–52)** looks similar to a sewing thread spool. The large circular parts of the valve are called the lands. There is a minimum of two lands per valve. Each land

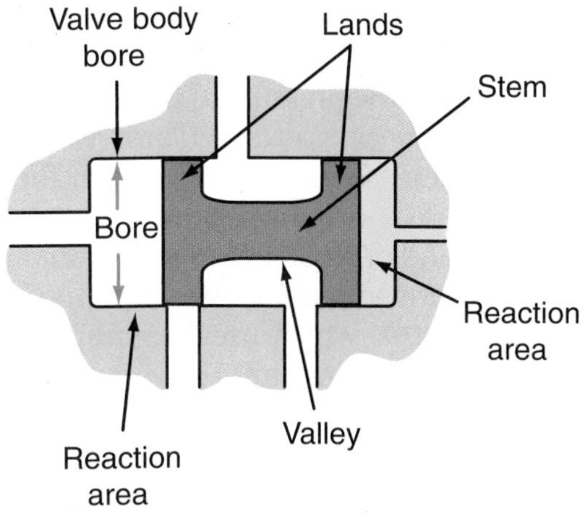

Figure 40-52 Components of a spool valve assembly.

Figure 40-53 A gear-type oil pump. *Courtesy of Transtar Industries*

of the assembly is connected by a stem. The space between the lands and stem is called the valley. Valleys form a fluid pressure chamber between the spools and valve body bore. Fluid flow can be directed into other passages depending on the spool valve and valve body design.

Precisely machined around the periphery of each valve, the land is the part of the valve that rides on a thin film of fluid in a bore of the valve body. The land must be treated very carefully because any damage, even a small score or scratch, can impair smooth valve operation. As the spool valve moves, the land covers (closes) or uncovers (opens) ports in the valve body.

The reaction area, also known as the face, is the space at the outside of the lands at the end of the valve. Forces acting against the reaction area that cause the valve to move include spring tension, fluid pressure, or mechanical linkage.

Oil Pump

The source of fluid flow through the transmission is the oil pump **(Figure 40–53)**. Three types of oil pumps are commonly used in automatic transmissions: gear type **(Figure 40–54A)**, rotor type, and vane type **(Figure 40–54B)**. Oil pumps are driven by the pump drive hub of the T/C or oil pump shaft and/or converter cover on transaxles. Therefore, whenever the T/C cover is rotating, the oil pump is driven. The oil pump creates fluid flow throughout the transmission.

Pressure Regulator Valve

Transmission pumps are capable of creating excessive amounts of pressure that may cause damage;

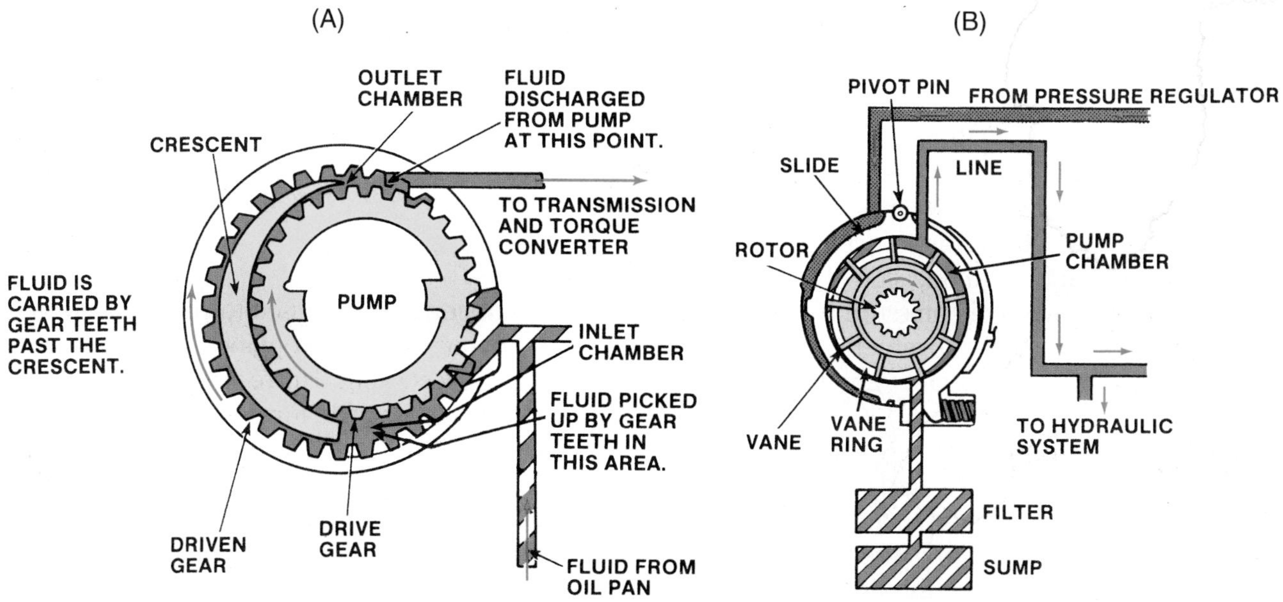

Figure 40-54 (A) Operation of a gear-type pump and (B) a vane-type pump.

therefore, the transmission is equipped with a pressure regulator valve, which is normally located in the valve body. Pressure regulating valves are typically spool valves that toggle back and forth in their bores to open and close an exhaust passage. By opening the exhaust passage, the valve decreases the pressure of the fluid. As soon as the pressure decreases to a predetermined amount, the spool valve moves to close off the exhaust port and pressure again begins to build. The action of the spool valve regulates the fluid pressure.

Many late-model transmissions use an electronic pressure control (EPC or PC) solenoid to regulate system pressure.

Governor Assembly

The **governor** assembly is driven by the transmission's output shaft, senses road speed, and sends a fluid pressure signal to the valve body to either upshift or downshift. When vehicle speed is increased, the pressure developed by the governor is directed to the shift valve. As the speed (and therefore the pressure) increases, the spring tension and throttle pressure on the shift valve are overcome and the valve moves. This action causes an upshift. Likewise, a decrease in speed results in a decrease in pressure and a downshift.

Although the governor sends a signal that will force an upshift, engine load may cause a delay in the shift. This allows for operation in a lower gear when there is a heavy load and the vehicle needs the gear reduction. During heavy-load operation, the governor pressure must be strong enough to overcome the high throttle pressure plus the spring tension on the shift valve before it can force an upshift. Because of this, the transmission will remain in a particular gear range until a higher-than-normal engine speed is reached.

PRESSURE BOOSTS

When the engine is operating under heavy load conditions, fluid pressure must be increased to increase the holding capacity of a hydraulic member. Increasing the fluid pressure holds the band and clutch control units tighter to reduce the chance of slipping while under heavy load. This is accomplished by sending pressurized fluid to one side of the pressure regulator's spool valve. This pressure works against the spool valve's normal movement to open the exhaust port and allows pressure to build to a higher point than normal.

Engine load can be monitored electronically by various electronic sensors (primarily the TP and MAP sensors) that send information to an electronic control unit, which in turn controls the pressure at the valve body. Load can also be monitored by throttle pressure. Throttle pedal movement moves a **throttle valve** in the valve body via a throttle cable. When the throttle plate is opened, the throttle valve opens and applies pressure to the pressure regulator. This delays the opening of the pressure regulator valve, which allows for an increase in pressure. When the driver lets off the throttle pedal, the pressure regulator valve is free to move and normal pressure is maintained.

Many early transmissions were equipped with a **vacuum modulator**, which uses engine vacuum to change transmission pressure. The vacuum modulator allows for an increase in pressure when vacuum is low and decreases it when vacuum is high.

MAP Sensor

Engine load can be monitored electronically through the use of various electronic sensors that send information to an electronic control unit, which in turn controls the pressure at the valve body. The most commonly used sensor is the MAP sensor. The MAP sensor senses air pressure in the intake manifold. The control unit uses this information as an indication of engine load. A pressure-sensitive ceramic or silicon element and electronic circuit in the sensor generates a voltage signal that changes in direct proportion to pressure. A MAP sensor measures manifold air pressure against a precalibrated absolute pressure; therefore, the readings from these sensors are not adversely affected by changes in operating altitudes or barometric pressures.

Kickdown Valve

The valve body is also fitted with a **kickdown** circuit, which provides a downshift when the driver requires additional power. When the throttle pedal is quickly opened wide, throttle pressure rapidly increases and directs a large amount of pressure onto the kickdown valve. This moves the kickdown valve, which opens a port and allows mainline pressure to flow against the shift valve. The spring tension on the shift valve, the kickdown pressure, and throttle pressure will push on the end of the shift valve, causing it to move to the downshift position and forcing a quick downshift.

SHIFT QUALITY

All transmissions are designed to change gears at the correct time according to engine and vehicle speed, load, and driver intent. However, transmissions are also designed to provide for positive change of gear ratios without jarring the driver or passengers. If a band or clutch is applied too quickly, a harsh shift will occur. **Shift feel** is controlled by the pressure at which each hydraulic member is applied or released, the rate at which each is pressurized or exhausted, and the relative timing of the apply and release of the members.

To improve shift feel during gear changes, a band is often released while a multiple-disc pack is being applied. The timing of these two actions must be just right or both components will be released or applied at the same time, which would cause engine flare-up or clutch and band slippage. Several other methods are used to smooth gear changes and improve shift feel.

Multiple friction disc packs sometimes contain a wavy spring-steel separator plate that helps smooth the application of the clutch. Shift feel can also be smoothed out by using a restricting **orifice** or an accumulator piston **(Figure 40–55)** in the band or clutch apply circuit. A restricting orifice or check ball in the passage to the apply piston restricts fluid flow and slows the pressure increase at the piston by limiting the quantity of fluid that can pass in a given time. An accumulator piston slows pressure buildup at the apply piston by diverting a portion of the pressure to a second spring-loaded piston in the same hydraulic circuit. This delays and smooths the application of a clutch or band.

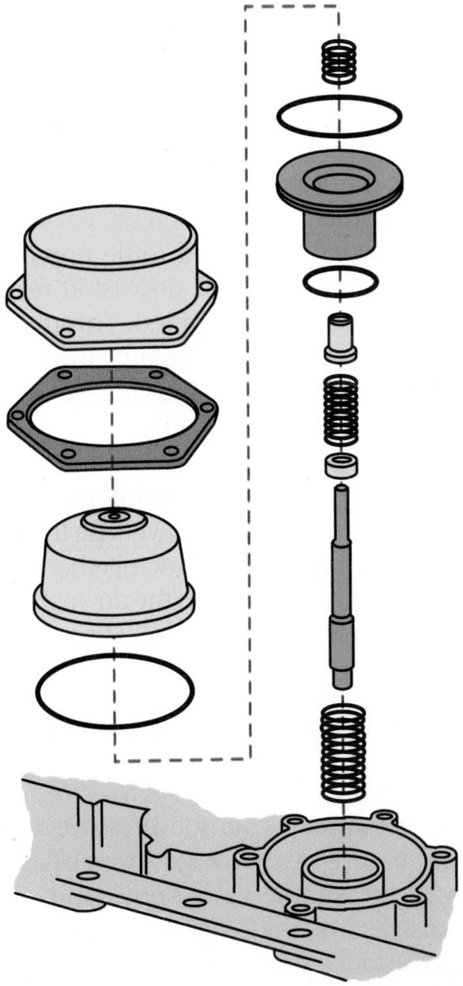

Figure 40-55 An accumulator assembly is used to control shift feel.

Manufacturers have also applied electronics to get the desired shift feel. One of the most common techniques is the pulsing (turning on and off) of the shift solenoids, which prevents the immediate engagement of a gear by allowing some slippage.

Shift Timing

Shift timing is determined by throttle pressure and governor pressure acting on opposite ends of the shift valve. When a vehicle is accelerating from a stop, throttle pressure is high and governor pressure is low. As vehicle speed increases, the throttle pressure decreases and the governor pressure increases. When governor pressure overcomes throttle pressure and the spring tension at the shift valve, the shift valve moves to direct pressure to the appropriate apply device and the transmission upshifts.

HYDRAULIC CIRCUITS

To provide a practical look at the operation of a valve body, a detailed look at a simple hydraulic circuit for a three-speed transaxle with computer-controlled lockup torque converter follows. This transaxle and valve body is not very complex, but it will serve as an example of the basic operation of a valve body and the control of various fluid pressures.

Flow Charts

The hydraulic circuits for each gear selector position is explained using flow charts, accompanied by the torque converter's lockup clutch controls and operation along with diagrams of the power flows through the transaxle.

While examining each flow chart, note the gear range, torque converter mode, engaged planetary controls, approximate vehicle speed, and throttle position. Check the direction of the fluid flow and the pressure chart to get a good understanding of the operation of the valve body and transaxle.

Gear Range: Neutral and park

Gear Selector Position: N and P

Throttle Position: 0 to 10 psi (0 to 70 kPa) (approximately closed)

Pump pressure leaves the transmission pump and is directed to the pressure regulator valve and manual valve (**Figure 40–56**). At the pressure regulator valve, pump pressure is regulated to become line pressure. Line pressure enters the pressure regulator valve and leaves as converter pressure, flowing to the switch valve. The switch valve allows line pressure to enter the torque converter. Converter pressure circulates from the switch valve to fill the torque converter and

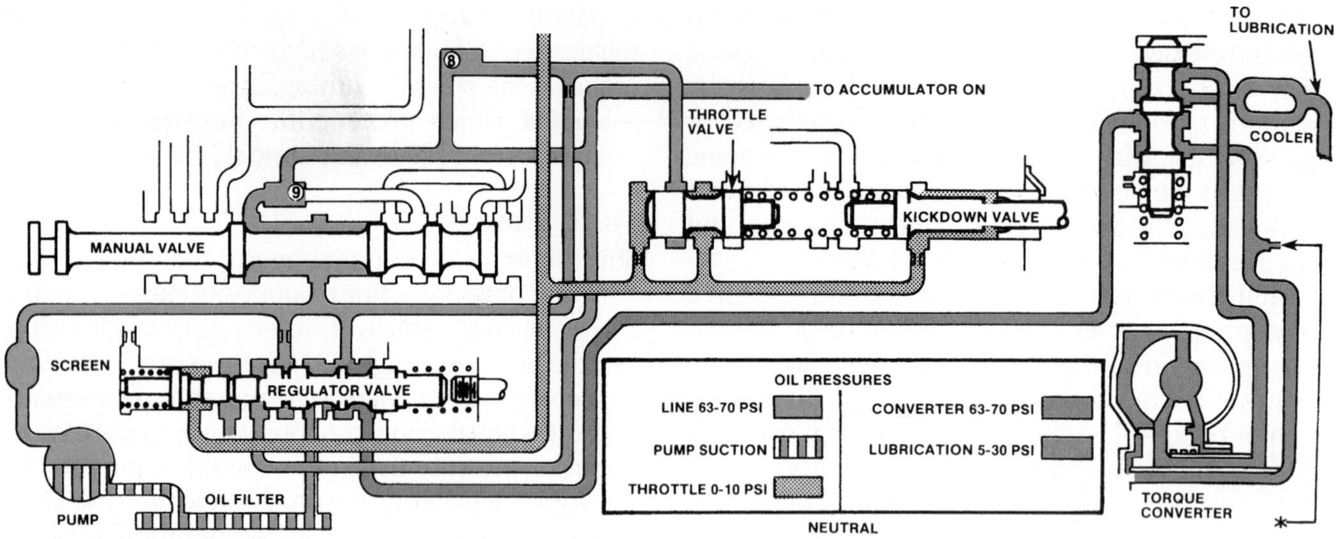

Figure 40-56 Fluid flows in the neutral gear range. *Courtesy of Chrysler LLC*

returns to the switch valve to become cooling and lubrication pressure.

From the pressure regulator valve, line pressure flows to the manual valve. From the manual valve, line pressure seats check ball 9 and flows around check ball 8 to stop at the land of the closed throttle valve. Throttle pressure is low because the throttle valve is not open. Line pressure flows to the accumulator to cushion the engagement of the planetary controls when the gear selector is moved to D or R ranges. The accumulator is basically a hydraulic shock absorber designed to absorb the shock of engaging planetary controls.

In neutral, line pressure is maintained by the pressure regulator valve and flows to the manual and throttle valves.

Gear Range: D first gear

Selector Position: D

Torque Converter Mode: Unlock

Planetary Controls Engaged: Rear clutch; overrunning clutch

Approximate Speed: 8 mph (13 km/h)

Throttle Position: Half throttle

In **Figure 40–57**, pressure between the manual valve and pressure regulator valve is considered to be line pressure. This line pressure circulates to the switch valve. Since the switch valve is held in the torque converter unlocked position, line pressure flows no further.

Line Pressure Beginning at the first manual valve outlet 1, line pressure seats #9 check ball and flows past #8 check ball to enter the throttle valve and establish throttle pressure. Line pressure also flows to the pressure regulator valve to regulate pressure. Line pressure moves the accumulator piston against coil spring tension, cushioning the engagement of the rear clutch. At outlet port 2 of the manual valve, line pressure fills the worm track, which engages the rear clutch and flows to the governor assembly. When the rear clutch is engaged and the governor assembly is filled with line pressure, the forward circuit is ready to drive the vehicle forward.

Throttle Pressure As line pressure passes through the valley of the throttle valve, it becomes throttle pressure. Throttle pressure circulates around the kickdown valve valley. With throttle pressure at the kickdown valve, a very quick downshift response to full-throttle operation is provided. Throttle pressure is directed to the pressure regulator throttle plug. It acts on the throttle plug, compressing the throttle plug spring. The result is that the pressure regulator valve closes the exhaust port, which results in a line pressure increase. Throttle pressure moves to act on the spring end of the 1-2 shift valve. The throttle pressure and coil spring tension work together to hold the shift valve and governor plug in the downshifted position against governor pressure. Throttle pressure passes check ball 5, which is acting on the spring end of the 2-3 shift valve. From the 2-3 shift valve, throttle pressure flows to hold the shuttle valve throttle plug against its stop in the valve body.

Governor Pressure Governor pressure developed from line pressure leaves the governor assembly terminating at the shuttle valve spool land. Governor pressure also acts on the 2-3 and 1-2 shift valve governor plugs. Because the vehicle is traveling at 8 mph (13 km/h), governor pressure is not strong enough to overcome throttle pressure and spring tension at the

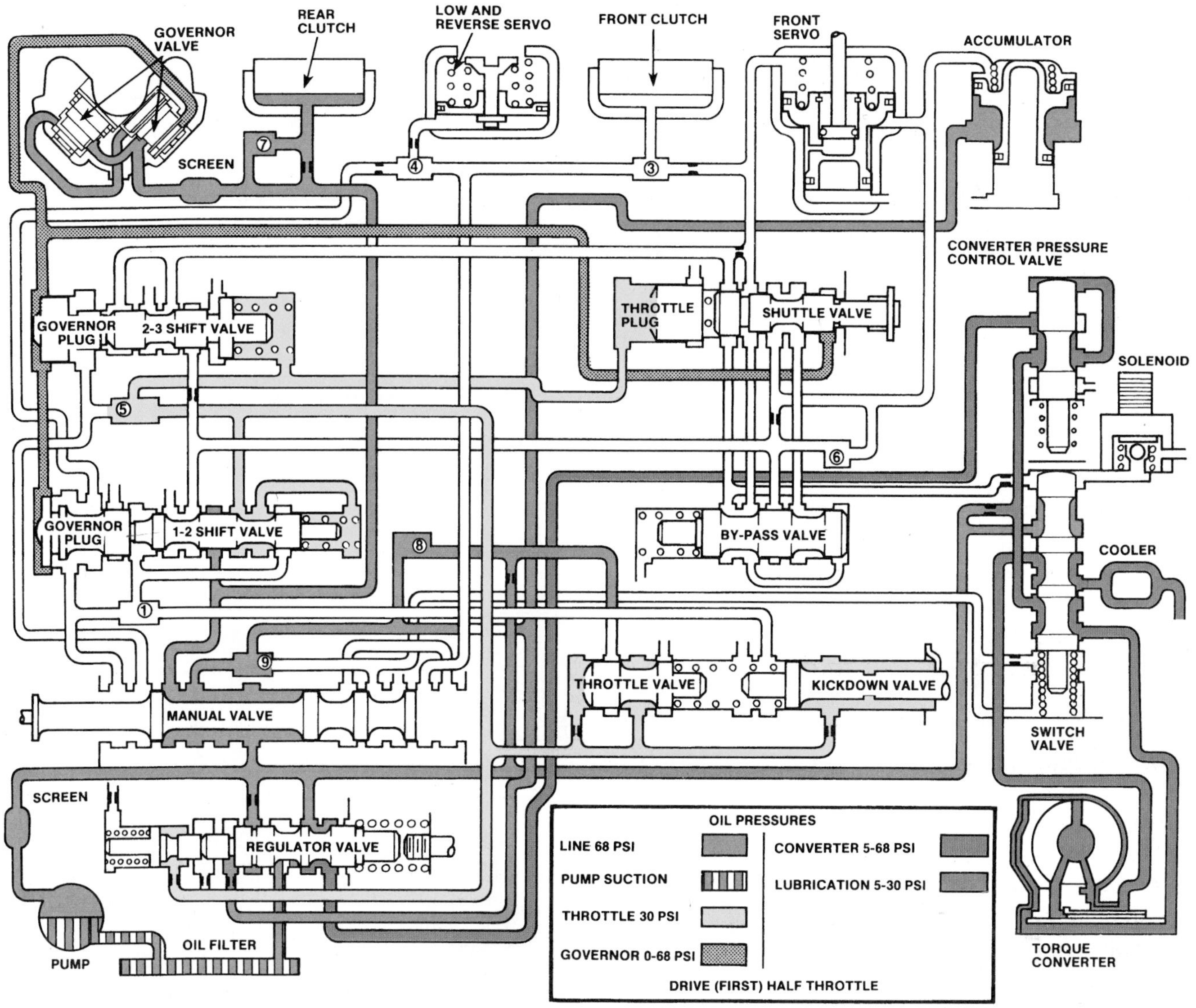

Figure 40-57 In drive range first gear, the rear clutch and overrunning clutch are engaged. *Courtesy of Chrysler LLC*

opposite end of the shift valves. Therefore, the transmission stays downshifted in drive range first gear.

Converter Pressure From the pressure regulator valve, converter pressure is directed to the converter pressure control valve. From the converter pressure control valve, converter pressure flows through the switch valve valley and enters the torque converter turbine shaft to keep the lockup piston disengaged. Converter pressure entering between the impeller and turbine fills the torque converter. Converter pressure flows back to the switch valve and enters the cooler to become cooler pressure. When cooler pressure returns to the transmission, it cools and lubricates transmission parts.

Drive Range: D second gear

Gear Selector Position: D

Torque Converter Mode: Unlock

Planetary Controls Engaged: Rear clutch; front kickdown band

Approximate Vehicle Speed: 15 mph (25 km/h)

Throttle Position: Half throttle

Line pressure leaving the area between the pressure regulator valve and manual valve develops throttle pressure and torque converter control pressure.

Referring to the manual valve (**Figure 40–58**), line pressure leaves outlet port 1, seating check ball 9. Line pressure passing check ball 8 moves to the throttle valve to become throttle valve pressure. From outlet 1, line pressure also flows to the pressure regulator valve, regulating line pressure. Line pressure from the same circuit flows to the accumulator, opposing line pressure and spring tension. The accumulator cushions the engagement of the intermediate band.

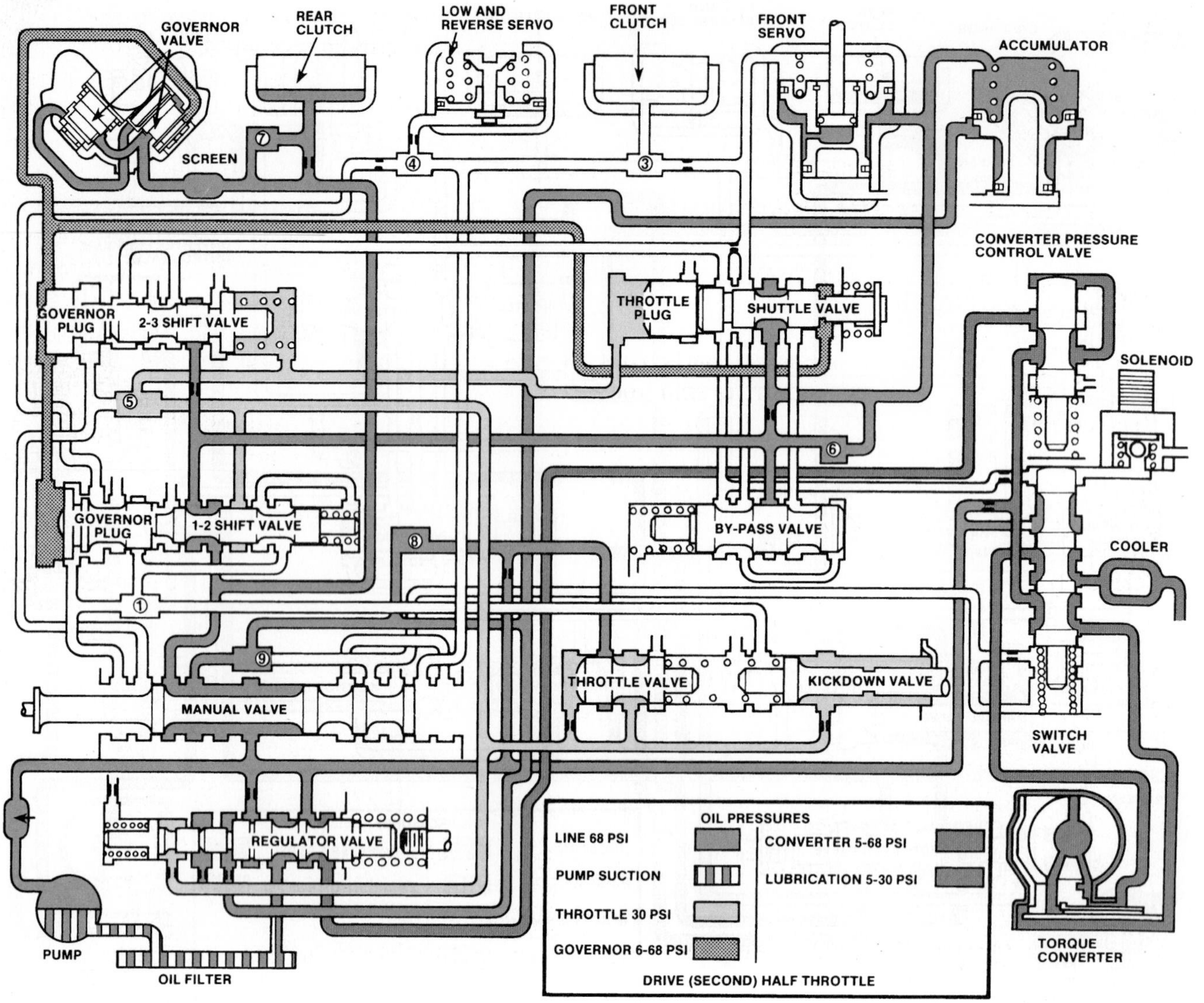

Figure 40-58 In drive range second gear, the rear clutch and front kickdown band are applied. *Courtesy of Chrysler LLC*

From manual valve outlet 2, fluid moves to the upshifted 1-2 shift valve. Line pressure flows to engage the rear clutch, then around check ball 7, through the governor screen to the governor assembly.

From the 1-2 shift valve, line pressure circulates to the shuttle valve and bypass valve through the restriction, or around check ball 6 to operate the front servo. When the front servo piston strokes in the cylinder, the front kickdown band engages around the front clutch drum.

Throttle Pressure The throttle pedal is in the half-open position, developing throttle pressure, which is directed to the kickdown valve. Throttle pressure is also directed to the throttle plug of the pressure regulator valve. Throttle pressure moving the throttle plug forces it against the throttle plug spring tension. The action of the throttle plug removes some of the opposition to the pressure regulator valve spring, resulting in a boost in line pressure. Throttle pressure moves to the 1-2 shift valve, where it is blocked by the upshifted valve's spool land. After seating check ball 5, throttle pressure and coil spring tension push on the 2-3 shift valve reaction area and move it to the downshifted position against governor pressure. Throttle pressure leaves the 2-3 shift valve to hold the shuttle valve throttle plug against its seat.

Governor Pressure Governor pressure leaves the rotating governor assembly and flows to the shuttle valve throttle plug. Governor pressure also acts on the governor plugs of the two shift valves. Vehicle speed and governor pressure are high enough to overcome throttle valve pressure and coil spring tension at the 1-2 shift valve. Governor pressure forces the 1-2 shift valve to move against throttle pressure. Therefore, throttle pressure is blocked from acting on the shift valve reaction area.

Converter Pressure Converter pressure flow is the same as in first gear.

Drive Range: D third gear

Gear Selector Position: D

Torque Converter Mode: Unlock

Planetary Controls Engaged: Rear clutch; front clutch

Approximate Vehicle Speed: 25 mph (40 km/h)

Throttle Position: Half throttle

Drive range third gear is covered in two parts. The first explains the hydraulic operation to shift the transmission into third gear. The second part introduces the sensors and controls affiliated with engine computer operation and torque converter lockup clutch control. (To this point, in drive range first and second gears, the torque converter clutch has been unlocked.)

Line Pressure Line pressure between the manual and pressure regulator valves is directed to the throttle valve and switch valve **(Figure 40–59)**. Start at manual valve outlet 1, where line pressure seats check ball 9 and circulates to the throttle valve. With the throttle valve open, throttle pressure is developed. In this circuit, pump pressure flows to the pressure regulator valve to develop line pressure, which, in this circuit, also operates the accumulator. From manual valve outlet 2, line pressure charges the forward circuit, engages the rear clutch, and enters the governor assembly to produce governor pressure.

Line pressure also flows from manual valve outlet 2 to circulate around the valley of the upshifted 1-2 shift valve and then around the valley of the 2-3 shift valve to the restriction above the shuttle valve. This restriction biases line pressure to the shuttle and by-pass valves. Line pressure leaves the shuttle

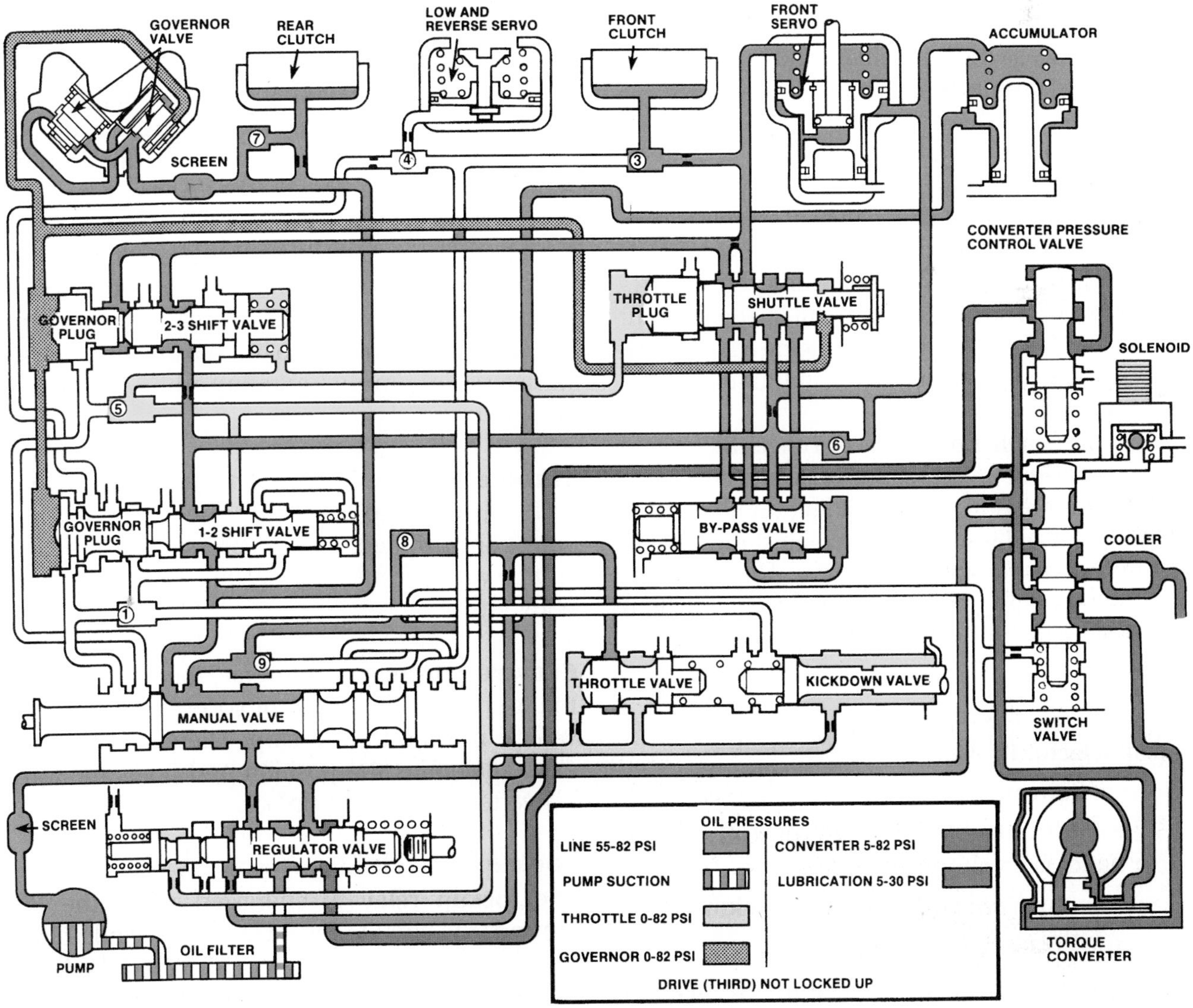

Figure 40–59 Drive range third gear with the lockup torque converter clutch disengaged and front and rear clutches engaged. *Courtesy of Chrysler LLC*

and by-pass valve area and moves to the release side of the front servo and the front clutch. The feed line to the front clutch has a restriction that causes pressure to build on the release side of the front servo, which disengages the front kickdown band. Line pressure flowing through the front clutch feed restriction engages the front clutch for third gear direct drive operation.

Throttle Pressure With the throttle valve open, throttle pressure flows to the kickdown valve, the throttle plug of the pressure regulator valve, 1-2 and 2-3 shift valves, and the shuttle valve throttle plug. At the 2-3 shift valve, throttle pressure and coil spring tension are opposed by increasing governor pressure.

Governor Pressure Governor pressure leaves the governor to move the shuttle valve, opening line pressure circuits. The shuttle valve's movement buries the coil spring in the hollow shuttle valve throttle plug. The governor pressure has moved the 1-2 and 2-3 shift valves to the upshifted position, directing line pressure to engage the front clutch and hold the front servo released during third gear operation.

Torque Converter Controls and Pressure

The torque converter lockup clutch assembly is controlled by the PCM. When the PCM receives signals from the different sensors confirming the requirements for lockup have been met, clutch engagement begins. These sensors include an engine coolant sensor, vehicle speed sensor, engine vacuum sensor, and throttle position sensor.

When the clutch is disengaged, a check ball is held off its seat by fluid pressure. The unseated check ball prevents line pressure from building until it is high enough to move the switch valve against the tension of the spring inside the switch valve. The converter clutch is never engaged in reverse gear and its action during reverse is similar to other times when it is not engaged.

Drive Range: Reverse gear

Gear Selector Position: R

Torque Converter Mode: Unlock

Planetary Controls Engaged: Low and reverse rear band; front clutch

Approximate Vehicle Speed: 5 mph (8 km/h)

Throttle Position: Part throttle

Line pressure from the outlet of the manual valve outlet that was not used before circulates through a bypass around the manual valve. Line pressure circulates to the low and reverse servo and front clutch (**Figure 40–60**). During the process of engaging the low and reverse servo and front clutch, the numbers 4 and 3 check balls are seated by line pressure.

Line Pressure Line pressure from between the pressure regulator and manual valve circulates around the pressure regulator valve. After flowing through a restriction to the seat number 8 check ball, line pressure enters the throttle valve, which produces line-to-throttle pressure. With the throttle valve open, throttle pressure charges the kickdown valve and strokes the throttle plug to its extreme left position at the pressure regulator valve. Line pressure from the manual valve does not flow to the pressure regulator valve to oppose spring tension. The pressure regulator valve coil spring pushes the pressure regulator valve over to close the exhaust port. Line pressure builds to approximately 200 to 300 psi (1,380 to 2,070 kPa).

You may wonder why line pressure must be increased so much in reverse gear. The planetary control units engaged in reverse are the front clutch and the low and reverse rear band. The front clutch, unlike the rear clutch, does not have a Belleville spring. The Belleville spring multiplies clutch piston apply force on the clutch pack, which reduces possible slippage. Therefore, to keep the front clutch from slipping when moving the vehicle from a stop, reverse line pressure is increased. The concept of increasing line pressure in reverse is common in many automatic transaxles and transmissions.

Throttle Pressure Throttle pressure circulates to the shift valve area to keep both the 1-2 and 2-3 shift valves downshifted. Throttle pressure also keeps the shuttle valve throttle plug against its seat in the valve body.

Converter Pressure In reverse, the transaxle needs the torque multiplication of vortex flow to start the vehicle moving from a stop. Therefore, the switch valve maintains the position and holds the torque converter piston in the unlocked position.

Lockup Clutch Engagement

The **lockup relay** is energized when the PCM grounds the circuit. The lockup relay sends 12 volts to energize the lockup solenoid. When the **lockup solenoid** is energized, it seats the check ball, which stops the exhaust of line pressure. Line pressure builds up on the reaction area of the switch valve

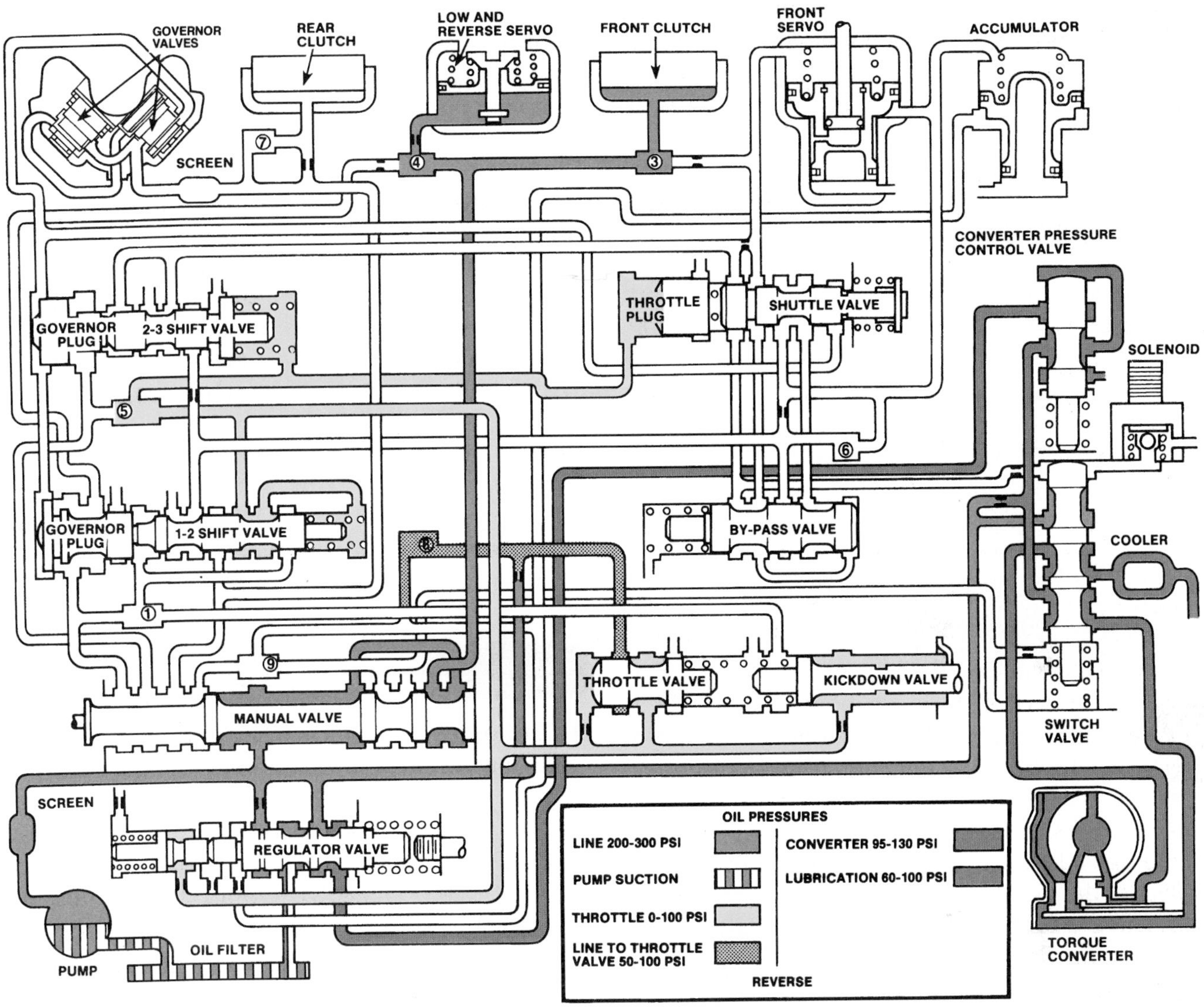

Figure 40-60 Reverse line pressure is increased to keep the front clutch, which has no Belleville spring, from slipping. The front clutch and the low reverse bands are engaged. *Courtesy of Chrysler LLC*

and moves the valve against spring tension to begin lockup engagement.

Drive Range: D third gear

Gear Selector Position: D

Torque Converter Mode: Locked

Planetary Controls Engaged: Rear clutch; front clutch

Approximate Vehicle Speed: 40 mph

Throttle Position: Half throttle

In drive range third gear lockup, the transaxle operates in the same manner as third gear unlock. The focus of attention is on the torque converter lockup clutch control system.

The coolant sensor reports to the computer that the engine has reached a temperature of at least 150°F (66°C). The vehicle's speed is monitored by a VSS. When the vehicle has a speed of more than 40 mph (65 km/h), it is at the desired engagement speed. Before engagement is initiated, the vacuum transducer must report to the PCM that engine vacuum is between 4 and 22 in. Hg (102 to 560 mm Hg). Based on these inputs, the PCM energizes the clutch relay and lockup solenoid to move the solenoid check ball onto its seat. The check ball stops the exhausting of line pressure. The increasing line pressure forces the switch valve to move against spring tension. Line pressure from the switch valve is directed to the pump drive hub and stator support to fill the torque converter with fluid. Fluid in the torque converter during lockup operation resides there to become the cooling and lubricating pressure. Line pressure flows from the impeller and turbine to fill the space behind the torque converter clutch piston and force engagement.

USING SERVICE INFORMATION

Before beginning to service a transaxle or transmission, have the service manual and latest service bulletins on the bench for ready reference. Many service manual publishers have complete volumes dedicated to automatic transmissions and transaxles. These volumes contain detailed information on domestic and imported transmissions and transaxles.

CASE STUDY

A fast-lube business has recently been faced with numerous complaints from customers who had their transmission fluid and filters changed. All of these customers owned late-model Chrysler products and have a common complaint of harsh shifting.

Concerned about these complaints, the business owner reviewed the files of each of these customers and their vehicles. He found something else the cars have in common. All of the vehicles were refilled with ATF-3 transmission fluid. The owner wondered if the fluid was faulty. After doing some research, he found that late-model Chrysler products call for ATF-4 fluid, which has special friction modifiers in it to provide smoother shifts.

This could be the cause of the complaints. The owner contacted one of the customers and requested that the vehicle be brought in. The fluid was drained and the correct fluid installed. The vehicle was then returned to the customer because the transmission now shifted normally. All of the other customers were contacted and the same service was performed to their vehicles. The shop owner quickly realized that the money saved by buying ATF-3 fluid from a surplus supplier was more than lost. Not only did he lose what he paid for the fluid, he also risked losing his customers. The correct type of fluid should always be used.

KEY TERMS

Annulus
Band
Belleville spring
Butt-end seals
Check ball valve
Continuously variable transmission (CVT)
Coupling point
Flow-directing valves
Governor
Hook-end seals
Impeller
Kickdown
Lip seals
Lockup relay
Lockup solenoid
Multiple-disc clutch
Orifice
Overrunning clutch
Planetary carrier
Planetary pinions
Pressure regulating valves
Ravigneaux geartrain
Rotary oil flow
Servo
Shift feel
Simpson geartrain
Spool valve
Square-cut seal
Stator
Sun gear
Throttle valve
Torque converter clutch (TCC)
Torrington bearing
Turbine
Two-mode hybrid system
Vacuum modulator
Valve body
Vortex oil flow

SUMMARY

- The torque converter is a fluid clutch used to transfer engine torque from the engine to the transmission. It automatically engages and disengages power transfer from the engine to the transmission in relation to engine rpm. It consists of three elements: the impeller (input), turbine (output), and stator (torque multiplier).
- Two types of oil flow take place inside the torque converter: rotary and vortex flow.
- An overrunning clutch keeps the stator assembly from rotating in one direction and permits overrunning when turned in the opposite direction.
- A lockup torque converter eliminates the 10% slip that takes place between the impeller and turbine at the coupling stage of operation.
- Planetary gear sets transfer power and alter the engine's torque. Compound gear sets combine two simple planetary gear sets so that load can be spread over a greater number of teeth for strength and also so the largest number of gear ratios possible can be obtained in a compact area. A simple planetary gear set consists of a sun gear, a carrier with planetary pinions mounted to it, and an internally toothed ring gear.
- Planetary gear controls include transmission bands, servos, and clutches. A band is a braking assembly positioned around a drum. There are two types: single wrap and double wrap. Simple and compound servos are used to engage bands. Transmission clutches, either overrunning or

multiple disc, are capable of both holding and driving members.

- There are two common designs of compound gear sets, the Simpson gear set, in which two planetary gear sets share a common sun gear, and the Ravigneaux gear set, which has two sun gears, two sets of planet gears, and a common ring gear.
- Most Honda transaxles do not use a planetary gear set; rather, constant-mesh helical and square-cut gears are used in a manner similar to that of a manual transmission.
- The operation of most CVTs is based on a steel belt linking two variable pulleys.
- Many hybrid vehicles have CVTs that rely on a planetary gear set and two electric motors rather than pulleys and belts.
- An automatic transmission uses ATF pressure to control the action of the planetary gear sets. This fluid pressure is regulated and directed to change gears automatically through the use of various pressure regulators and control valves. To form a complete working hydraulic system, the following elements are needed: fluid reservoir (transmission oil pan), pressure source (oil pump), control valving (valve control body), and output devices (servos and clutches).
- The transmission's pump is driven by the torque converter shell at engine speed. The purpose of the pump is to create fluid flow and pressure in the system. Excessive pump pressure is controlled by the pressure regulator valve. Three common types of oil pumps are installed in automatic transmissions: the gear, rotor, and vane.
- The valve body is the control center of the automatic transmission. It is made of two or three main parts. Internally, the valve body has many fluid passages called worm tracks.
- The purpose of a valve is to start, stop, or to direct and regulate fluid flow. Generally, in most valve bodies used in automatic transmissions, three types of valves are used: check ball, poppet, and, most common, the spool.
- To prevent stalling, the automatic transmission pump has a pressure regulator valve normally located in the valve body. It maintains a basic fluid pressure. The valve's movement to the exhaust position is controlled by calibrated coil spring tension. The three stages of pressure regulation are charging the torque converter, exhausting fluid pressure, and establishing a balanced position. There are times when fluid pressure must be increased above its baseline pressure to accomplish these stages.
- The purpose of the governor assembly is to sense vehicle road speed and send a fluid pressure signal to the transmission valve body to either upshift or permit the transmission to downshift. Throttle pressure delays the transmission upshift and forces the downshift.
- Bearings that take radial loads are called bushings and those that take axial loads are called thrust washers.
- The gaskets and seals of an automatic transmission help to contain the fluid within the transmission and prevent the fluid from leaking out of the various hydraulic circuits. Different types of seals are used in automatic transmissions; they can be made of rubber, metal, or Teflon.
- Three major types of rubber seals are used in automatic transmissions: the O-ring, the lip seal, and the square-cut seal.
- Three types of sealing rings are used in automatic transmissions: butt-end seals, open-end seals, and hook-end seals.
- There are four common configurations used as the final drives on FWD vehicles: helical gear, planetary gear, hypoid gear, and chain drive.

REVIEW QUESTIONS

1. Explain the difference between rotary and vortex fluid flow in a torque converter.
2. What component keeps the stator assembly from rotating when driven in one direction and permits rotation when turned in the opposite direction?
3. When a transmission is described as having two planetary gear sets in tandem, what does this mean?
4. The four common configurations used as the final drives on FWD vehicles are the ______ gear, ______ gear, ______ gear, and ______ ______.
5. Three major types of rubber seals are used in automatic transmissions: the ______, the ______, and the ______ seal.
6. *True or False?* The vent in a transmission housing is designed to allow fluid to escape when there is excessive pressure in the system.
7. To achieve a slow overdrive in a simple planetary gear set, the ______.
   a. sun gear must be the input member
   b. ring gear must be the input member

c. planetary carrier must be the input member
d. ring gear must be held

8. In a simple planetary gear set, when the planetary carrier is held, the gear set produces a ________.
   a. reverse
   b. direct drive
   c. fast overdrive
   d. forward reduction
9. Overrunning clutches are capable of ________.
   a. holding a planetary gear member stationary
   b. driving a planetary gear member
   c. both a and b
   d. neither a nor b
10. In the coupling stage of conventional torque converter operation, ________.
   a. both the impeller and turbine are turning at essentially the same speed
   b. both the stator and turbine are stationary
   c. both the stator and impeller are stationary
   d. the stator turns with the turbine and impeller
11. How can shift feel be controlled?
12. What determines the timing of the shifts in an automatic transmission?
13. Why must hydraulic line pressures increase when there is an increase of load on the engine?
14. What is necessary for torque converter clutch lockup to take place?
   a. The lockup solenoid check ball must be off its seat and the switch valve held up by spring tension.
   b. The lockup solenoid must be energized with the check ball on its seat and the switch valve held down by line pressure.
   c. The lockup check ball must be off its seat and the switch valve held down by line pressure.
   d. The solenoid must have the check ball seated and the switch valve held up by spring tension.
15. The three types of metal seals used in automatic transmissions are the ________, ________, and ________ seals.

ASE-STYLE REVIEW QUESTIONS

1. Technician A says that bearings that take radial loads are called Torrington washers. Technician B says that bearings that take axial loads are called thrust washers. Who is correct?
   a. Technician A
   b. Technician B
   c. Both A and B
   d. Neither A nor B
2. Technician A says that rotary oil flow is the oil flow around the circumference of the torque converter caused by the rotation of the torque converter on its axis. Technician B says that the centrifugal lockup clutch is the type installed in most automatic transmissions. Who is correct?
   a. Technician A
   b. Technician B
   c. Both A and B
   d. Neither A nor B
3. Technician A says that a stator aids in directing the flow of fluid from the impeller to the turbine. Technician B says that a stator is equipped with a one way clutch that allows it to remain stationary when necessary. Who is correct?
   a. Technician A
   b. Technician B
   c. Both A and B
   d. Neither A nor B
4. Technician A says that a Simpson gear set is two planetary gear sets that share a common sun gear. Technician B says that a Ravigneaux gear set has two sun gears, two sets of planet gears, and a common ring gear. Who is correct?
   a. Technician A
   b. Technician B
   c. Both A and B
   d. Neither A nor B
5. Technician A says that that one of the primary purposes of the pressure regulator valve is to fill the torque converter with fluid. Technician B says that the pressure regulator valve directly controls throttle pressure. Who is correct?
   a. Technician A
   b. Technician B
   c. Both A and B
   d. Neither A nor B
6. Technician A says that the purpose of a valve body is to increase oil pressures in response to increases in engine speed. Technician B says that valve bodies are normally fitted with three different types of valves, which start, stop, or use movable parts to regulate and direct the flow of fluid throughout the transmission. Who is correct?
   a. Technician A
   b. Technician B
   c. Both A and B
   d. Neither A nor B
7. Technician A says that changes in engine load cause changes in hydraulic pressure inside the transmission. Technician B says that engine load is monitored by throttle pedal movement or by engine vacuum. Who is correct?
   a. Technician A
   b. Technician B
   c. Both A and B
   d. Neither A nor B

8. Technician A says that the rear hub of the torque converter is bolted to the flexplate. Technician B says that the flexplate is designed to be flexible enough to allow the front of the converter to move forward or backward if it expands or contracts because of heat or pressure. Who is correct?
   a. Technician A
   b. Technician B
   c. Both A and B
   d. Neither A nor B
9. While discussing a Lepelletier system: Technician A says that two simple planetary gear sets are connected in tandem. Technician B says that these systems rely on various combinations of the planetary gears to obtain many forward gears. Who is correct?
   a. Technician A
   b. Technician B
   c. Both A and B
   d. Neither A nor B
10. While discussing the CVT transaxle used in Toyota hybrid vehicles: Technician A says that the transaxle contains two electric motor/generators, a differential, and a simple planetary gear set. Technician B says that the planetary gear set adjusts the diameter of the pulleys to provide the various speed ratios. Who is correct?
    a. Technician A
    b. Technician B
    c. Both A and B
    d. Neither A nor B

CHAPTER 41

ELECTRONIC AUTOMATIC TRANSMISSIONS

OBJECTIVES

■ Explain the advantages of using electronic controls for transmission shifting. ■ Briefly describe what determines the shift characteristics of each selector lever position. ■ Identify the input and output devices in a typical electronic control system and briefly describe the function of each. ■ Diagnose electronic control systems and determine needed repairs. ■ Conduct preliminary checks on the EAT systems and determine needed repairs or service. ■ Perform converter clutch system tests and determine needed repairs or service. ■ Inspect, test, and replace electrical/electronic sensors. ■ Inspect, test, bypass, and replace actuators.

Electronic transmission control provides automatic gear changes when certain operating conditions are met. Through the use of electronics, transmissions have better shift timing and quality. As a result, the transmissions contribute to improved fuel economy, lower exhaust emission levels, and improved driver comfort. Although these transmissions function in the same way as earlier hydraulically based transmissions, a computer determines their shift points. The computer uses inputs from different sensors and matches the information to a predetermined schedule.

Hydraulically controlled transmissions relied on signals from a governor and throttle pressure device to force a shift in gears. Electronically controlled transmissions typically do not have governors nor throttle pressure devices. Hydraulically controlled transmissions relied on pressure differentials at the sides of a shift valve to hold or change a gear. Electronic transmissions do too. However, the pressure differential is caused by the action of shift solenoids that allow for changes in pressure on the side of a shift valve **(Figure 41–1)**. The computer controls

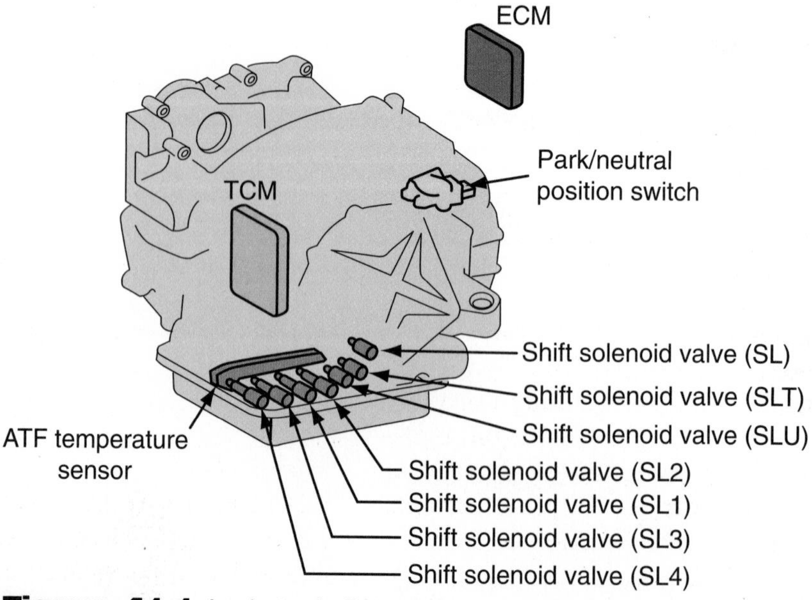

Figure 41-1 An electronically controlled transaxle.

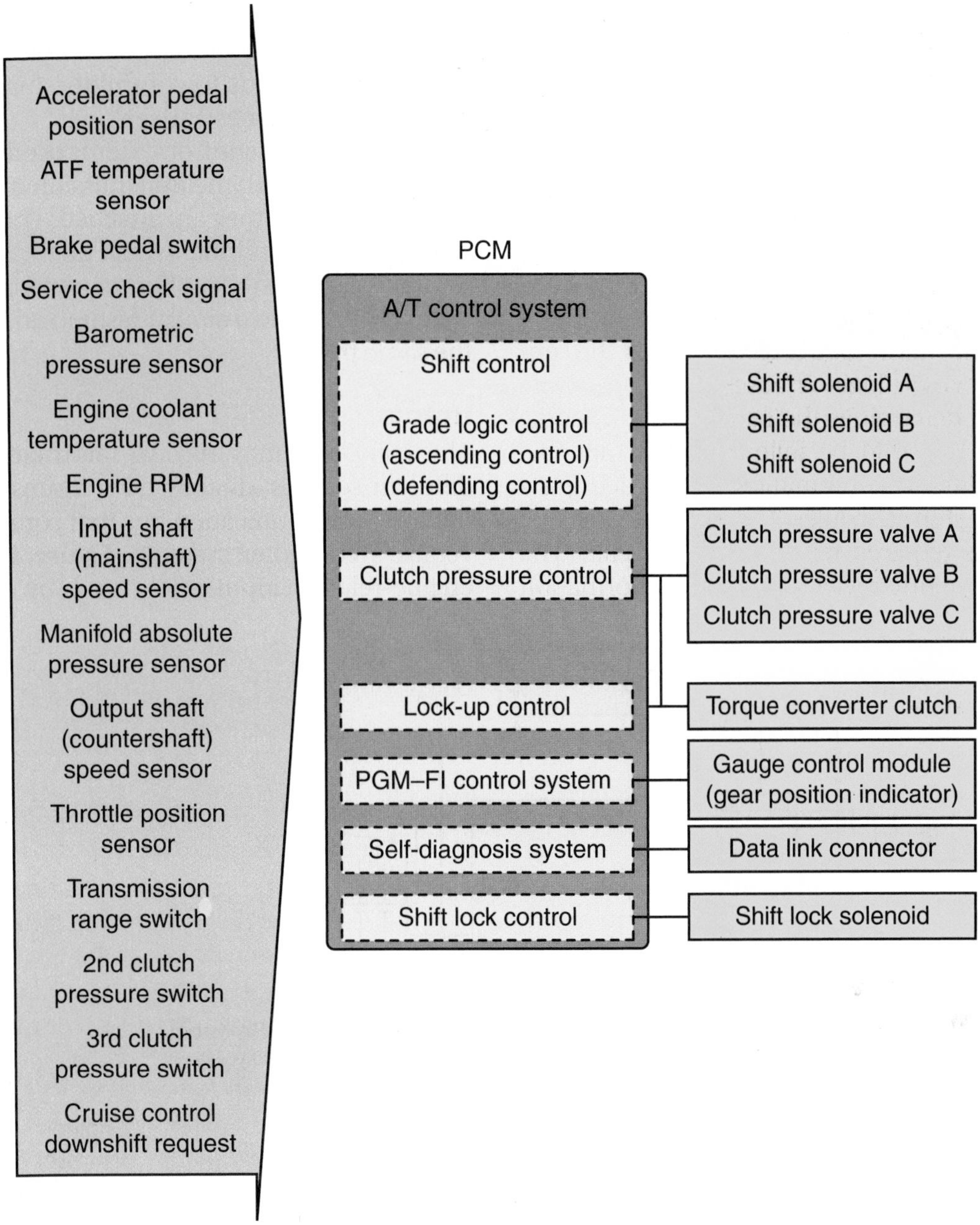

Figure 41–2 The different inputs, modules, and outputs in an EAT circuit.

these solenoids. The solenoids do not directly control the transmission's clutches and bands. These are engaged or disengaged in the same way as hydraulically controlled units. The solenoids simply control the fluid pressures and do not perform a mechanical function.

In an electronic automatic transmission (EAT) system, there is a central processing unit, inputs, and outputs (**Figure 41–2**). Often the central processing unit is a separate computer designated for transmission control. The computer receives information from various inputs and controls two to five (perhaps even more!) solenoids that control hydraulic pressure and fluid flow to the apply devices and to the clutch of the torque converter. This computer may be the **transmission control module (TCM)** or the powertrain control module (PCM). When the transmission is not controlled by the PCM, a TCM is used. The TCM is tied into the serial (CAN) bus and shares information with the PCM. The TCM may be a separate unit or may be part of another control module, such as the body control module (BCM).

TRANSMISSION CONTROL MODULE

A TCM is programmed to provide gear changes at the optimum time. The TCM controls shift timing, shift feel, and the operation of the torque converter clutch. To determine the best operating strategy for the transmission, the TCM uses inputs from some engine-related and driver-controlled sensors. The TCM also receives input signals from specific sensors connected to the transmission. By monitoring all of these inputs, the TCM can determine when to shift gears or when to apply or release the torque converter clutch.

The decision to shift or not to shift is based on the shift schedules and logic. A **shift schedule** contains the actual shift points to be used by the computer according to the input data it receives from the sensors. Shift schedule logic chooses the appropriate gear and then determines the correct shift schedule or pattern that should be followed. Each possible engine/transmission combination for a vehicle has a different set of shift schedules. The shift schedules set the conditions required for a change in gears. The computer frequently reviews the input information and can make quick adjustments to the schedule, if needed and *as* needed. The TCM controls the action of various solenoids to control the torque converter clutch and shift times and feel.

The electronic control systems used by the manufacturers differ with the transmission models and the engines to which they are attached. The components in each system and the overall operation of the system also vary with the different transmissions. However, all operate in a similar fashion and use basically the same parts.

Inputs

The computer may receive information from two different sources: directly from a sensor or through the CAN communication bus that connects all of the vehicle's computer systems (**Figure 41–3**). Normal engine-related inputs are used by the TCM to

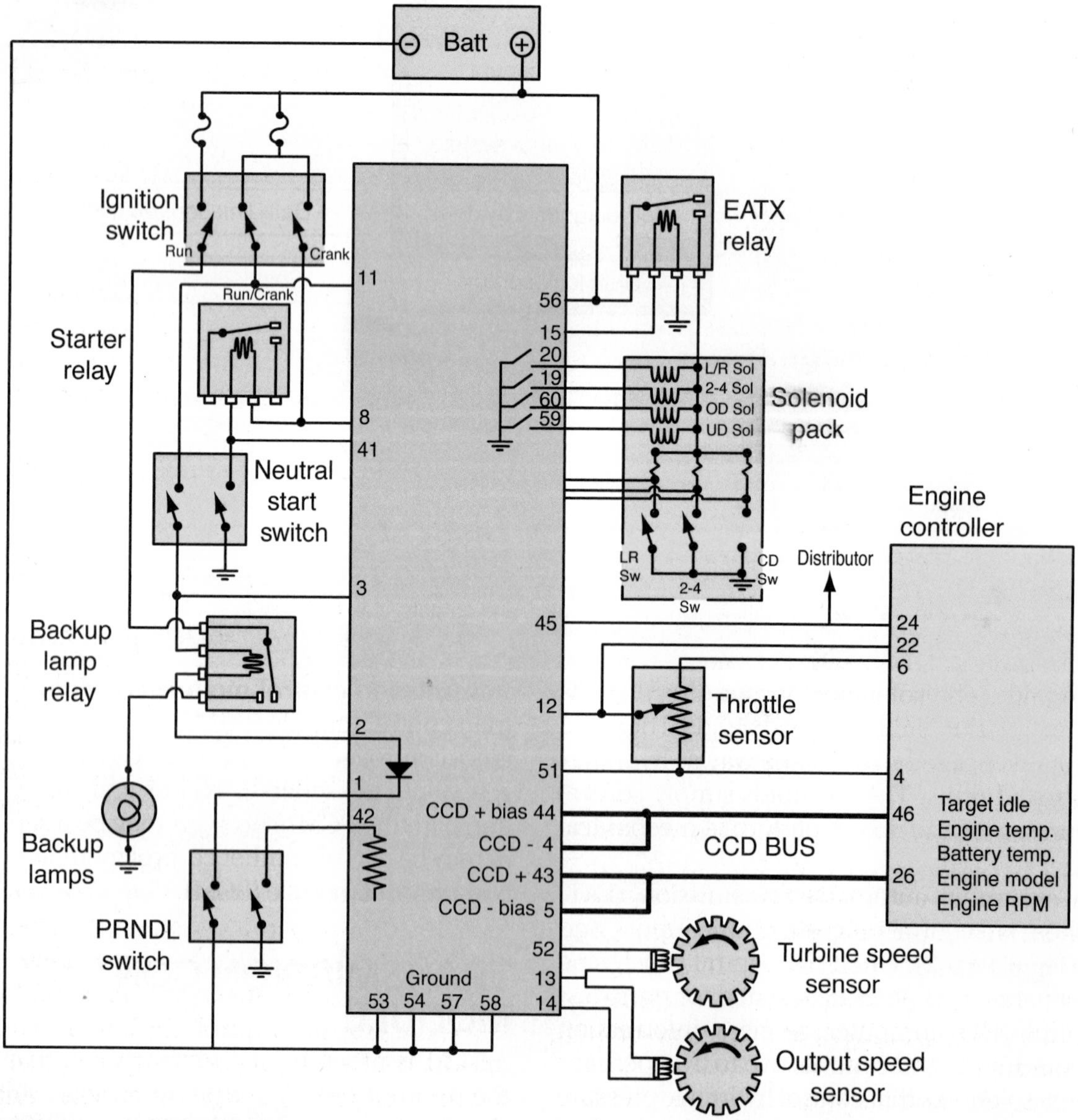

Figure 41–3 The electric circuit for a typical electronically controlled transmission. Note: The CCD data BUS is the data source for other inputs in this multiplexed circuit.

determine the best shift points. Many of these inputs are available at the common data bus. Other information, such as engine and body identification, the TCM's target idle speed, and speed control operation are not the result of monitoring by sensors; rather, these have been calculated or determined by the TCM and made available on the bus.

Typical bus inputs used by the TCM are from the throttle position (TP), manifold absolute pressure (MAP), mass airflow (MAF), intake air temperature (IAT), barometric pressure (BARO), engine coolant temperature (ECT), and crankshaft position (CKP) sensors. These provide the TCM with information about the operating condition of the engine. Through these, the TCM is able to control shifting and torque converter clutch (TCC) operation according to the temperature, speed, and load of the engine.

Inputs that are tied directly to the TCM are typically not available on the bus circuit. Many of these sensors produce an analog signal that must be changed to a digital signal before the TCM can respond. This conversion is handled by an analog-to-digital (A/D) converter, the PCM, or a digital radio adapter controller (DRAC). These typically convert an analog AC signal to a digital 5-volt square wave.

The TCM constantly monitors what is happening inside the transmission with various speed and gear range sensors that tell it if the gears are shifting correctly and at what speeds. Once the TCM commands a gear change, solenoids are energized to hydraulically control the engagement of the proper clutch or band.

On-Off Switches Several simple switches are used in the control of an EAT. The number and purpose of each depends on the system. A brake pedal position (BPP) switch tells the TCM when the brakes are applied. At that time, the torque converter clutch disengages. The BPP switch closes when the brakes are applied and opens when they are released. Its input has little to do with the up-and-down shifting of gears, except that in some systems it signals a need for engine braking.

The transmission control (TC) switch is a momentary contact switch located on the selector lever that allows the driver to cancel operation of overdrive. When the TC switch is depressed, a signal is sent to the PCM to disengage overdrive operation. At the same time, the PCM illuminates the transmission control indicator lamp (TCIL) to notify the driver that overdrive is canceled.

An A/C request switch tells the TCM that the A/C has been turned on. The TCM then changes the line pressure and shift timing to accommodate the extra engine load created by the A/C system. When the A/C clutch is engaged, the electronic pressure control (EPC) is adjusted by the TCM to compensate for the additional load on the engine.

EATs also have a cruise control switch that informs the TCM when cruise control is active. In this mode, shift patterns are altered to reduce excessive and harsh shifting. There may also be a four-wheel-drive low (4WDL) range switch that lets the TCM know when the four-wheel-drive transfer case gear is in its low range.

Digital Transmission Range (TR) Sensor The first input that the TCM looks at is the position of the gear shift lever. All shift schedules are based on the gear selected by the driver. These schedules are coded by the position of the gear selector and the current gear range and use throttle angle and vehicle speed as primary determining factors. A digital transmission range (TR) sensor informs the TCM of the gear selected by the driver. This sensor normally also contains the neutral safety switch and the reverse light switch. A TR sensor is typically a multiple-pole-type on-off switch **(Figure 41–4)**. The digital TR sensor may be located on the outside of the transmission at the manual lever or it may be part of the TCM.

Throttle Position (TP) Sensor The TP sensor sends a voltage signal to the TCM in response to TP. Not only is this signal used to inform the TCM of the driver's intent, it is also used in place of the hydraulic throttle pressure linkage. The TP sensor is very important to the operation of an EAT and is used for shift scheduling, EPC, and TCC control.

If the TP signal is wrong, it can affect transmission kickdown shifts during acceleration as well as upshifts and downshifts. When the TCM is unable to get a good TP signal, it may substitute a "calculated" TP based on other inputs from the data bus. Regardless, the TCM always calculates a TP based on its programming and inputs. This calculation is one of the most important inputs for shift pattern logic. A very low TP angle causes upshifts to occur early, and very high TP angles cause early downshifts.

Mass Airflow (MAF) Sensor The MAF sensor measures the mass of air flowing into the engine and is primarily used to calculate injector pulse width. It also is used to calculate engine load and to regulate electronic pressure control (EPC) and shift and TCC scheduling. Depending on the type of fuel injection system used (speed density or airflow), engine load may be determined by the TP sensor, MAP sensor, and/or a vane airflow VAF sensor or MAF sensor.

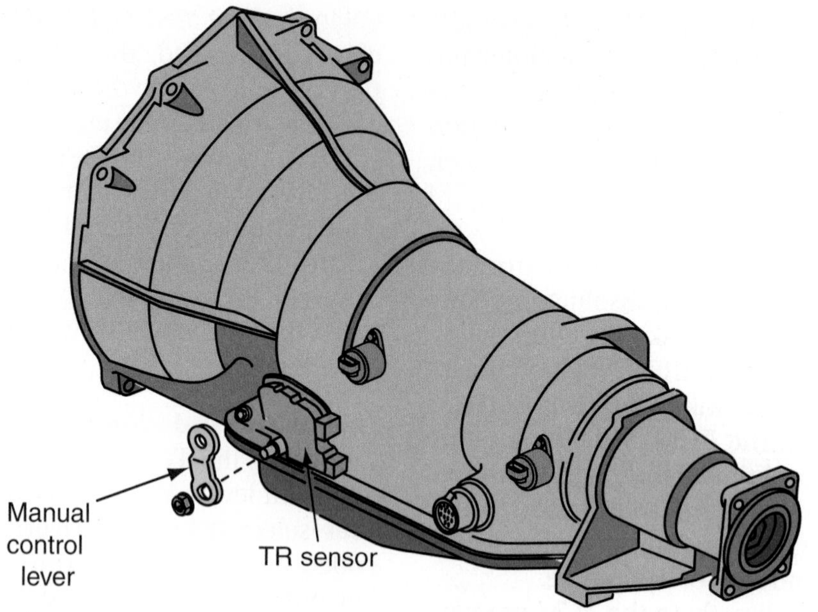

Figure 41–4 A TR sensor.

Signals from a BARO sensor are used to adjust line pressures according to changes in altitude. This sensor input may not be used; its use depends on the type of intake air monitoring system the vehicle is equipped with. On those vehicles using the BARO sensor as an input, the sensor may be integrated in the PCM or it may be mounted externally.

Temperature Sensors Shift schedules are also influenced by the engine's temperature. Sometimes the engine's temperature is raised by delaying gear shifts. When the engine is overheating, shifts will occur sooner. The computer may also engage the converter clutch in second or third gear if the coolant temperature rises. Engine temperatures are often tied to ATF temperatures in the computer and are critical to the operation of a transmission.

The IAT sensor provides the fuel injection system with mixture temperature information. The IAT sensor is used both as a density corrector for airflow calculation and to proportion cold enrichment fuel flow. It is installed in the air cleaner outlet tube. The IAT sensor also is used in determining line pressures.

The TCM may also use the signal from the IAT to calculate the temperature of the battery. The TCM then uses this temperature to estimate transmission fluid temperature.

Engine Speed To ensure that the transmission shifts at the correct time, the TCM must receive an engine speed input. This can be done through the ignition module and serial data bus or through a direct connection from a dedicated circuit between the CKP sensor and TCM. With the direct connection, all time delays at the bus circuit are eliminated and the TCM always knows the current engine speed. This input is used to determine shift timing, wide-open throttle (WOT) shift control, TCC control, and EPC pressure. Also, to prevent the engine from running at too high of a speed, the TCM will order an upshift.

Transmission Fluid Temperature (TFT) Sensor The transmission fluid temperature (TFT) sensor is normally located on the valve body **(Figure 41–5)**. Signals from the sensor tell the TCM what the temperature of the ATF is. The sensor is a thermistor whose output signal varies with ATF temperature. The TCM uses this signal to control shift timing, shift feel, and TCC engagement. When the signal indicates normal operating temperature, the transmission is shifted normally; however, when the fluid is too cold or too hot, shifting is altered.

When the fluid temperature is too high, the TCM will operate the transmission in a way to allow it to cool. Shifts will occur sooner than normal. When the fluid is too cold, shifting is delayed to help warm the fluid. Fluid temperature is used, along with other

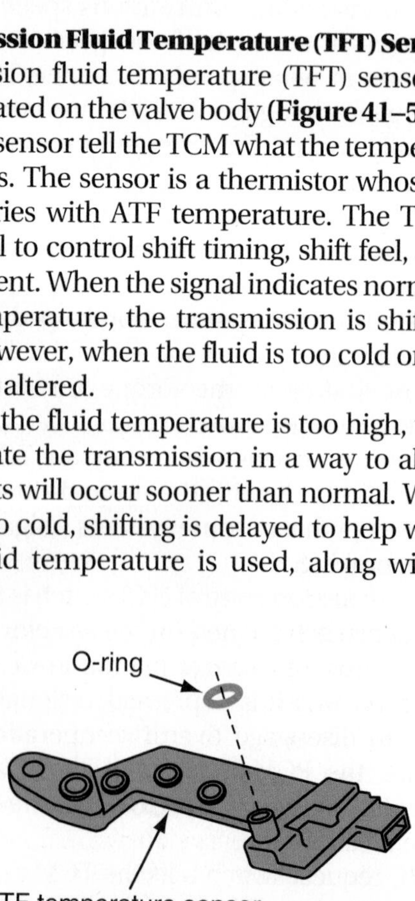

Figure 41–5 An example of an ATF temperature sensor.

Temperature Range	Gear Operation	TCC Operation
Below −16°F (−27°C)	Operates only in Park, Neutral, Reverse, 1st, and 3rd gears	NO
Between −12°F (−24°C) and 10°F (12°C)	Upshifts from 2nd to 3rd and 3rd to 4th are delayed and downshifts from 4th to 3rd and 3rd to 2nd are early	NO
Between 10°F (12°C) and 36°F (2°C)	All shifts are delayed	NO
Between 40°F (4°C) and 80°F (27°C)	Normal shifting	NO
Between 80°F (27°C) and 240°F (116°C)	Normal shifting	YES, normal
More than 240°F (116°C), or if the engine is overheating	2nd to 3rd and 3rd to 4th shifts are delayed	YES, but longer
Above 260°F (127°C)	2nd to 3rd and 3rd to 4th shifts are greatly delayed	YES, nearly full time

Figure 41–6 The operation of a Chrysler 45RFE transmission with the ATF at different temperatures.

inputs, to control TCC clutch engagement. When the fluid is cold, the TCM prevents TCC engagement until the fluid reaches a specific temperature.

Some transmissions operate with distinct shift schedules based on fluid temperature. **Figure 41–6** shows the operating characteristics of a Chrysler 45RFE transmission when the ATF is at different temperatures.

If the TFT sensor fails or the TCM determines temperature signals are incorrect, the TCM will look at engine temperature to estimate the temperature of the ATF.

Transmission Pressure Switches Various transmission pressure switches can be used to keep the TCM informed as to which hydraulic circuits are pressurized and which clutches and brakes are applied. These input signals can serve as verification to other inputs and as self-monitoring or feedback signals. The most commonly used pressure switch is the transmission fluid pressure (TFP) switch, which monitors fluid pressure to determine when a clutch or band is applied or released.

Voltage-Generating Sensors A variety of speed sensors, most commonly Hall-effect or PM generator sensors, are used in today's EATs. These sensors serve as the governor signal and monitor the operation of the transmission. Most TCMs receive a signal from a vehicle speed sensor (VSS) and use this to determine the correct time to shift. Some EATs have an output shaft speed (OSS) sensor. This sensor may be used with the VSS or provide a vehicle speed reference for the TCM. When a vehicle has both sensors, the OSS signal is used as a verification signal for the VSS, or vice versa. The OSS is typically used for control of

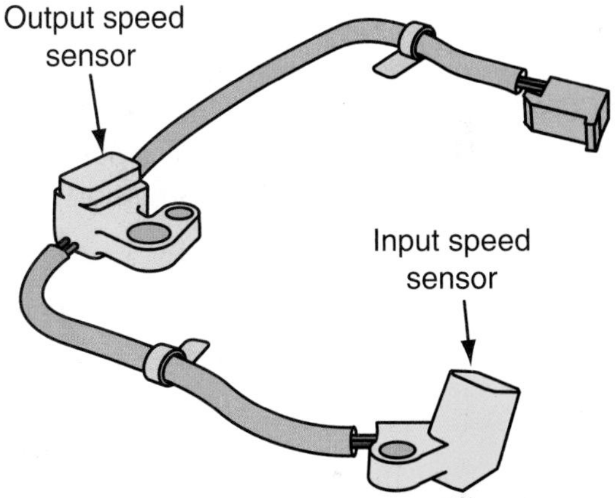

Figure 41–7 This wiring harness contains the input speed sensor (ISS) and the output speed sensor (OSS).

torque converter clutch operation, shift timing, and fluid pressure control.

Some transmissions have an input speed sensor (ISS) or a turbine speed (TSS) sensor. This sensor is identical to the OSS and its signal is used to calculate converter turbine speed (**Figure 41–7**). It is also used, along with an engine speed input, to determine the amount of T/C slip by providing the TCM with the difference between engine speed and transmission input shaft speed. This is used to determine the amount of pressure applied to the TCC. Four-wheel-drive vehicles may have an additional speed sensor in the transfer case.

Outputs

Common outputs used with EATs are indicator lamps and solenoids. The indicator lamps can show the selected gear range in the instrument panel. They may also be a MIL designated for the transmission, or this warning light may be incorporated into the

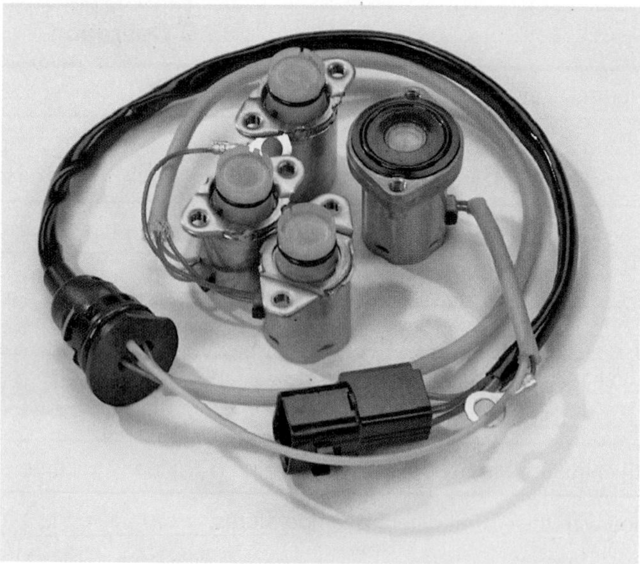

Figure 41-8 The solenoids for a transmission and their wiring harness. *Courtesy of Transtar Industries*

circuit for the engine's MIL. Shift, pressure control, and TCC solenoids are used in all modern EATs. All of these are controlled by the TCM. The solenoids are typically located inside the transmission and are mounted to the valve body.

Shift Solenoids Shift solenoids **(Figure 41–8)** are used to regulate shift timing and feel by controlling the delivery of fluid to the manual shift valve. These solenoids are on-off solenoids that are normally off and in the open position. When open, line pressure is present at the manual or selected shift valve. When the shift solenoids are energized, they block line pressure and allow pressure to exhaust from around the shift valve. This allows the valve to move.

The number and purpose of each depends on the model of transmission. A typical four-speed unit has two shift solenoids. The two solenoids offer four possible on-off combinations to control fluid to the various shift valves. This provides the engagement of the four forward gears **(Figure 41–9)**. Transmissions with additional forward gears rely on additional shift solenoids. The on-off combinations of these provide the additional gears. It is important to note that the TCM will stop current flow to a shift solenoid if it detects a fault in the solenoid or its circuit.

Pressure Control Solenoids The pressure control solenoid replaces the conventional TV cable setup to provide changes in line pressure in response to engine running conditions and engine load. The action of the solenoid controls the operating hydraulic pressures to the clutches and brakes to provide smooth and precise shifting. The pressure control solenoid is installed on the valve body.

In most cases, the pressure control solenoid is called an **electronic pressure control (EPC) solenoid** and is duty cycle controlled by the TCM. Most of these solenoids are **variable force solenoids (VFS)** or pulse width modulated (PWM) solenoids and contain a spool valve or plunger and a spring. To control fluid pressure, the PCM sends a varying signal to the solenoid. This changes the amount the solenoid will cause

Lever Position	Commanded Gear	Shift Solenoid A	Shift Solenoid B	TCC Solenoid
P/R/N	1	ON	OFF	Disabled hydraulically
D	1	ON	OFF	Disabled hydraulically
D	2	OFF	OFF	Controlled electronically
D	3	OFF	ON	Controlled electronically
D	4	ON	ON	Controlled electronically
O/D switched off				
1	1	ON	OFF	Disabled hydraulically
2	2	OFF	OFF	Controlled electronically
3	3	OFF	ON	Controlled electronically
Manual 2	2	OFF	OFF	Controlled electronically
Manual 1	1	ON	OFF	Disabled hydraulically
Manual 1	2	OFF	OFF	Controlled electronically

Figure 41-9 The action of the solenoids in a typical four-speed transmission.

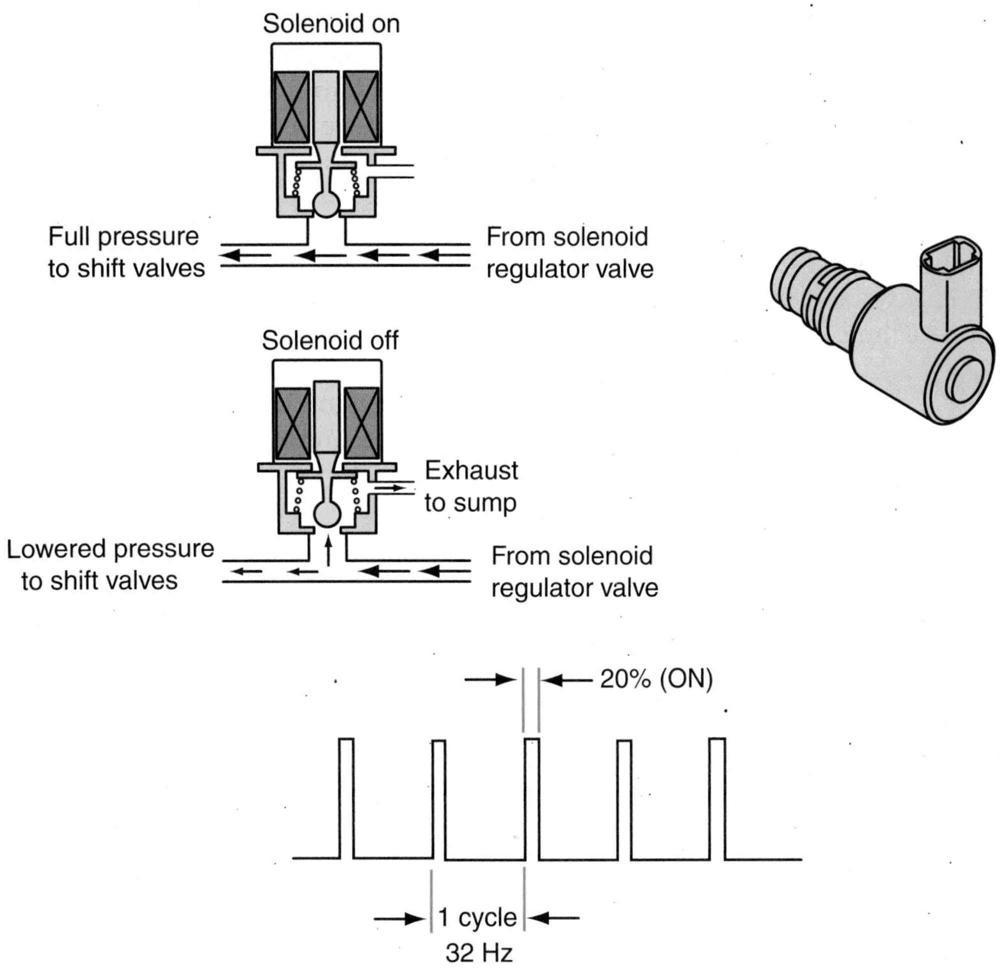

Figure 41–10 (Top) A typical PWM solenoid. (Bottom) The signal representing the control or ordered duty cycle from the computer.

the spool valve to move. When the solenoid is off, spring tension keeps the valve in place to maintain maximum pressure. When the solenoid is energized it moves the spool valve. The movement of the valve uncovers the exhaust port around the valve, thereby causing a decrease in pressure **(Figure 41–10)**.

TCC Solenoid The TCC solenoid is used to control the application, modulation, and release of the TCC. The operation of the converter clutch is also totally controlled by the TCM. The exception to this is when the clutch is hydraulically disabled to prevent engagement regardless of the commands from the computer. The converter clutch is hydraulically applied and electrically controlled through a PWM solenoid controlled by the TCM. When the solenoid is off, TCC signal fluid exhausts and the converter clutch remains released. Once the solenoid is energized, TCC signal fluid passes to the TCC regulator valve and the clutch, engaging it. Modulating the pressure to the clutch allows for smooth engagement and disengagement and also allows for partial engagement of the clutch.

Whenever the TCM detects a problem in any of the solenoids in a transmission, it will disable the TCC solenoid.

Adaptive Controls

Most late-model EATs have systems that allow the TCM to change transmission behavior based on the current condition of the transmission and engine, current operating conditions, and the habits of the driver. When systems are capable of doing this, they have **adaptive learning** capabilities. Adaptive learning provides consistent quality shifting and increases the durability of the transmission.

These transmissions have a line pressure control system that compensates for the normal wear of a transmission. As parts wear or conditions change, the time required to apply a clutch or brake changes. The TCM monitors the input and output speeds of the transmission during commanded shifts to determine if a shift is occurring too fast (harsh) or too slow (soft) and adjusts the line pressure to maintain the desired shift feel. Adaptive learning takes place as the TCM reads input and output speeds more than 140 times per second.

The system may also monitor the condition of the engine and compensate for any changes in the engine's performance. It also monitors and memorizes the typical driving style of the driver and the operating conditions of the vehicle. With this information, the computer adjusts the timing of shifts and converter clutch engagement to provide good shifting at the appropriate time.

SHOP TALK

EATs with adaptive learning may display harsh and/or abrupt shifting when they are new. This is normal because the TCM has not yet received enough information to learn about the vehicle or driver. This shifting behavior may also occur after the vehicle's battery has been disconnected for a long period.

Limp-In Mode

When the TCM detects a serious transmission problem or a problem in its circuit (or in some cases an engine control or data bus problem), it may switch to a default, limp-in, or fail-safe mode. Limp-in may also be initiated if the TCM loses its battery power feed. This mode allows for limited driving capabilities and is designed to prevent further transmission damage while allowing the driver to drive with decreased power and efficiency to a service facility for repairs.

The capabilities of a transmission while it is in limp-in mode depend on the extent of the fault, the manufacturer, and the model of transmission. When the TCM moves into the limp-in mode, a DTC is set and the transmission will only operate in this mode until the problem is corrected. Examples of operating characteristics during limp-in include:

- The transmission locks in third gear when the gear selector is in the drive position or second gear when it is in a lower position.
- The transmission will remain in whatever gear it was in but will shift into third or second gear and stay there as soon as the vehicle slows down.
- The transmission will only use first and third gears while in the drive position.
- The transmission will operate only in park, neutral, reverse, and second gears and will not upshift or downshift.

Operational Modes

With electronic controls, automatic transmissions can be programmed to operate in different modes. The desired mode is selected by the driver. The mode selection switch can be located on the center console or the instrument panel. Most transmissions with this feature have two selective modes, normally called "Normal" and "Power." During the normal mode, the transmission operates according to the shift schedule and logic set for normal or regular operation. In the power mode, the TCM uses different logic and shift schedules to provide for better acceleration and performance with heavy loads. Normally this means delaying upshifts.

If three modes are available, the third mode is called the "Auto" mode. The auto mode is a mixture between the normal and power modes. While in this mode, the TCM will control the shifts in a normal way. However, if the throttle is quickly opened, the shift pattern will switch to the power mode.

Manual Shifting

One of the most publicized features of electronically controlled transmissions is the availability of manual shift controls. Although not all electronic transmissions have this feature, they all could. Basically, these systems allow the driver to manually upshift and downshift the transmission at will much like a manual transmission. Unlike a manual transmission, though, the driver does not need to depress a clutch pedal nor is there a clutch assembly on the flywheel. The driver simply moves the gear selector or hits a button and the transmission changes gears. If the driver does not change gears and engine speed is high, the transmission shifts on its own. If the driver elects to let the transmission shift automatically, a switch disconnects the manual control and the transmission operates automatically.

Marketed as a sport option and a combination of a manual and an automatic transmission, these transmissions are still based on an automatic transmission with a torque converter. Therefore, performance numbers are not quite as good as if the vehicle were equipped with a manual transmission. In fact, manual control of an automatic transmission often results in slower acceleration times than when the transmission shifts by itself.

Not all manually shifted automatics behave in the same way, nor do they control the same things. Actually, the behavior of the transmission depends on the car in which it is installed. Some of these cars are pure high-performance cars, whereas others are moderate-performance family sedans. What follows are basic descriptions of some of the manually shifted automatic transmissions available. There are many other similar systems available; these are mentioned to serve only as examples.

Figure 41–11 The gear selector for a manual shifting BMW automatic transmission. *Courtesy of BMW of North America, LLC*

BMW's Steptronic Steptronic systems are based on five- or six-speed transmissions and offer the option of shifting gears manually using the selector lever **(Figure 41–11)** or through steering wheel mounted buttons. The driver can select to shift the gears manually or let the transmission shift automatically. The system relies on shift-by-wire technology, which replaced mechanical linkages to the transmission. These transmissions also are equipped with adaptive transmission controls (ATCs), which responds to the driver's style as well as the operating conditions. The ATC looks at the travel and movement patterns of the accelerator, deceleration rates during braking, and lateral acceleration in curves. The ATC then selects the most suitable shift schedule. Upshifts are delayed on uphill stretches to allow better use of the engine's power. On downhill grades, the ATC downshifts when the driver is forced to brake to counteract undesired acceleration.

In the manual mode, the ATC works to avoid overspeeding the engine by upshifting just before the engine reaches its automatic cutoff speed. At low speeds, it downshifts automatically without input from the driver. In the kickdown mode, the system downshifts to the lowest gear possible without overrevving the engine.

Chrysler Autostick This is the most familiar system. Manual shifting is performed by moving a control on the console. Moving the selector to the right provides for an upshift. A movement to the left allows for a downshift. The transaxle is not modified for this option; rather, it is fitted with a special gear selector and switch assembly. The driver can either manually shift the gears or allow the transaxle to shift automatically. The selected gear is displayed on the instrument panel to keep the driver informed of the selected gear.

Although the driver has control of the shifting, the TCM will override the controls during some conditions. Sometimes, the TCM will force upshifts or prevent manual shifting. These conditions relate to the projected engine speed if a gear change does or does not take place. The Autostick feature will also be deactivated if the TCM senses problems and/or sets a trouble code that relates to the TR sensor or Autostick switch or when there is a high engine and transmission temperature code.

Honda's Sequential Sportshift Manual shifting is performed by moving a control on the console. Moving the selector forward provides for an upshift. A movement down allows for a downshift. This transmission is unique in that it will not automatically upshift if the driver brings the engine's speed too high. All other transmissions of this design will upshift automatically at a predetermined engine speed to prevent damage to the engine. Honda vehicles use a rev limiter that prevents the engine from overspeeds. When some Acura models are operated in the manual shifting mode, they will start out in first gear and shift automatically to second gear. The transmission can only be shifted manually between second and fifth gears.

Tiptronic This is a five- or six-speed transmission available from a few European manufacturers, such as Porsches and Audis. Although the concept of all Tiptronic units is the same, driver control varies quite a bit. Manual shifting on the Audi is performed by moving a control on the console. Moving the selector forward provides for an upshift. A movement down allows for a downshift. To shift Porsche's transmission, the driver moves the gear selector into the manual gate, next to the automatic ranges, and depresses buttons on the steering wheel. These systems also are typically controlled by a computer that controls the change of gears and tries to mimic the action of a manual transmission. While operating in the manual shift mode, these systems study the driving habits of the driver and select the optimum driving range, reducing the driver's workload, particularly in stop-and-go traffic.

Toyota/Lexus Systems Toyota has a series of high-performance cars that feature five-speed transmissions that can operate in either of two modes, providing fully automatic shifting or electronic manual control. The TCM is programmed to allow for rapid shifts in response to the driver's commands. It will also prevent shifting during conditions that may cause engine or transmission failure. The transmission may also be

Figure 41–12 Fingertip controls for manually shifting Lexus' automatic/manual transmission. *Courtesy of Toyota Motor Sales, U.S.A., Inc.*

shifted by its gated console-mounted shift lever. The shift lever allows the driver to select individual gear ranges as well as the full-automatic mode.

Manual shifting may also be controlled by fingertip shifting buttons located on both horizontal spokes of the steering wheel **(Figure 41–12)**. Touching a button on the front of the steering wheel triggers downshifts. Contacting the buttons on the backside of the steering wheel controls upshifts. The buttons are located so that either thumb can be used to downshift and either index finger can be used to upshift.

CVT Controls

The electronic control system for Honda's CVT consists of a TCM, various sensors, linear solenoids, and an inhibitor solenoid. Pulley ratios are always controlled by the control system. Input from the various sensors determines which linear solenoid the TCM will activate **(Figure 41–13)**. Activating the shift

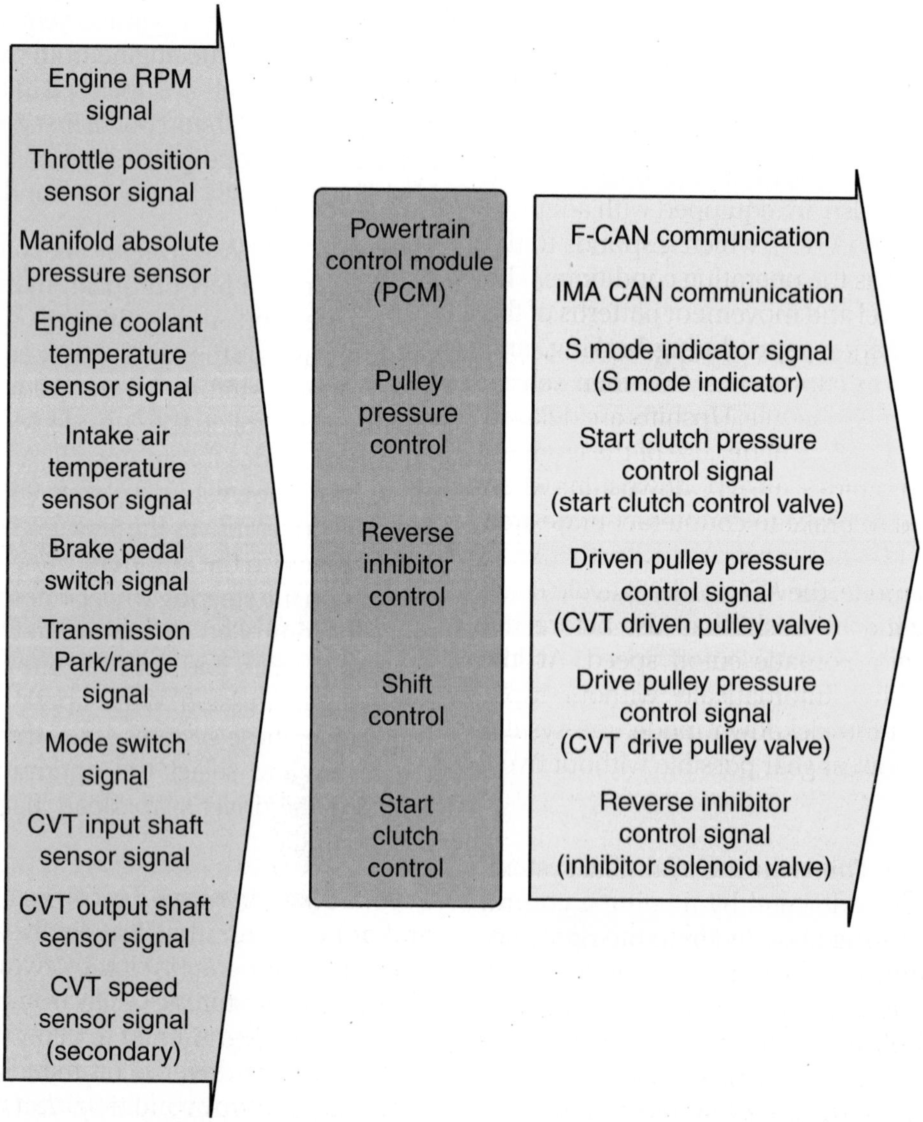

Figure 41–13 The electronic control system for a Honda CVT.

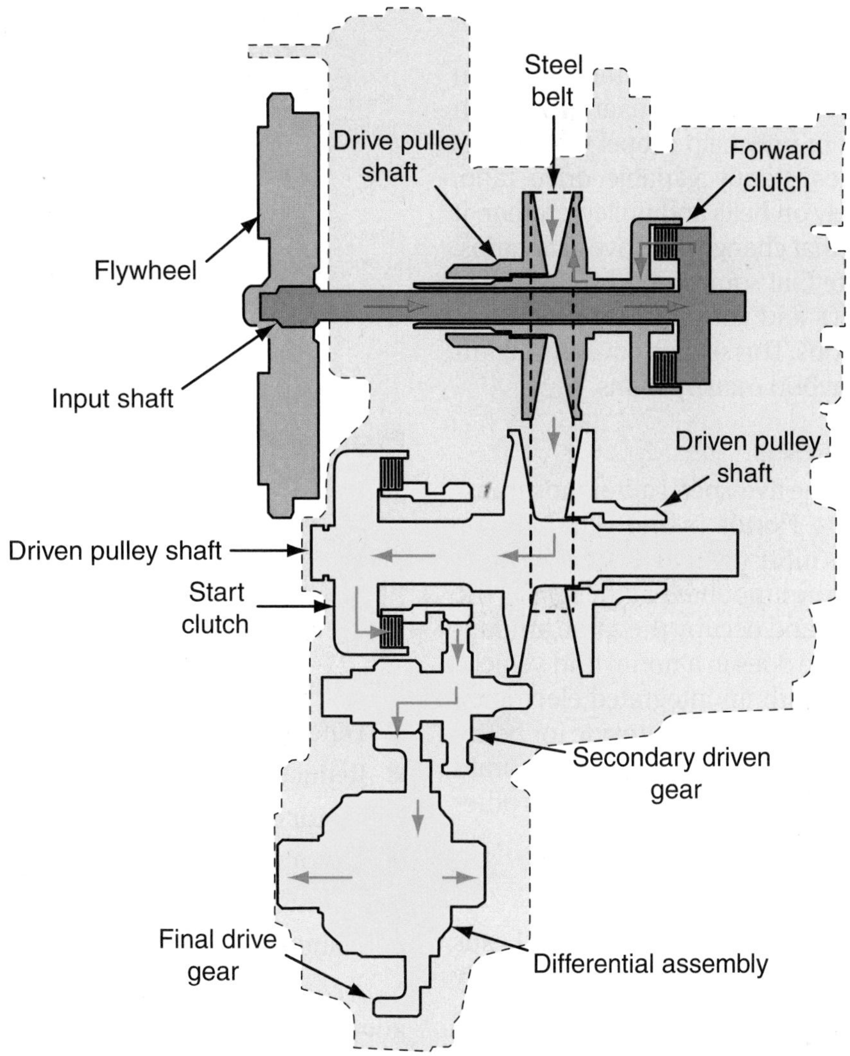

Figure 41–14 Components of a CVT assembly.

control solenoid changes the shift control valve pressure, causing the shift valve to move. This changes the pressure applied to the driven and drive pulleys, which change the effective pulley ratio. Activating the start clutch control solenoid moves the start clutch valve. This valve allows or disallows pressure to the start clutch assembly. When pressure is applied to the clutch, power is transmitted from the pulleys to the final drive gear set **(Figure 41–14)**.

The start clutch allows for smooth starting. Because this transaxle does not have a torque converter, the start clutch is designed to slip just enough to get the car moving without stalling or straining the engine. The slippage is controlled by the hydraulic pressure applied to the start clutch. To compensate for engine loads, the TCM monitors the engine's vacuum and compares it to the measured vacuum of the engine while the transaxle was in park or neutral.

The TCM controls the pulley ratios to reduce engine speed and maintain ideal engine temperatures during acceleration. If the car is continuously driven at full throttle acceleration, the TCM causes an increase in pulley ratio. This reduces engine speed and maintains normal engine temperature while not adversely affecting acceleration. After the car has been driven at a lower speed or not accelerated for a while, the TCM lowers the pulley ratio. When the gear selector is placed into reverse, the TCM sends a signal to the PCM. The PCM then turns off the car's air conditioning and causes a slight increase in engine speed.

Audi's stepless Multitronic CVT is based on what it refers to as a variator. The variator is made of vanadium-hardened steel that is encased in oil and offers more durability than belt-driven CVTs. A manual gear selection mode is available and six simulated gear ratios can be selected. In the automatic mode, Multitronic calculates the optimum gear ratio with the aid of a dynamic regulating program, according to engine load, the driver's preferences, and driving conditions.

HYBRID TRANSMISSIONS

Perhaps the most complex EATs are those used in most hybrid vehicles. The transmissions are fitted with electric motors that not only help propel the vehicle, but also provide a constantly variable drive ratio. These CVTs do not rely on belts and pulleys; rather, it is the electric motors that change the drive gear ratios. It is important to note that some hybrid vehicles rely on conventional CVTs and conventional manual or automatic transmissions. This section covers the common nontraditional hybrid transmissions.

Honda Hybrid Models

A modified version of the five-speed automatic transaxle or CVT is used in Honda hybrid vehicles. The five-speed transaxle is more compact so that it can fit behind the electric motor mounted at the rear of the engine **(Figure 41–15)** and occupy the same amount of space as the transaxle does in a nonhybrid vehicle. The transaxle is fitted with an integrated electric oil pump and different gear ratios that provide for better acceleration, fuel economy, and regenerative braking. These transaxles operate in the same way as other Honda units.

Toyota and Lexus Hybrids

The power-split device used in Toyota and Lexus hybrids **(Figure 41–16)** operates as a continuously variable transaxle, although it does not use the belts and pulleys normally associated with CVTs. The variability of this transaxle depends on the action of a motor/generator, referred to as MG1, and the torque supplied by another motor/generator, referred to as MG2, and/or the engine. The transaxle assembly contains the following:

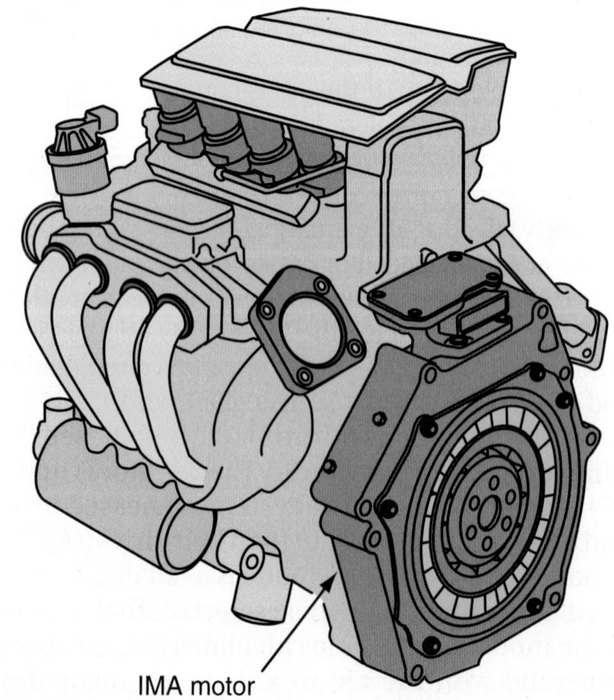

Figure 41–15 The IMA motor in Honda hybrids is mounted between the engine and transaxle.

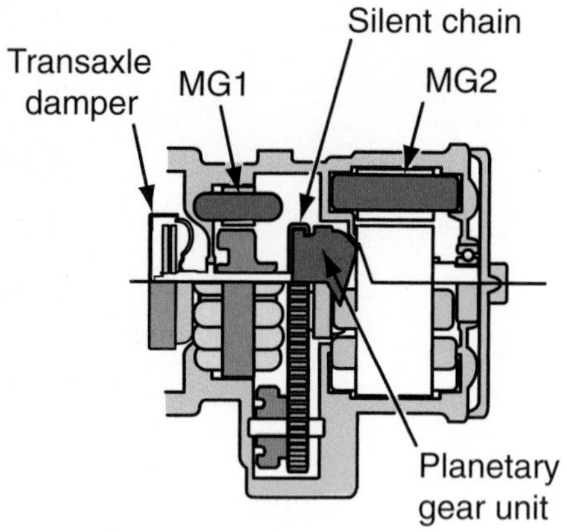

Figure 41–16 A typical power-split transaxle for a Toyota hybrid.

- Differential assembly
- Reduction unit
- Motor generator 1 (MG1)
- Motor generator 2 (MG2)
- Transaxle damper
- Planetary gear set

A conventional differential unit is used to allow for good handling when the car is making a turn. The reduction unit increases the final drive ratio so that ample torque is available to the drive wheels. The reduction unit is a gear set linked by a chain. MG1, which generates energy and serves as the engine starter, is connected to the planetary gear set, and so is MG2, which also is connected to the differential unit by the drive chain. This transaxle does not have a torque converter or clutch. Rather, a damper is used to cushion engine vibration and the power surges that result from the sudden engagement of power to the transaxle.

The engine, MG1, and MG2 are mechanically connected at the planetary gear set. The gear set can transfer power between the engine, MG1, MG2, and drive wheels in nearly any combination of these. The unit splits power from the engine to different paths: to drive MG1 or drive the car's wheels, or both. MG2 can drive the wheels or be driven by them.

In the planetary gear set used in the power-split device, the sun gear is attached to MG1. The ring gear is connected to MG2 and the final drive unit in the transaxle. The planetary carrier is connected to the engine's output shaft **(Figure 41–17)**. The key to understanding how this system splits power is to

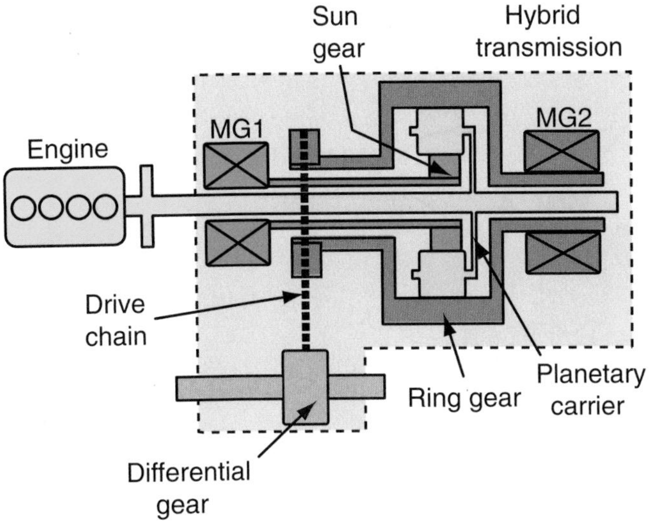

Figure 41-17 The arrangement of the electric motors and planetary gears in a power-split transaxle.

realize that when there are two sources of input power, they rotate in the same direction but not at the same speed. Therefore, one can assist the rotation of the other, slow down the rotation of the other, or work together. Also, keep in mind that the rotational speed of MG2 largely depends on the power generated by MG1. Therefore, MG1 basically controls the continuously variable transmission function of the transaxle. Here is a summary of the action of the planetary gear set during different operating conditions:

- To start the engine, MG1 is energized and the sun gear becomes the drive member of the gear set. Current is sent to MG2 to lock or hold the ring gear. The carrier is driven by the sun gear and walks around the inside of the ring gear to crank the engine at a speed higher than that of the sun gear.
- After the engine is started, MG1 becomes a generator. The ring gear remains locked by MG2 and the carrier now drives the sun gear, which spins MG1.
- When the car is driven solely by MG2, the carrier is held because the engine is not running. The ring gear rotates by the power of MG2 and drives the sun gear in an opposite direction. This causes MG1 to slowly spin in the opposite direction without generating electricity.
- If more drive torque is needed while running with MG2 only, MG1 is activated to start the engine. There are now two inputs to the gear set: the ring gear (MG2) and the sun gear (MG1). The carrier is driven by the sun gear and walks around the inside of the rotating ring gear. This cranks the engine at a faster speed than when the ring gear is held.
- After the engine is started, MG2 continues to rotate the ring gear and the engine rotates the carrier to drive the sun gear and MG1, which is now a generator.
- When the car is operating under light acceleration and the engine is running, some engine power is used to drive the sun gear and MG1 and the rest is rotating the ring gear to move the car. The energy produced by MG1 is fed to MG2. MG2 is also causing the ring gear to rotate and the power of the engine and MG1 combine to move the vehicle.
- This condition continues until the load on the engine or the condition of the battery changes. When the load decreases, such as during low-speed cruising, the HV ECU increases the generation ability of MG1, which now supplies more energy to MG2. The increased power at the ring gear allows the engine to do less work while driving the car's wheels and do more work driving the sun gear and MG1.
- During full throttle acceleration, battery power is sent to MG2 in addition to the power generated by MG1. This additional electrical energy allows MG2 to produce more torque. This torque is added to the high output of the engine at the carrier.
- When the car is decelerating and the transmission is in drive, the engine is shut off, which effectively holds the carrier. MG2 is now driven by the wheels and acts as a generator to charge the battery pack. The sun rotates slowly in the opposite direction and MG1 does not generate electricity. If the car is decelerating from a high speed, the engine is kept running to prevent damage to the gear set. The engine, however, merely keeps the carrier rotating within the ring gear.
- When the car is decelerating and the transmission is moved into the B range, MG2 acts as a generator to charge the battery pack and to supply energy to MG1. MG1 rotates the engine, which is not running at this time, to offer some engine braking.
- During normal deceleration with the brake pedal depressed, the engine is off and the skid control ECU calculates the required amount of regenerative brake force and sends a signal to the HV ECU. The HV ECU, in turn, controls the generative action of MG2 to provide a load on the ring gear. This load helps to slow down and stop the car. The hydraulic brake system does the rest of the braking.
- When reverse gear is selected, only MG2 powers the car. MG2 and the ring gear rotate in the reverse direction. Because the engine is not running, the carrier is effectively being held. The sun gear is rotating in its normal rotational direction but

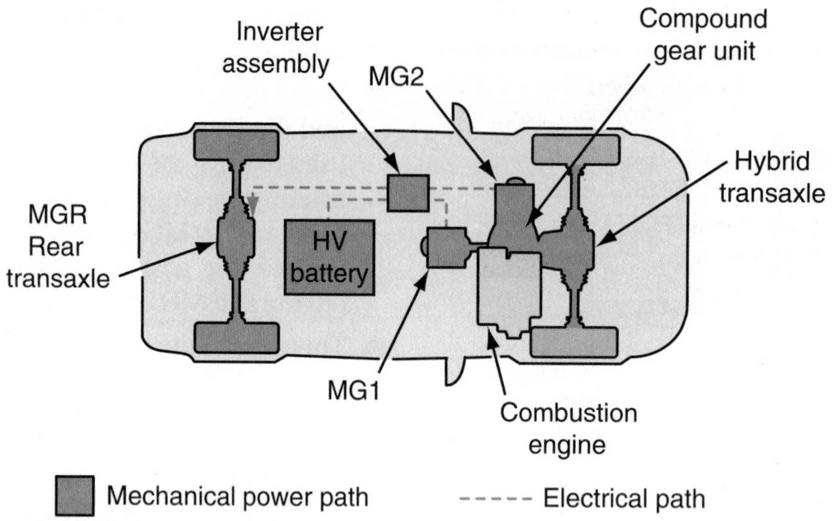

Figure 41–18 The layout of components for a Toyota 4WD vehicle.

slowly, and MG1 is not acting as a generator. Therefore, the only load on MG2 is the drive wheels.

- It is important to remember that at any time the car is powered only by MG2, the engine may be started to correct an unsatisfactory condition, such as low battery SOC, high battery temperature, and heavy electrical loads.

4WD Hybrids On Toyota and Lexus hybrid SUVs, the front transaxle assembly has been modified to include a speed reduction unit. This unit is a planetary gear set coupled to the power-split planetary gear set. This compound gear set has a common or shared ring gear that drives the vehicle's wheels. The sun gear of the power-split unit is driven by MG1 and the carrier is driven by the engine. In the reduction gear set, the carrier is held and the sun gear is driven by MG2. Because the sun gear is driving a larger gear (the ring gear), its output speed is reduced and its torque output is increased proportionally. High torque is available because MG2 can rotate at very high speeds. The rotational speed of MG1 essentially controls the overall gear ratio of the transaxle. The torque of the engine and MG2 moves from their designated gears to the common ring gear to the final drive gear and differential unit.

The transaxle has three distinct shafts: a main shaft that turns with MG1, MG2, and the compound gear unit; a shaft for the counter driven gear and final gear; and a third shaft for the differential. Because a clutch or torque converter is not used, a coil spring damper is used to absorb torque shocks from the engine and the initiation of MG2 to the driveline.

At the rear axle, an additional motor/generator (MGR) is placed in its own transaxle assembly to rotate the rear drive wheels. Unlike conventional 4WD vehicles, there is no physical connection between the front and rear axles (**Figure 41–18**). The aluminum housing of the rear transaxle contains the MGR, a counter drive gear, counter driven gear, and a differential. The unit has three shafts: MGR and the counter drive gear are located on the main shaft (MGR drives the counter drive gear), the counter driven gear and the differential drive pinion gear are located on the second shaft, and the third shaft holds the differential.

Ford Motor Company Hybrids

Ford's hybrids are equipped with an electronically controlled continuously variable transmission (eCVT). Based on a simple planetary gearset like the Toyotas, the overall gear ratios are determined by the motor/generator. Ford's transaxle is different in construction from that found in a Toyota. In a Ford transaxle, the traction motor is not directly connected to the ring gear of the gear set. Rather, it is connected to the transfer gear assembly. The transfer gear assembly is comprised of three gears: one connected to the ring gear of the planetary set, a counter gear, and the drive gear of the traction motor.

The effective gear ratios are determined by the speed of the members in the planetary gear set. This means that the speed or the motor/generator, engine, and traction motor determines the torque that moves to the final drive unit in the transaxle. All three of these power plants are controlled by the VSC through the TCM. Based on commands from the VSC and information from a variety of inputs, the TCM calculates the amount of torque required for the current operating conditions. A motor/generator control unit then sends commands to the inverter. The inverter, in turn, sends phased AC to the stator of the motors. The timing of the phased AC is critical to the operation of the motors as is the amount of voltage applied to each stator winding.

Angle sensors (resolvers) at the motors' stator track the position of the rotor within the stator. The signals from the resolvers also are used for the calculation of rotor speed. These calculations are shared with other control modules through CAN communications. The TCM, through these sensors, monitors the activity of the inverter and constantly checks for an open circuit, excessive current, and out-of-phase cycling. The TCM also monitors the temperature of the inverter and transaxle fluid.

4WD Hybrids Unlike the Toyotas with 4WD, the Escape and Mariner do not have a separate motor to drive the rear wheels. Rather, these wheels are driven in a conventional way with a transfer case, rear drive shaft, and a rear axle assembly. This 4WD system is fully automatic and has a computer-controlled clutch that engages the rear axle when traction and power at the rear are needed. The system relies on inputs from sensors located at each wheel and the accelerator pedal. The system calculates how much torque should be sent to the rear wheels. By monitoring these inputs, the control unit can predict and react to wheel slippage. It also can make adjustments to torque distribution when the vehicle is making a tight turn; this eliminates any driveline shudder that can occur when a 4WD vehicle is making a turn.

Two-Mode GM, Chrysler, and BMW Hybrids

GM, BMW, and Chrysler have developed a two-mode full hybrid system that can be used with gasoline or diesel engines. The two-mode hybrid system relies on advanced hybrid, transmission, and electronic technologies to improve fuel economy and overall vehicle performance. It is claimed that the fuel consumption of a full-size truck or SUV is decreased by at least 25% when it is equipped with this hybrid system.

The system fits into a standard transmission housing and is basically three planetary gearsets coupled to two electronically controlled electric motors, which are powered by a 300-volt battery pack. This combination results in four forward speeds plus continuously variable gear ratios at low speeds and motor/generators for hybrid operation **(Figure 41–19)**.

Operation One motor is used to restart the engine after it shuts down at a traffic light or stop sign. It also assists the engine during low-speed acceleration. The other motor provides all propulsion power when reverse gear is selected and assists the engine during low speeds with a heavy load and when cruising at high speeds. During light-load operation up to 30 mph (48 km/h), the motor can propel the vehicle without the assistance of the engine. Both motors are used as generators to charge the battery pack when the vehicle is decelerating and braking.

Figure 41–19 The electric motors and planetary gear sets in a GM, Chrysler, and BMW two-mode hybrid transmission. *Courtesy of Chrysler LLC*

The transmission uses clutch-to-clutch technology. The variable gear ratios are available through a mixing of power from the electric motors with the engine's power. When the motors are not providing power, the transmission operates like a conventional four-speed automatic.

The hybrid system has two distinct modes of operation. It operates in the first mode during low-speed and low-load conditions, and the second mode is used while cruising at highway speeds. The system can operate solely on electric or engine power or by a combination of the two. Typically, when one or both of the motors are not providing propulsion power, they are working as generators driven by the engine or by the drive wheels for regenerative braking.

The first mode of operation is called the input split, and the second is the compound split **(Figure 41–20)**. During the input split mode, the vehicle can be propelled by battery power or engine power, or both. This is the normal mode of operation when the vehicle is slowly accelerating from a stop and when it is cruising at slow speeds. When the control unit determines that battery power is sufficient for the current conditions, the engine shuts off. During this time, one motor is working to move the vehicle, while the other

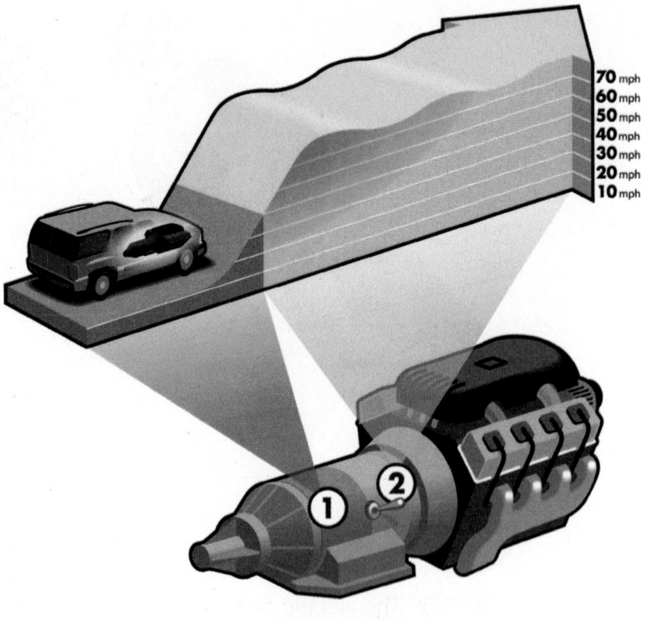

Figure 41-20 The two operating modes for a two-mode hybrid system. *Courtesy of Chrysler LLC*

may be working as a generator to supply power for the traction motor or to recharge the battery. If the engine is commanded to start, the traction motor may shut down but the second motor can continue to operate as a generator if needed.

During normal driving under light loads, the vehicle is powered solely by the engine. Depending on the load and other conditions, the engine may switch off some of its cylinders. When vehicle speed increases or when a heavy load is introduced to the vehicle, such as hard acceleration, climbing a hill, or towing, the system switches to the compound split mode.

In this mode, the control unit can order both motors to supply assist to the engine or require one of them to operate as a generator. The goal of the control unit is to maximize fuel economy while meeting the needs of the current operating conditions. The control unit also works with engine controls to determine if the other fuel savings features, such as cylinder deactivation and late intake valve closing (Atkinson cycle), should be initiated. It is important to realize that the deactivation of cylinders and the initiation of the Atkinson cycle reduce the power output of the engine. These fuel savings features do not hurt the performance of the vehicle because the engine's output is supplemented by the electric motor. This feature distinguishes a two-mode hybrid from other full hybrids. Typically, electric assist is available only when there is a high demand for power. In the two-mode system, the motors make it possible to reduce the work of the engine, even during light and moderate loads.

BASIC EAT TESTING

One of the first tasks during diagnosis of an EAT is to determine if the problem is caused by the transmission or by electronics. To determine this, the transmission must be observed to see if it responds to commands given by the computer. Identifying whether the problem is the transmission or electrical will determine what steps need to be followed to diagnose the cause of the problem.

EATs work only as well as the commands they receive from the computer, even if the hydraulic and mechanical parts of the transmission are fine. All diagnostics should begin with a scan tool to check for trouble codes in the system's computer. After the received codes are addressed, you can begin a more detailed diagnosis of the system and transmission. Your next step may be manually activating the shift solenoids by connecting a jumper wire to them or by using a transmission tester that allows you to manually activate the solenoids. Prior to doing this, study the wiring to the solenoids to determine if the computer activates them by supplying voltage to them or by completing the ground circuit. In addition, you need to know in which gear certain solenoids are activated. This information can be found in the service manual.

The best way to diagnose an electronically controlled transmission is to approach solving the problem in a logical way.

PROCEDURE

The recommended procedure for troubleshooting an EAT involves seven distinct steps that should be followed according to the order given:

STEP 1 Verify the customer's complaint. Pay attention to the conditions that exist when the problem occurs.

STEP 2 Check for any related symptoms, such as engine overheating, a lit MIL, and other driveability problems.

STEP 3 Conduct preliminary inspections and checks.

STEP 4 Check all service information for information that may apply to the complaint, including service bulletins, symptom charts, and recall notices.

STEP 5 Interpret and respond to all diagnostic codes.

STEP 6 Follow the diagnostic routines given by the manufacturer to define and isolate the cause of the problem.

STEP 7 Fix the problem and verify the repair.

Because many EAT problems are caused by the basics, it is wise to conduct all of the preliminary checks required for a nonelectronically controlled transmission. Also, thoroughly inspect the electronic system. This inspection should include a check of the MIL and the retrieval of diagnostic codes. Doing this will also allow you to pull engine-related codes as well as transmission codes. Whenever diagnosing a transmission, remember that an engine problem can and will cause the transmission to act abnormally.

Scan Tool Checks

Go To **Chapter 25 for details on connecting a scan tool and retrieving DTCs.**

Using a scan tool is one of the first steps of EAT diagnostics. Prior to retrieving the DTCs, pay attention to the MIL. The MIL is basically an engine malfunction light, but if the TCM detects a problem that may affect emissions, it will send a request over the data bus to the PCM to turn on the MIL lamp. Therefore, if the MIL is lit, the engine or transmission can have a problem. Remember, the MIL does not light after all DTCs are set; it only is lit when there is a fault that could affect emission levels.

The scan tool will display any trouble codes in the system. DTCs are designed to help technicians identify and locate problems in the transmission's system. It is very important that you refer to the service manual when interpreting DTCs. When diagnosing an EAT, make sure there are no engine-related codes that could affect the operation of the transmission. If there is an engine problem, fix it before continuing with your diagnosis.

DTCs that relate to transmission faults can be caused by engine or transmission input and/or output devices. These codes may seem to indicate a problem with an input or output circuit but may actually be caused by an internal transmission problem. Remember, codes are set by out-of-range values. Therefore, when the TCM is receiving a too low or high input signal, the cause is not necessarily a bad sensor. The sensor can be fine and a mechanical or hydraulic transmission problem is causing the abnormal signals. Not only can internal transmission problems cause codes to be set, but so can basic electrical problems. Problems such as loose connections, broken wires, corrosion, and poor grounds will affect the signals.

If the TCM is unable to communicate with the PCM, there is a data bus problem. These problems will normally result in poor operation as well as the inability to retrieve DTCs from the TCM. The PCM constantly monitors the data bus and if it is unable to establish communication, it will order a data bus DTC.

Although the first steps in diagnosis include retrieving DTCs, there are problems that will not be evident by a code. These problems are solved with further testing, symptom charts, or pure logic. This logic must be based on an understanding of the transmission and its controls. It is possible to pinpoint the exact cause of a transmission problem: Monitor the serial data with a scan tool **(Figure 41–21)**. The serial data stream allows you to monitor system activity during operation. Comparing the observed values to the manufacturer's specifications will greatly help in diagnostics. However, it is possible that the data displayed by a scan tool is not the actual value. Most computer

PID Name	Description of PID
EPC	Commanded Electronic Pressure Control Pressure—in psi
GEAR	Commanded Gear—not actual
LINEDSD	Commanded Line Pressure—in psi
OSS	Input from Output Shaft Speed Sensor—in rpm
RPM	Input from Engine Speed Sensor—in rpm
SSA	Commanded State of Shift Solenoid No. 1—ON or OFF
SSB	Commanded State of Shift Solenoid No. 2—ON or OFF
SS1F	Shift Solenoid No. 1 Circuit Fault—YES or NO
SS2F	Shift Solenoid No. 2 Circuit Fault—YES or NO
TCCACT	Slippage of Torque Converter Clutch—in rpm
TCCCMD	Commanded State of Torque Converter Clutch Solenoid—in %
TCCF	Torque Converter Clutch Solenoid Circuit Fault—YES or NO
TFT	Transmission Fluid Temperature—in voltage or degrees
TP	Throttle Position—in voltage
TR	Transmission Range Sensor—by position
TRANRAT	Actual Transmission Gear Ratio—by position
TSS	Input from Turbine Shaft Speed Sensor—in rpm

Figure 41–21 Common transmission PIDs for Ford products.

systems will disregard inputs that are well out of range and rely on a default value held in its memory. These default values are hard to recognize and do little for diagnostics; this is why the use of basic electrical troubleshooting equipment, such as wiring diagrams, diagnostic charts, DMMs, lab scopes, and special transmission testers, to check the system is common.

Preliminary EAT Checks

Critical to proper diagnosis of EAT and TCC control systems is a road test. The road test should be conducted in the same way as one for a nonelectronic transmission except that a scan tool should also be connected to the circuit to monitor engine and transmission operation.

During the road test, the vehicle should be driven in the normal manner. Pay close attention to all gear changes. Also, the various computer inputs should be monitored and the readings recorded for future reference. Some scan tools have the capability of printing out a report of the test drive. Critical information from the inputs includes engine speed, vehicle speed, manifold vacuum, operating gear, and the time it took to shift gears. If the scan tool does not have the ability to give a summary of the road test, you should record this same information after each gear or change in operating condition.

Often, accurately defining the problem and locating related information in TSBs and other materials can identify the cause of the problem. When a manufacturer recognizes common occurrences of a problem, a bulletin will be issued regarding the fix of the problem. Also, for many DTCs and symptoms, service manuals give a simple diagnostic chart or path for identifying the cause of the problem. These are designed to be followed step by step and will lead to a conclusion if you follow the path matched to the symptom. Check all available information before moving on in your diagnostics.

Sometimes the symptom will not match any of those described in the service manual. This does not mean that it is time to guess; rather, it is time to clearly identify what is working right. By eliminating those circuits and components that are working correctly from a list of possible causes, you can identify what may be causing the problem and what should be tested further.

Common problems that affect shift timing and quality as well as the timing and quality of TCC engagement are incorrect battery voltage, a blown fuse, poor connections, a defective TP sensor or VSS, defective solenoids, wires to the solenoids or sensors crossed, corrosion at an electrical terminal, or faulty installation of an accessory.

Often computer-controlled transmissions will start off in the wrong gear. This can happen for several reasons, such as either internal transmission problems or external control system problems. Internal transmission problems can be faulty solenoids or stuck valves. External problems can be the result of a complete loss of power or ground to the control circuit or the transmission is operating in its fail-safe mode. Typically, the default gear is simply the gear that is applied when the shift solenoids are off.

Electronic Defaults

While diagnosing a problem in an electronically controlled transmission, always refer to the appropriate service manual to identify the normal "default" operation of the transmission. You could spend time tracing the wrong problem by not recognizing that the transmission is operating in default.

Whenever the computer sees a potential problem that may increase wear and/or damage to the transmission, the system also defaults to limp-in mode. Minor slipping can be sensed by the computer through its input and output sensors. This slipping will cause premature wear and may cause the computer to move into its default mode; some systems may increase fluid pressure to compensate for the problem. A totally burnt clutch assembly will cause limp-in operation as will some internal pressure leaks that may not be apparent until pressure tests are run.

Guidelines for Diagnosing EATs

1. Make sure the battery has at least 12.6 volts before troubleshooting the transmission.
2. Check all fuses and identify the cause of any blown fuses.
3. Check the physical condition of all sensors and the wiring going to them **(Figure 41–22)**.
4. Compare the wiring to all suspected components with the wire colors given in the service manual.
5. When testing electronic circuits, always use a high-impedance testlight or DMM.
6. If an output device is not working properly, check the power circuit to it.
7. If an input device is not sending the correct signal back to the computer, check the reference voltage it is receiving and the voltage it is sending back to the computer.
8. Compare the voltages in and out of a sensor with the voltages the computer is sending out and receiving.
9. Before replacing a computer, check the solenoid isolation diodes according to the procedures outlined in the service manual.

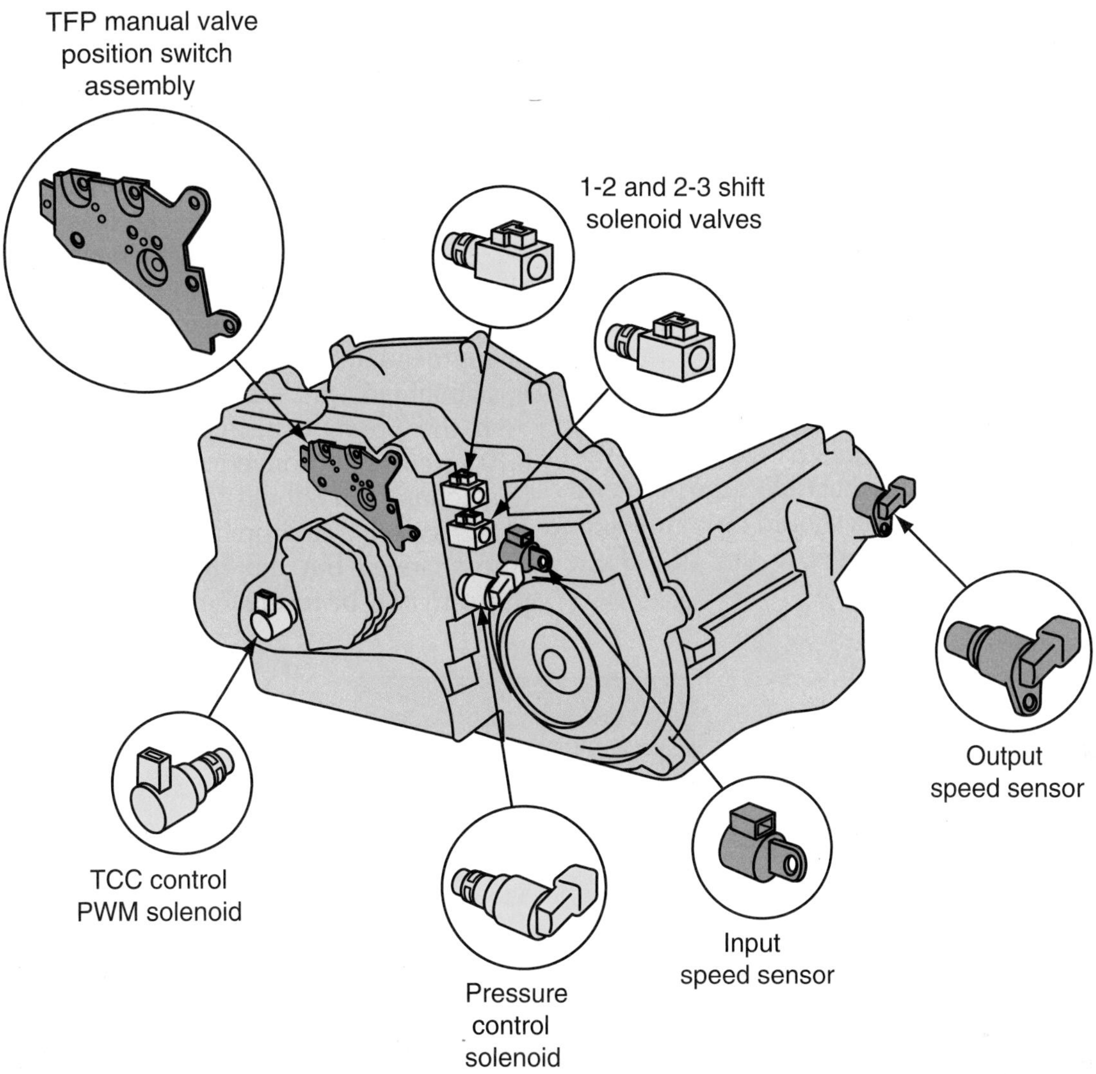

Figure 41-22 The typical location of the electronic components in or on a transaxle.

10. Make sure the computer wiring harnesses do not run parallel with any high-current wires or harnesses. The magnetic field created by the high current may induce a voltage in the computer harness. You should also be aware that antenna cables and CB radios could cause interference.
11. Take necessary precautions to prevent the possibility of static discharge while working with electronic systems.
12. While checking individual components, always check the voltage drop of the ground circuits. This becomes more and more important as cars are made of less material that conducts electricity well.
13. Make sure the ignition is off when you disconnect or connect an electronic component.
14. Check all sensors in cold and hot conditions.
15. Check all wire terminals and connections for tightness and cleanliness.
16. Use a TV-tuner cleaning spray to clean all connectors and terminals.
17. Use dielectric grease at all connections to prevent future corrosion.
18. If you must break through the insulation of a wire to take an electrical measurement, make sure you tightly tape over the area after you are finished testing.

WARNING!

Static electricity can destroy or render an electronic part useless. When handling any electronic part, do whatever is possible to reduce the chances of electrostatic buildup on your body and the inadvertent discharge to the electronic part.

CONVERTER CLUTCH CONTROL DIAGNOSTICS

To properly diagnose converter clutch problems, you must know when the TCC should engage and disengage and understand the function of the various controls involved with the system **(Figure 41–23)**. Although the actual controls for a TCC vary with the different models of transmissions, they all have certain operating conditions that must be met before the clutch can be engaged.

Diagnosis of a TCC circuit should be conducted in the same way as any other computer system. The computer will recognize problems within the system and store trouble codes that reflect the problem area of the circuit. A road test should be conducted to see if the problem is related to the TCC **(Figure 41–24)**.

On early electronically controlled systems with a TCC solenoid, the clutch is typically applied when oil flow through the torque converter is reversed. This change can be observed with a pressure gauge. Connect a pressure gauge to the fluid line from the transmission to the cooler. Position the gauge so that it is easily seen from the driver's seat. Then raise the vehicle on a hoist with the drive wheels off the ground and able to spin freely. Operate the vehicle until the transmission shifts into high gear. Then maintain a speed of approximately 55 mph (88 km/h). Once the speed is maintained, watch the pressure gauge.

If the pressure decreases 5 to 10 psi (0.35 to 0.70 kg/cm^2), the converter clutch was applied. With this action, you should feel the engagement of the clutch as well as a drop in engine speed. If the pressure changed but the clutch did not engage, the problem may be inside the converter or at the end of

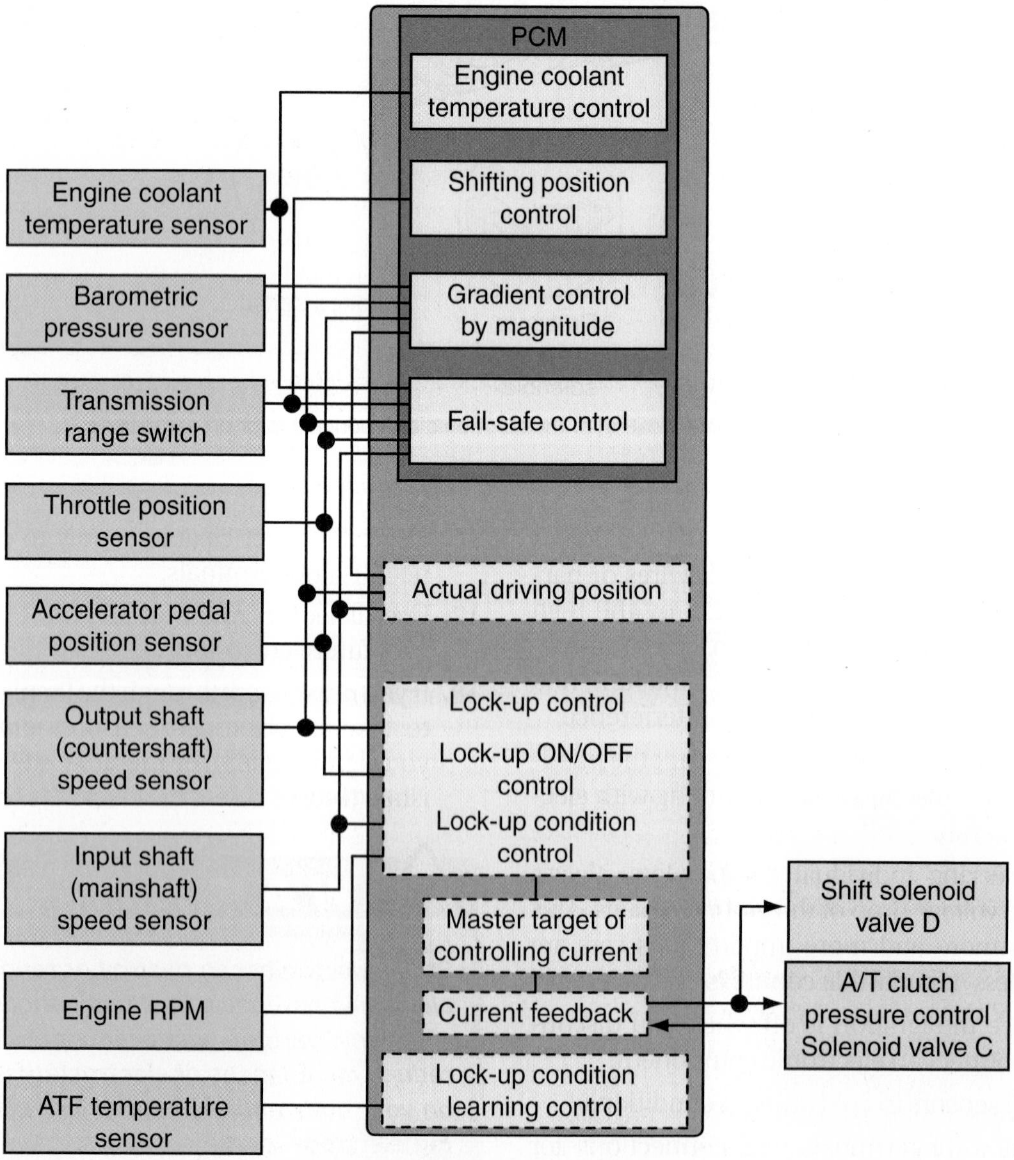

Figure 41-23 The electronic control system for a TCC.

Step	Findings	Rememdy
1. Does the TCC engage and disengage?	Yes No	Go to step #2 Go to symptoms chart—diagnose the system
2. Describe the vibration or shudder during a 3-4 or 4-3 shift.	Light Medium Heavy	Go to step #3 Go to step #3 Go to symptoms chart—is not TCC related
3. Is the vibration or shudder vehicle speed related and not gear related?	Yes No	Go to symptoms chart—is not TCC related Go to step #4
4. Is the vibration or shudder engine speed related and not gear related?	Yes No	Go to symptoms chart—is not TCC related Go to step #5
5. Does the vibration or shudder occur in coast, cruise, or reverse gear?	Yes No	Go to symptoms chart—is not TCC related Go to step #6
6. Does the vibration or shudder occur during long periods of light braking?	Yes No	Go to symptoms chart—is not TCC related Go to step #7
7. Did the vibration or shudder only occur in step #2?	Yes No	There is a probable TCC problem—diagnose the system Go to symptoms chart—is not TCC related

Figure 41-24 An evaluation form to use while doing a road test to check the torque converter.

the input shaft. If the input shaft end is worn or the O-ring at the end is cut or worn, there will be pressure loss at the converter clutch. This loss in pressure will prevent full engagement of the clutch. If the pressure did not change and the clutch did not engage, suspect a faulty clutch valve or control solenoid or a fault in the solenoid control circuit.

If the clutch does not engage, check for power to the solenoid. If power is available, make sure the ground of the circuit is good. If there is power available and the ground is good, check the voltage drop across the solenoid. The solenoids should drop very close to source voltage. If less than that is measured, check the voltage drop across the power and ground sides of the circuit. If the voltage drop testing results are good, remove the solenoid and test it with an ohmmeter. Suspect clutch material, dirt, or other material plugging up the solenoid valve passages if the solenoid checks out fine with the ohmmeter. Attempt to flush the valve with clean ATF if blockage is found. If the solenoid has a filter assembly, replace the filter after cleaning the fluid passages. Replace the solenoid if the blockage cannot be removed.

The TCC in late-model transmissions is controlled by the PCM or TCM. The computer turns on the converter clutch solenoid, which opens a valve and allows fluid pressure to engage the clutch. When the computer turns the solenoid off, the clutch disengages.

A malfunctioning converter clutch can cause a wide variety of driveability problems. Normally, the application of the clutch should feel like a smooth engagement into another gear. It should not feel harsh, nor should there be any noises related to the application of the clutch.

If the clutch engages at the wrong time, a sensor or switch in the circuit is probably the cause. If clutch engagement occurs at the wrong speed, check all speed-related sensors. A faulty temperature sensor may cause the clutch not to engage. If the sensor is not reading the correct temperature, the PCM may never realize that the temperature is suitable for engagement. Checking the appropriate sensors can be done with a scan tool, DMM, and/or lab scope.

Engagement Quality

If the clutch engages prematurely or is not applied with full pressure, a shudder or vibration results from the rapid grabbing and slipping of the clutch. This symptom can feel like an engine misfire or vibration. The clutch begins to engage then slips because it cannot hold the engine's torque. The torque capacity of the clutch is determined by the oil pressure applied to it and the condition of the frictional surfaces of the clutch assembly.

If the shudder is only noticeable during the engagement of the clutch, the problem is typically in the converter. When the shudder is only evident after the engagement of the clutch, the cause of the shudder is the engine, transmission, or another component of the driveline. If the shudder is caused by the clutch, the converter must be replaced to correct the problem.

A faulty clutch solenoid or its return spring may cause low apply pressure. The valve controlled by the solenoid is normally held in position by a coil-type return spring. If the spring loses tension, the clutch will engage too soon. Because there is insufficient pressure to hold the clutch, shudder occurs as the clutch begins to grab and then slip. If the solenoid valve and/or return spring are faulty, they should be replaced, as should the torque converter. If the TCC fails to release, it can cause the engine to jerk and stall when the vehicle is stopping.

An out-of-round torque converter prevents full clutch engagement, which will also cause shudder, as will contaminated clutch frictional material. The frictional material can become contaminated by metal particles circulating through the torque converter and collecting on the clutch. Broken or worn clutch dampener springs will also cause shudder.

DETAILED TESTING OF INPUTS

There are many different designs of sensors that are part of the control system for an EAT. The transmission will not work properly if it receives bad information from its sensors or from the CAN bus. The transmission may shift at the wrong speeds, shift harshly, or operate only in the limp-in mode.

Some sensors are nothing more than a switch that completes a circuit. Others are complex devices that react to chemical reactions and generate their own voltage during certain conditions. If the preliminary tests pointed to a possible problem in an input circuit, the circuit should be tested. Make sure to check all suspect circuits for resistance problems; conduct voltage drop tests on those circuits. Often the manufacturers will give specific testing procedures for specific sensors; always follow them.

Go To *Chapter 26 for details on specific tests of input devices.*

Testing Switches

Many different switches are used as inputs or control devices for EATs. Most of the switches are either mechanically or hydraulically controlled. The operation of these switches can be easily checked with an ohmmeter. With the meter connected across the switch's leads, there should be continuity or low resistance when the switch is closed and there should be infinite resistance across the switch when it is open. A testlight can also be used. When the switch is closed, power should be present at both sides of the switch. When the switch is open, power should be present at only one side.

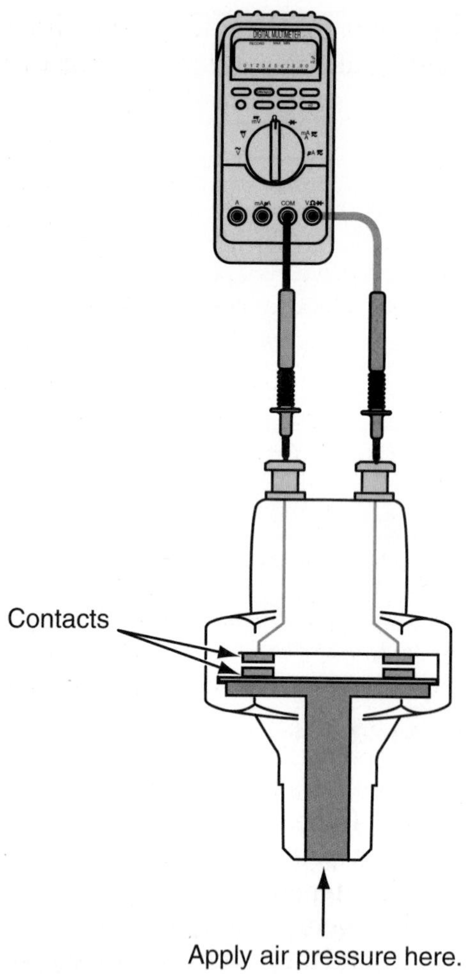
Contacts
Apply air pressure here.

Figure 41-25 Testing a typical normally open pressure switch. Apply air to the bottom fluid passage and check for continuity across the terminals.

Pressure switches can be checked by applying air pressure to the part that would normally be exposed to fluid pressure (**Figure 41–25**). When applying air pressure to these switches, check them for leaks. Although a malfunctioning electrical switch will probably not cause a shifting problem, it will if it leaks. If the switch leaks off the applied pressure in a hydraulic circuit to a holding device, the holding member may not be able function properly. When possible, you should check pressure switches when they are installed and controlled by the vehicle.

Throttle Position (TP) Sensor

Another type of switch is a potentiometer. Rather than open and close a circuit, a potentiometer controls the circuit by varying its resistance in response to something. A TP sensor is a potentiometer. A bad TP sensor can cause the following problems: no upshifts, quick upshifts, delayed shifts, line-pressure problems with transmissions that have a line pressure control solenoid, and erratic converter clutch engagement.

SHOP TALK

A faulty TP sensor may not cause a DTC to be set. The diagnostic capabilities of the PCM must be able to determine if the sensor is working correctly or not. If it does not have this capability, it may not set a code. The PCM must be able to look at the input from the TP sensor and compare it to other inputs, such as engine speed, MAP inputs, and airflow. If the PCM determines that the TP signal does not reflect a true value based on engine speed and load inputs, it will set a code.

A TP sensor can be checked with an ohmmeter or a voltmeter. If checked with an ohmmeter, you should be able to watch the resistance of the sensor change as the throttle is opened and closed. Often there will be a resistance specification given in the service manual. Compare your reading to this.

With a voltmeter, you will be able to measure the reference voltage and the output voltage. Both of these should be within specified amounts. If the reference voltage is lower than normal, check the voltage drop across the reference voltage circuit from the computer to the TP sensor. Replace the TP sensor if it is defective.

Testing with a lab scope is a good way to watch the sweep of the resistor. The waveform is a DC signal that moves up as the voltage increases. Most potentiometers in computer systems are fed a reference voltage of 5 volts. Therefore, the voltage output of these sensors will typically range from 0.5 to 4.5 volts. The change in voltage should be smooth. Look for glitches in the signal. These can be caused by changes in resistance or an intermittent open.

Mass Airflow Sensor

When a MAF sensor fails or sends faulty signals, the engine runs roughly and tends to stall as soon as you put the transmission into gear. This sensor can be checked with a multimeter set to the Hz frequency range. Check the service manual for specific values. Normally, at idle, 30 Hz is measured and the frequency will increase as the throttle opens.

A scan tool may also be used to test this sensor; most have a test mode that monitors MAF sensors. The output of some MAFs can be observed with a DMM; their output is variable DC voltage. While diagnosing these systems, keep in mind that cold air is denser than warm air.

Test connections	Specified value
E2 - THO1	90Ω to 156 kΩ
E2 - ground	156 kΩ or higher
THO1 - ground	156 kΩ or higher

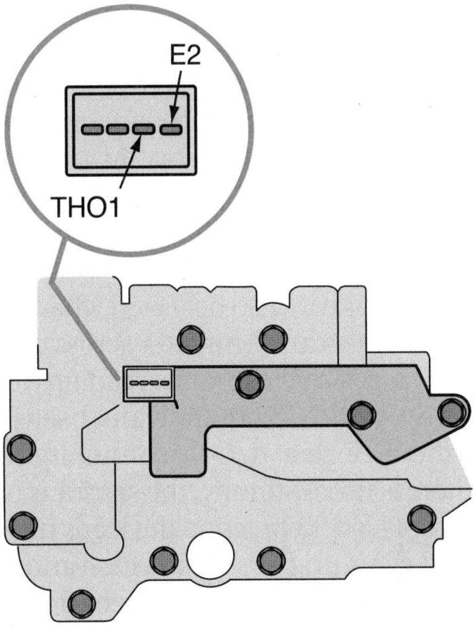

Figure 41-26 Checking an ATF temperature sensor.

Temperature Sensors

Temperature sensors can be checked with an ohmmeter. To do so, disconnect the sensor. In most cases, the sensor can be checked at room temperature **(Figure 41–26)**. Determine the temperature of the sensor and measure the resistance across it. Compare your reading to the chart of normal resistance for that temperature, which is given in the service information.

Thermistor activity can be monitored with a lab scope. Connect the scope across to the output of the thermistor or temperature sensor. Run the engine and watch the waveform. As the temperature increases, there should be a smooth increase or decrease in voltage. Look for glitches in the signal. These can be caused by changes in resistance or an intermittent open.

Speed Sensors

Speed sensors negate the need for hydraulic signals from a governor. When this sensor fails or sends faulty readings, it can cause complaints similar to those caused by a bad TP sensor. The most common complaints are no overdrive, no converter clutch engagement, and no upshifts.

The operation of a PM generator-type speed sensor can be tested with a DMM set to measure AC voltage. Raise the vehicle on a lift. Allow the wheels to be suspended so they can rotate freely. Connect the meter to the speed sensor. Start the engine and put the transmission in gear. Slowly increase the engine's speed until the vehicle is at approximately 20 mph, and then measure the voltage at the speed sensor. Slowly increase the engine's speed and observe the voltmeter. The voltage should increase smoothly and precisely with an increase in speed.

Magnetic pulse generator speed sensors can be tested with a lab scope. Connect the lab scope's leads across the sensor's terminals. The expected pattern is an AC signal, which should be a perfect sine wave when the speed is constant. When the speed is changing, the AC signal should change in amplitude and frequency. If the readings are not steady and do not smoothly change with speed, suspect a faulty connector, wiring harness, or sensor.

A speed sensor can also be tested when it is out of the vehicle. Connect an ohmmeter across the sensor's terminals. The desired resistance readings across the sensor will vary with every individual sensor; however, you should expect to have continuity across the leads. If there is no continuity, the sensor is open and should be replaced. Reposition the leads of the meter so that one lead is on the sensor's case and the other to a terminal. There should be no continuity in this position. If there is any measurable amount of resistance, the sensor is shorted.

DETAILED TESTING OF ACTUATORS

If you were unable to identify the cause of a transmission problem through the previous checks, you should continue your diagnostics with testing the solenoids. This will allow you to determine if the shifting problem is the solenoids or their control circuit, or if it is a hydraulic or mechanical problem.

Go To ***Chapter 26 for details on specific tests of output devices, including solenoids.***

Before continuing, however, you must first determine if the solenoids are case grounded and fed voltage by the computer or if they always have power applied to them and the computer merely supplies the ground. While looking in the service manual to find this, also find the section that tells you which solenoids are on and which are off for each of the different gears.

To begin this test you should collect the tools and/or equipment necessary to manually activate the solenoids. Switch panels that connect into the solenoid assembly are available and allow the technician to switch gears by depressing or flicking a switch.

SHOP TALK

This type of tester is easily made. Get a wiring harness for the transmission you want to test. Connect the harness to simple switches. Follow the solenoid/gear pattern when doing this. To change gears, all you will need to do is turn off one switch and turn on the next.

With the tester, the solenoids will be energized in the correct pattern and observe the action of the solenoids. To totally test the transmission, you should shift gears under light, half, and full throttle. If the transmission shifts fine with the movement of the switches, you know that the transmission is fine. Any shifting problem must therefore be caused by something electrical. If the transmission did not respond to the switch movements, the problem is probably in the transmission.

At times, a solenoid will work fine during light throttle operation but may not allow the valve to exhaust enough fluid when pressure increases. To verify that the valve is not exhausting, activate the solenoid then increase engine speed while pulling on the throttle cable. If the valve cannot exhaust, the transmission will downshift. Restricted solenoids are a common cause of rough shifting under heavy loads or full throttle but good shifting under light throttle.

Testing Actuators with a Lab Scope

You will be able to watch the actuator's electrical activity by observing its action on a lab scope. Normally, if there is a mechanical fault, this will affect its electrical activity as well. Some actuators are controlled by pulse width modulated signals **(Figure 41–27)**. These devices are controlled by varying the pulse width, signal frequency, and/or voltage levels. By watching the control signal, you can see the turning on and off of the solenoid **(Figure 41–28)**. All

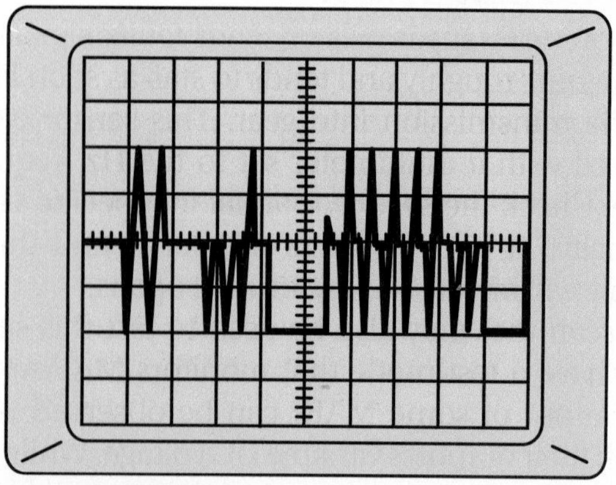

Figure 41-27 A typical control signal for a pulse width modulated solenoid.

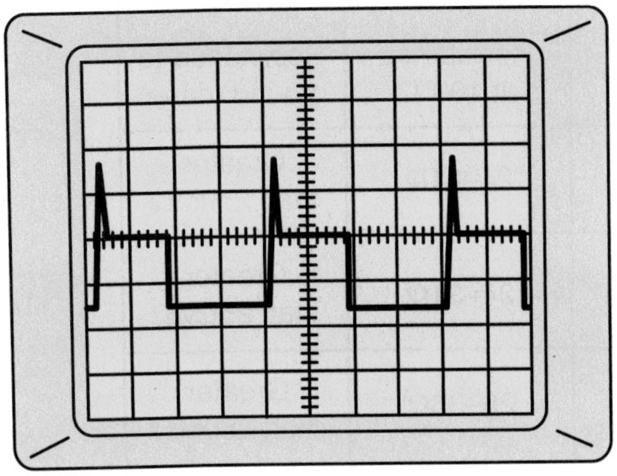

Figure 41-28 The activity of a solenoid as it cycles on and off.

waveforms should be checked for amplitude, time, and shape. You should also observe changes to the pulse width as operating conditions change. A bad waveform will have noise, glitches, or rounded corners. You should be able to see evidence that the actuator immediately turns off and on according to the commands of the computer.

Testing Actuators with an Ohmmeter

Solenoids can be checked for circuit resistance and shorts to ground. This can typically be done without removing the oil pan. The test can be conducted at the transmission case connector. Individual solenoids can be checked with an ohmmeter by identifying the proper pins in the connector. Remember, lower-than-normal resistance indicates a short, whereas higher-than-normal resistance indicates a problem of high resistance. If you get an infinite reading across the solenoid, the solenoid windings are open. The ohmmeter can also be used to check for shorts to ground. Simply connect one lead of the ohmmeter to one end of the solenoid windings and the other lead to ground **(Figure 41–29)**. The reading should be infinite. If there is any measurable resistance, the winding is shorted to ground.

Solenoids can also be tested on a bench. Resistance values are typically given in service manuals for each application **(Figure 41–30)**. A solenoid may be electrically fine but still may fail mechanically or hydraulically. A solenoid's check valve may fail to seat or its porting can be plugged. This is not an electrical problem; rather, it could be caused by the magnetic field collecting metal particles in the ATF and clogging the port or check valve. These would cause erratic shifting, no shift conditions, wrong gear starts, no limited passing (kickdown) gear, or binding shifts. When a solenoid affected in this way is activated, it will make a slow, dull thud. A good solenoid tends to snap when activated.

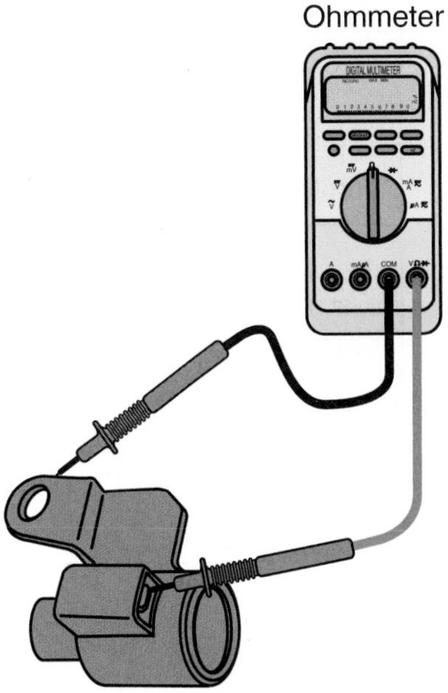

Figure 41-29 The meter connections for checking a solenoid for an open or short as well as resistance value.

CAUTION!

When servicing the transmission of a Honda hybrid, be careful of the high-voltage motor that is sandwiched between the engine and the transmission. Always disconnect the high voltage before working on or near the transmission.

CASE STUDY

A customer came into the shop with a 2003 Hyundai Elantra with an F4A42 EAT. A tow truck delivered the car because the transmission would not shift gears. The technician assigned to the car assumed the worse due to the fact that the car had very high mileage. He pulled the transmission and overhauled it. After installing an overhaul kit and inspecting all other transmission parts, the technician reassembled the transmission and installed it back in the car. The transmission still would not shift out of third gear.

The technician got serious and began to check the electronics involved with the transmission. He finally used a scan tool and retrieved DTC P1723, which indicated that the voltage from the EAT control relay was low. After a little more research in the service

Components	Pass thru pins	Resistance at 20°C	Resistance at 100°C	Resistance to ground (case)
1-2 shift solenoid valve	A, E	19–24Ω	24–31Ω	Greater than 250MΩ
2-3 shift solenoid valve	B, E	19–24Ω	24–31Ω	Greater than 250MΩ
TCC solenoid valve	T, E	21–24Ω	26–33Ω	Greater than 250MΩ
TCC PWM solenoid valve	U, E	10–11Ω	13–15Ω	Greater than 250MΩ
3-2 shift solenoid valve assembly	S, E	20–24Ω	29–32Ω	Greater than 250MΩ
Pressure control solenoid valve	C, D	3–5Ω	4–7Ω	Greater than 250MΩ
Transmission fluid temp. (TFT) sensor	M, L	3,088–3,942Ω	159–198Ω	Greater than 10MΩ
Vehicle speed sensor	A, B Vss conn	1,420Ω @ 25°C	2,140Ω @ 150°C	Greater than 10MΩ

IMPORTANT: The resistance of this device is necessarily temperature dependent and will therefore vary far more than that of any other device. Refer to transmission fluid temp (TFT) sensor specifications.

Figure 41-30 The appropriate service information lists the resistance values and test points for various transmission solenoids and sensors.

manual, he found that the transmission was in the fail-safe mode. With a voltmeter, he then checked the relay and found its output to the TCM to be normal. This certainly was not the problem. He then went back to the service information and followed the sequence given for that DTC. Checking the relay was step one. The additional checks were for the connectors and wiring harnesses. Finally, while checking the voltage from the control relay to the TCM, he found very low voltage. This was caused by heavy corrosion on the terminals of the computer and the wiring harness. The technician then cleaned up the area and replaced the computer. The transmission now shifted well. The lesson he learned was obvious—guessing can lead to much frustration and a big bunch of wasted time and money.

KEY TERMS

Adaptive learning
Electronic pressure control (EPC) solenoid
Shift schedule
Transmission control module (TCM)
Variable force solenoid (VFS)

SUMMARY

- A TCM is a separate computer designated for transmission operation or is part of the PCM. The PCM can be a multifunction computer that controls all engine and transmission operations.
- A shift schedule contains the actual shift points to be used by the computer according to the input data it receives from the sensors. Its logic chooses the proper shift schedule for the current conditions of the transmission.

- In most electronically controlled transmission systems, the hydraulically operated clutches are controlled by the transaxle controller.
- The typical output devices are solenoids and motors, which cause something mechanical or hydraulic to change.
- The computer may receive information from two different sources: directly from a sensor or through a twisted-pair bus circuit, which connects all of the vehicle computer systems.
- Fluid flow to the apply devices is directly controlled by the solenoids.
- Pressure switches give input to the transmission computer; they are all located within the solenoid assembly.
- Normally, the shift solenoids receive voltage through the ignition switch and are grounded through the PCM.
- The TCM is programmed to adjust its operating parameters in response to changes within the system, such as component wear. As component wear and shift overlap times increase, the TCM adjusts the line pressure to maintain proper shift timing calibrations. This is called adaptive learning.
- The pulse width modulated solenoid is a normally closed valve installed in the valve body. It controls the position of the TCC apply valve.
- If the TCM loses source voltage, the transmission will enter into limp-in mode. The transmission will also enter into default if the computer senses a transmission failure. While in the default mode, the transmission will operate only in park, neutral, reverse, and one forward gear. This allows the vehicle to be operated, although its efficiency and performance are hurt.
- The basic shift logic of the computer allows the releasing apply device to slip slightly during the engagement of the engaging apply device.
- The EPC solenoid provides changes in pressure in response to engine load.
- Toyota, Ford, and Nissan hybrid systems rely on electric motors to determine the overall gear ratio in their CVTs.
- The two-mode hybrid system fits into a standard automatic transmission housing and is comprised of two planetary gear sets coupled to two electric motor/generators. This arrangement allows for two distinct modes of hybrid drive operation.
- If a computer's input signals are correct and its output signals are incorrect, the computer must be replaced.
- Input devices are critical to the operation of an EAT and should be checked with a scan tool, DMM, or lab scope.
- Solenoid valves can be checked by measuring their resistance or by applying a current to it and listening and feeling for its movement. They can also be checked with a lab scope or DMM.

REVIEW QUESTIONS

1. Although computers receive different information from a variety of sensors, the decisions for shifting are actually based on more than the inputs. What are they based on?
2. What happens if the TCM determines that the signal from the TFT sensor are incorrect?
3. How can air pressure be used to check an electrical switch?
4. *True or False?* On most late-model EAT systems, separate procedures must be followed to retrieve the DTCs from the transmission control system and the TCC system.
5. Some transmissions receive information through multiplexing. How does this work?
6. List five things that could cause incorrect shift times and poor shifting quality.
7. Most late-model transmission control systems have adaptive learning. What does this mean?
8. What inputs are of prime importance to a computer in deciding when to shift gears?
9. What are the advantages of using electronic controls rather than relying on conventional hydraulic controls in a transmission?
10. Voltage generation devices are typically used to monitor ______ ______.
11. Which of the following is *not* a voltage-generating type sensor?
    a. vehicle speed sensor
    b. OSS
    c. MAP
    d. ISS
12. *True or False?* The neutral safety switch and reverse lamp switch are typically part of a digital transmission range sensor.
13. A glitch appears in the waveform of a vehicle speed sensor. Which of the following is *not* a probable cause of the problem?
    a. a loose connector
    b. a damaged wire

c. a poorly mounted sensor
d. a damaged magnet in the sensor

14. In a Toyota hybrid CVT, which of the following parts effectively controls the overall gear ratio of the transaxle?
    a. MG1
    b. MG2
    c. reduction unit
    d. transaxle damper
15. Which of the following would probably not be caused by a plugged filter screen at a solenoid?
    a. erratic shifting
    b. no shift conditions
    c. a solenoid that will not energize
    d. wrong gear starts

ASE-STYLE REVIEW QUESTIONS

1. Technician A says that some systems use a special modulated shift control solenoid. Technician B says that some systems use a special modulated converter clutch control solenoid. Who is correct?
   a. Technician A   c. Both A and B
   b. Technician B   d. Neither A nor B
2. Technician A says that shift solenoids direct fluid flow to and away from the apply devices in the transmission. Technician B says that shift solenoids are used to mechanically apply a friction brake or multiple-disc clutch assembly. Who is correct?
   a. Technician A   c. Both A and B
   b. Technician B   d. Neither A nor B
3. Technician A says that throttle position is an important input in most electronic shift control systems. Technician B says that vehicle speed is an important input for most electronic shift control systems. Who is correct?
   a. Technician A   c. Both A and B
   b. Technician B   d. Neither A nor B
4. Technician A says that a faulty TP sensor can cause delayed shifts. Technician B says that delayed shifts can be caused by an open shift solenoid. Who is correct?
   a. Technician A   c. Both A and B
   b. Technician B   d. Neither A nor B
5. Technician A says that some shift solenoids can be activated by providing a ground for the solenoid. Technician B says that some shift solenoids can be activated by applying hydraulic pressure to their valve. Who is correct?
   a. Technician A   c. Both A and B
   b. Technician B   d. Neither A nor B
6. An electronically controlled transmission has erratic shifting: Technician A says that a poor PCM ground will cause this problem. Technician B says that a bad AC generator-to-battery circuit will cause erratic performance of a transmission or transaxle. Who is correct?
   a. Technician A   c. Both A and B
   b. Technician B   d. Neither A nor B
7. While discussing transmission adaptive learning: Technician A says that the transmission learns to respond according to the current condition of the engine and transmission. Technician B says that the TCM learns to respond to the driver's driving habits. Who is correct?
   a. Technician A   c. Both A and B
   b. Technician B   d. Neither A nor B
8. While discussing valve body assemblies in late-model transmissions: Technician A says that the valve body is no longer needed in some electronically controlled transmissions. Technician B says that an EPC solenoid maintains constant fluid pressure through the valve body regardless of the vehicle's operating condition. Who is correct?
   a. Technician A   c. Both A and B
   b. Technician B   d. Neither A nor B
9. A hot-wire MAF has low resistance across it when the wire is hot: Technician A says that this condition will cause the engine to stall as soon as you put the transmission into gear. Technician B says that this condition will cause the transmission to shift late. Who is correct?
   a. Technician A   c. Both A and B
   b. Technician B   d. Neither A nor B
10. While diagnosing the cause of TCC shudder: Technician A checks the fluid pressure sent to the clutch. Technician B says that because the shudder occurs only after the clutch is engaged, the problem is the engine, transmission, or another component of the driveline. Who is correct?
    a. Technician A   c. Both A and B
    b. Technician B   d. Neither A nor B

AUTOMATIC TRANSMISSION AND TRANSAXLE SERVICE

CHAPTER 42

OBJECTIVES

- Listen to the driver's complaint, road test the vehicle, and then verify the repair.
- Diagnose unusual fluid usage, level, and condition problems.
- Replace automatic transmission fluid and filters.
- Diagnose noise and vibration problems.
- Diagnose hydraulic and vacuum control systems.
- Perform oil pressure tests and determine needed repairs.
- Inspect and adjust external linkages.
- Describe the basic steps for overhauling a transmission.

Because of the many similarities between a transmission and a transaxle, most diagnostic and service procedures are similar. Therefore, all references to a transmission apply equally to a transaxle unless otherwise noted. Whenever you are diagnosing or repairing a transaxle or transmission, make sure you refer first to the appropriate service manual before you begin.

Transmissions are strong and typically trouble-free units that require little maintenance. Normal maintenance usually includes fluid checks, scheduled linkage adjustments, and oil and filter changes.

IDENTIFICATION

Always make sure you know exactly which transmission you are working on. This ensures you are following the correct procedures and specifications and are installing the correct parts. Proper identification can be difficult because transmissions cannot be accurately identified by the way they look. The only exception to this is the shape of the oil pan, which can sometimes be used for identification. However, this is not foolproof.

The only positive way to identify a transmission is by its identification numbers. **Transmission identification numbers** are found on stickers on the transmission **(Figure 42–1)**, stamped numbers in the case, or on a metal tag held by a bolt. Also, all vehicles have a vehicle certification label that gives transmission information in a space marked TR. Use a service manual to decipher the identification number. Most identification numbers include the model, manufacturer, and build date. Most late-model transmissions have labels with bar codes that can be scanned for transmission identification.

BASIC DIAGNOSTICS

Automatic transmission problems are commonly caused by poor engine performance, problems in the hydraulic system, abuse resulting in overheating, mechanical malfunctions, electronic failures, and/or improper adjustments. Diagnosis of transmission problems should begin with checking the condition and level of the fluid, conducting a thorough visual inspection, checking the various linkage adjustments, retrieving all DTCs, and checking basic engine operation.

Engine performance can affect torque converter clutch operation. If the engine is running too poorly to maintain a constant speed, the converter clutch will engage and disengage at higher speeds. The customer complaint may be that the vehicle vibrates.

If the vehicle has an engine performance problem, the cause should be found and corrected before any conclusions on the transmission are made. A quick way to identify if the engine is causing shifting problems is to connect a vacuum gauge to the engine and take a reading while the engine is running. The gauge should be connected to intake manifold vacuum. A normal vacuum gauge reading is steady and at 17 in. Hg (431.8 mm Hg). The rougher the engine runs, the more the gauge readings will fluctuate.

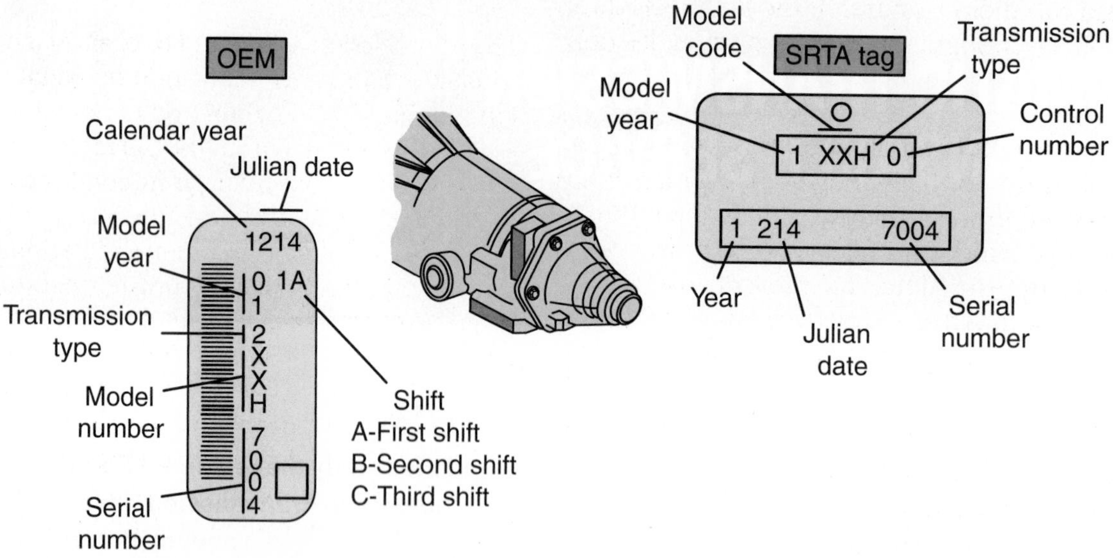

Figure 42-1 The location of and information contained on a GM transaxle ID plate.

In order to properly diagnose a problem, you must totally understand the customer's concern or complaint. Make sure you know the conditions that exist when the problem occurs.

Fluid Check

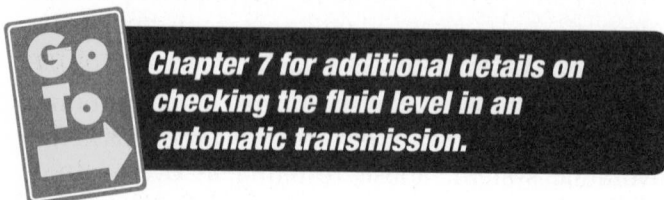

Go To *Chapter 7 for additional details on checking the fluid level in an automatic transmission.*

When checking the fluid level, make sure the vehicle is on a level surface. Check the level and condition of the fluid. If the transmission has a dipstick, wipe all dirt off the protective disc and the dipstick handle. On most automobiles, the ATF level can be checked accurately only when the transmission is at operating temperature, the transmission is in a specific gear, and the engine is running. Remove the dipstick and wipe it clean with a lint-free white cloth or paper towel. Reinsert the dipstick, remove it again, and note the reading. Markings on a dipstick indicate ADD levels and on some models indicate FULL levels for cool, warm, or hot fluid.

Low fluid levels can cause a variety of problems. Air can be drawn into the oil pump's inlet circuit and mix with the fluid. This will result in aerated fluid, which causes slow pressure buildup, and low pressures, which cause slippage between shifts. Air in the pressure regulator valve will cause a buzzing noise when the valve tries to regulate pump pressure. If the fluid level is low, the problem could be external fluid leaks. Check the transmission case, oil pan, and cooler lines for signs of leaks.

Excessively high fluid levels can also cause **aeration**. As the planetary gears rotate in high fluid levels, air can be forced into the fluid. Aerated fluid can foam, overheat, and oxidize. All of these problems can interfere with normal valve, clutch, and servo operation. Foaming may be evident by fluid leakage from the transmission's vent.

Examine the fluid carefully. The normal color of ATF is red. If the fluid has a dark brownish or blackish color and/or a burned odor, the fluid has been overheated. A milky color indicates that engine coolant has been leaking into the transmission's cooler in the radiator. If there is any question about the condition of the fluid, drain out a sample for closer inspection.

Synthetic ATF is normally a darker red than petroleum-based fluid. Synthetic fluids tend to look and smell burnt after normal use; therefore, the appearance and smell of these fluids is not a good indicator of the fluid's condition.

After checking the ATF level and color, wipe the dipstick on absorbent white paper and look at the stain left by the fluid. Dark particles are normally band and/or clutch material, whereas silvery metal particles are normally caused by the wearing of the transmission's metal parts. If the dipstick cannot be wiped clean, it is probably covered with varnish, which results from fluid oxidation. Varnish will cause the spool valves to stick, causing shifting malfunction. Varnish or other heavy deposits indicate the need to change the transmission's fluid and filter.

Contaminated fluid can sometimes be felt better than be seen. Place a few drops of fluid between two fingers and rub them together. If the fluid feels dirty or gritty, it is contaminated with burned friction material.

Transmissions without a Dipstick Many late-model transmissions do not have a dipstick, and the fluid level is checked in the same way as a manual transmission. The dipstick and filler tube were removed from these transmissions to prevent overfilling. Research has found that many transmission failures were caused by overfilling and/or using the wrong fluid. Without a dipstick, it is difficult to check the fluid level and condition. These transmissions have a vent/fill cap typically located on the side of the transmission. Some also have a drain plug in the bottom of the pan. In addition, these transmissions are fitted with a fluid level sensor that will inform the driver when the fluid level is dangerously low.

To check the fluid level, the transmission must be warm and the vehicle must be on a level surface or raised on a lift. It is important that all four wheels are raised, because just lifting the front will put the vehicle on a slant and you will be unable to get an accurate reading. Find the fill plug and remove it. A small amount of fluid should leak out of the vent/fill opening if the fluid is at the correct level. If fluid does not come out, add fluid until it does.

Some vehicles and the types of transmissions without a dipstick include:

- 2004 + Acura TL, RL, RSX w/5-speed
- 2004–05 Cadillac Catera w/5L40/5L50E
- 2004–05 Chevrolet Aveo w/Aisin 81-40LE
- 1997 + Chevrolet Cavalier, Cobalt w/4T40/45E
- 2005 + Chevrolet Equinox w/AF33
- 2005 + Chrysler 300 3.5L w/42RLE or NAG-1
- 2005 + Ford cars and light trucks w/5R55N/S/W
- 2005 + Lincoln Navigator 5.4L w/ZF-6SHP-26
- 2004 Mazda MPV w/5F31J
- 2004 Mazda Miata w/N4AEL
- 2004–05 Saturn Ion w/AF23
- 2004–05 Saturn Vue w/4 or 5-speed

SHOP TALK

Make sure you always install the correct fluid in a transmission. Often the manufacturer recommends a special fluid for special applications. This is especially true of CVTs. Not only can the wrong fluid result in poor shifting; it can damage the transmission.

Fluid Changes

The transmission's fluid and filter should be changed whenever there is an indication of oxidation or contamination. Periodic fluid and filter changes, as well as fluid flushes are part of the preventive program for most vehicles. The mileage interval recommended depends on the type of transmission.

Change the fluid only when the engine and transmission are at normal operating temperatures. Photo Sequence 42 shows a typical procedure for changing a transmission's fluid and filter. On most transmissions, you must remove the oil pan to drain the fluid. A filter or screen is normally attached to the bottom of the valve body. Filters are made of paper or fabric and held in place by screws, clips, or bolts, or by the pressure on the pickup tube seal **(Figure 42–2)**. Filters should be replaced, not cleaned.

Some late-model Saturn and a few other transmissions do not have a filter in the pan. Rather, they are fitted with a spin-on filter **(Figure 42–3)**. This filter looks like an engine filter. They are serviced in the same way as an engine oil filter. Other transmissions may have an inline filter in the fluid lines leading to the cooler **(Figure 42–4)**.

Check the bottom of the pan for deposits and metal particles. Slight contamination, such as blackish deposits from clutches and bands, is normal. Other contaminants should be of concern. Clean the oil pan and its magnet **(Figure 42–5)**.

Remove the filter and inspect it. Use a magnet to determine if metal particles are steel or aluminum. Steel particles will be attracted to the magnet and indicate severe internal transmission wear or damage. If the metal particles are aluminum, they may be part of the torque converter stator. Some torque converters use phenolic plastic stators; therefore, aluminum particles found in these transmissions must be from the transmission itself.

Remove any traces of the old pan gasket on the case and oil pan. Then install a new filter and gasket and tighten the retaining bolts to the specified torque. If the filter is sealed with an O-ring, make sure it is properly installed. Reinstall the pan using the gasket or sealant recommended by the manufacturer. Tighten the bolts to the specified torque. The required torque is often given in inch-pounds. You can easily break the bolts or damage something if you tighten the bolts to foot-pounds.

Pour a little less than the required amount of fluid into the transmission through the filler tube or fill point. Start the engine and allow it to idle for at least 1 minute. Then, with the parking and service brakes applied, move the gear selector lever momentarily to each position, ending in the park. Recheck the fluid

PHOTO SEQUENCE 42

Changing Automatic Transmission Fluid and Filter

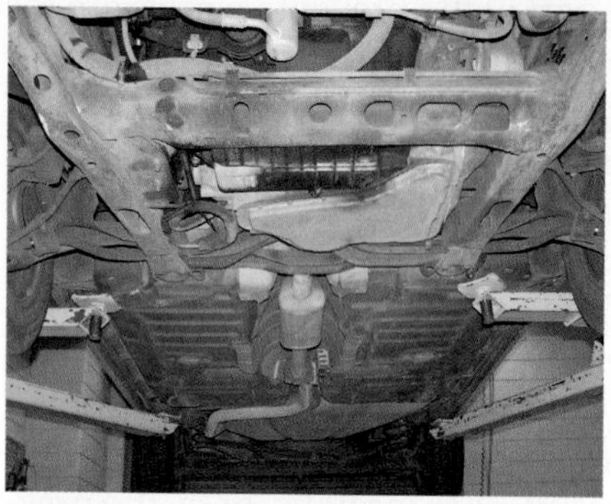

P42-1 *Raise the vehicle to a good working height and safely positioned on the lift.*

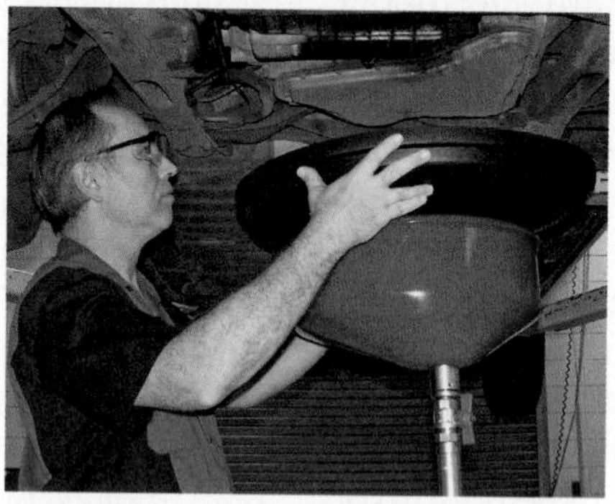

P42-2 *Place a large-diameter oil drain pan under the transmission pan.*

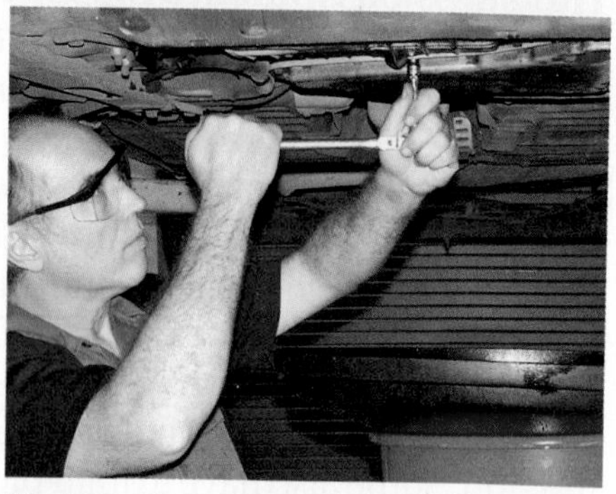

P42-3 *Loosen all of the pan bolts except three at one end. This will allow some fluid to drain out.*

P42-4 *Support the pan with one hand and remove the remaining bolts to remove the pan. Pour the fluid in the pan into the drain pan.*

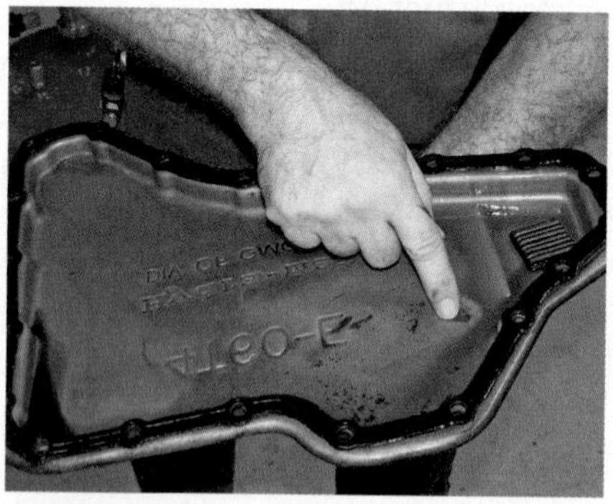

P42-5 *Inspect the residue in the pan for indications of transmission problems. Then remove the old pan gasket and wipe the pan clean with a lint-free rag.*

P42-6 *Unbolt the filter from the transmission.*

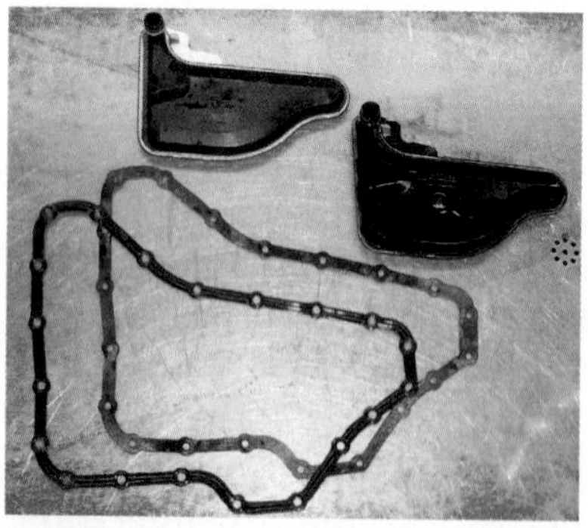

P42-7 *Compare the replacement gasket and filter with the old ones to make sure the replacements are the right ones for this application.*

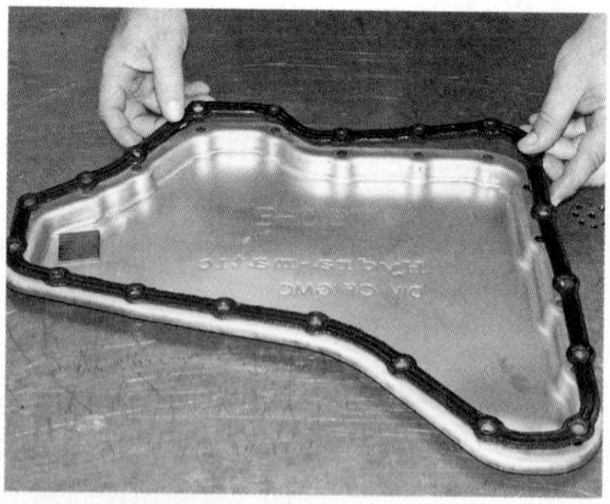

P42-8 *Install the new filter and tighten the attaching bolts to specifications. Then lay the new gasket over the sealing surface of the pan.*

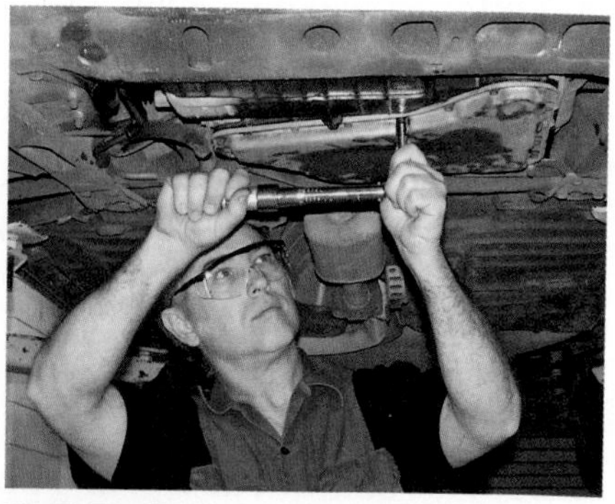

P42-9 *Install the pan onto the transmission. Install and tighten the bolts to specifications. Then lower the vehicle and pour new fluid into the transmission. Run the engine to circulate the new fluid, then turn it off and raise the vehicle. Check for fluid leaks.*

Figure 42-2 Transmission fluid filters are attached to the transmission case by screws, bolts, retaining clips, and/or by the pickup tube.

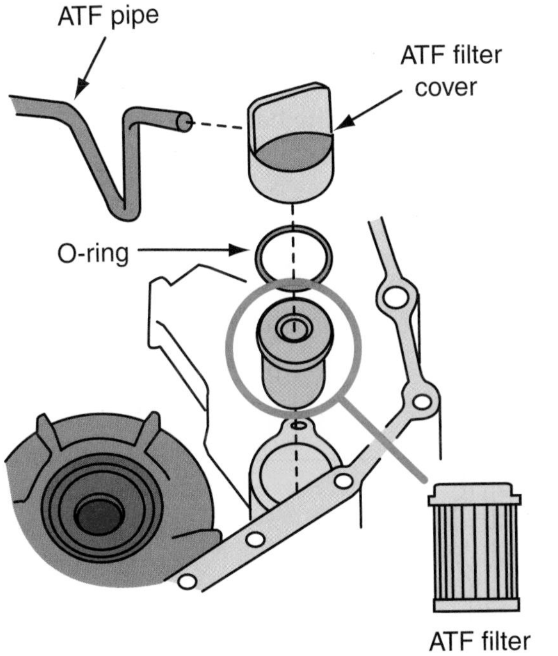

Figure 42-3 An engine oil filter-style transmission filter.

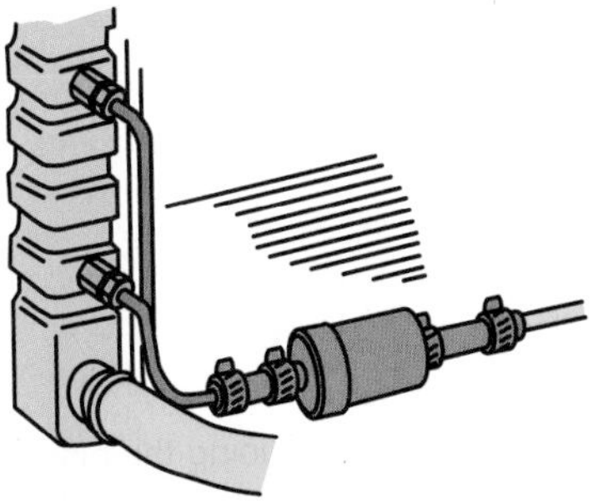

Figure 42-4 An inline filter for a fluid cooler.

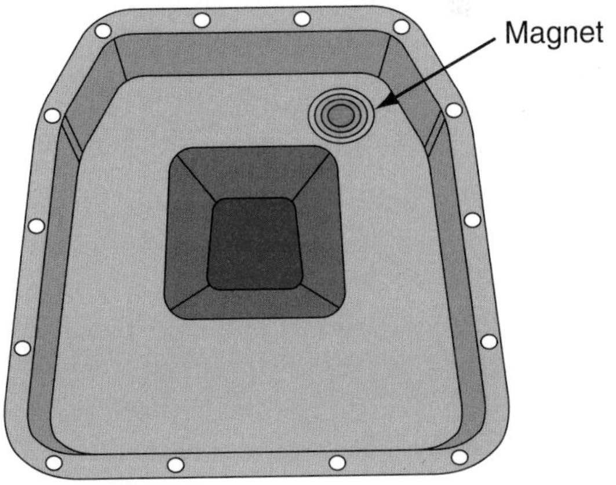

Figure 42-5 The magnet in the oil pan should be cleaned and the material it gathered should be analyzed.

level and add a sufficient amount of fluid to bring the level to about ⅛ inch below the FULL or MAX mark.

SHOP TALK

Always check the service information for the correct conditions for checking fluid level. Although most manufacturers recommend checking it with the gear selector in the park position, some must be checked with the selector in neutral. In those transmissions, the fluid does not flow to the cooler when the transmission is in park.

Run the engine until it reaches normal operating temperature. Then recheck the fluid level and correct it if necessary. Make sure the dipstick is fully seated into the dipstick tube opening. This will prevent dirt from entering the transmission.

Parking Pawl

Any time you have the oil pan off, you should inspect all of the exposed parts, especially the parking pawl assembly. This component is typically not hydraulically activated; rather, the gearshift linkage moves the pawl into position to lock the output shaft of the transmission. Unless the customer's complaint indicates a problem with the parking mechanism, no test will detect a problem here.

Check the pawl assembly for excessive wear and other damage. Also, check to see how firmly the pawl is in place when the gear selector is shifted into the PARK mode. If the pawl can be moved out easily, it should be repaired or replaced.

FLUID LEAKS

Continue your diagnostics with a quick and careful visual inspection. Check all drivetrain parts for looseness and leaks. If the transmission fluid was low or there was no fluid, raise the vehicle and carefully inspect the transmission for signs of leakage. Leaks are often caused by defective gaskets or seals. Common sources of leaks are the oil pan seal, rear cover and final drive cover (on transaxles), extension housing, speedometer gear assembly, and electrical switches mounted into the housing **(Figure 42–6)**. The housing may have a porosity problem, allowing fluid to seep through the metal. Case porosity may be repaired with an epoxy sealer.

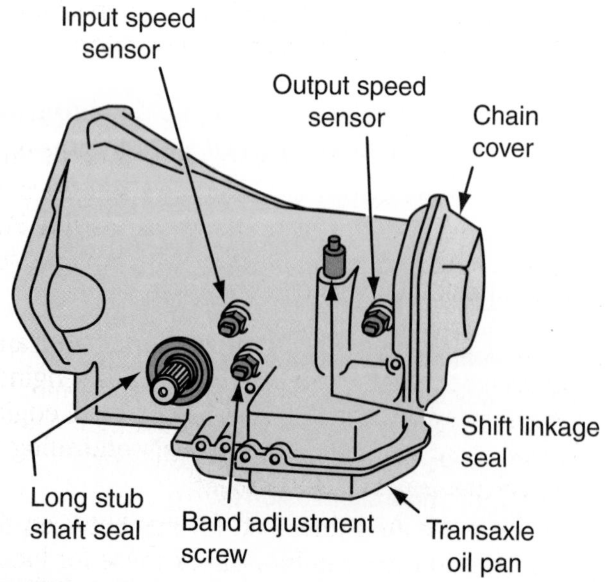

Figure 42-6 Possible sources of fluid leaks on this transaxle.

Oil Pan

A common source of fluid leakage is between the oil pan and the transmission housing. If fluid is present around the rim of the pan, retorquing the pan bolts may correct the problem. If it does not correct the problem, the pan must be removed and a new gasket installed. Make sure the sealing surface of the pan is flat and capable of providing a seal before reinstalling it.

Torque Converter

Torque converter problems can be caused by a leaking converter **(Figure 42–7)**. This type of problem may be the cause of customer complaints of slippage and a lack of power. To check the converter for leaks, remove the converter access cover and examine the area around the torque converter shell. An engine oil leak may be falsely diagnosed as a converter leak. The color of engine oil is different from that of transmission fluid and may help identify the true source of the leak. However, if the oil or fluid has absorbed much dirt, both will look the same. An engine leak typically leaves an oil film on the front of the converter shell, whereas a converter leak will cause the entire shell to be wet. If the transmission's oil pump seal is leaking, only the back side of the shell will be wet. If the converter is leaking or damaged, it should be replaced.

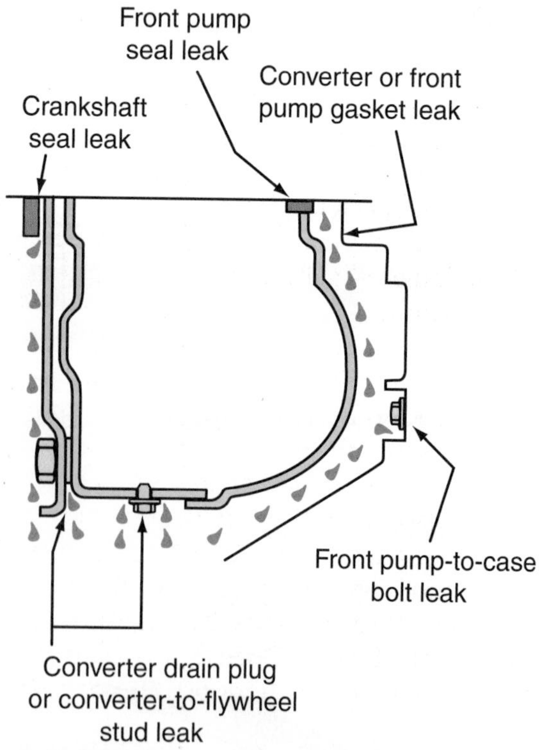

Figure 42-7 By determining the direction of fluid travel, the cause of a fluid leak around the torque converter can be identified.

Extension Housing

An oil leak stemming from the mating surfaces of the extension housing and the transmission case may be caused by loose bolts. To correct this problem, tighten the bolts to the specified torque. Also, check for signs of leakage at the rear of the extension housing. Fluid leaks from the seal of the extension housing can be corrected with the transmission in the car. Often, the cause for the leakage is a worn extension housing bushing, which supports the sliding yoke of the drive shaft. When the drive shaft is installed, the clearance between the sliding yoke and the bushing should be minimal. If the clearance is satisfactory, a new oil seal will correct the leak. If the clearance is excessive, the repair requires that a new seal and a new bushing be installed. If the seal is faulty, the transmission vent should be checked for blockage.

Speedometer Drive

The vehicle's speedometer can be purely electronic, which requires no mechanical hookup to the transmission, or it can be driven, via a cable, off the output shaft. If the transmission is equipped with a VSS, the bore and sensor can be a source of leaks. The sensor is retained in the bore with a retaining nut or bolt. An oil leak at the speedometer cable or VSS can be corrected by replacing the O-ring seal. While replacing the seal, inspect the drive gear for chips and missing teeth. Always lubricate the O-ring and gear prior to installation.

Electrical Connections

Check all electrical connections to the transmission. Faulty connectors or wires can cause harsh or delayed and missed shifts. On transaxles, the connectors can normally be inspected through the engine compartment, whereas they can only be seen from under the vehicle on longitudinally mounted transmissions. To check the connectors, release the locking tabs and disconnect them, one at a time, from the transmission. Carefully examine them for signs of corrosion, distortion, moisture, and transmission fluid. Carefully check the weather seals on all connectors. A connector or wiring harness may deteriorate if ATF reaches it. Also, check the connector at the transmission. Using a small mirror and flashlight may help you get a good look at the inside of the connectors. Inspect the entire transmission wiring harness for tears and other damage. Road debris can damage the wiring and connectors mounted underneath the vehicle.

Because the operation of the engine and transmission are integrated through the control computer, a faulty engine sensor or connector may affect the operation of the engine and the transmission. The various sensors and their locations can be identified by referring to the appropriate service manual. The engine control sensors that are the most likely to cause shifting problems are the TP, MAP, and VSS.

Checking Transmission and Transaxle Mounts

The engine and transmission mounts on FWD cars are important to the operation of the transaxle. Any engine movement may change the effective length of the shift and throttle cables or wiring harnesses and therefore may affect the engagement of the gears. Delayed or missed shifts may result from linkage changes as the engine pivots on its mounts.

Visually inspect the mounts for looseness and cracks. With a prybar, pull up and push down on the transaxle case while watching the mount. If the mount's rubber separates from the metal plate or if the case moves up but not down, replace the mount. If there is movement between the metal plate and its attaching point on the frame, tighten the attaching bolts.

Then, from the driver's seat, apply the foot brake, set the parking brake, and start the engine. Put the transmission into a forward gear and gradually increase the engine speed to about 1,500 to 2,000 rpm. Watch the torque reaction of the engine on its mounts. Repeat the check in reverse. If the engine's reaction to the torque appears to be excessive, broken or worn drivetrain mounts may be the cause.

If it is necessary to replace the transaxle mount, make sure you follow the manufacturer's recommendations for maintaining the alignment of the driveline. Failure to do so may result in poor gear shifting, vibrations, and/or broken cables. Some manufacturers recommend that a holding fixture or special bolt be used to keep the unit in its proper location.

Transmission Cooler and Line Inspection

Transmission coolers are a possible source of fluid leaks. The efficiency of the coolers is also critical to the operation and longevity of the transmission. Follow these steps when inspecting the transmission cooler and associated lines and fittings:

1. Check the engine's cooling system. The transmission cooler cannot be efficient if the engine's cooling system is defective. Repair all engine cooling system problems before continuing to check the transmission cooler.
2. Inspect the fluid lines and fittings between the cooler and transmission. Check these for looseness, damage, signs of leakage, and wear. Replace any damaged lines and fittings.

3. Inspect the engine's coolant for traces of transmission fluid. If ATF is present in the coolant, the transmission cooler leaks.
4. Check the transmission's fluid for signs of engine coolant. Water or coolant will cause the fluid to appear milky with a pink tint. This milky appearance is also an indication that the transmission cooler leaks and is allowing engine coolant to enter into the transmission fluid.

The cooler can be checked for leaks by disconnecting and plugging the transmission to cooler lines at the radiator. Then remove the radiator cap to relieve any pressure in the system. Tightly plug one of the ATF line fittings at the radiator. Using the shop air supply with a pressure regulator, apply 50 to 70 psi (3.52 to 4.92 kg/cm^2) of air pressure into the cooler at the other cooler line fitting. Look into the radiator. If bubbles are observed, the cooler leaks.

ROAD TESTING THE VEHICLE

Critical to proper diagnosis of an automatic transmission is a road test. If the vehicle has an EAT, connect a scan tool with recording capabilities, if possible, before taking the drive. Before beginning your road test, find and duplicate, from a service manual, the chart **(Figure 42–8)** that shows the band and clutch application for different gear selector positions. Using these charts will greatly simplify your diagnosis of automatic transmission problems. It is also wise to have a notebook or piece of paper to jot down notes about the operation of the transmission.

Begin the road test with a drive at normal speeds to warm the engine and transmission. If a problem appears only when starting and/or when the engine and transmission are cold, record this symptom on the chart or in your notebook.

During the road test, the transmission should be operated in all possible modes and its operation noted. Check for proper gear engagement as the selector lever is moved to each gear position, including park. There should be no hesitation or roughness as the gears are engaging. Check for proper operation in all forward ranges, especially the upshifts and converter clutch engagement during light throttle operation. These shifts should be smooth and positive and occur at the correct speeds. These same shifts should feel firmer under medium to heavy throttle pressures. Transmissions equipped with a torque converter clutch should be brought to the specified apply speed and their engagement noted. Again, record the operation of the transmission in these different modes in your notebook or on the diagnostic chart. Also, the various computer inputs should be monitored and the readings recorded.

Force the transmission to kick down and pay attention to the quality of this shift. Manual downshifts should also be made at a variety of speeds. The reaction of the transmission should be noted, as

Range		Gear ratio	Clutch		Low and reverse brake	Lockup	Band servo		One-way clutch	Parking pawl
			High-reverse clutch (front)	Forward clutch (rear)			Operation	Release		
Park										on
Reverse		2.364	on		on					
Neutral										
Drive	D_1 Low	2.826		on					on	
	D_2 Second	1.543		on			on			
	D_3 Top (3rd)	1.000	on	on		on	(on)	on		
2	2_1 Low	2.856		on					on	
	2_2 Second	1.543		on			on			
1	1_1 Low	2.826		on	on				on	
	1_2 Second	1.543		on			on			

Figure 42-8 A typical band and clutch application chart. This chart should be referred to during a road test and when determining the cause of any shifting problems. *Copyright, Nissan (2008). Nissan and the Nissan logo are registered trademarks of Nissan.*

should all abnormal noises, and the gears and speeds at which they occur.

All transmission complaints should be verified by road testing the vehicle and attempting to duplicate the customer's concern. Knowing the exact conditions that cause the symptom will allow you to accurately diagnose problems. Diagnosis becomes easy if you think about what is happening in the transmission when the problem occurs. If there is a shifting problem, think about the parts that are being engaged and disengaged.

SHOP TALK

Always refer to the service information to identify the particulars of the transmission you are diagnosing. Also check for any technical service bulletins that may be related to the customer's complaint.

After the road test, check the transmission for signs of leakage. Any new leaks and their probable cause should be noted. Then compare your written notes from the road test to the information given in the service information to identify the cause of the malfunction. This information usually has a diagnostic chart to aid you in this process.

Diagnosis of Noise and Vibration Problems

Many noise and vibration problems that appear to be transmission related may be caused by problems in the engine, drive shaft, U- or CV joints, wheel bearings, wheel/tire imbalance, or other conditions. Problems in those areas can lead customers and, unfortunately, some technicians, to mistakenly suspect that the problems are caused by the transmission or torque converter. The entire driveline should be checked before assuming that the noise is transmission related.

Abnormal transmission noises and vibrations can be caused by faulty bearings, damaged gears, worn or damaged clutches and bands, or a bad oil pump as well as contaminated fluid or improper fluid levels.

Most vibration problems are caused by an unbalanced torque converter assembly, the loosely mounted to the flexplate, a loose or cracked flexplate, torque converter, or a faulty output shaft. The key to determining the cause of the vibration is to pay particular attention to the vibration in relationship to engine and vehicle speed. If the vibration changes with a change in engine speed, the cause of the problem is most probably the torque converter. If the vibration changes with vehicle speed, the cause is probably the output shaft or the driveline connected to it. The latter can be a bad extension housing bushing or universal joint, which would become worse at higher speeds.

To determine if the problem is caused by the transmission or the driveline, put the transmission in gear and apply the foot brakes. If the noise is no longer evident, the problem can be in the driveline or the output of the transmission. If the noise is still present, the problem must be in the transmission or torque converter.

Noise problems are also best diagnosed by paying a great deal of attention to the speed and the conditions at which the noise occurs. If the noise is engine speed related and is present in all gears, including park and neutral, the oil pump is the most probable source because it rotates whenever the engine is running. However, if the noise is engine speed related and is present in all gears except park and neutral, the most probable sources of the noise are those parts that rotate in all gears, such as the drive chain, the input shaft, and torque converter.

Noises that only occur when a particular gear is operating must be related to those parts responsible for providing that gear, such as a brake or clutch. Often the exact cause of noise and vibration problems can only be identified through a careful inspection of a disassembled transmission.

CHECKING THE TORQUE CONVERTER

Many transmission problems are related to the operation of the torque converter. Normally, torque converter problems will cause abnormal noises, poor acceleration in all gears, normal acceleration but poor high-speed performance, or transmission overheating.

If the vehicle lacks power during acceleration, it has a restricted exhaust or the torque converter's one-way stator clutch is slipping. To determine which of these problems is causing the power loss, test for a restricted exhaust first. Other possible causes of this problem include a restricted air or fuel filter and a defective fuel pump.

If there is no evidence of a restricted exhaust, it can be assumed that the torque converter's stator clutch is slipping and not allowing any torque multiplication to take place in the converter. To correct this problem, the torque converter should be replaced.

If the engine's speed flares up during acceleration in drive and does not have normal acceleration, the clutches or bands in the transmission are slipping.

This symptom is similar to the slipping of a clutch in a manual transmission. Often this problem is mistakenly blamed on the torque converter.

Complaints of thumping or grinding noises are often thought to be caused by the torque converter (T/C), when they are actually caused by bad thrust washers or damaged gears and bearings in the transmission. Nontransmission problems, such as bad CV joints and wheel bearings, also cause these noises.

A common TCC problem is the presence of a shudder when it is being engaged or disengaged. This shudder also can be caused by an out-of-round torque converter and contaminated clutch frictional material. An out-of-round converter prevents full clutch engagement, which is a common cause of clutch shudder.

Engine-related problems can also cause the clutch to engage early or not at all. Vacuum leaks and bad electrical sensors can prevent the TCC from engaging at the correct time.

Go To Chapter 41 for more information on checking a TCC.

Stall Test

To test the torque converter, many technicians perform a stall test. The stall test checks the holding capacity of the converter's stator overrunning clutch assembly as well as the clutches and bands in the transmission. Some manufacturers do not recommend the stall test. Rather, diagnosis should be based on the symptoms of the problem.

SHOP TALK

Stall testing places extreme stress on the transmission and should only be conducted if recommended by the manufacturer.

CAUTION!

If a stall test is not correctly conducted, the converter and/or transmission can be damaged.

To conduct a stall test, connect a tachometer to the engine and position it so that it can be easily read from the driver's seat. Set the parking brake, raise the hood, and place blocks in front of the vehicle's nondriving tires. Conduct the test outdoors, if possible, especially if it is a cold day. If the test is conducted indoors, place a large fan in front of the vehicle to keep the engine cool. With the engine running, press and hold the brake pedal. Then move the gear selector to the drive position and press the throttle pedal to the floor. Hold the throttle down for 2 seconds then note the tachometer reading and immediately let off the throttle pedal and allow the engine to idle. Compare the measured stall speed to specifications.

WARNING!

Make sure no one is around the engine or the front of the vehicle while a stall test is being conducted. A lot of stress is put on the engine, transmission, and brakes during the test. If something lets go, somebody can get seriously hurt.

CAUTION!

To prevent serious damage to the transmission, follow these guidelines while conducting a stall test:

1. *Never conduct a stall test if there is an engine problem.*
2. *Check the fluid levels in the engine and transmission before conducting the test.*
3. *Ensure that the engine is at normal operating temperature during the test.*
4. *Never hold the throttle wide open for more than 5 seconds during the test.*
5. *Do not perform the test in more than two gear ranges without driving the vehicle a few miles to allow the engine and transmission to cool down.*
6. *After the test, allow the engine to idle for a few minutes to cool the transmission fluid before shutting off the ignition.*

The engine will reach a specific speed if the torque converter and transmission are functioning properly. If the tachometer indicates a speed above or below specifications, a possible problem exists in the transmission or torque converter. If a torque converter is suspected as being faulty, it should be removed and the one-way clutch checked on the bench.

A restricted exhaust or slipping stator clutch is indicated if the stall speed is below the specifications. If the stator's one-way clutch is not holding, ATF leaving the turbine of the converter works against the rotation of the impeller and slows down the engine. Both of these problems cause poor acceleration. If the stall speed is only slightly below normal, the engine is probably not producing enough power and should be diagnosed and repaired.

If the vehicle has poor acceleration but had good results from the stall test, suspect a seized one-way clutch. Excessively hot ATF in the transmission is a good indication that the clutch is seized. However, other problems can cause these same symptoms; therefore, be careful during your diagnosis.

If the stall speed is above specifications, the bands or clutches in the transmission may be slipping and not holding properly.

A stall test will generate a lot of noise, most of which is normal. If you hear any metallic noises during the test, however, diagnose the source of these noises. Operate the vehicle at low speeds on a hoist with the drive wheels free to rotate. If the noises are still present, the source of the noises is probably the torque converter.

TC-Related Cooler Problems

Vehicles equipped with a converter clutch may stall when the transmission is shifted into reverse gear. The cause of this problem may be plugged transmission cooler lines, or the cooler itself may be plugged. Fluid normally flows from the torque converter through the transmission cooler. If the cooler passages are blocked, fluid is unable to exhaust from the torque converter and the converter clutch piston remains engaged. When the clutch is engaged, there is no vortex flow in the converter and therefore little torque multiplication is taking place in the converter.

To verify that the transmission cooler is plugged, disconnect the cooler return line from the radiator or cooler **(Figure 42–9)**. Connect a short piece of hose to the outlet of the cooler and allow the other end of the hose to rest inside an empty container. Start the engine and measure the amount of fluid that flows into the container after twenty seconds. Normally, 1 quart of fluid should flow into the container. If less than that filled the container, a plugged cooler is indicated.

To correct a plugged transmission cooler, disconnect the cooler lines at the transmission and the radiator. Blow air through the cooler, one end at a time, then through the cooler lines. The air will clear large pieces of debris from the transmission cooler. Always use low air pressure, no more than 50 psi (3.5 kg/cm^2). Higher pressures may damage the cooler. If there is little airflow through the cooler, the radiator or external cooler must be removed and flushed or replaced.

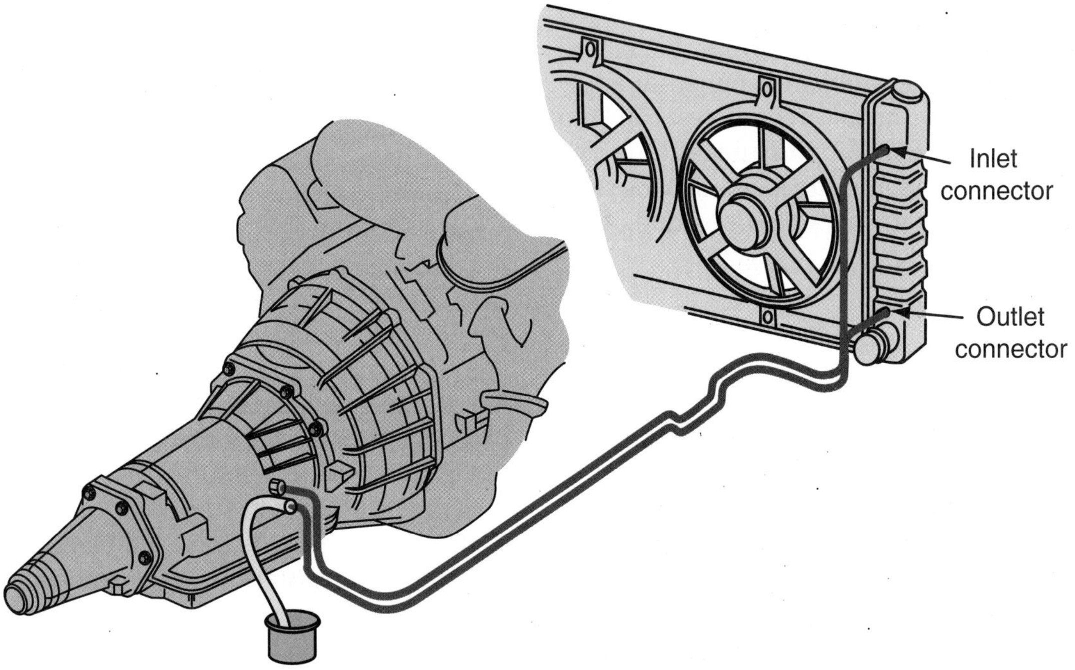

Figure 42-9 To check for a plugged cooler, disconnect the return line at the transmission and measure the flow of fluid out of the cooler.

DIAGNOSING HYDRAULIC AND VACUUM CONTROL SYSTEMS

The best way to identify the exact cause of the problem is to use the results of the road test, logic, and the oil circuit charts for the transmission being worked on. Before doing this, however, you should always check all sources for information about the symptom first. Also, make sure you check the basics: trouble codes in the computer, fluid level and condition, leaks, and mechanical and electrical connections. Using the oil circuits, you can trace problems to specific valves, servos, clutches, and bands.

The basic oil flow is the same for all transmissions. The oil pump creates the fluid flow used throughout the transmission. Fluid from the pump always goes to the pressure-regulating valve and torque converter. From there, the fluid is directed to the manual shift valve. When the gear selector is moved, the manual valve direct's the fluid to other valves and to the apply devices. By following the flow of the fluid on the oil circuit chart, you can identify which valves and apply devices should be operating in each particular gear selector position. Through a process of elimination, you can identify the most probable cause of the problem.

In most cases, the transmission or transaxle is removed to repair or replace the items causing the problem. However, some transmissions allow for a limited amount of service to the apply devices and control valves.

Mechanical and/or vacuum controls can also contribute to shifting problems. The condition and adjustment of the various linkages and cables should be checked whenever there is a shifting problem. If all checks indicate that the problem is either an apply device or in the valving, an air pressure test can help identify the exact problem. Air pressure tests are also performed during disassembly to locate leaking seals and during reassembly to check the operation of the clutches and servos.

Pressure Tests

If you cannot identify the cause of a transmission problem from your inspection or road test, a pressure test should be conducted. This test measures the fluid pressure of the different transmission circuits during the various operating gears and gear selector positions (**Figure 42–10**). The number of hydraulic circuits that can be tested varies with the different makes and models of transmissions. However, most transmissions are equipped with pressure taps, which allow the pressure test equipment to be connected to the transmission's hydraulic circuits (**Figure 42–11**).

Before conducting a pressure test on an electronic automatic transmission, check and correct all trouble codes retrieved from the system. Also make sure the transmission fluid level and condition is okay and that the shift linkage is in good order and properly adjusted.

The test is best conducted with three pressure gauges, but two will work. Two of the gauges should read up to 400 psi (28 kg/cm^2) and the other to 100 psi (7 kg/cm^2). The two 400 psi (28 kg/cm^2) gauges are usually used to check mainline and an individual circuit, such as mainline and direct or forward circuits. If a circuit is 15 psi (1 kg/cm^2) lower than mainline pressure when they are both tested at exactly the same time, a leak is indicated. A 100 psi (7 kg/cm^2) gauge may be used be used on TV and governor circuits.

The pressure gauges are connected to the pressure taps in the transmission housing and routed so that the gauges can be seen by the driver. The vehicle is then road tested and the gauge readings observed during the following operational modes: slow idle, fast idle, and WOT.

During the road test, observe the starting pressures and the steadiness of the increases that occur with slight increases in load. The pressure drops as the transmission shifts from one gear to another also should be noted. The pressure should not drop more than 15 psi (1 kg/cm^2) between shifts.

Any pressure reading not within the specifications indicates a problem (**Figure 42–12**). Typically, when the fluid pressures are low, there is an internal leak, clogged filter, low oil pump output, or faulty pressure regulator valve. If the pressure increased at the wrong time or the pressure was not high enough, sticking valves or leaking seals are indicated. If the pressure drop between shifts was greater than approximately 15 psi (1 kg/cm^2), an internal leak at a servo or clutch seal is indicated. Always check the manufacturer's specifications for maximum drop off.

On transmissions equipped with an electronic pressure control (EPC) solenoid, if the line pressure is not within specifications, the EPC pressure needs to be checked. To do this, connect the pressure gauge to the EPC tap. Start the engine and check EPC pressure, then compare it to specifications. If the pressure is not within specifications (**Figure 42–13**), follow the procedures for testing the EPC. If the pressure is okay, there is a mainline pressure problem.

If the pressure tests suggest a governor problem, it should be removed, disassembled, cleaned, and inspected. Some transmissions require that the transmission be removed to service the governor. In other transmissions, it can be serviced by removing the extension housing or oil pan, or by detaching an external retaining clamp and then removing the unit.

PRELIMINARY CHECK PROCEDURE

Check transmission oil level.
Check and adjust TV cable.
Check and adjust outside manual linkage.
Check engine tune.
Install oil pressure gauge and tachometer.
Check oil pressures in the following order.

Minimum TV line pressure check
Set TV cable to specifications; with brakes applied, take pressure readings in the ranges and rpm shown below.

Full TV line pressure check
Hold the TV cable to the full extent of its travel; with brakes applied, take pressure readings in the ranges and rpm shown below.

CAUTION Brakes must be applied at all times

NOTICE Total running time must not exceed 2 minutes

RANGE	MODEL	MIN. TV		MAX. TV	
		kPa	PSI	kPa	PSI
Park @ 1,000 rpm	7BPC	459–507	66–74	459–507	66–74
	7HLC	511–581	74–84	511–581	74–84
Reverse @ 1,000 rpm	7BPC	804–887	117–129	1630–1847	236–268
	7HLC	895–1018	130–148	1721–1978	250–287
Neutral @ 1,000 rpm	7BPC	459–507	67–74	931–1055	135–153
	7HLC	511–581	74–84	983–1130	143–164
Intermediate LO @ 1,000 rpm	7BPC	788–869	114–126	788–869	114–126
	7HLC	877–998	127–145	877–998	127–145

Figure 42-10 A typical manufacturer's line pressure chart. This chart should be constantly referred to when checking oil pressure to diagnose a problem.

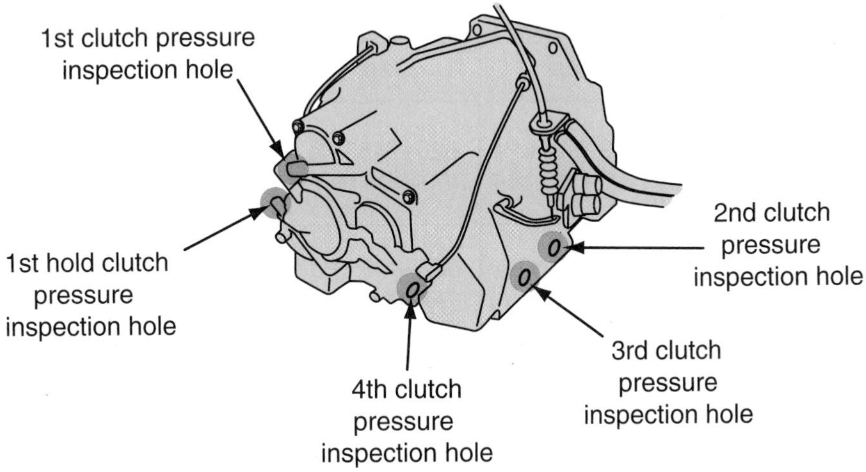

Figure 42-11 Pressure taps on the outside of a typical Honda transaxle case.

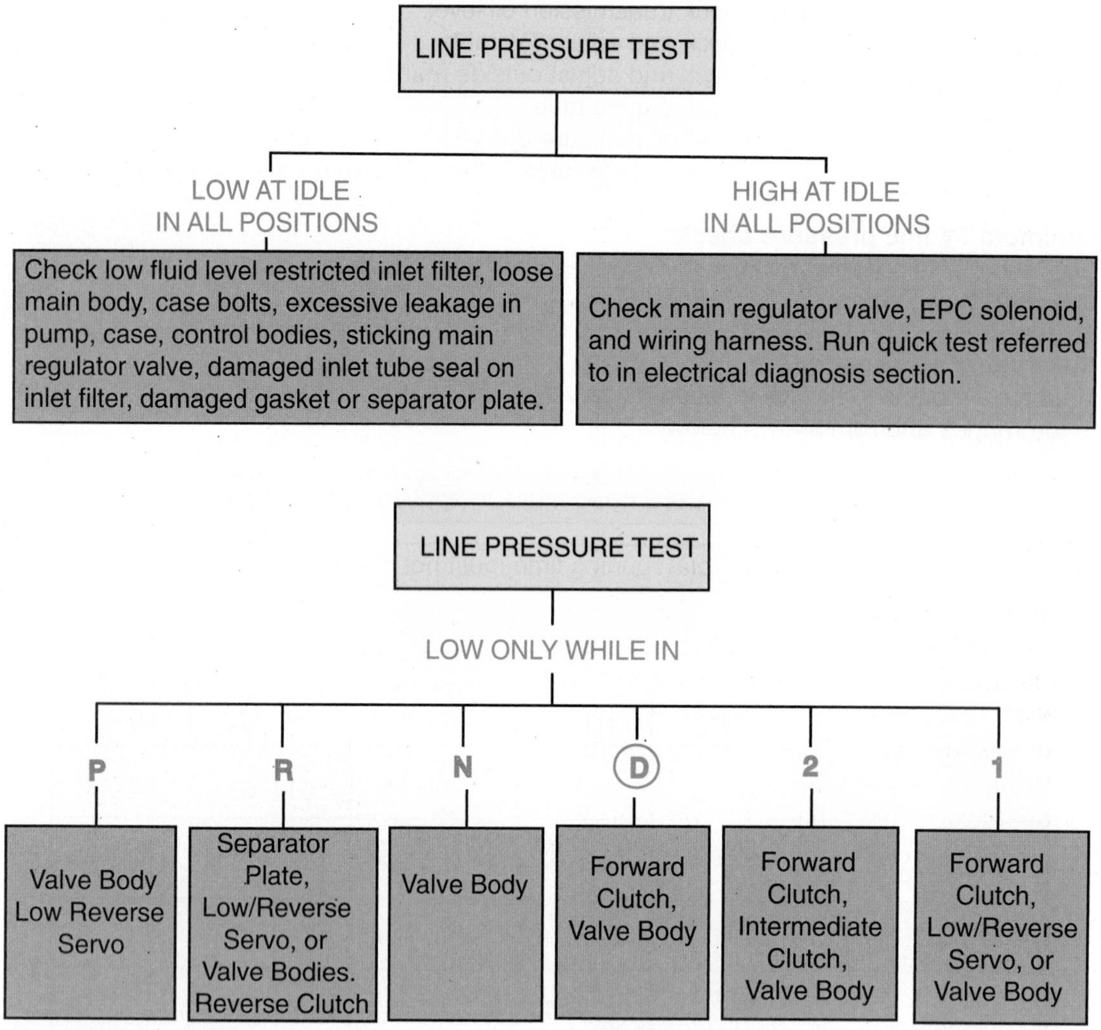

Figure 42-12 A troubleshooting chart for abnormal line pressure.

Transmission Pressure with TP at 1.5 Volts and Vehicle Speed above 8 km/h (5 mph)					
Gear	EPC Tap	Line Pressure Tap	Forward Clutch Tap	Intermediate Clutch Tap	Direct Clutch Tap
1	276–345 kPa (40–50 psi)	689–814 kPa (100–118 psi)	620–745 kPa (90–108 psi)	641–779 kPa (93–113 psi)	0–34 kPa (0–5 psi)
2	310–345 kPa (45–50 psi)	731–869 kPa (106–126 psi)	662–800 kPa (96–116 psi)	689–827 kPa (100–120 psi)	655–800 kPa (95–116 psi)
3	341–310 kPa (35–45 psi)	620–758 kPa (90–110 psi)	0–34 kPa (0–5 psi)	586–724 kPa (85–105 psi)	551–689 kPa (80–100 psi)

Figure 42-13 A pressure chart for a transmission equipped with an EPC solenoid.

Improper shift points are typically caused by a faulty governor or governor drive gear system. However, most electronically controlled transmissions do not rely on the hydraulic signals from a governor; rather, they rely on the electrical signals from sensors. Sensors, such as speed and load sensors, signal to the TCM when gears should be shifted. Faulty electrical components and/or loose connections can also cause improper shift points.

COMMON PROBLEMS

The following problems and their causes are given as examples. The actual causes of these types of problems vary with the different models of transmissions. Refer to the appropriate Band and Clutch Application Chart while diagnosing shifting problems. Doing so allows you to identify the cause of the shifting problems through the process of elimination.

Normally, if the shift for all forward gears is delayed, the clutch that is applied in all forward gears may be slipping. Likewise, if the slipping occurs in one or more gears but not all, suspect the clutch that is applied only during those gear ranges.

It is important to remember that delayed shifts or slippage may also be caused by leaking hydraulic circuits or sticking spool valves in the valve body. Since the application of bands and clutches is controlled by the hydraulic system, improper pressures will cause shifting problems. Other components of the transmission can also contribute to shifting problems. For example, on transmissions equipped with a vacuum modulator, if upshifts do not occur at the specified speeds or do not occur at all, the modulator may be faulty or the vacuum supply line may be leaking.

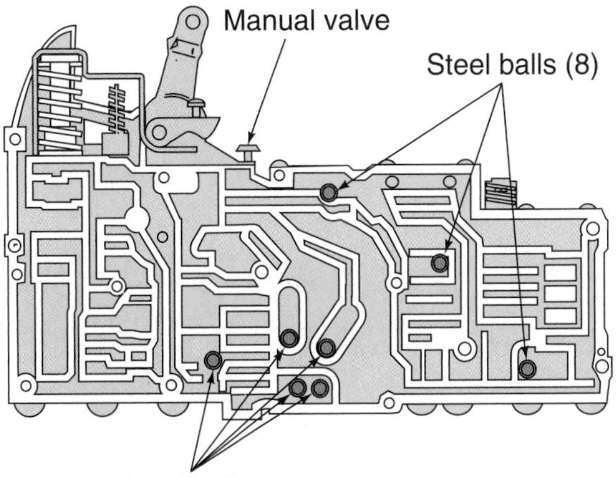

Figure 42-14 Location of the manual shift valve and check balls in a typical valve body. *Courtesy of Chrysler LLC*

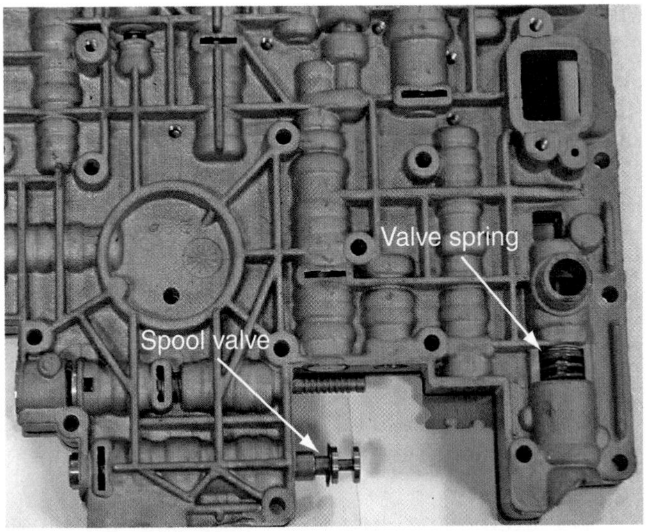

Figure 42-15 Examples of a spool valve and return spring in a valve body.

Valve Body

If the pressure problem was associated with the valve body, a thorough disassembly, cleaning in fresh solvent, careful inspection, and the freeing and polishing of the valves may correct the problem. Disconnect the lever and detent assemblies attached to the valve body, and then remove the valve body screws. Before lowering the valve body and separating the assembly, hold the assembly with the valve body on the bottom and the transfer and separator plates on top. Holding the assembly in this way reduces the chances of dropping the steel balls located in the valve body. Lower the valve body and note where these steel balls are located in the valve body **(Figure 42–14)**, then remove them and set them aside along with the various screws.

After all of the valves and springs **(Figure 42–15)** have been removed from the valve body, soak the valve body, separator, and transfer plates in mineral spirits for a few minutes. Thoroughly clean all parts and make sure all passages within the valve body are clear and free of debris. Carefully blow-dry each part individually with dry compressed air. Never wipe the parts of a valve body with a wiping rag or paper towel. Lint from either will collect in the valve body passages and cause shifting problems. As the parts of the valve body are dried, place them in a clean container.

Examine each valve for nicks, burrs, and scratches. Check that each valve properly fits into its respective bore. If a valve cannot be cleaned enough to move freely in its bore, the valve body is normally replaced. Individual valve body parts are available, as well as bore reamers. Care must be taken when rebuilding valve bodies.

During reassembly of the valve body, lube the valves with fresh ATF. Check the valve body gasket (if used) to make sure it is the correct one by laying it over the separator plate and holding it up to the light. No oil holes should be blocked. Then install the bolts

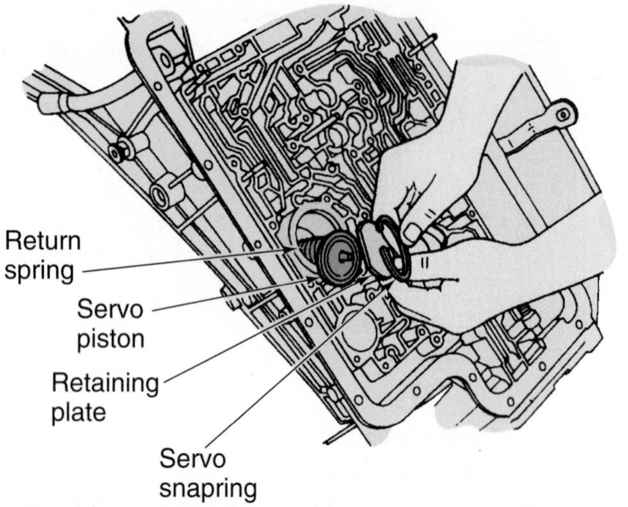

Figure 42–16 The servos in some transmissions are serviceable while the transmission is in the vehicle. The servos are contained in their own bores. *Courtesy of Ford Motor Company*

to hold valve body sections together and the valve body to the case. Tighten the bolts to the torque specifications to prevent valve body warpage and possible leakover.

Servo Assemblies

On some transmissions, the servo assemblies are serviceable with the transmission in the vehicle **(Figure 42–16)**. Others require the complete disassembly of the transmission. Internal leaks at the servo or clutch seal will cause excessive pressure drops during gear changes.

When removing the servo, check both the inner and outer parts of the seal for wet oil that means leakage. When removing the seal, inspect the sealing surface, or lips, before washing. Look for unusual wear, warping, cuts and gouges, or particles embedded in the seal.

The servo piston, spring, piston rod, and guide should be cleaned and dried. Then check the sealing rings to make sure they are able to turn freely in the groove of the piston ring. These seal rings are not typically replaced unless they are damaged, so carefully inspect them. Check the servo piston for cracks, burrs, scores, and wear. Inspect the servo cylinder for scores or other damage. Move the piston rod through the piston rod guide and check for freedom of movement. If all of the parts are in good condition, the servo assembly can be reassembled.

Lubricate the sealing ring with ATF and carefully install it on the piston rod. Lubricate and install the piston rod guide with its snap ring into the servo piston. Then, install the servo piston assembly, return spring, and piston guide into the servo cylinder. Some servos are fitted with rubber lip seals that should be replaced. Lubricate and install the new lip seal. On lip seals, make sure the spring is seated around the lip and that the lip is not damaged.

LINKAGES

Many transmission problems are caused by improper adjustment of the linkages. All transmissions have either a cable or a rod-type gear selector linkage. Some transmissions also have a throttle valve linkage, while others use an electric switch connected to the throttle to control forced downshifts.

Normal operation of a neutral safety switch provides a quick check for the adjustment of the gear selector linkage. To do this, move the selector lever slowly until it clicks into the park position. Turn the ignition key to the start position; if the starter operates, the park position is correct. After checking the park position, move the lever slowly toward the neutral position until the lever drops at the end of the N stop in the selector gate. If the starter also operates at this point, the gearshift linkage is properly adjusted. This quick test also tests the adjustment of the neutral safety switch. If the engine does not start in either or both of these positions, the neutral safety switch or the gear selector linkage needs adjustment or repair.

Gear Selector Linkage

A worn or misadjusted gear selection linkage will affect transmission operation. The transmission's manual shift valve must completely engage the selected gear **(Figure 42–17)**. Partial manual shift

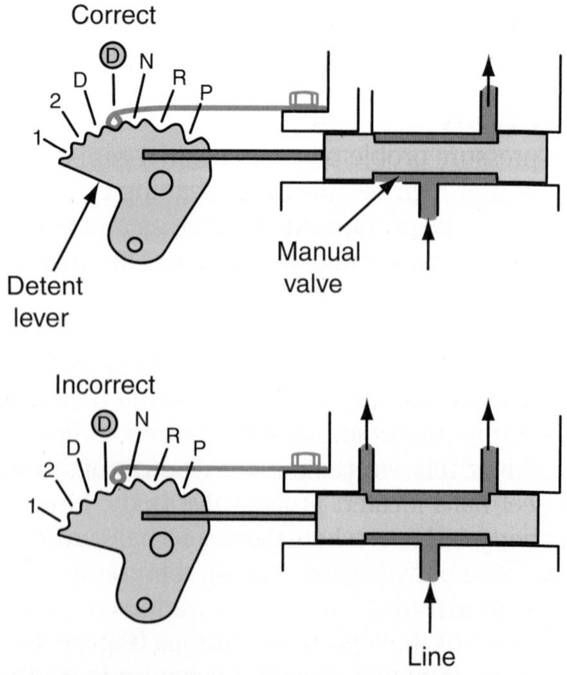

Figure 42–17 Incorrect linkage adjustments may cause the manual shift valve to be positioned improperly in its bore and cause slipping during gear changes.

valve engagement will not allow the proper amount of fluid pressure to reach the rest of the valve body. If the linkage is misadjusted, poor gear engagement, slipping, and excessive wear can result. The gear selector linkage should be adjusted so the manual shift valve detent position in the transmission matches the selector level detent and position indicator.

To check the adjustment of the linkage, move the shift lever from park to the lowest drive gear. Detents should be felt at each of these positions. If the detent cannot be felt, the linkage needs to be adjusted. While moving the shift lever, pay attention to the gear position indicator. Although the indicator will move with an adjustment of the linkage, the pointer may need to be adjusted so that it shows the exact gear after the linkage has been adjusted.

Throttle Valve Linkages

The throttle valve (TV) cable connects the movement of the throttle pedal movement to the throttle valve in the transmission's valve body. On some transmissions, the TV linkage may control both the downshift valve and the TV. Others use a vacuum modulator to control the TV and a throttle linkage to control the downshift valve. Late-model transmissions may not have a throttle cable. Instead, they rely on electronic sensors and switches to monitor engine load and throttle plate opening. The action of the TV produces throttle pressure. Throttle pressure is used as an indication of engine load and influences the speed at which automatic shifts take place.

A misadjusted TV linkage may also result in throttle pressure that is too low in relation to the amount the throttle plates are open, causing early upshifts. Throttle pressure that is too high can cause harsh and delayed upshifts, and downshifts will occur earlier than normal. When adjusting the TV linkage, always follow the manufacturer's recommended procedures. An adjustment as small as a half turn can make a big difference in shift timing and feel.

Kickdown Switch Adjustment

Some transmissions have a **kickdown switch** located at the upper post of the throttle pedal. On other transmissions, kickdown control is based on signals from the TP sensor. In both designs, the movement of the throttle pedal to the wide-open position signals to the transmission that the driver desires a forced downshift.

To check the operation of the switch, fully depress the throttle pedal and listen for a click that should be heard just before the pedal reaches its travel stop. If the click is not heard, loosen the locknut and extend the switch until the pedal lever makes contact with the switch. If the pedal contacts the switch too early, the transmission may downshift during part-throttle operation.

If you feel and hear the click of the switch but the transmission still does not kick down, check the continuity of the switch when it is depressed. An open switch will prevent forced downshifting, whereas a shorted switch can cause upshift problems. Defective switches should be replaced.

Band Adjustment

On some transmissions, slippage during shifting can be corrected by adjusting the holding bands. To help identify if a band adjustment will correct the problem, refer to the written results of your road test. Compare your results with the Clutch and Band Application Chart in the service manual. If slippage occurs when there is a gear change that requires the holding by a band, the problem may be corrected by tightening the band.

On some vehicles, the bands can be adjusted externally with a torque wrench. On others, the transmission fluid must be drained and the oil pan removed. Locate the band-adjusting nut **(Figure 42–18)**, then clean off all dirt on and around the nut. Now, loosen the band-adjusting bolt locknut and back it off approximately five turns. Use a calibrated pound-inch torque wrench to tighten the adjusting bolt to the specified torque. Then back off the adjusting screw the specified number of turns and tighten the adjusting bolt locknut while holding the adjusting stem stationary. Reinstall the oil pan with a new gasket and refill the transmission with fluid. If the transmission problem still exists, an oil pressure test or transmission teardown must be done.

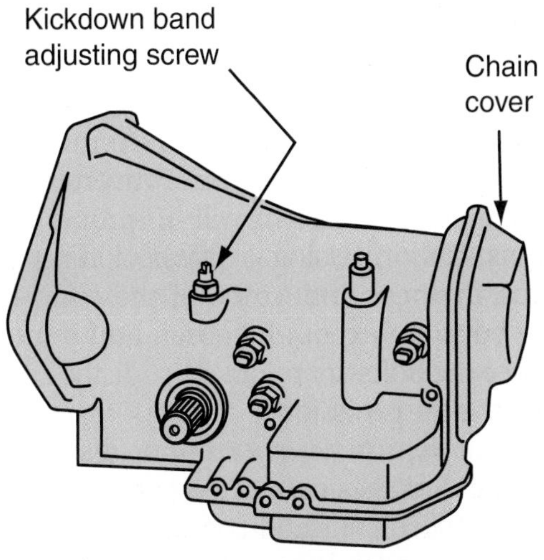

Figure 42–18 An example of the location of an external band adjusting screw.

WARNING!

Do not excessively back off the adjusting stem as the anchor block may fall out of place. It will then be necessary to remove and disassemble the transmission to fit it back in place.

Obviously, in order to rebuild or service a transmission, it must be removed. The procedure for this is very similar to removing a manual transmission or transaxle. The exact procedure for removing a transmission will vary with each year, make, and model of vehicle. Always refer to the service manual for the correct procedure.

Go To **Chapter 38 for details on removing and installing a transmission.**

CAUTION!

Always begin by disconnecting the battery ground cable. This is a safety-related precaution to help avoid any electrical surprises when removing starters or wiring harnesses. It is also possible to send voltage spikes, which may kill the PCM if wiring is disconnected when the battery is still connected.

Before removing the transmission, remove the torque converter access cover. Check for loose torque converter bolts. Then rotate the engine and watch the movement of the T/C. If it wobbles, this may be caused by a damaged flexplate or converter.

Place an index mark on the converter and the flexplate to ensure the two will be properly mated during installation. Using a flywheel turning tool, rotate the flywheel until some of the converter-to-flexplate bolts are exposed. Loosen and remove the bolts. Once the bolts are removed, slide the converter back into the transmission.

Once the vehicle is in position, disconnect all transmission linkages connected to the engine. Also remove the ATF dipstick. Plug the cooler line fittings on the housing and the lines. Then proceed to remove the transmission.

Inspecting the Torque Converter

After the transmission has been removed:

- Inspect the flexplate for evidence of cracking or other damage.
- Check the condition of the starter ring gear teeth and make sure the gear is firmly attached to the flexplate.
- Inspect the drive studs or lugs used to attach the converter to the flexplate.
- Check the shoulder area around the lugs and studs for cracked welds or other damage.
- Inspect the converter attaching bolts or nuts and replace them if they are damaged.
- Check the T/C for ballooning; this is caused by excessive pressure. If the converter is ballooned, it should be replaced and the cause of the high pressure corrected.
- Check the converter's balance weights to make sure they are still firmly attached to the unit.
- Check the pilot of the converter for wear and other damage.
- Check the area around the pilot for cracks.
- Check the drive hub of the torque converter for wear and other damage.
- Check for excessive runout of the flexplate and the converter's hub with a dial indicator.

In general, a torque converter should be replaced if there is fluid leakage from its seams or welds, loose drive studs, worn drive stud shoulders, stripped drive stud threads, heavily grooved hub, or excessive hub runout. The following additional tests of the T/C can be conducted to determine its condition:

- Stator one-way clutch check
- Internal interference check
- End play check
- Converter leakage tests

Transmission Overhaul

The exact procedure for overhauling a transmission depends solely on the specific transmission as well as the problems that the transmission may have. Photo Sequence 43 outlines the procedure for a commonly used transaxle. Always refer to the service information for the procedures for a specific transmission/transaxle.

The following guidelines will help you service any transmission:

- For the most part, torque converters are replaced if their condition is in doubt. Only specialists rebuild them.

P43-1 *Remove the torque converter. Then place the transaxle in a holding fixture. Remove the speedometer sensor and governor assembly.*

P43-2 *Remove the bottom oil pan, oil filter, and modulator valve.*

P43-3 *Remove the accumulator cover with governor feed and return pipes, accumulator pistons, gaskets, retainers, and pipes from their bores.*

P43-4 *Remove the reverse servo cover by applying pressure to it and removing the retaining ring. Then remove the servo assembly from the case.*

P43-5 *Apply pressure to the 1-2 servo cover and remove the cover's retaining ring. Then remove the cover and servo assembly.*

P43-6 *Remove the side cover and gaskets. Disconnect the wiring harness to the pressure switches, solenoid, and case connector. Unplug solenoids and switches.*

P43-7 *Remove the pump assembly cover. Remove the servo pipe retainer bolt, retainer plate, bolts, and valve body.*

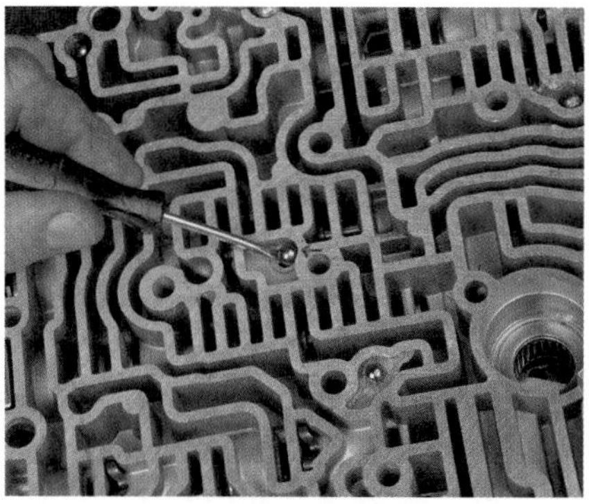

P43-8 *Mark the location of and remove all of the check balls between the spacer plate and the valve body and between the channel plate and the spacer plate.*

P43-9 *Disconnect the manual valve link from the manual valve. Place the detent lever in the park position and remove the retaining clip. Then remove the channel plate with its gaskets.*

P43-10 *Remove the oil pump drive shaft. Remove the input clutch accumulator and converter clutch piston assemblies (if so equipped). Remove the plates from the fourth clutch and the thrust bearing, hub, and shaft.*

P43-11 *Rotate the final drive unit until both ends of the output shaft retaining rings are showing.*

P43-12 *With a C-ring removal tool, loosen the ring from the output shaft. Rotate the shaft 180 degrees and remove the ring. Then remove the output shaft.*

P43-13 *Remove and discard the O-ring from the input shaft. Then remove the drive sprocket, driven sprocket, and chain as an assembly with the selective thrust washers.*

P43-14 *Remove the driven sprocket support and thrust washer from between the sprockets and the channel plate. Remove the scavenging scoop and driven sprocket with the second clutch's thrust washer.*

P43-15 *Using the correct tool, remove the second clutch and input shaft clutch housings as an assembly. Remove the reverse band.*

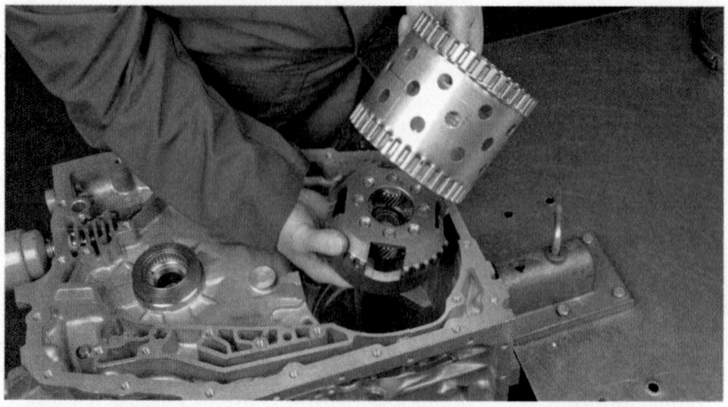

P43-16 *Measure the end play of the input shaft clutch housing. Select the correct thickness of thrust washer and set it aside for reassembly. Remove the input clutch sprag assembly, the third clutch assembly, and the input sun gear. Then remove the reverse band, reverse reaction drum input carrier assembly and thrust washer.*

P43-17 *Remove the reaction carrier, reaction sun gear/drum assembly, and forward band.*

P43-18 *Remove the reaction sun gear thrust bearing and final drive sun gear shaft.*

P43-19 *Using the proper tool, remove the final drive assembly and selective thrust washers and bearings.*

P43-20 *Check the final drive end play and select the correct size thrust washers for the unit.*

P43-21 *Clean and inspect the transaxle case and components. Then position the correct thrust washers and bearings onto the final drive and install the unit. Use petroleum jelly to hold the bearings in place.*

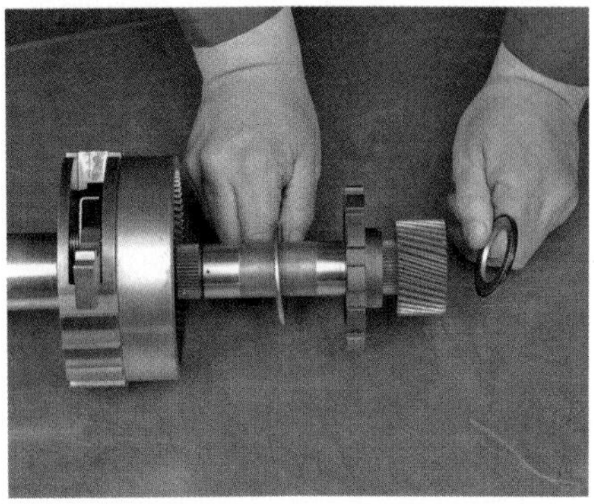

P43-22 *Install the final drive sun gear shaft through the final drive ring gear. The splines must engage with the parking gear and the final drive sun gear.*

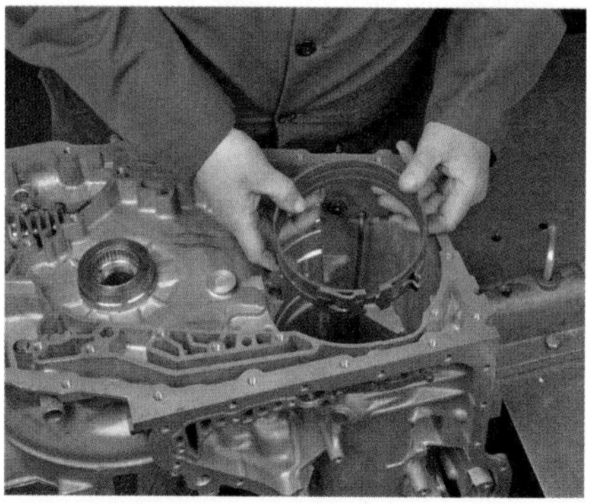

P43-23 *Install the forward band into the case, making sure the band is aligned with the anchor pin.*

P43-24 *Install the reaction sun gear to final drive ring gear thrust bearing. Then assemble the reaction sun gear and drum assembly onto the final drive ring gear.*

P43-25 *Check the end play of the carrier in the reaction gear set. Then install the thrust washer and carrier assembly into the case. Rotate the carrier until the pinions engage with the reaction sun gear.*

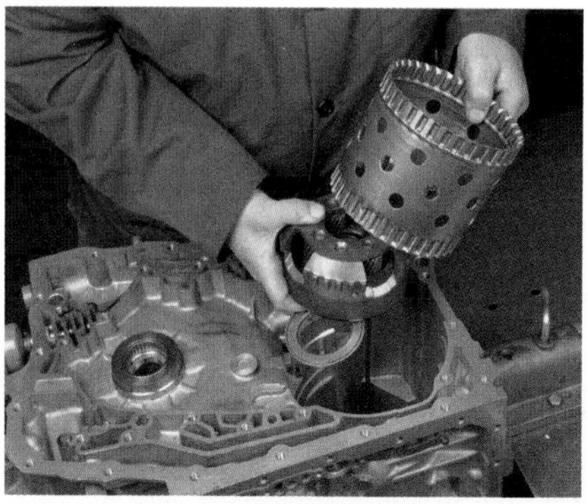

P43-26 *Install the input carrier with its thrust bearing in the case. Then install the reverse reaction drum, making sure its spline engages with the input carrier.*

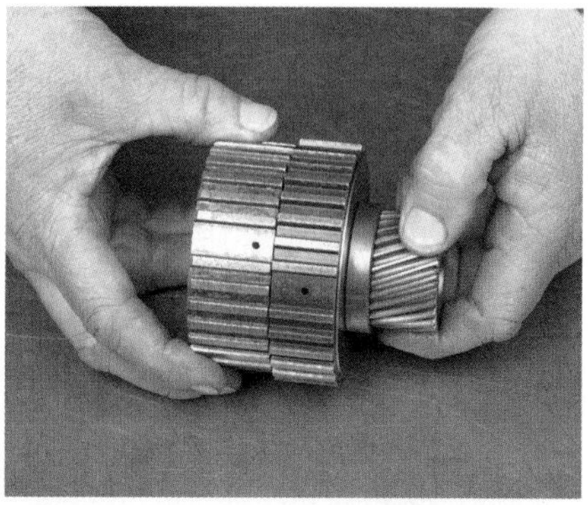

P43-27 *Install the spacer, input sprag retainer, sprag assembly, and roller clutch onto the sun gear. Hold the input sun gear and make sure the sprag and roller clutch hold and freewheel in opposite directions.*

P43-28 *Lubricate the inner seal on the input clutch's piston. Then install the seal and assemble the input piston into the input housing. Install a new O-ring onto the input shaft.*

P43-29 *Install the spring retainer and guide into the piston. Then install the third clutch piston housing into the input housing. Using the proper compressor, install the retaining snapring.*

P43-30 *Install the inner seal for the third clutch. Then install the third clutch piston into the housing. Compress the spring retainer and install the retaining snapring.*

P43-31 *Install the wave plate, the correct number of clutch plates (in the correct sequence), backing plate, and retaining snap ring into the input clutch housing. Air check its operation.*

P43-32 *Assemble the second clutch piston in the housing. Install the apply ring, return spring, snapring, and wave plate. Then install the correct number of plates (in the correct sequence), backing plate, and retaining snapring. Air check its operation.*

P43-33 *Install the thrust washers. Using the correct tool, install the second clutch and input shaft clutch housings as an assembly. Install the reverse band.*

P43-34 *Install the scavenging scoop and driven sprocket support with the second clutch thrust washer. Install the driven sprocket support and thrust washer between the sprockets and the channel plate.*

P43-35 *Install the drive sprocket, driven sprocket, and chain as an assembly.*

P43-36 *Install the output shaft. Start the C-ring onto the output shaft, then rotate the shaft 180 degrees and fully seat the ring onto the shaft.*

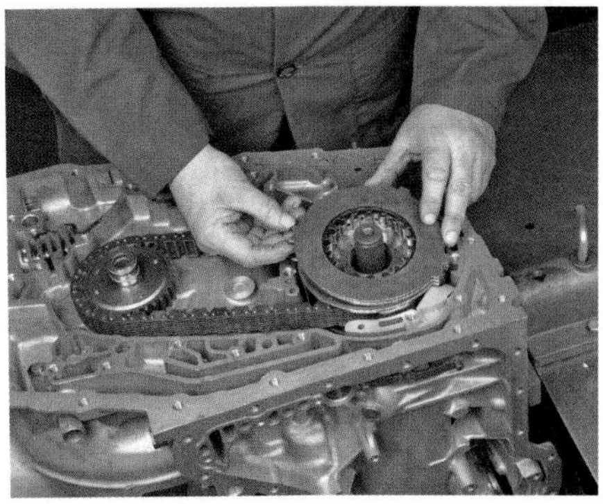

P43-37 *Install the fourth clutch's thrust bearing, hub, and shaft. Then install the fourth clutch's plates and the apply plate. Install the input clutch accumulator and converter clutch piston assemblies. Install the oil pump drive shaft.*

P43-38 *Install the channel plate with new gaskets. Connect the manual valve link to the manual valve. Place the detent lever in the park position and install the retaining clip.*

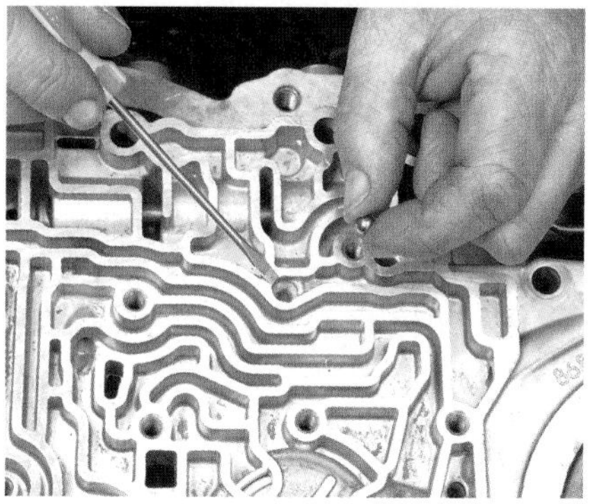

P43-39 *Install the oil reservoir weir and the check balls in their proper location between the spacer plate and the valve body and between the channel plate and the spacer plate.*

P43-40 *Install the servo pipe retainer bolt, retainer plate, mounting bolts, and valve body. Install the pump assembly cover. Tighten the bolts to specifications. Install the TV lever, linkage, and bracket onto the valve body.*

P43-41 *Connect and install the wiring harness to the pressure switches, solenoids, and case connector. Then install the side covers and gaskets. Tighten the bolts to the specified torque.*

P43-42 *Install the 1-2 servo cover, servo assembly, and retaining ring.*

P43-43 *Install the reverse servo cover, servo assembly, and retaining ring.*

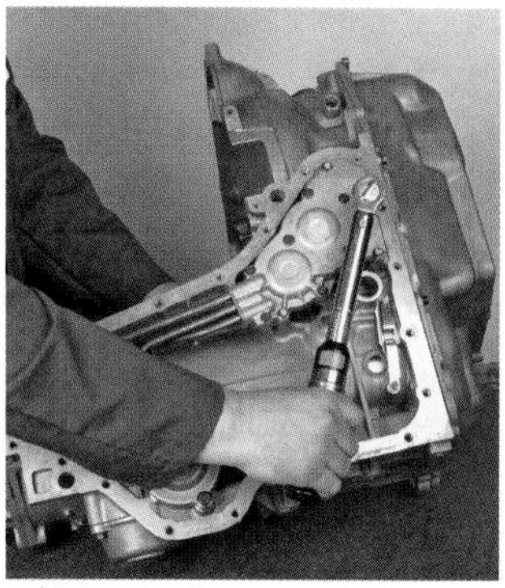

P43-44 *Install the accumulator cover with governor feed and return pipes and the accumulator pistons, gaskets, retainers, and pipes into their bores.*

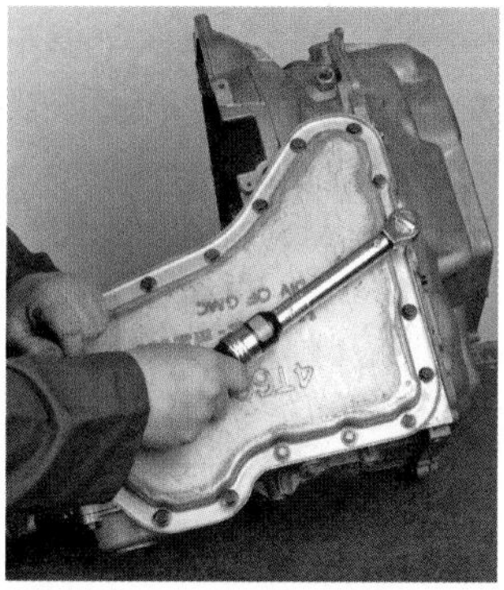

P43-45 *Install the bottom oil pan, oil filter, modulator, and modulator valve.*

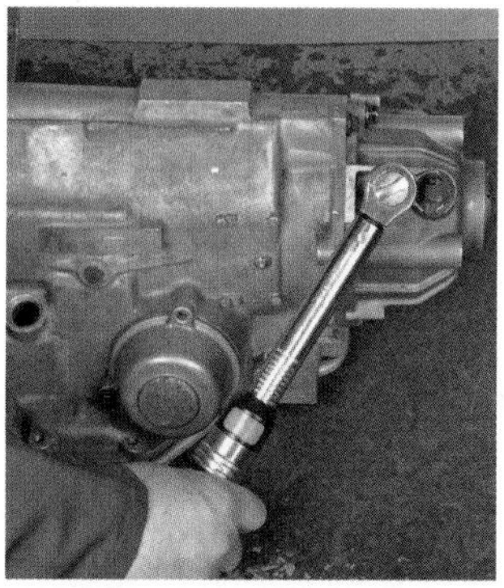

P43-46 *Install the torque converter, speedometer sensor, and governor assembly.*

- If the transaxle is fitted with a drive chain, inspect it for side play and stretch.
- Check all pumps, valve bodies, and cases for warpage and ensure that they are flat filed to take off any high spots or burrs prior to reassembly.
- Keep every part absolutely clean and air dry all parts. Only use lint-free rags to wipe parts off; lint can collect and damage the transmission.
- Make sure that the correct size thrust washers are used throughout the transmission. Also make sure that the thrust washers, bearings, and bushings are well lubricated before you install them.
- Soak all friction materials in clean ATF for at least 30 minutes prior to installing them.
- When tightening any fastener that directly or indirectly involves a rotating shaft, rotate the shaft during and after tightening to ensure that the part does not bind.
- Always use new gaskets and seals throughout the transmission.
- Make sure that all electrical wiring to the solenoids and sensors are connected and that all the harness clamps hold the wiring where it should.
- Always flush out the transmission cooling system before using a rebuilt or new transmission. The cooling system is a good place for debris to collect. Some manufacturers require replacement of the cooler rather than flushing it.

Installation

Transmission installation is generally the reverse of the removal procedure. A quick check of the following will greatly simplify your installation and reduce the chances of destroying something during installation.

- Make sure that the block alignment pins (dowels) are in their appropriate bores and are in good shape and that the alignment holes in the bell housing are not damaged.
- Make sure that the pilot hole in the crankshaft is smooth and not out of round. This will allow the converter to move in and out of the flexplate.
- Make sure that the pilot hub of the converter is smooth and that you cover it with a light coating of chassis lubricant to prevent chafing or rust.
- Make sure that the converter's drive hub is smooth and that you coat it with trans gel or equivalent lubrication.
- Secure all wiring harnesses out of the way to prevent their being pinched between the bell housing and engine block.
- Flush out the converter. It is recommended that clutch-type converters be replaced, because it is not possible to tell how much debris is in the unit.
- Always perform an end play check and check the overall height before reinstalling a torque converter or installing a fresh unit out of the box.
- Pour 1 quart of the recommended fluid into the converter before mounting the converter to the transmission. This will ensure that all parts in the converter have some lubrication before startup.

SHOP TALK

If the transmission is an EAT, check the service manual before taking it on its initial road test. Most EATs require that a "Learning Procedure" be followed. This includes a variety of driving conditions. Because you need to road test and teach the transmission, you may as well do both at the same time.

CASE STUDY

A customer with a late-model Ford pickup equipped with a 4R55E transmission had complained that every time he hit a bump, the transmission would downshift. Then when driven on smooth surfaces, the transmission would begin to cycle on its own between third and fourth gears.

The technician originally thought that the problem was caused by a bad fourth gear clutch that could not hold the planetary gear set in overdrive or by a problem in the torque converter clutch circuit. Beginning the diagnosis with a visual inspection of the transmission and the many sensors that provide information to the transmission's TCM, the technician found faulty signals from the TR sensor. These signals explained the erratic shifting, because the computer uses the sensor to determine what gear the transmission should be in and how much modified line pressure to supply.

The technician used the required special tool to realign the sensor with the gear selector but still found that the resistance readings were out of specifications. The sensor was replaced and the problem of erratic shifting solved.

KEY TERMS

Aeration
Kickdown switch
Transmission identification number

SUMMARY

- The ATF level should be checked at regular mileage and time intervals. Typically, when the fluid is checked, the vehicle should be level and running and the transmission should be at operating temperature.
- Both low fluid levels and high fluid levels can cause aeration of the fluid, which, in turn, can cause a number of transmission problems.
- Uncontaminated ATF is red in color and has no dark or metallic particles suspended in it.
- The fluid should be changed when the engine and transmission or transaxle are at normal operating temperatures. After draining the fluid, the pan should be inspected and the filter replaced.
- If ATF is leaking from the pump seal, the transmission must be removed from the vehicle so the seal can be replaced. Other worn or defective gaskets or seals can be replaced without removing the transmission. Case porosity may be repaired using an epoxy-type sealer.
- Slippage during shifting can indicate the need for band adjustment.
- Improper shift points can be caused by a malfunction in the governor or governor drive gear system or a misadjusted throttle linkage.
- The road test gives the technician the opportunity to check the transaxle or transmission operation for slipping, harshness, incorrect upshift speeds, and incorrect downshift.
- Accurate diagnosis depends on knowing what planetary controls are applied in a particular gear range.
- A pressure test checks hydraulic pressures in the transmission by using gauges attached to the transmission. Pressure readings reveal possible problems in the oil pump, governor, and throttle circuits.
- Proper adjustment of the gear selector or manual linkage is important to have the manual valve fluid inlet and outlets properly aligned in the valve body. If the manual valve does not align with the inlet and outlets, line pressure could be lost to an open circuit.

REVIEW QUESTIONS

1. What is the most probable cause of a low fluid level?
2. What does milky colored ATF indicate?
3. What do varnish or gum deposits on the dipstick indicate?
4. Typically during a pressure test, the pressure should not drop more than ______ psi between shifts.
5. What should you do if a valve does not move freely in its bore in the valve body?
6. Which of the following is the most likely cause for a shudder during the engagement of a lock-up torque converter?
   a. a bad converter
   b. worn or damaged CV or U-joints
   c. a worn front planetary gearset
   d. loose flexplate
7. If a transmission does not have a dipstick, how do you check the level of the fluid?
8. What is checked during a stall test?
9. When should the ATF level be checked on most vehicles?
   a. when the engine is cool
   b. when the engine is at operating temperature and the engine is off
   c. when the engine is at operating temperature and the engine is on
   d. it does not matter
10. List five reasons for replacing a torque converter.
11. Pressure readings reveal possible problems in which of the following?
    a. oil pump
    b. governor
    c. apply circuits
    d. all of the above
12. Which of the following is a probable cause for a converter's stall speed being below specifications?
    a. restricted exhaust
    b. slipping transmission clutches
    c. slipping transmission bands
    d. worn planetary gear set
13. When rebuilding an automatic transmission, which of the following is *not* true?

a. All pumps, valve bodies, and cases should be checked for warpage and should be flat filed to take off any high spots or burrs prior to reassembly.
b. Only use lint-free rags to wipe parts off.
c. Lubricate all thrust washers, bearings, and bushings before installing them.
d. Soak all friction materials in clean engine oil prior to installing them.

14. Explain why oil circuit diagrams are invaluable tools for diagnosing a transmission problem.
15. Explain how a plugged fluid cooler can cause a vehicle to stall when reverse is selected.

ASE-STYLE REVIEW QUESTIONS

1. Technician A says that a prerequisite to accurate road-testing analysis is knowing what planetary controls are applied in a particular gear range. Technician B says that all slipping conditions can be traced to a leaking hydraulic circuit. Who is correct?
   a. Technician A
   b. Technician B
   c. Both A and B
   d. Neither A nor B
2. While discussing proper band adjustment procedures: Technician A says that on some vehicles the bands can be adjusted externally with a torque wrench. Technician B says that a calibrated inch-pound torque wrench is normally used to tighten the band adjusting bolt to a specified torque. Who is correct?
   a. Technician A
   b. Technician B
   c. Both A and B
   d. Neither A nor B
3. Technician A says that if the shift for all forward gears is delayed, a slipping forward clutch is normally indicated. Technician B says that a bad forward clutch is indicated when there is a slip when the transmission shifts into any forward gear. Who is correct?
   a. Technician A
   b. Technician B
   c. Both A and B
   d. Neither A nor B
4. Technician A says that the only positive way to identify the exact design of the transmission is by the shape of its oil pan. Technician B says that identification numbers only identify the manufacturer and assembly date of the transmission. Who is correct?
   a. Technician A
   b. Technician B
   c. Both A and B
   d. Neither A nor B
5. Technician A says that delayed shifting can be caused by worn planetary gear set members. Technician B says that delayed shifts or slippage may be caused by leaking hydraulic circuits or sticking spool valves in the valve body. Who is correct?
   a. Technician A
   b. Technician B
   c. Both A and B
   d. Neither A nor B
6. While checking the condition of a car's ATF: Technician A says that if the fluid has a dark brownish or blackish color and/or a burned odor, the fluid has been overheated. Technician B says that if the fluid has a milky color, this indicates that engine coolant has been leaking into the transmission's cooler. Who is correct?
   a. Technician A
   b. Technician B
   c. Both A and B
   d. Neither A nor B
7. While discussing the results of an oil pressure test: Technician A says that when the fluid pressures are high, internal leaks, a clogged filter, low oil pump output, or a faulty pressure regulator valve are indicated. Technician B says that if the fluid pressure increased at the wrong time, an internal leak at the servo or clutch seal is indicated. Who is correct?
   a. Technician A
   b. Technician B
   c. Both A and B
   d. Neither A nor B
8. While discussing a pressure test: Technician A says that this test is the most valuable diagnostic check for slippage when a gear is engaged. Technician B says that the test can identify the cause of late or harsh shifting. Who is correct?
   a. Technician A
   b. Technician B
   c. Both A and B
   d. Neither A nor B
9. While checking the engine and transmission mounts on a FWD car: Technician A says that any engine movement may change the effective length of the shift and throttle cables and therefore may affect the engagement of the gears. Technician B says that delayed or missed shifts are caused by hydraulic problems, not linkage or control problems. Who is correct?
   a. Technician A
   b. Technician B
   c. Both A and B
   d. Neither A nor B
10. While discussing the cause of aerated fluid in a transmission: Technician A says that this can be caused by too much ATF in the transmission. Technician B says that this can be caused by too little ATF in the transmission. Who is correct?
    a. Technician A
    b. Technician B
    c. Both A and B
    d. Neither A nor B

FOUR- AND ALL-WHEEL DRIVE

CHAPTER 43

OBJECTIVES

■ Identify the advantages of four- and all-wheel drive. ■ Name the major components of a conventional four-wheel-drive system. ■ Name the components of a transfer case. ■ State the difference between the transfer, open, and limited slip differentials. ■ State the major purpose and operation of locking/unlocking hubs. ■ Name the five shift lever positions on a typical 4WD vehicle. ■ Understand the difference between four- and all-wheel drive. ■ Know the purpose of a viscous clutch in all-wheel drive.

With the popularity of SUVs **(Figure 43–1)**, pickup trucks, and crossover vehicles, the need for technicians who can diagnose and service four-wheel-drive systems has increased drastically. Although all-wheel-drive passenger cars are available, most prospective buyers for all-wheel and four-wheel-drive vehicles are opting for truck-based SUVs and pickups.

Four-wheel-drive (4WD) and **all-wheel-drive (AWD)** systems can dramatically increase a vehicle's traction and handling ability in rain, snow, and off-road driving. Considering that the vehicle's only contact with the road is the small areas of the tires, driving and handling are improved when the work load is evenly spread out to four tires rather than two.

Figure 43-1 A 4WD sport utility vehicle. *Courtesy of Chrysler LLC*

The increased traction also makes it possible to apply greater amounts of energy through the drive system. Vehicles with 4WD and AWD can maintain control while transmitting levels of power that would cause two wheels to spin either on takeoff or while rounding a curve. The improved traction of 4WD and AWD systems allows the use of tires that are narrower than those used on similar 2WD vehicles. These narrow tires tend to cut through snow and water rather than hydroplane over them. Of course, wider and larger tires are often used in off-the-road adventures.

Both 4WD and AWD systems add initial cost and weight. With most passenger cars, the weight problem is minor. A typical 4WD system adds approximately 170 pounds to a passenger car. An AWD system adds even less weight. The additional weight in larger 4WD trucks can be as much as 400 pounds or more.

The systems also add initial cost to the vehicle. Vehicles equipped with 4WD and AWD require special service and maintainance not performed on 2WD vehicles. However, the slight disadvantages of 4WD and AWD are heavily outweighed by the traction and performance these systems offer. Many vehicles are equipped with 4WD or AWD, therefore technicians must be prepared to diagnose and repair these systems.

TYPES OF FOUR-WHEEL DRIVES

Because of the many names manufacturers give their drive systems, it is often difficult to clearly define the difference between 4WD and AWD. Both have the

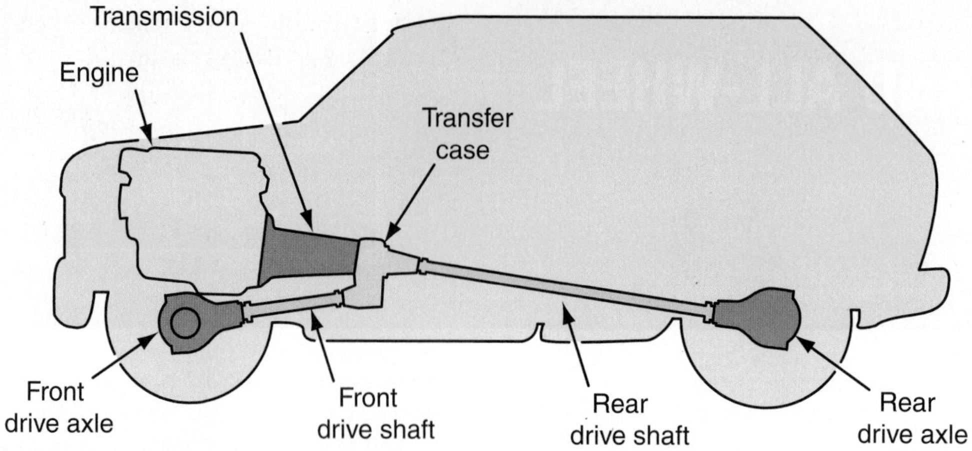

Figure 43-2 A typical arrangement of 4WD components.

ability to send torque to all four wheels. To do this, these vehicles have a transfer case and a differential unit at the front and rear axles **(Figure 43–2)**. A **transfer case** splits the power from the transmission between the front and rear axles. For clarity, the primary difference between the two is that the transfer case in a 4WD vehicle offers two speed ratios in four-wheel drive: high and low. These systems are mostly found in pickups and large SUVs. An AWD vehicle does not offer this. AWD vehicles are smaller SUVs or passenger cars **(Figure 43–3)**.

Four-Wheel-Drive Systems

Four-wheel-drive systems can have several names, including 4×4, and can be divided into separate types: part time and full time. These terms describe when power is sent to the four wheels. Both, however, have two-speed transfer cases. The transfer case is switched with a shifter or electrical switch. The high range could be called the normal range because the torque from the transmission is not altered by the transfer case. This gear selection is used to provide 4WD traction on roads covered with ice or packed snow. The low range provides extra torque at very low output speed. This allows drivers to slowly and smoothly climb very steep hills or pull heavy loads.

Figure 43-3 A 4WD passenger car moving through the dirt. *Courtesy of Subaru of America Inc.*

Part-Time 4WD Part-time systems are often found on 4WD pickups and older SUVs. These vehicles are basically RWD vehicles fitted with a two-speed transfer case and axle connects/disconnects, normally called locking hubs. The transfer case in part-time systems does two things in response to the driver's commands: it engages/disengages 4WD and adjusts the overall gear ratio for the final drive. Normally, the system operates in 2WD until the driver selects 4WD.

A selector switch or shifter **(Figure 43–4)** controls the transfer case so that power is directed to the axles as the driver desires. Normally, the driver can select 2WD, 4WD HI, 4WD LO. The shift control will either

Figure 43-4 A 4WD-mode selector switch.

physically move a gear in the transfer case or activate an electrically operated solenoid or clutch pack to send torque to the front axle. In addition, most mechanical shifters offer a neutral position that is used as a stopping point before 4WD LO can be selected.

The transmission is connected directly to the transfer case. In turn, the transfer case is connected to the front and rear axle assemblies. Two drive shafts extend from the transfer case: One sends torque to the front axle and the other to the rear axle **(Figure 43–5)**. When the system is operating in 4WD, all four wheels are powered by the engine. The front drive shaft is locked to the rear drive shaft; therefore, each axle receives half of the torque coming from the transmission and the axles rotate at the same speed. With this split of torque, the tires need to slip on the surface when the vehicle makes a turn. This is one of the reasons 4WD in part-time systems should not be selected when the vehicle is traveling on dry surfaces. Four-wheel drives should only be selected during situations where the tires can easily slip, such as mud or other slippery surfaces. When operated on dry surfaces, the tires have a difficult time slipping and jerking in turns and excessive tire wear will result.

Part-time systems also have **locking hubs**, which can be manually or automatically controlled. The hubs are designed to stop the rotation of the front axles and differential when 4WD is not selected. This increases fuel economy and decreases the wear on those parts. Newer systems have automatic locking hubs that engage when the driver switches into 4WD **(Figure 43–6)**. Older vehicles had manual locking hubs that required the driver to turn a knob on the front wheels **(Figure 43–7)**. Most manual or automatic hubs use a sliding collar that locks the front drive axles to the hub.

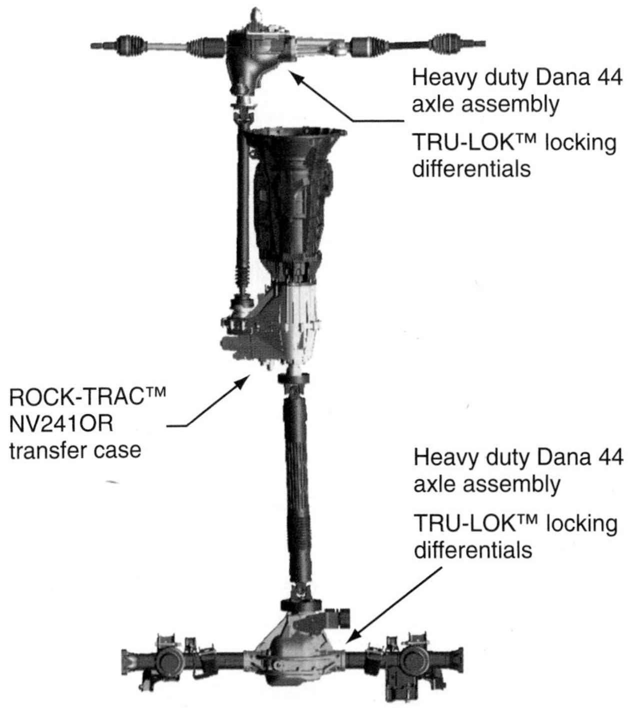

Figure 43-5 The drivetrain layout for a part-time heavy-duty 4WD vehicle. *Courtesy of Chrysler LLC*

Full-Time 4WD In a full-time 4WD system, all four wheels have power delivered to them all of the time. They do well off the road and can be used on all surfaces, including dry pavement. These systems have a center or **interaxle differential** or **viscous clutch** that allows for a difference in speeds between the front and rear axles. These units decrease the amount of slip the wheels experience when the vehicle is turning a corner. The center differential is located between the front and rear drive shafts and may be an integral part of the transfer case. Like all 4WD systems, the transfer case has a low- and high-speed range. However, the driver does not have the option of shifting into 2WD. Also, these systems have no need for locking hubs at the front wheels.

All-Wheel Drive

AWD systems are basically the same as 4WD systems, except there is no low gear option. Many AWD vehicles are based on FWD vehicles. They may not have a separate transfer case; rather, a viscous clutch, center differential **(Figure 43–8)**, or transfer clutch is used to transfer power from the transaxle to a rear driveline and rear axle assembly.

Full-Time AWD These systems are very similar to the full-time 4WD and power the four wheels all of the time. They are not designed for heavy off-the-road driving; rather, they enhance a vehicle's stability and performance during normal driving conditions. The engine's torque is divided according to operating conditions. When the vehicle is moving straight on a level road, each wheel receives the same amount of torque. During a turn, the front wheels receive less torque; this prevents wheel slip.

Some AWD vehicles have an electromagnetic clutch in the transfer case. During normal driving, the AWD control module keeps the clutch working at a minimum level. This allows for a slight difference between speeds of the front and rear drive shafts, enabling the vehicle to negotiate turns. When the system detects wheel slippage, the control module fully activates the clutch to send more torque to the front

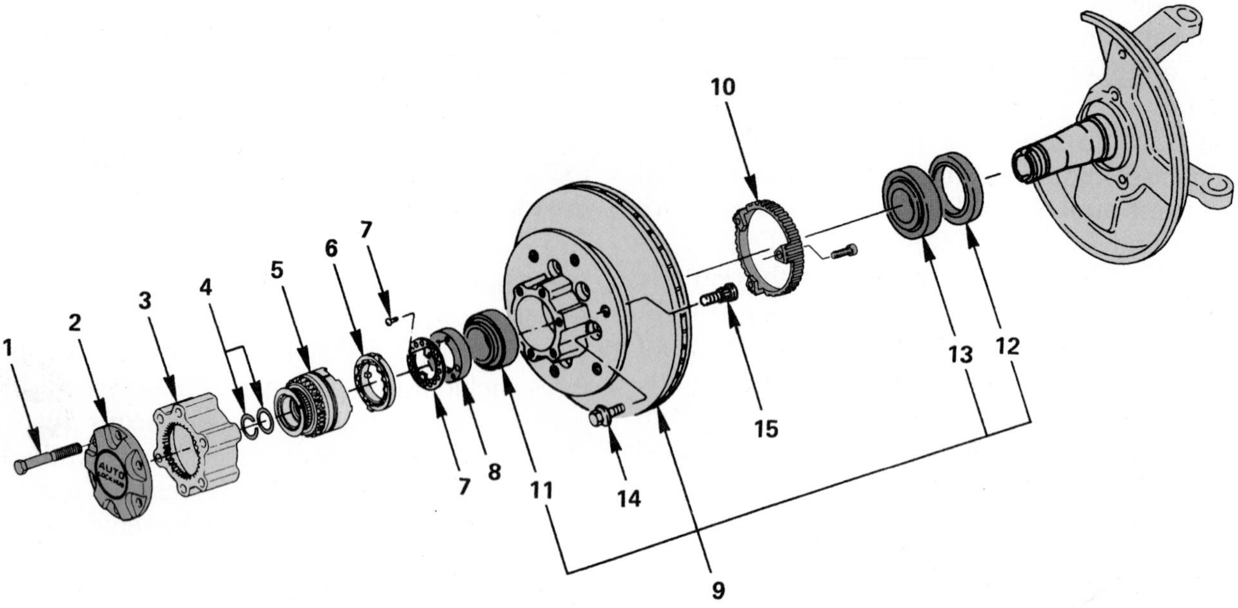

Disassembly steps
1. Bolt
2. Hub cap
3. Housing
4. Snap ring and shim
5. Drive clutch assembly
6. Inner cam assembly
7. Lock washer and lock screw
8. Hub nut
9. Hub and disc assembly
10. ABS sensor ring
11. Outer bearing outer race
12. Oil seal
13. Inner bearing outer race
14. Bolt
15. Wheel pin

Reassembly steps

15. Wheel pin
14. Bolt
13. Inner bearing outer race
12. Oil seal
11. Outer bearing outer race
10. ABS sensor ring
9. Hub and disc assembly
8. Hub nut
7. Lock washer and lock screw
6. Inner cam assembly
5. Drive clutch assembly
4. Snap ring and shim
3. Housing
2. Hub cap
1. Bolt

Figure 43-6 Automatic locking front hub components. *Courtesy of American Isuzu Motors Inc.*

or rear wheels **(Figure 43–9)**. Wheel slip is monitored by inputs from wheel speed sensors that are available on the CAN bus.

Some systems allow the driver to select an operating mode that locks the front and rear drive shafts. This causes the drive shafts to rotate at the same speed and torque. This mode is intended for very slippery conditions and should not be used on dry payment.

Automatic AWD These systems operate in 2WD most of the time. Power to four wheels only occurs when the conditions dictate the need. These systems were designed to enhance vehicle stability and safety and are not intended for off-the-road usage.

During normal driving, one axle receives all of the output from the transmission. When that axle experiences some slippage, the AWD control unit allows up to 50% of the torque to move to the other axle. This power split can be accomplished hydraulically, mechanically, or electrically, depending on the system. As soon as the axle that was slipping is no longer slipping, all torque is sent to that axle. Under normal conditions one axle gets 100% of the torque—meaning you are driving in 2WD.

FOUR-WHEEL-DRIVE SYSTEMS

The heart of conventional 4WD vehicles is the transfer case, which may be integrated into the rear of the transmission **(Figure 43–10)** or mounted to the back of the transmission **(Figure 43–11)**. A chain or gear drive within the case receives the power from the transmission and transfers it to the front and rear axles through two separate drive shafts.

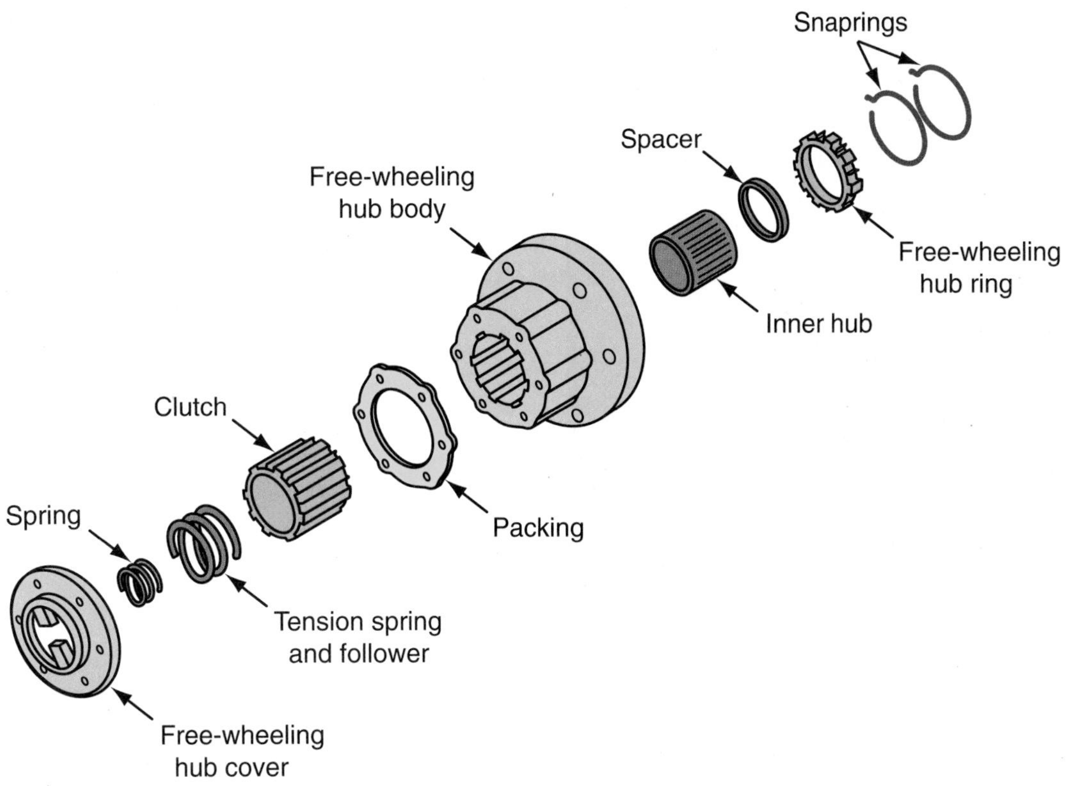

Figure 43-7 A disassembled manual locking hub.

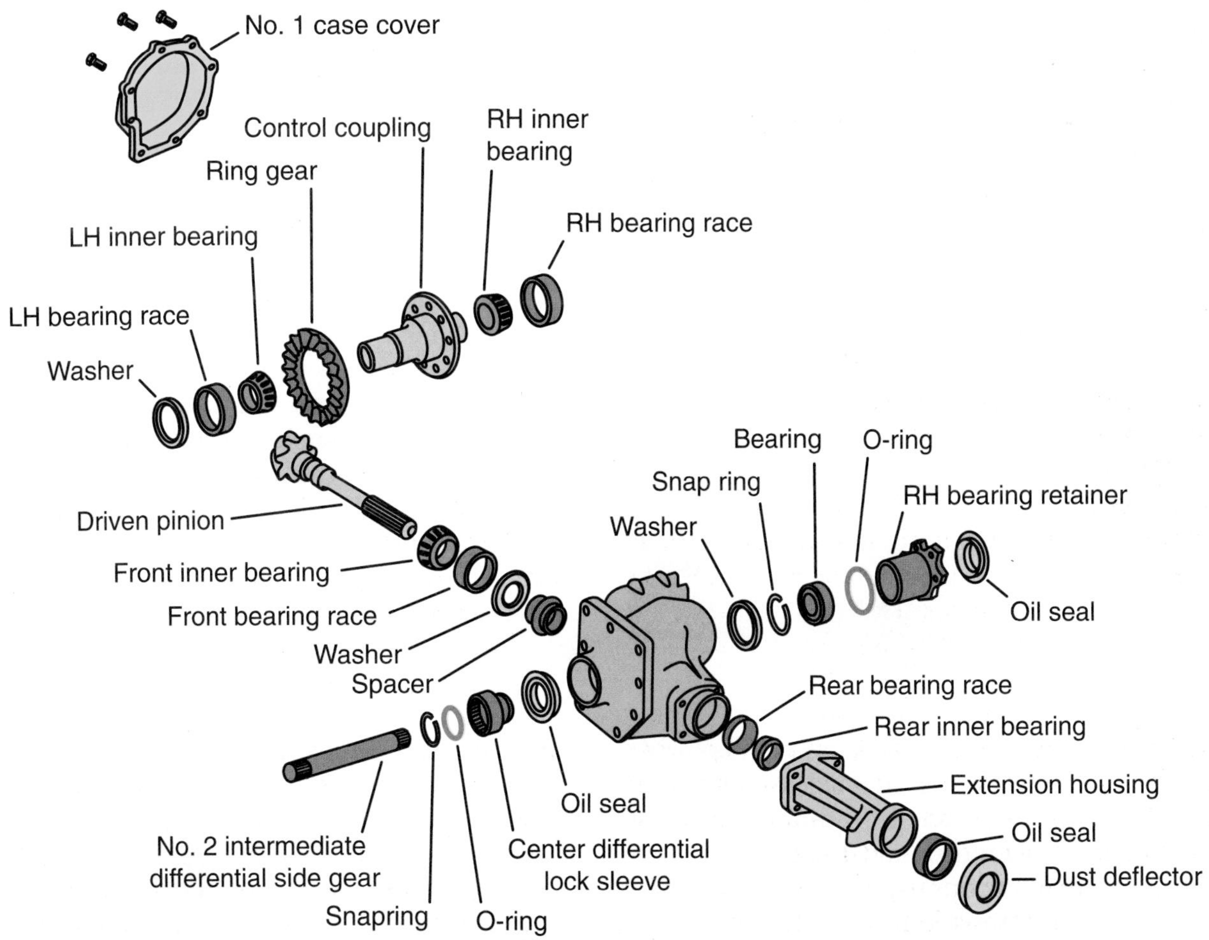

Figure 43-8 An exploded view of a center differential assembly for a full-time AWD vehicle.

Figure 43-9 The differential for the rear axle of a Honda with AWD. *Courtesy of American Honda Motor Co., Inc.*

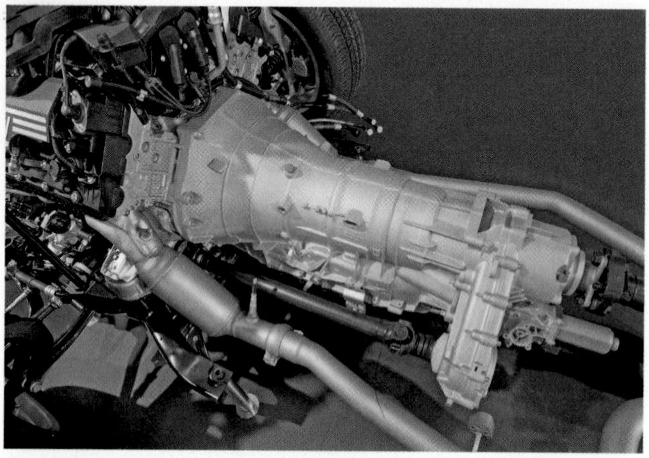

Figure 43-10 A transfer case integrated into the rear of a transmission. *Courtesy of BMW of North America, LLC*

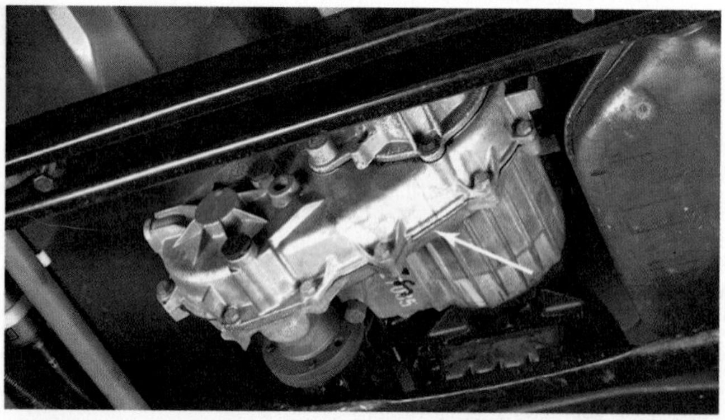

Figure 43-11 A typical bolt-on transfer case.

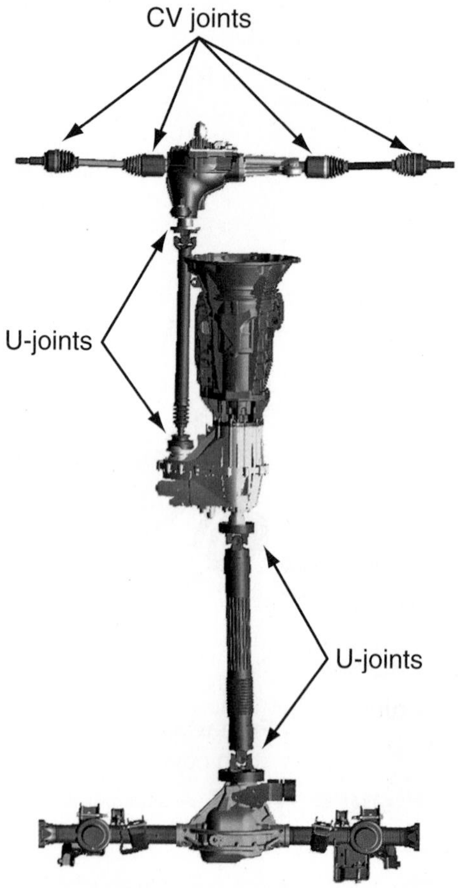

Figure 43-12 The location on the CV joints and U-joints on a typical 4WD vehicle. *Courtesy of Chrysler LLC*

The driveline from the transfer case shafts run to differentials at the front and rear axles. As on two-wheel-drive vehicles, these axle differentials are used to compensate for road and driving conditions by adjusting the rpm to opposing wheels. For example, the outer wheel must roll faster than the inner wheel during a turn because it has more ground to cover. To permit this action, the differential cuts back the power delivered to the inner wheel and boosts the amount of power delivered to the outer wheel.

U-joints are used to couple the driveline shafts with the differentials and transfer cases on all these vehicles. U-joints can also be used on some vehicles to connect the rear axle and wheels. Normally, however, rear axles are simply bolted to the wheel hubs.

The coupling between front wheels and axles is normally done with U-joints or CV (constant velocity) joints **(Figure 43–12)**. Generally, half axles or half shafts with CV joints are found on 4WD passenger cars. They can also be found on a number of passenger vans and on mini pickups and trucks.

To provide independent front suspension, some vehicles have one half shaft and one solid axle for the front drive axle **(Figure 43–13)**. The half shaft is able to move independently of the solid axle, thereby giving the vehicle the ride characteristics desired from an independent front suspension.

On 4WD systems adapted from front-wheel-drive systems, a separate front differential and driveline are not needed. The front wheels are driven by the

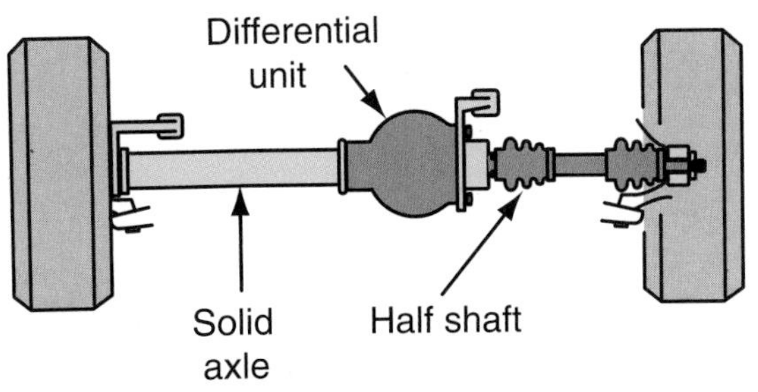

Figure 43–13 A front suspension for a front drive axle that relies on one solid axle and a half shaft to provide independent suspension.

transaxle differential of the base model. A **power takeoff** is added to the transaxle to transmit power to the rear wheels in four-wheel drive. This takeoff gearing is housed in the transaxle housing. The gearing connects to a rear driveline and rear axle assembly that includes the rear differential.

TRANSFER CASE

In a 4WD vehicle, as mentioned earlier, the transfer case delivers power to both the front and rear assemblies. Two drive shafts normally operate from the transfer case, one to each drive axle.

The transfer case **(Figure 43–14)** itself is constructed similarly to a standard transmission. It uses shift forks to select the operating mode, plus splines, gears, shims, bearings, and other components found in manual and automatic transmissions. The outer case of the unit is made of cast iron, magnesium, or aluminum. It is filled with lubricant (oil) that cuts friction on all moving parts.

INTERAXLE DIFFERENTIALS

The front and rear axle differentials may have different ratios. This allows for different speeds at the two axles but can result in a phenomenon called **driveline windup**. Driveline windup can be explained by associating the driveline to a torsion bar. The driveline twists up when both driving axles are rotating at different speeds, pushing and pulling the vehicle on hard, dry pavement.

Figure 43–14 A transfer case for a full-time 4WD system. *Courtesy of Chrysler LLC*

Driveline windup can cause handling problems, particularly when rounding turns on dry pavement. This is because the front axle wheels must travel farther than the rear axle wheels when rounding a curve. On wet or slippery roads, the front and rear wheels slide enough to prevent damage to the driveline components. However, this is not the case on dry surfaces.

Full-time 4WD and AWD systems require a device that allows the front and rear axles to rotate at different speeds. These units work to eliminate driveline windup. The most common setup has an interaxle differential located in the transfer case **(Figure 43–15)**. The front and rear drivelines are connected to the interaxle differential. Just as a drive axle differential allows for different left and right drive axle shaft speeds, the interaxle differential allows for different front and rear driveline shaft speeds.

While the interaxle differential solves the problem of driveline windup during turns, it also lowers performance in poor traction conditions. This is because the interaxle differential will tend to deliver more power to the wheels with the least traction. The result is increased slippage, which is the exact opposite of what is desired.

To counteract this problem, some interaxle differentials are designed much like a limited slip differential. They use a multiple-disc clutch pack to maintain a predetermined amount of torque transfer before the differential action begins to take effect. Other systems such as the one shown in Figure 43–15 use a cone braking system rather than a clutch pack, but the end result is the same. Power is supplied to both axles regardless of the traction encountered.

Many systems also give the driver the option of locking the interaxle differential in certain operating modes. This eliminates the differential action altogether. However, the interaxle differential should only be locked while driving in slippery conditions and can only be activated at low speeds.

Viscous Clutch

Some systems use a viscous clutch in the transfer case, outside the transfer case **(Figure 43–16)**, or as a separate unit to split the power to the front and rear axles. A viscous clutch is used to drive the axle with low tractive effort, taking the place of the interaxle differential. In existence for several years, the viscous clutch is installed to improve mobility under difficult

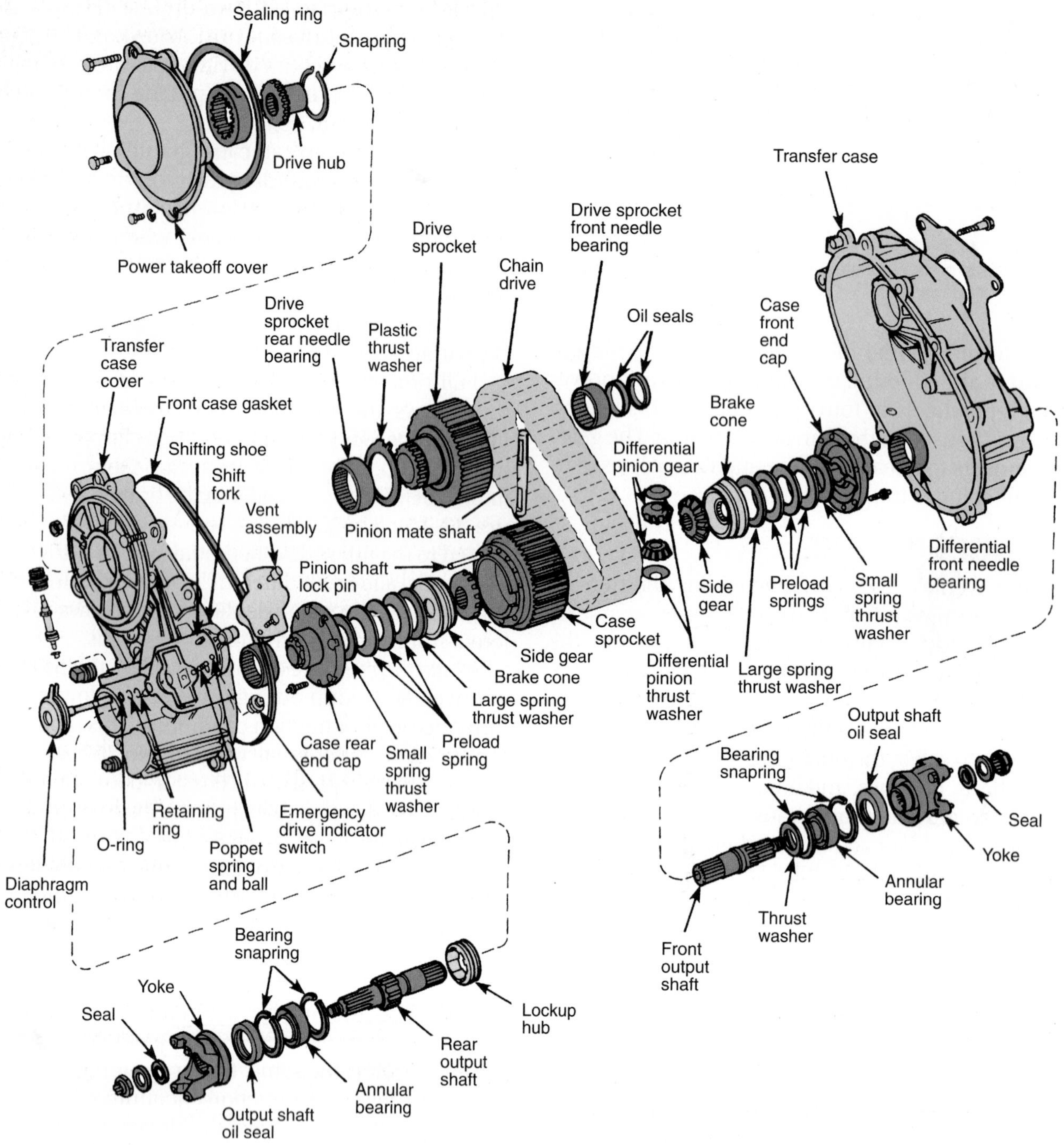

Figure 43–15 Four-wheel-drive transfer case with integral differential and cone brakes for limited slip.

driving conditions. Viscous clutches operate automatically while constantly transmitting power to the axle assembly as soon as it becomes necessary to improve driving wheel traction. This action is also known as biasing driving torque to the axle with tractive effort. The viscous clutch assembly is designed similarly to a multiple-disc clutch with alternating driven and driving plates **(Figure 43–17)**.

The viscous clutch parts fit inside a drum that is completely sealed. The clutch pack is made up of alternating steel and friction plates. One set of steel plates is splined internally to the clutch assembly hub. The second set of clutch plates is splined externally to the clutch drum. The clutch housing is filled with a small quantity of air and special silicone fluid with the purpose of transmitting force from the driving plates to the driven plates.

When a difference in speed of 8% exists between the input shaft driven by the driving axle with tractive effort, the clutch plates begin shearing (cutting) the

Figure 43-16 A viscous clutch assembly.

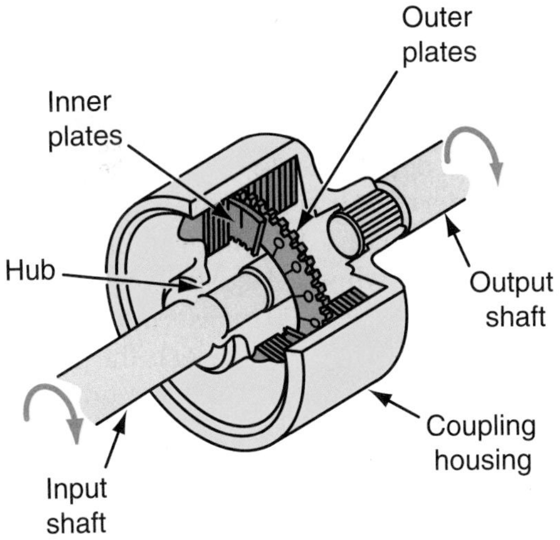

Figure 43-17 Inside a typical viscous clutch assembly.

special silicone fluid. The shearing action causes heat to build within the housing very rapidly, which results in the silicone fluid stiffening. The stiffening action causes a locking action between the clutch plates to take place within approximately 1/10 second. A locking action results from the stiff silicone fluid, making it very hard for the plates to shear. The stiff silicone fluid transfers power from the driving to the driven plates. The driving shaft is then connected to the driven shaft through the clutch plates and stiff silicone fluid.

The viscous clutch has a self-regulating control. When the clutch assembly locks up, there is very little, if any, relative movement between the clutch plates. Because there is little relative movement, silicone fluid temperature drops, which reduces pressure within the clutch housing. As speed fluctuates between the driving and driven members, heat increases, causing the silicone fluid to stiffen. Speed differences between the driving and driven members regulate the amount of slip in a viscous clutch driveline. The viscous clutch takes the place of the interaxle differential, biasing driving torque to the normally undriven axle during difficult driving conditions.

Haldex Clutch

Some AWD vehicles are equipped with a **Haldex clutch** (**Figure 43–18**), which serves as a center differential. This clutch unit distributes the drive force variably between two axles. In a typical application, the Haldex unit mounts in front of the rear differential and receives torque from the front axle.

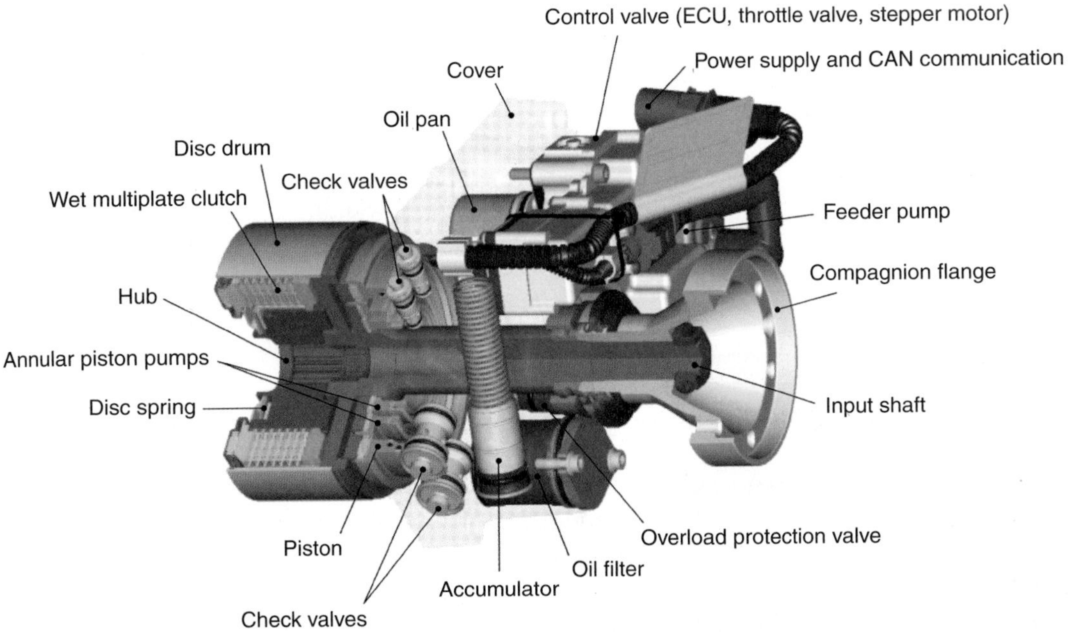

Figure 43-18 A Haldex clutch assembly. *Courtesy of Haldex Traction AB*

The Haldex unit has three main parts: the hydraulic pump driven by the slip between the axles or wheels, a wet multidisc clutch, and an electronically controlled valve. The unit is much like a hydraulic pump in which the housing and a piston are connected to one shaft and a piston actuator connected to the other.

When a front wheel slips, the input shaft to the Haldex unit spins faster than its output shaft. This causes the pump to immediately generate oil flow. The oil flow and pressure engages the multidisc clutch to send power to the rear wheels. This happens extremely quickly because an electric pump and accumulator keep the circuit primed.

The oil from the pump flows to the clutch's piston to compress the clutch pack. The oil returns to the reservoir through a controllable valve, which adjusts the oil pressure and the force on the clutch pack. An electronic control module controls the valve **(Figure 43–19)**.

In high slip conditions, a high pressure is delivered to the clutch pack: In tight curves or at high speeds, a much lower pressure is provided. When there is no difference in speed between the front and rear axles, the pump does not supply pressure to the clutch pack.

Volkswagen 4MOTION Haldex System The Volkswagen 4MOTION AWD system uses a Haldex coupling as a center differential. The coupling is mounted in front of the rear axle differential and is part of the rear differential case. The input shaft of the Haldex center differential is driven by a drive shaft from the transaxle. In a Haldex differential, the input shaft is totally separated from the output shaft, which is connected to the rear final drive gears. Therefore, power is only sent to the rear wheels when the Haldex clutch is engaged.

The Haldex unit is controlled by an ECU that receives inputs from a variety of sensors. This means the system can respond to other driving conditons and not just wheel slip. When there is no slip, understeer, or oversteer, the vehicle operates as a 2WD vehicle. It only distributes power to the rear axle when it is needed.

Saab XWD Saab's active AWD system is called a cross-wheel drive (XWD) system. It relies on computer-controlled center and rear differentials to control the stability of the vehicle. It is a fully automatic, on-demand system that can send up to 100% of the engine's torque to the front or rear wheels. It is called a cross system because it also can apply power to the opposite corner of the car from the wheel that has lost traction. It also can vary the amount of torque applied to each of the rear wheels. Adjustments to power distribution can be made in 80 milliseconds.

A three-piece drive shaft connects the power take-off at the transaxle to the rear drive module (RDM). The RDM is comprised of a torque transfer device (TTD) and an electronically controlled limited slip differential (eLSD). Both of these units are wet, multiple-plate Haldex clutches. The TTD regulates the amount of torque delivered to the rear drive axle. The eLSD

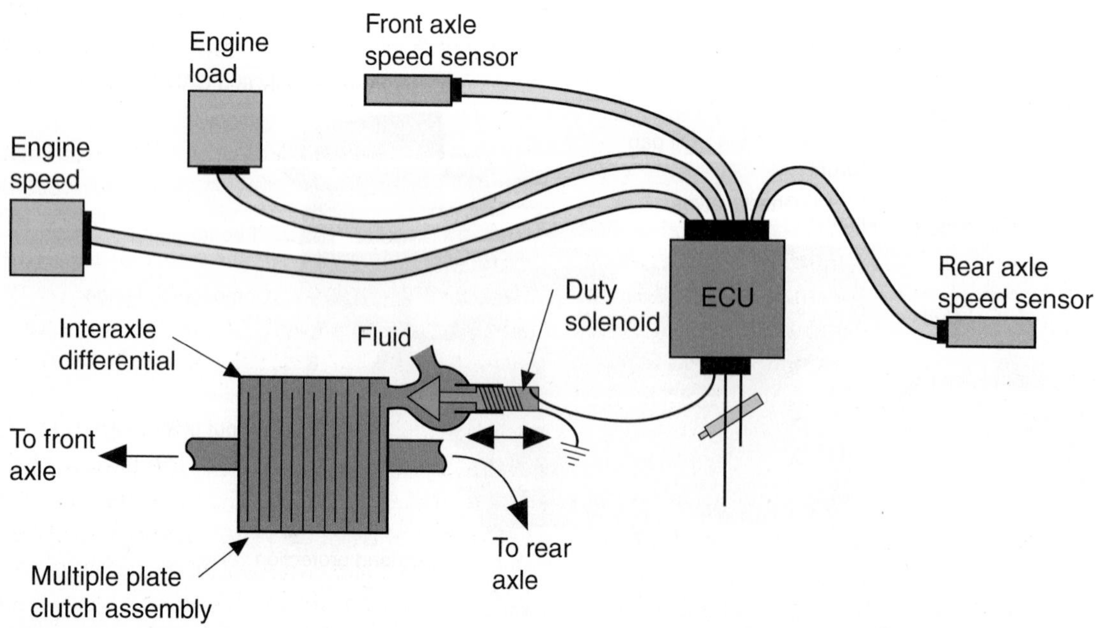

Figure 43–19 The oil flow to the Haldex unit is controlled by the ECU, which monitors wheel slippage.

controls the amount of torque applied to each of the rear wheels. The system's operation is controlled by a separate ECU that receives inputs from and shared information with the engine, transmission, and ABS/ESP ECUs. Based on the inputs, the system can control wheel slip, improve overall vehicle handling, counteract oversteer and understeer, and improve stability.

Other Systems

Manufacturers are constantly developing new 4WD systems. Many of these new systems are modifications of previously used systems. The following are some examples of these.

BMW xDrive BMW has an intelligent, permanent AWD system called the xDrive. Ford also uses a similar system. These systems rely on electronics to vary the distribution of engine power between the front and rear axles and to each wheel on the rear axle. The drive system is tied to the stability control system, the active steering system, and the ABS. It helps steer the vehicle by sending torque to either of the rear wheels, which not only prevents wheel slip but also eliminates oversteer and understeer. The system is designed to anticipate slippage and handling concerns by monitoring many inputs on the CAN bus.

Normally the system provides a 40/40 split of power between the front and rear drive axles but is capable of sending 100% of the engine's power to either the front or rear axles. The system is capable of responding to needs within 100 milliseconds, which is much quicker than conventional AWD systems; this is due to the advanced electronics involved.

In addition to the electronics, the system has a rear differential that contains two planetary gear sets and two electrically activated clutch packs that are capable of multiplying torque to each rear wheel. In addition, there is a power divider that splits the power from the front and rear axles; this unit also uses an electrically controlled clutch pack.

The system works to counteract any oversteer or understeer during turns by moving power to the front or rear. When the possibility of understeer is detected, the system reduces power to the front wheels. When oversteering is detected, more power is sent to the front axle.

Ford Ford also uses a system called Control Trac II. This system uses a simple power takeoff on the front transaxle to transfer torque through a two-piece drive shaft to a coupling at the rear axle. The coupling has a rotary blade pump comprised of a disc, a piston, and a chamber filled with silicone fluid. The rotation of the blades creates fluid pressure that engages the clutch. Torque to the rear wheels increases as the pressure on the clutch plates increases. When the front wheels lose traction, the drive shaft to the rear axle spins faster, causing the coupling's rotary blade to rotate faster. This increases the hydraulic pressure on the piston, which increases the torque to the rear axle.

GM Similar to the Ford system, GM's Versatrak uses a gerotor pump in the rear axle coupling to build oil pressure against a piston and the clutch pack to transfer torque from one side of the rear axle to the other. When the front wheels have good traction, the drive shaft to the rear axle axle and the rear ring gear spin freely. The clutches in the coupling are disengaged and no torque is transferred to the rear wheels. When the front wheels lose traction, the speed of the drive shaft and ring gear increases and they rotate at a greater speed than the rear drive axles. This causes the coupling's oil pump to turn faster and produce a pressure to engage the clutch packs. As the clutch packs engage, they become locked to the differential housing and transfer torque to the rear drive axles.

Honda The Acura MDX has an electronically controlled rear drive axle. In the rear axle there are two electromagnetic multidisc clutch packs—one at each rear axle shaft. These clutch packs are controlled by an ECU that monitors the powertrain and the ABS. The action of the clutch packs is controlled by varying the current sent to them. The system controls wheel spin and can transfer torque to the rear wheels during acceleration or to reduce torque steer. Like Ford's Control Trac II, this system has a driver-controlled switch that engages both clutch packs to lock the system if the vehicle gets stuck.

Mitsubishi S-AWC Mitsubishi's Super All Wheel Control (S-AWC) is a full-time 4WD system used in the Lancer Evolution. The system is integrated with the vehicle's active center differential (ACD), active yaw control (AYC), active stability control (ASC), and sports ABS components. This provides for regulation of torque and braking force at each wheel.

The ACD is an electronically controlled hydraulic multiplate clutch that limits the action of the center differential gears. The ACD regulates the torque split between the front and rear drive axles. The AYC acts like a limited slip differential by reducing rear-wheel slip to improve traction. It controls rear-wheel torque to limit the yaw of the vehicle to improve the vehicle's cornering performance. The ASC regulates engine power and the braking force at each wheel. This system improves vehicle stability and improves traction during acceleration. The system relies on two ECUs: one controls the ACD and AYC and the other controls ASC and ABS. The ECUs communicate to each other via the CAN bus.

LOCKING HUBS

Many 4WD vehicles have front-wheel locking hubs. These hubs connect the front wheels to the front drive axles when they are in the locked position. Manual locking hubs require that a lever or knob be turned by hand to the 2WD or 4WD position. Automatic locking hubs can be locked by shifting to 4WD and moving forward slowly. They are unlocked by slowly backing up the vehicle. On certain late-model 4WD systems, a front axle lock is used in place of individual locking hubs.

To reduce wear on the driveline of the front axle, the hubs are unlocked when the vehicle is in 2WD, which allows the wheels to rotate independently of the drive axles. Locking hubs are also called freewheeling hubs because their purpose is to disengage the front axle and allow it to free wheel while the vehicle is operating in 2WD. If the hubs remain in the 4WD or locked position when the vehicle is in 2WD, the driver will notice a drop in fuel economy and excessive noise will come from the front axle.

A locking hub is a type of clutch that engages or disengages the outer ends of the front axle shafts from the wheel hub. When the hub is in the locked position, the ring of the clutch is set onto the splines of the axle shaft. When the hub is in the unlocked position, spring pressure forces the clutch ring away from the axle shaft, thereby disconnecting the wheel hub and the axle.

Although automatic hubs are more convenient for the driver, they do have a disadvantage. Many self-locking hub designs are unlocked when the vehicle is moved in reverse. Therefore, if the vehicle is stuck and needs to back out of the trouble spot, only RWD will be available to move it. Other automatic hubs unlock immediately when 4WD is disengaged without the need to back up. On these systems, the hubs are automatically locked, regardless of the direction the vehicle is moving.

Locking hubs are not needed with full-time 4WD. The wheels and hubs are always engaged with the axle shafts. The interaxle differential or transfer case prevents damage and undue wear of the parts of the powertrain.

Some 4WD vehicles use a vacuum motor or mechanical linkage to move a splined collar to connect or disconnect the front drive axle (**Figure 43–20**). With this system, locking hubs are not needed. When 2WD is selected, one axle is disconnected from the front differential. As a result, all engine torque move to the side of the differential that has the disconnected axle. This is due to normal open differential action. When the vehicle is shifted into 4WD, the shift collar connects the two sections of the axle shaft together.

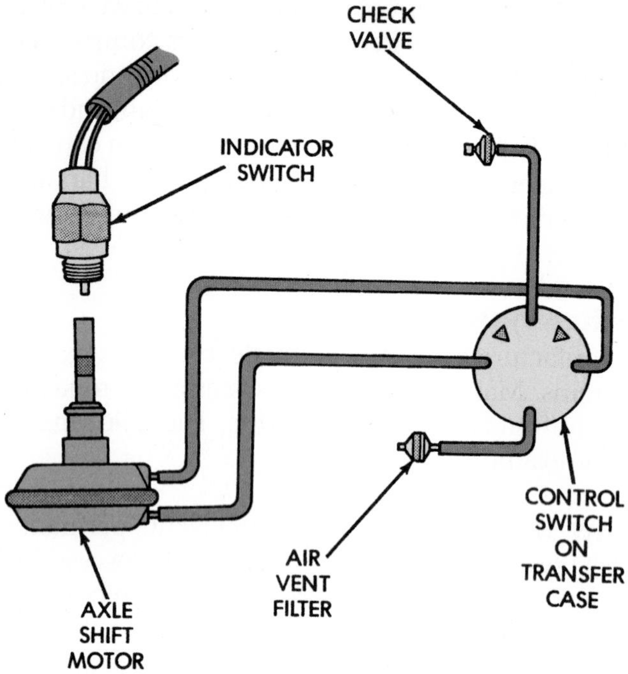

Figure 43-20 Axle disconnect system using a vacuum-operated shifter. *Courtesy of Chrysler LLC*

Other axle disconnects are operated electrically. An electric motor can be used to connect and disconnect the axle (**Figure 43–21**). This system allows for a smooth transition from 2WD to 4WD. GM uses a system whereby selecting 4WD on the selector switch energizes a heating element in the axle disconnect. The heating element heats a gas, causing the plunger to operate the shift mechanism.

FOUR-WHEEL-DRIVE PASSENGER CARS

While most 4WD trucks and utility vehicles are design variations of most basic rear-wheel-drive vehicles (**Figure 43–22**), most passenger cars featuring 4WD were developed from a front-wheel-drive base model. They feature a transaxle and differential that drive the front wheels, plus some type of mechanism for connecting the transaxle to a rear driveline. In many cases this is a simple clutch-type engagement.

Four-wheel-drive passenger cars normally differ from heavier-duty 4WD trucks and vehicles in several other ways. First, there is no separate transfer case;

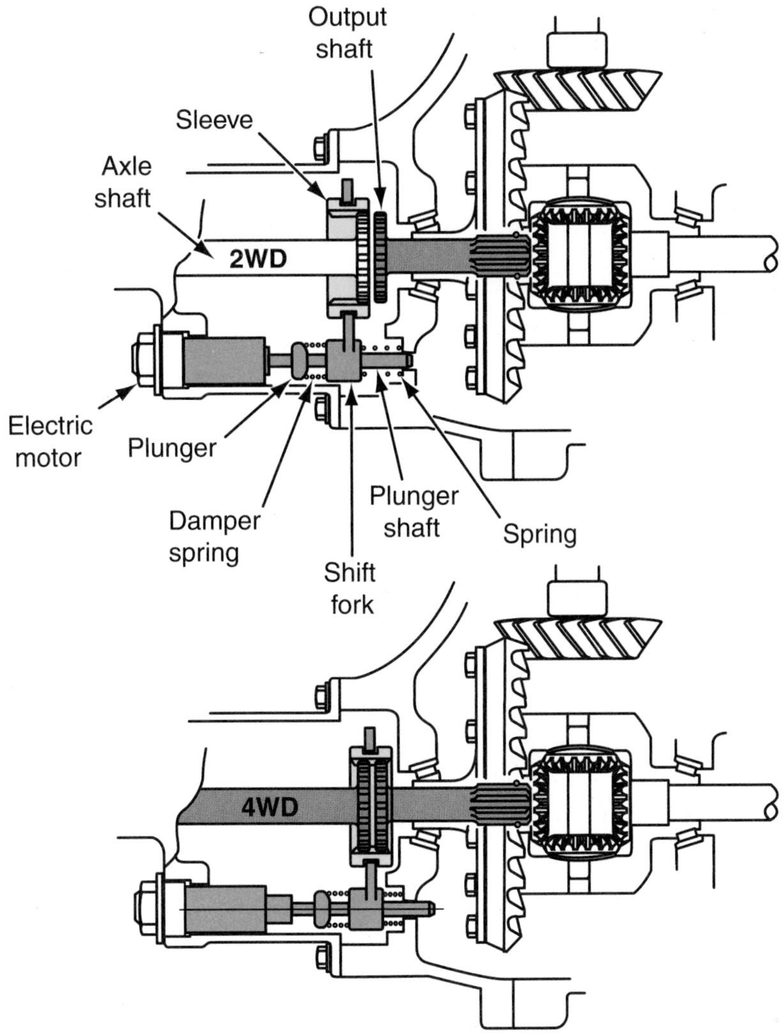

Figure 43-21 An electric motor disconnects the axles in this system.

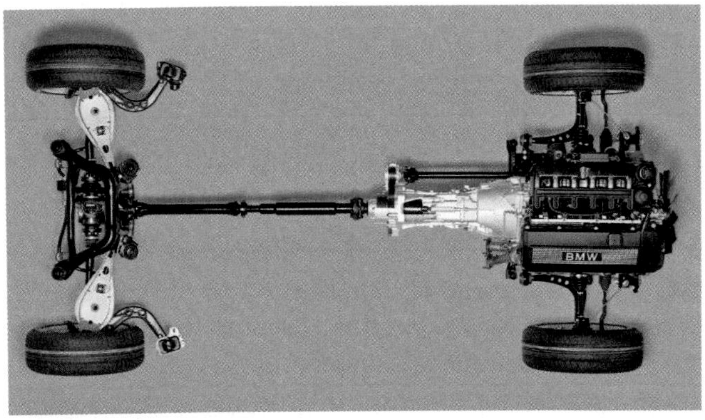

Figure 43-22 An all-wheel-drive system based on a RWD platform with a transfer case and three differentials. *Courtesy of BMW of North America, LLC*

Figure 43-23 This clutch assembly eliminates the need for a center differential and operates mechanically, not electronically. *Courtesy of American Honda Motor Co., Inc.*

any gearing needed to transfer power to the rear driveline is usually contained in the transaxle housing or small bolt-on extension housing.

Some AWD vehicles do not have a center differential; rather, a multiple-disc clutch is placed at the input of the rear differential **(Figure 43–23)**. In these systems, all torque is normally sent to the front wheels. If one of the front wheels begins to slip, the clutch at the rear differential can send torque to the rear wheels.

Most passenger cars with 4WD are not designed for the rigors of off-road driving, in which clearance over rocks and debris is needed. For simple road driving there are no clearance or durability problems.

USING SERVICE INFORMATION

Because of the wide variety of AWD and 4WD systems on the road today, make sure you locate the specific system in the service manual before beginning to service it. Normally, four-wheel-drive systems are covered in two distinct sections: transfer cases and propeller shafts.

Customer Care

Advise your customer that wear on 4WD systems is much greater than on a two-wheel-drive transaxle or transmission, especially if drivers leave the four-wheel drive engaged on dry pavement. Some car makers warn that abuse of the system is not covered under the warranty.

Customer Care

It is very important to remind your customers that the fluid level in a transfer case must be checked at recommended time intervals. (Check the service manual.) It is wise to show your customers where the transfer case fill plug is located and how to remove it. Also tell them the lubricant should be almost even with the fill hole. Always refer to the service manual for recommended transfer case lubricants. Many transfer cases require extreme pressure (EP) lubricants as used in differentials and in some manual transmissions.

ALL-WHEEL-DRIVE SYSTEMS

All-wheel-drive systems do not give the driver the option of 2WD or 4WD. They always drive all wheels. All-wheel-drive vehicles are usually passenger cars that are not designed for off-road operation. The all-wheel-drive-vehicle is designed to increase vehicle performance in poor traction situations, such as icy or snowy roads, and in emergencies. All-wheel drive gives the vehicle operator maximum control in adverse operating conditions by biasing the driving torque to the axle with driving traction. The advantage of all-wheel drive can be compared to walking on snowshoes. Snowshoes prevent the user from sinking into the snow by spreading the body weight over a large surface. All-wheel-drive vehicles spread the driving force over four wheels when needed rather than two wheels. When a vehicle travels over the road, the driving wheels transmit a tractive force to the road's surface. The ability of each tire to transmit tractive force is a result of vehicle weight pressing the tire into the road's surface and the coefficient of friction between the tire and the road. If the road's surface is dry and the tire is dry, the coefficient of friction is high and four driving wheels are not needed. If the road's surface is wet and slippery, the coefficient of friction between the tire and road is low. The tire loses its coefficient of friction on slippery road surfaces, which could result in loss of control by the operator. Unlike a two-wheel-drive vehicle, an all-wheel-drive vehicle spreads the tractive effort to all four driving wheels. In addition to spreading the driving torque, the all-wheel-drive vehicle biases the driving torque to the axle that has the traction only when it is needed.

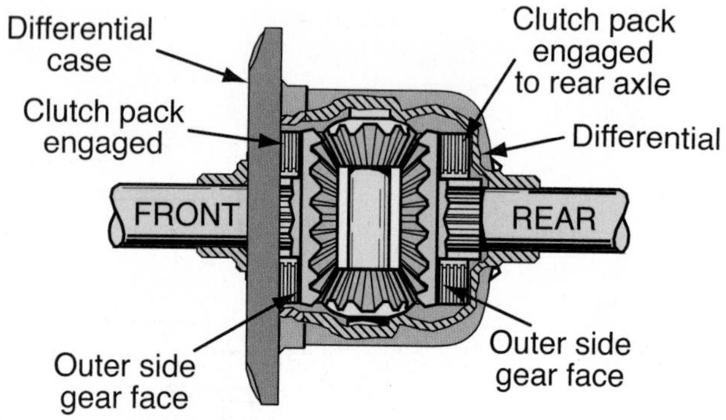

Figure 43-24 A cutaway of a center differential.

Center Differential All-Wheel Drive

Recent AWD designs feature a center differential to split the power between the front and rear axles **(Figure 43–24)**. On the manual transmission model, the driver can lock the center differential with a switch. On the automatic transmission model, the center differential locks automatically, depending on which transaxle range the driver selects and whether there is any slippage between front and rear wheels.

DIAGNOSING 4WD AND AWD SYSTEMS

Current 4WD and AWD systems are controlled by the PCM or a separate control module. Like all computer-monitored and computer-controlled systems, faults

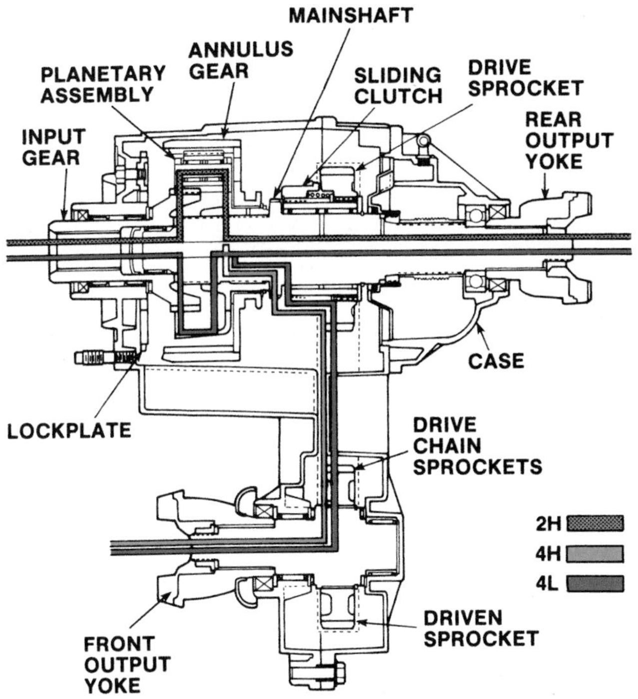

Figure 43-25 The power flow through transfer case in 2H, 4H, and 4L modes. *Courtesy of Chrysler LLC*

have to be identified by DTCs. It is also valuable to understand the power flow within the transfer case (**Figure 43–25**). This can help you identify what could be going wrong internally in the transfer case.

PROCEDURE

To diagnose late-model 4WD systems, follow these steps:

STEP 1 Verify the customer's complaint.

STEP 2 Visually inspect for obvious signs of mechanical or electrical damage. Check the condition of the following (some systems will not have all of these):

- Half shafts
- Locking hubs
- Drive shaft and U-joints
- Shift lever and/or linkage
- Mode switch
- Shift motor
- Electromagnetic clutch
- Vacuum and fluid lines
- Matching tire sizes
- Transfer case
- System fuses
- Wiring harnesses and connectors

STEP 3 If the cause was not visually evident, connect the scan tool to the DLC.

STEP 4 Note the DTCs retrieved.

STEP 5 Conduct the self-test for the 4WD control module.

STEP 6 If the DTCs retrieved are related to the concern, interpret the codes and follow the specific pinpoint tests for those codes.

STEP 7 If the DTCs are not related to the concern or there are no DTCs, go to the manufacturer's symptom chart (**Figure 43–26**).

STEP 8 Identify and repair the cause of the problem.

STEP 9 Verify the repair.

Basic System Diagnosis

Computer-controlled 4WD systems are diagnosed just like other computer-controlled systems. Make sure the scan tool can communicate with the vehicle by completing the test provided by the scan tool. If the system has a separate self-diagnostic routine for the 4WD system, run that and retrieve and record all DTCs. Always follow the procedures outlined by the manufacturer when connecting a scan tool to the system. Also always use the manufacturer's specifications when interpreting data.

Component Testing Individual inputs should be tested according to the procedures recommended by the manufacturer. The speed sensors are typically Hall-effect or magnetic pulse units. These are best checked with a lab scope. Often, the manufacturer will recommend testing the sensors with a voltmeter or ohmmeter. Normally, these tests involve taking readings across a multiple-pin connector and at specific terminals.

The selector switch can be checked by connecting an ohmmeter across the terminals of the switch. As the switch is moved to various positions, the circuit through the switch should be either open or closed depending on the position. By referring to the wiring diagram, you can easily identify what should happen in the different switch positions. If the switch does not function as it should, it must be replaced.

When the transfer case shift motor is suspected as being faulty, begin diagnosis with a careful inspection. Then follow the manufacturer's procedures for testing it. Often, testing involves checking the motor for opens, shorts, and high resistance. If the motor is found to be defective, it must be replaced. Where placing the motor, make sure it is

Condition	Possible Cause
No communication with the control module (PCM)	Scan tool Data link connector (DLC) Control module (PCM) Circuitry
The 4WD indicators do not operate correctly or do not operate	Indicator lamp Circuitry Control module (PCM) Ignition switch
The vehicle does not shift between modes correctly	Mode select switch Transfer case Transfer case clutch Control module (PCM) Circuitry Locking hubs Ignition switch Transfer case shift lever Mode indication switch Ignition switch
4WD does not engage correctly at speed	Transfer case clutch coil Control module (PCM) Locking hubs Ignition switch
The front axle does not engage or disengage correctly or makes noise in 2WD under heavy throttle	Mode select switch Locking hubs Vacuum leaks Control module (PCM) Front half shaft Ignition switch
The transfer case jumps out of gear	Transfer case Vacuum solenoid vent Mode select switch Transfer case Shift lever
Driveline wind-up while moving straight	Tire inflation pressure Tire and wheel size Tire wear Axle ratio
Grinding noise during 4WD engagement, especially at high speeds	The front half shafts are not turning at the same speed
Flashing 4WD indicators	Loss of CAN communication between 4WD control module and instrument cluster Loss of high-speed (HS-CAN) communication between 4WD control module and PCM Ignition switch
The transfer case makes noise	Tire inflation pressure Unmatched tire and wheel size Tire tread wear Internal components Fluid level
The vehicle binds in turns, resists turning, or pulsates or shudders while moving straight	Unmatched tire sizes Unequal amounts of tire wear Unequal tire inflation pressures Unmatched front and rear axle ratios

Figure 43-26 A typical symptom chart.

mounted correctly and its retaining bolts are tightened to specifications.

If the system uses a magnetic clutch, it can be checked with an ohmmeter. Noise from the clutch can be caused by a faulty clutch or a clutch that needs adjustment.

Shift-on-the-Fly Systems Shift-on-the-fly systems switch between 2WD and 4WD electrically by a switch on the instrument panel. Many of these systems use electrically controlled vacuum motors to connect and disconnect the wheels from the axle and an electrical motor or electromagnetic clutch at the transfer case to transfer power to the additional axle. The 4WD indicator light will illuminate when 4WD is engaged. If there is a problem with the system or if 4WD cannot be engaged, the indicator lamp will blink to notify the driver of a problem. Diagnosis begins with watching the frequency of the blinking lamp. Check the service information for an interpretation of the blinking lamp.

If the vehicle will not engage into 4WD, there may be a faulty transfer case actuator motor or faulty vacuum solenoid assembly. If the motor in the transfer case is a possible cause, remove it and check its operation. Check the transfer case and repair or replace any faulty components if the motor is not the cause of the problem. If no problem is found in the transfer case, it is likely that the electronic control unit is faulty.

Check for engine vacuum at the actuator and/or solenoid before checking anything else if the vacuum actuator is suspect. If no vacuum is present when the engine is running, check the transfer position switch. A voltage signal should be sent to the vacuum solenoid when 4WD is selected. If no signal is being sent, then the switch or switch circuit is faulty. However, if the solenoid is receiving a voltage signal, suspect a faulty solenoid assembly. Check the vacuum circuit for leaks if there are no electrical problems.

Vacuum solenoids can be checked by connecting jumper leads from the battery to the terminals of the solenoid. Remove the vacuum solenoids and connect the battery to the terminals of each solenoid. Connect a hand-held vacuum pump to the inlet port of the solenoid and a vacuum gauge to the port for axle engagement. With the battery connected to the solenoid, there should be vacuum at the outlet port to the axle. Move the vacuum gauge to the other outlet port. Vacuum should not be present there. Now disconnect the battery from the solenoid. Recheck the vacuum at the outlet ports. A good solenoid will have vacuum only at the disengagement port when vacuum is applied and the solenoid is not connected to the battery.

On-Demand Systems On-demand systems operate at the discretion of the PCM or override controls selected by the driver. The division of power between the front and rear axles is controlled by the output clutch at the transfer case. The activity of the clutch is controlled by regulating its duty cycle. During normal operation, the duty cycle is low. This allows for a slight speed difference between the front and rear drive shafts, which normally occurs when the vehicle is moving through a curve. When slip is detected at the rear wheels, the duty cycle to the clutch is increased until the difference between the front and rear drive axles is reduced.

To vary the torque split, a computer monitors many things, especially the rotational speeds of the front and rear drive axles. Some systems rely on wheel-speed sensors for this, whereas others have additional speed sensors on the front and rear output shafts from the transfer case. Many of the inputs to the electronic module for 4WD are shared and work with other systems, and they are available on CAN. The best place to start when diagnosing these systems is the service information. Go to the section on the system and identify the components involved in the various modes of operation.

AWD Systems Most AWD systems have a viscous coupling. Problems with AWD systems normally occur with the controls that provide for braking efficiency and AWD in reverse gear. Chrysler's AWD system uses a vacuum solenoid setup to provide AWD in reverse. The solenoids are used to bypass the overrunning clutch. If AWD is not available in reverse gear, suspect the vacuum solenoids, their controls, or the dog-clutch assembly.

An overrunning clutch is used to prevent any feedback of front-wheel braking torque to the rear wheels. This allows the brake system to control braking as if it were a 2WD vehicle. The controls for this feature vary with the manufacturer, and you should always follow the troubleshooting charts provided by the manufacturer of the vehicle.

Axle Hub Diagnosis

Front hubs may make a ratcheting sound when water or dirt has entered the hub and contaminated the lubricant. This prevents free movement of the components in the hub. A ratcheting sound from an automatic locking hub may indicate that the hub on the opposite side of the axle is not disengaging.

Locking hubs can be quickly checked by rotating the brake drum or rotor slightly and turning the hub selector into the lock position. A click should be heard when the hub engages the axle, and the axle should now turn with the hub. Next, with the hub still turning, turn the selector to the free position. The axle should now be free of the hub. If both of these events do not happen, the hub assembly needs to be repaired or replaced.

Wheel Bearings

To check the adjustment of the wheel bearings, raise the front of the vehicle. Grasp each front tire at the front and rear and push the wheel inward and outward. If any free play is noticed between the hub, rotor, and front spindle, adjust the wheel bearings.

Wheel bearings should be disassembled and serviced any time the hubs have been submerged in water. During normal operation the bearings get warm and when they are quickly cooled off by the splash of water, their lubricant breaks down and the bearings can be destroyed. Always replace the bearings if they are worn or damaged.

Another frequent cause of wheel bearing failure is the use of oversized tires mounted on wheels with substantial offset. These switch the load from the large inner bearing to the small outer wheel bearing, which was never intended to do much more than stabilize the wheel.

SERVICING 4WD VEHICLES

The components of 4WD vehicles are serviced in basically the same way as the same parts of a 2WD vehicle. Also it is important that the U-joints, slip joints, or CV joints in the drivetrain be lubed on a regular basis. To service the drive shafts and U-joints on a 4WD vehicle, use the general instructions given for a cross U-joint. A four-wheel-drive simply uses two drive shafts instead of one.

Go To ***Chapter 39 for a detailed look at servicing joints and drive shafts.***

Servicing the Transfer Case

Be sure to check the manufacturer's service manual for specific transfer case repair and overhaul procedures. It gives details for the particular make and model of transfer case to be worked on.

CAUTION!

When removing and working on a transfer case, be sure to support it on a transmission jack or safety stands. The unit is heavy and, if dropped, could cause part damage and personal injury.

When removing the transfer case, disconnect and remove all drive shaft assemblies. Be sure to mark the parts and their relative positions on their yokes so that the proper driveline balance can be maintained when reassembled. Disconnect the linkage to the transfer case shift lever. Also, disconnect all wires connected to the transfer case. Remove all fasteners holding the case and move it away from the transaxle or transmission.

Once the transfer case has been removed from the vehicle and safely supported, inspect it for signs of oil leaks. Remove the case cover and then carefully loosen and drive out the pins that hold the shift forks in place. Remove the front output shafts and chain drive or gear sets from the case. Keep in mind that some cases use chain drives, whereas others use spur or helical cut gear sets to transfer torque from the transaxle or transmission to the output shafts. Planetary gear sets provide the necessary gear reductions in some transfer cases. Photo Sequence 44 shows the procedure for disassembling a Warner 13-56 transfer case, which is a commonly used unit.

Clean and carefully inspect all parts for damage and wear. Check the slack in the chain drive according to the procedure given in the service manual. Replace any defective parts. It may be necessary to measure the shaft assembly end play. If excessive, new snaprings and shims may be used to correct the situation.

When reassembling the transfer case, the procedure is essentially the reverse of the removal. Be sure to use new gaskets between the covers when reassembling the unit. Photo Sequence 45 shows the procedure for reassembling a Warner 13-56 transfer case.

P44-1 *If not previously drained, remove the drain plug and allow the oil to drain, then reinstall the plug. Loosen the flange nuts.*

P44-2 *Remove two output shaft yoke nuts, washers, rubber seals, and output yokes from the case.*

P44-3 *Remove the four-wheel-drive indicator switch from the cover.*

P44-4 *Remove the wires from the electronic shift harness connector. Record the exact location of each disconnected wire.*

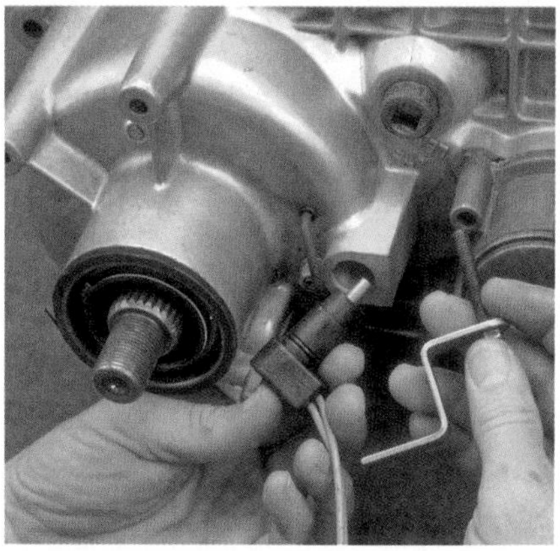

P44-5 *Remove the speed sensor retaining bracket screw, bracket, and sensor.*

P44-6 *Remove the bolts securing the electric shift motor and remove the motor.*

P44-7 *Note the location of the triangular shaft in the case and the triangular slot in the electric motor.*

P44-8 *Loosen and remove the front case to rear case retaining bolts.*

P44-9 *Separate the two halves by prying between the pry bosses on the case.*

PHOTO SEQUENCE 44

Typical Procedure for Disassembling a Warner 13-56 Transfer Case *(continued)*

P44-10 *Remove the shift rail for the electric motor.*

P44-11 *Pull the clutch coil off the mainshaft.*

P44-12 *Pull the 2WD/4WD shift fork and lockup assembly off the mainshaft.*

P44-13 *Remove the chain, driven sprocket, and drive sprocket as a unit.*

P44-14 *Remove the mainshaft with the oil pump assembly.*

P44-15 *Slip the high-low range shift fork out of the inside track of the shift cam.*

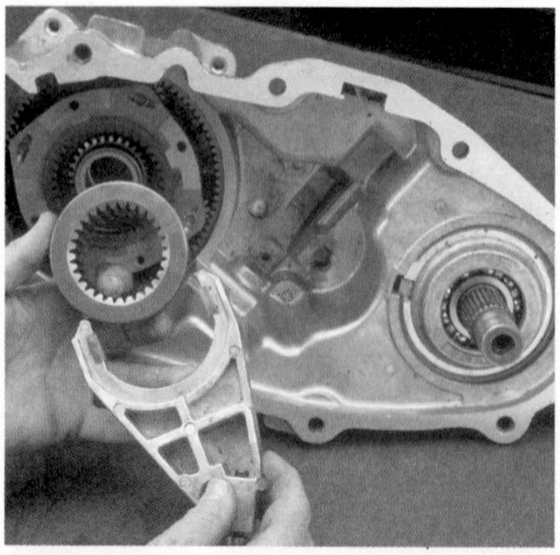

P44-16 *Remove the high-low shift collar from the shift fork.*

P44-17 *Unbolt and remove the planetary gear mounting plate from the case.*

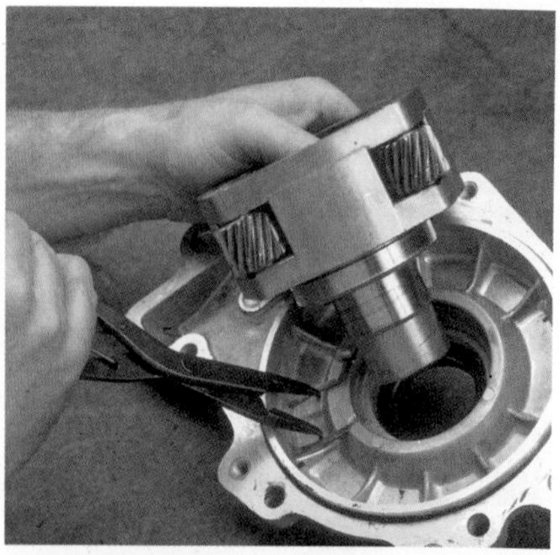

P44-18 *Pull the planetary gear set out of the mounting plate.*

P45-1 *Install the input shaft and front output shaft bearings into the case.*

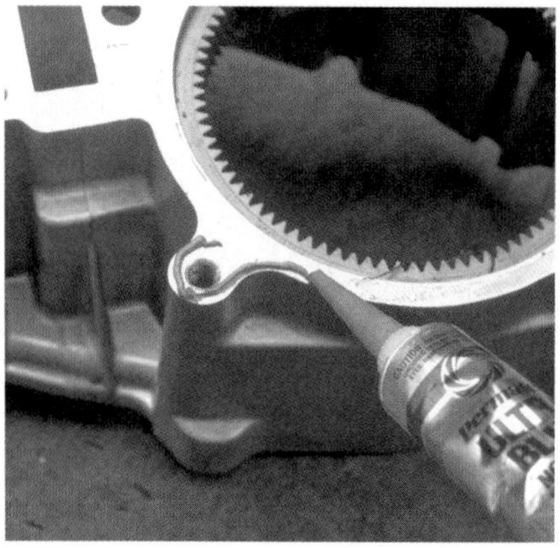

P45-2 *Apply a thin bead of sealer around the ring gear housing.*

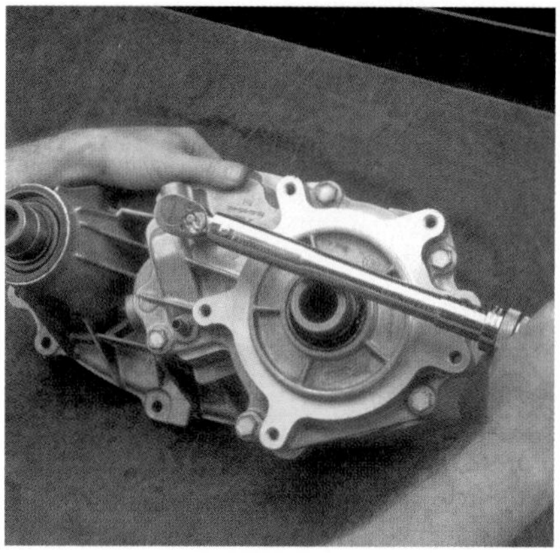

P45-3 *Install the input shaft with planetary gear set and tighten the retaining bolts to specifications.*

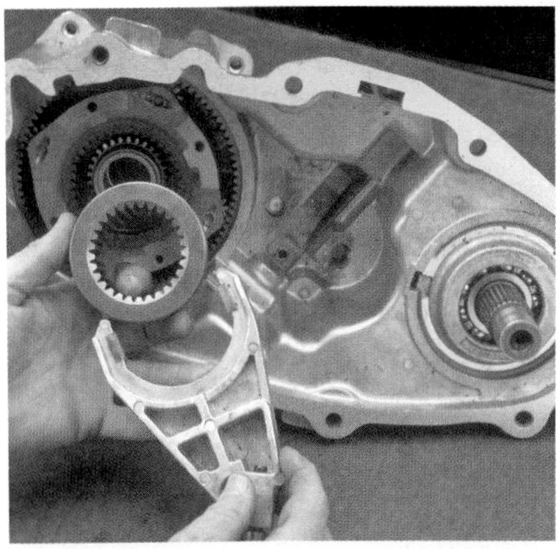

P45-4 *Install the high-low shift collar into the shift fork.*

P45-5 *Install the high-low shift assembly into the case.*

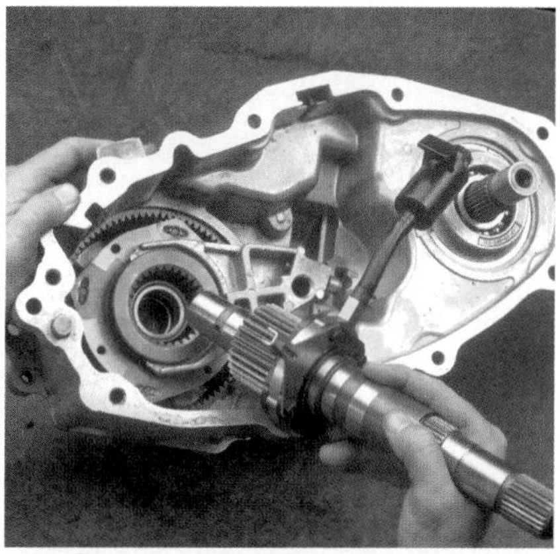

P45-6 *Install the mainshaft with oil pump assembly into the case.*

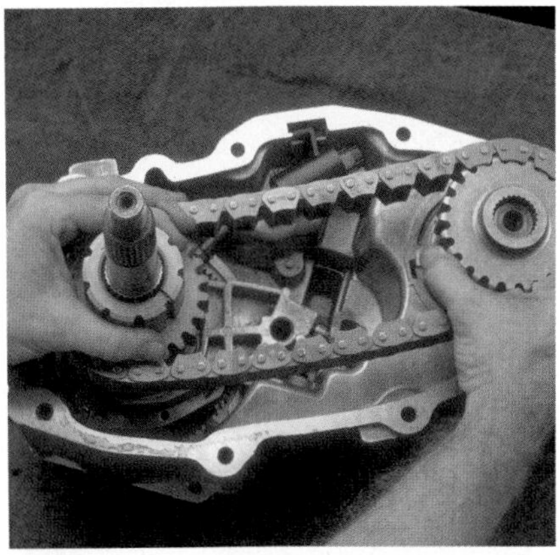

P45-7 *Install the drive and driven sprockets and chain into position in the case.*

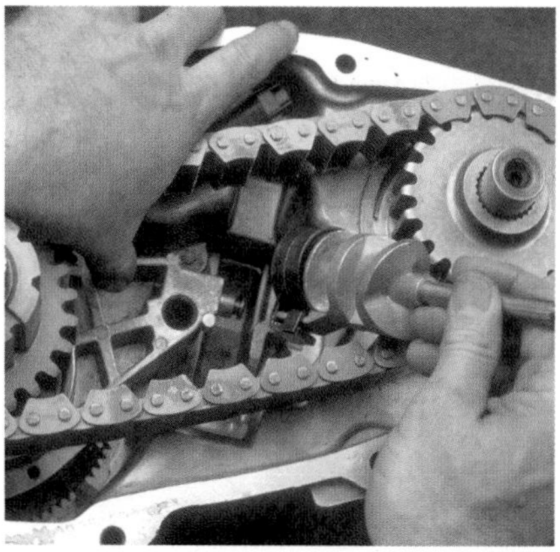

P45-8 *Install the shift rails.*

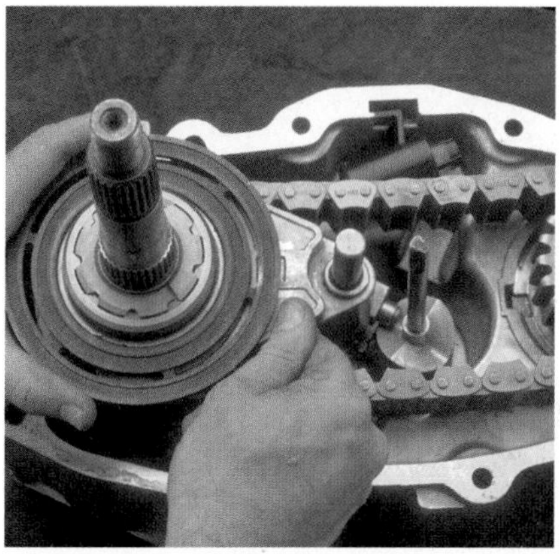

P45-9 *Install the 2WD/4WD shift fork and lockup assembly onto the mainshaft.*

PHOTO SEQUENCE 45 Typical Procedure for Reassembling a Warner 13-56 Transfer Case *(continued)*

P45-10 *Install the clutch coil onto the mainshaft.*

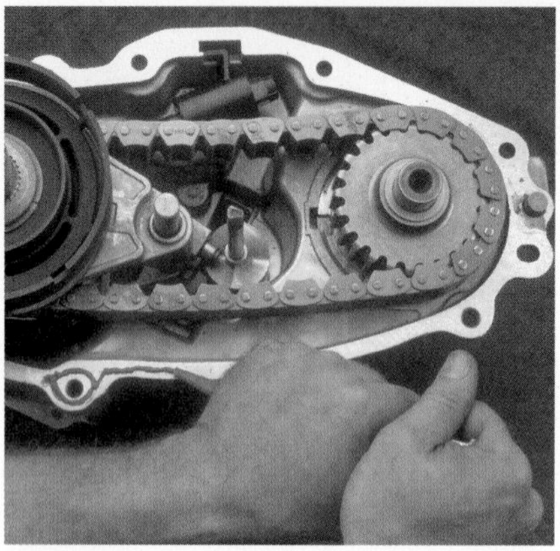

P45-11 *Clean the mating surface of the case. Apply the recommended sealant on the areas where the case halves join.*

P45-12 *Position the shafts and tighten the case halves together. Tighten attaching bolts to specifications.*

P45-13 *Apply a thin bead of sealer to the mating surface of the electric shift motor.*

P45-14 *Align the triangular shaft with the motor's triangular slot.*

P45-15 *Install the motor over the shaft, and wiggle the motor to make sure it is fully seated on the shaft.*

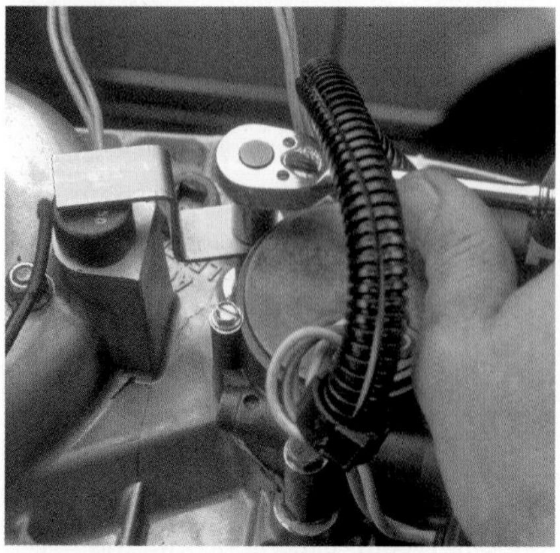

P45-16 *Tighten the motor's retaining bolts to specifications.*

P45-17 *Reinstall the wires into the connector and connect all electric sensors.*

P45-18 *Install the companion flanges' seal, washer, and nut. Then tighten the nut to specifications.*

Front Axles and Hubs

The axles and locking hubs must be removed before the front differential and final drive assembly can be removed from the axle housing. Removing these components requires special procedures, and each type of 4WD system requires its own specific procedure.

Removal of the axle shafts begins with the removal of the hubs. Normally, the procedures for replacing manual hubs are somewhat different from those for replacing automatic hubs. Locking hubs are not serviceable; therefore, service of the hubs consists of merely replacing them. Also, the procedure for the replacement of the hubs varies with the type of hub and the manufacturer. The following are general procedures for the removal and installation of both manual and automatic locking hubs. Always follow the specific procedures for the type of hub you are servicing.

Manual Locking Hub To remove a manual hub, set the hub's control knob to the "Free" or unlocked position. Then remove the cover bolts and the outer snapring. If hubs are equipped with shims, remove them as well. Then remove the drive flange or gear from the axle. Now remove the bolts that attach the hub body to the hub. The hub assembly can now be pulled from the axle shaft. After the hub assembly has been removed, inspect the splines of the axle shaft for nicks and burrs.

To install a manual hub, separate the base and handle units of the hub lock assembly. Apply a light coat of grease to the axle shaft splines and the hub lock base. Also apply a light coating of lubricant to the O-ring. Make sure the gasket surface is smooth and clean, then place a new hub gasket onto the hub. Set the control handle to the unlocked position and install the hub assembly and the snapring onto the axle shaft. Tighten the attaching bolts to the specified torque. Install the hub cover with a new gasket onto the hub and tighten the remaining attaching bolts to specifications. Check the control handle for ease of operation.

Automatic Locking Hub There are basically two designs of automatic locking hubs: internally and externally retained. The procedure for servicing automatic hubs depends on how they are retained. Remove the bolts from the outer cap assembly and pull the cap off the body of the hub. Then remove the axle shaft bolt, lock washer, and axle shaft stop. In the wheel hub there is a groove that retains a lock ring; remove it and slide the hub assembly off the wheel's hub. Then loosen the set screws in the spindle locknut until their heads are flush with the face of the locknut. Remove the spindle locknut.

To install an automatic hub, adjust the wheel bearings and install the spindle locknut and tighten it to specifications. Firmly tighten the locking set screws. Apply multipurpose grease to the inner splines of the hub; do not pack the hub with grease. Slide the assembly into the wheel hub. Push firmly on the body until it seats in the hub. Then install the lock ring in the groove of the wheel hub. Place the lock washer and axle shaft stop onto the axle bolt. Install the bolt and tighten it to specifications. Apply a small amount of lube to the seal on the cap assembly. Do not grease the cap assembly. Install the cap assembly over the body assembly and into the wheel hub. Install the attaching bolts and tighten them to specifications. Firmly turn the control from stop to stop to check the assembly. Then set both hub controls in the same position: "auto" or "lock."

Wheel Bearings

The front-wheel bearings on a 4WD vehicle should be disassembled, cleaned, inspected, and lubricated on a regular basis. Upon reassembly, it is very important that the bearings be properly adjusted. The procedure for doing this is similar to other wheel bearings. However, because of the load on the bearings, the adjustment is more critical.

To adjust the wheel bearings, begin by removing the snapring at the end of the spindle, the axle shaft spacer, needle thrust washer, bearing spacer, outer wheel bearing locknut, and bearing washer. Then loosen the inner bearing locknut and tighten it fully to seat the bearings. The wheel bearing locknuts on most 4WD vehicles require the use of a four- or six-pronged spanner wrench. Spin the brake rotor on the spindle, then loosen the inner bearing locknut approximately ¼ turn. Install the outer bearing locknut and tighten it to the specified amount, usually 70 to 90 ft.-lb (95 to 122 N-m). Assemble the remaining parts of the hub assembly, making sure that the snapring is fully seated in its groove on the axle.

CASE STUDY

A four-wheel-drive pickup truck is brought into the shop. There is a severe vibration through the truck when it is traveling at road speeds. New wheels and tires were recently installed on the truck. The new wheels and tires are drastically oversized. Lift kits were installed by the owner to provide the necessary tire clearance. The owner would like the alignment checked and the wheels balanced.

The technician carefully balances and aligns the wheels. He takes the truck for a test drive to verify the repair. The truck still has a harsh vibration. It seems to get worse as speed increases. A bent wheel or bad wheel bearing or hub is suspected.

After returning to the shop, he measures the runout and checks the play of each wheel. Finding nothing wrong, he carefully checks the roundness of the tires. While doing so on the rear tires, he discovers the cause of the problem. When the lift kit was installed, the operating angle of the rear U-joint was affected. As the drive shaft turns, it hesitates slightly. This hesitation is caused by a binding of the U-joint.

The technician informs the owner of the cause and suggests that the lift kits be removed or the drive shaft operating angle be corrected.

KEY TERMS

All-wheel drive (AWD)
Driveline windup
Four-wheel drive (4WD)
Haldex clutch
Interaxle differential
Locking hubs
Power takeoff
Transfer case
Viscous clutch

SUMMARY

- The importance of excellent traction and the benefits of four-wheel- and all-wheel-drive systems become readily apparent in snow or heavy rain.
- The heart of most conventional 4WD systems is the transfer case.
- The interaxle is placed in the transfer case to operate in the same fashion as the differentials at the axles. The only difference is that the transfer differential controls the speed of the drivelines instead of the axles.
- Many vehicles require that the front hubs be in a locked condition to operate as 4WD vehicles. This lockup may be made either manually or automatically.
- On 4WD vehicles with on-the-fly shifting, it is not necessary to come to a complete stop when changing operational modes.
- Components of 4WD vehicles can be serviced in basically the same manner as the identical components on a 2WD vehicle.
- All-wheel-drive vehicles may use a viscous clutch, rather than a transfer case, to drive the axle with low tractive effort.
- Many AWD vehicles have a third differential, called an interaxle differential, instead of a transfer case.
- All-wheel-drive passenger cars are typically based on FWD platforms.

REVIEW QUESTIONS

1. Name the three main driveline components that are added to a RWD vehicle to make it a 4WD vehicle.
2. Describe the purpose of a viscous clutch.
3. What is the purpose of the interaxle differential?
4. Briefly explain how a Haldex clutch works.
5. What is the primary purpose of a transfer case?
6. What are the primary differences between a full-time and a part-time 4WD system?
7. When the plates of a viscous coupling (clutch) rotate at different speeds, the plates ______ the fluid.
8. How does a viscous clutch assembly work? Why is it used in AWD systems?
9. What results from having different axle ratios on the front and rear of a 4WD vehicle when it is operating in 4WD?
   a. poor handling on dry surfaces
   b. driveline windup
   c. mechanical damage to the driveline
   d. all of the above

10. The transfer clutch in the all-wheel-drive automatic transaxle takes the place of the ______.
    a. transmission
    b. reduction gears
    c. torque converter
    d. interaxle differential
11. In a viscous clutch, when the silicone fluid is heated, it ______.
    a. becomes a very thin fluid
    b. boils to a vapor
    c. thickens to a solid mass
    d. stiffens
12. When servicing transfer cases, all of the following are correct, *except* ______.
    a. it must be supported on a transmission jack or safety stands
    b. the chain drive and a planetary gear set must be adjusted to specifications
    c. visual inspections for leaks and damage are necessary
    d. follow the manufacturer's recommendations for lubricants
13. *True or False?* The basic difference between 4WD and AWD systems is that 4WD systems have a two-speed transfer case.
14. Which of the following would *not* cause noise while driving in 2WD under a heavy throttle?
    a. faulty ignition switch
    b. worn transfer case clutch
    c. inoperative locking hubs
    d. bad front half shaft
15. Which of the following could cause the 4WD indicators not to operate correctly?
    a. a faulty control module (PCM)
    b. inoperative locking hubs
    c. worn transfer case clutch
    d. damaged transfer case shift lever

ASE-STYLE REVIEW QUESTIONS

1. Technician A says that viscous couplings are electronically controlled. Technician B says that Haldex clutches are electronically controlled. Who is correct?
    a. Technician A   c. Both A and B
    b. Technician B   d. Neither A nor B
2. Technician A says that some AWD systems have a center differential. Technician B says that some AWD vehicles have a viscous clutch. Who is correct?
    a. Technician A   c. Both A and B
    b. Technician B   d. Neither A nor B
3. While discussing AWD systems: Technician A says that some systems are electronically controlled and use an electromagnetic clutch to engage and disengage the drive axles. Technician B says that some cars do not use a transfer case, but they are equipped with a third differential unit. Who is correct?
    a. Technician A   c. Both A and B
    b. Technician B   d. Neither A nor B
4. Technician A says that the 4WD mode of part-time 4WD systems is designed to be used only when driving off the road or on slippery surfaces. Technician B says that AWD systems are intended for off-the-road use only. Who is correct?
    a. Technician A   c. Both A and B
    b. Technician B   d. Neither A nor B
5. While diagnosing the cause of noise from a transfer case: Technician A says that a probable cause is incorrect tire size. Technician B says that the most probable cause is a faulty bearing or gear. Who is correct?
    a. Technician A   c. Both A and B
    b. Technician B   d. Neither A nor B
6. Technician A says that a seized shift linkage may cause the transfer case to operate only in 4WD. Technician B says that a worn and loose linkage may cause a transfer case to operate only in 2WD. Who is correct?
    a. Technician A   c. Both A and B
    b. Technician B   d. Neither A nor B
7. While diagnosing the cause of the transfer case jumping out of gear: Technician A says that the most probable cause is worn or broken synchronizers. Technician B says that a possible cause is an improperly adjusted shift linkage. Who is correct?
    a. Technician A   c. Both A and B
    b. Technician B   d. Neither A nor B
8. While servicing an AWD vehicle: Technician A checks the service manual to identify the correct lubricant for the center differential. Technician B says that the same lubricant is used for the

viscous coupling, transfer case, and differentials on most vehicles. Who is correct?

a. Technician A
b. Technician B
c. Both A and B
d. Neither A nor B

9. While discussing the various speed gear positions of a transfer case: Technician A says that when the transfer case is in low, the overall gear ratio is numerically increased. Technician B says that when the transfer case is in the high position, the vehicle operates in an overdrive mode due to the decrease in torque multiplication. Who is correct?

a. Technician A
b. Technician B
c. Both A and B
d. Neither A nor B

10. While discussing automatic AWD systems: Technician A says that the systems operate in 2WD nearly all of the time. Technician B says that on many systems, when the PCM receives inputs from the CAN bus, it indicates that wheel slippage power is sent to the other drive axle. Who is correct?

a. Technician A
b. Technician B
c. Both A and B
d. Neither A nor B

SUSPENSION AND STEERING SYSTEMS

Section 7

CHAPTER 44

TIRES AND WHEELS

OBJECTIVES

■ Describe basic wheel and hub design. ■ Recognize the basic parts of a tubeless tire. ■ Explain the differences between the three types of tire construction in use today. ■ Explain the tire ratings and designations in use today. ■ Describe why certain factors affect tire performance, including inflation pressure, tire rotation, and tread wear. ■ Describe the operation, diagnosis, and service for a tire pressure monitor. ■ Remove and install a wheel and tire assembly. ■ Dismount and remount a tire. ■ Repair a damaged tire. ■ Describe the differences between static balance and dynamic balance. ■ Balance wheels both on and off a vehicle. ■ Describe the three popular types of wheel hub bearings.

Tire and wheel assemblies provide the only connection between the road and the vehicle. Tire design has improved dramatically during the past few years. Modern tires require increased attention to achieve their full potential of extended service and correct ride control. Tire wear that is uneven or premature is usually a good indicator of steering and suspension system problems. Tires, therefore, become not only a good diagnostic aid to a technician, but can also be clear evidence to the customer that there is need for service.

WHEELS

Wheels are made of either stamped or pressed steel discs riveted or welded together. They are also available in the form of aluminum or magnesium rims that are die-cast or forged (**Figure 44–1**). Magnesium wheels are commonly referred to as mag wheels, although they are usually made of an aluminum alloy. Aluminum wheels are lighter in weight when compared with the stamped steel type. This weight savings is important because the wheels and tires on a vehicle are unsprung weight.

SHOP TALK

Sprung weight represents the weight of the vehicle that is supported by the suspension. The suspension and wheels are unsprung weight. Lower amounts of unsprung weight make the vehicle handle better, primarily on irregular surfaces. It also gives a better ride. This is because when a tire hits a bump, the shock is moved through the tire and wheel to the suspension. The shock that is not absorbed by the suspension then moves to the rest of the vehicle. When the unsprung weight is high, that weight and the shock from the road must be absorbed by the suspension. This means more shock will be passed to the rest of the vehicle.

Near the center of the wheel are mounting holes that are tapered to fit tapered mounting nuts (lug nuts) that center the wheel over the hub. The rim has a hole for the tire's valve stem and a **drop center** area designed to allow for easy tire removal and

Figure 44-1 An alloy wheel on a late-model car.

installation. **Wheel offset** is the vertical distance between the centerline of the rim and the mounting face of the wheel. The offset is considered positive if the centerline of the rim is inboard of the mounting face and negative if outboard of the mounting face. The amount and type of offset is critical because changing the wheel offset changes the front suspension loading as well as the scrub radius.

The wheel is bolted to a **hub**, either by lug bolts that pass through the wheel and thread into the hub, or by studs that protrude from the hub. In the case of studs, special lug nuts are required. A few vehicles have left-hand threads (which turn counterclockwise to tighten) on the driver's side and right-hand threads (which turn clockwise to tighten) on the passenger's side. All other vehicles use right-hand threads on both sides.

Wheel size is designated by rim width and rim diameter (**Figure 44–2**). Rim width is determined by measuring across the rim between the flanges. Rim diameter is measured across the bead seating areas

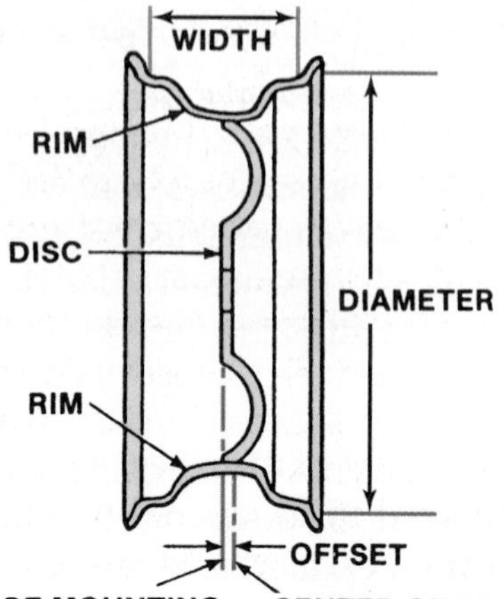

Figure 44-2 Wheel dimensions are important when replacing tires.

from the top to the bottom of the wheel. Some rims have safety ridges near their lips. In the event of a tire blowout, these ridges tend to keep the tire from moving into the dropped center and from coming off the wheel.

Replacement wheels must be equal to the original equipment wheels in load capacity, diameter, width, offset, and mounting configuration. An incorrect wheel can affect wheel and bearing life, ground and tire clearance, or speedometer and odometer calibrations. A wrong size wheel can also affect the antilock brake system. Using the wrong size tire or wheel or improperly inflated tires will affect the rotational speed of the tire, and the ABS will be unable to operate correctly.

Some performance-oriented cars are equipped with different-sized wheels at the front and rear. It is important that the specific size wheel be at the specific axle. The wider or larger wheels are designed to use different-sized tires. For example, a Porsche 911 Turbo has 8.5 × 19 front wheels that are fitted with 235/35 ZR 19 tires. The rear is equipped with 11 × 19 wheels that carry 305/30 ZR 19 tires.

TIRES

The primary purpose of tires is to provide traction. Tires also help the suspension absorb road shocks, but this is a side benefit. They must perform under a variety of conditions. The road might be wet or dry or paved with asphalt, concrete, or gravel, or there might be no road at all. The car might be traveling slowly on a straight road, or moving quickly through curves or over hills. All of these conditions call for special requirements that must be present, at least to some degree, in all tires.

In addition to providing good traction, tires are also designed to carry the weight of the vehicle, to withstand side thrust over varying speeds and conditions, and to transfer braking and driving torque to the road. As a tire rolls on the road, friction is created between the tire and the road. This friction gives the tire its traction. Although good traction is desirable, it must be limited. Too much traction means there is much friction. Too much friction means there is a lot of rolling resistance. Rolling resistance wastes engine power and fuel; therefore, it must be kept to a minimal level. This dilemma is a major concern in the design of today's tires.

Tube and Tubeless Tires

Early vehicle tires were solid rubber. These were replaced with pneumatic tires, which are filled with air.

There are two basic types of pneumatic tires: those that use inner tubes and those that do not. The latter are called tubeless tires and are about the only type used on passenger cars today. A tubeless tire has a soft inner lining that keeps air from leaking between the tire and rim. On some tires, this inner lining can form a seal around a nail or other object that punctures the tread. A self-sealing tire holds in air even after the object is removed. The key to this sealing is a lining of sticky rubber compound on the inside of the tread area that will seal a hole up to 3⁄16 inch (4.76 mm).

A tubeless tire air valve has a central core that is spring-loaded to allow air to pass inward only, unless the pin is depressed. If the core becomes defective, it can be unscrewed and replaced. The airtight cap on the end of the valve provides extra protection against valve leakage. A tubeless tire is mounted on a special rim that retains air between the rim and the tire casing when the tire is inflated.

Figure 44–3 shows a cutaway view of a typical tubeless tire. The basic parts are shown. The cord body or casing consists of layers of rubber-impregnated cords, called **plies**, that are bonded into a solid unit. Typically tires are made of 2, 4, or 8 plies; thus, the reference to 2-, 4-, and 8-ply tires. The plies determine a tire's strength, handling, ride, amount of road noise, traction, and resistance to fatigue, heat, and bruises. The **bead** is the portion of the tire that helps keep it in contact with the rim of the wheel. It also provides the air seal on tubeless tires. The bead is constructed of a heavy band of steel wire wrapped into the inner circumference of the tire's ply structure. The **tread**, or crown, is the portion of the tire that comes in contact with the road surface. It is a pattern of grooves and ribs that provides traction. The grooves are designed to drain off water, while the ribs grip the road surface. Tread thickness varies with tire quality. On some tires, small cuts, called sipes, are molded into the ribs of the tread. These sipes open as the tire flexes on the road, offering additional gripping action, especially on wet road surfaces. The **sidewalls** are the sides of the tire's body. They are constructed of thinner material than the tread to offer greater flexibility.

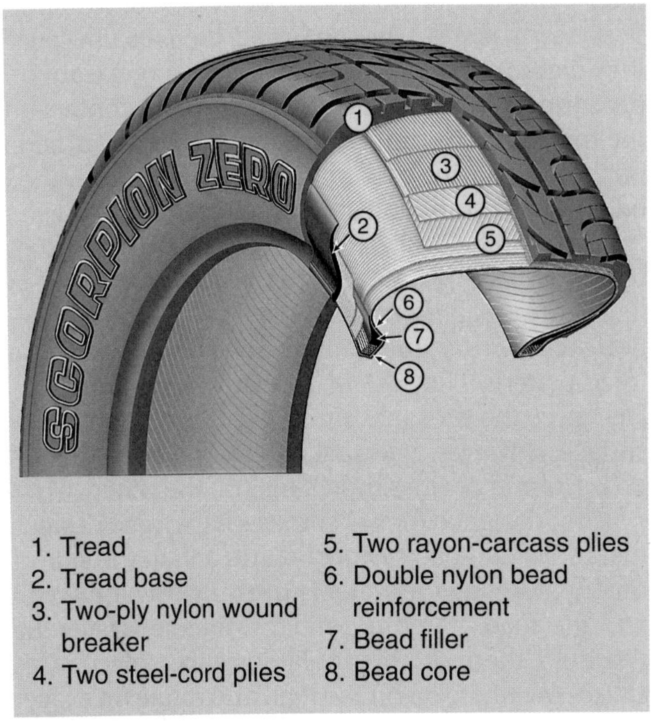

Figure 44-3 A typical tubeless tire. *Courtesy of Pirelli Tire of North America, LLC*

The tire body and belt material can be made of rayon, nylon, polyester, fiberglass, steel, amarid, or kevlar. Each has its advantages and disadvantages. For instance, rayon and cord tires are low in cost and give a good ride, but do not have the inherent strength needed to cope with long high-speed runs or extended periods of abusive use on rough roads. Nylon-cord tires generally give a slightly harder ride than rayon—especially for the first few miles after the car has been parked—but offer greater toughness and resistance to road damage. Polyester and fiberglass tires offer many of the best qualities of rayon and nylon, but without the disadvantages. They run as smoothly as rayon tires but are much tougher. They are almost as tough as nylon, but give a much smoother ride. Steel is tougher than fiberglass or polyester, but it gives a slightly rougher ride because the steel cord does not give under impact, as do fabric plies. Amarid and kevlar cords are lighter than steel cords and, pound for pound, stronger than steel.

Types of Tire Construction

There are three basic types of tire construction: bias ply, belted bias, and radial ply **(Figure 44–4)**. Bias ply and belted bias tires are only used on heavy equipment, trailers, and older cars. Today, nearly all vehicles are fitted with radial tires.

Bias ply tires have fabric plies that run alternatively and form a crisscross design. The angle varies from 30 to 38 degrees with the centerline of the tire. **Belted bias ply** tires are similar to bias ply tires, except that two or more belts run the circumference of the tire under the tread. This construction gives strength to the sidewall and greater stability to the tread.

Radial ply tires have body cords that extend from bead to bead at an angle of about 90 degrees or "radial" to the circumferential centerline of the tire—plus two or more layers of relatively inflexible belts under the tread. The construction of various combinations of rayon, nylon, fiberglass, and steel gives

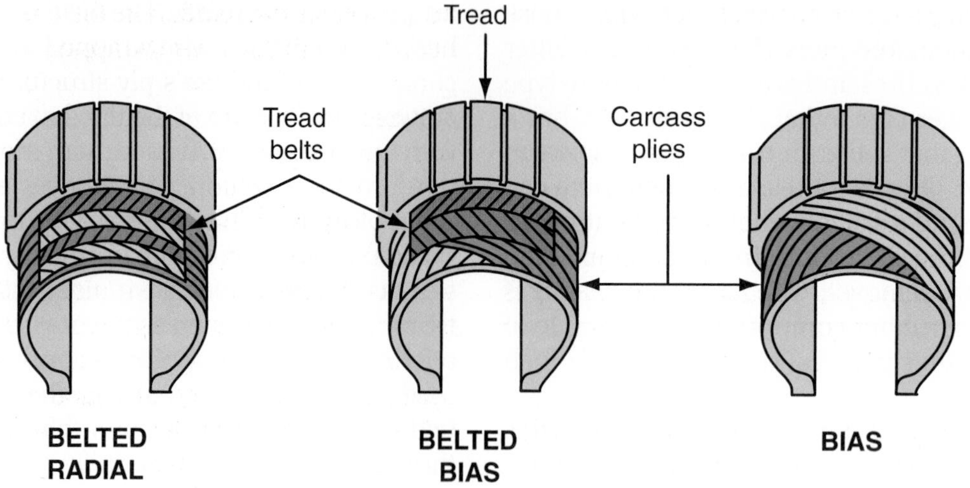

Figure 44-4 The construction of the three basic types of tires.

greater strength to the tread area and flexibility to the sidewall. The belts restrict tread motion during contact with the road, thus improving tread life and traction. Radial ply tires also offer greater fuel economy, increased skid resistance, and more positive braking.

Specialty Tires

Specialty tires reflect the advances made in tire development. These tires are designed for specific road conditions or applications. All-season tires are designed to perform well on all types of road conditions, but they are not excellent performers on all road surfaces. To provide traction in the snow and mud, at least 25% of the tread area is void **(Figure 44–5)**. This leaves open areas for the snow or mud to move into as the tire rotates. The open spaces also give the tires some bite as they move. The remaining tread area is designed to provide good traction on normal surfaces.

Tires can also be designed for heavy snow. These are commonly called "snow tires." The tread area is much more aggressive, with larger voids than all-season tires. Because of the decreased contact with the road, these tires do not provide good traction during normal conditions.

Figure 44-5 An all-season radial tire.

SHOP TALK

Snow tires are designed to increase traction in the snow. They should be installed on all wheels and replaced with normal tires once the season changes.

Studded tires provide superior traction on ice but are slowly disappearing from the tire market because their performance on dry surfaces is poor. In addition, many states have outlawed their use because they damage the road and can be a safety hazard on dry surfaces. Because the studs have more contact on the road than the rubber, it is easy for the tire to slide on the road during cornering and stopping. The studs offer much less friction than the rubber tire tread.

Tires can be designed to be great on dry surfaces or wet surfaces. However, it is nearly impossible to have a tire that performs extremely well on both. Tires designed for dry and smooth roads do not need a tread pattern. They can be "slicks." The smooth surface gives the tires maximum grip on a smooth, dry surface. However, when a slick hits a wet spot, there is no traction. The tire simply slides on the water.

Tires designed for wet surfaces have tread designs that move the road's water behind and to the side of the tire. Moving the water is the only way a tire can grip the road. When too much water separates the tire from the road, hydroplaning takes place. This causes the tire to lift off the road and rotate on a layer of water. Traction is reduced and this can create an unsafe condition. Needless to say, when a tire has many directed and open channels for water in its tread, less rubber meets the road.

Run-Flat Tires

There are several types of run-flat tires available. Vehicles equipped with these have no spare tire or jack in the luggage compartment. Run-flats can be divided into three categories: self-sealing, self-supporting, and auxiliary supported systems. Each of these uses different ways to allow a vehicle to be driven after a tire is punctured.

Self-Sealing Self-sealing tires are designed to quickly and permanently seal most tread-area punctures. The tires are constructed like other tires but there is an additional lining inside the tire under the tread area. The lining is coated with a sealant that can permanently seal most punctures up to 3⁄16 inch (4.76 mm) in diameter. The lining seals the area around the puncture and can fill in the hole once the object is removed from the tire **(Figure 44–6)**. There is no provision for reinflating the tire, so the tire will need to be inflated after the repair.

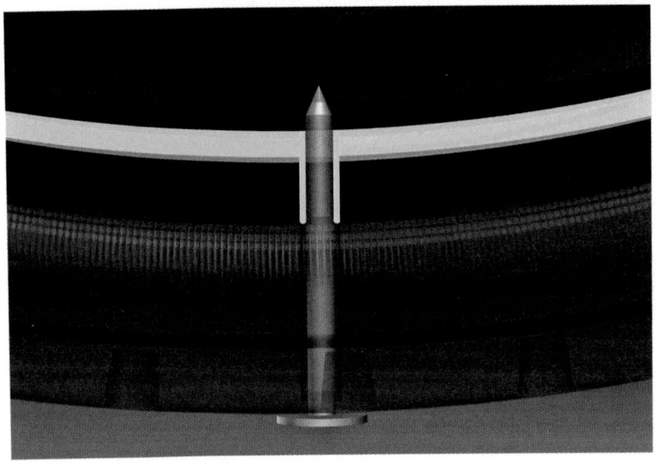

Figure 44-6 The action of a self-sealing tire.
Courtesy of The Goodyear Tire & Rubber Company

Self-Supporting When a self-supporting tire loses all of its air pressure, it is able to temporarily carry the weight of the vehicle and allow the vehicle to be driven. This is a result of the tire's construction. These tires have reinforced sidewalls and special beads **(Figure 44–7)**. The first "run-flat" tire available on a regular production vehicle was a self-supporting tire that was offered as an option on the 1994 Chevrolet Corvette. Today, many tire manufacturers offer self-supporting tires. Typically, a self-supporting tire can be driven for 50 miles (80 km) at speeds up to 55 mph (89 km/h) after it has lost air pressure.

Auxiliary Supported Run-Flat Systems Auxiliary supported systems are much different from other run-flat tires. They are systems that have special tires and wheels. The basis of these systems is a solid supporting ring that allows a flat tire's tread to rest on a support ring attached to the wheel when the tire loses pressure **(Figure 44–8)**. This support ring allows the tire to behave as it would when it was inflated. The wheel and tire are designed to prevent the tire from coming loose from the wheel when air pressure is lost. The most common system is Michelin's PAX system, which was introduced in 1996. This system allows the driver to drive the vehicle up to 125 miles at 55 mph before it needs service.

Tread Designs

The real purpose of a tire is to get a grip on the road. The ideal tire is one that wears little, holds the road well to provide sure handling and braking, and provides a cushion from road shock. The ideal tire should also provide maximum grip on dry roads, wet roads,

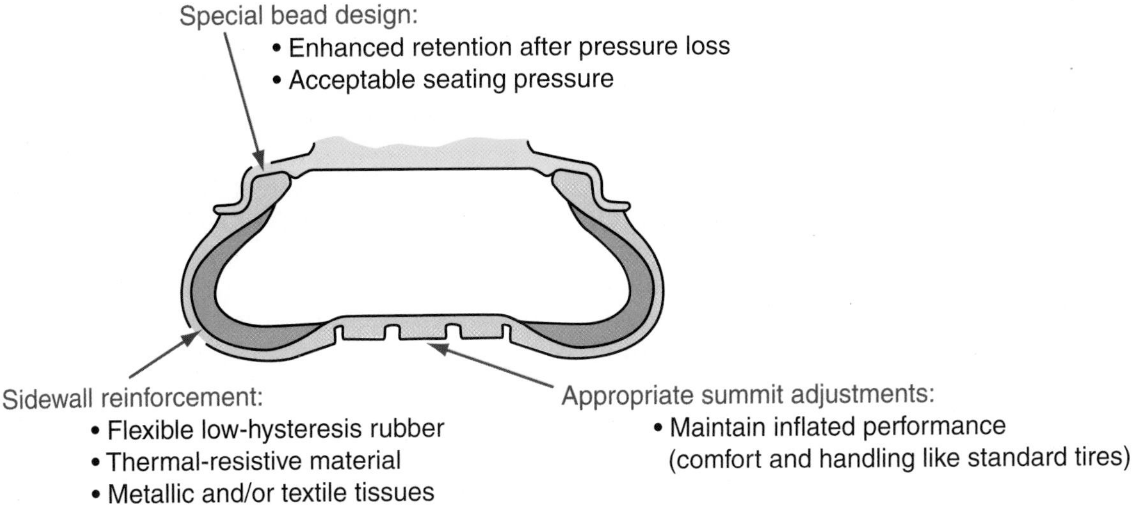

Figure 44-7 Features of a run-flat tire.

Figure 44-8 A cutaway of a run-flat tire with an insert for support in case the tire goes very flat. *Courtesy of The Goodyear Tire & Rubber Company*

Figure 44-9 There are many different tread designs available for today's vehicles. *Courtesy of The Goodyear Tire & Rubber Company*

and snow and ice, and operate quietly at any speed. This is a tall order, so tire manufacturers compromise on one or two of these qualities for the sake of excelling at another. A tire's tread design dictates what the tire will excel at.

There are basically three categories of tread patterns: directional, nondirectional, and symmetric and asymmetric.

A directional tire is mounted so that it revolves in a particular direction. These tires have an arrow on the sidewalls that show the designed direction of travel. A directional tire offers good performance only when it is rotating in the direction in which it was designed to rotate **(Figure 44–9)**. A nondirectional tire has the same handling qualities in either direction of rotation. A symmetric tire has the same tread pattern on both sides of the tire. An asymmetrical tire has a tread design that is different from one side to the other. Asymmetrical tires are typically designed to provide good grip when traveling straight (the inside half) and good grip in turns (the outside half of the tread). Most asymmetric tires are also directional tires.

The number and size of the blocks, sipes, and grooves on a tire's tread not only determine how much rubber contacts the road and how much water can be displaced; they also determine how quiet the tire will be during travel. The more aggressive the tread, the more noise it will make. This statement is especially true if the tire's tread is made of a hard compound. Softer tires typically make less noise but wear more quickly. Soft tires also adhere to the road better.

Channels are cut into a tire's tread to allow water to move away from the tire's direction of travel. The deeper the channel, the more water the tire can move. The disadvantage of these channels is decreased road contact.

SHOP TALK

Whenever a customer wants a better handling tire, make sure that he or she knows that a better gripping tire may make more noise and not wear as long as other tires. Knowing what design of tire will meet a customer's needs is a science. Always consult with a tire specialist before recommending one tire or another.

Spare Tires

Nearly all vehicles are equipped with a spare tire to be used in case one of the vehicle's tires loses air and goes flat. A spare tire can be a tire that matches the tires on the vehicle or can be a compact spare.

Compact spares are designed to reduce weight and storage space but still provide the driver with a tire in the case of an emergency. Compact tires are typically one of three types: high-pressure mini spare, space-saver spare, and lightweight skin spare.

A high-pressure mini-spare tire is a temporary tire. It should not be used for extended mileage or for speeds above 50 mph. A space-saver spare must be blown up with a compressor that operates from the cigarette lighter or a built-in air compressor. A skin spare is a normal bias ply type tire with a reduced tread depth.

Some cars do not have spare tires, even if they are not fitted with run-flat tires. Instead, these cars have a high-pressure air pump and a can of tire sealant in the trunk. Normally the sealant will allow the car to be driven until the tire can be properly repaired.

Customer Care

Make sure you warn your customer that a mini, space-saver, or similar type compact spare should be used only as a temporary tire. It should never be used as a regular tire. Any continuous use of a temporary spare will result in tire failure, which can cause loss of control and injury to the vehicle's occupants. It can also cause differential gear wear and will affect ABS and ATC operation.

TIRE RATINGS AND DESIGNATIONS

The construction of a tire depends on its application. Needless to say, there are many different tires. These differences are based on not only size, but their construction to meet intended driving conditions. There are also standards that tire manufacturers must meet to ensure that the tire will be safe, not wear rapidly, and offer good road isolation for the passengers in the vehicle. The uniqueness of each tire is represented by information given on the sidewall of every tire produced. In fact, everything you need to know about a tire is imprinted on the tire **(Figure 44–10)**.

Tire Size

The best way to describe and explain the information given on the sidewall of a tire is to look at an example. Look at the tire size designation of P215/65 R15 89H and see what it tells.

On a **P**215/65 R15 89H tire, the *P* represents the application of the tire; in this case P = passenger car. If the tire had an "LT" designation, the tire would be for a light truck.

The *215* in P**215**/65 R15 89H represents the width of the tire measured in millimeters from sidewall to sidewall. This tire width is 215 millimeters. This is also called section width, and it varies with the wheel (rim) on which the tire is mounted. A wide rim increases

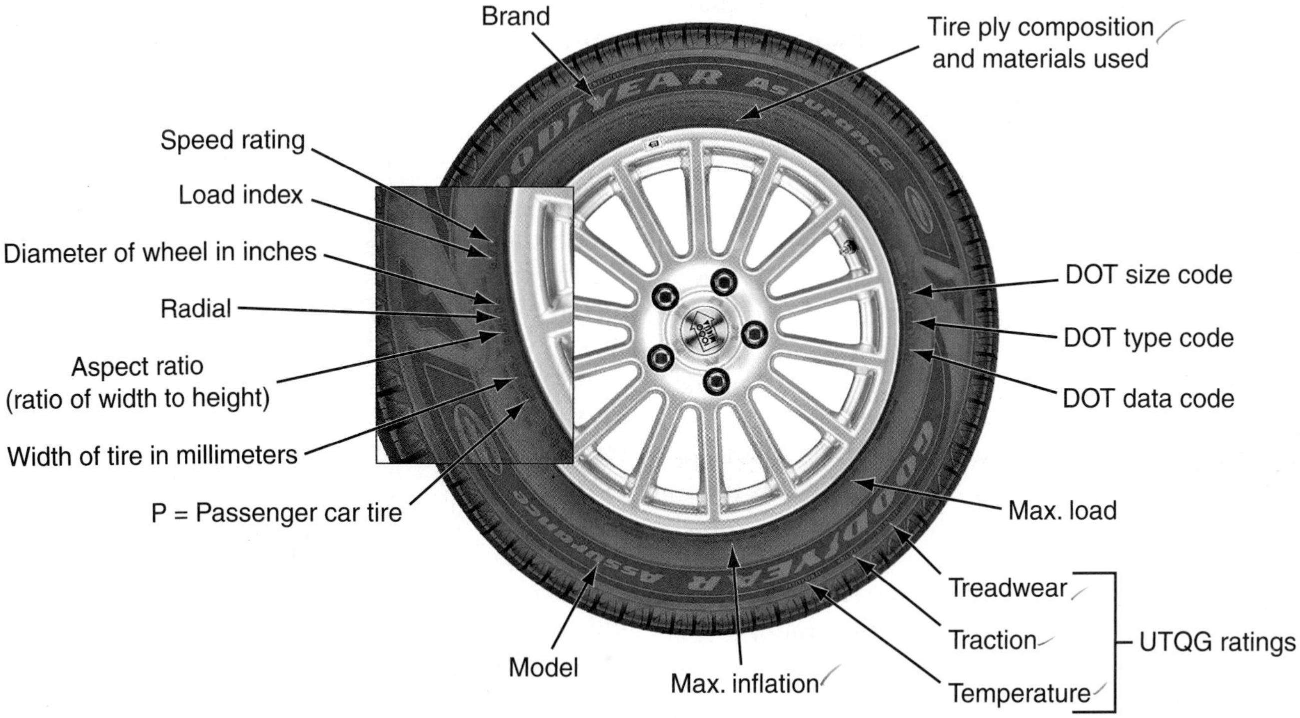

Figure 44–10 There is a lot of information about a tire on its sidewall. *Courtesy of The Goodyear Tire & Rubber Company*

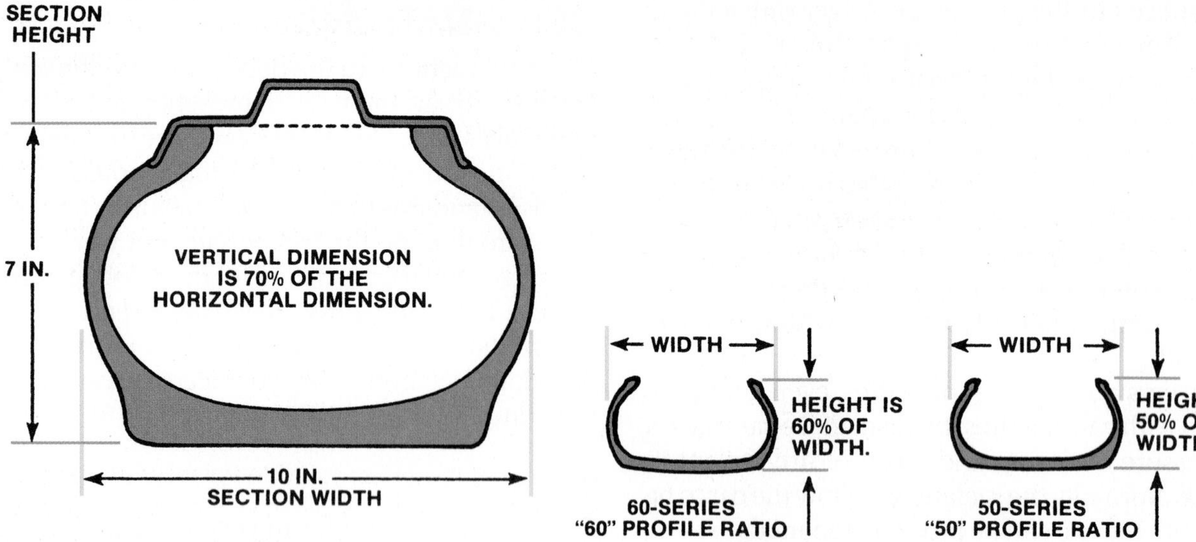

Figure 44–11 The aspect ratio (profile) of a tire is its cross-sectional height compared to its cross-sectional width expressed in a percentage.

the section width, whereas a narrow one decreases it. The measurement given on the tire was taken with the tire on a specific rim.

The *65* in P215/**65** R15 89H indicates the **aspect ratio** or profile (serios) of the tire **(Figure 44–11)**. A tire's aspect ratio is the relationship of its cross-sectional height (from tread to bead) to its cross-sectional width (from sidewall to sidewall). In our example, the tire's height is equal to 65% of its width (the width equals 215 mm × 65% or 140 mm). The aspect ratio determines a tire's performance characteristics. Higher aspect ratios provide a softer ride because they will deflect more over irregular surfaces and under heavy loads. Shorter sidewall heights demand stiffer sidewalls. Therefore, tires with a low aspect ratio have a harsher ride. However, they provide a larger contact area with the road and therefore better traction.

For a tire rated as P215/65 **R**15 89H the *R* represents the basic ply construction of the tire. This letter can be an "R" for radial construction, a "B" for belted-bias construction, or a "D" for bias ply ("bias" means the plies are set diagonally or at a slant).

The diameter of the wheel is indicated by the *15* after the R. The diameter of the wheel for this tire is 15 inches. Wheel diameter is the height of the wheel from one end to the other.

Following the size notation is the load and speed ratings. These are expressed by a number and a letter; in this case the ratings are given as 89H. The *89* is the load index and the *H* is the speed rating.

The maximum load rating lists the maximum amount of weight the tire can carry at the recommended tire pressure. Typically the load rating and the number of tread and sidewall plies are proportional. In most cases, the more plies a tire has, the more weight it can support. The higher the tire's load index number, the greater its load carrying capacity. The load ratings for passenger car and light truck tires range from 70 to 110. Following are some examples of load ratings and the weight they represent:

71 = 761 lb (345 kg)

79 = 963 lb (437 kg)

89 = 1,279 lb (580 kg)

99 = 1,709 lb (775 kg)

109 = 2,271 lb (1,030 kg)

So in our example, the tire can carry 1,279 lb (580 kg).

The speed rating indicates the maximum speed at which the tire should be used. In this case, the *H* means the tire has been tested to be safe at speeds up to 130 mph (210 km/h). The speed rating of a tire is really nothing more than an expression of how well the tire will withstand the temperatures of high speed. This does not necessarily mean that a high-speed-rated tire will perform better at low speeds than a lower-rated tire.

Table 44–1 lists the various letters used to designate the speed rating of a tire and the maximum speed at which the tire was designed to safely operate. Driving a vehicle at speeds greater than the speed rating of the tires is risky. The heat generated can cause the tire to come apart. If this happens at high speeds, it will be close to impossible for the driver to maintain control of the vehicle.

TABLE 44-1 SPEED RATINGS

Symbol	Max. Speed
Q	99 mph (160 km/h)
S	112 mph (180 km/h)
T	118 mph (190 km/h)
U	124 mph (200 km/h)
H	130 mph (210 km/h)
V	149 mph (240 km/h)
Z	Above 149 mph (240+ km/h)
Z-W	168 mph (270 km/h)
Z-Y	186 mph (300 km/h)

Other Information

The sidewall of a tire also has a DOT safety code, tire identification or serial number, UTQG ratings, and maximum inflation values. The DOT code indicates that the tire has met all of the applicable safety standards established by the U.S. Department of Transportation (DOT). Next to the DOT code is a tire identification or serial number. This is a combination of numbers and letters that identify the tire manufacturer, where it was made, the tire design and size, and the week and year the tire was manufactured.

UTQG stands for **Uniform Tire Quality Grading**, a rating system developed by the DOT. This rating is comprised of three factors: tread wear, traction, and temperature resistance. All tires, except snow tires, have these ratings.

Tread Wear The tread wear grade is a rating based on a tire's wear rate when tested under controlled conditions on a specified government test track. Tread wear is listed as a number: The higher the number, the longer the tread will last. A rating of 100 is considered normal, whereas ratings lower than 100 mean poor tread wear. Ratings above 100 mean the tire has better-than-normal tread wear. These ratings should be used to compare the anticipated wear of tires from the same manufacturer and not to compare wear between manufacturers.

Traction Tire traction ratings are based on a tire's ability to stop on wet concrete and asphalt. It is not an indication of how well a tire will handle. The traction rating is given as AA, A, B, or C. A tire rated as C will provide less traction than one rated with an A.

Temperature Resistance This rating is an indication of how well a tire will dissipate heat and how it works when it is heated. The temperature rating applies only to a properly inflated tire that is not overloaded. Heat builds up when a tire is underinflated or overloaded. Temperature also increases with excessive speeds. Temperature resistance rating is given as A, B, or C. A rating of C means the tire is acceptable. A tire with an A temperature rating will be able to withstand high temperatures better than one rated B or C.

Additional Ratings Some tires carry additional markings related to their intended service. An M&S or M+S designation means the tire has been rated by the manufacturer as suitable for mud and snow use. This is the designation for many all-season tires. Some tires may also have a mountain/snowflake symbol, which indicates that these tires are suitable for severe snow conditions; these are commonly called winter tires.

Maximum Cold Inflation and Load

The sidewalls of all passenger tires are marked to indicate the tires' maximum load capacity and maximum cold inflation pressure. It is important to remember that the maximum inflation number on a tire is its maximum inflation, not its recommended inflation. Tires should never be inflated beyond their maximum rating.

Tire Placard

The tire placard, or safety compliance certification label, is generally found on the driver's door jamb. It includes recommended maximum vehicle load, tire size (including spare), and the *correct* cold tire inflation for each tire of the vehicle **(Figure 44–12)**. Never use this information for other cars.

As a general rule, tires should be replaced with the same size designation or an approved optional size as recommended by the auto or tire manufacturer. Also, always follow the manufacturer's recommendations for tire type, inflation pressures, and rotation patterns.

SHOP TALK

Tires with a larger or smaller diameter than originally installed will affect the operation of the antilock brake system and the accuracy of the speedometer. It might be necessary to recalibrate the ABS and ATC computers and change the speedometer drive gears when tire size has been changed. Check the vehicle's service manual for details.

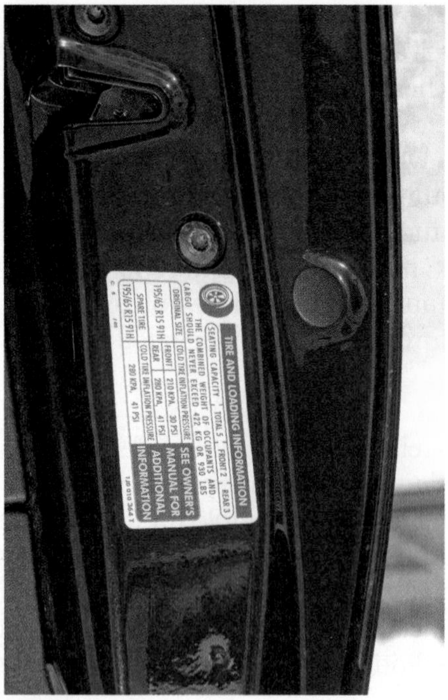

Figure 44-12 A tire placard on a door jamb.
Courtesy of The Goodyear Tire & Rubber Company

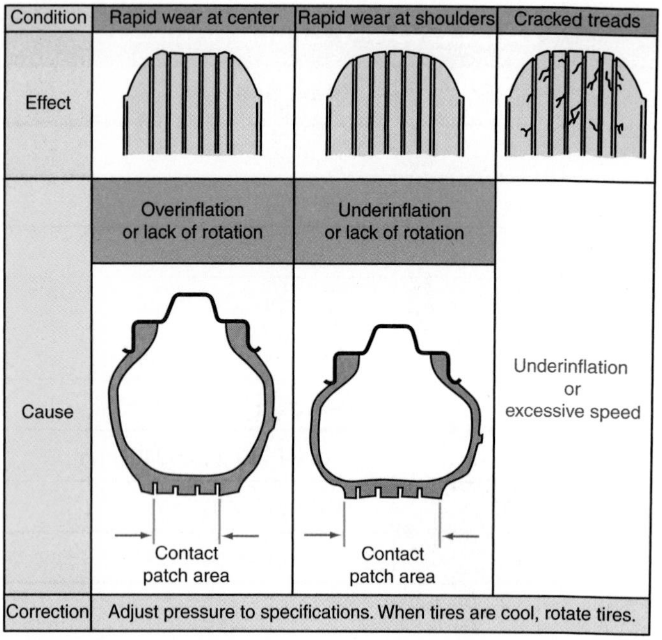

Figure 44-13 Effects of inflation on tread contact and wear.

Tire Care

To maximize tire performance, inspect for signs of improper inflation and uneven wear, which can indicate a need for balancing, rotation, or wheel alignment. Tires should also be checked frequently for cuts, stone bruises, abrasions, and blisters, and for objects that might have become imbedded in the tread. More frequent inspections are recommended when rapid or extreme temperature changes occur, or where road surfaces are rough or occasionally littered with debris.

To clean tires, use a mild soap and water solution only. Rinse thoroughly with clear water. Do not use any caustic solutions or abrasive materials. Never use steel wool or wire brushes. Avoid gasoline, paint thinner, and similar materials having a mineral oil base. These materials will cause premature drying of the tire's rubber. As the rubber dries, it gets harder and the tire will lose some of its performance characteristics.

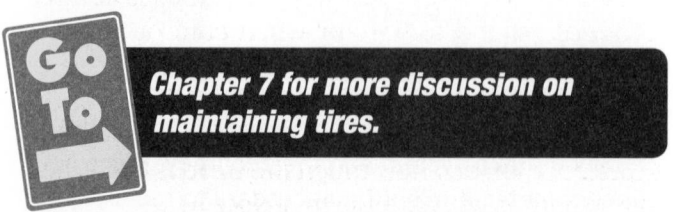

Go To
Chapter 7 for more discussion on maintaining tires.

Inflation Pressure A properly inflated tire gives the best tire life, riding comfort, handling stability, and fuel economy. Too little air pressure can result in tire squeal, hard steering, excessive tire heat, abnormal tire wear, and increased fuel consumption by as much as 10%. An underinflated tire shows maximum wear on the outside edges of the tread; there is little or no wear in the center (**Figure 44–13**).

Conversely, an overinflated tire shows its wear in the center of the tread and little wear on the outside edges. Too much air can also cause a hard ride and tire bruising.

Many inflation pressures listed for imported vehicles are given in kilopascals (kPa) rather than psi. **Table 44–2** converts kPa to psi.

TABLE 44-2 INFLATION PRESSURE CONVERSION (KILOPASCALS TO psi)

kPa	psi	kPa	psi
140	20	215	31
145	21	220	32
155	22	230	33
160	23	235	34
165	24	240	35
170	25	250	36
180	26	275	40
185	27	310	45
190	28	345	50
200	29	380	55
205	30	415	60

Conversion: 6.9 kPa = 1 psi.

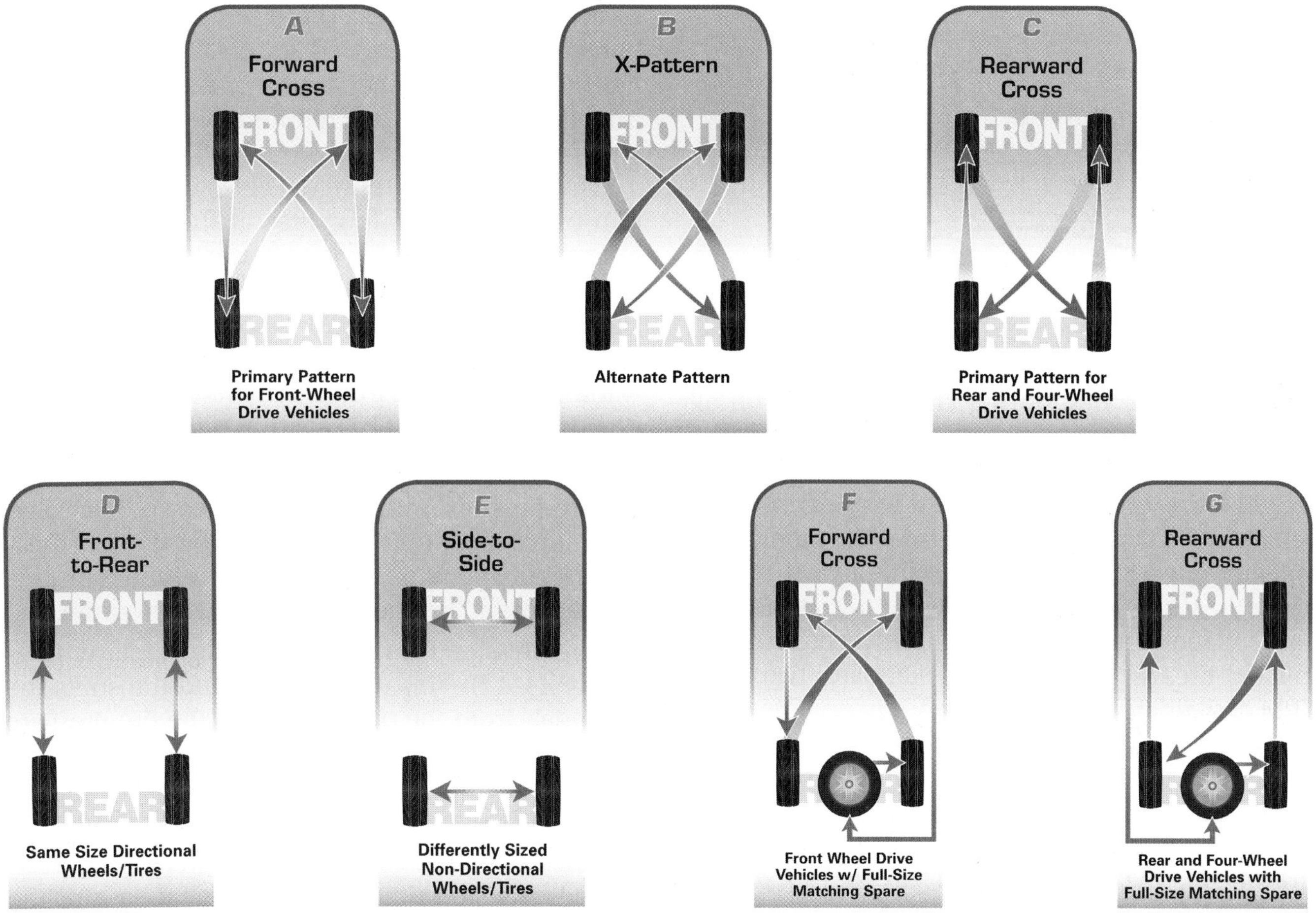

Figure 44–14 On FWD cars, rotate the tires in a forward cross pattern (A) or the alternative pattern (B). On RWD or 4WD vehicles, rotate the tires in a rearward cross pattern (C) or the alternative pattern (B). The front-to-rear (D) pattern may be used for vehicles equipped with the same-size directional wheels and/or directional tires. The side-to-side (E) pattern may be used for vehicles equipped with different-sized nondirectional tires and wheels on the front axle compared to the rear axle. On FWD cars with a full-size matching spare, rotate the tires in a forward cross pattern (F). On RWD or 4WD vehicles with a full-size matching spare, rotate the tires in a rearward cross pattern (G). *Courtesy of The Tire Rack*

SHOP TALK

The temperature of a tire affects its actual inflation. Pressure increases with an increase in temperature. The opposite is also true. Keep in mind that a 10°F (5.5°C) change in temperature will change the air pressure by 1 psi (6.9 kPa). Always adhere to the cold inflation pressures given on the tire placard.

Tire Rotation

Part of a preventive maintenance program is the rotation of the tires. Doing this on a regular basis can preserve balanced handling and traction and even out tire wear. Most manufacturers recommend this be done at least every 6,000 miles. It is important to remember that tire rotation will not correct wear problems due to misalignment, worn mechanical parts, or incorrect inflation pressures.

The industry has described seven rotation patterns that cover the procedure for most cars and light trucks **(Figure 44–14)**.

Tread Wear

Most tires used today have built-in tread wear indicators (wear bars) to show when they need replacement. These indicators appear as $\frac{1}{2}$-inch wide bands when the tire tread depth wears to $\frac{1}{16}$ inch. When the indicators appear in two or more adjacent grooves at three locations around the tire or when cord or fabric is exposed, tire replacement is recommended.

If the tires do not have tread wear indicators, a tread depth indicator **(Figure 44–15)** quickly shows in 32nds of an inch how much tire tread is left. When only $\frac{2}{32}$ inch is left, it is time to replace a tire.

Tire Pressure Monitor (TPM)

The U.S. DOT National Highway Traffic Safety Administration (NHTSA) has developed a federal motor

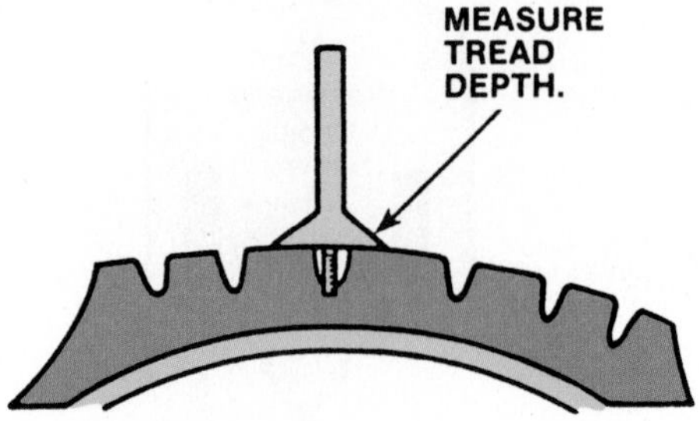

Figure 44–15 Checking tread depth.

vehicle safety standard that requires the installation of **tire pressure monitoring (TPM) systems** on all 2008 and newer passenger cars, trucks, multipurpose passenger vehicles, and buses with a gross vehicle weight rating of 10,000 pounds or less, except those vehicles with dual wheels on an axle. The monitoring systems must illuminate a warning light if one or more tires is at least 25% below the recommended cold-inflation pressure to warn the driver of an underinflated tire. As a result of this standard, two basic TPM designs are being used.

The most commonly used system is referred to as a direct system. In this system an air pressure sensor is strapped around the drop center of each wheel, or the sensor is attached to a special tire valve **(Figure 44–16)**. The pressure sensor measures the tire's inflation pressure and relays this information to the vehicle via radio waves. These signals are picked up by separate body-mounted antennas for each wheel. A central electronic control unit processes the signals from the four wheels and reports any variations to the system.

Figure 44–16 A tire pressure monitor and air valve assembly. *Courtesy of Adaptive Activity Network BV, www .tiresmonitor.com*

The tire pressure monitor (TPM) checks the inflation pressures in all four tires at frequent periodic intervals. The TPM sensors keep track of the tire pressures both when the vehicle is moving and when it is stationary. When the TPM detects changes in any tire's inflation pressure, it responds by triggering a warning lamp on the instrument panel.

A typical direct TPM system has the following components:

- Tire pressure warning transmitter and air valve. This is a single unit with a built-in battery that measures tire pressure and temperature and transmits a signal and ID number for that particular tire.
- Tire pressure warning antenna and receiver **(Figure 44–17)**. This unit receives and transmits the signals from the transmitters to the tire pressure warning control unit.
- Tire pressure warning control unit. This unit receives the signal from the receiver. If the measured air pressure is equal to or lower than a specified value, this unit transmits a signal, causing the air pressure warning light to illuminate.
- Tire pressure warning light. Located in the instrument cluster, this unit informs the driver of low tire pressure or a problem in the system.
- Tire pressure warning reset switch. This unit is used after sensor, tire, or wheel replacement. It is used to allow the control unit to relearn the system.

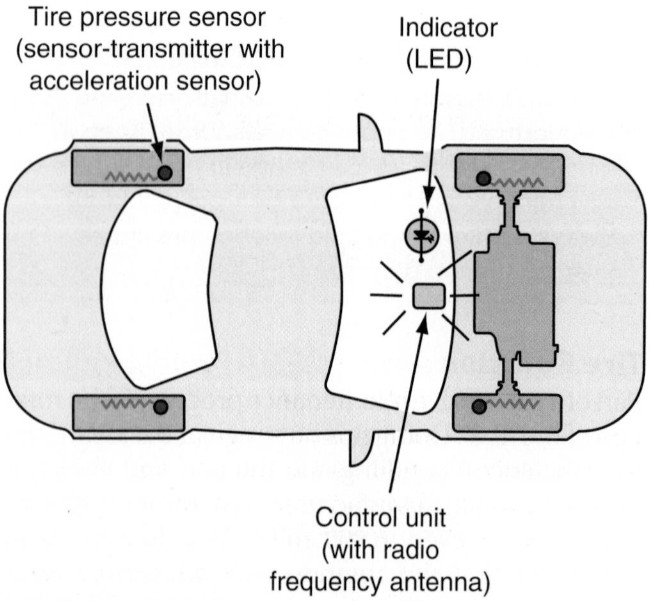

Figure 44–17 Basic components for a direct TPM system.

In an attempt to meet the TPM standard without the cost of a direct system, some manufacturers use an indirect system. These systems do not use pressure sensors; rather they rely on the inputs from wheel speed sensors. These signals have been used for ABS or other systems, With indirect TPM, the PCM is reprogrammed to use those signals to identify when a tire has lost air pressure. Indirect systems are also used on some older vehicles with run-flat tires. This was necessary because run-flat tires without air appear normal. The driver needs to be alerted to the loss of air.

The input signals from the wheel speed sensors are used to compare the rotational speeds of the four tires. When a tire loses or gains air pressure, it will roll at a slightly different number of revolutions per mile than the other three tires. When the computer senses this difference, a warning lamp will light.

Indirect systems are not as effective as direct systems, however. These systems cannot tell the driver which tire has low pressure. They also are not capable of informing the driver when all four tires are losing air pressure. This commonly happens when the outside temperature drops.

Warning System The TPM warning system can vary from a warning lamp to a graphic display that shows which tire is low on pressure. Some systems allow the driver to monitor the current air pressure of each tire. If the PCM detects a problem with the TPM system, a warning lamp or message will appear in the instrument cluster. After tires have been rotated or replaced, the system must relearn the tires at each position. During this time, a message or warning lamp will be illuminated. Once the relearning process is completed, the message will turn off.

If the system is working correctly, the TPM lamp and the low tire pressure lamp should illuminate for about 2 seconds when the ignition is turned on. If the lamps do not turn off, there is a problem in the system. The system will typically set a DTC when there is a problem with the system or when it detects low pressure in any of the four tires.

Testing a TPM System

The TPM system in most vehicles is tied directly to the PCM; therefore, faults cause DTCs to set. These can be retrieved with a scan tool. Special tools are required to accurately test and locate the problem tire(s). A TPM sensor tool **(Figure 44–18)** is a wireless tool that may be used with a scan tool for diagnosing sensors and allowing the system to relearn when a part has been replaced. Although trouble codes can identify a system problem, the DTCs do not indicate the exact

Figure 44–18 A TPM tester. *Courtesy of SPX Service Solutions*

location of the troubled tire. The TPM sensor tool is used to find the tire or sensor that is causing the problem.

The TPM sensor tester is used to reset the system, which is typically needed after tires are rotated, tires or wheels are replaced, and repairs are made to the system, and when the vehicle's battery was low or replaced. The tester activates the sensors and the transmitted data from them can be observed.

TIRE/WHEEL RUNOUT

A tire that is off center is said to run out. This is known as **radial runout** or eccentricity. One that wobbles side to side is said to have **lateral runout**. If a tire with some built-in runout is mismatched with a wheel's runout, the resulting total runout can exceed the ability of the balance weights to correct the problem. For this reason, part of a tire/wheel inspection should be for excessive runout. Sometimes tires or wheels can be remounted to lessen or correct runout problems.

To avoid false readings caused by temporary flat spots in the tires, check runout only after the vehicle has been driven. Visually inspect the tire for abnormal bulges or distortions. The extent of runout should be measured with a dial indicator. All measurements should be made on the vehicle with the tires inflated to recommended load inflation pressures and with the wheel bearing adjusted to specification.

Measure tire radial runout at the center and outside ribs of the tread face. Measure tire lateral runout just above the buffing rib on the sidewall **(Figure 44–19)**. Mark the high points of lateral and radial runout for future references. On bias or belted bias tires, radial runout must not exceed 0.06 inch

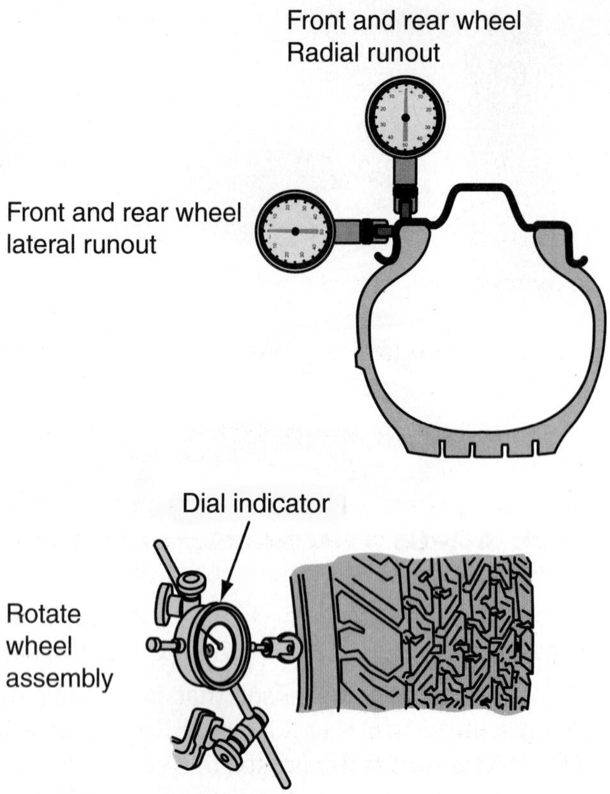

Figure 44-19 Checking wheel runout.

(1.5 mm) and lateral runout must not exceed 0.045 inch (1.1 mm). On radial ply tires, radial runout must not exceed 0.081 inch (2.06 mm) and lateral runout must not exceed 0.099 inch (2.51 mm).

If the total radial or lateral runout of the tire exceeds the specified limits, it is necessary to check wheel runout to determine whether the wheel or tire is at fault. Wheel radial runout is measured at the wheel rim just inside the wheel cover retaining nibs. Wheel lateral runout is measured at the wheel rim bead flange just inside the curved lip of the flange. Wheel radial runout should not exceed 0.035 inch (0.89 mm) and wheel lateral runout should not exceed 0.040 inch (1.02 mm). Mark the high points of radial and lateral runout for future reference.

If the total tire runout, either lateral or radial, exceeds the specified limit but wheel runout is within the specified limit, it might be possible to reduce runout to an acceptable level. This is done by changing the position of the tire on the wheel so that the previously marked high points are 180 degrees apart.

TIRE REPLACEMENT

Tires should be replaced when they are worn or heavily damaged. Replacement tires should match the tires that were on the vehicle originally. This is the preferred choice unless a change in appearance or handling is desired. Vehicle manufacturers know how important the right tire is. They spend a great deal of time developing a suspension system that works in the way they believe it should. Part of that development is tire design. However, the tire choice by manufacturers is a compromise—a compromise between characteristics. Manufacturers choose the characteristics they believe owners of that model vehicle want. If the owner is happy with the vehicle, recommend that the replacement tires match the ones the vehicle was originally sold with.

Replacing One Tire

In some cases, only one tire needs to be replaced. The usual causes of this are that the tire was damaged due to an accident, a road hazard, or vandalism. Replacing one tire is recommended only if the other tires have a satisfactory amount of tread left. Make sure the replacement tire is the same brand, type, size, and speed rating as the other tires. If the replacement tire is different from the rest, the vehicle can exhibit unsafe handling problems. The replacement tire should be mounted on the rear axle and the tire (of the remaining three) with the most tread depth should be mounted on the opposite side of the axle.

Replacing Two Tires

If there is a need to replace two tires and the other two have good treads, the replacement pair should be mounted on the front axle. The replacement tires should match the remaining pair of tires as closely as possible.

Changing Tire and/or Wheel Size

An owner may want more emphasis on handling or fuel economy and may desire a different type of tire and/or wheel. There are several other factors that may dictate a change in tire size, and the customer may come to you for advice. Perhaps one of the most important considerations is that the tire must be able to carry the weight of the vehicle. The load-carrying capacity of a tire must be the same as or higher than the OEM tire. Changing tires from one aspect ratio to another also changes the sectional width, which relates to the load-carrying capacity of the tire.

Most tire width changes affect the overall diameter of the tire. A change in the tire's outside diameter will cause a change in the overall gear ratio and will affect the accuracy of the speedometer and odometer. A change in tire diameter or aspect ratio may also affect overall driveability. This is due to false readings from the vehicle or wheel speed sensors. On passenger cars and mini-vans, a 3% or less (¾-inch or less) change in

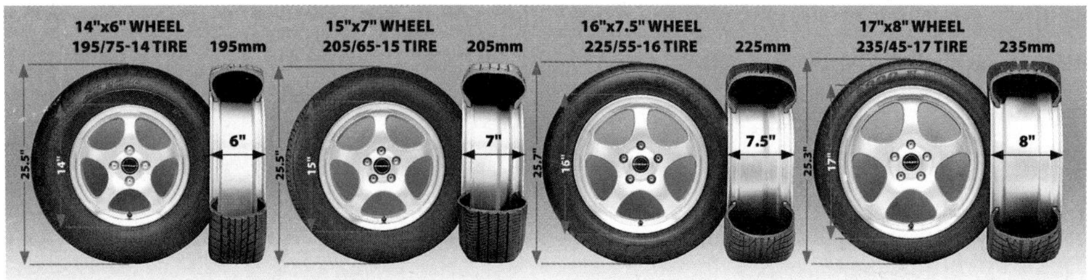

Figure 44–20 The effects of plus sizing. *Courtesy of The Tire Rack*

tire diameter is acceptable. Most SUVs and pickups can handle a change as much as 15%. The overall diameter of a tire can be calculated.

Plus Sizing A way to change the contact area of a tire without seriously affecting its overall diameter is using the plus sizing system. This system is based on the overall diameters of a combination of different-sized tires and wheels. The system requires much research but it is the best way to achieve the desired results. For example, the customer wants to have a wider tread area **(Figure 44–20)**. The OEM tire was a 195/75-14 tire mounted on a 14 × 6 inch wheel. The overall diameter of this assembly is 25.5 inches (647.7 mm). There are three available wheel/tire combinations that closely match that diameter. If a 205/65-15 tire is mounted on a 15 × 7 inch wheel, the width increases by 0.39 inches (10 mm), whereas the diameter stays the same. If a 16 × 7.5 wheel is used along with a 225/55-16 tire, the width increases by 1.18 inches (30 mm) and the overall diameter increases by only 0.2 inch (5.1 mm). Going even wider, if a 235/45-17 tire is mounted on a 17 × 8 wheel, the tire's width increases by 1.57 inches (40 mm) and the overall diameter decreases by 0.2 inch (5.1 mm). Although the latter two changes do not match the OEM diameter, they are certainly within the 3% rule.

Additional Points Here are some additional important points to consider:

- Handling improvement typically comes from more tire contact on the road, and fuel economy increases with less.
- Tires of different sizes, constructions, and wear may affect handling, stability, and fuel economy.
- Too wide of a tire may rub against the body or suspension.
- Radial tires should never be mixed with another type of tire on the same vehicle.
- All tires on the vehicle should be the same size, construction, and speed rating unless the vehicle was otherwise equipped by the OEM.
- Tires should be replaced with ones of the same or higher speed rating. Speed ratings should not be downgrades from original equipment ratings.
- A hard tread will provide long wear and low rolling resistance but will also have poorer traction.
- An aggressive tread pattern may provide resistance to hydroplaning or better traction in the snow, but they are noisier on dry surfaces.
- A tire with stiff sidewalls will increase high-speed stability and improved handling, but it will make the overall ride rougher.
- A replacement rim should provide the same overall tire diameter as the original.
- A narrower rim pulls the beads of the tire closer together, causing the sidewalls to curve. This allows the sidewalls to flex more, which results in a softer ride but reduces tire life.
- A wide rim increases the distance between the beads, which stiffens the sidewalls and provides a harsher ride and shorter tire life. However, it will improve the handling of the vehicle.

Calculating Tire Dimensions When replacing tires with other than the original tire size, you may need to calculate the dimensions of a desired replacement tire. Like everything else, there are formulas to make these calculations.

To determine the section height of a tire, multiply its aspect ratio by the sectional width.

$$\text{Width} \times \text{Aspect Ratio} = \text{Section Height}$$

To determine the overall diameter of a tire, multiply the sectional height by 2 (this is called the combined sectional height because there are two), then add the diameter of the wheel.

$$\text{Combined Section Height (Sectional Height} \times 2) + \text{Wheel Diameter} = \text{Tire Diameter}$$

TIRE REPAIR

The most common tire problem besides wear is a puncture. When properly repaired, the tire can be put back in service without the fear of an air leak recurring. Punctures in the tread area are the only ones that should be repaired or even attempted to be repaired. Never attempt to service punctures in the tire's shoulders or sidewalls. In addition, do not service any tire that has sustained the following damage:

- Bulges or blisters
- Ply separation
- Broken or cracked beads
- Fabric cracks or cuts
- Wear to the fabric or visible wear indicators
- Punctures larger than ¼-inch (6 mm) diameter

Some car owners attempt to seal punctures with tire sealants. These sealants are injected into the tire through the valve stem. Sometimes the chemicals in the sealant do a great job sealing the hole, other times they fail. The sealants should never be used and will not work on sidewall punctures. Some of the sealants are very flammable and carry a warning that the tire should be marked so that the next technician knows the sealant has been used.

WARNING!

Tire sealants injected through the valve stem can produce wheel rust and tire imbalance.

To locate a puncture in a tire, inflate it to the maximum inflation pressure indicated on its sidewall. Then submerge the tire/wheel assembly in a tank of water or sponge it with a soapy water solution. Bubbles will identify the location of any air leakage.

Mark the location of the leak with a tire crayon so it can be easily found once the tire is removed from the wheel. Also use the crayon to mark the location of the valve stem so that original tire and wheel balance can be maintained after the tire is put back on the wheel.

The proper procedure for dismounting and remounting a tire is illustrated in Photo Sequence 46. Do not use hand tools or tire irons alone to change a tire because they might damage the beads or wheel rim. When mounting or dismounting tires on vehicles using aluminum or wire spoke wheels, it is recommended that the tire changer manufacturer be contacted about the accessories that are required to protect the wheel's finish.

TPM Sensors

If the wheel/tire assembly has a direct TPM system, the sensor can be removed after the air pressure has been released from the tire. This should be done before the tire is removed from the rim. Unbolt the air valve assembly and allow it to drop into the tire. Before servicing the tire or wheel, remove the sensor. After tire and/or wheel repairs have been made, install the sensor with a new rubber O-ring or seal and aluminum retaining nut. The retaining nut must be torqued to specifications.

SHOP TALK

Some air valves for TPM systems are made of brass and have an aluminum valve stem. Over time, the valves will experience galvanic corrosion and will seize within the brass valve core. If the tire you are working on shows these problems, replace this unit with a nickel-plated valve core.

Repair Methods

Once the tire is off the wheel and the cause of the puncture is removed and the location marked, the tire can be repaired from the inside using a service plug and a vulcanized patch. Although the repair kit's instructions should always be followed, there are some general guidelines that help make a good, permanent patch of the puncture. The following methods are the most common methods used to repair a tire.

Customer Care

The repair of a Michelin PAX system requires special equipment, replacement parts, and specialized training. If you attempt to repair one of these systems without the necessary equipments, parts, and training, the tire warranty will become void. Make sure your customer understands this. Also, the use of plugs and/or sealants to repair the tire will void the warranty.

Plug Repair The head-type plug (**Figure 44–21**) is commonly used. A plug that is slightly larger than the size of the puncture is inserted into the hole from the inside of the tire with an insertion tool. Before doing this, insert the plug into the eye of the tool and coat the hole, plug, and tool with vulcanizing fluid.

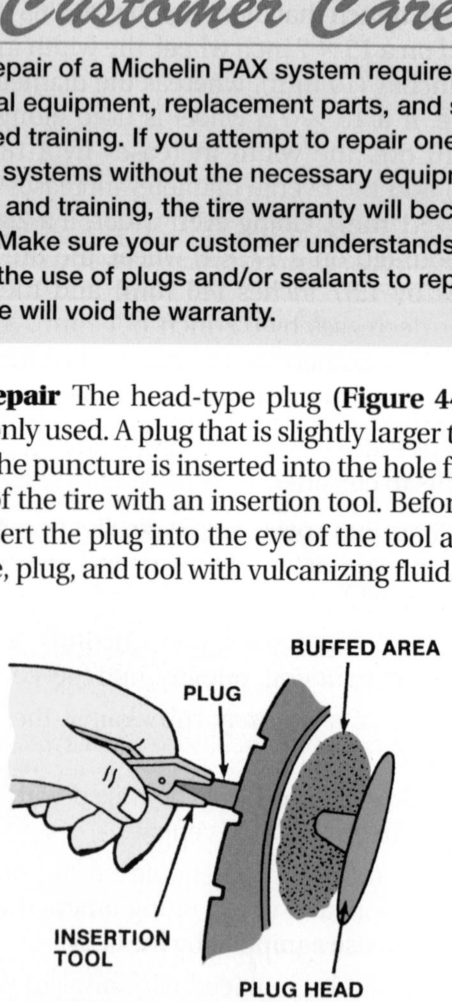

Figure 44-21 A plug for a radial tire.

P46-1 *Dismounting the tire from the wheel begins with releasing the air, removing the valve stem core, and unseating the tire from its rim. The machine does the unseating. The technician merely guides the operating lever.*

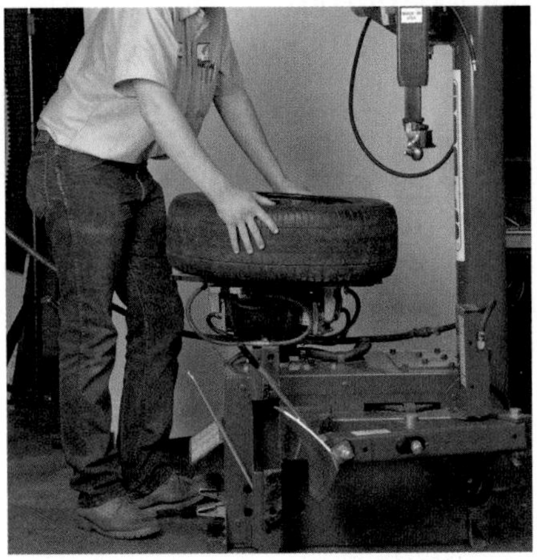

P46-2 *Once both sides of the tire are unseated, place the tire and wheel onto the machine. Then depress the pedal that clamps the wheel to the tire machine.*

P46-3 *Lower the machine's arm into position on the tire and wheel assembly.*

P46-4 *Insert the tire iron between the upper bead of the tire and the wheel. Depress the pedal that causes the wheel to rotate. Do the same with the lower bead.*

P46-5 *After the tire is totally free from the rim, remove the tire.*

P46-6 *Prepare the wheel for the mounting of the tire by using a wire brush to remove all dirt and rust from the sealing surface. Apply rubber compound to the bead area of the tire.*

P46-7 *Place the tire onto the wheel and lower the arm into place. As the machine rotates the wheel, the arm will force the tire over the rim. After the tire is completely over the rim, install the air ring over the tire. Activate it to seat the tire against the wheel.*

P46-8 *Reinstall the valve stem core and inflate the tire to the recommended inflation.*

While holding and stretching the long end of the plug, insert it into the hole. The plug must extend above both the tread and inner liner surface. If the plug pops through, throw it away and insert a new plug. Once the plug is in place, remove the tool and trim off the plug 1⁄32 inch (0.7 mm) above the inner surface. Be careful not to pull on the plug while cutting it.

Cold Patch Repair When using a cold patch, carefully remove the backing from the patch. Spread vulcanizing fluid on the punctured area. Let it dry, then center the patch base over the punctured area. Run a stitching tool over the patch to help bind it to the tire.

WARNING!

When repairing radial tires, use only a patch specially approved for that application. These special patches have arrows that must be lined up parallel to the radial plies.

Hot Patch Repair A hot tire patch application is similar to a cold patch. The difference is that the hot patch is clamped over the puncture and heat is applied to the patch to make it adhere.

Wheel Inspection

The wheels should be carefully inspected each time a tire is to be mounted on it. The major causes of wheel failure are improper maintenance, overloading, age, and accidents, including pothole damage. Wheels must be replaced when they are bent, dented, or heavily rusted; have leaks or elongated bolt holes; and have excessive lateral or radial runout. Wheels with a lateral or radial runout greater than specifications can cause high-speed vibrations. Wobble or shimmy caused by a damaged wheel eventually damages the wheel bearings. Stones wedged between the wheel and disc brake rotor or drum can unbalance the wheel.

SHOP TALK

When installing new tires, always install new air valve stems. The life of tire rubber is close to the life of the valve stem rubber. Most stems are the snap-in type. These are installed from inside the wheel with a pulling tool. Make sure that the stem is properly seated. Another style of stem has a retaining nut that must be removed when pulling off the old stem. Be sure to completely tighten the new nut.

Inflation

After the tire is mounted to the wheel, inflate the tires to the recommended pressure. Also check to see if any air is leaking from the beads or the point of repair.

Nitrogen Tire Inflation Many tire experts recommend that tires be filled with nitrogen rather than compressed air. Other experts say there is no need to do this if owners watch the air pressure in their vehicle's tires. The idea behind using nitrogen is simple: Nitrogen molecules are larger than air molecules. Therefore, it is less likely to leak out of a tire. Those in favor of using nitrogen claim that nitrogen-filled tires stay inflated about three times longer than air-filled tires. Nitrogen also helps keep tires cooler while traveling on the highway. This means the air pressure stays more constant and is less likely to leak out. The supposed result of these advantages is safer and longer lasting tires. The idea of using nitrogen in tires is not new. Race cars, commercial airliners, and trucks have used nitrogen-filled tires for many years.

INSTALLATION OF TIRE/WHEEL ASSEMBLY ON THE VEHICLE

Before reinstalling a tire/wheel assembly on a vehicle, inspect the wheel bearings as described later in this chapter, then clean the axle/rotor flange and wheel bore with a wire brush or steel wool. Coat the axle pilot flange with disc brake caliper slide grease or an equivalent.

Place the wheel on the hub. Make sure the wheel is seated on the hub. A common mounting problem is caused by improperly positioning the wheel on the wheel hub or by improperly tightening the lug nuts.

Install the locking wheelcover pedestal (if used) and lug nuts, and tighten them alternately to draw the wheel evenly against the hub. They should be tightened to a specified torque and sequence **(Figure 44–22)** to avoid distortion. The best way to do this is to snug up the lug nuts. Then when the car is lowered to the floor, they use a torque wrench for the final

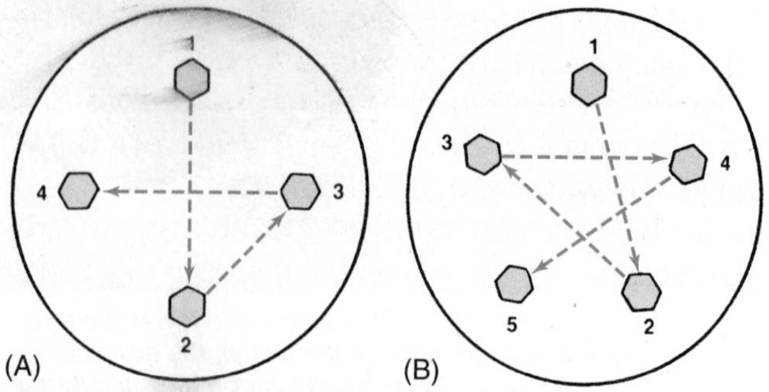

Figure 44-22 The lug nut tightening sequence for (A) a four-lug wheel and (B) a five-lug wheel.

tightening. Loose lug nuts can cause shimmy and vibration and can also distort the stud holes in the wheels. Once the vehicle is on the ground, check and adjust the air pressure in all tires.

> **WARNING!**
>
> *Overtorqueing and uneven tightening of the lug nuts are the most common cause of disc brake rotor distortion. Also, an overtorqued lug distorts the threads of the lug and could lead to premature failure.*

Mounting Run-Flat Tires

Mounting and dismounting a run-flat tire requires patience and skill. Most shops have rim-clamp tire changers. Some of these use rollers to loosen the beads, whereas others have side-shovel bead looseners. As the tire revolves on the changer, the rollers automatically loosen the beads. When using side-shovel bead looseners, the tire must be manuallly placed into position, rotated on the changer, and flipped. Roller bead looseners are preferred for run-flat tires.

The bead looseners must be positioned on the sidewall near the bead. If the loosener is placed too far away from the bead, it may damage the sidewall inserts in the tire. The rotation of the bead loosener should begin with the bead on the back of the tire and only be inserted enough to free the bead from the wheel. Tire lubricant should be applied to the tire and wheel as the bead moves away from the rim. The process is then repeated on the front side of the tire until the outside bead is loose. Care should be taken not to damage the TPMS sensor or the inner wheel ring (if so equipped). The tire should then be removed from the wheel, following the instructions of the tire changer.

TIRE/WHEEL ASSEMBLY SERVICE

For most tire/wheel service, the assembly must first be removed from the vehicle. The wheel and the tire must be separated whenever tires are replaced or repaired. The rear-wheel drum or disc brake rotor is usually attached to studs on the rear axle shaft's hub flange. The wheel and tire mount on the same studs and are held against the hub and drum or rotor by the wheel nuts.

Tire/Wheel Balance

Proper wheel alignment allows the tires to roll straight without excessive tread wear. The wheels can go out of alignment from striking raised objects or potholes. Misalignment subjects the tires to uneven and/or irregular wear. An out-of-balance condition can also cause increased wear on the ball joints, as well as deterioration of shock absorbers and other suspension components.

Should an inspection show uneven or irregular tire wear, wheel alignment and balance service is a must. Wheel balancing distributes weights along the wheel rim, which counteract heavy spots in the wheels and tires and allow them to roll smoothly without vibration. The wheel weights are adhered to the wheel or are clipped over the edge of the wheel's rim. There are two types of wheel imbalance: static and dynamic.

Static Balance **Static balance** is the equal distribution of weight around the wheel. Wheels that are statically unbalanced cause a bouncing action called **wheel tramp**. This condition eventually causes uneven tire wear. As the name implies, static balance means balancing a wheel at rest. This is done by adding a compensating weight. A statically unbalanced wheel tends to rotate by itself until the heavy portion is down. A bubble balancer is used to statically balance a tire and wheel. When it is placed on the balancer, any imbalance moves the bubble off center.

Many equipment manufacturers recommend static balancing a wheel at equal distances from the center of the light area. Balance weights are normally hammered on with their holding tabs between the tire bead and rim (**Figure 44–23**). Wheel weights are not normally hammered onto alloy or mag wheels; rather, special tape weights are adhered to the wheels to balance them.

Figure 44-23 A typical wheel weight attached to a wheel.

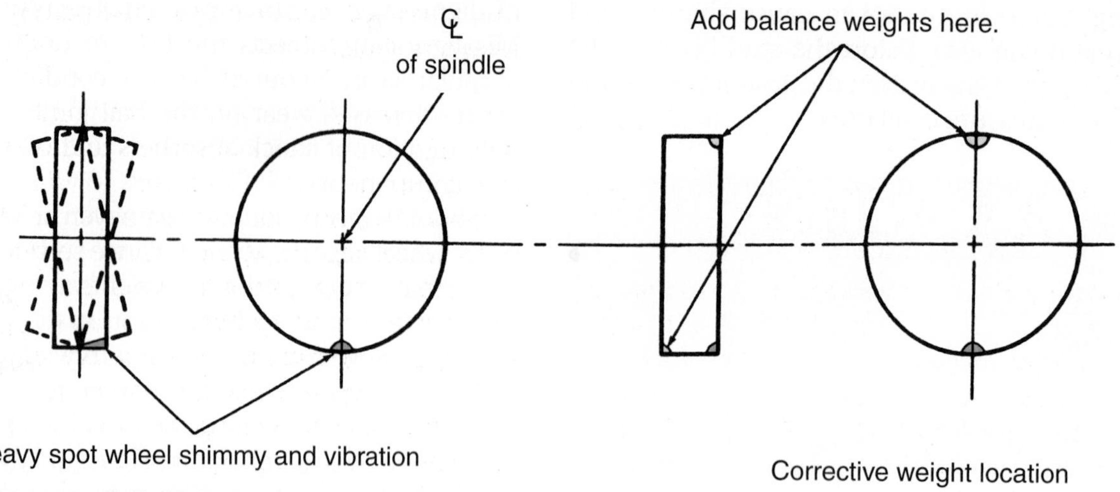

Figure 44-24 Dynamic wheel balancing calls for a weight to be attached to the wheel to compensate for a heavy spot on the wheel. *Courtesy of Chrysler LLC*

Dynamic Balance Dynamic balance is the equal distribution of weight on each side of the centerline. When the balanced tire spins, there is no tendency for the assembly to move from side to side. Wheels that are dynamically unbalanced can cause **wheel shimmy** and a wear pattern **(Figure 44–24)**. Dynamic balance, simply stated, means balancing a wheel in motion. Once a wheel starts to rotate and is in motion, the static weights try to reach the true plane of rotation of the wheel because of the action of centrifugal force. In an attempt to reach the true plane of rotation when there is an imbalance, the static weights force the spindle to one side.

At 180 degrees of wheel rotation, static weights kick the spindle in the opposite direction. The resultant side thrusts cause the wheel assembly to wobble or wiggle. When the imbalance is severe enough, as already mentioned, it causes vibration and front-wheel shimmy.

To correct dynamic unbalance, equal weights are placed 180 degrees opposite each other, one on the inside of the wheel and one on the outside, at the point of unbalance. This corrects the couple action or wiggle of the wheel assembly. Also, note that dynamic balance is obtained, while static balance remains unaffected.

The most commonly used dynamic wheel balancer requires that the tire/wheel assembly be taken off and mounted on the balancer's spindle **(Figure 44–25)**. The machine spins the entire assembly to indicate the heavy spot with a strobe light or other device. Two tests must be done, one for static and one for dynamic imbalance. One set of weights is placed to correct for static imbalance, and others are placed to correct for dynamic imbalance. Sometimes proper positioning of the static balance weights also corrects dynamic imbalance.

There are several electronic dynamic/static balancer units available that will permit balancing while the wheel and tire are on the car. A switch on the console sets the machine for either static or dynamic balancing. When the wheel balancing assembly is mounted for static balancing, it rotates until the heavy spot falls to the bottom. Weights are added to balance the assembly.

In the dynamic balance mode, the wheel assembly is rotated at high speed. Observing the balance scale, the operator reads out the amount of weight that has to be added and the location where the weights should be placed.

Road Force Measurements The wheel balancer shown in Figure 44–25 is designed to also eliminate all causes of vibration due to the wheel and tire assembly. Like many other current wheel balancers, this unit can simulate a road test with a load roller. This roller applies a heavy force on the tire while it is rotating on the balancer. The roller also measures the deflection of the tire as it rolls under pressure. The machine then makes recommendations for the service required to remove all runout and ensure vibration-free operation.

WHEEL BEARINGS

The purpose of all bearings is to allow a shaft to rotate smoothly in a housing or to allow the housing to rotate smoothly around a shaft. Wheel and axle bearings do this for a vehicle's wheels. Typically, on driving axles, the wheel is mounted to the hub of an axle shaft and the shaft rotates within a housing on an axle bearing. Wheel bearings are used on nondriving axles. The wheel's hub rotates on a shaft called the spindle. Axle bearings are typically serviced with the drive axle. Wheel bearings, however, require periodic

Figure 44-25 A computerized tire balancer.
Courtesy of Hunter Engineering Company

maintenance service and are often serviced with suspension and brake work. Although there is a distinction between axle and wheel bearings, the bearings for the front wheels on a FWD and 4WD vehicle are commonly called wheel bearings. Regardless of what they are called, bad bearings cause handling and tire wear problems.

Tapered Roller Bearing Troubleshooting

Bearings rarely fail suddenly. Rather, they deteriorate slowly because of dirt, lack of lubrication, and improper adjustment. Bearing wear and failure are almost always accompanied by noise and/or vibration.

Normal bearing sounds should be uniform as the wheel spins on its spindle. An uneven rumble or a grinding sound indicates possible bearing problems. While rotating the wheel, try to move it in and out on the spindle and note the amount of movement. Worn or damaged bearings or bearings that need adjustment will have a noticeable amount of end play. Also grasp the top and bottom of the tire and try to wobble the wheel and tire back and forth. You should feel little or no wobble.

End play can be measured with a dial indicator placed against the wheel hub. Set the indicator to zero, move the wheel in and out on the spindle, and note the reading. Compare your readings to specifications.

Inspect the wheel and the brake drum or rotor for grease that may be leaking past a bad seal. Bearing grease can contaminate brake linings, and a leaking grease seal can let dirt into the bearing. A leaking seal must be replaced, but the bearings also must be cleaned, inspected, and repacked to be sure they have not been damaged. Whether installing a used bearing or installing a new one, always install a new grease seal.

Front Wheel Hubs

Often the front wheel hub bearing assembly for driven and nondriven wheels has two tapered bearings **(Figure 44–26)** facing each other. Each of the

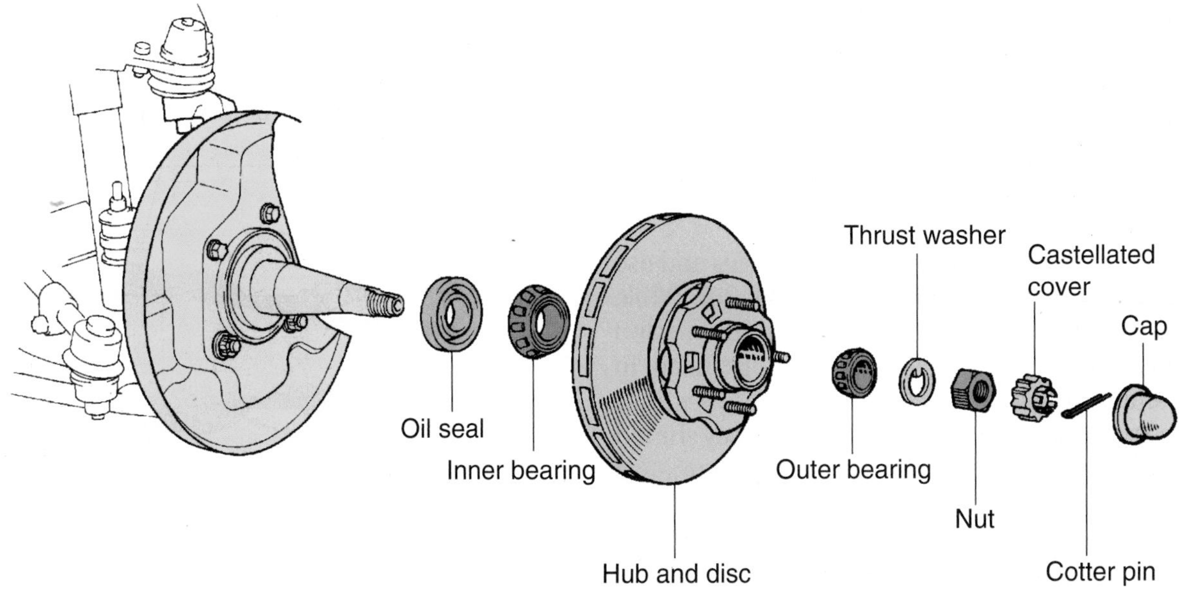

Figure 44-26 An exploded view of a typical front wheel bearing assembly for a RWD vehicle.

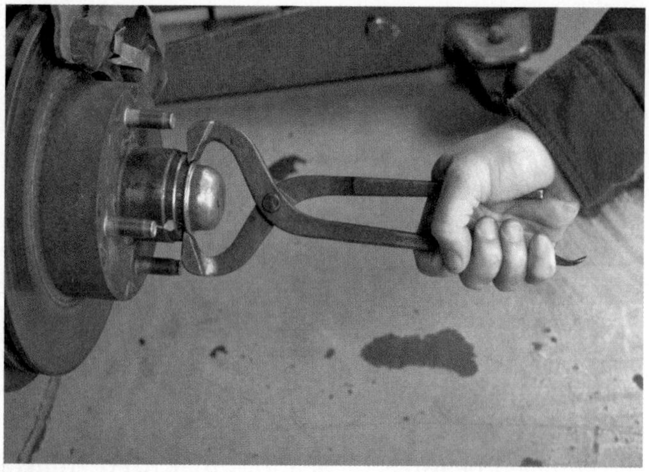

Figure 44-27 A special tool for removing a dust cap.

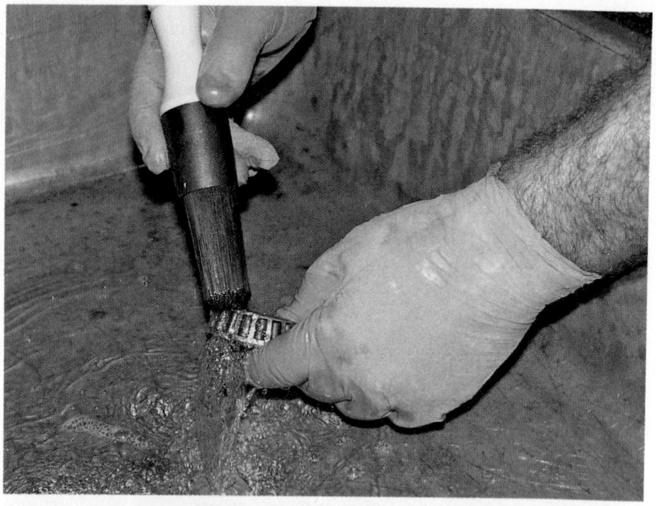

Figure 44-28 Thoroughly clean the bearings and races and then carefully inspect them before reusing them.

bearings rides in its own race. Some front wheel bearings are sealed units and are lubricated for life. They should be replaced and serviced as an assembly. Others are serviceable and require periodic lubrication and adjustment.

Except when making slight adjustments to the bearings, the bearing assembly must be removed for all service work. This is done with the vehicle lifted and the wheel assembly removed. In the center of the hub there is a dust (grease) cap. Using a special dust cap removal tool **(Figure 44–27)**, wiggle the cap out of its recess in the hub. Now remove the cotter pin and locknut from the end of the spindle. Loosen the spindle nut while supporting the brake assembly and hub. On many vehicles the brake caliper must be removed to remove the brake disc and hub. Once the hub is free from the spindle, remove the spindle nut and the washer behind the nut. Move the hub slightly forward, and then push it back. This should free the outer bearing so you can remove the hub assembly.

A grease seal located on the back of the hub normally keeps the inner bearing from falling out when the hub is removed. To remove the bearing assembly, the grease must be removed. In most cases, the seal can be pried out of the hub. The inner bearing should then fall out. Keep the outer bearing and inner bearing separated if you plan on reusing them.

Wipe the grease off the bearings and races and use a parts cleaner to clean them **(Figure 44–28)**. While doing this, pay close attention to the condition and movement of the bearings. The bearings need to rotate smoothly. Also visually inspect the bearings and races; any noticeable damage means they should be replaced. Also inspect the spindle. If it is damaged or excessively worn, the steering knuckle assembly should be replaced.

Whenever a bearing is replaced, its race must be replaced with it. Races are pressed in and out of the hub. Typically, the old race can be driven out with a large drift and a hammer. Once the race has been removed, wipe all grease from the inside of the hub. The new race should be installed with the proper driver.

During assembly, the bearings and hub assembly must be thoroughly lubricated **(Figure 44–29)**. Care must be taken not to get grease on the brake disc or on any part that will directly contact the disc. Always use the recommended grease. The grease must be able to withstand much heat and friction. If the wrong grease is used, it may not offer the correct protection or it may liquefy from the heat and leak out of the seals.

The bearings should be packed with grease. It is important that the grease be forced into and around all of the rollers in the bearing. Merely coating the outside of the bearing with grease will not do the job. A bearing packer does the best job at packing in the grease. If one is not available, force grease into the bearing with your hand.

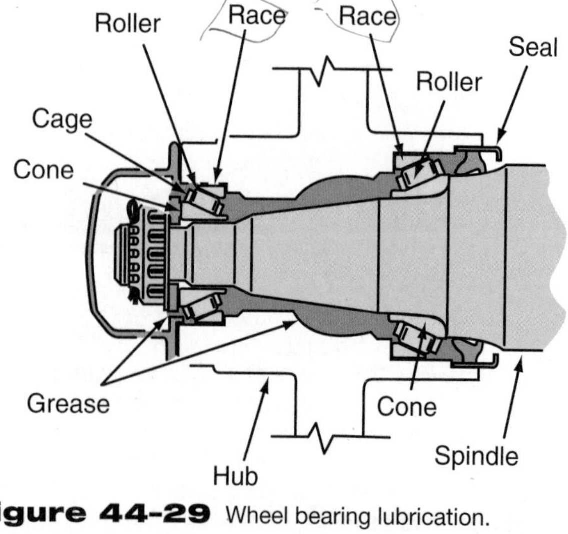

Figure 44-29 Wheel bearing lubrication.

Install the greased inner bearing into the hub. Install a new grease seal into the hub. To avoid damaging the seal, use the correct size driver to press the seal into the hub. Lubricate the spindle, and then slip the hub over the spindle. Install the outer bearing, washer, and locknut.

WARNING!

Throughout this entire process, your hands will have grease on them. Be very careful not to touch the brake assembly with your greasy hands. Clean them before handling the brake parts or use a clean rag to hold the brake assembly.

Wheel Bearing Adjustment The locknut should be adjusted to the exact specifications given by the manufacturer. Often it is tightened and loosened before it is adjusted. The initial tightening seats the bearings into their races.

PROCEDURE

To properly adjust front tapered bearings, follow these steps:

STEP 1 Support the rotor or drum with one hand and install the outer bearing and thrust washer into the hub.

STEP 2 Finger tighten the bearing adjusting nut against the thrust washer.

STEP 3 Then adjust the bearings by one of the following methods:

a. Rotate the drum or rotor and snug up the adjusting nut with a wrench to seat the bearings. While continuing to rotate the drum or rotor, back off the nut ¼ to ½ turn or until it is barely loose. Then tighten the nut by hand to a snug fit.

b. Rotate the drum or rotor and tighten the adjusting nut with a torque wrench to the specified torque. Then back off the nut ⅓ turn and retorque it to the specified value while continuing to rotate the drum or rotor. Final torque is usually in inch-pounds.

c. Rotate the drum or rotor and tighten the adjusting nut with a torque wrench to 12 to 25 foot-pounds. Then back off the nut ¼ to ½ turn or until it is just loose. Mount the base of a dial indicator as close as possible to the center of the hub. Locate the tip of the indicator's plunger on the tip of the spindle. Set the indicator to zero. Move the drum or rotor in and out and note the indicator reading. Turn the adjusting nut as necessary to obtain the specified end play, which is usually 0.001 to 0.005 inch (0.025 to 0.125 mm).

STEP 4 After adjustment, install the locknut over the top of the adjusting nut so the slots in the locknut align with the cotter pin hole in the spindle.

STEP 5 Install a new cotter pin through the spindle and locknut and bend its ends to secure it. Reinstall the dust cap in the hub.

STEP 6 If the axle has drum brakes that were backed off to allow removal of the drum, readjust the brakes.

STEP 7 If the axle has disc brakes, reinstall the caliper.

STEP 8 Reinstall the wheel and tire and lower the vehicle to the ground.

FWD and 4WD Vehicles

The front bearing arrangement often found on FWD and 4WD vehicles is often nonserviceable. To replace these bearings, they must be pressed in and out of the hub. To do this, the axle or half shaft is removed, as is the steering knuckle and hub assembly. The bearings may be sealed and require no additional lubrication or they may need to be packed with grease when they are reassembled. In most cases, the bearings are not adjusted. A heavily torqued axle nut is used to hold the assembly in place on the axle. This nut is typically replaced after it has been removed and is staked in place after it is tightened.

SHOP TALK

Because an axle nut is heavily torqued, it is wise to loosen the nut before the vehicle is raised for service. The weight of the vehicle on the tires will stop the hub from rotating while the nut is being loosened. The same holds true for final torquing. Tighten the nut as tight as possible with the vehicle raised, then lower the vehicle and torque the nut to specifications.

Rear Hubs

The rear bearings on a FWD vehicle are serviced in the same way as the nondriving front wheel bearings. Most RWD axle bearings are of the straight roller bearing design, in which the drive axle tube serves as the bearing race. Some rear wheel axle bearings are of the ball or tapered roller bearing type.

Wheel Bearing Grease Specification

The grease for wheel bearings should be smooth textured, consist of soaps and oils, and be free of filler and abrasives. Recommended are lithium complex (or equivalent) soaps, or solvent-refined petroleum oils. Additives could be added to inhibit corrosion and oxidation. The grease should be noncorrosive to bearing materials with no chance of separating during storage or use.

Using the correct amount of lube is also essential. Failure to maintain proper lubrication might result in bearing damage, causing a wheel to lock. Greases are classified by the National Lubricating Grease Institute (NLGI) to indicate their application.

Go To ***Chapter 7 for a detailed discussion and chart of NLGI lubricants.***

CASE STUDY

A customer complains of a bearing noise from the right front of his late-model 2WD pickup. The customer says this bearing has already been replaced twice in the last year and he is not very happy about paying for it again. Since not many miles have expired between bearing replacements, something must be causing the premature bearing failure.

The technician removes the right front wheel and hub. The outer bearing and race are heavily scored. After he cleans the bearing for a closer look at the wear, he notices that the bearing has an uneven wear pattern, which indicates poor bearing alignment. The same pattern is visible on the bearing race. Both must be replaced. After he removes the race from the hub, the technician inspects the hub only to find a small metal burr behind the bearing race. This small burr is the cause of the misalignment of the bearing. The burr is removed with a fine file, inner and outer bearing assemblies are installed, and the truck taken for a test drive.

There was no noise from the bearing during the test drive. The truck was returned to the customer. After 6 months, the technician called the customer to find out if the bearing noise had reappeared and found that it had not. Isn't it funny how the little things, like a small metal burr, can make life difficult? In this case the burr had frustrated the technician and the customer.

When servicing, replacing, or installing wheel bearings, always follow the procedure given in the service manual.

KEY TERMS

Aspect ratio
Bead
Belted bias ply
Bias ply
Drop center
Dynamic balance
Hub
Lateral runout
Plies
Radial ply
Radial runout
Sidewall
Sipes
Static balance
Tire pressure monitoring (TPM) system
Tread
Uniform Tire Quality Grading (UTQG)
Wheel offset
Wheel shimmy
Wheel tramp

SUMMARY

- Wheels are made of either stamped or pressed steel discs riveted or welded into a circular shape or are die-cast or forged aluminum or magnesium rims.
- The primary purpose of tires is to provide traction. They are also designed to carry the weight of the vehicle, to withstand side thrust over varying speeds and conditions, to transfer braking and driving torque to the road, and to absorb much of the rock shock from surface irregularities.
- Pneumatic tires are of two types: those that use inner tubes and those that do not. The latter are called tubeless tires and are the only type used on passenger cars today.
- There are three types of tire construction on the road today: bias ply, belted bias ply, and radial ply.
- Tires are rated by their profile, ratio, size, and load range.
- An ideal tire is one that wears little, holds the road well to provide sure handling and braking, and provides a cushion from road shock. It should also provide maximum grip on dry roads, wet roads, and snow and ice, and operate quietly at any speed.
- The number and size of the blocks, sipes, and grooves on a tire's tread determines how much rubber contacts the road, how much water can be displaced, and how quiet the tire will be during travel.

- To maximize tire performance, inspect for signs of improper inflation and uneven wear, which can indicate a need for balancing, rotation, or alignment. Tires should also be checked frequently for cuts, bruises, abrasions, and blisters, and for stones or other objects that might have become imbedded in the tread.
- A properly inflated tire gives the best tire life, riding comfort, handling stability, and even gas mileage during normal driving conditions.
- To equalize tire wear, most car and tire manufacturers recommend that the tires be rotated. It must be remembered that front and rear tires perform different jobs and can wear differently, depending on driving habits and the type of vehicle.
- Most tires used today have built-in tread wear indicators to show when tires need replacement.
- There are three popular methods of tire repair: head-type plug, cold patch repair, and hot patch repair.
- There are two types of wheel balancing: static balance and dynamic balance.
- The bearings on an axle that drives the wheels are called axle bearings. Wheel bearings are used on nondriving axles. Axle bearings are typically serviced with the drive axle. Wheel bearings require periodic maintenance service and are often serviced with suspension and brake work.
- Bad axle and wheel bearings will cause handling and tire wear problems.
- The front wheel hubs on ball or tapered roller bearings are lubricated by wheel bearing grease.
- Rear wheels are bolted to integral or detachable hubs.

REVIEW QUESTIONS

1. List five things that could cause premature bearing failure.
2. Explain the purpose of the wheel rim drop center and safety ridges.
3. Why is tire rotation recommended by most manufacturers?
4. Define dynamic and static wheel balance.
5. Describe the procedure for using a plug to seal a puncture in a tire.
6. The rim offset is the vertical distance between the rim centerline and the _______ _______ of the disc.
7. To calculate the tire aspect ratio, the tire section height is divided by the _______.
8. *True or False?* All TPMS have a pressure sensor mounted inside the tire and wheel assembly.
9. A tire that wobbles from side to side is said to have _______.
   a. radial runout
   b. lateral runout
   c. eccentric runout
   d. none of the above
10. A front tire has excessive wear on both edges of the tire tread. The most likely cause of this problem is _______.
    a. overinflation
    b. underinflation
    c. improper static balance
    d. improper dynamic balance
11. All of the following statements are correct *except* _______.
    a. belts are reinforcing materials that encircle the tire under the tread
    b. the carcass of a tire is made up of plies, layers of cloth, and rubber
    c. most tread patterns are designed to work well on both wet and dry roads
    d. the bead is the decorative pattern at the outer edge of the tread
12. All of the following statements are correct *except* _______.
    a. tire inflation pressure directly affects traction
    b. the recommended tire pressure for front and rear tires may be different
    c. the recommended tire pressures are often lower than maximum pressures
    d. the recommended tire pressure is molded into the sidewall of the tire
13. Which of the following statements about sidewall markings is correct?
    a. The aspect ratio, or profile, is a rating for tread wear.
    b. The traction and temperature ratings are based on the speed rating of the tire.
    c. The tire's maximum safe inflation pressure and load are indicated.
    d. Section width is the distance between the tread and the bead.
14. All of these statements about improper wheel balance are true *except* _______.
    a. dynamic imbalance may cause wheel shimmy
    b. dynamic imbalance may cause steering pull in either direction

c. static imbalance causes wheel tramp
d. static imbalance causes rapid wear on suspension components

15. Most information about a car's tires can be found on the ______.
a. engine block
b. tire placard
c. VIN tag
d. certification label

ASE-STYLE REVIEW QUESTIONS

1. Technician A says that the wheel bearings on a typical FWD car are pressed in and out of the steering knuckle assembly. Technician B says that the rear wheel bearings on a typical FWD car are serviced in the same way as the front bearings for a RWD car. Who is correct?
a. Technician A
b. Technician B
c. Both A and B
d. Neither A nor B

2. Technician A says that dynamic wheel unbalance causes wheel shimmy. Technician B says that static wheel unbalance causes wheel tramp. Who is correct?
a. Technician A
b. Technician B
c. Both A and B
d. Neither A nor B

3. Technician A says that front bearing assembly locknuts for RWD vehicles are typically heavily torqued to maintain the bearing adjustment. Technician B says that the axle nuts for a FWD wheel bearing assembly is often staked in place to maintain its specified torque. Who is correct?
a. Technician A
b. Technician B
c. Both A and B
d. Neither A nor B

4. Technician A says that replacement wheel rims should be the same as the original equipment wheels in load capacity, offset, width, diameter, and mounting configuration. Technician B says that if wheels are installed that have a different stock width or diameter, make sure to use a tire that has the same overall width as the original tires. Otherwise the vehicle's wheel speed sensors, speedometer, antilock brake system, and many other systems will be affected. Who is correct?
a. Technician A
b. Technician B
c. Both A and B
d. Neither A nor B

5. Technician A says that underinflation can increase the rolling resistance of a tire. Technician B says that underinflation can cause hard steering and a 10% decrease in fuel economy. Who is correct?
a. Technician A
b. Technician B
c. Both A and B
d. Neither A nor B

6. While choosing the correct tire for a vehicle: Technician A says that a tire with a temperature resistance rating of B will dissipate more heat than one rated with an A. Technician B says that the temperature rating indicates how well a tire will perform when it is hot. Who is correct?
a. Technician A
b. Technician B
c. Both A and B
d. Neither A nor B

7. Technician A says that on most vehicles, the wheel/tire assembly is considered sprung weight. Technician B says that low unsprung weight makes the vehicle handle better on irregular surfaces. Who is correct?
a. Technician A
b. Technician B
c. Both A and B
d. Neither A nor B

8. While discussing run-flat tires: Technician A says that some are self-sealing tires and are designed to quickly and permanently seal most tread area punctures. Technician B says that most of these tires have reinforced sidewalls that are able to support the vehicle at any speed without air in the tire. Who is correct?
a. Technician A
b. Technician B
c. Both A and B
d. Neither A nor B

9. Technician A says that the aspect ratio of a tire represents the relationship between the tire's cross-sectional height to its cross-sectional width. Technician B says that low aspect ratios provide a softer ride because they will deflect more over irregular surfaces and under heavy loads. Who is correct?
a. Technician A
b. Technician B
c. Both A and B
d. Neither A nor B

10. While discussing the effects of loose wheel bearings: Technician A says that the vehicle will handle poorly. Technician B says that the bearing may overheat. Who is correct?
a. Technician A
b. Technician B
c. Both A and B
d. Neither A nor B

SUSPENSION SYSTEMS

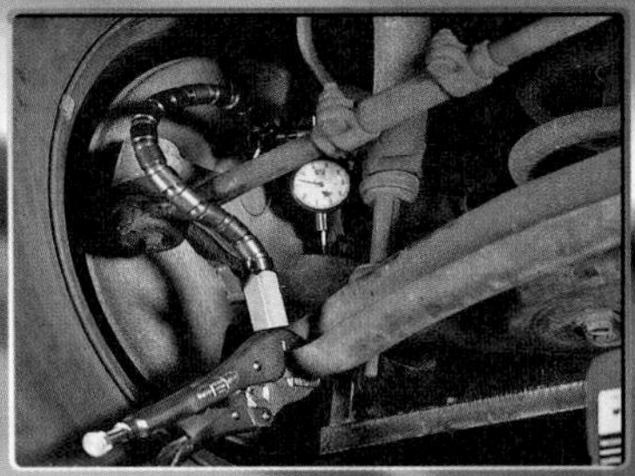

CHAPTER 45

OBJECTIVES

■ Name the different types of springs and how they operate. ■ Name the advantages of ball joint suspensions. ■ Explain the important differences between sprung and unsprung weight with regard to suspension control devices. ■ Identify the functions of shock absorbers and struts and describe their basic construction. ■ Identify the components of a MacPherson strut system and describe their functions. ■ Identify the functions of bushings and stabilizers. ■ Perform a general front suspension inspection. ■ Check chassis height measurements to specifications. ■ Identify the three basic types of rear suspensions and know their effects on traction and tire wear. ■ Identify the various types of springs, their functions, and their locations in the rear axle housing. ■ Describe the advantages and operation of the three basic electronically controlled suspension systems: level control, adaptive, and active. ■ Explain the function of electronic suspension components including air compressors, sensors, control modules, air shocks, electronic shock absorbers, and electronic struts. ■ Explain the basic towing, lifting, jacking, and service precautions that must be followed when servicing air springs and other electronic suspension components.

Like the other systems on cars and light trucks, the suspension system has become more advanced through the years. These advances have been made to provide better and safer handling and a better ride. Today, front and rear suspensions have many parts and can be quite complex **(Figure 45–1)**.

As a vehicle moves, the suspension and tires must react to the current driving conditions.

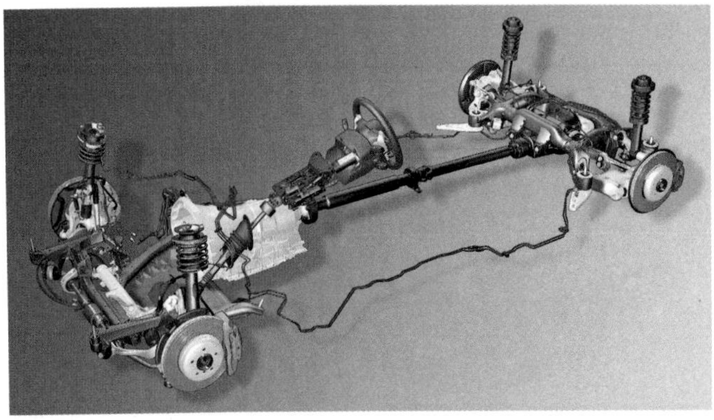

Figure 45-1 The suspension system for a late-model car. The front uses a strut setup, while the rear has a multilink system. *Courtesy of BMW of North America, LLC*

Specifically, the suspension system:

- Supports the weight of the vehicle
- Keeps the tires in contact with the road
- Controls the direction of the vehicle's travel
- Attempts to maintain the correct vehicle ride height
- Maintains proper wheel alignment
- Reduces the effect of shock forces as the vehicle travels on an irregular surface

FRAMES

To provide a rigid structural foundation for the vehicle body and a solid anchorage for the suspension system, a frame of some type is essential. There are two basic frames in common use today.

Conventional Frame Construction

In the conventional body-over-frame construction, the frame is the vehicle's foundation. The body and all major parts of a vehicle are attached to the frame. It must provide the support and strength needed by the assemblies and parts attached to it. In other words, the frame is an independent, separate component because it is not welded to any of the major units of the body shell.

Unibody Construction

Unibody construction has no separate frame. The body is constructed in such a manner that the body parts themselves supply the rigidity and strength required to maintain the structural integrity of the car. The unibody design significantly lowers the base weight of the car, and that, in turn, increases gas mileage capabilities.

SUSPENSION SYSTEM COMPONENTS

Nearly all automotive suspensions have the same basic components and they operate similarly. The basic differences between the suspensions found on various vehicles are the construction and placement of the parts.

Springs

A spring is the core of all suspension systems. Springs carry the weight of the vehicle and absorb shock forces while maintaining correct riding height. They are compressible links between the vehicle's frame and body and the tires. Doing this, they dampen road shock and provide a comfortable ride. If a spring is worn or damaged, other suspension parts will shift out of their proper positions and will experience increased wear.

Various types of springs are used in suspension systems (**Figure 45–2**)—coil, torsion bar, leaf (both mono- and multileaf types), and air springs. Springs are mounted in rubber or nylon to reduce road shock and noise.

Automotive springs are generally classified by the amount they compress under a specific load. This is referred to as the **spring rate**. A force (weight) applied to a spring causes it to compress in direct proportion to the force applied. When that force is removed, the spring returns to its original position if it is not overloaded. This is why a heavy vehicle needs stiffer springs than a lightweight car.

The springs take care of two fundamental vertical actions: jounce and rebound. **Jounce**, or compression, occurs when a wheel hits a bump and moves up (**Figure 45–3A**). When this happens, the suspension system acts to pull in the top of the wheel, maintaining an equal distance between the two wheels and preventing a sideways scrubbing action as the wheel moves up and down. **Rebound**, or extension, occurs when the wheel hits a dip or hole and moves downward (**Figure 45–3B**). In this case, the suspension system acts to move the wheel in at both the top and bottom equally, while maintaining an equal distance between the wheels.

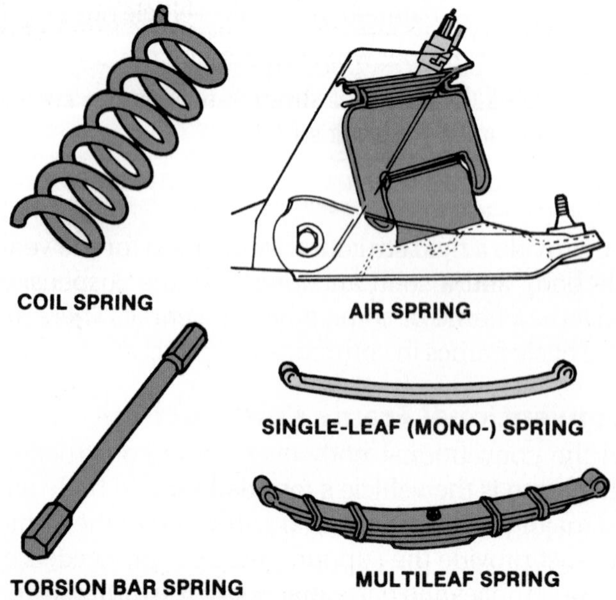

Figure 45-2 Various types of automotive springs.

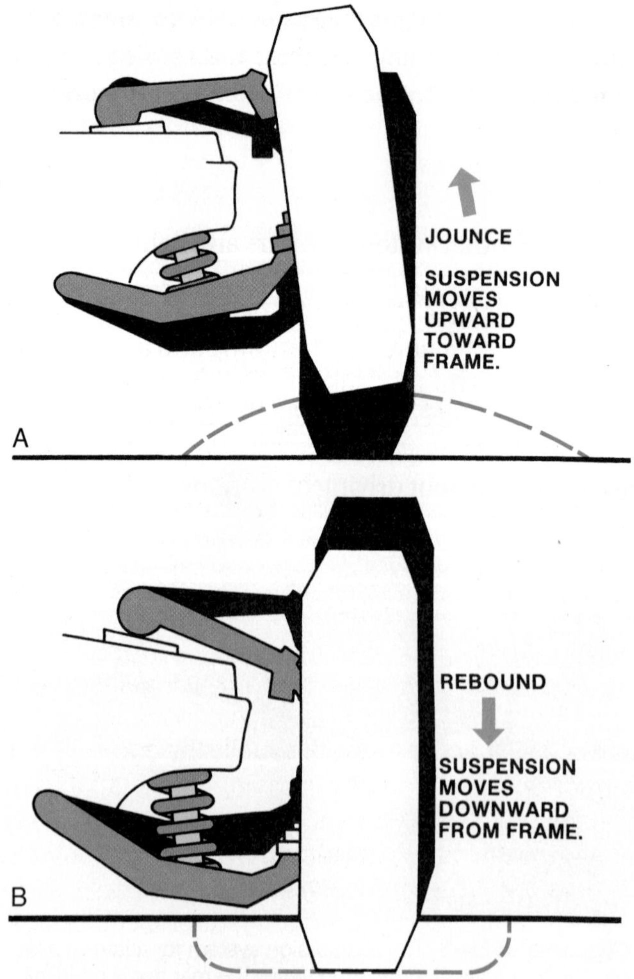

Figure 45-3 (A) Upward and (B) downward suspension movement.

Figure 45-4 The different designs of coil springs.

When the spring experiences compression or extension, it stores energy. This energy forces the spring to return to its normal shape. The spring oscillates between jounce and rebound until all energy has moved from the spring. Each oscillation becomes smaller until it stops. A **shock absorber** is added to each suspension to dampen and stop the motion of the spring after jounce.

Coil Springs Two basic designs of coil springs are used: linear rate and variable rate **(Figure 45–4)**. **Linear rate** springs characteristically have one basic shape and a consistent wire diameter. All linear springs are wound from a steel rod into a cylindrical shape with even spacing between the coils. As the load is increased, the spring is compressed and the coils twist (deflect). As the load is removed, the coils flex (unwind) back to the normal position. The amount of load necessary to deflect the spring 1 inch (25.4 mm) is the spring rate. On linear rate springs this is a constant rate, no matter how much the spring is compressed. For example, 250 pounds (112 kg) compress the spring 1 inch (25.4 mm) and 750 pounds (340 kg) compress the spring 3 inches (76.2 mm). Spring rates for linear rate springs are normally calculated between 20% and 60% of the total spring deflection.

Variable rate spring designs are characterized by a combination of wire sizes and shapes. The most commonly used variable rate springs have a consistent wire diameter, are wound in a cylindrical shape, and have unequally spaced coils. This type of spring is called a progressive rate coil spring.

The design of the coil spacing gives the spring three functional ranges of coils: inactive, transitional, and active. Inactive coils are usually the end coils and introduce force into the spring. Transitional coils become inactive as they are compressed to their point of maximum load-bearing capacity. Active coils work throughout the entire range of spring loading. Theoretically in this type of design, at stationary loads the inactive coils are supporting all of the vehicle's weight. As the loads are increased, the transitional coils take over until they reach maximum capacity. Finally, the active coils carry the remaining overload. This allows for automatic load adjustment while maintaining vehicle height.

Another common variable rate design uses tapered wire to achieve this same type of progressive rate action. In this design, the active coils have a large wire diameter and the inactive coils have a small wire diameter.

Later designs of variable rate springs deviate from the old cylindrical shape. These include the truncated cone, the double cone, and the barrel spring. The major advantage of these designs is the ability of the coils to nest, or bottom out, within each other without touching, which lessens the amount of space needed to store the springs in the vehicle.

Unlike a linear rate spring, a variable rate spring has no predictable standard spring rate. Instead, it has an average spring rate based on the load of a predetermined spring deflection. This makes it impossible to compare a linear rate spring to a variable rate spring. Variable rate springs, however, handle a load of up to 30% over standard rate springs in some applications.

Leaf Springs Although leaf springs were the first type of suspension spring used on automobiles, today they are generally found only on light-duty trucks, vans, and some passenger cars. There are three basic types of leaf springs: multiple leaf, monoleaf, and fiber composite.

Multiple-Leaf Springs **Multiple-leaf** springs consist of a series of flat steel leaves that are bundled together and held with clips or by a bolt placed slightly ahead of the center of the bundle. One leaf, called the **main leaf**, runs the entire length of the spring. The next leaf is a little shorter and attaches to the main leaf. The next leaf is shorter yet and attaches to the second leaf, and so on. This system allows almost any number of leaves to be used to support the vehicle's weight **(Figure 45–5)**. It also gives a progressively stiffer spring.

Figure 45-5 To provide more ground clearance, a lift kit that included ten-leaf rear springs was installed on this pickup. *Courtesy of Superlift Suspension Systems*

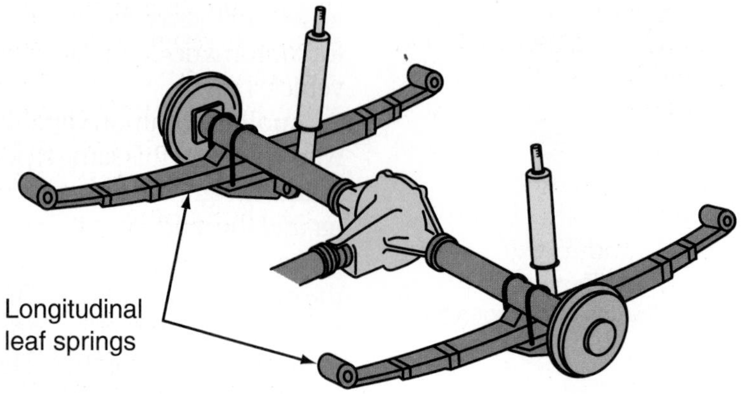

Figure 45–6 These leaf springs support and locate the drive axle.

The spring easily flexes over small distances for minor bumps. The farther the spring is deflected, the stiffer it gets. The more leaves and the thicker and shorter the leaves, the stronger the spring. It must be remembered that as the spring flexes, the ends of the leaves slide over one another. This sliding could be a source of noise and can also produce friction. These problems are reduced by interleaves of zinc and plastic placed between the spring's leaves. As the multiple leaves slide, friction produces a harsh ride as the spring flexes. This friction also dampens the spring motion.

Multiple-leaf springs have a curve in them. This curve, if doubled, forms an ellipse. Thus, leaf springs are sometimes called semielliptical or quarter-elliptical. The semi or quarter refers to how much of the ellipse the spring actually describes. The vast majority of leaf springs are semielliptical.

Leaf springs are typically mounted at right angles to the axle **(Figure 45–6)**. However, on late-model Toyota trucks, the leaf springs are angled outward toward the front, so the distance between the front ends of the two springs is greater than the distance between the rears of the springs. In addition to absorbing road shock, leaf springs also serve as a mount for the drive axle. A centering pin is often used to keep the axle properly located on the springs. If a spring is broken or not in its proper position, the drive axle may be sitting at an angle; this will cause handling problems.

Some vehicles have a transversely mounted leaf spring. The center of the spring is mounted to the vehicle's chassis and the outer ends are fastened to the ends of the axle housing or wheel spindles.

The front eye of the main leaf at either end of the axle is attached to a bracket on the frame of the vehicle with a bolt and bushing connection. The rear eye of the main leaf is secured to the frame with a shackle, which permits some fore and aft movement **(Figure 45–7)** in response to physical forces of acceleration, deceleration, and braking.

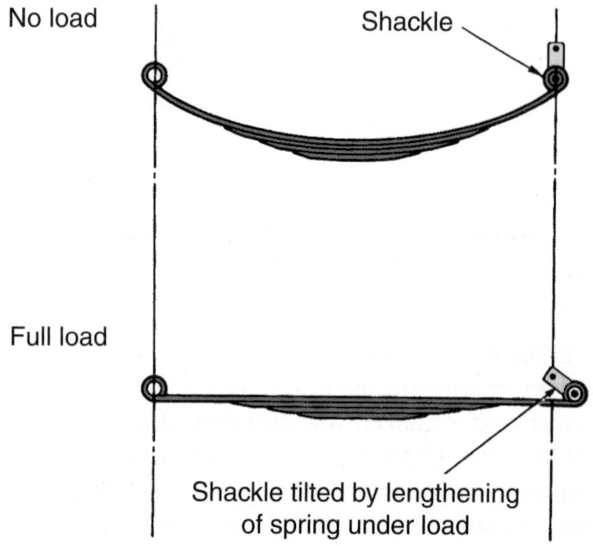

Figure 45–7 The action of a leaf spring as it compresses.

Monoleaf Springs **Monoleaf** or single-leaf springs are usually the tapered plate type with a heavy or thick center section tapering off at both ends. This provides a variable spring rate for a smooth ride and good load-carrying ability. In addition, single-leaf springs do not have the noise and static friction characteristic of multiple-leaf springs.

Fiber Composite Springs While most leaf springs are still made of steel, **fiber composite** types are increasing in popularity **(Figure 45–8)**. Some automotive people call them plastic springs in spite of the fact that

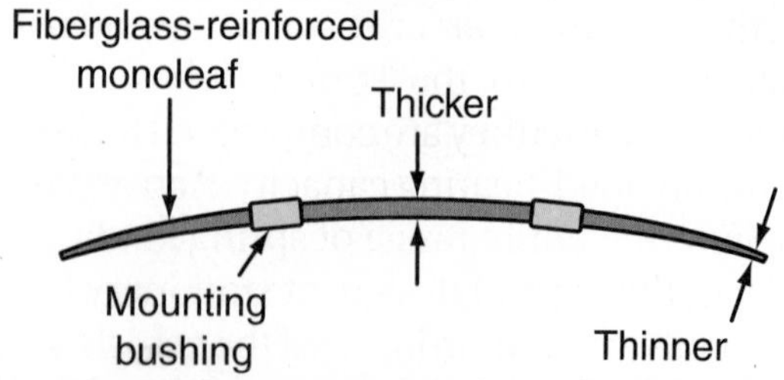

Figure 45–8 The construction of a fiberglass-reinforced monoleaf spring.

the springs contain no plastic at all. They are made of fiberglass, laminated and bonded together by tough polyester resins. The long strands of fiberglass are saturated with resin and bundled together by wrapping (a process called filament winding) or squeezed together under pressure (compression molding).

Fiber composite leaf springs are incredibly lightweight and possess some unique ride control characteristics. Conventional monoleaf steel springs are real heavyweights, tipping the scale at anywhere from 25 to 45 pounds (11 to 20 kg) apiece. Some multiple-leaf springs can weigh almost twice as much. A fiber composite leaf spring is a featherweight by comparison, weighing a mere 8 to 10 pounds (3.6 to 4.5 kg). As every performance enthusiast knows, springs are dead weight. Reducing the weight of the suspension not only reduces the overall weight of the vehicle, but also reduces the sprung mass of the suspension itself. This reduces the spring effort and amount of shock control that is required to keep the wheels in contact with the road. The result is a smoother riding, better handling, and faster responding suspension, which is exactly the sort of thing every performance enthusiast wants.

Air Springs Another type of spring, an air spring, is used in an air-operated microprocessor-controlled system that replaces the conventional coil springs with air springs to provide a comfortable ride and automatic front and rear load leveling. This system, fully described later in this chapter, uses four air springs to carry the vehicle's weight. The air springs are located in the same positions where coil springs are usually found. Each spring consists of a reinforced rubber bag pressurized with air. The bottom of each air bag is attached to an inverted pistonlike mount that reduces the interior volume of the air bag during jounce (**Figure 45–9**). This has the effect of increasing air pressure inside the spring as it is compressed, making it progressively stiffer. A vehicle equipped with an electronic air suspension system is able to provide a comfortable street ride, about a third softer than conventional coil springs. At the same time, its variable spring rate helps absorb bumps and protect against bottoming.

Torsion Bar Suspension System

Torsion bars serve the same function as coil springs. In fact, they are often described as straightened-out coil springs. Instead of compressing like coil springs, a torsion bar twists and straightens out on the recoil. That is, as the bar twists, it resists up-and-down movement. One end of the bar—made of heat-treated alloy spring steel—is attached to the vehicle frame. The other end is attached to the lower control arm (**Figure 45–10**). When the wheel moves up and down, the lower control arm is raised and lowered. This twists the torsion bar, which causes it to absorb road shocks. The bar's natural resistance to twisting quickly restores it to its original position, returning the wheel to the road.

When torsion bars are manufactured, they are prestressed to give them fatigue strength. Because of directional prestressing, torsion bars are directional. The torsion bar is marked either right or left to identify on which side it is to be used.

Because the torsion bar is connected to the lower control arm, the lower ball joint is the load carrier. A shock absorber is connected between the lower control arm and the frame to damp the twisting motion of the torsion bar.

Many late-model pickups and SUVs use torsion bars in their front suspensions. They are primarily used in this type vehicle because they can be mounted low and out of the way of the driveline components.

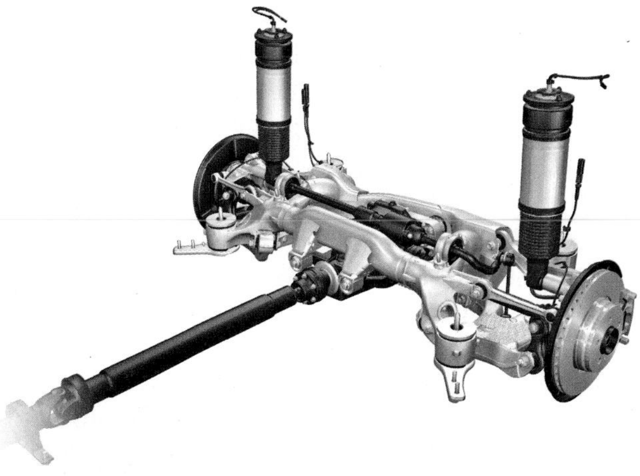

Figure 45–9 A rear suspension setup with air springs. *Courtesy of BMW of North America, LLC*

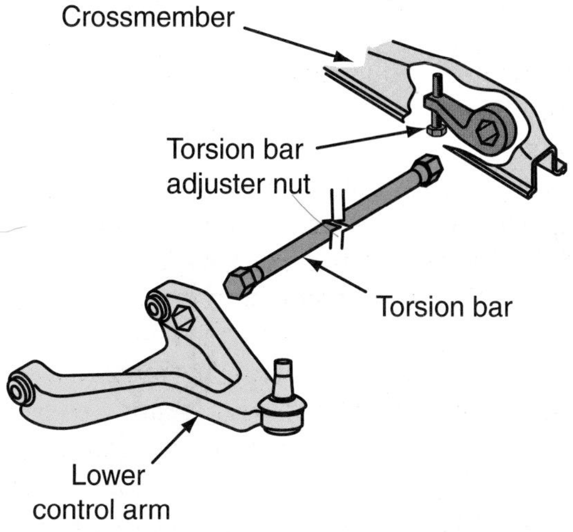

Figure 45–10 A torsion bar setup.

Shock Absorbers

Shock absorbers damp or control motion in a vehicle. If unrestrained, springs continue expanding and contracting after a blow until all the energy is absorbed. Not only would this lead to a rough and unstable—perhaps uncontrollable—ride after consecutive shocks, it would also create a great deal of wear on the suspension and steering systems. Shock absorbers prevent this. Despite their name, they actually dampen spring movement instead of absorbing shock. As a matter of fact, in England and almost everywhere else but the United States, shock absorbers are referred to as **dampers**.

Today's conventional shock absorber is a velocity-sensitive hydraulic damping device. The faster it moves, the more resistance it has to the movement **(Figure 45–11)**. This allows it to automatically adjust to road conditions. A shock absorber works on the principle of fluid displacement on both its compression (jounce) and extension (rebound) cycles. A typical car shock has more resistance during its extension cycle than its compression cycle. The extension cycle controls motions of the vehicle body spring weight. The compression cycle controls the same motions of the unsprung weight. This motion energy is converted into heat energy and dissipated into the atmosphere.

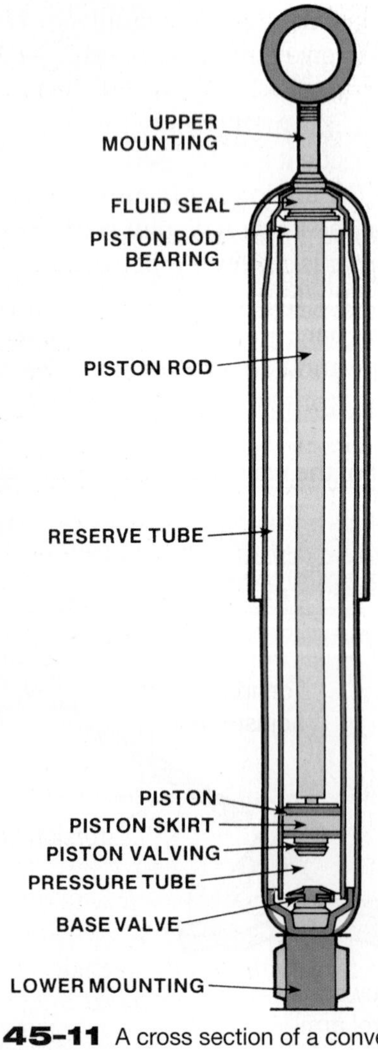

Figure 45–11 A cross section of a conventional shock absorber.

Shock absorbers can be mounted either vertically or at an angle. Angle mounting of shock absorbers improves vehicle stability and dampens accelerating and braking torque.

Conventional hydraulic shocks are available in two styles: single-tube and double-tube. The vast majority of domestic shocks are double tubed. While they are a little heavier and run hotter than the single-tubed type, they are easier to make. The double-tube shock has an outer tube that completely covers the inner tube. The area between the tubes is the oil reservoir. A compression valve at the bottom of the inner tube allows oil to flow between the two tubes. The piston moves up and down inside the inner tube.

In a single **monoshock**, there is a second floating piston near the bottom of the tube. When the fluid volume increases or decreases, the second piston moves up and down, compressing the reservoir. The fluid does not move back and forth between a reservoir and the main chamber. There are no other valves in a single-tube shock besides those in the main piston. The second piston prevents the oil from splashing around too much and getting air bubbles in it. Air in the shock oil is detrimental. Air, unlike oil, is compressible and slips past the piston easily. When this happens, the result is a shock that offers poor vehicle control on bumpy roads.

In addition to these conventional hydraulic shocks, there are a number of others the technician may encounter.

Gas-Charged Shock Absorbers On rough roads, the passage of fluid from chamber to chamber becomes so rapid that foaming can occur. Foaming is simply the mixing of the fluid with any available air. Since aeration can cause a skip in the shock's action, engineers have sought methods of eliminating it. One is the spiral groove reservoir, the shape of which breaks up bubbles. Another is a gas-filled cell or bag (usually nitrogen) that seals air out of the reservoir so the shock fluid can only contact the gas.

A gas-charged shock absorber **(Figure 45–12)** operates on the same hydraulic fluid principle as conventional shocks. It uses a piston and oil chamber similar to other shock absorbers. Instead of a double tube with a reserve chamber, it has a dividing piston that separates the oil chamber from the gas chamber. The oil chamber contains a special

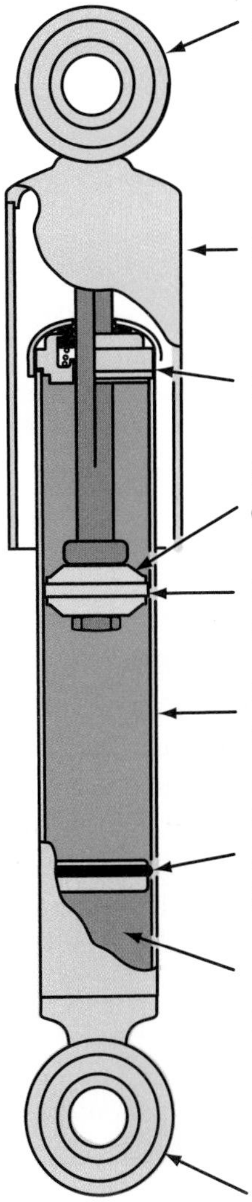

Dust shield

Rising rate valve system
dampens all bumps.
Working piston with valves

separates oil and nitrogen
gas to eliminate foaming.

principle gives better

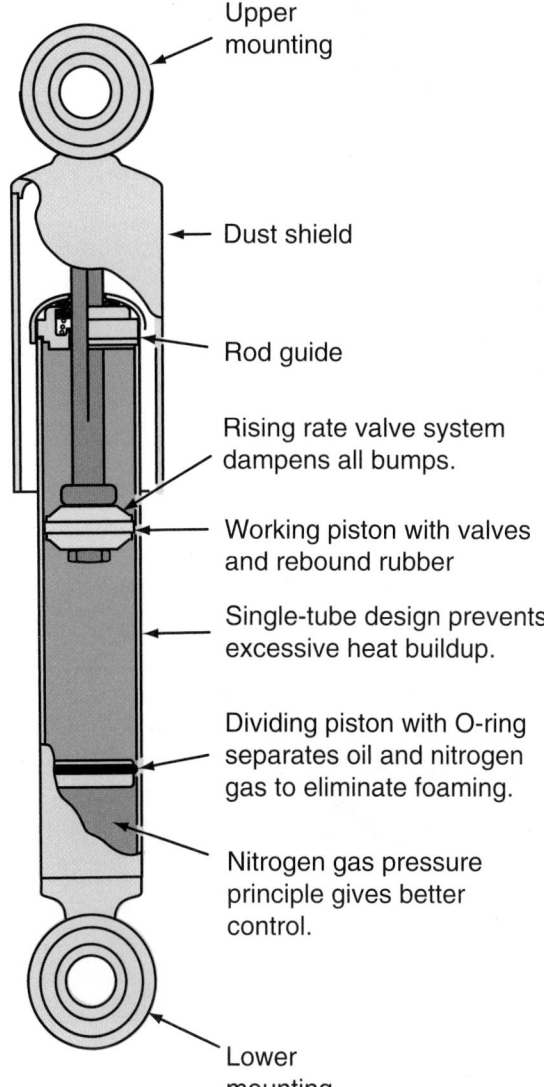

Figure 45-12 Gas-pressure damped shocks operate like conventional oil-filled shocks. Gas is used to keep oil pressurized, which reduces oil foaming and increases efficiency under seven conditions.

hydraulic oil, and the gas chamber contains nitrogen gas under pressure equal to approximately 25 times atmospheric pressure.

As the piston rod moves downward in the shock absorber, oil is displaced, just as it is in a double-tube shock. This oil displacement causes the divided piston to press on the gas chamber. The gas is compressed and the chamber reduces in size. When the piston rod returns, the gas pressure returns the dividing piston to its starting piston. Whenever the static pressure of the oil column is held at approximately 100 to 360 psi (690 to 2.482 kPa) (depending on the design), the pressure decreases behind the piston and so cannot be high enough for the gas to escape from the oil column. As a result, a gas-filled shock absorber operates without aeration.

SHOP TALK

Some high-pressure gas-charged shocks are monotube shocks with fluid and gas in separate chambers. The gas is charged to 360 psi (2.482 kPa). Its basic design does not allow the valving range needed for a more responsive ride over a broad range of road conditions. The high-pressure gas charge can provide a harsh ride under normal driving conditions and is usually found on small trucks.

Air Shock Systems There are two basic adjustable air shock systems: manual fill and automatic load-leveling. The manual fill system can be ordered on new vehicles or can be installed on almost any vehicle manufactured without it.

There are several different types of manual fill air shock systems available. One common manual fill air shock system uses a high-speed, direct current (DC) motor to transfer a command signal that is manually selected from the driver's seat. In another manual air system, the units are inflated through air valves mounted at the rear of the vehicle. Air lines run between the shocks and the valve. A tire air pressure pump is used to fill the shocks to bring the rear of the vehicle to the desired height.

Shock Absorber Ratio Most shock absorbers are valved to offer roughly equal resistance to suspension movement upward (jounce) and downward (rebound). The proportion of a shock absorber's ability to resist these movements is indicated by a numerical formula. The first number indicates jounce resistance. The second indicates rebound resistance. For example, passenger cars with normal suspension requirements use shock absorbers valued at 50/50 (50% jounce/50% rebound). Drag racers, on the other hand, use shocks valued at about 90/10. Small vehicles, because of their light weight and soft springs, require more control in both jounce and rebound in the shock absorbers. Damping rates within the shock absorbers are controlled by the size of the piston, the size of the orifices, and the closing force of the valves.

It is important to keep in mind that the shock absorber ratio only describes what percentage of the shock absorber's total control is compression and what percentage is extension. Two shocks with the same ratio can differ greatly in their control capacity. This is one reason the technician must be sure correct replacement shocks are installed on the vehicle.

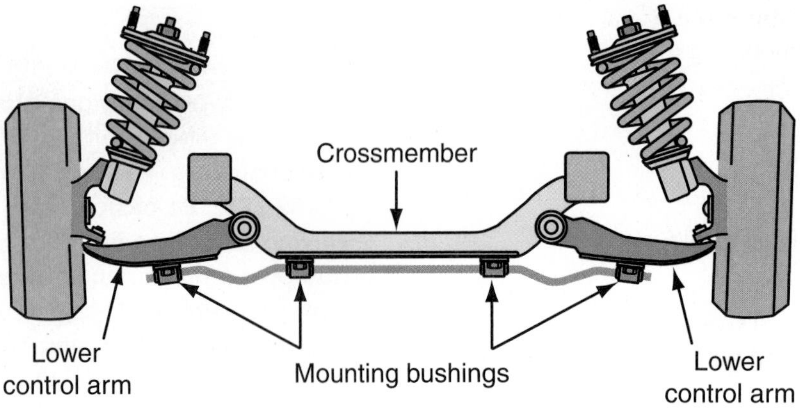

Figure 45–13 The typical location of a stabilizer bar.

Stabilizer Bars

Nearly all suspension systems have a **sway bar**, which is also known as the **antisway bar** or stabilizer. This bar, like the shock absorbers, provides directional stability by reducing body roll. It is a metal rod running between the opposite lower or upper control arms **(Figure 45–13)**. As the suspension at one wheel responds to the road surface, the sway bar transfers a similar movement to the suspension at the other wheel. For example, if the right wheel is drawn down by a dip in the road surface, the sway bar is drawn with it, creating a downward draw on the left wheel as well. In this way, a more level ride is produced. Sway or lean during cornering is also reduced. Depending on its thickness, the antisway bar can reduce vehicle roll or sway by up to 15%.

If both wheels go into a jounce, the sway bar simply rotates in its bushings. When only one wheel goes into jounce, the bar twists like a torsion bar to lift the frame and the opposite side of the suspension. This reduces body roll.

The sway bar is typically a one-piece, U-shaped rod connected to the control arms with rubber bushings, or it can be attached to each control arm by a separate sway bar link **(Figure 45–14)**. The arm is held to the links with nuts and rubber bushings and is also mounted to the frame in the center with rubber bushings. If it is too large, the sway bar causes the vehicle to wander. If it is too small, it has little effect on stability.

Strut rods are used on models that do not use the sway bar. Strut bars are attached to the lower control arm and frame with bushings, allowing the arm-limited forward and backward movement. Strut rods are directly affected by braking forces and road shocks, and their failure can quickly lead to failure of the entire suspension system.

Figure 45–14 A sway bar link connects the sway bar to the lower control arm.

Figure 45–15 The center section of a sway bar rides in bushings.

Bushings

Bushings are used at the stabilizer bars **(Figure 45–15)**, control arms, radius arms, and strut rods. They make good suspension system pivots, minimize the number of lubrication points, and allow for slight assembly misalignments. Bushings are able to absorb some of the road shock before the force is transferred to the vehicle's frame, or body.

Suspension bushings are typically made of a rubber material, commonly an elastomer. Elastomers are capable of compressing in response to a force. When the force is removed, elastomers return to their original shape. They also allow movement or shifting of the parts they are between. The amount of movement depends on the design of the bushing. For example, control arms are attached to the frame of the vehicle with rubber elastomeric bushings. The bushings become the pivoting point for the control arms. During suspension travel, the bushings twist as the control arm moves. The bushings, acting like a spring, attempt to untwist and push the control arm back into its original position. This action provides some resistance to suspension movement while the bushings absorb some of the road shock.

This twisting and untwisting of the bushings generate heat. Rough road conditions and/or bad shock absorbers will cause the suspension to move more than normal. This causes more heat to build up in the bushings, shortening their life. Excessive heat tends to harden the rubber and as the bushings become harder, they break, crack, or fall apart.

Worn suspension bushings may allow suspension parts to change positions. This can lead to vibrations, wheel alignment problems, tire wear, and poor ride and handling. Often, a clunking noise when traveling on a rough surface will be an indication of a worn bushing. Worn or damaged bushings should be replaced.

Noise may also result from dry bushings and this may be corrected by lubricating them. Only rubber lubricant or a silicone-based lubricant should be used. Rubber bushings should not be lubricated with petroleum-based lubricants. These will cause the bushings to deteriorate.

Performance TIP *Many technicians replace stock rubber bushings with harder bushings made of high-grade polyurethane materials. These bushings do not have the give that rubber bushings do and tend to improve handling, steering response, and ride control. These bushings can also help to reduce torque steer on FWD vehicles.*

MACPHERSON STRUT SUSPENSION COMPONENTS

The **MacPherson strut** suspension is dramatically different in appearance from the traditional independent front suspension **(Figure 45–16)**, but similar components operate in the same way to meet suspension demands.

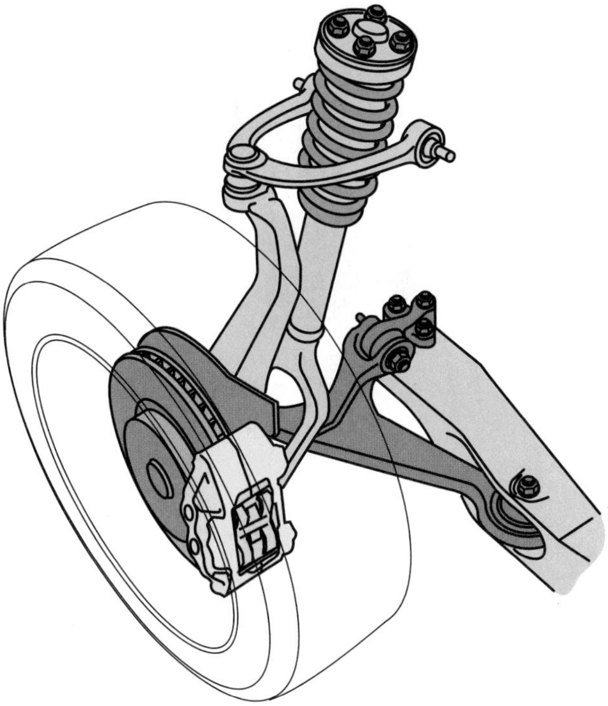

Figure 45–16 A complete MacPherson strut front suspension.

The MacPherson strut suspension's most distinctive feature is the combination of the main elements into a single assembly. It typically includes the spring, upper suspension locator, and shock absorber. It is mounted vertically between the top arm of the steering knuckle and the inner fender panel.

Struts have taken two forms: a concentric coil spring around the strut itself **(Figure 45–17)** and a spring located between the lower control arm and the frame **(Figure 45–18)**. The location of the spring on the lower control arm, not on the strut as in a conventional MacPherson strut system, allows minor road vibrations to be absorbed through the chassis rather than be fed back to the driver through the steering system. This system is called modified MacPherson suspension.

Struts

The core element of this type of suspension is the strut. With its cylindrical shape and protruding piston rod, it looks quite similar to the conventional shock absorber. In fact, the strut provides the damping function of the shock absorber, in addition to serving to locate the spring and to fix the position of the suspension.

The shock-damping function is accomplished differently on various types of struts. None of them uses a separate shock absorber as the traditional front

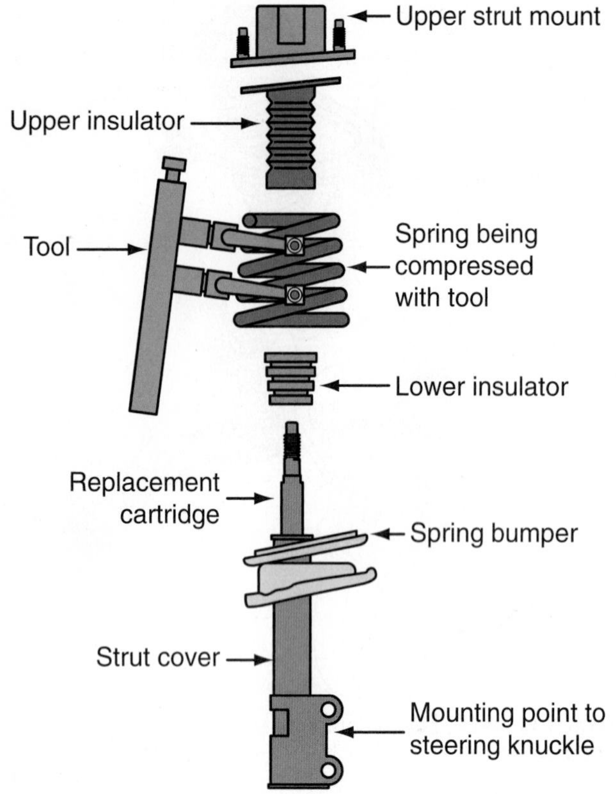

Figure 45-17 A MacPherson strut with a replaceable shock absorber cartridge.

suspension does. Some versions are designed so the damper can be independently serviced.

Struts fall into two broad categories: sealed and serviceable units. A sealed strut is designed so the top closure of the strut assembly is permanently sealed. There is no access to the shock absorber cartridge inside the strut housing and no means of replacing the cartridge. Therefore, it is necessary to replace the entire strut unit. A serviceable strut is designed so the cartridge inside the housing, which provides the shock-absorbing function, can be replaced with a new cartridge. Serviceable struts use a threaded body nut in place of a sealed cap to retain the cartridge.

The shock absorber device inside a serviceable strut is generally wet. This means the shock absorber contains oil that contacts and lubricates the inner wall of the strut body. The oil is sealed inside the strut by the body nut, O-ring, and piston rod seal. Servicing a wet strut with the equivalent components involves a thorough cleaning of the inside of the strut body, absolute cleanliness, and great care in reassembly (including replenishing the strut with oil).

Cartridge inserts were developed to simplify servicing wet struts. The insert is a factory-sealed replacement for the strut shock absorber. The replacement cartridge is simply substituted for the original shock absorber cartridge and retained with the body nut.

Most OE domestic struts are serviced by replacement of the entire unit. There is no strut cartridge to replace. Sealed OE units can also be serviced by replacement with an aftermarket unit that permits future servicing by cartridge replacement.

The use of the strut reduces suspension space and weight requirements. By mounting the bottom of the

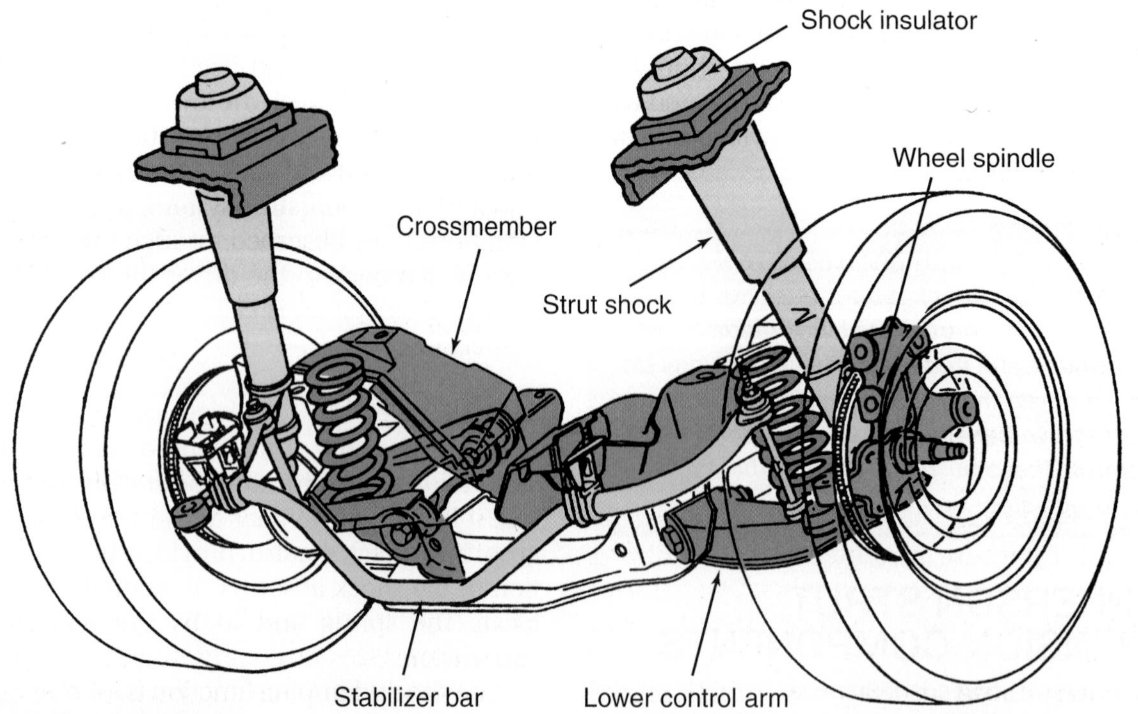

Figure 45-18 A modified MacPherson suspension has the spring mounted separately from the strut. *Courtesy of Ford Motor Company*

strut assembly to the steering knuckle, the upper control arm and ball joint of the traditional suspension are eliminated. In place of the ball joint, the upper mount, which is bolted to the fender panel, is the load-carrying member on MacPherson suspensions.

Strut Mounts

A MacPherson strut has a mount between the top of the strut to the chassis where the strut is supported. These mounts are designed to dampen vibrations as well as secure the strut in position. Often the mounts include a bearing, although some use a bushing. Bearings are most commonly found in front suspensions because the suspension turns or pivots between left and right. Some strut suspension use bushings, although they can also just have a bushing. Every application is different; however, most applications fall into one of three groupings **(Figure 45–19)**.

Spacer Bushing The spacer bushing design is used mostly by Volkswagen, Toyota, Mazda, and Mitsubishi, and it was used in early Chrysler vehicles. This design has a bearing centered in the mount and a separate inner bushing. The bearing is pressed into the strut mount. The bearing, bushing, and upper plate support the strut rod. If the bushing is cracked or torn or if the bearing is binding or seized, the strut mount must be replaced.

Inner Plate The inner plate design is used by GM and Ford. It has a rubber-encased inner plate placed between upper and lower surface plates. The plate prevents the strut piston rod from pushing through the upper or lower surface plate if the inner plate fails. The bearing is at the bottom of the strut mount and is not serviceable. If the bearing is bad, the mount must be replaced.

Center Sleeve The center sleeve design is widely used by Chrysler. It has a center sleeve molded to a rubber bushing. The stem of the strut passes through the sleeve. The bearing is not a part of the strut mount; rather it is a separate unit. To prevent the strut rod from pushing through the mount, upper and lower retainer washers are used. If there are cracks, tears, or other damage to the bushing, the mount should be replaced.

Mount Diagnosis and Service Worn or damaged strut mounts can cause the strut and the strut tower to move independently of one another. This can lead to abnormal noise, a bent or damaged strut, damage to the strut tower, and poor handling. A bad mount may cause a creaking or popping noise. This is caused by too much movement of the strut within the mount. Often the mount is replaced with the strut.

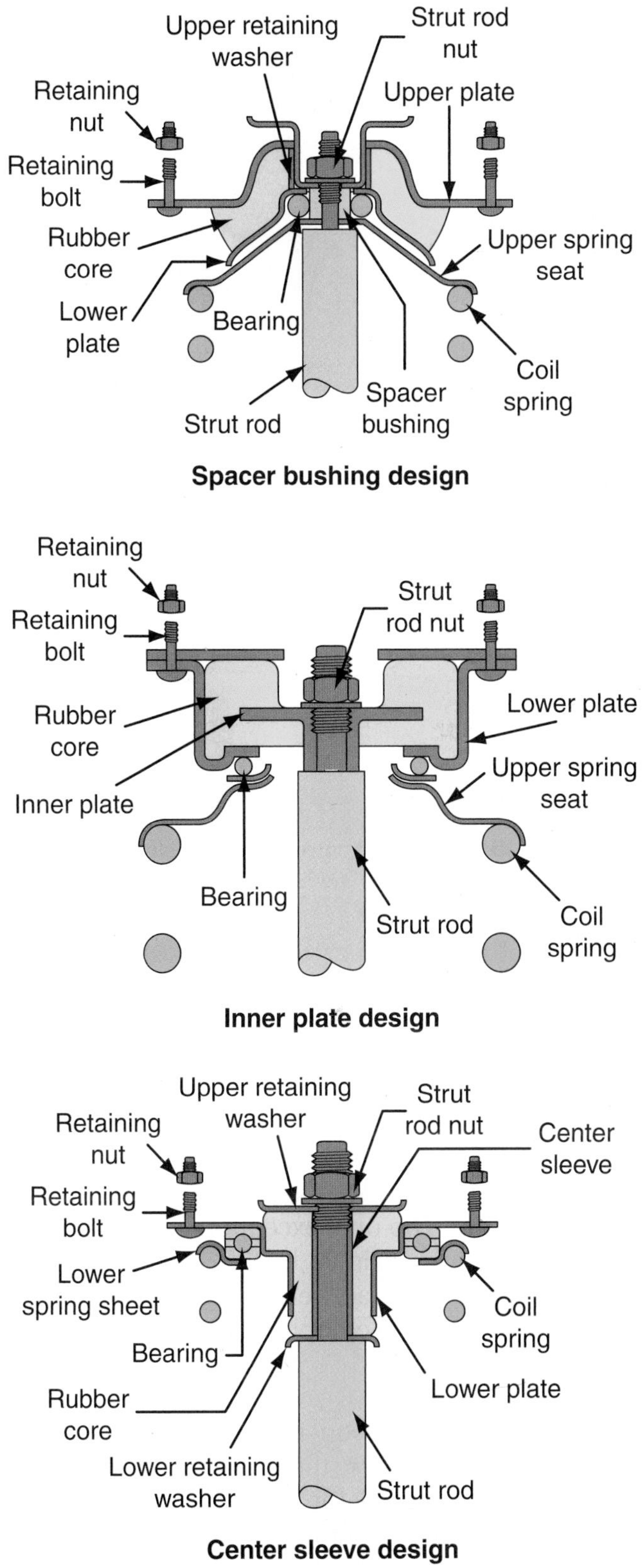

Figure 45–19 The different types of upper mounts for a strut.

Lower Suspension Components

The suspension's lower mounting position continues to be the frame, as on the traditional suspension, because the lower control arm and ball joint are retained **(Figure 45–20)**. As on those suspensions,

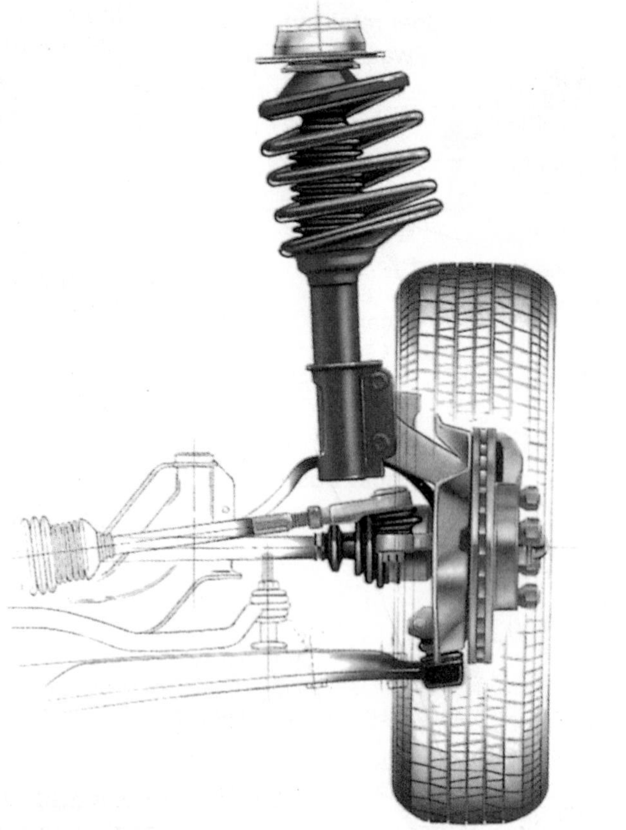

Figure 45-20 A front MacPherson strut assembly. *Courtesy of Ford Motor Company*

the control arm serves as the lower locator for the suspension.

MacPherson strut suspensions continue to use sway or stabilizer bars. On models with single-bushing control arms, strut rods or the sway bar can be fastened to the control arm to provide lateral stability.

The lower **ball joint** is a friction or steering ball joint and is used to stabilize the steering and to retard shimmy. The only exception is on modified MacPherson suspensions. In this design, the ball joint becomes the load bearer; the upper mount becomes the steering component.

Springs

Coil springs are used on all strut suspensions. A mounting plate welded to the strut serves as the lower spring seat. The upper seat is bolted to the strut piston rod. A bearing or rubber bushing in the upper mount permits the spring and strut to turn with the motion of the wheel as it is steered.

INDEPENDENT FRONT SUSPENSION

Front suspension systems are fairly complex. They have somewhat contradictory jobs. They must keep the wheels rigidly positioned and at the same time allow them to steer right and left. In addition, because of weight transfer during braking, the front suspension system absorbs most of the braking torque. While accomplishing this, it must provide good ride and stability characteristics.

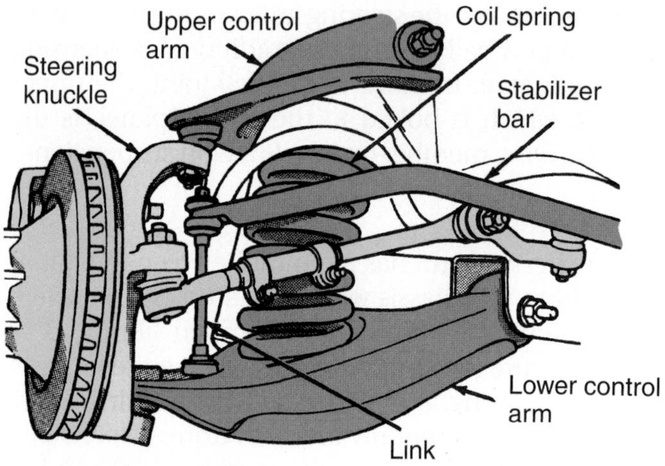

Figure 45-21 A typical SLA front suspension. *Courtesy of Chrysler LLC*

Short-Long Arm Suspension

The unequal length control arm or **short-long arm (SLA)** suspension system has been common on domestic-made vehicles for many years **(Figure 45–21)**. Each wheel is independently connected to the frame by a steering knuckle, ball joint assemblies, and short upper and longer lower control arms. Because the upper arm pivots in a shorter arc, the top of the wheel moves in and out slightly but the tire's road contact remains constant **(Figure 45–22)**.

One design of an SLA uses a narrow lower control arm, shaped like an "I" **(Figure 45–23)**. A strut rod is used to hold the control arm in place. The strut rod is

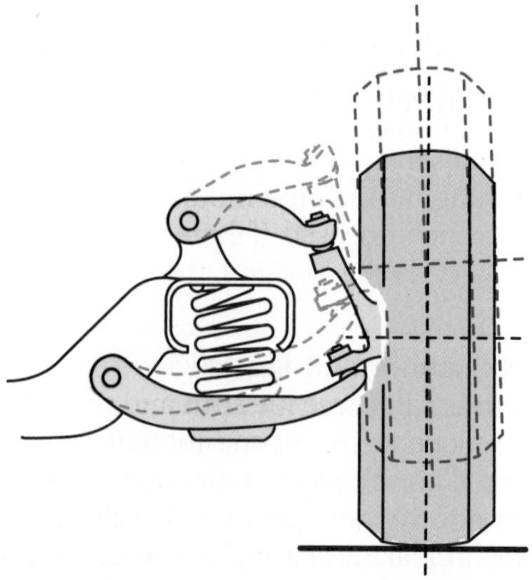

Figure 45-22 The movement of the wheel as a short-long arm suspension system moves up and down.

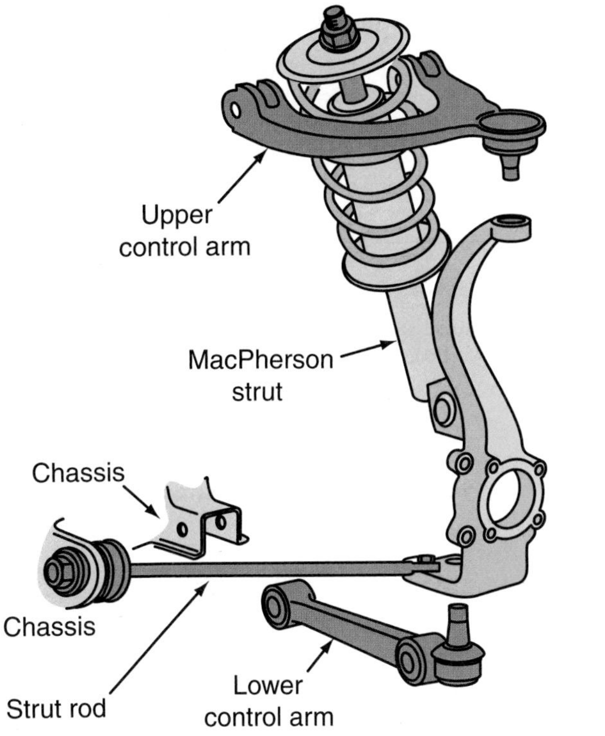

Figure 45-23 A FWD wishbone suspension system with a narrow lower control arm mounted on a single pivot.

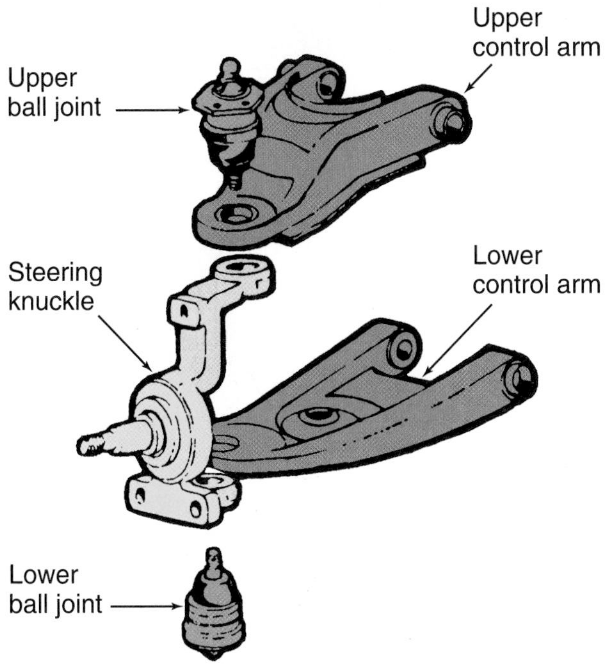

Figure 45-24 Ball joint locations in an SLA suspension. *Courtesy of Federal-Mogul Corporation*

attached to the control arm close to the steering knuckle and to the frame in front of the wheel assembly. Rubber bushings at the frame mounting allow the strut rod to move a little when the tire hits a bump. The bushing dampens the shock and prevents it from transmitting through the vehicle's frame.

The essential components of SLA systems are the wheel spindle assembly, control arms, ball joints, shock absorbers, and springs, among others.

Wheel Spindle A **wheel spindle** assembly consists of a wheel spindle and a steering knuckle. A wheel spindle is connected to the wheel through wheel bearings and is the point at which the wheel hub and wheel bearings are connected. A **steering knuckle** is connected to control arms. In most cases, a steering knuckle and wheel spindle are forged to form a single piece.

Control Arms The upper and lower **control arms** on the traditional **independent front suspension (IFS)** function primarily as locators. They fix the position of the system and its components relative to the vehicle and are attached to the frame with bushings that permit the wheel assemblies to move up and down separately in response to irregularities in the road surface. The outer ends are connected to the wheel assembly with ball joints **(Figure 45–24)** inserted through each arm into the steering knuckle.

There are two types of control arms: the wishbone, or double-pivot, control arm and the single-pivot, or single-bushing, control arm **(Figure 45–25)**. The wishbone offers greater lateral stability than the single-pivot arm, which is lighter and requires less space than the wishbone but also requires modifications in suspension design to compensate for the reduced lateral stability. Those modifications are discussed further later in this chapter.

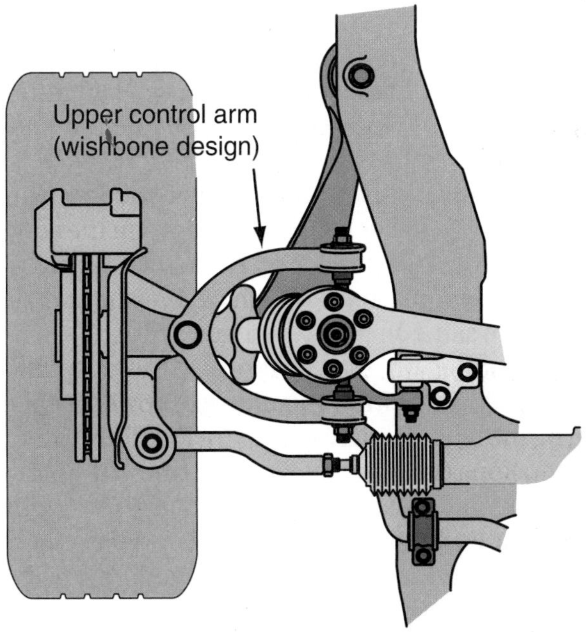

Figure 45-25 A front suspension system with an upper control arm shaped like a "wishbone," which is what this type of suspension is called.

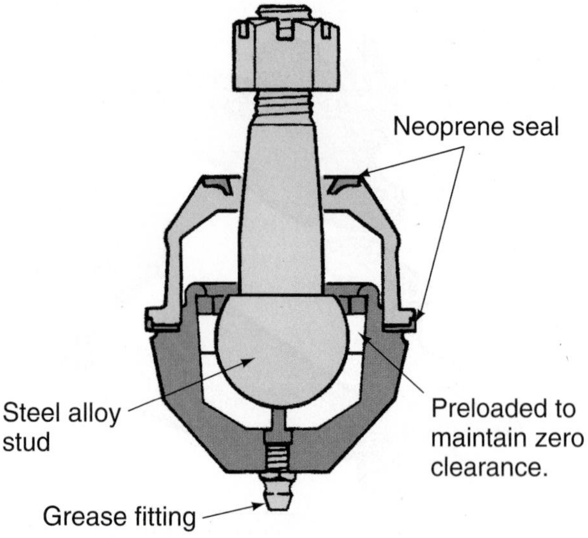

Figure 45-26 A typical ball joint.

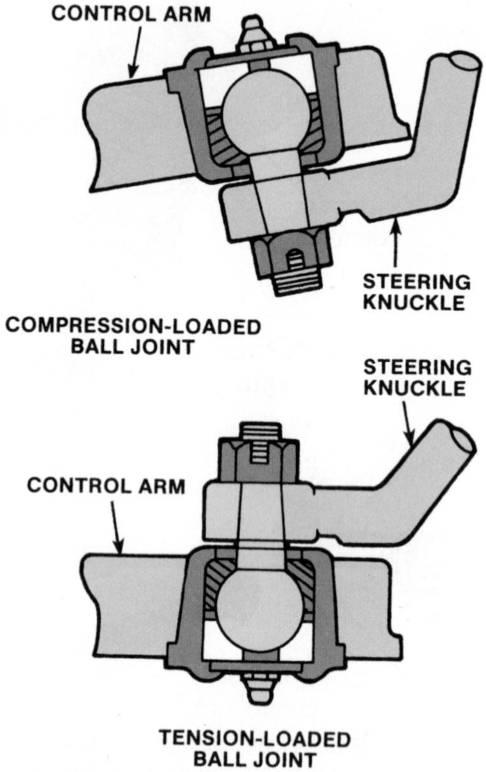

Figure 45-27 Two basic types of load-carrying ball joints.

Ball Joints A ball joint **(Figure 45–26)** connects the steering knuckle to the control arm, allowing it to pivot on the control arm during steering. Ball joints also permit up-and-down movement of the control arm as the suspension reacts to road conditions. The ball joint stud protrudes from its socket through a rubber seal that keeps lubricating grease in the housing and keeps dirt out. Some ball joints require periodic lubrication, while most do not. These maintenance-free ball joints move in a prelubricated nylon bearing.

Ball joints are either load carrying or are followers. A load-carrying ball joint supports the car's weight and is generally in the control arm that holds or seats the spring. Load-carrying joints can be called tension-loaded or compression-loaded ball joints **(Figure 45–27)**. The correct term depends on whether the force of the load tends to push the ball into the socket (compression) or pull it out of the socket (tension).

Follower ball joints are often called friction-loaded ball joints. A follower ball joint mounts on the control arm that does not provide a seat for the spring. The follower does not support vehicle weight and does not get the same stress as the load carrier.

Depending on the location of the suspension system's spring, either the upper or lower ball joint will be the load-carrying joint. In a MacPherson strut suspension, there is usually only one ball joint on each side and it is typically a follower. In modified strut suspensions, the ball joint is a load-carrying joint because the spring is positioned between the frame crossmember and the lower control arm.

Some ball joints have wear indicators. As the joint wears, the grease fitting of the joint recedes into the housing. When the shoulder of the fitting is flush with the housing, the joint needs to be replaced **(Figure 45–28)**.

A ball joint is nothing more than a ball-in-socket joint. As long as the ball is firmly in the socket and the ball and/or socket is not worn, the joint will provide a solid connection. Once the ball or socket is worn, the connection becomes sloppy. How the ball is kept in its socket depends on the type of ball joint it is. Load-carrying ball joints rely on the vehicle's weight to keep the ball in the socket **(Figure 45–29)**. As weight is removed from the joint, the ball relaxes in the socket and will feel loose. Follower ball joints are held in place by friction inside the joint. A spring inside the joint typically keeps the ball tight in the socket but allows for some flexibility. This type of joint should never have any play.

Four-Link Front Suspension

A four-link front suspension fixes the wheel with four rod-type control arms and the tie rod **(Figure 45–30)**. The suspension strut supports the vehicle weight against the body via the load-bearing link. By separating wheel attachment and suspension elements, this suspension optimizes ride quality and movement. The influence of drive forces on the steering system is minimal.

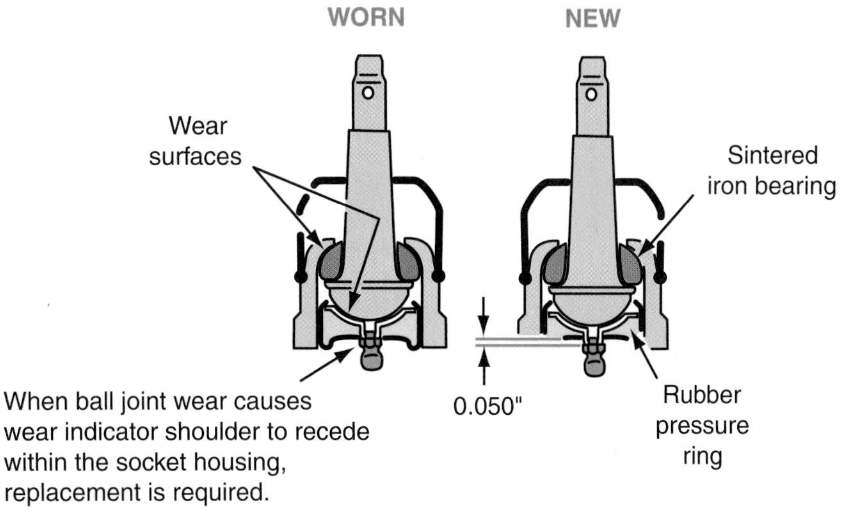

Figure 45-28 A wear indicator on a ball joint.

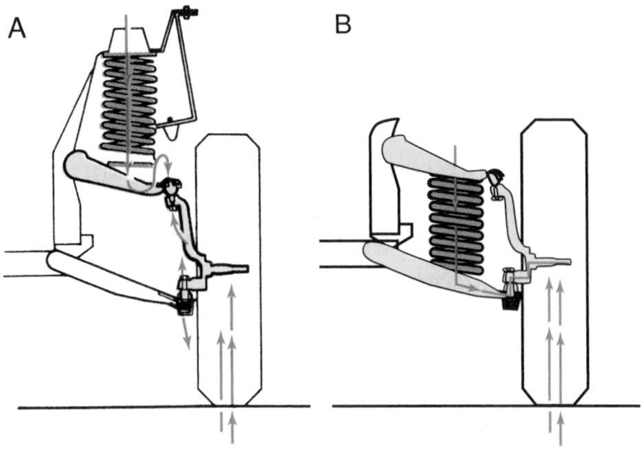

Figure 45-29 (A) The upper ball joint is the load-carrying joint. (B) The lower ball joint is the load-carriying joint in this system.

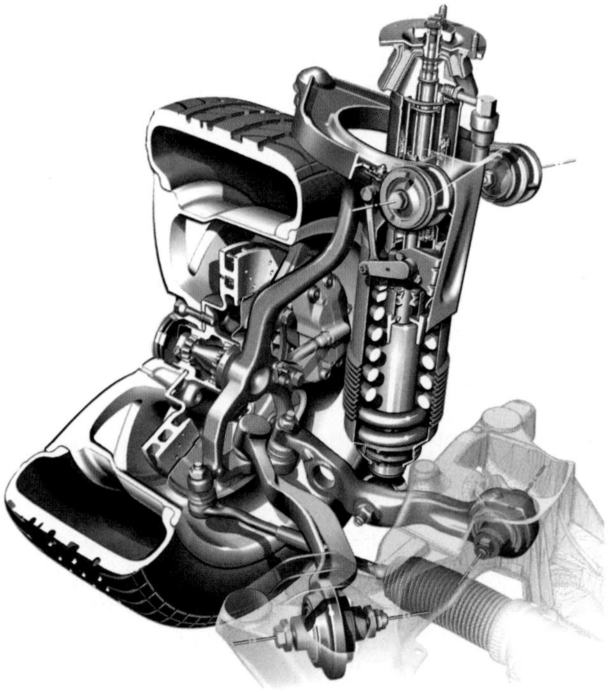

Figure 45-30 A four-link front suspension system.
Courtesy of Chrysler LLC

BASIC FRONT-SUSPENSION DIAGNOSIS

Diagnosis of suspension problems should follow a logical sequence. The following procedure can be used on most vehicles; however, it is also best to follow the sequence given by the manufacturer for a specific vehicle.

PROCEDURE

To diagnose a suspension system:

STEP 1 Take the vehicle on a road test and verify the customer's concern.

STEP 2 Inspect the tires. Check their condition and air pressure. Also make sure that the tires and wheels are the correct size.

STEP 3 Inspect the chassis and underbody. Remove any excessive accumulation of mud, dirt, or road deposits. Then:

- Inspect all parts to identify any aftermarket modifications that may have been made.
- Check vehicle attitude for evidence of overloading or sagging. Be sure the chassis height is correct.
- Raise the vehicle off the floor. Grasp the upper and lower surfaces of the tire and shake each front wheel to check for worn wheel bearings.
- Look for loose or damaged front and rear suspension parts.
- Check loose, damaged, or missing suspension bolts.
- Check the ball joints for looseness and wear.
- Check the condition of the struts' upper mounts.
- Check the shock absorbers and struts for signs of fluid leakage **(Figure 45–31)** and damage.
- Check all of the mountings for the shocks and struts.
- Check all suspension bushings for looseness, splits, cracks, misplacement, and noises.

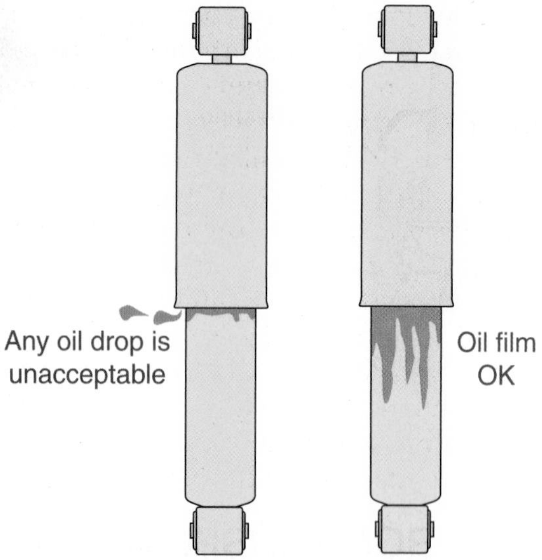

Figure 45-31 Check the shock absorbers for signs of fluid leakage.

- Check the steering mounts, linkages, and all connections for looseness, binding, or damage.
- Check for damaged or sagging springs.
- Check the drive axles for damage and looseness.

STEP 4 If the cause of the customer's concern was found, make the repair as necessary and verify that the repair fixed the problem.

STEP 5 If the cause of the concern was not found, refer to the symptom chart given in the service information and conduct all applicable checks. Then make the repair as necessary and verify that the repair fixed the problem.

Shock Absorber or Strut Bounce Test

With the vehicle on the ground, a quick check, called the bounce test, can be performed to check the operation of the shocks and struts. When the bounce test is performed, the bumper is pushed two or three times downward with considerable weight applied on each corner of the vehicle. The bumper is released after each push and the vehicle should oscillate about 1½ cycles and then settle. One free upward bounce should stop the vertical chassis movement if the shock absorber or strut provides proper spring control. If the vehicle's bumper does more than 1½ free upward bounces, the shock absorber or strut is defective.

Noises

Abnormal noises from the suspension system can be caused by a number of problems. Tire noise varies with road surface conditions, whereas differential noise is not affected when various road surfaces are encountered. Uneven tread surfaces may cause tire noises that seem to originate elsewhere in the vehicle. These noises may be confused with differential noise.

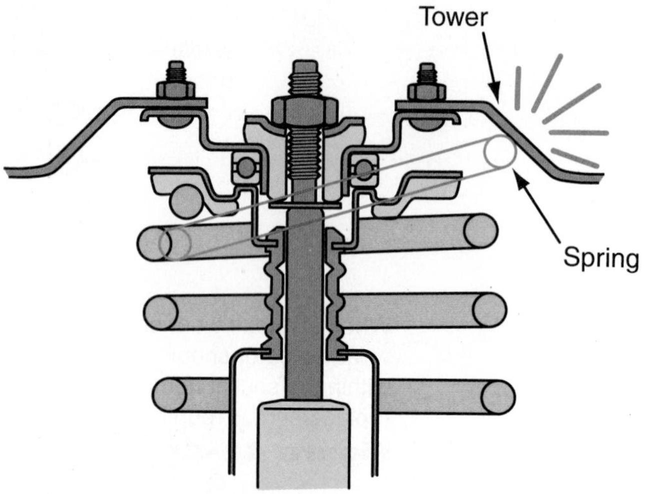

Figure 45-32 Coil spring to strut tower interference.

Differential noise usually varies with acceleration and deceleration, whereas tire noise remains more constant in relation to these forces. Tire noise is most pronounced on smooth asphalt road surfaces at speeds of 15 to 45 mph.

Rattling on road irregularities can be caused by worn shock absorber bushings or grommets, worn spring insulators, a broken coil spring or broken spring insulators, worn control arm bushings, worn stabilizer bar bushings, and worn strut rod grommets, worn leaf spring shackles and bushings, and worn torsion bars, anchors, and bushings. Dry or worn control arm bushings may cause a squeaking noise on irregular road surfaces.

Chatter while cornering can be caused by worn upper strut mounts. Front strut noise on sharp turns or during suspension jounce may be caused by interference between the coil spring and the strut tower **(Figure 45–32)** or between the coil spring and the upper mount.

SHOP TALK

Before visual inspection or suspension height measurements can be performed, the vehicle must be on a level surface. Tires must be at recommended pressures; gas tank must be full; and there should be no passenger or luggage compartment load. Beginning at the rear bumper, jounce the car up and down several times. Proceed to the front bumper and repeat, releasing during same cycle as rear jounce.

Chassis Height Specifications

A quick overall visual inspection detects any obvious sag from front to rear or from side to side. Under the car, at the level of the two ends of the control arms, check for out-of-level, damaged, or worn rubber bumpers, or shiny or worn spring coils. All indicate weak coil springs.

A more accurate inspection reveals less obvious problems by measuring heights at specific points on each side of the suspension system.

USING SERVICE INFORMATION

For the most accurate measurement of chassis height, use the service manual to check against the manufacturer's recommendations for the specific model. Photo Sequence 47 shows the typical procedure for checking vehicle ride height. Be careful. The measurement points vary from one model to another even if manufactured by the same company. When coil spring wear is suspected, it might be necessary to load the vehicle to the manufacturer's suggested capacities and measure at the designated points.

FRONT-SUSPENTION COMPONENT SERVICING

Each major component of the suspension system needs to be carefully checked. Each has its own procedure for doing this as well as the service procedures. The only maintenance required on suspension systems is a periodic chassis lubrication. If the owner has failed to have this done, many problems can result.

Go To *Chapter 7 for details on how to lubricate the chassis.*

Coil Springs

Coil springs require no adjustments and are basically trouble-free. However, the constant compression and extension of the spring can lead to a loss of elasticity and cause spring sag. Coil springs then need to be replaced. A weak spring upsets vehicle trim height, which can cause incorrect wheel alignment, abnormal steering, poor headlight aim, poor braking, increased tire wire, and a decrease in the service life of U-joints and shocks.

Coil springs can also break. This is often due to vehicles being overloaded or rusted. When a vehicle carries more weight than it was designed to carry, the springs can break from being overly compressed.

A vital step in replacing a coil spring is to identify the correct replacement. Begin by looking for the OEM part number, which is normally on a tag wrapped around a coil. Often, this tag falls off before replacement is necessary. If a set of aftermarket springs has been installed, the part number might be stamped on the end of the coil. Next, determine what types of ends the coils springs have. There are three types of ends used in automotive applications: full wire open, tapered wire closed, and pigtail. Springs with full wire open ends are cut straight off and sometimes flattened or ground into a D or square shape. Tapered wire closed ends are wound to ensure squareness and ground into a taper at the ends. Pigtail ends are wound into a smaller diameter at the ends.

The final step is to check all available service information. To do this, it is necessary to know the make, year, model, body style, and engine size, and if the vehicle is equipped with air conditioning. In some cases, it is also good to know the type of transmission, seating capacity, and other specifics that add extra weight to the vehicle. In most parts catalogs, springs are listed by vehicle and VIN in two sections: front and rear. Springs should always be replaced in pairs.

CAUTION!

The coil spring exerts a tremendous force on the control arm. Before you disconnect either control arm from the knuckle for any service operation, contain the spring with a spring compressor to prevent it from flying out and causing injury.

Removing a Spring To remove a coil spring, raise and support the vehicle by its frame. Let the control arm hang free. Remove wheels, shock absorbers, and stabilizer links. Disconnect the outer tie-rod ends from their respective arms.

Unload the ball joints with a roll-around floor jack. Jack under the lower control arm from the opposite side of the vehicle. This allows the jack to roll back when the control arm is lowered. Position the jack as close to the lower ball joint as possible for maximum leverage against the spring.

The spring is ready for the installation of the spring compressor (**Figure 45–33**). There are many different types of spring compressors. One type uses a threaded compression rod that fits through two plates, an upper and lower ball nut, a thrust washer, and a forcing nut. The two plates are positioned at either end of the spring. The compression rod fits through the

P47-1 *Check the trunk for extra weight.*

P47-2 *Check the tires for normal inflation pressure.*

P47-3 *Park the car on a level shop floor or alignment rack.*

P47-4 *Find the vehicle manufacturer's specified curb riding height measurement locations in the service manual.*

P47-5 *Measure and record the right front curb riding height.*

P47-6 *Measure and record the left front curb riding height.*

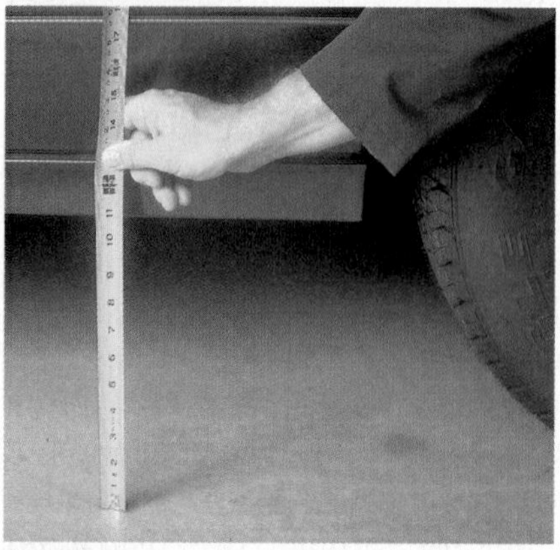

P47-7 *Measure and record the right rear curb riding height.*

P47-8 *Measure and record the left rear curb riding height.*

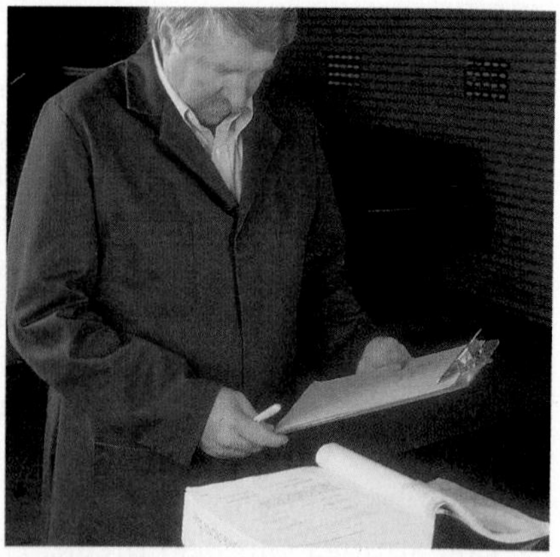

P47-9 *Compare the measurement results to the specified curb riding height in the service manual.*

Figure 45-33 A coil spring compressor is used to compress the spring before disconnecting some suspension parts.

plates with a ball nut at either end. The upper ball nut is pinned to the rod. The thrust washer and forcing nut are threaded onto the end of the rod. Turning the forcing nut draws the two plates together and compresses the spring.

In some cases, it is necessary to break the tapers of both upper and lower ball joints so the steering knuckle can be moved to one side (**Figure 45–34**). If the vehicle is equipped with a strut rod, this must be disconnected at the lower control arm. Push the control arm down until the spring can be removed. If necessary, a pry bar can be used to remove the spring from its lower seat. Remove the spring and compressor.

If the same spring is to be reinstalled, leave the compressor in position. If a new spring is to be used, slowly release the pressure on the tool by backing off the forcing nut. Compress the new spring prior to installing it.

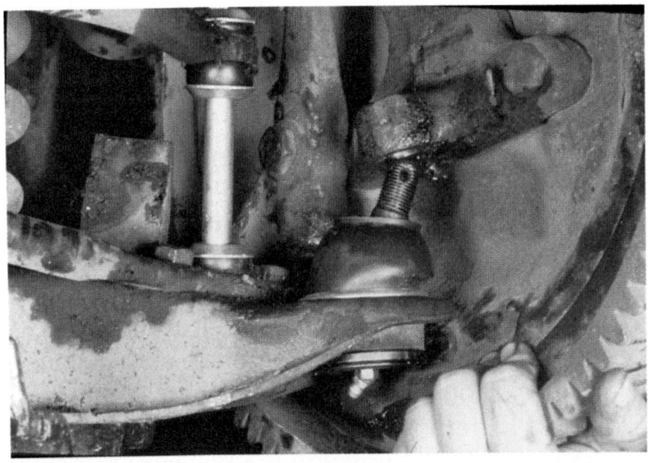

Figure 45-34 At times the steering knuckle must be disconnected from the lower and/or ball joint in order to remove a spring.

Torsion Bars

Torsion bars are subjected to many of the same conditions affecting coil springs. Periodic adjustment of the torsion bars is necessary to maintain the proper height. Replacement is sometimes necessary because of breakage. It should be noted that the bars are not interchangeable from side to side.

Height inspection and measurements for vehicles with torsion bar suspensions are the same for coil springs. However, sagging can usually be corrected by adjusting the bars. Procedures for adjusting torsion bars are given in the service manual.

Ball Joints

Begin your inspection of a ball joint by checking to see if the ball joint has a wear indicator on it. If it does, check the placement of the grease fitting. If it is recessed, the ball joint is worn and should be replaced. On some vehicles it is recommended that you check to see if the grease fitting can wiggle in the ball joint. If it does, the ball joint should be replaced. Always check the service manual when checking ball joints.

Look carefully at the joint's boot. A damaged boot or joint seal will allow lubricant to leak out and allow dirt to enter and contaminate the lubricant. If the boot is damaged, the ball joint should be replaced.

If no boot damage is evident, gently squeeze the boot. If the boot is filled with grease, it will feel somewhat firm. If the joint has a grease fitting and appears not to be filled with grease, use a grease gun and refill the joint. Fill the joint until fresh grease is seen flowing out of the boot's vent. If too much grease is forced into the joint or it is forced in too quickly, the boot can unseat or tear.

Ball joints should be checked for excessive wear. Load-carrying joints will have some slop when the weight of the vehicle is taken off them. Follower joints should never have play. To check a load-carrying joint, it must be unloaded.

When the coil spring is on the lower control arm, raise the vehicle by jacking under the control arm as close to the ball joint as possible. This gives the maximum amount of leverage against the spring. The ball joint is unloaded when the upper strike out bumper is not in contact with the control arm or frame. A quick check for looseness can be made by using a pry bar between the tire and the ground. To find out if the ball joint is loose beyond manufacturer's specifications, use an accurate measuring device. The following checking procedures demonstrate the use of a dial indicator. The dial indicator is a precision instrument and should be handled carefully to prevent damage. The mounting procedure for the checking tool might vary depending on the style of ball joint used on the vehicle. Manufacturer's tolerances can be axial (vertical), radial (horizontal), or both. To conduct these checks, follow these procedures.

Typical Radial Check For a radial check, attach a dial indicator to the control arm of the ball joint being checked. Position and adjust the plunger of the dial indicator against the edge of the wheel rim nearest to the ball joint being checked. Slip the dial ring to the zero marking. Move the wheel in and out and note the amount of ball joint radial looseness registered on the dial **(Figure 45–35)**. The procedure for checking the radial movement of a lower ball joint on a MacPherson strut front suspension is shown in Photo Sequence 48.

Typical Axial Check For an axial check, first fasten the dial indicator to the control arm, then clean off the flat on the spindle next to the ball joint stud nut. Position the dial indicator plunger on the flat of the spindle and depress the plunger approximately 0.350 inch. Turn the lever to tighten the indicator in place. Pry the bar between the floor and tire. Record the reading **(Figure 45–36)**.

If the ball joint looseness reading on the dial indicator exceeds manufacturer's specifications, the ball joint should be replaced.

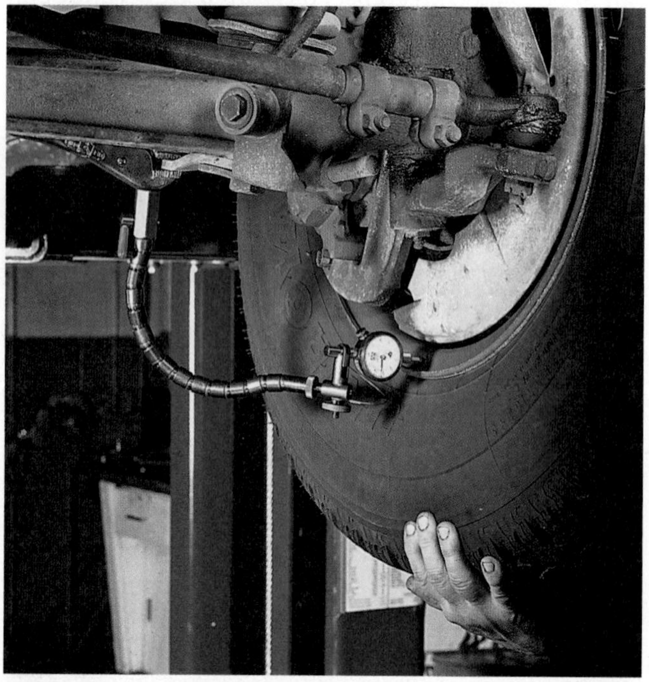

Figure 45–35 A typical mounting of a dial indicator for a radial check. *Courtesy of Federal-Mogul Corporation*

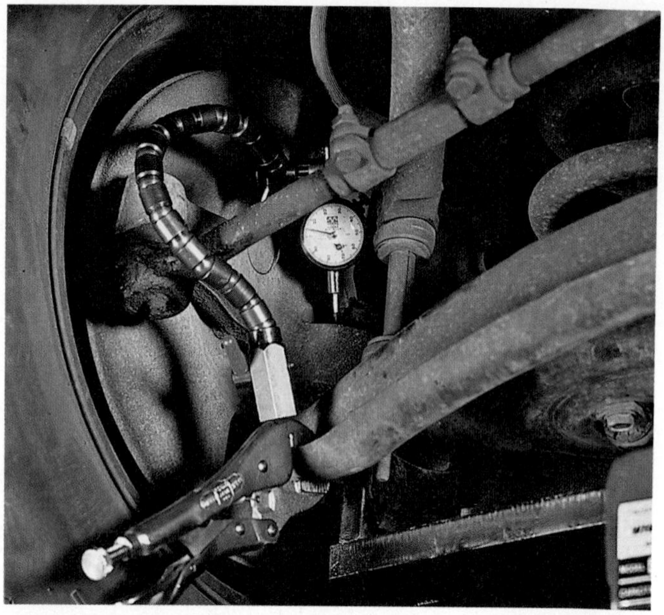

Figure 45–36 A typical mounting of a dial indicator for an axial check. *Courtesy of Federal-Mogul Corporation*

When the load-carrying ball joints are on the upper control arm (spring mounted on the upper arm), raise the vehicle by its frame using support tools to unload the ball joints and hold them in their normal position. To determine the condition of the non-load-carrying (or follower) ball joint, vigorously push and pull on the tire, while watching the ball joint for signs of movement. Refer to the manufacturer's specifications for tolerances.

Inspection of Wear Indicators Wear-indicator-type ball joints must remain loaded to check for wear. The vehicle should be checked with the suspension at curb height. The most common type has a small-diameter boss, which protrudes from the center of the lower housing. As wear occurs internally, this boss recedes very gradually into the housing. When it is flush with the housing, the ball joint should be replaced. To remove and install a ball joint, follow the procedure given in the service manual.

Ball joints are mounted to the control arm in one of four basic ways: rivets, bolts, press fit, and threaded. The most common method is press fit. Some manufacturers require you to replace the entire control arm assembly if a ball joint is to be replaced. In these cases, the ball joint and control arm are made as a single assembly and individual parts are not available.

Press-fit ball joints are removed and installed using special tools and a hydraulic press. The control arm must be removed before attempting to remove the ball joint. While pressing the ball joint out of or into the control arm, make sure you do not damage the arm.

PHOTO SEQUENCE 48

Measuring the Lower Ball Joint Radial Movement on a MacPherson Strut Front Suspension

P48-1 *Raise the front suspension with a floor jack and place jack stands under the chassis at the vehicle's lift points.*

P48-2 *Grasp the front tire at the top and bottom and rock the tire inward and outward while a coworker visually checks for movement in the front wheel bearing. If there is movement, adjust or replace the wheel bearing.*

P48-3 *Position a dial indicator against the inner edge of the rim at the bottom. Preload and zero the dial indicator.*

P48-4 *Grasp the bottom of the tire and push outward.*

P48-5 *With the tire held outward, read the dial indicator.*

P48-6 *Pull the bottom of the tire inward and be sure the dial indicator reading is zero. Adjust the dial indicator as required.*

P48-7 *Grasp the bottom of the tire and push outward.*

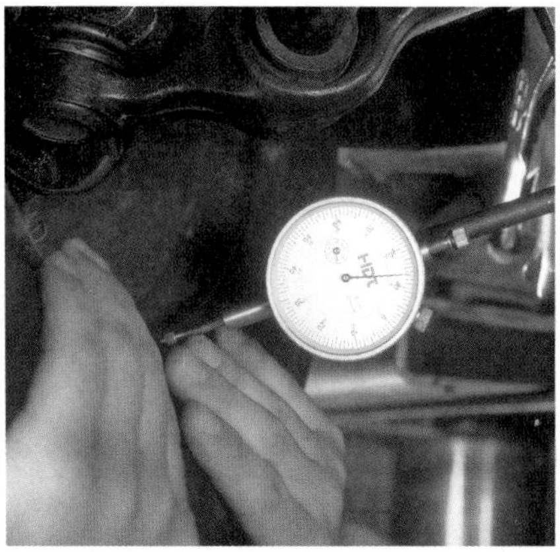

P48-8 *With the tire held in this position, read the dial indicator.*

P48-9 *If the dial indicator reading is more than specified, replace the lower ball joint. Before doing this, make sure that the wheel bearing and hub are good.*

WARNING!

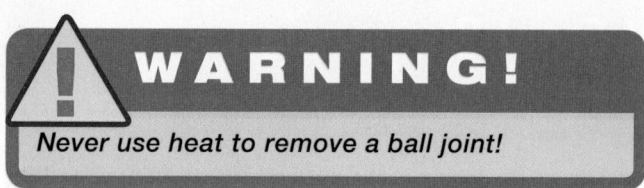

Never use heat to remove a ball joint!

Control Arm Bushings

Visually inspect each rubber bushing for signs of distortion, movement, off-center condition, and presence of heavy cracking. Check metal bushings for noise and loose seals.

To remove the control arm bushings, raise the vehicle and support the frame on safety jack stands. Remove the wheel assembly. Install a spring compressor on the coil spring.

Disconnect the ball joint studs from the steering knuckle as described previously. Remove the bolts attaching the control arm assembly to the frame and remove the control arm.

Bushings are pressed in and out of their bores by using special tools. A special tool is installed over the bushing **(Figure 45–37)** after the correct size adapter for the tool has been selected. Tightening the tool pushes the bushing out of the control arm. The same process is used to press a new bushing into the control arm. As the special tool is tightened, the bushing moves into its bore. When installing new bushings, make sure they stay straight while they are being pressed in.

Once the new bushings are started into the control arm, measure and mark the center between mounting holes and center the control arm. Now, alternately press in the bushings on each side, keeping the reference marks aligned. This ensures the shaft is not off center, causing binding. End cap nuts or bolts should not be torqued until the vehicle is at curb height and the suspension has been bounced and allowed to settle out.

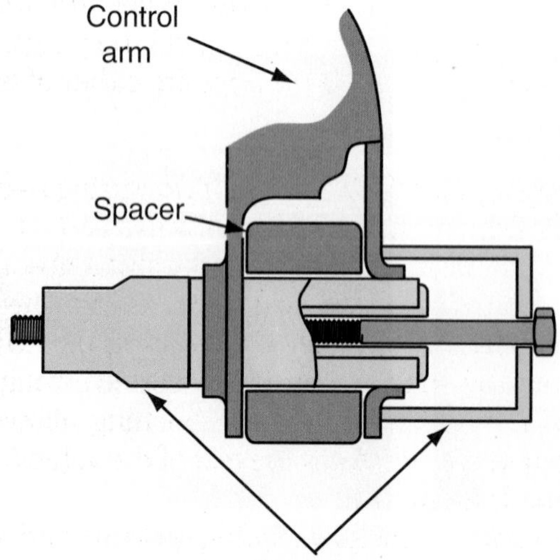

Figure 45–37 Removing a control-arm bushing.

Rebolt the control arm and tighten the bolts to specifications, then install the coil spring into position. Install the ball joint studs into the control arms. Remove the coil spring compressor. Install the wheel assembly and lower the car. Road test the car, retighten all bolts, and set wheel alignment.

Strut Rod Bushings

Except in the case of accidental damage, the strut rod itself is rarely replaced. Rather, it is the bushing that wears, deteriorates, and needs replacement.

Sway Bar Bushings

These bushings anchor the sway bar securely to the vehicle frame and the control arms on each side. The condition of the bushings affects the performance of the bar. Visual inspection of mounting bushings indicates if the bushings are worn, have taken a permanent set, or are possibly missing. Also check the sway bar links. Any damage indicates that they should be replaced.

Shock Absorbers

A shock absorber that is functioning properly ensures vehicle stability, handling, and rideability. Most motorists fail to notice gradual changes in the operation of their cars as a result of worn shock absorbers. Some common indications of shock absorber failure follow:

- Steering and handling are more difficult.
- Braking is not smooth.
- Bouncing is excessive after stops.
- Tire wear patterns are unusual, especially cupping.
- Springs are bottoming out.

WARNING!

Gas-pressurized shock absorbers will extend on their own and can create a dangerous situation. Never apply heat or flame to the shock absorbers during removal; the heat will cause the gas to expand and the shock will quickly extend.

Shock absorbers should be inspected for loose mounting bolts and worn mounting bushings. If these components are loose, they will rattle, and replacement of the bushings and bolts is necessary. The upper mounts of struts should also be carefully checked.

In some shock absorbers, the bushing is permanently mounted in the shock, and the complete unit must be replaced if the bushing is worn. When the mounting bushings are worn, the shock absorber will not provide proper spring control.

Vibrations set up by a worn shock absorber can cause premature wear in many of the undercar systems. They can cause wear in the front and rear component parts of the suspension system, the linkage component parts of the steering system, and the U-joints and motor or transmission mounts of the driveline. Also, vibrations can cause unnatural wear patterns on the tires.

A shock absorber can be bench tested. First, turn it up in the same direction it occupies in the vehicle. Then extend it fully. Next, turn it upside down and fully compress it. Repeat this operation several times. Install a new shock absorber if a lag or skip occurs near mid-stroke of the shaft's change in travel direction, or if the shaft seizes at any point in its travel, except at the ends. Also, install a new shock absorber if noise, other than a switch or click, is encountered when the stroke is reversed rapidly, if there are any leaks, or if action remains erratic after purging air.

When removing and installing shock absorbers, be sure to follow the aftermarket manufacturer's instructions or those given in the service manual.

MacPherson Strut Suspension

The MacPherson strut suspension system is based on a triangular design. The strut shaft is a structural member that does away with the upper control arm bushings and the upper ball joint. Since this shaft is also the shock absorber shaft, it receives a tremendous amount of force vertically and horizontally. Therefore, this assembly should be inspected very closely for leakage, bent shaft, and poor damping.

To remove and replace the MacPherson strut, proceed as shown in Photo Sequence 49.

During the disassembly of the strut, make sure you check the strut pivot bearing **(Figure 45–38)**. Move the bearing with your hand. If the bearing is hard to move or seems to bind, it must be replaced. When replacing the bearing, make sure the correct side is up. Manufacturers normally mark the up side with paint or some other marking. Make sure you check all rubber insulators for deterioration and other damage and replace them if necessary. Also make sure you mark the eccentric camber bolts before loosening them. Returning the bolts in the same location will help maintain the correct camber angle after reassembly. Also replace all cotter pins in the suspension and align the wheels after everything is reassembled.

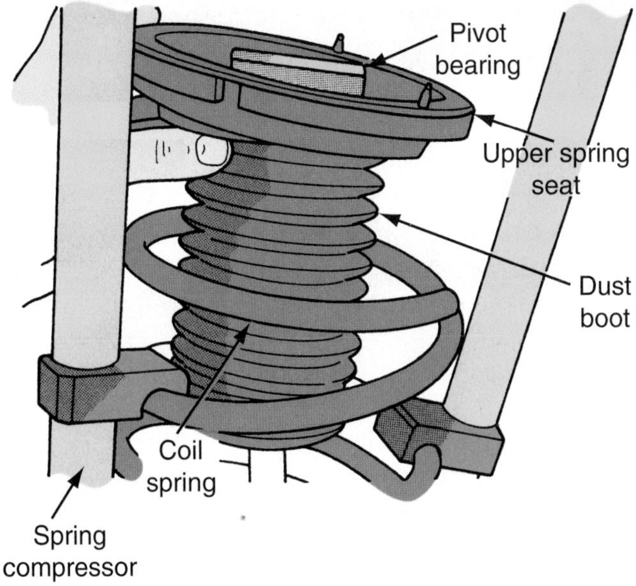

Figure 45-38 Check the strut's pivot bearing for free movement. *Courtesy of Chrysler LLC*

REAR-SUSPENSION SYSTEMS

There are three basic types of rear suspensions: live-axle, semi-independent, and independent. There are distinct designs of each, but the types of components and the principles involved are the same as on front suspension systems described earlier in this chapter. Live-axle suspensions are found on rear-wheel-drive (RWD) trucks, vans, and many four-wheel-drive (4WD) passenger cars. Semi-independent systems are used on front-wheel-drive (FWD) vehicles. Independent suspensions can be found on both RWD and FWD vehicles, as well as 4WD cars.

Live-Axle Rear-Suspension Systems

This traditional rear-suspension system consists of springs used in conjunction with a live-axle (one in which the differential axle, wheel bearings, and brakes act as a unit). The springs are either of leaf or coil type.

Leaf Spring Live-Axle System Two springs—either multiple-leaf or monoleaf—are mounted at right angles to the axle and along with the shock absorbers, are positioned below the rear axle housing. The front of the two springs is attached to brackets on the vehicle's frame by a bolt and bushing inserted through the eyes of the springs. While the bushing allows the spring to move, it isolates the rest of the vehicle from noisy road vibrations.

The center of each leaf spring is connected to the rear axle housing with U-bolts. Rubber bumpers are located between the rear axle housing and frame or unit body to dampen severe shocks. The rear eye

PHOTO SEQUENCE 49 Removing and Replacing a MacPherson Strut

P49-1 *The top of the strut assembly is mounted directly to the chassis of the car.*

P49-2 *Prior to loosening the strut chassis bolts, scribe alignment marks on the strut bolts and the chassis.*

P49-3 *With the top strut bolts or nuts removed, raise the car to a working height. It is important that the car be supported on its frame and not on its suspension components.*

P49-4 *Remove the wheel assembly. The strut is accessible from the wheel well after the wheel is removed.*

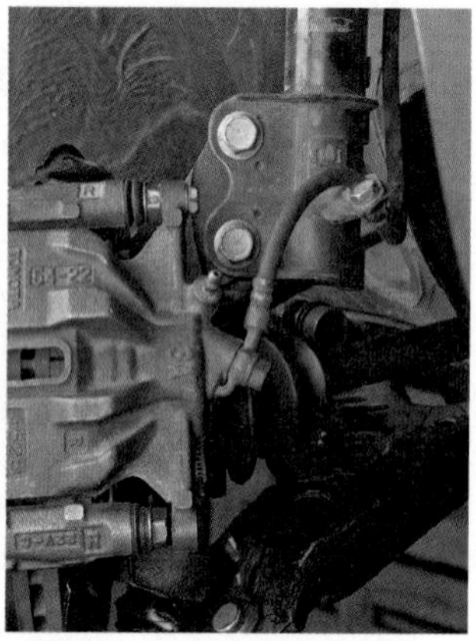

P49-5 *Remove the bolt that fastens the brake line or hose to the strut assembly.*

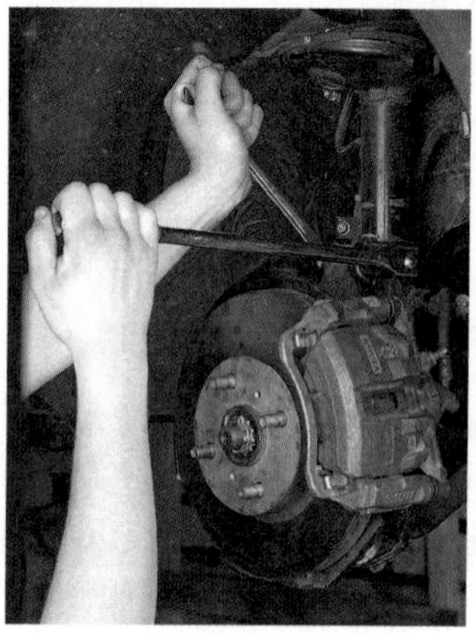

P49-6 *Remove the strut's two steering knuckle bolts.*

P49-7 *Support the steering knuckle with wire and remove the strut assembly from the car.*

P49-8 *Install the strut assembly into the proper type of spring compressor. Then compress the spring until it is possible to safely loosen the retaining bolts.*

P49-9 *Remove the old strut assembly from the spring and install the new strut. Compress the spring to allow for reassembly and tighten the retaining bolts.*

P49-10 *Reinstall the strut assembly into the car. Make sure all bolts are properly tightened and in the correct locations.*

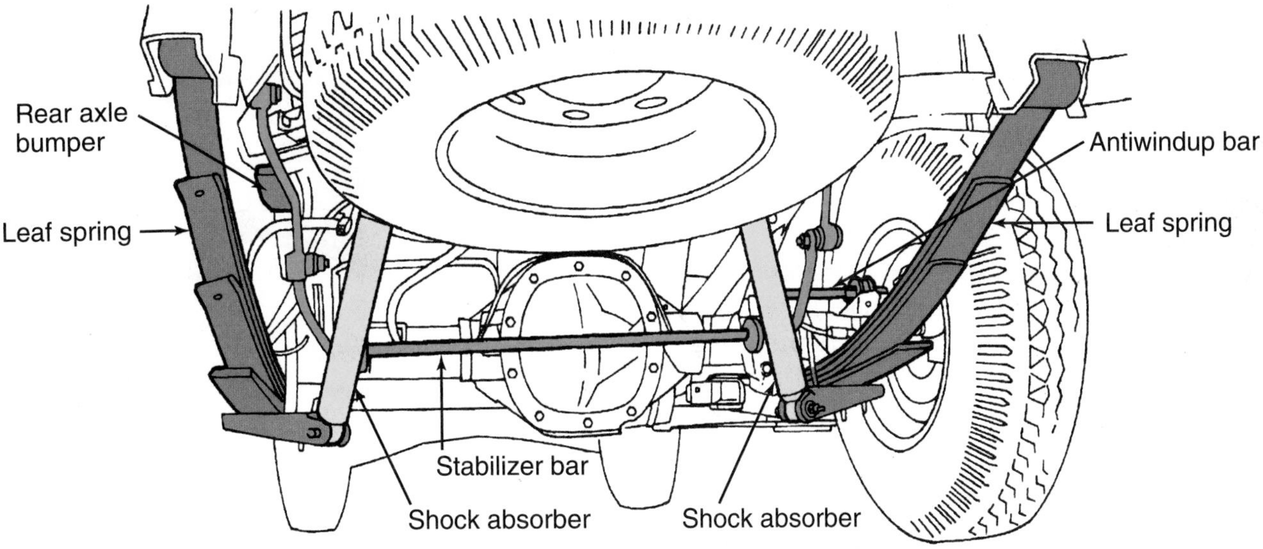

Figure 45–39 A typical rear suspension with a live axle. *Courtesy of Ford Motor Company*

pivot bushings are held to the frame with shackles, which attach to the springs by a bolt and bushing **(Figure 45–39)**.

There are some disadvantages to the live-axle suspension system. First, this design has a large amount of unsprung weight. Another drawback is the instability caused by the use of a solid axle. Since both rear wheels are connected to the same axle, movement up or down by one wheel affects the other. Consequently, poor traction results because both wheels are pushed out of alignment with the road. Under severe acceleration, this type of suspension is subject to **axle tramp**, a rapid up-and-down jumping of the rear axle due to the torque absorption of the leaf springs. This condition can break spring mounts and shock absorbers and cause premature wear of wheel bearings. Axle tramp is reduced by mounting shock absorbers on the opposing sides (front and back) of the axle. Some heavy-duty vehicles have two-stage springs that allow the vehicle to ride comfortably with both a light or heavy load.

Coil Spring Live-Axle System Some vehicles use two coil springs at the rear with a live rear axle. Because coil springs can only support weight and have little axle-locating capability, such vehicles need forward and lateral control arms or links. This type of suspension is called the link-type rigid axle.

The coil springs, located between the brackets on the axle housing and the vehicle body or frame, are held in place by the weight of the vehicle and sometimes by the shock absorbers **(Figure 45–40)**. The control arms are usually made of channeled steel and mounted with rubber bushings. Accelerating, driving, and braking torque are transmitted through three or

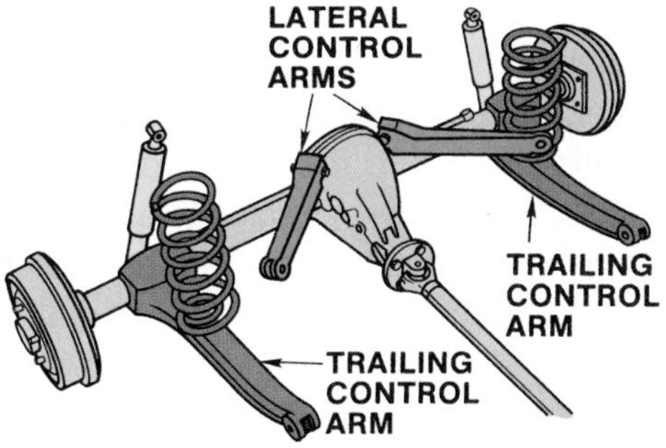

Figure 45–40 A typical live-axle suspension with coil springs.

four control arms, depending on the design. Two forward links are always used, but either one or two lateral links can be found on individual models. **Trailing arms** mount to the underside of the axle, and run forward at a 90-degree angle to the axle to brackets on the car frame. Rubber bushings are used at mounting locations to permit up-and-down movement of the arms and to reduce noise and the effect of shock.

Some rear-axle assemblies are connected to the body by two lower control arms and a tracking bar. A single torque arm is used in place of upper control arms and is rigidly mounted to the rear-axle housing at the rear and through a rubber bushing to the transmission at the front.

Live-Axle Suspension System Servicing Typical service to both coil-spring and leaf-spring systems include the replacement of shock absorbers or springs. Bushings, shackles, or control arms do not need replacement frequently. Always follow the procedures outlined in the service manual whenever servicing the rear suspension.

SEMI-INDEPENDENT SUSPENSION

A semi-independent suspension system is used on many front-wheel-drive models. On some, the suspension position is fixed by an axle beam, or crossmember, running between two trailing arms. Although there is a solid connection between the two halves of the suspension because of the axle beam, the beam twists as the wheel assemblies move up and down. The twisting action not only permits semi-independent suspension movement, but it also acts as a stabilizer. Frequently, a separate shock and spring trailing arm system is also used. In either an integrated or separate shock system, each rear wheel is independently suspended by a coil spring.

A coil spring and shock absorber–strut assembly are ordinarily used with this suspension system. The bottom of the strut is mounted to the rear end of the trailing arm. The top is mounted to the reinforced inner fender panel. Braking torque is transmitted through the trailing control arms and struts. The arms and struts also maintain the fore and aft, and lateral positioning of the wheels. A tracking bar is also used on some trailing arm suspension systems. The tracking bar helps to reduce sideways movement of the axle.

Semi-Independent Suspension System Servicing

As in most rear-system servicing, the first step is to remove the shock absorber. It is important to remember not to remove both shock absorbers at one time. Suspending the rear axle at full length could result in damage to the brake lines and hoses. The servicing of a semi-independent suspension system usually involves the removal and reinstallation of shock absorbers, springs, insulators, and control arm bushings. Follow the procedures given in the vehicle's service manual.

WARNING!

When removing the rear springs, do not use a twin-post hoist. The swing arc tendency of the rear-axle assembly when certain fasteners are removed might cause it to slip from the hoist. Perform this operation on the floor if necessary.

Independent Suspension

Independent suspensions can be found in large numbers on both FWD and RWD vehicles. The introduction of independent rear suspensions was brought about by the same concerns for improved traction ride that prompted the introduction of independent front suspensions. If the wheels can move separately on the road, traction and ride are improved.

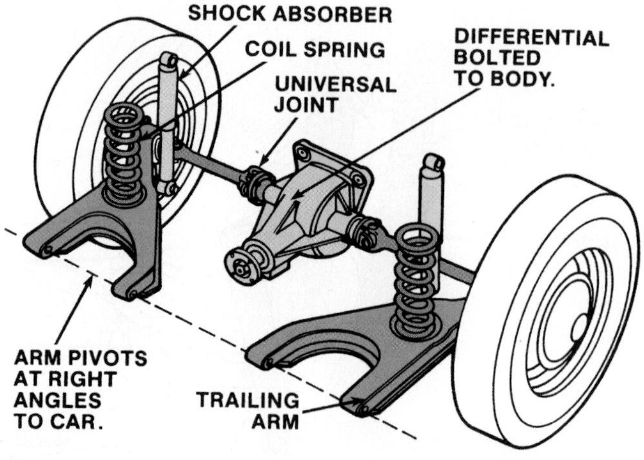

Figure 45–41 Trailing arms are often used with independent rear suspensions.

Independent coil-spring rear suspensions can have several control arm arrangements. For example, A-shaped control arms are sometimes employed. When the wide bottom of a control arm is toward the front of the car and the point turns in to meet the upright, they are called trailing arms **(Figure 45–41)**. When the entire A-shaped control arms are mounted at an angle, they are known as semitrailing control arms or multilink suspensions. Coil springs are used between the control arm and the vehicle body. The control arms pivot on a crossmember and are attached at the other end to a spindle. A shock absorber is attached to the spindle or control arm.

Some vehicles use a rear-suspension system that uses a lower control arm and open driving axles. A crossmember supports the control arms, while the tops of the shock absorbers are mounted to the body. The springs are set in seats at the bottom and top of the crossmember.

A few cars use only lower control arms, but substitute a wishbone-shaped subframe for the upper control arms. Two torque arms transfer the rear-end torque to the subframe. In fact, many cars are now featuring rear **double-wishbone** suspension **(Figure 45–42)**. Torque loads create bushing and control arm deflection during braking, cornering, acceleration, and deceleration. It is interesting to note this rear-suspension system allows for a small amount of toe-in change to enhance straight line stability. The toe-in change during cornering leads to quicker and more responsive turning. The rear-suspension system can also be tuned to ensure minimal dive under braking and minimal squat under acceleration.

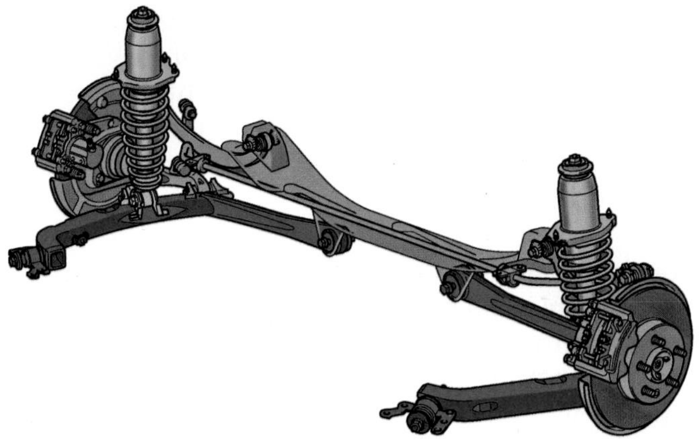

Figure 45–42 A double wishbone rear suspension. *Courtesy of Toyota Motor Sales, U.S.A., Inc.*

Currently, struts are replacing conventional shock absorbers in rear independent-suspension systems. One of the latest strut rear-suspension designs used by car manufacturers is shown in **Figure 45–43**.

On this type of system, the spindle is used to secure the strut, the outer ends of two of the four control arms, the rear ends of the tie-rods, and a rear wheel. The control arms contain bushings of different sizes at their outer ends. The ends with the smaller bushings attach near the body centerline. The ends with the larger bushings attach at the spindle. (When replacing control arms, it is mandatory that offsets at their outer ends and the flanges on the arms face in the direction prescribed in the manufacturer's service manual.) This system is also called the nonmodified MacPherson strut system.

The modified MacPherson strut rear suspension is common for vehicles with front-wheel drive. The major components on each side of the vehicle are a modified MacPherson shock strut, lower control arm, tie-rod, and wheel spindle. A coil spring mounts between the lower control arm and the body crossmember/side rail. The spindle, in addition to supporting the rear wheel, is used as an attaching location for the outer end of the control arm and the rear end of the tie-rod. The inner end of the control arm attaches to the crossmember. The forward end of the tie-rod attaches to the side rail.

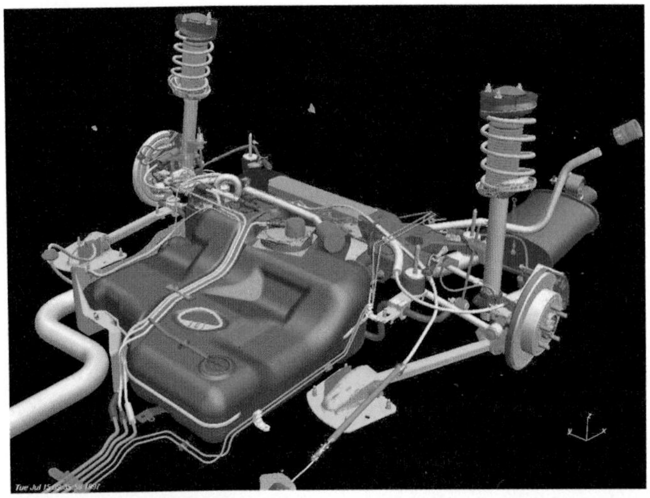

Figure 45–43 A strut-based rear-suspension system. *Courtesy of Chrysler LLC*

Another rear strut design uses a **Chapman strut** that is similar to the modified MacPherson strut. The difference between the two struts is that the MacPherson strut is involved directly in the car's steering system. The Chapman strut is not. In addition, the Chapman strut can be used with conventional springs, often a leaf-type spring. It frees the Chapman strut of load-carrying duties so it can concentrate on providing exact wheel location and shock absorbing functions.

Rear leaf-spring suspension systems are used on many vehicles with conventional rear drives. These leaf springs are generally mounted longitudinally in the same manner as described earlier in this chapter for live-axle systems. A few leaf-spring systems, however, employ springs mounted transversely. Both multiple-leaf and monoleaf, or single-leaf, can be used. The transverse-leaf spring is mounted to the differential housing rather than the vehicle frame as in the longitudinal installation. The transversely mounted spring's eyes are connected to the wheel spindle assemblies.

Rear shock absorbers or shock struts have the same service limitations as those used on the front of the vehicle. They cannot be adjusted, refilled, or repaired. The procedure for inspecting rear shocks or shock struts is similar to previously mentioned front-end parts inspection. Repetition is not necessary.

Multilink Rear Suspension

A multilink rear suspension uses several control arms to guide the wheel **(Figure 45–44)**. Different models feature different types of multilink rear suspensions

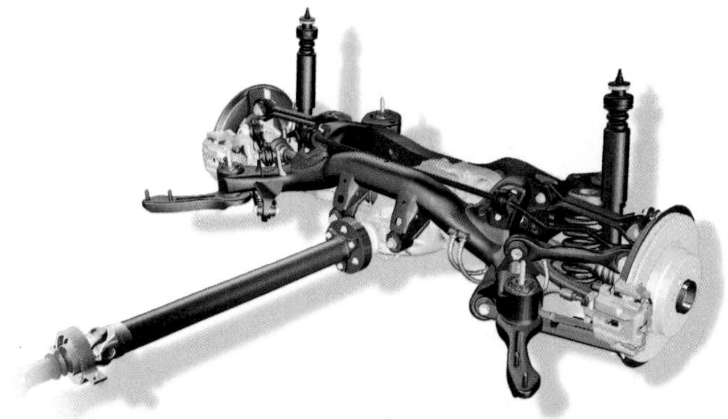

Figure 45–44 A five-link rear suspension. *Courtesy of BMW of North America, LLC*

that satisfy the varying demands of vehicle dynamics, ride comfort, and space requirements. These include the double-wishbone rear suspension, trailing-link double-wishbone rear suspension, and trapezoidal-link rear suspension.

In the double-wishbone suspension, the wheel is guided by two triangulated lateral control arms (the wishbones) and a tie rod. The suspension strut is attached to the lower wishbone to provide vertical support.

The trailing-link double-wishbone suspension has a trailing link that also carries the wheel and upper and lower wish bones. The spring is located on the trailing link ahead of the center of the wheel; the shock absorber is behind it.

The trapezoidal-link rear suspension permits excellent performance, handling, and comfort. The rear wheel is fixed by an upper lateral control arm and a trapezoidal lower link with a tie rod behind it. For reduced weight, the trapezoidal link and upper control are hollow aluminum castings.

Servicing Independent Suspension Systems

Most of the servicing techniques for rear independent suspension systems—except coil, control arm, and strut removal and installation—are similar to other front- and rear-suspension parts. They have been covered earlier in this chapter. Of course, check the service manual for all inspection and repair techniques of the vehicle's independent rear system.

Servicing Rear Coil Spring Raise the vehicle on a frame contact hoist or position jack stands under the frame forward of the rear-axle assembly. This allows the shock absorbers to fully extend. Place a floor jack under the center of the rear-axle housing and support the weight of the rear axle, but do not lift the vehicle off the jack stands. Disconnect the lower end of the shock absorber. Then, lower the floor jack until all of the coil spring force is relieved. If a coil spring positioner is used, remove it from the center of the coil spring. The coil spring can usually be removed from the vehicle at this time by lifting it from its spring seat. If the springs are to be used again, mark or tag each one so it can be returned to its original location. When a replacement is needed, always replace coil springs in pairs. This ensures equal height.

To install a spring, place the insulator on top of the coil spring and position the spring on the spring seat. The end of the top coil must be positioned to line up with the recess in the spring seat. Jack up the rear-axle housing so the spring is properly seated at the lower end and the shock absorbers line up. Reconnect the shock absorbers.

There are some definite advantages to working on one spring at a time. First of all, the assembled side of the vehicle helps support the disassembled side. It also keeps the parts aligned and eliminates the possibility of putting the parts on the wrong side of the vehicle.

Servicing Rear Control Arms To remove the upper rear control arms from the vehicle, remove the bolts passing through the control arms at the frame and at the axle ends. Usually the rear coil spring does not have to be removed for this. Service one side of the vehicle at a time. This simplifies realigning the parts during assembly. On a serviceable control arm, replace the control arm bushings by removing the defective bushing with an appropriate puller. Properly position the new bushing and press it into place in the same manner as is done on front suspensions. Position the repaired control arm on the vehicle and loosely install the bolts. Repeat the service on the other control arm if necessary. Properly torque the nuts and bolts once the vehicle's entire weight is on the springs again.

The coil springs must be removed to service the lower rear control arms. Again, one side of the vehicle should be serviced at a time. Once the vehicle is properly supported and the springs are dismantled, remove the nuts and bolts that pass through the control arm. Remove the control arm from the vehicle and service it in the same way as the upper control arm.

Check the service manual to see if there is an adjustment for the driveline working angle. If none is specified, torque the control arm bolts to specification while the full vehicle weight is on the rear axle. This sets neutral bushing tension at normal curb height.

When there is a driveline working angle adjustment, adjust the angle before torquing the control arm bolts. After the rear suspension has been serviced, always check the working angle of the universal joints on the drive shaft. This minimizes the possibility of driveline vibration.

Some independent rear-suspension systems have ball joints that perform a function similar to the front ball joints, and they should be inspected in the same way.

Although very few cars have rear wheels that steer, some independent suspension systems have components that would normally be seen only on vehicles with four-wheel steering. Components such as tie-rod ends may appear to serve the same purpose as if they were on the front suspension. They are used to adjust the angle of the wheels for stable straight ahead performance. These suspension and steering items are covered in greater detail in Chapters 46 and 47 of this book.

ELECTRONICALLY CONTROLLED SUSPENSIONS

All of the suspension systems covered up to this point are known as passive systems. Vehicle height and damping depend on fixed nonadjustable coil springs, shock absorbers, or MacPherson struts. When weight is added, the vehicle lowers as the springs are compressed. Air-adjustable shock absorbers may provide some amount of flexibility in ride height and ride firmness, but there is no way to vary this setting during operation. Passive systems can be set to provide a soft, firm, or compromise ride. Vehicle body motion and tire traction vary due to road conditions and turning and braking forces. Passive systems have no way of adjusting to these changes.

Advances in electronic sensor and computer control technology have led to a new generation of suspension systems. The simplest systems are level control systems that use electronic height sensors to control an air compressor linked to air-adjustable shock absorbers.

More advanced adaptive suspensions are capable of altering shock damping and ride height continuously. Electronic sensors **(Figure 45–45)** provide input data to a computer. The computer adjusts air spring and shock damping settings to match road and driving conditions.

The most advanced computer-controlled suspension systems are true active suspensions. These

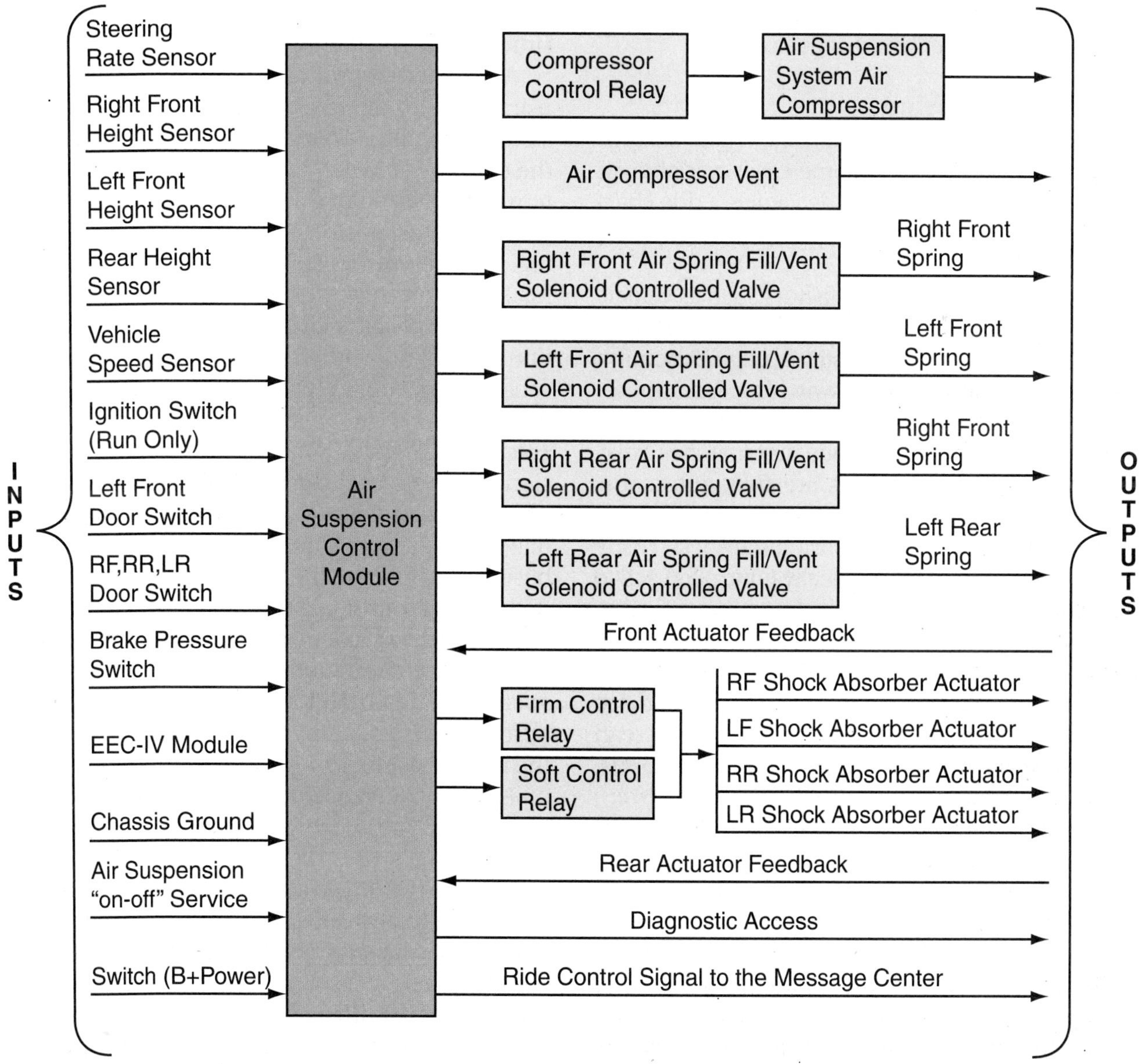

Figure 45-45 The various inputs and outputs for an electronic suspention system.

systems are hydraulically, rather than air, controlled. They use high pressure hydraulic actuators to carry the vehicle's weight rather than conventional springs or air springs.

The unique feature of an active suspension is that it can be programmed to respond almost perfectly to various operating conditions. For example, by raising the height of the outside actuators and lowering the inside actuators when going around a curve, the vehicle can be made to lean into a curve, much like a motorcycle. Active systems using hydraulic actuators are presently used on a limited number of high-performance vehicles. Most manufacturers are introducing various adaptive suspension systems that rely on pneumatically actuated air springs and dampers.

Some late-model pickups and SUVs offer air suspension systems. These systems are added to existing leaf-spring suspensions. The air spring is positioned between the center of the leaf spring and the frame of the truck. The air spring serves as an adjustable and additional spring at each end of the axle.

Adaptive Suspensions

Adaptive suspensions use electronic shock absorbers with variable valving. In some cases, variable air spring rates are used to adapt the vehicle's ride characteristics to the prevailing road conditions or driver demands.

Electronic sensors monitor factors such as vehicle height, vehicle speed, steering angle, braking force, door position, shock damping status, engine vacuum, throttle position, and ignition switching. A computer is used to analyze this input and switch the suspension into a preset operating mode that matches existing conditions. Some systems are fully automatic. Others allow the driver to select the ride mode.

At present, adaptive suspensions are less costly and complicated than hydraulically controlled active suspensions. However, they do have some limitations. Although they can reduce body roll, adaptive suspensions cannot eliminate it like true active systems. Adaptive systems also experience a slight delay in their reaction time, although some systems can change shock valving in as little as 150 microseconds.

System Components There are many different designs and components used by the manufacturers to accomplish the same task. Some systems use adjustable shocks, while others use air springs at each side of the axles. The air spring membrane is similar to a tire in construction. A solenoid valve and filter assembly allows clean air to be added or released from the air spring. Adding or removing air changes the ride height of the vehicle.

The airflow to the springs is controlled by the interaction of the air compressor, system sensors, computer control module, and solenoid valves. All of the air-operated parts of the system are connected by nylon tubing.

Compressor The compressor supplies the air pressure for operating the entire system. It is often a positive displacement single piston pump powered by a 12-volt DC motor. A regenerative air dryer is attached to the compressor output to remove moisture from the air before it is delivered to the air springs. The compressor is operated through the use of an electric relay controlled by the computer module.

Sensors Vehicle height sensors can be rotary Hall-effect sensors that enable the computer to more accurately measure ride height as well as to compensate for road variations. This prevents the vehicle from bottoming out when crossing over railroad tracks or similar road irregularities.

Advanced systems also read the steering angle by using a photo diode and shutter location inside the steering column. This allows the system to firm up the suspension when the vehicle is turning. The system also reads engine vacuum or throttle position to stiffen the suspension when the vehicle is accelerating. A brake sensor allows the system to compensate for front nose dive during hard braking. Some systems use a special G-sensor to sense sudden acceleration or braking. Other adaptive systems use a yaw sensor to pick up body roll when cornering.

Electronic Shock Absorbers Many adaptive suspension systems use electronically controlled shock absorbers that feature variable shock damping. The degree of damping is controlled by the computer based on input from the vehicle speed, steering angle, and braking sensors. As explained earlier in this chapter, variable shock damping is accomplished by varying the size of the metering orifices inside the shock. A small actuating motor mounted on top of the shock absorber rotates a control rod that alters the size of the metering orifices.

A recent advancement in adaptive suspension technology is the use of real-time shock damping. These systems use solenoid-actuated shocks rather than the motor driven shocks. Solenoids allow almost instantaneous valving changes. This means the suspension can react to bumps and body motions as they happen. Real-time adaptive systems deliver most of the handling advantages of a full active suspension without increased vehicle weight and power drain. With these systems, changes to shock valving

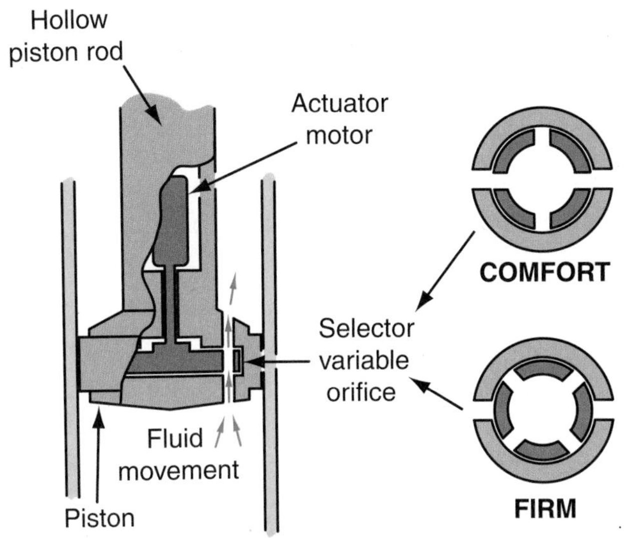

Figure 45–46 Computer command ride strut. Four electronically controlled struts are used on many adaptive suspension systems. Based on input from the computer, the valving selector shifts the variable by-pass orifice to a comfort, normal, or firm setting.

in as little as 10 milliseconds are possible when bumps are encountered.

Electronic Struts Some systems use an electronically controlled strut in place of the air spring and shock absorber **(Figure 45–46)**. Design and operation is similar to electronic shock absorbers. A valve selector or variable orifice located inside the strut controls fluid pressure in the suspension system based on input from many sensors and commands from the system's control module.

Some variable damping suspension systems use air or gas rather than a fluid. At speeds up to 40 mph (65 km/h), the orifice is fully open and provides full flow. From 40 to 60 mph (65 to 100 km/h), the orifice is in the normal position and flow is restricted. At speeds more than 60 mph (100 km/h), or when the vehicle is accelerating or braking, the variable orifice is shifted to the firm position.

The use of a variable orifice in the damper control, coupled with the deflected disc valving, provides optimum fluid flow control for both rebound and jounce strokes. In the comfort mode, the selector is set to allow fluid flow primarily through the large selector orifice to achieve minimum damping forces. While in the normal mode, the unit is set to balance fluid flows between the small selector orifice and the deflected disc valving to provide moderate damping forces. Under conditions in which the firm mode is needed, the selector is rotated to its firm or blocked position and fluid flows entirely through the deflected disc valving.

The damper control also can raise or lower the vehicle's height. This action also improves the car's aerodynamic characteristics at highway speed. As speed increases, the suspension reduces the vehicle's height and the front end angles downward. This action tends to reduce wind resistance for greater stability and better gas mileage. As the vehicle slows, the suspension brings the body up to its normal height and level position.

Computer Control Module A microcomputer-based module controls the air compressor motor (through a relay), the compressor vent solenoid, and the four air spring solenoids. The computer module also controls operation of electronic shock absorber actuating motors and electronic strut valving selectors. The control module receives input from all system sensors.

The computer module also has the capability of performing diagnostic tests on the system. It has a preprogrammed routine for properly fitting air springs after servicing. The module also controls the dash-mounted system warning light.

Electrical power to operate the basic air suspension system is distributed by the main body wiring harness. Each wiring harness involved has a special function in the typical air suspension system.

WARNING!

The compressor relay, compressor vent solenoid, and all air spring solenoids have internal diodes for electrical noise suppression and are polarity sensitive. Care must be taken when servicing these components not to switch the battery feed and ground circuits, or component damage results. When charging the battery, the ignition switch must be in the off position if the air suspension switch is on, or damage to the air compressor relay or motor may occur. However, use of a battery charger while performing the diagnostic test or air spring fill option is acceptable. Set to a rate to maintain but not damage the vehicle battery.

Electronic Leveling Control Adaptive suspension systems are capable of adjusting the suspension system during operation. Less complicated electronic level-control systems are used on many large and mid-size vehicles.

These systems do not use a computer module. In most cases, height sensors are the only types of sensors used. These height sensors sense when passenger

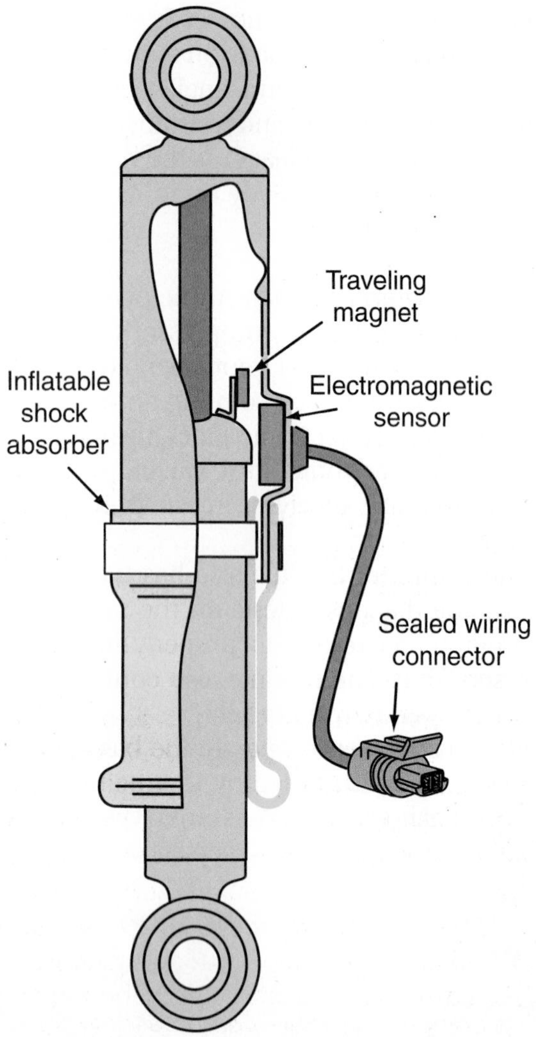

Figure 45–47 A load-sensing shock absorber.

weight or cargo is added to or removed from the vehicle (**Figure 45–47**). The height sensors control two basic circuits. The compressor relay coil grounds circuits that activate the compressor. The exhaust solenoid coil grounds circuits that vent air from the system.

To prevent falsely actuating the compressor relay or exhaust solenoid circuits during normal ride motions, the sensor circuitry provides an 8- to 15-second delay before either circuit can be completed.

In addition, the typical sensor electronically limits compressor run time or exhaust solenoid energized time to a maximum of approximately 3½ minutes. This time limit function is necessary to prevent continuous compressor operation in a case of a solenoid malfunction. Turning the ignition off and on resets the electronic timer circuit to renew the 3½-minute maximum run time. The height sensor is mounted to the frame crossmember in the rear. The sensor actuator arm is attached to the rear upper control arm by a link. The link should be attached to the metal arm when making any trim adjustment.

When the air line is attached to the shock absorber fittings or compressor dryer fitting, the retainer clip snaps into a groove in the fitting, locking the air line in position. To remove the air line, spread the retainer clip, release it from the groove, and pull on the air line.

Adjustable Pneumatic Suspension Adjustable pneumatic suspension at the front and rear wheels is a feature on some AWD vehicles. By varying the vehicle's ground clearance, it can be used off-road but also performs and handles well on the highway. There are four ride-height positions that can be selected either manually or automatically, with a total range of over 8 inches (20 mm) of ground clearance. At highway speeds, the vehicle's ground clearance is 5.6 inches (14 mm). Urban mode raises it a full inch. For moderate off-road and local driving, ground clearance is 7.6 inches (19 mm). For severe off-road conditions at speeds under 25 mph (40 km/h), maximum clearance is 8.2 inches (21 mm). The vehicle will adjust to the desired height based on vehicle speed, or the driver can temporarily override the setting by depressing a button.

MagneRide

MagneRide is a semiactive suspension system, which features shocks or struts with no electromechanical valves or small moving parts. Instead of valve-controlled orifices, MagneRide regulates the flow of fluid by a variable magnetic field produced by a small electric coil mounted in the shock (**Figure 45–48**). The shocks are filled with magneto-rheological (MR) fluid. MR fluid consists of magnetically soft particles, such as iron, suspended in synthetic hydrocarbon fluid.

The action of the shock forces the MR fluid through a magnetized opening in each shock. When the shock

Figure 45–48 A magneto-rheological fluid-based strut. *Courtesy of the Delphi Corporation*

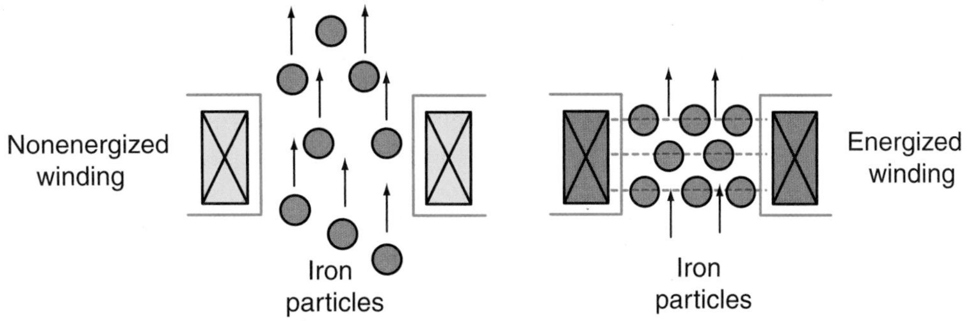

Figure 45–49 The iron particles in the MR fluid align themselves when they pass through the magnetic field, causing the fluid to stiffen.

is in its off state, the fluid is not magnetized and flows freely through the orifice. When current is sent to the coil, the fluid becomes magnetized and its viscosity changes instantly **(Figure 45–49)**.

The material changes from a fluid state to a semi-solid state that is directly proportional to the magnetic field applied to it. With little or no electrical current, the iron particles are randomly distributed, and the fluid passes freely through the piston orifice. When a strong electrical current is applied to the coil, the resulting magnetic field aligns the iron particles so that the fluid stiffens and the flow is resisted. This condition causes heavy damping. The resulting damping force is proportional to the viscosity, which is proportional to the strength of the magnetic field.

Sensors monitoring wheel position, lateral acceleration, vehicle speed, steering wheel angle, and brake pedal angle are inputs to the control module that sends current to the coil in the shocks. This system provides extremely quick response time, typically about 5 ms, and the fluid is capable of reacting 30,000 times per second.

SERVICING ELECTRONIC SUSPENSION COMPONENTS

Most electronic suspension servicing requires the removal and replacement of a component. The correct procedures for doing this are given in the manufacturer's service information. Serviceable items include the air compressor, charger, mounting brackets, height sensors, air springs, air lines and connections, gas struts, strut mounts, control arm components, shock absorbers, and stabilizer bars.

Attention to fasteners is extremely important when serving all suspension systems, especially electronic systems. Failure to adhere to the advice given about fasteners can result in a sudden failure of the air spring or suspension system. Suspension fasteners can affect performance of vital components and systems or could result in additional service expenses. All fasteners must be replaced with fasteners of the same part number or with an equivalent part. Never use a replacement part of lesser quality or substitute design. All fasteners must be tightened to the specified torque. New fasteners must be installed whenever the originals are loosened or removed and when new components are installed.

Diagnosis

A scan tool and/or a special electronic tester is used to diagnose most electronic suspension system. These can only retrieve DTCs; they may also be able to activate various actuators in the system. The exact procedures and available data from the vehicle's computer will vary with manufacturer and the system found on the vehicle. Always refer to the correct service information when diagnosing electronic systems.

Diagnosis should begin with gathering as much information as possible from the customer. Make sure you know exactly what the concern is and the conditions at the time the malfunction occurred. Verify the customer's concern by attempting to duplicate these conditions during a road test.

Check the voltage at the battery. If the voltage is below 11 volts, recharge or replace the battery before continuing with diagnostics. Check the fuses, connectors, and wiring harnesses for the suspension system and repair them as necessary. Start the engine and allow it to warm up. Then connect the scan tool or tester. Retrieve all DTCs from the system. If the scan tool or tester is unable to communicate with the vehicle's computer, diagnose the cause of this before proceeding.

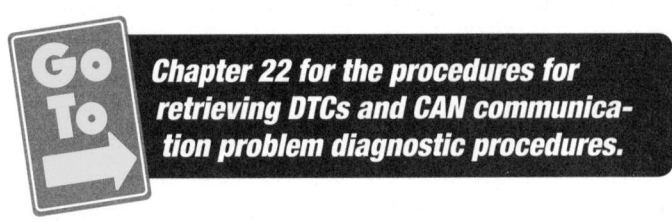

Go To ***Chapter 22 for the procedures for retrieving DTCs and CAN communication problem diagnostic procedures.***

Check the DTC charts to see if the DTCs match the symptoms exhibited by the vehicle. If they do, proceed to follow the troubleshooting chart related to each DTC. If the DTCs do not match the symptoms, clear them and recheck for trouble codes after your diagnosis of the suspension system has been completed.

If there are no DTCs related to the problem, record the PIDs and compare them to specifications **(Figure 45–50)**. If the cause of the problem is still not evident, refer to the symptoms chart provided by the manufacturer. Check those areas identified as possible causes of the problem. If the tester has the capability of activating the system's actuators, do this now. In many cases, the tester can cause action at one wheel at a time. The suspension of one wheel can be manually energized and the vehicle will tilt. The actuator can then be de-energized and the vehicle will return to its original position. This procedure should be completed at each wheel. If the suspension does not respond correctly to these commands, the corner of the suspension that did not should be thoroughly checked. If all corners responded as they should, the problem is most likely not caused by something in the electronic system, and diagnosis should continue with checks of the conventional suspension parts.

PID	DESCRIPTION	EXPECTED VALUES
4X4HIGH	4X4 High Input	IN, OUT
4X4_LOW	4X4 Low Input	IN, OUT
AS_COMP	Compressor Relay Status	ON---, ONO--, ON-B-,ON--G, OFF---, OFFO--, OFF-B-, OFF--G
AS_GATE	Front Gate Solenoid Status	ON---, ONO--, ON-B-,ON--G, OFF---, OFFO--, OFF-B-, OFF--G
AS_VENT	Vent Solenoid Status	ON---, ONO--, ON-B-,ON--G, OFF---, OFFO--, OFF-B-, OFF--G
BOO_ARC	Brake Pedal Position Switch Input	ON, OFF
CCNTARC	Number of Continuous DTCs Counted by the ARC Module	one count per bit
DR_OPEN	Door Ajar Input	OPEN, CLOSED
F_FILL	Front Fill Solenoid Status	ON---, ONO--, ON-B-,ON--G, OFF---, OFFO--, OFF-B-, OFF--G
FHGTSEN	Front Height Sensor	#, ## VDC
HGTSENS	Height Sensor	ON, OFF
IGN_RUN	Detection of Ignition Switch in the RUN Position	RUN, not RUN
LFSHK_E	Left Front Shock Encoder Status	SOFT, FIRM
LRSHK_E	Left Rear Shock Encoder Status	SOFT, FIRM
OFFROAD	Vehicle Off Road Status	ON, OFF
OPSTRAT	Operational Strategy	ARC
PCM_ACC	Acceleration Signal From the Powertrain Control Module (PCM)	YES, NO
R_FILL	Rear Fill Solenoid Status	ON---, ONO--, ON-B-,ON--G, OFF---, OFFO--, OFF-B-, OFF--G
RASGATE	Rear Gate Solenoid Status	ON---, ONO--, ON-B-,ON--G, OFF---, OFFO--, OFF-B-, OFF--G
RFSHK_E	Right Front Shock Encoder Status	SOFT, FIRM
RHGTSEN	Rear Height Sensor	#, ## VDC
RRSHK_E	Right Rear Shock Encoder Status	SOFT, FIRM
STEER_A	Steering Rotation Sensor A	LOW,HIGH

Figure 45-50 Typical PIDs for an electronic suspension system.

Once the cause is identified it should be replaced or repaired; then the repair ought to be confirmed. Drive the vehicle under the same conditions during which the concern was present. Also recheck the system for any new DTCs.

WARNING!

Do not remove an air spring under any circumstances when there is pressure in the air spring. Do not remove any components supporting an air spring without either exhausting the air or providing support for the air spring. Power to the air system must be shut off by turning the air suspension switch (in the luggage compartment) off or by disconnecting the battery when servicing any air suspension components. Most air suspension systems are equipped with a warning light. The light comes on if there is a problem, or when servicing the system.

Customer Care

Because the technician is seldom present when a vehicle requires towing, it is important to advise the customer of proper procedures so the tow operator does not damage the electronic suspension system. You must also know the proper hoist lifting and jacking restrictions. While it is necessary to check the service manual for specific instructions, the following are the basics for electronic suspension.

When towing, it must be remembered that when the ignition is off, the automatic leveling suspension is still on. Before lifting the vehicle, be sure the ignition switch is turned off and the trunk switch deactivated. When towed from the front, towing should not exceed a speed of 35 mph (60 km/h) or a distance of 50 miles (80 km/h). When the car is towed from the rear, speeds should not exceed 50 mph (80 km/h) (or 35 mph [60 km/h] on bumpy pavement).

A body hoist is usually the only type of lift recommended. In most service manuals, manufacturers warn against using a suspension hoist. The proper sequence is to position the car over the lift, shut off the ignition, then deactivate the system.

If a body hoist is not available, a floor jack and jack stands will do. Lift the car by the front cross member and the rear jacking points that are just in front of the rear wheel wells. Jack stands should be used to support the car.

In all situations, the lifting theory is the same. The suspension should be free to hang down while the car is in the air. This allows the wheels to be supported by the struts in the front and the shocks at the rear, both in their full extension (rebound) positions. Thus, the membrane of each air spring retains its proper shape while the car is in the air.

Vehicle Alignment

Aligning a vehicle with an electronic suspension system is essentially the same as the aligning procedure described in Chapter 47—with one notable exception—curb height.

Curb height is an important dimension because it affects the other alignment angles. Caster is the most obvious one that is affected, but front camber and toe can also be included. Curb height is especially critical when checking rear camber and toe on independent rear suspensions. With electronic suspension, the ride can vary depending on various circumstances. The only way to guarantee the suspension at curb height is to preset it.

ACTIVE SUSPENSIONS

Some of the advanced adaptive suspension systems may be called **active suspensions**. In this text, active suspensions refer to those controlled by double-acting hydraulic cylinders or solenoids (usually called actuators) that are mounted at each wheel. Each actuator maintains a sort of hydraulic equilibrium with the others to carry the vehicle's weight, while maintaining the desired body attitude. At the same time, each actuator serves as its own shock absorber, eliminating the need for yet another traditional suspension component.

In other words, each hydraulic actuator acts as both a spring (with variable-rate damping characteristics) and a variable-rate shock absorber. This is accomplished in an active suspension system by varying the hydraulic pressure within each cylinder and the rate at which it increases or decreases. By bleeding or adding hydraulic pressure from the individual actuators, each wheel can react independently to changing road conditions.

The components that make such a system possible are the actuator control valves, various sensors, and the chassis computer (**Figure 45–51**). Feeding information to a computer are a number of specialized sensors. Each actuator has a linear displacement sensor

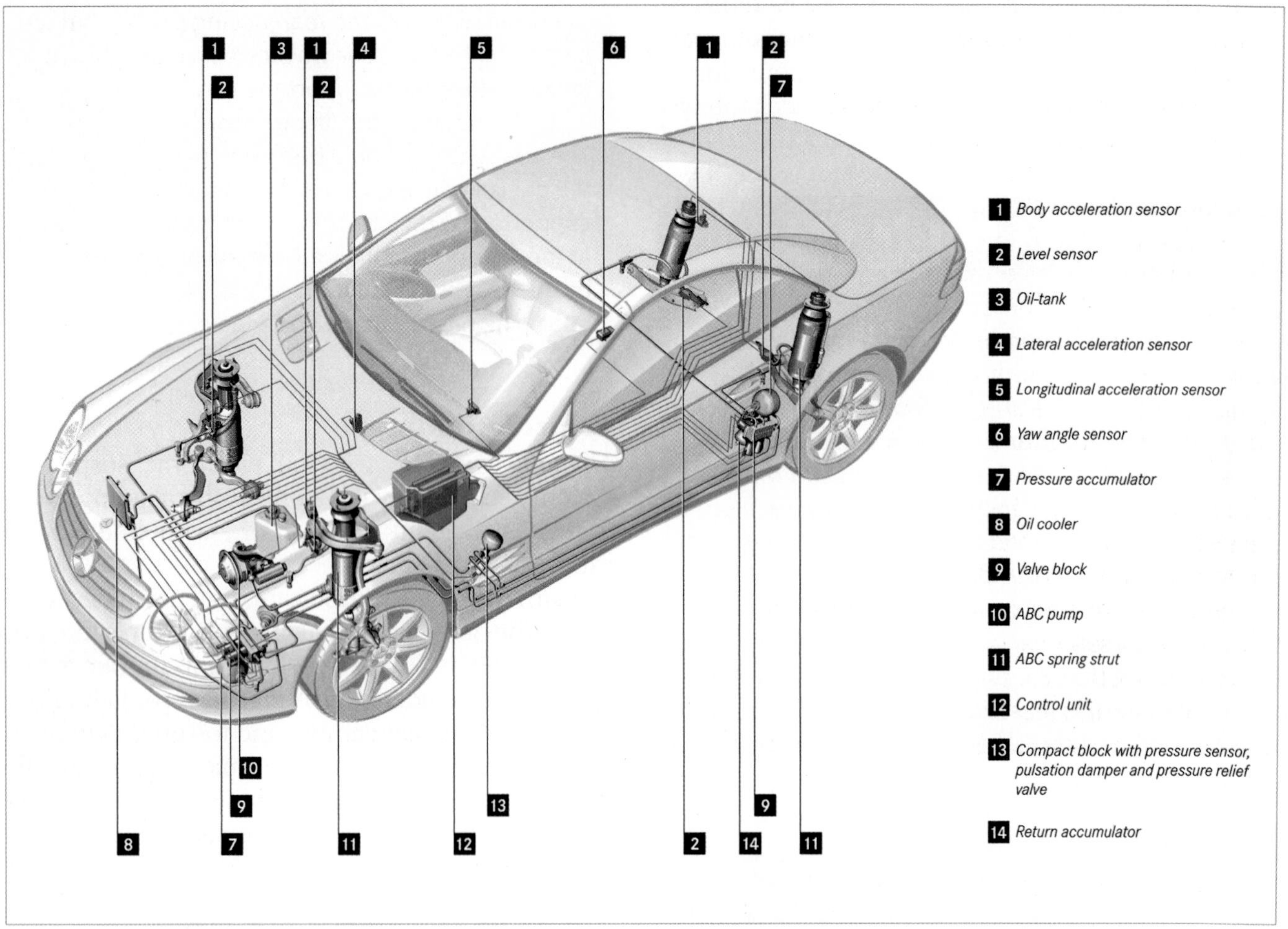

Figure 45-51 An active suspension system. *Courtesy of Chrysler LLC*

and an acceleration sensor to keep the computer informed about the actuator's relative position. This enables the computer to track the extension and compression of each actuator, and to know when each wheel is undergoing jounce or rebound. There are also load sensors and hub acceleration sensors in each wheel to measure how heavily each wheel is loaded.

A steering angle sensor is used to signal the computer when the vehicle is turning. To monitor body motions, a roll sensor and lateral acceleration and G-sensors are used. The computer also monitors hydraulic pressure within the system and the speed of the pump monitor.

Once it has all the necessary inputs, the computer can then regulate the flow of hydraulic pressure within each individual actuator according to any number of variables and its own built-in program. Another nice feature of a suspension such as this is that it can be programmed to behave in a variety of unique and currently impossible ways: leaning or rolling into turns, for example, or even raising a flat tire on command in order to change the tire without using a separate jack.

When the wheel of an active suspension hits a bump, the sensors detect the sudden upward deflection of the wheel. The computer recognizes the change as a bump, and instantly opens a control valve to bleed pressure from the hydraulic actuator. The rate at which pressure is bled from the actuator determines the cushioning of the bump and the relative harshness or softness of the ride. The rate can be varied at any point during jounce or rebound to produce a variable spring rate effect. In other words, the feel of the suspension can be programmed to respond in an almost indefinite variety of ways. Once the bump has been absorbed by the actuator, pressure is forced back into it to keep the wheel in contact with the road and to maintain the suspension's desired ride height.

During hard braking with a conventional suspension system, there is a tendency for a vehicle to make a dive. The weight of the vehicle seemingly pushes the front of the car downward and the back upward. During hard braking, the active suspension increases air pressure in the front actuators and reduces air pressure in the rear actuators. These actions minimize dive to keep the vehicle level and make it easier for the driver to control. After braking, valves operate to equalize air pressures in front and rear air actuators and level the vehicle again.

Frequently, when a driver depresses the accelerator quickly during hard acceleration, the front end of the vehicle tends to lift up, while the rear end lowers. The action is known as **squat**. With an active suspension system, squat is controlled by the operating valve's solenoids, which increase the air pressure in rear wheel actuators and reduce air pressure in front wheel actuators. When the vehicle is no longer accelerating quickly, the control system operates valves to equalize air pressures and level the vehicle. Thus, an active suspension changes the height of the front, rear, or either side of the vehicle to counteract tilting, rolling, and leaning. These active attitude control functions improve vehicle stability and increase tire traction and driver control.

The power required for a totally active system is only 3 to 5 horsepower (about the same as a typical power steering pump). Power consumption is lowest when the system is least active, as when driving on a smooth road. Rough roads and hard maneuvers, on the other hand, put more of a demand on the system. The hydraulic pump works harder and thus requires more power.

Power consumption can be reduced by going with a semiactive suspension that uses small springs with the hydraulic actuators. The springs help to support the vehicle's weight, which reduces the load on the actuators. Smaller actuators that require less hydraulic power can then be used, which reduces the bulk and weight of the system. The addition of springs also adds a certain margin of safety to the system to keep it from going flat should the hydraulics spring a leak.

Although not as widely used as electronic leveling or adaptive suspension systems, hydraulic active suspensions are sure to become more common.

CASE STUDY

An 8-year-old vehicle is brought to the shop. The customer states that the vehicle runs well, but the ride has become so bouncy he wants the shock absorbers replaced.

The work order is written and the technician replaces the shock absorbers that afternoon.

The following day, the customer returns, complaining of only a slight improvement in ride and handling. He states that the shocks must be defective or mismatched to his vehicle. Faced with an angry customer, the technician now takes the time to do what he should have done the previous day. He takes the vehicle on a road test and performs a complete inspection. On checking the ride height of the vehicle, he finds the vehicle is riding extremely low. This is a sure sign the coil springs are severely weakened.

Showing the customer the ride height measurement and the factory specification for that measurement convinces the customer of the need for new coil springs.

With new coil springs and shock absorbers, ride and handling improve dramatically.

KEY TERMS

Active suspension
Adaptive suspension
Antisway bar
Axle tramp
Ball joint
Chapman strut
Control arm
Dampers
Double wishbone
Fiber composite
Independent front suspension (IFS)
Jounce
Linear rate
MacPherson strut
Main leaf
Monoleaf
Monoshock
Multiple-leaf
Rebound
Shock absorber
Short-long arm (SLA)
Spring rate
Squat
Steering knuckle
Strut rod
Sway bar
Torsion bar
Trailing arm
Variable rate
Wheel spindle

SUMMARY

- Four types of springs are used in suspension systems: coil, leaf, torsion bar, and air.
- Springs take care of two fundamental wheel actions: jounce and rebound.
- Common coil spring materials include carbon steel, carbon boron, steel, and alloy steels. Alloy steels, such as those containing chromium and silicon, improve the coil's resistance to relaxation.
- Two basic designs of coil springs are used in vehicles: linear rate and variable rate.
- Leaf springs are made of steel or a fiber composite.

- In torsion suspension, the bar may run either from front to rear or side to side across the chassis.
- Air springs are generally only used in computer-controlled suspension systems.
- Shock absorbers damp or control motion in a vehicle. A conventional shock absorber is a velocity-sensitive hydraulic damping device. The faster it moves, the more resistance it has to the movement.
- Shock absorbers can be mounted either vertically or at an angle. Angle mounting of shock absorbers improves vehicle stability and dampens accelerating and braking torque.
- There are two basic adjustable air shock systems: the manual fill type and the automatic or electronic load-leveling type.
- MacPherson struts provide the damping function of a shock absorber. In addition, they serve to locate the spring and to fix the position of the suspension.
- Domestic struts have taken two forms: a concentric coil spring around the strut itself and a spring located between the lower control arm and the frame.
- Independent front suspension (IFS) must keep the wheels rigidly positioned and at the same time allow them to steer right and left. In addition, because of weight transfer during braking, the front suspension system absorbs most of the braking torque. When accomplishing this, it must provide good ride and stability characteristics.
- The unequal length arm or short-long arm (SLA) suspension system is most commonly used on domestic vehicles.
- Live-axle is the traditional rear suspension system and consists of springs used in conjunction with a live-axle (one in which the differential axle, wheel bearings, and brakes act as a unit). The springs are either of leaf or coil type.
- Semi-independent suspension is used on many front-wheel-drive models.
- Three strut designs are frequently used in IFS systems: the conventional MacPherson strut, the modified MacPherson strut, and the Chapman strut.
- The two basic types of computer suspension systems are adaptive and active.
- Electronically controlled suspensions can be either simple load-leveling systems, adaptive systems, or fully active systems. Adaptive and active suspension systems are computer controlled.
- Adaptive suspensions can alter vehicle ride height and shock absorber damping while the vehicle is in motion. Such systems use air springs and electronic shock absorbers or struts.
- Active suspensions are hydraulically operated actuators to control up-and-down and side-to-side movement. They can be programmed to respond to certain road conditions and turning forces.

REVIEW QUESTIONS

1. How does a stabilizer bar work?
2. Explain the difference between sprung and unsprung weight.
3. What is the purpose of an air spring?
4. Explain the action of the conventional shock absorber on both compression (jounce) and rebound strokes.
5. Describe the action of the independent front wheel suspension system.
6. The core of any suspension system is the ______.
7. What occurs when a wheel hits a dip or hole and moves downward?
   a. jounce
   b. free length
   c. deflection
   d. rebound
8. Which of the following is part of the sprung weight of a vehicle?
   a. steering linkage
   b. tires
   c. engine
   d. all of the above
9. What occurs when a wheel hits a dip or hole and moves upward?
   a. jounce
   b. free length
   c. deflection
   d. rebound
10. Describe uncontrolled spring action without a shock absorber.
11. The modified MacPherson rear strut suspension is very common in ______.
    a. FWD vehicles
    b. RWD vehicles
    c. pickup trucks
    d. station wagons

12. What controls the movement of a vehicle as it moves down a bumpy road?
    a. struts
    b. shock absorbers
    c. both a and b
    d. neither a nor b
13. The coil springs of the vehicle ______.
    a. support the weight of the vehicle
    b. provide axle location
    c. stabilize the up-and-down movement
    d. all of the above
14. The two SLA systems in common use today are the ______.
    a. coil spring and strut suspension
    b. coil spring and torsion bar suspension
    c. coil spring and single control arm suspension
    d. single and double control arm suspension
15. When the bottom of a control arm is toward the front of the car and the point turns in to meet the upright, it is called a(n) ______.
    a. trailing arm
    b. semitrailing arm
    c. wishbone arm
    d. A-shaped arm

ASE-STYLE REVIEW QUESTIONS

1. Technician A says that the use of firmer, urethane bushings in the suspension improves the vehicle's road-holding ability and handling. Technician B says that firmer bushings help eliminate torque steer in some FWD vehicles. Who is correct?
   a. Technician A c. Both A and B
   b. Technician B d. Neither A nor B
2. Technician A says that load-carrying ball joints should always have some play in them. Technician B says that follower ball joints should never have some play in them. Who is correct?
   a. Technician A c. Both A and B
   b. Technician B d. Neither A nor B
3. Technician A says that a weak suspension spring can cause a loss of traction during acceleration. Technician B says that a weak suspension spring can cause poor braking power. Who is correct?
   a. Technician A c. Both A and B
   b. Technician B d. Neither A nor B
4. Technician A says that leaf spring-type rear suspensions are subject to wheel tramp. Technician B says that an antisway bar is designed to limit wheel tramp. Who is correct?
   a. Technician A c. Both A and B
   b. Technician B d. Neither A nor B
5. Technician A says that shock absorbers should be inspected for loose mounting bolts and worn mounting bushings. Technician B says that shock absorbers and struts should be inspected for oil leakage. Who is correct?
   a. Technician A c. Both A and B
   b. Technician B d. Neither A nor B
6. While discussing types of coil springs: Technician A says that linear-rate coil springs have equal spacing between the coils. Technician B says that a variable-rate spring may have a cylindrical shape with unequally spaced coils. Who is correct?
   a. Technician A c. Both A and B
   b. Technician B d. Neither A nor B
7. When the front wheels are turned on a vehicle equipped with front struts, the left front coil spring provides a chattering action and noise: Technician A says that the strut has internal defects and strut replacement is necessary. Technician B says that the upper strut bearing and mount are defective. Who is correct?
   a. Technician A c. Both A and B
   b. Technician B d. Neither A nor B
8. Technician A says that the suspension switch must be turned off before raising any corner of a car with an electronic air suspension. Technician B says that the ignition switch must not be turned on while any corner of a car with electronic air suspension is raised. Who is correct?
   a. Technician A c. Both A and B
   b. Technician B d. Neither A nor B
9. While conducting a bounce test: Technician A says that the bumper should be pushed two or three times downward with considerable weight applied on each corner of the vehicle. The bumper should be released after each push and the vehicle should oscillate about 1½ cycles and then settle. Technician B says that one free upward bounce should stop the vertical chassis

movement if the shock absorber or strut provides proper spring control. If the vehicle's bumper does more than 1½ free upward bounces, the shock absorber or strut is defective. Who is correct?

a. Technician A
b. Technician B
c. Both A and B
d. Neither A nor B

10. Technician A says that rattling on road irregularities can be caused by worn shock absorber bushings or grommets, worn spring insulators, a broken coil spring or broken spring insulators, worn control arm bushings, worn stabilizer bar bushings, worn strut rod grommets, worn leaf spring shackles and bushings, and worn torsion bars, anchors, and bushings. Technician B says that dry or worn control arm bushings may cause a squeaking noise on irregular road surfaces. Who is correct?

a. Technician A
b. Technician B
c. Both A and B
d. Neither A nor B

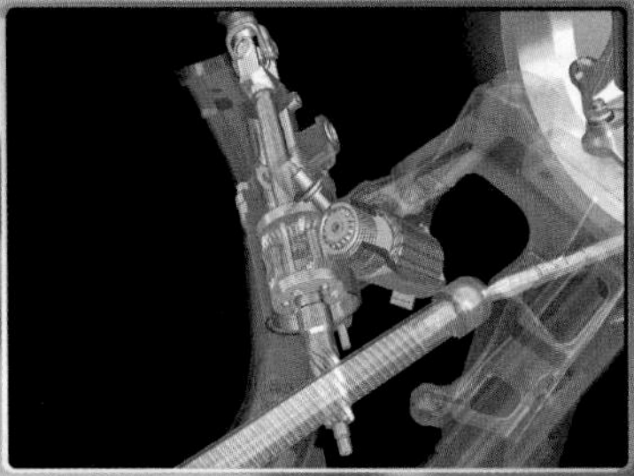

STEERING SYSTEMS

CHAPTER 46

OBJECTIVES

■ Describe the similarities and differences between parallelogram, worm and roller, and rack and pinion steering linkage systems. ■ Identify the typical manual-steering system components and their functions. ■ Name the five basic types of steering linkage systems. ■ Identify the components in a parallelogram steering linkage arrangement and describe the function of each. ■ Identify the components in a manual rack and pinion steering arrangement and describe the function of each. ■ Describe the function and operation of a manual-steering gearbox and the steering column. ■ Explain the various manual-steering service procedures. ■ Describe the service to the various power-steering designs. ■ Perform general power-steering system checks. ■ Describe the common four-wheel steering systems.

The purpose of the steering system is to turn the front wheels. In some cases, it also turns the rear wheels. The wheels constantly change direction while switching lanes, rounding sharp turns, and when avoiding roadway obstacles **(Figure 46–1)**. Few older vehicles had some sort of power assist to turn the wheels; today all vehicles have power-assist steering. This point is important, because in most cases, the power-assist systems work with manual-steering systems. Therefore, it is important to look at manual systems and then power-assist systems to understand how today's steering systems work.

Figure 46-1 Allowing a vehicle to do something other than go straight is the job of the steering system. *Courtesy of Toyota Motor Sales, U.S.A., Inc.*

MANUAL-STEERING SYSTEMS

The steering system is composed of three major subsystems: the steering linkage, steering gear, and steering column and wheel. As the steering wheel is turned by the driver, the steering gear transfers this motion to the steering linkage. The steering linkage turns the wheels to control the vehicle's direction. Although there are many variations to this system, these three major assemblies are in all steering systems.

Steering Linkage

The term **steering linkage** is applied to the system of pivots and connecting parts placed between the steering gear and the steering arms that are attached to the front or rear wheels that control the direction of vehicle travel. The steering linkage transfers the motion of the steering gear output shaft to the steering arms, turning the wheels to maneuver the vehicle.

The type of front-wheel suspension (independent wheel suspension as compared with a solid front axle) greatly influences steering geometry. Most passenger cars and many light trucks and recreational vehicles have independent front-wheel suspension systems.

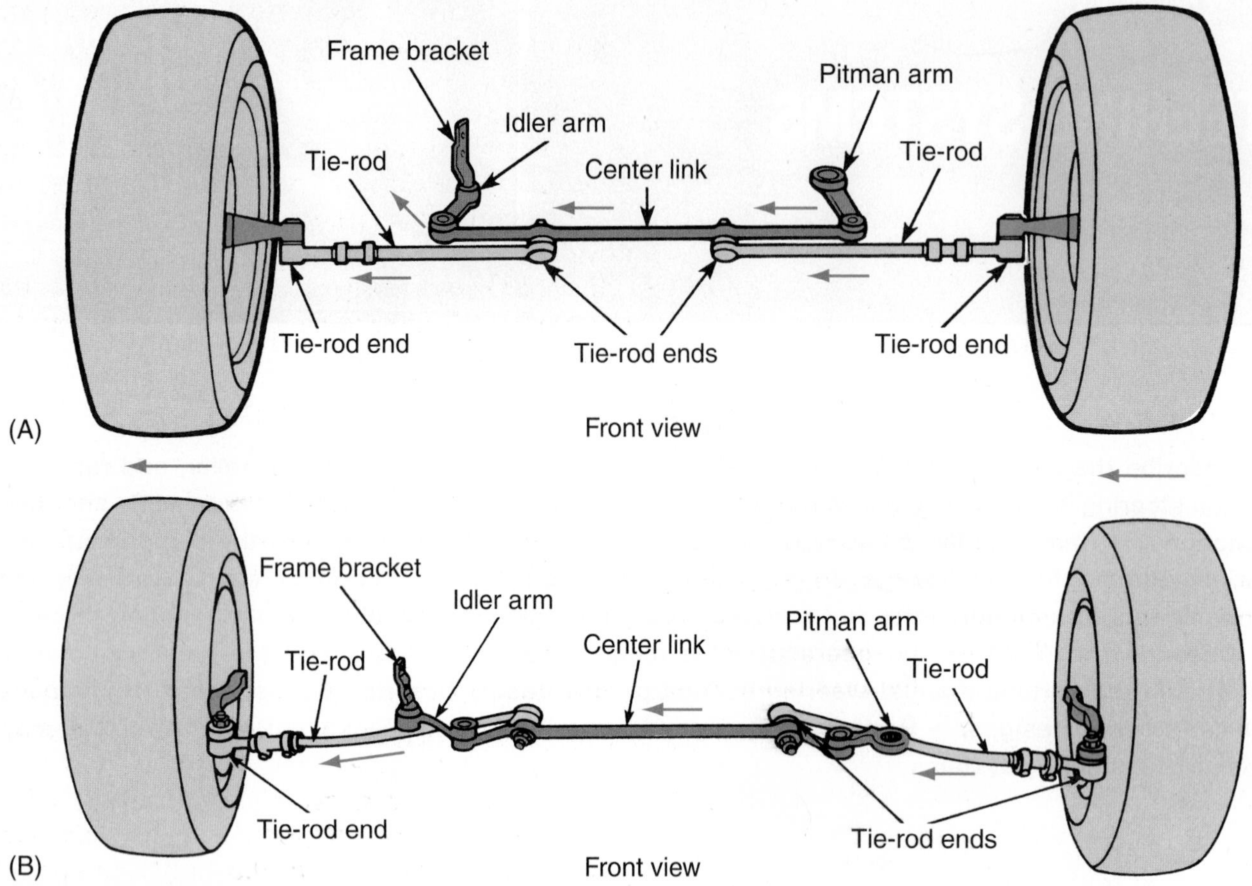

Figure 46-2 Parallelogram steering system mounts (A) behind the front suspension, and (B) ahead of the front suspension.

Therefore, a steering linkage arrangement that tolerates relatively large wheel movement must be used.

Parallelogram Steering Linkage

A parallelogram type of steering linkage arrangement was at one time the most common type used on passenger cars. Now it is found mostly on larger cars, pickups, and larger SUVs. It is used with the recirculating ball steering gear and can be classified into two distinct configurations: parallelogram steering linkage placed behind the front-wheel suspension **(Figure 46–2A)** and parallelogram steering linkage placed ahead of the front-wheel suspension **(Figure 46–2B)**. This type of steering linkage is most often used where motor and chassis components would interfere with normal operation of the steering linkage.

These designs are the basic steering systems used in conjunction with independent front-wheel suspensions. This type of linkage also provides good steering and suspension geometry. Road vibrations and impact forces are transmitted to the linkage from the tires, causing wear and looseness in the system, which permits intermittent changes in the toe setting of the front wheels, allowing further tire wear.

In a parallelogram steering linkage, the tie rods have ball socket assemblies at each end. One end is attached to the wheel's steering arm and the other end to the center link.

The components in a parallelogram steering linkage arrangement are the pitman arm, idler arm, links, and tie rods.

Pitman Arm The **pitman arm (Figure 46–3)** connects the linkage to the steering column through a steering gear located at the base of the column. It

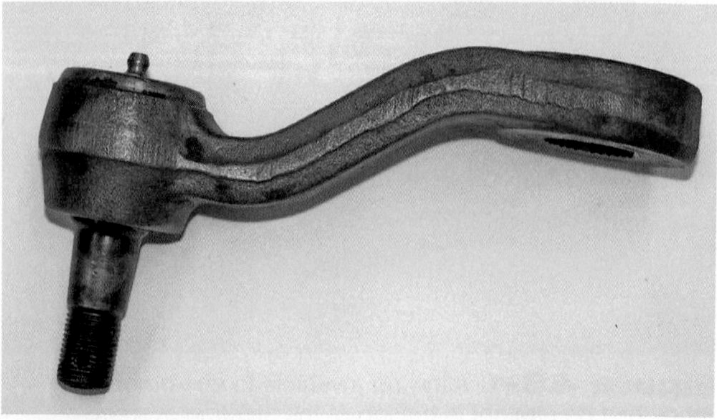

Figure 46-3 A typical pitman arm. It connects the steering column to the center link.

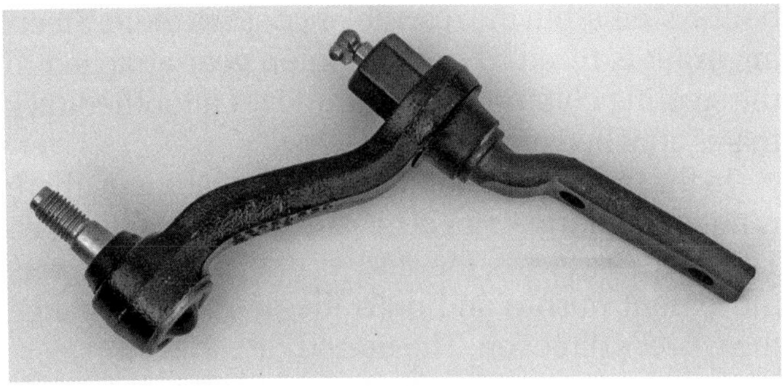

Figure 46-4 A typical idler arm. It supports the center link and it is mounted to the frame.

transmits the motion it receives from the gear to the linkage, causing the linkage to move left or right to turn the wheels in the appropriate direction. It also serves to maintain the height of the center link. This ensures that the tie-rods are able to be parallel to the control arm movement and avoid unsteady toe settings or **bump steer**. *Toe,* a critical alignment factor, is a term that defines how well the tires point to the direction of the vehicle.

Idler Arm The **idler arm** or idler arm assembly (**Figure 46–4**) is normally attached, on the opposite side of the center link, from the pitman arm and to the car frame, supporting the center link at the correct height. A pivot built into the arm or assembly permits sideways movement of the linkage. On some linkages, such as those on a few light-duty trucks, two idler arms are used.

Idler arms normally wear more than pitman arms because of this pivot function, with wear usually showing up at the swivel point of the arm or assembly. Worn bushings or stud assemblies on idler arms permit excessive vertical movement in the idler arms.

Links Links, depending on the design application, can be referred to as **center links** or steering links (**Figure 46–5**). Their purpose is to control sideways linkage movement, which changes the wheel's direction. Because they usually are also mounting locations for tie-rods, they are very important for maintaining correct toe settings. If they are not mounted at the correct height, toe is unstable and a condition known as the toe change or bump steer is produced. Center links and drag links can be used either alone or in conjunction with each other, depending on the particular steering design.

Center links with stud or bushing ends are likely to become unserviceable from the effects of normal operation and should be inspected periodically. Links with open tapers usually need to be replaced only if they have been damaged in an accident or through excessive tolerance at the mounting position of the idler or pitman arms.

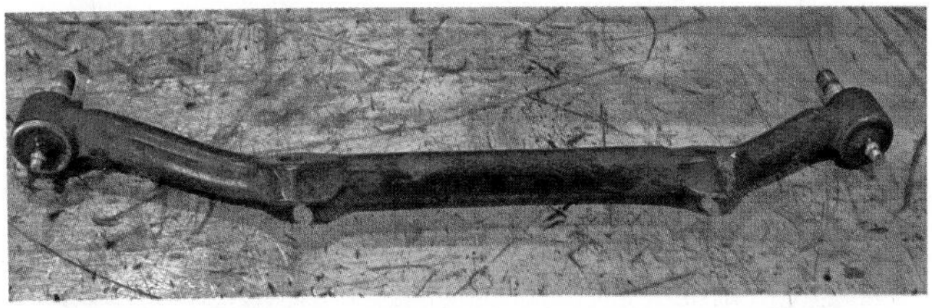

Figure 46-5 A typical center link.

SHOP TALK

If a center link is nonwear, the pitman arm is normally a wear arm. If the link is wear, the pitman arm is usually nonwear. Idler arms or assemblies are always subject to wear.

Tie-Rods **Tie-rods** are actually assemblies that make the final connections between the steering linkage and steering knuckles. They consist of inner tie-rod ends, which are connected to the opposite sides of the center link; outer tie-rod ends, which connect to the steering knuckles; and adjusting sleeves or bolts, which join the inner and outer tie-rod ends, permitting the tie-rod length to be adjusted for correct toe settings (**Figure 46–6**).

Tie-rods are subject to wear and damage, particularly if the rubber or plastic dust boots covering the ball stud have been damaged or are missing. Contaminants such as dirt and moisture can enter and cause rapid part failure. A special bonded ball stud, in which no boot is used, is available for use on certain light-duty two-wheel-drive and four-wheel-drive trucks. An elastomer bushing bonded to the stud ball

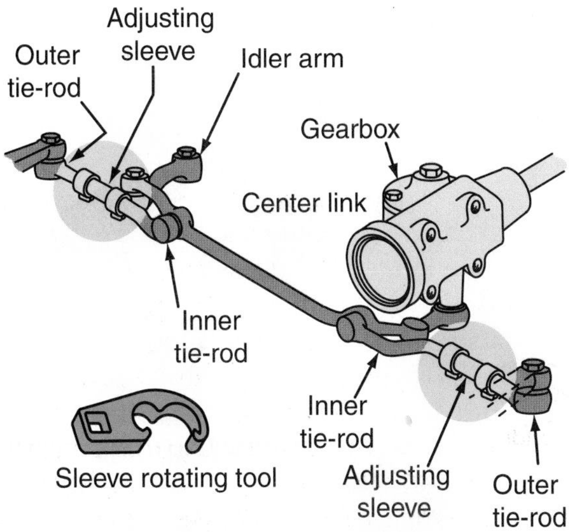

Figure 46-6 A tie-rod assembly.

provides strong shock absorption and steering return in downsized vehicles.

Rack and Pinion Steering Linkage

Rack and pinion is lighter in weight and has fewer components than parallelogram steering **(Figure 46–7)**. Tie rods are used in the same fashion on both systems, but the resemblance stops there. Steering input is received from a pinion gear attached to the steering column. This gear moves a toothed rack that is attached to the tie-rods.

Figure 46-7 A rack and pinion steering system.

In the rack and pinion steering arrangement, there is no pitman arm, idler arm assembly, or center link. The **rack** performs the task of the center link. Its movement pushes and pulls the tie-rods to change the wheel's direction. The tie-rods are the only steering linkage parts used in a rack and pinion system.

Most rack and pinion constructions **(Figure 46–8)** are composed of a tube in which the steering rack can slide. The rack is a rod with gear teeth cut along one end-spur and helical. The other end is fitted with two balls to which the ends of the divided track rods are attached. The rack meshes with the teeth of a small pinion at the end of the steering column. The two inner tie-rod ends, which are attached to the rack, are covered by rubber bellows boots that protect the rack from contamination. The inner tie-rods connect to

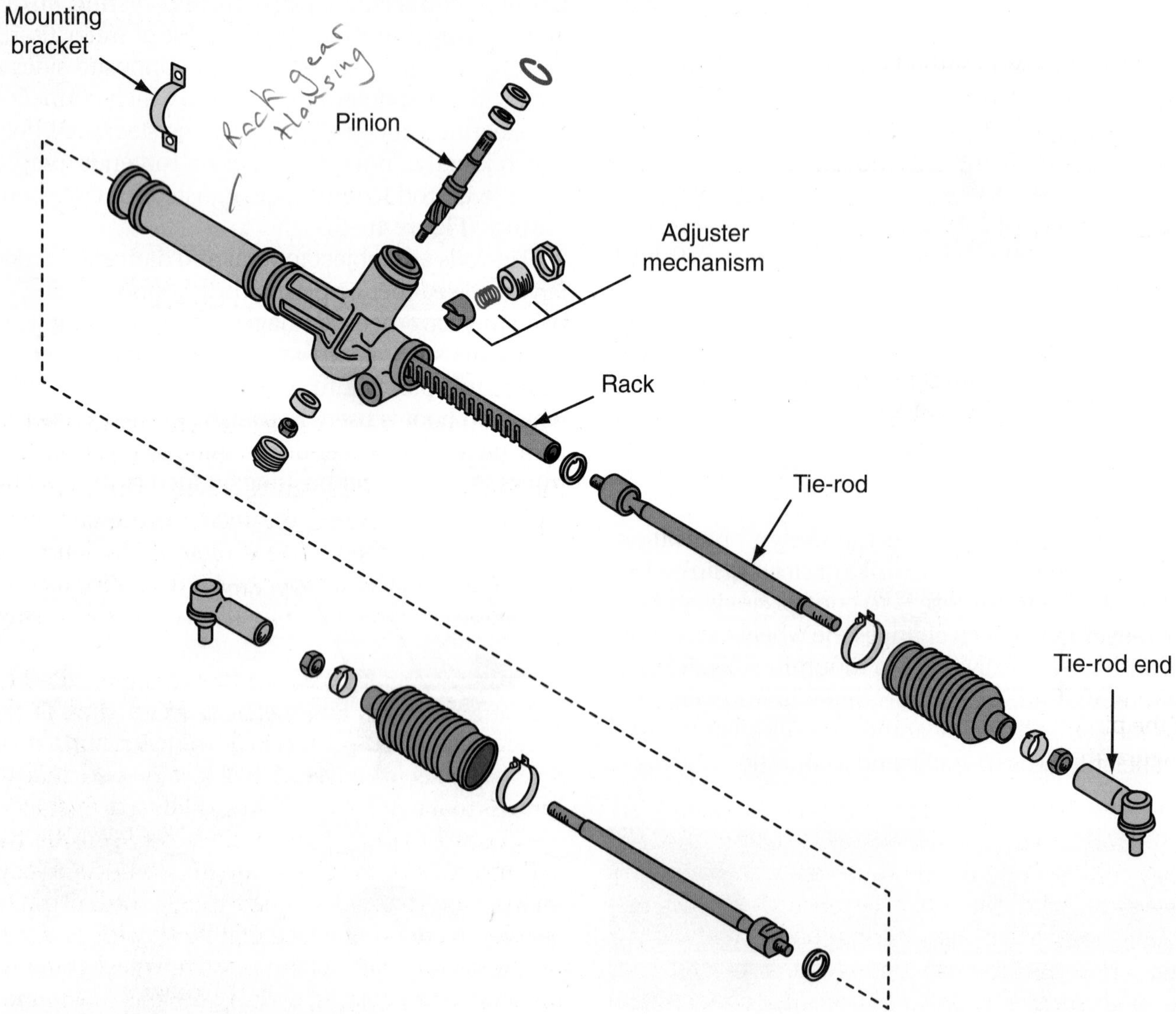

Figure 46-8 A disassembled view of a manual rack and pinion steering gear.

outer tie-rod ends, which connect to the steering arms. The rack and pinion housing is fastened to the vehicle at two or three points. Like the parallelogram linkage, it can be mounted in front of or behind the suspension.

In some cases, the rack and pinion steering gear on unibody cars is bolted directly to a body panel, like a cowl. When this is done, the body panel must hold the steering gear in its correct location. The unibody structure must maintain the proper relationship of the steering and suspension parts to each other. Along with other advantages, the rack and pinion steering system combined with the MacPherson strut suspension system is found in most front-wheel-drive unibody vehicles because of their weight- and space-saving feature.

The driver gets a greater feeling of the road with rack and pinion because there are fewer friction points. This means a higher probability of car owners with steering complaints. Fewer friction points can reduce the system's total ability to isolate and dampen vibrations.

Rack The rack is a toothed bar contained in a metal housing. The rack maintains the correct height of the steering components so that the tie-rod movement is able to parallel control arm movement.

The rack is similar to the parallelogram center link in that its sideways movement in the housing is what pulls or pushes the tie-rods to change wheel directions.

Pinion The pinion is a toothed or worm gear mounted at the base of the steering column assembly where it is moved by the steering wheel. The pinion gear meshes with the teeth in the rack so that the rack is propelled sideways in response to the turning of the pinion.

Yoke Adjustment The rack-to-pinion lash, or preload, affects steering harshness, feedback, and noise. It is set according to the manufacturer's specifications. An adjustment screw, plug, or shim pack are located on the outside of the housing at the junction of the pinion and rack to correct or set the **yoke lash** (**Figure 46–9**).

Tie-Rods Tie-rods in rack and pinion systems are very similar to those used on parallelogram systems. They consist of inner and outer ends and adjusting sleeves or bolts. The inner tie-rod ends on rack and pinion units are usually spring-loaded ball sockets that screw onto the rack ends (**Figure 46–10**). They are preloaded and protected against contaminant entry by rubber bellows or boots.

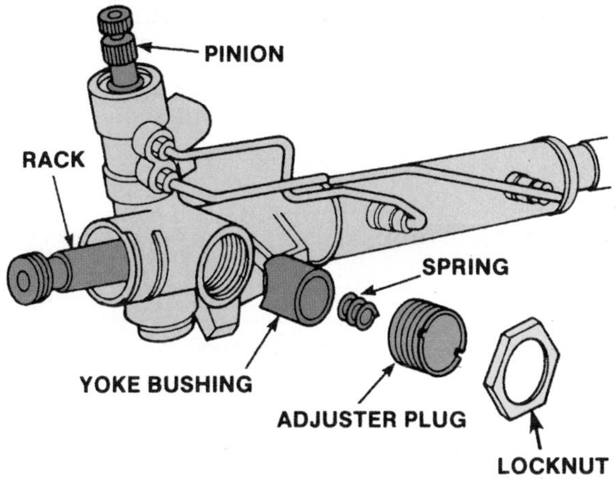

Figure 46–9 The rack preload (yoke lash) is adjusted by a screw, plug, or shim pack.

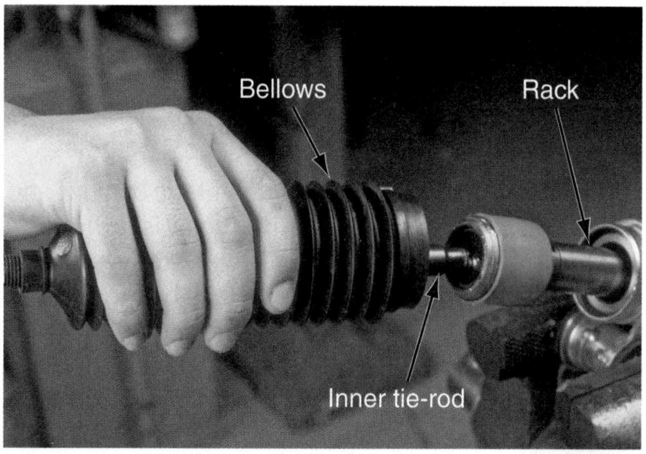

Figure 46–10 The inner tie-rod is a spring-loaded ball socket in a rack and pinion steering box.

Manual-Steering Gear

The purpose of the steering gear is to change the rotational motion of the steering wheel to a reciprocating motion to move the steering linkage. There are three styles currently in use: the recirculating ball, worm and roller, and the rack and pinion. The latter gear assembly incorporates the already described rack and pinion linkage system and steering gear as a single unit.

The recirculating ball, as shown in **Figure 46–11**, is generally found in larger cars. A sector shaft is supported by needle bearings in the housing and a bushing in the sector cover. A ball nut is used that has threads that mate to the threads of the wormshaft via continuous rows or ball bearings between the two. Ball bearings recirculate through two outside loops, referred to as ball return guide tubes. The ball nut has gear teeth cut on one face that mesh with gear teeth on the sector shaft. As the steering wheel is rotated, the wormshaft rotates, causing the ball nut to move up or down the wormshaft. Since the gear teeth on

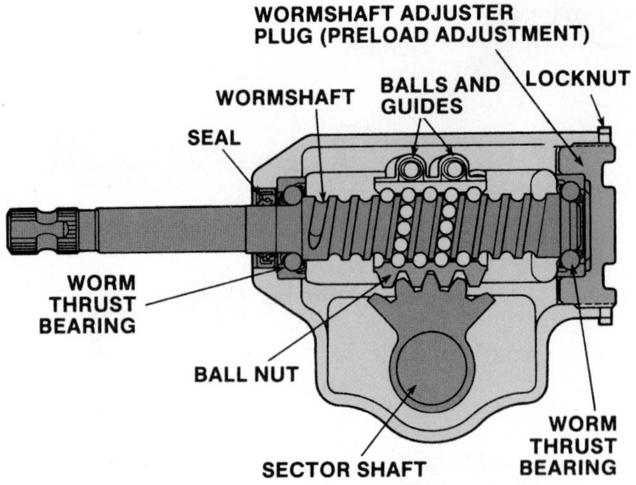

Figure 46–11 A top cutaway of a manual recirculating ball steering gear.

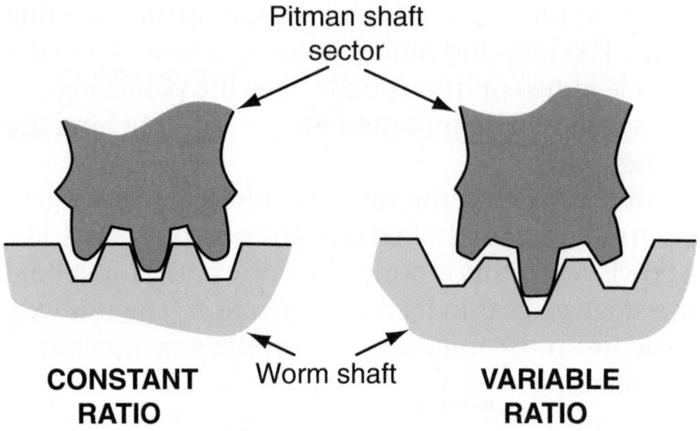

Figure 46–12 Constant and variable ratio steering gears.

the ball nut are meshed with the gear teeth on the sector shaft, the movement of the nut causes the sector shaft to rotate and swing the pitman arm.

The design of two separate circuits of balls results in an almost friction-free operation of the ball nut and the wormshaft. When the steering wheel is turned, the ball bearings roll in the ball thread grooves of the wormshaft and ball nut. When the ball bearings reach the end of their respective circuit, they enter the guide tubes and are returned to the other end of the circuits.

The teeth on the sector shaft and the ball nut are designed so that an interference fit exists between the two when the front wheels are straight ahead. This interference fit eliminates gear tooth lash for a positive feel when driving straight ahead. Proper mesh engagement between the sector and ball nut is obtained by an adjusting screw that moves the sector shaft axially.

CAUTION!

Proper adjustment between the worm gear and the pitman gear is important for adequate steering response. Refer to the vehicle manual for specific adjustment procedures.

The worm thrust bearing adjuster can be turned to provide proper preloading of the worm thrust bearings. Worm bearing preload eliminates worm end play and is necessary to prevent steering free play and vehicle wander.

Variable Steering The number of input turns per output turn of the steering gearbox is called the gearbox ratio. Steering gears can have a constant or a variable ratio. The sector teeth in a constant ratio unit are identical in size and shape, while the sector of a variable ratio unit has larger center teeth (**Figure 46–12**). This makes the steering faster in turns than in a straight direction. Variable ratio is normally used only in power-steering units.

Worm and Roller The **worm and roller gearbox** is similar to the recirculating ball except a single roller replaces the balls and ball nut. This reduces internal friction, making it ideal for smaller cars. The steering linkage used with a worm and roller gearbox typically includes a pitman arm, center link, idler arm, and two tie-rod assemblies. The function of these components is the same as in the parallelogram steering linkage described earlier in this chapter.

In operation, the steering shaft rotates the worm gear. It, in turn, engages the roller, causing the roller shaft to turn. The shaft moves the pitman arm left or right to steer the vehicle.

It must be noted that the steering gear does not cause the vehicle to pull to one side, nor does it cause road wheel shimmy.

Steering Wheel and Column

The purpose of the steering wheel and column is to produce the necessary force to turn the steering gear. The exact type of steering wheel and column depends on the year and the car manufacturer. The steering column, also called a steering shaft, relays the movement of the steering wheel to the steering gear.

Major parts of the steering wheel and column are shown in **Figure 46–13**. The steering wheel is used to produce the turning effort. The lower and upper covers conceal parts. The universal joints rotate at angles. Support brackets are used to hold the steering column in place. Assorted screws, nuts, bolt pins, and seals are used to make the steering wheel and column perform correctly. Since 1968, all steering columns

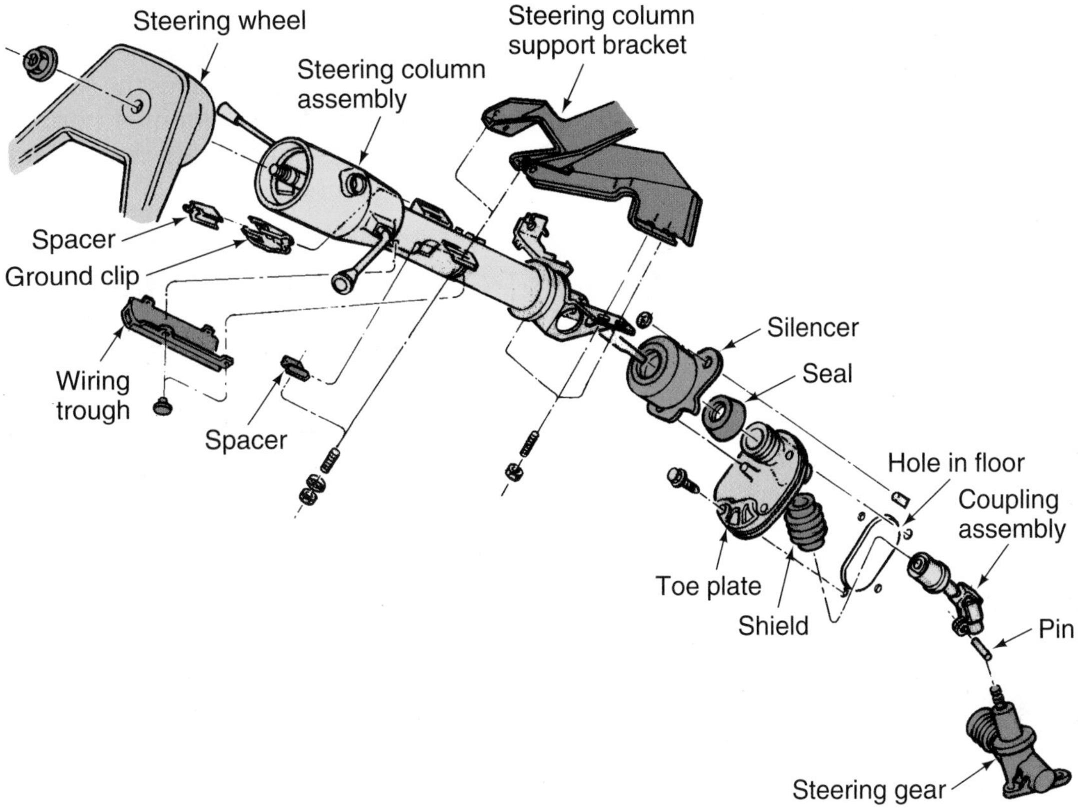

Figure 46-13 Typical steering column components. The steering wheel is splined to the shaft that extends through the column and down to the steering gearbox. *Courtesy of Chrysler LLC*

have a collapsible feature that allows the column to fold into itself on impact. This feature prevents injury to the driver.

In most vehicles equipped with a driver's side air bag, the air bag assembly is contained in the center portion of the steering wheel. This assembly must be disarmed and removed before the steering wheel can be removed.

WARNING!

Always disconnect the negative battery cable and wait one full minute before beginning to remove an air bag assembly. Failure to do this may result in accidental air bag deployment. (See Chapter 24 of this textbook.) Air bag service precautions vary depending on the vehicle. Always follow the manufacturer's service precautions and recommendations when working with an air bag system.

Differences in steering wheel and column designs include fixed column, telescoping column, tilt column, manual transmission, floor shift, and automatic transmission column shift. The tilt columns **(Figure 46–14)** feature at least five driving positions

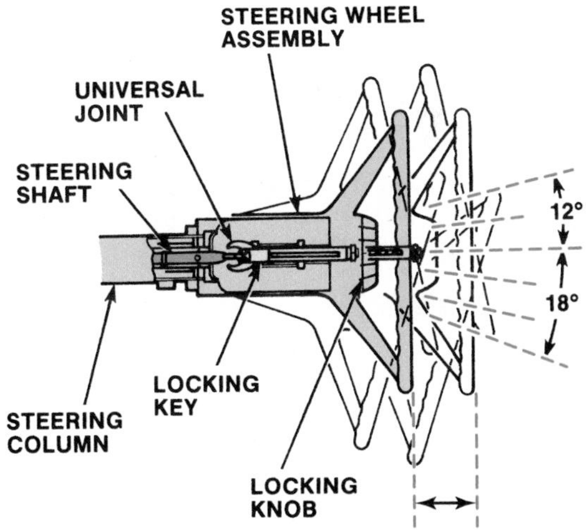

Figure 46-14 Tilt steering column operation.

(two up, two down, and a center position). Both fixed and tilt columns may house an emergency warning flasher control, a turn signal switch, ignition key, lights (high/low beams), horn, windshield wipers and washers, and an antitheft device that locks the steering system. On automatic-transmission-equipped vehicles, the transmission linkage locks also.

Methods used to lock the shaft to the tube include a breakaway plastic capsule or a series of inserts or steel balls held in a plastic retainer that allow the shaft to roll forward inside the tube. There are also

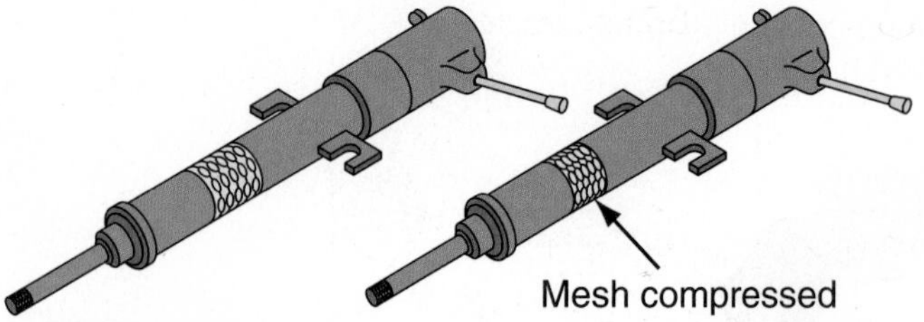

Figure 46–15 This figure shows the condition of the collapsible mesh of a steering column before and after an accident.

collapsible steel mesh **(Figure 46–15)** or accordion-pleated devices that give way under pressure. After the vehicle has been in an accident, the steering column should be checked for evidence of collapse. Although the car can be steered with a collapsed column that has been pulled back, the collapsed portion must be replaced. All service manuals provide explicit instructions for doing this.

The steering wheel is usually held in place on the steering column by either a bolt or nut.

Steering Damper

The purpose of a steering damper is simply to reduce the amount of road shock that is transmitted up through the steering column. Steering dampers are found mostly on 4WD, especially those fitted with large tires. The damper serves the same function as a shock absorber but is mounted horizontally to the steering linkage—one end to the center link and the other to the frame.

POWER-STEERING SYSTEMS

The power-steering unit is designed to reduce the amount of effort required to turn the steering wheel. It also reduces driver fatigue on long drives and makes it easier to steer the vehicle at slow road speeds, particularly during parking.

Power steering can be broken down into two design arrangements: conventional and nonconventional or electronically controlled. In the conventional arrangement, hydraulic power is used to assist the driver. In the nonconventional arrangement, an electric motor and electronic controls provide power assistance in steering.

There are several power-steering systems in use on passenger cars and light-duty trucks. The most common ones are the integral-piston, and power-assisted rack and pinion system **(Figure 46–16)**.

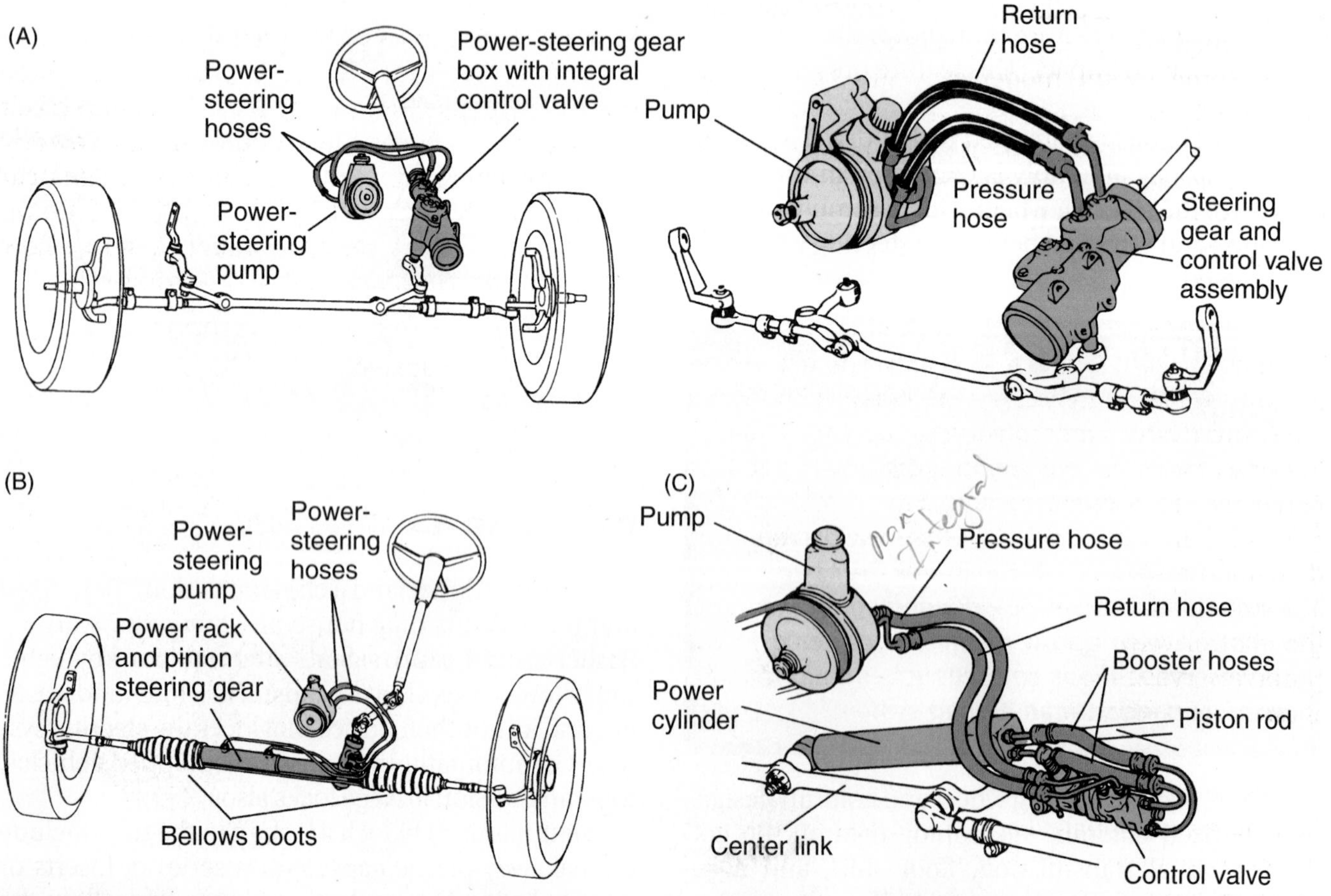

Figure 46–16 The three common power-steering systems: (A) Integral-piston linkage. (B) Rack and pinion. (C) External-piston linkage. *Courtesy of Federal-Mogul Corporation*

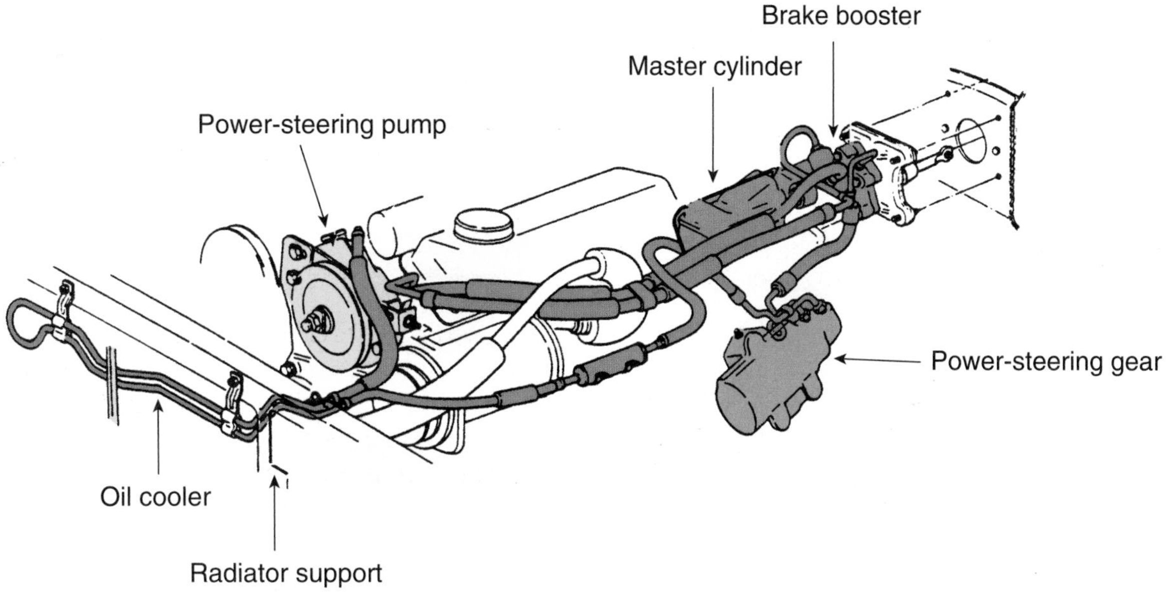

Figure 46-17 A typical hydro-boost system that uses the power-steering pump to power assist brake applications. *Courtesy of Ford Motor Company*

Integral Piston System

The **integral piston** system is the most common conventional power-steering systems in use today. It consists of a power-steering pump and reservoir, power-steering pressure and return hose, and steering gear. The power cylinder and the control valve are in the same housing as the steering gear.

On some recent model cars and light trucks, instead of the conventional vacuum-assist brake booster, the hydraulic fluid from the power-steering pump is also used to actuate the brake booster. This brake system is called the hydro-boost system **(Figure 46–17)**.

Power-Assisted Rack and Pinion System

The power-assisted rack and pinion system is similar to the integral system because the power cylinder and the control valve are in the same housing. The rack housing acts as the cylinder and the power piston is part of the rack. Control valve location is in the pinion housing. Turning the steering wheel moves the valve, directing pressure to either end of the back piston. The system utilizes a pressure hose from the pump to the control valve housing and a return line to the pump reservoir. This type of steering system is common in front-wheel-drive vehicles.

Components

Several of the manual-steering parts described earlier in this chapter, such as the steering linkage, are used in conventional power-steering systems. The components that have been added for power steering provide the hydraulic power that drives the system. They are the power-steering pump, flow control and pressure relief valves, reservoir, spool valves and power pistons, hydraulic hose lines, and gearbox or assist assembly on the linkage.

Power-Steering Pump The steering pump is used to develop hydraulic flow, which provides the force needed to operate the steering gear. The pump is belt driven from the engine crankshaft, providing flow any time the engine is running. It is usually mounted near the front of the engine **(Figure 46–18)**. The pump assembly includes a reservoir and an internal flow control valve. The drive pulley is normally pressed onto the pump's shaft.

Figure 46-18 A power-steering pump. *Courtesy of Visteon Corporation™*

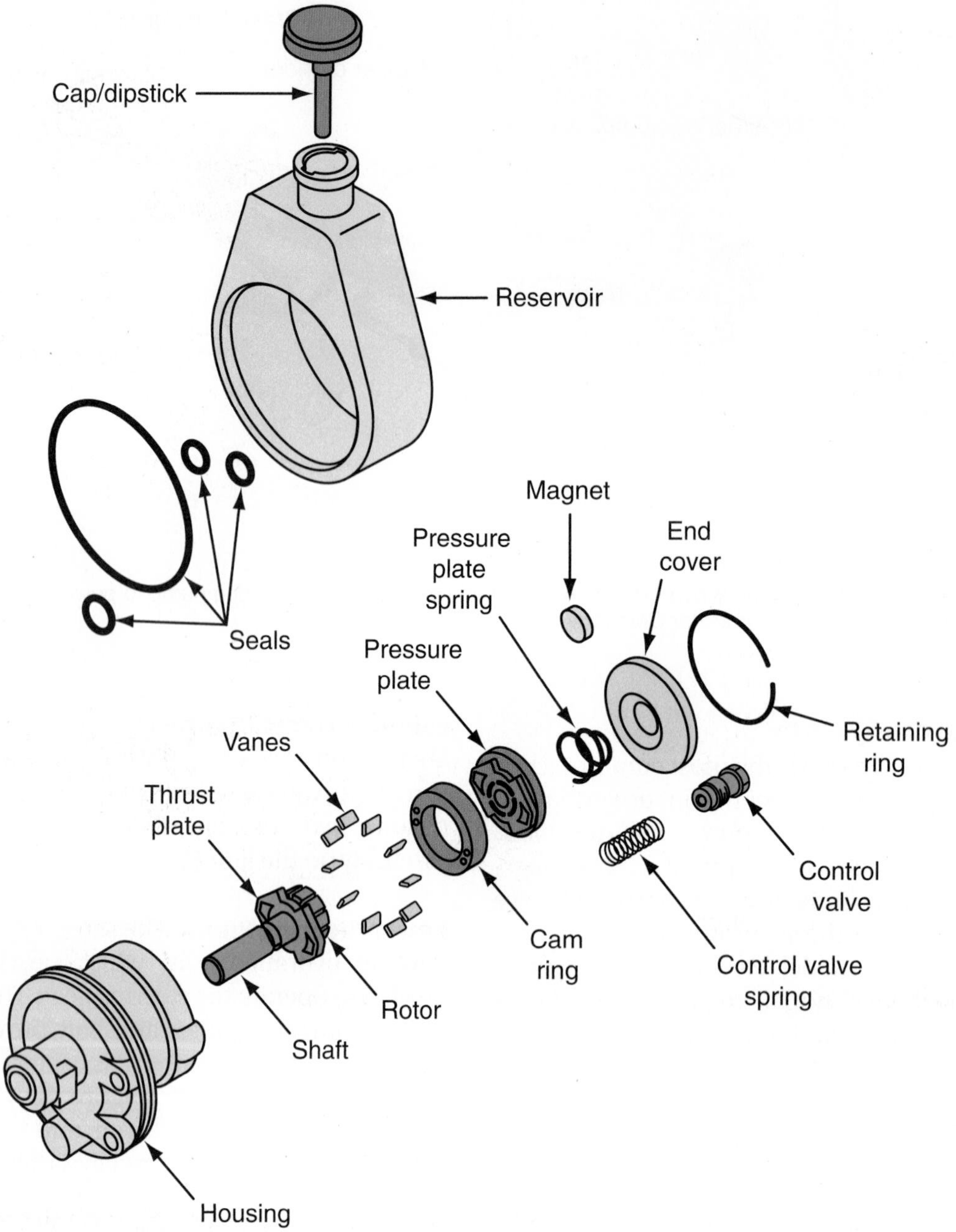

Figure 46–19 A vane-type power-steering pump.

There are four general types of power-steering pumps: roller, vane **(Figure 46–19)**, slipper, and gear. Functionally, all pumps operate in the same basic manner. Hydraulic fluid for the power-steering pump is stored in a reservoir. Fluid is routed to and from the pump by hoses and lines. Excessive pressure is controlled by a relief valve.

Power-Steering Pump Drive Belts Many power-steering pumps are driven by a belt that connects the crankshaft pulley to the power-steering pump pulley. Nearly all late-model vehicles use a serpentine belt. This belt may be used to drive all the belt-driven components. Most serpentine belts have a spring-loaded automatic belt tensioner that eliminates periodic belt tension adjustments. The smooth backside of the drive belt may also be used to drive some of the components. Older vehicles used a V-belt to drive the accessories, including the power-steering pump. Regardless of the type of belt, the belt tension is critical. A power-steering pump will never develop full pressure if the belt is slipping.

Electric Power Steering Many vehicles use a 12- or 42-volt electric motor mounted to or in the steering gear **(Figure 46–20)**. The motor replaces the conventional pump and its belts and hoses. These systems are discussed in more detail later in this chapter.

Flow Control and Pressure Relief Valves A pressure relief valve controls the pressure output from the

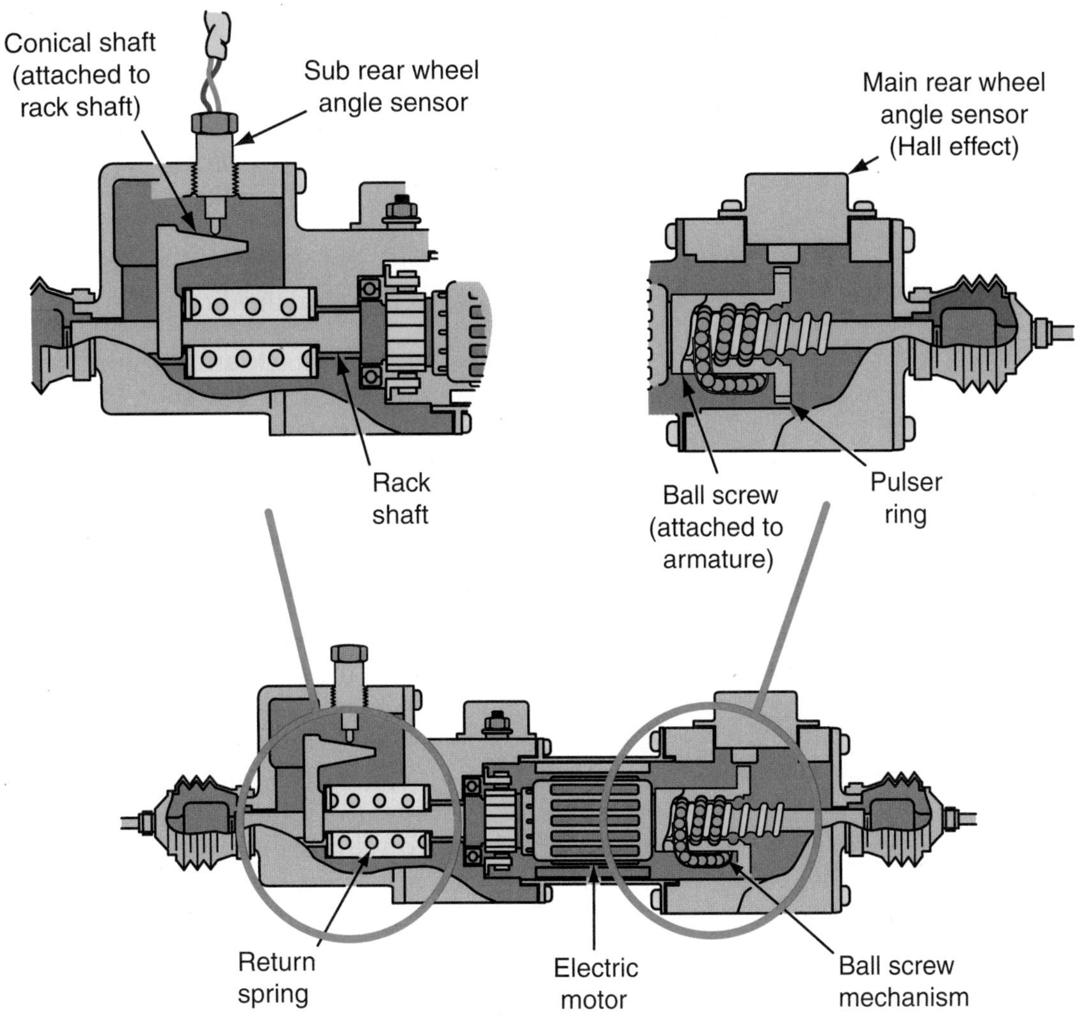

Figure 46-20 An electric/electronic rack and pinion system.

pump. This valve is necessary because of the variations in engine rpm and the need for consistent steering ability in all ranges from idle to highway speeds. It is positioned in a chamber that is exposed to pump outlet pressure at one end and supply hose pressure at the other. A spring is used at the supply pressure end to help maintain a balance.

As the fluid leaves the pump rotor, it passes the end of the flow control valve and is forced through an orifice that causes a slight drop in pressure. This reduced pressure, aided by the springs, holds the flow control valve in the closed position. All pump flow is sent to the steering gear.

When engine speed increases, the pump can deliver more flow than is required to operate the system. Since the outlet orifice restricts the amount of fluid leaving the pump, the difference in pressure at the two ends of the valve becomes greater until pump outlet pressure overcomes the combined force of supply line pressure and spring force. The valve is pushed down against the spring, opening a passage that returns the excess flow back to the inlet side of the pump.

A spring and ball contained inside the flow control valve are used to relieve pump outlet pressure. This is done to protect the system from damage due to excessive pressure when the steering wheel is held against the stops. Since flow in the system is severely restricted, the pump would continue to build pressure until a hose ruptured or the pump destroyed itself.

When outlet pressure reaches a preset level, the pressure relief ball is forced off its seat, creating a greater pressure differential at the two ends of the flow control valve. This allows the flow control valve to open wider, permitting more pump pressure to flow back to the pump inlet and pressure is held at a safe level.

Power-Steering Gearbox A power-steering gearbox is basically the same as a manual recirculating ball gearbox with the addition of a hydraulic assist. A power-steering gearbox is filled with hydraulic fluid and uses a control valve.

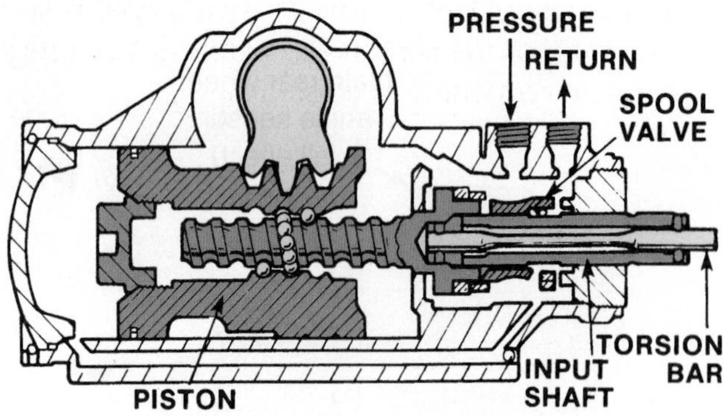

Figure 46-21 A torsion bar moves the spool valve to direct the oil flow to the piston.

In a power rack and pinion gear, the movement of the rack is assisted by hydraulic pressure. When the wheel is turned, the rotary valve changes hydraulic flow to create a pressure differential on either side of the rack. The unequal pressure causes the rack to move toward the lower pressure, reducing the effort required to turn the wheels.

The integral power steering has the spool valve and a power piston integrated with the gearbox. The spool valve directs the oil pressure to the left or right power chamber to steer the vehicle. The spool valve is actuated by a lever or a small torsion bar **(Figure 46–21)**.

In linkage systems, the control valve is connected directly to the steering center link and the pitman arm on the steering gear. Any movement of the steering wheel and the pitman arm compresses the centering spring and moves the valve spool. This opens and closes a series of ports directing fluid under pressure from the pump to one side or the other of the power cylinder piston.

The power cylinder in the linkage system is attached to the steering center link and the piston shaft is attached to the frame. As fluid under pressure is directed to one side of the piston by the control valve, power assist is provided to aid the driver in moving the steering linkage and road wheels. Two lines connect the cylinder to the control valve. Each one functions as both the return line or the supply line, depending on the direction of the turn.

Power-Assisted Rack and Pinion Steering Power-assisted rack and pinion components are basically the same as for manual rack and pinion steering **(Figure 46–22)**, except for the hydraulic control housing. As mentioned earlier, the power rack and pinion steering unit may be classified as integral. The rack functions as the power piston and the spool valve is connected to the pinion gear.

In a power rack and pinion gear, the piston is mounted on the rack, inside the rack housing. The rack housing is sealed on either side of the rack piston to form two separate hydraulic chambers for the left and right turn circuits. When the wheel is turned to the right, the rotary valve creates a pressure differential on either side of the rack piston. This causes the rack to move toward the lower pressure and reduces the total effort required to turn the wheels.

Power-Steering Hoses The primary purpose of power-steering hoses is to transmit power (fluid under pressure) from the pump to the steering gearbox, and to return the fluid ultimately to the pump reservoir. Hoses also, through material and construction, function as additional reservoirs and act as sound and vibration dampers.

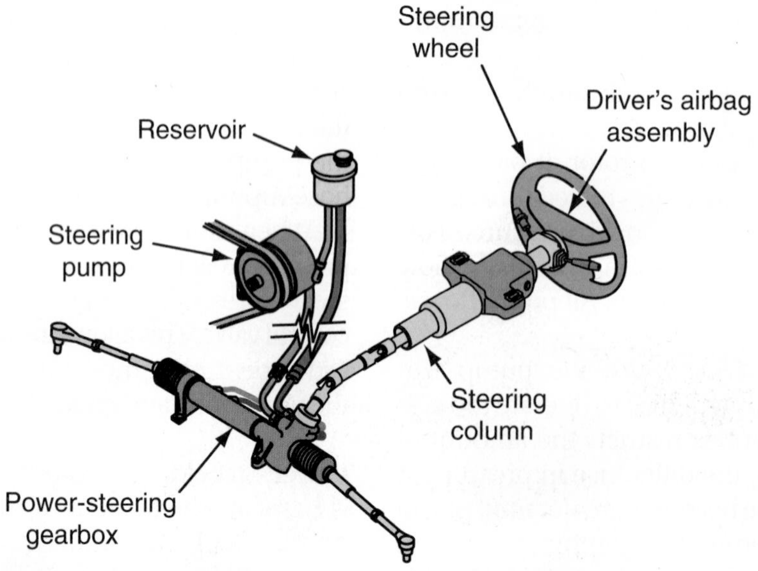

Figure 46-22 A complete power rack and pinion steering system.

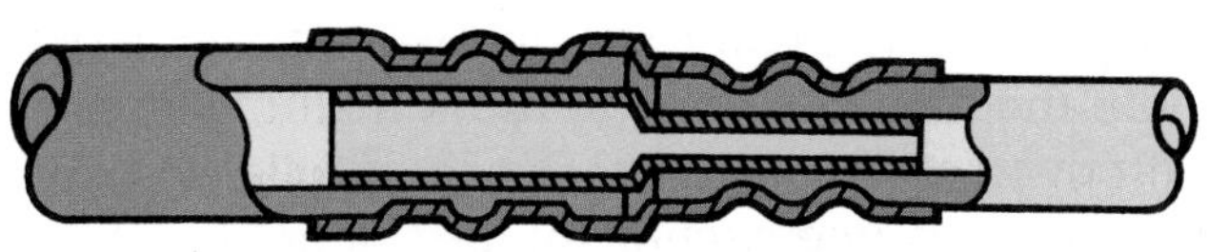

Figure 46-23 Power-steering hoses may have two internal diameters.

Hoses are generally a reinforced synthetic rubber (neoprene) material coupled to metal tubing at the connecting points. The pressure side must be able to handle pressures up to 1,500 psi (10,342 kPa). For that reason, wherever there is a metal tubing to a rubber connection, the connection is crimped. Pressure hoses are also subject to surges in pressure and pulsations from the pump. The reinforced construction permits the hose to expand slightly and absorb changes in pressure.

Two internal diameters of hose **(Figure 46–23)** may be used on the pressure side; the larger diameter or pressure hose is at the pump end. It acts as a reservoir and as an accumulator, absorbing pulsations. The smaller diameter or return hose reduces the effects of kickback from the gear itself. By restricting fluid flow, it also maintains constant backpressure on the pump, which reduces pump noise. If the hose is of one diameter, the gearbox is performing the damping functions internally.

Because of working fluid temperature and adjacent engine temperatures, these hoses must be able to withstand temperatures up to 300°F (150°C). Due to various weather conditions, they must also tolerate subzero temperatures as well. Hose material is specially formulated to resist breakdown or deterioration due to oil or temperature conditions.

CAUTION!

Hoses must be carefully routed away from engine manifolds. Power-steering fluid is very flammable. If it comes in contact with hot engine parts, it could start an underhood fire.

ELECTRONICALLY CONTROLLED POWER-STEERING SYSTEMS

The object of power steering is to make steering easier at low speeds, especially while parking. However, higher steering efforts are desirable at higher speeds in order to provide improved down-the-road feel. The electronically controlled power-steering (EPS) systems **(Figure 46–24)** provide both of these benefits. The hydraulic boost of these systems is tapered off by electronic control as road speed increases. Thus, these systems require well under 1 pound (4.4 N) of steering effort at low road speeds and 3 pounds plus (13.2 N) of steering effort at higher road speeds to enable the driver to maintain control of the steering wheel for improved high-speed handling.

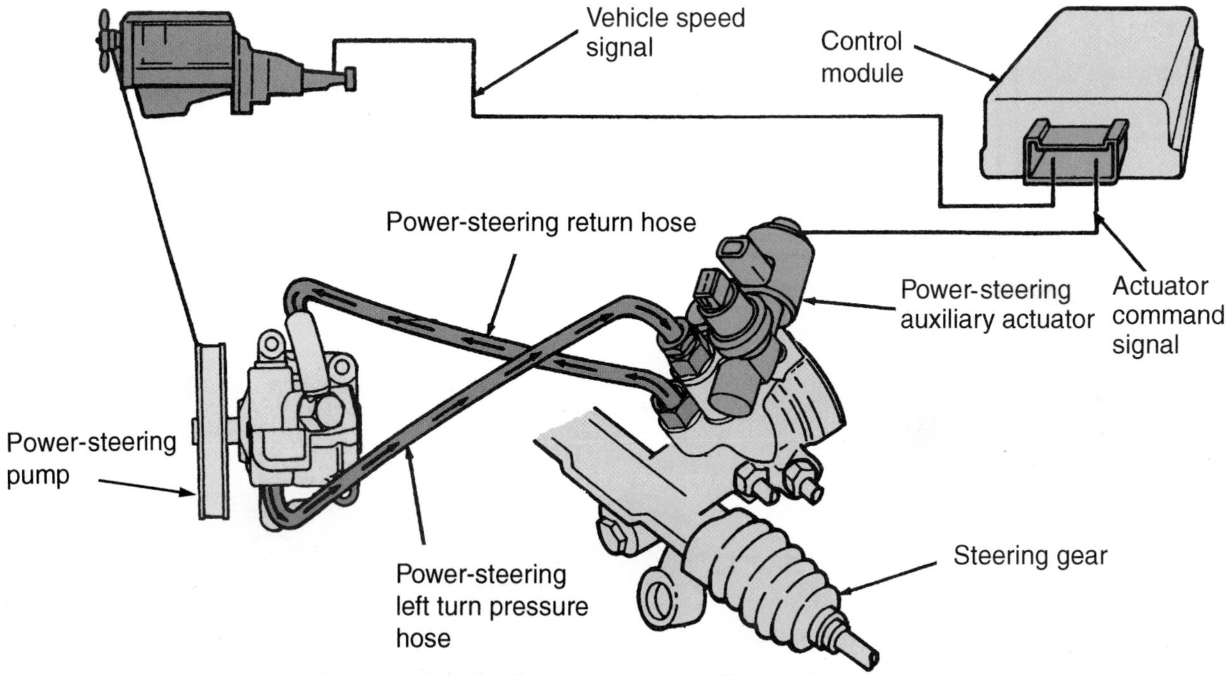

Figure 46-24 A variable-assist power-steering system. *Courtesy of Ford Motor Company*

A rotary valve electronic power-steering system consists of the power-steering gearbox, power-steering oil pump, pressure hose, and the return hose. The amount of hydraulic fluid flow (pressure) used to boost steering is controlled by a solenoid valve that is identified as its PCV (pressure control valve).

SHOP TALK

The PCV (pressure control valve) is not to be confused with the PCV (positive crankcase ventilator) used with emission controls systems.

The electronic power-steering system's PCV **(Figure 46–25)** is exposed to spring tension on the top and plunger force on the bottom. The plunger slips inside an electromagnet. By varying the electrical current to the electromagnet, the upward force exerted by the plunger can be varied as it works against the opposing spring. Current flow to the electromagnet is variable with vehicle road speed and, therefore, provides steering to match the vehicle's road speed.

General Motors' variable effort steering (VES) system relies on an input signal from the vehicle speed sensor to the VES controller to control the amount of power assist. The controller, in turn, supplies a pulse width modulated voltage to the actuator solenoid in the power-steering pump. The controller also provides a ground connection for the solenoid.

When the vehicle is operating at low speeds, the controller supplies a signal to cycle the solenoid faster so it allows high pump pressure. This provides for maximum power assist during cornering and parking. As the vehicle's speed increases, the solenoid cycles less and the pump provides a lower amount of assist. This gives the driver better road feel during high speeds.

Active Steering

Active steering improves vehicle stability by turning the wheels more or less sharply than commanded by the turn of the steering wheel during some situations. Through inputs and computer programming, this system can adjust the steering to respond quickly to the threat of skidding **(Figure 46–26)**. The system also allows for a variable steering ratio dependent on vehicle speed.

Current active steering systems are not true steer-by-wire systems. There is still a mechanical connection between the steering wheel and vehicle's wheels

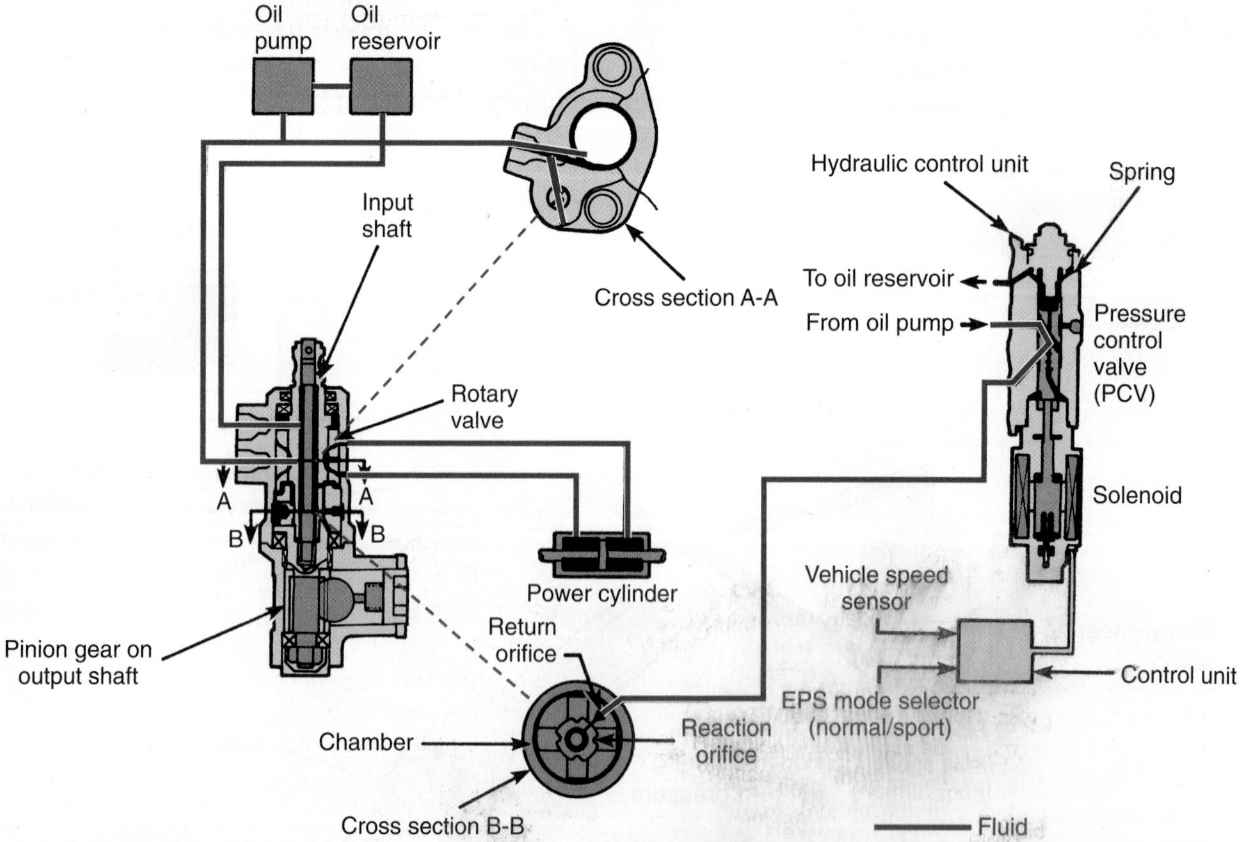

Figure 46-25 An outline of electronic power-steering components. The EPS PCV is exposed to spring tension and plunger force. *Courtesy of Mitsubishi Motors North America, Inc.*

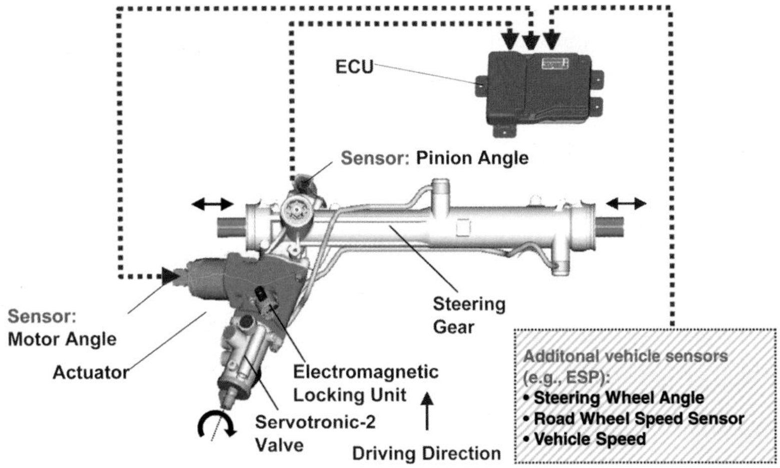

Figure 46-26 The main components and circuits of an active steering system. *Courtesy of Robert Bosch GmbH, www.bosch-presse.de*

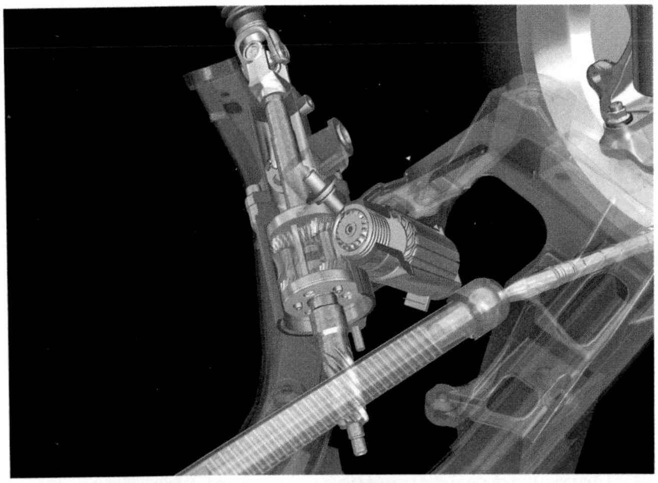

Figure 46-27 The planetary gear set and electric motor that turn the road wheels when the steering wheel is turned. *Courtesy of BMW of North America, LLC*

(**Figure 46–27**). The systems have an overriding drive built into the steering column. This drive is controlled by an electric motor, which is controlled by the system's computer. The computer determines whether the steering angle needs to be changed and by how much. If the system fails, the planetary gear unit will rotate directly with the steering wheel.

Electric/Electronic Rack and Pinion System

The electric/electronic rack and pinion unit replaces the hydraulic pump, hoses, and fluid associated with conventional power-steering systems with electronic controls and an electric motor located concentric to the rack itself. The design features a DC motor armature with a follow shaft to allow passage of the rack through it. The outboard housing and rack are designed so that the rotary motion of the armature can be transferred to linear movement of the rack through a ball nut with thrust bearings. The armature is mechanically connected to the ball nut through an internal/external spline arrangement.

The basis of system operation is its ability to change the rotational direction of the electric motor while being able to deliver the necessary amount of current to meet torque requirements at the same time. The system can deliver up to 75 amperes to the motor. The higher the current, the greater the force exerted on the rack. The direction of the turn is controlled by changing the polarity of the signal to the motor.

The field assembly houses permanent ceramic magnets while providing structural integrity for the gear system. In essence, the electronic/electric rack design allows for a direct power source to the rack and steering linkage. The system monitors steering wheel movement through a sensor mounted on the input shaft of the rack and pinion steering gear. After receiving directional and load information from the sensor, an electronic controller activates the motor to provide power assistance.

These units are readily retrofitted to conventionally equipped vehicles. As for servicing, there are currently no replacement parts available; therefore, if the rack should become faulty, the entire unit should be replaced. Rebuild kits, with complete installation instructions, are available.

Unlike conventional power steering, electric/electronic units provide power assistance even when the engine stalls, since the power source is the battery rather than the engine-driven pump. The feel of the steering can also be adjusted to match the particular driving characteristics of cars and drivers, from high performance to luxury touring cars. It also eliminates hydraulic oil, which means no leaks.

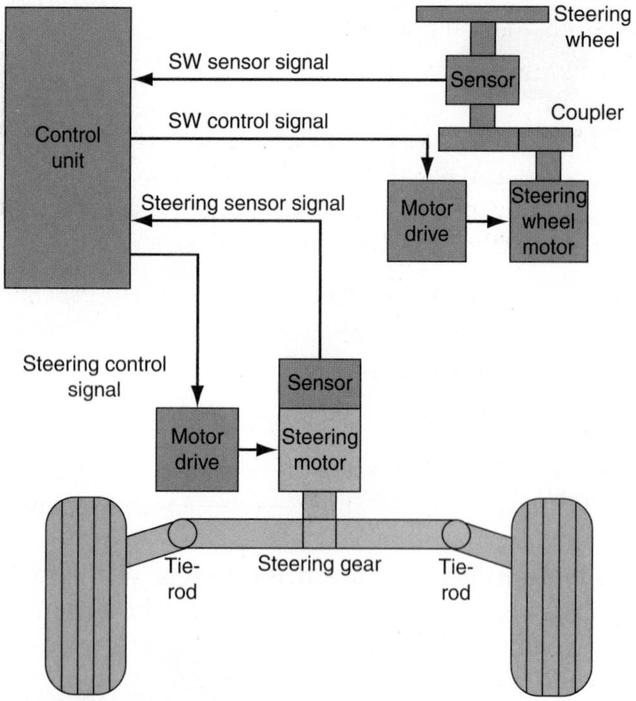

Figure 46-28 The basic layout for a steer-by-wire system.

Steer-by-Wire System

Steer-by-wire systems **(Figure 46–28)** are not found on any production vehicles today. They are being tested and have appeared on many concept cars. These systems do not use a steering column or shaft to connect the steering wheel to the steering gear. The system is totally electronic. The turning of the steering wheel is monitored by a sensor. The sensor sends an input signal to a controller. The controller, in turn, sends commands to an electric motor in the steering gear. The commands from the controller are also based on inputs from a variety of other inputs, such as vehicle speed.

These systems also have a small motor attached to the mount for the steering wheel. This motor is controlled by a steering controller. This motor provides the correct steering feel for the current conditions. The driver needs this feel to maintain control of the vehicle.

Steer-by-wire systems allow total customization of steering performance and can provide a constantly variable steering ratio. The absence of a steering column opens up space in the vehicle's interior and engine compartment. The systems are also lighter than conventional steering systems.

STEERING SYSTEM DIAGNOSIS

It is important to realize that many steering complaints are caused by problems in areas other than the steering system. A good diagnosis is one that finds the exact cause of the customer's complaint. Although customers may describe the problem in different ways, the most common complaints and their typical causes are discussed next.

Common Complaints

Excessive Steering-Wheel Play Excessive play in the steering wheel is apparent when there is too much steering-wheel movement before the wheels begin to turn. A small amount of play is normal.

This problem can be caused by the following:

- Loose, worn, or damaged steering linkages or tie-rod ends
- Worn ball joints
- Loose, worn, or damaged steering column U-joints
- Loose, worn, or damaged steering column bearings
- Damaged or worn steering gear
- Aerated fluid
- Loose steering gear bolts
- Faulty strut bearing or plate

Feedback When the driver feels the surface of the road through the steering wheel it is called feedback.

This problem can be caused by the following:

- Loose, worn, or damaged steering linkages or tie-rod ends
- Loose, worn, or damaged steering column U-joints
- Loose or damaged steering gear mounting bolts
- Damaged or worn steering column bearings
- Loose suspension bushings, fasteners, or ball joints

Hard Steering Obviously, a complaint of hard steering results when extra effort is needed to turn the steering wheel. This problem may be simply an absence of power assist. Hard steering problems can occur whenever the steering wheel is turned or just when it has been turned close to its limit.

This problem can be caused by the following:

- A faulty power-steering pump
- Damaged or faulty steering column bearings
- Seized steering column U-joints
- Steering gear set too tight or is binding
- Faulty suspension components
- Stuck flow control valve
- Inadequately inflated tires
- Restricted power-steering lines or hoses

Nibble This feeling is similar to a shimmy. Nibble results from the interaction of the tires with the road's surface. The customer's complaint may describe the nibble problem as slight rotational oscillations of the steering wheel.

This problem can be caused by the following:

- Loose, worn, or damaged steering linkages or tie-rod ends
- Loose, worn, or damaged suspension parts

Pulling or Drifting A pull is a tugging sensation felt at the steering wheel. The driver must push the steering wheel in the opposite direction of the pull to keep the vehicle going straight. Drifting is a condition in which the vehicle slowly moves to one side of the road when the driver's hands are taken off the steering wheel.

These problems can be caused by the following:

- Improper frame alignment
- Brake system problem
- Worn or binding suspension components, especially springs
- Poor wheel alignment
- Unevenly loaded or overloaded vehicle
- Loose, worn, or damaged steering linkages or tie-rod ends
- Out of balance steering gear valve
- Torque steer
- Tire problems
- Binding strut bearing

Shimmy When the wheels shimmy, the driver will feel large, consistent, rotational oscillations at the steering wheel. These motions are caused by the lateral movement of the tires.

This problem can be caused by the following:

- Loose, worn, or damaged steering linkages or tie-rod ends
- Loose, worn, or damaged suspension parts
- Out-of-balance tires
- Excessive wheel runout
- A bad tire
- Loose wheel bearings

Sticking Steering or Poor Return Poor returnability and sticky steering describes the steering wheel's resistance to return to center after a turn.

This problem can be caused by the following:

- Binding steering column U-joints
- Loose, worn, or damaged steering linkages or tie-rod ends
- Steering gear set too tight or is binding
- Loose, damaged, or worn suspension parts
- Poor wheel alignment
- Binding steering column bearings

Wandering When a vehicle wanders, the driver must constantly turn the steering wheel to the left and right to keep the vehicle going straight on a level road.

This problem can be caused by the following:

- Loose or worn suspension components
- Poor wheel alignment
- Unevenly loaded or overloaded vehicle
- Loose or damaged steering gear bolts
- Loose steering column U-joint bolts
- Loose, worn, or binding steering linkages or tie-rod ends
- Improper steering gear preload adjustment
- Leaking rack pistons

DIAGNOSIS

As with the diagnosis of any problem, your diagnosis should begin with trying to duplicate the customer's complaint. For steering problems, this is done on a road test; make sure you drive carefully and cautiously, especially because the vehicle has a control problem. It is very important that during the road test the vehicle is driven under conditions similar to the owner's normal driving. Try to duplicate the conditions on which the customer's concern is based. Before going on the road test, do a thorough safety inspection of the vehicle, including the tires.

Once the road test has been completed and it has been determined that there is an abnormal condition, use the symptom to identify the possible trouble area. Then check the parts in that area.

Noise Diagnosis

Often customers complain of abnormal noises or vibrations coming from the steering system. Pay attention to these during the road test. There can be many causes for these; some are not the steering system. The cause of these noises is best identified by paying close attention to where the noise is coming from. Some noises may be caused by tires or interference between the steering wheel and the steering column covers. Others can result from a faulty power-steering pump or system. **Figure 46–29** features a list of common noises and the possible problem areas.

A handy tool for identifying the exact cause of a noise in called the ChassisEar™ **(Figure 46–30)**. This is a wireless electronic device that can identify the

Symptom	Possible Cause
Drive belt squeal or chirp when moving the steering wheel from stop to stop	Loose or worn drive belt
Noise during a cold start	Blockage in the power-steering fluid reservoir Air in the steering hydraulic system
Steering grunt, growl, or shudder when turning into or out of a turn at low speeds	Air in the steering hydraulic system Restricted power-steering hoses
Steering system clunk	Air in the steering hydraulic system Steering column U-joints Steering gear
Steering gear squeak	Incorrect fluid in system Steering gear rotary seal Steering column components
Power-steering hiss or whistle	Steering column shaft or binding or misaligned Damaged or worn steering gear input shaft and valve Power-steering pump low relief pressure Restricted power-steering lines
Power-steering pump moan when the steering wheel is rotated to the stop position	Low fluid Air in the hydraulic system Power-steering fluid reservoir or screen is blocked or damaged Power-steering pump brackets loose or misaligned Bad steering gear isolators
Power-steering pump whine noise	Aerated fluid Damaged power-steering pump
Rattle, chuckle, or knocking noise or roughness is felt in the steering wheel when the vehicle is driven over rough surfaces	Steering column shaft/coupling joints damaged or worn Loose, damaged, or worn tie-rod ends Steering gear insulators or mounting bolts loose or damaged Steering column shaft/coupling bolts are loose Steering column damaged or worn Loose suspension bushings, bolts, or ball joints
Steering column rattle	Loose bolts or attaching brackets Loose, worn or insufficiently lubricated column bearings Steering shaft insulators damaged or worn Steering column shaft/coupling compressed or extended
Steering column squeak, cracks, or grinds	Poorly lubricated steering shaft bushings Loose or misaligned steering column shrouds Steering wheel rubbing against steering column shrouds Upper or lower bearing sleeves out of position

Figure 46-29 Common causes of steering noises.

source of a noise during a road test. The unit relies on inductive pickups that clamp on or near the component you want to listen to. The device also has a control unit with adjustable volume and headphones. The value of this tool is that it simply is hard to identify the source of a noise while you are driving.

Visual Inspection

Begin your visual inspection of the steering system by inspecting the tires. Check for correct pressure, construction, size, wear, and damage, and for defects that include ply separations, sidewall knots, concentricity problems, and force problems. Keep in mind that tire wear patterns are good indicators of steering and

Figure 46-30 The ChassisEar™ tool is valuable for identifying the source of steering and suspension noises. *Courtesy of JS Products, Inc.*

suspension problems (**Figure 46-31**). Tire wear is also a great indicator for wheel alignment problems, which is covered in Chapter 47.

Also check the tire and wheel assemblies for radial and lateral runout, static, and dynamic imbalance. Check the adjustment of the wheel bearings.

Continue your inspection by checking the drive belt for the power steering, if so equipped. Power-steering belt condition and tension are extremely important for satisfactory power-steering pump operation. A loose belt causes low pump pressure and hard steering. A loose, dry, or worn belt may cause squealing and chirping noises, especially during engine acceleration and cornering. The power-steering pump belt should be checked for tension, cracks, oil soaking, worn or glazed edges, tears, and splits. If any of these conditions are present, the belt should be replaced and adjusted properly.

Go To **Chapter 7 for procedures on checking, replacing, and adjusting belt tension.**

Also check the fluid level and condition. The fluid is checked at the pump reservoir with the dipstick attached to the reservoir cap. Before checking the fluid, allow the engine to run at idle for 2 or 3 minutes and cycle the steering wheel from lock to lock several times. This warms the fluid to its normal operating temperature and gives a more accurate reading.

Examine the condition of the fluid carefully. Check for evidence of contamination such as solid particles or water. If either of these conditions is present or the fluid has a burnt odor, the system should be flushed.

Also check the fluid for evidence of air trapped in the system. If the fluid looks foamy, it is likely that air is in the system. To verify this, run the engine until it reaches normal operating temperature. Then turn the steering wheel to the left and to the right several times without hitting the stops. If air is in the system, bubbles will appear in the fluid reservoir. Bleeding the system will remove the air.

With the ignition off, wipe off the outside of the power-steering pump, pressure hose, return hose,

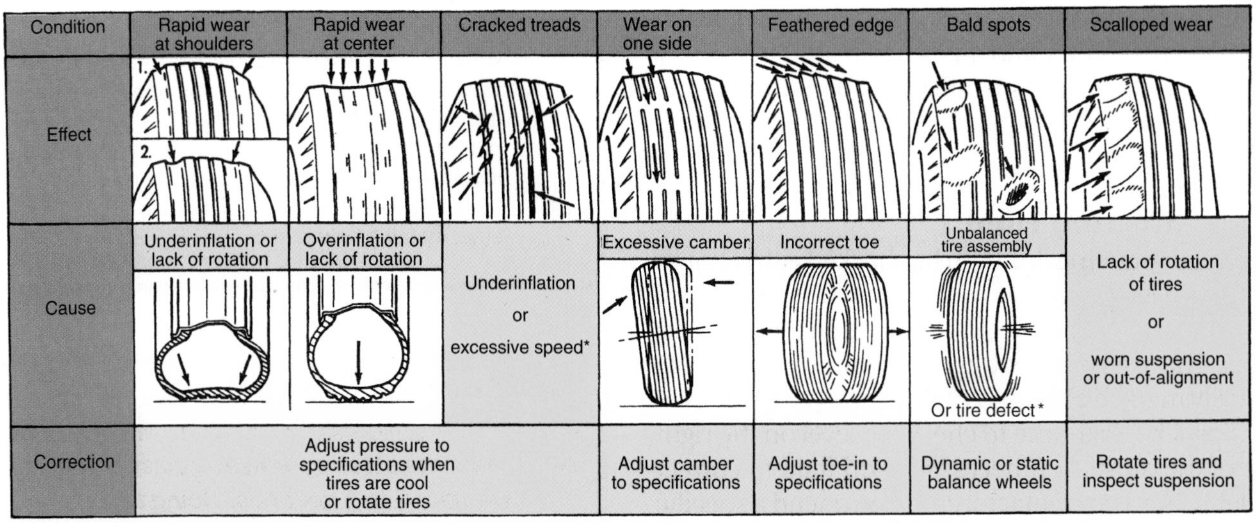

Condition	Rapid wear at shoulders	Rapid wear at center	Cracked treads	Wear on one side	Feathered edge	Bald spots	Scalloped wear
Effect	1. 2.						
Cause	Underinflation or lack of rotation	Overinflation or lack of rotation	Underinflation or excessive speed*	Excessive camber	Incorrect toe	Unbalanced tire assembly Or tire defect*	Lack of rotation of tires or worn suspension or out-of-alignment
Correction	Adjust pressure to specifications when tires are cool or rotate tires			Adjust camber to specifications	Adjust toe-in to specifications	Dynamic or static balance wheels	Rotate tires and inspect suspension

*Have tire inspected for further use.

Figure 46-31 Tire wear patterns. *Courtesy of Chrysler LLC*

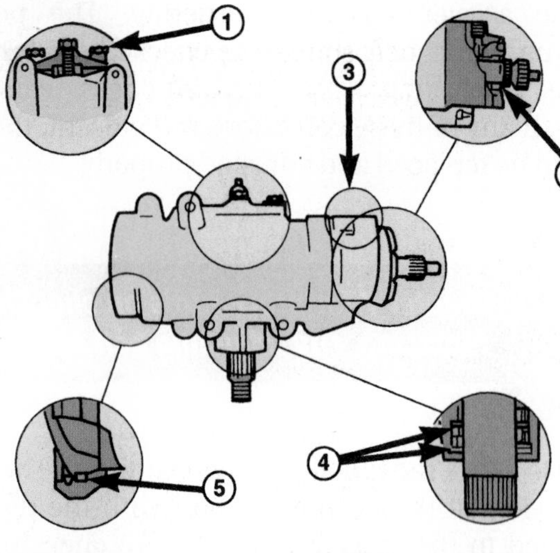
1
2
3
4
5
1. SIDE COVER LEAK - TORQUE SIDE COVER BOLTS TO SPECIFICATION. REPLACE THE SIDE COVER SEAL IF THE LEAKAGE PERSISTS.
2. ADJUSTER PLUG SEAL - REPLACE THE ADJUSTER PLUG SEALS.
3. PRESSURE LINE FITTING - TORQUE THE HOSE FITTING NUT TO SPECIFICATIONS. IF LEAKAGE PERSISTS, REPLACE THE SEAL.
4. PITMAN SHAFT SEALS - REPLACE THE SEALS.
5. TOP COVER SEAL - REPLACE THE SEAL.

Figure 46-32 Points for fluid leaks at a steering gear. *Courtesy of Chrysler LLC*

fluid cooler, and the steering gear **(Figure 46–32)**. Start the engine and turn the steering wheel several times from stop to stop. Check for leaks **(Figure 46–33)**. Fluid leakage will cause abnormal noises and may result in unequal and abnormal steering efforts. If there are no signs of leakage initially, repeat the test several times. While doing this, look at the hoses for signs of swelling. Always replace power-steering hoses with an exact replacement hose. Never attempt to patch or seal a leak in a hose or the hose's fittings.

On all systems, carefully check all mechanical parts of the steering and suspension systems. Many suspension problems can affect the operation of the steering system. If any part is found to be defective it should be replaced.

Power-Steering Fluid

The power-steering fluid can be checked either hot or cold. Fluid level will vary with temperature, however, and a more accurate check is done when the engine is warm. The reservoir cap has a dipstick **(Figure 46–34)** typically marked HOT and COLD on opposite sides of the dipstick. Make sure to check the level on the right side of the dipstick. If necessary, add fluid to correct the level. Some manufacturers recommend a specific fluid for use in a power-steering system. However, most recommend a specific type of ATF. Always check the service information before installing fluid to the system.

SPECIFIC CHECKS

If preliminary checks did not reveal the cause of the customer's concern, detailed checks of the various subsystems and parts should be done.

Power-Steering Pressure Checks

Many steering problems are caused by incorrect pressures in the power-steering system. Most late-model vehicles have a power-steering pressure sensor that can be monitored with a scan tool. In others you must connect a power-steering pressure gauge to the system.

PROCEDURE

Follow these steps to check the pressure in a power-steering system using a scan tool:

STEP 1 Place a thermometer in the power-steering fluid reservoir.

STEP 2 Connect the scan tool.

STEP 3 Set the tool to monitor the power-steering pressure sensor PID.

STEP 4 Start the engine.

STEP 5 With the engine at idle, raise the power-steering fluid temperature to the specified amount by rotating the steering wheel fully to the left and right several times.

STEP 6 With the steering wheel in the straight-ahead position and the engine speed at idle record the pressure reading.

STEP 7 Compare the readings to specifications.

If the pressure reading is higher than specifications, check the power-steering lines and hoses for restrictions.

STEP 8 Turn the steering wheel to the left and right stops. Record the pressure readings at each stop. The pressure reading at both stops should be almost the same. If the pressure reaches the maximum pump pressure at one stop but not the other, install a new steering gear. If the pressure does not reach the maximum pump pressure at either side, install a new pump.

CAUTION!

Never hold the steering wheel against its stops for more than 3 to 5 seconds. Doing this can damage the pump.

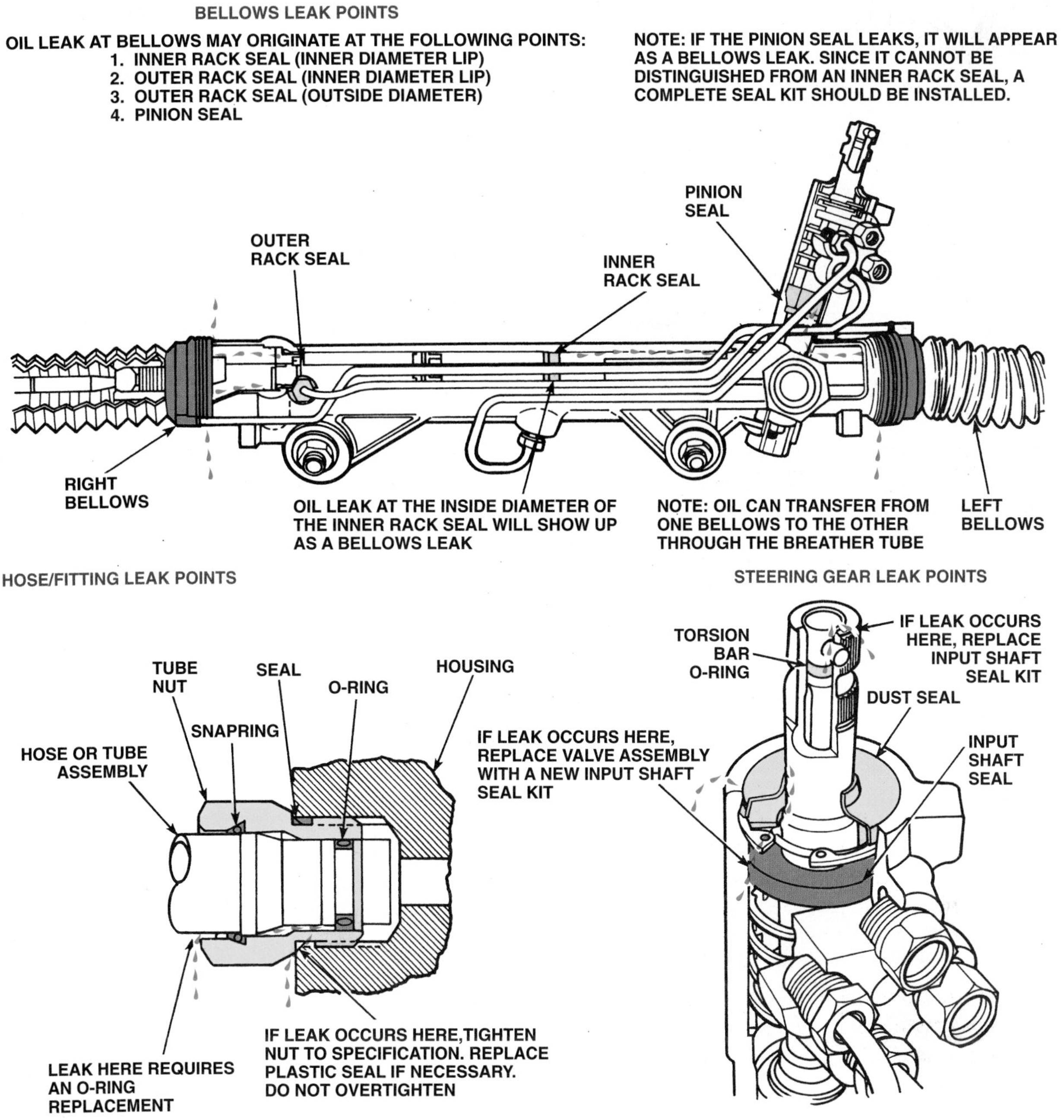
BELLOWS LEAK POINTS
OIL LEAK AT BELLOWS MAY ORIGINATE AT THE FOLLOWING POINTS:
1. INNER RACK SEAL (INNER DIAMETER LIP)
2. OUTER RACK SEAL (INNER DIAMETER LIP)
3. OUTER RACK SEAL (OUTSIDE DIAMETER)
4. PINION SEAL
NOTE: IF THE PINION SEAL LEAKS, IT WILL APPEAR AS A BELLOWS LEAK. SINCE IT CANNOT BE DISTINGUISHED FROM AN INNER RACK SEAL, A COMPLETE SEAL KIT SHOULD BE INSTALLED.
PINION SEAL
OUTER RACK SEAL
INNER RACK SEAL
RIGHT BELLOWS
OIL LEAK AT THE INSIDE DIAMETER OF THE INNER RACK SEAL WILL SHOW UP AS A BELLOWS LEAK
NOTE: OIL CAN TRANSFER FROM ONE BELLOWS TO THE OTHER THROUGH THE BREATHER TUBE
LEFT BELLOWS
HOSE/FITTING LEAK POINTS
TUBE NUT
SEAL
O-RING
HOUSING
SNAPRING
HOSE OR TUBE ASSEMBLY
LEAK HERE REQUIRES AN O-RING REPLACEMENT
IF LEAK OCCURS HERE,TIGHTEN NUT TO SPECIFICATION. REPLACE PLASTIC SEAL IF NECESSARY. DO NOT OVERTIGHTEN
STEERING GEAR LEAK POINTS
TORSION BAR O-RING
IF LEAK OCCURS HERE, REPLACE INPUT SHAFT SEAL KIT
DUST SEAL
IF LEAK OCCURS HERE, REPLACE VALVE ASSEMBLY WITH A NEW INPUT SHAFT SEAL KIT
INPUT SHAFT SEAL

Figure 46-33 Possible leakage points on power-steering systems.

Testing Systems with a Pressure Tester With the engine off, place a drain pan under the vehicle in order to catch any power-steering fluid. Disconnect the pressure hose at the pump. Install the pressure gauge between the pump and the steering gear and bleed the system.

Run the engine for about 2 minutes, and then stop the engine and add fluid to the power-steering pump if necessary. Restart the engine and allow it to idle. Turn the steering wheel and briefly hold it against the steering stops in order to release any trapped air from the system. Observe the pressure reading. The readings should be about 30 to 80 psi (200 to 550 kPa). If the pressure is low, the pump may be faulty. If the pressure is too high, the problem may be restricted hoses.

Continue testing by closing the shut-off valve on the gauge and observe the pressure reading **(Figure 46–35)**. When the valve is closed, the pressure should increase. If the pressure is too high, there may be a faulty flow control valve. If the pressure is too low, the pump may be bad. If the flow control valve is suspect, remove and inspect it. If there are any burrs or scratches on the valve, replace it. Do *not* attempt to clean it. Also, inspect the valve's bore. If there are any burrs or scratches, replace the power-steering pump.

After testing and verifying the repairs, remove the gauge, reconnect the lines, refill the system, bleed it, and refill it again.

Figure 46-34 Check the fluid level with the dipstick attached to the cap.

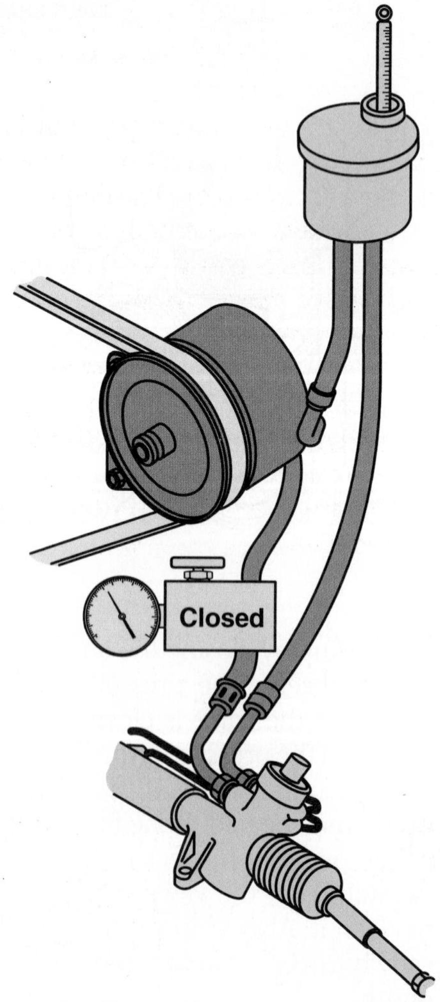

Figure 46-35 The final step when checking power-steering pump pressure is to close the tester's valve and observe the pressure in the system. Never keep the valve closed for more than 5 seconds.

Electric Power-Steering Systems

Many vehicles, including all hybrid vehicles, have electric power steering. These systems do not have a power-steering pump and require no fluid services. Because these systems are electronically controlled, diagnosis is performed in the typical way. There is a warning lamp on the dash that comes on when a problem is detected. A DTC will also be set. In most cases, when the warning lamp is lit, there will be no power assist available. However, the system will allow for manual steering.

The system is constantly (from key on to key off) monitored by the control module. If a problem is detected, the EPS lamp may stay lit after the ignition is turned on, or it may come on while it is being driven. To identify the reason for the illumination of the lamp, interview the customer to find out when and where it first came on. Try to duplicate the situation during a road test and then retrieve any and all DTCs. If the problem cannot be duplicated, do a careful inspection of all associated wiring and connectors.

Steering problems in an EPS can be caused by the same things as a conventional steering system. However, power assist is provided by electronics, not hydraulics. Therefore, power-steering problems are related to the electronic/electrical components. If the driver complains about how hard it is to turn the steering wheel, the problem can be caused by power-steering motor, speed sensor, power-steering control unit, or electrical circuit problems.

If the steering effort does not change with vehicle speed or is greater when turning left or right, suspect a faulty power-steering motor, speed sensor, or power-steering control unit. A defective power-steering motor may be evident by a high-pitched sound that occurs when the vehicle is stopped and the steering wheel is turned slowly.

On some models, if the power-steering motor gets too hot, the control module will decrease the current flow to it. This can cause increased steering effort. The systems do this to prevent permanent damage to the motor. As soon as the motor's temperature drops, normal current flow resumes. Therefore, a temporary loss of full assist may be normal.

Pitman Arm

Because of its function, the pitman arm is the most heavily stressed point in the system. To inspect the pitman arm, grasp it and vigorously shake it to detect any looseness. Check the socket to reveal any damage or looseness. Either condition must be corrected by replacing the worn part. Their removal normally requires the use of a special puller **(Figure 46–36)**.

Figure 46–36 A pitman arm puller. *Courtesy of Federal-Mogul Corporation*

Idler Arm

A worn or damaged idler arm can cause steering instability, uneven tire wear, front-end shimmy, hard steering, excessive play in steering, or poor returnability. Because an idler arm is the weakest link in a parallelogram steering system, it wears more quickly than the rest of the system.

The procedure is simple for checking an idler arm for looseness or wear. The suspension should be normally loaded on the ground or on an alignment rack. When raised by a frame contact hoist, the vehicle's steering linkage is allowed to hang, and proper testing cannot be done. Check the idler arm ends for worn sockets or deteriorated bushings. Grasp the center link firmly with your hand at the idler arm end. Push up with approximately a 25-pound (110-N) load. Pull down with the same load. The allowable movement of the idler arm and support assembly in one direction is ⅛ inch (3 mm), for a total acceptable movement of ¼ inch (6 mm). The load can be accurately measured by using a dial indicator or pull-spring scale located as near the center link end of the idler arm as possible. Keep in mind that the test forces should not exceed 25 pounds (110 N), as even a new idler arm might be forced to show movement due to steel flexing when excessive pressure is applied. It is also necessary that a scale or ruler be rested against the frame and used to determine the amount of movement. Observers tend to overestimate the actual movement when a scale is not used. The idler arm should always be replaced if it fails this test. Jerking the right front wheel and tire assembly back and forth (causing an up-and-down movement in the idler arm) is not an acceptable method of checking, as there is no control on the amount of force being applied.

Center Link

Worn or bent center links can cause front-end shimmy, vehicle pull to one side, or change in the toe setting, causing excessive tire wear.

When inspecting the center link, look closely to insure it has not been bent or damaged. Grasp the center link firmly and try moving it in all directions. Any movement, or sign of damage, is reason for replacement. Tapered openings seldom wear but should be checked for enlargement caused by a loose connection. If necessary, replace the center link.

Tie-Rod Assembly

Worn tie-rod ends result in incorrect toe-in settings, scalloped and scuffed tires, wheel shimmy, understeering, or front-end noise and tire squeal on turns.

Tie-rod end and center link inspections are similar. Grasp the tie-rod end firmly. Push vertically with the stud, and inspect for movement at the joint with the steering knuckle. Any movement over ⅛ inch (3 mm) or observation of damaged or missing parts, such as seals, is sufficient evidence that replacement is necessary.

Adjusting sleeves resemble a piece of internally threaded pipe. They have a slot or separation that runs either their entire length or just part way. Adjusting sleeves also have two crimping or squeezing clamps located at each end to lock the toe adjustment. Badly rusted, worn, or damaged adjusting sleeves should be replaced.

An additional check of the tie-rods can be made by rotating each tie-rod end to feel for roughness or binding, which could indicate that the socket has probably rusted internally. A special puller is often required to separate a tie-rod end from the steering knuckle (**Figure 46–37**).

Steering Damper

The steering dampers found in some steering linkage designs are generally nonadjustable, nonrefillable, and not repairable. At each inspection interval, inspect the mountings and check the assembly for damage (such as being bent) and fluid leaks. A light film of fluid is evidence of fluid leakage. However, a light film of fluid is permissible on the body of the damper near the shaft seal. A dripping damper should be replaced. A bad steering damper may cause wheel shimmy even though the rest of the suspension and steering system is fine.

Dry Park Check

An excellent overall check for worn or loose conventional steering components is the **dry park check**.

Figure 46-37 A tie-rod end-separating tool. *Courtesy of Federal-Mogul Corporation*

With the full weight of the vehicle on the wheels, have an assistant rock the steering wheel back and forth. Start your inspection from one side to the other side. Note any looseness in tie-rod, center link, idler arm, or pitman arm sockets **(Figure 46–38)**. If a second person is not available, reach up under the vehicle and grasp the flexible coupling on the steering shaft. Rock the linkage.

Before assembling any steering linkage parts, thoroughly check all tapered holes for out-of-roundness and wear. Thoroughly clean all bores that the stud tapers mount in. On new and reused parts, firmly install the tapered stud into its tapered hole. The stud must seat firmly without rocking. Only thread should protrude from the hole. If the parts do not meet these requirements, the mating part is worn and must be replaced, or the correct parts are not being used. Always follow the manufacturer's stud and mounting bolt torque specifications when installing chassis parts.

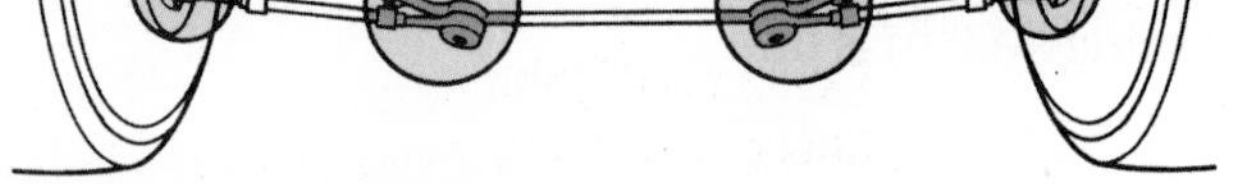

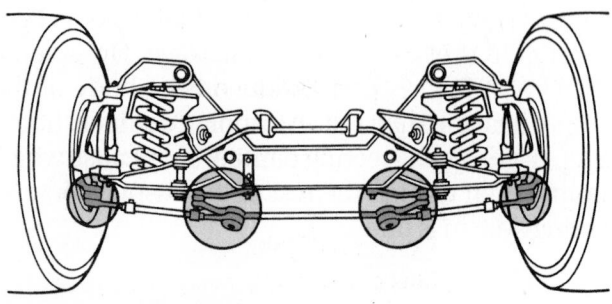

Figure 46-38 The circled areas indicate where a dry park check of steering linkage should be made.

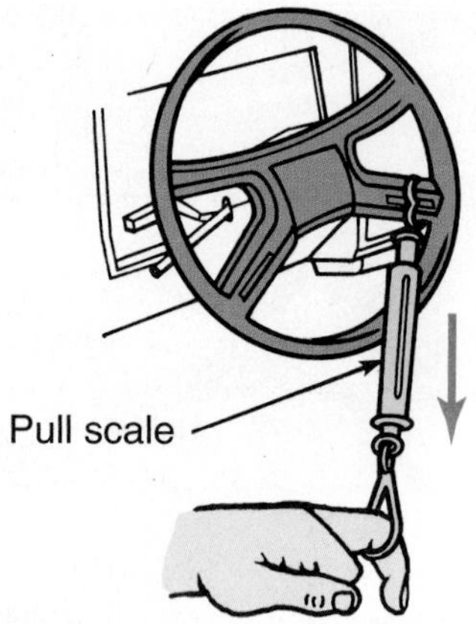

Figure 46-39 A pull scale is used to measure steering effort. *Courtesy of Toyota Motor Sales, U.S.A., Inc.*

Turning Effort

If an owner's concern indicates excessive turning effort, a pull scale should be used to read the actual force required to turn the wheel **(Figure 46–39)**. Compare the test results to the specifications in the service manual. If the effort exceeds the maximum, carefully inspect the entire steering system before performing a pressure test.

Tie-Rod Articulation Effort

The effort required to move the tie-rod or its inner ball socket should be checked with a pull scale if excessive steering effort or looseness is noted during the road test. If the effort is not within the specified limits, the tie-rod must be replaced.

Worm and Roller Steering

Since the worm and roller steering linkage components are almost identical to those of a parallelogram linkage, the same methods of inspection are used.

Rack and Pinion Steering

A rack and pinion system has no idler or pitman arms and no center link. Instead, they are replaced with a rack. Because the rack has no wear points, the number of wear points on rack and pinion systems is reduced to four—each of the tie-rod ends. Tie-rod ends are also wear points on the parallelogram steering system. Power rack and pinion assemblies should be carefully checked for leaks. If leaks cause the pump to run out of fluid, the pump will be damaged.

In order to solve customer complaints, a very thorough inspection of the entire system is needed. Everything, including ball joints, tires, outer tie-rods, bellows boots, inner tie-rods, rack-mounting bushings, mounting bolts, steering couplings, and gearbox adjustment must be checked. Rack and pinion steering inspection must be very thorough because of the system's sensitivity.

PROCEDURE

Rack and Pinion Steering Inspection

STEP 1 Check all working components **(Figure 46–40)** of the systems. Inspect the flexible steering coupling or the universal joints for wear or looseness. If any play is found, recommend replacement. Universal joints can also seize or bind. They should be checked closely.

STEP 2 Grasp the pinion gear shaft at the flexible steering coupling and try to move it in and out of the gear. If there is movement, the pinion bearing preload might need adjustment. If there is no adjustment, internal components have to be replaced.

STEP 3 Carefully inspect the rack housing. In most cases, the rack and pinion steering assemblies are mounted in rubber bushings. As the vehicle gets older, these mounting bushings deteriorate from heat, age, and oil leakage from the engine. When this happens, the housing moves within its mounting and causes loose and erratic steering. Also, be alert for excessive movement of the rack housing. Stiffness in steering can be caused by a bent rack assembly, tight yoke bearing adjustment, loose power-steering belt, weak pump, internal leaks in the power-steering system, and damaged CV joints in front-wheel-drive vehicles.

STEP 4 Check the inner tie-rod socket assemblies located inside the bellows. The most foolproof way of checking these sockets is to loosen the inner bellows clamp and pull the bellows back, giving a clear view of the socket. During the dry park check, observe any looseness. The inner tie-rod socket can also be checked by squeezing the bellows boot until the inner socket can be felt. Push and pull on the tire. If looseness is found in the tie rod, it should be replaced. On some vehicles the boot might be made of hard plastic. For this type of boot, lock the steering wheel and push and pull on the tire. Watch for in-and-out movement of the tie rod. If movement is observed, replace the inner tie rod.

One fact to keep in mind is that the condition of the bellows boot determines the life of the inner socket. The bellows boot protects the rack from contamination. It might also contain fluid that helps keep the rack lubricated. If any cracks, splits, or leaks exist, the boot should be replaced. Also, be sure that clamps for the bellows are in their proper place and fastened tightly.

STEP 5 Inspect the outer tie-rod ends. In addition to the dry park check, grab each end and rotate to feel for any roughness that would indicate internal rusting. Be sure to check for bent or damaged forgings and studs, split or deteriorated seals, and damaged, out-of-round, or loose tapers. If any of these conditions exist, the parts should be replaced.

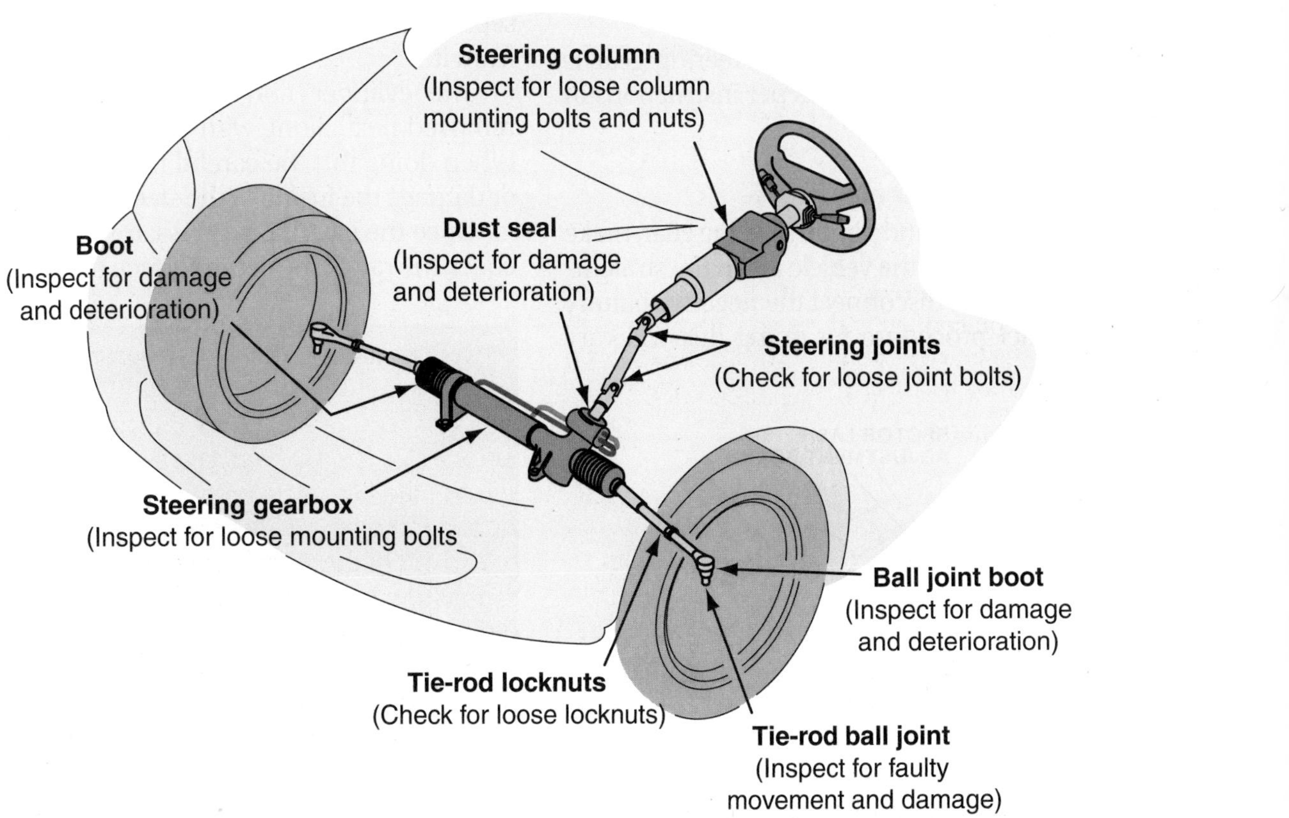

Figure 46–40 All steering parts should be carefully checked during diagnosis.

STEERING SYSTEM SERVICING

When a steering system component is found to be faulty, it is replaced. Most often part replacement is quite straightforward but you should always refer to the service manual before proceeding. At times, diagnosis will indicate a need to adjust the steering gear or inspect and repair the steering column.

Steering Gear Adjustments

Before any adjustments are made or servicing procedures performed to the steering gear, a careful check should be made of front-end alignment, shock absorbers, wheel balance, and tire pressure for possible steering system problems.

Before adjusting or servicing a manual steering gear, the technician must disconnect the battery ground cable. Raise the vehicle with the front wheels in the straight-ahead position. Remove the pitman arm nut. Mark the relationship of the pitman shaft. Remove the pitman arm with a pitman arm puller. Loosen the steering gear adjuster plug lock nut and back the adjuster plug off one-quarter turn (**Figure 46–41**). Remove the horn shroud or button cap. Turn the steering wheel gently in one direction until stopped by the gear; then turn back one-half turn. Measure and record bearing drag by applying a torque wrench with a socket on the steering wheel nut and rotating through a 90-degree arc. Check the service manual for the correct amount of drag.

Once these steps are taken, the steering gear is ready for adjusting or servicing as per instructions in the vehicle's service manual.

Rack and Pinion Service

When removing a rack and pinion steering gear, make sure the front wheels of the vehicle are in the straight-ahead position. Also, disconnect the negative battery cable. The exact procedure for doing this will vary with the model and make of the vehicle, always refer to the procedures outlined by the manufacturer. A typical procedure includes removing the wheels and disconnecting all linkages, hydraulic lines, and electrical connectors from the rack. In some cases, the engine and/or transaxle must be removed.

To disassemble and inspect the assembly, the rack assembly should be secured in a vise or the special tool recommended by the manufacturer. Once it is secured, all tubes can be removed, along with the tie-rod ends and boots. Check the boots for cracks or signs of leakage; replace them, if damaged.

WARNING!

It is easy to damage the boots and the rack while disassembling the unit, so be careful and adhere to all of the manufacturer's recommendations.

Make an index mark on the outer tie-rod end, jam nut, and tie-rod (**Figure 46–42**). Remove the rack guide spring cap, compression spring, and rack guide subassembly. Then remove the O-ring from the rack guide spring cap and the dust cover. Now remove the control valve assembly and subassembly. In some cases the subassembly is removed by pressing it away from the main assembly. Once the subassembly is separated, remove the snaprings, seals, and spacers from it.

The cylinder and "stoppers" in the rack are removed next, along with the steering rack bushing. When doing this, be careful not to drop the bushing or damage the inside of the steering gear housing.

Once the steering gear assembly is disassembled, check the rack. Check the surface of the rack for wear

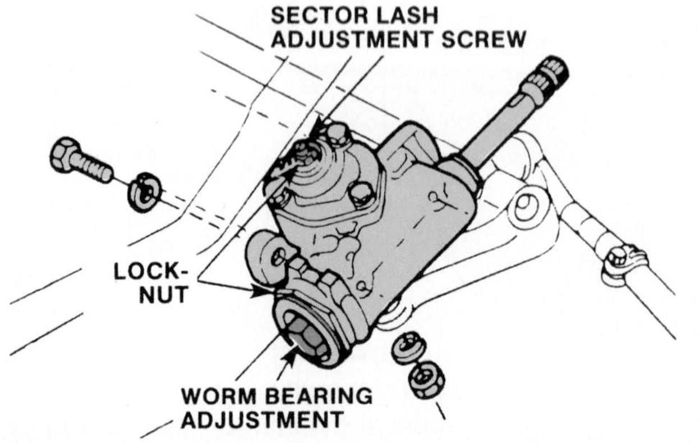

Figure 46-41 Typical steering gear adjustment points.

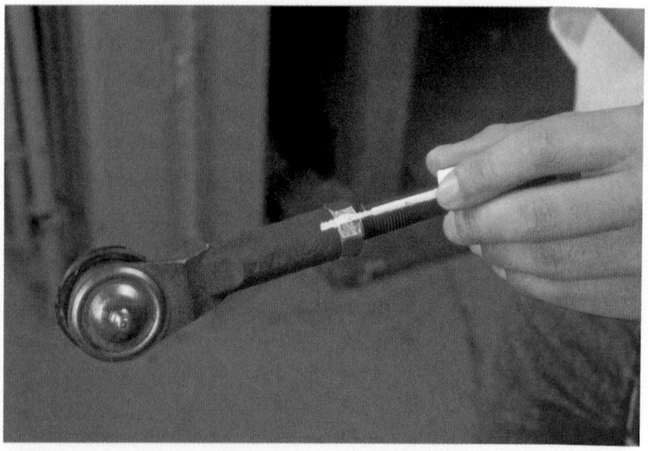

Figure 46-42 Before disassembling the steering gear, make an index mark on the outer tie-rod end, jam nut, and tie-rod.

and damage. Then place the rack on V-blocks. Using a dial indicator, check for runout of the steering rack. Compare the measured runout to the specifications. If the runout is greater than the maximum allowable, replace the power-steering link assembly. Then check the movement of the tie-rod ends. This check is normally done with a torque wrench. All parts of the assembly should be inspected and replaced as necessary.

Also check the following:

- *Inner Tie-Rod Sockets.* If the inner tie-rod ends were found to be loose during the inspection or if they require too much effort to move, replace them.
- *Pinion and Bearing Assembly.* If the pinion bearing is loose on the shaft, replace the pinion and bearing assembly. A pinion shaft with worn or chipped teeth must be replaced. Inspect the pilot bearing contact area on the pinion shaft. Wear, pitting, or scoring in this area indicates that a new pinion shaft is required. If the pinion bearing is bad, replace the pinion shaft and bearing assembly.
- *Pilot Bearing.* Check the pilot bearing in the steering gear housing. If this bearing is worn or scored, replace the pilot bearing and the pinion shaft and bearing assembly.
- *Rack Bushing.* If the rack bushing is worn, bushing replacement is necessary. Remove the bushing-retaining ring prior to bushing removal. Position the puller fingers behind the bushing, and operate the slide hammer on the puller to remove the bushing. If a puller is not available, an appropriate bushing driver or socket may be used to drive the bushing out of the housing.
- *Mounting Bushings.* If the mounting bushings are loose, replace the bushings. Always replace the bushings in pairs. If the bushings are in satisfactory condition, do not disturb them.

SHOP TALK

Some manufactures actually recommend different lengths for the torque wrench that is used to measure, test, and tighten things. This is a reason for always checking the manufacturer's recommendations before proceeding with a repair or service. For example, Toyota recommends using a torque wrench with a different fulcrum length for different applications. Torque wrenches with a fulcrum length of 9.84 in. (250 mm) or 11.81 in. (300 mm) are recommended to measure interference and tightening. Always check the service information for the correct procedures.

When installing new parts to the rack assembly or when reassembling the unit, all moving parts should be lubricated with power-steering fluid, molybdenum disulfide lithium-based grease, MP grease, or silicon grease. Also install new bellows, clamps, bushings, and seals. There are certain parts that need to be preloaded; again follow the manufacturer's recommendations. Reassembly is the reverse of the disassembly procedure. Once the unit is reassembled, check for poor sealing by connecting a vacuum gauge to the housing. If the unit cannot hold a vacuum, a seal(s) is not sealing and should be replaced.

Installing a new or rebuilt rack and pinion steering gear follows the opposite procedure as when removing it. Make sure everything is tightened to the torque specified by the manufacturer with the wheels in a straight forward position. Also make sure to check the fluid level when finished and before the vehicle is driven. The vehicle will also need to have its wheels aligned.

Steering Columns

To perform service procedures on the steering column upper end components, it is not necessary to remove the column from the vehicle. The steering wheel, horn components, directional signal switch, ignition switch, and lock cylinder can be removed with the column remaining in the vehicle.

To determine if the energy-absorbing steering column components are functioning as designed, or if repairs are required, a close inspection should be made. An inspection is called for in all cases where damage is evident or whenever the vehicle is being repaired due to a front-end collision. If damage is evident, the affected parts must be replaced. Because of the differences in the steering column styles and various components, consult the service manual for more explicit inspection and servicing procedures.

WARNING!

Set the parking brake before removing the steering column. Also, remove the battery cable from the negative terminal. Remember that special precautions must be observed before beginning disassembly and during assembly to ensure the correct fitting together of the steering column shaft and steering gear shaft connections.

Steering Wheels

At times, a customer may complain about an uncentered steering wheel. When the steering wheel is in its centered position, the front wheels should be pointing straight ahead. If the wheels are not in the straight-ahead position, this can be corrected by adjusting the toe of the vehicle. However, this adjustment should only be made if the steering wheel index mark is aligned with the steering column index marks. As a rule, indexing teeth or mating flats on the wheel hub and steering shaft prevent misindexing of these components. One way to verify an incorrectly positioned steering wheel is to turn it from stop to stop and count the number of turns it took. Then take that number and divide it by 2; the result represents the center or straight-ahead position. Now turn the steering wheel to a stop and then the number of turns that represents the center. Look at the front wheels; if they are not in the straight-ahead position, either the steering was installed wrong or the wheels need to have their toe adjusted.

At times the steering wheel must be removed, such as when servicing the multifunctional switch, horn, or the steering column covers. The steering wheel is very unlikely to cause a steering problem.

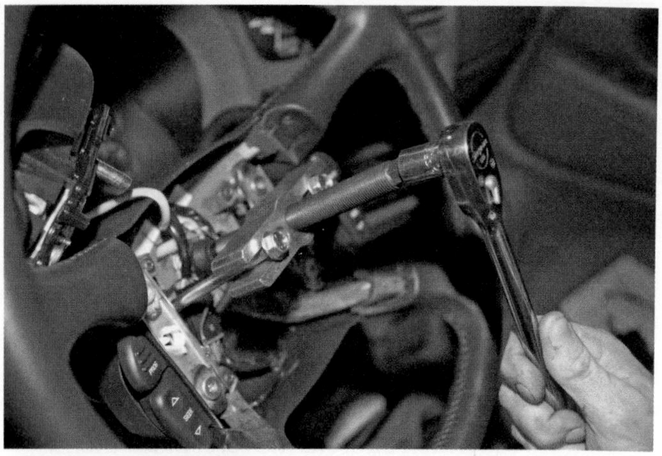

Figure 46-43 Use the correct puller to remove the steering wheel assembly.

WARNING!

The center of the steering wheel contains the driver side air bag. Failure to disarm it can cause serious personal injury.

PROCEDURE

The following is a typical procedure for removing and reinstalling a steering wheel:

To remove:

STEP 1 Place the front wheels facing straight ahead.

STEP 2 Disconnect the cable from the negative battery terminal.

STEP 3 Remove the steering pad.

STEP 4 Remove the steering wheel assembly set nut.

STEP 5 Put alignment marks on the steering wheel and the steering main shaft.

STEP 6 Disconnect the connectors from the spiral cable.

STEP 7 Using the correct puller, remove the steering wheel assembly **(Figure 46–43)**.

To install:

STEP 1 Slip the steering wheel over the main shaft.

STEP 2 Align the alignment marks on the steering wheel assembly and steering main shaft **(Figure 46–44)**.

STEP 3 Install the steering wheel assembly set nut. Torque the nut to specifications.

STEP 4 Connect the connectors to the clockspring (spiral cable) subassembly.

STEP 5 Connect the cable to the negative battery terminal.

STEP 6 Check the SRS warning light.

Go To ***Chapter 24 for specific precautions that should be followed when working around or near an air bag.***

POWER-STEERING SYSTEM SERVICING

Vehicles with power-steering systems have the same type of steering linkage as manual steering. The power-steering linkage is checked and serviced as previously described. Actually, the only difference is the servicing of the hydraulic components, such as the hoses, pump, and power-steering gear, that are fully covered in the vehicle's service manual. One of the common procedures that is recommended by manufacturers as part of a preventive maintenance program is flushing the hydraulic system.

Bleeding the System

Often the procedure for bleeding the power-steering system is called "purging" the system, because it moves air that may be trapped in the fluid. Purging the system must be done after the replacement of any part of the power-steering hydraulic system or when there is a problem that indicates there may be aerated fluid, such as a whining noise.

The method for bleeding the system depends on the type of power steering the vehicle is equipped

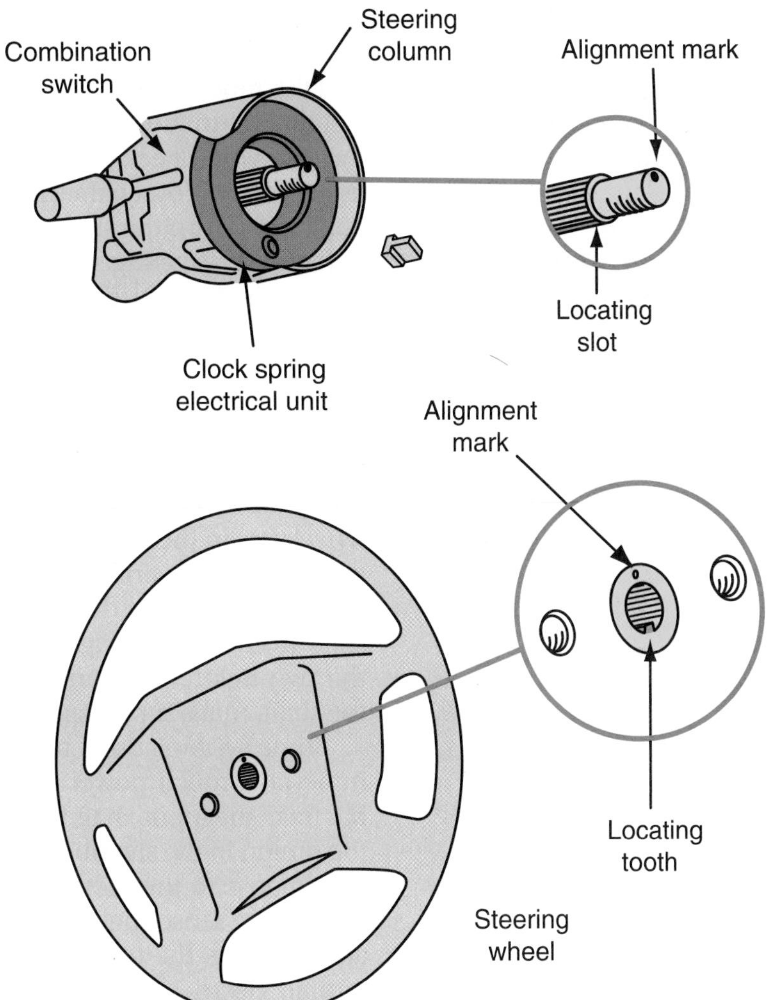

Figure 46-44 The steering wheel is splined to the steering column and must be aligned to the reference marks.

with. Follow the procedures given in the service manual. If the system is not purged correctly or if air is allowed to remain in the system, the power-steering pump can fail prematurely. What follows is a typical procedure.

PROCEDURE

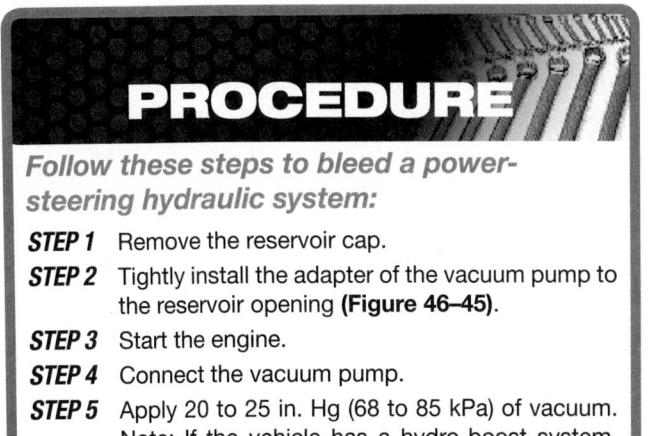

Follow these steps to bleed a power-steering hydraulic system:

STEP 1 Remove the reservoir cap.

STEP 2 Tightly install the adapter of the vacuum pump to the reservoir opening **(Figure 46–45)**.

STEP 3 Start the engine.

STEP 4 Connect the vacuum pump.

STEP 5 Apply 20 to 25 in. Hg (68 to 85 kPa) of vacuum. Note: If the vehicle has a hydro-boost system, depress the brake pedal two times.

STEP 6 Fully cycle the steering wheel from stop to stop ten times.

STEP 7 Turn off the engine.

STEP 8 Release the vacuum and remove the adapter from the reservoir.

STEP 9 Fill the reservoir with the correct fluid.

STEP 10 Tightly install the adapter of the vacuum pump to the reservoir opening.

STEP 11 Start the engine.

STEP 12 Apply 20 to 25 in. Hg (68 to 85 kPa) of vacuum. Note: If the vehicle has a hydro-boost system, depress the brake pedal two times.

STEP 13 Turn off the engine.

STEP 14 Release the vacuum and remove the adapter from the reservoir.

STEP 15 Fill the reservoir with the correct fluid.

STEP 16 Check the system for signs of leaks. Make repairs as necessary.

STEP 17 Install the reservoir cap.

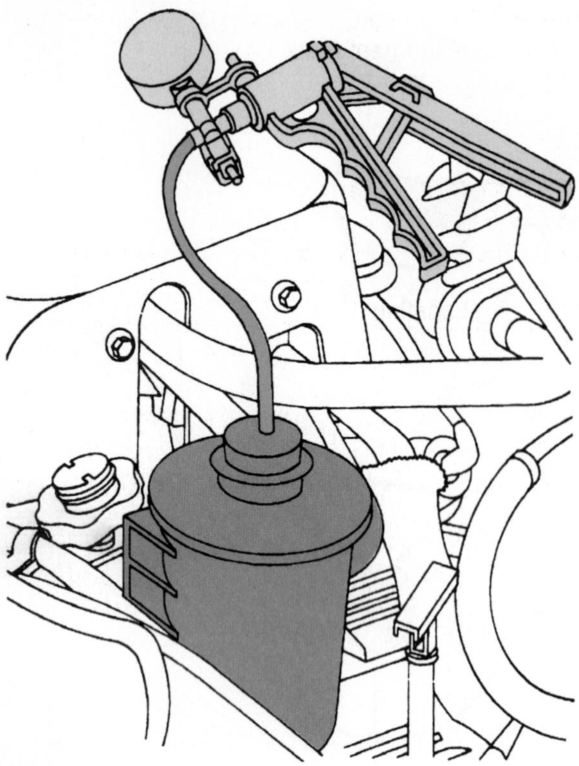

Figure 46-45 To bleed the system, a vacuum pump is connected to the fluid's reservoir. *Courtesy of Ford Motor Company*

Flushing the System

The reason for flushing the system should be obvious. Hydraulic fluid is easily contaminated by moisture and dirt. Flushing removes the old fluid with its contaminants and new fluid is added to the system. It is also wise to flush the system after you have replaced or repaired a part in the system. Before beginning to flush the system, you should disable the engine's ignition. Then disconnect the power-steering return hose and plug the reservoir. Attach an extension hose between the power-steering return hose and an empty container. Raise the vehicle's front wheels off the ground. Fill the reservoir with the correct type of fluid.

WARNING!

Never mix oil types. Any mixture or any unapproved oil can lead to seal deterioration and leaks.

Turn the steering wheel from stop to stop while cranking the engine until the fluid leaving the return hose is clean. Never crank the engine for more than 5 seconds at a time. Add fluid to the reservoir to make sure it does not empty. Once the fluid is clear, fill the reservoir to its full mark and lower the vehicle.

Disconnect the extension hose from the power-steering return hose and reconnect the return hose to the reservoir. Check the fluid level again and add fluid as necessary. Now enable the ignition system. Start the engine and turn the steering wheel from stop to stop. If the power-steering system is noisy and bubbles are forming in the fluid, the system must be purged of air.

Electric Power-Steering Systems

When working on a vehicle with EPS, remember that the electric motor adds the assist. It must be positioned properly in order to provide assist equally to both sides. Therefore, it is important that the front wheels are in their straight-ahead position whenever any part of the steering linkage is removed or installed. If it is necessary to disconnect the steering column shaft from the steering gear, mark the alignment of the two before disconnecting it. When reattaching the shaft, make sure the marks are aligned.

Some systems use a torque sensor to help determine how much power assist to provide. A torsion bar links the input shaft from the steering column to the pinion in the steering gear. The torsion bar twists in response to the movement of the steering wheel. The torque sensor detects the twist of the torsion bar and converts the torque applied to the torsion bar into an electrical signal. The torque sensor must be recalibrated to its zero point whenever the steering gear, steering wheel, steering column, or steering control module has been removed or replaced. This calibration must also be completed if there is a difference in steering effort when turning right or left. Follow the manufacturer's directions for calibrating this sensor. If the sensor is faulty, steering effort in both directions may increase or the steering wheel may not return properly.

CAUTION!

The power-steering unit on some hybrid vehicles is powered by high voltage. Make sure you disarm the high-voltage circuit before working on or near all high-voltage systems.

Hoses and Lines

Hoses and lines should also be carefully inspected for leaks, dents, sharp bends, cracks, and swelling. Always replace the power-steering hoses with an exact replacement hose. Never attempt to patch or seal a leak in a hose or the hose's fittings. Lines and hoses must not rub against other components. This can wear a hole in the line or hose. Many high-pressure lines are made of high-pressure steel-braided hose with molded steel fittings on each end.

PROCEDURE

When power steering hose replacement is required, follow these steps:

STEP 1 With the engine stopped, remove the return hose at the power-steering gear, and allow the fluid to drain into a drain pan.

STEP 2 Loosen and remove all hose fittings from the pump and steering gear.

STEP 3 Remove all hose-to-chassis clips.

STEP 4 Remove the hoses from the chassis, and cap the pump and steering gear fittings.

STEP 5 If O-rings are used on the hose ends, install new O-rings. Some lines have gaskets. The old gasket must be pried out of the fitting before installing the new lines. Lubricate the O-rings with power-steering fluid.

STEP 6 Install the new hose by reversing the steps for removal. Make sure to tighten all fittings to the specified torque. Be sure all hose-to-chassis clips are in place. Do not position hoses where they rub on other components.

STEP 7 Fill the pump reservoir to the full mark with the manufacturer's recommended fluid. Bleed air from the power-steering system, then check the fluid level in the reservoir and add fluid as required.

Power-Steering Pump

Although power-steering pumps are not typically rebuilt by technicians, some parts are replaceable. The actual parts that can be replaced depend on the make of the pump. The common parts that can be replaced are discussed next.

Power-Steering Pump Pulley Replacement If the pulley wobbles while it is rotating, it is undoubtedly bent and should be replaced. Worn pulley grooves and/or cracks also indicate that the pulley should be replaced. A pulley that is loose on the pump's shaft must be replaced. Never hammer on the pump's drive shaft during pulley removal or installation. This will damage the internal parts of the pump.

If the pulley is pressed onto the pump's shaft, a special puller is required to remove it and a pulley installation tool is used to install the pulley.

If the power-steering pump pulley is retained with a nut, mount the pump in a vise. Always tighten the vise on one of the pump's mounting bolt surfaces. Do not tighten the vise with excessive force. Use a special holding tool to keep the pulley from turning, and loosen the nut with a box-end wrench. Remove the nut, pulley, and woodruff key. Inspect the pulley, shaft, and woodruff key for wear. Replace all worn components.

Remove and Replace the Flow Control Valve and End Cover To replace the flow control valve and end plate, remove the retaining ring with a slotted screwdriver and punch. Then remove the flow control valve, end cover, spring, and magnet. Check the flow control valve for burrs. Remove minor burrs with crocus cloth, and clean the flow control valve in solvent. Damaged or worn flow control valves must be replaced. Inspect the end cover sealing surface for damage. Also check the pump's drive shaft for corrosion and damage. Remove any corrosion with crocus cloth. Clean all parts and lubricate the end cover with power-steering fluid. Make sure to clean the magnet with a shop towel. To reassemble, install the flow control valve, end cover, retaining ring, and related components.

Remove and Replace the Pressure Relief Valve Follow this procedure to service the pressure relief valve:

1. Wrap a shop towel around the land end of the flow control valve and clamp this end in a soft-jawed vise. Be very careful not to damage the valve lands.
2. Remove the hex-head ball seat. Clean the components in solvent. A worn or damaged pressure relief ball, spring, guide, or seat must be replaced.
3. Reinstall the new or cleaned components, and then install the ball seat.

FOUR-WHEEL STEERING SYSTEMS

A few manufacturers have offered four-wheel steering systems in which the rear wheels also help to turn the car by electrical, hydraulic, or mechanical means. Although they certainly are not very common, you should be aware of how they work.

Production-built cars tend to understeer or, in a few instances, oversteer. If a car could automatically compensate for an understeer/oversteer problem, the driver would enjoy nearly neutral steering under varying operating conditions. **Four-wheel steering (4WS)** is a serious effort on the part of automotive design engineers to provide near-neutral steering with the following advantages:

- The vehicle's cornering behavior becomes more stable and controllable at high speeds as well as on wet or slippery road surfaces **(Figure 46–46)**.

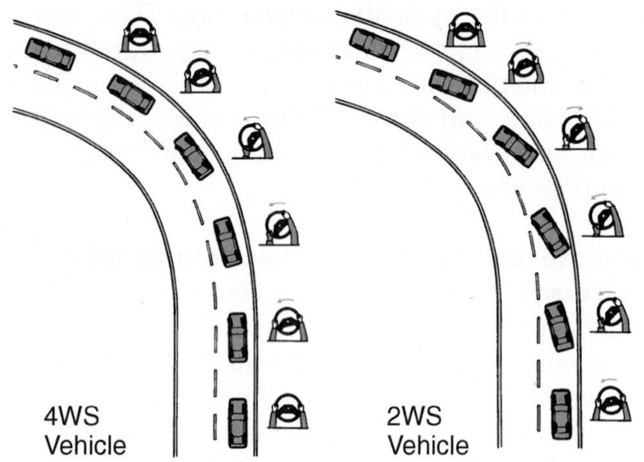

Figure 46-46 A comparison of 2WS and 4WS vehicle behavior during cornering.

- The vehicle's response to steering input becomes quicker and more precise throughout the vehicle's entire speed range.
- The vehicle's straight-line stability at high speeds is improved. Negative effects of road irregularities and crosswinds on the vehicle's stability are minimized.
- Stability in lane changing at high speeds is improved. High-speed slalom-type operations become easier. The vehicle is less likely to go into a spin even in situations in which the driver must make a sudden and relatively large change of direction.
- By steering the rear wheels in the direction opposite the front wheels at low speeds, the vehicle's turning circle is greatly reduced. Therefore, vehicle maneuvering on narrow roads and during parking becomes easier.

To understand the advantages of four-wheel steering, it is wise to review the dynamics of typical steering maneuvers with a conventional front-steered vehicle. The tires are subject to the forces of grip, momentum, and steering input when making a movement other than straight-ahead driving. These forces compete with each other during steering maneuvers. With a front-steered vehicle, the rear end is always trying to catch up to the directional changes of the front wheels. This causes the vehicle to sway. As a normal part of operating a vehicle, the driver learns to adjust to these forces without thinking about them.

When turning, the driver is putting into motion a complex series of forces. Each of these must be balanced against the others. The tires are subjected to road grip and slip angle. Grip holds the car's wheels to the road, and momentum moves the car straight ahead. Steering input causes the front wheels to turn. The car momentarily resists the turning motion, causing a tire slip angle to form. Once the vehicle begins to respond to the steering input, cornering forces are generated. The vehicle sways as the rear wheels attempt to keep up with the cornering forces already generated by the front tires. This is referred to as rear-end lag because there is a time delay between steering input and vehicle reaction. When the front wheels are turned back to a straight-ahead position, the vehicle must again try to adjust by reversing the same forces developed by the turn. As the steering is turned, the vehicle body sways as the rear wheels again try to keep up with the cornering forces generated by the front wheels.

The idea behind four-wheel steering is that a vehicle requires less driver input for any steering maneuver if all four wheels are steering the vehicle. As with two-wheel-steer vehicles, tire grip holds the four wheels on the road. However, when the driver turns the wheel slightly, all four wheels react to the steering input, causing slip angles to form at all four wheels. The entire vehicle moves in one direction rather than the rear half attempting to catch up to the front. There is also less sway when the wheels are turned back to a straight-ahead position. The vehicle responds more quickly to steering input because rear wheel lag is eliminated.

Because each 4WS system is unique in its construction and repair needs, the vehicle's service manual must be followed for proper diagnosis, repair, and alignment of a four-wheel system.

Mechanical 4WS

In a straight-mechanical type of 4WS, two steering gears are used—one for the front and the other for the rear wheels. A steel shaft connects the two steering gearboxes and terminates at an eccentric shaft that is fitted with an offset pin (**Figure 46–47**). This pin engages a second offset pin that fits into a planetary gear.

The planetary gear meshes with the matching teeth of an internal gear that is secured in a fixed position to the gearbox housing. This means that the planetary gear can rotate but the internal gear cannot. The eccentric pin of the planetary gear fits into a hole in a slider for the steering gear.

A 120-degree turn of the steering wheel rotates the planetary gear to move the slider in the same direction that the front wheels are headed. Proportionately, the rear wheels turn the steering wheel about 1.5 to 10 degrees. Further rotation of the steering wheel, past the 120-degree point, causes the rear wheels to start straightening out due to the double-crank action (two eccentric pins) and rotation of the planetary gear. Turning the steering wheel to a greater angle, about 230 degrees, finds the rear wheels in a neutral position regarding the front wheels. Further rotation of

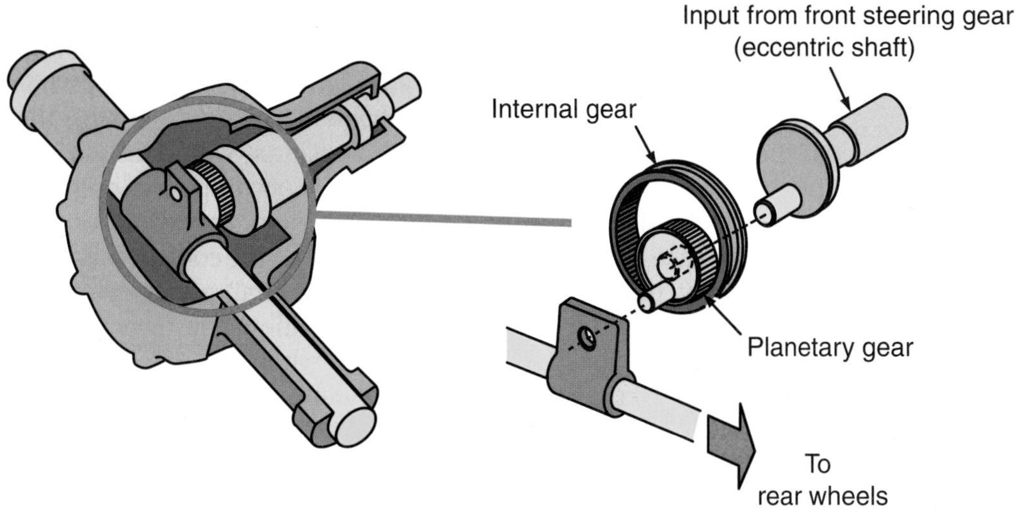

Figure 46-47 Inside a rear-steering gearbox is a simple planetary gear setup.

the steering wheel results in the rear wheels going counterphase with regard to the front wheels. About 5.3 degrees maximum counter phase rear steering is possible.

Mechanical 4WS is steering angle sensitive. It is not sensitive to vehicle road speed.

Hydraulic 4WS

The hydraulically operated 4WS system shown in **Figure 46–48** is a simple design, both in components and operation. The rear wheels turn only in the same direction as the front wheels. They also turn no more than 1½ degrees. The system only activates at speeds above 30 mph (50 km/h) and does not operate when the vehicle moves in reverse.

A two-way hydraulic cylinder mounted on the rear stub frame turns the wheels. Fluid for this cylinder is supplied by a rear steering pump that is driven by the differential. The pump only operates when the front wheels are turning. A tank in the engine compartment supplies the rear steering pump with fluid.

When the steering wheel is turned, the front steering pump sends fluid under pressure to the rotary valve in the front rack and pinion unit. This forces fluid into the front power cylinder, and the front wheels turn in the direction steered. The fluid pressure varies with the turning of the steering wheel. The faster and farther the steering wheel is turned, the greater the fluid pressure.

The fluid is also fed under the same pressure to the control valve, where it opens a spool valve in the control valve housing. As the spool valve moves, it allows fluid from the rear steering pump to move through and operate the rear power cylinder. The higher the pressure on the spool, the farther it moves. The farther it moves, the more fluid it allows through to move the rear wheels. As mentioned earlier, this system limits rear wheel movement to 1½ degrees in either the left or right direction.

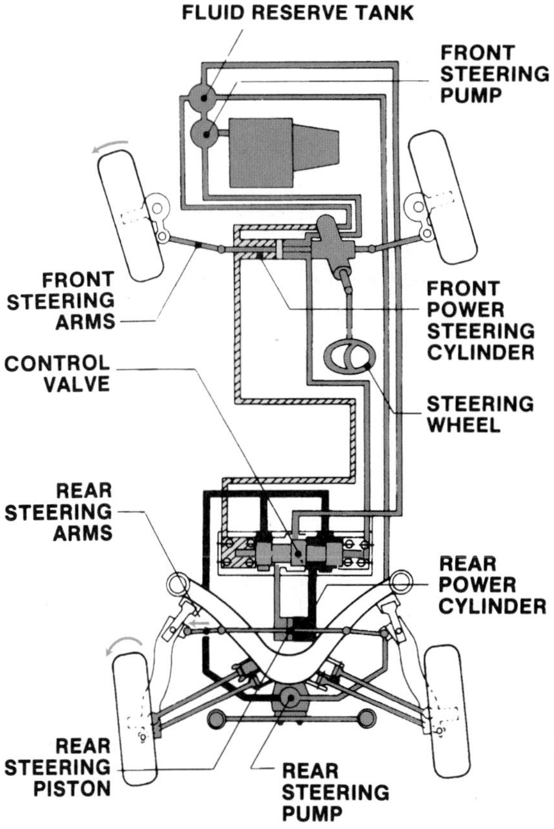

Figure 46-48 A simple hydraulic 4WS system. *Courtesy of Mitsubishi Motors North America, Inc.*

Electrohydraulic 4WS

Several 4WS systems combine computer electronic controls with hydraulics to make the system sensitive to both steering angle and road speeds. In this design, a speed sensor and steering wheel angle sensor feed information to the electronic control unit (ECU). By processing the information received, the ECU commands the hydraulic system to steer the rear wheels. At low road speed, the rear wheels of this system are not considered a dynamic factor in the steering process.

At moderate road speeds, the rear wheels are steered momentarily counterphase, through neutral, then in phase with the front wheels. At high road speeds, the rear wheels turn only in phase with the front wheels. The ECU must know not only road speed, but also how much and how quickly the steering wheel is turned. These three factors—road speed, amount of steering wheel turn, and the quickness of the steering wheel turn—are interpreted by the ECU to maintain continuous and desired steering angle of the rear wheels.

Another electrohydraulic 4WS system is shown in **Figure 46–49**. The basic working elements of the design are the control unit, a stepper motor, a swing arm, a set of beveled gears, a control rod, and a control valve with an output rod. Two electronic sensors tell the ECU how fast the car is going.

The yoke is a major mechanical component of this electrohydraulic design. The position of the control yoke varies with vehicle road speed. For example, at speeds below 33 mph (53 km/h), the yoke is in its downward position, which results in the rear wheels steering in the counterphase (opposite front wheels) direction. As road speeds approach and exceed 22 mph (35 km/h), the control yoke swings up through a neutral (horizontal) position to an up position. In the neutral position, the rear wheels steer in phase with the front wheels.

The stepper motor moves the control yoke. A swing arm is attached to the control yoke. The position of the yoke determines the arc of the swing rod. The arc of the swing arm is transmitted through a control arm that passes through a large bevel gear. Stepper motor action eventually causes a push-or-pull movement of its output shaft to steer the rear wheels up to a maximum of 5 degrees in either direction.

The electronically controlled 4WS system regulates the angle and direction of the rear wheels in

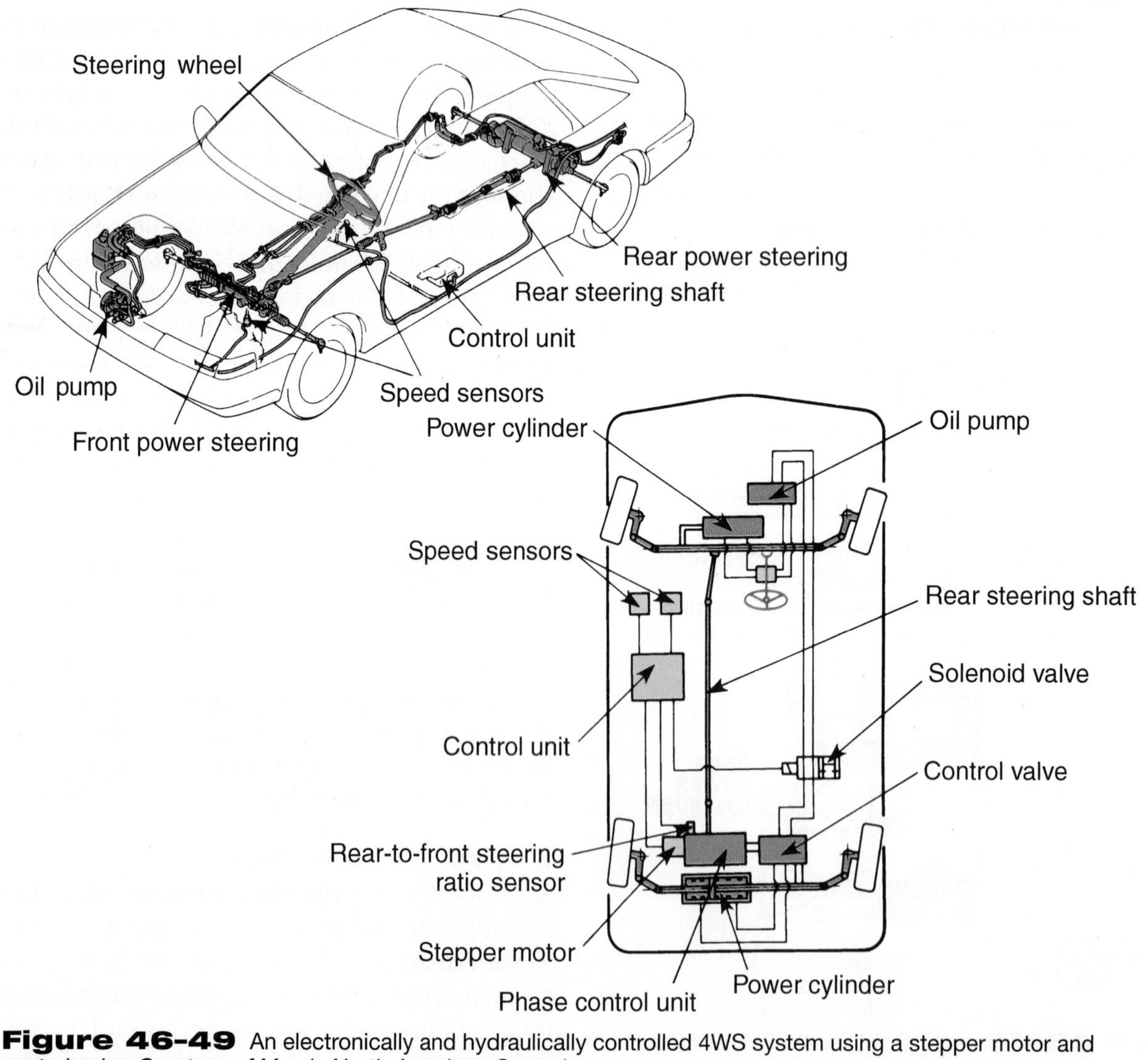

Figure 46–49 An electronically and hydraulically controlled 4WS system using a stepper motor and control yoke. *Courtesy of Mazda North American Operations*

response to speed and driver's steering. This speed-sensing system optimizes the vehicle's dynamic characteristics at any speed, thereby producing enhanced stability and, within certain parameters, agility.

The actual 4WS system consists of a rack and pinion front steering that is hydraulically powered by a main twin-tandem pump. The system also has a rear-steering mechanism, hydraulically powered by the main pump. The rear-steering shaft extends from the rack bar of the front-steering assembly to the rear-steering-phase control unit.

The rear steering is comprised of the input end of the rear-steering shaft, vehicle speed sensors, and steering-phase control unit (deciding direction and degree), a power cylinder, and an output rod. A centering lock spring is incorporated that locks the rear system in a neutral (straight-ahead) position in the event of hydraulic failure. Additionally, a solenoid valve that disengages the hydraulic boost (thereby activating the centering lock spring in case of an electrical failure) is included.

All 4WS systems have fail-safe measures. For example, with the electrohydraulic setup, the system automatically counteracts possible causes of failure, both electronic and hydraulic, and converts the entire steering system to a conventional two-wheel steering type. Specifically, if a hydraulic defect should reduce pressure level (by a movement malfunction or a broken driving belt), the rear-wheel-steering mechanism is automatically locked in a neutral position, activating a low-level warning light.

An electrical failure would be detected by a self-diagnostic circuit integrated in the four-wheel-steering control unit. The control unit stimulates a solenoid valve, which neutralizes hydraulic pressure, thereby alternating the system to two-wheel steering. The failure would be indicated by the system's warning light in the main instrument display.

On any 4WS system, there must be near-perfect compliance between the position of the steering wheel, the position of the front wheels, and the position of the rear wheels. It is usually recommended that the car be driven about 20 feet (6 meters) in a dead-straight line. Then, the position of the front/rear wheels is checked with respect to steering wheel position. The base reference point is a strip of masking tape on the steering wheel hub and the steering column. When the wheel is positioned dead center, draw a line down the tape. Run the car a short distance straight ahead to see if the reference line holds. If not, corrections are needed, such as repositioning the steering wheel.

Even severe imbalance of a rear wheel on a speed-sensitive 4WS system can cause problems and make basic troubleshooting a bit frustrating.

Quadrasteer

Quadrasteer is a 4WS system that improves low-speed maneuverability (decreasing the turning radius), high-speed stability, and trailering capabilities for full-size pickups, vans, and SUVs. The system combines normal front-wheel steering with an electrically powered and electronically controlled rear-wheel steering system. Besides the mechanical part, the system uses wheel position and vehicle speed sensors and a central control module **(Figure 46–50)**.

At low speeds, the rear wheels turn in the opposite direction from the front wheels **(Figure 46–51)**. At moderate speeds, the rear wheels remain straight. At high speeds, the rear wheels turn in the same direction as the front wheels. If the system were to fail, the truck would default to normal two-wheel steering.

Figure 46–50 An exploded view of the Quadrasteer setup. *Courtesy of the Delphi Corporation*

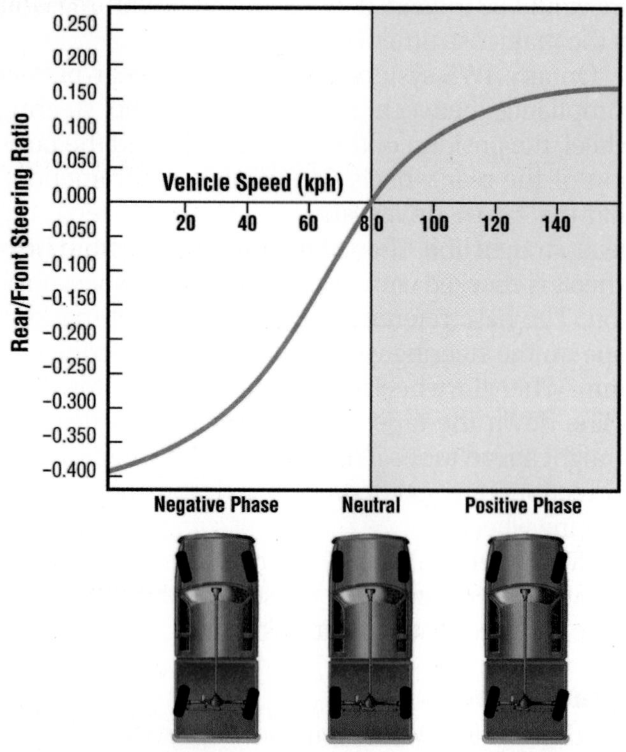

Figure 46-51 The different operational modes of a Quadrasteer system. *Courtesy of the Delphi Corporation*

CASE STUDY

A customer complained of excessive steering effort on his 1998 Oldsmobile Aurora. The technician took the car out for a road test to verify the customer's complaint, but found no evidence of hard steering. Further questioning of the customer revealed that the problem only occurred when the engine was cold. The service writer informed the customer that the car would need to be parked overnight at the shop so the problem could be duplicated. The customer agreed to this.

The next morning prior to starting the engine, the technician connected a power-steering pressure tester in series with the high-pressure hose. When the engine was first started, the technician found the steering effort to be overly hard and it felt as if there was no power assist. But the power-steering pump was working and was putting out a higher than normal pressure. Once the steering wheel was turned in one direction, the steering effort quickly returned to normal. Since the power-steering pump pressure was higher than specified, the technician concluded the pump and drive belt must be in fine shape. He also reasoned that other causes of hard steering, such as binding in the steering column, would be always present. They would not cause the problem to occur only when the system was cold. Therefore, he decided the problem must be in the steering gear.

He then removed and inspected the steering gear. He found that the rack cylinder was severely ridged and scored in the center area, and the Teflon ring on the rack piston was severely worn. A new steering gear assembly was installed, the system flushed, and the car road tested. To verify the repair, the car was left out overnight again and the steering checked the next morning. Everything worked fine.

KEY TERMS

Bump steer
Center link
Dry park check
Four-wheel steering (4WS)
Idler arm
Integral piston
Pitman arm
Rack
Steering linkage
Tie-rod
Worm and roller gearbox
Yoke lash

SUMMARY

- The components of a manual-steering system include the steering linkage, steering gear, and the steering column and wheel.
- The term *steering linkage* is applied to the system of pivots and connecting parts placed between the steering gear and the steering arms that are attached to the front wheels, controlling the direction of vehicle travel. The steering linkage transfers the motion of the steering gear output shaft to the steering arms, turning the wheels to maneuver the vehicle.
- Basic components of a parallelogram steering linkage system include the pitman arm, idler arm, links, tie-rods, and, in some designs, a steering damper.
- The worm and roller steering components are basically the same as found in the parallelogram system.

- In rack and pinion steering linkage, steering input is received from a pinion gear attached to the steering column. This gear moves a toothed rack that is attached to the tie-rods that move the wheels.
- There are three types of manual steering gears in use today: recirculating ball, worm and roller, and rack and pinion.
- The steering wheel and column produce the necessary force to turn the steering gear.
- The power-steering unit is designed to reduce the amount of effort required to turn the steering wheel. It also reduces driver fatigue on long drives and makes it easier to steer the vehicle at slow road speeds, particularly during parking.
- There are several power-steering systems in use on passenger cars and light-duty trucks. The most common ones are the integral, linkage, hydro-boost, and power-assisted rack and pinion systems.
- The major components of a conventional power-steering system are the steering linkage, power-steering pump, flow control and pressure relief valves, reservoir, spool valves and power pistons, hydraulic hose lines, and gearbox or assist assembly on the linkage.
- Electronic rack and pinion systems replace the hydraulic pump, hoses, and fluid associated with conventional power-steering systems with electronic controls and an electric motor located concentric to the rack itself.
- Four-wheel steering (4WS) advantages include cornering capability, steering response, straight-line stability, lane changing, and low-speed maneuverability.

REVIEW QUESTIONS

1. Describe how a rack and pinion steering, a parallelogram steering, and a worm and roller system operate.
2. A power-steering hose transmits fluid under pressure from the ______ to the ______ ______ ______.
3. What is an integral power-steering system?
4. Define the term *gearbox ratio.*
5. What are the basic features of all four-wheel steering systems?
6. List the four main components in a parallelogram steering linkage and explain the purpose of each component.
7. What is the primary purpose of power-steering hoses?
   a. to lubricate the pump
   b. to relieve pressure
   c. to transmit power through fluid under pressure
   d. none of the above
8. In a rack and pinion steering system, what protects the rack from contamination?
   a. the inner tie-rod socket
   b. the outer tie-rod socket
   c. grommets
   d. the bellows boot
9. The main job of the idler arm is to ______.
   a. support the left side of the center link
   b. support the right side of the center link
   c. support the pitman arm
   d. keep both ends of the steering system level
10. Which of the following is *true* of the steering linkage system?
    a. The power cylinder can be integrated with the steering gear.
    b. The power cylinder can have four hydraulic lines connected to it.
    c. The power cylinder may be attached to the steering center link.
    d. All of the above.
11. Rack and pinion steering ______.
    a. is lighter in weight and has fewer components than parallelogram steering
    b. does not provide as much feel for the road as parallelogram steering
    c. does not use tie-rods in the same fashion as parallelogram steering
    d. all of the above
12. If an electrical defect occurs in the electronic power-steering (EPS) system, ______.
    a. the system continues to operate normally, but the EPS warning light is illuminated
    b. the system continues to operate with a slightly reduced power-steering assist
    c. the EPS control unit shuts the system down by opening the fail-safe and power relays.
    d. the EPS control unit locks the armature in the steering gear to prevent armature and screw shaft damage.
13. Wandering or poor straight-ahead tracking may be caused by all of the following *except* ______.

a. worn or loose rack mounting bushings
b. rack piston seal leaks
c. off-center steering gear
d. loose or worn tie-rod ends

14. A rack and pinion steering gear ______.
    a. has tie-rods that connect the rack directly to the steering arms
    b. is a gear in which the rack needs an idler arm to change the direction of the steering arms ______.
    c. has inner tie-rod ends that are pressed onto the rack
    d. has more friction points compared to a parallelogram steering linkage

15. In a manual recirculating ball steering gear ______.
    a. the interference fit between the sector shaft teeth and ball nut teeth is present when the front wheels are straight ahead
    b. the proper interference fit between the sector shaft teeth and ball nut teeth is obtained by radial sector shaft movement
    c. the interference fit between the sector shaft teeth and ball nut teeth becomes tighter when the front wheels are turned to the right or left
    d. tightening the worm shaft adjusting plug tightens the interference fit between the sector shaft teeth and ball nut teeth

ASE-STYLE REVIEW QUESTIONS

1. Technician A says that the manual-steering gear is the probable cause of a shimmy. Technician B says that it could be loose steering linkage. Who is correct?
   a. Technician A c. Both A and B
   b. Technician B d. Neither A nor B
2. Technician A says that adjusting the steering gear too tightly can cause hard steering. Technician B says that adjusting the steering gear too tightly can cause poor returnability. Who is correct?
   a. Technician A c. Both A and B
   b. Technician B d. Neither A nor B
3. An owner's concern indicates excessive turning effort: Technician A checks the alignment of the vehicle's frame. Technician B uses a pull scale. Who is correct?
   a. Technician A c. Both A and B
   b. Technician B d. Neither A nor B
4. While discussing steering problems: Technician A says that a "jerky" steering wheel and a "clunking" noise could indicate worn steering column U-joints. Technician B says that lack of assist and a "growling" noise in a fluid-filled steering pump could indicate a hose or pump internal restriction. Who is correct?
   a. Technician A c. Both A and B
   b. Technician B d. Neither A nor B
5. While discussing the fail-safe function: Technician A says that the 4WS indicator light is illuminated during the fail-safe function. Technician B says that the rear wheels steer normally when the 4WS control unit enters the fail-safe mode. Who is correct?
   a. Technician A c. Both A and B
   b. Technician B d. Neither A nor B
6. Technician A says that electronically controlled power-steering systems allow for power assistance even when an engine stalls. Technician B says that an electronic rack and pinion system provides power assistance even when the engine stalls. Who is correct?
   a. Technician A c. Both A and B
   b. Technician B d. Neither A nor B
7. A vehicle has continual excessive steering effort, but there is no noise and the fluid level in the reservoir is correct: Technician A says that the cause of the problem could be a stuck flow control valve. Technician B says that the cause of the problem could be air in the power-steering fluid. Who is correct?
   a. Technician A c. Both A and B
   b. Technician B d. Neither A nor B
8. While diagnosing a loose feeling when steering a vehicle with power steering: Technician A says that the problem could be caused by air in the system. Technician B says that the problem could be caused by tight worm shaft thrust bearing preload adjustment. Who is correct?
   a. Technician A c. Both A and B
   b. Technician B d. Neither A nor B
9. While diagnosing the problems that can be caused by worn tie-rod ends: Technician A says that they can cause scalloped and scuffed tires.

Technician B says that they can cause the vehicle to pull to one side. Who is correct?

a. Technician A
b. Technician B
c. Both A and B
d. Neither A nor B

10. While diagnosing the cause of wheel shimmy: Technician A says that this problem can be caused by loose wheel bearings. Technician B says that this can be caused by binding steering column U-joints. Who is correct?

a. Technician A
b. Technician B
c. Both A and B
d. Neither A nor B

CHAPTER 47

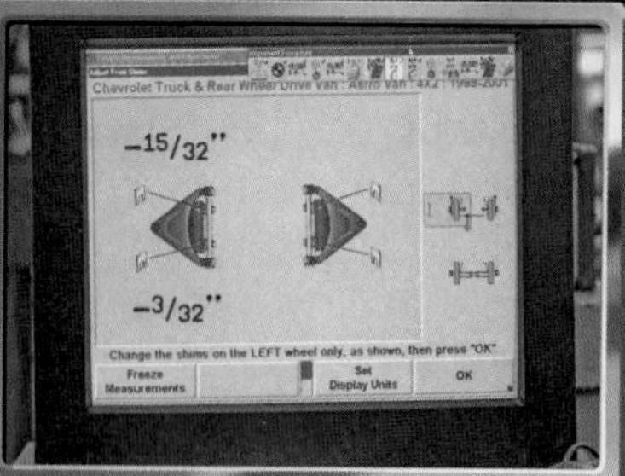

WHEEL ALIGNMENT

OBJECTIVES

■ Explain the benefits of accurate wheel alignment. ■ Explain the importance of correct wheel alignment angles. ■ Describe the different functions of camber and caster with regard to the vehicle's suspension. ■ Identify the purposes of steering axis inclination. ■ Explain why toe is the most critical tire wear factor of all the alignment angles. ■ Identify the purposes of turning radius or toe-out. ■ Explain the conditions known as tracking and thrust angle. ■ Perform a prealignment inspection. ■ Describe the various types of equipment that can be used to align the wheels of a vehicle. ■ Describe how alignment angles can be changed on a vehicle. ■ Understand the importance of rear-wheel alignment. ■ Know the difference between two-wheel and four-wheel alignment procedures.

A vehicle's wheels, tires, suspension system, and steering system are all designed to work together to provide safe, stable, and reliable handling. This is the purpose of a wheel alignment **(Figure 47–1)**. During a wheel alignment, the angles of the wheels are measured and adjusted to place the tires perpendicular to the ground and parallel to each other and to the geometric centerline of the vehicle. These angles are adjusted by changing the position of various steering and suspension parts. The desired angles are those set by the vehicle's manufacturer.

Correct wheel alignment allows the wheels to roll without scuffing, dragging, or slipping on different types of road surfaces. Proper alignment of both the front and rear wheels ensures greater safety, easier steering, longer tire life, reduction in fuel consumption, and less strain on the steering and suspension systems.

The alignment of the wheels should be checked whenever new tires or steering and suspension parts are installed. The wheels also should be aligned whenever the tires are wearing abnormally. A vehicle's wheels become out-of-alignment because of worn suspension parts, a change in ride height, or driving hard into a pothole or curb. All of these can affect the angle of the wheels.

A wheel alignment restores the geometry of the suspension to the angles that were determined to properly locate the vehicle's weight on the tires and to facilitate steering.

Figure 47–1 Technicians who specialize in wheel alignment are always in demand. *Courtesy of Hunter Engineering Company*

Types of Wheel Alignment

There are two basic types of wheel alignment performed today: two wheel and four wheel. In a two-wheel alignment, only the angles of the front wheels are measured and adjusted. This does not take into account the relationship between the front and rear axles. Two-wheel alignment was common before suspension and steering systems became more complex.

Four-wheel alignment measures the angles at the four wheels. On some vehicles, adjustments are made only to the front wheels. This is primarily due to the fact that there is no way to make adjustments to the rear wheels. However, by adjusting the front wheels so they are rotating in the same direction as the rear wheels, the vehicle will tend to move straight. Many vehicles have provisions for adjusting the rear wheels. When this is the case, the rear wheels are adjusted first and then the fronts are aligned to the vehicle's centerline.

Road Crown

Most roads are not designed to be flat. They are paved at a slight angle. This allows water to flow off the road's surface rather than allowing it to accumulate. The angle is called **road crown** and can cause a vehicle to tend to pull toward the right of the road. To compensate for this, different angles are set at each side of the vehicle.

ALIGNMENT GEOMETRY

The proper alignment of a suspension/steering system centers on the accuracy of the following angles.

Caster

Caster is the angle of the steering axis of a wheel from the vertical, as viewed from the side of the vehicle. The forward or rearward tilt from the vertical line **(Figure 47–2)** is caster. Caster is most often the first angle adjusted during an alignment. Tilting the wheel forward is negative caster. Tilting backward is positive caster.

Caster is designed to provide steering stability. The caster angle for each wheel on an axle should be equal. Unequal caster angles cause the vehicle to steer toward the side with less caster. Too much negative caster can cause the vehicle to have sensitive steering at high speeds. The vehicle might wander as a result of negative caster. Caster is not considered to be a tire wearing angle.

Caster is affected by worn or loose ball joints, strut rods, and control arm bushings. Caster adjustments are not possible on some strut suspension systems without an aftermarket service kit. Often if these vehicles have a caster problem, something is worn or bent and those parts must be replaced or repaired. Where adjustment points are provided, they can be made at the top or bottom mount of the strut assembly.

Camber

Camber is the angle represented by the tilt of either the front or rear wheels inward or outward from the vertical as viewed from the front of the car **(Figure 47–3)**. Camber is designed into the vehicle to compensate for road crown, passenger weight, and vehicle weight. Camber is usually set equally for each wheel. Equal camber means each wheel is tilted outward or inward the same amount. Unequal camber causes tire wear and causes the vehicle to pull toward the side that is more positive.

Camber angle changes, due to the travel of the suspension system, are controlled by the suspension's pivots. Camber is affected by weak or broken springs and worn or loose ball joints, control arm bushings, and wheel bearings. Anything that changes riding height also affects camber.

Camber is adjustable at the control arms on most vehicles. Some vehicles with a strut suspension include a camber adjustment at the spindle assembly. Camber adjustments are also provided on some strut

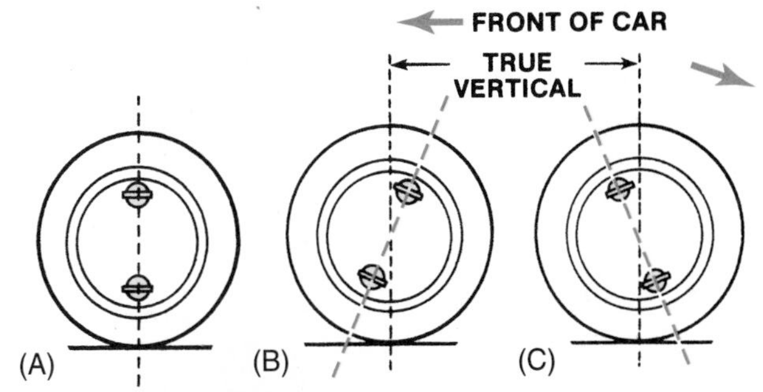

Figure 47–2 Three types of caster: (A) zero, (B) positive, and (C) negative.

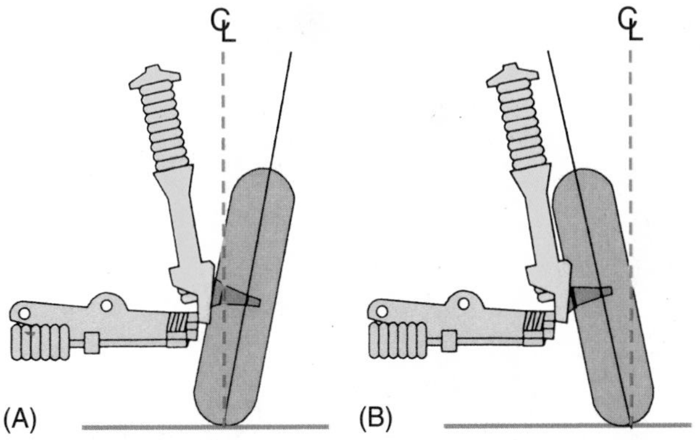

Figure 47–3 (A) Positive and (B) negative camber.

suspension systems at the top mounting of the strut. Very little adjustment of camber (or caster) is required on strut suspensions if the tower and lower control arm locations are correct. If serious camber error has occurred and the suspension mounting positions have not been damaged, it is an indication of bent suspension parts. Damaged parts should be replaced.

Toe

Toe is the distance comparison between the leading edge and trailing edge of the front tires. If the leading edge distance is less, then there is toe-in. If it is greater, there is toe-out **(Figure 47–4)**. Actually, toe is critical as a tire-wearing angle. Wheels that do not track straight ahead have to drag as they travel forward. Excessive toe measurements (in or out) cause a sawtooth edge on the tread surface from dragging the tire sideways. Excessive toe-in will cause tire wear on the outside edge of the tire. Toe-out causes wear on the inside edge.

Toe adjustments are made at the tie-rod. They must be set equally on both sides of the car. If the toe settings are not equal, the car may tend to pull due to the steering wheel being off-center. An off-center steering wheel and steering pull should be corrected by making the toe adjustments equal on both sides of the car with the steering wheel centered.

Toe will change with vehicle speed. As the vehicle moves, friction forces the tires to move straight ahead or have zero toe. However, aerodynamic forces on the vehicle cause a change in its riding height. This will also change the toe as well as camber. Therefore, most toe specifications anticipate these changes and are set to provide zero toe at highway speeds.

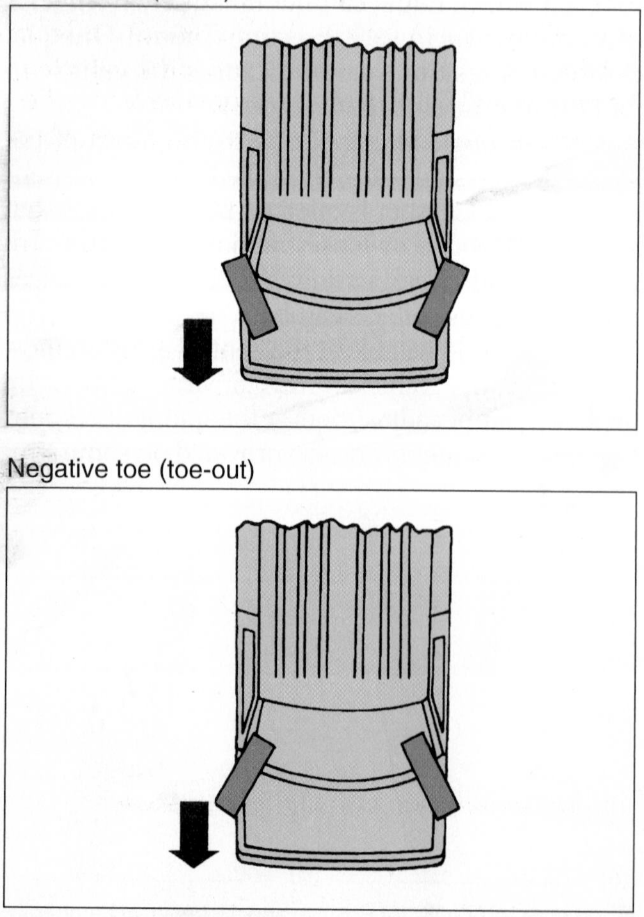

Figure 47–4 (Top) Toe-in. (Bottom) Toe-out. *Courtesy of Ford Motor Company*

Thrust Line Alignment

A main consideration in any alignment is to make sure the vehicle runs straight down the road, with the rear tires tracking directly behind the front tires when the steering wheel is in the straight-ahead position. The geometric centerline of the vehicle should parallel the road direction. This is the case when rear toe is parallel to the vehicle's geometric centerline in the straight-ahead position. If rear toe does not parallel the vehicle centerline, a thrust direction to the left or right is created **(Figure 47–5)**. This difference of rear toe from the geometric centerline is called the **thrust angle.**

Any time the centerline of the front axle is not parallel to the rear axle, handling will be affected. This is because the vehicle will tend to travel according to the angle of the rear axle. This causes the vehicle to pull in the opposite direction as the thrust angle. If the thrust line is to the right, the vehicle will pull to the left; when the thrust line is to the left, the vehicle will pull to the right.

This problem can cause tire wear and poor directional stability on ice, snow, or wet pavement. It can also make a vehicle pull during braking or hard acceleration. Also, if the thrust angle is not zero, the steering wheel will not be centered. Nonparallel axles, or thrust line deviations, are usually caused by the shifting of the rear axle on its spring supports, rear wheel misalignment, or damage from an accident.

Steering Axis Inclination (SAI)

Steering axis inclination (SAI) locates the vehicle weight to the inside or outside of the vertical centerline

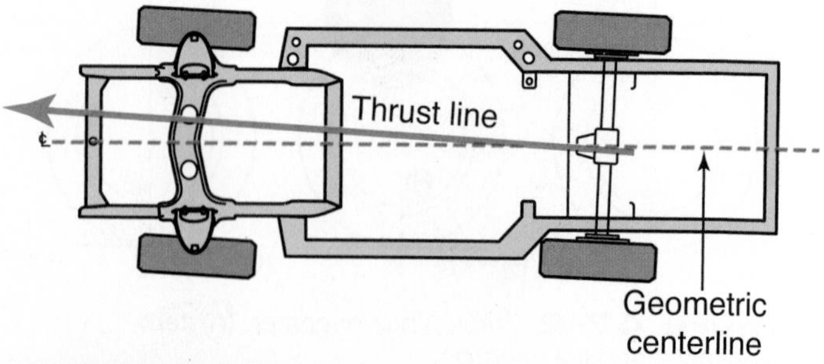

Figure 47–5 The thrust line or driving direction of the rear wheels.

of the tire. The SAI is the angle between true vertical and a line drawn between the steering pivots as viewed from the front of the vehicle **(Figure 47–6)**. It is an engineering angle designed to project the weight of the vehicle to the road surface for stability. The SAI helps the vehicle's steering system return to straight ahead after a turn.

If the vehicle has 0 (zero) SAI, the upper and lower ball joints (or strut pivot points) would be located directly over one another. Problems associated with this simple relationship include tire scrub in turns, lack of control, and increased effort during turn recovery. If the SAI is tilted, a triangle is formed between ball joints and spindle. An arc is then formed when turning. There is a high point at straight-ahead position and a drop downward turning to each side. This motion travels through the control arms to the springs and, finally, to the weight of the vehicle. The forces generated in a turn are actually trying to lift the vehicle. The tilting and loading effect of SAI offsets the lifting forces and helps to pull the tires back to straight ahead when the turn is finished.

Front-wheel-drive vehicles with strut suspensions typically have a higher SAI angle (12 to 18 degrees) than a short-long arm rear-wheel-drive suspension (6 to 8 degrees). This is because the extra leverage provided by a larger angle helps directional stability.

If the SAI angles are unequal side-to-side, torque steer, brake pull, and bump steer (jerking from side to side) can occur even if static camber angles are within specifications.

Checking the SAI angle can help locate various problems that affect wheel alignment. For example, an SAI angle that varies from side to side may indicate an out-of-position upper strut tower, a bowed lower control arm, or a shifted center crossmember.

On a short-long arm suspension, SAI is the angle between true vertical and a line drawn from the upper ball joint through the lower ball joint. In a strut-equipped vehicle this line is drawn through the center of the strut's upper mount down through the center of the lower ball joint.

Included Angle When the camber angle is added to the SAI angle, the sum of the two is called the **included angle (Figure 47–7)**. This angle is not measured by an alignment machine; it is simply obtained by adding the camber and SAI on one side of the vehicle together. The included angle must be the same on each side of

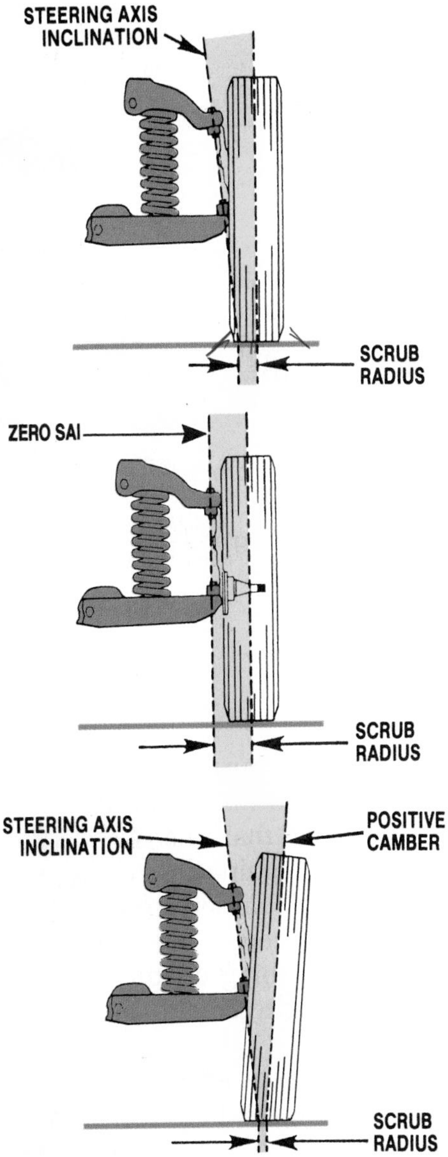

Figure 47–6 The effects of steering axis inclination changes.

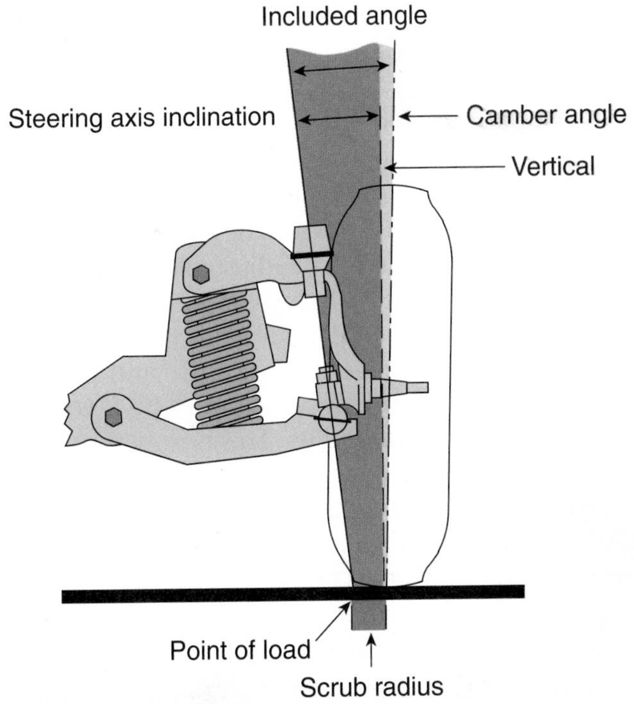

Figure 47–7 The included angle is the sum of the camber and SAI angles.

TABLE 47-1 ALIGNMENT ANGLE DIAGNOSTIC CHART

Suspension Systems	SAI	Camber	Included Angle	Probable Cause
Short Arm/Long Arm Suspension	Correct	Less	Less	Bent knuckle
	Less	Greater	Correct	Bent lower control arm
	Greater	Less	Correct	Bent upper control arm
	Less	Greater	Greater	Bent knuckle
MacPherson Strut Suspension	Correct	Less	Less	Bent knuckle and/or bent strut
	Correct	Greater	Greater	Bent knuckle and/or bent strut
	Less	Greater	Correct	Bent control arm or strut tower (out at top)
	Greater	Less	Correct	Strut tower (in at top)
	Greater	Greater	Greater	Strut tower (in at top) and spindle and/or bent strut
	Less	Greater	Greater	Bent control arm or strut tower (out at top) plus bent knuckle and/or bent strut
	Less	Less	Less	Strut tower (out at top) and knuckle and/or strut bent or bent control arm
Twin I-Beam Suspension	Correct	Greater	Greater	Bent knuckle
	Greater	Less	Correct	Bent I beam
	Less	Greater	Correct	Bent I beam
	Less	Greater	Greater	Bent knuckle

the vehicle even if the camber on each side is different. If it is not, the vehicle will pull.

Comparing the SAI, included, and camber angles can help identify damaged or worn components. For example, if the SAI reading is correct but the camber and included angles are less than specifications, the steering knuckle or strut tower may be bent. **Table 47–1** summarizes the various angle combinations used to troubleshoot short-long arm, strut, and twin I-beam suspension system alignment problems.

Scrub Radius Scrub radius is the distance between the center of the tire and where the SAI angle intersects the ground. The scrub radius must be equal on both sides of the vehicle. Otherwise the vehicle will pull to one side. Scrub radius is not adjustable or measured; rather, it is observed. Scrub radius is part of the suspension's design. There is positive scrub when the tire's contact patch is outside the SAI angle and negative when the patch is inside the angle. Most FWD vehicles have a negative scrub radius. This is done to reduce torque steer. To the contrary, most SLA suspensions have a positive scrub radius. If a vehicle pulls after it has been properly aligned, look for offset wheels or a problem that would affect SAI.

Turning Radius

Turning radius relates to the amount of toe-out present in turns **(Figure 47–8)**. This is also called "toe-out on turns" or "turning angle." As a car goes around a corner, the inside tire must travel in a smaller radius circle than the outside tire. This is accomplished by designing the steering geometry to turn the inside wheel sharper than the outside wheel. The result can be seen as toe-out in turns. This eliminates tire scrubbing on the road surface by keeping the tires pointed in the direction they have to move.

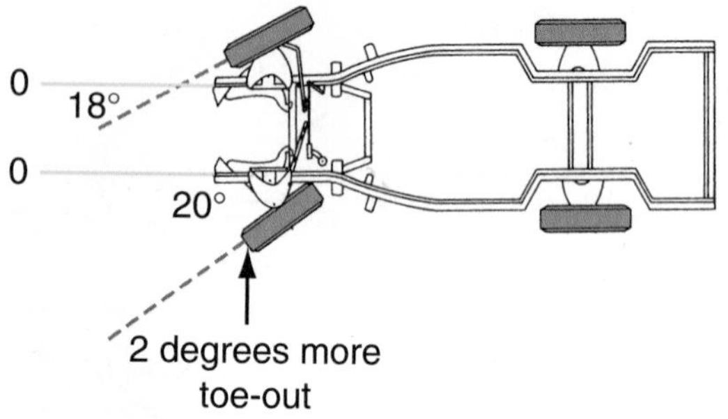

Figure 47-8 Turning angle is affected by toe-out on turns.

Turning radius is not an adjustable angle. If the angle is not correct, the tie-rods, steering arm, or steering knuckle are damaged and will need to be replaced.

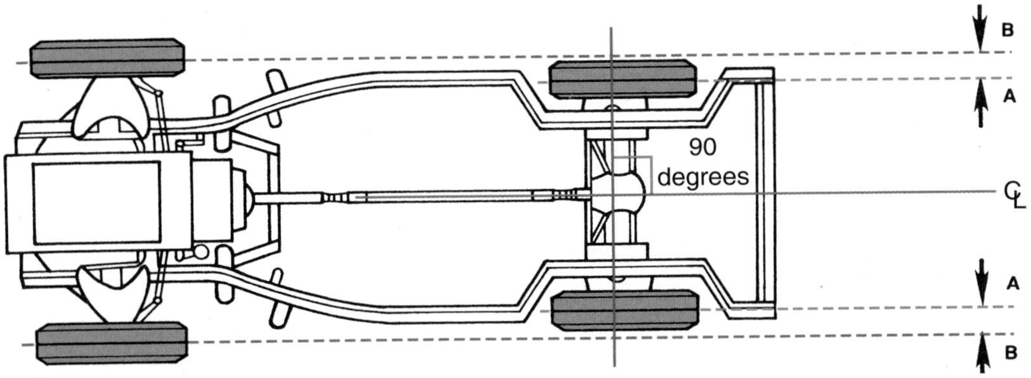

Figure 47-9 When a car is tracking correctly, its rear wheels are the same distance from the front wheels on both sides.

Thrust Line

All vehicles are built around a geometric centerline that runs through the center of the chassis from the back to the front. The **thrust line** is the direction the rear axle would travel if unaffected by the front wheels. This condition is also called **tracking**. Correct tracking results from having all suspension parts in their correct location, in good condition, and aligned so that the rear wheels follow directly behind the front wheels while moving in a straight line **(Figure 47–9)**. For this to occur, the axles and wheels must be parallel with one another and the centerlines through the axles and spindles must be at 90-degree angles to the vehicle's centerline. Simply stated, all four wheels should form a perfect rectangle.

An offset thrust line affects handling by causing a pull in the direction away from the thrust line and can cause tire wear similar in pattern to that of incorrect toe settings. As a general rule, minor variations between the thrust line and centerline are not noticeable and do not cause handling problems as long as the front wheels are aligned parallel with the thrust line.

Load Distribution

Load distribution refers to the load placed on each wheel. Every vehicle is engineered to operate at a designed curb height (also called trim height). At this height, each wheel must carry the correct amount of weight. Excessive loading to the front, rear, or one side of the vehicle changes the curb height, upsetting vehicle balance and steering geometry.

In correct alignment, sagging springs and bent suspension parts can also change this condition, upsetting geometry and placing excessive load on only one or two wheels.

All of these elements—springs, shocks, suspension, and geometry—are engineered to work together as a balanced team to provide safe and comfortable riding and handling. Quite naturally, if one wheel is running under a different condition of weight load and steering geometry, the vehicle does not ride and handle as it is capable of doing.

PREALIGNMENT INSPECTION

Before beginning to align a vehicle, make sure you know why the vehicle needs to be aligned or why the customer has requested an alignment. Often the symptoms will lead to identifying the reason the wheels need to be aligned. Follow the customer interview with a test drive. While driving the car, check to see that the steering wheel is straight. Feel for vibrations in the steering wheel as well as in the floor or seats. Notice any pulling or abnormal handling problems, such as hard steering, tire squeal while cornering, or mechanical pops or clunks. This helps find problems that must be corrected before proceeding with the alignment.

An extremely important part of a wheel alignment is the prealignment inspection. During this inspection, if any parts are found to be defective, they should be replaced before proceeding with the alignment. The inspection should include a careful look at the tires, wheel, suspension system, and steering system. The procedures for inspecting these are detailed in the previous three chapters. A typical alignment inspection report is shown in **Figure 47–10**.

It is important that all abnormal loads be removed before taking any measurements. Obviously, added weight will affect the vehicle's ride height and, therefore, the alignment angles. If the vehicle is normally used to carry heavy objects, such toolboxes, leave them in the vehicle.

Ride Height

Check the vehicle's ride height. Every vehicle is designed to ride at a specific curb height. Curb height specifications and the specific measuring points are given in service manuals. Proper alignment is impossible if the ride height is incorrect. This is especially true for camber because as the height of a vehicle changes, so does the camber of the wheels.

PREALIGNMENT INSPECTION CHECKLIST

Owner ______________________ Phone ______________ Date __________

Address ______________________________ VIN ____________________

Make ________ Model ________ Year _____ Lic. number ________ Mileage ________

1. Road test results	Yes	No	Right	Left
Above 30 MPH				
Below 30 MPH				
Bump steer				
When braking				
Steering wheel movement				
Stopping from 2-3 MPH (Front)				
Vehicle steers hard				
Strg wheel returnability normal				
Strg wheel position				
Vibration	Yes	No	Frnt	Rear

2. Tire pressure Specs Frnt ____ Rear ____

Record pressure found

RF _____ LF _______ RR _______ LR _______

3. Chassis height Specs Frnt ____ Rear ____

Record height found

RF _____ LF _______ RR _______ LR _______

	Yes	No
Springs sagged		
Torsion bars adjusted		

4. Rubber bushings	OK
Upper control arm	
Lower control arm	
Sway bar/stabilizer link	
Strut rod	
Rear bushing	

5. Shock absorbers/struts	Frnt	Rear

6. Steering linkage	Frnt OK	Rear OK
Tie-rod ends		
Idler arm		
Center link		
Sector shaft		
Pitman arm		
Gearbox/rack adjustment		
Gearbox/rack mounting		

7. Ball joints	OK
Load bearings	
Specs: Right ____ Left ____ / Readings: Right ____ Left ____	
Follower	
Upper strut bearing mount	
Rear	

8. Power steering	OK
Belt tension	
Fluid level	
Leaks/hose fittings	
Spool valve centered	

9. Tires/wheels	OK
Wheel runout	
Condition	
Equal tread depth	
Wheel bearing	

10. Brakes operating properly	

11. Alignment	Spec		Initial reading		Adjusted reading	
	R	L	R	L	R	L
Camber						
Caster						
Toe						

Bump steer	Toe change right wheel		Toe change left wheel	
	Amount	Direction	Amount	Direction
Chassis down 3"				
Chassis up 3"				

	Spec		Initial reading		Adjusted reading	
	R	L	R	L	R	L
Toe-out on turns						
SAI						
Rear camber						
Rear total toe						
Rear indiv. toe						
Wheel balance						
Radial tire pull						

Figure 47-10 A checklist for a prealignment inspection and wheel alignment.

Figure 47-11 The sensors used to measure ride height with a wheel aligner. *Courtesy of Hunter Engineering Company*

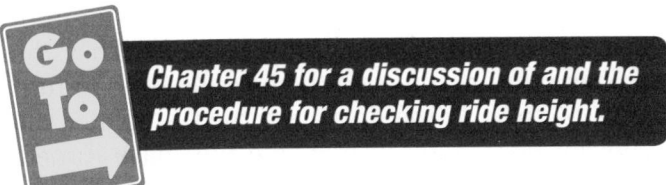

Go To *Chapter 45 for a discussion of and the procedure for checking ride height.*

Some current alignment machines have adapters that allow the unit to measure ride height. This is done without a tape measure; the machine's sensors are installed on the fenders and the measurements are instantly transmitted to the alignment machine (**Figure 47–11**).

Results of Poor Alignment

If no problems were found during the inspection, a good alignment may take care of the customer's concerns. Poor wheel alignment will cause many different things. Use the chart in **Table 47–2** when checking your work after the alignment.

WHEEL ALIGNMENT EQUIPMENT

There are many different ways to measure the alignment angles on a vehicle. The most common way is the use of an alignment machine or rack. The equipment used for checking alignment angles has

TABLE 47-2 EFFECTS OF INCORRECT ALIGNMENT

Problem	Effect
Incorrect camber setting	Tire wear Ball joint/wheel bearing wear Pull to side of most positive/ least negative camber
Too much positive caster	Hard steering Excessive road shock Wheel shimmy
Too much negative caster	Wander Weave Instability at high speeds
Unequal caster	Pull to side most negative/ least positive caster
Incorrect SAI	Instability Poor return Pull to side of lesser inclination Hard steering
Incorrect toe setting	Tire wear
Incorrect turning radius	Tire wear Squeal in turns

evolved from string and measuring tapes to computerized machines. Today, computerized machines are the most commonly used (**Figure 47–12**). A typical computerized system gives information on a CRT screen to guide the technician step-by-step through the alignment process.

Alignment Machine Care

The alignment machine is a precise piece of equipment; therefore, it needs to be taken care of. Failure to

Figure 47-12 A computer-based wheel alignment machine.

do so will lead to incorrect measurements and adjustments. Alignment machines should be periodically calibrated. This can be done by a technician from the manufacturer, or the shop may purchase a calibration fixture for its own use.

The turn tables and slip plates on the rack should be checked for dirt buildup and wear. Both of these can cause them to bind, which will cause incorrect settings. Also, never use the console as a workbench, even for small parts. Extra care should be paid to the alignment heads. They may fall while connecting and disconnecting them. Delicate electronics are contained in the heads; they can be easily damaged. If the heads are dropped, they should be recalibrated before they are used again.

The heads on some newer alignment machines **(Figure 47–13)** do not contain electronic parts. Rather, the heads serve as targets **(Figure 47–14)** for high-imaging cameras mounted above the console **(Figure 47–15)**. Therefore, dropping these will have no effect on their operation.

Figure 47–13 A wheel alignment machine that relies on high-imaging cameras and targets to gather measurements.

Figure 47–14 These targets are in the machine shown in Figure 47–13 in place of the traditional wheel heads.

Figure 47–15 These high-imaging cameras are positioned on a beam above the console of the machine shown in Figure 47–13.

Turning Radius Gauges

Turning radius gauges measure how many degrees the front wheels are turned. They are commonly used to measure camber, caster, and toe-out on turns. Turning radius gauges (sometimes called turn tables) may be portable but are commonly found as part of an alignment rack. To use these gauges, the front wheels are centered on the gauge plates. Then the locking pins are removed to allow the plate to turn with the tires. As the tires are turned, a pointer will indicate how many degrees the tires have turned. To check toe-out on turns, turn one of the tires to 20 degrees. Then look at the gauge on the other tire.

Miscellaneous Tools

Figure 47–16 shows an assortment of the special tools required for wheel alignment and other steering and suspension system work.

ALIGNMENT MACHINES

There are many varieties of alignment machines that have been used through the years. Some are equipped to measure alignment angles at all four wheels of the vehicle, and others measure the angles at only two wheels. Some alignment machines simply display the angle readings, whereas most current models give much more **(Figure 47–17)**, such as:

- A model-specific inspection screen with photos and illustrations
- Caster, camber, SAI, included angle, setback, and toe readings

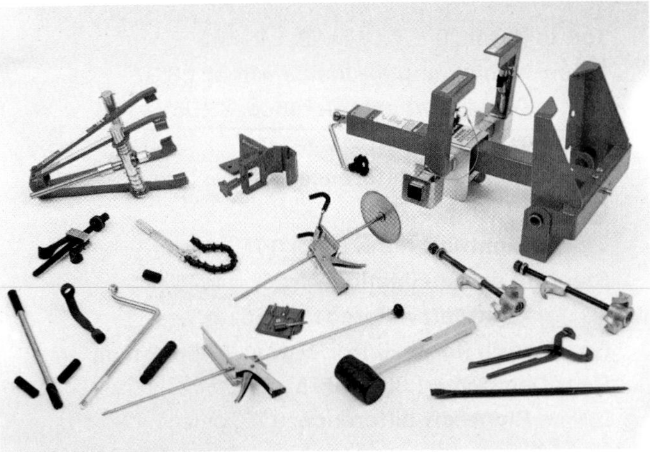

Figure 47-16 An assortment of steering and suspension tools. *Reproduced under license from Snap-on Incorporated. All of the marks are marks of their owners.*

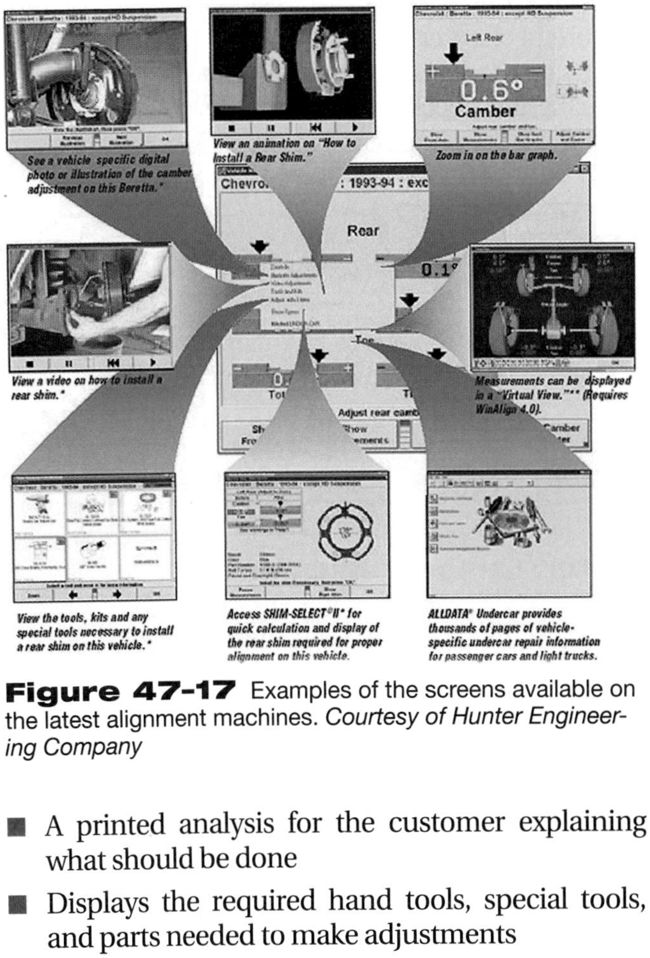

Figure 47-17 Examples of the screens available on the latest alignment machines. *Courtesy of Hunter Engineering Company*

- A printed analysis for the customer explaining what should be done
- Displays the required hand tools, special tools, and parts needed to make adjustments
- Photos and videos showing how to make the adjustment
- The steps required to correct the alignment, including:
  - The correct size and placement of shims (front and rear)
  - How much to turn an eccentric
  - How much to move an arm mounted in the slot
  - How much and in what direction to adjust the tie rod
- Prints a summary and the results of the work

Normally an alignment rack is part of the alignment machine's package. The rack is best described as a limited-purpose vehicle hoist (lift) equipped with turning radius plates.

> **Customer Care**
>
> **Many of the alignment machines have printers that allow the customer to see a before-and-after picture of what the technician did. Along with these readings are the specifications. This is a great tool to gain a customer's confidence in your service, so offer him or her a copy when you have completed an alignment.**

Two-Wheel Alignment

Two-wheel alignment equipment aligns the front wheels to the geometric centerline, assuming that the rear wheels are square with respect to the centerline. If this is true, this type of alignment produces satisfactory results. If not, steering and tracking might be a problem.

Rear-wheel alignment can be checked on a two-wheel machine by simply backing the vehicle onto the rack. These machines do not have the capability of setting the front axle parallel to the rear or aligning the wheels to the vehicle's centerline or thrust line.

Four-Wheel Alignment

Four-wheel (or **total wheel**) **alignment** sets the alignment angles on all four wheels so they are positioned straight ahead with the steering wheel centered. The wheels and axles must also be parallel to one another and perpendicular to a common centerline. More than 85% of all new vehicles require that all four wheels be aligned.

Four-wheel alignment is the best approach in terms of both accuracy and completeness for wheel alignment because it references the front wheels to the rear wheels. The first step during an alignment is adjusting rear camber and toe or bringing the rear axle or wheels into square with the chassis. The front caster, camber, and toe settings can then be adjusted.

PERFORMING AN ALIGNMENT

All wheel alignment angles are interrelated. Regardless of the make of a car or the type of suspension, the same adjustment order—caster, camber, toe—should be followed. Some MacPherson suspensions do not provide for caster or camber adjustments. However, there are some aftermarket kits that allow for these adjustments. Additionally, adjustment methods vary from model to model and, occasionally, even among different model years.

After the vehicle is properly placed on the rack and turn tables, vehicle information is keyed into the machine and the wheel units (heads) are installed. On some machines, the wheel units or heads must be compensated for wheel runout. When compensation is complete, alignment measurements are instantly displayed. Also displayed are the specifications for that vehicle. In addition to the normal alignment specifications, the CRT may display asymmetric tolerances, different left- and right-side specifications, and cross specifications (the difference allowed between the left and right sides). Graphics and text on the screen show the technician where and how to make adjustments **(Figure 47–18)**. As the adjustments are made on the vehicle, the technician can observe the center block slide toward the target. When the block aligns with the target, the adjustment is within half the specified tolerance. A typical procedure for checking the alignment of all four wheels with a computerized alignment machine is shown in Photo Sequence 50.

Specifications

All angles and measurements should be set to the manufacturer's specifications. These specifications usually list a preferred angle for camber, caster, and toe. They are also listed with a minimum and maximum specification, sometimes listed as a plus or minus **(Figure 47–19)**. When making adjustments, you should attempt to achieve the preferred angle. If this is not possible, make sure the measurements are within the minimum and maximum range. If you cannot make an adjustment within that range, some-

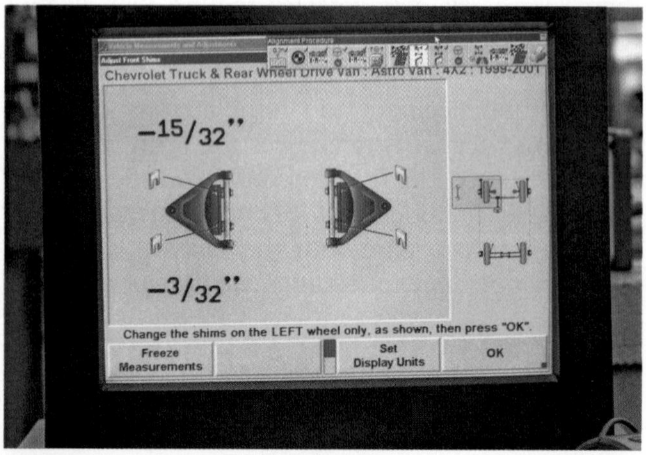

Figure 47-18 On many wheel alignment machines, the screen shows how and where to make adjustments.

Toe-in (total): 0 ± 0.08 in. (0 ± 2 mm)
Wheel turning angle – Inside wheel: 38.37° ± 2°
Outside wheel reference: 33.55°
Camber: −0.67° ± 0.75°
Right-left difference: 0.75° or less
Caster: 3.00° ± 0.75°
Right-left difference: 0.75° or less
Steering axis inclination: 12.25° ± 0.75°
Right-left difference: 0.75° or less
Rear Toe-in (total): 0.16 ± 0.08 in. (−4 ± 2 mm)
Rear Camber: −1.30° ± 0.75°
Right-left difference: 0.75° or less

Figure 47-19 An example of alignment specifications.

thing is wrong and the suspension and frame should be careful checked.

In most cases, the bolts or nuts need to be loosened to make the necessary adjustments. Make sure that all of these are retightened to the required torque.

Performance TIP *If you want to improve the vehicle's handling, for the street only, you can do this by aligning the wheels to improve the tire's road contact, while staying within the manufacturer's acceptable ranges. To do this, set negative camber to the maximum allowable, the maximum positive caster, and the preferred toe settings.*

Caster/Camber Adjustment

Caster affects steering stability and steering wheel returnability. Zero (0) caster is present when the upper ball joint or top strut bearing and lower ball joint are in the same plane as viewed from the side of the vehicle. Positive caster exists when the upper ball joint or top strut bearing is toward the rear of the vehicle in relationship to the lower ball joint. When the upper ball joint or top strut bearing is toward the front of the vehicle in relationship to the lower ball joint, negative caster is present. If the caster at both wheels is not equal, the vehicle will tend to drift toward the side with the lowest caster.

Camber is the inward or outward tilt of the top of the wheel. Adjusting camber centers the vehicle's weight on the tire. Proper camber adjustment minimizes tire wear. Zero (0) camber is present when the wheel is at a perfectly vertical position. The tires have positive camber when the top of the tire is tilted out, or away from the engine. When the top of the tire is tilted in, there is negative camber. Incorrect camber will

P50-1 *Position the vehicle on the alignment rack.*

P50-2 *Make sure the front tires are positioned properly on the turn tables.*

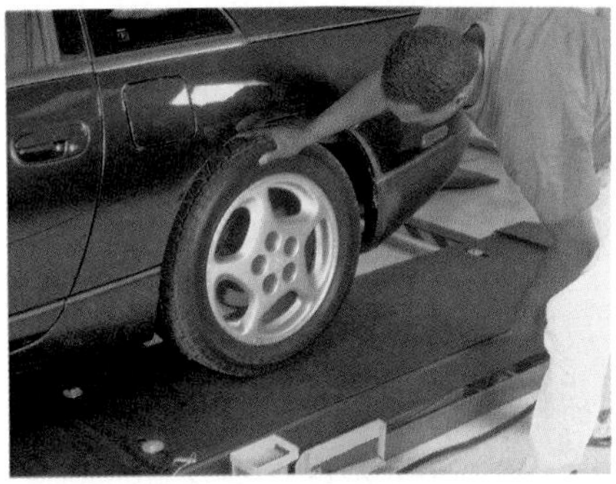

P50-3 *Position the rear wheels on the slip plates.*

P50-4 *Attach the wheel units. Make sure the wheel is raised before attaching an electronic wheel unit.*

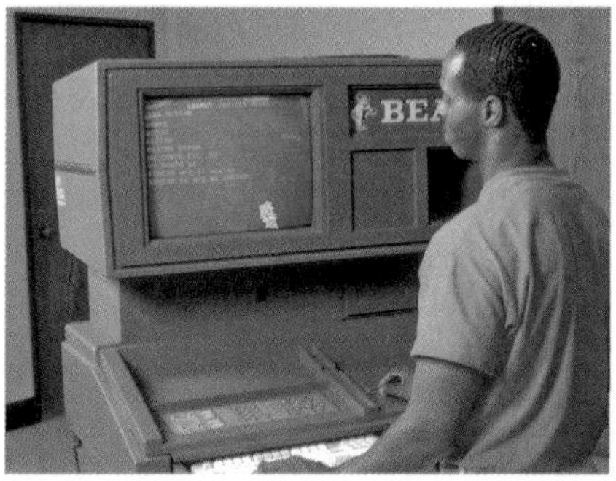

P50-5 *Select the vehicle make and model year.*

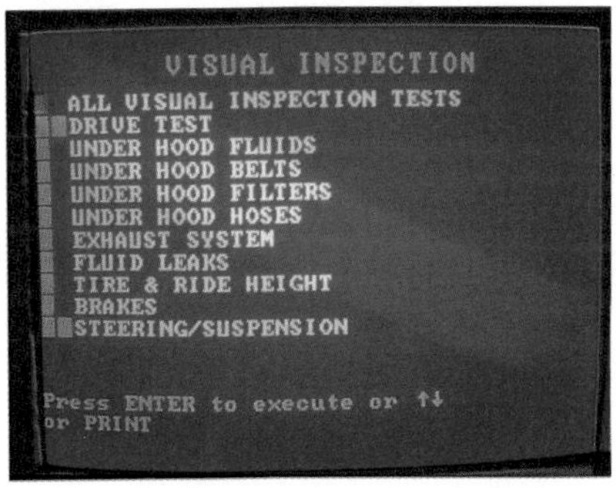

P50-6 *Check the items on the screen before the preliminary inspection.*

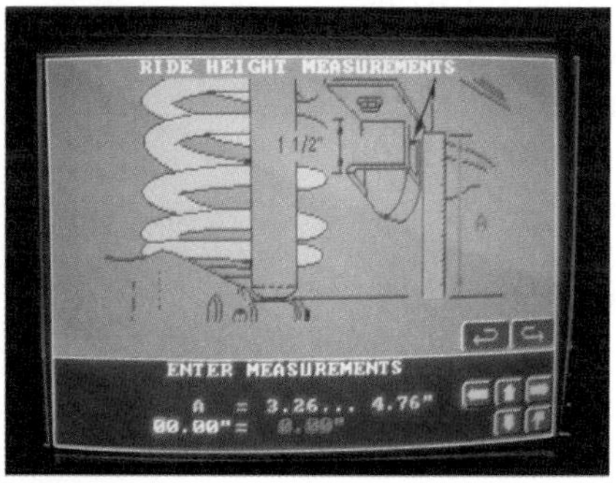

P50-7 *Display the ride height screen. Check the tire condition for each tire on the tire condition screen.*

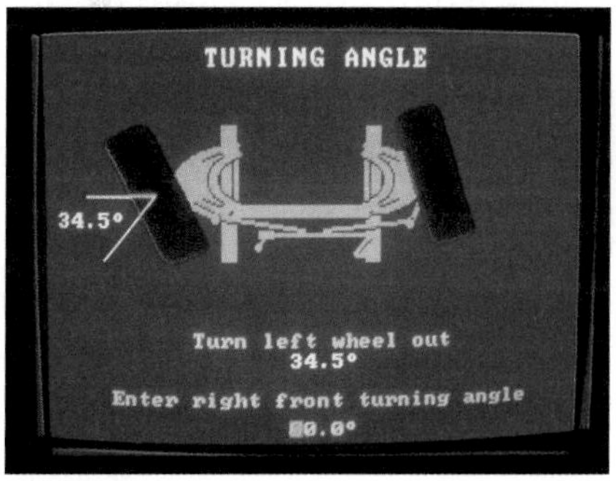

P50-8 *Measure the turning angle.*

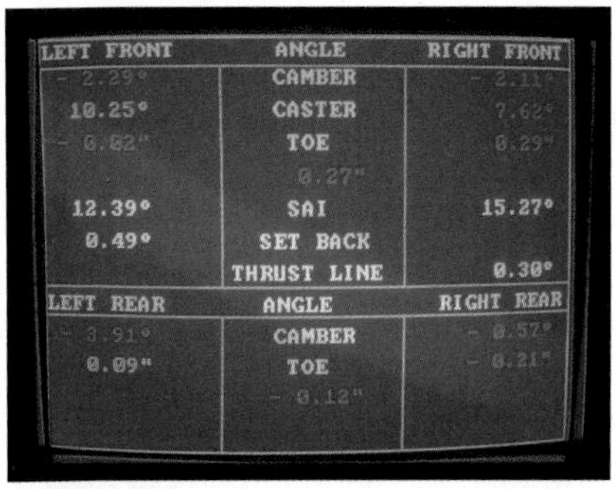

P50-9 *Display the front- and rear-wheel alignment angle screen.*

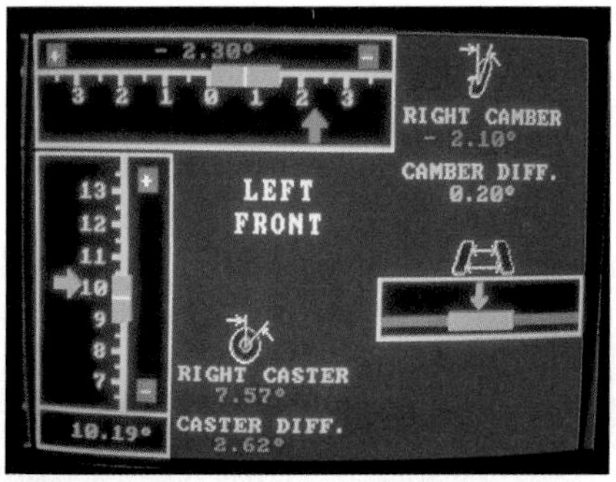

P50-10 *Display the adjustment screen.*

cause excessive stress and wear on suspension parts. Too much negative camber will cause wear on the inside tread of the tire, whereas too much positive camber will cause tire wear on the outside tread. If camber is not the same on both wheels, the vehicle will pull toward the side with the most positive camber.

SHOP TALK

Sometimes it is impossible to adjust the camber to specifications. When this happens, suspect a poorly seated wheel hub and hearings, a bent frame, weak springs, or a bent control arm.

Adjusting for Road Crown To compensate for road crown, most alignment specifications allow for slightly different caster and camber specifications on each side of the vehicle. This difference causes the vehicle to naturally pull slightly to the left to overcome the natural pull of the road to the right.

SHOP TALK

When the vehicle has been aligned to compensate for road crown, it will pull to the left when it is traveling on a flat road or on a road that is angled to the left.

Typically, specifications call for slightly more negative camber (approximately 1/4 degree) on the right side of the vehicle. Or the caster on the left should be set with slightly more negative caster (approximately 1/4 degree).

Adjusting Caster and Camber

Several methods are used to adjust caster and camber; always check the service information to determine how the alignment angles can be changed. Most current wheel alignment machines feature illustrations of what should happen and where.

Camber on nearly all front suspensions is adjustable. On strut-equipped vehicles, camber can be adjusted by moving the top of the strut mount or by adjusting an eccentric bolt located where the strut attaches to the steering knuckle. Caster is adjustable on some strut-equipped vehicles. This is done by again moving the top of the strut. Typically, there is no provision for adjusting caster. Vehicles that have no simple means for adjusting camber or caster require the installation of special kits from the aftermarket to obtain the correct angles. These kits basically contain an adjustable strut mount that permits the top of the strut to move front to back and left to right.

On other vehicles, the upper or lower control arm is used to adjust camber and caster. This is done by adding or removing shims between the control arm and the frame or by rotating an eccentric shaft or eccentric washers. Two bolts attach the control arm to the frame. An equal amount of shims is placed behind or in front of both bolts to correct camber. To gain more negative camber, the lower control arm must be moved outward or the upper arm moved inward. The opposite is true for gaining positive camber.

Caster is adjustable on all vehicles with an upper and a lower control arm. This is done by rotating an eccentric bushing at one of the pivot points for the control arm or adding or subtracting shims between a control arm and the frame.

Rear-wheel camber may be adjustable. When it is, camber is adjusted by eccentric bushings at the control arm pivot point or by an eccentric on the lower mounting of the strut. Solid rear axles seldom have a provision for camber adjustment.

SHOP TALK

A feature of some alignment machines makes the job of adjusting camber and caster much easier and quicker. The feature has many names; the most common one is "jack and hold." This allows for caster and camber adjustments with the wheels off the ground. The advantage here is simply more room to work, and the vehicle's weight is off the adjusting points. Although camber and caster will change when the wheel is lifted, the machine will display the reading before the wheel was raised. This allows you to make adjustments and monitor them just as if the vehicle was sitting flat. Of course all measurements should be verified after the wheels are back on the ground.

Shims Many cars use shims for adjusting caster and camber **(Figure 47–20)**. The shims can be located between the control arm pivot shaft and the inside of the frame. Both caster and camber can be adjusted in one operation requiring the loosening of the shim bolts just once. Caster is changed by adding or subtracting shims from one end of the pivot shaft only. Then, camber is adjusted by adding or subtracting an equal amount of shims from the front and rear bolts. This procedure allows camber to change without affecting the caster setting.

Some cars use shims located between the control arm pivot shaft and the outside of the frame. The adjustment procedure is the same as just described. Always

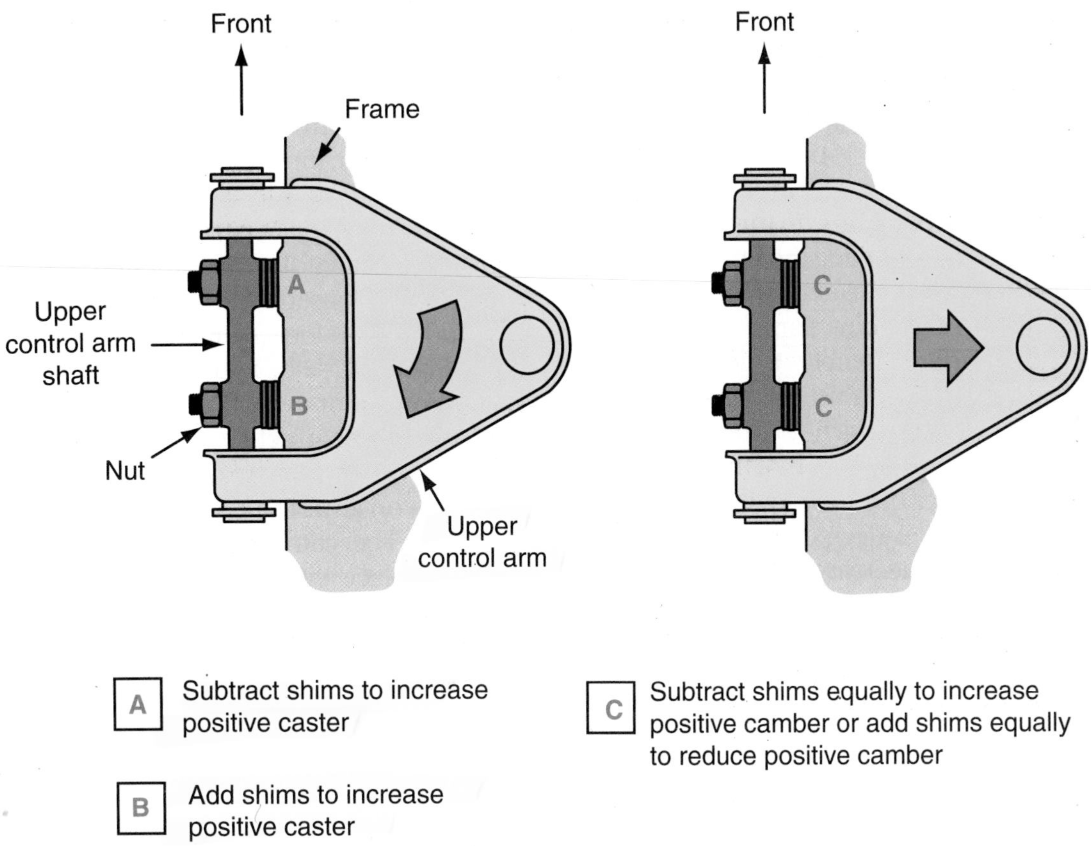

Figure 47-20 Adding and subtracting shims between the control arm and the frame will change caster and camber.

look at the shim arrangements to determine the desired direction of change before loosening the bolts.

Eccentrics and Shims Eccentrics and shims are used on some vehicles to adjust caster and camber. In some designs, an eccentric bolt and cam on the upper control arm adjust both caster and camber. To adjust, the nuts on the upper control arm are loosened first. Then, one eccentric bolt at a time is turned to set caster. Both bolts are turned equally to set camber.

The **eccentric bolt and cam** assembly **(Figure 47–21)** can be located on the inner lower or upper control arm. Unlike other designs, camber is adjusted first. Some car models have a camber eccentric between the steering knuckle and the upper control arm. The camber eccentric is rotated to set camber **(Figure 47–22)**. Caster is set with an adjustable strut rod.

Slotted Frame The slotted frame adjustment has slotted holes under the control arm inner shaft that allow the shaft to be repositioned to the correct caster and camber settings. Caster and camber adjusting tools help in making adjustments. One end of the shaft is moved for caster adjustment. Both ends of the shaft are moved for camber adjustment. Turning a

Figure 47-21 Eccentric bolt and cam, shown on an upper control arm.

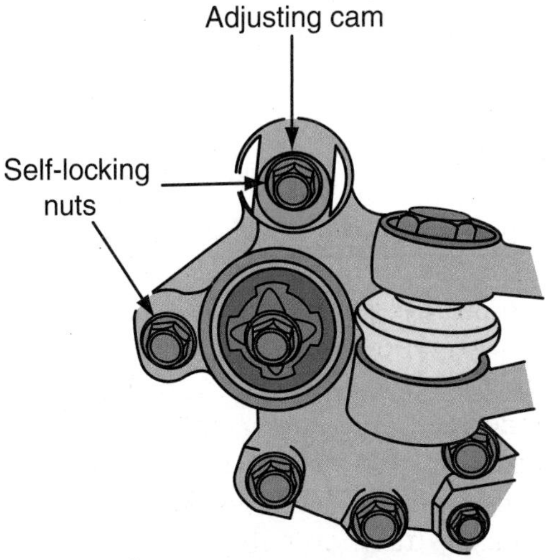

Figure 47-22 Graduated cam for adjusting camber.

nut on one end of the rod changes its length and adjusts caster. Camber is set by an eccentric at the inner end of the lower control arm, or by a camber eccentric in the steering knuckle of the upper support arm, as described earlier.

Rotating Ball Joint and Washers In this design, camber is increased by disconnecting the upper ball joint, rotating it 180 degrees, and reconnecting. This positions the flat of the ball joint flange inboard and increases camber approximately 1 degree. Caster angle is changed with two washers, one 0.12 inch (3 mm) thick and one 0.36 inch (9 mm) thick. The washers are placed at opposite ends of the upper control arm pivot. Placement of the large washer at the rear increases caster by 1 degree. Placement of the large washer at the front decreases caster by 1 degree.

Ball Joint Stud Bushings Some suspension systems have an eccentric bushing at the top of the steering knuckle. This bushing can be used to adjust camber and camber. The bore for the ball joint stud through the bushing is off-center. Rotating the bushing moves the wheel's geometry.

MacPherson Suspension Adjustments

Caster/camber adjustments are made only on certain models with MacPherson suspensions. There are two general OEM procedures for doing this, although aftermarket kit adapters are available for some models. Service information must be consulted for an accurate listing of models on which adjustments can be made.

In one version, a cam bolt at the base of the strut assembly is used to adjust camber **(Figure 47–23)**. On different models, this bolt can be either the upper or the lower of the two bolts connecting the strut assembly to the steering knuckle. Both bolts must be loosened to make the adjustment, and the wheel assembly must be centered. Turn the cam bolt to reach the correct alignment, then retighten the bolts to the appropriate torque specifications. There is no caster adjustment on this version.

In the other form of kit adapter, both caster and camber are adjustable at the strut upper mount. Slots in the mounting plate permit the strut assembly to be shifted to reach the alignment specifications. To adjust caster, loosen the three locknuts on the mounting studs and relocate the plate. Do not remove the nuts **(Figure 47–24)**. Loosen the center locknut and slide it toward or away from the engine as needed to adjust camber correctly.

While caster cannot be adjusted on many MacPherson strut front suspensions, camber can be adjusted. The camber is such that, although it is locked in place with a pop rivet, it can be adjusted by removing the rivet from the camber plate and

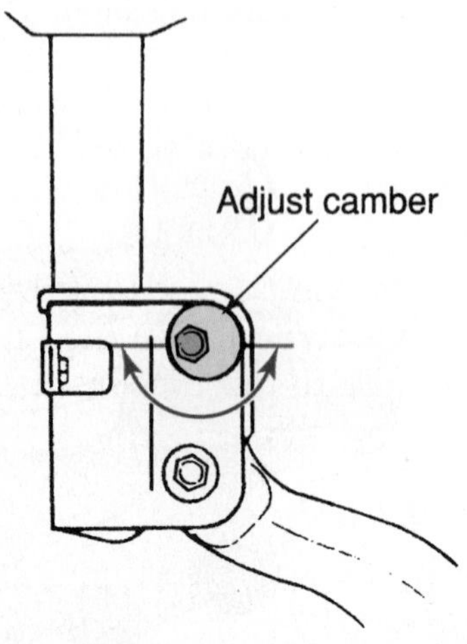
Adjust camber

Figure 47–23 Some MacPherson suspensions use cam bolts at the connection to the steering knuckle for camber adjustments. *Courtesy of Hunter Engineering Company*

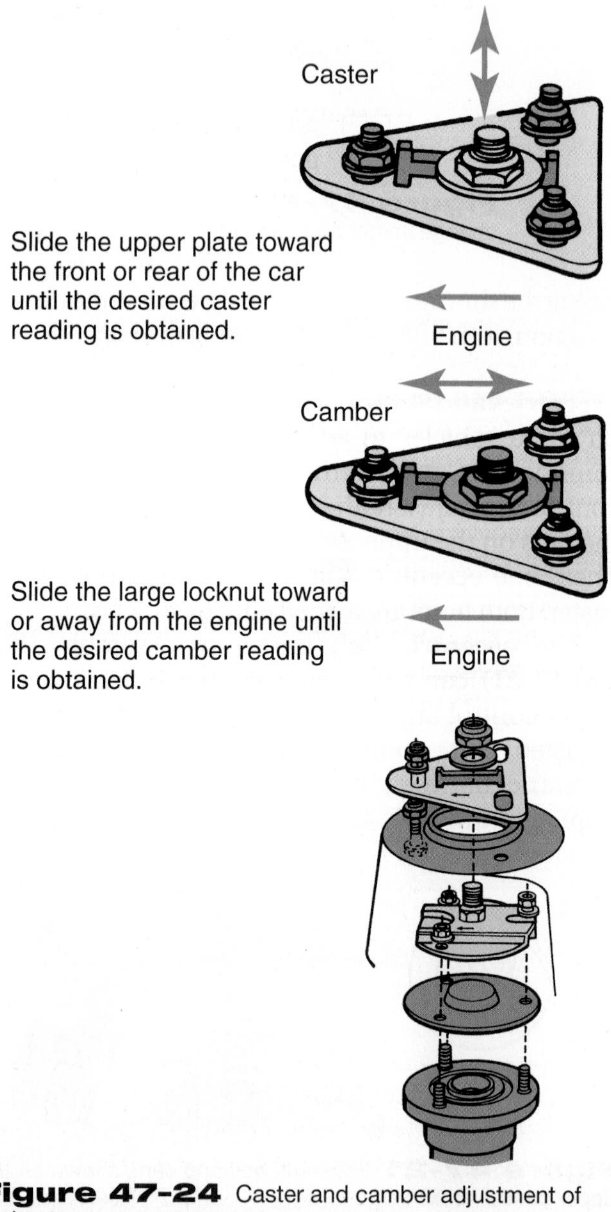
Caster
Slide the upper plate toward the front or rear of the car until the desired caster reading is obtained.
Engine
Camber
Slide the large locknut toward or away from the engine until the desired camber reading is obtained.
Engine

Figure 47–24 Caster and camber adjustment of locknuts.

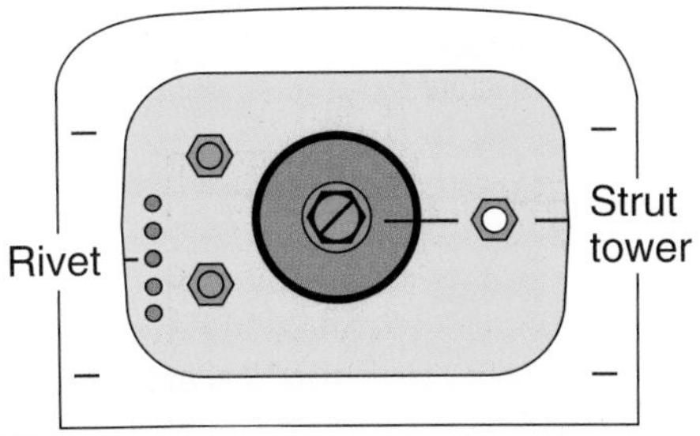

Figure 47–25 An upper strut mount with a camber plate. Note the location of the rivet and the attaching bolts. *Courtesy of Hunter Engineering Company*

loosening the three nuts that hold the plate to the body apron **(Figure 47–25)**. Camber is changed by moving the top of the shock strut to the position in which the desired camber setting is achieved. The nuts are then tightened to specifications. (It is not necessary to install a new pop rivet.)

Rear-Wheel Camber Adjustments

Like front camber, rear camber affects both tire wear and handling. The ideal situation is to have zero running camber on all four wheels to keep the tread in full contact with the road for optimum traction and handling.

Camber is not a static angle. It changes as the suspension moves up and down. Camber also changes as the vehicle is loaded and the suspension sags under the weight.

To compensate for loading, most vehicles with independent rear suspension often call for a slight amount of positive camber. A collapsed or mislocated strut tower, bent strut, collapsed upper control arm bushing, bent upper control arm, sagging spring, or an overloaded suspension can cause the rear wheels to have negative camber. A bent spindle or strut or bowed lower control arm can cause too much positive camber. Even rigid rear axle housings in rear-wheel-drive vehicles can become bowed by excessive torque, severe overloading, or road damage.

CAUTION!

Never jack up or lift a FWD vehicle on its rear axle. The weight of the vehicle may cause the axle to bend and result in misalignment of the rear wheels. Always lift the vehicle at the recommended lifting points.

Besides wearing the tires unevenly across the tread, uneven side-to-side camber (as when one wheel leans in and the other does not) creates a steering pull just like it does when the camber readings on the front wheels do not match. It is like leaning on a bicycle. A vehicle always pulls toward a wheel with the most positive camber. If the mismatch is at the rear wheels, the rear axle pulls toward the side with the greatest amount of positive camber. If the rear axle pulls to the right, the front of the car drifts to the left, and the result is a steering pull even though the front wheels may be perfectly aligned.

The methods used to adjust rear suspensions vary. On some semi-independent suspensions, camber and toe are adjusted by inserting different sizes of shims or a full contact shim between the rear spindle and the spindle mounting **(Figure 47–26)**. The shim thickness is changed between the top or bottom of the spindle to adjust camber. Many shims are now available that are round but have different thicknesses through their diameters.

On others, a camber adjustment can be made by installing a wedge spacer between the top of the knuckle and the strut. Many rear suspensions have eccentric bolts and cams at the mounting points for the control and/or trailing arms **(Figure 47–27)**.

Toe Adjustment

Toe is the last alignment angle to be set. The same procedure is followed on all vehicles, except those with bonded ball stud sockets. Correct toe will minimize tire wear and rolling friction.

To adjust toe, start by being sure the steering wheel is centered **(Figure 47–28)** when the front wheels point straight ahead. Using a steering wheel holder, secure the steering wheel in that position. Many steering wheel holders are installed between the steering wheel and the top of the front seat. Then loosen the retaining bolts on the tie-rod adjusting sleeves. Turn the sleeves to move the tie-rod ends **(Figure 47–29)**.

Some manufacturers of alignment machines recommend that a steering wheel holder not be used. Instead, a steering wheel angle gauge is used to ensure that the steering wheel is centered.

On many rack and pinion systems, the tie-rod locknut must be loosened and the tie rod rotated to adjust toe at each wheel **(Figure 47–30)**. Before rotating the tie-rod, the small outer bellows clamp must be loosened to prevent the boot from twisting.

Other rack and pinion tie-rod ends have internal threads and a threaded adjuster. One end of the adjuster has right-hand threads, the other has left-hand threads. As the adjuster is turned, it changes the overall length of the tie-rod, thereby changing toe.

SHOP TALK

If you are unable to adjust the toe to specifications, check for a bent lower control arm, worn or stripped tie-rod sleeves, or a bent frame.

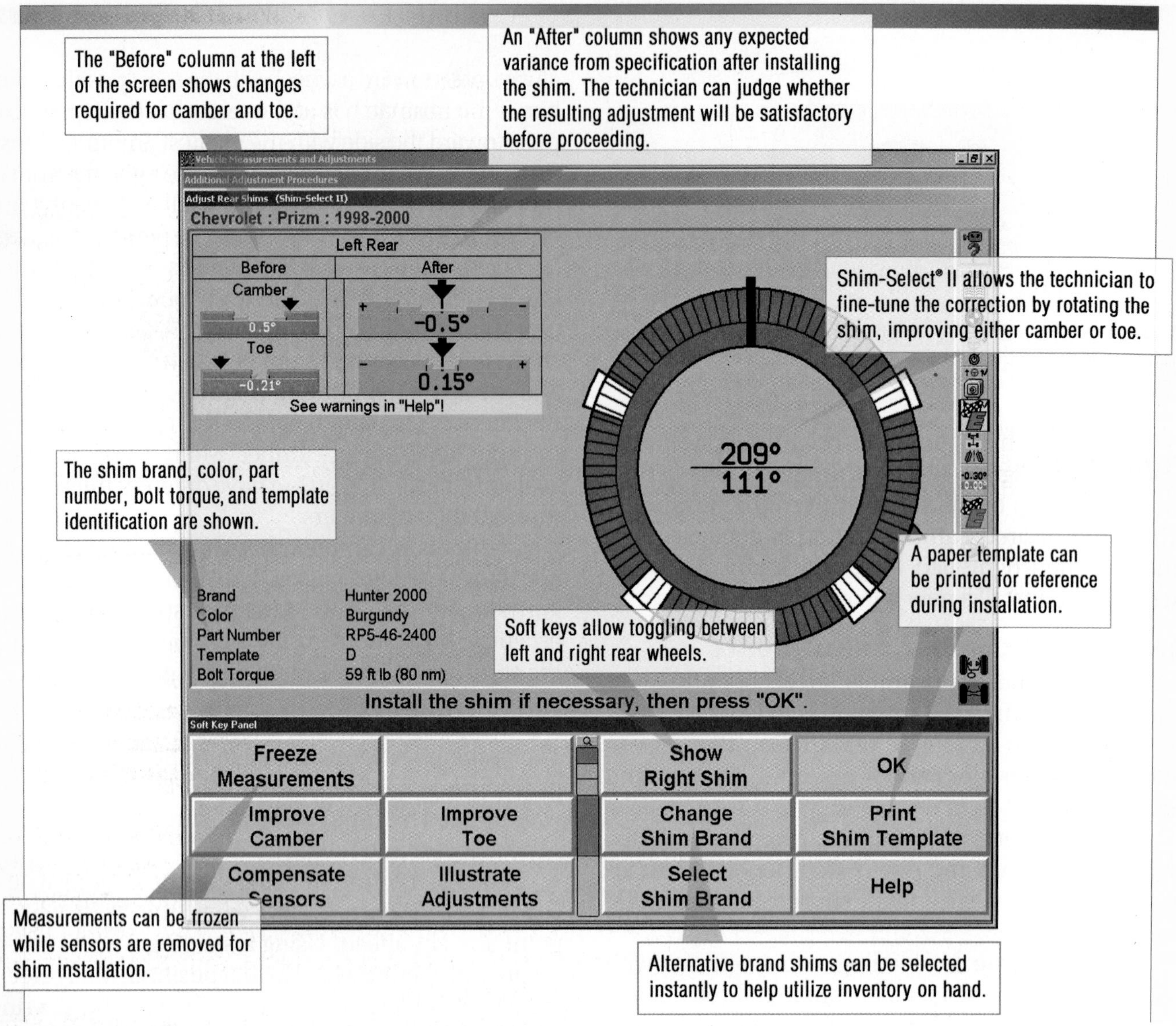
The "Before" column at the left of the screen shows changes required for camber and toe.
An "After" column shows any expected variance from specification after installing the shim. The technician can judge whether the resulting adjustment will be satisfactory before proceeding.
Vehicle Measurements and Adjustments
Additional Adjustment Procedures
Adjust Rear Shims (Shim-Select II)
Chevrolet : Prizm : 1998-2000
Left Rear
Before
After
Camber
0.5°
-0.5°
Toe
-0.21°
0.15°
See warnings in "Help"!
209°
111°
Shim-Select® II allows the technician to fine-tune the correction by rotating the shim, improving either camber or toe.
The shim brand, color, part number, bolt torque, and template identification are shown.
A paper template can be printed for reference during installation.
Brand
Hunter 2000
Color
Burgundy
Part Number
RP5-46-2400
Template
D
Bolt Torque
59 ft lb (80 nm)
Soft keys allow toggling between left and right rear wheels.
Install the shim if necessary, then press "OK".
Soft Key Panel
Freeze Measurements
Show Right Shim
OK
Improve Camber
Improve Toe
Change Shim Brand
Print Shim Template
Compensate Sensors
Illustrate Adjustments
Select Shim Brand
Help
Measurements can be frozen while sensors are removed for shim installation.
Alternative brand shims can be selected instantly to help utilize inventory on hand.

Figure 47-26 A computer screen showing how the adjusting shim should be installed to correct camber and toe at the rear of the vehicle. *Courtesy of Hunter Engineering Company*

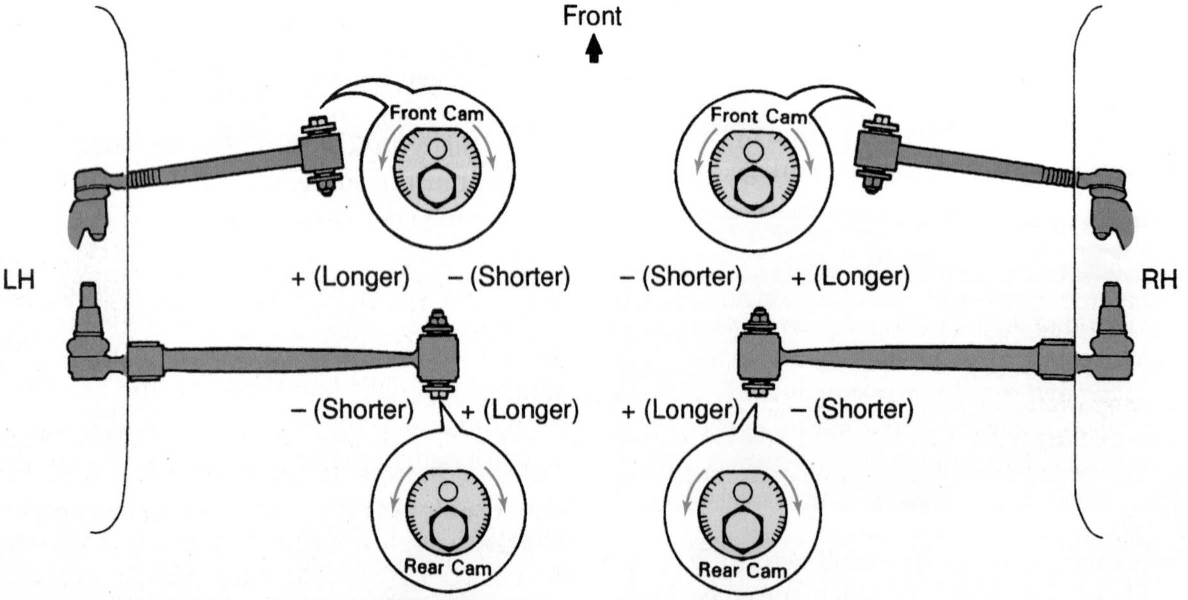
Front
Front Cam
Front Cam
LH
+ (Longer)
– (Shorter)
– (Shorter)
+ (Longer)
RH
– (Shorter)
+ (Longer)
+ (Longer)
– (Shorter)
Rear Cam
Rear Cam

Figure 47-27 An independent rear suspension may have eccentric cams at all of its control arms, one for camber adjustment and another for toe adjustment. *Courtesy of Toyota Motor Sales, U.S.A., Inc.*

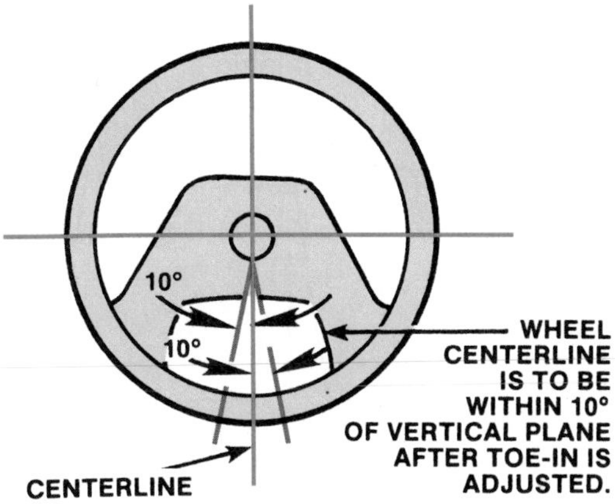

Figure 47-28 A typical acceptable steering wheel position—measured from a normal spoke angle.

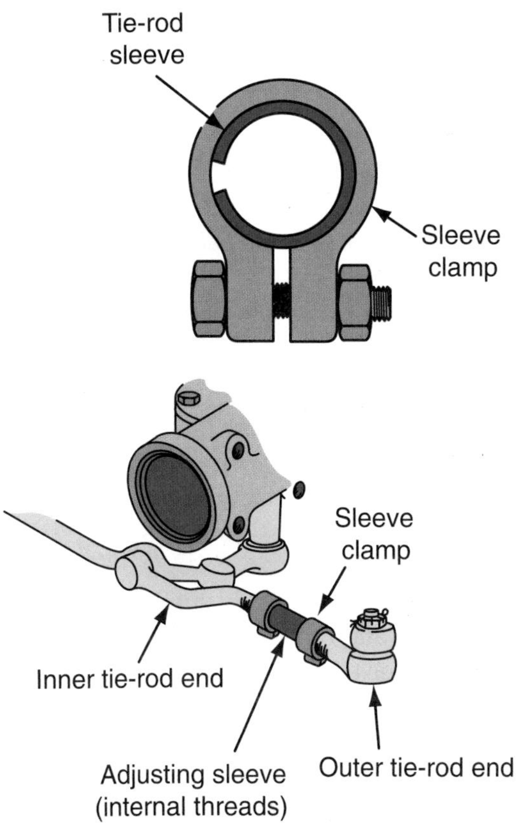

Figure 47-29 To adjust toe, the sleeve clamps on the tie-rod assembly are loosened and the sleeve is rotated.

An ideal toe condition is both wheels exactly straight ahead, which would minimize tire wear. This, however, is not possible because of the many factors affecting alignment. As a result of these numerous conditions dealing with both tire wear and handling, all suspensions are designed with a slight toe-in or toe-out.

Any misalignment of the steering linkage pivot point or control arm pivot point (such as the center link or rack and pinion out of place) causes the condition known as **toe-change**. Toe-change involves turning

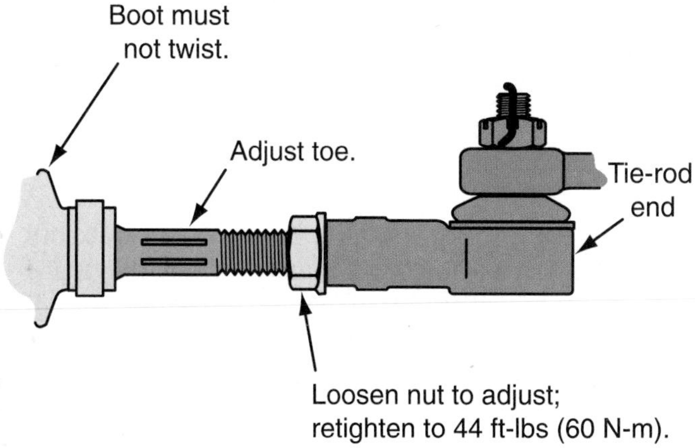

Figure 47-30 Rotating the tie rod to adjust the toe on a rack and pinion steering gear.

the wheels from their straight-ahead position as the suspension moves up and down.

The change might be only one wheel, both wheels in the same direction, or both wheels in the opposite direction. Regardless of the condition, any change of one or more wheels is a toe-change condition. The results are tire wear and a hard-to-handle vehicle. The poor handling effects can get to the point that the vehicle is dangerous to drive.

Toe change is not a specification; it is a condition in which the toe setting constantly varies. It must be determined by equipment or a method that measures individual wheel toe at all suspension heights. There must be a change in suspension heights for any changes to occur.

Lightweight front-wheel-drive vehicles can be affected greatly by toe-change. With these vehicles, the front wheels are no longer being pushed. They actually pull the vehicle forward and, as a result, if the wheels are not maintaining a straight-ahead position, they affect directional control. Adverse road conditions, such as wet or icy conditions, can also increase the handling effects created by toe-change in the front-wheel-drive car.

Rear Toe Rear toe, like front toe, is a critical tire wear angle. If toed-in or toed-out, the rear tires scuff just like the front ones. Either condition can also contribute to steering instability as well as reduced braking effectiveness. Keep this in mind with antilock brake systems.

Like camber, rear toe is not a static alignment angle. It changes as the suspension goes through jounce and rebound. It also changes in response to rolling resistance and the application of engine torque. With FWD vehicles, the front wheels tend to toe-in under power while the rear wheels toe-out in response to rolling resistance and suspension compliance.

With RWD vehicles, the opposite happens: the front wheels toe-out while the rear wheels on an independent suspension try to toe-in as they push the vehicle ahead.

If rear toe is not within specifications, it affects tire wear and steering stability just as much as front toe. A total toe reading that is within specifications does not necessarily mean the wheels are properly aligned—especially when it comes to rear toe measurements. If one rear wheel is toed-in while the other is toed-out by an equal amount, total toe would be within specifications. However, the vehicle would have a steering pull because the rear wheels would not be parallel to center.

Remember, the ideal situation is to have all four wheels at zero running toe when the car is traveling down the road. This is especially true with antilock brakes, where improper toe can affect brakes; such a condition can affect brake balance when braking on slick or wet surfaces, causing the antilock brakes to cycle on and off to prevent a skid. Without antilock brakes, this condition may upset traction enough to cause an uncontrollable skid.

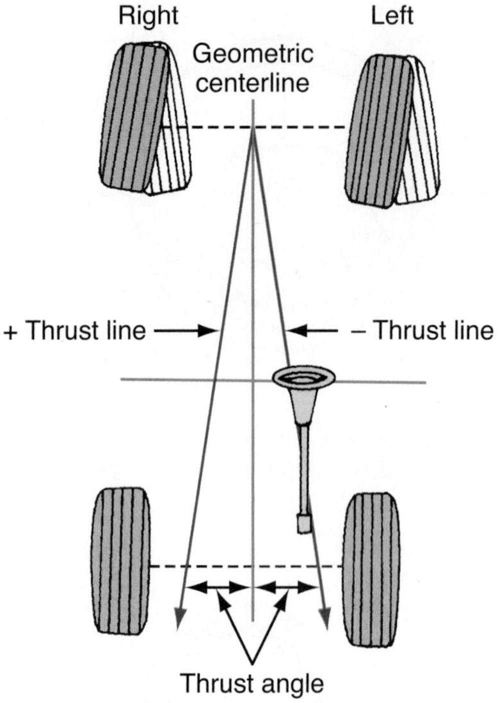

Note:
A + thrust line will cause the vehicle to pull to the left
A – thrust line will cause the vehicle to pull to the right

Figure 47–31 The thrust angle is the angle between the geometric centerline of the vehicle and the thrust line.

Thrust Line

If both rear wheels are square to one another and the rest of the vehicle, the thrust line is perpendicular to the rear axle and coincides with the vehicle's centerline. But if one or both rear wheels are toed in or out, or one is set back slightly with respect to the other, the thrust line is thrown off-center.

The thrust angle **(Figure 47–31)** is eliminated by aligning the rear axle with the geometric centerline of the vehicle and then adjusting the toe of the front wheels. This is normally done during a four-wheel alignment if the rear toe is adjustable. In most cases, doing this will also center the steering wheel. If the thrust angle is ignored and front toe is set to the geometric centerline, the steering wheel will be off-center.

Four-wheel-alignment machines check the toe at each wheel. An incorrect thrust line may be caused by unequal amounts of toe at the rear wheels. Therefore, each rear wheel should be adjusted to the same toe specification. On most FWD vehicles and those with adjustable rear toe, this can be easily done by using the factory-provided toe adjustments, by placing toe/camber shims between the rear spindles and axle, or by using eccentric bushing kits.

On RWD vehicles that have a solid rear axle, changing rear toe is not as easy. Sometimes the floor pan or frame rails are misaligned from the factory or from collision damage. Short of straightening the chassis on a collision bench to restore the correct control arm or spring-mount geometry, the only other options are to try some type of offset trailing arm bushing with the coil springs, or to reposition the spring shackles or U-bolts with leaf springs.

If rear toe cannot be easily changed, the next best alternative is to align the front wheels to the rear axle thrust line rather than the vehicle centerline. Doing this puts the steering wheel back on center and eliminates the steering pull—but it does not eliminate tracking.

Setback

Setback is a condition when one wheel on an axle is set behind the other **(Figure 47–32)**. This means the distance between the centers of the tires on one side of the vehicle will be different than on the other side. Like the thrust angle, setback will cause an off-centered steering wheel. Most alignment machines measure setback. Excessive setback is evident by a reading of more than ¼ inch (6.35 mm). Setback is typically caused by a bent suspension part, problems with the upper strut mount, or a misaligned lower control arm. It can also be caused by a misaligned front cradle. If the setback is slight, the difference will be compensated during a good four-wheel alignment.

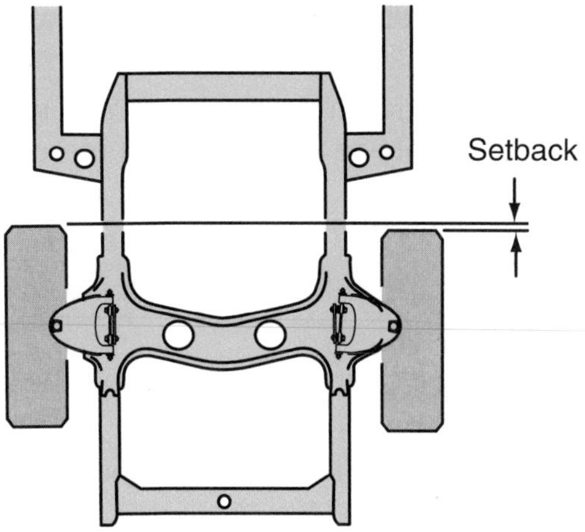

Figure 47-32 Setback is a condition when one wheel on an axle is set behind the other.

FOUR-WHEEL-DRIVE VEHICLE ALIGNMENT

With front-wheel-drive and full-time 4WD vehicles, the front wheels are also driving wheels. As the front wheels pull the vehicle, the wheels tend to toe-in when torque is applied. To offset this tendency, the front wheels usually need less static toe-in to produce zero running toe. In fact, the preferred toe alignment specifications in this instance can be zero to slightly toed-out (1⁄16-inch [1.5 mm] toe-out).

It is important to note that when the front wheels of a part-time 4WD system are freewheeling, they behave the same as the front wheels in a rear-wheel-drive vehicle. That is, they roll rather than pull. The wheels tend to-toe-out, so the static toe setting would have to toe-in to achieve zero running toe when driving in the two-wheel mode.

The tires suffer in proportion to toe misalignment. For a tire that is only 1⁄8 inch (3 mm) off (1⁄4 degree), the tire is scrubbed sideways 12 feet (3.6 meters) for every mile traveled. That may not sound like much, but 12 feet (3.6 meters) of sideways scrub every mile can cut a tire's life in half.

If rapid tire wear seems to be the problem, look for the telltale feathered wear pattern. If the wheels are running toe-in, the feathered wear pattern leaves sharp edges on the inside edges of the tread. If the wheels are running toe-out, the sharp edges are toward the outside of the tread. It is usually easier to feel the feathered wear pattern than to see it. To tell which way the wear pattern runs, rub your fingers sideways across the tread.

On most 4WD vehicles, caster is not adjustable. Aftermarket companies do provide caster adjustment kits for some pickups. These kits may contain shims or eccentric cam and bolt. For some pickups, the aftermarket cam kit will also provide for camber adjustments.

On other 4WD vehicles, camber is adjusted by installing adjustment shims between the spindle and the steering knuckle or by installing and/or adjusting an eccentric bushing at the upper ball joint. Most aftermarket parts manufacturers have camber adjustment shims available in various thicknesses and diameters. Never stack the shims. Only one shim per side should be used.

CASE STUDY

Two days after having a complete alignment performed on her late-model vehicle, a customer returns to the shop complaining of improper service. Since having the alignment work done, the vehicle now consistently pulls to the right.

The technician who performed the work road tests the vehicle and verifies that the car is drifting to the right. The technician rechecks the alignment against factory specifications. All settings appear to be within range. A check of the suspension system finds nothing wrong. Desperately trying to figure out why the vehicle is drifting, the technician reviews some notes she kept from an alignment class attended several months before. Here, she finds the answer. When setting the right wheel, she forgot to allow for road crown. The technician readjusts the right wheel to account for road crown angle and test drives the vehicle. The drifting has been eliminated.

KEY TERMS

Camber
Caster
Eccentric bolt and cam
Four-wheel alignment
Included angle
Road crown
Setback
Steering axis inclination (SAI)
Thrust angle
Thrust line
Toe
Toe-change
Total wheel alignment
Tracking

SUMMARY

- Caster is the angle of the steering axis of a wheel from the vertical, as viewed from the side of the vehicle. Tilting the wheel forward is negative caster. Tilting backward is positive.
- Camber is the angle represented by the tilt of either the front or rear wheels inward or outward from the vertical as viewed from the front of the car.
- Toe is the distance comparison between the leading edge and trailing edge of the front tires. If the edge distance is less, then there is toe-in. If it is greater, there is toe-out.
- The difference of rear toe from the geometric centerline of the vehicle is called the thrust angle. The vehicle tends to travel in the direction of the thrust line, rather than straight ahead.
- Steering axis inclination (SAI) angles locate the vehicle weight to the inside or outside of the vertical centerline of the tire. The SAI is the angle between the true vertical and a line drawn between the steering pivots as viewed from the front of the vehicle.
- Turning radius or cornering angle is the amount of toe-out present on turns.
- In correct tracking, all suspensions and wheels are in their correct locations and conditions and are aligned so that the rear wheels follow directly behind the front wheels while moving in a straight line.
- It is important to remember that approximately 85% of today's vehicles not only undergo front-end alignment but require rear-wheel alignment as well.
- The primary objective of four-wheel or total-wheel alignment, whether front or rear drive, solid axle, or independent rear suspension, is to align all four wheels so the vehicle drives and tracks straight with the steering wheel centered. To accomplish this, the wheels must be parallel to one another and perpendicular to a common centerline.

REVIEW QUESTIONS

1. Define positive camber.
2. What tire wear pattern will result from excessive negative camber?
3. What can be caused by excessive negative caster?
4. Describe the difference between toe-in and toe-out.
5. Describe thrust angle.
6. How will a vehicle handle if it has too much toe-out?
7. In what direction must the bottom of a tire and wheel assembly be moved to add more positive camber?
8. What parts provide for toe-out on turns?
9. Define the term *tracking* as it applies to vehicle handling.
10. Describe how caster adjustment may be used to compensate for road crown.
11. Which of the following is a good definition of SAI?
    a. forward tilt of the top of the steering knuckle
    b. inward tilt of the spindle steering arm
    c. inward tilt of the top of the ball joint or strut
    d. outward tilt of the top of the ball joint or strut
12. What is the correct sequence for adjusting alignment angles during a four-wheel alignment?
13. A 3-degree difference in the SAI angle on each side of the front suspension may cause a FWD vehicle to have increased ______ during hard acceleration.
14. Unequal SAI angles on the left and right sides of the front suspension may cause______.
    a. tread wear on the front tires
    b. brake pull during sudden stops
    c. ball joint wear
    d. steering wander while driving straight ahead
15. While driving straight, a FWD car pulls to the right. The most likely cause is______.
    a. more positive camber on the left front wheel compared to the right front wheel
    b. sagged front springs and improper front-wheel toe setting
    c. less positive caster on the right front wheel compared to the left front wheel
    d. the SAI on the right front wheel is 1½ degrees more than the SAI on the left.

ASE-STYLE REVIEW QUESTIONS

1. Technician A says that positive caster provides directional stability. Technician B says that positive caster results from excessive positive camber. Who is correct?
    a. Technician A
    b. Technician B
    c. Both A and B
    d. Neither A nor B

2. Technician A says that the purpose of the steering wheel holder is to make sure that the steering wheel is centered after the alignment. Technician B says that if the steering wheel is not centered after an alignment, it should be pulled off the column and positioned correctly. Who is correct?
   - a. Technician A
   - b. Technician B
   - c. Both A and B
   - d. Neither A nor B
3. Technician A says that the presence of a thrust angle can cause poor directional stability on ice, snow, or a wet pavement. Technician B says that it can also increase tire wear because the front wheels fight the rear ones for steering control. Who is correct?
   - a. Technician A
   - b. Technician B
   - c. Both A and B
   - d. Neither A nor B
4. Technician A says that camber changes as the suspension moves up and down. Technician B says that camber changes as the vehicle is loaded and the suspension sags under the weight. Who is correct?
   - a. Technician A
   - b. Technician B
   - c. Both A and B
   - d. Neither A nor B
5. Technician A says that camber can be adjusted on nearly all front suspensions. Technician B says that SAI can only be adjusted on strut suspensions. Who is correct?
   - a. Technician A
   - b. Technician B
   - c. Both A and B
   - d. Neither A nor B
6. Technician A says that on FWD vehicles, the front wheels toe-out while the rear wheels on an independent suspension try to toe-in as the vehicle moves ahead. Technician B says that on RWD vehicles, the front wheels tend to toe-in while the rear wheels toe-out in response to rolling resistance and suspension compliance. Who is correct?
   - a. Technician A
   - b. Technician B
   - c. Both A and B
   - d. Neither A nor B
7. Technician A says that if caster at both wheels is not equal, the vehicle will tend to drift toward the side with the most positive caster. Technician B says that if camber is not the same on both wheels, the vehicle will pull toward the side with the most positive camber. Who is correct?
   - a. Technician A
   - b. Technician B
   - c. Both A and B
   - d. Neither A nor B
8. While discussing front-wheel caster: Technician A says that excessive positive caster increases the steering effort and causes harsh riding. Technician B says that excessive negative caster causes front-wheel shimmy. Who is correct?
   - a. Technician A
   - b. Technician B
   - c. Both A and B
   - d. Neither A nor B
9. While performing a prealignment inspection: Technician A says that improper front-wheel bearing adjustment may affect wheel alignment angles. Technician B says that worn ball joints have no effect on wheel alignment angles. Who is correct?
   - a. Technician A
   - b. Technician B
   - c. Both A and B
   - d. Neither A nor B
10. While performing a prealignment inspection: Technician A says that a prealignment inspection should include checking the vehicle interior for heavy items. Technician B says that tools and other items normally carried in the vehicle should be included during an alignment. Who is correct?
    - a. Technician A
    - b. Technician B
    - c. Both A and B
    - d. Neither A nor B

Section 8 BRAKES

CHAPTER 48

BRAKE SYSTEMS

OBJECTIVES

■ Explain the basic principles of braking, including kinetic and static friction, friction materials, application pressure, and heat dissipation. ■ Describe the components of a hydraulic brake system and their operation, including brake lines and hoses, master cylinders, system control valves, and safety switches. ■ Perform both manual and pressure bleeding of the hydraulic system. ■ Briefly describe the operation of drum and disc brakes. ■ Inspect and service hydraulic system components. ■ Describe the operation and components of both vacuum-assist and hydraulic-assist braking units.

It is commonly believed that the purpose of a brake system is to slow or halt the motion of a vehicle. However, that is really not true. The friction of the tires against the road is what slows down and stops a vehicle. The brake system slows or stops the rotation of the wheels. This is a minor point but one that extends the responsibility for braking to the tires as well as the brake system.

The brake system converts the momentum of the vehicle into heat by slowing and stopping the vehicle's wheels. This is done by causing friction at the wheels. The application of the friction units is controlled by a hydraulic system. This chapter looks at the basics of all brake systems and gives a detailed look at the hydraulic systems required to stop a vehicle.

Go To Chapter 8 for a detailed discussion on friction and its effects.

FRICTION

There are two basic types of friction that explain how brake systems work: kinetic, or moving, and **static**, or stationary **(Figure 48–1)**. The amount of friction, or resistance to movement, depends on the type of materials in contact, the smoothness of their rubbing surfaces, and the pressure holding them together (often gravity or weight). Friction always converts moving, or kinetic, energy into heat. The greater the friction between two moving surfaces, the greater the amount of heat produced.

As the brakes on a moving automobile are applied, rough-textured pads or shoes are pressed against rotating parts of the vehicle—either rotors (discs) or drums. The kinetic energy, or momentum, of the vehicle is then converted into heat energy by the kinetic friction of rubbing surfaces and the car or truck slows down.

When the vehicle comes to a stop, it is held in place by static friction. The friction between the surfaces of the brakes and between the tires and the road resists any movement.

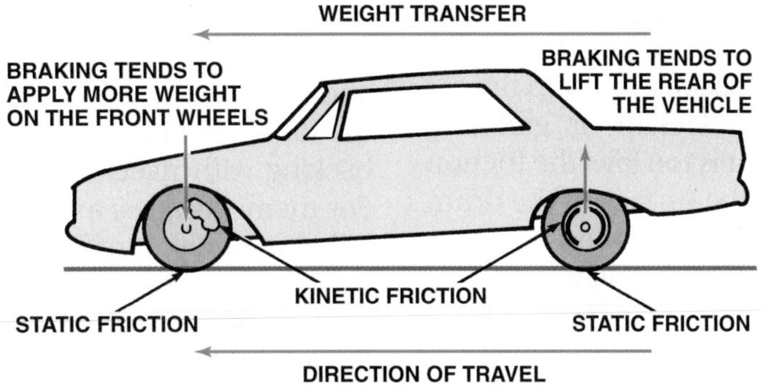

Figure 48-1 Braking action creates kinetic friction in the brakes and static friction between the tire and road to slow the vehicle. When brakes are applied, the vehicle's weight is transferred to the front wheels and is unloaded on the rear wheels.

Factors Governing Braking

Four basic factors determine the braking power of a system. The first three factors govern the generation of friction: pressure, coefficient of friction, and frictional contact surface. The fourth factor is a result of friction. It is heat or, more precisely, heat dissipation.

An additional factor influences how well a vehicle will stop when the brakes are applied, that being weight transfer. When the brakes are applied while the vehicle is moving forward, the weight of the vehicle shifts forward. This causes the front of the vehicle to drop or "nose dive." It also means that the front brakes will need the most stopping power. If the vehicle is overloaded or if the front suspension is weak, more weight will be thrown forward and the brakes will need to work harder.

Pressure The amount of friction generated between moving surfaces in contact with each other depends in part on the pressure exerted on the surfaces. For example, if you slowly increase the downward pressure on the palm of your hand as you move it across a desk, you will feel a gradual increase in friction.

In a brake system, hydraulic systems provide application pressure. Hydraulic force is used to move brake pads or brake shoes against spinning rotors or drums mounted to the wheels. The amount of pressure is determined by the pressure on the brake pedal and the design of the brake system.

Coefficient of Friction The amount of friction generated between two surfaces is expressed as a **coefficient of friction (COF)**. The COF is determined by dividing the force required to pull an object across a surface by the weight of the object **(Figure 48–2)**. For example, if it requires 100 pounds (455 N) of pull to slide a 100-pound (45.4 kg) metal part across a concrete floor, the COF is 100 ÷ 100 or 1. To pull a 100-pound (45.4 kg) block of ice across the same surface may require only 2 pounds (9 N) of pull. The COF then would be only 0.02.

As it applies to automotive brakes, the COF expresses the frictional relationship between pads and rotors or shoes and drums. The required COF depends on the vehicle and other factors and is carefully chosen by the manufacturer to ensure safe and reliable braking. Therefore, when replacing

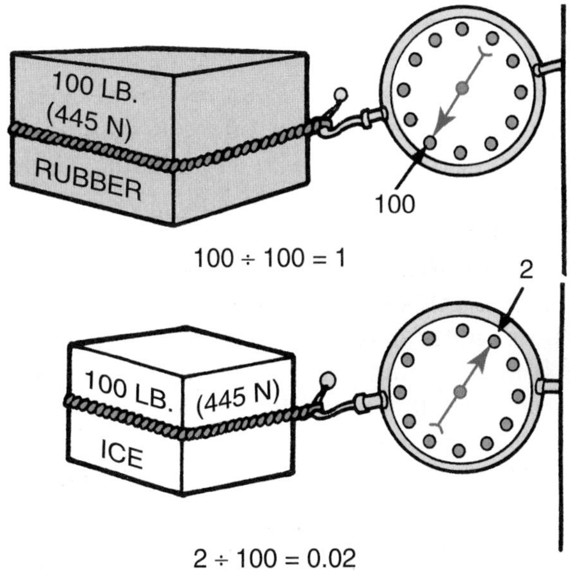

Figure 48-2 Coefficient of friction is equal to the pounds of pull divided by the weight of the object.

pads or shoes, it is important to use replacement parts with similar COF. If, for example, the COF is too high, the brakes will be too sticky to stop the car smoothly. Premature wheel lockup or grabbing would result. If the coefficient is too low, the friction material tends to slide over the surface of the drum or rotor rather than slowing it down. Most automotive friction materials are engineered with a COF of between 0.25 and 0.55.

Frictional Contact Surface The third factor is the amount of surface area that is in contact. Simply put, bigger brakes stop a car more quickly than smaller brakes used on the same car. For the most part, the vehicle's weight and potential speed determines the size of the friction surface areas. Also, the greater the surface areas of the wheel brake units, the faster heat can be dissipated.

Heat Dissipation Any braking system must be able to effectively handle the heat created by friction within the system. The tremendous heat created by the rubbing brake surfaces must be conducted away from the pad and rotor (or shoe and drum) and be absorbed by the air. Brakes that do not effectively dissipate heat experience brake fade during hard, continuous braking.

Brake fade is a condition where the stopping power of the brakes has been drastically reduced. This is commonly caused by excessive heat buildup. With brake fade, the brake pedal seems normal but there is reduced stopping ability. Brake fade may become worse as heat builds up; this may be due to outgassing. As the shoes or pads become extremely hot, they can generate a gas. This gas can become an air bearing between the frictional material and the rotor or drum. Rather than clamp on the wheel brake, the friction elements will slip on the air (gas buildup). Fade can also be caused by overheating the brake fluid because gases form in the fluid.

The friction materials must be able to dissipate heat and the system must be designed to allow the material to get rid of its heat. This may be done by allowing ample airflow past the brake units. Another way is to ventilate the rotors. Ventilated rotors have internal vanes that move the hot air from the disc to the outside. Some rotors are cross-drilled or slotted. Both of these designs allow the rotor to run cooler and reduce the chances of gas buildup.

Heat can also cause the linings of the pads and shoes to become glazed and harden the rotor and drum. Therefore, the COF is reduced and excessive foot pressure must be applied to the brake pedal to produce the desired braking effect.

Brake Lining Friction Materials

Brake linings are made up relatively soft but tough and heat-resistant material with a high coefficient of friction. The lining is typically attached to a metal backing with rivets or high-temperature adhesives. For many decades, asbestos was the standard brake lining material. It offers good friction qualities, long wear, and low noise. But new materials are being used because of the health hazards of breathing asbestos dust. Asbestos has not been used in brake linings or pads since 2003. Many different materials are used as lining material and the type of lining is defined by its composition. Each type has different heat dissipation, fade resitance, rotor wear, noise generation, and braking force characteristics.

Nonasbestos Organic **Nonasbestos organic (NAO)** linings are installed on many vehicles by the OEM. Organic linings are made of nonmetallic fibers bonded together to form a composite material. Today's organic brake linings contain the following types of materials:

- Friction materials and friction modifiers. Some common examples of these are graphite, powdered metals, and even nut shells.
- Fillers. Fillers are secondary materials added for noise reduction, heat transfer, and other purposes.
- Binders. Binders are glues that hold the other materials together.
- Curing agents. These accelerate the chemical reaction of the binders and other materials.

Organic linings have a high COF, and they are economical, quiet, wear slowly, and are only mildly abrasive to drums. However, organic linings fade more quickly than other materials and do not operate well at high temperatures. High-temperature organic linings are available for high-performance use but they do not work as well at low temperatures. They also wear faster than regular organic linings.

Metallic Linings Fully metallic materials were used for many years in racing. **Metallic lining** is made of powdered metal that is formed into blocks by heat and pressure. These materials provide excellent resistance to brake fade but require high brake pedal pressure; they create the most wear on rotors and drums. Metallic linings work very poorly until they are fully warmed. Improved high-temperature organic linings and semimetallic materials have made metallic linings almost obsolete for late-model automotive use. Metallic linings are also extremely noisy, which is something that must be considered by customers when choosing the type of brake lining to install on their vehicle.

Semimetallic Linings Semimetallic materials are made of a mixture of organic or synthetic fibers and certain metals molded together. **Semimetallic linings** are harder and more fade resistant than organic materials but require higher brake pedal effort.

Most semimetallic linings contain about 50% iron and steel fibers. Copper also has been used in some semimetallic linings and, in smaller amounts, in organic linings. Concerns about copper contamination of the nation's water systems has led to its reduced use in brake linings, however.

Semimetallic linings operate best above 200°F (93.3°C) and actually must be warmed up to bring them into full efficiency. Consequently, they are typically less efficient than organic linings at low temperatures.

Semimetallic linings were sometimes used on older heavy or high-performance vehicles with four-wheel drum brakes. Currently, semimetallic linings are used only on the front disc brakes of passenger cars and light trucks. The lighter braking loads on rear brakes, particularly on FWD cars, may never heat semimetallic linings to their required operating efficiency. Semimetallic linings also have a lower static COF than organic linings, which makes them less efficient with parking brakes.

Synthetic Linings The goals of improved braking performance and the disadvantages of the other lining materials have led to the development of **synthetic lining** materials. They are classified as synthetic because they are made of nonorganic, nonmetallic, and nonasbestos materials. Two types of synthetic materials are commonly used as brake linings for drum brakes: fiberglass and **aramid fibers.**

Fiberglass was introduced as a brake lining material to help eliminate asbestos. Like asbestos, it has good heat resistance, good COF, and excellent structural strength. The disadvantages of fiberglass are its higher cost and its reduced friction at very high temperatures. Overall, fiberglass linings perform similarly to organic linings and are used primarily in rear drum brakes.

Aramid fibers are a family of synthetic materials that are five times stronger than steel, pound for pound, but weigh little more than half what an equal volume of fiberglass weighs. Friction materials made with aramid fibers are made similarly to organic and fiberglass linings. Aramid fibers have a COF similar to semimetallic linings when cold and close to that of organic linings when hot. Overall, the performance of aramid linings is somewhere between organic and semimetallic materials but with much better wear resistance and longevity than organic materials.

Carbon-Metallic/Ceramic Carbon-metallic and ceramic linings are found on many FWD vehicles, because they have high heat resistance. Most ceramic pads are made of a ceramic material mixed with copper fibers. These pads are quiet and produce little dust. Carbon is mixed with metals to make a lining that has a good COF and high heat resistance. Carbon linings are also able to withstand very high temperatures without causing brake fade. A few aftermarket companies offer linings made of carbon, Kevlar, and various other materials. These also have ceramic heat shields that reduce the amount of heat that can transfer from the linings to the rest of the brake system. Some high-performance cars are fitted with carbon-ceramic pads. These pads are comprised of a ceramic composite of carbon fiber reinforced with silicon carbide. These pads offer excellent braking performance and are extremely lightweight. They also provide a consistent COF through a wide range of temperatures and weather conditions.

PRINCIPLES OF HYDRAULIC BRAKE SYSTEMS

A hydraulic system **(Figure 48–3)** uses a brake fluid to transfer pressure from the brake pedal to the pads or shoes. This transfer of pressure is reliable and consistent because liquids are not compressible. That is, pressure applied to a liquid in a closed system is transmitted by that liquid equally to every other part of that system. Apply a force of 5 pounds (35 kPa) per square inch (psi) through the master cylinder and you can measure 5 psi (35 kPa) anywhere in the lines and at each wheel where the brakes operate.

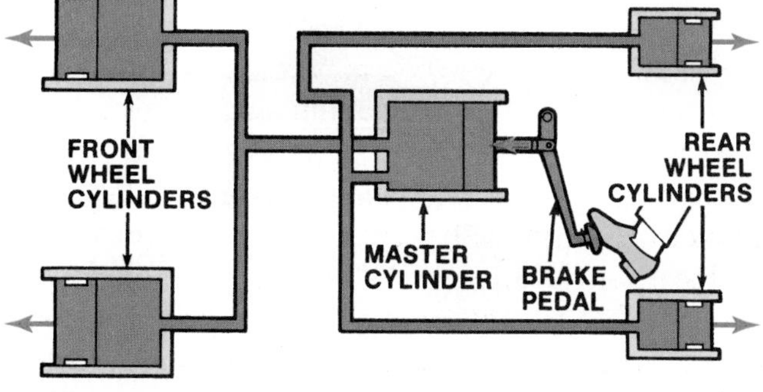

Figure 48–3 A schematic of a basic automotive hydraulic brake system.

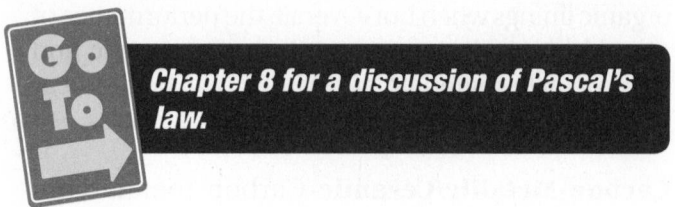

Chapter 8 for a discussion of Pascal's law.

The force can be increased at output (that is, at the wheel) by increasing the size of the wheel's piston, though piston travel decreases. The force at output can be decreased by decreasing the size of the wheel piston, but the piston travel increases.

Thus, to double the output force of the 5 psi (35 kPa) at the master cylinder to 10 psi (69 kPa) at the wheels, simply use a wheel cylinder piston with a surface area of 2 square inches (13 sq. cm). To triple the force of 100 psi (690 kPa), use a piston with 3 square inches (20 sq. cm) and 300 pounds (2,068 kPa) of output result **(Figure 48–4)**. No matter what the fluid pressure is, the output force can be increased with a larger piston, though piston travel decreases proportionately. In actual practice, however, fluid movement in an automotive hydraulic brake system is very slight. In an emergency, when the pedal goes all the way to the floor, the volume of fluid displaced amounts to only about 20 cubic centimeters. About 15 cubic centimeters goes to the front discs and 5 cubic centimeters goes to the rear drums. Even under these conditions, the wheel cylinder and caliper pistons move only slightly.

Of course, the hydraulic system does not stop the car all by itself. In fact, it really just transmits the action of the driver's foot on the brake pedal out to the wheels. In the wheels, sets of friction pads are forced against rotors or drums to slow their turning and bring the car to a stop. Mechanical force (the driver stepping on the brake pedal) is changed into hydraulic pressure, which is changed back into mechanical force (brake shoes and disc pads contacting the drums and rotors). The amount of force acting on the friction pads and shoes is equivalent to the psi applied to the pedal multiplied by the area of the piston affected. A force of 25 pounds applied to the pedal times 4 square inches of piston area equals 100 pounds (psi) of pressure in the system.

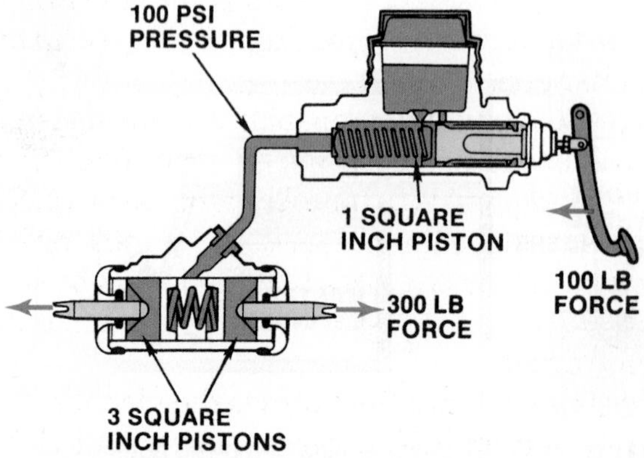

Figure 48–4 Output force increases with piston size.

Dual Braking Systems

Since 1967, federal law has required that all cars be equipped with two separate brake systems. If one circuit fails, the other provides enough braking power to safely stop the car.

The dual system differs from the single system by employing a tandem master cylinder, which is essentially two master cylinders formed by installing two separate pistons and fluid reservoirs into one cylinder bore. Each piston applies hydraulic pressure to two wheels **(Figure 48–5)**.

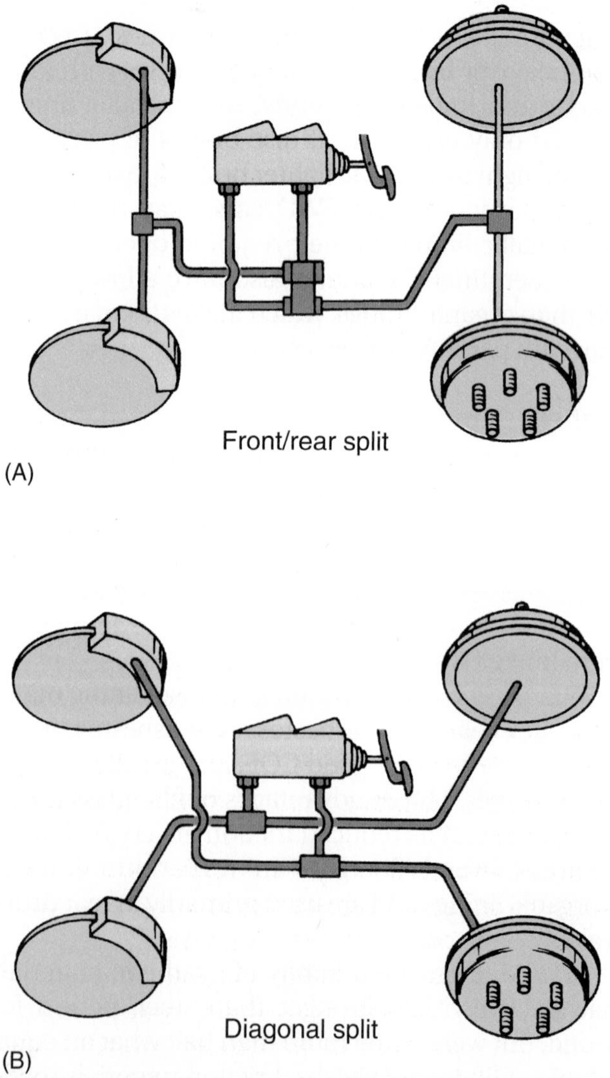

Figure 48–5 (A) A front/rear split of the braking system. (B) A diagonal split. The master cylinder is also split to allow pressure only to its designated wheel units. *Courtesy of Ford Motor Company*

Front/Rear Split System In early dual systems, the hydraulic circuits were separated front and rear. Both front wheels were on one hydraulic circuit and both rear wheels on another. If a failure occurred in one system, the other system was still available to stop the vehicle. However, the front brakes do approximately 70% of the braking work. A failure in the front brake system would only leave 20% to 40% braking power. This problem was somewhat reduced with the development of diagonally split systems.

Diagonally Split System The **diagonally split system** operates on the same principles as the front and rear split system.

The hydraulic brake lines on this system, however, have been diagonally split front to rear (left front to right rear and right front to left rear). The circuit split can occur within the master cylinder or externally at a proportioning valve or pressure differential switch.

In the event of a system failure, the remaining good system would do all the braking on one front wheel and the opposite rear wheel, thus maintaining 50% of the total braking force.

HYDRAULIC BRAKE SYSTEM COMPONENTS

The following sections describe the major components of a hydraulic brake system, including power-assisted systems and antilock braking systems.

Brake Fluid

Brake fluid is the lifeblood of any hydraulic brake system. It is what makes the system operate properly. Brake fluid is specially blended to perform a variety of functions. Brake fluid must be able to flow freely at extremely high temperatures (500°F [260°C]) and at very low temperatures (−104°F [−75°C]). Brake fluid also serves as a lubricant for many parts to ensure smooth and even operation. In addition, brake fluid must fight corrosion and rust in the brake lines and various assemblies and components. Another important property of brake fluid is that it must resist evaporation.

All brake fluids are hygroscopic; that is, they readily absorb water. Moisture can enter the fluid when the fluid is exposed to the atmosphere and while it is in the brake system. For this reason, brake fluid should always be kept in a sealed container and should only be exposed to outside air for limited periods. Moisture also builds up in the fluid due to condensation. The fluid gets hot because of brake applications, and then when at rest it cools. This change in temperature causes condensation. Today's vehicles are more prone to hot brake fluid because there is little airflow under the hood to cool the fluid and braking system.

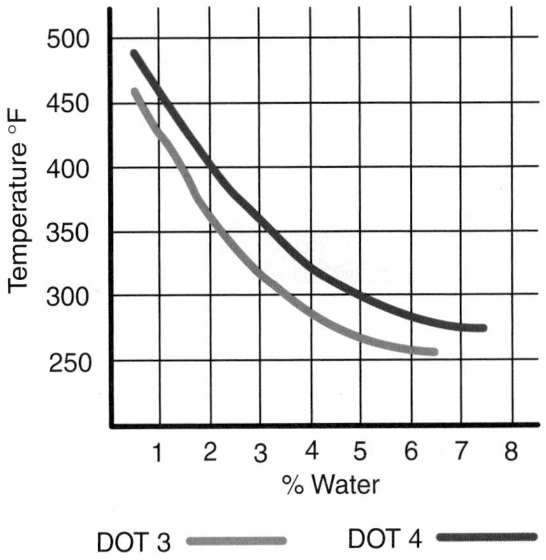

Figure 48-6 Moisture affects the boiling point of brake fluid.

The performance of brake fluid is affected by moisture. As the amount of water in the fluid increases, the boiling point of the fluid decreases **(Figure 48–6)**. This can cause vapor to build in the system, which could lead to sudden brake failure or an unpredictable spongy pedal. Vapor is a gas and therefore is compressible. When pressure is applied to the brake fluid, it will compress the vapor before moving on through the system if it is able to do so.

The viscosity of the fluid at low temperatures is also affected by the amount of moisture. The viscosity increases, which means the fluid will have a harder time moving through the system, when it is cold. This means poor cold weather braking.

Moisture may cause corrosion to build on internal parts of the system. This also decreases the efficiency of the brake system.

Tests have shown that within 1 year of service in a typical vehicle, the water content of the fluid is about 2%. It takes approximately 2 years for the fluid to have moisture levels that lower the fluid's boiling point to a dangerous level.

Brake fluid must be compatible with the materials used within the brake system to avoid damage to them. It must provide a controlled amount of swell to the brake system cups and seals. There must be just enough swell to form a good seal. However, the swell cannot be too great. If it is, drag and poor brake response occur.

Every can of brake fluid carries the identification letters of SAE and DOT. These letters (and corresponding numbers) indicate the nature, blend, and performance characteristics of that particular brand of brake fluid. Always use the fluid recommended by the manufacturer.

Go To Chapter 7 for details on the types of brake fluids.

WARNING!

Use only approved brake fluid in a brake system. Any other lubricant that has a petroleum base must never be used. Petroleum-based fluids attack the rubber components in the brake system and cause them to swell and disintegrate.

Most vehicles have brake fluid level sensors that provide the driver with an early warning message when the brake fluid in the master cylinder reservoir has dropped below the normal level.

As the brake fluid in the master cylinder reservoir drops below the designated level, the sensor closes the warning message circuit. About 15 seconds later the message "Brake fluid low" appears on the instrument panel. At this time, the master cylinder reservoir should be checked and filled to the correct level with the specified brake fluid.

Brake Pedal

The brake pedal is where the brake's hydraulic system gets its start. When the brake pedal is depressed, force is applied to the master cylinder. On a basic hydraulic brake system (where there is no power assist), the force applied is transmitted mechanically. As the pedal pivots, the force applied to it is multiplied mechanically. The force that the pushrod applies to the master cylinder piston is, therefore, much greater than the force applied to the brake pedal.

MASTER CYLINDERS

The master cylinder **(Figure 48–7)** transmits the pressure on the brake pedal to each of the four wheel brakes to stop the vehicle. It changes the driver's mechanical pressure on the pedal to hydraulic force, which is changed back to mechanical force at the wheel brake units. The master cylinder uses the fact that fluids are not compressible to transmit the pedal movement to the wheel brake units.

Figure 48-7 A brake master cylinder.

The master cylinder also uses hydraulics to increase the pedal force applied by the driver. A 100-pound force on the brake pedal can be used to push on a 1 sq. in. master cylinder piston to create a 100 psi pressure in the hydraulic system. This 100 psi can be used to push on 4 sq. in. output pistons at a wheel brake. The result is a 400-pound force at the 4 sq. in. output pistons. The driver's 100-pound force has been multiplied to a force of 400 pounds.

Dual-Piston Master Cylinders

Figure 48–8 is a simplified illustration of a master cylinder. A pushrod is connected to a piston inside the cylinder, and hydraulic fluid is in front of the piston. When the pedal is pressed, the piston is pushed forward. The fluid transmits the force of the piston to all the inner surfaces of the system. Only the pistons in the drum brake wheel cylinders and/or disc brake calipers can move, and they move outward to force the brake shoes or pads against the rotating brake drums and/or rotors.

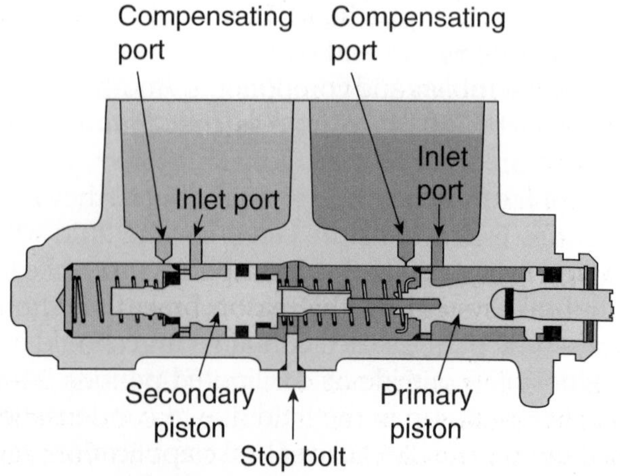

Figure 48-8 The basic components of a dual master cylinder.

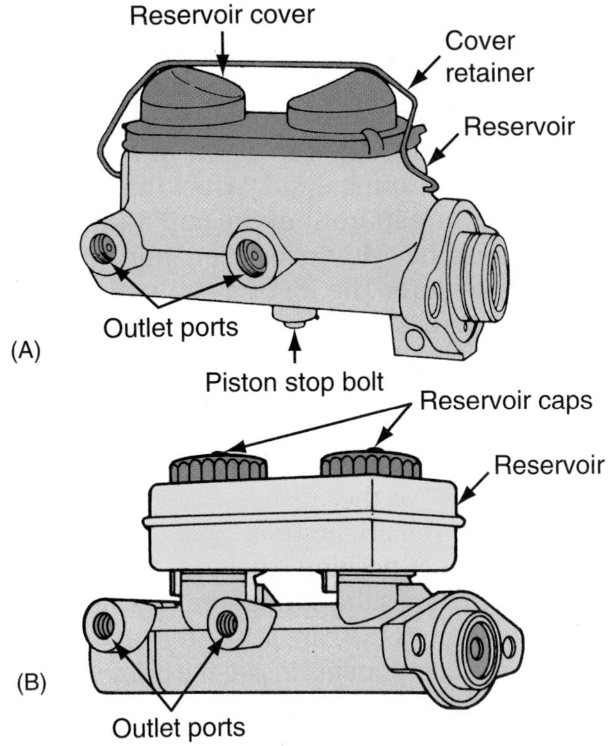

Figure 48–9 (A) A typical cast-iron dual master cylinder and (B) a typical aluminum/composite dual master cylinder.

Master Cylinder Reservoir The reservoir may be cast as one piece with the cylinder body, or it may be a separate molded nylon or plastic container **(Figure 48–9)**. The one-piece body and reservoir casting is usually made of cast iron. The cylinder is directly below the reservoir. All reservoirs have a removable cover so that brake fluid can be added to the system. One-piece reservoirs typically have a single cover that is held on the reservoir with a retainer bail. Nylon or plastic reservoirs typically have two screw caps on top of the reservoir. Separate reservoirs may be clamped or bolted to the cylinder body, or they may be pressed into holes at the top of the body and sealed with grommets or O-rings.

The caps or covers are vented to prevent a vacuum lock as the fluid level drops in the reservoir. A flexible rubber diaphragm at the top of the reservoir is incorporated in the caps or covers. The diaphragm separates the brake fluid from the air above it while remaining free to move up and down with changes in fluid level. The diaphragm keeps the moisture and air from entering the brake fluid in the reservoir.

If a vehicle with front disc and rear drum brakes has a hydraulic system split front to rear, the reservoir chamber for the disc brakes is larger than the chamber for the drum brakes. As disc pads wear, the caliper pistons move out farther in their bores. More fluid is then required to keep the system full in the master cylinder. With drum brake wheel cylinder pistons, the volume of fluid does not increase much with lining wear because the pistons always fully retract into the cylinders regardless of brake lining wear. Vehicles with four-wheel disc brakes or diagonally split hydraulic systems usually have master cylinders with equally sized reservoirs because each circuit of the hydraulic system requires the same volume of fluid.

Plastic reservoirs are often translucent so that fluid level can be seen without removing the cover. Although this feature allows a quick check of fluid level without opening the system to the air, you should not rely on it for thorough brake fluid inspection. Stains inside the reservoir can give a false indication of fluid level, and contamination cannot be seen without removing the reservoir caps or cover.

Master Cylinder Ports Different names have been used for the ports in the master cylinder. This text refers to the forward port as the "vent" port and the rearward port as the "replenishing" port. These are the names established by SAE Standard J1153. The **vent port** has been called a compensating port or a replenishing port. To further confuse the issue, the **replenishing port** has been described by many manufacturers as the compensating port, as well as the vent port, the bypass port or hole, the filler port, or the intake port **(Figure 48–10)**. The vent ports and replenishing ports let fluid pass between each pressure chamber and its fluid reservoir during operation. The names of these ports are not important as long as you understand their purposes and operations.

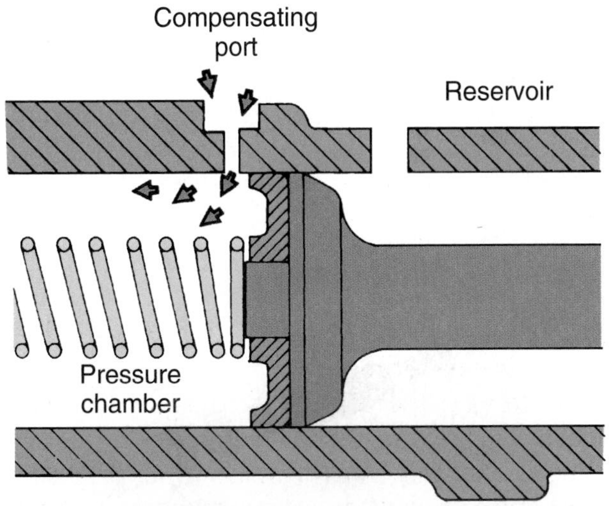

Figure 48–10 Fluid from the reservoir fills the cylinder through the compensating port. *Courtesy of Ford Motor Company*

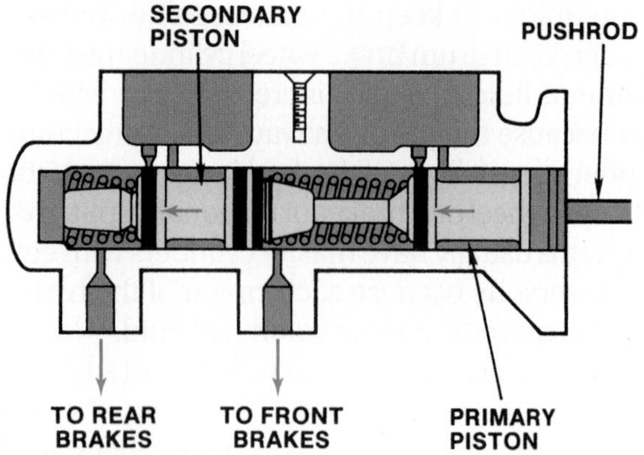

Figure 48-11 The position of the pistons in a master cylinder when the brakes are applied. Notice that the seals stop fluid flow throughout the master cylinder.

Master Cylinder Construction

A single cylinder bore contains two piston assemblies **(Figure 48–11)**. The piston assembly at the rear is the primary piston, and the one at the front is the secondary piston. Each piston has a return spring in front of it. There is a **cup seal** in front of each piston and a cup or seal at the rear of each piston. The seals retain fluid in the cylinders and prevent seepage between the cylinders.

Inside the cylinder are two spool-shaped pistons. The piston has a head on one end and a groove for an O-ring seal on the other end. The seal seats against the cylinder wall and keeps fluid from leaking past the piston. The smaller diameter center of the piston is the valley or spool area, which lets fluid get behind the head of the piston.

Each master cylinder piston works with a rubber cup seal, which fits in front of the piston head. The cup has flexible lips that fit against the cylinder walls to seal fluid pressure ahead of the piston head. The cup lip also can bend to let fluid get around the cup from behind. When the brakes are applied, pressure in front of the cup forces the lip tightly against the cylinder wall, enabling it to hold very high pressure. The lip of a cup seal is always installed toward the pressure to be contained or away from the body of the piston. The cup seals in only one direction. If pressure behind the lip exceeds the pressure in front of it, the higher pressure will force the lip away from the cylinder wall and let fluid bypass the cup.

Pistons have small coil springs that return the pistons to the proper position when the brake pedal is released. Sometimes the springs are attached to the pistons; sometimes they are separate parts. A snapring holds the components inside the cylinder, and a rubber boot fits around the rear of the cylinder and pushrod to keep dirt from entering the cylinder **(Figure 48–12)**.

A two-piece master cylinder has an aluminum body. Because aluminum can be nicked or gouged easily, the bore of the aluminum cylinder is anodized to protect it from wear and damage. They are fitted with a removable nylon or plastic reservoir. Because the master cylinder is made of two materials, it is often called a composite master cylinder. The pistons, cups, and springs used in a composite master cylinder are essentially the same and work the same way as those in a one-piece master cylinder.

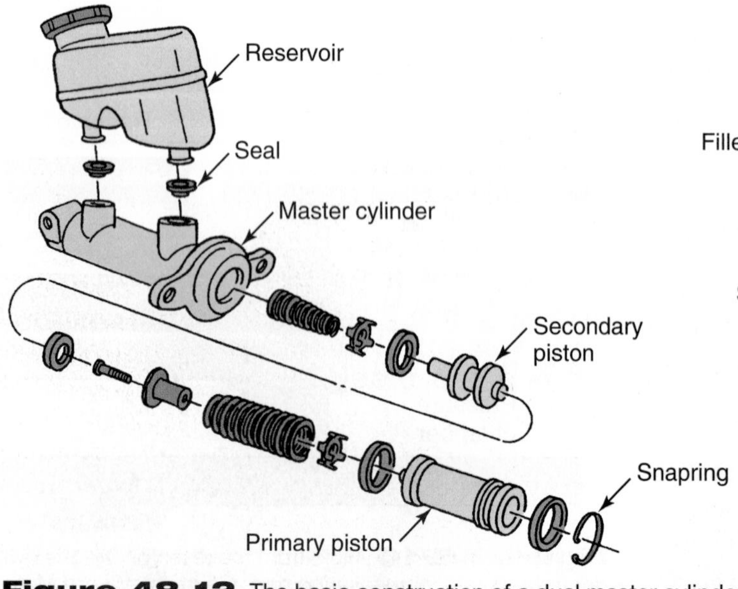

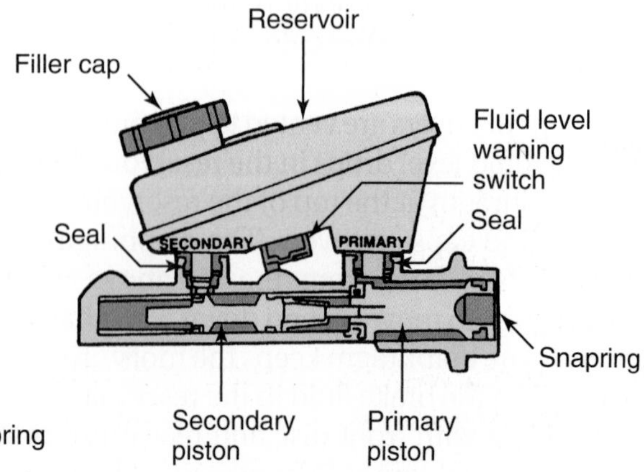

Figure 48-12 The basic construction of a dual master cylinder. *Courtesy of Ford Motor Company*

MASTER CYLINDER OPERATION

The vent port in the bottom of the reservoir is located just ahead of the piston cup. Fluid flows from the reservoir into the pressure chamber in front of the piston cup. The replenishing port is located above the valley area of the piston behind the piston head. The O-ring seal on the piston keeps the fluid from leaking out the rear of the cylinder. The return spring in front of the piston and cup returns the piston when the brakes are released.

When the driver depresses the brake pedal, the pushrod pushes the piston forward. As it moves forward, the piston pushes the cup past the vent port. As soon as the vent port is covered, fluid is trapped ahead of the cup. The fluid, which is under pressure, goes through the outlet lines to the wheel brake units to apply the brakes.

When the driver releases the brake pedal, the return spring forces the piston back to its released position. As the piston moves back, it pulls away from the fluid faster than the fluid can flow back from the brake lines to the pressure chamber. This creates a low pressure ahead of the piston.

The piston must rapidly move back to the released position so it can be ready for another forward stroke if necessary. The low-pressure area must be filled with fluid as the piston moves back. A path for fluid flow is provided by the valley area, past the primary cup protector washer and through several small holes in the head of the piston, or by having enough clearance between the piston head and the cylinder bore. Fluid flows through the piston or around the lip of the cup and into the chamber ahead of the piston. This flow quickly relieves the low-pressure condition.

The fluid that flows from the valley area to the pressure chamber must be replaced. When the piston is fully returned to its released position, the space in front of it is full of fluid. The piston cup again seals off the head of the piston. In the meantime, the fluid from the rest of the system has begun to flow back to the high-pressure chamber. If this pressure is not released, the brakes would not release. The returning fluid flows back to the reservoir through the vent port. The vent port is covered by the piston cup at all times, except when the piston is released.

Residual Pressure Check Valve

The pressure chamber in a master cylinder for some drum brake systems may have an additional part called a **residual pressure check valve**. This valve **(Figure 48–13)** can be installed in the pressure chamber or the outlet line of the master cylinder. A residual pressure check valve is the oldest type of pressure control valve used in a brake system. It can be found in older four-wheel drum brake systems as well as in some late-model systems.

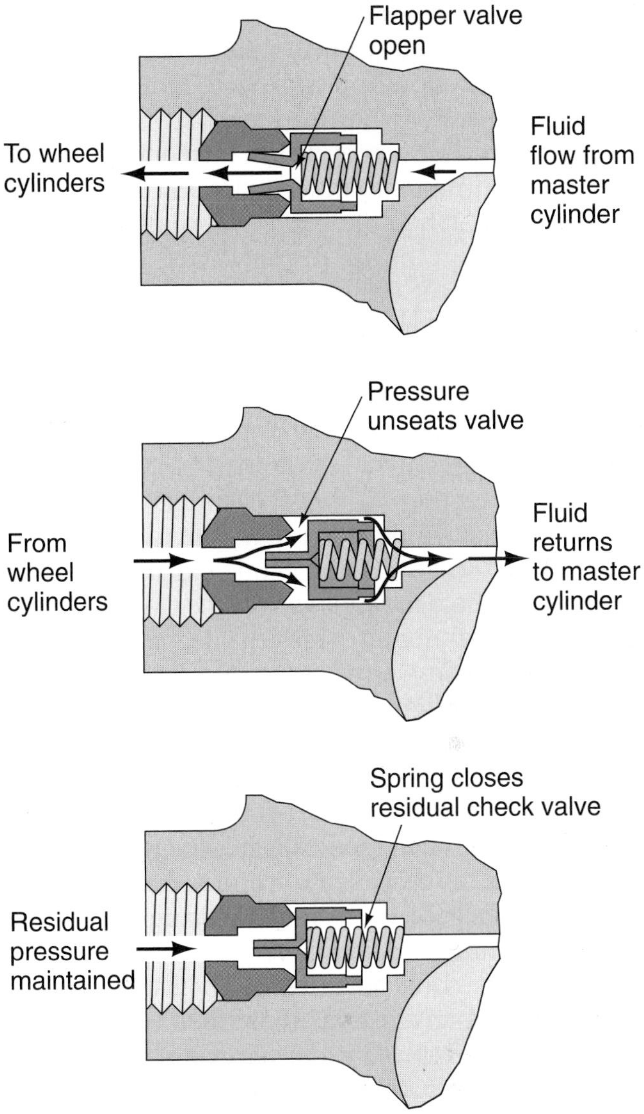

Figure 48–13 The operation of a master cylinder's residual valve.

This valve is usually installed in the master cylinder outlet port to drum brakes. It maintains a residual pressure of 6 to 25 psi in the brake lines when the pedal is released. This residual line pressure maintains slight pressure on the wheel cylinder pistons to keep the sealing lips of the piston cups forced outward against the cylinder walls. When the pedal is released, the retracting master cylinder piston creates a pressure drop in the lines. If pressure were to drop low enough, the piston cups could be pulled away from the wheel cylinder walls and draw air into the system. The slight residual pressure prevents this but is not high enough to overcome brake shoe spring tension. Maintaining pressure on the piston cups also maintains their fluid-sealing integrity and helps prevent fluid leakage. Disc brake systems do not use this valve because residual pressure would cause the pads to drag on the rotor when the brakes are released.

The residual valve may be installed under the tubing seat in the master cylinder outlet port or inside the cylinder bore. When the brakes are applied, master cylinder pressure opens the valve and allows fluid flow to the wheel cylinders. When the brakes are released, pressure in the lines unseats the valve in the reverse direction to allow fluid return flow to the master cylinder. When line pressure drops below the pressure of the check valve spring, the valve closes to hold residual pressure in the lines.

With the redesign of wheel cylinders, many late-model master cylinders do not have a residual pressure check valve. Piston cup expanders were developed to hold the cups against the wheel cylinder walls. Cup expanders are simpler, cheaper, and more reliable than check valves.

Diagonally split brake systems are another reason for the elimination of residual pressure check valves. A diagonally split system usually pairs one disc brake with one drum brake for half of the hydraulic system. Disc brakes rely on the action of the square-cut piston seal to retract the caliper piston and remove pressure from the brake pad. Any residual pressure at all would cause brake drag.

Although residual pressure check valves have been eliminated in many systems, they are still used in others. When you replace a master cylinder, it is very important to verify whether or not the vehicle requires a residual pressure check valve. Installing the wrong cylinder will cause improper brake operation and possible system failure.

Split Hydraulic Systems

Most late-model vehicles have a diagonally split hydraulic system. If there is a hydraulic failure in the brake lines served by the master cylinder's secondary piston, both pistons will move forward when the brakes are applied, but there is nothing to resist piston travel except the secondary piston spring. This lets the primary piston build up only a small amount of pressure until the secondary piston bottoms in the cylinder bore. Then the primary piston will build enough hydraulic pressure to operate the brakes served by this half of the system.

In case of a hydraulic failure in the brake system served by the primary piston, the piston will move forward when the brakes are applied but will not build up hydraulic pressure. Very little force is transferred to the secondary piston through the primary piston spring until the piston extension screw comes in contact with the secondary piston. Then, pushrod force is transmitted directly to the secondary piston and enough pressure is built up to operate its brakes.

Fast-Fill and Quick Take-Up Master Cylinders

Several manufacturers use fast-fill, or quick take-up, master cylinders. These cylinders fill the hydraulic system quickly to take up the slack in the caliper pistons of low-drag disc brakes. Low-drag calipers retract the pistons and pads farther from the rotor than traditional calipers. This reduces friction and brake drag and improves fuel mileage.

If a conventional master cylinder were used with low-drag calipers, excessive pedal travel would be needed on the first stroke to fill the lines and calipers with fluid and take up the slack in the pads. To overcome this, fast-fill, or **quick take-up**, master cylinders provide a large volume of fluid on the first stroke of the brake pedal.

You can recognize a fast-fill, or quick take-up, master cylinder by the bulge, or larger diameter, on the outside of the casting **(Figure 48–14)**. The cylinder has a larger diameter bore for the rear primary piston than the front primary piston. Inside the cylinder, a fast-fill, or quick take-up, valve replaces the conventional vent and replenishing ports for the primary piston. Some

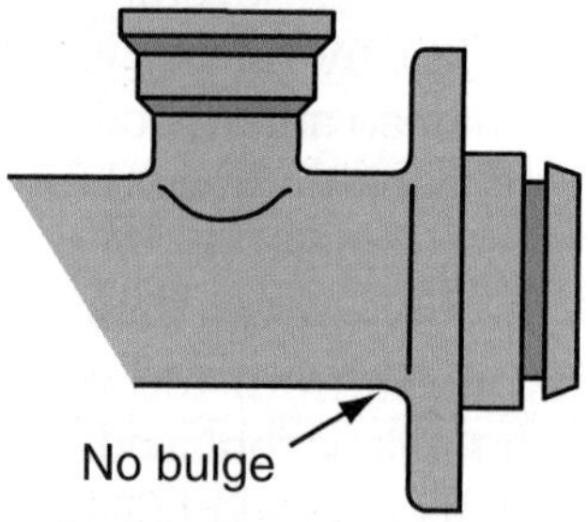

Standard Master Cylinder

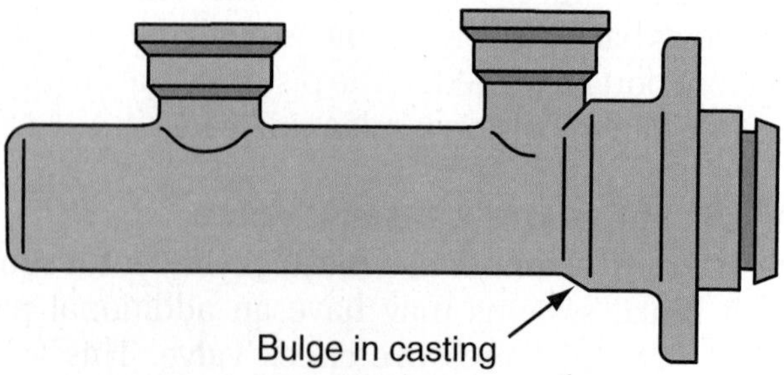

Quick Take-Up Master Cylinder

Figure 48-14 You can recognize a quick take-up master cylinder by the bulge or step in the casting.

master cylinders for four-wheel disc brakes also have a quick take-up valve for the secondary piston.

The quick take-up valve contains a spring-loaded check ball that has a small by-pass groove cut in the edge of its seat. The outer circumference of the valve is sealed to the cylinder body with a lip seal. Several holes around the edge of the hole let fluid bypass the lip seal under certain conditions. Some valves (those more often called "fast-fill" valves) are pressed into the cylinder body and sealed tightly by an O-ring. A rubber flapper-type check valve under the fast-fill valve performs the same functions as a lip seal of a quick take-up valve.

Brakes Not Applied When the brakes are off, both master cylinder pistons are retracted, and all vent and replenishing ports are open. However, fluid to both ports of the primary piston must flow through the groove in the check ball seat.

Brakes Applied As the brakes are applied, the primary piston moves forward in its bore. Remember that the diameter of the primary chamber is larger than the diameter of the secondary. As the secondary piston moves forward, the volume is reduced. This causes hydraulic pressure to instantly rise in the low-pressure chamber. The higher pressure forces the large volume of fluid into the low-pressure chamber past the cup seal of the primary piston. This provides the extra volume of fluid to take up the slack in the caliper pistons.

The lip seal of the quick take-up valve keeps fluid from flowing from the low-pressure chamber back to the reservoir. Initially a small amount of fluid bypasses the check ball through the by-pass groove, but this is not enough to affect quick take-up operation.

As brake application continues, pressure in the low-pressure chamber rises to about 70 to 100 psi. The check ball in the quick take-up valve then opens to let excess fluid return to the reservoir. Pressures in both chambers of the primary piston equalize, and the piston moves forward to actuate the secondary piston.

All of the actions apply to the primary piston if it is serving front disc brakes and the secondary piston is serving drum brakes. If the hydraulic system is diagonally split, or if the car has four-wheel, low-drag discs, the quick take-up fluid volume must be available to both pistons. Some master cylinders have a second quick take-up valve for the secondary piston. Others provide the needed fluid volume through the design of the cylinder itself. As long as the primary quick take-up valve stays closed, the fluid bypassing the primary piston cup causes the secondary piston to move farther. This provides equal fluid displacement from both pistons and maintains equal pressure in the system. When the quick take-up valve opens, both pistons move together just as in any other master cylinder.

Brakes Released When the driver releases the brake pedal, the return springs force the primary and secondary pistons to move back. Pressure drops in the high-pressure chambers, and fluid bypasses the piston cup seals from the low-pressure chambers. Low pressure is created in the low-pressure chamber, which allows atmospheric pressure to move past the lip seal of the quick take-up valve. Fluid from the reservoir then flows through both the vent and replenishing ports to equalize pressure in the pressure chambers and valley areas.

On the return stroke, fluid flows to the secondary piston through the replenishing port unless the secondary piston also has a quick take-up valve. If a secondary quick take-up valve is installed, it works in the same way as a primary quick take-up valve.

Central-Valve Master Cylinders

Some antilock brake systems (ABS) use master cylinders that have central check valves in the tops of the pistons. These valves are designed to prevent seal damage and pedal vibration. If the master cylinder provides pressure during antilock operation and the system also has a motor-driven pump, the master cylinder's pistons may shift back and forth rapidly during antilock operation. This will cause excessive pedal vibration and—more importantly—wear on the piston cups where they pass over the vent ports.

When the brakes are released, fluid flows from the replenishing ports to the low-pressure chambers, through the open central check valves, and into the high-pressure chambers. As the brakes are applied, the central valves close to hold fluid in the high-pressure chambers. When the brakes are released again, the check valves open to let fluid flow back through the pistons to the low-pressure chambers and the reservoir.

The central check valves provide supplementary fluid passages to let fluid move rapidly back and forth between the high- and low-pressure chambers during antilock operation. This is not much different in principle from non-ABS fluid flow, but the extra passages reduce piston and pedal vibration and cup seal wear.

HYDRAULIC TUBES AND HOSES

Steel tubing and flexible synthetic rubber hosing serve as the arteries and veins of the hydraulic brake system. These brake lines transmit brake fluid pressure (the blood) from the master cylinder (the heart) to the wheel cylinders and calipers (the muscles and working parts) of the drum and disc brakes.

Fluid transfer from the driver-actuated master cylinder is usually routed through one or more valves and then into the steel tubing and hoses **(Figure 48–15)**. The design of the brake lines offers quick fluid transfer response with very little friction loss. Engineering and installing the brake lines so they do not wrap around sharp curves is very important in maintaining this good fluid transfer.

SHOP TALK

Never use copper tubing as a replacement for defective lines or hoses. This type of tubing is subject to fatigue, cracking, and corrosion. If a section of the brake tubing is damaged, the entire section must be installed with a new tube of the same type, size, shape, and length. Also, when installing tubing, hoses, or connectors, tighten all connections to specifications. Install a new flexible brake hose if the hose shows signs of softening, cracking, or other damage. When installing a new hose, position the hose to avoid contact with other vehicle components. After installation, bleed the brake system.

Brake Line Tubing

Most brake line tubing consists of copper-fused double-wall steel tubing in diameters ranging from ⅛ to ⅜ inch (3 mm to 9 mm). Some OEM brake tubing is manufactured with soft steel strips, sheathed with copper. The strips are rolled into a double-wall assembly and then bonded in a furnace at extremely high temperatures. Corrosion protection is often added by tin-plating the tubing.

Fittings

Assorted fittings are used to connect steel tubing to junction blocks or other tubing sections. The most common fitting is the double or inverted flare style. Double flaring is important to maintain the strength and safety of the system. Single flare or sleeve compression fittings may not hold up in the rigorous operating environment of a standard vehicle brake system.

Fittings are constructed of steel or brass. The 37-degree inverted flare or standard flare fitting is the most commonly used coupling. Newer vehicles may use the ISO or metric bubble flare fitting.

Never change the style of fitting being used on the vehicle. Replace ISO fittings only with ISO fittings. Replace standard fittings with standard fittings.

The metal composition of the fittings must also match exactly. Using an aluminum-alloy fitting with

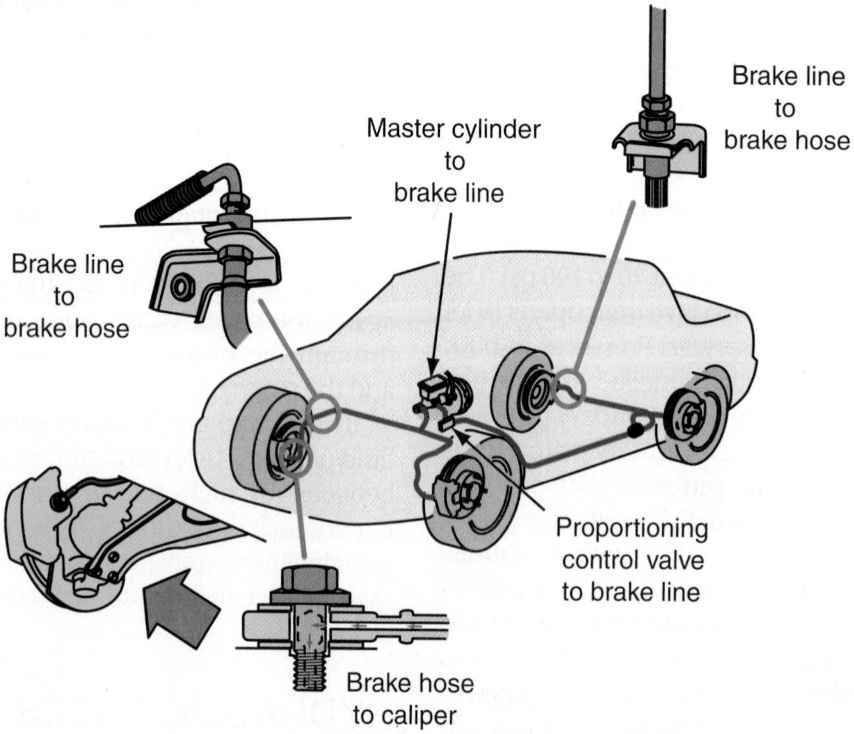

Figure 48-15 A typical layout of the hoses and tubes for the brake system.

steel tubing may provide a good initial seal, but the dissimilar metals create a corrosion cell that eats away the metal and reduces the connection's service life.

Brake Line Hoses

Brake line hoses offer flexible connections to wheel units so steering and suspension members can operate without damaging the brake system. Typical brake hoses range from 10 to 30 inches (25 to 76 mm) in length and are constructed of multiple layers of fabric impregnated with a synthetic rubber. Brake hose material must offer high heat resistance and withstand harsh operating conditions.

SHOP TALK

Many brake hose failures can be traced to errors made in the original installation or repair of the hose. Hoses twisted into place become stressed and are prime candidates for leaks and bursting. Most manufacturers now print a natural lay indicator or line on the hose. By making sure this line is not spiralled after fittings are tightened, you can ensure the hose is not overly stressed. Also, always use a hose of the same length and diameter as the original during servicing to maintain brake balance at all wheels.

HYDRAULIC SYSTEM SAFETY SWITCHES AND VALVES

Switches and valves are installed in the brake system hydraulic lines to act as warning devices to pressure control devices.

Pressure Differential (Warning Light) Switches

A pressure differential valve is used to operate a warning light switch. Its main purpose is to tell the driver if pressure is lost in either of the two hydraulic systems. Since each brake hydraulic system functions independently, it is possible the driver might not notice immediately that pressure and braking are lost. When a pressure loss occurs, brake pedal travel increases and a more-than-usual effort is needed for braking. Should the driver not notice the extra effort needed, the warning light is actuated by the hydraulic system safety switch.

Under normal conditions, the hydraulic pressure on each side of the pressure differential valve piston is balanced. The piston is located at its center point, so the spring-loaded warning switch plunger fits into the tapered groove of the piston. This leaves the contacts of the warning switch open. The brake warning light stays off.

If there is a leak in the front or rear braking system, the hydraulic pressure in the two systems is unequal. For example, if there is a leak in the system supplying the front brakes, there is lower pressure in the front system when the brake pedal is applied. The hydraulic pressure in the rear system then pushes the piston toward the front side, where the pressure is lower. As the piston moves, the plunger is pushed out **(Figure 48–16)**. This closes the switch and illuminates the brake warning light.

While all brake warning light switches serve the same function, there are three common variations in the design of these switches. These variations include switches with centering springs, without centering springs, and with centering springs and two pistons.

Metering and Proportioning Valves

Metering and proportioning valves are used to balance the braking characteristics of disc and drum brakes.

The braking response of the disc brakes is immediate when the brake pedal is applied. It is directly proportionate to the effort applied at the pedal. Drum brake response is delayed while rear brake hydraulic pressure moves the wheel cylinder pistons to overcome the force of their return springs and force the brake shoes to contact the drum. Their actions are self-energizing and tend to multiply the pedal effort.

Metering Valve A **metering valve** (hold-off valve) in the front brake line holds off pressure going from the master cylinder to the front disc calipers. This delay allows pressure to build up in the rear drums first. When the rear brakes begin to take hold, the hydraulic pressure builds to the level needed to open the metering valve. When the metering valve opens, line pressure is high enough to operate the front discs. This process provides for better balance of the front and rear brakes. It also prevents lockup of the front brakes by keeping pressure from them until the rear brakes have started to operate. The metering valve has the most effect at the start of each brake operation and all during light braking conditions.

Proportioning Valve The self-energizing action of the delayed response rear drum brakes can cause them to lock the rear wheels at a lower hydraulic pressure than the front brakes. The **proportioning valve**

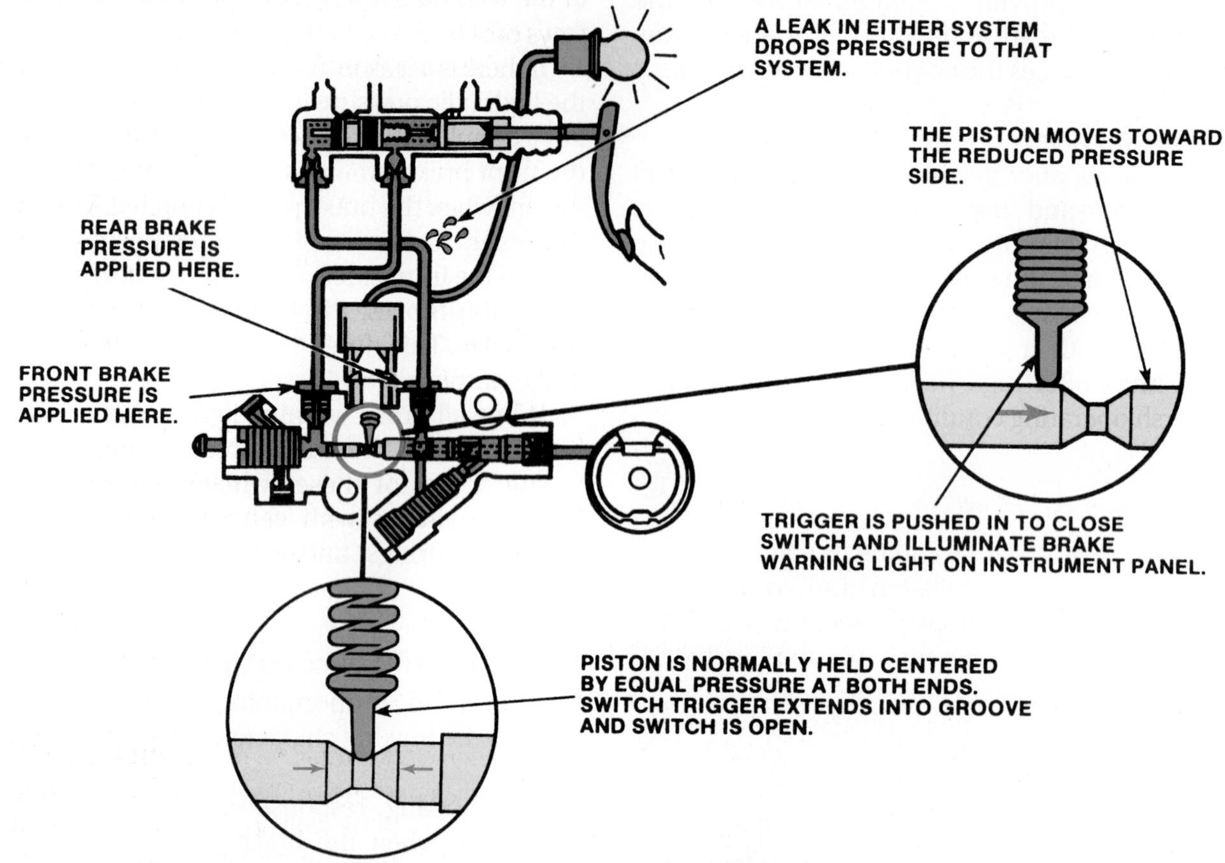
A LEAK IN EITHER SYSTEM DROPS PRESSURE TO THAT SYSTEM.
THE PISTON MOVES TOWARD THE REDUCED PRESSURE SIDE.
REAR BRAKE PRESSURE IS APPLIED HERE.
FRONT BRAKE PRESSURE IS APPLIED HERE.
TRIGGER IS PUSHED IN TO CLOSE SWITCH AND ILLUMINATE BRAKE WARNING LIGHT ON INSTRUMENT PANEL.
PISTON IS NORMALLY HELD CENTERED BY EQUAL PRESSURE AT BOTH ENDS. SWITCH TRIGGER EXTENDS INTO GROOVE AND SWITCH IS OPEN.

Figure 48-16 A pressure differential valve under normal conditions.

(balance valve) **(Figure 48–17)** is used to control rear brake pressures, particularly during hard stops. When the pressure to the rear brakes reaches a specified level, the proportioning valve overcomes the force of its spring-loaded piston, stopping the flow of fluid to the rear brakes. By doing so, it regulates rear brake system pressure and adjusts for the difference in pressure between front and rear brake systems. This keeps front and rear braking forces in balance.

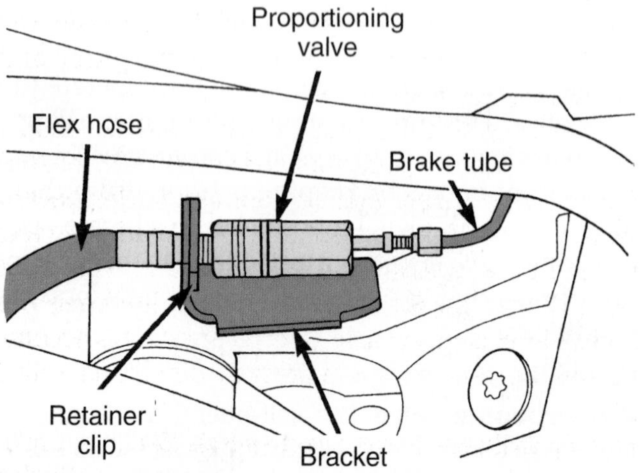
Proportioning valve
Flex hose
Brake tube
Retainer clip
Bracket

Figure 48-17 A proportioning valve. *Courtesy of Chrysler LLC*

Height-Sensing Proportional Valve The height-sensing proportional valve provides two different brake balance modes to the rear brakes based on vehicle load. This is accomplished by turning the valve on or off. When the vehicle is not loaded, hydraulic pressure is reduced to the rear brakes. When the vehicle is carrying a full load, the actuator lever moves up to change the valve's setting. The valve now allows full hydraulic pressure to the rear brakes. The valve contains a plunger, cam, torsional clutch spring, and an actuator shaft **(Figure 48–18)**.

The valve is mounted to the frame above the rear axle and has an actuator lever connected by a link to the lower shock absorber bracket. The valve is turned on and off as the axle-to-frame height changes due to load in the vehicle. The torsional clutch spring attached to the valve shaft is used as an override. Once the valve is positioned during braking, the spring prevents the valve from changing position if the vehicle goes over a bump or moves off the road.

Height-sensing proportional valves are replaced when defective and are not adjustable.

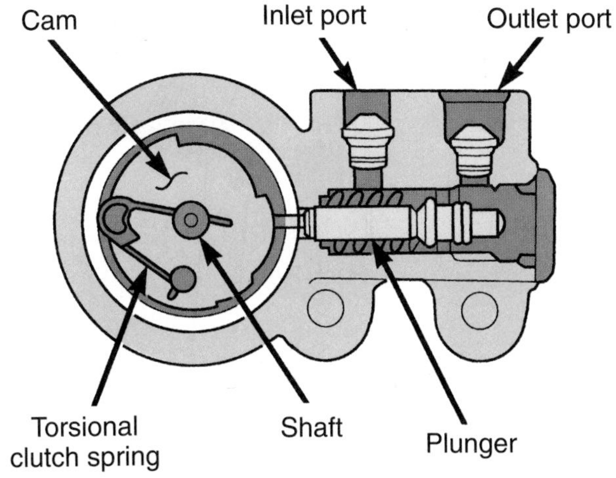

Figure 48-18 A height-sensing proportional valve. *Courtesy of Chrysler LLC*

SHOP TALK

Distribution and weight on some vehicles, such as station wagons, is such that a rear-to-front weight transfer does not present a similar problem. For this reason, standard proportioning valves are not required on all station wagon models.

Combination Valves Most newer cars have a **combination valve** (**Figure 48–19**) in their hydraulic system. This valve is simply a single unit that combines the metering and proportioning valves with the pressure differential valve. Combination valves are described as three-function or two-function valves, depending on the number of functions they perform in the hydraulic system.

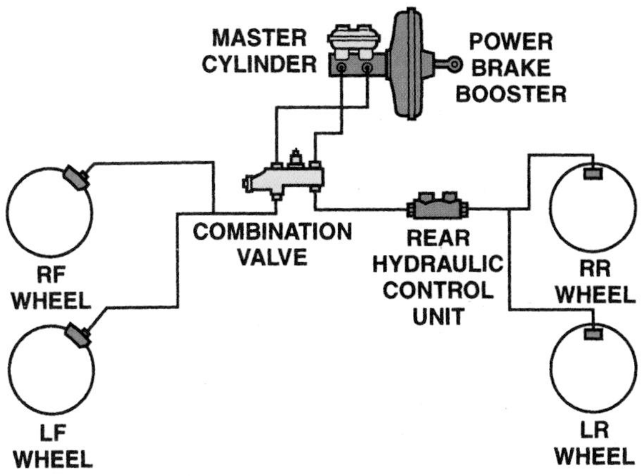

Figure 48-19 The hydraulic circuit for a pickup with rear-wheel antilock brakes and a combination valve. *Courtesy of Chrysler LLC*

Three-Function Valve This type of valve performs the functions of the metering valve, brake warning light switch, and proportioning valve.

Two-Function Valves There are two variations of the two-function combination valve. One variation works the proportioning valve and brake warning light switch functions. The other performs the metering valve and brake warning light switch functions.

If any one of its several operations fail, the entire combination valve must be replaced, because these units are not repairable.

Warning Lights

A wide variety of electrical and electronic components are found in a brake system, especially with ABS. These include the warning lamp switch operation of a pressure differential valve and the electrical switches to operate the failure warning lamp and the stop lamps as well as sensors to indicate low brake fluid level.

Failure Warning Lamp Switch A pressure differential valve is a hydraulically operated switch that controls the brake failure warning lamp on the instrument panel. Each side of the pressure differential valve is connected to half of the hydraulic system (one chamber of the master cylinder). Each master cylinder piston provides pressure to a separate hydraulic system. If one of the circuits fails, the brake pedal travel will increase and more brake pedal effort will be required to stop the car. The driver might not notice a problem, however, but the lamp on the instrument panel will provide a warning in case of hydraulic failure.

Failure in one half of the hydraulic system causes a pressure loss on one side of the pressure differential valve. Pressure on the other side moves the valve's plunger into contact with the switch terminal. This closes the circuit and the warning lamp is illuminated. All pressure differential valves work in this basic way but differ in the details of the shape of the piston and the use of centering springs.

The pressure differential valve on most late-model vehicles is part of a combination valve or built into the body of the master cylinder. Some vehicles have a switch in the float assembly of the fluid reservoir instead of a pressure differential valve. This float switch turns on the brake warning lamp when the fluid level changes to a dangerous point. This accomplishes the same thing as a pressure differential valve.

Master Cylinder Fluid Level Switch Because brake fluid level is important to safe braking, many vehicles have a fluid level switch that causes illumination of the instrument panel's red brake warning lamp when the fluid level is too low. This warning system is similar to the pressure differential valve because fluid level in the reservoir will drop from a leak caused by hydraulic failure. Therefore, a fluid level switch has replaced the pressure differential valve on many vehicles. An added advantage of a fluid level switch is that it will alert the driver of a dangerous fluid level caused by inattention and poor maintenance practices.

Fluid level sensors are built into the reservoir body or cap. One type has a float with a pair of switch contacts on a rod that extends above the float. If the fluid level drops too low, the float will drop and cause the rod-mounted contacts to touch a set of fixed contacts and close the lamp circuit. Another type of switch uses a magnet in a movable float. If the float drops low enough, the magnet pulls a set of switch contacts together to close the lamp circuit. The contacts typically provide a ground path for the brake warning lamp.

Parking Brake Switch The parking brake should only be applied to hold a vehicle stationary. If the parking brake is even partially applied while the vehicle is moving, it will produce enough heat to glaze friction materials, expand drum dimensions, and increase pedal travel. On rear disc brake systems with integral parking brake actuators, driving with the parking brake applied will distort the brake rotors and reduce pad life.

A normally closed, single-pole, single-throw switch is used to ground the circuit of the red brake warning lamp in the instrument cluster. This switch is located within the parking brake handle or pedal assembly and is designed to turn on the light whenever the parking brake is applied.

Some vehicles with daytime running lights (DRL) use the parking brake switch to complete a circuit that prevents the headlights from coming on if the parking brake is applied when the engine is started. When the parking brake is released, the DRLs operate normally.

Stop Lamps

Stop lamps are included in the right and left tail lamp assemblies. Vehicles built since 1986 also have a center high-mounted stop lamp (CHMSL).

Brake stop lamp switches are operated hydraulically or mechanically. Hydraulic switches were used on older vehicles and were installed in the master cylinder's high-pressure chamber; they were activated by system pressure. A mechanical switch is mounted on the bracket for the brake pedal and activated by the movement of the pedal lever **(Figure 48–20)**.

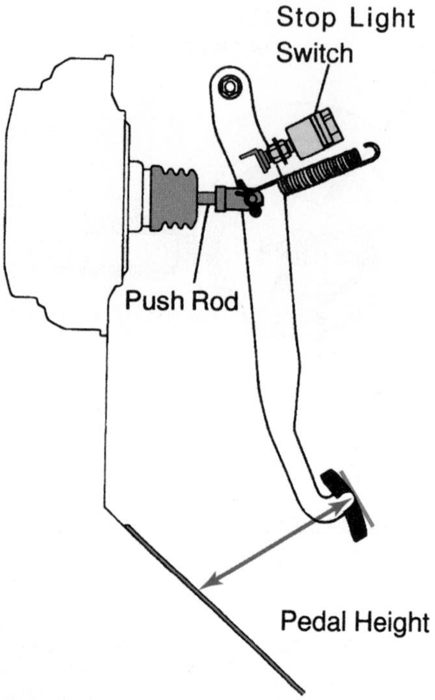

Figure 48-20 A mechanically activated stoplight switch. *Courtesy of Toyota Motor Sales, U.S.A., Inc.*

Mechanical switches are found on today's vehicles because they can be adjusted to illuminate the stop lamps with the slightest pedal movement. Stop lamp switches may be single-function or multifunction units. Single-function switches have only one set of switch contacts that control electric current to the stop lamps at the rear of the vehicle. Multifunction switches have one set of switch contacts for the stop lamps and at least one additional set of contacts for the torque converter clutch, the cruise control, or ABS. Some multifunction switches have contacts for all of these functions.

The brake lamp switch contacts are often connected to the brake lamps through the turn signal and hazard flasher switch. If a vehicle has antilock brakes, there is a connection or a separate switch for the ABS control unit to sense when the brakes are being applied. Wiring diagrams are essential for accurate identification of brake pedal switches and their functions.

DRUM AND DISC BRAKE ASSEMBLIES

Although drum and disc brakes are explained in great detail in later chapters, a brief explanation of their components and operating principles is essential at this point.

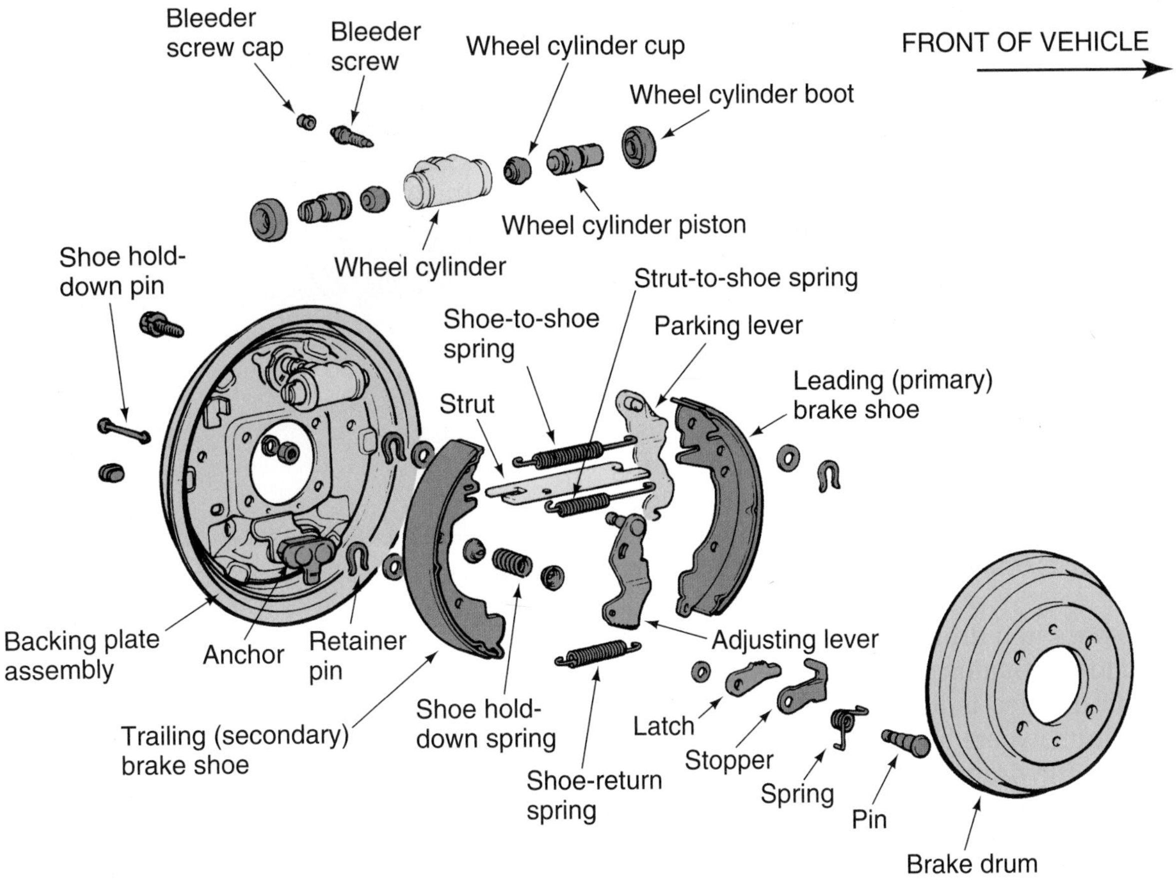

Figure 48-21 Typical nonservo drum brake assembly.

Drum Brakes

A drum brake assembly consists of a cast-iron drum, which is bolted to and rotates with the vehicle's wheel, and a fixed backing plate to which the shoes, wheel cylinders, automatic adjusters, and linkages are attached **(Figure 48–21)**. Additionally, there might be some extra hardware for parking brakes. The shoes are surfaced with frictional linings, which contact the inside of the drum when the brakes are applied. The shoes are forced outward by pistons located inside the wheel cylinder. They are actuated by hydraulic pressure. As the drum rubs against the shoes, the energy of the moving drum is transformed into heat. This heat energy is passed into the atmosphere. When the brake pedal is released, hydraulic pressure drops and the pistons are pulled back to their unapplied position by return springs.

Disc Brakes

Disc brakes resemble the brakes on a bicycle: the friction elements are in the form of pads, which are squeezed or clamped about the edge of a rotating wheel. With automotive disc brakes, this wheel is a separate unit, called a **rotor**, inboard of the vehicle wheel **(Figure 48–22)**. The rotor is made of cast iron. Since the pads clamp against both sides of it, both sides are machined smooth. Usually the two surfaces are separated by a finned center section for better cooling (such rotors are called **ventilated rotors**). The pads are attached to a metal backing that is actuated

Figure 48-22 The caliper of a disc brake assembly.

by pistons the same as with drum brakes. The pistons are contained within a caliper assembly, a housing that wraps around the edge of the rotor. The caliper is kept from rotating by way of bolts holding it to the car's suspension framework.

The **caliper** is a housing containing the pistons and related seals, springs, and boots as well as the cylinders and fluid passages necessary to force the friction linings or pads against the rotor. The caliper resembles a hand in the way it wraps around the edge of the rotor. It is attached to the steering knuckle. Some models use spring pressure to keep the pads close against the rotor. In other caliper designs this is achieved by a square-cut seal that distorts during brake application and returns to its original position as the brakes are released. This assists in retracting the piston into its cylinder and moves the pad off the rotor.

Unlike shoes in a drum brake, the pads act perpendicular to the rotation of the disc when the brakes are applied. This effect is different from that produced in a brake drum, where frictional drag actually pulls the shoe into the drum. Disc brakes are said to be non–self-energizing and so require more force to achieve the same braking effort. For this reason, they are ordinarily used in conjunction with a power brake unit.

HYDRAULIC SYSTEM SERVICE

Hydraulic system service is relatively uncomplicated, but it is vital to the vehicle's safe operation.

Brake Fluid Inspection

WARNING!

Clean the cover before removal to avoid dropping dirt into the reservoir.

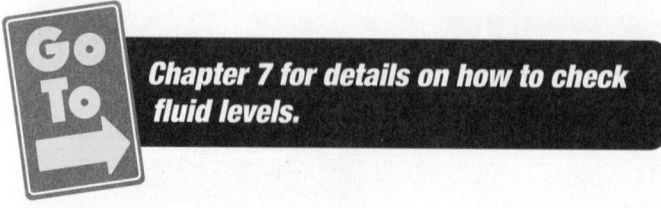

Go To
Chapter 7 for details on how to check fluid levels.

Brake Fluid Inspection

There are many ways to check for contaminated fluid. The color of the fluid is a good indication of its condition. Most brake fluids should be clear to amber in color; if the fluid is cloudy or dark, it is contaminated. The fluid can also be checked by placing a small amount of fluid in a clear glass jar. If the fluid is dirty or separates into layers, it is contaminated.

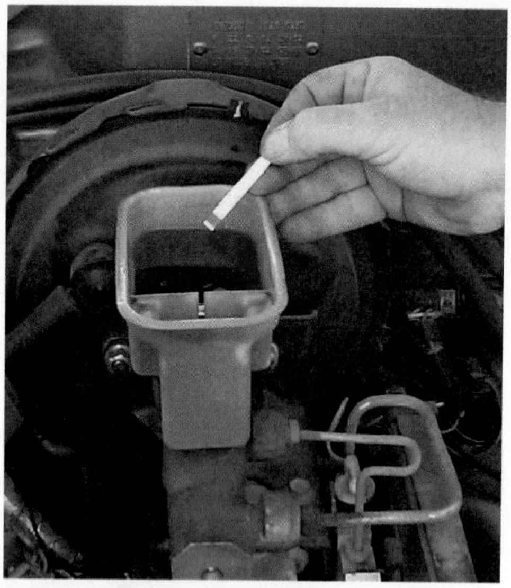

Figure 48-23 A test strip can be used to check for copper contaminants in the brake fluid. *Courtesy of Phoenix Systems LLC*

Test strips are also available. A strip of treated paper is dipped into the reservoir (**Figure 48–23**). The paper will change colors corresponding to the condition of the fluid. The resulting color is then matched to the color chart that accompanies the test strips.

Special brake fluid testers are also available. These measure the fluid's boiling point, which is an indication of moisture content.

Contaminated brake fluid can damage rubber parts and cause leaks. To rid the system of contaminated brake fluid, the system should be flushed and refilled with new fluid.

Master Cylinder Inspection

Master cylinder problems are quite common but not always readily evident. However, there are times when the master cylinder is suspect and the problem lies elsewhere. Accurate and logical troubleshooting is the only way to truly determine if the master cylinder is working properly. Although brake pedal response and reservoir fluid levels are strong indicators of problems with the master cylinder or hydraulic system, other tests can be performed to help pinpoint the problem.

Check the master cylinder housing for cracks and damage (**Figure 48–24**). Look for drops of brake fluid around the master cylinder. If a reservoir chamber is cracked, it may be completely empty and the surrounding area may be dry. This is because the fluid

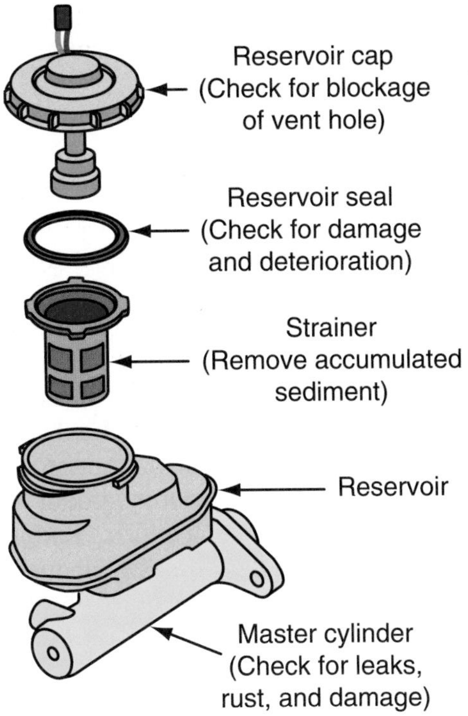

Figure 48-24 When inspecting a master cylinder, make sure to carefully check these items.

drained very quickly and has had time to evaporate or wash away. But with only one-half of the brake system operational, the brake warning lamp should be lit and a test drive should reveal the loss of braking power.

Refill the master cylinder reservoir section that is empty and apply the brakes several times. Wait 5 to 10 minutes and check for leakage or fluid level drop in the reservoir.

Hydraulic brake system leaks can be internal or external. Most internal leaks are actually fluid bypassing the cups in the master cylinder. If the cups lose their ability to seal the pistons, brake fluid leaks past the cups and the pistons cannot develop system pressure.

Internal and external rubber parts wear with use or can deteriorate with age or fluid contamination. Moisture or dirt in the hydraulic system can cause corrosion or deposits to form in the bore, resulting in the wear of the cylinder bore or its parts. Although internal leaks do not cause a loss of brake fluid, they can result in a loss of brake performance. Internal leakage will cause sinking pedal complaints and can be hard to pinpoint.

If the primary piston cup seal is leaking, the fluid will bypass the seal and move between the vent and replenishing ports for that reservoir or, in some cases, between reservoirs.

If there are no signs of external leakage, but the brake warning lamp is lit, the master cylinder may have an internal leak. To check for an internal leak in the master cylinder, remove the master cylinder cover and be sure the reservoirs are at least half full. Watch the fluid levels in the reservoirs while a helper slowly presses the brake pedal and then quickly releases it. If the fluid level rises slightly under steady pressure, the piston cups are probably leaking. Fluid level rising in one reservoir and falling in the other as the brake pedal is pressed and released also can indicate that fluid is bypassing the piston cups. Replace or rebuild the master cylinder if there is evidence of leakage.

Another quick test for internal leakage is to hold pressure on the brake pedal for about 1 minute. If the pedal drops but no sign of external leakage exists, fluid is probably bypassing the piston cups.

Quick Take-Up Valve Checks The quick take-up valve is used to provide a high volume of fluid on the first pedal stroke. This action takes up the slack in the low-drag caliper pistons. No direct test method exists for a quick take-up valve, but excessive pedal travel on the first stroke may indicate that fluid is bypassing the valve. If this symptom exists, check for a damaged or unseated valve. If the pedal returns slowly when the brakes are released, the quick take-up valve may be clogged so that fluid flow from the cylinder to the reservoir is delayed.

Air Entrapment Test

Poor pedal feel or action may be caused by trapped air in the system. Trapped air can be the result of a worn or defective master cylinder or other parts of the system. Air will also enter the system if there is a fluid leak. If there are no signs of leaks, check the system for trapped air.

CAUTION!

This test may result in brake fluid bubbling or spraying out of the master cylinder reservoir. Wear safety goggles. Cover the master cylinder reservoirs with clear plastic wrap or other suitable cover to keep brake fluid off the vehicle's paint.

To check for entrapped air, remove the cover of the master cylinder and make sure the reservoirs are filled to the proper level. Hold the cover and gasket against the reservoir top but do not secure it with its clamp or screws. Then have an assistant pump the brake pedal ten to twenty times rapidly and maintain pressure after the last pedal application. Remove the cover and have the assistant quickly release pedal pressure. Watch for a squirt of brake fluid from the reservoirs. If air is compressed in the system, it will force fluid back through the compensating ports faster than normal and cause fluid to squirt in the reservoir. If a fluid squirt appears in one side of the reservoir but not the other, that side of the split hydraulic system contains the trapped air. If there is air trapped in the system, bleed the system and recheck.

System Flushing

Currently, more than a dozen manufacturers specify periodic brake fluid changes for some, or all, of their models built during the past 12 years. Change intervals vary from as often as every 12 months or 15,000 miles to as infrequently as every 60,000 miles. All brake systems accumulate sludge over some period of time. Flushing the system can remove this sludge and any moisture, but once you have disturbed the sludge, you want to be sure you get it *all* out of the system. Stirring up sludge from the master cylinder reservoir may cause it to get into ABS valves and pumps if you do not get it all out of the system.

Brake hoses for disc brakes usually enter the caliper near the top of the caliper body. The bleeder valve is also located at the top of the caliper bore. If sludge accumulates in the caliper bore, it collects at the bottom. A quick, superficial bleeding of the caliper will not flush out the sludge and all of the old fluid. To flush a caliper thoroughly, pump several ounces of fluid through it. On some vehicles during brake pad replacement, you may want to remove the caliper from its mounts and retract the piston to force out all the old fluid. Then reinstall it and thoroughly flush it with fresh fluid.

Flushing should be done at each bleeder screw in the same manner as bleeding. Open the bleeder screw approximately 1½ turns and force fluid through the system until the fluid emerges clear and uncontaminated. Do this at each bleeder screw in the system. After all lines have been flushed, bleed the system using one of the common bleeding procedures. All contaminated fluid should be drawn out of the master cylinder reservoir before bleeding. Make sure you dispose of the old brake fluid in the proper manner.

Brake Line Inspection Check all tubing, hoses, and connections from under the hood to the wheels for leaks and damage. Also check for weak areas where the hose may collapse. Wheels and tires should also be inspected for signs of brake fluid leaks. Check all hoses for flexibility, bulges, and cracks. Check parking brake linkage, cable, and connections for damage and wear. Replace parts where necessary.

Brake Pedal Inspection Depress and release the brake pedal several times (engine running for power brakes). Check for friction and noise. Pedal movement should be smooth, with no squeaks from the pedal or brakes. The pedal should return quickly when it is released.

When operating the engine, be sure the transmission lever is in neutral or park. Be sure the area is properly ventilated for the exhaust to escape.

Apply heavy foot pressure to the brake pedal (engine running for power brakes). Check for a spongy pedal and pedal reserve. Spongy pedal action is springy. Pedal action should feel firm. **Pedal reserve** is the distance between the brake pedal and the floor after the pedal has been depressed fully. The pedal should not go lower than 1 or 2 inches (25 or 50 mm) above the floor.

With the engine off, hold light foot pressure on the pedal for about 15 seconds. There should be no pedal movement during this time. Pedal movement indicates a leak. Repeat the procedure using heavy pedal pressure (engine running for power brakes).

If there is pedal movement but the fluid level is not low, the master cylinder has internal leakage. It must be rebuilt or replaced. If the fluid level is low, there is an external leak somewhere in the brake system. The leak must be repaired.

Depress the pedal and check for proper stoplight operation.

To check power brake operation, depress and release the pedal several times while the engine is stopped. This eliminates vacuum from the system. Hold the brake down with moderate foot pressure and start the engine. If the power unit is operating properly, the brake pedal moves downward when the engine is started.

SHOP TALK

Master cylinders are seldom rebuilt; rather, they are replaced as a unit. The following overview for rebuilding a master cylinder is only intended to give you an understanding of what may be done. Always refer to the service information when attempting to rebuild a master cylinder.

Master Cylinder Rebuilding A master cylinder is rebuilt to replace leaking seals or gaskets. If a more serious problem exists, the master cylinder should be replaced.

To remove a master cylinder, disconnect the brake lines at the master cylinder. Install plugs in the brake lines and master cylinder to prevent dirt from entering. Remove the nuts that attach the master cylinder to the fire wall power brake unit, and remove the cylinder.

Remove the cover and seal. Drain the master cylinder and carefully mount it in a vise. Remove the piston assembly and seals according to the manufacturer's instructions. New pistons, pushrods, and seals are usually included in rebuilding kits.

Clean master cylinder parts only with brake fluid, brake cleaning solvent, or alcohol. Do not use a solvent containing mineral oil, such as gasoline. Mineral oil is very harmful to rubber seals.

Inspect the master cylinder. Damage, cracks, porous leaks, and worn piston bores mean the master cylinder must be replaced. Check very carefully for pitting or roughness in the bore. If any are present, the cylinder must be replaced.

Reassemble, install, and bleed the master cylinder according to the manufacturer's directions. Photo Sequence 51 is a typical procedure for bench bleeding a master cylinder.

WARNING!

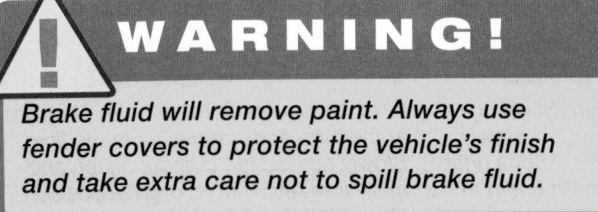

Brake fluid will remove paint. Always use fender covers to protect the vehicle's finish and take extra care not to spill brake fluid.

Hydraulic System Bleeding

Fluids cannot be compressed, whereas gases are compressible. Any air in the brake hydraulic system is compressed as the pressure increases. This action reduces the amount of force that can be transmitted by the fluid, therefore it is very important to keep all air out of the hydraulic system. To do this, air must be bled from brakes. This procedure is called bleeding the brake system.

Bleeding is a process of forcing fluid through the brake lines and out through a bleeder valve or bleeder screw **(Figure 48–25)**. The fluid eliminates any air that might be in the system. Bleeder screws and valves are fastened to the wheel cylinders or calipers. The bleeder must be cleaned. A drain hose then is connected from the bleeder to a glass jar **(Figure 48–26)**.

Two types of brake bleeding procedures are used: manual bleeding (vacuum bleeding) and pressure bleeding. On some antilock brake systems, a scan tool can be used to help bleed the brakes. Always follow the manufacturer's recommendations when bleeding brakes. The sequence in which bleeding is performed can be critical. To remove vacuum, the engine must be off. Pump the brake pedal several times.

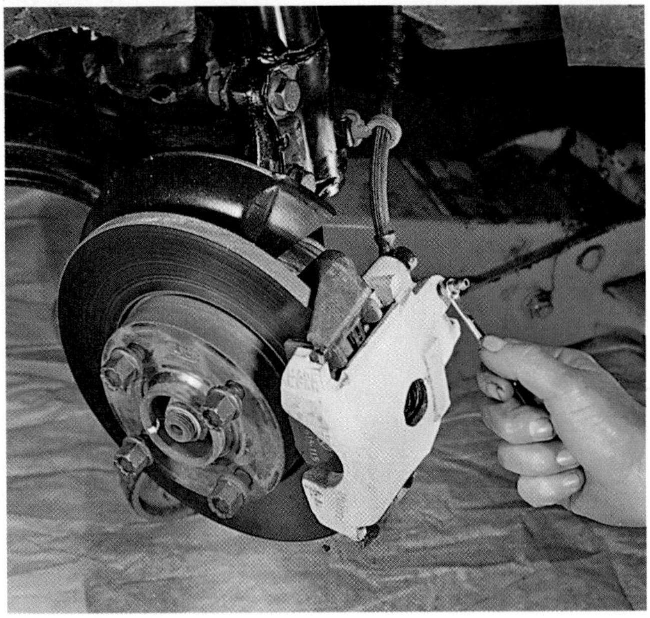

Figure 48–25 The wheel brake hydraulic units are fitted with bleeder screws.

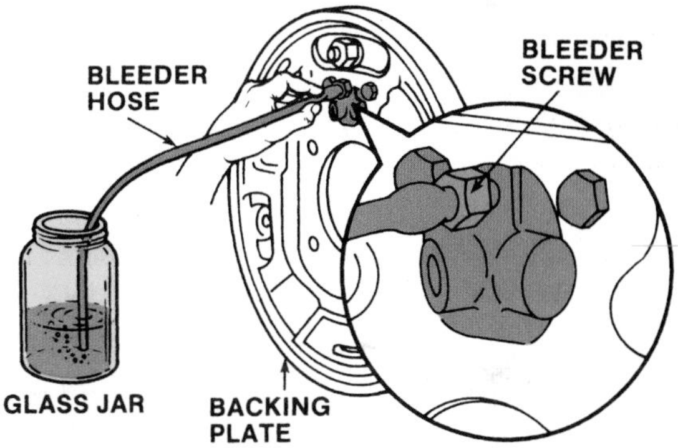

Figure 48–26 While bleeding the brake system, releasing the fluid into a glass jar will help to determine when the system is free of air.

WARNING!

Always use fresh brake fluid when bleeding the system. Do not use fluid that has been drained. Drained fluid may be contaminated and can damage the system.

Bleeding Sequence All manufacturers recommend a specific sequence to follow when bleeding a vehicle's brakes. These recommendations can be found in service manuals and should be followed. If the

P51-1 *Mount the master cylinder firmly in a vise, being careful not to apply excessive pressure to the casting. Position the master cylinder so the bore is horizontal.*

P51-2 *Connect short lengths of tubing to the outlet ports, making sure the connections are tight.*

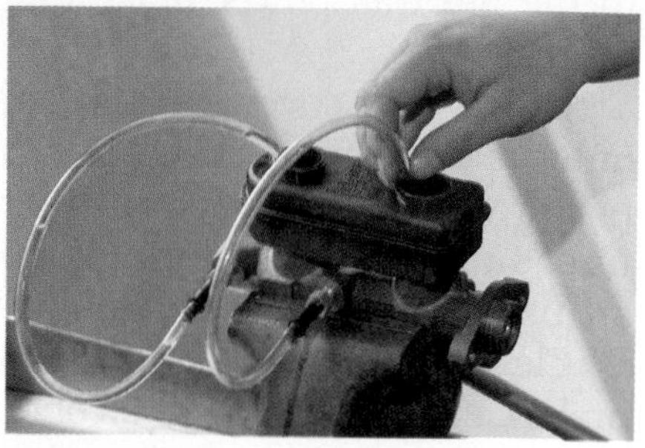

P51-3 *Bend the tubing lines so that the ends are in each chamber of the master cylinder reservoir.*

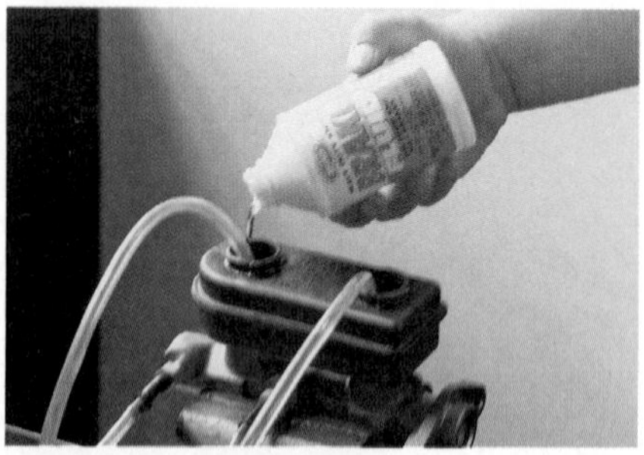

P51-4 *Fill the reservoirs with fresh brake fluid until the level is above the ends of the tubes.*

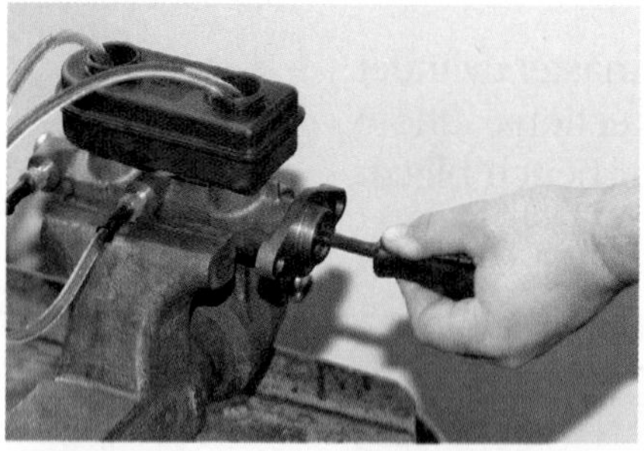

P51-5 *Using a wooden dowel or the blunt end of a drift or punch, slowly push on the master cylinder pistons until both are completely bottomed out in their bore.*

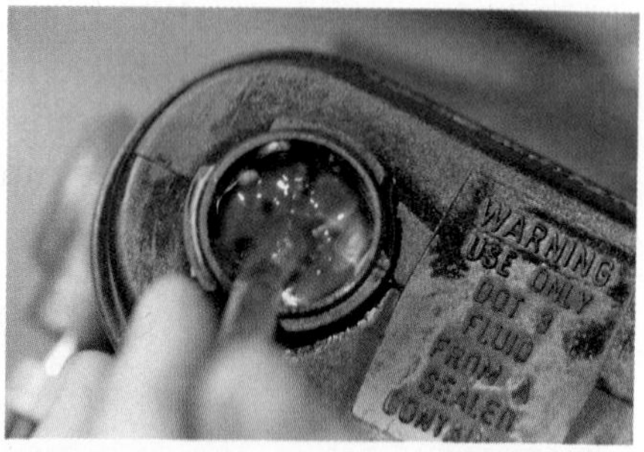

P51-6 *Watch for bubbles to appear at the tube ends immersed in the fluid. Slowly release the cylinder piston and allow it to return to its original position. On quick take-up master cylinders, wait 15 seconds before pushing in the piston again. On other units, repeat the stroke as soon as the piston returns to its original position. Slow piston return is normal for some master cylinders.*

manufacturer's recommendations are not available, the following sequence will work on most vehicles:

1. Master cylinder
2. Combination valve or proportioning valve (if fitted with bleeder screws)
3. Right rear
4. Left rear
5. Right front
6. Left front
7. Height-sensing proportioning valve (if there is a bleeder screw)

This sequence is based on the principles of starting at the highest point in the system and working downward, then starting at the wheel farthest from the master cylinder and working to the closest. A couple of more general rules also are worth remembering.

If the brake system is split between the front and rear wheels, the rear wheels (which are farthest from the master cylinder) usually are bled first. If the brake system is split diagonally, the most common sequence is: RR-LF-LR-RF **(Figure 48–27)**. This sequence also applies to most systems with a quick take-up master cylinder. If you bleed a quick take-up system in any other sequence, you may chase air throughout the system.

Exceptions to the general rules exist, however. Chrysler, for example, recommends bleeding both rear brakes before the front brakes, regardless of how the hydraulic system is split.

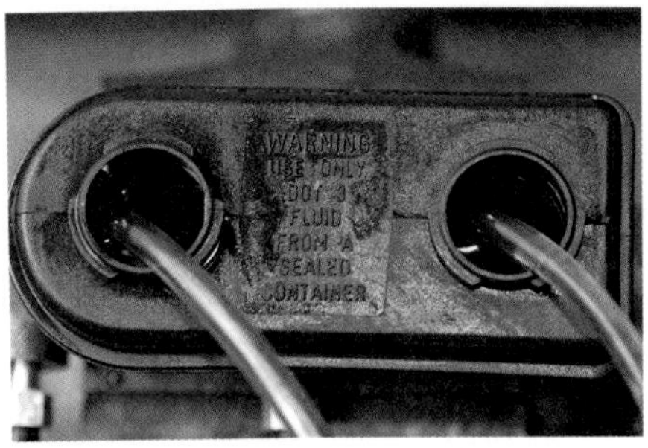

P51-7 *Pump the cylinder piston until no bubbles appear in the fluid.*

P51-8 *Remove the tubes from the outlet ports and plug the openings with temporary plugs or your fingers. Keep the ports covered until you install the master cylinder on the vehicle.*

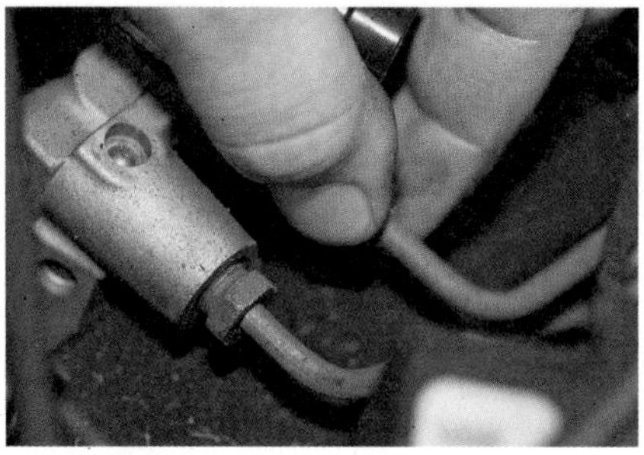

P51-9 *Install the master cylinder on the vehicle. Attach the lines, but do not tighten the tube connections.*

P51-10 *Slowly depress the pedal several times to force out any air that might be trapped in the connections each time the pedal is depressed. Before releasing the pedal, tighten the nut slightly and loosen it before depressing the pedal each time. Soak up the fluid with a rag to avoid damaging the car finish.*

P51-11 *When there are no air bubbles in the fluid, tighten the connections to the manufacturer's specifications. Make sure the master cylinder reservoirs are adequately filled with brake fluid.*

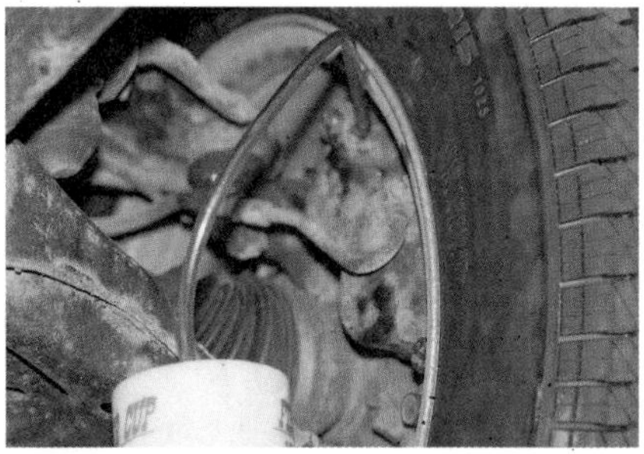

P51-12 *After reinstalling the master cylinder, bleed the entire brake system on the vehicle.*

Manual Bleeding A manual bleeding procedure requires two people. One person operates the bleeder; the other, the brake pedal. Bleed only one wheel at a time.

WARNING!

Be sure the bleeder hose is below the surface of the liquid in the jar at all times. Do not allow the master cylinder to run out of fluid at any time. If these precautions are not followed, air can enter the system, and it must be bled again. The master cylinder cover must be kept in place.

Place the bleeder hose and jar in position. Have a helper pump the brake pedal several times and then hold it down with moderate pressure. Slowly open the bleeder valve. After fluid/air has stopped flowing, close the bleeder valve. Have the helper slowly release the pedal. Repeat this procedure until fluid that flows from the bleeder is clear and free of bubbles.

Discard all used brake fluid. Fill the master cylinder reservoir. Check the brakes for proper operation.

WARNING!

Clean the master cylinder and cover before adding fluid. This is important for preventing dirt from entering the reservoir.

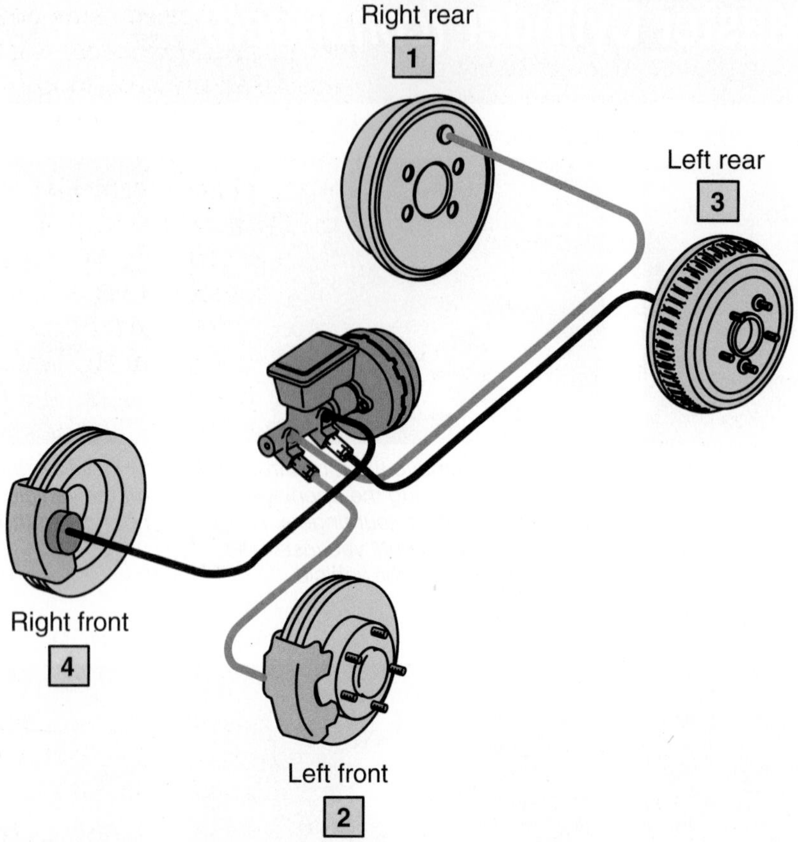

Figure 48-27 The recommended bleeding sequence for a diagonally split brake system.

Pressure Bleeding A pressure bleeding procedure can be done by one person. Pressure bleeding equipment uses pressurized fluid that flows through a special adapter fitted into the master cylinder **(Figure 48–28)**.

The use of pressure bleeding equipment varies with different automobiles and different equipment makers. Always follow the automobile manufacturer's recommendations when using pressure bleeding equipment.

On automobiles with metering valves, the valve must be held open during pressure bleeding. A special tool is used to hold open the metering section of a combination valve.

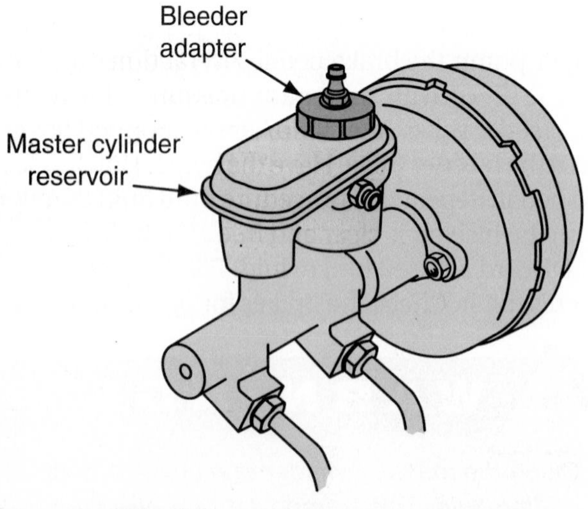

Figure 48-28 An adapter that installs onto the reservoir's fill opening is required to pressure the brake system.

USING SERVICE INFORMATION

Consult the service manual to be sure the proper bleeding sequence is followed. If a vehicle requiring a special sequence is bled in the conventional manner, air might be chased throughout the system.

Open the bleeder valves one at a time until clear, air-free fluid is flowing. Progress from the wheel cylinder farthest from the master cylinder to the cylinder closest.

Do not exceed recommended pressure while bleeding the brakes. Always release air pressure after bleeding. Clean and fill the master cylinder after pressure bleeding. Check the brakes for proper operation. Be sure to remove the special tool used to hold the metering valve.

POWER BRAKES

Power brakes are nothing more than a standard hydraulic brake system with a booster unit located between the brake pedal and the master cylinder to help activate the brakes.

Two basic types of power-assist mechanisms are used. The first is **vacuum assist**. These systems use engine vacuum, or sometimes vacuum pressure developed by an external vacuum pump, to help apply the brakes. The second type of power assist is **hydraulic assist**. It is normally found on larger vehicles. This system uses hydraulic pressure developed by the power steering pump or other external pump to help apply the brakes.

Both vacuum and hydraulic assist act to multiply the force exerted on the master cylinder pistons by the driver. This increases the hydraulic pressure delivered to the wheel cylinders or calipers while decreasing driver foot pressure.

Vacuum-Assist Power Brakes

All vacuum-assisted units are similar in design. They generate application energy by opposing engine vacuum to atmospheric pressure. A piston and cylinder, flexible diaphragm, or bellows use this energy to provide braking assistance.

All modern vacuum-assist units are vacuum-suspended systems. This means the diaphragm inside the unit is balanced using engine vacuum until the brake pedal is depressed. Applying the brake allows atmospheric pressure to unbalance the diaphragm and allows it to move generating application pressure.

Atmospheric pressure is normally between 14 and 15 psi (96.5 and 103 kPa). If the diameter of the diaphragm is 12 inches (305 mm), the area of the diaphragm is about 113 square inches (72,907 sq. mm). Since the vacuum to the booster is typically 17 inches of mercury (432 mm Hg) or about 7 psi (48 kPa), the pressure differential on the diaphragm is 7.7 psi (53 kPa). Therefore, the resulting force on the diaphragm would be 870 pounds (3870 N) (7.7×113).

Vacuum boosters may be single diaphragm or tandem diaphragm. The unit consists of three basic elements combined into a single power unit **(Figure 48–29)**.

The three basic elements of the single diaphragm follow:

1. A vacuum power section that includes a front and rear shell, a power diaphragm, a return spring, and a pushrod.
2. A control valve built as an integral part of the power diaphragm and connected through a valve rod to the brake pedal. It controls the degree of brake application or release in accordance with the pressure applied to the brake pedal.

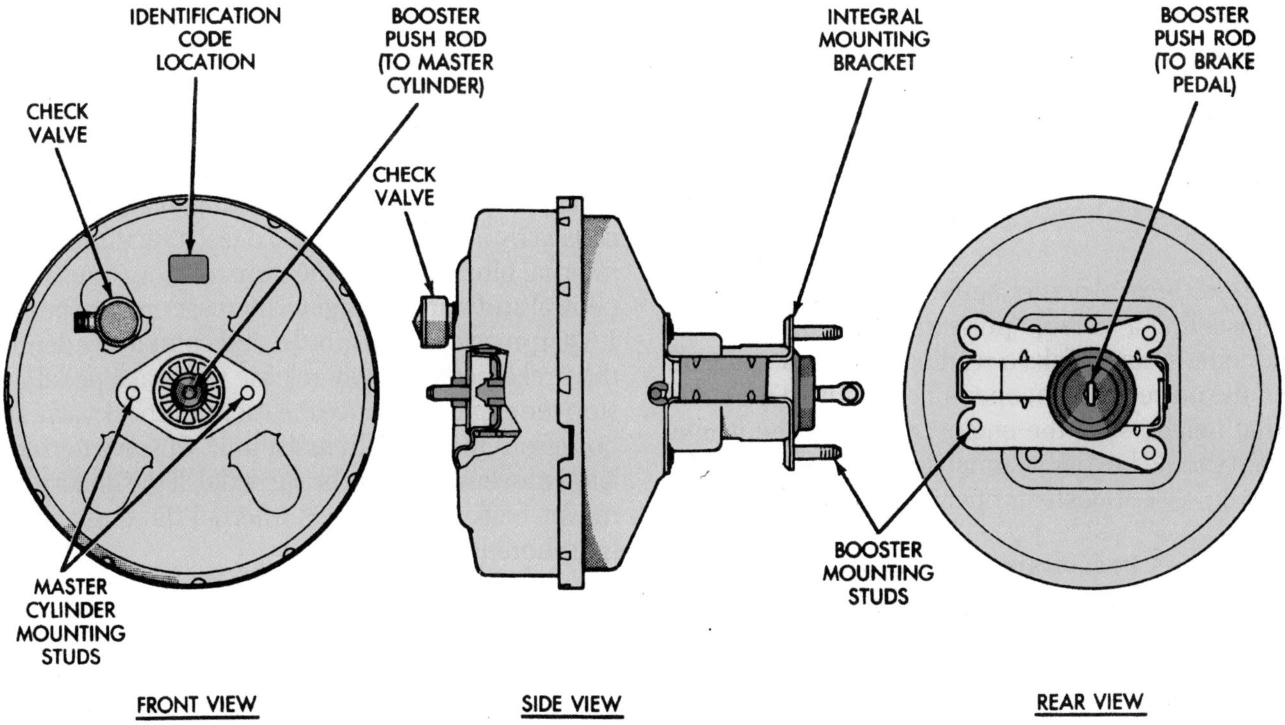

Figure 48-29 A typical vacuum brake booster. *Courtesy of Chrysler LLC*

3. A hydraulic master cylinder, attached to the vacuum power section that contains all the elements of the conventional brake master cylinder except for the pushrod. It supplies fluid under pressure to the pressure applied by the brake booster.

Operation When the brakes are applied, the valve rod and plunger move to the left in the power diaphragm. This action closes the control valve's vacuum port and opens the atmospheric port to admit air through the valve at the rear diaphragm chamber. With vacuum in the rear chamber, a force develops that moves the power diaphragm, hydraulic pushrod, and hydraulic piston or pistons to close the compensating port or ports and force fluid under pressure through the residual check valve or valves and lines into the front and rear brake assemblies.

As pressure develops in the master cylinder, a counterforce acts through the hydraulic pushrod and reaction disc against the vacuum power diaphragm and valve plunger. This force tends to close the atmospheric port and reopen the vacuum port. Since this force is in opposition to the force applied to the brake pedal by the operator, it gives the operator a feel for the amount of brake applied.

Servicing Vacuum-Assist Booster Units The fact that a vehicle's brakes still operate when the vacuum-assist unit fails indicates that the hydraulic brake system and the vacuum-assist system are two separate systems. This means you should always check for faults in the hydraulic system first. If it checks out satisfactorily, start inspecting the vacuum-assist circuit.

For a fast check of vacuum-assist operation, press the brake pedal firmly and then start the engine. The pedal should fall away slightly and less pressure should be needed to maintain the pedal in any position.

Pressure Check Another simple check can be made by installing a suitable pressure gauge in the brake hydraulic system. Take a reading with the engine off and the power unit not operating. Maintain the same pedal height, start the engine, and read the gauge. There should be a substantial pressure increase if the vacuum-assist booster is operating correctly.

Pedal Travel **Pedal travel** and total travel are critical on vacuum-assisted vehicles. Pedal travel should be kept strictly to specifications listed in the vehicle's service manual.

Vacuum Reading If the power unit is not giving sufficient assistance, take a manifold vacuum reading. If manifold vacuum level is below specifications, tune the engine and retest the unit. Loose or damaged vacuum lines and clogged air intake filters reduce braking assistance. Most units have a check valve that retains some vacuum in the system when the engine is off. A vacuum gauge check of this valve indicates if it is restricted or stays open.

Release Problems Failure of the brakes to release is often caused by a tight or misaligned connection between the power unit and the brake linkage. Broken pistons, diaphragms, bellows, or return springs can also cause this problem.

To help pinpoint the problem, loosen the connection between the master cylinder and the brake booster. If the brakes release, the problem is caused by internal binding in the vacuum unit. If the brakes do not release, look for a crimped or restricted brake line or similar problem in the hydraulic system.

Hard Pedal Power brakes that have a hard pedal may have collapsed or leaking vacuum lines of insufficient manifold vacuum. Punctured diaphragms or bellows and leaky piston seals all lead to weak power unit operation and hard pedal. A steady hiss when the brake is held down indicates a vacuum leak that causes poor operation.

Grabbing Brakes First, look for all the usual causes of brake grab, such as greasy linings, or scored rotors or drums. If the trouble appears to be in the power unit, check for a damaged reaction control. The reaction control is made up of a diaphragm, spring, and valve that tend to resist pedal action. It is put into the system to give the driver more brake pedal feel.

Check of Internal Binding Release problems, hard pedal, and dragging (slow releasing) brakes can all be caused by internal binding. To test a vacuum unit for internal binding, place the transmission/transaxle in neutral and start the engine. Increase engine speed to 1,500 rpm, close the throttle, and completely depress the brake pedal. Slowly release the brake pedal and stop the engine. Remove the vacuum check valve and hose from the vacuum assist unit. Observe for backward movement of the brake pedal. If the brake pedal moves backward, there is internal binding and the unit should be replaced.

PUSHROD ADJUSTMENT

Proper adjustment of the master cylinder pushrod is necessary to ensure proper operation of the power brake system. A pushrod that is too long causes the master cylinder piston to close off the vent port, preventing hydraulic pressure from being released and resulting in brake drag. A pushrod that is too short causes excessive brake pedal travel and causes groaning noises to come from the booster when the brakes are applied. A properly adjusted pushrod that remains assembled to the booster with which it was matched during production should not require service adjustment. However, if the booster, master cylinder, or pushrod are replaced, the pushrod might require adjustment.

Two methods can be used to check for proper pushrod length and installation: the gauge method and the air method.

Gauge Method

In most vacuum power units, the master cylinder pushrod length is fixed, and length is usually checked only after the unit has been overhauled or replaced. A typical adjustment using the gauge method is shown in **Figure 48–30**.

Air Method

The air-testing method uses compressed air applied to the hydraulic outlet of the master cylinder. Air pressure is regulated to a value of approximately 5 psi (35 kPa) to prevent brake fluid spraying from the master cylinder.

If air passes through the replenishing port, which is the smaller of the two holes in the bottom of the master cylinder reservoir, the adjustment is satisfactory. If air does not flow through the replenishing port, adjust the pushrod as required, either by means of the adjustment screw (if provided) or by adding shims between the master cylinder and power unit shell until the air flows freely.

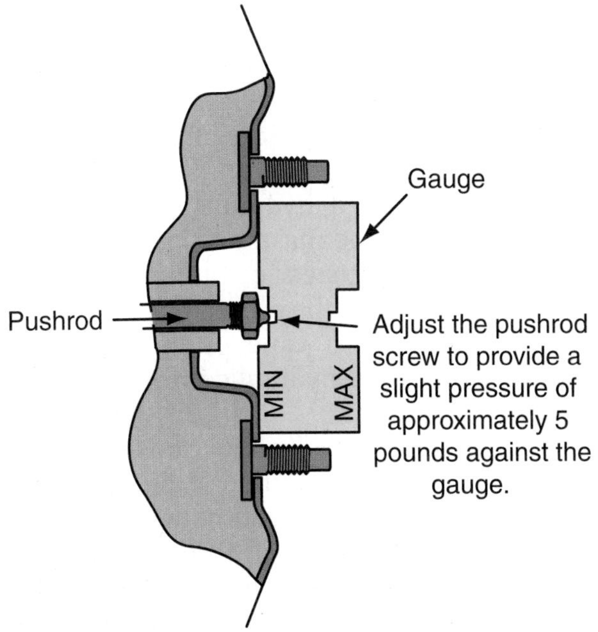

Figure 48-30 A gauge for measuring pushrod length.

HYDRAULIC BRAKE BOOSTERS

Decreases in engine size, plus the continued use of engine vacuum to operate other engine systems, such as emission control devices, led to the development of hydraulic-assist power brakes. These systems use fluid pressure, not vacuum pressure, to help apply the brakes. They are mostly found on diesel engines and other engines that have low vacuum.

Fluid pressure from the power-steering pump provides the power assist to the brakes. The power brake booster is located between the cowl and the master cylinder. Hoses connect the power-steering pump to the booster assembly.

The power-steering pump provides a continuous flow of fluid to the brake booster whenever the engine is running. Three flexible hoses route the power-steering fluid to the booster. One hose supplies pressurized fluid from the pump. Another hose routes the pressurized fluid from the booster to the power-steering gear assembly. The third hose returns fluid from the booster to the power-steering pump.

The hydraulic pressure in the hydraulic booster should not be confused with the hydraulic pressure in the brake lines. Remember that they are two separate systems and require two different types of fluid: power-steering fluid for the pump and brake fluid for the brake system. Never put power-steering fluid in the brake reservoir. If the brake fluid becomes contaminated by power-steering fluid, the brake system must be flushed.

Some systems have a nitrogen charged pneumatic accumulator on the booster to provide reserve power-assist pressure. If power-steering pump pressure is not available due to belt failure or similar problems, the accumulator pressure is used to provide brake assist.

The booster assembly **(Figure 48–31)** consists of an open center spool valve and sleeve assembly, a lever assembly, an input rod assembly, a power piston, an output pushrod, and the accumulator. The booster assembly is mounted on the vehicle in much the same manner as a vacuum booster. The pedal rod is connected at the booster input rod end.

Power-steering fluid flow in the booster unit is controlled by a hollow center spool valve. The spool valve has lands, annular grooves, and drilled passages. These mate with grooves and lands in the valve bore. The flow pattern of the fluid depends on the alignment of the valve in the bore.

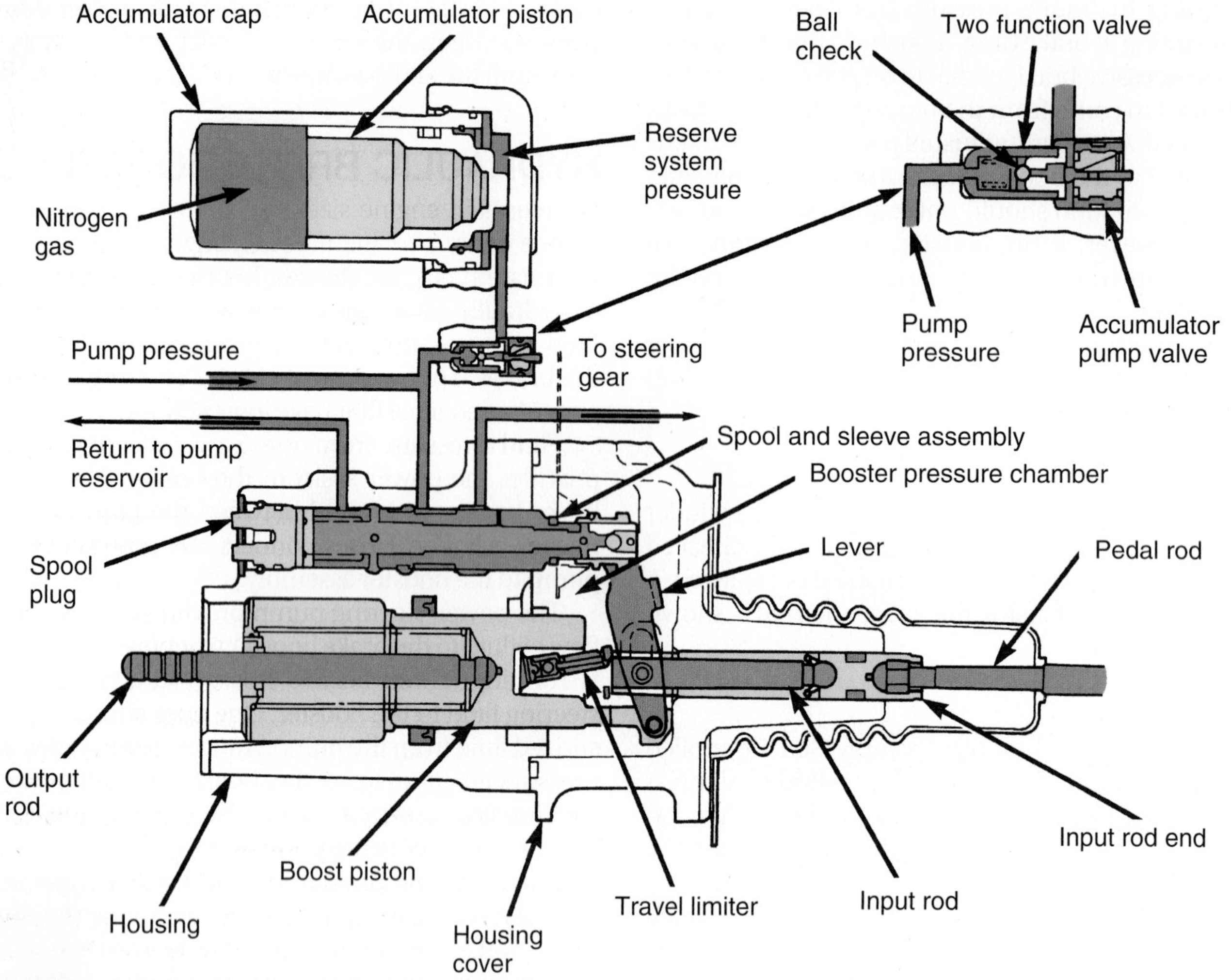

Figure 48-31 A hydraulic brake booster. *Courtesy of Chrysler LLC*

Operation

When the brake pedal is depressed, the pedal's pushrod moves the master cylinder's primary piston forward. This causes the lever assembly of the booster to move a sleeve forward to close off the holes leading to the open center of the spool valve. A small additional lever movement moves the spool valve into the spool valve bore. The spool valve then diverts some hydraulic fluid into the cavity behind the booster piston building up hydraulic pressure that moves the piston and a pushrod forward. The output pushrod moves the primary and secondary master cylinder pistons that apply pressure to the brake system. When the brake pedal is released, the spool and sleeve assemblies return to their normal positions. Excess fluid behind the piston returns to the power-steering pump reservoir through the return line. After the brakes have been released, pressurized fluid from the power-steering pump flows into the booster through the open center of the spool valve and back to the power-steering pump.

There have been many variations of and names given to hydraulic brake booster systems. The most common names are the hydro-boost system produced by Bendix and the Powermaster system produced by General Motors. The Powermaster system is a little different from the rest in that it uses a self-contained hydraulic booster that is built directly onto the master cylinder. Instead of relying on the power-steering pump for hydraulic pressure as is done in the other systems, the Powermaster has its own vane pump and electric motor to provide the hydraulic pressure required for booster operation.

Any investigation of a hydraulic boost complaint should begin with an inspection of the power-steering pump belt, fluid level, and hose condition and connections. Hydraulic boost systems do not work properly if they are not supplied with a continuous supply of clean, bubble-free power-steering fluid at the proper pressure.

WARNING!

Always depressurize the accumulator of any hydraulic boost system before disconnecting any brake lines or hoses. This is usually done by turning the engine off and depressing and releasing the brake pedal up to ten times.

Basic Operational Test

The basic operational test of these systems is as follows. With the engine off, pump the brake pedal numerous times to bleed off the residual hydraulic pressure that is stored in the accumulator. Hold firm pressure on the brake pedal and start the engine. The brake pedal should move downward, then push up against the foot.

Accumulator Test

To be sure the accumulator is performing properly, rotate the steering wheel with the engine running until it stops and hold it in that position for no more than 5 seconds. Return the steering wheel to the center position and shut off the engine. Pump the brake pedal. You should feel two to three power-assisted strokes. Now repeat the steps. That pressurizes the accumulator. Wait one hour, then pump the brake pedal. There should be two or three power-assisted strokes. If the system does not perform as just described, the accumulator is leaking and should be replaced.

Noise Troubleshooting

The booster is also part of another major subsystem of the vehicle, the power-steering system. Problems or malfunctions in the steering system may affect brake-assist operation. The following are some common troubleshooting tips.

Moan or low-frequency hum usually accompanied by a vibration in the pedal or steering column might be encountered during parking or other very low-speed maneuvers. This can be caused by a low fluid level in the power-steering pump, or by air in the power-steering fluid due to holding the pump at relief pressure (steering wheel held all the way in one direction) for an excessive amount of time (more than 5 seconds). Check the fluid level and add fluid if necessary. Allow the system to sit for 1 hour with the cap removed to eliminate the air. If the condition persists, it might be a sign of excessive pump wear. Check the pump according to the vehicle manufacturer's recommended procedure.

At or near power runout (brake pedal near fully depressed position), a high-speed fluid noise (like a faucet can make) might occur. This is a normal condition and will not be heard, except in emergency braking conditions.

Whenever the accumulator pressure is used, a slight hiss is noticed. It is the sound of the hydraulic fluid escaping through the accumulator valve and is completely normal.

After the accumulator has been emptied and the engine is started again, another hissing sound might be heard during the first brake application or the first steering maneuver. This sound is caused by the fluid rushing through the accumulator charging orifice. It is normal and will only be heard once after the accumulator is emptied. However, if this sound continues even though no apparent accumulator pressure assist was made, it could indicate that the accumulator is not holding pressure. Check for this possibility using the accumulator test discussed previously.

After bleeding, a gulping sound might be present during brake applications, as noted in the bleeding instructions. This sound is normal and should disappear with normal driving and braking.

Diagnosis and testing of the Powermaster unit requires the use of a special adapter and test gauge or aftermarket equivalents. The Powermaster pressure switch is removed and the adapter and test gauge is installed in its port. The unit can then be energized, and the switch's high-pressure cut-off and low-pressure turn-on points observed and checked against specifications. Follow service manual instructions for the connection and operation of the test gauge and all system test procedures.

ELECTRIC PARKING BRAKES

Electrically operated parking brakes are becoming more common and are replacing mechanical systems. These systems operate as a conventional hydraulic brake for normal braking and as an electric brake for parking. With electric parking brakes, there is no need for a parking brake lever or pedal. This frees up space in the interior.

Electric parking brakes are seen as the first step toward brake-by-wire systems. Two different techniques are currently being used by manufacturers. Some systems have an electric motor mounted on the rear brake calipers and others use an undercar motor to pull on the parking brake cables.

When the caliper is fitted with a motor **(Figure 48–32)**, there is no need for parking brake cables

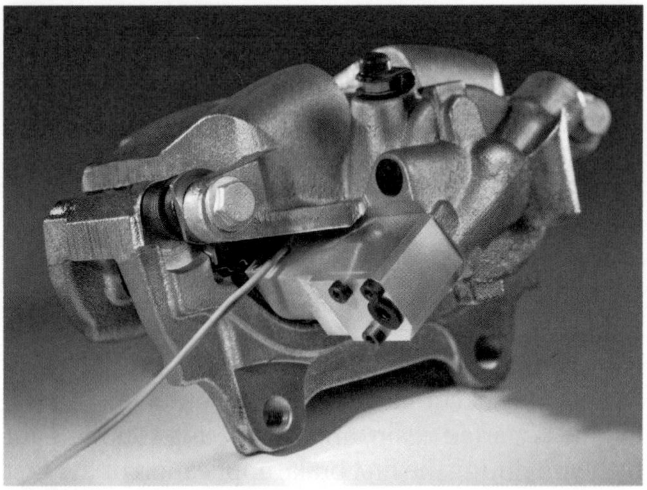

Figure 48-32 This caliper is fitted with an electronically controlled motor that controls the application of the parking brake. *Courtesy of Robert Bosch GmbH, www.bosch-presse.de*

and linkages. The motor is controlled by the PCM. The system interfaces with the vehicle's controller area network (CAN) for continuous monitoring and feedback. This allows the system to do many things besides apply the parking brake, such as:

- Provide some control during emergency braking.
- Help stop the car if the hydraulic system fails.
- Automatically release the parking brakes when the throttle is opened.
- Automatically engage when the ignition is turned off.
- Automatically engage when the driver's door is opened.
- Keep the vehicle from rolling backward when stopped on a hill by applying the rear brakes until the driver operates the clutch or throttle pedals.

CASE STUDY

A customer brings her late-model vehicle to a brake specialty shop, complaining of occasional rear brake lockup when the brakes are lightly applied. She is nervous because she feels the vehicle is unsafe to drive. The vehicle is equipped with front disc and rear drum brakes.

A thorough road test by the technician does not verify the complaint. Talking to the customer, the technician learns the problem most often occurs in damp or cold weather or when the vehicle has been sitting for a length of time.

The technician performs a complete visual and operational inspection. He checks the action of the brake pedal, the power brake booster, and the combination valve. All systems are working fine. He then removes the rear drums and inspects the shoes and drums. Nothing appears out of order or broken. The technician turns to his supervisor for a second opinion.

After listening to a summary of the inspection and checks made to date, the shop supervisor checks the file on factory service bulletins for that model vehicle. The answer is found in a bulletin issued about a year after the vehicle was introduced. The OEM brake shoes were found to have a tendency to swell when subjected to cold and wet. The factory recommended that another type of shoe be installed to correct the problem. If the technician had the experience to check the service bulletin file on the vehicle, valuable diagnostic time would have been saved.

The correct shoe is installed, and the vehicle is road tested. The rear brakes no longer lock up.

KEY TERMS

Aramid fibers
Brake fade
Caliper
Coefficient of friction (COF)
Combination valve
Cup seal
Diagonally split system
Hydraulic assist
Metallic linings
Metering valve
Nonasbestos organic linings (NAO)
Pedal reserve
Pedal travel
Proportioning valve
Quick take-up
Replenishing port
Residual pressure check valve
Semimetallic lining
Static friction
Synthetic lining
Vacuum assist
Vent port
Ventilated rotor

SUMMARY

- The four factors that determine a vehicle's braking power are pressure, which is provided by the hydraulic system; coefficient of friction, which represents the frictional relationship between pads and rotors or shoes and drums; frictional contact surface, which means bigger brakes stop a car more quickly than smaller brakes; and head dissipation, which is necessary to prevent brake fade.

- Today's brake linings are fully metallic, semimetallic, nonasbestos organic, synthetic, carbon, or ceramic.
- Since 1967, all cars have been required to have two separate brake systems. The dual brake system uses a tandem master cylinder, which is two master cylinders with two separated pistons and fluid reservoirs in one cylinder bore.
- The brake lines transmit brake fluid pressure from the master cylinder to the wheel cylinders and calipers of drum and disc brakes. Brake hoses offer flexible connections to wheel units and must offer high heat resistance.
- A pressure differential valve is used in all dual brake systems to operate a warning light switch that alerts the driver if pressure is lost in either hydraulic system.
- The metering valve, located in the front brake line, provides for better balance of the front and rear brakes while also preventing lockup of the front brakes.
- The proportioning valve controls rear brake pressure, particularly during hard stops.
- Bleeding removes air from the hydraulic system.
- Flushing removes old brake fluid from the system and replaces it with fresh new fluid. This should be done about every 2 years.
- Power brakes can be either vacuum assist or hydraulic assist. Vacuum-assisted units use engine vacuum or vacuum developed by an external pump to help apply the brakes. Hydraulic-assisted units use fluid pressure.

REVIEW QUESTIONS

1. Explain why bleeding air out of a hydraulic system is so important.
2. Explain why modern hydraulic braking systems are dual designs, and why this is important.
3. Describe the functions of the hydraulic system combination valve.
4. When the brakes are applied on a moving car, the frictional parts (brake ______ or ______) are forced against the rotating parts of the car (brake ______ or ______). The friction causes the rotating parts to slow down and stop. Just as the energy of the rotating parts is called kinetic ______, the friction used to stop them is called kinetic ______.
5. What is the purpose of the master cylinder vent port?
6. What is the purpose of the master cylinder replenishing port?
7. Explain why a height-sensing proportioning valve is required on some vehicles.
8. A three-function combination valve has a brake system failure switch, a ______ valve, and a ______ valve.
9. Explain how vacuum is used to provide a power assist.
10. The purpose of the master cylinder is to ______.
    a. generate the hydraulic pressure needed to apply the brake mechanisms
    b. automatically pump the brakes during panic stops
    c. apply braking power when wheel slippage occurs
    d. all of the above
11. Which of the following can lead to brake hose failure?
    a. improperly matched fittings
    b. stressing the hose during installations
    c. deterioration from heat and contaminants
    d. all of the above
12. *True or False?* Metering and proportioning valves balance the braking characteristics of disc and drum brakes.
13. Which type of brake requires greater application force and is commonly used with power-boost units?
    a. drum
    b. disc
    c. parking
    d. none of the above
14. Which of the following is *not* a factor in determining the effectiveness of a brake system?
    a. heat dissipation
    b. lubricant
    c. pressure
    d. frictional contact area
15. Which of the following could cause an extremely hard brake pedal?
    a. air in the system
    b. excessively worn brake pads
    c. use of the wrong fluid
    d. a leaking diaphragm in the vacuum power booster

ASE-STYLE REVIEW QUESTIONS

1. A vehicle's power brakes are grabbing: Technician A says that the most likely cause is the power brake booster. Technician B says that a likely cause is greasy linings or scored drums. Who is correct?
   a. Technician A c. Both A and B
   b. Technician B d. Neither A nor B
2. While discussing what affects the amount of pressure exerted by the brakes: Technician A says that the shorter the line, the more pressure there will be. Technician B says that braking force will increase if the size of the pistons in a master cylinder are increased. Who is correct?
   a. Technician A c. Both A and B
   b. Technician B d. Neither A nor B
3. The metering valve portion of a combination valve fails: Technician A says that this means the entire combination valve must be replaced. Technician B says that the metering part can be repaired. Who is correct?
   a. Technician A c. Both A and B
   b. Technician B d. Neither A nor B
4. While discussing quick take-up master cylinders: Technician A says that this design allows for increased braking power. Technician B says that this design is only used on drum brake systems. Who is correct?
   a. Technician A c. Both A and B
   b. Technician B d. Neither A nor B
5. While bleeding a brake system: Technician A loosens the brake line fitting at the master cylinder to bleed the system if a bleeder screw is seized cannot be loosened. Technician B uses shop air to push the fluid and air from the wheel units to the master cylinder. Who is correct?
   a. Technician A c. Both A and B
   b. Technician B d. Neither A nor B
6. The basic frictional parts of a brake system are being discussed: Technician A says that the harder the frictional parts are pushed together, the higher the friction. Technician B says that the harder the frictional parts are pushed together, the more heat is developed. Who is correct?
   a. Technician A c. Both A and B
   b. Technician B d. Neither A nor B
7. Vehicle dynamics during braking are being discussed: Technician A says that the rear of the car rises during braking. Technician B says that the front of the car lowers during braking. Who is correct?
   a. Technician A c. Both A and B
   b. Technician B d. Neither A nor B
8. Pressure bleeding is being discussed: Technician A says that metering and combination valves must be held open using a special tool during the bleeding operation to ensure good results. Technician B says that the pressure bleeder requires special adapters to connect it to the master cylinder reservoir. Who is correct?
   a. Technician A c. Both A and B
   b. Technician B d. Neither A nor B
9. Technician A says that the master cylinder should be bled before any individual wheel assembly. Technician B says that the bleeder screw should be closed before the brake pedal is released during manual bleeding of the system. Who is correct?
   a. Technician A c. Both A and B
   b. Technician B d. Neither A nor B
10. The hydraulic system of the hydro-boost and Powermaster is being discussed: Technician A says that the hydro-boost uses a power steering pump. Technician B says that the Powermaster uses an electric motor-driven vane pump. Who is correct?
    a. Technician A c. Both A and B
    b. Technician B d. Neither A nor B

DRUM BRAKES

CHAPTER 49

OBJECTIVES

■ Explain how drum brakes operate. ■ Identify the major components of a typical drum brake and describe their functions. ■ Explain the difference between duo-servo and nonservo drum brakes. ■ Perform a cleaning and inspection of a drum brake assembly. ■ Recognize conditions that adversely affect the performance of drums, shoes, linings, and related hardware. ■ Reassemble a drum brake after servicing. ■ Explain how typical drum parking brakes operate. ■ Adjust a typical drum parking brake.

For many years, drum brakes **(Figure 49–1)** were used on all four wheels on virtually every vehicle on the road. Today, disc brakes have replaced drum brakes on the front wheels of most vehicles and some models are equipped with both front and rear disc brakes. One reason for their continued use is that drum brakes can easily handle the 20% to 40% of total braking load placed on the rear wheels. Another is that drum brakes can also be built with a simple parking brake mechanism.

DRUM BRAKE OPERATION

Drum brake operation is fairly simple. The most important feature contributing to the effectiveness of the braking force supplied by the drum brake is the brake shoe pressure or force directed against the drum **(Figure 49–2)**. With the vehicle moving in either the forward or reverse direction with the brakes on, the applied force of the brake shoe pressing against the brake drum increasingly multiplies itself (called self-energizing) because the brake's anchor pin acts as a brake shoe stop and prohibits the brake shoe from its tendency to follow the movement of the

Figure 49-1 A drum brake assembly.

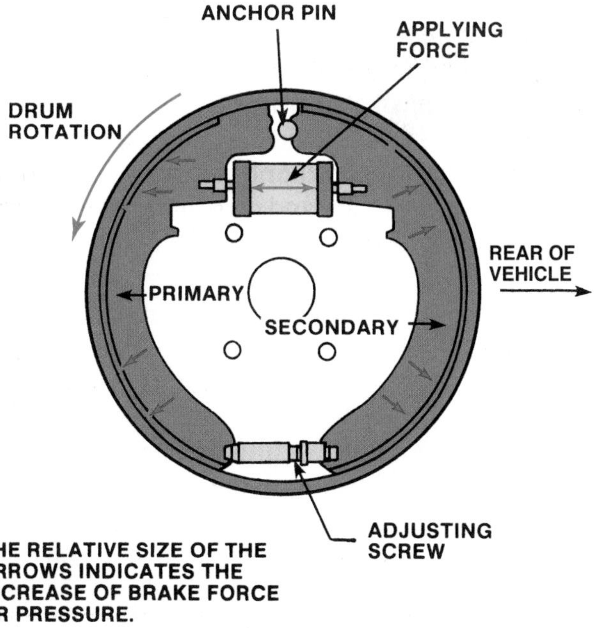

Figure 49-2 The wheel cylinder pushes the primary and secondary shoes against the inside surface of the rotating brake drum.

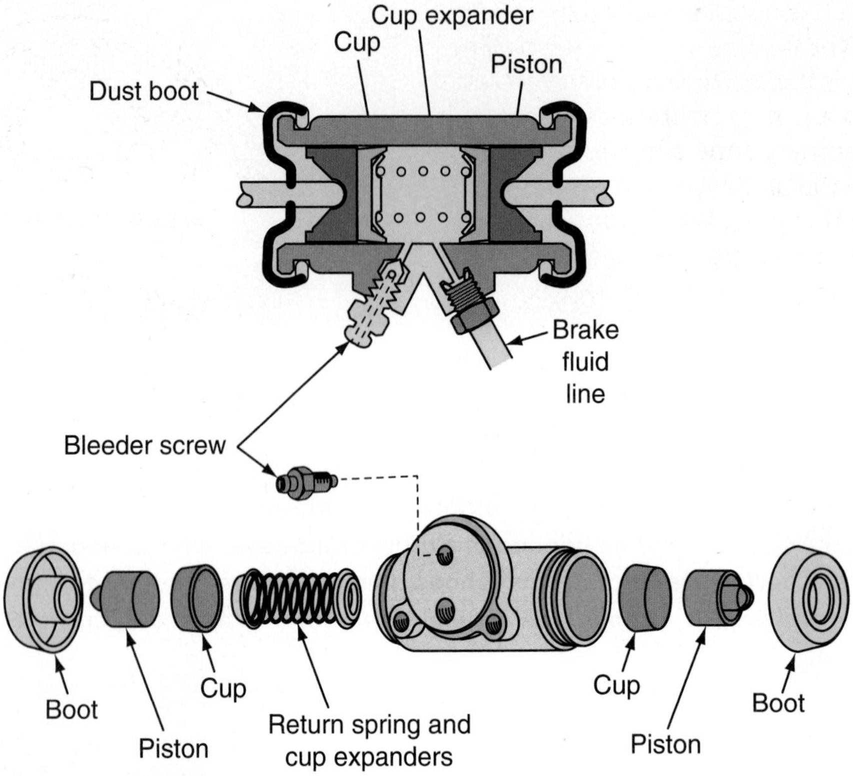

Figure 49-3 The exploded view of a typical wheel cylinder.

rotating drum. The result is a wedging action between the brake shoe and brake drum. The wedging action combined with the applied brake force creates a self-multiplied brake force.

DRUM BRAKE COMPONENTS

The **backing plate** provides a foundation for the brake shoes and associated hardware. The plate is secured and bolted to the axle flange or spindle. The wheel cylinder, under hydraulic pressure, forces the brake's shoes against the drum. There are also two linked brake shoes attached to the backing plate. Brake shoes are the backbone of a drum brake. They must support the lining and carry it into the drum so the pressure is distributed across the lining surface during brake application. Shoe return springs and shoe holddown parts maintain the correct shoe position and clearance. Some drum brakes are self-adjusting. Others require manual adjustment mechanisms. Brake drums provide the rubbing surface area for the linings. Drums must withstand high pressures without excessive flexing and must also dissipate large quantities of heat generated during brake application. Finally, the rear drum brakes on most vehicles include the parking brakes.

Wheel Cylinders

Wheel cylinders convert hydraulic pressure from the master cylinder into a mechanical force at the brakes (**Figure 49–3**). The wheel cylinder bore is filled with fluid. When the brake pedal is depressed, additional brake fluid is forced into the cylinder. The additional fluid moves the cups and pistons outward. This piston movement forces the brake shoes outward to the contact drum and thus applies the brakes. Piston stops prevent the fluid leakage or air from getting into the system when the pistons move to the end of their bores.

Brake Shoes and Linings

In the same brake shoe sizes, there can be differences in **web** thickness, shape of web cutouts, and positions of any reinforcements.

The shoe rim is welded to the web to provide a stable surface for the lining. The web thickness might differ to provide the stiffness or flexibility needed for a specific application. Many shoes have nibs or indented places along the edge of the rim. These nibs rest against shoe support ledges on the backing plate and keep the shoe from hanging up.

Each drum in the drum braking system contains a set of shoes. In a servo system, the **primary shoe** (or leading shoe) is the one that is toward the front of the vehicle. The friction between the primary shoe and the brake drum forces the primary shoe to shift slightly in the direction that the drum is turning. (An **anchor pin** permits just limited movement.) The shifting of the primary shoe forces it against the bottom of the secondary shoe, which causes the secondary shoe to contact the drum. The **secondary shoe** (or trailing shoe) is the one that is toward the rear of the vehicle. It comes into contact as a result of the movement and pressure from the primary shoe and wheel cylinder piston and increases the braking action.

The brake shoe lining provides friction against the drum to stop the car. It contains heat-resistant fibers. The lining is molded with a high-temperature synthetic bonding agent.

The two general methods of attaching the lining to the shoe are riveting and bonding. Regardless of the method of attachment, brake shoes are usually held in a position by spring tension. They are either held against the anchor by the shoe return springs or against the support plate pads by shoe **holddown springs**. The shoe webs are linked together at the end opposite the anchor by an adjuster and a spring. The adjuster holds them apart. The spring holds them against the adjuster ends.

Mechanical Components

In the unapplied position, the shoes are held against the anchor pin by the return springs. The shoes are held to the backing plate by holddown springs or spring clips. Opposite the anchor pin, a star wheel adjuster links the shoe webs and provides a threaded adjustment that permits the shoes to be expanded or contracted. The shoes are held against the star wheel by a spring.

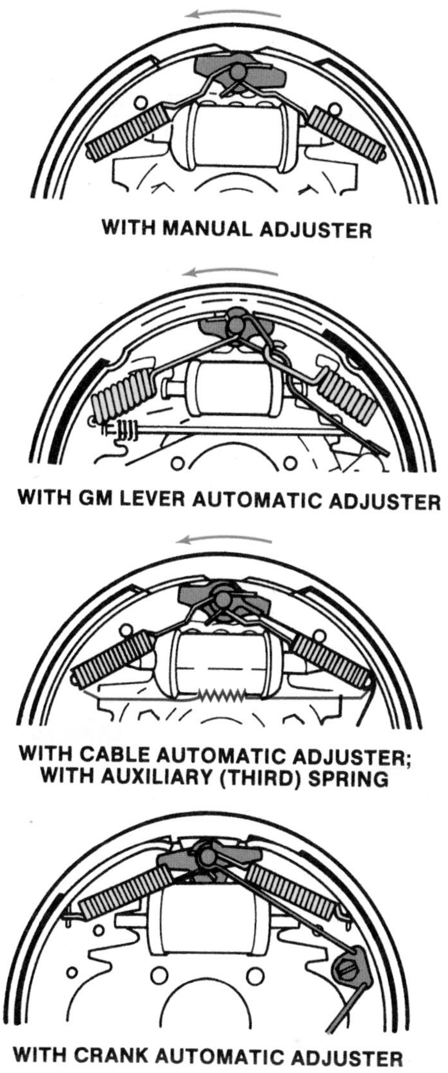

Figure 49–5 Typical brake shoe return spring anchoring points.

Shoe Return Springs **Return springs** can be separately hooked into a link or a guide **(Figure 49–4)** or strung between the shoes. Springs are normally installed on the anchor in the order shown under each category listed in **Figure 49–5**.

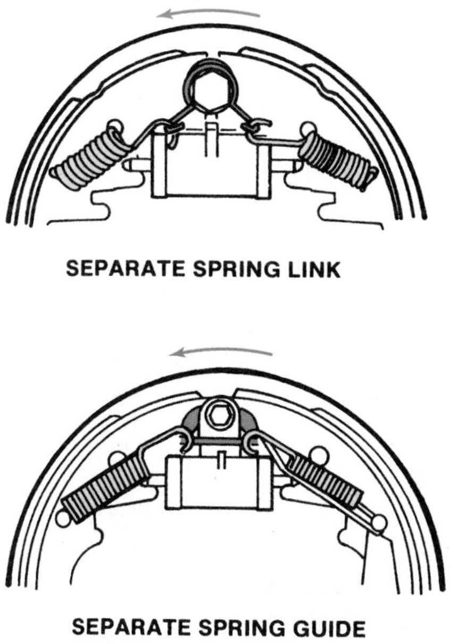

Figure 49–4 Typical brake shoe return spring alignments.

While shoe brake springs look the same, they are usually not interchangeable. Sometimes to help distinguish between them, they are color coded. Pay close attention to the colors and the way they are hooked up.

Shoe Holddowns Various shoe holddowns are illustrated in **Figure 49–6**. To unlock or lock the straight pin holddowns, depress the locking cup and coil spring or the spring clip, and rotate the pin or lock 90 degrees. On General Motors' lever adjusters, the inner (bottom) cup has a sleeve that aligns the adjuster lever.

Shoe Anchors There are various types of **shoe anchors** such as the fixed nonadjustable type, self-centering shoe sliding type, or on some earlier models, adjustable fixed-type providing either an eccentric or

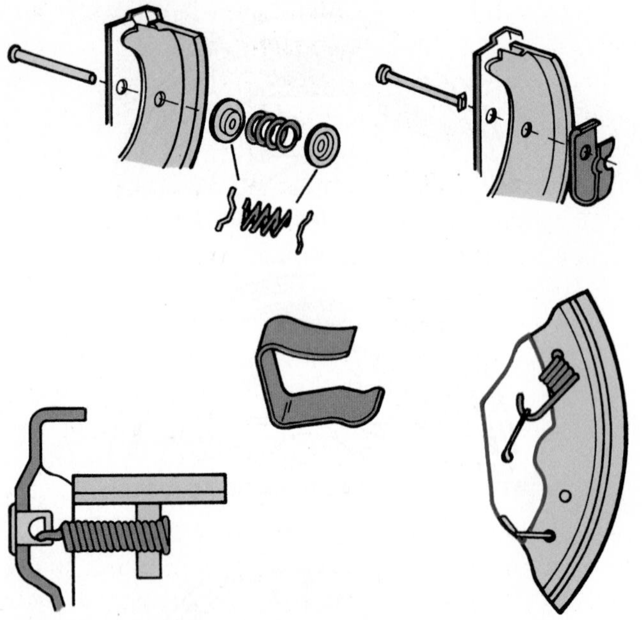

Figure 49-6 Types of brake shoe holddowns.

a slotted adjustment. On some front brakes, fixed anchors are threaded into or are bolted through the steering knuckle and also support the wheel cylinder.

On adjustable anchors, when it is necessary to recenter the shoes in the drum or drum gauge, loosen the locknut enough to permit the anchor to slip but not so much that it can tilt.

Drums

Modern automotive brake drums are made of heavy cast iron (some are aluminum with an iron or steel sleeve or liner) with a machined surface inside against which the linings on the brake shoes generate friction when the brakes are applied. This results in the creation of a great deal of heat. The inability of drums to dissipate as much heat as disc brakes is one of the main reasons discs have replaced drums at the front of all late-model cars and light trucks, and at the rear of some sports and luxury cars.

Sometimes the rear drums of FWD cars are integral with the hub and cannot be removed without disassembling the wheel bearing. The rear drums of other FWD and most RWD cars are held in place by the wheel lugs so they can be removed without tampering with the wheel bearings.

DRUM BRAKE DESIGNS

There are two brake designs in common use. They are **duo-servo** (or self-energizing) **drum brakes** and **nonservo** (or leading-trailing) **drum brakes**.

Most large American cars use the duo-servo design of brake. However, the nonservo type has become popular as the size of cars has become smaller. Because the smaller cars are lighter, this type of brake helps reduce rear brake lockup without reducing braking ability.

Duo-Servo Drum Brakes

The name duo-servo drum brake is derived from the fact that the **self-energizing force** is transferred from one shoe to the other with the wheel rotating in either direction. Both the primary (front) and secondary (rear) brake shoes are actuated by a double-piston wheel cylinder. The upper end of each shoe is held against a single anchor by a heavy coil return spring. An adjusting screw assembly and spring connect the lower ends of the shoes.

The wheel cylinder is mounted on the backing plate at the top of the brake. When the brakes are applied, hydraulic pressure behind the wheel cylinder cups forces both pistons outward causing the brakes to be applied.

When the brake shoes contact the rotating drum in either direction of rotation, they tend to move with the drum until one shoe contacts the anchor and the other shoe is stopped by the star wheel adjuster link **(Figure 49–7)**. With forward rotation, frictional forces between the lining and the drum of the primary shoe result in a force acting on the adjuster link to apply the secondary shoe. This adjuster link force into the secondary shoe is many times greater than the wheel cylinder input force acting on the primary shoe. The force of the adjuster link into the secondary shoe is again multiplied by the frictional forces between the secondary lining and rotating drum, and all of the resultant force is taken on the anchor pin. In normal forward braking, the friction developed by the secondary lining is greater than the primary lining. Therefore, the secondary brake lining is usually thicker and has more surface area. The roles of the

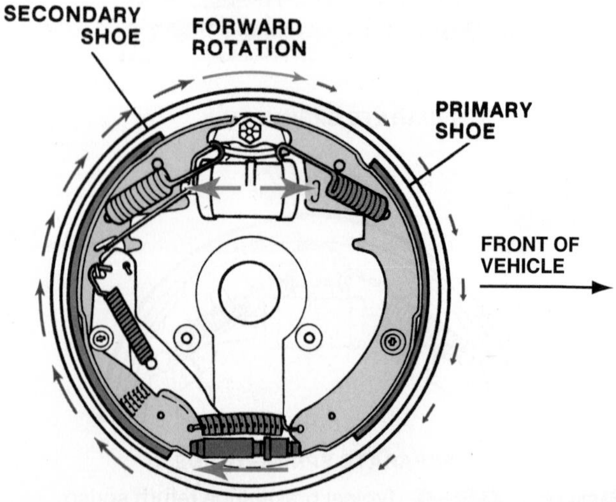

Figure 49-7 Duo-servo braking forces.

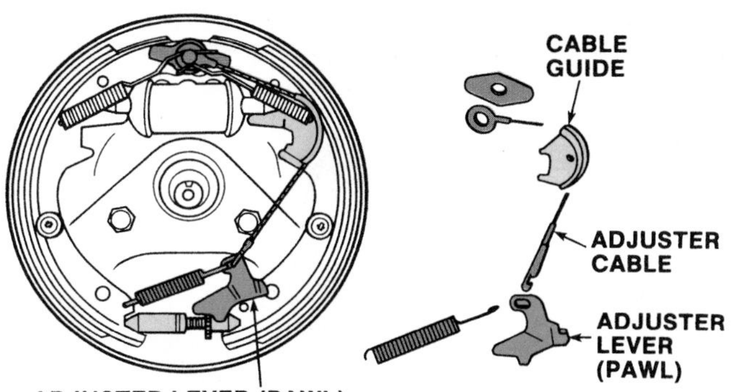

Figure 49-8 Cable self-adjusters.

primary and secondary linings are reversed in braking the vehicle when backing up.

Automatically Adjusted Servo Brakes

Since the early 1960s, automatic drum brake adjusters have been used on all American and most import vehicles. There are several variations of automatic adjusters used with servo brakes. The more common types available follow.

Basic Cable **Figure 49–8** shows a typical automatic adjusting system. Adjusters, whether cable, crank, or lever, are installed on one shoe and operated whenever the shoe moves away from its anchor. The upper link, or cable eye, is attached to the anchor. As the shoe moves, the cable pulls over a guide mounted on the shoe web (the crank or lever pivots on the shoe web) and operates a lever (pawl), which is attached to the shoe so it engages a star wheel tooth. The pawl is located on the outer side of the star whe
ferent styles, slightly above or below th
terline so it serves as a ratchet lock, which
the adjustment from backing off. However, whenev
lining wears enough to permit sufficient shoe movement, brake application pulls the pawl high enough to engage the next tooth. As the brake is released, the adjuster spring returns the pawl, thus advancing the star wheel one notch.

On most vehicles, the adjuster system is installed on the secondary shoe and operates when the brakes are applied as the vehicle is backing up. On a few models, it is located on the primary shoe and operates when the brakes are applied as the vehicle is moving forward. Left-hand and right-hand threaded star wheels are used on opposite sides of the car, so parts should be kept separated. If the wrong star wheel thread is installed, the system does not adjust at all or will unadjust with every brake application.

Another system uses a cable and pawl, with the left brake having right-hand threads and the right brake, left-hand threads. The first cable guide is usually retained on the shoe web by the secondary shoe return spring, and the lever-pawl engages a hole in the shoe web. The adjuster operates in either direction of vehicle movement.

Cable with Overtravel Spring **Figure 49–9** shows a system with an upstroke pawl advance. The left brake has left-hand threads, and the right brake has right-hand threads. The lever (pawl) is installed on a

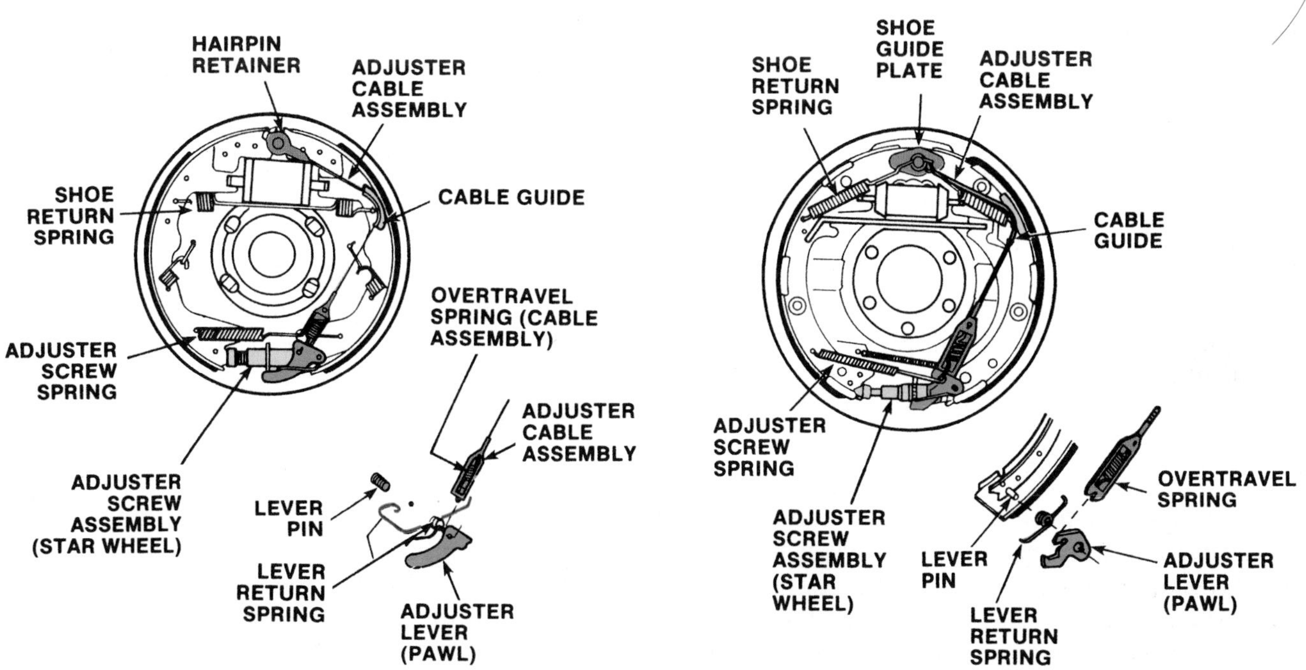

Figure 49-9 Cable automatic adjustment with overtravel springs.

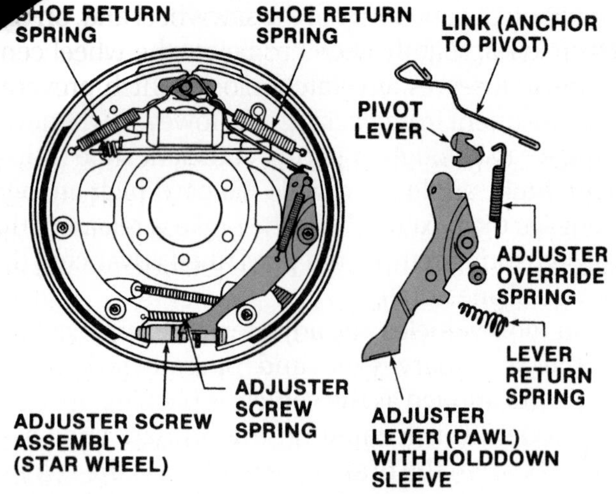

Figure 49–10 An adjusting lever with a pivot and override spring.

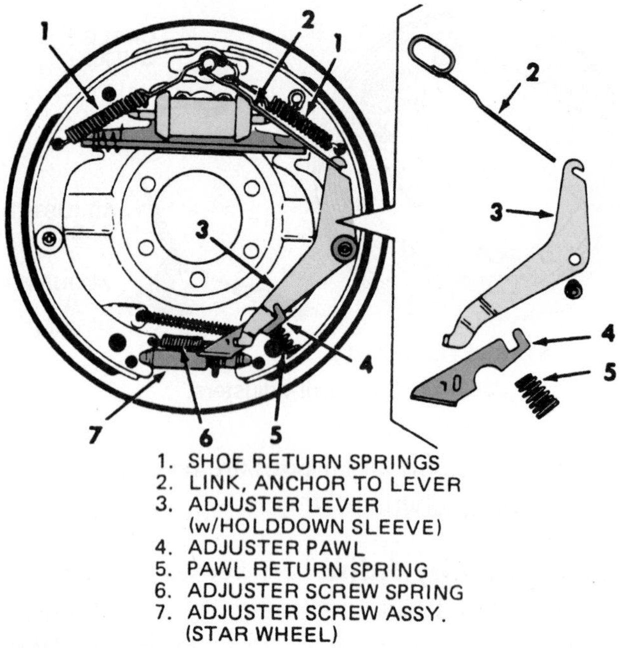

Figure 49–11 Lever and pawl automatic adjustment. *Courtesy of Federal-Mogul Corporation*

web pin with an additional pawl return mousetrap spring. The cable is hooked to the lever (pawl) by means of an **overtravel spring** installed in the cable hook. The overtravel spring dampens movements and prevents unnecessary adjustment should sudden hard braking cause excessive drum deflection and shoe movement.

Lever with Override The system illustrated in **Figure 49–10** uses a downstroke pawl advance. The left brake has right-hand threads, and the right brake has left-hand threads.

The lever (pawl) is mounted on a shoe holddown, pivoting on a cup sleeve. It has a separate lever-pawl return spring located between the lever and the shoe table. A pivot lever and an override spring assembled to the upper end of the main lever dampen movement, preventing unnecessary adjustment in the event of excessive drum deflection.

Lever and Pawl The system illustrated in **Figure 49–11** uses a downstroke pawl advance. The left brake has right-hand threads, and the right brake has left-hand threads. The lever is mounted on a shoe holddown, pivoting on a cup sleeve, and engages the pawl. A separate pawl return spring is located between the pawl and the shoe.

Nonservo Drum Brakes

The nonservo (or as it is better known today as the leading-trailing shoe) drum brake is often used on small cars. The basic difference between this type and the duo-servo brake is that both brake shoes are held against a fixed anchor at the bottom by a retaining spring **(Figure 49–12)**. Nonservo brakes have no servo action.

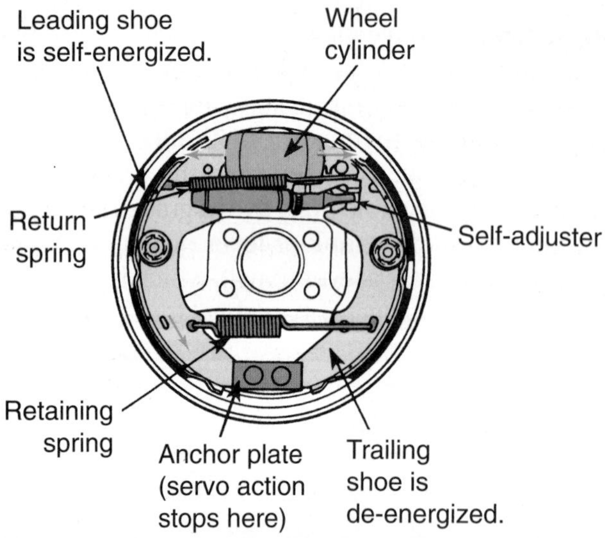

Figure 49–12 A typical nonservo drum brake.

On a forward brake application, the forward (leading) shoe friction forces are developed by wheel cylinder fluid pressure forcing the lining into contact with the rotating brake drum. The shoe's friction forces work against the anchor pin at the bottom of the shoe. The trailing shoe is also actuated by wheel cylinder pressure but can only support a friction force equal to the wheel cylinder piston forces. The trailing shoe anchor pin supports no friction load. The leading shoe in this brake is energized and does most of the braking in comparison to the nonenergized trailing shoe. In reverse braking, the leading and trailing brake shoes switch functions.

SHOP TALK

It is important for the technician to remember that on nonservo drum systems the forward shoe is called the leading shoe and the rear one is known as the trailing shoe (when the vehicle is moving in the forward direction). On duo-servo designs, the forward shoe is the primary, and the rear is the secondary.

Automatically Adjusted Nonservo Brakes

While some standard automatic adjusters similar to the one already discussed are employed on small cars, some of the automatic adjuster mechanisms are unique and varied, using expanding struts between the shoes, or special ratchet adjusting mechanisms. Among the more common of these designs are automatic cam, ratchet automatic, and semiautomatic adjusters.

Automatic Cam Adjusters This rear nonservo drum brake is for use with front disc brakes and has one forward acting (leading) and one reverse acting (trailing) shoe. Shoes rest against the wheel cylinder pistons at the top and are held against the anchor plate by a shoe-to-shoe pull-back spring. The anchor plate and retaining plate are riveted to the backing plate. Adjustment of the brake shoes takes place automatically as needed when the brakes are applied. The automatic cam adjusters are attached to each shoe by a pin through a slot in the shoe webbing. As the shoes move outward during application, the pin in the slot moves the cam adjuster, rotating it outward. Shoes always return enough to provide proper clearance because the pin diameter is smaller than the width of the slot.

Ratchet Automatic Adjuster These brakes are a leading-trailing shoe design with a ratchet self-adjusting mechanism. The shoes are held to the backing plate by spring and pin holddowns, and are held against the anchors at the top by a shoe-to-shoe spring. At the bottom, the shoe webs are held against the wheel cylinder piston ends by a return spring **(Figure 49–13)**.

The self-adjusting mechanism consists of a spacer strut and a pair of toothed ratchets attached to the secondary brake shoe. The parking brake actuating lever is pivoted on the spacer strut.

The self-adjusting mechanism automatically senses the correct lining-to-drum clearance. As the linings wear, the clearance is adjusted by increasing the effective length of the spacer strut. This strut has projections to engage the inner edge of the secondary shoe via the hand brake lever and the inner edge of the large ratchet on the secondary shoe. As wear on the linings increases, the movement of the shoes to bring them in contact with the drums becomes greater than the gap. The spacer strut, bearing on the shoe web, is moved together with the primary shoe to close the gap. Further movement causes the large ratchet behind the secondary shoe to rotate inward against the spring-loaded small ratchet, and the serrations on the mating edges maintain this new setting until further wear on the shoe results in another adjustment. On releasing brake pedal pressure, the return springs cause the shoes to move into contact with the

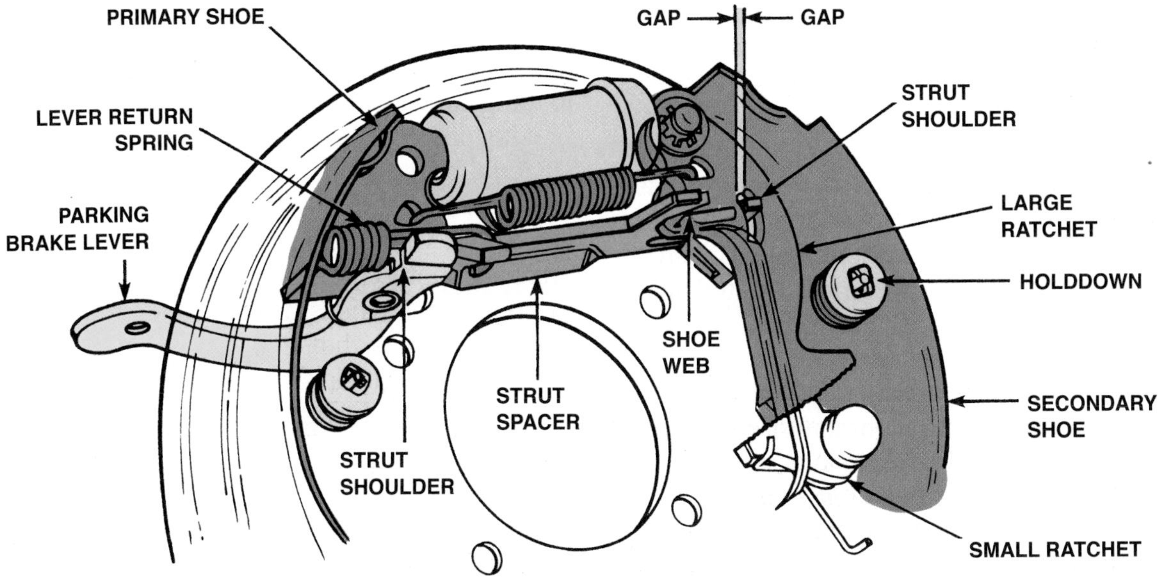

Figure 49-13 A typical nonservo self-adjusting mechanism.

shoulders of the spacer strut/hand brake actuating lever. This restores the clearance between the linings and the drum proportionate to the gap.

Inspection and Service

The first rule of quality brake service is to perform a complete job. For example, if new linings are installed without regard to the condition of the hydraulic system, the presence of a leaking wheel cylinder quickly ruins the new linings. Braking power and safety are also compromised.

Problems such as spongy pedal, excessive pedal travel, pedal pulsation, poor braking ability, brake drag, lock, or pulling to one side, and braking noises can be caused by trouble in the hydraulic system or the mechanical components of the brake assembly. To aid in doing a complete inspection and diagnosis, a form like the one shown in **Figure 49–14** is very helpful. Working with such a form helps the technician avoid missing any brake test and components that may cause problems.

Brake Noise

All customer complaints related to brake performance must be carefully considered. The number one customer complaint is brake noise. Noise is often the first indication of wear or problems within the braking system, particularly in the mechanical components. Rattles, clicking, grinding, and hammering from the wheels when the brake is in the unapplied position should be carefully investigated. Be sure the noise is not caused by the bearings or various suspension parts. If the noise is coming from the brake assembly, it is most likely caused by worn, damaged, or missing brake hardware, or the poor fastening or mounting of brake components. Grinding noises usually occur when a stone or other object becomes trapped between the lining material and the rotor or drum.

When the brakes are applied, a clicking noise usually indicates play or hardware failure in the attachment of the pad or shoe. On recent systems, the noise could be caused by the lining tracking cutting tool marks in the rotor or drum. A nondirectional finish on rotors eliminates this and so does a less pointed tip on the cutting tool used to refinish drums.

Grinding noises on application can mean metal-to-metal contact, either from badly worn pads or shoes, or from a serious misalignment of the caliper, rotor, wheel cylinder, or backing plate. Wheel cylinders and calipers that are frozen due to internal corrosion can also cause grinding or squealing noises.

Other noise problems and their solutions are covered later in this chapter.

ROAD TESTING BRAKES

Road testing allows the brake technician to evaluate brake performance under actual driving conditions. Whenever practical, perform the road test before beginning any work on the brake system. In every case, road test the vehicle after any brake work to make sure the brake system is working safely and properly.

WARNING!

Before test driving any car, first check the fluid level in the master cylinder. Depress the brake pedal to be sure there is adequate pedal reserve. Make a series of low-speed stops to be sure the brakes are safe for road testing. Always make a preliminary inspection of the brake system in the shop before taking the vehicle on the road.

Brakes should be road tested on a dry, clean, reasonably smooth, and level roadway. A true test of brake performance cannot be made if the roadway is wet, greasy, or covered with loose dirt. All tires do not grip the road equally. Testing is also adversely affected if the roadway is crowned so as to throw the weight of the vehicle toward the wheels on one side, or if the roadway is so rough that wheels tend to bounce.

Test brakes at different speeds with both light and heavy pedal pressure. Avoid locking the wheels and sliding the tires on the roadway. There are external conditions that affect brake road-test performance. Tires having unequal contact and grip on the road cause unequal braking. Tires must be equally inflated and the tread pattern of right and left tires must be approximately equal. When the vehicle has unequal loading, the most heavily loaded wheels require more braking power than others and a heavily loaded vehicle requires more braking effort. Misalignment of the front end causes the brakes to pull to one side. Also, a loose front-wheel bearing could permit the disc to tilt and have spotty contact with brake shoe linings, causing pulsations when the brakes are applied. Faulty shock absorbers that do not prevent the car from bouncing on quick stops can give the erroneous impression that the brakes are too severe.

DRUM BRAKE INSPECTION

Place the vehicle in neutral, release the parking brake, and raise the vehicle on the hoist. Once the wheels are removed, mark the wheel-to-drum and

PRE-BRAKE-JOB INSPECTION CHECKLIST

Owner ________________ (LAST) ________________ (FIRST) Phone ____________ Date ___/___/___

Address ________________________________ License No. ____________

Make __________ Model __________ Mileage __________ Serial No. __________ Year __________

Special Key for Hubcaps/Wheels Location ______________ Owner Use Parking Brake Yes ☐ No ☐

4 Drum ☐ 4 Disc ☐ Disc/Drum ☐ P/B No ☐ Yes ☐ Vacuum ☐ Hydro ☐ ABS ☐

Owner Comments ________________________________

1. CHECKS BEFORE ROAD TEST	Safe	Unsafe
Stoplight Operation		
Brake Warning Light Operation		
Master Cylinder Checks		
Fluid Level		
Fluid Contamination		
Under Hood Fluid Leaks		
Under Dash Fluid Leaks (No Power)		
Bypassing		

BRAKE PEDAL HEIGHT AND FEEL				
Check One		Check One		
Low		Spongy		
Med		Firm		
High				

Power Brake Unit Checks					
VACUUM	Safe	Unsafe	HYDRO	Safe	Unsafe
Vacuum Unit			Hydro Unit		
Engine Vacuum			P/S Fluid		
Vacuum Hose			P/S Belt Tension		
Unit Check Valve			P/S Belt Condition		
Reserve Braking			P/S Fluid Leaks		
			Reserve Braking		

Tire Pressure Specs	Front	Rear
Record Pressure Found		
RF ______ LF ______ RR ______ LR ______		
Tire Condition		
RF ______ LF ______ RR ______ LR ______		

2. ROAD TEST	Yes	No	RF	LF	RR	LR
Brake Pull						
Brake Clunk						
Brake Scraping						
Brake Squeal						
Brake Grabby						
Brakes Lock Prematurely						
Wheel Bearing Noise						
Vehicle Vibrates						

STEERING WHEEL MOVEMENT WHEN STOPPING FROM 2-3 MPH YES/NO/RGT/LFT

Does ABS Work	YES	NO
Pedal Pulsation when Braking	YES	NO
Steering Wheel Oscillation when Braking	YES	NO
No Stopping Power	YES	NO
Warning Light Comes on when Braking	YES	NO
Difference in Pedal Height after Cornering	YES	NO
Nose Dive	YES	NO

3. In Shop Checks On Hoist	Yes	No	RF	LF	RR	LR
Brake Drag						
Intermittent Brake Drag						
Brake Pedal Linkage Binding						
Wheel Bearing Looseness						
Missing or Broken Wheel Fasteners						
Suspension Looseness						

	Front					Rear				
Mark Wheels and Remove										
Caliper/Piston Stuck RF LF RR LR	Right			Left		Right			Left	
Mark Drums and Remove	Spec	Safe	Unsafe	Safe	Unsafe	Spec	Safe	Unsafe	Safe	Unsafe
Measure Rotor Thickness or Drum Diameter.										
Measure Rotor Thickness Variation.										
Measure Rotor Runout.										
Lining Thickness										
Tubes and Hoses										
Fluid Leaks										
Broken Bleeders										
Leaky Seals										
Self-Adjuster Operation										
									Safe	Unsafe
Parking Brake Cables and Linkage										

Figure 49-14 A sample of pre-brake-job inspection checklist. *Courtesy of Hennessy Industries, Inc.*

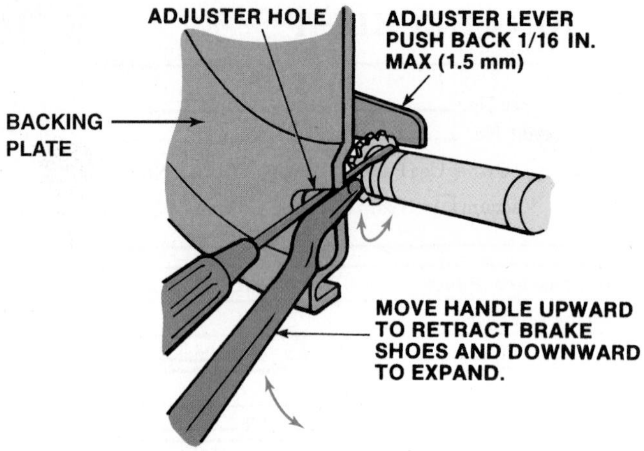

Figure 49-15 Backing off the self-adjusters in order to remove the brake drum.

drum-to-axle positions so the components can be correctly reassembled.

Drum Removal

Drum removal procedures are different for fixed and floating drums. In all cases, however, you may need to back off the parking brake adjustment or manually retract the self-adjusters **(Figure 49–15)** to have enough shoe-to-drum clearance to remove the drum. Wear on the friction surface of the drum creates a ridge at the edge of the drum's rim. As the self-adjusters move the shoes outward to take up clearance, the shoe diameter becomes larger than the ridge diameter. If the adjuster is not retracted, the drum's ridge may jam on the shoes and prevent drum removal. Trying to force the drum over the shoes may damage brake parts. To do this, reach through the adjusting slot with a thin screwdriver (or similar tool) and carefully push the self-adjusting lever away from the star wheel a maximum of 1⁄16 inch. While holding the lever back, insert a brake adjusting tool into the slot and turn the star wheel in the proper direction until the brake drum can be removed. On vehicles that have the adjusting slot in the drum rather than in the backing plate, reach through the slot with a thin wire hook and pull the adjuster lever away from the star wheel.

Before you remove a drum, mark it "L" or "R" for left or right so that it gets reinstalled on the same side of the vehicle from which it was removed.

> **WARNING!**
>
> *Do not step on the brake pedal while a brake drum is off. This will cause the piston in the wheel cylinder to overextend or pop apart.*

Brake drums that are made as a one-piece unit with the wheel hub are common as rear drums on FWD cars and on the front wheels of older vehicles with four-wheel drum brakes. The hub contains the wheel bearings and is held onto the spindle by a single large nut. This nut also is used to adjust the wheel bearings. To remove this type of drum, remove the dust cap from the center of the hub. Then remove the cotter pin from the castellated nut or nut lock on the spindle. Next, remove the spindle nut and washer. Pull the drum outward to slide it off the spindle.

Floating drums do not have a built-in hub. In most cases, the drums are held in place by studs on the axle flange and the wheel and lug nuts. On many floating drums, push nuts or speed nuts are used during vehicle assembly to hold the drum onto two or three studs. Typically the push nuts do not need to be reinstalled after service. However, on some vehicles, the push nuts are used to hold the drum squarely against the axle or hub flange.

Floating drums are pulled off the hub or axle flange **(Figure 49–16)**. If the brake drum is rusted or corroded to the axle flange and cannot be removed, lightly tap the axle flange to the drum mounting surface with a ball-peen hammer. Penetrating oil may help in loosening a stuck drum. If the drum is stuck to its flange, use a large scribe or center punch to score around the joint at the drum and flange and break the surface tension. Remember that if the drum is worn, the brake shoe adjustment has to be backed off for the drums to clear the brake shoes. Do not force the drum or distort it. Do not allow the drum to drop.

After the drum is removed, inspect the grease in the hub and on the bearings. If the grease is dirty or dried out and hard, it is a clue to possible bearing damage. Also inspect the rear axle gaskets and wheel

Figure 49-16 Pull the drum away from the axle flange or hub, being careful not to drop it.

Figure 49–17 Before disassembling the brakes, use an OSHA-approved washer to make sure all asbestos dust is removed from the parts.

seals for leaks. Replace worn components as needed. Set the drum and all bearing parts aside for cleaning and close inspection. If the grease seems to be in good condition, place the drum on a bench with the open side down. Cover the outer bearing opening with a shop cloth to keep dirt out.

CAUTION!

*When servicing wheel brake parts, do not create dust by cleaning with a dry brush or with compressed air. Asbestos fibers can become airborne if dust is created during servicing. Breathing dust containing asbestos fibers can cause serious bodily harm. To clean away asbestos from brake surfaces, use an OSHA-approved washer (**Figure 49–17**). Follow the manufacturer's instructions when using the washer.*

Drum Inspection

One of the most important parts that need to be inspected is the brake drum (**Figure 49–18**). Thoroughly clean the drums with a water-dampened cloth or a water-based solution. If the drums have been exposed to leaking oil or grease, thoroughly clean them with a non-oil base solvent after washing to remove dust and dirt. It is important to determine the source of the oil or grease leak and correct the problem before reinstalling the drums.

Brake drums act as a heat sink. They absorb heat and dissipate it into the air. As drums wear from normal use or are machined, their cooling surface area is reduced and their operating temperatures increase. Their structural strength is also reduced. This leads to distortion, which causes some of the drum conditions shown in **Figure 49–19**.

Figure 49–18 Carefully check the inside surface of the brake drum.

Also take a look at the brake shoes while they are still mounted. Their condition can often reveal defects in the drums. If the linings on one wheel are worn more than the others, it might indicate a rough drum. Uneven wear from side to side on any one set of shoes can be caused by a tapered drum. If some linings are worn badly at the toe or heel, it might indicate an out-of-round drum.

Scored Drum Surface The most common cause of this condition is buildup of brake dust and dirt between the brake lining and drum. A glazed brake lining, hardened by high heat or in some cases by very hard inferior grade brake lining, can also groove the drum surface. Excessive lining wear that exposes the rivet head or shoe steel will score the drum surface. If the grooves are not too deep, the drum can be turned.

Bell-Mouthed Drum This distortion is due to extreme heat and braking pressure. It occurs mostly on wide drums and is caused by poor support at the

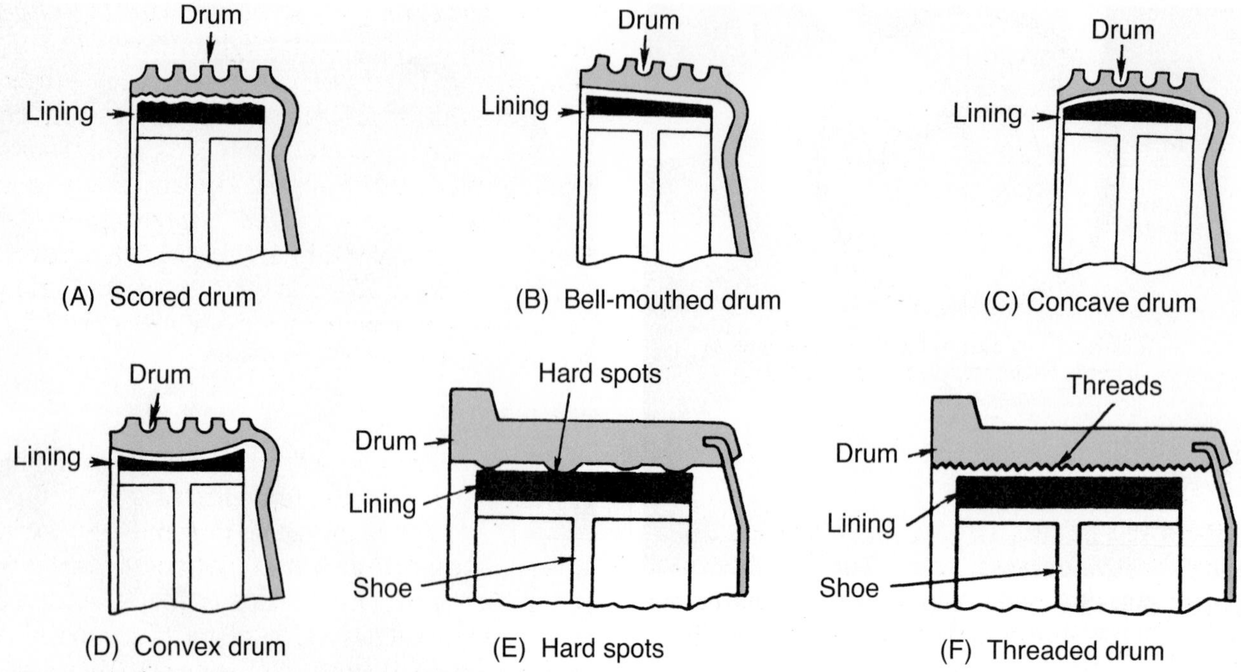

Figure 49-19 Drum wear conditions.

outside of the drum. Full drum-to-lining contact cannot be achieved and fading can be expected. Drums must be turned.

Concave Drum This is an excessive wear pattern in the center area of the drum brake surface. Extreme braking pressure can distort the shoe platform so braking pressure is concentrated at the center of the drum.

Convex Drum This wear pattern is greater at the closed end of the drum. It is the result of excessive heat or an oversized drum, which allows the open end of the drum to distort.

Hard Spots on the Drum This condition in the cast-iron surface, sometimes called chisel spots or islands of steel, results from a change in metallurgy caused by braking heat. Chatter, pulling, rapid wear, hard pedal, and noise can occur. These spots can be removed by grinding. However, only the raised surfaces are removed, and they can reappear when heat is applied. If this condition reappears, the drum must be replaced.

Threaded Drum Surface An extremely sharp or chipped tool bit or a lathe that turns too fast can result in a threaded drum surface. This condition can cause a snapping sound during brake application as the shoes ride outward on the thread, then snap back. To avoid this, recondition drums using a rounded tool and proper lathe speed. Check the edge of the drum surface around the mounting flange side for tool marks indicating a previous machining. If the drum has been machined, it might have worn too thin for use. Check the diameter.

Heat Checks **Heat checks** are visible, unlike hard spots that do not appear until the machining of the drum **(Figure 49–20)**. Extreme operating temperatures are the major cause. The drum might also show a bluish/gold tint, which is a sign of high temperatures. Hardened carbide lathe bits or special grinding attachments are available through lathe manufacturers to service these conditions. Excessive damage by heat checks or hard spots requires drum replacement.

Cracked Drum Cracks in the cast-iron drum are caused by excessive stress. They can be anywhere but usually are in the vicinity of the bolt circle or at the outside of the flange. Fine cracks in the drums are often hard to see and, unfortunately, often do not show up until after machining. Nevertheless, should

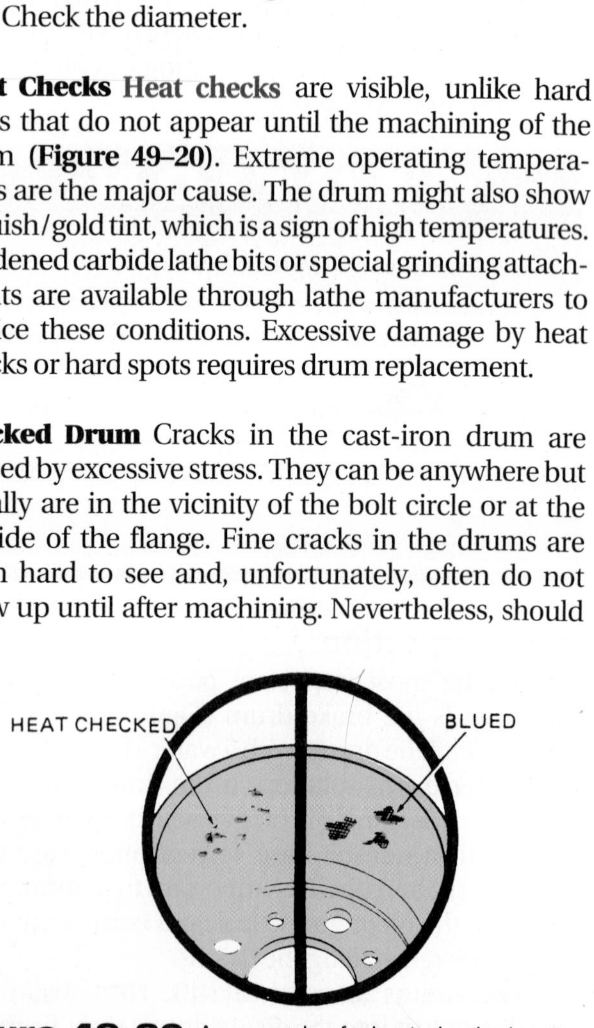

Figure 49-20 An example of a heat-checked and overheated brake drum. *Courtesy of Federal-Mogul Corporation*

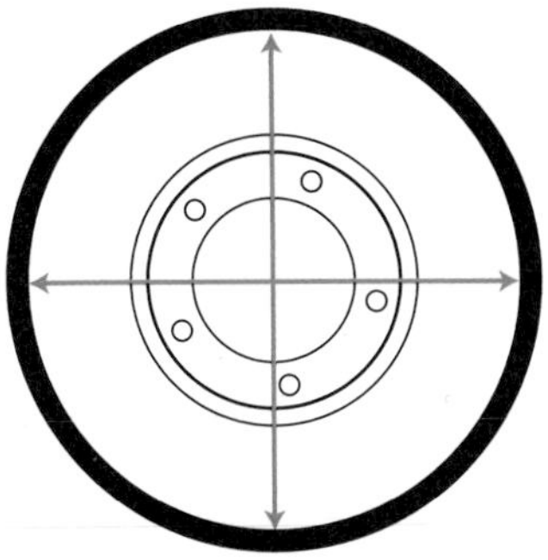

Figure 49-21 Measure the inside diameter of the drum in several spots to determine out-of-roundness.

any cracks appear, no matter how small, the drum must be replaced.

Out-of-Round Drums Drums with eccentric distortion might appear fine to the eye but can cause pulling, grabbing, and pedal vibration or pulsation. An out-of-round or egg-shaped condition **(Figure 49–21)** is often caused by heating and cooling during normal brake operation. Out-of-round drums can be detected before the drum is removed by adjusting the brake to a light drag and feeling the rotation of the drum by hand. After removing the drum, gauge it to determine the amount of eccentric distortion. Drums with this defect should be machined or replaced.

Drum Measurements

Measure every drum with a drum micrometer **(Figure 49–22)**, even if the drum passed a visual inspection, to make sure that it is within the safe oversize limits. If the drum is within safe limits, even though the surface appears smooth, it should be turned to ensure a true drum surface and to remove any possible contamination in the surface from previous brake linings, road dust, and so forth. Remember that if too much metal is removed from a drum, unsafe conditions can result.

Figure 49-22 Measuring the inside diameter with a drum micrometer.

Take measurements at the open and closed edges of the friction surface and at right angles to each other. Drums with taper or out-of-roundness exceeding 0.006 inch (0.152 mm) are unfit for service and should be turned or replaced. If the maximum diameter reading (measured from the bottom of any grooves that might be present) exceeds the new drum diameter by more than 0.060 inch (1.5 mm), the drum cannot be reworked. If the drums are smooth and true but exceed the new diameter by 0.090 inch (2.2 mm) or more, they must be replaced.

Drum Refinishing

Brake drums can be refinished by either turning or grinding on a **brake lathe** **(Figure 49–23)**.

Only enough metal should be removed to obtain a true, smooth friction surface. When one drum must be machined to remove defects, the other drum on the same axle set must also be machined in the same manner and to the same diameter (±0.010 inches) so braking is equal.

Brake drums are stamped with a discard dimension **(Figure 49–24)**. This is the allowable wear dimension and not the allowable machining dimension. There must be 0.030 inch (0.762 mm) left for wear after turning the drums. Some states have laws about measuring the limits of a brake drum.

Figure 49-23 Brake drums can be resurfaced by grinding or turning them on a brake lathe. *Courtesy of Hennessy Industries, Inc.*

Figure 49-24 The drum's discard diameter is stamped on the drum.

Machining or grinding brake drums increases the inside diameter of the drum and changes the lining-to-drum fit. When remachining a drum, follow the equipment instructions for the specific tool you are using.

USING SERVICE INFORMATION

Service manuals list the standard brake drum inside diameter alng with the discard dimension. They also state the standard and minimum lining thickness. Manual illustrations should be used to accurately identify all components plus the disassemble/reassembly procedure. Tightening torques for backing plate nuts and other components should always be followed.

Cleaning Newly Refaced Drums

The friction surface of a newly refaced drum contains millions of tiny metal particles. These particles not only remain free on the surface, they always lodge themselves in the open pores of the newly machined surface. If the metal particles are allowed to remain in the drum, they become embedded in the brake lining. Once the brake lining gets contaminated in this manner, it acts as a fine grinding stone and scores the drum.

PROCEDURE

Mechanical Component Service of Duo-Servo Drum Brakes

STEP 1 Disconnect the cable from the parking brake lever.

STEP 2 If required, install wheel cylinder clamps on the wheel cylinders to prevent fluid leakage or air from getting into the system while the shoes are removed. Some brakes have wheel cylinder stops; therefore, wheel cylinder clamps are not required. Regardless of whether the clamps are needed, do not press down on the brake pedal after shoe return springs have been removed. To prevent this, block up the brake pedal so it cannot be depressed.

STEP 3 Remove the brake shoe return springs. Use a brake spring removal and installation tool to unhook the springs from the anchor pin or anchor plate **(Figure 49–25)**.

STEP 4 Remove the shoe retaining or holddown cups and springs. Special tools are available **(Figure 49–26)**, but the holddown springs can be removed by using pliers to compress the spring and rotating the cup with relation to the pin.

STEP 5 Self-adjuster parts can now be removed. Lift off the actuating link, lever and pivot assembly, sleeve (through lever), and return spring. No advantage is gained by disassembling the lever and pivot assembly unless one of the parts is damaged.

STEP 6 Spread the shoes slightly to free the parking brake strut and remove the strut with its spring. Disconnect the parking brake lever from the secondary shoe. It can be attached with a retaining clip, bolt, or simply hooked into the shoe.

STEP 7 Slip the anchor plate off the pin. No advantage is gained by removing the plate if it is bolted on or riveted. Spread the anchor ends of the shoes and disengage them from the wheel cylinder links, if used. Remove the shoes connected at the bottom by the adjusting screw and spring, as an assembly.

STEP 8 Overlap the anchor end of the shoes to relieve spring tension. Unhook the adjusting screw spring, and remove the adjusting screw assembly.

SHOP TALK

Keep the adjusting screws and automatic adjuster parts for left and right brakes separate. These parts usually are different. For example, on some automatic adjusters, the adjusting screws on the right brakes have left-hand threads and the adjusting screws on the left brakes have right-hand threads.

Figure 49-25 A brake spring tool.

Figure 49-26 A holddown spring tool.

PROCEDURE

Disassembling Nonservo or Leading-Trailing Brakes

STEP 1 Install the wheel cylinder clamp. Then unhook the adjuster spring from the parking brake strut and reverse shoe.

STEP 2 Unhook the upper shoe-to-shoe spring from the shoes and unhook the antinoise spring from the spring bracket.

STEP 3 Remove the parking brake strut and disengage the shoe webs from the flat, clamp shoe holddown clips.

STEP 4 Unhook the lower shoe-to-shoe spring and remove the forward shoe. Disconnect the parking brake cable, then remove the reserve shoe.

STEP 5 Remove the shoe holddown clips from the backing plate.

STEP 6 Press off the C-shaped retainers from the pins and remove the parking brake lever, automatic adjuster lever, and adjuster latch.

STEP 7 Remove the parking brake lever.

SHOP TALK

Mark the shoe positions if shoes and linings are to be reused. When disassembling an unfamiliar brake assembly, work on one wheel at a time and use the other wheel as a reference.

PROCEDURE

Cleaning and Inspecting Brake Parts

STEP 1 Clean the backing plates, struts, levers, and other metal parts to be reused using a water-dampened cloth or a water-based solution. Equipment is commercially available to perform washing functions of brake parts. Wet cleaning methods must be used to prevent asbestos fibers from becoming airborne.

STEP 2 Carefully examine the raised shoe support pads on the backing plate to make sure they are free from corrosion or other surface defects that might prevent the shoes from sliding freely. Use fine emery cloth to remove surface defects, if necessary. Clean them thoroughly.

STEP 3 Check to make sure that the backing plates are not cracked or bent. If so, they must be replaced. Make sure backing plate bolts and bolted-on anchor pins are torqued to specifications.

STEP 4 If replacement of the wheel cylinder is needed, it should be done at this time. To determine wheel cylinder condition, carefully inspect the boots. If they are cut, torn, heat-cracked, or show evidence of leakage, the wheel cylinders should be replaced. If more than a drop of fluid spills out, leakage is excessive and indicates that replacement is necessary.

STEP 5 Disassemble the adjusting screw assembly **(Figure 49–27)** and clean the parts in a suitable solvent. Make sure the adjusting screw threads into the pivot nut over its complete length without sticking or binding. Check that none of the adjusting screw teeth are damaged. Lubricate the adjusting screw threads with brake lubricant.

STEP 6 Examine the shoe anchor, support plate, and small parts for signs of looseness, wear, or damage that could cause faulty shoe alignment. Check springs for spread or collapsed coils, twisted or nicked shanks, and severe discoloration **(Figure 49–28)**. Operate star wheel automatic adjusters by prying

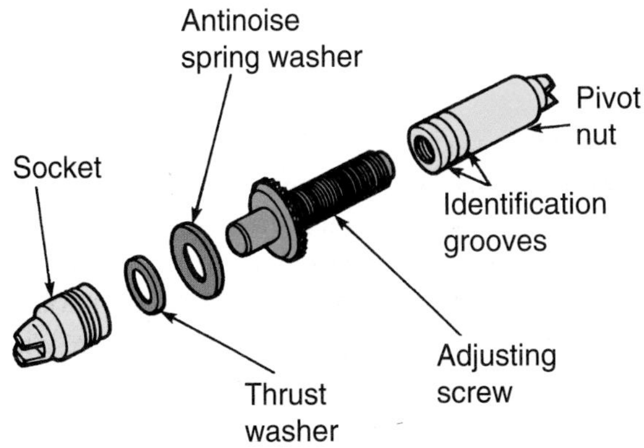

Figure 49-27 An exploded view of a brake adjuster assembly.

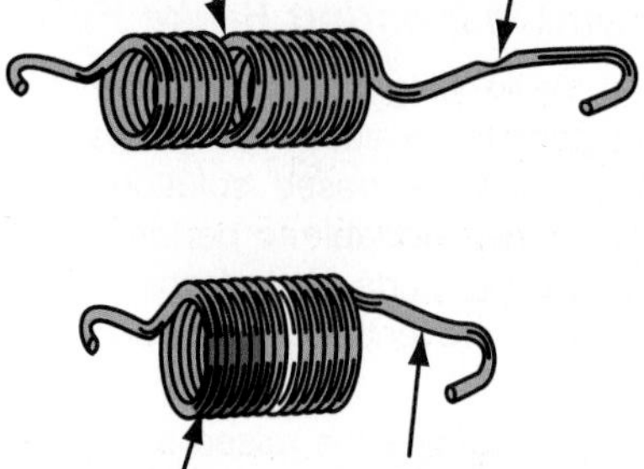

Figure 49-28 All springs should be checked for distortion and damage.

the shoe lightly away from its anchor or by pulling the cable to make sure the adjuster advances easily, one notch at a time. Adjuster cables tend to stretch, and star wheels and pawls become blunted after a long period of use. For rear-axle parking brakes, pull on the cable and shoe linkage to make sure no binding condition is present that could cause the shoes to drag when the parking brake is released.

These metal particles must be removed by washing or cleaning the drum. Do not blow out the drum with air pressure. Either of the following methods is recommended to clean a newly refaced brake drum. The first method involves washing the brake drum thoroughly with hot water and wiping with a lint-free rag. Then use the air pressure to thoroughly dry it. If the front hub and drums are being cleaned, be very careful to avoid contaminating the wheel bearing grease. Or, completely remove all the old grease, then regrease and repack the wheel bearing after the drum has been cleaned and dried. The wheel bearings and the grease seals must be removed from the drum before cleaning. The second method involves wiping the inside of the brake drum (especially the newly machined surface) with a lint-free white cloth dipped in one of the many available brake cleaning solvents that do not leave a residue. This operation should be repeated until dirt is no longer apparent on the wiping cloth. Allow the drum to dry before reinstalling it on the vehicle.

Both of these procedures are also good for cleaning disc brake rotors.

Cleaning, Inspecting, and Lubricating Brake Parts

To complete the drum brake inspection, examine wheel bearings and hub grease seals for signs of damage. Service or replace, if necessary.

BRAKE SHOES AND LININGS

Lining materials influence braking operation. The use of a lining with a friction value that is too high can result in a severe grabbing condition. A friction value that is too low can make stopping difficult because of a hard pedal.

Overheating a lining accelerates wear and can result in dangerous lining heat fade—a friction-reducing condition that hardens the pedal and lengthens the stopping distance. Continual overheating eventually pushes the lining beyond the point of recovery into a permanent fade condition. In addition to fade, overheating can cause squeal.

Overheating is indicated by a lining that is charred or has a glass-hard glazed surface, or if severe, random cracking of the surface is present.

CAUTION!

Automotive friction materials often contain substantial amounts of asbestos. Studies indicate that exposure to excessive amounts of asbestos dust can be a potential health hazard. It is important that anyone handling brake linings understands this and takes the necessary precautions to avoid injury.

Inspect the linings for uneven wear, embedded foreign material, loose rivets, and to see if they are oil soaked. If linings are oil soaked, replace them.

If linings are otherwise serviceable, tighten or replace loose rivets, remove imbedded foreign material, and clean the rivet counterbores.

If linings at any wheel show a spotty wear pattern or an uneven contact with the brake drum, it is an indication that the linings are not centered in the drums. Linings should be circle ground to provide better contact with the drum.

Brake Shoe Replacement

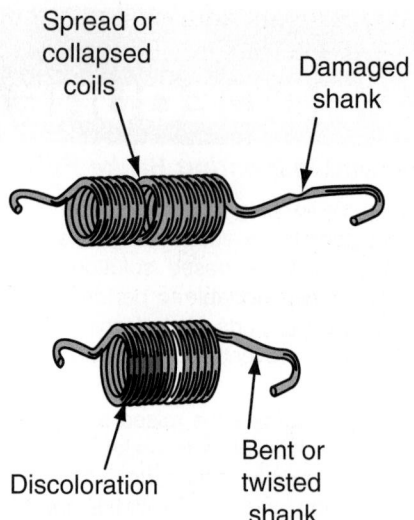

Brake linings that are worn to within 1⁄32 inch (0.79 mm) of a rivet head or that have been contaminated with brake fluid, grease, or oil must be replaced (**Figure 49–29**). Failure to replace worn linings results in a scored drum. When it is necessary to replace brake shoes, they must also be replaced on the wheel on the

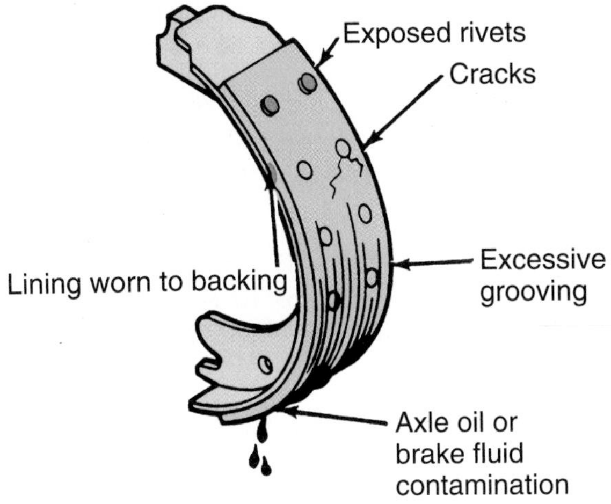

Figure 49-29 Potential brake shoe problems.

opposite side of the vehicle. Inspect brake shoes for distortion, cracks, or looseness. If these conditions exist, the shoe must be discarded.

Do not let brake fluid, oil, or grease touch the brake lining.

The two general methods of attaching the linings to the brake shoes are bonding and riveting. The **bonded linings** are fastened (glued) with a special adhesive to the shoe, clamped in place, then cured in an oven. Instead of using an adhesive, some linings are riveted to the shoe. Riveted linings allow for better heat transfer than bonded linings.

Selecting Replacement Linings

Identification codes, called the automotive friction material edge codes, are printed on the edges of drum brake linings **(Figure 49–30)** and disc brake pads. The letters and numbers identify the manufacturer of the lining material and the material used, and the last two letters identify the cold and hot coefficients of friction (COF).

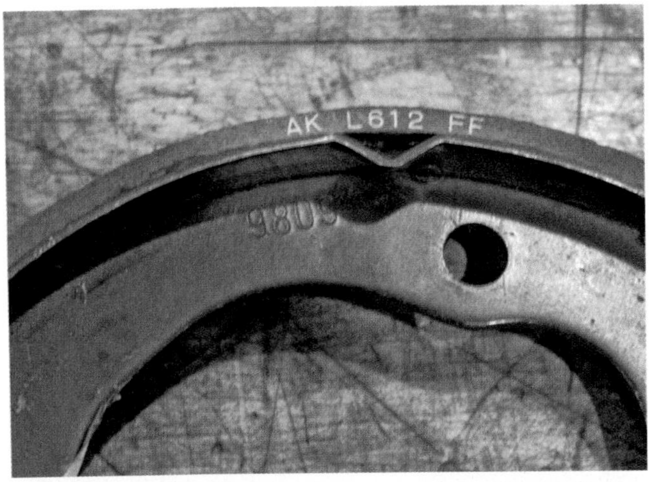

Figure 49-30 Identification codes, called the automotive friction material edge codes, are printed on the edges of drum brake linings.

These codes do not address lining quality or its hardness. From a service standpoint, the COF codes are the most important and are coded as follows:

C = not over 0.15
D = over 0.15 but not over 0.25
E = over 0.25 but not over 0.35
F = over 0.35 but not over 0.45
G = over 0.45 but not over 0.55
H = over 0.55

It is also important to use the recommended friction material when replacing brake shoes. The incorrect type of friction material can affect the stopping characteristics of the car.

Hard and *soft* are terms applied to linings within a general category of material. Thus, any particular organic lining may be considered as a hard or a soft organic material. Overall, organic linings are considered softer than semimetallic linings, and semimetallic linings are considered softer than fully metallic linings. A hard lining usually has a low COF but resists fade better and lasts longer than a soft lining. A soft lining has a higher COF but fades sooner and wears faster than a hard lining. A soft lining is less abrasive on drum surfaces and operates more quietly than a hard lining. It also is common to use linings with a lower COF on the rear brakes than on the front to minimize rear brake lockup.

Sizing New Linings

Modern brake shoes are usually supplied with what is known as cam, offset, contour, or eccentric shape, which is ground in at the factory. That is, the full thickness of the lining is only present at the heel and toe of the shoe, and is ground down slightly at the center. The diameter of the circle the shoes make is slightly smaller than that of the drum. This compensates for the minor tolerance variations of drums and brake mountings and promotes proper wearing-in of the linings to match the drum.

SHOP TALK

On duo-servo shoe designs, the forward shoe is the primary and the rear, the secondary. The secondary shoe lining is longer.

Lining Adjustment

New eccentric-ground linings tolerate a closer new lining clearance adjustment than concentric

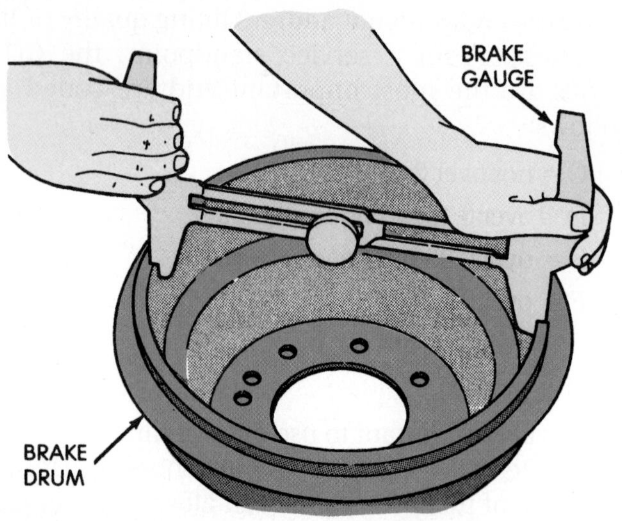

Figure 49-31 Using a brake shoe set gauge to match the diameter of the drum to the brake shoes so further shoe adjustment is limited. *Courtesy of Chrysler LLC*

Figure 49-32 Using the other side of the brake gauge to set the brake shoes.

ground linings. With manual adjusters, the shoes should be expanded into the drums until the linings are at the point of drag but not dragging heavily against the drum. With star wheel automatic adjusters, a drum/shoe gauge **(Figure 49–31)** provides a convenient means of making the preliminary adjustment. This type of gauge, when set at actual drum diameter, automatically provides the working clearance of the shoes **(Figure 49–32)**. If new linings have been concentrically ground, the initial clearance adjustment must be backed off an amount that provides sufficient working clearance.

SHOP TALK

Some brake technicians check brake spring tension by the drop method. This method is not overly scientific and the results are not always correct. Drop the brake spring on a clean concrete floor. If it bounces with a chunky sound, it is good. If the bounced spring gives off a tinny sound, it is tired and should be replaced.

Drum Shoe and Brake Installation

Before installing the shoes, sand the inner edge of the shoe to smoothen any metal nicks and burrs that could interfere with the sliding on the support pads.

A support (backing) plate must be tight on its mount and not bent. Remove any burrs or grooves on the shoe support pads that could cause the shoes to bind or hang up.

Using an approved lubricant, lightly coat the support pads **(Figure 49–33)** and the threads of servo star wheel adjusters. On rear axle parking brakes, lubricate any point of potential binding in the linkage and the cable.

Reassemble the brakes in the reverse order of disassembly. Make sure all parts are in their proper locations and that both brake shoes are properly positioned in either end of the adjuster. Also, both brake shoes should correctly engage the wheel cylinder pushrods and parking brake links. They should be centered on the backing plate. Parking brake links and levers should be in place on the rear brakes. With all of the parts in place, try the fit of the brake drum over the new shoes. If not slightly snug, pull it off and turn the star wheel until a slight drag is felt when sliding on the drum. A brake preset gauge makes this job easy and final brake adjustment simple. Then install the brake drum and wheel/tire assemblies, and make the final brake adjustments as specified in individual

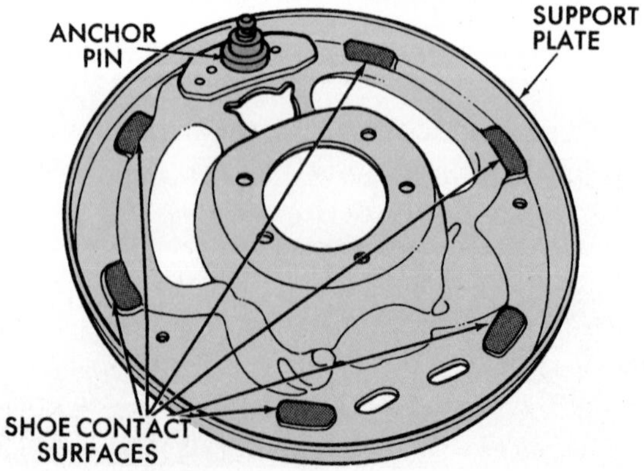

Figure 49-33 The areas or pads where the brake shoe will rub or contact the backing plate. *Courtesy of Chrysler LLC*

instructions in the vehicle's service manual. Torque the spindle and lug nuts to specifications.

WHEEL CYLINDER INSPECTION AND SERVICING

Wheel cylinders might need replacement when the brake shoes are replaced or when they begin to leak.

Inspecting and Replacing Wheel Cylinders

Wheel cylinder leaks reveal themselves in several ways: (1) fluid can be found when the dust boot is peeled back; (2) the cylinder, linings, and backing plate, or the inside of a tire might be wet; or (3) there might be a drop in the level of fluid in the master cylinder reservoir.

Such leaks can cause the brakes to grab or fail and should be immediately corrected. Note the amount of fluid present when the dust boot is pulled back. A small amount of fluid seepage dampening the interior of the boot is normal. A dripping boot is not.

WARNING!

Hydraulic system parts should not be allowed to come in contact with oil or grease. They should not be handled with greasy hands. Even a trace of any petroleum-based product is sufficient to cause damage to the rubber parts.

Cylinder binding can be caused by rust deposits, swollen cups due to fluid contamination, or by a cup wedged into an excessive piston clearance. If the clearance between the pistons and the bore wall exceeds allowable values, a condition called heel drag might exist. It can result in rapid cup wear and can cause the piston to retract very slowly when the brakes are released.

Care must be taken when installing new or reconditioned wheel cylinders on cars equipped with wheel cylinder piston stops. The rubber dust boots and the pistons must be squeezed into the cylinder before it is tightened to the backing plate. If this is not done, the pistons jam against the stops causing hydraulic fluid leaks and erratic brake performance.

SHOP TALK

Wheel cylinders are seldom rebuilt; rather, they are replaced as a unit. The time and risk involved with rebuilding them is not worth it. However, if it is necessary to rebuild one, refer to the service information first.

PROCEDURE

Replacing a Wheel Cylinder

STEP 1 Because brake hoses are an important link in the hydraulic system, it is recommended that they be replaced when a new cylinder is to be installed or when the old cylinder is to be reconditioned. Remove the brake shoe assemblies from the backing plate before proceeding. The smallest amount of brake fluid contaminates the friction surface of the brake lining.

STEP 2 Use the appropriate tubing wrench and disconnect the hydraulic line where it enters the wheel cylinder. Care must be taken while removing this steel line. It might bend and be difficult to reinstall.

STEP 3 Remove the plates, shims, and bolts that hold the wheel cylinder to the backing plate. Some later design wheel cylinders are held to the backing plate with a retaining ring that can be removed with two small picks.

STEP 4 Remove the wheel cylinder from the backing plate and clean the area with a proper cleaning solvent.

STEP 5 Install the new wheel cylinder. Care must be taken when installing wheel cylinders on cars equipped with wheel cylinder piston stops. The rubber dust boots and pistons must be squeezed into the cylinder before it is tightened to the backing plate. If this is not done, the pistons will jam against the stops, causing fluid leaks and erratic brake performance.

STEP 6 Thread the brake line into the cylinder before attaching the wheel cylinder to the backing plate. Once the cylinder's mounting bolts are tightened to specifications, tighten the brake line. Then reassemble the brake unit and bleed the system.

DRUM PARKING BRAKES

The parking brake keeps a vehicle from rolling while it is parked. It is important to remember that the parking brake is not part of the vehicle's hydraulic braking system. It works mechanically, using a lever assembly connected through a cable system to the rear drum service brakes.

Types of Parking Brake Systems

Parking brakes can be either hand or foot operated. In general, downsized cars and light trucks use hand-operated self-adjusting lever systems **(Figure 49–34)**. Full-size vehicles normally use a foot-operated parking brake pedal **(Figure 49–35A)**. The pedal or lever assembly is designed to latch into an applied position and is released by pulling a brake release handle or pushing a release button.

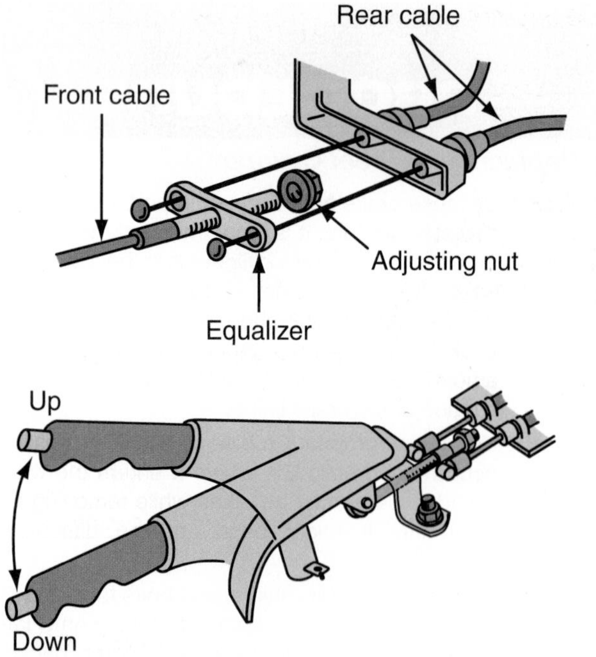

Figure 49-34 A typical setup for a center-mounted hand-operated parking brake.

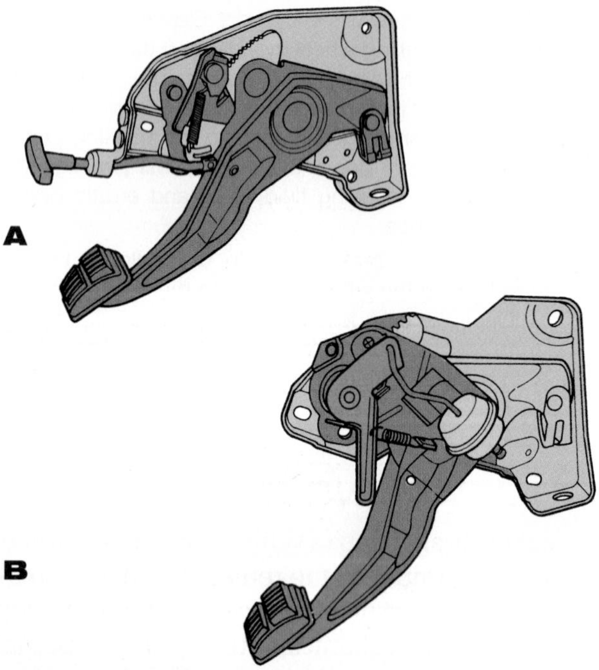

Figure 49-35 Typical pedal-operated parking brakes: (A) mechanical release and (B) vacuum release.

On some vehicles, a vacuum power unit **(Figure 49–35B)** is connected by a rod to the upper end of the release lever. The vacuum motor is actuated to release the parking brake whenever the engine is running and the transmission is in forward driving gear. The lower end of the release lever extends down for alternate manual release in the event of vacuum power failure or for optional manual release at any time. Hoses connect the power unit and the engine manifold to a vacuum release valve on the steering column.

The starting point of a typical parking brake cable and lever system is the foot pedal or hand lever. This assembly is a variable ratio lever mechanism that converts input effort of the operator and pedal/lever travel into output force with less travel. Tensile force from the front cable is transmitted through the car's brake cable system to the rear brakes. This tension pulls the flexible steel cables attached to each of the rear brakes. It operates the internal lever and strut mechanism of each rear brake, expanding the brake shoes against the drum. Springs return the shoes to the unapplied position when the parking brake pedal is released and tensile forces in the cable system are relaxed.

An electronic switch, triggered when the brake pedal is applied, lights the brake indicator in the instrument panel when the ignition is turned on. The light goes out when either the pedal or control is released or the ignition is turned off.

The cable/lever routing system in a typical parking brake arrangement uses a three-lever setup to multiply the physical effort of the operator. First is the pedal assembly or hand grip. When moved, it multiplies the operator's effect and pulls the front cable. The front cable, in turn, pulls the equalizer lever.

The **equalizer lever** multiplies the effort of the pedal assembly, or hand grip, and pulls the rear cables. This pulling effort passes through an equalizer, which ensures equal pull on both rear cables. The equalizer functions by allowing the rear brake cables to slip slightly to balance out small differences in cable length or adjustment.

Figure 49–36 shows a typical parking brake system. When the parking brake pedal is applied, the cables and equalizer exert a balanced pull on the parking brake levers of both rear brakes. The levers and the parking brake struts move the shoes outward against the brake drums. The shoes are held in this position until the parking brake pedal is released.

The rear cable enters each rear brake through a conduit **(Figure 49–37)**. The cable end engages the lower end of the parking brake lever. This lever is hinged to the web of the secondary shoe and linked with the primary shoe by means of a strut. The lever and strut expand both shoes away from the anchor and wheel cylinder and into contact with the drum as the cable and lever are drawn forward. The shoe return springs reposition the shoes when the cable is slacked.

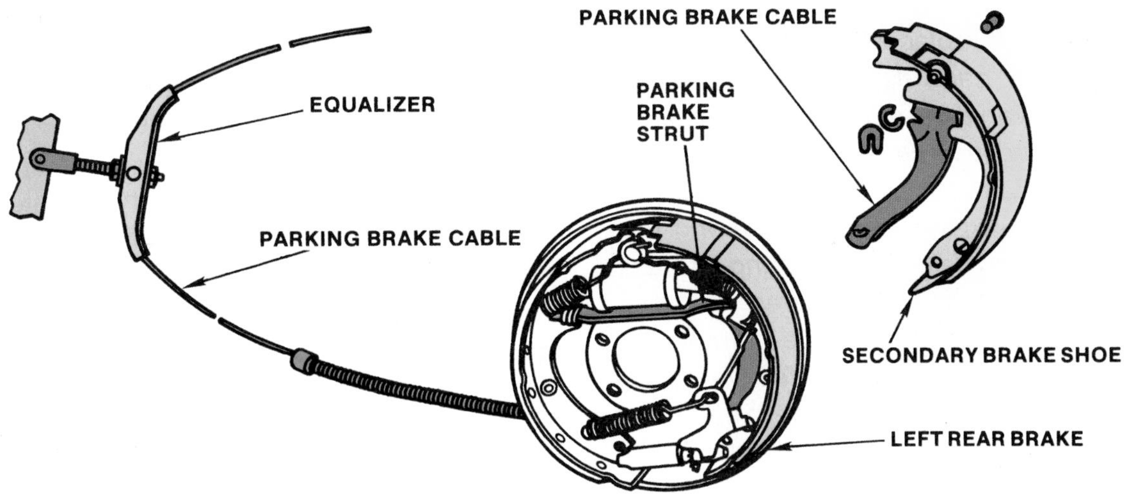

Figure 49-36 Parking brake components.

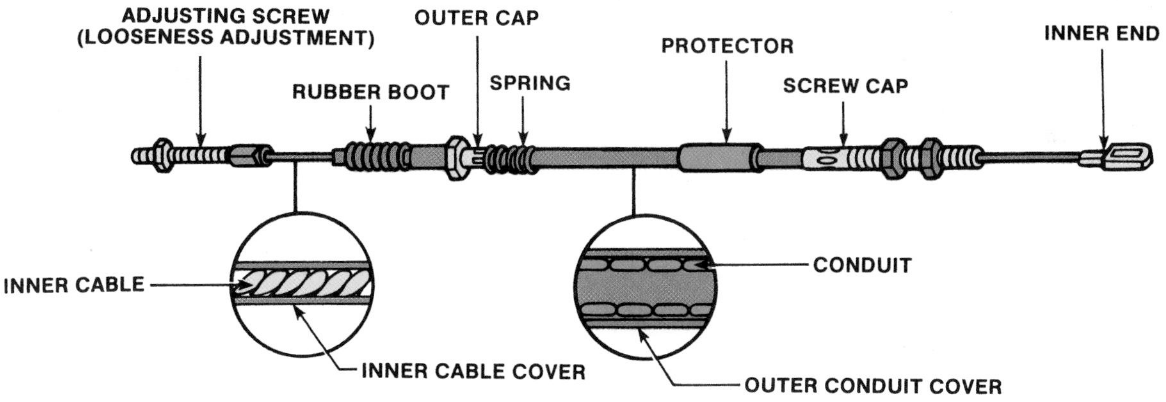

Figure 49-37 Rear cable and conduit details.

To remove and replace the brake shoes, it might be necessary to relieve the parking brake cable tension by backing off the adjusting nuts at the equalizer.

Adjusting and Replacing Parking Brakes

Regular wheel brake service should be completed before adjusting the parking brake. Then check the parking brake for free movement of the parking brake cables in the conduits. If the cables drag, replace them. Check for worn equalizer and linkage parts. Replace any defective parts. Finally, check for broken strands in the cables. Replace any cable that has broken strands or shows signs of wear.

Testing Test the parking brake by parking the vehicle facing up on an incline of 30 degrees or less. Set the parking brake fully and place the transmission in neutral. The vehicle should hold steady. Reverse the vehicle position so it is facing down the incline and repeat the test. If the vehicle creeps or rolls in either case, the parking brake requires adjustment.

CASE STUDY

Shortly after new brake shoes and hardware were installed on a vehicle, it is returned to the shop on the rear of a tow truck. An extremely irate and distressed customer is heard throughout the entire service bay area. It seems when the customer applied the brakes while traveling at a high speed, there was a sudden loss of braking power and a fine metallic grinding noise was heard. Luckily, the customer was able to control the vehicle and avoid an accident, but the situation was potentially disastrous.

The vehicle is immediately placed on a hoist and the drum brakes are disassembled. Although all the parts are new, they had failed. The shoes are pulled away from the backing plate and the holddown pins and retainers are broken.

The technician replaces all broken components with new parts from the shop's parts department. Everything appears to be installed correctly and in good working order.

The technician is reluctant to release the vehicle to the owner without uncovering the cause of the first failure. Inspecting the assembly, the technician compares the new shoes with the failed shoes. He notices the holddown pin bores on the defective shoes do not line up perfectly with the pins. More comparisons between the new and failed parts uncover slight differences. The shoes and hardware that had failed were the wrong parts for the vehicle. Although similar in many ways, these slight differences were enough to cause a major failure. The technician who had performed the original replacement should have visually compared the new parts with the old parts.

KEY TERMS

Anchor pin
Backing plate
Bonded linings
Brake lathe
Duo-servo drum brake
Equalizer lever
Floating drum
Heat checks
Holddown spring
Nonservo drum brake
Overtravel spring
Primary shoe
Return spring
Secondary shoe
Self-energizing force
Shoe anchor
Web

SUMMARY

- Drum brakes are still used on the rear wheels of many cars and light trucks.
- The drum is mounted to the wheel hub. When the brakes are applied, a wheel cylinder uses hydraulic power to press two brake shoes against the inside surface of the drum. The resulting friction between the shoe's lining and drum slows the drum and wheel.
- The brake's anchor pin acts as a brake shoe stop, keeping the shoes from following the rotating drum. This creates a wedging action that multiplies braking force.
- The shoes and wheel cylinder are mounted on a backing plate. Hardware, such as shoe return springs, holddown parts, and linkages are also mounted on the backing plate.
- The primary or leading shoe is toward the front of the vehicle while the secondary or trailing shoe is toward the rear of the vehicle.
- Brake lining can be attached to the shoes by riveting or a special adhesive bonding process.
- Brake drums act as a heat sink to dissipate the heat of braking friction. Drums can be refinished on a brake lathe provided the inside diameter is not increased above a safe limit (discard dimension).
- Servicing brakes requires performing a complete system inspection. Partial replacement of worn or damaged parts does not solve the braking problems and may ruin the new parts installed.
- When servicing brakes, extreme care must be taken to avoid generating asbestos dust.
- Wheel cylinders should be replaced if they show any signs of hydraulic fluid leakage or component wear.
- Drum brakes allow for the use of a simple parking brake mechanism that can be activated with a hand lever or foot pedal. This is a mechanical system, completely separate from the service brake hydraulic system.

REVIEW QUESTIONS

1. Name the two methods of attaching brake lining materials to the brake shoes.
2. Explain how drum brakes create a self-multiplying brake force.
3. List at least five separate types of wear and distortion to look for when inspecting brake drums.
4. What is the job of wheel cylinder stops?
5. Explain the operation of an integral drum brake parking brake.
6. In a typical drum brake, which component provides a foundation for the brake shoes and associated hardware?
   a. wheel cylinder
   b. drum
   c. backing plate
   d. lining
7. *True or False?* The name *duo-servo drum brake* is derived from the fact that the self-energizing

force is transferred from one shoe to another with the wheel rotating forward.

8. Explain how a duo-servo brake assembly works to provide great braking ability.
9. Which of the following statements about drum brake shoes is *not* true?
   a. Web thickness varies only with changes in shoe size.
   b. Many shoes have indented places along the edge of the rim.
   c. The shoe rim is welded to the web.
   d. The primary/leading shoe is typically the one that is positioned toward the front of the vehicle.
10. Brake linings should be replaced when ______.
    a. linings are worn to within 1/32 inch of a rivet head
    b linings are contaminated with oil or grease
    c. linings are contaminated with brake fluid
    d. all of the above
11. In the unapplied position, drum brake shoes are held against the anchor pin by the ______.
    a. holddown springs
    b. star wheel adjuster
    c. shoe holddown
    d. return springs
12. Duo-servo drum brakes are also known as what type of brake assembly?
    a. leading-trailing brakes
    b. self-energizing brakes
    c. nonservo brakes
    d. none of the above
13. On most vehicles, the automatic adjuster cables or levers are ______.
    a. installed on the secondary shoe
    b. set up to operate when the brakes are applied as the vehicle moves forward
    c. installed on the primary shoe
    d. set up to operate when the brakes are not applied
14. Backing plates, struts, levers, and other metal brake parts should be ______.
    a. wet-cleaned using water or a water-based solution
    b. wet-cleaned using an alcohol-based solvent
    c. wet-cleaned using gasoline
    d. dry-cleaned only
15. A buildup of brake dust and dirt between the lining and the drum is the most common cause of a ______.
    a. concave/barrel-shaped drum
    b. convex/tapered drum
    c. threaded drum surface
    d. scored drum surface

ASE-STYLE REVIEW QUESTIONS

1. Technician A says that an out-of-round drum can cause a pulsating brake pedal. Technician B says that an out-of-round drum can cause the brakes to grab. Who is correct?
   a. Technician A
   b. Technician B
   c. Both A and B
   d. Neither A nor B
2. Technician A says that a grinding noise from a drum brake when it is not applied can be caused by a bad wheel bearing. Technician B says that a grinding noise from a drum brake when it is not applied can be caused by worn brake hardware. Who is correct?
   a. Technician A
   b. Technician B
   c. Both A and B
   d. Neither A nor B
3. It has been determined that chatter and brake pull are being caused by hard spots on the brake drum: Technician A says that the problem can be solved by grinding off the hard spots. Technician B says that the drum must be replaced. Who is correct?
   a. Technician A
   b. Technician B
   c. Both A and B
   d. Neither A nor B
4. Technician A says that the discard dimension of a brake drum is the drum's allowable machining dimension. Technician B says that the discard dimension is the allowable wear dimension. There must be 0.030 inch left for wear after machining. Who is correct?
   a. Technician A
   b. Technician B
   c. Both A and B
   d. Neither A nor B
5. After resurfacing a brake drum: Technician A cleans it using hot water and a lint-free cloth. He then uses compressed air to thoroughly dry it. Technician B cleans the drum using a lint-free cloth dipped in a special brake cleaning solvent. She then allows the drum to dry before reinstallation. Who is correct?

a. Technician A c. Both A and B
b. Technician B d. Neither A nor B

6. Drum linings are badly worn at their heel and toe: Technician A says that the problem is an out-of-round drum. Technician B says that the problem is a tapered drum. Who is correct?
 a. Technician A c. Both A and B
 b. Technician B d. Neither A nor B

7. When machining brake drums: Technician A tries to remove only enough metal to obtain a true, smooth surface. Technician B cuts the drum on the other side of the axle to the same diameter as the one that was cut first. Who is correct?
 a. Technician A c. Both A and B
 b. Technician B d. Neither A nor B

8. While discussing what would happen if too much metal is removed from a drum by machining: Technician A says that noise can result from the thin drum vibrating when the brakes are applied. Technician B says that the brakes could fade because the thin drum is unable to absorb heat during braking. Who is correct?
 a. Technician A c. Both A and B
 b. Technician B d. Neither A nor B

9. Technician A checks the surface of the drum for scoring by running a fingernail across the surface. Technician B replaces any drum that is scored. Who is correct?
 a. Technician A c. Both A and B
 b. Technician B d. Neither A nor B

10. While discussing pull during braking: Technician A says that this can be caused by one tire being underinflated. Technician B says that a frozen wheel cylinder on one side of the car can cause this. Who is correct?
 a. Technician A c. Both A and B
 b. Technician B d. Neither A nor B

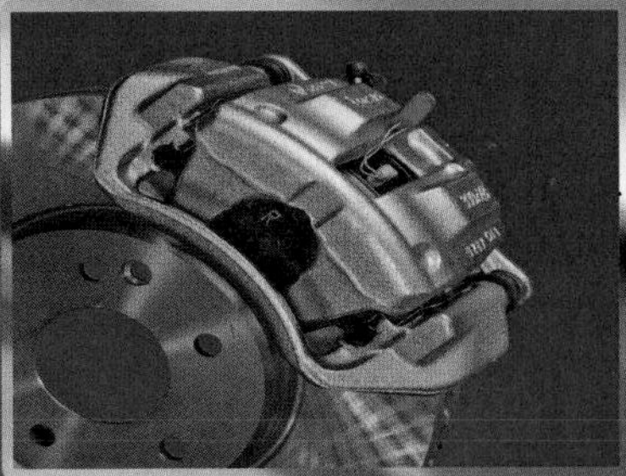

CHAPTER 50

DISC BRAKES

OBJECTIVES

■ List the advantages of disc brakes. ■ List disc brake components and describe their functions. ■ Explain the difference between the three types of calipers commonly used on disc brakes. ■ Describe the two types of parking brake systems used with disc brakes. ■ Describe the causes of common disc brake problems. ■ Explain what precautions should be taken when servicing disc brake systems. ■ Describe the general procedure involved in replacing disc brake pads. ■ List and describe five typical disc brake rotor problems.

Disc brakes resemble the brakes on a bicycle. The friction elements are in the form of pads, which are squeezed or clamped about the edge of a rotating wheel. With automotive disc brakes, this wheel is a separate unit mounted to the wheel, called the rotor (**Figure 50–1**). The rotor is typically made of cast iron. Because the pads clamp against both sides of a rotor, both sides are machined smooth. The pads are attached to metal backings, which are actuated by pistons. The pistons are contained within a caliper assembly, which is a housing that wraps around the edge of the rotor. The caliper is mounted to the steering knuckle to stop it from rotating. The caliper contains the pistons and related seals, springs, bleeder screws, and boots as well as the cylinder(s) and fluid passages necessary to force the pads against the rotor.

Figure 50-1 A typical disc brake assembly. *Courtesy of BMW of North America, LLC*

Disc brakes offer four major advantages over drum brakes. Disc brakes are more resistant to heat fade during high-speed brake stops or repeated stops. The design of the disc brake rotor exposes more surface to the air and thus dissipates heat more efficiently. They are also resistant to water fade because the rotation of the rotor tends to throw off moisture. The squeeze of the sharp edges of the pads clears the surface of water. Disc brakes perform more straight-line stops. Due to their clamping action, disc brakes are less apt to pull. Finally, disc brakes automatically adjust as pads wear.

DISC BRAKE COMPONENTS AND THEIR FUNCTIONS

The disc brakes used today are typically of two basic designs: **fixed caliper** or **floating caliper**. There is also a **sliding caliper**, but its design is very similar to the floating caliper (**Figure 50–2**). The only difference is that sliding calipers slide on surfaces that have been machined smooth for this purpose, and floating calipers slide on special pins or bolts. The disc brake, regardless of its design, consists of a hub and rotor assembly, a caliper assembly, and a brake pad assembly.

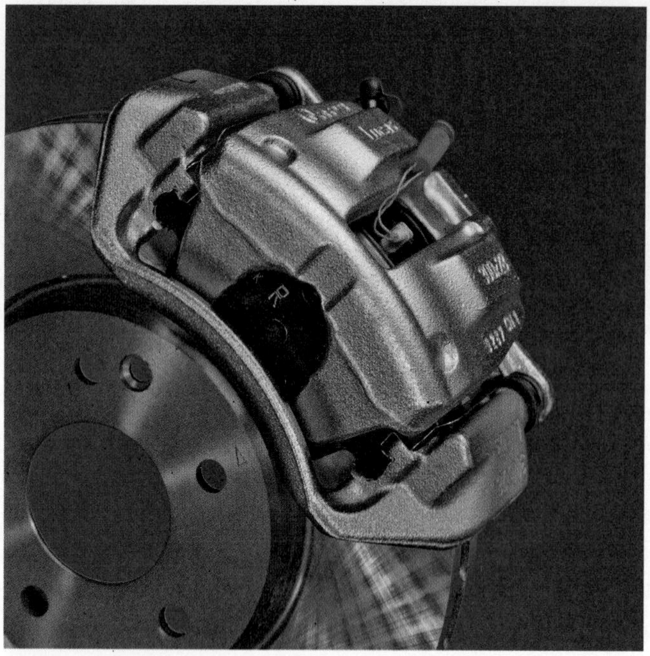

Figure 50-2 A floating caliper assembly. *Courtesy of Chrysler LLC*

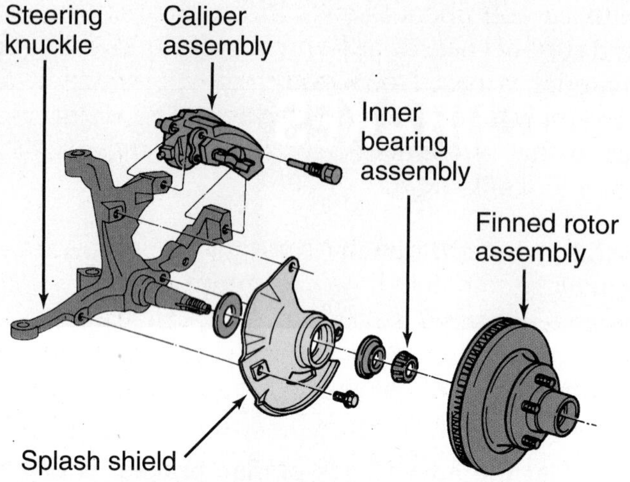

Figure 50-3 A steering knuckle, splash shield, bearings, and finned rotor for a disc brake assembly. *Courtesy of Ford Motor Company*

Rotors

The disc brake rotor has two main parts: the hub and the braking surface. The hub is where the wheel is mounted and contains the wheel bearings. The braking surface is the machined surface on both sides of the rotor. It is carefully machined to provide a friction surface for the brake pads. The entire rotor is usually made of cast iron, which provides an excellent friction surface.

The size of the rotor braking surface is determined by the diameter of the rotor. Large cars, which require more braking energy, have large rotors. Smaller, lighter cars can use smaller rotors. Generally, manufacturers want to keep parts as small and light as possible while maintaining efficient braking ability.

The rotor is protected from water and dirt due to road splash by a **splash shield** bolted to the steering knuckle **(Figure 50–3)**. The outboard side is shielded by the vehicle's wheel. The splash shield and wheel also are important in directing air over the rotor to aid cooling.

Fixed and Floating Rotors Rotors are classified by their hub design. A **fixed rotor** has the hub and the rotor cast as a single unit. **Floating rotors** and hubs are made as two separate parts. The hub is a conventional casting and is mounted on wheel bearings or on the axle. The wheel studs are mounted in the hub and pass through the rotor center section. One advantage of this design is that the rotor is less expensive and can be replaced easily and economically.

Composite Rotors The need to reduce vehicle weight led to the development of **composite rotors**. Composite rotors are made of different materials, usually cast iron and steel, to reduce weight. The friction surfaces and the hubs are cast iron, but the supporting parts of the rotor are made of lighter steel stampings. The steel and iron sections are bonded to each under heat and high pressure to form a one-piece finished assembly. Composite rotors may be fixed or floating rotors. Because the friction surfaces of composite rotors are cast iron, the wear standards are generally the same as they are for other rotors.

Ceramic Rotors In the late 1990s, Porsche first offered a carbon-ceramic brake option on the 911 GT2 and then on its 911 Turbo in 2000. Today, ceramic brakes are an option on all Ferraris and most Lamborghinis, Porsches, and Bentleys. Soon they will be standard or optional equipment on many more vehicles.

Ceramic brakes are costly but weigh about one-half of a conventional rotor. This means they allow for lower unsprung weight that helps ride quality and handling and improves fuel economy. They also last four times longer than steel discs. Brake pads also last about three times longer. The brake pads designed to be used with ceramic discs contain a ceramic powder mixed with metal wires or particles. The pads have heat shields to prevent the heat from traveling through the system.

Ceramic brakes have excellent fade resistance and stopping power. Also, the vehicle's wheels stay cleaner because no black brake dust is released.

The disc assembly is a two-piece unit: a ceramic ring and a steel center piece or hub. The ring is bolted to the hub. The ring of the rotor is made of ceramic with carbon fibers arranged to strengthen the disc and conduct heat away from the surface. The ceramic material is based on silicon carbide, which is an extremely hard material with a crystal structure similar to that of diamond. The finished surface of the rotor looks like stone.

Solid and Ventilated Rotors A rotor may be solid or it may be ventilated. A solid rotor is simply a solid piece of metal with a friction surface on each side. A solid rotor is light, simple, cheap, and easy to manufacture. Because they do not have the cooling capacity of a ventilated rotor, solid rotors usually are used on small cars of moderate performance and the rear brakes of performance-oriented vehicles.

A ventilated rotor has cooling fins cast between the braking surfaces to increase the cooling area of the rotor. When the wheel is in motion, the rotation of these fins in the rotor also increases air circulation and brake cooling. Although ventilated rotors are larger and heavier than solid rotors, these disadvantages are more than offset by their better cooling ability and heat dissipation.

Some ventilated rotors have cooling fins that are curved or formed at an angle to the hub center (**Figure 50–4**). These fins increase the centrifugal force on the rotor airflow and increase the air volume that removes heat. Such rotors are called unidirectional rotors because the fins only work properly when the rotor rotates in one direction. Therefore, unidirectional rotors cannot be interchanged from the right side to the left side on the car.

Drilled vs. Slotted Rotors Many high-performance vehicles are fitted with cross-drilled rotors (**Figure 50–5**). The idea behind having holes through the rotor is simply to allow heat, gases, and dirt to escape. In addition, the edges of the holes give a place for the pads to grab. They, however, also decrease the overall surface area of the rotor, which reduces the thermal capacity of the discs and the discs have a poor service life. The latest trend is to cut a series of tangential slots or channels into the surface. These slots do the same thing as the holes without the disadvantages.

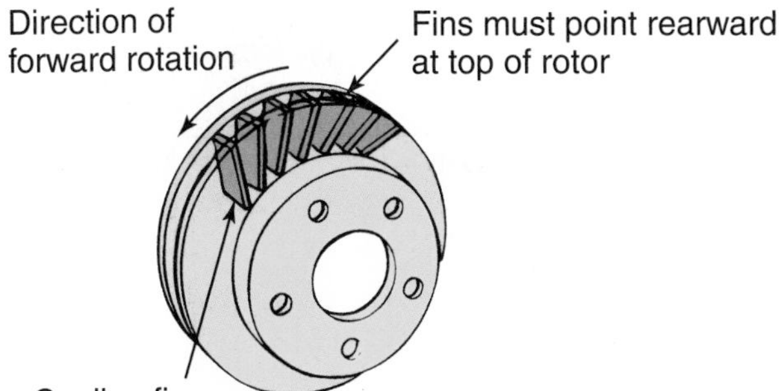

Figure 50-4 Some ventilated rotors have curved cooling fins and are directional. *Courtesy of Federal-Mogul Corporation*

Figure 50-5 A cross-drilled brake rotor.

Rotor Hubs and Wheel Bearings

Tapered roller bearings, which are installed in the wheel hubs, are the most common bearings used on the front wheels of RWD and the rear wheels of FWD vehicles. The tapered roller bearing has two main parts: the inner bearing cone and the outer bearing cup. The bearing cone is an assembly that contains steel tapered rollers. The rollers ride on an inner cone-shaped race and are held together by a bearing cage. The bearing fits into the outer cup, or race, which is pressed into the hub. This provides two surfaces, an inner cone and outer cup, for the rollers to ride on. The bearings are held in place with a thrust washer, nut, locknut, and cotter pin. A dust cap fits over the assembly to keep dirt out and lubricant in. A seal on the inboard side prevents lubricant from escaping at this end.

Caliper Assembly

A brake caliper converts hydraulic pressure into mechanical force. The caliper housing is usually a one-piece construction of cast iron or aluminum and has an inspection hole in the top to allow for lining wear inspection. The housing contains the cylinder bore(s). In the cylinder bore is a groove that seats a square-cut seal. This groove is tapered toward the bottom of the bore to increase the compression on the edge of the seal that is nearest hydraulic pressure. The top of the cylinder bore is also grooved as a seat for the dust boot. A fluid inlet hole is machined into the bottom of the cylinder bore and a bleeder valve is located near the top of the casting.

A caliper can contain one, two, or four cylinder bores and pistons that provide uniform pressure distribution against the brake's friction pads. The pistons are relatively large in diameter and short in stroke to provide high pressure on the friction pad assemblies with a minimum of fluid displacement.

Basically, the hydraulics of disc brakes are the same as for drum brakes, in that the master cylinder piston forces the brake fluid into the wheel cylinders and against the wheel pistons.

The disc brake piston is made of steel, aluminum, or fiberglass-reinforced **phenolic resin**. Steel pistons are usually nickel-chrome plated for improved durability and smoothness. The top of the pistons is grooved to accept the **dust boot** that seats in a groove at the top of the cylinder bore and also in a groove in the piston. The dust boot prevents moisture and road contamination from entering the bore.

A piston hydraulic (square-cut) seal prevents fluid leakage between the cylinder bore wall and the piston. This rubber sealing ring also acts as a retracting mechanism for the piston when hydraulic pressure is released, causing the piston to return in its bore **(Figure 50–6)**. When hydraulic pressure is diminished, the seal functions as a return spring to retract the piston.

In addition, as the disc brake pads wear, the seal allows the piston to move farther out to adjust automatically for the wear without allowing fluid to leak. Since the brake pads need to retract only slightly after they have been applied, the piston moves back only slightly into its bore. The additional brake fluid in the caliper bore keeps the piston out and ready to clamp the surface of the rotor.

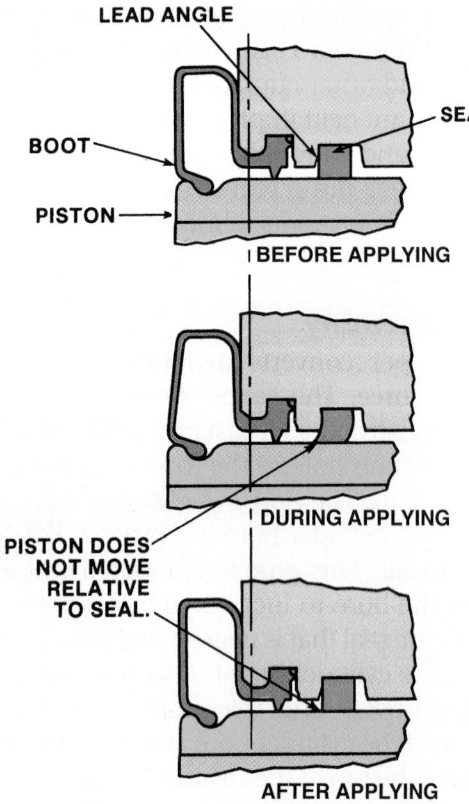

Figure 50-6 Action of the piston's hydraulic seal in the caliper's cylinder.

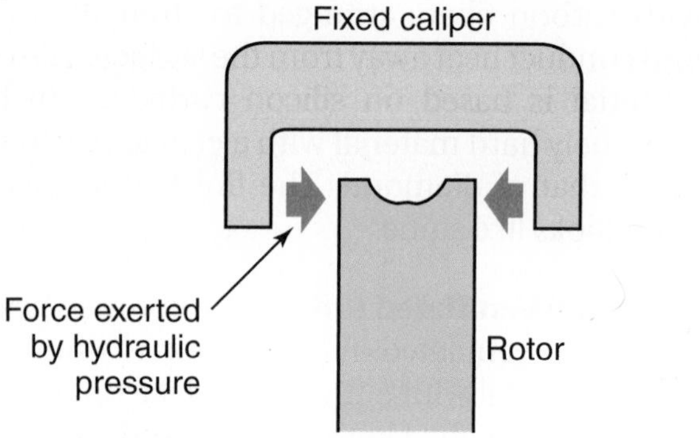

Figure 50-7 Operation of a fixed caliper. *Courtesy of Federal-Mogul Corporation*

Fixed Caliper Disc Brakes Fixed caliper disc brakes have a caliper assembly that is bolted in a fixed position and does not move when the brakes are applied. The pistons in both sides of the caliper come inward to force the pads against the rotor **(Figure 50–7)**.

Floating Caliper Disc Brakes A typical floating caliper disc brake is a one-piece casting that has one hydraulic cylinder and a single piston. The caliper is attached to the **spindle anchor plate** with two threaded locating pins. A Teflon sleeve separates the caliper housing from each pin and the caliper slides back and forth on the pins as the brakes are actuated. When the brakes are applied, hydraulic pressure builds in the cylinder behind the piston and seal. Because hydraulic pressure exerts equal force in all directions, the piston moves evenly out of its bore.

The piston presses the inboard pad against the rotor. As the pad contacts the revolving rotor, greater resistance to outward movement is increased, forcing pressure to push the caliper away from the piston. This action forces the outboard pad against the rotor **(Figure 50–8)**. However, both pads are applied with equal pressure.

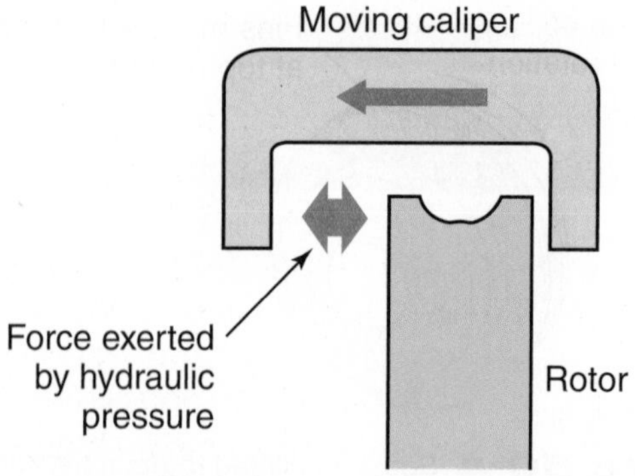

Figure 50-8 Operation of a floating caliper. *Courtesy of Federal-Mogul Corporation*

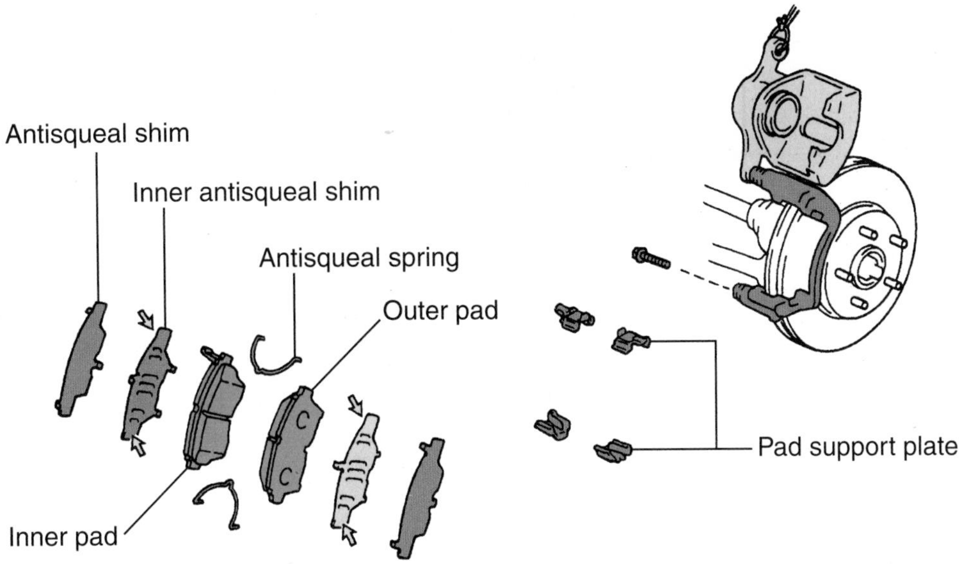

Figure 50–9 A sliding caliper with support plates (keys). *Courtesy of Toyota Motor Sales, U.S.A., Inc.*

Sliding Caliper Disc Brakes With a sliding caliper assembly, the caliper slides or moves sideways when the brakes are applied. As mentioned previously, in operation, these brakes are almost identical to the floating type. But unlike the floating caliper, the sliding caliper does not float on pins or bolts attached to the anchor plate. It has angular machined surfaces at each end that slide in mating machined surfaces on the anchor plate. This is where the caliper slides back and forth.

Some sliding calipers use a support key to locate and support the caliper in the anchor plate (**Figure 50–9**). The caliper support key is inserted between the caliper and the anchor plate. A worn support key may cause tapered brake pad wear. Always inspect the support keys when replacing brake pads. Also make sure they are lubricated when reassembling the unit.

Brake Pad Assembly

Brake pads are metal plates with the linings either riveted or bonded to them. Pads are placed at each side of the caliper and straddle the rotor. The linings are made of semimetallic or other nonasbestos material.

Disc Pad Wear Sensors Some brake pads have wear sensing indicators. The three most common design wear sensors are audible, visual, and tactile.

Audible sensors are thin, spring steel tabs that are riveted to or installed onto the edge of the pad's backing plate and are bent to contact the rotor when the lining wears down to a point that replacement is necessary. At that point, the sensor causes a high-pitched squeal whenever the wheel is turning, except when the brakes are applied. Then the noise goes away. The noise gives a warning to the driver that brake service is needed and perhaps saves the rotor from destruction (**Figure 50–10**).

Some vehicles have systems with an electronic wear indicator in the disc brake pads. As the pad wears to a predetermined point, a warning light in the instrument panel is illuminated by the wear sensors. In some systems, a small pellet is contained in the brake pad's friction material. The pellets are wired in series or in parallel to the red brake warning lamp circuit (**Figure 50–11**) and complete the lamp circuit when the pellets contact the rotor.

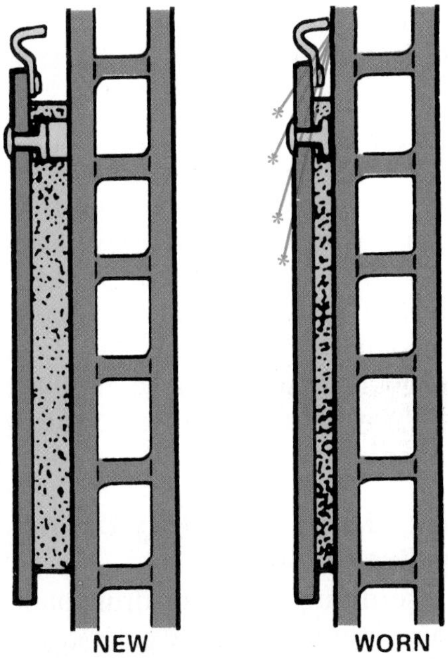

Figure 50–10 Operation of the wear indicator. *Courtesy of Federal-Mogul Corporation*

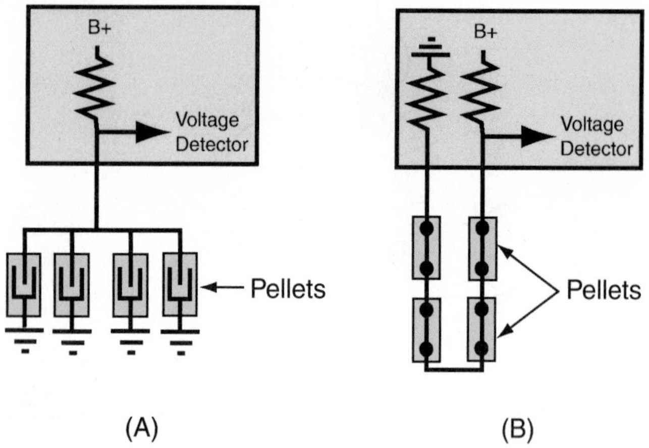

Figure 50–11 Electronic pad wear sensors can be wired in parallel or in series with the warning lamp. The pellets are embedded in the brake pads.

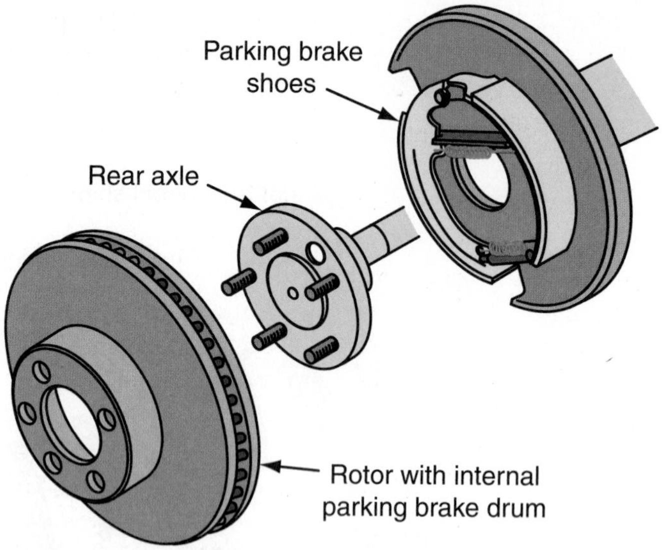

Figure 50–12 A few rear disc brakes have a rotor with an internal drum and a small set of brake shoes to serve as the parking brake.

Customer Care

Excessive heat liquefies the resin binder that holds the brake pad material together. Once liquefied, the binder rises to the surface of the pad to form a glaze. A glazed pad may cause squealing because more heat is needed to achieve an amount of friction equal to a pad in good condition. One common cause of pad glazing is the improper break-in of the pads. Remember to inform customers that unnecessary hard braking during the first 200 miles (320 km) of a pad's life can generate enough heat to glaze the pads and ruin the quality of the work just performed. There are pads now available that require no break-in time.

REAR-WHEEL DISC BRAKES

Rear-wheel disc brake calipers may be fixed, floating, or sliding, and all of these designs work in the same way as when they are used at the front wheels. The only difference between a front and rear disc brake caliper is that the rear disc brake caliper needs a parking brake. Four-wheel disc brake installations must have some way to apply the rear brakes when the parking brake is set.

Rear Disc/Drum (Auxiliary Drum) Parking Brake

The rear disc/drum or auxiliary drum parking brake arrangement is found on some vehicles with fixed or sliding calipers. On these brakes, the inside of each rear wheel hub and rotor assembly is used as the parking brake drum (**Figure 50–12**). A pair of small brake shoes is mounted on a backing plate that is bolted to the axle housing or the hub carrier. These parking brake shoes operate independently of the service brakes. They are applied by linkage and cables from the control pedal or lever. The cable at each wheel operates a lever and strut that apply the shoes in the same way that rear drum parking brakes work.

The assembly (often called the drum-in-heat system) is a smaller version of a drum brake and is serviced much like any other drum brake. However, they do not have self-adjusters. The parking brakes must be adjusted manually with star wheels that are accessible through the backing plate or through the outboard surface of the drum (**Figure 50–13**).

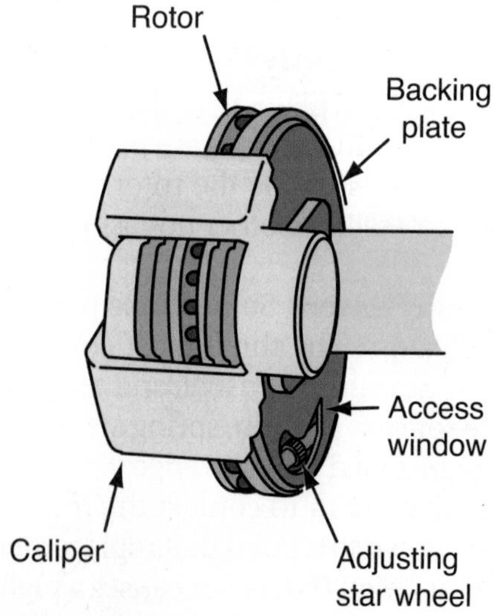

Figure 50–13 The auxiliary drum parking brakes are adjusted through a window in the backing plate.

Late-model GM trucks use an expandable, single metal band covered with friction material inside the parking brake drum that is machined on the inside of the rear brake disc.

Caliper-Actuated Parking Brakes

Most floating or sliding caliper rear disc brakes mechanically apply the calipers' pistons to lock the pads against the rotors for parking. All caliper-actuated parking brakes have a lever that protrudes from the inboard side of the caliper. These levers are operated by linkage and cables from the control pedal or lever.

The two most common types of caliper-actuated parking brakes are the screw-and-nut type and the ball-and-ramp type. A few imported cars have a third type that uses an eccentric shaft and a rod to apply the caliper piston. An eccentric acts like a cam. One portion of the shaft is oval-shaped. As the shaft rotates, the high part of the oval pushes the operating rod out to apply the brakes.

General Motors' floating caliper rear disc brakes are the most common example of the screw-and-nut parking brake mechanism **(Figure 50–14)**. The caliper lever is attached to an actuator screw inside the caliper that is threaded into a large nut. The nut, in turn, is splined to the inside of a large cone that fits inside the caliper piston. When the parking brake is applied, the caliper lever rotates the actuator screw. Because the nut is splined to the inside of the cone, it cannot rotate, so it forces the cone outward against the inside of the piston, forcing it outward. Similarly, the piston cannot rotate because it is keyed to the brake pad, which is fixed in the caliper. The piston then applies the inboard brake pad, and the caliper slides as it does for service brake operation and forces the outboard pad against the rotor. An adjuster spring inside the nut and cone rotates the nut outward when the parking brakes are released to provide self-adjustment. Rotation of the nut takes up clearance as the brake pads wear.

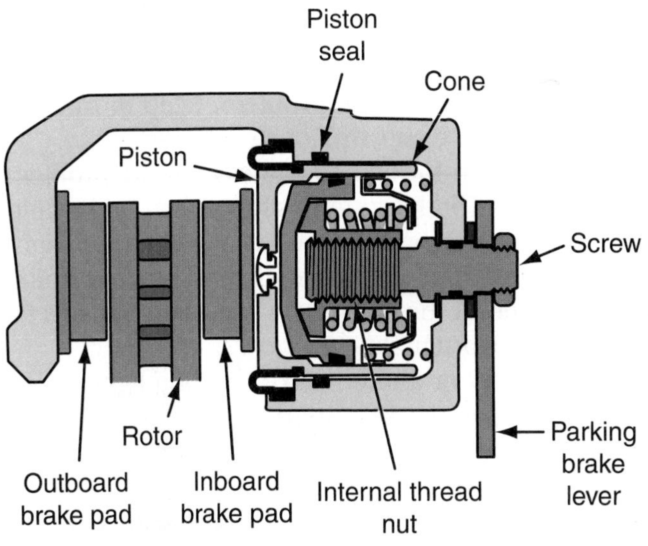

Figure 50-14 A GM screw-and-nut parking brake mechanism for a disc brake.

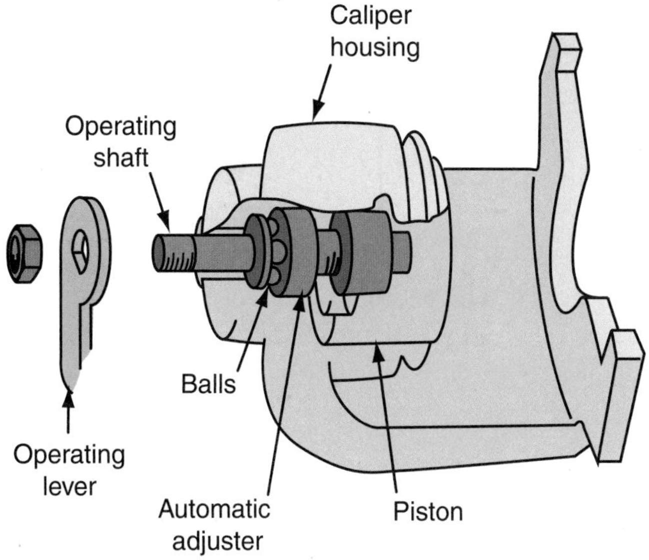

Figure 50-15 A Ford ball-and-ramp parking mechanism for a disc brake.

Ford's floating caliper rear disc brakes are the most common example of the ball-and-ramp parking brake mechanism **(Figure 50–15)**. The caliper lever is attached to a shaft inside the caliper that has a small plate on the other end. Another plate is attached to a thrust screw inside the caliper piston. The two plates face each other, and three steel balls separate them. When the parking brake is applied, the caliper lever rotates the shaft and plate. Ramps in the surface of the plate force the balls outward against similar ramps in the other plate. As the plates move farther apart, the thrust screw forces the piston outward. The thrust screw cannot rotate because it is keyed to the caliper. The piston then applies the inboard brake pad, and the caliper slides as it does for service brake operation and forces the outboard pad against the rotor. When the piston moves away from the thrust screw, an adjuster nut inside the piston rotates on the screw to take up clearance and provide self-adjustment. A drive ring on the nut keeps it from rotating backward.

Another way to tighten the pads against the rotor when the parking brake is applied is to use a threaded, spring-loaded pushrod **(Figure 50–16)**. As the parking brakes are applied, a mechanism rotates or unscrews the pushrod, which in turn pushes the piston out.

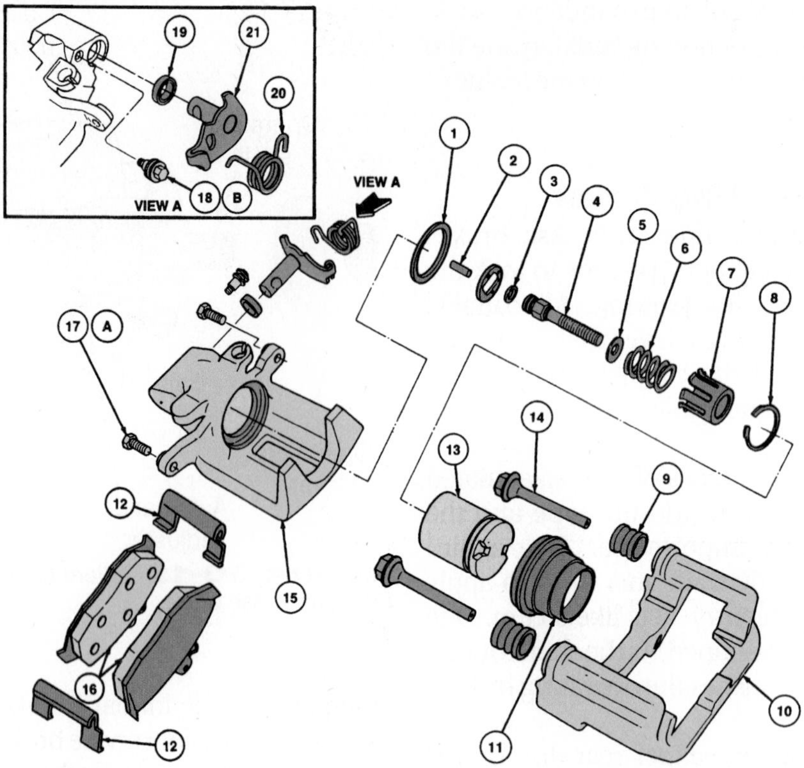

1. Piston seal
2. Pin
3. O-ring
4. Threaded pushrod
5. Flat washer
6. Spring
7. Spring cage
8. Snapring
9. Slider pin boot seal
10. Rear support
11. Piston dust boot
12. Antirattle clip
13. Piston
14. Locating pin
15. Caliper
16. Brake pads
17. Pin retainer
18. Limiting bolt
19. Parking brake shaft seal
20. Parking brake lever return spring
21. Parking brake lever

Figure 50-16 A rear caliper with a threaded drive for the piston.
Courtesy of Ford Motor Company

DISC BRAKE DIAGNOSIS

Many problems experienced on vehicles with disc brakes are the same as those evident with drum brake systems. Some problems occur only with disc brakes. Before covering the typical complaints, it is important to remind you to get as much information as possible about the complaint from the customer. Then road test the vehicle to verify the complaint. A complete inspection of the rotor, caliper, and pads **(Figure 50–17)** should be done any time you are working on the brakes.

What follows is a brief discussion of common complaints and their typical causes.

Warning Lights

Today's vehicles are normally equipped with more than one brake warning light on the instrument panel. Regardless of what warning light is lit, it is an indication of warning to the driver. You need to understand what would cause the different lights to illuminate in order to take care of the problem. Keep in mind, a vehicle may have one, two, or all of these lights.

The red warning light indicates there is a problem in the regular brake system, such as low brake fluid levels or that the parking brake is on. A low fluid light may be present in addition to the red brake warning light. Whenever the fluid is low, you should suspect a leak or very worn brake pads.

The yellow or amber brake warning light is tied into the antilock brake system (ABS). This light turns on for two reasons: the ABS is performing a self-test or there is a fault in the ABS.

A blue or yellow warning light lets the driver know the wheels are slipping because of poor road conditions.

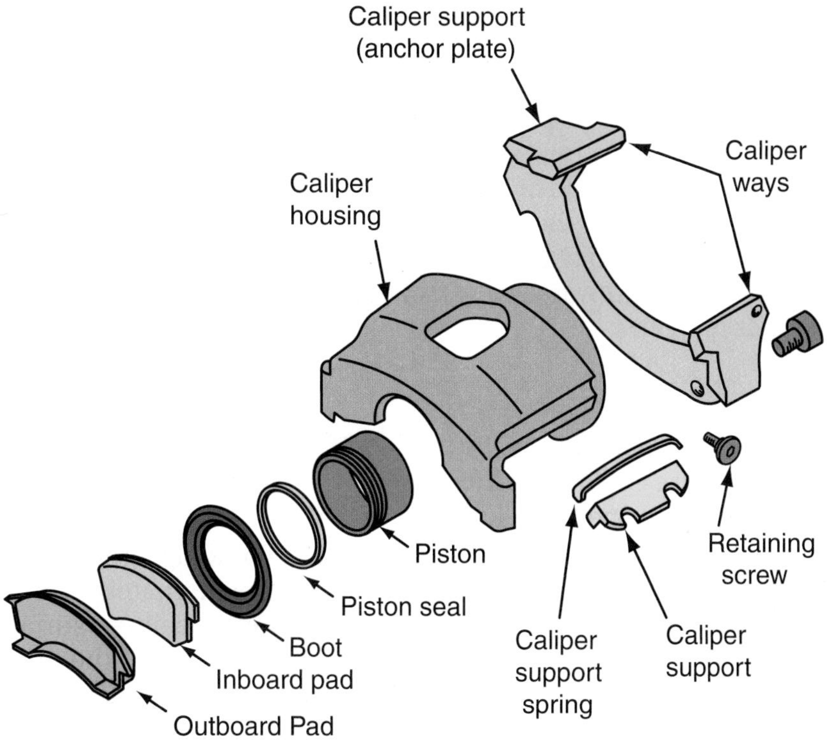

Figure 50-17 Whenever you are doing brake work, thoroughly inspect the entire brake assembly.

Pulsating Pedal

Customers will feel a vibration or pulsation in the brake pedal when the brakes are applied if a brake rotor is warped. If this symptom exists, check the rotors for runout and parallelism. A warped rotor should be replaced and is often caused by improper tightening of the wheel lug nuts. In fact, uneven lug nut torque can cause a pulsating brake pedal. You should be aware that pedal pulsation is normal on vehicles with ABS when the ABS is working.

Spongy Pedal

With a spongy pedal, the customer will probably feel the need to pump the brake pedal to get good stopping ability. The complaint may also be described as a soft pedal. This problem is caused by air in the hydraulic system. Although bleeding the system may remove the air, you should always question how the air got in there. Check for leaks and for proper master cylinder operation.

Hard Pedal

The driver's complaint of a hard pedal normally indicates a problem with the power brake booster. However, it can also be caused by a restricted brake line or hose. Carefully check the lines and hoses for damage. Feel the brake hoses. If they seem to have lost their rigidity, the hose may have collapsed on the inside, and this is causing the restriction. Restrictions can also be caused by frozen caliper or wheel cylinder pistons.

Dragging Brakes

Dragging brakes make the vehicle feel as if it has lost or is losing power as it drives down the road. The problem also wastes a lot of fuel and generates destructive amounts of heat that can cause serious brake damage and brake failure. While trying to find the cause of this problem, check the parking brake first. Make sure it is off. Check the rear wheels to make sure the parking brakes are released when they should be. If the problem is not in the parking brakes, check for sticky or seized pistons at the hoses to the calipers and wheel cylinders.

Grabbing Brakes

When the brakes seem to be overly sensitive to pedal pressure, they are grabbing. Normally this problem is caused by contaminated brake linings. If the linings are covered or saturated with oil, find the source of the oil and repair it. Then replace the pads and refinish or replace the rotor.

Noise

If the customer's complaint is noisy brakes, verify during the road test that the problem is in the brakes.

If the noise is caused by the brakes, pay attention to the type of noise and let that lead to the source of the problem. Remember, some brake pads have wear sensors that are designed to make a high-pitched squeal when the pads are worn. Other causes could be the rotor rubbing against the splash shield or that something has become wedged between the rotor and another part of the vehicle. Noise may also be caused by failure to install all of the hardware when placing a caliper or brake pads in service.

Pulling

When a vehicle drifts or pulls to one side while cruising or when braking, the cause could be in the brake system or in the steering and suspension system. Check the inflation of the tires, the tires' tread condition, and verify that the tires on each axle are the same size. Check the operation of the brakes. If only one front wheel is actually doing the braking, the vehicle will seem to stumble or pivot on that one wheel. If no problems are found in the brake system, suspect an alignment or suspension problem.

SERVICE GUIDELINES

The following general service guidelines apply to all disc brake systems and should always be followed:

- Be sure the vehicle is properly centered and secured on stands or a hoist.
- If the vehicle has antilock brakes, depressurize the system according to the procedures given in the service manual.
- Disconnect the battery ground cable.
- Before any service is performed, check the following:
  - Tires for excessive wear or improper inflation
  - Wheels for bent or warped rims
  - Wheel bearings for looseness or wear
  - Suspension components to see if they are worn or broken
  - Brake fluid level
  - Master cylinder, brake lines or hoses, and each wheel for leaks
- Before you remove a brake hydraulic part, use a pedal depressor to slightly depress the brake pedal. This closes the master cylinder's ports and prevents fluid from draining. It also makes the bleeding process easier after the system is reassembled.
- Before beginning brake work, remove about two-thirds of the brake fluid from the master cylinder's reservoir. If this is not done, the fluid could overflow and spill when the pistons are forced back into the caliper bore. You also can open the caliper bleeder screw and run a hose to a container to catch the fluid that is expelled. This prevents dirty brake fluid from being forced back into the ABS control unit.
- If the bleeder screws are frozen tight with corrosion, it is sometimes possible to free them using a torch and penetrating oil. Of course, the caliper must be removed from the car, taken to a bench, and worked on there. If the bleeder screws cannot be loosened, the entire caliper is probably in bad shape so it is best to replace the entire unit.
- During servicing, grease, oil, brake fluid, or any other foreign material must be kept off the brake linings, caliper, surfaces of the disc, and external surfaces of the hub. Handle the brake disc and caliper in such a way to avoid damage to the disc.
- When a hydraulic hose is disconnected, plug it to prevent foreign material from entering.
- Never permit the caliper assembly to hang with its weight on the brake hose. Support it on the suspension or hang it by away from the assembly.
- When using compressed air to remove the caliper pistons, avoid high pressures. A safe pressure to use is 30 psi (207 kPa).
- Clean the hydraulic brake components in either denatured alcohol or clean brake fluid. Do not use mineral-based cleaning solvent such as gasoline, kerosene, carbon tetrachloride, acetone, or paint thinner to clean the caliper. It causes rubber parts to become soft and swollen in an extremely short time.
- With the recommended lubricant, lubricate all moving parts, such as the caliper housing or mounting bracket to ensure a free-moving action.
- Before the brake pads are installed, apply a disc brake noise suppressor to the back of the pads to prevent brake squeal. For best results, follow the directions on the container.
- Obtain a firm brake pedal after servicing the brakes and before moving the vehicle. Be sure to road test the vehicle.
- Always torque the lug nuts when installing a wheel on a vehicle with disc brakes. Never use an impact gun to tighten the lug nuts. Warpage of the rotor could result if an impact gun is used.

WARNING!

When using the propane torch to loosen a bleeder screw, use it with extreme care.

GENERAL CALIPER INSPECTION AND SERVICING

Frequently, caliper service involves only the removal and installation of the brake pads. However, since the new pads are thicker than the worn-out set they replace, they locate the piston farther back in the bore where dirt and corrosion might cause the seals to leak. For this reason, it is often good practice to carefully inspect the calipers whenever installing new pads. Of course, it is also good practice to true-up or replace the rotors when replacing brake pads.

When bench testing or servicing a caliper assembly, use a vise that is equipped with protector jaws. Excessive vise pressure causes bore and piston distortion.

Caliper Removal

To be able to replace brake pads, service the rotor, or to replace the caliper, the caliper must be removed. The procedure for doing this varies according to caliper design. Always follow the specific procedures given in a service manual. Use the following as an example of these procedures:

1. Remove the brake fluid from the master cylinder.
2. Raise the vehicle and remove the wheel and tire assembly.
3. On a sliding or floating caliper, install a C-clamp with the solid end of the clamp on the caliper housing and the screw end on the metal portion of the outboard brake pad. Tighten the clamp until the piston bottoms in the caliper bore **(Figure 50–18)**, then remove the clamp. Bottoming the piston allows room for the brake pad to slide over the ridge of rust that accumulates on the edge of the rotor.
4. On threaded-type rear calipers, the piston must be rotated to depress it, which requires a special tool **(Figure 50–19)**.
5. Disconnect the brake hose from the caliper and remove the copper gaskets or washer and cap the end of the brake hose. If only the brake pads are to be replaced, do not disconnect the brake hose.
6. Remove the two mounting brackets to the steering knuckle bolts. Support the caliper when removing the second bolt to prevent the caliper from falling.
7. On a sliding caliper, remove the top bolts, retainer clip, and **antirattle springs** **(Figure 50–20)**. On a floating caliper, remove the two special pins that hold the caliper to the anchor plate **(Figure 50–21)**. On a fixed caliper, remove the bolts holding it to the steering knuckle. On all three types, get the caliper off by prying it straight up and lifting it clear of the rotor.

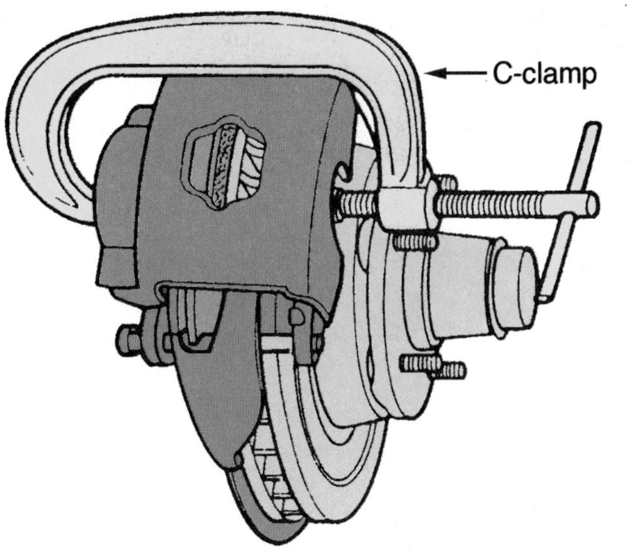

Figure 50–18 Bottoming the piston in the caliper's bore.

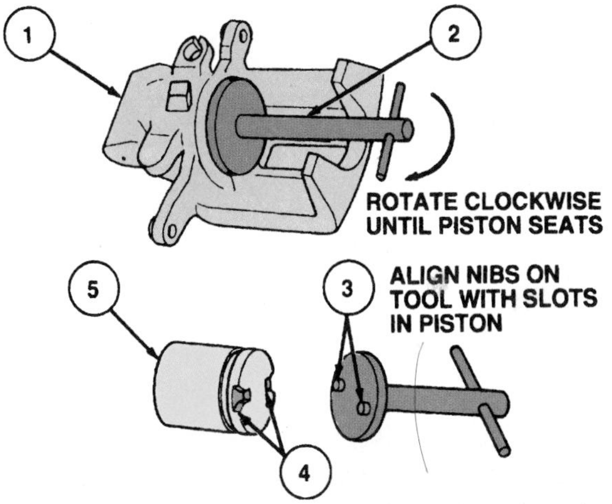

Item	Description
1	Caliper Housing
2	Rear Caliper Piston Adjuster Tool
3	Nibs
4	Slots
5	Rear Disc Brake Piston and Adjuster

Figure 50–19 A special tool is required to move a threaded piston into its bore. *Courtesy of Ford Motor Company*

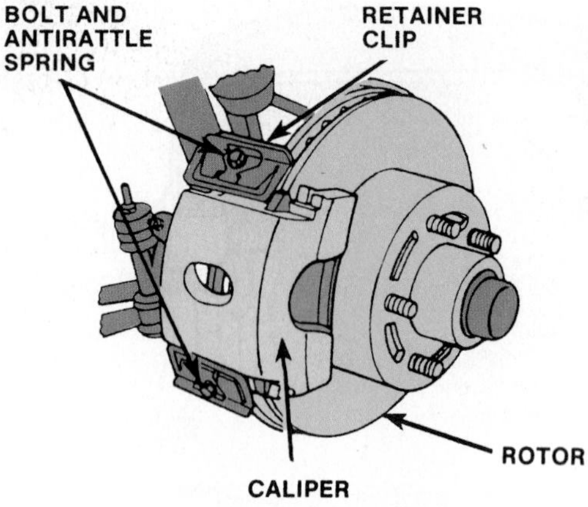

Figure 50-20 Sliding caliper removal.

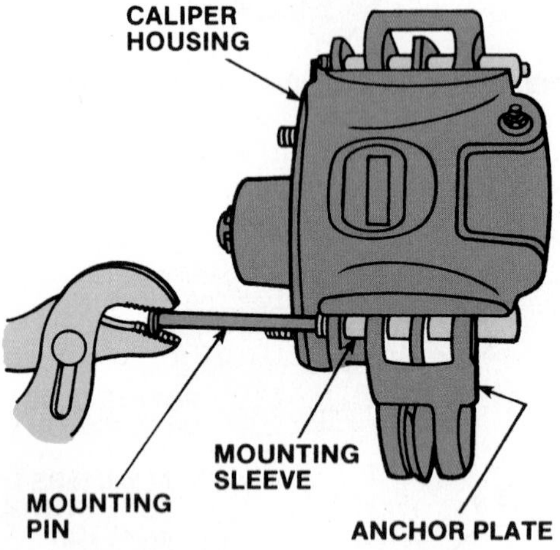

Figure 50-21 Floating caliper removal.

Brake Pad Removal

Sliding or floating calipers must always be lifted off the rotor for pad replacement. Fixed calipers might have pads that can be replaced by removing the retaining pins or clips instead of having to lift the entire caliper off. Brake pads may be held in position by retaining pins, guide pins, or a support key. Note the position of the shims, antirattle clips, keys, bushings, or pins during disassembly. A typical procedure for replacing brake pads is outlined in Photo Sequence 52.

If you are only replacing the pads, lift the caliper off the rotor and hang it up by a wire. Remove the pads. Remove the old sleeves and bushings and install new ones. Replace the rusty pins to ensure free movement. You may need to transfer the shoe retainers onto the new pads.

Brake Pad Inspection

Disc brake pads should be checked periodically. Some calipers have inspection holes in the caliper body. If they do not, the pads can be inspected from the outer ends of the caliper.

If you are not sure the pads are worn enough to warrant replacement, measure them at the thinnest part of the pad. Compare this measurement to the minimum brake pad lining thickness listed in the service manual, and replace the pads if needed. Typically, if the friction material remaining on the backing plate is less than ⅛ inch (3.175 mm), the pads should be replaced.

When a pad on one side of the rotor has worn more than the other side, there is uneven wear. Uneven pad wear often means the caliper is sticking and not giving equal pressure to both pads. On a sliding caliper, the problem could be caused by poor lubrication or deformation of the machined sliding areas on the caliper and/or anchor plate.

SHOP TALK

Most often, calipers are replaced rather than rebuilt. The old caliper is sent back to the manufacturer as a core. The following overview for rebuilding is intended only to give you an understanding of what may be done. Always refer to the appropriate service manual when rebuilding a caliper.

Caliper Disassembly

If the caliper must be rebuilt, it should be taken to the workbench for servicing. Drain any brake fluid from the caliper by way of bleeder screws. Remove the bleeder valve protector, if so equipped.

On a floating caliper, examine the mounting pins for rust that could limit travel. Most manufacturers recommend that these pins and their bushings be replaced each time the caliper is removed. This is a good idea because the pins are inexpensive and a good insurance against costly comebacks. On a fixed caliper, check the pistons for sticking and rebuild the caliper if this problem is found.

To disassemble the caliper, the piston and dust boot must first be removed. Place the caliper face down on a workbench (**Figure 50–22**). Insert the used outer pad or a block of wood into the caliper. Place a folded shop towel on the face of the lining to cushion the piston. Apply low air pressure (NEVER MORE THAN 30 PSI) to the fluid inlet port of the caliper to force the piston from the caliper housing.

CAUTION!

Wear safety glasses while disassembling the caliper to protect your eyes from spraying brake fluid.

P52-1 *Front brake pad replacement begins with removing brake fluid from the master cylinder reservoir.*

P52-2 *Raise the car. Make sure it is safely positioned on the lift. Remove its wheel assemblies.*

P52-3 *Inspect the brake assembly. Look for signs of fluid leaks, broken or cracked lines, or a damaged brake rotor. If any problem is found, correct it before installing the new brake pads.*

P52-4 *Loosen the bolts and remove the pad locator pins.*

P52-5 *Lift and rotate the caliper assembly from the rotor.*

P52-6 *Remove the brake pads from the caliper assembly.*

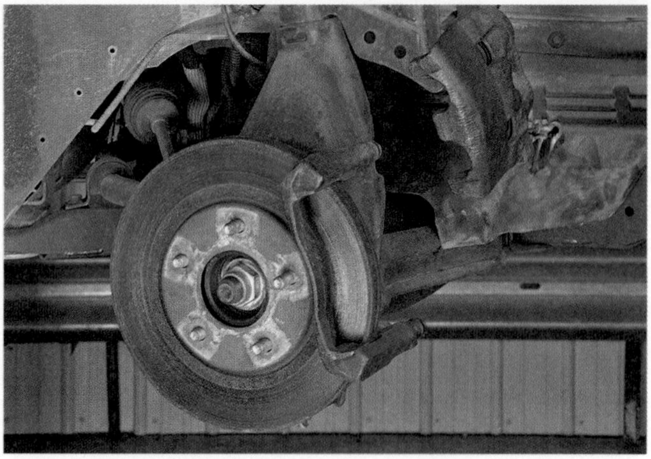

P52-7 *Fasten a piece of wire to the car's frame and support the caliper with the wire.*

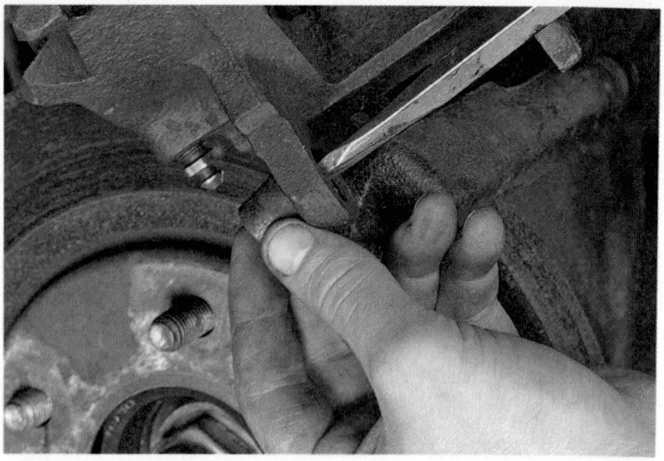

P52-8 *Check the condition of the locating pin insulators and sleeves.*

P52-9 *Place a piece of wood over the caliper's piston and install a C-clamp over the wood and caliper. Tighten the clamp to force the piston back into its bore.*

P52-10 *Remove the clamp and install new locating pin insulators and sleeves, if necessary.*

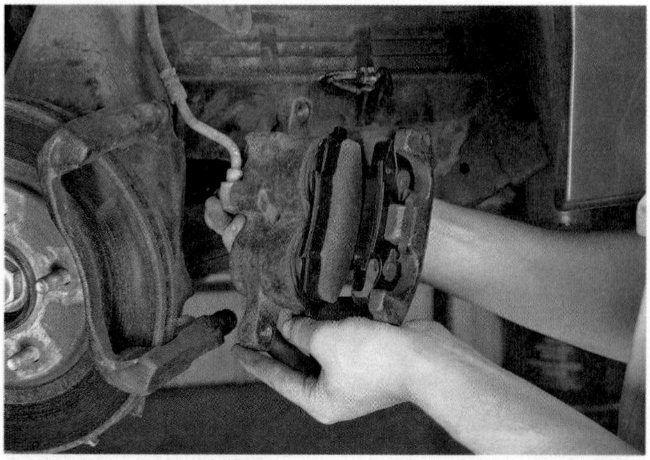

P52-11 *Install the new pads into the caliper.*

P52-12 *Set caliper with pads over the rotor and install the locating pins. After the assembly is in the proper position, torque the pins according to specifications.*

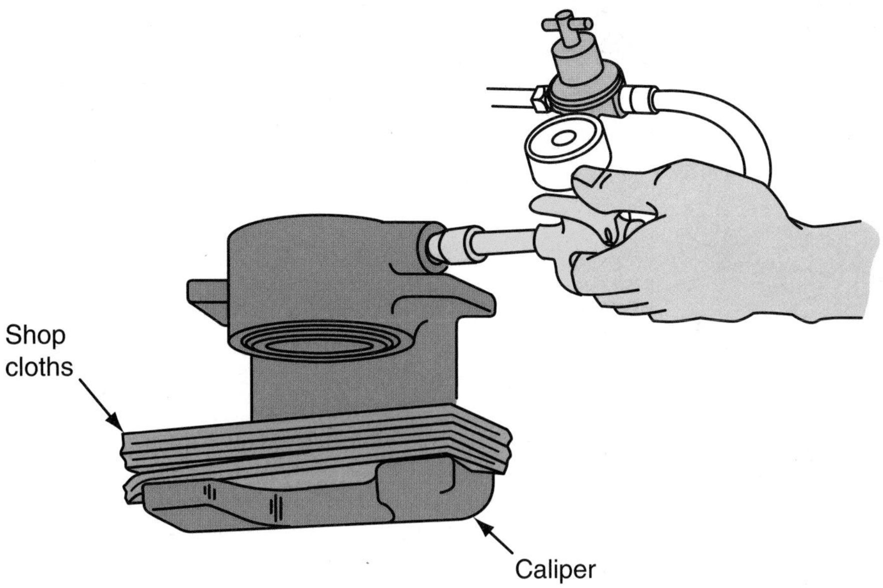

Figure 50-22 Using air to remove a piston.

WARNING!

Be careful to apply air pressure very gradually. Be sure there are enough cloths to catch the piston when it comes out of the bore. Never place your fingers in front of the piston for any reason when applying compressed air. Personal injury could occur if the piston is popped out of the bore.

If a piston is frozen, release air pressure and tap the piston into its bore with a soft-headed hammer or mallet. Reapply air pressure. Frozen phenolic (plastic) pistons can be broken into pieces with a chisel and hammer. Be careful not to damage the cylinder bore while doing this. Internal expanding pliers are sometimes used to remove pistons from caliper bores.

Inspect phenolic pistons for cracks, chips, or gouges. Replace the piston if any of these conditions are evident. If the plated surface of a steel piston is worn, pitted, scored, or corroded, it also should be replaced.

Dust boots vary in design depending on the type of piston and seal, but they all fit into one groove in the piston and another groove in the cylinder. One type comes out with the piston and peels off. Another type stays in place and the piston comes out through the boot, and then is removed from the cylinder **(Figure 50–23)**. In either case, peel the boot from its groove. In some cases it might be necessary to pry it out, but be careful not to scratch the cylinder bore while doing so. The old boot can be discarded since it must be replaced along with the seal.

Figure 50-23 The dust boot is peeled away from the piston and caliper.

Remove the piston's and the cylinder's seal by prying them out with a wooden or plastic tool **(Figure 50–24)**. Do not use a screwdriver or other metal tool. Any of these could nick the metal in the caliper bore and cause a leak. Inspect the bore for pitting or scoring. A bore that shows light scratches or corrosion can usually be cleaned with crocus cloth. However, a bore that has deep scratches or scoring normally indicates that the caliper should be replaced. In some cases, the cylinder can be honed. Check the service

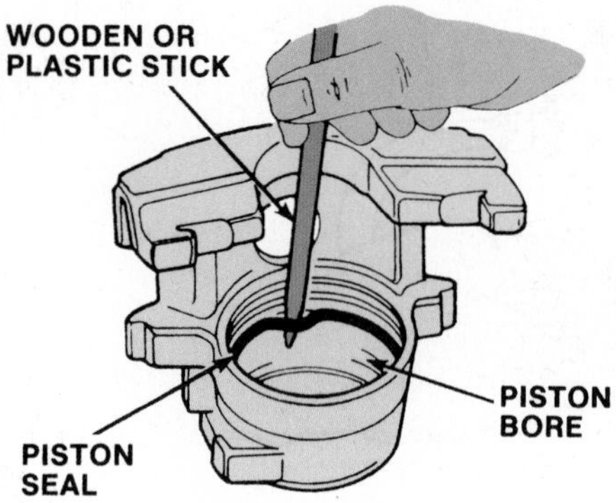

Figure 50-24 Removing a piston seal with a wooden or plastic stick.

manual before doing this. If there is no mention of honing the bore, the manufacturer probably does not recommend it. Black stains on the bore walls are caused by piston seals. They do no harm.

When using a hone, be sure to install the hone baffle before honing the bore. The baffle protects the hone stones from damage. Use extreme care in cleaning the caliper after honing. Remove all dust and grit by flushing the caliper with alcohol. Wipe it dry with a clean lint-free cloth and then clean the caliper a second time in the same manner.

Loaded Calipers

Rather than overhaul a caliper, many shops install loaded calipers **(Figure 50–25)**. Loaded calipers are complete units with friction pads and mounting hardware included. Besides the convenience and the savings of installation time, preassembled calipers also reduce the odds of errors during caliper overhaul.

Mistakes that are frequently made when replacing calipers include forgetting to bend the pad locating tabs that prevent pad vibration and noise, leaving off antirattle clips and pad insulators, and reusing corroded caliper mounting hardware that can cause a floating caliper to bind up and wear the pads unevenly.

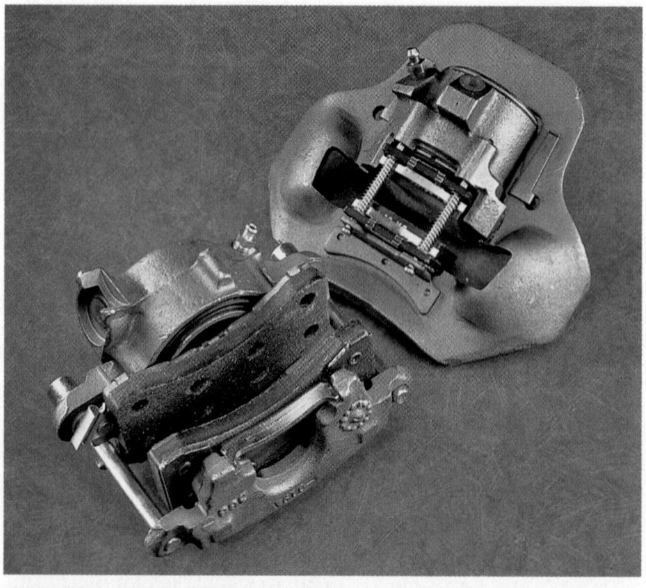

Figure 50-25 A pair of loaded brake calipers. *Courtesy of Honeywell International Inc.*

SHOP TALK

Avoid mismatching friction materials from side to side. When one caliper is bad, both calipers should be replaced using the same friction material.

Caliper Reassembly

Before assembling the caliper, clean the phenolic piston (if so equipped) and all metal parts to be reused in clean denatured alcohol or brake fluid. Then, clean out and dry the grooves and passageways with compressed air. Make sure that the caliper bore and component parts are thoroughly clean.

To replace a typical piston seal, dust boot, and piston, first lubricate the new piston seal with clean brake fluid or assembly lubricant (usually supplied with the caliper rebuild kit). Make sure the seal is not distorted. Insert it into the groove in the cylinder bore so it does not become twisted or rolled. Install a new dust boot by setting the flange squarely in the outer groove of the caliper bore. Next, coat the piston with brake fluid or assembly lubricant and install it in the cylinder bore. Be sure to use a wood block or other flat stock when installing the piston back into the piston bore. Never apply a C-clamp directly to a phenolic piston, and be sure the pistons are not cocked. Spread the dust boot over the piston as it is installed. Seat the dust boot in the piston groove.

With some types of boot/piston arrangements, the procedure for installation is slightly different from that already described. That is, the new dust boot is pulled over the end of the piston **(Figure 50–26)**. Lubricate

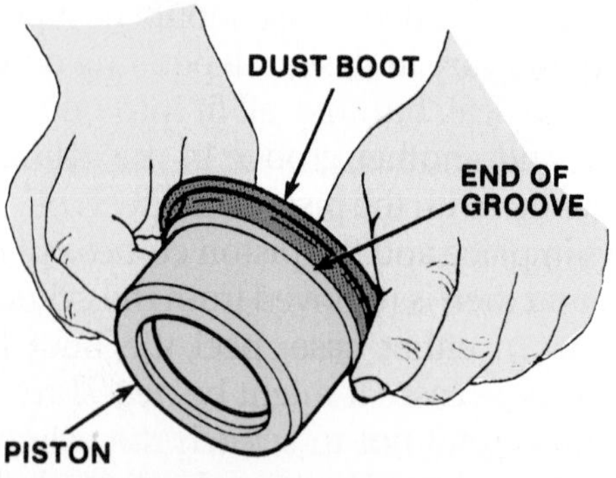

Figure 50-26 Some installation procedures require the dust boot to be pulled over the end of the piston.

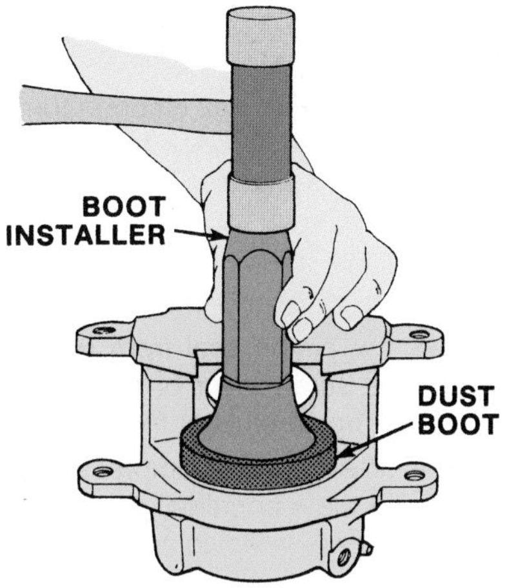

Figure 50-27 Seating a dust boot with a boot installer.

the piston with brake fluid before installing it in the caliper. Then by hand, slip the piston carefully into the cylinder bore, pushing it straight, so the piston seal is not damaged during installation. Use an installation tool or wooden block to seat the new dust boot (**Figure 50–27**).

Another point to keep in mind is that some caliper designs have a slot cut in the face of the pistons that must align with an **antisqueal shim**. Make sure that the piston and shim align. It might be necessary to turn the piston to achieve proper alignment. To complete the caliper assembly job, install the bleeder screw.

CAUTION!

On fixed calipers, bridge bolts are used to hold the two caliper halves together. These are high-tensile bolts ordered only by specific part number. They require accurate torque tightness to prevent leakage. Do not attempt to use standard bolts in place of bridge bolts.

Brake Pad Installation

It is a good practice to replace disc brake hardware when replacing disc brake pads. Replacement of the hardware ensures proper caliper movement and brake pad retention. It also aids in preventing brake noise and uneven brake pad wear.

One of the major causes of premature brake wear is rust. It causes improper slider and piston operation that leads to uneven pad wear. Tests have shown that when only the pads are replaced, the new pads can wear out in half the mileage they should if the calipers are corroded. Therefore, if the calipers are corroded, replace them.

Fixed Caliper Brake Pads The designs of fixed caliper disc brakes vary slightly. Generally, to replace the pads, insert new pads and plates in the caliper with the metal plates against the end of the pistons. Be sure that the plates are properly seated in the caliper. Spread the pads apart and slide the caliper into position on the rotor. With some pads, mounting bolts are used to hold them in place. These bolts are usually tightened 80 to 90 foot-pounds (108 to 122 N-m). On some fixed disc brakes, the pads are held in place by retaining clips and/or retaining pins. Reinstall the antirattle spring/clips and other hardware (if so equipped).

Sliding Caliper Brake Pads Push the piston carefully back into the bore until it bottoms. Lightly lubricate the sliding surfaces of the caliper and the caliper anchor. Slide a new outer pad into the recess of the caliper. No free play between the brake pad flanges and caliper fingers should exist. If free play is found, remove the pad from the caliper and bend the flanges to eliminate all vertical free play. Install the pad.

Place the inner pad into position on the caliper anchor with the pad's flange on the machined sliding area. Fit the caliper over the rotor. Align the caliper to the anchor and slide it into position. Be careful not to pull the dust boot from its groove when the piston and boot slide over the inboard pad. Install the antirattle springs (if so equipped) on top of the retainer plate and tighten the retaining screws to specification.

On some calipers, especially those used as rear brakes, there is a notch or groove in the piston and a tab on the rear of the inner pad. During installation of the pad, the tab must fit into the groove in the piston (**Figure 50–28**).

Floating Caliper Brake Pads For floating or pin caliper disc brakes, compress the flanges of the outer

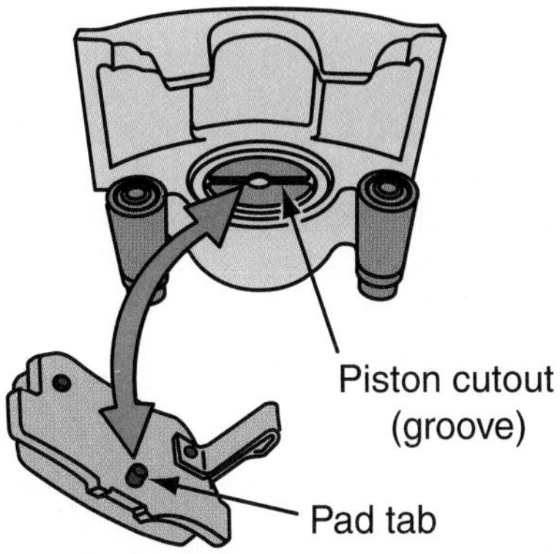

Figure 50-28 On some rear brake assemblies with rotating piston parking brake, there is a tab on the back of the pad that must line up with the groove in the piston.

bushing in the caliper fingers and work them into position in the hole from the outer side of the caliper. Compress the flanges of the inner guide pin bushings and install them.

Slide the new pad and lining assemblies into position in the adapter and caliper. Be sure that the metal portion of the pad is fully recessed in the caliper and adapter and that the proper pad is on the outer side of the caliper.

Hold the outer pad and carefully slide the caliper into position on the anchor and over the disc. Align the guide pin holes of the anchor with those of the inner and outer pads. Lightly lubricate and install the guide pins through the bushings, caliper, anchor, and inner and outer pads into the outer bushings in the caliper and antirattle spring.

When installing any type of caliper, follow these guidelines:

- Make sure the correct caliper is installed on the correct anchor plate. Also make sure that the bolts are tightened to specifications.
- Lubricate the rubber insulators (if so equipped) with silicone dielectric compound.
- After the caliper assembly is in its mounting brackets, connect the brake hose to the caliper. If copper washers or gaskets are used, be sure to use new ones—the old ones might have taken a set and might not form a tight seal if reused.
- Fill the master cylinder reservoirs and bleed the hydraulic system.
- Check for fluid leaks under maximum pedal pressure.
- Lower the vehicle and road test it.

REAR DISC BRAKE CALIPERS

Rear disc brakes calipers with some type of parking brake mechanism have different inspection and overhaul procedures than front brake calipers.

PROCEDURE

To overhaul a typical rear wheel caliper when there is no auxiliary drum parking brake follow these steps:

STEP 1 Unbolt the caliper.

STEP 2 Disconnect the parking brake cable from the lever on the caliper.

STEP 3 Disconnect the brake hose from the caliper, remove the caliper mounting bolts, and lift the caliper off its support.

STEP 4 Remove the brake pads and all pad shims and retainers.

STEP 5 Remove the piston from its bore; turn the locknut counterclockwise to back the piston out of the bore. When the piston is free, remove the piston boot.

STEP 6 Carefully inspect the piston for wear. Replace the piston if it is worn or damaged in any way.

STEP 7 Remove the piston seal from the caliper using the tip of a screwdriver or a wooden or plastic scraper, being careful not to scratch the bore.

STEP 8 Install a brake spring compressor between the caliper and the rear caliper guide. Turn the shaft of the compressor tool to compress the adjusting spring and use snapring pliers to remove the circlip holding the spring in place.

STEP 9 After removing the circlip, remove the spring cover, adjusting spring, spacer, bearing, adjusting bolt, and cup.

STEP 10 Next remove the sleeve piston, O-ring, and the rod from the cam.

STEP 11 Remove the return spring, the parking lever and cam assembly, and the cam boot from the caliper body. Do not loosen the nut on the parking lever and cam assembly with the cam installed in the caliper. If the lever and shaft must be separated, secure the lever in a vise before loosening the parking nut.

STEP 12 Begin reassembly by packing all cavities of the needle bearing with the specified lubricant. Coat the new cam boot with assembly lubricant and install it in the caliper. Apply lubricant to the area on the pin that contacts the cam. Install the cam and lever assembly into the caliper body. Then install the return spring. If the cam and lever were separated, reassemble them before installing the cam in the caliper body.

STEP 13 Install the rod in the cam, followed by a new O-ring on the sleeve piston. Then install the sleeve piston so the hole in the bottom of the piston is aligned with the rod in the cam and the two pins on the piston are aligned with the holes in the caliper.

STEP 14 Install a new cup with its groove facing the bearing side of the adjusting bolt. Fit the bearing, spacer, adjusting spring, and spring cover on the adjusting bolt then install it in the caliper bore.

STEP 15 Install the brake spring compressor and compress the spring until the tool bottoms out.

STEP 16 Make sure that the flared end of the spring cover is below the circlip groove. Then install the circlip. Make sure the circlip is properly seated in the groove.

STEP 17 Coat the new piston seal and piston boot with silicone grease and install them in the caliper.

STEP 18 Coat the outside of the piston with brake fluid and install it on the adjusting bolt while rotating it clockwise with the locknut wrench.

STEP 19 Install the new brake pads, pad shims, retainers, and springs onto the caliper bracket.

STEP 20 Reinstall the caliper and the splash shield and tighten the caliper bolts to torque specifications.

STEP 21 Reconnect the brake hose to the caliper with new sealing washers and tighten the banjo bolt to specifications.

STEP 22 Then reconnect the parking brake cable to the arm on the caliper and reinstall the caliper shield.

STEP 23 Top off the master cylinder reservoir and bleed the brake system. Adjust the parking brake as needed. Before making adjustments, be sure the parking brake arm on the caliper touches the pin.

ROTOR INSPECTION

The rotors should be inspected whenever brake pads are replaced and when the wheels are removed for other services. They should be carefully checked to determine if they can be reused or machined or if they should be replaced. When inspecting the rotor, make sure you look at the sensor wheel for the wheel speed sensor **(Figure 50–29)**.

If a good look at the surface is impossible because of dirt, clean the surfaces with a shop cloth dampened in brake cleaning solvent or alcohol. If the surface is rusted, remove it with medium-grit sandpaper or emery cloth and then clean it with brake cleaner or alcohol.

Most brake rotors have a discard thickness dimension cast into them. If you cannot find this dimension on the rotor or if it is hard to read, check a service manual for thickness specifications. Rotor discard thickness dimensions are given in two or three decimal points (hundredths or thousandths of an inch or hundredths of a millimeter), such as 1.25 inches, 1.375 inches, 0.750 inch, or 24.75 mm. If you resurface the rotor, it must be 0.015- to 0.030-inch (0.38 to 0.76 mm) thicker than the discard dimension after machining to allow for wear. If a rotor is already below the minimal thickness spec, replace it. It is always wise to replace both rotors on the same axle.

New rotors come with a protective coating on the friction surfaces. To remove this coating, use carburetor cleaner, brake cleaner, or the solvent recommended by the manufacturer.

Figure 50-29 Check the wheel speed sensor's rotor when servicing a rotor.

SHOP TALK

Cross-drilled or slotted brake rotors may not be able to be machined. Therefore, if the rotor is scored or otherwise damaged, it should be replaced.

Thickness and Parallelism

To measure rotor thickness, place a brake disc micrometer about 1 inch in from the outer edge of the rotor and measure the thickness **(Figure 50–30)**. Compare the measurement to specifications. Repeat the measurement at about eight points equidistant (45 degrees) around the surface of the rotor and compare each measurement to specifications. Take all measurements at the same distance from the edge so that rotor taper does not affect the measurements. If the rotor is thinner than the minimum thickness at any point or if thickness variations exceed limits, it must be replaced. Also check the service manual for an allowable thickness variation. Many manufacturers hold tolerances on thickness variations as close as 0.0005 inch (0.013 mm).

Rotor **parallelism** refers to thickness variations in the rotor from one measurement point to another around the rotor surface. If the rotor is out of parallel, it can cause excessive pedal travel; front end

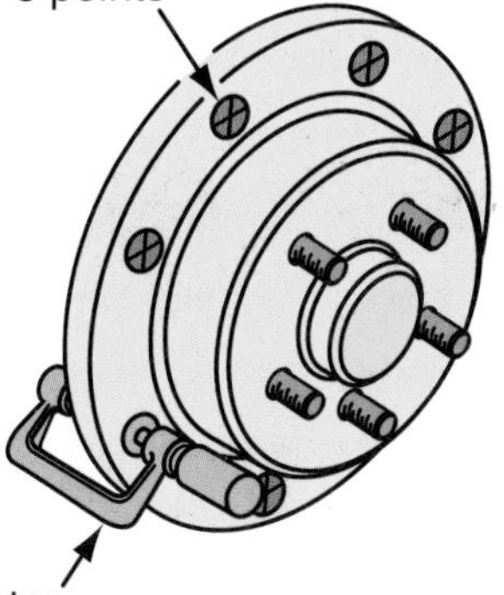

Figure 50-30 To check a rotor's thickness and parallelism, measure the rotor at eight different spots.

vibration; pedal pulsation; chatter; and, on occasion, grabbing of the brakes. The rotor then must be resurfaced or replaced.

Lateral Runout

Excessive lateral runout is the wobbling of a rotor from side to side when it rotates. This wobble knocks the pads farther back than normal, causing the pedal to pulse and vibrate during braking. Chatter can also result. Lateral runout also causes excessive pedal travel because the pistons have farther to travel to reach the rotor. If runout exceeds specifications, the rotor must be turned or replaced.

For the best braking performance, lateral runout should be less than 0.003 inch (0.08 mm) for most vehicles. Some manufacturers, however, specify runout limits as small as 0.002 inch (0.05 mm) or as great as 0.008 inch (0.20 mm).

Runout measurements are taken only on the outboard surface of the rotor, using a dial indicator and suitable mounting adapters **(Figure 50–31)**. If the rotor is mounted on adjustable wheel bearings, readjust the bearings to remove bearing end play. Do not overtighten the bearings. On rotors bolted solidly to the axles of FWD vehicles, bearing end play is not a factor in rotor runout measurement. If there is excessive bearing end play, the bearing assembly must be replaced. Bearing end play is best checked with a dial indicator.

Clamp the dial indicator support to the steering knuckle or other suspension part that will hold it securely as you turn the rotor. Position the dial indicator so that its tip contacts the rotor at 90 degrees. Place the indicator tip on the friction surface about 1 inch in from the outer edge of the rotor. Do not place the dial indicator on a dirty, rusted, grooved, or scored area. Rotate the rotor until the lowest reading appears on the dial indicator; then set the indicator to zero. Turn the rotor through one complete revolution and compare the lowest to the highest reading. This is the maximum runout of the rotor.

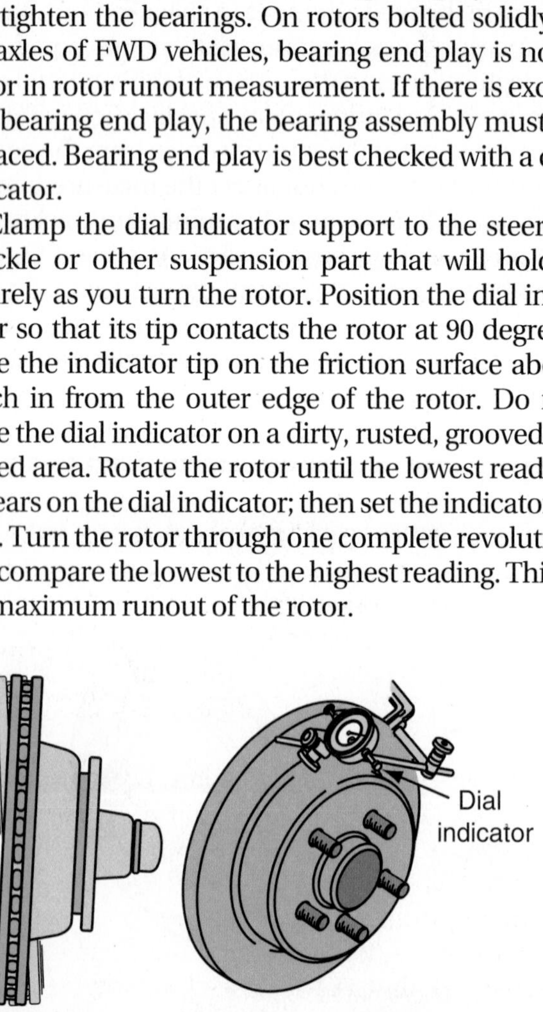

Figure 50-31 Checking the lateral runout of a brake rotor.

Additional Checks

The following are some of the typical rotor conditions that warrant disc replacement or machining.

Grooves and Scoring Inspect both rotor surfaces for scoring and grooving. Scoring or small grooves up to 0.010 inch (0.25 mm) deep are usually acceptable for proper braking performance. Scoring can be caused by linings that are worn through to the rivets or backing plate or by friction material that is harsh or unkind to the mating surface. Rust, road dirt, and other contamination could also cause rotor scoring. Any rotor having score marks more than 0.15 inch should be refinished or replaced.

If the rotor is deeply grooved, it must be thick enough to allow the grooves to be completely removed without machining the rotor to less than its minimum thickness. Measure rotor thickness at the bottom of the deepest groove. If rotor thickness at the bottom of the deepest grooves is at or near the discard dimension, replace the rotor.

SHOP TALK

Some rotors, particularly on GM vehicles, have a single deep groove manufactured into each surface. This groove helps to keep the pads from moving outward and also reduces operating noise.

Cracks Check the rotor thoroughly for cracks or broken edges. Replace any rotor that is cracked or chipped, but do not mistake small surface checks in the rotor for structural cracks. Surface checks will normally disappear when a rotor is resurfaced. Structural cracks, however, will be more visible when surrounded by a freshly turned rotor surface.

Bluing or Heat Checking Inspect the rotor surfaces for heat checking and hard spots **(Figure 50–32)**. Heat checking appears as many small interlaced cracks on the surface. Heat checking lowers the heat dissipation ability and friction coefficient of the rotor surface. Heat checking does not disappear with resurfacing. Therefore, a rotor with heat checks should be replaced.

Hard spots appear as round, shiny, bluish areas on the friction surface. Hard spots on the surface of a rotor usually results from a change in the metallurgy caused by brake heat. Pulling, rapid wear, hard pedal, and noise occur. These spots can be removed by machining. However, only the raised surfaces are removed, and they could reappear when heat is again encountered. The rotor should be replaced.

Figure 50-32 Some of the typical conditions that should be looked for on a brake rotor.

Rust If the vehicle has not been driven for a period of time, the discs will rust in the area not covered by the lining and cause noise and chatter. This also can result in excessive wear and scoring of the discs and pads. Wear ridges on the discs can cause temporary improper pad contact if the ridges are not removed before the installation of new pads. Rusted rotors should be cleaned before any measurements are taken.

Inspect the fins of vented rotors for cracks and rust. Rust near the fins can cause the rotor to expand and lead to rotor thickness variations and excessive runout problems. Machining the rotor may remove runout and thickness variations, but rotor expansion due to rust may cause these problems to reappear soon. Rusted rotors should be replaced.

ROTOR SERVICE

If the thickness of the rotor is below or close to the minimal allowable thickness or is badly distorted, it must be replaced. If the thickness is greater than the minimum specifications, it can be trued and smoothened with a brake lathe. Rotors that have minor imperfections or are slightly unparallel can be turned true and smooth with a brake lathe.

Removing a Rotor

To remove a rotor, raise the vehicle and remove the wheel. Then remove the caliper from the rotor and suspend it with wire from the suspension of the vehicle. Before you remove a rotor, mark it "L" or "R" for left or right so that it gets reinstalled on the same side of the vehicle from which it was removed.

If the rotor is a two-piece floating rotor, remove it from the hub by pulling it off the hub studs. If you cannot pull the rotor off by hand, apply penetrating oil on the front and rear rotor-to-hub mating surfaces. Strike the rotor between the studs using a ball-peen hammer. If this does not free the rotor, attach a three-jaw puller to the rotor and pull it off.

Whenever you separate a floating rotor from the hub flange, clean any rust or dirt from the mating surfaces of the hub and rotor. Neglecting to clean rust and dirt from the rotor and hub mounting surfaces before installing the rotor will result in increased rotor lateral runout, leading to premature brake pulsation and other problems.

If the rotor and hub are a one-piece assembly, remove the outer wheel bearing and lift the rotor and hub off the spindle.

SHOP TALK

New rotors have the correct surface finish, which may be disturbed by them turning on a lathe. Clean any oil film off a new rotor with brake cleaning solvent or alcohol and let the rotor air dry before installing it on the vehicle.

Brake Lathes

A brake lathe cuts metal away to achieve the desired surface finish. There are basically two types of brake disc lathes used by the industry. The first one, a bench brake lathe, has the capability of resurfacing brake drums and brake discs after they have been removed from the vehicle. The lathe rotates the disc as cutting tools work their way across the braking surface of the disc. The second type is an on-vehicle brake lathe. This type of brake lathe is a time saver because the rotor does not need to be removed from the vehicle. Special fixtures are used to straddle the rotor so the cutting tools can precisely cut both sides of the rotor. An electric motor is used to rotate the disc and hub assembly during cutting.

Whenever you refinish a rotor, remove the least amount of metal possible to achieve the proper finish. This helps to ensure the longest service life from the rotor. Never turn the rotor on one side of the vehicle without turning the rotor on the other side. Left- and right-side rotors should be the same thickness, generally within 0.002 inch to 0.003 inch. Similarly, equal amounts of metal should be cut off both surfaces of a rotor.

CAUTION!

Do not attempt to use a brake lathe without proper training. You can be seriously injured if the lathe is not properly set up or is operated improperly.

Bench Lathes On a bench, off-vehicle lathe, the rotor is mounted on the lathe's arbor and turned at a controlled speed while a cutting bit passes across the rotor surface to remove a few thousandths of an inch of metal **(Figure 50–33)**. The lathe turns the rotor perpendicularly to the cutting bits so that the entire rotor surface is refinished. Most rotor cutting assemblies have two cutting bits. The rotor mounts between the bits and is pinched between them. As the cut is made, the same amount of surface material should be cut from both sides of the rotor.

On-Vehicle Brake Lathes The advantage of an on-vehicle lathe **(Figure 50–34)** is that the rotor does not need to be removed. On-vehicle lathes also are ideal for rotors with excessive runout problems.

To install the lathe, remove the wheel and then remove the caliper. If any end play is present in an adjustable tapered roller bearing, carefully tighten the adjusting nut by hand just enough to remove the end play before installing the lathe. After turning the rotor, readjust the bearing. To hold a two-piece floating rotor to its hub, reinstall the wheel nuts with flat washers or adapters against the rotor. Carefully follow the manufacturer's mounting instructions and

Figure 50-33 A typical bench lathe.

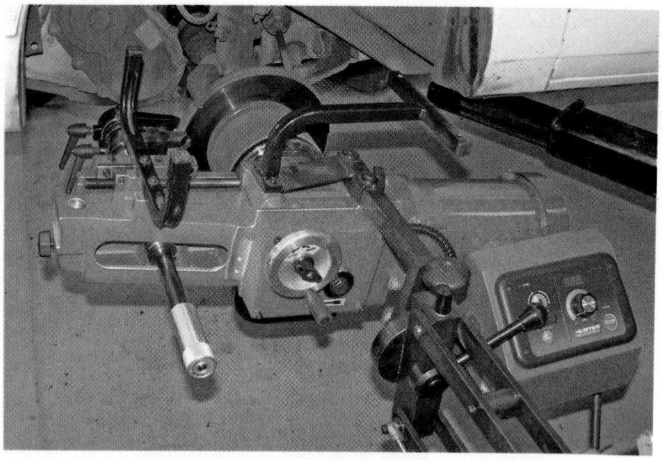

Figure 50-34 An on-the-vehicle brake lathe.

attach the lathe to the rotor. Some on-vehicle lathes mount on the caliper support; others are supported on a separate stand.

Some on-vehicle lathes use the vehicle's power to rotate the rotor; others use an electric motor. If the engine is used to turn the rotor, the lathe can be used only on drive wheels. However, this presents a problem because the differential gearing in the transaxle transmits the power to the opposite wheel, not to the rotor to be resurfaced. To prevent that opposite wheel from turning, that wheel can be lowered to the floor or the brake on the opposite side can be applied. This is done by removing and plugging the brake hose to the caliper on the rotor you are refinishing. Apply the brakes; this will lock the other wheel.

Self-powered on-vehicle lathes are more popular. These have the advantage of being able to machine on nondriving wheels, and rotor speed can be controlled more exactly.

An on-car lathe may be mounted on the brake caliper support or on its own stand and indexed to the hub and the wheel studs. Each lathe has its own operating instructions, which you must follow carefully.

Installing a Rotor

If the rotor is a two-piece floating rotor, make sure all mounting surfaces are clean. Apply a small amount of antiseize compound to the pilot diameter of the disc brake rotor before installing the rotor on the hub. Reinstall the caliper. If the rotor is a fixed, one-piece assembly with the hub that contains the wheel bearings, clean and repack the bearings and install the rotor.

Install the wheel and tire on the rotor and torque the wheel nuts to specifications, following the recommended tightening pattern. Failure to tighten in the correct pattern may result in increased lateral runout, brake roughness, or pulsation as well as damage to the wheels.

After lowering the vehicle to the ground, pump the brake pedal several times before moving the vehicle. This positions the brake linings against the rotor.

SHOP TALK

Special cutters are required to resurface composite rotors. Make sure you have the correct machine and cutting tools before attempting to true up a composite brake rotor.

CASE STUDY

An obviously embarrassed customer explains his problem to the brake shop technician. He just finished installing new disc pads on his light truck, but the disc brakes are dragging. The young man insists he performed the work correctly and that the rest of the braking system is in fine working order.

The technician inspects the front brakes and finds that the owner cleaned all components well and installed the pads correctly. The technician checks the action of the caliper. Its piston shows no signs of sticking, but the caliper appears to be binding. Removal and inspection of the caliper locates the problem. Although the owner had carefully cleaned the slides of the caliper, he had failed to apply any lubricant to the surface. This was causing the calipers to bind and the brakes to drag.

Omitting the simplest of tasks when performing repairs can often lead to failure and wasted time.

KEY TERMS

Antirattle spring
Antisqueal shim
Audible sensor
Composite rotor
Dust boot
Fixed caliper
Fixed rotor
Floating caliper
Floating rotor
Parallelism
Phenolic resin
Sliding caliper
Spindle anchor plate
Splash shield

SUMMARY

- Disc brakes offer four major advantages over drum brakes: resistance to heat fade, resistance to water fade, increased straight-line stopping ability, and automatic adjustment.
- The typical rotor is attached to and rotates with the wheel hub assembly. Heavier vehicles generally use ventilated rotors. Splash shields protect the rotors and pads from road moisture and dirt.
- The caliper assembly includes cylinder bores and pistons, dust boots, and piston hydraulic seals.
- Brake pads are placed in each side of the caliper and together straddle the rotor. Some brake pads have wear sensors.
- Fixed caliper disc brakes do not move when the brakes are applied. Floating caliper disc brakes slide back and forth on pins or bolts. Sliding calipers slide on surfaces that have been machined smooth for this purpose.
- In a rear disc brake system the inside of each rear wheel hub and rotor assembly is used as the parking brake drum.
- Rear disc parking brakes have a mechanism that forces the pads against the rotor mechanically.
- The general procedures involved in a complete caliper overhaul include tasks such as: caliper removal, brake pad removal, caliper disassembly, caliper assembly, brake pad installation, and caliper installation.
- The first step in proper caliper service is to remove the caliper assembly from the vehicle.
- Disc brake pads should be checked periodically or whenever the wheels are removed. They should be replaced if they fail to exceed minimum lining thickness as listed in the service manual.
- To disassemble the caliper, the piston and dust boot must first be removed. Compressed air is used to pop the piston out of the bore.
- Before assembling the caliper, all metal parts and the phenolic piston are cleaned in denatured alcohol or brake fluid. The grooves and passageways of the caliper are cleaned out and dried with compressed air.
- It is a good practice to replace disc brake hardware when replacing disc brake pads.
- Disc brake rotor conditions that must be corrected include lateral runout, lack of parallelism, scoring, blueing or heat checking, and rusty rotors.

REVIEW QUESTIONS

1. Name the three major assemblies that make up a disc brake.
2. Name the three types of calipers used on disc brakes.
3. What type of brake uses the inside of each rear wheel hub and rotor assembly as a parking brake drum?
4. *True or False?* Disc brakes are not as likely to fade during heavy braking as are drum brakes.
5. What are the two basic types of brake rotors used on today's vehicles?
6. *True or False?* All calipers have at least one piston that pushes the brake pad against the rotor?
7. What is the difference between floating and sliding calipers?
8. What is rotor parallelism and how is it checked?
9. Why is brake fluid removed from the master cylinder prior to working on disc brakes?
10. Describe the procedure for using compressed air to remove a piston from a brake caliper.
11. List three conditions that dictate that a rotor should be refinished.
12. Give two main advantages of an on-vehicle lathe versus a bench lathe.
13. What should be done to remove rust, corrosion, pitting, and scratches from the piston bore?
14. Which term refers to variations in thickness of the rotor?
    a. torque
    b. lateral runout
    c. parallelism
    d. pedal pulsation
15. Which of the following is not likely to cause a pulsating brake pedal?
    a. loose wheel bearings
    b. worn brake pad linings
    c. excessive lateral runout
    d. nonparallel rotors

ASE-STYLE REVIEW QUESTIONS

1. Technician A says that a hard brake pedal on a vehicle with disc brakes can be caused by air in the hydraulic system. Technician B says that a pulsating pedal on a vehicle with disc brakes can be caused by a restricted brake line. Who is correct?
   a. Technician A c. Both A and B
   b. Technician B d. Neither A nor B
2. When replacing brake pads on a vehicle: Technician A works on one wheel before beginning work on another. Technician B uses a minimum of 50 psi of air pressure to force the piston from the caliper housing. Who is correct?
   a. Technician A c. Both A and B
   b. Technician B d. Neither A nor B
3. When examining disc brakes: Technician A visually inspects the rotor and says that the rotor can be reused if it is not damaged or scored. Technician B says that it is normal for the inboard pad to be slightly more worn than the outside pad. Who is correct?
   a. Technician A c. Both A and B
   b. Technician B d. Neither A nor B
4. Technician A cleans brake components in denatured alcohol. Technician B cleans brake components in clean brake fluid. Who is correct?
   a. Technician A c. Both A and B
   b. Technician B d. Neither A nor B
5. While discussing how to remove the piston from a brake caliper: Technician A says that the dust boot should be removed, then a large dull screwdriver should be inserted into the piston groove to pry the piston out. Technician B says that air pressure should be injected into the bleeder screw's bore to force the piston out of the caliper. Who is correct?
   a. Technician A c. Both A and B
   b. Technician B d. Neither A nor B
6. The functions of a splash shield are being discussed: Technician A says that disc brakes will function normally without a splash shield in place. Technician B says that a splash shield helps to direct cooling air over the rotor. Who is correct?
   a. Technician A c. Both A and B
   b. Technician B d. Neither A nor B
7. Technician A says that the piston seal retracts the caliper piston when hydraulic pressure is released. Technician B says that a return spring is used to retract a caliper piston. Who is correct?
   a. Technician A c. Both A and B
   b. Technician B d. Neither A nor B

8. Technician A says that fixed calipers use a piston on each side of the rotor to apply the brakes. Technician B says that sliding calipers typically use only one piston or pistons on one side of the rotor. Who is correct?
   a. Technician A
   b. Technician B
   c. Both A and B
   d. Neither A nor B
9. While installing new brake pads: Technician A coats the inside diameter of the bushings with silicone grease before installing the mounting bolts and sleeves of floating calipers. Technician B lubricates the caliper ways on the caliper support and the mating parts of the sliding caliper housings with the recommended lubricant. Who is correct?
   a. Technician A
   b. Technician B
   c. Both A and B
   d. Neither A nor B
10. Loaded calipers are being discussed: Technician A says that loaded calipers are replacement calipers that come with pads and hardware already installed. Technician B says that loaded calipers should always be installed in axle sets. Who is correct?
    a. Technician A
    b. Technician B
    c. Both A and B
    d. Neither A nor B

CHAPTER

51

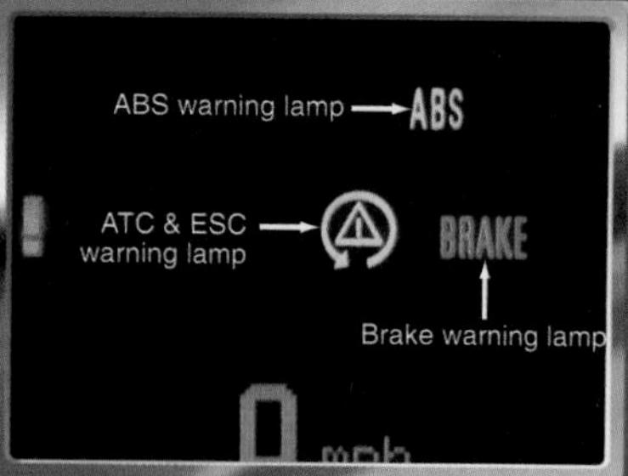

ANTILOCK BRAKE, TRACTION CONTROL, AND STABILITY CONTROL SYSTEMS

OBJECTIVES

■ Explain how antilock brake systems work to bring a vehicle to a controlled stop. ■ Describe the differences between an integrated and a nonintegrated antilock brake system. ■ Briefly describe the major components of a two-wheel antilock brake system. ■ Briefly describe the major components of a four-wheel antilock brake system. ■ Describe the operation of the major components of an antilock brake system. ■ Describe the operation of the major components of automatic traction and stability control systems. ■ Explain the best procedure for finding ABS faults. ■ List the precautions that should be followed whenever working on an antilock brake system.

Antilock brake systems (ABS) and traction and stability control systems are rapidly gaining popularity. ABS is now standard equipment on most vehicles and is available on others. These systems add yet another group of electronically controlled systems to the increasingly complex modern vehicle.

ANTILOCK BRAKES

Modern antilock brake systems (**Figure 51–1**) can be thought of as electronic/hydraulic pumping of the brakes for straight-line stopping under panic conditions. Good drivers have always pumped the brake pedal during panic stops to avoid wheel lockup and the loss of steering control. Antilock brake systems simply get the pumping job done much faster and in a much more precise manner than the fastest human foot. Keep in mind that a tire on the verge of slipping produces more friction with respect to the road than one that is locked and skidding. Once a tire loses its grip, friction is reduced and the vehicle takes longer to stop.

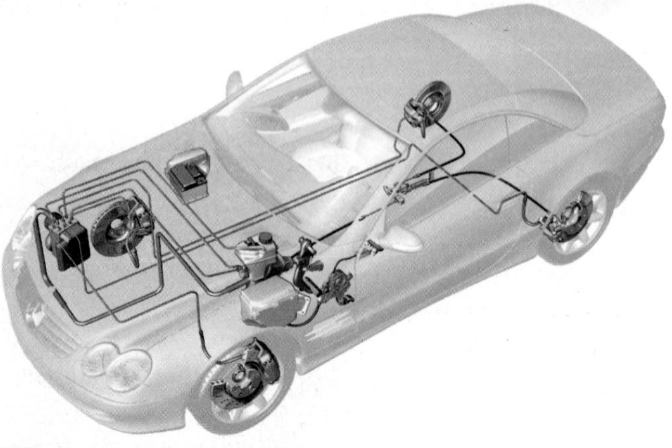

Figure 51–1 A common four-wheel antilock brake system. *Courtesy of Chrysler LLC*

Pressure Modulation

When the driver quickly and firmly applies the brakes and holds the pedal down, the brakes of a vehicle not equipped with ABS will almost immediately lock the wheels. The vehicle slides rather than rolls to a stop. During this time, the driver also has a very difficult time keeping the vehicle straight and the vehicle will skid out of control. The skidding and lack of control was caused by the locking of the wheels. If the driver was able to release the brake pedal just before the wheels locked up then reapply the brakes, the skidding could be avoided.

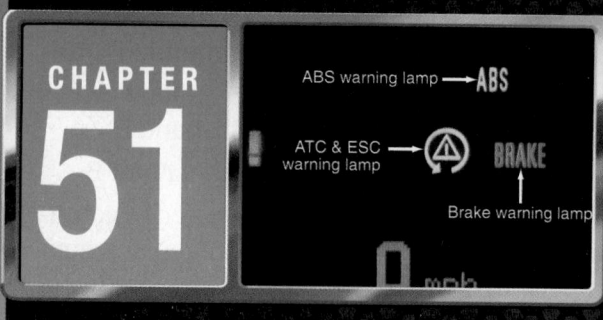

This release and apply of the brake pedal is exactly what an antilock system does. When the brake pedal is pumped or pulsed, pressure is quickly applied and released at the wheels. This is called **pressure modulation**. Pressure modulation works to prevent wheel locking. Antilock brake systems can modulate the pressure to the brakes as often as fifteen times per second. By modulating the pressure to the brakes, friction between the tires and the road is maintained and the vehicle is able to come to a controllable stop.

The only time reduced friction aids in braking is when a tire is on loose snow. A locked tire allows a small wedge of snow to build up ahead of it, which allows it to stop in a shorter distance than a rolling tire.

Steering is another important consideration. As long as a tire does not slip, it goes only in the direction in which it is turned. But once it skids, it has little or no directional stability. One of the big advantages of ABS, therefore, is the ability to keep control of the vehicle under all conditions.

Slip Rate

The maneuverability of the vehicle is reduced if the front wheels are locked, and the stability of the vehicle is reduced if the rear wheels are locked. A locked tire skids on pavement and has poor traction. This condition allows for 100% tire slip, whereas a tire rolling freely has a slip of nearly 0%. Slip is the difference between the actual speed of the vehicle and the speed of the tire's tread as it rotates on the pavement. Antilock brake systems control the slip rate **(Figure 51–2)** of the wheels to ensure maximum grip force, or traction, at the tires. It is the traction of the tires that actually stops the vehicle; therefore, ABS can improve braking and handling by controlling the brake fluid pressure at each wheel to attain the target slip rate at that wheel.

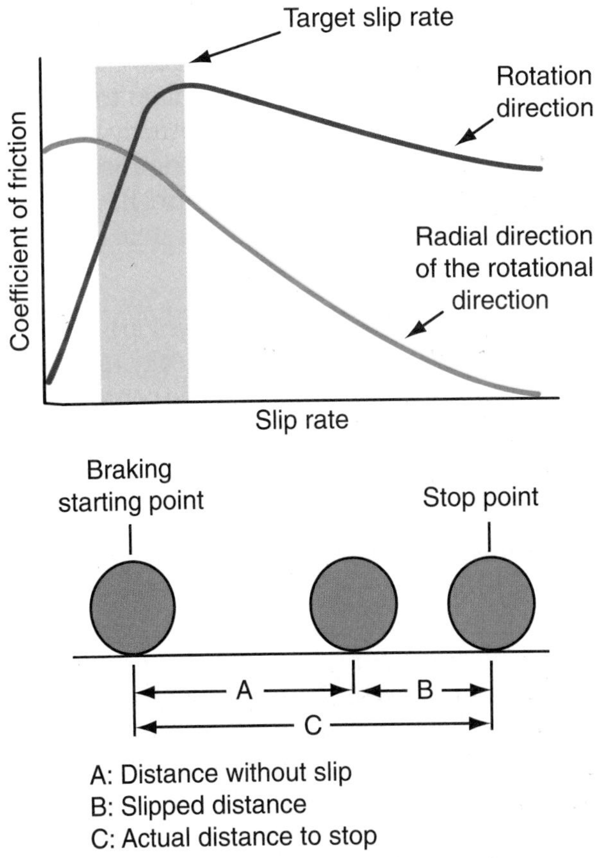

$$\text{SLIP RATE} = \frac{B}{C} = \frac{\text{Vehicle speed} - \text{Wheel speed}}{\text{Vehicle speed}}$$

Figure 51-2 Defining slip rate.

Although ABS prevents complete wheel lockup, it allows some wheel slip in order to achieve the best braking possible. During ABS operation, the target slip rate can be from 10% to 30%. A slip rate of 25% means the velocity of a wheel is 25% less than that of a free rolling wheel at the same vehicle speed. Many things are considered when determining the target slip rate for a particular vehicle. For some the range is very low—5% to 10%—while on others it is high—20% to 30%.

> **Customer Care**
>
> **Remind your customers that pumping the brake pedal while stopping will prevent the ABS from activating. They should always keep firm steady pressure on the brake pedal during braking.**

Pedal Feel

The brake pedal on a vehicle equipped with ABS has a different feel than that of a conventional braking system. When the ABS is activated, a small bump followed by rapid pedal pulsations will continue until the vehicle comes to a stop or the ABS turns off. These pulsations are the result of the modulation of pressure to the brakes and are felt more on some systems than on others. This is due to the use of damping valves in some modulation units. If pedal feel is of concern during diagnosis of a brake problem, compare the brake pedal feel with that of a similar vehicle with a normal operating antilock brake system. With ABS, the brake pedal effort and pedal feel during normal braking are similar to that of a conventional power brake system.

ABS COMPONENTS

Many different designs of antilock brake systems are found on today's vehicles. These designs vary in their basic layout, operation, and components. There are also variations based on the type of power-assist used and on whether the system is integral. The ABS components that may be found on a vehicle can be divided into two categories: hydraulic and electrical/electronic components. Keep in mind that no one system uses all of the parts discussed here. Normal or conventional brake

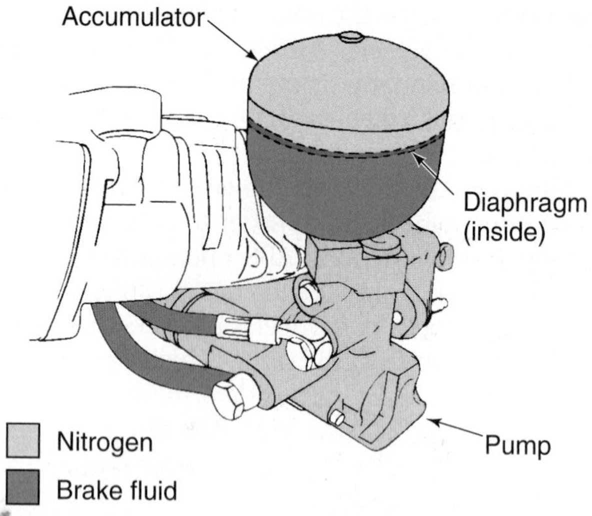

Figure 51-3 Pressure in an accumulator. *Courtesy of Federal-Mogul Corporation*

parts are part of the overall brake system but are not in the following discussion.

Hydraulic Components

Accumulator An accumulator is used to store hydraulic fluid to maintain high pressure in the brake system and to provide residual pressure for power-assisted braking. Normally, the accumulator is charged with nitrogen gas **(Figure 51–3)** and is an integral part of the modulator unit. This unit is typically found on vehicles with a hydraulically assisted brake system.

Antilock Hydraulic Control Valve Assembly This assembly controls the release and application of brake system pressure to the wheel brake assemblies. It may be of the integral type, meaning this unit is combined with the power-boost and master cylinder units into one assembly **(Figure 51–4)**. The nonintegral type is mounted externally from the master cylinder/power booster unit and is located between the master cylinder and wheel brake assemblies. Both types generally contain solenoid valves that control the releasing, the holding, and the applying of brake system pressure.

Booster Pump The booster pump is an assembly of an electric motor and pump. The booster pump is used to provide pressurized hydraulic fluid for the ABS. The pump's motor is controlled by the system's control unit. The booster pump is also called the electric pump and motor assembly.

Booster/Master Cylinder Assembly The booster/master cylinder assembly **(Figure 51–5)**, sometimes referred to as the hydraulic unit, contains the valves and pistons needed to modulate hydraulic pressure in the wheel circuits during ABS operation. Power brake-assist is provided by pressurized brake fluid supplied by a hydraulic pump.

Fluid Accumulators Different than a pressure accumulator, fluid accumulators temporarily store brake fluid removed from the wheel brake units during an ABS cycle. This fluid is then used by the pump to build pressure for the brake hydraulic system. There are normally two fluid accumulators in a hydraulic control unit, one each for the primary and secondary hydraulic circuits.

Hydraulic Control Unit This assembly contains the solenoid valves, fluid accumulators, pump, and an electric motor. This is actually a combination unit of many individual components found separately in some systems. The unit may have one pump and one motor or it will have one motor and two pumps: one pump for half of the hydraulic system and the other for the other half.

Main Valve This two-position valve is also controlled by the ABS control module and is open only in the ABS mode. When open, pressurized brake fluid from the booster circuit is directed into the master cylinder (front brake) circuits to prevent excessive pedal travel.

Modulator Unit (Figure 51–6) The modulator unit controls the flow of pressurized brake fluid to the individual wheel circuits. Normally the modulator is made up of solenoids that open and close valves, several valves that control the flow of fluid to the wheel brake units, and electrical relays that activate or deactivate the solenoids through the commands of the control module. This unit may also be called the hydraulic actuator, hydraulic power unit, or the electrohydraulic control valve.

Solenoid Valves The solenoid valves are located in the modulator unit and are electrically operated by signals from the control module. The control module switches the solenoids on or off to increase, decrease, or maintain the hydraulic pressure to the individual wheel units.

Valve Block Assembly The valve block assembly attaches to the side of the booster/master cylinder and contains the hydraulic wheel circuit solenoid valves. The control module controls the position of these solenoid valves. The valve block is serviceable separate from the booster/master cylinder but should not be disassembled. An electrical connector links the valve block to the ABS control module.

Wheel Circuit Valves Two solenoid valves are used to control each circuit or channel. One controls the inlet valve of the circuit, the other controls the outlet valve. When inlet and outlet valves of a circuit are used in combination, pressure can be increased, decreased,

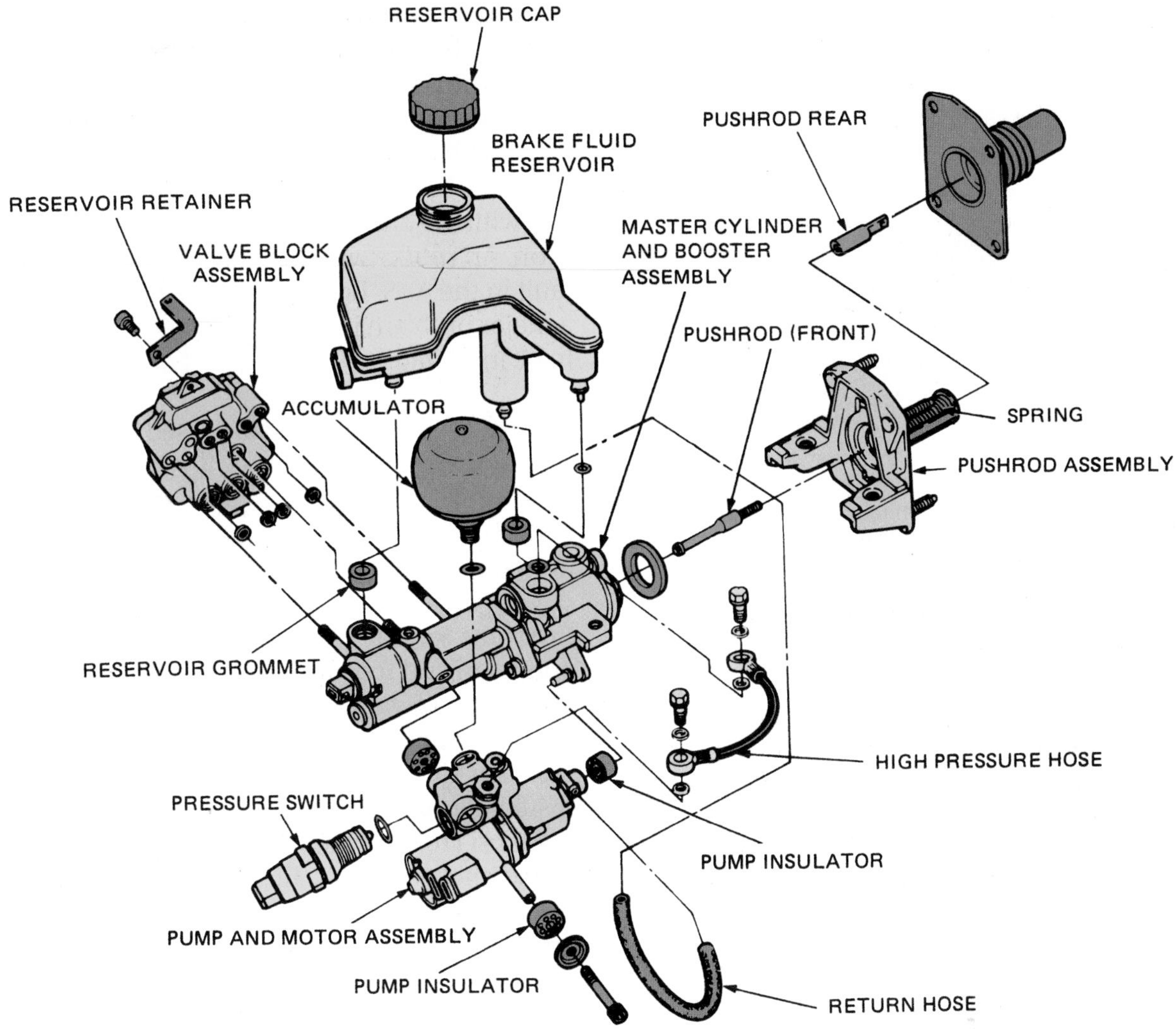

Figure 51–4 A hydraulic unit for a Teves antilock brake system. *Courtesy of Federal-Mogul Corporation*

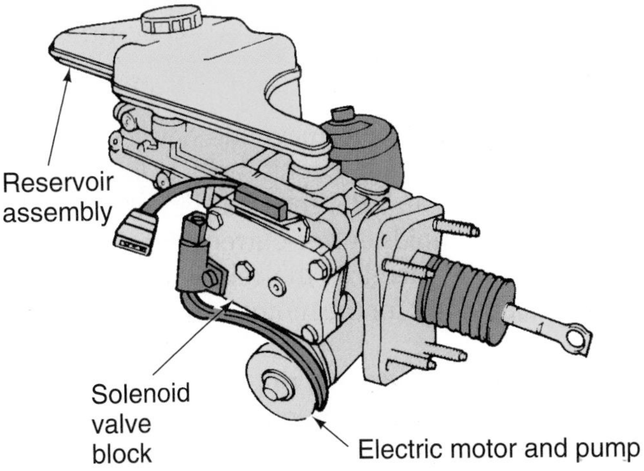

Figure 51–5 A master brake cylinder in this typical integrated system uses an electric pump for power boost. *Courtesy of Ford Motor Company*

or held steady in the circuit. The position of each valve is determined by the control module. Outlet valves are normally closed, and inlet valves are normally open. Valves are activated when the ABS control module switches 12 volts to the circuit solenoids. During normal driving, the circuits are not activated.

Figure 51–6 A rear wheel ABS modulator assembly.

Electrical/Electronic Components

ABS Control Module This small control computer is normally mounted inside the trunk on the wheel housing, mounted to the master cylinder **(Figure 51–7)**, or is part of the hydraulic control unit. It monitors

Figure 51-7 On the left is an old-style hydraulic modulator with its control unit. On the right is the smaller, new-generation modulator and control module. *Courtesy of Robert Bosch GmbH, www.bosch-presse.de*

system operation and controls antilock function when needed. The module relies on inputs from the wheel speed sensors and feedback from the hydraulic unit to determine if the antilock brake system is operating correctly and to determine when the antilock mode is required. The module has a self-diagnostic function including numerous trouble codes. This module may also be called the ECU (electronic control unit), EBCM (electronic brake control module), the antilock brake controller, or the ECM (electronic control module). The name used depends on the manufacturer and year of the vehicle.

Brake Pedal Sensor The antilock brake pedal sensor switch is normally closed. When the brake pedal travel exceeds the antilock brake pedal sensor switch setting during an antilock stop, the antilock brake control module senses that the antilock brake pedal sensor switch is open and grounds the pump motor relay coil. This energizes the relay and turns the pump motor on. When the pump motor is running, the hydraulic reservoir is filled with high-pressure brake fluid, and the brake pedal will be pushed up until the antilock brake pedal sensor switch closes. When the antilock brake pedal sensor switch closes, the pump motor is turned off and the brake pedal will drop some with each ABS control cycle until the antilock brake pedal sensor switch opens and the pump motor is turned on again. This minimizes pedal feedback during ABS cycling.

Data Link Connector (DLC). The DLC provides access and/or control of vehicle information, operating conditions, and diagnostic information.

Diagnostic Trouble Code (DTC). These trouble codes are numeric identifiers for fault conditions identified by the ABS's internal diagnostic system.

Indicator Lights Most ABS-equipped vehicles are fitted with two different brake warning lights. One of the warning lights is tied directly to the ABS, whereas the other lamp is part of the base brake system. All vehicles have a *red* warning light. This lamp lights when the brake fluid level is low, when there is a problem with the brake system, or when the parking brake is on. An *amber* warning lamp lights when there is a fault in the ABS. Both lamps will illuminate if there is a major problem in the base system, causing the ABS to be inhibited.

Lateral Acceleration Sensor Used on some vehicles with stability control, this switch monitors the sideward movement of the vehicle while it is turning a corner. This information is sent to the control module to ensure proper braking during turns.

Pressure Switch This switch controls pump motor operation and the low pressure warning light circuit. The pressure switch grounds the pump motor relay coil circuit, activating the pump when accumulator pressure drops below 2,030 psi (14,000 kPa). The switch cuts off the motor when the pressure reaches 2,610 psi (18,000 kPa). The pressure switch also contains switches to activate the dash-mounted warning light if accumulator pressure drops below 1,500 psi (10,343 kPa). This unit is typically found on vehicles with a hydraulically assisted brake system.

Pressure Differential Switch The pressure differential switch is located in the modulator unit. This switch sends a signal to the control module whenever there is an undesirable difference in hydraulic pressures within the brake system.

Relays Relays are electromagnetic devices used to control a high-current circuit with a low-current switching circuit. In ABS, relays are used to switch motors and solenoids. A low-current signal from the control module energizes the relays that complete the electrical circuit for the motor or solenoid.

Toothed Ring The toothed ring, also called a toner ring, can be located on an axle shaft, differential gear, or a wheel's hub. This ring is used in conjunction with the wheel speed sensor. The ring has a number of teeth around its circumference. The number of teeth varies by manufacturer and vehicle model. As the ring rotates and each tooth passes by the wheel speed sensor, an AC voltage signal is generated between the sensor and the tooth. As the tooth moves away from the sensor, the signal is broken until the next tooth comes close to the sensor. The end result is a pulsing

Figure 51–8 A wheel-speed sensor on a FWD drive axle. *Courtesy of Chrysler LLC*

signal that is sent to the control module. The control module translates the signal into wheel speed. The toothed ring may also be called the reluctor, tone ring, or gear pulser.

Wheel-Speed Sensor (Figure 51–8) The wheel-speed sensors are mounted near the different toothed rings. As the ring's teeth rotate past the sensor, an AC voltage is generated. As the teeth move away from the sensor, the signal is broken until the next tooth comes close to the sensor. The end result is a pulsing signal that is sent to the control module. The control module translates the signal into wheel speed. The sensor is normally a small coil of wire with a permanent magnet in its center.

Multiplexing

The electronic circuit and wiring of an ABS is tied into the vehicle's CAN network. This allows the ABS control unit to communicate with other control modules and share input devices with them. Every computer in the vehicle has access to all of the data in the CAN network, but individual computers use only the data that applies to their function. CAN communications is especially important when ABS is modified with additional features, such as traction and stability control.

Basic Operation

The control unit processes inputs and controls the operation of **isolation/dump valves** in the hydraulic modulator unit. The isolation/dump valves block off or isolate the master cylinder from certain brakes. As long as the brakes are applied and the vehicle is moving, the master cylinder remains isolated so additional fluid cannot be directed to those brakes. At the same time, the dump valve opens and allows a very small amount of fluid from the brake lines to enter an accumulator. This reduces the hydraulic pressure delivered to the brake and it is slightly released to allow the wheels to turn. If the wheels speed up too much, the dump valve reverses and the accumulator forces a small amount of fluid back into the brake. This constant dump/recharge is what causes the pulsation of the brake pedal during a panic or ABS stop. Most systems have a dedicated isolation/dump valve for each wheel.

TYPES OF ANTILOCK BRAKE SYSTEMS

The ABSs found on today's vehicles are manufactured by one of many different companies. Each manufacturer has a unique way to accomplish the same thing—vehicle control during braking. When working with ABS, it is important that you identify the exact system you are working with and follow the specific service procedures for that system. Often the system is initially identified by the manufacturer and then its model number; for example, a Teves Mark 20 system was manufactured by Teves and this model 20 is found on 1997 and later Chrysler vehicles. Keep in mind that there have been nearly 50 different ABSs used by the industry in recent years.

The exact manner in which hydraulic pressure is controlled depends on the ABS design. A great majority of the earlier ABSs were integrated or **integral** systems. They combine the master cylinder, hydraulic booster, and ABS hydraulic circuitry into a single hydraulic assembly.

Nearly all of today's systems are **nonintegral**. They use a conventional vacuum-assist booster and master cylinder. The ABS hydraulic control unit is a separate mechanism. In some nonintegrated systems, the master cylinder supplies brake fluid to the hydraulic unit. Although the hydraulic unit is a separate assembly, it still uses a high-pressure pump/motor, an accumulator, and fast-acting solenoid valves to control hydraulic pressure to the wheels.

Both integral and nonintegral systems operate in much the same way; therefore, an understanding of one system will lend itself to the understanding of the other systems.

General Motors' electromagnetic ABS is a different type of nonintegral system that uses a conventional

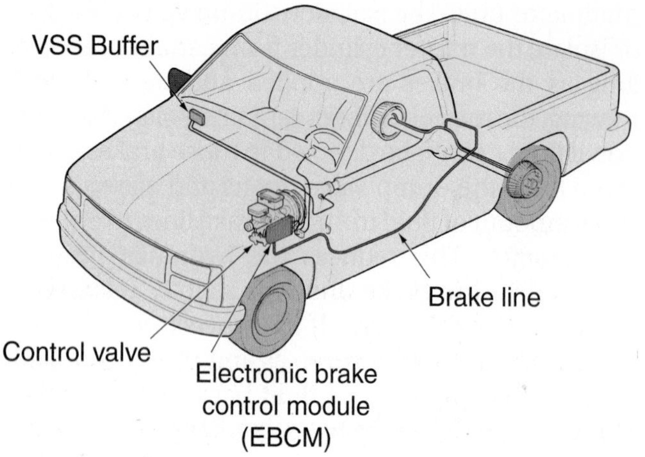

Figure 51-9 The main components of a rear wheel ABS.

vacuum power booster and master brake cylinder. But it does not use a high-pressure pump/motor, an accumulator, and fast-acting solenoid valves to control hydraulic pressure. Instead, it uses motors in a hydraulic modulator.

In addition to being classified as integral and nonintegral ABSs, systems can be broken down into the level of control they provide. ABSs can be one-, two-, three-, or four-channel, two- or four-wheel systems. A channel is merely a hydraulic circuit to the brakes.

Two-Wheel Systems

These basic systems offer antilock brake performance to the rear wheels only. They do not provide antilock performance to the steering wheels. Two-wheel systems are most often found on light trucks and some sport utility vehicles **(Figure 51–9)**.

These systems can be either one- or two-channel systems. In **one-channel systems**, the rear brakes on both sides of the vehicle are modulated at the same time to control skidding. These systems rely on the input from a centrally located speed sensor. The speed sensor is normally positioned on the ring gear in the differential unit **(Figure 51–10)**, transmission, or transfer case.

Although not commonly found, a **two-channel system** can be used to modulate the pressure to each of the rear wheels independently of each other. Modulation is controlled by the speed variances recorded by speed sensors located at each wheel.

Two-channel systems may be found on some diagonally split brake systems. These systems use two speed sensors to provide wheel speed data for the regulation of all four wheels. One sensor has input that controls the right front wheel; the other sensor performs identically for the left front wheel.

Brake hydraulic pressure to the opposite rear wheel is controlled simultaneously with its diagonally located front wheel. For example, the right rear wheel receives the same pumping instructions as the left front wheel. This system is an upgrade from the two-wheel system since it does provide steering control. However, it can have shortcomings under certain operating conditions.

Full (Four-Wheel) Systems

Some hydraulic systems that are split from front to rear use a **three-channel system** and are called four-wheel antilock brake systems. These systems have individual hydraulic circuits to each of the two front wheels, and a single circuit to the two rear wheels.

The most effective and most common ABS available is a **four-channel system**, in which sensors monitor each of the four wheels. With this continuous information, the ABS control module ensures that each wheel receives the exact braking force it needs to maintain both antilock and steering control.

ABS OPERATION

The exact operation of an antilock brake system depends on its design and manufacturer. It would take many pages to try to explain the operation of each, and as soon as you read the explanations there would be two or more new systems that would have to be explained. The exact operation of any system can be easily understood if you understand the basic operation of a few. The primary difference in operation between them all is based on the components used by the system. Therefore, the following systems were chosen as examples of how certain systems operate with the components they have.

Two-Wheel Systems (Nonintegral)

These systems are used to prevent rear wheel lockup on pickup trucks and SUVs, especially under light payload conditions. They consist of a standard power brake system, an electronic control unit (control module), and an isolation/dump valve assembly. The valve assembly is attached to the master cylinder at the rear brake line. Both rear wheel brake assemblies are controlled by the valve assembly under ABS conditions.

Under normal braking, pressure will pass through the valve assembly. The control module receives a signal from the brake switch when brakes are applied and begins to monitor the vehicle speed sensor (VSS) signal at speeds over 5 miles per hour (8 km/h). If the control module detects a deceleration rate from the VSS that would indicate probable rear wheel lockup, it activates the isolation valve, which stops the buildup of pressure to the rear wheels. If further deceleration occurs that would indicate lockup, the control module will rapidly

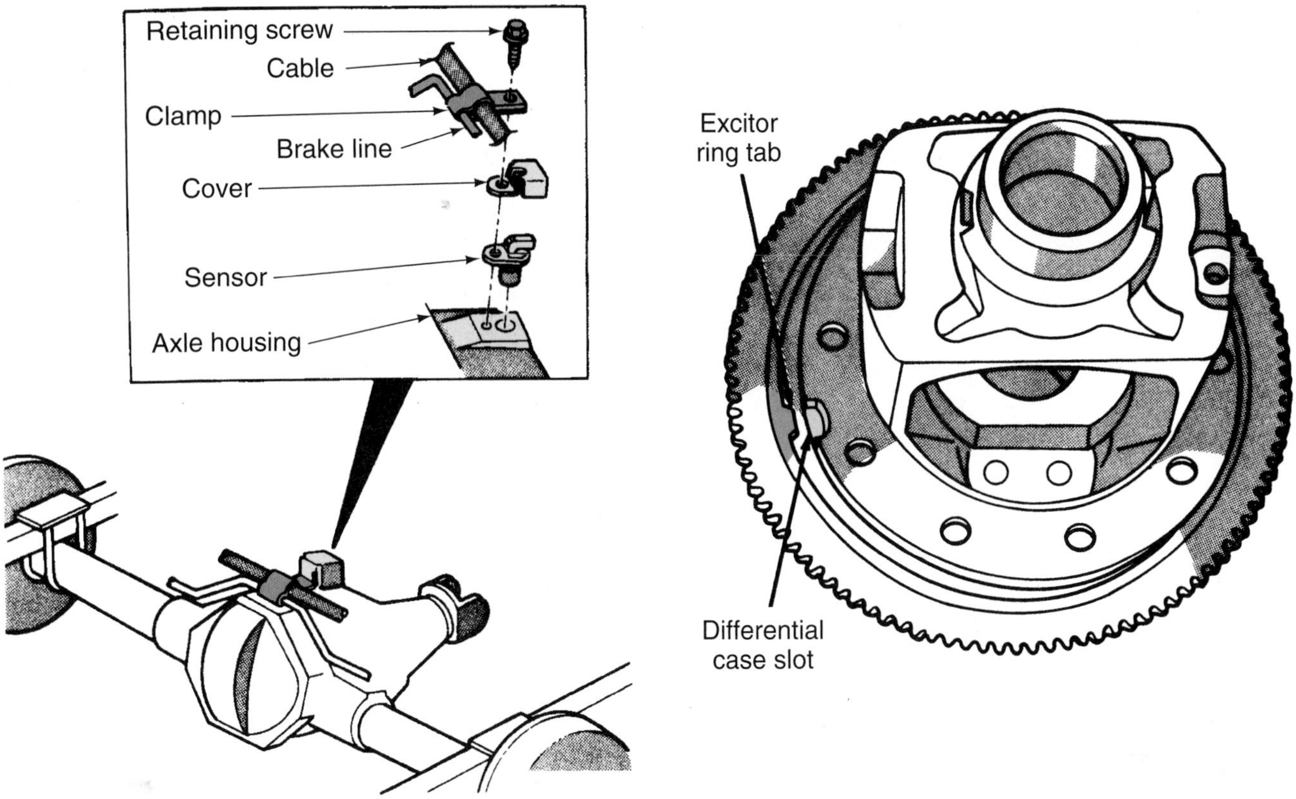

Figure 51–10 The speed sensor for both rear wheels is located on the differential unit. *Courtesy of Chrysler LLC*

pulse the dump valve to release brake pressure into the accumulator. The control module continues to pulse the dump valve until rear wheel deceleration matches the vehicle's deceleration rate or the desired slip rate. When wheel speed picks up, the control module will turn off the isolation valve, allowing the fluid in the accumulator to return to the master cylinder and normal braking control to resume.

The control unit has three distinct functions: it performs self-test diagnostics, it monitors the ABS action and system, and it controls the ABS solenoid valves.

When the ignition switch is turned on, the control module checks its ROM and RAM. If an error is detected, a DTC is set in memory. A DTC will also be set if the control module senses a problem during ABS operation.

The control module continuously monitors the speed of the differential ring gear through signals from the rear wheel speed sensor. The control module also receives signals from the brake light switch, brake warning lamp switch, reset switch, and the 4WD switch.

Preventing wheel lockup is the primary responsibility of the control module. It does this by controlling the operation of the isolation and dump solenoid valves. To check the effectiveness of the system, the dump valve can only be cycled a predetermined number of times during one stop before a DTC is set.

This system is disabled on four-wheel drive vehicles when in the four-wheel drive mode due to transfer case operation. Switching the transfer case into two-wheel drive mode will re-enable the ABS.

Four-Wheel Systems (Nonintegral)

The hydraulic circuit for this system is an independent four-channel type (**Figure 51–11**). The hydraulic control unit is a separate unit. In the hydraulic control unit there are two valves per wheel; therefore, a total of eight valves are used. Some systems have three channels, one for each of the front wheels and one for the rear axle. Obviously these systems have only three pairs of solenoids (**Figure 51–12**).

The system prevents wheel lockup during an emergency stop by modulating brake pressure. It allows the driver to maintain steering control and stop the vehicle in the shortest possible distance under most conditions. During ABS operation, the driver will sense a pulsation in the brake pedal and a clicking sound.

Operation The ABS control module calculates the slip rate of the wheels and controls the brake fluid pressure to certain wheel brakes to reach the target slip rate. If the control module senses that a wheel is about to lock, based on input sensor data, it pulses the normally open inlet solenoid valve closed for that circuit. This prevents any more fluid from entering

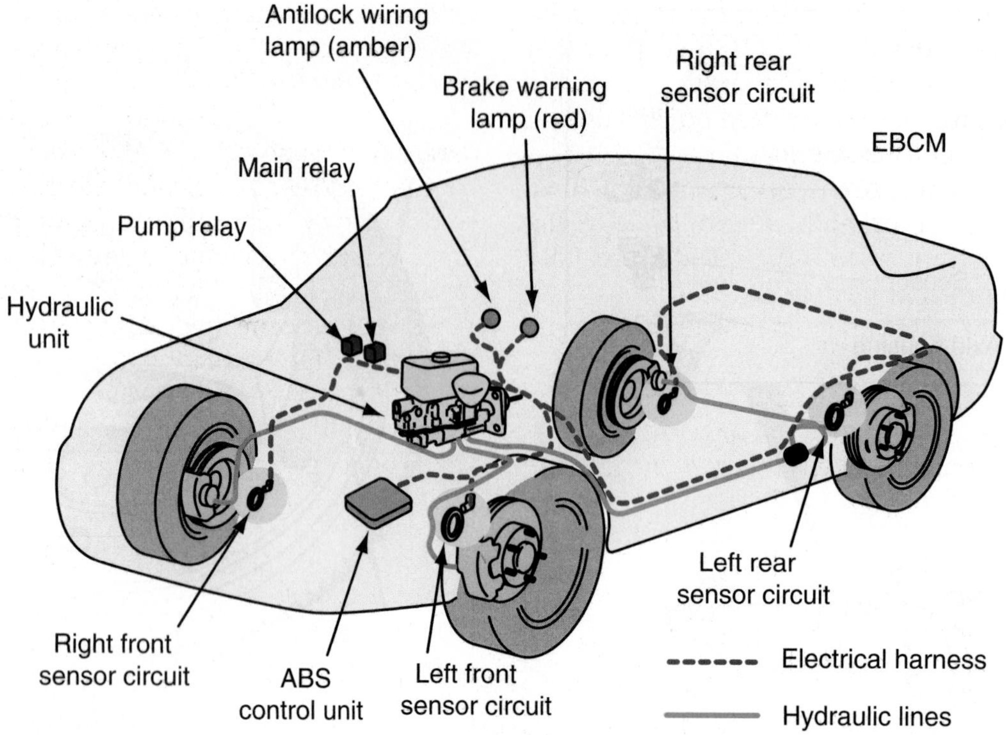

Figure 51-11 The basic electrical and hydraulic components of a four-wheel antilock brake system.

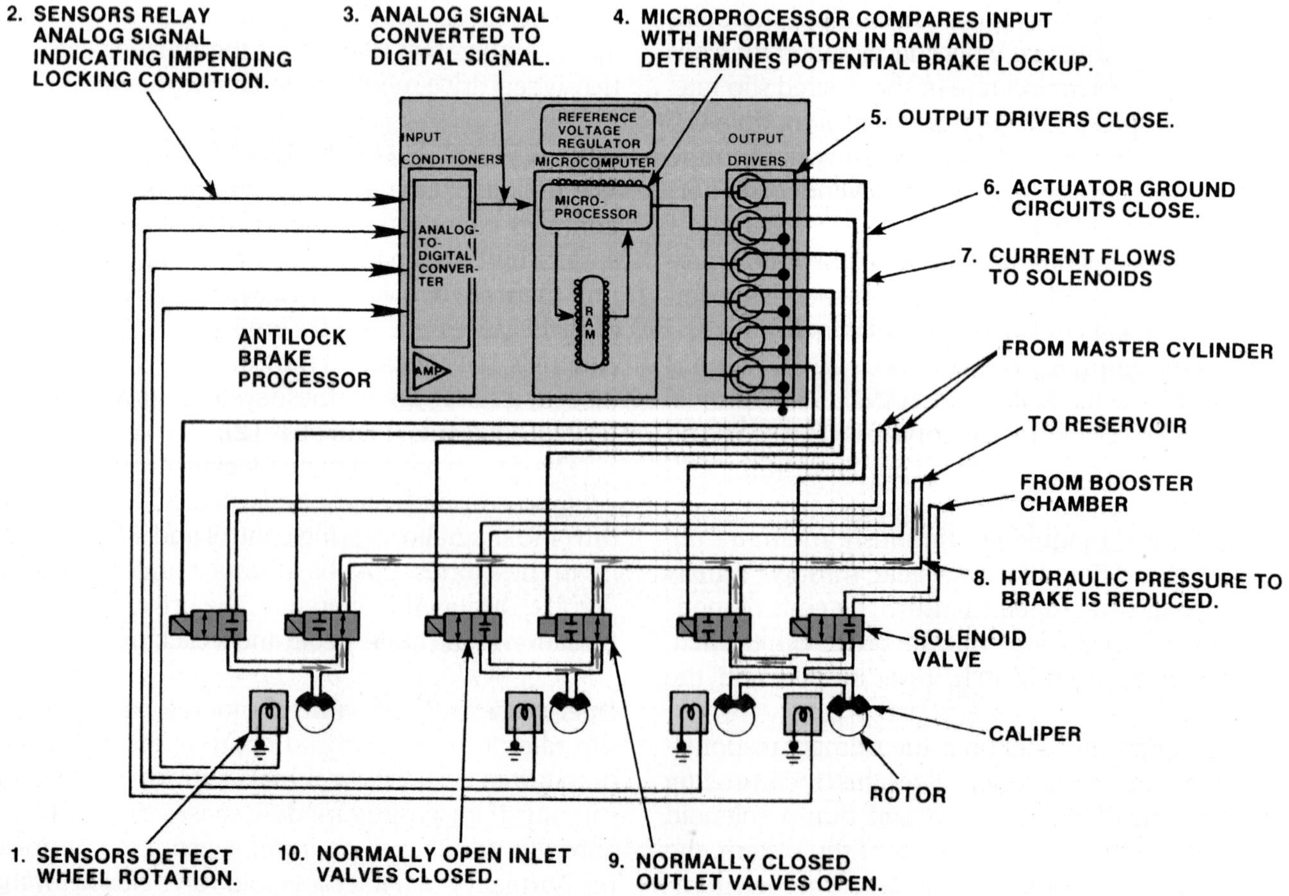

Figure 51-12 ABS operation—potential brake lock condition.

that circuit. The ABS control module then looks at the sensor signal from the affected wheel again. If that wheel is still decelerating faster than the other three wheels, it opens the normally closed outlet solenoid valve for that circuit. This dumps any pressure that is trapped between the closed inlet valve and the brake back to the master cylinder reservoir. Once the affected wheel returns to the same speed as the other wheels, the control module returns the valves to their normal condition, allowing fluid flow to the affected brake.

Wheel speed at each wheel is measured by variable-reluctance sensors or PM generators. As the teeth on the gear pulser or tone wheel rotate past the sensor, AC current is generated. The AC frequency and amplitude change in accordance with the wheel speed (**Figure 51–13**).

Modulator Assembly The ABS modulator assembly consists of the inlet solenoid valve, outlet solenoid valve, reservoir, pump, pump motor, and the damping chamber. The hydraulic control has three modes: pressure reduction (decrease), pressure retaining (hold), and pressure intensifying (increase).

While in the pressure reduction decrease mode (**Figure 51–14**), the inlet valve is closed and the outlet valve is open. During this mode, fluid pressure to the wheel brake is blocked and the existing fluid in the

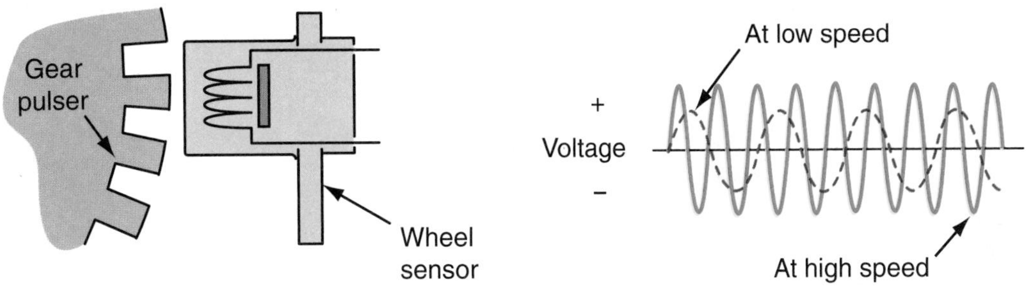

Figure 51-13 A wheel-speed sensor.

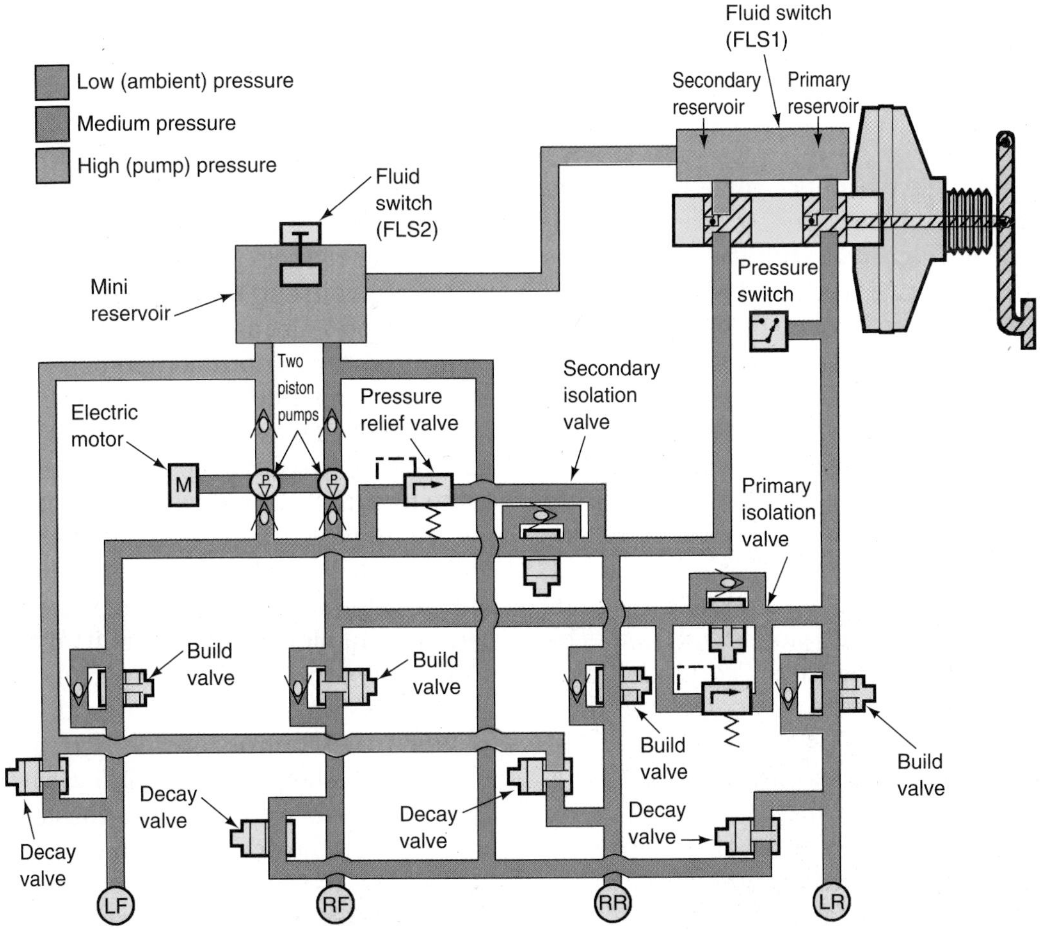

Figure 51-14 Antilock braking, pressure reduction. *Courtesy of Chrysler LLC*

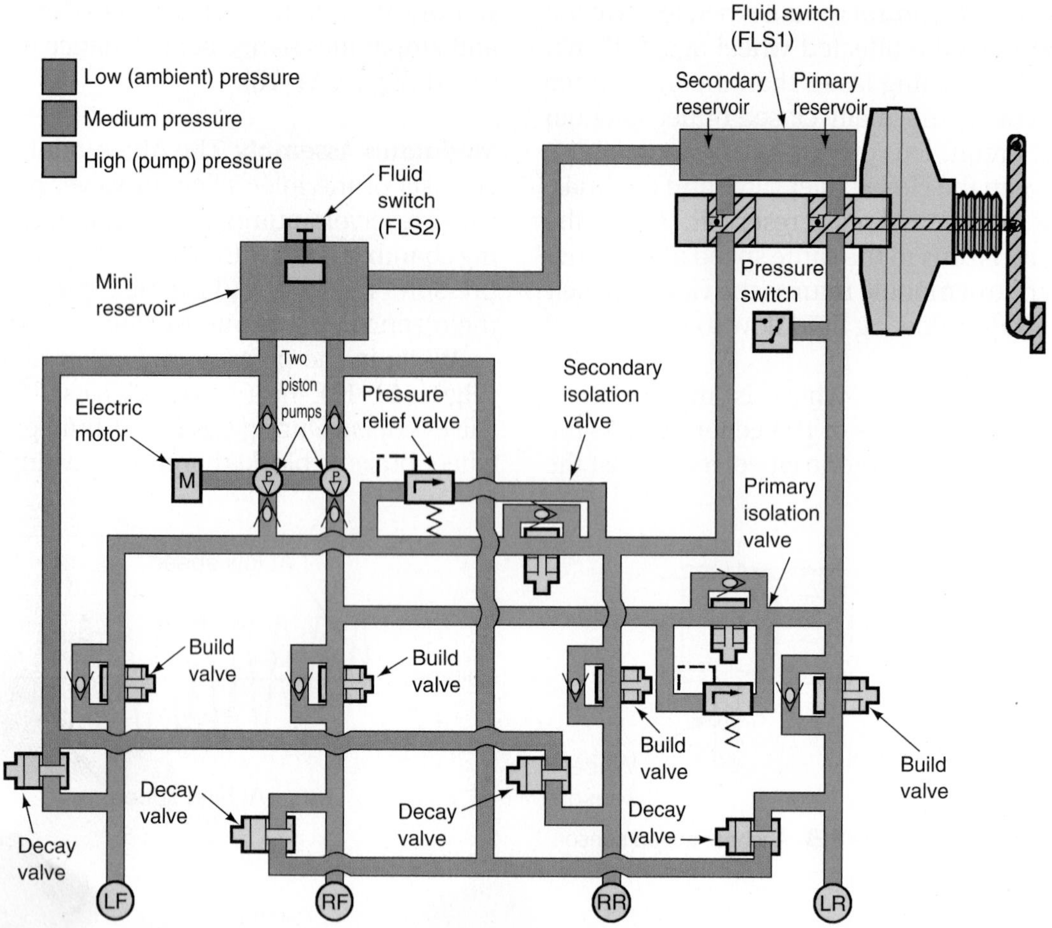

Figure 51-15 Antilock braking, pressure increase. *Courtesy of Chrysler LLC*

caliper flows through the outlet valve back to the master cylinder reservoir. During the pressure-intensifying mode **(Figure 51–15)**, the inlet valve is open and the outlet valve is closed. Pressurized fluid is pumped to the caliper. To keep the pressure at the caliper during the pressure-retaining mode, the inlet and outlet valves are closed.

The pump/motor provides the extra fluid required during an ABS stop. The pump is supplied fluid that is released to the accumulators when the outlet valve is open during an ABS stop. The accumulators provide temporary fluid storage for use during an ABS stop. The pump also drains the accumulator circuits after the ABS stop is complete. The pump is run by an electric motor controlled by a relay that is controlled by the ABS control module. The pump is continuously on during an ABS stop and remains on for about 5 seconds after the stop is complete.

Keep in mind that the activity of the solenoid valves changes rapidly, several times each second. This means the fluid under pressure must be redirected quickly; this is the primary job of the pump.

Integral Four-Wheel Systems

When the brakes are released, the piston in the master cylinder retracts. The booster chamber is vented to the reservoir, and the fluid in the chamber is at the same low pressure as the reservoir. When the brakes are applied, under normal conditions, the brake pedal actuates a pushrod **(Figure 51–16)**. This moves a lever, which moves a spool valve. When the spool valve moves, it closes the port from the booster chamber to the reservoir and partially opens the port from the accumulator in proportion to the pressure on the brake pedal. This allows hydraulic fluid under pressure from the accumulator to enter the booster chamber. As hydraulic pressure enters, it pushes the booster piston forward, providing hydraulic assist to the mechanical thrust from the pushrod.

When the control module determines that the wheels are locking up, it opens a valve that supplies one chamber between the two master cylinder pistons and another chamber between the retraction sleeve and the first master cylinder piston. The hydraulic pressure on the retraction sleeve retracts the pushrod, pushing back the brake pedal. In effect,

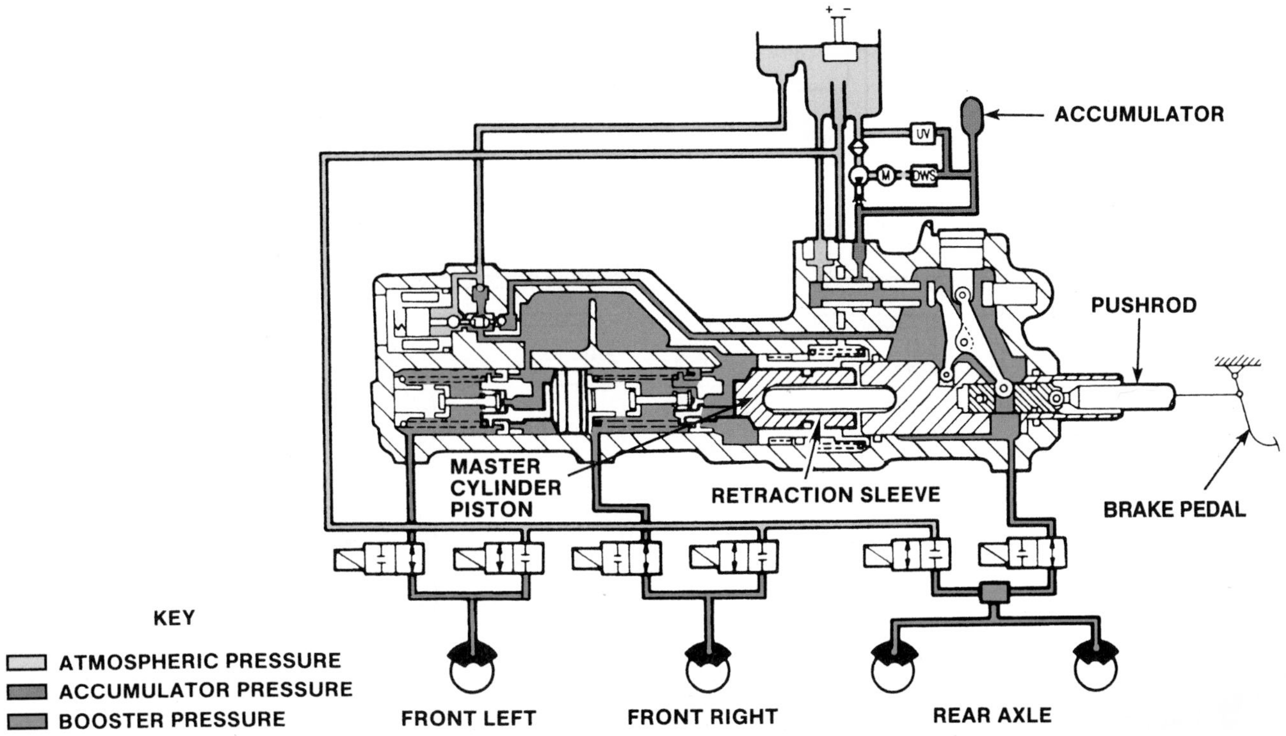

Figure 51–16 Normal braking with antilock system.

the hydraulic pressure to the wheels is now supplied by the accumulator, not by the brake pedal action. The control module also opens and closes the solenoid valves to cycle the brakes on the wheels that have been locking up.

When the solenoid valves are open, the master cylinder pistons supply hydraulic fluid to the front brakes, and the boost pressure chamber provides hydraulic pressure to the rear. When the solenoid valves are closed, the hydraulic fluid from the master cylinder pistons and booster pressure chamber is cut off. The hydraulic fluid is returned from the brakes to the reservoir.

General Motors' Electromagnetic Antilock Brake Systems

Beginning in 1991, General Motors began equipping certain small and midsize vehicles with an antilock braking system called the ABS-VI. This system is an add-on system that uses a conventional vacuum power booster and master brake cylinder. It does not use a high-pressure pump/motor and accumulator and fast-acting solenoid valves to control hydraulic pressure. Instead, it uses a hydraulic modulator **(Figure 51–17)** that operates using a principle called electromagnetic braking.

As in integrated systems, wheel speed is monitored using individual speed sensors. When one wheel begins to decelerate faster than the others while braking, the control module signals the hydraulic modulator assembly to reduce pressure to the affected brake.

The ABS-VI modulator contains three small screw plungers—one for each front brake circuit and one for the rear brake circuit **(Figure 51–18)**—that are driven by electric motors. At the top of each plunger cavity is a check ball that controls hydraulic pressure within the brake circuit.

The hydraulic valve body (modulator) assembly and motor pack is mounted to the master cylinder. The hydraulic brake circuit for each front wheel is controlled by a motor, gear-driven ball screw, solenoid, piston, and check valve. The rear wheel circuit is controlled by check balls and a single motor; therefore, both rear brakes are modulated together.

The motors are high-speed, bidirectional motors that quickly and precisely position the ball screws. Each motor has a brake that allows its ball screw to maintain its position against hydraulic pressures. The front motors have an electromagnetic brake (EMB) and the rear motor uses an expansion spring brake (ESB).

During normal braking conditions, each plunger is all the way up. The check ball at the top is unseated and the bypass solenoid is normally open, which allows brake pressure from the master cylinder and vacuum power booster to apply the brakes during normal stopping.

During panic stops, the ABS mode operates. In this situation, brake pressure must be reduced to

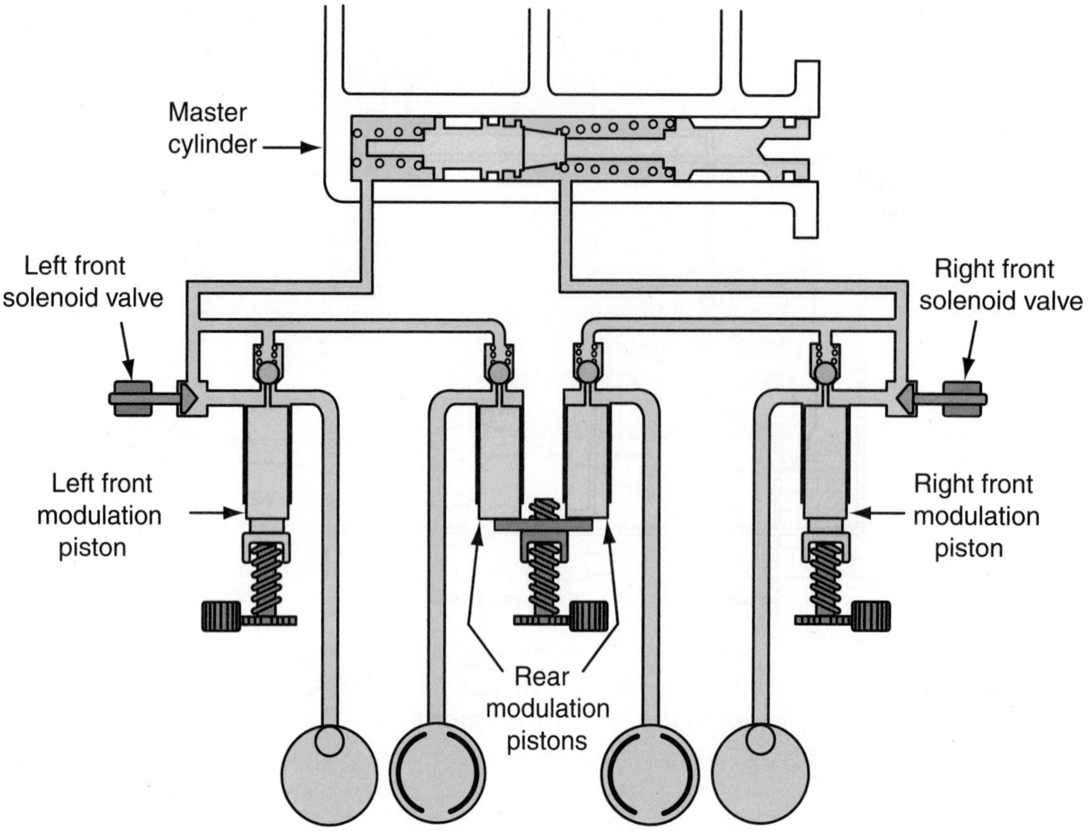

Figure 51-17 The hydraulic layout for GM's ABS-VI system.

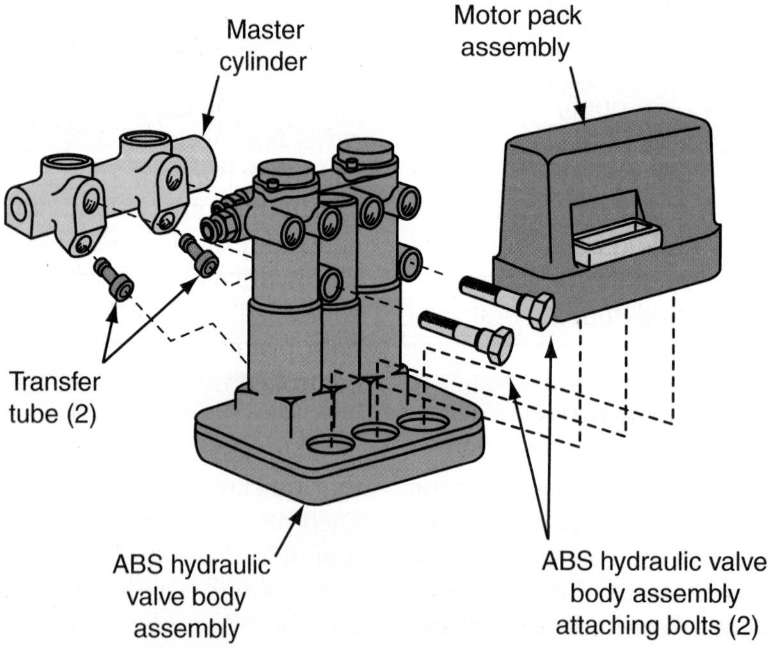

Figure 51-18 The hydraulic modulator is an add-on component to the conventional master cylinder. The electric motors can be replaced separately, as can the two solenoids.

prevent wheel lockup. This is done by closing the normally open solenoid to isolate the circuit and then turning the plunger down to reduce braking pressure. As the plunger turns down, it increases the volume within the brake circuit. This causes a drop in pressure that keeps the wheel from locking. The amount of pressure applied is controlled by running the plunger up and down as required. To decrease pressure further, the plunger is run down. To reapply pressure, the plunger moves back up.

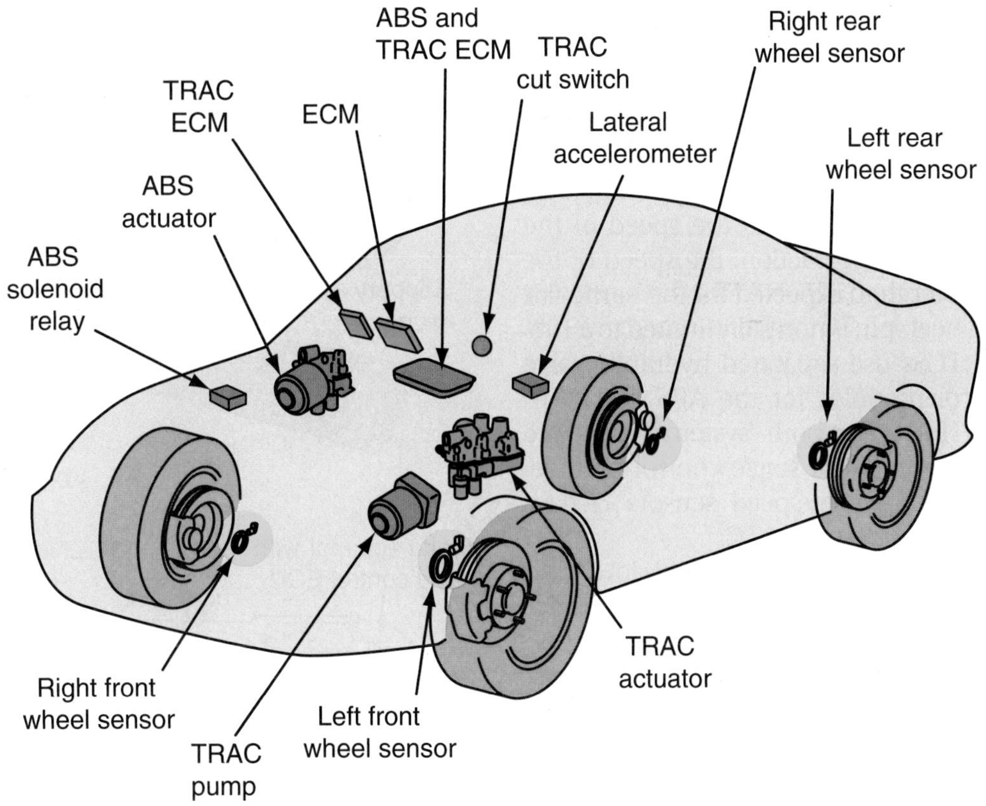

Figure 51-19 A typical ATC system.

The system can cycle the brakes seven times per second. Because the system does not have a high-pressure pump and accumulator, it cannot increase brake pressure above what can be provided by the master cylinder and vacuum-assist booster.

Other Brake System Controls

A few automobiles are now equipped with an electronic brake-assist system that can recognize emergency braking and automatically apply full-power brake force for shorter stopping distances. This system is activated only in emergency braking situations and does not affect normal brake operation.

The system recognizes emergency braking by the speed at which it was depressed and the brakes are automatically applied under full power. The system is driver-adaptive as it learns the driver's braking habits by using sensors to monitor every movement of the brake pedal. When the sensors detect an emergency stop, an electronic valve at the power brake booster is turned on. This supplies full braking power to the wheels. The ABS prevents the wheels from locking in spite of the full-power braking force.

AUTOMATIC TRACTION CONTROL

Automakers use the technology and hardware of ABSs to control tire traction and vehicle stability. As explained earlier, an ABS pumps the brakes when a braking wheel attempts to go into a locked condition. **Automatic traction control** (**ATC**) systems apply the brakes when a drive wheel attempts to spin and lose traction (**Figure 51–19**). Manufacturers have used various basic designs for these systems and they are referred to as ATC, traction control systems (TCS), and acceleration slip reduction (ASR) systems.

Controlling wheel slip is the goal of both ABS and ATC. ABS controls negative wheel slip by modulating the hydraulic pressure to the wheel, or wheels, that is skidding. An ATC system controls positive wheel spin by modulating hydraulic pressure at the wheel that is spinning to slow down the wheel. Many systems use other methods before applying brake pressure.

ATC is most helpful on four-wheel or all-wheel drive vehicles where loss of traction at one wheel could hamper driver control. It is also desirable on high-powered front-wheel-drive vehicles for the same reason. Often if traction control is fitted to a FWD vehicle, the ABS modified system is a three-channel system because ATC is not needed at the rear wheels. On RWD and 4WD vehicles, the system is based on a four-channel ABS.

In order to use the brakes to control wheel spin, the ABS must have a pump to develop hydraulic pressure and an accumulator to store reserve pressure. During braking, the driver's foot applies pressure to the brake pedal. During acceleration, the driver's foot is nowhere near the brake pedal; so if brake pressure is to be used to stop wheel spin, the brake system must have an independent source of hydraulic pressure.

During operation, the ATC system monitors the wheel-speed sensors. If a wheel enters a loss-of-traction situation, the module applies braking force to the wheel in trouble. Loss of traction is identified by comparing the vehicle's speed to the speed of the wheel. If there is a loss of traction, the speed of the wheel will be greater than expected for the particular vehicle speed. Wheel spin is normally limited to a 10% slippage. Some TCSs use separated hydraulic valve units and control modules for the ABS and ATC, whereas others integrate both systems into one hydraulic control unit and a single control module. The pulse rings and wheel-speed sensors remain unchanged from the ABS to the ATC.

Some ATC systems function only at low road speeds of 5 to 25 miles per hour. These systems are designed to reduce wheel slip and maintain traction at the drive wheels when the road is wet or snow covered. If during acceleration the module detects drive wheel slip and the brakes are not applied, the control module enters into the traction control mode. The inlet and outlet solenoid valves are pulsed and allow the brake to be quickly applied and released. The pump/motor assembly is turned on and supplies pressurized fluid to the slipping wheel's brake.

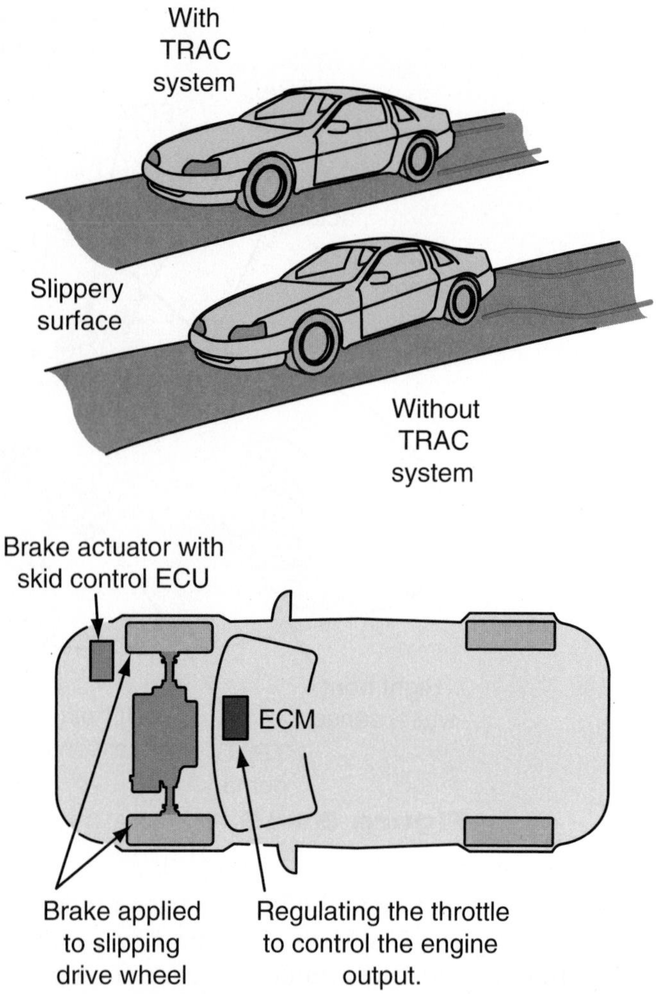

Figure 51–20 An ATC (TRAC) system applies the appropriate brakes and regulates engine output to help prevent wheel slippage when accelerating on slippery surfaces.

Engine Controls

More advanced systems work at higher speeds and integrate some engine control functions into the control loop. Most ATC systems rely on inputs available on the CAN bus and compare front wheel speeds to rear wheel speeds to determine if drive wheels lose traction.

When drive wheel slip is detected while the brake is not applied, the electronic brake control module (EBCM) will enter into the traction control mode. At that time, the PCM will initiate an engine torque reduction routine to slow down the drive wheels. The following shows how the PCM reduces torque to the drive wheels:

- By retarding spark timing
- By decreasing the opening of the throttle plate
- By reducing or cutting off fuel injection pulses to one or more cylinders
- By increasing exhaust gas recirculation (EGR) flow
- By momentarily upshifting the transmission to a higher gear

If the engine torque reduction does not eliminate drive wheel slip, the EBCM will gradually apply the brakes at the driving wheels **(Figure 51–20)**. The master cylinder's isolation valve closes to isolate the cylinder from the rest of the hydraulic system. Then prime valve opens to allow the pump to accumulate brake fluid and build hydraulic pressure. The drive wheel inlet and outlet solenoid valves then open and close and pass through stages of pressure hold, pressure increase, and pressure decrease.

Driver Controls and Indicators

Most TCSs have two warning lights. However, some vehicles display the status of the system with messages on the instrument cluster. Normally, an amber lamp will illuminate or a service message will appear as any of the ABS-disabling DTCs is set. When this occurs, the TCS is automatically disabled by the control unit. When the system is actively controlling wheel spin, a green lamp lights or a message is displayed saying the system is active.

There may also be a manual cut-off switch so the driver can turn off the TCS. This will cause the amber lamp to light or a message displayed, stating the system is off.

AUTOMATIC STABILITY CONTROL

Various stability control systems are found on today's vehicles. Like TCSs, stability controls are based on and linked to the ABS **(Figure 51–21)**. On some vehicles, the stability control system is also linked to the electronic suspension system. Most often, the stability control system is called an **electronic stability control (ESC)** system, although many other names are used. ESC helps prevent skids, swerves, and rollover accidents. Basically the system applies the brakes at one or more wheels to help correct the steering. In some cases, power to the drive wheels is also reduced.

It is important to remember that a vehicle's tendency to roll is influenced by its height, track width, and the stiffness of its suspension. ESC cannot override a car's physical limits nor can it increase traction. If the vehicle is pushed beyond its traction limits, ESC may not be able to correct the vehicle's movement. ESC simply helps the driver maintain control using the available traction.

ESC systems can control the vehicle during acceleration, braking, and coasting. If the brakes are applied but oversteer or understeer is occurring, the fluid pressure to the appropriate brake is increased **(Figure 51–22)**. Understeer is a condition where the vehicle is slow to respond to steering changes. When the system senses understeer in a turn, the brake at the inside rear wheel is applied to regain vehicle stability. Oversteer occurs when the rear wheels try to swing around or fishtail. When this occurs, the ESC system will apply the brake at the outer rear or front wheel in an attempt to neutralize the oversteering.

The control unit, normally the EBCM, receives signals from the wheel-speed sensors, a steering angle sensor (typically part of the combination switch body

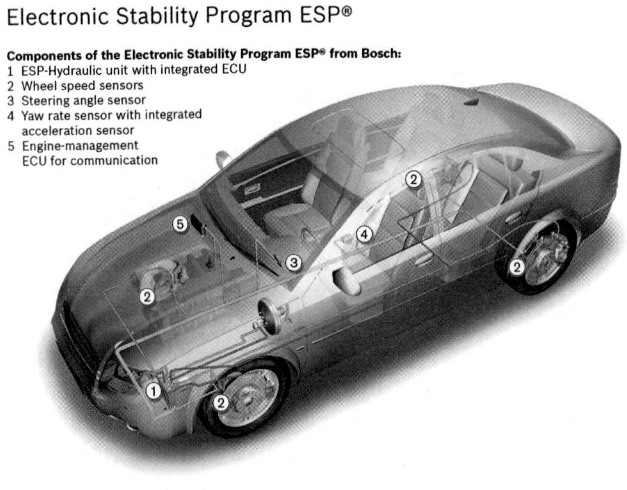

Figure 51-21 The components of a typical vehicle stability control system. *Courtesy of Robert Bosch GmbH, www.bosch-presse.de*

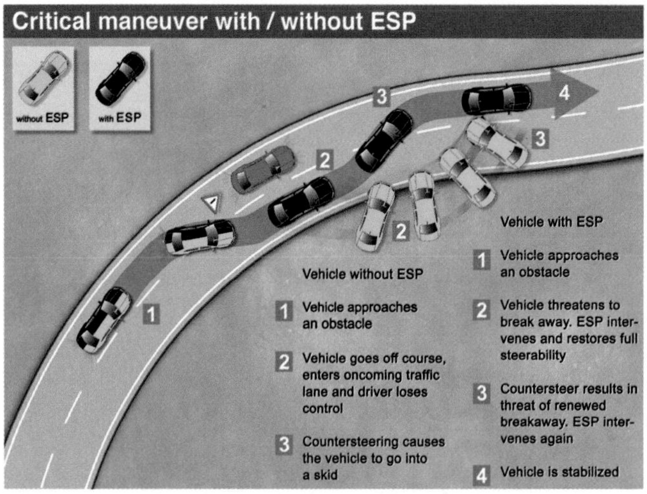

Figure 51-22 The effects of a stability control (ESP = Electronic Stability Program) system. *Courtesy of Robert Bosch GmbH, www.bosch-presse.de*

behind the steering wheel), a lateral-acceleration sensor, a yaw sensor, roll sensors, and a brake-pressure sensor. It also communicates with other control units through the CAN bus **(Figure 51–23)**. The sensors basically let the control unit know the current status of the vehicle.

The ESC control unit compares the driver's intended direction (by monitoring steering angle) to the vehicle's actual direction (by measuring lateral acceleration, yaw, and individual wheel speeds). If there is a difference between the two, the control unit intervenes by modulating individual front or rear wheels and/or reducing engine power output.

ESC continuously monitors key inputs such as yaw rate and wheel speed. **Yaw** is defined as the natural tendency of a vehicle to rotate on its vertical center axis or twist during a turn. A vehicle may also rotate naturally on its horizontal axis; this movement is called roll and pitch.

A yaw rate sensor is a gyroscopic sensor that measures the side-to-side twist of the vehicle. Two types of yaw rate sensors are used: micromechanical and piezoelectric. A micromechanical sensor relies on an oscillating element. The movement of this element is changed in response to yaw and speed. During a turn, the vehicle tends to yaw and the output from the sensor changes. The control unit uses those signals to determine how much yaw is occurring.

A piezoelectric sensor has a vibration-type gyroscope shaped like a tuning fork **(Figure 51–24)**. The device is divided into two sections: upper and lower. Both sections have piezoelectric elements attached to them. As current flows through the piezoelectric materials, the sections oscillate from one side to the other. When the vehicle is making a turn, the movement of the vehicle causes the upper elements to move away

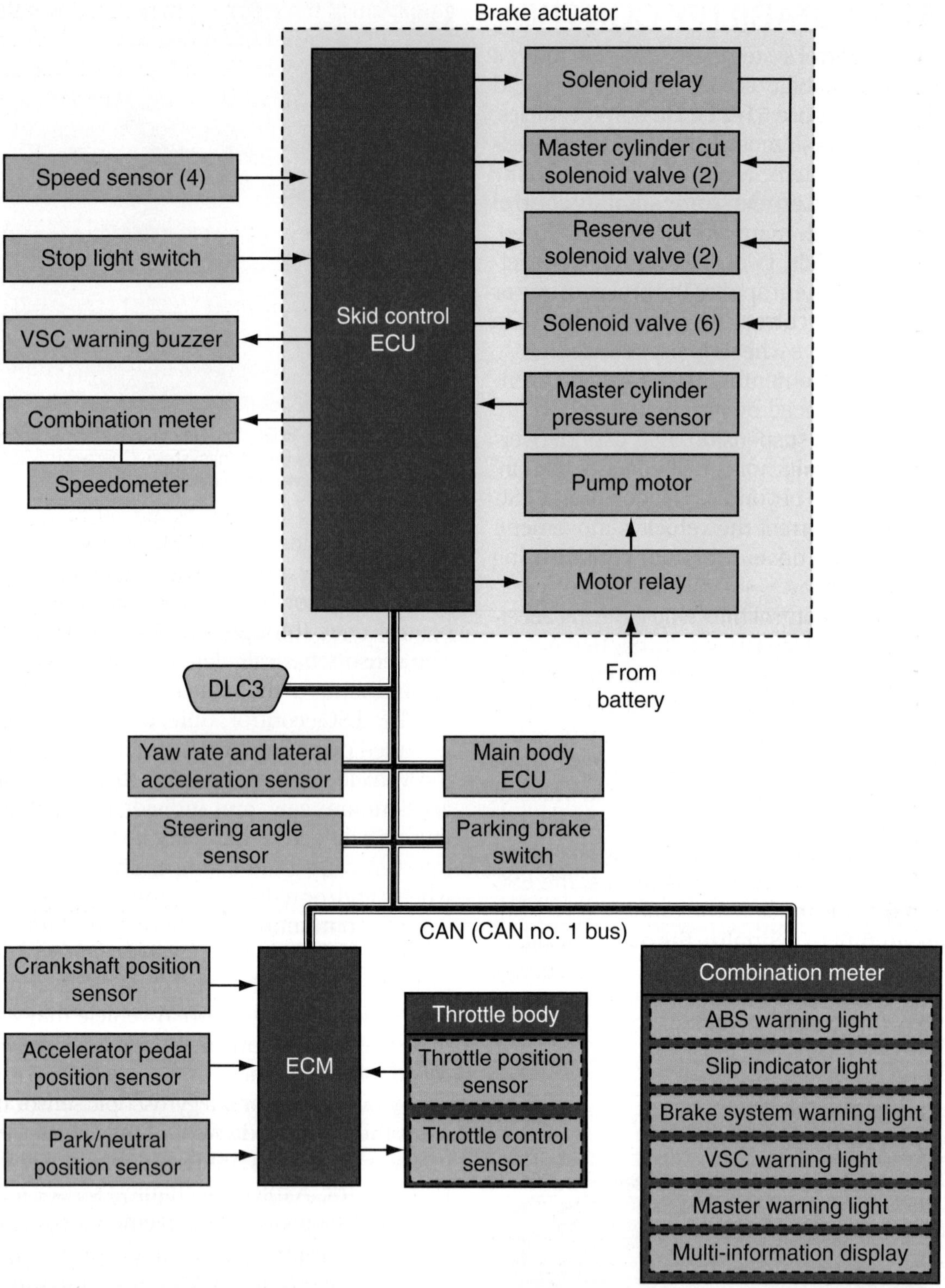

Figure 51-23 A typical system diagram for an ESC system.

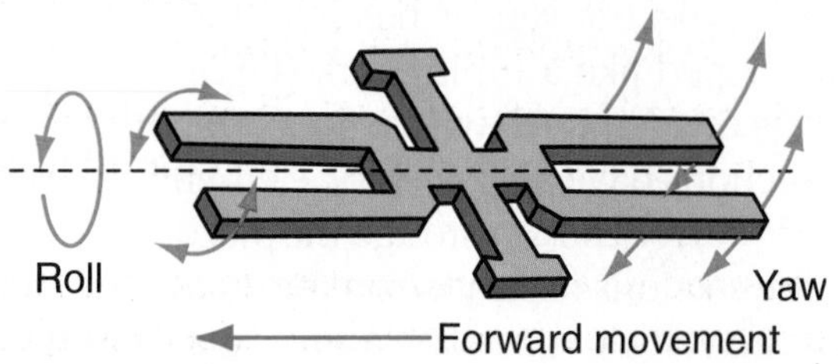

Figure 51-24 The basis of a yaw rate sensor.

from the lower elements. This action generates an AC voltage signal that represents the vehicle's speed and yaw rate. In a right turn, the signal voltage increases and decreases when the vehicle is turning right. The output signals range from 0.25 to 4.75 volts. When the vehicle has 0 yaw, the output signal will be 2.5 volts.

The control unit looks at the actual yaw rate and compares it to the calculated desired rate. It responds

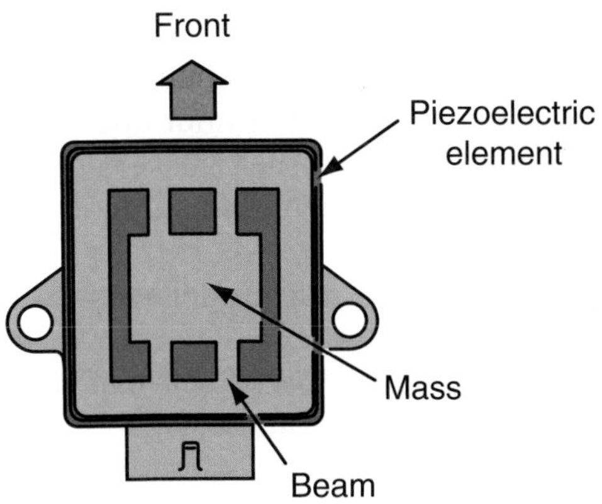

Figure 51-25 An acceleration/deceleration sensor. The semiconductor material rests on a plate that is supported by four beams that flex as thrust is applied in any direction.

to the difference between the two. This difference represents the amount of understeer or oversteer that is occurring. To correct the yaw, the system applies the brake at the appropriate wheel.

Typically the yaw rate sensor and lateral accelerometer share the same housing. They are mounted in the center of the vehicle. The lateral accelerometer monitors acceleration, deceleration, and cornering forces. These sensors are commonly Hall-effect or piezoelectric units. Semiconductor materials are placed on a plate and are set 45 degrees away from the centerline of the vehicle. The plate is supported by four beams **(Figure 51–25)**. The beams are designed to be able to flex in response to the movement of the vehicle. The amount of flex determines the output signal from the sensor. The signal can range from 0.25 to 4.75 volts depending on the G-forces the vehicle is experiencing.

G-force is a measurement of a vehicle's acceleration. The term is based on the normal acceleration rate due to gravity (32.174 ft./s^2 or 9.80665 m/s^2). A G-force value of 1 means the acceleration rate of the vehicle is the same as the acceleration rate of gravity. A value of less than 1 means the rate is lower than that of gravity. Values greater than 1 indicate that the forces acting on the vehicle are greater than the normal acceleration rate of gravity. Most lateral accelerometer sensors have an operating range of -1.5 to $+1.5$ g. Zero lateral acceleration provides a 2.5-volt output signal. The voltage of the signal increases with an increase in lateral acceleration.

Stability Control System Indicators

ESC systems use an indicator light or the message center on the dash to tell the driver when the system is active (i.e., it has detected and corrected yaw rates). A lamp or the message center will also inform the driver when the system is turned off. If the control unit detects a problem in the system, it may shut down the system and alert the driver of needed service. Also, many ESC systems have an "off" switch so the driver can disable ESC, for example, when stuck in mud or snow. However, ESC defaults to "on" when the ignition is restarted. When the system has been turned off by the driver, a warning lamp or message is displayed.

ANTILOCK BRAKE SYSTEM SERVICE

Most of the service done to the brakes of an antilock brake system are identical to those in a conventional brake system. There are, however, some important differences. Always refer to the recommended procedures in the appropriate service manual before attempting to service the brakes on an ABS-equipped vehicle.

Many ABS components are simply remove-and-replace items. Normal brake repairs, such as replacing brake pads, caliper replacement, rotor machining or replacement, brake hose replacement, master cylinder or power booster replacement, or parking brake repair, can all be performed as usual. In other words, brake service on an ABS-equipped vehicle is similar to brake service on a conventional system with a few exceptions. Before beginning any service, check the service manual. It may be necessary to depressurize the accumulator to prevent personal injury from high-pressure fluid.

Safety Precautions

- When replacing brake lines and/or hoses, always use lines and hoses designed for and specifically labeled for use on ABS vehicles.
- Never use silicone brake fluids in ABS vehicles. Use only the brake fluid type recommended by the manufacturer (normally DOT-3).
- Never begin to bleed the hydraulic brake system on a vehicle equipped with ABS until you have checked a service manual for the proper procedure.
- Never open a bleeder screw or loosen a hydraulic brake line or hose while the ABS is pressurized.
- Disconnect all vehicle computers, including the ABS control unit, before doing any electrical welding on the vehicle.
- Never disconnect or reconnect electrical connectors while the ignition switch is on.
- Never install a telephone or CB antenna close to the ABS control unit or any other control module or computer.
- Check the wheel-speed sensor to sensor ring air gap after any parts of the wheel-speed circuit have been replaced.

- Keep the wheel sensor clean. Never cover the sensor with grease unless the manufacturer specifies doing so; in those cases, use only the recommended type.
- When replacing speed (toothed) rings, never beat them on with a hammer. These rings should be pressed onto their flange. Hammering may result in a loss of polarization or magnetization.

Relieving Accumulator Pressure

Some services require that brake tubing or hoses be disconnected. Many ABSs use hydraulic pressures as high as 2,800 psi (19,300 kPa) and an accumulator to store this pressurized fluid. Before disconnecting any lines or fittings in many systems, the accumulator must be fully depressurized. A common method of depressurizing the ABS follows:

1. Turn the ignition switch to the off position.
2. Pump the brake pedal between twenty-five and fifty times.
3. The pedal should be noticeably harder when the accumulator is discharged.

Procedures for depressurizing the system vary with the design of the system. Always refer to the service manual.

DIAGNOSIS AND TESTING

Always follow the vehicle manufacturer's procedures when diagnosing an ABS. In general, ABS diagnostics requires three to five different types of testing that must be performed in the specified order listed in the service manual. Types of testing may include the following:

1. Prediagnostic inspections and test drive
2. Warning light symptom troubleshooting
3. On-board ABS control module testing (trouble code reading)
4. Individual trouble code or component troubleshooting

CAUTION!

Following the wrong sequence or bypassing steps may lead to unnecessary replacement of parts or incorrect resolution of the symptom. The information and procedures given in this chapter are typical of the various antilock systems on the market. For specific instructions, consult the vehicle's service manual.

Prediagnostic Inspection

Before undertaking any actual checks, take just a few minutes to talk with the customer about his or her ABS complaint. The customer is a very good source of information, especially when diagnosing intermittent problems. Make sure you find out what symptoms are present and under what conditions they occur.

All ABSs have some sort of self-test. This test is activated each time the ignition switch is turned on. You should begin all diagnosis with this simple test.

Warning Lamps Place the ignition switch in the start position while observing both the red brake system light and amber ABS indicator lights **(Figure 51–26)**. Both lights should turn on. Start the vehicle. The red brake system light should quickly turn off. This lamp will stay illuminated if the brake fluid level is low, the parking brake switch is closed, or the bulb test switch section of the ignition switch is closed, or when certain ABS trouble codes are set.

With the ignition switch in the run position, the antilock brake control module will perform a preliminary self-check on the antilock electrical system. The self-check takes 3 to 6 seconds, during which time the amber antilock indicator light remains on. Once the self-check is complete, the ABS indicator light should turn off. If any malfunction is detected during this test, the amber lamp will either flash or light continuously to alert the driver of the problem. In some systems, a flashing ABS indicator lamp indicates that the control unit detected a problem but has not suspended ABS operation. However, a flashing ABS indicator lamp is a signal that repairs must be made to the system as soon as possible.

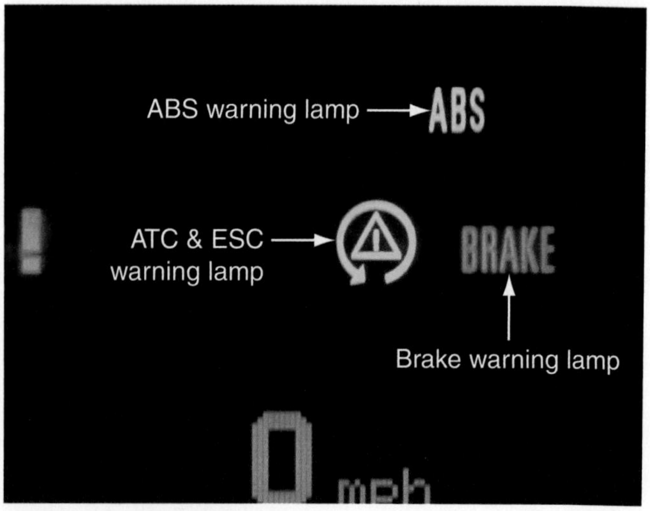

Figure 51–26 The warning lights for a typical ABS-, ATC-, and ESC-equipped vehicle.

A solid ABS indicator lamp indicates that a problem has been detected that affects the operation of ABS. No antilock braking will be available, but normal, nonantilock brake performance will remain. In order to regain ABS braking ability, the ABS system must be serviced.

If both brake indicators stay on and the parking brake is fully released, the front-to-rear braking distribution system may be shut down due to a pressure loss.

If the lamp does not light when the ignition is turned on, the computer probably will not go into its self-test mode. The problem may be as simple as a burned-out bulb, or it may be a problem with the computer itself. To identify the cause for the bulb not illuminating, begin by checking the bulb. If the bulb is good, you will need to make some voltage tests at a diagnostic connector. Follow the testing procedures given by the manufacturer. Nearly all diagnostic connectors have a ground terminal that is used for one or more test modes. Use a voltmeter or ohmmeter to check the continuity between the diagnostic ground terminal and the battery negative terminal. High ground resistance or an open circuit can keep the computer out of the self-test mode and may be a clue to other system problems.

Various other terminals on the diagnostic connector may have other levels of voltage applied to them at different times. Some may have battery (system) voltage under certain conditions, whereas others may have 5 volts, 7 volts, or a variable voltage applied to them.

SHOP TALK

Dirty or damaged wheel-speed sensors and damaged sensor wiring harnesses are leading triggers to turn on ABS warning lamps. Do not rush to condemn the ABS computer or hydraulic module before checking the speed sensors and tone wheels.

Visual Inspection

The prediagnosis inspection consists of a quick visual check of system components. Problems can often be spotted during this inspection, which can eliminate the need to conduct other more time-consuming procedures. This inspection should include the following:

1. Check the master cylinder fluid level.
2. Inspect all brake hoses, lines, and fittings for signs of damage, deterioration, and leakage. Inspect the hydraulic modulator unit for any leaks or wiring damage.
3. Inspect the brake components at all four wheels. Make sure that no brake drag exists and that all brakes react normally when they are applied.
4. Inspect for worn or damaged wheel bearings that may allow a wheel to wobble.
5. Check the alignment and operation of the outer CV joints.
6. Make sure the tires meet the legal tread depth requirements and that they are the correct size.
7. Inspect all electrical connections for signs of corrosion, damage, fraying, and disconnection.
8. Inspect the wheel-speed sensors and their wiring. Check the air gaps between the sensor and ring, and make sure these gaps are within the specified range. Also check the mounting of the sensors and the condition of the toothed ring and wiring to the sensor.

SHOP TALK

Remember that faulty base brake system components may cause the ABS to shut down. Do not condemn the ABS too quickly.

Test Drive

After the visual inspection is completed, test drive the vehicle to evaluate the performance of the entire brake system. Begin the test drive with a feel of the brake pedal while the vehicle is sitting still. Then accelerate to a speed of about 20 mph (32 km/h). Bring the vehicle to a stop using normal braking procedures. Look for any signs of swerving or improper operation. Next, accelerate the vehicle to about 25 mph (40 km/h) and apply the brakes with firm and constant pressure. You should feel the pedal pulsate if the antilock brake system is working properly.

During the test drive, both brake warning lights should remain off. If either of the lights turns on, take note of the condition that may have caused it. After you have stopped the vehicle, place the gear selector into park or neutral and observe the warning lights. They should both be off.

If the control module detects a problem with the system, the amber ABS indicator lamp will either flash or light continuously to alert the driver of the problem. In some systems, a flashing ABS indicator lamp indicates that the control unit detected a problem but has not suspended ABS operation. However, a flashing ABS indicator lamp is a signal that repairs must be made to the system as soon as possible.

A solid ABS indicator lamp indicates that a problem has been detected that affects the operation of ABS. No antilock braking will be available, but normal, nonantilock, brake performance will remain. In order to regain ABS braking ability, the ABS must be serviced.

The red brake warning lamp will be illuminated when the brake fluid level is low, the parking brake switch is closed, the bulb test switch section of the ignition switch is closed, or when certain ABS trouble codes are set.

Intermittent problems can often be identified during the test drive. If the scan tool is equipped with a "snap-shot" feature, use it to capture system performance during normal acceleration, stopping, and turning maneuvers. If this does not reproduce the malfunction, perform an antilock stop on a low coefficient surface such as gravel, from approximately 30 to 50 mph (48 to 80 km/h) while triggering the snapshot mode on the scan tool.

Self-Diagnostics

The control module monitors the electromechanical components of the system. A malfunction of the system will cause the control module to shut off or inhibit the system. However, normal power-assisted braking remains. Malfunctions are indicated by a warning indicator in the instrument cluster. The system is self-monitoring. When the ignition switch is placed in the run position, the ABS control module will perform a preliminary self-check on its electrical system indicated by a second illumination of the amber ABS indicator in the instrument cluster. During vehicle operation, the control module monitors all electrical ABS functions and some hydraulic functions during normal and antilock braking. With most malfunctions of the ABS, the amber ABS indicator will be illuminated and a DTC recorded.

The electronic control system of most ABSs includes sophisticated on-board diagnostics that, when accessed with the proper scan tool, can identify the source of a problem within the system. Each of the DTCs represents a specific possible problem in the system (**Figure 51–27**). The service manual contains a detailed step-by-step troubleshooting chart for each DTC.

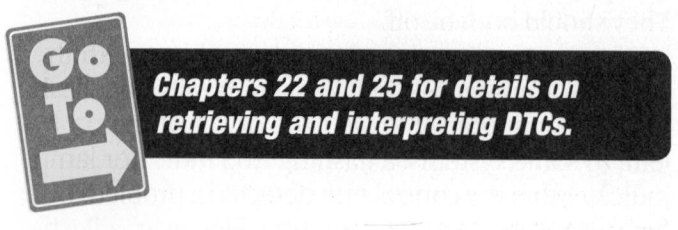

Go To
Chapters 22 and 25 for details on retrieving and interpreting DTCs.

Each system has its own self-diagnostic capabilities. For example, an ABS-VI control module has five separate diagnostic modes. Data available for troubleshooting includes wheel-speed sensor readings, vehicle speed, battery voltage, individual motor and solenoid command status, warning light status, and brake switch status. Numerous trouble codes are programmed into the control module to help pinpoint problems. Other diagnostic modes store past trouble codes. This data can help technicians determine if an earlier fault code, such as an intermittent wheel-speed sensor, is linked to the present problem, such as a completely failed wheel sensor. Another mode enables testing of individual system components.

Testers and Scanning Tools

Different vehicle manufacturers provide ABS test and scan tools with varying capabilities (**Figure 51–28**). Some testers are used simply to access the digital trouble codes. Others may also provide functional test modes for checking wheel sensor circuits, pump operation, solenoid testing, and so forth. Current ABSs are tied to the GAN bus and do not require special testers.

On some vehicles, the amber ABS light and red brake light flash out the digital trouble codes. As you can see, it is important to research the capabilities and proper use of the test equipment the vehicle manufacturer provides. Misuse of test equipment can be dangerous. For example, connecting test equipment during a test drive that is not designed for this use may lead to loss of braking ability.

Once all system malfunctions have been corrected, clear the ABS' DTCs. Codes cannot be erased until all codes have been retrieved, all faults have been corrected, and the vehicle has been driven above a set speed (usually 18 to 25 mph [30 to 40 km/h]). It may be necessary to disconnect a fuse for several seconds to clear the codes on some systems. After service work is performed on the ABS, repeat the previous test procedure to confirm that all codes have been erased.

Testing Components with ABS Scan Tools

ABS scan tools and testers can often be used to monitor and/or trigger input and output signals in the ABS. This allows you to confirm the presence of a suspected problem with an input sensor, switch, or output solenoid in the system. You can also check that the repair has been successful before driving the vehicle. Manual control of components and automated functional tests are also available when using many diagnostic testers. Details of typical functional tests follow.

DTC	Definition
C1211	ABS indicator signal circuit high
C1214	System relay contact or coil circuit open
C1217	BPMV pump motor control circuit shorted
C1218	Pump motor voltage
C1221–C1224	Wheel speed sensor input signal is zero
C1225–C1228	Excessive wheel speed sensor variation
C1232–C1235	Wheel speed sensor circuit open or shorted
C1236	Low system voltage
C1237	High system voltage
C1238	Drive wheel brake rotor excessive temperature
C1242	BPMV pump motor shorted to ground
C1243	BPMV pump motor stalled
C1245	Tire inflation monitor
C1246	Brake lining wear circuit open
C1248	Dynamic rear proportioning control system
C1251	RSS indicated malfunction
C1252	LF normal force malfunction
C1253	RF normal force malfunction
C1254	Checksum error
C1255	EBTCM internal malfunction (ABS/TCS disabled)
C1256	EBTCM internal malfunction
C1261	LF inlet solenoid valve malfunction
C1262	LF outlet solenoid malfunction
C1263	RF inlet solenoid valve malfunction
C1264	RF outlet solenoid malfunction
C1265	LR inlet solenoid valve malfunction
C1266	LR outlet solenoid malfunction
C1266	RR inlet solenoid valve malfunction
C1267	RR outlet solenoid malfunction
C1271	LF TCS master cylinder isolation valve malfunction
C1272	LF TCS prime valve malfunction
C1273	RF TCS master cylinder isolation valve malfunction
C1274	RF TCS prime valve malfunction
C1276	Delivered torque signal circuit malfunction
C1277	Requested torque signal circuit malfunction
C1278	TCS temporarily inhibited by PCM
C1281	Steering sensor uncorrelated malfunction
C1282	Yaw rate sensor bias circuit malfunction
C1283	Excessive time to center steering
C1284	Lateral accelerometer self-test malfunction
C1285	Lateral accelerometer circuit malfunction
C1286	Steering sensor bias malfunction
C1287	Steering sensor rate malfunction
C1288	Steering sensor circuit malfunction
C1291	Open brake switch contacts during deceleration
C1293	DTC C1291 set in previous ignition cycle
C1294	Brake light switch circuit always active
C1295	Brake light switch circuit open
C1298	Class 2 serial data link malfunction
P1571	Requested torque signal circuit malfunction
P1644	Delivered torque signal voltage invalid
P1689	Delivered torque signal voltage invalid

Figure 51-27 Examples of some of the DTCs that can be set by an ABS, ATC, ESC system.

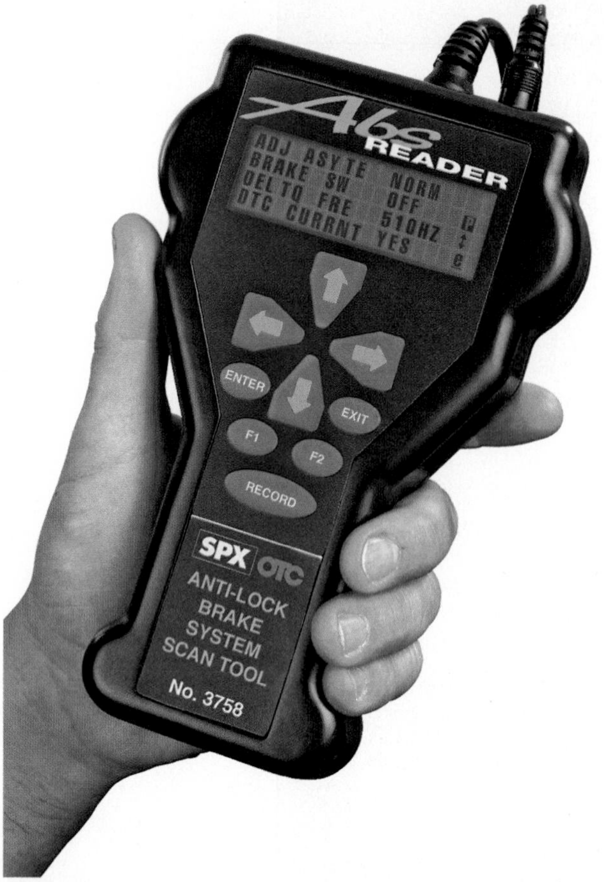

Figure 51–28 A hand-held antilock brake system scan tool. *Courtesy of SPX Service Solutions*

Solenoid Leak Test

PROCEDURE

Checking for Solenoid Leaks in the Hydraulic Modulator

STEP 1 Disconnect the inspection connector from the connector cover and connect the inspection connector to the tester.

STEP 2 Remove the modulator reservoir filter, then fill the reservoir to the max level with fresh fluid.

STEP 3 Bleed the high-pressure fluid from the maintenance bleeder connection. Often this procedure requires the use of a special tool.

STEP 4 Start the engine and release the parking brake.

STEP 5 Set the tester to the proper mode and press the start-test button.

STEP 6 While the ABS pump is running, place your finger over the top of the solenoid return tube in the modulator reservoir. If you can feel brake fluid coming from the return tube, one of the solenoids is leaking. Go to step 7. If you cannot feel brake fluid coming from the return tube, the solenoids are OK. Reinstall the modular reservoir filter and refill the reservoir to the max level.

STEP 7 Bleed the high-pressure fluid from the maintenance bleeder, then run through steps 3 and 6 with the tester. Repeat this procedure three or four times.

STEP 8 Repeat steps 5 and 6. If the solenoid leakage has stopped, reinstall the modulator reservoir filter and refill the reservoir to the max level. If one of the solenoids is leaking, the entire hydraulic modulator may require replacement. In some cases it is possible to remove, inspect, and replace individual solenoids.

WARNING!

Certain components of the ABS are not intended to be serviced individually. Do not attempt to remove or disconnect these components. Only those components with approved removal and installation procedures in the manufacturer's service manual should be serviced.

Testing Components with a Lab Scope

Like most electrical/electronic systems, antilock brake traction control and stability control system components can be tested with a lab scope. A lab scope offers one distinct advantage over many other testing tools. You can watch the component's activity over time. For example, all antilock-based systems rely on the cycling of solenoid valves. **Figure 51–29** is a waveform of a solenoid valve showing the solenoid's activity immediately after the brake pedal was depressed hard. When looking at the solenoid on the lab scope you should see a change in the waveform as soon as ABS operation begins. When a wheel tries to slip, the ABS control module should begin pulsing the solenoid to that wheel.

A critical input to the antilock brake, traction control, and stability control systems is from the wheel sensors. These too can be monitored on a lab scope **(Figure 51–30)**. As the wheel begins to spin, the waveform of the sensor's output should begin to oscillate above and below zero volts. The oscillations should get taller as speed increases. If the wheel's speed is kept constant, the waveform should also stay constant.

Component Replacement

A typical antilock braking system consists of a conventional hydraulic brake system (the base system) plus a number of antilock components. The base brake system consists of a vacuum power booster, master cylinder, front disc brakes, rear drum or disc brakes,

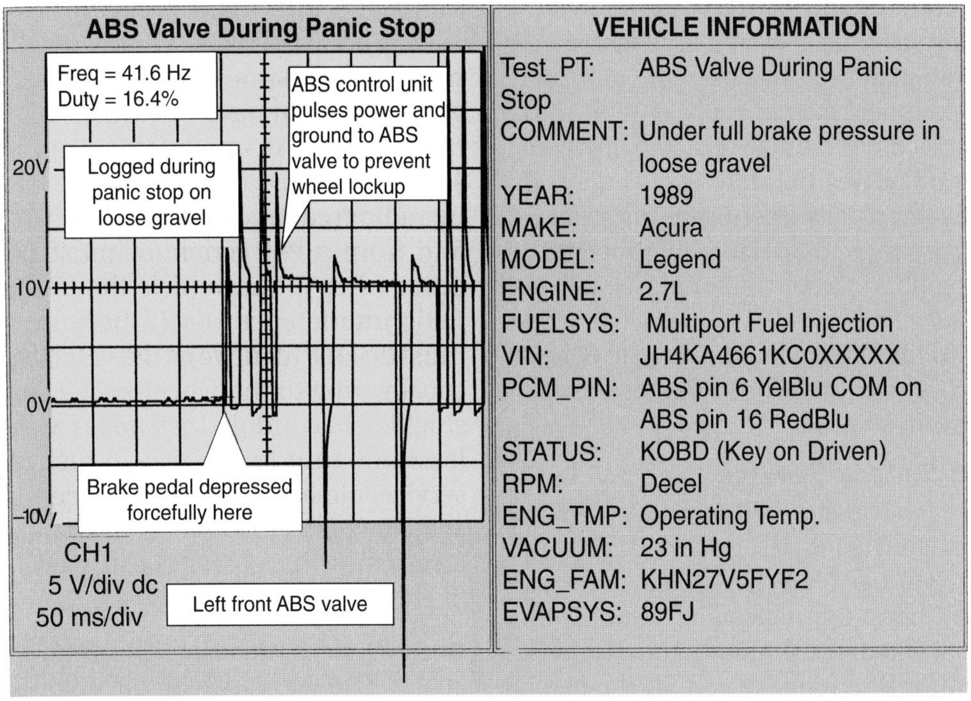

Figure 51-29 The waveform of a solenoid valve before and after the pedal was depressed hard and quickly. *Courtesy of Progressive Diagnostics—WaveFile AutoPro*

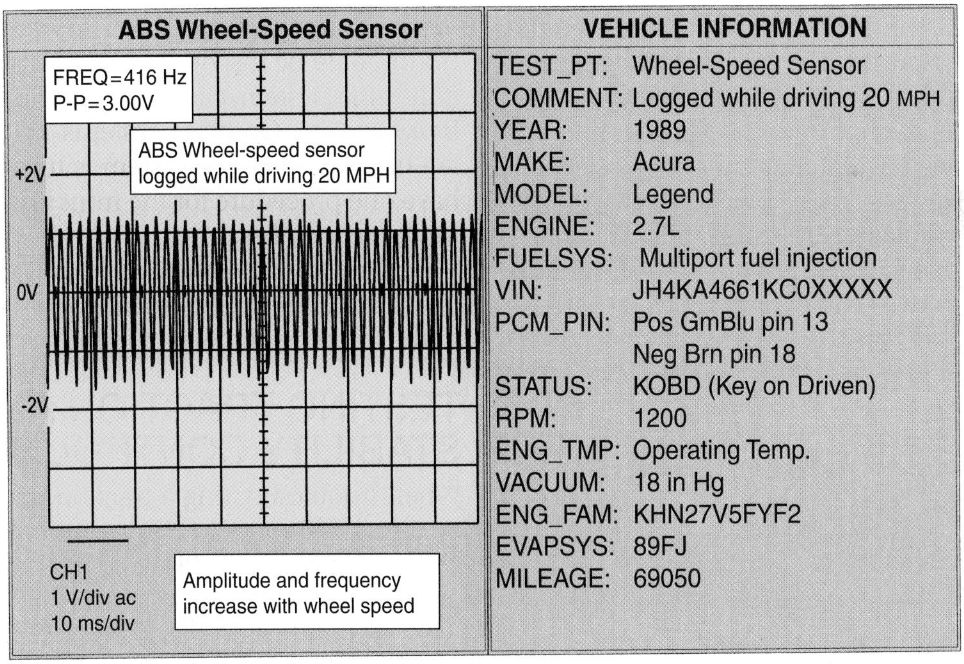

Figure 51-30 The waveform from a wheel-speed sensor. *Courtesy of Progressive Diagnostics—WaveFile AutoPro*

interconnecting hydraulic tubing and hoses, a low fluid sensor, and a red brake system warning light.

Antilock components are added to this base system to provide antilocking braking ability. Most ABSs use the same operational principles, but the major components may be configured and/or named differently.

The electrical components of the ABS are generally very stable. Common electrical system failures are usually caused by poor or broken connections. Other common faults can be caused by malfunction of the wheel-speed sensors, pump and motor assembly, or the hydraulic module assembly.

Many of the components of ABSs are serviced by replacement only. On some systems, wheel-speed sensors must be adjusted. Normal brake repairs, such as replacing brake pads, caliper replacement, rotor machining or replacement, brake hose replacement, master cylinder or power booster replacement, or parking brake repair, can all be performed as usual. In other words, brake service on an ABS-equipped vehicle is similar to service on a conventional system with a few exceptions.

Always refer to the service manual for the correct adjustment and replacements procedures for any and all brake parts in an ABS.

Wheel-Speed Sensor Service

Visually inspect each wheel-speed sensor pulsers for chipped or damaged teeth. Use a feeler gauge to measure the air gap between the sensor and rotor **(Figure 51–31)**. Check the gap while rotating the drive shaft, wheel, or rear hub unit by hand. If there is a specification on this gap, make certain the gap is within the required specification (typically 0.02 to 0.04 inch or 0.4 mm to 1.0 mm). If the gap exceeds specifications, the problem is likely a distorted knuckle that should be replaced.

Also check for a buildup of metal and rust on the sensor. Some sensors have a very strong magnet that will attract metal from all around the area. This buildup can cause the sensor to have inaccurate readings. If the collected material cannot be cleaned, replace the sensor.

Sensors are replaced by simply disconnecting the wiring at the sensor and unbolting the fasteners. Be careful not to twist the wiring cables or harness when installing the sensors. Many vehicles have a jumper harness made of highly flexible twisted pair wiring between each wheel-speed sensor and the main wiring harness. Wheel-speed sensors are PM generators that have AC voltage outputs. Radio signals are also AC-modulated signals; therefore, the signal produced by PM generators can be affected by radio frequency interference (RFI). The wires to and from a PM generator must be shielded from other wires in the vehicle that radiate electromagnetic interference (EMI). Because the suspension must be able to move as the vehicle travels, the electrical wiring to the wheel-speed sensors cannot be shielded in conduit. Instead the wires are twisted at least one turn for every 1.75 inches. These are not serviceable and must be replaced if they are damaged or corroded. Never attempt to solder, splice, or crimp these harnesses because eventual failure will likely result.

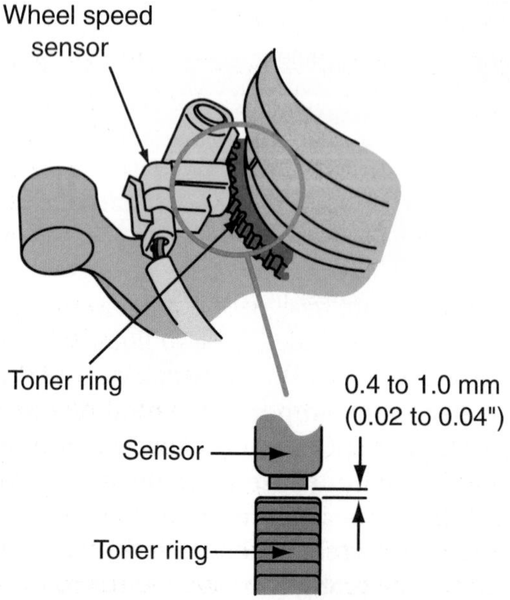

Figure 51-31 Wheel sensor gap measurement.

Brake System Bleeding This is one service area that is most likely to vary with the different designs of the ABS. Always refer to the service manual before attempting to bleed the brake system on a vehicle with ABS.

Some systems require that the accumulator be depressurized and the power source to the ABS control module be disconnected. After these have been done, the system can be bled like a conventional brake system. On other systems, the bleeder screws are opened while the system is turned on, and they have one procedure for the front brakes and another for the rear brakes.

Failure to follow the correct procedure can result in personal injury, destruction of ABS components, or the trapping of more air in the system.

TESTING TRACTION AND STABILITY CONTROL SYSTEMS

When troubleshooting a traction or stability control system, it is important to remember that the systems will be automatically disabled by the control unit if a fault occurs. Because TCS and ESC are combined with ABS and share many input signals and output functions, TCS and ESC will be disabled if an ABS fault occurs.

Traction and stability control systems add very few, if any, extra components to a vehicle. The operation of these systems is principally based on altering the control programs of the brake and engine control systems. It follows then that TCS and ESC testing should be based on retrieving DTCs and then testing individual components. Scan tools can read trouble codes and operating system data. The scan tool will also identify any communications errors that may be present. Proper communications is important for both traction and stability control systems. As with other electronic tests, you should refer to the manufacturer's troubleshooting procedures for the vehicle you are servicing.

NEW TRENDS

The widespread use of ABS has allowed engineers to pursue many different features for an automobile. Not only have ABS, traction control, and stability control systems advanced, but the entire brake system is in a state of evolution.

Electric Calipers

A recent development is the use of electrically operated calipers **(Figure 51–32)**. Doing this eliminates the need to run hydraulic lines to the wheels and provides better brake control. This system is computer controlled and can evenly distribute braking power at the four wheels. At the rear wheels, this system can also be used as the parking brake, thereby eliminating the need to have cables at the rear wheels.

Brake-by-Wire

In a brake-by-wire system, the braking command of the brake pedal is electronically processed. A control module receives sensor input from the brake pedal and activates the master cylinder **(Figure 51–33)**. For safety purposes, the normal linkage from the pedal to the master cylinder may still exist. The advantage of brake-by-wire is simply that the pressure on the pedal can be immediately and directly relayed to the master cylinder and wheel brakes. In a true brake-by-wire system, the hydraulic system has been totally eliminated; these systems are currently being analyzed.

Electronic Wedge Brake (EWB)

Much development is being done with electronic wedge brakes **(Figure 51–34)**. This system is totally electrically operated and electronically controlled. There is no hydraulic system. This development opens the door for brake-by-wire systems. One of the obstacles to brake-by-wire systems has been providing enough clamping power on the rotor to safely stop the vehicle. The electronic wedge brake uses metal wedges with a series of interlocking triangular teeth between the caliper and the disc.

The wedges are designed to react to the speed of the wheels; as speed increases, so does the clamping pressure of the brakes. When the pad is pushed against the rotor, the momentum of the rotating disc draws the pads farther up the interlocking series of wedges, applying greater braking pressure and increasing braking efficiency. This allows for the use of a low-power 12-volt electric motor to initially push the brake pad toward the rotor. The motor turns roller screws that press the wedge and pads against the rotor.

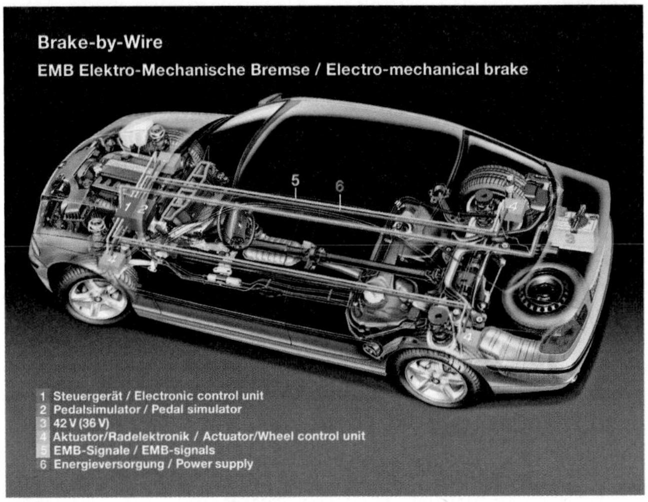

Figure 51–33 The layout for a brake-by-wire system. *Courtesy of BWW of North America, LLC*

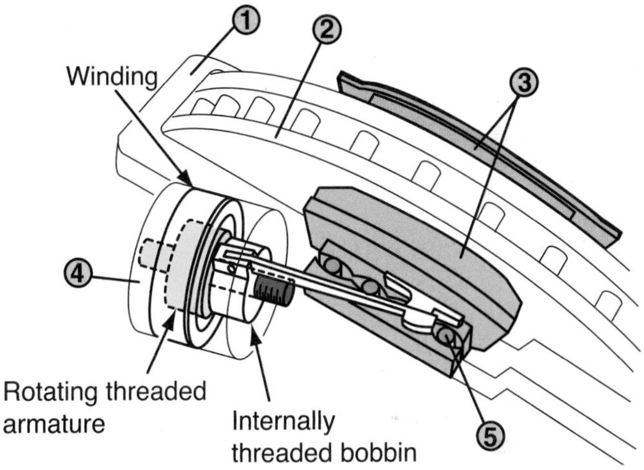

Figure 51–34 The layout of an electronic wedge brake. The brake caliper (1) spans the brake disc (2) on two sides. The brake disc is braked by a pad (3) that is moved by an electric motor (4) via several rollers (5) along wedge-shaped faces.

Figure 51–32 An electric brake caliper. *Courtesy of the Delphi Corporation*

The system relies on the vehicle's battery for power but also has a built-in backup battery. This is important because one of the fears of brake-by-wire systems is the loss of power. Because the system uses a low-power motor, a small backup battery can power the brakes if needed. Also, as a safety feature, each wheel brake in the system is independent of the others. This means if one brake fails, there are three others that can be used to stop the vehicle. The system can also quickly retract the pads when the brake pedal is released; this reduces brake drag and allows for ABS cycling to occur up to six times faster.

It is claimed that EWB systems greatly shorten stopping distances, require much less energy than hydraulic brakes, allow for much better control in ATC and ESC systems, are light, and are more reliable and durable than hydraulic brake systems.

Several suppliers, such as Siemens VDO, Continental AG, Robert Bosch LLC, Delphi Corporation, and TRW Automotive, also are working on electromechanical brake programs.

Wet Brakes

Some vehicles are equipped with rain-sense windshield wipers. These systems have been modified by a few manufacturers (notably BMW) to include removal of water from the brake rotors. When the windshield rain sensor sends a signal to the ABS, the brakes are lightly applied to wipe water off the discs and pads.

Hold Feature

Some vehicles with manual transmissions have a brake hold feature. The purpose of this system is to apply the brakes while the vehicle is stopped on a hill. This prevents the vehicle from rolling. To enable the system, a driver presses a button while the brakes are applied. This locks the brakes until the throttle pedal is depressed.

CASE STUDY

A customer brought in a late-model General Motors vehicle because her brake warning light came on and stayed on. Her expressed concern was she paid extra for ABS and is sure the system is not working, which is why the light is on. It seems she read her in the owner's manual that when the ABS light remains on, there is a problem with the system and the ABS is inoperative.

The technician verified the complaint and found the red brake warning lamp to be lit at all times. He then referred to the test procedures for that ABS. After the preliminary tests were completed, he used a scan tool to retrieve any trouble codes from the system. No DTCs were found and the brake warning light remained on.

Confused and frustrated, the technician sought help from a fellow certified technician. He summarized the problem and the tests he conducted. The certified technician went over to the car to see what he could find and immediately identified the problem. The red warning light was on, not the amber one. The amber lamp will light when there is an ABS problem. It will also indicate that the ABS is inoperative. The red light, however, will light when there is a problem of low brake fluid, a closed parking brake switch, or if there is a hydraulic problem in the base brake system.

Through a basic inspection, the technician found the brake fluid level to be low. Once the fluid level was corrected, the light went out. It seems the technician was sidetracked by the customer's concern over the light and her desire to have ABS. The moral of the story is simply to make sure you understand the function of all warning lights before you jump to any conclusions.

KEY TERMS

Automatic traction control (ATC)
Electronic stability control (ESC)
Four-channel system
Integral antilock brake system
Isolation/dump valves
Nonintegral antilock brake system
One-channel system
Pressure modulation
Three-channel system
Two-channel system
Yaw

SUMMARY

- Brake fluid is the lifeblood of any hydraulic brake system. DOT 3 is recommended for most antilock brake systems and some power brake systems.

- Modern antilock brake systems provide electronic/hydraulic pumping of the brakes for straight-line stopping under panic conditions.
- In addition to being classified as integral and nonintegral, antilock brake systems can be divided into the level of control they provide. They can be one-, two-, three-, or four-channel, two- or four-wheel systems.
- Pressure modulation works to prevent wheel locking. Antilock brake systems can modulate the pressure to the brakes as often as fifteen times per second.
- Integrated antilock brake systems combine the master cylinder, hydraulic booster, and hydraulic circuitry into a single assembly. The great majority of antilock brake systems are of this type.
- On a nonintegrated ABS, the master cylinder and hydraulic valve unit are separate assemblies and a vacuum boost is used. In some nonintegrated systems, the master cylinder supplies brake fluid to the hydraulic unit.
- Automatic traction control (ATC) is a system that applies the brakes when a drive wheel attempts to spin and loses traction.
- Automatic stability systems correct oversteer and understeer by applying one wheel brake.
- Malfunction of the ABS causes the electronic control module to shut off or inhibit the system. However, normal power-assisted braking remains. Malfunctions are indicated by one or two warning lights inside the vehicle.
- Loss of hydraulic fluid or power booster pressure disables the antilock brake system.

REVIEW QUESTIONS

1. What is the primary difference between an automatic traction control system and a stability control system?
2. Briefly describe the proper steps and testing needed to accurately diagnose antilock braking systems.
3. An ABS modulates brake pressure. What does this mean?
4. Explain the difference between oversteer and understeer.
5. List the various methods used by traction control systems to eliminate wheel spin.
6. Besides indicating ABS faults for some systems, the red brake warning lamp on the instrument panel can indicate failure of the ______ brake system, application of the ______ brake, or low fluid level in the ______.
7. What are the functions of an isolation/dump valve?
8. Describe the normal pedal feel when the ABS is activated during hard braking.
9. Which of the following is a *true* statement?
   a. In some antilock brake systems, the power brake assist is provided by pressurized brake fluid supplied by the hydraulic accumulator.
   b. The accumulator is a small, sealed chamber mounted to the pump/motor assembly.
   c. The accumulator holds a highly pressurized nitrogen gas that is used to generate a charging pressure.
   d. All of the statements are true.
10. What is the name for the gas-filled pressure chamber that is part of the antilock braking system's pump and motor assembly?
    a. control module
    b. accumulator
    c. sensor
    d. reservoir
11. What four things should be checked before replacing the computer in an ABS?
12. When inspecting wheel-speed sensors, check for all of the following *except* ______.
    a. buildup of metal or rust on the sensor
    b. proper contact between the pole piece and tone ring
    c. secure sensor mounting
    d. condition of the tone ring teeth
13. Define the difference between an integrated and a nonintegrated antilock braking system.
14. What is the basic difference between a Delco ABS VI and other typical systems?
15. Why are the wires leading to some wheel-speed sensors twisted?

ASE-STYLE REVIEW QUESTIONS

1. Technician A says that a malfunction of the ABS causes the control module to shut off or inhibit the system. Technician B says that a loss of

hydraulic fluid or power booster pressure disables the antilock brake system. Who is correct?

a. Technician A
b Technician B
c. Both A and B
d. Neither A nor B

2. While road testing a car equipped with an ABS: Technician A says that during heavy braking, several pulses may be felt through the brake pedal. Technician B says that a spongy pedal during normal braking is normal. Who is correct?

a. Technician A
b Technician B
c. Both A and B
d. Neither A nor B

3. Technician A says that the accumulator in an ABS is used to store hydraulic fluid to provide residual pressure for power-assist braking. Technician B says that the booster pump in an antilock system provides pressurized hydraulic fluid for the ABS. Who is correct?

a. Technician A
b Technician B
c. Both A and B
d. Neither A nor B

4. Technician A says that it is normal for the amber brake lamp to light when the ABS is activated during braking. Technician B says that it is normal for the red brake lamp to be on whenever the ignition is on and the engine is off. Who is correct?

a. Technician A
b Technician B
c. Both A and B
d. Neither A nor B

5. Technician A says that some antilock brake systems use a lateral acceleration sensor in addition to the wheel-speed sensors. Technician B says that traction control systems apply one wheel brake to get the vehicle going in the direction in which it is being steered. Who is correct?

a. Technician A
b Technician B
c. Both A and B
d. Neither A nor B

6. Technician A says that wheel-speed sensors send an AC voltage signal to the control module. Technician B says that wheel speed signals create an AC signal voltage by altering a reference voltage received from the control module. Who is correct?

a. Technician A
b Technician B
c. Both A and B
d. Neither A nor B

7. Technician A says that an ABS problem can cause the electronic control module to shut off or inhibit the system. Technician B says that a loss of hydraulic fluid or power booster pressure will disable the antilock brake system. Who is correct?

a. Technician A
b Technician B
c. Both A and B
d. Neither A nor B

8. Technician A says that each ABS has its own diagnostic procedure involving use of the brake and antilock warning lamps, special testers, breakout boxes, scan tools, troubleshooting charts, and wiring diagrams. Technician B says that ABSs vary in the number of codes they can store and whether they will retain the code in memory when the ignition is turned off. Who is correct?

a. Technician A
b Technician B
c. Both A and B
d. Neither A nor B

9. When removing a vehicle's ABS electronic control module: Technician A relieves hydraulic pressure in the system. Technician B disconnects the battery ground cable. Who is correct?

a. Technician A
b Technician B
c. Both A and B
d. Neither A nor B

10. Technician A says that some traction control systems eliminate wheel spin by applying the brakes on the driving wheels. Technician B says that some traction control systems eliminate wheel spin by retarding ignition timing and cutting off fuel injection. Who is correct?

a. Technician A
b Technician B
c. Both A and B
d. Neither A nor B

CHAPTER 52

HEATING AND AIR-CONDITIONING

OBJECTIVES

■ Identify the purpose of a ventilation system. ■ Identify the common parts of a heating system. ■ Compare the vacuum and mechanical controls of a heating system. ■ Diagnose temperature control problems in the heater/ventilation system. ■ Remove, inspect, and reinstall the heater control valve(s) and heater core. ■ Describe how an automotive air-conditioning system operates. ■ Explain why R-134a is the current refrigerant of choice. ■ Locate, identify, and describe the function of the various air-conditioning components. ■ Describe the operation of the types of air-conditioning control systems.

The first automobiles had few features that provided for driver and passenger comfort. In winter, heavy coats and blankets were all that allowed passengers to survive behind a woefully inadequate windscreen. In summer, the breeze generated by 15 mph (25 km/h) travel was the only thing that cooled the passengers.

Today's vehicles have ventilation, heating, and air-conditioning systems to provide passenger comfort. Ventilation and heating systems are standard equipment on all passenger vehicles and air-conditioning is standard on most and available for nearly all. These systems move heat to warm or cool the passengers.

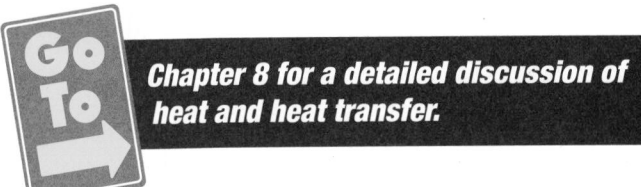

Go To ***Chapter 8 for a detailed discussion of heat and heat transfer.***

VENTILATION SYSTEM

The ventilation system on most vehicles is designed to supply outside air to the passenger compartment through upper or lower vents or both. Several systems are used to vent air into the passenger compartment, the most common of which is the flow-through system (**Figure 52–1**). In this arrangement, a supply of outside air, called ram air, flows into the car when it is moving. When the car is not moving, a steady flow of outside air can be produced by the heater fan. In operation, ram air is forced through an inlet grille. The pressurized air then circulates throughout the passenger and trunk compartment. From there the air is forced outside the vehicle through an exhaust area.

On certain older vehicles, air is admitted by opening or closing two vent knobs under the dashboard. The left knob controls air through the left inlet. The

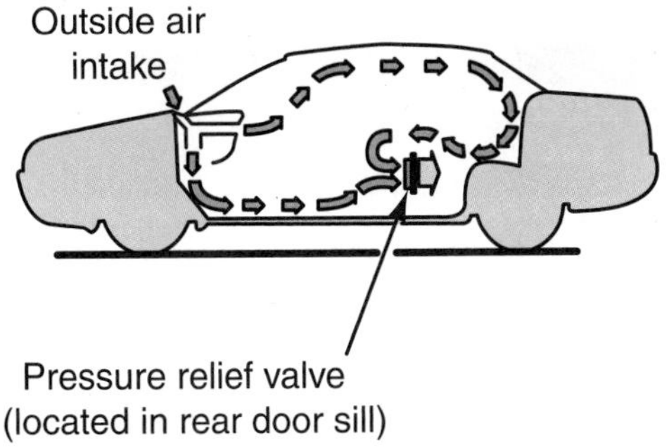

Figure 52–1 A flow-through ventilation system.

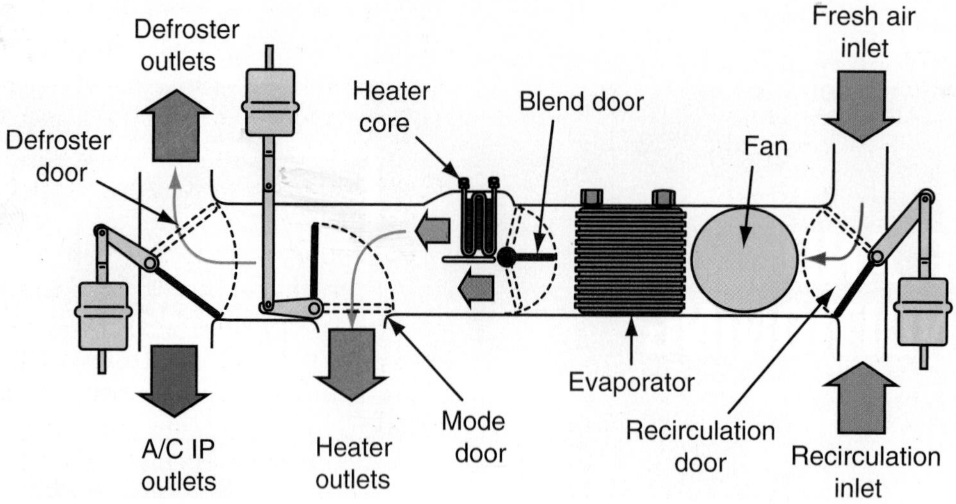

Figure 52-2 A typical HVAC airflow for a heater and air-conditioning.

right knob controls air through the right inlet. The air is still considered ram air and is circulated through the passenger compartment.

Rather than using ram air (especially if the vehicle is stopped), a ventilation fan can be used. It can be accessible from under the dashboard or from inside the engine compartment. A blower assembly is placed inside the blower housing. As the squirrel cage blower rotates, it produces a strong suction on the intake. A pressure is also created on the output. When the fan motor is energized by using the temperature controls on the dashboard, air is moved through the passenger compartment. Most vehicles have a cabin filtration system that cleans the air before it enters the passenger compartment.

AUTOMOTIVE HEATING SYSTEMS

The automotive heating system has been designed to work hand in hand with the cooling system to maintain proper temperatures inside the car. The heating system's primary job is to provide a comfortable passenger compartment temperature and to keep car windows clear of fog or frost.

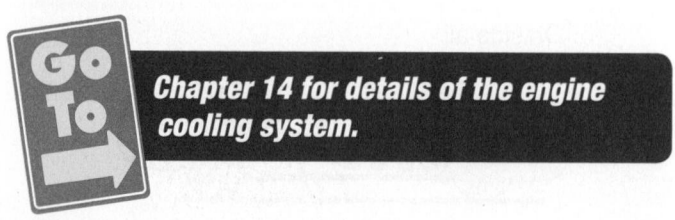

Go To Chapter 14 for details of the engine cooling system.

To meet federal safety standards, all vehicles must be equipped with passenger compartment heating and windshield defrosting systems. The main components of an automotive heating system are the heater core, the heater control valve, the blower motor and the fan, and the heater and defroster ducts **(Figure 52–2)**. The heating system works with the engine's cooling system and converts the heat from the coolant circulating inside the engine to hot air, which is blown into the passenger compartment.

SHOP TALK

In most vehicles, the air-conditioning will turn on when the defroster is selected by the driver and the ambient temperature is more than 35°F (1.67°C). This is done to remove moisture from the vehicle's interior.

In the liquid-cooling system, heat from the coolant circulating inside the engine is converted to hot air, which is blown into the passenger compartment. Hot coolant from the engine is transferred by a heater hose to the heater control valve and then to the heater core inlet **(Figure 52–3)**. As the coolant circulates through the core, heat is transferred from the coolant to the tubes and fins of the core. Air blown through the core by the blower motor and fan then picks up the heat from the surfaces of the core and transfers it to the passenger compartment of the car. After giving up its heat, the coolant is then pumped out through the heater core outlet, where it is returned to the engine to be heated again.

Heater Core

The **heater core** is generally designed and constructed much like a miniature radiator **(Figure 52–4)**. It features an inlet or outlet tube and a tube and fin core to facilitate coolant flow between them.

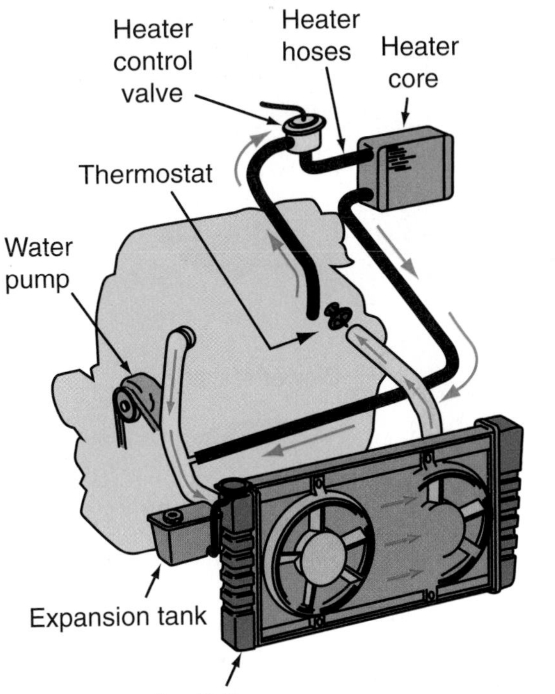

Figure 52-3 Hot coolant from the engine is sent from the upper portion of the engine to the heater core.

Figure 52-4 A heater core.

Although all heater cores basically function in the same manner, several variations in design and materials are used by different automakers to achieve the same results. Although the core construction varies, this type has an aluminum core with plastic tanks.

Heater Control Valve

The **heater control valve** (sometimes called the water flow valve) controls the flow of coolant into the heater core from the engine. In a closed position, the valve allows no flow of hot coolant to the heater core, keeping it cool. In an open position, the valve allows heated coolant to circulate through the heater core, maximizing heater efficiency. Heater control valves are operated in three basic ways: by cable, thermostat, or vacuum. Some vehicles do not use a heater control valve; rather, a heater door controls how much heat is released into the passenger compartment from the heater core.

Cable-operated valves are controlled directly from the heater control lever on the dashboard. Thermostatically controlled valves feature a liquid-filled capillary tube located in the discharge air stream off the heater core. This tube senses air temperature, and the valve modulates the flow of coolant to maintain a constant temperature, regardless of engine speed or temperature.

Most heater valves utilized on today's car are vacuum operated. These valves are normally located in the heater hose line or mounted directly in the engine block. When a vacuum signal reaches the valve, a diaphragm inside the valve is raised, either opening or closing the valve against an opposing spring. When the temperature selection on the dashboard is changed, vacuum to the valve is vented and the valve returns to its original position. Vacuum-actuated heater control valves are either normally open or normally closed designs. Some vehicles do not use a heater control valve; rather, a heater door controls how much heat is released into the passenger compartment from the heater core.

On late-model vehicles, heater control valves are typically made of plastic for corrosion resistance and light weight (**Figure 52–5**). These valves feature few internal working parts and no external working parts. With the reduced weight of these valves, external mounting brackets are not required.

Systems without a Control Valve Some systems do not have a heater control valve; rather, heat inside the passenger compartment is controlled by changing

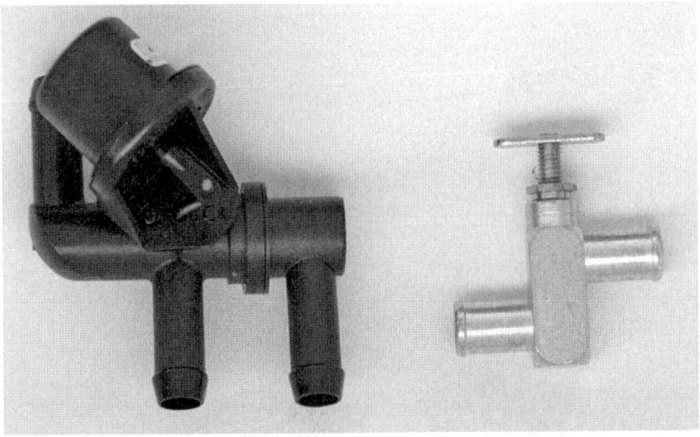

Figure 52-5 Two types of heater control valves: (left) a vacuum-operated plastic unit, and (right) a metal cable-operated valve.

the airflow over the heater core. In these systems, hot coolant always flows through the heater core. When heat is needed, air is allowed to flow through the heater core. The amount of heat is determined by the amount of air allowed to pass through the core.

Airflow is controlled by a **blend door** inside the blower housing. The movement of this door can be controlled by a cable, vacuum, or electrically with a motor.

PTC Heaters Today's engines are designed to reduce the amount of waste heat they emit; therefore, the amount of heat available for the heating system may be limited. This is especially true with hybrid vehicles with the stop-start feature. Also, in conventional vehicles, interior heat is not available until the coolant warms up and transfers heat through the heater core.

Some late-model vehicles, especially hybrids, have an auxiliary positive temperature coefficient (PTC) unit **(Figure 52–6)** or a PTC heating element in the heater core. These heaters are able to provide warm air before the engine's coolant warms up. PTC heaters are electronically controlled and work independently of the cooling system.

When current passes through a PTC element, heat is produced and the air that passes through the heater core is warmed. Some vehicles have an additional PTC element installed in an air duct from the blower housing; this helps to increase the air temperature in the ducts. PTC elements react quickly to changes in current. The elements are small ceramic stones. These are typically barium titanate polycrystalline ceramics doped with metals.

Blower Motor

The blower motor is usually located in the heater housing assembly. It ensures that air is circulated through the system **(Figure 52–7)**. Its speed is controlled by a multiposition switch in the control panel. The switch works in connection with a resistor block that is usually located on the heater housing.

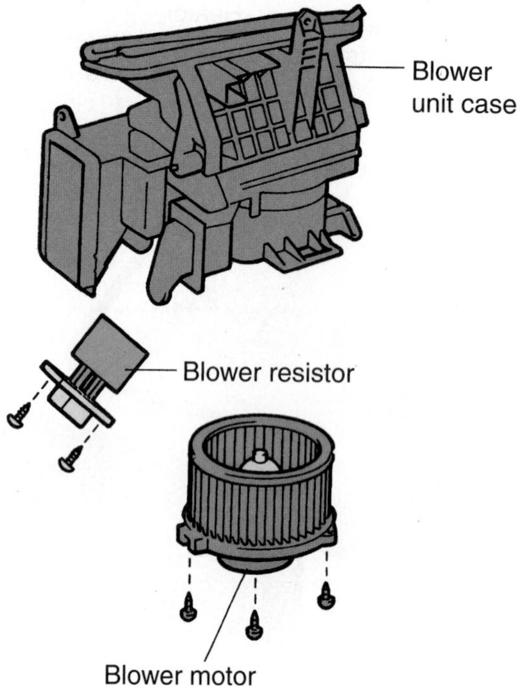

Figure 52-7 A blower motor assembly. *Courtesy of Toyota Motor Sales, U.S.A., Inc.*

On some vehicles, when the engine is running, the blower motor is in constant operation at low speed. On automatic temperature control systems, the blower motor is activated only when the engine reaches a predetermined temperature. The blower motor circuit is protected by a fuse located in the fuse panel.

The blower motor resistor block is used to control the blower motor speed. The typical resistor block is composed of three or four wire resistors in series with the blower motor that control its voltage and current. The speed of the motor is determined by the control panel switch, which puts the resistors in series. Increasing the resistance in the system slows the blower speed.

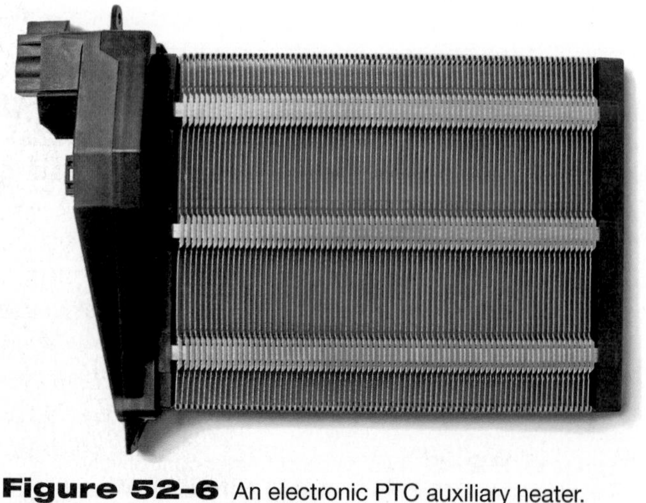

Figure 52-6 An electronic PTC auxiliary heater. *Courtesy of BERU Corporation*

Heater and Defroster Duct Hoses

Transferring heated air from the heater core to the passenger compartment heater and defroster outlets is the job of the heater and defroster ducts. The ducts are typical parts of a large plastic shell that connects to the necessary inside and outside vents. This ductwork also has mounting points for the evaporator and heater core assemblies. Contained inside the duct are the mode doors required to direct air to the floor, dash, and/or windshield. Sometimes the duct

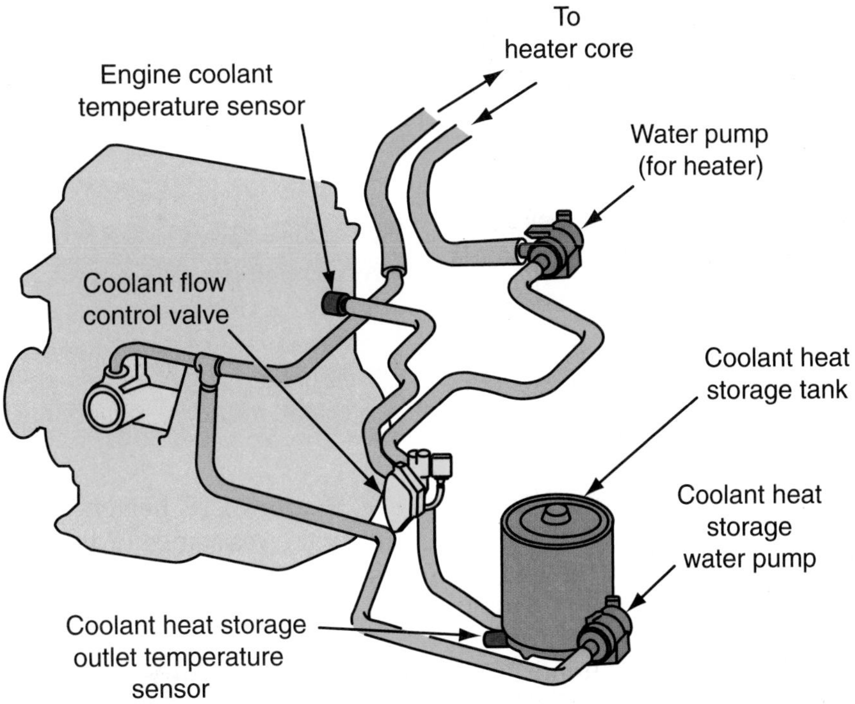

Figure 52-8 The coolant heat storage tank is a large vacuum-insulated container that is capable of storing hot coolant for a long period.

is connected directly to the vents, whereas other times hoses are used.

Recirc/Fresh Door Many vehicles have a recirc/fresh mode. This feature allows the driver to recirculate (recirc) the air inside the vehicle. This mode is normally selected when outside air is entering the vehicle and it has unpleasant odors. Outside or fresh air normally enters the interior through the normal heating, ventilation, and A/C systems. While operating in the recirc mode, the system uses the air from the interior for those systems. A recirculation door closes off the outside airflow and opens the intake for interior air.

Heaters in Hybrid Vehicles

In a hybrid electric vehicle, the engine can be used to supply the heat, so heating and defrosting systems are similar to those used in conventional vehicles. However, some hybrids have additional electrical heaters. These keep the passenger compartment warm when the engine is off.

Other systems have a 12- or 42-volt auxiliary electric water pump. This pump is used to circulate coolant through the engine to maintain heater performance during an idle stop. A heater control module operates the pump based on an idle stop signal from the PCM. If the engine's temperature drops below a predetermined level during an idle stop, the engine will restart. The system also has a temperature sensor at the heater core. This sensor monitors the activity of the electric water pump. If it senses that the heater core is losing heat quickly, the control module will assume that the pump has failed and will start the engine and set a DTC.

The cooling systems used in some hybrids feature coolant storage tanks. Hot coolant is stored in a container **(Figure 52–8)** where it can remain hot for quite a long period. The hot coolant is circulated through the engine immediately after startup. The fluid also may circulate through the engine many hours after it is shut off. The stored coolant can provide heat for the passenger compartment and allows the engine to warm up quickly, thereby reducing emission levels during startup.

HEATING SYSTEM SERVICE

When doing system checks and services, always follow the procedures recommended by the manufacturer. In most cases, problems with the heating system are problems with the engine's cooling system. Therefore, most service work and diagnosis are done to the cooling system. Problems that pertain specifically to the heater are few, these being the heater control valve and the heater core. Most often if these two items are faulty, the engine's cooling system will be negatively affected. Both of these items are replaced rather than repaired. Some problems will pertain only to the heater controls. In some cases, it is possible to make repairs to vacuum hose and electrical connections without removing the heater assembly. If it is necessary to remove the heater assembly, the cooling system must be drained before removing the heater core.

Go To Chapter 14 for details on diagnosing and servicing engine cooling systems.

Basic Heater Inspection and Checks

When there is a problem of insufficient heat, begin your diagnosis with a visual inspection and a check of the coolant level. If the level is correct, turn the heater controls on and run the engine until it reaches normal operating temperature. Then measure the temperature of the upper radiator hose. The temperature can be measured with a pyrometer. If one is not available, gently touch the hose. You should not be able to hold the hose long because of the heat. While doing this, make sure you stay clear of the area around the cooling fan. A spinning fan can chop off your hand. If the temperature of the hose is not within specifications, suspect a faulty thermostat.

If the hose is the correct temperature, check the temperature of the two heater hoses. They should both be hot. If only one of the hoses is hot, suspect the problem to be the heater control valve or a plugged heater core.

Heater Core Service

Like the radiator, heater core tanks, tubes, and fins can become clogged over time by rust, scale, and mineral deposits circulated by the coolant. Heater core failures are generally caused by leakage or clogging. Feel the heater inlet and outlet hoses while the engine is idling and warm with the heater temperature control on hot. If the hose downstream of the heater valve does not feel hot, the valve is not opening.

If the heater core appears to be plugged, the inlet hose may feel hot up to the core but the outlet hose remains cool. Reverse flushing the core with a power flusher may open up the blockage, but usually the core has to be removed for cleaning or replacement. Air pockets in the heater core can also interfere with proper coolant circulation. Air pockets form when the coolant level is low or when the cooling system is not properly filled after draining.

When the heater core leaks and must be repaired or replaced, it is a very difficult and time-consuming job, primarily because of the core's location deep within the bulkhead of the car. For this reason always leak test a replacement heater core before installation. Also flush the cooling system and replace the coolant seasonally.

SHOP TALK

A common procedure for servicing a cooling system is reverse flushing it. However, some manufacturers do not recommend reverse flushing their heater cores. Always check with the service information before reverse flushing a system.

PTC Heaters PTC heaters can be checked by measuring the resistance of the heating element. If the resistance does not meet specifications **(Figure 52–9)**, the assembly should be replaced.

Heater Control Valve Service

When there is a problem with the control valve, it is typically caused by the controls or the valve itself. When the valve is bad, it should be replaced. If the heater control valve or its controls prevent the valve from opening all the way, heating will be reduced in the passenger compartment. If the valve does not completely close, the compartment will not cool properly when the A/C is turned on.

With cable-operated control valves, check the cable for sticking, slipping (loose mounting bracket), or misadjustment. With valves that are vacuum operated, there should be no vacuum to the valve when the heater is on (except for those that are normally closed and need vacuum to open).

The thermostat, which helps regulate the coolant temperature in the cooling system, plays a large part in the heating system. A malfunctioning thermostat

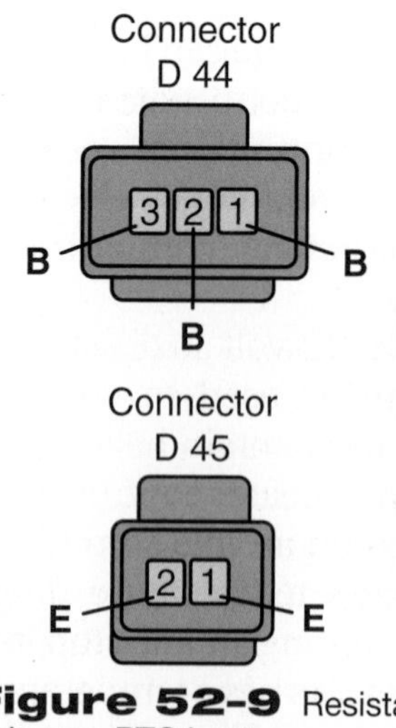

Connections for testing
D44-2 (B) to D45-1 (E)
D44-2 (B) to D45-2 (E)
D44-3 (B) to D45-2 (E)
Value for all: 0.5 to 2.5Ω

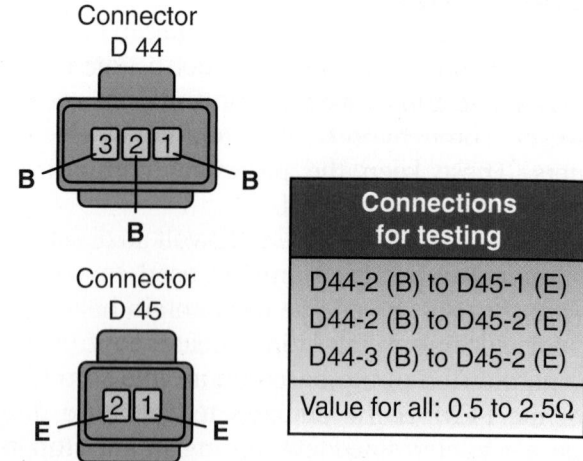

Figure 52-9 Resistance testing points at the terminals on a PTC heater.

can cause the engine to overheat or not reach normal operating temperature, or it can be the cause of poor heater performance.

SHOP TALK

Poor heating can be caused by many things. A low coolant level can starve the heater, resulting in little or no heat output. Check the coolant level in the radiator (not the overflow tank) to see if it is low. Air pockets can form in long heater hoses or where the heater is mounted higher than the radiator. To vent the trapped air, some vehicles have bleeder valves on the hoses. Opening the valves allows air to escape as the system is filled.

Blower Motor Service

If the blower motor does not operate, use a DMM to make sure there is voltage on both sides of the fuse. Then check to see if there is voltage at the motor. If the blower motor is getting voltage, the problem is either a burned-out blower motor or a bad ground on the motor. In situations where no voltage is available at the motor, backtrack to check for an open resistor. Check also for proper relay operation and for burned or corroded connections at the blower relays or in the bulkhead connectors. Inline fuses are another potential problem.

Heater and Defroster Duct Service

If the blower motor runs but no air comes out of the ducts, the problem is either a stuck or an inoperative airflow control valve or blend door. This can also affect the operation of the defrosters. These doors may be cable, electrically, or vacuum operated. To further diagnose the problem, change the position of the temperature selector knob, sliding it from hot to cold. If you do not hear the sound of doors opening and closing, it means the control cables have slipped loose from the dash switch or door arm, which can sometimes occur, rendering the door inoperative. If this is the case, you will feel little or no resistance when sliding the temperature control knob. A kinked or rusted cable can also prevent a door from working. If this is the case, you will feel resistance when trying to move the control knob. In either case, it is necessary to get under the dash, find the cable, and then replace, reroute, or reconnect it. Doors can also be jammed by objects that have fallen down the defroster ducts. Remove the obstruction from the plenum by fishing through the heater outlet with a coat hanger or magnet, or remove the plenum. Be careful when fishing for the obstruction; some heater cores are easily pierced.

Most current HVAC systems use electric servomotors to move the various doors. Each motor is fitted with a potentiometer that sends a signal to the A/C control unit regarding the actual position of the door. If the actual position is not the same as the commanded position, most systems will set a DTC, indicating that the door is not working properly. Before proceeding with detailed diagnostics of the electrical system, make sure there are no linkage problems or that the door is not physically stuck. Normally, if the problem is electrical, the A/C control unit, servomotor, or wiring harness is faulty.

With vacuum-controlled doors, the most common causes of failure are leaky or loose vacuum hoses or defective diaphragms in the vacuum motors that move the doors. Check the vacuum by starting the engine and disconnecting the hose that goes to one of the door's vacuum motor. If you feel a vacuum or hear a hissing sound when trying different temperature settings, the vacuum source is good. Apply vacuum to the motor with a hand-held pump; if it moves and holds vacuum for about 1 minute then bleeds off, the problem is a bad vacuum motor. If it does not, check for leaky vacuum hose connections, a defective temperature control switch, or a leaky vacuum reservoir or check valve under the dash or in the engine compartment.

THEORY OF AUTOMOTIVE AIR-CONDITIONING

Air-conditioning basically supplies cool air into the passenger compartment. It also removes moisture from the air. An A/C system cools the air by moving heat from the confined space of the passenger compartment to the atmosphere.The operation of all A/C systems is based on fundamental laws of nature.

Go To *Chapter 8 for a detailed discussion of these laws.*

Heat Flow

An air-conditioning system is designed to pump heat from one point to another. All materials or substances, as cold as −459°F (−273°C), have heat in them. Also, heat always flows from a warmer object to a colder one. For example, if one object is at 30°F (−1.1°C) and another object is 80°F (27°C), heat flows from the warmer object (80°F [27°C]) to the colder one (30°F [−1.1°C]). The greater the temperature difference between the objects, the greater the amount of heat flow.

Heat Absorption

Objects can be in one of three forms: solid, liquid, or gas. When objects change from one state to another, large amounts of heat can be transferred. For example, when water temperature goes below 32°F (0°C), water changes from a liquid to a solid (ice). If the temperature of water is raised to 212°F (100°C), the liquid turns into a gas (steam). But an interesting thing occurs when water, or any matter, changes from a solid to a liquid and then from a liquid to a gas. Additional heat is necessary to change the state of the substance, even though this heat does not register on a thermometer. For example, ice at 32°F (0°C) requires heat to change into water, which will also be at 32°F (0°C). Additional heat raises the temperature of the water until it reaches the boiling point of 212°F (100°C). More heat is required to change water into steam. But if the temperature of the steam were measured, it would also be 212°F (100°C). The amount of heat necessary to change the state of a substance is called latent heat—or hidden heat—because it cannot be measured with a thermometer. This hidden heat is the basic principle behind all air-conditioning systems.

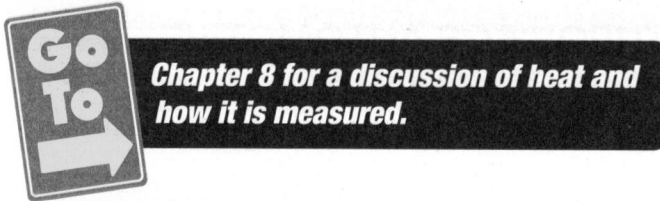

Go To Chapter 8 for a discussion of heat and how it is measured.

Pressure and Boiling Points

Pressure also plays an important part in air-conditioning. Pressure on a substance, such as a liquid, changes its boiling point. The greater the pressure on a liquid, the higher the boiling point. If pressure is placed on a vapor, the vapor condenses at a higher-than-normal temperature. In addition, as the pressure on a substance is reduced, the boiling point can also be reduced. For example, the boiling point of water is 212°F (100°C). The boiling point can be increased by increasing the pressure on the fluid. It can also be decreased by reducing the pressure or placing the fluid in a vacuum.

Relative Humidity

The amount of moisture that air can hold is directly related to the temperature of the air. The warmer the air, the more moisture it can hold. Therefore, lowering the temperature of the air extracts the moisture from the air and lowers the air's relative humidity. Low relative humidity is more comfortable than high humidity.

REFRIGERANTS

An air conditioning (A/C) system is designed to move heat from one point to another. In an automobile, heat is removed from the passenger compartment and moved to outside the vehicle. The substance used to move the heat is called the **refrigerant**.

Before 1994, most automotive A/C systems used a refrigerant called Refrigerant-12 (commonly referred to as R-12 and Freon). R-12 is dichlorodifluoromethane (CCl_2F_2). By law, R-12 is no longer used in A/C systems. This resulted from studies showing that the earth's ozone layer was being depleted by the **chlorofluorocarbons (CFCs)** found in R-12. The ozone layer is the earth's outermost shield of protection. This delicate layer protects against harmful effects of the sun's ultraviolet rays. Because A/C systems with R-12 are susceptible to leaks, further damage to the ozone layer could be avoided by not using R-12 in A/C units.

R-134a

Of the many chemicals that could have replaced R-12, automobile manufacturers decided to use R-134a, which contains no chlorine. this refrigerant may also be referred to as SUVA. R-134a is tetrafluoroethane (CH_3CF_3) and considered a **hydrofluorocarbon (HFC)** that causes less damage to the ozone layer when released to the atmosphere. Although R-134a air conditioners operate in the same way and with the same basic components as R-12 systems, the two refrigerants are not interchangeable.

R-134a systems operate at higher pressures and are designed to handle these pressures **(Figure 52–10)**. R-134a systems also require different service techniques and equipment. All R-134a systems are identified by an underhood decal **(Figure 52–11)** and by the hoses and fittings used in the system. Service equipment for R-134a is also different. Retrofit kits are available to convert older R-12 systems to R-134a.

Although R-134a is less likely to have an adverse effect on the ozone layer, it still has the capability of contributing to the "greenhouse effect" when released into the air. The recovery and recycling of R-12 and R-134a is mandatory by law.

In 1987, many countries worldwide agreed to the Montreal Protocol agreement. This agreement began the phasing out of R-12 refrigerant. This agreement was followed up by 1990 U.S. Clean Air

Ambient Temp	High side PSIG, R-134a	Low side PSIG, R-134a	High side PSIG, R-12	Low side PSIG, R-12
60°F (15.6°C)	120–170	7–15	120–150	5–15
70°F (21.1°C)	150–250	8–16	140–180	8–16
80°F (26.7°C)	190–280	10–20	160–250	10–18
90°F (32.2°C)	220–330	15–25	200–280	12–25
100°F(37.8°C)	250–350	20–30	220–300	15–30

Figure 52-10 A chart comparing the pressures of R-134a and R-12 at various temperatures.

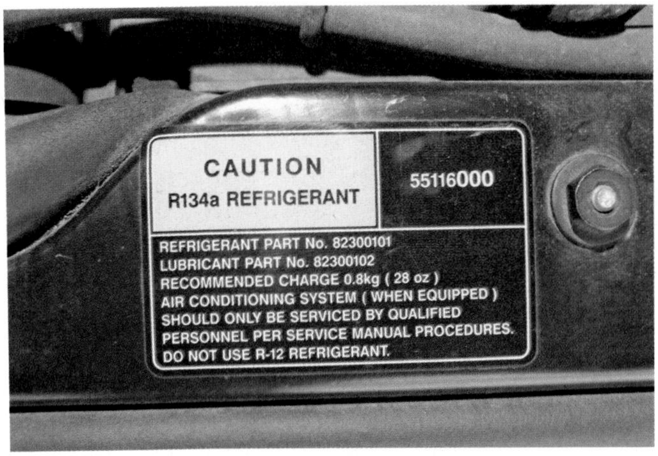

Figure 52-11 A typical decal informing the technician that the system uses R-134a.

Act that mandated that all refrigerants were to be recovered and recycled. Section 609 of the act requires that all service technicians who work on A/C systems must be certified in proper refrigerant recovery and recycling. This section also mandates that all recovery and recycling equipment must be certified. Section 609 is a United States requirement and is not valid in other countries. Some states have stricter laws regarding refrigerant and these must be followed.

The Kyoto Protocol, a worldwide environmental program, has set standards for the reduced use of greenhouse gases, including R-134a. Several countries have signed on to the program; so far the United States is not one of them. As a result, European and Japanese manufacturers will eliminate R-134a A/C systems by 2011. This means those companies will be shipping vehicles to the United States with an alternative refrigerant and U.S. service facilities will need to be equipped to service R-134a systems and the variety of alternative refrigerants that will be used.

Alternative Refrigerants

There are different refrigerants that can be used in automotive A/C systems; these are typically called alternative refrigerants. The EPA has a list of approved alternative refrigerants; however, the OEMs say that only R-134a should be used when retrofitting an R-12 system. The use of any refrigerant that contains flammable substances, such as propane and butane, is illegal and dangerous.

Finding an alternative to R-134a is the focus of much research. The replacement cannot be a greenhouse gas. The safety and functionality of several new refrigerants is currently being tested and some of these will be found in automobiles before 2011. A rating scale has been developed to specify a refrigerant's global warming potential (GWP). R-134a has a GWP of 1,430. An alternative refrigerant is R-152a, which has a GWP level of 124. R-152a is a hydrofluorocarbon variant.

CO_2 Systems

To meet the standards set up by the Kyoto Protocol, auto manufacturers need to find an alternative to R-134a. BMW and other European manufacturers have chosen CO_2 as the refrigerant for their A/C systems in the future. CO_2 is nontoxic when used as a refrigerant. It is known as R-744 and has a GWP of 1. The extremely low GWP means there is little environmental concern with its use. In fact, this refrigerant will not need to be recovered and recycled. If CO_2 leaks from an A/C system, the effect on the environment is very small. CO_2 is abundantly available in our air; therefore, it does not need to be manufactured.

It is claimed that CO_2-based systems will be up to 25% more energy efficient than the best of today's R-134a systems. In addition, CO_2 can be used for heat pump systems. Heat pump systems can supply cool and warm air and can be used in hybrid vehicles. Also, CO_2 systems need a smaller amount of gas

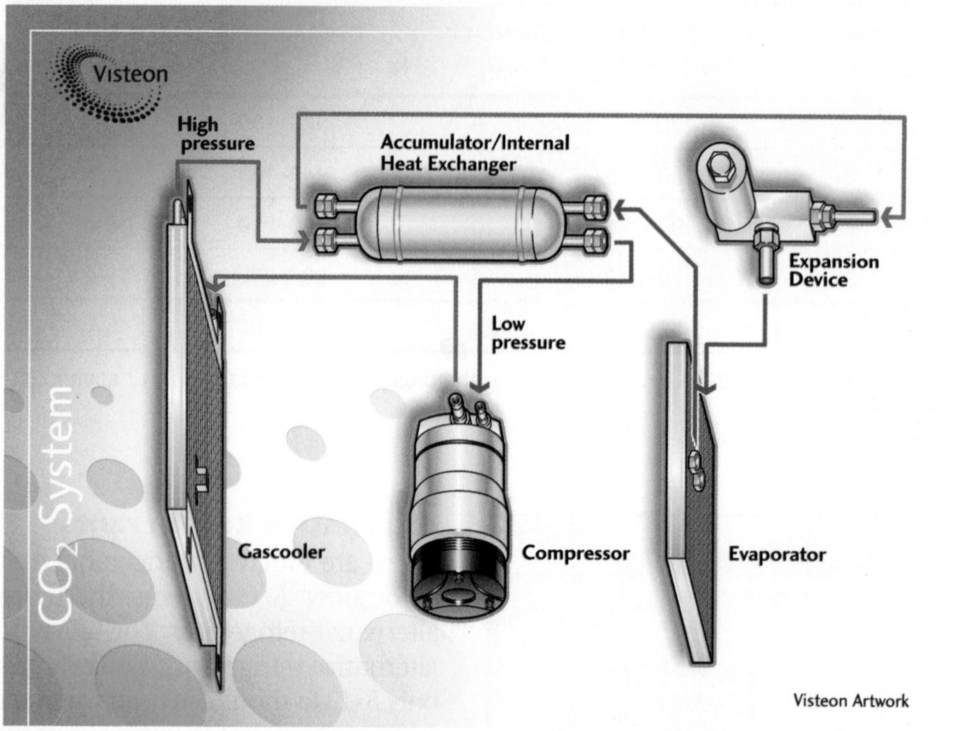

Figure 52-12 The layout of a CO_2 A/C system. *Courtesy of Visteon Corporation™*

than conventional systems, which means the size of the system is substantially less than a conventional system.

The operation of a CO_2 system is the same as that of any other A/C system but the components are different due to the higher pressures. CO_2 systems operate at pressures that are nearly ten times that of an R-134a system. CO_2 also has a critical temperature that is much lower than R-134a. Critical temperature is the temperature above which a substance cannot exist in the liquid state regardless of the pressure. Therefore, these systems need an internal heat exchanger (IHX) and accumulator. The materials used for hoses and gaskets must also be slightly different. The IHX is located between the condenser and the evaporator. The accumulator is part of the IHX **(Figure 52–12)**.

BASIC OPERATION OF AN AIR-CONDITIONING SYSTEM

Refrigerants are used to carry heat from the inside of the vehicle to the outside of the vehicle. Automotive refrigerants have a low boiling point (the point at which **evaporation** occurs). For example, at any temperature above −15.34°F (−26.3°C), liquid R-134a can become a vapor. As a refrigerant changes state, it absorbs a large amount of heat. Because the heat that it absorbs is from the inside of the vehicle, passengers are cooler.

To understand how a refrigerant is used to cool the interior of a vehicle, the effects of pressure and temperature must be understood first. If the pressure of the refrigerant is high, so is its temperature. Likewise, if the pressure is low, so is its temperature. Therefore, the temperature of the refrigerant can be changed by changing its pressure. As the pressure on a liquid increases, its boiling point also increases. Likewise, as the pressure decreases, the boiling point of the liquid also decreases. Figure 52–6 compares the pressures of R-12 and R-134a at various temperatures.

To absorb heat, the temperature and pressure of the refrigerant are kept low. To dissipate heat, the temperature and pressure are high. As the refrigerant absorbs heat, it changes from a liquid to a vapor. As it dissipates heat, it changes from a vapor to a liquid. The change from a vapor to a liquid is called **condensation**. These two changes of state—evaporation and condensation—occur continuously as the refrigerant circulates through the system.

Refrigeration Cycle

In a basic A/C system, the heat is absorbed and transferred in the following steps **(Figure 52–13)**.

1. When the system is off, the refrigerant occupies the system as a vapor and its pressure is the same throughout the system.

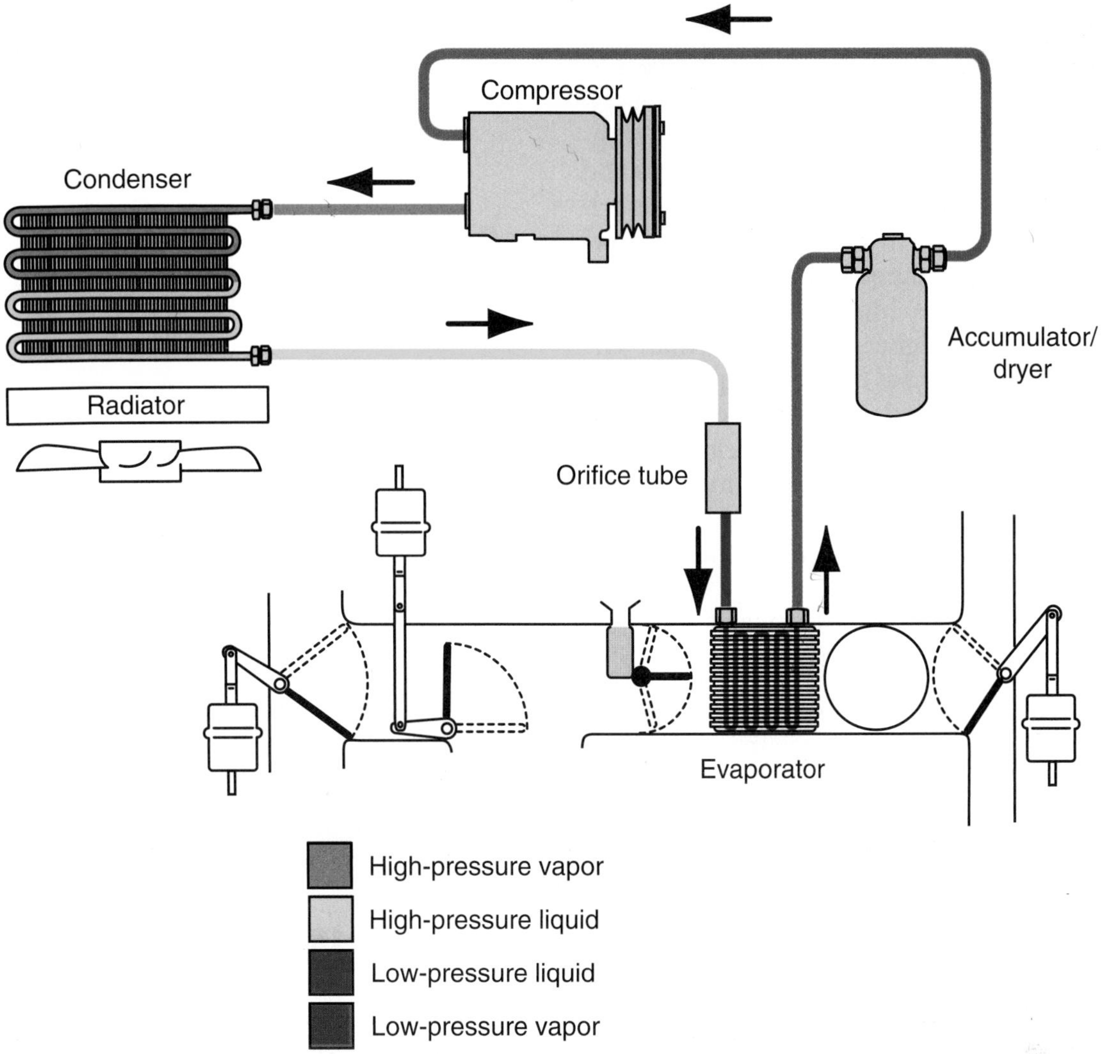

Figure 52-13 The basic refrigerant flow cycle.

2. When the A/C compressor is turned on, it increases pressure on the refrigerant and raises its temperature.
3. The refrigerant is pumped out of the compressor as a high-pressure, high-temperature vapor.
4. The high-pressure gas is sent to the condenser. Heat in the refrigerant is moved to the outside air by conduction and convection. The removal of the heat causes the refrigerant vapor to condense into a liquid under high pressure.
5. The high-pressure, high-temperature liquid leaves the bottom of the condenser and enters a receiver/dryer or accumulator. These devices remove moisture and contaminants and store clean refrigerant until it is needed.
6. The refrigerant then flows to the inlet side of the evaporator core orifice or expansion valve. These control the flow of refrigerant into the evaporator. The restriction causes the pressure and boiling point of the refrigerant to drop.
7. As the liquid refrigerant leaves the restriction, it is at its lowest pressure and temperature.
8. The refrigerant then passes through the evaporator where it absorbs heat, through convection, from the air inside the passenger compartment. The additional heat causes the refrigerant to boil and change back to a vapor.
9. The refrigerant then returns to the compressor as a low-pressure, low-temperature vapor and the cycle continues.

An automotive A/C system is a closed, pressurized system. It consists of a compressor, condenser, receiver/dryer or accumulator, expansion valve or orifice tube, and an evaporator. To understand the operation of these components, remember that an A/C system is divided into two sides: the high side and the low side. **High side** refers to the side of the system that is under high pressure and high temperature. **Low side** refers to the low-pressure, low-temperature side of the system.

COMPRESSORS

The compressor is the heart of the automotive A/C system. It separates the high and low sides of the system. The compressor is designed to pump only refrigerant vapor; liquid refrigerant will not compress and its presence in the compressor can damage it. Its primary purpose is to draw the low-pressure and low-temperature vapor from the evaporator and compress it into high-temperature, high-pressure vapor. The refrigerant then has a higher temperature than surrounding air and condenses back to a liquid form at the condenser. The secondary purpose of the compressor is to circulate or pump the refrigerant through the A/C system under the different pressures required for proper operation. The compressor is located on the engine and is typically driven by the engine's crankshaft via a drive belt.

Although there are numerous types of compressors in use today **(Figure 52–14)**, they are usually based on one of these designs.

Piston Compressor

This type of compressor **(Figure 52–15)** can have its pistons arranged in an inline, axial, radial, or V design. It is designed to have an intake stroke and a compression stroke for each cylinder. On the intake stroke, the refrigerant from the low side of the system (evaporator) is drawn into the compressor. The intake of refrigerant occurs through intake reed valves **(Figure 52–16)**. These one-way valves control the flow of refrigerant vapors into the cylinder. During the compression stroke, the refrigerant vapor is compressed. This increases both the pressure and the temperature of the refrigerant. The outlet or **discharge side** reed valves then open to allow the refrigerant to move to the condenser. The outlet reed valves are the beginning of the high side of the system. Reed valves are made of spring steel, which can weaken or break if improper charging procedures are used, such as liquid charging with the engine running.

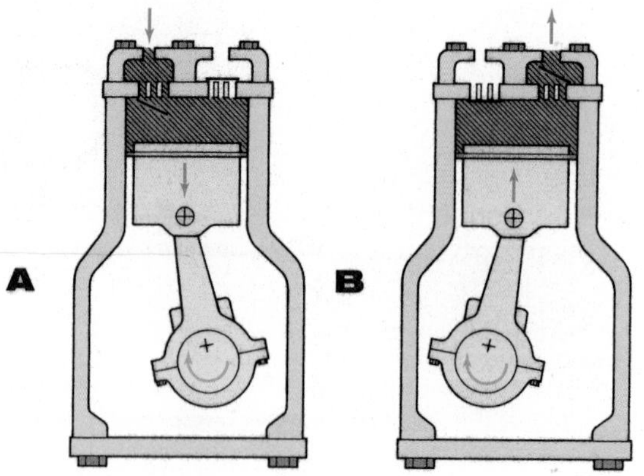

Figure 52–15 (A) A piston on the downstroke (intake) pulls low-pressure refrigerant into the cylinder cavity. The intake (suction) valve is open, and the discharge valve is closed. (B) A piston on the upstroke (discharge) compresses refrigerant vapor and forces it out through the discharge valve. The intake (suction) valve is closed, and the discharge valve is open.

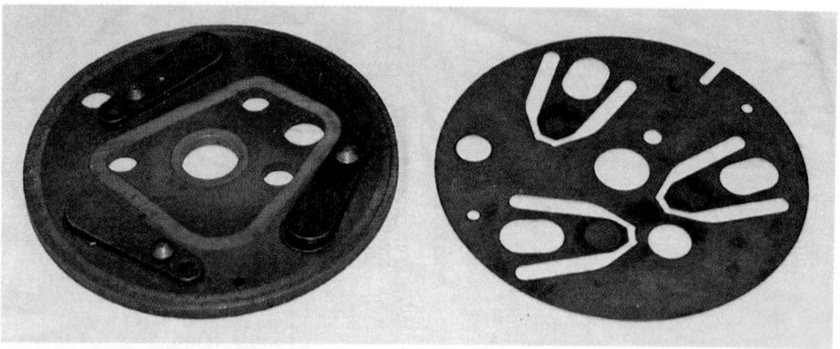

Figure 52–16 The reed valves for an A/C compressor.

Figure 52–14 Various A/C compressors.

Variable Displacement Compressor Today nearly all manufacturers use variable displacement compressors **(Figure 52–17)**. These compressors are normally axial compressors, with the pistons arranged around and parallel to the drive shaft. The pistons are driven by a wobble plate or a swash plate. When a wobble plate is used, short pushrods connect the pistons to the plate **(Figure 52–18)**. As the compressor's drive shaft rotates, the plate wobbles and the pistons move in their bores. When a swash plate is used, it is set at an angle to the drive shaft. As the shaft rotates, the pistons move back and forth in their bores.

The angle of the wobble plate or swash plate determines the stroke of the pistons. When the stroke of the pistons is increased, more refrigerant is being pumped and there is increased cooling. The angle is controlled by the difference in pressure between the outlet and inlet of the compressor. When the pressure inside the compressor increases, that pressure works on the bottom of the pistons and moves them closer to the cylinder head. This action shortens their

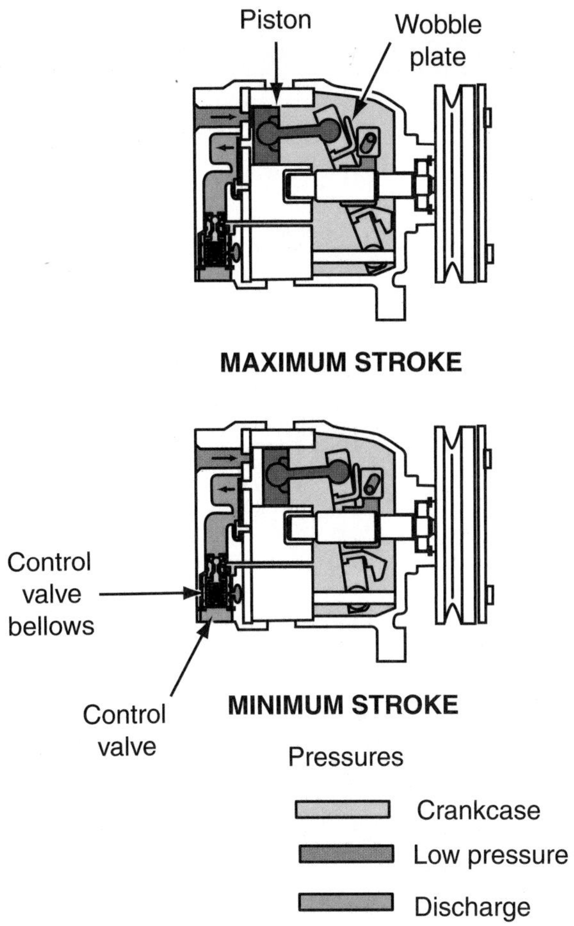

Figure 52-17 A V5 variable displacement compressor.

Figure 52-18 The pistons are driven by a short pushrod that connects the pistons to the wobble plate.

stroke, thereby decreasing the displacement of the pump. As pressure inside the compressor decreases, spring tension overcomes the pressure and the pistons move down in their bores. This causes the plate angle to increase, resulting in increased stroke and displacement.

Some swash plate compressors use a solenoid control valve that opens and closes to adjust the inlet to the compressor. Controlling the suction side of the compressor changes the volume capacity and pressure. The change in pressure also affects the angle of the swash plate.

On some vehicles, the compressor does not have a clutch, which means the compressor runs constantly. These units have an electrically controlled rotary valve at the rear of the compressor to control the flow of refrigerant into a special chamber that changes the angle of the swash plate. This regulates the flow of refrigerant by changing the stroke of the pistons.

Rotary Vane Compressor

The rotary vane compressor does not have pistons. It has a rotor with several vanes and a carefully shaped housing. The sliding vanes seal against the housing at both ends. As the compressor shaft rotates, the vanes and housing form chambers. As the rotor turns, the size of the chambers changes.

The refrigerant is drawn into the chambers through the suction port. The discharge port is located at the point where the gas is completely compressed. No sealing rings are used in a vane compressor. The vanes are sealed against the housing by centrifugal force and lubricating oil. The oil sump is located on the discharge side, so the high pressure tends to force it around the vanes into the low-pressure side. This action ensures continuous lubrication. Because this type of compressor depends on a good oil supply, it is subject to damage if the system charge is lost. A protection device is used to disengage the clutch if pressure drops too low.

Scroll-Type Compressor

The scroll-type compressor has a movable scroll and a fixed or nonmovable scroll that provide an eccentric-like motion. As the compressor's crankshaft rotates, the movable scroll forces the refrigerant against the fixed scroll and toward the center of the compressor. This motion pressurizes the refrigerant. The action of a scroll-type compressor can be compared to that of a tornado. The pressure of air moving in a circular pattern increases as it moves toward the center of the circle **(Figure 52–19)**. A delivery port is positioned at the center of the compressor and allows the high-pressure refrigerant to flow into the A/C system. These compressors do not have a suction valve. They are smaller and operate more smoothly than other designs.

Compressor Clutches

Most compressors are equipped with an electromagnetic clutch as part of the compressor pulley assembly **(Figure 52–20)**. It is designed to engage the pulley

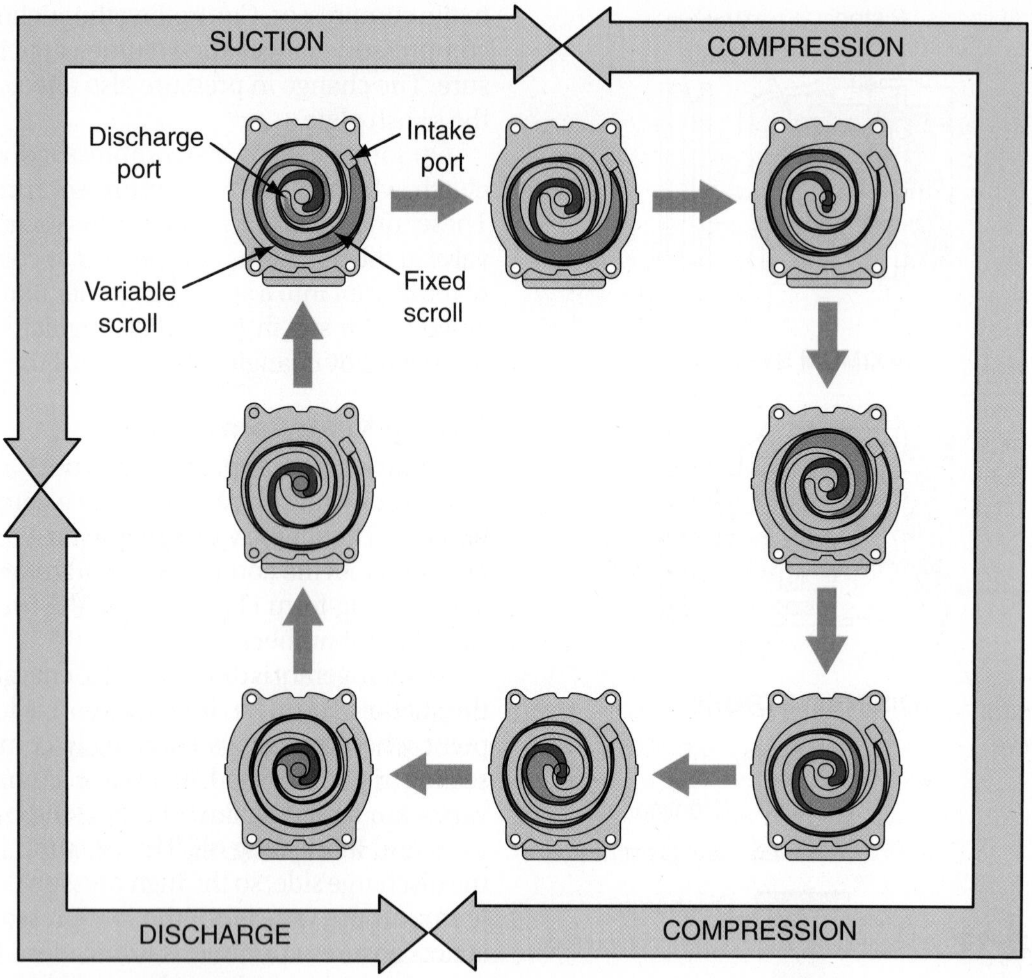

Figure 52-19 A scroll compressor passes through three distinct phases: suction, compression, and discharge.

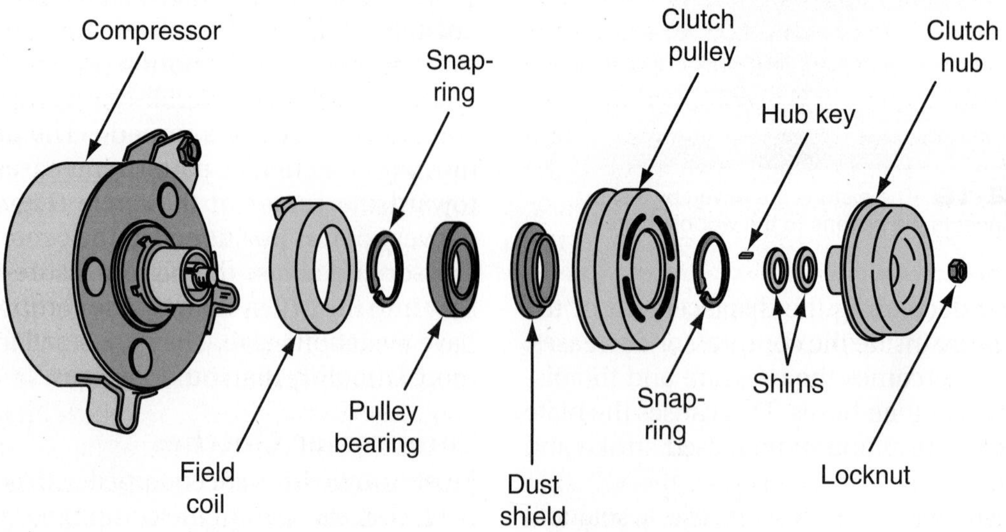

Figure 52-20 An A/C compressor clutch assembly.

to the compressor shaft when the clutch coil is energized. The purpose of the clutch is to transmit power from the engine to the compressor and to provide a means of engaging and disengaging the refrigeration system from engine operation. The clutch is driven by power from the engine's crankshaft, which is transmitted through one or more belts (a few use gears) to the pulley, which is in operation whenever the engine is running. When the clutch is engaged, power is transmitted from the pulley to the compressor shaft by the clutch drive plate. When the clutch is not engaged, the compressor shaft does not rotate, and the pulley freewheels.

The clutch allows the A/C system to be controlled by an electric circuit. The clutch relay is controlled by a temperature signal from the evaporator and a pressure switch in the refrigerant line. In most systems, the compressor clutch cycles on and off periodically to allow the evaporator to warm up during periods of high cooling demand.

The clutch is engaged by a magnetic field and disengaged by springs when the magnetic field is broken. When the controls call for compressor operation, the electrical circuit to the clutch is completed, the magnetic clutch is energized, and the clutch engages the compressor. When the electrical circuit is opened, the clutch disengages the compressor.

Two types of electromagnetic clutches have been in use for many years. Early-model A/C systems use a rotating coil clutch. The magnetic coil is mounted within the pulley and rotates with it. Electrical connections for the clutch operation are made through a stationary brush assembly and rotating slip rings, which are part of the field coil assembly. This older rotating coil clutch, now in limited use, has been largely replaced by the stationary coil clutch.

With the stationary coil, wear has been measurably reduced, efficiency increased, and serviceability made much easier. The clutch coil does not rotate. When the system is turned on, the pulley assembly is magnetized by the stationary coil on the compressor body, thus engaging the clutch to the clutch hub attached to the compressor shaft. This activates the A/C system. Depending on the system, the magnetic clutch is usually pressure controlled to cycle the operation of the compressor (depending on system temperature or pressure). In some system designs, the clutch might operate continually when the system is turned on.

Electric Drive Compressors

One of the features of hybrid vehicles is the idle-stop or stop-start system. When the vehicle is sitting at a light, the system shuts down the engine. This means there is no power to drive the A/C compressor. Therefore, most hybrid vehicles have an electric drive for the compressor. An electric motor is built into the compressor and is powered by the vehicle's high-voltage system **(Figure 52–21)**. This means the A/C system can operate when the engine is not running.

Figure 52–21 An electric A/C compressor. *Courtesy of DENSO Corporation*

A/C compressors in hybrid vehicles are typically identical to those used in a conventional vehicle, except for the portion that is actuated by the electric motor. Because the compressor is energized by electricity, the control module can control its speed. This enables the system to provide ideal cooling while using a minimum of energy to run the system.

In most cases, the compressor's motor is driven by AC voltage from the vehicle's inverter. The system controls the voltage to rotate the compressor at a desired speed. This speed is calculated by the control module and is based on a target evaporator temperature and the actual evaporator temperature.

The Accord hybrid is also equipped with a dual-scroll "hybrid" A/C compressor **(Figure 52–22)**. The A/C system uses two compressors built into a single housing; one compressor is driven by the engine and the other is driven by an electric motor powered by the high-voltage battery. A driver circuit in the control module provides switched high voltage to the motor. The action of the driver is determined by inputs from the climate control settings and the CAN bus. The mechanically driven side uses a normal electric clutch controlled by the climate control module. The electrically driven side uses a high-voltage, brushless, three-phase motor driven by a controller. When full cooling is needed, the A/C unit relies on both power sources to provide maximum cooling. During normal cooling,

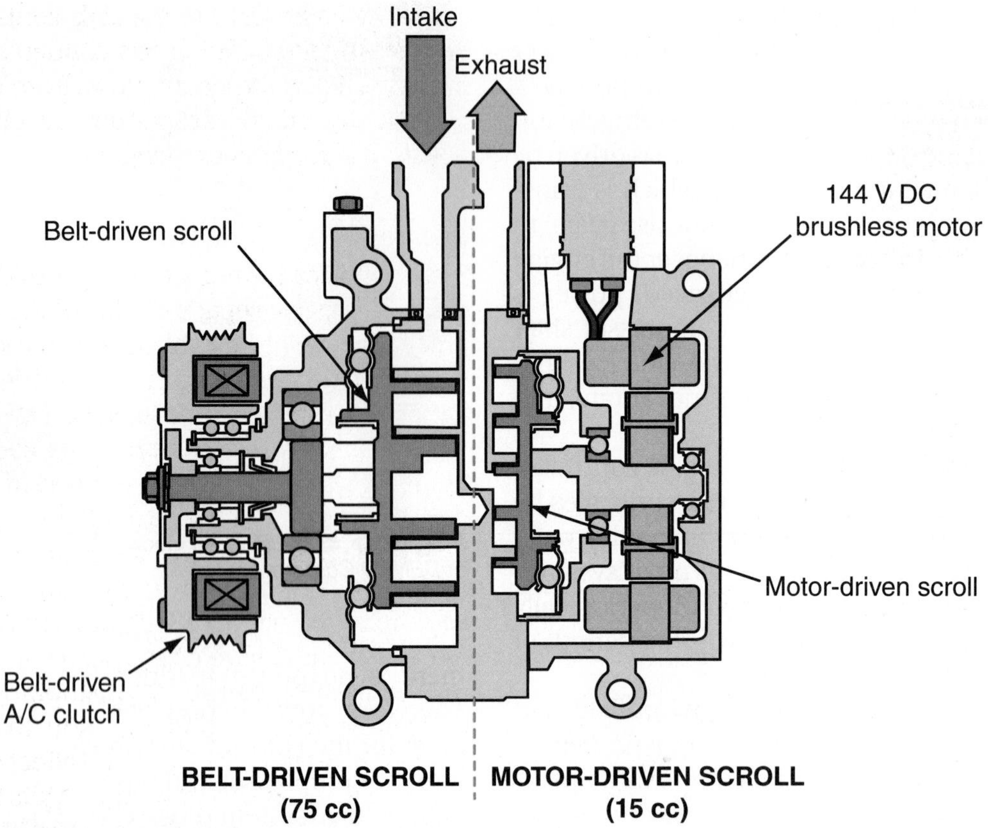

Figure 52-22 A dual-scroll A/C compressor powered by a drive belt and/or an electric motor.

the A/C is powered by either the belt-driven compressor or the electric motor-driven compressor.

Refrigerant Oils

Normally the only source of lubrication for a compressor is the oil mixed with the refrigerant. A/C systems carry oil through the system with the refrigerant to lubricate the parts of the system, including the compressor. Because of the loads and speeds at which the compressor operates, proper lubrication is a must for long compressor life.

The refrigerant oil required by the system depends on a number of things, but it is primarily dictated by the refrigerant used in the system. R-12 systems use a mineral oil. Mineral oil mixes well with R-12 without breaking down. Mineral oil, however, cannot be used with R-134a. R-134a systems require that a synthetic oil, **polyalkaline glycol (PAG)** or POE (polyester) oil. Most manufacturers use PAG oil in R-134a systems. Aftermarket companies, on the other hand, often choose ester oils for lubrication with R-134a because they tend to attract less moisture than PAG oils. Most often when a system has been converted from R-12 to R-134a, ester oil will be recommended. Ester oil mixes well with mineral oil because it is hydrocarbon based.

There are a number of different blends of refrigerant oil; always use the one recommended by the vehicle manufacturer or compressor manufacturer. Failure to use the correct oil will cause damage to the compressor.

WARNING!

Hybrid vehicles often have an electrically driven compressor. Because the electric motor is inside the compressor case and is in contact with the oil in the compressor, only the specified oil should be used in the compressor. This oil has electrical insulating qualities that protect you from dangerous electrical shocks. Also, if you use the wrong oil, the A/C unit will be contaminated and this may result in a need to replace the compressor, condenser, evaporator, and/or all of the refrigerant lines.

CONDENSER

The condenser **(Figure 52–23)** consists of coiled refrigerant tubing mounted in a series of thin cooling fins to provide maximum heat transfer in a minimum

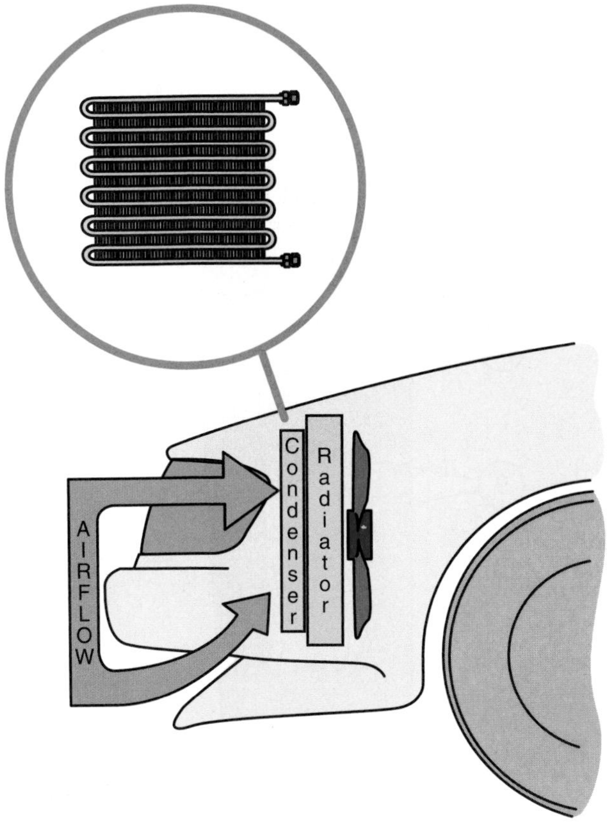

Figure 52-23 A typical condenser.

amount of space. The condenser is normally mounted just in front of the vehicle's radiator. It receives the full flow of ram air from the movement of the vehicle or airflow from the radiator fan when the vehicle is standing still.

The purpose of the condenser is to condense or liquefy the high-pressure, high-temperature vapor coming from the compressor. To do so, it must give up its heat. The condenser receives very hot (normally 200 to 400°F), high-pressure refrigerant vapor from the compressor through its discharge hose. The refrigerant vapor enters the inlet at the top of the condenser and as the hot vapor passes down through the condenser coils, heat (following its natural tendencies) moves from the hot refrigerant into the cooler air as it flows across the condenser coils and fins. This process causes a large quantity of heat to be transferred to the outside air and the refrigerant to change from a high-pressure hot vapor to a high-pressure warm liquid. This high-pressure warm liquid flows from the outlet at the bottom of the condenser through a line to the receiver/dryer or to the refrigerant metering device if an accumulator is used instead of a dryer.

In an A/C system which is operating under an average heat load, the condenser has a combination of hot refrigerant vapor in the upper two-thirds of its coils. The lower third of the coils contains the warm liquid refrigerant, which has condensed. This high-pressure, liquid refrigerant flows from the condenser and on toward the evaporator. In effect, the condenser is a true heat exchanger.

Subcoolers

Some vehicles have a subcooler built into the condenser or have a separate subcooler. Subcooling is a process by which sensible heat is removed from liquid refrigerant, resulting in lower refrigerant temperatures. The separate subcooler is located between the condenser and evaporator. A subcooler is a heat exchanger that allows the refrigerant to lose additional heat after it becomes a liquid. The subcooler increases the efficiency of the system by cooling the refrigerant and prevents premature vaporization or flash off as the refrigerant passes through the expansion valve and before it reaches the evaporator. Premature flash off can result in stopping some of the refrigerant from evaporating, and that part of the refrigerant would have no useful effect on the cooling of the vehicle. The heat exchange causes the liquid to be subcooled to a level that ensures little or no flash gas on its way to the evaporator.

Many vehicles have a condenser with a subcool chamber **(Figure 52–24)**. In these condensers, refrigerant enters at the top as a high-pressure gas. It then passes through a receiver/dryer or modulator to separate the liquid from the gaseous refrigerant. The modulator contains a desiccant and filter to remove the moisture and foreign material from the refrigerant. After passing through the modulator, the refrigerant flows into the bottom subcooler chamber to further cool the liquid. This results in complete liquidization of the refrigerant and better A/C performance. These condensers have thin tubes and low fin height, which improve the heat exchange rate. Because of this two-step approach, the refrigerant sent to the evaporator is almost completely liquefied.

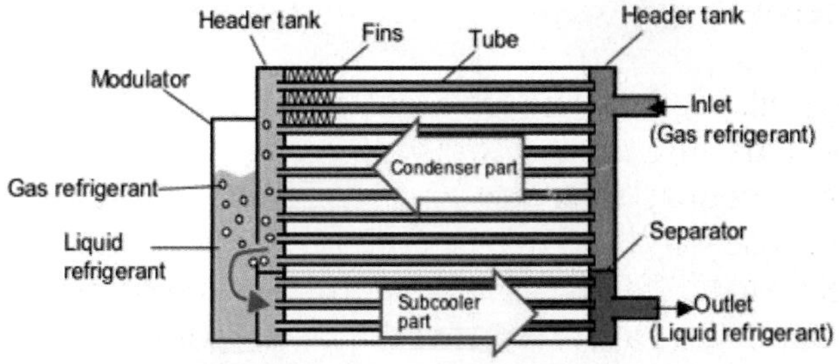

Figure 52-24 A condenser with a built-in subcooler. *Courtesy of DENSO Corporation*

RECEIVER/DRYER

Used on many early systems, the receiver/dryer is a storage tank for the liquid refrigerant from the condenser, which flows into the upper portion of the receiver tank containing a bag of desiccant (moisture-absorbing material such as silica alumina or silica gel). As the refrigerant flows through an opening in the lower portion of the receiver, it is filtered through a mesh screen attached to a baffle at the bottom of the receiver. The purpose of the desiccant in this assembly is to absorb any moisture present that might enter the system during assembly. These features of the assembly prevent obstruction to the valves or damage to the compressor.

Depending on the manufacturer, the receiver/dryer may be known by other names such as filter or dehydrator. Regardless of its name, the function is the same. Included in many receiver/dryers are additional features such as a high-pressure fitting, a pressure relief valve, and a sight glass for determining the state and condition of the refrigerant in the system.

Accumulator

Most late-model systems are not equipped with a receiver/dryer; rather, they use an accumulator to accomplish the same thing **(Figure 52–25)**. The accumulator is connected into the low side at the outlet of the evaporator. The accumulator also contains a desiccant and is designed to store excess refrigerant and to filter and dry the refrigerant **(Figure 52–26)**. If liquid refrigerant flows out of the evaporator, it will be collected by and stored in the accumulator. The main purpose of an accumulator is to prevent liquid from entering the compressor.

Figure 52-25 Two different accumulator designs.

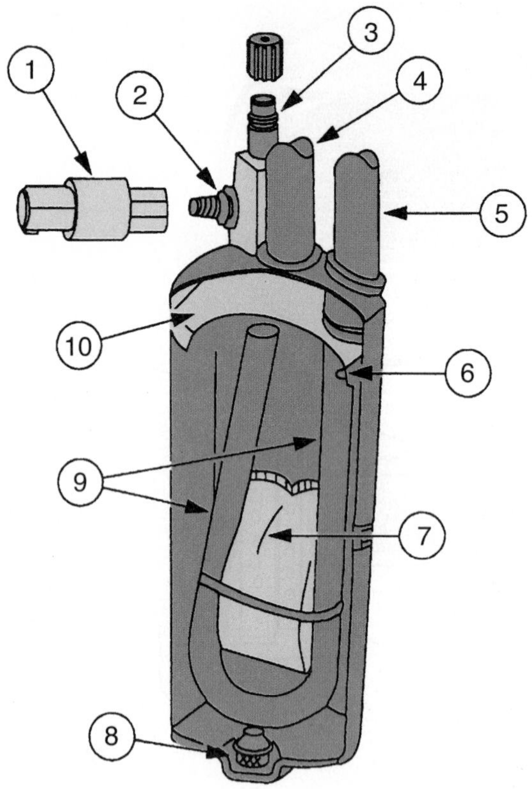

1. Cycling switch
2. O-ring seal
3. Low-side port
4. From evaporator
5. To compressor
6. Antisiphon hole
7. Desiccant bag
8. Oil return filter
9. Vapor return tube
10. Accumulator/dryer dome

Figure 52-26 An accumulator/dryer. *Courtesy of Ford Motor Company*

THERMOSTATIC EXPANSION VALVE/ORIFICE TUBE

The refrigerant flow to the evaporator must be controlled to obtain maximum cooling while ensuring complete evaporation of the liquid refrigerant within the evaporator. This is accomplished by a thermostatic expansion valve (TEV or TXV) or a fixed orifice tube **(Figure 52–27)**.

The TEV is mounted at the inlet to the evaporator and separates the high-pressure side of the system from the low-pressure side. The TEV regulates refrigerant flow to the evaporator to prevent evaporator flooding or starving. In operation, the TEV regulates the refrigerant flow to the evaporator by balancing the inlet flow to the outlet temperature.

Both externally and internally equalized TEVs are used in A/C systems. The only difference between the two valves is that the external TEV uses an equalizer line connected to the evaporator outlet line as a means of sensing evaporator outlet pressure. The internal TEV senses evaporator inlet pressure through

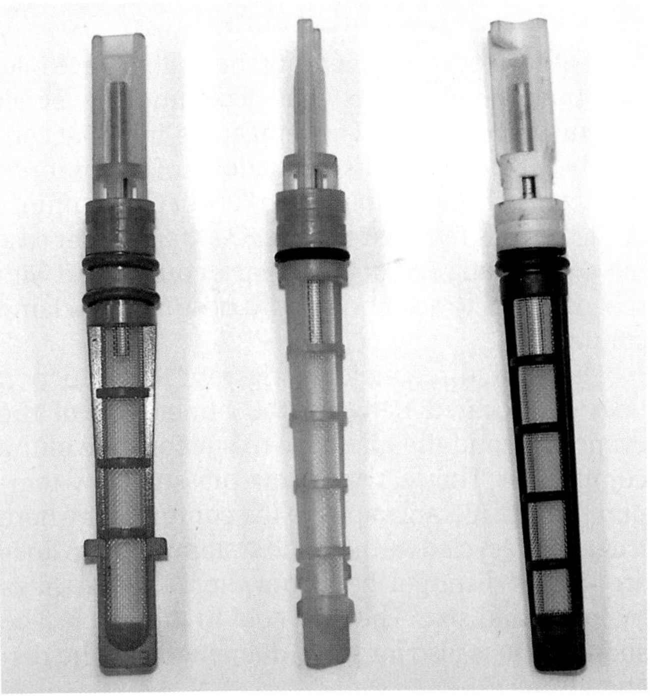

Figure 52-27 Various orifice expansion tube assemblies.

an internal equalizer passage. Both valves have a capillary tube to sense evaporator outlet temperature. The tube is filled with a gas, which, if allowed to escape due to careless handling, will ruin the TXV.

During stabilized conditions, the pressure on the bottom of the expansion valve diaphragm becomes equal to the pressure on the top of the diaphragm. This allows the valve spring to close the valve. When the system is started, the pressure on the bottom of the diaphragm drops rapidly, allowing the valve to open and meter liquid refrigerant to the lower evaporator tubes where it begins to vaporize.

Compressor suction draws the vaporized refrigerant out of the top of the evaporator at the top tube, where it passes by the sealed sensing bulb. The bottom of the valve diaphragm internally senses the evaporator pressure through the internal equalization passage around the sealed sensing bulb. As evaporator pressure is increased, the diaphragm flexes upward, pulling the pushrod away from the ball seat of the expansion valve. The expansion valve spring forces the ball onto the tapered seat and the liquid refrigerant flow is reduced.

As the pressure is reduced due to restricted refrigerant flow, the diaphragm flexes downward again, opening the expansion valve to provide the required controlled pressure and refrigerant flow condition. As the cool refrigerant passes by the body of the sensing bulb, the gas above the diaphragm contracts and allows the expansion valve spring to close the expansion valve. When heat from the passenger compartment is absorbed by the refrigerant, it causes the gas to expand. The pushrod again forces the expansion valve to open, allowing more refrigerant to flow so that more heat can be absorbed.

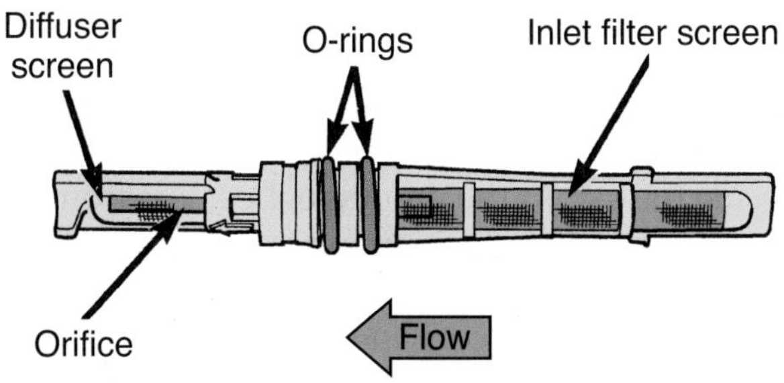

Figure 52-28 A typical orifice expansion tube. *Courtesy of Chrysler LLC*

Orifice Tube

Like the TEV, the orifice tube is the dividing point between the high- and low-pressure parts of the system. However, its metering or flow rate control does not depend on comparing evaporator pressure and temperature. It is a fixed orifice **(Figure 52–28)**. The flow rate is determined by pressure difference across the orifice and by subcooling at the bottom of the condenser.

EVAPORATOR

The evaporator, like the condenser, consists of a refrigerant coil mounted in a series of thin cooling fins **(Figure 52–29)**. It provides a maximum amount of heat transfer in a minimum amount of space. The evaporator is usually located beneath the dashboard or instrument panel.

Upon receiving the low-pressure, low-temperature liquid refrigerant from the TEV or orifice tube in the form of an atomized (or droplet) spray, the evaporator serves as a boiler or vaporizer. This regulated flow

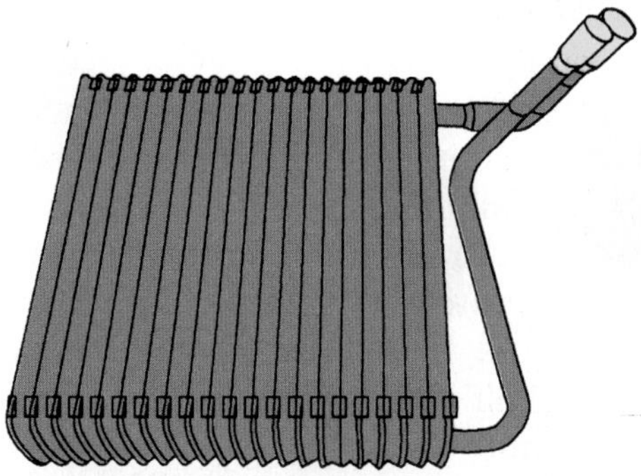

Figure 52-29 A typical evaporator. *Courtesy of Ford Motor Company*

of refrigerant boils immediately. Heat from the core surface is lost to the boiling and vaporizing refrigerant, which is cooler than the core, thereby cooling the core. The air passing over the evaporator loses its heat to the cooler surface of the core, thereby cooling the air inside the car. As the process of heat loss from the air to the evaporator core surface is taking place, any moisture (humidity) in the air condenses on the outside of the evaporator core and is drained off as water. A drain tube in the bottom of the evaporator housing leads the water outside the vehicle. This dehumidification of air is an added feature of the A/C system that adds to passenger comfort. It also is used as a means of controlling fogging of the vehicle windows. Under certain conditions, however, too much moisture can accumulate on the evaporator coils. An example would be when humidity is extremely high and the maximum cooling mode is selected. The evaporator temperature might become so low that moisture would freeze on the evaporator coils before it can drain off.

Through the metering, or controlling, action of the TEV or orifice tube, greater or lesser amounts of refrigerant are provided in the evaporator to adequately cool the car under all heat load conditions. If too much refrigerant is allowed to enter, the evaporator floods. This results in poor cooling due to the higher pressure (and temperature) of the refrigerant. The refrigerant can neither boil away rapidly nor vaporize. On the other hand, if too little refrigerant is metered, the evaporator starves. Poor cooling again results because the refrigerant boils away or vaporizes too quickly before passing through the evaporator.

The temperature of the refrigerant vapor at the evaporator outlet will be approximately 4 to 16°F higher than the temperature of the liquid refrigerant at the evaporator inlet. This temperature differential ensures that the vapor will not contain any droplets of liquid refrigerant that would be harmful to the compressor.

Blower Motor/Fan

The blower motor/fan assembly is located in the evaporator housing. Its purpose is to increase airflow in the passenger compartment. The blower, which is basically the same type as those used in heater systems, draws warm air from the passenger compartment, forces it over the coils and fins of the evaporator, and blows the cooled, cleaned, and dehumidified air into the passenger compartment. The blower motor is controlled by a fan switch. On some systems, blower speed is regulated by the speed of the compressor.

REFRIGERANT LINES

All of the major components of the system have inlet and outlet connections that accommodate either flare or O-ring fittings. The refrigerant lines that connect between these units are made up of an appropriate length of hose or tubing with flare or O-ring fittings at each end as required **(Figure 52–30)**. In either case the hose or tube end of the fitting is constructed with sealing beads to accommodate a hose or tube clamp connection.

There are three major refrigerant lines. Suction lines are located between the outlet side of the evaporator and the inlet side or suction side of the compressor. They carry the low-pressure, low-temperature refrigerant vapor to the compressor where it again is recycled through the system. Suction lines are always distinguished from the discharge lines by touch and size. They are cold to the touch. The suction line is also larger in diameter than the discharge line.

Beginning at the discharge outlet on the compressor, the discharge or high-pressure line connects the compressor to the condenser. The liquid lines connect the condenser to the receiver/dryer and the receiver/dryer to the inlet side of the expansion valve. Through these lines, the refrigerant travels in its path from a gas state (compressor outlet) to a liquid state (condenser outlet) and then to the inlet side of the expansion valve, where it vaporizes on entry to the evaporator. Discharge and liquid lines are always very warm to the touch and easily distinguishable from the suction lines.

Aluminum tubing is commonly used to connect A/C components where flexibility is not required. Where the line is subjected to vibrations, special

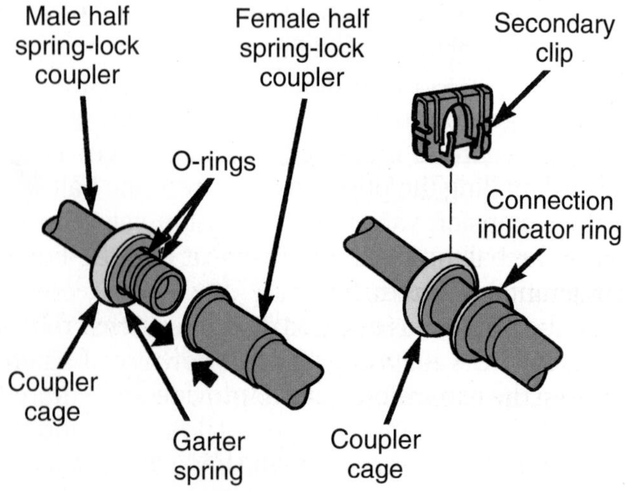

Figure 52-30 Spring-lock line coupler with O-rings. *Courtesy of Chrysler LLC*

Figure 52-31 The location of a typical sight glass.

rubber hoses are used. Typically the compressor outlet and inlet lines are rubber hoses with aluminum ends and fittings.

R-134a systems are required to be fitted with quick-disconnect service fittings throughout the system. These also have hoses specially made for R-134a. They have an additional layer of rubber that serves as a barrier to prevent the refrigerant from escaping through the pores of the hose. Some late-model R-12 systems also use these barrier hoses to prevent the loss of refrigerant through the walls of the hoses.

Sight Glass

The **sight glass** allows the technician to see the flow of refrigerant in the lines. A sight glass is normally found on systems using R-12 and a TEV. It is rare to find a sight glass on R-134a systems **(Figure 52–31)**. However, some OEMs have a sight glass on all of their A/C systems, regardless of the refrigerant used. It can be located on the receiver/dryer or inline between the receiver/dryer and the expansion valve or tube.

AIR-CONDITIONING SYSTEMS AND CONTROLS

There are two basic types of automotive air-conditioning systems. They are classified according to the method used in obtaining temperature control and are known as cycling clutch systems or evaporator pressure or temperature control systems.

Evaporator Pressure Control System

Evaporator controls maintain a backpressure in the evaporator. Because of the refrigerant temperature/pressure relationship, the effect is to regulate evaporator temperature. The temperature is controlled to a point that provides effective air cooling but prevents the freezing of moisture that condenses on the evaporator.

In this type of system the compressor operates continually when dash controls are in the air-conditioning position. Evaporator outlet air temperature is automatically controlled by an evaporator pressure control valve. This type of valve throttles the flow of refrigerant out of the evaporator as required to establish a minimum evaporator pressure and thereby prevent freezing of condensation on the evaporator core.

Cycling Clutch System

In every **cycling clutch** system, the compressor is run intermittently by means of controlling the application and release of its clutch through a thermostatic or pressure switch. The thermostatic switch senses the evaporator's outlet air temperature through a capillary tube that is part of the switch assembly. With a high sensing temperature, the thermostatic switch is closed and the compressor clutch is energized. As the evaporator outlet temperature drops to a preset level, the thermostatic switch opens the circuit to the compressor clutch. The compressor then ceases to operate until such time as the evaporator temperature rises above the switch setting. From this on-and-off operation is derived the term *cycling clutch.* In effect, the thermostatic switch is calibrated to allow the lowest possible evaporator outlet temperature that would prevent the freezing of condensation that might form on the evaporator.

Variations of the cycling clutch system include a system with a thermostatic expansion valve and a system with an orifice tube.

Cycling Clutch System with Thermostatic Expansion Valve Some factory installations utilize a cycling clutch system that incorporates a TEV and receiver/dryer, as do some add-on units. The evaporator and control components are either in the engine compartment or an integral part of the cowl. In such cases there is a common blower and duct work for both heating and air-conditioning purposes. Also in these installations, the thermostatic switch has no temperature control knob and is usually mounted on the evaporator or its case. Temperature control is accomplished by using fresh or recirculating air and by reheating the cooled air in the heater core. The clutch cycles only to prevent evaporator icing.

A common form of the cycling clutch system is the field-installed (add-on) unit. With this installation, the evaporator, the thermostatic expansion valve, and the thermostatic switch are self-contained in an underdash assembly. An installation of this type operates solely on passenger compartment recirculated air. Temperature control depends on intermittent operation of the compressor. The thermostatic switch has a control knob to cycle at higher temperatures when less cooling is desired.

Cycling Clutch System with Orifice Tube (CCOT) A typical **CCOT system** is illustrated in **Figure 52–32**. The system is factory installed and can use a thermostatic clutch cycling switch mounted on the evaporator case or a pressure cycling switch located on the accumulator. An expansion (orifice) tube is used in place of the TEV. Also, the system has an accumulator in the evaporator outlet. The accumulator is used primarily to separate vapor from liquid refrigerant before it enters the compressor. It also contains a drying agent or desiccant to remove moisture. The CCOT system has no receiver/dryer or sight glass. This system does have a special orifice, the oil bleed orifice, that allows refrigerant oil to return to the compressor rather than to collect in the accumulator.

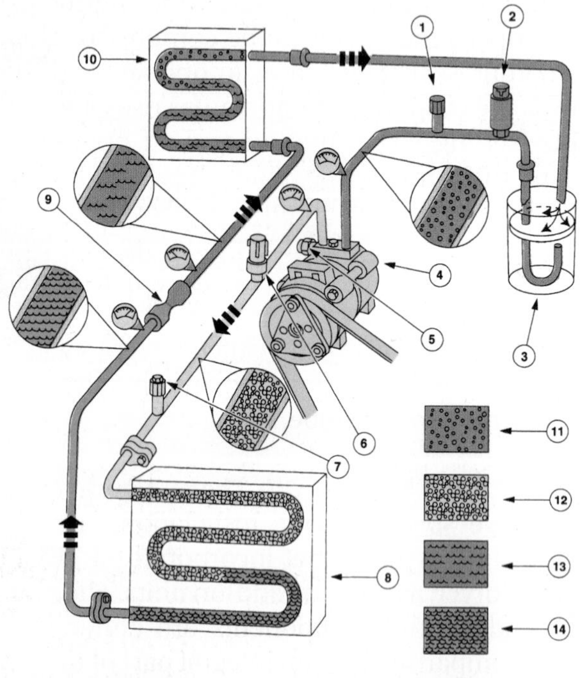

Item	Description	Item	Description
1	Low side port	8	Condenser
2	Cycling switch	9	Orifice tube
3	Accumulator/drier	10	Evaporator
4	Compressor	11	Low-pressure vapor
5	Pressure relief valve	12	High-pressure vapor
6	Pressure cutoff switch	13	Low-pressure liquid
7	High side port	14	High-pressure liquid

Figure 52-32 A typical cycling clutch system with an expansion (orifice) tube. *Courtesy of Ford Motor Company*

Compressor Controls

Many controls are used to monitor and trigger the compressor during its operational cycle. Each of these represents the most common protective control devices designed to ensure safe and reliable operation of the compressor.

Ambient Temperature Switch This switch senses outside air temperature and is designed to prevent compressor clutch engagement when air-conditioning is not required or when compressor operation might cause internal damage to seals and other parts.

The switch is in series with the compressor clutch electrical circuit and closes at about 37°F (2.7°C). At all lower temperatures, the switch is open, preventing clutch engagement.

On some vehicles, the ambient switch is located in the air inlet duct of air-conditioning systems regulated by evaporator pressure controls; other makes have it installed near the radiator. It is not required on systems with a thermostatic or pressure switch.

Thermostatic Switch In cycling clutch systems, the thermostatic switch is placed in series with the compressor clutch circuit so it can turn the clutch on or off. It has two purposes. It de-energizes the clutch and stops the compressor if the evaporator is at the freezing point **(Figure 52–33)**. On hang-on units or systems without reheat temperature control, it also controls the air temperature by turning the compressor on and off intermittently. For this purpose, it has a control knob to change the switch setting.

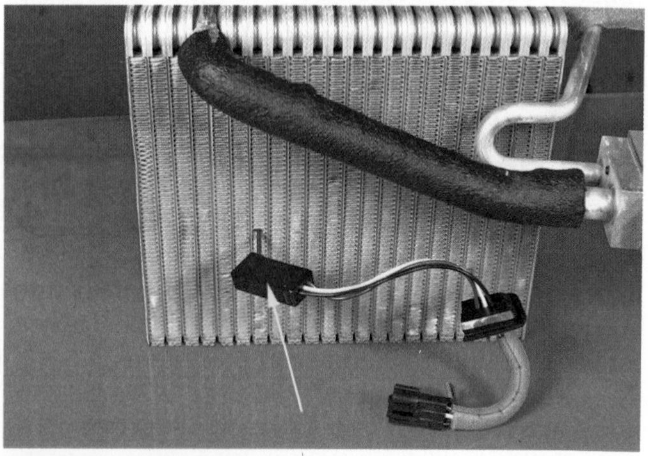

Figure 52-33 A temperature sensor mounted onto an evaporator.

When the temperature of the evaporator approaches the freezing point (or the low setting of the switch), the thermostatic switch opens the circuit and disengages the compressor clutch. The compressor remains inoperative until the evaporator temperature rises to the preset temperature, at which time the switch closes and compressor operation resumes.

Pressure Cycling Switch This switch is electrically connected in series with the compressor electromagnetic clutch. Like the thermostatic switch, the turning on and off of the pressure cycling switch controls the operation of the compressor.

Low-Pressure Cutoff or Discharge Pressure Switch This switch is located on the high side of the system and senses any low-pressure conditions. It is tied into the compressor clutch circuit, allowing it to immediately disengage the clutch when the pressure falls too low.

High-Pressure Cutout Switch This switch, normally located in the vicinity of the compressor or discharge (high side) muffler, is wired with the compressor clutch (in series). Designed to open (cut out) and disengage the clutch at 350 to 375 psi (2,400 to 2,600 kPa), it again closes and normally reengages the clutch when pressure returns to 250 psi (1,700 kPa) (higher if the system uses R-134a).

High-Pressure Relief Valve A high-pressure relief valve is incorporated into many air-conditioning systems. This valve may be installed on the receiver/dryer, compressor, or elsewhere in the high side of the system. It is a high-pressure protection device that opens (normally at 440 psi [3,033 kPa]) to bleed off excessive pressure that might occur in the system.

Compressor Control Valve This valve regulates the crankcase pressure in some compressors (commonly General Motors' V5 compressor). It has a pressure-sensitive bellows exposed to the suction side that acts on a ball and pin valve, which is exposed to high-side pressure. The bellows also controls a bleed port that is also exposed to the low side. The control valve is continuously modulating—changing the displacement of the compressor according to pressure or temperature.

Electronic Cycling Clutch Switch (ECCS) The ECCS prevents evaporator freeze-up by sending a signal to the engine control computer. The computer, in turn, cycles the compressor on and off by monitoring suction line temperature. If the temperature gets too low, the ECCS will open the input circuit to the computer, which causes the A/C clutch relay to open, disengaging the compressor clutch. Often this switch is the thermostatic switch at the evaporator.

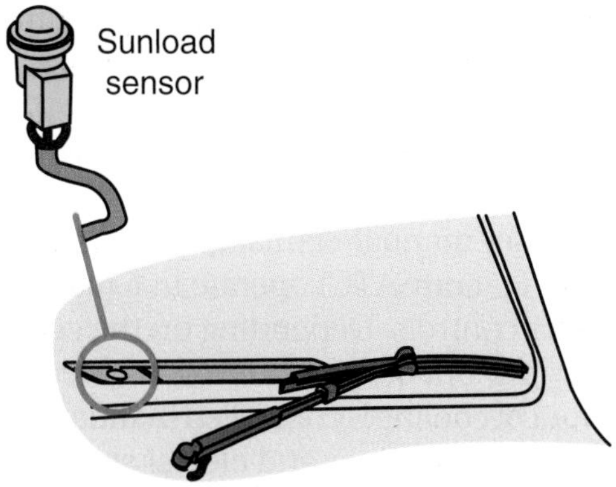

Figure 52–34 A solar (sunload) sensor.

Solar "Sunload" Sensor

More than half of the heat that is in a vehicle's passenger compartment comes from solar radiation. Many A/C systems have a solar sensor, also called the sunload sensor, that anticipates the amount of heat that will result from the sunlight. The solar sensor is usually located on top of the instrument panel **(Figure 52–34)**. The solar sensor is one of the main controls in many automatic climate control systems.

The solar sensor is a photo diode that receives a 5-volt reference signal from the control module. The diode normally blocks the flow of current in both directions except in the presence of light. The signal voltage from the solar sensor varies with the amount of sunlight on the sensor. When the sunlight increases, the output voltage increases, and as the sunlight decreases, the output voltage decreases. Bright sunlight causes the vehicle's interior temperature to increase; therefore, the system adds more cool air to the passenger compartment to overcome the increased heat.

Some vehicles have a solar sensor that measures sunlight at two separate angles. This allows the system to react differently on the driver and passenger sides of the vehicle. Other dual-zone systems have left and right solar sensors.

TEMPERATURE CONTROL SYSTEMS

Temperature control systems for air conditioners usually are connected with heater controls. Most heater and air-conditioning systems use the same plenum chamber for air distribution. Two types of air-conditioning controls are used: manual/semiautomatic and automatic.

Manual/Semiautomatic Temperature Controls

Air conditioner manual/semiautomatic temperature controls (MTC and SATC) operate in a manner similar to heater controls. Depending on the control setting, doors are opened and closed to direct airflow. The amount of cooling is controlled manually through the use of control settings and blower speed.

Automatic Temperature Control

An automatic or electronic temperature control system **(Figure 52–35)** maintains a specific temperature automatically inside the passenger compartment. To maintain a selected temperature, heat sensors send signals to a computer unit that controls compressor, heater valve, blower, and plenum door operation. A typical electronic control system might contain a coolant temperature sensor, in-car temperature sensor **(Figure 52–36)**, outside temperature sensor, high-side temperature switch, low-side temperature switch, low-pressure switch, vehicle speed sensor, throttle position sensor, sunload sensor, and power-steering cutout switch.

The control panel is found in the instrument panel at a convenient location for both driver and front-seat passenger access. Three types of control panels may be found: manual, push-button, or touch pad. All serve the same purpose. They provide operator input control for the air-conditioning and heating system. Some control panels have features that other panels do not have, such as provisions to display in-car and outside air temperature in degrees.

Figure 52-35 Typical automatic climate control panel and selectors.

Provisions are made on the control panel for operator selection of an in-car temperature between 65 and 85°F (18 and 29°C) in one-degree increments. Some have an override feature that provides for a setting of either 60 or 90°F (15 or 32°C). Either of these two settings overrides all in-car temperature control circuits to provide maximum cooling or heating conditions.

Usually, a microprocessor is located in the control head to input data to the programmer, based on operator-selected conditions. When the ignition switch is turned off, a memory circuit remembers the previous setting. These conditions are restored the next time the ignition switch is turned on. If the battery is disconnected, however, the memory circuit is cleared and must be reprogrammed.

Many automotive electronic temperature control systems have self-diagnostic test provisions in which an on-board microprocessor-controlled subsystem displays a code. This code (number, letter, or alphanumeric) is displayed to tell the technician the cause of the malfunction. Some systems also display a code to indicate which computer detected the malfunction. Manufacturers' specifications must be followed to identify the malfunction display codes, because they differ from car to car. For example, in some General Motors' car lines ".7,0" indicates no malfunction if "no trouble" codes are stored in the computer. In some Ford car lines the no trouble code is 888.

Case and Duct Systems

A typical automotive heater/air conditioner/case and duct system is shown in **Figure 52–37**. The purpose of the system is twofold: It is used to house the heater core and the air conditioner evaporator and to direct the selected supply air through these components into the passenger compartment of the vehicle. The supply air selected can be either fresh (outside) or recirculated air, depending on the system mode. After the air is heated or cooled, it is delivered to the floor outlet, dash panel outlets **(Figure 52–38)**, or the defrost outlets.

In domestic vehicles, there are two basic duct systems employed. In the stacked-core-reheat system the basic control is in the water valve. For maximum air, the water valve is completely closed. All air enters the vehicle compartment through the heater core.

The access door, which is activated by a cable, controls only fresh or recirculated air. Recirculated air is used during maximum cold operation. The air-conditioning unit is not operative and the evaporator will not be cold. The evaporator is used only in the max air or maximum cold position. As the control level inside the car is moved, it controls the water

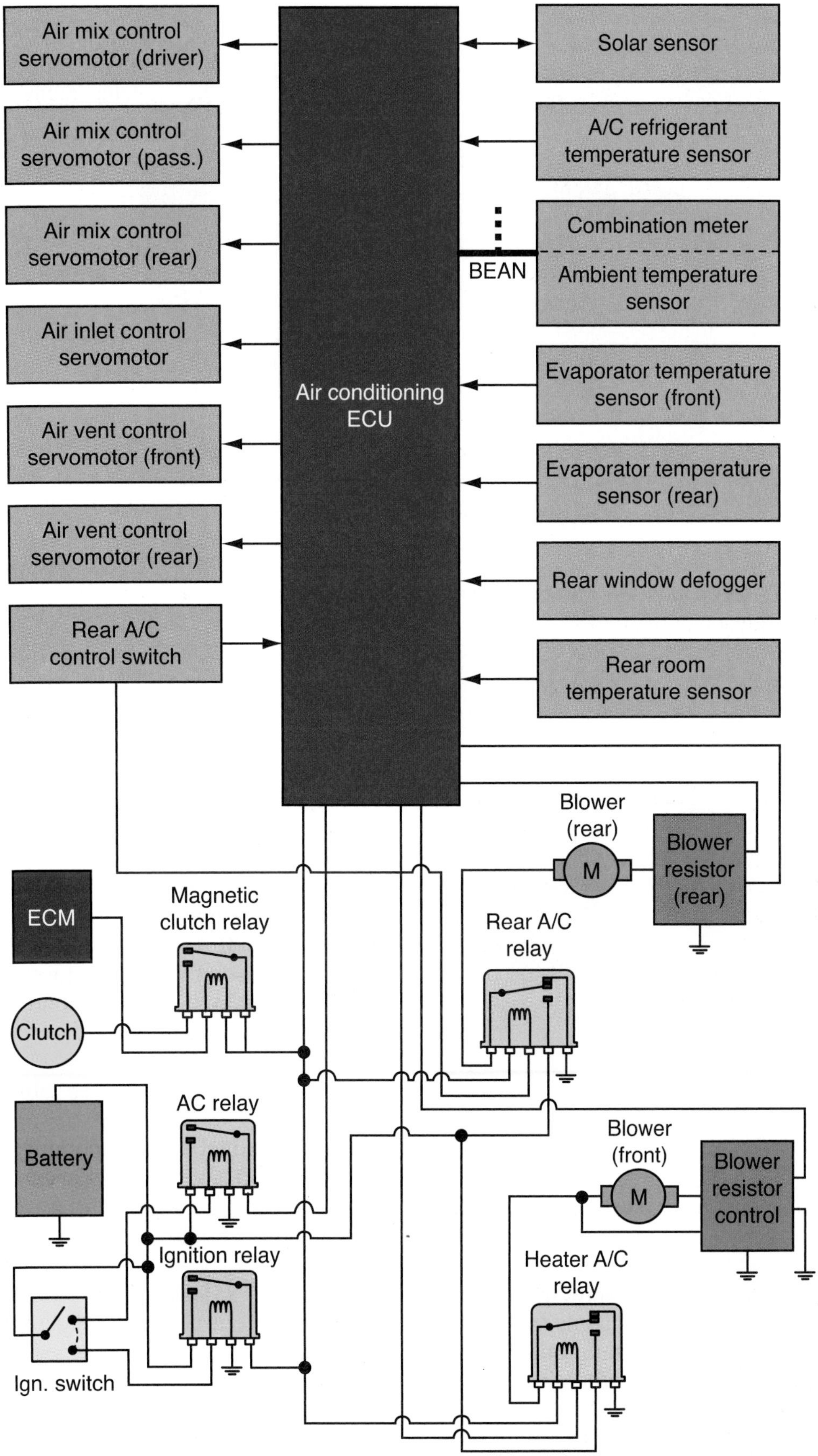

Figure 52-36 The layout of a computer-controlled automatic climate control system.

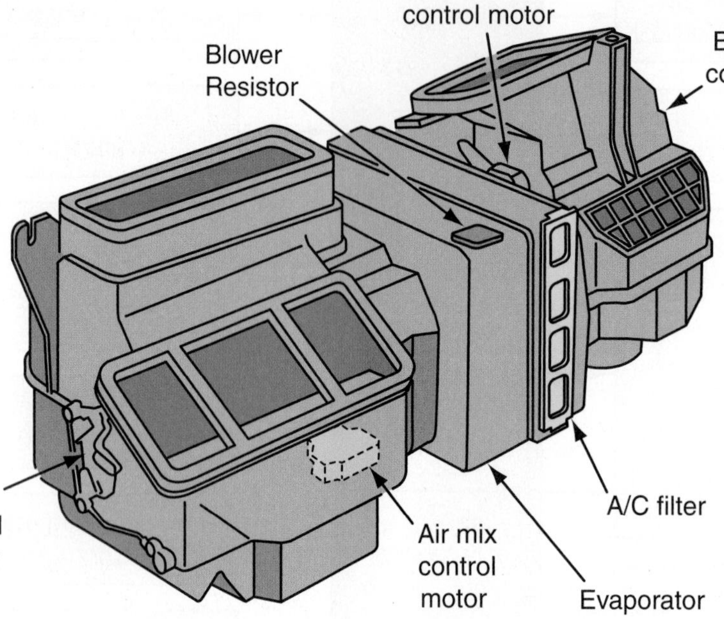

Figure 52-37 Typical heater/air conditioner ducts.

Figure 52-38 Air moving out from the dash vents to create comfort zones for the driver and passengers. *Courtesy of Chrysler LLC*

valve by means of a vacuum or a cable to control the amount of hot water entering the heater core and the temperature of the air at the unit outlet.

A blend-air-reheat mode door is found on General Motors and Ford vehicles and some truck units with factory-controlled heater system units. During heater-only operation, the air-conditioning unit is shut off, and the evaporator performs no function in air distribution or temperature control. During maximum air or extreme cold air, the air-conditioning system operates, the evaporator is cold, and the blend-air-door damper is completely closed. Only cold air enters the car.

As the control lever is moved in the vehicle from max air toward heat with the air conditioner on, the blend-air-door is moving. In maximum cold, it is completely shut. On maximum hot, it is completely open. The water valve on this unit is a vacuum on/off unit to regulate water flow. Normal position would be open. This type of blend-air system is extremely popular and can be used with or without a water valve.

To check the proper functioning of the duct work, move the temperature control lever to see if any change occurs. If it does not, shut off the air conditioner and turn on the heater. Move the temperature control arm again to see if any change occurs. If not, check the cable and the flap door connected to the temperature control lever. You might be able to reach under the dash to reconnect the cable or free a stuck flap.

If no substantial airflow is coming out of the registers, check the fuses in the blower circuit. Remove the fan switch and test it. Check the blower motor by hot-wiring it directly to the battery with jumper cables.

Dual-Zone Systems

Some climate control systems offer separate temperature settings for the driver's seat and passenger's seat **(Figure 52–39)** as well as for rear-seat passengers. The temperature on each side of the vehicle is controlled by the doors in the ductwork for the heating and A/C system. Each side of the vehicle also has an inside air temperature sensor and a solar sensor. The temperature setting for either side of the vehicle is made at the control panel. If the passenger's control is turned off,

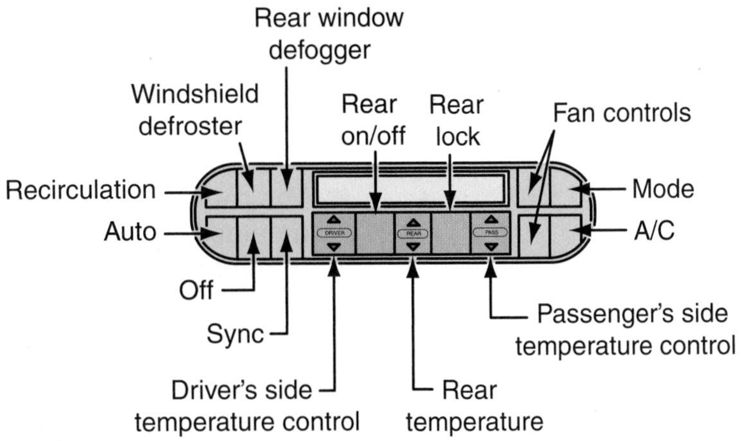

Figure 52-39 The control panel for a vehicle with dual zone and a rear heating and A/C system.

the climate control system will maintain both sides of the vehicle according to the setting on the driver's side. When the passenger's control is turned on, the temperature on that side of the vehicle can be adjusted independently from the driver's setting.

On some systems, the climate control system is tied into the navigation system. In these cases, the A/C control module calculates the direction of the vehicle, the time, longitude, latitude, intensity of the sunlight, ambient temperature, and other information to determine the ideal amount of heat or cooling for each side of the vehicle. This system automatically controls the temperature in response to all of these inputs.

Rear Systems

Some vehicles have a separate rear A/C system to provide comfort and temperature control for the passengers in the rear of the vehicle. These vehicles have a separate evaporator mounted in the rear of the vehicle **(Figure 52–40)**. A single conventional compressor is used to move refrigerant through the front and rear systems. There are also independent controls for the front and rear of the vehicle. Refrigerant lines connect the rear system to the compressor.

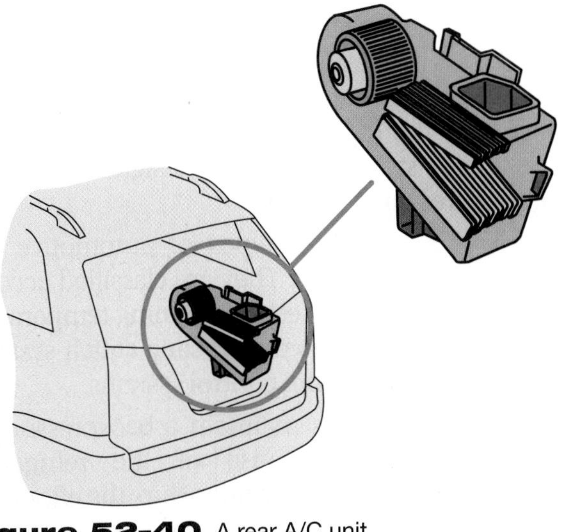

Figure 52-40 A rear A/C unit.

CASE STUDY

A customer brings in a car, recently purchased from a used car dealer, for service with a complaint of a water leak. The situation is that water spills out from under the dash when she is turning right. She also complains that the passenger side floor mat seems to always be wet.

The technician verifies the wet floor mat and suspects that the heater core is leaking. However, there is no evidence of engine coolant on the floor mat. Further discussion with the customer verifies that the car is not overheating nor does it have a heater problem. While talking to the customer, the technician notices that there are not the familiar drips on the shop floor usually experienced when a car's air-conditioning is running.

He inspects the drain tube for the air conditioner's evaporator and notices that the car was recently undercoated. He further notices that the outside drain opening is plugged with undercoating. With no place to go, the moisture pulled out of the air in the car is backing up into the evaporator housing and spilling out. Cleaning the drain tube takes care of the problem.

KEY TERMS

Blend door
CCOT system
Chlorofluorocarbon (CFC)
Condensation
Cycling clutch
Discharge side
Evaporation
Heater control valve
Heater core
Hydrofluorocarbon (HFC)
High side
Low side
Polyalkaline glycol (PAG)
Refrigerant
Sight glass

SUMMARY

- Ventilation, heating, and air-conditioning provide for the comfort of the vehicle's passengers.
- The ventilation system on most vehicles is designed to supply outside air to the passenger compartment through upper or lower vents or both. Several systems are used to vent air into the passenger compartment. The most common is the flow-through system. In this arrangement, a supply of outside air, called ram air, flows into the car when it is moving.
- Automotive heating systems have been designed to work with the cooling system to maintain proper temperature inside the car. The heating system's primary job is to provide a comfortable passenger compartment temperature and to keep car windows clear of fog or frost.
- The main components of an automotive heating system are the heater control valve, the heater core, the blower motor and fan, and heater and defroster ducts.
- All air-conditioning systems are based on three fundamental laws of nature: heat flow, heat absorption, and pressure and boiling points.
- The major components of an air-conditioning system are compressor, condenser, receiver/dryer or accumulator, expansion valve or orifice tube, and evaporator.
- The compressor is the heart of an automotive air-conditioning system. It separates the high-pressure and low-pressure sides of the system. The primary purpose of the unit is to draw the low-pressure vapor from the evaporator and compress this vapor into high-temperature, high-pressure vapor. This action results in the refrigerant having a higher temperature than surrounding air, enabling the condenser to condense the vapor back to liquid.
- The secondary purpose of the compressor is to circulate or pump the refrigerant through the condenser under the different pressures required for proper operation. The compressor is located in the engine compartment.
- The condenser consists of a refrigerant coil tube mounted in a series of thin cooling fins to provide maximum heat transfer in a minimum amount of space. The condenser is normally mounted just in front of the vehicle's radiator.
- The receiver/dryer is a storage tank for the liquid refrigerant.
- The refrigerant flow to the evaporator must be controlled to obtain maximum cooling, while ensuring complete evaporation of the liquid refrigerant within the evaporator. This is accomplished by a thermostatic expansion valve or a fixed orifice tube.
- The evaporator, like the condenser, consists of a refrigerant coil mounted in a series of thin cooling fins. The evaporator is usually located beneath the dashboard or instrument panel.
- An amendment to the United States' Clean Air Act and international agreements have mandated that R-12 be phased out as the refrigerant of choice. This came as a result of research that found the earth's ozone layer was being deteriorated by the chemicals found in R-12.
- Of the many chemicals that could have been used in place of R-12, the automobile manufacturers have decided to use R-134a, a hydrofluorocarbon (HFC) that causes less damage to the ozone layer when released to the atmosphere.
- Although R-134a air conditioners operate in the same way and with the same basic components as R-12 systems, the two refrigerants are not interchangeable. Because it is less efficient than R-12, R-134a operates at higher pressures to make up for the loss of performance and requires new service techniques and system component designs. Basically, the higher system pressures of R-134a mean the system must be designed for those higher pressures.
- There are two basic types of automotive air-conditioning systems. They are classified according to the method used in obtaining temperature control and are known as cycling clutch systems or evaporator pressure control systems.
- Evaporator controls maintain a backpressure in the evaporator. Because of the refrigerant temperature/pressure relationship, the effect is to regulate evaporator temperature.

- Temperature control systems for air conditioners are usually connected with heater controls. Most heater and air conditioner systems use the same plenum chamber for air distribution.
- Two types of air conditioner controls are used: manual/semiautomatic and automatic.

REVIEW QUESTIONS

1. What is the amount of heat necessary to change the state of a substance called?
2. What type of electromagnetic clutch is used in late-model vehicles?
3. Describe the different ways heat is supplied in a hybrid vehicle when the engine is not running.
4. What is the purpose of the thermostatic switch?
5. What does "change of state" mean? And why is it important to air-conditioning units?
6. What state is the refrigerant in when it leaves the condenser?
7. What causes condensed water to leak from the air-conditioning system?
8. On which of the following laws of nature is the air-conditioning system based?
   a. heat flow
   b. heat absorption
   c. pressure and boiling points
   d. all of the above
9. Which of the following statements is *true*?
   a. Refrigerant leaves the compressor as a high-pressure, high-temperature liquid.
   b. Refrigerant leaves the condenser as a low-pressure, low-temperature liquid.
   c. Refrigerant returns to the compressor as a low-pressure, high-temperature vapor.
   d. None of the above.
10. Which of the following statements is *false*?
    a. The condenser is normally mounted just in front of the radiator.
    b. The receiver/dryer is a storage tank for the liquid refrigerant from the condenser.
    c. An accumulator is not used in any system with a receiver/dryer.
    d. All of the above.
11. Which of the following statements about PTC heaters is *not* true?
    a. Some vehicles have a PTC heating element in the heater core.
    b. Some vehicles have a PTC element installed in an air duct from the blower housing to increase the air temperature in the ducts.
    c. PTC heaters are electronically controlled and work with the cooling system to provide hot coolant.
    d. PTC elements are small ceramic stones that react quickly to changes in current.
12. Which of the following is electrically connected in series with the compressor electromagnetic clutch?
    a. ambient temperature switch
    b. thermostatic switch
    c. pressure cycling switch
    d. all of the above
13. How can you identify an R-134a air-conditioning system?
14. Which of the following oils is used in an original equipment R-134a system?
    a. mineral oil
    b. CCOT
    c. ester oil
    d. PAG
15. A heating system includes all of the following parts except a ______.
    a. heater core
    b. ventilation system
    c. receiver/dryer
    d. distribution plenum

ASE-STYLE REVIEW QUESTIONS

1. Technician A says that a great amount of heat is transferred when a liquid boils or a vapor condenses. Technician B says that a change in pressure does not change the boiling point of a substance. Who is correct?
   a. Technician A   c. Both A and B
   b. Technician B   d. Neither A nor B
2. Technician A says that suction lines become very warm during A/C operation. Technician B says that discharge lines connect the compressor to the condenser. Who is correct?
   a. Technician A   c. Both A and B
   b. Technician B   d. Neither A nor B
3. While discussing the reasons the use of R-134a is being discontinued in some countrie

Technician A says that the chemical composition of the refrigerant has been proven to be harmful to the earth's ozone layer. Technician B says that R-134a leaks contribute to the undesirable greenhouse effect. Who is correct?

a. Technician A
b. Technician B
c. Both A and B
d. Neither A nor B

4. While discussing the reaction by a refrigerant to heat: Technician A says that to absorb heat, the temperature and pressure of the refrigerant are kept low. Technician B says that to dissipate heat, the temperature and pressure are kept high. Who is correct?

a. Technician A
b. Technician B
c. Both A and B
d. Neither A nor B

5. While discussing solar sensors: Technician A says that a solar sensor is a photo diode that receives a 5-volt reference signal from the control module. Technician B says that when the intensity of sunlight increases, the output voltage increases, and as the sunlight decreases, the output voltage decreases. Who is correct?

a. Technician A
b. Technician B
c. Both A and B
d. Neither A nor B

6. While discussing the future use of CO_2 as a refrigerant: Technician A says that CO_2 is nontoxic and presents little threat to the environment. Technician B says that only slight changes from conventional systems will be required because CO_2 systems operate at low pressures and temperatures. Who is correct?

a. Technician A
b. Technician B
c. Both A and B
d. Neither A nor B

7. Technician A says that when the A/C system is off, the refrigerant occupies the system as a liquid and its pressure is the same throughout the system. Technician B says that refrigerant is pumped out of the compressor as a high-pressure, high-temperature vapor and is sent to the condenser. Who is correct?

a. Technician A
b. Technician B
c. Both A and B
d. Neither A nor B

8. Technician A says that when the refrigerant is pumped out of the compressor and sent to the condenser, heat in the refrigerant is moved to the outside air by conduction and convection. Technician B says that the refrigerant leaves the bottom of the condenser as a high-pressure, high-temperature liquid. Who is correct?

a. Technician A
b. Technician B
c. Both A and B
d. Neither A nor B

9. A heater does not supply enough heat and the coolant level and flow are correct: Technician A says that a misadjusted heater control could be the cause. Technician B says that a bad thermostat could be the cause. Who is correct?

a. Technician A
b. Technician B
c. Both A and B
d. Neither A nor B

10. While discussing relative humidity: Technician A says that the amount of moisture that air can hold is directly related to the temperature of the air. Technician B says that the colder the air is, the more moisture it can hold. Who is correct?

a. Technician A
b. Technician B
c. Both A and B
d. Neither A nor B

AIR-CONDITIONING DIAGNOSIS AND SERVICE

CHAPTER 53

OBJECTIVES

■ Understand the special handling procedures for automotive refrigerants. ■ Explain the concerns and precautions regarding retrofitting an air-conditioning (A/C) system. ■ Describe how to connect a manifold gauge set and a recovery/recycling machine to a system. ■ Describe methods used to check refrigerant leaks. ■ Use approved methods and equipment to discharge, reclaim/recycle, evacuate, and recharge an automotive A/C system. ■ Perform a performance test on an A/C system. ■ Interpret pressure readings as an aid to diagnose A/C problems. ■ Diagnose and repair A/C control systems.

Air-conditioning (A/C) service, in many ways, is different from service to other parts of the vehicle. Although there are few parts in the system (**Figure 53–1**), each component has a specific purpose and service procedure. That by itself is not a big deal. The challenge with A/C is how it operates. The system operates on changes of refrigerant pressure. There are many things that can cause the pressure to change; some are part of the system, some are part of the environment, and some of them are faults or bad components in the system.

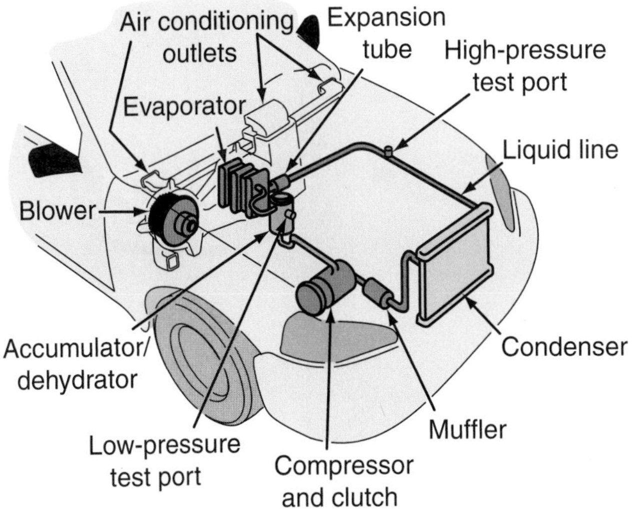

Figure 53-1 A late-model air-conditioning system.

SERVICE PRECAUTIONS

A/C systems are extremely sensitive to moisture and dirt. Therefore, clean working conditions are very important. The smallest particle of foreign matter in an A/C system contaminates the refrigerant, causing rust, ice, or damage to the compressor. For this reason, all replacement parts are sold in vacuum-sealed containers and should not be opened until they are ready to be installed in the system. If, for any reason, a part has been removed from its container for any length of time, the part must be completely flushed using only the recommended solvent to remove any dust or moisture that might have accumulated during storage. If the system has been open for more than 2 minutes, the entire system must be completely evacuated and a new accumulator or desiccant bag must be installed.

SHOP TALK

It is important to remember that just one drop of water added to the refrigerant will start chemical changes that can result in corrosion and eventual breakdown of the chemicals in the system. The smallest amount of moist air in the refrigerant system might start reactions that can cause malfunctions.

Whenever it is necessary to disconnect a refrigerant line, wipe away any dirt or oil at and near the connection to eliminate the possibility of dirt entering the system. Both sides of the connection should be immediately capped or plugged to prevent the entrance of dirt, foreign material, and moisture. It must be remembered that all air contains moisture. Air that enters any part of the system carries moisture with it.

Keep all tools clean and dry. This includes the gauge set and replacement parts. Be careful not to overtighten any connection. Overtightening can result in distortion and a system leak.

When adding oil, the container and the transfer tube through which the oil will flow should be exceptionally clean and dry. Refrigerant oil quickly absorbs any moisture it contacts. For this reason, the oil container should not be opened until it is time to use it and should be capped immediately after use.

When it is necessary to open a system, have everything needed immediately available so as little time as possible is required to perform the operation. Do not leave the system open any longer than necessary.

Any time the system has been opened for repairs, it must be properly evacuated after the repair. Also, before disconnecting or removing any part, the refrigerant needs to be recovered and not allowed to escape the atmosphere.

REFRIGERANT SAFETY PRECAUTIONS

- Always work in a well-ventilated and clean area. Refrigerants are colorless and invisible. Refrigerant is heavier than oxygen and will displace it in a confined area. Avoid breathing the refrigerant vapors. Exposure to refrigerant may irritate your eyes, nose, and throat.
- Refrigerant evaporates quickly when it is exposed to the atmosphere. It will freeze anything it contacts. If liquid refrigerant gets in your eyes or on your skin, it can cause frostbite. Never rub your eyes or skin if refrigerant has contacted these areas. Immediately flush the exposed areas with cool water for 15 minutes and seek medical help. Also check the MSDS for the refrigerant to identify other safety-related procedures.
- An A/C system's high pressure can cause severe injury to your eyes and/or skin if a hose were to burst. Always wear eye protection when working around the A/C system and refrigerant. It is also advisable to wear protective gloves and clothing.
- Never use R-134a in combination with compressed air for leak testing. Pressurized R-134a in the presence of oxygen may form a combustible mixture. Never introduce compressed air into R-134a containers (empty or full ones), A/C systems, or A/C service equipment.
- Be careful when handling refrigerant containers. Never drop, strike, puncture, or burn the containers. Always use DOT-approved refrigerant containers.
- Never expose A/C system components to high temperatures. Heat will cause the refrigerant's pressure to increase. Never expose refrigerant to an open flame.
- Never overfill refrigerant containers. The filling level of the container should never exceed 60% of the container's gross weight rating. Always store refrigerant containers in temperatures below 125°F (52°C) and keep them out of direct sunlight.
- Refrigerant comes in 30- and 50-pound cylinders. Keep the drums in an upright position. Make sure that valves are protected by safety caps when the drums are not in use. Avoid dropping the drums. Handle them carefully.
- R-12 should be stored and sold in white containers, whereas R-134a should be stored in light blue containers **(Figure 53–2)**. R-12 and R-134a should never be mixed. If the two refrigerants are mixed, contamination will occur and may result in A/C system failure. Separate service equipment should be used for the different refrigerants.
- To prevent cross-contamination, identify whether the A/C system being worked on uses R-12 or R-134a. Check the fittings in the system; all

Figure 53-2 R-12 should be stored and sold in white containers, whereas R-134a should be stored in light blue containers.

R-134a-based systems use ½-inch 16 ACME threaded fittings and quick-disconnect service couplings. Most R-134a systems can be identified by underhood labels clearly stating that R-134a is used. Most manufacturers identify the type of refrigerant used by labeling the compressor. Also look for a label with the words, "CAUTION—SYSTEM TO BE SERVICED BY QUALIFIED PERSONNEL." This label or plate can be found under the hood near a component of the system. This label also indicates what kind refrigerant is used the required quantity to fill the system, and the type of refrigerant oil.

Special Precautions for Hybrid Vehicles

Hybrid vehicles have high-voltage systems. Most often, the A/C compressor is powered by high voltage. Careless handling of some components can lead to serious injury, including death. Always follow and adhere to the precautions given by the manufacturer. These precautions are clearly labeled in their service manuals. All service procedures should be followed exactly as defined by the manufacturer. Being careless and/or not following the procedures can cause serious injury and can cause the battery to explode! Following is a list of commonsense items to consider when working on a hybrid vehicle:

- Before doing any service on a hybrid vehicle, refer to the service manual for that specific vehicle. All hybrids have similar operation but have different systems and components; this is true for vehicles made by the same manufacturer.
- All high-voltage wires and harnesses are wrapped in orange-colored insulation. Respect the color and stay away from it unless the system is depowered.
- Warning and/or caution labels are attached to all high-voltage parts. Be careful not to touch these cables and parts without the correct protective gear, such as safety gloves.
- Make sure that the high-voltage system is shut down and isolated from the vehicle before working near or with any high-voltage component.
- When working on or near the high-voltage system, even when it is depowered, always use insulated tools.
- Never leave tools or loose parts under the hood or close to the battery pack. These can easily cause a short.
- Never wear anything metallic, such as rings, necklaces, watches, and earrings, when working on a hybrid vehicle.

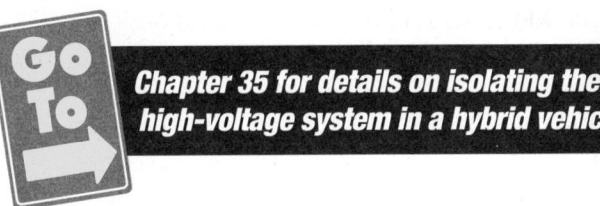

Go To ***Chapter 35 for details on isolating the high-voltage system in a hybrid vehicle.***

GUIDELINES FOR CONVERTING (RETROFITTING) R-12 SYSTEMS TO R-134A

The following guidelines should be followed when converting an older A/C system to R-134a. These guidelines should allow you to provide the customer with a cool vehicle and to meet current legislative mandates. The guidelines are listed in order and reflect the necessary steps for making this conversion.

1. Visually inspect all A/C and heater system components.
2. Use a refrigerant identifier to make sure the system only contains R-12.
3. Check the system for leaks.
4. Run a performance test and record the temperature and pressure readings.
5. Remove all R-12 from the system with a recycling machine.
6. Make sure all old refrigerant and oil are removed from the system.
7. If the system uses a compressor with an oil sump, remove the compressor and drain all oil from it. Measure the amount of oil drained out.
8. Remove and inspect the expansion valve or the orifice tube; replace it if necessary. If either is contaminated, flush the condenser.
9. Remove the filter/dryer or accumulator; drain it and measure the amount of oil in it.
10. Before converting, make all necessary system changes, such as the compressors, hoses, gaskets, and seals.
11. Install R-134a compatible oil into the system; put in the same amount you took out.
12. Install a new filter/dryer or accumulator with the correct desiccant.
13. Permanently install conversion fittings using a thread locking chemical. Unique service fittings for R-134a should be used to minimize the risk of cross-contamination.
14. Install a high-pressure cutoff switch if the system does not have one.
15. Install R-134a conversion labels and remove the R-12 label (**Figure 53–3**).
16. Connect R-12 system evacuation equipment and evacuate for at least 30 minutes.

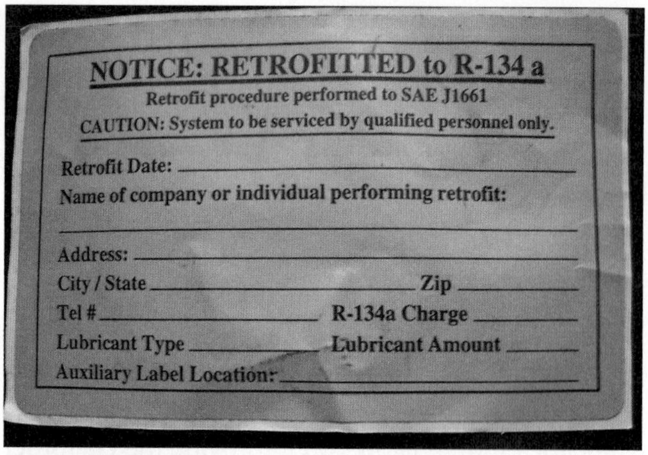

Figure 53-3 A label showing that an R-12 A/C system was converted to R-134a. This must be attached to the vehicle whenever the conversion has taken place.

17. Recharge the system with R-134a to 80% of the original R-12 charge.
18. Run a performance check and compare readings with those taken before the conversion.
19. Leak check the system.

Go To *Chapter 5 for a complete description of the various tools used to diagnose and service A/C systems.*

WARNING!

Under no circumstances, because of environmental and health concerns, should refrigerants or refrigerant oils be mixed. Never add one type of refrigerant to a system that has or has had another type refrigerant in it. Also make sure that the vehicle is clearly labeled as to the type of refrigerant it was converted to use.

SHOP TALK

In some states and provinces, you must be certified by an EPA-approved program to buy refrigerant. To become certified, a technician must have and use approved refrigerant recycling equipment and pass an exam on refrigerant recovery and recycling. A common test for this certification is administered by the ASE. MACS worldwide training/testing and some states also offer the test.

INITIAL SYSTEM CHECKS

The first step in A/C diagnosis, as in all automotive diagnostic work, is to get the customer's story. Is the problem no cold air, the air from the ducts never gets cold enough, or the unit only cools at certain times of the day? The complaint could also be that the system does not work at all, or that the air blows out of the wrong ducts. An accurate description of the customer's problem helps pinpoint whether the problem is refrigerant, mechanical, vacuum, or electrical related. It also reduces diagnosis time and, most important, satisfies the customer.

Because of the many construction and operational variations that exist, there is no uniform or standard diagnostic procedure applicable to all automotive A/C systems except the performance test. For complete specific diagnostic information on a given air conditioner, check the manufacturer's service manual.

A quick verification of the customer's concern can be done by starting the engine and allowing it to warm up. Move the temperature control to its full heat position. Check the amount of airflow and heat in each of the fan positions. The amount of heat coming out the system should change with a change in blower speed. Also pay attention to all smells and noises. Turn the temperature control to the defrost mode. Make sure there is adequate airflow coming out of the defroster vents. Now turn the temperature control to cool or to the A/C position. Pay attention to how the engine reacts when the A/C compressor clutch engages. Also pay attention to any abnormal noises or smells. The air leaving the vents should be cool. While doing these checks, attempt to duplicate the customer's concerns and the conditions that exist when the problem occurs. If the system does not respond as expected, the system should be inspected and performance tested.

Inspection

A visual check of the A/C and cooling systems can result in an immediate diagnosis. Begin by checking the condition of the compressor's drive belt. Check the tension of the belt with a belt tension gauge (if the compressor is driven by a serpentine belt, check the belt tensioner markings). Carefully look at the compressor clutch for signs of oil leakage and belt slippage. Belt slippage may be evident by large amounts of black dust around the clutch plate. The presence of oil on or around the clutch normally indicates a bad compressor seal.

When the system is low on refrigerant charge, the performance of the system suffers. Low charge is

Figure 53-4 The oily film by this fitting is evidence of a refrigerant leak.

typically the result of system leaks. The area of leakage can often be identified with a thorough inspection. Because refrigerant oil leaks out with the refrigerant, there will be an oily film at the point of leakage **(Figure 53–4)**. This film will collect dirt, and a buildup of dirt around a fitting is a good indication of a leak. Check the following refrigerant hoses and fittings for signs of leakage or damage and bends or distortion:

- Compressor to the condenser
- Receiver/dryer to the accumulator
- Condenser to the evaporator
- Evaporator to the compressor

Visually check the condenser and the area between the condenser and radiator for a buildup of dirt, leaves, and other debris **(Figure 53–5)**. Also check the fins of the condenser for damage. If the airflow through the condenser is blocked, clean the condenser and the area around it. If the fins are damaged, they may be able to be straightened. In most cases, a damaged condenser is replaced.

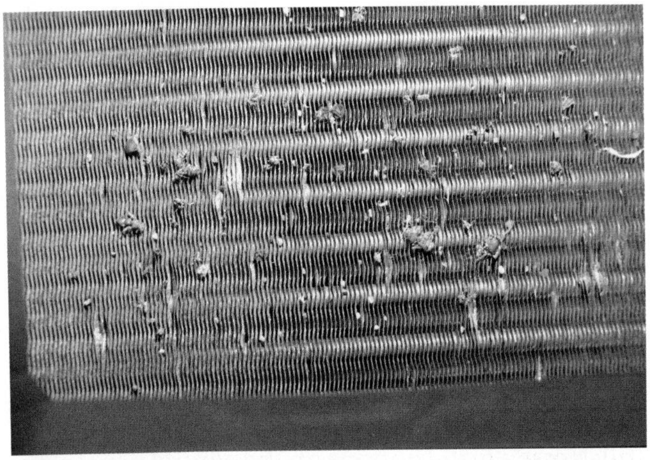

Figure 53-5 The condenser should be checked for damage and a buildup of debris that could cause a restriction to airflow.

A/C systems rely on many switches; these can be the cause of several operating problems. Look at the wiring and connectors to all sensors and switches carefully. Make sure all connections are secure and clean. Sometimes a bad connection or corrosion can cause intermittent problems.

In a well-ventilated area, start the engine, place the transmission in park (if the vehicle has a manual transmission, put it in neutral), set the parking brake, and turn on the A/C system to MAX cooling. On most vehicles, the electric engine cooling fan will turn on when the A/C system is turned on. Check to see if the fan(s) came on.

Listen to the system and record any unusual sounds. Be sure to listen to the system while the compressor clutch cycles on and off. You should hear a click or a slight drop in engine speed when the clutch engages. Of course, if the compressor runs at all times, the clutch will not engage or disengage with a change of the A/C controls.

Simple Checks Feel the air coming out of the vents. Cold or slightly cool air means the system has some cooling ability and the compressor is working. If the air is not cool, this means the compressor is not working or the system is not capable of providing conditioned air. If the air is warm, there is probably a problem with the air duct system or the controls.

Move the temperature control from hot to cool. Feel the air from the vents. If the air temperature does not change or changes little, there is a problem with the blend door or doors in the air distribution case.

Operate the blower control through all of its speed positions; the amount of air should change. If not, there is a problem with the switch or the resistor block for the blower.

Feel the discharge line from the compressor to the condenser. The line should be hot and the temperature should be the same along its full length. Any change is a sign of restriction, and the line should be flushed or replaced. Check the condenser by feeling up and down the face or along the return bends. There should be a gradual change from hot to warm as you move from the top to the bottom. Any abrupt change indicates a restriction, and the condenser has to be flushed or replaced.

If the system has a receiver/dryer, check the inlet and outlet lines. They should have the same temperature. Any temperature difference or frost on the lines or receiver tank are signs of a restriction. The receiver/dryer must be replaced. Also feel the liquid line from the receiver/dryer to the expansion valve. The entire length of the line should be warm. Typically the formation of frost on the outside of a line or component means there is a restriction to refrigerant flow. A restriction can cause the refrigerant to stay longer in the condenser. This is called condenser flooding and will cause a starved evaporator.

The expansion valve should be free of frost, and there should be a sharp temperature difference between its inlet and outlet. On vehicles equipped with the orifice tube, feel the liquid line from the condenser outlet to the evaporator inlet. A restriction is indicated by any temperature change in the liquid line before the orifice tube. Flush the line or replace the orifice tube if restricted.

Carefully place your hand on the inlet line of the evaporator **(Figure 53–6)**. Do the same on the outlet. Both tubes should be colder than ambient temperature and both should be about the same temperature. If the inlet is significantly colder than the outlet, the system is probably low on refrigerant. If both lines are not colder than ambient temperature, the system needs further diagnosis.

The suction line to the compressor should be cool to the touch from the evaporator to the compressor. If it is covered with thick frost, this might indicate that the expansion valve is flooding the evaporator. The accumulator should also be cool to the touch.

Sight Glass The sight glass allows a technician to see the flow of refrigerant in the lines. A sight glass is normally found on systems using R-12 and a thermal expansion valve. It is not commonly found on R-134a systems. It can be located on the receiver/dryer, between the receiver/dryer and the expansion valve or tube, or in the liquid line.

Figure 53-6 Feeling the temperature of the pressure lines in an A/C system is part of a basic visual inspection of the system.

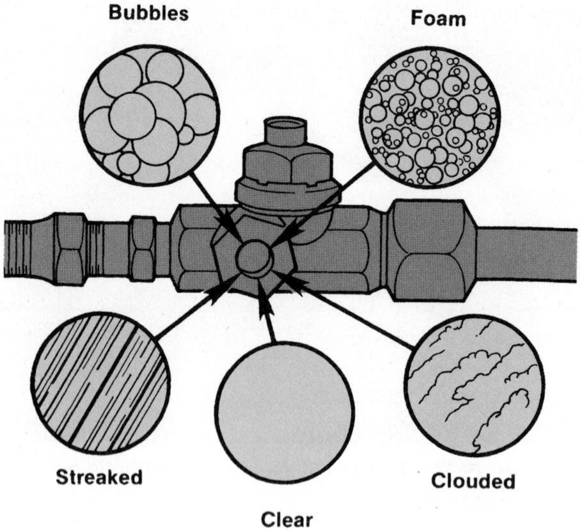

Figure 53-7 The appearance of the refrigerant in the sight glass can give an indication of system problems.

To check the refrigerant, set the controls for maximum cooling and set the blower on its highest speed. Let the system run for about 5 minutes. Be sure the vehicle is in a well-ventilated area or connected to an exhaust gas ventilation system.

Check the sight glass while the engine is running **(Figure 53–7)**. If oil streaking is seen, this indicates that the system is empty. Bubbles, or foam, indicate that the refrigerant is low. A sufficient level of refrigerant is indicated by what looks like a flow of clear water with no bubbles. A clouded sight glass is an indication of desiccant contamination with subsequent infiltration and circulation through the system.

Accumulator systems do not have a receiver/dryer to separate the gas from the liquid as it flows from the condenser. The liquid line will always have a certain amount of bubbles in it. R-134a systems can also be checked by viewing the sight glass if the system has one. The normal appearance of R-134a tends to be milky. This is why R-134a systems and systems with an accumulator normally do not have a sight glass. Pressure and performance testing are the only ways to identify low-refrigerant levels.

Trouble Codes

Some late-model A/C systems have a self-diagnostic capability. An indicator lamp on the control panel may be used to flash codes. On some vehicles, part of the A/C system can be checked with a scan tool connected to the vehicle's BCM; this is especially true for those vehicles with automatic climate control. Normally the buttons on the control panel must be depressed in a certain sequence to activate the self-test.

PROCEDURE

To access DTCs from a system, follow these steps:

Using a Scan Tool

STEP 1 Using the service manual or a component locator for this vehicle, locate where the diagnostic connector is for the body control module (BCM).

STEP 2 Connect the scan tool.

STEP 3 Turn the ignition on.

STEP 4 Program the scan tool for that vehicle.

STEP 5 From the menu on the scan tool, select the BCM.

STEP 6 Retrieve and record all DTCs.

STEP 7 To observe the activity of the inputs and outputs, select DATA. Pay attention to any data that is outside its normal operating range, such as the ambient temperature, evaporator temperature, and low- and high-side pressures.

STEP 8 Diagnose any abnormal operating perimeters.

STEP 9 Turn the ignition off.

STEP 10 Disconnect the scan tool.

Using the Control Panel

STEP 1 Refer to the service information for the correct procedure to enter into the HVAC's self-diagnostic mode. Normally this is done by depressing and holding in buttons on the panel.

STEP 2 Determine what display on the panel should be observed to retrieve the DTCs and data.

STEP 3 Turn the ignition on.

STEP 4 Press and hold the designated buttons for the designated time. Make sure you hold the buttons down for the required period.

STEP 5 Record all DTCs that were displayed and describe what is indicated by each.

STEP 6 If there are no DTCs detected and the system is still not operating properly, check the sensor input to the HVAC control unit. NOTE: The procedure for checking sensor inputs will vary with the model of vehicle.

STEP 7 Turn the ignition switch off.

STEP 8 Depress and hold the designated buttons.

STEP 9 Start the engine and release the buttons after the engine starts.

STEP 10 The control panel should begin to display the data for the sensors. Compare all readings to specifications.

STEP 11 Diagnose and correct all problems noted.

STEP 12 Cancel the self-diagnostic function.

STEP 13 After completing all repairs to the system, run the self-diagnostic test again to make sure there are no other problems.

DIAGNOSIS

Often diagnosis of an A/C system can be done by observing the operation of the system, including listening for noises. When abnormal noises are heard, they can lead to the problem area.

Noise

Often a customer's concern is the noise emitted by the A/C system. The following discussion covers the common abnormal noises and their causes.

Clutch Noises The clutch normally makes a clicking noise when it engages and disengages. This noise will become louder as the clutch wears. Once the clutch is severely worn, it will make a squealing noise when it is engaged. This noise can also result from oil on the clutch. A very loud screech or squeal can indicate a seized compressor. A bent drive pulley also can cause a growling or rubbing noise.

Hose Noise The change in pressures in the suction/discharge hoses can set up vibrations that cause sounds to appear from the inside of the vehicle. The noise is typically caused by a hose contacting another hose or part in the engine. Check the routing of the hoses to make sure they are not in contact. Also look for abrasions on the hoses. These can be caused by the contact.

Compressor Noise If the mounts for the compressor are loose or damaged, there will be a rattling or groaning noise. The noise is normally random. At times, loose mounts will also cause premature wear of the drive belt.

Hissing or Whistling This is a common and normal noise. When the A/C is turned off, a high-pitched whistle may be heard. This sound is caused by the pressures that are equalizing in the system. The high pressure moves to the low-pressure side through the metering device. This movement can result in a whistle or hiss.

Odors

A some what common concern of customers is a smell emitting from the A/C system. This moldy and musty smell is due to a buildup of moisture on the evaporator or the cabin filter. Moisture buildup is caused by the evaporator doing its job. As the temperature inside the vehicle is lowered, moisture is separated from the air; the moisture should be able to drain but sometimes the drain is plugged or dislocated. When a customer complains about the smell, check the evaporator drain. Also, the cabin air filter needs to be periodically changed; if it has not been changed, there will be odors that offend the passengers. Products are available to disinfect the evaporator and ductwork. These kill any odor-causing bacteria.

Cabin Filters Cabin air filters are designed to capture soot, dirt, pollen, and other pollutants that enter a vehicle through its heating, A/C, and defrost systems. Today more than 80% of all new vehicles sold in the United States have a cabin air filter or a slot where one can be installed. Cabin air filters are typically located behind the glove compartment, under the dash, and under the hood **(Figure 53–8)**. Some cars have the filter in the HVAC case between the blower motor and evaporator core. A cabin filter is a critical part of the HVAC system. If the filter becomes dirty or clogged, less air will be able to pass through the filter. This will adversely affect the operation of the HVAC system.

Many Toyotas have an ion generator called the Plasmacluster generator **(Figure 53–9)** that improves interior air quality. It uses a high-voltage device that emits a slight sound. With these systems, dust will accumulate at the driver-side air vent. Clean it with a cloth; never spray solvent into the air vent.

Replacement of the filters is part of a vehicle's PM program; normally a filter should be replaced every 12,000 to 15,000 miles (19,300 to 24,140 km) or at least once a year. This interval really depends on where the vehicle is typically driven. The procedure for replacing the filters varies with make and model; always check with the manufacturer.

There are two basic types of cabin air filters: particle-trapping and one that has an additional charcoal layer for odor absorption. Some particulate filters have a section that traps larger particles and another that is electrostatically charged to attract and hold smaller particles.

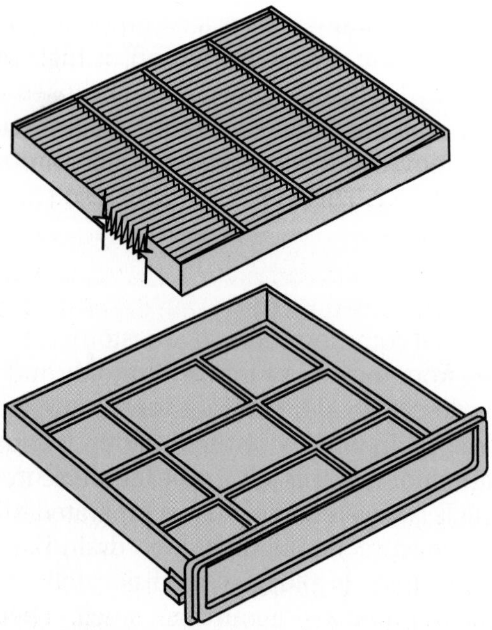

Figure 53-8 A typical cabin air filter and housing.

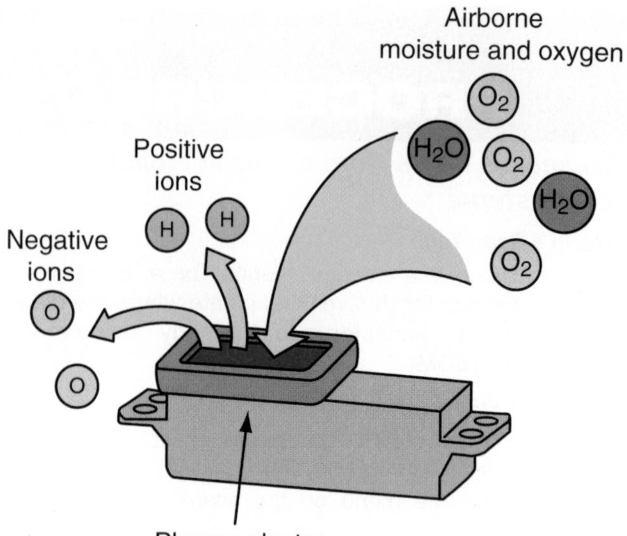

Figure 53-9 Many Toyotas have an ion generator called the Plasmacluster generator that improves interior air quality.

A charcoal type will hold a lot of odorant, and under some conditions it may release trapped odors. This is why these filters should be replaced on a regular basis. There are two alternatives to the charcoal filter: a filter element impregnated with baking soda and a filter with a biocidal cartridge. Both prevent the growth of bacteria, mold, mildew, algae, and yeast.

Some vehicles are also equipped with an air quality sensor in the main inlet duct of the HVAC system or in front of the condenser. The sensor detects the presence of undesirable gases in the incoming air, specifically carbon monoxide and nitrogen dioxide. Within seconds of detection, the ventilation system closes the outside air inlet and recirculates the air until the pollutants are no longer at an unacceptable level.

PERFORMANCE TESTING

Performance testing provides a measure of A/C system operating efficiency. A manifold pressure gauge set is used to determine both high and low pressures in the refrigeration system. The desired pressure readings will vary according to temperature. Use temperature/pressure charts as a guide to determine the proper pressures. At the same time, a thermometer is used to determine air discharge temperature into the passenger compartment **(Figure 53–10)**.

Service Valves

System service valves provide the attachment point for the manifold pressure gauge set. They are located

Figure 53-10 A thermometer is used to determine air discharge temperature into the passenger compartment.

in the low and high sides of most A/C systems. Two basic types of service valves were used on R-12 systems: stem and Schrader. R-134a systems have quick-connect fittings. The fittings for the gauges are connected to the system's fittings. The gauge's fitting has a knob that is turned in the clockwise direction to open the valve. No refrigerant should leak out when the gauge set is connected.

Stem Valves The **stem valve** was sometimes used on two-cylinder reciprocating-piston compressors. The service valves are mounted on the compressor head. These valves can be used to isolate the rest of the A/C system from the compressor when the compressor is being serviced. These valves have a stem under a cap with the hose connection directly opposite it.

Schrader Valves R-12 systems without a stem service valve have **Schrader service valves (Figure 53–11)**. Closely resembling a tire valve, Schrader valves are usually located in the high-pressure line and in the low-pressure line. All test hoses have a Schrader core depressor in them. As the hose is threaded into the service port, the pin in the center of the valve is depressed, allowing refrigerant to flow to the manifold gauge set. The valve closes automatically when the hose is removed.

CAUTION!

Always wear safety goggles when working on or around A/C systems.

Purity Test

When you are not sure of the refrigerant used in a system or if you suspect that a mixing of refrigerants has occurred, you should run a purity test and/or use a refrigerant identifier **(Figure 53–12)**. Knowing what refrigerant is in the system, or what condition it is in, will help you determine what steps you need to take to properly service the system. It is recommended that the system be tested for sealant identification and/or contamination before conducting the purity test.

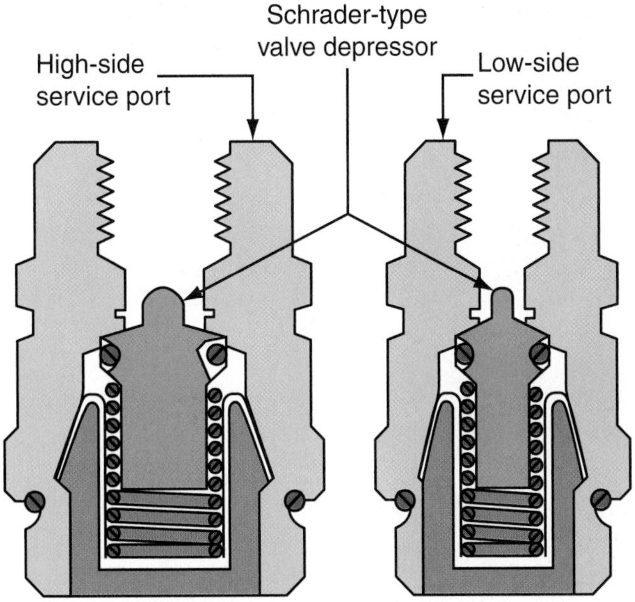

Figure 53-11 Schrader service valves for a R-134a system.

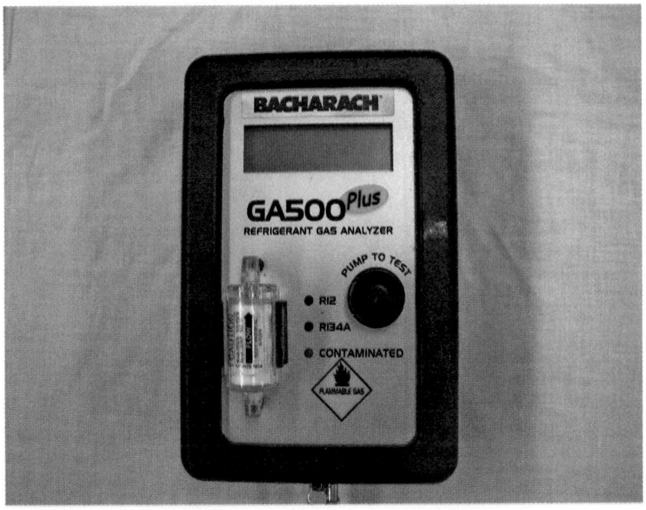

Figure 53-12 A refrigerant identifier.

Manifold Pressure Gauge

The manifold gauge set **(Figure 53–13)** is one of the most important A/C tools. It is used when discharging, charging, and evacuating the system; it also is used for diagnosing the system. With the new legislation on handling refrigerants, all gauge sets are required to have a valve device to close off the end of the hose so that the fitting not in use is automatically shut **(Figure 53–14)**.

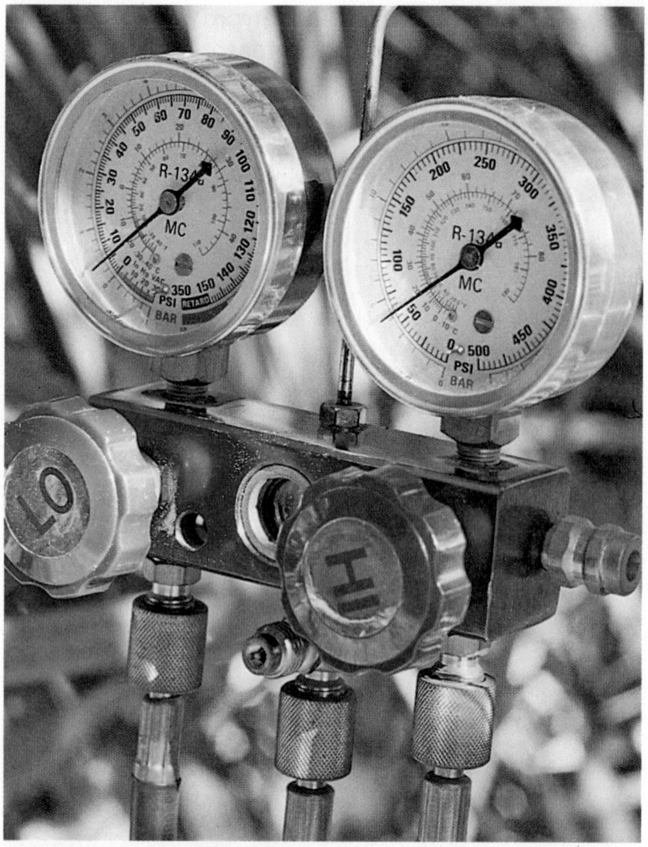

Figure 53-13 A manifold gauge set for R-134a.

Figure 53-14 The test connectors with shutoff valves for an R-134a system.

The low-pressure gauge is graduated into pounds of pressure from 1 to 120 (with cushion to 250) in 1-pound graduations, and, in the opposite direction, in inches of vacuum from 0 to 30. This is the gauge that should always be used in checking pressure on the low-pressure side of the system. The gauge at the right is graduated from 0 to 500 pounds pressure in 10-pound graduations. This is the high-pressure gauge that is used for checking pressure on the high-pressure side of the system.

The center manifold fitting is common to both the low and the high sides and is for removing or adding refrigerant to the system. When this fitting is not being used, it should be capped. A test hose connected to the fitting directly under the low-side gauge is used to connect the low side of the test manifold to the low side of the system. A similar connection is found on the high side.

The gauge set has hand valves that isolate the low and high sides from the central portion of the manifold. During all tests, both the low- and high-side hand valves are in the closed position (turned inward until the valve is seated).

Because R-134a is not interchangeable with R-12, separate sets of hoses, gauges, and other equipment are required to service vehicles. Manifold gauge sets for R-134a can be identified by one or all of the following: Labeled "FOR USE WITH R-134a," labeled "HFC-134" or "R-134a," and/or have a light blue color on the face of the gauges. Also, R-134a service hoses have a black stripe along their length and are clearly labeled "SAE J2196/R-134a." The low-pressure hose is blue with a black stripe. The high-pressure hose is red with black stripe, and the center service hose is yellow with a black stripe. Service hoses for one type of refrigerant will not easily connect into the wrong system, because the fittings for an R-134a system are different from those used in an R-12 system.

Connecting the Gauge Set Identify the type of refrigerant used in the systems and obtain the appropriate gauge set. Locate the high- and low-side service valves. Make sure the valves on the gauge set are fully closed. Remove the protective cap from the low-side service valve and connect the low-side service hose to the service valve. Then remove the protective cap from the high-side service valve and connect the high-side service hose to the valve.

Testing Procedures

PROCEDURE

To conduct a performance test, follow these steps:

STEP 1 With the manifold gauge set connected, start the engine in neutral or park with the parking brake on and allow the engine to run for about 5 minutes.

STEP 2 Increase engine speed to 1,500 to 2,000 rpm.

STEP 3 Keeping your hands away from the cooling fan, measure and record the temperature and humidity of the air in front of the condenser **(Figure 53–15)**.

STEP 4 Place a high-volume fan in front of the radiator grille to ensure an adequate supply of airflow across the condenser.

STEP 5 Adjust the A/C controls to maximum cooling and high blower position.

STEP 6 Place the thermometer in the center air duct in the dash. Measure and record the temperature.

STEP 7 Read the high and low pressures. If the vehicle uses a cycling clutch system, note the pressures at which the clutch engages and disengages.

STEP 8 Compare your readings to specifications.

STEP 9 Return the engine to a normal idle speed.

STEP 10 Close the service hose valves, if so equipped.

STEP 11 Disconnect the service hoses.

STEP 12 After disconnecting the gauge lines, check the valve areas to be sure that the service valves are correctly seated and that the Schrader valves are not leaking.

STEP 13 Reinstall the protective caps on the service valves.

STEP 14 Turn off the A/C and the engine.

Figure 53-15 Measure the temperature of the air entering the condenser.

Always refer to the manufacturer's pressure charts when using observed pressures for diagnostics **(Figure 53–16)**. It is normal for the gauge readings to fluctuate as the clutch cycles and the heat load changes, but they should not greatly fluctuate. Operating pressures will vary with humidity as well as with outside air temperature. Accordingly, on more humid days, operating pressures will be on the high side of the acceptable pressure range. On less humid days, the operating pressures will read toward the lower side. If operating pressures are found to be within the normal range, the refrigeration portion of the A/C system is functioning properly. This is further confirmed with a check of evaporator outlet air temperatures.

Here are some guidelines to help you interpret abnormal readings:

- If the observed pressures are below the normal ranges, this indicates an undercharged system.
- If both pressures are above normal pressures, the system is overcharged or there is insufficient air flowing through the condenser.
- If there is normal low-side pressure but higher-than-normal high-side pressure, this indicates air in the system.
- If both gauges show equally low pressures but a medium amount of pressure, this indicates a fully charged system with an inoperative compressor.
- If the high-side pressure is too high, suspect air in the system, too much refrigerant in the system, a restriction in the high-side of the system, and poor airflow across the condenser.
- If the high-side pressure is too low, suspect a low refrigerant level or defective compressor.
- If the low-side pressure is higher than normal, suspect refrigerant overcharge, a defective compressor, or a faulty metering device.
- If the low-side pressure is lower than normal, suspect a faulty metering device, poor airflow across the evaporator, a restriction in the low side of the system, or a system that is undercharged with refrigerant.

Evaporator Outlet Temperature Evaporator outlet air temperature also varies according to outside (ambient) air and humidity conditions. Further variations depend on whether the system is controlled by a cycling clutch compressor or an evaporator pressure control valve. Because of these variations, it is difficult to pinpoint what the evaporator outlet air temperature should be for all applications. In general, with low air temperatures and humidity, the evaporator outlet air temperature should be from 35 to 40°F (1.7 to 4.4°C). On the other extreme of high temperatures and humidity, the evaporator air outlet temperature might be in the 55 to 60°F (12.8 to 15.6°C) range.

LEAK TESTING

Testing the refrigerant system for leaks is one of the most important phases of troubleshooting. Over a period of time, all A/C systems lose or leak some refrigerant. In systems that are in good condition, R-12 losses of up to ½ pound (8 oz.) per year are considered normal (R-134a systems lose about half as much, or 4 oz.). Higher loss rates signal a need to locate and repair the leaks.

Leaks are often evident by low-pressure gauge readings. Normally, if there is some pressure in the system, the leak is a small one. Large leaks typically will empty the system.

Ambient Air Temperature	Relative Humidity	Low-Side Pressure	High-Side Pressure
55–65°F (13–18°C)	0–100%	22–49 psi (51–337 kPa)	112–167 psi (771–1,150 kPa)
66–75°F (19–24°C)	Below 40%	22–34 psi (151–234 kPa)	107–160 psi (737–1,102 kPa)
	Above 40%	24–37 psi (165–254 kPa)	128–173 psi (881–1,191 kPa)
76–85°F (25–29°C)	Below 35%	24–37 psi (165–254 kPa)	120–176 psi (826–1,212 kPa)
	35–50%	28–38 psi (192–261 kPa)	127–177 psi (875–1,219 kPa)
	Above 50%	26–40 psi (179–275 kPa)	132–179 psi (909–1,233 kPa)
86–95°F (30–35°C)	Below 30%	26–41 psi (179–282 kPa)	152–199 psi (1,047–1,371 kPa)
	30–50%	31–42 psi (213–289 kPa)	164–193 psi (1,129–1,329 kPa)
	Above 50%	32–44 psi (220–303 kPa)	165–186 psi (1,136–1,281 kPa)
96–105°F (36–41°C)	Below 20%	35–46 psi (241–316 kPa)	193–228 psi (1,329–1,570 kPa)
	20–40%	35–46 psi (241–316 kPa)	190–216 psi (1,309–1,488 kPa)
	Above 40%	36–47 psi (248–323 kPa)	184–205 psi (1,267–1,412 kPa)
106–115°F (42–46°C)	Below 20%	40–50 psi (275–344 kPa)	224–251 psi (1,543–1,729 kPa)
	Above 20%	40–50 psi (275–344 kPa)	216–253 psi (1,488–1,743 kPa)
116–120°F (47–49°C)	Below 30%	44–54 psi (303–372 kPa)	239–285 psi (1,646–1,963 kPa)

Figure 53-16 A typical pressure chart.

Leaks are most often found at the compressor hose connections and at the various fittings and joints in the system. Refrigerant can be lost through hose permeation. Leaks can also be traced to pinholes in the evaporator caused by acid, which forms when water and refrigerant mix. Because oil and refrigerant leak out together, oily spots on hoses, fittings, and components mean there is a leak.

When the source of leakage is not obvious, a leak detector can be used to locate it. However, to use a leak detector, there must be some refrigerant in the system. Most manufacturers say that a leak detector will only work if the high-side pressure is above 60 psig. If the pressure is below this, add no more than 1 pound (453.6 grams) of refrigerant to the system and then check for leaks.

Be sure to check the entire system, including inside the passenger compartment around the evaporator. If a leak is found at a connection, tighten the connection carefully and recheck. If the leak is still apparent, the affected components should be replaced. After leak testing, and before opening the system for repairs, use the recovery station to remove all refrigerant from the system.

Stop Leak Products Products that can be added to seal leaks in the system are sold. Many OEMs do not recommend their use. Sealants may cause buildups in the system, causing unwanted restrictions. They may also cause compressor clutch problems if there is a leak around the compressor shaft. These manufacturers also will void the warranty on the system if a sealant has been installed. Always check with the manufacturer before adding a sealant to the system.

Electronic Leak Detector

This is the preferred method of leak detection; it is safe, effective, and can be used with all types of refrigerants. The hand-held battery-operated electronic leak detector **(Figure 53–17)** contains a test probe that is moved under the areas of suspected leaks. (Remember that refrigerant gas is heavier than air, so the probe should be positioned below the test point.) An alarm or a buzzer on the detector indicates the presence of a leak **(Figure 53–18)**. On some models, a light flashes to establish the leak. When using an electronic leak detector, keep the following mind:

- Many leak detectors have the sensing unit inside the tool and not at the tip.
- When checking the system, move slowly, about an inch per second, to allow the detector to get a sample of the air.
- Normally, small leaks can only be found when the A/C and engine are turned off and there are no cooling fans operating.
- Low-side leaks may be easier to find when the system is turned off.
- High-side leaks may be easier to find when the system is running.

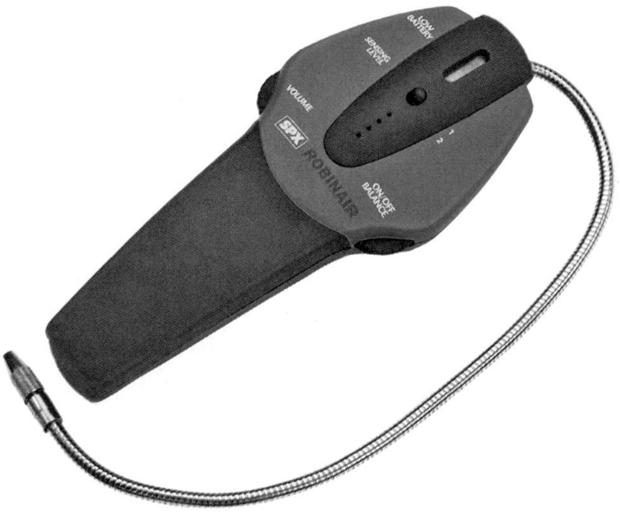

Figure 53–17 An electronic leak detector. *Courtesy of SPX Service Solutions*

Fluorescent Dye Leak Detection

A common method of leak detection is using a fluorescent dye. The dye can be injected into the system for testing. Many new vehicles have R-134a fluorescent dye installed in the refrigerant system when the vehicle is produced. The fluorescent dye mixes and flows with the PAG oil throughout the A/C system. In these systems, additional dye should not added unless more than 50% of the refrigerant oil has been lost because of bad fittings, hose rupture, or other major leakage. There is no need to add dye to the system after the system has been flushed. Part of the flushing process is the replacement of the accumulator or receiver/dryer. New accumulators or receiver/dryers for these systems have a fluorescent dye element in the desiccant bag. The element will dissolve while the system runs, adding the correct amount of dye to the system.

Using a UV lamp, and in some cases special glasses or goggles, system leaks will be apparent by a luminous yellow-green **(Figure 53–19)** trace. When checking

Figure 53–18 When the probe of the leak detector is near the area of a leak, its light will flash.

Figure 53–19 A fluorescent dye is added to the system before checking for leaks with the black light. *Courtesy of Tracer Products*

for leaks, inspect all components, lines, and fittings of the refrigerant system. This includes the evaporator drain tube. The latter is done to check the evaporator for leaks. If a leak is found, continue checking the rest of the system because there may be more than one leak. It is important to remember that PAG oil is water soluble. Therefore, condensation on the evaporator or the refrigerant lines may wash the oil and fluorescent dye away from where the leak is.

If a leak is found, recover the refrigerant. Then correct the problem. Next, evacuate and charge the system. Thoroughly clean all traces of fluorescent dye from any area where leaks were found. Run the system for a short time and recheck the system for leaks to verify the repair.

Leak Detector Fluid Leaks can also be located by applying leak detector fluid around areas to be tested. Detection fluids are available for R-12 and R-134a systems; make sure you use the correct one for the refrigerant you are looking for. If a leak is present, it will form clusters of bubbles around the source **(Figure 53–20)**. A very small leak will cause white foam to form around the leak source within several seconds to a minute. Adequate lighting over the entire surface being tested is necessary for an accurate diagnosis.

Adding Fluorescent Dye Dye may need to be added to a system that was produced with dye in the system or added to other systems for leak detection. This can be done by injecting it into the system or by pouring it directly into a removed component. It is important to note that all fluorescent dyes are not compatible with PAG oil. Before adding dye to the system, make sure it is the right type. Make sure the right quantity of dye is installed; never overcharge the system with dye. Also, the dye needs time to properly circulate through the system before it can be useful. This can be anywhere from 15 minutes to several days, depending on the size of the leak.

Figure 53-20 The bubbles show the location of a leak.

SHOP TALK

Some OEMs, such as Toyota, do not recommend the use of leak-detecting dyes. Always check with the manufacturer's recommendations before adding dyes to the A/C system.

There are two basic ways to inject dye into the system. One uses an A/C charging station and a manifold gauge set. This should only be used when the system is not fully charged. A small tank of dye and injector is attached to the center hose on the gauge set. The dye is added with some refrigerant to the low side of the system.

When the system is fully charged, a special kit must be used. The kit contains a reservoir for the dye and is connected between the high-pressure and low-pressure service valves. With the A/C turned on, the high-side valve is opened and remains open until the dye has left the reservoir. This method does not add refrigerant to the system.

EMPTYING THE SYSTEM

Before the system is opened for service, all refrigerant in the system needs to be recovered. However, prior to doing this, the refrigerant needs to be positively identified. This is necessary to prevent mixing of refrigerants. Although the system may be clearly marked as using a particular refrigerant, there are many different refrigerants available that may have been added to the system. Some of these are blends and not pure. Any refrigerant that is not pure is considered contaminated. Also, the passage of a refrigerant that the recovery machine is not designed for will contaminate the machine and the storage containers. Identifying refrigerant should also take place before using a container of recycled refrigerant just to be safe.

The composition of refrigerant is analyzed by a refrigerant identifier. This tool takes a sample of the refrigerant from the low side of the system and analyzes it. The amount of R-12 or R-134a is displayed on the tool. All refrigerants that are less than 98% pure will fail the tests and should not be recycled. This should be recovered into a storage container labeled as contaminated. This container should only be used for contaminated refrigerant.

There are many different identifiers available; always follow the specific instructions for the tool you will be using.

Recovery

There are currently two types of refrigerant recovery machines: the single pass and the multipass. Both have the ability to draw the refrigerant from the vehicle, filter and separate the oil from it, remove moisture and air from it, and store the refrigerant until it is reused.

In a single-pass system, the refrigerant goes through each stage before being stored. In multipass systems, the refrigerant may go through all stages or some of the stages before being stored. Either system is acceptable if it has the UL-approved label.

To minimize the amount of refrigerant released to the atmosphere when A/C systems are serviced, always follow these steps.

1. The recovery/recycling equipment **(Figure 53–21)** must have shutoff valves within 12 inches of the hoses' service ends. With the valves closed, connect the hoses to the vehicle's A/C service fittings.
2. Always follow the equipment manufacturer's procedures for use. Recover the refrigerant from the vehicle and continue the process until the vehicle's system shows vacuum instead of pressure. Turn off the recovery/recycling unit for at least 5 minutes. If the system still has pressure, repeat the recovery process to remove any remaining refrigerant. Continue until the A/C system holds a stable vacuum for 2 minutes.
3. Close the valves in the recovery/recycling unit's service lines and disconnect them from the system's service fittings. On recovery/recycling stations with automatic shutoff valves, make sure they work properly.
4. You may now make repairs and/or replace parts in the system.

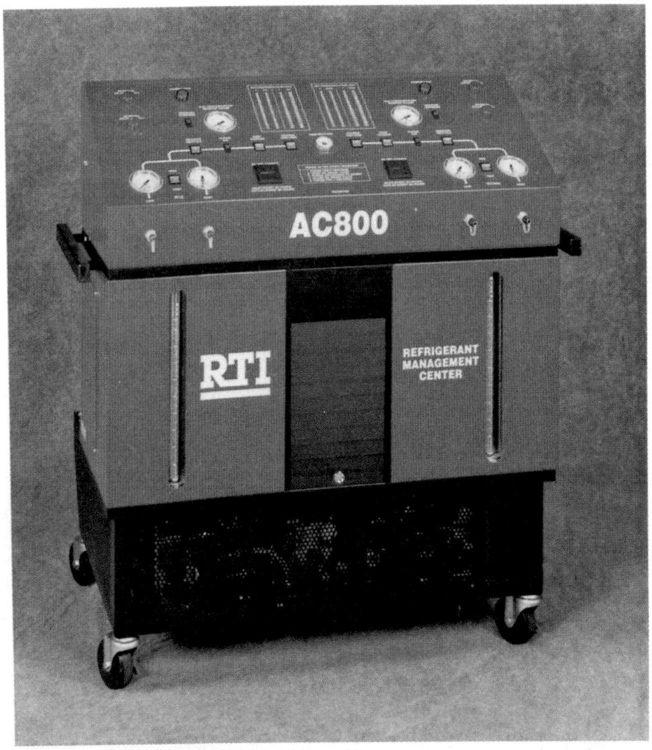

Figure 53-21 A dual refrigerant management center that is capable of doing all refrigerant services. *Courtesy of RTI Technologies, Inc.*

Recycling

Recycling collects old refrigerant from the system being serviced and cleans it up and prepares it to be used in the future. All recycled refrigerant must be safely stored in DOT CFR Title 49 or UL-approved containers. Containers specifically made for R-134a should be so marked. Before any container of recycled refrigerant can be used, it must be checked for noncondensable gases.

The SAE has set standards for recycled R-12 and R-134a to make sure the refrigerants provide proper system performance and longevity. J1991 and J2099 are purity standards for recycled refrigerant and specifies a limit in parts per million (ppm) by weight. These limits are placed on three potential contaminants. There can be no more than 15 ppm of moisture in recycled R-12 and R-134a. The limit for the amount of refrigerant oil in R-12 is 4,000 ppm and for R-134a, 500 ppm. The additional contaminant looked at by the SAE is air or noncondensable gases. There should be no more than 330 ppm in R-12 and no more than 150 ppm in R-134a.

GENERAL SERVICE

All refrigerant must be discharged from the system before repair or replacement of any component (except for compressors with stem service valves). The refrigerant must be recovered and recycled. This means the refrigerant cannot be released into the atmosphere. According to Section 609 of the Federal Clean Air Act, no one repairing or servicing a motor vehicle's A/C system can do so without the proper refrigerant recovery recycling equipment. It further states that no one can do this service unless that person is properly trained and certified.

Before replacing any part in the system, recover the refrigerant first. When replacing a part in an A/C system make sure it is the correct replacement part. Also make sure it is properly installed and the line connections are secure. After a component has been replaced it may be necessary to add refrigerant oil to the system to replace the oil that settled in the replaced component. Typically the amount of oil can be estimated by considering the component replaced. Some parts, such as an orifice tube, require special tools for removal and installation.

Refrigerant oil

Normally the only source of lubrication for a compressor and other moving parts in the A/C system is the oil mixed with the refrigerant. The oil required for the system depends on the refrigerant used in it.

Generally, compressor oil level is checked only where there is evidence of a major loss of system oil that could be caused by a broken refrigerant hose, severe hose fitting leak, badly leaking compressor seal, or collision damage to the system's components.

When replacing refrigerant oil, it is important to use the specific type and quantity of oil recommended by the manufacturer. If there is a surplus of oil in the system, too much will circulate with the refrigerant, causing the cooling capacity of the system to be reduced. Too little oil results in poor lubrication. When there has been excessive leakage or it is necessary to replace a part of the A/C system, certain procedures must be followed to ensure that the total oil charge in the system is correct after leak repair or that the new part is on the car. Most A/C recovery/recycling equipment will display how much oil has been removed with the refrigerant **(Figure 53–22).**

When the compressor is running, oil gradually leaves the compressor and is circulated through the system with the refrigerant. Eventually a balanced condition is reached in which a certain amount of oil is retained in the compressor and a certain amount is continually circulated. If a component of the system is replaced after the system has been operated, some oil goes with it. Always add the specific amount and the correct type of oil when replacing a part **(Figure 53–23)** and do so according to the manufacturer's recommendations.

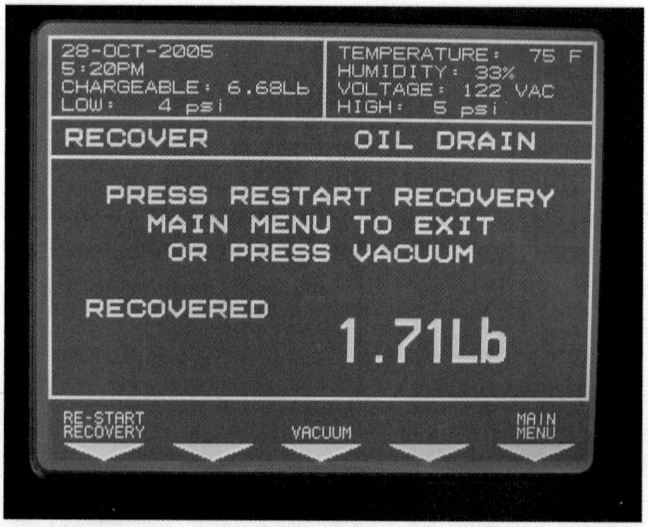

Figure 53-22 Most A/C recovery/recycling equipment will display how much oil has been removed with the refrigerant.

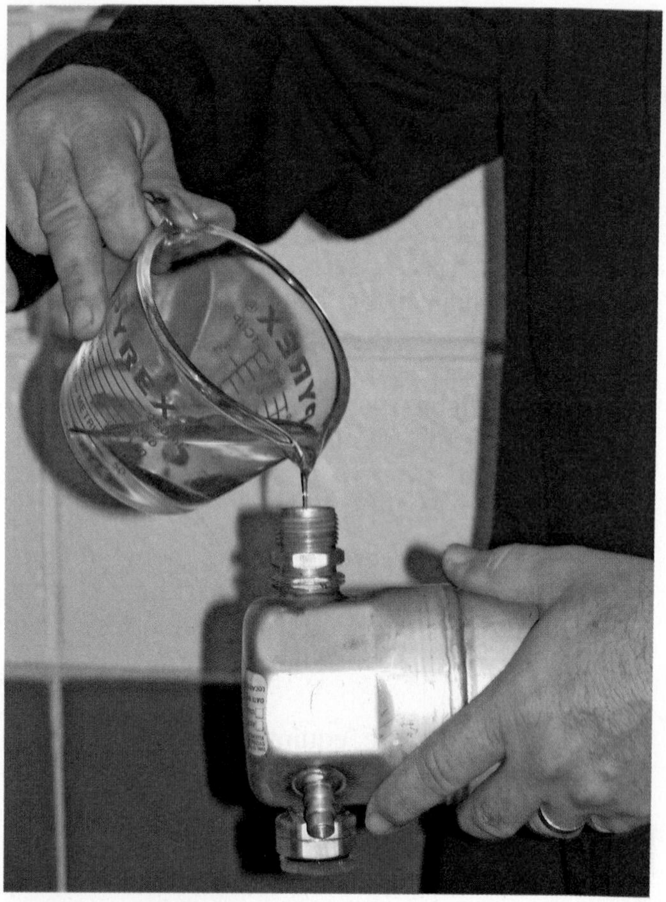

Figure 53-23 When a part is replaced, make sure to add the correct amount and type of refrigerant oil for that part to the system.

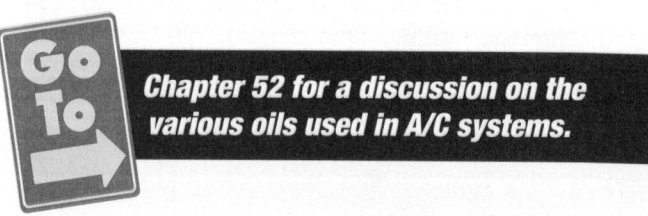

Go To *Chapter 52 for a discussion on the various oils used in A/C systems.*

Clutch Service

The compressor clutch is engaged by a magnetic field and disengaged by springs when the magnetic field is broken. When the controls call for compressor operation, the electrical circuit to the clutch is completed, the magnetic clutch is energized, and the clutch engages the compressor. When the electrical circuit is opened, the clutch disengages the compressor.

The clutch assembly should be carefully inspected for discoloration, peeling, or other damage. If there is damage, replace the clutch assembly. Also check the play and drag of the compressor pulley bearing by rotating the pulley by hand. Replace the clutch

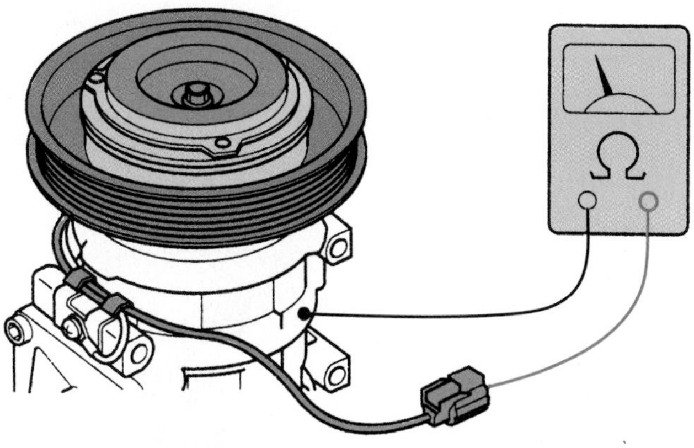

Figure 53-24 Testing a compressor clutch with an ohmmeter. *Courtesy of Toyota Motor Sales, U.S.A., Inc.*

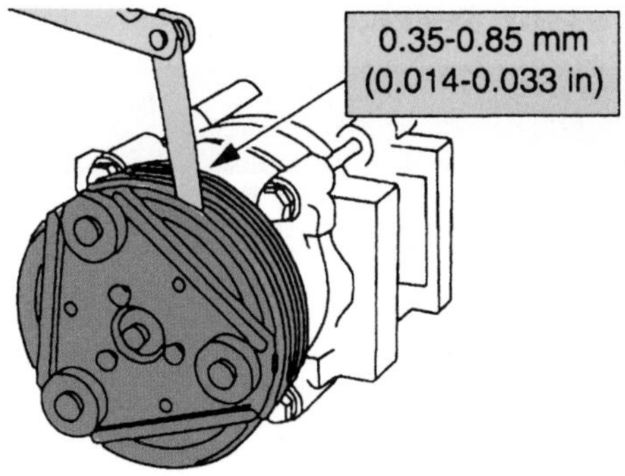

Figure 53-25 Checking the gap of a compressor clutch assembly. *Courtesy of Ford Motor Company*

assembly if it is noisy or has excessive play or drag. The field coil for the clutch can be checked with an ohmmeter (**Figure 53–24**). The exact testing points and acceptable resistance readings are given in the vehicle's service information. If resistance is not within specifications, replace the field coil. Also, check the clutch clamping diode.

If the clutch is not operating, check the electrical connections to it. Make sure they are secure and corrosion free. Then, check for power to the clutch with a testlight or DMM. If there is no power, locate and repair the problem. If there is power to the clutch, check the ground circuit with a DMM. If there is power and a good ground, the clutch is defective and must be replaced.

Clutch Clearance Nearly all clutch assemblies have a clearance spec for the distance between the clutch and the pressure plate. This clearance is measured with a feeler gauge. If the clearance is too great, the clutch may slip and cause a scraping or squealing noise. If the clearance is insufficient, the compressor may run when not electrically activated and the clutch may chatter at all times. As the clutch assembly wears, the clearance increases and therefore should be checked and adjusted whenever symptoms suggest doing so. Always follow the specific procedures given by the manufacturer for measuring and correcting the gap.

Measure the clearance between the rotor pulley and the armature plate (**Figure 53–25**). Do this at more than one location. If the clearance is not within the specified limits, remove the armature plate and add or remove shims as needed to increase or decrease the clearance. The shims are available in different thickness and it is recommended that no more three shims be installed to correct the clearance.

Some clutches are press-fit to the compressor shaft. The gap of these clutches is adjusted by tightening the retaining bolt. With a feeler gauge placed between the clutch plate and pulley, tighten the bolt until the feeler gauge of the specified size has a slight drag.

Clutch and Pulley Replacement On some vehicles, the clutch can be serviced on the vehicle. On others, the compressor assembly must be removed to service the clutch. Refer to the service manual on your specific vehicle to determine what must be done.

Normally, to remove the clutch assembly, the magnetic clutch hub must be held while the retaining bolt is loosened. After the bolt is removed, the clutch hub and shims can be removed. This will give access to a snapring around the compressor shaft. Using snapring pliers, expand and remove the snapring and then remove the rotor of the magnetic clutch. Be careful not to damage the compressor seals while doing this. Now, remove the snapring for the clutch stator and then the stator.

When installing a new clutch assembly, begin by aligning the protrusion on the stator with the notch on the compressor. Then install a new snapring with its chamfered side facing up. Install the clutch rotor with a new snapring. Now install the clutch shims and hub. Hold the hub while tightening the center bolt to specifications. While installing the clutch assembly, make sure all parts are kept clean and free of oil or grease.

Clutch Pulley Bearing Replacement The clutch pulley can be a source of unwanted noise and can be replaced on most systems. To replace the bearing, remove the pulley. Use a bearing driver or press to push the bearing out of the front of the pulley. To install a new bearing, align the pulley and press it into place.

Compressor Service

PROCEDURE

To remove and install a compressor, follow these steps:

STEP 1 Disconnect the negative battery cable.

STEP 2 Identify and disconnect all electrical connections to the compressor.

STEP 3 Discharge the system using a recovery/recycling machine.

STEP 4 Disconnect the refrigerant lines at the compressor. Immediately cap or seal the ends of the lines or hoses.

STEP 5 Remove the drive belt for the compressor.

STEP 6 Loosen and remove the compressor mounting brackets. Note the location of the bolts because they are typically a different length.

STEP 7 Remove the compressor.

STEP 8 To install a compressor, reverse the removal procedure. Make sure the drive belt is tightened to the proper tension.

Shaft Seal Replacement The seal at the compressor shaft is a common source of leaks. To replace the seal, remove the clutch assembly. Then remove the internal snapring that holds the seal in place. Install the special seal remover/installer tool against the face of the seal. Twist it to expand the jaws of the tool on the seal. With a gentle twisting and pulling motion, remove the seal. Then remove and discard the O-ring seal that is located between the housing and the shaft seal.

Lubricate and install a new O-ring into its groove. Coat the new shaft seal with refrigerant oil. Then place the seal into the jaws of the seal remover/installer tool. Install a seal protector over the threads of the compressor's shaft (**Figure 53–26**). With a gentle twisting motion, slide the seal over the protector and into the groove in the compressor. Release the installer from the seal. Then reinstall the snapring and clutch assembly. The system will need to be evacuated before it is recharged.

Figure 53-26 Install the protector over the threads of the compressor shaft. Place the new seal on the installer and install the seal with a gentle twisting motion.

Flushing Compressor failure causes foreign material to pass into the system. The condenser must be flushed and the receiver/dryer or accumulator replaced. Filter screens are sometimes located in the suction side of the compressor and in the receiver/dryer. The compressor inlet screen should be replaced whenever the compressor is replaced. These screens confine foreign material to the compressor, condenser, receiver/dryer, and connecting hoses. If a screen becomes clogged, it will block the flow of refrigerant. Use only recommended flushing solvents. Never use CFCs or methylchloroform for flushing. Some manufacturers recommend replacing the clogged components and installing a liquid line (inline) filter just ahead of the expansion valve or orifice tube instead of flushing the system. If flushing is recommended by the manufacturer, use only the recommended flushing agent and follow the specified procedures. After the system has been flushed, be sure to oil all components that require it.

Inline Filters The most effective way to collect debris that may be in the system is to install an inline filter (**Figure 53–27**). These filters should be installed whenever the compressor seizes or has severe damage and needs to be replaced. A filter should also be installed if the expansion valve or orifice tube is plugged with debris. Filters can contain an orifice and these should be installed in a different location than filters without an orifice. If the filter does not have an orifice, it should be installed in the liquid line between the condenser outlet and the evaporator inlet. If the filter has an orifice, it is installed between the low-pressure side of the system beyond

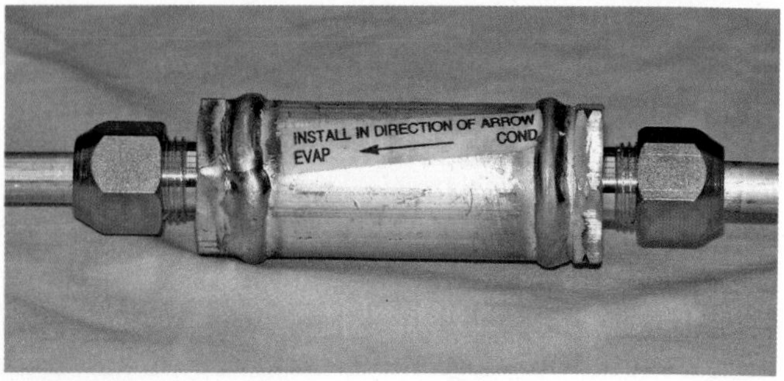

Figure 53-27 An inline filter.

the expansion tube. If the filter has a built-in orifice, remove the original expansion tube. In either case, the filter should be inserted at a point where the refrigerant line is straight.

To install the filter, cut the line at the desired filter location. Make sure the cut is smooth and straight. Slide the ferrules, cones, and seals for the filter onto each end of the cut line. Lubricate the seals and insert the filter and tighten the ferrules. Make sure the lines are properly sealed, and then evacuate and charge the system. When finished, check the installation for leaks.

Compressor Controls

The ambient temperature sensor or the engine's IAT measures outside air temperature to prevent compressor clutch engagement when A/C is not required or when compressor operation might cause internal damage to the seals and other parts. The switch is in series with the compressor clutch electrical circuit and closes at about 37°F (2.8°C). At all lower temperatures, the switch is open, preventing clutch engagement.

In cycling clutch systems, a thermostatic switch is placed in series with the compressor clutch circuit so it can turn the clutch on or off. It de-energizes the clutch and stops the compressor if the evaporator is at the freezing point. When the temperature of the evaporator approaches the freezing point, the thermostatic switch opens the circuit and disengages the compressor clutch. The compressor remains inoperative until the evaporator temperature rises to the preset temperature, at which time the switch closes and compressor operation resumes.

A pressure cycling switch is electrically connected in series with the compressor electromagnetic clutch. Like the thermostatic switch, the turning on and off of the pressure cycling switch controls the operation of the compressor.

The low-pressure cutoff or discharge pressure switch is located on the low side of the system and senses any low-pressure conditions. It is tied into the compressor clutch circuit, allowing it to immediately disengage the clutch when the pressure falls too low.

The electronic cycling clutch switch (ECCS) prevents evaporator freeze-up by sending a signal to the engine control computer. The computer, in turn, cycles the compressor on and off by monitoring suction line temperature. If the temperature gets too low, the ECCS will open the input circuit to the computer, which causes the A/C clutch relay to open, which disengages the compressor clutch. Often this switch is the thermostatic switch at the evaporator.

The high-pressure relief valve is used to keep system pressures from reaching a point that may cause compressor lockup or other component damage because of excessive high pressures. When system pressures exceed a predetermined point, the pressure relief valve opens, reducing the system's pressure.

The A/C high-pressure switch is used for additional A/C system pressure control. It is normally closed and its high-pressure contacts open at a predetermined A/C pressure. This results in the A/C turning off, preventing the A/C pressure from rising to a level that would open the A/C high-pressure relief valve. The A/C pressure transducer sensor is located in the high-pressure (discharge) side of the A/C system. The ACP transducer sensor provides a voltage signal to the PCM that is proportional to the A/C pressure. The PCM uses this information for A/C clutch control, fan control, and idle speed control.

The evaporator temperature sensor measures the temperature of the cool air immediately after the evaporator. This input to the A/C control module is used to detect evaporator freeze-up.

PROCEDURE

To check the various control devices, follow these steps:

STEP 1 Using the service manual, locate the various pressure and temperature sensors in the system that controls the operation of the compressor.

STEP 2 Check the manual to determine if each of these is normally closed or normally open.

STEP 3 Checking one sensor at a time, complete the following steps:

- ***a.*** Disconnect the wires at the sensor or switch.
- ***b.*** Connect the DMM across the switch terminals **(Figure 53–28)**.
- ***c.*** Set the meter for resistance or continuity checks. If the switch is normally closed, the reading should be zero ohms. If the switch is normally open, you should get an infinite reading. Any reading other than these is an indication of a faulty switch.

STEP 4 If a switch needs to be replaced, the system may need to be discharged. Check the service manual before proceeding. Some systems have a Schrader-type disconnect below the switch. switch replacement in these systems does not require system evacuation.

STEP 5 To remove the defective switch, loosen it and remove it.

STEP 6 To install the new switch, carefully thread it into place and tighten it. Then reconnect the wires leading to the switches.

Figure 53-28 All sensors and switches can be checked with an ohmmeter.

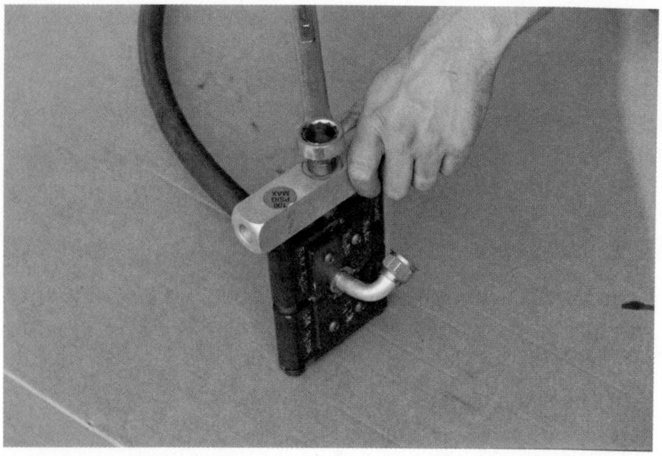

Figure 53-29 The ferrule is crimped with a special crimping tool.

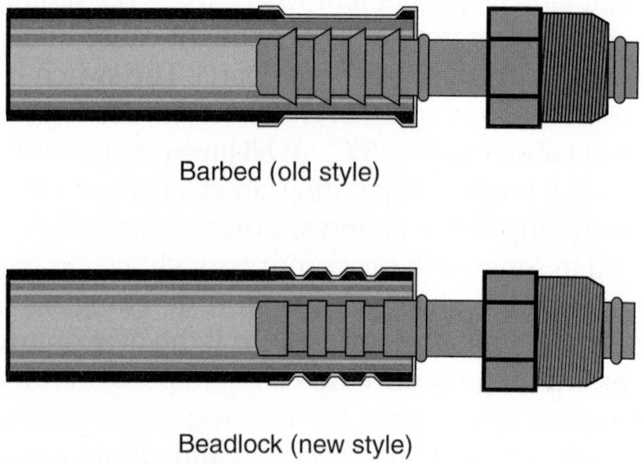

Figure 53-30 A comparison of a barbed fitting and the new-style beadlock fitting.

Hoses and Fittings

Although total hose replacement is the preferred way to correct for a hose leak, there are several accepted ways to repair refrigerant hoses and fittings. Using insert barb fittings and a length of replacement hose, you can fabricate an acceptable replacement for an original equipment hose. Insert barb fittings can also be used to replace bad original fittings or to replace a section of a hose.

PROCEDURE

To fabricate a replacement A/C pressure hose, follow these steps:

STEP 1 Measure and mark the required length of replacement high-pressure hose.

STEP 2 Using the razor blade, cut the hose to the desired length. Make sure the cut is square and flat.

STEP 3 Apply clean refrigerant oil to the inside of the hose.

STEP 4 Install the correct ferrule onto the end of the hose.

STEP 5 Carefully inspect the new fitting to make sure it is free of nicks and other damage.

STEP 6 Coat the fitting with refrigerant oil and insert it into the hose.

STEP 7 Using the crimping tool designed for the ferrule, crimp the ferrule **(Figure 53–29)**.

When using an insert barb fitting **(Figure 53-30)** to replace a bad original fitting, the original fitting must be removed without damaging too much of the hose. Remove the hose from the vehicle. Set the hose in a vise while cutting through the fitting's ferrule with a hacksaw. Make this cut in the direction of the hose, not through the hose. With the ferrule cut, use pliers to peel or pull the ferrule off. Cut the hose at a point just beyond the insert of the original fitting. Make sure the cut is square and flat. Apply clean refrigerant oil to the inside of the hose and to the fitting's insert. Insert the fitting into the hose. Position the hose clamp over the barb closest to the fitting before tightening it.

If the hose or line has spring lock fittings, a special tool is required to separate the sections. Put the tool over the coupling. Close the tool and push the lines into the tool to release the female fitting from the garter spring of the coupling. Then pull the male and female fittings apart. Remove the tool. When joining two fittings with a spring lock coupling, lubricate new O-rings and put them in their proper location on the male fitting. Insert the male fitting into the female fitting. Firmly push them together until they are secured by the spring lock coupling.

Rigid Line Repair Like pressure hoses, solid or rigid refrigeration lines are replaced rather than repaired if they leak. However, repairs can be made with

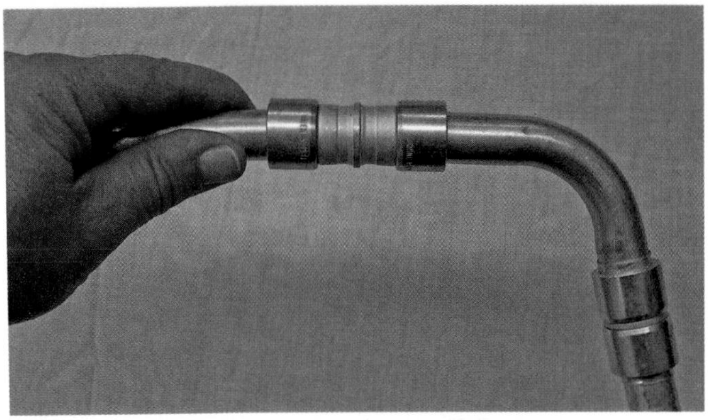

Figure 53-31 Repairs to rigid lines can be made with special collars.

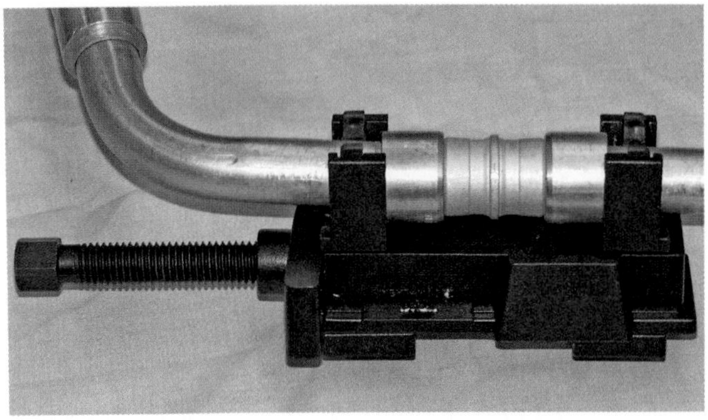

Figure 53-32 A designated tool is used to press the collar into the tube.

special collars **(Figure 53–31)** inserted into the line to correct a kink or leak. The damaged section of the tube is cleanly cut out. The collar or fitting is inserted into the line after sealant has been applied to it. Using a special tool, the collar is pressed into the ends of the existing line **(Figure 53–32)**. The line is then crimped around the insert. This provides for a permanent seal.

Receiver-Dryer/Accumulator

The receiver-dryer/accumulator is often neglected when the A/C system is serviced or repaired. Failure to replace it can lead to poor system performance or replacement part failure. It is recommended that the receiver-dryer/accumulator and/or its desiccant be changed whenever a component is replaced, the system has lost the refrigerant charge, or the system has been open to the atmosphere for any length of time.

The desiccant draws moisture like a magnet and it can become contaminated in less than 5 minutes if it is exposed to the atmosphere. Keep it sealed. To replace a receiver-dryer/accumulator, disconnect the electrical connector to the low-pressure switch, if there is one. Then disconnect the inlet and outlet hoses or lines at the receiver-dryer/accumulator. Remove the mounting bolts and brackets from the receiver-dryer/accumulator and lift it out. The procedure for installation is the reverse of removal.

Accumulator Most late-model systems do not have a receiver/dryer; rather, they use an accumulator. The accumulator is connected to the low side at the outlet of the evaporator. The accumulator also contains a desiccant and is designed to store excess refrigerant and to filter and dry the refrigerant. If liquid refrigerant flows out of the evaporator, it will be collected by and stored in the accumulator. An accumulator/dryer should be replaced if there is excessive moisture or debris in the A/C system or when the accumulator leaks. To replace an accumulator, disconnect the lines at the accumulator's inlet and outlet fittings. Then loosen and/or remove the accumulator mounting bolts or screws. Lift the accumulator out. The procedure for installation is the reverse of removal. As soon as the accumulator is installed, immediately evacuate the system.

Evaporator

The evaporator and its controls often need service. If the evaporator is clogged, leaking, or damaged, it should be replaced.

PROCEDURE

To remove an evaporator and install a new one, follow these steps:

STEP 1 Recover the refrigerant using an approved recovery/recycling equipment.

STEP 2 Disconnect the negative battery cable.

STEP 3 If the evaporator and heater core are a combined unit, drain the engine's coolant.

STEP 4 Disconnect and label all electrical connectors, cables, and vacuum hoses that are connected to the evaporator.

STEP 5 Disconnect the refrigerant hoses at the evaporator and plug or cap the hose ends to prevent dirt and moisture from entering the system.

STEP 6 Unbolt and remove the evaporator.

STEP 7 Drain the oil from the evaporator into a graduated container. Record the amount of oil removed.

STEP 8 Check the oil for dirt. If the oil is contaminated, replace the accumulator or receiver/dryer.

STEP 9 Add the above amount of new refrigerant oil or the amount specified in the service manual to the new evaporator.

STEP 10 Coat the new O-rings for the evaporator and the line fittings on the evaporator with clean refrigerant oil.

STEP 11 Install the new evaporator and tighten the fittings. Also install a new receiver-dryer/accumulator or its desiccant bag.

STEP 12 Immediately evacuate the system.

STEP 13 Reinstall and reconnect all parts, wires, cables, and vacuum hoses that were disconnected during removal. Add coolant if needed.

STEP 14 Connect the negative battery cable.

STEP 15 Evacuate and recharge the system.

STEP 16 Perform a leak test and correct any problems.

Evaporator Water Drain Because the evaporator is also responsible for controlling the humidity in the vehicle, a check of its drain is necessary. This is especially true if the customer had concerns about odor or water on the carpet. To check the drain, raise the vehicle on a lift. Locate the evaporator case drain tube. Place the drain pan under the evaporator case. Disconnect the tube from the case. If no water comes out, carefully clean out the drain hole and tube at the evaporator case. You may need to insert a rod through the tube to clean it out; do this carefully because the tube is accordion shaped and one of its bends may feel like a restriction. If no water comes out while trying to clear the tube, the entire evaporator case must be cleared. If water comes out of the case, the outside drain tube should be cleared by inserting the air nozzle into an end of the tube. Low air pressure should remove any restrictions in the tube.

Expansion Devices Basically there are two types of expansion devices. On late-model vehicles, the orifice tube is the most common. To replace an orifice tube assembly, always use the tools and procedures recommended by the manufacturer. Begin by disconnecting the inlet line at the evaporator. Then pour a small amount of clean refrigerant oil into the orifice tube. Insert the orifice removal tool. Turn the handle of the tool just enough for it to engage onto the tabs of the orifice tube. Hold the handle of the tool in position while turning the tool's outer sleeve clockwise to remove the orifice tube **(Figure 53–33)**. Coat the new orifice tube with clean refrigerant oil. Place it into the evaporator line inlet. Push it in until it stops. Install a new O-ring on the refrigerant line **(Figure 53–34)** and reconnect it.

To replace a thermostatic expansion valve, disconnect the inlet and outlet lines at the TXV. Then remove whatever is used to keep the remote sensing bulb secure. Loosen and/or remove the mounting clamp for the TXV and remove the TXV from the evaporator. The procedure for installation is the reverse of removal.

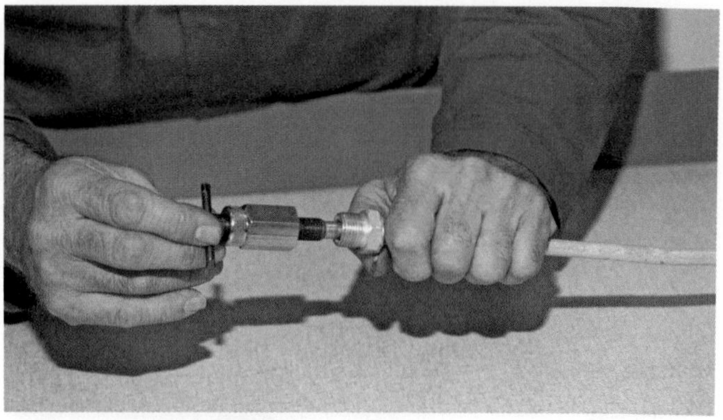

Figure 53-33 The tool designed to remove and replace an orifice tube.

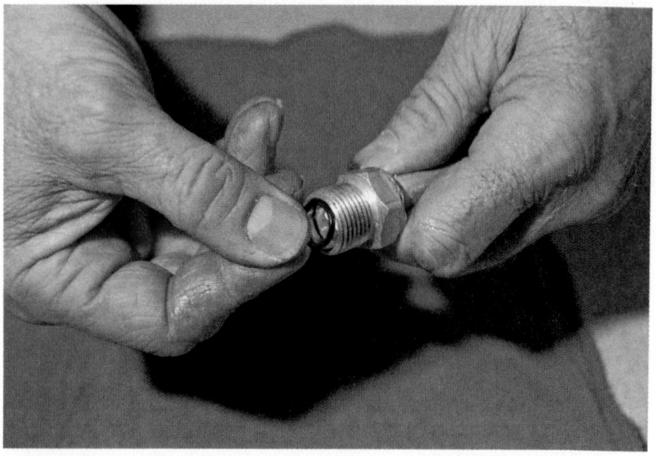

Figure 53-34 Always install a new O-ring when replacing an orifice tube.

Condenser

The condenser is quite reliable and only needs replacement when there are leaks due to damage. To replace a condenser, disconnect the negative battery cable. On many vehicles, the radiator, cooling fan, and/or shroud must be removed to gain access to the condenser. If the radiator must be removed, drain the coolant. Disconnect and label all electrical connectors that are connected to the condenser and all those that may be in the way. Disconnect the refrigerant hoses at the condenser and plug or cap the hose ends to prevent dirt and moisture from entering the system. Unbolt and remove the condenser. If the receiver-dryer/accumulator mounts to the condenser, disconnect it and remove it with the condenser. Drain and measure the oil from the condenser into a graduated container. Check the oil for dirt. If the oil is contaminated with moisture, replace the accumulator or receiver/dryer. Add the above amount of new refrigerant oil or the amount specified in the service manual to the new condenser. Install the new condenser and loosely tighten the mounting bolts. Coat the new O-rings for the condenser and the line fittings on the condenser with clean refrigerant oil.

Tighten the fittings, and then tighten the mounting bolts. Then immediately evacuate the system. Reinstall and reconnect all parts, wires, cables, and vacuum hoses that were disconnected or removed during removal. Connect the negative battery cable. Check for coolant and refrigerant leaks.

RECHARGING THE SYSTEM

All of the refrigerant in the system must be recovered prior to evacuation. Evacuation is the name given to the process that pulls all traces of air and moisture from the system. This is done by creating a vacuum in the system. A vacuum pump is connected to the system to do this **(Figure 53–35)**. The vacuum pump should remain on and connected to the system for at least 30 minutes after 26 to 29 inches of mercury (in. Hg) is reached.

CAUTION!

Always wear safety goggles when working with refrigerant containers or servicing A/C systems.

Any air or moisture that is left inside an A/C system reduces the system's efficiency and eventually leads to major problems, such as compressor failure.

Air causes excessive pressure within the system, restricting the refrigerant's ability to change its state from gas to liquid within the refrigeration cycle, which drastically reduces its heat absorbing and transferring ability. Moisture, on the other hand, can cause freeze-up at the orifice tube or expansion valve, which restricts refrigerant flow or blocks it completely. Both of these problems result in intermittent cooling or no cooling at all. Moisture mixed with R-134a forms hydrochloric acid, causing internal corrosion, which is especially dangerous to the compressor. The vacuum pump reduces system pressure, vaporizes the moisture, and then exhausts it with the air.

The main responsibility of the vacuum pump is to remove the contaminating air and moisture from the system. The vacuum pump reduces system pressure in order to vaporize the moisture and then exhausts the vapor along with all remaining air. The pump's ability to clean the system is directly related to its ability to reduce pressure—create a vacuum—low enough to boil off all the contaminating moisture.

An electronic thermistor vacuum gauge is designed to work with the vacuum pump to measure the last, most critical inch of mercury during evacuation. It constantly monitors and visually indicates the vacuum level so you know when the system has a full vacuum and will be moisture free. After the system is evacuated, it can be recharged. If the system will not pull down to a good vacuum, there is probably a leak somewhere in the system. Photo Sequence 53 goes through a common way to evacuate and recharge a system.

Charging

Refilling the system with refrigerant is called charging the system. Refrigerant is added through the system's service ports **(Figure 53–36)**. Depending on the

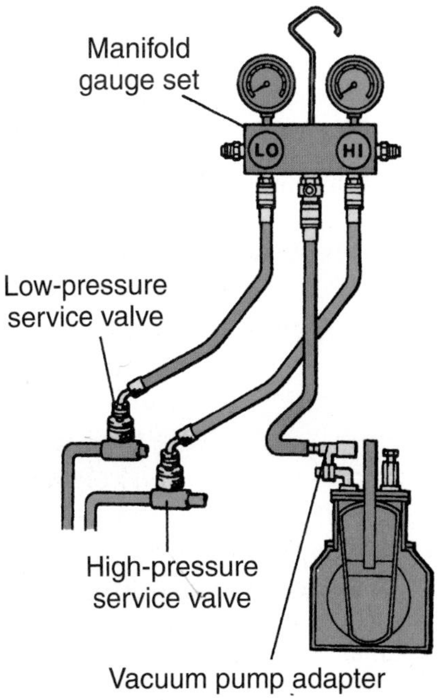

Figure 53–35 A vacuum pump connected to an A/C system through a gauge set. *Courtesy of Toyota Motor Sales, U.S.A., Inc.*

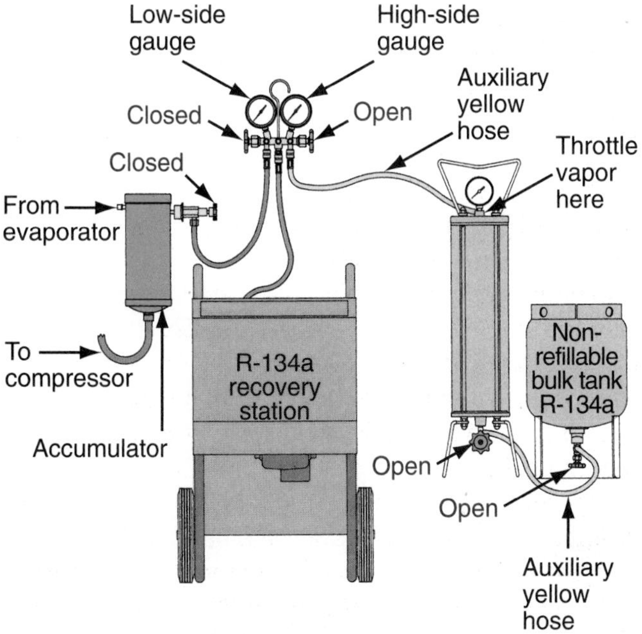

Figure 53–36 The connections for filling a system with refrigerant using a charging cylinder. *Courtesy of SPX Service Solutions*

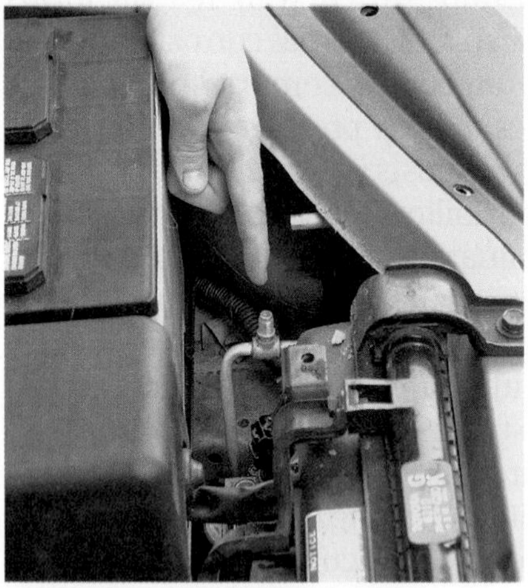

P53-1 *Physically locate the pressure fittings of the system you will be working with. Make sure you have the proper adapters for the fittings prior to starting any service work on air-conditioning system.*

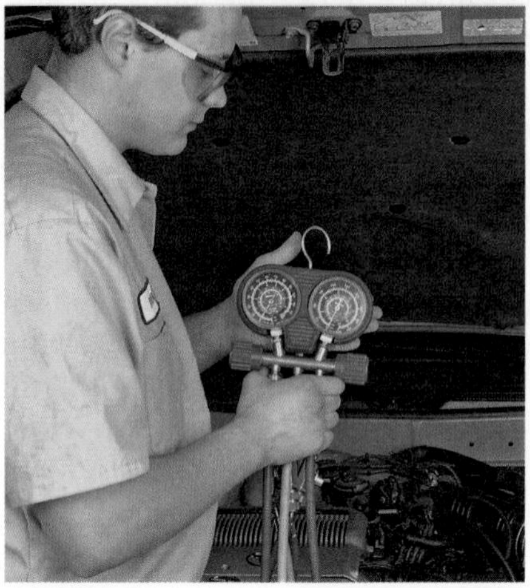

P53-2 *Connect a gauge set to the system.*

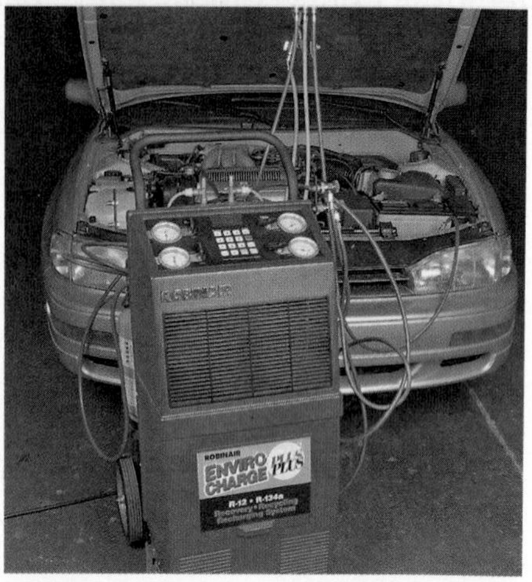

P53-3 *Typical refrigerant recovery/reclaiming/recycling machine.*

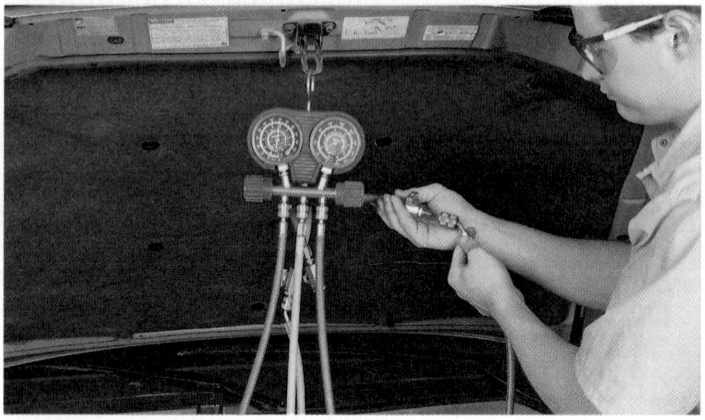

P53-4 *Connect the reclaiming unit to the gauge set.*

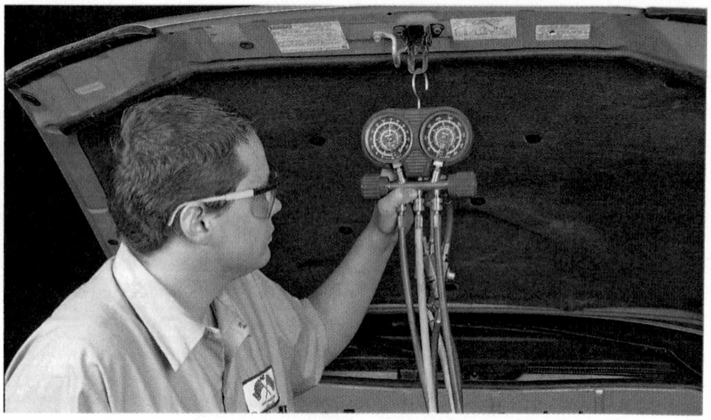

P53-5 *Activate the reclaimer to discharge the system, then begin evacuation. Allow enough time for the machine to draw a good vacuum (at least 29 in. Hg) in the system. Observe the gauge set to determine when a vacuum is created.*

P53-6 *Close the valves at the gauge set and disconnect the reclaiming machine. Connect the charging station and open the gauge set valves. Set the unit to deliver the required amount of refrigerant for that system.*

P53-7 *After the system has been recharged, check the gauge readings while the system is in operation to make sure it is working properly.*

method and source of new refrigerant, refrigerant is introduced through the low or high side of the system. When the system is operating, refrigerant is charged through the low side. On some vehicles with the system off, charging is done through the high side. Always refer to the service manual for the correct procedure.

Always use the same type of refrigerant and refrigerant oil for the system. The importance of the correct charge cannot be stressed enough. The efficient operation of the A/C system greatly depends on the correct amount of refrigerant in the system. A low charge results in inadequate cooling under high heat loads due to a lack of reserve refrigerant and can cause the clutch cycling switch to cycle faster than normal. An overcharge can cause inadequate cooling because of a high liquid refrigerant level in the condenser. Refrigerant controls will not operate properly and compressor damage can result. In general, an overcharge of refrigerant will cause higher-than-normal gauge readings and noisy compressor operation.

The charging cylinder is designed to meter out a desired amount of a specific refrigerant by weight. Compensation for temperature variations is accomplished by reading the pressure on the gauge of the cylinder and dialing the plastic shroud. The calibrated chart on the shroud contains corresponding pressure readings for the refrigerant being used.

When charging an A/C system with refrigerant, the pressure in the system often reaches a point at which it is equal to the pressure in the cylinder from which the system is being charged. To get more refrigerant into the system to complete the charge, heat must be applied to the cylinder.

The A/C system may only be charged through the high side with the system off **(Figure 53–37)**.

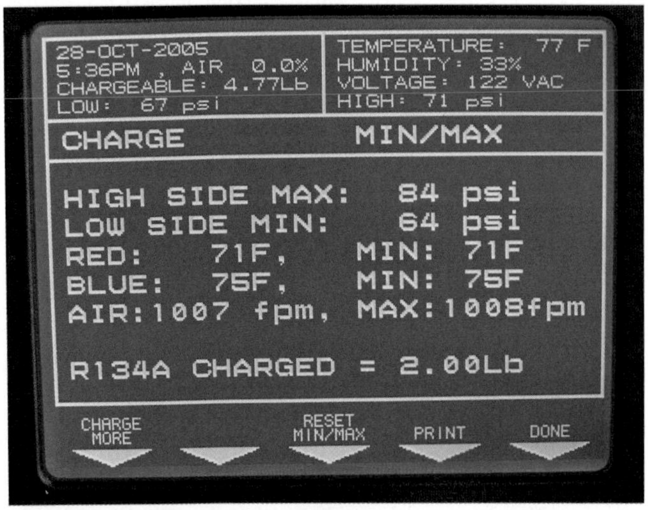

Figure 53-37 The recharging equipment will display how much refrigerant was added to the system.

Inverting the refrigerant container disperses liquid refrigerant. As a general practice, the system is vapor charged through the low side while the system is running. If the system is charged through the high side (with the system off), the compressor should be turned a few times by hand afterward to ensure there is no liquid refrigerant on top of the piston. If liquid refrigerant enters the compressor, this component may be damaged. The vehicle should always be charged from the low-pressure side if the engine is on. If the engine is off, it can be charged from the high-pressure side.

Never open the high-side hand valve with the system operating and a refrigerant can connected to the center hose. The refrigerant will flow out of the system under high pressure into the can. High-side pressure is between 150 and 300 psi and will cause the refrigerant tank to burst. The only occasion for opening both hand valves at the same time would be when evacuating the system or when reclaiming refrigerant with the system off.

Charging Cylinder

With an increase in temperature in any cylinder filled with refrigerant, there is a corresponding increase in pressure and a change in the volume of liquid refrigerant in the cylinder. To measure an accurate charge according to the weight of a cylinder, it is absolutely necessary to compensate for liquid volume variations caused by temperature variations. These temperature variations are directly related to pressure variations and accurate measurements by weight can be calibrated in relation to pressure.

The charging cylinder is designed to meter out a desired amount of a specific refrigerant by weight. Compensation for temperature variations is accomplished by reading the pressure on the gauge of the cylinder. A calibrated chart on the machine contains the corresponding pressure readings for the refrigerant being used.

When charging an A/C system with refrigerant, the pressure in the system often reaches a point at which it is equal to the pressure in the cylinder from which the system is being charged. To get more refrigerant into the system to complete the charge, the cylinder can be placed on a heating plate.

Testing for Noncondensable Gases Whenever using recycled refrigerant, it should be checked for noncondensable gases. Install a calibrated pressure gauge on the container and measure the pressure in the container. Measure the temperature of the air 4 inches away from the container's surface. Compare the measured pressure and temperature to the manufacturer's pressure/temperature chart. Determine if the recycled refrigerant has excessive noncondensable gases. If the refrigerant has excessive noncondensable gases, it should be recycled again and then retested before it is used.

CLIMATE CONTROL SYSTEMS

When diagnosing climate control systems, it is essential to have accurate information about the system before beginning. This information can be obtained from a number of sources: manufacturer service manuals, manufacturer service information Web sites, CD- or Internet-based computerized information systems, technical service bulletins, and aftermarket printed manuals.

Self-Diagnosis

Most climate control systems are computer controlled and have a self-diagnostic feature. Manufacturers' specifications must be followed to identify the malfunction display codes, because they differ from car to car. To check the system, refer to the service manual and determine if the automatic climate control system relies on the PCM, BCM, or a separate computer for control.

If the unit is part of the PCM/BCM system, connect the scan tool and retrieve any codes that may be present. If the unit is controlled by its own computer system, use your service manual and the procedure for retrieving trouble codes. If the system is a semiautomatic system, check the service manual for the proper diagnostic procedures for checking evaporator and heating controls. Test the components or subsystems identified by the trouble codes.

The DTCs from the system can lead a technician to the following problems:

- An open or short in the air mix control motor circuit
- A problem with the air mix control linkage, door, or motor circuit
- An open or short in the mode control circuit
- A problem with the mode control linkage, doors, or motor circuit
- A problem in the blower motor circuit
- A problem with the HVAC control unit
- An open or short in the evaporator temperature sensor circuit

Blower Motor

The blower motor is usually located in the heater housing assembly. It ensures that air is circulated through the system. Its speed is controlled by a multiposition switch in the control panel. The switch works in connection with a resistor block that is usually located on the heater housing.

On some vehicles, when the engine is running, the blower motor is in constant operation at low speed. On automatic temperature control systems, the blower motor is activated only when the engine reaches a predetermined temperature. The blower motor circuit is protected by a fuse located in the fuse panel. The fuse rating is usually 20 to 30 amperes.

The blower motor resistor block is used to control the blower motor speed. The typical resistor block is composed of three- or four-wire resistors in series with the blower motor, which control its voltage and current. The speed of the motor is determined by the control panel switch, which puts the resistors in series. Increasing the resistance in the system slows the blower speed.

If the blower does not operate, use a testlight to make sure there is voltage on both sides of the fuse. Check the wiring diagram and identify whether the blower circuit is controlled by a ground side switch or an insulated (power side) switch. Then check the ground of the circuit. Check to see if current is arriving at the motor. On cars where the blower motor is behind the inner fender shell, hunt out the wiring and check for current. If the blower motor is getting current, the problem is either a burned-out blower motor or a bad ground on the motor. In situations where no current is available at the motor, backtrack to check for an open resistor. Check also for burned or corroded connections in the blower relays or bulkhead connectors. An open blower motor ground would cause the motor not to run at all. If the correct amount of voltage is available to the motor and the ground is good, suspect a bad motor.

If the blower works at some speeds but not all, check the voltage to the blower motor at the various switch positions. If the voltage does not change when a new position is selected, check the circuit from the switch to the resistor block. If there was zero voltage in a switch position, check for an open in the resistor block.

If the motor runs when it should not, there is probably a short in the circuit. If a ground side switch controls the circuit, check for a short to ground in the control circuit. The exact problem can be isolated by disconnecting portions of the circuit until the motor stops. The short is in that part of the circuit that was disconnected last. If the circuit is controlled by an insulated switch, check for a wire-to-wire short. Check other circuits of the vehicle to identify what circuit is involved in this problem. That circuit will also experience a lack of control, or when that circuit is turned off, the blower motor will turn off. The exact problem can be isolated by disconnecting portions of the circuit until the motor stops. The short is in that part of the circuit that was disconnected last.

Doors and Ductwork

To check the proper functioning of the ductwork, move the temperature control lever to see if any change occurs. If it does not, shut off the air conditioner and turn on the heater. Move the temperature control arm again to see if any change occurs. If not, check the cable and the flap door connected to the temperature control lever. You might be able to reach under the dash to reconnect the cable or free a stuck flap.

If no substantial airflow is coming out of the registers, check the fuses in the blower circuit. Remove the fan switch and test it. Check the blower motor by hot-wiring it directly to the battery with jumper wires.

Controls

Most systems use electric motors to control the blend and mode doors; however, some use vacuum. To check these systems, identify and inspect the hoses and components for vacuum motors and doors. Look for disconnected or broken hoses, broken connectors, misrouted vacuum lines, and loose or disconnected electrical connectors at vacuum switches and controls. Identify the vacuum source for the various doors and switches of the duct system and make sure vacuum is available when the engine is running. Disconnect that vacuum source. Connect a vacuum gauge to the inlet for the defroster door motor(s) and record the vacuum available there when the master control switch is in the following positions: max, norm, bilevel, vent, heat, blend, defrost, and off. If they do not move with the engine's vacuum, use a hand-operated vacuum pump to check the component's activity and its ability to hold a vacuum. Use the vacuum pump to check all one-way check valves. Also, check all mechanical or cable linkages and controls. Make sure the cables are properly attached to the levers of the control switch. If the cable(s) is equipped with an automatic adjuster, make sure the adjusting mechanism operates freely and is not damaged.

Control Panel At times the control is the root of problems **(Figure 53–38)**. Inspect the connectors to the control unit, look for damage and corrosion. With on ohmmeter, check the continuity through the blower motor switch in all of its positions. Compare your results with the wiring diagram for the switch. Check the resistance across the temperature control switch. If the mode selector switch is electrical, check for continuity across the terminals as you move the switch through the various mode selections. If the mode selector switch is a vacuum switch, apply a vacuum to the inlet of the switch and feel for a vacuum at the various tube connectors on the switch while you move the selector through its various positions. Compare your results to the information in the service manual.

Figure 53-38 The control must be removed to test it and what it controls.

ATC control panels are not repaired; rather, they are replaced as a unit. Care needs to be taken when replacing these units. Like the other electronic circuits on a vehicle, the control panel may be part of the BCM circuit. Follow all precautions to eliminate static and voltage spikes at the BCM.

Solar Sensor

The solar sensor detects sunlight. It controls the climate control system on many vehicles. The output voltage from the sensor varies with the amount of sunlight. When the sunlight increases, the output voltage increases. As the sunlight decreases, the output voltage decreases. The sensor is monitored by the A/C control module. DTCs will indicate a perceived problem. The systems typically look for an open or short in the solar sensor circuit.

CASE STUDY

A customer complained about his air-conditioning not working. Questioning the customer revealed that the system had not worked since the end of last summer. "It was going to get cool in a few weeks and I wouldn't need the air conditioner, so I put it off until spring."

A visual inspection of the system did not reveal any oil spots or ruptured lines and hoses; therefore, no leaks were indicated. The technician then connected the manifold set to the system. He found the pressure to be equal on both sides. Further, the temperature/pressure chart indicated that the system pressure was within acceptable limits for the ambient temperature.

The technician noticed that the lead wire to the clutch coil had been disconnected. Assuming that it had been unintentionally disconnected, he reconnected it. Further questioning of the customer, however, revealed no knowledge of the disconnected wire.

Shortly after starting the engine and turning the A/C on, cool air was coming out of the dash vents. The manifold gauges showed normal readings. A thermometer was placed in the vents and the output temperature was within specifications.

Further discussion with the customer revealed that the car broke down while he was on vacation. While traveling, a belt needed to be replaced. What undoubtedly happened was the technician who installed the belt must have unplugged the wire. Because the weather was cool, the customer had no reason to turn on the air conditioner and therefore could never link the problem to the belt replacement. Whenever you are working on a car, make sure everything is connected before returning it to the customer.

KEY TERMS

Schrader service valves **Stem valve**

SUMMARY

- Air-conditioning (A/C) systems are extremely sensitive to moisture and dirt. Therefore, clean working conditions are extremely important. The smallest particle of foreign matter in an A/C system contaminates the refrigerant, causing rust, ice, or damage to the compressor.
- Air conditioner testing and servicing equipment includes a manifold gauge set, service valves, vacuum pumps, charging station, charging cylinder, recovery/recycling systems, and leak-detecting devices.
- Refrigerant evaporates quickly when it is exposed to the atmosphere. It will freeze anything it contacts. If liquid refrigerant gets in your eyes or on your skin, it can cause frostbite. Never rub your eyes or skin if refrigerant has contacted you. Immediately flush the exposed areas with cool water for 15 minutes and seek medical help.
- Never expose A/C system components to high temperatures. Heat will cause the refrigerant's pressure to increase. Never expose refrigerant to an open flame.
- Refrigerants should never be mixed. Their oils and desiccants are not compatible. If refrigerants are mixed, contamination will occur and may result in A/C system failure. Separate service equipment, including recovery/recycling machines and service gauges, should be used for the different refrigerants.
- The manifold gauge set is used when discharging, charging, and evacuating the system; it is also used for diagnosing trouble in the system. Because R-134a is not interchangeable with R-12, separate sets of hoses, gauges, and other equipment are required to service vehicles.
- Performance testing provides a measure of A/C system operating efficiency.
- A manifold pressure gauge set is used to determine both high and low pressures in the refrigeration system. The desired pressure readings will vary according to temperature.
- Using an electronic leak detector is the preferred method of leak detection.
- A common leak detection method includes a fluorescent dye and a black light glow gun to locate the source of leaks.
- All refrigerant must be discharged from the system before repair or replacement of any component, except for compressors with stem service valves.
- Recycling collects old refrigerant from the system being serviced and cleans it up and prepares it to be used in the future.
- R-12-based systems use mineral oil, whereas R-134a systems use synthetic polyalkylene glycol (PAG) oils. Using a mineral oil with R-134a will result in A/C compressor failure because of poor lubrication. Use only the oil specified for the system.

- While evacuating the system, the vacuum pump should remain on and connected to the system for at least 30 minutes after 26 to 29 inches of mercury is reached.

REVIEW QUESTIONS

1. What should you do if you suspect that an alternative refrigerant had been used in the A/C system and it is not performing properly?
2. Explain the purpose of a blower motor resistor block and how it basically works.
3. What is the purpose of the thermostatic switch in a cycling clutch system?
4. Why should the probe of an electronic leak detector be placed below the area being checked?
5. What will happen if moisture and air have contaminated a refrigerant?
6. Which of the following statements best defines the term *evacuation*?
   a. A process in which the refrigerant in a system is released
   b. A condition that exists when a compressor runs until there is no refrigerant left in the system
   c. The process that pulls all traces of air and moisture from the system
   d. A process that uses air to force moisture and dirt out of the system
7. Which of the following statements is *not* true about orifice tube systems?
   a. The suction line must be cool to the touch from the evaporator outlet to the compressor.
   b. The liquid line from the condenser outlet to the evaporator inlet should be cool.
   c. The evaporator should feel very cold.
   d. The accumulator must be cool to the touch.
8. To charge an air-conditioning system while it is running, the refrigerant should be added to ______.
   a. the high side
   b. the low side
   c. both the high and low sides
   d. either the high or the low side
9. All of the following are true statements about flushing an A/C system after a compressor failure, *except* ______.
   a. the condenser must be flushed and the receiver/dryer replaced
   b. filter screens are sometimes located in the suction side of the compressor and in the receiver/dryer. These screens confine foreign material to the compressor, condenser, receiver/dryer, and connecting hoses
   c. use only CFCs for flushing. After the system has been flushed, be sure to oil all components that require it
   d. some manufacturers recommend replacing the clogged components and installing a liquid line (inline) filter just ahead of the expansion valve or orifice tube instead of flushing the system
10. A very important part of a performance test on an A/C system is a pressure test. Pressure also is the key to the operation of the system. Which of the following statements is *not* true?
    a. Pressure on a substance, such as a liquid, changes its boiling point.
    b. The greater the pressure on a liquid, the higher the boiling point.
    c. If pressure is placed on a liquid, the liquid freezes at a higher-than-normal temperature.
    d. The boiling point can be increased by increasing the pressure on the fluid. It can also be decreased by reducing the pressure or placing the fluid in a vacuum.
11. While conducting a pressure test on an A/C system, the ambient temperature is 80°F, the low-side gauge reads low (8 psi), and the high side also reads low (85 psi). Which of these could cause these readings?
    a. low refrigerant level
    b. normal operation
    c. bad compressor
    d. a high-side restriction
12. The high-side pressure on a system with a cycling clutch and pressure cycling switch is 245 psi, the low-side pressure is 28 psi, and the ambient temperature is 78°F. What do these gauge readings indicate?
    a. The system is undercharged.
    b. The system is overcharged.
    c. The system is normal.
    d. The evaporator pressure regulator is bad.
13. When checking an A/C system, the evaporator's inlet and outlet tubes feel like they are at the same temperature. Which of these could cause this?
    a. a plugged evaporator
    b. correct charge of refrigerant in the system

c. restricted orifice tube
d. leaking condenser

14. All of these are good reasons for replacing an accumulator unit, *except*______.
a. the accumulator/dryer is punctured
b. the accumulator/dryer is saturated with water
c. the A/C system has been open to the atmosphere for 2 or more hours
d. the outer shell of the accumulator/dryer is dented

15. In an A/C system, if the low-side pressure is lower than normal, which of the following is the least likely cause?
a. a faulty metering device
b. poor airflow across the evaporator
c. a restriction in the low side of the system
d. the system being overcharged with refrigerant

ASE-STYLE REVIEW QUESTIONS

1. Technician A says that evacuating (pumping down) an air-conditioning system removes air and moisture from the system. Technician B says that evacuating (pumping down) an air-conditioning system removes dirt particles from the system. Who is correct?
a. Technician A
b. Technician B
c. Both A and B
d. Neither A nor B

2. Technician A says that some refrigerant leaks can also be traced to pinholes in the evaporator caused by acid, which forms when water and refrigerant mix. Technician B says that leaks are most often found at the compressor hose connections and at the various fittings and joints in the system. Who is correct?
a. Technician A
b. Technician B
c. Both A and B
d. Neither A nor B

3. Technician A says that the presence of oil around a fitting of an air-conditioning line may indicate an oil leak but not a refrigerant leak. Technician B says that using an electronic leak detector is the best way to find the source of a refrigerant leak. Who is correct?
a. Technician A
b. Technician B
c. Both A and B
d. Neither A nor B

4. Technician A says that a faulty accumulator will allow contaminants to enter the A/C system. Technician B says that a damaged compressor will allow contaminants to enter the A/C system. Who is correct?
a. Technician A
b. Technician B
c. Both A and B
d. Neither A nor B

5. Technician A says that air in the system can cause excessive pressure within the system, restricting the refrigerant's ability to change its state from gas to liquid within the refrigeration cycle. Technician B says that moisture can cause freeze-up at the cap tube or expansion valve, which restricts refrigerant flow or blocks it completely. Who is correct?
a. Technician A
b. Technician B
c. Both A and B
d. Neither A nor B

6. Technician A feels up and down the face or along the return bends of the condenser for a temperature change and says that there should be a gradual change from hot to warm from the bottom to the top. Technician B says that a restriction in the condenser will be evident by a sudden change in temperature. Who is correct?
a. Technician A
b. Technician B
c. Both A and B
d. Neither A nor B

7. Technician A says that a manifold gauge set can be used during the evacuation of the system. Technician B says that a manifold gauge set can be used to diagnose a problem in the A/C system. Who is correct?
a. Technician A
b. Technician B
c. Both A and B
d. Neither A nor B

8. While discussing an inoperative blend air door: Technician A says that to check the proper functioning of the ductwork, move the temperature control lever to see if any change at the doors occurs. If it does not, shut off the air conditioner and turn on the heater. Move the temperature control arm again to see if any change occurs. If not, check the cable and the flap door connected to the temperature control lever. Technician B says that in some cases you can reach under the dash to reconnect the cable or free a stuck flap. Who is right?
a. Technician A
b. Technician B
c. Both A and B
d. Neither A nor B

9. Two technicians were discussing how comparing the temperature of a high-pressure line to that of a low-pressure line would help evaluate an A/C system when the system does not have a sight glass. Technician A says that the high-

pressure line will feel warmer than the low-pressure line if the system is operating normally. Technician B says that both lines should feel cold after the system has been run for a while if the system is working normally. Who is right?

a. Technician A
b. Technician B
c. Both A and B
d. Neither A nor B

10. While conducting an A/C system performance test: Technician A inserts a thermometer into the air duct at the center of the dash and monitors discharge temperature. Technician B follows this test with a complete visual inspection. Who is right?

a. Technician A
b. Technician B
c. Both A and B
d. Neither A nor B

APPENDIX A

DECIMAL AND METRIC EQUIVALENTS

DECIMAL AND METRIC EQUIVALENTS					
Fractions	**Decimal (in.)**	**Metric (mm)**	**Fractions**	**Decimal (in.)**	**Metric (mm)**
1/64	0.015625	0.397	33/64	0.515625	13.097
1/32	0.03125	0.794	17/32	0.53125	13.494
3/64	0.046875	1.191	35/64	0.546875	13.891
1/16	0.0625	1.588	9/16	0.5625	14.288
5/64	0.078125	1.984	37/64	0.578125	14.684
3/32	0.09375	2.381	19/32	0.59375	15.081
7/64	0.109375	2.778	39/64	0.609375	15.478
1/8	0.125	3.175	5/8	0.625	15.875
9/64	0.140625	3.572	41/64	0.640625	16.272
5/32	0.15625	3.969	21/32	0.65625	16.669
11/64	0.171875	4.366	43/64	0.671875	17.066
3/16	0.1875	4.763	11/16	0.6875	17.463
13/64	0.203125	5.159	45/64	0.703125	17.859
7/32	0.21875	5.556	23/32	0.71875	18.256
15/64	0.234275	5.953	47/64	0.734375	18.653
1/4	0.250	6.35	3/4	0.750	19.05
17/64	0.265625	6.747	49/64	0.765625	19.447
9/32	0.28125	7.144	25/32	0.78125	19.844
19/64	0.296875	7.54	51/64	0.796875	20.241
5/16	0.3125	7.938	13/16	0.8125	20.638
21/64	0.328125	8.334	53/64	0.828125	21.034
11/32	0.34375	8.731	27/32	0.84375	21.431
23/64	0.359375	9.128	55/64	0.859375	21.828
3/8	0.375	9.525	7/8	0.875	22.225
25/64	0.390625	9.922	57/64	0.890625	22.622
13/32	0.40625	10.319	29/32	0.90625	23.019
27/64	0.421875	10.716	59/64	0.921875	23.416
7/16	0.4375	11.113	15/16	0.9375	23.813
29/64	0.453125	11.509	61/64	0.953125	24.209
15/32	0.46875	11.906	31/32	0.96875	24.606
31/64	0.484375	12.303	63/64	0.984375	25.003
1/2	0.500	12.7	1	1.00	25.4

APPENDIX B

GENERAL TORQUE SPECIFICATIONS

The values in this chart should only be used when manufacturer's specifications are *not* available. Also, the values are only valid when SAE 10 oil is used to lubricate the threads of the bolt.

Bolt Diameter (in inches)	**Torque (pounds-feet)**		
	SAE 2	**SAE 5**	**SAE 8**
1/4	7	10	14
5/16	14	21	30
3/8	24	37	52
7/16	39	60	84
1/2	59	90	128
9/16	85	130	184
5/8	117	180	255
3/4	205	320	450
7/8	200	515	730
1	300	775	1,090

Bolt Diameter (in millimeters)	**TORQUE: kg. cm* kg. m Property Class:**									
	4.6	**4.8**	**5.6**	**5.8**	**6.6**	**6.8**	**6.9**	**8.8**	**10.9**	**12.9**
6	49*	63*	61*	79*	74*	95*	103*	126*	172*	206*
8	119*	153*	148*	178*	178*	230*	250*	306*	417*	500*
10	235*	303*	294*	379*	353*	455*	495*	606*	8.2	10
12	411*	529*	427*	662*	616*	7.9	8.6	10.5	14	17
14	654*	8.4	8.2	10.5	10	12	13	17	23	27
16	10	13	12	16	15	20	21	26	36	43
18	14	18	17	23	21	27	30	36	49	59
22	27	35	34	44	41	52	57	70	95	114

GLOSSARY

Note: Terms are highlighted in color, followed by Spanish translation in bold.

Abrasion Wearing or rubbing away of a part.

Abrasión El desgaste o consumo por rozamiento de una parte.

Abrasive cleaning Cleaning that relies on physical abrasion, such as wire brushing and glass bead blasting.

Limpieza abrasiva La limpieza que cuenta con un abrasión físico, tal como el limpiar con cepillo de alambre o con un chorro de perlitas de vidrio.

Absorbed glass mat (AGM) battery A design for sealed lead-acid batteries. The electrolyte is absorbed in a matrix of glass fibers, which holds the electrolyte next to the plate and immobilizes it, preventing spills.

Batería con malla de fibra de vidrio absorbente (AGM) Diseño de baterías selladas de ácido-plomo. El electrolito se absorbe en una matriz de fibra de vidrio, lo cual retiene al electrolito junto al plato, y lo inmoviliza, previniendo así derrames.

Acceleration The rate of increase in speed.

Aceleración El índice de incremento en velocidad.

Accumulator A container that stores hydraulic fluid under pressure. It can be used as a fluid shock absorber or as an alternate pressure source. A spring or compressed gas behind a sealed diaphragm provides the accumulator pressure. They are commonly used in automatic transmissions, air-conditioning systems, and antilock brake systems.

Acumulador Contenedor que almacena fluido hidráulico bajo presión. Puede usarse como fluido para amortiguadores o como fuente de presión alterna. Resorte o gas comprimido detrás de un diafragma sellado proporciona depression al acumulador. Por lo general, se usan en transmisiones automáticas, sistemas de aire acondicionado, y sistemas de frenos antibloqueo.

Acid A compound that has an excess of H ions and breaks into hydrogen (H^+) ions and another compound when placed in an aqueous (water) solution.

Ácido Un compuesto que tiene un exceso de iones de H, que se rompe en los iones de hidrógeno y un otro compuesto cuando está colocado en una solución acuosa (del agua).

Acidity In lubrication, acidity denotes the presence of acid-type chemicals, which are identified by the acid number. Acidity within oil causes corrosion, sludges, and varnish to increase.

Acidez En la lubricación, la acidez se refiere a la presencia de los químicos de tipo ácido, que se identifican por un número de ácido. La acidez en los aceites causa aumentos en la corrosión, la grasa y el barnís.

Active restraint A manual seat and shoulder belt assembly that must be activated by an occupant.

Sujeción activo Una asamblea manual de cinturón de seguridad que debe ser implementado por el ocupante.

Actuator A control device that delivers mechanical action in response to an electrical signal.

Solenoide Un dispositivo de control que entrega una acción mecánica como respuesta a un señal eléctrico.

A/D converter An electronic device that changes analog signals to digital signals.

Convertidor de A/D Un dispositivo electrónico que cambia los señales análogos a los señales digitales.

Adaptive learning An operational mode of a control computer that adjusts operation parameters according to various conditions.

Operación de adaptación Un modo de operación de una computadora de control que ajusta los parámetros de operación según varias condiciones.

Additive In automotive oils, a material added to the oil to give it certain properties; for example, a material added to engine oil to lessen its tendency to congeal or thicken at low temperatures.

Aditamento En los aceites automotrices, una materia añadida al aceite para que ésta tenga ciertas propiedades; por ejemplo, una materia agregada al aceite de motor para disminuir su tendencia a solidificarse o ponerse espeso en las temperaturas bajas.

Adhesion The property of lubricating oil that causes it to stick or cling to a bearing surface.

Adhesión La propriedad del aceite de lubricación que causa que se pega o se adhiera a una superficie portante.

Adhesive Substance that causes adjoining bodies to stick together.

Adhesivo Una substancia que causa que dos cuerpos contiguos se pegan.

Aeration The process of mixing air with a liquid. Aeration occurs in a shock absorber from rapid fluctuation in movement.

Aeración El proceso de mezclar el aire con un líquido. La aeración ocurre en un amortiguador de barquinazos por causa de una variación rápida en el movimiento.

Aerobic Typically refers to a sealant or adhesive that dries in the presence of oxygen.

Aeróbico Típicamente se refiere a un sellante o un adhesivo que se seca cuando esta presente el oxígeno.

Aerodynamics The study of the effects of air on a moving object.

Aerodinámica El estudio de los efectos del aire en un objeto móvil.

Air bag system System that uses impact sensors, vehicle's on-board computer, an inflation module, and a nylon bag in the steering column and dash to protect the driver and passenger during a head-on collision.

Sistema de bolsa de aire Un sistema que utiliza sensores de impacto, la computadora del vehículo, un módulo inflador y bolsas de nilón en la columna de dirección y el tablero de instrumentos, para proteger al conductor y al pasajero en caso de un choque de frente.

Air-conditioning The process of adjusting and regulating, by heating or refrigerating, the quality, quantity, temperature, humidity, and circulation of air in a space or enclosure; to condition the air.

Aire acondicionado El proceso de ajustar, regular calentando o enfriando, la calidad, la cantidad, la temperatura, la humedad, y la circulación de aire en un espacio o espacio cerrado; para acondicionar el aire.

Air-conditioning clutch An electromechanical device mounted on the air-conditioning compressor used to start and stop compressor action, thereby controlling refrigerant circulating through the system.

Embrague de climatizador (aire acondicionado) Dispositivo electromecánico montado en el compresor del equipo climatizador y que se utiliza para poner en marcha y detener dicho compresor, controlando de ese modo el refrigerante que circula por el sistema.

Air-conditioning compressor A vapor pump that pumps vapor (refrigerant or air) from one pressure level to a higher pressure level.

Compresor de aire acondicionado Dispositivo del climatizador que bombea vapor (refrigerante o aire) de un nivel de presión a otro más elevado.

Air ducts Tubes, channels, or other tubular structures used to carry air to a specific location.

Conductos de aire Los tubos, las canalizaciones, u otras estructuras tubulares que sirven para traer el aire a una posición específica.

Air filter A filter that removes dust, dirt, and particles from the air passing through it.

Filtro de aire Un filtro que retira polvo, suciedad y partículas del aire que pasa por él.

Air gap The space between spark plug electrodes, motor and generator armatures, and field shoes.

Entrehierro El espacio entre los electrodos de las bujías, el enducido del motor y del generador, y las piezas inductoras.

Air injection The introduction of fresh air into an exhaust manifold for additional burning and to provide oxygen to a catalytic converter.

Inyección por aire La introducción del aire fresco al múltiple de escape para mejor combustión y para proveer el oxígeno al convertidor catalítico.

Air Injection Reactor (AIR) An emission control system responsible for adding outside air to the catalytic converter or exhaust stream.

Reactor de inyección de aire (AIR) Sistema de control de emisión responsable de agregar aire externo al convertidor catalítico o flujo de escape.

Air-fuel ratio (A/F) sensor Similar to an oxygen sensor, this sensor is placed in the exhaust stream and sends signals to the PCM is response to very slight changes in exhaust oxygen content.

Sensor de proporción aire/combustible Similar a un sensor de oxígeno, este sensor se coloca en el flujo de escape y envía señales al PCM, es responsable de ligeros cambios

Alignment An adjustment to a line or to bring into a line.

Alineación Un ajuste que se efectúa en una linea o alinear.

Alloy A mixture of different metals such as solder, which is an alloy consisting of lead and tin.

Aleación Una mezcla de metales distinctos como la soldadura, que es una aleación que consiste del plomo y el estaño.

All-wheel drive (AWD) System of driving all four wheels only when traction conditions dictate.

Tracción total (todoterreno) Un sistema de tracción que maneja las cuatro ruedas sólo cuando indican las condiciones de la tracción.

Alternating current Electrical current that changes direction between positive and negative.

Corriente alterna Corriente eléctrica que recorre un circuito ya sea en dirección positiva o en dirección negativa.

Alternator An AC (alternating current) generator that produces electrical current and forces it into the battery to recharge it.

Alternador Un generador de ca (corriente alterna) que produce corriente eléctrica y la dirige a una batería para cargarla.

Ambient temperature Temperature of air surrounding an object.

Temperatura del ambiente La temperatura en la atmósfera alrededor de un objeto.

American wire gauge (AWG) A standard method of denoting the diameter of electrically conducting wire.

Calibre Americano (AWG) Método estándar que indica el diámetro de cable conductor de electricidad.

Ammeter The instrument used to measure electrical current flow in a circuit.

Amperímetro El instrumento que se usa para medir los corrientes eléctricos de un circuito.

Ampere The unit for measuring electrical current; usually called an amp.

Amperio La unidad para medir el corriente eléctrico; por lo común se refiere como "amp" en inglés.

Ampere-hour (AH) rating The rating that is based on the total number of amperes the battery can supply in a 20-hour period at a fixed rate of discharge. If a battery is rated at 200 ampere/hours, it can supply 10 amperes per hour for 20 hours.

Tasa ampere-hora (AH) Esta tasa se basa en el total del número de amperes que una batería puede suministrar en un período de 20 horas a una tasa fija de descarga. Si una batería se tasa a 200 amperes/horas puede suministrar 10 amperes durante 20 horas.

Amplifier A circuit or device used to increase the voltage or current of a signal.

Amplificador Un circuito o un dispositivo que se usa para aumentar el voltage o el corriente de un señal.

Amplify To enlarge or strengthen original characteristics; a term associated with electronics.

Amplificar Aumentar o fortalecer las características originales; un término que se asocia con lo electrónico.

Amplitude A measurement of a vibration's intensity.

Amplitud Una medida de intensidad de una vibración.

Anaerobic Typically refers to a sealant or adhesive that dries without being exposed to oxygen.

Anaerobio Típicamente se refiere a un sellante o un adhesivo que se seca sin ser expuesto al oxígeno.

Anaerobic sealer Liquid or gel that operates to bond two parts together in the absence of air.

Sellante anaerobio Un líquido o un gel cuyo propósito es adherir dos partes en la ausencia del aire.

Analog A nondigital measuring method that uses a needle to indicate readings. A typical dashboard gauge with a moving needle is an analog instrument.

Análogo Un método no numérico de medir que usa una aguja para indicar las lecturas. Un indicador típico de un tablero de intrumentos con una aguja que se mueve es un instrumento análogo.

Anodize An electrochemical process that coats and hardens the surface of aluminum.

Anodizar Un proceso electroquímico que cubre y endurece la superficie del aluminio.

Antifreeze A material, such as alcohol or glycerin, added to water to lower its freezing point.

Anticongelante Una materia, tal como el alcohol o la glicerina, que se agrega al agua para disminuir su punto de congelación.

Antiknock Index (AKI) The octane rating required by law and the one displayed on gasoline pumps. It is the average of RON and MON and is stated as (R1M)/2.

Índice antidetonante (AKI) Proporción de octano se requiere por ley y se muestra en las bombas de gasolina. Es el promedio entre RON y MON y se escribe como (R+M)/2.

Antilock brake system (ABS) A series of sensing devices at each wheel that control braking action to prevent wheel lockup.

Sistema de frenos antideslizante Una seria de dispositivos sensibles en cada rueda que controla la acción del enfrenamiento y previene que las ruedas se bloquean.

Antiseize compounds A type of lubrication that prevents dissimilar metals from reacting with one another and seizing.

Compuestas antiagarrotamiento Un tipo de lubricación que previene que los metales desemejantes se reaccionan y se agarran.

Antisway bar A suspension geometry that resists a vehicle's tendency to drop or squat on the rear springs when accelerating.

Barra de anticabeceo Una geometría de suspensión que resiste la tendencia de un vehículo a desplomarse o apoyarse sobre los muelles posteriores al acelerar.

Aqueous A solution that is mostly water.

Acuosa Una solución acuosa es una solución que es agua en su mayor parte.

Aramid fibers Refer to a family of synthetic materials that are stronger than steel but weigh little more than half of what an equal volume of fiberglass would weigh.

Fibras de aramida Se refieren a una familia de materiales sintéticos que son más fuertes que el acero pero pesan poco más de la mitad de lo que un volumen igual de fibra de vidrio pesaría.

Asynchronous motor A type of AC motor in which the rotor rotates at a slower speed than that of the rotating magnetic field.

Motor asíncrono Tipo de motor CA en el cual el rotor gira a una velocidad menor que la rotación del campo magnetico.

Aspect ratio The height of a tire, from bead to tread, expressed as a percentage of the tire's section width.

Relación de aspecto La altura de un neumático, de la ceja a la banda, que se expresa como un porcentaje de lo ancho de una sección del neumático.

Asymmetric Unequal surfaces or sizes.

Asimétrico Desigual en las superficies o en los tamaños.

ATF Automatic transmission fluid.

ATF Fluido para transmisión automática.

Atkinson cycle The cycle during which the intake valves are kept open for a while during the compression stroke, reducing the actual displacement and the power output of the engine.

Ciclo Atkinson Durante el ciclo Atkinson, las válvulas de entrada se mantienen abiertas por un tiempo durante el golpe de compresión, esto reduce el desplazamiento y la salida de energía del motor.

Atmospheric pressure The weight of the air at sea level (about 14.7 pounds per square inch or less at higher altitudes).

Presión atmosférica El peso del aire a nivel del mar (aproximadamente unas 14.7 libras por pulgada cuadrada o menos en las altitudes más altas).

Atom The smallest particle of an element in which all the chemical characteristics of the element are present.

Átomo La partícula más pequeña de un elemento en el la cual todas las características químicas del elemento estén presentes.

Atomization The stage in which the metered air-fuel emulsion is drawn into the airstream in the form of tiny droplets.

Atomización La etapa en la cual la emulsión calibrada del aire-combustible se introduce al chorro del aire en la forma de pequeñas gotas.

Automotive aftermarket The network of businesses that supplies replacement parts and services to independent service shops, specialty repair shops, car and truck dealerships, fleet and industrial operations, and the general buying public.

Mercado de repuestas no originales La cadena de negocios que proveen las repuestas y los servicios a los talleres de servicio independientes, los talleres de refacción especializadas, los comerciantes de coches y camiones, las operaciones de flete e industrias, y las compras del público común.

Average responding A type of voltage reading in which the average voltage peak is displayed.

Respuesta promedio Tipo de lectura de voltaje en el cual se muestra el pico promedio de voltaje.

Axial Having the same direction or being parallel to the axis or rotation.

Axial Teniendo el mismo dirección o siendo paralelo al eje o la rotación.

Axial load A type of load placed on a bearing that is parallel to the axis of the rotating shaft.

Carga axial Un tipo de carga puesto en un cojinete paralelo al eje de una flecha que gira.

Axial play Movement that is parallel to the axis or rotation.

Holgura axial El movimiento que es paralelo al eje o a la rotación.

Back pressure Pressure created by restriction in an exhaust system.

Contrapresión La presión creada por una restricción en un sistema de escape.

Backlash The clearance or play between two parts, such as meshed gears.

Juego La cantidad de holgura o juego entre dos partes, tal como los engranajes endentados.

Balancing coil gauge A type of gauge that utilizes coils to create magnetic fields instead of a permanent magnet.

Bobina equilibrada Un tipo de manómetro que utiliza las bobinas para crear un campo magnético en vez de un imán permanente.

Ball bearing An antifriction bearing that uses a series of steel balls held between inner and outer bearing races.

Cojinete de bolas Un cojinete de antifricción que usa una serie de bolas del acero sujetados entre pistas interiores y exteriores.

Ball joint A pivot point for turning a front wheel to the right or left. Ball joints can be considered either nonloaded or loaded when carrying the car's weight.

Unión esférica Un punto de pivote en el cual gira una rueda delantera hacia la derecha o a la izquierda. Las uniones esféricas se pueden considerar como sin carga o con carga al sostener el peso del coche.

Band A steel band with an inner lining of friction material; a device used to hold a clutch drum at certain times during automatic transmission operation.

Banda Banda de acero con un revestimiento de material de fricción. Dispositivo que se usa para sostener un tambor del embrague en ciertos momentos durante la operación de transmisión automática.

Barometric pressure A sensor or its signal circuit that sends a varying frequency signal to the processor relating actual barometric pressure.

Presión barométrico Un sensor o su circuito de señal que manda un señal de frecuencia variada al procesor referiendo la presión barométrico actual.

Base A solution that has an excess of OH ions, also called an alkali. Also the center layer of a bipolar transistor.

Base Una solución que tiene un exceso de iones del OH, también llamada un álcali. También la capa central de un transistor bipolar.

Battery A device for storing energy in chemical form so it can be released as electricity.

Batería Un dispositivo para almacenar energía en forma química para que se pueda liberar como electricidad.

Battery cable A heavy electrical conductor used to connect a vehicle's battery to the starter motor or chassis ground.

Cable de batería Un conductor eléctrico sólido que se usa para conectar la batería de un vehículo al motor de arranque o la masa del chasís (tierra).

Battery cell That part of a storage battery made from two dissimilar metals and an acid solution. A cell stores chemical energy for use later as electrical energy.

Célula de batería Aquella parte de una batería acumuladora hecha de dos metales desemejantes y una solución de ácido. Una célula almacena la energía química para usarse después como energía eléctrica.

Baud rate The speed of communication and is equal to the number of bits per second that a computer can process.

Velocidad de baudios La velocidad de comunicación y es igual al número de bits por segundo que puede procesar una computadora.

BDC Bottom dead center, position of the piston.

BDC Punto muerto exterior, posición del pistón.

Bead The edge of a tire's sidewall, usually made of steel wires wrapped in rubber, used to hold the tire to the wheel.

Ceja La extremidad del refuerzo lateral de un neumático, suele ser hecho de alambres de acero cubiertos en caucho, que sujeta el neumático a la rueda.

Bearing Soft metallic shells used to reduce friction created by rotational forces.

Cojinete Una pieza hueca de metal blanda que sirve para reducir la fricción creada por las fuerzas giratorias.

Bearing clearance The amount of space left between a shaft and the bearing surface for lubricating oil to enter.

Holgura del cojinete La cantidad del espacio que se deja entre la flecha y la superficie portante por la cual entra el lubricante.

Bearing crush The process of compressing a bearing into place as the bearing cap is tightened.

Aplastamiento del cojinete El proceso de colocar un cojinete comprimiéndolo cuando se aprieta la tapa del cojinete.

Bearing race The machined circular surface of a bearing against which the roller or ball bearings ride.

Pista del cojinete La superficie circular maquinada del cojinete en la cual ruedan los rodillos o bolas.

Bearing spread The condition in which the distance across the outside parting edges of the bearing insert is slightly greater than the diameter of the housing bore.

Aplastamiento del cojinete La condición en la cual la distancia de las extremidades exteriores de la pieza inserta del cojinete es un poco más grande que el diámetro de la caja.

Bell housing A term often used for clutch housing.

Cárter del embrague Un término que se usa con frecuencia para la caja del embrague.

Belleville spring A round, slightly cone-shaped disc used to return a hydraulic piston in a clutch assembly to a static, unapplied position.

Resorte Belleville o resorte de presión Disco redondo, ligeramente cónico, usado para regresar un pistón hidráulico en un ensamble de embrague a la posición estática, posición no aplicada.

Belt alternator starter (BAS) A combination motor/generator that is driven by the engine's crankshaft via a drive belt. It replaces both the engine's alternator and starter motor.

Arrancador de alternador de banda (BAS) Combinación de motor/generador que maneja el cigüeñal, por medio de la banda de transmisión. Reemplaza tanto al alternador del motor como al motor del arrancador.

Bias A diagonal line of direction. In relationship to tires, bias means that belts and plies are laid diagonally or crisscrossing each other.

Bies Una línea de dirección diagonal. Perteneciente a los neumáticos, bies quiere decir que las bandas y los pliegues se colocan diagonalmente y se cruzán entre sí.

Binary code A series of numbers represented by 1s and 0s or offs and ons.

Código binario Una serie de números representados por los 1s y los 0s o por apagado y prendido.

Biodiesel fuels A biodegradable fuel for use in diesel engines. It is produced with organically derived oils or fats. It may be used as a replacement for, or as a component of, diesel fuel.

Combustibles biodiésel Combustible biodegradable para usar en motores diésel. Se produce a partir de derivados del petróleo o grasas en forma organic. Puede utilizarse como reemplazo o como un componente del combustible diésel.

Bit One character of binary code.

Bit Un carácter del código binario.

Bloodborne pathogens Pathogenic microorganisms that are present in human blood and can cause disease. These pathogens include, but are not limited to, hepatitis B virus (HBV) and human immunodeficiency virus (HIV).

Patógenos transmisibles por sangre Microorganismos patógenos que están presentes en la sangre humana y pueden causar enfermedades. Estos patógenos incluyen, pero no están limitados a, virus de hepatitis B (HBV) y virus de inmunodeficiencia humana (HIV).

Blowby The unburned fuel and products of combustion that leak past the piston rings and into the crankcase during the last part of the combustion stroke.

Fuga El combustible no consumido y los productos de la combustión que se escapan alrededor de los anillos de los pistones y entran al cárter en las últimas etapas de la carrera de combustión.

Body-over-frame Type of vehicle construction in which the frame is the foundation, with the body and all major parts of the vehicle attached to it.

Carrocería sobre armazón Un tipo de construcción de vehículos en el cual el armazón es la fundación, al cual se conectan la carrocería y los componentes principales del vehículo.

Bolt diameter The measurement across the major diameter of a bolt's threaded area or across the bolt shank.

Diámetro del perno La medida del diámetro mayor de la parte fileteada de un perno o atravéz del asta del perno.

Bolt head The part of a bolt that the socket or wrench fits over in order to torque or tighten the bolt.

Cabeza del perno La parte del perno sobre la cual se pone el dado o la llave para apretar o torcer al perno.

Boot Rubber protective cover with accordion pleats used to contain lubricants and exclude contaminating dirt, water, and grime. Located at each end of rack and pinion assembly and front-wheel-drive CV joints.

Protectores de caucho Una cubierta protectora hecho de caucho que tiene pliegues estilo acordión que sirve para contener las lubricantes y prevenir la entrada de los contaminantes como el lodo, el agua y el mugre. Se ubican en cada extremidad de la asamblea de piñon y cremallera.

Bore A dimension of cylinder size representing the diameter of the cylinder.

Diámetro del orificio Una dimensión de tamaño del cilindro que representa el diámetro del cilindro.

Brake band A circle-shaped part lined with friction material that acts as a brake or holding device to stop and hold a rotating drum that has a gear train member connected to it.

Banda de freno Una parte circular recubierta con materia fricativa que sirve para frenar o como un dispositivo de asir para detener y sostener un tambor giratorio al cual está conectado un miembro del tren de engrenaje.

Brake drum A bowl-shaped cast-iron housing against which the brake shoes press to stop its rotation.

Tambor de freno Un alojamiento de hierro colado en forma de tazón contra el que se oprimen las zapatas de los frenos para detener su rotación.

Brake fade Occurs when friction surfaces become hot enough to cause the coefficient of friction to drop to a point where the application of severe pedal pressure results in little actual braking.

Amortiguamiento de frenar Ocurre cuando la superficie de fricción se sobrecalienta al punto de causar que caiga el coeficiente de fricción a tal punto que la aplicación de presión riguroso en el pedal de frenos resulta en muy poco enfrenamiento.

Brake fluid A hydraulic fluid used to transmit force through brake lines. Brake fluid must be noncorrosive to both the metal and rubber components of the brake system.

Líquido para frenos Un líquido hidráulico que se usa para transmitir potencia por las líneas de frenos. Dicho líquido debe ser no corrosivo para los componentes tanto metálicos como de caucho del sistema de frenos.

Brake pads The part of a disc brake system that holds the linings.

Tacos de presión Las piezas de un sistema de frenado que sostienen las guarniciones (balatas).

Brake rotor Disc-shaped component that revolves with hub and wheel. The lining pads are forced against the rotor to provide a friction surface for the brake system, so as to slow or stop a vehicle.

Rotor de freno Componente en forma de disco que gira con el cubo y la rueda.

Brake shoe The metal assembly onto which the frictional lining is attached for drum brake systems.

Zapata de freno El ensamblaje metálico sobre el que se sujeta la guarnición de fricción (balata) para los sistemas de freno de tambor.

British thermal unit (Btu) A measurement of the amount of heat required to raise the temperature of 1 pound of water 1 degree Fahrenheit.

Unidad térmica británica Una medida de la cantidad del calor que se requiere para aumentar la temperatura de 1 libra de agua por 1 grado fahrenheit.

BTDC Before top dead center, position of the piston.

BTDC Antes del punto muerto interior Una posición del pistón.

Bump steer Erratic steering that is caused from rolling over bumps, cornering, or heavy braking; same as orbital steer and roll steer.

Dirección de choques La dirección errática que se causa al pasar por encima de los topes, a dar la vuelta, o en el enfrenamiento violento; quiere decir lo mismo que la dirección orbital y la dirección de inclinación.

Burnish To smooth or polish with a sliding tool under pressure.

Bruñir Pulir o suavizar por medio de una herramienta deslizandola bajo presión.

Byte A group of individual characters of binary code.

Byte Grupode caracteres individuales del código binario.

CAFE standards Law requiring automakers to not only manufacture clean-burning engines but also to equip vehicles with engines that burn gasoline efficiently.

Normas CAFE La ley requeriendo que los fabricantes de automóviles no sólo fabrican los motores de combustión limpia sino que tambien equipan los vehículos con los motores que consumen la gasolina en una manera eficiente.

Caliper Major component of a disc brake system. Houses the piston(s) and supports the brake pads.

Articulación Uno de los componentes principales de un sistema de freno de disco. Contiene el o los pistones y sostiene los tacos de presión.

Camber The attitude of a wheel and tire assembly when viewed from the front of a car. If it leans outward, away from the car at the top, the wheel is said to have positive camber. If it leans inward, it is said to have negative camber.

Camber (comba) La actitud de una asamblea de la rueda y el neumático al verse desde la frente del coche. Si se inclina hacia afuera, se dice que la rueda tiene una comba positiva. Si se inclina hacia el interior, se dice que tiene una comba negativa.

Camshaft The component in the engine that opens and closes the valves.

Árbol de levas El componente en el motor que abre y cierra las válvulas.

CAN (Controller Area Network) Bus A commonly used multiplexing protocol for serial communication. The communication wire is a twisted-pair wire.

CAN (Controlador de área de red) Bus Protocolo multiplexado comúnmente usado para comunicación serial. El cable de comunicación es un cable conductor doble retorcido.

Candlepower A measurement of the brightness of light.

Unidad de intensidad luminosa Una medida del brillantez de la luz.

Capacitor A device for holding and storing a surge of current.

Capacitor Un dispositivo que sostiene y almacena una sobretensión de corriente.

Carbon dioxide (CO_2) Compressed into solid form, this material is known as dry ice and remains at a temperature of −109°. It goes directly from a solid to a vapor state.

Anhídrido carbónico Comprimido para formar un sólido, esta materia se conoce como hielo seco y sostiene una temperatura de −109°F. Va directamente del estado sólido a un vapor.

Carbon monoxide (CO) Poisonous gas formed in engine exhaust.

Óxido de carbono Un gas venenoso producido en el escape de un motor.

Carburizing A method used to surface-harden steel by heat or mechanical means to increase the hardness of the outer surface while leaving the core relatively soft.

Carburación Un método usado para endurecer la superficie del acero por el calor o medios mecánicos para aumentar la dureza de la superficie externa mientras que deja la base relativamente suave.

Case harden To harden the surface of steel.

Cementar Endurecer la superficie del acero.

Caster Angle formed between the kingpin axis and a vertical axis as viewed from the side of the vehicle. Caster is considered positive when the top of the kingpin axis is behind the vertical axis.

Ángulo de caster El ángulo formado entre la linea de pivote y un eje vertical al verse del lado del vehículo. El caster se considera positivo cuando la extremidad superior de la línea de pivote es atrás del eje vertical.

Catalyst A compound or substance that can speed up or slow down the reaction of other substances without being consumed itself. In an automatic catalytic converter, special metals (for example, platinum or palladium) are used to promote more complete combustion of unburned hydrocarbons and a reduction of carbon monoxide.

Agente catalítico Una compuesta o una substancia que puede acelerar o retardar la reacción de otras substancias sin ser consumida. En un convertidor catalítico automático, se usan metales especiales (por ejemplo, el platino y el paladio) para promover la combustión completa del hidrocarburo sobrante y para reducir el óxido de carbono.

Catalytic converter An emission device located in front of the muffler in the exhaust system. It looks very much like a heavy muffler and contains catalysts to clean up an engine's emissions before they leave the end of the exhaust pipe.

Convertidor catalítico Un dispositivo de emisiones ubicado en la parte delantera del silenciador en el sistema del escape. Se parece mucho a un silenciador de trabajo pesado y contiene loscatalizadores para limpiar las emisiones del motor antes de que salgan del tubo de escape.

Caustic Something that causes corrosion.

Cáustico Algo que causa la corrosión.

CC-ing The process of measuring a volume in cubic centimeters, commonly used to measure combustion chamber size.

Medida en CC El proceso de medir un volumen en centímetros cúbicos, suelen usarse al medir el tamaño de la cámara de combustión.

Center link A steering linkage component connected between the pitman and idler arm.

Varilla central Un componente de enlace de dirección que conecta entre el brazo Pitman y el brazo loco.

Center of gravity The point about which the weight of a car is evenly distributed; the point of balance.

Centro de gravedad El punto en el cual el peso del coche es distribuido en una manera uniforme; el punto del equilibrio.

Centrifugal force A force tending to pull an object outward when it is rotating rapidly around a center.

Fuerza centrífuga Una fuerza que tiene la tendencia de jalar un objeto hacia afuera cuando éste gira rápidamente alrededor de un punto central.

Centripetal force A force that acts on something and keeps it in a circular motion by pulling it toward the center of the circle.

Fuerza centrípeta Una fuerza que actua sobre algo y lo mantiene moviendo circularmente tirandolo hacia el centro del círculo.

Cetane The rating used to measure a diesel fuel's ignition quality. The higher the cetane number, the shorter the ignition lag time from the point the fuel enters the combustion chamber until it ignites.

Cetano Usado para medir la calidad de arranque de un combustible diesel. Mientras más alto el índice de cetano, más corto es el tiempo de demora entre el momento en que el combustible entra a la cámara de combustión y el momento en que enciende.

CFC Chlorofluorocarbon; CFCs are depleting the protective ozone layer through a chemical reaction.

Clorofluorocarburo Compuesto que está agotando la capa protectora de ozono mediante una reacción química.

Chamfer A bevel or taper at the edge of a hole.

Chaflán Un bisel o ahusamiento en la orilla de un hoyo.

Chamfering The process of removing the sharp edges around a bore or hole.

Achaflanar El proceso de quitar las orillas afiladas alrededor de un orificio o un hoyo.

Charcoal canister A small plastic or steel container filled with activated charcoal that can store gasoline vapors until the right time for them to be drawn into the engine and burned.

Bote de carbón vegetal Un pequeño recipiente de plástico o acero, lleno de carbón activado, que puede almacenar vapores de gasolina hasta el momento adecuado para su introducción al motor y su combustión.

Chase To straighten or repair damaged threads.

Embutir Enderezar o reparar los filetes dañados.

Chassis ground The use of the vehicle's frame and/or body as a common connection to the negative terminal of a battery.

Tierra de chasis el uso del marco y/o el cuerpo del vehículo como una conexión común a la terminal negativa de la batería.

Check valve A gate or valve that allows passage of gas or fluid in one direction only.

Válvula de seguridad Una válvula que permite fluir la presión en un sólo sentido.

Chemical cleaning Cleaning that relies primarily on some type of chemical action to remove dirt, grease, scale, paint, or rust.

Limpieza química La limpieza que cuenta primariamente con la acción química para remover el lodo, la grasa, las incrustaciones, la pintura, o los óxidos.

Clamping diode A special diode used to prevent voltage spikes. It is typically installed in parallel to a coil, creating a bypass for the electrons during the time the circuit is opened.

Diodo de fijación Diodo especial usado para prevenir picos de voltaje. Por lo general, se instala paralelo a un serpentín creando así una desviación para los electrones durante el tiempo que durante el que el circuito permanece abierto.

Closed loop An electronic feedback system in which the sensors provide constant information on what is taking place in the engine.

Bucle cerrado Un sistema de reacción electrónico en el cual los detectores proveen información constante de lo que está pasando en el interior del motor.

Clutch An electromechanical device mounted on the air-conditioning compressor used to start and stop compressor action, thereby controlling refrigerant circulating through the system.

Embrague Un dispositivo electromecánico montado en el compresor del aire acondicionado, utilizado para comenzar y detener la acción del compresor, regulando así la circulación del refrigerante a través del sistema.

Clutch disc The part of a clutch that receives the driving motion from the flywheel and pressure plate assembly and transmits that motion to the transmission input shaft.

Disco del embrague La parte de un embrague que recibe el movimiento propulsor del volante y la asamblea del plato opresor y transmite esa acción a la flecha de entrada de la transmisión.

Clutch fork A forked lever that moves the clutch release bearing and hub back and forth.

Horquilla del embrague Una palanca bifurcada que mueve al balero collarín del embrague hacia afrente y hacia atrás.

Clutch release bearing A sealed, prelubricated ball bearing that moves the pressure plate release levers or diaphragm spring through the engagement and disengagement process.

Collarín del embrague Un cojinete esférico, prelubricado y sellado, que mueve las palancas de liberación del plato opresor o el resorte diafragma por medio del proceso del enganche y el desenganche.

Coefficient of friction A relative measurement of the friction developed between two objects in contact with each other.

Coeficiente de la fricción La medida relativa de la fricción que ocurre entre dos superficies que estan en contacto la una con la otra.

Coil pack The name commonly used to describe the assembly of multiple ignition coils on an engine with distributorless ignition.

Paquete de bobinas Término utilizado normalmente para describir el conjunto de varias bobinas en un motor con encendido sin distribuidor.

Cold cranking amps (CCA) A common method of rating most automotive starting batteries. This rating is based on the load in amperes that a battery is able to deliver for 30 seconds at 0°F (−17.7°C) without its voltage dropping below a predetermined level.

Amperes de arranque en frío (AAF) Método común para medir la mayoría de las baterías de arranque. Esta medición se basa en la carga, en amperes, que puede entregar una batería durante 30 segundos a 0°F (−17.7 °C) sin que su voltaje caiga más abajo de un nivel preestablecido.

Collector The portion of a bipolar transistor that receives the majority of electrical current.

Colector La porción de un transistor bipolar que recibe la mayoría del corriente eléctrico.

Combination valve An H-valve, used in some early air-conditioning systems, combining a suction throttling valve and an expansion valve.

Válvula combinada Una válvula en H utilizada en algunos sistemas iniciales de climatización (aire acondicionado), que combina una válvula reguladora de succión y otra de expansión.

Combustion Rapid oxidation with the release of energy in the form of heat and light.

Combustión Oxidación rápida, con liberación de energía en la forma de calor y luz.

Combustion chamber The space between the top of a piston and the cylinder head where the engine's combustion takes place.

Cámara de combustión El espacio entre la parte superior del pistón y la cabeza del cilindro en donde se lleva acabo la combustión.

Compound A mixture of two or more ingredients.

Compuesta Una mezcla de dos o más ingredientes.

Compression The act of reducing volume by pressure.

Compresión La acción de reducir el volumen por efectos de la presión.

Compression ratio The ratio of the volume in the cylinder above the piston when the piston is at bottom dead center to the volume in the cylinder above the piston when the piston is at top dead center.

Relación índice de compresión La relación del volumen en el cilindro arriba del pistón cuando el pistón está en el punto muerto inferior al volumen en el cilindro arriba del pistón cuando el pistón está en el punto muerto superior.

Compression stroke The second stroke of the four-stroke engine cycle, in which the piston moves from bottom dead center and the intake valve closes. This traps and compresses the air-fuel mixture in the cylinder.

Carrera de compresión La segunda carrera de un ciclo de un motor de cuatro tiempos, en la cual el pistón mueve del punto muerto inferior y se cierre la válvula de admisión. Esto encierre y comprime la mezcla de aire-combustible en el cilindro.

Concentric Two or more circles having a common center.

Concéntrico Dos círculos o más que comparten un centro común.

Condensation The process of a vapor becoming a liquid; the reverse of evaporation.

Condensación El proceso por el cual un vapor se convierte en un líquido; lo opuesto de la evaporación.

Condense To cool a vapor to below its boiling point. The vapor condenses into a liquid.

Condensar El proceso de enfriar un vapor a una temperatura más baja de su punto de ebullición. El vapor condensa, formando un líquido.

Condenser A capacitor made from two sheets of metal foil separated by an insulator.

Condensador Es un capacitador construido de dos láminas de metal separadas por un aislante.

Conductance test A test that measures conductance and provides a reliable indication of a battery's condition and is correlated to battery capacity. Conductance can be used to detect cell defects, shorts, normal aging, and open circuits, which can cause the battery to fail.

Prueba de conductancia Prueba que mide la conductancia y proporciona una indicación confiable de la condición de una batería y se correlaciona a la capacidad de la batería. La conductancia puede ser usada para detectar defectos, cortocircuitos, edad normal, y circuitos abiertos, que pueden hacer que la batería falle.

Conduction The movement of heat through a material.

Conducción El movimiento de calor a través de un material.

Conductor A device that readily allows for current flow.

Conductor Un dispositivo que permite fácilmente el flujo del corriente.

Connecting rod The link between the piston and crankshaft.

Biela La acoplación entre el pistón y la cigüeñal.

Continuously variable transmission (CVT) A transmission that automatically changes torque and speed ranges without requiring a change in engine speed. A CVT is a transmission without fixed forward speeds.

Transmisión continua variable (TCV) Transmisiones que automáticamente cambian de rangos de torque y velocidad sin requerir un cambio en la velocidad del motor. Una TCV es una transmisión sin velocidades delanteras fijas.

Continuity A term used to describe the presence of a completed circuit between two points.

Continuidad Término usado para describir la presencia de un circuito completo entre dos puntos.

Contraction A reduction in mass or dimension; the opposite of expansion.

Contracción Una reducción en la masa o en la dimensión; lo opuesto de la expansión.

Control arms Suspension parts that control coil spring action as a wheel is affected by road conditions.

Palanca de comando Las partes de suspención que controlan la acción del muelle de control al afectarse una rueda por las condiciones de la carretera.

Convection The transfer of heat by the movement of a heated object.

Convección El transferimiento de calor a través del movimiento de un objeto que ha sido calentado.

Coolant A mixture of water and ethylene glycol-based antifreeze that circulates through the engine to help maintain proper temperatures.

Fluido refrigerante Una mezcla del agua con el anticongelante a base de glicol etileno que circula por el motor y ayuda en mantener las temperaturas indicadas.

Cords The inner materials running through the plies that produce strength in the tire. Common cord materials are fiberglass and steel.

Cuerdas Las materiales interiores recorriendo por los pliegues que producen la solidez del neumático. Las materiales más comunes para las cuerdas son la fibra de vidrio y el acero.

Core The center of the radiator, made of tubes and fins, used to transfer heat from the coolant to the air.

Núcleo El centro de un radiador, hecho de tubos y aletas, que sirve para transferir el calor del líquido refrigerante a la atmósfera.

Core plug A plug used to seal openings where necessary for casting the object in a cast structure.

Tapón del núcleo Un tapón diseñado para sellar las aperturas necesarias para fundar el objeto en un molde.

Corrosivity The characteristic of a material that enables it to dissolve metals and other materials or burn the skin.

Corrosidad La característica de una materia que permite que disuelva a los metales u otras materias o quema la piel.

Counter EMF (CEMF) A force that is created in a rotating armature. The faster the armature spins, the more induced voltage is present in the armature. The induced voltage opposes, or is counter to, the battery's voltage. This limits the current following through the armature windings.

Contra EMF (CEMF) Fuerza que se crea en una armadura giratoria. Mientras más rápido gira la armadura, mas voltaje inducido está presente en la armadura. El voltaje inducido se opone, o es contrario al voltaje de la batería. Esto limita la corriente a través de los devanados de la armadura.

Counterweight Weight forged or cast into the crankshaft to reduce vibration.

Contrapeso Un peso que se ha forjado o colado al cigueñal para reducir la vibración.

Crank pin The machined, offset area of a crankshaft where the connecting rod journals are machined.

Espiga de manívela La superficie maquinada, descentrada de un cigueñal en donde los muñones de la biela se maquinan.

Crank throw The distance from the crankshaft main bearing centerline to the connecting rod journal centerline. The stroke of any engine is the crank throw.

Carrera La distancia del eje de quilla del cojinete principal del cigueñal al eje de quilla del muñon de la biela. Una carrera del pistón de cualquier motor es una carrera.

Cranking amps (CA) rating A method of rating automotive starting batteries. This rating is based on the load in amperes that a battery is able to deliver for 30 seconds at 32°F (0°C) without its voltage dropping below a predetermined level.

Tasa de amperes de arranque en frío (CA) Método que consiste en medir baterías automotrices de arranque. Esta tasa se basa en la carga, en amperes, que una batería puede entregar durante 30 segundos a 32°F (0°C) sin que su voltaje caiga más abajo de un nivel predeterminado.

Crankshaft A rotating component mounted in the lower side of the block that changes vertical piston motion to rotary motion.

Cigüeñal Un componente giratorio montado en la parte inferior del bloque que convierte el movimiento vertical del pistón en movimiento giratorio.

Crimp The use of pressure to force a thin holding part to clamp to, or conform to the shape of, a part so it cannot move.

Engarzar El uso de la presión en forzar una parte delgada de sujeción para agarrar, o conformar a la forma de, una parte sujetada.

Cross counts The number of times that the O_2 voltage signal changes above or below 0.45 volt in a second.

Cuentas cruzadas Número de veces que la señal O_2 del voltaje cambia arriba o debajo de 0.45 voltios en un segundo.

Crude oil The natural state of oil as it is pulled from the earth and before it is refined.

Aceite crudo El estado natural del aceite cuando sale de la tierra y antes de que sea refinado.

Curb weight The weight of the vehicle when it is not loaded with passengers or cargo.

Peso en vacío Peso del vehículo cuando no está cargado con pasajeros o carga.

Current The number of electrons flowing past a given point in a given amount of time.

Corriente El número de los electronos que fluyen al través de un punto específico en un dado periódo del tiempo.

Cutpoint The report from an I/M 240 test that shows the amount of gases emitted during the different speeds and loads of the test. It also shows the average output for each of the gases and the cutpoint, which is the maximum allowable amount of each gas.

Punto de corte Informe de una prueba I/M 240 que indica la cantidad de gases emitida durante las diferentes velocidades y cargas de la prueba. También indica el promedio de salida para cada uno de los gases y el punto de corte, la cantidad máxima permitida de cada gas.

CV joint Constant velocity joint. A flexible coupling between two shafts that permits each shaft to maintain the same driving or driven speed regardless of operating angle, allowing for a smooth transfer of power. The CV joint consists of an inner and outer housing with balls in between or a tripod and yoke assembly.
Junta homocinética Junta de velocidad constante: un acoplamiento flexible entre dos ejes que permite que la velocidad entre ellos sea constant en todo momentosin importar el ángulo de operación, lo cual permite una transferencia regulada de la energía. La velocidad constante consiste de una carcasa interna y externa con balines en el medio o un ensamble de trípode y horquilla.
Cylinder A circular tubelike opening in a compressor block or casting in which the piston moves up and down or back and forth; a circular drum used to store refrigerant.
Cilindro Una apertura circular parecida a un tubo en un bloque del compresor o una pieza en los que el pistón se mueve de arriba hacia abajo o de un lado a otro; un tambor circular utilizado para el almacenaje de anticongelante.
Cylinder head On most engines the cylinder head contains the valves, valve seats, valve guides, valve springs, and the upper portion of the combustion chamber.
Cabeza de los cilindros En la mayoría de los motores, la cabeza de los cilindros contiene las válvulas, asientos de las válvulas, sus guías, sus resortes y la parte superior de la cámara de combustión.
Cylindrical cell A type of electrochemical cell in which the electrodes are rolled together and fit into a metal cylinder. A separator soaked in an electrolyte is placed between the plates.
Célula cilíndrica Un tipo de célula electroquímica en la cual los electrodos están enrollados juntos y caben en un cilindro de metal. Un separador empapado en un electrolito se coloca entre las placas.
Dampen To slow or reduce oscillations or movement.
Amortiguar Retardar o reducir las vibraciones o el movimiento.
Dead axle An axle that does not rotate but merely forms a base on which to attach the wheels.
Eje muerto Un eje que no gira sino solamente forma una plataforma en la cual se puede conectar las ruedas.
Deceleration The rate of decrease in speed.
Desaceleración El índice de la disminución de la velocidad.
Deck The top of the engine block where the cylinder head mounts.
Cubierta del monoblock La parte superior de un bloque del motor en donde se conecta la culata.
Deflection Bending or movement away from normal due to loading.
Desviación Curvación o movimiento fuera de lo normal debido a la carga.
Deflection angle The angle at which the oil is deflected inside the torque converter during operation. The greater the angle of deflection, the greater the amount of torque applied to the output shaft.
Ángulo de desviación El ángulo en el que se desvia el aceite en el convertidor de par durante su operación. Cuanto más el ángulo de desviación, más torsión se aplica en la flecha de potencia.
Density Compactness; relative mass of matter in a given volume.
Densidad La firmeza; la masa relativa de la materia en un volumen indicado.
Desiccant A special substance that absorbs moisture.
Desicante Una substancia especial que absorbe la humedad.
Detergent A compound of soaplike nature used in engine oil to remove engine deposits and hold them in suspension in the oil.
Detergente Una compuesta parecida al jabón que se usa en el aceite de motor para quitar los depósitos del motor y mantenerlos suspendidos en el aceite.
Detonation As used in an automobile, indicates a hasty burning or explosion of the mixture in the engine cylinders. It becomes audible through a vibration of the combustion chamber walls and is sometimes confused with a ping or spark knock.
Detonación Aplicado al automóvil, indica una combustión apresurada o una explosión de la mezcla en los cilindros del motor. Se oye por medio de las vibraciones de los paredes de la cámara de combustión y a veces se confunde con un golpeteo o una detonación de las bujías.
Diagnosis A way of looking at systems that are not functioning properly and finding out why. It is based on an understanding of the purpose and operation of the system that is not working properly.
Diagnóstico Forma de observar los sistemas que no están funcionando adecuadamente, y averiguar el porqué. Se basa en la comprensión del propósito y la operación del sistema que no está funcionando adecuadamente.
Dial caliper Versatile measuring instrument capable of taking inside, outside, depth, and step measurements.
Calibre de carátula (pie de rey) Un instrumento de medir versátil que es capable de tomar las medidas interiores, exteriores, de profundidad, y de paso a paso.
Dial indicator A measuring tool used to adjust small clearances up to 0.001 inch. The clearance is read on a dial.
Calibrador de carátula Una herramienta de medida que sirve para ajustar las pequeñas holguras hasta el 0.001 de una pulgada. La holgura se lee en un cuadrante.
Diaphragm A flexible, impermeable membrane on which pressure acts to produce mechanical movement.
Diafragma Una membrana flexible e impermeable sobre la cual tiene un efecto la presión para producir un movimiento mecánico.
Dielectric An insulator material.
Dieléctrico Una material de aislador.
Diesel fuel Fuel for diesel engines obtained from the distillation of crude oil. It is composed of hydrocarbons and its efficiency is measured by cetane number.
Combustible Diésel Combustible para motores diesel que se obtiene a partir de destilación del petróleo crudo. Está compuesto por hidrocarburos y su eficiencia se mide en número de cetano.
Differential A gear assembly that transmits power from the drive shaft to the wheels and allows two opposite wheels to turn at different speeds for cornering and traction.
Diferencial Una asamblea de engrenajes que transmite la potencia del árbol de mando a las ruedas y permite que dos ruedas opuestas giran a velocidades distinctos para ejecutar las vueltas y la tracción.
Diffusion The random movement of gas particles; also ensures that any two gases sharing the same container will totally mix.
Difusión El movimiento al azar de las partículas de gas también asegura que cualquier dos gases que comparten el mismo envase se mezclarán totalmente.
Digital A voltage signal that is pulsed on or off.
Digital Un señal del voltaje que pulsa prendida o apagada.
Dilution A thinner or weaker solution. Oil is diluted by the addition of fuel and water droplets.
Dilución Desleir o debilitar. El aceite se diluye por la agregación de las gotas del combustible o del agua.
Diode A simple semiconductor device that permits flow of electricity in one direction but not in the opposite direction.
Diodo Un dispositivo sencillo de semiconductor que permite el flujo de la electricidad en una dirección pero no en la dirección opuesta.
Direct current (DC) A type of electrical power used in mobile applications. A unidirectional current of substantially constant value.
Corriente continua Un tipo de potencia eléctrica utilizada en aplicaciones movibles. Una corriente que fluye en un solo sentido de un valor substancialmente constante.
Direct drive The downward gear engagement in which the input shaft and output shaft are locked together.
Transmisión directa El enganchamiento en descenso del engrenaje en el cual la flecha de entrada y la flecha de producción se enclavan.
Direct injection A diesel fuel injection system in which fuel is injected directly onto the top of the piston.
Inyección directa Un sistema de inyección de diesel en el cual el combustible se inyecta directamente arriba del pistón.
Direct ignition system (DIS) A distributorless ignition system in which spark distribution is controlled by the vehicle's computer.
Encendido directo Un sistema del encendido sin distribuidor en el cual la distribución de la chispa se controla por medio de la computador del vehículo.
Directional stability The ability of a car to travel in a straight line with a minimum of driver control.
Estabilidad direccional La abilidad de un coche de viajar en una linea recta con control mínimo del conductor.
Disc brakes Brakes in which the frictional forces act on the faces of a disc.
Frenos de disco Frenos en los que las fuerzas de fricción se aplican sobre las caras de un disco.
Discharge line Connects the compressor outlet to the condenser inlet.
Línea de descarga La que conecta la salida del compresor a la entrada del condensador.
Displacement The volume the cylinder holds between the top dead center and bottom dead center positions of the piston.
Desplazamiento (cilindrada) El volumen que contiene el cilindro entre la posiciones del punto muerto superior y el punto muerto inferior del pistón.
Distributor The mechanism within the ignition system that controls the primary circuit and directs the secondary voltage to the correct spark plug.

Distribuidor El mecanismo dentro del sistema de ignición que controla el circuito primario y dirige el voltaje secundario a las bujías correctas.

Distributor ignition (DI) system SAE J1930 terminology for an ignition system with a distributor.

Sistema de la ignición con distribuidor El término utilizado por la SAE J1930 para referirse a un sistema de la ignición que tiene un distribuidor.

DOHC Dual overhead camshaft.

DOHC Arbol de levas doble superior.

Dome Typically refers to the shape of the top of a piston.

Fondo hemisférico Típicamente se refiere al contorno de la parte superior de un pistón.

DOT U.S. Department of Transportation.

DOT Departamento de Transportes de los Estados Unidos de América.

Dowel A pin extending from one part to fit into a hole in an attached part; used for both location and retention.

Espiga Una clavija que extiende de una parte para quedarse en el hoyo de otra parte conjunta; se usa para la localización y la retención.

Drive cycle The operating conditions that must exist before OBD-II self-diagnosis can take place. It includes an engine start and operation that brings the vehicle into closed loop and includes whatever specific operating conditions are necessary, either to initiate and complete a specific monitoring sequence or to verify a symptom or repair.

Ciclo de manejo Las condiciones de operación que deben existir antes de que se lleve a cabo el auto diagnóstico OBD-II, Incluye un arranque y operación de motor que trae al vehículo a una curva cerrada e incluye las condiciones de operación específicas necesarias ya sea para encender o terminar una secuencia de monitoreo completa o para verificar un síntoma o una reparación.

Drive member A gear that drives, or provides power for, other gears in a planetary gearset.

Miembro de ataque Un engrenaje que impulsa, o provee la potencia para los otros engrenajes en un conjunto planetario.

Driveability The degree to which a vehicle operates properly. Includes starting, running smoothly, accelerating, and delivering reasonable fuel mileage.

Manejo El punto en el cual un vehículo opera correctamente. Incluye el encendido, marchar sin averías, la aceleración y la entregada de un kilometraje económico.

Drive shaft A hollow metal tube that has a universal joint at each end.

Eje motriz Un tubo metálico hueco que tiene una junta universal en cada extremo.

Dry sump An oil pan or sump that does not store oil. It merely seals the bottom of the crankcase.

Cárter seco Colector o recipiente que no guarda aceite. Simplemente sella la parte inferior del cárter.

Duo-servo A drum brake design with increased stopping power due to the servo or self-energizing effect of the brake.

Frenos tipo servo Un diseño de freno de tambor que tiene más potencia de enfrenar debido al servo o al efecto autoenergético del freno.

Duty cycle The percentage of on-time to total cycle time of fuel injectors.

Ciclo del trabajo El porcentaje del tiempo de trabajar al tiempo del ciclo total de los inyectores del combustible.

Dwell time The degree of crankshaft rotation during which the primary circuit is on.

Ángulo de cierre El grado de rotación del cigueñal en el cual esta encendido el circuito primario.

Dynamic Refers to balance when the object is in motion.

Dinámico Refiere al balance mientras que un objeto esté en movimiento.

Dynamic pressure The pressure of a fluid while it is in motion.

Presión dinámica La presión de un líquido mientras que está en movimiento.

Dynamometer An instrument for measuring mechanical power.

Dinamómetro Un instrumento para medir el poder mecánico.

Eccentric The part of a camshaft that operates the fuel pump.

Eccéntrico La parte de un árbol de levas que opera la bomba del combustible.

Efficiency A ratio of the amount of energy put into an engine as compared to the amount of energy coming out of the engine; a measure of the quality of how well a particular machine works.

Eficiencia Una relación de la cantidad de energía que consume un motor comparado con la cantidad de energía que produce el motor; una medida de la calidad de la máquina y su marcha.

EGR Exhaust gas recirculation. The system that allows a small amount of exhaust gas to be routed into the incoming air-fuel mixture to reduce NO_x emissions.

EGR Sistema de recirculación de gas del escape. Permite que se dirija una cantidad pequeña del gas del escape a la mezcla de admisión de aire y combustible para reducir descargas (emisiones) de NO_x.

EI The J1930 acronym for a distributorless ignition system.

EI La abreviación J1930 para un sistema de la ignición sin distrubuidor.

Elasticity The principle by which a bolt can be stretched a certain amount. Each time the stretching load is reduced, the bolt returns back to its exact, original, normal size.

Elasticidad El principio por el cual un perno puede extenderse por un cierta cantidad. Cada vez que la carga tensor se reduce, el perno regresa exactamente a su tamaño normal y original.

Electrochemical The chemical action of two dissimilar materials in a chemical solution.

Electroquímico La acción química de dos materiales desemejantes en una solución química.

Electrochemical degradation (ECD) An electrochemical attack on rubber cooling system hoses. It occurs because the hose, engine coolant, and the engine/radiator fittings form a galvanic cell. This chemical reaction causes very small cracks in the hose, allowing the coolant to attack and weaken the reinforcement in the hose.

Degradación electroquímica Ataque electroquímico en mangueras de hule del sistema de enfriamiento. Ocurre porque la manguera, el enfriador del motor, y los conectores del motor/radiador forman una célula galvánica. Esta reacción química causa grietas muy pequeñas en la manguera, lo que permite que el enfriador ataque y debilite el refuerzo de la manguera.

Electrode The firing terminals found in a spark plug.

Electrodo Los terminales de chispeo que se encuentran en una bujía.

Electrolysis A chemical and electrical decomposition process that can damage metals, such as brass, copper, and aluminum, in the cooling system.

Electrólisis (deposición) Un proceso de decomposición química y eléctrica que puede dañar a los metales, tales como el latón, el cobre y el aluminio, en el sistema de enfriamiento.

Electrolyte A material that has atoms that become ionized, or electrically charged, in solution. Automobile battery electrolyte is a mixture of sulfuric acid and water.

Electrólito Una materia que tiene átomos que se han ionizado, o que tienen una carga eléctrica, en una solución. El electrólito para las baterías automotrices es una mezcla del ácido sulfúrico y el agua.

Electromagnet A magnet formed by electrical flow through a conductor.

Electroimán Un imán que se produce por un corriente eléctrico fluyendo por un conductor.

Electromagnetic induction Moving a wire through a magnetic field to create current flow in the wire.

Inducción electromagnético Crear un corriente eléctrico en un alambre al moverlo por un campo magnético.

Electromagnetism A form of magnetism that occurs when current flows through a conductor.

Electromagnetismo Una forma del magnetismo que ocurre al fluir un corriente por un conductor.

Electromechanical Refers to a device that incorporates both electronic and mechanical principles together in its operation.

Electromecánico Refiere a un dispositivo que incorpora ambos principios del electrónico y mecánico en su operación.

Electromotive force (EMF) A technical name for voltage.

Fuerza electromotriz (FEM) Nombre técnico para voltaje.

Electronic Pertaining to the control of systems or devices by the use of small electrical signals and various semiconductor devices and circuits.

Electrónico Perteneciente al control de los sistemas o de los dispositivos por medio de los pequeños señales eléctricos y por varios de los dispositivos semiconductores y los circuitos.

Electronic fuel injection (EFI) A generic term applied to various types of fuel injection systems.

Inyección electrónica de combustible Un término general aplicado a varios sistemas de inyección de combustible.

Element A substance with only one type of atom.

Elemento Una sustancia con solamente un tipo de átomo.

Embedability The ability of the bearing lining material to absorb dirt.

Incrustabilidad La abilidad que tiene una material de forro de un cojinete en absorber la suciedad.

Emitter A portion of a transistor from which electrons are emitted, or forced out.

Emisor Una porción de un transistor de la cual se emite o se expulsan los electrones.

Enable criteria The operating conditions that must be met to run some OBD-II monitors.

Criterios de activación Condiciones de operación requeridas para el funcionamiento de algunos monitores OBD-II.

End play The amount of axial or end-to-end movement in a shaft due to clearance in the bearings.

Juego axial La cantidad del movimiento axial o de una extremidad a otra de una flecha debido a la holgura de los cojinetes.

Energy The ability to do work.

Energía La abilidad de trabajar.

Engine block The main structure of the engine that houses the pistons and crankshaft. Most other engine components attach to the engine block.

Monobloque La estructura principal del motor que contiene los pistones y el cigüeñal. La mayoría de los otros componentes del motor se conectan al monobloque.

Engine efficiency A measure of the relationship between the amount of energy put into the engine and the amount of energy available from the engine.

Eficacia del motor Una medida de la relación entre la cantidad de energía puesta en el motor y la cantidad de energía disponible del motor.

EP toxicity The characteristic of a material that enables it to leach one or more of eight heavy metals in concentrations greater than 100 times standard drinking water concentrations.

Toxicidad EP La característica de una materia que permite que disuelva uno o más de los ocho metales pesados en concentraciones mayores de 100 veces la concentración del agua potable.

Equilibrium Exists when the applied forces on an object are balanced and there is no overall resultant force.

Equilibrio Existe cuando las fuerzas aplicadas en un objeto es equilibradas y no hay fuerza resultante total.

Ethanol A widely used gasoline additive known for its abilities as an octane enhancer.

Etanol Un aditivo muy común que se reconoce por su capacidad de elevar los niveles del octano.

Evacuate The process of applying vacuum to a closed refrigeration system to remove air and moisture.

Vaciar El proceso de aplicar un vacío a un sistema cerrado de refrigeración para quitarle el aire y la humedad.

Evaporate Action of atoms or molecules when they break free from the body of the liquid to become gas particles.

Evaporar Acción de los átomos o las moléculas cuando se escapan del cuerpo de un líquido para convertirse en partículas del gas.

Evaporation A natural process in which the moisture contained by an object leaves and enters the atmosphere.

Evaporación Un proceso natural en el cual la humedad deja un objeto y entra la atmósfera.

Evaporator The component of an air-conditioning system that conditions the air.

Evaporador El componente en un sistema de aire acondicionado que acondiciona el aire.

Exhaust manifold A component that collects and then directs engine exhaust gases from the cylinders.

Múltiple de escape Un componente que colecciona y luego dirige los gases de escape del motor desde los cilindros.

Exhaust valve An engine part that controls the expulsion of spent gases and emissions out of the cylinder.

Válvula de escape Una parte del motor que controla la expulsión de los gases consumidos y las emisiones afuera del cilindro.

Expansion An increase in size. For example, when a metal rod is heated, it increases in length and perhaps also in diameter; expansion is the opposite of contraction.

Expansión Un aumento en el tamaño. Por ejemplo, al calentarse una varilla de metal, aumenta en longitud y quizá en su diámetro tambien; la expansión es lo contrario de la contracción.

Expansion tank A small tank sometimes located inside the main gas tank that allows for expansion of fuel in a full tank on a hot day.

Tanque de dilatación Un tanque pequeño que suele estar al interior del depósito principal de gasolina y que permite que se dilate el combustible cuando el depósito esté lleno en días calurosos.

Expansion valve A component in an air-conditioning system used to create a pressure on one side and reduce the pressure on the other side.

Válvula de expansión Un componente de un sistema de aire acondicionado que sirve para crear presión en un lado y reducirla en el otro.

Fail-safe circuit The circuit that is found in some hydraulic systems that allows limited operation when a component or components have failed.

Circuito de seguridad en falla Circuito que se encuentra en algunos sistemas hidráulicos, que permite operación limitada cuando uno o varios componentes han fallado.

Fan clutch A device used on engine-driven fans to limit their terminal speed, reduce power requirements, and lower noise levels.

Embrague del ventilador Un dispositivo utilizado en los ventiladores accionados por motores para limitar su velocidad terminal, disminuir los requisitos de potencia, y bajar los niveles de ruido.

Fatigue Deterioration of a bearing metal under excessive intermittent loads or prolonged operation.

Fatiga La deterioración de un metal de apoyo bajo las cargas intermitentes o durante la operación prolongada.

Feedback Normally refers to the process in which computer commands and the results of such are monitored by the computer.

Retroalimentación Normalmente se refiere al proceso en el cual la computadora manda y los resultados se regulan por la computadora.

Ferrous metal A metal that contains iron or steel and is, therefore, subject to rust.

Metal férreo Un metal que contiene el hierro o el acero y que, por lo tanto, puede oxidarse.

Field coil A coil of wire on an alternator rotor or starter motor frame that produces a magnetic field when energized.

Bobina inductora Una bobina de alambre en un rotor de un alternador o un armazón de un motor de arranque que produce un campo magnético al establecer el corriente.

Fillets The smooth curve where the shank flows into the bolt head.

Filetes La curva lisa en donde el asta se convierte a la cabeza del tornillo.

Final drive The final set of reduction gears the engine's power passes through on its way to the drive wheels.

Engrane o velocidad final El ultimo juego de reductores por el cual pasa la fuerza del motor en camino a las ruedas motrices.

Firing order The order in which the cylinders of an engine move through the power stroke.

Orden de encendido El orden en el cual mueven los pistones de un motor en la carrera encendido-expansión.

Flange A projecting rim or collar on an object that keeps it in place.

Collarín Una orilla sobresaliente o un collar en un objeto que lo sujeta en su lugar.

Flare An expanded, shaped end on a metal tube or pipe.

Abocinado (abocardado) Una extremidad extendida o formada en un tubo de metal.

Flat-rate manuals Literature containing figures dealing with the length of time specific repairs are supposed to require. Flat-rate manuals often contain a parts list with prices as well.

Manuales de valuación La literatura que se trata de los detalles del tiempo de obra requirido por varias reparaciones específicas. Los manuales de valuación suelen incluir tambien una lista de repuestas con sus precios.

Flexible fuel vehicles (FFV) A vehicle that can operate on either methanol or ethanol and regular gasoline or any combination of the two from the same tank.

Vehículos de combustible flexible (VCF) Un vehículo que puede operar con metanol o etanol y gasolina común o cualquier combinación de ambos en el mismo tanque.

Flexplate A round, flywheel-like disc mounted to an engine's crankshaft when the vehicle is equipped with an automatic transmission.

Placa de flexión Una placa redonda parecida al volante montada al cigueñal de un motor de un vehículo que ha sido equipado con una transmisión automática.

Flooding A condition in which excess, unvaporized fuel in the intake manifold prevents the engine from starting.

Ahogo Una condición en la cual un exceso de combustible no vaporizado se queda en el múltiple de admisión y previene que arranque el motor.

Fluid Something that does not have a definite shape; therefore, liquids and gases are fluids.

Fluido Algo que no tiene una forma definida, por lo tanto los líquidos y los gases son fluidos.

Flux density The number of flux lines per square centimeter.

Densidad de flujo El número de lineas de flujo por centímetro cuadrado.

Flux field The magnetic field formed by magnetic lines of force.

Campo de flujo El campo magnético formado por líneas de fuerza magnética.

Flywheel A heavy circular component located on the rear of the crankshaft that keeps the crankshaft rotating during nonproductive strokes.

Volante del motor Un componente pesado redondo, ubicado en la parte trasera del cigüeñal que mantiene el cigüeñal girando durante las carreras no productivas.

Foot-pound A unit of measurement for torque. One foot-pound is the torque obtained by a force of 1 pound applied to a wrench handle 12 inches (one foot) long.

Pie-libra Una unidad de medida de torsión. Un pie-libra es la torsión que se obtiene por la fuerza de una libra que se aplica a una llave de tuercas cuyo mango mide 12 pulgadas en longitud.

Force A push or pushing effort measured in pounds.

Fuerza Un esfuerzo de jalar o empujar que se mide en libras.

Forge The process of shaping metal by stamping it into a desired shape.

Forgar El proceso de formar el metal al imprimirlo en una forma requerida.

Forward bias A positive voltage that is applied to the P-material and a negative voltage that is applied to the N-material in a semiconductor.

Polarización directa Un voltaje positivo que se aplica al cristal P y un voltaje negativo que se aplica al cristal N de un semiconductor.

Free play Looseness in a linkage between the start of application and the actual movement of the device, such as the movement in the steering wheel before the wheels start to turn.

Juego libre Flojedad en una biela que aparece entre el tiempo en que se comienza usar a una aplicación y el movimiento actual del dispositivo, tal como el movimiento en un volante de dirección antes de que comienzan a girar las ruedas.

Free travel The distance a clutch pedal moves before it begins to take up slack in the clutch linkage.

Juego libre La distancia que mueve el pedal de embrague antes de que comienza accionar la biela del embrague.

Freewheeling A mechanical device that engages the driving member to impart motion to a driven member in one direction but not the other. Also known as an overrunning clutch.

Embrague de volante libre Un dispositivo mecánico que engrana el miembro de impulso con el miembro de tracción y confiere el movimiente en una dirección, pero no en la otra. Tambien se conoce como un embrague de sobremarcha.

Frequency The rate at which something occurs.

Frecuencia La velocidad en la cual algo ocurre.

Friction The resistance to motion that occurs when two objects rub against each other.

Fricción La resistencia al movimiento que ocurre al frotarse dos objetos.

Fuel cell stack An assembly of several hundred fuel cells connected in series and layered or stacked next to each other. Each fuel cell produces electricity and the combined output of the cells is used to power a vehicle.

Pila de células de combustible Ensamble de varios cientos de células de combustible en una serie, acomodados o apilados uno junto a otro. Cada célula de combustible produce electricidad y el resultado de las células se usa para suministrar energía a un vehículo.

Fuel pressure regulator A device designed to limit the amount of pressure buildup in a fuel delivery system.

Regulador de presión del combustible Un dispositivo diseñado para limitar la cantidad de acumulación de presión en un sistema de suministro de combustible.

Fuel pump A mechanical or electrical device used to move fuel from the fuel tank to the carburetor or injectors.

Bomba de combustible Un dispositivo mecánico o eléctrico que se utiliza para llevar combustible del depósito al carburador o los inyectores.

Fuel rail A metal or plastic pipe in which the upper ends of the injectors are installed in port injection systems.

Carril del combustible Un tubo de metal o plástico en el cual se instalan las extremidades superiores de los inyectores en los sistemas de inyección de puerto/de lumbreras.

Fulcrum The support or point of rest on which a lever rests; also called the pivot point.

Punto de apoyo El soporte o punto de descanso en el cual descansa una palanca; tambien se llama el punto de pivote.

Full-floating A type of live axle arrangement in which the weight of the vehicle is not supported by the axle.

Flotante Un tipo de dispositivo de eje motor en el cual el peso del vehículo no se soporta por el eje.

Fuse An electrical device used to protect a circuit against accidental overload or unit malfunction.

Fusible Un dispositivo eléctrico utilizado para proteger un circuito contra una sobrecarga imprevista o un mal funcionamiento de la unidad.

Fusible link A type of fuse made of a special wire that melts to open a circuit when current draw is excessive.

Fusible térmico El tipo de fusible fabricado de un alambre especial que se funde para abrir un circuito cuando ocurre una sobrecarga del circuito.

Galling wear Uniting two solid surfaces that are in rubbing contact; used to describe the normal wear of valve lifters.

Desgaste por rozamiento Uniendo a dos superficies sólidos que están en contacto frotativo; se usa para describir el gasto normal de las levantaválvulas.

Ganged A type of switch in which several circuits are controlled by moving one control or lever.

Acoplados Un tipo de interruptor en el cual varios circuitos se controlan al mover un control o una palanca.

Gasket A thin layer of material or composition that is placed between two machined surfaces to provide a leakproof seal between them.

Junta Una capa delgada de material o compuesto que se coloca entre dos superficies rectificadas para proveer una junta hermética para evitar fugas entre ellas.

Gateway A module that allows for data exchange between different buses. It translates a message on one bus and transfers that message to another bus without changing the message. The gateway interacts with each bus according to the protocol of that bus.

Puerta de enlace Módulo que permite el intercambio de información entre diferentes camiones. Traduce un mensaje en un camión y lo transfiere a otro camión sin cambiar el mensaje. La puerta de enlace interactúa con cada camión según el protocolo de dicho camión.

Gear A wheel with external or internal teeth that serves to transmit or change motion.

Engranaje Rueda con dientes externos o internos que sirve para transmitir o modificar un movimiento.

Gear pitch The number of teeth per given unit of pitch diameter. Gear pitch is determined by dividing the number of teeth by the pitch diameter of the gear.

Paso del engrenaje El número de los dientes por una unidad dada de un diámetro de paso. El paso del engrenaje se determine al dividir el número de los dientes por el diámetro del paso del engrenaje.

Gear ratio An expression of gear size and tooth count of gears that are meshed together. The ratio reflects torque multiplication.

Relación de engranes Una expresión del tamaño del engrenaje y el número de los dientes que se engranan. La relación refleja la multiplicación de par.

Generator An electrical device that produces alternating current that is rectified to DC current by the brushes and commutator.

Generador Un dispositivo eléctrico que produce corriente alterna que se rectifica como corriente continua mediante las escobillas y el conmutador.

Glaze A thin residue on the cylinder walls formed by a combination of heat, engine oil, and piston movement.

Porcelana (barniz) Un residuo en los paredes de los cilindros que se forma por una combinación del calor, el aceite del motor y el movimiento de los pistones.

Glitches Abnormal, slight movements of a waveform on a lab scope. These can be caused by circuit problems or noise in the circuit.

Fallos transitorios Pequeños movimientos, anormales, de una forma de onda en un ámbito de laboratorio. Estos los pueden causar problemas de circuito o ruido en el circuito.

Grade markings Marks on fasteners that indicate strength.

Marcos de grado Las marcas en las fijaciones que indican su fuerza.

Graphite Very fine carbon dust with a slippery texture used as a lubricant.

Grafito Polvo de carbón muy fino con textura resbalosa usado como lubricante.

Grease A combination of oil and a thickening agent.

Grasa Una combinación de aceite y un agente de espesamiento.

Greenhouse gas A name given to the effect that certain gases have on global warming.

Gas de invernadero Nombre que se le da al efecto que ciertos gases tienen en el calentamiento global.

Ground The negatively charged side of a circuit. A ground can be a wire, the negative side of the battery, or even the vehicle chassis.

Tierra El lado de un circuito que tiene una carga negativa. Una tierra puede ser un alambre, el lado negativo de una batería, o hasta el chasis del vehículo.

Gum In automotive fuels, gum refers to oxidized petroleum products that accumulate in the fuel system, carburetor, or engine parts.

Sarro (goma) En los combustibles automotrices, el sarro se refiere a los productos petróleos oxidados que acumulan en el sistema de combustible, el carburador, o en las partes del motor.

Haldex clutch Found as the center differential in some all-wheel-drive vehicles, this unit distributes the drive force variably between two axles. It consists of a hydraulic pump, a wet multidisc clutch, and an electronically controlled valve.

Embrague Haldex Se encuentra como el diferencial central en algunos vehículos de tracción en todas las llantas, esta unidad distribuye la fuerza de manejo variada entre los dos ejes. Consiste en una bomba hidráulica, un embrague húmedo de muchos discos, y una válvula controlada electrónicamente.

Half shaft Either of the two drive shafts that connect the transaxle to the wheel hubs in FWD cars. Half shafts have constant velocity joints attached to each end to allow for suspension motions and steering. The shafts may be of solid or tubular steel and may be of different lengths. Balance is not critical, as half shafts turn at roughly one-third the speed of RWD drive shafts.

Semieje cualquiera De los dos ejes motores que conectan el dispositivo de acoplamiento intermedio a los cubos de las ruedas en los vehículos de tracción en las cuatro ruedas. Los semiejes tienen juntas de velocidad constante fijas a cada extremo para permitir los movimientos de la suspensión y la dirección. Los ejes pueden ser de acero tubular o sólido y tener longitudes diferentes. El equilibrio no es crítico, ya que los semiejes giran a aproximadamente un tercio de la velocidad de los ejes motores de los vehículos de tracción trasera.

Hall effect The consequence of moving current through a thin conductor that is exposed to a magnetic field; as a result of this, voltage is produced.

Efecto Hall La consecuencia de pasar un corriente por un conductor delgado que se expone a un campo magnético; por consecuencia, se produce el voltaje.

Halogen The term used to define a group of chemically related nonmetallic elements.

Halógeno El termino para describir un grupo de elementos no metálicos del mismo género químico.

Hand tap A tool used for hand cutting internal threads.

Macho de roscar de mano Una herramienta que sirve para cortar a mano las roscas interiores.

Hardening A process that increases the hardness of a metal, deliberately or accidentally, by hammering, rolling, carburizing, heat treating, tempering, or other physical processes.

Endurecer Un proceso que aumenta la dureza de un metal, deliberadamente o accidentalmente, por el martilleo, rodar, la carburación, el calor, templar, u otros procesos físicos.

Hard spots Areas in the friction surface of a brake drum or rotor that have become harder than the surrounding metal; sometimes called islands of steel.

Regiones endurecidos Las averías en la superficie de fricción del tambor o el rotor de un freno que se han endurecido más que el metal que los rodea; a veces se llaman islas de acero.

Harmonics Periods of vibration that occur when valve springs are opened and closed rapidly.

Armónico Los períodos de la vibración que ocurren cuando los resortes de las válvulas abren y cierran rápidamente.

Hazardous waste Any material found on the Environmental Protection Agency list of known harmful materials. Waste is also considered hazardous if it has one or more of the following characteristics: ignitability, corrosivity, reactivity, and EP toxicity.

Desechos tóxicos Cualquier material que se nombra en la lista de materiales dañosos de la agencia de protección del medio ambiente (EPA). Los desechos tambien se consideran tóxicos si tiene uno o más de las características siguientes: la inflamabilidad, la calidad corrosiva, la reactividad y la toxicidad EP.

Head gasket Gasket used to prevent compression pressures, gases, and fluids from leaking. It is located on the connection between the cylinder head and the engine block.

Junta de la cabeza Una junta que se emplea para prevenir que se escapen las presiones, los gases y los fluidos de la compresión. Se ubica en la conexión entre la cabeza de los cilindros y el monobloque.

Heat A form of energy caused by the movement of atoms and molecules.

Calor Una forma de energía causada por el movimiento de átomos y de moléculas.

Heat range A rating used to express how well a spark dissipates heat.

Rango de calor Una medida que expresa la abilidad de una chispa de dispersar el calor.

Heat shield An assembly designed to restrict the transfer of high heat from one component to another.

Pantalla térmica Un ensamblaje diseñado para restringir la transferencia de calor elevado de un componente a otro.

Heat sink A device used to dissipate heat and protect parts.

Fuente fría Un dispositivo que sirve para dispersar el calor y protejer las partes.

Heat treating The changing of the properties of a metal by using heat.

Tratamiento por calor El cambiar de las características de un metal usando calor.

Heater control valve A manual or automatic valve in the heater hose used for opening or closing, providing coolant flow control to the heater core.

Válvula de control del calentador Una válvula manual o automática en la manguera del calentador que se usa para abrir o cerrar, proporcionando control de flujo de refrigerante al núcleo del calentador.

Heater core A small heat exchanger in the passenger compartment through which engine coolant is circulated.

Núcleo del calentador Un pequeño intercambiador térmico en el compartimiento del pasajero por el que se hace circular el refrigerante del motor.

Helical gear A gear with teeth that are cut at an angle or are spiral to the gear's axis of rotation.

Engranaje helicoidal Un engranaje cuyos dientes han sido cortados en un ángulo o que son un espiral al eje de rotación del engrenaje.

Heptane A standard reference fuel with an octane number of zero, meaning that it knocks severely in an engine.

Eptano Un combustible de norma cuyo índice de octano es el cero, lo que significa que causa que el motor golpetea severamente.

Hertz (Hz) The unit by which frequency is most often expressed, equal to one cycle per second.

Hertzio (Hz) La unidad de medida que se utiliza para expresar frecuencia y es igual a un ciclo por segundo.

High-intensity discharge (HID) headlamps Headlamps that are also called Xenon headlamps. These are high-voltage lamps that produce about twice as much light as conventional headlamps.

Faros de descarga de alta intensidad Estos faros también se llaman faros de xénon. Son faros de alto voltaje que producen como el doble de luz que los faros convencionales.

High tension High voltage. In automotive ignition systems, voltages (up to 40 kilovolts) in the secondary circuit of the system as contrasted with the low, primary circuit voltage (nominally 6 or 12 volts).

Alta tensión El voltaje alto. En los sistemas del encendido, los voltajes (hasta los 40 kilovoltios) en el circuito secundario del sistema en contraste del voltaje bajo del circuito principal (de 6 o 12 voltios nominalmente).

Hone An abrasive for correcting small irregularities of differences in diameter in a cylinder, such as an engine cylinder or brake cylinder.

Piedra de rectificar Un abrasivo que sirve para corregir las pequeñas irregularidades en los diámetros de un cilindro, tal como un cilindro de un motor o un cilindro de freno.

Horsepower The rate at which torque is produced.

Caballo de fuerza El índice de la cual se produce el esfuerzo de torsión.

Hot spot Refers to a comparatively thin section of area of the wall between the inlet and exhaust manifold of an engine, the purpose being to allow the hot exhaust gases to heat the comparatively cool incoming mixture. Also, used to designate local areas of the cooling system that have attained above average temperatures.

Región de calor Refiere a una sección delgada en el muro entre la admisión y el escape del múltiple de un motor, su propósito es permitir que los gases calientes del escape calientan a la mezcla entrando que es comparativamente fría. Este término tambien sirve para designar las areas locales del sistema del enfriamento que han alcanzado las temperaturas más altas que las normales.

Hunting gear set A differential gear set in which one drive pinion gear tooth contacts every ring gear tooth after several rotations.

Conjunto de engranaje con diente suplementario Un conjunto de engranajes del diferencial en el cual cada diente del piñon de mando se engrena con cada diente del engranaje de corona después de varias rotaciones.

Hybrid electric vehicle (HEV) A vehicle with two distinct power sources, typically an electric motor and an internal combustion engine.

Vehículo híbrido eléctrico (HEV) Vehículo con dos fuentes distintas de energía. Por lo general, consta de un motor eléctrico y un motor interno de combustión.

Hydraulic hybrid Similar to a hybrid electric vehicle, except energy for the alternative power source is stored in tanks of hydraulic fluid under pressure. This vehicle has a hydraulic propulsion system.

Hibrido hidráulico Parecido a los vehículos híbridos eléctricos, excepto que energía para la fuente alterna se almacena en tanques de líquido hidráulico bajo presión. Estos vehículos tienen un sistema de propulsión hidráulica.

Hydrocarbons Particles of gasoline present in the exhaust and in crankcase vapors that have not been fully burned.

Hidrocarburos Las partículas de la gasolina presentes en los gases de escape y en los vapores de la caja de cigueñal que no se han quemado completamente.

Hydrofluorocarbon (HFC) The chemical family currently used as the source of refrigerant (R-134a) for air-conditioning systems. It contains no chlorine and causes less damage to the ozone layer when released to the atmosphere.

Hidrofluorocarbono (HFC) Familia química usada actualmente como fuente del refrigerante (R-134a) para sistemas de aire acondicionado. Contiene cloro y provoca menos daño a la capa de ozono cuando se libera a la atmosfera.

Hypoid gears A type of spiral, beveled ring and pinion gear set in a differential. Hypoid gears mesh below the ring gear centerline.

Engrenajes hipoíde Un conjunto de engrenajes de un diferencial de anillo y piñon de forma espiral y biselado. Los engrenajes hipoídes se engrenan abajo de la linea central de la corona.

Idler arm A steering linkage component fastened to the car frame that supports the right end of the center link.

Brazo intermedio Componente de enlace de dirección montado en el armazón del vehículo que sostiene el extremo derecho del enlace central.

Idler pulley A pulley used to tension or torque the belt(s).

Polea tensora Una polea utilizada para proveer tensión o par de torsión a la(s) banda(s).

Ignitability The characteristic of a solid that enables it to spontaneously ignite. Any liquid with a flash point below 140°F (60°C) is also said to possess ignitability.

Inflamabilidad La característica de un sólido que lo permite encenderse espontáneamente. Cualquier líquido con una temperatura de inflamabilidad menos del 140°F (60°C) posee la característica de la inflamabilidad.

Ignition coil A transformer containing a primary and secondary winding that acts to boost the battery voltage of 12 volts to as much as 30,000 volts to fire the spark plugs.

Bobina de encendido Un transformador contiene un devanado primario y otro secundario que actúan para reforzar el voltaje de la batería de 12 voltios a hasta 30,000 voltios para activar las bujías.

Ignition system The system that creates and distributes a timed spark to the cylinder.

Sistema de encendido Crea y distribuye una chispa sincronizada al cilindro.

Ignition timing The point at which ignition occurs in a cylinder.

Tiempo de encendido El punto en el que se produce el encendido en el cilindro.

Impedance Refers to the operating resistance of a component or piece of equipment. The higher the impedance, the lower the operating amperage.

Impedancia Se refiere a la resistencia de operación de un componente o parte de equipo. Mientras más alta la impedancia, menor el amperaje de operación.

Impeller The part of a torque converter that is driven by the engine's crankshaft.

Impulsor Una parte de un convertidor del par que es accionada por el cigueñal del motor.

Impermeable Characterístic of materials that adsorb fluids.

Impermeable Una característica de los materiales que fijan los líquidos por adsorción.

Inboard joint The CV joint that connects the drive axle to the transaxle of the differential assembly.

Junta interior La junta de velocidad constante que conecta el eje motor al árbol intermedio del conjunto del diferencial.

Included angle The sum of the angle of camber and steering axis inclination.

Ángulo comprendido La suma del ángulo del camber y la inclinación del eje de la dirección.

Induction The process of producing electricity through magnetism rather than direct flow through a conductor.

Inducción El proceso de producción de la electricidad por el magnetismo en vez de un corriente directo por un conductor.

Inductive reluctance A characteristic of a material that defines how easily it can be magnetized.

Renuencia inductiva Característica de un material que define qué tan fácilmente se puede magnetizar.

Inertia The constant moving force applied to carry the crankshaft from one firing stroke to the next.

Inercia La fuerza del movimiento constante aplicada a un cigueñal para progresar de una carrera a la próxima.

Inertia switch A switch that automatically shuts off the fuel pump if the vehicle is involved in a collision or rolls over.

Conmutador inercia Un interruptor que automáticamente apaga a la bomba de combustible si el vehículo se involucra en un choque o si se voltea.

Insert bearing A bearing made as a self-contained part and then inserted into the bearing housing.

Cojinete inserta Un cojinete fabricado como una parte autónoma y que se inserta en la caja del cojinete.

Installed spring height The distance from the valve spring seat to the underside of the retainer when it is assembled with keepers and held in place.

Altura del resorte instalado La distancia del asiento del resorte de la válvula a la parte inferior del retén cuando ha sido instalado con las cuñas y se fija en su lugar.

Installed stem height The distance from the valve spring seat and to the stem tip.

Altura del vástago instalado La distancia del asiento del muelle de la válvula al punto extremo del vástago.

Insulated circuit A circuit that includes all of the high-current cables and connections from the battery to the starter motor.

Circuito aislado Un circuito que incluye a todos los cables de alta tensión y las conexiones de la batería al motor de arranque.

Insulated gate bipolar transistor (IGBT) A solid-state device that converts DC voltage into three-phase AC voltage to power an electrical motor or other device.

Transistor Bipolar de compuerta aislada (IGBT) Dispositivo en estado sólido que convierte voltaje de CD a voltaje de CA trifásico para activar un motor eléctrico u otro dispositivo.

Insulator A material that does not allow for good current flow.

Aislador Una materia que no permite un flujo del corriente.

Intake valve The control passage of the air-fuel mixture entering the cylinder.

Válvula de admisión El pasaje de control para la mezcla de aire-combustible entrado en el cilindro.

Integral Made in one piece.

Integral Hecho en una sóla pieza.

Integrated circuit A large number of diodes, transistors, and other electronic components, all mounted on a single piece of semiconductor material and able to perform numerous functions.

Circuito integrado La cantidad de diodos, transistores u otros componentes electrónicos, todos montados en una sola pieza de material semiconductor y capaz de ejecutar varias funciones.

Integrated motor assist (IMA) The motor/generator assembly used in Honda hybrids and fits between the engine and the transmission.

Asistencia integrada de motor (IMA) el ensamble del motor/generador usado en los híbridos de Honda y cabe entre el motor y la transmisión.

Integrated starter alternator damper (ISAD) Similar to Honda's IMA, this unit replaces the flywheel, generator, and starter motor. It is placed between the engine and transmission in some hybrid vehicles.

Arrancador regulador de alternador integrado (ISAD) Parecida al IMA de Honda, esta unidad remplaza el volante, generador, y arrancador. Se coloca entre el motor y la transmisión en algunos vehículos híbridos .

Intercooler A device used on some turbocharged engines to cool the compressed air.

Refrigerante intermedio Un dispositivo que se usa en algunos motores turbocargados para enfriar al aire comprimido.

Inverter A device that converts AC electricity to DC electricity and DC to AC.

Inversor Dispositivo que convierte electricidad CA a electricidad CD , y CD a CA.

Isooctane A standard reference fuel with an octane number of 100, meaning that it does not knock in an engine.

Iso-octano Un combustible de norma con un índice de octano de cien, lo que significa que resulta en que el motor no golpetea.

Jack stands Safety stands, used to hold a vehicle up while using a floor jack.

Torres Los soportes de seguridad. Sirven para sostener un vehículo en alto mientras que se usa un gato de piso.

Jounce Upward suspension movement.

Sacudo Un movimiento hacia arriba de la suspensión.

Keep-alive memory (KAM) A series of vehicle battery-powered memory locations in the microcomputer that allows the microcomputer to store information on input failure identified in normal operations for use in diagnostic routines. KAM adopts some calibration parameters to compensate for changes in the vehicle system.

Memoria de retención Una seria de localizaciones de memoria mantenidas por la batería del vehículo que permiten que la microcomputador registra la información de fallos de entrada, que se identifican en las operaciones normales para usarse en las pruebas diagnósticos. La memoria de retención utiliza algunos de los parámetros calibrados para compensar por los cambios en el sistema del vehículo.

Kickdown Normally refers to an automatic transmission shifting into a lower gear to meet the demands of the driver or a heavier load.

Disparo descendente Normalmente se refiere a una transmisión automática que efectúa un cambio descendente para acomodar al chofer o para acomodar a una carga más pesada.

Kinetic balance Balance of the radial forces on a spinning tire determined by an electronic wheel balancer.

Balanceo cinético El balanceo dinámico de las fuerzas radiales en una llanta que gira. Se determine por un balanceador de ruedas electrónico.

Kinetic energy Energy in motion.

Energía cinético La energía en movimiento.

Knurling A technique used for restoring the inside diameter dimensions of a worn valve guide by plowing tiny furrows through the surface of the metal.

Moletear Un método para restorar las dimensiones del diámetro interior de una guía de la válvula gastada partiendo al metal de la superficie en pequeños surcos.

KOEO The acronym for a test condition in which the key is on and the engine is not running.

KOEO Acrónimo para una condición de prueba en la cual la llave está puesta y el motor no está encendido.

KOER The acronym for a test condition in which the key is on and the engine is running.

KOER Acrónimo para una condición de prueba en la cual la llave está puesta y el motor está encendido.

Lambda sensor A form of oxygen sensor used on some European cars.

Sensor lambda Un tipo de detector del oxígeno que se usa en algunos coches europeos.

Lamination Thin layers of soft metal used as the core for a magnetic circuit.

Láminas Las hojas delgadas del metal blando que sirven de alma de un circuito magnético.

Land The areas on a piston between the grooves.

Meseta del pared Las areas de un pistón entre las muescas.

Lash The operational gap between two objects.

Juego La holgura que permite operar dos objetos.

Latent heat The heat required to change a mass's state of matter.

Calor latente El calor requerido para cambiar una masa a un estado differente.

Leakdown The relative movement of the plunger with respect to the hydraulic valve lifter body after the check valve is seated by pressurized oil. A small amount of oil leakdown is necessary for proper hydraulic valve lifter operation.

Fuga por abajo El movimiento relativo del émbolo con respeto al cuerpo de la levantaválvula hidráulica después de que la válvula de un sólo paso se asienta por el aceite a presión. Una pequeña cantidad de fuga es necesario para asegurar una operación correcta de la levantaválvula hidráulica.

Lean An air-fuel mixture that has more air than is required for a stoichiometric mixture.

Mezcla pobre Una mezcla de combustible y aire que tiene más aire de lo que requiere una mezcla estoquiométrica.

Lepelletier gears A compound planetary gear set that is based on a simple planetary gear set connected to a Ravigneaux compound gear set.

Engranaje Lepelletier Un conjunto de engranaje central compuesto que se basa en un simple conjunto de engranaje central conectado a un conjunto compuesto de engranaje Ravigneaux.

Lever A device made up of a bar turning about a fixed point, called the fulcrum, that uses a force applied at one point to move a mass on the other end of the bar.

Palanca Un dispositivo compuesto de una barra que da la vuelta sobre un punto fijo, llamado el fulcro, que utiliza una fuerza aplicada en un punto para mover una masa en el otro extremo de la barra.

Light-emitting diode (LED) A type of digital electronic display used as either single indicator lights or grouped to show a set of letters or numbers.

Diodo emisor de luz (LED) Un tipo de indicador electrónico digital que tiene luces indicadores individuales o en grupos para formar un conjunto de letras o números.

Limited-slip differential (LSD) A type of differential that allows more torque to be sent to the wheel with the most traction.

Deslizamiento limitado Un tipo de diferencial que permite que más energía del par se envía a la rueda con más tracción.

Line boring An engine block machining operation in which the main bearing housing bores are rebored to standard size and in perfect alignment.

Rectificación en linea Una operación maquinaria del bloque del motor en la cual los taladros principales de la caja se taladran de nuevo para asegurar un tamaño uniforme y una alineación perfecta.

Line pressure The hydraulic pressure that operates apply devices and is the source of all other pressures in an automatic transmission. It is developed by pump pressure and regulated by the pressure regulator.

Presión de línea Presión hidráulica que opera dispositivos de aplicación y es la fuente de todas las otras presiones en una transmisión automática. Se desarrolla por presión de bomba y se regula por regulador de presión.

Linear rate springs A coil spring with equal spacing between the coils, one basic shape, and constant wire diameter having a constant deflection rate regardless of load.

Resortes de variación lineal Resortes en espiral de espaciamiento igual entre espiras, una forma básica y un diámetro de alambre constante, con una tasa constante de desviación, sea cual sea la carga.

Lip-type seal An assembly consisting of a metal or plastic casing, a sealing element made of rubber, and a garter spring to help hold the seal against a turning shaft.

Sellos con borde Una asamblea que consiste de un cárter de metal o de plástico, un elemento sellante hecho de caucho, y un resorte de liga que sirve para apretar al sello contra una flecha que gira.

Lithium-ion (Li-Ion) A battery in which lithium is used as the electrochemically active material and the electrolyte is a liquid that conducts lithium ions.

Ion de litio (Li-Ion) Batería que funciona con litio usada como material electroquímicamente activo y el electrolito es un líquido que conduce iones de litio.

Live axle An axle on which the wheels are firmly affixed. The axle drives the wheels.

Eje motor Un eje en el cual las ruedas se montan rígidamente. El eje acciona las ruedas.

Load The work an engine must do, under which it operates more slowly and less efficiently. The load could be that of driving up a hill or pulling extra weight.

Carga El trabajo que debe ejecutar el motor, bajo el cual opera más lentamente y con menos eficiencia. La carga puede ser el de viajar cuesta arriba o de arrastrar el peso extra.

Lobe The part of the camshaft that raises the lifter.

Lóbulo La parte del árbol de levas que alza el buzo.

Lockup The point at which braking power overcomes the traction of the vehicle's tires and skidding occurs. The most efficient stopping occurs just before lockup is reached. Locked wheels cause loss of control, long stopping distances, and flat-spotting of the tires.

Enclavamiento El punto en el cual el enfrenamiento supera la tracción de los neumáticos del vehículo y ocurre el patinaje. El enfrenamiento más eficiente se efectúa justo antes de que ocurre el enclavamiento. Las ruedas enclavadas causan una pérdida del control, el enfrenamiento dilatorio y el gasto irregular de los neumáticos.

Lockup torque converter A converter with a friction disc that locks the impeller and turbine together.

Convertidor de par de torsión de bloqueo Un convertidor con un disco de fricción que bloquea al impulsor y la turbina.

Logic gate Electronic circuit that acts as a gate to output voltage signals depending on different combinations of input signals.

Compuerta lógico El circuito electrónico que sirve como una puerta a las señales del voltaje de salida según las combinaciones distinctas de señales de entrada.

Look-up tables The part of a microcomputer's memory that indicates in the form of calibrations and specifications how an engine should perform.

Obtención de datos de una tabla La parte de una memoria del microcomputador que indica en la forma de las calibraciones y las especificaciones como debe funcionar el motor.

LP gas Liquified petroleum gas, often referred to as propane, which burns clean in the engine and can be precisely controlled.

Licuados del petróleo El gas de petróleo en forma líquido, típicamente llamado el propano, que se quema completamente en el motor y que puede ser controlado precisamente.

Lubrication The process of reducing friction between the moving parts of an engine.

Lubricación El proceso de reducir la fricción entre dos partes que mueven en un motor.

MacPherson strut suspension A suspension system in which the strut is connected from the steering knuckle to an upper strut mount, and the strut replaces the shock absorber.

Suspensión del puntal de tipo MacPherson Un sistema de suspensión en donde el puntal se conecta al muñón de la dirección a un montaje superior para el puntal, y el puntal reemplaza el amortiguador.

Magnet Any body with the property of attracting iron or steel.

Imán Cualquier cuerpo que tiene la propriedad de atraer al hierro o al acero.

Magnetic field The area surrounding the poles of a magnet that is affected by its attraction or repulsion forces.

Campo magnético El área alrededor de los polos de un imán que se afectan por las fuerzas de atracción o repulsión.

Magnetic pulse generator An engine position sensor used to monitor the position of the crankshaft and control the flow of current to the center base terminal of the switching transistor.

Generador de pulsos magnéticos Un detector de la posición del motor que sirve para regular la posición de la cigueñal y controlar el flujo del corriente al terminal de la base central del transistor interruptor.

Magnetism A force between two poles of opposite potential caused by the alignment of electrons.

Magnetismo Fuerza entre dos polos de potencial opuesto, provocada por la alineación de los electrones.

Magnitude Normally refers to the height of an electrical signal.

Magnitúd Normalmente se refiere a la intensidad de un señal eléctrico.

Mainline pressure Pressure that is regulated in an automatic transmission hydraulic system.

Presión de la línea procedente de la bomba La presión que se regula en un sistema de transmisión hidráulica automática.

Malleable Able to be shaped.

Maleable Capaz de ser formado.

Manifold absolute pressure A measure of the degree of vacuum or pressure within an intake manifold; used to measure air volume flow.

Presión absoluta en el múltiple Una medida del grado del vacío o de la presión dentro de un múltiple de admisión; sirve para medir el volumen del flujo del aire.

MAP sensor The sensor that measures changes in the intake manifold pressure that result from changes in engine load and speed.

Sensor MAP El detector que mide los cambios de presión en el múltiple de admisión que resultan según los cambios de la carga y la velocidad del motor.

Margin The area between the valve face and the head of the valve.

Margen El área entre la cara de la válvula y la cabeza de la válvula.

Mass The amount of matter in an object.

Masa La cantidad de materia en un objeto.

Mass airflow (MAF) sensor An EFI air intake sensor that measures the mass, not the volume, of the air flowing into the intake manifold.

Sensor de la masa del aire Un detector del admisión del aire de inyección de combustible electrónico que mide la masa, en vez del volumen, del aire que fluye al múltiple de admisión.

Master cylinder The liquid-filled cylinder in the hydraulic brake system or clutch where hydraulic pressure is developed when the driver depresses a foot pedal.

Cilindro maestro El cilindro lleno de líquido del sistema de freno hidráulico o embrague donde se desarrolla la presión hidráulica cuando el conductor oprime un pedal.

Material safety data sheets Information sheets containing chemical composition and precautionary information for all products that can present a health or safety hazard.

Folletos de información de materiales y precauciones El folleto que contiene la información pertinente a la composición química y la indicaciones precavidos para todos los productos que pueden presentar peligros a la salud o la seguridad.

Matter Anything that occupies space.

Materia Cualquier cosa que ocupa el espacio.

Mechanical advantage The result of a machine or system that increases the force available to do work.

Ventaja mecánica El resultado de una máquina o de un sistema que aumente la fuerza disponible para hacer el trabajo.

Mechanical efficiency (engine) The ratio between the indicated horsepower and the brake horsepower of an engine.

Eficiencia mecánica (motor) La relación entre la potencia indicada en caballos y el caballo indicado al freno de un motor.

Memory The part of a computer that stores, or holds, the programs and other data.

Memoria La parte de una computador que archiva, o guarda, a las aplicaciones u otros datos.

Mesh To fit together, as gear teeth.

Engrenar Ajustarse bien, tal como los dientes de un engrenaje.

Metered To control the amount of fuel passing into an injector. Fuel is metered to obtain the correct measured quantity.

Medir Controlar la cantidad del combustible que pasa al inyector. El combustible se mide para asegurar una cantidad medida correcta.

Metering valve A component that momentarily delays the application of front disc brakes until the rear drum brakes begin to move. Helps to provide balanced braking.

Válvula dosificadora Un componente que retrasa momentáneamente la aplicación de los frenos de disco delanteros hasta que comienzan a moverse los frenos de tambor traseros. Contribuye a asegurar un frenado más equilibrado.

Methanol The lightest and simplest of the alcohols; also known as wood alcohol.

Metano El más ligero y más sencillo del grupo de alcohol; tambien llamado el alcohol piroleñoso.

Microprocessor The portion of a microcomputer that receives sensor input and handles all calculations.

Microprocesor La parte de un microcomputador que recibe la información de los detectores y se encarga de todas las calculaciones.

Millisecond One thousandth of a second.

Milisegundo Una milésima de un segundo.

Miscibility The ability of one chemical to mix well with another chemical.

Miscible La característica de una química a mezclarse bien con otra química.

Misfiring Failure of an explosion to occur in one or more cylinders while the engine is running; can be continuous or intermittent failure.

Fallo de encendido El fallo en que una detonación ocurre en uno o más cilindros mientras que este en marcha el motor; puede ser un fallo continuo o intermitente.

Mode Manner or state of existence of a thing; for example, heat or cool.

Modo Manera o estado de existencia de una cosa; por ejempto, calor o fresco.

Modulation Pulsing.

Modulación Pulsante.

Molecule The smallest particle of an element or compound that can exist in the free state and still retain the characteristics of the element or compound.

Molécula La partícula más pequeña de un elemento o un compuesto que puede existir en el estado libre y todavía conservar las características del elemento o del compuesto.

Momentum A type of mechanical energy that is the product of an object's weight times its speed.

Momento Un tipo de energía mecánica que es el producto del peso de un objeto multiplicado por su velocidad.

Monolith A single body shaped like a pillar or long tubular structure used as a catalyst in a catalytic converter.

Monolito Un cuerpo sólo en forma de columna o una estructura larga y tubular que sirve de catalizador en un convertidor catalítico.

Muffler (1) A hollow, tubular device used in the lines of some air conditioners to minimize the compressor noise or surges transmitted to the inside of the car. (2) A device in the exhaust system used to reduce noise.

Silenciador (1) Un dispositivo tubular hueco que se usa en las líneas de algunos equipos de aire acondicionado para reducir al mínimo el ruido del compresor o las sobrecargas transmitidas al interior del vehículo. (2) Un dispositivo del sistema de escape que sirve para reducir el ruido.

Multimeter A tool that combines the voltmeter, ohmmeter, and ammeter together in one diagnostic instrument.

Multímetro Una herramienta que combina las funciones del voltiometro, del ohmímetro, y del amperímetro en un instrumento diagnóstico.

Multiplexing A means of transmitting information between computers. A system in which electrical signals are transmitted by a peripheral serial bus instead of common wires, allowing several devices to share signals on a common conductor.

Multiplexar Una manera de transferir la información entre las computadoras. Un sistema en el cual los señales eléctricos se transmiten por medio de un bus serial periférico en vez de por los alambres comunes, así permitiendo que varios dispositivos comparten los señales en un conductor común.

Multiviscosity oil A chemically modified oil that has been tested for viscosity at cold and hot temperatures.

Aceite de viscosidad de grado múltiple Un aceite modificado químicamente cuya viscosidad se ha probado en temperaturas bajas y altas.

Newton-meter (N-m) The metric measurement of torque or twisting force.

Newton metro (Nm) Medida métrica de torque o fuerza de torsión.

Nickel-metal hydride (NiMH) A battery made of nickel hydroxide and hydride alloys. The electrolyte is potassium hydroxide.

Níquel metal hidruro (NiMH) Batería hecha de hidróxido de níquel y aleación de hidruro. El electrolito es hidróxido de potasio.

Nodular iron A metal used in pressure plates that contains graphite, which acts as a lubricating agent.

Hierro nodular Un metal usado en los platos opresores que contiene el grafito, que funciona como un agente de lubricación.

Nonasbestos organic linings (NAO) Organic brake linings that are made of nonmetallic fibers bonded together to form a composite material.

Revestimientos orgánicos no de asbestos (NAO) estos revestimientos parafrenos se hacen con fibras no metálicas unidas para formar un material compuesto.

Nonhunting gear set A differential gear set in which one drive pinion gear tooth contacts only three ring gear teeth after several rotations.

Conjunto de engrenaje sin diente suplementario Un conjunto de engrenaje en el cual un diente del engrenaje de piñón engrane solamente con tres dientes del engrenaje del anillo después de varias rotaciones.

Normally aspirated The method by which an internal combustion engine draws air into the combustion chamber. As the piston moves downward in the cylinder, it creates a vacuum that draws air into the combustion chamber through the intake manifold.

Aspirado normalmente El método por el cual un motor de combustión interno hace entrar al aire en la cámara de combustión. Al moverse hacia abajo el pistón en el cilindro crea un vacío que hace entrar el aire a la cámara de combustión a través del múltiple de admisión.

Octane number A unit of measurement on a scale intended to indicate the tendency of a fuel to detonate or knock.

Índice de octano Una unidad de medida en una gama cuya intención es de indicar la tendencia de un combustible de causar una detonación o los golpeteos.

OEM parts Parts made by the original vehicle manufacturer.

Partes EOF Las partes hechas por el fabricante original del vehículo.

Offset Placed off center. Also, the measurement between the center of the rim and the point where a wheel's center is mounted.

Descentrado Ubicado fuera del centro. Tambien, la medida entre el centro del ancho del rim y el punto en que se monta la rueda.

Ohm A unit of measured electrical resistance.

Ohmio Una unidad de medida de la resistencia eléctrica.

Ohm's law A basic law of electricity expressing the relationship between current, resistance, and voltage in any electrical circuit. It states that the voltage in the circuit is equal to the current (in amperes) multiplied by the resistance (in ohms).

Ley de los ohmios Una ley básica de la electricidad que describe la relación entre el corriente, la resistencia, y el voltaje en cualquier circuito eléctrico. Declara que el voltaje en un circuito iguala al corriente (en amperios) multiplicado por la resistencia (en ohmios).

Open circuit An electrical circuit that has a break in the wire.

Circuito abierto Un circuito eléctrico que tiene una quebradura en el alambre.

Open loop An electronic control system in which sensors provide information, the microcomputer gives orders, and the output actuators obey the orders without feedback to the microcomputer.

Bucle abierto Un sistema de control electrónico en el cual los detectores proveen la información, el microcomputador da los ordenes, y los actuadores de salida obedecen a éstas ordenes sin realimentación a la microcomputadora.

Orifice A precisely sized hole that controls fluid flow. Also, an opening.

Orificio Un hoyo de tamaño preciso que controla el flujo de los líquidos. Tambien, una apertura.

Oscillation Any single swing of an object back and forth between the extremes of its travel.

Oscilación Cualquier movimiento único de un objeto hacia adelante y hacia atrás entre los extremos de su recorrido.

Occupational Safety and Health Administration (OSHA) A government agency charged with ensuring safe work environments for all workers.

OSHA Administración de Salud y Seguridad Ocupacional, la cual es una agencia gubernamental encargada de asegurar ambientes de trabajo seguros para todos los trabajadores.

Out-of-round An inside or outside diameter, designed to be perfectly round, having varying diameters when measured at different points across its diameter.

Ovulado Un diámetro interior o exterior, por diseño perfectamente redondo, que tiene varios diámetros cuando se mide en puntos distintos a través de su diámetro.

Outboard joint The CV joint that connects the drive axle to the wheel spindle.

Junta exterior La junta de velocidad constante que conecta el eje motor al husillo de la rueda.

Output driver An electronic on/off switch located in the processor and operated by the digital commands of the computer used to control the ground circuit of a specific actuator.

Controlador de salida Ubicado en el procesador y operado por comandos digitales de la computadora, un controlador de salida es un interruptor electrónico de encendido/apagado usado para controlar el circuito de tierra de un activador específico.

Overdrive Normally used to express a gear ratio that allows the driven gear to rotate faster than the drive gear.

Sobremarcha Normalemente se usa para describir una relación de engrenaje que permite que el engrenaje arrastrado gira más rápidamente que el engrenaje propulsor.

Overlap The period during which the intake and exhaust valves are open at the same time.

Temporización El periódo en el cual las válvulas de admisión y de escape están abiertas a la misma vez.

Override clutch The mechanism that disengages the starter from the engine as soon as the engine turns more rapidly than the starter has cranked it.

Embrague de invalidación El mecanismo que desembraga el motor de arranque en el momento que el motor gira más rápidamente que el motor de arranque.

Oversquare A term used to describe an engine that has a larger bore than stroke.

Oversquare Usado para describir un motor que tenga un cilindro con un diámetro interior más grande que la longitud del movimiento del pistón.

Oxidation The combination of a substance with oxygen to produce an oxygen-containing compound. Also, the chemical breakdown of a substance or compound caused by its combination with oxygen.

Oxidación La combinación de una sustancia con el oxígeno para producir una compuesta que tenga oxígeno. Tambien, una descomposición química de una sustancia o una compuesta causada por su combinación con el oxígeno.

Oxidation inhibitor Gasoline additives used to promote gasoline stability by controlling gum and deposit formation and staleness.

Inhibitor del oxígeno Los aditivos de la gasolina que promuevan la estabilidad de la gasolina por medio de controlar la formación del sarro y los depósitos y el rancidez.

Oxidation rate Speed by which oxygen is taken into a substance.

Tasa de oxidación La velocidad en que el oxígeno se absorbe por una sustancia.

Oxides of nitrogen Various compounds of oxygen and nitrogen that are formed in the cylinders during combustion and are part of the exhaust gas.

Óxido de nitrógeno Varias compuesta del oxígeno y el nitrógeno que se forman en los cilindros durante la combustión y son parte del vapor del escape.

Oxygen (O_2) sensor An input sensor that sends a voltage signal to the computer in relation to the amount of oxygen in the exhaust stream.

Sensor de oxígeno (O_2) Es un sensor de entrada que le envía una señal de voltaje a la computadora referente a la cantidad de oxígeno en la pipa del escape.

Oxygenates Compounds such as alcohols and ethers that contain oxygen that are added to gasoline to reduce emissions and increase its octane rating.
Oxigenadores Compuestos como alcoholes y éteres que contienen oxígeno que se agregan a la gasolina para reducir emisiones y aumentar su índice de octanaje.
Parallel circuit In this type of circuit, there is more than one path for the current to follow.
Circuito paralelo En este tipo de circuito, hay más que una senda en que puede viajar el corriente.
Parameter identification (PID) A scan tool display for OBD-II systems that shows current emission-related data values of inputs and outputs, calculated values, and system status information.
Identificación de parámetros (PID) Un despliegue de herramienta de escaneo para sistemas OBD II que muestra los valores de la información relacionada a la emisión actual de entradas, salidas, valores calculados e información del estado del sistema.
Parasitic load An electrical load that is still present when the ignition switch is off.
Carga parásita Una carga eléctrica que siga presente cuando el interruptor de encendido esta apagado.
Particulate filter A diesel vehicle emission control device that traps and incinerates diesel particulate emissions after they are exhausted but before they are expelled into the atmosphere.
Filtro de partículas (PM) Dispositivo de control de emisiones de vehículos diésel que atrapa e incinera partículas de emisiones de diésel después que han sido quemadas pero antes que sean lanzadas a la atmósfera.
PCM Power control module.
PCM Módulo de control de la potencia.
PCV valve Positive crankcase ventilation valve. A valve that delivers crankcase vapors into the intake manifold rather than allowing them to escape to the atmosphere.
Válvula de ventilación positiva del cárter Una válvula que conduce vapores del cigüeñal al múltiple de admisión, en lugar de dejar que se descarguen a la atmósfera.
Peen To stretch or clinch over by pounding with the rounded end of a hammer.
Martillazo Estirar o remachar con la extremidad redondeado de un martillo de bola.
Permeable A characteristic of materials that absorb fluids.
Permeable Una característica de los materiales que absorben los líquidos.
Petroleum Oil as it is found in its natural state under the ground.
Petróleo Petróleo crudo como se encuentra en su estado natural bajo la tierra.
pH scale A scale used to measure how acidic or basic a solution is.
Escala del pH Un sistema que mide el grado de acidez de una solución.
Phase Refers to the rotational positions of the various elements of a driveline.
Fase Se refiere a las posiciones giratorias de los varios elementos de una flecha motríz.
Pickup coil A weak permanent magnet and wire assembly that in combination with a reluctor forms a position sensor.
Bobina de captación Una asamblea de alambre y un imán débil permanente que en combinación con un reluctor forma un detector de posición.
Piezoresistive A characteristic of something that changes resistances in relationship to changes in pressure.
Piezoresistivo Una característica de algo que cambia la resistencia relativamente con los cambios de la presión.
Pilot bushing A plain bearing fitted in the end of a crankshaft. The primary purpose is to support the input shaft of the transmission.
Buje piloto Chumacera simple que se adapta al extremo de un cigüeñal. Su finalidad primordial es la de soportar al eje de entrada de la transmisión.
Pinion gear The smaller of two meshing gears.
Piñón diferencial El más pequeño de dos engranajes enganchados uno al otro.
Pinning Cold crack repair process involving the installation of tapered plugs in the crack or on either side of the crack.
Chavetear Un proceso de reparar las grietas sin calor que involucra la instalación de los espárragos cónicos en las grietas o a un lado de la grieta.
Pintle The center pin used to control a fluid passing through a hole; a small pin or pointed shaft used to open or close a passageway.
Clavija La aguja central que controla al flúido que pasa por un hoyo; una pequeña clavija o flecha puntiaguda que sirve para abrir o tapar un pasillo.
Piston An engine component in the form of a hollow cylinder that is enclosed at the top and open at the bottom. Combustion forces are applied to the top of the piston to force it down. The piston, when assembled to the connecting rod, is designed to transmit the power produced in the combustion chamber to the crankshaft.
Pistón Un componente del motor que consiste de un cilindro hueco cerrado en la parte de arriba y abierto en la parte de abajo. Las fuerzas de combustión se aplican en la parte superior del pistón para forzarlo hacia abajo. El pistón, al conectarse a la biela, es diseñado para transmitir la fuerza producida en la cámara de combustión al cigüeñal.
Piston rings Components that seal the compression and expansion gases and prevent oil from entering the combustion chamber.
Anillos (aros) del pistón Los componentes que sellan los gases de la compresión y la expansión y previenen que entre el aceite en la cámara de combustión.
Pitch The angle of the valve spring twist. A variable pitch valve spring has unevenly spaced coils.
Paso El ángulo de las espiras de un resorte de válvula. Un resorte de válvula con paso variable tiene separaciones iniguales entre las espiras.
Pitch gauge A tool used to measure the thread pitch of a bolt.
Comprobador de paso de rosca Una herramienta que sirve para medir el paso de las roscas de un perno.
Pitman arm A steering linkage component that connects the steering gear to the linkage at the left end of the center link.
Brazo de dirección (biela de mando) Componente de enlace del sistema de dirección que conecta el mecanismo al extremo izquierdo del enlace central.
Pitting Surface irregularities resulting from corrosion.
Picadura Las irregularidades de la superficie causadas por la corrosión.
Planetary gear set A group of gears named after the solar system because of their arrangement and action. This unit consists of a center (sun) gear around which pinion (planet) gears revolve. The assembly is placed inside a ring gear having internal teeth. All gears mesh constantly. Planetary gear sets may be used to increase or decrease torque and/or obtain neutral, low, intermediate, high, or reverse.
Juego de engranajes planetarios (piñones satélites) Un grupo de engranajes cuyo nombre se inspira en el sistema solar por su disposición y su modo de actuar. Esta unidad consiste en un engranaje central (sol) en torno al que giran piñones diferenciales (planetas). El conjunto se coloca al interior de un engranaje anular con dientes internos. Todos los engranajes están enganchados continuamente. Estos juegos se pueden utilizar para hacer que aumente o disminuya el par de torsión y/u obtener velocidades neutra, baja, intermedia, alta o de retroceso.
Plasma An ionized gas and the fourth state of matter after solid, liquid, and gas.
Plasma Un gas ionizado y el estado cuarto después de sólido, de líquido, y del gas.
Plastigage A special wax used to measure bearing clearances.
Plastigage (calibrador de plástico) Una cera especial que sirve para medir las holguras.
Play Movement between two parts.
Juego El movimiento entre dos partes.
Plug-in hybrid electric vehicles (PHEVs) Full hybrids with larger batteries and the ability to recharge from an electric power grid. They are equipped with a power socket that allows the batteries to be recharged when the engine is not running.
Vehículo híbrido eléctrico de conectar (PHEV) Los híbridos totales con baterías más grandes y la habilidad de recargar de una matriz energética eléctrica. Están equipados con un interruptor de energía que permite que las baterías se recarguen cuando el motor no está funcionando.
Pneumatic Operated by compressed air.
Neumático Operado por el aire comprimido.
Polarity The particular state, either positive or negative, with reference to the two poles or to electrification.
Polaridad Un estado particular, sea positivo o negativo, en referencia a los dos polos o a la electrificación.
Polyglycol Polyalkaline-glycol-ether brake fluids that meet specifications for DOT 3 and DOT 4 brake fluids.
Poliglicol Líquido de frenos polialcalino-glicol-éter que cumple con las especificaciones de DOT 3 y DOT 4 para líquidos de frenos.
Poppet valve A valve consisting of a round head with a tapered face, an elongated stem that guides the valve, and a machined slot at the top of the stem for the valve spring retainer.

Válvula champiñón Una válvula que consiste de una cabeza redonda de cara cónica, un vástago alargado que guía a la válvula, y una muesca maquinada en la parte superior del vástago para el retenedor del resorte de la válvula.

Porosity Tiny holes in a casting caused by air bubbles.

Porosidad Los pequeños agujeros en una pieza moldeada causados por las burbujas del aire.

Port fuel injection A fuel injection system that uses one injector at each cylinder, thus making fuel distribution exactly equal among all of the cylinders.

Inyección de combustible por aperturas Un sistema de inyección de combustible que usa un inyector en cada cilindro, así proporcionando una distribución del combustible exactamente igual en todos los cilindros.

Positive displacement pumps Oil pump through which a fixed volume of oil passes with each revolution of its drive shaft.

Bombas de desplazamiento positivo Una bomba de aceite por el cual pasa un volumen fijo de aceite con cada revolución del eje propulsor.

Postcombustion control systems Emission control systems that clean up the exhaust gases after the fuel has been burned.

Controles de escape poscombustión Los sistemas de control de emisiones que purifican los vapores de gas después de que se haya quemado el combustible.

Potential energy Stored energy.

Energía potencial La energía almacenada.

Potentiometer A variable resistor that acts as a circuit divider to provide accurate voltage drop reading in response to the movement of an object.

Potenciómetro Un resistor variable que sirve de divisor de tensión para proveer una lectura precisa de una caída del voltaje según el movimiento de un objeto.

Power A measure of work being done.

Potencia Una medida del trabajo que se efectúa.

Power brake booster Used to increase pedal pressure applied to a brake master cylinder.

Reforzador de freno mecánico Se usa para incrementar la presión de pedal que se aplica al cilindro maestro del freno.

Power split device The basic name for the CVT transmission used in Ford and Toyota hybrid vehicles. This device is based on planetary gears and divides the output of the engine and the electric motors to drive the wheels or the generator.

Dispositivo divisor de energía El nombre básico para la transmisión CVT usada en los vehículos híbridos Ford y Honda. Este dispositivo conta de engranajes planetarios y divide la salida del motor y de los motores eléctricos al manejar las ruedas o el generador.

Power steering pump A hydraulic pump driven by a belt from a crankshaft pulley to provide up to 1,300 psi (8,964 kPa) "boost" pressure necessary to operate the power-steering system.

Bomba de dirección hidráulica (asistida) Una bomba hidráulica impulsada por una banda a partir de una polea de cigüeñal para proporcionar hasta 8,964 kPa (1,300 lbs/pulgada^2) de presión "de refuerzo" necesaria para manejar el sistema de dirección hidráulica (asistida).

Precombustion control system Emission control systems that prevent emissions from being created in the engine, either during or before the combustion cycle.

Control de escape precombustión Los sistemas de control de emisión que previenen que se crean las emisiones en el motor, sea durante o antes del ciclo de combustión.

Preheating The application of heat as a preliminary step to some further thermal or mechanical treatment.

Precalentamiento La aplicación del calor como un paso preliminario de un tratamiento termal o mecánico.

Preignition The process of a glowing spark or deposit igniting the air-fuel mixture before the spark plug.

Detonación El proceso en que una chispa o un depósito caliente enciende la mezcla de aire-combustible antes de que lo haga la bujía.

Preload A thrust load applied to bearings that support a rotating part to eliminate axial play or movement.

Carga previa Una carga de empuje aplicada a los cojinetes que soportan una parte giratoria con el fin de eliminar el juego o movimiento axial.

Pressure The exertion of force upon a body in contact with it. Pressure is developed within the cooling system and is measured in pounds per square inch on a gauge.

Presión El esfuerzo de una fuerza sobre un cuerpo con el que esta en contacto. La presión se desarrolla dentro del sistema de enfriamiento y se mide en libras por pie cuadrado con un calibre.

Preventive maintenance Normally scheduled maintenance designed to prevent vehicle failures.

Mantenimiento preventivo El mantenimiento periódico con el fin de prevenir los fallos del vehículo.

Primary circuit The low-voltage circuit of an ignition system.

Circuito primario El circuito de bajo voltaje de un sistema de encendido.

Prismatic cells Voltaic cells with flat electrodes placed into a box with separators placed between them.

Células prismáticas Células voltaicas con electrodos planos colocados en la caja con separadores entre ellos.

Program A set of instructions or procedures that a computer must follow when controlling a system.

Programa Un conjunto de instrucciones o procedimientos que debe cumplir una computadora en el control de un sistema.

Proportioning valve A pressure reduction valve used in the rear brake circuit.

Válvula dosificadora Una válvula de reducción de presión que se usa en un circuito trasero de los frenos.

Pulley A wheel with a grooved rim in which a rope, belt, or chain runs to raise something by pulling on the other end of the rope, belt, or chain.

Polea Una rueda con un borde acanalado en el cual mueve una cuerda, una correa, o una cadena. Es posible levantar un objeto pesado tirando en el otro extremo de la cuerda, de la correa, o de la cadena.

Pulse width The length of time in milliseconds that an injector is energized.

Anchura de impulso La cantidad del tiempo en milisegundos en la cual se acciona un inyector.

Pumping loss A term used to describe the difficulty that a piston has in moving air into the cylinder and moving it out on the exhaust stroke. Pumping losses are a major reason that engines consume a disproportionately large amount of fuel in city driving.

Pérdida de bombeo Término utilizado para describir la dificultad que tiene un pistón para mover aire hacia el cilindro y mover aire hacia fuera en el tiempo de escape. Las pérdidas de bombeo son una razón muy importante de porqué los motores consumen una cantidad inmensa de combustible cuando se conduce en una ciudad.

Purge To separate or clean by carrying off gasoline fumes. The carbon canister has a purge line to remove impurities.

Purgar Separar o limpiar al vaciar los vapores de la gasolina. El recipiente lleno de carbón tiene una línea de purgar para quitarle las impurezas.

Purge control valve A valve in some evaporative emission-control system charcoal canisters to limit the flow of vapor and air to the carburetor during idle.

Válvula de control de purga Una válvula de algunos recipientes de carbón del sistema de control de emisiones de evaporación para limitar el flujo de vapor y aire al carburador durante la marcha de vacío o en punto muerto.

Quad driver A group of transistors in a computer that controls specific outputs.

Ejecutor cuarteto Un grupo de transistores en una computadora que controla las salidas específicas.

Quenching The cooling of gases by pressing them into a thin area.

Templar El enfriamiento de los gases al oprimirlos en una area pequeña.

R-12 An air-conditioning system refrigerant that contains chlorofluorocarbons. When released into the atmosphere, this refrigerant is harmful to the earth's ozone layer.

R-12 Un refrigerante del sistema de aire acondicionado que contiene los clorofluorocarbonos. Al descargarse este refrigerante a la atmósfera, causa daños a la capa de la ozona de la tierra.

R-134a An air-conditioning system refrigerant that replaces R-12. R-134a has little effect on the earth's ozone layer.

R-134a Un refrigerante del sistema del aire acondicionado que reemplaza el R-12. R-134a tiene po efecto sobre la tierra capa ozono.

Race A channel in the inner or outer ring of an antifriction bearing in which the balls or rollers operate.

Pista Un canal en el anillo interior o exterior de un cojinete antifricción en el cual operan las bolas o los rodillos.

Rack and pinion steering A steering system in which the end of the steering shaft has a pinion gear that meshes with a rack gear.

Dirección de piñón y cremallera El extremo del eje de dirección tiene un piñón que se endenta con un engranaje de cremallera.

Radial Perpendicular to the shaft or bearing bore.

Radial Perpendicular a la flecha o al taladro del cojinete.

Radial load A load that is applied at 90 degrees to an axis of rotation.
Carga radial Una carga aplicada a 90 grados del eje de la rotación.
Radiation The transfer of heat by rays, such as heat from the sun.
Radiación La transferencia del calor por medio de los rayos, tal como el calor del sol.
Radiator A coolant-to-air heat exchanger. The device that removes heat from coolant passing through it.
Radiador Un intercambiador de calor hacia el aire fresco. El dispositivo que remueve el calor del refrigerante que pasa atraves de él.
Random-access memory (RAM) A type of memory used to store information temporarily.
Memoria de acceso aleatorio Un tipo de memoria que sirve para archivar la información temporalmente.
Ratio The relation of proportion that one number bears to another.
Relación La relación entre la proporción de un número a otro.
Ravigneaux geartrain A compound gear set that combines two planetary units with a common ring gear.
Tren de engranaje Ravigneaux Conjunto compuesto de engranajes que combina dos unidades planetarias con una corona de engranaje común.
Reach A design feature of a spark plug. It is the length of the threaded portion of a spark plug.
Alcance Un rasgo del diseño de una bujía. Se trata de la longitud de la porción fileteada de una bujía.
Reactivity The characteristic of a material that enables it to react violently with water or other materials. Materials that release cyanide gas, hydrogen sulfide gas, or similar gases when exposed to low pH acid solutions are also said to possess reactivity, as are materials that generate toxic mists, fumes, vapors, and flammable gases.
Reactividad La característica de una materia que facilita que reacciona violentamente con el agua o con otras materias. Las materias que descargan el gas cianuro, el gas sulfhídrico, o los gases semejantes al exponerse a las soluciones de bajo pH se consideran tener la propriedad de reactividad, así como son las materiales que crean la toxicidad en forma de la llovizna, el humo y el vapor y los gases inflamables.
Read-only memory (ROM) A type of memory in an automotive microcomputer used to store information permanently.
Memoria de sólo lectura Un tipo de la memoria en una computador automotríz que sirve para archivar la información permanentemente.
Reaming A technique used to repair worn valve guides either by increasing the guide hole size to take an oversize valve stem or by restoring the guide to its original diameter.
Escariado Un método para reparar las guías de las válvulas gastadas sea por aumentar el tamaño del hoyo de la guía para que accepta un vástago más grande o por restaurar el diámetro original de la guía.
Rebound An expansion of a suspension spring after it has been compressed as the result of jounce.
Repercusión Una expansión de un muelle de suspensión después de que se haya comprimido por razón de un sacudo.
Receiver/dryer A refrigerant reservoir and dryer that supplies liquid refrigerant to the expansion valve.
Colector desecador Un depósito desecador de refrigerante líquido que suministra éste último a la válvula de dilatación (expansión).
Recess A shaped hollow space on a part.
Muesca Un espacio hueco formado en una parte.
Recirculating ball steering gear A popular low friction steering gear box. A sector gear meshes with a ball nut that rides on ball bearings on the worm shaft to provide a smooth steering feel.
Engranaje de dirección de rótula de recirculación Un cárter de dirección de baja fricción muy usado. Un sector dentado se engancha en una tuerca esférica que va sobre cojinetes de bolas (baleros) en el eje sinfín para proporcionar una sensación de manejo suave.
Reciprocating An up-and-down or back-and-forth motion.
Reciprocante Un movimiento de arriba y abajo o de un lado a otro.
Recovery tank See Expansion tank.
Depósito de recuperación Vea Depósito de expansión.
Rectifier An electrical device used to convert alternating current to direct current.
Convertidor estático (rectificador) Un dispositivo eléctrico que se usa para convertir la corriente alterna en continua.
Rectify To change one type of voltage to another.
Rectificar Cambiar un tipo del voltaje por otro.
Reduction The removal of oxygen from exhaust gases.
Reducción Remover el oxígeno de los gases de escape.
Reference voltage A voltage provided by a voltage regulator to operate potentiometers and other sensors at a constant level.
Voltaje de referencia Un voltaje proporcionado por un regulador de voltaje para operar a los potenciómetros u otros detectores en un nivel constante.
Reformer An in-vehicle device that extracts hydrogen from a fossil fuel to provide fuel for a fuel cell or hydrogen-fueled engine.
Reformador Dispositivo interno del vehículo que extrae el hidrogeno de un combustible fósil para proporcionar combustible a una celda de combustible o motor movido por hidrógeno.
Reformulated gasoline (RFG) Gasolines that have had their compositions and/or characteristics altered to reduce vehicular emissions of pollutants, particularly pursuant to the EPA regulations under the CAA.
Gasolina reformulada (RFG) Gasolinas a las que se les han alterado sus composiciones y/o características para reducir las emisiones vehiculares contaminantes, particularmente siguiendo las regulaciones de la EPA bajo la CAA.
Refrigerant The chemical compound used in a refrigeration system to produce the desired cooling.
Refrigerante El compuesto químico utilizado en un sistema de refrigeración para producir el enfriamiento deseado.
Regenerative braking A method that captures a vehicle's kinetic energy while it is slowing down or is being stopped. This captured energy is used to charge batteries and/or an ultracapacitor.
Frenado regenerativo Método que captura la energía cinética mientras disminuye o se detiene la velocidad. Esta energía capturada se usa para cargar baterías y/o un ultra capacitador.
Relay An electrical switching device that uses a low-current circuit to control a high-current circuit.
Relé Un dispositivo interruptor eléctrico que usa un circuito de bajo corriente para controlar un circuito de alta corriente.
Relief The amount one surface is set below or above another surface.
Relieve La cantidad por el cual una superficie queda arriba o abajo de otra superficie.
Reluctance A term used to indicate a material's resistance to the passage of magnetic lines of flux.
Reluctancia Un término que indica la resistencia de una materia al pasaje de las líneas de flujo magnético.
Renewable fuels Fuels derived from nonfossil sources and produced from plant or animal products or wastes (biomass).
Combustibles renovables Combustibles derivados de fuentes no fósiles y producidos con plantas o productos o desechos de animales (biomasa).
Residual Remaining or leftover pressure.
Residual La presión que queda o sobra.
Residue Surplus or what remains after a separation takes place.
Residuo Lo que sobra o se queda después de que se efectúa una separación.
Resilience Elastic or rebound action.
Resiliencia La acción elástica o de repercusión.
Resistance The opposition offered by a substance or body to the passage of electric current through it.
Resistencia La oposición que ofrece una sustancia o un cuerpo al pasaje de la corriente eléctrica que lo atraviesa.
Resonator A second muffler in line with the other muffler.
Resonador Un silenciador secundario en linea con el primario.
Reverse bias A positive voltage applied to N-material and a negative voltage applied to P-material in a semiconductor.
Polarización inversa Un voltaje positivo aplicado a un cristal N y un voltaje negativo aplicado al cristal P en un semiconductor.
Rheostat A two-terminal variable resistor used to regulate electrical current.
Reóstato Un resistor variable con dos terminales que sirve para regular al corriente eléctrico.
Rich An air-fuel mixture that has more fuel than is required for a stoichiometric mixture.
Rica Una mezcla de aire-combustible que contiene más combustible de lo que requiere una mezcla estoquiométrica.
Right-To-Know laws Laws requiring employers to provide employees with a safe workplace as it relates to hazardous materials.
Leyes de derechos de los empleados Las leyes que requieren que los patrones proveen sus empleados con un ambiente del trabajo libre de los peligros asociados con las materiales tóxicas.
Ring gear (1) The gear around the edge of a flywheel. (2) A large, circular gear such as that in the final drive assembly.
Engranaje anular (anillo dentado) (1) El engranaje en torno al borde de un volante. (2) Un gran engranaje circular como el del ensamblaje de transmisión final.

Ring lands The high parts of a piston between the grooves.

Tierras de las coronas Las partes altas de un pistón entre las ranuras.

Road crown The slant of a road that allows water to drain off. Road crown causes a vehicle steering to drift to the right.

Corona de la carretera Inclinación transversal de la carretera que permite el escurrimiento del agua. La corona de la carretera provoca que la dirección de un vehículo se mueva hacia la derecha.

Rolling resistance A term used to describe the amount of resistance a tire has to rolling on the road. Tires that have a lower rolling resistance usually get better gas mileage. Typically, radial tires have lower rolling resistance.

Resistencia a la rotación Un término que describe la cantidad de resistencia que tiene un neumático a rodar en el camino. Los neumáticos que tienen una resistencia a rodar más baja suelen proporcionar un kilometraje más económico.

Rotary A circular motion.

Rotativo Un movimiento circular.

Rotary oil flow Torque converter oil flow associated with the coupling stage of operation.

Flujo rotativo El flujo del aceite del convertidor de par que se asocia con la etapa de acoplamiento de su operación.

Rotor The rotating or freewheeling portion of a clutch; the belt slides on the rotor.

Rotor La parte giratoria o con marcha libre de un embrague; la correa/banda se desliza sobre el rotor.

Rotor-type oil pump A type of oil pump that utilizes two rotors, a four-lobe inner, and a five-lobe outer; output per revolution depends on rotor diameter and thickness.

Bomba de tipo rotor Un tipo de bomba de aceite que utiliza dos rotores, uno interior con cuatro resaltes, y uno exterior con cinco rebajes; la producción por revolución depende del diámetro y espesor del rotor.

RTV Room temperature vulcanizing. A formed-in-place gasket product used in place of conventional paper, cork, and cork/rubber gaskets.

Sellador RTV Un empaque que se forma en sitio que se usa en vez de los sellos más convencionales del papel, del caucho o de caucho con corcho.

Runout Out-of-round or wobble.

Corrimiento Ovulado o con una oscilación irregular.

Sampling The act of periodically collecting information, as from a sensor. A microcomputer samples input from various sensors in the process of controlling a system.

Muestreo Coleccionar la información periódicamente, tal como de un detector. Una microcomputadora toma las muestras de varios detectores en el proceso de controlar a un sistema.

Saturation The point reached when current flowing through a coil or wire has built up the maximum magnetic field.

Saturación El punto en que el corriente que fluye por una bobina o un alambre llega a lo máximo de un campo magnético.

Scale A flaky deposit occurring on steel or iron. Ordinarily used to describe the accumulation of minerals and metals in an automobile cooling system.

Incrustación Un depósito que ocurre en el acero o el hierro. Típicamente se usa para describir la acumulación de los minerales y los metales en un sistema de enfriamiento automotríz.

Scan tool A microprocessor designed to communicate with a vehicle's on-board computer to perform diagnosis and troubleshooting.

Herramienta de exploración Un microprocesador diseñado para comunicar con una computadora a bordo de un vehículo para efectuar los prognósticos y localización de fallas.

Schematics Wiring diagrams used to show how circuits are constructed.

Esquemáticos Los dibujos de los conexiones que sirven para enseñar cómo se han construídos los circuitos.

Score A scratch, ridge, or groove marring a finished surface.

Rayo Un rasguño, una arruga, o una muesca que echa a perder una superficie acabada.

Scuffing Scraping and heavy wear from the piston on the cylinder walls.

Rozamiento La raspadura y el gasto extremo de un pistón en los paredes del cilíndro.

Seal Generally refers to a compressor shaft oil seal; matching shaft-mounted seal face and front head-mounted seal seat to prevent refrigerant and/or oil from escaping. May also refer to any gasket or O-ring used between two mating surfaces for the same purpose.

Junta hermética Generalmente se refiere a la junta hermética en el eje del compresor; el eje hace juego con el montaje de la cara del sello y el sello instalado con la cabeza mirando hacia adelante se asienta para prevenir que el refrigerante y/o aceite se escape. Puede referirse también a cualquier junta o sello de tipo anillo O utilizada entre dos superficies que hacen juego para el mismo propósito.

Sealant Material used as a seal between two mating objects.

Sellante Una material que sirve de empaque entre dos objetos de contacto.

Seat A surface, usually machined, on which another part rests or seats; for example, the surface on which a valve face rests.

Asiento Una superficie, típicamente maquinada, sobre la cual otra parte descanza o se asienta; por ejemplo, una superficie en la que descansa la cara de la válvula.

Secondary circuit A circuit in the ignition system that uses 20,000 or more volts to operate. It includes the secondary coil windings, the rotor, distributor cap, coil and spark plug wires, and spark plugs.

Circuito secundario Circuito del sistema de encendido que utiliza 20,000 voltios o más para funcionar. Incluye devanados secundarios de la bobina, rotor, tapa del distribuidor, bobina, bujías y sus cables.

Seize When one surface moving on another scratches. An example is a piston score or abrasion in a cylinder due to a lack of lubrication or overexpansion.

Agarrotar Cuando una superficie moviendo en otra la raya. Otro ejemplo sería un rayo o abrasión en un cilindro debido a la falta de lubricación o la sobreexpansión.

Selective catalytic reduction (SCR) To clean diesel exhaust, an ammonia-like substance is injected into the exhaust stream.

Reducción catalítica selectiva (RCS) Para limpiar el escape de diesel, una sustancia como amoníaco se inyecta al caudal del escape.

Semifloating axle A design of live axle in which only part of the vehicle's weight is supported by the axles.

Semiflotante Un diseño de un eje motor en el que sólo una porción del peso del vehículo se soporta por los ejes.

Sending unit A device that sends a signal to a gauge regarding oil pressure or coolant temperature.

Unidad emisora (detectora) Un dispositivo que envía una señal a un medidor sobre la temperatura del refrigerante o la presión del aceite.

Sensor Any device that provides an input to the computer.

Detector Cualquier dispositivo que provee una entrada para la computadora.

Serial data bus The communications to and from the computer.

Información serial Comunicación desde y hacia la computadora.

Serpentine belts Multiple-ribbed belts used to drive water pumps, power-steering pumps, air-conditioning compressors, alternators, and emission control pumps.

Correa serpentín Las bandas dentadas que sirven para propulsar las bombas de agua, las bombas de dirección hidráulica, los compresores acondicionadores, las alternadores y las bombas de control de emisiones.

Servo The part of a cruise control system that maintains the desired car speed by receiving a controlled amount of vacuum from the transducer.

Servo La parte de un sistema de control crucero que mantiene la velocidad deseada por medio de una cantidad de vacío que recibe y se controla por el transconductor.

Shift forks Semicircular castings connected to the shift rails that help control the movement of the synchronizer.

Horquillas de cambio de velocidades Partes moldeadas de forma semicircular conectadas a las palancas de cambio para ayudar en el control del movimiento del sincronizador.

Shift rails The parts of a transmission shift linkage that transfer motion from the driver-controlled gear shift lever to the shift forks.

Palancas de cambio Las partes de una biela de desembrague de una transmisión que transfieren el movimiento de la palanca del desembrague manual a las horquillas de desembrague.

Shift schedule Best described as a three-dimensional graph that plots engine speed and load as well as other operating conditions. Certain parts of the graph have designated gear ranges. When the conditions fall into a range, the computer causes the transmission to shift into that gear.

Programa de cambios Se describe mejor como una grafica tridimensional que marca la velocidad y carga del motor así como otras condiciones de operación. Ciertas partes de la grafica tienen rangos de enganajes designados. Cuando las condiciones caen en un rango, la computadora provoca que la transmisión cambie a esa velocidad.

Shim Thin sheets, usually metal, which are used as spacers between two parts, such as the two halves of a journal bearing.

Chapa Las hojas delgadas, por lo regular del metal, que sirven de arandelas entre dos partes, tal como las dos mitades de un cojinete de contacto plano.

Shock absorber A device that dampens spring oscillations by converting the energy from spring movement into heat energy.

Amortiguador Un dispositivo que absorbe las oscilaciones de los muelles, convirtiendo la energía del movimiento de estos últimos en energía térmica.

Short Of brief duration; for example, short cycling. Also refers to an intentional or unintentional grounding of an electrical circuit.

Breve/corto De una duración breve; funcionamiento cíclico breve. Se refiere también a una aplicación a tierra previsto o imprevista de un circuito eléctrico.

Short and long arm suspension (SLA) A suspension system using an upper and lower control arm. The upper arm is shorter than the lower arm. This is done to allow the wheel to deflect in a vertical direction with a minimum change in camber.

Suspensión de brazo corto y largo Un sistema de suspensíon que usa un brazo de control superior e inferior. El brazo superior es más corto que el brazo inferior. Este arreglo permite que la rueda se desvía en una dirección vertical con un cambio mínimo del camber.

Shrink fit The shaft or part is slightly larger than the hole in which it is to be inserted. The outer part is heated above its normal operating temperature or the inner part chilled below its normal operating temperature and assembled in this condition. Upon cooling, an exceptionally tight fit is obtained.

Calado por contracción La flecha o la parte que se debe insertar es un poquito más grande que el orificio. La parte exterior se calienta a una temperatura más alta de lo normal al funcionar, o la parte interior se enfría a una temperatura más baja de lo normal al funcionar y se asamblea en estas condiciones. Al regresar a la temperatura normal, se obtiene un ajuste extremadamente apretado.

Shudder Momentary shake or quiver; can sometimes be severe.

Estremecimiento Un sacudo o temblor momentáneo; a veces puede ser severo.

Shunt More than one path for current to flow, such as a parallel part of a circuit.

Shunt (en derivación) Cuando hay más que una senda en que puede fluir un corriente, tal como una parte paralela de un circuito.

Shunt circuits The branches of the parallel circuit.

Circuitos en desviación Las ramas del circuito paralelo.

Simpson geartrain A compound gear set in which two planetary units work together through a common sun gear.

Tren de engranajes Simpson Conjunto de engranajes compuesto en el cual dos unidades planetarias trabajan a través de un engranaje planetario.

Sine wave Basically, this wave is a circle drawn over time. It appears as a wave with equal positive and negative magnitudes and amplitudes.

Onda senoidal Basicamente, esta onda es un círculo que se dibuja a través del tiempo. Aparece como una onda con magnitudes y amplitudes positivos y negativos iguales.

Sleeving A means of reconditioning an engine by boring the cylinder oversize and installing a thin metal liner called a sleeve. The inside diameter of the sleeve is then bored, usually to the original or standard piston size.

Restaurar con manguito Un metodo de recondicionar a un motor por medio de taladrar más grande al cilindro e instalar un forro de metal delgado llamado un manguito. Luego se taladra al diámetro interior del manguito al tamaño normal o indicado según el pistón.

Sliding fit Where sufficient clearance has been allowed between the shaft and journal to permit free running without overheating.

Ajuste deslizante Cuando una holgura suficiente grande se ha dejado entro la flecha y el muñón para permitir la operación sín que se sobrecalienta.

Slip A condition caused when a driving part rotates faster than a driven part.

Deslizamiento Una condición causada cuando una parte propulsor gira más rapidamente que la parte arrastrada.

Slip yoke A component having internal splines that slide on the transmission outputshaft external splines, allowing the driveline to adjust for variations in length as the rear axle assembly moves.

Horqueta deslizante Un componente con ranuras externas que se desliza sobre las acanaladuras externas del eje de salida de la transmisión, permitiendo que la línea de propulsión se adapte a las variaciones de longitud del conjunto del eje posterior al moverse.

Sludge As used in connection with automobile engines, it indicates a composition of oxidized petroleum products along with an emulsion formed by the mixture of oil and water. This forms a pasty substance and clogs oil lines and passages and interferes with engine lubrication.

Sebo (grasa) En referencia con los motores automotrices, indica una composición de productos petróleos oxidados junta con una emulsión formada de una mezcla del aceite y el agua. Esto forma una sustancia pastosa y causa las obstrucciones en las lineas y pasajes del aceite preveniendo la lubricación.

Smog Air pollution created by the reaction of nitrogen oxides to sunlight.

Smog La contaminación del aire producido por la reacción de los óxidos de nitrógeno a la luz del sol.

Solenoid An electromagnetic switch with a movable core.

Solenoide Un interruptor electromagnético con un núcleo portátil.

Solution Formed when a solid dissolves into a liquid; its particles break away from this structure and mix evenly in the liquid.

Solución Se forma cuando un sólido disuelve en un líquido. Las partículas del sólido se separan y se mezclan uniformemente en el líquido.

Solvent The liquid in a solution.

Solvente El líquido en una solución.

Spark plug A device used to plug a hole in the combustion chamber while allowing a spark to enter the chamber.

Bujía Un dispositivo que sirve para tapar un hoyo en la cámara de combustión mientras que permite pasar una chispa a la cámara.

Specific gravity The weight of a given volume of a liquid divided by the weight of the same volume of water.

Gravedad específica El peso de un volumen dado de un líquido dividido por el peso del mismo volumen del agua.

Speed The distance an object travels in a set amount of time. Speed is the relationship between the distance traveled and the time it takes to travel it.

Velocidad La distancia que viaja un objeto dividido por la medida del tiempo usada para viajar esa distancia. La velocidad es la relación entre el espacio recorrido y el tiempo empleado en recorrerlo.

Speed ratio Comparison of the difference in speed between two moving parts, such as impeller speed and turbine speed.

Relación de la velocidad La comparación de la diferencia en velocidad entre dos partes en movimiento, tal como la velocidad del impulsor y la velocidad de la turbina.

Splay To spread or move outward from a central point.

Achaflanar Desplegar o moverse hacia afuera de un punto central.

Splice To join. Electrical wires can be joined by soldering or by using crimped connectors.

Empalmar Juntar. Los alambres eléctricos se pueden unir por soldado o por medio de los conectores de presión.

Splines External or internal teeth cut into a shaft that are used to keep a pulley or hub secured on a rotating shaft.

Espárragos Los dientes exteriores o interiores cortados en una flecha que sirven para fijar a una polea o un cubo en una flecha rotativa.

Sponginess A feel of a soft brake pedal.

Esponjoso Una sensación blanda en un pedal de freno.

Spontaneous combustion Process by which a combustible material ignites by itself and starts a fire.

Combustión espontánea Un proceso por el cual una material combustible se encienda sí mismo y causa un fuego.

Spool valve A cylindrical sliding valve that uses lands and valleys around its circumference to control the flow of hydraulic fluid through the valve body.

Válvula de carrete Válvula cilíndrica móvil que usa planos y valles alrededor de su circunferencia para controlar el flujo de fluido hidráulico a través del cuerpo de la válvula.

Spring A device that changes shape when it is stretched or compressed but returns to its original shape when the force is removed.

Resorte Dispositivo que cambia de forma cuando se extiende o contrae, pero regresa a su forma original cuando se le retira la fuerza.

Spur gear A gear with teeth that are cut straight across the gear.

Engrenaje recto Un engrenaje cuyos dientes atraviesan el engranaje en una linea recta.

Square wave Typically a digital waveform that shows the cycling of a circuit or device.

Onda cuadrada Típicamente una forma de onda digital que indica el ciclado de un circuito o un dispositivo.

Squib The igniter for an air bag.

Estopín Detonador de la bolsa de aire.

Squirm To wiggle or twist about a body. When applied to tires, squirm is the wiggle or movement of the tread against the road surface. Squirm increases tire wear.

Encorvadura Torcerse o doblar alrededor de un cuerpo. Al aplicarse a los neumáticos, la encorvadura se refiere a la torción o el movimiento de la banda contra la superficie del camino. La encorvadura aumenta el desgaste de los neumáticos.

Stainless steel An iron-carbon alloy with a minimum of 10.5% chromium content.

Acero inoxidable Aleación de carbón y hierro con un mínimo de 10.5% de contenido de cromo.

Stall speed With the engine operating at full throttle, the gear selection in D range, and the vehicle stationary, this speed can be read on a tachometer.

Velocidad de calar Al estar el motor en marcha a todo gas, el engrenaje selecionado en el rango D, el vehículo estacionario, este velocidad se lee en el taquímetro.

Stamping A piece of sheet metal cut and formed into the desired shape with the use of dies.

Estampar Una pieza de metal de hoja que se corta y se forma a lo deseado con los matrices.

Starter relay A magnetic switch, generally operated by the ignition switch, that uses low current to close a circuit to control the flow of very high current to the starter.

Relé (relevador) del motor de arranque Un interruptor magnético, activado generalmente por el interruptor de encendido, que usa una corriente baja para cerrar un circuito y controlar el flujo de corriente muy alta al motor de arranque.

Static balance Balance at rest; still balance. It is the equal distribution of weight of the wheel and tire around the axis of rotation such that the wheel assembly has no tendency to rotate by itself regardless of its position.

Balanceo estático El equilibreo en descanso; el balanceo inmóvil. Es la distribución equilibrada del peso de la rueda y el neumático alrededor del eje de rotación para que la asamblea de la rueda no tenga tendencia a girar por sí mismo, sin que importa su posición.

Static pressure The pressure inside the hydraulic system.

Presión estática La presión dentro del sistema hydráulico.

Stator The name used to describe a part in a torque converter or the stationary windings of an AC generator.

Estátor El nombre que describe una parte en un convertidor de par o el devanado estático de un alternador.

Steering axis inclination (SAI) The angle of a line through the center of the upper strut mount and lower ball joint in relation to the true vertical centerline of the tire viewed from the front of the vehicle.

Inclinación del eje de dirección (SAI) El ángulo de una línea a través del centro de la montura superior de puntal y unión inferior de la junta esférica en relación con la línea central vertical real de la rueda vista del frente del vehículo.

Stellite An alloy of nickel, chromium, and tungsten and is nonmagnetic. It is a hard facing material that is welded to valve faces and stems.

Estelita una aleación de níquel, cromo, y tungsteno y no es magnética. Es un material de reforzado que se solda a la caras de las válvula y a los vástagos.

Step-up transformer A transformer in which the voltage created in a secondary coil is greater than the voltage in the primary, or first, coil.

Transformador elevador Una transformador en la cual el voltaje creado en la bobina secundaria es más alta que el voltaje en la bobina primaria.

Stoichiometric Chemically correct. An air-fuel mixture is considered stoichiometric when it is neither too rich nor too lean; stoichiometric ratio is 14.7 parts of air for every part of fuel.

Estequiométrica Lo que es correcto químicamente. Una mezlca de aire-combustible se considera estequiométrica cuando no es demasiada rica o pobre; la relación estequiométrica es 14.7 partes del aire por cada parte del combustible.

Stress The force of strain to which a material is subjected.

Esfuerzo La fuerza de la tensión a la cual se somete una materia.

Stroke A term used to describe cylinder size that represents the amount of movement the piston has inside the bore.

Carrera Un término que describe el tamaño de un cilindro, representa la cantidad del movimiento que tiene el pistón dentro del orificio.

Strut Components connected from the top of the steering knuckle to the upper strut mount that maintain the knuckle position and act as shock absorbers to control spring action in a vehicle's suspension system. Used on most front-wheel drive cars and some rear-wheel drive cars.

Riostras Componentes conectados de la parte superior de la charnela de dirección a la montura de tirante superior para mantener la posición de la charnela y actuar como amortiguadores para controlar la acción de resorte del sistema de suspensión de un vehículo. Se utiliza en la mayoría de los automóviles de tracción delantera y en muchos de tracción posterior.

Substrate A ceramic honeycomb grid structure coated with catalyst materials.

Sustrato Una estructura cerámica de rejilla agrietada cubierta con materiales de catalizador.

Suction Suction exists in a vessel when the pressure is lower than the atmospheric pressure.

Succión La succión existe en una vasija cuando la presión es más baja que la presión atmosférica.

Sulfation A potential condition of a lead-acid battery in which the battery operates only in a partially discharged condition. This condition results from excessive stop-and-go driving or a fault in the charging system, which causes the sulfate normally formed in the plates to become dense, hard, and chemically irreversible.

Sulfatación Estado potencial de una batería de ácido-plomo en la cual la batería opera en estado de descarga parcial. Este estado resulta de manejar cuando se efectúan excesivas paradas y arranques o existe una falla en el sistema de carga, provocando que el sulfato que normalmente se forma en las placas se haga denso, duro y químicamente irreversible.

Supercharger A belt-driven pump, also called a blower.

Sobrealimentador Una bomba de banda que se denomina también aventador.

Surface tension The result of forces of attraction, which pull on the particles of a liquid in all directions. These forces create liquid bubbles and drops in a spherical shape.

Tensión de superficie El resultado de las fuerzas de atracción que halan las partículas de un líquido en todas las direcciones. Estas fuerzas crean burbujas y gotas líquidas en una forma esférica.

Surging A condition in which the engine speeds up and slows down with the throttle held steady.

Operación pulsatorio Una condición en la cual el motor acelera y desacelera mientras que la presión de admisión se mantiene fijo.

Sway bar Also called a stabilizer bar. It prevents the vehicle's body from diving into turns.

Barra de oscilación lateral Llamada también barra estabilizadora. Impide que la carrocería del vehículo se clave durante las curvas.

Swirl combustion A swirling of the air-fuel mixture in a corkscrew pattern. The swirling effect improves combustion.

Combustión remolino Un movimiento circular de la mezcla aire-combustible en el patrón de un remolino. El efecto de este movimiento es de amejorar la combustión.

Synchronizer An assembly used in manual transmissions to bring components that are rotating at different speeds to one synchronized speed.

Sincronizador Ensamble usado en las transmisiones manuales para traer los componentes que están rotando a diferentes velocidades a una velocidad sincronizada.

Tailpipe The part of the exhaust system that allows the exhaust gases to leave the rear of the vehicle and go into the atmosphere.

Tubo de escape La parte del sistema de escape que permite que los gases se descarguen a la atmósfera por la parte trasera del vehículo.

Tap To cut threads in a hole with a tapered, fluted, threaded tool.

Roscar con macho Cortar las roscas en un agujero con una herramienta cónica, acanalada y fileteada.

Taper The difference in diameter between the cylinder bore at the bottom of the hole and the bore at the top of the hole, just below the ridge.

Ahuso La diferencia en el diámetro del orificio del cilindro en la parte inferior y del orificio en la parte superior de la apertura, justo abajo del reborde.

Tapered roller bearing A bearing containing tapered roller bearings mounted between the inner and outer races.

Balero de rodillos cónicos Los baleros que contienen rodillos cónicos montados entre las pistas interiores y exteriores.

Telescoping gauges Tools used for measuring bore diameters and other clearances; also known as snap gauges.

Calibradores telescópicos Las herramientas que sirven para medir los diámetros de los orificios y otras holguras; tambien se conocen como calibres de ensarte.

Temperature An indication of an object's kinetic energy.

Temperatura La temperatura es el grado de calor, relacionado con la energía cinética de las moléculas de los mismos.

Tempering The heat treatment of metal alloys, particularly steel, to result in specific properties.

Templar El tratar de las aleaciones del metal, particularmente de acero, con calor para dar ciertas características al metal.

Tensile strength The amount of pressure per square inch the bolt can withstand just before breaking when being pulled apart.

Resistencia a la tracción La cantidad de presión por pulgada cuadrada que puede resistir un perno al estirarse justo antes de que se quiebra.

Tension Effort that elongates or stretches a material.

Tracción Un esfuerzo que alarga o estira a una materia.

Thermal cleaning Cleaning that relies exclusively on heat to bake off or oxidize surface contaminants.

Limpieza termal La limpieza que cuenta exclusivamente con el calor para quemar u oxidar los contaminantes de una superficie.

Thermal contraction A decrease in the size of a mass due to the movement of atoms and molecules as heat moves out of a mass.

Contracción termal Una contracción termal es una disminución del volumen de una masa debido al movimiento de átomos y moléculas cuando el calor se mueve de una masa.

Thermal expansion An increase in the size of a mass due to the movement of atoms and molecules as heat moves into a mass.

Expansión termal Una expansión termal es un aumento en el volumen de una masa debido al movimiento de átomos y moléculas cuando el calor entra a una masa.

Thermistor A solid-state variable resistor made from semiconductor material that changes resistance in response to changes in temperature.

Termistor Un resistor variable de estado sólido hecho de material semiconductor que cambia su resistencia según los cambios de la temperatura.

Thermo efficiency A gallon of fuel contains a certain amount of potential energy in the form of heat when burned in the combustion chamber. Some of this heat is lost and some is converted into power. The thermal efficiency is the ratio of work accomplished compared to the total quantity of heat contained in the fuel.

Rendimiento térmico Un galón del combustible contiene una cierta cantidad de energía potencial en la forma de calor cuando se quema en la cámara de combustión. Algo de éste calor se pierde y algo se convierte a la fuerza. El rendimiento térmico es la relación del trabajo que se efectúa comparado con la cantidad total del calor que contiene el combustible.

Thermostat A device used to cycle the clutch to control the rate of refrigerant flow as a means of temperature control. The driver has control of the temperature desired.

Termostato Un dispositivo utilizado para ciclar el embrague para regular la proporción del flujo de refrigerante como medio de regulación de temperatura. El accionador puede regular la temperatura deseada.

Thread pitch The number of threads in 1 inch of threaded bolt length. In the metric system, thread pitch is the distance in millimeters between two adjacent threads.

Paso de rosca El número de las cuerdas por pulgada de un perno con rosca. En el sistema métrico, el paso de las roscas es la medida de la distancia entre dos pasos contiguos en milímetros.

Three-quarter floating axle A drive axle arrangement in which the axle shaft supports 25% of the vehicle's weight. The bearing is outside the axle housing.

Eje flotante de tres cuartos Disposición del eje de tracción en la que el eje axial soporta el 25% del peso del vehículo. El balero está fuera de la carcasa del eje.

Throw With reference to an automobile engine, usually the distance from the center of the crankshaft main bearing to the center of the connecting rod journal.

Recorrido Al referirse a un motor de automóvil, típicamente es la distancia del cojinete principal del cigueñal al centro de los muñones de la biela.

Throwout bearing In the clutch, the bearing that can be moved inward to the release levers and clutch-pedal action to cause declutching, which disengages the engine crankshaft from the transmission. A common name for a clutch release bearing.

Cojinete de desembrague En el embrague, el cojinete que puede moverse hacia adentro hasta las palancas de liberación, por medio de la acción del pedal del embrague, para logar el desembrague, desenganchando el cigüeñal del motor de la transmisión. Es un nombre común para designar una chumacera de liberación del embrague.

Thrust line A line that divides the total toe angle of the rear wheels.

Línea de empuje Una línea que divide el ángulo total de la convergencia de las ruedas traseras.

Thrust load Load placed on a part that is parallel to the center of the axis.

Carga de empuje Una carga en una parte que queda paralela al centro del eje.

Tie-rod A rod connected from the steering arm to the rack or center link, depending on the type of steering linkage.

Barra de acoplamiento Una varilla conectada desde el brazo de la dirección a la cremallera o a la barra central, dependiendo del tipo de varilla de la dirección.

Tolerance A permissible variation between the two extremes of a specification or dimension.

Tolerancia Una variación permitible entre dos extremos de una especificación o una dimensión.

Torque A twisting force applied to a shaft or bolt.

Par de ajuste Una fuerza torcedura que se aplica a una flecha o un perno.

Torque converter A turbine device utilizing a rotary pump, one or more reactors (stators), and a driven circular turbine or vane whereby power is transmitted from a driving to a driven member by hydraulic action. It provides varying drive ratios; with a speed reduction, it increases torque.

Convertidor del par de torsión o de torque Una turbina que utiliza una bomba giratoria, uno o más reactores (estator) y una turbina circular accionada o paleta; la fuerza se transmite desde el mecanismo de accionamiento al mecanismo accionado mediante una acción hidráulica. Provee relaciones de accionamiento variadas; con una reducción de velocidad, aumenta el par de torsión.

Torque steer A twisting axle movement in front-wheel-drive automobiles that causes a pulling action under acceleration toward the side with the longer driving axle.

Dirección de torsión Un movimiento de torsión del eje en un automóvil de tracción delantera que causa que jale mientras que accelera hacia el lado con un eje motor más largo.

Torque-to-yield (TTY) Bolts that are intentionally torqued just barely into a yield condition, although not far enough to distort the bolt. This type of bolt will provide 100% of its intended strength compared to 75% when torqued to normal values.

Torque a rendimiento (TTY) Los tornillos son enroscados intencionalmente justamente a condición de entrada, aunque no lo suficiente como para distorsionar el tornillo. Este tipo de tornillo proporcionará el 100% de su fuerza intencionada, comparada con el 75% cuando se enrosca a valores normales.

Torrington bearing A thrust washer fitted with roller bearings.

Rodamiento Torrington Arandela de empuje con rodamientos rodantes.

Torsion bar A steel bar connected from the chassis to the lower control arm. As the vehicle's weight pushes the chassis downward, the torsion bar twists to support this weight. Torsion bars are used in place of coil springs.

Barra de torsión Una barra de acero conectada del chasis al brazo de control inferior de la suspensión. Mientras el peso del vehículo presiona el chasis hacia abajo, la barra de torsión se tuerce para soportar este peso. Las barras de torsión se utilizan en lugar de los resortes espirales.

Total indicator reading (TIR) The total deflection of the dial indicator as the object is rotated through one complete rotation. It is equal to the amount greater than zero plus the amount less than zero.

Lectura total del indicador (TIR) La deflexión total del indicador de carátula conforme el objeto es rotado a una rotación completa. Es igual a la cantidad mayor de cero más la cantidad menor de cero.

Trace A lab scope converts electrical signals to a visual image representing voltage changes over a specific period. This information is displayed in the form of a continuous voltage line called a trace.

Rastro Un ámbito de laboratorio convierte las señales eléctricas a una imagen visual representando cambios durante un período específico de tiempo. Esta información se despliega en la forma de una línea continua de voltaje llamado rastro.

Tracking The travel of the rear wheels in a parallel path with the front wheels.

Seguimiento El progreso de las ruedas traseras en una senda paralela con las ruedas delanteras.

Traction A tire's ability to hold or grip the road surface.

Tracción La abilidad de un neumático para agarrar o mantenerse en la superficie del camino.

Traction motor An AC or DC electric motor designed for vehicle propulsion.

Motor de tracción Motor eléctrico de CA o CD diseñado para vehículos a propulsión.

Tractive effort Pushing force exerted by the vehicle's driving wheels against the road's surface.

Esfuerzo de tracción Una fuerza de empujo que resulta de tracción de las ruedas contra la superficie del camino.
Tramp Wheel hop caused by static balance.
Pisoteo Un brinquito de la rueda causado por el balanceo estático.
Transaxle A unit that combines a transmission and differential.
Eje de transmisión Una unidad que combina una transmisión y un diferencial.
Transfer case An additional component added to a four-wheel-drive vehicle in which the power from the engine is divided between both the front and rear wheels equally.
Caja de transferencia Componente adicional que se agrega a un vehículo de tracción en las cuatro ruedas, donde la potencia del motor se divide por igual entre las ruedas delanteras y las traseras.
Transistor An electronic device produced by joining three sections of semiconductor materials. A transistor is very useful as a switching device, functioning as either a conductor or an insulator.
Transistor Un dispositivo electrónico producido al unir tres secciones de materiales de semiconductor. Un transistor es muy útil como dispositivo interruptor, funciona como un conductor o como aislador.
Transverse Perpendicular, or at right angles, to a front-to-back centerline.
Transverso Perpendicular, o en ángulo recto, a una línea central de delantero a trasero.
Trip An OBD-II drive cycle that includes all of the conditions (enable criteria) required for a monitor to run.
Viaje Ciclo de transmisión OBD II que incluye todas las condiciones (habilitar criterios) requeridos para que un monitor funcione.
Trouble codes Output of the self-diagnostics program in the form of a numbered code that indicates faulty circuits or components. Trouble codes are two- or three- digit characters that are displayed in the diagnostic display if the testing and failure requirements are both met.
Códigos indicadores de fallas Los datos del programa autodiagnóstico en forma de código numerado que indica los circuitos o los componentes defectuosos. Los códigos de problemas son compuestos de dos o tres caracteres digitales que se muestran en el despliegue de diagnóstico si se llenan los requisitos de prueba y de falla.
Turbo boost The positive pressure increase created by a turbocharger.
Reforzar de turbo Un aumento del presión positivo creado por un turbocargador.
Turbocharger A small radial fan pump driven by the energy of the exhaust flow.
Turbosobrealimentador Un pequeño ventilador radial impulsado por la energía del flujo del escape.
Turbulence A very rapid movement of gases in a combustion chamber that causes better combustion because the mixture is thoroughly mixed.
Turbulencia Movimiento rápido de gases en una cámara de combustión que provoca mejor combustión porque la mezcla es totalmente homogénea.
Turning torque Amount of torque required to keep a shaft or gear rotating; measured with a torque wrench.
Torsión giratoria La cantidad de torsión requerida para mantener girando a una flecha o un engrenaje; se mide con una llave de torsión.
Two-mode hybrid system A hybrid transmission that fits into a standard housing and is basically two planetary gear sets coupled to two electric motors. This results in a continuously variable transmission and motor/generator for hybrid operation. This also allows for two distinct modes of hybrid drive operation: low speed/low load and cruising at highway speeds.
Sistema híbrido de dos modos Transmisión híbrida que cabe en una carcasa estándar y es básicamente dos engranajes planetarios unidos a dos motores eléctricos. Esto trae como resultado una transmisión continuamente variable y motores/generadores para operación híbrida. También permite que dos distintos modos de operación híbrida de manejo: velocidad baja/carga baja y manejo a las velocidades de carretera.
Ultracapacitors Used in hybrid vehicles and in some fuel cell electric vehicles, they are capacitors with a large electrode surface area and a very small distance between the electrodes, which gives them very high capacitance.
Capacitadores ultra Usado en vehículos híbridos y en algunos vehículos con combustible de pila eléctrica, son capacitadores con área de superficie grande de electrodos y muy pequeña distancia entre los electrodos, lo cual les da muy alta capacidad.
Ultrasonic cleaning Method of cleaning that utilizes high-frequency sound waves to create microscopic bubbles that work to loosen soil from parts.
Limpieza ultrasónica Un método de limpieza que usa las ondas de sonido de alta frecuencia para crear las burbújas microscópicas que hacen el trabajo de aflojar la suciedad de las partes.
Undersquare A term used to describe an engine that has a larger stroke than bore.
Undersquare Usado para describir un motor que tenga un cilindro con un diámetro interior más pequeño que la longitud del movimiento del pistón.
Unibody A stressed hull body structure that eliminates the need for a separate frame.
Monocasco Una estructura de casco arriostrada que elimina la necesidad de un chassis aparte.
Union A hydraulic coupling used to connect two brake lines.
Unión Un acoplador hidráulico que sirve para conectar dos lineas de freno.
Universal joint A joint that allows the driveshaft to transmit torque at different angles as the suspension moves up and down.
Junta universal Una junta que permite que la flecha motríz transmite el par en los ángulos diferentes mientras que la suspensión mueve hacia arriba y hacia abajo.
Urea An ammonia-like substance that is injected in the exhaust stream of a diesel engine to reduce emissions.
Urea Sustancia similar al amoníaco que se inyecta en el caudal de escape de un motor diésel para reducir las emisiones.
Vacuum The absence of atmospheric pressure; commonly used to refer to any pressure less than atmospheric pressure.
Vacio La ausencia de la presión atmosférica; palabra que se utiliza comúnmente para referir a cualquier presión que es menos que la presión atmosférica.
Valve Device that controls the flow of gases into and out of the engine cylinder.
Válvula Un dispositivo que controla el flujo de los gases entrando y saliendo del cilindro del motor.
Valve body The hydraulic control assembly of the transmission.
Cuerpo de válvula El conjunto del control hidráulico de la transmisión.
Valve lifter A cylindrically shaped hydraulic or mechanical device in the valve train that rides on the camshaft lobe to lift the valve off its seat.
Filtro de válvula Un dispositivo hidráulico o mecánico de forma cilíndrica, en el tren de la válvula, que va sobre el lóbulo del árbol de levas para levantar la válvula a partir de su asiento.
Valve seat Machined surface of the cylinder head that provides the mating surface for the valve face. The valve seat can be either machined into the cylinder head or a separate component that is pressed into the cylinder head.
Asiento de la válvula La superficie rectificada a máquina en la cabeza del cilindro que provee una superficie de contacto para la cara de la válvula. El asiento puede ser rectificado a máquina en la cabeza del cilindro o puede ser un componente aparte para prensarlo en la cabeza de la válvula.
Valve spring A coil of specially constructed metal used to force the valve closed, providing a positive seal between the valve face and seat.
Resorte de la válvula Es un espiral de metal de construcción especial que provee la fuerza para mantener cerrada la válvula, así proveyendo un sello positivo entre la cara de la válvula y el asiento.
Vapor A substance in a gaseous state. Liquid becomes a vapor when brought above the boiling point.
Vapor Una substancia en un estado gaseoso. El líquido se convierte en un vapor al superar el punto de ebullición.
Vapor lock A condition wherein the fuel boils in the fuel system, forming bubbles that retard or stop the flow of fuel to the carburetor.
Tapón de vapor Una condicion en la cual el combustible hierve en el sistema de combustible, formando unas burbújas que retrasan o detienen al flujo del combustible al carburador.
Vaporization The last stage of carburetion in which a fine mist of fuel is created below the venturi in the bore.
Vaporización La última etapa del carburacíon en la cual una llovizna muy fina del combustible se crea en el orificio abajo del venturi.
Variable-rate coil spring Rather than having a standard spring deflection rate, these springs have an average spring rate based on load at a predetermined deflection.
Resorte espiral de capacidad variable En vez de tener una capacidad de desviación de resorte estándar, estos resortes tienen un valor

promedio de elasticidad basado en la carga a una desviación predeterminada.

Variable resistor A resistor that allows for a change in resistance based on the physical movement of a control. The control can be moved by an individual or a component.

Resistor variable Resistor que permite un cambio de resistencia basado en el movimiento físico de un control. El control puede moverse a través de una persona o un componente.

Varnish A deposit in an engine lubrication system resulting from oxidation of the motor oil. Varnish is similar to, but softer than, lacquer.

Barníz Un depósito en el sistema de lubricación de un motor que resulta de una oxidación del aceite del motor. El barníz parece a la laca, pero es más blando.

V-belt A rubberlike continuous loop placed between the engine crankshaft pulley and accessories to transfer rotary motion of the crankshaft to the accessories.

Correa/banda en V Un ciclo de caucho continuo parecido a una liga de caucho ubicada entre la polea del cigüeñal del motor y los accesorios para transferir el movimiento giratorio del cigüeñal a los accesorios.

Vehicle identification number (VIN) The number that is made up of seventeen characters and contains all pertinent information about the vehicle.

Número de identificación de vehículo Este número se compone de 17 caracteres y contiene toda la información pertinente del vehículo.

Velocity The speed of an object in a particular direction.

Velocidad Velocidad es el índice de movimiento de un objeto en una dirección particular.

Ventilation The act of supplying fresh air to an enclosed space such as the inside of an automobile.

Ventilación El proceso de suministrar el aire fresco a un espacio cerrado, como por ejemplo al interior de un automóvil.

Viscosity The resistance to flow that is exhibited by a liquid. Thick oil has greater viscosity than thin oil.

Viscosidad Resistencia al flujo mostrada por un líquido. El aceite pesado tiene mayor viscosidad que aceite ligero.

Viscous clutch A clutch assembly that is enclosed in a drum-filled thick fluid. It houses one set of the steel discs connected to the front wheels and another set connected to the rear. A viscous coupling splits torque according to the needs of each axle.

Embrague viscoso Ensamble de embrague dentro de un tibor lleno de fluido pesado que alberga un juego de discos de acero conectados a las llantas anteriores y otro conjunto conectado a la parte posterior. Un acoplamiento viscoso divide el torque según las necesidades de cada eje.

VNT Variable nozzle turbine turbocharger. Designed to reduce turbo lag time.

VNT La unidad de turbina con boquilla variante diseñada para permitir que un turbocargador se acelerara rápidamente, así disminuyendo el tiempo del retraso.

Volatile liquid A liquid that vaporizes very quickly.

Líquido volátil Un líquido que se vaporiza muy rápidamente.

Volatility The tendency for a fluid to evaporate rapidly or pass off in the form of vapor. For example, gasoline is more volatile than kerosene because it evaporates at a lower temperature.

Volatilidad La tendencia de un flúido a evaporarse rápidamente o escaparse en la forma de un vapor. Por ejemplo, la gasolina es más volátil que el kerosén porque se evapora en las temperatura más bajas.

Volt A unit of measurement of electromotive force. One volt of electromotive force applied steadily to a conductor of 1-ohm resistance produces a current of 1 ampere.

Voltio Una unidad de medida de la fuerza electromotor. Un voltio de la fuerza electromotor aplicado constantemente a un conductor que tiene una resistancia de un ohmio produce una corriente de un amperio.

Voltage drop Voltage lost by the passage of electrical current through resistance.

Caída de voltaje El voltaje que se pierde al pasarse un corriente eléctrico por la resistencia.

Voltage-generating sensors Sensors that are capable of producing their own input voltage signal. This varying voltage signal enables the computer to monitor and adjust for changes in the computerized control system.

Sensores que general voltaje Sensores capaces de producir su propia señal de voltaje. Este voltaje variante le permite a la computadora controlar y ajustar cambios en el sistema computarizado de control.

Voltage regulator A device that controls the strength of the electromagnetic field in the rotor of an alternator.

Regulador del voltaje Un dispositivo que controla la fuerza del campo electromagnético en el rotor de un alternador.

Voltmeter A tool used to measure the voltage available at any point in an electrical system.

Voltímetro Una herramienta que sirve para medir el voltaje disponible en cualquier punto del sistema eléctrico.

Volume The measure of space expressed as cubic inches, cubic feet, and so forth.

Volumen La medida del espacio expresado como pulgadas cúbicas, pies cúbicos, etc.

Volumetric efficiency A measure of how well air flows in and out of an engine.

Rendimiento volumétrico Una medida de lo fácil que circula el aire en un motor.

Vortex flow A swirling, twisting motion of fluid.

Flujo vórtice Un movimiento como remolino o giratorio de un flúido.

V-ribbed belt A belt that has multiple ribs on one side and is flat on the other side.

Banda encostillada en V Una banda que tiene varias nervaduras en un lado y esta plana en el otro.

Vulcanized A process of heating rubber under pressure to mold it into a special shape.

Vulcanizado Un proceso de calentar al caucho bajo presión para moldearlo en una forma especial.

VVT Variable valve timing. A system that automatically changes the valve timing according to operating conditions.

Warpage Bending.

Alabeo Doblando.

Waste spark A spark occurring during the exhaust stroke on a computerized ignition system.

Chispa de desperdicio Chispa que se produce durante el ciclo de escape de un sistema de encendido computarizado.

Water pump A device, usually located on the front of the engine and driven by one of the accessory drive belts, that circulates the coolant by causing it to move from the lower radiator-outlet section into the engine by centrifugal action of a finned impeller on the pump shaft.

Bomba de agua Un dispositivo, situado por lo común en la parte delantera del motor e impulsado por una de las bandas para accesorios, que hace circular el refrigerante, haciéndolo pasar de la sección de salida inferior del radiador al motor, mediante la acción centrífuga de un propulsor de aletas en el eje de la bomba.

Water soluble A type of contaminant that is the easiest to clean and that includes dirt, dust, and mud.

Soluble en agua Tipo de contaminante que es el más fácil de limpiar, tal como la tierra, el polvo y el lodo.

Watt A unit of measure of electric power.

Watt Unidad de medida de energía eléctrica

Watt's law A basic law of electricity used to find the power of an electrical circuit expressed in watts. It states that power equals the voltage multiplied by the current (in amperes).

Ley de Watt Una ley básica de la electricidad que sirve para expresar la potencia de un circuito eléctrico en vatios. Declara que la potencia iguala al voltaje multiplicado por el corriente (en ampérios).

Wave Any single swing of an object back and forth between the extremes of its travel through matter or space.

Onda Cualquier movimiento único de un objeto hacia adelante y hacia atrás entre los extremos de su recorrido a través de materia o de espacio.

Waveform (Trace) A lab scope converts electrical signals to a visual image representing voltage changes over a specific period of time. This information is displayed in the form of a continuous voltage line called a waveform.

Rastro – Ondas Un ámbito de laboratorio convierte señales eléctricas a imágenes visuales que representan cambios en el voltaje durante un período específico de tiempo. Esta información se despliega en la forma de una línea continua de voltaje llamada una forma de onda.

Wavelength The distance between each compression of a sound or electrical wave.

Longitud de onda La distancia entre cada compresión de un sonido u onda eléctrica.

Weight A force exerted on a mass by the gravitational force.

Peso El peso es una fuerza ejercida en una masa por la gravedad.

Wheatstone bridge A series-parallel arrangement of resistors between an input terminal and ground.

Puente Wheatstone Un arreglo en series-paralelo de los resistores entre un terminal de entrada y la tierra.

Wheelbase The distance between the center of a front wheel and the center of a rear wheel.

Batalla (empate) La distancia entre el centro de una rueda delantera y el centro de una rueda trasera.

Wheel cylinder A device used to convert hydraulic fluid pressure to mechanical force for brake applications.

Cilindro de rueda Un dispositivo que sirve para convertir la presión de flúido hidráulico en una fuerza mecánica para aplicaciones de frenos.

Wheel spindle The short shaft on the front wheel upon which the wheel bearings ride and to which the wheel is attached.

Husillo de rueda El eje corto de la rueda delantera sobre el que se asientan los cojinetees y al que va acoplada la rueda.

Work What is accomplished when a force moves a certain mass a specific distance.

Trabajo Lo que se logra cuando una fuerza mueve una cierta masa una distancia específica.

Yaw A swinging motion to the left or right of the vertical centerline or rotation around the vertical centerline.

Desviación o derrape Se define como un movimiento a la izquierda o derecha de la línea vertical del centro o rotación alrededor de la línea vertical del centro.

Yield Commonly refers to the point where a bolt is stretched to its limit and is unable to return to its original shape.

Rendimiento Se refiere comúnmente al punto donde un tornillo se extiende a su límite y no puede regresar a su forma original.

Zener diode A diode that allows reverse current to flow above a set voltage limit.

Diodo zener Un diodo que permite que el corriente en reverso fluye en cantidades más altas del límite de voltaje predeterminado.

INDEX

C

D

E

F

I

M

N

Q

R

T

W

X

Y

Z